YAWS HANDBOOK OF PROPERTIES FOR ENVIRONMENTAL AND GREEN ENGINEERING

Adsorption capacity, water solubility, Henry's law constant, octanol-water partition coefficient, bioconcentration factor, soil sorption coefficient, threshold limit value, and permissible exposure limit for environmental and green engineering. Coverage for more than 4000 hydrocarbons and chemicals.

CARL L. YAWS

PROFESSOR OF CHEMICAL ENGINEERING
LAMAR UNIVERSITY
BEAUMONT, TEXAS

2008

Gulf Publishing Company
Houston, Texas

YAWS HANDBOOK OF PROPERTIES FOR ENVIRONMENTAL AND GREEN ENGINEERING

Adsorption capacity, water solubility, Henry's law constant, octanol-water partition coefficient, bioconcentration factor, soil sorption coefficient, threshold limit value, and permissible exposure limit for environmental and green engineering. Coverage for more than 4000 hydrocarbons and chemicals.

CARL L. YAWS

PROFESSOR OF CHEMICAL ENGINEERING
LAMAR UNIVERSITY
BEAUMONT, TEXAS

2008

Gulf Publishing Company
Houston, Texas

THE YAWS HANDBOOK OF PROPERTIES FOR ENVIRONMENTAL AND GREEN ENGINEERING

Gulf Publishing Company
2 Greenway Plaza, Suite 1020
Houston, TX 77046

10 9 8 7 6 5 4 3 2 1

Library of Congress Cataloging-in-Publication Data

Yaws, Carl L.
The Yaws handbook of properties for environmental and green engineering: adsorption capacity, water solubility, Henry's law constant ... / Carl L. Yaws.
p. cm.
Includes bibliographical references and index.
ISBN 978-1-933762-15-9 (alk. paper)
1. Chemicals--Handbooks, manuals, etc. 2. Hydrocarbons--Handbooks, manuals, etc. 3. Chemicals--Tables. 4. Hydrocarbons--Tables. 5. Environmental chemistry--Industrial applications--Handbooks, manuals, etc. 6. Environmental engineering--Handbooks, manuals, etc. I. Title. II. Title: Handbook of properties for environmental and green engineering.

TP200.Y394 2009
660--dc22

2008039449

Printed in the United States of America
Printed on acid-free paper. ∞

CONTENTS

CONTRIBUTORS

Daniel H. Chen	Professor, Chemical Engineering Department, Lamar University, P.O. Box 10053, Beaumont, Texas 77710, USA
Helen Lou	Associate Professor, Chemical Engineering Department, Lamar University, P.O. Box 10053, Beaumont, Texas 77710, USA
Avanindra Shishir Jaiswal	Graduate Student, Chemical Engineering Department, Lamar University, P.O. Box 10053, Beaumont, Texas 77710, USA
Akhila Kodali	Graduate Student, Chemical Engineering Department, Lamar University, P.O. Box 10053, Beaumont, Texas 77710, USA
Prasad K. Narasimhan	Senior Software Development Engineer, Invensys Process Systems, 10900 Equity Drive, Houston, TX 77041, USA
Satish K. Padala	Graduate Student, Chemical Engineering Department, Lamar University, P.O. Box 10053, Beaumont, Texas 77710, USA
Ralph W. Pike	Professor, Chemical Engineering Department, Louisiana State University, Baton Rouge, Louisiana 70803, USA
Swetha Pingali	Graduate Student, Chemical Engineering Department, Lamar University, P.O. Box 10053, Beaumont, Texas 77710, USA
Kayzad J. Vajifdar	Graduate Student, Chemical Engineering Department, Lamar University, P.O. Box 10053, Beaumont, Texas 77710, USA
Meenakshi S. Yadav	Graduate Student, Mechanical Engineering Department, Lamar University, P.O. Box 10053, Beaumont, Texas 77710, USA
Preeti S. Yadav	Graduate Student, Chemical Engineering Department, Lamar University, P.O. Box 10053, Beaumont, Texas 77710, USA
Carl L. Yaws	Professor, Chemical Engineering Department, Lamar University, P.O. Box 10053, Beaumont, Texas 77710, USA

ACKNOWLEDGMENTS

The author wishes to acknowledge special appreciation to his wife (Annette) and family (Kent, Michele, Chelsea, Brandon, Lindsay, Rebecca, Chloe, Trey, and Sarah).

Carl L. Yaws, Lamar University, Beaumont, Texas

DISCLAIMER

Chapter 1

ADSORPTION CAPACITY OF ACTIVATED CARBON - GAS

Carl L. Yaws, Daniel H. Chen, and Helen Lou
Lamar University, Beaumont, Texas

Tabulation Results

The results for adsorption capacity of activated carbon are given in Table 1. For the tabulation, the Freundlich equation is used for adsorption capacity as a function of concentration in gas:

$$\log_{10} Q = \log_{10} K + 1/n \log_{10} y \qquad (1)$$

where Q = adsorption capacity at equilibrium, g of compound/100 g of carbon
y = concentration in gas at 25 C and 1 atm, parts per million by volume, ppmv
K, 1/n = regression coefficients

The tabulation is arranged by carbon number (C, C2, C3, ...) to provide ease of use in quickly locating the data by using the chemical formula. The compound name, CAS No (Chemical Abstracts Registry Number), and regression coefficients are provided in the adjacent columns. The range of application is denoted by minimum and maximum concentrations (ymin and ymax). Concentrations outside the range of application should not be used. The next column provides the code for the tabulation which is based on both experimental data and estimated values.

In preparing the tabulation, a literature search was conducted to identify data source publications for organics (1-39). Both experimental values for the property under consideration and parameter values for estimation of the property are included in the source publications. The publications were screened and copies of appropriate data were made. These data were then keyed into the computer to provide a database of values for compounds for which experimental data are available. The database also served as a basis to check the accuracy of the estimation methods. Upon completion of data collection, estimation of the values for the remaining compounds was performed. The numerous point values were processed using a computer program for minimum deviation.

The compilations of Calgon (1-2) were used primarily for experimental data. Estimates were primarily based on an empirical equation developed by Calgon (BPL activated carbon, 4 x 10 mesh) for equilibrium adsorption capacity as a fifth order polynominal function (6,34) of adsorption potential. Data (vapor pressure, liquid molar volume, and refractive index) for estimation parameters were from compilations of Daubert and Danner (7), TAMU (26-27), and Yaws (34-39).

Experimental data for adsorption capacity are very limited. Data are especially meager for large size molecules (C8 and higher) and substituted organics containing halogens, nitrogen, oxygen, and sulfur. In the absence of data, the estimates for large size and substituted compounds should be considered as rough approximations. If initial analysis using the compilation is favorable, follow-up experimental determination of equilibrium adsorption capacity is recommended.

A comparison of calculated and data values is shown in Figure 1 for a representative compound. The graph discloses favorable agreement of equation and data.

Example

Air from an industrial process contains 1,000 ppmv of 1,2 dichloroethane (C2H4Cl2). Calculate the adsorption capacity of activated carbon for removing the compound at 25 C and 1 atm.

Substitution of the regression coefficients from the table into the equation for adsorption capacity yields:

$$\log_{10} Q = \log_{10}(4.9266) - 0.2532 \log_{10}(1000) = 1.45209$$

$$Q = 10^{1.45209} = 28.32 \text{ g of compound/100 g of carbon}$$

References

1. Calgon Adsorption Isotherms, personal communication to Carl L. Yaws, Calgon Carbon Corporation, Pittsburg, PA(1994).
2. CALGON CARBON ADSORPTION HANDBOOK, Calgon Carbon Corporation, Pittsburg, PA (1994).

3. Cheremisnoff, P. N. and F. Ellerbusch, CARBON ADSORPTION HANDBOOK, Ann Arbor Science Publishers, Ann Arbor, MI (1980).
4. Cook, W. H. and D. Basmadjian, Can. J. Chem. Eng., 42, 146 (1964).
5. Copper, C. D. and F. C. Alley, AIR POLLUTION CONTROL, 2nd ed., Waveland Press, Prospects Heights, IL (1994).
6. Damie, A. S. and T. N. Rogers, AIR STRIPPER DESIGN MANUAL, EPA-450/1-90-003, U. S. Environmental Protection Agency (May,1990).
7. Daubert, T. E. and R. P. Danner, DATA COMPILATION OF PROPERTIES OF PURE COMPOUNDS, Parts 1, 2, 3, and 4, Supplements 1 and 2, DIPPR Project, AIChE, New York, NY (1985-1992).
8. Graham, J. R. and M. Ramaratnam, Chem. Eng., 100, 6 (1993).
9. Grant, R. J. and M. Manes, Ind. Eng. Chem. Fundam., 3, 221 (1964).
10. Grant, R. J. and M. Manes, Ind. Eng. Chem. Fundam., 5, 490 (1966).
11. Grant, R. J., M. Manes, and S. B. Smith, AIChE J., 8, 403 (1962).
12. Howard, P. H. and W. M. Meylan, eds., HANDBOOK OF PHYSICAL PROPERTIES OF ORGANIC CHEMICALS, CRC Press, Boca Raton, FL (1997).
13. Kaul, B. K., Ind. Eng. Chem. Res., 26, 928 (1987).
14. Kuro-Oka, M., T. Suzuki, T. Nitta, and T. Katayama, J. Chem. Eng. Japan, 17, 588 (1984).
15. Laukhuf, W. L. S. and C. A. Plank, J. Chem. Eng. Data, 14, 48 (1969).
16. Lewis, W. K., E. R. Gilliand, B. Chertow, and W. P. Cadogan, Ind. Eng. Chem., 42, 1326 (1950).
17. Lewis, W. K., E. R. Gilliland, B. Chertow, and W. Milliken, J. Am. Chem. Soc., 72, 1157 (1950).
18. Lide, D. R. and H. V. Kehianian, CRC HANDBOOK OF THERMOPHYSICAL AND THERMOCHEMICAL DATA, CRC Press, Boca Raton, FL (1994).
19. Maslan, F. D., M. Altman, and E. R. Aberth, J. Phys. Chem., 57, 106 (1953).
20. Payne, H. K., G. A. Studervant, and T. W. Leland, Ind. Eng. Chem. Fundam., 7, 363 (1968).
21. Ray, G. C. and E. O. Box, Ind. Eng. Chem., 42, 1315 (1950).
22. Reich, R., W. T. Zeigler, and K. A. Rogers, Ind. Eng. Chem. Proc. Des. Dev., 19, 336 (1980).
23. Reich, R., W. T. Ziegler, and K. A. Rogers, Ind. Eng. Chem. Process Des. Dev., 19, 336 (1980).
24. Ritter, J. A. and R. T. Yang, Ind. Eng. Chem. Res., 26, 1679 (1987).
25. Schenz, T. W. and M. Manes, J. Phy. Chem., 79, 604 (1975).
26. SELECTED VALUES OF PROPERTIES OF CHEMICAL COMPOUNDS, Thermodynamics Research Center, TAMU, College Station, TX (1997).
27. SELECTED VALUES OF PROPERTIES OF HYDROCARBONS AND RELATED COMPOUNDS, Thermodynamics Research Center, TAMU, College Station, TX (1997).
28. Szepesy, L. and V. Illes, Acta Chim. Hung., 35, 37 (1963).
29. Valenzuela, D. P. and A. L. Myers, ADSORPTION EQUILIBRIUM DATA HANDBOOK, Prentice Hall, Engle Cliffs, NJ (1989).
30. Yang, R. T., GAS SEPARATION BY ADSORPTION PROCESSES, Butterworth Publishers, Boston, MA (1987).
31. Yaws, C. L. and others, Pollution Engineering, 27 (2), 34 (1995).
32. Yaws, C. L. and others, Environmental Engineering World, 1 (3), 16 (May-June, 1995).
33. Yaws, C. L. and others, Oil & Gas Journal, 93 (7), 64 (Feb. 13, 1995).
34. Yaws, C. L., CHEMICAL PROPERTIES HANDBOOK, McGraw-Hill, New York, NY (1999).
35. Yaws, C. L., MATHESON GAS DATA BOOK, 7th ed., Matheson Tri-Gas (Parsippany, NJ), McGraw-Hill, New York, NY (2001).
36. Yaws, C. L., YAWS HANDBOOK OF THERMODYNAMIC AND PHYSICAL PROPERTIES FOR CHEMICALS, electronic edition, www.knovel.com, knovel, Norvich, NY (2003).
37. Yaws, C. L., YAWS HANDBOOK OF ANTOINE COEFFICIENTS FOR VAPOR PRESSURE OF ORGANIC COMPOUNDS, electronic edition, www.knovel.com, knovel, Norvich, NY (2005).
38. Yaws, C. L., YAWS HANDBOOK OF PHYSICAL PROPERTIES FOR HYDROCARBONS AND CHEMICALS, Gulf Publishing Co., Houston, TX (2005).
39. Yaws, C. L., YAWS HANDBOOK OF ANTOINE COEFFICIENTS FOR VAPOR PRESSURE, Gulf Publishing Co., Houston, TX (2007).

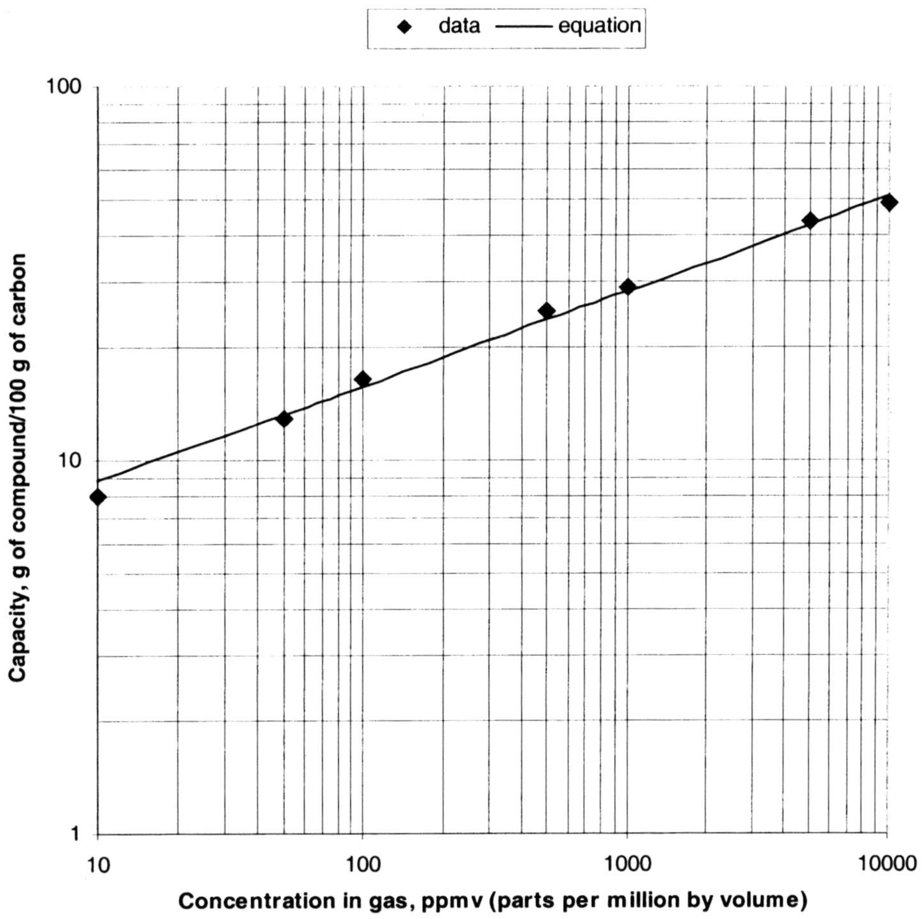

Figure 1 Adsorption Capacity for 1,2-Dichloroethane

Table 1 Adsorption Capacity of Activated Carbon - Gas

$\log_{10} Q = \log_{10} K + 1/n \log_{10} y$ (Q - g of compound/100 g of carbon, y - ppmv)

NO	ID	FORMULA	NAME	CAS No	K	1/n	ymin	ymax	code	Q @ymin	Q @10ppm	Q @ymax
1	16	CBrCl3	bromotrichloromethane	75-62-7	36.4852	0.1078	10	10000	2	46.76	46.76	98.44
2	17	CBrF3	bromotrifluoromethane	75-63-8	0.0402	0.5268	10	10000	1,2	0.14	0.14	5.15
3	23	CBr2F2	dibromodifluoromethane	75-61-6	8.1256	0.2317	10	10000	1,2	13.85	13.85	68.67
4	34	CClF3	chlorotrifluoromethane	75-72-9	0.0057	0.7571	10	10000	2	0.03	0.03	6.06
5	42	CCl2F2	dichlorodifluoromethane	75-71-8	1.0396	0.3251	10	10000	1,2	2.20	2.20	20.76
6	44	CCl2O	phosgene	75-44-5	0.3528	0.4419	10	10000	1,2	0.98	0.98	20.65
7	47	CCl3F	trichlorofluoromethane	75-69-4	1.9786	0.3030	10	10000	1,2	3.97	3.97	32.23
8	50	CCl4	carbon tetrachloride	56-23-5	16.6393	0.1582	10	10000	1,2	23.95	23.95	71.45
9	96	CHBr3	bromoform	75-25-2	74.2917	0.0805	10	7238	1,2	89.41	89.41	151.88
10	101	CHCl2F	dichlorofluoromethane	75-43-4	0.6575	0.4000	10	10000	2	1.65	1.65	26.17
11	104	CHCl3	chloroform	67-66-3	6.5813	0.2370	10	10000	2	11.36	11.36	58.40
12	140	CH2BrCl	bromochloromethane	74-97-5	7.0307	0.2597	10	10000	2	12.79	12.79	76.89
13	144	CH2Br2	dibromomethane	74-95-3	22.4639	0.1815	10	10000	2	34.12	34.12	119.55
14	149	CH2Cl2	dichloromethane	75-09-2	1.1915	0.3683	10	10000	2	2.78	2.78	35.42
15	161	CH2I2	diiodomethane	75-11-6	114.8041	0.0549	10	1583	2	130.27	130.27	172.03
16	175	CH2O	formaldehyde	50-00-0	0.0035	0.6708	10	10000	2	0.02	0.02	1.67
17	176	CH2O2	formic acid	64-18-6	0.0428	0.7494	10	10000	1,2	0.24	0.24	42.60
18	191	CH3Br	methyl bromide	74-83-9	0.1309	0.4853	10	10000	2	0.40	0.40	11.44
19	196	CH3Cl	methyl chloride	74-87-3	0.0131	0.5907	10	10000	1,2	0.05	0.05	3.02
20	210	CH3Cl3Si	methyl trichlorosilane	75-79-6	16.1048	0.1359	10	10000	2	22.02	22.02	56.32
21	222	CH3I	methyl iodide	74-88-4	6.6925	0.2575	10	10000	1,2	12.11	12.11	71.72
22	232	CH3NO	formamide	75-12-7	20.6755	0.2586	10	76	2	37.51	37.51	63.37
23	235	CH3NO2	nitromethane	75-52-5	1.1393	0.4151	10	10000	2	2.96	2.96	52.12
24	251	CH4	methane	74-82-8	0.0001	0.7447	10	10000	1,2	0.0003	0.0003	0.05
25	254	CH4Cl2Si	methyl dichlorosilane	75-54-7	7.3852	0.1896	10	10000	2	11.43	11.43	42.34
26	272	CH4O	methyl alcohol	67-56-1	0.0133	0.7453	10	10000	2	0.07	0.07	12.69
27	278	CH4S	methyl mercaptan	74-93-1	0.1028	0.4918	10	10000	1,2	0.32	0.32	9.53
28	284	CH5ClSi	methyl chlorosilane	993-00-0	2.3860	0.2550	10	10000	2	4.29	4.29	24.99
29	286	CH5N	methylamine	74-89-5	0.0136	0.5896	10	10000	2	0.05	0.05	3.10
30	363	CN4O8	tetranitromethane	509-14-8	41.2048	0.0780	10	10000	2	49.31	49.31	84.51
31	372	CO	carbon monoxide	630-08-0	0.0000	0.8273	10	10000	1,2	0.0001	0.0001	0.02
32	373	COS	carbonyl sulfide	463-58-1	0.0371	0.5121	10	10000	1,2	0.12	0.12	4.15
33	375	CO2	carbon dioxide	124-38-9	0.0002	0.7840	10	10000	1,2	0.0014	0.0014	0.32
34	382	CS2	carbon disulfide	75-15-0	0.8060	0.3904	10	10000	1,2	1.98	1.98	29.37
35	413	C2Br2F4	1,2-dibromotetrafluoroethane	124-73-2	19.0355	0.1664	10	10000	2	27.92	27.92	88.15
36	423	C2ClF3	chlorotrifluoroethylene	79-38-9	1.8821	0.2842	10	10000	2	3.62	3.62	25.78
37	427	C2ClF5	chloropentafluoroethane	76-15-3	1.4760	0.2745	10	10000	1,2	2.78	2.78	18.50
38	441	C2Cl2F4	1,2-dichloro-1,1,2,2-tetrafluoroethane	76-14-2	3.1670	0.2697	10	10000	2	5.89	5.89	37.96
39	442	C2Cl2F4	1,1-dichloro-1,2,2,2-tetrafluoroethane	374-07-2	4.0324	0.2503	10	10000	2	7.18	7.18	40.42
40	449	C2Cl3F3	1,1,2-trichloro-1,2,2-trifluoroethane	76-13-1	22.5386	0.1196	10	10000	1,2	29.68	29.68	67.82
41	450	C2Cl3F3	1,1,1-trichloro-2,2,2-trifluoroethane	354-58-5	12.5153	0.1749	10	10000	2	18.72	18.72	62.67
42	455	C2Cl4	tetrachloroethylene	127-18-4	34.7500	0.0940	10	10000	1,2	43.14	43.14	82.56
43	456	C2Cl4F2	1,1,2,2-tetrachloro-1,2-difluoroethane	76-12-0	29.3346	0.0966	10	10000	1,2	36.64	36.64	71.39
44	459	C2Cl4O	trichloroacetyl chloride	76-02-8	37.9003	0.0846	10	10000	2	46.05	46.05	82.60
45	511	C2HBrClF3	halothane	151-67-7	14.1193	0.1817	10	10000	2	21.46	21.46	75.28
46	539	C2HCl3	trichloroethylene	79-01-6	15.4014	0.1612	10	10000	1,2	22.32	22.32	67.97
47	540	C2HCl3F2	1,1,1-trichloro-2,2-difluoroethane	354-12-1	16.8804	0.1545	10	10000	2	24.09	24.09	70.07
48	541	C2HCl3F2	1,1,2-trichloro-1,2-difluoroethane	354-15-4	16.3958	0.1575	10	10000	2	23.56	23.56	69.93
49	542	C2HCl3F2	1,2,2-trichloro-1,1-difluoroethane	354-21-2	16.0155	0.1588	10	10000	2	23.08	23.08	69.12
50	543	C2HCl3O	dichloroacetyl chloride	79-36-7	26.2945	0.1168	10	10000	2	34.41	34.41	77.09
51	544	C2HCl3O	trichloroacetaldehyde	75-87-6	22.6281	0.1282	10	10000	2	30.40	30.40	73.71
52	549	C2HCl4F	1,1,1,2-tetrachloro-2-fluoroethane	354-11-0	39.3169	0.0820	10	10000	2	47.49	47.49	83.69
53	550	C2HCl4F	1,1,2,2-tetrachloro-1-fluoroethane	354-14-3	38.8422	0.0831	10	10000	2	47.03	47.03	83.50
54	551	C2HCl5	pentachloroethane	76-01-7	54.8483	0.0549	10	4828	2	62.23	62.23	87.34
55	556	C2HF3O2	trifluoroacetic acid	76-05-1	1.1190	0.4173	10	10000	2	2.93	2.93	52.25
56	571	C2H2	acetylene	74-86-2	0.0095	0.6401	10	10000	1,2	0.04	0.04	3.44
57	592	C2H2Br4	1,1,2,2-tetrabromoethane	79-27-6	136.2172	0.0307	1	26	2	136.22	146.20	150.55
58	593	C2H2Br4	1,1,1,2-tetrabromoethane	630-16-0	136.2172	0.0307	1	26	2	136.22	146.20	150.55
59	604	C2H2ClF3	1-chloro-1,1,2-trifluoroethane	421-04-5	0.6532	0.3954	10	10000	2	1.62	1.62	24.91
60	605	C2H2ClF3	1-chloro-1,2,2-trifluoroethane	431-07-2	0.6896	0.3923	10	10000	2	1.70	1.70	25.57
61	606	C2H2ClF3	2-chloro-1,1,1-trifluoroethane	75-88-7	0.7223	0.3897	10	10000	2	1.77	1.77	26.16
62	615	C2H2Cl2	1,1-dichloroethylene	75-35-4	4.2365	0.2368	10	10000	2	7.31	7.31	37.51
63	616	C2H2Cl2	cis-1,2-dichloroethylene	156-59-2	5.3173	0.2415	10	10000	2	9.27	9.27	49.15
64	617	C2H2Cl2	trans-1,2-dichloroethylene	156-60-5	4.3660	0.2517	10	10000	1,2	7.79	7.79	44.34
65	620	C2H2Cl2F2	1,1-dichloro-1,2-difluoroethane	1842-05-3	5.9952	0.2285	10	10000	2	10.15	10.15	49.17
66	621	C2H2Cl2F2	1,1-dichloro-2,2-difluoroethane	471-43-2	7.3942	0.2162	10	10000	2	12.17	12.17	54.18
67	622	C2H2Cl2F2	1,2-dichloro-1,1-difluoroethane	1649-08-7	6.3284	0.2254	10	10000	2	10.64	10.64	50.48
68	623	C2H2Cl2F2	1,2-dichloro-1,2-difluoroethane	431-06-1	6.2226	0.2264	10	10000	2	10.48	10.48	50.06
69	625	C2H2Cl2O	chloroacetyl chloride	79-04-9	14.7077	0.1734	10	10000	2	21.93	21.93	72.64
70	627	C2H2Cl2O2	dichloroacetic acid	79-43-6	55.5967	0.0682	10	234	2	65.04	65.04	80.64
71	630	C2H2Cl3F	1,1,1-trichloro-2-fluoroethane	2366-36-1	17.7489	0.1564	10	10000	2	25.44	25.44	74.94
72	631	C2H2Cl3F	1,1,2-trichloro-1-fluoroethane	811-95-0	20.2594	0.1311	10	10000	2	27.40	27.40	67.79
73	632	C2H2Cl3F	1,1,2-trichloro-2-fluoroethane	359-28-4	22.0084	0.1231	10	10000	2	29.22	29.22	68.37
74	635	C2H2Cl4	1,1,1,2-tetrachloroethane	630-20-6	37.8217	0.0809	10	10000	2	45.57	45.57	79.68
75	636	C2H2Cl4	1,1,2,2-tetrachloroethane	79-34-5	44.2164	0.0730	10	6073	1,2	52.31	52.31	83.54
76	688	C2H3Br	vinyl bromide	593-60-2	2.0306	0.3158	10	10000	2	4.20	4.20	37.21
77	697	C2H3Br3	1,1,1-tribromoethane	2311-14-0	99.1116	0.0432	10	563	2	109.47	109.47	130.29
78	698	C2H3Br3	1,1,2-tribromoethane	78-74-0	109.8610	0.0381	10	198	2	119.95	119.95	134.42
79	701	C2H3Cl	vinyl chloride	75-01-4	0.1946	0.4307	10	10000	1,2	0.52	0.52	10.28
80	707	C2H3ClO	acetyl chloride	75-36-5	1.5757	0.3353	10	10000	2	3.41	3.41	34.57

Table 1 Adsorption Capacity of Activated Carbon - Gas

$\log_{10} Q = \log_{10} K + 1/n \log_{10} y$ (Q - g of compound/100 g of carbon, y - ppmv)

NO	ID	FORMULA	NAME	CAS No	K	1/n	ymin	ymax	code	Q @ymin	Q @10ppm	Q @ymax
81	711	C2H3ClO2	methyl chloroformate	79-22-1	4.2102	0.2686	10	10000	2	7.81	7.81	49.96
82	714	C2H3Cl2F	1,1-dichloro-1-fluoroethane	1717-00-6	2.5470	0.2988	10	10000	2	5.07	5.07	39.91
83	715	C2H3Cl2F	1,1-dichloro-2-fluoroethane	430-53-5	4.4097	0.2516	10	10000	2	7.87	7.87	44.75
84	716	C2H3Cl2F	1,2-dichloro-1-fluoroethane	430-57-9	7.9014	0.2178	10	10000	2	13.05	13.05	58.75
85	720	C2H3Cl3	1,1,1-trichloroethane	71-55-6	13.1793	0.1635	10	10000	1,2	19.20	19.20	59.42
86	721	C2H3Cl3	1,1,2-trichloroethane	79-00-5	22.3051	0.1285	10	10000	1,2	29.99	29.99	72.88
87	728	C2H3F	vinyl fluoride	75-02-5	0.0047	0.7501	10	10000	2	0.03	0.03	4.74
88	743	C2H3I	vinyl iodide	593-66-8	13.9593	0.1941	10	10000	2	21.82	21.82	83.39
89	753	C2H3N	acetonitrile	75-05-8	0.2347	0.4938	10	10000	2	0.73	0.73	22.17
90	754	C2H3N	methyl isocyanide	593-75-9	0.0472	0.6573	10	10000	2	0.21	0.21	20.08
91	755	C2H3NO	methyl isocyanate	624-83-9	0.1780	0.5141	10	10000	2	0.58	0.58	20.27
92	787	C2H4	ethylene	74-85-1	0.0067	0.5375	10	10000	1,2	0.02	0.02	0.94
93	798	C2H4Br2	1,1-dibromoethane	557-91-5	36.2088	0.1137	10	10000	2	47.05	47.05	103.19
94	799	C2H4Br2	1,2-dibromoethane	106-93-4	43.2681	0.1041	10	10000	2	54.99	54.99	112.90
95	801	C2H4ClF	1-chloro-1-fluoroethane	1615-75-4	0.6128	0.3986	10	10000	2	1.53	1.53	24.09
96	802	C2H4ClF	1-chloro-2-fluoroethane	762-50-5	1.5734	0.3445	10	10000	2	3.48	3.48	37.56
97	812	C2H4Cl2	1,1-dichloroethane	75-34-3	5.2472	0.2337	10	10000	2	8.99	8.99	45.17
98	813	C2H4Cl2	1,2-dichloroethane	107-06-2	4.9266	0.2532	10	10000	1,2	8.83	8.83	50.73
99	816	C2H4Cl2O	bis(chloromethyl) ether	542-88-1	15.2511	0.1585	10	10000	2	21.97	21.97	65.65
100	844	C2H4I2	1,1-diiodoethane	594-02-5	106.7495	0.0423	10	973	2	117.67	117.67	142.82
101	884	C2H4O	acetaldehyde	75-07-0	0.0975	0.4930	10	10000	1,2	0.30	0.30	9.14
102	885	C2H4O	ethylene oxide	75-21-8	0.0069	0.7278	10	10000	2	0.04	0.04	5.61
103	888	C2H4OS	thioacetic-acid	507-09-5	7.5117	0.2085	10	10000	2	12.14	12.14	51.28
104	889	C2H4O2	acetic acid	64-19-7	2.1642	0.3539	10	10000	2	4.89	4.89	56.34
105	890	C2H4O2	methyl formate	107-31-3	0.1328	0.5165	10	10000	2	0.44	0.44	15.45
106	901	C2H4S	ethylene sulfide	420-12-2	1.7520	0.3200	10	10000	2	3.66	3.66	33.39
107	910	C2H5Br	bromoethane	74-96-4	3.2779	0.2684	10	10000	1,2	6.08	6.08	38.83
108	914	C2H5Cl	ethyl chloride	75-00-3	0.4285	0.3851	10	10000	1,2	1.04	1.04	14.88
109	917	C2H5ClO	2-chloroethanol	107-07-3	12.4563	0.1829	10	9446	2	18.98	18.98	66.47
110	948	C2H5I	ethyl iodide	75-03-6	16.2093	0.1784	10	10000	2	24.44	24.44	83.84
111	954	C2H5N	ethyleneimine	151-56-4	0.1855	0.5214	10	10000	2	0.62	0.62	22.59
112	957	C2H5NO	N-methylformamide	123-39-7	23.9945	0.1374	10	332	2	32.93	32.93	53.28
113	958	C2H5NO2	nitroethane	79-24-3	5.5793	0.2462	10	10000	2	9.84	9.84	53.88
114	985	C2H6	ethane	74-84-0	0.0052	0.5764	10	10000	1,2	0.02	0.02	1.06
115	1065	C2H6O	ethyl alcohol	64-17-5	0.5976	0.4319	10	10000	1,2	1.62	1.62	31.93
116	1067	C2H6OS	dimethyl sulfoxide	67-68-5	28.6060	0.1057	10	808	2	36.49	36.49	58.04
117	1072	C2H6O2	ethylene glycol	107-21-1	28.3262	0.1133	10	121	2	36.77	36.77	48.77
118	1083	C2H6O4S	dimethyl sulfate	77-78-1	28.4341	0.1129	10	890	2	36.87	36.87	61.20
119	1092	C2H6S	ethyl mercaptan	75-08-1	1.3229	0.3070	10	10000	1,2	2.68	2.68	22.37
120	1093	C2H6S	dimethyl sulfide	75-18-3	4.6030	0.2229	10	10000	2	7.69	7.69	35.87
121	1094	C2H6S2	dimethyl disulfide	624-92-0	8.8896	0.1987	10	10000	2	14.05	14.05	55.40
122	1115	C2H7N	ethylamine	75-04-7	0.1416	0.4838	10	10000	2	0.43	0.43	12.19
123	1116	C2H7N	dimethylamine	124-40-3	0.0967	0.4620	10	10000	1,2	0.28	0.28	6.81
124	1117	C2H7NO	monoethanolamine	141-43-5	23.4661	0.1355	10	485	2	32.06	32.06	54.24
125	1159	C2H8N2	ethylenediamine	107-15-3	8.1021	0.1934	10	10000	2	12.65	12.65	48.12
126	1284	C3F6	hexafluoropropylene	116-15-4	16.1811	0.1176	10	10000	2	21.21	21.21	47.79
127	1402	C3H3Cl	propargyl chloride	624-65-7	2.8758	0.2773	10	10000	2	5.45	5.45	36.97
128	1462	C3H3N	acrylonitrile	107-13-1	2.0045	0.3095	10	10000	1,2	4.09	4.09	34.67
129	1463	C3H3NO	oxazole	288-42-6	6.4549	0.1724	10	10000	2	9.60	9.60	31.59
130	1492	C3H4	methylacetylene	74-99-7	0.0731	0.5704	10	10000	1,2	0.27	0.27	13.98
131	1493	C3H4	allene	463-49-0	0.1872	0.4064	10	10000	2	0.48	0.48	7.90
132	1519	C3H4Br4	1,2,2,3-tetrabromopropane	54268-02-9	133.4144	0.0259	1	3.9	2	133.41	---	138.20
133	1533	C3H4Cl2	2,3-dichloropropene	78-88-6	14.0969	0.1517	10	10000	2	19.99	19.99	56.99
134	1565	C3H4Cl4	1,1,1,2-tetrachloropropane	812-03-3	48.7170	0.0540	10	4108	2	55.17	55.17	76.34
135	1566	C3H4Cl4	1,1,1,3-tetrachloropropane	1070-78-6	51.3305	0.0536	10	2529	2	58.07	58.07	78.13
136	1567	C3H4Cl4	1,1,2,2-tetrachloropropane	13116-60-4	50.2798	0.0546	10	3237	2	57.01	57.01	78.15
137	1568	C3H4Cl4	1,1,2,3-tetrachloropropane	18495-30-2	57.1400	0.0451	10	928	2	63.39	63.39	77.77
138	1569	C3H4Cl4	1,1,3,3-tetrachloropropane	1653-17-4	54.1353	0.0493	10	1611	2	60.64	60.64	77.89
139	1570	C3H4Cl4	1,2,2,3-tetrachloropropane	13116-53-5	52.9174	0.0498	10	1985	2	59.34	59.34	77.21
140	1578	C3H4F4	1,1,1,2-tetrafluoropropane	421-48-7	0.3514	0.4402	10	10000	2	0.97	0.97	20.26
141	1579	C3H4F4	1,1,1,3-tetrafluoropropane	460-36-6	0.8334	0.3881	10	10000	2	2.04	2.04	29.73
142	1580	C3H4F4	1,1,2,2-tetrafluoropropane	40723-63-5	0.7788	0.3919	10	10000	2	1.92	1.92	28.77
143	1581	C3H4F4	1,1,2,3-tetrafluoropropane	------	0.7796	0.3918	10	10000	2	1.92	1.92	28.79
144	1582	C3H4F4	1,1,3,3-tetrafluoropropane	------	0.7796	0.3918	10	10000	2	1.92	1.92	28.79
145	1583	C3H4F4	1,2,2,3-tetrafluoropropane	813-75-2	0.7791	0.3919	10	10000	2	1.92	1.92	28.78
146	1621	C3H4O	acrolein	107-02-8	0.7290	0.3600	10	10000	2	1.67	1.67	20.07
147	1622	C3H4O	propargyl alcohol	107-19-7	4.3617	0.2662	10	10000	2	8.05	8.05	50.64
148	1626	C3H4O2	acrylic acid	79-10-7	12.6304	0.1780	10	5164	2	19.03	19.03	57.86
149	1627	C3H4O2	beta-propiolactone	57-57-8	12.8524	0.2215	10	2243	2	21.40	21.40	70.99
150	1628	C3H4O2	vinyl formate	692-45-5	1.1347	0.3504	10	10000	2	2.54	2.54	28.60
151	1634	C3H4O3	pyruvic acid	127-17-3	23.2971	0.1453	10	1678	2	32.55	32.55	68.51
152	1642	C3H5Br	cis-1-bromo-1-propene	590-13-6	9.0370	0.2028	10	10000	2	14.42	14.42	58.51
153	1643	C3H5Br	trans-1-bromo-1-propene	590-15-8	10.0046	0.1945	10	10000	2	15.66	15.66	59.98
154	1644	C3H5Br	2-bromo-1-propene	557-93-7	7.6940	0.2109	10	10000	2	12.50	12.50	53.65
155	1645	C3H5Br	3-bromo-1-propene	106-95-6	10.9883	0.1845	10	10000	2	16.80	16.80	60.10
156	1663	C3H5Br3	1,1,1-tribromopropane	62127-61-1	101.8943	0.0376	10	50	2	111.10	111.10	118.02
157	1664	C3H5Br3	1,1,2-tribromopropane	14602-62-1	102.5127	0.0349	10	113	2	111.10	111.10	120.91
158	1665	C3H5Br3	1,1,3-tribromopropane	23511-78-6	103.8039	0.0349	10	82	2	112.49	112.49	121.07
159	1666	C3H5Br3	1,2,2-tribromopropane	14476-30-3	108.9922	0.0432	1	10	2	108.99	120.40	120.40
160	1667	C3H5Br3	1,2,3-tribromopropane	96-11-7	110.7880	0.0333	1	30	2	110.79	119.62	124.07

Table 1 Adsorption Capacity of Activated Carbon - Gas

$\log_{10} Q = \log_{10} K + 1/n \log_{10} y$ (Q - g of compound/100 g of carbon, y - ppmv)

NO	ID	FORMULA	NAME	CAS No	K	1/n	ymin	ymax	code	Q @ymin	Q @10ppm	Q @ymax
161	1668	C3H5Cl	cis-1-chloro-1-propene	16136-84-8	2.3475	0.2730	10	10000	2	4.40	4.40	29.02
162	1669	C3H5Cl	trans-1-chloro-1-propene	16136-85-9	2.5287	0.2693	10	10000	2	4.70	4.70	30.21
163	1670	C3H5Cl	2-chloro-1-propene	557-98-2	2.1069	0.2734	10	10000	2	3.95	3.95	26.13
164	1671	C3H5Cl	3-chloro-1-propene	107-05-1	2.8003	0.2647	10	10000	1,2	5.15	5.15	32.07
165	1676	C3H5ClO	alpha-epichlorohydrin	106-89-8	12.1662	0.1760	10	10000	1,2	18.24	18.24	61.51
166	1688	C3H5ClO2	methyl chloroacetate	96-34-4	20.4959	0.1278	10	9959	2	27.51	27.51	66.45
167	1689	C3H5ClO2	ethyl chloroformate	541-41-3	14.5785	0.1489	10	10000	2	20.54	20.54	57.45
168	1703	C3H5Cl3	1,1,1-trichloropropane	7789-89-1	27.3701	0.0937	10	10000	2	33.96	33.96	64.85
169	1704	C3H5Cl3	1,1,2-trichloropropane	598-77-6	36.3959	0.0751	10	9273	2	43.27	43.27	72.28
170	1705	C3H5Cl3	1,1,3-trichloropropane	20395-25-9	39.2441	0.0704	10	4646	2	46.15	46.15	71.10
171	1706	C3H5Cl3	1,2,2-trichloropropane	3175-23-3	32.4370	0.0813	10	10000	2	39.12	39.12	68.61
172	1707	C3H5Cl3	1,2,3-trichloropropane	96-18-4	39.8360	0.0709	10	4842	2	46.90	46.90	72.68
173	1711	C3H5F	cis-1-fluoro-1-propene	19184-10-2	0.5229	0.3627	10	10000	2	1.21	1.21	14.76
174	1712	C3H5F	trans-1-fluoro-1-propene	20327-65-5	0.5229	0.3627	10	10000	2	1.21	1.21	14.76
175	1713	C3H5F	2-fluoro-1-propene	1184-60-7	0.3200	0.3893	10	10000	2	0.78	0.78	11.54
176	1714	C3H5F	3-fluoro-1-propene	818-92-8	0.3230	0.3888	10	10000	2	0.79	0.79	11.59
177	1721	C3H5F3	1,1,1-trifluoropropane	421-07-8	0.1255	0.5111	10	10000	2	0.41	0.41	13.89
178	1722	C3H5F3	1,1,2-trifluoropropane	------	0.1892	0.4743	10	10000	2	0.56	0.56	14.93
179	1723	C3H5F3	1,1,3-trifluoropropane	------	0.1897	0.4741	10	10000	2	0.57	0.57	14.94
180	1724	C3H5F3	1,2,2-trifluoropropane	811-94-9	0.1764	0.4793	10	10000	2	0.53	0.53	14.59
181	1725	C3H5F3	1,2,3-trifluoropropane	------	0.1897	0.4741	10	10000	2	0.57	0.57	14.94
182	1734	C3H5I	cis-1-iodo-1-propene	7796-36-3	29.9474	0.1159	10	10000	2	39.11	39.11	87.08
183	1735	C3H5I	trans-1-iodo-1-propene	7796-54-5	29.9474	0.1159	10	10000	2	39.11	39.11	87.08
184	1736	C3H5I	2-iodo-1-propene	4375-96-6	32.1589	0.1105	10	10000	2	41.48	41.48	88.99
185	1737	C3H5I	3-iodo-1-propene	556-56-9	32.6892	0.1108	10	10000	2	42.19	42.19	90.69
186	1752	C3H5N	propionitrile	107-12-0	2.1882	0.3026	10	10000	2	4.39	4.39	35.53
187	1753	C3H5N	ethyl isocyanide	624-79-3	1.9726	0.3002	10	10000	2	3.94	3.94	31.34
188	1756	C3H5NO	hydracrylonitrile	109-78-4	31.6055	0.1200	10	105	2	41.66	41.66	55.24
189	1757	C3H5NO	lactonitrile	78-97-7	32.3304	0.0910	10	157	2	39.87	39.87	51.22
190	1809	C3H6	cyclopropane	75-19-4	0.0519	0.5337	10	10000	2	0.18	0.18	7.08
191	1810	C3H6	propylene	115-07-1	0.2048	0.3682	10	10000	1,2	0.48	0.48	6.08
192	1824	C3H6Br2	1,1-dibromopropane	598-17-4	50.7008	0.0779	10	10000	2	60.66	60.66	103.91
193	1825	C3H6Br2	1,2-dibromopropane	78-75-1	51.7245	0.0726	10	10000	2	61.13	61.13	100.93
194	1826	C3H6Br2	1,3-dibromopropane	109-64-8	70.2582	0.0550	10	959	2	79.73	79.73	102.46
195	1827	C3H6Br2	2,2-dibromopropane	594-16-1	47.0475	0.0754	10	10000	2	55.97	55.97	94.26
196	1844	C3H6Cl2	1,1-dichloropropane	78-99-9	13.5531	0.1482	10	10000	2	19.07	19.07	53.09
197	1845	C3H6Cl2	1,2-dichloropropane	78-87-5	14.2636	0.1472	10	10000	1,2	20.02	20.02	55.32
198	1846	C3H6Cl2	1,3-dichloropropane	142-28-9	19.8675	0.1214	10	10000	2	26.27	26.27	60.76
199	1847	C3H6Cl2	2,2-dichloropropane	594-20-7	10.4129	0.1670	10	10000	2	15.30	15.30	48.49
200	1865	C3H6F2	1,1-difluoropropane	430-61-5	0.3010	0.4340	10	10000	2	0.82	0.82	16.40
201	1866	C3H6F2	1,2-difluoropropane	62126-90-3	0.4498	0.4066	10	10000	2	1.15	1.15	19.02
202	1867	C3H6F2	1,3-difluoropropane	462-39-5	0.8646	0.3753	10	10000	2	2.05	2.05	27.40
203	1868	C3H6F2	2,2-difluoropropane	420-45-1	0.2285	0.4513	10	10000	2	0.65	0.65	14.59
204	1871	C3H6I2	1,1-diiodopropane	10250-52-9	100.2014	0.0357	10	918	2	108.79	108.79	127.84
205	1872	C3H6I2	1,2-diiodopropane	598-29-8	105.3931	0.0356	10	610	2	114.39	114.39	132.39
206	1873	C3H6I2	1,3-diiodopropane	627-31-6	113.3690	0.0331	10	102	2	122.35	122.35	132.13
207	1874	C3H6I2	2,2-diiodopropane	630-13-7	97.1284	0.0377	10	1510	2	105.94	105.94	128.02
208	1920	C3H6O	allyl alcohol	107-18-6	4.4625	0.2450	10	10000	2	7.84	7.84	42.60
209	1921	C3H6O	methyl vinyl ether	107-25-5	0.8478	0.3223	10	10000	2	1.78	1.78	16.51
210	1922	C3H6O	propionaldehyde	123-38-6	2.1577	0.2618	10	10000	1,2	3.94	3.94	24.06
211	1923	C3H6O	acetone	67-64-1	1.0039	0.3506	10	10000	1,2	2.25	2.25	25.36
212	1924	C3H6O	1,2-propylene oxide	75-56-9	0.5613	0.3886	10	10000	2	1.37	1.37	20.12
213	1925	C3H6O	1,3-propylene oxide	503-30-0	0.6960	0.3886	10	10000	2	1.70	1.70	24.95
214	1930	C3H6O2	propanoic acid	79-09-4	12.4291	0.1741	10	4872	2	18.56	18.56	54.53
215	1931	C3H6O2	ethyl formate	109-94-4	1.8856	0.3051	10	10000	2	3.81	3.81	31.31
216	1932	C3H6O2	methyl acetate	79-20-9	1.8793	0.3095	10	10000	2	3.83	3.83	32.50
217	1942	C3H6O2S	3-mercaptopropionic acid	107-96-0	48.7584	0.0610	10	65	2	56.11	56.11	62.89
218	1948	C3H6O3	lactic acid	50-21-5	39.8760	0.0989	10	106	2	50.08	50.08	63.25
219	1949	C3H6O3	methoxyacetic acid	625-45-6	41.1862	0.0935	10	71	2	51.08	51.08	61.35
220	1967	C3H6S	thiacyclobutane	287-27-4	8.5222	0.1878	10	10000	2	13.13	13.13	48.08
221	1968	C3H6S	2-methylthiacyclopropane	1072-43-1	6.2515	0.2035	10	10000	2	9.99	9.99	40.74
222	1978	C3H7Br	1-bromopropane	106-94-5	10.4870	0.1861	10	10000	2	16.10	16.10	58.21
223	1979	C3H7Br	2-bromopropane	75-26-3	9.4737	0.1870	10	10000	2	14.57	14.57	53.04
224	1986	C3H7Cl	1-chloropropane	540-54-5	3.5301	0.2364	10	10000	2	6.08	6.08	31.14
225	1987	C3H7Cl	2-chloropropane	75-29-6	2.7445	0.2531	10	10000	2	4.92	4.92	28.24
226	2020	C3H7F	1-fluoropropane	460-13-9	0.1087	0.5066	10	10000	2	0.35	0.35	11.55
227	2021	C3H7F	2-fluoropropane	420-26-8	0.1451	0.4674	10	10000	2	0.43	0.43	10.75
228	2025	C3H7I	1-iodopropane	107-08-4	29.9035	0.1138	10	10000	2	38.86	38.86	85.30
229	2026	C3H7I	2-iodopropane	75-30-9	26.7657	0.1198	10	10000	2	35.27	35.27	80.67
230	2030	C3H7N	allylamine	107-11-9	2.2033	0.2732	10	10000	2	4.13	4.13	27.29
231	2031	C3H7N	propyleneimine	75-55-8	1.8234	0.3003	10	10000	2	3.64	3.64	28.99
232	2036	C3H7NO	N,N-dimethylformamide	68-12-2	14.8415	0.1455	10	5220	1,2	20.75	20.75	51.57
233	2045	C3H7NO2	1-nitropropane	108-03-2	14.2364	0.1436	10	10000	2	19.81	19.81	53.43
234	2046	C3H7NO2	2-nitropropane	79-46-9	11.6077	0.1611	10	10000	2	16.82	16.82	51.17
235	2094	C3H8	propane	74-98-6	0.2290	0.3599	10	10000	1,2	0.52	0.52	6.30
236	2129	C3H8O	propyl alcohol	71-23-8	5.0369	0.2288	10	10000	2	8.53	8.53	41.44
237	2130	C3H8O	isopropyl alcohol	67-63-0	3.2275	0.2639	10	10000	2	5.93	5.93	36.69
238	2131	C3H8O	methyl ethyl ether	540-67-0	0.7375	0.3291	10	10000	2	1.57	1.57	15.28
239	2138	C3H8O2	2-methoxyethanol	109-86-4	10.8492	0.1712	10	10000	1,2	16.09	16.09	52.50
240	2139	C3H8O2	methylal	109-87-5	2.0536	0.2828	10	10000	2	3.94	3.94	27.77

Table 1 Adsorption Capacity of Activated Carbon - Gas

$\log_{10} Q = \log_{10} K + 1/n \log_{10} y$ (Q - g of compound/100 g of carbon, y - ppmv)

NO	ID	FORMULA	NAME	CAS No	K	1/n	ymin	ymax	code	Q @ymin	Q @10ppm	Q @ymax
241	2140	C3H8O2	1,2-propanediol (propylene glycol)	57-55-6	30.3914	0.1159	10	170	1,2	39.69	39.69	55.12
242	2141	C3H8O2	1,3-propanediol	504-63-2	37.9166	0.0920	10	57	2	46.86	46.86	55.00
243	2158	C3H8S	propyl mercaptan	107-03-9	5.3863	0.1949	10	10000	1,2	8.44	8.44	32.44
244	2159	C3H8S	isopropyl mercaptan	75-33-2	5.2235	0.1974	10	10000	2	8.23	8.23	32.17
245	2160	C3H8S	methyl ethyl sulfide	624-89-5	6.4337	0.1855	10	10000	2	9.86	9.86	35.52
246	2200	C3H9N	propylamine	107-10-8	1.3727	0.2817	10	10000	1,2	2.63	2.63	18.38
247	2201	C3H9N	isopropylamine	75-31-0	1.4982	0.2858	10	10000	2	2.89	2.89	20.83
248	2202	C3H9N	methylethylamine	624-78-2	1.6535	0.2793	10	10000	2	3.15	3.15	21.67
249	2203	C3H9N	trimethylamine	75-50-3	0.8462	0.3075	10	10000	2	1.72	1.72	14.37
250	2204	C3H9NO	1-amino-2-propanol	78-96-6	27.4217	0.0948	10	617	2	34.11	34.11	50.43
251	2205	C3H9NO	3-amino-1-propanol	156-87-6	34.0775	0.0872	10	101	2	41.66	41.66	50.96
252	2206	C3H9NO	methylethanolamine	109-83-1	22.7094	0.1079	10	1421	2	29.11	29.11	49.69
253	2241	C3H9O4P	trimethyl phosphate	512-56-1	39.9536	0.0655	10	984	2	46.45	46.45	62.73
254	2255	C3H10N2	1,2-propanediamine	78-90-0	13.5941	0.1308	10	10000	2	18.37	18.37	45.36
255	2318	C4Cl6	hexachloro-1,3-butadiene	87-68-3	70.2139	0.0229	10	291	2	74.02	74.02	79.97
256	2336	C4F8	octafluorocyclobutane	115-25-3	1.3233	0.3363	10	10000	2	2.87	2.87	29.29
257	2374	C4H2	biacetylene	460-12-8	0.6460	0.3440	10	10000	2	1.43	1.43	15.36
258	2510	C4H4	vinylacetylene	689-97-4	0.9277	0.3084	10	10000	2	1.89	1.89	15.89
259	2592	C4H4O	furan	110-00-9	1.5194	0.3102	10	10000	2	3.10	3.10	26.45
260	2598	C4H4O2	diketene	674-82-8	13.9427	0.1514	10	10000	2	19.76	19.76	56.23
261	2613	C4H4S	thiophene	110-02-1	9.5144	0.1773	10	10000	1,2	14.31	14.31	48.71
262	2634	C4H5Cl	chloroprene	126-99-8	7.9529	0.1749	10	10000	2	11.90	11.90	39.83
263	2695	C4H5N	trans-crotonitrile	627-26-9	9.2773	0.1640	10	10000	2	13.53	13.53	42.01
264	2696	C4H5N	cis-crotonitrile	1190-76-7	6.8858	0.1927	10	10000	2	10.73	10.73	40.60
265	2697	C4H5N	methacrylonitrile	126-98-7	4.8914	0.2169	10	10000	2	8.06	8.06	36.05
266	2698	C4H5N	vinylacetonitrile	109-75-1	7.8887	0.1843	10	10000	2	12.06	12.06	43.07
267	2701	C4H5N	pyrrole	109-97-7	13.4874	0.1473	10	10000	2	18.94	18.94	52.39
268	2713	C4H5NO2	methyl cyanoacetate	105-34-0	40.8280	0.0676	10	180	2	47.71	47.71	58.00
269	2754	C4H6	dimethylacetylene	503-17-3	1.1229	0.3067	10	10000	2	2.28	2.28	18.93
270	2755	C4H6	ethylacetylene	107-00-6	1.0935	0.2789	10	10000	1,2	2.08	2.08	14.27
271	2756	C4H6	1,2-butadiene	590-19-2	1.5586	0.2649	10	10000	2	2.87	2.87	17.88
272	2757	C4H6	butadiene (1,3 butadiene)	106-99-0	1.1218	0.2771	10	10000	1,2	2.12	2.12	14.40
273	2806	C4H6Cl2	1,3-dichloro-trans-2-butene	7415-31-8	27.7975	0.0836	10	10000	2	33.70	33.70	60.04
274	2807	C4H6Cl2	1,4-dichloro-cis-2-butene	1476-11-5	33.7256	0.0717	10	5382	2	39.78	39.78	62.42
275	2808	C4H6Cl2	1,4-dichloro-trans-2-butene	110-57-6	34.4269	0.0705	10	4502	2	40.50	40.50	62.32
276	2809	C4H6Cl2	3,4-dichloro-1-butene	760-23-6	24.1169	0.0958	10	10000	2	30.07	30.07	58.30
277	2940	C4H6O	trans-crotonaldehyde	123-73-9	8.4921	0.1741	10	10000	2	12.68	12.68	42.20
278	2941	C4H6O	methacrolein	78-85-3	4.2565	0.2273	10	10000	2	7.18	7.18	34.53
279	2942	C4H6O	2,5-dihydrofuran	1708-29-8	3.5163	0.2559	10	10000	2	6.34	6.34	37.11
280	2944	C4H6O	divinyl ether	109-93-3	3.1668	0.2237	10	10000	2	5.30	5.30	24.87
281	2965	C4H6O2	gamma-butyrolactone	96-48-0	31.0983	0.1015	10	592	2	39.28	39.28	59.43
282	2966	C4H6O2	cis-2-butenoic acid	503-64-0	29.8798	0.0885	10	773	2	36.63	36.63	53.83
283	2967	C4H6O2	trans-2-butenoic acid	107-93-7	29.1568	0.0946	10	500	2	36.26	36.26	52.50
284	2968	C4H6O2	vinylacetic acid	625-38-7	28.0388	0.0950	10	828	2	34.90	34.90	53.09
285	2969	C4H6O2	methacrylic acid	79-41-4	26.2306	0.0999	10	1282	2	33.02	33.02	53.63
286	2970	C4H6O2	methyl acrylate	96-33-3	3.8682	0.2122	10	10000	1,2	6.31	6.31	27.31
287	2971	C4H6O2	vinyl acetate	108-05-4	5.9941	0.2068	10	10000	1,2	9.65	9.65	40.27
288	2990	C4H6O3	acetic anhydride	108-24-7	21.7595	0.1157	10	4608	2	28.40	28.40	57.75
289	3047	C4H7Br	cis-1-bromo-1-butene	31849-78-2	19.9989	0.1253	10	10000	2	26.69	26.69	63.42
290	3048	C4H7Br	trans-1-bromo-1-butene	32620-08-9	21.9424	0.1180	10	10000	2	28.79	28.79	65.04
291	3049	C4H7Br	2-bromo-1-butene	23074-36-4	18.7351	0.1301	10	10000	2	25.28	25.28	62.08
292	3050	C4H7Br	3-bromo-1-butene	22037-73-6	25.3971	0.1058	10	10000	2	32.40	32.40	67.28
293	3051	C4H7Br	4-bromo-1-butene	5162-44-7	23.3911	0.1125	10	10000	2	30.31	30.31	65.95
294	3052	C4H7Br	1-bromo-cis-2-butene	39616-19-8	25.8974	0.1035	10	10000	2	32.87	32.87	67.19
295	3053	C4H7Br	1-bromo-trans-2-butene	29576-14-5	25.8974	0.1035	10	10000	2	32.87	32.87	67.19
296	3054	C4H7Br	2-bromo-cis-2-butene	3017-68-3	19.8968	0.1260	10	10000	2	26.59	26.59	63.50
297	3055	C4H7Br	2-bromo-trans-2-butene	3017-71-8	21.8603	0.1190	10	10000	2	28.75	28.75	65.41
298	3056	C4H7Br	1-bromo-2-methyl-1-propene	3017-69-4	21.4438	0.1193	10	10000	2	28.22	28.22	64.32
299	3057	C4H7Br	3-bromo-2-methyl-1-propene	1458-98-6	23.8463	0.1084	10	10000	2	30.61	30.61	64.74
300	3083	C4H7Br3	1,1,1-tribromobutane	62127-62-2	99.1718	0.0333	1	25	2	99.17	107.08	110.40
301	3084	C4H7Br3	1,1,2-tribromobutane	3675-68-1	99.5122	0.0313	1	49	2	99.51	106.95	112.40
302	3085	C4H7Br3	1,1,3-tribromobutane	------	101.5746	0.0249	10	51	2	107.56	107.56	112.01
303	3086	C4H7Br3	1,1,4-tribromobutane	------	101.5659	0.0249	10	50	2	107.56	107.56	111.97
304	3087	C4H7Br3	1,2,2-tribromobutane	3675-69-2	100.6628	0.0304	1	39	2	100.66	107.96	112.51
305	3088	C4H7Br3	1,2,3-tribromobutane	632-05-3	100.6628	0.0304	1	39	2	100.66	107.96	112.51
306	3089	C4H7Br3	1,2,4-tribromobutane	38300-67-3	99.4929	0.0313	1	47	2	99.49	106.94	112.25
307	3090	C4H7Br3	1,3,3-tribromobutane	62127-46-2	96.7651	0.0310	10	105	2	103.93	103.93	111.79
308	3091	C4H7Br3	2,3,3-tribromobutane	62127-47-3	96.7651	0.0310	10	105	2	103.93	103.93	111.79
309	3092	C4H7Br3	1,1,1-tribromo-2-methylpropane	------	100.0267	0.0290	10	51	2	106.92	106.92	112.09
310	3093	C4H7Br3	1,1,2-tribromo-2-methylpropane	15331-16-5	100.0267	0.0290	10	51	2	106.92	106.92	112.09
311	3094	C4H7Br3	1,1,3-tribromo-2-methylpropane	------	100.0267	0.0290	10	51	2	106.92	106.92	112.09
312	3095	C4H7Br3	1,2,3-tribromo-2-methylpropane	631-28-7	100.0267	0.0290	10	51	2	106.92	106.92	112.09
313	3096	C4H7Br3	1,3-dibromo-2-(bromomethyl)propane	62127-48-4	98.8770	0.0302	1	35	2	98.88	106.00	110.10
314	3098	C4H7Cl	cis-1-chloro-1-butene	7611-86-1	7.7760	0.1752	10	10000	2	11.64	11.64	39.06
315	3099	C4H7Cl	trans-1-chloro-1-butene	7611-87-2	8.2885	0.1713	10	10000	2	12.30	12.30	40.14
316	3100	C4H7Cl	2-chloro-1-butene	2211-70-3	7.3583	0.1781	10	10000	2	11.09	11.09	37.95
317	3101	C4H7Cl	3-chloro-1-butene	563-52-0	8.0940	0.1702	10	10000	2	11.98	11.98	38.82
318	3102	C4H7Cl	4-chloro-1-butene	927-73-1	9.3399	0.1611	10	10000	2	13.54	13.54	41.20
319	3103	C4H7Cl	1-chloro-cis-2-butene	4628-21-1	11.0101	0.1512	10	10000	2	15.59	15.59	44.31
320	3104	C4H7Cl	1-chloro-trans-2-butene	4894-61-5	11.1385	0.1488	10	10000	2	15.69	15.69	43.85

Table 1 Adsorption Capacity of Activated Carbon - Gas

$\log_{10} Q = \log_{10} K + 1/n \log_{10} y$ (Q - g of compound/100 g of carbon, y - ppmv)

NO	ID	FORMULA	NAME	CAS No	K	1/n	ymin	ymax	code	Q @ymin	Q @10ppm	Q @ymax
321	3105	C4H7Cl	2-chloro-cis-2-butene	2211-69-0	8.3758	0.1701	10	10000	2	12.39	12.39	40.11
322	3106	C4H7Cl	2-chloro-trans-2-butene	2211-68-9	7.8921	0.1737	10	10000	2	11.77	11.77	39.07
323	3107	C4H7Cl	1-chloro-2-methyl-1-propene	513-37-1	8.5396	0.1686	10	10000	2	12.59	12.59	40.34
324	3108	C4H7Cl	3-chloro-2-methyl-1-propene	563-47-3	9.1183	0.1643	10	10000	2	13.31	13.31	41.42
325	3156	C4H7Cl3	1,1,1-trichlorobutane	13279-85-1	37.6843	0.0589	10	8884	2	43.16	43.16	64.37
326	3157	C4H7Cl3	1,1,2-trichlorobutane	------	40.7481	0.0548	10	4585	2	46.22	46.22	64.65
327	3158	C4H7Cl3	1,1,3-trichlorobutane	13279-87-3	41.2286	0.0549	10	4585	2	46.78	46.78	65.49
328	3159	C4H7Cl3	1,1,4-trichlorobutane	------	41.9084	0.0556	10	4348	2	47.63	47.63	66.75
329	3160	C4H7Cl3	1,2,2-trichlorobutane	------	45.5194	0.0503	10	1876	2	51.11	51.11	66.51
330	3161	C4H7Cl3	1,2,3-trichlorobutane	18338-40-4	46.6523	0.0508	10	1824	2	52.44	52.44	68.32
331	3162	C4H7Cl3	1,2,4-trichlorobutane	1790-22-3	46.7174	0.0475	10	1645	2	52.12	52.12	66.42
332	3163	C4H7Cl3	1,3,3-trichlorobutane	------	41.5812	0.0558	10	4865	2	47.28	47.28	66.77
333	3164	C4H7Cl3	2,3,3-trichlorobutane	10403-60-8	39.9481	0.0565	10	6123	2	45.50	45.50	65.41
334	3165	C4H7Cl3	1,1,1-trichloro-2-methylpropane	------	39.8465	0.0572	10	6459	2	45.45	45.45	65.80
335	3166	C4H7Cl3	1,1,2-trichloro-2-methylpropane	29559-52-2	40.4380	0.0570	10	5709	2	46.10	46.10	66.18
336	3167	C4H7Cl3	1,1,3-trichloro-2-methylpropane	62108-65-0	40.4307	0.0572	10	5816	2	46.12	46.12	66.37
337	3168	C4H7Cl3	1,2,3-trichloro-2-methylpropane	1871-58-5	45.5700	0.0515	10	2151	2	51.31	51.31	67.66
338	3169	C4H7Cl3	1,3-dichloro-2-(chloromethyl)-propane	------	38.7778	0.0611	10	9889	2	44.64	44.64	68.05
339	3177	C4H7F	cis-1-fluoro-1-butene	66675-34-1	2.7885	0.2391	10	10000	2	4.84	4.84	25.22
340	3178	C4H7F	trans-1-fluoro-1-butene	66675-35-2	2.7885	0.2391	10	10000	2	4.84	4.84	25.22
341	3179	C4H7F	2-fluoro-1-butene	430-44-4	1.5465	0.2922	10	10000	2	3.03	3.03	22.81
342	3180	C4H7F	3-fluoro-1-butene	------	1.6097	0.2812	10	10000	2	3.08	3.08	21.46
343	3181	C4H7F	4-fluoro-1-butene	------	1.6175	0.2810	10	10000	2	3.09	3.09	21.52
344	3182	C4H7F	cis-1-fluoro-2-butene	------	1.8130	0.2761	10	10000	2	3.42	3.42	23.05
345	3183	C4H7F	trans-1-fluoro-2-butene	------	1.8130	0.2761	10	10000	2	3.42	3.42	23.05
346	3184	C4H7F	cis-2-fluoro-2-butene	66675-38-5	2.7563	0.2430	10	10000	2	4.82	4.82	25.85
347	3185	C4H7F	trans-2-fluoro-2-butene	66675-39-6	2.7563	0.2430	10	10000	2	4.82	4.82	25.85
348	3186	C4H7F	1-fluoro-2-methyl-1-propene	------	3.2188	0.2320	10	10000	2	5.49	5.49	27.27
349	3187	C4H7F	3-fluoro-2-methyl-1-propene	------	3.2194	0.2320	10	10000	2	5.49	5.49	27.27
350	3192	C4H7F3	1,1,1-trifluorobutane	460-34-4	2.0646	0.2813	10	10000	2	3.95	3.95	27.54
351	3193	C4H7F3	1,1,2-trifluorobutane	------	1.9942	0.2827	10	10000	2	3.82	3.82	26.95
352	3194	C4H7F3	1,1,3-trifluorobutane	------	1.9942	0.2827	10	10000	2	3.82	3.82	26.95
353	3195	C4H7F3	1,1,4-trifluorobutane	------	2.0045	0.2825	10	10000	2	3.84	3.84	27.04
354	3196	C4H7F3	1,2,2-trifluorobutane	------	1.9279	0.2841	10	10000	2	3.71	3.71	26.40
355	3197	C4H7F3	1,2,3-trifluorobutane	------	1.9942	0.2827	10	10000	2	3.82	3.82	26.95
356	3198	C4H7F3	1,2,4-trifluorobutane	------	2.0045	0.2825	10	10000	2	3.84	3.84	27.04
357	3199	C4H7F3	1,3,3-trifluorobutane	------	1.9279	0.2841	10	10000	2	3.71	3.71	26.40
358	3200	C4H7F3	2,3,3-trifluorobutane	------	1.9191	0.2843	10	10000	2	3.69	3.69	26.33
359	3201	C4H7F3	1,1,1-trifluoro-2-methylpropane	------	1.9191	0.2843	10	10000	2	3.69	3.69	26.33
360	3202	C4H7F3	1,1,2-trifluoro-2-methylpropane	------	1.9191	0.2843	10	10000	2	3.69	3.69	26.33
361	3203	C4H7F3	1,1,3-trifluoro-2-methylpropane	------	1.9942	0.2827	10	10000	2	3.82	3.82	26.95
362	3204	C4H7F3	1,3-difluoro-2-(fluoromethyl)-propane	------	2.0045	0.2825	10	10000	2	3.84	3.84	27.04
363	3209	C4H7I	cis-1-iodo-1-butene	54068-75-6	56.8917	0.0542	10	1773	2	64.45	64.45	85.34
364	3210	C4H7I	trans-1-iodo-1-butene	62154-92-1	45.4701	0.0687	10	10000	2	53.26	53.26	85.61
365	3211	C4H7I	2-iodo-1-butene	24308-61-0	45.0934	0.0694	10	10000	2	52.91	52.91	85.47
366	3212	C4H7I	3-iodo-1-butene	62154-74-9	45.0626	0.0695	10	10000	2	52.88	52.88	85.46
367	3213	C4H7I	4-iodo-1-butene	7766-51-0	45.3235	0.0690	10	10000	2	53.13	53.13	85.56
368	3214	C4H7I	1-iodo-cis-2-butene	53121-23-6	46.9129	0.0661	10	9765	2	54.62	54.62	86.10
369	3215	C4H7I	1-iodo-trans-2-butene	38169-04-9	46.9129	0.0661	10	9765	2	54.62	54.62	86.10
370	3216	C4H7I	2-iodo-cis-2-butene	24298-09-7	41.4559	0.0766	10	10000	2	49.46	49.46	83.97
371	3217	C4H7I	2-iodo-trans-2-butene	24298-08-6	41.4559	0.0766	10	10000	2	49.46	49.46	83.97
372	3218	C4H7I	1-iodo-2-methyl-1-propene	20687-01-8	41.4559	0.0766	10	10000	2	49.46	49.46	83.97
373	3219	C4H7I	3-iodo-2-methyl-1-propene	3756-30-7	37.1769	0.0858	10	10000	2	45.30	45.30	81.97
374	3238	C4H7N	butyronitrile	109-74-0	7.9445	0.1771	10	10000	2	11.95	11.95	40.61
375	3239	C4H7N	isobutyronitrile	78-82-0	6.3976	0.1931	10	10000	2	9.98	9.98	37.89
376	3240	C4H7N	propyl isocyanide	627-36-1	7.3189	0.1778	10	10000	2	11.02	11.02	37.65
377	3241	C4H7N	isopropyl isocyanide	598-45-8	5.8920	0.1934	10	10000	2	9.20	9.20	34.99
378	3246	C4H7NO	acetone cyanohydrin	75-86-5	23.7589	0.1000	10	1455	2	29.91	29.91	49.20
379	3248	C4H7NO	3-methoxypropionitrile	110-67-8	21.2883	0.1067	10	2598	2	27.21	27.21	49.24
380	3249	C4H7NO	2-pyrrolidone	616-45-5	45.1616	0.0747	1	26	2	45.16	53.63	57.60
381	3318	C4H8	methylcyclopropane	594-11-6	1.0089	0.2938	10	10000	2	1.98	1.98	15.11
382	3319	C4H8	cyclobutane	287-23-0	0.7390	0.3332	10	10000	2	1.59	1.59	15.91
383	3320	C4H8	1-butene	106-98-9	1.4676	0.2480	10	10000	1,2	2.60	2.60	14.41
384	3322	C4H8	cis-2-butene	590-18-1	1.3806	0.2675	10	10000	2	2.56	2.56	16.22
385	3323	C4H8	trans-2-butene	624-64-6	1.6254	0.2507	10	10000	2	2.89	2.89	16.35
386	3324	C4H8	isobutene	115-11-7	1.5592	0.2489	10	10000	2	2.77	2.77	15.44
387	3333	C4H8Br2	1,1-dibromobutane	62168-25-6	60.2643	0.0532	10	3437	2	68.11	68.11	92.91
388	3334	C4H8Br2	1,2-dibromobutane	533-98-2	60.2894	0.0520	10	4093	2	67.96	67.96	92.94
389	3335	C4H8Br2	1,3-dibromobutane	107-80-2	69.7696	0.0426	10	692	2	76.96	76.96	92.18
390	3336	C4H8Br2	1,4-dibromobutane	110-52-1	77.3066	0.0362	10	198	2	84.03	84.03	93.64
391	3337	C4H8Br2	2,2-dibromobutane	50341-35-0	59.2875	0.0520	10	3894	2	66.82	66.82	91.11
392	3338	C4H8Br2	DL-2,3-dibromobutane	598-71-0	65.6977	0.0465	10	1574	2	73.12	73.12	92.51
393	3339	C4H8Br2	meso-2,3-dibromobutane	5780-13-2	64.4448	0.0475	10	1856	2	71.89	71.89	92.11
394	3340	C4H8Br2	1,1-dibromo-2-methylpropane	33693-78-6	61.4532	0.0502	10	3146	2	68.98	68.98	92.05
395	3341	C4H8Br2	1,2-dibromo-2-methylpropane	594-34-3	61.3895	0.0492	10	3027	2	68.75	68.75	91.06
396	3342	C4H8Br2	1,3-dibromo-2-methylpropane	28148-04-1	70.2125	0.0431	10	692	2	77.54	77.54	93.07
397	3360	C4H8Cl2	1,1-dichlorobutane	541-33-3	23.1028	0.0937	10	10000	2	28.67	28.67	54.77
398	3361	C4H8Cl2	1,2-dichlorobutane	616-21-7	27.2583	0.0821	10	10000	2	32.93	32.93	58.05
399	3362	C4H8Cl2	1,3-dichlorobutane	1190-22-3	29.2656	0.0761	10	8797	2	34.87	34.87	58.41
400	3363	C4H8Cl2	1,4-dichlorobutane	110-56-5	32.0313	0.0723	10	5427	2	37.84	37.84	59.66

Table 1 Adsorption Capacity of Activated Carbon - Gas

$\log_{10} Q = \log_{10} K + 1/n \log_{10} y$ (Q - g of compound/100 g of carbon, y - ppmv)

NO	ID	FORMULA	NAME	CAS No	K	1/n	ymin	ymax	code	Q @ymin	Q @10ppm	Q @ymax
401	3364	C4H8Cl2	2,2-dichlorobutane	4279-22-5	21.5711	0.0984	10	10000	2	27.05	27.05	53.38
402	3365	C4H8Cl2	DL-2,3-dichlorobutane	2211-67-8	26.0170	0.0856	10	10000	2	31.68	31.68	57.21
403	3366	C4H8Cl2	meso-2,3-dichlorobutane	4028-56-2	25.1547	0.0883	10	10000	2	30.83	30.83	56.74
404	3367	C4H8Cl2	1,1-dichloro-2-methylpropane	598-76-5	22.4983	0.0979	10	10000	2	28.19	28.19	55.45
405	3368	C4H8Cl2	1,2-dichloro-2-methylpropane	594-37-6	23.0120	0.0946	10	10000	2	28.62	28.62	55.02
406	3369	C4H8Cl2	1,3-dichloro-2-methylpropane	616-19-3	29.4957	0.0749	10	8200	2	35.05	35.05	57.93
407	3389	C4H8F2	1,1-difluorobutane	2358-38-5	2.8881	0.2576	10	10000	2	5.23	5.23	30.97
408	3390	C4H8F2	1,2-difluorobutane	686-65-7	3.4271	0.2507	10	10000	2	6.10	6.10	34.48
409	3391	C4H8F2	1,3-difluorobutane	691-42-9	3.8029	0.2486	10	10000	2	6.74	6.74	37.55
410	3392	C4H8F2	1,4-difluorobutane	372-90-7	5.6512	0.2215	10	10000	2	9.41	9.41	43.45
411	3393	C4H8F2	2,2-difluorobutane	353-81-1	2.3494	0.2687	10	10000	2	4.36	4.36	27.92
412	3394	C4H8F2	2,3-difluorobutane	666-21-7	3.0471	0.2615	10	10000	2	5.56	5.56	33.86
413	3395	C4H8F2	1,1-difluoro-2-methylpropane	62126-91-4	2.4302	0.2731	10	10000	2	4.56	4.56	30.07
414	3396	C4H8F2	1,2-difluoro-2-methylpropane	62126-92-5	2.5045	0.2717	10	10000	2	4.68	4.68	30.57
415	3397	C4H8F2	1,3-difluoro-2-methylpropane	62126-93-6	4.0021	0.2455	10	10000	2	7.04	7.04	38.39
416	3401	C4H8I2	1,1-diiodobutane	66587-65-3	99.6779	0.0274	10	299	2	106.18	106.18	116.55
417	3402	C4H8I2	1,2-diiodobutane	53161-72-1	92.5482	0.0326	10	1608	2	99.75	99.75	117.69
418	3403	C4H8I2	1,3-diiodobutane	18371-24-9	100.2798	0.0275	10	299	2	106.83	106.83	117.27
419	3404	C4H8I2	1,4-diiodobutane	628-21-7	103.6555	0.0274	10	271	2	110.40	110.40	120.82
420	3405	C4H8I2	2,2-diiodobutane	29443-50-3	99.0468	0.0283	10	400	2	105.71	105.71	117.34
421	3406	C4H8I2	2,3-diiodobutane	66587-66-4	99.8860	0.0277	10	329	2	106.47	106.47	117.30
422	3407	C4H8I2	1,1-diiodo-2-methylpropane	10250-55-2	99.8860	0.0277	10	329	2	106.47	106.47	117.30
423	3408	C4H8I2	1,2-diiodo-2-methylpropane	66794-23-8	99.0468	0.0283	10	400	2	105.71	105.71	117.34
424	3409	C4H8I2	1,3-diiodo-2-methylpropane	17616-44-3	100.2798	0.0275	10	299	2	106.83	106.83	117.27
425	3474	C4H8O	1,2-epoxybutane	106-88-7	3.4835	0.2426	10	10000	2	6.09	6.09	32.54
426	3476	C4H8O	ethyl vinyl ether	109-92-2	2.8729	0.2378	10	10000	2	4.97	4.97	25.68
427	3478	C4H8O	butyraldehyde	123-72-8	4.3943	0.2227	10	10000	2	7.34	7.34	34.17
428	3479	C4H8O	isobutyraldehyde	78-84-2	3.6547	0.2336	10	10000	2	6.26	6.26	31.43
429	3480	C4H8O	methyl ethyl ketone	78-93-3	4.4215	0.2245	10	10000	1,2	7.41	7.41	34.97
430	3481	C4H8O	tetrahydrofuran	109-99-9	2.5023	0.2722	10	10000	1,2	4.68	4.68	30.69
431	3504	C4H8O2	butyric acid	107-92-6	25.1951	0.0974	10	1244	2	31.53	31.53	50.44
432	3505	C4H8O2	isobutyric acid	79-31-2	21.7251	0.1083	10	2388	2	27.88	27.88	50.44
433	3506	C4H8O2	propyl formate	110-74-7	7.0419	0.1908	10	10000	2	10.93	10.93	40.83
434	3507	C4H8O2	isopropyl formate	625-55-8	6.0786	0.1978	10	10000	2	9.58	9.58	37.57
435	3508	C4H8O2	ethyl acetate	141-78-6	6.4470	0.1980	10	10000	2	10.17	10.17	39.95
436	3509	C4H8O2	methyl propanoate	554-12-1	6.8197	0.1949	10	10000	2	10.68	10.68	41.06
437	3510	C4H8O2	1,4-dioxane	123-91-1	7.2967	0.1971	10	10000	1,2	11.49	11.49	44.81
438	3511	C4H8O2	cis-2-butene-1,4-diol	6117-80-2	47.5712	0.0742	1	8.6	2	47.57	---	55.81
439	3512	C4H8O2	trans-2-butene-1,4-diol	821-11-4	47.5712	0.0742	1	8.6	2	47.57	---	55.81
440	3536	C4H8O2S	sulfolane	126-33-0	59.8823	0.0015	10	10000	1,2	60.08	60.08	60.69
441	3581	C4H8S	tetrahydrothiophene	110-01-0	15.3941	0.1211	10	10000	1,2	20.34	20.34	46.96
442	3583	C4H8S	2-ethylthiacyclopropane	3195-86-6	14.3871	0.1256	10	10000	2	19.21	19.21	45.75
443	3584	C4H8S	2,2-dimethylthiacyclopropane	3772-13-2	11.5680	0.1419	10	10000	2	16.04	16.04	42.74
444	3585	C4H8S	2,cis-3-dimethylthiacyclopropane	5954-71-2	9.5450	0.1559	10	10000	2	13.67	13.67	40.11
445	3586	C4H8S	2,trans-3-dimethylthiacyclopropane	5955-98-6	9.5450	0.1559	10	10000	2	13.67	13.67	40.11
446	3598	C4H9Br	1-bromobutane	109-65-9	21.1866	0.1171	10	10000	2	27.74	27.74	62.31
447	3599	C4H9Br	2-bromobutane	78-76-2	19.6892	0.1217	10	10000	2	26.05	26.05	60.37
448	3600	C4H9Br	1-bromo-2-methylpropane	78-77-3	20.9067	0.1175	10	10000	2	27.40	27.40	61.69
449	3601	C4H9Br	2-bromo-2-methylpropane	507-19-7	17.1856	0.1288	10	10000	2	23.12	23.12	56.27
450	3610	C4H9Cl	1-chlorobutane	109-69-3	9.2568	0.1589	10	10000	2	13.35	13.35	40.01
451	3611	C4H9Cl	1-chloro-2-methylpropane	513-36-0	8.5769	0.1639	10	10000	2	12.51	12.51	38.80
452	3612	C4H9Cl	2-chlorobutane	78-86-4	8.0634	0.1677	10	10000	2	11.86	11.86	37.78
453	3613	C4H9Cl	2-chloro-2-methylpropane	507-20-0	6.6509	0.1769	10	10000	2	9.99	9.99	33.91
454	3657	C4H9F	1-fluorobutane	2366-52-1	2.1572	0.2641	10	10000	2	3.96	3.96	24.55
455	3658	C4H9F	2-fluorobutane	359-01-3	1.9082	0.2686	10	10000	2	3.54	3.54	22.66
456	3659	C4H9F	1-fluoro-2-methylpropane	359-00-2	1.5213	0.2821	10	10000	2	2.91	2.91	20.44
457	3660	C4H9F	2-fluoro-2-methylpropane	353-61-7	1.5002	0.2794	10	10000	2	2.85	2.85	19.67
458	3666	C4H9I	1-iodobutane	542-69-8	42.5308	0.0724	10	10000	2	50.25	50.25	82.89
459	3667	C4H9I	2-iodobutane	513-48-4	47.1167	0.0629	10	8452	2	54.47	54.47	83.25
460	3668	C4H9I	1-iodo-2-methylpropane	513-38-2	46.9726	0.0639	10	8274	2	54.42	54.42	83.59
461	3669	C4H9I	2-iodo-2-methylpropane	558-17-8	34.8265	0.0842	10	10000	2	42.27	42.27	75.60
462	3675	C4H9N	pyrrolidine	123-75-1	6.8242	0.1921	10	10000	2	10.62	10.62	40.04
463	3681	C4H9NO	N,N-dimethylacetamide	127-19-5	23.6611	0.0941	10	2631	2	29.39	29.39	49.65
464	3682	C4H9NO	morpholine	110-91-8	16.9727	0.1235	10	10000	2	22.56	22.56	52.95
465	3697	C4H9NO2	1-nitrobutane	627-05-4	23.2987	0.0914	10	5678	2	28.76	28.76	51.34
466	3698	C4H9NO2	2-nitrobutane	600-24-8	19.4108	0.1065	10	10000	2	24.80	24.80	51.74
467	3699	C4H9NO2	1-nitro-2-methylpropane	625-74-1	23.6797	0.0896	10	4971	2	29.11	29.11	50.77
468	3700	C4H9NO2	2-nitro-2-methylpropane	594-70-7	23.6797	0.0896	10	4971	2	29.11	29.11	50.77
469	3783	C4H10	butane	106-97-8	1.3597	0.2562	10	10000	1,2	2.45	2.45	14.40
470	3784	C4H10	isobutane	75-28-5	1.1621	0.2657	10	10000	1,2	2.14	2.14	13.42
471	3865	C4H10O	butanol	71-36-3	13.5388	0.1281	10	9276	2	18.18	18.18	43.63
472	3866	C4H10O	isobutanol	78-83-1	11.9479	0.1380	10	10000	2	16.42	16.42	42.57
473	3867	C4H10O	sec-butanol	78-92-2	10.0131	0.1548	10	10000	2	14.30	14.30	41.65
474	3868	C4H10O	tert-butanol	75-65-0	10.0131	0.1548	10	10000	2	14.30	14.30	41.65
475	3869	C4H10O	diethyl ether	60-29-7	2.3917	0.2388	10	10000	1,2	4.15	4.15	21.58
476	3870	C4H10O	methyl propyl ether	557-17-5	3.0824	0.2298	10	10000	2	5.23	5.23	25.60
477	3871	C4H10O	methyl isopropyl ether	598-53-8	2.9905	0.2279	10	10000	2	5.05	5.05	24.41
478	3881	C4H10O2	t-butyl hydroperoxide	75-91-2	18.5978	0.1052	10	7205	2	23.69	23.69	47.33
479	3882	C4H10O2	1,2-dimethoxyethane	110-71-4	8.4582	0.1688	10	10000	2	12.48	12.48	40.03
480	3883	C4H10O2	2-ethoxyethanol	110-80-5	18.7799	0.1095	10	6982	2	24.17	24.17	49.51

Table 1 Adsorption Capacity of Activated Carbon - Gas

$\log_{10} Q = \log_{10} K + 1/n \log_{10} y$ (Q - g of compound/100 g of carbon, y - ppmv)

NO	ID	FORMULA	NAME	CAS No	K	1/n	ymin	ymax	code	Q @ymin	Q @10ppm	Q @ymax
481	3884	C4H10O2	1,2-butanediol	584-03-2	37.1571	0.0630	10	191	2	42.96	42.96	51.73
482	3885	C4H10O2	1,3-butanediol	107-88-0	41.8051	0.0670	1	26	2	41.81	48.78	52.01
483	3886	C4H10O2	1,4-butanediol	110-63-4	43.2491	0.0784	1	13	2	43.25	51.80	52.88
484	3887	C4H10O2	DL-2,3-butanediol	6982-25-8	35.0233	0.0677	10	281	2	40.93	40.93	51.30
485	3888	C4H10O2	meso-2,3-butanediol	5341-95-7	33.6608	0.0722	10	394	2	39.75	39.75	51.82
486	3889	C4H10O2	2-methyl-1,2-propanediol	558-43-0	32.8321	0.0740	10	460	2	38.93	38.93	51.69
487	3890	C4H10O2	1-methyl-1,3-propanediol	------	39.2780	0.0635	10	58	2	45.46	45.46	50.82
488	3918	C4H10O3	diethylene glycol	111-46-6	50.9952	0.0626	1	7.5	2	51.00	---	57.85
489	3937	C4H10O4S	diethyl sulfate	64-67-5	47.2131	0.0443	10	268	2	52.29	52.29	60.50
490	3942	C4H10S	butyl mercaptan	109-79-5	13.3589	0.1214	10	10000	2	17.67	17.67	40.88
491	3943	C4H10S	isobutyl mercaptan	513-44-0	12.3738	0.1259	10	10000	2	16.53	16.53	39.45
492	3944	C4H10S	sec-butyl mercaptan	513-53-1	11.9271	0.1281	10	10000	2	16.02	16.02	38.82
493	3945	C4H10S	tert-butyl mercaptan	75-66-1	9.6343	0.1404	10	10000	2	13.31	13.31	35.10
494	3947	C4H10S	diethyl sulfide	352-93-2	13.0785	0.1215	10	10000	2	17.30	17.30	40.03
495	3948	C4H10S	methyl propyl sulfide	3877-15-4	13.5009	0.1196	10	10000	2	17.78	17.78	40.62
496	3949	C4H10S	methyl isopropyl sulfide	1551-21-9	12.0485	0.1271	10	10000	2	16.14	16.14	38.84
497	3950	C4H10S2	diethyl disulfide	110-81-6	32.2374	0.0544	10	5627	2	36.54	36.54	51.56
498	3977	C4H11N	butylamine	109-73-9	6.7194	0.1749	10	10000	2	10.05	10.05	33.64
499	3978	C4H11N	isobutylamine	78-81-9	5.8885	0.1827	10	10000	2	8.97	8.97	31.67
500	3979	C4H11N	sec-butylamine	13952-84-6	5.4691	0.1863	10	10000	2	8.40	8.40	30.40
501	3980	C4H11N	tert-butylamine	75-64-9	4.3072	0.1972	10	10000	2	6.78	6.78	26.49
502	3982	C4H11N	methyl-propylamine	627-35-0	5.9740	0.1784	10	10000	2	9.01	9.01	30.88
503	3983	C4H11N	methylisopropylamine	4747-21-1	6.5834	0.1714	10	10000	2	9.77	9.77	31.92
504	3984	C4H11N	diethylamine	109-89-7	4.8232	0.1934	10	10000	1,2	7.53	7.53	28.63
505	3985	C4H11N	dimethylethylamine	598-56-1	3.9674	0.1991	10	10000	2	6.27	6.27	24.82
506	3988	C4H11NO	dimethylethanolamine	108-01-0	22.1181	0.0896	10	4183	2	27.19	27.19	46.70
507	4080	C4H12N2O	N-aminoethyl ethanolamine	111-41-1	51.0733	0.0380	1	2	2	51.07	---	52.44
508	4110	C4H12Si	tetramethylsilane	75-76-3	6.3768	0.1490	10	10000	2	8.99	8.99	25.16
509	4118	C4H13N3	diethylene triamine	111-40-0	37.5885	0.0474	10	307	2	41.92	41.92	49.30
510	4169	C5Cl6	hexachlorocyclopentadiene	77-47-4	78.7762	0.0242	10	78	2	83.30	83.30	87.55
511	4327	C5H4O2	furfural	98-01-1	29.7325	0.0900	10	2907	2	36.58	36.58	60.96
512	4391	C5H5N	pyridine	110-86-1	13.2088	0.1448	10	10000	1,2	18.44	18.44	50.14
513	4447	C5H6	cyclopentadiene	542-92-7	2.4197	0.2547	10	10000	1,2	4.35	4.35	25.27
514	4448	C5H6	2-methyl-1-butene-3-yne	78-80-8	3.4029	0.2148	10	10000	2	5.58	5.58	24.62
515	4449	C5H6	1-pentene-3-yne	646-05-9	5.1975	0.1907	10	10000	2	8.06	8.06	30.10
516	4450	C5H6	1-pentene-4-yne	871-28-3	3.4203	0.2217	10	10000	2	5.70	5.70	26.35
517	4490	C5H6N2	glutaronitrile	544-13-8	43.4046	0.0695	1	10	2	43.40	50.93	50.93
518	4534	C5H6O	2-methylfuran	534-22-5	6.4812	0.1933	10	10000	2	10.11	10.11	38.44
519	4535	C5H6O	3-methylfuran	930-27-8	6.8764	0.1879	10	10000	2	10.60	10.60	38.80
520	4548	C5H6O2	furfuryl alcohol	98-00-0	35.5850	0.0757	10	800	2	42.36	42.36	59.01
521	4587	C5H6S	2-methylthiophene	554-14-3	20.1199	0.1014	10	10000	2	25.41	25.41	51.21
522	4588	C5H6S	3-methylthiophene	616-44-4	20.5983	0.0997	10	10000	2	25.91	25.91	51.59
523	4658	C5H7N	N-methylpyrrole	96-54-8	14.8477	0.1227	10	10000	2	19.70	19.70	45.98
524	4677	C5H7NO2	ethyl cyanoacetate	105-56-6	42.4913	0.0583	10	80	2	48.60	48.60	54.86
525	4736	C5H8	spiropentane	157-40-4	3.5392	0.2189	10	10000	2	5.86	5.86	26.57
526	4737	C5H8	cyclopentene	142-29-0	3.5628	0.2243	10	10000	2	5.97	5.97	28.12
527	4738	C5H8	isoprene	78-79-5	4.3143	0.1913	10	10000	2	6.70	6.70	25.13
528	4739	C5H8	3-methyl-1,2-butadiene	598-25-4	4.5137	0.1906	10	10000	2	7.00	7.00	26.11
529	4740	C5H8	1,2-pentadiene	591-95-7	4.6370	0.1905	10	10000	2	7.19	7.19	26.80
530	4741	C5H8	cis-1,3-pentadiene	1574-41-0	5.1008	0.1811	10	10000	2	7.74	7.74	27.04
531	4742	C5H8	trans-1,3-pentadiene	2004-70-8	5.0207	0.1798	10	10000	2	7.59	7.59	26.29
532	4743	C5H8	1,4-pentadiene	591-93-5	3.3430	0.2081	10	10000	2	5.40	5.40	22.72
533	4744	C5H8	2,3-pentadiene	591-96-8	5.0420	0.1843	10	10000	2	7.71	7.71	27.53
534	4745	C5H8	1-pentyne	627-19-0	3.3636	0.2181	10	10000	2	5.56	5.56	25.06
535	4746	C5H8	2-pentyne	627-21-4	4.4820	0.2005	10	10000	2	7.11	7.11	28.40
536	4747	C5H8	3-methyl-1-butyne	598-23-2	2.9356	0.2216	10	10000	2	4.89	4.89	22.61
537	4848	C5H8O	cyclopentanone	120-92-3	15.1559	0.1296	10	10000	2	20.43	20.43	50.01
538	4849	C5H8O	methyl isopropenyl ketone	814-78-8	10.9580	0.1438	10	10000	2	15.26	15.26	41.19
539	4882	C5H8O2	acetylacetone	123-54-6	24.8497	0.0843	10	5458	2	30.17	30.17	51.32
540	4883	C5H8O2	allyl acetate	591-87-7	13.8236	0.1297	10	10000	2	18.64	18.64	45.66
541	4888	C5H8O2	ethyl acrylate	140-88-5	4.1509	0.2218	10	10000	1,2	6.92	6.92	32.02
542	4889	C5H8O2	methyl methacrylate	80-62-6	17.8360	0.0734	10	10000	1,2	21.12	21.12	35.05
543	4890	C5H8O2	2-methylenebutanoic acid	3586-58-1	35.8456	0.0611	10	370	2	41.26	41.26	51.46
544	4891	C5H8O2	trans-2-methyl-2-butenoic acid	80-59-1	31.9731	0.0721	10	799	2	37.75	37.75	51.76
545	4892	C5H8O2	cis-2-methyl-2-butenoic acid	565-63-9	31.9731	0.0721	10	799	2	37.75	37.75	51.76
546	4893	C5H8O2	2-methyl-3-butenoic acid	53774-20-2	31.9731	0.0721	10	799	2	37.75	37.75	51.76
547	4894	C5H8O2	3-methyl-2-butenoic acid	541-47-9	32.5128	0.0695	10	792	2	38.16	38.16	51.71
548	4895	C5H8O2	3-methyl-3-butenoic acid	1617-31-8	32.5128	0.0695	10	792	2	38.16	38.16	51.71
549	4896	C5H8O2	trans-2-pentenoic acid	13991-37-2	38.0589	0.0530	10	227	2	43.00	43.00	50.73
550	4897	C5H8O2	cis-2-pentenoic acid	16666-42-5	38.0589	0.0530	10	227	2	43.00	43.00	50.73
551	4898	C5H8O2	trans-3-pentenoic acid	1617-32-9	36.1000	0.0591	10	327	2	41.36	41.36	50.83
552	4899	C5H8O2	cis-3-pentenoic acid	33698-87-2	36.1000	0.0591	10	327	2	41.36	41.36	50.83
553	4900	C5H8O2	4-pentenoic acid	591-80-0	34.7983	0.0623	10	418	2	40.16	40.16	50.67
554	4927	C5H8O3	2-hydroxyethyl acrylate	818-61-1	42.6475	0.0473	10	68	2	47.55	47.55	52.06
555	4929	C5H8O3	methyl acetoacetate	105-45-3	34.3524	0.0693	10	1175	2	40.29	40.29	56.06
556	4964	C5H9Br	cis-1-bromo-1-pentene	31849-75-9	31.6866	0.0754	10	10000	2	37.69	37.69	63.46
557	4965	C5H9Br	trans-1-bromo-1-pentene	31849-76-0	31.6866	0.0754	10	10000	2	37.69	37.69	63.46
558	4966	C5H9Br	2-bromo-1-pentene	31844-95-8	29.7359	0.0802	10	10000	2	35.76	35.76	62.21
559	4967	C5H9Br	3-bromo-1-pentene	53045-71-9	31.4973	0.0740	10	10000	2	37.35	37.35	62.27
560	4968	C5H9Br	4-bromo-1-pentene	31950-56-8	30.2233	0.0802	10	10000	2	36.35	36.35	63.26

Table 1 Adsorption Capacity of Activated Carbon - Gas

$\log_{10} Q = \log_{10} K + 1/n \log_{10} y$ (Q - g of compound/100 g of carbon, y - ppmv)

NO	ID	FORMULA	NAME	CAS No	K	1/n	ymin	ymax	code	Q @ymin	Q @10ppm	Q @ymax
561	4969	C5H9Br	5-bromo-1-pentene	1119-51-3	35.2883	0.0678	10	9710	2	41.25	41.25	65.78
562	4970	C5H9Br	1-bromo-cis-2-pentene	7348-78-9	34.6248	0.0686	10	10000	2	40.55	40.55	65.10
563	4971	C5H9Br	1-bromo-trans-2-pentene	7348-71-2	34.6248	0.0686	10	10000	2	40.55	40.55	65.10
564	4972	C5H9Br	2-bromo-cis-2-pentene	54653-29-1	30.6806	0.0817	10	10000	2	37.03	37.03	65.13
565	4973	C5H9Br	2-bromo-trans-2-pentene	54653-30-4	30.6806	0.0817	10	10000	2	37.03	37.03	65.13
566	4974	C5H9Br	3-bromo-cis-2-pentene	23068-94-2	31.0038	0.0798	10	10000	2	37.26	37.26	64.67
567	4975	C5H9Br	3-bromo-trans-2-pentene	54653-28-0	31.0038	0.0798	10	10000	2	37.26	37.26	64.67
568	4976	C5H9Br	4-bromo-cis-2-pentene	56535-63-8	32.4962	0.0759	10	10000	2	38.70	38.70	65.37
569	4977	C5H9Br	4-bromo-trans-2-pentene	23068-95-3	32.4962	0.0759	10	10000	2	38.70	38.70	65.37
570	4978	C5H9Br	5-bromo-cis-2-pentene	50273-84-2	34.8102	0.0699	10	10000	2	40.88	40.88	66.24
571	4979	C5H9Br	5-bromo-trans-2-pentene	7515-62-0	34.8102	0.0699	10	10000	2	40.88	40.88	66.24
572	4980	C5H9Br	cis-1-bromo-2-methyl-1-butene	36668-55-0	31.9709	0.0752	10	10000	2	38.01	38.01	63.88
573	4981	C5H9Br	trans-1-bromo-2-methyl-1-butene	54265-17-7	31.9709	0.0752	10	10000	2	38.01	38.01	63.88
574	4982	C5H9Br	3-bromo-2-methyl-1-butene	51872-48-1	28.9783	0.0825	10	10000	2	35.04	35.04	61.97
575	4983	C5H9Br	4-bromo-2-methyl-1-butene	20038-12-4	29.2355	0.0818	10	10000	2	35.29	35.29	62.10
576	4984	C5H9Br	cis-1-bromo-3-methyl-1-butene	16416-44-7	30.3840	0.0788	10	10000	2	36.42	36.42	62.75
577	4985	C5H9Br	trans-1-bromo-3-methyl-1-butene	16416-41-4	30.3840	0.0788	10	10000	2	36.42	36.42	62.75
578	4986	C5H9Br	2-bromo-3-methyl-1-butene	31844-96-9	27.7151	0.0869	10	10000	2	33.85	33.85	61.69
579	4987	C5H9Br	3-bromo-3-methyl-1-butene	865-58-7	27.4782	0.0876	10	10000	2	33.62	33.62	61.55
580	4988	C5H9Br	4-bromo-3-methyl-1-butene	31950-55-7	28.5032	0.0836	10	10000	2	34.55	34.55	61.55
581	4989	C5H9Br	1-bromo-2-methyl-cis-2-butene	57253-29-9	35.1389	0.0670	10	10000	2	41.00	41.00	65.16
582	4990	C5H9Br	1-bromo-2-methyl-trans-2-butene	57253-30-2	35.1389	0.0670	10	10000	2	41.00	41.00	65.16
583	4991	C5H9Br	1-bromo-3-methyl-2-butene	870-63-3	32.6496	0.0753	10	10000	2	38.83	38.83	65.35
584	4992	C5H9Br	2-bromo-3-methyl-2-butene	3017-70-7	33.0449	0.0756	10	10000	2	39.32	39.32	66.27
585	4993	C5H9Br	3-bromo-2-ethyl-1-propene	------	29.5877	0.0848	10	10000	2	35.97	35.97	64.63
586	5023	C5H9Cl	cis-1-chloro-1-pentene	66213-69-2	15.2526	0.1145	10	10000	2	19.85	19.85	43.79
587	5024	C5H9Cl	trans-1-chloro-1-pentene	66213-70-5	15.2526	0.1145	10	10000	2	19.85	19.85	43.79
588	5025	C5H9Cl	2-chloro-1-pentene	42131-85-1	15.1784	0.1147	10	10000	2	19.76	19.76	43.64
589	5026	C5H9Cl	3-chloro-1-pentene	24356-00-1	16.0392	0.1092	10	10000	2	20.62	20.62	43.85
590	5027	C5H9Cl	4-chloro-1-pentene	10524-08-0	16.4102	0.1059	10	10000	2	20.94	20.94	43.53
591	5028	C5H9Cl	5-chloro-1-pentene	928-50-7	17.9223	0.1013	10	10000	2	22.63	22.63	45.58
592	5029	C5H9Cl	1-chloro-cis-2-pentene	6261-19-4	19.7129	0.0913	10	10000	2	24.33	24.33	45.72
593	5030	C5H9Cl	1-chloro-trans-2-pentene	6261-25-2	19.5492	0.0945	10	10000	2	24.30	24.30	46.69
594	5031	C5H9Cl	2-chloro-cis-2-pentene	42131-98-6	16.8773	0.1042	10	10000	2	21.45	21.45	44.06
595	5032	C5H9Cl	2-chloro-trans-2-pentene	42132-00-3	16.8773	0.1042	10	10000	2	21.45	21.45	44.06
596	5033	C5H9Cl	3-chloro-cis-2-pentene	26423-60-9	15.7325	0.1104	10	10000	2	20.28	20.28	43.48
597	5034	C5H9Cl	3-chloro-trans-2-pentene	26423-61-0	15.7325	0.1104	10	10000	2	20.28	20.28	43.48
598	5035	C5H9Cl	4-chloro-cis-2-pentene	26423-63-2	17.1811	0.1028	10	10000	2	21.77	21.77	44.29
599	5036	C5H9Cl	4-chloro-trans-2-pentene	18610-33-8	17.1811	0.1028	10	10000	2	21.77	21.77	44.29
600	5037	C5H9Cl	5-chloro-cis-2-pentene	53543-44-5	18.9050	0.0956	10	10000	2	23.56	23.56	45.62
601	5038	C5H9Cl	5-chloro-trans-2-pentene	10524-07-9	19.1608	0.0942	10	10000	2	23.80	23.80	45.61
602	5039	C5H9Cl	cis-1-chloro-2-methyl-1-butene	7611-88-3	14.9149	0.1168	10	10000	2	19.52	19.52	43.72
603	5040	C5H9Cl	trans-1-chloro-2-methyl-1-butene	7611-89-4	14.9149	0.1168	10	10000	2	19.52	19.52	43.72
604	5041	C5H9Cl	3-chloro-2-methyl-1-butene	5166-35-8	16.3236	0.1084	10	10000	2	20.95	20.95	44.32
605	5042	C5H9Cl	4-chloro-2-methyl-1-butene	10523-96-3	17.8864	0.1003	10	10000	2	22.53	22.53	45.06
606	5043	C5H9Cl	cis-1-chloro-3-methyl-1-butene	66213-67-0	16.4453	0.1072	10	10000	2	21.05	21.05	44.12
607	5044	C5H9Cl	trans-1-chloro-3-methyl-1-butene	66213-68-1	16.4453	0.1072	10	10000	2	21.05	21.05	44.12
608	5045	C5H9Cl	2-chloro-3-methyl-1-butene	17773-64-7	15.9393	0.1096	10	10000	2	20.52	20.52	43.75
609	5046	C5H9Cl	3-chloro-3-methyl-1-butene	2190-48-9	14.0022	0.1180	10	10000	2	18.37	18.37	41.50
610	5047	C5H9Cl	4-chloro-3-methyl-1-butene	10524-01-3	16.3153	0.1078	10	10000	2	20.91	20.91	44.03
611	5048	C5H9Cl	1-chloro-2-methyl-cis-2-butene	23009-74-7	19.4314	0.0959	10	10000	2	24.23	24.23	47.01
612	5049	C5H9Cl	1-chloro-2-methyl-trans-2-butene	23009-73-6	18.7999	0.0983	10	10000	2	23.58	23.58	46.50
613	5050	C5H9Cl	1-chloro-3-methyl-2-butene	503-60-6	19.9023	0.0933	10	10000	2	24.67	24.67	46.99
614	5051	C5H9Cl	2-chloro-3-methyl-2-butene	17773-65-8	16.4391	0.1078	10	10000	2	21.07	21.07	44.38
615	5052	C5H9Cl	3-chloro-2-ethyl-1-propene	------	16.5988	0.1055	10	10000	2	21.16	21.16	43.86
616	5100	C5H9Cl3	1,1,1-trichloropentane	3922-27-8	43.4986	0.0431	10	2982	2	48.03	48.03	61.40
617	5109	C5H9F	cis-1-fluoro-1-pentene	66213-83-0	6.8650	0.1693	10	10000	2	10.14	10.14	32.65
618	5110	C5H9F	trans-1-fluoro-1-pentene	66213-84-1	6.8650	0.1693	10	10000	2	10.14	10.14	32.65
619	5111	C5H9F	2-fluoro-1-pentene	66213-89-6	6.4338	0.1728	10	10000	2	9.58	9.58	31.61
620	5112	C5H9F	3-fluoro-1-pentene	66702-88-3	6.4103	0.1730	10	10000	2	9.55	9.55	31.55
621	5113	C5H9F	4-fluoro-1-pentene	66702-91-8	6.4103	0.1730	10	10000	2	9.55	9.55	31.55
622	5114	C5H9F	5-fluoro-1-pentene	407-79-4	6.9071	0.1756	10	10000	2	10.35	10.35	34.81
623	5115	C5H9F	1-fluoro-cis-2-pentene	66213-85-2	7.0056	0.1676	10	10000	2	10.31	10.31	32.81
624	5116	C5H9F	1-fluoro-trans-2-pentene	66213-86-3	7.0056	0.1676	10	10000	2	10.31	10.31	32.81
625	5117	C5H9F	2-fluoro-cis-2-pentene	66213-87-4	6.9835	0.1678	10	10000	2	10.28	10.28	32.76
626	5118	C5H9F	2-fluoro-trans-2-pentene	66213-88-5	6.9835	0.1678	10	10000	2	10.28	10.28	32.76
627	5119	C5H9F	3-fluoro-cis-2-pentene	66213-90-9	6.9835	0.1678	10	10000	2	10.28	10.28	32.76
628	5120	C5H9F	3-fluoro-trans-2-pentene	66702-87-2	6.9835	0.1678	10	10000	2	10.28	10.28	32.76
629	5121	C5H9F	4-fluoro-cis-2-pentene	66702-89-4	6.9570	0.1681	10	10000	2	10.24	10.24	32.71
630	5122	C5H9F	4-fluoro-trans-2-pentene	66702-90-7	6.9570	0.1681	10	10000	2	10.24	10.24	32.71
631	5123	C5H9F	5-fluoro-cis-2-pentene	66702-92-9	7.0056	0.1676	10	10000	2	10.31	10.31	32.81
632	5124	C5H9F	5-fluoro-trans-2-pentene	66702-93-0	7.0056	0.1676	10	10000	2	10.31	10.31	32.81
633	5125	C5H9F	cis-1-fluoro-2-methyl-1-butene	66213-72-7	6.9835	0.1678	10	10000	2	10.28	10.28	32.76
634	5126	C5H9F	trans-1-fluoro-2-methyl-1-butene	66213-74-9	6.9835	0.1678	10	10000	2	10.28	10.28	32.76
635	5127	C5H9F	3-fluoro-2-methyl-1-butene	53731-24-1	6.3951	0.1732	10	10000	2	9.53	9.53	31.52
636	5128	C5H9F	4-fluoro-2-methyl-1-butene	66213-81-8	6.4338	0.1728	10	10000	2	9.58	9.58	31.61
637	5129	C5H9F	cis-1-fluoro-3-methyl-1-butene	66213-73-8	6.9570	0.1681	10	10000	2	10.24	10.24	32.71
638	5130	C5H9F	trans-1-fluoro-3-methyl-1-butene	66213-75-0	6.9570	0.1681	10	10000	2	10.24	10.24	32.71
639	5131	C5H9F	2-fluoro-3-methyl-1-butene	66213-78-3	6.3951	0.1732	10	10000	2	9.53	9.53	31.52
640	5132	C5H9F	3-fluoro-3-methyl-1-butene	66213-80-7	6.2059	0.1749	10	10000	2	9.28	9.28	31.09

Table 1 Adsorption Capacity of Activated Carbon - Gas

$\log_{10} Q = \log_{10} K + 1/n \log_{10} y$ (Q - g of compound/100 g of carbon, y - ppmv)

NO	ID	FORMULA	NAME	CAS No	K	1/n	ymin	ymax	code	Q @ymin	Q @10ppm	Q @ymax
641	5133	C5H9F	4-fluoro-3-methyl-1-butene	66213-82-9	6.4103	0.1730	10	10000	2	9.55	9.55	31.55
642	5134	C5H9F	1-fluoro-2-methyl-cis-2-butene	66213-76-1	6.6624	0.1745	10	10000	2	9.96	9.96	33.24
643	5135	C5H9F	1-fluoro-2-methyl-trans-2-butene	66213-77-2	6.6624	0.1745	10	10000	2	9.96	9.96	33.24
644	5136	C5H9F	1-fluoro-3-methyl-2-butene	23425-81-2	6.9835	0.1678	10	10000	2	10.28	10.28	32.76
645	5137	C5H9F	2-fluoro-3-methyl-2-butene	66213-79-4	6.9618	0.1680	10	10000	2	10.25	10.25	32.72
646	5138	C5H9F	3-fluoro-2-ethyl-1-propene	------	6.4338	0.1728	10	10000	2	9.58	9.58	31.61
647	5144	C5H9F3	1,1,1-trifluoropentane	406-82-6	6.5805	0.1930	10	10000	2	10.26	10.26	38.92
648	5147	C5H9I	cis-1-iodo-1-pentene	66703-02-4	52.5014	0.0492	10	4417	2	58.79	58.79	79.32
649	5148	C5H9I	trans-1-iodo-1-pentene	66703-03-5	52.5014	0.0492	10	4417	2	58.79	58.79	79.32
650	5149	C5H9I	2-iodo-1-pentene	66688-59-3	48.7137	0.0533	10	10000	2	55.08	55.08	79.62
651	5150	C5H9I	3-iodo-1-pentene	66688-61-7	48.9307	0.0543	10	10000	2	55.45	55.45	80.68
652	5151	C5H9I	4-iodo-1-pentene	59967-14-5	49.4677	0.0546	10	10000	2	56.10	56.10	81.81
653	5152	C5H9I	5-iodo-1-pentene	7766-48-5	54.1275	0.0489	10	4166	2	60.57	60.57	81.33
654	5153	C5H9I	1-iodo-cis-2-pentene	66703-04-6	50.7922	0.0521	10	7424	2	57.26	57.26	80.78
655	5154	C5H9I	1-iodo-trans-2-pentene	66688-56-0	50.7922	0.0521	10	7424	2	57.26	57.26	80.78
656	5155	C5H9I	2-iodo-cis-2-pentene	66688-57-1	50.5675	0.0523	10	7771	2	57.04	57.04	80.79
657	5156	C5H9I	2-iodo-trans-2-pentene	66688-58-2	50.5675	0.0523	10	7771	2	57.04	57.04	80.79
658	5157	C5H9I	3-iodo-cis-2-pentene	35895-39-7	50.5675	0.0523	10	7771	2	57.04	57.04	80.79
659	5158	C5H9I	3-iodo-trans-2-pentene	66688-60-6	50.5675	0.0523	10	7771	2	57.04	57.04	80.79
660	5159	C5H9I	4-iodo-cis-2-pentene	66688-62-8	50.4672	0.0524	10	7931	2	56.94	56.94	80.80
661	5160	C5H9I	4-iodo-trans-2-pentene	66688-63-9	50.4672	0.0524	10	7931	2	56.94	56.94	80.80
662	5161	C5H9I	5-iodo-cis-2-pentene	66688-64-0	50.9567	0.0517	10	7424	2	57.40	57.40	80.77
663	5162	C5H9I	5-iodo-trans-2-pentene	56399-98-5	50.9567	0.0517	10	7424	2	57.40	57.40	80.77
664	5163	C5H9I	cis-1-iodo-2-methyl-1-butene	52812-57-4	50.5675	0.0523	10	7771	2	57.04	57.04	80.79
665	5164	C5H9I	trans-1-iodo-2-methyl-1-butene	66702-95-2	50.5675	0.0523	10	7771	2	57.04	57.04	80.79
666	5165	C5H9I	3-iodo-2-methyl-1-butene	66703-00-2	50.1720	0.0527	10	8421	2	56.65	56.65	80.82
667	5166	C5H9I	4-iodo-2-methyl-1-butene	53750-52-0	50.3919	0.0525	10	8053	2	56.87	56.87	80.80
668	5167	C5H9I	cis-1-iodo-3-methyl-1-butene	64245-25-6	50.4672	0.0524	10	7931	2	56.94	56.94	80.80
669	5168	C5H9I	trans-1-iodo-3-methyl-1-butene	66702-96-3	50.4672	0.0524	10	7931	2	56.94	56.94	80.80
670	5169	C5H9I	2-iodo-3-methyl-1-butene	32442-47-0	48.5187	0.0551	10	10000	2	55.08	55.08	80.58
671	5170	C5H9I	3-iodo-3-methyl-1-butene	59128-91-5	47.5638	0.0567	10	10000	2	54.20	54.20	80.22
672	5171	C5H9I	4-iodo-3-methyl-1-butene	66703-01-3	48.7516	0.0547	10	10000	2	55.29	55.29	80.66
673	5172	C5H9I	1-iodo-2-methyl-cis-2-butene	66702-97-4	50.5675	0.0523	10	7771	2	57.04	57.04	80.79
674	5173	C5H9I	1-iodo-2-methyl-trans-2-butene	66702-98-5	50.5675	0.0523	10	7771	2	57.04	57.04	80.79
675	5174	C5H9I	1-iodo-3-methyl-2-butene	41004-19-7	50.5675	0.0523	10	7771	2	57.04	57.04	80.79
676	5175	C5H9I	2-iodo-3-methyl-2-butene	66702-99-6	50.3477	0.0526	10	8126	2	56.82	56.82	80.81
677	5176	C5H9I	3-iodo-2-ethyl-1-propene	------	48.8621	0.0545	10	10000	2	55.39	55.39	80.70
678	5183	C5H9N	valeronitrile	110-59-8	16.2278	0.1048	10	9596	2	20.66	20.66	42.44
679	5184	C5H9N	2-methylbutanenitrile	18936-17-9	15.3043	0.1086	10	10000	2	19.65	19.65	41.62
680	5185	C5H9N	3-methylbutanenitrile	625-28-5	15.9668	0.1064	10	8989	2	20.40	20.40	42.05
681	5186	C5H9N	2,2-dimethylpropanenitrile	630-18-2	11.6130	0.1310	10	10000	2	15.70	15.70	38.82
682	5187	C5H9N	butyl isocyanide	2769-64-4	15.7690	0.1030	10	10000	2	19.99	19.99	40.71
683	5188	C5H9N	isobutyl isocyanide	590-94-3	13.0291	0.1194	10	10000	2	17.15	17.15	39.14
684	5189	C5H9N	sec-butyl isocyanide	14069-89-7	13.3075	0.1181	10	10000	2	17.47	17.47	39.50
685	5190	C5H9N	tert-butyl isocyanide	7188-38-7	10.1228	0.1389	10	10000	2	13.94	13.94	36.40
686	5197	C5H9NO	butyl isocyanate	111-36-4	17.1832	0.1046	10	10000	2	21.86	21.86	45.04
687	5198	C5H9NO	N-methyl-2-pyrrolidone	872-50-4	36.7372	0.0609	10	447	2	42.26	42.26	53.26
688	5280	C5H10	1,1-dimethylcyclopropane	1630-94-0	2.8786	0.2192	10	10000	2	4.77	4.77	21.68
689	5281	C5H10	1,cis-2-dimethylcyclopropane	930-18-7	3.5157	0.2129	10	10000	2	5.74	5.74	24.97
690	5282	C5H10	1,trans-2-dimethylcyclopropane	2402-06-4	3.1446	0.2159	10	10000	2	5.17	5.17	22.98
691	5283	C5H10	ethylcyclopropane	1191-96-4	3.4770	0.2122	10	10000	2	5.67	5.67	24.54
692	5284	C5H10	methylcyclobutane	598-61-8	3.2818	0.2161	10	10000	2	5.40	5.40	24.02
693	5285	C5H10	cyclopentane	287-92-3	3.9498	0.2149	10	10000	2	6.48	6.48	28.59
694	5286	C5H10	1-pentene	109-67-1	3.5948	0.2007	10	10000	2	5.71	5.71	22.82
695	5288	C5H10	cis-2-pentene	627-20-3	4.0303	0.1952	10	10000	2	6.32	6.32	24.32
696	5289	C5H10	trans-2-pentene	646-04-8	3.9890	0.1949	10	10000	2	6.25	6.25	24.01
697	5290	C5H10	2-methyl-1-butene	563-46-2	3.6980	0.2001	10	10000	2	5.86	5.86	23.34
698	5291	C5H10	3-methyl-1-butene	563-45-1	3.2076	0.2052	10	10000	2	5.15	5.15	21.24
699	5292	C5H10	2-methyl-2-butene	513-35-9	4.1369	0.1943	10	10000	2	6.47	6.47	24.77
700	5297	C5H10Br2	1,1-dibromopentane	13320-56-4	65.3177	0.0380	10	1195	2	71.29	71.29	85.50
701	5298	C5H10Br2	1,2-dibromopentane	3234-49-9	69.9794	0.0335	10	431	2	75.59	75.59	85.76
702	5299	C5H10Br2	1,3-dibromopentane	42474-20-4	70.9169	0.0323	10	313	2	76.39	76.39	85.39
703	5300	C5H10Br2	1,4-dibromopentane	626-87-9	73.5581	0.0308	10	183	2	78.96	78.96	86.34
704	5301	C5H10Br2	1,5-dibromopentane	111-24-0	76.3305	0.0347	1	49	2	76.33	82.67	87.35
705	5302	C5H10Br2	2,2-dibromopentane	54653-26-8	65.5637	0.0369	10	1045	2	71.37	71.37	84.72
706	5303	C5H10Br2	2,3-dibromopentane	5398-25-4	68.8417	0.0352	10	621	2	74.66	74.66	86.34
707	5304	C5H10Br2	2,4-dibromopentane	19398-53-9	68.8055	0.0349	10	517	2	74.56	74.56	85.56
708	5305	C5H10Br2	3,3-dibromopentane	54653-27-9	65.4249	0.0382	10	1219	2	71.44	71.44	85.81
709	5306	C5H10Br2	1,1-dibromo-2-methylbutane	62127-58-6	66.8627	0.0367	10	868	2	72.76	72.76	85.73
710	5307	C5H10Br2	1,1-dibromo-3-methylbutane	62127-59-7	66.8627	0.0367	10	868	2	72.76	72.76	85.73
711	5308	C5H10Br2	1,2-dibromo-2-methylbutane	10428-64-5	66.9278	0.0363	10	895	2	72.76	72.76	85.64
712	5309	C5H10Br2	1,2-dibromo-3-methylbutane	10288-13-8	67.4091	0.0365	10	868	2	73.31	73.31	86.27
713	5310	C5H10Br2	1,3-dibromo-2-methylbutane	49623-50-9	70.9125	0.0413	10	868	2	77.98	77.98	93.76
714	5311	C5H10Br2	1,3-dibromo-3-methylbutane	24443-15-0	66.3416	0.0374	10	1059	2	72.31	72.31	86.08
715	5312	C5H10Br2	1,4-dibromo-2-methylbutane	69498-28-8	68.8537	0.0368	10	791	2	74.95	74.95	88.03
716	5313	C5H10Br2	2,2-dibromo-3-methylbutane	62127-60-0	66.5195	0.0378	10	1156	2	72.57	72.57	86.83
717	5314	C5H10Br2	2,3-dibromo-2-methylbutane	594-51-4	57.7161	0.0309	10	1791	2	61.97	61.97	72.75
718	5315	C5H10Br2	1,1-dibromo-2,2-dimethylpropane	2443-91-6	68.9735	0.0346	10	533	2	74.70	74.70	85.72
719	5316	C5H10Br2	1,3-dibromo-2,2-dimethylpropane	5434-27-5	69.4090	0.0356	10	573	2	75.34	75.34	87.01
720	5317	C5H10Br2	1,3-dibromo-2-ethylpropane	------	66.7858	0.0385	10	1059	2	72.99	72.99	87.36

Table 1 Adsorption Capacity of Activated Carbon - Gas

$\log_{10} Q = \log_{10} K + 1/n \log_{10} y$ (Q - g of compound/100 g of carbon, y - ppmv)

NO	ID	FORMULA	NAME	CAS No	K	1/n	ymin	ymax	code	Q @ymin	Q @10ppm	Q @ymax
721	5330	C5H10Cl2	1,1-dichloropentane	820-55-3	31.4536	0.0608	10	9431	2	36.18	36.18	54.85
722	5331	C5H10Cl2	1,2-dichloropentane	1674-33-5	34.8917	0.0561	10	4621	2	39.70	39.70	56.00
723	5332	C5H10Cl2	1,3-dichloropentane	30122-12-4	36.4201	0.0539	10	3155	2	41.23	41.23	56.22
724	5333	C5H10Cl2	1,4-dichloropentane	626-92-6	37.5784	0.0508	10	2254	2	42.24	42.24	55.62
725	5334	C5H10Cl2	1,5-dichloropentane	628-76-2	39.7300	0.0492	10	1484	2	44.49	44.49	56.90
726	5335	C5H10Cl2	2,2-dichloropentane	34887-14-4	30.1544	0.0629	10	10000	2	34.85	34.85	53.83
727	5336	C5H10Cl2	2,3-dichloropentane	600-11-3	33.2856	0.0591	10	7089	2	38.13	38.13	56.19
728	5337	C5H10Cl2	2,4-dichloropentane	625-67-2	34.4101	0.0562	10	4434	2	39.16	39.16	55.14
729	5338	C5H10Cl2	3,3-dichloropentane	21571-91-5	31.5457	0.0600	10	10000	2	36.22	36.22	54.83
730	5339	C5H10Cl2	1,1-dichloro-2-methylbutane	23010-05-1	30.8030	0.0622	10	10000	2	35.54	35.54	54.61
731	5340	C5H10Cl2	1,1-dichloro-3-methylbutane	625-66-1	30.8569	0.0617	10	10000	2	35.57	35.57	54.47
732	5341	C5H10Cl2	1,2-dichloro-2-methylbutane	23010-04-0	32.4684	0.0604	10	8272	2	37.31	37.31	55.98
733	5342	C5H10Cl2	1,2-dichloro-3-methylbutane	600-10-2	34.1649	0.0578	10	5879	2	39.03	39.03	56.41
734	5343	C5H10Cl2	1,3-dichloro-2-methylbutane	23010-07-3	34.6120	0.0579	10	5390	2	39.55	39.55	56.92
735	5344	C5H10Cl2	1,3-dichloro-3-methylbutane	624-96-4	34.0680	0.0574	10	5712	2	38.88	38.88	55.97
736	5345	C5H10Cl2	1,4-dichloro-2-methylbutane	623-34-7	35.6813	0.0556	10	5117	2	40.56	40.56	57.38
737	5346	C5H10Cl2	2,2-dichloro-3-methylbutane	17773-66-9	34.8126	0.0565	10	6000	2	39.65	39.65	56.90
738	5347	C5H10Cl2	2,3-dichloro-2-methylbutane	507-45-9	34.8126	0.0565	10	6000	2	39.65	39.65	56.90
739	5348	C5H10Cl2	1,3-dichloro-2-ethylpropane	------	35.3995	0.0555	10	5117	2	40.22	40.22	56.86
740	5349	C5H10Cl2	1,1-dichloro-2,2-dimethylpropane	29559-54-4	34.8126	0.0565	10	6000	2	39.65	39.65	56.90
741	5350	C5H10Cl2	1,3-dichloro-2,2-dimethylpropane	29559-55-5	34.4518	0.0577	10	5415	2	39.35	39.35	56.60
742	5358	C5H10F2	1,1-difluoropentane	62127-40-6	8.2543	0.1713	10	10000	2	12.25	12.25	40.00
743	5359	C5H10F2	1,2-difluoropentane	62126-94-7	9.0231	0.1666	10	10000	2	13.24	13.24	41.87
744	5360	C5H10F2	1,3-difluoropentane	62126-95-8	9.4635	0.1647	10	10000	2	13.83	13.83	43.15
745	5361	C5H10F2	1,4-difluoropentane	62126-96-9	9.9111	0.1644	10	10000	2	14.47	14.47	45.04
746	5362	C5H10F2	1,5-difluoropentane	373-17-1	16.9739	0.1140	10	10000	2	22.07	22.07	48.51
747	5363	C5H10F2	2,2-difluoropentane	371-65-3	7.4362	0.1762	10	10000	2	11.16	11.16	37.69
748	5364	C5H10F2	2,3-difluoropentane	66688-46-8	6.9446	0.1864	10	10000	2	10.67	10.67	38.65
749	5365	C5H10F2	2,4-difluoropentane	66688-47-9	6.9446	0.1864	10	10000	2	10.67	10.67	38.65
750	5366	C5H10F2	3,3-difluoropentane	358-03-2	7.5171	0.1764	10	10000	2	11.28	11.28	38.16
751	5367	C5H10F2	1,1-difluoro-2-methylbutane	66688-67-3	5.9109	0.1910	10	10000	2	9.18	9.18	34.32
752	5368	C5H10F2	1,1-difluoro-3-methylbutane	53731-22-9	5.9109	0.1910	10	10000	2	9.18	9.18	34.32
753	5369	C5H10F2	1,2-difluoro-2-methylbutane	66688-68-4	5.7371	0.1927	10	10000	2	8.94	8.94	33.84
754	5370	C5H10F2	1,2-difluoro-3-methylbutane	66688-69-5	5.9109	0.1910	10	10000	2	9.18	9.18	34.32
755	5371	C5H10F2	1,3-difluoro-2-methylbutane	66688-43-5	5.9109	0.1910	10	10000	2	9.18	9.18	34.32
756	5372	C5H10F2	1,3-difluoro-3-methylbutane	66688-44-6	5.7371	0.1927	10	10000	2	8.94	8.94	33.84
757	5373	C5H10F2	1,4-difluoro-2-methylbutane	66688-45-7	5.9527	0.1906	10	10000	2	9.23	9.23	34.43
758	5374	C5H10F2	2,2-difluoro-3-methylbutane	51891-58-8	5.6999	0.1931	10	10000	2	8.89	8.89	33.74
759	5375	C5H10F2	2,3-difluoro-2-methylbutane	53731-25-2	5.6999	0.1931	10	10000	2	8.89	8.89	33.74
760	5376	C5H10F2	1,3-difluoro-2-ethylpropane	------	5.9527	0.1906	10	10000	2	9.23	9.23	34.43
761	5377	C5H10F2	1,1-difluoro-2,2-dimethylpropane	53731-23-0	5.6999	0.1931	10	10000	2	8.89	8.89	33.74
762	5378	C5H10F2	1,3-difluoro-2,2-dimethylpropane	66688-66-2	5.7371	0.1927	10	10000	2	8.94	8.94	33.84
763	5380	C5H10I2	1,1-diiodopentane	66688-35-5	96.5450	0.0253	10	94	2	102.34	102.34	108.31
764	5381	C5H10I2	1,2-diiodopentane	998-76-5	96.5879	0.0256	10	94	2	102.46	102.46	108.52
765	5382	C5H10I2	1,3-diiodopentane	66688-36-6	96.9814	0.0258	10	94	2	102.91	102.91	109.04
766	5383	C5H10I2	1,4-diiodopentane	55930-45-5	97.3741	0.0259	10	94	2	103.37	103.37	109.55
767	5384	C5H10I2	1,5-diiodopentane	628-77-3	100.0551	0.0250	10	84	2	105.99	105.99	111.78
768	5385	C5H10I2	2,2-diiodopentane	66688-37-7	95.1118	0.0262	10	130	2	101.03	101.03	108.06
769	5386	C5H10I2	2,3-diiodopentane	66688-38-8	97.0023	0.0262	10	105	2	103.03	103.03	109.58
770	5387	C5H10I2	2,4-diiodopentane	66719-29-7	95.0771	0.0286	10	105	2	101.55	101.55	108.61
771	5388	C5H10I2	3,3-diiodopentane	66688-39-9	95.1118	0.0262	10	130	2	101.03	101.03	108.06
772	5389	C5H10I2	1,1-diiodo-2-methylbutane	66688-50-4	96.4219	0.0254	10	105	2	102.23	102.23	108.53
773	5390	C5H10I2	1,1-diiodo-3-methylbutane	66688-51-5	96.4219	0.0254	10	105	2	102.23	102.23	108.53
774	5391	C5H10I2	1,2-diiodo-2-methylbutane	66688-52-6	96.4714	0.0262	10	130	2	102.48	102.48	109.60
775	5392	C5H10I2	1,2-diiodo-3-methylbutane	66688-53-7	97.2042	0.0257	10	105	2	103.14	103.14	109.57
776	5393	C5H10I2	1,3-diiodo-2-methylbutane	66688-54-8	97.2042	0.0257	10	105	2	103.14	103.14	109.57
777	5394	C5H10I2	1,3-diiodo-3-methylbutane	66688-55-9	96.4714	0.0262	10	130	2	102.48	102.48	109.60
778	5395	C5H10I2	1,4-diiodo-2-methylbutane	66688-32-2	97.5687	0.0255	10	94	2	103.46	103.46	109.54
779	5396	C5H10I2	2,2-diiodo-3-methylbutane	66688-33-3	96.4919	0.0266	10	144	2	102.59	102.59	110.14
780	5397	C5H10I2	2,3-diiodo-2-methylbutane	66688-34-4	96.4919	0.0266	10	144	2	102.59	102.59	110.14
781	5398	C5H10I2	1,3-diiodo-2-ethylpropane	------	98.9423	0.0255	10	94	2	104.92	104.92	111.09
782	5399	C5H10I2	1,1-diiodo-2,2-dimethylpropane	2443-89-2	97.5751	0.0262	10	144	2	103.65	103.65	111.16
783	5400	C5H10I2	1,3-diiodo-2,2-dimethylpropane	66688-49-1	97.8294	0.0262	10	130	2	103.92	103.92	111.15
784	5445	C5H10O	pentanal	110-62-3	11.4104	0.1359	10	10000	2	15.60	15.60	39.90
785	5446	C5H10O	2-methylbutanal	96-17-3	10.2886	0.1437	10	10000	2	14.32	14.32	38.64
786	5447	C5H10O	3-methylbutanal	590-86-3	10.4540	0.1416	10	10000	2	14.48	14.48	38.53
787	5448	C5H10O	2,2-dimethylpropanal	630-19-3	7.9046	0.1627	10	10000	2	11.50	11.50	35.37
788	5450	C5H10O	methyl propyl ketone	107-87-9	11.0045	0.1388	10	10000	2	15.15	15.15	39.53
789	5451	C5H10O	diethyl ketone	96-22-0	11.0091	0.1399	10	10000	2	15.19	15.19	39.94
790	5452	C5H10O	methyl isopropyl ketone	563-80-4	9.8525	0.1485	10	10000	2	13.87	13.87	38.69
791	5453	C5H10O	2-methyltetrahydrofuran	96-47-9	8.4324	0.1666	10	10000	2	12.37	12.37	39.10
792	5454	C5H10O	3-methyltetrahydrofuran	13423-15-9	20.1581	0.1006	10	10000	2	25.41	25.41	50.91
793	5492	C5H10O2	pentanoic acid	109-52-4	34.8530	0.0573	10	308	2	39.77	39.77	48.41
794	5493	C5H10O2	2-methylbutyric acid	116-53-0	31.8335	0.0653	10	645	2	37.00	37.00	48.57
795	5494	C5H10O2	3-methylbutanoic acid	503-74-2	5.5285	0.2011	10	578	1,2	8.78	8.78	19.86
796	5497	C5H10O2	butyl formate	592-84-7	14.9109	0.1186	10	10000	2	19.59	19.59	44.46
797	5498	C5H10O2	isobutyl formate	542-55-2	13.6319	0.1246	10	10000	2	18.16	18.16	42.96
798	5499	C5H10O2	sec-butyl formate	589-40-2	13.0713	0.1284	10	10000	2	17.57	17.57	42.63
799	5500	C5H10O2	tert-butyl formate	762-75-4	11.8151	0.1360	10	10000	2	16.16	16.16	41.35
800	5501	C5H10O2	propyl acetate	109-60-4	14.1775	0.1227	10	10000	2	18.81	18.81	43.90

Table 1 Adsorption Capacity of Activated Carbon - Gas

$\log_{10} Q = \log_{10} K + 1/n \log_{10} y$ (Q - g of compound/100 g of carbon, y - ppmv)

NO	ID	FORMULA	NAME	CAS No	K	1/n	ymin	ymax	code	Q @ymin	Q @10ppm	Q @ymax
801	5502	C5H10O2	isopropyl acetate	108-21-4	8.3317	0.1647	10	10000	2	12.17	12.17	37.98
802	5503	C5H10O2	ethyl propanoate	105-37-3	13.7358	0.1256	10	10000	2	18.34	18.34	43.69
803	5504	C5H10O2	methyl butanoate	623-42-7	14.2565	0.1235	10	10000	2	18.95	18.95	44.48
804	5506	C5H10O2	tetrahydrofurfuryl alcohol	97-99-4	33.0130	0.0729	10	1058	2	39.05	39.05	54.87
805	5537	C5H10O2S	3-methyl sulfolane	872-93-5	55.2550	0.0471	1	9.5	2	55.25	---	61.44
806	5543	C5H10O3	diethyl carbonate	105-58-8	21.7591	0.0920	10	10000	2	26.89	26.89	50.76
807	5544	C5H10O3	ethyl lactate	97-64-3	27.8852	0.0780	10	4933	2	33.37	33.37	54.11
808	5593	C5H10S	thiacyclohexane	1613-51-0	25.8000	0.0753	10	10000	2	30.69	30.69	51.63
809	5594	C5H10S	2,2,3-trimethylthiacyclopropane	53971-47-4	17.6066	0.0992	10	10000	2	22.13	22.13	43.92
810	5595	C5H10S	2-methylthiacyclopentane	1795-09-1	23.6337	0.0801	10	10000	2	28.42	28.42	49.44
811	5596	C5H10S	3-methylthiacyclopentane	4740-00-5	24.6724	0.0772	10	10000	2	29.48	29.48	50.26
812	5597	C5H10S	cyclopentanethiol	1679-07-8	28.1925	0.0617	10	10000	2	32.49	32.49	49.74
813	5603	C5H11Br	1-bromopentane	110-53-2	32.3780	0.0721	10	10000	2	38.23	38.23	62.90
814	5604	C5H11Br	2-bromopentane	107-81-3	31.6121	0.0733	10	10000	2	37.42	37.42	62.09
815	5605	C5H11Br	3-bromopentane	1809-10-5	32.0539	0.0724	10	10000	2	37.87	37.87	62.45
816	5606	C5H11Br	1-bromo-2-methylbutane	10422-35-2	32.5213	0.0719	10	10000	2	38.37	38.37	63.05
817	5607	C5H11Br	1-bromo-3-methylbutane	107-82-4	32.3433	0.0712	10	10000	2	38.11	38.11	62.33
818	5608	C5H11Br	2-bromo-2-methylbutane	507-36-8	29.1143	0.0810	10	10000	2	35.09	35.09	61.40
819	5609	C5H11Br	2-bromo-3-methylbutane	18295-25-5	31.1708	0.0756	10	10000	2	37.10	37.10	62.55
820	5610	C5H11Br	1-bromo-2,2-dimethylpropane	630-17-1	28.4191	0.0818	10	10000	2	34.31	34.31	60.37
821	5620	C5H11Cl	1-chloropentane	543-59-9	17.1793	0.1015	10	10000	2	21.70	21.70	43.75
822	5621	C5H11Cl	2-chloropentane	625-29-6	15.5834	0.1082	10	10000	2	19.99	19.99	42.20
823	5622	C5H11Cl	3-chloropentane	616-20-6	16.5163	0.1062	10	10000	2	21.09	21.09	43.91
824	5623	C5H11Cl	1-chloro-2-methylbutane	616-13-7	16.2241	0.1062	10	10000	2	20.72	20.72	43.13
825	5624	C5H11Cl	1-chloro-3-methylbutane	107-84-6	15.5458	0.1081	10	10000	2	19.94	19.94	42.06
826	5625	C5H11Cl	2-chloro-2-methylbutane	594-36-5	14.1761	0.1149	10	10000	2	18.47	18.47	40.85
827	5626	C5H11Cl	2-chloro-3-methylbutane	631-65-2	15.6961	0.1068	10	10000	2	20.07	20.07	41.96
828	5627	C5H11Cl	1-chloro-2,2-dimethylpropane	753-89-9	14.5052	0.1133	10	10000	2	18.83	18.83	41.18
829	5649	C5H11F	1-fluoropentane	592-50-7	6.7375	0.1743	10	10000	2	10.06	10.06	33.55
830	5650	C5H11F	2-fluoropentane	590-87-4	6.3032	0.1768	10	10000	2	9.47	9.47	32.11
831	5651	C5H11F	3-fluoropentane	41909-29-9	6.1276	0.1782	10	10000	2	9.24	9.24	31.63
832	5652	C5H11F	1-fluoro-2-methylbutane	10086-64-3	6.0804	0.1817	10	10000	2	9.24	9.24	32.41
833	5653	C5H11F	1-fluoro-3-methylbutane	407-06-7	5.8975	0.1828	10	10000	2	8.98	8.98	31.75
834	5654	C5H11F	2-fluoro-2-methylbutane	661-53-0	5.1825	0.1908	10	10000	2	8.04	8.04	30.04
835	5655	C5H11F	2-fluoro-3-methylbutane	62108-95-6	5.5054	0.1859	10	10000	2	8.45	8.45	30.50
836	5656	C5H11F	1-fluoro-2,2-dimethylpropane	59006-05-2	5.0470	0.1911	10	10000	2	7.84	7.84	29.33
837	5660	C5H11I	1-iodopentane	628-17-1	50.4475	0.0505	10	6409	2	56.67	56.67	78.56
838	5661	C5H11I	2-iodopentane	637-97-8	54.8406	0.0451	10	2418	2	60.85	60.85	77.95
839	5662	C5H11I	3-iodopentane	1809-05-8	55.4546	0.0445	10	2167	2	61.43	61.43	78.03
840	5663	C5H11I	1-iodo-2-methylbutane	29394-58-9	55.8608	0.0446	10	2051	2	61.90	61.90	78.50
841	5664	C5H11I	1-iodo-3-methylbutane	541-28-6	55.8707	0.0436	10	1818	2	61.77	61.77	77.51
842	5665	C5H11I	2-iodo-2-methylbutane	594-38-7	50.7563	0.0486	10	5818	2	56.76	56.76	77.34
843	5666	C5H11I	2-iodo-3-methylbutane	18295-27-7	53.4558	0.0461	10	3028	2	59.44	59.44	77.36
844	5667	C5H11I	1-iodo-2,2-dimethylpropane	15501-33-4	51.4457	0.0485	10	4465	2	57.52	57.52	77.32
845	5669	C5H11N	N-methylpyrrolidine	120-94-5	10.0157	0.1421	10	10000	2	13.89	13.89	37.08
846	5670	C5H11N	piperidine	110-89-4	13.7302	0.1233	10	10000	2	18.24	18.24	42.75
847	5675	C5H11NO	tert-butylformamide	2425-74-3	34.0986	0.0529	10	359	2	38.52	38.52	46.56
848	5696	C5H11NO2	1-nitropentane	628-05-7	34.1563	0.0545	10	845	2	38.72	38.72	49.32
849	5697	C5H11NO2	2-nitropentane	4609-89-6	29.2301	0.0648	10	2966	2	33.93	33.93	49.08
850	5698	C5H11NO2	3-nitropentane	551-88-2	29.2417	0.0649	10	2966	2	33.95	33.95	49.13
851	5699	C5H11NO2	1-nitro-2-methylbutane	3457-57-6	31.2577	0.0602	10	1787	2	35.91	35.91	49.08
852	5700	C5H11NO2	1-nitro-3-methylbutane	627-67-8	31.2577	0.0602	10	1787	2	35.91	35.91	49.08
853	5701	C5H11NO2	2-nitro-2-methylbutane	595-42-6	28.8296	0.0664	10	3523	2	33.60	33.60	49.60
854	5702	C5H11NO2	2-nitro-3-methylbutane	2625-35-6	29.1643	0.0652	10	3183	2	33.89	33.89	49.35
855	5703	C5H11NO2	1-nitro-2,2-dimethylpropane	34715-98-5	29.1312	0.0649	10	3471	2	33.83	33.83	49.45
856	5772	C5H12	pentane	109-66-0	3.9440	0.1931	10	10000	1,2	6.15	6.15	23.35
857	5773	C5H12	isopentane	78-78-4	3.5818	0.1981	10	10000	1,2	5.65	5.65	22.21
858	5774	C5H12	neopentane	463-82-1	3.3638	0.1924	10	10000	1,2	5.24	5.24	19.79
859	5831	C5H12O	1-pentanol	71-41-0	18.4317	0.1001	10	3250	2	23.21	23.21	41.40
860	5832	C5H12O	2-pentanol	6032-29-7	18.6622	0.0922	10	8030	2	23.07	23.07	42.74
861	5833	C5H12O	3-pentanol	584-02-1	17.6342	0.0977	10	10000	2	22.09	22.09	43.39
862	5834	C5H12O	2-methyl-1-butanol	137-32-6	21.1526	0.0853	10	4116	2	25.74	25.74	43.01
863	5837	C5H12O	3-methyl-1-butanol	123-51-3	20.9258	0.0862	10	4187	2	25.52	25.52	42.93
864	5838	C5H12O	tert-pentyl-alcohol	75-85-4	14.9414	0.1109	10	10000	2	19.29	19.29	41.50
865	5839	C5H12O	3-methyl-2-butanol	598-75-4	17.3263	0.0988	10	10000	2	21.75	21.75	43.03
866	5840	C5H12O	2,2-dimethyl-1-propanol	75-84-3	17.3263	0.0988	10	10000	2	21.75	21.75	43.03
867	5841	C5H12O	methyl butyl ether	628-28-4	8.7028	0.1466	10	10000	2	12.20	12.20	33.57
868	5842	C5H12O	methyl isobutyl ether	625-44-5	8.3226	0.1446	10	10000	2	11.61	11.61	31.53
869	5843	C5H12O	methyl sec-butyl ether	6795-87-5	7.4316	0.1573	10	10000	2	10.68	10.68	31.65
870	5844	C5H12O	methyl tert-butyl ether	1634-04-4	6.9291	0.1624	10	10000	2	10.07	10.07	30.91
871	5845	C5H12O	ethyl propyl ether	628-32-0	7.7323	0.1528	10	10000	2	10.99	10.99	31.60
872	5846	C5H12O	ethyl isopropyl ether	625-54-7	6.9595	0.1593	10	10000	2	10.04	10.04	30.19
873	5857	C5H12O2	ethylene glycol monopropyl ether	2807-30-9	25.3219	0.0760	10	4104	2	30.17	30.17	47.67
874	5859	C5H12O2	1,2-pentanediol	5343-92-0	39.8728	0.0535	10	72	2	45.10	45.10	50.12
875	5861	C5H12O2	1,3-pentanediol	3174-67-2	41.4269	0.0548	1	38	2	41.43	47.00	50.57
876	5862	C5H12O2	1,4-pentanediol	626-95-9	41.6369	0.0530	1	33	2	41.64	47.05	50.12
877	5863	C5H12O2	1,5-pentanediol	111-29-5	47.5614	0.0500	1	4.6	2	47.56	---	51.34
878	5866	C5H12O2	DL-erythro-2,3-pentanediol	61828-35-1	38.1231	0.0582	10	116	2	43.59	43.59	50.28
879	5867	C5H12O2	DL-threo-2,3-pentanediol	61828-36-2	38.0980	0.0584	10	116	2	43.58	43.58	50.29
880	5868	C5H12O2	DL-2,4-pentanediol	1825-14-5	38.0980	0.0584	10	116	2	43.58	43.58	50.29

Table 1 Adsorption Capacity of Activated Carbon - Gas

$\log_{10} Q = \log_{10} K + 1/n \log_{10} y$ (Q - g of compound/100 g of carbon, y - ppmv)

NO	ID	FORMULA	NAME	CAS No	K	1/n	ymin	ymax	code	Q @ymin	Q @10ppm	Q @ymax
881	5869	C5H12O2	meso-2,4-pentanediol	3950-21-8	38.0980	0.0584	10	116	2	43.58	43.58	50.29
882	5870	C5H12O2	2-methyl-1,2-butanediol	41051-72-3	37.8938	0.0503	10	253	2	42.55	42.55	50.06
883	5871	C5H12O2	2-methyl-1,3-butanediol	684-84-4	38.3940	0.0491	10	219	2	42.99	42.99	50.01
884	5872	C5H12O2	2-methyl-1,4-butanediol	2938-98-9	38.4310	0.0491	10	214	2	43.03	43.03	50.02
885	5873	C5H12O2	2-methyl-2,3-butanediol	5396-58-7	34.4367	0.0596	10	592	2	39.50	39.50	50.38
886	5874	C5H12O2	3-methyl-1,2-butanediol	50468-22-9	38.2933	0.0589	10	128	2	43.85	43.85	50.95
887	5875	C5H12O2	3-methyl-1,3-butanediol	2568-33-4	37.7837	0.0591	10	150	2	43.29	43.29	50.81
888	5876	C5H12O2	2-ethyl-1,3-propanediol	2612-29-5	39.0023	0.0584	10	117	2	44.61	44.61	51.50
889	5899	C5H12O3	2-(2-methoxyethoxy)ethanol	111-77-3	40.6785	0.0467	10	236	2	45.30	45.30	52.50
890	5921	C5H12S	pentyl mercaptan	110-66-7	21.7392	0.0746	10	10000	2	25.81	25.81	43.20
891	5922	C5H12S	2-pentanethiol	2084-19-7	19.4057	0.0831	10	10000	2	23.50	23.50	41.71
892	5923	C5H12S	3-pentanethiol	616-31-9	19.6862	0.0828	10	10000	2	23.82	23.82	42.22
893	5924	C5H12S	2-methyl-1-butanethiol	1878-18-8	20.5649	0.0798	10	10000	2	24.71	24.71	42.87
894	5925	C5H12S	3-methyl-1-butanethiol	541-31-1	20.3599	0.0794	10	10000	2	24.44	24.44	42.31
895	5926	C5H12S	2-methyl-2-butanethiol	1679-09-0	17.3690	0.0913	10	10000	2	21.43	21.43	40.25
896	5927	C5H12S	3-methyl-2-butanethiol	2084-18-6	19.0773	0.0854	10	10000	2	23.22	23.22	41.88
897	5928	C5H12S	2,2-dimethyl-1-propanethiol	1679-08-9	18.0713	0.0886	10	10000	2	22.16	22.16	40.87
898	5930	C5H12S	methyl butyl sulfide	628-29-5	21.3667	0.0760	10	10000	2	25.45	25.45	43.02
899	5931	C5H12S	ethyl propyl sulfide	4110-50-3	20.6112	0.0783	10	10000	2	24.68	24.68	42.39
900	5932	C5H12S	3-methyl-2-thiapentane	10359-64-5	19.4795	0.0825	10	10000	2	23.56	23.56	41.65
901	5933	C5H12S	4-methyl-2-thiapentane	5008-69-5	19.5536	0.0825	10	10000	2	23.64	23.64	41.79
902	5934	C5H12S	2-methyl-3-thiapentane	5145-99-3	18.8465	0.0843	10	10000	2	22.89	22.89	40.98
903	5935	C5H12S	3,3-dimethyl-2-thiabutane	6163-64-0	17.5476	0.0902	10	10000	2	21.60	21.60	40.28
904	5947	C5H13N	pentylamine	110-58-7	13.8417	0.1084	10	10000	2	17.77	17.77	37.57
905	5948	C5H13N	1-methylbutylamine	625-30-9	12.7472	0.1133	10	10000	2	16.55	16.55	36.21
906	5949	C5H13N	2-methylbutylamine	96-15-1	13.4856	0.1104	10	10000	2	17.39	17.39	37.28
907	5950	C5H13N	3-methylbutylamine	107-85-7	13.6004	0.1092	10	10000	2	17.49	17.49	37.20
908	5951	C5H13N	1,1-dimethylpropylamine	594-39-8	10.7820	0.1255	10	10000	2	14.40	14.40	34.26
909	5952	C5H13N	1,2-dimethylpropylamine	598-74-3	11.9264	0.1206	10	10000	2	15.74	15.74	36.23
910	5953	C5H13N	2,2-dimethylpropylamine	5813-64-9	10.7757	0.1256	10	10000	2	14.39	14.39	34.27
911	5955	C5H13N	methyl-butylamine	110-68-9	12.6789	0.1128	10	10000	2	16.44	16.44	35.84
912	5956	C5H13N	methylisobutylamine	14610-37-8	11.0948	0.1216	10	10000	2	14.68	14.68	33.99
913	5957	C5H13N	methyl-sec-butylamine	7713-69-1	11.2786	0.1203	10	10000	2	14.88	14.88	34.16
914	5958	C5H13N	methyl-tert-butylamine	625-43-4	10.2176	0.1302	10	10000	2	13.79	13.79	33.89
915	5959	C5H13N	ethylpropylamine	20193-20-8	11.3236	0.1212	10	10000	2	14.97	14.97	34.58
916	5960	C5H13N	ethylisopropylamine	19961-27-4	10.6634	0.1254	10	10000	2	14.23	14.23	33.83
917	5961	C5H13N	methyldiethylamine	616-39-7	9.4746	0.1316	10	10000	2	12.83	12.83	31.85
918	5962	C5H13N	dimethyl-propylamine	926-63-6	9.1608	0.1333	10	10000	2	12.45	12.45	31.27
919	5963	C5H13N	dimethylisopropylamine	996-35-6	9.3541	0.1341	10	10000	2	12.74	12.74	32.16
920	6080	C6F6	hexafluorobenzene	392-56-3	19.7066	0.1447	10	10000	2	27.50	27.50	74.69
921	6305	C6H3Cl3	1,2,4-trichlorobenzene	120-82-1	53.1649	0.0527	10	566	2	60.03	60.03	74.26
922	6306	C6H3Cl3	1,2,3-trichlorobenzene	87-61-6	53.1649	0.0527	10	566	2	60.03	60.03	74.26
923	6307	C6H3Cl3	1,3,5-trichlorobenzene	108-70-3	53.1649	0.0527	10	566	2	60.03	60.03	74.26
924	6367	C6H3N3O6	1,3,5-trinitrobenzene	99-35-4	53.1649	0.0527	10	566	1,2	60.03	60.03	74.26
925	6411	C6H4Br2	m-dibromobenzene	108-36-1	83.1526	0.0319	10	353	2	89.49	89.49	100.27
926	6469	C6H4Cl2	o-dichlorobenzene	95-50-1	40.1808	0.0623	10	1783	1,2	46.38	46.38	64.05
927	6470	C6H4Cl2	m-dichlorobenzene	541-73-1	40.6056	0.0597	10	2821	1,2	46.59	46.59	65.23
928	6471	C6H4Cl2	p-dichlorobenzene	106-46-7	40.6056	0.0597	10	2821	1,2	46.59	46.59	65.23
929	6524	C6H4F2	m-difluorobenzene	372-18-9	14.5654	0.1451	10	10000	2	20.34	20.34	55.43
930	6525	C6H4F2	o-difluorobenzene	367-11-3	15.7034	0.1369	10	10000	2	21.52	21.52	55.40
931	6526	C6H4F2	p-difluorobenzene	540-36-3	14.6813	0.1439	10	10000	2	20.45	20.45	55.25
932	6656	C6H5Br	bromobenzene	108-86-1	46.3882	0.0598	10	5593	2	53.24	53.24	77.73
933	6692	C6H5Cl	chlorobenzene	108-90-7	17.4945	0.1238	10	10000	1,2	23.27	23.27	54.72
934	6720	C6H5ClO	m-chlorophenol	108-43-0	40.5285	0.0592	10	3329	2	46.45	46.45	65.52
935	6721	C6H5ClO	o-chlorophenol	95-57-8	40.5285	0.0592	10	3329	2	46.45	46.45	65.52
936	6722	C6H5ClO	p-chlorophenol	106-48-9	40.5285	0.0592	10	3329	2	46.45	46.45	65.52
937	6769	C6H5F	fluorobenzene	462-06-6	12.1131	0.1486	10	10000	2	17.06	17.06	47.61
938	6795	C6H5I	iodobenzene	591-50-4	70.2047	0.0407	10	1397	2	77.11	77.11	94.29
939	6819	C6H5NO2	nitrobenzene	98-95-3	47.9852	0.0440	10	328	2	53.10	53.10	61.92
940	6876	C6H6	benzene	71-43-2	9.2103	0.1593	10	10000	1,2	13.29	13.29	39.94
941	6923	C6H6ClN	m-chloroaniline	108-42-9	52.3568	0.0397	10	86	2	57.36	57.36	62.47
942	6924	C6H6ClN	o-chloroaniline	95-51-2	50.3221	0.0385	10	268	2	54.99	54.99	62.42
943	6925	C6H6ClN	p-chloroaniline	106-47-8	52.3568	0.0397	10	86	2	57.36	57.36	62.47
944	6979	C6H6N2	cis-dicyano-1-butene	2141-58-4	44.8468	0.0568	1	37	2	44.85	51.11	55.05
945	6980	C6H6N2	trans-dicyano-1-butene	2141-59-5	44.1217	0.0570	1	43	2	44.12	50.31	54.67
946	6981	C6H6N2	trans-1,4-dicyano-2-butene	1119-85-3	44.1217	0.0570	1	43	2	44.12	50.31	54.67
947	7057	C6H6O	phenol	108-95-2	33.5662	0.0444	10	10000	2	37.18	37.18	50.53
948	7121	C6H6S	phenyl mercaptan	108-98-5	37.6581	0.0518	10	1988	2	42.42	42.42	55.79
949	7172	C6H7N	aniline	62-53-3	37.4016	0.0534	10	645	2	42.29	42.29	52.83
950	7173	C6H7N	2-methylpyridine	109-06-8	21.5533	0.0893	10	10000	2	26.47	26.47	49.05
951	7174	C6H7N	3-methylpyridine	108-99-6	24.2898	0.0810	10	7967	2	29.27	29.27	50.28
952	7175	C6H7N	4-methylpyridine	108-89-4	24.4800	0.0801	10	7760	2	29.44	29.44	50.15
953	7267	C6H8	1,3-cyclohexadiene	592-57-4	10.2395	0.1439	10	10000	2	14.26	14.26	38.55
954	7268	C6H8	methylcyclopentadiene	26519-91-5	9.3611	0.1473	10	10000	2	13.14	13.14	36.34
955	7317	C6H8N2	methylglutaronitrile	4553-62-2	44.6771	0.0498	1	6.7	2	44.68	---	49.12
956	7321	C6H8N2	phenylhydrazine	100-63-0	48.7392	0.0419	1	33	2	48.74	53.67	56.43
957	7381	C6H8O	2-ethylfuran	3208-16-0	14.4995	0.1198	10	10000	2	19.11	19.11	43.71
958	7382	C6H8O	3-ethylfuran	67363-95-5	14.4778	0.1196	10	10000	2	19.07	19.07	43.55
959	7383	C6H8O	2,3-dimethylfuran	14920-89-9	14.8798	0.1173	10	10000	2	19.49	19.49	43.83
960	7384	C6H8O	2,4-dimethylfuran	3710-43-8	14.7842	0.1181	10	10000	2	19.40	19.40	43.87

Table 1 Adsorption Capacity of Activated Carbon - Gas

$\log_{10} Q = \log_{10} K + 1/n \log_{10} y$ (Q - g of compound/100 g of carbon, y - ppmv)

NO	ID	FORMULA	NAME	CAS No	K	1/n	ymin	ymax	code	Q @ymin	Q @10ppm	Q @ymax
961	7385	C6H8O	2,5-dimethylfuran	625-86-5	14.7437	0.1173	10	10000	2	19.32	19.32	43.43
962	7386	C6H8O	3,4-dimethylfuran	20843-07-6	12.8372	0.1280	10	10000	2	17.24	17.24	41.72
963	7446	C6H8O4	dimethyl maleate	624-48-6	47.3827	0.0410	10	227	2	52.07	52.07	59.19
964	7467	C6H8S	2-ethylthiophene	872-55-9	27.6241	0.0674	10	10000	2	32.26	32.26	51.39
965	7468	C6H8S	3-ethylthiophene	1795-01-3	28.0685	0.0665	10	10000	2	32.71	32.71	51.78
966	7469	C6H8S	2,3-dimethylthiophene	632-16-6	29.1779	0.0635	10	9644	2	33.77	33.77	52.25
967	7470	C6H8S	2,4-dimethylthiophene	638-00-6	28.6428	0.0646	10	10000	2	33.24	33.24	51.93
968	7471	C6H8S	2,5-dimethylthiophene	638-02-8	28.0925	0.0650	10	10000	2	32.63	32.63	51.11
969	7472	C6H8S	3,4-dimethylthiophene	632-15-5	29.7931	0.0628	10	8360	2	34.43	34.43	52.53
970	7619	C6H10	1-methylcyclopentene	693-89-0	9.3060	0.1446	10	10000	2	12.98	12.98	35.24
971	7620	C6H10	3-methylcyclopentene	1120-62-3	8.4186	0.1501	10	10000	2	11.90	11.90	33.56
972	7621	C6H10	4-methylcyclopentene	1759-81-5	8.9704	0.1464	10	10000	2	12.57	12.57	34.54
973	7622	C6H10	cyclohexene	110-83-8	10.1744	0.1414	10	10000	2	14.09	14.09	37.42
974	7623	C6H10	1,2-hexadiene	592-44-9	10.3655	0.1252	10	10000	2	13.83	13.83	32.83
975	7624	C6H10	1,cis-3-hexadiene	14596-92-0	10.6737	0.1200	10	10000	2	14.07	14.07	32.22
976	7625	C6H10	1,trans-3-hexadiene	20237-34-7	10.6737	0.1200	10	10000	2	14.07	14.07	32.22
977	7627	C6H10	1,cis-4-hexadiene	7318-67-4	9.0610	0.1336	10	10000	2	12.33	12.33	31.03
978	7628	C6H10	1,trans-4-hexadiene	7319-00-8	9.0610	0.1336	10	10000	2	12.33	12.33	31.03
979	7629	C6H10	1,5-hexadiene	592-42-7	8.1515	0.1412	10	10000	2	11.28	11.28	29.92
980	7630	C6H10	2,3-hexadiene	592-49-4	8.7442	0.1351	10	10000	2	11.93	11.93	30.34
981	7632	C6H10	cis-2,trans-4-hexadiene	5194-50-3	12.2719	0.1108	10	10000	2	15.84	15.84	34.06
982	7633	C6H10	cis-2,cis-4-hexadiene	6108-61-8	11.5931	0.1153	10	10000	2	15.12	15.12	33.53
983	7634	C6H10	trans-2,trans-4-hexadiene	5194-51-4	12.0501	0.1111	10	10000	2	15.56	15.56	33.53
984	7635	C6H10	3-methyl-1,2-pentadiene	7417-48-3	9.6464	0.1307	10	10000	2	13.03	13.03	32.16
985	7636	C6H10	4-methyl-1,2-pentadiene	13643-05-5	9.7063	0.1291	10	10000	2	13.07	13.07	31.87
986	7638	C6H10	2-methyl-1,cis-3-pentadiene	1501-60-6	11.0232	0.1193	10	10000	2	14.51	14.51	33.07
987	7639	C6H10	2-methyl-1,trans-3-pentadiene	926-54-5	11.0232	0.1193	10	10000	2	14.51	14.51	33.07
988	7640	C6H10	3-methyl-1,cis-3-pentadiene	2787-45-3	11.1256	0.1208	10	10000	2	14.69	14.69	33.85
989	7641	C6H10	3-methyl-1,trans-3-pentadiene	2787-43-1	11.1256	0.1208	10	10000	2	14.69	14.69	33.85
990	7642	C6H10	4-methyl-1,3-pentadiene	926-56-7	11.2678	0.1171	10	10000	2	14.76	14.76	33.14
991	7643	C6H10	2-methyl-1,4-pentadiene	763-30-4	7.5816	0.1461	10	10000	2	10.61	10.61	29.11
992	7644	C6H10	3-methyl-1,4-pentadiene	1115-08-8	7.9151	0.1434	10	10000	2	11.01	11.01	29.66
993	7645	C6H10	2-methyl-2,3-pentadiene	3043-33-2	9.8708	0.1283	10	10000	2	13.26	13.26	32.19
994	7646	C6H10	2-ethyl-1,3-butadiene	3404-63-5	10.9362	0.1196	10	10000	2	14.40	14.40	32.91
995	7647	C6H10	2,3-dimethyl-1,3-butadiene	513-81-5	9.9011	0.1295	10	10000	2	13.34	13.34	32.63
996	7648	C6H10	1-hexyne	693-02-7	8.5610	0.1434	10	10000	2	11.91	11.91	32.07
997	7649	C6H10	2-hexyne	764-35-2	10.2042	0.1312	10	10000	2	13.80	13.80	34.17
998	7650	C6H10	3-hexyne	928-49-4	9.9396	0.1318	10	10000	2	13.46	13.46	33.46
999	7651	C6H10	3-methyl-1-pentyne	922-59-8	7.4256	0.1518	10	10000	2	10.53	10.53	30.05
1000	7652	C6H10	4-methyl-1-pentyne	7154-75-8	7.7333	0.1489	10	10000	2	10.90	10.90	30.49
1001	7653	C6H10	4-methyl-2-pentyne	21020-27-9	9.0725	0.1380	10	10000	2	12.46	12.46	32.33
1002	7654	C6H10	3,3-dimethyl-1-butyne	917-92-0	5.9426	0.1616	10	10000	2	8.62	8.62	26.34
1003	7760	C6H10O	cyclohexanone	108-94-1	19.8144	0.0978	10	5690	1,2	24.82	24.82	46.17
1004	7761	C6H10O	mesityl oxide	141-79-7	20.8346	0.0821	10	10000	2	25.17	25.17	44.40
1005	7813	C6H10O2	epsilon-caprolactone	502-44-3	41.7432	0.0505	10	248	2	46.89	46.89	55.15
1006	7814	C6H10O2	ethyl methacrylate	97-63-2	20.5388	0.0880	10	10000	2	25.15	25.15	46.19
1007	7815	C6H10O2	propyl acrylate	925-60-0	22.0032	0.0810	10	10000	2	26.51	26.51	46.39
1008	7881	C6H10O3	ethylacetoacetate	141-97-9	36.7241	0.0533	10	1025	2	41.52	41.52	53.15
1009	7882	C6H10O3	propionic anhydride	123-62-6	34.5760	0.0570	10	1486	2	39.43	39.43	52.44
1010	7918	C6H10O4	diethyl oxalate	95-92-1	41.7609	0.0452	10	543	2	46.34	46.34	55.51
1011	7919	C6H10O4	ethylene glycol diacetate	111-55-7	46.3033	0.0447	10	101	2	51.32	51.32	56.90
1012	7920	C6H10O4	ethylidene diacetate	542-10-9	40.0600	0.0489	10	765	2	44.83	44.83	55.41
1013	7979	C6H11Br	cis-1-bromo-1-hexene	13154-12-6	39.6735	0.0512	10	5464	2	44.64	44.64	61.66
1014	7980	C6H11Br	trans-1-bromo-1-hexene	13154-13-7	39.6735	0.0512	10	5464	2	44.64	44.64	61.66
1015	8006	C6H11Cl	cis-1-chloro-1-hexene	50586-18-0	24.4451	0.0695	10	10000	2	28.68	28.68	46.35
1016	8007	C6H11Cl	trans-1-chloro-1-hexene	50586-19-1	24.4451	0.0695	10	10000	2	28.68	28.68	46.35
1017	8066	C6H11Cl3	1,1,1-trichlorohexane	17760-40-6	47.4633	0.0324	10	921	2	51.14	51.14	59.21
1018	8071	C6H11F	cis-1-fluoro-1-hexene	66225-46-5	12.9106	0.1155	10	10000	2	16.84	16.84	37.41
1019	8072	C6H11F	trans-1-fluoro-1-hexene	66225-47-6	12.9106	0.1155	10	10000	2	16.84	16.84	37.41
1020	8077	C6H11F3	1,1,1-trifluorohexane	------	13.9283	0.1279	10	10000	2	18.70	18.70	45.25
1021	8080	C6H11I	cis-1-iodo-1-hexene	16538-47-9	57.0419	0.0368	10	1536	2	62.09	62.09	74.74
1022	8081	C6H11I	trans-1-iodo-1-hexene	16644-98-7	57.0419	0.0368	10	1536	2	62.09	62.09	74.74
1023	8088	C6H11N	hexanenitrile	628-73-9	23.4453	0.0710	10	3746	2	27.61	27.61	42.04
1024	8089	C6H11N	2-methylpentanenitrile	6339-13-5	22.8849	0.0716	10	3998	2	26.98	26.98	41.43
1025	8090	C6H11N	2-ethylbutanenitrile	617-80-1	22.6936	0.0731	10	4407	2	26.85	26.85	41.89
1026	8091	C6H11N	3-methylpentanenitrile	21101-88-2	24.3615	0.0682	10	2982	2	28.50	28.50	42.03
1027	8092	C6H11N	4-methylpentanenitrile	542-54-1	24.2949	0.0685	10	2840	2	28.44	28.44	41.88
1028	8093	C6H11N	2,2-dimethylbutanenitrile	20654-46-0	19.4111	0.0807	10	10000	2	23.38	23.38	40.82
1029	8094	C6H11N	2,3-dimethylbutanenitrile	20654-44-8	19.6703	0.0820	10	10000	2	23.76	23.76	41.86
1030	8095	C6H11N	3,3-dimethylbutanenitrile	3302-16-7	19.6703	0.0820	10	10000	2	23.76	23.76	41.86
1031	8200	C6H12	1-methyl-1-ethylcyclopropane	53778-43-1	7.6922	0.1476	10	10000	2	10.81	10.81	29.97
1032	8201	C6H12	1-methyl-cis-2-ethylcyclopropane	19781-68-1	8.5595	0.1418	10	10000	2	11.86	11.86	31.59
1033	8202	C6H12	1-methyl-trans-2-ethylcyclopropane	19781-69-2	7.7821	0.1459	10	10000	2	10.89	10.89	29.83
1034	8203	C6H12	1,1,2-trimethylcyclopropane	4127-45-1	7.3747	0.1494	10	10000	2	10.40	10.40	29.19
1035	8204	C6H12	1,cis-2,cis-3-trimethylcyclopropane	4806-58-0	8.4529	0.1431	10	10000	2	11.75	11.75	31.57
1036	8205	C6H12	1,cis-2,trans-3-trimethylcyclopropane	4806-59-1	7.7344	0.1472	10	10000	2	10.86	10.86	30.02
1037	8206	C6H12	propylcyclopropane	2415-72-7	8.7671	0.1397	10	10000	2	12.09	12.09	31.75
1038	8207	C6H12	isopropylcyclopropane	3638-35-5	7.7621	0.1468	10	10000	2	10.88	10.88	29.99
1039	8208	C6H12	1,1-dimethylcyclobutane	------	7.7535	0.1481	10	10000	2	10.91	10.91	30.34
1040	8209	C6H12	1,cis-2-dimethylcyclobutane	15679-01-3	8.6440	0.1439	10	10000	2	12.04	12.04	32.53

Table 1 Adsorption Capacity of Activated Carbon - Gas

$\log_{10} Q = \log_{10} K + 1/n \log_{10} y$ (Q - g of compound/100 g of carbon, y - ppmv)

NO	ID	FORMULA	NAME	CAS No	K	1/n	ymin	ymax	code	Q @ymin	Q @10ppm	Q @ymax
1041	8210	C6H12	1,trans-2-dimethylcyclobutane	15679-02-4	7.9942	0.1462	10	10000	2	11.19	11.19	30.74
1042	8211	C6H12	1,cis-3-dimethylcyclobutane	2398-09-6	8.0112	0.1459	10	10000	2	11.21	11.21	30.71
1043	8212	C6H12	1,trans-3-dimethylcyclobutane	2398-10-9	7.7757	0.1467	10	10000	2	10.90	10.90	30.04
1044	8213	C6H12	ethylcyclobutane	4806-61-5	8.9798	0.1400	10	10000	2	12.39	12.39	32.59
1045	8214	C6H12	methylcyclopentane	96-37-7	8.9628	0.1431	10	10000	2	12.46	12.46	33.48
1046	8215	C6H12	cyclohexane	110-82-7	6.6035	0.1727	10	10000	1,2	9.83	9.83	32.39
1047	8216	C6H12	1-hexene	592-41-6	8.6122	0.1340	10	10000	2	11.72	11.72	29.58
1048	8218	C6H12	cis-2-hexene	7688-21-3	9.2414	0.1308	10	10000	2	12.49	12.49	30.82
1049	8219	C6H12	trans-2-hexene	4050-45-7	9.1442	0.1301	10	10000	2	12.34	12.34	30.31
1050	8221	C6H12	cis-3-hexene	7642-09-3	9.0418	0.1311	10	10000	2	12.23	12.23	30.25
1051	8222	C6H12	trans-3-hexene	13269-52-8	9.1224	0.1302	10	10000	2	12.31	12.31	30.26
1052	8223	C6H12	2-methyl-1-pentene	763-29-1	8.5660	0.1353	10	10000	2	11.70	11.70	29.79
1053	8224	C6H12	3-methyl-1-pentene	760-20-3	7.7982	0.1405	10	10000	2	10.78	10.78	28.44
1054	8225	C6H12	4-methyl-1-pentene	691-37-2	7.7804	0.1401	10	10000	2	10.74	10.74	28.27
1055	8227	C6H12	4-methyl-cis-2-pentene	691-38-3	8.1146	0.1376	10	10000	2	11.14	11.14	28.82
1056	8228	C6H12	4-methyl-trans-2-pentene	674-76-0	8.3428	0.1354	10	10000	2	11.40	11.40	29.04
1057	8229	C6H12	2-methyl-2-pentene	625-27-4	9.2873	0.1296	10	10000	2	12.52	12.52	30.65
1058	8230	C6H12	3-methyl-cis-2-pentene	922-62-3	9.2271	0.1315	10	10000	2	12.49	12.49	30.97
1059	8232	C6H12	3-methyl-trans-2-pentene	616-12-6	9.5453	0.1294	10	10000	2	12.86	12.86	31.45
1060	8233	C6H12	2,3-dimethyl-1-butene	563-78-0	8.0180	0.1397	10	10000	2	11.06	11.06	29.03
1061	8234	C6H12	2,3-dimethyl-2-butene	563-79-1	11.4619	0.1132	10	10000	2	14.87	14.87	32.50
1062	8235	C6H12	3,3-dimethyl-1-butene	558-37-2	6.8759	0.1469	10	10000	2	9.64	9.64	26.60
1063	8236	C6H12	2-ethyl-1-butene	760-21-4	8.8340	0.1345	10	10000	2	12.04	12.04	30.48
1064	8245	C6H12Br2	1,1-dibromohexane	58133-26-9	67.3227	0.0292	10	416	2	72.01	72.01	80.31
1065	8273	C6H12Cl2	1,1-dichlorohexane	62017-16-7	37.1987	0.0443	10	3113	2	41.20	41.20	53.13
1066	8298	C6H12F2	1,1-difluorohexane	62127-41-7	16.3499	0.1077	10	10000	2	20.95	20.95	44.09
1067	8305	C6H12I2	1,1-diiodohexane	66225-50-1	94.7503	0.0221	1	28	2	94.75	99.69	101.98
1068	8394	C6H12O	butyl vinyl ether	111-34-2	14.9263	0.1009	10	10000	2	18.83	18.83	37.81
1069	8400	C6H12O	cyclohexanol	108-93-0	32.5623	0.0623	10	993	2	37.58	37.58	50.04
1070	8401	C6H12O	hexanal	66-25-1	19.4477	0.0841	10	10000	2	23.60	23.60	42.20
1071	8402	C6H12O	2-methylpentanal	123-15-9	18.1610	0.0894	10	10000	2	22.31	22.31	41.38
1072	8403	C6H12O	3-methylpentanal	15877-57-3	18.9390	0.0854	10	10000	2	23.05	23.05	41.57
1073	8404	C6H12O	4-methylpentanal	1119-16-0	19.0526	0.0845	10	10000	2	23.15	23.15	41.51
1074	8405	C6H12O	2-ethylbutanal	97-96-1	18.2269	0.0900	10	10000	2	22.42	22.42	41.74
1075	8406	C6H12O	2,2-dimethylbutanal	2094-75-9	16.0868	0.0992	10	10000	2	20.22	20.22	40.12
1076	8407	C6H12O	2,3-dimethylbutanal	2109-98-0	17.5343	0.0928	10	10000	2	21.71	21.71	41.20
1077	8408	C6H12O	3,3-dimethylbutanal	2987-16-8	16.1748	0.0996	10	10000	2	20.34	20.34	40.48
1078	8409	C6H12O	2-hexanone	591-78-6	19.5014	0.0836	10	10000	2	23.64	23.64	42.13
1079	8410	C6H12O	3-hexanone	589-38-8	18.5059	0.0887	10	10000	2	22.70	22.70	41.90
1080	8411	C6H12O	3-methyl-2-pentanone	565-61-7	17.4354	0.0928	10	10000	2	21.59	21.59	40.99
1081	8412	C6H12O	4-methyl-2-pentanone	108-10-1	17.1620	0.0936	10	10000	2	21.29	21.29	40.64
1082	8413	C6H12O	ethyl isopropyl ketone	565-69-5	17.5184	0.0930	10	10000	2	21.70	21.70	41.26
1083	8414	C6H12O	3,3-dimethyl-2-butanone	75-97-8	15.4390	0.1030	10	10000	2	19.57	19.57	39.86
1084	8416	C6H12O	2-ethyltetrahydrofuran	1003-30-1	16.9575	0.1013	10	10000	2	21.41	21.41	43.12
1085	8417	C6H12O	3-ethyltetrahydrofuran	93716-28-0	21.0647	0.0898	10	10000	2	25.91	25.91	48.18
1086	8418	C6H12O	2,2-dimethyltetrahydrofuran	1003-17-4	14.2406	0.1141	10	10000	2	18.52	18.52	40.72
1087	8419	C6H12O	2,3-dimethyltetrahydrofuran	------	14.9042	0.1102	10	10000	2	19.21	19.21	41.14
1088	8420	C6H12O	2,4-dimethyltetrahydrofuran	64265-26-5	15.0567	0.1093	10	10000	2	19.37	19.37	41.22
1089	8421	C6H12O	2,5-dimethyltetrahydrofuran	1003-38-9	13.7121	0.1172	10	10000	2	17.96	17.96	40.37
1090	8422	C6H12O	3,3-dimethyltetrahydrofuran	15833-75-7	15.3889	0.1074	10	10000	2	19.71	19.71	41.38
1091	8423	C6H12O	3,4-dimethyltetrahydrofuran	32970-37-9	16.9922	0.0987	10	10000	2	21.33	21.33	42.17
1092	8460	C6H12O2	hexanoic acid	142-62-1	39.7343	0.0443	10	57	2	44.00	44.00	47.53
1093	8461	C6H12O2	2-ethyl butyric acid	88-09-5	37.1154	0.0445	10	246	2	41.12	41.12	47.43
1094	8468	C6H12O2	pentyl formate	638-49-3	24.7452	0.0696	10	8095	2	29.04	29.04	46.28
1095	8469	C6H12O2	1-methylbutyl formate	58368-66-4	22.3564	0.0774	10	10000	2	26.72	26.72	45.62
1096	8470	C6H12O2	2-methylbutyl formate	35073-27-9	22.2245	0.0780	10	10000	2	26.59	26.59	45.57
1097	8471	C6H12O2	isopentyl formate	110-45-2	21.7883	0.0792	10	10000	2	26.15	26.15	45.19
1098	8472	C6H12O2	1-ethylpropyl formate	58368-67-5	22.2739	0.0778	10	10000	2	26.64	26.64	45.59
1099	8473	C6H12O2	1,1-dimethylpropyl formate	757-88-0	20.7994	0.0843	10	10000	2	25.26	25.26	45.22
1100	8474	C6H12O2	1,2-dimethylpropyl formate	66794-46-5	22.3989	0.0780	10	10000	2	26.80	26.80	45.92
1101	8475	C6H12O2	2,2-dimethylpropyl formate	23361-67-3	21.5565	0.0813	10	10000	2	25.99	25.99	45.57
1102	8476	C6H12O2	butyl acetate	123-86-4	22.3285	0.0774	10	10000	2	26.69	26.69	45.56
1103	8477	C6H12O2	isobutyl acetate	110-19-0	20.4219	0.0844	10	10000	2	24.81	24.81	44.45
1104	8478	C6H12O2	sec-butyl acetate	105-46-4	19.6087	0.0878	10	10000	2	24.00	24.00	44.01
1105	8479	C6H12O2	tert-butyl acetate	540-88-5	16.9473	0.0990	10	10000	2	21.29	21.29	42.19
1106	8480	C6H12O2	propyl propanoate	106-36-5	21.4898	0.0809	10	10000	2	25.89	25.89	45.26
1107	8481	C6H12O2	isopropyl propanoate	637-78-5	20.4187	0.0833	10	10000	2	24.73	24.73	43.96
1108	8482	C6H12O2	ethyl butanoate	105-54-4	20.6846	0.0839	10	10000	2	25.09	25.09	44.79
1109	8483	C6H12O2	ethyl isobutanoate	97-62-1	19.1816	0.0889	10	10000	2	23.54	23.54	43.52
1110	8484	C6H12O2	cyclohexyl peroxide	766-07-4	44.9180	0.0456	1	29	2	44.92	49.89	52.38
1111	8485	C6H12O2	diacetone alcohol	123-42-2	30.6037	0.0603	10	2247	2	35.16	35.16	48.75
1112	8531	C6H12O3	2-ethoxyethyl acetate	111-15-9	32.1621	0.0561	10	3081	2	36.59	36.59	50.45
1113	8533	C6H12O3	paraldehyde	123-63-7	26.3052	0.0722	10	10000	2	31.06	31.06	51.15
1114	8621	C6H12S	thiacycloheptane	4753-80-4	27.3153	0.0408	10	9862	2	30.00	30.00	39.74
1115	8622	C6H12S	2-ethylthiacyclopentane	1551-32-2	30.3239	0.0556	10	5574	2	34.47	34.47	48.99
1116	8623	C6H12S	3-ethylthiacyclopentane	62184-67-2	31.5041	0.0538	10	4006	2	35.66	35.66	49.24
1117	8624	C6H12S	2,2-dimethylthiacyclopentane	5161-75-1	25.7281	0.0655	10	10000	2	29.92	29.92	47.04
1118	8625	C6H12S	2,cis-3-dimethylthiacyclopentane	5161-77-3	25.0020	0.0679	10	10000	2	29.23	29.23	46.73
1119	8626	C6H12S	2,trans-3-dimethylthiacyclopentane	5161-78-4	25.0020	0.0679	10	10000	2	29.23	29.23	46.73
1120	8627	C6H12S	2,cis-4-dimethylthiacyclopentane	5161-79-5	25.0020	0.0679	10	10000	2	29.23	29.23	46.73

Table 1 Adsorption Capacity of Activated Carbon - Gas

$\log_{10} Q = \log_{10} K + 1/n \log_{10} y$ (Q - g of compound/100 g of carbon, y - ppmv)

NO	ID	FORMULA	NAME	CAS No	K	1/n	ymin	ymax	code	Q @ymin	Q @10ppm	Q @ymax
1121	8628	C6H12S	2,trans-4-dimethylthiacyclopentane	5161-80-8	25.0020	0.0679	10	10000	2	29.23	29.23	46.73
1122	8629	C6H12S	2,cis-5-dimethylthiacyclopentane	5161-13-7	27.9987	0.0586	10	9879	2	32.04	32.04	47.98
1123	8630	C6H12S	2,trans-5-dimethylthiacyclopentane	5161-14-8	27.6585	0.0593	10	10000	2	31.70	31.70	47.74
1124	8631	C6H12S	3,3-dimethylthiacyclopentane	5161-76-2	25.6223	0.0656	10	10000	2	29.80	29.80	46.87
1125	8632	C6H12S	3,cis-4-dimethylthiacyclopentane	5161-11-5	24.8988	0.0679	10	10000	2	29.12	29.12	46.55
1126	8633	C6H12S	3,trans-4-dimethylthiacyclopentane	5161-12-6	24.8988	0.0679	10	10000	2	29.12	29.12	46.55
1127	8634	C6H12S	2-methylthiacyclohexane	5161-16-0	29.6413	0.0567	10	6996	2	33.78	33.78	48.99
1128	8635	C6H12S	3-methylthiacyclohexane	5258-50-4	30.3920	0.0557	10	5714	2	34.55	34.55	49.20
1129	8636	C6H12S	4-methylthiacyclohexane	5161-17-1	30.4873	0.0555	10	5540	2	34.64	34.64	49.18
1130	8647	C6H13Br	1-bromohexane	111-25-1	39.3365	0.0508	10	5394	2	44.21	44.21	60.85
1131	8648	C6H13Br	2-bromohexane	3377-86-4	39.6498	0.0497	10	4666	2	44.46	44.46	60.34
1132	8649	C6H13Br	3-bromohexane	3377-87-5	39.3681	0.0506	10	5293	2	44.23	44.23	60.76
1133	8650	C6H13Br	1-bromo-2-methylpentane	25346-33-2	40.0273	0.0499	10	4643	2	44.90	44.90	60.98
1134	8651	C6H13Br	1-bromo-3-methylpentane	51116-73-5	40.5267	0.0500	10	3822	2	45.47	45.47	61.23
1135	8652	C6H13Br	1-bromo-4-methylpentane	626-88-0	40.4059	0.0485	10	4013	2	45.18	45.18	60.44
1136	8653	C6H13Br	2-bromo-2-methylpentane	4283-80-1	36.5851	0.0555	10	10000	2	41.58	41.58	61.02
1137	8654	C6H13Br	2-bromo-3-methylpentane	62168-41-6	37.2511	0.0536	10	10000	2	42.14	42.14	61.02
1138	8655	C6H13Br	2-bromo-4-methylpentane	30310-22-6	36.9445	0.0532	10	9119	2	41.76	41.76	60.03
1139	8656	C6H13Br	3-bromo-2-methylpentane	4283-83-4	37.2511	0.0536	10	10000	2	42.14	42.14	61.02
1140	8657	C6H13Br	3-bromo-3-methylpentane	25346-31-0	36.5782	0.0558	10	10000	2	41.59	41.59	61.16
1141	8658	C6H13Br	1-bromo-2,2-dimethylbutane	62168-42-7	38.5546	0.0518	10	6564	2	43.44	43.44	60.80
1142	8659	C6H13Br	1-bromo-2,3-dimethylbutane	30540-31-9	39.4852	0.0516	10	6289	2	44.46	44.46	61.99
1143	8660	C6H13Br	1-bromo-3,3-dimethylbutane	1647-23-0	38.0998	0.0512	10	6853	2	42.87	42.87	59.89
1144	8661	C6H13Br	2-bromo-2,3-dimethylbutane	594-52-5	36.6833	0.0558	10	10000	2	41.71	41.71	61.33
1145	8662	C6H13Br	2-bromo-3,3-dimethylbutane	26356-06-9	37.4494	0.0533	10	9604	2	42.34	42.34	61.03
1146	8663	C6H13Br	1-bromo-2-ethylbutane	3814-34-4	39.4660	0.0512	10	5656	2	44.40	44.40	61.43
1147	8669	C6H13Cl	1-chlorohexane	544-10-5	25.1967	0.0641	10	10000	2	29.20	29.20	45.47
1148	8670	C6H13Cl	2-chlorohexane	638-28-8	24.4483	0.0659	10	10000	2	28.46	28.46	44.86
1149	8671	C6H13Cl	3-chlorohexane	2346-81-8	24.3991	0.0658	10	10000	2	28.39	28.39	44.72
1150	8672	C6H13Cl	1-chloro-2-methylpentane	14753-05-0	24.2290	0.0668	10	10000	2	28.26	28.26	44.84
1151	8673	C6H13Cl	1-chloro-3-methylpentane	62016-93-7	25.6160	0.0631	10	10000	2	29.62	29.62	45.79
1152	8674	C6H13Cl	1-chloro-4-methylpentane	62016-94-8	24.9971	0.0642	10	10000	2	28.98	28.98	45.14
1153	8675	C6H13Cl	2-chloro-2-methylpentane	4325-48-8	22.2170	0.0731	10	10000	2	26.29	26.29	43.57
1154	8676	C6H13Cl	2-chloro-3-methylpentane	24319-09-3	22.1241	0.0744	10	10000	2	26.26	26.26	43.91
1155	8677	C6H13Cl	2-chloro-4-methylpentane	25346-32-1	22.5064	0.0719	10	10000	2	26.56	26.56	43.62
1156	8678	C6H13Cl	3-chloro-2-methylpentane	38384-05-3	22.8651	0.0717	10	10000	2	26.97	26.97	44.25
1157	8679	C6H13Cl	3-chloro-3-methylpentane	918-84-3	23.2752	0.0705	10	10000	2	27.38	27.38	44.57
1158	8680	C6H13Cl	1-chloro-2-ethylbutane	4737-41-1	25.2278	0.0654	10	10000	2	29.33	29.33	46.08
1159	8681	C6H13Cl	1-chloro-2,2-dimethylbutane	6366-35-4	23.1662	0.0716	10	10000	2	27.32	27.32	44.78
1160	8682	C6H13Cl	1-chloro-2,3-dimethylbutane	600-06-6	24.2667	0.0588	10	10000	2	27.79	27.79	41.73
1161	8683	C6H13Cl	1-chloro-3,3-dimethylbutane	2855-08-5	23.1896	0.0700	10	10000	2	27.24	27.24	44.18
1162	8684	C6H13Cl	2-chloro-2,3-dimethylbutane	594-57-0	22.6035	0.0653	10	10000	2	26.27	26.27	41.23
1163	8685	C6H13Cl	2-chloro-3,3-dimethylbutane	5750-00-5	22.2743	0.0743	10	10000	2	26.43	26.43	44.17
1164	8705	C6H13F	1-fluorohexane	373-14-8	13.6455	0.1124	10	10000	2	17.68	17.68	38.42
1165	8706	C6H13F	2-fluorohexane	372-54-3	12.7985	0.1167	10	10000	2	16.74	16.74	37.48
1166	8707	C6H13F	3-fluorohexane	52688-75-2	12.5053	0.1188	10	10000	2	16.44	16.44	37.35
1167	8708	C6H13F	1-fluoro-2-methylpentane	62127-27-9	12.6271	0.1163	10	10000	2	16.50	16.50	36.84
1168	8709	C6H13F	1-fluoro-3-methylpentane	62127-28-0	13.1686	0.1131	10	10000	2	17.08	17.08	37.31
1169	8710	C6H13F	1-fluoro-4-methylpentane	62127-29-1	9.0727	0.1509	10	10000	2	12.84	12.84	36.40
1170	8711	C6H13F	2-fluoro-2-methylpentane	62127-30-4	11.9418	0.1204	10	10000	2	15.76	15.76	36.21
1171	8712	C6H13F	2-fluoro-3-methylpentane	62127-31-5	12.0170	0.1200	10	10000	2	15.84	15.84	36.28
1172	8713	C6H13F	2-fluoro-4-methylpentane	62127-32-6	12.0170	0.1200	10	10000	2	15.84	15.84	36.28
1173	8714	C6H13F	3-fluoro-2-methylpentane	62127-33-7	11.5324	0.1230	10	10000	2	15.31	15.31	35.80
1174	8715	C6H13F	3-fluoro-3-methylpentane	19031-61-9	10.9780	0.1265	10	10000	2	14.69	14.69	35.21
1175	8716	C6H13F	1-fluoro-2,2-dimethylbutane	62127-34-8	12.0863	0.1195	10	10000	2	15.92	15.92	36.35
1176	8717	C6H13F	1-fluoro-2,3-dimethylbutane	62127-35-9	12.3603	0.1179	10	10000	2	16.21	16.21	36.60
1177	8718	C6H13F	1-fluoro-3,3-dimethylbutane	371-64-2	11.8974	0.1203	10	10000	2	15.69	15.69	36.02
1178	8719	C6H13F	2-fluoro-2,3-dimethylbutane	354-09-6	10.9780	0.1265	10	10000	2	14.69	14.69	35.21
1179	8720	C6H13F	2-fluoro-3,3-dimethylbutane	92089-78-6	11.2194	0.1250	10	10000	2	14.96	14.96	35.47
1180	8721	C6H13F	1-fluoro-2-ethylbutane	------	10.2390	0.1314	10	10000	2	13.86	13.86	34.35
1181	8724	C6H13I	1-iodohexane	638-45-9	54.9757	0.0383	10	2466	2	60.05	60.05	74.16
1182	8725	C6H13I	2-iodohexane	59654-13-6	59.3358	0.0322	10	563	2	63.91	63.91	72.78
1183	8726	C6H13I	3-iodohexane	31294-91-4	59.4653	0.0325	10	596	2	64.09	64.09	73.20
1184	8727	C6H13I	1-iodo-2-methylpentane	31294-94-7	59.4653	0.0325	10	596	2	64.09	64.09	73.20
1185	8728	C6H13I	1-iodo-3-methylpentane	24346-53-0	59.5384	0.0308	10	425	2	63.91	63.91	71.74
1186	8729	C6H13I	1-iodo-4-methylpentane	6196-80-1	60.3189	0.0318	10	447	2	64.90	64.90	73.24
1187	8730	C6H13I	2-iodo-2-methylpentane	31294-95-8	56.2992	0.0361	10	1554	2	61.17	61.17	73.38
1188	8731	C6H13I	2-iodo-3-methylpentane	24319-07-1	57.0080	0.0353	10	1264	2	61.83	61.83	73.34
1189	8732	C6H13I	2-iodo-4-methylpentane	31294-96-9	57.2332	0.0350	10	1183	2	62.04	62.04	73.33
1190	8733	C6H13I	3-iodo-2-methylpentane	31294-97-0	57.0080	0.0353	10	1264	2	61.83	61.83	73.34
1191	8734	C6H13I	3-iodo-3-methylpentane	24319-05-2	56.2992	0.0361	10	1554	2	61.17	61.17	73.38
1192	8735	C6H13I	1-iodo-2,2-dimethylbutane	31294-98-1	57.3284	0.0349	10	1149	2	62.13	62.13	73.32
1193	8736	C6H13I	1-iodo-2,3-dimethylbutane	31295-00-8	58.4145	0.0337	10	828	2	63.13	63.13	73.26
1194	8737	C6H13I	1-iodo-3,3-dimethylbutane	15672-88-5	58.3308	0.0338	10	849	2	63.05	63.05	73.26
1195	8738	C6H13I	2-iodo-2,3-dimethylbutane	594-59-2	56.8970	0.0366	10	1554	2	61.91	61.91	74.48
1196	8739	C6H13I	2-iodo-3,3-dimethylbutane	24556-56-7	56.9094	0.0354	10	1301	2	61.74	61.74	73.35
1197	8740	C6H13I	1-iodo-2-ethylbutane	------	58.8992	0.0342	10	844	2	63.72	63.72	74.15
1198	8742	C6H13N	cyclohexylamine	108-91-8	22.1213	0.0772	10	10000	2	26.43	26.43	45.05
1199	8745	C6H13N	hexamethyleneimine	111-49-9	23.2479	0.0740	10	10000	2	27.57	27.57	45.98
1200	8790	C6H13NO2	1-nitrohexane	646-14-0	35.7424	0.0433	10	748	2	39.49	39.49	47.60

Table 1 Adsorption Capacity of Activated Carbon - Gas

$\log_{10} Q = \log_{10} K + 1/n \log_{10} y$ (Q - g of compound/100 g of carbon, y - ppmv)

NO	ID	FORMULA	NAME	CAS No	K	1/n	ymin	ymax	code	Q @ymin	Q @10ppm	Q @ymax
1201	8791	C6H13NO2	2-nitrohexane	14255-44-8	35.7424	0.0433	10	748	2	39.49	39.49	47.60
1202	8792	C6H13NO2	3-nitrohexane	7286-40-0	35.9751	0.0443	10	748	2	39.84	39.84	48.23
1203	8793	C6H13NO2	1-nitro-2-methylpentane	------	35.8412	0.0451	10	881	2	39.77	39.77	48.68
1204	8794	C6H13NO2	1-nitro-3-methylpentane	------	35.8412	0.0451	10	881	2	39.77	39.77	48.68
1205	8795	C6H13NO2	1-nitro-4-methyl pentane	------	35.8412	0.0451	10	881	2	39.77	39.77	48.68
1206	8796	C6H13NO2	2-nitro-2-methylpentane	------	35.8412	0.0451	10	881	2	39.77	39.77	48.68
1207	8797	C6H13NO2	2-nitro-3-methylpentane	42202-42-6	35.5337	0.0450	10	1095	2	39.42	39.42	48.70
1208	8798	C6H13NO2	2-nitro-4-methylpentane	------	35.5116	0.0451	10	1103	2	39.40	39.40	48.70
1209	8799	C6H13NO2	3-nitro-2-methylpentane	------	35.5116	0.0451	10	1103	2	39.40	39.40	48.70
1210	8800	C6H13NO2	3-nitro-3-methylpentane	------	35.5116	0.0451	10	1103	2	39.40	39.40	48.70
1211	8801	C6H13NO2	1-nitro-2,2-dimethylbutane	2625-29-8	35.5116	0.0451	10	1103	2	39.40	39.40	48.70
1212	8802	C6H13NO2	1-nitro-2,3-dimethylbutane	66553-34-2	36.7970	0.0443	10	806	2	40.75	40.75	49.48
1213	8803	C6H13NO2	1-nitro-3,3-dimethylbutane	------	36.7970	0.0443	10	806	2	40.75	40.75	49.48
1214	8804	C6H13NO2	2-nitro-2,3-dimethylbutane	34075-28-0	34.9993	0.0460	10	1333	2	38.91	38.91	48.73
1215	8805	C6H13NO2	2-nitro-3,3-dimethylbutane	599-02-0	34.9993	0.0460	10	1333	2	38.91	38.91	48.73
1216	8806	C6H13NO2	1-nitro-2-ethylbutane	------	36.0350	0.0440	10	927	2	39.88	39.88	48.66
1217	8877	C6H14	hexane	110-54-3	9.0303	0.1291	10	10000	1,2	12.16	12.16	29.64
1218	8878	C6H14	2,2-dimethylbutane	75-83-2	7.5646	0.1394	10	10000	2	10.43	10.43	27.32
1219	8879	C6H14	2,3-dimethylbutane	79-29-8	8.2045	0.1358	10	10000	2	11.22	11.22	28.65
1220	8880	C6H14	2-methylpentane	107-83-5	8.3886	0.1328	10	10000	2	11.39	11.39	28.52
1221	8881	C6H14	3-methylpentane	96-14-0	8.6409	0.1325	10	10000	2	11.72	11.72	29.27
1222	8970	C6H14O	1-hexanol	111-27-3	25.1506	0.0676	10	1220	1,2	29.39	29.39	40.66
1223	8971	C6H14O	2-hexanol	626-93-7	25.2952	0.0637	10	3271	2	29.29	29.29	42.37
1224	8972	C6H14O	3-hexanol	623-37-0	23.2126	0.0695	10	6551	2	27.24	27.24	42.77
1225	8973	C6H14O	2-methyl-1-pentanol	105-30-6	26.4200	0.0628	10	2522	2	30.53	30.53	43.22
1226	8974	C6H14O	3-methyl-1-pentanol	589-35-5	27.8744	0.0581	10	1551	2	31.87	31.87	42.72
1227	8975	C6H14O	4-methyl-1-pentanol	626-89-1	27.9586	0.0568	10	1373	2	31.86	31.86	42.14
1228	8976	C6H14O	4-methyl-2-pentanol	108-11-2	22.8040	0.0699	10	6902	2	26.78	26.78	42.28
1229	8977	C6H14O	2-methyl-2-pentanol	590-36-3	22.0235	0.0725	10	8886	2	26.03	26.03	42.59
1230	8978	C6H14O	3-methyl-2-pentanol	565-60-6	24.5922	0.0671	10	4510	2	28.70	28.70	43.23
1231	8979	C6H14O	2-methyl-3-pentanol	565-67-3	22.9476	0.0709	10	7232	2	27.02	27.02	43.08
1232	8980	C6H14O	3-methyl-3-pentanol	77-74-7	21.4592	0.0759	10	10000	2	25.56	25.56	43.18
1233	8981	C6H14O	2-ethyl-1-butanol	97-95-0	27.2840	0.0606	10	2013	2	31.37	31.37	43.26
1234	8982	C6H14O	2,2-dimethyl-1-butanol	1185-33-7	24.6347	0.0668	10	4505	2	28.73	28.73	43.22
1235	8983	C6H14O	2,3-dimethyl-1-butanol	19550-30-2	26.7666	0.0616	10	2289	2	30.85	30.85	43.11
1236	8985	C6H14O	3,3-dimethyl-1-butanol	624-95-3	25.3677	0.0636	10	3150	2	29.37	29.37	42.35
1237	8986	C6H14O	2,3-dimethyl-2-butanol	594-60-5	21.4046	0.0754	10	10000	2	25.46	25.46	42.87
1238	8987	C6H14O	3,3-dimethyl-2-butanol	464-07-3	21.9190	0.0729	10	9827	2	25.92	25.92	42.84
1239	8993	C6H14O	dipropyl ether	111-43-3	13.9576	0.1021	10	10000	2	17.66	17.66	35.76
1240	8994	C6H14O	propyl isopropyl ether	627-08-7	13.0407	0.1064	10	10000	2	16.66	16.66	34.75
1241	8995	C6H14O	diisopropyl ether	108-20-3	11.3542	0.1157	10	10000	2	14.82	14.82	32.95
1242	8996	C6H14O	methyl pentyl ether	628-80-8	15.6433	0.0948	10	10000	2	19.46	19.46	37.46
1243	8997	C6H14O	methyl 1-methylbutyl ether	6795-88-6	14.4487	0.1006	10	10000	2	18.22	18.22	36.50
1244	8998	C6H14O	methyl 2-methylbutyl ether	62016-48-2	14.4411	0.0999	10	10000	2	18.18	18.18	36.24
1245	8999	C6H14O	methyl 3-methylbutyl ether	626-91-5	14.3177	0.1009	10	10000	2	18.06	18.06	36.28
1246	9000	C6H14O	methyl 1-ethylpropyl ether	36839-67-5	14.0120	0.1031	10	10000	2	17.76	17.76	36.20
1247	9001	C6H14O	methyl 1,2-dimethylpropyl ether	62016-49-3	13.4325	0.1069	10	10000	2	17.18	17.18	35.96
1248	9002	C6H14O	methyl tert-pentyl ether	994-05-8	13.5825	0.1075	10	10000	2	17.40	17.40	36.56
1249	9003	C6H14O	methyl 2,2-dimethylpropyl ether	1118-00-9	13.3351	0.1075	10	10000	2	17.08	17.08	35.88
1250	9004	C6H14O	ethyl butyl ether	628-81-9	14.5075	0.0996	10	10000	2	18.25	18.25	36.32
1251	9005	C6H14O	ethyl isobutyl ether	627-02-1	13.1801	0.1056	10	10000	2	16.81	16.81	34.86
1252	9006	C6H14O	ethyl sec-butyl ether	2679-87-0	13.2163	0.1061	10	10000	2	16.87	16.87	35.11
1253	9007	C6H14O	ethyl tert-butyl ether	637-92-3	12.1269	0.1123	10	10000	2	15.71	15.71	34.12
1254	9020	C6H14O2	acetal	105-57-7	18.5399	0.0868	10	10000	2	22.64	22.64	41.25
1255	9021	C6H14O2	2-butoxyethanol	111-76-2	32.4446	0.0512	10	1147	2	36.50	36.50	46.53
1256	9022	C6H14O2	hexylene glycol	107-41-5	42.3107	0.0395	1	16	2	42.31	46.34	47.21
1257	9023	C6H14O2	1,2-hexanediol	6920-22-5	43.8164	0.0400	1	17	2	43.82	48.05	49.08
1258	9024	C6H14O2	1,3-hexanediol	21531-91-9	43.8164	0.0400	1	17	2	43.82	48.05	49.08
1259	9025	C6H14O2	1,4-hexanediol	16432-53-4	46.3955	0.0422	1	6	2	46.40	---	50.04
1260	9026	C6H14O2	1,5-hexanediol	928-40-5	43.8532	0.0410	1	21	2	43.85	48.19	49.68
1261	9027	C6H14O2	1,6-hexanediol	629-11-8	43.8532	0.0410	1	21	2	43.85	48.19	49.68
1262	9028	C6H14O2	2,3-hexanediol	617-30-1	42.2978	0.0442	1	36	2	42.30	46.83	49.56
1263	9030	C6H14O2	2,5-hexanediol	2935-44-6	42.2978	0.0442	1	36	2	42.30	46.83	49.56
1264	9031	C6H14O2	3,4-hexanediol	922-17-8	42.2978	0.0442	1	36	2	42.30	46.83	49.56
1265	9032	C6H14O2	2-methyl-1,2-pentanediol	20667-05-4	41.7955	0.0414	10	58	2	45.98	45.98	49.46
1266	9033	C6H14O2	2-methyl-1,3-pentanediol	149-31-5	42.2189	0.0452	1	43	2	42.22	46.85	50.04
1267	9034	C6H14O2	2-methyl-1,5-pentanediol	42856-62-2	42.6771	0.0440	1	36	2	42.68	47.23	49.96
1268	9035	C6H14O2	2-methyl-2,3-pentanediol	7795-80-4	42.5301	0.0444	1	30	2	42.53	47.10	49.46
1269	9036	C6H14O2	3-methyl-1,3-pentanediol	33879-72-0	40.8147	0.0455	10	84	2	45.32	45.32	49.93
1270	9037	C6H14O2	3-methyl-1,5-pentanediol	4457-71-0	45.4131	0.0447	1	8.9	2	45.41	---	50.08
1271	9038	C6H14O2	3-methyl-2,3-pentanediol	63521-37-9	42.9597	0.0438	1	30	2	42.96	47.52	49.86
1272	9039	C6H14O2	3-methyl-2,4-pentanediol	5683-44-3	44.0552	0.0478	1	12	2	44.06	49.18	49.61
1273	9040	C6H14O2	4-methyl-1,4-pentanediol	1462-10-8	41.6617	0.0455	1	47	2	41.66	46.26	49.64
1274	9041	C6H14O2	DL-erythro-4-methyl-2,3-pentanediol	6702-10-9	41.6617	0.0455	1	47	2	41.66	46.26	49.64
1275	9042	C6H14O2	DL-threo-4-methyl-2,3-pentanediol	6464-40-0	41.6617	0.0455	1	47	2	41.66	46.26	49.64
1276	9043	C6H14O2	2-ethyl-1,2-butanediol	66553-16-0	40.9363	0.0450	10	86	2	45.40	45.40	50.01
1277	9044	C6H14O2	2-ethyl-1,3-butanediol	66553-17-1	40.9363	0.0450	10	86	2	45.40	45.40	50.01
1278	9045	C6H14O2	2-ethyl-1,4-butanediol	57716-79-7	41.4028	0.0445	10	86	2	45.87	45.87	50.47
1279	9046	C6H14O2	2,2-dimethyl-1,3-butanediol	76-35-7	39.3298	0.0484	10	136	2	43.96	43.96	49.88
1280	9047	C6H14O2	2,2-dimethyl-1,4-butanediol	32812-23-0	40.0424	0.0459	10	109	2	44.51	44.51	49.66

Table 1 Adsorption Capacity of Activated Carbon - Gas

$\log_{10} Q = \log_{10} K + 1/n \log_{10} y$ (Q - g of compound/100 g of carbon, y - ppmv)

NO	ID	FORMULA	NAME	CAS No	K	1/n	ymin	ymax	code	Q @ymin	Q @10ppm	Q @ymax
1281	9048	C6H14O2	2,3-dimethyl-1,2-butanediol	66553-15-9	40.0424	0.0459	10	109	2	44.51	44.51	49.66
1282	9049	C6H14O2	2,3-dimethyl-1,3-butanediol	24893-35-4	40.0963	0.0465	10	109	2	44.62	44.62	49.86
1283	9050	C6H14O2	DL-2,3 -dimethyl-1,4-butanediol	------	42.8454	0.0444	1	36	2	42.85	47.46	50.23
1284	9051	C6H14O2	meso-2,3-dimethyl-1,4-butanediol	------	42.8338	0.0445	1	36	2	42.83	47.45	50.24
1285	9052	C6H14O2	2,3-dimethyl-2,3-butanediol	76-09-5	42.8338	0.0445	1	36	2	42.83	47.45	50.24
1286	9053	C6H14O2	3,3-dimethyl-1,2-butanediol	59562-82-2	42.8338	0.0445	1	36	2	42.83	47.45	50.24
1287	9054	C6H14O2	2-propyl-1,3-propanediol	2612-28-4	40.1928	0.0456	10	105	2	44.64	44.64	49.69
1288	9055	C6H14O2	2-isopropyl-1,3-propanediol	2612-27-3	42.7951	0.0447	1	36	2	42.80	47.44	50.23
1289	9056	C6H14O2	2-methyl-2-ethyl-1,3-propanediol	77-84-9	42.7951	0.0447	1	36	2	42.80	47.44	50.23
1290	9075	C6H14O3	diethylene glycol dimethyl ether	111-96-6	31.0965	0.0553	10	3910	2	35.32	35.32	49.11
1291	9076	C6H14O3	dipropylene glycol	25265-71-8	44.9744	0.0419	1	41	2	44.97	49.53	52.54
1292	9077	C6H14O3	2-(2-ethoxyethoxy)ethanol	111-90-0	41.9846	0.0367	10	166	2	45.69	45.69	50.64
1293	9092	C6H14O4	triethylene glycol	112-27-6	56.6079	0.0319	1	1.7	2	56.61	---	57.57
1294	9107	C6H14S	hexyl mercaptan	111-31-9	28.1557	0.0506	10	5576	2	31.64	31.64	43.58
1295	9108	C6H14S	2-hexanethiol	1679-06-7	26.3003	0.0541	10	10000	2	29.79	29.79	43.30
1296	9112	C6H14S	dipropyl sulfide	111-47-7	26.9630	0.0527	10	8493	2	30.44	30.44	43.43
1297	9113	C6H14S	diisopropyl sulfide	625-80-9	23.2999	0.0635	10	10000	2	26.97	26.97	41.81
1298	9114	C6H14S	methyl pentyl sulfide	1741-83-9	25.9129	0.0563	10	10000	2	29.50	29.50	43.51
1299	9115	C6H14S	ethyl butyl sulfide	638-46-0	27.1276	0.0524	10	7991	2	30.60	30.60	43.42
1300	9116	C6H14S	3-methyl-2-thiahexane	13286-91-4	26.6481	0.0537	10	10000	2	30.16	30.16	43.71
1301	9117	C6H14S	4-methyl-2-thiahexane	15013-37-3	26.7583	0.0535	10	9705	2	30.26	30.26	43.72
1302	9118	C6H14S	5-methyl-2-thiahexane	13286-90-3	26.2720	0.0552	10	10000	2	29.83	29.83	43.69
1303	9119	C6H14S	2-methyl-3-thiahexane	5008-73-1	25.1856	0.0568	10	10000	2	28.71	28.71	42.51
1304	9120	C6H14S	4-methyl-3-thiahexane	5008-72-0	25.5325	0.0566	10	10000	2	29.08	29.08	42.99
1305	9121	C6H14S	5-methyl-3-thiahexane	1613-45-2	25.5043	0.0562	10	10000	2	29.03	29.03	42.79
1306	9122	C6H14S	3-ethyl-2-thiapentane	57093-84-2	25.4527	0.0564	10	10000	2	28.98	28.98	42.77
1307	9123	C6H14S	3,3-dimethyl-2-thiapentane	13286-92-5	24.5572	0.0580	10	10000	2	28.07	28.07	41.91
1308	9124	C6H14S	3,4-dimethyl-2-thiapentane	53897-51-1	24.7287	0.0575	10	10000	2	28.23	28.23	41.98
1309	9125	C6H14S	4,4-dimethyl-2-thiapentane	6079-57-8	24.5572	0.0580	10	10000	2	28.07	28.07	41.91
1310	9126	C6H14S	2,2-dimethyl-3-thiapentane	14290-92-7	23.4428	0.0620	10	10000	2	27.04	27.04	41.51
1311	9130	C6H14S2	dipropyl disulfide	629-19-6	40.2650	0.0305	10	720	2	43.20	43.20	49.22
1312	9135	C6H15Al	triethyl aluminum	97-93-8	36.8043	0.0351	10	80	2	39.90	39.90	42.92
1313	9173	C6H15N	hexylamine	111-26-2	21.2954	0.0677	10	10000	2	24.89	24.89	39.73
1314	9174	C6H15N	2-aminohexane	5329-79-3	22.4048	0.0640	10	8479	2	25.96	25.96	39.99
1315	9175	C6H15N	3-aminohexane	16751-58-9	23.1872	0.0631	10	6448	2	26.81	26.81	40.32
1316	9176	C6H15N	1-amino-2-methylpentane	13364-16-4	21.5505	0.0674	10	9879	2	25.17	25.17	40.06
1317	9177	C6H15N	1-amino-3-methylpentane	42245-37-4	21.9775	0.0657	10	9879	2	25.57	25.57	40.24
1318	9178	C6H15N	1-amino-4-methylpentane	5344-20-7	21.1084	0.0702	10	10000	2	24.81	24.81	40.30
1319	9179	C6H15N	2-amino-2-methylpentane	53310-02-4	17.8298	0.0808	10	10000	2	21.48	21.48	37.53
1320	9180	C6H15N	2-amino-3-methylpentane	35399-81-6	22.4875	0.0630	10	6934	2	26.00	26.00	39.27
1321	9181	C6H15N	2-amino-4-methylpentane	108-09-8	18.6178	0.0775	10	10000	2	22.26	22.26	38.02
1322	9182	C6H15N	3-amino-2-methylpentane	54287-41-1	22.4875	0.0630	10	6934	2	26.00	26.00	39.27
1323	9183	C6H15N	3-amino-3-methylpentane	3495-46-3	21.0156	0.0691	10	10000	2	24.64	24.64	39.71
1324	9184	C6H15N	1-amino-2,2-dimethylbutane	41781-17-3	19.3042	0.0764	10	10000	2	23.02	23.02	39.02
1325	9185	C6H15N	1-amino-2,3-dimethylbutane	66553-05-7	18.0707	0.0811	10	10000	2	21.78	21.78	38.15
1326	9186	C6H15N	1-amino-3,3-dimethylbutane	15673-00-4	22.0577	0.0656	10	8214	2	25.66	25.66	39.86
1327	9187	C6H15N	2-amino-2,3-dimethylbutane	4358-75-2	17.5667	0.0843	10	10000	2	21.33	21.33	38.18
1328	9188	C6H15N	2-amino-3,3-dimethyl butane	3850-30-4	17.5871	0.0842	10	10000	2	21.35	21.35	38.20
1329	9189	C6H15N	1-amino-2-ethylbutane	617-79-8	22.7279	0.0639	10	6448	2	26.33	26.33	39.81
1330	9191	C6H15N	methylpentylamine	25419-06-1	20.1638	0.0701	10	10000	2	23.70	23.70	38.46
1331	9192	C6H15N	methyl-1-methylbutylamine	51932-19-5	18.9703	0.0755	10	10000	2	22.57	22.57	38.02
1332	9193	C6H15N	methyl-2-methylbutylamine	90023-96-4	18.9703	0.0755	10	10000	2	22.57	22.57	38.02
1333	9194	C6H15N	methyl-3-methylbutylamine	52317-98-3	18.2389	0.0788	10	10000	2	21.87	21.87	37.69
1334	9195	C6H15N	methyl-1,1-dimethylpropylamine	------	18.3368	0.0784	10	10000	2	21.96	21.96	37.73
1335	9196	C6H15N	methyl-1,2-dimethylpropylamine	34317-39-0	18.9703	0.0755	10	10000	2	22.57	22.57	38.02
1336	9197	C6H15N	methyl-2,2-dimethylpropylamine	26153-91-3	18.3368	0.0784	10	10000	2	21.96	21.96	37.73
1337	9198	C6H15N	ethylbutylamine	13360-63-9	18.6103	0.0761	10	10000	2	22.17	22.17	37.50
1338	9199	C6H15N	ethyl-sec-butylamine	21035-44-9	16.9840	0.0831	10	10000	2	20.56	20.56	36.50
1339	9200	C6H15N	ethylisobutylamine	13205-60-2	16.8888	0.0837	10	10000	2	20.48	20.48	36.51
1340	9201	C6H15N	ethyl-tert-butylamine	4432-77-3	14.0501	0.0986	10	10000	2	17.63	17.63	34.83
1341	9202	C6H15N	di-propylamine	142-84-7	18.2511	0.0782	10	10000	2	21.85	21.85	37.49
1342	9203	C6H15N	diisopropylamine	108-18-9	14.2273	0.0949	10	10000	2	17.70	17.70	34.11
1343	9204	C6H15N	propylisopropylamine	21968-17-2	16.8446	0.0830	10	10000	2	20.39	20.39	36.19
1344	9206	C6H15N	methylethylpropylamine	4458-32-6	16.0499	0.0859	10	10000	2	19.56	19.56	35.40
1345	9207	C6H15N	methylethylisopropylamine	39198-07-7	15.9349	0.0870	10	10000	2	19.47	19.47	35.51
1346	9208	C6H15N	dimethylbutylamine	927-62-8	15.7502	0.0873	10	10000	2	19.26	19.26	35.19
1347	9209	C6H15N	dimethyl-sec-butylamine	921-04-0	16.2228	0.0875	10	10000	2	19.84	19.84	36.30
1348	9210	C6H15N	dimethylisobutylamine	7239-24-9	14.2873	0.0961	10	10000	2	17.82	17.82	34.61
1349	9211	C6H15N	dimethyl-tert-butylamine	918-02-5	15.6333	0.0908	10	10000	2	19.27	19.27	36.08
1350	9212	C6H15N	triethylamine	121-44-8	14.7619	0.0934	10	10000	2	18.30	18.30	34.90
1351	9239	C6H15NO3	triethanolamine	102-71-6	14.5539	0.0866	10	10000	2	17.77	17.77	32.31
1352	9247	C6H15N3	N-aminoethyl piperazine	140-31-8	43.4416	0.0354	10	77	2	47.13	47.13	50.65
1353	9268	C6H15O4P	triethyl phosphate	78-40-0	43.6612	0.0349	10	748	2	47.32	47.32	55.01
1354	9356	C6H18N3OP	hexamethyl phosphoramide	680-31-9	47.0980	0.0262	10	60	2	50.03	50.03	52.44
1355	9362	C6H18OSi2	hexamethyldisiloxane	107-46-0	24.8486	0.0456	10	10000	2	27.60	27.60	37.83
1356	9378	C6H19NSi2	hexamethyldisilazane	999-97-3	28.3206	0.0360	10	10000	2	30.77	30.77	39.47
1357	9508	C7H3ClF3NO2	4-chloro-3-nitrobenzotrifluoride	121-17-5	67.8613	0.0305	10	78	2	72.80	72.80	77.52
1358	9684	C7H4ClF3	p-chlorobenzotrifluoride	98-56-6	39.9800	0.0509	10	10000	2	44.95	44.95	63.88
1359	9736	C7H4Cl2O	m-chlorobenzoyl chloride	618-46-2	62.4611	0.0313	10	183	2	67.14	67.14	73.54
1360	9799	C7H4F3NO2	3-nitrobenzotrifluoride	98-46-4	59.3548	0.0367	10	338	2	64.59	64.59	73.51

Table 1 Adsorption Capacity of Activated Carbon - Gas

$\log_{10} Q = \log_{10} K + 1/n \log_{10} y$ (Q - g of compound/100 g of carbon, y - ppmv)

NO	ID	FORMULA	NAME	CAS No	K	1/n	ymin	ymax	code	Q @ymin	Q @10ppm	Q @ymax
1361	9927	C7H5ClO	benzoyl chloride	98-88-4	47.2682	0.0413	10	821	2	51.98	51.98	62.36
1362	9967	C7H5Cl3	benzotrichloride	98-07-7	58.1803	0.0300	10	616	2	62.34	62.34	70.53
1363	10003	C7H5F3	benzotrifluoride	98-08-8	23.1537	0.1001	10	10000	2	29.16	29.16	58.23
1364	10045	C7H5N	benzonitrile	100-47-0	35.8347	0.0539	10	1009	2	40.57	40.57	52.01
1365	10049	C7H5NO	phenyl isocyanate	103-71-9	36.0291	0.0564	10	3384	2	41.03	41.03	57.00
1366	10252	C7H6Cl2	benzyl dichloride	98-87-3	49.9745	0.0390	10	672	2	54.67	54.67	64.41
1367	10253	C7H6Cl2	2,4-dichlorotoluene	95-73-8	51.4461	0.0349	10	602	2	55.75	55.75	64.33
1368	10404	C7H6O	benzaldehyde	100-52-7	35.8890	0.0559	10	1553	2	40.82	40.82	54.14
1369	10410	C7H6O2	salicylaldehyde	90-02-8	44.3453	0.0458	10	780	2	49.28	49.28	60.17
1370	10492	C7H7Cl	benzyl chloride	100-44-7	32.1450	0.0718	10	1716	1,2	37.92	37.92	54.85
1371	10493	C7H7Cl	o-chlorotoluene	95-49-8	35.6662	0.0537	10	4637	2	40.36	40.36	56.13
1372	10494	C7H7Cl	p-chlorotoluene	106-43-4	36.1541	0.0519	10	3670	2	40.74	40.74	55.33
1373	10495	C7H7Cl	m-chlorotoluene	108-41-8	35.8165	0.0529	10	4170	2	40.45	40.45	55.65
1374	10555	C7H7F	p-fluorotoluene	352-32-9	22.1611	0.0892	10	10000	2	27.22	27.22	50.42
1375	10626	C7H7NO2	m-nitrotoluene	99-08-1	48.6503	0.0354	10	272	2	52.79	52.79	59.34
1376	10627	C7H7NO2	o-nitrotoluene	88-72-2	49.1750	0.0351	10	242	2	53.32	53.32	59.63
1377	10628	C7H7NO2	p-nitrotoluene	99-99-0	48.6503	0.0354	10	272	2	52.79	52.79	59.34
1378	10660	C7H7NO3	o-nitroanisole	91-23-6	60.9581	0.0320	1	4.6	2	60.96	---	64.01
1379	10724	C7H8	1,3,5-cycloheptatriene	544-25-2	20.6843	0.0836	10	10000	2	25.08	25.08	44.68
1380	10725	C7H8	toluene	108-88-3	17.5566	0.0983	10	10000	1,2	22.02	22.02	43.41
1381	10850	C7H8O	anisole	100-66-3	30.6442	0.0620	10	4749	2	35.34	35.34	51.78
1382	10851	C7H8O	benzyl alcohol	100-51-6	39.1390	0.0630	10	107	2	45.25	45.25	52.54
1383	10852	C7H8O	m-cresol	108-39-4	42.0043	0.0478	10	148	2	46.89	46.89	53.34
1384	10853	C7H8O	o-cresol	95-48-7	42.0043	0.0478	10	148	2	46.89	46.89	53.34
1385	10854	C7H8O	p-cresol	106-44-5	42.0043	0.0478	10	148	2	46.89	46.89	53.34
1386	10918	C7H8S	o-methylthiophenol	137-06-4	40.7968	0.0395	10	1008	2	44.68	44.68	53.61
1387	10919	C7H8S	m-methylthiophenol	108-40-7	41.2064	0.0391	10	901	2	45.08	45.08	53.75
1388	10920	C7H8S	p-methylthiophenol	106-45-6	40.7968	0.0395	10	1008	2	44.68	44.68	53.61
1389	10921	C7H8S	methyl phenyl sulfide	100-68-5	41.6418	0.0390	10	977	2	45.55	45.55	54.46
1390	10953	C7H9N	benzylamine	100-46-9	37.0652	0.0468	10	857	2	41.28	41.28	50.83
1391	10954	C7H9N	N-methylaniline	100-61-8	38.7904	0.0421	10	596	2	42.74	42.74	50.76
1392	10955	C7H9N	m-toluidine	108-44-1	39.9172	0.0404	10	398	2	43.81	43.81	50.85
1393	10956	C7H9N	o-toluidine	95-53-4	40.6866	0.0397	10	338	2	44.59	44.59	51.28
1394	10957	C7H9N	p-toluidine	106-49-0	39.9172	0.0404	10	398	2	43.81	43.81	50.85
1395	10958	C7H9N	2,3-dimethylpyridine	583-61-9	30.5328	0.0579	10	3775	2	34.88	34.88	49.17
1396	10959	C7H9N	2,4-dimethylpyridine	108-47-4	29.7518	0.0583	10	4263	2	34.02	34.02	48.42
1397	10960	C7H9N	2,5-dimethylpyridine	589-93-5	29.4861	0.0584	10	4656	2	33.73	33.73	48.29
1398	10961	C7H9N	2,6-dimethylpyridine	108-48-5	27.9528	0.0610	10	7151	2	32.17	32.17	48.03
1399	10962	C7H9N	3,4-dimethylpyridine	583-58-4	33.3088	0.0533	10	1777	2	37.66	37.66	49.62
1400	10963	C7H9N	3,5-dimethylpyridine	591-22-0	31.8283	0.0550	10	2424	2	36.12	36.12	48.84
1401	10964	C7H9N	2-ethylpyridine	100-71-0	28.2310	0.0617	10	6576	2	32.54	32.54	48.57
1402	10965	C7H9N	3-ethylpyridine	536-78-7	30.4090	0.0580	10	3603	2	34.75	34.75	48.90
1403	10966	C7H9N	4-ethylpyridine	536-75-4	30.8179	0.0570	10	3237	2	35.14	35.14	48.86
1404	11225	C7H10S	2-propylthiophene	1551-27-5	33.6038	0.0476	10	4414	2	37.50	37.50	50.10
1405	11226	C7H10S	3-propylthiophene	1518-75-8	33.9856	0.0472	10	3967	2	37.89	37.89	50.24
1406	11227	C7H10S	2-isopropylthiophene	4095-22-1	32.8742	0.0489	10	5577	2	36.79	36.79	50.12
1407	11228	C7H10S	3-isopropylthiophene	29488-27-5	33.2876	0.0486	10	5050	2	37.23	37.23	50.39
1408	11229	C7H10S	2-methyl-3-ethylthiophene	53119-51-0	33.4740	0.0484	10	4760	2	37.42	37.42	50.42
1409	11230	C7H10S	2-methyl-4-ethyl thiophene	13678-54-1	34.3767	0.0465	10	3687	2	38.26	38.26	50.36
1410	11231	C7H10S	2-methyl-5-ethyl thiophene	40323-88-4	33.8193	0.0468	10	4172	2	37.67	37.67	49.97
1411	11232	C7H10S	3-methyl-2-ethylthiophene	31805-48-8	34.2600	0.0474	10	4015	2	38.21	38.21	50.77
1412	11233	C7H10S	3-methyl-4-ethyl thiophene	------	33.3983	0.0489	10	5360	2	37.38	37.38	50.82
1413	11234	C7H10S	3-methyl-5-ethylthiophene	------	33.3983	0.0489	10	5360	2	37.38	37.38	50.82
1414	11235	C7H10S	2,3,4-trimethylthiophene	1795-04-6	36.9793	0.0435	10	1915	2	40.87	40.87	51.36
1415	11236	C7H10S	2,3,5-trimethylthiophene	1795-05-7	34.5906	0.0461	10	3501	2	38.47	38.47	50.40
1416	11275	C7H11NO	cyclohexyl isocyanate	3173-53-3	40.0875	0.0460	10	1340	2	44.57	44.57	55.84
1417	11323	C7H12	1-ethylcyclopentene	2146-38-5	16.8967	0.0923	10	10000	2	20.90	20.90	39.55
1418	11324	C7H12	3-ethylcyclopentene	694-35-9	15.5680	0.0973	10	10000	2	19.48	19.48	38.16
1419	11325	C7H12	4-ethylcyclopentene	3742-38-9	15.6224	0.0970	10	10000	2	19.53	19.53	38.18
1420	11326	C7H12	1,2-dimethylcyclopentene	765-47-9	16.8596	0.0921	10	10000	2	20.84	20.84	39.38
1421	11327	C7H12	1,3-dimethylcyclopentene	62184-82-1	14.7963	0.0989	10	10000	2	18.58	18.58	36.81
1422	11328	C7H12	1,4-dimethylcyclopentene	19550-48-2	14.8855	0.0993	10	10000	2	18.71	18.71	37.15
1423	11329	C7H12	1,5-dimethylcyclopentene	16491-15-9	16.0251	0.0944	10	10000	2	19.92	19.92	38.23
1424	11330	C7H12	3,3-dimethylcyclopentene	58049-91-5	14.0732	0.1041	10	10000	2	17.88	17.88	36.71
1425	11331	C7H12	3,cis-4-dimethylcyclopentene	56039-55-5	14.7027	0.1010	10	10000	2	18.55	18.55	37.27
1426	11332	C7H12	3,trans-4-dimethylcyclopentene	53225-40-4	14.5725	0.1019	10	10000	2	18.43	18.43	37.26
1427	11333	C7H12	3,cis-5-dimethylcyclopentene	30213-29-7	14.5725	0.1019	10	10000	2	18.43	18.43	37.26
1428	11334	C7H12	3,trans-5-dimethylcyclopentene	61394-27-2	14.5725	0.1019	10	10000	2	18.43	18.43	37.26
1429	11335	C7H12	4,4-dimethylcyclopentene	19037-72-0	14.0732	0.1041	10	10000	2	17.88	17.88	36.71
1430	11336	C7H12	1-methylcyclohexene	591-49-1	17.6196	0.0901	10	10000	2	21.68	21.68	40.42
1431	11337	C7H12	3-methylcyclohexene	591-48-0	16.4190	0.0949	10	10000	2	20.43	20.43	39.35
1432	11338	C7H12	4-methylcyclohexene	591-47-9	16.3676	0.0950	10	10000	2	20.37	20.37	39.28
1433	11340	C7H12	1-heptyne	628-71-7	14.6510	0.0961	10	10000	2	18.28	18.28	35.49
1434	11341	C7H12	2-heptyne	1119-65-9	17.1886	0.0846	10	10000	2	20.88	20.88	37.45
1435	11342	C7H12	3-heptyne	2586-89-2	16.5438	0.0864	10	10000	2	20.19	20.19	36.66
1436	11343	C7H12	3-methyl-1-hexyne	40276-93-5	13.4390	0.1012	10	10000	2	16.97	16.97	34.14
1437	11344	C7H12	4-methyl-1-hexyne	52713-81-2	14.2607	0.0983	10	10000	2	17.88	17.88	35.27
1438	11345	C7H12	5-methyl-1-hexyne	2203-80-7	14.3289	0.0972	10	10000	2	17.92	17.92	35.08
1439	11346	C7H12	4-methyl-2-hexyne	20198-49-6	15.5492	0.0916	10	10000	2	19.20	19.20	36.16
1440	11347	C7H12	5-methyl-2-hexyne	53566-37-3	15.9065	0.0896	10	10000	2	19.55	19.55	36.32

Table 1 Adsorption Capacity of Activated Carbon - Gas

$\log_{10} Q = \log_{10} K + 1/n \log_{10} y$ (Q - g of compound/100 g of carbon, y - ppmv)

NO	ID	FORMULA	NAME	CAS No	K	1/n	ymin	ymax	code	Q @ymin	Q @10ppm	Q @ymax
1441	11348	C7H12	2-methyl-3-hexyne	36566-80-0	14.9766	0.0928	10	10000	2	18.54	18.54	35.20
1442	11349	C7H12	3-ethyl-1-pentyne	21020-26-8	13.3971	0.1026	10	10000	2	16.97	16.97	34.46
1443	11350	C7H12	3,3-dimethyl-1-pentyne	918-82-1	31.9589	0.0107	10	10000	2	32.75	32.75	35.26
1444	11351	C7H12	3,4-dimethyl-1-pentyne	61064-08-2	12.9170	0.1049	10	10000	2	16.45	16.45	33.95
1445	11352	C7H12	4,4-dimethyl-1-pentyne	13361-63-2	12.4233	0.1066	10	10000	2	15.88	15.88	33.15
1446	11353	C7H12	4,4-dimethyl-2-pentyne	999-78-0	13.4537	0.1003	10	10000	2	16.95	16.95	33.88
1447	11450	C7H12O2	butyl acrylate	141-32-2	28.3215	0.0563	10	7171	2	32.24	32.24	46.68
1448	11451	C7H12O2	isobutyl acrylate	106-63-8	27.0476	0.0581	10	10000	2	30.92	30.92	46.18
1449	11452	C7H12O2	propyl methacrylate	2210-28-8	28.0761	0.0567	10	8386	2	31.99	31.99	46.85
1450	11530	C7H12O4	diethyl malonate	105-53-3	43.7717	0.0362	10	354	2	47.58	47.58	54.13
1451	11569	C7H13Br	cis-1-bromo-1-heptene	39924-57-7	44.7373	0.0378	10	1737	2	48.81	48.81	59.32
1452	11570	C7H13Br	trans-1-bromo-1-heptene	53434-74-5	44.7373	0.0378	10	1737	2	48.81	48.81	59.32
1453	11587	C7H13Cl	cis-1-chloro-1-heptene	53268-66-9	31.2410	0.0473	10	4092	2	34.84	34.84	46.30
1454	11588	C7H13Cl	trans-1-chloro-1-heptene	53268-67-0	31.2410	0.0473	10	4092	2	34.84	34.84	46.30
1455	11612	C7H13Cl3	1,1,1-trichloroheptane	3922-26-7	49.8543	0.0256	10	302	2	52.88	52.88	57.69
1456	11613	C7H13F	cis-1-fluoro-1-heptene	28028-91-3	19.4224	0.0790	10	10000	2	23.30	23.30	40.22
1457	11614	C7H13F	trans-1-fluoro-1-heptene	28028-92-4	19.4224	0.0790	10	10000	2	23.30	23.30	40.22
1458	11618	C7H13F3	1,1,1-trifluoroheptane	------	22.3542	0.0833	10	10000	2	27.08	27.08	48.13
1459	11619	C7H13I	cis-1-iodo-1-heptene	63318-29-6	59.4914	0.0287	10	543	2	63.55	63.55	71.26
1460	11620	C7H13I	trans-1-iodo-1-heptene	60595-37-1	84.2412	0.0614	10	543	2	97.03	97.03	123.98
1461	11622	C7H13N	heptanenitrile	629-08-3	31.4828	0.0443	10	562	2	34.86	34.86	41.67
1462	11696	C7H14	ethylcyclopentane	1640-89-7	16.1125	0.0923	10	10000	2	19.93	19.93	37.71
1463	11697	C7H14	1,1-dimethylcyclopentane	1638-26-2	14.1197	0.1012	10	10000	2	17.82	17.82	35.85
1464	11699	C7H14	cis-1,2-dimethylcyclopentane	1192-18-3	15.6772	0.0951	10	10000	2	19.52	19.52	37.65
1465	11700	C7H14	trans-1,2-dimethylcyclopentane	822-50-4	14.5497	0.0985	10	10000	2	18.25	18.25	36.05
1466	11702	C7H14	cis-1,3-dimethylcyclopentane	2532-58-3	14.3890	0.0985	10	10000	2	18.05	18.05	35.66
1467	11703	C7H14	trans-1,3-dimethylcyclopentane	1759-58-6	14.4995	0.0984	10	10000	2	18.19	18.19	35.90
1468	11704	C7H14	methylcyclohexane	108-87-2	15.6065	0.0955	10	10000	1,2	19.44	19.44	37.60
1469	11705	C7H14	cycloheptane	291-64-5	18.8610	0.0841	10	10000	2	22.89	22.89	40.90
1470	11706	C7H14	1-heptene	592-76-7	14.8814	0.0889	10	10000	2	18.26	18.26	33.76
1471	11708	C7H14	cis-2-heptene	6443-92-1	15.5118	0.0867	10	10000	2	18.94	18.94	34.49
1472	11709	C7H14	trans-2-heptene	14686-13-6	15.4427	0.0863	10	10000	2	18.84	18.84	34.19
1473	11712	C7H14	cis-3-heptene	7642-10-6	15.2786	0.0872	10	10000	2	18.67	18.67	34.10
1474	11713	C7H14	trans-3-heptene	14686-14-7	15.2759	0.0867	10	10000	2	18.65	18.65	33.93
1475	11714	C7H14	2-methyl-1-hexene	6094-02-6	14.8100	0.0898	10	10000	2	18.21	18.21	33.87
1476	11715	C7H14	3-methyl-1-hexene	3404-61-3	13.8377	0.0938	10	10000	2	17.17	17.17	32.83
1477	11716	C7H14	4-methyl-1-hexene	3769-23-1	14.2107	0.0926	10	10000	2	17.59	17.59	33.36
1478	11717	C7H14	5-methyl-1-hexene	3524-73-0	14.0473	0.0928	10	10000	2	17.39	17.39	33.01
1479	11718	C7H14	2-methyl-2-hexene	2738-19-4	15.4830	0.0867	10	10000	2	18.90	18.90	34.41
1480	11719	C7H14	3-methyl-cis-2-hexene	10574-36-4	15.6876	0.0868	10	10000	2	19.16	19.16	34.88
1481	11720	C7H14	3-methyl-trans-2-hexene	20710-38-7	15.4597	0.0878	10	10000	2	18.92	18.92	34.69
1482	11721	C7H14	4-methyl-cis-2-hexene	3683-19-0	14.3396	0.0919	10	10000	2	17.72	17.72	33.43
1483	11722	C7H14	4-methyl-trans-2-hexene	3683-22-5	14.4787	0.0907	10	10000	2	17.84	17.84	33.39
1484	11723	C7H14	5-methyl-cis-2-hexene	13151-17-2	14.6667	0.0904	10	10000	2	18.06	18.06	33.72
1485	11724	C7H14	5-methyl-trans-2-hexene	7385-82-2	14.4812	0.0902	10	10000	2	17.82	17.82	33.22
1486	11725	C7H14	2-methyl-cis-3-hexene	15840-60-5	14.3249	0.0912	10	10000	2	17.67	17.67	33.18
1487	11726	C7H14	2-methyl-trans-3-hexene	692-24-0	14.2898	0.0907	10	10000	2	17.61	17.61	32.95
1488	11727	C7H14	3-methyl-cis-3-hexene	4914-89-0	15.6888	0.0862	10	10000	2	19.13	19.13	34.70
1489	11728	C7H14	3-methyl-trans-3-hexene	3899-36-3	15.3877	0.0874	10	10000	2	18.82	18.82	34.41
1490	11729	C7H14	2-ethyl-1-pentene	3404-71-5	14.9729	0.0899	10	10000	2	18.42	18.42	34.26
1491	11730	C7H14	3-ethyl-1-pentene	4038-04-4	13.8931	0.0942	10	10000	2	17.26	17.26	33.08
1492	11731	C7H14	2,3-dimethyl-1-pentene	3404-72-6	14.1161	0.0940	10	10000	2	17.53	17.53	33.56
1493	11732	C7H14	2,4-dimethyl-1-pentene	2213-32-3	13.6530	0.0951	10	10000	2	16.99	16.99	32.77
1494	11733	C7H14	3,3-dimethyl-1-pentene	3404-73-7	13.3456	0.0974	10	10000	2	16.70	16.70	32.72
1495	11734	C7H14	3,4-dimethyl-1-pentene	7385-78-6	13.6853	0.0955	10	10000	2	17.05	17.05	32.97
1496	11735	C7H14	4,4-dimethyl-1-pentene	762-62-9	12.5782	0.0998	10	10000	2	15.83	15.83	31.54
1497	11736	C7H14	3-ethyl-2-pentene	816-79-5	15.6294	0.0877	10	10000	2	19.12	19.12	35.04
1498	11737	C7H14	2,3-dimethyl-2-pentene	10574-37-5	15.9858	0.0866	10	10000	2	19.51	19.51	35.48
1499	11738	C7H14	2,4-dimethyl-2-pentene	625-65-0	14.1677	0.0919	10	10000	2	17.51	17.51	33.04
1500	11739	C7H14	3,4-dimethyl-cis-2-pentene	4914-91-4	14.8336	0.0910	10	10000	2	18.29	18.29	34.28
1501	11740	C7H14	3,4-dimethyl-trans-2-pentene	4914-92-5	15.1234	0.0898	10	10000	2	18.60	18.60	34.58
1502	11741	C7H14	4,4-dimethyl-cis-2-pentene	762-63-0	13.6253	0.0958	10	10000	2	16.99	16.99	32.91
1503	11742	C7H14	4,4-dimethyl-trans-2-pentene	690-08-4	13.2836	0.0962	10	10000	2	16.58	16.58	32.23
1504	11743	C7H14	3-methyl-2-ethyl-1-butene	7357-93-9	14.3447	0.0932	10	10000	2	17.78	17.78	33.84
1505	11744	C7H14	2,3,3-trimethyl-1-butene	594-56-9	13.2467	0.0988	10	10000	2	16.63	16.63	32.92
1506	11751	C7H14Br2	1,1-dibromoheptane	59104-79-9	67.0104	0.0263	10	148	2	71.20	71.20	76.43
1507	11763	C7H14Cl2	1,1-dichloroheptane	821-25-0	41.2172	0.0333	10	990	2	44.50	44.50	51.86
1508	11772	C7H14F2	1,1-difluoroheptane	407-96-5	23.9434	0.0703	10	10000	2	28.15	28.15	45.76
1509	11774	C7H14I2	1,1-diiodoheptane	66675-46-5	92.2641	0.0204	1	8.6	2	92.26	---	96.40
1510	11827	C7H14O	1-heptanal	111-71-7	26.2995	0.0566	10	4648	2	29.96	29.96	42.42
1511	11829	C7H14O	2-heptanone	110-43-0	25.9651	0.0574	10	4995	2	29.63	29.63	42.34
1512	11830	C7H14O	3-heptanone	106-35-4	26.0686	0.0576	10	4858	2	29.76	29.76	42.50
1513	11831	C7H14O	4-heptanone	123-19-3	24.2028	0.0618	10	9030	2	27.90	27.90	42.49
1514	11832	C7H14O	3-methyl-2-hexanone	2550-21-2	23.1854	0.0663	10	10000	2	27.01	27.01	42.71
1515	11833	C7H14O	4-methyl-2-hexanone	105-42-0	24.5589	0.0604	10	8115	2	28.22	28.22	42.30
1516	11834	C7H14O	5-methyl-2-hexanone	110-12-3	24.9775	0.0595	10	6843	2	28.65	28.65	42.24
1517	11835	C7H14O	2-methyl-3-hexanone	7379-12-6	23.4371	0.0644	10	10000	2	27.18	27.18	42.40
1518	11836	C7H14O	4-methyl-3-hexanone	17042-16-9	22.4616	0.0679	10	10000	2	26.26	26.26	41.99
1519	11837	C7H14O	5-methyl-3-hexanone	623-56-3	22.4227	0.0676	10	10000	2	26.20	26.20	41.79
1520	11838	C7H14O	3-ethyl-2-pentanone	6137-03-7	22.7453	0.0671	10	10000	2	26.55	26.55	42.22

Table 1 Adsorption Capacity of Activated Carbon - Gas

$\log_{10} Q = \log_{10} K + 1/n \log_{10} y$ (Q - g of compound/100 g of carbon, y - ppmv)

NO	ID	FORMULA	NAME	CAS No	K	1/n	ymin	ymax	code	Q @ymin	Q @10ppm	Q @ymax
1521	11839	C7H14O	3,3-dimethyl-2-pentanone	20669-04-9	21.6475	0.0708	10	10000	2	25.48	25.48	41.56
1522	11840	C7H14O	3,4-dimethyl-2-pentanone	565-78-6	21.9717	0.0700	10	10000	2	25.82	25.82	41.88
1523	11841	C7H14O	4,4-dimethyl-2-pentanone	590-50-1	20.7084	0.0733	10	10000	2	24.51	24.51	40.67
1524	11842	C7H14O	2,2-dimethyl-3-pentanone	564-04-5	20.7543	0.0741	10	10000	2	24.61	24.61	41.05
1525	11843	C7H14O	2,4-dimethyl-3-pentanone	565-80-0	21.3067	0.0864	10	10000	2	26.00	26.00	47.22
1526	11844	C7H14O	1-methylcyclohexanol	590-67-0	32.5441	0.0530	10	1836	2	36.77	36.77	48.48
1527	11845	C7H14O	cis-2-methylcyclohexanol	7443-70-1	32.5441	0.0530	10	1836	2	36.77	36.77	48.48
1528	11846	C7H14O	trans-2-methylcyclohexanol	7443-52-9	32.7358	0.0516	10	1575	2	36.87	36.87	47.86
1529	11847	C7H14O	cis-3-methylcyclohexanol	5454-79-5	35.3768	0.0438	10	697	2	39.13	39.13	47.13
1530	11848	C7H14O	trans-3-methylcyclohexanol	7443-55-2	34.8598	0.0469	10	753	2	38.83	38.83	47.55
1531	11849	C7H14O	cis-4-methylcyclohexanol	7731-28-4	36.3370	0.0431	10	429	2	40.12	40.12	47.17
1532	11850	C7H14O	trans-4-methylcyclohexanol	7731-29-5	35.6341	0.0442	10	511	2	39.45	39.45	46.95
1533	11884	C7H14O2	heptanoic acid	111-14-8	43.6369	0.0382	1	7.3	2	43.64	---	47.08
1534	11891	C7H14O2	hexyl formate	629-33-4	30.7977	0.0489	10	2747	2	34.47	34.47	45.37
1535	11892	C7H14O2	pentyl acetate	628-63-7	29.1619	0.0525	10	4610	2	32.91	32.91	45.42
1536	11893	C7H14O2	1-methylbutyl acetate	626-38-0	27.2285	0.0560	10	8029	2	30.98	30.98	45.05
1537	11894	C7H14O2	2-methylbutyl acetate	624-41-9	28.3921	0.0544	10	5741	2	32.18	32.18	45.47
1538	11895	C7H14O2	isopentyl acetate	123-92-2	27.7960	0.0546	10	7362	1,2	31.52	31.52	45.21
1539	11896	C7H14O2	1-ethylpropyl acetate	620-11-1	27.0759	0.0564	10	8349	2	30.83	30.83	45.07
1540	11897	C7H14O2	tert-pentyl acetate	625-16-1	25.9185	0.0604	10	10000	2	29.79	29.79	45.22
1541	11898	C7H14O2	1,2-dimethylpropyl acetate	5343-96-4	26.3067	0.0581	10	10000	2	30.07	30.07	44.92
1542	11899	C7H14O2	neopentyl acetate	926-41-0	25.9184	0.0586	10	10000	2	29.66	29.66	44.47
1543	11901	C7H14O2	butyl propanoate	590-01-2	28.8949	0.0533	10	4853	2	32.67	32.67	45.44
1544	11902	C7H14O2	isobutyl propanoate	540-42-1	27.5571	0.0561	10	6926	2	31.35	31.35	45.24
1545	11903	C7H14O2	sec-butyl propanoate	591-34-4	29.5673	0.0510	10	3498	2	33.25	33.25	44.82
1546	11904	C7H14O2	tert-butyl propanoate	20487-40-5	30.1692	0.0506	10	2819	2	33.90	33.90	45.09
1547	11905	C7H14O2	propyl butanoate	105-66-8	27.4137	0.0562	10	7716	2	31.20	31.20	45.32
1548	11906	C7H14O2	isopropyl butanoate	638-11-9	26.7086	0.0562	10	8897	2	30.40	30.40	44.51
1549	11909	C7H14O2	ethyl isovalerate	108-64-5	26.3540	0.0583	10	10000	2	30.14	30.14	45.10
1550	11931	C7H14O3	ethyl-3-ethoxypropionate	763-69-9	35.2761	0.0442	10	1690	2	39.05	39.05	48.98
1551	11960	C7H14S	thiacyclooctane	6572-99-2	37.5727	0.0395	10	1061	2	41.15	41.15	49.48
1552	11961	C7H14S	2-ethylthiacyclohexane	------	32.2171	0.0457	10	6416	2	35.79	35.79	48.10
1553	11962	C7H14S	3-ethylthiacyclohexane	------	32.2171	0.0457	10	6416	2	35.79	35.79	48.10
1554	11963	C7H14S	4-ethylthiacyclohexane	------	32.2171	0.0457	10	6416	2	35.79	35.79	48.10
1555	11964	C7H14S	2,2-dimethylthiacyclohexane	------	31.7838	0.0444	10	6786	2	35.21	35.21	47.04
1556	11965	C7H14S	2,cis-3-dimethylthiacyclohexane	------	31.2316	0.0454	10	8414	2	34.67	34.67	47.07
1557	11966	C7H14S	2,trans-3-dimethylthiacyclohexane	------	31.2316	0.0454	10	8414	2	34.67	34.67	47.07
1558	11967	C7H14S	2,cis-4-dimethylthiacyclohexane	------	31.2316	0.0454	10	8414	2	34.67	34.67	47.07
1559	11968	C7H14S	2,trans-4-dimethylthiacyclohexane	------	31.2316	0.0454	10	8414	2	34.67	34.67	47.07
1560	11969	C7H14S	2,cis-5-dimethylthiacyclohexane	------	31.2316	0.0454	10	8414	2	34.67	34.67	47.07
1561	11970	C7H14S	2,trans-5-dimethylthiacyclohexane	------	31.2316	0.0454	10	8414	2	34.67	34.67	47.07
1562	11971	C7H14S	2,cis-6-dimethylthiacyclohexane	------	31.2316	0.0454	10	8414	2	34.67	34.67	47.07
1563	11972	C7H14S	2,trans-6-dimethylthiacyclohexane	------	31.2316	0.0454	10	8414	2	34.67	34.67	47.07
1564	11973	C7H14S	3,3-dimethylthiacyclohexane	------	31.7838	0.0444	10	6786	2	35.21	35.21	47.04
1565	11974	C7H14S	3,cis-4-dimethylthiacyclohexane	------	31.2316	0.0454	10	8414	2	34.67	34.67	47.07
1566	11975	C7H14S	3,trans-4-dimethylthiacyclohexane	------	31.2316	0.0454	10	8414	2	34.67	34.67	47.07
1567	11976	C7H14S	3,cis-5-dimethylthiacyclohexane	------	31.2316	0.0454	10	8414	2	34.67	34.67	47.07
1568	11977	C7H14S	3,trans-5-dimethylthiacyclohexane	------	31.2316	0.0454	10	8414	2	34.67	34.67	47.07
1569	11978	C7H14S	4,4-dimethylthiacyclohexane	------	31.7838	0.0444	10	6786	2	35.21	35.21	47.04
1570	11982	C7H15Br	1-bromoheptane	629-04-9	44.4282	0.0375	10	1673	2	48.44	48.44	58.69
1571	11988	C7H15Cl	1-chloroheptane	629-06-1	30.9399	0.0458	10	4079	2	34.38	34.38	45.28
1572	12015	C7H15F	1-fluoroheptane	661-11-0	20.8768	0.0730	10	10000	2	24.70	24.70	40.89
1573	12020	C7H15I	1-iodoheptane	4282-40-0	57.5011	0.0302	10	983	2	61.64	61.64	70.78
1574	12024	C7H15N	N-methylcyclohexylamine	100-60-7	27.7805	0.0562	10	5619	2	31.62	31.62	45.14
1575	12112	C7H16	heptane	142-82-5	15.3837	0.0847	10	10000	1,2	18.69	18.69	33.55
1576	12113	C7H16	2-methylhexane	591-76-4	14.3894	0.0888	10	10000	2	17.65	17.65	32.60
1577	12114	C7H16	3-methylhexane	589-34-4	14.6500	0.0888	10	10000	2	17.97	17.97	33.19
1578	12115	C7H16	3-ethylpentane	617-78-7	14.9226	0.0888	10	10000	2	18.31	18.31	33.81
1579	12116	C7H16	2,2-dimethylpentane	590-35-2	13.1569	0.0957	10	10000	2	16.40	16.40	31.76
1580	12117	C7H16	2,3-dimethylpentane	565-59-3	14.4653	0.0907	10	10000	2	17.82	17.82	33.34
1581	12118	C7H16	2,4-dimethylpentane	108-08-7	13.3375	0.0938	10	10000	2	16.55	16.55	31.66
1582	12119	C7H16	3,3-dimethylpentane	562-49-2	14.0059	0.0925	10	10000	2	17.33	17.33	32.84
1583	12120	C7H16	2,2,3-trimethylbutane	464-06-2	13.4214	0.0959	10	10000	2	16.74	16.74	32.46
1584	12164	C7H16O	1-heptanol	111-70-6	34.1348	0.0379	10	283	2	37.24	37.24	42.27
1585	12165	C7H16O	2-heptanol	543-49-7	29.7862	0.0473	10	1620	2	33.21	33.21	42.24
1586	12166	C7H16O	3-heptanol	589-82-2	30.3119	0.0464	10	1364	2	33.73	33.73	42.36
1587	12167	C7H16O	4-heptanol	589-55-9	29.4714	0.0480	10	1810	2	32.92	32.92	42.26
1588	12168	C7H16O	2-methyl-1-hexanol	624-22-6	31.6618	0.0440	10	846	2	35.04	35.04	42.60
1589	12169	C7H16O	3-methyl-1-hexanol	13231-81-7	33.8410	0.0391	10	341	2	37.03	37.03	42.50
1590	12170	C7H16O	4-methyl-1-hexanol	818-49-5	33.6801	0.0389	10	352	2	36.83	36.83	42.30
1591	12171	C7H16O	5-methyl-1-hexanol	627-98-5	32.6187	0.0403	10	507	2	35.79	35.79	41.94
1592	12172	C7H16O	2-methyl-2-hexanol	625-23-0	27.9050	0.0511	10	3180	2	31.39	31.39	42.14
1593	12173	C7H16O	3-methyl-2-hexanol	2313-65-7	28.2718	0.0515	10	3106	2	31.83	31.83	42.77
1594	12174	C7H16O	4-methyl-2-hexanol	2313-61-3	31.3006	0.0440	10	938	2	34.64	34.64	42.30
1595	12175	C7H16O	5-methyl-2-hexanol	627-59-8	28.5499	0.0497	10	2486	2	32.01	32.01	42.10
1596	12176	C7H16O	2-methyl-3-hexanol	617-29-8	28.8661	0.0500	10	2428	2	32.39	32.39	42.62
1597	12177	C7H16O	3-methyl-3-hexanol	597-96-6	27.9849	0.0520	10	3438	2	31.54	31.54	42.74
1598	12178	C7H16O	4-methyl-3-hexanol	615-29-2	26.8820	0.0544	10	5068	2	30.47	30.47	42.75
1599	12179	C7H16O	5-methyl-3-hexanol	623-55-2	27.6191	0.0542	10	3848	2	31.29	31.29	43.20
1600	12180	C7H16O	2-ethyl-1-pentanol	27522-11-8	29.7171	0.0488	10	1981	2	33.25	33.25	43.06

Table 1 Adsorption Capacity of Activated Carbon - Gas

				$\log_{10} Q = \log_{10} K + 1/n \log_{10} y$			(Q - g of compound/100 g of carbon, y - ppmv)					
NO	ID	FORMULA	NAME	CAS No	K	1/n	ymin	ymax	code	Q @ymin	Q @10ppm	Q @ymax
1601	12181	C7H16O	3-ethyl-1-pentanol	66225-51-2	29.2030	0.0505	10	2266	2	32.81	32.81	43.15
1602	12182	C7H16O	2,2-dimethyl-1-pentanol	2370-12-9	27.3484	0.0544	10	4631	2	30.99	30.99	43.27
1603	12183	C7H16O	2,3-dimethyl-1-pentanol	10143-23-4	32.4099	0.0432	10	754	2	35.80	35.80	43.15
1604	12184	C7H16O	2,4-dimethyl-1-pentanol	6305-71-1	28.1512	0.0505	10	3309	2	31.62	31.62	42.39
1605	12185	C7H16O	3,3-dimethyl-1-pentanol	19264-94-9	32.5789	0.0422	10	642	2	35.91	35.91	42.81
1606	12186	C7H16O	3,4-dimethyl-1-pentanol	6570-87-2	30.6159	0.0465	10	1401	2	34.08	34.08	42.89
1607	12187	C7H16O	4,4-dimethyl-1-pentanol	3121-79-7	27.7379	0.0515	10	3451	2	31.23	31.23	42.21
1608	12188	C7H16O	3-ethyl-2-pentanol	609-27-8	27.6110	0.0540	10	4460	2	31.27	31.27	43.47
1609	12189	C7H16O	2,3-dimethyl-2-pentanol	4911-70-0	27.8915	0.0529	10	3774	2	31.50	31.50	43.12
1610	12190	C7H16O	2,4-dimethyl-2-pentanol	625-06-9	26.2870	0.0547	10	5625	2	29.81	29.81	42.14
1611	12191	C7H16O	3,3-dimethyl-2-pentanol	19781-24-9	26.3335	0.0553	10	7051	2	29.91	29.91	43.00
1612	12192	C7H16O	3,4-dimethyl-2-pentanol	64502-86-9	27.6813	0.0535	10	4596	2	31.31	31.31	43.46
1613	12193	C7H16O	4,4-dimethyl-2-pentanol	6144-93-0	27.2985	0.0521	10	3944	2	30.78	30.78	42.02
1614	12194	C7H16O	3-ethyl-3-pentanol	597-49-9	28.0649	0.0536	10	3988	2	31.75	31.75	43.76
1615	12195	C7H16O	2,2-dimethyl-3-pentanol	3970-62-5	26.7376	0.0549	10	5458	2	30.34	30.34	42.87
1616	12196	C7H16O	2,3-dimethyl-3-pentanol	595-41-5	27.7384	0.0540	10	4293	2	31.41	31.41	43.57
1617	12197	C7H16O	2,4-dimethyl-3-pentanol	600-36-2	27.2957	0.0538	10	4632	2	30.90	30.90	43.00
1618	12198	C7H16O	2-methyl-2-ethyl-1-butanol	18371-13-6	31.6900	0.0439	10	872	2	35.06	35.06	42.65
1619	12199	C7H16O	3-methyl-2-ethyl-1-butanol	32444-34-1	30.8180	0.0468	10	1356	2	34.33	34.33	43.20
1620	12200	C7H16O	2,2,3-trimethyl-1-butanol	55505-23-2	28.0558	0.0537	10	4217	2	31.75	31.75	43.93
1621	12201	C7H16O	2,3,3-trimethyl-1-butanol	36794-64-6	28.0986	0.0514	10	3724	2	31.63	31.63	42.89
1622	12202	C7H16O	2,3,3-trimethyl-2-butanol	594-83-2	24.3761	0.0627	10	10000	2	28.16	28.16	43.43
1623	12211	C7H16O	isopropyl-tert-butyl-ether	17348-59-3	18.1833	0.0768	10	10000	2	21.70	21.70	36.88
1624	12224	C7H16O2	1,2-heptanediol	3710-31-4	44.2888	0.0418	1	14	2	44.29	48.76	49.45
1625	12225	C7H16O2	1,3-heptanediol	23433-04-7	45.3453	0.0362	1	8.3	2	45.35	---	48.96
1626	12226	C7H16O2	1,4-heptanediol	40646-07-9	47.2871	0.0324	1	2.8	2	47.29	---	48.89
1627	12228	C7H16O2	1,6-heptanediol	13175-27-4	42.8887	0.0358	10	56	2	46.57	46.57	49.53
1628	12229	C7H16O2	1,7-heptanediol	629-30-1	46.5979	0.0348	1	4.8	2	46.60	---	49.21
1629	12230	C7H16O2	2,3-heptanediol	21508-07-6	44.4372	0.0353	1	19	2	44.44	48.20	49.31
1630	12231	C7H16O2	2,4-heptanediol	20748-86-1	44.3784	0.0358	1	19	2	44.38	48.19	49.31
1631	12232	C7H16O2	2,6-heptanediol	5969-12-0	44.5110	0.0347	1	19	2	44.51	48.21	49.30
1632	12233	C7H16O2	2-methyl-1,2-hexanediol	56255-50-6	42.6672	0.0387	1	45	2	42.67	46.65	49.45
1633	12234	C7H16O2	2-methyl-2,4-hexanediol	66225-35-2	44.7883	0.0347	1	16	2	44.79	48.51	49.31
1634	12235	C7H16O2	2-methyl-2,5-hexanediol	29044-06-2	44.7883	0.0347	1	16	2	44.79	48.51	49.31
1635	12236	C7H16O2	3-methyl-1,6-hexanediol	4089-71-8	42.6993	0.0359	10	56	2	46.37	46.37	49.33
1636	12237	C7H16O2	3-methyl-2,4-hexanediol	------	46.0520	0.0367	1	6.8	2	46.05	---	49.41
1637	12238	C7H16O2	3-methyl-3,4-hexanediol	18938-47-1	44.7035	0.0338	1	16	2	44.70	48.33	49.10
1638	12239	C7H16O2	4-methyl-1,5-hexanediol	66225-37-4	44.7041	0.0343	1	19	2	44.70	48.37	49.45
1639	12240	C7H16O2	4-methyl-2,4-hexanediol	38836-25-8	44.8735	0.0348	1	16	2	44.87	48.61	49.41
1640	12241	C7H16O2	5-methyl-1,5-hexanediol	1462-11-9	42.5964	0.0398	1	45	2	42.60	46.68	49.56
1641	12242	C7H16O2	5-methyl-2,4-hexanediol	54877-00-8	46.0804	0.0376	1	6.8	2	46.08	---	49.52
1642	12243	C7H16O2	2-ethyl-1,3-pentanediol	29887-11-4	43.3880	0.0374	1	34	2	43.39	47.29	49.51
1643	12244	C7H16O2	2-ethyl-1,5-pentanediol	14189-13-0	42.9675	0.0356	10	56	2	46.63	46.63	49.58
1644	12245	C7H16O2	3-ethyl-2,3-pentanediol	66225-32-9	41.4193	0.0388	10	104	2	45.29	45.29	49.59
1645	12246	C7H16O2	3-ethyl-2,4-pentanediol	66225-33-0	41.6123	0.0378	10	89	2	45.40	45.40	49.31
1646	12247	C7H16O2	2,2-dimethyl-1,3-pentanediol	2157-31-5	42.6259	0.0384	1	45	2	42.63	46.57	49.34
1647	12248	C7H16O2	2,2-dimethyl-1,5-pentanediol	3121-82-2	42.6259	0.0384	1	45	2	42.63	46.57	49.34
1648	12249	C7H16O2	2,3-dimethyl-1,3-pentanediol	66225-52-3	44.8143	0.0337	1	16	2	44.81	48.43	49.20
1649	12250	C7H16O2	2,3-dimethyl-2,3-pentanediol	6931-70-0	44.5509	0.0399	1	13	2	44.55	48.84	49.35
1650	12251	C7H16O2	DL-2,4-dimethyl-1,5-pentanediol	54630-82-9	44.1971	0.0343	1	19	2	44.20	47.83	48.89
1651	12252	C7H16O2	meso-2,4-dimethyl-1,5-pentanediol	2121-69-9	43.7807	0.0339	1	19	2	43.78	47.33	48.38
1652	12253	C7H16O2	2,4-dimethyl-2,3-pentanediol	66225-53-4	39.2601	0.0342	10	214	2	42.48	42.48	47.17
1653	12254	C7H16O2	2,4-dimethyl-2,4-pentanediol	24892-49-7	39.2601	0.0342	10	214	2	42.48	42.48	47.17
1654	12255	C7H16O2	3,3-dimethyl-1,5-pentanediol	53120-74-4	42.1336	0.0370	1	45	2	42.13	45.88	48.51
1655	12256	C7H16O2	3,4-dimethyl-1,4-pentanediol	63521-36-8	44.4643	0.0335	1	16	2	44.46	48.03	48.79
1656	12257	C7H16O2	4,4-dimethyl-1,2-pentanediol	66225-54-5	44.4617	0.0343	1	16	2	44.46	48.12	48.90
1657	12258	C7H16O2	erythro-4,4-dimethyl-2,3-pentanediol	23646-57-3	44.4617	0.0343	1	16	2	44.46	48.12	48.90
1658	12259	C7H16O2	threo-4,4-dimethyl-2,3-pentanediol	23646-58-4	44.4617	0.0343	1	16	2	44.46	48.12	48.90
1659	12260	C7H16O2	2-propyl-1,4-butanediol	62946-68-3	42.7201	0.0355	10	56	2	46.35	46.35	49.27
1660	12261	C7H16O2	2-isopropyl-1,4-butanediol	39497-66-0	43.8510	0.0337	1	19	2	43.85	47.39	48.43
1661	12262	C7H16O2	2-methyl-3-ethyl-1,4-butanediol	------	43.8510	0.0337	1	19	2	43.85	47.39	48.43
1662	12263	C7H16O2	2-sec-butyl-1,3-propanediol	33673-01-7	43.8510	0.0337	1	19	2	43.85	47.39	48.43
1663	12264	C7H16O2	2-methyl-2-propyl-1,3-propanediol	78-26-2	43.8510	0.0337	1	19	2	43.85	47.39	48.43
1664	12265	C7H16O2	2,2-diethyl-1,3-propanediol	115-76-4	43.8510	0.0337	1	19	2	43.85	47.39	48.43
1665	12300	C7H16S	heptyl mercaptan	1639-09-4	32.8161	0.0375	10	1718	2	35.77	35.77	43.38
1666	12301	C7H16S	2-heptanethiol	1628-00-2	30.9569	0.0404	10	3544	2	33.97	33.97	43.06
1667	12302	C7H16S	methyl hexyl sulfide	20291-60-5	30.6486	0.0415	10	4645	2	33.72	33.72	43.51
1668	12303	C7H16S	ethyl pentyl sulfide	26158-99-6	30.6486	0.0415	10	4645	2	33.72	33.72	43.51
1669	12304	C7H16S	propyl butyl sulfide	1613-46-3	30.6486	0.0415	10	4645	2	33.72	33.72	43.51
1670	12315	C7H17N	heptylamine	111-68-2	27.1134	0.0478	10	3604	2	30.27	30.27	40.12
1671	12322	C7H17N	methylhexylamine	35161-70-7	26.0691	0.0485	10	4659	2	29.15	29.15	39.26
1672	12323	C7H17N	ethylpentylamine	17839-26-8	24.7998	0.0508	10	7240	2	27.88	27.88	38.94
1673	12324	C7H17N	dimethylpentylamine	26153-88-8	22.1607	0.0574	10	10000	2	25.29	25.29	37.61
1674	12325	C7H17N	dimethyl-2-pentylamine	57303-85-2	20.7377	0.0611	10	10000	2	23.87	23.87	36.42
1675	12326	C7H17N	dimethyl-3-pentylamine	18636-94-7	22.1796	0.0599	10	10000	2	25.46	25.46	38.50
1676	12327	C7H17N	dimethyl-2-methylbutylamine	66225-39-6	22.0869	0.0604	10	10000	2	25.38	25.38	38.52
1677	12328	C7H17N	dimethyl-3-methylbutylamine	2315-43-7	21.9179	0.0610	10	10000	2	25.22	25.22	38.45
1678	12329	C7H17N	dimethyl-1,1-dimethylpropylamine	57757-60-5	22.0025	0.0580	10	10000	2	25.15	25.15	37.54
1679	12330	C7H17N	dimethyl-1,2-dimethylpropylamine	66225-38-5	22.1531	0.0574	10	10000	2	25.29	25.29	37.60
1680	12331	C7H17N	dimethyl-2,2-dimethylpropylamine	10076-31-0	21.8843	0.0585	10	10000	2	25.04	25.04	37.51

Table 1 Adsorption Capacity of Activated Carbon - Gas

$\log_{10} Q = \log_{10} K + 1/n \log_{10} y$ (Q - g of compound/100 g of carbon, y - ppmv)

NO	ID	FORMULA	NAME	CAS No	K	1/n	ymin	ymax	code	Q @ymin	Q @10ppm	Q @ymax
1681	12332	C7H17N	methylethylbutylamine	66225-40-9	22.2080	0.0572	10	10000	2	25.34	25.34	37.62
1682	12333	C7H17N	methylethyl-sec-butylamine	66225-41-0	19.9975	0.0639	10	10000	2	23.16	23.16	36.01
1683	12334	C7H17N	methylethylisobutylamine	60247-14-5	20.6453	0.0632	10	10000	2	23.88	23.88	36.94
1684	12335	C7H17N	methylethyl-tert-butylamine	52841-28-8	21.2431	0.0625	10	10000	2	24.53	24.53	37.77
1685	12336	C7H17N	methyldipropylamine	3405-42-3	22.0551	0.0603	10	10000	2	25.34	25.34	38.42
1686	12337	C7H17N	methylpropylisopropylamine	66225-42-1	21.7888	0.0613	10	10000	2	25.09	25.09	38.31
1687	12338	C7H17N	methyldiisopropylamine	10342-97-9	21.5812	0.0620	10	10000	2	24.89	24.89	38.19
1688	12339	C7H17N	diethylpropylamine	4458-31-5	21.5678	0.0606	10	10000	2	24.80	24.80	37.69
1689	12340	C7H17N	diethylisopropylamine	6006-15-1	21.0605	0.0627	10	10000	2	24.33	24.33	37.51
1690	12713	C8H6	ethynylbenzene	536-74-3	28.3428	0.0567	10	7666	2	32.29	32.29	47.05
1691	13237	C8H8	styrene	100-42-5	28.0564	0.0574	10	8044	1,2	32.02	32.02	47.01
1692	13238	C8H8	1,3,5,7-cyclooctatetraene	629-20-9	27.7215	0.0583	10	9950	2	31.71	31.71	47.42
1693	13391	C8H8O	acetophenone	98-86-2	41.1034	0.0403	10	521	2	45.10	45.10	52.89
1694	13392	C8H8O	p-tolualdehyde	104-87-0	42.4030	0.0369	10	333	2	46.17	46.17	52.55
1695	13412	C8H8O2	methyl benzoate	93-58-3	44.1534	0.0382	10	505	2	48.21	48.21	56.00
1696	13445	C8H8O3	methyl salicylate	119-36-8	52.9557	0.0354	1	45	2	52.96	57.45	60.59
1697	13789	C8H10	ethylbenzene	100-41-4	25.4997	0.0616	10	10000	2	29.38	29.38	44.95
1698	13790	C8H10	o-xylene	95-47-6	27.0996	0.0578	10	8721	2	30.96	30.96	45.78
1699	13791	C8H10	m-xylene	108-38-3	25.6457	0.0609	10	10000	1,2	29.51	29.51	44.95
1700	13792	C8H10	p-xylene	106-42-3	25.7331	0.0599	10	10000	2	29.54	29.54	44.67
1701	13912	C8H10O	phenetole	103-73-1	35.2505	0.0454	10	2054	2	39.14	39.14	49.85
1702	13913	C8H10O	2-phenylethanol	60-12-8	43.6532	0.0388	10	113	2	47.73	47.73	52.43
1703	13918	C8H10O	o-ethylphenol	90-00-6	42.7050	0.0358	10	283	2	46.37	46.37	52.27
1704	13919	C8H10O	m-ethylphenol	620-17-7	42.1515	0.0362	10	320	2	45.81	45.81	51.93
1705	13920	C8H10O	p-ethylphenol	123-07-9	42.1515	0.0362	10	320	2	45.81	45.81	51.93
1706	13921	C8H10O	2,3-xylenol	526-75-0	43.1917	0.0395	10	134	2	47.31	47.31	52.42
1707	13922	C8H10O	2,4-xylenol	105-67-9	43.1917	0.0395	10	134	2	47.31	47.31	52.42
1708	13923	C8H10O	2,5-xylenol	95-87-4	43.1917	0.0395	10	134	2	47.31	47.31	52.42
1709	13924	C8H10O	2,6-xylenol	576-26-1	43.1917	0.0395	10	134	2	47.31	47.31	52.42
1710	13925	C8H10O	3,4-xylenol	95-65-8	43.1917	0.0395	10	134	2	47.31	47.31	52.42
1711	13926	C8H10O	3,5-xylenol	108-68-9	43.1917	0.0395	10	134	2	47.31	47.31	52.42
1712	14039	C8H10S	o-ethylthiophenol	4500-58-7	44.1285	0.0306	10	417	2	47.35	47.35	53.09
1713	14040	C8H10S	m-ethylthiophenol	62154-77-2	44.3406	0.0305	10	399	2	47.57	47.57	53.24
1714	14041	C8H10S	p-ethylthiophenol	4946-13-8	44.3406	0.0305	10	399	2	47.57	47.57	53.24
1715	14042	C8H10S	2,3-dimethylthiophenol	------	44.8003	0.0298	10	320	2	47.99	47.99	53.21
1716	14043	C8H10S	2,4-dimethylthiophenol	13616-82-5	44.0170	0.0311	10	464	2	47.28	47.28	53.26
1717	14044	C8H10S	2,5-dimethylthiophenol	4001-61-0	43.7254	0.0315	10	530	2	47.01	47.01	53.28
1718	14045	C8H10S	2,6-dimethylthiophenol	118-72-9	44.8003	0.0298	10	320	2	47.99	47.99	53.21
1719	14046	C8H10S	3,4-dimethylthiophenol	18800-53-8	44.8003	0.0298	10	320	2	47.99	47.99	53.21
1720	14047	C8H10S	3,5-dimethylthiophenol	38360-81-5	44.8003	0.0298	10	320	2	47.99	47.99	53.21
1721	14048	C8H10S	(1-thiapropyl)-benzene	622-38-8	43.0578	0.0311	10	540	2	46.26	46.26	52.38
1722	14049	C8H10S	2-methyl-(1-thiaethyl)-benzene	------	43.7742	0.0309	10	446	2	47.00	47.00	52.85
1723	14050	C8H10S	3-methyl-(1-thiaethyl)-benzene	4886-77-5	43.9035	0.0304	10	446	2	47.08	47.08	52.84
1724	14051	C8H10S	4-methyl-(1-thiaethyl)-benzene	623-13-2	44.4324	0.0294	10	323	2	47.54	47.54	52.65
1725	14082	C8H11N	N,N-dimethylaniline	121-69-7	38.1993	0.0365	10	931	2	41.54	41.54	49.01
1726	14083	C8H11N	o-ethylaniline	578-54-1	42.2522	0.0321	10	224	2	45.49	45.49	50.27
1727	14096	C8H11N	2,4,6-trimethylpyridine	108-75-8	33.3190	0.0447	10	2601	2	36.93	36.93	47.37
1728	14124	C8H11NO	p-phenetidine	156-43-4	49.8365	0.0344	1	13	2	49.84	53.94	54.43
1729	14228	C8H12	1,5-cyclooctadiene	111-78-4	27.9247	0.0564	10	6515	2	31.80	31.80	45.84
1730	14231	C8H12	vinylcyclohexene	100-40-3	23.1135	0.0656	10	10000	2	26.88	26.88	42.30
1731	14380	C8H12O4	diethyl maleate	141-05-9	43.2004	0.0276	10	136	2	46.03	46.03	49.47
1732	14463	C8H14	1-propylcyclopentene	3074-61-1	23.4759	0.0611	10	10000	2	27.02	27.02	41.20
1733	14466	C8H14	1-ethylcyclohexene	1453-24-3	24.5871	0.0593	10	10000	2	28.18	28.18	42.45
1734	14467	C8H14	3-ethylcyclohexene	2808-71-1	23.6095	0.0615	10	10000	2	27.20	27.20	41.59
1735	14468	C8H14	4-ethylcyclohexene	3742-42-5	23.7539	0.0610	10	10000	2	27.34	27.34	41.66
1736	14469	C8H14	1,2-dimethylcyclohexene	1674-10-8	24.7146	0.0592	10	10000	2	28.32	28.32	42.64
1737	14470	C8H14	1,3-dimethylcyclohexene	2808-76-6	22.8406	0.0633	10	10000	2	26.43	26.43	40.93
1738	14471	C8H14	1,4-dimethylcyclohexene	2808-79-9	22.8138	0.0634	10	10000	2	26.40	26.40	40.92
1739	14472	C8H14	1,5-dimethylcyclohexene	2808-77-7	22.9914	0.0630	10	10000	2	26.58	26.58	41.07
1740	14473	C8H14	1,6-dimethylcyclohexene	1759-64-4	23.5887	0.0620	10	10000	2	27.21	27.21	41.74
1741	14474	C8H14	3,3-dimethylcyclohexene	695-28-3	21.8315	0.0673	10	10000	2	25.49	25.49	40.58
1742	14475	C8H14	3,cis-4-dimethylcyclohexene	------	22.2220	0.0672	10	10000	2	25.94	25.94	41.27
1743	14476	C8H14	3,trans-4-dimethylcyclohexene	------	22.2220	0.0672	10	10000	2	25.94	25.94	41.27
1744	14477	C8H14	3,cis-5-dimethylcyclohexene	------	22.2220	0.0672	10	10000	2	25.94	25.94	41.27
1745	14478	C8H14	3,trans-5-dimethylcyclohexene	------	22.2220	0.0672	10	10000	2	25.94	25.94	41.27
1746	14479	C8H14	3,cis-6-dimethylcyclohexene	------	22.2220	0.0672	10	10000	2	25.94	25.94	41.27
1747	14480	C8H14	3,trans-6-dimethylcyclohexene	3685-00-5	22.2220	0.0672	10	10000	2	25.94	25.94	41.27
1748	14481	C8H14	4,4-dimethylcyclohexene	14072-86-7	21.3793	0.0689	10	10000	2	25.05	25.05	40.33
1749	14482	C8H14	4,cis-5-dimethylcyclohexene	------	22.9255	0.0632	10	10000	2	26.51	26.51	41.01
1750	14483	C8H14	4,trans-5-dimethylcyclohexene	------	22.3180	0.0654	10	10000	2	25.94	25.94	40.75
1751	14484	C8H14	1-octyne	629-05-0	21.8183	0.0609	10	10000	2	25.10	25.10	38.23
1752	14485	C8H14	2-octyne	2809-67-8	23.7890	0.0549	10	9985	2	26.99	26.99	39.42
1753	14486	C8H14	3-octyne	15232-76-5	23.1374	0.0563	10	10000	2	26.34	26.34	38.86
1754	14599	C8H14O2	butyl methacrylate	97-88-1	33.1554	0.0420	10	2675	2	36.52	36.52	46.18
1755	14659	C8H14O3	butyric anhydride	106-31-0	40.6406	0.0332	10	372	2	43.87	43.87	49.46
1756	14688	C8H14O4	diethyl succinate	123-25-1	47.1677	0.0302	10	57	2	50.56	50.56	53.29
1757	14736	C8H15Br	cis-1-bromo-1-octene	42843-49-2	47.3308	0.0298	10	723	2	50.70	50.70	57.61
1758	14737	C8H15Br	trans-1-bromo-1-octene	51751-87-2	47.3308	0.0298	10	723	2	50.70	50.70	57.61
1759	14749	C8H15Cl	cis-1-chloro-1-octene	64531-23-3	35.9347	0.0347	10	1090	2	38.92	38.92	45.79
1760	14750	C8H15Cl	trans-1-chloro-1-octene	59871-24-8	35.9347	0.0347	10	1090	2	38.92	38.92	45.79

Table 1 Adsorption Capacity of Activated Carbon - Gas

$\log_{10} Q = \log_{10} K + 1/n \log_{10} y$ (Q - g of compound/100 g of carbon, y - ppmv)												
NO	ID	FORMULA	NAME	CAS No	K	1/n	ymin	ymax	code	Q @ymin	Q @10ppm	Q @ymax
1761	14765	C8H15Cl3	1,1,1-trichlorooctane	4905-79-7	50.6155	0.0237	10	101	2	53.45	53.45	56.46
1762	14766	C8H15F	cis-1-fluoro-1-octene	32814-16-7	25.3239	0.0548	10	10000	2	28.73	28.73	41.94
1763	14767	C8H15F	trans-1-fluoro-1-octene	32814-17-8	25.3239	0.0548	10	10000	2	28.73	28.73	41.94
1764	14770	C8H15F3	1,1,1-trifluorooctane	53392-86-2	29.0662	0.0573	10	10000	2	33.17	33.17	49.29
1765	14772	C8H15I	cis-1-iodo-1-octene	52356-93-1	60.6098	0.0231	10	196	2	63.92	63.92	68.48
1766	14773	C8H15I	trans-1-iodo-1-octene	42599-17-7	60.6098	0.0231	10	196	2	63.92	63.92	68.48
1767	14775	C8H15N	octanenitrile	124-12-9	35.1514	0.0324	10	187	2	37.88	37.88	41.65
1768	14861	C8H16	propylcyclopentane	2040-96-2	22.8265	0.0605	10	10000	2	26.24	26.24	39.85
1769	14862	C8H16	isopropylcyclopentane	3875-51-2	22.0792	0.0632	10	10000	2	25.54	25.54	39.53
1770	14863	C8H16	1-methyl-1-ethylcyclopentane	16747-50-5	21.5303	0.0659	10	10000	2	25.06	25.06	39.49
1771	14864	C8H16	1-methyl-cis-2-ethylcyclopentane	930-89-2	22.5151	0.0626	10	10000	2	26.01	26.01	40.08
1772	14865	C8H16	1-methyl-trans-2-ethylcyclopentane	930-90-5	21.3635	0.0651	10	10000	2	24.82	24.82	38.90
1773	14866	C8H16	1-methyl-cis-3-ethylcyclopentane	2613-66-3	21.2831	0.0653	10	10000	2	24.73	24.73	38.82
1774	14867	C8H16	1-methyl-trans-3-ethylcyclopentane	2613-65-2	21.1915	0.0656	10	10000	2	24.65	24.65	38.78
1775	14868	C8H16	1,1,2-trimethylcyclopentane	4259-00-1	20.4402	0.0691	10	10000	2	23.96	23.96	38.62
1776	14869	C8H16	1,1,3-trimethylcyclopentane	4516-69-2	19.0067	0.0721	10	10000	2	22.44	22.44	36.93
1777	14870	C8H16	1,cis-2,cis-3-trimethylcyclopentane	2613-69-6	21.6717	0.0651	10	10000	2	25.18	25.18	39.48
1778	14871	C8H16	1,cis-2,trans-3-trimethylcyclopentane	15890-40-1	20.8595	0.0672	10	10000	2	24.35	24.35	38.74
1779	14872	C8H16	1,trans-2,cis-3-trimethylcyclopentane	19374-46-0	19.6820	0.0699	10	10000	2	23.12	23.12	37.48
1780	14873	C8H16	1,cis-2,cis-4-trimethylcyclopentane	2613-72-1	20.8639	0.0674	10	10000	2	24.37	24.37	38.83
1781	14874	C8H16	1,cis-2,trans-4-trimethylcyclopentane	4850-28-6	20.6166	0.0674	10	10000	2	24.08	24.08	38.34
1782	14875	C8H16	1,trans-2,cis-4-trimethylcyclopentane	16883-48-0	19.4898	0.0700	10	10000	2	22.90	22.90	37.15
1783	14877	C8H16	ethylcyclohexane	1678-91-7	23.0406	0.0609	10	10000	2	26.51	26.51	40.37
1784	14878	C8H16	1,1-dimethylcyclohexane	590-66-9	21.2220	0.0669	10	10000	2	24.76	24.76	39.31
1785	14880	C8H16	cis-1,2-dimethylcyclohexane	2207-01-4	22.8295	0.0626	10	10000	2	26.37	26.37	40.64
1786	14881	C8H16	trans-1,2-dimethylcyclohexane	6876-23-9	21.5860	0.0649	10	10000	2	25.07	25.07	39.26
1787	14883	C8H16	cis-1,3-dimethylcyclohexane	638-04-0	21.1143	0.0656	10	10000	2	24.55	24.55	38.62
1788	14884	C8H16	trans-1,3-dimethylcyclohexane	2207-03-6	22.0211	0.0644	10	10000	2	25.54	25.54	39.84
1789	14886	C8H16	cis-1,4-dimethylcyclohexane	624-29-3	21.9212	0.0645	10	10000	2	25.43	25.43	39.71
1790	14887	C8H16	trans-1,4-dimethylcyclohexane	2207-04-7	20.8716	0.0662	10	10000	2	24.31	24.31	38.40
1791	14888	C8H16	cyclooctane	292-64-8	26.7681	0.0542	10	7108	2	30.33	30.33	43.29
1792	14890	C8H16	1-octene	111-66-0	20.9874	0.0594	10	10000	2	24.06	24.06	36.26
1793	14892	C8H16	cis-2-octene	7642-04-8	21.5394	0.0582	10	10000	2	24.63	24.63	36.83
1794	14893	C8H16	trans-2-octene	13389-42-9	21.3380	0.0585	10	10000	2	24.41	24.41	36.57
1795	14894	C8H16	cis-3-octene	14850-22-7	21.2225	0.0591	10	10000	2	24.31	24.31	36.57
1796	14895	C8H16	trans-3-octene	14919-01-8	21.1603	0.0585	10	10000	2	24.21	24.21	36.26
1797	14897	C8H16	cis-4-octene	7642-15-1	21.2442	0.0589	10	10000	2	24.33	24.33	36.55
1798	14898	C8H16	trans-4-octene	14850-23-8	21.0477	0.0588	10	10000	2	24.10	24.10	36.17
1799	14899	C8H16	2-methyl-1-heptene	15870-10-7	20.8321	0.0606	10	10000	2	23.95	23.95	36.39
1800	14900	C8H16	3-methyl-1-heptene	4810-09-7	19.7079	0.0730	10	10000	2	23.31	23.31	38.59
1801	14901	C8H16	4-methyl-1-heptene	13151-05-8	20.1629	0.0628	10	10000	2	23.30	23.30	35.96
1802	14902	C8H16	5-methyl-1-heptene	13151-04-7	20.1595	0.0628	10	10000	2	23.29	23.29	35.94
1803	14903	C8H16	6-methyl-1-heptene	5026-76-6	20.0448	0.0627	10	10000	2	23.16	23.16	35.71
1804	14904	C8H16	2-methyl-2-heptene	627-97-4	21.3407	0.0589	10	10000	2	24.44	24.44	36.71
1805	14905	C8H16	3-methyl-cis-2-heptene	22768-19-0	21.3976	0.0593	10	10000	2	24.53	24.53	36.94
1806	14906	C8H16	3-methyl-trans-2-heptene	22768-20-3	21.3976	0.0593	10	10000	2	24.53	24.53	36.94
1807	14907	C8H16	4-methyl-cis-2-heptene	66225-16-9	20.2658	0.0622	10	10000	2	23.39	23.39	35.95
1808	14908	C8H16	4-methyl-trans-2-heptene	66225-17-0	20.2658	0.0622	10	10000	2	23.39	23.39	35.95
1809	14909	C8H16	5-methyl-cis-2-heptene	24608-84-2	20.8097	0.0609	10	10000	2	23.94	23.94	36.47
1810	14910	C8H16	5-methyl-trans-2-heptene	24608-85-3	20.8097	0.0609	10	10000	2	23.94	23.94	36.47
1811	14912	C8H16	6-methyl-cis-2-heptene	66225-18-1	20.6263	0.0610	10	10000	2	23.74	23.74	36.18
1812	14913	C8H16	6-methyl-trans-2-heptene	51065-65-7	20.6263	0.0610	10	10000	2	23.74	23.74	36.18
1813	14916	C8H16	2-methyl-cis-3-heptene	20488-34-0	19.9495	0.0622	10	10000	2	23.02	23.02	35.37
1814	14917	C8H16	2-methyl-trans-3-heptene	692-96-6	19.9923	0.0620	10	10000	2	23.06	23.06	35.39
1815	14919	C8H16	3-methyl-cis-3-heptene	22768-17-8	21.2624	0.0597	10	10000	2	24.40	24.40	36.85
1816	14920	C8H16	3-methyl-trans-3-heptene	22768-18-9	21.2624	0.0597	10	10000	2	24.40	24.40	36.85
1817	14922	C8H16	4-methyl-cis-3-heptene	14255-24-4	21.3046	0.0591	10	10000	2	24.41	24.41	36.73
1818	14923	C8H16	4-methyl-trans-3-heptene	13714-85-7	21.3046	0.0591	10	10000	2	24.41	24.41	36.73
1819	14925	C8H16	5-methyl-cis-3-heptene	50422-80-5	20.0815	0.0625	10	10000	2	23.19	23.19	35.72
1820	14926	C8H16	5-methyl-trans-3-heptene	53510-18-2	20.0815	0.0625	10	10000	2	23.19	23.19	35.72
1821	14927	C8H16	6-methyl-cis-3-heptene	66225-19-2	20.3537	0.0614	10	10000	2	23.45	23.45	35.84
1822	14928	C8H16	6-methyl-trans-3-heptene	66225-20-5	20.2845	0.0617	10	10000	2	23.38	23.38	35.81
1823	14929	C8H16	2-ethyl-1-hexene	1632-16-2	20.9821	0.0608	10	10000	2	24.13	24.13	36.73
1824	14930	C8H16	3-ethyl-1-hexene	3404-58-8	19.8019	0.0641	10	10000	2	22.95	22.95	35.75
1825	14931	C8H16	4-ethyl-1-hexene	16746-85-3	20.2005	0.0639	10	10000	2	23.40	23.40	36.39
1826	14932	C8H16	2,3-dimethyl-1-hexene	16746-86-4	19.9817	0.0641	10	10000	2	23.16	23.16	36.06
1827	14933	C8H16	2,4-dimethyl-1-hexene	16746-87-5	20.0527	0.0636	10	10000	2	23.22	23.22	36.04
1828	14934	C8H16	2,5-dimethyl-1-hexene	6975-92-4	20.0463	0.0633	10	10000	2	23.19	23.19	35.90
1829	14935	C8H16	3,3-dimethyl-1-hexene	3404-77-1	19.1851	0.0665	10	10000	2	22.36	22.36	35.40
1830	14936	C8H16	3,4-dimethyl-1-hexene	16745-94-1	20.2087	0.0635	10	10000	2	23.39	23.39	36.27
1831	14937	C8H16	3,5-dimethyl-1-hexene	7423-69-0	19.0886	0.0662	10	10000	2	22.23	22.23	35.11
1832	14938	C8H16	4,4-dimethyl-1-hexene	1647-08-1	19.6198	0.0654	10	10000	2	22.81	22.81	35.84
1833	14939	C8H16	4,5-dimethyl-1-hexene	16106-59-5	19.9700	0.0650	10	10000	2	23.19	23.19	36.34
1834	14940	C8H16	5,5-dimethyl-1-hexene	7116-86-1	18.9601	0.0668	10	10000	2	22.11	22.11	35.09
1835	14941	C8H16	3-ethyl-cis-2-hexene	36880-72-5	21.5307	0.0597	10	10000	2	24.70	24.70	37.30
1836	14942	C8H16	3-ethyl-trans-2-hexene	66225-15-8	21.5307	0.0597	10	10000	2	24.70	24.70	37.30
1837	14943	C8H16	4-ethyl-cis-2-hexene	54616-49-8	20.2644	0.0635	10	10000	2	23.45	23.45	36.37
1838	14944	C8H16	4-ethyl-trans-2-hexene	19781-63-6	20.2644	0.0635	10	10000	2	23.45	23.45	36.37
1839	14945	C8H16	2,3-dimethyl-2-hexene	7145-20-2	21.6939	0.0587	10	10000	2	24.83	24.83	37.24
1840	14946	C8H16	2,4-dimethyl-2-hexene	14255-23-3	20.0125	0.0639	10	10000	2	23.19	23.19	36.06

Table 1 Adsorption Capacity of Activated Carbon - Gas

$\log_{10} Q = \log_{10} K + 1/n \log_{10} y$ (Q - g of compound/100 g of carbon, y - ppmv)

NO	ID	FORMULA	NAME	CAS No	K	1/n	ymin	ymax	code	Q @ymin	Q @10ppm	Q @ymax
1841	14947	C8H16	2,5-dimethyl-2-hexene	3404-78-2	20.3345	0.0624	10	10000	2	23.47	23.47	36.12
1842	14948	C8H16	3,4-dimethyl-cis-2-hexene	19550-81-3	20.8046	0.0628	10	10000	2	24.04	24.04	37.10
1843	14949	C8H16	3,4-dimethyl-trans-2-hexene	19550-82-4	20.8046	0.0628	10	10000	2	24.04	24.04	37.10
1844	14950	C8H16	3,5-dimethyl-cis-2-hexene	66225-31-8	20.3162	0.0631	10	10000	2	23.49	23.49	36.32
1845	14951	C8H16	3,5-dimethyl-trans-2-hexene	66225-12-5	20.3162	0.0631	10	10000	2	23.49	23.49	36.32
1846	14952	C8H16	4,4-dimethyl-cis-2-hexene	19550-83-5	19.6398	0.0655	10	10000	2	22.84	22.84	35.90
1847	14953	C8H16	4,4-dimethyl-trans-2-hexene	------	19.6398	0.0655	10	10000	2	22.84	22.84	35.90
1848	14955	C8H16	4,5-dimethyl-cis-2-hexene	65036-71-7	20.0202	0.0644	10	10000	2	23.22	23.22	36.23
1849	14956	C8H16	4,5-dimethyl-trans-2-hexene	66225-14-7	20.0202	0.0644	10	10000	2	23.22	23.22	36.23
1850	14957	C8H16	5,5-dimethyl-cis-2-hexene	39761-61-0	19.6343	0.0648	10	10000	2	22.79	22.79	35.66
1851	14958	C8H16	5,5-dimethyl-trans-2-hexene	39782-43-9	19.1347	0.0657	10	10000	2	22.26	22.26	35.03
1852	14959	C8H16	3-ethyl-3-hexene	16789-51-8	20.8027	0.0616	10	10000	2	23.97	23.97	36.69
1853	14960	C8H16	2,2-dimethyl-cis-3-hexene	690-92-6	19.2540	0.0658	10	10000	2	22.41	22.41	35.31
1854	14961	C8H16	2,2-dimethyl-trans-3-hexene	690-93-7	18.8066	0.0665	10	10000	2	21.92	21.92	34.71
1855	14962	C8H16	2,3-dimethyl-cis-3-hexene	59643-75-3	20.5239	0.0627	10	10000	2	23.71	23.71	36.56
1856	14963	C8H16	2,3-dimethyl-trans-3-hexene	66225-30-7	20.5239	0.0627	10	10000	2	23.71	23.71	36.56
1857	14964	C8H16	2,4-dimethyl-cis-3-hexene	37549-89-6	19.8951	0.0638	10	10000	2	23.04	23.04	35.81
1858	14965	C8H16	2,4-dimethyl-trans-3-hexene	61847-78-7	19.7640	0.0638	10	10000	2	22.89	22.89	35.58
1859	14966	C8H16	2,5-dimethyl-cis-3-hexene	10557-44-5	18.7870	0.0676	10	10000	2	21.95	21.95	35.02
1860	14967	C8H16	2,5-dimethyl-trans-3-hexene	692-70-6	18.9968	0.0667	10	10000	2	22.15	22.15	35.12
1861	14968	C8H16	3,4-dimethyl-cis-3-hexene	19550-87-9	21.8902	0.0595	10	10000	2	25.10	25.10	37.85
1862	14969	C8H16	3,4-dimethyl-trans-3-hexene	19550-88-0	21.8902	0.0595	10	10000	2	25.10	25.10	37.85
1863	14970	C8H16	2-propyl-1-pentene	15918-08-8	20.7507	0.0613	10	10000	2	23.90	23.90	36.50
1864	14971	C8H16	2-isopropyl-1-pentene	16746-02-4	20.2581	0.0634	10	10000	2	23.44	23.44	36.32
1865	14972	C8H16	3-methyl-2-ethyl-1-pentene	3404-67-9	20.2719	0.0640	10	10000	2	23.49	23.49	36.53
1866	14973	C8H16	4-methyl-2-ethyl-1-pentene	3404-80-6	19.9255	0.0641	10	10000	2	23.09	23.09	35.96
1867	14974	C8H16	2-methyl-3-ethyl-1-pentene	19780-66-6	19.8815	0.0656	10	10000	2	23.12	23.12	36.37
1868	14975	C8H16	3-methyl-3-ethyl-1-pentene	6196-60-7	20.4162	0.0633	10	10000	2	23.62	23.62	36.59
1869	14976	C8H16	4-methyl-3-ethyl-1-pentene	61847-80-1	19.6290	0.0654	10	10000	2	22.82	22.82	35.86
1870	14977	C8H16	2,3,3-trimethyl-1-pentene	560-23-6	20.0223	0.0656	10	10000	2	23.29	23.29	36.64
1871	14978	C8H16	2,3,4-trimethyl-1-pentene	565-76-4	19.9104	0.0653	10	10000	2	23.14	23.14	36.34
1872	14979	C8H16	2,4,4-trimethyl-1-pentene	107-39-1	18.6247	0.0689	10	10000	2	21.83	21.83	35.13
1873	14980	C8H16	3,3,4-trimethyl-1-pentene	560-22-5	19.5741	0.0667	10	10000	2	22.83	22.83	36.19
1874	14981	C8H16	3,4,4-trimethyl-1-pentene	564-03-4	19.4076	0.0660	10	10000	2	22.59	22.59	35.65
1875	14982	C8H16	2-methyl-3-ethyl-2-pentene	19780-67-7	21.3573	0.0606	10	10000	2	24.56	24.56	37.33
1876	14983	C8H16	4-methyl-3-ethyl-cis-2-pentene	42067-48-1	21.0599	0.0618	10	10000	2	24.28	24.28	37.20
1877	14984	C8H16	4-methyl-3-ethyl-trans-2-pentene	42067-49-2	20.7429	0.0626	10	10000	2	23.96	23.96	36.91
1878	14985	C8H16	2,3,4-trimethyl-2-pentene	565-77-5	21.2084	0.0616	10	10000	2	24.44	24.44	37.42
1879	14986	C8H16	2,4,4-trimethyl-2-pentene	107-40-4	19.4891	0.0658	10	10000	2	22.68	22.68	35.72
1880	14987	C8H16	3,4,4-trimethyl-cis-2-pentene	39761-64-3	20.6298	0.0635	10	10000	2	23.88	23.88	37.02
1881	14988	C8H16	3,4,4-trimethyl-trans-2-pentene	39761-57-4	20.6235	0.0635	10	10000	2	23.87	23.87	37.02
1882	14991	C8H16	3-methyl-2-isopropyl-1-butene	111823-35-9	19.2388	0.0674	10	10000	2	22.47	22.47	35.80
1883	14992	C8H16	3,3-dimethyl-2-ethyl-1-butene	18231-53-3	20.1147	0.0643	10	10000	2	23.33	23.33	36.38
1884	14994	C8H16Br2	1,1-dibromooctane	62168-26-7	67.1658	0.0221	10	52	2	70.67	70.67	73.29
1885	15005	C8H16Cl2	1,1-dichlorooctane	20395-24-8	43.7486	0.0261	10	334	2	46.46	46.46	50.92
1886	15016	C8H16F2	1,1-difluorooctane	61350-03-6	30.0796	0.0489	10	6772	2	33.67	33.67	46.32
1887	15017	C8H16I2	1,1-diiodooctane	66225-22-7	90.7410	0.0173	1	2.5	2	90.74	---	92.19
1888	15062	C8H16O	2-ethylhexanal	123-05-7	30.0778	0.0439	10	2585	2	33.28	33.28	42.48
1889	15063	C8H16O	1-octanal	124-13-0	31.2152	0.0412	10	1549	2	34.32	34.32	42.24
1890	15066	C8H16O	2-octanone	111-13-7	32.8000	0.0379	10	723	2	35.79	35.79	42.09
1891	15156	C8H16O2	octanoic acid	124-07-2	44.3360	0.0308	1	4.5	2	44.34	---	46.44
1892	15159	C8H16O2	heptyl formate	112-23-2	35.1870	0.0366	10	901	2	38.28	38.28	45.14
1893	15160	C8H16O2	hexyl acetate	142-92-7	33.4590	0.0395	10	1741	2	36.64	36.64	44.92
1894	15167	C8H16O2	pentyl propanoate	624-54-4	33.6375	0.0392	10	1580	2	36.82	36.82	44.91
1895	15168	C8H16O2	1-methylbutyl propanoate	54004-43-2	34.6057	0.0368	10	931	2	37.66	37.66	44.50
1896	15169	C8H16O2	2-methylbutyl propanoate	2438-20-2	33.9844	0.0381	10	1257	2	37.10	37.10	44.61
1897	15170	C8H16O2	isopentyl propanoate	105-68-0	32.5513	0.0411	10	2379	2	35.78	35.78	44.81
1898	15171	C8H16O2	1-ethylpropyl propanoate	------	34.7919	0.0368	10	887	2	37.87	37.87	44.67
1899	15172	C8H16O2	1,1-dimethylpropyl propanoate	34949-22-9	34.7047	0.0370	10	923	2	37.79	37.79	44.68
1900	15173	C8H16O2	1,2-dimethylpropyl propanoate	66576-70-3	34.8931	0.0366	10	846	2	37.96	37.96	44.67
1901	15174	C8H16O2	2,2-dimethylpropyl propanoate	3581-69-9	34.2281	0.0379	10	1146	2	37.35	37.35	44.71
1902	15175	C8H16O2	butyl butanoate	109-21-7	32.5447	0.0413	10	2383	2	35.79	35.79	44.86
1903	15176	C8H16O2	isobutyl butanoate	539-90-2	30.6624	0.0444	10	4810	2	33.96	33.96	44.68
1904	15177	C8H16O2	isobutyl isobutyrate	97-85-8	30.0572	0.0447	10	5700	2	33.32	33.32	44.25
1905	15178	C8H16O2	sec-butyl butanoate	819-97-6	31.6748	0.0425	10	3150	2	34.93	34.93	44.62
1906	15179	C8H16O2	tert-butyl butanoate	2308-38-5	29.6321	0.0464	10	7226	2	32.97	32.97	44.74
1907	15232	C8H16O4	diethylene glycol ethyl ether acetate	112-15-2	44.6760	0.0278	10	194	2	47.63	47.63	51.71
1908	15243	C8H16S	thiacyclononane	6007-54-1	40.4641	0.0305	10	426	2	43.40	43.40	48.66
1909	15248	C8H17Br	1-bromooctane	111-83-1	47.1960	0.0293	10	637	2	50.49	50.49	57.03
1910	15252	C8H17Cl	1-chlorooctane	111-85-3	34.9982	0.0345	10	1353	2	37.89	37.89	44.89
1911	15273	C8H17F	1-fluorooctane	463-11-6	26.8730	0.0498	10	7924	2	30.14	30.14	42.01
1912	15276	C8H17I	1-iodooctane	629-27-6	58.7801	0.0245	10	402	2	62.19	62.19	68.09
1913	15354	C8H18	octane	111-65-9	14.0500	0.1191	10	10000	1,2	18.48	18.48	42.07
1914	15355	C8H18	2-methylheptane	592-27-8	20.2957	0.0601	10	10000	2	23.31	23.31	35.31
1915	15356	C8H18	3-methylheptane	589-81-1	20.5855	0.0597	10	10000	2	23.62	23.62	35.67
1916	15357	C8H18	4-methylheptane	589-53-7	20.4710	0.0618	10	10000	2	23.60	23.60	36.18
1917	15358	C8H18	3-methyl-3-ethylpentane	1067-08-9	20.6258	0.0623	10	10000	2	23.81	23.81	36.61
1918	15359	C8H18	3-ethylhexane	619-99-8	20.6828	0.0603	10	10000	2	23.77	23.77	36.05
1919	15360	C8H18	3-ethyl-2-methylpentane	609-26-7	20.4480	0.0611	10	10000	2	23.54	23.54	35.90
1920	15361	C8H18	2,2-dimethylhexane	590-73-8	19.0157	0.0648	10	10000	2	22.08	22.08	34.54

Table 1 Adsorption Capacity of Activated Carbon - Gas

$\log_{10} Q = \log_{10} K + 1/n \log_{10} y$ (Q - g of compound/100 g of carbon, y - ppmv)

NO	ID	FORMULA	NAME	CAS No	K	1/n	ymin	ymax	code	Q @ymin	Q @10ppm	Q @ymax
1921	15362	C8H18	2,3-dimethylhexane	584-94-1	20.2714	0.0617	10	10000	2	23.36	23.36	35.78
1922	15363	C8H18	2,4-dimethylhexane	589-43-5	19.3750	0.0634	10	10000	2	22.42	22.42	34.73
1923	15364	C8H18	2,5-dimethylhexane	592-13-2	19.2512	0.0636	10	10000	2	22.29	22.29	34.58
1924	15365	C8H18	3,3-dimethylhexane	563-16-6	19.7326	0.0637	10	10000	2	22.85	22.85	35.50
1925	15366	C8H18	3,4-dimethylhexane	583-48-2	20.6071	0.0614	10	10000	2	23.74	23.74	36.27
1926	15367	C8H18	2,2,3-trimethylpentane	564-02-3	19.5754	0.0650	10	10000	2	22.73	22.73	35.61
1927	15368	C8H18	2,2,4-trimethylpentane	540-84-1	18.0589	0.0687	10	10000	1,2	21.16	21.16	34.02
1928	15369	C8H18	2,3,3-trimethylpentane	560-21-4	20.2034	0.0637	10	10000	2	23.39	23.39	36.32
1929	15370	C8H18	2,3,4-trimethylpentane	565-75-3	20.0502	0.0636	10	10000	2	23.21	23.21	36.01
1930	15371	C8H18	2,2,3,3-tetramethylbutane	594-82-1	19.1242	0.0754	10	1000	2	22.75	22.75	32.19
1931	15430	C8H18O	1-octanol	111-87-5	36.2117	0.0341	10	103	2	39.17	39.17	42.41
1932	15431	C8H18O	2-octanol	123-96-6	34.8379	0.0328	10	317	2	37.57	37.57	42.07
1933	15433	C8H18O	3-octanol	589-98-0	32.8278	0.0372	10	872	2	35.76	35.76	42.23
1934	15434	C8H18O	4-octanol	589-62-8	33.3234	0.0358	10	675	2	36.19	36.19	42.08
1935	15436	C8H18O	2-methyl-1-heptanol	60435-70-3	31.7982	0.0370	10	1143	2	34.63	34.63	41.26
1936	15437	C8H18O	3-methyl-1-heptanol	1070-32-2	32.5425	0.0337	10	633	2	35.17	35.17	40.45
1937	15438	C8H18O	4-methyl-1-heptanol	817-91-4	33.0074	0.0353	10	717	2	35.80	35.80	41.62
1938	15439	C8H18O	5-methyl-1-heptanol	7212-53-5	34.4827	0.0331	10	385	2	37.22	37.22	42.00
1939	15440	C8H18O	6-methyl-1-heptanol	1653-40-3	35.5842	0.0312	10	212	2	38.23	38.23	42.05
1940	15441	C8H18O	2-methyl-2-heptanol	625-25-2	31.8540	0.0380	10	1212	2	34.77	34.77	41.73
1941	15442	C8H18O	3-methyl-2-heptanol	31367-46-1	30.6593	0.0417	10	2335	2	33.75	33.75	42.37
1942	15443	C8H18O	4-methyl-2-heptanol	56298-90-9	32.2496	0.0361	10	921	2	35.04	35.04	41.25
1943	15444	C8H18O	5-methyl-2-heptanol	54630-50-1	31.8807	0.0382	10	1259	2	34.81	34.81	41.86
1944	15445	C8H18O	6-methyl-2-heptanol	4730-22-7	32.8448	0.0355	10	710	2	35.64	35.64	41.46
1945	15446	C8H18O	2-methyl-3-heptanol	18720-62-2	32.6433	0.0377	10	1028	2	35.60	35.60	42.40
1946	15447	C8H18O	3-methyl-3-heptanol	5582-82-1	32.1021	0.0392	10	1423	2	35.14	35.14	42.68
1947	15448	C8H18O	4-methyl-3-heptanol	14979-39-6	28.6024	0.0432	10	4673	2	31.59	31.59	41.20
1948	15449	C8H18O	5-methyl-3-heptanol	18720-65-5	29.0038	0.0449	10	4384	2	32.17	32.17	42.28
1949	15450	C8H18O	6-methyl-3-heptanol	18720-66-6	29.8593	0.0426	10	3016	2	32.93	32.93	42.00
1950	15451	C8H18O	2-methyl-4-heptanol	21570-35-4	32.4408	0.0372	10	937	2	35.34	35.34	41.83
1951	15452	C8H18O	3-methyl-4-heptanol	1838-73-9	30.9139	0.0429	10	2419	2	34.12	34.12	43.17
1952	15453	C8H18O	4-methyl-4-heptanol	598-01-6	31.8849	0.0391	10	1438	2	34.89	34.89	42.38
1953	15454	C8H18O	2-ethyl-1-hexanol	104-76-7	36.3722	0.0309	10	178	2	39.05	39.05	42.68
1954	15455	C8H18O	3-ethyl-1-hexanol	41065-95-6	32.8133	0.0381	10	1169	2	35.82	35.82	42.95
1955	15456	C8H18O	4-ethyl-1-hexanol	------	32.6397	0.0380	10	1169	2	35.63	35.63	42.69
1956	15457	C8H18O	2,2-dimethyl-1-hexanol	2370-13-0	35.0371	0.0327	10	323	2	37.78	37.78	42.33
1957	15458	C8H18O	2,3-dimethyl-1-hexanol	------	31.8249	0.0393	10	1651	2	34.84	34.84	42.58
1958	15459	C8H18O	2,4-dimethyl-1-hexanol	3965-59-1	31.9324	0.0391	10	1571	2	34.94	34.94	42.57
1959	15460	C8H18O	2,5-dimethyl-1-hexanol	6886-16-4	33.9867	0.0313	10	1323	2	36.53	36.53	42.56
1960	15461	C8H18O	3,3-dimethyl-1-hexanol	10524-70-6	32.5814	0.0391	10	1345	2	35.65	35.65	43.19
1961	15462	C8H18O	3,4-dimethyl-1-hexanol	------	32.5214	0.0387	10	1221	2	35.55	35.55	42.81
1962	15463	C8H18O	3,5-dimethyl-1-hexanol	13501-73-0	32.5227	0.0385	10	1113	2	35.53	35.53	42.59
1963	15464	C8H18O	4,4-dimethyl-1-hexanol	6481-95-4	32.5056	0.0381	10	1289	2	35.49	35.49	42.70
1964	15465	C8H18O	4,5-dimethyl-1-hexanol	------	32.7112	0.0377	10	1170	2	35.68	35.68	42.69
1965	15466	C8H18O	5,5-dimethyl-1-hexanol	------	32.5056	0.0381	10	1289	2	35.49	35.49	42.70
1966	15467	C8H18O	3-ethyl-2-hexanol	24448-19-9	31.4184	0.0403	10	2124	2	34.47	34.47	42.77
1967	15468	C8H18O	4-ethyl-2-hexanol	------	31.4041	0.0403	10	2138	2	34.46	34.46	42.77
1968	15469	C8H18O	2,3-dimethyl-2-hexanol	19550-03-9	33.0949	0.0377	10	1037	2	36.09	36.09	42.99
1969	15470	C8H18O	2,4-dimethyl-2-hexanol	42328-76-7	30.4629	0.0406	10	2352	2	33.45	33.45	41.74
1970	15471	C8H18O	2,5-dimethyl-2-hexanol	3730-60-7	31.7999	0.0388	10	1298	2	34.77	34.77	41.99
1971	15472	C8H18O	3,3-dimethyl-2-hexanol	22025-20-3	30.5515	0.0438	10	3534	2	33.79	33.79	43.69
1972	15473	C8H18O	3,4-dimethyl-2-hexanol	19550-05-1	31.8098	0.0408	10	2025	2	34.94	34.94	43.39
1973	15474	C8H18O	3,5-dimethyl-2-hexanol	66576-27-0	30.7246	0.0429	10	3232	2	33.92	33.92	43.46
1974	15475	C8H18O	4,4-dimethyl-2-hexanol	------	30.7111	0.0429	10	3250	2	33.90	33.90	43.46
1975	15476	C8H18O	4,5-dimethyl-2-hexanol	------	30.9094	0.0425	10	2987	2	34.09	34.09	43.45
1976	15477	C8H18O	5,5-dimethyl-2-hexanol	31841-77-7	30.7164	0.0439	10	2732	2	33.98	33.98	43.46
1977	15478	C8H18O	3-ethyl-3-hexanol	597-76-2	31.5272	0.0414	10	1962	2	34.68	34.68	43.16
1978	15479	C8H18O	4-ethyl-3-hexanol	19780-44-0	32.1542	0.0396	10	1545	2	35.22	35.22	42.99
1979	15480	C8H18O	2,2-dimethyl-3-hexanol	4209-90-9	29.0744	0.0460	10	5308	2	32.32	32.32	43.12
1980	15481	C8H18O	2,3-dimethyl-3-hexanol	4166-46-5	30.9004	0.0424	10	2664	2	34.07	34.07	43.18
1981	15482	C8H18O	2,4-dimethyl-3-hexanol	13432-25-2	29.9931	0.0433	10	3769	2	33.14	33.14	42.85
1982	15483	C8H18O	2,5-dimethyl-3-hexanol	19550-07-3	31.5588	0.0394	10	1549	2	34.55	34.55	42.14
1983	15484	C8H18O	3,4-dimethyl-3-hexanol	19550-08-4	31.2917	0.0414	10	2419	2	34.42	34.42	43.21
1984	15485	C8H18O	3,5-dimethyl-3-hexanol	4209-91-0	29.2539	0.0448	10	4714	2	32.43	32.43	42.73
1985	15486	C8H18O	4,4-dimethyl-3-hexanol	19550-09-5	31.3667	0.0409	10	2229	2	34.46	34.46	42.99
1986	15487	C8H18O	4,5-dimethyl-3-hexanol	------	30.0703	0.0432	10	3855	2	33.21	33.21	42.95
1987	15488	C8H18O	5,5-dimethyl-3-hexanol	66576-31-6	30.5751	0.0428	10	2789	2	33.74	33.74	42.93
1988	15489	C8H18O	2-propyl-1-pentanol	58175-57-8	35.3735	0.0330	10	288	2	38.16	38.16	42.63
1989	15490	C8H18O	2-isopropyl-1-pentanol	------	34.2288	0.0354	10	522	2	37.13	37.13	42.71
1990	15491	C8H18O	2-methyl-2-ethyl-1-pentanol	5970-63-8	34.4552	0.0343	10	514	2	37.29	37.29	42.69
1991	15492	C8H18O	3-methyl-2-ethyl-1-pentanol	------	32.2487	0.0390	10	1451	2	35.28	35.28	42.83
1992	15493	C8H18O	4-methyl-2-ethyl-1-pentanol	106-67-2	33.9038	0.0356	10	615	2	36.80	36.80	42.62
1993	15494	C8H18O	2-methyl-3-ethyl-1-pentanol	------	32.0000	0.0400	10	1581	2	35.08	35.08	42.95
1994	15495	C8H18O	3-methyl-3-ethyl-1-pentanol	------	31.7927	0.0404	10	1735	2	34.89	34.89	42.96
1995	15496	C8H18O	4-methyl-3-ethyl-1-pentanol	------	32.0000	0.0400	10	1581	2	35.08	35.08	42.95
1996	15497	C8H18O	2,2,3-trimethyl-1-pentanol	57409-53-7	34.3117	0.0347	10	588	2	37.17	37.17	42.81
1997	15498	C8H18O	2,2,4-trimethyl-1-pentanol	123-44-4	32.8876	0.0386	10	1118	2	35.94	35.94	43.11
1998	15499	C8H18O	2,3,3-trimethyl-1-pentanol	------	32.0001	0.0409	10	1809	2	35.16	35.16	43.49
1999	15500	C8H18O	2,3,4-trimethyl-1-pentanol	6570-88-3	33.2213	0.0387	10	1216	2	36.32	36.32	43.74
2000	15501	C8H18O	2,4,4-trimethyl-1-pentanol	16325-63-6	32.5883	0.0387	10	1184	2	35.63	35.63	42.86

Table 1 Adsorption Capacity of Activated Carbon - Gas

$\log_{10} Q = \log_{10} K + 1/n \log_{10} y$ (Q - g of compound/100 g of carbon, y - ppmv)

NO	ID	FORMULA	NAME	CAS No	K	1/n	ymin	ymax	code	Q @ymin	Q @10ppm	Q @ymax
2001	15502	C8H18O	3,3,4-trimethyl-1-pentanol	------	31.2721	0.0415	10	2230	2	34.40	34.40	43.05
2002	15503	C8H18O	3,4,4-trimethyl-1-pentanol	16325-64-7	31.2721	0.0415	10	2230	2	34.40	34.40	43.05
2003	15504	C8H18O	2-methyl-3-ethyl-2-pentanol	19780-63-3	32.6573	0.0390	10	1284	2	35.73	35.73	43.18
2004	15505	C8H18O	3-methyl-3-ethyl-2-pentanol	66576-22-5	31.2030	0.0429	10	3534	2	34.44	34.44	44.29
2005	15506	C8H18O	4-methyl-3-ethyl-2-pentanol	66576-23-6	31.0710	0.0427	10	2727	2	34.28	34.28	43.55
2006	15507	C8H18O	2,3,3-trimethyl-2-pentanol	23171-85-9	30.3047	0.0410	10	3298	2	33.30	33.30	42.23
2007	15508	C8H18O	2,3,4-trimethyl-2-pentanol	66576-26-9	29.5854	0.0412	10	4112	2	32.53	32.53	41.68
2008	15509	C8H18O	2,4,4-trimethyl-2-pentanol	690-37-9	29.5458	0.0436	10	4163	2	32.67	32.67	42.50
2009	15510	C8H18O	3,3,4-trimethyl-2-pentanol	19411-41-7	31.1570	0.0435	10	3079	2	34.44	34.44	44.19
2010	15511	C8H18O	3,4,4-trimethyl-2-pentanol	10575-56-1	31.1570	0.0435	10	3079	2	34.44	34.44	44.19
2011	15512	C8H18O	2-methyl-3-ethyl-3-pentanol	597-05-7	30.4085	0.0419	10	3512	2	33.49	33.49	42.82
2012	15513	C8H18O	2,2,3-trimethyl-3-pentanol	7294-05-5	30.3119	0.0442	10	3904	2	33.56	33.56	43.70
2013	15514	C8H18O	2,2,4-trimethyl-3-pentanol	5162-48-1	30.8285	0.0421	10	2601	2	33.96	33.96	42.91
2014	15515	C8H18O	2,3,4-trimethyl-3-pentanol	3054-92-0	29.8272	0.0455	10	4856	2	33.12	33.12	43.89
2015	15516	C8H18O	3-methyl-2-isopropyl-1-butanol	18593-92-5	32.0903	0.0407	10	1860	2	35.24	35.24	43.59
2016	15517	C8H18O	2,2-diethyl-1-butanol	13023-60-4	32.2473	0.0399	10	1888	2	35.35	35.35	43.58
2017	15518	C8H18O	2,3-dimethyl-2-ethyl-1-butanol	------	31.8942	0.0410	10	2051	2	35.05	35.05	43.61
2018	15519	C8H18O	3,3-dimethyl-2-ethyl-1-butanol	66576-56-5	31.9153	0.0411	10	2051	2	35.08	35.08	43.66
2019	15520	C8H18O	2,2,3,3-tetramethyl-1-butanol	66576-24-7	31.5596	0.0439	10	1000	2	34.92	34.92	42.73
2020	15527	C8H18O	dibutyl ether	142-96-1	26.2649	0.0461	10	7920	2	29.20	29.20	39.72
2021	15528	C8H18O	di-sec-butyl ether	6863-58-7	23.6660	0.0536	10	10000	2	26.77	26.77	38.76
2022	15529	C8H18O	di-tert-butyl ether	6163-66-2	21.9139	0.0597	10	10000	2	25.15	25.15	37.99
2023	15544	C8H18O2	1,2-octanediol	1117-86-8	46.7244	0.0284	1	3.9	2	46.72	---	48.57
2024	15545	C8H18O2	1,3-octanediol	23433-05-8	46.7244	0.0284	1	3.9	2	46.72	---	48.57
2025	15546	C8H18O2	1,4-octanediol	------	46.7244	0.0284	1	3.9	2	46.72	---	48.57
2026	15547	C8H18O2	1,8-octanediol	629-41-4	46.7244	0.0284	1	3.9	2	46.72	---	48.57
2027	15556	C8H18O2	di-t-butyl peroxide	110-05-4	24.7156	0.0518	10	10000	2	27.85	27.85	39.84
2028	15578	C8H18O3	diethylene glycol diethyl ether	112-36-7	37.9158	0.0316	10	684	2	40.78	40.78	46.60
2029	15579	C8H18O3	diethylene glycol monobutyl ether	112-34-5	45.9342	0.0301	1	8.6	2	45.93	---	49.01
2030	15591	C8H18O4	triethylene glycol dimethyl ether	112-49-2	45.2509	0.0271	10	53	2	48.16	48.16	50.39
2031	15602	C8H18S	octyl mercaptan	111-88-6	36.0376	0.0289	10	558	2	38.51	38.51	43.26
2032	15603	C8H18S	2-octanethiol	3001-66-9	34.5259	0.0310	10	1147	2	37.08	37.08	42.95
2033	15604	C8H18S	tert-octyl mercaptan	141-59-3	31.5326	0.0368	10	6493	2	34.32	34.32	43.55
2034	15606	C8H18S	dibutyl sulfide	544-40-1	34.0502	0.0323	10	1776	2	36.68	36.68	43.36
2035	15607	C8H18S	methyl heptyl sulfide	20291-61-6	34.0395	0.0324	10	1776	2	36.67	36.67	43.36
2036	15608	C8H18S	ethyl hexyl sulfide	7309-44-6	34.0395	0.0324	10	1776	2	36.67	36.67	43.36
2037	15609	C8H18S	propyl pentyl sulfide	42841-80-5	34.0395	0.0324	10	1776	2	36.67	36.67	43.36
2038	15614	C8H18S2	dibutyl disulfide	629-45-8	43.5818	0.0224	10	74	2	45.89	45.89	48.00
2039	15631	C8H19N	octylamine	111-86-4	31.1314	0.0359	10	1272	2	33.82	33.82	40.24
2040	15635	C8H19N	methylheptylamine	36343-05-2	30.3966	0.0360	10	1458	2	33.03	33.03	39.52
2041	15636	C8H19N	ethylhexylamine	20352-67-4	29.3859	0.0376	10	2184	2	32.04	32.04	39.24
2042	15637	C8H19N	dibutylamine	111-92-2	28.3984	0.0397	10	3401	2	31.11	31.11	39.21
2043	15638	C8H19N	diisobutylamine	110-96-3	25.8855	0.0436	10	9559	2	28.62	28.62	38.59
2044	15641	C8H19N	dimethylhexylamine	4385-04-0	26.8158	0.0416	10	6331	2	29.51	29.51	38.60
2045	15642	C8H19N	diethylbutylamine	4444-68-2	26.7184	0.0424	10	6941	2	29.46	29.46	38.87
2046	15756	C8H24O4Si4	octamethylcyclotetrasiloxane	556-67-2	43.1649	0.0171	10	1284	2	44.89	44.89	48.77
2047	15903	C9H6N2O2	toluene diisocyanate	584-84-9	33.7354	0.0973	10	18	1,2	42.21	42.21	44.70
2048	16010	C9H7N	isoquinoline	119-65-3	48.8313	0.0319	10	79	2	52.55	52.55	56.12
2049	16011	C9H7N	quinoline	91-22-5	48.8313	0.0319	10	79	2	52.55	52.55	56.12
2050	16090	C9H8	indene	95-13-6	37.9039	0.0421	10	1428	2	41.76	41.76	51.47
2051	16185	C9H8O	2-methylbenzofuran	4265-25-2	42.8188	0.0366	10	640	2	46.59	46.59	54.25
2052	16508	C9H10	indane	496-11-7	35.5592	0.0442	10	2027	2	39.37	39.37	49.77
2053	16509	C9H10	cis-1-propenylbenzene	766-90-5	34.3134	0.0403	10	2205	2	37.65	37.65	46.81
2054	16510	C9H10	trans-1-propenylbenzene	873-66-5	33.7091	0.0408	10	2991	2	37.03	37.03	46.73
2055	16511	C9H10	2-propenylbenzene	300-57-2	31.3410	0.0456	10	4744	2	34.81	34.81	46.12
2056	16512	C9H10	alpha-methylstyrene	98-83-9	33.1046	0.0428	10	3459	2	36.54	36.54	46.94
2057	16513	C9H10	m-methylstyrene	100-80-1	34.0447	0.0413	10	2502	2	37.44	37.44	47.04
2058	16514	C9H10	o-methylstyrene	611-15-4	34.1803	0.0409	10	2434	2	37.56	37.56	47.03
2059	16515	C9H10	p-methylstyrene	622-97-9	34.3308	0.0416	10	2404	2	37.78	37.78	47.46
2060	16676	C9H10O2	ethyl benzoate	93-89-0	44.6927	0.0312	10	351	2	48.02	48.02	53.65
2061	16680	C9H10O2	benzyl acetate	140-11-4	45.9567	0.0287	10	233	2	49.10	49.10	53.74
2062	17004	C9H12	cumene	98-82-8	27.7773	0.0486	10	5984	1,2	31.06	31.06	42.38
2063	17005	C9H12	m-ethyltoluene	620-14-4	31.1059	0.0435	10	3994	2	34.38	34.38	44.62
2064	17006	C9H12	o-ethyltoluene	611-14-3	31.9987	0.0432	10	3432	2	35.35	35.35	45.49
2065	17007	C9H12	p-ethyltoluene	622-96-8	31.0407	0.0434	10	3939	2	34.30	34.30	44.46
2066	17008	C9H12	1,2,3-trimethylbenzene	526-73-8	33.5867	0.0412	10	2221	2	36.93	36.93	46.15
2067	17009	C9H12	1,2,4-trimethylbenzene	95-63-6	32.2640	0.0422	10	2976	2	35.55	35.55	45.20
2068	17011	C9H12	mesitylene	108-67-8	31.6885	0.0424	10	3260	2	34.94	34.94	44.64
2069	17012	C9H12	propylbenzene	103-65-1	30.6801	0.0446	10	4502	2	34.00	34.00	44.65
2070	17077	C9H12O	benzyl ethyl ether	539-30-0	37.6372	0.0366	10	1216	2	40.95	40.95	48.83
2071	17101	C9H12O	2-propylphenol	644-35-9	45.0619	0.0332	10	88	2	48.64	48.64	52.28
2072	17102	C9H12O	3-propylphenol	621-27-2	45.7769	0.0325	1	45	2	45.78	49.33	51.80
2073	17103	C9H12O	4-propylphenol	645-56-7	45.7769	0.0325	1	45	2	45.78	49.33	51.80
2074	17104	C9H12O	2-isopropylphenol	88-69-7	43.6170	0.0332	10	124	2	47.08	47.08	51.18
2075	17105	C9H12O	3-isopropylphenol	618-45-1	43.6170	0.0332	10	124	2	47.08	47.08	51.18
2076	17106	C9H12O	4-isopropylphenol	99-89-8	43.6170	0.0332	10	124	2	47.08	47.08	51.18
2077	17107	C9H12O	2-methyl-3-ethylphenol	1123-73-5	45.0428	0.0311	10	61	2	48.39	48.39	51.19
2078	17108	C9H12O	2-methyl-4-ethylphenol	2219-73-0	45.0428	0.0311	10	61	2	48.39	48.39	51.19
2079	17109	C9H12O	2-methyl-5-ethylphenol	1687-65-6	45.0428	0.0311	10	61	2	48.39	48.39	51.19
2080	17110	C9H12O	2-methyl-6-ethylphenol	1687-64-5	45.0428	0.0311	10	61	2	48.39	48.39	51.19

Table 1 Adsorption Capacity of Activated Carbon - Gas

			$\log_{10} Q = \log_{10} K + 1/n \log_{10} y$				(Q - g of compound/100 g of carbon, y - ppmv)					
NO	ID	FORMULA	NAME	CAS No	K	1/n	ymin	ymax	code	Q @ymin	Q @10ppm	Q @ymax
2081	17111	C9H12O	3-methyl-4-ethylphenol	1123-94-0	45.0428	0.0311	10	61	2	48.39	48.39	51.19
2082	17112	C9H12O	3-methyl-5-ethylphenol	698-71-5	45.0428	0.0311	10	61	2	48.39	48.39	51.19
2083	17113	C9H12O	3-methyl-6-ethylphenol	1687-61-2	45.0428	0.0311	10	61	2	48.39	48.39	51.19
2084	17114	C9H12O	4-methyl-2-ethylphenol	3855-26-3	45.0428	0.0311	10	61	2	48.39	48.39	51.19
2085	17115	C9H12O	4-methyl-3-ethylphenol	6161-67-7	45.0428	0.0311	10	61	2	48.39	48.39	51.19
2086	17138	C9H12O2	cumene hydroperoxide	80-15-9	50.6283	0.0291	1	7.4	2	50.63	---	53.67
2087	17213	C9H12S	propyl phenyl sulfide	874-79-3	44.4653	0.0253	10	256	2	47.13	47.13	51.16
2088	17214	C9H12S	isopropyl phenyl sulfide	3019-20-3	42.7352	0.0269	10	487	2	45.47	45.47	50.49
2089	17215	C9H12S	3-methyl-(1-thiapropyl)-benzene	34786-24-8	44.3657	0.0253	10	272	2	47.03	47.03	51.13
2090	17216	C9H12S	4-methyl-(1-thiapropyl)-benzene	622-63-9	44.4289	0.0254	10	260	2	47.11	47.11	51.18
2091	17217	C9H12S	2-ethyl-(1-thiaethyl)-benzene	20760-06-9	46.1478	0.0246	10	182	2	48.84	48.84	52.46
2092	17218	C9H12S	3-ethyl-(1-thiaethyl)-benzene	------	45.7610	0.0278	10	147	2	48.78	48.78	52.56
2093	17219	C9H12S	4-ethyl-(1-thiaethyl)-benzene	------	45.7610	0.0278	10	147	2	48.78	48.78	52.56
2094	17220	C9H12S	2,3-dimethyl-(1-thiaethyl)-benzene	------	46.1256	0.0272	10	120	2	49.11	49.11	52.55
2095	17221	C9H12S	2,4-dimethyl-(1-thiaethyl)-benzene	------	45.9413	0.0268	10	120	2	48.87	48.87	52.24
2096	17222	C9H12S	2,5-dimethyl-(1-thiaethyl)-benzene	------	45.9413	0.0268	10	120	2	48.87	48.87	52.24
2097	17223	C9H12S	2,6-dimethyl-(1-thiaethyl)-benzene	------	45.9413	0.0268	10	120	2	48.87	48.87	52.24
2098	17224	C9H12S	3,4-dimethyl-(1-thiaethyl)-benzene	------	45.9413	0.0268	10	120	2	48.87	48.87	52.24
2099	17225	C9H12S	3,5-dimethyl-(1-thiaethyl)-benzene	------	45.9413	0.0268	10	120	2	48.87	48.87	52.24
2100	17436	C9H14O	isophorone	78-59-1	38.4417	0.0331	10	575	2	41.48	41.48	47.43
2101	17513	C9H14O6	glyceryl triacetate	102-76-1	57.7926	0.0243	1	3.3	2	57.79	---	59.50
2102	17590	C9H16	1-butylcyclopentene	2423-01-0	28.6687	0.0438	10	5217	2	31.71	31.71	41.71
2103	17593	C9H16	1-nonyne	3452-09-3	26.1389	0.0448	10	7895	2	28.98	28.98	39.07
2104	17594	C9H16	2-nonyne	19447-29-1	28.0801	0.0415	10	4040	2	30.90	30.90	39.64
2105	17595	C9H16	3-nonyne	20184-89-8	27.3875	0.0421	10	4951	2	30.18	30.18	39.19
2106	17785	C9H17Br	cis-1-bromo-1-nonene	39924-58-8	49.6943	0.0233	10	194	2	52.44	52.44	56.20
2107	17786	C9H17Br	trans-1-bromo-1-nonene	53434-75-6	49.6943	0.0233	10	194	2	52.44	52.44	56.20
2108	17791	C9H17Cl	cis-1-chloro-1-nonene	66688-72-0	38.9778	0.0267	10	323	2	41.45	41.45	45.47
2109	17792	C9H17Cl	trans-1-chloro-1-nonene	66688-73-1	38.9778	0.0267	10	323	2	41.45	41.45	45.47
2110	17804	C9H17Cl3	1,1,1-trichlorononane	1071-84-7	51.4647	0.0210	1	34	2	51.46	54.01	55.42
2111	17805	C9H17F	cis-1-fluoro-1-nonene	66688-74-2	30.1110	0.0393	10	7770	2	32.96	32.96	42.82
2112	17806	C9H17F	trans-1-fluoro-1-nonene	66688-75-3	30.1110	0.0393	10	7770	2	32.96	32.96	42.82
2113	17808	C9H17F3	1,1,1-trifluorononane	55757-34-1	34.4128	0.0421	10	6526	2	37.92	37.92	49.81
2114	17809	C9H17I	cis-1-iodo-1-nonene	27370-94-1	60.5038	0.0218	10	68	2	63.61	63.61	66.32
2115	17810	C9H17I	trans-1-iodo-1-nonene	26143-08-8	60.5038	0.0218	10	68	2	63.61	63.61	66.32
2116	17813	C9H17N	nonanenitrile	2243-27-8	36.9665	0.0296	10	65	2	39.58	39.58	41.84
2117	17859	C9H18	butylcyclopentane	2040-95-1	28.0564	0.0430	10	5154	2	30.98	30.98	40.53
2118	17860	C9H18	isobutylcyclopentane	3788-32-7	27.0632	0.0447	10	7666	2	30.00	30.00	40.36
2119	17861	C9H18	sec-butylcyclopentane	4850-32-2	28.0462	0.0439	10	5866	2	31.03	31.03	41.04
2120	17862	C9H18	tert-butylcyclopentane	3875-52-3	26.7576	0.0462	10	9921	2	29.76	29.76	40.93
2121	17863	C9H18	1-methyl-1-propylcyclopentane	16631-63-3	27.3038	0.0457	10	8620	2	30.34	30.34	41.33
2122	17864	C9H18	1-methyl-cis-2-propylcyclopentane	932-43-4	27.6873	0.0444	10	6646	2	30.67	30.67	40.93
2123	17865	C9H18	1-methyl-trans-2-propylcyclopentane	932-44-5	26.6880	0.0452	10	8586	2	29.62	29.62	40.20
2124	17866	C9H18	1-methyl-cis-3-propylcyclopentane	2443-04-1	26.8389	0.0453	10	8029	2	29.79	29.79	40.33
2125	17867	C9H18	1-methyl-trans-3-propylcyclopentane	2443-03-0	26.8389	0.0453	10	8029	2	29.79	29.79	40.33
2126	17868	C9H18	1-methyl-1-isopropylcyclopentane	61828-00-0	27.3130	0.0458	10	8458	2	30.35	30.35	41.33
2127	17869	C9H18	1-methyl-cis-2-isopropylcyclopentane	61868-01-7	27.1281	0.0455	10	8554	2	30.12	30.12	40.96
2128	17870	C9H18	1-methyl-trans-2-isopropylcyclopentane	61828-01-1	26.1413	0.0468	10	10000	2	29.12	29.12	40.23
2129	17871	C9H18	1-methyl-cis-3-isopropylcyclopentane	61828-02-2	26.1469	0.0471	10	10000	2	29.14	29.14	40.33
2130	17872	C9H18	1-methyl-trans-3-isopropylcyclopentane	61828-03-3	26.1469	0.0471	10	10000	2	29.14	29.14	40.33
2131	17873	C9H18	1,1-diethylcyclopentane	2721-38-2	27.8217	0.0451	10	7159	2	30.86	30.86	41.50
2132	17874	C9H18	1,cis-2-diethylcyclopentane	932-39-8	27.8744	0.0444	10	6382	2	30.88	30.88	41.13
2133	17875	C9H18	1,trans-2-diethylcyclopentane	932-40-1	26.9404	0.0452	10	8186	2	29.90	29.90	40.50
2134	17877	C9H18	1,cis-3-diethylcyclopentane	62016-59-5	27.2520	0.0450	10	7393	2	30.23	30.23	40.68
2135	17878	C9H18	1,trans-3-diethylcyclopentane	62016-60-8	27.2660	0.0451	10	7393	2	30.25	30.25	40.74
2136	17879	C9H18	1,1-dimethyl-2-ethylcyclopentane	54549-80-3	26.0296	0.0481	10	10000	2	29.08	29.08	40.54
2137	17880	C9H18	1,1-dimethyl-3-ethylcyclopentane	62016-61-9	25.0428	0.0493	10	10000	2	28.06	28.06	39.45
2138	17881	C9H18	1,cis-2-dimethyl-1-ethylcyclopentane	62016-63-1	26.9454	0.0465	10	10000	2	29.99	29.99	41.34
2139	17882	C9H18	1,trans-2-dimethyl-1-ethylcyclopentane	62016-62-0	26.8396	0.0465	10	10000	2	29.87	29.87	41.18
2140	17883	C9H18	1,cis-2-dimethyl-cis-3-ethylcyclopentane	------	27.5210	0.0452	10	7359	2	30.54	30.54	41.15
2141	17884	C9H18	1,cis-2-dimethyl-trans-3-ethylcyclopentane	------	26.6741	0.0462	10	9776	2	29.67	29.67	40.77
2142	17885	C9H18	1,trans-2-dimethyl-cis-3-ethylcyclopentane	------	25.4333	0.0478	10	10000	2	28.39	28.39	39.49
2143	17886	C9H18	1,trans-2-dimethyl-trans-3-ethylcyclopentane	------	26.5208	0.0463	10	9776	2	29.50	29.50	40.56
2144	17887	C9H18	1,cis-2-dimethyl-cis-4-ethylcyclopentane	62016-64-2	26.9913	0.0464	10	9015	2	30.03	30.03	41.19
2145	17888	C9H18	1,cis-2-dimethyl-trans-4-ethylcyclopentane	62016-65-3	26.5948	0.0459	10	9015	2	29.56	29.56	40.40
2146	17889	C9H18	1,trans-2-dimethyl-cis-4-ethylcyclopentane	62016-66-4	25.5695	0.0475	10	10000	2	28.53	28.53	39.62
2147	17890	C9H18	1,cis-3-dimethyl-1-ethylcyclopentane	62016-68-6	25.3559	0.0488	10	10000	2	28.37	28.37	39.73
2148	17891	C9H18	1,trans-3-dimethyl-1-ethylcyclopentane	62016-67-5	25.4692	0.0489	10	10000	2	28.50	28.50	39.94
2149	17892	C9H18	1,cis-3-dimethyl-cis-2-ethylcyclopentane	19903-00-5	27.3567	0.0449	10	7359	2	30.33	30.33	40.79
2150	17893	C9H18	1,cis-3-dimethyl-trans-2-ethylcyclopentane	19902-98-8	25.2776	0.0480	10	10000	2	28.23	28.23	39.35
2151	17894	C9H18	1,trans-3-dimethyl-cis-2-ethylcyclopentane	19902-99-9	26.4607	0.0463	10	10000	2	29.44	29.44	40.55
2152	17895	C9H18	1,cis-3-dimethyl-cis-4-ethylcyclopentane	------	26.6515	0.0464	10	9776	2	29.66	29.66	40.83
2153	17896	C9H18	1,cis-3-dimethyl-trans-4-ethylcyclopentane	------	25.2412	0.0482	10	10000	2	28.20	28.20	39.35
2154	17897	C9H18	1,trans-3-dimethyl-cis-4-ethylcyclopentane	------	30.0287	0.0377	10	1478	2	32.76	32.76	39.55
2155	17898	C9H18	1,trans-3-dimethyl-trans-4-ethylcyclopentane	------	26.3427	0.0461	10	9776	2	29.30	29.30	40.25
2156	17899	C9H18	1,1,2,2-tetramethylcyclopentane	52688-89-8	25.5042	0.0505	10	10000	2	28.65	28.65	40.59
2157	17900	C9H18	1,1,cis-2,cis-3-tetramethylcyclopentane	62016-70-0	25.5464	0.0487	10	10000	2	28.58	28.58	40.00
2158	17901	C9H18	1,1,cis-2,trans-3-tetramethylcyclopentane	------	24.4607	0.0508	10	10000	2	27.50	27.50	39.07
2159	17902	C9H18	1,1,cis-2,cis-4-tetramethylcyclopentane	62016-72-2	24.6316	0.0511	10	10000	2	27.71	27.71	39.43
2160	17903	C9H18	1,1,cis-2,trans-4-tetramethylcyclopentane	------	24.3518	0.0508	10	10000	2	27.37	27.37	38.87

Table 1 Adsorption Capacity of Activated Carbon - Gas

$\log_{10} Q = \log_{10} K + 1/n \log_{10} y$ (Q - g of compound/100 g of carbon, y - ppmv)												
NO	ID	FORMULA	NAME	CAS No	K	1/n	ymin	ymax	code	Q @ymin	Q @10ppm	Q @ymax
2161	17904	C9H18	1,1,3,3-tetramethylcyclopentane	50876-33-0	22.8536	0.0545	10	10000	2	25.91	25.91	37.76
2162	17905	C9H18	1,1,cis-3,cis-4-tetramethylcyclopentane	20309-77-7	24.7125	0.0501	10	10000	2	27.73	27.73	39.19
2163	17906	C9H18	1,1,cis-3,trans-4-tetramethylcyclopentane	------	23.1492	0.0533	10	10000	2	26.17	26.17	37.81
2164	17907	C9H18	1,2,2,cis-3-tetramethylcyclopentane	62016-73-3	25.6031	0.0487	10	10000	2	28.64	28.64	40.11
2165	17908	C9H18	1,2,2,trans-3-tetramethylcyclopentane	------	25.8294	0.0489	10	10000	2	28.91	28.91	40.52
2166	17909	C9H18	1,cis-2,cis-3,cis-4-tetramethylcyclopentane	2532-65-2	26.4291	0.0465	10	10000	2	29.42	29.42	40.57
2167	17910	C9H18	1,cis-2,cis-3,trans-4-tetramethylcyclopentane	2532-69-6	25.3446	0.0485	10	10000	2	28.34	28.34	39.63
2168	17911	C9H18	1,cis-2,trans-3,cis-4-tetramethylcyclopentane	2532-68-5	24.6172	0.0510	10	10000	2	27.68	27.68	39.36
2169	17912	C9H18	1,cis-2,trans-3,trans-4-tetramethylcyclopentane	2532-64-1	25.4113	0.0487	10	10000	2	28.43	28.43	39.79
2170	17913	C9H18	1,trans-2,cis-3,trans-4-tetramethylcyclopentane	2532-67-4	23.6566	0.0514	10	10000	2	26.63	26.63	37.99
2171	17914	C9H18	1,trans-2,trans-3,cis-4-tetramethylcyclopentane	19907-40-5	24.4918	0.0507	10	10000	2	27.52	27.52	39.05
2172	17916	C9H18	propylcyclohexane	1678-92-8	28.2262	0.0433	10	5512	2	31.19	31.19	41.00
2173	17917	C9H18	isopropylcyclohexane	696-29-7	28.1714	0.0441	10	6307	2	31.18	31.18	41.44
2174	17918	C9H18	1-methyl-1-ethylcyclohexane	4926-90-3	27.5855	0.0457	10	8528	2	30.65	30.65	41.72
2175	17919	C9H18	1-methyl-cis-2-ethylcyclohexane	4923-77-7	28.6185	0.0440	10	5651	2	31.67	31.67	41.85
2176	17920	C9H18	1-methyl-trans-2-ethylcyclohexane	4923-78-8	27.8138	0.0441	10	6749	2	30.79	30.79	41.03
2177	17921	C9H18	1-methyl-cis-3-ethylcyclohexane	19489-10-2	27.0577	0.0449	10	8338	2	30.00	30.00	40.58
2178	17922	C9H18	1-methyl-trans-3-ethylcyclohexane	4926-76-5	27.8697	0.0440	10	6731	2	30.84	30.84	41.08
2179	17923	C9H18	1-methyl-cis-4-ethylcyclohexane	4926-78-7	28.1648	0.0438	10	5931	2	31.16	31.16	41.22
2180	17924	C9H18	1-methyl-trans-4-ethylcyclohexane	6236-88-0	26.5523	0.0459	10	10000	2	29.51	29.51	40.52
2181	17925	C9H18	1,1,2-trimethylcyclohexane	7094-26-0	26.6745	0.0474	10	10000	2	29.75	29.75	41.29
2182	17926	C9H18	1,1,3-trimethylcyclohexane	3073-66-3	25.4895	0.0487	10	10000	2	28.52	28.52	39.93
2183	17927	C9H18	1,1,4-trimethylcyclohexane	7094-27-1	24.8715	0.0501	10	10000	2	27.91	27.91	39.44
2184	17929	C9H18	1,cis-2,cis-3-trimethylcyclohexane	1839-88-9	26.4335	0.0480	10	10000	2	29.53	29.53	41.15
2185	17930	C9H18	1,cis-2,trans-3-trimethylcyclohexane	7667-55-2	26.7632	0.0479	10	10000	2	29.88	29.88	41.60
2186	17931	C9H18	1,trans-2,cis-3-trimethylcyclohexane	1678-81-5	25.6231	0.0485	10	10000	2	28.65	28.65	40.07
2187	17933	C9H18	1,cis-2,cis-4-trimethylcyclohexane	1678-80-4	25.7623	0.0488	10	10000	2	28.83	28.83	40.38
2188	17934	C9H18	1,cis-2,trans-4-trimethylcyclohexane	7667-58-5	25.9582	0.0482	10	10000	2	29.00	29.00	40.46
2189	17935	C9H18	1,trans-2,cis-4-trimethylcyclohexane	7667-59-6	25.7384	0.0490	10	10000	2	28.81	28.81	40.41
2190	17936	C9H18	1,trans-2,trans-4-trimethylcyclohexane	7667-60-9	25.0073	0.0495	10	10000	2	28.03	28.03	39.46
2191	17938	C9H18	cis,cis-1,3,5-trimethylcyclohexane	1795-27-3	26.5043	0.0447	10	9258	2	29.38	29.38	39.86
2192	17939	C9H18	cis,trans-1,3,5-trimethylcyclohexane	1795-26-2	25.9957	0.0396	10	8712	2	28.48	28.48	37.23
2193	17941	C9H18	cyclononane	293-55-0	31.9548	0.0405	10	2433	2	35.08	35.08	43.83
2194	17942	C9H18	1-nonene	124-11-8	25.9968	0.0417	10	7085	2	28.62	28.62	37.62
2195	17946	C9H18	2-methyl-1-octene	4588-18-5	25.9319	0.0422	10	7976	2	28.58	28.58	37.88
2196	17955	C9H18Br2	1,1-dibromononane	62168-27-8	66.5868	0.0198	1	20	2	66.59	69.70	70.66
2197	17959	C9H18Cl2	1,1-dichlorononane	821-88-5	44.7764	0.0243	10	120	2	47.35	47.35	50.30
2198	17964	C9H18F2	1,1-difluorononane	62127-42-8	34.5751	0.0369	10	2329	2	37.64	37.64	46.03
2199	17997	C9H18O	1-nonanal	124-19-6	34.8240	0.0310	10	487	2	37.40	37.40	42.19
2200	17998	C9H18O	diisobutyl ketone	108-83-8	31.3702	0.0362	10	2214	2	34.10	34.10	41.47
2201	17999	C9H18O	2-nonanone	821-55-6	34.4248	0.0316	10	576	2	37.03	37.03	42.09
2202	18052	C9H18O2	nonanoic acid	112-05-0	45.7957	0.0246	1	1.5	2	45.80	---	46.26
2203	18058	C9H18O2	octyl formate	112-32-3	36.3837	0.0313	10	801	2	39.10	39.10	44.85
2204	18059	C9H18O2	heptyl acetate	112-06-1	36.8594	0.0301	10	561	2	39.50	39.50	44.59
2205	18061	C9H18O2	hexyl propanoate	2445-76-3	36.6531	0.0307	10	588	2	39.34	39.34	44.59
2206	18062	C9H18O2	pentyl butanoate	540-18-1	35.9392	0.0314	10	852	2	38.64	38.64	44.43
2207	18063	C9H18O2	1-methylbutyl butanoate	60415-61-4	37.1298	0.0291	10	418	2	39.71	39.71	44.27
2208	18064	C9H18O2	2-methylbutyl butanoate	51115-64-1	36.1674	0.0309	10	727	2	38.83	38.83	44.33
2209	18065	C9H18O2	3-methylbutyl butanoate	106-27-4	34.4594	0.0339	10	1768	2	37.26	37.26	44.41
2210	18066	C9H18O2	1-ethylpropyl butanoate	------	38.4881	0.0267	10	176	2	40.93	40.93	44.19
2211	18067	C9H18O2	1,1-dimethylpropyl butanoate	2050-00-2	38.4238	0.0268	10	183	2	40.87	40.87	44.19
2212	18068	C9H18O2	1,2-dimethylpropyl butanoate	88736-68-9	38.4630	0.0268	10	178	2	40.91	40.91	44.19
2213	18069	C9H18O2	2,2-dimethylpropyl butanoate	23361-69-5	37.9989	0.0276	10	241	2	40.49	40.49	44.22
2214	18071	C9H18O2	butyl valerate	591-68-4	36.2012	0.0310	10	775	2	38.88	38.88	44.48
2215	18111	C9H18S	thiacyclodecane	6048-83-5	42.1083	0.0246	10	192	2	44.57	44.57	47.93
2216	18115	C9H19Br	1-bromononane	693-58-3	49.0429	0.0236	10	226	2	51.78	51.78	55.72
2217	18117	C9H19Cl	1-chlorononane	2473-01-0	37.7753	0.0271	10	465	2	40.20	40.20	44.61
2218	18127	C9H19F	1-fluorononane	463-18-3	31.2246	0.0375	10	2684	2	34.04	34.04	41.98
2219	18129	C9H19I	1-iodononane	4282-42-2	59.3087	0.0206	10	169	2	62.18	62.18	65.90
2220	18176	C9H20	nonane	111-84-2	21.6208	0.0485	10	5855	2	24.18	24.18	32.94
2221	18177	C9H20	2-methyloctane	3221-61-2	25.3194	0.0416	10	8191	2	27.87	27.87	36.84
2222	18178	C9H20	3-methyloctane	2216-33-3	25.5142	0.0418	10	8256	2	28.09	28.09	37.21
2223	18179	C9H20	4-methyloctane	2216-34-4	25.3271	0.0421	10	9015	2	27.91	27.91	37.17
2224	18180	C9H20	3-ethylheptane	15869-80-4	25.5478	0.0423	10	8938	2	28.16	28.16	37.53
2225	18181	C9H20	4-ethylheptane	2216-32-2	25.4012	0.0427	10	9758	2	28.02	28.02	37.59
2226	18182	C9H20	2,2-dimethylheptane	1071-26-7	24.1092	0.0448	10	10000	2	26.73	26.73	36.42
2227	18183	C9H20	2,3-dimethylheptane	3074-71-3	25.2184	0.0430	10	10000	2	27.84	27.84	37.47
2228	18184	C9H20	2,4-dimethylheptane	2213-23-2	24.3083	0.0446	10	10000	2	26.94	26.94	36.67
2229	18185	C9H20	2,5-dimethylheptane	2216-30-0	24.6092	0.0439	10	10000	2	27.23	27.23	36.87
2230	18186	C9H20	2,6-dimethylheptane	1072-05-5	24.3727	0.0438	10	10000	2	26.96	26.96	36.49
2231	18187	C9H20	3,4-dimethylheptane	922-28-1	25.3585	0.0431	10	10000	2	28.00	28.00	37.72
2232	18188	C9H20	3,5-dimethylheptane	926-82-9	24.7697	0.0441	10	10000	2	27.42	27.42	37.18
2233	18189	C9H20	4,4-dimethylheptane	1068-19-5	24.5829	0.0449	10	10000	2	27.26	27.26	37.19
2234	18190	C9H20	3-ethyl-2-methylhexane	16789-46-1	25.1730	0.0439	10	10000	2	27.85	27.85	37.73
2235	18191	C9H20	4-ethyl-2-methylhexane	3074-75-7	24.5504	0.0448	10	10000	2	27.22	27.22	37.09
2236	18192	C9H20	3-ethyl-3-methylhexane	3074-76-8	25.4530	0.0441	10	10000	2	28.17	28.17	38.19
2237	18193	C9H20	3-ethyl-4-methylhexane	3074-77-9	25.5743	0.0435	10	10000	2	28.27	28.27	38.18
2238	18194	C9H20	2,2,3-trimethylhexane	16747-25-4	24.5311	0.0455	10	10000	2	27.24	27.24	37.32
2239	18195	C9H20	2,2,4-trimethylhexane	16747-26-5	23.5172	0.0476	10	10000	2	26.24	26.24	36.45
2240	18196	C9H20	2,2,5-trimethylhexane	3522-94-9	23.1000	0.0483	10	10000	2	25.82	25.82	36.05

Table 1 Adsorption Capacity of Activated Carbon - Gas

$\log_{10} Q = \log_{10} K + 1/n \log_{10} y$ (Q - g of compound/100 g of carbon, y - ppmv)

NO	ID	FORMULA	NAME	CAS No	K	1/n	ymin	ymax	code	Q @ymin	Q @10ppm	Q @ymax
2241	18197	C9H20	2,3,3-trimethylhexane	16747-28-7	25.0720	0.0449	10	10000	2	27.80	27.80	37.90
2242	18198	C9H20	2,3,4-trimethylhexane	921-47-1	25.2818	0.0443	10	10000	2	28.00	28.00	38.03
2243	18199	C9H20	2,3,5-trimethylhexane	1069-53-0	24.2028	0.0459	10	10000	2	26.90	26.90	36.92
2244	18200	C9H20	2,4,4-trimethylhexane	16747-30-1	23.9940	0.0468	10	10000	2	26.72	26.72	36.91
2245	18201	C9H20	3,3,4-trimethylhexane	16747-31-2	25.5100	0.0443	10	10000	2	28.25	28.25	38.35
2246	18202	C9H20	3,3-diethylpentane	1067-20-5	26.1682	0.0434	10	9590	2	28.92	28.92	38.96
2247	18203	C9H20	2,2-dimethyl-3-ethylpentane	16747-32-3	24.3349	0.0474	10	10000	2	27.14	27.14	37.64
2248	18204	C9H20	2,3-dimethyl-3-ethylpentane	16747-33-4	26.1936	0.0432	10	10000	2	28.93	28.93	39.00
2249	18205	C9H20	2,4-dimethyl-3-ethylpentane	1068-87-7	25.0099	0.0451	10	10000	2	27.75	27.75	37.88
2250	18206	C9H20	2,2,3,3-tetramethylpentane	7154-79-2	25.7222	0.0450	10	10000	2	28.53	28.53	38.92
2251	18207	C9H20	2,2,3,4-tetramethylpentane	1186-53-4	24.5407	0.0467	10	10000	2	27.33	27.33	37.73
2252	18208	C9H20	2,2,4,4-tetramethylpentane	1070-87-7	23.0378	0.0495	10	10000	2	25.82	25.82	36.33
2253	18209	C9H20	2,3,3,4-tetramethylpentane	16747-38-9	25.5902	0.0430	10	10000	2	28.25	28.25	38.03
2254	18236	C9H20O	1-nonanol	143-08-8	38.8510	0.0276	1	23	2	38.85	41.40	42.36
2255	18237	C9H20O	2-nonanol	628-99-9	37.0422	0.0291	10	88	2	39.61	39.61	42.20
2256	18239	C9H20O	3-nonanol	624-51-1	34.8884	0.0304	10	573	2	37.42	37.42	42.32
2257	18241	C9H20O	4-nonanol	5932-79-6	34.7533	0.0308	10	617	2	37.31	37.31	42.36
2258	18243	C9H20O	5-nonanol	623-93-8	34.7782	0.0304	10	563	2	37.30	37.30	42.16
2259	18244	C9H20O	2-methyl-1-octanol	818-81-5	35.3469	0.0294	10	457	2	37.82	37.82	42.32
2260	18245	C9H20O	3-methyl-1-octanol	38514-02-2	35.4800	0.0296	10	457	2	37.98	37.98	42.53
2261	18246	C9H20O	4-methyl-1-octanol	38514-03-3	35.3517	0.0294	10	457	2	37.83	37.83	42.32
2262	18247	C9H20O	5-methyl-1-octanol	38514-04-4	35.5164	0.0296	10	457	2	38.02	38.02	42.58
2263	18248	C9H20O	6-methyl-1-octanol	38514-05-5	36.0389	0.0288	10	348	2	38.51	38.51	42.66
2264	18249	C9H20O	7-methyl-1-octanol	2430-22-0	36.0389	0.0288	10	348	2	38.51	38.51	42.66
2265	18250	C9H20O	2-methyl-2-octanol	628-44-4	33.0543	0.0334	10	1425	2	35.69	35.69	42.12
2266	18251	C9H20O	3-methyl-2-octanol	27644-49-1	34.0409	0.0331	10	1147	2	36.74	36.74	42.99
2267	18252	C9H20O	5-methyl-2-octanol	66793-81-5	33.6800	0.0326	10	1147	2	36.30	36.30	42.36
2268	18253	C9H20O	7-methyl-2-octanol	66793-83-7	33.7725	0.0329	10	1147	2	36.43	36.43	42.57
2269	18254	C9H20O	2-methyl-3-octanol	26533-34-6	34.0200	0.0324	10	1001	2	36.65	36.65	42.55
2270	18255	C9H20O	3-methyl-3-octanol	5340-36-3	34.3434	0.0321	10	889	2	36.98	36.98	42.70
2271	18256	C9H20O	4-methyl-3-octanol	66793-80-4	34.8909	0.0324	10	923	2	37.59	37.59	43.53
2272	18257	C9H20O	6-methyl-3-octanol	40225-75-0	34.5558	0.0317	10	923	2	37.17	37.17	42.91
2273	18258	C9H20O	7-methyl-3-octanol	66793-84-8	34.3347	0.0323	10	923	2	36.99	36.99	42.81
2274	18259	C9H20O	2-methyl-4-octanol	40575-41-5	33.6703	0.0321	10	1001	2	36.26	36.26	42.04
2275	18260	C9H20O	3-methyl-4-octanol	26533-35-7	33.8105	0.0331	10	1191	2	36.49	36.49	42.73
2276	18261	C9H20O	4-methyl-4-octanol	23418-37-3	33.5984	0.0330	10	1253	2	36.25	36.25	42.51
2277	18262	C9H20O	5-methyl-4-octanol	59734-23-5	33.5365	0.0325	10	1099	2	36.14	36.14	42.10
2278	18263	C9H20O	6-methyl-4-octanol	66793-82-6	33.6256	0.0328	10	1099	2	36.26	36.26	42.31
2279	18264	C9H20O	7-methyl-4-octanol	33933-77-6	33.3186	0.0324	10	1099	2	35.90	35.90	41.79
2280	18265	C9H20O	2-ethyl-1-heptanol	817-60-7	35.1977	0.0297	10	477	2	37.69	37.69	42.28
2281	18266	C9H20O	3-ethyl-1-heptanol	3525-25-5	36.1831	0.0290	10	333	2	38.68	38.68	42.81
2282	18267	C9H20O	5-ethyl-1-heptanol	998-65-2	36.1644	0.0306	1C	477	2	38.80	38.80	43.68
2283	18268	C9H20O	2,2-dimethyl-1-heptanol	14250-79-4	35.0945	0.0323	10	782	2	37.80	37.80	43.52
2284	18269	C9H20O	4,6-dimethyl-1-heptanol	820-05-3	34.4101	0.0337	10	1099	2	37.18	37.18	43.56
2285	18270	C9H20O	6,6-dimethyl-1-heptanol	65769-10-0	34.3547	0.0328	10	1208	2	37.05	37.05	43.35
2286	18271	C9H20O	2-ethyl-2-heptanol	------	34.1082	0.0329	10	1208	2	36.80	36.80	43.09
2287	18272	C9H20O	3-ethyl-2-heptanol	19780-39-3	34.1359	0.0333	10	1099	2	36.85	36.85	43.09
2288	18273	C9H20O	2,3-dimethyl-2-heptanol	66794-00-1	33.9910	0.0331	10	1316	2	36.68	36.68	43.10
2289	18274	C9H20O	2,4-dimethyl-2-heptanol	65822-93-7	33.6013	0.0335	10	1316	2	36.30	36.30	42.74
2290	18275	C9H20O	2,5-dimethyl-2-heptanol	------	33.5342	0.0338	10	1316	2	36.25	36.25	42.75
2291	18276	C9H20O	2,6-dimethyl-2-heptanol	13254-34-7	32.2246	0.0350	10	2083	2	34.93	34.93	42.12
2292	18277	C9H20O	4,6-dimethyl-2-heptanol	51079-52-8	35.7231	0.0328	10	711	2	38.52	38.52	44.30
2293	18278	C9H20O	5,6-dimethyl-2-heptanol	58795-24-7	34.9523	0.0302	10	774	2	37.47	37.47	42.72
2294	18279	C9H20O	3-ethyl-3-heptanol	19780-41-7	34.1494	0.0332	10	1190	2	36.86	36.86	43.20
2295	18280	C9H20O	2,2-dimethyl-3-heptanol	19549-70-3	33.2203	0.0337	10	1559	2	35.90	35.90	42.55
2296	18281	C9H20O	2,3-dimethyl-3-heptanol	19549-71-4	33.2156	0.0345	10	1835	2	35.96	35.96	43.03
2297	18282	C9H20O	2,6-dimethyl-3-heptanol	19549-73-6	32.7947	0.0341	10	1606	2	35.48	35.48	42.20
2298	18283	C9H20O	3,5-dimethyl-3-heptanol	19549-74-7	32.9889	0.0338	10	1626	2	35.66	35.66	42.35
2299	18284	C9H20O	3,6-dimethyl-3-heptanol	1573-28-0	33.0154	0.0338	10	1626	2	35.69	35.69	42.40
2300	18285	C9H20O	4-ethyl-4-heptanol	597-90-0	33.6816	0.0335	10	1365	2	36.38	36.38	42.90
2301	18286	C9H20O	2,2-dimethyl-4-heptanol	66793-99-5	32.8204	0.0350	10	1835	2	35.57	35.57	42.68
2302	18287	C9H20O	2,4-dimethyl-4-heptanol	19549-77-0	32.5544	0.0347	10	2048	2	35.26	35.26	42.42
2303	18288	C9H20O	2,6-dimethyl-4-heptanol	108-82-7	34.8778	0.0292	10	398	2	37.30	37.30	41.54
2304	18289	C9H20O	3,3-dimethyl-4-heptanol	19549-78-1	33.8146	0.0368	10	1686	2	36.81	36.81	44.46
2305	18290	C9H20O	3,5-dimethyl-4-heptanol	19549-79-2	34.8938	0.0342	10	960	2	37.75	37.75	44.13
2306	18291	C9H20O	2-propyl-1-hexanol	817-46-9	36.4041	0.0279	10	741	2	38.82	38.82	43.78
2307	18292	C9H20O	3-propyl-1-hexanol	66793-85-9	35.4133	0.0323	10	741	2	38.14	38.14	43.83
2308	18293	C9H20O	1-methyl-2-ethyl-1-hexanol	------	35.3262	0.0322	10	810	2	38.05	38.05	43.83
2309	18294	C9H20O	1-methyl-3-ethyl-1-hexanol	------	35.5496	0.0311	10	707	2	38.19	38.19	43.61
2310	18295	C9H20O	3-methyl-2-ethyl-1-hexanol	66794-04-5	35.1762	0.0311	10	681	2	37.79	37.79	43.09
2311	18296	C9H20O	4-methyl-2-ethyl-1-hexanol	66794-06-7	34.9029	0.0307	10	620	2	37.46	37.46	42.51
2312	18297	C9H20O	5-methyl-2-ethyl-1-hexanol	66794-07-8	34.5993	0.0308	10	681	2	37.15	37.15	42.31
2313	18298	C9H20O	3,3,5-trimethyl-1-hexanol	1484-87-3	34.3328	0.0313	10	826	2	36.90	36.90	42.38
2314	18299	C9H20O	3,4,4-trimethyl-1-hexanol	66793-73-5	34.1764	0.0317	10	894	2	36.76	36.76	42.39
2315	18300	C9H20O	3,5,5-trimethyl-1-hexanol	3452-97-9	34.3661	0.0316	10	821	2	36.96	36.96	42.49
2316	18301	C9H20O	4,5,5-trimethyl-1-hexanol	66793-75-7	35.2743	0.0298	10	561	2	37.78	37.78	42.60
2317	18302	C9H20O	2-methyl-3-ethyl-2-hexanol	66794-02-3	33.5308	0.0340	10	1549	2	36.26	36.26	43.04
2318	18303	C9H20O	2,3,4-trimethyl-2-hexanol	21102-13-6	34.4374	0.0314	10	906	2	37.02	37.02	42.64
2319	18304	C9H20O	2,4,4-trimethyl-2-hexanol	66793-91-7	34.7954	0.0324	10	1002	2	37.49	37.49	43.54
2320	18305	C9H20O	2,4,5-trimethyl-2-hexanol	66793-93-9	34.3104	0.0321	10	906	2	36.95	36.95	42.70

Table 1 Adsorption Capacity of Activated Carbon - Gas

$\log_{10} Q = \log_{10} K + 1/n \log_{10} y$ (Q - g of compound/100 g of carbon, y - ppmv)												
NO	ID	FORMULA	NAME	CAS No	K	1/n	ymin	ymax	code	Q @ymin	Q @10ppm	Q @ymax
2321	18306	C9H20O	2,5,5-trimethyl-2-hexanol	66793-71-3	34.1154	0.0333	10	1002	2	36.83	36.83	42.93
2322	18307	C9H20O	2-methyl-3-ethyl-3-hexanol	66794-03-4	34.4110	0.0334	10	1197	2	37.16	37.16	43.59
2323	18308	C9H20O	2-methyl-4-ethyl-3-hexanol	33943-21-4	33.5589	0.0328	10	1361	2	36.19	36.19	42.53
2324	18309	C9H20O	3-methyl-4-ethyl-3-hexanol	66794-05-6	35.4810	0.0370	10	1494	2	38.63	38.63	46.49
2325	18310	C9H20O	4-methyl-3-ethyl-3-hexanol	51200-80-7	34.1623	0.0348	10	1494	2	37.01	37.01	44.06
2326	18311	C9H20O	5-methyl-3-ethyl-3-hexanol	597-77-3	33.0531	0.0354	10	1997	2	35.86	35.86	43.26
2327	18312	C9H20O	2,2,3-trimethyl-3-hexanol	5340-41-0	33.1036	0.0354	10	2259	2	35.92	35.92	43.52
2328	18313	C9H20O	2,2,4-trimethyl-3-hexanol	66793-89-3	32.3059	0.0349	10	2452	2	35.01	35.01	42.44
2329	18314	C9H20O	2,2,5-trimethyl-3-hexanol	3970-60-3	30.6432	0.0343	10	3571	2	33.16	33.16	40.58
2330	18315	C9H20O	2,3,4-trimethyl-3-hexanol	66793-90-6	32.1780	0.0346	10	2797	2	34.84	34.84	42.33
2331	18316	C9H20O	2,3,5-trimethyl-3-hexanol	65927-60-8	32.0706	0.0358	10	2797	2	34.83	34.83	42.61
2332	18317	C9H20O	2,4,4-trimethyl-3-hexanol	66793-92-8	33.1797	0.0356	10	2256	2	36.02	36.02	43.68
2333	18318	C9H20O	2,5,5-trimethyl-3-hexanol	66793-72-4	32.1683	0.0355	10	2468	2	34.91	34.91	42.44
2334	18319	C9H20O	3,4,4-trimethyl-3-hexanol	66793-74-6	32.4151	0.0348	10	3026	2	35.12	35.12	42.86
2335	18320	C9H20O	3,5,5-trimethyl-3-hexanol	66810-87-5	32.1422	0.0362	10	3045	2	34.94	34.94	42.98
2336	18321	C9H20O	4-methyl-2-propyl-1-pentanol	54004-41-0	38.0596	0.0158	10	701	2	39.47	39.47	42.22
2337	18322	C9H20O	4-methyl-2-isopropyl-1-pentanol	55505-24-3	36.7429	0.0215	10	967	2	38.60	38.60	42.59
2338	18323	C9H20O	2,2-dimethyl-3-ethyl-1-pentanol	66793-95-1	34.0929	0.0325	10	1021	2	36.75	36.75	42.72
2339	18324	C9H20O	2,4-dimethyl-2-ethyl-1-pentanol	66793-98-4	34.5292	0.0323	10	1015	2	37.19	37.19	43.18
2340	18325	C9H20O	3,3-diethyl-2-pentanol	66793-94-0	34.8333	0.0317	10	1021	2	37.47	37.47	43.38
2341	18326	C9H20O	2,3-dimethyl-3-ethyl-2-pentanol	66793-97-3	34.4309	0.0325	10	1231	2	37.11	37.11	43.40
2342	18327	C9H20O	4,4-dimethyl-3-ethyl-2-pentanol	21102-09-0	34.5505	0.0325	10	1116	2	37.23	37.23	43.39
2343	18328	C9H20O	2,3,3,4-tetramethyl-2-pentanol	66793-86-0	33.6731	0.0338	10	1862	2	36.40	36.40	43.44
2344	18329	C9H20O	2,3,4,4-tetramethyl-2-pentanol	66793-87-1	33.6272	0.0340	10	1862	2	36.37	36.37	43.44
2345	18330	C9H20O	3,3,4,4-tetramethyl-2-pentanol	66793-88-2	33.6272	0.0340	10	1862	2	36.37	36.37	43.44
2346	18331	C9H20O	2-methyl-3-isopropyl-3-pentanol	------	34.1541	0.0347	10	1694	2	37.00	37.00	44.22
2347	18332	C9H20O	2,2-dimethyl-3-ethyl-3-pentanol	66793-96-2	33.5247	0.0356	10	2177	2	36.39	36.39	44.07
2348	18333	C9H20O	2,4-dimethyl-3-ethyl-3-pentanol	3970-59-0	34.1044	0.0347	10	1691	2	36.94	36.94	44.13
2349	18334	C9H20O	2,2,3,4-tetramethyl-3-pentanol	62185-29-9	33.3681	0.0358	10	2342	2	36.24	36.24	44.06
2350	18335	C9H20O	2,2,4,4-tetramethyl-3-pentanol	14609-79-1	33.3681	0.0358	10	2342	2	36.24	36.24	44.06
2351	18339	C9H20O2	1,2-nonanediol	42789-13-9	47.1799	0.0270	1	3.3	2	47.18	---	48.72
2352	18340	C9H20O2	1,3-nonanediol	------	47.5527	0.0232	1	1.9	2	47.55	---	48.27
2353	18341	C9H20O2	1,4-nonanediol	2430-73-1	47.5527	0.0232	1	1.9	2	47.55	---	48.27
2354	18342	C9H20O2	1,9-nonanediol	3937-56-2	47.5527	0.0232	1	1.9	2	47.55	---	48.27
2355	18365	C9H20S	nonyl mercaptan	1455-21-6	38.3767	0.0230	10	172	2	40.46	40.46	43.19
2356	18366	C9H20S	2-nonanethiol	13281-11-3	36.9200	0.0248	10	415	2	39.09	39.09	42.88
2357	18367	C9H20S	methyl octyl sulfide	3698-95-1	37.6402	0.0241	10	299	2	39.79	39.79	43.18
2358	18368	C9H20S	ethyl heptyl sulfide	24768-44-3	37.6402	0.0241	10	299	2	39.79	39.79	43.18
2359	18369	C9H20S	propyl hexyl sulfide	24768-43-2	37.6402	0.0241	10	299	2	39.79	39.79	43.18
2360	18370	C9H20S	butyl pentyl sulfide	24768-42-1	37.6402	0.0241	10	299	2	39.79	39.79	43.18
2361	18388	C9H21N	nonylamine	112-20-9	34.3551	0.0274	10	366	2	36.59	36.59	40.39
2362	18389	C9H21N	methyloctylamine	2439-54-5	33.5041	0.0279	10	472	2	35.73	35.73	39.78
2363	18390	C9H21N	ethylheptylamine	66793-76-8	32.7431	0.0289	10	677	2	35.00	35.00	39.53
2364	18391	C9H21N	dimethylheptylamine	5277-11-2	30.8715	0.0313	10	1622	2	33.18	33.18	38.91
2365	18392	C9H21N	diethylpentylamine	2162-91-6	30.3221	0.0333	10	2562	2	32.74	32.74	39.39
2366	18393	C9H21N	tripropylamine	102-69-2	30.4898	0.0322	10	1988	2	32.83	32.83	38.93
2367	18601	C10H7Br	1-bromonaphthalene	90-11-9	71.2607	0.0256	1	12	2	71.26	75.60	75.95
2368	18606	C10H7Cl	1-chloronaphthalene	90-13-1	55.4224	0.0251	1	26	2	55.42	58.73	60.15
2369	18669	C10H8	naphthalene	91-20-3	93.1819	0.0242	10	10000	2	98.52	98.52	116.43
2370	18849	C10H9N	2-methylquinoline	91-63-4	50.5392	0.0288	1	12	2	50.54	54.00	54.28
2371	18850	C10H9N	3-methylquinoline	612-58-8	50.5392	0.0288	1	12	2	50.54	54.00	54.28
2372	18851	C10H9N	4-methylquinoline	491-35-0	51.3531	0.0296	1	13	2	51.35	54.97	55.40
2373	18852	C10H9N	5-methylquinoline	7661-55-4	51.3531	0.0296	1	13	2	51.35	54.97	55.40
2374	18853	C10H9N	6-methylquinoline	91-62-3	51.3531	0.0296	1	13	2	51.35	54.97	55.40
2375	18854	C10H9N	7-methylquinoline	612-60-2	51.3531	0.0296	1	13	2	51.35	54.97	55.40
2376	18855	C10H9N	8-methylquinoline	611-32-5	51.3531	0.0296	1	13	2	51.35	54.97	55.40
2377	18958	C10H10	m-divinylbenzene	108-57-6	38.9969	0.0303	10	761	2	41.81	41.81	47.67
2378	19195	C10H10O4	dimethyl phthalate	131-11-3	59.0361	0.0245	1	4	2	59.04	---	61.07
2379	19359	C10H12	1,2,3,4-tetrahydronaphthalene	119-64-2	40.8558	0.0321	10	483	2	43.99	43.99	49.83
2380	19360	C10H12	cis-(1-butenyl)benzene	1560-09-4	37.1466	0.0325	10	1186	2	40.04	40.04	46.77
2381	19361	C10H12	trans-(1-butenyl)benzene	1005-64-7	37.7658	0.0305	10	777	2	40.52	40.52	46.27
2382	19362	C10H12	cis-(1-methyl-1-propenyl)benzene	767-99-7	34.9422	0.0356	10	2435	2	37.92	37.92	46.10
2383	19363	C10H12	trans-(1-methyl-1-propenyl)benzene	768-00-3	37.7640	0.0321	10	1001	2	40.66	40.66	47.13
2384	19364	C10H12	(2-methyl-1-propenyl)benzene	768-49-0	36.6283	0.0323	10	1342	2	39.46	39.46	46.23
2385	19365	C10H12	cis-1-methyl-2-(1-propenyl)benzene	2077-33-0	37.1802	0.0316	10	1036	2	39.99	39.99	46.32
2386	19366	C10H12	cis-1-methyl-3-(1-propenyl)benzene	2077-31-8	36.9901	0.0312	10	1036	2	39.75	39.75	45.95
2387	19367	C10H12	cis-1-methyl-4-(1-propenyl)benzene	2077-29-4	37.0981	0.0306	10	886	2	39.80	39.80	45.65
2388	19368	C10H12	trans-1-methyl-4-(1-propenyl)benzene	2077-30-7	38.0693	0.0304	10	712	2	40.83	40.83	46.48
2389	19369	C10H12	1-methyl-2-(1-methylethenyl)benzene	7399-49-7	34.7039	0.0356	10	2371	2	37.67	37.67	45.77
2390	19370	C10H12	1-methyl-3-(1-methylethenyl)benzene	1124-20-5	36.6264	0.0331	10	1358	2	39.53	39.53	46.51
2391	19371	C10H12	1-methyl-4-(1-methylethenyl)benzene	1195-32-0	36.5813	0.0327	10	1300	2	39.44	39.44	46.24
2392	19372	C10H12	o-ethylstyrene	7564-63-8	36.8747	0.0326	10	1263	2	39.75	39.75	46.53
2393	19373	C10H12	m-ethylstyrene	7525-62-4	36.7326	0.0318	10	1120	2	39.53	39.53	45.93
2394	19374	C10H12	p-ethylstyrene	3454-07-7	36.8989	0.0313	10	1015	2	39.65	39.65	45.81
2395	19375	C10H12	1-ethenyl-2,4-dimethylbenzene	2234-20-0	38.1356	0.0302	10	670	2	40.88	40.88	46.43
2396	19376	C10H12	1-ethenyl-3,5-dimethylbenzene	5379-20-4	37.0625	0.0315	10	1023	2	39.85	39.85	46.11
2397	19377	C10H12	2-ethenyl-1,3-dimethylbenzene	2039-90-9	37.2242	0.0320	10	1023	2	40.07	40.07	46.47
2398	19378	C10H12	2-ethenyl-1,4-dimethylbenzene	2039-89-6	37.3898	0.0316	10	996	2	40.22	40.22	46.52
2399	19379	C10H12	4-ethenyl-1,2-dimethylbenzene	27831-13-6	37.5526	0.0315	10	1023	2	40.38	40.38	46.73
2400	19380	C10H12	2,3-dimethylstyrene	40243-75-2	36.9487	0.0313	10	1023	2	39.71	39.71	45.90

Table 1 Adsorption Capacity of Activated Carbon - Gas

$\log_{10} Q = \log_{10} K + 1/n \log_{10} y$ (Q - g of compound/100 g of carbon, y - ppmv)

NO	ID	FORMULA	NAME	CAS No	K	1/n	ymin	ymax	code	Q @ymin	Q @10ppm	Q @ymax
2401	19381	C10H12	3,4-dimethylstyrene	------	37.5526	0.0315	10	1023	2	40.38	40.38	46.73
2402	19382	C10H12	2,6-dimethylstyrene	------	37.2242	0.0320	10	1023	2	40.07	40.07	46.47
2403	19383	C10H12	3,5-dimethylstyrene	------	37.0625	0.0315	10	1023	2	39.85	39.85	46.11
2404	19387	C10H12	(1-methylenepropyl)benzene	2039-93-2	35.6981	0.0335	10	1709	2	38.56	38.56	45.80
2405	19484	C10H12O	anethole	104-46-1	45.2440	0.0261	10	73	2	48.04	48.04	50.60
2406	19675	C10H12O4	diallyl maleate	999-21-3	52.2290	0.0251	1	9	2	52.23	---	55.19
2407	19865	C10H14	butylbenzene	104-51-8	34.8030	0.0335	10	1346	2	37.59	37.59	44.29
2408	19866	C10H14	isobutylbenzene	538-93-2	33.2458	0.0354	10	2533	2	36.07	36.07	43.89
2409	19867	C10H14	sec-butylbenzene	135-98-8	33.7415	0.0353	10	2304	2	36.60	36.60	44.35
2410	19868	C10H14	tert-butylbenzene	98-06-6	33.5068	0.0361	10	2809	2	36.41	36.41	44.63
2411	19870	C10H14	m-cymene	535-77-3	33.8086	0.0350	10	2268	2	36.64	36.64	44.29
2412	19871	C10H14	o-cymene	527-84-4	34.6583	0.0347	10	1974	2	37.54	37.54	45.11
2413	19872	C10H14	p-cymene	99-87-6	33.9645	0.0343	10	1927	2	36.75	36.75	44.01
2414	19873	C10H14	1-methyl-2-propylbenzene	1074-17-5	34.3165	0.0350	10	2222	2	37.20	37.20	44.95
2415	19874	C10H14	1-methyl-3-propylbenzene	1074-43-7	33.6451	0.0352	10	2473	2	36.49	36.49	44.30
2416	19875	C10H14	1-methyl-4-propylbenzene	1074-55-1	33.8131	0.0347	10	2174	2	36.63	36.63	44.15
2417	19878	C10H14	o-diethylbenzene	135-01-3	35.5090	0.0334	10	1384	2	38.35	38.35	45.22
2418	19879	C10H14	m-diethylbenzene	141-93-5	34.7741	0.0335	10	1489	2	37.56	37.56	44.40
2419	19880	C10H14	p-diethylbenzene	105-05-5	34.8424	0.0332	10	1387	2	37.61	37.61	44.29
2420	19881	C10H14	3-ethyl-o-xylene	933-98-2	37.0328	0.0316	10	817	2	39.83	39.83	45.79
2421	19882	C10H14	4-ethyl-o-xylene	934-80-5	36.0258	0.0320	10	983	2	38.78	38.78	44.93
2422	19883	C10H14	2-ethyl-m-xylene	2870-04-4	36.6367	0.0322	10	965	2	39.45	39.45	45.70
2423	19884	C10H14	4-ethyl-m-xylene	874-41-9	35.7350	0.0326	10	1170	2	38.52	38.52	45.00
2424	19885	C10H14	5-ethyl-m-xylene	934-74-7	35.0363	0.0330	10	1355	2	37.80	37.80	44.44
2425	19886	C10H14	2-ethyl-p-xylene	1758-88-9	35.6506	0.0328	10	1245	2	38.45	38.45	45.05
2426	19887	C10H14	1,2,3,4-tetramethylbenzene	488-23-3	38.5005	0.0301	10	498	2	41.26	41.26	46.40
2427	19888	C10H14	1,2,3,5-tetramethylbenzene	527-53-7	37.4402	0.0308	10	653	2	40.19	40.19	45.71
2428	19889	C10H14	1,2,4,5-tetramethylbenzene	95-93-2	37.4402	0.0308	10	653	2	40.19	40.19	45.71
2429	19982	C10H14O	3-butylphenol	4074-43-5	44.9250	0.0268	10	57	2	47.79	47.79	50.08
2430	19983	C10H14O	4-butylphenol	1638-22-8	46.1734	0.0265	1	21	2	46.17	49.07	50.05
2431	19984	C10H14O	2-isobutylphenol	4167-75-3	45.1801	0.0283	1	41	2	45.18	48.22	50.18
2432	19985	C10H14O	3-isobutylphenol	30749-25-8	45.1801	0.0283	1	41	2	45.18	48.22	50.18
2433	19986	C10H14O	4-isobutylphenol	4167-74-2	45.1801	0.0283	1	41	2	45.18	48.22	50.18
2434	19987	C10H14O	2-sec-butylphenol	89-72-5	44.4804	0.0271	10	61	2	47.34	47.34	49.71
2435	19988	C10H14O	3-sec-butylphenol	3522-86-9	44.4804	0.0271	10	61	2	47.34	47.34	49.71
2436	19989	C10H14O	4-sec-butylphenol	99-71-8	44.4804	0.0271	10	61	2	47.34	47.34	49.71
2437	19990	C10H14O	2-tert-butylphenol	88-18-6	44.2556	0.0280	10	95	2	47.20	47.20	50.26
2438	19991	C10H14O	3-tert-butylphenol	585-34-2	44.2556	0.0280	10	95	2	47.20	47.20	50.26
2439	19992	C10H14O	4-tert-butylphenol	98-54-4	44.2556	0.0280	10	95	2	47.20	47.20	50.26
2440	19993	C10H14O	2-methyl-3-propylphenol	------	46.9247	0.0286	1	11	2	46.92	50.11	50.25
2441	19994	C10H14O	2-methyl-4-propylphenol	18441-56-0	46.9247	0.0286	1	11	2	46.92	50.11	50.25
2442	19995	C10H14O	2-methyl-5-propylphenol	------	46.9247	0.0286	1	11	2	46.92	50.11	50.25
2443	19996	C10H14O	2-methyl-6-propylphenol	------	46.9247	0.0286	1	11	2	46.92	50.11	50.25
2444	19997	C10H14O	3-methyl-2-propylphenol	------	46.9247	0.0286	1	11	2	46.92	50.11	50.25
2445	19998	C10H14O	3-methyl-4-propylphenol	30388-40-0	46.9247	0.0286	1	11	2	46.92	50.11	50.25
2446	19999	C10H14O	3-methyl-5-propylphenol	36186-96-6	46.6400	0.0291	1	13	2	46.64	49.87	50.25
2447	20000	C10H14O	4-methyl-2-propylphenol	4074-46-8	46.6400	0.0291	1	13	2	46.64	49.87	50.25
2448	20001	C10H14O	4-methyl-3-propylphenol	------	46.6400	0.0291	1	13	2	46.64	49.87	50.25
2449	20002	C10H14O	5-methyl-2-propylphenol	------	46.6400	0.0291	1	13	2	46.64	49.87	50.25
2450	20003	C10H14O	2-methyl-3-isopropylphenol	4371-48-6	45.9430	0.0267	1	26	2	45.94	48.85	50.12
2451	20004	C10H14O	2-methyl-4-isopropylphenol	1740-97-2	47.4816	0.0272	1	7.5	2	47.48	---	50.15
2452	20005	C10H14O	2-methyl-5-isopropylphenol	499-75-2	45.2100	0.0280	1	41	2	45.21	48.22	50.16
2453	20006	C10H14O	2-methyl-6-isopropylphenol	3228-04-4	45.9430	0.0267	1	26	2	45.94	48.85	50.12
2454	20007	C10H14O	3-methyl-2-isopropylphenol	3228-01-1	45.9430	0.0267	1	26	2	45.94	48.85	50.12
2455	20008	C10H14O	3-methyl-4-isopropylphenol	3228-02-2	45.9430	0.0267	1	26	2	45.94	48.85	50.12
2456	20009	C10H14O	3-methyl-5-isopropylphenol	3228-03-3	45.9430	0.0267	1	26	2	45.94	48.85	50.12
2457	20010	C10H14O	4-methyl-2-isopropylphenol	4427-56-9	45.9430	0.0267	1	26	2	45.94	48.85	50.12
2458	20011	C10H14O	5-methyl-2-isopropylphenol	89-83-8	45.9430	0.0267	1	26	2	45.94	48.85	50.12
2459	20037	C10H14O	2-butylphenol	3180-09-4	44.9250	0.0268	10	57	2	47.79	47.79	50.08
2460	20195	C10H15N	N,N-dietyhlaniline	91-66-7	42.2470	0.0238	10	185	2	44.62	44.62	47.82
2461	20196	C10H15N	2,6-diethylaniline	579-66-8	44.6473	0.0231	1	5	2	44.65	---	46.34
2462	20316	C10H16	D-limonene	5989-27-5	32.8227	0.0354	10	2604	2	35.61	35.61	43.38
2463	20319	C10H16	alpha-phellandrene	99-83-2	33.4272	0.0348	10	1995	2	36.22	36.22	43.56
2464	20320	C10H16	beta-phellandrene	555-10-2	33.5306	0.0332	10	2085	2	36.20	36.20	43.23
2465	20321	C10H16	alpha-pinene	80-56-8	31.1487	0.0407	10	6250	2	34.21	34.21	44.44
2466	20323	C10H16	beta-pinene	127-91-3	32.8005	0.0380	10	3849	2	35.80	35.80	44.89
2467	20325	C10H16	alpha-terpinene	99-86-5	33.2505	0.0335	10	1987	2	35.91	35.91	42.87
2468	20326	C10H16	gamma-terpinene	99-85-4	34.1739	0.0336	10	1430	2	36.92	36.92	43.63
2469	20327	C10H16	terpinolene	586-62-9	35.8065	0.0314	10	848	2	38.49	38.49	44.24
2470	20593	C10H18	1-decyne	764-93-2	30.4982	0.0335	10	2043	2	32.94	32.94	39.36
2471	20594	C10H18	2-decyne	2384-70-5	31.0502	0.0324	10	1475	2	33.45	33.45	39.33
2472	20595	C10H18	3-decyne	2384-85-2	30.8049	0.0325	10	1852	2	33.20	33.20	39.34
2473	20601	C10H18	cis-decahydronaphthalene	493-01-6	36.4913	0.0338	10	1034	2	39.44	39.44	46.14
2474	20602	C10H18	trans-decahydronaphthalene	493-02-7	34.6467	0.0350	10	1603	2	37.55	37.55	44.84
2475	20603	C10H18	1-pentylcyclopentene	4291-98-9	32.4386	0.0336	10	1834	2	35.05	35.05	41.76
2476	20856	C10H19Br	cis-1-bromo-1-decene	66291-72-3	50.3258	0.0218	10	65	2	52.92	52.92	55.12
2477	20857	C10H19Br	trans-1-bromo-1-decene	66291-73-4	50.3258	0.0218	10	65	2	52.92	52.92	55.12
2478	20868	C10H19Cl	cis-1-chloro-1-decene	66291-47-2	40.4526	0.0245	10	102	2	42.80	42.80	45.31
2479	20869	C10H19Cl	trans-1-chloro-1-decene	66291-48-3	40.4526	0.0245	10	102	2	42.80	42.80	45.31
2480	20876	C10H19Cl3	1,1,1-trichlorodecane	62108-56-9	51.8783	0.0199	1	12	2	51.88	54.31	54.51

Table 1 Adsorption Capacity of Activated Carbon - Gas

$\log_{10} Q = \log_{10} K + 1/n \log_{10} y$ (Q - g of compound/100 g of carbon, y - ppmv)												
NO	ID	FORMULA	NAME	CAS No	K	1/n	ymin	ymax	code	Q @ymin	Q @10ppm	Q @ymax
2481	20877	C10H19F	cis-1-fluoro-1-decene	66291-49-4	33.6192	0.0306	10	2537	2	36.07	36.07	42.74
2482	20878	C10H19F	trans-1-fluoro-1-decene	66291-50-7	33.6192	0.0306	10	2537	2	36.07	36.07	42.74
2483	20883	C10H19I	cis-1-iodo-1-decene	66291-51-8	60.4840	0.0195	1	26	2	60.48	63.26	64.45
2484	20884	C10H19I	trans-1-iodo-1-decene	66291-52-9	60.4840	0.0195	1	26	2	60.48	63.26	64.45
2485	20886	C10H19N	decanenitrile	1975-78-6	38.7060	0.0256	1	23	2	38.71	41.05	41.94
2486	20922	C10H20	pentylcyclopentane	3741-00-2	31.9912	0.0325	10	1541	2	34.48	34.48	40.62
2487	20923	C10H20	butylcyclohexane	1678-93-9	32.1390	0.0330	10	1728	2	34.68	34.68	41.10
2488	20942	C10H20	cyclodecane	293-96-9	35.4862	0.0315	10	940	2	38.16	38.16	44.03
2489	20943	C10H20	1-decene	872-05-9	29.7417	0.0320	10	2199	2	32.02	32.02	38.05
2490	20947	C10H20	2-methyl-1-nonene	2980-71-4	29.5761	0.0327	10	2704	2	31.89	31.89	38.28
2491	20962	C10H20Cl2	1,1-dichlorodecane	3162-62-7	45.9347	0.0215	1	41	2	45.93	48.26	49.75
2492	20966	C10H20F2	1,1-difluorodecane	62127-43-9	37.6632	0.0295	10	794	2	40.31	40.31	45.86
2493	20992	C10H20O	decanal	112-31-2	36.9034	0.0277	10	135	2	39.33	39.33	42.27
2494	20993	C10H20O	2-decanone	693-54-9	36.6739	0.0254	10	235	2	38.88	38.88	42.13
2495	21043	C10H20O2	nonyl formate	5451-92-3	39.1264	0.0237	10	195	2	41.32	41.32	44.34
2496	21044	C10H20O2	2-ethylhexyl acetate	103-09-3	38.6984	0.0252	10	301	2	41.01	41.01	44.68
2497	21045	C10H20O2	octyl acetate	112-14-1	39.0484	0.0240	10	199	2	41.27	41.27	44.34
2498	21049	C10H20O2	heptyl propanoate	2216-81-1	38.8477	0.0246	10	226	2	41.11	41.11	44.38
2499	21050	C10H20O2	hexyl butanoate	2639-63-6	38.3129	0.0251	10	312	2	40.60	40.60	44.26
2500	21052	C10H20O2	isopentyl isovalerate	659-70-1	35.9782	0.0286	10	1164	2	38.42	38.42	44.01
2501	21088	C10H20S	thiacycloundecane	408-32-2	42.4544	0.0236	10	102	2	44.83	44.83	47.35
2502	21090	C10H21Br	1-bromodecane	112-29-8	49.6694	0.0221	10	82	2	52.27	52.27	54.76
2503	21093	C10H21Cl	1-chlorodecane	1002-69-3	39.6914	0.0221	10	170	2	41.76	41.76	44.46
2504	21097	C10H21F	1-fluorodecane	334-56-5	34.3993	0.0292	10	929	2	36.79	36.79	42.00
2505	21100	C10H21I	1-iododecane	2050-77-3	58.9861	0.0197	10	72	2	61.73	61.73	64.18
2506	21133	C10H22	decane	124-18-5	29.8049	0.0307	10	1877	1,2	31.99	31.99	37.56
2507	21134	C10H22	2-methylnonane	871-83-0	29.0771	0.0320	10	2479	2	31.30	31.30	37.33
2508	21135	C10H22	3-methylnonane	5911-04-6	29.2198	0.0322	10	2602	2	31.47	31.47	37.65
2509	21136	C10H22	4-methylnonane	17301-94-9	28.9217	0.0327	10	3055	2	31.19	31.19	37.61
2510	21137	C10H22	5-methylnonane	15869-85-9	29.0335	0.0326	10	2914	2	31.30	31.30	37.66
2511	21138	C10H22	3-ethyloctane	5881-17-4	29.3234	0.0326	10	2859	2	31.61	31.61	38.02
2512	21139	C10H22	4-ethyloctane	15869-86-0	29.0104	0.0330	10	3340	2	31.30	31.30	37.93
2513	21140	C10H22	2,2-dimethyloctane	15869-87-1	27.9444	0.0340	10	4790	2	30.22	30.22	37.28
2514	21141	C10H22	2,3-dimethyloctane	7146-60-3	29.0212	0.0330	10	3299	2	31.31	31.31	37.93
2515	21142	C10H22	2,4-dimethyloctane	4032-94-4	28.0982	0.0339	10	4546	2	30.38	30.38	37.38
2516	21143	C10H22	2,5-dimethyloctane	15869-89-3	28.3595	0.0336	10	4165	2	30.64	30.64	37.53
2517	21144	C10H22	2,6-dimethyloctane	2051-30-1	28.2612	0.0335	10	4204	2	30.53	30.53	37.37
2518	21145	C10H22	2,7-dimethyloctane	1072-16-8	28.3353	0.0331	10	3715	2	30.58	30.58	37.21
2519	21146	C10H22	3,3-dimethyloctane	4110-44-5	28.7567	0.0336	10	4023	2	31.07	31.07	37.99
2520	21147	C10H22	3,4-dimethyloctane	15869-92-8	29.1619	0.0334	10	3519	2	31.49	31.49	38.29
2521	21148	C10H22	3,5-dimethyloctane	15869-93-9	28.5967	0.0338	10	4141	2	30.91	30.91	37.89
2522	21149	C10H22	3,6-dimethyloctane	15869-94-0	28.7293	0.0334	10	3765	2	31.03	31.03	37.83
2523	21150	C10H22	4,4-dimethyloctane	15869-95-1	28.3331	0.0340	10	4773	2	30.64	30.64	37.80
2524	21151	C10H22	4,5-dimethyloctane	15869-96-2	29.1122	0.0336	10	3776	2	31.46	31.46	38.40
2525	21152	C10H22	4-propylheptane	3178-29-8	28.5528	0.0338	10	4180	2	30.86	30.86	37.84
2526	21153	C10H22	4-isopropylheptane	52896-87-4	28.5740	0.0340	10	4428	2	30.90	30.90	38.00
2527	21154	C10H22	3-ethyl-2-methylheptane	14676-29-0	28.9242	0.0337	10	3957	2	31.26	31.26	38.25
2528	21155	C10H22	4-ethyl-2-methylheptane	52896-88-5	28.3117	0.0342	10	4843	2	30.63	30.63	37.85
2529	21156	C10H22	5-ethyl-2-methylheptane	13475-78-0	28.5773	0.0337	10	4107	2	30.89	30.89	37.84
2530	21157	C10H22	3-ethyl-3-methylheptane	17302-01-1	29.1998	0.0337	10	3838	2	31.56	31.56	38.56
2531	21158	C10H22	4-ethyl-3-methylheptane	52896-89-6	29.1590	0.0338	10	3921	2	31.52	31.52	38.56
2532	21159	C10H22	3-ethyl-5-methylheptane	52896-90-9	28.6688	0.0340	10	4358	2	31.00	31.00	38.10
2533	21160	C10H22	3-ethyl-4-methylheptane	52896-91-0	29.2553	0.0337	10	3761	2	31.62	31.62	38.61
2534	21161	C10H22	4-ethyl-4-methylheptane	17302-04-4	29.0100	0.0341	10	4368	2	31.38	31.38	38.62
2535	21162	C10H22	2,2,3-trimethylheptane	52896-92-1	28.4913	0.0344	10	4942	2	30.84	30.84	38.17
2536	21163	C10H22	2,2,4-trimethylheptane	14720-74-2	27.3647	0.0354	10	7171	2	29.69	29.69	37.47
2537	21164	C10H22	2,2,5-trimethylheptane	20291-95-6	27.5800	0.0349	10	6397	2	29.89	29.89	37.46
2538	21165	C10H22	2,2,6-trimethylheptane	1190-83-6	27.3854	0.0350	10	6589	2	29.68	29.68	37.26
2539	21166	C10H22	2,3,3-trimethylheptane	52896-93-2	28.8824	0.0342	10	4530	2	31.25	31.25	38.52
2540	21167	C10H22	2,3,4-trimethylheptane	52896-95-4	28.9078	0.0342	10	4417	2	31.28	31.28	38.52
2541	21168	C10H22	2,3,5-trimethylheptane	20278-85-7	28.7586	0.0342	10	4398	2	31.11	31.11	38.31
2542	21169	C10H22	2,3,6-trimethylheptane	4032-93-3	28.2514	0.0343	10	4910	2	30.57	30.57	37.80
2543	21170	C10H22	2,4,4-trimethylheptane	4032-92-2	27.7646	0.0351	10	6686	2	30.10	30.10	37.83
2544	21171	C10H22	2,4,5-trimethylheptane	20278-84-6	28.4187	0.0344	10	5031	2	30.76	30.76	38.12
2545	21172	C10H22	2,4,6-trimethylheptane	2613-61-8	27.3120	0.0351	10	6724	2	29.61	29.61	37.21
2546	21173	C10H22	2,5,5-trimethylheptane	1189-99-7	28.0870	0.0350	10	5942	2	30.45	30.45	38.08
2547	21174	C10H22	3,3,4-trimethylheptane	20278-87-9	29.1985	0.0342	10	4344	2	31.59	31.59	38.88
2548	21175	C10H22	3,3,5-trimethylheptane	7154-80-5	28.3113	0.0348	10	5586	2	30.67	30.67	38.23
2549	21176	C10H22	3,4,4-trimethylheptane	20278-88-0	29.1473	0.0344	10	4530	2	31.55	31.55	38.94
2550	21177	C10H22	3,4,5-trimethylheptane	20278-89-1	29.3154	0.0340	10	4016	2	31.70	31.70	38.88
2551	21178	C10H22	3-isopropyl-2-methylhexane	62016-13-1	29.0619	0.0338	10	3976	2	31.41	31.41	38.46
2552	21179	C10H22	3,3-diethylhexane	17302-02-2	29.6930	0.0337	10	3594	2	32.09	32.09	39.13
2553	21180	C10H22	3,4-diethylhexane	19398-77-7	29.2149	0.0339	10	3760	2	31.59	31.59	38.61
2554	21181	C10H22	3-ethyl-2,2-dimethylhexane	20291-91-2	28.5191	0.0349	10	5563	2	30.91	30.91	38.54
2555	21182	C10H22	4-ethyl-2,2-dimethylhexane	52896-99-8	27.4672	0.0356	10	7711	2	29.82	29.82	37.79
2556	21183	C10H22	3-ethyl-2,3-dimethylhexane	52897-00-4	29.5576	0.0342	10	4122	2	31.98	31.98	39.29
2557	21184	C10H22	4-ethyl-2,3-dimethylhexane	52897-01-5	29.1610	0.0343	10	4377	2	31.56	31.56	38.89
2558	21185	C10H22	3-ethyl-2,4-dimethylhexane	7220-26-0	29.0633	0.0344	10	4566	2	31.46	31.46	38.84
2559	21186	C10H22	4-ethyl-2,4-dimethylhexane	52897-03-7	29.1380	0.0343	10	4494	2	31.53	31.53	38.89
2560	21187	C10H22	3-ethyl-2,5-dimethylhexane	52897-04-8	28.2092	0.0349	10	5669	2	30.57	30.57	38.13

Table 1 Adsorption Capacity of Activated Carbon - Gas

$\log_{10} Q = \log_{10} K + 1/n \log_{10} y$ (Q - g of compound/100 g of carbon, y - ppmv)

NO	ID	FORMULA	NAME	CAS No	K	1/n	ymin	ymax	code	Q @ymin	Q @10ppm	Q @ymax
2561	21188	C10H22	4-ethyl-3,3-dimethylhexane	52897-05-9	29.4790	0.0344	10	4307	2	31.91	31.91	39.30
2562	21189	C10H22	3-ethyl-3,4-dimethylhexane	52897-06-0	29.4012	0.0345	10	4491	2	31.83	31.83	39.30
2563	21190	C10H22	2,2,3,3-tetramethylhexane	13475-81-5	29.2469	0.0348	10	5085	2	31.69	31.69	39.36
2564	21191	C10H22	2,2,3,4-tetramethylhexane	52897-08-2	28.7592	0.0350	10	5373	2	31.17	31.17	38.85
2565	21192	C10H22	2,2,3,5-tetramethylhexane	52897-09-3	27.6266	0.0356	10	7420	2	29.99	29.99	37.94
2566	21193	C10H22	2,2,4,4-tetramethylhexane	51750-65-3	27.9694	0.0355	10	7501	2	30.35	30.35	38.41
2567	21194	C10H22	2,2,4,5-tetramethylhexane	16747-42-5	27.4442	0.0358	10	7946	2	29.80	29.80	37.84
2568	21195	C10H22	2,2,5,5-tetramethylhexane	1071-81-4	26.2580	0.0370	10	10000	2	28.59	28.59	36.92
2569	21196	C10H22	2,3,3,4-tetramethylhexane	52897-10-6	29.6977	0.0343	10	4229	2	32.14	32.14	39.56
2570	21197	C10H22	2,3,3,5-tetramethylhexane	52897-11-7	28.2633	0.0354	10	6441	2	30.66	30.66	38.55
2571	21198	C10H22	2,3,4,4-tetramethylhexane	52897-12-8	29.6972	0.0360	10	4812	2	32.27	32.27	40.31
2572	21199	C10H22	2,3,4,5-tetramethylhexane	52897-15-1	28.6237	0.0348	10	5357	2	31.01	31.01	38.59
2573	21200	C10H22	3,3,4,4-tetramethylhexane	5171-84-6	30.4189	0.0341	10	3729	2	32.90	32.90	40.27
2574	21201	C10H22	2,4-dimethyl-3-isopropylpentane	13475-79-1	28.8240	0.0350	10	5636	2	31.25	31.25	39.01
2575	21202	C10H22	3,3-diethyl-2-methylpentane	52897-16-2	30.4447	0.0338	10	3361	2	32.91	32.91	40.06
2576	21203	C10H22	3-ethyl-2,2,3-trimethylpentane	52897-17-3	30.4593	0.0337	10	3853	2	32.91	32.91	40.21
2577	21204	C10H22	3-ethyl-2,2,4-trimethylpentane	52897-18-4	28.5641	0.0356	10	6210	2	31.00	31.00	38.97
2578	21205	C10H22	3-ethyl-2,3,4-trimethylpentane	52897-19-5	30.2230	0.0340	10	3660	2	32.69	32.69	39.96
2579	21206	C10H22	2,2,3,3,4-pentamethylpentane	16747-44-7	30.0098	0.0347	10	4541	2	32.50	32.50	40.18
2580	21207	C10H22	2,2,3,4,4-pentamethylpentane	16747-45-8	28.9971	0.0354	10	6321	2	31.46	31.46	39.54
2581	21224	C10H22O	1-decanol	112-30-1	39.8465	0.0262	1	11	2	39.85	42.33	42.43
2582	21225	C10H22O	isodecanol	25339-17-7	39.5850	0.0253	1	27	2	39.59	41.96	43.02
2583	21226	C10H22O	2-decanol	1120-06-5	38.3877	0.0259	1	40	2	38.39	40.75	42.23
2584	21228	C10H22O	3-decanol	1565-81-7	36.7680	0.0250	10	275	2	38.94	38.94	42.31
2585	21229	C10H22O	4-decanol	2051-31-2	36.6534	0.0249	10	274	2	38.82	38.82	42.15
2586	21230	C10H22O	5-decanol	5205-34-5	36.1064	0.0259	10	408	2	38.32	38.32	42.18
2587	21231	C10H22O	2-methyl-1-nonanol	40589-14-8	37.5675	0.0237	10	161	2	39.67	39.67	42.37
2588	21232	C10H22O	4-methyl-1-nonanol	1489-47-0	37.2408	0.0245	10	231	2	39.40	39.40	42.55
2589	21233	C10H22O	5-methyl-1-nonanol	2768-16-3	37.3166	0.0246	10	231	2	39.49	39.49	42.66
2590	21234	C10H22O	7-methyl-1-nonanol	33234-93-4	37.2408	0.0245	10	231	2	39.40	39.40	42.55
2591	21235	C10H22O	3-methyl-1-nonanol	22663-64-5	37.3050	0.0246	10	231	2	39.48	39.48	42.66
2592	21236	C10H22O	2-methyl-2-nonanol	10297-57-1	36.9211	0.0254	10	285	2	39.14	39.14	42.62
2593	21237	C10H22O	3-methyl-2-nonanol	60671-32-1	37.2215	0.0248	10	255	2	39.41	39.41	42.71
2594	21238	C10H22O	5-methyl-2-nonanol	66731-95-1	37.1353	0.0250	10	255	2	39.34	39.34	42.66
2595	21239	C10H22O	6-methyl-2-nonanol	66256-60-8	37.1821	0.0246	10	255	2	39.35	39.35	42.61
2596	21240	C10H22O	7-methyl-2-nonanol	66256-61-9	36.9611	0.0250	10	255	2	39.15	39.15	42.46
2597	21241	C10H22O	8-methyl-2-nonanol	14779-92-1	36.7858	0.0248	10	255	2	38.95	38.95	42.20
2598	21242	C10H22O	2-methyl-3-nonanol	26533-33-5	36.6051	0.0254	10	324	2	38.81	38.81	42.40
2599	21243	C10H22O	3-methyl-3-nonanol	21078-72-8	36.6265	0.0255	10	354	2	38.84	38.84	42.53
2600	21244	C10H22O	5-methyl-3-nonanol	66719-43-5	36.6264	0.0253	10	317	2	38.82	38.82	42.37
2601	21245	C10H22O	2-methyl-4-nonanol	26533-31-3	36.4409	0.0253	10	317	2	38.63	38.63	42.16
2602	21246	C10H22O	4-methyl-4-nonanol	23418-38-4	36.5504	0.0256	10	354	2	38.77	38.77	42.48
2603	21247	C10H22O	5-methyl-4-nonanol	66719-44-6	36.7859	0.0254	10	317	2	39.00	39.00	42.57
2604	21248	C10H22O	7-methyl-4-nonanol	26981-98-6	36.4776	0.0254	10	317	2	38.67	38.67	42.21
2605	21249	C10H22O	2-methyl-5-nonanol	29843-62-7	36.3180	0.0253	10	317	2	38.49	38.49	42.01
2606	21250	C10H22O	3-methyl-5-nonanol	66779-42-4	36.4905	0.0253	10	317	2	38.68	38.68	42.21
2607	21251	C10H22O	5-methyl-5-nonanol	33933-78-7	36.5088	0.0256	10	354	2	38.73	38.73	42.43
2608	21252	C10H22O	2-ethyl-1-octanol	20592-10-3	36.7209	0.0248	10	288	2	38.88	38.88	42.26
2609	21253	C10H22O	3-ethyl-1-octanol	66719-36-6	36.6837	0.0248	10	288	2	38.84	38.84	42.21
2610	21254	C10H22O	2,2-dimethyl-1-octanol	2370-14-1	36.5257	0.0256	10	367	2	38.74	38.74	42.48
2611	21255	C10H22O	2,6-dimethyl-1-octanol	62417-08-7	36.7119	0.0251	10	304	2	38.89	38.89	42.37
2612	21256	C10H22O	3,7-dimethyl-1-octanol	106-21-8	37.1731	0.0249	10	276	2	39.37	39.37	42.77
2613	21257	C10H22O	4,5-dimethyl-1-octanol	66719-32-2	36.9511	0.0245	10	278	2	39.09	39.09	42.41
2614	21258	C10H22O	4,6-dimethyl-1-octanol	66719-33-3	36.9153	0.0246	10	278	2	39.07	39.07	42.41
2615	21259	C10H22O	4,7-dimethyl-1-octanol	66719-34-4	36.9033	0.0247	10	278	2	39.06	39.06	42.41
2616	21260	C10H22O	7,7-dimethyl-1-octanol	66719-35-5	36.7558	0.0250	10	310	2	38.93	38.93	42.41
2617	21261	C10H22O	2,4-dimethyl-2-octanol	18675-20-2	36.4205	0.0255	10	343	2	38.62	38.62	42.27
2618	21262	C10H22O	2,6-dimethyt-2-octanol	18479-57-7	36.4530	0.0258	10	343	2	38.68	38.68	42.38
2619	21263	C10H22O	2,7-dimethyl-2-octanol	42007-73-8	36.4665	0.0257	10	343	2	38.69	38.69	42.38
2620	21264	C10H22O	3,7-dimethyl-2-octanol	15340-96-2	36.7204	0.0252	10	307	2	38.91	38.91	42.42
2621	21265	C10H22O	3-ethyl-3-octanol	2051-32-3	36.3443	0.0268	10	541	2	38.66	38.66	43.03
2622	21266	C10H22O	6-ethyl-3-octanol	19781-27-2	36.2778	0.0263	10	492	2	38.55	38.55	42.71
2623	21267	C10H22O	2,2-dimethyl-3-octanol	19841-72-6	35.5880	0.0277	10	740	2	37.93	37.93	42.74
2624	21268	C10H22O	2,3-dimethyl-3-octanol	19781-10-3	35.1617	0.0278	10	907	2	37.49	37.49	42.49
2625	21269	C10H22O	2,7-dimethyl-3-octanol	66719-55-9	35.0499	0.0269	10	666	2	37.29	37.29	41.75
2626	21270	C10H22O	3,5-dimethyl-3-octanol	56065-42-0	35.8183	0.0279	10	772	2	38.19	38.19	43.11
2627	21271	C10H22O	3,6-dimethyl-3-octanol	151-19-9	35.5171	0.0275	10	801	2	37.84	37.84	42.69
2628	21272	C10H22O	3,7-dimethyl-3-octanol	78-69-3	35.6117	0.0272	10	675	2	37.92	37.92	42.53
2629	21273	C10H22O	3-ethyl-4-octanol	63126-48-7	35.6153	0.0265	10	640	2	37.85	37.85	42.26
2630	21274	C10H22O	4-ethyl-4-octanol	38395-42-5	35.8580	0.0275	10	706	2	38.20	38.20	42.94
2631	21275	C10H22O	2,2-dimethyl-4-octanol	66719-52-6	34.8435	0.0279	10	872	2	37.15	37.15	42.08
2632	21276	C10H22O	2,4-dimethyl-4-octanol	33933-79-8	35.0149	0.0277	10	841	2	37.32	37.32	42.18
2633	21277	C10H22O	2,5-dimethyl-4-octanol	66719-53-7	34.8494	0.0275	10	763	2	37.12	37.12	41.82
2634	21278	C10H22O	2,6-dimethyl-4-octanol	66719-54-8	34.8987	0.0272	10	763	2	37.16	37.16	41.81
2635	21279	C10H22O	2,7-dimethyl-4-octanol	19781-11-4	34.8659	0.0274	10	763	2	37.14	37.14	41.82
2636	21280	C10H22O	3,4-dimethyl-4-octanol	66719-30-0	35.2292	0.0271	10	841	2	37.50	37.50	42.28
2637	21281	C10H22O	3,6-dimethyl-4-octanol	66719-31-1	35.8847	0.0269	10	763	2	38.18	38.18	42.89
2638	21282	C10H22O	4,6-dimethyl-4-octanol	56065-43-1	35.3413	0.0279	10	841	2	37.69	37.69	42.65
2639	21283	C10H22O	4,7-dimethyl-4-octanol	19781-13-6	35.2542	0.0274	10	801	2	37.55	37.55	42.33
2640	21284	C10H22O	2-propyl-1-heptanol	10042-59-8	37.4747	0.0241	10	193	2	39.61	39.61	42.54

Table 1 Adsorption Capacity of Activated Carbon - Gas

$\log_{10} Q = \log_{10} K + 1/n \log_{10} y$ (Q - g of compound/100 g of carbon, y - ppmv)

NO	ID	FORMULA	NAME	CAS No	K	1/n	ymin	ymax	code	Q @ymin	Q @10ppm	Q @ymax
2641	21285	C10H22O	3-isopropyl-1-heptanol	38514-15-7	37.6121	0.0247	10	240	2	39.81	39.81	43.07
2642	21286	C10H22O	2,4,6-trimethyl-2-heptanol	66256-47-1	35.4492	0.0280	10	845	2	37.81	37.81	42.81
2643	21287	C10H22O	2,5,6-trimethyl-2-heptanol	66256-48-2	35.3294	0.0278	10	840	2	37.66	37.66	42.60
2644	21288	C10H22O	4,6,6-trimethyl-2-heptanol	51079-79-9	35.3972	0.0282	10	845	2	37.77	37.77	42.81
2645	21289	C10H22O	2-methyl-3-ethyl-3-heptanol	66719-37-7	35.9916	0.0281	10	767	2	38.39	38.39	43.37
2646	21290	C10H22O	4-methyl-5-ethyl-3-heptanol	66731-94-0	36.9792	0.0273	10	700	2	39.38	39.38	44.23
2647	21291	C10H22O	5-methyl-3-ethyl-3-heptanol	------	35.6379	0.0277	10	772	2	37.99	37.99	42.85
2648	21292	C10H22O	2,2,3-trimethyl-3-heptanol	29772-40-5	35.2374	0.0291	10	1240	2	37.68	37.68	43.35
2649	21293	C10H22O	2,2,6-trimethyl-3-heptanol	66256-43-7	34.8294	0.0281	10	959	2	37.16	37.16	42.25
2650	21294	C10H22O	2,3,6-trimethyl-3-heptanol	58046-40-5	35.4446	0.0284	10	959	2	37.84	37.84	43.07
2651	21295	C10H22O	3,5,5-trimethyl-3-heptanol	66256-50-6	36.3068	0.0282	10	832	2	38.74	38.74	43.89
2652	21296	C10H22O	4-propyl-4-heptanol	2198-72-3	35.6519	0.0273	10	672	2	37.96	37.96	42.58
2653	21297	C10H22O	4-isopropyl-4-heptanol	51200-82-9	35.7725	0.0283	10	872	2	38.18	38.18	43.33
2654	21298	C10H22O	2,2,4-trimethyl-4-heptanol	57233-31-5	34.4570	0.0294	10	1527	2	36.87	36.87	42.75
2655	21299	C10H22O	2,2,5-trimethyl-4-heptanol	66256-42-6	34.6589	0.0296	10	1344	2	37.10	37.10	42.90
2656	21300	C10H22O	2,2,6-trimethyl-4-heptanol	66256-44-8	34.2896	0.0299	10	1344	2	36.74	36.74	42.54
2657	21301	C10H22O	2,4,5-trimethyl-4-heptanol	66256-46-0	34.5636	0.0288	10	1344	2	36.93	36.93	42.53
2658	21302	C10H22O	2,4,6-trimethyl-4-heptanol	60836-07-9	34.1366	0.0288	10	1335	2	36.48	36.48	42.01
2659	21303	C10H22O	3,3,6-trimethyl-4-heptanol	------	34.9957	0.0293	10	1344	2	37.43	37.43	43.20
2660	21304	C10H22O	4-methyl-2-propyl-1-hexanol	66256-62-0	36.6248	0.0254	10	329	2	38.83	38.83	42.42
2661	21305	C10H22O	4-methyl-2-isopropyl-1-hexanol	66719-41-3	36.1331	0.0264	10	496	2	38.39	38.39	42.56
2662	21306	C10H22O	5-methyl-2-isopropyl-1-hexanol	2051-33-4	36.9612	0.0251	10	292	2	39.16	39.16	42.62
2663	21307	C10H22O	5,5-dimethyl-2-ethyl-1-hexanol	------	36.9797	0.0258	10	329	2	39.24	39.24	42.94
2664	21308	C10H22O	2,3,4,4-tetramethyl-2-hexanol	66256-66-4	35.4981	0.0282	10	1147	2	37.88	37.88	43.29
2665	21309	C10H22O	3,3,5,5-tetramethyl-2-hexanol	66256-69-7	35.4165	0.0285	10	1154	2	37.82	37.82	43.29
2666	21310	C10H22O	2-methyl-3-isopropyl-3-hexanol	51200-81-8	35.9857	0.0282	10	882	2	38.40	38.40	43.58
2667	21311	C10H22O	2,2-dimethyl-4-ethyl-3-hexanol	66719-47-9	35.0639	0.0284	10	1081	2	37.44	37.44	42.77
2668	21312	C10H22O	2,4-dimethyl-4-ethyl-3-hexanol	66719-48-0	36.0407	0.0291	10	1088	2	38.54	38.54	44.17
2669	21313	C10H22O	5,5-dimethyl-3-ethyl-3-hexanol	5340-62-5	35.1233	0.0288	10	1197	2	37.54	37.54	43.09
2670	21314	C10H22O	2,2,3,4-tetramethyl-3-hexanol	66256-63-1	36.0149	0.0288	10	1057	2	38.48	38.48	44.01
2671	21315	C10H22O	2,2,3,5-tetramethyl-3-hexanol	66256-64-2	37.5288	0.0273	10	145	2	39.96	39.96	42.98
2672	21316	C10H22O	2,2,4,4-tetramethyl-3-hexanol	66256-65-3	35.7672	0.0290	10	1147	2	38.23	38.23	43.86
2673	21317	C10H22O	2,2,5,5-tetramethyl-3-hexanol	55073-86-4	35.7672	0.0290	10	1147	2	38.23	38.23	43.86
2674	21318	C10H22O	2,3,4,4-tetramethyl-3-hexanol	66256-67-5	37.2401	0.0281	10	730	2	39.73	39.73	44.81
2675	21319	C10H22O	2,3,5,5-tetramethyl-3-hexanol	5396-09-8	35.8440	0.0276	10	705	2	38.19	38.19	42.95
2676	21320	C10H22O	2,4,4,5-tetramethyl-3-hexanol	66256-68-6	36.5635	0.0269	10	635	2	38.90	38.90	43.50
2677	21321	C10H22O	3,4,4,5-tetramethyl-3-hexanol	66256-39-1	37.3074	0.0279	10	701	2	39.79	39.79	44.80
2678	21322	C10H22O	3,4,5,5-tetramethyl-3-hexanol	66256-40-4	36.3799	0.0285	10	934	2	38.85	38.85	44.21
2679	21323	C10H22O	4-methyl-2-isobutyl-1-pentanol	22417-45-4	36.5909	0.0254	10	435	2	38.79	38.79	42.69
2680	21324	C10H22O	4,4-dimethyl-3-isopropyl-1-pentanol	66719-51-5	35.5253	0.0276	10	1288	2	37.85	37.85	43.28
2681	21325	C10H22O	3,4-dimethyl-3-isopropyl-2-pentanol	66719-50-4	35.3837	0.0278	10	1410	2	37.72	37.72	43.29
2682	21326	C10H22O	2,4-dimethyl-3-propyl-3-pentanol	------	35.5867	0.0293	10	1227	2	38.07	38.07	43.82
2683	21327	C10H22O	2,4-dimethyl-3-isopropyl-3-pentanol	51200-83-0	36.4998	0.0285	10	869	2	38.97	38.97	44.26
2684	21328	C10H22O	2,2,4-trimethyl-3-ethyl-3-pentanol	66256-41-5	36.0890	0.0290	10	1101	2	38.58	38.58	44.22
2685	21335	C10H22O	di-pentyl ether	693-65-2	32.9959	0.0281	10	1127	2	35.20	35.20	40.19
2686	21380	C10H22O5	tetraethylene glycol dimethyl ether	143-24-8	50.5696	0.0199	1	2.9	2	50.57	---	51.65
2687	21385	C10H22S	decyl mercaptan	143-10-2	39.6083	0.0226	1	48	2	39.61	41.73	43.24
2688	21386	C10H22S	2-decanethiol	13402-60-3	38.6401	0.0206	10	154	2	40.51	40.51	42.86
2689	21387	C10H22S	dipentyl sulfide	872-10-6	39.7057	0.0213	10	52	2	41.70	41.70	43.19
2690	21388	C10H22S	methyl nonyl sulfide	59973-07-8	39.7021	0.0213	10	52	2	41.70	41.70	43.19
2691	21389	C10H22S	ethyl octyl sulfide	3698-94-0	39.7021	0.0213	10	52	2	41.70	41.70	43.19
2692	21390	C10H22S	propyl heptyl sulfide	24768-46-5	39.7021	0.0213	10	52	2	41.70	41.70	43.19
2693	21391	C10H22S	butyl hexyl sulfide	16967-04-7	39.7021	0.0213	10	52	2	41.70	41.70	43.19
2694	21394	C10H22S2	dipentyl disulfide	112-51-6	45.5046	0.0185	1	6.8	2	45.50	---	47.14
2695	21400	C10H23N	decylamine	2016-57-1	35.9020	0.0255	10	137	2	38.07	38.07	40.70
2696	21401	C10H23N	methylnonylamine	39093-27-1	35.7010	0.0224	10	159	2	37.59	37.59	40.00
2697	21402	C10H23N	dimethyloctylamine	7378-99-6	33.2072	0.0255	10	692	2	35.22	35.22	39.23
2698	21403	C10H23N	diethylhexylamine	44979-90-0	33.1528	0.0260	10	790	2	35.20	35.20	39.43
2699	21404	C10H23N	ethyloctylamine	4088-36-2	35.1467	0.0231	10	216	2	37.06	37.06	39.78
2700	21405	C10H23N	dipentylamine	2050-92-2	35.1915	0.0231	10	199	2	37.11	37.11	39.77
2701	21601	C11H10	1-methylnaphthalene	90-12-0	46.4482	0.0265	10	89	2	49.37	49.37	52.32
2702	21602	C11H10	2-methylnaphthalene	91-57-6	46.4482	0.0265	10	89	2	49.37	49.37	52.32
2703	21705	C11H11N	2,3-dimethylquinoline	1721-89-7	49.7648	0.0240	1	24	2	49.76	52.60	53.72
2704	21706	C11H11N	2,4-dimethylquinoline	1198-37-4	49.7648	0.0240	1	24	2	49.76	52.60	53.72
2705	21707	C11H11N	2,6-dimethylquinoline	877-43-0	49.7648	0.0240	1	24	2	49.76	52.60	53.72
2706	21708	C11H11N	2,7-dimethylquinoline	93-37-8	49.7648	0.0240	1	24	2	49.76	52.60	53.72
2707	21709	C11H11N	2,8-dimethylquinoline	1463-17-8	49.7648	0.0240	1	24	2	49.76	52.60	53.72
2708	21710	C11H11N	3,4-dimethylquinoline	2436-92-2	49.7648	0.0240	1	24	2	49.76	52.60	53.72
2709	21711	C11H11N	3,6-dimethylquinoline	20668-26-2	49.7648	0.0240	1	24	2	49.76	52.60	53.72
2710	21712	C11H11N	3,7-dimethylquinoline	20668-28-4	49.7648	0.0240	1	24	2	49.76	52.60	53.72
2711	21713	C11H11N	3,8-dimethylquinoline	20668-29-5	49.7648	0.0240	1	24	2	49.76	52.60	53.72
2712	21714	C11H11N	4,5-dimethylquinoline	59129-74-7	52.0146	0.0233	1	4	2	52.01	---	53.73
2713	21715	C11H11N	4,6-dimethylquinoline	826-77-7	52.0146	0.0233	1	4	2	52.01	---	53.73
2714	21716	C11H11N	4,8-dimethylquinoline	13362-80-6	52.0146	0.0233	1	4	2	52.01	---	53.73
2715	21717	C11H11N	5,6-dimethylquinoline	20668-30-8	52.0146	0.0233	1	4	2	52.01	---	53.73
2716	21718	C11H11N	5,7-dimethylquinoline	20150-89-4	52.0146	0.0233	1	4	2	52.01	---	53.73
2717	21719	C11H11N	6,7-dimethylquinoline	20668-33-1	52.0146	0.0233	1	4	2	52.01	---	53.73
2718	22025	C11H14	1-methyl-[1,2,3,4-tetrahydronaphthalene]	1559-81-5	42.0493	0.0269	10	321	2	44.74	44.74	49.11
2719	22026	C11H14	2-methyl-[1,2,3,4-tetrahydronaphthalene]	3877-19-8	41.3471	0.0269	10	330	2	43.99	43.99	48.33
2720	22113	C11H14O2	butyl benzoate	136-60-7	46.8421	0.0249	1	44	2	46.84	49.61	51.47

Table 1 Adsorption Capacity of Activated Carbon - Gas

$\log_{10} Q = \log_{10} K + 1/n \log_{10} y$ (Q - g of compound/100 g of carbon, y - ppmv)

NO	ID	FORMULA	NAME	CAS No	K	1/n	ymin	ymax	code	Q @ymin	Q @10ppm	Q @ymax
2721	22304	C11H16	pentylbenzene	538-68-1	37.0988	0.0269	10	576	2	39.47	39.47	44.01
2722	22305	C11H16	2-phenylpentane	2719-52-0	36.0346	0.0286	10	1112	2	38.49	38.49	44.04
2723	22306	C11H16	3-phenylpentane	1196-58-3	35.9636	0.0283	10	1065	2	38.38	38.38	43.80
2724	22307	C11H16	1-phenyl-2-methylbutane	3968-85-2	36.5238	0.0279	10	819	2	38.95	38.95	44.04
2725	22308	C11H16	1-phenyl-3-methylbutane	2049-94-7	36.5276	0.0277	10	753	2	38.93	38.93	43.88
2726	22309	C11H16	2-phenyl-2-methylbutane	2049-95-8	36.5324	0.0290	10	1211	2	39.06	39.06	44.89
2727	22310	C11H16	2-phenyl-3-methylbutane	4481-30-5	36.0636	0.0297	10	1325	2	38.62	38.62	44.65
2728	22311	C11H16	1-phenyl-2,2-dimethylpropane	1007-26-7	35.3447	0.0294	10	1588	2	37.82	37.82	43.88
2729	22312	C11H16	1-methyl-2-butylbenzene	1595-11-5	37.9591	0.0263	10	463	2	40.33	40.33	44.61
2730	22313	C11H16	1-methyl-3-butylbenzene	1595-04-6	37.2731	0.0265	10	529	2	39.61	39.61	44.00
2731	22314	C11H16	1-methyl-4-butylbenzene	1595-05-7	37.5332	0.0265	10	484	2	39.89	39.89	44.20
2732	22315	C11H16	1-methyl-2-sec-butylbenzene	1595-06-8	37.0550	0.0279	10	867	2	39.52	39.52	44.76
2733	22316	C11H16	1-methyl-3-sec-butylbenzene	1772-10-7	36.3224	0.0280	10	946	2	38.74	38.74	44.00
2734	22317	C11H16	1-methyl-4-sec-butylbenzene	1595-16-0	36.8376	0.0278	10	830	2	39.27	39.27	44.40
2735	22318	C11H16	1-methyl-2-isobutylbenzene	36301-29-8	36.7378	0.0278	10	867	2	39.17	39.17	44.35
2736	22319	C11H16	1-methyl-3-isobutylbenzene	5160-99-6	36.1611	0.0279	10	946	2	38.56	38.56	43.77
2737	22320	C11H16	1-methyl-4-isobutylbenzene	5161-04-6	36.2165	0.0277	10	867	2	38.60	38.60	43.67
2738	22321	C11H16	1-methyl-2-tert-butyl benzene	1074-92-6	37.7491	0.0280	10	870	2	40.26	40.26	45.63
2739	22322	C11H16	1-methyl-3-tert-butylbenzene	1075-38-3	35.9869	0.0291	10	1400	2	38.48	38.48	44.43
2740	22323	C11H16	1-methyl-4-tert-butylbenzene	98-51-1	36.0536	0.0287	10	1207	2	38.51	38.51	44.19
2741	22324	C11H16	1-ethyl-2-propylbenzene	16021-20-8	37.7771	0.0268	10	579	2	40.18	40.18	44.81
2742	22325	C11H16	1-ethyl-3-propylbenzene	20024-91-3	37.0872	0.0269	10	632	2	39.46	39.46	44.11
2743	22326	C11H16	1-ethyl-4-propylbenzene	20024-90-2	37.3011	0.0264	10	529	2	39.64	39.64	44.03
2744	22327	C11H16	1-ethyl-2-isopropylbenzene	18970-44-0	37.2687	0.0286	10	1126	2	39.80	39.80	45.56
2745	22328	C11H16	1-ethyl-3-isopropylbenzene	4920-99-4	36.2451	0.0281	10	1032	2	38.67	38.67	44.05
2746	22329	C11H16	1-ethyl-4-isopropylbenzene	4218-48-8	36.5534	0.0276	10	844	2	38.95	38.95	44.02
2747	22330	C11H16	1,2-dimethyl-3-propylbenzene	17059-44-8	38.8109	0.0260	10	416	2	41.20	41.20	45.40
2748	22331	C11H16	1,2-dimethyl-4-propylbenzene	3982-66-9	38.0680	0.0260	10	451	2	40.42	40.42	44.64
2749	22332	C11H16	1,3-dimethyl-2-propylbenzene	17059-45-9	38.5561	0.0264	10	478	2	40.97	40.97	45.37
2750	22333	C11H16	1,3-dimethyl-4-propylbenzene	61827-85-8	37.9391	0.0263	10	500	2	40.31	40.31	44.69
2751	22334	C11H16	1,3-dimethyl-5-propylbenzene	3982-64-7	37.1827	0.0266	10	607	2	39.53	39.53	44.10
2752	22335	C11H16	1,4-dimethyl-2-propylbenzene	3042-50-0	37.7634	0.0266	10	554	2	40.15	40.15	44.66
2753	22336	C11H16	1,2-dimethyl-3-isopropylbenzene	22539-65-7	38.9034	0.0241	10	658	2	41.12	41.12	45.48
2754	22337	C11H16	1,2-dimethyl-4-isopropylbenzene	4132-77-8	37.3717	0.0271	10	681	2	39.77	39.77	44.59
2755	22338	C11H16	1,3-dimethyl-2-isopropylbenzene	14411-75-7	37.9829	0.0276	10	770	2	40.47	40.47	45.63
2756	22339	C11H16	1,3-dimethyl-4-isopropylbenzene	4706-89-2	37.2978	0.0274	10	767	2	39.73	39.73	44.75
2757	22340	C11H16	1,3-dimethyl-5-isopropylbenzene	4706-90-5	36.5428	0.0278	10	937	2	38.96	38.96	44.20
2758	22341	C11H16	1,4-dimethyl-2-isopropylbenzene	4132-72-3	37.1352	0.0277	10	870	2	39.59	39.59	44.81
2759	22342	C11H16	1-methyl-2,3-diethylbenzene	13632-93-4	38.7301	0.0265	10	500	2	41.16	41.16	45.65
2760	22343	C11H16	1-methyl-2,4-diethylbenzene	1758-85-6	37.9555	0.0264	10	537	2	40.34	40.34	44.82
2761	22344	C11H16	1-methyl-2,5-diethylbenzene	13632-94-5	38.1419	0.0262	10	489	2	40.51	40.51	44.86
2762	22345	C11H16	1-methyl-2,6-diethyl benzene	13632-95-6	38.8726	0.0262	10	453	2	41.29	41.29	45.63
2763	22346	C11H16	1-methyl-3,4-diethylbenzene	13732-80-4	37.9341	0.0265	10	571	2	40.33	40.33	44.90
2764	22347	C11H16	1-methyl-3,5-diethylbenzene	2050-24-0	37.1799	0.0268	10	650	2	39.55	39.55	44.23
2765	22348	C11H16	1,2,3-trimethyl-4-ethylbenzene	61827-86-9	40.1226	0.0250	10	274	2	42.50	42.50	46.16
2766	22349	C11H16	1,2,3-trimethyl-5-ethylbenzene	31366-00-4	39.1566	0.0253	10	336	2	41.51	41.51	45.37
2767	22350	C11H16	1,2,4-trimethyl-3-ethylbenzene	61827-87-0	39.5611	0.0254	10	325	2	41.94	41.94	45.82
2768	22351	C11H16	1,2,4-trimethyl-5-ethylbenzene	------	38.8183	0.0256	10	381	2	41.18	41.18	45.21
2769	22352	C11H16	1,2,4-trimethyl-6-ethylbenzene	17851-27-3	39.1146	0.0257	10	381	2	41.50	41.50	45.56
2770	22353	C11H16	1,3,5-trimethyl-2-ethylbenzene	3982-67-0	38.7777	0.0257	10	392	2	41.14	41.14	45.21
2771	22665	C11H20	1-methyl-cis-decahydronaphthalene	------	40.3528	0.0251	10	161	2	42.75	42.75	45.84
2772	22666	C11H20	1-methyl-trans-decahydronaphthalene	------	39.8483	0.0260	10	227	2	42.30	42.30	45.87
2773	22667	C11H20	2-methyl-cis-decahydronaphthalene	------	38.5608	0.0281	10	507	2	41.14	41.14	45.94
2774	22668	C11H20	2-methyl-trans-decahydronaphthalene	4683-94-7	37.9838	0.0291	10	710	2	40.61	40.61	45.97
2775	22669	C11H20	4a-methyl-cis-decahydronaphthalene	2547-26-4	38.6317	0.0276	10	529	2	41.17	41.17	45.94
2776	22670	C11H20	4a-methyl-trans-decahydronaphthalene	2547-27-5	37.6586	0.0287	10	805	2	40.23	40.23	45.62
2777	22671	C11H20	1-hexylcyclopentene	4291-99-0	35.1370	0.0267	10	679	2	37.37	37.37	41.83
2778	22672	C11H20	1-undecyne	2243-98-3	33.2726	0.0265	10	708	2	35.36	35.36	39.59
2779	22673	C11H20	2-undecyne	60212-29-5	33.7445	0.0252	10	550	2	35.76	35.76	39.56
2780	22674	C11H20	3-undecyne	60212-30-8	33.4155	0.0258	10	708	2	35.46	35.46	39.58
2781	22701	C11H20O2	2-ethylhexyl acrylate	103-11-7	40.1026	0.0220	10	233	2	42.18	42.18	45.21
2782	22749	C11H21Br	cis-1-bromo-1-undecene	66142-59-4	50.5583	0.0200	1	26	2	50.56	52.94	53.97
2783	22750	C11H21Br	trans-1-bromo-1-undecene	66142-60-7	50.5583	0.0200	1	26	2	50.56	52.94	53.97
2784	22756	C11H21Cl	cis-1-chloro-1-undecene	66172-73-4	41.8500	0.0215	1	33	2	41.85	43.97	45.11
2785	22757	C11H21Cl	trans-1-chloro-1-undecene	66142-61-8	41.8500	0.0215	1	33	2	41.85	43.97	45.11
2786	22764	C11H21F	cis-1-fluoro-1-undecene	66172-74-5	36.2671	0.0245	10	800	2	38.37	38.37	42.71
2787	22765	C11H21F	trans-1-fluoro-1-undecene	66142-62-9	36.2671	0.0245	10	800	2	38.37	38.37	42.71
2788	22768	C11H21I	cis-1-iodo-1-undecene	66142-63-0	60.1089	0.0188	1	11	2	60.11	62.77	62.89
2789	22769	C11H21I	trans-1-iodo-1-undecene	66142-64-1	60.1089	0.0188	1	11	2	60.11	62.77	62.89
2790	22771	C11H21N	undecanenitrile	2244-07-7	40.0382	0.0239	1	7.9	2	40.04	---	42.07
2791	22791	C11H22	hexylcyclopentane	4457-00-5	34.8488	0.0256	10	474	2	36.96	36.96	40.80
2792	22792	C11H22	pentylcyclohexane	4292-92-6	35.0257	0.0258	10	523	2	37.17	37.17	41.16
2793	22794	C11H22	cycloundecane	294-41-7	37.7513	0.0258	10	451	2	40.06	40.06	44.19
2794	22795	C11H22	1-undecene	821-95-4	32.6487	0.0252	10	648	2	34.60	34.60	38.44
2795	22803	C11H22	2-methyl-1-decene	13151-27-4	32.3312	0.0261	10	940	2	34.33	34.33	38.65
2796	22807	C11H22Cl2	1,1-dichloroundecane	822-01-5	46.3839	0.0188	1	24	2	46.38	48.44	49.24
2797	22809	C11H22F2	1,1-difluoroundecane	62127-44-0	38.6545	0.0248	10	584	2	40.93	40.93	45.27
2798	22822	C11H22O	undecanal	112-44-7	38.4531	0.0231	10	62	2	40.55	40.55	42.29
2799	22823	C11H22O	2-undecanone	112-12-9	38.3317	0.0232	10	65	2	40.44	40.44	42.24
2800	22837	C11H22O2	nonyl acetate	143-13-5	40.6162	0.0230	1	42	2	40.62	42.83	44.27

Table 1 Adsorption Capacity of Activated Carbon - Gas

$\log_{10} Q = \log_{10} K + 1/n \log_{10} y$ (Q - g of compound/100 g of carbon, y - ppmv)												
NO	ID	FORMULA	NAME	CAS No	K	1/n	ymin	ymax	code	Q @ymin	Q @10ppm	Q @ymax
2801	22838	C11H22O2	octyl propanoate	142-60-9	40.6100	0.0231	1	44	2	40.61	42.83	44.33
2802	22839	C11H22O2	heptyl butanoate	5870-93-9	39.4689	0.0236	10	124	2	41.68	41.68	44.23
2803	22866	C11H22S	thiacyclododecane	294-65-5	43.0364	0.0202	10	56	2	45.08	45.08	46.68
2804	22870	C11H23Br	1-bromoundecane	693-67-4	50.3693	0.0199	1	30	2	50.37	52.73	53.89
2805	22872	C11H23Cl	1-chloroundecane	2473-03-2	40.7385	0.0208	10	62	2	42.74	42.74	44.40
2806	22875	C11H23F	1-fluoroundecane	506-05-8	36.6887	0.0236	10	328	2	38.73	38.73	42.05
2807	22876	C11H23I	1-iodoundecane	4282-44-4	58.8826	0.0181	1	31	2	58.88	61.38	62.65
2808	22890	C11H24	undecane	1120-21-4	33.3966	0.0282	10	542	2	35.64	35.64	39.89
2809	22891	C11H24	2-methyldecane	6975-98-0	31.8933	0.0253	10	769	2	33.81	33.81	37.73
2810	22892	C11H24	3-methyldecane	13151-34-3	32.2058	0.0253	10	742	2	34.14	34.14	38.07
2811	22893	C11H24	4-methyldecane	2847-72-5	32.0429	0.0255	10	813	2	33.98	33.98	38.01
2812	22894	C11H24	5-methyldecane	13151-35-4	31.9671	0.0257	10	883	2	33.92	33.92	38.06
2813	22895	C11H24	3-ethylnonane	17302-11-3	32.3365	0.0256	10	809	2	34.30	34.30	38.38
2814	22896	C11H24	4-ethylnonane	5911-05-7	32.0797	0.0260	10	963	2	34.06	34.06	38.35
2815	22897	C11H24	5-ethylnonane	17302-12-4	31.8400	0.0264	10	1146	2	33.84	33.84	38.35
2816	22898	C11H24	2,2-dimethylnonane	17302-14-6	31.1008	0.0266	10	1361	2	33.07	33.07	37.68
2817	22899	C11H24	2,3-dimethylnonane	2884-06-2	32.0403	0.0260	10	962	2	34.02	34.02	38.30
2818	22900	C11H24	2,4-dimethylnonane	17302-24-8	31.2350	0.0267	10	1356	2	33.21	33.21	37.86
2819	22901	C11H24	2,5-dimethylnonane	17302-27-1	31.3556	0.0266	10	1299	2	33.34	33.34	37.94
2820	22902	C11H24	2,6-dimethylnonane	17302-28-2	31.3668	0.0266	10	1299	2	33.35	33.35	37.96
2821	22903	C11H24	2,7-dimethylnonane	17302-29-3	31.4456	0.0265	10	1244	2	33.42	33.42	37.98
2822	22904	C11H24	2,8-dimethylnonane	17302-30-6	31.3037	0.0261	10	1094	2	33.24	33.24	37.58
2823	22905	C11H24	3,3-dimethylnonane	17302-15-7	31.6917	0.0266	10	1251	2	33.69	33.69	38.30
2824	22906	C11H24	3,4-dimethylnonane	17302-22-6	32.1304	0.0261	10	1004	2	34.12	34.12	38.49
2825	22907	C11H24	3,5-dimethylnonane	17302-25-9	31.6538	0.0266	10	1244	2	33.65	33.65	38.25
2826	22908	C11H24	3,6-dimethylnonane	17302-31-7	31.7797	0.0264	10	1142	2	33.77	33.77	38.26
2827	22909	C11H24	3,7-dimethylnonane	17302-32-8	31.9177	0.0262	10	1048	2	33.90	33.90	38.29
2828	22910	C11H24	4,4-dimethylnonane	17302-18-0	31.4434	0.0270	10	1482	2	33.46	33.46	38.29
2829	22911	C11H24	4,5-dimethylnonane	17302-23-7	31.9614	0.0264	10	1142	2	33.97	33.97	38.50
2830	22912	C11H24	4,6-dimethylnonane	17302-26-0	31.5873	0.0267	10	1299	2	33.59	33.59	38.25
2831	22913	C11H24	5,5-dimethylnonane	6414-96-6	31.3735	0.0271	10	1546	2	33.39	33.39	38.28
2832	22914	C11H24	4-propyloctane	17302-13-5	31.7070	0.0263	10	1146	2	33.69	33.69	38.16
2833	22915	C11H24	4-isopropyloctane	62016-15-3	31.7645	0.0269	10	1356	2	33.79	33.79	38.55
2834	22916	C11H24	2-methyl-3-ethyloctane	62016-16-4	32.0311	0.0265	10	1142	2	34.04	34.04	38.59
2835	22917	C11H24	2-methyl-4-ethyloctane	62016-17-5	31.3912	0.0270	10	1477	2	33.40	33.40	38.22
2836	22918	C11H24	2-methyl-5-ethyloctane	62016-18-6	31.5190	0.0268	10	1356	2	33.52	33.52	38.23
2837	22919	C11H24	2-methyl-6-ethyloctane	62016-19-7	31.7797	0.0264	10	1142	2	33.77	33.77	38.26
2838	22920	C11H24	3-methyl-3-ethyloctane	17302-16-8	32.2940	0.0263	10	1070	2	34.31	34.31	38.80
2839	22921	C11H24	3-methyl-4-ethyloctane	62016-20-0	32.2708	0.0265	10	1142	2	34.30	34.30	38.90
2840	22922	C11H24	3-methyl-5-ethyloctane	62016-21-1	31.5176	0.0270	10	1477	2	33.54	33.54	38.38
2841	22923	C11H24	3-methyl-6-ethyloctane	62016-22-2	32.0311	0.0265	10	1142	2	34.04	34.04	38.59
2842	22924	C11H24	4-methyl-3-ethyloctane	62016-23-3	32.3412	0.0264	10	1094	2	34.37	34.37	38.91
2843	22925	C11H24	4-methyl-4-ethyloctane	17302-19-1	32.0552	0.0269	10	1361	2	34.11	34.11	38.93
2844	22926	C11H24	4-methyl-5-ethyloctane	62016-24-4	32.1423	0.0267	10	1244	2	34.18	34.18	38.89
2845	22927	C11H24	4-methyl-6-ethyloctane	62016-25-5	31.8941	0.0266	10	1244	2	33.91	33.91	38.56
2846	22928	C11H24	2,2,3-trimethyloctane	62016-26-6	31.5871	0.0271	10	1540	2	33.62	33.62	38.55
2847	22929	C11H24	2,2,4-trimethyloctane	18932-14-4	30.6192	0.0277	10	2107	2	32.64	32.64	37.86
2848	22930	C11H24	2,2,5-trimethyloctane	62016-27-7	30.5943	0.0278	10	2152	2	32.62	32.62	37.86
2849	22931	C11H24	2,2,6-trimethyloctane	62016-28-8	30.7858	0.0275	10	1898	2	32.80	32.80	37.88
2850	22932	C11H24	2,2,7-trimethyloctane	62016-29-9	30.3272	0.0278	10	2244	2	32.33	32.33	37.58
2851	22933	C11H24	2,3,3-trimethyloctane	62016-30-2	32.0180	0.0269	10	1359	2	34.07	34.07	38.88
2852	22934	C11H24	2,3,4-trimethyloctane	62016-31-3	31.9796	0.0269	10	1350	2	34.02	34.02	38.83
2853	22935	C11H24	2,3,5-trimethyloctane	62016-32-4	31.5318	0.0271	10	1533	2	33.57	33.57	38.48
2854	22936	C11H24	2,3,6-trimethyloctane	62016-33-5	31.7294	0.0268	10	1350	2	33.75	33.75	38.50
2855	22937	C11H24	2,3,7-trimethyloctane	62016-34-6	31.4276	0.0269	10	1409	2	33.43	33.43	38.18
2856	22938	C11H24	2,4,4-trimethyloctane	62016-35-7	30.8014	0.0276	10	2152	2	32.82	32.82	38.07
2857	22939	C11H24	2,4,5-trimethyloctane	62016-36-8	31.4662	0.0272	10	1600	2	33.50	33.50	38.47
2858	22940	C11H24	2,4,6-trimethyloctane	62016-37-9	31.0958	0.0274	10	1742	2	33.12	33.12	38.14
2859	22941	C11H24	2,4,7-trimethyloctane	62016-38-0	30.5429	0.0275	10	1978	2	32.54	32.54	37.62
2860	22942	C11H24	2,5,5-trimethyloctane	62016-39-1	30.8858	0.0278	10	2064	2	32.92	32.92	38.18
2861	22943	C11H24	2,5,6-trimethyloctane	62016-14-2	31.6023	0.0270	10	1470	2	33.63	33.63	38.49
2862	22944	C11H24	2,6,6-trimethyloctane	54166-32-4	31.1484	0.0274	10	1746	2	33.17	33.17	38.21
2863	22945	C11H24	3,3,4-trimethyloctane	62016-40-4	32.2728	0.0270	10	1359	2	34.34	34.34	39.21
2864	22946	C11H24	3,3,5-trimethyloctane	62016-41-5	31.2537	0.0276	10	1898	2	33.31	33.31	38.51
2865	22947	C11H24	3,3,6-trimethyloctane	62016-42-6	31.4472	0.0273	10	1674	2	33.49	33.49	38.52
2866	22948	C11H24	3,4,4-trimethyloctane	62016-43-7	32.2054	0.0271	10	1417	2	34.28	34.28	39.21
2867	22949	C11H24	3,4,5-trimethyloctane	62016-44-8	32.2893	0.0269	10	1294	2	34.35	34.35	39.15
2868	22950	C11H24	3,4,6-trimethyloctane	62016-45-9	31.9091	0.0270	10	1409	2	33.96	33.96	38.81
2869	22951	C11H24	3,5,5-trimethyloctane	61868-94-8	31.0073	0.0281	10	2244	2	33.08	33.08	38.50
2870	22952	C11H24	4,4,5-trimethyloctane	61868-95-9	32.1347	0.0272	10	1477	2	34.21	34.21	39.19
2871	22953	C11H24	4-tert-butylheptane	60302-21-8	31.4840	0.0277	10	1898	2	33.56	33.56	38.82
2872	22954	C11H24	2-methyl-4-propylheptane	61868-96-0	31.2499	0.0272	10	1610	2	33.27	33.27	38.20
2873	22955	C11H24	3-methyl-4-propylheptane	61868-97-1	32.0713	0.0268	10	1299	2	34.12	34.12	38.88
2874	22956	C11H24	4-methyl-4-propylheptane	17302-20-4	31.9168	0.0272	10	1482	2	33.98	33.98	38.91
2875	22957	C11H24	2-methyl-3-isopropylheptane	6876-18-2	31.8271	0.0271	10	1451	2	33.87	33.87	38.76
2876	22958	C11H24	2-methyl-4-isopropylheptane	61868-98-2	31.1948	0.0277	10	1896	2	33.25	33.25	38.44
2877	22959	C11H24	3-methyl-4-isopropylheptane	61868-99-3	32.0198	0.0273	10	1533	2	34.10	34.10	39.12
2878	22960	C11H24	4-methyl-4-isopropylheptane	61869-00-9	32.3893	0.0273	10	1477	2	34.49	34.49	39.53
2879	22961	C11H24	3,3-diethylheptane	17302-17-9	32.8785	0.0264	10	1017	2	34.94	34.94	39.48
2880	22962	C11H24	3,4-diethylheptane	61869-01-0	32.3999	0.0268	10	1244	2	34.46	34.46	39.22

Table 1 Adsorption Capacity of Activated Carbon - Gas

$\log_{10} Q = \log_{10} K + 1/n \log_{10} y$ (Q - g of compound/100 g of carbon, y - ppmv)

NO	ID	FORMULA	NAME	CAS No	K	1/n	ymin	ymax	code	Q @ymin	Q @10ppm	Q @ymax
2881	22963	C11H24	3,5-diethylheptane	61869-02-1	32.0842	0.0268	10	1299	2	34.13	34.13	38.89
2882	22964	C11H24	4,4-diethylheptane	17302-21-5	32.3765	0.0271	10	1344	2	34.46	34.46	39.34
2883	22965	C11H24	2,2-dimethyl-3-ethylheptane	61869-03-2	31.8091	0.0278	10	1746	2	33.91	33.91	39.16
2884	22966	C11H24	2,2-dimethyl-4-ethylheptane	62016-46-0	30.6171	0.0282	10	2440	2	32.67	32.67	38.14
2885	22967	C11H24	2,2-dimethyl-5-ethylheptane	62016-47-1	30.8858	0.0278	10	2064	2	32.92	32.92	38.18
2886	22968	C11H24	2,3-dimethyl-3-ethylheptane	61868-21-1	32.4149	0.0273	10	1477	2	34.52	34.52	39.56
2887	22969	C11H24	2,3-dimethyl-4-ethylheptane	61868-22-2	32.1471	0.0271	10	1409	2	34.22	34.22	39.13
2888	22970	C11H24	2,3-dimethyl-5-ethylheptane	61868-23-3	31.8384	0.0271	10	1470	2	33.89	33.89	38.80
2889	22971	C11H24	2,4-dimethyl-3-ethylheptane	61868-24-4	32.1471	0.0271	10	1409	2	34.22	34.22	39.13
2890	22972	C11H24	2,4-dimethyl-4-ethylheptane	61868-25-5	31.2475	0.0282	10	2244	2	33.34	33.34	38.83
2891	22973	C11H24	2,4-dimethyl-5-ethylheptane	61868-26-6	31.7120	0.0273	10	1600	2	33.77	33.77	38.80
2892	22974	C11H24	2,5-dimethyl-3-ethylheptane	61868-27-7	32.4788	0.0240	10	1600	2	34.32	34.32	38.76
2893	22975	C11H24	2,5-dimethyl-4-ethylheptane	61868-28-8	31.6406	0.0274	10	1669	2	33.70	33.70	38.78
2894	22976	C11H24	2,5-dimethyl-5-ethylheptane	61868-29-9	31.7520	0.0273	10	1606	2	33.81	33.81	38.85
2895	22977	C11H24	2,6-dimethyl-3-ethylheptane	61868-30-2	31.4662	0.0272	10	1600	2	33.50	33.50	38.47
2896	22978	C11H24	2,6-dimethyl-4-ethylheptane	61868-31-3	30.8253	0.0278	10	2064	2	32.86	32.86	38.10
2897	22979	C11H24	3,3-dimethyl-4-ethylheptane	61868-32-4	32.3910	0.0273	10	1477	2	34.49	34.49	39.53
2898	22980	C11H24	3,3-dimethyl-5-ethylheptane	61868-33-5	31.5552	0.0276	10	1820	2	33.63	33.63	38.83
2899	22981	C11H24	3,4-dimethyl-3-ethylheptane	61868-34-6	32.9843	0.0269	10	1199	2	35.09	35.09	39.91
2900	22982	C11H24	3,4-dimethyl-4-ethylheptane	61868-35-7	32.1341	0.0269	10	1417	2	34.19	34.19	39.06
2901	22983	C11H24	3,4-dimethyl-5-ethylheptane	61868-36-8	32.6080	0.0269	10	1241	2	34.69	34.69	39.49
2902	22984	C11H24	3,5-dimethyl-3-ethylheptane	61868-37-9	32.1014	0.0271	10	1477	2	34.17	34.17	39.11
2903	22985	C11H24	3,5-dimethyl-4-ethylheptane	61868-38-0	32.5364	0.0270	10	1294	2	34.62	34.62	39.48
2904	22986	C11H24	4,4-dimethyl-3-ethylheptane	61868-39-1	32.1526	0.0277	10	1746	2	34.27	34.27	39.54
2905	22987	C11H24	2,2,3,3-tetramethylheptane	61868-40-4	32.0226	0.0278	10	1749	2	34.14	34.14	39.40
2906	22988	C11H24	2,2,3,4-tetramethylheptane	61868-41-5	31.6963	0.0278	10	1889	2	33.79	33.79	39.09
2907	22989	C11H24	2,2,3,5-tetramethylheptane	61868-42-6	31.3229	0.0279	10	2051	2	33.40	33.40	38.76
2908	22990	C11H24	2,2,3,6-tetramethylheptane	61868-43-7	30.8514	0.0282	10	2420	2	32.92	32.92	38.44
2909	22991	C11H24	2,2,4,4-tetramethylheptane	61868-44-8	30.9921	0.0283	10	2627	2	33.08	33.08	38.74
2910	22992	C11H24	2,2,4,5-tetramethylheptane	61868-45-9	30.7553	0.0283	10	2522	2	32.83	32.83	38.40
2911	22993	C11H24	2,2,4,6-tetramethylheptane	61868-46-0	29.7916	0.0289	10	3341	2	31.84	31.84	37.65
2912	22994	C11H24	2,2,5,5-tetramethylheptane	61868-47-1	30.2539	0.0287	10	3092	2	32.32	32.32	38.11
2913	22995	C11H24	2,2,5,6-tetramethylheptane	61868-48-2	30.5322	0.0282	10	2522	2	32.58	32.58	38.09
2914	22996	C11H24	2,2,6,6-tetramethylheptane	40117-45-1	29.6225	0.0289	10	3495	2	31.66	31.66	37.49
2915	22997	C11H24	2,3,3,4-tetramethylheptane	61868-49-3	32.6109	0.0274	10	1475	2	34.73	34.73	39.82
2916	22998	C11H24	2,3,3,5-tetramethylheptane	61868-50-6	31.6963	0.0278	10	1889	2	33.79	33.79	39.09
2917	22999	C11H24	2,3,3,6-tetramethylheptane	61868-51-7	31.3936	0.0278	10	1968	2	33.47	33.47	38.77
2918	23000	C11H24	2,3,4,4-tetramethylheptane	61868-52-8	32.1164	0.0275	10	1739	2	34.22	34.22	39.44
2919	23001	C11H24	2,3,4,5-tetramethylheptane	61868-53-9	32.2381	0.0274	10	1528	2	34.34	34.34	39.40
2920	23002	C11H24	2,3,4,6-tetramethylheptane	61868-54-0	31.4077	0.0277	10	1883	2	33.48	33.48	38.71
2921	23003	C11H24	2,3,5,5-tetramethylheptane	61868-55-1	31.1936	0.0281	10	2228	2	33.28	33.28	38.75
2922	23004	C11H24	2,3,5,6-tetramethylheptane	52670-32-3	31.4789	0.0276	10	1806	2	33.55	33.55	38.72
2923	23005	C11H24	2,4,4,5-tetramethylheptane	61868-56-2	31.5663	0.0280	10	2051	2	33.67	33.67	39.08
2924	23006	C11H24	2,4,4,6-tetramethylheptane	61868-57-3	30.2629	0.0285	10	3101	2	32.32	32.32	38.06
2925	23007	C11H24	2,4,5,5-tetramethylheptane	61868-58-4	31.6963	0.0278	10	1889	2	33.79	33.79	39.09
2926	23008	C11H24	3,3,4,4-tetramethylheptane	61868-59-5	32.9141	0.0275	10	1487	2	35.06	35.06	40.23
2927	23009	C11H24	3,3,4,5-tetramethylheptane	61868-60-8	32.5395	0.0275	10	1537	2	34.66	34.66	39.81
2928	23010	C11H24	3,3,5,5-tetramethylheptane	61868-61-9	32.1118	0.0275	10	1749	2	34.21	34.21	39.44
2929	23011	C11H24	3,4,4,5-tetramethylheptane	61868-62-0	33.0671	0.0272	10	1303	2	35.20	35.20	40.18
2930	23012	C11H24	2,2-dimethyl-3-isopropylhexane	61868-63-1	31.4260	0.0282	10	2228	2	33.54	33.54	39.06
2931	23013	C11H24	2,3-dimethyl-3-isopropylhexane	61868-64-2	32.7558	0.0277	10	1601	2	34.91	34.91	40.17
2932	23014	C11H24	2,4-dimethyl-3-isopropylhexane	61868-65-3	31.9630	0.0278	10	1806	2	34.07	34.07	39.37
2933	23015	C11H24	2,5-dimethyl-3-isopropylhexane	61868-66-4	31.1340	0.0281	10	2227	2	33.22	33.22	38.67
2934	23016	C11H24	2-methyl-3,3-diethylhexane	61868-67-5	33.3136	0.0269	10	1150	2	35.44	35.44	40.26
2935	23017	C11H24	2-methyl-3,4-diethylhexane	61868-68-6	32.4057	0.0272	10	1409	2	34.50	34.50	39.47
2936	23018	C11H24	2-methyl-4,4-diethylhexane	61868-69-7	37.1982	0.0076	10	1540	2	37.86	37.86	39.34
2937	23019	C11H24	3-methyl-3,4-diethylhexane	61868-70-0	33.3136	0.0269	10	1150	2	35.44	35.44	40.26
2938	23020	C11H24	3-methyl-4,4-diethylhexane	61868-71-1	33.8493	0.0265	10	973	2	35.98	35.98	40.63
2939	23021	C11H24	2,2,3-trimethyl-3-ethylhexane	61868-72-2	32.6658	0.0279	10	1749	2	34.83	34.83	40.22
2940	23022	C11H24	2,2,3-trimethyl-4-ethylhexane	61868-73-3	31.9468	0.0279	10	1889	2	34.07	34.07	39.43
2941	23023	C11H24	2,2,4-trimethyl-3-ethylhexane	61868-74-4	31.8749	0.0280	10	1968	2	34.00	34.00	39.41
2942	23024	C11H24	2,2,4-trimethyl-4-ethylhexane	61868-75-5	31.9839	0.0277	10	1897	2	34.09	34.09	39.43
2943	23025	C11H24	2,2,5-trimethyl-3-ethylhexane	61868-76-6	31.0508	0.0283	10	2420	2	33.14	33.14	38.72
2944	23026	C11H24	2,2,5-trimethyl-4-ethylhexane	61868-77-7	30.5567	0.0286	10	2855	2	32.64	32.64	38.38
2945	23027	C11H24	2,3,3-trimethyl-4-ethylhexane	61868-78-8	32.9337	0.0274	10	1415	2	35.07	35.07	40.16
2946	23028	C11H24	2,3,4-trimethyl-3-ethylhexane	61868-79-9	33.4667	0.0270	10	1200	2	35.62	35.62	40.54
2947	23029	C11H24	2,3,4-trimethyl-4-ethylhexane	61868-80-2	32.7558	0.0277	10	1601	2	34.91	34.91	40.17
2948	23030	C11H24	2,3,5-trimethyl-3-ethylhexane	61868-81-3	31.9468	0.0279	10	1889	2	34.07	34.07	39.43
2949	23031	C11H24	2,3,5-trimethyl-4-ethylhexane	61868-82-4	32.0946	0.0276	10	1661	2	34.20	34.20	39.38
2950	23032	C11H24	2,4,4-trimethyl-3-ethylhexane	61868-83-5	32.3368	0.0278	10	1739	2	34.47	34.47	39.78
2951	23033	C11H24	3,3,4-trimethyl-4-ethylhexane	61868-84-6	33.8566	0.0270	10	1166	2	36.03	36.03	40.98
2952	23034	C11H24	2,2,3,3,4-pentamethylhexane	61868-85-7	32.9962	0.0277	10	1614	2	35.17	35.17	40.50
2953	23035	C11H24	2,2,3,3,5-pentamethylhexane	61868-86-8	31.5427	0.0290	10	2773	2	33.72	33.72	39.70
2954	23036	C11H24	2,2,3,4,4-pentamethylhexane	61868-87-9	33.1161	0.0274	10	1489	2	35.27	35.27	40.45
2955	23037	C11H24	2,2,3,4,5-pentamethylhexane	61868-88-0	31.6399	0.0283	10	2216	2	33.77	33.77	39.34
2956	23038	C11H24	2,2,3,5,5-pentamethylhexane	14739-73-2	30.4993	0.0291	10	3323	2	32.61	32.61	38.61
2957	23039	C11H24	2,2,4,4,5-pentamethylhexane	60302-23-0	31.6798	0.0281	10	2224	2	33.80	33.80	39.34
2958	23040	C11H24	2,3,3,4,4-pentamethylhexane	61868-89-1	33.4717	0.0275	10	1431	2	35.66	35.66	40.89
2959	23041	C11H24	2,3,3,4,5-pentamethylhexane	52670-33-4	32.6176	0.0277	10	1666	2	34.77	34.77	40.07
2960	23042	C11H24	2,2,4-trimethyl-3-isopropylpentane	61868-90-4	31.3665	0.0287	10	2608	2	33.51	33.51	39.30

Table 1 Adsorption Capacity of Activated Carbon - Gas

$\log_{10} Q = \log_{10} K + 1/n \log_{10} y$ (Q - g of compound/100 g of carbon, y - ppmv)												
NO	ID	FORMULA	NAME	CAS No	K	1/n	ymin	ymax	code	Q @ymin	Q @10ppm	Q @ymax
2961	23043	C11H24	2,3,4-trimethyl-3-isopropylpentane	61868-91-5	33.4933	0.0274	10	1360	2	35.68	35.68	40.82
2962	23044	C11H24	2,2-dimethyl-3,3-diethylpentane	60302-28-5	33.7819	0.0271	10	1214	2	35.96	35.96	40.96
2963	23045	C11H24	2,4-dimethyl-3,3-diethylpentane	61868-92-6	33.8796	0.0269	10	1106	2	36.05	36.05	40.92
2964	23046	C11H24	2,2,3,4-tetramethyl-3-ethylpentane	61868-93-7	33.3969	0.0276	10	1489	2	35.59	35.59	40.87
2965	23047	C11H24	2,2,4,4-tetramethyl-3-ethylpentane	------	32.4477	0.0279	10	1894	2	34.60	34.60	40.05
2966	23048	C11H24	2,2,3,3,4,4-hexamethylpentane	60302-27-4	34.1947	0.0273	10	1295	2	36.41	36.41	41.57
2967	23055	C11H24O	1-undecanol	112-42-5	41.4398	0.0216	1	3.9	2	41.44	---	42.68
2968	23056	C11H24O	2-undecanol	1653-30-1	39.6935	0.0242	1	14	2	39.69	41.96	42.31
2969	23084	C11H24S	1-undecanethiol	5332-52-5	41.7363	0.0196	1	5.8	2	41.74	---	43.20
2970	23085	C11H24S	2-undecanethiol	62155-02-6	39.9520	0.0201	1	36	2	39.95	41.84	42.93
2971	23086	C11H24S	methyl decyl sulfide	22438-39-7	40.8872	0.0190	1	17	2	40.89	42.72	43.15
2972	23087	C11H24S	ethyl nonyl sulfide	59973-08-9	40.8872	0.0190	1	17	2	40.89	42.72	43.15
2973	23088	C11H24S	propyl octyl sulfide	3698-93-9	40.8872	0.0190	1	17	2	40.89	42.72	43.15
2974	23089	C11H24S	butyl heptyl sulfide	40813-84-1	40.8872	0.0190	1	17	2	40.89	42.72	43.15
2975	23090	C11H25N	undecylamine	7307-55-3	37.5625	0.0223	1	44	2	37.56	39.54	40.87
2976	23091	C11H25N	methyldecylamine	7516-82-7	37.5926	0.0213	1	27	2	37.59	39.48	40.33
2977	23092	C11H25N	ethylnonylamine	66563-84-6	37.3066	0.0215	1	29	2	37.31	39.20	40.10
2978	23093	C11H25N	dimethylnonylamine	17373-27-2	35.0226	0.0213	10	290	2	36.78	36.78	39.52
2979	23094	C11H25N	diethylheptylamine	26981-81-7	34.8457	0.0220	10	393	2	36.65	36.65	39.73
2980	23675	C12H12	1,2-dimethylnaphthalene	573-98-8	49.3071	0.0239	1	10	2	49.31	52.10	52.10
2981	23676	C12H12	1,3-dimethylnaphthalene	575-41-7	46.7976	0.0229	10	67	2	49.34	49.34	51.54
2982	23677	C12H12	1,4-dimethylnaphthalene	571-58-4	46.9961	0.0234	10	80	2	49.60	49.60	52.07
2983	23678	C12H12	1,5-dimethylnaphthalene	571-61-9	46.9961	0.0234	10	80	2	49.60	49.60	52.07
2984	23679	C12H12	1,6-dimethylnaphthalene	575-43-9	47.0655	0.0236	1	40	2	47.07	49.69	51.34
2985	23680	C12H12	1,7-dimethylnaphthalene	575-37-1	48.2539	0.0243	1	13	2	48.25	51.03	51.36
2986	23681	C12H12	1,8-dimethylnaphthalene	569-41-5	48.2539	0.0243	1	13	2	48.25	51.03	51.36
2987	23682	C12H12	2,3-dimethylnaphthalene	581-40-8	48.2539	0.0243	1	13	2	48.25	51.03	51.36
2988	23683	C12H12	2,6-dimethylnaphthalene	581-42-0	48.2539	0.0243	1	13	2	48.25	51.03	51.36
2989	23684	C12H12	2,7-dimethylnaphthalene	582-16-1	48.2539	0.0243	1	13	2	48.25	51.03	51.36
2990	23685	C12H12	1-ethylnaphthalene	1127-76-0	47.5684	0.0233	1	33	2	47.57	50.19	51.61
2991	23686	C12H12	2-ethylnaphthalene	939-27-5	46.3261	0.0226	10	57	2	48.80	48.80	50.76
2992	24111	C12H16	cyclohexylbenzene	827-52-1	43.9984	0.0233	10	52	2	46.42	46.42	48.24
2993	24112	C12H16	1-ethyl-[1,2,3,4-tetrahydronaphthalene]	13556-58-6	43.2338	0.0252	10	127	2	45.82	45.82	48.84
2994	24113	C12H16	2-ethyl-[1,2,3,4-tetrahydronaphthalene]	32367-54-7	42.9296	0.0249	10	111	2	45.46	45.46	48.26
2995	24357	C12H18	hexylbenzene	1077-16-3	39.3761	0.0215	10	158	2	41.37	41.37	43.90
2996	24361	C12H18	m-diisopropylbenzene	99-62-7	37.8325	0.0235	10	517	2	39.94	39.94	43.82
2997	24362	C12H18	p-diisopropylbenzene	100-18-5	38.4730	0.0226	10	323	2	40.53	40.53	43.84
2998	24364	C12H18	1,2,3-triethylbenzene	------	38.2689	0.0240	10	678	2	40.45	40.45	44.76
2999	24365	C12H18	1,2,4-triethylbenzene	877-44-1	39.2520	0.0225	10	329	2	41.34	41.34	44.71
3000	24366	C12H18	1,3,5-triethylbenzene	102-25-0	38.6821	0.0250	10	748	2	40.98	40.98	45.65
3001	24652	C12H22	bicyclohexyl	92-51-3	40.1976	0.0247	10	141	2	42.55	42.55	45.43
3002	24653	C12H22	1-dodecyne	765-03-7	35.2381	0.0218	10	272	2	37.05	37.05	39.82
3003	24654	C12H22	2-dodecyne	629-49-2	35.9059	0.0192	10	214	2	37.53	37.53	39.79
3004	24655	C12H22	3-dodecyne	6790-27-8	35.2640	0.0214	10	295	2	37.04	37.04	39.82
3005	24658	C12H22	4a-ethyl-cis-decahydronaphthalene	66553-61-5	40.1358	0.0226	10	224	2	42.28	42.28	45.36
3006	24659	C12H22	4a-ethyl-trans-decahydronaphthalene	66553-62-6	38.5446	0.0232	10	316	2	40.66	40.66	44.05
3007	24660	C12H22	1-ethyl,cis-decahydronaphthalene	------	39.9334	0.0231	10	69	2	42.11	42.11	44.04
3008	24661	C12H22	1-ethyl-trans-decahydronaphthalene	------	39.6714	0.0235	10	86	2	41.88	41.88	44.05
3009	24662	C12H22	2-ethyl-cis-decahydronaphthalene	------	40.1032	0.0225	10	206	2	42.24	42.24	45.21
3010	24663	C12H22	2-ethyl-trans-decahydronaphthalene	------	38.9385	0.0229	10	278	2	41.05	41.05	44.30
3011	24664	C12H22	1,4a-dimethyl-cis-decahydronaphthalene	------	39.3359	0.0241	10	442	2	41.58	41.58	45.54
3012	24665	C12H22	1,4a-dimethyl-trans-decahydronaphthalene	------	37.7789	0.0246	10	593	2	39.98	39.98	44.21
3013	24666	C12H22	1-heptylcyclopentene	4292-00-6	37.0730	0.0220	10	266	2	39.00	39.00	41.93
3014	24787	C12H23Br	cis-1-bromo-1-dodecene	66553-39-7	51.2545	0.0188	1	10	2	51.25	53.52	53.52
3015	24788	C12H23Br	trans-1-bromo-1-dodecene	66553-40-0	51.2545	0.0188	1	10	2	51.25	53.52	53.52
3016	24792	C12H23Cl	cis-1-chloro-1-dodecene	66553-41-1	42.8423	0.0203	1	11	2	42.84	44.89	44.98
3017	24793	C12H23Cl	trans-1-chloro-1-dodecene	66553-42-2	42.8423	0.0203	1	11	2	42.84	44.89	44.98
3018	24796	C12H23F	cis-1-fluoro-1-dodecene	66553-43-3	38.2680	0.0201	10	244	2	40.08	40.08	42.73
3019	24797	C12H23F	trans-1-fluoro-1-dodecene	66553-44-4	38.2680	0.0201	10	244	2	40.08	40.08	42.73
3020	24800	C12H23I	cis-1-iodo-1-dodecene	66553-45-5	60.0513	0.0163	1	4.5	2	60.05	---	61.54
3021	24801	C12H23I	trans-1-iodo-1-dodecene	66553-46-6	60.0513	0.0163	1	4.5	2	60.05	---	61.54
3022	24804	C12H23N	dicyclohexylamine	101-83-7	42.9666	0.0222	1	44	2	42.97	45.22	46.72
3023	24805	C12H23N	dodecanenitrile	2437-25-4	41.1183	0.0201	1	3.3	2	41.12	---	42.11
3024	24818	C12H24	heptylcyclopentane	5617-42-5	36.8421	0.0238	10	144	2	38.91	38.91	41.46
3025	24819	C12H24	hexylcyclohexane	4292-75-5	40.5862	0.0237	10	167	2	42.86	42.86	45.82
3026	24823	C12H24	1-dodecene	112-41-4	34.7662	0.0207	10	209	2	36.46	36.46	38.83
3027	24824	C12H24	2-methyl-1-undecene	18516-37-5	34.3858	0.0215	10	337	2	36.13	36.13	38.97
3028	24833	C12H24Cl2	1,1-dichlorododecane	62017-17-8	46.9517	0.0180	1	9.6	2	46.95	---	48.90
3029	24838	C12H24F2	1,1-difluorododecane	62127-45-1	40.2485	0.0207	10	270	2	42.22	42.22	45.21
3030	24852	C12H24O	dodecanal	112-54-9	39.9056	0.0204	1	20	2	39.91	41.82	42.41
3031	24854	C12H24O	2-dodecanone	6175-49-1	39.7706	0.0204	1	20	2	39.77	41.69	42.28
3032	24868	C12H24O2	undecyl formate	5454-24-0	42.3802	0.0211	1	11	2	42.38	44.49	44.58
3033	24869	C12H24O2	decyl acetate	112-17-4	40.8127	0.0210	1	44	2	40.81	42.83	44.18
3034	24870	C12H24O2	nonyl propanoate	53184-67-1	42.1148	0.0209	1	9.9	2	42.11	---	44.19
3035	24871	C12H24O2	octyl butanoate	110-39-4	40.7056	0.0212	1	47	2	40.71	42.74	44.16
3036	24892	C12H24S	thiacyclotridecane	295-05-6	43.2303	0.0185	1	30	2	43.23	45.11	46.04
3037	24894	C12H25Br	1-bromododecane	143-15-7	50.7918	0.0189	1	11	2	50.79	53.06	53.15
3038	24897	C12H25Cl	1-chlorododecane	112-52-7	41.8225	0.0188	1	22	2	41.82	43.67	44.32
3039	24899	C12H25F	1-fluorododecane	334-68-9	38.0232	0.0220	10	113	2	40.00	40.00	42.19
3040	24900	C12H25I	1-iodododecane	4292-19-7	58.6510	0.0175	1	13	2	58.65	61.06	61.34

Table 1 Adsorption Capacity of Activated Carbon - Gas

$\log_{10} Q = \log_{10} K + 1/n \log_{10} y$ (Q - g of compound/100 g of carbon, y - ppmv)

NO	ID	FORMULA	NAME	CAS No	K	1/n	ymin	ymax	code	Q @ymin	Q @10ppm	Q @ymax
3041	24914	C12H26	dodecane	112-40-3	33.1847	0.0222	10	179	2	34.93	34.93	37.24
3042	24915	C12H26	3-methylundecane	1002-43-3	34.2458	0.0208	10	258	2	35.93	35.93	38.45
3043	24916	C12H26	4-methylundecane	2980-69-0	34.1734	0.0209	10	270	2	35.86	35.86	38.42
3044	24917	C12H26	5-methylundecane	1632-70-8	34.0275	0.0211	10	308	2	35.73	35.73	38.41
3045	24918	C12H26	6-methylundecane	17302-33-9	34.0226	0.0211	10	308	2	35.72	35.72	38.41
3046	24919	C12H26	3-ethyldecane	17085-96-0	34.6532	0.0212	10	273	2	36.39	36.39	39.03
3047	24920	C12H26	4-ethyldecane	1636-44-8	34.1666	0.0214	10	336	2	35.89	35.89	38.69
3048	24921	C12H26	5-ethyldecane	17302-36-2	34.0527	0.0215	10	368	2	35.78	35.78	38.67
3049	24922	C12H26	2,2-dimethyldecane	17302-37-3	33.2941	0.0218	10	467	2	35.01	35.01	38.07
3050	24923	C12H26	2,3-dimethyldecane	17312-44-6	34.1248	0.0214	10	339	2	35.85	35.85	38.65
3051	24924	C12H26	2,4-dimethyldecane	2801-84-5	33.4121	0.0217	10	441	2	35.13	35.13	38.14
3052	24925	C12H26	2,5-dimethyldecane	17312-50-4	33.4539	0.0219	10	481	2	35.19	35.19	38.30
3053	24926	C12H26	2,6-dimethyldecane	13150-81-7	33.4576	0.0219	10	481	2	35.19	35.19	38.31
3054	24927	C12H26	2,7-dimethyldecane	17312-51-5	33.6550	0.0216	10	404	2	35.37	35.37	38.32
3055	24928	C12H26	2,8-dimethyldecane	17312-52-6	33.7728	0.0215	10	370	2	35.48	35.48	38.34
3056	24929	C12H26	2,9-dimethyldecane	1002-17-1	33.5248	0.0214	10	370	2	35.22	35.22	38.05
3057	24930	C12H26	3,3-dimethyldecane	17302-38-4	33.8893	0.0218	10	429	2	35.63	35.63	38.67
3058	24931	C12H26	3,4-dimethyldecane	17312-45-7	34.3181	0.0215	10	354	2	36.06	36.06	38.93
3059	24932	C12H26	3,5-dimethyldecane	17312-48-0	33.7961	0.0218	10	441	2	35.54	35.54	38.60
3060	24933	C12H26	3,6-dimethyldecane	17312-53-7	33.8461	0.0217	10	422	2	35.58	35.58	38.60
3061	24934	C12H26	3,7-dimethyldecane	17312-54-8	33.9069	0.0217	10	404	2	35.64	35.64	38.61
3062	24935	C12H26	3,8-dimethyldecane	17312-55-9	34.0698	0.0214	10	354	2	35.79	35.79	38.64
3063	24936	C12H26	4,4-dimethyldecane	17312-39-9	33.6597	0.0221	10	510	2	35.42	35.42	38.63
3064	24937	C12H26	4,5-dimethyldecane	17312-46-8	34.1488	0.0217	10	404	2	35.90	35.90	38.90
3065	24938	C12H26	4,6-dimethyldecane	17312-49-1	33.6804	0.0220	10	481	2	35.43	35.43	38.58
3066	24939	C12H26	4,7-dimethyldecane	17312-56-0	33.7361	0.0219	10	460	2	35.48	35.48	38.59
3067	24940	C12H26	5,5-dimethyldecane	17453-92-8	33.5476	0.0223	10	556	2	35.31	35.31	38.62
3068	24941	C12H26	5,6-dimethyldecane	1636-43-7	34.0925	0.0218	10	422	2	35.85	35.85	38.89
3069	24942	C12H26	4-propylnonane	6165-37-3	33.8908	0.0217	10	420	2	35.63	35.63	38.65
3070	24943	C12H26	5-propylnonane	998-35-6	33.7642	0.0219	10	459	2	35.51	35.51	38.61
3071	24944	C12H26	4-isopropylnonane	62184-71-8	33.9242	0.0220	10	481	2	35.69	35.69	38.87
3072	24945	C12H26	5-isopropylnonane	62184-72-9	33.9069	0.0220	10	481	2	35.67	35.67	38.85
3073	24946	C12H26	2-methyl-3-ethylnonane	62184-73-0	34.1099	0.0218	10	422	2	35.87	35.87	38.92
3074	24947	C12H26	2-methyl-4-ethylnonane	62184-37-6	33.6620	0.0220	10	481	2	35.41	35.41	38.56
3075	24948	C12H26	2-methyl-5-ethylnonane	62184-38-7	33.5678	0.0221	10	525	2	35.32	35.32	38.56
3076	24949	C12H26	2-methyl-6-ethylnonane	62184-39-8	33.6804	0.0220	10	481	2	35.43	35.43	38.58
3077	24950	C12H26	2-methyl-7-ethylnonane	62184-40-1	33.9069	0.0217	10	404	2	35.64	35.64	38.61
3078	24951	C12H26	3-methyl-3-ethylnonane	17302-39-5	34.4444	0.0218	10	411	2	36.22	36.22	39.27
3079	24952	C12H26	3-methyl-4-ethylnonane	62184-41-2	34.3463	0.0219	10	422	2	36.12	36.12	39.20
3080	24953	C12H26	3-methyl-5-ethylnonane	62184-42-3	33.7354	0.0222	10	549	2	35.51	35.51	38.81
3081	24954	C12H26	3-methyl-6-ethylnonane	62184-43-4	33.9800	0.0220	10	460	2	35.74	35.74	38.88
3082	24955	C12H26	3-methyl-7-ethylnonane	62184-44-5	34.0680	0.0219	10	441	2	35.83	35.83	38.92
3083	24956	C12H26	4-methyl-3-ethylnonane	62184-45-6	34.3138	0.0219	10	441	2	36.09	36.09	39.22
3084	24957	C12H26	4-methyl-4-ethylnonane	17312-40-2	34.1602	0.0222	10	510	2	35.95	35.95	39.23
3085	24958	C12H26	4-methyl-5-ethylnonane	62184-46-7	34.1777	0.0221	10	481	2	35.96	35.96	39.17
3086	24959	C12H26	4-methyl-6-ethylnonane	62184-47-8	33.8118	0.0222	10	525	2	35.58	35.58	38.85
3087	24960	C12H26	4-methyl-7-ethylnonane	62184-48-9	34.0408	0.0219	10	441	2	35.80	35.80	38.89
3088	24961	C12H26	5-methyl-3-ethylnonane	62184-49-0	33.9800	0.0220	10	460	2	35.74	35.74	38.88
3089	24962	C12H26	5-methyl-4-ethylnonane	1632-71-9	34.2302	0.0220	10	460	2	36.01	36.01	39.18
3090	24963	C12H26	5-methyl-5-ethylnonane	14531-16-9	34.0989	0.0223	10	532	2	35.89	35.89	39.21
3091	24964	C12H26	2,2,3-trimethylnonane	55499-04-2	33.7782	0.0222	10	535	2	35.55	35.55	38.83
3092	24965	C12H26	2,2,4-trimethylnonane	62184-50-3	32.8320	0.0228	10	786	2	34.60	34.60	38.22
3093	24966	C12H26	2,2,5-trimethylnonane	62184-51-4	32.8307	0.0228	10	786	2	34.60	34.60	38.22
3094	24967	C12H26	2,2,6-trimethylnonane	62184-52-5	32.7963	0.0229	10	820	2	34.57	34.57	38.24
3095	24968	C12H26	2,2,7-trimethylnonane	62184-53-6	33.0591	0.0225	10	662	2	34.82	34.82	38.26
3096	24969	C12H26	2,2,8-trimethylnonane	62184-54-7	32.7627	0.0224	10	662	2	34.49	34.49	37.89
3097	24970	C12H26	2,3,3-trimethylnonane	62184-55-8	37.5112	0.0067	10	492	2	38.10	38.10	39.10
3098	24971	C12H26	2,3,4-trimethylnonane	62184-56-9	34.1377	0.0221	10	484	2	35.92	35.92	39.13
3099	24972	C12H26	2,3,5-trimethylnonane	62184-57-0	33.6649	0.0223	10	575	2	35.44	35.44	38.80
3100	24973	C12H26	2,3,6-trimethylnonane	62184-58-1	33.7730	0.0222	10	527	2	35.54	35.54	38.81
3101	24974	C12H26	2,3,7-trimethylnonane	62184-59-2	33.9405	0.0220	10	463	2	35.70	35.70	38.84
3102	24975	C12H26	2,3,8-trimethylnonane	62184-60-5	33.6455	0.0220	10	484	2	35.39	35.39	38.54
3103	24976	C12H26	2,4,4-trimethylnonane	62184-61-6	33.0122	0.0229	10	820	2	34.80	34.80	38.51
3104	24977	C12H26	2,4,5-trimethylnonane	62184-62-7	33.6037	0.0224	10	600	2	35.38	35.38	38.79
3105	24978	C12H26	2,4,6-trimethylnonane	62184-10-5	33.1926	0.0226	10	683	2	34.97	34.97	38.47
3106	24979	C12H26	2,4,7-trimethylnonane	62184-11-6	33.3102	0.0224	10	627	2	35.08	35.08	38.49
3107	24980	C12H26	2,4,8-trimethylnonane	49542-74-7	32.8203	0.0226	10	714	2	34.57	34.57	38.08
3108	24981	C12H26	2,5,5-trimethylnonane	62184-12-7	33.0122	0.0229	10	820	2	34.80	34.80	38.51
3109	24982	C12H26	2,5,6-trimethylnonane	62184-13-8	33.6600	0.0224	10	575	2	35.44	35.44	38.80
3110	24983	C12H26	2,5,7-trimethylnonane	62184-14-9	33.3066	0.0224	10	627	2	35.07	35.07	38.49
3111	24984	C12H26	2,5,8-trimethylnonane	49557-09-7	32.9673	0.0226	10	683	2	34.73	34.73	38.20
3112	24985	C12H26	2,6,6-trimethylnonane	62184-15-0	33.0340	0.0229	10	820	2	34.83	34.83	38.53
3113	24986	C12H26	2,6,7-trimethylnonane	62184-16-1	33.8336	0.0221	10	505	2	35.60	35.60	38.83
3114	24987	C12H26	2,7,7-trimethylnonane	62184-17-2	33.4030	0.0224	10	608	2	35.17	35.17	38.56
3115	24988	C12H26	3,3,4-trimethylnonane	62184-18-3	34.4271	0.0222	10	492	2	36.23	36.23	39.50
3116	24989	C12H26	3,3,5-trimethylnonane	62184-19-4	33.2764	0.0230	10	820	2	35.09	35.09	38.83
3117	24990	C12H26	3,3,7-trimethylnonane	62184-21-8	33.5320	0.0226	10	662	2	35.32	35.32	38.83
3118	24991	C12H26	3,3,6-trimethylnonane	62184-20-7	33.6998	0.0224	10	583	2	35.48	35.48	38.86
3119	24992	C12H26	3,4,4-trimethylnonane	62184-22-9	34.3149	0.0223	10	535	2	36.13	36.13	39.48
3120	24993	C12H26	3,4,5-trimethylnonane	62184-23-0	34.3861	0.0221	10	484	2	36.19	36.19	39.43

Table 1 Adsorption Capacity of Activated Carbon - Gas

$\log_{10} Q = \log_{10} K + 1/n \log_{10} y$ (Q - g of compound/100 g of carbon, y - ppmv)

NO	ID	FORMULA	NAME	CAS No	K	1/n	ymin	ymax	code	Q @ymin	Q @10ppm	Q @ymax
3121	24994	C12H26	3,4,6-trimethylnonane	62184-24-1	33.9641	0.0223	10	550	2	35.76	35.76	39.10
3122	24995	C12H26	3,4,7-trimethylnonane	27802-85-3	34.1340	0.0221	10	484	2	35.91	35.91	39.13
3123	24996	C12H26	3,5,5-trimethylnonane	62184-25-2	33.3082	0.0229	10	786	2	35.11	35.11	38.81
3124	24997	C12H26	3,5,6-trimethylnonane	62184-26-3	33.9592	0.0223	10	550	2	35.75	35.75	39.10
3125	24998	C12H26	3,5,7-trimethylnonane	62184-27-4	33.6037	0.0224	10	600	2	35.38	35.38	38.79
3126	24999	C12H26	3,6,6-trimethylnonane	62184-28-5	33.2764	0.0230	10	820	2	35.09	35.09	38.83
3127	25000	C12H26	4,4,5-trimethylnonane	62184-29-6	34.2010	0.0225	10	583	2	36.02	36.02	39.47
3128	25001	C12H26	4,4,6-trimethylnonane	62184-30-9	33.3033	0.0229	10	786	2	35.11	35.11	38.80
3129	25002	C12H26	4,5,5-trimethylnonane	62184-31-0	34.1924	0.0225	10	583	2	36.01	36.01	39.46
3130	25003	C12H26	4,5,6-trimethylnonane	62211-85-2	34.2766	0.0223	10	527	2	36.08	36.08	39.42
3131	25004	C12H26	4-tert-butyloctane	62184-32-1	33.5510	0.0230	10	786	2	35.37	35.37	39.11
3132	25005	C12H26	2-methyl-4-propyloctane	62184-33-2	33.2881	0.0225	10	654	2	35.06	35.06	38.53
3133	25006	C12H26	2-methyl-5-propyloctane	62184-34-3	33.2504	0.0225	10	654	2	35.02	35.02	38.47
3134	25007	C12H26	3-methyl-4-propyloctane	62184-35-4	34.1163	0.0222	10	503	2	35.90	35.90	39.16
3135	25008	C12H26	3-methyl-5-propyloctane	62184-36-5	34.0654	0.0227	10	573	2	35.90	35.90	39.35
3136	25009	C12H26	4-methyl-4-propyloctane	17312-41-3	33.9278	0.0225	10	606	2	35.73	35.73	39.19
3137	25010	C12H26	4-methyl-5-propyloctane	62183-85-1	34.0545	0.0223	10	525	2	35.84	35.84	39.15
3138	25011	C12H26	2-methyl-3-isopropyloctane	13287-19-9	33.9078	0.0224	10	575	2	35.70	35.70	39.10
3139	25012	C12H26	2-methyl-4-isopropyloctane	62183-86-2	33.3180	0.0228	10	745	2	35.12	35.12	38.75
3140	25013	C12H26	2-methyl-5-isopropyloctane	62183-87-3	33.3803	0.0227	10	714	2	35.18	35.18	38.76
3141	25014	C12H26	3-methyl-4-isopropyloctane	62183-88-4	34.1005	0.0225	10	600	2	35.92	35.92	39.39
3142	25015	C12H26	3-methyl-5-isopropyloctane	62183-89-5	33.6781	0.0227	10	683	2	35.49	35.49	39.06
3143	25016	C12H26	4-methyl-4-isopropyloctane	62183-90-8	34.4520	0.0225	10	583	2	36.29	36.29	39.77
3144	25017	C12H26	4-methyl-5-isopropyloctane	62183-91-9	34.0382	0.0226	10	627	2	35.86	35.86	39.38
3145	25018	C12H26	3,3-diethyloctane	17302-40-8	34.6409	0.0217	10	393	2	36.42	36.42	39.45
3146	25019	C12H26	3,4-diethyloctane	62183-92-0	34.4829	0.0221	10	460	2	36.28	36.28	39.48
3147	25020	C12H26	3,5-diethyloctane	62183-93-1	34.1163	0.0222	10	503	2	35.90	35.90	39.16
3148	25021	C12H26	3,6-diethyloctane	62183-94-2	34.4304	0.0217	10	404	2	36.20	36.20	39.23
3149	25022	C12H26	4,4-diethyloctane	17312-42-4	34.5418	0.0223	10	488	2	36.36	36.36	39.64
3150	25023	C12H26	4,5-diethyloctane	1636-41-5	34.3680	0.0222	10	503	2	36.17	36.17	39.46
3151	25024	C12H26	2,2-dimethyl-3-ethyloctane	62183-95-3	33.8324	0.0226	10	635	2	35.64	35.64	39.14
3152	25025	C12H26	2,2-dimethyl-4-ethyloctane	62183-96-4	32.8350	0.0232	10	932	2	34.64	34.64	38.48
3153	25026	C12H26	2,2-dimethyl-5-ethyloctane	62183-97-5	32.9506	0.0230	10	856	2	34.74	34.74	38.49
3154	25027	C12H26	2,2-dimethyl-6-ethyloctane	62183-98-6	33.1826	0.0227	10	721	2	34.96	34.96	38.53
3155	25028	C12H26	2,3-dimethyl-3-ethyloctane	62183-99-7	34.6838	0.0222	10	492	2	36.51	36.51	39.81
3156	25029	C12H26	2,3-dimethyl-4-ethyloctane	62184-00-3	34.2110	0.0224	10	550	2	36.02	36.02	39.40
3157	25030	C12H26	2,3-dimethyl-5-ethyloctane	62184-01-4	33.7965	0.0226	10	627	2	35.60	35.60	39.08
3158	25031	C12H26	2,3-dimethyl-6-ethyloctane	62184-02-5	34.0768	0.0222	10	505	2	35.86	35.86	39.12
3159	25032	C12H26	2,4-dimethyl-3-ethyloctane	62184-03-6	34.2159	0.0224	10	550	2	36.03	36.03	39.41
3160	25033	C12H26	2,4-dimethyl-4-ethyloctane	62184-04-7	33.5510	0.0230	10	786	2	35.37	35.37	39.11
3161	25034	C12H26	2,4-dimethyl-5-ethyloctane	62184-05-8	33.7306	0.0226	10	654	2	35.54	35.54	39.07
3162	25035	C12H26	2,4-dimethyl-6-ethyloctane	62184-06-9	33.4372	0.0227	10	683	2	35.23	35.23	38.77
3163	25036	C12H26	2,5-dimethyl-3-ethyloctane	62184-07-0	33.7965	0.0226	10	627	2	35.60	35.60	39.08
3164	25037	C12H26	2,5-dimethyl-4-ethyloctane	62184-08-1	33.6781	0.0227	10	683	2	35.49	35.49	39.06
3165	25038	C12H26	2,5-dimethyl-5-ethyloctane	62184-09-2	33.6618	0.0228	10	721	2	35.48	35.48	39.12
3166	25039	C12H26	2,5-dimethyl-6-ethyloctane	62183-50-0	33.9592	0.0223	10	550	2	35.75	35.75	39.10
3167	25040	C12H26	2,6-dimethyl-3-ethyloctane	62183-51-1	33.9641	0.0223	10	550	2	35.76	35.76	39.10
3168	25041	C12H26	2,6-dimethyl-4-ethyloctane	62183-52-2	33.3216	0.0228	10	745	2	35.12	35.12	38.75
3169	25042	C12H26	2,6-dimethyl-5-ethyloctane	62183-53-3	33.8497	0.0225	10	600	2	35.65	35.65	39.09
3170	25043	C12H26	2,6-dimethyl-6-ethyloctane	62183-54-4	33.9462	0.0224	10	583	2	35.75	35.75	39.16
3171	25044	C12H26	2,7-dimethyl-3-ethyloctane	62183-55-5	33.6649	0.0223	10	575	2	35.44	35.44	38.80
3172	25045	C12H26	2,7-dimethyl-4-ethyloctane	62183-56-6	33.0744	0.0228	10	745	2	34.85	34.85	38.45
3173	25046	C12H26	3,3-dimethyl-4-ethyloctane	62183-57-7	34.4557	0.0225	10	583	2	36.29	36.29	39.78
3174	25047	C12H26	3,3-dimethyl-5-ethyloctane	62183-58-8	33.5510	0.0230	10	786	2	35.37	35.37	39.11
3175	25048	C12H26	3,3-dimethyl-6-ethyloctane	62183-59-9	33.8324	0.0226	10	635	2	35.64	35.64	39.14
3176	25049	C12H26	3,4-dimethyl-3-ethyloctane	62212-28-6	34.8580	0.0224	10	535	2	36.71	36.71	40.14
3177	25050	C12H26	3,4-dimethyl-4-ethyloctane	62183-60-2	34.8883	0.0224	10	513	2	36.73	36.73	40.11
3178	25051	C12H26	3,4-dimethyl-5-ethyloctane	62183-61-3	34.5227	0.0224	10	527	2	36.35	36.35	39.72
3179	25052	C12H26	3,4-dimethyl-6-ethyloctane	62183-62-4	34.2717	0.0223	10	527	2	36.08	36.08	39.42
3180	25053	C12H26	3,5-dimethyl-3-ethyloctane	62183-63-5	34.0825	0.0226	10	635	2	35.91	35.91	39.45
3181	25054	C12H26	3,5-dimethyl-4-ethyloctane	62183-64-6	34.4701	0.0224	10	550	2	36.30	36.30	39.72
3182	25055	C12H26	3,5-dimethyl-5-ethyloctane	62183-65-7	33.7549	0.0231	10	820	2	35.60	35.60	39.40
3183	25056	C12H26	3,5-dimethyl-6-ethyloctane	62183-66-8	34.2159	0.0224	10	550	2	36.03	36.03	39.41
3184	25057	C12H26	3,6-dimethyl-3-ethyloctane	62183-67-9	34.2536	0.0224	10	558	2	36.07	36.07	39.47
3185	25058	C12H26	3,6-dimethyl-4-ethyloctane	62183-68-0	34.1530	0.0225	10	575	2	35.97	35.97	39.39
3186	25059	C12H26	4,4-dimethyl-3-ethyloctane	62183-69-1	32.2091	0.0204	10	583	2	33.76	33.76	36.67
3187	25060	C12H26	4,4-dimethyl-5-ethyloctane	62183-70-4	34.3361	0.0227	10	635	2	36.18	36.18	39.75
3188	25061	C12H26	4,4-dimethyl-6-ethyloctane	62183-71-5	33.5510	0.0230	10	786	2	35.37	35.37	39.11
3189	25062	C12H26	4,5-dimethyl-3-ethyloctane	62183-72-6	34.5848	0.0223	10	505	2	36.41	36.41	39.73
3190	25063	C12H26	4,5-dimethyl-4-ethyloctane	62183-73-7	34.8262	0.0224	10	535	2	36.67	36.67	40.10
3191	25064	C12H26	2,2,3,3-tetramethyloctane	62183-74-8	34.3282	0.0227	10	649	2	36.17	36.17	39.77
3192	25065	C12H26	2,2,3,4-tetramethyloctane	62183-75-9	33.8672	0.0229	10	724	2	35.70	35.70	39.37
3193	25066	C12H26	2,2,3,5-tetramethyloctane	62183-76-0	33.4487	0.0231	10	822	2	35.27	35.27	39.05
3194	25067	C12H26	2,2,3,6-tetramethyloctane	62183-77-1	33.5719	0.0229	10	755	2	35.39	35.39	39.07
3195	25068	C12H26	2,2,3,7-tetramethyloctane	62183-78-2	33.3238	0.0228	10	755	2	35.12	35.12	38.77
3196	25069	C12H26	2,2,4,4-tetramethyloctane	62183-79-3	33.3481	0.0231	10	904	2	35.17	35.17	39.03
3197	25070	C12H26	2,2,4,5-tetramethyloctane	62183-80-6	32.9202	0.0234	10	1014	2	34.74	34.74	38.71
3198	25071	C12H26	2,2,4,6-tetramethyloctane	62183-81-7	32.5733	0.0235	10	1104	2	34.38	34.38	38.40
3199	25072	C12H26	2,2,4,7-tetramethyloctane	62183-82-8	36.2316	0.0068	10	1113	2	36.80	36.80	37.99
3200	25073	C12H26	2,2,5,5-tetramethyloctane	62183-83-9	32.4441	0.0238	10	1260	2	34.27	34.27	38.44

Table 1 Adsorption Capacity of Activated Carbon - Gas

$\log_{10} Q = \log_{10} K + 1/n \log_{10} y$ (Q - g of compound/100 g of carbon, y - ppmv)

NO	ID	FORMULA	NAME	CAS No	K	1/n	ymin	ymax	code	Q @ymin	Q @10ppm	Q @ymax
3201	25074	C12H26	2,2,5,6-tetramethyloctane	62183-84-0	33.0966	0.0232	10	894	2	34.91	34.91	38.74
3202	25075	C12H26	2,2,5,7-tetramethyloctane	62199-19-3	32.3427	0.0234	10	1104	2	34.14	34.14	38.11
3203	25076	C12H26	2,2,6,6-tetramethyloctane	62199-20-6	32.5175	0.0237	10	1209	2	34.34	34.34	38.48
3204	25077	C12H26	2,2,6,7-tetramethyloctane	62199-21-7	32.9164	0.0230	10	857	2	34.71	34.71	38.45
3205	25078	C12H26	2,2,7,7-tetramethyloctane	1071-31-4	32.0535	0.0235	10	1209	2	33.84	33.84	37.88
3206	25079	C12H26	2,3,3,4-tetramethyloctane	62199-22-8	34.7252	0.0225	10	563	2	36.57	36.57	40.05
3207	25080	C12H26	2,3,3,5-tetramethyloctane	62199-23-9	33.8066	0.0230	10	755	2	35.64	35.64	39.36
3208	25081	C12H26	2,3,3,6-tetramethyloctane	62199-24-0	33.9289	0.0228	10	694	2	35.76	35.76	39.39
3209	25082	C12H26	2,3,3,7-tetramethyloctane	62199-25-1	33.6762	0.0227	10	694	2	35.49	35.49	39.08
3210	25083	C12H26	2,3,4,4-tetramethyloctane	62199-26-2	34.1778	0.0229	10	694	2	36.03	36.03	39.69
3211	25084	C12H26	2,3,4,5-tetramethyloctane	62199-27-3	34.3061	0.0226	10	603	2	36.14	36.14	39.65
3212	25085	C12H26	2,3,4,6-tetramethyloctane	62199-28-4	33.9415	0.0227	10	656	2	35.76	35.76	39.33
3213	25086	C12H26	2,3,4,7-tetramethyloctane	62199-29-5	33.6400	0.0227	10	685	2	35.45	35.45	39.02
3214	25087	C12H26	2,3,5,5-tetramethyloctane	62199-30-8	33.5700	0.0228	10	724	2	35.38	35.38	39.01
3215	25088	C12H26	2,3,5,6-tetramethyloctane	62199-31-9	33.9464	0.0227	10	656	2	35.77	35.77	39.33
3216	25089	C12H26	2,3,5,7-tetramethyloctane	62199-32-0	33.1676	0.0230	10	813	2	34.97	34.97	38.69
3217	25090	C12H26	2,3,6,6-tetramethyloctane	62199-33-1	33.5140	0.0230	10	788	2	35.34	35.34	39.07
3218	25091	C12H26	2,3,6,7-tetramethyloctane	52670-34-5	33.7618	0.0226	10	629	2	35.56	35.56	39.04
3219	25092	C12H26	2,4,4,5-tetramethyloctane	62199-34-2	33.6416	0.0232	10	857	2	35.49	35.49	39.35
3220	25093	C12H26	2,4,4,6-tetramethyloctane	62199-35-3	32.8072	0.0236	10	1104	2	34.64	34.64	38.70
3221	25094	C12H26	2,4,4,7-tetramethyloctane	35866-96-7	32.5436	0.0236	10	1151	2	34.36	34.36	38.42
3222	25095	C12H26	2,4,5,5-tetramethyloctane	62199-36-4	33.6944	0.0231	10	822	2	35.54	35.54	39.35
3223	25096	C12H26	2,4,5,6-tetramethyloctane	62199-37-5	33.8844	0.0228	10	685	2	35.71	35.71	39.32
3224	25097	C12H26	2,4,5,7-tetramethyloctane	2217-17-6	32.9673	0.0233	10	952	2	34.79	34.79	38.68
3225	25098	C12H26	2,4,6,6-tetramethyloctane	62199-38-6	32.9237	0.0234	10	1014	2	34.75	34.75	38.71
3226	25099	C12H26	2,5,5,6-tetramethyloctane	62199-39-7	33.8138	0.0230	10	755	2	35.65	35.65	39.37
3227	25100	C12H26	2,5,6,6-tetramethyloctane	62199-40-0	33.9253	0.0228	10	694	2	35.75	35.75	39.38
3228	25101	C12H26	3,3,4,4-tetramethyloctane	62199-41-1	34.9669	0.0227	10	597	2	36.84	36.84	40.43
3229	25102	C12H26	3,3,4,5-tetramethyloctane	62199-42-2	34.5526	0.0228	10	638	2	36.41	36.41	40.03
3230	25103	C12H26	3,3,4,6-tetramethyloctane	62199-43-3	34.1814	0.0229	10	694	2	36.03	36.03	39.70
3231	25104	C12H26	3,3,5,5-tetramethyloctane	62199-44-4	34.1224	0.0228	10	735	2	35.96	35.96	39.67
3232	25105	C12H26	3,3,5,6-tetramethyloctane	62199-45-5	33.7482	0.0230	10	788	2	35.59	35.59	39.35
3233	25106	C12H26	3,3,6,6-tetramethyloctane	62199-46-6	33.1831	0.0234	10	995	2	35.02	35.02	39.01
3234	25107	C12H26	3,4,4,5-tetramethyloctane	62199-47-7	34.9849	0.0226	10	563	2	36.85	36.85	40.37
3235	25108	C12H26	3,4,4,6-tetramethyloctane	62185-19-7	34.1193	0.0229	10	724	2	35.97	35.97	39.68
3236	25109	C12H26	3,4,5,5-tetramethyloctane	62185-20-0	34.4916	0.0228	10	666	2	36.35	36.35	40.01
3237	25110	C12H26	3,4,5,6-tetramethyloctane	62185-21-1	34.6812	0.0225	10	553	2	36.53	36.53	39.98
3238	25111	C12H26	4,4,5,5-tetramethyloctane	62185-22-2	34.8480	0.0228	10	649	2	36.73	36.73	40.40
3239	25112	C12H26	2-methyl-4-tert-butylheptane	62185-23-3	33.0562	0.0236	10	1104	2	34.91	34.91	39.01
3240	25113	C12H26	3-methyl-4-tert-butylheptane	62185-24-4	33.8107	0.0233	10	894	2	35.68	35.68	39.62
3241	25114	C12H26	4-methyl-4-tert-butylheptane	62185-25-5	34.7245	0.0230	10	705	2	36.61	36.61	40.38
3242	25115	C12H26	3-ethyl-4-propylheptane	62185-26-6	34.3340	0.0223	10	525	2	36.15	36.15	39.49
3243	25116	C12H26	4-ethyl-4-propylheptane	17312-43-5	34.4308	0.0225	10	556	2	36.26	36.26	39.70
3244	25117	C12H26	3-ethyl-4-isopropylheptane	62185-27-7	34.3523	0.0226	10	600	2	36.19	36.19	39.70
3245	25118	C12H26	4-ethyl-4-isopropylheptane	62185-28-8	35.0865	0.0225	10	535	2	36.95	36.95	40.41
3246	25119	C12H26	2,2-dimethyl-4-propylheptane	------	33.1603	0.0211	10	1059	2	34.81	34.81	38.42
3247	25120	C12H26	2,3-dimethyl-4-propylheptane	62185-30-2	34.0492	0.0226	10	627	2	35.87	35.87	39.39
3248	25121	C12H26	2,4-dimethyl-4-propylheptane	62185-31-3	33.4400	0.0232	10	856	2	35.27	35.27	39.10
3249	25122	C12H26	2,5-dimethyl-4-propylheptane	62185-32-4	33.5579	0.0229	10	745	2	35.37	35.37	39.04
3250	25123	C12H26	2,6-dimethyl-4-propylheptane	62185-33-5	32.6189	0.0232	10	942	2	34.41	34.41	38.23
3251	25124	C12H26	3,3-dimethyl-4-propylheptane	62185-34-6	34.3349	0.0227	10	635	2	36.18	36.18	39.75
3252	25125	C12H26	3,4-dimethyl-4-propylheptane	62185-35-7	34.7298	0.0226	10	583	2	36.59	36.59	40.11
3253	25126	C12H26	3,5-dimethyl-4-propylheptane	62185-36-8	34.4041	0.0225	10	575	2	36.24	36.24	39.70
3254	25127	C12H26	2,2-dimethyl-3-isopropylheptane	62185-37-9	33.5614	0.0233	10	894	2	35.41	35.41	39.31
3255	25128	C12H26	2,2-dimethyl-4-isopropylheptane	62185-38-0	32.6226	0.0238	10	1252	2	34.46	34.46	38.66
3256	25129	C12H26	2,3-dimethyl-3-isopropylheptane	62185-39-1	34.9837	0.0226	10	563	2	36.85	36.85	40.37
3257	25130	C12H26	2,3-dimethyl-4-isopropylheptane	62185-40-4	34.0174	0.0230	10	746	2	35.87	35.87	39.61
3258	25131	C12H26	2,4-dimethyl-3-isopropylheptane	62185-41-5	34.0174	0.0230	10	746	2	35.87	35.87	39.61
3259	25132	C12H26	2,4-dimethyl-4-isopropylheptane	62185-42-6	33.9484	0.0232	10	822	2	35.81	35.81	39.67
3260	25133	C12H26	2,5-dimethyl-3-isopropylheptane	62185-43-7	33.6630	0.0231	10	813	2	35.50	35.50	39.30
3261	25134	C12H26	2,5-dimethyl-4-isopropylheptane	62185-44-8	33.5749	0.0232	10	848	2	35.42	35.42	39.25
3262	25135	C12H26	2,6-dimethyl-3-isopropylheptane	62185-45-9	33.3782	0.0229	10	785	2	35.19	35.19	38.89
3263	25136	C12H26	2,6-dimethyl-4-isopropylheptane	35866-89-8	32.7721	0.0236	10	1096	2	34.60	34.60	38.65
3264	25137	C12H26	3,3-dimethyl-4-isopropylheptane	62185-46-0	34.3010	0.0231	10	755	2	36.17	36.17	39.97
3265	25138	C12H26	3,4-dimethyl-4-isopropylheptane	62185-47-1	35.2421	0.0226	10	563	2	37.13	37.13	40.68
3266	25139	C12H26	3,5-dimethyl-4-isopropylheptane	62198-89-4	34.3716	0.0229	10	685	2	36.23	36.23	39.91
3267	25140	C12H26	2-methyl-3,3-diethylheptane	62198-90-7	35.2749	0.0223	10	471	2	37.13	37.13	40.46
3268	25141	C12H26	2-methyl-3,4-diethylheptane	62198-91-8	34.3536	0.0226	10	600	2	36.19	36.19	39.70
3269	25142	C12H26	2-methyl-3,5-diethylheptane	62198-92-9	34.1005	0.0225	10	600	2	35.92	35.92	39.39
3270	25143	C12H26	2-methyl-4,4-diethylheptane	62198-93-0	34.1976	0.0228	10	691	2	36.04	36.04	39.71
3271	25144	C12H26	2-methyl-4,5-diethylheptane	62198-94-1	33.9985	0.0227	10	654	2	35.82	35.82	39.40
3272	25145	C12H26	2-methyl-5,5-diethylheptane	62198-95-2	34.4740	0.0226	10	583	2	36.31	36.31	39.80
3273	25146	C12H26	3-methyl-3,4-diethylheptane	62198-96-3	35.1380	0.0224	10	513	2	37.00	37.00	40.41
3274	25147	C12H26	3-methyl-3,5-diethylheptane	62198-97-4	34.3861	0.0226	10	608	2	36.22	36.22	39.75
3275	25148	C12H26	3-methyl-4,4-diethylheptane	62198-98-5	35.5840	0.0222	10	451	2	37.45	37.45	40.77
3276	25149	C12H26	3-methyl-4,5-diethylheptane	62198-99-6	34.7982	0.0224	10	527	2	36.64	36.64	40.05
3277	25150	C12H26	3-methyl-5,5-diethylheptane	62199-00-2	34.7798	0.0225	10	558	2	36.63	36.63	40.11
3278	25151	C12H26	4-methyl-3,3-diethylheptane	62199-01-3	35.6740	0.0222	10	433	2	37.54	37.54	40.82
3279	25152	C12H26	4-methyl-3,4-diethylheptane	62199-02-4	35.1368	0.0224	10	513	2	37.00	37.00	40.41
3280	25153	C12H26	4-methyl-3,5-diethylheptane	62199-03-5	34.8987	0.0223	10	484	2	36.74	36.74	40.05

Table 1 Adsorption Capacity of Activated Carbon - Gas

$\log_{10} Q = \log_{10} K + 1/n \log_{10} y$ (Q - g of compound/100 g of carbon, y - ppmv)

NO	ID	FORMULA	NAME	CAS No	K	1/n	ymin	ymax	code	Q @ymin	Q @10ppm	Q @ymax
3281	25154	C12H26	2,2,3-trimethyl-3-ethylheptane	62199-04-6	34.8637	0.0229	10	649	2	36.75	36.75	40.43
3282	25155	C12H26	2,2,3-trimethyl-4-ethylheptane	62199-05-7	33.9992	0.0231	10	788	2	35.86	35.86	39.66
3283	25156	C12H26	2,2,3-trimethyl-5-ethylheptane	62199-06-8	33.6993	0.0231	10	822	2	35.54	35.54	39.36
3284	25157	C12H26	2,2,4-trimethyl-3-ethylheptane	62199-07-9	33.9484	0.0232	10	822	2	35.81	35.81	39.67
3285	25158	C12H26	2,2,4-trimethyl-4-ethylheptane	62199-08-0	33.9627	0.0231	10	832	2	35.82	35.82	39.66
3286	25159	C12H26	2,2,4-trimethyl-5-ethylheptane	62199-09-1	33.1599	0.0235	10	1014	2	35.00	35.00	39.01
3287	25160	C12H26	2,2,5-trimethyl-3-ethylheptane	62199-10-4	33.5138	0.0237	10	894	2	35.39	35.39	39.36
3288	25161	C12H26	2,2,5-trimethyl-4-ethylheptane	62199-11-5	33.1075	0.0236	10	1058	2	34.95	34.95	39.01
3289	25162	C12H26	2,2,5-trimethyl-5-ethylheptane	62199-12-6	33.1948	0.0235	10	1024	2	35.04	35.04	39.06
3290	25163	C12H26	2,2,6-trimethyl-3-ethylheptane	62199-13-7	33.3471	0.0232	10	894	2	35.18	35.18	39.05
3291	25164	C12H26	2,2,6-trimethyl-4-ethylheptane	62199-14-8	32.2778	0.0239	10	1363	2	34.10	34.10	38.35
3292	25165	C12H26	2,2,6-trimethyl-5-ethylheptane	62199-15-9	32.9132	0.0234	10	1014	2	34.73	34.73	38.70
3293	25166	C12H26	2,3,3-trimethyl-4-ethylheptane	62199-16-0	34.8833	0.0228	10	612	2	36.76	36.76	40.37
3294	25167	C12H26	2,3,3-trimethyl-5-ethylheptane	62199-17-1	34.0509	0.0230	10	755	2	35.90	35.90	39.66
3295	25168	C12H26	2,3,4-trimethyl-3-ethylheptane	62199-18-2	35.3792	0.0225	10	517	2	37.26	37.26	40.72
3296	25169	C12H26	2,3,4-trimethyl-4-ethylheptane	62198-55-4	34.7962	0.0228	10	638	2	36.67	36.67	40.32
3297	25170	C12H26	2,3,4-trimethyl-5-ethylheptane	62198-56-5	34.5627	0.0227	10	603	2	36.41	36.41	39.96
3298	25171	C12H26	2,3,5-trimethyl-3-ethylheptane	62198-57-6	34.5556	0.0227	10	652	2	36.41	36.41	40.03
3299	25172	C12H26	2,3,5-trimethyl-4-ethylheptane	62198-58-7	34.5097	0.0228	10	629	2	36.37	36.37	39.96
3300	25173	C12H26	2,3,5-trimethyl-5-ethylheptane	62198-59-8	34.0509	0.0230	10	755	2	35.90	35.90	39.66
3301	25174	C12H26	2,3,6-trimethyl-3-ethylheptane	62198-60-1	34.1887	0.0229	10	694	2	36.04	36.04	39.71
3302	25175	C12H26	2,3,6-trimethyl-4-ethylheptane	62198-61-2	33.6630	0.0231	10	813	2	35.50	35.50	39.30
3303	25176	C12H26	2,3,6-trimethyl-5-ethylheptane	62198-62-3	33.8149	0.0229	10	715	2	35.64	35.64	39.30
3304	25177	C12H26	2,4,4-trimethyl-3-ethylheptane	62198-63-4	34.0871	0.0228	10	755	2	35.93	35.93	39.66
3305	25178	C12H26	2,4,4-trimethyl-5-ethylheptane	62198-64-5	33.9484	0.0232	10	822	2	35.81	35.81	39.67
3306	25179	C12H26	2,4,5-trimethyl-3-ethylheptane	62198-65-6	34.5109	0.0228	10	629	2	36.37	36.37	39.96
3307	25180	C12H26	2,4,5-trimethyl-4-ethylheptane	62198-66-7	34.3891	0.0230	10	724	2	36.26	36.26	40.02
3308	25181	C12H26	2,4,5-trimethyl-5-ethylheptane	62198-67-8	34.4916	0.0228	10	666	2	36.35	36.35	40.01
3309	25182	C12H26	2,4,6-trimethyl-3-ethylheptane	62198-68-9	33.7139	0.0230	10	779	2	35.55	35.55	39.30
3310	25183	C12H26	2,4,6-trimethyl-4-ethylheptane	62198-69-0	33.0562	0.0236	10	1104	2	34.91	34.91	39.01
3311	25184	C12H26	2,5,5-trimethyl-3-ethylheptane	62198-70-3	33.5614	0.0233	10	894	2	35.41	35.41	39.31
3312	25185	C12H26	2,5,5-trimethyl-4-ethylheptane	62198-71-4	33.9484	0.0232	10	822	2	35.81	35.81	39.67
3313	25186	C12H26	3,3,4-trimethyl-4-ethylheptane	62198-72-5	35.6621	0.0226	10	527	2	37.57	37.57	41.08
3314	25187	C12H26	3,3,4-trimethyl-5-ethylheptane	62198-73-6	34.8833	0.0228	10	612	2	36.76	36.76	40.37
3315	25188	C12H26	3,3,5-trimethyl-4-ethylheptane	62198-74-7	34.7457	0.0229	10	666	2	36.63	36.63	40.32
3316	25189	C12H26	3,3,5-trimethyl-5-ethylheptane	62198-75-8	34.9413	0.0226	10	597	2	36.81	36.81	40.37
3317	25190	C12H26	3,4,4-trimethyl-3-ethylheptane	62198-76-9	35.7003	0.0226	10	527	2	37.61	37.61	41.13
3318	25191	C12H26	3,4,4-trimethyl-5-ethylheptane	62198-77-0	35.3290	0.0226	10	539	2	37.22	37.22	40.72
3319	25192	C12H26	3,4,5-trimethyl-3-ethylheptane	62198-78-1	35.3302	0.0226	10	539	2	37.22	37.22	40.72
3320	25193	C12H26	3,4,5-trimethyl-4-ethylheptane	62198-79-2	35.7391	0.0224	10	476	2	37.63	37.63	41.03
3321	25194	C12H26	2,2,3,3,4-pentamethylheptane	62198-80-5	34.9642	0.0231	10	710	2	36.87	36.87	40.69
3322	25195	C12H26	2,2,3,3,5-pentamethylheptane	62198-81-6	34.0816	0.0234	10	908	2	35.97	35.97	39.98
3323	25196	C12H26	2,2,3,3,6-pentamethylheptane	62198-82-7	33.7824	0.0235	10	946	2	35.66	35.66	39.67
3324	25197	C12H26	2,2,3,4,4-pentamethylheptane	62198-83-8	35.0447	0.0228	10	655	2	36.93	36.93	40.63
3325	25198	C12H26	2,2,3,4,5-pentamethylheptane	62198-84-9	34.1518	0.0233	10	825	2	36.03	36.03	39.92
3326	25199	C12H26	2,2,3,4,6-pentamethylheptane	62198-85-0	33.3138	0.0236	10	1058	2	35.17	35.17	39.27
3327	25200	C12H26	2,2,3,5,5-pentamethylheptane	62198-86-1	33.1951	0.0239	10	1210	2	35.07	35.07	39.33
3328	25201	C12H26	2,2,3,5,6-pentamethylheptane	62198-87-2	33.3479	0.0236	10	1024	2	35.21	35.21	39.26
3329	25202	C12H26	2,2,3,6,6-pentamethylheptane	62198-88-3	32.5708	0.0239	10	1313	2	34.41	34.41	38.66
3330	25203	C12H26	2,2,4,4,5-pentamethylheptane	62199-61-5	34.1642	0.0231	10	837	2	36.03	36.03	39.92
3331	25204	C12H26	2,2,4,4,6-pentamethylheptane	62199-62-6	32.8491	0.0238	10	1281	2	34.70	34.70	38.94
3332	25205	C12H26	2,2,4,5,5-pentamethylheptane	62199-63-7	33.1098	0.0239	10	1260	2	34.98	34.98	39.28
3333	25206	C12H26	2,2,4,5,6-pentamethylheptane	62199-64-8	32.8784	0.0238	10	1199	2	34.73	34.73	38.91
3334	25207	C12H26	2,2,4,6,6-pentamethylheptane	13475-82-6	31.8325	0.0243	10	1755	2	33.67	33.67	38.18
3335	25208	C12H26	2,2,5,5,6-pentamethylheptane	62199-65-9	32.9002	0.0239	10	1260	2	34.76	34.76	39.02
3336	25209	C12H26	2,3,3,4,4-pentamethylheptane	62199-66-0	35.4063	0.0229	10	628	2	37.33	37.33	41.04
3337	25210	C12H26	2,3,3,4,5-pentamethylheptane	62199-67-1	35.0370	0.0229	10	643	2	36.93	36.93	40.63
3338	25211	C12H26	2,3,3,4,6-pentamethylheptane	62199-68-2	34.2037	0.0232	10	791	2	36.08	36.08	39.92
3339	25212	C12H26	2,3,3,5,5-pentamethylheptane	62199-69-3	34.6519	0.0229	10	710	2	36.53	36.53	40.27
3340	25213	C12H26	2,3,3,5,6-pentamethylheptane	52670-35-6	33.8006	0.0234	10	896	2	35.67	35.67	39.62
3341	25214	C12H26	2,3,4,4,5-pentamethylheptane	62199-70-6	35.0370	0.0229	10	643	2	36.93	36.93	40.63
3342	25215	C12H26	2,3,4,4,6-pentamethylheptane	62199-71-7	33.6624	0.0235	10	974	2	35.53	35.53	39.57
3343	25216	C12H26	2,3,4,5,5-pentamethylheptane	62199-72-8	34.5930	0.0231	10	728	2	36.48	36.48	40.28
3344	25217	C12H26	2,3,4,5,6-pentamethylheptane	27574-98-7	34.2749	0.0230	10	718	2	36.14	36.14	39.86
3345	25218	C12H26	2,4,4,5,5-pentamethylheptane	62199-73-9	34.4704	0.0234	10	837	2	36.37	36.37	40.34
3346	25219	C12H26	3,3,4,4,5-pentamethylheptane	62199-74-0	35.8523	0.0227	10	556	2	37.78	37.78	41.40
3347	25220	C12H26	3,3,4,5,5-pentamethylheptane	62199-75-1	35.9793	0.0224	10	492	2	37.88	37.88	41.34
3348	25221	C12H26	2,2-dimethyl-3-tert-butylhexane	62199-76-2	34.4169	0.0232	10	837	2	36.31	36.31	40.23
3349	25222	C12H26	2-methyl-3-ethyl-3-isopropylhexane	62199-77-3	35.6018	0.0226	10	535	2	37.50	37.50	41.03
3350	25223	C12H26	2-methyl-4-ethyl-3-isopropylhexane	62199-78-4	34.2744	0.0230	10	745	2	36.14	36.14	39.92
3351	25224	C12H26	2,2,3-trimethyl-3-isopropylhexane	62199-79-5	35.2643	0.0231	10	682	2	37.19	37.19	40.99
3352	25225	C12H26	2,2,4-trimethyl-3-isopropylhexane	62199-80-8	33.8559	0.0237	10	1015	2	35.75	35.75	39.88
3353	25226	C12H26	2,2,5-trimethyl-3-isopropylhexane	62199-81-9	32.8676	0.0243	10	1477	2	34.76	34.76	39.23
3354	25227	C12H26	2,2,5-trimethyl-4-isopropylhexane	62199-82-0	32.6216	0.0242	10	1477	2	34.49	34.49	38.92
3355	25228	C12H26	2,3,4-trimethyl-3-isopropylhexane	62199-83-1	35.8266	0.0227	10	545	2	37.75	37.75	41.34
3356	25229	C12H26	2,3,5-trimethyl-3-isopropylhexane	62199-84-2	34.4073	0.0233	10	825	2	36.30	36.30	40.23
3357	25230	C12H26	2,3,5-trimethyl-4-isopropylhexane	62199-85-3	33.9848	0.0234	10	886	2	35.86	35.86	39.82
3358	25231	C12H26	2,4,4-trimethyl-3-isopropylhexane	62199-86-4	34.3070	0.0235	10	896	2	36.21	36.21	40.24
3359	25232	C12H26	3,3,4-triethylhexane	62199-87-5	35.9851	0.0221	10	415	2	37.87	37.87	41.12
3360	25233	C12H26	2,2-dimethyl,3,3-diethylhexane	62199-88-6	21.5831	0.1038	10	649	2	27.41	27.41	42.27

Table 1 Adsorption Capacity of Activated Carbon - Gas

$\log_{10} Q = \log_{10} K + 1/n \log_{10} y$ (Q - g of compound/100 g of carbon, y - ppmv)

NO	ID	FORMULA	NAME	CAS No	K	1/n	ymin	ymax	code	Q @ymin	Q @10ppm	Q @ymax
3361	25234	C12H26	2,2-dimethyl-3,4-diethylhexane	62199-89-7	34.0938	0.0234	10	894	2	35.98	35.98	39.98
3362	25235	C12H26	2,2-dimethyl-4,4-diethylhexane	62184-89-8	34.8023	0.0227	10	649	2	36.67	36.67	40.32
3363	25236	C12H26	2,3-dimethyl-3,4-diethylhexane	62184-90-1	37.3301	0.0150	10	496	2	38.64	38.64	40.97
3364	25237	C12H26	2,3-dimethyl-4,4-diethylhexane	62184-91-2	35.6912	0.0225	10	496	2	37.59	37.59	41.03
3365	25238	C12H26	2,4-dimethyl-3,3-diethylhexane	62184-92-3	35.7862	0.0229	10	587	2	37.72	37.72	41.40
3366	25239	C12H26	2,4-dimethyl-3,4-diethylhexane	62184-93-4	35.1878	0.0227	10	587	2	37.08	37.08	40.68
3367	25240	C12H26	2,5-dimethyl-3,3-diethylhexane	62184-94-5	34.3944	0.0235	10	894	2	36.30	36.30	40.34
3368	25241	C12H26	2,5-dimethyl-3,4-diethylhexane	62184-95-6	34.1720	0.0232	10	813	2	36.05	36.05	39.92
3369	25242	C12H26	3,3-dimethyl-4,4-diethylhexane	62184-96-7	36.5527	0.0223	10	429	2	38.48	38.48	41.84
3370	25243	C12H26	3,4-dimethyl-3,4-diethylhexane	62184-97-8	36.5527	0.0223	10	429	2	38.48	38.48	41.84
3371	25244	C12H26	2,2,3,3-tetramethyl-4-ethylhexane	62184-98-9	35.2643	0.0231	10	682	2	37.19	37.19	40.99
3372	25245	C12H26	2,2,3,4-tetramethyl-3-ethylhexane	62184-99-0	35.7163	0.0229	10	603	2	37.65	37.65	41.35
3373	25246	C12H26	2,2,3,4-tetramethyl-4-ethylhexane	62185-00-6	35.9347	0.0225	10	512	2	37.84	37.84	41.34
3374	25247	C12H26	2,2,3,5-tetramethyl-3-ethylhexane	62185-01-7	34.3263	0.0235	10	908	2	36.24	36.24	40.29
3375	25248	C12H26	2,2,3,5-tetramethyl-4-ethylhexane	62185-02-8	33.9565	0.0235	10	934	2	35.84	35.84	39.88
3376	25249	C12H26	2,2,4,4-tetramethyl-3-ethylhexane	62185-03-9	35.2044	0.0230	10	710	2	37.12	37.12	40.94
3377	25250	C12H26	2,2,4,5-tetramethyl-3-ethylhexane	62185-04-0	33.9064	0.0236	10	974	2	35.80	35.80	39.88
3378	25251	C12H26	2,2,4,5-tetramethyl-4-ethylhexane	62185-05-1	34.4656	0.0231	10	803	2	36.35	36.35	40.23
3379	25252	C12H26	2,2,5,5-tetramethyl-3-ethylhexane	62185-06-2	32.5300	0.0243	10	1613	2	34.40	34.40	38.93
3380	25253	C12H26	2,3,3,4-tetramethyl-4-ethylhexane	62185-07-3	36.2431	0.0227	10	512	2	38.19	38.19	41.75
3381	25254	C12H26	2,3,3,5-tetramethyl-4-ethylhexane	62185-08-4	34.8459	0.0231	10	728	2	36.75	36.75	40.59
3382	25255	C12H26	2,3,4,4-tetramethyl-3-ethylhexane	62185-09-5	36.2431	0.0227	10	512	2	38.19	38.19	41.75
3383	25256	C12H26	2,3,4,5-tetramethyl-3-ethylhexane	62185-10-8	35.3484	0.0228	10	617	2	37.26	37.26	40.94
3384	25257	C12H26	2,2,3,3,4,4-hexamethylhexane	62185-11-9	37.0134	0.0225	10	454	2	38.98	38.98	42.46
3385	25258	C12H26	2,2,3,3,4,5-hexamethylhexane	62185-12-0	34.9709	0.0234	10	809	2	36.91	36.91	40.90
3386	25259	C12H26	2,2,3,3,5,5-hexamethylhexane	60302-24-1	34.1118	0.0236	10	1046	2	36.02	36.02	40.19
3387	25260	C12H26	2,2,3,4,4,5-hexamethylhexane	62185-13-1	35.6415	0.0228	10	610	2	37.56	37.56	41.24
3388	25261	C12H26	2,2,3,4,5,5-hexamethylhexane	62185-14-2	33.9030	0.0239	10	1164	2	35.82	35.82	40.15
3389	25262	C12H26	2,3,3,4,4,5-hexamethylhexane	52670-36-7	35.9579	0.0229	10	610	2	37.91	37.91	41.66
3390	25263	C12H26	2,4-dimethyl-3-ethyl-3-isopropylpentane	62185-16-4	36.1949	0.0226	10	501	2	38.13	38.13	41.65
3391	25264	C12H26	2,2,4-trimethyl-3,3-diethylpentane	62185-15-3	36.1080	0.0228	10	556	2	38.05	38.05	41.70
3392	25265	C12H26	2,2,3,4-tetramethyl-3-isopropylpentane	62185-17-5	35.8107	0.0231	10	662	2	37.77	37.77	41.61
3393	25266	C12H26	2,2,4,4-tetramethyl-3-isopropylpentane	62185-18-6	34.3802	0.0236	10	990	2	36.30	36.30	40.45
3394	25267	C12H26	2,2,3,4,4-pentamethyl-3-ethylpentane	66576-21-4	37.0290	0.0227	10	492	2	39.02	39.02	42.62
3395	25268	C12H26	2-methylundecane	7045-71-8	33.9674	0.0208	10	258	2	35.63	35.63	38.12
3396	25277	C12H26O	1-dodecanol	112-53-8	41.0428	0.0200	1	4.9	2	41.04	---	42.37
3397	25278	C12H26O	2-dodecanol	10203-28-8	41.0428	0.0200	1	4.9	2	41.04	---	42.37
3398	25285	C12H26O	dihexyl ether	112-58-3	37.3543	0.0202	10	60	2	39.13	39.13	40.57
3399	25294	C12H26O3	diethylene glycol dibutyl ether	112-73-2	42.2973	0.0191	1	34	2	42.30	44.20	45.25
3400	25310	C12H26S	1-dodecanethiol	112-55-0	41.3409	0.0188	1	11	2	41.34	43.17	43.25
3401	25311	C12H26S	2-dodecanethiol	14402-50-7	40.9086	0.0191	1	13	2	40.91	42.74	42.96
3402	25312	C12H26S	dihexyl sulfide	6294-31-1	41.6261	0.0184	1	8	2	41.63	---	43.25
3403	25313	C12H26S	methyl undecyl sulfide	7289-44-3	41.6297	0.0183	1	8	2	41.63	---	43.25
3404	25314	C12H26S	ethyl decyl sulfide	19313-61-2	41.6297	0.0183	1	8	2	41.63	---	43.25
3405	25315	C12H26S	propyl nonyl sulfide	62103-66-6	41.6297	0.0183	1	8	2	41.63	---	43.25
3406	25316	C12H26S	butyl octyl sulfide	16900-07-5	41.6297	0.0183	1	8	2	41.63	---	43.25
3407	25327	C12H27BO3	tributyl borate	688-74-4	38.7732	0.0190	10	692	2	40.50	40.50	43.89
3408	25341	C12H27N	methylundecylamine	66553-53-5	38.3416	0.0209	1	15	2	38.34	40.23	40.57
3409	25342	C12H27N	ethyldecylamine	66553-52-4	38.2966	0.0187	1	16	2	38.30	39.99	40.34
3410	25343	C12H27N	dihexylamine	143-16-8	38.1699	0.0189	1	17	2	38.17	39.87	40.27
3411	25344	C12H27N	dimethyldecylamine	1120-24-7	36.1185	0.0203	10	123	2	37.85	37.85	39.83
3412	25345	C12H27N	diethyloctylamine	4088-37-3	36.1969	0.0189	10	185	2	37.80	37.80	39.94
3413	25346	C12H27N	tributylamine	102-82-9	36.2361	0.0186	10	152	2	37.82	37.82	39.78
3414	25739	C13H12	1-methylbiphenyl	------	47.5630	0.0213	10	52	2	49.95	49.95	51.74
3415	25740	C13H12	3-methylbiphenyl	643-93-6	48.5384	0.0209	1	23	2	48.54	50.93	51.83
3416	25741	C13H12	4-methylbiphenyl	644-08-6	48.5384	0.0209	1	23	2	48.54	50.93	51.83
3417	25903	C13H14	1-propylnaphthalene	2765-18-6	48.3788	0.0214	1	9.2	2	48.38	---	50.73
3418	25904	C13H14	2-propylnaphthalene	2027-19-2	47.8323	0.0210	1	8.2	2	47.83	---	49.99
3419	25905	C13H14	1-isopropylnaphthalene	6158-45-8	48.2101	0.0177	1	22	2	48.21	50.22	50.92
3420	25906	C13H14	2-isopropylnaphthalene	2027-17-0	46.7560	0.0205	1	25	2	46.76	49.01	49.94
3421	25924	C13H14	1,2,6-trimethylnaphthalene	3031-05-8	50.1479	0.0201	1	3.8	2	50.15	---	51.51
3422	25929	C13H14	1,3,7-trimethylnaphthalene	2131-38-6	48.5264	0.0204	1	18	2	48.53	50.86	51.47
3423	26158	C13H18	1-propyl-[1,2,3,4-tetrahydronaphthalene]	66324-83-2	44.4100	0.0211	10	54	2	46.62	46.62	48.31
3424	26159	C13H18	2-propyl-[1,2,3,4-tetrahydronaphthalene]	66324-84-3	43.9366	0.0222	1	43	2	43.94	46.24	47.76
3425	26283	C13H20	heptylbenzene	1078-71-3	40.5495	0.0210	1	40	2	40.55	42.56	43.82
3426	26402	C13H24	1-tridecyne	26186-02-7	36.5051	0.0206	10	100	2	38.28	38.28	40.14
3427	26403	C13H24	2-tridecyne	28467-75-6	36.6837	0.0203	10	83	2	38.44	38.44	40.13
3428	26404	C13H24	3-tridecyne	60186-78-9	36.3698	0.0208	10	114	2	38.15	38.15	40.14
3429	26405	C13H24	dicyclohexylmethane	3178-23-2	40.6592	0.0215	10	97	2	42.73	42.73	44.87
3430	26406	C13H24	1-octylcyclopentene	52315-44-3	38.1685	0.0209	10	106	2	40.05	40.05	42.08
3431	26439	C13H25Br	cis-1-bromo-1-tridecene	66324-50-3	51.6418	0.0166	1	3.9	2	51.64	---	52.82
3432	26440	C13H25Br	trans-1-bromo-1-tridecene	66324-51-4	51.6418	0.0166	1	3.9	2	51.64	---	52.82
3433	26443	C13H25Cl	cis-1-chloro-1-tridecene	66324-52-5	43.8063	0.0174	1	4.2	2	43.81	---	44.91
3434	26444	C13H25Cl	trans-1-chloro-1-tridecene	66324-53-6	43.8063	0.0174	1	4.2	2	43.81	---	44.91
3435	26446	C13H25F	cis-1-fluoro-1-tridecene	66324-54-7	39.5203	0.0188	10	72	2	41.27	41.27	42.83
3436	26447	C13H25F	trans-1-fluoro-1-tridecene	66324-55-8	39.5203	0.0188	10	72	2	41.27	41.27	42.83
3437	26449	C13H25I	cis-1-iodo-1-tridecene	66324-56-9	59.8559	0.0145	1	1.8	2	59.86	---	60.37
3438	26450	C13H25I	trans-1-iodo-1-tridecene	66324-57-0	59.8559	0.0145	1	1.8	2	59.86	---	60.37
3439	26465	C13H26	octylcyclopentane	1795-20-6	38.0549	0.0207	1	44	2	38.05	39.91	41.15
3440	26466	C13H26	heptylcyclohexane	5617-41-4	38.3651	0.0196	10	51	2	40.13	40.13	41.43

Table 1 Adsorption Capacity of Activated Carbon - Gas

$\log_{10} Q = \log_{10} K + 1/n \log_{10} y$ (Q - g of compound/100 g of carbon, y - ppmv)

NO	ID	FORMULA	NAME	CAS No	K	1/n	ymin	ymax	code	Q @ymin	Q @10ppm	Q @ymax
3441	26468	C13H26	cyclotridecane	295-02-3	40.0679	0.0209	10	103	2	42.04	42.04	44.14
3442	26469	C13H26	1-tridecene	2437-56-1	35.8797	0.0197	10	83	2	37.54	37.54	39.14
3443	26470	C13H26	2-methyl-1-dodecene	16435-49-7	35.6608	0.0203	10	124	2	37.36	37.36	39.32
3444	26474	C13H26Cl2	1,1-dichlorotridecane	821-93-2	47.4634	0.0158	1	3.9	2	47.46	---	48.49
3445	26475	C13H26F2	1,1-difluorotridecane	62127-02-0	41.0956	0.0201	10	116	2	43.05	43.05	45.22
3446	26486	C13H26O	1-tridecanal	10486-19-8	41.4167	0.0185	1	3.8	2	41.42	---	42.45
3447	26494	C13H26O2	dodecyl formate	28303-42-6	43.8797	0.0165	1	2	2	43.88	---	44.38
3448	26495	C13H26O2	undecyl acetate	1731-81-3	43.7591	0.0170	1	1.6	2	43.76	---	44.11
3449	26496	C13H26O2	decyl propanoate	5454-19-3	43.6394	0.0171	1	1.8	2	43.64	---	44.08
3450	26497	C13H26O2	nonyl butanoate	2639-64-7	41.8515	0.0180	1	17	2	41.85	43.62	44.04
3451	26502	C13H26O2	butyl nonanoate	50623-57-9	41.1078	0.0184	1	27	2	41.11	42.89	43.68
3452	26505	C13H26O2	methyl dodecanoate	111-82-0	51.3681	0.0228	1	5.4	2	51.37	---	53.38
3453	26515	C13H27Br	1-bromotridecane	765-09-3	51.1768	0.0165	1	4.9	2	51.18	---	52.53
3454	26516	C13H27Cl	1-chlorotridecane	822-13-9	42.5917	0.0180	1	9.1	2	42.59	---	44.31
3455	26517	C13H27F	1-fluorotridecane	1536-21-6	39.2551	0.0198	1	43	2	39.26	41.08	42.29
3456	26518	C13H27I	1-iodotridecane	35599-77-0	58.6180	0.0153	1	5.8	2	58.62	---	60.22
3457	26526	C13H28	tridecane	629-50-5	35.6380	0.0194	10	73	2	37.26	37.26	38.73
3458	26527	C13H28	2-methyldodecane	1560-97-0	35.3124	0.0195	10	85	2	36.94	36.94	38.52
3459	26528	C13H28	3-methyldodecane	17312-57-1	35.6416	0.0196	10	82	2	37.28	37.28	38.85
3460	26529	C13H28	2,2-dimethylundecane	17312-64-0	35.0251	0.0183	10	156	2	36.53	36.53	38.42
3461	26530	C13H28	2,3-dimethylundecane	17312-77-5	35.8571	0.0199	10	102	2	37.54	37.54	39.31
3462	26531	C13H28	2,4-dimethylundecane	17312-80-0	35.1579	0.0185	10	180	2	36.69	36.69	38.71
3463	26535	C13H28O	1-tridecanol	112-70-9	42.0410	0.0171	1	1.7	2	42.04	---	42.42
3464	26536	C13H28O	2-tridecanol	1653-31-2	42.0410	0.0171	1	1.7	2	42.04	---	42.42
3465	26549	C13H28S	1-tridecanethiol	19484-26-5	41.9195	0.0164	1	4.6	2	41.92	---	42.98
3466	26550	C13H28S	2-tridecanethiol	62155-03-7	41.9195	0.0164	1	4.6	2	41.92	---	42.98
3467	26551	C13H28S	methyl dodecyl sulfide	------	42.4103	0.0160	1	3.2	2	42.41	---	43.21
3468	26552	C13H28S	ethyl undecyl sulfide	66577-30-8	42.4103	0.0160	1	3.2	2	42.41	---	43.21
3469	26553	C13H28S	propyl decyl sulfide	66577-31-9	42.4103	0.0160	1	3.2	2	42.41	---	43.21
3470	26554	C13H28S	butyl nonyl sulfide	66577-32-0	42.4103	0.0160	1	3.2	2	42.41	---	43.21
3471	26556	C13H29N	methyldodecylamine	7311-30-0	39.1374	0.0184	1	9.5	2	39.14	---	40.79
3472	26557	C13H29N	ethylundecylamine	59570-04-6	38.9183	0.0185	1	10	2	38.92	40.61	40.61
3473	26558	C13H29N	dimethylundecylamine	17373-28-3	37.3647	0.0184	1	45	2	37.36	38.98	40.07
3474	26559	C13H29N	diethylnonylamine	45124-35-4	37.2896	0.0179	10	67	2	38.86	38.86	40.21
3475	26866	C14H12	cis-stilbene	645-49-8	49.5341	0.0201	1	10	2	49.53	51.88	51.88
3476	26867	C14H12	trans-stilbene	103-30-0	49.5341	0.0201	1	10	2	49.53	51.88	51.88
3477	26868	C14H12	1,1-diphenylethene	530-48-3	49.3649	0.0193	1	21	2	49.36	51.61	52.36
3478	27052	C14H14	1,1-diphenylethane	612-00-0	48.3103	0.0214	1	15	2	48.31	50.75	51.19
3479	27053	C14H14	1,2-diphenylethane	103-29-7	48.3103	0.0214	1	15	2	48.31	50.75	51.19
3480	27054	C14H14	2-methyldiphenylmethane	713-36-0	48.3895	0.0214	1	15	2	48.39	50.84	51.28
3481	27055	C14H14	3-methyldiphenylmethane	620-47-3	48.0448	0.0191	1	16	2	48.04	50.21	50.66
3482	27056	C14H14	4-methyldiphenylmethane	620-83-7	47.7752	0.0212	1	14	2	47.78	50.16	50.52
3483	27057	C14H14	2,2'-dimethylbiphenyl	605-39-0	47.3719	0.0203	1	39	2	47.37	49.64	51.02
3484	27058	C14H14	2,3'-trimethylbiphenyl	611-43-8	47.3719	0.0203	1	39	2	47.37	49.64	51.02
3485	27059	C14H14	2,4'-dimethlbiphenyl	611-61-0	47.3719	0.0203	1	39	2	47.37	49.64	51.02
3486	27060	C14H14	3,3'-dimethylbiphenyl	612-75-9	48.8428	0.0196	1	8.8	2	48.84	---	50.97
3487	27061	C14H14	3,4'-dimethylbiphenyl	7383-90-6	48.4319	0.0185	1	16	2	48.43	50.54	50.98
3488	27062	C14H14	4,4'-dimethylbiphenyl	613-33-2	48.4319	0.0185	1	16	2	48.43	50.54	50.98
3489	27063	C14H14	2-ethylbiphenyl	1812-51-7	47.6525	0.0198	1	30	2	47.65	49.88	50.98
3490	27064	C14H14	3-ethylbiphenyl	5668-93-9	48.8041	0.0201	1	10	2	48.80	51.11	51.11
3491	27065	C14H14	4-ethylbiphenyl	5707-44-8	48.8041	0.0201	1	10	2	48.80	51.11	51.11
3492	27267	C14H16	1-butylnaphthalene	1634-09-9	48.6254	0.0185	1	4.3	2	48.63	---	49.95
3493	27268	C14H16	2-butylnaphthalene	1134-62-9	46.7742	0.0183	1	19	2	46.77	48.79	49.37
3494	27269	C14H16	1-sec-butylnaphthalene	1680-58-6	46.9866	0.0185	1	20	2	46.99	49.03	49.67
3495	27270	C14H16	2-sec-butylnaphthalene	4614-03-3	46.9866	0.0185	1	20	2	46.99	49.03	49.67
3496	27271	C14H16	1-isobutylnaphthalene	16727-91-6	46.9866	0.0185	1	20	2	46.99	49.03	49.67
3497	27272	C14H16	2-isobutylnaphthalene	26490-07-3	46.9866	0.0185	1	20	2	46.99	49.03	49.67
3498	27273	C14H16	1-tert-butylnaphthalene	17085-91-5	48.0014	0.0187	1	20	2	48.00	50.11	50.76
3499	27274	C14H16	2-tert-butylnaphthalene	2876-35-9	46.7574	0.0186	1	22	2	46.76	48.80	49.53
3500	27285	C14H16	1,2-diethylnaphthalene	19182-11-7	47.2388	0.0195	1	42	2	47.24	49.41	50.81
3501	27286	C14H16	1,4-diethylnaphthalene	37796-58-0	47.2388	0.0195	1	42	2	47.24	49.41	50.81
3502	27287	C14H16	1,6-diethylnaphthalene	19182-13-9	47.2388	0.0195	1	42	2	47.24	49.41	50.81
3503	27288	C14H16	1,7-diethylnaphthalene	66577-37-5	47.2388	0.0195	1	42	2	47.24	49.41	50.81
3504	27289	C14H16	2,3-diethylnaphthalene	66610-91-1	47.2388	0.0195	1	42	2	47.24	49.41	50.81
3505	27290	C14H16	2,4-dimethyl-1-ethylnaphthalene	66577-14-8	48.3053	0.0178	1	17	2	48.31	50.33	50.81
3506	27291	C14H16	4,6-dimethyl-1-ethylnaphthalene	66577-15-9	48.3053	0.0178	1	17	2	48.31	50.33	50.81
3507	27292	C14H16	2,5-dimethyl-3-ethylnaphthalene	------	48.3053	0.0178	1	17	2	48.31	50.33	50.81
3508	27293	C14H16	1,2-dimethyl-4-ethylnaphthalene	66577-38-6	48.3053	0.0178	1	17	2	48.31	50.33	50.81
3509	27294	C14H16	1,3-dimethyl-5-ethylnaphthalene	66577-40-0	48.3053	0.0178	1	17	2	48.31	50.33	50.81
3510	27295	C14H16	1,4-dimethyl-5-ethylnaphthalene	66309-90-8	48.3053	0.0178	1	17	2	48.31	50.33	50.81
3511	27296	C14H16	1,3-dimethyl-6-ethylnaphthalene	39622-45-2	48.3053	0.0178	1	17	2	48.31	50.33	50.81
3512	27297	C14H16	1,4-dimethyl-6-ethylnaphthalene	66610-92-2	48.3053	0.0178	1	17	2	48.31	50.33	50.81
3513	27298	C14H16	1,2-dimethyl-7-ethylnaphthalene	66577-39-7	48.3053	0.0178	1	17	2	48.31	50.33	50.81
3514	27299	C14H16	1,2,3,4-tetramethylnaphthalene	3031-15-0	48.7134	0.0188	1	25	2	48.71	50.87	51.75
3515	27300	C14H16	1,2,3,6-tetramethylnaphthalene	66577-20-6	48.7134	0.0188	1	25	2	48.71	50.87	51.75
3516	27301	C14H16	1,2,4,6-tetramethylnaphthalene	66577-21-7	48.7134	0.0188	1	25	2	48.71	50.87	51.75
3517	27302	C14H16	1,2,4,7-tetramethylnaphthalene	16020-17-0	48.7134	0.0188	1	25	2	48.71	50.87	51.75
3518	27303	C14H16	1,2,4,8-tetramethylnaphthalene	66577-22-8	48.7134	0.0188	1	25	2	48.71	50.87	51.75
3519	27304	C14H16	1,2,5,6-tetramethylnaphthalene	2131-43-3	48.7134	0.0188	1	25	2	48.71	50.87	51.75
3520	27305	C14H16	1,2,5,8-tetramethylnaphthalene	3031-16-1	48.7134	0.0188	1	25	2	48.71	50.87	51.75

Table 1 Adsorption Capacity of Activated Carbon - Gas

$\log_{10} Q = \log_{10} K + 1/n \log_{10} y$ (Q - g of compound/100 g of carbon, y - ppmv)

NO	ID	FORMULA	NAME	CAS No	K	1/n	ymin	ymax	code	Q @ymin	Q @10ppm	Q @ymax
3521	27306	C14H16	1,2,6,8-tetramethylnaphthalene	66577-00-2	48.7134	0.0188	1	25	2	48.71	50.87	51.75
3522	27307	C14H16	1,3,5,8-tetramethylnaphthalene	14558-12-4	48.7134	0.0188	1	25	2	48.71	50.87	51.75
3523	27308	C14H16	1,3,6,7-tetramethylnaphthalene	7435-50-9	48.7134	0.0188	1	25	2	48.71	50.87	51.75
3524	27309	C14H16	1,4,5,8-tetramethylnaphthalene	2717-39-7	48.7134	0.0188	1	25	2	48.71	50.87	51.75
3525	27310	C14H16	1,4,6,7-tetramethylnaphthalene	13764-18-6	48.7134	0.0188	1	25	2	48.71	50.87	51.75
3526	27311	C14H16	2,3,6,7-tetramethylnaphthalene	1134-40-3	48.7134	0.0188	1	25	2	48.71	50.87	51.75
3527	27560	C14H20	1-butyl-[1,2,3,4-tetrahydronaphthalene]	38857-76-0	45.0287	0.0192	1	22	2	45.03	47.06	47.78
3528	27561	C14H20	2-butyl-[1,2,3,4-tetrahydronaphthalene]	36230-28-1	44.9259	0.0189	1	17	2	44.93	46.92	47.39
3529	27644	C14H22	octylbenzene	2189-60-8	41.5548	0.0199	1	14	2	41.55	43.51	43.80
3530	27645	C14H22	1,2,3,4-tetraethylbenzene	642-32-0	41.6615	0.0188	10	91	2	43.51	43.51	45.35
3531	27646	C14H22	1,2,3,5-tetraethylbenzene	38842-05-6	40.9215	0.0180	10	213	2	42.66	42.66	45.08
3532	27647	C14H22	1,2,4,5-tetraethylbenzene	635-81-4	40.7192	0.0181	10	228	2	42.45	42.45	44.93
3533	27777	C14H26	1,1-dicyclohexylethane	2319-61-1	42.5403	0.0196	1	39	2	42.54	44.51	45.71
3534	27779	C14H26	1-nonylcyclopentene	62184-74-1	39.2656	0.0190	1	43	2	39.27	41.02	42.18
3535	27780	C14H26	1-tetradecyne	765-10-6	37.5293	0.0182	10	59	2	39.14	39.14	40.43
3536	27781	C14H26	3-tetradecyne	60212-32-0	37.5687	0.0190	1	49	2	37.57	39.25	40.45
3537	27829	C14H27Cl	cis-1-chloro-1-tetradecene	66827-04-1	44.4565	0.0152	1	1.6	2	44.46	---	44.78
3538	27830	C14H27Cl	trans-1-chloro-1-tetradecene	66827-05-2	44.4565	0.0152	1	1.6	2	44.46	---	44.78
3539	27833	C14H27F	cis-1-fluoro-1-tetradecene	66827-06-3	40.7191	0.0168	1	21	2	40.72	42.33	42.86
3540	27834	C14H27F	trans-1-fluoro-1-tetradecene	66827-07-4	40.7191	0.0168	1	21	2	40.72	42.33	42.86
3541	27847	C14H28	nonylcyclopentane	2882-98-6	39.2791	0.0193	1	13	2	39.28	41.07	41.28
3542	27848	C14H28	octylcyclohexane	1795-15-9	39.5861	0.0176	1	16	2	39.59	41.22	41.56
3543	27850	C14H28	1-tetradecene	1120-36-1	37.4437	0.0173	1	19	2	37.44	38.97	39.40
3544	27851	C14H28	2-tetradecene	638-60-8	37.0788	0.0179	1	31	2	37.08	38.64	39.43
3545	27852	C14H28	2-methyl-1-tridecene	18094-01-4	37.1868	0.0179	1	32	2	37.19	38.75	39.57
3546	27856	C14H28Cl2	1,1-dichlorotetradecane	4168-40-5	47.8436	0.0141	1	1.6	2	47.84	---	48.16
3547	27857	C14H28F2	1,1-difluorotetradecane	62127-03-1	42.0145	0.0179	10	58	2	43.78	43.78	45.18
3548	27862	C14H28O	tetradecanal	124-25-4	41.2045	0.0175	1	5.9	2	41.20	---	42.50
3549	27877	C14H28O2	decyl butanoate	5454-09-1	42.2261	0.0179	1	10	2	42.23	44.00	44.00
3550	27895	C14H29Br	1-bromotetradecane	112-71-0	51.4842	0.0145	1	1.9	2	51.48	---	51.97
3551	27897	C14H29Cl	1-chlorotetradecane	2425-54-9	43.4004	0.0156	1	3.4	2	43.40	---	44.24
3552	27899	C14H29F	1-fluorotetradecane	73180-09-3	40.3982	0.0170	1	16	2	40.40	42.01	42.34
3553	27901	C14H29I	1-iodotetradecane	19218-94-1	58.3752	0.0138	1	2.8	2	58.38	---	59.21
3554	27912	C14H30	tetradecane	629-59-4	37.0277	0.0186	1	15	2	37.03	38.65	38.94
3555	27913	C14H30	2-methyltridecane	1560-96-9	36.5896	0.0176	1	29	2	36.59	38.11	38.83
3556	27914	C14H30	3-methyltridecane	6418-41-3	36.9167	0.0176	1	28	2	36.92	38.45	39.15
3557	27915	C14H30	2,2-dimethyldodecane	49598-54-1	36.1821	0.0174	10	53	2	37.67	37.67	38.78
3558	27916	C14H30	2,3-dimethyldodecane	6117-98-2	37.2075	0.0179	1	34	2	37.21	38.77	39.63
3559	27917	C14H30	2,4-dimethyldodecane	6117-99-3	36.3112	0.0177	10	64	2	37.82	37.82	39.08
3560	27928	C14H30O	diheptyl ether	629-64-1	38.6452	0.0171	1	28	2	38.65	40.20	40.91
3561	27946	C14H30S	1-tetradecanethiol	2079-95-0	42.7269	0.0144	1	1.5	2	42.73	---	42.98
3562	27947	C14H30S	2-tetradecanethiol	62155-04-8	42.7269	0.0144	1	1.5	2	42.73	---	42.98
3563	27959	C14H31N	diheptylamine	2470-68-0	39.4428	0.0167	1	7.1	2	39.44	---	40.76
3564	27960	C14H31N	dimethyldodecylamine	112-18-5	38.3769	0.0161	1	19	2	38.38	39.83	40.24
3565	27961	C14H31N	diethyldecylamine	6308-94-7	38.1848	0.0166	1	30	2	38.18	39.67	40.40
3566	28379	C15H16	1,1-diphenylpropane	1530-03-6	48.0137	0.0194	1	13	2	48.01	50.21	50.46
3567	28380	C15H16	1,2-diphenylpropane	5814-85-7	47.5681	0.0194	1	13	2	47.57	49.74	49.99
3568	28381	C15H16	1,3-diphenylpropane	1081-75-0	48.5429	0.0182	1	5.8	2	48.54	---	50.13
3569	28382	C15H16	2,2-diphenylpropane	778-22-3	48.0137	0.0194	1	13	2	48.01	50.21	50.46
3570	28389	C15H16	2-ethyldiphenylmethane	28122-25-0	48.7687	0.0187	1	8.6	2	48.77	---	50.77
3571	28390	C15H16	3-ethyldiphenylmethane	28122-24-9	48.1989	0.0186	1	8.3	2	48.20	---	50.13
3572	28391	C15H16	4-ethyldiphenylmethane	620-85-9	48.3464	0.0182	1	6.4	2	48.35	---	50.01
3573	28647	C15H22	1-pentyl-[1,2,3,4-tetrahydronaphthalene]	66359-06-6	45.5418	0.0183	1	9.3	2	45.54	---	47.44
3574	28648	C15H22	2-pentyl-[1,2,3,4-tetrahydronaphthalene]	66359-07-7	45.5183	0.0181	1	7.2	2	45.52	---	47.17
3575	28725	C15H24	nonylbenzene	1081-77-2	42.5144	0.0170	1	5.3	2	42.51	---	43.74
3576	28815	C15H28	1,1-dicyclohexylpropane	54934-91-7	43.3709	0.0173	1	19	2	43.37	45.13	45.63
3577	28819	C15H28	1-decylcyclopentene	62184-75-2	40.2297	0.0166	1	19	2	40.23	41.80	42.24
3578	28820	C15H28	1-pentadecyne	765-13-9	38.6098	0.0178	1	15	2	38.61	40.23	40.52
3579	28821	C15H28	2-pentadecyne	52112-25-1	38.6693	0.0177	1	14	2	38.67	40.28	40.52
3580	28822	C15H28	3-pentadecyne	61886-61-1	38.4994	0.0165	1	21	2	38.50	39.99	40.49
3581	28843	C15H29F	cis-1-fluoro-1-pentadecene	66291-42-7	41.7054	0.0159	1	6.1	2	41.71	---	42.92
3582	28844	C15H29F	trans-1-fluoro-1-pentadecene	66291-43-8	41.7054	0.0159	1	6.1	2	41.71	---	42.92
3583	28855	C15H30	decylcyclopentane	1795-21-7	40.4352	0.0165	1	4.3	2	40.44	---	41.42
3584	28856	C15H30	nonylcyclohexane	2883-02-5	40.6266	0.0166	1	5.1	2	40.63	---	41.74
3585	28859	C15H30	1-pentadecene	13360-61-7	38.5428	0.0164	1	6	2	38.54	---	39.69
3586	28860	C15H30	2-methyl-1-tetradecene	52254-38-3	37.9840	0.0160	1	19	2	37.98	39.41	39.82
3587	28900	C15H31F	1-fluoropentadecane	1555-17-5	41.1654	0.0164	1	6.5	2	41.17	---	42.45
3588	28904	C15H32	pentadecane	629-62-9	38.3334	0.0160	1	4.5	2	38.33	---	39.27
3589	28905	C15H32	2-methyltetradecane	1560-95-8	37.6314	0.0168	1	10	2	37.63	39.12	39.12
3590	28906	C15H32	3-methyltetradecane	18435-22-8	37.9554	0.0168	1	10	2	37.96	39.45	39.45
3591	28907	C15H32	2,2-dimethyltridecane	61869-04-3	37.2670	0.0160	1	19	2	37.27	38.66	39.06
3592	28908	C15H32	2,3-dimethyltridecane	18435-20-6	38.3011	0.0170	1	12	2	38.30	39.83	39.96
3593	28909	C15H32	2,4-dimethyltridecane	61868-05-1	37.4358	0.0161	1	23	2	37.44	38.85	39.38
3594	28934	C15H33N	dimethyltridecylamine	17373-29-4	39.1128	0.0156	1	9.2	2	39.11	---	40.49
3595	28935	C15H33N	diethylundecylamine	54334-64-4	38.9542	0.0161	1	13	2	38.95	40.43	40.60
3596	28936	C15H33N	tripentylamine	621-77-2	38.6298	0.0150	1	19	2	38.63	39.99	40.38
3597	29232	C16H16	1,1-diphenyl-1-butene	1726-14-3	49.1068	0.0168	1	7.9	2	49.11	---	50.84
3598	29342	C16H18	1,1-diphenylbutane	719-79-9	47.4912	0.0165	1	19	2	47.49	49.33	49.85
3599	29559	C16H24	1-hexyl-[1,2,3,4-tetrahydronaphthalene]	66325-11-9	46.0683	0.0161	1	4	2	46.07	---	47.11
3600	29560	C16H24	2-hexyl-[1,2,3,4-tetrahydronaphthalene]	66325-09-5	46.0655	0.0159	1	3.1	2	46.07	---	46.90

Table 1 Adsorption Capacity of Activated Carbon - Gas

$\log_{10} Q = \log_{10} K + 1/n \log_{10} y$ (Q - g of compound/100 g of carbon, y - ppmv)												
NO	ID	FORMULA	NAME	CAS No	K	1/n	ymin	ymax	code	Q @ymin	Q @10ppm	Q @ymax
3601	29595	C16H26	decylbenzene	104-72-3	43.3744	0.0148	1	1.7	2	43.37	---	43.72
3602	29640	C16H30	1-hexadecyne	629-74-3	39.8942	0.0152	1	3.7	2	39.89	---	40.69
3603	29641	C16H30	2-hexadecyne	629-75-4	39.5880	0.0156	1	6	2	39.59	---	40.71
3604	29642	C16H30	3-hexadecyne	61886-62-2	39.2703	0.0160	1	9.6	2	39.27	---	40.72
3605	29643	C16H30	1,1-dicyclohexylbutane	54890-00-5	43.7015	0.0170	1	11	2	43.70	45.44	45.52
3606	29646	C16H30	1-undecylcyclopentene	62184-76-3	40.9273	0.0161	1	8.6	2	40.93	---	42.37
3607	29693	C16H31F	cis-1-fluoro-1-hexadecene	66271-71-4	42.6080	0.0139	1	1.7	2	42.61	---	42.92
3608	29694	C16H31F	trans-1-fluoro-1-hexadecene	66271-72-5	42.6080	0.0139	1	1.7	2	42.61	---	42.92
3609	29707	C16H32	1-hexadecene	629-73-2	39.1479	0.0148	1	3.5	2	39.15	---	39.88
3610	29708	C16H32	2-methyl-1-pentadecene	29833-69-0	38.8153	0.0155	1	7.8	2	38.82	---	40.07
3611	29744	C16H33F	1-fluorohexadecane	408-38-8	41.9467	0.0145	1	2.4	2	41.95	---	42.48
3612	29754	C16H34	hexadecane	544-76-3	39.1472	0.0142	1	1.9	2	39.15	---	39.51
3613	29755	C16H34	2-methylpentadecane	1560-93-6	38.6095	0.0147	1	3.8	2	38.61	---	39.38
3614	29756	C16H34	3-methylpentadecane	2882-96-4	38.9059	0.0148	1	3.7	2	38.91	---	39.66
3615	29757	C16H34	2,2-dimethyltetradecane	59222-86-5	38.2137	0.0153	1	7	2	38.21	---	39.37
3616	29758	C16H34	2,3-dimethyltetradecane	18435-23-9	39.3291	0.0149	1	4.2	2	39.33	---	40.18
3617	29759	C16H34	2,4-dimethyltetradecane	61868-06-2	38.3298	0.0155	1	9.4	2	38.33	---	39.69
3618	29800	C16H35N	dioctylamine	1120-48-5	40.3334	0.0140	1	3.6	2	40.33	---	41.07
3619	29803	C16H35N	dimethyltetradecylamine	112-75-4	39.9479	0.0139	1	3.5	2	39.95	---	40.65
code: 1 - data, 2 - estimate												
BPL Activated Carbon, 4x10 mesh, Calgon												
Q - adsorption capacity at equilibrium, g of compound/100 g of carbon												
y - concentration of compound in gas, parts per million by volume, ppmv												
K and 1/n - coefficients of Freundlich equation												
T - temperature, K												
ymin - minimum concentration, ppmv												
ymax - maximum concentration, ppmv												

Chapter 2

ADSORPTION CAPACITY OF ACTIVATED CARBON – AQUEOUS LIQUID

Carl L. Yaws, Helen Lou, and Avanindra Shishir Jaiswal
Lamar University, Beaumont, Texas

Tabulation Results

The results for adsorption capacity of activated carbon are given in Table 2. For the tabulation, the Freundlich equation is used for adsorption capacity as a function of concentration in aqueous liquid :

$$\log_{10} Q = \log_{10} K + 1/n \log_{10} x \qquad (2)$$

where Q = adsorption capacity at equilibrium, g of compound/100 g of carbon
x = concentration in water at 25 C and 1 atm, parts per million by weight, ppm(wt)
K, $1/n$ = regression coefficients

The tabulation is arranged by carbon number (C, C2, C3, ...) to provide ease of use in quickly locating the data by using the chemical formula. The compound name, CAS No (Chemical Abstracts Registry Number), and regression coefficients are provided in the adjacent columns. The range of application is denoted by minimum and maximum concentrations (xmin and xmax). Concentrations outside the range of application should not be used. The next column provides the code for the tabulation which is based on both experimental data and estimated values.

In preparing the tabulation, a literature search was conducted to identify data source publications (1-15). Both experimental values for the property under consideration and parameter values for estimation of the property are included in the source publications. The publications were screened and copies of appropriate data were made. These data were then keyed into the computer to provide a database of values for compounds for which experimental data are available. The database also served as a basis to check the accuracy of the estimation methods. Upon completion of data collection, estimation of the values for the remaining compounds was performed. The numerous point values were processed using a computer program for minimum deviation.

The excellent compilations by Arbuckle (1); Sonthier, Crittenden and Summer (2); and Speth (3-4) were utilized for adsorption capacity. The compilations by Howard and Meylan (5); Mackay, Shiu, and Ma (6); Verschueren (7); Yalkowsky (8); and Yaws (12-14) were used for water solubility. Liquid molar volume was ascertained from data compilations by Yaws (12-15). Estimates were primarily based on a modified version by Yaws (9-11) of the equation originally provided by Arbuckle for adsorption capacity on activated carbon (F 400, Calgon) as a function of adsorption potential (Polyani theory). The modified version was developed for alcohols, ketones, aldehydes, acetates, benzenes, chloro alkanes, chloro olefins, and bromo alkanes.

In general, experimental data available in the literature are very limited for adsorption capacity of activated carbon for organic compounds in water at low concentrations of several hundred parts per million. The original work of Arbuckle (1) did involve concentrations in the range of several hundred parts per million for oxygenated petrochemicals (twenty-two compounds). In view of the very limited experimental data, estimates in the tabulation should be considered rough approximations useful for initial analysis. If initial analysis is favorable, follow-up experimental determination of equilibrium adsorption capacity is recommended.

A comparison of calculated and data values is shown in Figure 2 for a representative compound. The graph discloses favorable agreement of equation and data.

Example

Water from an industrial process contains 10 ppm(wt) of toluene (C7H8). Estimate the adsorption capacity of activated carbon for removing the compound from the water.

Substitution of the coefficients from the tabulation and concentration into the equation for adsorption capacity yields:

$$\log_{10} Q = \log_{10} (9.5958) + 0.3318 \log_{10} (10) = 1.3139$$

$$Q = 10^{1.3139}$$

$$Q = 20.60 \text{ g of toluene/100 g of carbon}$$

References

1. Arbuckle, W. B., Environ. Sci. Technol., 15(7), 812 (1981).
2. Sontheimer, H., Crittenden, J. C. and Summers, R. S., ACTIVATED CARBON FOR WATER TREATMENT, 2nd edition, DVGW – Forschungsstelle, FRG (1988).
3. Speth, T. F., Masters Thesis, Michigan Technological University, Houghton, MI (1986).
4. Speth, T. F. and Miller, J. R., Journal AWWA, 72 (February, 1990).
5. Howard, P. H. and W. M. Meylan, eds., HANDBOOK OF PHYSICAL PROPERTIES OF ORGANIC CHEMICALS, CRC Press, Boca Raton, FL (1997).
6. Mackay, D., W. Y. Shiu, and K. C. Ma, ILLUSTRATED HANDBOOK OF PHYSICAL-CHEMICAL PROPERTIES AND ENVIRONMENTAL FATE FOR ORGANIC CHEMICALS, Vols. 1, 2, 3, 4 and 5, Lewis Publishers, New York, NY (1992, 1992, 1993, 1995, 1997).
7. Verschueren, K., HANDBOOK OF ENVIRONMENTAL DATA ON ORGANIC CHEMICALS, 3rd and 4th eds., Van Nostrand Reinhold, New York, NY (1996, 2001).
8. Yalkowsky, S. H., AQUASOL DATABASE, University of Arizona, Tucson, AZ (1990-2004).
9. Yaws, C. L. and others, "Correlation calculates adsorption capacity of activated carbon for alkyl benzenes in water", Oil & Gas Journal, 101.18, 80-82 (May 5, 2003).
10. Yaws, C. L. and others, "Equation predicts adsorption of oxgenated petrochems on activated carbon", Oil & Gas Journal, 103.4, 51-53 (Jan 24, 2005).
11. Yaws, C. L. and others, "Adsorption Capacity of Activated Carbon for Water Purification", WATER ENCYCLOPEDIA, Volume 4, ISBN 0-471-73684-8, pages 561-564, John Wiley, New York, NY (2005).
12. Yaws, C. L., CHEMICAL PROPERTIES HANDBOOK, McGraw-Hill, New York, NY (1999).
13. Yaws, C. L., MATHESON GAS DATA BOOK, 7th ed., Matheson Tri-Gas (Parsippany, NJ), McGraw-Hill, New York, NY (2001).
14. Yaws, C. L., YAWS HANDBOOK OF THERMODYNAMIC AND PHYSICAL PROPERTIES FOR CHEMICALS, electronic edition, www.knovel.com, knovel, Norvich, NY (2003).
15. Yaws, C. L., YAWS HANDBOOK OF PHYSICAL PROPERTIES FOR HYDROCARBONS AND CHEMICALS, Gulf Publishing Co., Houston, TX (2005).

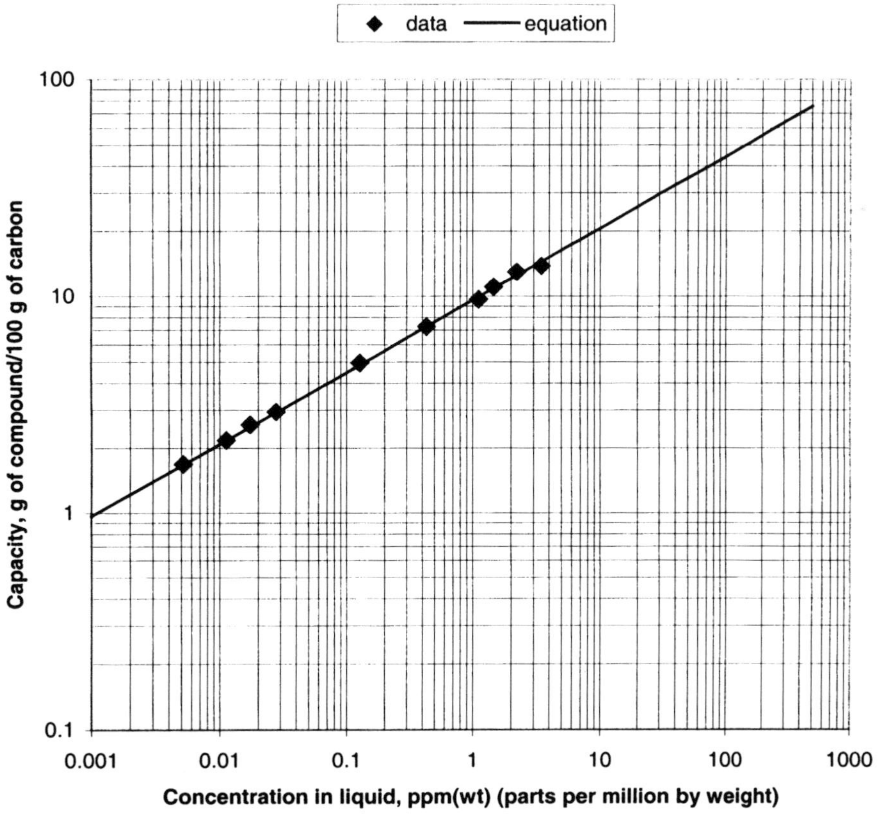

Figure 2 Adsorption Capacity for Toluene

Table 2 Adsorption Capacity of Activated Carbon - Aqueous Liquid

$\log_{10} Q = \log_{10} K + 1/n \log_{10} x$ (Q - g of compound/100 g of carbon, x - ppm(wt))												
NO	ID	FORMULA	NAME	CAS No	K	1/n	xmin	xmax	code	Q @xmin	Q @10ppm	Q @xmax
1	96	CHBr3	bromoform	75-25-2	5.0300	0.6754	0.1	90	1,2	1.06	23.82	105.08
2	104	CHCl3	chloroform	67-66-3	0.6861	0.7137	1	1000	1,2	0.69	3.55	94.95
3	149	CH2Cl2	dichloromethane	75-09-2	0.0591	0.8934	1	1000	2	0.059	0.46	28.28
4	175	CH2O	formaldehyde	50-00-0	0.0003	0.8843	1	1000	2	0.00026	0.0020	0.12
5	191	CH3Br	methyl bromide	74-83-9	0.0542	1.0392	1	1000	2	0.054	0.59	71.03
6	196	CH3Cl	methyl chloride	74-87-3	0.0547	1.0411	1	1000	2	0.055	0.60	72.68
7	272	CH4O	methyl alcohol	67-56-1	0.00026	0.8861	1	1000	2	0.00026	0.0020	0.12
8	455	C2Cl4	tetrachloroethylene	127-18-4	14.3503	0.3875	0.1	100	1,2	5.88	35.02	85.46
9	539	C2HCl3	trichloroethylene	79-01-6	4.0564	0.5291	0.1	70	1,2	1.20	13.72	38.41
10	615	C2H2Cl2	1,1-dichloroethylene	75-35-4	1.9121	0.5493	1	1000	2	1.91	6.77	85.00
11	616	C2H2Cl2	cis-1,2-dichloroethylene	156-59-2	1.4880	0.5180	1	1000	1,2	1.49	4.91	53.30
12	617	C2H2Cl2	trans-1,2-dichloroethylene	156-60-5	0.7820	0.6118	1	1000	2	0.78	3.20	53.52
13	618	C2H2Cl2	dichloroethylene	25323-30-2	0.8280	0.6314	1	1000	2	0.83	3.54	64.91
14	619	C2H2Cl2	1,2-dichloroethylene; mixture of cis and trans	540-59-0	0.8845	0.6236	1	1000	2	0.88	3.72	65.67
15	701	C2H3Cl	vinyl chloride	75-01-4	0.7768	0.6888	1	1000	2	0.78	3.79	90.53
16	799	C2H4Br2	1,2-dibromoethane	106-93-4	4.8722	0.4724	0.1	100	1,2	1.64	14.46	42.91
17	813	C2H4Cl2	1,2-dichloroethane	107-06-2	0.5576	0.7249	1	1000	2	0.56	2.96	83.38
18	884	C2H4O	acetaldehyde	75-07-0	0.0084	0.6338	1	1000	2	0.008	0.036	0.67
19	1065	C2H6O	ethyl alcohol	64-17-5	0.0106	0.6163	1	1000	2	0.011	0.044	0.75
20	----	C3H4Cl2O2	dalapon	------	2.3119	0.2240	0.076	2.73	1,2	1.30	----	2.90
21	----	C3H5Br2Cl	dibromochloropropane	------	22.0028	0.5010	0.0002	0.069	1,2	0.31	----	5.77
22	----	C3H8NO5P	glyphosate	------	19.9299	0.1190	0.0263	2.37	1,2	12.93	----	22.09
23	1668	C3H5Cl	cis-1-chloro-1-propene	16136-84-8	1.3711	0.5776	1	1000	2	1.37	5.18	74.09
24	1669	C3H5Cl	trans-1-chloro-1-propene	16136-85-9	1.3718	0.5775	1	1000	2	1.37	5.19	74.10
25	1670	C3H5Cl	2-chloro-1-propene	557-98-2	1.6184	0.5576	1	1000	2	1.62	5.84	76.17
26	1671	C3H5Cl	3-chloro-1-propene	107-05-1	1.3437	0.5800	1	1000	2	1.34	5.11	73.84
27	1672	C3H5Cl	1-chloro-1-propene	590-21-6	1.4297	0.5725	1	1000	2	1.43	5.34	74.61
28	1922	C3H6O	propionaldehyde	123-38-6	0.0806	0.4877	1	1000	1,2	0.08	0.25	2.34
29	1923	C3H6O	acetone	67-64-1	0.0515	0.4873	1	1000	1,2	0.05	0.16	1.49
30	1932	C3H6O2	methyl acetate	79-20-9	0.1964	0.4514	1	1000	2	0.20	0.56	4.44
31	2129	C3H8O	propyl alcohol	71-23-8	0.1051	0.4371	1	1000	1,2	0.11	0.29	2.15
32	2130	C3H8O	isopropyl alcohol	67-63-0	0.1001	0.4389	1	1000	1,2	0.10	0.28	2.08
33	3478	C4H8O	butyraldehyde	123-72-8	0.6566	0.3706	1	1000	1,2	0.66	1.54	8.49
34	3479	C4H8O	isobutyraldehyde	78-84-2	0.6076	0.3922	1	1000	2	0.61	1.50	9.13
35	3480	C4H8O	methyl ethyl ketone	78-93-3	0.2583	0.3717	1	1000	1,2	0.26	0.61	3.37
36	3508	C4H8O2	ethyl acetate	141-78-6	0.8774	0.3661	1	1000	2	0.88	2.04	11.00
37	3865	C4H10O	butanol	71-36-3	0.9607	0.3414	1	1000	1,2	0.96	2.11	10.16
38	3866	C4H10O	isobutanol	78-83-1	0.9274	0.3411	1	1000	1,2	0.93	2.03	9.78
39	3867	C4H10O	sec-butanol	78-92-2	0.4599	0.3918	1	1000	2	0.46	1.13	6.89
40	3868	C4H10O	tert-butanol	75-65-0	0.1709	0.4154	1	1000	2	0.17	0.44	3.01
41	3872	C4H10O	(R)-(-)-2-butanol	14898-79-4	0.4979	0.3913	1	1000	2	0.50	1.23	7.43
42	3873	C4H10O	(S)-(+)-2-butanol	4221-99-2	0.5031	0.3903	1	1000	2	0.50	1.24	7.46
43	3874	C4H10O	(<+->)-2-butanol	15892-23-6	0.5915	0.3755	1	1000	2	0.59	1.40	7.91
44	5445	C5H10O	pentanal	110-62-3	2.2561	0.3372	1	1000	2	2.26	4.90	23.17
45	5446	C5H10O	2-methylbutanal	96-17-3	1.8148	0.3342	1	1000	2	1.81	3.92	18.26
46	5447	C5H10O	3-methylbutanal	590-86-3	1.8653	0.3325	1	1000	2	1.87	4.01	18.55
47	5448	C5H10O	2,2-dimethylpropanal	630-19-3	1.4339	0.3280	1	1000	2	1.43	3.05	13.82
48	5449	C5H10O	2-methylbutanal, (±)	57456-98-1	1.8317	0.3342	1	1000	2	1.83	3.95	18.42
49	5450	C5H10O	methyl propyl ketone	107-87-9	1.3543	0.3359	1	1000	2	1.35	2.94	13.79
50	5451	C5H10O	diethyl ketone	96-22-0	1.9077	0.3019	1	1000	1,2	1.91	3.82	15.35
51	5452	C5H10O	methyl isopropyl ketone	563-80-4	1.2943	0.3372	1	1000	2	1.29	2.81	13.29
52	5501	C5H10O2	propyl acetate	109-60-4	3.0624	0.2840	1	1000	1,2	3.06	5.89	21.78
53	5502	C5H10O2	isopropyl acetate	108-21-4	2.4967	0.2874	1	1000	1,2	2.50	4.84	18.18
54	5831	C5H12O	1-pentanol	71-41-0	2.7169	0.2799	1	1000	1,2	2.72	5.18	18.78
55	5832	C5H12O	2-pentanol	6032-29-7	1.5559	0.3295	1	1000	2	1.56	3.32	15.15
56	5833	C5H12O	3-pentanol	584-02-1	1.4002	0.3348	1	1000	2	1.40	3.03	14.14
57	5834	C5H12O	2-methyl-1-butanol	137-32-6	1.7120	0.3331	1	1000	2	1.71	3.69	17.10
58	5835	C5H12O	2-methyl-1-butanol, (±)	34713-94-5	1.8265	0.3336	1	1000	2	1.83	3.94	18.30
59	5836	C5H12O	3-methyl-2-butanol, (±)	70116-68-6	1.3516	0.3328	1	1000	2	1.35	2.91	13.47
60	5837	C5H12O	3-methyl-1-butanol	123-51-3	1.7880	0.3323	1	1000	2	1.79	3.84	17.76
61	5838	C5H12O	tert-pentyl-alcohol	75-85-4	1.1590	0.3295	1	1000	2	1.16	2.47	11.28
62	5839	C5H12O	3-methyl-2-butanol	598-75-4	1.3906	0.3331	1	1000	2	1.39	2.99	13.89
63	5840	C5H12O	2,2-dimethyl-1-propanol	75-84-3	0.8353	0.3769	1	1000	2	0.84	1.99	11.28
64	5842	C5H12O	methyl isobutyl ether	625-44-5	2.8952	0.2681	1	1000	1,2	2.90	5.37	18.45
65	5849	C5H12O	(S)-(-)-2-methyl-1-butanol	1565-80-6	1.8739	0.3319	1	1000	2	1.87	4.02	18.56
66	5850	C5H12O	(S)-(+)-2-pentanol	26184-62-3	1.5394	0.3319	1	1000	2	1.54	3.31	15.24
67	5851	C5H12O	pentanol	30899-19-5	2.2089	0.3197	1	1000	2	2.21	4.61	20.10
68	5852	C5H12O	(R)-(-)-2-pentanol	31087-44-2	1.7048	0.3331	1	1000	2	1.70	3.67	17.03
69	5853	C5H12O	(S)-3-methyl-2-butanol	1517-66-4	1.1583	0.3197	1	1000	2	1.16	2.42	10.54
70	----	C6HCl5O	pentachlorophenol	------	44.3006	0.3390	0.0004	3.41	1,2	3.12	----	67.14
71	----	C6H3Cl3N2O2	pichloram	------	8.1136	0.1800	0.0002	3.07	1,2	1.75	----	9.93
72	6876	C6H6	benzene	71-43-2	3.9187	0.3995	1	1000	2	3.92	9.83	61.90
73	8401	C6H12O	hexanal	66-25-1	4.2705	0.2917	1	1000	2	4.27	8.36	32.04
74	8402	C6H12O	2-methylpentanal	123-15-9	4.1188	0.2894	1	1000	2	4.12	8.02	30.40
75	8403	C6H12O	3-methylpentanal	15877-57-3	4.4650	0.2885	1	1000	2	4.47	8.68	32.76
76	8404	C6H12O	4-methylpentanal	1119-16-0	4.4788	0.2881	1	1000	2	4.48	8.70	32.78
77	8405	C6H12O	2-ethylbutanal	97-96-1	4.0136	0.2920	1	1000	2	4.01	7.86	30.16
78	8406	C6H12O	2,2-dimethylbutanal	2094-75-9	3.4322	0.2885	1	1000	2	3.43	6.67	25.18
79	8407	C6H12O	2,3-dimethylbutanal	2109-98-0	3.8390	0.2907	1	1000	2	3.84	7.50	28.59
80	8408	C6H12O	3,3-dimethylbutanal	2987-16-8	3.3510	0.2910	1	1000	2	3.35	6.55	25.02

Table 2 Adsorption Capacity of Activated Carbon - Aqueous Liquid

$\log_{10} Q = \log_{10} K + 1/n \log_{10} x$ (Q - g of compound/100 g of carbon, x - ppm(wt))

NO	ID	FORMULA	NAME	CAS No	K	1/n	xmin	xmax	code	Q @xmin	Q @10ppm	Q @xmax
81	8409	C6H12O	2-hexanone	591-78-6	3.1614	0.2907	1	1000	2	3.16	6.17	23.54
82	8410	C6H12O	3-hexanone	589-38-8	3.2299	0.2917	1	1000	2	3.23	6.32	24.23
83	8411	C6H12O	3-methyl-2-pentanone	565-61-7	3.1873	0.2892	1	1000	2	3.19	6.20	23.50
84	8412	C6H12O	4-methyl-2-pentanone	108-10-1	3.1496	0.2867	1	1000	2	3.15	6.09	22.82
85	8413	C6H12O	ethyl isopropyl ketone	565-69-5	2.9253	0.2903	1	1000	2	2.93	5.71	21.73
86	8414	C6H12O	3,3-dimethyl-2-butanone	75-97-8	2.5993	0.2885	1	1000	2	2.60	5.05	19.08
87	8415	C6H12O	3-methyl-2-pentanone, (±)	55156-16-6	3.1324	0.2912	1	1000	2	3.13	6.12	23.41
88	8453	C6H12O	2,2-dimethyl-3-butanone	------	3.0920	0.2996	1	1000	2	3.09	6.16	24.48
89	8476	C6H12O2	butyl acetate	123-86-4	5.7536	0.2407	1	1000	1,2	5.75	10.02	30.35
90	8477	C6H12O2	isobutyl acetate	110-19-0	5.0073	0.2699	1	1000	2	5.01	9.32	32.30
91	8478	C6H12O2	sec-butyl acetate	105-46-4	5.0426	0.2696	1	1000	2	5.04	9.38	32.46
92	8479	C6H12O2	tert-butyl acetate	540-88-5	3.6047	0.2674	1	1000	2	3.60	6.67	22.86
93	8970	C6H14O	1-hexanol	111-27-3	5.7324	0.2370	1	1000	1,2	5.73	9.89	29.47
94	8971	C6H14O	2-hexanol	626-93-7	3.4613	0.2860	1	1000	2	3.46	6.69	24.96
95	8972	C6H14O	3-hexanol	623-37-0	3.3449	0.2875	1	1000	2	3.34	6.49	24.38
96	8973	C6H14O	2-methyl-1-pentanol	105-30-6	3.8348	0.2920	1	1000	2	3.83	7.51	28.82
97	8974	C6H14O	3-methyl-1-pentanol	589-35-5	4.4057	0.2895	1	1000	2	4.41	8.58	32.55
98	8975	C6H14O	4-methyl-1-pentanol	626-89-1	4.5025	0.2858	1	1000	2	4.50	8.70	32.43
99	8976	C6H14O	4-methyl-2-pentanol	108-11-2	3.4518	0.2842	1	1000	2	3.45	6.64	24.59
100	8977	C6H14O	2-methyl-2-pentanol	590-36-3	2.6539	0.2858	1	1000	2	2.65	5.13	19.11
101	8978	C6H14O	3-methyl-2-pentanol	565-60-6	3.1614	0.2912	1	1000	2	3.16	6.18	23.63
102	8979	C6H14O	2-methyl-3-pentanol	565-67-3	2.8041	0.2894	1	1000	2	2.80	5.46	20.70
103	8980	C6H14O	3-methyl-3-pentanol	77-74-7	2.5639	0.2909	1	1000	2	2.56	5.01	19.12
104	8981	C6H14O	2-ethyl-1-butanol	97-95-0	5.0970	0.2378	1	1000	1,2	5.10	8.81	26.35
105	8982	C6H14O	2,2-dimethyl-1-butanol	1185-33-7	3.3120	0.2912	1	1000	2	3.31	6.48	24.75
106	8983	C6H14O	2,3-dimethyl-1-butanol	19550-30-2	4.0859	0.2915	1	1000	2	4.09	7.99	30.60
107	8984	C6H14O	2,3-dimethyl-1-butanol, (±)-	20281-85-0	3.7843	0.2913	1	1000	2	3.78	7.40	28.31
108	8985	C6H14O	3,3-dimethyl-1-butanol	624-95-3	3.8747	0.2859	1	1000	2	3.87	7.48	27.92
109	8986	C6H14O	2,3-dimethyl-2-butanol	594-60-5	2.4403	0.2889	1	1000	2	2.44	4.75	17.95
110	8987	C6H14O	3,3-dimethyl-2-butanol	464-07-3	2.5447	0.2874	1	1000	2	2.54	4.93	18.52
111	8988	C6H14O	3,3-dimethyl-2-butanol, (±)	20281-91-8	2.5798	0.2868	1	1000	2	2.58	4.99	18.70
112	8989	C6H14O	(RS)-2-hexanol	20281-86-1	3.6594	0.2865	1	1000	2	3.66	7.08	26.48
113	8990	C6H14O	2-hexanol, (R)	26549-24-6	3.4728	0.2865	1	1000	2	3.47	6.72	25.14
114	8991	C6H14O	3-hexanol, (R)	13471-42-6	3.2949	0.2884	1	1000	2	3.29	6.40	24.15
115	8992	C6H14O	3-methyl-1-pentanol, (±)	20281-83-8	4.4560	0.2894	1	1000	2	4.46	8.68	32.90
116	8995	C6H14O	diisopropyl ether	108-20-3	4.0057	0.2457	1	1000	1,2	4.01	7.05	21.86
117	9008	C6H14O	(S)-(+)-2-hexanol	52019-78-0	3.5354	0.2874	1	1000	2	3.54	6.85	25.74
118	9009	C6H14O	tert-hexyl alcohol	26401-20-7	2.6641	0.2786	1	1000	2	2.66	5.06	18.26
119	10725	C7H8	toluene	108-88-3	9.5958	0.3318	0.001	500	1,2	0.97	20.60	75.44
120	----	C7H12ClN5	simazine	------	15.0157	0.2270	0.00002	0.017	1,2	1.29	----	5.91
121	----	C7H13N3O3S	oxamyl	------	41.6437	0.7930	0.0025	3	1,2	0.36	----	99.52
122	----	C7H14N2O2S	aldicarb	------	13.2894	0.4020	0.0013	0.274	1,2	0.92	----	7.90
123	11827	C7H14O	1-heptanal	111-71-7	8.0912	0.2569	1	1000	2	8.09	14.62	47.71
124	11828	C7H14O	3-methylhexanal	19269-28-4	7.6426	0.2578	1	1000	2	7.64	13.84	45.37
125	11829	C7H14O	2-heptanone	110-43-0	6.2232	0.2562	1	1000	2	6.22	11.23	36.53
126	11830	C7H14O	3-heptanone	106-35-4	7.2500	0.3000	20	1000	1,2	17.81	---	57.59
127	11831	C7H14O	4-heptanone	123-19-3	6.6813	0.2564	1	1000	2	6.68	12.06	39.27
128	11832	C7H14O	3-methyl-2-hexanone	2550-21-2	6.1132	0.2597	1	1000	2	6.11	11.12	36.76
129	11833	C7H14O	4-methyl-2-hexanone	105-42-0	6.2373	0.2554	1	1000	2	6.24	11.23	36.42
130	11834	C7H14O	5-methyl-2-hexanone	110-12-3	6.8257	0.2553	1	1000	2	6.83	12.29	39.81
131	11835	C7H14O	2-methyl-3-hexanone	7379-12-6	5.6126	0.2569	1	1000	2	5.61	10.14	33.09
132	11836	C7H14O	4-methyl-3-hexanone	17042-16-9	5.9017	0.2565	1	1000	2	5.90	10.65	34.71
133	11837	C7H14O	5-methyl-3-hexanone	623-56-3	5.9627	0.2553	1	1000	2	5.96	10.73	34.78
134	11838	C7H14O	3-ethyl-2-pentanone	6137-03-7	6.0459	0.2573	1	1000	2	6.05	10.93	35.76
135	11839	C7H14O	3,3-dimethyl-2-pentanone	20669-04-9	5.4208	0.2565	1	1000	2	5.42	9.79	31.89
136	11840	C7H14O	3,4-dimethyl-2-pentanone	565-78-6	5.4942	0.2575	1	1000	2	5.49	9.94	32.54
137	11841	C7H14O	4,4-dimethyl-2-pentanone	590-50-1	5.1335	0.2528	1	1000	2	5.13	9.19	29.42
138	11842	C7H14O	2,2-dimethyl-3-pentanone	564-04-5	5.0223	0.2551	1	1000	2	5.02	9.04	29.26
139	11843	C7H14O	2,4-dimethyl-3-pentanone	565-80-0	4.3934	0.2881	1	1000	2	4.39	8.53	32.15
140	11892	C7H14O2	pentyl acetate	628-63-7	8.7611	0.2416	1	1000	2	8.76	15.28	46.50
141	11893	C7H14O2	1-methylbutyl acetate	626-38-0	7.7162	0.2391	1	1000	2	7.72	13.38	40.24
142	11894	C7H14O2	2-methylbutyl acetate	624-41-9	8.2816	0.2416	1	1000	2	8.28	14.45	43.95
143	11895	C7H14O2	isopentyl acetate	123-92-2	8.5489	0.2403	1	1000	2	8.55	14.87	44.94
144	11896	C7H14O2	1-ethylpropyl acetate	620-11-1	7.5556	0.2401	1	1000	2	7.56	13.13	39.68
145	11897	C7H14O2	tert-pentyl acetate	625-16-1	6.5345	0.2450	1	1000	2	6.53	11.49	35.49
146	11898	C7H14O2	1,2-dimethylpropyl acetate	5343-96-4	7.3290	0.2383	1	1000	2	7.33	12.69	38.02
147	11899	C7H14O2	neopentyl acetate	926-41-0	7.3855	0.2366	1	1000	2	7.39	12.74	37.87
148	11900	C7H14O2	sec-pentyl acetate (R)	54638-10-7	8.4682	0.2421	1	1000	2	8.47	14.79	45.09
149	12164	C7H16O	1-heptanol	111-70-6	7.9440	0.2546	1	1000	2	7.94	14.28	46.11
150	12165	C7H16O	2-heptanol	543-49-7	6.8522	0.2527	1	1000	2	6.85	12.26	39.26
151	12166	C7H16O	3-heptanol	589-82-2	6.3961	0.2536	1	1000	2	6.40	11.47	36.88
152	12167	C7H16O	4-heptanol	589-55-9	6.2569	0.2527	1	1000	2	6.26	11.20	35.85
153	12168	C7H16O	2-methyl-1-hexanol	624-22-6	6.9223	0.2555	1	1000	2	6.92	12.47	40.44
154	12169	C7H16O	3-methyl-1-hexanol	13231-81-7	7.9058	0.2558	1	1000	2	7.91	14.25	46.28
155	12170	C7H16O	4-methyl-1-hexanol	818-49-5	8.0997	0.2546	1	1000	2	8.10	14.56	47.01
156	12171	C7H16O	5-methyl-1-hexanol	627-98-5	8.1281	0.2521	1	1000	2	8.13	14.52	46.37
157	12172	C7H16O	2-methyl-2-hexanol	625-23-0	5.2940	0.2515	1	1000	2	5.29	9.45	30.07
158	12173	C7H16O	3-methyl-2-hexanol	2313-65-7	5.8826	0.2552	1	1000	2	5.88	10.59	34.29
159	12174	C7H16O	4-methyl-2-hexanol	2313-61-3	5.8731	0.2536	1	1000	2	5.87	10.53	33.87
160	12175	C7H16O	5-methyl-2-hexanol	627-59-8	5.9850	0.2515	1	1000	2	5.98	10.68	34.00

Table 2 Adsorption Capacity of Activated Carbon - Aqueous Liquid

$\log_{10} Q = \log_{10} K + 1/n \log_{10} x$ (Q - g of compound/100 g of carbon, x - ppm(wt))

NO	ID	FORMULA	NAME	CAS No	K	1/n	xmin	xmax	code	Q @xmin	Q @10ppm	Q @xmax
161	12176	C7H16O	2-methyl-3-hexanol	617-29-8	5.3206	0.2546	1	1000	2	5.32	9.56	30.88
162	12177	C7H16O	3-methyl-3-hexanol	597-96-6	5.1308	0.2549	1	1000	2	5.13	9.23	29.84
163	12178	C7H16O	4-methyl-3-hexanol	615-29-2	5.6528	0.2546	1	1000	2	5.65	10.16	32.81
164	12179	C7H16O	5-methyl-3-hexanol	623-55-2	5.4319	0.2574	1	1000	2	5.43	9.82	32.14
165	12180	C7H16O	2-ethyl-1-pentanol	27522-11-8	7.1367	0.2574	1	1000	2	7.14	12.91	42.23
166	12181	C7H16O	3-ethyl-1-pentanol	66225-51-2	7.1194	0.2577	1	1000	2	7.12	12.89	42.22
167	12182	C7H16O	2,2-dimethyl-1-pentanol	2370-12-9	5.8463	0.2577	1	1000	2	5.85	10.58	34.67
168	12183	C7H16O	2,3-dimethyl-1-pentanol	10143-23-4	6.8396	0.2589	1	1000	2	6.84	12.42	40.91
169	12184	C7H16O	2,4-dimethyl-1-pentanol	6305-71-1	6.6532	0.2530	1	1000	2	6.65	11.91	38.20
170	12185	C7H16O	3,3-dimethyl-1-pentanol	19264-94-9	7.0468	0.2571	1	1000	2	7.05	12.74	41.61
171	12186	C7H16O	3,4-dimethyl-1-pentanol	6570-87-2	7.0640	0.2567	1	1000	2	7.06	12.76	41.62
172	12187	C7H16O	4,4-dimethyl-1-pentanol	3121-79-7	6.8216	0.2518	1	1000	2	6.82	12.18	38.84
173	12188	C7H16O	3-ethyl-2-pentanol	609-27-8	5.6967	0.2589	1	1000	2	5.70	10.34	34.07
174	12189	C7H16O	2,3-dimethyl-2-pentanol	4911-70-0	4.7428	0.2571	1	1000	2	4.74	8.57	28.00
175	12190	C7H16O	2,4-dimethyl-2-pentanol	625-06-9	4.5884	0.2508	1	1000	2	4.59	8.18	25.95
176	12191	C7H16O	3,3-dimethyl-2-pentanol	19781-24-9	5.4240	0.2558	1	1000	2	5.42	9.78	31.75
177	12192	C7H16O	3,4-dimethyl-2-pentanol	64502-86-9	5.7845	0.2589	1	1000	2	5.78	10.50	34.60
178	12193	C7H16O	4,4-dimethyl-2-pentanol	6144-93-0	4.9667	0.2505	1	1000	2	4.97	8.84	28.03
179	12194	C7H16O	3-ethyl-3-pentanol	597-49-9	4.8380	0.2608	1	1000	2	4.84	8.82	29.31
180	12195	C7H16O	2,2-dimethyl-3-pentanol	3970-62-5	4.6080	0.2552	1	1000	2	4.61	8.29	26.86
181	12196	C7H16O	2,3-dimethyl-3-pentanol	595-41-5	4.7065	0.2595	1	1000	2	4.71	8.56	28.27
182	12197	C7H16O	2,4-dimethyl-3-pentanol	600-36-2	4.7697	0.2561	1	1000	2	4.77	8.60	27.98
183	12198	C7H16O	2-methyl-2-ethyl-1-butanol	18371-13-6	6.3096	0.2558	1	1000	2	6.31	11.37	36.94
184	12199	C7H16O	3-methyl-2-ethyl-1-butanol	32444-34-1	6.6513	0.2586	1	1000	2	6.65	12.06	39.70
185	12200	C7H16O	2,2,3-trimethyl-1-butanol	55505-23-2	6.0072	0.2617	1	1000	2	6.01	10.97	36.63
186	12201	C7H16O	2,3,3-trimethyl-1-butanol	36794-64-6	6.6008	0.2558	1	1000	2	6.60	11.90	38.64
187	12202	C7H16O	2,3,3-trimethyl-2-butanol	594-83-2	4.0580	0.2602	1	1000	2	4.06	7.39	24.48
188	12203	C7H16O	2,4-dimethyl-1-pentanol, (±)	111768-02-6	7.1125	0.2449	1	1000	2	7.11	12.50	38.61
189	12204	C7H16O	2-heptanol, (±)	52390-72-4	6.6998	0.2522	1	1000	2	6.70	11.97	38.25
190	12205	C7H16O	3-heptanol, (S)	26549-25-7	6.4049	0.2540	1	1000	2	6.40	11.50	37.03
191	12206	C7H16O	2-methyl-1-hexanol, (±)	111768-04-8	7.0536	0.2550	1	1000	2	7.05	12.69	41.07
192	12207	C7H16O	3-methyl-1-hexanol, (±)	111768-08-2	7.7207	0.2550	1	1000	2	7.72	13.89	44.93
193	12208	C7H16O	4-methyl-1-hexanol, (±)	111768-05-9	8.1112	0.2544	1	1000	2	8.11	14.57	47.01
194	12209	C7H16O	2-methyl-3-hexanol, (±)	100295-82-7	4.9315	0.2595	1	1000	2	4.93	8.96	29.62
195	12210	C7H16O	5-methyl-3-hexanol, (±)	100295-83-8	5.6711	0.2508	1	1000	2	5.67	10.10	32.07
196	12220	C7H16O	(S)-(+)-2-heptanol	6033-23-4	6.1761	0.2524	1	1000	2	6.18	11.04	35.31
197	12221	C7H16O	(R)-(-)-2-heptanol	6033-24-5	5.5651	0.2540	1	1000	2	5.57	9.99	32.16
198	----	C8H5Cl3O3	2,4,5-trichlorophenoxyacetic acid	------	18.3429	0.2100	0.0208	1.79	1,2	8.13	----	20.73
199	----	C8H6Cl2O3	dicamba	------	9.1375	0.1470	0.0210	2.89	1,2	5.18	----	10.68
200	13789	C8H10	ethylbenzene	100-41-4	15.1828	0.2965	0.01	150	1,2	3.88	30.05	67.07
201	13790	C8H10	o-xylene	95-47-6	16.3327	0.2559	0.01	150	1,2	5.03	29.44	58.88
202	13791	C8H10	m-xylene	108-38-3	18.8246	0.2402	0.01	150	1,2	6.23	32.73	62.72
203	13792	C8H10	p-xylene	106-42-3	16.7289	0.2889	1	202	2	16.73	32.54	77.50
204	13793	C8H10	xylenes	1330-20-7	15.2621	0.2902	1	270	2	15.26	29.78	77.50
205	----	C8H14ClN5	atrazine	------	28.8876	0.2910	0.0042	0.908	1,2	5.88	----	28.09
206	----	C8H14N4OS	metribuzin	------	18.4726	0.1930	0.0400	2.83	1,2	9.92	----	22.58
207	15062	C8H16O	2-ethylhexanal	123-05-7	11.5385	0.2304	1	752	2	11.54	19.62	53.09
208	15063	C8H16O	1-octanal	124-13-0	13.6539	0.2296	1	370	2	13.65	23.17	53.09
209	15064	C8H16O	5-methylheptanal	75579-88-3	11.7767	0.2297	1	703	2	11.78	19.99	53.09
210	15065	C8H16O	2-propylpentanal	18295-59-5	11.2936	0.2325	1	777	2	11.29	19.29	53.09
211	15066	C8H16O	2-octanone	111-13-7	11.1567	0.2293	1	900	2	11.16	18.92	53.09
212	15067	C8H16O	2,5-dimethyl-3-hexanone	1888-57-9	8.4976	0.2327	1	1000	2	8.50	14.52	42.39
213	15068	C8H16O	2,2-dimethyl-3-hexanone	5405-79-8	8.6345	0.2281	1	1000	2	8.63	14.60	41.73
214	15069	C8H16O	3,3-dimethyl-2-hexanone	26118-38-7	8.5363	0.2312	1	1000	2	8.54	14.54	42.16
215	15070	C8H16O	3,4-dimethyl-2-hexanone	19550-10-8	9.8130	0.2327	1	1000	2	9.81	16.77	48.97
216	15071	C8H16O	4,4-dimethyl-3-hexanone	19550-14-2	8.6016	0.2320	1	1000	2	8.60	14.67	42.71
217	15072	C8H16O	3-ethyl-4-methyl-2-pentanone	71172-57-1	9.7325	0.2274	1	1000	2	9.73	16.43	46.82
218	15073	C8H16O	2-methyl-3-heptanone	13019-20-0	10.1111	0.2286	1	1000	2	10.11	17.12	49.04
219	15074	C8H16O	2-methyl-4-heptanone	626-33-5	9.6164	0.2281	1	1000	2	9.62	16.26	46.49
220	15075	C8H16O	3-methyl-2-heptanone	2371-19-9	10.8443	0.2301	1	994	2	10.84	18.42	53.09
221	15076	C8H16O	3-methyl-4-heptanone	15726-15-5	9.1491	0.2329	1	1000	2	9.15	15.64	45.72
222	15077	C8H16O	5-methyl-3-heptanone	541-85-5	9.6514	0.2429	1	1000	2	9.65	16.88	51.66
223	15078	C8H16O	6-methyl-2-heptanone	928-68-7	11.4358	0.2283	1	833	2	11.44	19.34	53.09
224	15079	C8H16O	6-methyl-3-heptanone	624-42-0	10.6663	0.2325	1	994	2	10.67	18.22	53.09
225	15080	C8H16O	3-octanone	106-68-3	11.2818	0.2313	1	809	2	11.28	19.22	53.09
226	15081	C8H16O	4-octanone	589-63-9	10.7677	0.2292	1	1000	2	10.77	18.25	52.45
227	15082	C8H16O	2,2,4-trimethyl-3-pentanone	5857-36-3	7.5767	0.2259	1	1000	2	7.58	12.75	36.06
228	15147	C8H16O	3-ethyl-3-methyl-2-pentanone	19780-65-5	8.5426	0.2429	1	1000	2	8.54	14.94	45.73
229	15148	C8H16O	5-methyl-2-heptanone	18217-12-4	9.9109	0.2429	1	1000	2	9.91	17.34	53.05
230	15160	C8H16O2	hexyl acetate	142-92-7	13.7087	0.2171	1	511	2	13.71	22.60	53.09
231	15161	C8H16O2	1,1-dimethylbutyl acetate	34859-98-8	10.8737	0.2268	1	1000	2	10.87	18.33	52.11
232	15162	C8H16O2	2-ethylbutyl acetate	10031-87-5	12.8634	0.2188	1	651	2	12.86	21.29	53.09
233	15163	C8H16O2	sec-hexyl acetate	108-84-9	10.6669	0.2203	1	1000	2	10.67	17.71	48.85
234	15164	C8H16O2	2-methyl-1-pentyl acetate	7789-99-3	13.0385	0.2176	1	634	2	13.04	21.52	53.09
235	15165	C8H16O2	2-methyl-3-pentyl acetate	35897-16-6	11.1736	0.2163	1	1000	2	11.17	18.39	49.78
236	15166	C8H16O2	3-methyl-3-pentyl acetate	10250-47-2	10.7890	0.2195	1	1000	2	10.79	17.89	49.15
237	15430	C8H18O	1-octanol	111-87-5	12.6489	0.2280	1	540	2	12.65	21.38	53.09
238	15431	C8H18O	2-octanol	123-96-6	10.4771	0.2263	1	1000	2	10.48	17.64	50.03
239	15432	C8H18O	2-octanol, (±)	4128-31-8	11.0009	0.2258	1	1000	2	11.00	18.50	52.33
240	15433	C8H18O	3-octanol	589-98-0	10.1956	0.2265	1	1000	2	10.20	17.18	48.74

Table 2 Adsorption Capacity of Activated Carbon - Aqueous Liquid

				$\log_{10} Q = \log_{10} K + 1/n \log_{10} x$			(Q - g of compound/100 g of carbon, x - ppm(wt))					
NO	ID	FORMULA	NAME	CAS No	K	1/n	xmin	xmax	code	Q @xmin	Q @10ppm	Q @xmax
241	15434	C8H18O	4-octanol	589-62-8	10.5039	0.2259	1	1000	2	10.50	17.67	50.00
242	15435	C8H18O	4-octanol, (±)	74778-22-6	10.4842	0.2256	1	1000	2	10.48	17.62	49.81
243	15436	C8H18O	2-methyl-1-heptanol	60435-70-3	10.7194	0.2212	1	1000	2	10.72	17.84	49.41
244	15437	C8H18O	3-methyl-1-heptanol	1070-32-2	12.6150	0.2173	1	745	2	12.61	20.81	53.09
245	15438	C8H18O	4-methyl-1-heptanol	817-91-4	11.6696	0.2234	1	881	2	11.67	19.52	53.09
246	15439	C8H18O	5-methyl-1-heptanol	7212-53-5	12.0180	0.2259	1	719	2	12.02	20.22	53.09
247	15440	C8H18O	6-methyl-1-heptanol	1653-40-3	12.1505	0.2265	1	673	2	12.15	20.47	53.09
248	15441	C8H18O	2-methyl-2-heptanol	625-25-2	8.2176	0.2236	1	1000	2	8.22	13.75	38.51
249	15442	C8H18O	3-methyl-2-heptanol	31367-46-1	9.0895	0.2265	1	1000	2	9.09	15.31	43.46
250	15443	C8H18O	4-methyl-2-heptanol	56298-90-9	10.1681	0.2213	1	1000	2	10.17	16.93	46.90
251	15444	C8H18O	5-methyl-2-heptanol	54630-50-1	9.9809	0.2243	1	1000	2	9.98	16.73	47.01
252	15445	C8H18O	6-methyl-2-heptanol	4730-22-7	10.1157	0.2225	1	1000	2	10.12	16.89	47.05
253	15446	C8H18O	2-methyl-3-heptanol	18720-62-2	9.2175	0.2273	1	1000	2	9.22	15.56	44.31
254	15447	C8H18O	3-methyl-3-heptanol	5582-82-1	8.3536	0.2286	1	1000	2	8.35	14.14	40.51
255	15448	C8H18O	4-methyl-3-heptanol	14979-39-6	8.3307	0.2200	1	1000	2	8.33	13.82	38.07
256	15449	C8H18O	5-methyl-3-heptanol	18720-65-5	7.7575	0.2256	1	1000	2	7.76	13.04	36.85
257	15450	C8H18O	6-methyl-3-heptanol	18720-66-6	8.3752	0.2264	1	1000	2	8.38	14.10	40.00
258	15451	C8H18O	2-methyl-4-heptanol	21570-35-4	9.2458	0.2243	1	1000	2	9.25	15.50	43.54
259	15452	C8H18O	3-methyl-4-heptanol	1838-73-9	8.6309	0.2307	1	1000	2	8.63	14.68	42.49
260	15453	C8H18O	4-methyl-4-heptanol	598-01-6	8.4723	0.2270	1	1000	2	8.47	14.29	40.64
261	15454	C8H18O	2-ethyl-1-hexanol	104-76-7	12.9341	0.1824	1	500	1,2	12.93	19.69	40.19
262	15455	C8H18O	3-ethyl-1-hexanol	41065-95-6	10.6448	0.2302	1	1000	2	10.64	18.09	52.21
263	15456	C8H18O	4-ethyl-1-hexanol	------	10.9332	0.2264	1	1000	2	10.93	18.41	52.22
264	15457	C8H18O	2,2-dimethyl-1-hexanol	2370-13-0	9.8113	0.2277	1	1000	2	9.81	16.57	47.30
265	15458	C8H18O	2,3-dimethyl-1-hexanol	------	10.0892	0.2264	1	1000	2	10.09	16.99	48.19
266	15459	C8H18O	2,4-dimethyl-1-hexanol	3965-59-1	10.2453	0.2264	1	1000	2	10.25	17.25	48.94
267	15460	C8H18O	2,5-dimethyl-1-hexanol	6886-16-4	10.6456	0.2285	1	1000	2	10.65	18.02	51.62
268	15461	C8H18O	3,3-dimethyl-1-hexanol	10524-70-6	10.7089	0.2313	1	1000	2	10.71	18.24	52.93
269	15462	C8H18O	3,4-dimethyl-1-hexanol	------	11.0808	0.2264	1	1000	2	11.08	18.66	52.93
270	15463	C8H18O	3,5-dimethyl-1-hexanol	13501-73-0	11.2615	0.2283	1	892	2	11.26	19.05	53.09
271	15464	C8H18O	4,4-dimethyl-1-hexanol	6481-95-4	11.2308	0.2264	1	955	2	11.23	18.91	53.09
272	15465	C8H18O	4,5-dimethyl-1-hexanol	------	11.2310	0.2264	1	955	2	11.23	18.91	53.09
273	15466	C8H18O	5,5-dimethyl-1-hexanol	------	11.2308	0.2264	1	955	2	11.23	18.91	53.09
274	15467	C8H18O	3-ethyl-2-hexanol	24448-19-9	9.3332	0.2264	1	1000	2	9.33	15.72	44.58
275	15468	C8H18O	4-ethyl-2-hexanol	------	9.3146	0.2264	1	1000	2	9.31	15.69	44.49
276	15469	C8H18O	2,3-dimethyl-2-hexanol	19550-03-9	8.1257	0.2305	1	1000	2	8.13	13.81	39.93
277	15470	C8H18O	2,4-dimethyl-2-hexanol	42328-76-7	7.6141	0.2233	1	1000	2	7.61	12.73	35.60
278	15471	C8H18O	2,5-dimethyl-2-hexanol	3730-60-7	7.6866	0.2249	1	1000	2	7.69	12.90	36.35
279	15472	C8H18O	3,3-dimethyl-2-hexanol	22025-20-3	7.9173	0.2332	1	1000	2	7.92	13.55	39.66
280	15473	C8H18O	3,4-dimethyl-2-hexanol	19550-05-1	9.4325	0.2303	1	1000	2	9.43	16.03	46.29
281	15474	C8H18O	3,5-dimethyl-2-hexanol	66576-27-0	8.3916	0.2264	1	1000	2	8.39	14.13	40.08
282	15475	C8H18O	4,4-dimethyl-2-hexanol	------	8.6004	0.2264	1	1000	2	8.60	14.48	41.08
283	15476	C8H18O	4,5-dimethyl-2-hexanol	------	8.6005	0.2264	1	1000	2	8.60	14.48	41.08
284	15477	C8H18O	5,5-dimethyl-2-hexanol	31841-77-7	9.0879	0.2264	1	1000	2	9.09	15.30	43.41
285	15478	C8H18O	3-ethyl-3-hexanol	597-76-2	7.9819	0.2308	1	1000	2	7.98	13.58	39.32
286	15479	C8H18O	4-ethyl-3-hexanol	19780-44-0	8.8501	0.2264	1	1000	2	8.85	14.90	42.27
287	15480	C8H18O	2,2-dimethyl-3-hexanol	4209-90-9	7.7327	0.2299	1	1000	2	7.73	13.13	37.85
288	15481	C8H18O	2,3-dimethyl-3-hexanol	4166-46-5	7.9011	0.2308	1	1000	2	7.90	13.44	38.90
289	15482	C8H18O	2,4-dimethyl-3-hexanol	13432-25-2	8.3916	0.2264	1	1000	2	8.39	14.13	40.08
290	15483	C8H18O	2,5-dimethyl-3-hexanol	19550-07-3	8.3286	0.2257	1	1000	2	8.33	14.00	39.59
291	15484	C8H18O	3,4-dimethyl-3-hexanol	19550-08-4	7.2449	0.2310	1	1000	2	7.24	12.33	35.74
292	15485	C8H18O	3,5-dimethyl-3-hexanol	4209-91-0	7.4377	0.2280	1	1000	2	7.44	12.57	35.93
293	15486	C8H18O	4,4-dimethyl-3-hexanol	19550-09-5	8.0420	0.2299	1	1000	2	8.04	13.65	39.37
294	15487	C8H18O	4,5-dimethyl-3-hexanol	------	7.9408	0.2264	1	1000	2	7.94	13.37	37.93
295	15488	C8H18O	5,5-dimethyl-3-hexanol	66576-31-6	7.6445	0.2264	1	1000	2	7.64	12.87	36.51
296	15489	C8H18O	2-propyl-1-pentanol	58175-57-8	10.8091	0.2264	1	1000	2	10.81	18.20	51.63
297	15490	C8H18O	2-isopropyl-1-pentanol	------	14.5279	0.2264	1	306	2	14.53	24.47	53.09
298	15491	C8H18O	2-methyl-2-ethyl-1-pentanol	5970-63-8	10.5369	0.2264	1	1000	2	10.54	17.75	50.33
299	15492	C8H18O	3-methyl-2-ethyl-1-pentanol	------	10.5019	0.2264	1	1000	2	10.50	17.69	50.16
300	15493	C8H18O	4-methyl-2-ethyl-1-pentanol	106-67-2	10.2710	0.2288	1	1000	2	10.27	17.40	49.90
301	15494	C8H18O	2-methyl-3-ethyl-1-pentanol	------	10.2248	0.2264	1	1000	2	10.22	17.22	48.84
302	15495	C8H18O	3-methyl-3-ethyl-1-pentanol	------	10.2247	0.2264	1	1000	2	10.22	17.22	48.84
303	15496	C8H18O	4-methyl-3-ethyl-1-pentanol	------	10.2248	0.2264	1	1000	2	10.22	17.22	48.84
304	15497	C8H18O	2,2,3-trimethyl-1-pentanol	57409-53-7	10.1767	0.2264	1	1000	2	10.18	17.14	48.61
305	15498	C8H18O	2,2,4-trimethyl-1-pentanol	123-44-4	9.0413	0.2310	1	1000	2	9.04	15.39	44.60
306	15499	C8H18O	2,3,3-trimethyl-1-pentanol	------	10.3619	0.2264	1	1000	2	10.36	17.45	49.50
307	15500	C8H18O	2,3,4-trimethyl-1-pentanol	6570-88-3	10.8048	0.2344	1	892	2	10.80	18.53	53.09
308	15501	C8H18O	2,4,4-trimethyl-1-pentanol	16325-63-6	9.4771	0.2296	1	1000	2	9.48	16.08	46.30
309	15502	C8H18O	3,3,4-trimethyl-1-pentanol	------	9.6933	0.2264	1	1000	2	9.69	16.32	46.30
310	15503	C8H18O	3,4,4-trimethyl-1-pentanol	16325-64-7	9.6933	0.2264	1	1000	2	9.69	16.32	46.30
311	15504	C8H18O	2-methyl-3-ethyl-2-pentanol	19780-63-3	7.8224	0.2313	1	1000	2	7.82	13.32	38.66
312	15505	C8H18O	3-methyl-3-ethyl-2-pentanol	66576-22-5	7.7055	0.2366	1	1000	2	7.71	13.29	39.49
313	15506	C8H18O	4-methyl-3-ethyl-2-pentanol	66576-23-6	8.8497	0.2264	1	1000	2	8.85	14.90	42.27
314	15507	C8H18O	2,3,3-trimethyl-2-pentanol	23171-85-9	8.4322	0.2258	1	1000	2	8.43	14.18	40.11
315	15508	C8H18O	2,3,4-trimethyl-2-pentanol	66576-26-9	8.4208	0.2227	1	1000	2	8.42	14.06	39.22
316	15509	C8H18O	2,4,4-trimethyl-2-pentanol	690-37-9	6.9642	0.2269	1	1000	2	6.96	11.74	33.38
317	15510	C8H18O	3,3,4-trimethyl-2-pentanol	19411-41-7	8.3132	0.2360	1	1000	2	8.31	14.31	42.44
318	15511	C8H18O	3,4,4-trimethyl-2-pentanol	10575-56-1	5.4616	0.2751	1	1000	2	5.46	10.29	36.53
319	15512	C8H18O	2-methyl-3-ethyl-3-pentanol	597-05-7	8.2254	0.2288	1	1000	2	8.23	13.93	39.96
320	15513	C8H18O	2,2,3-trimethyl-3-pentanol	7294-05-5	7.1078	0.2332	1	1000	2	7.11	12.16	35.60

Table 2 Adsorption Capacity of Activated Carbon - Aqueous Liquid

				$\log_{10} Q = \log_{10} K + 1/n \log_{10} x$			(Q - g of compound/100 g of carbon, x - ppm(wt))					
NO	ID	FORMULA	NAME	CAS No	K	1/n	xmin	xmax	code	Q @xmin	Q @10ppm	Q @xmax
321	15514	C8H18O	2,2,4-trimethyl-3-pentanol	5162-48-1	7.2505	0.2294	1	1000	2	7.25	12.30	35.36
322	15515	C8H18O	2,3,4-trimethyl-3-pentanol	3054-92-0	7.5613	0.2341	1	1000	2	7.56	12.96	38.09
323	15516	C8H18O	3-methyl-2-isopropyl-1-butanol	18593-92-5	9.4814	0.2332	1	1000	2	9.48	16.22	47.49
324	15517	C8H18O	2,2-diethyl-1-butanol	13023-60-4	9.9554	0.2264	1	1000	2	9.96	16.77	47.55
325	15518	C8H18O	2,3-dimethyl-2-ethyl-1-butanol	------	9.9551	0.2264	1	1000	2	9.96	16.77	47.55
326	15519	C8H18O	3,3-dimethyl-2-ethyl-1-butanol	66576-56-5	9.4425	0.2335	1	1000	2	9.44	16.17	47.39
327	15520	C8H18O	2,2,3,3-tetramethyl-1-butanol	66576-24-7	6.5608	0.2828	1	1000	2	6.56	12.58	46.27
328	15521	C8H18O	3,5-dimethyl-3-hexanol, (±)	19113-78-1	7.2650	0.2307	1	1000	2	7.26	12.36	35.76
329	15522	C8H18O	2-methyl-1-heptanol, (±)	111675-77-5	10.7304	0.2211	1	1000	2	10.73	17.85	49.42
330	15523	C8H18O	5-methyl-1-heptanol, (±)	111767-95-4	12.0180	0.2259	1	719	2	12.02	20.22	53.09
331	15524	C8H18O	2-methyl-3-heptanol, (±)	100296-26-2	9.2312	0.2269	1	1000	2	9.23	15.57	44.26
332	15525	C8H18O	6-methyl-3-heptanol, (±)	100295-85-0	9.4473	0.2265	1	1000	2	9.45	15.92	45.17
333	15526	C8H18O	3-methyl-4-heptanol, (R*,S*)-(±)	92737-91-2	8.3097	0.2309	1	1000	2	8.31	14.14	40.95
334	15535	C8H18O	isooctanol	26952-21-6	11.5615	0.2305	1	745	2	11.56	19.66	53.09
335	15536	C8H18O	(R)-(-)-2-octanol	5978-70-1	9.9262	0.2321	1	1000	2	9.93	16.94	49.34
336	15537	C8H18O	(S)-(+)-2-octanol	6169-06-8	10.5502	0.2277	1	1000	2	10.55	17.82	50.86
337	15538	C8H18O	octyl alcohol; mixed isomers	29063-28-3	13.4394	0.2264	1	432	2	13.44	22.63	53.09
338	17004	C9H12	cumene	98-82-8	28.4928	0.2558	0.10	50	2	15.81	51.35	77.50
339	17005	C9H12	m-ethyltoluene	620-14-4	24.0503	0.2558	0.10	97	2	13.35	43.34	77.50
340	17006	C9H12	o-ethyltoluene	611-14-3	23.7567	0.2608	0.10	93	2	13.03	43.31	77.50
341	17007	C9H12	p-ethyltoluene	622-96-8	24.2871	0.2549	0.10	95	2	13.50	43.68	77.50
342	17008	C9H12	1,2,3-trimethylbenzene	526-73-8	24.6661	0.2650	0.10	75	2	13.40	45.40	77.50
343	17009	C9H12	1,2,4-trimethylbenzene	95-63-6	26.9172	0.2594	0.10	59	2	14.81	48.91	77.50
344	17010	C9H12	trimethyl benzene	25551-13-7	29.1837	0.2554	0.10	46	2	16.21	52.54	77.50
345	17011	C9H12	mesitylene	108-67-8	24.0176	0.2561	0.10	97	2	13.32	43.31	77.50
346	17012	C9H12	propylbenzene	103-65-1	28.1863	0.2558	0.10	52	2	15.64	50.79	77.50
347	----	C9H13ClN6	cyanazine	------	24.3557	0.1260	0.0740	2.2	1,2	17.54	----	26.90
348	17997	C9H18O	1-nonanal	124-19-6	20.1870	0.2077	1	105	2	20.19	32.57	53.09
349	17998	C9H18O	diisobutyl ketone	108-83-8	10.6910	0.2034	1	1000	2	10.69	17.08	43.58
350	17999	C9H18O	2-nonanone	821-55-6	18.3929	0.2071	1	167	2	18.39	29.63	53.09
351	18000	C9H18O	di-tert-butyl ketone	815-24-7	10.9421	0.2076	1	1000	2	10.94	17.65	45.91
352	18001	C9H18O	3,5-dimethyl-4-heptanone	19549-84-9	12.3890	0.2073	1	1000	2	12.39	19.97	51.88
353	18002	C9H18O	3-methyl-2-octanone	6137-08-2	16.1000	0.2074	1	315	2	16.10	25.96	53.09
354	18003	C9H18O	3-methyl-4-octanone	20754-04-5	14.2729	0.2081	1	552	2	14.27	23.04	53.09
355	18004	C9H18O	4-methyl-3-octanone	6137-15-1	16.0495	0.2080	1	315	2	16.05	25.91	53.09
356	18005	C9H18O	5-methyl-2-octanone	58654-67-4	14.9738	0.2200	1	315	2	14.97	24.85	53.09
357	18006	C9H18O	7-methyl-3-octanone	5408-57-1	15.5036	0.2119	1	334	2	15.50	25.25	53.09
358	18007	C9H18O	7-methyl-4-octanone	20809-46-5	15.0010	0.2080	1	436	2	15.00	24.21	53.09
359	18008	C9H18O	3-nonanone	925-78-0	17.4095	0.2080	1	213	2	17.41	28.11	53.09
360	18009	C9H18O	4-nonanone	4485-09-0	16.8991	0.2077	1	247	2	16.90	27.26	53.09
361	18010	C9H18O	5-nonanone	502-56-7	15.6363	0.2074	1	363	2	15.64	25.21	53.09
362	18044	C9H18O	3,5,5-trimethylhexanal	5435-64-3	17.1006	0.2072	1	237	2	17.10	27.56	53.09
363	18045	C9H18O	2,6-dimethyl-3-heptanone	19549-83-8	12.8692	0.2200	1	627	2	12.87	21.36	53.09
364	18047	C9H18O	2-methyl-3-octanone	923-28-4	14.8520	0.2200	1	327	2	14.85	24.65	53.09
365	18048	C9H18O	2-methyloctanal	7786-29-0	15.9405	0.2201	1	237	2	15.94	26.46	53.09
366	18059	C9H18O2	heptyl acetate	112-06-1	20.3119	0.1973	1	130	2	20.31	32.00	53.09
367	18060	C9H18O2	4-heptyl acetate	5921-84-6	15.9933	0.1993	1	412	2	15.99	25.31	53.09
368	18236	C9H20O	1-nonanol	143-08-8	19.4770	0.2061	1	130	2	19.48	31.30	53.09
369	18237	C9H20O	2-nonanol	628-99-9	16.9779	0.2051	1	260	2	16.98	27.22	53.09
370	18238	C9H20O	2-nonanol, (±)	74683-66-2	14.4933	0.2107	1	474	2	14.49	23.55	53.09
371	18239	C9H20O	3-nonanol	624-51-1	15.2040	0.2054	1	440	2	15.20	24.40	53.09
372	18240	C9H20O	3-nonanol, (±)	74742-08-8	15.2748	0.2053	1	432	2	15.27	24.50	53.09
373	18241	C9H20O	4-nonanol	5932-79-6	14.8699	0.2056	1	488	2	14.87	23.87	53.09
374	18242	C9H20O	4-nonanol, (S)	52708-03-9	14.7313	0.2061	1	503	2	14.73	23.68	53.09
375	18243	C9H20O	5-nonanol	623-93-8	15.3528	0.2046	1	429	2	15.35	24.59	53.09
376	18244	C9H20O	2-methyl-1-octanol	818-81-5	16.0286	0.2076	1	321	2	16.03	25.85	53.09
377	18245	C9H20O	3-methyl-1-octanol	38514-02-2	16.1201	0.2066	1	321	2	16.12	25.94	53.09
378	18246	C9H20O	4-methyl-1-octanol	38514-03-3	16.2134	0.2056	1	321	2	16.21	26.03	53.09
379	18247	C9H20O	5-methyl-1-octanol	38514-04-4	16.0968	0.2068	1	321	2	16.10	25.92	53.09
380	18248	C9H20O	6-methyl-1-octanol	38514-05-5	17.3822	0.2073	1	218	2	17.38	28.02	53.09
381	18249	C9H20O	7-methyl-1-octanol	2430-22-0	12.7784	0.2645	1	218	2	12.78	23.49	53.09
382	18250	C9H20O	2-methyl-2-octanol	628-44-4	12.4965	0.2040	1	1000	2	12.50	19.99	51.15
383	18251	C9H20O	3-methyl-2-octanol	27644-49-1	12.5560	0.2083	1	1000	2	12.56	20.28	52.94
384	18252	C9H20O	5-methyl-2-octanol	66793-81-5	12.8195	0.2053	1	1000	2	12.82	20.57	52.94
385	18253	C9H20O	7-methyl-2-octanol	66793-83-7	12.6229	0.2076	1	1000	2	12.62	20.36	52.94
386	18254	C9H20O	2-methyl-3-octanol	26533-34-6	13.2358	0.2063	1	840	2	13.24	21.28	53.09
387	18255	C9H20O	3-methyl-3-octanol	5340-36-3	14.0134	0.2070	1	622	2	14.01	22.57	53.09
388	18256	C9H20O	4-methyl-3-octanol	66793-80-4	13.1201	0.2111	1	752	2	13.12	21.33	53.09
389	18257	C9H20O	6-methyl-3-octanol	40225-75-0	13.3813	0.2081	1	752	2	13.38	21.61	53.09
390	18258	C9H20O	7-methyl-3-octanol	66793-84-8	13.4279	0.2076	1	752	2	13.43	21.66	53.09
391	18259	C9H20O	2-methyl-4-octanol	40575-41-5	13.4585	0.2038	1	840	2	13.46	21.52	53.09
392	18260	C9H20O	3-methyl-4-octanol	26533-35-7	12.5334	0.2071	1	1000	2	12.53	20.19	52.39
393	18261	C9H20O	4-methyl-4-octanol	23418-37-3	12.7830	0.2060	1	1000	2	12.78	20.54	53.04
394	18262	C9H20O	5-methyl-4-octanol	59734-23-5	13.0889	0.2041	1	955	2	13.09	20.94	53.09
395	18263	C9H20O	6-methyl-4-octanol	66793-82-6	12.9993	0.2051	1	955	2	13.00	20.84	53.09
396	18264	C9H20O	7-methyl-4-octanol	33933-77-6	13.2243	0.2026	1	955	2	13.22	21.08	53.09
397	18265	C9H20O	2-ethyl-1-heptanol	817-60-7	15.8239	0.2076	1	341	2	15.82	25.52	53.09
398	18266	C9H20O	3-ethyl-1-heptanol	3525-25-5	17.5439	0.2081	1	205	2	17.54	28.33	53.09
399	18267	C9H20O	5-ethyl-1-heptanol	998-65-2	15.4126	0.2121	1	341	2	15.41	25.12	53.09
400	18268	C9H20O	2,2-dimethyl-1-heptanol	14250-79-4	14.5040	0.2076	1	519	2	14.50	23.39	53.09

Table 2 Adsorption Capacity of Activated Carbon - Aqueous Liquid

				$\log_{10} Q = \log_{10} K + 1/n \log_{10} x$			(Q - g of compound/100 g of carbon, x - ppm(wt))					
NO	ID	FORMULA	NAME	CAS No	K	1/n	xmin	xmax	code	Q @xmin	Q @10ppm	Q @xmax
401	18269	C9H20O	4,6-dimethyl-1-heptanol	820-05-3	12.7797	0.2076	1	955	2	12.78	20.61	53.09
402	18270	C9H20O	6,6-dimethyl-1-heptanol	65769-10-0	12.5608	0.2101	1	955	2	12.56	20.37	53.09
403	18271	C9H20O	2-ethyl-2-heptanol	------	12.7797	0.2076	1	955	2	12.78	20.61	53.09
404	18272	C9H20O	3-ethyl-2-heptanol	19780-39-3	12.7797	0.2076	1	955	2	12.78	20.61	53.09
405	18273	C9H20O	2,3-dimethyl-2-heptanol	66794-00-1	12.7796	0.2076	1	955	2	12.78	20.61	53.09
406	18274	C9H20O	2,4-dimethyl-2-heptanol	65822-93-7	12.8221	0.2071	1	955	2	12.82	20.66	53.09
407	18275	C9H20O	2,5-dimethyl-2-heptanol	------	12.8221	0.2071	1	955	2	12.82	20.66	53.09
408	18276	C9H20O	2,6-dimethyl-2-heptanol	13254-34-7	11.5040	0.2038	1	1000	2	11.50	18.39	47.02
409	18277	C9H20O	4,6-dimethyl-2-heptanol	51079-52-8	14.2271	0.2148	1	459	2	14.23	23.33	53.09
410	18278	C9H20O	5,6-dimethyl-2-heptanol	58795-24-7	14.5254	0.2073	1	519	2	14.53	23.41	53.09
411	18279	C9H20O	3-ethyl-3-heptanol	19780-41-7	12.6808	0.2093	1	935	2	12.68	20.53	53.09
412	18280	C9H20O	2,2-dimethyl-3-heptanol	19549-70-3	12.2939	0.2061	1	1000	2	12.29	19.76	51.04
413	18281	C9H20O	2,3-dimethyl-3-heptanol	19549-71-4	11.5390	0.2083	1	1000	2	11.54	18.64	48.65
414	18282	C9H20O	2,6-dimethyl-3-heptanol	19549-73-6	12.0264	0.2043	1	1000	2	12.03	19.25	49.33
415	18283	C9H20O	3,5-dimethyl-3-heptanol	19549-74-7	12.2321	0.2051	1	1000	2	12.23	19.61	50.43
416	18284	C9H20O	3,6-dimethyl-3-heptanol	1573-28-0	11.6536	0.2053	1	1000	2	11.65	18.70	48.13
417	18285	C9H20O	4-ethyl-4-heptanol	597-90-0	12.3152	0.2078	1	1000	2	12.32	19.87	51.75
418	18286	C9H20O	2,2-dimethyl-4-heptanol	66793-99-5	11.6041	0.2075	1	1000	2	11.60	18.71	48.67
419	18287	C9H20O	2,4-dimethyl-4-heptanol	19549-77-0	11.4303	0.2053	1	1000	2	11.43	18.34	47.21
420	18288	C9H20O	2,6-dimethyl-4-heptanol	108-82-7	13.1689	0.2018	1	1000	2	13.17	20.96	53.09
421	18289	C9H20O	3,3-dimethyl-4-heptanol	19549-78-1	11.8919	0.2076	1	1000	2	11.89	19.18	49.88
422	18290	C9H20O	3,5-dimethyl-4-heptanol	19549-79-2	13.0748	0.2138	1	702	2	13.07	21.39	53.09
423	18291	C9H20O	2-propyl-1-hexanol	817-46-9	13.9417	0.2076	1	628	2	13.94	22.48	53.09
424	18292	C9H20O	3-propyl-1-hexanol	66793-85-9	13.9417	0.2076	1	628	2	13.94	22.48	53.09
425	18293	C9H20O	1-methyl-2-ethyl-1-hexanol	------	13.9415	0.2076	1	628	2	13.94	22.48	53.09
426	18294	C9H20O	1-methyl-3-ethyl-1-hexanol	------	14.1444	0.2116	1	519	2	14.14	23.02	53.09
427	18295	C9H20O	3-methyl-2-ethyl-1-hexanol	66794-04-5	14.5511	0.2091	1	488	2	14.55	23.55	53.09
428	18296	C9H20O	4-methyl-2-ethyl-1-hexanol	66794-06-7	15.1785	0.2063	1	432	2	15.18	24.41	53.09
429	18297	C9H20O	5-methyl-2-ethyl-1-hexanol	66794-07-8	14.8651	0.2053	1	493	2	14.87	23.85	53.09
430	18298	C9H20O	3,3,5-trimethyl-1-hexanol	1484-87-3	14.6602	0.2076	1	493	2	14.66	23.64	53.09
431	18299	C9H20O	3,4,4-trimethyl-1-hexanol	66793-73-5	14.3224	0.2076	1	551	2	14.32	23.10	53.09
432	18300	C9H20O	3,5,5-trimethyl-1-hexanol	3452-97-9	14.8240	0.2061	1	488	2	14.82	23.82	53.09
433	18301	C9H20O	4,5,5-trimethyl-1-hexanol	66793-75-7	16.4795	0.2076	1	280	2	16.48	26.58	53.09
434	18302	C9H20O	2-methyl-3-ethyl-2-hexanol	66794-02-3	12.1128	0.2084	1	1000	2	12.11	19.57	51.11
435	18303	C9H20O	2,3,4-trimethyl-2-hexanol	21102-13-6	14.7274	0.2068	1	493	2	14.73	23.71	53.09
436	18304	C9H20O	2,4,4-trimethyl-2-hexanol	66793-91-7	14.3443	0.2111	1	493	2	14.34	23.32	53.09
437	18305	C9H20O	2,4,5-trimethyl-2-hexanol	66793-93-9	14.7046	0.2071	1	493	2	14.70	23.69	53.09
438	18306	C9H20O	2,5,5-trimethyl-2-hexanol	66793-71-3	14.6601	0.2076	1	493	2	14.66	23.64	53.09
439	18307	C9H20O	2-methyl-3-ethyl-3-hexanol	66794-03-4	12.8225	0.2112	1	835	2	12.82	20.85	53.09
440	18308	C9H20O	2-methyl-4-ethyl-3-hexanol	33943-21-4	12.4449	0.2061	1	1000	2	12.44	20.00	51.66
441	18309	C9H20O	3-methyl-4-ethyl-3-hexanol	66794-05-6	10.8979	0.2249	1	1000	2	10.90	18.29	51.54
442	18310	C9H20O	4-methyl-3-ethyl-3-hexanol	51200-80-7	11.8255	0.2133	1	1000	2	11.83	19.33	51.62
443	18311	C9H20O	5-methyl-3-ethyl-3-hexanol	597-77-3	11.1756	0.2093	1	1000	2	11.18	18.10	47.45
444	18312	C9H20O	2,2,3-trimethyl-3-hexanol	5340-41-0	11.2242	0.2106	1	1000	2	11.22	18.23	48.07
445	18313	C9H20O	2,2,4-trimethyl-3-hexanol	66793-89-3	10.9164	0.2075	1	1000	2	10.92	17.60	45.79
446	18314	C9H20O	2,2,5-trimethyl-3-hexanol	3970-60-3	10.7199	0.1963	1	1000	2	10.72	16.85	41.60
447	18315	C9H20O	2,3,4-trimethyl-3-hexanol	66793-90-6	10.5051	0.2075	1	1000	2	10.51	16.94	44.06
448	18316	C9H20O	2,3,5-trimethyl-3-hexanol	65927-60-8	10.6273	0.2061	1	1000	2	10.63	17.08	44.12
449	18317	C9H20O	2,4,4-trimethyl-3-hexanol	66793-92-8	10.8743	0.2113	1	1000	2	10.87	17.69	46.81
450	18318	C9H20O	2,5,5-trimethyl-3-hexanol	66793-72-4	11.0836	0.2053	1	1000	2	11.08	17.78	45.78
451	18319	C9H20O	3,4,4-trimethyl-3-hexanol	66793-74-6	10.5433	0.2073	1	1000	2	10.54	16.99	44.15
452	18320	C9H20O	3,5,5-trimethyl-3-hexanol	66810-87-5	10.4831	0.2078	1	1000	2	10.48	16.92	44.05
453	18321	C9H20O	4-methyl-2-propyl-1-pentanol	54004-41-0	14.6852	0.2056	1	519	2	14.69	23.57	53.09
454	18322	C9H20O	4-methyl-2-isopropyl-1-pentanol	55505-24-3	13.6231	0.2076	1	702	2	13.62	21.97	53.09
455	18323	C9H20O	2,2-dimethyl-3-ethyl-1-pentanol	66793-95-1	13.7682	0.2076	1	667	2	13.77	22.20	53.09
456	18324	C9H20O	2,4-dimethyl-2-ethyl-1-pentanol	66793-98-4	13.6368	0.2093	1	661	2	13.64	22.08	53.09
457	18325	C9H20O	3,3-diethyl-2-pentanol	66793-94-0	13.7682	0.2076	1	667	2	13.77	22.20	53.09
458	18326	C9H20O	2,3-dimethyl-3-ethyl-2-pentanol	66793-97-3	13.7680	0.2076	1	667	2	13.77	22.20	53.09
459	18327	C9H20O	4,4-dimethyl-3-ethyl-2-pentanol	21102-09-0	13.7680	0.2076	1	667	2	13.77	22.20	53.09
460	18328	C9H20O	2,3,3,4-tetramethyl-2-pentanol	66793-86-0	12.4678	0.2076	1	1000	2	12.47	20.11	52.29
461	18329	C9H20O	2,3,4,4-tetramethyl-2-pentanol	66793-87-1	12.4678	0.2076	1	1000	2	12.47	20.11	52.29
462	18330	C9H20O	3,3,4,4-tetramethyl-2-pentanol	66793-88-2	8.6233	0.2607	1	1000	2	8.62	15.72	52.21
463	18331	C9H20O	2-methyl-3-isopropyl-3-pentanol	------	11.6370	0.2141	1	1000	2	11.64	19.05	51.06
464	18332	C9H20O	2,2-dimethyl-3-ethyl-3-pentanol	66793-96-2	11.1320	0.2132	1	1000	2	11.13	18.19	48.55
465	18333	C9H20O	2,4-dimethyl-3-ethyl-3-pentanol	3970-59-0	11.6574	0.2136	1	1000	2	11.66	19.07	51.00
466	18334	C9H20O	2,2,3,4-tetramethyl-3-pentanol	62185-29-9	11.1661	0.2131	1	1000	2	11.17	18.24	48.68
467	18335	C9H20O	2,2,4,4-tetramethyl-3-pentanol	14609-79-1	7.3742	0.2551	1	1000	2	7.37	13.27	42.95
468	----	C10H12N2O5	dinoseb	------	20.8869	0.2790	0.0011	2.84	1,2	3.12	----	27.95
469	19865	C10H14	butylbenzene	104-51-8	42.5279	0.2285	0.10	14	2	25.13	71.98	77.50
470	19866	C10H14	isobutylbenzene	538-93-2	45.9412	0.2261	0.10	10	2	27.29	77.33	77.50
471	19867	C10H14	sec-butylbenzene	135-98-8	40.2419	0.2285	0.10	18	2	23.78	68.11	77.50
472	19868	C10H14	tert-butylbenzene	98-06-6	35.6010	0.2298	0.10	30	2	20.97	60.44	77.50
473	19869	C10H14	cymene	25155-15-1	34.0619	0.2302	0.10	36	2	20.05	57.87	77.50
474	19870	C10H14	m-cymene	535-77-3	33.7469	0.2283	0.10	38	2	19.95	57.08	77.50
475	19871	C10H14	o-cymene	527-84-4	34.6024	0.2325	0.10	32	2	20.26	59.10	77.50
476	19872	C10H14	p-cymene	99-87-6	34.7811	0.2269	0.10	34	2	20.63	58.65	77.50
477	19873	C10H14	1-methyl-2-propylbenzene	1074-17-5	37.8824	0.2316	0.10	22	2	22.22	64.58	77.50
478	19874	C10H14	1-methyl-3-propylbenzene	1074-43-7	36.8712	0.2282	0.10	26	2	21.80	62.36	77.50
479	19875	C10H14	1-methyl-4-propylbenzene	1074-55-1	36.2053	0.2276	0.10	28	2	21.44	61.14	77.50
480	19876	C10H14	(S)-(1-methylpropyl)benzene	5787-28-0	37.5722	0.2302	0.10	23	2	22.11	63.84	77.50

Table 2 Adsorption Capacity of Activated Carbon - Aqueous Liquid

$\log_{10} Q = \log_{10} K + 1/n \log_{10} x$ (Q - g of compound/100 g of carbon, x - ppm(wt))

NO	ID	FORMULA	NAME	CAS No	K	1/n	xmin	xmax	code	Q @xmin	Q @10ppm	Q @xmax
481	19877	C10H14	diethylbenzene	25340-17-4	36.4487	0.2314	0.10	26	2	21.39	62.11	77.50
482	19878	C10H14	o-diethylbenzene	135-01-3	36.9725	0.2333	0.10	24	2	21.61	63.27	77.50
483	19879	C10H14	m-diethylbenzene	141-93-5	36.3783	0.2291	0.10	27	2	21.47	61.64	77.50
484	19880	C10H14	p-diethylbenzene	105-05-5	37.1983	0.2285	0.10	25	2	21.98	62.96	77.50
485	19881	C10H14	3-ethyl-o-xylene	933-98-2	41.9863	0.2365	0.10	13	2	24.36	72.38	77.50
486	19882	C10H14	4-ethyl-o-xylene	934-80-5	40.2785	0.2320	0.10	17	2	23.61	68.72	77.50
487	19883	C10H14	2-ethyl-m-xylene	2870-04-4	39.9621	0.2360	0.10	17	2	23.21	68.81	77.50
488	19884	C10H14	4-ethyl-m-xylene	874-41-9	39.5618	0.2322	0.10	18	2	23.18	67.53	77.50
489	19885	C10H14	5-ethyl-m-xylene	934-74-7	37.5977	0.2293	0.10	23	2	22.17	63.75	77.50
490	19886	C10H14	2-ethyl-p-xylene	1758-88-9	38.7194	0.2325	0.10	20	2	22.67	66.14	77.50
491	19887	C10H14	1,2,3,4-tetramethylbenzene	488-23-3	48.0374	0.2400	0.10	7	2	27.64	---	77.50
492	19888	C10H14	1,2,3,5-tetramethylbenzene	527-53-7	44.2517	0.2362	0.10	11	2	25.69	76.24	77.50
493	19889	C10H14	1,2,4,5-tetramethylbenzene	95-93-2	52.7670	0.3082	0.10	3	2	25.95	---	77.50
494	20992	C10H20O	decanal	112-31-2	25.4945	0.1895	0.10	48	2	16.48	39.44	53.09
495	20993	C10H20O	2-decanone	693-54-9	24.4133	0.1893	0.10	61	2	15.79	37.75	53.09
496	20994	C10H20O	3-decanone	928-80-3	22.3565	0.1896	0.10	96	2	14.45	34.60	53.09
497	20995	C10H20O	4-decanone	624-16-8	23.3362	0.1894	0.10	77	2	15.09	36.09	53.09
498	20996	C10H20O	2-methyl-5-nonanone	22287-02-1	22.5785	0.1887	0.10	93	2	14.62	34.87	53.09
499	21019	C10H20O	5-decanone	820-29-1	21.6100	0.1995	0.10	90	2	13.65	34.21	53.09
500	21025	C10H20O	isodecanal	1321-89-7	21.2596	0.1914	1	119	2	21.26	33.03	53.09
501	21034	C10H20O	2-methyl-3-nonanone	5445-31-8	20.5473	0.1995	1	116	2	20.55	32.53	53.09
502	21044	C10H20O2	2-ethylhexyl acetate	103-09-3	23.2596	0.1820	0.10	93	2	15.30	35.37	53.09
503	21045	C10H20O2	octyl acetate	112-14-1	26.4695	0.1807	0.10	47	2	17.46	40.13	53.09
504	21046	C10H20O2	6-methyl-2-heptyl acetate	67952-57-2	21.3376	0.1767	1	174	2	21.34	32.05	53.09
505	21047	C10H20O2	6-methyl-3-heptyl acetate	32764-34-4	20.6538	0.1783	1	199	2	20.65	31.14	53.09
506	21048	C10H20O2	2-methylheptyl acetate, (±)	74112-36-0	22.7864	0.1789	1	113	2	22.79	34.40	53.09
507	21224	C10H22O	1-decanol	112-30-1	26.9203	0.1880	0.10	37	2	17.46	41.51	53.09
508	21225	C10H22O	isodecanol	25339-17-7	22.5593	0.1908	0.10	89	2	14.54	35.00	53.09
509	21226	C10H22O	2-decanol	1120-06-5	22.1251	0.1872	1	107	2	22.13	34.05	53.09
510	21227	C10H22O	2-decanol, (±)	74742-10-2	20.5736	0.1871	1	159	2	20.57	31.65	53.09
511	21228	C10H22O	3-decanol	1565-81-7	20.2437	0.1876	1	171	2	20.24	31.18	53.09
512	21229	C10H22O	4-decanol	2051-31-2	20.3515	0.1869	1	169	2	20.35	31.30	53.09
513	21230	C10H22O	5-decanol	5205-34-5	18.3005	0.1869	1	298	2	18.30	28.14	53.09
514	21231	C10H22O	2-methyl-1-nonanol	40589-14-8	23.3638	0.1890	0.10	77	2	15.12	36.11	53.09
515	21232	C10H22O	4-methyl-1-nonanol	1489-47-0	21.6990	0.1887	1	115	2	21.70	33.51	53.09
516	21233	C10H22O	5-methyl-1-nonanol	2768-16-3	21.0841	0.1892	1	132	2	21.08	32.59	53.09
517	21234	C10H22O	7-methyl-1-nonanol	33234-93-4	21.1310	0.1887	1	132	2	21.13	32.63	53.09
518	21235	C10H22O	3-methyl-1-nonanol	22663-64-5	21.0841	0.1892	1	132	2	21.08	32.59	53.09
519	21236	C10H22O	2-methyl-2-nonanol	10297-57-1	21.0989	0.1890	1	132	2	21.10	32.61	53.09
520	21237	C10H22O	3-methyl-2-nonanol	60671-32-1	21.0606	0.1894	1	132	2	21.06	32.57	53.09
521	21238	C10H22O	5-methyl-2-nonanol	66731-95-1	21.0990	0.1890	1	132	2	21.10	32.61	53.09
522	21239	C10H22O	6-methyl-2-nonanol	66256-60-8	21.1075	0.1889	1	132	2	21.11	32.61	53.09
523	21240	C10H22O	7-methyl-2-nonanol	66256-61-9	21.0990	0.1890	1	132	2	21.10	32.61	53.09
524	21241	C10H22O	8-methyl-2-nonanol	14779-92-1	21.2962	0.1871	1	132	2	21.30	32.77	53.09
525	21242	C10H22O	2-methyl-3-nonanol	26533-33-5	19.8591	0.1879	1	187	2	19.86	30.61	53.09
526	21243	C10H22O	3-methyl-3-nonanol	21078-72-8	19.9074	0.1885	1	182	2	19.91	30.73	53.09
527	21244	C10H22O	5-methyl-3-nonanol	66719-43-5	19.8520	0.1890	1	182	2	19.85	30.68	53.09
528	21245	C10H22O	2-methyl-4-nonanol	26533-31-3	20.0734	0.1869	1	182	2	20.07	30.87	53.09
529	21246	C10H22O	4-methyl-4-nonanol	23418-38-4	19.9311	0.1883	1	182	2	19.93	30.75	53.09
530	21247	C10H22O	5-methyl-4-nonanol	66719-44-6	19.8838	0.1887	1	182	2	19.88	30.71	53.09
531	21248	C10H22O	7-methyl-4-nonanol	26981-98-6	19.8520	0.1890	1	182	2	19.85	30.68	53.09
532	21249	C10H22O	2-methyl-5-nonanol	29843-62-7	20.1450	0.1862	1	182	2	20.14	30.93	53.09
533	21250	C10H22O	3-methyl-5-nonanol	66779-42-4	20.0496	0.1871	1	182	2	20.05	30.85	53.09
534	21251	C10H22O	5-methyl-5-nonanol	33933-78-7	18.3964	0.1880	1	280	2	18.40	28.36	53.09
535	21252	C10H22O	2-ethyl-1-octanol	20592-10-3	19.8522	0.1890	1	182	2	19.85	30.68	53.09
536	21253	C10H22O	3-ethyl-1-octanol	66719-36-6	20.0496	0.1871	1	182	2	20.05	30.85	53.09
537	21254	C10H22O	2,2-dimethyl-1-octanol	2370-14-1	19.7287	0.1883	1	192	2	19.73	30.43	53.09
538	21255	C10H22O	2,6-dimethyl-1-octanol	62417-08-7	20.2200	0.1878	1	171	2	20.22	31.16	53.09
539	21256	C10H22O	3,7-dimethyl-1-octanol	106-21-8	20.5640	0.1896	1	149	2	20.56	31.82	53.09
540	21257	C10H22O	4,5-dimethyl-1-octanol	66719-32-2	20.5882	0.1890	1	150	2	20.59	31.82	53.09
541	21258	C10H22O	4,6-dimethyl-1-octanol	66719-33-3	20.5882	0.1890	1	150	2	20.59	31.82	53.09
542	21259	C10H22O	4,7-dimethyl-1-octanol	66719-34-4	20.5882	0.1890	1	150	2	20.59	31.82	53.09
543	21260	C10H22O	7,7-dimethyl-1-octanol	66719-35-5	20.5881	0.1890	1	150	2	20.59	31.82	53.09
544	21261	C10H22O	2,4-dimethyl-2-octanol	18675-20-2	20.7620	0.1874	1	150	2	20.76	31.96	53.09
545	21262	C10H22O	2,6-dimethyl-2-octanol	18479-57-7	20.7146	0.1878	1	150	2	20.71	31.92	53.09
546	21263	C10H22O	2,7-dimethyl-2-octanol	42007-73-8	20.5879	0.1890	1	150	2	20.59	31.82	53.09
547	21264	C10H22O	3,7-dimethyl-2-octanol	15340-96-2	20.6910	0.1880	1	150	2	20.69	31.90	53.09
548	21265	C10H22O	3-ethyl-3-octanol	2051-32-3	17.5053	0.1905	1	338	2	17.51	27.15	53.09
549	21266	C10H22O	6-ethyl-3-octanol	19781-27-2	17.6300	0.1890	1	341	2	17.63	27.24	53.09
550	21267	C10H22O	2,2-dimethyl-3-octanol	19841-72-6	16.6359	0.1890	1	464	2	16.64	25.71	53.09
551	21268	C10H22O	2,3-dimethyl-3-octanol	19781-10-3	15.8567	0.1880	1	618	2	15.86	24.45	53.09
552	21269	C10H22O	2,7-dimethyl-3-octanol	66719-55-9	17.0971	0.1848	1	459	2	17.10	26.17	53.09
553	21270	C10H22O	3,5-dimethyl-3-octanol	56065-42-0	16.2682	0.1908	1	493	2	16.27	25.24	53.09
554	21271	C10H22O	3,6-dimethyl-3-octanol	151-19-9	16.2933	0.1889	1	519	2	16.29	25.17	53.09
555	21272	C10H22O	3,7-dimethyl-3-octanol	78-69-3	17.1327	0.1883	1	406	2	17.13	26.43	53.09
556	21273	C10H22O	3-ethyl-4-octanol	63126-48-7	16.4458	0.1890	1	493	2	16.45	25.41	53.09
557	21274	C10H22O	4-ethyl-4-octanol	38395-42-5	16.3373	0.1901	1	493	2	16.34	25.31	53.09
558	21275	C10H22O	2,2-dimethyl-4-octanol	66719-52-6	16.2045	0.1862	1	586	2	16.20	24.88	53.09
559	21276	C10H22O	2,4-dimethyl-4-octanol	33933-79-8	16.3133	0.1867	1	556	2	16.31	25.07	53.09
560	21277	C10H22O	2,5-dimethyl-4-octanol	66719-53-7	16.4787	0.1851	1	556	2	16.48	25.23	53.09

Table 2 Adsorption Capacity of Activated Carbon - Aqueous Liquid

$\log_{10} Q = \log_{10} K + 1/n \log_{10} x$ (Q - g of compound/100 g of carbon, x - ppm(wt))

NO	ID	FORMULA	NAME	CAS No	K	1/n	xmin	xmax	code	Q @xmin	Q @10ppm	Q @xmax
561	21278	C10H22O	2,6-dimethyl-4-octanol	66719-54-8	17.3789	0.1840	1	432	2	17.38	26.55	53.09
562	21279	C10H22O	2,7-dimethyl-4-octanol	19781-11-4	18.7062	0.1851	1	280	2	18.71	28.65	53.09
563	21280	C10H22O	3,4-dimethyl-4-octanol	66719-30-0	16.0724	0.1890	1	556	2	16.07	24.84	53.09
564	21281	C10H22O	3,6-dimethyl-4-octanol	66719-31-1	15.9876	0.1899	1	556	2	15.99	24.75	53.09
565	21282	C10H22O	4,6-dimethyl-4-octanol	56065-43-1	16.1032	0.1887	1	556	2	16.10	24.87	53.09
566	21283	C10H22O	4,7-dimethyl-4-octanol	19781-13-6	16.4566	0.1874	1	519	2	16.46	25.33	53.09
567	21284	C10H22O	2-propyl-1-heptanol	10042-59-8	22.2179	0.1887	1	101	2	22.22	34.31	53.09
568	21285	C10H22O	3-isopropyl-1-heptanol	38514-15-7	21.2519	0.1910	1	121	2	21.25	32.99	53.09
569	21286	C10H22O	2,4,6-trimethyl-2-heptanol	66256-47-1	16.4452	0.1890	1	493	2	16.45	25.41	53.09
570	21287	C10H22O	2,5,6-trimethyl-2-heptanol	66256-48-2	16.5281	0.1885	1	488	2	16.53	25.51	53.09
571	21288	C10H22O	4,6,6-trimethyl-2-heptanol	51079-79-9	16.4452	0.1890	1	493	2	16.45	25.41	53.09
572	21289	C10H22O	2-methyl-3-ethyl-3-heptanol	66719-37-7	16.1820	0.1919	1	488	2	16.18	25.17	53.09
573	21290	C10H22O	4-methyl-5-ethyl-3-heptanol	66731-94-0	15.7702	0.1958	1	493	2	15.77	24.75	53.09
574	21291	C10H22O	5-methyl-3-ethyl-3-heptanol	------	16.3836	0.1896	1	493	2	16.38	25.35	53.09
575	21292	C10H22O	2,2,3-trimethyl-3-heptanol	29772-40-5	14.9434	0.1917	1	745	2	14.94	23.23	53.09
576	21293	C10H22O	2,2,6-trimethyl-3-heptanol	66256-43-7	16.1067	0.1869	1	591	2	16.11	24.77	53.09
577	21294	C10H22O	2,3,6-trimethyl-3-heptanol	58046-40-5	15.7362	0.1905	1	591	2	15.74	24.40	53.09
578	21295	C10H22O	3,5,5-trimethyl-3-heptanol	66256-50-6	16.4549	0.1942	1	417	2	16.45	25.73	53.09
579	21296	C10H22O	4-propyl-4-heptanol	2198-72-3	16.7191	0.1885	1	459	2	16.72	25.80	53.09
580	21297	C10H22O	4-isopropyl-4-heptanol	51200-82-9	15.6493	0.1917	1	586	2	15.65	24.33	53.09
581	21298	C10H22O	2,2,4-trimethyl-4-heptanol	57233-31-5	14.3817	0.1889	1	1000	2	14.38	22.22	53.05
582	21299	C10H22O	2,2,5-trimethyl-4-heptanol	66256-42-6	14.4519	0.1896	1	955	2	14.45	22.36	53.09
583	21300	C10H22O	2,2,6-trimethyl-4-heptanol	66256-44-8	14.5127	0.1890	1	955	2	14.51	22.43	53.09
584	21301	C10H22O	2,4,5-trimethyl-4-heptanol	66256-46-0	14.5127	0.1890	1	955	2	14.51	22.43	53.09
585	21302	C10H22O	2,4,6-trimethyl-4-heptanol	60836-07-9	14.8659	0.1858	1	946	2	14.87	22.80	53.09
586	21303	C10H22O	3,3,6-trimethyl-4-heptanol	------	14.3169	0.1910	1	955	2	14.32	22.23	53.09
587	21304	C10H22O	4-methyl-2-propyl-1-hexanol	66256-62-0	19.7523	0.1880	1	192	2	19.75	30.45	53.09
588	21305	C10H22O	4-methyl-2-isopropyl-1-hexanol	66719-41-3	18.0800	0.1890	1	298	2	18.08	27.94	53.09
589	21306	C10H22O	5-methyl-2-isopropyl-1-hexanol	2051-33-4	20.8884	0.1889	1	139	2	20.89	32.27	53.09
590	21307	C10H22O	5,5-dimethyl-2-ethyl-1-hexanol	------	20.7095	0.1903	1	141	2	20.71	32.10	53.09
591	21308	C10H22O	2,3,4,4-tetramethyl-2-hexanol	66256-66-4	15.9164	0.1890	1	586	2	15.92	24.60	53.09
592	21309	C10H22O	3,3,5,5-tetramethyl-2-hexanol	66256-69-7	15.8892	0.1890	1	591	2	15.89	24.55	53.09
593	21310	C10H22O	2-methyl-3-isopropyl-3-hexanol	51200-81-8	15.9034	0.1928	1	519	2	15.90	24.79	53.09
594	21311	C10H22O	2,2-dimethyl-4-ethyl-3-hexanol	66719-47-9	15.3668	0.1892	1	702	2	15.37	23.76	53.09
595	21312	C10H22O	2,4-dimethyl-4-ethyl-3-hexanol	66719-48-0	14.7334	0.1953	1	708	2	14.73	23.10	53.09
596	21313	C10H22O	5,5-dimethyl-3-ethyl-3-hexanol	5340-62-5	15.2035	0.1905	1	708	2	15.20	23.58	53.09
597	21314	C10H22O	2,2,3,4-tetramethyl-3-hexanol	66256-63-1	15.7231	0.1946	1	519	2	15.72	24.61	53.09
598	21315	C10H22O	2,2,3,5-tetramethyl-3-hexanol	66256-64-2	29.6973	0.1903	0.10	21	2	19.16	46.03	53.09
599	21316	C10H22O	2,2,4,4-tetramethyl-3-hexanol	66256-65-3	15.4237	0.1940	1	586	2	15.42	24.11	53.09
600	21317	C10H22O	2,2,5,5-tetramethyl-3-hexanol	55073-86-4	9.2141	0.2316	1	1000	2	9.21	15.70	45.62
601	21318	C10H22O	2,3,4,4-tetramethyl-3-hexanol	66256-67-5	17.1498	0.1983	1	298	2	17.15	27.07	53.09
602	21319	C10H22O	2,3,5,5-tetramethyl-3-hexanol	5396-09-8	18.1526	0.1901	1	283	2	18.15	28.12	53.09
603	21320	C10H22O	2,4,4,5-tetramethyl-3-hexanol	66256-68-6	18.2611	0.1890	1	283	2	18.26	28.22	53.09
604	21321	C10H22O	3,4,4,5-tetramethyl-3-hexanol	66256-39-1	17.3631	0.1983	1	280	2	17.36	27.41	53.09
605	21322	C10H22O	3,4,5,5-tetramethyl-3-hexanol	66256-40-4	16.2029	0.1956	1	432	2	16.20	25.42	53.09
606	21323	C10H22O	4-methyl-2-isobutyl-1-pentanol	22417-45-4	18.7168	0.1892	1	247	2	18.72	28.93	53.09
607	21324	C10H22O	4,4-dimethyl-3-isopropyl-1-pentanol	66719-51-5	14.6770	0.1890	1	900	2	14.68	22.68	53.09
608	21325	C10H22O	3,4-dimethyl-3-isopropyl-2-pentanol	66719-50-4	14.6771	0.1890	1	900	2	14.68	22.68	53.09
609	21326	C10H22O	2,4-dimethyl-3-propyl-3-pentanol	------	14.4041	0.1937	1	840	2	14.40	22.50	53.09
610	21327	C10H22O	2,4-dimethyl-3-isopropyl-3-pentanol	51200-83-0	16.0817	0.1958	1	446	2	16.08	25.24	53.09
611	21328	C10H22O	2,2,4-trimethyl-3-ethyl-3-pentanol	66256-41-5	15.4494	0.1956	1	551	2	15.45	24.24	53.09
612	21329	C10H22O	2,2,3,4,4-pentamethyl-3-pentanol	5857-69-2	12.0534	0.2428	1	448	2	12.05	21.08	53.09
613	21330	C10H22O	2,3-dimethyl-2-tert-butyl-1-butanol	81931-81-9	11.9520	0.2405	1	493	2	11.95	20.79	53.09
614	21331	C10H22O	3,7-dimethyl-3-octanol, (±)	57706-88-4	12.7352	0.2389	1	394	2	12.74	22.07	53.09
615	21332	C10H22O	3,7-dimethyl-1-octanol, (±)	59204-02-3	20.9992	0.1866	1	144	2	21.00	32.27	53.09
616	21333	C10H22O	3-ethyl-6-methyl-3-heptanol	66719-40-2	17.7335	0.1896	1	325	2	17.73	27.44	53.09
617	21334	C10H22O	8-methyl-1-nonanol	55505-26-5	17.7902	0.1890	1	325	2	17.79	27.49	53.09
618	21338	C10H22O	(S)-(+)-2-decanol	33758-16-6	20.4316	0.1885	1	159	2	20.43	31.53	53.09
619	21339	C10H22O	(R)-3,7-dimethyl-1-octanol	1117-60-8	20.7654	0.1890	1	143	2	20.77	32.09	53.09
620	21340	C10H22O	s-3,7-dimethyl-1-octanol	68680-98-8	20.5885	0.1890	1	150	2	20.59	31.82	53.09
621	21341	C10H22O	decyl alcohol (mixed isomers)	85566-12-7	17.7902	0.1890	1	325	2	17.79	27.49	53.09
622	22304	C11H16	pentylbenzene	538-68-1	58.7255	0.2062	0.10	4	2	36.53	---	77.50
623	22305	C11H16	2-phenylpentane	2719-52-0	51.4841	0.2061	0.10	7	2	32.03	---	77.50
624	22306	C11H16	3-phenylpentane	1196-58-3	52.1837	0.2050	0.10	7	2	32.55	---	77.50
625	22307	C11H16	1-phenyl-2-methylbutane	3968-85-2	46.9962	0.2062	0.10	11	2	29.23	75.55	77.50
626	22308	C11H16	1-phenyl-3-methylbutane	2049-94-7	48.0820	0.2054	0.10	10	2	29.96	77.17	77.50
627	22309	C11H16	2-phenyl-2-methylbutane	2049-95-8	44.1617	0.2100	0.10	15	2	27.23	71.62	77.50
628	22310	C11H16	2-phenyl-3-methylbutane	4481-30-5	42.1227	0.2088	0.10	19	2	26.04	68.13	77.50
629	22311	C11H16	1-phenyl-2,2-dimethylpropane	1007-26-7	41.6112	0.2052	0.10	21	2	25.94	66.74	77.50
630	22312	C11H16	1-methyl-2-butylbenzene	1595-11-5	52.7374	0.2091	0.10	6	2	32.59	---	77.50
631	22313	C11H16	1-methyl-3-butylbenzene	1595-04-6	51.3290	0.2062	0.10	7	2	31.93	---	77.50
632	22314	C11H16	1-methyl-4-butylbenzene	1595-05-7	52.3563	0.2071	0.10	7	2	32.50	---	77.50
633	22315	C11H16	1-methyl-2-sec-butylbenzene	1595-06-8	46.0858	0.2095	0.10	12	2	28.45	74.66	77.50
634	22316	C11H16	1-methyl-3-sec-butylbenzene	1772-10-7	45.4692	0.2059	0.10	13	2	28.30	73.05	77.50
635	22317	C11H16	1-methyl-4-sec-butylbenzene	1595-16-0	46.8041	0.2079	0.10	11	2	29.00	75.53	77.50
636	22318	C11H16	1-methyl-2-isobutylbenzene	36301-29-8	46.3069	0.2076	0.10	12	2	28.71	74.69	77.50
637	22319	C11H16	1-methyl-3-isobutylbenzene	5160-99-6	45.5915	0.2049	0.10	13	2	28.44	73.08	77.50
638	22320	C11H16	1-methyl-4-isobutylbenzene	5161-04-6	46.6739	0.2044	0.10	12	2	29.15	74.73	77.50
639	22321	C11H16	1-methyl-2-tert-butyl benzene	1074-92-6	48.0267	0.2136	0.10	9	2	29.37	---	77.50
640	22322	C11H16	1-methyl-3-tert-butylbenzene	1075-38-3	42.8640	0.2078	0.10	17	2	26.56	69.17	77.50

Table 2 Adsorption Capacity of Activated Carbon - Aqueous Liquid

$\log_{10} Q = \log_{10} K + 1/n \log_{10} x$ (Q - g of compound/100 g of carbon, x - ppm(wt))												
NO	ID	FORMULA	NAME	CAS No	K	1/n	xmin	xmax	code	Q @xmin	Q @10ppm	Q @xmax
641	22323	C11H16	1-methyl-4-tert-butylbenzene	98-51-1	44.7436	0.2067	0.10	14	2	27.80	72.02	77.50
642	22324	C11H16	1-ethyl-2-propylbenzene	16021-20-8	49.8280	0.2099	0.10	8	2	30.73	---	77.50
643	22325	C11H16	1-ethyl-3-propylbenzene	20024-91-3	49.0827	0.2066	0.10	9	2	30.50	---	77.50
644	22326	C11H16	1-ethyl-4-propylbenzene	20024-90-2	51.3167	0.2063	0.10	7	2	31.91	---	77.50
645	22327	C11H16	1-ethyl-2-isopropylbenzene	18970-44-0	42.5843	0.2132	0.10	17	2	26.07	69.57	77.50
646	22328	C11H16	1-ethyl-3-isopropylbenzene	4920-99-4	44.4260	0.2062	0.10	15	2	27.64	71.42	77.50
647	22329	C11H16	1-ethyl-4-isopropylbenzene	4218-48-8	46.7976	0.2061	0.10	12	2	29.12	75.21	77.50
648	22330	C11H16	1,2-dimethyl-3-propylbenzene	17059-44-8	53.9561	0.2128	0.10	5	2	33.05	---	77.50
649	22331	C11H16	1,2-dimethyl-4-propylbenzene	3982-66-9	53.2416	0.2092	0.10	6	2	32.89	---	77.50
650	22332	C11H16	1,3-dimethyl-2-propylbenzene	17059-45-9	52.1627	0.2126	0.10	6	2	31.97	---	77.50
651	22333	C11H16	1,3-dimethyl-4-propylbenzene	61827-85-8	51.9042	0.2094	0.10	7	2	32.05	---	77.50
652	22334	C11H16	1,3-dimethyl-5-propylbenzene	3982-64-7	49.7584	0.2066	0.10	9	2	30.92	---	77.50
653	22335	C11H16	1,4-dimethyl-2-propylbenzene	3042-50-0	50.6206	0.2093	0.10	8	2	31.27	---	77.50
654	22336	C11H16	1,2-dimethyl-3-isopropylbenzene	22539-65-7	49.2641	0.2132	0.10	8	2	30.16	---	77.50
655	22337	C11H16	1,2-dimethyl-4-isopropylbenzene	4132-77-8	49.2803	0.2088	0.10	9	2	30.47	---	77.50
656	22338	C11H16	1,3-dimethyl-2-isopropylbenzene	14411-75-7	47.2290	0.2136	0.10	10	2	28.88	77.24	77.50
657	22339	C11H16	1,3-dimethyl-4-isopropylbenzene	4706-89-2	47.7338	0.2095	0.10	10	2	29.46	77.33	77.50
658	22340	C11H16	1,3-dimethyl-5-isopropylbenzene	4706-90-5	45.6127	0.2069	0.10	13	2	28.33	73.45	77.50
659	22341	C11H16	1,4-dimethyl-2-isopropylbenzene	4132-72-3	46.1661	0.2098	0.10	12	2	28.48	74.83	77.50
660	22342	C11H16	1-methyl-2,3-diethylbenzene	13632-93-4	51.4581	0.2139	0.10	7	2	31.44	---	77.50
661	22343	C11H16	1-methyl-2,4-diethylbenzene	1758-85-6	50.9373	0.2100	0.10	7	2	31.41	---	77.50
662	22344	C11H16	1-methyl-2,5-diethylbenzene	13632-94-5	52.1061	0.2102	0.10	7	2	32.11	---	77.50
663	22345	C11H16	1-methyl-2,6-diethyl benzene	13632-95-6	52.7426	0.2138	0.10	6	2	32.23	---	77.50
664	22346	C11H16	1-methyl-3,4-diethylbenzene	13732-80-4	50.1169	0.2103	0.10	8	2	30.88	---	77.50
665	22347	C11H16	1-methyl-3,5-diethylbenzene	2050-24-0	48.8598	0.2072	0.10	9	2	30.32	---	77.50
666	22348	C11H16	1,2,3-trimethyl-4-ethylbenzene	61827-86-9	59.5710	0.2165	0.10	3	2	36.18	---	77.50
667	22349	C11H16	1,2,3-trimethyl-5-ethylbenzene	31366-00-4	57.0151	0.2128	0.10	4	2	34.93	---	77.50
668	22350	C11H16	1,2,4-trimethyl-3-ethylbenzene	61827-87-0	57.3355	0.2149	0.10	4	2	34.96	---	77.50
669	22351	C11H16	1,2,4-trimethyl-5-ethylbenzene	------	55.3976	0.2120	0.10	5	2	34.00	---	77.50
670	22352	C11H16	1,2,4-trimethyl-6-ethylbenzene	17851-27-3	55.2539	0.2136	0.10	5	2	33.79	---	77.50
671	22353	C11H16	1,3,5-trimethyl-2-ethylbenzene	3982-67-0	55.0400	0.2120	0.10	5	2	33.78	---	77.50
672	22354	C11H16	pentamethylbenzene	700-12-9	63.1764	0.3008	0.10	2	2	31.60	---	77.50
673	22355	C11H16	1-ethyl-2,3,5-trimethylbenzene	18262-85-6	55.2538	0.2136	0.10	5	2	33.79	---	77.50
674	22356	C11H16	sec-pentylbenzene	29316-05-0	57.3598	0.2090	0.10	4	2	35.45	---	77.50
675	22822	C11H22O	undecanal	112-44-7	31.6967	0.1743	0.10	19	2	21.22	47.35	53.09
676	22823	C11H22O	2-undecanone	112-12-9	31.8635	0.1741	0.10	19	2	21.34	47.58	53.09
677	22824	C11H22O	3-undecanone	2216-87-7	31.5153	0.1745	0.10	20	2	21.09	47.10	53.09
678	22825	C11H22O	4-undecanone	14476-37-0	34.4671	0.1753	0.10	12	2	23.02	51.60	53.09
679	22826	C11H22O	5-undecanone	33083-83-9	31.5175	0.1745	0.10	20	2	21.09	47.10	53.09
680	22827	C11H22O	6-undecanone	927-49-1	31.8270	0.1752	0.10	19	2	21.26	47.65	53.09
681	22834	C11H22O	2-methyl-1-decanal	19009-56-4	31.8194	0.1775	0.10	18	2	21.14	47.89	53.09
682	22837	C11H22O2	nonyl acetate	143-13-5	27.6068	0.1668	0.10	50	2	18.80	40.53	53.09
683	22856	C11H22O2	3,5,5-trimethylhexyl acetate	58430-94-7	26.7633	0.1674	0.10	60	2	18.20	39.35	53.09
684	23055	C11H24O	1-undecanol	112-42-5	33.1654	0.1740	0.10	15	2	22.22	49.51	53.09
685	23056	C11H24O	2-undecanol	1653-30-1	29.3418	0.1724	0.10	31	2	19.73	43.64	53.09
686	23057	C11H24O	2-undecanol, (±)	113666-64-1	26.9649	0.1723	0.10	51	2	18.14	40.09	53.09
687	23058	C11H24O	3-undecanol, (R)-	107494-37-1	27.2155	0.1730	0.10	48	2	18.27	40.53	53.09
688	23059	C11H24O	5-undecanol	37493-70-2	27.2221	0.1729	0.10	48	2	18.28	40.53	53.09
689	23060	C11H24O	6-undecanol	23708-56-7	20.9259	0.2367	0.10	51	2	12.13	36.09	53.09
690	23061	C11H24O	2,8-dimethyl-5-nonanol	19780-96-2	28.6554	0.1721	0.10	36	2	19.28	42.59	53.09
691	23062	C11H24O	undecanol	30207-98-8	33.2622	0.1729	0.10	15	2	22.34	49.53	53.09
692	23063	C11H24O	5-ethyl-2-nonanol	103-08-2	25.9092	0.1729	0.10	63	2	17.40	38.58	53.09
693	----	C12H12N2	diquat	------	1.2026	0.2420	0.03	3.25	1,2	0.51	----	1.60
694	----	C12H15NO3	carbofuran	------	27.4691	0.4080	0.0029	0.111	1,2	2.53	----	11.20
695	24357	C12H18	hexylbenzene	1077-16-3	77.2397	0.1884	0.10	1	2	50.06	---	77.50
696	24358	C12H18	1,4-dipropylbenzene	4815-57-0	55.1827	0.1877	0.10	6	2	35.82	---	77.50
697	24359	C12H18	diisopropylbenzene	25321-09-9	50.6963	0.1895	0.10	9	2	32.77	---	77.50
698	24360	C12H18	1,2-diisopropylbenzene	577-55-9	52.3910	0.1909	0.10	8	2	33.76	---	77.50
699	24361	C12H18	m-diisopropylbenzene	99-62-7	52.3068	0.1877	0.10	8	2	33.95	---	77.50
700	24362	C12H18	p-diisopropylbenzene	100-18-5	56.1819	0.1879	0.10	6	2	36.45	---	77.50
701	24363	C12H18	triethylbenzene; (mixed isomers)	25340-18-5	59.8150	0.1911	0.10	4	2	38.52	---	77.50
702	24364	C12H18	1,2,3-triethylbenzene	------	59.7481	0.1917	0.10	4	2	38.43	---	77.50
703	24365	C12H18	1,2,4-triethylbenzene	877-44-1	59.7481	0.1917	0.10	4	2	38.43	---	77.50
704	24366	C12H18	1,3,5-triethylbenzene	102-25-0	58.5829	0.1954	0.10	4	2	37.36	---	77.50
705	24367	C12H18	1-tert-butyl-3,5-dimethylbenzene	98-19-1	54.0667	0.1902	0.10	7	2	34.90	---	77.50
706	24368	C12H18	(1,1-dimethylbutyl)benzene	1985-57-5	53.1359	0.1914	0.10	7	2	34.20	---	77.50
707	24369	C12H18	2,4-dimethyl-1-sec-butylbenzene	1483-60-9	54.9475	0.1901	0.10	6	2	35.47	---	77.50
708	24370	C12H18	(1-ethylbutyl)benzene	4468-42-2	65.0479	0.1815	0.10	3	2	42.83	---	77.50
709	24371	C12H18	(1-ethyl-1-methylpropyl)benzene	1985-97-3	61.8830	0.1917	0.10	3	2	39.80	---	77.50
710	24372	C12H18	(1-methylpentyl)benzene	6031-02-3	63.8919	0.1899	0.10	3	2	41.27	---	77.50
711	24373	C12H18	(3-methylpentyl)benzene	54410-69-4	71.7261	0.1887	0.10	2	2	46.45	---	77.50
712	24374	C12H18	(4-methylpentyl)benzene	4215-86-5	69.8018	0.1874	0.10	2	2	45.34	---	77.50
713	24375	C12H18	1,2,4-trimethyl-5-isopropylbenzene	10222-95-4	61.6393	0.1931	0.10	3	2	39.51	---	77.50
714	24376	C12H18	4-tert-butyl-o-xylene	7397-06-0	52.8167	0.1919	0.10	7	2	33.95	---	77.50
715	24377	C12H18	hexamethyldewarbenzene	7641-77-2	37.5145	0.1767	0.10	61	2	24.98	56.35	77.50
716	24378	C12H18	hexamethylbenzene	87-85-4	92.2215	0.3257	0.10	1	2	43.57	---	77.50
717	24852	C12H24O	dodecanal	112-54-9	37.8242	0.1617	0.10	8	2	26.07	---	53.09
718	24853	C12H24O	2-methylundecanal	110-41-8	36.6483	0.1615	0.10	10	2	25.27	---	53.09
719	24854	C12H24O	2-dodecanone	6175-49-1	41.2687	0.1612	0.10	5	2	28.47	---	53.09
720	24855	C12H24O	6-dodecanone	6064-27-3	40.9245	0.1618	0.10	5	2	28.19	---	53.09

Table 2 Adsorption Capacity of Activated Carbon - Aqueous Liquid

$\log_{10} Q = \log_{10} K + 1/n \log_{10} x$ (Q - g of compound/100 g of carbon, x - ppm(wt))

NO	ID	FORMULA	NAME	CAS No	K	1/n	xmin	xmax	code	Q @xmin	Q @10ppm	Q @xmax
721	24860	C12H24O	trimethyl nonanone	1331-50-6	28.9426	0.1599	0.10	44	2	20.03	41.83	53.09
722	24861	C12H24O	3-dodecanone	1534-27-6	40.9245	0.1618	0.10	5	2	28.19	---	53.09
723	24862	C12H24O	5-dodecanone	19780-10-0	40.9245	0.1618	0.10	5	2	28.19	---	53.09
724	24863	C12H24O	4-dodecanone	6137-26-4	40.9245	0.1618	0.10	5	2	28.19	---	53.09
725	24869	C12H24O2	decyl acetate	112-17-4	38.3561	0.1555	0.10	8	2	26.81	---	53.09
726	24880	C12H24O2	3,7-dimethyloctanyl acetate	20780-49-8	34.6200	0.1558	0.10	16	2	24.19	49.56	53.09
727	25277	C12H26O	1-dodecanol	112-53-8	42.4867	0.1607	0.10	4	2	29.35	---	53.09
728	25278	C12H26O	2-dodecanol	10203-28-8	37.5682	0.1597	0.10	9	2	26.01	---	53.09
729	25279	C12H26O	3-dodecanol	10203-30-2	26.5301	0.2222	0.10	23	2	15.91	44.25	53.09
730	25280	C12H26O	6-dodecanol	6836-38-0	21.2982	0.2207	0.10	63	2	12.81	35.40	53.09
731	25281	C12H26O	2-methyl-1-undecanol	10522-26-6	32.2710	0.1595	0.10	23	2	22.35	46.59	53.09
732	25282	C12H26O	2-methyl-3-undecanol	60671-36-5	32.1569	0.1606	0.10	23	2	22.22	46.54	53.09
733	25283	C12H26O	2-methyl-5-undecanol	33978-71-1	32.3252	0.1589	0.10	23	2	22.42	46.61	53.09
734	25284	C12H26O	2-butyl-1-octanol	3913-02-8	34.0215	0.1718	0.10	13	2	22.91	50.53	53.09
735	25286	C12H26O	2,6,8-trimethylnonanol-4	123-17-1	27.6171	0.1586	0.10	62	2	19.17	39.79	53.09
736	26283	C13H20	heptylbenzene	1078-71-3	88.2225	0.1730	0.01	0.47	2	39.78	---	77.50
737	26284	C13H20	1-methyl-2,4-diisopropylbenzene	1460-98-6	64.6228	0.1745	0.10	3	2	43.24	---	77.50
738	26285	C13H20	1-methyl-3,5-diisopropylbenzene	3055-14-9	60.5732	0.1751	0.10	4	2	40.48	---	77.50
739	26286	C13H20	[2-methyl-1-isopropylpropyl]benzene	21777-84-4	72.5156	0.1768	0.10	1	2	48.27	---	77.50
740	26288	C13H20	(2,4-dimethylpentyl)benzene	54518-00-2	71.5768	0.1733	0.10	2	2	48.03	---	77.50
741	26486	C13H26O	1-tridecanal	10486-19-8	44.0830	0.1504	0.10	3	2	31.18	---	53.09
742	26487	C13H26O	2-tridecanone	593-08-8	49.9240	0.2088	0.10	1	2	30.87	---	53.09
743	26488	C13H26O	3-tridecanone	1534-26-5	52.3773	0.2131	0.10	1	2	32.06	---	53.09
744	26489	C13H26O	7-tridecanone	462-18-0	48.5606	0.2093	0.10	2	2	29.99	---	53.09
745	26495	C13H26O2	undecyl acetate	1731-81-3	39.9108	0.1473	0.10	7	2	28.43	---	53.09
746	26535	C13H28O	1-tridecanol	112-70-9	49.6052	0.2137	0.10	1	2	30.33	---	53.09
747	26536	C13H28O	2-tridecanol	1653-31-2	46.8061	0.1488	0.10	2	2	33.22	---	53.09
748	26537	C13H28O	3-tridecanol	10289-68-6	37.2229	0.2107	0.10	5	2	22.91	---	53.09
749	26538	C13H28O	5-butyl-5-nonanol	597-93-3	41.1763	0.1508	0.10	5	2	29.10	---	53.09
750	----	C14H20ClNO2	alachor	------	48.2194	0.2570	0.0006	0.512	1,2	7.16	----	40.60
751	27644	C14H22	octylbenzene	2189-60-8	96.7882	0.1602	0.01	0.25	2	46.28	---	77.50
752	27645	C14H22	1,2,3,4-tetraethylbenzene	642-32-0	78.9555	0.1658	0.10	1	2	53.89	---	77.50
753	27646	C14H22	1,2,3,5-tetraethylbenzene	38842-05-6	78.7059	0.1649	0.10	1	2	53.84	---	77.50
754	27647	C14H22	1,2,4,5-tetraethylbenzene	635-81-4	78.4605	0.1643	0.10	1	2	53.74	---	77.50
755	27648	C14H22	1,4-di-tert-butylbenzene	1012-72-2	70.0521	0.2481	0.10	2	2	39.57	---	77.50
756	27649	C14H22	(1-methylheptyl)benzene	777-22-0	90.1696	0.1631	0.01	0.40	2	42.54	---	77.50
757	27650	C14H22	1,3-di-tert-butylbenzene	1014-60-4	78.8099	0.1613	0.10	1	2	54.36	---	77.50
758	27862	C14H28O	tetradecanal	124-25-4	48.8061	0.1978	0.10	2	2	30.95	---	53.09
759	27863	C14H28O	2-tetradecanone	2345-27-9	65.5462	0.1991	0.01	0.35	2	26.20	---	53.09
760	27864	C14H28O	7-ethyl-2-methyl-4-undecanone	6976-00-7	42.6666	0.1964	0.10	3	2	27.14	---	53.09
761	27865	C14H28O	2-methyl-3-tridecanone	40239-35-8	53.9341	0.1967	0.10	1	2	34.29	---	53.09
762	27866	C14H28O	3-tetradecanone	629-23-2	35.3187	0.1961	0.10	8	2	22.49	---	53.09
763	27875	C14H28O2	dodecyl acetate	112-66-3	46.6041	0.1357	0.10	3	2	34.09	---	53.09
764	27922	C14H30O	1-tetradecanol	112-72-1	67.4257	0.1986	0.01	0.30	2	27.02	---	53.09
765	27923	C14H30O	2-tetradecanol	4706-81-4	59.9591	0.1998	0.10	1	2	37.85	---	53.09
766	27924	C14H30O	3-tetradecanol	1653-32-3	54.7863	0.2006	0.10	1	2	34.52	---	53.09
767	27925	C14H30O	12-methyl-1-tridecanol	21987-21-3	54.2765	0.1411	0.10	1	2	39.22	---	53.09
768	27926	C14H30O	2-methyl-3-tridecanol	98930-89-3	50.7693	0.1407	0.10	1	2	36.72	---	53.09
769	27927	C14H30O	2-methyl-4-tridecanol	36691-77-7	54.7849	0.2005	0.10	1	2	34.53	---	53.09
770	27929	C14H30O	tetradecanol; mixed isomers	27196-00-5	66.1777	0.2012	0.01	0.33	2	26.20	---	53.09
771	----	C15H22ClNO2	metolachlor	------	23.2869	0.1250	0.0180	1.77	1,2	14.09	----	25.01
772	28725	C15H24	nonylbenzene	1081-77-2	101.7694	0.1490	0.01	0.16	2	51.23	---	77.50
773	28726	C15H24	1,2,4-triisopropylbenzene	948-32-3	75.6788	0.1500	0.10	1	2	53.58	---	77.50
774	28727	C15H24	1,3,5-triisopropylbenzene	717-74-8	72.9397	0.1489	0.10	2	2	51.77	---	77.50
775	28728	C15H24	3,5-di-tert-butyltoluene	15181-11-0	74.9512	0.2109	0.10	1	2	46.12	---	77.50
776	28873	C15H30O	pentadecanal	2765-11-9	56.0683	0.1924	0.10	1	2	36.00	---	53.09
777	28875	C15H30O	8-pentadecanone	818-23-5	81.7939	0.1874	0.01	0.10	2	34.51	---	53.09
778	28881	C15H30O2	tridecyl acetate	1072-33-9	52.7332	0.1279	0.10	1	2	39.28	---	53.09
779	28913	C15H32O	2-pentadecanol	1653-34-5	79.4590	0.1906	0.01	0.12	2	33.04	---	53.09
780	29595	C16H26	decylbenzene	104-72-3	103.1634	0.1395	0.01	0.13	2	54.27	---	77.50
781	29596	C16H26	pentaethylbenzene	605-01-6	93.1425	0.2049	0.01	0.41	2	36.25	---	77.50
782	29599	C16H26	(1-methylnonyl)benzene	4537-13-7	101.7733	0.1394	0.01	0.14	2	53.56	---	77.50
783	29717	C16H32O	hexadecanal	629-80-1	63.6166	0.1822	0.01	0.37	2	27.49	---	53.09
784	29724	C16H32O2	tetradecyl acetate	638-59-5	58.1553	0.1207	0.01	0.47	2	33.35	---	53.09
785	29768	C16H34O	3-hexadecanol	593-03-3	72.7536	0.1813	0.01	0.18	2	31.57	---	53.09
786	29769	C16H34O	2-methyl-1-pentadecanol	25354-98-7	65.8355	0.1238	0.01	0.18	2	37.23	---	53.09
787	29770	C16H34O	6-methyl-6-pentadecanol	108836-86-8	72.8948	0.1824	0.01	0.18	2	31.47	---	53.09
788	29773	C16H34O	2-hexyl-1-decanol	2425-77-6	72.9897	0.1831	0.01	0.18	2	31.40	---	53.09

code: 1 - data, 2 - estimate
F 400 Activated Carbon, Calgon
Q - adsorption capacity at equilibrium, g of compound/100 g of carbon
x - concentration of compound in liquid, parts per million by weight, ppm(wt)
K and 1/n - coefficients of Freundlich equation
T - temperature, K
xmin - minimum concentration, ppm(wt)
xmax - maximum concentration, ppm(wt)

Chapter 3

SOLUBILITY IN WATER – ORGANIC COMPOUNDS

Carl L. Yaws, Satish K. Padala, and Swetha Pingali
Lamar University, Beaumont, Texas

Tabulation Results

The results for solubility in water are given in Table 3. The tabulation is arranged by carbon number (C, C2, C3, ...) to provide ease of use in quickly locating the data by using the chemical formula. The compound name, CAS No (Chemical Abstracts Registry Number), and solubility values are provided in the adjacent columns. The last column provides the code for the tabulation which is based on both experimental data and estimated values.

In preparing the tabulation, a literature search was conducted to identify data source publications for organics (1-172). Both experimental values for the property under consideration and parameter values for estimation of the property are included in the source publications. The publications were screened and copies of appropriate data were made. These data were then keyed into the computer to provide a database of values for compounds for which experimental data are available. The database also served as a basis to check the accuracy of the estimation methods. Upon completion of data collection, estimation of the values for the remaining compounds was performed. The numerous point values were processed using a computer program for minimum deviation.

The excellent compilations by Horvath (56); Howard and Meylan (57); Mackay, Shiu, and Ma (91-95); Verschueren (141); Yalkowsky (147); and Yaws (149-171) were primarily used for experimental data for water solubility. Most of the estimates were based on the boiling point method as developed by Yaws (150-169). Parameter values for the estimates were from data compilations of Yaws (169-172).

For hydrocarbons and chemicals, experimental data available in the literature are very limited for isomers, large molecules (C10 and higher), and substituted organics containing halogens, nitrogen, oxygen, and sulfur. In the absence of data, the estimates for isomers, large molecules, and substituted organics should be considered as rough approximations useful for initial analysis. If initial analysis is favorable, follow-up experimental determination is recommended.

A comparison of calculated and data values is shown in Figure 3 for a representative chemical family. The graph discloses favorable agreement of equation and data.

Example

A chemical spill of pentane (C5H12) occurs into a body of water at ambient conditions (25 C). Determine the concentration in water at saturation.

Inspection of the tabulation for solubility in water yields:

S = 38.5 parts per million by weight, ppm(wt)

References

1. Acree, W. E., Jr., THERMODYNAMIC PROPERTIES OF NONELECTROLYTE SOLUTIONS, Academic Press, New York, NY (1984).
2. Akiyoshi, M., T. Deguchi and I. Sanemasa, Bull. Chem. Soc. Jap. (Nippon Kagakukai Bull.), 60, 3935 (1987).
3. Albert, A., Chem. & Ind. (London), 252, (1956).
4. Alexander, D. M., J. Phys. Chem., 63, 1021 (1959).
5. Allott, P. R., A. Steward, V. Flook and W. W. Mapleson, Brit. J. Anaesth., 45, 294 (1973).
6. Almgren, M., F. Grieser and J. K. Thomas, J. Am. Chem. Soc., 101, 279 (1979).
7. Amoore, J. E. and R. G. Buttery, Chem. Senses and Flavor, 3, 57 (1978).
8. Andrews, L. J. and R. M. Keefer, J. Am. Chem. Soc., 72, 3113 (1950).
9. Andrews, L. J. and R. M. Keefer, J. Am. Chem. Soc., 72, 5034 (1950).
10. Anthony, R. G. and J. J. McKetta, J. Chem. Eng. Data, 12, 17 (1967).
11. Anthony, R. G. and J. J. McKetta, J. Chem. Eng. Data, 12, 21 (1967).
12. Arnold, D. S., C. A. Plank and E. E. Erickson, Chem. Eng. Data. Ser., 3, 253 (1958).
13. Azarnoosh, A. and J. J. McKetta, J. Chem. Eng. Data, 4, 211 (1959).
14. Azarnoosh, A. and J. J. McKetta, Petrol. Refiner, 37[11], 275 (1958).
15. Azaz, E. and M. Donbrow, J. Colloid & Interface Sc., 57, 11 (1976).
16. Back, E. and B. Steenberg, Acta Chem. Scand., 4, 810 (1950).

17. Baker, E. G., FUNDAMENTAL ASPECTS OF PETROLEUM GEOCHEMISTRY, Nagy, B. and U. Colombo, eds., Chapter 7, 299, Elsevier Publishing Co., New York, NY (1967).
18. Baker, E. G., Science, 129, 871 (1959).
19. Baker, R. J., B. J. Donelan, L. J. Peterson, W. E. Acree, Jr. and C. C. Tsai, Phys. Chem. Liq., 16, 279 (1987).
20. Banerjee, S., S. H. Yalkowsky and S. C. Valvani, Env. Sc. & Tech., 14, 1227 (1980).
21. Banerjee, S., Env. Sc. & Tech., 18, 587 (1984).
22. Bergen, Jr., R. L., and F. A. Long, J. Phy. Chem., 60, 1131 (1956).
23. Bidner, M. S. and M. de Santiago, Chem. Eng. Sc., 26, 1484 (1971).
24. Bohon, R. L. and W. F. Claussen, J. Am. Chem. Soc., 73, 1571 (1951).
25. Bowman, M. C., J. R. King and C. L. Holder, Internat. J. Env. Anal. Chem., 4, 205 (1976).
26. Bradbury, E. J., D. McNulty, R. L. Savage and E. E. McSweeney, Ind. Eng. Chem., 44, 211 (1952).
27. Brooks, W. B. and J. J. McKetta, Petrol. Refiner, 34[2], 143 (1955).
28. Brooks, W. B., G. B. Gibbs and J. J. McKetta, Petrol. Refiner, 30[10], 118 (1951).
29. Brooks, W. B., J. E. Haughn and J. J. McKetta, Petrol. Refiner, 34[8], 129 (1955).
30. Burris, D. R. and W. G. MacIntyre, Env. Sc. & Tech., 20, 296 (1986).
31. Chiou, C. T. and D. W. Schmedding, Env. Sc. & Tech., 16, 4 (1982).
32. Clarke, G. A., T. R. Williams and R. W. Taft, J. Am. Chem. Soc., 84, 2292 (1962).
33. Conway, J. B. and J. J. Norton, Ind. & Eng. Chem., 43, 1433 (1951).
34. CRC HANDBOOK OF CHEMISTRY AND PHYSICS, 75th - 86th eds., CRC Press, Inc., Boca Raton, FL (1994-2006).
35. Culberson, O. L. and J. J. McKetta, Trans. AIME, 189, 319 (1950).
36. Culberson, O. L. and J. J. McKetta, Trans. AIME, 192, 233 (1951).
37. Culberson, O. L., A. B. Horn and J. J. McKetta, Trans. AIME, 189, 1 (1950).
38. Curme, G. O., Jr., ed., GLYCOLS, American Chemical Society Series, Reinhold Publishing Corporation, New York, NY (1953).
39. Davis, J. E. and J. J. McKetta, J. Chem. Eng. Data, 5, 374 (1960).
40. Davis, J. E. and J. J. McKetta, Petrol. Refiner, 39[3], 205 (1960).
41. Deno, N. C. and H. E. Berkheimer, J. Chem. & Eng. Data, 5, 1 (1960).
42. Dixon, M. R., C. E. Rehberg and C. H. Fisher, J. Am. Chem. Soc., 70, 3733 (1948).
43. Eckert, J. W., Phytopath., 52, 642 (1962).
44. Eganhouse, R. P. and J. A. Calder, Geochim. et Cosmochim. Acta, 40, 555 (1976).
45. Eik-Nes, K., J. A. Schellman, R. Lumry and L. T. Samuels, J. Bio. Chem., 206, 411 (1954).
46. Evans, B. K., K. C. James and D. K. Luscombe, J. Pharm. Sc., 67, 277 (1978).
47. Fendler, J. H., and others, J. Am. Chem. Soc., 97, 89 (1975).
48. Franks, F., M. Gent and H. H. Johnson, J. Chem. Soc., 2716 (1963).
49. Franks, F., Nature, 210, 87 (1966).
50. Freier, R. K., AQUEOUS SOLUTIONS - DATA FOR INORGANIC AND ORGANIC COMPOUNDS, Vols. 1 and 2, Walter de Gruyter, Berlin, Germany - New York, NY (1976).
51. Hafkenscheid, T. L. and E. Tomlinson, J. Chromatogr., 218, 409 (1981).
52. Hansch, C., J. E. Quinlan, and G. L. Lawrence, J. Org. Chem., 33(1), 347 (1968).
53. Hashimoto, Y., K. Tokura, K. Ozaki and W. M. J. Strachan, Chemosphere, 11, 991 (1982).
54. Hildebrand, J. L. and R. L. Scott, THE SOLUBILITY OF NONELECTROLYTES, 3rd ed., Reinhold Publishing Corporation, New York, NY (1950).
55. Hollifield, J. C., Bull. Env. Contamin. & Toxic., 23, 579 (1979).
56. Horvath, A. L., HALOGENATED HYDROCARBONS, Marcel Dekker, Inc., New York, NY (1982).
57. Howard, P. H. and W. M. Meylan, eds., HANDBOOK OF PHYSICAL PROPERTIES OF ORGANIC CHEMICALS, CRC Press, Boca Raton, FL (1997).
58. Ikeda, Y. and others, Yakugaku Zasshi (Tokyo), 102, 83 (1982).
59. Inga, R. F. and J. J. McKetta, J. Chem. Eng. Data, 6, 337 (1961).
60. Inga, R. F. and J. J. McKetta, Petrol. Refiner, 40(3),191 (1961).
61. Irmann, F., Chem. Ing. Tech. 37(8), 789 (1965).
62. Iwamoto, E., Y. Hiyama and Y. Yamamoto, J. Sol. Chem., 6, 371 (1977).
63. Kabadi, V. N. and R. P. Danner, Hydrocarbon Processing, 58(5), 245 (1979).
64. Kertes, A. S., O. Levy and G. Y. Markovits, EXPERIMENTAL THERMODYNAMICS OF NONPOLAR FLUIDS, Dack, M. R. J., ed., Vol. II, Chapter 15, Solubility, 725, Butterworth, London (1975).
65. Klevens, H. B., Chem. Rev., 47, 1 (1950).
66. Klevens, H. B., J. Phys. Colloid Chem., 54, 283 (1950).
67. Kobayashi, R. and D. Katz, Ind. Eng. Chem., 45, 440 (1953).
68. Krzyzanowska, T. and J. Szeliga, Nafta (Katowice, Poland), 34, 413 (1978).
69. Kudchadker, A. P. and J. J. McKetta, AICHE J., 7, 707 (1961).
70. Kudchadker, A. P. and J. J. McKetta, Petrol. Refiner, 40[9], 231 (1961).
71. Kudchadlker, A. P. and J. J. McKetta, Petrol. Refiner, 41[3], 191 (1962).
72. Lamouroux, F., Comptes Rendus Hebdoa. des Seances de l'Acad. des Sc., 128, 998 (1899).

73. LANGE'S HANDBOOK OF CHEMISTRY, 13th, 14th, and 15th eds., McGraw-Hill, New York, NY (1985, 1992, 1999).
74. LeBreton, J. G. and J. J. McKetta, Hydrocarbon Processing, 43[6], 136 (1964).
75. Leinonen, P. J. and D. Mackay, Can. J. Chem. Eng., 51[2], 230 (1973).
76. Leland, T. W., J. J. McKetta and K. A. Kobe, Ind, Eng. Chem., 47, 1265 (1955).
77. Leyder, F. and P. Boulanger, Bull. Env. Contamin. & Toxic., 30, 152 (1983).
78. Li, C. C. and J. J. McKetta, J. Chem. Eng. Data, 8[2], 271 (1963).
79. Li, C. C. and J. J. McKetta, Petrol. Refiner, 42[3], 135 (1963); 42[6], 107 (1963).
80. Lindstrom, R. E. and C. H. Lee, J. Pharm. & Pharmaco., 32, 245 (1980).
81. LIQUID-LIQUID EQUILIBRIUM DATA COLLECTION, Supplement 1, Vol. V, Part 4, Dechema Chemistry Data Series, 6000 Frankfurt/Main, Germany (1987).
82. LIQUID-LIQUID EQUILIBRIUM DATA COLLECTION, Vol. V, part 1, Dechema Chemistry Data Series, 6000 Frankfurt/Main, Germany (1979).
83. Lozano, F. J., J. Chem. & Eng. Data, 26, 131 (1981).
84. Lu, P-Y. and R. L. Metcalf, Env. Health Perspect., 10, 269 (1975).
85. Lu, P-Y., R. L. Metcalf and E. M. Carlson, Env. Health Perspect., 24, 201 (1978).
86. Lyman, W. J., W. F. Reehl, and D. H. Rosenblatt, HANDBOOK OF CHEMICAL PROPERTY ESTIMATION METHODS, American Chemical Society, Washington, DC (1990).
87. Mackay, D. and others, Env. Sc. & Tech., 19, 522 (1985).
88. Mackay, D. and W. Y. Shiu, J. Chem. & Eng. Data, 22, 399 (1977).
89. Mackay, D. and W. Y. Shiu, J. Phys. Chem. Ref. Data, 10(4), 1175 (1981).
90. Mackay, D., W. Y. Shiu and R. P. Sutherland, Environ. Sci. Technol., 13(3), 333 (1979).
91. Mackay, D., W. Y. Shiu, and K. C. Ma, ILLUSTRATED HANDBOOK OF PHYSICAL-CHEMICAL PROPERTIES AND ENVIRONMENTAL FATE FOR ORGANIC CHEMICALS, Volume 1 – Monoaromatic Hydrocarbons, Chlorobenzenes and PCBs, Lewis Publishers, New York, NY (1992).
92. Mackay, D., W. Y. Shiu, and K. C. Ma, ILLUSTRATED HANDBOOK OF PHYSICAL-CHEMICAL PROPERTIES AND ENVIRONMENTAL FATE FOR ORGANIC CHEMICALS, Volume 2 – Polynuclear Aromatic Hydrocarbons, Polychlorinated Dioxins, and Dibenzofurans, Lewis Publishers, New York, NY (1992).
93. Mackay, D., W. Y. Shiu, and K. C. Ma, ILLUSTRATED HANDBOOK OF PHYSICAL-CHEMICAL PROPERTIES AND ENVIRONMENTAL FATE FOR ORGANIC CHEMICALS, Volume 3 – Volatile Organic Compounds, Lewis Publishers, New York, NY (1993).
94. Mackay, D., W. Y. Shiu, and K. C. Ma, ILLUSTRATED HANDBOOK OF PHYSICAL-CHEMICAL PROPERTIES AND ENVIRONMENTAL FATE FOR ORGANIC CHEMICALS, Volume 4 – Oxygen, Nitrogen, and Sulfur Containing Compounds, Lewis Publishers, New York, NY (1995).
95. Mackay, D., W. Y. Shiu, and K. C. Ma, ILLUSTRATED HANDBOOK OF PHYSICAL-CHEMICAL PROPERTIES AND ENVIRONMENTAL FATE FOR ORGANIC CHEMICALS, Volume 5 – Pesticide Chemicals, Lewis Publishers, New York, NY (1997).
96. Marcus, Y., L. E. Asher, J. Hormadaly and E. Pross, Hydrometallurgy, 7, 27 (1981).
97. Markley, K. S., FATTY ACIDS, Interscience Publishers, Inc., New York, NY (1947).
98. Martin, H., PESTICIDE MANUAL, British Crop Protection Council (1977).
99. Massaldi, H. A. and C. J. King, J. Chem. & Eng. Data, 18, 393 (1973).
100. McAuliffe, C., J. Phy. Chem., 70, 1267 (1966).
101. McAuliffe, C., Nature (London), 200, 1092 (1963).
102. McAuliffe, C., Science, 158, 478 (1969).
103. McBain, J. W. and P. H. Richards, Ind. & Eng. Chem., 38, 642 (1946).
104. Merckel, J. H. C., Recueil des Travaux Chim. des Pays-Bas, 56, 811 (1937).
105. Merrill, E. J., J. Pharm. Sc., 54, 1670 (1965).
106. Miller, M. M., S. P. Waslk, G. L. Huang, W. Y. Shiu and D. Mackay, Environ. Sci. Technol., 19(6), 522 (1985).
107. Mitchell, A. G. and L. S. C. Wan, J. Pharm. Sc., 53, 1467 (1964).
108. Mitchell, A. G., L. S. C. Wan and S. G. Bjaastad, J. Pharm. & Pharmaco., 16, 632 (1964).
109. Montgomery, J. H., GROUNDWATER CHEMICALS, 3th ed., CRC Press, Boca Raton, FL (2000).
110. Natarajan, G. S. and K. A. Venkatachalam, J. Chem. Eng. Data, 17[3], 328 (1972).
111. Nathan, M. F., Chem. Eng. (New York), 85, 93 (1978).
112. Peake, E. and G. W. Hodgson, J. Amer. Oil Chem. Soc., 43, 215 (1966).
113. Peake, E. and G. W. Hodgson, J. Amer. Oil Chem. Soc., 44, 696 (1967).
114. Polak, J. and B. C. Y. Lu, Can. J. Chem., 51, 4018 (1973).
115. Price, L. C., Bull. Am. Assoc. Petrol. Geologists, 60(2), 213-244 (1976).
116. Reamer, H. H., B. H. Sage and W. N. Lacey, Ind. Eng. Chem., 44, 609 (1952).
117. Rebert, C. J. and K. E. Hayworth, AICHE J., 13[1], 118 (1967).
118. Rebert, C. J. and W. B. Kay, AIChE J., 5[3], 285 (1959).
119. Reed, C. D. and J. J. McKetta, J. Chem. Eng. Data, 4, 294 (1959).
120. Reed, C. D. and J. J. McKetta, Petrol. Refiner, 38[4], 159 (1959).

121. Sanchez, M. and H. Lentz, High Temperatures-High Pressures, 5, 689 (1973).
122. Schatzberg, P., J. Phys. Chem., 67, 776 (April, 1963).
123. Schwarz, F. P., Anal. Chem., 52, 10 (1980).
124. Schwarz, F. P., J. Chem. Eng. Data, 22, 273 (1977).
125. Simpson, L. B. and F. P. Lovell, J. Chem. & Eng. Data, 7, 498 (1962).
126. Simpson, L. B. and F. P. Lovell, J. Chem. Eng. Data, 7, 498 (1962).
127. SOLUBILITY DATA SERIES, IUPAC, Vol. 9, Ethane, Pergamon Press, New York, NY (1982).
128. SOLUBILITY DATA SERIES, IUPAC, Vol. 15, Alcohols with Water, Pergamon Press, Oxford, England (1984).
129. SOLUBILITY DATA SERIES, IUPAC, Vol. 24, Propane, Butane, and 2-Methylpropane, Pergamon Press, New York, NY (1986).
130. SOLUBILITY DATA SERIES, IUPAC, Vol. 27/28, Methane, Pergamon Press, New York, NY (1987).
131. SOLUBILITY DATA SERIES, IUPAC, Vol. 37, Hydrocarbons with Water and Seawater, Part 1 – Hydrocarbons C5 to C7, Pergamon Press, Oxford, England (1989).
132. SOLUBILITY DATA SERIES, IUPAC, Vol. 38, Hydrocarbons with Water and Seawater, Part 1 – Hydrocarbons C8 to C36, Pergamon Press, Oxford, England (1989).
133. SOLUBILITY DATA SERIES, IUPAC, Vol. 43, Carbon Monoxide, Pergamon Press, New York, NY (1990).
134. SOLUBILITY DATA SERIES, IUPAC, Vol. 48, Esters with Water, Part 1 – Esters C2 to C6, Pergamon Press, Oxford, England (1992).
135. SOLUBILITY DATA SERIES, IUPAC, Vol. 49, Esters with Water, Part 2 – Esters C7 to C32, Pergamon Press, Oxford, England (1992).
136. Speece, R. E. and N. N. Nirmalakhandan, Environ. Sci. Technol., 22(3), 328 (1988).
137. Stephen, H. and T. Stephen, SOLUBILITIES OF INORGANIC AND ORGANIC COMPOUNDS, Vol. 1, Part 1, Macmillan Company, New York, NY (1963).
138. Sutton, C. and J. A. Calder, Environ. Sci. Techn., 8(7), 654 (1974).
139. Suzuki, T., J. Computer Aided Molecular Design, 5, 149 (1991).
140. TECHNICAL DATA BOOK - PETROLEUM REFINING, Vols. I and II, American Petroleum Institute, Washington, DC (1972, 1977, 1982).
141. Verschueren, K., HANDBOOK OF ENVIRONMENTAL DATA ON ORGANIC CHEMICALS, 3rd and 4th eds., Van Nostrand Reinhold, New York, NY (1996, 2001).
142. Vesala, A., Acta Chem. Scand., Series A:Phy. & Inorg. Chem., 28, 839 (1974).
143. Wauchope, R. D. and F. W. Getzen, J. Chem. Eng. Data, 17[1], 38 (1972).
144. Wen, W. Y. and J. H. Hung, J. Phys. Chem., 74, 170 (1970).
145. Wilhelm, E., CRC Crit. Rev. Anal. Chem., 16(2), 129 (1985).
146. Yalkowsky, S. H. and S. Banerjee, AQUEOUS SOLUBILITY, Marcel Dekker, Inc., New York, NY (1992).
147. Yalkowsky, S. H., Aquasol Database, 5th edition, College of Pharmacy, University of Arizona, Tuscon, Az (1990).
148. Yalkowsky, S. H., S. Banerjee and R. S. Pearlman, J. Phys. Chem. Ref. Data, 13(2), 555 (1984).
149. Yaws, C. L. and others, "Hydrocarbons: Water Solubility Data", Chem. Eng., 97, 177 (April,1990).
150. Yaws, C. L. and others, "Organic Chemicals: Water Solubility Data", Chem. Eng., 97, 115 (July, 1990).
151. Yaws, C. L. and others, "Water Solubility Data for Organic Compounds", Pollution Engineering, 22, 70 (Oct., 1990).
152. Yaws, C. L. and others, "New Correlation Calculates Reliable Paraffin Solubilities", Oil & Gas Journal, 89, 79 (April 8, 1991).
153. Yaws, C. L. and others, "New Correlation Accurately Calculates Naphthene Water Solubilities", Oil & Gas Journal, 89(37), 86 (Sept. 16, 1991).
154. Yaws, C. L. and others, "Water Solubility Data for 151 Hydrocarbons", [temperature effect – alkanes], Chem. Eng., 100, 108 (February, 1993).
155. Yaws, C. L. and others, "The Water Solubility of Naphthenes", [temperature effect – cycloalkanes], Chem. Eng., 100 (10), 122 (Oct., 1993).
156. Yaws, C. L. and others, "New Correlation Accurately Calculates Water Solubilities of Aromatics", Oil & Gas Journal, 95 (35), 80 (Aug. 29, 1994).
157. Yaws, C. L. and others, "Calculate the Solubility of Aromatics", [temperature effect – benzenes], Chem. Eng., 102 (2), 113 (Feb., 1995).
158. Yaws, C. L. and others, "Solubility and Henry's Law Constant for Alcohols in Water", Waste Management, 4(8), 541(1997).
159. Yaws, C. L. and others, "Using Solubility and Henry's Law Constant for Ketones in Water", Pollution Engineering, 30 (2), 44 (1998).
160. Yaws, C. L. and others, "Calculating Solubility and Henry's Law Constant for Gases in Water", Chemical Engineering, 106 (6), 102 (June, 1999).
161. Yaws, C. L. and others, "Solubility and Henry's Law Constant for Amines in Water", Chem. Eng., 108 (8), 84 (Aug, 2001).
162. Yaws, C. L. and others, "Solubility and Henry's Law Constants for Sulfur Compounds in Water", Chem. Eng., 110 (8), 60-64 (August, 2003).

163. Yaws, C. L. and others, "Correlation Calculates Solubility, Henry's Law Constant for Olefins, Diolefins in Water", Oil & Gas Journal, 101.37, 76-78 (Sept. 29, 2003).
164. Yaws, C. L. and others, "Solubility and Henry's Law Constant for Chlorinated Compounds in Water", Chem. Eng., 112 (2), 50-56 (February, 2005).
165. Yaws, C. L. and others, "Solubility of Chemicals in Water", WATER ENCYCLOPEDIA, Volume 4, ISBN 0-471-73684-8, pages 555-559, John Wiley, New York, NY (2005).
166. Yaws, C. L. and others, "Solubility of Hydrocarbons in Salt Water", WATER ENCYCLOPEDIA, Volume 4, ISBN 0-471-73684-8, pages 559-561, John Wiley, New York, NY (2005).
167. Yaws, C. L. and others, "Solubility of Hydrocarbons and Sulfur Compounds in Water", WATER ENCYCLOPEDIA, Volume 4, ISBN 0-471-73684-8, pages 561-564, John Wiley, New York, NY (2005).
168. Yaws, C. L. and others, "Correlation Measures Properties of Brominated Petchems", [solubility and Henry's law constant in water], Oil & Gas Journal, 104.11, 53-56 (March 20, 2006).
169. Yaws, C. L., CHEMICAL PROPERTIES HANDBOOK, McGraw-Hill, New York, NY (1999).
170. Yaws, C. L., MATHESON GAS DATA BOOK, 7th ed., Matheson Tri-Gas (Parsippany, NJ), McGraw-Hill, New York, NY (2001).
171. Yaws, C. L., YAWS HANDBOOK OF THERMODYNAMIC AND PHYSICAL PROPERTIES OF CHEMICAL COMPOUNDS, electronic edition, www.knovel.com, knovel, Norwich, NY (2003).
172. Yaws, C. L., YAWS HANDBOOK OF PHYSICAL PROPERTIES FOR HYDROCARBONS AND CHEMICALS, gulf Publishing Co., Houston, TX (2005).

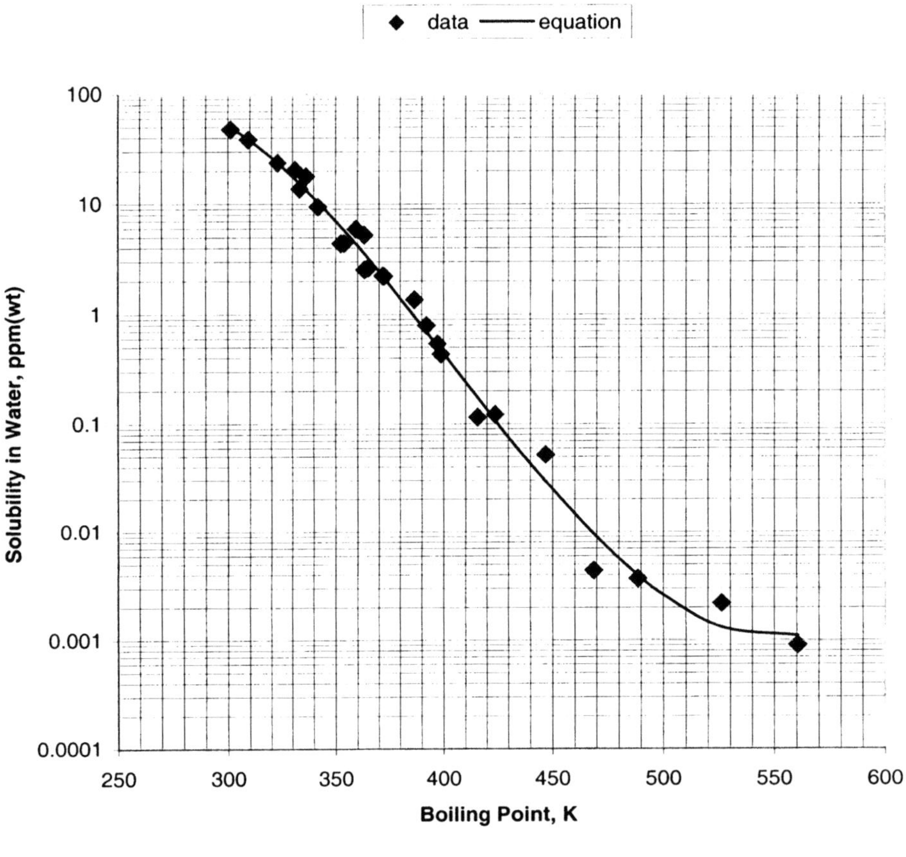

Figure 3 Solubility of Alkanes in Water

Table 3 Solubility in Water - Organic Compounds

NO	ID	FORMULA	NAME	CAS No	Solubility in Water, S (ppm - parts per million, wt - weight, g - gram) T, C	S@T, ppm(wt)	S@T, ppm(mol)	S@T, g/1000 g of H2O	S@T, mol/1000 g of H2O	code
1	13	CBrClF2	bromochlorodifluoromethane	353-59-3	2	9.9100E+02	1.0806E+02	9.9198E-01	5.9988E-03	1
2	17	CBrF3	bromotrifluoromethane	75-63-8	25	3.2242E+02	3.9017E+01	3.2252E-01	2.1659E-03	1
3	29	CBr4	carbon tetrabromide	558-13-4	30	2.4000E+02	1.3040E+01	2.4006E-01	7.2388E-04	1
4	34	CClF3	chlorotrifluoromethane	75-72-9	25	9.0929E+01	1.5683E+01	9.0937E-02	8.7055E-04	1
5	39	CClN	cyanogen chloride	506-77-4	25	3.0000E+04	8.9825E+03	3.0928E+01	5.0313E-01	1
6	42	CCl2F2	dichlorodifluoromethane	75-71-8	25	3.0340E+02	4.5216E+01	3.0349E-01	2.5100E-03	1
7	44	CCl2O	phosgene	75-44-5	25	6.8255E+03	1.2501E+03	6.8724E+00	6.9478E-02	1
8	47	CCl3F	trichlorofluoromethane	75-69-4	30	1.0800E+03	1.4177E+02	1.0812E+00	7.8707E-03	1
9	50	CCl4	carbon tetrachloride	56-23-5	25	7.8570E+02	9.2082E+01	7.8632E-01	5.1119E-03	1
10	74	CF4	carbon tetrafluoride	75-73-0	25	1.8227E+01	3.7312E+00	1.8227E-02	2.0712E-04	1
11	96	CHBr3	bromoform	75-25-2	25	3.1100E+03	2.2233E+02	3.1197E+00	1.2344E-02	1
12	97	CHClF2	chlorodifluoromethane	75-45-6	25	2.7760E+03	5.7963E+02	2.7837E+00	3.2194E-02	1
13	101	CHCl2F	dichlorofluoromethane	75-43-4	25	1.8691E+04	3.3228E+03	1.9047E+01	1.8506E-01	1
14	104	CHCl3	chloroform	67-66-3	25	7.8400E+03	1.1911E+03	7.9020E+00	6.6194E-02	1
15	113	CHF3	fluoroform	75-46-7	25	7.2028E+02	1.8543E+02	7.2080E-01	1.0295E-02	1
16	117	CHI3	iodoform	75-47-8	25	1.0000E+02	4.5759E+00	1.0001E-01	2.5401E-04	1
17	122	CHN	hydrogen cyanide	74-90-8	25	total	total	total	total	1
18	140	CH2BrCl	bromochloromethane	74-97-5	25	1.5000E+04	2.1159E+03	1.5228E+01	1.1770E-01	1
19	144	CH2Br2	dibromomethane	74-95-3	25	1.1000E+04	1.1513E+03	1.1122E+01	6.3982E-02	1
20	145	CH2ClF	chlorofluoromethane	593-70-4	25	1.0014E+04	2.6540E+03	1.0115E+01	1.4772E-01	1
21	149	CH2Cl2	dichloromethane	75-09-2	25	1.9400E+04	4.1788E+03	1.9784E+01	2.3294E-01	1
22	158	CH2F2	difluoromethane	75-10-5	25	4.3880E+03	1.5239E+03	4.4073E+00	8.4718E-02	1
23	161	CH2I2	diiodomethane	75-11-6	25	1.2400E+03	8.3501E+01	1.2415E+00	4.6354E-03	1
24	175	CH2O	formaldehyde	50-00-0	25	total	total	total	total	1
25	176	CH2O2	formic acid	64-18-6	100	total	total	total	total	1
26	191	CH3Br	methyl bromide	74-83-9	25	1.4576E+04	2.7989E+03	1.4792E+01	1.5580E-01	1
27	196	CH3Cl	methyl chloride	74-87-3	25	5.1457E+03	1.8422E+03	5.1723E+00	1.0245E-01	1
28	214	CH3F	methyl fluoride	593-53-3	25	1.9266E+03	1.0207E+03	1.9303E+00	5.6719E-02	1
29	222	CH3I	methyl iodide	74-88-4	25	2.6070E+04	3.3859E+03	2.6768E+01	1.8859E-01	1
30	232	CH3NO	formamide	75-12-7	20	total	total	total	total	1
31	235	CH3NO2	nitromethane	75-52-5	25	1.0000E+04	2.9723E+03	1.0101E+01	1.6548E-01	1
32	251	CH4	methane	74-82-8	25	2.1949E+01	2.4647E+01	2.1949E-02	1.3682E-03	1
33	272	CH4O	methyl alcohol	67-56-1	25	total	total	total	total	1
34	278	CH4S	methyl mercaptan	74-93-1	25	2.3300E+04	8.8541E+03	2.3856E+01	4.9587E-01	1
35	286	CH5N	methylamine	74-89-5	25	total	total	total	total	1
36	372	CO	carbon monoxide	630-08-0	25	2.5937E+01	1.6682E+01	2.5938E-02	9.2600E-04	1
37	373	COS	carbonyl sulfide	463-58-1	25	1.2373E+03	3.7135E+02	1.2388E+00	2.0621E-02	1
38	375	CO2	carbon dioxide	124-38-9	25	1.4429E+03	5.9114E+02	1.4450E+00	3.2833E-02	1
39	382	CS2	carbon disulfide	75-15-0	25	1.8790E+03	4.4520E+02	1.8825E+00	2.4724E-02	1
40	427	C2ClF5	chloropentafluoroethane	76-15-3	25	5.8589E+01	6.8334E+00	5.8592E-02	3.7932E-04	1
41	441	C2Cl2F4	1,2-dichloro-1,1,2,2-tetrafluoroethane	76-14-2	25	1.3871E+02	1.4622E+01	1.3873E-01	8.1166E-04	1
42	449	C2Cl3F3	1,1,2-trichloro-1,2,2-trifluoroethane	76-13-1	25	1.7000E+02	1.6347E+01	1.7003E-01	9.0743E-04	1
43	455	C2Cl4	tetrachloroethylene	127-18-4	25	1.5000E+02	1.6297E+01	1.5002E-01	9.0467E-04	1
44	456	C2Cl4F2	1,1,2,2-tetrachloro-1,2-difluoroethane	76-12-0	25	1.2000E+02	1.0607E+01	1.2001E-01	5.8880E-04	1
45	463	C2Cl6	hexachloroethane	67-72-1	25	8.0000E+00	6.0878E-01	8.0001E-03	3.3793E-05	1
46	488	C2F4	tetrafluoroethylene	116-14-3	25	1.5193E+02	2.7369E+01	1.5195E-01	1.5193E-03	1
47	494	C2F6	hexafluoroethane	76-16-4	25	7.8970E+00	1.0308E+00	7.8971E-03	5.7220E-05	1
48	511	C2HBrClF3	halothane	151-67-7	24	3.7200E+03	3.4068E+02	3.7339E+00	1.8917E-02	1
49	539	C2HCl3	trichloroethylene	79-01-6	25	1.1000E+03	1.5097E+02	1.1012E+00	8.3814E-03	1
50	551	C2HCl5	pentachloroethane	76-01-7	25	5.0000E+02	4.4547E+01	5.0025E-01	2.4729E-03	1
51	571	C2H2	acetylene	74-86-2	25	1.0504E+03	7.2698E+02	1.0515E+00	4.0384E-02	1
52	592	C2H2Br4	1,1,2,2-tetrabromoethane	79-27-6	30	6.5100E+02	3.3950E+01	6.5142E-01	1.8846E-03	1
53	593	C2H2Br4	1,1,1,2-tetrabromoethane	630-16-0	25	6.5100E+02	3.3950E+01	6.5142E-01	1.8846E-03	2
54	615	C2H2Cl2	1,1-dichloroethylene	75-35-4	25	3.3450E+03	6.2330E+02	3.3562E+00	3.4621E-02	1
55	616	C2H2Cl2	cis-1,2-dichloroethylene	156-59-2	25	3.5000E+03	6.5227E+02	3.5123E+00	3.6231E-02	1
56	617	C2H2Cl2	trans-1,2-dichloroethylene	156-60-5	25	6.3000E+03	1.1768E+03	6.3399E+00	6.5399E-02	1
57	618	C2H2Cl2	dichloroethylene	25323-30-2	25	4.3817E+03	8.1717E+02	4.4010E+00	4.5397E-02	2
58	619	C2H2Cl2	1,2-dichloroethylene, cis and trans	540-59-0	25	4.3817E+03	8.1717E+02	4.4010E+00	4.5397E-02	2
59	627	C2H2Cl2O2	dichloroacetic acid	79-43-6	20	total	total	total	total	1
60	635	C2H2Cl4	1,1,1,2-tetrachloroethane	630-20-6	25	1.1000E+03	1.1818E+02	1.1012E+00	6.5608E-03	1
61	636	C2H2Cl4	1,1,2,2-tetrachloroethane	79-34-5	25	2.9000E+03	3.1206E+02	2.9084E+00	1.7328E-02	1
62	642	C2H2F2	1,1-difluoroethylene	75-38-7	25	1.5968E+02	4.4928E+01	1.5971E-01	2.4940E-03	1
63	643	C2H2F2	cis-1,2-difluoroethene	1630-77-9	25	1.5968E+02	4.4928E+01	1.5971E-01	2.4940E-03	2
64	644	C2H2F2	trans-1,2-difluoroethene	1630-78-0	25	1.5968E+02	4.4928E+01	1.5971E-01	2.4940E-03	2
65	650	C2H2F4	1,1,1,2-tetrafluoroethane	811-97-2	25	6.6005E+02	1.1660E+02	6.6049E-01	6.4733E-03	1
66	651	C2H2F4	1,1,2,2-tetrafluoroethane	359-35-3	25	6.6005E+02	1.1660E+02	6.6049E-01	6.4733E-03	2
67	684	C2H2O4	oxalic acid	144-62-7	20	8.3000E+04	1.7788E+04	9.0513E+01	1.0053E+00	1
68	701	C2H3Cl	vinyl chloride	75-01-4	25	2.3906E+03	6.9026E+02	2.3963E+00	3.8342E-02	1
69	702	C2H3ClF2	1-chloro-1,1-difluoroethane	75-68-3	25	1.0178E+03	1.8261E+02	1.0188E+00	1.0138E-02	---
70	720	C2H3Cl3	1,1,1-trichloroethane	71-55-6	25	1.0000E+03	1.3516E+02	1.0010E+00	7.5036E-03	1
71	721	C2H3Cl3	1,1,2-trichloroethane	79-00-5	25	4.4200E+03	5.9917E+02	4.4396E+00	3.3280E-02	1
72	753	C2H3N	acetonitrile	75-05-8	25	total	total	total	total	1
73	787	C2H4	ethylene	74-85-1	25	1.3646E+02	8.7633E+01	1.3648E-01	4.8649E-03	1
74	798	C2H4Br2	1,1-dibromoethane	557-91-5	25	1.4454E+04	1.4044E+03	1.4666E+01	7.8067E-02	2
75	799	C2H4Br2	1,2-dibromoethane	106-93-4	25	4.1700E+03	4.0140E+02	4.1875E+00	2.2290E-02	1
76	812	C2H4Cl2	1,1-dichloroethane	75-34-3	25	5.1000E+03	9.3232E+02	5.1261E+00	5.1801E-02	1
77	813	C2H4Cl2	1,2-dichloroethane	107-06-2	20	8.7000E+03	1.5951E+03	8.7764E+00	8.8687E-02	1
78	814	C2H4Cl2	dichloroethane	1300-21-6	25	6.5238E+03	1.1940E+03	6.5666E+00	6.6357E-02	2
79	834	C2H4F2	1,1-difluoroethane	75-37-6	25	2.6802E+03	7.3244E+02	2.6874E+00	4.0687E-02	1
80	835	C2H4F2	1,2-difluoroethane	624-72-6	25	2.6800E+03	7.3238E+02	2.6872E+00	4.0684E-02	2

Table 3 Solubility in Water - Organic Compounds

					Solubility in Water, S (ppm - parts per million, wt - weight, g - gram)						
NO	ID	FORMULA	NAME	CAS No	T, C	S@T, ppm(wt)	S@T, ppm(mol)	S@T, g/1000 g of H2O	S@T, mol/1000 g of H2O	code
81	884	C2H4O	acetaldehyde	75-07-0	99	total	total	total	total	1
82	885	C2H4O	ethylene oxide	75-21-8	25	total	total	total	total	1
83	889	C2H4O2	acetic acid	64-19-7	100	total	total	total	total	1
84	890	C2H4O2	methyl formate	107-31-3	20	3.0400E+05	1.1585E+05	4.3678E+02	7.2733E+00	1
85	910	C2H5Br	bromoethane	74-96-4	20	9.0000E+03	1.4992E+03	9.0817E+00	8.3345E-02	1
86	914	C2H5Cl	ethyl chloride	75-00-3	20	5.7000E+03	1.5982E+03	5.7327E+00	8.8859E-02	1
87	939	C2H5F	ethyl fluoride	353-36-6	25	2.0889E+03	7.8403E+02	2.0933E+00	4.3555E-02	1
88	948	C2H5I	ethyl iodide	75-03-6	20	3.9000E+03	4.5203E+02	3.9153E+00	2.5103E-02	1
89	954	C2H5N	ethyleneimine	151-56-4	25	total	total	total	total	1
90	956	C2H5NO	acetamide	60-35-5	25	7.0500E+05	4.2159E+05	2.3898E+03	4.0459E+01	1
91	958	C2H5NO2	nitroethane	79-24-3	25	5.0140E+04	1.2510E+04	5.2787E+01	7.0319E-01	1
92	985	C2H6	ethane	74-84-0	25	5.5018E+01	3.2963E+01	5.5021E-02	1.8298E-03	1
93	1065	C2H6O	ethyl alcohol	64-17-5	25	total	total	total	total	1
94	1066	C2H6O	dimethyl ether	115-10-6	18	6.5180E+04	2.6542E+04	6.9725E+01	1.5135E+00	1
95	1072	C2H6O2	ethylene glycol	107-21-1	25	total	total	total	total	1
96	1083	C2H6O4S	dimethyl sulfate	77-78-1	18	2.8000E+04	4.0974E+03	2.8807E+01	2.2838E-01	1
97	1084	C2H6O4S	ethyl sulfate	540-82-9	25	2.8000E+04	4.0974E+03	2.8807E+01	2.2838E-01	2
98	1092	C2H6S	ethyl mercaptan	75-08-1	25	1.5000E+04	4.3958E+03	1.5228E+01	2.4508E-01	1
99	1093	C2H6S	dimethyl sulfide	75-18-3	25	2.2000E+04	6.4797E+03	2.2495E+01	3.6203E-01	1
100	1115	C2H7N	ethylamine	75-04-7	20	total	total	total	total	1
101	1116	C2H7N	dimethylamine	124-40-3	20	total	total	total	total	1
102	1117	C2H7NO	monoethanolamine	141-43-5	25	total	total	total	total	1
103	1159	C2H8N2	ethylenediamine	107-15-3	20	total	total	total	total	1
104	1210	C2N2	cyanogen	460-19-5	20	9.4870E+03	3.3049E+03	9.5779E+00	1.8406E-01	1
105	1284	C3F6	hexafluoropropylene	116-15-4	25	1.8972E+02	2.2786E+01	1.8976E-01	1.2648E-03	1
106	1299	C3F8	octafluoropropane	76-19-7	25	1.1580E+01	1.1095E+00	1.1580E-02	6.1590E-05	1
107	1377	C3H2N2	malononitrile	109-77-3	20	1.1760E+05	3.5068E+04	1.3327E+02	2.0174E+00	1
108	1462	C3H3N	acrylonitrile	107-13-1	20	7.3500E+04	2.6226E+04	7.9331E+01	1.4950E+00	1
109	1492	C3H4	methylacetylene	74-99-7	25	2.6344E+03	1.1863E+03	2.6414E+00	6.5927E-02	1
110	1533	C3H4Cl2	2,3-dichloropropene	78-88-6	25	2.1500E+03	3.4966E+02	2.1546E+00	1.9416E-02	1
111	1534	C3H4Cl2	1,1-dichloropropene	563-58-6	25	2.1500E+03	3.4966E+02	2.1546E+00	1.9416E-02	2
112	1535	C3H4Cl2	cis-1,2-dichloropropene	6923-20-2	25	2.1500E+03	3.4966E+02	2.1546E+00	1.9416E-02	2
113	1536	C3H4Cl2	trans-1,2-dichloropropene	7069-38-7	25	2.1500E+03	3.4966E+02	2.1546E+00	1.9416E-02	2
114	1537	C3H4Cl2	cis-1,3-dichloropropene	10061-01-5	25	2.1500E+03	3.4966E+02	2.1546E+00	1.9416E-02	2
115	1538	C3H4Cl2	trans-1,3-dichloropropene	10061-02-6	25	2.1500E+03	3.4966E+02	2.1546E+00	1.9416E-02	2
116	1539	C3H4Cl2	3,3-dichloropropene	563-57-5	25	2.1500E+03	3.4966E+02	2.1546E+00	1.9416E-02	2
117	1540	C3H4Cl2	1,3-dichloropropene; (cis+trans)	542-75-6	25	2.1500E+03	3.4966E+02	2.1546E+00	1.9416E-02	2
118	1541	C3H4Cl2	1,2-dichloropropene	563-54-2	25	2.1500E+03	3.4966E+02	2.1546E+00	1.9416E-02	2
119	1542	C3H4Cl2	dichloropropylene	26952-23-8	25	2.1500E+03	3.4966E+02	2.1546E+00	1.9416E-02	2
120	1621	C3H4O	acrolein	107-02-8	20	2.1100E+05	7.9132E+04	2.6743E+02	4.7700E+00	1
121	1622	C3H4O	propargyl alcohol	107-19-7	20	total	total	total	total	1
122	1626	C3H4O2	acrylic acid	79-10-7	100	total	total	total	total	1
123	1628	C3H4O2	vinyl formate	692-45-5	25	1.4003E+05	3.9114E+04	1.6283E+02	2.2596E+00	2
124	1634	C3H4O3	pyruvic acid	127-17-3	20	total	total	total	total	1
125	1642	C3H5Br	cis-1-bromo-1-propene	590-13-6	25	3.8220E+03	5.7100E+02	3.8367E+00	3.1714E-02	2
126	1643	C3H5Br	trans-1-bromo-1-propene	590-15-8	25	3.8220E+03	5.7100E+02	3.8367E+00	3.1714E-02	2
127	1644	C3H5Br	2-bromo-1-propene	557-93-7	25	3.8220E+03	5.7100E+02	3.8367E+00	3.1714E-02	2
128	1645	C3H5Br	3-bromo-1-propene	106-95-6	25	3.8220E+03	5.7100E+02	3.8367E+00	3.1714E-02	1
129	1647	C3H5Br	1-bromo-1-propene; (cis+trans)	590-14-7	25	3.8220E+03	5.7100E+02	3.8367E+00	3.1714E-02	2
130	1668	C3H5Cl	cis-1-chloro-1-propene	16136-84-8	25	4.0000E+03	9.4454E+02	4.0161E+00	5.2480E-02	2
131	1669	C3H5Cl	trans-1-chloro-1-propene	16136-85-9	25	4.0000E+03	9.4454E+02	4.0161E+00	5.2480E-02	2
132	1670	C3H5Cl	2-chloro-1-propene	557-98-2	25	4.0000E+03	9.4454E+02	4.0161E+00	5.2480E-02	2
133	1671	C3H5Cl	3-chloro-1-propene	107-05-1	25	4.0000E+03	9.4454E+02	4.0161E+00	5.2480E-02	1
134	1672	C3H5Cl	1-chloro-1-propene	590-21-6	25	4.0000E+03	9.4454E+02	4.0161E+00	5.2480E-02	2
135	1703	C3H5Cl3	1,1,1-trichloropropane	7789-89-1	25	1.9000E+03	2.3256E+02	1.9036E+00	1.2912E-02	1
136	1704	C3H5Cl3	1,1,2-trichloropropane	598-77-6	25	1.9000E+03	2.3256E+02	1.9036E+00	1.2912E-02	1
137	1705	C3H5Cl3	1,1,3-trichloropropane	20395-25-9	25	1.8500E+03	2.2643E+02	1.8534E+00	1.2572E-02	2
138	1706	C3H5Cl3	1,2,2-trichloropropane	3175-23-3	25	1.8500E+03	2.2643E+02	1.8534E+00	1.2572E-02	2
139	1707	C3H5Cl3	1,2,3-trichloropropane	96-18-4	25	1.7500E+03	2.1417E+02	1.7531E+00	1.1891E-02	1
140	1752	C3H5N	propionitrile	107-12-0	25	9.3382E+04	3.2591E+04	1.0300E+02	1.8700E+00	1
141	1755	C3H5NO	acrylamide	79-06-1	30	6.8300E+05	3.5320E+05	2.1546E+03	3.0312E+01	1
142	1796	C3H5N3O9	nitroglycerine	55-63-0	20	1.3800E+03	1.0962E+02	1.3819E+00	6.0853E-03	1
143	1809	C3H6	cyclopropane	75-19-4	25	4.7881E+02	2.0504E+02	4.7904E-01	1.1384E-02	1
144	1810	C3H6	propylene	115-07-1	25	2.6979E+02	1.1552E+02	2.6986E-01	6.4130E-03	1
145	1824	C3H6Br2	1,1-dibromopropane	598-17-4	25	4.1121E+03	3.6831E+02	4.1291E+00	2.0452E-02	2
146	1825	C3H6Br2	1,2-dibromopropane	78-75-1	25	1.4380E+03	1.2848E+02	1.4401E+00	7.1330E-03	1
147	1826	C3H6Br2	1,3-dibromopropane	109-64-8	20	1.6900E+03	1.5104E+02	1.6929E+00	8.3851E-03	1
148	1827	C3H6Br2	2,2-dibromopropane	594-16-1	25	1.0302E+04	9.2799E+02	1.0409E+01	5.1560E-02	2
149	1844	C3H6Cl2	1,1-dichloropropane	78-99-9	25	3.2134E+03	5.1375E+02	3.2238E+00	2.8533E-02	2
150	1845	C3H6Cl2	1,2-dichloropropane	78-87-5	25	2.7400E+03	4.3789E+02	2.7475E+00	2.4317E-02	1
151	1846	C3H6Cl2	1,3-dichloropropane	142-28-9	25	2.8000E+03	4.4750E+02	2.8079E+00	2.4851E-02	1
152	1847	C3H6Cl2	2,2-dichloropropane	594-20-7	25	6.7861E+03	1.0882E+03	6.8324E+00	6.0472E-02	2
153	1848	C3H6Cl2	1,2-dichloropropane, (±)	26198-63-0	25	2.2410E+03	3.5799E+02	2.2460E+00	1.9879E-02	2
154	1849	C3H6Cl2	dichloropropane	26638-19-7	25	2.2211E+03	3.5481E+02	2.2261E+00	1.9702E-02	2
155	1920	C3H6O	allyl alcohol	107-18-6	25	total	total	total	total	1
156	1922	C3H6O	propionaldehyde	123-38-6	25	4.0450E+05	1.7402E+05	6.7926E+02	1.1695E+01	1
157	1923	C3H6O	acetone	67-64-1	25	total	total	total	total	1
158	1924	C3H6O	1,2-propylene oxide	75-56-9	20	4.0500E+05	1.7432E+05	6.8067E+02	1.1720E+01	1
159	1925	C3H6O	1,3-propylene oxide	503-30-0	25	4.0500E+05	1.7432E+05	6.8067E+02	1.1720E+01	2
160	1927	C3H6O	(S)-(-)-propylene oxide	16088-62-3	25	4.0500E+05	1.7432E+05	6.8067E+02	1.1720E+01	2

Table 3 Solubility in Water - Organic Compounds

					Solubility in Water, S (ppm - parts per million, wt - weight, g - gram)					
NO	ID	FORMULA	NAME	CAS No	T, C	S@T, ppm(wt)	S@T, ppm(mol)	S@T, g/1000 g of H2O	S@T, mol/1000 g of H2O	code
161	1928	C3H6O	(R)-(+)-propylene oxide	15448-47-2	25	4.0500E+05	1.7432E+05	6.8067E+02	1.1720E+01	2
162	1930	C3H6O2	propanoic acid	79-09-4	100	total	total	total	total	1
163	1931	C3H6O2	ethyl formate	109-94-4	18	1.0000E+05	2.6310E+04	1.1111E+02	1.4999E+00	1
164	1932	C3H6O2	methyl acetate	79-20-9	20	2.4370E+05	7.2666E+04	3.2223E+02	4.3497E+00	1
165	1949	C3H6O3	methoxyacetic acid	625-45-6	20	total	total	total	total	1
166	1950	C3H6O3	trioxane	110-88-3	25	1.7500E+05	4.0696E+04	2.1212E+02	2.3548E+00	1
167	1978	C3H7Br	1-bromopropane	106-94-5	20	2.3400E+03	3.4343E+02	2.3455E+00	1.9070E-02	1
168	1979	C3H7Br	2-bromopropane	75-26-3	20	3.2000E+03	4.6999E+02	3.2103E+00	2.6101E-02	1
169	1986	C3H7Cl	1-chloropropane	540-54-5	25	2.5000E+03	5.7453E+02	2.5063E+00	3.1910E-02	1
170	1987	C3H7Cl	2-chloropropane	75-29-6	20	3.0400E+03	6.9892E+02	3.0493E+00	3.8824E-02	1
171	2020	C3H7F	1-fluoropropane	460-13-9	25	2.9129E+03	8.4695E+02	2.9214E+00	4.7053E-02	1
172	2021	C3H7F	2-fluoropropane	420-26-8	25	2.8386E+03	8.2530E+02	2.8467E+00	4.5850E-02	1
173	2025	C3H7I	1-iodopropane	107-08-4	24	1.0700E+03	1.1350E+02	1.0711E+00	6.3011E-03	1
174	2026	C3H7I	2-iodopropane	75-30-9	20	1.4000E+03	1.4855E+02	1.4020E+00	8.2472E-03	1
175	2030	C3H7N	allylamine	107-11-9	20	total	total	total	total	1
176	2045	C3H7NO2	1-nitropropane	108-03-2	20	1.3810E+04	2.8235E+03	1.4003E+01	1.5717E-01	1
177	2046	C3H7NO2	2-nitropropane	79-46-9	20	1.6720E+04	3.4265E+03	1.7004E+01	1.9086E-01	1
178	2094	C3H8	propane	74-98-6	25	6.5123E+01	2.6606E+01	6.5127E-02	1.4769E-03	1
179	2129	C3H8O	propyl alcohol	71-23-8	25	total	total	total	total	1
180	2130	C3H8O	isopropyl alcohol	67-63-0	25	total	total	total	total	1
181	2131	C3H8O	methyl ethyl ether	540-67-0	25	1.9795E+05	6.8888E+04	2.4680E+02	4.1068E+00	2
182	2139	C3H8O2	methylal	109-87-5	20	2.3000E+05	6.6045E+04	2.9870E+02	3.9254E+00	1
183	2140	C3H8O2	1,2-propanediol (propylene glycol)	57-55-6	25	total	total	total	total	1
184	2141	C3H8O2	1,3-propanediol	504-63-2	25	total	total	total	total	2
185	2148	C3H8O3	glycerol	56-81-5	25	total	total	total	total	1
186	2158	C3H8S	propyl mercaptan	107-03-9	25	3.6899E+03	8.7525E+02	3.7036E+00	4.8627E-02	2
187	2159	C3H8S	isopropyl mercaptan	75-33-2	25	6.1810E+03	1.4690E+03	6.2195E+00	8.1661E-02	2
188	2160	C3H8S	methyl ethyl sulfide	624-89-5	25	8.6976E+03	2.0710E+03	8.7739E+00	1.1520E-01	2
189	2200	C3H9N	propylamine	107-10-8	25	total	total	total	total	1
190	2201	C3H9N	isopropylamine	75-31-0	20	total	total	total	total	1
191	2202	C3H9N	methylethylamine	624-78-2	25	total	total	total	total	2
192	2203	C3H9N	trimethylamine	75-50-3	30	total	total	total	total	2
193	2204	C3H9NO	1-amino-2-propanol	78-96-6	25	total	total	total	total	1
194	2241	C3H9O4P	trimethyl phosphate	512-56-1	25	5.0000E+05	1.1395E+05	1.0000E+03	7.1390E+00	1
195	2318	C4Cl6	hexachloro-1,3-butadiene	87-68-3	25	3.2300E+00	2.2315E-01	3.2300E-03	1.2387E-05	1
196	2336	C4F8	octafluorocyclobutane	115-25-3	25	5.1621E+01	4.6492E+00	5.1624E-02	2.5808E-04	1
197	2374	C4H2	biacetylene	460-12-8	25	9.4510E+03	3.4218E+03	9.5412E+00	1.9060E-01	1
198	2510	C4H4	vinylacetylene	689-97-4	25	1.9161E+03	6.6368E+02	1.9198E+00	3.6865E-02	1
199	2562	C4H4N2	succinonitrile	110-61-2	20	1.2690E+05	3.1658E+04	1.4534E+02	1.8148E+00	1
200	2592	C4H4O	furan	110-00-9	25	9.9010E+03	2.6394E+03	1.0000E+01	1.4690E-01	1
201	2605	C4H4O4	fumaric acid	110-17-8	25	7.0000E+03	1.0929E+03	7.0493E+00	6.0732E-02	1
202	2606	C4H4O4	maleic acid	110-16-7	25	4.4070E+05	1.0897E+05	7.8795E+02	6.7884E+00	1
203	2613	C4H4S	thiophene	110-02-1	25	3.0200E+03	6.4813E+02	3.0291E+00	3.6001E-02	1
204	2754	C4H6	dimethylacetylene	503-17-3	25	3.3534E+03	1.1194E+03	3.3647E+00	6.2204E-02	2
205	2755	C4H6	ethylacetylene	107-00-6	25	3.9964E+03	1.3345E+03	4.0124E+00	7.4178E-02	1
206	2756	C4H6	1,2-butadiene	590-19-2	25	6.4695E+02	2.1556E+02	6.4737E-01	1.1968E-02	2
207	2757	C4H6	butadiene (1,3 butadiene)	106-99-0	25	7.3500E+02	2.4491E+02	7.3554E-01	1.3598E-02	1
208	2940	C4H6O	trans-crotonaldehyde	123-73-9	20	1.5000E+05	4.3389E+04	1.7647E+02	2.5177E+00	1
209	2941	C4H6O	methacrolein	78-85-3	25	6.4000E+04	1.7271E+04	6.8376E+01	9.7553E-01	1
210	2944	C4H6O	divinyl ether	109-93-3	37	3.8410E+03	9.9005E+02	3.8558E+00	5.5011E-02	1
211	2956	C4H6O	crotonaldehyde	4170-30-3	25	1.0700E+05	2.9877E+04	1.1982E+02	1.7095E+00	2
212	2963	C4H6O2	2-butyne-1,4-diol	110-65-6	25	total	total	total	total	2
213	2965	C4H6O2	gamma-butyrolactone	96-48-0	20	total	total	total	total	1
214	2966	C4H6O2	cis-2-butenoic acid	503-64-0	20	total	total	total	total	1
215	2970	C4H6O2	methyl acrylate	96-33-3	23	5.0000E+04	1.0894E+04	5.2632E+01	6.1135E-01	1
216	2971	C4H6O2	vinyl acetate	108-05-4	25	2.5530E+04	5.4524E+03	2.6199E+01	3.0432E-01	1
217	2972	C4H6O2	allyl formate	1838-59-1	25	3.7765E+04	8.1458E+03	3.9247E+01	4.5588E-01	2
218	2990	C4H6O3	acetic anhydride	108-24-7	25	1.5100E+05	3.0430E+04	1.7786E+02	1.7422E+00	1
219	3006	C4H6O4	succinic acid	110-15-6	25	8.3230E+04	1.3661E+04	9.0786E+01	7.6879E-01	1
220	3024	C4H6O5	malic acid	6915-15-7	26	5.9200E+05	1.6314E+05	1.4510E+03	1.0821E+01	1
221	3033	C4H6O6	tartaric acid	87-69-4	20	1.7100E+05	2.4161E+04	2.0627E+02	1.3743E+00	1
222	3238	C4H7N	butyronitrile	109-74-0	25	3.1950E+04	8.5304E+03	3.3004E+01	4.7759E-01	1
223	3239	C4H7N	isobutyronitrile	78-82-0	25	3.1950E+04	8.5304E+03	3.3004E+01	4.7759E-01	2
224	3249	C4H7NO	2-pyrrolidone	616-45-5	20	total	total	total	total	1
225	3320	C4H8	1-butene	106-98-9	25	2.2200E+02	7.1290E+01	2.2205E-01	3.9576E-03	1
226	3321	C4H8	2-butene; (cis+trans)	107-01-7	25	2.2671E+02	7.2804E+01	2.2676E-01	4.0416E-03	2
227	3322	C4H8	cis-2-butene	590-18-1	25	2.2352E+02	7.1779E+01	2.2357E-01	3.9847E-03	2
228	3323	C4H8	trans-2-butene	624-64-6	25	2.2995E+02	7.3843E+01	2.3000E-01	4.0993E-03	2
229	3324	C4H8	isobutene	115-11-7	25	3.1083E+02	9.9822E+01	3.1093E-01	5.5416E-03	1
230	3333	C4H8Br2	1,1-dibromobutane	62168-25-6	25	1.1317E+03	9.4521E+01	1.1330E+00	5.2473E-03	2
231	3334	C4H8Br2	1,2-dibromobutane	533-98-2	25	7.2256E+02	6.0327E+01	7.2308E-01	3.3489E-03	2
232	3335	C4H8Br2	1,3-dibromobutane	107-80-2	25	4.5054E+02	3.7607E+01	4.5074E-01	2.0876E-03	2
233	3336	C4H8Br2	1,4-dibromobutane	110-52-1	25	3.5000E+02	2.9212E+01	3.5012E-01	1.6216E-03	1
234	3337	C4H8Br2	2,2-dibromobutane	50341-35-0	25	2.3832E+03	1.9928E+02	2.3889E+00	1.1064E-02	2
235	3338	C4H8Br2	DL-2,3-dibromobutane	598-71-0	25	9.8920E+02	8.2609E+01	9.9018E-01	4.5860E-03	2
236	3339	C4H8Br2	meso-2,3-dibromobutane	5780-13-2	25	1.1751E+03	9.8147E+01	1.1764E+00	5.4486E-03	2
237	3340	C4H8Br2	1,1-dibromo-2-methylpropane	33693-78-6	25	2.0347E+03	1.7008E+02	2.0388E+00	9.4427E-03	2
238	3341	C4H8Br2	1,2-dibromo-2-methylpropane	594-34-3	25	1.8302E+03	1.5296E+02	1.8336E+00	8.4921E-03	2
239	3342	C4H8Br2	1,3-dibromo-2-methylpropane	28148-04-1	25	4.5054E+02	3.7607E+01	4.5074E-01	2.0876E-03	2
240	3343	C4H8Br2	2,3-dibromobutane	5408-86-6	25	9.6289E+02	8.0410E+01	9.6381E-01	4.4639E-03	2

Table 3 Solubility in Water - Organic Compounds

					Solubility in Water, S (ppm - parts per million, wt - weight, g - gram)					
NO	ID	FORMULA	NAME	CAS No	T, C	S@T, ppm(wt)	S@T, ppm(mol)	S@T, g/1000 g of H2O	S@T, mol/1000 g of H2O	code
241	3360	C4H8Cl2	1,1-dichlorobutane	541-33-3	25	5.0100E+02	7.1091E+01	5.0125E-01	3.9465E-03	1
242	3361	C4H8Cl2	1,2-dichlorobutane	616-21-7	25	6.0663E+02	8.6088E+01	6.0700E-01	4.7791E-03	2
243	3362	C4H8Cl2	1,3-dichlorobutane	1190-22-3	25	3.7503E+02	5.3210E+01	3.7517E-01	2.9538E-03	2
244	3363	C4H8Cl2	1,4-dichlorobutane	110-56-5	25	1.2858E+02	1.8240E+01	1.2860E-01	1.0125E-03	2
245	3364	C4H8Cl2	2,2-dichlorobutane	4279-22-5	25	1.6630E+03	2.3621E+02	1.6658E+00	1.3115E-02	2
246	3365	C4H8Cl2	DL-2,3-dichlorobutane	2211-67-8	20	5.6200E+02	7.9751E+01	5.6232E-01	4.4272E-03	1
247	3366	C4H8Cl2	meso-2,3-dichlorobutane	4028-56-2	20	5.6200E+02	7.9751E+01	5.6232E-01	4.4272E-03	1
248	3367	C4H8Cl2	1,1-dichloro-2-methylpropane	598-76-5	25	1.5164E+03	2.1536E+02	1.5187E+00	1.1957E-02	2
249	3368	C4H8Cl2	1,2-dichloro-2-methylpropane	594-37-6	25	1.3815E+03	1.9618E+02	1.3834E+00	1.0892E-02	2
250	3369	C4H8Cl2	1,3-dichloro-2-methylpropane	616-19-3	25	3.4720E+02	4.9261E+01	3.4732E-01	2.7346E-03	2
251	3370	C4H8Cl2	2,3-dichlorobutane, dl and meso	7581-97-7	25	8.1796E+02	1.1610E+02	8.1863E-01	6.4453E-03	2
252	3371	C4H8Cl2	mixo-dichlorobutane	26761-81-9	25	7.1156E+02	1.0099E+02	7.1207E-01	5.6063E-03	2
253	3478	C4H8O	butyraldehyde	123-72-8	25	8.3720E+04	2.2318E+04	9.1369E+01	1.2671E+00	1
254	3479	C4H8O	isobutyraldehyde	78-84-2	25	8.9000E+04	2.3826E+04	9.7695E+01	1.3549E+00	1
255	3480	C4H8O	methyl ethyl ketone	78-93-3	25	total	total	total	total	1
256	3481	C4H8O	tetrahydrofuran	109-99-9	20	total	total	total	total	1
257	3504	C4H8O2	butyric acid	107-92-6	100	total	total	total	total	1
258	3505	C4H8O2	isobutyric acid	79-31-2	20	2.0000E+05	4.8631E+04	2.5000E+02	2.8375E+00	1
259	3506	C4H8O2	propyl formate	110-74-7	22	2.2000E+04	4.5784E+03	2.2495E+01	2.5532E-01	1
260	3507	C4H8O2	isopropyl formate	625-55-8	25	2.0700E+04	4.3034E+03	2.1138E+01	2.3991E-01	1
261	3508	C4H8O2	ethyl acetate	141-78-6	25	8.0000E+04	1.7469E+04	8.6957E+01	9.8695E-01	1
262	3509	C4H8O2	methyl propanoate	554-12-1	25	6.2400E+04	1.3425E+04	6.6553E+01	7.5537E-01	1
263	3510	C4H8O2	1,4-dioxane	123-91-1	25	total	total	total	total	1
264	3511	C4H8O2	cis-2-butene-1,4-diol	6117-80-2	25	total	total	total	total	2
265	3512	C4H8O2	trans-2-butene-1,4-diol	821-11-4	25	total	total	total	total	2
266	3581	C4H8S	tetrahydrothiophene	110-01-0	25	4.0277E+02	8.2317E+01	4.0293E-01	4.5697E-03	2
267	3598	C4H9Br	1-bromobutane	109-65-9	25	8.6800E+02	1.1421E+02	8.6875E-01	6.3404E-03	1
268	3599	C4H9Br	2-bromobutane	78-76-2	25	7.7446E+02	1.0189E+02	7.7506E-01	5.6566E-03	2
269	3600	C4H9Br	1-bromo-2-methylpropane	78-77-3	18	5.1000E+02	6.7083E+01	5.1026E-01	3.7240E-03	1
270	3601	C4H9Br	2-bromo-2-methylpropane	507-19-7	25	1.6115E+03	2.1217E+02	1.6141E+00	1.1780E-02	2
271	3602	C4H9Br	2-bromobutane, (±)	5787-31-5	25	7.7179E+02	1.0154E+02	7.7239E-01	5.6371E-03	2
272	3610	C4H9Cl	1-chlorobutane	109-69-3	25	8.7200E+02	1.6982E+02	8.7276E-01	9.4283E-03	1
273	3611	C4H9Cl	1-chloro-2-methylpropane	513-36-0	20	9.2400E+02	1.7996E+02	9.2485E-01	9.9911E-03	1
274	3612	C4H9Cl	2-chlorobutane	78-86-4	25	1.0000E+03	1.9477E+02	1.0010E+00	1.0814E-02	1
275	3613	C4H9Cl	2-chloro-2-methylpropane	507-20-0	15	2.8700E+03	5.5984E+02	2.8783E+00	3.1093E-02	1
276	3614	C4H9Cl	2-chlorobutane	53178-20-4	25	1.0218E+03	1.9903E+02	1.0229E+00	1.1050E-02	2
277	3675	C4H9N	pyrrolidine	123-75-1	25	total	total	total	total	1
278	3681	C4H9NO	N,N-dimethylacetamide	127-19-5	25	total	total	total	total	1
279	3682	C4H9NO	morpholine	110-91-8	20	total	total	total	total	1
280	3783	C4H10	butane	106-97-8	25	7.0137E+01	2.1740E+01	7.0142E-02	1.2068E-03	1
281	3784	C4H10	isobutane	75-28-5	25	5.1857E+01	1.6073E+01	5.1860E-02	8.9223E-04	1
282	3865	C4H10O	butanol	71-36-3	25	7.4600E+04	1.9216E+04	8.0614E+01	1.0876E+00	1
283	3866	C4H10O	isobutanol	78-83-1	25	8.1000E+04	2.0972E+04	8.8139E+01	1.1891E+00	1
284	3867	C4H10O	sec-butanol	78-92-2	25	1.8350E+05	5.1792E+04	2.2474E+02	3.0320E+00	1
285	3868	C4H10O	tert-butanol	75-65-0	25	total	total	total	total	1
286	3869	C4H10O	diethyl ether	60-29-7	25	6.0880E+04	1.5511E+04	6.4827E+01	8.7458E-01	1
287	3870	C4H10O	methyl propyl ether	557-17-5	25	3.0520E+04	7.5931E+03	3.1481E+01	4.2471E-01	1
288	3871	C4H10O	methyl isopropyl ether	598-53-8	25	6.5070E+04	1.6634E+04	6.9599E+01	9.3897E-01	1
289	3872	C4H10O	(R)-(-)-2-butanol	14898-79-4	25	1.5210E+05	4.1777E+04	1.7939E+02	2.4201E+00	2
290	3873	C4H10O	(S)-(+)-2-butanol	4221-99-2	25	1.5263E+05	4.1941E+04	1.8012E+02	2.4300E+00	2
291	3874	C4H10O	(<+->)-2-butanol	15892-23-6	25	1.5913E+05	4.3973E+04	1.8925E+02	2.5532E+00	2
292	3883	C4H10O2	2-ethoxyethanol	110-80-5	20	total	total	total	total	1
293	3884	C4H10O2	1,2-butanediol	584-03-2	25	total	total	total	total	2
294	3885	C4H10O2	1,3-butanediol	107-88-0	25	total	total	total	total	2
295	3886	C4H10O2	1,4-butanediol	110-63-4	20	total	total	total	total	1
296	3887	C4H10O2	DL-2,3-butanediol	6982-25-8	20	total	total	total	total	2
297	3888	C4H10O2	meso-2,3-butanediol	5341-95-7	20	total	total	total	total	2
298	3889	C4H10O2	2-methyl-1,2-propanediol	558-43-0	20	total	total	total	total	2
299	3890	C4H10O2	1-methyl-1,3-propanediol	---	20	total	total	total	total	2
300	3918	C4H10O3	diethylene glycol	111-46-6	25	total	total	total	total	1
301	3942	C4H10S	butyl mercaptan	109-79-5	25	6.0000E+02	1.1991E+02	6.0036E-01	6.6567E-03	1
302	3943	C4H10S	isobutyl mercaptan	513-44-0	25	1.6210E+03	3.2421E+02	1.6236E+00	1.8002E-02	2
303	3944	C4H10S	sec-butyl mercaptan	513-53-1	25	1.8786E+03	3.7581E+02	1.8822E+00	2.0869E-02	2
304	3945	C4H10S	tert-butyl mercaptan	75-66-1	25	4.1849E+03	8.3873E+02	4.2025E+00	4.6596E-02	2
305	3946	C4H10S	2-butanethiol	91840-99-2	25	1.8771E+03	3.7550E+02	1.8806E+00	2.0852E-02	2
306	3947	C4H10S	diethyl sulfide	352-93-2	25	3.2000E+03	6.4083E+02	3.2103E+00	3.5595E-02	1
307	3948	C4H10S	methyl propyl sulfide	3877-15-4	25	2.7391E+03	5.4833E+02	2.7467E+00	3.0454E-02	2
308	3949	C4H10S	methyl isopropyl sulfide	1551-21-9	25	4.3550E+03	8.7294E+02	4.3740E+00	4.8498E-02	2
309	3977	C4H11N	butylamine	109-73-9	25	total	total	total	total	1
310	3978	C4H11N	isobutylamine	78-81-9	20	total	total	total	total	1
311	3979	C4H11N	sec-butylamine	13952-84-6	25	total	total	total	total	1
312	3980	C4H11N	tert-butylamine	75-64-9	25	total	total	total	total	2
313	3981	C4H11N	2-butanamine, (.+/-.)-	33966-50-6	25	total	total	total	total	2
314	3982	C4H11N	methyl-propylamine	627-35-0	25	total	total	total	total	1
315	3983	C4H11N	methylisopropylamine	4747-21-1	25	total	total	total	total	2
316	3984	C4H11N	diethylamine	109-89-7	25	total	total	total	total	1
317	3985	C4H11N	dimethylethylamine	598-56-1	25	total	total	total	total	2
318	3986	C4H11N	(R)-(-)-sec-butylamine	13250-12-9	25	total	total	total	total	2
319	3987	C4H11N	(S)-(+)-sec-butylamine	513-49-5	25	total	total	total	total	2
320	3988	C4H11NO	dimethylethanolamine	108-01-0	20	total	total	total	total	1

Table 3 Solubility in Water - Organic Compounds

					Solubility in Water, S (ppm - parts per million, wt - weight, g - gram)					
NO	ID	FORMULA	NAME	CAS No	T, C	S@T, ppm(wt)	S@T, ppm(mol)	S@T, g/1000 g of H2O	S@T, mol/1000 g of H2O	code
321	4006	C4H11NO2	diethanolamine	111-42-2	20	total	total	total	total	1
322	4118	C4H13N3	diethylene triamine	111-40-0	20	total	total	total	total	1
323	4327	C5H4O2	furfural	98-01-1	25	7.9000E+04	1.5828E+04	8.5776E+01	8.9271E-01	1
324	4391	C5H5N	pyridine	110-86-1	25	total	total	total	total	1
325	4548	C5H6O2	furfuryl alcohol	98-00-0	25	total	total	total	total	1
326	4576	C5H6O4	citraconic acid	498-23-7	25	7.8300E+05	3.3317E+05	3.6083E+03	2.7735E+01	1
327	4577	C5H6O4	itaconic acid	97-65-4	20	7.6800E+04	1.1388E+04	8.3189E+01	6.3942E-01	1
328	4587	C5H6S	2-methylthiophene	554-14-3	25	6.1097E+02	1.1218E+02	6.1135E-01	6.2275E-03	2
329	4588	C5H6S	3-methylthiophene	616-44-4	25	4.0000E+02	7.3428E+01	4.0016E-01	4.0763E-03	1
330	4677	C5H7NO2	ethyl cyanoacetate	105-56-6	25	2.0000E+04	3.2397E+03	2.0408E+01	1.8042E-01	1
331	4737	C5H8	cyclopentene	142-29-0	25	5.3500E+02	1.4154E+02	5.3529E-01	7.8582E-03	1
332	4738	C5H8	isoprene	78-79-5	25	6.4200E+02	1.6987E+02	6.4241E-01	9.4308E-03	1
333	4739	C5H8	3-methyl-1,2-butadiene	598-25-4	25	4.0952E+02	1.0834E+02	4.0969E-01	6.0143E-03	2
334	4740	C5H8	1,2-pentadiene	591-95-7	25	3.5458E+02	9.3798E+01	3.5470E-01	5.2071E-03	2
335	4741	C5H8	cis-1,3-pentadiene	1574-41-0	25	3.6502E+02	9.6561E+01	3.6515E-01	5.3606E-03	2
336	4742	C5H8	trans-1,3-pentadiene	2004-70-8	25	3.9300E+02	1.0396E+02	3.9315E-01	5.7716E-03	2
337	4743	C5H8	1,4-pentadiene	591-93-5	25	5.5800E+02	1.4763E+02	5.5831E-01	8.1962E-03	1
338	4744	C5H8	2,3-pentadiene	591-96-8	25	3.1195E+02	8.2519E+01	3.1205E-01	4.5810E-03	2
339	4745	C5H8	1-pentyne	627-19-0	25	1.5700E+03	4.1569E+02	1.5725E+00	2.3084E-02	1
340	4746	C5H8	2-pentyne	627-21-4	25	1.2100E+03	3.2028E+02	1.2114E+00	1.7784E-02	2
341	4747	C5H8	3-methyl-1-butyne	598-23-2	25	3.1710E+03	8.4058E+02	3.1811E+00	4.6700E-02	2
342	4751	C5H8	1,3-pentadiene; (cis+trans)	504-60-9	25	3.9467E+02	1.0441E+02	3.9483E-01	5.7962E-03	2
343	4845	C5H8N4O12	pentaerythritol tetranitrate	78-11-5	25	2.0000E+00	1.1397E-01	2.0000E-03	6.3263E-06	1
344	4849	C5H8O	methyl isopropenyl ketone	814-78-8	25	4.7000E+04	1.0452E+04	4.9318E+01	5.8630E-01	1
345	4882	C5H8O2	acetylacetone	123-54-6	20	1.0900E+05	2.1539E+04	1.2233E+02	1.2219E+00	1
346	4883	C5H8O2	allyl acetate	591-87-7	25	2.0000E+04	3.6588E+03	2.0408E+01	2.0384E-01	2
347	4884	C5H8O2	isopropenyl acetate	108-22-5	25	2.0000E+04	3.6588E+03	2.0408E+01	2.0384E-01	2
348	4885	C5H8O2	vinyl propionate	105-38-4	25	2.0000E+04	3.6588E+03	2.0408E+01	2.0384E-01	2
349	4886	C5H8O2	methyl cis-2-butenoate	4358-59-2	25	2.0000E+04	3.6588E+03	2.0408E+01	2.0384E-01	2
350	4887	C5H8O2	methyl trans-2-butenoate	623-43-8	25	2.0000E+04	3.6588E+03	2.0408E+01	2.0384E-01	2
351	4888	C5H8O2	ethyl acrylate	140-88-5	25	2.0000E+04	3.6588E+03	2.0408E+01	2.0384E-01	1
352	4889	C5H8O2	methyl methacrylate	80-62-6	25	2.0000E+04	3.6588E+03	2.0408E+01	2.0384E-01	2
353	4918	C5H8O2	methyl 3-butenoate	3724-55-8	25	2.0000E+04	3.6588E+03	2.0408E+01	2.0384E-01	2
354	4945	C5H8O4	glutaric acid	110-94-1	65	5.3270E+05	1.3453E+05	1.1400E+03	8.6284E+00	1
355	5245	C5H9NO4	L-glutamic acid	56-86-0	25	8.6000E+03	1.0610E+03	8.6746E+00	5.8958E-02	1
356	5285	C5H10	cyclopentane	287-92-3	25	1.5600E+02	4.0075E+01	1.5602E-01	2.2246E-03	1
357	5286	C5H10	1-pentene	109-67-1	25	1.4800E+02	3.8020E+01	1.4802E-01	2.1105E-03	1
358	5287	C5H10	2-pentene; (cis+trans)	109-68-2	25	1.7688E+02	4.5440E+01	1.7691E-01	2.5224E-03	2
359	5288	C5H10	cis-2-pentene	627-20-3	25	2.0300E+02	5.2151E+01	2.0304E-01	2.8950E-03	1
360	5289	C5H10	trans-2-pentene	646-04-8	25	2.0300E+02	5.2151E+01	2.0304E-01	2.8950E-03	1
361	5290	C5H10	2-methyl-1-butene	563-46-2	25	1.9019E+02	4.8860E+01	1.9023E-01	2.7123E-03	2
362	5291	C5H10	3-methyl-1-butene	563-45-1	25	1.3000E+02	3.3396E+01	1.3002E-01	1.8538E-03	1
363	5292	C5H10	2-methyl-2-butene	513-35-9	25	1.5044E+02	3.8648E+01	1.5047E-01	2.1454E-03	2
364	5297	C5H10Br2	1,1-dibromopentane	13320-56-4	25	3.4306E+02	2.6886E+01	3.4317E-01	1.4924E-03	2
365	5298	C5H10Br2	1,2-dibromopentane	3234-49-9	25	2.7582E+02	2.1615E+01	2.7589E-01	1.1998E-03	2
366	5299	C5H10Br2	1,3-dibromopentane	42474-20-4	25	1.9889E+02	1.5585E+01	1.9893E-01	8.6512E-04	2
367	5300	C5H10Br2	1,4-dibromopentane	626-87-9	25	1.1556E+02	9.0547E+00	1.1557E-01	5.0262E-04	2
368	5301	C5H10Br2	1,5-dibromopentane	111-24-0	25	3.5202E+01	2.7580E+00	3.5203E-02	1.5309E-04	2
369	5302	C5H10Br2	2,2-dibromopentane	54653-26-8	25	5.9141E+02	4.6360E+01	5.9176E-01	2.5735E-03	2
370	5303	C5H10Br2	2,3-dibromopentane	5398-25-4	25	3.7230E+02	2.9178E+01	3.7244E-01	1.6197E-03	2
371	5304	C5H10Br2	2,4-dibromopentane	19398-53-9	25	3.0761E+02	2.4106E+01	3.0770E-01	1.3382E-03	2
372	5305	C5H10Br2	3,3-dibromopentane	54653-27-9	25	6.9603E+02	5.4566E+01	6.9651E-01	3.0291E-03	2
373	5306	C5H10Br2	1,1-dibromo-2-methylbutane	62127-58-6	25	5.2788E+02	4.1377E+01	5.2816E-01	2.2969E-03	2
374	5307	C5H10Br2	1,1-dibromo-3-methylbutane	62127-59-7	25	5.2788E+02	4.1377E+01	5.2816E-01	2.2969E-03	2
375	5308	C5H10Br2	1,2-dibromo-2-methylbutane	10428-64-5	25	5.0238E+02	3.9377E+01	5.0263E-01	2.1859E-03	2
376	5309	C5H10Br2	1,2-dibromo-3-methylbutane	10288-13-8	25	5.2788E+02	4.1377E+01	5.2816E-01	2.2969E-03	2
377	5310	C5H10Br2	1,3-dibromo-2-methylbutane	49623-50-9	25	5.2788E+02	4.1377E+01	5.2816E-01	2.2969E-03	2
378	5311	C5H10Br2	1,3-dibromo-3-methylbutane	24443-15-0	25	5.9983E+02	4.7020E+01	6.0019E-01	2.6102E-03	2
379	5312	C5H10Br2	1,4-dibromo-2-methylbutane	69498-28-8	25	5.1539E+02	4.0398E+01	5.1566E-01	2.2426E-03	2
380	5313	C5H10Br2	2,2-dibromo-3-methylbutane	62127-60-0	25	6.1434E+02	4.8158E+01	6.1472E-01	2.6734E-03	2
381	5314	C5H10Br2	2,3-dibromo-2-methylbutane	594-51-4	25	5.6440E+02	4.4242E+01	5.6472E-01	2.4559E-03	2
382	5315	C5H10Br2	1,1-dibromo-2,2-dimethylpropane	2443-91-6	25	3.4306E+02	2.6886E+01	3.4317E-01	1.4924E-03	2
383	5316	C5H10Br2	1,3-dibromo-2,2-dimethylpropane	5434-27-5	25	2.9128E+02	2.2827E+01	2.9136E-01	1.2671E-03	2
384	5317	C5H10Br2	1,3-dibromo-2-ethylpropane	---	25	5.9983E+02	4.7020E+01	6.0019E-01	2.6102E-03	2
385	5330	C5H10Cl2	1,1-dichloropentane	820-55-3	25	2.7077E+02	3.4594E+01	2.7085E-01	1.9204E-03	2
386	5331	C5H10Cl2	1,2-dichloropentane	1674-33-5	25	2.8600E+02	3.6540E+01	2.8608E-01	2.0284E-03	1
387	5332	C5H10Cl2	1,3-dichloropentane	30122-12-4	25	1.2123E+02	1.5487E+01	1.2125E-01	8.5967E-04	2
388	5333	C5H10Cl2	1,4-dichloropentane	626-92-6	25	8.3203E+01	1.0628E+01	8.3210E-02	5.8998E-04	2
389	5334	C5H10Cl2	1,5-dichloropentane	628-76-2	25	3.1287E+01	3.9964E+00	3.1288E-02	2.2184E-04	2
390	5335	C5H10Cl2	2,2-dichloropentane	34887-14-4	25	5.2191E+02	6.6695E+01	5.2219E-01	3.7024E-03	2
391	5336	C5H10Cl2	2,3-dichloropentane	600-11-3	25	2.8600E+02	3.6540E+01	2.8608E-01	2.0284E-03	1
392	5337	C5H10Cl2	2,4-dichloropentane	625-67-2	25	1.6692E+02	2.1324E+01	1.6695E-01	1.1837E-03	2
393	5338	C5H10Cl2	3,3-dichloropentane	21571-91-5	25	4.0497E+02	5.1745E+01	4.0513E-01	2.8725E-03	2
394	5339	C5H10Cl2	1,1-dichloro-2-methylbutane	23010-05-1	25	5.2587E+02	6.7201E+01	5.2615E-01	3.7305E-03	2
395	5340	C5H10Cl2	1,1-dichloro-3-methylbutane	625-66-1	25	4.5762E+02	5.8475E+01	4.5783E-01	3.2461E-03	2
396	5341	C5H10Cl2	1,2-dichloro-2-methylbutane	23010-04-0	25	3.6553E+02	4.6704E+01	3.6566E-01	2.5926E-03	2
397	5342	C5H10Cl2	1,2-dichloro-3-methylbutane	600-10-2	25	2.2907E+02	2.9265E+01	2.2912E-01	1.6245E-03	2
398	5343	C5H10Cl2	1,3-dichloro-2-methylbutane	23010-07-3	25	2.0785E+02	2.6553E+01	2.0789E-01	1.4740E-03	2
399	5344	C5H10Cl2	1,3-dichloro-3-methylbutane	624-96-4	25	1.8853E+02	2.4085E+01	1.8856E-01	1.3370E-03	2
400	5345	C5H10Cl2	1,4-dichloro-2-methylbutane	623-34-7	25	5.8516E+01	7.4747E+00	5.8520E-02	4.1492E-04	2

Table 3 Solubility in Water - Organic Compounds

NO	ID	FORMULA	NAME	CAS No	Solubility in Water, S (ppm - parts per million, wt - weight, g - gram)					
					T, C	S@T, ppm(wt)	S@T, ppm(mol)	S@T, g/1000 g of H2O	S@T, mol/1000 g of H2O	code
401	5346	C5H10Cl2	2,2-dichloro-3-methylbutane	17773-66-9	25	2.0785E+02	2.6553E+01	2.0789E-01	1.4740E-03	2
402	5347	C5H10Cl2	2,3-dichloro-2-methylbutane	507-45-9	25	2.8600E+02	3.6540E+01	2.8608E-01	2.0284E-03	1
403	5348	C5H10Cl2	1,3-dichloro-2-ethylpropane	---	25	2.0785E+02	2.6553E+01	2.0789E-01	1.4740E-03	2
404	5349	C5H10Cl2	1,1-dichloro-2,2-dimethylpropane	29559-54-4	25	2.0785E+02	2.6553E+01	2.0789E-01	1.4740E-03	2
405	5350	C5H10Cl2	1,3-dichloro-2,2-dimethylpropane	29559-55-5	25	1.9563E+02	2.4992E+01	1.9566E-01	1.3873E-03	2
406	5351	C5H10Cl2	dichloropentane	30586-10-8	25	4.4841E+02	5.7298E+01	4.4862E-01	3.1808E-03	2
407	5445	C5H10O	pentanal	110-62-3	25	1.1700E+04	2.4699E+03	1.1839E+01	1.3744E-01	1
408	5446	C5H10O	2-methylbutanal	96-17-3	25	2.4361E+04	5.1953E+03	2.4970E+01	2.8989E-01	2
409	5447	C5H10O	3-methylbutanal	590-86-3	25	2.3632E+04	5.0368E+03	2.4204E+01	2.8100E-01	2
410	5448	C5H10O	2,2-dimethylpropanal	630-19-3	25	6.0632E+04	1.3320E+04	6.4545E+01	7.4936E-01	2
411	5449	C5H10O	2-methylbutanal, (±)	57456-98-1	25	2.3752E+04	5.0629E+03	2.4330E+01	2.8247E-01	2
412	5450	C5H10O	methyl propyl ketone	107-87-9	25	5.5360E+04	1.2109E+04	5.8604E+01	6.8039E-01	1
413	5451	C5H10O	diethyl ketone	96-22-0	25	3.4000E+04	7.3076E+03	3.5197E+01	4.0863E-01	1
414	5452	C5H10O	methyl isopropyl ketone	563-80-4	25	6.0800E+04	1.3359E+04	6.4736E+01	7.5157E-01	1
415	5492	C5H10O2	pentanoic acid	109-52-4	25	2.4000E+04	4.3187E+03	2.4590E+01	2.4077E-01	1
416	5493	C5H10O2	2-methylbutyric acid	116-53-0	25	2.4929E+04	4.4893E+03	2.5566E+01	2.5032E-01	2
417	5494	C5H10O2	3-methylbutanoic acid	503-74-2	20	4.0700E+04	7.4280E+03	4.2427E+01	4.1541E-01	1
418	5495	C5H10O2	2,2-dimethylpropanoic acid	75-98-9	25	4.8652E+04	8.9398E+03	5.1140E+01	5.0072E-01	2
419	5496	C5H10O2	2-methylbutanoic acid	600-07-7	25	2.4929E+04	4.4893E+03	2.5566E+01	2.5032E-01	2
420	5497	C5H10O2	butyl formate	592-84-7	27	7.5580E+03	1.3415E+03	7.6156E+00	7.4565E-02	1
421	5498	C5H10O2	isobutyl formate	542-55-2	25	1.0200E+04	1.8144E+03	1.0305E+01	1.0090E-01	1
422	5499	C5H10O2	sec-butyl formate	589-40-2	25	1.0320E+04	1.8359E+03	1.0428E+01	1.0210E-01	2
423	5500	C5H10O2	tert-butyl formate	762-75-4	25	1.7872E+04	3.1995E+03	1.8197E+01	1.7817E-01	2
424	5501	C5H10O2	propyl acetate	109-60-4	25	2.0000E+04	3.5868E+03	2.0408E+01	1.9982E-01	1
425	5502	C5H10O2	isopropyl acetate	108-21-4	20	3.0950E+04	5.6020E+03	3.1938E+01	3.1271E-01	1
426	5503	C5H10O2	ethyl propanoate	105-37-3	25	2.2000E+04	3.9521E+03	2.2495E+01	2.2025E-01	1
427	5504	C5H10O2	methyl butanoate	623-42-7	25	1.5000E+04	2.6789E+03	1.5228E+01	1.4910E-01	1
428	5505	C5H10O2	methyl isobutanoate	547-63-7	25	2.8513E+04	5.1503E+03	2.9350E+01	2.8737E-01	2
429	5506	C5H10O2	tetrahydrofurfuryl alcohol	97-99-4	25	total	total	total	total	1
430	5530	C5H10O2	(S)-(+)-2-methylbutyric acid	1730-91-2	25	2.0365E+01	3.5921E+00	2.0365E-02	1.9940E-04	2
431	5543	C5H10O3	diethyl carbonate	105-58-8	20	1.8800E+04	2.9134E+03	1.9160E+01	1.6219E-01	1
432	5544	C5H10O3	ethyl lactate	97-64-3	20	total	total	total	total	1
433	5603	C5H11Br	1-bromopentane	110-53-2	25	1.2700E+02	1.5149E+01	1.2702E-01	8.4091E-04	1
434	5604	C5H11Br	2-bromopentane	107-81-3	25	2.3142E+02	2.7606E+01	2.3147E-01	1.5325E-03	2
435	5605	C5H11Br	3-bromopentane	1809-10-5	25	2.1823E+02	2.6033E+01	2.1828E-01	1.4451E-03	2
436	5606	C5H11Br	1-bromo-2-methylbutane	10422-35-2	25	1.9877E+02	2.3711E+01	1.9881E-01	1.3162E-03	2
437	5607	C5H11Br	1-bromo-3-methylbutane	107-82-4	17	2.0000E+02	2.3858E+01	2.0004E-01	1.3244E-03	1
438	5608	C5H11Br	2-bromo-2-methylbutane	507-36-8	25	3.6307E+02	4.3316E+01	3.6320E-01	2.4045E-03	2
439	5609	C5H11Br	2-bromo-3-methylbutane	18295-25-5	25	2.5628E+02	3.0573E+01	2.5635E-01	1.6972E-03	2
440	5610	C5H11Br	1-bromo-2,2-dimethylpropane	630-17-1	25	3.9868E+02	4.7566E+01	3.9884E-01	2.6405E-03	2
441	5611	C5H11Br	1-bromo-2-methylbutane, (S)	534-00-9	25	1.8825E+02	2.2456E+01	1.8829E-01	1.2466E-03	2
442	5620	C5H11Cl	1-chloropentane	543-59-9	25	2.0100E+02	3.3976E+01	2.0104E-01	1.8860E-03	1
443	5621	C5H11Cl	2-chloropentane	625-29-6	25	2.5100E+02	4.2429E+01	2.5106E-01	2.3553E-03	1
444	5622	C5H11Cl	3-chloropentane	616-20-6	25	2.5100E+02	4.2429E+01	2.5106E-01	2.3553E-03	1
445	5623	C5H11Cl	1-chloro-2-methylbutane	616-13-7	25	2.6895E+02	4.5464E+01	2.6903E-01	2.5238E-03	2
446	5624	C5H11Cl	1-chloro-3-methylbutane	107-84-6	25	2.9478E+02	4.9831E+01	2.9486E-01	2.7662E-03	2
447	5625	C5H11Cl	2-chloro-2-methylbutane	594-36-5	25	3.2900E+02	5.5618E+01	3.2911E-01	3.0875E-03	1
448	5626	C5H11Cl	2-chloro-3-methylbutane	631-65-2	25	3.7952E+02	6.4161E+01	3.7967E-01	3.5618E-03	2
449	5627	C5H11Cl	1-chloro-2,2-dimethylpropane	753-89-9	25	5.4461E+02	9.2083E+01	5.4491E-01	5.1120E-03	2
450	5628	C5H11Cl	1-chloro-2-methylbutane, (±)	114180-21-1	25	2.7687E+02	4.6803E+01	2.7694E-01	2.5981E-03	2
451	5629	C5H11Cl	2-chloropentane, (+)	29882-57-3	25	3.1538E+02	5.3315E+01	3.1548E-01	2.9597E-03	2
452	5772	C5H12	pentane	109-66-0	25	3.8500E+01	9.6132E+00	3.8501E-02	5.3363E-04	1
453	5773	C5H12	isopentane	78-78-4	25	4.7800E+01	1.1935E+01	4.7802E-02	6.6254E-04	1
454	5774	C5H12	neopentane	463-82-1	25	4.1450E+01	1.0350E+01	4.1452E-02	5.7452E-04	1
455	5831	C5H12O	1-pentanol	71-41-0	25	2.2000E+04	4.5762E+03	2.2495E+01	2.5519E-01	1
456	5832	C5H12O	2-pentanol	6032-29-7	25	4.5000E+04	9.5381E+03	4.7120E+01	5.3455E-01	1
457	5833	C5H12O	3-pentanol	584-02-1	25	5.2000E+04	1.1086E+04	5.4852E+01	6.2226E-01	1
458	5834	C5H12O	2-methyl-1-butanol	137-32-6	25	3.0000E+04	6.2810E+03	3.0928E+01	3.5086E-01	1
459	5835	C5H12O	2-methyl-1-butanol, (±)	34713-94-5	25	2.4335E+04	5.0716E+03	2.4942E+01	2.8295E-01	2
460	5836	C5H12O	3-methyl-2-butanol, (±)	70116-68-6	25	6.1684E+04	1.3257E+04	6.5739E+01	7.4576E-01	2
461	5837	C5H12O	3-methyl-1-butanol	123-51-3	25	2.7000E+04	5.6391E+03	2.7749E+01	3.1480E-01	1
462	5838	C5H12O	tert-pentyl-alcohol	75-85-4	25	1.1000E+05	2.4637E+04	1.2360E+02	1.4021E+00	1
463	5839	C5H12O	3-methyl-2-butanol	598-75-4	25	5.6000E+04	1.1978E+04	5.9322E+01	6.7297E-01	1
464	5840	C5H12O	2,2-dimethyl-1-propanol	75-84-3	25	6.0882E+04	1.3076E+04	6.4829E+01	7.3544E-01	2
465	5841	C5H12O	methyl butyl ether	628-28-4	25	7.0194E+03	1.4426E+03	7.0690E+00	8.0193E-02	2
466	5842	C5H12O	methyl isobutyl ether	625-44-5	25	1.2665E+04	2.6146E+03	1.2827E+01	1.4552E-01	2
467	5843	C5H12O	methyl sec-butyl ether	6795-87-5	25	1.2379E+04	2.5551E+03	1.2535E+01	1.4220E-01	2
468	5844	C5H12O	methyl tert-butyl ether	1634-04-4	25	5.1260E+04	1.0921E+04	5.4030E+01	6.1293E-01	1
469	5845	C5H12O	ethyl propyl ether	628-32-0	25	9.6779E+03	1.9932E+03	9.7725E+00	1.1086E-01	2
470	5846	C5H12O	ethyl isopropyl ether	625-54-7	25	1.6776E+04	3.4750E+03	1.7063E+01	1.9357E-01	2
471	5847	C5H12O	sec-butyl methyl ether	116783-23-4	25	1.2317E+04	2.5421E+03	1.2470E+01	1.4147E-01	2
472	5848	C5H12O	sec-butyl methyl ether (d)	66610-39-7	25	1.1358E+04	2.3424E+03	1.1488E+01	1.3033E-01	2
473	5849	C5H12O	(S)-(-)-2-methyl-1-butanol	1565-80-6	25	2.3738E+04	4.9447E+03	2.4315E+01	2.7584E-01	2
474	5850	C5H12O	(S)-(+)-2-pentanol	26184-62-3	25	4.2928E+04	9.0834E+03	4.4854E+01	5.0883E-01	2
475	5851	C5H12O	pentanol	30899-19-5	25	2.0845E+04	4.3319E+03	2.1289E+01	2.4151E-01	2
476	5852	C5H12O	(R)-(-)-2-pentanol	31087-44-2	25	3.0380E+04	6.3626E+03	3.1332E+01	3.5544E-01	2
477	5853	C5H12O	(S)-3-methyl-2-butanol	1517-66-4	25	1.5700E+05	3.6666E+04	1.8624E+02	2.1128E+00	2
478	5858	C5H12O2	neopentyl glycol	126-30-7	20	total	total	total	total	2
479	5859	C5H12O2	1,2-pentanediol	5343-92-0	20	total	total	total	total	2
480	5861	C5H12O2	1,3-pentanediol	3174-67-2	20	total	total	total	total	2

Table 3 Solubility in Water - Organic Compounds

					Solubility in Water, S (ppm - parts per million, wt - weight, g - gram)					
NO	ID	FORMULA	NAME	CAS No	T, C	S@T, ppm(wt)	S@T, ppm(mol)	S@T, g/1000 g of H2O	S@T, mol/1000 g of H2O	code
481	5862	C5H12O2	1,4-pentanediol	626-95-9	20	total	total	total	total	2
482	5863	C5H12O2	1,5-pentanediol	111-29-5	20	total	total	total	total	1
483	5899	C5H12O3	2-(2-methoxyethoxy)ethanol	111-77-3	20	total	total	total	total	1
484	5912	C5H12O4	pentaerythritol	115-77-5	20	6.0000E+04	8.3752E+03	6.3830E+01	4.6883E-01	1
485	5921	C5H12S	pentyl mercaptan	110-66-7	25	2.7257E+02	4.7127E+01	2.7264E-01	2.6161E-03	2
486	5922	C5H12S	2-pentanethiol	2084-19-7	25	5.4872E+02	9.4895E+01	5.4902E-01	5.2681E-03	2
487	5923	C5H12S	3-pentanethiol	616-31-9	25	5.1059E+02	8.8299E+01	5.1085E-01	4.9019E-03	2
488	5924	C5H12S	2-methyl-1-butanethiol	1878-18-8	25	3.9851E+02	6.8910E+01	3.9867E-01	3.8254E-03	2
489	5925	C5H12S	3-methyl-1-butanethiol	541-31-1	25	4.1147E+02	7.1152E+01	4.1164E-01	3.9499E-03	2
490	5926	C5H12S	2-methyl-2-butanethiol	1679-09-0	25	1.0171E+03	1.7597E+02	1.0182E+00	9.7697E-03	2
491	5927	C5H12S	3-methyl-2-butanethiol	2084-18-6	25	6.2217E+02	1.0760E+02	6.2256E-01	5.9737E-03	2
492	5928	C5H12S	2,2-dimethyl-1-propanethiol	1679-08-9	25	8.2658E+02	1.4298E+02	8.2726E-01	7.9379E-03	2
493	5929	C5H12S	2-methyl-1-butanethiol, (+)	20089-07-0	25	3.9652E+02	6.8566E+01	3.9668E-01	3.8063E-03	2
494	5930	C5H12S	methyl butyl sulfide	628-29-5	25	7.3497E+02	1.2713E+02	7.3551E-01	7.0575E-03	2
495	5931	C5H12S	ethyl propyl sulfide	4110-50-3	25	9.3766E+02	1.6221E+02	9.3854E-01	9.0057E-03	2
496	5932	C5H12S	3-methyl-2-thiapentane	10359-64-5	25	1.2843E+03	2.2224E+02	1.2859E+00	1.2339E-02	2
497	5933	C5H12S	4-methyl-2-thiapentane	5008-69-5	25	1.2539E+03	2.1698E+02	1.2555E+00	1.2047E-02	2
498	5934	C5H12S	2-methyl-3-thiapentane	5145-99-3	25	1.5982E+03	2.7663E+02	1.6007E+00	1.5360E-02	2
499	5935	C5H12S	3,3-dimethyl-2-thiabutane	6163-64-0	25	2.3600E+03	4.0875E+02	2.3656E+00	2.2699E-02	2
500	5947	C5H13N	pentylamine	110-58-7	25	total	total	total	total	1
501	5948	C5H13N	1-methylbutylamine	625-30-9	25	total	total	total	total	2
502	5949	C5H13N	2-methylbutylamine	96-15-1	25	total	total	total	total	2
503	5950	C5H13N	3-methylbutylamine	107-85-7	25	total	total	total	total	2
504	5951	C5H13N	1,1-dimethylpropylamine	594-39-8	25	total	total	total	total	2
505	5952	C5H13N	1,2-dimethylpropylamine	598-74-3	25	total	total	total	total	2
506	5953	C5H13N	2,2-dimethylpropylamine	5813-64-9	25	total	total	total	total	2
507	5954	C5H13N	3-pentanamine	616-24-0	25	high	high	high	high	2
508	5955	C5H13N	methyl-butylamine	110-68-9	25	total	total	total	total	1
509	5956	C5H13N	methylisobutylamine	14610-37-8	25	total	total	total	total	2
510	5957	C5H13N	methyl-sec-butylamine	7713-69-1	25	total	total	total	total	2
511	5958	C5H13N	methyl-tert-butylamine	625-43-4	25	total	total	total	total	2
512	5959	C5H13N	ethylpropylamine	20193-20-8	25	total	total	total	total	2
513	5960	C5H13N	ethylisopropylamine	19961-27-4	25	total	total	total	total	1
514	5961	C5H13N	methyldiethylamine	616-39-7	25	total	total	total	total	2
515	5962	C5H13N	dimethyl-propylamine	926-63-6	25	total	total	total	total	1
516	5963	C5H13N	dimethylisopropylamine	996-35-6	25	total	total	total	total	1
517	5964	C5H13N	(S)-(-)-2-methylbutylamine	20626-52-2	25	high	high	high	high	---
518	5965	C5H13N	N,N-dimethyl-2-propanamine	996-35-0	25	total	total	total	total	2
519	6067	C6Cl6	hexachlorobenzene	118-74-1	25	6.0000E-03	3.7955E-04	6.0000E-06	2.1069E-08	1
520	6281	C6H3ClN2O4	1-chloro-2,4-dinitrobenzene	97-00-7	15	8.0000E+00	7.1152E-01	8.0001E-03	3.9496E-05	1
521	6305	C6H3Cl3	1,2,4-trichlorobenzene	120-82-1	25	3.4570E+01	3.4324E+00	3.4571E-02	1.9053E-04	1
522	6306	C6H3Cl3	1,2,3-trichlorobenzene	87-61-6	25	3.4570E+01	3.4324E+00	3.4571E-02	1.9053E-04	2
523	6307	C6H3Cl3	1,3,5-trichlorobenzene	108-70-3	25	3.4570E+01	3.4324E+00	3.4571E-02	1.9053E-04	2
524	6308	C6H3Cl3	trichlorobenzene	12002-48-1	25	3.4570E+01	3.4324E+00	3.4571E-02	1.9053E-04	2
525	6367	C6H3N3O6	1,3,5-trinitrobenzene	99-35-4	15	2.7800E+02	2.3507E+01	2.7808E-01	1.3049E-03	1
526	6411	C6H4Br2	m-dibromobenzene	108-36-1	25	9.8350E+01	7.5112E+00	9.8360E-02	4.1694E-04	1
527	6412	C6H4Br2	o-dibromobenzene	583-53-9	25	9.8350E+01	7.5112E+00	9.8360E-02	4.1694E-04	2
528	6413	C6H4Br2	p-dibromobenzene	106-37-6	25	9.8350E+01	7.5112E+00	9.8360E-02	4.1694E-04	2
529	6414	C6H4Br2	dibromobenzene	26249-12-7	25	9.8350E+01	7.5112E+00	9.8360E-02	4.1694E-04	2
530	6443	C6H4ClNO2	m-chloronitrobenzene	121-73-3	20	2.7290E+02	3.1211E+01	2.7297E-01	1.7326E-03	1
531	6444	C6H4ClNO2	o-chloronitrobenzene	88-73-3	20	4.4120E+02	5.0467E+01	4.4139E-01	2.8015E-03	1
532	6445	C6H4ClNO2	p-chloronitrobenzene	100-00-5	20	2.2480E+02	2.5709E+01	2.2485E-01	1.4271E-03	1
533	6469	C6H4Cl2	o-dichlorobenzene	95-50-1	25	9.2320E+01	1.1315E+01	9.2329E-02	6.2807E-04	1
534	6470	C6H4Cl2	m-dichlorobenzene	541-73-1	25	1.2300E+02	1.5075E+01	1.2302E-01	8.3682E-04	1
535	6471	C6H4Cl2	p-dichlorobenzene	106-46-7	25	8.0000E+01	9.8046E+00	8.0006E-02	5.4425E-04	1
536	6524	C6H4F2	m-difluorobenzene	372-18-9	25	1.1400E+02	1.8002E+01	1.1401E-01	9.9928E-04	1
537	6525	C6H4F2	o-difluorobenzene	367-11-3	25	1.1400E+03	1.8017E+02	1.1413E+00	1.0003E-02	1
538	6526	C6H4F2	p-difluorobenzene	540-36-3	25	1.2230E+03	1.9330E+02	1.2245E+00	1.0732E-02	1
539	6580	C6H4N2O4	m-dinitrobenzene	99-65-0	25	5.7490E+02	6.1639E+01	5.7523E-01	3.4218E-03	1
540	6581	C6H4N2O4	o-dinitrobenzene	528-29-0	25	1.3300E+02	1.4254E+01	1.3302E-01	7.9126E-04	1
541	6582	C6H4N2O4	p-dinitrobenzene	100-25-4	25	6.8760E+01	7.3689E+00	6.8765E-02	4.0905E-04	1
542	6656	C6H5Br	bromobenzene	108-86-1	25	4.1000E+02	4.7060E+01	4.1017E-01	2.6124E-03	1
543	6692	C6H5Cl	chlorobenzene	108-90-7	25	3.9070E+02	6.2552E+01	3.9085E-01	3.4724E-03	1
544	6720	C6H5ClO	m-chlorophenol	108-43-0	20	2.5300E+04	3.6242E+03	2.5957E+01	2.0191E-01	1
545	6721	C6H5ClO	o-chlorophenol	95-57-8	25	1.1350E+04	1.6062E+03	1.1480E+01	8.9301E-02	1
546	6722	C6H5ClO	p-chlorophenol	106-48-9	20	2.6300E+04	3.7707E+03	2.7010E+01	2.1010E-01	1
547	6769	C6H5F	fluorobenzene	462-06-6	25	1.5500E+03	2.9092E+02	1.5524E+00	1.6153E-02	1
548	6795	C6H5I	iodobenzene	591-50-4	25	2.0070E+02	1.7726E+01	2.0074E-01	9.8397E-04	1
549	6819	C6H5NO2	nitrobenzene	98-95-3	25	1.9360E+03	2.8377E+02	1.9398E+00	1.5756E-02	1
550	6876	C6H6	benzene	71-43-2	25	1.7550E+03	4.0530E+02	1.7581E+00	2.2507E-02	1
551	6923	C6H6ClN	m-chloroaniline	108-42-9	20	5.4420E+03	7.7209E+02	5.4718E+00	4.2891E-02	1
552	6924	C6H6ClN	o-chloroaniline	95-51-2	20	3.7650E+03	5.3339E+02	3.7792E+00	2.9624E-02	1
553	6925	C6H6ClN	p-chloroaniline	106-47-8	20	2.7550E+03	3.8996E+02	2.7626E+00	2.1655E-02	1
554	6996	C6H6N2O2	m-nitroaniline	99-09-2	25	9.0000E+02	1.1747E+02	9.0081E-01	6.5216E-03	1
555	6997	C6H6N2O2	o-nitroaniline	88-74-4	25	1.2600E+03	1.6451E+02	1.2616E+00	9.1336E-03	1
556	6998	C6H6N2O2	p-nitroaniline	100-01-6	25	6.0000E+02	7.8295E+01	6.0036E-01	4.3465E-03	1
557	7000	C6H6N2O2	N-nitroaniline	645-55-6	25	9.2000E+02	1.2009E+02	9.2085E-01	6.6667E-03	2
558	7057	C6H6O	phenol	108-95-2	25	8.0190E+04	1.6414E+04	8.7181E+01	9.2634E-01	1
559	7071	C6H6O2	resorcinol	108-46-3	30	6.9700E+05	2.7344E+05	2.3003E+03	2.0891E+01	1
560	7072	C6H6O2	p-hydroquinone	123-31-9	25	7.0000E+04	1.2165E+04	7.5269E+01	6.8356E-01	1

Table 3 Solubility in Water - Organic Compounds

NO	ID	FORMULA	NAME	CAS No	Solubility in Water, S (ppm - parts per million, wt - weight, g - gram)					code
					T, C	S@T, ppm(wt)	S@T, ppm(mol)	S@T, g/1000 g of H2O	S@T, mol/1000 g of H2O	
561	7085	C6H6O3	1,2,3-benzenetriol	87-66-1	25	3.8000E+05	8.0504E+04	6.1290E+02	4.8600E+00	1
562	7172	C6H7N	aniline	62-53-3	25	3.4160E+04	6.7952E+03	3.5368E+01	3.7978E-01	1
563	7173	C6H7N	2-methylpyridine	109-06-8	25	total	total	total	total	1
564	7174	C6H7N	3-methylpyridine	108-99-6	25	total	total	total	total	1
565	7175	C6H7N	4-methylpyridine	108-89-4	20	total	total	total	total	1
566	7267	C6H8	1,3-cyclohexadiene	592-57-4	25	7.0000E+02	1.5746E+02	7.0049E-01	8.7420E-03	1
567	7269	C6H8	1,4-cyclohexadiene	628-41-1	25	7.0000E+02	1.5746E+02	7.0049E-01	8.7420E-03	2
568	7318	C6H8N2	m-phenylenediamine	108-45-2	25	3.5100E+05	8.2648E+04	5.4083E+02	5.0011E+00	1
569	7319	C6H8N2	o-phenylenediamine	95-54-5	35	4.1500E+04	7.1609E+03	4.3297E+01	4.0036E-01	1
570	7320	C6H8N2	p-phenylenediamine	106-50-3	24	3.8000E+04	6.5372E+03	3.9501E+01	3.6526E-01	1
571	7464	C6H8O7	citric acid	77-92-9	20	5.4100E+05	9.9520E+04	1.1786E+03	6.1348E+00	1
572	7467	C6H8S	2-ethylthiophene	872-55-9	25	2.9200E+02	4.6897E+01	2.9209E-01	2.6034E-03	1
573	7468	C6H8S	3-ethylthiophene	1795-01-3	25	1.8975E+02	3.0473E+01	1.8979E-01	1.6916E-03	2
574	7469	C6H8S	2,3-dimethylthiophene	632-16-6	25	1.4184E+02	2.2778E+01	1.4186E-01	1.2644E-03	2
575	7470	C6H8S	2,4-dimethylthiophene	638-00-6	25	1.4867E+02	2.3875E+01	1.4869E-01	1.3253E-03	2
576	7471	C6H8S	2,5-dimethylthiophene	638-02-8	25	1.8301E+02	2.9391E+01	1.8305E-01	1.6315E-03	2
577	7472	C6H8S	3,4-dimethylthiophene	632-15-5	25	1.1866E+02	1.9055E+01	1.1868E-01	1.0578E-03	2
578	7622	C6H10	cyclohexene	110-83-8	25	2.1310E+02	4.6742E+01	2.1315E-01	2.5947E-03	1
579	7623	C6H10	1,2-hexadiene	592-44-9	25	9.0596E+01	1.9870E+01	9.0604E-02	1.1030E-03	2
580	7624	C6H10	1,cis-3-hexadiene	14596-92-0	25	1.0503E+02	2.3036E+01	1.0504E-01	1.2787E-03	2
581	7625	C6H10	1,trans-3-hexadiene	20237-34-7	25	1.0503E+02	2.3036E+01	1.0504E-01	1.2787E-03	2
582	7626	C6H10	1,4-hexadiene	592-45-0	25	1.5341E+02	3.3647E+01	1.5343E-01	1.8678E-03	2
583	7627	C6H10	1,cis-4-hexadiene	7318-67-4	25	1.5341E+02	3.3647E+01	1.5343E-01	1.8678E-03	2
584	7628	C6H10	1,trans-4-hexadiene	7319-00-8	25	1.5341E+02	3.3647E+01	1.5343E-01	1.8678E-03	2
585	7629	C6H10	1,5-hexadiene	592-42-7	25	1.6900E+02	3.7068E+01	1.6903E-01	2.0577E-03	1
586	7630	C6H10	2,3-hexadiene	592-49-4	25	1.3346E+02	2.9271E+01	1.3347E-01	1.6249E-03	2
587	7631	C6H10	2,4-hexadiene	592-46-1	25	7.4053E+01	1.6241E+01	7.4058E-02	9.0155E-04	2
588	7632	C6H10	cis-2,trans-4-hexadiene	5194-50-3	25	6.1830E+01	1.3560E+01	6.1834E-02	7.5274E-04	2
589	7633	C6H10	cis-2,cis-4-hexadiene	6108-61-8	25	7.4053E+01	1.6241E+01	7.4058E-02	9.0155E-04	2
590	7634	C6H10	trans-2,trans-4-hexadiene	5194-51-4	25	6.7175E+01	1.4733E+01	6.7179E-02	8.1781E-04	2
591	7635	C6H10	3-methyl-1,2-pentadiene	7417-48-3	25	1.2139E+02	2.6625E+01	1.2141E-01	1.4780E-03	2
592	7636	C6H10	4-methyl-1,2-pentadiene	13643-05-5	25	1.2139E+02	2.6625E+01	1.2141E-01	1.4780E-03	2
593	7637	C6H10	3-methyl-1,3-pentadiene	4549-74-0	25	8.6183E+01	1.8902E+01	8.6191E-02	1.0492E-03	2
594	7638	C6H10	2-methyl-1,cis-3-pentadiene	1501-60-6	25	9.0596E+01	1.9870E+01	9.0604E-02	1.1030E-03	2
595	7639	C6H10	2-methyl-1,trans-3-pentadiene	926-54-5	25	9.0596E+01	1.9870E+01	9.0604E-02	1.1030E-03	2
596	7640	C6H10	3-methyl-1,cis-3-pentadiene	2787-45-3	25	8.6183E+01	1.8902E+01	8.6191E-02	1.0492E-03	2
597	7641	C6H10	3-methyl-1,trans-3-pentadiene	2787-43-1	25	8.6183E+01	1.8902E+01	8.6191E-02	1.0492E-03	2
598	7642	C6H10	4-methyl-1,3-pentadiene	926-56-7	25	8.9252E+01	1.9575E+01	8.9260E-02	1.0866E-03	2
599	7643	C6H10	2-methyl-1,4-pentadiene	763-30-4	25	2.9136E+02	6.3911E+01	2.9144E-01	3.5479E-03	2
600	7644	C6H10	3-methyl-1,4-pentadiene	1115-08-8	25	2.3783E+02	5.2168E+01	2.3789E-01	2.8960E-03	2
601	7645	C6H10	2-methyl-2,3-pentadiene	3043-33-2	25	1.1026E+02	2.4183E+01	1.1027E-01	1.3424E-03	2
602	7646	C6H10	2-ethyl-1,3-butadiene	3404-63-5	25	9.5204E+01	2.0880E+01	9.5213E-02	1.1591E-03	2
603	7647	C6H10	2,3-dimethyl-1,3-butadiene	513-81-5	25	1.2864E+02	2.8214E+01	1.2865E-01	1.5662E-03	2
604	7648	C6H10	1-hexyne	693-02-7	25	3.6000E+02	7.8972E+01	3.6013E-01	4.3841E-03	1
605	7649	C6H10	2-hexyne	764-35-2	25	3.1195E+02	6.8429E+01	3.1205E-01	3.7987E-03	2
606	7650	C6H10	3-hexyne	928-49-4	25	5.5860E+02	1.2256E+02	5.5891E-01	6.8039E-03	1
607	7651	C6H10	3-methyl-1-pentyne	922-59-8	25	1.1289E+03	2.4779E+02	1.1302E+00	1.3758E-02	2
608	7652	C6H10	4-methyl-1-pentyne	7154-75-8	25	9.7028E+02	2.1295E+02	9.7122E-01	1.1823E-02	2
609	7653	C6H10	4-methyl-2-pentyne	21020-27-9	25	5.5534E+02	1.2184E+02	5.5565E-01	6.7643E-03	2
610	7654	C6H10	3,3-dimethyl-1-butyne	917-92-0	25	2.4243E+03	5.3266E+02	2.4301E+00	2.9583E-02	2
611	7657	C6H10	1,3-hexadiene, cis and trans	592-48-3	25	1.0300E+02	2.2590E+01	1.0301E-01	1.2540E-03	2
612	7658	C6H10	2-methyl-1,3-pentadiene	1118-58-7	25	9.3337E+01	2.0471E+01	9.3346E-02	1.1363E-03	2
613	7659	C6H10	methylpentadiene, isomers	54363-49-4	25	9.0596E+01	1.9870E+01	9.0604E-02	1.1030E-03	2
614	7660	C6H10	hexadiene	42296-74-2	25	1.1303E+02	2.4791E+01	1.1305E-01	1.3762E-03	2
615	7760	C6H10O	cyclohexanone	108-94-1	25	9.3190E+04	1.8514E+04	1.0277E+02	1.0471E+00	1
616	7761	C6H10O	mesityl oxide	141-79-7	20	2.8000E+04	5.2598E+03	2.8807E+01	2.9351E-01	1
617	7881	C6H10O3	ethylacetoacetate	141-97-9	17	1.1100E+05	1.6990E+04	1.2486E+02	9.5940E-01	1
618	7917	C6H10O4	adipic acid	124-04-9	34	2.9880E+04	3.7824E+03	3.0800E+01	2.1075E-01	1
619	7919	C6H10O4	ethylene glycol diacetate	111-55-7	22	1.2460E+05	1.7243E+04	1.4233E+02	9.7394E-01	1
620	8214	C6H12	methylcyclopentane	96-37-7	25	4.2640E+01	9.1275E+00	4.2642E-02	5.0667E-04	1
621	8215	C6H12	cyclohexane	110-82-7	25	5.6100E+01	1.2009E+01	5.6103E-02	6.6661E-04	1
622	8216	C6H12	1-hexene	592-41-6	25	5.0000E+01	1.0703E+01	5.0003E-02	5.9413E-04	1
623	8217	C6H12	2-hexene; (cis+trans)	592-43-8	25	4.5325E+01	9.7023E+00	4.5327E-02	5.3858E-04	2
624	8218	C6H12	cis-2-hexene	7688-21-3	25	4.3482E+01	9.3079E+00	4.3484E-02	5.1668E-04	2
625	8219	C6H12	trans-2-hexene	4050-45-7	25	4.5603E+01	9.7618E+00	4.5605E-02	5.4187E-04	2
626	8220	C6H12	hex-3-ene	592-47-2	25	4.7942E+01	1.0263E+01	4.7945E-02	5.6968E-04	2
627	8221	C6H12	cis-3-hexene	7642-09-3	25	4.8729E+01	1.0431E+01	4.8731E-02	5.7902E-04	2
628	8222	C6H12	trans-3-hexene	13269-52-8	25	4.7299E+01	1.0125E+01	4.7301E-02	5.6203E-04	2
629	8223	C6H12	2-methyl-1-pentene	763-29-1	25	7.8000E+01	1.6697E+01	7.8006E-02	9.2686E-04	1
630	8224	C6H12	3-methyl-1-pentene	760-20-3	25	8.3573E+01	1.7890E+01	8.3580E-02	9.9310E-04	2
631	8225	C6H12	4-methyl-1-pentene	691-37-2	25	4.8020E+01	1.0279E+01	4.8022E-02	5.7060E-04	1
632	8226	C6H12	4-methyl-2-pentene	4461-48-7	25	7.3611E+01	1.5758E+01	7.3617E-02	8.7471E-04	2
633	8227	C6H12	4-methyl-cis-2-pentene	691-38-3	25	7.6222E+01	1.6317E+01	7.6228E-02	9.0574E-04	2
634	8228	C6H12	4-methyl-trans-2-pentene	674-76-0	25	6.9312E+01	1.4837E+01	6.9317E-02	8.2362E-04	2
635	8229	C6H12	2-methyl-2-pentene	625-27-4	25	4.6837E+01	1.0026E+01	4.6839E-02	5.5654E-04	2
636	8230	C6H12	3-methyl-cis-2-pentene	922-62-3	25	4.5968E+01	9.8400E+00	4.5970E-02	5.4622E-04	2
637	8231	C6H12	3-methyl-2-pentene, cis and trans	922-61-2	25	4.5539E+01	9.7480E+00	4.5541E-02	5.4111E-04	2
638	8232	C6H12	3-methyl-trans-2-pentene	616-12-6	25	4.0351E+01	8.6375E+00	4.0352E-02	4.7946E-04	2
639	8233	C6H12	2,3-dimethyl-1-butene	563-78-0	25	7.8738E+01	1.6855E+01	7.8744E-02	9.3563E-04	2
640	8234	C6H12	2,3-dimethyl-2-butene	563-79-1	25	3.5325E+01	7.5616E+00	3.5326E-02	4.1974E-04	2

Table 3 Solubility in Water - Organic Compounds

					Solubility in Water, S (ppm - parts per million, wt - weight, g - gram)					
NO	ID	FORMULA	NAME	CAS No	T, C	S@T, ppm(wt)	S@T, ppm(mol)	S@T, g/1000 g of H2O	S@T, mol/1000 g of H2O	code
641	8235	C6H12	3,3-dimethyl-1-butene	558-37-2	25	1.3715E+02	2.9361E+01	1.3717E-01	1.6298E-03	2
642	8236	C6H12	2-ethyl-1-butene	760-21-4	25	5.2895E+01	1.1323E+01	5.2898E-02	6.2853E-04	2
643	8245	C6H12Br2	1,1-dibromohexane	58133-26-9	25	1.0372E+02	7.6598E+00	1.0373E-01	4.2519E-04	2
644	8246	C6H12Br2	1,2-dibromo-2,3-dimethylbutane	29916-45-8	25	5.7888E+01	4.2747E+00	5.7891E-02	2.3729E-04	2
645	8247	C6H12Br2	1,4-dibromo-2,3-dimethylbutane	54462-70-3	25	5.7888E+01	4.2747E+00	5.7891E-02	2.3729E-04	2
646	8248	C6H12Br2	2,3-dibromo-2,3-dimethylbutane	594-81-0	25	5.7888E+01	4.2747E+00	5.7891E-02	2.3729E-04	2
647	8249	C6H12Br2	1,2-dibromohexane	624-20-4	25	5.7888E+01	4.2747E+00	5.7891E-02	2.3729E-04	2
648	8250	C6H12Br2	1,4-dibromohexane	25118-28-9	25	5.7888E+01	4.2747E+00	5.7891E-02	2.3729E-04	2
649	8251	C6H12Br2	1,5-dibromohexane	627-96-3	25	5.7888E+01	4.2747E+00	5.7891E-02	2.3729E-04	2
650	8252	C6H12Br2	1,6-dibromohexane	629-03-8	25	1.0844E+01	8.0077E-01	1.0845E-02	4.4450E-05	2
651	8253	C6H12Br2	2,3-dibromohexane	6423-02-5	25	1.4353E+02	1.0600E+01	1.4355E-01	5.8839E-04	2
652	8254	C6H12Br2	2,5-dibromohexane, (R*,R**)-(±)-	54462-68-9	25	5.7888E+01	4.2747E+00	5.7891E-02	2.3729E-04	2
653	8255	C6H12Br2	1,2-dibromo-3,3-dimethylbutane	640-21-1	25	5.7888E+01	4.2747E+00	5.7891E-02	2.3729E-04	2
654	8256	C6H12Br2	3,4-dibromohexane	16230-28-7	25	5.7888E+01	4.2747E+00	5.7891E-02	2.3729E-04	2
655	8257	C6H12Br2	2,5-dibromohexane	24774-58-1	25	7.5107E+01	5.5464E+00	7.5112E-02	3.0788E-04	2
656	8258	C6H12Br2	1,5-dibromo-3-methylpentane	4457-72-1	25	5.7888E+01	4.2747E+00	5.7891E-02	2.3729E-04	2
657	8273	C6H12Cl2	1,1-dichlorohexane	62017-16-7	25	7.4679E+01	8.6765E+00	7.4684E-02	4.8163E-04	2
658	8274	C6H12Cl2	1,2-dichlorohexane	2162-92-7	25	4.5817E+01	5.3231E+00	4.5819E-02	2.9548E-04	2
659	8275	C6H12Cl2	1,6-dichlorohexane	2163-00-0	25	8.4925E+00	9.8663E-01	8.4926E-03	5.4767E-05	2
660	8276	C6H12Cl2	2,2-dichlorohexane	42131-89-5	25	4.9511E+01	5.7523E+00	4.9514E-02	3.1931E-04	2
661	8277	C6H12Cl2	2,3-dichlorohexane	54305-87-2	25	7.6726E+01	8.9143E+00	7.6732E-02	4.9483E-04	2
662	8278	C6H12Cl2	2,5-dichlorohexane, (R*,R*)-(±)-	41761-12-0	25	3.6847E+01	4.2809E+00	3.6848E-02	2.3763E-04	2
663	8279	C6H12Cl2	1,1-dichloro-3,3-dimethylbutane	6130-96-7	25	1.7647E+02	2.0505E+01	1.7650E-01	1.1382E-03	2
664	8280	C6H12Cl2	hexane, 2,5-dichloro-, (R*,S*)-	41761-11-9	25	4.9511E+01	5.7523E+00	4.9514E-02	3.1931E-04	2
665	8281	C6H12Cl2	2,2-dichloro-3,3-dimethylbutane	594-84-3	25	4.9511E+01	5.7523E+00	4.9514E-02	3.1931E-04	2
666	8400	C6H12O	cyclohexanol	108-93-0	25	3.8160E+04	7.0852E+03	3.9674E+01	3.9610E-01	1
667	8401	C6H12O	hexanal	66-25-1	30	5.6440E+03	1.0199E+03	5.6760E+00	5.6669E-02	1
668	8402	C6H12O	2-methylpentanal	123-15-9	25	6.8659E+03	1.2419E+03	6.9134E+00	6.9023E-02	2
669	8403	C6H12O	3-methylpentanal	15877-57-3	25	5.3297E+03	9.6281E+02	5.3582E+00	5.3496E-02	2
670	8404	C6H12O	4-methylpentanal	1119-16-0	25	5.3297E+03	9.6281E+02	5.3582E+00	5.3496E-02	2
671	8405	C6H12O	2-ethylbutanal	97-96-1	25	6.9359E+03	1.2546E+03	6.9843E+00	6.9731E-02	2
672	8406	C6H12O	2,2-dimethylbutanal	2094-75-9	25	1.3265E+04	2.4121E+03	1.3443E+01	1.3422E-01	2
673	8407	C6H12O	2,3-dimethylbutanal	2109-98-0	25	8.4081E+03	1.5228E+03	8.4794E+00	8.4658E-02	2
674	8408	C6H12O	3,3-dimethylbutanal	2987-16-8	25	1.3265E+04	2.4121E+03	1.3443E+01	1.3422E-01	2
675	8409	C6H12O	2-hexanone	591-78-6	25	1.6400E+04	2.9899E+03	1.6673E+01	1.6647E-01	1
676	8410	C6H12O	3-hexanone	589-38-8	25	1.4700E+04	2.6762E+03	1.4919E+01	1.4895E-01	1
677	8411	C6H12O	3-methyl-2-pentanone	565-61-7	25	1.6734E+04	3.0517E+03	1.7019E+01	1.6992E-01	2
678	8412	C6H12O	4-methyl-2-pentanone	108-10-1	25	1.9000E+04	3.4715E+03	1.9368E+01	1.9337E-01	1
679	8413	C6H12O	ethyl isopropyl ketone	565-69-5	25	2.1686E+04	3.9711E+03	2.2166E+01	2.2131E-01	2
680	8414	C6H12O	3,3-dimethyl-2-butanone	75-97-8	25	3.4716E+04	6.4271E+03	3.5965E+01	3.5907E-01	2
681	8415	C6H12O	3-methyl-2-pentanone, (±)	55156-16-6	25	1.6627E+04	3.0319E+03	1.6908E+01	1.6881E-01	2
682	8453	C6H12O	2,2-dimethyl-3-butanone	---	25	1.3244E+04	2.4082E+03	1.3422E+01	1.3400E-01	2
683	8460	C6H12O2	hexanoic acid	142-62-1	20	9.6800E+03	1.5136E+03	9.7746E+00	8.4148E-02	1
684	8461	C6H12O2	2-ethyl butyric acid	88-09-5	25	1.0644E+04	1.6657E+03	1.0758E+01	9.2617E-02	2
685	8462	C6H12O2	tert-butylacetic acid	1070-83-3	25	1.2903E+04	2.0232E+03	1.3072E+01	1.1253E-01	2
686	8463	C6H12O2	2,2-dimethylbutanoic acid	595-37-9	25	1.5802E+04	2.4838E+03	1.6055E+01	1.3822E-01	2
687	8464	C6H12O2	2,3-dimethylbutanoic acid	14287-61-7	25	1.1839E+04	1.8546E+03	1.1980E+01	1.0314E-01	2
688	8465	C6H12O2	2-methylpentanoic acid, (±)	22160-39-0	25	9.7163E+03	1.5193E+03	9.8116E+00	8.4466E-02	2
689	8466	C6H12O2	3-methylpentanoic acid, (±)	22160-40-3	25	8.8247E+03	1.3789E+03	8.9033E+00	7.6647E-02	2
690	8467	C6H12O2	4-methylpentanoic acid	646-07-1	25	7.5805E+03	1.1832E+03	7.6384E+00	6.5758E-02	2
691	8468	C6H12O2	pentyl formate	638-49-3	25	1.1692E+03	1.8150E+02	1.1705E+00	1.0077E-02	2
692	8469	C6H12O2	1-methylbutyl formate	58368-66-4	25	2.3340E+03	3.6269E+02	2.3395E+00	2.0140E-02	2
693	8470	C6H12O2	2-methylbutyl formate	35073-27-9	25	2.3340E+03	3.6269E+02	2.3395E+00	2.0140E-02	2
694	8471	C6H12O2	isopentyl formate	110-45-2	25	3.5000E+03	5.4442E+02	3.5123E+00	3.0237E-02	1
695	8472	C6H12O2	1-ethylpropyl formate	58368-67-5	25	2.3340E+03	3.6269E+02	2.3395E+00	2.0140E-02	2
696	8473	C6H12O2	1,1-dimethylpropyl formate	757-88-0	25	3.6119E+03	5.6187E+02	3.6250E+00	3.1207E-02	2
697	8474	C6H12O2	1,2-dimethylpropyl formate	66794-46-5	25	2.3900E+03	3.7140E+02	2.3957E+00	2.0624E-02	2
698	8475	C6H12O2	2,2-dimethylpropyl formate	23361-67-3	25	2.7123E+03	4.2161E+02	2.7197E+00	2.3413E-02	2
699	8476	C6H12O2	butyl acetate	123-86-4	20	4.3000E+03	6.6931E+02	4.3186E+00	3.7178E-02	1
700	8477	C6H12O2	isobutyl acetate	110-19-0	25	6.3000E+03	9.8228E+02	6.3399E+00	5.4579E-02	1
701	8478	C6H12O2	sec-butyl acetate	105-46-4	20	6.2000E+03	9.6661E+02	6.2387E+00	5.3708E-02	1
702	8479	C6H12O2	tert-butyl acetate	540-88-5	25	2.3363E+04	3.6963E+03	2.3922E+01	2.0594E-01	2
703	8480	C6H12O2	propyl propanoate	106-36-5	25	5.3000E+03	8.2566E+02	5.3282E+00	4.5870E-02	1
704	8481	C6H12O2	isopropyl propanoate	637-78-5	25	5.9500E+03	9.2743E+02	5.9856E+00	5.1529E-02	1
705	8482	C6H12O2	ethyl butanoate	105-54-4	22	6.1680E+03	9.6159E+02	6.2063E+00	5.3429E-02	1
706	8483	C6H12O2	ethyl isobutanoate	97-62-1	25	1.1084E+04	1.7352E+03	1.1208E+01	9.6489E-02	2
707	8485	C6H12O2	diacetone alcohol	123-42-2	25	total	total	total	total	1
708	8497	C6H12O2	methyl 2,2-dimethylpropanoate	598-98-1	25	1.7754E+04	2.7953E+03	1.8074E+01	1.5560E-01	2
709	8514	C6H12O2	(±)-3-methylvaleric acid	105-43-1	25	9.5455E+03	1.4924E+03	9.6375E+00	8.2968E-02	2
710	8515	C6H12O2	2-methylvaleric acid	97-61-0	25	5.2626E+04	8.5414E+03	5.5549E+01	4.7821E-01	2
711	8531	C6H12O3	2-ethoxyethyl acetate	111-15-9	20	2.3000E+05	3.9124E+04	2.9870E+02	2.2602E+00	1
712	8533	C6H12O3	paraldehyde	123-63-7	30	1.1200E+05	1.6902E+04	1.2613E+02	9.5435E-01	1
713	8647	C6H13Br	1-bromohexane	111-25-1	25	2.5800E+01	2.8157E+00	2.5801E-02	1.5630E-04	1
714	8648	C6H13Br	2-bromohexane	3377-86-4	25	6.0179E+01	6.5679E+00	6.0183E-02	3.6458E-04	2
715	8649	C6H13Br	3-bromohexane	3377-87-5	25	6.8963E+01	7.5266E+00	6.8968E-02	4.1780E-04	2
716	8650	C6H13Br	1-bromo-2-methylpentane	25346-33-2	25	5.9864E+01	6.5335E+00	5.9867E-02	3.6267E-04	2
717	8651	C6H13Br	1-bromo-3-methylpentane	51116-73-5	25	4.8475E+01	5.2905E+00	4.8477E-02	2.9367E-04	2
718	8652	C6H13Br	1-bromo-4-methylpentane	626-88-0	25	5.1109E+01	5.5779E+00	5.1111E-02	3.0963E-04	2
719	8653	C6H13Br	2-bromo-2-methylpentane	4283-80-1	25	6.4766E+01	7.0685E+00	6.4770E-02	3.9237E-04	2
720	8654	C6H13Br	2-bromo-3-methylpentane	62168-41-6	25	1.3119E+02	1.4319E+01	1.3121E-01	7.9483E-04	2

Table 3 Solubility in Water - Organic Compounds

					Solubility in Water, S (ppm - parts per million, wt - weight, g - gram)					
NO	ID	FORMULA	NAME	CAS No	T, C	S@T, ppm(wt)	S@T, ppm(mol)	S@T, g/1000 g of H2O	S@T, mol/1000 g of H2O	code
721	8655	C6H13Br	2-bromo-4-methylpentane	30310-22-6	25	1.1738E+02	1.2811E+01	1.1739E-01	7.1114E-04	2
722	8656	C6H13Br	3-bromo-2-methylpentane	4283-83-4	25	1.3119E+02	1.4319E+01	1.3121E-01	7.9483E-04	2
723	8657	C6H13Br	3-bromo-3-methylpentane	25346-31-0	25	1.2350E+02	1.3480E+01	1.2352E-01	7.4827E-04	2
724	8658	C6H13Br	1-bromo-2,2-dimethylbutane	62168-42-7	25	8.6919E+01	9.4864E+00	8.6926E-02	5.2659E-04	2
725	8659	C6H13Br	1-bromo-2,3-dimethylbutane	30540-31-9	25	7.3804E+01	8.0550E+00	7.3809E-02	4.4713E-04	2
726	8660	C6H13Br	1-bromo-3,3-dimethylbutane	1647-23-0	25	8.6247E+01	9.4131E+00	8.6254E-02	5.2252E-04	2
727	8661	C6H13Br	2-bromo-2,3-dimethylbutane	594-52-5	25	1.0597E+02	1.1566E+01	1.0599E-01	6.4205E-04	2
728	8662	C6H13Br	2-bromo-3,3-dimethylbutane	26356-06-9	25	1.1154E+02	1.2174E+01	1.1155E-01	6.7576E-04	2
729	8663	C6H13Br	1-bromo-2-ethylbutane	3814-34-4	25	6.2105E+01	6.7781E+00	6.2109E-02	3.7625E-04	2
730	8669	C6H13Cl	1-chlorohexane	544-10-5	25	6.4300E+01	9.6038E+00	6.4304E-02	5.3311E-04	1
731	8670	C6H13Cl	2-chlorohexane	638-28-8	25	8.7685E+01	1.3097E+01	8.7693E-02	7.2701E-04	2
732	8671	C6H13Cl	3-chlorohexane	2346-81-8	25	9.2165E+01	1.3766E+01	9.2173E-02	7.6415E-04	2
733	8672	C6H13Cl	1-chloro-2-methylpentane	14753-05-0	25	1.0692E+02	1.5970E+01	1.0693E-01	8.8648E-04	2
734	8673	C6H13Cl	1-chloro-3-methylpentane	62016-93-7	25	6.8189E+01	1.0185E+01	6.8194E-02	5.6535E-04	2
735	8674	C6H13Cl	1-chloro-4-methylpentane	62016-94-8	25	8.3410E+01	1.2458E+01	8.3417E-02	6.9156E-04	2
736	8675	C6H13Cl	2-chloro-2-methylpentane	4325-48-8	25	1.6144E+02	2.4115E+01	1.6147E-01	1.3386E-03	2
737	8676	C6H13Cl	2-chloro-3-methylpentane	24319-09-3	25	2.0109E+02	3.0038E+01	2.0113E-01	1.6675E-03	2
738	8677	C6H13Cl	2-chloro-4-methylpentane	25346-32-1	25	1.5027E+02	2.2445E+01	1.5029E-01	1.2460E-03	2
739	8678	C6H13Cl	3-chloro-2-methylpentane	38384-05-3	25	1.5877E+02	2.3716E+01	1.5880E-01	1.3165E-03	2
740	8679	C6H13Cl	3-chloro-3-methylpentane	918-84-3	25	1.3001E+02	1.9420E+01	1.3003E-01	1.0780E-03	2
741	8680	C6H13Cl	1-chloro-2-ethylbutane	4737-41-1	25	7.9331E+01	1.1849E+01	7.9337E-02	6.5774E-04	2
742	8681	C6H13Cl	1-chloro-2,2-dimethylbutane	6366-35-4	25	1.2384E+02	1.8498E+01	1.2386E-01	1.0268E-03	2
743	8682	C6H13Cl	1-chloro-2,3-dimethylbutane	600-06-6	25	9.6857E+01	1.4467E+01	9.6867E-02	8.0306E-04	2
744	8683	C6H13Cl	1-chloro-3,3-dimethylbutane	2855-08-5	25	1.2384E+02	1.8498E+01	1.2386E-01	1.0268E-03	2
745	8684	C6H13Cl	2-chloro-2,3-dimethylbutane	594-57-0	25	1.5764E+02	2.3546E+01	1.5766E-01	1.3071E-03	2
746	8685	C6H13Cl	2-chloro-3,3-dimethylbutane	5750-00-5	25	1.6534E+02	2.4697E+01	1.6536E-01	1.3709E-03	2
747	8877	C6H14	hexane	110-54-3	25	9.4700E+00	1.9797E+00	9.4701E-03	1.0989E-04	1
748	8878	C6H14	2,2-dimethylbutane	75-83-2	25	2.3820E+01	4.9796E+00	2.3821E-02	2.7641E-04	1
749	8879	C6H14	2,3-dimethylbutane	79-29-8	25	2.0540E+01	4.2939E+00	2.0540E-02	2.3835E-04	1
750	8880	C6H14	2-methylpentane	107-83-5	25	1.3800E+01	2.8849E+00	1.3800E-02	1.6014E-04	1
751	8881	C6H14	3-methylpentane	96-14-0	25	1.7910E+01	3.7441E+00	1.7910E-02	2.0783E-04	1
752	8970	C6H14O	1-hexanol	111-27-3	25	5.8750E+03	1.0409E+03	5.9097E+00	5.7838E-02	1
753	8971	C6H14O	2-hexanol	626-93-7	25	1.4000E+04	2.4972E+03	1.4199E+01	1.3896E-01	1
754	8972	C6H14O	3-hexanol	623-37-0	25	1.4975E+04	2.6732E+03	1.5202E+01	1.4878E-01	2
755	8973	C6H14O	2-methyl-1-pentanol	105-30-6	25	8.1000E+03	1.4377E+03	8.1661E+00	7.9922E-02	1
756	8974	C6H14O	3-methyl-1-pentanol	589-35-5	25	5.4156E+03	9.5912E+02	5.4451E+00	5.3291E-02	2
757	8975	C6H14O	4-methyl-1-pentanol	626-89-1	25	5.6111E+03	9.9390E+02	5.6427E+00	5.5225E-02	2
758	8976	C6H14O	4-methyl-2-pentanol	108-11-2	27	1.5000E+04	2.6778E+03	1.5228E+01	1.4904E-01	1
759	8977	C6H14O	2-methyl-2-pentanol	590-36-3	25	3.5663E+04	6.4781E+03	3.6982E+01	3.6194E-01	2
760	8978	C6H14O	3-methyl-2-pentanol	565-60-6	25	1.6101E+04	2.8769E+03	1.6364E+01	1.6015E-01	2
761	8979	C6H14O	2-methyl-3-pentanol	565-67-3	25	2.5882E+04	4.6628E+03	2.6570E+01	2.6004E-01	2
762	8980	C6H14O	3-methyl-3-pentanol	77-74-7	25	3.3476E+04	6.0696E+03	3.4636E+01	3.3898E-01	2
763	8981	C6H14O	2-ethyl-1-butanol	97-95-0	25	1.0000E+04	1.7778E+03	1.0101E+01	9.8858E-02	1
764	8982	C6H14O	2,2-dimethyl-1-butanol	1185-33-7	25	1.3754E+04	2.4528E+03	1.3946E+01	1.3649E-01	2
765	8983	C6H14O	2,3-dimethyl-1-butanol	19550-30-2	25	6.6228E+03	1.1741E+03	6.6669E+00	6.5249E-02	2
766	8984	C6H14O	2,3-dimethyl-1-butanol, (±)-	20281-85-0	25	8.6559E+03	1.5371E+03	8.7315E+00	8.5455E-02	2
767	8985	C6H14O	3,3-dimethyl-1-butanol	624-95-3	25	9.4680E+03	1.6824E+03	9.5585E+00	9.3549E-02	2
768	8986	C6H14O	2,3-dimethyl-2-butanol	594-60-5	25	4.2654E+04	7.7943E+03	4.4555E+01	4.3605E-01	2
769	8987	C6H14O	3,3-dimethyl-2-butanol	464-07-3	25	3.9006E+04	7.1055E+03	4.0589E+01	3.9724E-01	2
770	8988	C6H14O	3,3-dimethyl-2-butanol, (±)	20281-91-8	25	3.8025E+04	6.9211E+03	3.9528E+01	3.8686E-01	2
771	8989	C6H14O	(RS)-2-hexanol	20281-86-1	25	1.1336E+04	2.0176E+03	1.1466E+01	1.1222E-01	2
772	8990	C6H14O	2-hexanol, (R)	26549-24-6	25	1.3588E+04	2.4229E+03	1.3776E+01	1.3482E-01	2
773	8991	C6H14O	3-hexanol, (R)	13471-42-6	25	1.5344E+04	2.7399E+03	1.5583E+01	1.5251E-01	2
774	8992	C6H14O	3-methyl-1-pentanol, (±)	20281-83-8	25	5.2271E+03	9.2559E+02	5.2546E+00	5.1426E-02	2
775	8993	C6H14O	dipropyl ether	111-43-3	25	3.8200E+03	6.7564E+02	3.8346E+00	3.7530E-02	1
776	8994	C6H14O	propyl isopropyl ether	627-08-7	25	4.2727E+03	7.5599E+02	4.2910E+00	4.1996E-02	2
777	8995	C6H14O	diisopropyl ether	108-20-3	25	1.1250E+04	2.0021E+03	1.1378E+01	1.1136E-01	1
778	8996	C6H14O	methyl pentyl ether	628-80-8	25	1.6319E+03	2.8812E+02	1.6346E+00	1.5998E-02	2
779	8997	C6H14O	methyl 1-methylbutyl ether	6795-88-6	25	2.4474E+03	4.3237E+02	2.4534E+00	2.4011E-02	2
780	8998	C6H14O	methyl 2-methylbutyl ether	62016-48-2	25	2.5745E+03	4.5489E+02	2.5812E+00	2.5262E-02	2
781	8999	C6H14O	methyl 3-methylbutyl ether	626-91-5	25	2.5745E+03	4.5489E+02	2.5812E+00	2.5262E-02	2
782	9000	C6H14O	methyl 1-ethylpropyl ether	36839-67-5	25	2.8490E+03	5.0350E+02	2.8572E+00	2.7963E-02	2
783	9001	C6H14O	methyl 1,2-dimethylpropyl ether	62016-49-3	25	3.6703E+03	6.4908E+02	3.6838E+00	3.6053E-02	2
784	9002	C6H14O	methyl tert-pentyl ether	994-05-8	25	3.1053E+03	5.4890E+02	3.1149E+00	3.0486E-02	2
785	9003	C6H14O	methyl 2,2-dimethylpropyl ether	1118-00-9	25	3.6703E+03	6.4908E+02	3.6838E+00	3.6053E-02	2
786	9004	C6H14O	ethyl butyl ether	628-81-9	25	2.3030E+03	4.0682E+02	2.3083E+00	2.2592E-02	2
787	9005	C6H14O	ethyl isobutyl ether	627-02-1	25	4.0411E+03	7.1488E+02	4.0575E+00	3.9711E-02	2
788	9006	C6H14O	ethyl sec-butyl ether	2679-87-0	25	4.0207E+03	7.1125E+02	4.0369E+00	3.9509E-02	2
789	9007	C6H14O	ethyl tert-butyl ether	637-92-3	25	6.1532E+03	1.0904E+03	6.1913E+00	6.0594E-02	2
790	9008	C6H14O	(S)-(+)-2-hexanol	52019-78-0	25	1.2410E+04	2.2105E+03	1.2565E+01	1.2298E-01	2
791	9009	C6H14O	tert-hexyl alcohol	26401-20-7	25	4.6071E+04	8.4432E+03	4.8296E+01	4.7267E-01	2
792	9020	C6H14O2	acetal	105-57-7	25	4.4000E+04	6.9673E+03	4.6025E+01	3.8946E-01	1
793	9021	C6H14O2	2-butoxyethanol	111-76-2	20	total	total	total	total	1
794	9076	C6H14O3	dipropylene glycol	25265-71-8	25	total	total	total	total	1
795	9077	C6H14O3	2-(2-ethoxyethoxy)ethanol	111-90-0	20	total	total	total	total	1
796	9092	C6H14O4	triethylene glycol	112-27-6	20	total	total	total	total	1
797	9100	C6H14O6	sorbitol	50-70-4	25	total	total	total	total	1
798	9107	C6H14S	hexyl mercaptan	111-31-9	25	7.0459E+01	1.0735E+01	7.0464E-02	5.9593E-04	2
799	9108	C6H14S	2-hexanethiol	1679-06-7	25	1.4553E+02	2.2175E+01	1.4555E-01	1.2309E-03	2
800	9109	C6H14S	2,3-dimethyl-2-butanethiol	1639-01-6	25	1.8638E+02	2.8401E+01	1.8641E-01	1.5765E-03	2

Table 3 Solubility in Water - Organic Compounds

					Solubility in Water, S (ppm - parts per million, wt - weight, g - gram)					
NO	ID	FORMULA	NAME	CAS No	T, C	S@T, ppm(wt)	S@T, ppm(mol)	S@T, g/1000 g of H2O	S@T, mol/1000 g of H2O	code
801	9110	C6H14S	3-hexanethiol	1633-90-5	25	1.8638E+02	2.8401E+01	1.8641E-01	1.5765E-03	2
802	9111	C6H14S	2-methyl-2-pentanethiol	1633-97-2	25	1.8638E+02	2.8401E+01	1.8641E-01	1.5765E-03	2
803	9112	C6H14S	dipropyl sulfide	111-47-7	25	2.7188E+02	4.1432E+01	2.7196E-01	2.3000E-03	2
804	9113	C6H14S	diisopropyl sulfide	625-80-9	25	8.6935E+02	1.3255E+02	8.7011E-01	7.3586E-03	2
805	9114	C6H14S	methyl pentyl sulfide	1741-83-9	25	5.8408E+02	8.9032E+01	5.8442E-01	4.9425E-03	2
806	9115	C6H14S	ethyl butyl sulfide	638-46-0	25	2.5248E+02	3.8475E+01	2.5254E-01	2.1358E-03	2
807	9116	C6H14S	3-methyl-2-thiahexane	13286-91-4	25	3.3242E+02	5.0660E+01	3.3253E-01	2.8122E-03	2
808	9117	C6H14S	4-methyl-2-thiahexane	15013-37-3	25	3.1556E+02	4.8090E+01	3.1566E-01	2.6696E-03	2
809	9118	C6H14S	5-methyl-2-thiahexane	13286-90-3	25	3.5014E+02	5.3361E+01	3.5026E-01	2.9622E-03	2
810	9119	C6H14S	2-methyl-3-thiahexane	5008-73-1	25	4.7567E+02	7.2501E+01	4.7590E-01	4.0248E-03	2
811	9120	C6H14S	4-methyl-3-thiahexane	5008-72-0	25	4.3825E+02	6.6795E+01	4.3844E-01	3.7080E-03	2
812	9121	C6H14S	5-methyl-3-thiahexane	1613-45-2	25	4.2561E+02	6.4867E+01	4.2579E-01	3.6010E-03	2
813	9122	C6H14S	3-ethyl-2-thiapentane	57093-84-2	25	4.8007E+02	7.3171E+01	4.8030E-01	4.0620E-03	2
814	9123	C6H14S	3,3-dimethyl-2-thiapentane	13286-92-5	25	5.8437E+02	8.9077E+01	5.8471E-01	4.9450E-03	2
815	9124	C6H14S	3,4-dimethyl-2-thiapentane	53897-51-1	25	5.8437E+02	8.9077E+01	5.8471E-01	4.9450E-03	2
816	9125	C6H14S	4,4-dimethyl-2-thiapentane	6079-57-8	25	5.8437E+02	8.9077E+01	5.8471E-01	4.9450E-03	2
817	9126	C6H14S	2,2-dimethyl-3-thiapentane	14290-92-7	25	8.5364E+02	1.3015E+02	8.5437E-01	7.2256E-03	2
818	9173	C6H15N	hexylamine	111-26-2	25	1.2000E+04	2.1576E+03	1.2146E+01	1.2003E-01	1
819	9174	C6H15N	2-aminohexane	5329-79-3	25	high	high	high	high	2
820	9175	C6H15N	3-aminohexane	16751-58-9	25	high	high	high	high	2
821	9176	C6H15N	1-amino-2-methylpentane	13364-16-4	25	high	high	high	high	2
822	9177	C6H15N	1-amino-3-methylpentane	42245-37-4	25	high	high	high	high	2
823	9178	C6H15N	1-amino-4-methylpentane	5344-20-7	25	high	high	high	high	2
824	9179	C6H15N	2-amino-2-methylpentane	53310-02-4	25	high	high	high	high	2
825	9180	C6H15N	2-amino-3-methylpentane	35399-81-6	25	high	high	high	high	2
826	9181	C6H15N	2-amino-4-methylpentane	108-09-8	25	high	high	high	high	2
827	9182	C6H15N	3-amino-2-methylpentane	54287-41-1	25	high	high	high	high	2
828	9183	C6H15N	3-amino-3-methylpentane	3495-46-3	25	high	high	high	high	2
829	9184	C6H15N	1-amino-2,2-dimethylbutane	41781-17-3	25	high	high	high	high	2
830	9185	C6H15N	1-amino-2,3-dimethylbutane	66553-05-7	25	high	high	high	high	2
831	9186	C6H15N	1-amino-3,3-dimethylbutane	15673-00-4	25	high	high	high	high	2
832	9187	C6H15N	2-amino-2,3-dimethylbutane	4358-75-2	25	high	high	high	high	2
833	9188	C6H15N	2-amino-3,3-dimethyl butane	3850-30-4	25	high	high	high	high	2
834	9189	C6H15N	1-amino-2-ethylbutane	617-79-8	25	high	high	high	high	2
835	9190	C6H15N	2-hexanamine, (±)	68107-05-1	25	high	high	high	high	2
836	9191	C6H15N	methylpentylamine	25419-06-1	25	high	high	high	high	2
837	9192	C6H15N	methyl-1-methylbutylamine	51932-19-5	25	high	high	high	high	2
838	9193	C6H15N	methyl-2-methylbutylamine	90023-96-4	25	high	high	high	high	2
839	9194	C6H15N	methyl-3-methylbutylamine	52317-98-3	25	high	high	high	high	2
840	9195	C6H15N	methyl-1,1-dimethylpropylamine	---	25	high	high	high	high	2
841	9196	C6H15N	methyl-1,2-dimethylpropylamine	34317-39-0	25	high	high	high	high	2
842	9197	C6H15N	methyl-2,2-dimethylpropylamine	26153-91-3	25	high	high	high	high	2
843	9198	C6H15N	ethylbutylamine	13360-63-9	30	3.6664E+04	6.7300E+03	3.8059E+01	3.7611E-01	1
844	9199	C6H15N	ethyl-sec-butylamine	21035-44-9	30	6.2352E+04	1.1700E+04	6.6498E+01	6.5715E-01	1
845	9200	C6H15N	ethylisobutylamine	13205-60-2	25	high	high	high	high	2
846	9201	C6H15N	ethyl-tert-butylamine	4432-77-3	25	high	high	high	high	2
847	9202	C6H15N	di-propylamine	142-84-7	36	2.9100E+04	5.3076E+03	2.9972E+01	2.9619E-01	1
848	9203	C6H15N	diisopropylamine	108-18-9	25	high	high	high	high	2
849	9204	C6H15N	propylisopropylamine	21968-17-2	25	high	high	high	high	2
850	9205	C6H15N	N-ethyl-2-butanamine, (±)	116724-10-8	25	high	high	high	high	2
851	9206	C6H15N	methylethylpropylamine	4458-32-6	25	high	high	high	high	2
852	9207	C6H15N	methylethylisopropylamine	39198-07-7	25	high	high	high	high	2
853	9208	C6H15N	dimethylbutylamine	927-62-8	25	high	high	high	high	2
854	9209	C6H15N	dimethyl-sec-butylamine	921-04-0	25	high	high	high	high	2
855	9210	C6H15N	dimethylisobutylamine	7239-24-9	25	high	high	high	high	2
856	9211	C6H15N	dimethyl-tert-butylamine	918-02-5	25	high	high	high	high	2
857	9212	C6H15N	triethylamine	121-44-8	25	7.3670E+04	1.3961E+04	7.9529E+01	7.8592E-01	1
858	9230	C6H15NO2	diisopropanolamine	110-97-4	25	total	total	total	total	1
859	9239	C6H15NO3	triethanolamine	102-71-6	25	total	total	total	total	1
860	9268	C6H15O4P	triethyl phosphate	78-40-0	25	5.0000E+05	8.9998E+04	1.0000E+03	5.4898E+00	1
861	9292	C6H16N2	hexamethylenediamine	124-09-4	25	total	total	total	total	1
862	9359	C6H18N4	triethylene tetramine	112-24-3	25	total	total	total	total	1
863	9932	C7H5ClO2	o-chlorobenzoic acid	118-91-2	25	2.1000E+03	2.4208E+02	2.1044E+00	1.3441E-02	1
864	10003	C7H5F3	benzotrifluoride	98-08-8	25	4.5000E+02	5.5505E+01	4.5020E-01	3.0812E-03	1
865	10045	C7H5N	benzonitrile	100-47-0	100	1.0000E+04	1.7615E+03	1.0101E+01	9.7950E-02	1
866	10049	C7H5NO	phenyl isocyanate	103-71-9	20	7.3499E-01	1.1115E-01	7.3499E-04	6.1700E-06	1
867	10118	C7H5N3O6	2,4,6-trinitrotoluene	118-96-7	20	1.2000E+02	9.5188E+00	1.2001E-01	5.2839E-04	1
868	10363	C7H6N2O4	2,4-dinitrotoluene	121-14-2	22	2.7000E+02	2.6712E+01	2.7007E-01	1.4828E-03	1
869	10404	C7H6O	benzaldehyde	100-52-7	25	6.5700E+03	1.1214E+03	6.6135E+00	6.2318E-02	1
870	10408	C7H6O2	benzoic acid	65-85-0	25	3.3900E+03	5.0152E+02	3.4015E+00	2.7853E-02	1
871	10409	C7H6O2	p-hydroxybenzaldehyde	123-08-0	30	1.2900E+04	1.9241E+03	1.3069E+01	1.0701E-01	1
872	10410	C7H6O2	salicylaldehyde	90-02-8	86	1.7000E+04	2.5446E+03	1.7294E+01	1.4161E-01	1
873	10420	C7H6O3	salicylic acid	69-72-7	25	2.2280E+03	2.9116E+02	2.2330E+00	1.6167E-02	1
874	10492	C7H7Cl	benzyl chloride	100-44-7	25	1.0630E+02	1.5130E+01	1.0631E-01	8.3984E-04	2
875	10493	C7H7Cl	o-chlorotoluene	95-49-8	25	1.0630E+02	1.5130E+01	1.0631E-01	8.3984E-04	2
876	10494	C7H7Cl	p-chlorotoluene	106-43-4	25	1.0630E+02	1.5130E+01	1.0631E-01	8.3984E-04	1
877	10495	C7H7Cl	m-chlorotoluene	108-41-8	25	1.0630E+02	1.5130E+01	1.0631E-01	8.3984E-04	2
878	10626	C7H7NO2	m-nitrotoluene	99-08-1	30	5.0000E+02	6.5710E+01	5.0025E-01	3.6478E-03	1
879	10628	C7H7NO2	p-nitrotoluene	99-99-0	15	4.0000E+01	5.2547E+00	4.0002E-02	2.9169E-04	1
880	10724	C7H8	1,3,5-cycloheptatriene	544-25-2	25	6.2010E+02	1.2130E+02	6.2048E-01	6.7341E-03	1

Table 3 Solubility in Water - Organic Compounds

					Solubility in Water, S (ppm - parts per million, wt - weight, g - gram)					
NO	ID	FORMULA	NAME	CAS No	T, C	S@T, ppm(wt)	S@T, ppm(mol)	S@T, g/1000 g of H2O	S@T, mol/1000 g of H2O	code
881	10725	C7H8	toluene	108-88-3	25	5.4240E+02	1.0609E+02	5.4269E-01	5.8899E-03	1
882	10850	C7H8O	anisole	100-66-3	25	1.4000E+02	2.3325E+01	1.4002E-01	1.2948E-03	1
883	10851	C7H8O	benzyl alcohol	100-51-6	25	4.2900E+04	7.4117E+03	4.4823E+01	4.1449E-01	1
884	10852	C7H8O	m-cresol	108-39-4	20	2.1800E+04	3.6989E+03	2.2286E+01	2.0608E-01	1
885	10853	C7H8O	o-cresol	95-48-7	20	2.4500E+04	4.1665E+03	2.5115E+01	2.3225E-01	1
886	10854	C7H8O	p-cresol	106-44-5	20	1.9400E+04	3.2850E+03	1.9784E+01	1.8295E-01	1
887	10855	C7H8O	cresol mixture	1319-77-3	25	2.1900E+04	3.7161E+03	2.2390E+01	2.0705E-01	2
888	10866	C7H8O2	guaiacol	90-05-1	37	1.3160E+03	1.9119E+02	1.3177E+00	1.0615E-02	1
889	10867	C7H8O2	p-methoxyphenol	150-76-5	37	4.9660E+02	7.2097E+01	4.9685E-01	4.0023E-03	1
890	10953	C7H9N	benzylamine	100-46-9	25	3.1400E+03	5.2928E+02	3.1499E+00	2.9396E-02	1
891	10954	C7H9N	N-methylaniline	100-61-8	25	5.6240E+03	9.4995E+02	5.6558E+00	5.2781E-02	1
892	10955	C7H9N	m-toluidine	108-44-1	20	1.5030E+04	2.5588E+03	1.5259E+01	1.4240E-01	1
893	10956	C7H9N	o-toluidine	95-53-4	25	1.5000E+04	2.5537E+03	1.5228E+01	1.4212E-01	1
894	10957	C7H9N	p-toluidine	106-49-0	15	6.5000E+03	1.0987E+03	6.5425E+00	6.1056E-02	1
895	11225	C7H10S	2-propylthiophene	1551-27-5	25	5.7817E+01	8.2522E+00	5.7820E-02	4.5808E-04	2
896	11226	C7H10S	3-propylthiophene	1518-75-8	25	5.0533E+01	7.2126E+00	5.0536E-02	4.0037E-04	2
897	11227	C7H10S	2-isopropylthiophene	4095-22-1	25	7.7635E+01	1.1081E+01	7.7641E-02	6.1511E-04	2
898	11228	C7H10S	3-isopropylthiophene	29488-27-5	25	6.2669E+01	8.9449E+00	6.2673E-02	4.9653E-04	2
899	11229	C7H10S	2-methyl-3-ethylthiophene	53119-51-0	25	6.2669E+01	8.9449E+00	6.2673E-02	4.9653E-04	2
900	11230	C7H10S	2-methyl-4-ethyl thiophene	13678-54-1	25	4.5361E+01	6.4743E+00	4.5363E-02	3.5939E-04	2
901	11231	C7H10S	2-methyl-5-ethyl thiophene	40323-88-4	25	5.3045E+01	7.5712E+00	5.3048E-02	4.2027E-04	2
902	11232	C7H10S	3-methyl-2-ethylthiophene	31805-48-8	25	5.0533E+01	7.2126E+00	5.0536E-02	4.0037E-04	2
903	11233	C7H10S	3-methyl-4-ethyl thiophene	---	25	3.9622E+01	5.6552E+00	3.9623E-02	3.1392E-04	2
904	11234	C7H10S	3-methyl-5-ethylthiophene	---	25	7.2813E+01	1.0393E+01	7.2819E-02	5.7691E-04	2
905	11235	C7H10S	2,3,4-trimethylthiophene	1795-04-6	25	2.6799E+01	3.8250E+00	2.6800E-02	2.1232E-04	2
906	11236	C7H10S	2,3,5-trimethylthiophene	1795-05-7	25	4.1826E+01	5.9698E+00	4.1828E-02	3.3138E-04	2
907	11340	C7H12	1-heptyne	628-71-7	25	9.4000E+01	1.7609E+01	9.4009E-02	9.7750E-04	1
908	11341	C7H12	2-heptyne	1119-65-9	25	1.6349E+02	3.0629E+01	1.6352E-01	1.7002E-03	1
909	11342	C7H12	3-heptyne	2586-89-2	25	8.9892E+01	1.6840E+01	8.9900E-02	9.3478E-04	2
910	11343	C7H12	3-methyl-1-hexyne	40276-93-5	25	3.0420E+02	5.6997E+01	3.0429E-01	3.1641E-03	2
911	11344	C7H12	4-methyl-1-hexyne	52713-81-2	25	2.2102E+02	4.1408E+01	2.2107E-01	2.2986E-03	2
912	11345	C7H12	5-methyl-1-hexyne	2203-80-7	25	2.1109E+02	3.9547E+01	2.1113E-01	2.1953E-03	2
913	11346	C7H12	4-methyl-2-hexyne	20198-49-6	25	1.3825E+02	2.5900E+01	1.3827E-01	1.4377E-03	2
914	11347	C7H12	5-methyl-2-hexyne	53566-37-3	25	1.1736E+02	2.1987E+01	1.1738E-01	1.2205E-03	2
915	11348	C7H12	2-methyl-3-hexyne	36566-80-0	25	1.7311E+02	3.2432E+01	1.7314E-01	1.8003E-03	1
916	11349	C7H12	3-ethyl-1-pentyne	21020-26-8	25	3.2054E+02	6.0059E+01	3.2064E-01	3.3341E-03	2
917	11350	C7H12	3,3-dimethyl-1-pentyne	918-82-1	25	6.4594E+02	1.2106E+02	6.4636E-01	6.7209E-03	2
918	11351	C7H12	3,4-dimethyl-1-pentyne	61064-08-2	25	3.9404E+02	7.3836E+01	3.9420E-01	4.0989E-03	2
919	11352	C7H12	4,4-dimethyl-1-pentyne	13361-63-2	25	4.8015E+02	8.9977E+01	4.8038E-01	4.9950E-03	2
920	11353	C7H12	4,4-dimethyl-2-pentyne	999-78-0	25	3.3766E+02	6.3269E+01	3.3778E-01	3.5122E-03	2
921	11358	C7H12	1,2-heptadiene	2384-90-9	25	1.8569E+01	3.4785E+00	1.8570E-02	1.9309E-04	2
922	11359	C7H12	1,4-heptadiene	5675-22-9	25	3.7267E+01	6.9811E+00	3.7269E-02	3.8752E-04	2
923	11360	C7H12	1,5-heptadiene	1541-23-7	25	3.5288E+01	6.6103E+00	3.5289E-02	3.6694E-04	2
924	11361	C7H12	1,6-heptadiene	3070-53-9	25	2.9925E+01	5.6056E+00	2.9926E-02	3.1117E-04	2
925	11362	C7H12	2,4-heptadiene	628-72-8	25	1.6102E+01	3.0162E+00	1.6102E-02	1.6743E-04	2
926	11363	C7H12	2-methyl-1,5-hexadiene	4049-81-4	25	4.8527E+01	9.0904E+00	4.8529E-02	5.0460E-04	2
927	11364	C7H12	2-methyl-2,4-hexadiene	28823-41-8	25	1.3168E+01	2.4667E+00	1.3168E-02	1.3692E-04	2
928	11365	C7H12	2,4-dimethyl-1,3-pentadiene	1000-86-8	25	3.6863E+01	6.9054E+00	3.6865E-02	3.8332E-04	2
929	11369	C7H12	2,4-dimethyl-2,3-pentadiene	1000-87-9	25	5.5981E+01	1.0487E+01	5.5984E-02	5.8212E-04	2
930	11450	C7H12O2	butyl acrylate	141-32-2	20	1.6000E+03	2.2520E+02	1.6026E+00	1.2503E-02	1
931	11451	C7H12O2	isobutyl acrylate	106-63-8	23	2.0000E+03	2.8159E+02	2.0040E+00	1.5635E-02	1
932	11452	C7H12O2	propyl methacrylate	2210-28-8	25	1.8000E+03	2.5339E+02	1.8032E+00	1.4069E-02	2
933	11453	C7H12O2	isopropyl methacrylate	4655-34-9	25	1.8000E+03	2.5339E+02	1.8032E+00	1.4069E-02	2
934	11454	C7H12O2	allyl butanoate	2051-78-7	25	1.8000E+03	2.5339E+02	1.8032E+00	1.4069E-02	2
935	11480	C7H12O2	ethyl 3-methyl-2-butenoate	638-10-8	25	1.8000E+03	2.5339E+02	1.8032E+00	1.4069E-02	2
936	11481	C7H12O2	ethyl trans-2-methyl-2-butenoate	5837-78-5	25	1.8000E+03	2.5339E+02	1.8032E+00	1.4069E-02	2
937	11483	C7H12O2	methyl 3-hexenoate	2396-78-3	25	1.8000E+03	2.5339E+02	1.8032E+00	1.4069E-02	2
938	11484	C7H12O2	methyl 2-methyl-2-pentenoate	10478-12-3	25	1.8000E+03	2.5339E+02	1.8032E+00	1.4069E-02	2
939	11485	C7H12O2	methyl 3-methyl-3-pentenoate	2258-58-4	25	1.8000E+03	2.5339E+02	1.8032E+00	1.4069E-02	2
940	11486	C7H12O2	methyl cis-3-methyl-2-pentenoate	17447-00-6	25	1.8000E+03	2.5339E+02	1.8032E+00	1.4069E-02	2
941	11487	C7H12O2	methyl trans-3-methyl-2-pentenoate	17447-01-7	25	1.8000E+03	2.5339E+02	1.8032E+00	1.4069E-02	2
942	11488	C7H12O2	tert-butyl acrylate	1663-39-4	25	1.8000E+03	2.5339E+02	1.8032E+00	1.4069E-02	2
943	11491	C7H12O2	methyl trans-2-methyl-2-pentenoate	1567-14-2	25	1.8000E+03	2.5339E+02	1.8032E+00	1.4069E-02	2
944	11492	C7H12O2	4-pentenyl acetate	1576-85-8	25	1.8000E+03	2.5339E+02	1.8032E+00	1.4069E-02	2
945	11497	C7H12O2	3,3-dimethylallyl acetate	1191-16-8	25	1.8000E+03	2.5339E+02	1.8032E+00	1.4069E-02	2
946	11500	C7H12O2	cis-3-hexenyl formate	33467-73-1	25	1.8000E+03	2.5339E+02	1.8032E+00	1.4069E-02	2
947	11530	C7H12O4	diethyl malonate	105-53-3	37	2.3220E+04	2.6666E+03	2.3772E+01	1.4842E-01	1
948	11696	C7H14	ethylcyclopentane	1640-89-7	25	9.5450E+00	1.7513E+00	9.5451E-03	9.7212E-05	2
949	11697	C7H14	1,1-dimethylcyclopentane	1638-26-2	25	2.2530E+01	4.1337E+00	2.2530E-02	2.2946E-04	2
950	11698	C7H14	1,2-dimethylcyclopentane	2452-99-5	25	1.3822E+01	2.5360E+00	1.3822E-02	1.4077E-04	2
951	11699	C7H14	cis-1,2-dimethylcyclopentane	1192-18-3	25	1.1910E+01	2.1852E+00	1.1910E-02	1.2130E-04	2
952	11700	C7H14	trans-1,2-dimethylcyclopentane	822-50-4	25	1.8155E+01	3.3310E+00	1.8155E-02	1.8490E-04	2
953	11701	C7H14	1,3-dimethylcyclopentane	2453-00-1	25	1.8076E+01	3.3165E+00	1.8076E-02	1.8410E-04	2
954	11702	C7H14	cis-1,3-dimethylcyclopentane	2532-58-3	25	1.9267E+01	3.5351E+00	1.9267E-02	1.9623E-04	2
955	11703	C7H14	trans-1,3-dimethylcyclopentane	1759-58-6	25	1.8293E+01	3.3563E+00	1.8293E-02	1.8631E-04	2
956	11704	C7H14	methylcyclohexane	108-87-2	25	1.4020E+01	2.5723E+00	1.4020E-02	1.4279E-04	1
957	11705	C7H14	cycloheptane	291-64-5	25	3.0000E+01	5.5044E+00	3.0001E-02	3.0554E-04	1
958	11706	C7H14	1-heptene	592-76-7	25	1.8200E+01	3.3393E+00	1.8200E-02	1.8536E-04	1
959	11707	C7H14	2-heptene	592-77-8	25	9.4812E+00	1.7396E+00	9.4813E-03	9.6562E-05	2
960	11708	C7H14	cis-2-heptene	6443-92-1	25	9.3972E+00	1.7242E+00	9.3973E-03	9.5707E-05	2

Table 3 Solubility in Water - Organic Compounds

					Solubility in Water, S (ppm - parts per million, wt - weight, g - gram)					
NO	ID	FORMULA	NAME	CAS No	T, C	S@T, ppm(wt)	S@T, ppm(mol)	S@T, g/1000 g of H2O	S@T, mol/1000 g of H2O	code
961	11709	C7H14	trans-2-heptene	14686-13-6	25	9.6405E+00	1.7688E+00	9.6406E-03	9.8185E-05	2
962	11710	C7H14	3-heptene	592-78-9	25	1.0828E+01	1.9866E+00	1.0828E-02	1.1028E-04	2
963	11711	C7H14	3-heptene (mixed isomers)	25339-56-4	25	1.4458E+01	2.6527E+00	1.4458E-02	1.4725E-04	2
964	11712	C7H14	cis-3-heptene	7642-10-6	25	1.0887E+01	1.9976E+00	1.0888E-02	1.1088E-04	2
965	11713	C7H14	trans-3-heptene	14686-14-7	25	1.0935E+01	2.0064E+00	1.0936E-02	1.1137E-04	2
966	11714	C7H14	2-methyl-1-hexene	6094-02-6	25	1.3480E+01	2.4733E+00	1.3480E-02	1.3729E-04	2
967	11715	C7H14	3-methyl-1-hexene	3404-61-3	25	2.0566E+01	3.7734E+00	2.0566E-02	2.0946E-04	2
968	11716	C7H14	4-methyl-1-hexene	3769-23-1	25	1.7724E+01	3.2520E+00	1.7725E-02	1.8052E-04	2
969	11717	C7H14	5-methyl-1-hexene	3524-73-0	25	1.9103E+01	3.5049E+00	1.9103E-02	1.9455E-04	2
970	11718	C7H14	2-methyl-2-hexene	2738-19-4	25	1.1093E+01	2.0353E+00	1.1093E-02	1.1298E-04	2
971	11719	C7H14	3-methyl-cis-2-hexene	10574-36-4	25	1.0016E+01	1.8378E+00	1.0017E-02	1.0201E-04	2
972	11720	C7H14	3-methyl-trans-2-hexene	20710-38-7	25	1.1234E+01	2.0612E+00	1.1234E-02	1.1442E-04	2
973	11721	C7H14	4-methyl-cis-2-hexene	3683-19-0	25	1.8122E+01	3.3250E+00	1.8123E-02	1.8457E-04	2
974	11722	C7H14	4-methyl-trans-2-hexene	3683-22-5	25	1.6960E+01	3.1118E+00	1.6961E-02	1.7274E-04	2
975	11723	C7H14	5-methyl-cis-2-hexene	13151-17-2	25	1.5293E+01	2.8058E+00	1.5293E-02	1.5575E-04	2
976	11724	C7H14	5-methyl-trans-2-hexene	7385-82-2	25	1.6472E+01	3.0222E+00	1.6472E-02	1.6776E-04	2
977	11725	C7H14	2-methyl-cis-3-hexene	15840-60-5	25	1.8421E+01	3.3799E+00	1.8422E-02	1.8762E-04	2
978	11726	C7H14	2-methyl-trans-3-hexene	692-24-0	25	1.8519E+01	3.3977E+00	1.8519E-02	1.8861E-04	2
979	11727	C7H14	3-methyl-cis-3-hexene	4914-89-0	25	1.1098E+01	2.0363E+00	1.1099E-02	1.1303E-04	2
980	11728	C7H14	3-methyl-trans-3-hexene	3899-36-3	25	1.2288E+01	2.2546E+00	1.2288E-02	1.2515E-04	2
981	11729	C7H14	2-ethyl-1-pentene	3404-71-5	25	1.1985E+01	2.1989E+00	1.1985E-02	1.2206E-04	2
982	11730	C7H14	3-ethyl-1-pentene	4038-04-4	25	2.0342E+01	3.7323E+00	2.0342E-02	2.0718E-04	2
983	11731	C7H14	2,3-dimethyl-1-pentene	3404-72-6	25	2.0162E+01	3.6993E+00	2.0162E-02	2.0534E-04	2
984	11732	C7H14	2,4-dimethyl-1-pentene	2213-32-3	25	2.3158E+01	4.2489E+00	2.3158E-02	2.3586E-04	2
985	11733	C7H14	3,3-dimethyl-1-pentene	3404-73-7	25	2.8573E+01	5.2426E+00	2.8574E-02	2.9102E-04	2
986	11734	C7H14	3,4-dimethyl-1-pentene	7385-78-6	25	2.4142E+01	4.4295E+00	2.4143E-02	2.4588E-04	2
987	11735	C7H14	4,4-dimethyl-1-pentene	762-62-9	25	3.6518E+01	6.7004E+00	3.6520E-02	3.7194E-04	2
988	11736	C7H14	3-ethyl-2-pentene	816-79-5	25	1.0733E+01	1.9692E+00	1.0733E-02	1.0931E-04	2
989	11737	C7H14	2,3-dimethyl-2-pentene	10574-37-5	25	9.9391E+00	1.8236E+00	9.9392E-03	1.0123E-04	2
990	11738	C7H14	2,4-dimethyl-2-pentene	625-65-0	25	2.1219E+01	3.8931E+00	2.1219E-02	2.1611E-04	2
991	11739	C7H14	3,4-dimethyl-cis-2-pentene	4914-91-4	25	1.5499E+01	2.8437E+00	1.5499E-02	1.5785E-04	2
992	11740	C7H14	3,4-dimethyl-trans-2-pentene	4914-92-5	25	1.3731E+01	2.5192E+00	1.3731E-02	1.3984E-04	2
993	11741	C7H14	4,4-dimethyl-cis-2-pentene	762-63-0	25	2.4604E+01	4.5142E+00	2.4604E-02	2.5058E-04	2
994	11742	C7H14	4,4-dimethyl-trans-2-pentene	690-08-4	25	2.9653E+01	5.4408E+00	2.9654E-02	3.0201E-04	2
995	11743	C7H14	3-methyl-2-ethyl-1-butene	7357-93-9	25	1.8070E+01	3.3154E+00	1.8070E-02	1.8404E-04	2
996	11744	C7H14	2,3,3-trimethyl-1-butene	594-56-9	25	2.7990E+01	5.1356E+00	2.7991E-02	2.8507E-04	2
997	11745	C7H14	3,4-dimethylpent-2-ene	24910-63-2	25	1.6880E+01	3.0971E+00	1.6881E-02	1.7192E-04	2
998	11751	C7H14Br2	1,1-dibromoheptane	59104-79-9	25	3.5758E+01	2.4970E+00	3.5760E-02	1.3860E-04	2
999	11752	C7H14Br2	1,2-dibromoheptane	42474-21-5	25	2.6178E+01	1.8280E+00	2.6179E-02	1.0147E-04	2
1000	11753	C7H14Br2	1,5-dibromoheptane	1622-10-2	25	1.5996E+01	1.1170E+00	1.5997E-02	6.2003E-05	2
1001	11754	C7H14Br2	1,7-dibromoheptane	4549-31-9	25	4.7404E+00	3.3101E-01	4.7404E-03	1.8374E-05	2
1002	11755	C7H14Br2	2,3-dibromoheptane	21266-88-6	25	1.5996E+01	1.1170E+00	1.5997E-02	6.2003E-05	2
1003	11756	C7H14Br2	3,4-dibromoheptane	21266-90-0	25	1.5996E+01	1.1170E+00	1.5997E-02	6.2003E-05	2
1004	11763	C7H14Cl2	1,1-dichloroheptane	821-25-0	25	2.1359E+01	2.2756E+00	2.1360E-02	1.2632E-04	2
1005	11764	C7H14Cl2	1,2-dichloro-4,4-dimethylpentane	6065-90-3	25	4.1090E+01	4.3778E+00	4.1091E-02	2.4301E-04	2
1006	11765	C7H14Cl2	1,5-dichloro-3,3-dimethylpentane	62496-53-1	25	2.9627E+01	3.1565E+00	2.9627E-02	1.7521E-04	2
1007	11766	C7H14Cl2	2,4-dichloro-2,4-dimethylpentane	33553-93-4	25	2.9627E+01	3.1565E+00	2.9627E-02	1.7521E-04	2
1008	11767	C7H14Cl2	1,2-dichloroheptane	10575-87-8	25	2.9627E+01	3.1565E+00	2.9627E-02	1.7521E-04	2
1009	11768	C7H14Cl2	1,7-dichloroheptane	821-76-1	25	2.9627E+01	3.1565E+00	2.9627E-02	1.7521E-04	2
1010	11769	C7H14Cl2	2,2-dichloroheptane	65786-09-6	25	2.9627E+01	3.1565E+00	2.9627E-02	1.7521E-04	2
1011	11770	C7H14Cl2	4,4-dichloroheptane	89796-76-9	25	2.9627E+01	3.1565E+00	2.9627E-02	1.7521E-04	2
1012	11827	C7H14O	1-heptanal	111-71-7	30	1.5160E+03	2.3948E+02	1.5183E+00	1.3297E-02	1
1013	11828	C7H14O	3-methylhexanal	19269-28-4	25	1.8395E+03	2.9066E+02	1.8429E+00	1.6139E-02	2
1014	11829	C7H14O	2-heptanone	110-43-0	25	4.3000E+03	6.8086E+02	4.3186E+00	3.7820E-02	1
1015	11830	C7H14O	3-heptanone	106-35-4	25	2.6454E+03	4.1829E+02	2.6524E+00	2.3229E-02	2
1016	11831	C7H14O	4-heptanone	123-19-3	25	3.2395E+03	5.1249E+02	3.2501E+00	2.8463E-02	2
1017	11832	C7H14O	3-methyl-2-hexanone	2550-21-2	25	4.1179E+03	6.5192E+02	4.1349E+00	3.6211E-02	2
1018	11833	C7H14O	4-methyl-2-hexanone	105-42-0	25	4.3737E+03	6.9257E+02	4.3929E+00	3.8471E-02	2
1019	11834	C7H14O	5-methyl-2-hexanone	110-12-3	25	3.0884E+03	4.8852E+02	3.0980E+00	2.7131E-02	2
1020	11835	C7H14O	2-methyl-3-hexanone	7379-12-6	25	6.2972E+03	9.9879E+02	6.3371E+00	5.5498E-02	2
1021	11836	C7H14O	4-methyl-3-hexanone	17042-16-9	25	5.2446E+03	8.3110E+02	5.2723E+00	4.6172E-02	2
1022	11837	C7H14O	5-methyl-3-hexanone	623-56-3	25	5.2446E+03	8.3110E+02	5.2723E+00	4.6172E-02	2
1023	11838	C7H14O	3-ethyl-2-pentanone	6137-03-7	25	4.6460E+03	7.3586E+02	4.6677E+00	4.0877E-02	2
1024	11839	C7H14O	3,3-dimethyl-2-pentanone	20669-04-9	25	7.2969E+03	1.1583E+03	7.3506E+00	6.4373E-02	2
1025	11840	C7H14O	3,4-dimethyl-2-pentanone	565-78-6	25	6.6952E+03	1.0623E+03	6.7403E+00	5.9028E-02	2
1026	11841	C7H14O	4,4-dimethyl-2-pentanone	590-50-1	25	1.0332E+04	1.6443E+03	1.0440E+01	9.1426E-02	2
1027	11842	C7H14O	2,2-dimethyl-3-pentanone	564-04-5	25	1.0332E+04	1.6443E+03	1.0440E+01	9.1426E-02	2
1028	11843	C7H14O	2,4-dimethyl-3-pentanone	565-80-0	25	5.7000E+03	9.0361E+02	5.7327E+00	5.0204E-02	1
1029	11844	C7H14O	1-methylcyclohexanol	590-67-0	25	3.5000E+04	5.6896E+03	3.6269E+01	3.1763E-01	2
1030	11845	C7H14O	cis-2-methylcyclohexanol	7443-70-1	25	3.5000E+04	5.6896E+03	3.6269E+01	3.1763E-01	2
1031	11846	C7H14O	trans-2-methylcyclohexanol	7443-52-9	25	3.5000E+04	5.6896E+03	3.6269E+01	3.1763E-01	2
1032	11847	C7H14O	cis-3-methylcyclohexanol	5454-79-5	25	3.5000E+04	5.6896E+03	3.6269E+01	3.1763E-01	2
1033	11848	C7H14O	trans-3-methylcyclohexanol	7443-55-2	25	3.5000E+04	5.6896E+03	3.6269E+01	3.1763E-01	2
1034	11849	C7H14O	cis-4-methylcyclohexanol	7731-28-4	25	3.5000E+04	5.6896E+03	3.6269E+01	3.1763E-01	2
1035	11850	C7H14O	trans-4-methylcyclohexanol	7731-29-5	20	3.5000E+04	5.6896E+03	3.6269E+01	3.1763E-01	1
1036	11851	C7H14O	trans-2-methylcyclohexanol, (±)	615-39-4	25	3.5000E+04	5.6896E+03	3.6269E+01	3.1763E-01	2
1037	11877	C7H14O	methylcyclohexanol	25639-42-3	25	3.5000E+04	5.6896E+03	3.6269E+01	3.1763E-01	2
1038	11878	C7H14O	2-methylcyclohexanol; (cis+trans)	583-59-5	25	3.5000E+04	5.6896E+03	3.6269E+01	3.1763E-01	2
1039	11879	C7H14O	4-methylcyclohexanol; (cis+trans)	589-91-3	25	3.5000E+04	5.6896E+03	3.6269E+01	3.1763E-01	2
1040	11880	C7H14O	3-methylcyclohexanol; (cis+trans)	591-23-1	25	3.5000E+04	5.6896E+03	3.6269E+01	3.1763E-01	2

Table 3 Solubility in Water - Organic Compounds

NO	ID	FORMULA	NAME	CAS No	Solubility in Water, S (ppm - parts per million, wt - weight, g - gram) T, C	S@T, ppm(wt)	S@T, ppm(mol)	S@T, g/1000 g of H2O	S@T, mol/1000 g of H2O	code
1041	11884	C7H14O2	heptanoic acid	111-14-8	20	2.4400E+03	3.3835E+02	2.4460E+00	1.8788E-02	1
1042	11885	C7H14O2	2-ethyl-2-methylbutanoic acid	19889-37-3	25	5.4538E+03	7.5825E+02	5.4837E+00	4.2122E-02	2
1043	11886	C7H14O2	2-ethylpentanoic acid, (±)	116908-83-9	25	4.8787E+03	6.7795E+02	4.9026E+00	3.7658E-02	2
1044	11887	C7H14O2	2-methylhexanoic acid, (±)	22160-12-9	25	3.5457E+03	4.9215E+02	3.5583E+00	2.7332E-02	2
1045	11888	C7H14O2	3-methylhexanoic acid	3780-58-3	25	4.0244E+03	5.5882E+02	4.0407E+00	3.1037E-02	2
1046	11889	C7H14O2	4-methylhexanoic acid, (±)	22160-41-4	25	3.2041E+03	4.4460E+02	3.2144E+00	2.4690E-02	2
1047	11890	C7H14O2	5-methylhexanoic acid	628-46-6	25	3.4570E+03	4.7980E+02	3.4690E+00	2.6646E-02	2
1048	11891	C7H14O2	hexyl formate	629-33-4	25	3.5670E+02	4.9375E+01	3.5683E-01	2.7409E-03	2
1049	11892	C7H14O2	pentyl acetate	628-63-7	20	1.7000E+03	2.3559E+02	1.7029E+00	1.3080E-02	1
1050	11893	C7H14O2	1-methylbutyl acetate	626-38-0	25	3.1871E+03	4.4224E+02	3.1973E+00	2.4560E-02	2
1051	11894	C7H14O2	2-methylbutyl acetate	624-41-9	25	2.1864E+03	3.0312E+02	2.1912E+00	1.6831E-02	2
1052	11895	C7H14O2	isopentyl acetate	123-92-2	25	2.0000E+03	2.7723E+02	2.0040E+00	1.5393E-02	1
1053	11896	C7H14O2	1-ethylpropyl acetate	620-11-1	25	3.3634E+03	4.6678E+02	3.3748E+00	2.5923E-02	2
1054	11897	C7H14O2	tert-pentyl acetate	625-16-1	25	5.1741E+03	7.1919E+02	5.2011E+00	3.9951E-02	2
1055	11898	C7H14O2	1,2-dimethylpropyl acetate	5343-96-4	25	4.0609E+03	5.6391E+02	4.0774E+00	3.1320E-02	2
1056	11899	C7H14O2	neopentyl acetate	926-41-0	25	4.1717E+03	5.7935E+02	4.1892E+00	3.2178E-02	2
1057	11900	C7H14O2	sec-pentyl acetate (R)	54638-10-7	25	1.9632E+03	2.7212E+02	1.9671E+00	1.5109E-02	2
1058	11901	C7H14O2	butyl propanoate	590-01-2	20	1.5000E+03	2.0784E+02	1.5023E+00	1.1539E-02	1
1059	11902	C7H14O2	isobutyl propanoate	540-42-1	25	2.6256E+03	3.6415E+02	2.6325E+00	2.0221E-02	2
1060	11903	C7H14O2	sec-butyl propanoate	591-34-4	25	3.1871E+03	4.4224E+02	3.1973E+00	2.4560E-02	2
1061	11904	C7H14O2	tert-butyl propanoate	20487-40-5	25	2.0276E+03	2.8107E+02	2.0318E+00	1.5607E-02	2
1062	11905	C7H14O2	propyl butanoate	105-66-8	17	1.6140E+03	2.2365E+02	1.6166E+00	1.2418E-02	1
1063	11906	C7H14O2	isopropyl butanoate	638-11-9	25	3.5495E+03	4.9268E+02	3.5621E+00	2.7362E-02	2
1064	11907	C7H14O2	isopropyl isobutanoate	617-50-5	25	6.1802E+03	8.5978E+02	6.2186E+00	4.7767E-02	2
1065	11908	C7H14O2	propyl isobutanoate	644-49-5	25	2.7858E+03	3.8642E+02	2.7936E+00	2.1458E-02	2
1066	11910	C7H14O2	ethyl 2,2-dimethylpropanoate	3938-95-2	25	6.9948E+03	9.7380E+02	7.0441E+00	5.4108E-02	2
1067	11911	C7H14O2	ethyl 2-methylbutanoate, (+)	10307-61-6	25	3.3634E+03	4.6678E+02	3.3748E+00	2.5923E-02	2
1068	11913	C7H14O2	methyl 2-ethylbutanoate	816-11-5	25	2.7118E+03	3.7613E+02	2.7192E+00	2.0887E-02	2
1069	11925	C7H14O2	2-methylhexanoic acid	4536-23-6	25	4.8051E+03	6.6768E+02	4.8283E+00	3.7087E-02	2
1070	11926	C7H14O2	neoheptanoic acid	33113-10-9	25	2.0365E+01	2.8181E+00	2.0365E-02	1.5643E-04	2
1071	11931	C7H14O3	ethyl-3-ethoxypropionate	763-69-9	25	3.7970E+02	4.6807E+01	3.7984E-01	2.5984E-03	1
1072	11982	C7H15Br	1-bromoheptane	629-04-9	25	6.6800E+00	6.7192E-01	6.6800E-03	3.7298E-05	1
1073	11983	C7H15Br	2-bromoheptane	1974-04-5	25	1.8457E+01	1.8566E+00	1.8458E-02	1.0306E-04	2
1074	11984	C7H15Br	4-bromoheptane	998-93-6	25	2.4187E+01	2.4329E+00	2.4187E-02	1.3505E-04	2
1075	11988	C7H15Cl	1-chloroheptane	629-06-1	25	1.3600E+01	1.8196E+00	1.3600E-02	1.0101E-04	1
1076	11989	C7H15Cl	2-chloro-2,3-dimethylpentane	59889-45-1	25	3.8197E+01	5.1106E+00	3.8198E-02	2.8369E-04	2
1077	11990	C7H15Cl	2-chloro-2,4-dimethylpentane	35951-33-8	25	7.1728E+01	9.5973E+00	7.1733E-02	5.3275E-04	2
1078	11991	C7H15Cl	3-chloro-2,3-dimethylpentane	595-38-0	25	7.1728E+01	9.5973E+00	7.1733E-02	5.3275E-04	2
1079	11992	C7H15Cl	4-chloro-2,2-dimethylpentane	33429-72-0	25	7.1728E+01	9.5973E+00	7.1733E-02	5.3275E-04	2
1080	11993	C7H15Cl	3-chloro-3-ethylpentane	994-25-2	25	3.2253E+01	4.3153E+00	3.2254E-02	2.3954E-04	2
1081	11994	C7H15Cl	2-chloroheptane	1001-89-4	25	3.8197E+01	5.1106E+00	3.8198E-02	2.8369E-04	2
1082	11995	C7H15Cl	3-chloroheptane	999-52-0	25	3.1417E+01	4.2035E+00	3.1418E-02	2.3333E-04	2
1083	11996	C7H15Cl	4-chloroheptane	998-95-8	25	3.1417E+01	4.2035E+00	3.1418E-02	2.3333E-04	2
1084	11997	C7H15Cl	1-chloro-3-methylhexane	101257-63-0	25	2.1695E+01	2.9027E+00	2.1695E-02	1.6112E-04	2
1085	11998	C7H15Cl	2-chloro-2-methylhexane	4398-65-6	25	5.0186E+01	6.7148E+00	5.0189E-02	3.7274E-04	2
1086	11999	C7H15Cl	2-chloro-5-methylhexane	58766-17-9	25	4.2977E+01	5.7503E+00	4.2979E-02	3.1920E-04	2
1087	12000	C7H15Cl	3-chloro-3-methylhexane	43197-78-0	25	5.0186E+01	6.7148E+00	5.0189E-02	3.7274E-04	2
1088	12001	C7H15Cl	2-chloro-2,3,3-trimethylbutane	918-07-0	25	3.8197E+01	5.1106E+00	3.8198E-02	2.8369E-04	2
1089	12112	C7H16	heptane	142-82-5	25	2.2400E+00	4.0272E-01	2.2400E-03	2.2354E-05	1
1090	12113	C7H16	2-methylhexane	591-76-4	25	2.5400E+00	4.5665E-01	2.5400E-03	2.5348E-05	1
1091	12114	C7H16	3-methylhexane	589-34-4	25	2.6400E+00	4.7463E-01	2.6400E-03	2.6346E-05	1
1092	12115	C7H16	3-ethylpentane	617-78-7	25	2.9458E+00	5.2960E-01	2.9458E-03	2.9398E-05	2
1093	12116	C7H16	2,2-dimethylpentane	590-35-2	25	4.4000E+00	7.9105E-01	4.4000E-03	4.3911E-05	1
1094	12117	C7H16	2,3-dimethylpentane	565-59-3	25	5.2500E+00	9.4387E-01	5.2500E-03	5.2393E-05	1
1095	12118	C7H16	2,4-dimethylpentane	108-08-7	25	4.4100E+00	7.9285E-01	4.4100E-03	4.4010E-05	1
1096	12119	C7H16	3,3-dimethylpentane	562-49-2	25	5.9400E+00	1.0679E+00	5.9400E-03	5.9279E-05	1
1097	12120	C7H16	2,2,3-trimethylbutane	464-06-2	25	5.7410E+00	1.0321E+00	5.7411E-03	5.7294E-05	2
1098	12121	C7H16	3-methylhexane, (S)-	6131-24-4	25	3.1910E+00	5.7369E-01	3.1910E-03	3.1845E-05	2
1099	12164	C7H16O	1-heptanol	111-70-6	25	1.7400E+03	2.7015E+02	1.7430E+00	1.5000E-02	1
1100	12165	C7H16O	2-heptanol	543-49-7	30	3.3000E+03	5.1303E+02	3.3109E+00	2.8492E-02	1
1101	12166	C7H16O	3-heptanol	589-82-2	25	4.2029E+03	6.5389E+02	4.2206E+00	3.6321E-02	2
1102	12167	C7H16O	4-heptanol	589-55-9	25	4.7283E+03	7.3597E+02	4.7508E+00	4.0883E-02	2
1103	12168	C7H16O	2-methyl-1-hexanol	624-22-6	25	2.9022E+03	4.5103E+02	2.9106E+00	2.5048E-02	2
1104	12169	C7H16O	3-methyl-1-hexanol	13231-81-7	25	1.7101E+03	2.6549E+02	1.7130E+00	1.4741E-02	2
1105	12170	C7H16O	4-methyl-1-hexanol	818-49-5	25	1.6122E+03	2.5028E+02	1.6148E+00	1.3897E-02	2
1106	12171	C7H16O	5-methyl-1-hexanol	627-98-5	25	1.7101E+03	2.6549E+02	1.7130E+00	1.4741E-02	2
1107	12172	C7H16O	2-methyl-2-hexanol	625-23-0	25	9.5821E+03	1.4976E+03	9.6748E+00	8.3257E-02	2
1108	12173	C7H16O	3-methyl-2-hexanol	2313-65-7	25	5.5451E+03	8.6371E+02	5.5761E+00	4.7985E-02	2
1109	12174	C7H16O	4-methyl-2-hexanol	2313-61-3	25	5.8829E+03	9.1658E+02	5.9177E+00	5.0925E-02	2
1110	12175	C7H16O	5-methyl-2-hexanol	627-59-8	25	5.8829E+03	9.1658E+02	5.9177E+00	5.0925E-02	2
1111	12176	C7H16O	2-methyl-3-hexanol	617-29-8	25	8.4015E+03	1.3118E+03	8.4726E+00	7.2912E-02	2
1112	12177	C7H16O	3-methyl-3-hexanol	597-96-6	25	9.5821E+03	1.4976E+03	9.6748E+00	8.3257E-02	2
1113	12178	C7H16O	4-methyl-3-hexanol	615-29-2	25	6.6228E+03	1.0325E+03	6.6669E+00	5.7373E-02	2
1114	12179	C7H16O	5-methyl-3-hexanol	623-55-2	25	7.0277E+03	1.0960E+03	7.0774E+00	6.0905E-02	2
1115	12180	C7H16O	2-ethyl-1-pentanol	27522-11-8	25	2.4334E+03	3.7803E+02	2.4393E+00	2.0992E-02	2
1116	12181	C7H16O	3-ethyl-1-pentanol	66225-51-2	25	2.4334E+03	3.7803E+02	2.4393E+00	2.0992E-02	2
1117	12182	C7H16O	2,2-dimethyl-1-pentanol	2370-12-9	25	5.2271E+03	8.1395E+02	5.2546E+00	4.5219E-02	2
1118	12183	C7H16O	2,3-dimethyl-1-pentanol	10143-23-4	25	2.7367E+03	4.2525E+02	2.7442E+00	2.3615E-02	2
1119	12184	C7H16O	2,4-dimethyl-1-pentanol	6305-71-1	25	3.6710E+03	5.7089E+02	3.6845E+00	3.1708E-02	2
1120	12185	C7H16O	3,3-dimethyl-1-pentanol	19264-94-9	25	2.5806E+03	4.0094E+02	2.5873E+00	2.2265E-02	2

Table 3 Solubility in Water - Organic Compounds

					Solubility in Water, S (ppm - parts per million, wt - weight, g - gram)					
NO	ID	FORMULA	NAME	CAS No	T, C	S@T, ppm(wt)	S@T, ppm(mol)	S@T, g/1000 g of H2O	S@T, mol/1000 g of H2O	code
1121	12186	C7H16O	3,4-dimethyl-1-pentanol	6570-87-2	25	2.5806E+03	4.0094E+02	2.5873E+00	2.2265E-02	2
1122	12187	C7H16O	4,4-dimethyl-1-pentanol	3121-79-7	25	3.4615E+03	5.3820E+02	3.4735E+00	2.9891E-02	2
1123	12188	C7H16O	3-ethyl-2-pentanol	609-27-8	25	5.5451E+03	8.6371E+02	5.5761E+00	4.7985E-02	2
1124	12189	C7H16O	2,3-dimethyl-2-pentanol	4911-70-0	25	1.2041E+04	1.8858E+03	1.2187E+01	1.0488E-01	2
1125	12190	C7H16O	2,4-dimethyl-2-pentanol	625-06-9	25	1.7336E+04	2.7276E+03	1.7642E+01	1.5182E-01	2
1126	12191	C7H16O	3,3-dimethyl-2-pentanol	19781-24-9	25	7.4580E+03	1.1635E+03	7.5140E+00	6.4663E-02	2
1127	12192	C7H16O	3,4-dimethyl-2-pentanol	64502-86-9	25	5.2271E+03	8.1395E+02	5.2546E+00	4.5219E-02	2
1128	12193	C7H16O	4,4-dimethyl-2-pentanol	6144-93-0	25	1.2790E+04	2.0045E+03	1.2956E+01	1.1149E-01	2
1129	12194	C7H16O	3-ethyl-3-pentanol	597-49-9	25	9.7558E+03	1.5250E+03	9.8519E+00	8.4782E-02	2
1130	12195	C7H16O	2,2-dimethyl-3-pentanol	3970-62-5	25	1.4438E+04	2.2660E+03	1.4650E+01	1.2607E-01	2
1131	12196	C7H16O	2,3-dimethyl-3-pentanol	595-41-5	25	1.1336E+04	1.7745E+03	1.1466E+01	9.8674E-02	2
1132	12197	C7H16O	2,4-dimethyl-3-pentanol	600-36-2	25	1.2187E+04	1.9090E+03	1.2337E+01	1.0617E-01	2
1133	12198	C7H16O	2-methyl-2-ethyl-1-butanol	18371-13-6	25	4.1293E+03	6.4241E+02	4.1464E+00	3.5682E-02	2
1134	12199	C7H16O	3-methyl-2-ethyl-1-butanol	32444-34-1	25	3.0777E+03	4.7839E+02	3.0872E+00	2.6568E-02	2
1135	12200	C7H16O	2,2,3-trimethyl-1-butanol	55505-23-2	25	4.1293E+03	6.4241E+02	4.1464E+00	3.5682E-02	2
1136	12201	C7H16O	2,3,3-trimethyl-1-butanol	36794-64-6	25	3.4615E+03	5.3820E+02	3.4735E+00	2.9891E-02	2
1137	12202	C7H16O	2,3,3-trimethyl-2-butanol	594-83-2	25	1.9600E+04	3.0897E+03	1.9991E+01	1.7204E-01	2
1138	12203	C7H16O	2,4-dimethyl-1-pentanol, (±)	111768-02-6	25	3.6710E+03	5.7089E+02	3.6845E+00	3.1708E-02	2
1139	12204	C7H16O	2-heptanol, (±)	52390-72-4	25	3.6710E+03	5.7089E+02	3.6845E+00	3.1708E-02	2
1140	12205	C7H16O	3-heptanol, (S)	26549-25-7	25	4.1293E+03	6.4241E+02	4.1464E+00	3.5682E-02	2
1141	12206	C7H16O	2-methyl-1-hexanol, (±)	111768-04-8	25	2.7367E+03	4.2525E+02	2.7442E+00	2.3615E-02	2
1142	12207	C7H16O	3-methyl-1-hexanol, (±)	111768-08-2	25	1.9236E+03	2.9871E+02	1.9273E+00	1.6586E-02	2
1143	12208	C7H16O	4-methyl-1-hexanol, (±)	111768-05-9	25	1.6122E+03	2.5028E+02	1.6148E+00	1.3897E-02	2
1144	12209	C7H16O	2-methyl-3-hexanol, (±)	100295-82-7	25	9.4680E+03	1.4797E+03	9.5585E+00	8.2257E-02	2
1145	12210	C7H16O	5-methyl-3-hexanol, (±)	100295-83-8	25	7.4580E+03	1.1635E+03	7.5140E+00	6.4663E-02	2
1146	12211	C7H16O	isopropyl-tert-butyl-ether	17348-59-3	25	5.0000E+02	7.7548E+01	5.0025E-01	4.3050E-03	1
1147	12212	C7H16O	butyl propyl ether	3073-92-5	25	6.2015E+02	9.6193E+01	6.2054E-01	5.3401E-03	2
1148	12213	C7H16O	2-ethoxy-2-methylbutane	919-94-8	25	1.4018E+03	2.1759E+02	1.4038E+00	1.2081E-02	2
1149	12214	C7H16O	ethyl isopentyl ether	628-04-6	25	8.2357E+02	1.2777E+02	8.2425E-01	7.0931E-03	2
1150	12215	C7H16O	ethyl pentyl ether	17952-11-3	25	6.3606E+02	9.8662E+01	6.3647E-01	5.4772E-03	2
1151	12216	C7H16O	hexyl methyl ether	4747-07-3	25	4.1352E+02	6.4131E+01	4.1369E-01	3.5601E-03	2
1152	12217	C7H16O	1-(1-methylethoxy)butane	1860-27-1	25	1.0344E+03	1.6051E+02	1.0355E+00	8.9111E-03	2
1153	12218	C7H16O	2-methyl-1-propoxypropane	15268-49-2	25	1.2042E+03	1.8688E+02	1.2057E+00	1.0375E-02	2
1154	12219	C7H16O	2-methyl-2-propoxypropane	29072-93-3	25	1.5513E+03	2.4082E+02	1.5537E+00	1.3371E-02	2
1155	12220	C7H16O	(S)-(+)-2-heptanol	6033-23-4	25	5.0304E+03	7.8319E+02	5.0558E+00	4.3508E-02	2
1156	12221	C7H16O	(R)-(-)-2-heptanol	6033-24-5	25	7.1962E+03	1.1225E+03	7.2484E+00	6.2377E-02	2
1157	12300	C7H16S	heptyl mercaptan	1639-09-4	25	1.8959E+01	2.5823E+00	1.8960E-02	1.4334E-04	2
1158	12301	C7H16S	2-heptanethiol	1628-00-2	25	3.9137E+01	5.3307E+00	3.9139E-02	2.9590E-04	2
1159	12302	C7H16S	methyl hexyl sulfide	20291-60-5	25	7.5363E+01	1.0265E+01	7.5369E-02	5.6981E-04	2
1160	12303	C7H16S	ethyl pentyl sulfide	26158-99-6	25	7.5363E+01	1.0265E+01	7.5369E-02	5.6981E-04	2
1161	12304	C7H16S	propyl butyl sulfide	1613-46-3	25	1.2253E+02	1.6690E+01	1.2254E-01	9.2645E-04	2
1162	12315	C7H17N	heptylamine	111-68-2	25	8.1707E+03	1.2864E+03	8.2380E+00	7.1498E-02	2
1163	12316	C7H17N	2,4-dimethyl-2-pentanamine	64379-30-2	25	2.2629E+05	4.3730E+04	2.9247E+02	2.5384E+00	2
1164	12317	C7H17N	2-heptanamine	123-82-0	25	2.1900E+04	3.4887E+03	2.2391E+01	1.9433E-01	2
1165	12318	C7H17N	4-heptanamine	16751-59-0	25	2.6459E+04	4.2314E+03	2.7178E+01	2.3588E-01	2
1166	12319	C7H17N	3-methyl-1-hexanamine	65530-93-0	25	1.3472E+04	2.1306E+03	1.3656E+01	1.1852E-01	2
1167	12320	C7H17N	4-methyl-2-hexanamine	105-41-9	25	4.8463E+04	7.9005E+03	5.0932E+01	4.4204E-01	2
1168	12321	C7H17N	4-methyl-1-hexylamine	34263-68-8	25	1.0409E+04	1.6419E+03	1.0519E+01	9.1292E-02	2
1169	12322	C7H17N	methylhexylamine	35161-70-7	25	high	high	high	high	2
1170	12323	C7H17N	ethylpentylamine	17839-26-8	25	high	high	high	high	2
1171	12324	C7H17N	dimethylpentylamine	26153-88-8	25	high	high	high	high	2
1172	12325	C7H17N	dimethyl-2-pentylamine	57303-85-2	25	high	high	high	high	2
1173	12326	C7H17N	dimethyl-3-pentylamine	18636-94-7	25	high	high	high	high	2
1174	12327	C7H17N	dimethyl-2-methylbutylamine	66225-39-6	25	high	high	high	high	2
1175	12328	C7H17N	dimethyl-3-methylbutylamine	2315-43-7	25	high	high	high	high	2
1176	12329	C7H17N	dimethyl-1,1-dimethylpropylamine	57757-60-5	25	high	high	high	high	2
1177	12330	C7H17N	dimethyl-1,2-dimethylpropylamine	66225-38-5	25	high	high	high	high	2
1178	12331	C7H17N	dimethyl-2,2-dimethylpropylamine	10076-31-0	25	high	high	high	high	2
1179	12332	C7H17N	methylethylbutylamine	66225-40-9	25	high	high	high	high	2
1180	12333	C7H17N	methylethyl-sec-butylamine	66225-41-0	25	high	high	high	high	2
1181	12334	C7H17N	methylethylisobutylamine	60247-14-5	25	high	high	high	high	2
1182	12335	C7H17N	methylethyl-tert-butylamine	52841-28-8	25	high	high	high	high	2
1183	12336	C7H17N	methyldipropylamine	3405-42-3	25	high	high	high	high	2
1184	12337	C7H17N	methylpropylisopropylamine	66225-42-1	25	high	high	high	high	2
1185	12338	C7H17N	methyldiisopropylamine	10342-97-9	25	high	high	high	high	2
1186	12339	C7H17N	diethylpropylamine	4458-31-5	25	high	high	high	high	2
1187	12340	C7H17N	diethylisopropylamine	6006-15-1	25	high	high	high	high	2
1188	12341	C7H17N	N-tert-butylisopropylamine	7515-80-2	25	high	high	high	high	2
1189	12342	C7H17N	(S)-2-heptylamine	44745-29-1	25	2.0360E+04	3.2390E+03	2.0783E+01	1.8038E-01	2
1190	12343	C7H17N	(R)-2-heptylamine	6240-90-0	25	2.0360E+04	3.2390E+03	2.0783E+01	1.8038E-01	2
1191	12344	C7H17N	N-propylbutylamine	20193-21-9	25	high	high	high	high	2
1192	12345	C7H17N	3-aminoheptane	28292-42-4	25	2.3592E+04	3.7636E+03	2.4162E+01	2.0970E-01	2
1193	12346	C7H17N	5-methyl-2-hexylamine	28292-43-5	25	1.3775E+05	2.4371E+04	1.5976E+02	1.3866E+00	2
1194	12594	C8H4O3	phthalic anhydride	85-44-9	25	6.0000E+03	7.3362E+02	6.0362E+00	4.0753E-02	1
1195	12713	C8H6	ethynylbenzene	536-74-3	25	4.5620E+02	8.0496E+01	4.5641E-01	4.4686E-03	1
1196	12934	C8H6O4	isophthalic acid	121-91-5	25	1.3000E+02	1.4098E+01	1.3002E-01	7.8261E-04	1
1197	12935	C8H6O4	phthalic acid	88-99-3	25	1.4150E+03	1.5363E+02	1.4170E+00	8.5293E-03	1
1198	12936	C8H6O4	terephthalic acid	100-21-0	20	1.5000E+01	1.6266E+00	1.5000E-02	9.0290E-05	1
1199	12947	C8H6S	benzothiophene	95-15-8	20	1.3018E+02	1.7477E+01	1.3020E-01	9.7016E-04	1
1200	13107	C8H7N	indole	120-72-9	20	1.8744E+03	2.8870E+02	1.8779E+00	1.6030E-02	1

Table 3 Solubility in Water - Organic Compounds

					Solubility in Water, S (ppm - parts per million, wt - weight, g - gram)					
NO	ID	FORMULA	NAME	CAS No	T, C	S@T, ppm(wt)	S@T, ppm(mol)	S@T, g/1000 g of H2O	S@T, mol/1000 g of H2O	code
1201	13237	C8H8	styrene	100-42-5	25	3.2160E+02	5.5642E+01	3.2170E-01	3.0888E-03	1
1202	13391	C8H8O	acetophenone	98-86-2	25	6.8420E+03	1.0319E+03	6.8891E+00	5.7337E-02	1
1203	13392	C8H8O	p-tolualdehyde	104-87-0	25	2.2710E+03	3.4116E+02	2.2762E+00	1.8944E-02	1
1204	13412	C8H8O2	methyl benzoate	93-58-3	25	1.9720E+03	2.6138E+02	1.9759E+00	1.4513E-02	1
1205	13413	C8H8O2	o-toluic acid	118-90-1	25	1.1850E+03	1.5696E+02	1.1864E+00	8.7139E-03	1
1206	13414	C8H8O2	p-toluic acid	99-94-5	25	4.0000E+02	5.2945E+01	4.0016E-01	2.9391E-03	1
1207	13445	C8H8O3	methyl salicylate	119-36-8	30	7.0000E+02	8.2933E+01	7.0049E-01	4.6040E-03	1
1208	13446	C8H8O3	vanillin	121-33-5	25	1.1020E+04	1.3176E+03	1.1143E+01	7.3236E-02	1
1209	13671	C8H9NO	acetanilide	103-84-4	25	6.1000E+03	8.1733E+02	6.1374E+00	4.5407E-02	1
1210	13789	C8H10	ethylbenzene	100-41-4	25	1.6510E+02	2.8019E+01	1.6513E-01	1.5553E-03	1
1211	13790	C8H10	o-xylene	95-47-6	25	2.2080E+02	3.7473E+01	2.2085E-01	2.0802E-03	1
1212	13791	C8H10	m-xylene	108-38-3	25	1.7400E+02	2.9529E+01	1.7403E-01	1.6392E-03	1
1213	13792	C8H10	p-xylene	106-42-3	25	2.0170E+02	3.4231E+01	2.0174E-01	1.9002E-03	1
1214	13793	C8H10	xylenes	1330-20-7	25	2.6997E+02	4.5820E+01	2.7004E-01	2.5436E-03	2
1215	13913	C8H10O	2-phenylethanol	60-12-8	25	4.9000E+03	7.2560E+02	4.9241E+00	4.0307E-02	1
1216	13918	C8H10O	o-ethylphenol	90-00-6	25	5.2797E+03	7.8207E+02	5.3077E+00	4.3446E-02	2
1217	13919	C8H10O	m-ethylphenol	620-17-7	25	5.2797E+03	7.8207E+02	5.3077E+00	4.3446E-02	2
1218	13920	C8H10O	p-ethylphenol	123-07-9	25	5.2797E+03	7.8207E+02	5.3077E+00	4.3446E-02	2
1219	13921	C8H10O	2,3-xylenol	526-75-0	25	4.5690E+03	6.7639E+02	4.5900E+00	3.7571E-02	1
1220	13922	C8H10O	2,4-xylenol	105-67-9	25	7.8680E+03	1.1681E+03	7.9304E+00	6.4914E-02	1
1221	13923	C8H10O	2,5-xylenol	95-87-4	25	3.5430E+03	5.2404E+02	3.5556E+00	2.9104E-02	1
1222	13924	C8H10O	2,6-xylenol	576-26-1	25	6.0470E+03	8.9633E+02	6.0838E+00	4.9799E-02	1
1223	13925	C8H10O	3,4-xylenol	95-65-8	25	4.7640E+03	7.0538E+02	4.7868E+00	3.9183E-02	1
1224	13926	C8H10O	3,5-xylenol	108-68-9	25	4.8870E+03	7.2366E+02	4.9110E+00	4.0199E-02	1
1225	13934	C8H10O	dimethyl phenol	1300-71-6	25	5.2797E+03	7.8207E+02	5.3077E+00	4.3446E-02	2
1226	14082	C8H11N	N,N-dimethylaniline	121-69-7	25	1.1050E+03	1.6442E+02	1.1062E+00	9.1286E-03	1
1227	14083	C8H11N	o-ethylaniline	578-54-1	25	1.1050E+03	1.6442E+02	1.1062E+00	9.1286E-03	2
1228	14084	C8H11N	N,2-dimethylaniline	611-21-2	25	1.1050E+03	1.6442E+02	1.1062E+00	9.1286E-03	2
1229	14085	C8H11N	N,3-dimethylaniline	696-44-6	25	1.1050E+03	1.6442E+02	1.1062E+00	9.1286E-03	2
1230	14086	C8H11N	N,4-dimethylaniline	623-08-5	25	1.1050E+03	1.6442E+02	1.1062E+00	9.1286E-03	2
1231	14087	C8H11N	2,3-dimethylaniline	87-59-2	25	1.1050E+03	1.6442E+02	1.1062E+00	9.1286E-03	2
1232	14088	C8H11N	2,4-dimethylaniline	95-68-1	25	1.1050E+03	1.6442E+02	1.1062E+00	9.1286E-03	2
1233	14089	C8H11N	2,5-dimethylaniline	95-78-3	25	1.1050E+03	1.6442E+02	1.1062E+00	9.1286E-03	2
1234	14090	C8H11N	2,6-dimethylaniline	87-62-7	25	1.1050E+03	1.6442E+02	1.1062E+00	9.1286E-03	2
1235	14091	C8H11N	3,4-dimethylaniline	95-64-7	25	1.1050E+03	1.6442E+02	1.1062E+00	9.1286E-03	2
1236	14092	C8H11N	3,5-dimethylaniline	108-69-0	25	1.1050E+03	1.6442E+02	1.1062E+00	9.1286E-03	2
1237	14093	C8H11N	m-ethylaniline	587-02-0	25	1.1050E+03	1.6442E+02	1.1062E+00	9.1286E-03	2
1238	14094	C8H11N	p-ethylaniline	589-16-2	25	1.1050E+03	1.6442E+02	1.1062E+00	9.1286E-03	2
1239	14095	C8H11N	N-ethylaniline	103-69-5	25	1.1050E+03	1.6442E+02	1.1062E+00	9.1286E-03	2
1240	14231	C8H12	vinylcyclohexene	100-40-3	25	5.0000E+01	8.3265E+00	5.0003E-02	4.6220E-04	1
1241	14379	C8H12O4	1,4-cyclohexanedicarboxylic acid	619-82-9	17	8.0000E+02	8.3763E+01	8.0064E-01	4.6500E-03	1
1242	14410	C8H12S	2-butylthiophene	1455-20-5	25	1.6590E+01	2.1310E+00	1.6590E-02	1.1829E-04	2
1243	14411	C8H12S	3-butylthiophene	34722-01-5	25	1.6143E+01	2.0736E+00	1.6144E-02	1.1511E-04	2
1244	14412	C8H12S	2-(1,1-dimethylethyl)thiophene	1689-78-7	25	4.2973E+01	5.5201E+00	4.2975E-02	3.0642E-04	2
1245	14413	C8H12S	3-(1,1-dimethylethyl)thiophene	1689-79-8	25	3.2771E+01	4.2095E+00	3.2772E-02	2.3367E-04	2
1246	14484	C8H14	1-octyne	629-05-0	25	2.4000E+01	3.9235E+00	2.4001E-02	2.1779E-04	1
1247	14485	C8H14	2-octyne	2809-67-8	25	1.5052E+01	2.4607E+00	1.5052E-02	1.3659E-04	2
1248	14486	C8H14	3-octyne	15232-76-5	25	1.9731E+01	3.2256E+00	1.9732E-02	1.7905E-04	2
1249	14487	C8H14	4-octyne	1942-45-6	25	2.1608E+01	3.5324E+00	2.1608E-02	1.9608E-04	2
1250	14488	C8H14	3-ethyl-3-methyl-1-pentyne	919-12-0	25	1.2388E+02	2.0253E+01	1.2389E-01	1.1243E-03	2
1251	14493	C8H14	1,7-octadiene	3710-30-3	25	1.0445E+01	1.7076E+00	1.0445E-02	9.4786E-05	2
1252	14494	C8H14	2,6-octadiene	4974-27-0	25	6.1698E+00	1.0086E+00	6.1698E-03	5.5988E-05	2
1253	14495	C8H14	3-methyl-1,5-heptadiene	4894-62-6	25	1.3553E+01	2.2157E+00	1.3553E-02	1.2299E-04	2
1254	14496	C8H14	2,5-dimethyl-1,5-hexadiene	627-58-7	25	1.1199E+01	1.8308E+00	1.1199E-02	1.0163E-04	2
1255	14497	C8H14	2,5-dimethyl-2,4-hexadiene	764-13-6	25	3.4222E+00	5.5946E-01	3.4222E-03	3.1055E-05	2
1256	14861	C8H16	propylcyclopentane	2040-96-2	25	2.0400E+00	3.2750E-01	2.0400E-03	1.8179E-05	1
1257	14862	C8H16	isopropylcyclopentane	3875-51-2	25	2.5249E+00	4.0534E-01	2.5249E-03	2.2500E-05	2
1258	14863	C8H16	1-methyl-1-ethylcyclopentane	16747-50-5	25	3.3671E+00	5.4055E-01	3.3671E-03	3.0006E-05	2
1259	14864	C8H16	1-methyl-cis-2-ethylcyclopentane	930-89-2	25	2.2936E+00	3.6821E-01	2.2936E-03	2.0439E-05	2
1260	14865	C8H16	1-methyl-trans-2-ethylcyclopentane	930-90-5	25	3.4308E+00	5.5079E-01	3.4308E-03	3.0574E-05	2
1261	14866	C8H16	1-methyl-cis-3-ethylcyclopentane	2613-66-3	25	3.4510E+00	5.5402E-01	3.4510E-03	3.0754E-05	2
1262	14867	C8H16	1-methyl-trans-3-ethylcyclopentane	2613-65-2	25	3.4510E+00	5.5402E-01	3.4510E-03	3.0754E-05	2
1263	14868	C8H16	1,1,2-trimethylcyclopentane	4259-00-1	25	5.3034E+00	8.5142E-01	5.3035E-03	4.7262E-05	2
1264	14869	C8H16	1,1,3-trimethylcyclopentane	4516-69-2	25	3.7300E+00	5.9882E-01	3.7300E-03	3.3240E-05	1
1265	14870	C8H16	1,cis-2,cis-3-trimethylcyclopentane	2613-69-6	25	3.0871E+00	4.9560E-01	3.0871E-03	2.7510E-05	2
1266	14871	C8H16	1,cis-2,trans-3-trimethylcyclopentane	15890-40-1	25	4.2594E+00	6.8380E-01	4.2594E-03	3.7957E-05	2
1267	14872	C8H16	1,trans-2,cis-3-trimethylcyclopentane	19374-46-0	25	6.5020E+00	1.0438E+00	6.5021E-03	5.7943E-05	2
1268	14873	C8H16	1,cis-2,cis-4-trimethylcyclopentane	2613-72-1	25	4.3854E+00	7.0403E-01	4.3854E-03	3.9080E-05	2
1269	14874	C8H16	1,cis-2,trans-4-trimethylcyclopentane	4850-28-6	25	4.4546E+00	7.1515E-01	4.4546E-03	3.9697E-05	2
1270	14875	C8H16	1,trans-2,cis-4-trimethylcyclopentane	16883-48-0	25	6.8511E+00	1.0999E+00	6.8511E-03	6.1053E-05	2
1271	14876	C8H16	1-methyl-3-ethylcyclopentane	3726-47-4	25	3.4611E+00	5.5565E-01	3.4611E-03	3.0844E-05	2
1272	14877	C8H16	ethylcyclohexane	1678-91-7	25	6.3000E+00	1.0114E+00	6.3000E-03	5.6143E-05	1
1273	14878	C8H16	1,1-dimethylcyclohexane	590-66-9	25	5.9890E+00	9.6147E-01	5.9890E-03	5.3371E-05	2
1274	14879	C8H16	1,2-dimethylcyclohexane, (cis+trans)	583-57-3	25	4.1258E+00	6.6235E-01	4.1258E-03	3.6767E-05	2
1275	14880	C8H16	cis-1,2-dimethylcyclohexane	2207-01-4	25	6.0000E+00	9.6324E-01	6.0000E-03	5.3469E-05	1
1276	14881	C8H16	trans-1,2-dimethylcyclohexane	6876-23-9	25	5.2000E+00	8.3481E-01	5.2000E-03	4.6340E-05	1
1277	14882	C8H16	1,3-dimethylcyclohexane, cis and trans	591-21-9	25	5.8504E+00	9.3924E-01	5.8505E-03	5.2136E-05	2
1278	14883	C8H16	cis-1,3-dimethylcyclohexane	638-04-0	25	5.8027E+00	9.3157E-01	5.8027E-03	5.1711E-05	2
1279	14884	C8H16	trans-1,3-dimethylcyclohexane	2207-03-6	25	4.4905E+00	7.2090E-01	4.4905E-03	4.0017E-05	2
1280	14885	C8H16	1,4-dimethylcyclohexane, (cis+trans)	589-90-2	25	4.9650E+00	7.9708E-01	4.9650E-03	4.4246E-05	2

Table 3 Solubility in Water - Organic Compounds

					Solubility in Water, S (ppm - parts per million, wt - weight, g - gram)					
NO	ID	FORMULA	NAME	CAS No	T, C	S@T, ppm(wt)	S@T, ppm(mol)	S@T, g/1000 g of H2O	S@T, mol/1000 g of H2O	code
1281	14886	C8H16	cis-1,4-dimethylcyclohexane	624-29-3	25	4.5276E+00	7.2686E-01	4.5276E-03	4.0348E-05	2
1282	14887	C8H16	trans-1,4-dimethylcyclohexane	2207-04-7	25	3.8400E+00	6.1648E-01	3.8400E-03	3.4220E-05	1
1283	14888	C8H16	cyclooctane	292-64-8	25	7.9000E+00	1.2683E+00	7.9001E-03	7.0401E-05	1
1284	14890	C8H16	1-octene	111-66-0	25	4.1000E+00	6.5822E-01	4.1000E-03	3.6537E-05	1
1285	14891	C8H16	oct-2-ene	111-67-1	25	2.0467E+00	3.2858E-01	2.0467E-03	1.8239E-05	2
1286	14892	C8H16	cis-2-octene	7642-04-8	25	1.9595E+00	3.1458E-01	1.9595E-03	1.7462E-05	2
1287	14893	C8H16	trans-2-octene	13389-42-9	25	2.0347E+00	3.2665E-01	2.0347E-03	1.8132E-05	2
1288	14894	C8H16	cis-3-octene	14850-22-7	25	2.3019E+00	3.6955E-01	2.3019E-03	2.0514E-05	2
1289	14895	C8H16	trans-3-octene	14919-01-8	25	2.2485E+00	3.6098E-01	2.2485E-03	2.0038E-05	2
1290	14896	C8H16	oct-4-ene	592-99-4	25	2.0527E+00	3.2955E-01	2.0527E-03	1.8293E-05	2
1291	14897	C8H16	cis-4-octene	7642-15-1	25	2.3511E+00	3.7745E-01	2.3511E-03	2.0952E-05	2
1292	14898	C8H16	trans-4-octene	14850-23-8	25	2.3900E+00	3.8370E-01	2.3901E-03	2.1299E-05	2
1293	14899	C8H16	2-methyl-1-heptene	15870-10-7	25	2.8425E+00	4.5634E-01	2.8425E-03	2.5331E-05	2
1294	14900	C8H16	3-methyl-1-heptene	4810-09-7	25	4.6030E+00	7.3898E-01	4.6031E-03	4.1020E-05	2
1295	14901	C8H16	4-methyl-1-heptene	13151-05-8	25	4.1487E+00	6.6604E-01	4.1487E-03	3.6971E-05	2
1296	14902	C8H16	5-methyl-1-heptene	13151-04-7	25	4.0304E+00	6.4704E-01	4.0304E-03	3.5917E-05	2
1297	14903	C8H16	6-methyl-1-heptene	5026-76-6	25	4.0538E+00	6.5080E-01	4.0538E-03	3.6125E-05	2
1298	14904	C8H16	2-methyl-2-heptene	627-97-4	25	2.3428E+00	3.7612E-01	2.3428E-03	2.0878E-05	2
1299	14905	C8H16	3-methyl-cis-2-heptene	22768-19-0	25	2.4268E+00	3.8960E-01	2.4268E-03	2.1626E-05	2
1300	14906	C8H16	3-methyl-trans-2-heptene	22768-20-3	25	2.4268E+00	3.8960E-01	2.4268E-03	2.1626E-05	2
1301	14907	C8H16	4-methyl-cis-2-heptene	66225-16-9	25	3.8702E+00	6.2133E-01	3.8702E-03	3.4489E-05	2
1302	14908	C8H16	4-methyl-trans-2-heptene	66225-17-0	25	3.8702E+00	6.2133E-01	3.8702E-03	3.4489E-05	2
1303	14909	C8H16	5-methyl-cis-2-heptene	24608-84-2	25	3.0668E+00	4.9234E-01	3.0668E-03	2.7329E-05	2
1304	14910	C8H16	5-methyl-trans-2-heptene	24608-85-3	25	3.0668E+00	4.9234E-01	3.0668E-03	2.7329E-05	2
1305	14911	C8H16	6-methyl-2-heptene	73548-72-8	25	3.2509E+00	5.2190E-01	3.2509E-03	2.8970E-05	2
1306	14912	C8H16	6-methyl-cis-2-heptene	66225-18-1	25	3.2509E+00	5.2190E-01	3.2509E-03	2.8970E-05	2
1307	14913	C8H16	6-methyl-trans-2-heptene	51065-65-7	25	3.2509E+00	5.2190E-01	3.2509E-03	2.8970E-05	2
1308	14914	C8H16	6-methyl-3-heptene	3404-57-7	25	3.6520E+00	5.8630E-01	3.6521E-03	3.2545E-05	2
1309	14915	C8H16	2-methyl-3-heptene	17618-76-7	25	4.3450E+00	6.9755E-01	4.3450E-03	3.8721E-05	2
1310	14916	C8H16	2-methyl-cis-3-heptene	20488-34-0	25	4.3450E+00	6.9755E-01	4.3450E-03	3.8721E-05	2
1311	14917	C8H16	2-methyl-trans-3-heptene	692-96-6	25	4.3450E+00	6.9755E-01	4.3450E-03	3.8721E-05	2
1312	14918	C8H16	3-methyl-3-heptene	7300-03-0	25	2.5733E+00	4.1312E-01	2.5733E-03	2.2932E-05	2
1313	14919	C8H16	3-methyl-cis-3-heptene	22768-17-8	25	2.5733E+00	4.1312E-01	2.5733E-03	2.2932E-05	2
1314	14920	C8H16	3-methyl-trans-3-heptene	22768-18-9	25	2.5733E+00	4.1312E-01	2.5733E-03	2.2932E-05	2
1315	14921	C8H16	4-methyl-3-heptene	4485-16-9	25	2.5733E+00	4.1312E-01	2.5733E-03	2.2932E-05	2
1316	14922	C8H16	4-methyl-cis-3-heptene	14255-24-4	25	2.4268E+00	3.8960E-01	2.4268E-03	2.1626E-05	2
1317	14923	C8H16	4-methyl-trans-3-heptene	13714-85-7	25	2.4268E+00	3.8960E-01	2.4268E-03	2.1626E-05	2
1318	14924	C8H16	5-methyl-3-heptene	13172-91-3	25	4.3450E+00	6.9755E-01	4.3450E-03	3.8721E-05	2
1319	14925	C8H16	5-methyl-cis-3-heptene	50422-80-5	25	4.3450E+00	6.9755E-01	4.3450E-03	3.8721E-05	2
1320	14926	C8H16	5-methyl-trans-3-heptene	53510-18-2	25	4.3450E+00	6.9755E-01	4.3450E-03	3.8721E-05	2
1321	14927	C8H16	6-methyl-cis-3-heptene	66225-19-2	25	3.6520E+00	5.8630E-01	3.6521E-03	3.2545E-05	2
1322	14928	C8H16	6-methyl-trans-3-heptene	66225-20-5	25	3.6520E+00	5.8630E-01	3.6521E-03	3.2545E-05	2
1323	14929	C8H16	2-ethyl-1-hexene	1632-16-2	25	2.7285E+00	4.3803E-01	2.7285E-03	2.4315E-05	2
1324	14930	C8H16	3-ethyl-1-hexene	3404-58-8	25	4.7924E+00	7.6937E-01	4.7924E-03	4.2707E-05	2
1325	14931	C8H16	4-ethyl-1-hexene	16746-85-3	25	4.8758E+00	7.8276E-01	4.8758E-03	4.3450E-05	2
1326	14932	C8H16	2,3-dimethyl-1-hexene	16746-86-4	25	4.7375E+00	7.6056E-01	4.7375E-03	4.2218E-05	2
1327	14933	C8H16	2,4-dimethyl-1-hexene	16746-87-5	25	4.5503E+00	7.3051E-01	4.5503E-03	4.0550E-05	2
1328	14934	C8H16	2,5-dimethyl-1-hexene	6975-92-4	25	4.4465E+00	7.1385E-01	4.4465E-03	3.9625E-05	2
1329	14935	C8H16	3,3-dimethyl-1-hexene	3404-77-1	25	6.8668E+00	1.1024E+00	6.8669E-03	6.1194E-05	2
1330	14936	C8H16	3,4-dimethyl-1-hexene	16745-94-1	25	4.3450E+00	6.9755E-01	4.3450E-03	3.8721E-05	2
1331	14937	C8H16	3,5-dimethyl-1-hexene	7423-69-0	25	6.8668E+00	1.1024E+00	6.8669E-03	6.1194E-05	2
1332	14938	C8H16	4,4-dimethyl-1-hexene	1647-08-1	25	5.7244E+00	9.1900E-01	5.7244E-03	5.1013E-05	2
1333	14939	C8H16	4,5-dimethyl-1-hexene	16106-59-5	25	5.1640E+00	8.2903E-01	5.1640E-03	4.6019E-05	2
1334	14940	C8H16	5,5-dimethyl-1-hexene	7116-86-1	25	7.4740E+00	1.1999E+00	7.4740E-03	6.6605E-05	2
1335	14941	C8H16	3-ethyl-cis-2-hexene	36880-72-5	25	2.5733E+00	4.1312E-01	2.5733E-03	2.2932E-05	2
1336	14942	C8H16	3-ethyl-trans-2-hexene	66225-15-8	25	2.5733E+00	4.1312E-01	2.5733E-03	2.2932E-05	2
1337	14943	C8H16	4-ethyl-cis-2-hexene	54616-49-8	25	4.1010E+00	6.5838E-01	4.1010E-03	3.6546E-05	2
1338	14944	C8H16	4-ethyl-trans-2-hexene	19781-63-6	25	4.1010E+00	6.5838E-01	4.1010E-03	3.6546E-05	2
1339	14945	C8H16	2,3-dimethyl-2-hexene	7145-20-2	25	2.4597E+00	3.9489E-01	2.4597E-03	2.1920E-05	2
1340	14946	C8H16	2,4-dimethyl-2-hexene	14255-23-3	25	4.7103E+00	7.5620E-01	4.7103E-03	4.1976E-05	2
1341	14947	C8H16	2,5-dimethyl-2-hexene	3404-78-2	25	4.2951E+00	6.8954E-01	4.2951E-03	3.8276E-05	2
1342	14948	C8H16	3,4-dimethyl-cis-2-hexene	19550-81-3	25	3.4458E+00	5.5319E-01	3.4458E-03	3.0707E-05	2
1343	14949	C8H16	3,4-dimethyl-trans-2-hexene	19550-82-4	25	3.4458E+00	5.5319E-01	3.4458E-03	3.0707E-05	2
1344	14950	C8H16	3,5-dimethyl-cis-2-hexene	66225-31-8	25	4.3450E+00	6.9755E-01	4.3450E-03	3.8721E-05	2
1345	14951	C8H16	3,5-dimethyl-trans-2-hexene	66225-12-5	25	4.3450E+00	6.9755E-01	4.3450E-03	3.8721E-05	2
1346	14952	C8H16	4,4-dimethyl-cis-2-hexene	19550-83-5	25	6.1298E+00	9.8408E-01	6.1298E-03	5.4626E-05	2
1347	14953	C8H16	4,4-dimethyl-trans-2-hexene	---	25	6.1298E+00	9.8408E-01	6.1298E-03	5.4626E-05	2
1348	14954	C8H16	4,5-dimethyl-2-hexene	73548-71-7	25	4.8758E+00	7.8276E-01	4.8758E-03	4.3450E-05	2
1349	14955	C8H16	4,5-dimethyl-cis-2-hexene	65036-71-7	25	4.8758E+00	7.8276E-01	4.8758E-03	4.3450E-05	2
1350	14956	C8H16	4,5-dimethyl-trans-2-hexene	66225-14-7	25	4.8758E+00	7.8276E-01	4.8758E-03	4.3450E-05	2
1351	14957	C8H16	5,5-dimethyl-cis-2-hexene	39761-61-0	25	5.8233E+00	9.3487E-01	5.8233E-03	5.1894E-05	2
1352	14958	C8H16	5,5-dimethyl-trans-2-hexene	39782-43-9	25	6.8280E+00	1.0962E+00	6.8281E-03	6.0848E-05	2
1353	14959	C8H16	3-ethyl-3-hexene	16789-51-8	25	3.4458E+00	5.5319E-01	3.4458E-03	3.0707E-05	2
1354	14960	C8H16	2,2-dimethyl-cis-3-hexene	690-92-6	25	6.3318E+00	1.0165E+00	6.3318E-03	5.6426E-05	2
1355	14961	C8H16	2,2-dimethyl-trans-3-hexene	690-93-7	25	8.2004E+00	1.3165E+00	8.2004E-03	7.3078E-05	2
1356	14962	C8H16	2,3-dimethyl-cis-3-hexene	59643-75-3	25	3.8702E+00	6.2133E-01	3.8702E-03	3.4489E-05	2
1357	14963	C8H16	2,3-dimethyl-trans-3-hexene	66225-30-7	25	3.8702E+00	6.2133E-01	3.8702E-03	3.4489E-05	2
1358	14964	C8H16	2,4-dimethyl-cis-3-hexene	37549-89-6	25	5.1640E+00	8.2903E-01	5.1640E-03	4.6019E-05	2
1359	14965	C8H16	2,4-dimethyl-trans-3-hexene	61847-78-7	25	5.5950E+00	8.9823E-01	5.5951E-03	4.9860E-05	2
1360	14966	C8H16	2,5-dimethyl-cis-3-hexene	10557-44-5	25	7.6874E+00	1.2341E+00	7.6875E-03	6.8507E-05	2

Table 3 Solubility in Water - Organic Compounds

NO	ID	FORMULA	NAME	CAS No	Solubility in Water, S (ppm - parts per million, wt - weight, g - gram) T, C	S@T, ppm(wt)	S@T, ppm(mol)	S@T, g/1000 g of H2O	S@T, mol/1000 g of H2O	code
1361	14967	C8H16	2,5-dimethyl-trans-3-hexene	692-70-6	25	7.6874E+00	1.2341E+00	7.6875E-03	6.8507E-05	2
1362	14968	C8H16	3,4-dimethyl-cis-3-hexene	19550-87-9	25	2.4268E+00	3.8960E-01	2.4268E-03	2.1626E-05	2
1363	14969	C8H16	3,4-dimethyl-trans-3-hexene	19550-88-0	25	2.4268E+00	3.8960E-01	2.4268E-03	2.1626E-05	2
1364	14970	C8H16	2-propyl-1-pentene	15918-08-8	25	3.1209E+00	5.0103E-01	3.1209E-03	2.7812E-05	2
1365	14971	C8H16	2-isopropyl-1-pentene	16746-02-4	25	4.1010E+00	6.5838E-01	4.1010E-03	3.6546E-05	2
1366	14972	C8H16	3-methyl-2-ethyl-1-pentene	3404-67-9	25	4.4722E+00	7.1798E-01	4.4723E-03	3.9854E-05	2
1367	14973	C8H16	4-methyl-2-ethyl-1-pentene	3404-80-6	25	4.7924E+00	7.6937E-01	4.7924E-03	4.2707E-05	2
1368	14974	C8H16	2-methyl-3-ethyl-1-pentene	19780-66-6	25	4.8758E+00	7.8276E-01	4.8758E-03	4.3450E-05	2
1369	14975	C8H16	3-methyl-3-ethyl-1-pentene	6196-60-7	25	4.3450E+00	6.9755E-01	4.3450E-03	3.8721E-05	2
1370	14976	C8H16	4-methyl-3-ethyl-1-pentene	61847-80-1	25	5.6271E+00	9.0338E-01	5.6271E-03	5.0146E-05	2
1371	14977	C8H16	2,3,3-trimethyl-1-pentene	560-23-6	25	5.3723E+00	8.6247E-01	5.3723E-03	4.7875E-05	2
1372	14978	C8H16	2,3,4-trimethyl-1-pentene	565-76-4	25	5.4685E+00	8.7791E-01	5.4685E-03	4.8732E-05	2
1373	14979	C8H16	2,4,4-trimethyl-1-pentene	107-39-1	25	7.9333E+00	1.2736E+00	7.9334E-03	7.0698E-05	2
1374	14980	C8H16	3,3,4-trimethyl-1-pentene	560-22-5	25	6.4884E+00	1.0416E+00	6.4884E-03	5.7821E-05	2
1375	14981	C8H16	3,4,4-trimethyl-1-pentene	564-03-4	25	6.8668E+00	1.1024E+00	6.8669E-03	6.1194E-05	2
1376	14982	C8H16	2-methyl-3-ethyl-2-pentene	19780-67-7	25	3.2509E+00	5.2190E-01	3.2509E-03	2.8970E-05	2
1377	14983	C8H16	4-methyl-3-ethyl-cis-2-pentene	42067-48-1	25	3.4458E+00	5.5319E-01	3.4458E-03	3.0707E-05	2
1378	14984	C8H16	4-methyl-3-ethyl-trans-2-pentene	42067-49-2	25	3.8035E+00	6.1061E-01	3.8035E-03	3.3895E-05	2
1379	14985	C8H16	2,3,4-trimethyl-2-pentene	565-77-5	25	3.3941E+00	5.4488E-01	3.3941E-03	3.0246E-05	2
1380	14986	C8H16	2,4,4-trimethyl-2-pentene	107-40-4	25	6.5216E+00	1.0470E+00	6.5216E-03	5.8117E-05	2
1381	14987	C8H16	3,4,4-trimethyl-cis-2-pentene	39761-64-3	25	4.3450E+00	6.9755E-01	4.3450E-03	3.8721E-05	2
1382	14988	C8H16	3,4,4-trimethyl-trans-2-pentene	39761-57-4	25	4.3450E+00	6.9755E-01	4.3450E-03	3.8721E-05	2
1383	14989	C8H16	3,4,4-trimethyl-2-pentene	598-96-9	25	4.3450E+00	6.9755E-01	4.3450E-03	3.8721E-05	2
1384	14990	C8H16	2,4,4-trimethylpentene	25167-70-8	25	7.2662E+00	1.1665E+00	7.2662E-03	6.4753E-05	2
1385	14991	C8H16	3-methyl-2-isopropyl-1-butene	111823-35-9	25	6.8668E+00	1.1024E+00	6.8669E-03	6.1194E-05	2
1386	14992	C8H16	3,3-dimethyl-2-ethyl-1-butene	18231-53-3	25	4.8758E+00	7.8276E-01	4.8758E-03	4.3450E-05	2
1387	14994	C8H16Br2	1,1-dibromooctane	62168-26-7	25	1.2885E+01	8.5334E-01	1.2885E-02	4.7369E-05	2
1388	14995	C8H16Br2	1,2-dibromooctane	6269-92-7	25	1.3541E+01	8.9678E-01	1.3541E-02	4.9780E-05	2
1389	14996	C8H16Br2	1,4-dibromooctane	70690-24-3	25	8.1752E+00	5.4142E-01	8.1753E-03	3.0054E-05	2
1390	14997	C8H16Br2	1,5-dibromooctane	17912-17-3	25	8.1752E+00	5.4142E-01	8.1753E-03	3.0054E-05	2
1391	14998	C8H16Br2	1,8-dibromooctane	4549-32-0	25	3.3193E+00	2.1983E-01	3.3194E-03	1.2202E-05	2
1392	14999	C8H16Br2	4,5-dibromooctane	61539-75-1	25	8.1752E+00	5.4142E-01	8.1753E-03	3.0054E-05	2
1393	15005	C8H16Cl2	1,1-dichlorooctane	20395-24-8	25	6.8498E+00	6.7388E-01	6.8499E-03	3.7406E-05	2
1394	15006	C8H16Cl2	2,5-dichloro-2,5-dimethylhexane	6223-78-5	25	3.9538E+00	3.8897E-01	3.9538E-03	2.1591E-05	2
1395	15007	C8H16Cl2	1,8-dichlorooctane	2162-99-4	25	1.2350E+00	1.2149E-01	1.2350E-03	6.7440E-06	2
1396	15008	C8H16Cl2	2,3-dichlorooctane	21948-47-0	25	7.6260E+00	7.5024E-01	7.6261E-03	4.1645E-05	2
1397	15062	C8H16O	2-ethylhexanal	123-05-7	25	7.5232E+02	1.0577E+02	7.5288E-01	5.8721E-03	2
1398	15063	C8H16O	1-octanal	124-13-0	30	3.7030E+02	5.2046E+01	3.7044E-01	2.8892E-03	1
1399	15064	C8H16O	5-methylheptanal	75579-88-3	25	7.0296E+02	9.8830E+01	7.0345E-01	5.4865E-03	2
1400	15065	C8H16O	2-propylpentanal	18295-59-5	25	7.7750E+02	1.0932E+02	7.7810E-01	6.0688E-03	2
1401	15066	C8H16O	2-octanone	111-13-7	20	9.0000E+02	1.2655E+02	9.0081E-01	7.0258E-03	1
1402	15067	C8H16O	2,5-dimethyl-3-hexanone	1888-57-9	25	2.6297E+03	3.7033E+02	2.6367E+00	2.0565E-02	2
1403	15068	C8H16O	2,2-dimethyl-3-hexanone	5405-79-8	25	2.8752E+03	4.0499E+02	2.8835E+00	2.2490E-02	2
1404	15069	C8H16O	3,3-dimethyl-2-hexanone	26118-38-7	25	2.7091E+03	3.8153E+02	2.7164E+00	2.1187E-02	2
1405	15070	C8H16O	3,4-dimethyl-2-hexanone	19550-10-8	25	1.4142E+03	1.9895E+02	1.4162E+00	1.1046E-02	2
1406	15071	C8H16O	4,4-dimethyl-3-hexanone	19550-14-2	25	2.5528E+03	3.5947E+02	2.5593E+00	1.9961E-02	2
1407	15072	C8H16O	3-ethyl-4-methyl-2-pentanone	71172-57-1	25	1.7379E+03	2.4455E+02	1.7409E+00	1.3578E-02	2
1408	15073	C8H16O	2-methyl-3-heptanone	13019-20-0	25	1.4142E+03	1.9895E+02	1.4162E+00	1.1046E-02	2
1409	15074	C8H16O	2-methyl-4-heptanone	626-33-5	25	1.7899E+03	2.5188E+02	1.7931E+00	1.3985E-02	2
1410	15075	C8H16O	3-methyl-2-heptanone	2371-19-9	25	9.9408E+02	1.3979E+02	9.9507E-01	7.7610E-03	2
1411	15076	C8H16O	3-methyl-4-heptanone	15726-15-5	25	1.8987E+03	2.6722E+02	1.9023E+00	1.4837E-02	2
1412	15077	C8H16O	5-methyl-3-heptanone	541-85-5	25	1.1187E+03	1.5733E+02	1.1199E+00	8.7347E-03	2
1413	15078	C8H16O	6-methyl-2-heptanone	928-68-7	25	8.3349E+02	1.1720E+02	8.3419E-01	6.5062E-03	2
1414	15079	C8H16O	6-methyl-3-heptanone	624-42-0	25	9.9408E+02	1.3979E+02	9.9507E-01	7.7610E-03	2
1415	15080	C8H16O	3-octanone	106-68-3	25	8.0937E+02	1.1380E+02	8.1002E-01	6.3177E-03	2
1416	15081	C8H16O	4-octanone	589-63-9	25	1.0542E+03	1.4826E+02	1.0553E+00	8.2309E-03	2
1417	15082	C8H16O	2,2,4-trimethyl-3-pentanone	5857-36-3	25	5.5397E+03	7.8209E+02	5.5705E+00	4.3447E-02	2
1418	15147	C8H16O	3-ethyl-3-methyl-2-pentanone	19780-65-5	25	1.8489E+03	2.6020E+02	1.8524E+00	1.4447E-02	2
1419	15148	C8H16O	5-methyl-2-heptanone	18217-12-4	25	1.0029E+03	1.4103E+02	1.0039E+00	7.8297E-03	2
1420	15156	C8H16O2	octanoic acid	124-07-2	20	6.8000E+02	8.4995E+01	6.8046E-01	4.7184E-03	1
1421	15157	C8H16O2	2-ethylhexanoic acid	149-57-5	25	1.8823E+03	2.3553E+02	1.8859E+00	1.3077E-02	2
1422	15158	C8H16O2	3-ethylhexanoic acid	41065-91-2	25	5.0147E+04	6.5518E+03	5.2795E+01	3.6609E-01	2
1423	15159	C8H16O2	heptyl formate	112-23-2	25	1.0565E+02	1.3199E+01	1.0566E-01	7.3268E-04	2
1424	15160	C8H16O2	hexyl acetate	142-92-7	25	5.1050E+02	6.3799E+01	5.1076E-01	3.5417E-03	1
1425	15161	C8H16O2	1,1-dimethylbutyl acetate	34859-98-8	25	1.0858E+03	1.3577E+02	1.0870E+00	7.5375E-03	2
1426	15162	C8H16O2	2-ethylbutyl acetate	10031-87-5	25	6.5108E+02	8.1378E+01	6.5150E-01	4.5176E-03	2
1427	15163	C8H16O2	sec-hexyl acetate	108-84-9	20	1.3000E+03	1.6258E+02	1.3017E+00	9.0261E-03	1
1428	15164	C8H16O2	2-methyl-1-pentyl acetate	7789-99-3	25	6.3378E+02	7.9215E+01	6.3419E-01	4.3975E-03	2
1429	15165	C8H16O2	2-methyl-3-pentyl acetate	35897-16-6	25	1.3468E+03	1.6843E+02	1.3486E+00	9.3512E-03	2
1430	15166	C8H16O2	3-methyl-3-pentyl acetate	10250-47-2	25	1.4213E+03	1.7776E+02	1.4233E+00	9.8692E-03	2
1431	15167	C8H16O2	pentyl propanoate	624-54-4	25	8.1000E+02	1.0126E+02	8.1066E-01	5.6212E-03	1
1432	15168	C8H16O2	1-methylbutyl propanoate	54004-43-2	25	5.0907E+02	6.3620E+01	5.0933E-01	3.5318E-03	2
1433	15169	C8H16O2	2-methylbutyl propanoate	2438-20-2	25	5.0907E+02	6.3620E+01	5.0933E-01	3.5318E-03	2
1434	15170	C8H16O2	isopentyl propanoate	105-68-0	25	7.6520E+02	9.5652E+01	7.6579E-01	5.3101E-03	2
1435	15171	C8H16O2	1-ethylpropyl propanoate	---	25	5.0907E+02	6.3620E+01	5.0933E-01	3.5318E-03	2
1436	15172	C8H16O2	1,1-dimethylpropyl propanoate	34949-22-9	25	5.9158E+02	7.3937E+01	5.9193E-01	4.1045E-03	2
1437	15173	C8H16O2	1,2-dimethylpropyl propanoate	66576-70-3	25	5.2127E+02	6.5146E+01	5.2154E-01	3.6165E-03	2
1438	15174	C8H16O2	2,2-dimethylpropyl propanoate	3581-69-9	25	5.9158E+02	7.3937E+01	5.9193E-01	4.1045E-03	2
1439	15175	C8H16O2	butyl butanoate	109-21-7	20	5.0000E+02	6.2487E+01	5.0025E-01	3.4688E-03	1
1440	15176	C8H16O2	isobutyl butanoate	539-90-2	25	8.8018E+02	1.1004E+02	8.8095E-01	6.1087E-03	2

Table 3 Solubility in Water - Organic Compounds

					Solubility in Water, S (ppm - parts per million, wt - weight, g - gram)					
NO	ID	FORMULA	NAME	CAS No	T, C	S@T, ppm(wt)	S@T, ppm(mol)	S@T, g/1000 g of H2O	S@T, mol/1000 g of H2O	code
1441	15177	C8H16O2	isobutyl isobutyrate	97-85-8	25	5.7000E+02	7.1239E+01	5.7033E-01	3.9547E-03	1
1442	15178	C8H16O2	sec-butyl butanoate	819-97-6	25	1.1459E+03	1.4329E+02	1.1472E+00	7.9549E-03	2
1443	15179	C8H16O2	tert-butyl butanoate	2308-38-5	25	2.7118E+03	3.3956E+02	2.7192E+00	1.8855E-02	2
1444	15180	C8H16O2	sec-butyl butanoate, (S)	116836-55-6	25	1.1459E+03	1.4329E+02	1.1472E+00	7.9549E-03	2
1445	15181	C8H16O2	tert-butyl 2-methylpropanoate	16889-72-8	25	4.4741E+03	5.6110E+02	4.4942E+00	3.1164E-02	2
1446	15185	C8H16O2	ethyl hexanoate	123-66-0	25	5.7000E+02	7.1239E+01	5.7033E-01	3.9547E-03	2
1447	15186	C8H16O2	ethyl 4-methylpentanoate	25415-67-2	25	5.7000E+02	7.1239E+01	5.7033E-01	3.9547E-03	2
1448	15191	C8H16O2	isopropyl 3-methylbutanoate	32665-23-9	25	1.9632E+03	2.4566E+02	1.9671E+00	1.3640E-02	2
1449	15192	C8H16O2	isopropyl pentanoate	18362-97-5	25	5.7000E+02	7.1239E+01	5.7033E-01	3.9547E-03	2
1450	15193	C8H16O2	propyl 3-methylbutanoate	557-00-6	25	9.2887E+02	1.1613E+02	9.2973E-01	6.4469E-03	2
1451	15194	C8H16O2	propyl pentanoate	141-06-0	25	5.7000E+02	7.1239E+01	5.7033E-01	3.9547E-03	2
1452	15195	C8H16O2	methyl heptanoate	106-73-0	25	5.7000E+02	7.1239E+01	5.7033E-01	3.9547E-03	2
1453	15204	C8H16O2	2-methylheptanoic acid	1188-02-9	25	5.0147E+04	6.5518E+03	5.2795E+01	3.6609E-01	2
1454	15205	C8H16O2	2-propylpentanoic acid	99-66-1	25	2.6835E+03	3.3601E+02	2.6907E+00	1.8658E-02	2
1455	15207	C8H16O2	butyl isobutyrate	---	25	5.7000E+02	7.1239E+01	5.7033E-01	3.9547E-03	2
1456	15232	C8H16O4	diethylene glycol ethyl ether acetate	112-15-2	20	total	total	total	total	1
1457	15248	C8H17Br	1-bromooctane	111-83-1	25	1.6700E+00	1.5578E-01	1.6700E-03	8.6472E-06	1
1458	15249	C8H17Br	2-bromooctane, (±)-	60251-57-2	25	5.4211E+00	5.0569E-01	5.4211E-03	2.8070E-05	2
1459	15250	C8H17Br	2-ethylhexyl bromide	18908-66-2	25	3.8794E+00	3.6188E-01	3.8795E-03	2.0088E-05	2
1460	15252	C8H17Cl	1-chlorooctane	111-85-3	25	4.8900E+00	5.9252E-01	4.8900E-03	3.2891E-05	1
1461	15253	C8H17Cl	2-chloro-2,5-dimethylhexane	29342-44-7	25	8.0968E+00	9.8110E-01	8.0969E-03	5.4460E-05	2
1462	15254	C8H17Cl	3-chloro-2,3-dimethylhexane	101654-30-2	25	8.0968E+00	9.8110E-01	8.0969E-03	5.4460E-05	2
1463	15255	C8H17Cl	2-chloro-2-methylheptane	4325-49-9	25	8.0968E+00	9.8110E-01	8.0969E-03	5.4460E-05	2
1464	15256	C8H17Cl	2-chloro-6-methylheptane	2350-19-8	25	8.0968E+00	9.8110E-01	8.0969E-03	5.4460E-05	2
1465	15257	C8H17Cl	3-chloro-3-methylheptane	5272-02-6	25	8.0968E+00	9.8110E-01	8.0969E-03	5.4460E-05	2
1466	15258	C8H17Cl	4-chloro-4-methylheptane	61764-94-1	25	8.0968E+00	9.8110E-01	8.0969E-03	5.4460E-05	2
1467	15259	C8H17Cl	3-(chloromethyl)heptane	123-04-6	25	6.9930E+00	8.4735E-01	6.9930E-03	4.7036E-05	2
1468	15260	C8H17Cl	2-chlorooctane	628-61-5	25	6.9930E+00	8.4735E-01	6.9930E-03	4.7036E-05	2
1469	15261	C8H17Cl	2-chloro-2,4,4-trimethylpentane	6111-88-2	25	6.9930E+00	8.4735E-01	6.9930E-03	4.7036E-05	2
1470	15262	C8H17Cl	(S)-2-chlorooctane	16844-08-9	25	8.0968E+00	9.8110E-01	8.0969E-03	5.4460E-05	2
1471	15263	C8H17Cl	(<+->)-2-chlorooctane	51261-14-4	25	6.9740E+00	8.4504E-01	6.9740E-03	4.6908E-05	2
1472	15264	C8H17Cl	2-ethylhexyl-6-chloride	2350-24-5	25	8.0968E+00	9.8110E-01	8.0969E-03	5.4460E-05	2
1473	15354	C8H18	octane	111-65-9	25	4.3100E-01	6.7972E-02	4.3100E-04	3.7731E-06	1
1474	15355	C8H18	2-methylheptane	592-27-8	25	7.4739E-01	1.1787E-01	7.4739E-04	6.5428E-06	2
1475	15356	C8H18	3-methylheptane	589-81-1	25	7.9200E-01	1.2490E-01	7.9200E-04	6.9333E-06	1
1476	15357	C8H18	4-methylheptane	589-53-7	25	7.4478E-01	1.1746E-01	7.4478E-04	6.5199E-06	2
1477	15358	C8H18	3-methyl-3-ethylpentane	1067-08-9	25	7.2083E-01	1.1368E-01	7.2083E-04	6.3103E-06	2
1478	15359	C8H18	3-ethylhexane	619-99-8	25	7.0956E-01	1.1190E-01	7.0956E-04	6.2116E-06	2
1479	15360	C8H18	3-ethyl-2-methylpentane	609-26-7	25	8.3923E-01	1.3235E-01	8.3923E-04	7.3468E-06	2
1480	15361	C8H18	2,2-dimethylhexane	590-73-8	25	1.3953E+00	2.2004E-01	1.3953E-03	1.2214E-05	2
1481	15362	C8H18	2,3-dimethylhexane	584-94-1	25	8.4167E-01	1.3274E-01	8.4167E-04	7.3681E-06	2
1482	15363	C8H18	2,4-dimethylhexane	589-43-5	25	1.2030E+00	1.8972E-01	1.2030E-03	1.0531E-05	2
1483	15364	C8H18	2,5-dimethylhexane	592-13-2	25	1.2253E+00	1.9324E-01	1.2253E-03	1.0727E-05	2
1484	15365	C8H18	3,3-dimethylhexane	563-16-6	25	1.0393E+00	1.6391E-01	1.0393E-03	9.0983E-06	2
1485	15366	C8H18	3,4-dimethylhexane	583-48-2	25	7.4391E-01	1.1732E-01	7.4391E-04	6.5123E-06	2
1486	15367	C8H18	2,2,3-trimethylpentane	564-02-3	25	1.1743E+00	1.8520E-01	1.1743E-03	1.0280E-05	2
1487	15368	C8H18	2,2,4-trimethylpentane	540-84-1	25	2.1425E+00	3.3789E-01	2.1425E-03	1.8756E-05	2
1488	15369	C8H18	2,3,3-trimethylpentane	560-21-4	25	8.8376E-01	1.3938E-01	8.8376E-04	7.7366E-06	2
1489	15370	C8H18	2,3,4-trimethylpentane	565-75-3	25	2.3020E+00	3.6304E-01	2.3020E-03	2.0152E-05	1
1490	15371	C8H18	2,2,3,3-tetramethylbutane	594-82-1	25	1.4266E+00	2.2499E-01	1.4266E-03	1.2489E-05	2
1491	15372	C8H18	octane, mixture	---	25	1.7282E+00	2.7255E-01	1.7282E-03	1.5129E-05	2
1492	15430	C8H18O	1-octanol	111-87-5	25	5.4000E+02	7.4734E+01	5.4029E-01	4.1487E-03	1
1493	15431	C8H18O	2-octanol	123-96-6	25	1.3000E+03	1.8003E+02	1.3017E+00	9.9953E-03	1
1494	15432	C8H18O	2-octanol, (±)	4128-31-8	25	1.0658E+03	1.4756E+02	1.0669E+00	8.1924E-03	2
1495	15433	C8H18O	3-octanol	589-98-0	25	1.4584E+03	2.0200E+02	1.4606E+00	1.1215E-02	2
1496	15434	C8H18O	4-octanol	589-62-8	25	1.3036E+03	1.8053E+02	1.3053E+00	1.0023E-02	2
1497	15435	C8H18O	4-octanol, (±)	74778-22-6	25	1.3269E+03	1.8377E+02	1.3287E+00	1.0203E-02	2
1498	15436	C8H18O	2-methyl-1-heptanol	60435-70-3	25	1.3830E+03	1.9154E+02	1.3849E+00	1.0634E-02	2
1499	15437	C8H18O	3-methyl-1-heptanol	1070-32-2	25	7.4511E+02	1.0314E+02	7.4566E-01	5.7257E-03	2
1500	15438	C8H18O	4-methyl-1-heptanol	817-91-4	25	8.8094E+02	1.2196E+02	8.8172E-01	6.7705E-03	2
1501	15439	C8H18O	5-methyl-1-heptanol	7212-53-5	25	7.1876E+02	9.9489E+01	7.1928E-01	5.5231E-03	2
1502	15440	C8H18O	6-methyl-1-heptanol	1653-40-3	25	6.7279E+02	9.3122E+01	6.7324E-01	5.1696E-03	2
1503	15441	C8H18O	2-methyl-2-heptanol	625-25-2	25	4.2029E+03	5.8350E+02	4.2206E+00	3.2409E-02	2
1504	15442	C8H18O	3-methyl-2-heptanol	31367-46-1	25	2.4191E+03	3.3534E+02	2.4250E+00	1.8621E-02	2
1505	15443	C8H18O	4-methyl-2-heptanol	56298-90-9	25	1.7508E+03	2.4256E+02	1.7539E+00	1.3467E-02	2
1506	15444	C8H18O	5-methyl-2-heptanol	54630-50-1	25	1.7201E+03	2.3830E+02	1.7231E+00	1.3231E-02	2
1507	15445	C8H18O	6-methyl-2-heptanol	4730-22-7	25	1.7201E+03	2.3830E+02	1.7231E+00	1.3231E-02	2
1508	15446	C8H18O	2-methyl-3-heptanol	18720-62-2	25	2.2151E+03	3.0701E+02	2.2200E+00	1.7047E-02	2
1509	15447	C8H18O	3-methyl-3-heptanol	5582-82-1	25	3.2639E+03	4.5278E+02	3.2746E+00	2.5145E-02	2
1510	15448	C8H18O	4-methyl-3-heptanol	14979-39-6	25	4.5372E+03	6.3011E+02	4.5579E+00	3.4999E-02	2
1511	15449	C8H18O	5-methyl-3-heptanol	18720-65-5	25	5.0452E+03	7.0096E+02	5.0708E+00	3.8937E-02	2
1512	15450	C8H18O	6-methyl-3-heptanol	18720-66-6	25	3.4921E+03	4.8453E+02	3.5044E+00	2.6909E-02	2
1513	15451	C8H18O	2-methyl-4-heptanol	21570-35-4	25	2.4191E+03	3.3534E+02	2.4250E+00	1.8621E-02	2
1514	15452	C8H18O	3-methyl-4-heptanol	1838-73-9	25	2.6265E+03	3.6415E+02	2.6334E+00	2.0221E-02	2
1515	15453	C8H18O	4-methyl-4-heptanol	598-01-6	25	3.2448E+03	4.5012E+02	3.2554E+00	2.4997E-02	2
1516	15454	C8H18O	2-ethyl-1-hexanol	104-76-7	20	6.0000E+02	8.3042E+01	6.0036E-01	4.6100E-03	1
1517	15455	C8H18O	3-ethyl-1-hexanol	41065-95-6	25	1.0753E+03	1.4889E+02	1.0765E+00	8.2658E-03	2
1518	15456	C8H18O	4-ethyl-1-hexanol	---	25	1.0753E+03	1.4889E+02	1.0765E+00	8.2658E-03	2
1519	15457	C8H18O	2,2-dimethyl-1-hexanol	2370-13-0	25	1.6604E+03	2.3002E+02	1.6632E+00	1.2771E-02	2
1520	15458	C8H18O	2,3-dimethyl-1-hexanol	---	25	1.5334E+03	2.1240E+02	1.5358E+00	1.1793E-02	2

Table 3 Solubility in Water - Organic Compounds

					Solubility in Water, S (ppm - parts per million, wt - weight, g - gram)					
NO	ID	FORMULA	NAME	CAS No	T, C	S@T, ppm(wt)	S@T, ppm(mol)	S@T, g/1000 g of H2O	S@T, mol/1000 g of H2O	code
1521	15459	C8H18O	2,4-dimethyl-1-hexanol	3965-59-1	25	1.4328E+03	1.9845E+02	1.4349E+00	1.1018E-02	2
1522	15460	C8H18O	2,5-dimethyl-1-hexanol	6886-16-4	25	1.1309E+03	1.5659E+02	1.1322E+00	8.6938E-03	2
1523	15461	C8H18O	3,3-dimethyl-1-hexanol	10524-70-6	25	1.0133E+03	1.4029E+02	1.0143E+00	7.7886E-03	2
1524	15462	C8H18O	3,4-dimethyl-1-hexanol	---	25	1.0133E+03	1.4029E+02	1.0143E+00	7.7886E-03	2
1525	15463	C8H18O	3,5-dimethyl-1-hexanol	13501-73-0	25	8.9152E+02	1.2342E+02	8.9231E-01	6.8518E-03	2
1526	15464	C8H18O	4,4-dimethyl-1-hexanol	6481-95-4	25	9.5475E+02	1.3218E+02	9.5567E-01	7.3383E-03	2
1527	15465	C8H18O	4,5-dimethyl-1-hexanol	---	25	9.5475E+02	1.3218E+02	9.5567E-01	7.3383E-03	2
1528	15466	C8H18O	5,5-dimethyl-1-hexanol	---	25	9.5475E+02	1.3218E+02	9.5567E-01	7.3383E-03	2
1529	15467	C8H18O	3-ethyl-2-hexanol	24448-19-9	25	2.1636E+03	2.9986E+02	2.1683E+00	1.6650E-02	2
1530	15468	C8H18O	4-ethyl-2-hexanol	---	25	2.1828E+03	3.0252E+02	2.1876E+00	1.6798E-02	2
1531	15469	C8H18O	2,3-dimethyl-2-hexanol	19550-03-9	25	3.4412E+03	4.7744E+02	3.4531E+00	2.6515E-02	2
1532	15470	C8H18O	2,4-dimethyl-2-hexanol	42328-76-7	25	5.9883E+03	8.3266E+02	6.0243E+00	4.6259E-02	2
1533	15471	C8H18O	2,5-dimethyl-2-hexanol	3730-60-7	25	5.3837E+03	7.4821E+02	5.4129E+00	4.1564E-02	2
1534	15472	C8H18O	3,3-dimethyl-2-hexanol	22025-20-3	25	3.4921E+03	4.8453E+02	3.5044E+00	2.6909E-02	2
1535	15473	C8H18O	3,4-dimethyl-2-hexanol	19550-05-1	25	1.8137E+03	2.5129E+02	1.8170E+00	1.3952E-02	2
1536	15474	C8H18O	3,5-dimethyl-2-hexanol	66576-27-0	25	3.4615E+03	4.8026E+02	3.4735E+00	2.6672E-02	2
1537	15475	C8H18O	4,4-dimethyl-2-hexanol	---	25	3.1050E+03	4.3067E+02	3.1146E+00	2.3916E-02	2
1538	15476	C8H18O	4,5-dimethyl-2-hexanol	---	25	3.1050E+03	4.3067E+02	3.1146E+00	2.3916E-02	2
1539	15477	C8H18O	5,5-dimethyl-2-hexanol	31841-77-7	25	2.4334E+03	3.3732E+02	2.4393E+00	1.8731E-02	2
1540	15478	C8H18O	3-ethyl-3-hexanol	597-76-2	25	3.6710E+03	5.0943E+02	3.6845E+00	2.8293E-02	2
1541	15479	C8H18O	4-ethyl-3-hexanol	19780-44-0	25	2.7367E+03	3.7946E+02	2.7442E+00	2.1072E-02	2
1542	15480	C8H18O	2,2-dimethyl-3-hexanol	4209-90-9	25	4.3540E+03	6.0456E+02	4.3730E+00	3.3579E-02	2
1543	15481	C8H18O	2,3-dimethyl-3-hexanol	4166-46-5	25	3.8478E+03	5.3405E+02	3.8627E+00	2.9660E-02	2
1544	15482	C8H18O	2,4-dimethyl-3-hexanol	13432-25-2	25	3.4615E+03	4.8026E+02	3.4735E+00	2.6672E-02	2
1545	15483	C8H18O	2,5-dimethyl-3-hexanol	19550-07-3	25	3.6710E+03	5.0943E+02	3.6845E+00	2.8293E-02	2
1546	15484	C8H18O	3,4-dimethyl-3-hexanol	19550-08-4	25	5.5451E+03	7.7075E+02	5.5761E+00	4.2817E-02	2
1547	15485	C8H18O	3,5-dimethyl-3-hexanol	4209-91-0	25	5.5451E+03	7.7075E+02	5.5761E+00	4.2817E-02	2
1548	15486	C8H18O	4,4-dimethyl-3-hexanol	19550-09-5	25	3.6710E+03	5.0943E+02	3.6845E+00	2.8293E-02	2
1549	15487	C8H18O	4,5-dimethyl-3-hexanol	---	25	4.4185E+03	6.1356E+02	4.4382E+00	3.4079E-02	2
1550	15488	C8H18O	5,5-dimethyl-3-hexanol	66576-31-6	25	5.2271E+03	7.2634E+02	5.2546E+00	4.0348E-02	2
1551	15489	C8H18O	2-propyl-1-pentanol	58175-57-8	25	1.1309E+03	1.5659E+02	1.1322E+00	8.6938E-03	2
1552	15490	C8H18O	2-isopropyl-1-pentanol	---	25	3.0614E+02	4.2361E+01	3.0624E-01	2.3515E-03	2
1553	15491	C8H18O	2-methyl-2-ethyl-1-pentanol	5970-63-8	25	1.2656E+03	1.7527E+02	1.2672E+00	9.7306E-03	2
1554	15492	C8H18O	3-methyl-2-ethyl-1-pentanol	---	25	1.2845E+03	1.7788E+02	1.2861E+00	9.8758E-03	2
1555	15493	C8H18O	4-methyl-2-ethyl-1-pentanol	106-67-2	25	1.3113E+03	1.8160E+02	1.3131E+00	1.0083E-02	2
1556	15494	C8H18O	2-methyl-3-ethyl-1-pentanol	---	25	1.4456E+03	2.0022E+02	1.4477E+00	1.1116E-02	2
1557	15495	C8H18O	3-methyl-3-ethyl-1-pentanol	---	25	1.4456E+03	2.0022E+02	1.4477E+00	1.1116E-02	2
1558	15496	C8H18O	4-methyl-3-ethyl-1-pentanol	---	25	1.4456E+03	2.0022E+02	1.4477E+00	1.1116E-02	2
1559	15497	C8H18O	2,2,3-trimethyl-1-pentanol	57409-53-7	25	1.4758E+03	2.0440E+02	1.4779E+00	1.1349E-02	2
1560	15498	C8H18O	2,2,4-trimethyl-1-pentanol	123-44-4	25	2.1258E+03	2.9461E+02	2.1304E+00	1.6358E-02	2
1561	15499	C8H18O	2,3,3-trimethyl-1-pentanol	---	25	1.3627E+03	1.8873E+02	1.3645E+00	1.0478E-02	2
1562	15500	C8H18O	2,3,4-trimethyl-1-pentanol	6570-88-3	25	8.9152E+02	1.2342E+02	8.9231E-01	6.8518E-03	2
1563	15501	C8H18O	2,4,4-trimethyl-1-pentanol	16325-63-6	25	1.8137E+03	2.5129E+02	1.8170E+00	1.3952E-02	2
1564	15502	C8H18O	3,3,4-trimethyl-1-pentanol	---	25	1.8298E+03	2.5352E+02	1.8332E+00	1.4076E-02	2
1565	15503	C8H18O	3,4,4-trimethyl-1-pentanol	16325-64-7	25	1.8298E+03	2.5352E+02	1.8332E+00	1.4076E-02	2
1566	15504	C8H18O	2-methyl-3-ethyl-2-pentanol	19780-63-3	25	3.9394E+03	5.4681E+02	3.9550E+00	3.0369E-02	2
1567	15505	C8H18O	3-methyl-3-ethyl-2-pentanol	66576-22-5	25	3.4921E+03	4.8453E+02	3.5044E+00	2.6909E-02	2
1568	15506	C8H18O	4-methyl-3-ethyl-2-pentanol	66576-23-6	25	2.7367E+03	3.7946E+02	2.7442E+00	2.1072E-02	2
1569	15507	C8H18O	2,3,3-trimethyl-2-pentanol	23171-85-9	25	3.4615E+03	4.8026E+02	3.4735E+00	2.6672E-02	2
1570	15508	C8H18O	2,3,4-trimethyl-2-pentanol	66576-26-9	25	3.8934E+03	5.4039E+02	3.9086E+00	3.0013E-02	2
1571	15509	C8H18O	2,4,4-trimethyl-2-pentanol	690-37-9	25	7.7290E+03	1.0763E+03	7.7892E+00	5.9811E-02	2
1572	15510	C8H18O	3,3,4-trimethyl-2-pentanol	19411-41-7	25	2.5806E+03	3.5777E+02	2.5873E+00	1.9867E-02	2
1573	15511	C8H18O	3,4,4-trimethyl-2-pentanol	10575-56-1	25	3.8934E+03	5.4039E+02	3.9086E+00	3.0013E-02	2
1574	15512	C8H18O	2-methyl-3-ethyl-3-pentanol	597-05-7	25	3.4615E+03	4.8026E+02	3.4735E+00	2.6672E-02	2
1575	15513	C8H18O	2,2,3-trimethyl-3-pentanol	7294-05-5	25	5.5451E+03	7.7075E+02	5.5761E+00	4.2817E-02	2
1576	15514	C8H18O	2,2,4-trimethyl-3-pentanol	5162-48-1	25	5.8829E+03	8.1794E+02	5.9177E+00	4.5440E-02	2
1577	15515	C8H18O	2,3,4-trimethyl-3-pentanol	3054-92-0	25	4.1293E+03	5.7325E+02	4.1464E+00	3.1839E-02	2
1578	15516	C8H18O	3-methyl-2-isopropyl-1-butanol	18593-92-5	25	1.6122E+03	2.2333E+02	1.6148E+00	1.2400E-02	2
1579	15517	C8H18O	2,2-diethyl-1-butanol	13023-60-4	25	1.6265E+03	2.2532E+02	1.6292E+00	1.2510E-02	2
1580	15518	C8H18O	2,3-dimethyl-2-ethyl-1-butanol	---	25	1.6265E+03	2.2532E+02	1.6292E+00	1.2510E-02	2
1581	15519	C8H18O	3,3-dimethyl-2-ethyl-1-butanol	66576-56-5	25	1.6265E+03	2.2532E+02	1.6292E+00	1.2510E-02	2
1582	15520	C8H18O	2,2,3,3-tetramethyl-1-butanol	66576-24-7	25	1.6265E+03	2.2532E+02	1.6292E+00	1.2510E-02	2
1583	15521	C8H18O	3,5-dimethyl-3-hexanol, (±)	19113-78-1	25	5.5451E+03	7.7075E+02	5.5761E+00	4.2817E-02	2
1584	15522	C8H18O	2-methyl-1-heptanol, (±)	111675-77-5	25	1.3830E+03	1.9154E+02	1.3849E+00	1.0634E-02	2
1585	15523	C8H18O	5-methyl-1-heptanol, (±)	111767-95-4	25	7.1876E+02	9.9489E+01	7.1928E-01	5.5231E-03	2
1586	15524	C8H18O	2-methyl-3-heptanol, (±)	100296-26-2	25	2.2282E+03	3.0882E+02	2.2331E+00	1.7148E-02	2
1587	15525	C8H18O	6-methyl-3-heptanol, (±)	100295-85-0	25	2.0401E+03	2.8271E+02	2.0443E+00	1.5698E-02	2
1588	15526	C8H18O	3-methyl-4-heptanol, (R*,S*)-(±)	92737-91-2	25	3.0777E+03	4.2688E+02	3.0872E+00	2.3706E-02	2
1589	15527	C8H18O	dibutyl ether	142-96-1	25	3.0000E+02	4.1510E+01	3.0009E-01	2.3043E-03	1
1590	15528	C8H18O	di-sec-butyl ether	6863-58-7	25	5.3407E+02	7.3913E+01	5.3436E-01	4.1032E-03	2
1591	15529	C8H18O	di-tert-butyl ether	6163-66-2	25	1.0745E+03	1.4877E+02	1.0756E+00	8.2595E-03	2
1592	15530	C8H18O	butyl isobutyl ether	17071-47-5	25	1.1714E+02	1.6206E+01	1.1715E-01	8.9958E-04	2
1593	15531	C8H18O	tert-butyl isobutyl ether	33021-02-2	25	5.2679E+01	7.2875E+00	5.2682E-02	4.0453E-04	2
1594	15532	C8H18O	di-sec-butyl ether, (±)	17226-28-7	25	5.4916E+02	7.6002E+01	5.4946E-01	4.2192E-03	2
1595	15533	C8H18O	ethyl hexyl ether	5756-43-4	25	1.7567E+02	2.4305E+01	1.7570E-01	1.3492E-03	2
1596	15534	C8H18O	1-methoxyheptane	629-32-3	25	1.1714E+02	1.6206E+01	1.1715E-01	8.9958E-04	2
1597	15535	C8H18O	isooctanol	26952-21-6	25	7.4511E+02	1.0314E+02	7.4566E-01	5.7257E-03	2
1598	15536	C8H18O	(R)-(-)-2-octanol	5978-70-1	25	1.3708E+03	1.8984E+02	1.3726E+00	1.0540E-02	2
1599	15537	C8H18O	(S)-(+)-2-octanol	6169-06-8	25	1.2071E+03	1.6715E+02	1.2085E+00	9.2801E-03	2
1600	15538	C8H18O	octyl alcohol; mixed isomers	29063-28-3	25	4.3212E+02	5.9798E+01	4.3230E-01	3.3195E-03	2

Table 3 Solubility in Water - Organic Compounds

					Solubility in Water, S (ppm - parts per million, wt - weight, g - gram)						
NO	ID	FORMULA	NAME	CAS No	T, C	S@T, ppm(wt)	S@T, ppm(mol)	S@T, g/1000 g of H2O	S@T, mol/1000 g of H2O	code
1601	15539	C8H18O	isobutyl ether	628-55-7	25	4.9125E+02	6.7984E+01	4.9149E-01	3.7740E-03	2
1602	15578	C8H18O3	diethylene glycol diethyl ether	112-36-7	20	total	total	total	total	1
1603	15579	C8H18O3	diethylene glycol monobutyl ether	112-34-5	20	total	total	total	total	1
1604	15598	C8H18O5	tetraethylene glycol	112-60-7	20	total	total	total	total	1
1605	15602	C8H18S	octyl mercaptan	111-88-6	25	5.6934E+00	7.0109E-01	5.6935E-03	3.8917E-05	2
1606	15603	C8H18S	2-octanethiol	3001-66-9	25	1.1319E+01	1.3938E+00	1.1319E-02	7.7368E-05	2
1607	15604	C8H18S	tert-octyl mercaptan	141-59-3	25	5.9408E+01	7.3159E+00	5.9412E-02	4.0610E-04	2
1608	15605	C8H18S	2-octanethiol, (±)	10435-81-1	25	1.1319E+01	1.3938E+00	1.1319E-02	7.7368E-05	2
1609	15606	C8H18S	dibutyl sulfide	544-40-1	25	3.3037E+01	4.0682E+00	3.3038E-02	2.2583E-04	2
1610	15607	C8H18S	methyl heptyl sulfide	20291-61-6	25	2.3116E+01	2.8465E+00	2.3116E-02	1.5801E-04	2
1611	15608	C8H18S	ethyl hexyl sulfide	7309-44-6	25	2.5359E+01	3.1228E+00	2.5360E-02	1.7335E-04	2
1612	15609	C8H18S	propyl pentyl sulfide	42841-80-5	25	2.5359E+01	3.1228E+00	2.5360E-02	1.7335E-04	2
1613	15610	C8H18S	di-sec-butyl sulfide	626-26-6	25	8.3309E+01	1.0259E+01	8.3316E-02	5.6950E-04	2
1614	15611	C8H18S	di-tert-butyl sulfide	107-47-1	25	1.9552E+02	2.4081E+01	1.9556E-01	1.3367E-03	2
1615	15612	C8H18S	diisobutyl sulfide	592-65-4	25	6.0157E+01	7.4081E+00	6.0160E-02	4.1122E-04	2
1616	15631	C8H19N	octylamine	111-86-4	25	2.0000E+02	2.7882E+01	2.0004E-01	1.5477E-03	1
1617	15632	C8H19N	2-ethylhexylamine	104-75-6	20	2.5000E+03	3.4922E+02	2.5063E+00	1.9391E-02	1
1618	15633	C8H19N	6-methyl-2-heptanamine, (±)	5984-58-7	25	1.0502E+04	1.4772E+03	1.0614E+01	8.2122E-02	2
1619	15634	C8H19N	2-octanamine, (±)	44855-57-4	25	6.0651E+03	8.4983E+02	6.1022E+00	4.7214E-02	2
1620	15635	C8H19N	methylheptylamine	36343-05-2	25	3.5685E+03	4.9893E+02	3.5813E+00	2.7709E-02	2
1621	15636	C8H19N	ethylhexylamine	20352-67-4	25	5.3429E+03	7.4816E+02	5.3716E+00	4.1561E-02	2
1622	15637	C8H19N	dibutylamine	111-92-2	25	3.5000E+03	4.8932E+02	3.5123E+00	2.7175E-02	1
1623	15638	C8H19N	diisobutylamine	110-96-3	25	1.7527E+04	2.4804E+03	1.7840E+01	1.3803E-01	2
1624	15639	C8H19N	di-sec-butylamine	626-23-3	25	2.5429E+04	3.6238E+03	2.6093E+01	2.0189E-01	2
1625	15640	C8H19N	N-methyl-2-heptanamine	540-43-2	25	6.3771E+03	8.9379E+02	6.4181E+00	4.9658E-02	2
1626	15641	C8H19N	dimethylhexylamine	4385-04-0	25	2.9811E+03	4.1659E+02	2.9900E+00	2.3134E-02	2
1627	15642	C8H19N	diethylbutylamine	4444-68-2	25	4.9885E+03	6.9832E+02	5.0135E+00	3.8790E-02	2
1628	15643	C8H19N	N-ethyl-N-isopropyl-2-propanamine	7087-68-5	25	8.8012E+03	1.2361E+03	8.8793E+00	6.8701E-02	2
1629	15644	C8H19N	N-ethyl-N-propyl-1-propanamine	20634-92-8	25	4.4426E+03	6.2161E+02	4.4624E+00	3.4526E-02	2
1630	15645	C8H19N	2-amino-6-methylheptane	543-82-8	25	6.3771E+03	8.9379E+02	6.4181E+00	4.9658E-02	2
1631	15646	C8H19N	(S)-2-amino-6-methylheptane	70419-10-2	25	6.3771E+03	8.9379E+02	6.4181E+00	4.9658E-02	2
1632	15647	C8H19N	(R)-2-amino-6-methylheptane	70419-11-3	25	6.3771E+03	8.9379E+02	6.4181E+00	4.9658E-02	2
1633	15648	C8H19N	(S)-2-aminooctane	34566-04-6	25	6.0651E+03	8.4983E+02	6.1022E+00	4.7214E-02	2
1634	15649	C8H19N	(R)-2-aminooctane	34566-05-7	25	6.0651E+03	8.4983E+02	6.1022E+00	4.7214E-02	2
1635	15650	C8H19N	2-aminooctane	693-16-3	25	5.7175E+03	8.0088E+02	5.7504E+00	4.4492E-02	2
1636	15651	C8H19N	N-isopropyl-N-methyl-tert-butylamine	85523-00-8	25	8.5344E+03	1.1984E+03	8.6079E+00	6.6601E-02	2
1637	15652	C8H19N	tert-octylamine	107-45-9	25	3.0144E+04	4.3135E+03	3.1081E+01	2.4048E-01	2
1638	15747	C8H23N5	tetraethylenepentamine	112-57-2	20	total	total	total	total	1
1639	16010	C9H7N	isoquinoline	119-65-3	20	4.5206E+03	6.3298E+02	4.5411E+00	3.5159E-02	1
1640	16011	C9H7N	quinoline	91-22-5	20	6.7164E+03	9.4223E+02	6.7618E+00	5.2352E-02	1
1641	16018	C9H7NO	8-hydroxyquinoline	148-24-3	20	5.5560E+02	6.8986E+01	5.5591E-01	3.8296E-03	1
1642	16508	C9H10	indane	496-11-7	25	1.0926E+02	1.6657E+01	1.0927E-01	9.2464E-04	1
1643	16512	C9H10	alpha-methylstyrene	98-83-9	25	8.9000E+01	1.3568E+01	8.9008E-02	7.5317E-04	2
1644	16513	C9H10	m-methylstyrene	100-80-1	25	8.9000E+01	1.3568E+01	8.9008E-02	7.5317E-04	1
1645	16514	C9H10	o-methylstyrene	611-15-4	25	8.9000E+01	1.3568E+01	8.9008E-02	7.5317E-04	2
1646	16515	C9H10	p-methylstyrene	622-97-9	25	8.9000E+01	1.3568E+01	8.9008E-02	7.5317E-04	1
1647	16519	C9H10	methylstyrene	25013-15-4	25	8.9000E+01	1.3568E+01	8.9008E-02	7.5317E-04	2
1648	16520	C9H10	3(4)-methylstyrene, isomers	39294-88-7	25	8.9000E+01	1.3568E+01	8.9008E-02	7.5317E-04	2
1649	16676	C9H10O2	ethyl benzoate	93-89-0	25	7.2000E+02	8.6425E+01	7.2052E-01	4.7978E-03	1
1650	17004	C9H12	cumene	98-82-8	25	5.0000E+01	7.4944E+00	5.0003E-02	4.1601E-04	1
1651	17005	C9H12	m-ethyltoluene	620-14-4	25	9.7000E+01	1.4540E+01	9.7009E-02	8.0711E-04	1
1652	17006	C9H12	o-ethyltoluene	611-14-3	25	9.3050E+01	1.3948E+01	9.3059E-02	7.7424E-04	1
1653	17007	C9H12	p-ethyltoluene	622-96-8	25	9.4850E+01	1.4217E+01	9.4859E-02	7.8921E-04	1
1654	17008	C9H12	1,2,3-trimethylbenzene	526-73-8	25	7.5200E+01	1.1272E+01	7.5206E-02	6.2570E-04	1
1655	17009	C9H12	1,2,4-trimethylbenzene	95-63-6	25	5.9000E+01	8.8435E+00	5.9003E-02	4.9090E-04	1
1656	17010	C9H12	trimethyl benzene	25551-13-7	25	4.5810E+01	6.8664E+00	4.5812E-02	3.8115E-04	2
1657	17011	C9H12	mesitylene	108-67-8	25	9.7000E+01	1.4540E+01	9.7009E-02	8.0711E-04	1
1658	17012	C9H12	propylbenzene	103-65-1	25	5.2160E+01	7.8182E+00	5.2163E-02	4.3399E-04	1
1659	17436	C9H14O	isophorone	78-59-1	25	1.2000E+04	1.5806E+03	1.2146E+01	8.7879E-02	1
1660	17520	C9H14S	2-isopentylthiophene	26963-33-7	25	7.5325E+00	8.7958E-01	7.5325E-03	4.8825E-05	2
1661	17593	C9H16	1-nonyne	3452-09-3	25	7.2000E+00	1.0441E+00	7.2001E-03	5.7959E-05	1
1662	17594	C9H16	2-nonyne	19447-29-1	25	3.7002E+00	5.3660E-01	3.7002E-03	2.9786E-05	2
1663	17595	C9H16	3-nonyne	20184-89-8	25	4.8645E+00	7.0544E-01	4.8645E-03	3.9158E-05	2
1664	17596	C9H16	4-nonyne	20184-91-2	25	5.8145E+00	8.4320E-01	5.8145E-03	4.6806E-05	2
1665	17597	C9H16	7-methyl-3-octyne	37050-06-9	25	8.2617E+00	1.1981E+00	8.2618E-03	6.6506E-05	2
1666	17598	C9H16	2,6-dimethyl-3-heptyne	19549-97-4	25	2.2921E+01	3.3240E+00	2.2921E-02	1.8451E-04	2
1667	17599	C9H16	5,5-dimethyl-3-heptyne	23097-98-5	25	8.2617E+00	1.1981E+00	8.2618E-03	6.6506E-05	2
1668	17605	C9H16	1,8-nonadiene	4900-30-5	25	2.1364E+00	3.0982E-01	2.1364E-03	1.7198E-05	2
1669	17606	C9H16	2,7-nonadiene	51333-70-1	25	1.2629E+00	1.8314E-01	1.2629E-03	1.0166E-05	2
1670	17607	C9H16	4-methyl-3,5-octadiene	36903-95-4	25	1.4185E+00	2.0571E-01	1.4185E-03	1.1419E-05	2
1671	17608	C9H16	7-methyl-2,4-octadiene	2216-70-8	25	1.4605E+00	2.1179E-01	1.4605E-03	1.1756E-05	2
1672	17609	C9H16	2,4-dimethyl-2,4-heptadiene	74421-05-9	25	2.7841E+00	4.0374E-01	2.7841E-03	2.2412E-05	2
1673	17610	C9H16	2,6-dimethyl-1,3-heptadiene	2436-84-2	25	2.3334E+00	3.3838E-01	2.3334E-03	1.8783E-05	2
1674	17611	C9H16	2,6-dimethyl-1,5-heptadiene	6709-39-3	25	2.0746E+00	3.0086E-01	2.0746E-03	1.6700E-05	2
1675	17612	C9H16	3,5-dimethyl-2,4-heptadiene	101935-28-8	25	2.4748E+00	3.5888E-01	2.4748E-03	1.9921E-05	2
1676	17613	C9H16	3-ethyl-2-methyl-1,5-hexadiene	73398-15-9	25	1.8450E+00	2.6756E-01	1.8450E-03	1.4852E-05	2
1677	17628	C9H16	7-methyl-1,6-octadiene	42152-47-6	25	2.0146E+00	2.9216E-01	2.0146E-03	1.6217E-05	2
1678	17755	C9H16O4	azelaic acid	123-99-9	20	2.4470E+03	2.3472E+02	2.4530E+00	1.3032E-02	1
1679	17859	C9H18	butylcyclopentane	2040-95-1	25	4.3077E-01	6.1472E-02	4.3077E-04	3.4122E-06	2
1680	17860	C9H18	isobutylcyclopentane	3788-32-7	25	7.1134E-01	1.0151E-01	7.1134E-04	5.6348E-06	2

Table 3 Solubility in Water - Organic Compounds

NO	ID	FORMULA	NAME	CAS No	Solubility in Water, S (ppm - parts per million, wt - weight, g - gram) T, C	S@T, ppm(wt)	S@T, ppm(mol)	S@T, g/1000 g of H2O	S@T, mol/1000 g of H2O	code
1681	17861	C9H18	sec-butylcyclopentane	4850-32-2	25	4.9061E-01	7.0011E-02	4.9061E-04	3.8863E-06	2
1682	17862	C9H18	tert-butylcyclopentane	3875-52-3	25	8.5273E-01	1.2169E-01	8.5273E-04	6.7548E-06	2
1683	17863	C9H18	1-methyl-1-propylcyclopentane	16631-63-3	25	7.9720E-01	1.1376E-01	7.9720E-04	6.3149E-06	2
1684	17864	C9H18	1-methyl-cis-2-propylcyclopentane	932-43-4	25	5.4346E-01	7.7553E-02	5.4346E-04	4.3049E-06	2
1685	17865	C9H18	1-methyl-trans-2-propylcyclopentane	932-44-5	25	7.8013E-01	1.1133E-01	7.8013E-04	6.1797E-06	2
1686	17866	C9H18	1-methyl-cis-3-propylcyclopentane	2443-04-1	25	7.0927E-01	1.0121E-01	7.0927E-04	5.6183E-06	2
1687	17867	C9H18	1-methyl-trans-3-propylcyclopentane	2443-03-0	25	7.0927E-01	1.0121E-01	7.0927E-04	5.6183E-06	2
1688	17868	C9H18	1-methyl-1-isopropylcyclopentane	61828-00-0	25	7.0927E-01	1.0121E-01	7.0927E-04	5.6183E-06	2
1689	17869	C9H18	1-methyl-cis-2-isopropylcyclopentane	61868-01-7	25	7.0927E-01	1.0121E-01	7.0927E-04	5.6183E-06	2
1690	17870	C9H18	1-methyl-trans-2-isopropylcyclopentan	61828-01-1	25	1.0690E+00	1.5255E-01	1.0690E-03	8.4679E-06	2
1691	17871	C9H18	1-methyl-cis-3-isopropylcyclopentane	61828-02-2	25	1.0080E+00	1.4384E-01	1.0080E-03	7.9845E-06	2
1692	17872	C9H18	1-methyl-trans-3-isopropylcyclopentan	61828-03-3	25	1.0080E+00	1.4384E-01	1.0080E-03	7.9845E-06	2
1693	17873	C9H18	1,1-diethylcyclopentane	2721-38-2	25	6.1316E-01	8.7500E-02	6.1316E-04	4.8571E-06	2
1694	17874	C9H18	1,cis-2-diethylcyclopentane	932-39-8	25	5.1351E-01	7.3279E-02	5.1351E-04	4.0677E-06	2
1695	17875	C9H18	1,trans-2-diethylcyclopentane	932-40-1	25	7.2900E-01	1.0403E-01	7.2900E-04	5.7746E-06	2
1696	17876	C9H18	1,3-diethylcyclopentane	19398-75-5	25	6.3125E-01	9.0081E-02	6.3125E-04	5.0003E-06	2
1697	17877	C9H18	1,cis-3-diethylcyclopentane	62016-59-5	25	6.3125E-01	9.0081E-02	6.3125E-04	5.0003E-06	2
1698	17878	C9H18	1,trans-3-diethylcyclopentane	62016-60-8	25	6.3125E-01	9.0081E-02	6.3125E-04	5.0003E-06	2
1699	17879	C9H18	1,1-dimethyl-2-ethylcyclopentane	54549-80-3	25	1.2755E+00	1.8202E-01	1.2755E-03	1.0104E-05	2
1700	17880	C9H18	1,1-dimethyl-3-ethylcyclopentane	62016-61-9	25	1.6148E+00	2.3044E-01	1.6148E-03	1.2791E-05	2
1701	17881	C9H18	1,cis-2-dimethyl-1-ethylcyclopentane	62016-63-1	25	8.9630E-01	1.2790E-01	8.9630E-04	7.0999E-06	2
1702	17882	C9H18	1,trans-2-dimethyl-1-ethylcyclopentane	62016-62-0	25	8.9630E-01	1.2790E-01	8.9630E-04	7.0999E-06	2
1703	17883	C9H18	1,cis-2-dimethyl-cis-3-ethylcyclopentan	---	25	5.6203E-01	8.0203E-02	5.6203E-04	4.4520E-06	2
1704	17884	C9H18	1,cis-2-dimethyl-trans-3-ethylcyclopent	---	25	8.4527E-01	1.2062E-01	8.4527E-04	6.6957E-06	2
1705	17885	C9H18	1,trans-2-dimethyl-cis-3-ethylcyclopent	---	25	1.2755E+00	1.8202E-01	1.2755E-03	1.0104E-05	2
1706	17886	C9H18	1,trans-2-dimethyl-trans-3-ethylcyclope	---	25	8.4527E-01	1.2062E-01	8.4527E-04	6.6957E-06	2
1707	17887	C9H18	1,cis-2-dimethyl-cis-4-ethylcyclopentan	62016-64-2	25	7.5192E-01	1.0730E-01	7.5192E-04	5.9562E-06	2
1708	17888	C9H18	1,cis-2-dimethyl-trans-4-ethylcyclopent	62016-65-3	25	7.5192E-01	1.0730E-01	7.5192E-04	5.9562E-06	2
1709	17889	C9H18	1,trans-2-dimethyl-cis-4-ethylcyclopent	62016-66-4	25	1.1338E+00	1.6179E-01	1.1338E-03	8.9811E-06	2
1710	17890	C9H18	1,cis-3-dimethyl-1-ethylcyclopentane	62016-68-6	25	1.4351E+00	2.0479E-01	1.4351E-03	1.1368E-05	2
1711	17891	C9H18	1,trans-3-dimethyl-1-ethylcyclopentane	62016-67-5	25	1.4351E+00	2.0479E-01	1.4351E-03	1.1368E-05	2
1712	17892	C9H18	1,cis-3-dimethyl-cis-2-ethylcyclopentan	19903-00-5	25	5.6203E-01	8.0203E-02	5.6203E-04	4.4520E-06	2
1713	17893	C9H18	1,cis-3-dimethyl-trans-2-ethylcyclopent	19902-98-8	25	1.3529E+00	1.9307E-01	1.3529E-03	1.0717E-05	2
1714	17894	C9H18	1,trans-3-dimethyl-cis-2-ethylcyclopent	19902-99-9	25	8.9630E-01	1.2790E-01	8.9630E-04	7.0999E-06	2
1715	17895	C9H18	1,cis-3-dimethyl-cis-4-ethylcyclopentan	---	25	8.4527E-01	1.2062E-01	8.4527E-04	6.6957E-06	2
1716	17896	C9H18	1,cis-3-dimethyl-trans-4-ethylcyclopent	---	25	1.3529E+00	1.9307E-01	1.3529E-03	1.0717E-05	2
1717	17897	C9H18	1,trans-3-dimethyl-cis-4-ethylcyclopent	---	25	1.2132E+00	1.7313E-01	1.2132E-03	9.6102E-06	2
1718	17898	C9H18	1,trans-3-dimethyl-trans-4-ethylcyclope	---	25	8.4527E-01	1.2062E-01	8.4527E-04	6.6957E-06	2
1719	17899	C9H18	1,1,2,2-tetramethylcyclopentane	52688-89-8	25	1.7129E+00	2.4444E-01	1.7129E-03	1.3569E-05	2
1720	17900	C9H18	1,1,cis-2,cis-3-tetramethylcyclopentane	62016-70-0	25	1.2755E+00	1.8202E-01	1.2755E-03	1.0104E-05	2
1721	17901	C9H18	1,1,cis-2,trans-3-tetramethylcyclopenta	---	25	2.0445E+00	2.9176E-01	2.0445E-03	1.6195E-05	2
1722	17902	C9H18	1,1,cis-2,cis-4-tetramethylcyclopentane	62016-72-2	25	2.0445E+00	2.9176E-01	2.0445E-03	1.6195E-05	2
1723	17903	C9H18	1,1,cis-2,trans-4-tetramethylcyclopenta	---	25	2.0445E+00	2.9176E-01	2.0445E-03	1.6195E-05	2
1724	17904	C9H18	1,1,3,3-tetramethylcyclopentane	50876-33-0	25	4.1466E+00	5.9173E-01	4.1466E-03	3.2847E-05	2
1725	17905	C9H18	1,1,cis-3,cis-4-tetramethylcyclopentane	20309-77-7	25	1.7129E+00	2.4444E-01	1.7129E-03	1.3569E-05	2
1726	17906	C9H18	1,1,cis-3,trans-4-tetramethylcyclopenta	---	25	3.3513E+00	4.7824E-01	3.3513E-03	2.6547E-05	2
1727	17907	C9H18	1,2,2,cis-3-tetramethylcyclopentane	62016-73-3	25	1.2755E+00	1.8202E-01	1.2755E-03	1.0104E-05	2
1728	17908	C9H18	1,2,2,trans-3-tetramethylcyclopentane	---	25	1.2755E+00	1.8202E-01	1.2755E-03	1.0104E-05	2
1729	17909	C9H18	1,cis-2,cis-3,cis-4-tetramethylcyclopent	2532-65-2	25	8.4527E-01	1.2062E-01	8.4527E-04	6.6957E-06	2
1730	17910	C9H18	1,cis-2,cis-3,trans-4-tetramethylcyclope	2532-69-6	25	1.2755E+00	1.8202E-01	1.2755E-03	1.0104E-05	2
1731	17911	C9H18	1,cis-2,trans-3,cis-4-tetramethylcyclope	2532-68-5	25	1.9274E+00	2.7505E-01	1.9274E-03	1.5268E-05	2
1732	17912	C9H18	1,cis-2,trans-3,trans-4-tetramethylcyclo	2532-64-1	25	1.2755E+00	1.8202E-01	1.2755E-03	1.0104E-05	2
1733	17913	C9H18	1,trans-2,cis-3,trans-4-tetramethylcyclo	2532-67-4	25	2.3831E+00	3.4008E-01	2.3831E-03	1.8878E-05	2
1734	17914	C9H18	1,trans-2,trans-3,cis-4-tetramethylcyclo	19907-40-5	25	1.9274E+00	2.7505E-01	1.9274E-03	1.5268E-05	2
1735	17915	C9H18	1-methyl-2-isopropylcyclopentane	89223-57-4	25	9.5047E-01	1.3563E-01	9.5047E-04	7.5290E-06	2
1736	17916	C9H18	propylcyclohexane	1678-92-8	25	6.7726E-01	9.6646E-02	6.7726E-04	5.3648E-06	2
1737	17917	C9H18	isopropylcyclohexane	696-29-7	25	7.5939E-01	1.0837E-01	7.5939E-04	6.0154E-06	2
1738	17918	C9H18	1-methyl-1-ethylcyclohexane	4926-90-3	25	8.8254E-01	1.2594E-01	8.8254E-04	6.9909E-06	2
1739	17919	C9H18	1-methyl-cis-2-ethylcyclohexane	4923-77-7	25	7.0302E-01	1.0032E-01	7.0302E-04	5.5688E-06	2
1740	17920	C9H18	1-methyl-trans-2-ethylcyclohexane	4923-78-8	25	9.0639E-01	1.2934E-01	9.0639E-04	7.1798E-06	2
1741	17921	C9H18	1-methyl-cis-3-ethylcyclohexane	19489-10-2	25	1.1241E+00	1.6041E-01	1.1241E-03	8.9045E-06	2
1742	17922	C9H18	1-methyl-trans-3-ethylcyclohexane	4926-76-5	25	9.4397E-01	1.3471E-01	9.4398E-04	7.4775E-06	2
1743	17923	C9H18	1-methyl-cis-4-ethylcyclohexane	4926-78-7	25	9.4397E-01	1.3471E-01	9.4398E-04	7.4775E-06	2
1744	17924	C9H18	1-methyl-trans-4-ethylcyclohexane	6236-88-0	25	1.1241E+00	1.6041E-01	1.1241E-03	8.9045E-06	2
1745	17925	C9H18	1,1,2-trimethylcyclohexane	7094-26-0	25	1.3241E+00	1.8895E-01	1.3241E-03	1.0488E-05	2
1746	17926	C9H18	1,1,3-trimethylcyclohexane	3073-66-3	25	1.7700E+00	2.5258E-01	1.7700E-03	1.4021E-05	1
1747	17927	C9H18	1,1,4-trimethylcyclohexane	7094-27-1	25	2.4127E+00	3.4430E-01	2.4127E-03	1.9112E-05	2
1748	17928	C9H18	1,2,3-trimethylcyclohexane	1678-97-3	25	1.5975E+00	2.2797E-01	1.5975E-03	1.2655E-05	2
1749	17929	C9H18	1,cis-2,cis-3-trimethylcyclohexane	1839-88-9	25	8.4064E-01	1.1996E-01	8.4064E-04	6.6590E-06	2
1750	17930	C9H18	1,cis-2,trans-3-trimethylcyclohexane	7667-55-2	25	9.4397E-01	1.3471E-01	9.4398E-04	7.4775E-06	2
1751	17931	C9H18	1,trans-2,cis-3-trimethylcyclohexane	1678-81-5	25	1.4205E+00	2.0271E-01	1.4205E-03	1.1253E-05	2
1752	17932	C9H18	1,2,4-trimethylcyclohexane, isomers	2234-75-5	25	1.6069E+00	2.2931E-01	1.6069E-03	1.2729E-05	2
1753	17933	C9H18	1,cis-2,cis-4-trimethylcyclohexane	1678-80-4	25	1.1241E+00	1.6041E-01	1.1241E-03	8.9045E-06	2
1754	17934	C9H18	1,cis-2,trans-4-trimethylcyclohexane	7667-58-5	25	1.1241E+00	1.6041E-01	1.1241E-03	8.9045E-06	2
1755	17935	C9H18	1,trans-2,cis-4-trimethylcyclohexane	7667-59-6	25	1.3397E+00	1.9117E-01	1.3397E-03	1.0612E-05	2
1756	17936	C9H18	1,trans-2,trans-4-trimethylcyclohexane	7667-60-9	25	1.6725E+00	2.3867E-01	1.6725E-03	1.3248E-05	2
1757	17937	C9H18	1,3,5,-trimethylcyclohexane	1839-63-0	25	2.0395E+00	2.9104E-01	2.0395E-03	1.6155E-05	2
1758	17938	C9H18	cis,cis-1,3,5-trimethylcyclohexane	1795-27-3	25	1.9617E+00	2.7994E-01	1.9617E-03	1.5539E-05	2
1759	17939	C9H18	cis,trans-1,3,5-trimethylcyclohexane	1795-26-2	25	1.7397E+00	2.4826E-01	1.7397E-03	1.3781E-05	2
1760	17940	C9H18	1-ethyl-2-methylcyclohexane	3728-54-9	25	8.7288E-01	1.2456E-01	8.7288E-04	6.9143E-06	2

Table 3 Solubility in Water - Organic Compounds

					Solubility in Water, S (ppm - parts per million, wt - weight, g - gram)					
NO	ID	FORMULA	NAME	CAS No	T, C	S@T, ppm(wt)	S@T, ppm(mol)	S@T, g/1000 g of H2O	S@T, mol/1000 g of H2O	code
1761	17942	C9H18	1-nonene	124-11-8	25	1.1200E+00	1.5983E-01	1.1200E-03	8.8719E-06	1
1762	17943	C9H18	trans-2-nonene	6434-78-2	25	6.4524E-01	9.2077E-02	6.4524E-04	5.1111E-06	2
1763	17944	C9H18	trans-3-nonene	20063-92-7	25	5.4136E-01	7.7254E-02	5.4136E-04	4.2883E-06	2
1764	17945	C9H18	4-nonene; (cis+trans)	2198-23-4	25	5.9444E-01	8.4828E-02	5.9444E-04	4.7088E-06	2
1765	17946	C9H18	2-methyl-1-octene	4588-18-5	25	6.3959E-01	9.1271E-02	6.3959E-04	5.0664E-06	2
1766	17947	C9H18	3-methyl-2-octene	113426-22-5	25	6.2661E-01	8.9419E-02	6.2661E-04	4.9636E-06	2
1767	17948	C9H18	7-methyl-3-octene	86668-33-9	25	7.4722E-01	1.0663E-01	7.4722E-04	5.9190E-06	2
1768	17949	C9H18	2-methyl-4-octene	64501-77-5	25	8.9146E-01	1.2721E-01	8.9146E-04	7.0615E-06	2
1769	17950	C9H18	2,3-dimethyl-2-heptene	3074-64-4	25	6.1931E-01	8.8377E-02	6.1931E-04	4.9057E-06	2
1770	17951	C9H18	2,6-dimethylhept-3-ene	2738-18-3	25	2.4771E+00	3.5349E-01	2.4771E-03	1.9622E-05	2
1771	17955	C9H18Br2	1,1-dibromononane	62168-27-8	25	5.4386E+00	3.4252E-01	5.4387E-03	1.9013E-05	2
1772	17956	C9H18Br2	1,2-dibromononane	73642-91-8	25	3.1101E+00	1.9587E-01	3.1101E-03	1.0873E-05	2
1773	17957	C9H18Br2	1,9-dibromononane	4549-33-1	25	1.8494E+00	1.1647E-01	1.8494E-03	6.4653E-06	2
1774	17959	C9H18Cl2	1,1-dichlorononane	821-88-5	25	2.3875E+00	2.1816E-01	2.3875E-03	1.2110E-05	2
1775	17960	C9H18Cl2	2,6-dichloro-2,6-dimethylheptane	35951-36-1	25	1.0646E+00	9.7282E-02	1.0646E-03	5.4001E-06	2
1776	17961	C9H18Cl2	1,9-dichlorononane	821-99-8	25	4.9601E-01	4.5325E-02	4.9601E-04	2.5159E-06	2
1777	17997	C9H18O	1-nonanal	124-19-6	30	1.0510E+02	1.3312E+01	1.0511E-01	7.3896E-04	1
1778	17998	C9H18O	diisobutyl ketone	108-83-8	24	2.6400E+03	3.3513E+02	2.6470E+00	1.8609E-02	1
1779	17999	C9H18O	2-nonanone	821-55-6	25	1.6685E+02	2.1135E+01	1.6688E-01	1.1732E-03	2
1780	18000	C9H18O	di-tert-butyl ketone	815-24-7	25	2.0142E+03	2.5556E+02	2.0183E+00	1.4189E-02	2
1781	18001	C9H18O	3,5-dimethyl-4-heptanone	19549-84-9	25	1.1180E+03	1.4173E+02	1.1192E+00	7.8685E-03	2
1782	18002	C9H18O	3-methyl-2-octanone	6137-08-2	25	3.1484E+02	3.9886E+01	3.1494E-01	2.2141E-03	2
1783	18003	C9H18O	3-methyl-4-octanone	20754-04-5	25	5.5211E+02	6.9959E+01	5.5241E-01	3.8836E-03	2
1784	18004	C9H18O	4-methyl-3-octanone	6137-15-1	25	3.1484E+02	3.9886E+01	3.1494E-01	2.2141E-03	2
1785	18005	C9H18O	5-methyl-2-octanone	58654-67-4	25	3.1484E+02	3.9886E+01	3.1494E-01	2.2141E-03	2
1786	18006	C9H18O	7-methyl-3-octanone	5408-57-1	25	3.3364E+02	4.2268E+01	3.3375E-01	2.3464E-03	2
1787	18007	C9H18O	7-methyl-4-octanone	20809-46-5	25	4.3588E+02	5.5226E+01	4.3607E-01	3.0657E-03	2
1788	18008	C9H18O	3-nonanone	925-78-0	25	2.1274E+02	2.6948E+01	2.1278E-01	1.4959E-03	2
1789	18009	C9H18O	4-nonanone	4485-09-0	25	2.4734E+02	3.1333E+01	2.4741E-01	1.7393E-03	2
1790	18010	C9H18O	5-nonanone	502-56-7	25	3.6270E+02	4.5951E+01	3.6283E-01	2.5508E-03	1
1791	18044	C9H18O	3,5,5-trimethylhexanal	5435-64-3	25	2.3676E+02	2.9992E+01	2.3682E-01	1.6649E-03	2
1792	18045	C9H18O	2,6-dimethyl-3-heptanone	19549-83-8	25	6.2668E+02	7.9413E+01	6.2707E-01	4.4085E-03	2
1793	18047	C9H18O	2-methyl-3-octanone	923-28-4	25	3.2675E+02	4.1395E+01	3.2686E-01	2.2979E-03	2
1794	18048	C9H18O	2-methyloctanal	7786-29-0	25	2.3676E+02	2.9992E+01	2.3682E-01	1.6649E-03	2
1795	18052	C9H18O2	nonanoic acid	112-05-0	20	2.6000E+02	2.9607E+01	2.6007E-01	1.6435E-03	1
1796	18053	C9H18O2	2-ethylheptanoic acid	3274-29-1	25	1.7702E+04	2.0474E+03	1.8021E+01	1.1388E-01	2
1797	18054	C9H18O2	3-ethyl-3-methylhexanoic acid	50902-82-4	25	1.7702E+04	2.0474E+03	1.8021E+01	1.1388E-01	2
1798	18055	C9H18O2	2-ethyl-3-methylhexanoic acid	74581-94-5	25	1.6170E+03	1.8435E+02	1.6196E+00	1.0235E-02	2
1799	18056	C9H18O2	3-methyloctanoic acid	6061-10-5	25	1.7702E+04	2.0474E+03	1.8021E+01	1.1388E-01	2
1800	18057	C9H18O2	2-methyloctanoic acid	3004-93-1	25	1.7702E+04	2.0474E+03	1.8021E+01	1.1388E-01	2
1801	18058	C9H18O2	octyl formate	112-32-3	25	3.4663E+01	3.9463E+00	3.4664E-02	2.1906E-04	2
1802	18059	C9H18O2	heptyl acetate	112-06-1	25	1.3017E+02	1.4821E+01	1.3019E-01	8.2271E-04	2
1803	18060	C9H18O2	4-heptyl acetate	5921-84-6	25	4.1199E+02	4.6920E+01	4.1216E-01	2.6046E-03	2
1804	18061	C9H18O2	hexyl propanoate	2445-76-3	25	1.4812E+02	1.6866E+01	1.4815E-01	9.3621E-04	2
1805	18062	C9H18O2	pentyl butanoate	540-18-1	20	6.0000E+01	6.8311E+00	6.0004E-02	3.7919E-04	1
1806	18063	C9H18O2	1-methylbutyl butanoate	60415-61-4	25	2.2787E+02	2.5947E+01	2.2792E-01	1.4403E-03	2
1807	18064	C9H18O2	2-methylbutyl butanoate	51115-64-1	25	2.6781E+02	3.0496E+01	2.6788E-01	1.6929E-03	2
1808	18065	C9H18O2	3-methylbutyl butanoate	106-27-4	25	2.6781E+02	3.0496E+01	2.6788E-01	1.6929E-03	2
1809	18066	C9H18O2	1-ethylpropyl butanoate	---	25	1.4852E+02	1.6911E+01	1.4855E-01	9.3873E-04	2
1810	18067	C9H18O2	1,1-dimethylpropyl butanoate	2050-00-2	25	1.7260E+02	1.9652E+01	1.7263E-01	1.0909E-03	2
1811	18068	C9H18O2	1,2-dimethylpropyl butanoate	88736-68-9	25	1.5208E+02	1.7316E+01	1.5211E-01	9.6124E-04	2
1812	18069	C9H18O2	2,2-dimethylpropyl butanoate	23361-69-5	25	1.7260E+02	1.9652E+01	1.7263E-01	1.0909E-03	2
1813	18070	C9H18O2	isopentyl 2-methylpropanoate	2050-01-3	25	4.7135E+02	5.3683E+01	4.7157E-01	2.9801E-03	2
1814	18072	C9H18O2	butyl 2-methylbutanoate	15706-73-7	25	2.6781E+02	3.0496E+01	2.6788E-01	1.6929E-03	2
1815	18073	C9H18O2	butyl 3-methylbutanoate	109-19-3	25	2.0728E+02	2.3603E+01	2.0733E-01	1.3102E-03	2
1816	18075	C9H18O2	isobutyl 3-methylbutanoate	589-59-3	25	4.7135E+02	5.3683E+01	4.7157E-01	2.9801E-03	2
1817	18082	C9H18O2	3,5,5-trimethylhexanoic acid	3302-10-1	25	1.7702E+04	2.0474E+03	1.8021E+01	1.1388E-01	2
1818	18083	C9H18O2	7-methyloctanoic acid	693-19-6	25	1.7702E+04	2.0474E+03	1.8021E+01	1.1388E-01	2
1819	18087	C9H18O2	4-octanecarboxylic acid	3274-28-0	25	2.0365E+01	2.3185E+00	2.0365E-02	1.2870E-04	2
1820	18115	C9H19Br	1-bromononane	693-58-3	25	9.2687E-01	8.0605E-02	9.2687E-04	4.4743E-06	2
1821	18117	C9H19Cl	1-chlorononane	2473-01-0	25	1.1495E+00	1.2728E-01	1.1495E-03	7.0651E-06	2
1822	18118	C9H19Cl	3-chloro-3-ethyl-2,2-dimethylpentane	86661-53-2	25	1.7339E+00	1.9198E-01	1.7339E-03	1.0657E-05	2
1823	18119	C9H19Cl	3-chloro-3-ethylheptane	28320-89-0	25	1.7339E+00	1.9198E-01	1.7339E-03	1.0657E-05	2
1824	18120	C9H19Cl	3-chloro-3-methyloctane	28320-88-9	25	1.7339E+00	1.9198E-01	1.7339E-03	1.0657E-05	2
1825	18121	C9H19Cl	4-chloro-4-methyloctane	36903-89-6	25	1.7339E+00	1.9198E-01	1.7339E-03	1.0657E-05	2
1826	18122	C9H19Cl	2-chlorononane	2216-36-6	25	2.6218E+00	2.9030E-01	2.6218E-03	1.6114E-05	2
1827	18123	C9H19Cl	5-chlorononane	28123-70-8	25	1.7339E+00	1.9198E-01	1.7339E-03	1.0657E-05	2
1828	18124	C9H19Cl	3-chloro-2,2,3-trimethylhexane	102449-95-6	25	1.7339E+00	1.9198E-01	1.7339E-03	1.0657E-05	2
1829	18176	C9H20	nonane	111-84-2	25	1.2200E-01	1.7136E-02	1.2200E-04	9.5121E-07	1
1830	18177	C9H20	2-methyloctane	3221-61-2	25	1.6550E-01	2.3246E-02	1.6550E-04	1.2904E-06	2
1831	18178	C9H20	3-methyloctane	2216-33-3	25	1.5653E-01	2.1986E-02	1.5653E-04	1.2204E-06	2
1832	18179	C9H20	4-methyloctane	2216-34-4	25	1.1500E-01	1.6153E-02	1.1500E-04	8.9663E-07	1
1833	18180	C9H20	3-ethylheptane	15869-80-4	25	1.6628E-01	2.3356E-02	1.6628E-04	1.2965E-06	2
1834	18181	C9H20	4-ethylheptane	2216-32-2	25	1.8690E-01	2.6252E-02	1.8690E-04	1.4572E-06	2
1835	18182	C9H20	2,2-dimethylheptane	1071-26-7	25	3.0880E-01	4.3374E-02	3.0880E-04	2.4077E-06	2
1836	18183	C9H20	2,3-dimethylheptane	3074-71-3	25	1.9476E-01	2.7356E-02	1.9476E-04	1.5185E-06	2
1837	18184	C9H20	2,4-dimethylheptane	2213-23-2	25	3.0500E-01	4.2840E-02	3.0500E-04	2.3780E-06	2
1838	18185	C9H20	2,5-dimethylheptane	2216-30-0	25	2.5388E-01	3.5660E-02	2.5388E-04	1.9795E-06	2
1839	18186	C9H20	2,6-dimethylheptane	1072-05-5	25	2.6615E-01	3.7383E-02	2.6615E-04	2.0751E-06	2
1840	18187	C9H20	3,4-dimethylheptane	922-28-1	25	1.9362E-01	2.7195E-02	1.9362E-04	1.5096E-06	2

Table 3 Solubility in Water - Organic Compounds

					Solubility in Water, S (ppm - parts per million, wt - weight, g - gram)					
NO	ID	FORMULA	NAME	CAS No	T, C	S@T, ppm(wt)	S@T, ppm(mol)	S@T, g/1000 g of H2O	S@T, mol/1000 g of H2O	code
1841	18188	C9H20	3,5-dimethylheptane	926-82-9	25	2.5388E-01	3.5660E-02	2.5388E-04	1.9795E-06	2
1842	18189	C9H20	4,4-dimethylheptane	1068-19-5	25	2.6615E-01	3.7383E-02	2.6615E-04	2.0751E-06	2
1843	18190	C9H20	3-ethyl-2-methylhexane	16789-46-1	25	2.2565E-01	3.1694E-02	2.2565E-04	1.7593E-06	2
1844	18191	C9H20	4-ethyl-2-methylhexane	3074-75-7	25	2.8906E-01	4.0601E-02	2.8906E-04	2.2537E-06	2
1845	18192	C9H20	3-ethyl-3-methylhexane	3074-76-8	25	1.9362E-01	2.7195E-02	1.9362E-04	1.5096E-06	2
1846	18193	C9H20	3-ethyl-4-methylhexane	3074-77-9	25	1.9591E-01	2.7517E-02	1.9591E-04	1.5275E-06	2
1847	18194	C9H20	2,2,3-trimethylhexane	16747-25-4	25	2.9266E-01	4.1107E-02	2.9266E-04	2.2818E-06	2
1848	18195	C9H20	2,2,4-trimethylhexane	16747-26-5	25	4.4376E-01	6.2330E-02	4.4376E-04	3.4599E-06	2
1849	18196	C9H20	2,2,5-trimethylhexane	3522-94-9	25	5.1255E-01	7.1993E-02	5.1255E-04	3.9963E-06	2
1850	18197	C9H20	2,3,3-trimethylhexane	16747-28-7	25	2.2994E-01	3.2297E-02	2.2994E-04	1.7928E-06	2
1851	18198	C9H20	2,3,4-trimethylhexane	921-47-1	25	2.1224E-01	2.9811E-02	2.1224E-04	1.6548E-06	2
1852	18199	C9H20	2,3,5-trimethylhexane	1069-53-0	25	3.3420E-01	4.6942E-02	3.3420E-04	2.6057E-06	2
1853	18200	C9H20	2,4,4-trimethylhexane	16747-30-1	25	3.4809E-01	4.8892E-02	3.4809E-04	2.7140E-06	2
1854	18201	C9H20	3,3,4-trimethylhexane	16747-31-2	25	1.9522E-01	2.7420E-02	1.9522E-04	1.5221E-06	2
1855	18202	C9H20	3,3-diethylpentane	1067-20-5	25	1.3955E-01	1.9602E-02	1.3955E-04	1.0881E-06	2
1856	18203	C9H20	2,2-dimethyl-3-ethylpentane	16747-32-3	25	2.8855E-01	4.0529E-02	2.8855E-04	2.2497E-06	2
1857	18204	C9H20	2,3-dimethyl-3-ethylpentane	16747-33-4	25	1.5219E-01	2.1376E-02	1.5219E-04	1.1866E-06	2
1858	18205	C9H20	2,4-dimethyl-3-ethylpentane	1068-87-7	25	2.4347E-01	3.4198E-02	2.4347E-04	1.8983E-06	2
1859	18206	C9H20	2,2,3,3-tetramethylpentane	7154-79-2	25	1.9730E-01	2.7712E-02	1.9730E-04	1.5383E-06	2
1860	18207	C9H20	2,2,3,4-tetramethylpentane	1186-53-4	25	3.0267E-01	4.2513E-02	3.0267E-04	2.3599E-06	2
1861	18208	C9H20	2,2,4,4-tetramethylpentane	1070-87-7	25	5.6969E-01	8.0019E-02	5.6970E-04	4.4418E-06	2
1862	18209	C9H20	2,3,3,4-tetramethylpentane	16747-38-9	25	1.8299E-01	2.5702E-02	1.8299E-04	1.4267E-06	2
1863	18210	C9H20	3,3-dimethylheptane	4032-86-4	25	2.3529E-01	3.3048E-02	2.3529E-04	1.8345E-06	2
1864	18236	C9H20O	1-nonanol	143-08-8	25	1.2980E+02	1.6211E+01	1.2982E-01	8.9990E-04	1
1865	18237	C9H20O	2-nonanol	628-99-9	15	2.5966E+02	3.2434E+01	2.5973E-01	1.8004E-03	1
1866	18238	C9H20O	2-nonanol, (±)	74683-66-2	25	4.7357E+02	5.9164E+01	4.7379E-01	3.2844E-03	2
1867	18239	C9H20O	3-nonanol	624-51-1	25	4.4012E+02	5.4984E+01	4.4031E-01	3.0523E-03	2
1868	18240	C9H20O	3-nonanol, (±)	74742-08-8	25	4.3212E+02	5.3984E+01	4.3230E-01	2.9968E-03	2
1869	18241	C9H20O	4-nonanol	5932-79-6	25	4.8821E+02	6.0995E+01	4.8845E-01	3.3860E-03	2
1870	18242	C9H20O	4-nonanol, (S)	52708-03-9	25	5.0329E+02	6.2879E+01	5.0355E-01	3.4906E-03	2
1871	18243	C9H20O	5-nonanol	623-93-8	25	4.2948E+02	5.3654E+01	4.2967E-01	2.9785E-03	2
1872	18244	C9H20O	2-methyl-1-octanol	818-81-5	25	3.2054E+02	4.0040E+01	3.2064E-01	2.2227E-03	2
1873	18245	C9H20O	3-methyl-1-octanol	38514-02-2	25	3.2054E+02	4.0040E+01	3.2064E-01	2.2227E-03	2
1874	18246	C9H20O	4-methyl-1-octanol	38514-03-3	25	3.2054E+02	4.0040E+01	3.2064E-01	2.2227E-03	2
1875	18247	C9H20O	5-methyl-1-octanol	38514-04-4	25	3.2054E+02	4.0040E+01	3.2064E-01	2.2227E-03	2
1876	18248	C9H20O	6-methyl-1-octanol	38514-05-5	25	2.1810E+02	2.7241E+01	2.1814E-01	1.5122E-03	2
1877	18249	C9H20O	7-methyl-1-octanol	2430-22-0	25	2.1810E+02	2.7241E+01	2.1814E-01	1.5122E-03	2
1878	18250	C9H20O	2-methyl-2-octanol	628-44-4	25	1.2000E+03	1.5001E+02	1.2014E+00	8.3282E-03	2
1879	18251	C9H20O	3-methyl-2-octanol	27644-49-1	25	1.0133E+03	1.2665E+02	1.0143E+00	7.0312E-03	2
1880	18252	C9H20O	5-methyl-2-octanol	66793-81-5	25	1.0133E+03	1.2665E+02	1.0143E+00	7.0312E-03	2
1881	18253	C9H20O	7-methyl-2-octanol	66793-83-7	25	1.0133E+03	1.2665E+02	1.0143E+00	7.0312E-03	2
1882	18254	C9H20O	2-methyl-3-octanol	26533-34-6	25	8.3986E+02	1.0496E+02	8.4056E-01	5.8268E-03	2
1883	18255	C9H20O	3-methyl-3-octanol	5340-36-3	25	6.2211E+02	7.7732E+01	6.2250E-01	4.3152E-03	2
1884	18256	C9H20O	4-methyl-3-octanol	66793-80-4	25	7.5184E+02	9.3952E+01	7.5240E-01	5.2157E-03	2
1885	18257	C9H20O	6-methyl-3-octanol	40225-75-0	25	7.5184E+02	9.3952E+01	7.5240E-01	5.2157E-03	2
1886	18258	C9H20O	7-methyl-3-octanol	66793-84-8	25	7.5184E+02	9.3952E+01	7.5240E-01	5.2157E-03	2
1887	18259	C9H20O	2-methyl-4-octanol	40575-41-5	25	8.3986E+02	1.0496E+02	8.4056E-01	5.8268E-03	2
1888	18260	C9H20O	3-methyl-4-octanol	26533-35-7	25	1.0658E+03	1.3322E+02	1.0669E+00	7.3958E-03	2
1889	18261	C9H20O	4-methyl-4-octanol	23418-37-3	25	1.0043E+03	1.2553E+02	1.0053E+00	6.9687E-03	2
1890	18262	C9H20O	5-methyl-4-octanol	59734-23-5	25	9.5475E+02	1.1933E+02	9.5567E-01	6.6247E-03	2
1891	18263	C9H20O	6-methyl-4-octanol	66793-82-6	25	9.5475E+02	1.1933E+02	9.5567E-01	6.6247E-03	2
1892	18264	C9H20O	7-methyl-4-octanol	33933-77-6	25	9.5475E+02	1.1933E+02	9.5567E-01	6.6247E-03	2
1893	18265	C9H20O	2-ethyl-1-heptanol	817-60-7	25	3.4101E+02	4.2599E+01	3.4113E-01	2.3647E-03	2
1894	18266	C9H20O	3-ethyl-1-heptanol	3525-25-5	25	2.0471E+02	2.5569E+01	2.0475E-01	1.4193E-03	2
1895	18267	C9H20O	5-ethyl-1-heptanol	998-65-2	25	3.4101E+02	4.2599E+01	3.4113E-01	2.3647E-03	2
1896	18268	C9H20O	2,2-dimethyl-1-heptanol	14250-79-4	25	5.1882E+02	6.4820E+01	5.1909E-01	3.5983E-03	2
1897	18269	C9H20O	4,6-dimethyl-1-heptanol	820-05-3	25	9.5475E+02	1.1933E+02	9.5567E-01	6.6247E-03	2
1898	18270	C9H20O	6,6-dimethyl-1-heptanol	65769-10-0	25	9.5475E+02	1.1933E+02	9.5567E-01	6.6247E-03	2
1899	18271	C9H20O	2-ethyl-2-heptanol	---	25	9.5475E+02	1.1933E+02	9.5567E-01	6.6247E-03	2
1900	18272	C9H20O	3-ethyl-2-heptanol	19780-39-3	25	9.5475E+02	1.1933E+02	9.5567E-01	6.6247E-03	2
1901	18273	C9H20O	2,3-dimethyl-2-heptanol	66794-00-1	25	9.5475E+02	1.1933E+02	9.5567E-01	6.6247E-03	2
1902	18274	C9H20O	2,4-dimethyl-2-heptanol	65822-93-7	25	9.5475E+02	1.1933E+02	9.5567E-01	6.6247E-03	2
1903	18275	C9H20O	2,5-dimethyl-2-heptanol	---	25	9.5475E+02	1.1933E+02	9.5567E-01	6.6247E-03	2
1904	18276	C9H20O	2,6-dimethyl-2-heptanol	13254-34-7	25	1.8137E+03	2.2686E+02	1.8170E+00	1.2596E-02	2
1905	18277	C9H20O	4,6-dimethyl-2-heptanol	51079-52-8	25	4.5934E+02	5.7387E+01	4.5956E-01	3.1857E-03	2
1906	18278	C9H20O	5,6-dimethyl-2-heptanol	58795-24-7	25	5.1882E+02	6.4820E+01	5.1909E-01	3.5983E-03	2
1907	18279	C9H20O	3-ethyl-3-heptanol	19780-41-7	25	9.3506E+02	1.1687E+02	9.3594E-01	6.4880E-03	2
1908	18280	C9H20O	2,2-dimethyl-3-heptanol	19549-70-3	25	1.2107E+03	1.5135E+02	1.2121E+00	8.4026E-03	2
1909	18281	C9H20O	2,3-dimethyl-3-heptanol	19549-71-4	25	1.5199E+03	1.9006E+02	1.5222E+00	1.0552E-02	2
1910	18282	C9H20O	2,6-dimethyl-3-heptanol	19549-73-6	25	1.4328E+03	1.7916E+02	1.4349E+00	9.9468E-03	2
1911	18283	C9H20O	3,5-dimethyl-3-heptanol	19549-74-7	25	1.2845E+03	1.6059E+02	1.2861E+00	8.9155E-03	2
1912	18284	C9H20O	3,6-dimethyl-3-heptanol	1573-28-0	25	1.6122E+03	2.0162E+02	1.6148E+00	1.1194E-02	2
1913	18285	C9H20O	4-ethyl-4-heptanol	597-90-0	25	1.1309E+03	1.4137E+02	1.1322E+00	7.8485E-03	2
1914	18286	C9H20O	2,2-dimethyl-4-heptanol	66793-99-5	25	1.5199E+03	1.9006E+02	1.5222E+00	1.0552E-02	2
1915	18287	C9H20O	2,4-dimethyl-4-heptanol	19549-77-0	25	1.7715E+03	2.2157E+02	1.7747E+00	1.2302E-02	2
1916	18288	C9H20O	2,6-dimethyl-4-heptanol	108-82-7	25	1.0000E+03	1.2499E+02	1.0010E+00	6.9390E-03	1
1917	18289	C9H20O	3,3-dimethyl-4-heptanol	19549-78-1	25	1.3507E+03	1.6887E+02	1.3525E+00	9.3756E-03	2
1918	18290	C9H20O	3,5-dimethyl-4-heptanol	19549-79-2	25	7.0170E+02	8.7683E+01	7.0220E-01	4.8677E-03	2
1919	18291	C9H20O	2-propyl-1-hexanol	817-46-9	25	6.2776E+02	7.8439E+01	6.2816E-01	4.3544E-03	2
1920	18292	C9H20O	3-propyl-1-hexanol	66793-85-9	25	6.2776E+02	7.8439E+01	6.2816E-01	4.3544E-03	2

Table 3 Solubility in Water - Organic Compounds

					Solubility in Water, S (ppm - parts per million, wt - weight, g - gram)					
NO	ID	FORMULA	NAME	CAS No	T, C	S@T, ppm(wt)	S@T, ppm(mol)	S@T, g/1000 g of H2O	S@T, mol/1000 g of H2O	code
1921	18293	C9H20O	1-methyl-2-ethyl-1-hexanol	---	25	6.2776E+02	7.8439E+01	6.2816E-01	4.3544E-03	2
1922	18294	C9H20O	1-methyl-3-ethyl-1-hexanol	---	25	5.1882E+02	6.4820E+01	5.1909E-01	3.5983E-03	2
1923	18295	C9H20O	3-methyl-2-ethyl-1-hexanol	66794-04-5	25	4.8821E+02	6.0995E+01	4.8845E-01	3.3860E-03	2
1924	18296	C9H20O	4-methyl-2-ethyl-1-hexanol	66794-06-7	25	4.3212E+02	5.3984E+01	4.3230E-01	2.9968E-03	2
1925	18297	C9H20O	5-methyl-2-ethyl-1-hexanol	66794-07-8	25	4.9269E+02	6.1554E+01	4.9293E-01	3.4170E-03	2
1926	18298	C9H20O	3,3,5-trimethyl-1-hexanol	1484-87-3	25	4.9269E+02	6.1554E+01	4.9293E-01	3.4170E-03	2
1927	18299	C9H20O	3,4,4-trimethyl-1-hexanol	66793-73-5	25	5.5126E+02	6.8875E+01	5.5157E-01	3.8235E-03	2
1928	18300	C9H20O	3,5,5-trimethyl-1-hexanol	3452-97-9	25	4.8821E+02	6.0995E+01	4.8845E-01	3.3860E-03	2
1929	18301	C9H20O	4,5,5-trimethyl-1-hexanol	66793-75-7	25	2.8040E+02	3.5026E+01	2.8048E-01	1.9443E-03	2
1930	18302	C9H20O	2-methyl-3-ethyl-2-hexanol	66794-02-3	25	1.2000E+03	1.5001E+02	1.2014E+00	8.3282E-03	2
1931	18303	C9H20O	2,3,4-trimethyl-2-hexanol	21102-13-6	25	4.9269E+02	6.1554E+01	4.9293E-01	3.4170E-03	2
1932	18304	C9H20O	2,4,4-trimethyl-2-hexanol	66793-91-7	25	4.9269E+02	6.1554E+01	4.9293E-01	3.4170E-03	2
1933	18305	C9H20O	2,4,5-trimethyl-2-hexanol	66793-93-9	25	4.9269E+02	6.1554E+01	4.9293E-01	3.4170E-03	2
1934	18306	C9H20O	2,5,5-trimethyl-2-hexanol	66793-71-3	25	4.9269E+02	6.1554E+01	4.9293E-01	3.4170E-03	2
1935	18307	C9H20O	2-methyl-3-ethyl-3-hexanol	66794-03-4	25	8.3485E+02	1.0433E+02	8.3555E-01	5.7921E-03	2
1936	18308	C9H20O	2-methyl-4-ethyl-3-hexanol	33943-21-4	25	1.1410E+03	1.4263E+02	1.1423E+00	7.9187E-03	2
1937	18309	C9H20O	3-methyl-4-ethyl-3-hexanol	66794-05-6	25	1.1410E+03	1.4263E+02	1.1423E+00	7.9187E-03	2
1938	18310	C9H20O	4-methyl-3-ethyl-3-hexanol	51200-80-7	25	1.1410E+03	1.4263E+02	1.1423E+00	7.9187E-03	2
1939	18311	C9H20O	5-methyl-3-ethyl-3-hexanol	597-77-3	25	1.7101E+03	2.1387E+02	1.7130E+00	1.1874E-02	2
1940	18312	C9H20O	2,2,3-trimethyl-3-hexanol	5340-41-0	25	1.6028E+03	2.0043E+02	1.6053E+00	1.1128E-02	2
1941	18313	C9H20O	2,2,4-trimethyl-3-hexanol	66793-89-3	25	2.0401E+03	2.5523E+02	2.0443E+00	1.4171E-02	2
1942	18314	C9H20O	2,2,5-trimethyl-3-hexanol	3970-60-3	25	3.4615E+03	4.3359E+02	3.4735E+00	2.4078E-02	2
1943	18315	C9H20O	2,3,4-trimethyl-3-hexanol	66793-90-6	25	2.4549E+03	3.0723E+02	2.4610E+00	1.7060E-02	2
1944	18316	C9H20O	2,3,5-trimethyl-3-hexanol	65927-60-8	25	2.4549E+03	3.0723E+02	2.4610E+00	1.7060E-02	2
1945	18317	C9H20O	2,4,4-trimethyl-3-hexanol	66793-92-8	25	1.8137E+03	2.2686E+02	1.8170E+00	1.2596E-02	2
1946	18318	C9H20O	2,5,5-trimethyl-3-hexanol	66793-72-4	25	2.0582E+03	2.5750E+02	2.0625E+00	1.4297E-02	2
1947	18319	C9H20O	3,4,4-trimethyl-3-hexanol	66793-74-6	25	2.4334E+03	3.0453E+02	2.4393E+00	1.6910E-02	2
1948	18320	C9H20O	3,5,5-trimethyl-3-hexanol	66810-87-5	25	2.4549E+03	3.0723E+02	2.4610E+00	1.7060E-02	2
1949	18321	C9H20O	4-methyl-2-propyl-1-pentanol	54004-41-0	25	5.1882E+02	6.4820E+01	5.1909E-01	3.5983E-03	2
1950	18322	C9H20O	4-methyl-2-isopropyl-1-pentanol	55505-24-3	25	7.0170E+02	8.7683E+01	7.0220E-01	4.8677E-03	2
1951	18323	C9H20O	2,2-dimethyl-3-ethyl-1-pentanol	66793-95-1	25	6.6674E+02	8.3312E+01	6.6719E-01	4.6250E-03	2
1952	18324	C9H20O	2,4-dimethyl-2-ethyl-1-pentanol	66793-98-4	25	6.6075E+02	8.2563E+01	6.6119E-01	4.5834E-03	2
1953	18325	C9H20O	3,3-diethyl-2-pentanol	66793-94-0	25	6.6674E+02	8.3312E+01	6.6719E-01	4.6250E-03	2
1954	18326	C9H20O	2,3-dimethyl-3-ethyl-2-pentanol	66793-97-3	25	6.6674E+02	8.3312E+01	6.6719E-01	4.6250E-03	2
1955	18327	C9H20O	4,4-dimethyl-3-ethyl-2-pentanol	21102-09-0	25	6.6674E+02	8.3312E+01	6.6719E-01	4.6250E-03	2
1956	18328	C9H20O	2,3,3,4-tetramethyl-2-pentanol	66793-86-0	25	1.0753E+03	1.3441E+02	1.0765E+00	7.4621E-03	2
1957	18329	C9H20O	2,3,4,4-tetramethyl-2-pentanol	66793-87-1	25	1.0753E+03	1.3441E+02	1.0765E+00	7.4621E-03	2
1958	18330	C9H20O	3,3,4,4-tetramethyl-2-pentanol	66793-88-2	25	1.0658E+03	1.3322E+02	1.0669E+00	7.3958E-03	2
1959	18331	C9H20O	2-methyl-3-isopropyl-3-pentanol	---	25	1.2000E+03	1.5001E+02	1.2014E+00	8.3282E-03	2
1960	18332	C9H20O	2,2-dimethyl-3-ethyl-3-pentanol	66793-96-2	25	1.5199E+03	1.9006E+02	1.5222E+00	1.0552E-02	2
1961	18333	C9H20O	2,4-dimethyl-3-ethyl-3-pentanol	3970-59-0	25	1.2071E+03	1.5090E+02	1.2085E+00	8.3777E-03	2
1962	18334	C9H20O	2,2,3,4-tetramethyl-3-pentanol	62185-29-9	25	1.5021E+03	1.8783E+02	1.5044E+00	1.0428E-02	2
1963	18335	C9H20O	2,2,4,4-tetramethyl-3-pentanol	14609-79-1	25	2.2946E+03	2.8713E+02	2.2999E+00	1.5943E-02	2
1964	18336	C9H20O	ethyl heptyl ether	1969-43-3	25	5.4790E+01	6.8425E+00	5.4793E-02	3.7983E-04	2
1965	18365	C9H20S	nonyl mercaptan	1455-21-6	25	1.8771E+00	2.1092E-01	1.8771E-03	1.1708E-05	2
1966	18366	C9H20S	2-nonanethiol	13281-11-3	25	3.4756E+00	3.9054E-01	3.4756E-03	2.1678E-05	2
1967	18367	C9H20S	methyl octyl sulfide	3698-95-1	25	8.5797E+00	9.6407E-01	8.5798E-03	5.3515E-05	2
1968	18368	C9H20S	ethyl heptyl sulfide	24768-44-3	25	8.5797E+00	9.6407E-01	8.5798E-03	5.3515E-05	2
1969	18369	C9H20S	propyl hexyl sulfide	24768-43-2	25	8.5797E+00	9.6407E-01	8.5798E-03	5.3515E-05	2
1970	18370	C9H20S	butyl pentyl sulfide	24768-42-1	25	8.5797E+00	9.6407E-01	8.5798E-03	5.3515E-05	2
1971	18371	C9H20S	tert-nonyl mercaptan, isomers	25360-10-5	25	1.0373E+01	1.1656E+00	1.0373E-02	6.4703E-05	2
1972	18388	C9H21N	nonylamine	112-20-9	25	7.9564E+02	1.0011E+02	7.9627E-01	5.5577E-03	2
1973	18389	C9H21N	methyloctylamine	2439-54-5	25	1.1877E+03	1.4950E+02	1.1891E+00	8.2998E-03	2
1974	18390	C9H21N	ethylheptylamine	66793-76-8	25	1.6514E+03	2.0795E+02	1.6542E+00	1.1546E-02	2
1975	18391	C9H21N	dimethylheptylamine	5277-11-2	25	7.4822E+02	9.4142E+01	7.4878E-01	5.2262E-03	2
1976	18392	C9H21N	diethylpentylamine	2162-91-6	25	1.7964E+03	2.2624E+02	1.7997E+00	1.2561E-02	2
1977	18393	C9H21N	tripropylamine	102-69-2	25	7.4790E+02	9.4102E+01	7.4846E-01	5.2240E-03	1
1978	18394	C9H21N	tert-amyl-tert-butylamine	2085-66-7	25	1.4307E+04	1.8218E+03	1.4515E+01	1.0131E-01	2
1979	18395	C9H21N	N-methyldibutylamine	3405-45-6	---	1.4730E+03	---	1.4752E+00	1.0296E-02	2
1980	18598	C10H6O8	pyromellitic acid	89-05-4	16	1.4000E+04	1.0054E+03	1.4199E+01	5.5867E-02	1
1981	18601	C10H7Br	1-bromonaphthalene	90-11-9	21	9.3180E+00	8.1067E-01	9.3181E-03	4.5000E-05	1
1982	18606	C10H7Cl	1-chloronaphthalene	90-13-1	25	2.2400E+01	2.4815E+00	2.2401E-02	1.3775E-04	1
1983	18607	C10H7Cl	2-chloronaphthalene	91-58-7	25	2.2400E+01	2.4815E+00	2.2401E-02	1.3775E-04	2
1984	18608	C10H7Cl	monochloronaphthalene	25586-43-0	25	2.2400E+01	2.4815E+00	2.2401E-02	1.3775E-04	2
1985	18669	C10H8	naphthalene	91-20-3	25	3.2050E+01	4.5048E+00	3.2051E-02	2.5006E-04	1
1986	19195	C10H10O4	dimethyl phthalate	131-11-3	20	4.2920E+03	3.9973E+02	4.3105E+00	2.2198E-02	1
1987	19359	C10H12	1,2,3,4-tetrahydronaphthalene	119-64-2	28	4.6700E+01	6.3638E+00	4.6702E-02	3.5326E-04	1
1988	19484	C10H12O	anethole	104-46-1	25	1.1100E+02	1.3494E+01	1.1101E-01	7.4905E-04	1
1989	19865	C10H14	butylbenzene	104-51-8	25	1.3820E+01	1.8549E+00	1.3820E-02	1.0297E-04	1
1990	19866	C10H14	isobutylbenzene	538-93-2	25	1.0100E+01	1.3556E+00	1.0100E-02	7.5250E-05	1
1991	19867	C10H14	sec-butylbenzene	135-98-8	25	1.7600E+01	2.3623E+00	1.7600E-02	1.3113E-04	1
1992	19868	C10H14	tert-butylbenzene	98-06-6	25	2.9500E+01	3.9596E+00	2.9501E-02	2.1979E-04	1
1993	19869	C10H14	cymene	25155-15-1	25	3.5549E+01	4.7714E+00	3.5550E-02	2.6486E-04	2
1994	19870	C10H14	m-cymene	535-77-3	25	3.8184E+01	5.1252E+00	3.8185E-02	2.8450E-04	2
1995	19871	C10H14	o-cymene	527-84-4	25	3.2072E+01	4.3048E+00	3.2073E-02	2.3896E-04	2
1996	19872	C10H14	p-cymene	99-87-6	25	3.4150E+01	4.5837E+00	3.4151E-02	2.5444E-04	1
1997	19873	C10H14	1-methyl-2-propylbenzene	1074-17-5	25	2.1981E+01	2.9503E+00	2.1981E-02	1.6377E-04	2
1998	19874	C10H14	1-methyl-3-propylbenzene	1074-43-7	25	2.5916E+01	3.4785E+00	2.5917E-02	1.9309E-04	2
1999	19875	C10H14	1-methyl-4-propylbenzene	1074-55-1	25	2.8346E+01	3.8047E+00	2.8347E-02	2.1120E-04	2
2000	19876	C10H14	(S)-(1-methylpropyl)benzene	5787-28-0	25	2.3216E+01	3.1162E+00	2.3217E-02	1.7298E-04	2

Table 3 Solubility in Water - Organic Compounds

					Solubility in Water, S (ppm - parts per million, wt - weight, g - gram)					
NO	ID	FORMULA	NAME	CAS No	T, C	S@T, ppm(wt)	S@T, ppm(mol)	S@T, g/1000 g of H2O	S@T, mol/1000 g of H2O	code
2001	19877	C10H14	diethylbenzene	25340-17-4	25	2.6033E+01	3.4941E+00	2.6033E-02	1.9396E-04	2
2002	19878	C10H14	o-diethylbenzene	135-01-3	25	2.3859E+01	3.2023E+00	2.3859E-02	1.7776E-04	2
2003	19879	C10H14	m-diethylbenzene	141-93-5	25	2.7164E+01	3.6460E+00	2.7165E-02	2.0239E-04	2
2004	19880	C10H14	p-diethylbenzene	105-05-5	25	2.4830E+01	3.3327E+00	2.4831E-02	1.8500E-04	1
2005	19881	C10H14	3-ethyl-o-xylene	933-98-2	25	1.3351E+01	1.7920E+00	1.3351E-02	9.9470E-05	2
2006	19882	C10H14	4-ethyl-o-xylene	934-80-5	25	1.6797E+01	2.2544E+00	1.6797E-02	1.2514E-04	2
2007	19883	C10H14	2-ethyl-m-xylene	2870-04-4	25	1.6557E+01	2.2223E+00	1.6558E-02	1.2336E-04	2
2008	19884	C10H14	4-ethyl-m-xylene	874-41-9	25	1.8088E+01	2.4278E+00	1.8088E-02	1.3476E-04	2
2009	19885	C10H14	5-ethyl-m-xylene	934-74-7	25	2.3436E+01	3.1457E+00	2.3437E-02	1.7461E-04	2
2010	19886	C10H14	2-ethyl-p-xylene	1758-88-9	25	1.9776E+01	2.6544E+00	1.9777E-02	1.4735E-04	2
2011	19887	C10H14	1,2,3,4-tetramethylbenzene	488-23-3	25	7.3385E+00	9.8498E-01	7.3386E-03	5.4675E-05	2
2012	19888	C10H14	1,2,3,5-tetramethylbenzene	527-53-7	25	1.0720E+01	1.4388E+00	1.0720E-02	7.9865E-05	2
2013	19889	C10H14	1,2,4,5-tetramethylbenzene	95-93-2	25	3.4800E+00	4.6708E-01	3.4800E-03	2.5927E-05	1
2014	20089	C10H14O2	p-tert-butylcatechol	98-29-3	25	2.0000E-07	2.1676E-08	2.0000E-10	1.2032E-12	1
2015	20195	C10H15N	N,N-dietyhlaniline	91-66-7	12	1.4400E+04	1.7606E+03	1.4610E+01	9.7901E-02	1
2016	20196	C10H15N	2,6-diethylaniline	579-66-8	27	6.7000E+02	8.0927E+01	6.7045E-01	4.4925E-03	1
2017	20197	C10H15N	2-sec-butylaniline	55751-54-7	25	7.5350E+03	9.1565E+02	7.5922E+00	5.0874E-02	2
2018	20198	C10H15N	2-tert-butylaniline	6310-21-0	25	7.5350E+03	9.1565E+02	7.5922E+00	5.0874E-02	2
2019	20199	C10H15N	4-butylaniline	104-13-2	25	7.5350E+03	9.1565E+02	7.5922E+00	5.0874E-02	2
2020	20200	C10H15N	4-sec-butylaniline	30273-11-1	25	7.5350E+03	9.1565E+02	7.5922E+00	5.0874E-02	2
2021	20201	C10H15N	4-tert-butylaniline	769-92-6	25	7.5350E+03	9.1565E+02	7.5922E+00	5.0874E-02	2
2022	20202	C10H15N	N-sec-butylaniline	6068-69-5	25	7.5350E+03	9.1565E+02	7.5922E+00	5.0874E-02	2
2023	20203	C10H15N	N-tert-butylaniline	937-33-7	25	7.5350E+03	9.1565E+02	7.5922E+00	5.0874E-02	2
2024	20204	C10H15N	N-butylaniline	1126-78-9	25	7.5350E+03	9.1565E+02	7.5922E+00	5.0874E-02	2
2025	20205	C10H15N	4-isobutylaniline	30090-17-6	25	7.5350E+03	9.1565E+02	7.5922E+00	5.0874E-02	2
2026	20206	C10H15N	N-isobutylaniline	588-47-6	25	7.5350E+03	9.1565E+02	7.5922E+00	5.0874E-02	2
2027	20207	C10H15N	4-methyl-N-isopropylaniline	10436-75-6	25	7.5350E+03	9.1565E+02	7.5922E+00	5.0874E-02	2
2028	20208	C10H15N	2-methyl-5-isopropylaniline	2051-53-8	25	7.5350E+03	9.1565E+02	7.5922E+00	5.0874E-02	2
2029	20209	C10H15N	4-methyl-N-propylaniline	54837-90-0	25	7.5350E+03	9.1565E+02	7.5922E+00	5.0874E-02	2
2030	20210	C10H15N	2,3,4,5-tetramethylaniline	2217-45-0	25	7.5350E+03	9.1565E+02	7.5922E+00	5.0874E-02	2
2031	20211	C10H15N	2,3,4,6-tetramethylaniline	488-71-1	25	7.5350E+03	9.1565E+02	7.5922E+00	5.0874E-02	2
2032	20212	C10H15N	N,N,2,6-tetramethylaniline	769-06-2	25	7.5350E+03	9.1565E+02	7.5922E+00	5.0874E-02	2
2033	20230	C10H15N	3,5,N,N-tetramethylaniline	4913-13-7	25	7.5350E+03	9.1565E+02	7.5922E+00	5.0874E-02	2
2034	20231	C10H15N	aniline, p-isopropyl-N-methyl-,	6950-79-4	25	7.5350E+03	9.1565E+02	7.5922E+00	5.0874E-02	2
2035	20316	C10H16	D-limonene	5989-27-5	25	1.3800E+01	1.8248E+00	1.3800E-02	1.0130E-04	1
2036	20317	C10H16	limonene	138-86-3	25	1.3800E+01	1.8248E+00	1.3800E-02	1.0130E-04	2
2037	20318	C10H16	L-limonene	5989-54-8	25	1.3800E+01	1.8248E+00	1.3800E-02	1.0130E-04	2
2038	20423	C10H16O	camphor	76-22-2	25	1.5680E+03	1.8581E+02	1.5705E+00	1.0316E-02	1
2039	20540	C10H16S	3-hexylthiophene	1693-86-3	25	1.3048E+00	1.3967E-01	1.3048E-03	7.7528E-06	2
2040	20593	C10H18	1-decyne	764-93-2	25	1.8898E+00	2.4625E-01	1.8898E-03	1.3669E-05	2
2041	20594	C10H18	2-decyne	2384-70-5	25	1.0795E+00	1.4067E-01	1.0795E-03	7.8083E-06	2
2042	20595	C10H18	3-decyne	2384-85-2	25	1.4234E+00	1.8548E-01	1.4234E-03	1.0296E-05	2
2043	20596	C10H18	4-decyne	2384-86-3	25	2.2892E+00	2.9830E-01	2.2892E-03	1.6558E-05	2
2044	20597	C10H18	5-decyne	1942-46-7	25	1.6087E+00	2.0962E-01	1.6087E-03	1.1636E-05	2
2045	20598	C10H18	2,2-dimethyl-3-octyne	19482-57-6	25	2.2892E+00	2.9830E-01	2.2892E-03	1.6558E-05	2
2046	20599	C10H18	8-methyl-4-nonyne	70732-43-3	25	2.2892E+00	2.9830E-01	2.2892E-03	1.6558E-05	2
2047	20600	C10H18	5-ethyl-5-methyl-3-heptyne	61228-10-2	25	2.2892E+00	2.9830E-01	2.2892E-03	1.6558E-05	2
2048	20601	C10H18	cis-decahydronaphthalene	493-01-6	25	8.8900E-01	1.1584E-01	8.8900E-04	6.4302E-06	1
2049	20602	C10H18	trans-decahydronaphthalene	493-02-7	25	8.8900E-01	1.1584E-01	8.8900E-04	6.4302E-06	1
2050	20608	C10H18	1,3-decadiene	2051-25-4	25	4.6729E-01	6.0890E-02	4.6729E-04	3.3800E-06	2
2051	20609	C10H18	1,9-decadiene	1647-16-1	25	5.2216E-01	6.8040E-02	5.2216E-04	3.7768E-06	2
2052	20610	C10H18	2,6-dimethyl-2,5-octadiene	116668-48-5	25	4.3022E-01	5.6060E-02	4.3022E-04	3.1118E-06	2
2053	20611	C10H18	2,7-dimethyl-2,6-octadiene	16736-42-8	25	5.5212E-01	7.1944E-02	5.5212E-04	3.9936E-06	2
2054	20612	C10H18	3,6-dimethyl-2,6-octadiene	116668-49-6	25	1.0927E+00	1.4239E-01	1.0927E-03	7.9038E-06	2
2055	20613	C10H18	3,7-dimethyl-2,4-octadiene	56523-26-3	25	5.6779E-01	7.3985E-02	5.6779E-04	4.1069E-06	2
2056	20614	C10H18	4,5-dimethyl-2,6-octadiene	18476-57-8	25	1.1247E+00	1.4656E-01	1.1247E-03	8.1354E-06	2
2057	20615	C10H18	cis-2,6-dimethyl-2,6-octadiene	2492-22-0	25	4.9391E-01	6.4359E-02	4.9391E-04	3.5725E-06	2
2058	20636	C10H18	decahydronaphthalene; (cis+trans)	91-17-8	25	8.8900E-01	1.1584E-01	8.8900E-04	6.4302E-06	2
2059	20637	C10H18	3,7-dimethyl-1,6-octadiene	2436-90-0	25	9.4893E-01	1.2365E-01	9.4893E-04	6.8637E-06	2
2060	20820	C10H18O4	sebacic acid	111-20-6	20	1.0000E+03	8.9154E+01	1.0010E+00	4.9493E-03	1
2061	20922	C10H20	pentylcyclopentane	3741-00-2	25	1.1500E-01	1.4770E-02	1.1500E-04	8.1985E-07	1
2062	20923	C10H20	butylcyclohexane	1678-93-9	25	1.7779E-01	2.2833E-02	1.7779E-04	1.2675E-06	2
2063	20924	C10H20	tert-butylcyclohexane	3178-22-1	25	2.9568E-01	3.7974E-02	2.9568E-04	2.1079E-06	2
2064	20925	C10H20	sec-butylcyclohexane	7058-01-7	25	1.9424E-01	2.4946E-02	1.9424E-04	1.3847E-06	2
2065	20926	C10H20	isobutylcyclohexane	1678-98-4	25	2.9894E-01	3.8393E-02	2.9894E-04	2.1312E-06	2
2066	20927	C10H20	1-methyl-1-propylcyclohexane	4258-93-9	25	2.8770E-01	3.6949E-02	2.8770E-04	2.0510E-06	2
2067	20928	C10H20	1-methyl-2-propylcyclohexane	4291-79-6	25	2.3161E-01	2.9746E-02	2.3161E-04	1.6512E-06	2
2068	20929	C10H20	1-methyl-3-propylcyclohexane	4291-80-9	25	4.3606E-01	5.6004E-02	4.3606E-04	3.1087E-06	2
2069	20930	C10H20	1-methyl-4-propylcyclohexane	4291-81-0	25	2.3792E-01	3.0557E-02	2.3792E-04	1.6962E-06	2
2070	20931	C10H20	1-methyl-2-isopropylcyclohexane	16580-23-7	25	3.0390E-01	3.9030E-02	3.0390E-04	2.1665E-06	2
2071	20932	C10H20	1-methyl-3-isopropylcyclohexane	16580-24-8	25	3.8986E-01	5.0071E-02	3.8986E-04	2.7794E-06	2
2072	20933	C10H20	1-methyl-4-isopropylcyclohexane	99-82-1	25	3.0894E-01	3.9678E-02	3.0894E-04	2.2025E-06	2
2073	20934	C10H20	cis-1-methyl-4-isopropylcyclohexane	6069-98-3	25	2.8770E-01	3.6949E-02	2.8770E-04	2.0510E-06	2
2074	20935	C10H20	trans-1-methyl-4-isopropylcyclohexane	1678-82-6	25	3.1065E-01	3.9897E-02	3.1065E-04	2.2147E-06	2
2075	20936	C10H20	1,1,3,4-tetramethylcyclohexane	24612-75-7	25	5.6244E-01	7.2235E-02	5.6244E-04	4.0097E-06	2
2076	20937	C10H20	trans-1,1,3,5-tetramethylcyclohexane	50876-31-8	25	6.9120E-01	8.8773E-02	6.9120E-04	4.9277E-06	2
2077	20938	C10H20	cis-1,1,3,5-tetramethylcyclohexane	50876-32-9	25	8.6533E-01	1.1114E-01	8.6533E-04	6.1691E-06	2
2078	20939	C10H20	diethylcyclohexane; (mixed isomers)	1331-43-7	25	2.5801E-01	3.3137E-02	2.5801E-04	1.8394E-06	2
2079	20943	C10H20	1-decene	872-05-9	25	1.4530E-01	1.8661E-02	1.4530E-04	1.0359E-06	2
2080	20944	C10H20	4-decene	19689-18-0	25	1.4530E-01	1.8661E-02	1.4530E-04	1.0359E-06	2

Table 3 Solubility in Water - Organic Compounds

NO	ID	FORMULA	NAME	CAS No	Solubility in Water, S (ppm - parts per million, wt - weight, g - gram) T, C	S@T, ppm(wt)	S@T, ppm(mol)	S@T, g/1000 g of H2O	S@T, mol/1000 g of H2O	code
2081	20945	C10H20	cis-5-decene	7433-78-5	25	1.4214E-01	1.8256E-02	1.4214E-04	1.0134E-06	2
2082	20946	C10H20	trans-5-decene	7433-56-9	25	1.4214E-01	1.8256E-02	1.4214E-04	1.0134E-06	2
2083	20947	C10H20	2-methyl-1-nonene	2980-71-4	25	1.6406E-01	2.1071E-02	1.6406E-04	1.1696E-06	2
2084	20948	C10H20	2-methyl-3-nonene	53966-53-3	25	2.4853E-01	3.1920E-02	2.4853E-04	1.7718E-06	2
2085	20949	C10H20	5-methyl-4-nonene	15918-07-7	25	1.7927E-01	2.3023E-02	1.7927E-04	1.2780E-06	2
2086	20950	C10H20	3,7-dimethyl-1-octene	4984-01-4	25	3.7112E-01	4.7663E-02	3.7112E-04	2.6457E-06	2
2087	20951	C10H20	2,6-dimethyl-2-octene	4057-42-5	25	2.2193E-01	2.8503E-02	2.2193E-04	1.5822E-06	2
2088	20952	C10H20	4-propyl-3-heptene	4485-13-6	25	2.5569E-01	3.2839E-02	2.5569E-04	1.8229E-06	2
2089	20953	C10H20	2,4,6-trimethyl-3-heptene, isomer	123434-07-1	25	1.7927E-01	2.3023E-02	1.7927E-04	1.2780E-06	2
2090	20954	C10H20	2,2,5,5-tetramethylhex-3-ene	22808-06-6	25	2.1202E+00	2.7230E-01	2.1202E-03	1.5115E-05	2
2091	20955	C10H20	cis-1,2-di-tert-butylethene	692-47-7	25	1.7927E-01	2.3023E-02	1.7927E-04	1.2780E-06	2
2092	20956	C10H20	(E)-2,2,5,5-tetramethylhex-3-ene	692-48-8	25	2.0109E+00	2.5827E-01	2.0109E-03	1.4336E-05	2
2093	20958	C10H20Br2	1,1-dibromodecane	59104-80-2	25	2.5670E+00	1.5411E-01	2.5670E-03	8.5546E-06	2
2094	20959	C10H20Br2	1,10-dibromodecane	4101-68-2	25	2.5670E+00	1.5411E-01	2.5670E-03	8.5546E-06	2
2095	20960	C10H20Br2	5,6-dibromodecane	77928-86-0	25	2.5670E+00	1.5411E-01	2.5670E-03	8.5546E-06	2
2096	20962	C10H20Cl2	1,1-dichlorodecane	3162-62-7	25	9.1917E-01	7.8413E-02	9.1917E-04	4.3527E-06	2
2097	20963	C10H20Cl2	1,10-dichlorodecane	2162-98-3	25	9.1917E-01	7.8413E-02	9.1917E-04	4.3527E-06	2
2098	20992	C10H20O	decanal	112-31-2	25	4.7940E+01	5.5269E+00	4.7943E-02	3.0680E-04	2
2099	20993	C10H20O	2-decanone	693-54-9	25	6.0624E+01	6.9892E+00	6.0627E-02	3.8797E-04	2
2100	20994	C10H20O	3-decanone	928-80-3	25	9.5683E+01	1.1031E+01	9.5692E-02	6.1236E-04	2
2101	20995	C10H20O	4-decanone	624-16-8	25	7.6755E+01	8.8491E+00	7.6760E-02	4.9121E-04	2
2102	20996	C10H20O	2-methyl-5-nonanone	22287-02-1	25	9.2731E+01	1.0691E+01	9.2740E-02	5.9346E-04	2
2103	21019	C10H20O	5-decanone	820-29-1	25	9.0432E+01	1.0426E+01	9.0440E-02	5.7875E-04	2
2104	21025	C10H20O	isodecanal	1321-89-7	25	1.1932E+02	1.3757E+01	1.1933E-01	7.6363E-04	2
2105	21034	C10H20O	2-methyl-3-nonanone	5445-31-8	25	1.1643E+02	1.3424E+01	1.1645E-01	7.4517E-04	2
2106	21040	C10H20O2	decanoic acid	334-48-5	20	1.5000E+02	1.5688E+01	1.5002E-01	8.7087E-04	1
2107	21041	C10H20O2	2,2-dimethyloctanoic acid	29662-90-6	25	7.5505E+03	7.9497E+02	7.6079E+00	4.4163E-02	2
2108	21042	C10H20O2	3-methylnonanoic acid	35205-79-9	25	7.5505E+03	7.9497E+02	7.6079E+00	4.4163E-02	2
2109	21043	C10H20O2	nonyl formate	5451-92-3	25	1.5416E+01	1.6121E+00	1.5416E-02	8.9489E-05	2
2110	21044	C10H20O2	2-ethylhexyl acetate	103-09-3	25	8.7536E+01	9.1548E+00	8.7543E-02	5.0818E-04	2
2111	21045	C10H20O2	octyl acetate	112-14-1	25	4.7053E+01	4.9208E+00	4.7056E-02	2.7315E-04	2
2112	21046	C10H20O2	6-methyl-2-heptyl acetate	67952-57-2	25	1.7409E+02	1.8208E+01	1.7412E-01	1.0108E-03	2
2113	21047	C10H20O2	6-methyl-3-heptyl acetate	32764-34-4	25	1.9917E+02	2.0832E+01	1.9921E-01	1.1564E-03	2
2114	21048	C10H20O2	2-methylheptyl acetate, (±)	74112-36-0	25	1.1317E+02	1.1836E+01	1.1318E-01	6.5700E-04	2
2115	21049	C10H20O2	heptyl propanoate	2216-81-1	25	5.0465E+01	5.2776E+00	5.0467E-02	2.9296E-04	2
2116	21050	C10H20O2	hexyl butanoate	2639-63-6	25	6.2591E+01	6.5459E+00	6.2595E-02	3.6336E-04	2
2117	21051	C10H20O2	1-methylpentyl butanoate, (S)	116723-96-7	25	8.3807E+01	8.7648E+00	8.3814E-02	4.8653E-04	2
2118	21053	C10H20O2	1,1-dimethylpropyl 3-methylbutanoate	542-37-0	25	3.6011E+02	3.7670E+01	3.6024E-01	2.0911E-03	2
2119	21066	C10H20O2	8-methylnonanoic acid	5963-14-4	25	7.5505E+03	7.9497E+02	7.6079E+00	4.4163E-02	2
2120	21067	C10H20O2	neodecanoic acid	26896-20-8	25	4.7940E+02	5.0155E+01	4.7963E-01	2.7842E-03	2
2121	21068	C10H20O2	tert-decanoic acid	52627-73-3	25	7.5505E+03	7.9497E+02	7.6079E+00	4.4163E-02	2
2122	21074	C10H20O2	2-(4-isobutylphenyl)butyric acid	55837-18-8	25	7.5505E+03	7.9497E+02	7.6079E+00	4.4163E-02	2
2123	21077	C10H20O2	4-nonanecarboxylic acid	31080-39-4	25	7.5505E+03	7.9497E+02	7.6079E+00	4.4163E-02	2
2124	21090	C10H21Br	1-bromodecane	112-29-8	25	3.4697E-01	2.8261E-02	3.4697E-04	1.5687E-06	2
2125	21091	C10H21Br	2-bromodecane	39563-53-6	25	3.4697E-01	2.8261E-02	3.4697E-04	1.5687E-06	2
2126	21093	C10H21Cl	1-chlorodecane	1002-69-3	25	3.8530E-01	3.9276E-02	3.8530E-04	2.1802E-06	2
2127	21094	C10H21Cl	decyl chloride (mixed isomers)	28519-06-4	25	3.8530E-01	3.9276E-02	3.8530E-04	2.1802E-06	2
2128	21133	C10H22	decane	124-18-5	25	5.2000E-02	6.5838E-03	5.2000E-05	3.6546E-07	1
2129	21134	C10H22	2-methylnonane	871-83-0	25	4.2345E-02	5.3614E-03	4.2345E-05	2.9761E-07	2
2130	21135	C10H22	3-methylnonane	5911-04-6	25	4.0502E-02	5.1280E-03	4.0502E-05	2.8465E-07	2
2131	21136	C10H22	4-methylnonane	17301-94-9	25	4.5533E-02	5.7650E-03	4.5533E-05	3.2001E-07	2
2132	21137	C10H22	5-methylnonane	15869-85-9	25	4.6957E-02	5.9453E-03	4.6957E-05	3.3002E-07	2
2133	21138	C10H22	3-ethyloctane	5881-17-4	25	4.3518E-02	5.5099E-03	4.3518E-05	3.0585E-07	2
2134	21139	C10H22	4-ethyloctane	15869-86-0	25	5.1087E-02	6.4682E-03	5.1087E-05	3.5904E-07	2
2135	21140	C10H22	2,2-dimethyloctane	15869-87-1	25	7.4992E-02	9.4949E-03	7.4992E-05	5.2706E-07	2
2136	21141	C10H22	2,3-dimethyloctane	7146-60-3	25	4.9197E-02	6.2289E-03	4.9197E-05	3.4576E-07	2
2137	21142	C10H22	2,4-dimethyloctane	4032-94-4	25	7.9380E-02	1.0050E-02	7.9380E-05	5.5790E-07	2
2138	21143	C10H22	2,5-dimethyloctane	15869-89-3	25	6.8388E-02	8.6588E-03	6.8388E-05	4.8064E-07	2
2139	21144	C10H22	2,6-dimethyloctane	2051-30-1	25	6.1438E-02	7.7788E-03	6.1438E-05	4.3180E-07	2
2140	21145	C10H22	2,7-dimethyloctane	1072-16-8	25	6.3247E-02	8.0079E-03	6.3247E-05	4.4451E-07	2
2141	21146	C10H22	3,3-dimethyloctane	4110-44-5	25	5.8642E-02	7.4248E-03	5.8642E-05	4.1215E-07	2
2142	21147	C10H22	3,4-dimethyloctane	15869-92-8	25	5.1782E-02	6.5562E-03	5.1782E-05	3.6393E-07	2
2143	21148	C10H22	3,5-dimethyloctane	15869-93-9	25	6.4964E-02	8.2252E-03	6.4964E-05	4.5657E-07	2
2144	21149	C10H22	3,6-dimethyloctane	15869-94-0	25	5.9989E-02	7.5953E-03	5.9989E-05	4.2161E-07	2
2145	21150	C10H22	4,4-dimethyloctane	15869-95-1	25	7.2415E-02	9.1686E-03	7.2415E-05	5.0894E-07	2
2146	21151	C10H22	4,5-dimethyloctane	15869-96-2	25	5.5632E-02	7.0437E-03	5.5632E-05	3.9099E-07	2
2147	21152	C10H22	4-propylheptane	3178-29-8	25	7.2415E-02	9.1686E-03	7.2415E-05	5.0894E-07	2
2148	21153	C10H22	4-isopropylheptane	52896-87-4	25	6.6843E-02	8.4632E-03	6.6843E-05	4.6979E-07	2
2149	21154	C10H22	3-ethyl-2-methylheptane	14676-29-0	25	5.8642E-02	7.4248E-03	5.8642E-05	4.1215E-07	2
2150	21155	C10H22	4-ethyl-2-methylheptane	52896-88-5	25	7.8023E-02	9.8786E-03	7.8023E-05	5.4836E-07	2
2151	21156	C10H22	5-ethyl-2-methylheptane	13475-78-0	25	6.3862E-02	8.0858E-03	6.3862E-05	4.4884E-07	2
2152	21157	C10H22	3-ethyl-3-methylheptane	17302-01-1	25	5.0628E-02	6.4102E-03	5.0628E-05	3.5582E-07	2
2153	21158	C10H22	4-ethyl-3-methylheptane	52896-89-6	25	5.5412E-02	7.0159E-03	5.5412E-05	3.8945E-07	2
2154	21159	C10H22	3-ethyl-5-methylheptane	52896-90-9	25	6.9571E-02	8.8085E-03	6.9571E-05	4.8896E-07	2
2155	21160	C10H22	3-ethyl-4-methylheptane	52896-91-0	25	5.2964E-02	6.7058E-03	5.2964E-05	3.7224E-07	2
2156	21161	C10H22	4-ethyl-4-methylheptane	17302-04-4	25	5.9989E-02	7.5953E-03	5.9989E-05	4.2161E-07	2
2157	21162	C10H22	2,2,3-trimethylheptane	52896-92-1	25	7.2001E-02	9.1163E-03	7.2001E-05	5.0604E-07	2
2158	21163	C10H22	2,2,4-trimethylheptane	14720-74-2	25	1.2330E-01	1.5611E-02	1.2330E-04	8.6656E-07	2
2159	21164	C10H22	2,2,5-trimethylheptane	20291-95-6	25	1.0660E-01	1.3497E-02	1.0660E-04	7.4920E-07	2
2160	21165	C10H22	2,2,6-trimethylheptane	1190-83-6	25	1.1885E-01	1.5048E-02	1.1885E-04	8.3531E-07	2

Table 3 Solubility in Water - Organic Compounds

					Solubility in Water, S (ppm - parts per million, wt - weight, g - gram)					
NO	ID	FORMULA	NAME	CAS No	T, C	S@T, ppm(wt)	S@T, ppm(mol)	S@T, g/1000 g of H2O	S@T, mol/1000 g of H2O	code
2161	21166	C10H22	2,3,3-trimethylheptane	52896-93-2	25	6.2070E-02	7.8589E-03	6.2070E-05	4.3624E-07	2
2162	21167	C10H22	2,3,4-trimethylheptane	52896-95-4	25	6.3139E-02	7.9942E-03	6.3139E-05	4.4375E-07	2
2163	21168	C10H22	2,3,5-trimethylheptane	20278-85-7	25	6.0331E-02	7.6386E-03	6.0331E-05	4.2401E-07	2
2164	21169	C10H22	2,3,6-trimethylheptane	4032-93-3	25	7.8925E-02	9.9929E-03	7.8925E-05	5.5470E-07	2
2165	21170	C10H22	2,4,4-trimethylheptane	4032-92-2	25	1.0537E-01	1.3341E-02	1.0537E-04	7.4054E-07	2
2166	21171	C10H22	2,4,5-trimethylheptane	20278-84-6	25	7.6690E-02	9.7099E-03	7.6690E-05	5.3899E-07	2
2167	21172	C10H22	2,4,6-trimethylheptane	2613-61-8	25	1.2844E-01	1.6262E-02	1.2844E-04	9.0269E-07	2
2168	21173	C10H22	2,5,5-trimethylheptane	1189-99-7	25	9.4926E-02	1.2019E-02	9.4926E-05	6.6716E-07	2
2169	21174	C10H22	3,3,4-trimethylheptane	20278-87-9	25	5.6361E-02	7.1360E-03	5.6361E-05	3.9612E-07	2
2170	21175	C10H22	3,3,5-trimethylheptane	7154-80-5	25	8.0345E-02	1.0173E-02	8.0345E-05	5.6467E-07	2
2171	21176	C10H22	3,4,4-trimethylheptane	20278-88-0	25	5.8976E-02	7.4671E-03	5.8976E-05	4.1449E-07	2
2172	21177	C10H22	3,4,5-trimethylheptane	20278-89-1	25	5.4480E-02	6.8979E-03	5.4480E-05	3.8290E-07	2
2173	21178	C10H22	3-isopropyl-2-methylhexane	62016-13-1	25	4.3035E-02	5.4488E-03	4.3035E-05	3.0246E-07	2
2174	21179	C10H22	3,3-diethylhexane	17302-02-2	25	4.4006E-02	5.5717E-03	4.4006E-05	3.0928E-07	2
2175	21180	C10H22	3,4-diethylhexane	19398-77-7	25	5.0344E-02	6.3742E-03	5.0344E-05	3.5383E-07	2
2176	21181	C10H22	3-ethyl-2,2-dimethylhexane	20291-91-2	25	7.8472E-02	9.9356E-03	7.8472E-05	5.5152E-07	2
2177	21182	C10H22	4-ethyl-2,2-dimethylhexane	52896-99-8	25	1.3302E-01	1.6842E-02	1.3302E-04	9.3489E-07	2
2178	21183	C10H22	3-ethyl-2,3-dimethylhexane	52897-00-4	25	5.0914E-02	6.4464E-03	5.0914E-05	3.5783E-07	2
2179	21184	C10H22	4-ethyl-2,3-dimethylhexane	52897-01-5	25	5.9649E-02	7.5523E-03	5.9649E-05	4.1922E-07	2
2180	21185	C10H22	3-ethyl-2,4-dimethylhexane	7220-26-0	25	6.2424E-02	7.9037E-03	6.2424E-05	4.3873E-07	2
2181	21186	C10H22	4-ethyl-2,4-dimethylhexane	52897-03-7	25	5.8976E-02	7.4671E-03	5.8976E-05	4.1449E-07	2
2182	21187	C10H22	3-ethyl-2,5-dimethylhexane	52897-04-8	25	8.8054E-02	1.1149E-02	8.8054E-05	6.1886E-07	2
2183	21188	C10H22	4-ethyl-3,3-dimethylhexane	52897-05-9	25	5.3263E-02	6.7438E-03	5.3263E-05	3.7434E-07	2
2184	21189	C10H22	3-ethyl-3,4-dimethylhexane	52897-06-0	25	5.5727E-02	7.0557E-03	5.5727E-05	3.9166E-07	2
2185	21190	C10H22	2,2,3,3-tetramethylhexane	13475-81-5	25	6.1648E-02	7.8054E-03	6.1648E-05	4.3327E-07	2
2186	21191	C10H22	2,2,3,4-tetramethylhexane	52897-08-2	25	6.7226E-02	8.5116E-03	6.7226E-05	4.7248E-07	2
2187	21192	C10H22	2,2,3,5-tetramethylhexane	52897-09-3	25	1.2258E-01	1.5520E-02	1.2258E-04	8.6152E-07	2
2188	21193	C10H22	2,2,4,4-tetramethylhexane	51750-65-3	25	8.9593E-02	1.1344E-02	8.9593E-05	6.2967E-07	2
2189	21194	C10H22	2,2,4,5-tetramethylhexane	16747-42-5	25	1.2636E-01	1.5998E-02	1.2636E-04	8.8806E-07	2
2190	21195	C10H22	2,2,5,5-tetramethylhexane	1071-81-4	25	2.3281E-01	2.9476E-02	2.3281E-04	1.6362E-06	2
2191	21196	C10H22	2,3,3,4-tetramethylhexane	52897-10-6	25	4.8429E-02	6.1317E-03	4.8429E-05	3.4036E-07	2
2192	21197	C10H22	2,3,3,5-tetramethylhexane	52897-11-7	25	9.3293E-02	1.1812E-02	9.3293E-05	6.5568E-07	2
2193	21198	C10H22	2,3,4,4-tetramethylhexane	52897-12-8	25	5.7327E-02	7.2583E-03	5.7327E-05	4.0291E-07	2
2194	21199	C10H22	2,3,4,5-tetramethylhexane	52897-15-1	25	7.8023E-02	9.8786E-03	7.8023E-05	5.4836E-07	2
2195	21200	C10H22	3,3,4,4-tetramethylhexane	5171-84-6	25	3.5840E-02	4.5378E-03	3.5840E-05	2.5189E-07	2
2196	21201	C10H22	2,4-dimethyl-3-isopropylpentane	13475-79-1	25	7.4350E-02	9.4136E-03	7.4350E-05	5.2254E-07	2
2197	21202	C10H22	3,3-diethyl-2-methylpentane	52897-16-2	25	3.6438E-02	4.6135E-03	3.6438E-05	2.5609E-07	2
2198	21203	C10H22	3-ethyl-2,2,3-trimethylpentane	52897-17-3	25	3.6842E-02	4.6647E-03	3.6842E-05	2.5893E-07	2
2199	21204	C10H22	3-ethyl-2,2,4-trimethylpentane	52897-18-4	25	8.2168E-02	1.0404E-02	8.2168E-05	5.7749E-07	2
2200	21205	C10H22	3-ethyl-2,3,4-trimethylpentane	52897-19-5	25	3.6964E-02	4.6802E-03	3.6964E-05	2.5979E-07	2
2201	21206	C10H22	2,2,3,3,4-pentamethylpentane	16747-44-7	25	4.4625E-02	5.6501E-03	4.4625E-05	3.1363E-07	2
2202	21207	C10H22	2,2,3,4,4-pentamethylpentane	16747-45-8	25	6.5372E-02	8.2769E-03	6.5372E-05	4.5945E-07	2
2203	21224	C10H22O	1-decanol	112-30-1	25	3.7020E+01	4.2135E+00	3.7021E-02	2.3389E-04	1
2204	21225	C10H22O	isodecanol	25339-17-7	25	8.8776E+01	1.0105E+01	8.8784E-02	5.6092E-04	2
2205	21226	C10H22O	2-decanol	1120-06-5	25	1.0725E+02	1.2207E+01	1.0726E-01	6.7762E-04	2
2206	21227	C10H22O	2-decanol, (±)	74742-10-2	25	1.5853E+02	1.8045E+01	1.5855E-01	1.0017E-03	2
2207	21228	C10H22O	3-decanol	1565-81-7	25	1.7068E+02	1.9429E+01	1.7071E-01	1.0785E-03	2
2208	21229	C10H22O	4-decanol	2051-31-2	25	1.6905E+02	1.9243E+01	1.6908E-01	1.0682E-03	2
2209	21230	C10H22O	5-decanol	5205-34-5	25	2.9844E+02	3.3975E+01	2.9853E-01	1.8860E-03	2
2210	21231	C10H22O	2-methyl-1-nonanol	40589-14-8	25	7.6862E+01	8.7486E+00	7.6868E-02	4.8563E-04	2
2211	21232	C10H22O	4-methyl-1-nonanol	1489-47-0	25	1.1454E+02	1.3037E+01	1.1455E-01	7.2370E-04	2
2212	21233	C10H22O	5-methyl-1-nonanol	2768-16-3	25	1.3182E+02	1.5005E+01	1.3184E-01	8.3292E-04	2
2213	21234	C10H22O	7-methyl-1-nonanol	33234-93-4	25	1.3182E+02	1.5005E+01	1.3184E-01	8.3292E-04	2
2214	21235	C10H22O	3-methyl-1-nonanol	22663-64-5	25	1.3182E+02	1.5005E+01	1.3184E-01	8.3292E-04	2
2215	21236	C10H22O	2-methyl-2-nonanol	10297-57-1	25	1.3182E+02	1.5005E+01	1.3184E-01	8.3292E-04	2
2216	21237	C10H22O	3-methyl-2-nonanol	60671-32-1	25	1.3182E+02	1.5005E+01	1.3184E-01	8.3292E-04	2
2217	21238	C10H22O	5-methyl-2-nonanol	66731-95-1	25	1.3182E+02	1.5005E+01	1.3184E-01	8.3292E-04	2
2218	21239	C10H22O	6-methyl-2-nonanol	66256-60-8	25	1.3182E+02	1.5005E+01	1.3184E-01	8.3292E-04	2
2219	21240	C10H22O	7-methyl-2-nonanol	66256-61-9	25	1.3182E+02	1.5005E+01	1.3184E-01	8.3292E-04	2
2220	21241	C10H22O	8-methyl-2-nonanol	14779-92-1	25	1.3182E+02	1.5005E+01	1.3184E-01	8.3292E-04	2
2221	21242	C10H22O	2-methyl-3-nonanol	26533-33-5	25	1.8727E+02	2.1317E+01	1.8730E-01	1.1833E-03	2
2222	21243	C10H22O	3-methyl-3-nonanol	21078-72-8	25	1.8196E+02	2.0713E+01	1.8200E-01	1.1498E-03	2
2223	21244	C10H22O	5-methyl-3-nonanol	66719-43-5	25	1.8196E+02	2.0713E+01	1.8200E-01	1.1498E-03	2
2224	21245	C10H22O	2-methyl-4-nonanol	26533-31-3	25	1.8196E+02	2.0713E+01	1.8200E-01	1.1498E-03	2
2225	21246	C10H22O	4-methyl-4-nonanol	23418-38-4	25	1.8196E+02	2.0713E+01	1.8200E-01	1.1498E-03	2
2226	21247	C10H22O	5-methyl-4-nonanol	66719-44-6	25	1.8196E+02	2.0713E+01	1.8200E-01	1.1498E-03	2
2227	21248	C10H22O	7-methyl-4-nonanol	26981-98-6	25	1.8196E+02	2.0713E+01	1.8200E-01	1.1498E-03	2
2228	21249	C10H22O	2-methyl-5-nonanol	29843-62-7	25	1.8196E+02	2.0713E+01	1.8200E-01	1.1498E-03	2
2229	21250	C10H22O	3-methyl-5-nonanol	66779-42-4	25	1.8196E+02	2.0713E+01	1.8200E-01	1.1498E-03	2
2230	21251	C10H22O	5-methyl-5-nonanol	33933-78-7	25	2.8040E+02	3.1922E+01	2.8048E-01	1.7720E-03	2
2231	21252	C10H22O	2-ethyl-1-octanol	20592-10-3	25	1.8196E+02	2.0713E+01	1.8200E-01	1.1498E-03	2
2232	21253	C10H22O	3-ethyl-1-octanol	66719-36-6	25	1.8196E+02	2.0713E+01	1.8200E-01	1.1498E-03	2
2233	21254	C10H22O	2,2-dimethyl-1-octanol	2370-14-1	25	1.9210E+02	2.1867E+01	1.9214E-01	1.2139E-03	2
2234	21255	C10H22O	2,6-dimethyl-1-octanol	62417-08-7	25	1.7068E+02	1.9429E+01	1.7071E-01	1.0785E-03	2
2235	21256	C10H22O	3,7-dimethyl-1-octanol	106-21-8	25	1.4863E+02	1.6918E+01	1.4865E-01	9.3912E-04	2
2236	21257	C10H22O	4,5-dimethyl-1-octanol	66719-32-2	25	1.5007E+02	1.7083E+01	1.5010E-01	9.4827E-04	2
2237	21258	C10H22O	4,6-dimethyl-1-octanol	66719-33-3	25	1.5007E+02	1.7083E+01	1.5010E-01	9.4827E-04	2
2238	21259	C10H22O	4,7-dimethyl-1-octanol	66719-34-4	25	1.5007E+02	1.7083E+01	1.5010E-01	9.4827E-04	2
2239	21260	C10H22O	7,7-dimethyl-1-octanol	66719-35-5	25	1.5007E+02	1.7083E+01	1.5010E-01	9.4827E-04	2
2240	21261	C10H22O	2,4-dimethyl-2-octanol	18675-20-2	25	1.5007E+02	1.7083E+01	1.5010E-01	9.4827E-04	2

Table 3 Solubility in Water - Organic Compounds

					Solubility in Water, S (ppm - parts per million, wt - weight, g - gram)					
NO	ID	FORMULA	NAME	CAS No	T, C	S@T, ppm(wt)	S@T, ppm(mol)	S@T, g/1000 g of H2O	S@T, mol/1000 g of H2O	code
2241	21262	C10H22O	2,6-dimethyt-2-octanol	18479-57-7	25	1.5007E+02	1.7083E+01	1.5010E-01	9.4827E-04	2
2242	21263	C10H22O	2,7-dimethyl-2-octanol	42007-73-8	25	1.5007E+02	1.7083E+01	1.5010E-01	9.4827E-04	2
2243	21264	C10H22O	3,7-dimethyl-2-octanol	15340-96-2	25	1.5007E+02	1.7083E+01	1.5010E-01	9.4827E-04	2
2244	21265	C10H22O	3-ethyl-3-octanol	2051-32-3	25	3.3786E+02	3.8465E+01	3.3798E-01	2.1353E-03	2
2245	21266	C10H22O	6-ethyl-3-octanol	19781-27-2	25	3.4101E+02	3.8824E+01	3.4113E-01	2.1552E-03	2
2246	21267	C10H22O	2,2-dimethyl-3-octanol	19841-72-6	25	4.6357E+02	5.2782E+01	4.6378E-01	2.9301E-03	2
2247	21268	C10H22O	2,3-dimethyl-3-octanol	19781-10-3	25	6.1837E+02	7.0418E+01	6.1875E-01	3.9091E-03	2
2248	21269	C10H22O	2,7-dimethyl-3-octanol	66719-55-9	25	4.5934E+02	5.2301E+01	4.5956E-01	2.9034E-03	2
2249	21270	C10H22O	3,5-dimethyl-3-octanol	56065-42-0	25	4.9269E+02	5.6100E+01	4.9293E-01	3.1142E-03	2
2250	21271	C10H22O	3,6-dimethyl-3-octanol	151-19-9	25	5.1882E+02	5.9076E+01	5.1909E-01	3.2795E-03	2
2251	21272	C10H22O	3,7-dimethyl-3-octanol	78-69-3	25	4.0644E+02	4.6275E+01	4.0660E-01	2.5688E-03	2
2252	21273	C10H22O	3-ethyl-4-octanol	63126-48-7	25	4.9269E+02	5.6100E+01	4.9293E-01	3.1142E-03	2
2253	21274	C10H22O	4-ethyl-4-octanol	38395-42-5	25	4.9269E+02	5.6100E+01	4.9293E-01	3.1142E-03	2
2254	21275	C10H22O	2,2-dimethyl-4-octanol	66719-52-6	25	5.8565E+02	6.6691E+01	5.8600E-01	3.7022E-03	2
2255	21276	C10H22O	2,4-dimethyl-4-octanol	33933-79-8	25	5.5629E+02	6.3345E+01	5.5660E-01	3.5165E-03	2
2256	21277	C10H22O	2,5-dimethyl-4-octanol	66719-53-7	25	5.5629E+02	6.3345E+01	5.5660E-01	3.5165E-03	2
2257	21278	C10H22O	2,6-dimethyl-4-octanol	66719-54-8	25	4.3212E+02	4.9200E+01	4.3230E-01	2.7312E-03	2
2258	21279	C10H22O	2,7-dimethyl-4-octanol	19781-11-4	25	2.8040E+02	3.1922E+01	2.8048E-01	1.7720E-03	2
2259	21280	C10H22O	3,4-dimethyl-4-octanol	66719-30-0	25	5.5629E+02	6.3345E+01	5.5660E-01	3.5165E-03	2
2260	21281	C10H22O	3,6-dimethyl-4-octanol	66719-31-1	25	5.5629E+02	6.3345E+01	5.5660E-01	3.5165E-03	2
2261	21282	C10H22O	4,6-dimethyl-4-octanol	56065-43-1	25	5.5629E+02	6.3345E+01	5.5660E-01	3.5165E-03	2
2262	21283	C10H22O	4,7-dimethyl-4-octanol	19781-13-6	25	5.1882E+02	5.9076E+01	5.1909E-01	3.2795E-03	2
2263	21284	C10H22O	2-propyl-1-heptanol	10042-59-8	25	1.0106E+02	1.1503E+01	1.0107E-01	6.3851E-04	2
2264	21285	C10H22O	3-isopropyl-1-heptanol	38514-15-7	25	1.2070E+02	1.3739E+01	1.2072E-01	7.6267E-04	2
2265	21286	C10H22O	2,4,6-trimethyl-2-heptanol	66256-47-1	25	4.9269E+02	5.6100E+01	4.9293E-01	3.1142E-03	2
2266	21287	C10H22O	2,5,6-trimethyl-2-heptanol	66256-48-2	25	4.8821E+02	5.5590E+01	4.8845E-01	3.0859E-03	2
2267	21288	C10H22O	4,6,6-trimethyl-2-heptanol	51079-79-9	25	4.9269E+02	5.6100E+01	4.9293E-01	3.1142E-03	2
2268	21289	C10H22O	2-methyl-3-ethyl-3-heptanol	66719-37-7	25	4.8821E+02	5.5590E+01	4.8845E-01	3.0859E-03	2
2269	21290	C10H22O	4-methyl-5-ethyl-3-heptanol	66731-94-0	25	4.9269E+02	5.6100E+01	4.9293E-01	3.1142E-03	2
2270	21291	C10H22O	5-methyl-3-ethyl-3-heptanol	---	25	4.9269E+02	5.6100E+01	4.9293E-01	3.1142E-03	2
2271	21292	C10H22O	2,2,3-trimethyl-3-heptanol	29772-40-5	25	7.4511E+02	8.4860E+01	7.4566E-01	4.7109E-03	2
2272	21293	C10H22O	2,2,6-trimethyl-3-heptanol	66256-43-7	25	5.9099E+02	6.7298E+01	5.9134E-01	3.7359E-03	2
2273	21294	C10H22O	2,3,6-trimethyl-3-heptanol	58046-40-5	25	5.9099E+02	6.7298E+01	5.9134E-01	3.7359E-03	2
2274	21295	C10H22O	3,5,5-trimethyl-3-heptanol	66256-50-6	25	4.1653E+02	4.7425E+01	4.1670E-01	2.6326E-03	2
2275	21296	C10H22O	4-propyl-4-heptanol	2198-72-3	25	4.5934E+02	5.2301E+01	4.5956E-01	2.9034E-03	2
2276	21297	C10H22O	4-isopropyl-4-heptanol	51200-82-9	25	5.8565E+02	6.6691E+01	5.8600E-01	3.7022E-03	2
2277	21298	C10H22O	2,2,4-trimethyl-4-heptanol	57233-31-5	25	1.0043E+03	1.1440E+02	1.0053E+00	6.3512E-03	2
2278	21299	C10H22O	2,2,5-trimethyl-4-heptanol	66256-42-6	25	9.5475E+02	1.0876E+02	9.5567E-01	6.0377E-03	2
2279	21300	C10H22O	2,2,6-trimethyl-4-heptanol	66256-44-8	25	9.5475E+02	1.0876E+02	9.5567E-01	6.0377E-03	2
2280	21301	C10H22O	2,4,5-trimethyl-4-heptanol	66256-46-0	25	9.5475E+02	1.0876E+02	9.5567E-01	6.0377E-03	2
2281	21302	C10H22O	2,4,6-trimethyl-4-heptanol	60836-07-9	25	9.4627E+02	1.0779E+02	9.4716E-01	5.9839E-03	2
2282	21303	C10H22O	3,3,6-trimethyl-4-heptanol	---	25	9.5475E+02	1.0876E+02	9.5567E-01	6.0377E-03	2
2283	21304	C10H22O	4-methyl-2-propyl-1-hexanol	66256-62-0	25	1.9210E+02	2.1867E+01	1.9214E-01	1.2139E-03	2
2284	21305	C10H22O	4-methyl-2-isopropyl-1-hexanol	66719-41-3	25	2.9844E+02	3.3975E+01	2.9853E-01	1.8860E-03	2
2285	21306	C10H22O	5-methyl-2-isopropyl-1-hexanol	2051-33-4	25	1.3931E+02	1.5857E+01	1.3933E-01	8.8023E-04	2
2286	21307	C10H22O	5,5-dimethyl-2-ethyl-1-hexanol	---	25	1.4067E+02	1.6012E+01	1.4069E-01	8.8884E-04	2
2287	21308	C10H22O	2,3,4,4-tetramethyl-2-hexanol	66256-66-4	25	5.8565E+02	6.6691E+01	5.8600E-01	3.7022E-03	2
2288	21309	C10H22O	3,3,5,5-tetramethyl-2-hexanol	66256-69-7	25	5.9099E+02	6.7298E+01	5.9134E-01	3.7359E-03	2
2289	21310	C10H22O	2-methyl-3-isopropyl-3-hexanol	51200-81-8	25	5.1882E+02	5.9076E+01	5.1909E-01	3.2795E-03	2
2290	21311	C10H22O	2,2-dimethyl-4-ethyl-3-hexanol	66719-47-9	25	7.0170E+02	7.9914E+01	7.0220E-01	4.4363E-03	2
2291	21312	C10H22O	2,4-dimethyl-4-ethyl-3-hexanol	66719-48-0	25	7.0805E+02	8.0637E+01	7.0856E-01	4.4765E-03	2
2292	21313	C10H22O	5,5-dimethyl-3-ethyl-3-hexanol	5340-62-5	25	7.0805E+02	8.0637E+01	7.0856E-01	4.4765E-03	2
2293	21314	C10H22O	2,2,3,4-tetramethyl-3-hexanol	66256-63-1	25	5.1882E+02	5.9076E+01	5.1909E-01	3.2795E-03	2
2294	21315	C10H22O	2,2,3,5-tetramethyl-3-hexanol	66256-64-2	25	2.1165E+01	2.4089E+00	2.1165E-02	1.3372E-04	2
2295	21316	C10H22O	2,2,4,4-tetramethyl-3-hexanol	66256-65-3	25	5.8565E+02	6.6691E+01	5.8600E-01	3.7022E-03	2
2296	21317	C10H22O	2,2,5,5-tetramethyl-3-hexanol	55073-86-4	25	1.9236E+03	2.1931E+02	1.9273E+00	1.2177E-02	2
2297	21318	C10H22O	2,3,4,4-tetramethyl-3-hexanol	66256-67-5	25	2.9844E+02	3.3975E+01	2.9853E-01	1.8860E-03	2
2298	21319	C10H22O	2,3,5,5-tetramethyl-3-hexanol	5396-09-8	25	2.8304E+02	3.2222E+01	2.8312E-01	1.7887E-03	2
2299	21320	C10H22O	2,4,4,5-tetramethyl-3-hexanol	66256-68-6	25	2.8304E+02	3.2222E+01	2.8312E-01	1.7887E-03	2
2300	21321	C10H22O	3,4,4,5-tetramethyl-3-hexanol	66256-39-1	25	2.8040E+02	3.1922E+01	2.8048E-01	1.7720E-03	2
2301	21322	C10H22O	3,4,5,5-tetramethyl-3-hexanol	66256-40-4	25	4.3212E+02	4.9200E+01	4.3230E-01	2.7312E-03	2
2302	21323	C10H22O	4-methyl-2-isobutyl-1-pentanol	22417-45-4	25	2.4740E+02	2.8164E+01	2.4746E-01	1.5634E-03	2
2303	21324	C10H22O	4,4-dimethyl-3-isopropyl-1-pentanol	66719-51-5	25	8.9953E+02	1.0246E+02	9.0034E-01	5.6881E-03	2
2304	21325	C10H22O	3,4-dimethyl-3-isopropyl-2-pentanol	66719-50-4	25	8.9953E+02	1.0246E+02	9.0034E-01	5.6881E-03	2
2305	21326	C10H22O	2,4-dimethyl-3-propyl-3-pentanol	---	25	8.3986E+02	9.5659E+01	8.4056E-01	5.3105E-03	2
2306	21327	C10H22O	2,4-dimethyl-3-isopropyl-3-pentanol	51200-83-0	25	4.4553E+02	5.0728E+01	4.4573E-01	2.8160E-03	2
2307	21328	C10H22O	2,2,4-trimethyl-3-ethyl-3-pentanol	66256-41-5	25	5.5126E+02	6.2772E+01	5.5157E-01	3.4847E-03	2
2308	21329	C10H22O	2,2,3,4,4-pentamethyl-3-pentanol	5857-69-2	25	4.4826E+02	5.1039E+01	4.4846E-01	2.8333E-03	2
2309	21330	C10H22O	2,3-dimethyl-2-tert-butyl-1-butanol	81931-81-9	25	4.9269E+02	5.6100E+01	4.9293E-01	3.1142E-03	2
2310	21331	C10H22O	3,7-dimethyl-3-octanol, (±)	57706-88-4	25	3.9415E+02	4.4876E+01	3.9431E-01	2.4911E-03	2
2311	21332	C10H22O	3,7-dimethyl-1-octanol, (±)	59204-02-3	25	1.4390E+02	1.6379E+01	1.4392E-01	9.0923E-04	2
2312	21333	C10H22O	3-ethyl-6-methyl-3-heptanol	66719-40-2	25	3.2504E+02	3.7005E+01	3.2514E-01	2.0542E-03	2
2313	21334	C10H22O	8-methyl-1-nonanol	55505-26-5	25	3.2504E+02	3.7005E+01	3.2514E-01	2.0542E-03	2
2314	21335	C10H22O	di-pentyl ether	693-65-2	25	1.9151E+01	2.1797E+00	1.9152E-02	1.2100E-04	2
2315	21336	C10H22O	diisopentyl ether	544-01-4	25	3.9418E+01	4.4865E+00	3.9420E-02	2.4905E-04	2
2316	21337	C10H22O	ethyl octyl ether	929-61-3	25	1.9593E+01	2.2300E+00	1.9593E-02	1.2379E-04	2
2317	21338	C10H22O	(S)-(+)-2-decanol	33758-16-6	25	1.5853E+02	1.8045E+01	1.5855E-01	1.0017E-03	2
2318	21339	C10H22O	(R)-3,7-dimethyl-1-octanol	1117-60-8	25	1.4343E+02	1.6326E+01	1.4345E-01	9.0629E-04	2
2319	21340	C10H22O	s-3,7-dimethyl-1-octanol	68680-98-8	25	1.5007E+02	1.7083E+01	1.5010E-01	9.4827E-04	2
2320	21341	C10H22O	decyl alcohol (mixed isomers)	85566-12-7	25	3.2504E+02	3.7005E+01	3.2514E-01	2.0542E-03	2

Table 3 Solubility in Water - Organic Compounds

					Solubility in Water, S (ppm - parts per million, wt - weight, g - gram)					
NO	ID	FORMULA	NAME	CAS No	T, C	S@T, ppm(wt)	S@T, ppm(mol)	S@T, g/1000 g of H2O	S@T, mol/1000 g of H2O	code
2321	21380	C10H22O5	tetraethylene glycol dimethyl ether	143-24-8	20	total	total	total	total	1
2322	21385	C10H22S	decyl mercaptan	143-10-2	25	6.9284E-01	7.1589E-02	6.9285E-04	3.9739E-06	2
2323	21386	C10H22S	2-decanethiol	13402-60-3	25	1.2056E+00	1.2457E-01	1.2056E-03	6.9150E-06	2
2324	21387	C10H22S	dipentyl sulfide	872-10-6	25	2.9837E+00	3.0829E-01	2.9837E-03	1.7113E-05	2
2325	21388	C10H22S	methyl nonyl sulfide	59973-07-8	25	2.9837E+00	3.0829E-01	2.9837E-03	1.7113E-05	2
2326	21389	C10H22S	ethyl octyl sulfide	3698-94-0	25	2.9837E+00	3.0829E-01	2.9837E-03	1.7113E-05	2
2327	21390	C10H22S	propyl heptyl sulfide	24768-46-5	25	2.9837E+00	3.0829E-01	2.9837E-03	1.7113E-05	2
2328	21391	C10H22S	butyl hexyl sulfide	16967-04-7	25	2.9837E+00	3.0829E-01	2.9837E-03	1.7113E-05	2
2329	21392	C10H22S	bis(2-methylbutyl) sulfide	96034-00-3	25	8.3309E+01	8.6087E+00	8.3316E-02	4.7787E-04	2
2330	21393	C10H22S	diisopentyl sulfide	544-02-5	25	5.2529E+00	5.4276E-01	5.2529E-03	3.0128E-05	2
2331	21400	C10H23N	decylamine	2016-57-1	25	5.5000E+02	6.3020E+01	5.5030E-01	3.4984E-03	1
2332	21401	C10H23N	methylnonylamine	39093-27-1	25	4.3330E+02	4.9643E+01	4.3349E-01	2.7558E-03	2
2333	21402	C10H23N	dimethyloctylamine	7378-99-6	25	2.9415E+02	3.3697E+01	2.9424E-01	1.8705E-03	2
2334	21403	C10H23N	diethylhexylamine	44979-90-0	25	5.6252E+02	6.4455E+01	5.6283E-01	3.5781E-03	2
2335	21404	C10H23N	ethyloctylamine	4088-36-2	25	5.6790E+02	6.5073E+01	5.6823E-01	3.6124E-03	2
2336	21405	C10H23N	dipentylamine	2050-92-2	25	5.3797E+02	6.1641E+01	5.3826E-01	3.4219E-03	2
2337	21406	C10H23N	diisopentylamine	544-00-3	25	1.2167E+03	1.3950E+02	1.2182E+00	7.7446E-03	2
2338	21407	C10H23N	N,N-diisopropylisobutylamine	44976-81-0	25	-9.1484E+04	-9.6923E+03	-8.3816E+01	-5.3285E-01	2
2339	21601	C11H10	1-methylnaphthalene	90-12-0	25	2.8040E+01	3.5524E+00	2.8041E-02	1.9719E-04	1
2340	21602	C11H10	2-methylnaphthalene	91-57-6	25	2.5400E+01	3.2179E+00	2.5401E-02	1.7863E-04	1
2341	21603	C11H10	methylnaphthalene	1321-94-4	25	2.6720E+01	3.3852E+00	2.6721E-02	1.8791E-04	2
2342	22304	C11H16	pentylbenzene	538-68-1	25	3.8400E+00	4.6664E-01	3.8400E-03	2.5903E-05	1
2343	22305	C11H16	2-phenylpentane	2719-52-0	25	7.2769E+00	8.8429E-01	7.2770E-03	4.9086E-05	2
2344	22306	C11H16	3-phenylpentane	1196-58-3	25	6.8868E+00	8.3688E-01	6.8868E-03	4.6455E-05	2
2345	22307	C11H16	1-phenyl-2-methylbutane	3968-85-2	25	1.1315E+01	1.3750E+00	1.1315E-02	7.6328E-05	2
2346	22308	C11H16	1-phenyl-3-methylbutane	2049-94-7	25	1.0212E+01	1.2409E+00	1.0212E-02	6.8882E-05	2
2347	22309	C11H16	2-phenyl-2-methylbutane	2049-95-8	25	1.4557E+01	1.7690E+00	1.4557E-02	9.8197E-05	2
2348	22310	C11H16	2-phenyl-3-methylbutane	4481-30-5	25	1.8534E+01	2.2523E+00	1.8534E-02	1.2502E-04	2
2349	22311	C11H16	1-phenyl-2,2-dimethylpropane	1007-26-7	25	2.0710E+01	2.5167E+00	2.0710E-02	1.3970E-04	2
2350	22312	C11H16	1-methyl-2-butylbenzene	1595-11-5	25	6.3046E+00	7.6613E-01	6.3046E-03	4.2527E-05	2
2351	22313	C11H16	1-methyl-3-butylbenzene	1595-04-6	25	7.3773E+00	8.9649E-01	7.3774E-03	4.9764E-05	2
2352	22314	C11H16	1-methyl-4-butylbenzene	1595-05-7	25	6.6421E+00	8.0715E-01	6.6422E-03	4.4805E-05	2
2353	22315	C11H16	1-methyl-2-sec-butylbenzene	1595-06-8	25	1.1946E+01	1.4517E+00	1.1946E-02	8.0584E-05	2
2354	22316	C11H16	1-methyl-3-sec-butylbenzene	1772-10-7	25	1.3322E+01	1.6189E+00	1.3322E-02	8.9862E-05	2
2355	22317	C11H16	1-methyl-4-sec-butylbenzene	1595-16-0	25	1.1315E+01	1.3750E+00	1.1315E-02	7.6328E-05	2
2356	22318	C11H16	1-methyl-2-isobutylbenzene	36301-29-8	25	1.1946E+01	1.4517E+00	1.1946E-02	8.0584E-05	2
2357	22319	C11H16	1-methyl-3-isobutylbenzene	5160-99-6	25	1.3322E+01	1.6189E+00	1.3322E-02	8.9862E-05	2
2358	22320	C11H16	1-methyl-4-isobutylbenzene	5161-04-6	25	1.1946E+01	1.4517E+00	1.1946E-02	8.0584E-05	2
2359	22321	C11H16	1-methyl-2-tert-butyl benzene	1074-92-6	25	9.3957E+00	1.1418E+00	9.3958E-03	6.3379E-05	2
2360	22322	C11H16	1-methyl-3-tert-butylbenzene	1075-38-3	25	1.7286E+01	2.1006E+00	1.7286E-02	1.1660E-04	2
2361	22323	C11H16	1-methyl-4-tert-butylbenzene	98-51-1	25	1.4257E+01	1.7325E+00	1.4257E-02	9.6172E-05	2
2362	22324	C11H16	1-ethyl-2-propylbenzene	16021-20-8	25	8.2007E+00	9.9655E-01	8.2007E-03	5.5318E-05	2
2363	22325	C11H16	1-ethyl-3-propylbenzene	20024-91-3	25	9.1231E+00	1.1086E+00	9.1232E-03	6.1540E-05	2
2364	22326	C11H16	1-ethyl-4-propylbenzene	20024-90-2	25	7.3773E+00	8.9649E-01	7.3774E-03	4.9764E-05	2
2365	22327	C11H16	1-ethyl-2-isopropylbenzene	18970-44-0	25	1.6594E+01	2.0165E+00	1.6594E-02	1.1193E-04	2
2366	22328	C11H16	1-ethyl-3-isopropylbenzene	4920-99-4	25	1.4864E+01	1.8063E+00	1.4864E-02	1.0027E-04	2
2367	22329	C11H16	1-ethyl-4-isopropylbenzene	4218-48-8	25	1.1563E+01	1.4052E+00	1.1564E-02	7.8001E-05	2
2368	22330	C11H16	1,2-dimethyl-3-propylbenzene	17059-44-8	25	5.4826E+00	6.6625E-01	5.4826E-03	3.6983E-05	2
2369	22331	C11H16	1,2-dimethyl-4-propylbenzene	3982-66-9	25	6.0166E+00	7.3114E-01	6.0166E-03	4.0585E-05	2
2370	22332	C11H16	1,3-dimethyl-2-propylbenzene	17059-45-9	25	6.4373E+00	7.8226E-01	6.4374E-03	4.3423E-05	2
2371	22333	C11H16	1,3-dimethyl-4-propylbenzene	61827-85-8	25	6.7826E+00	8.2422E-01	6.7826E-03	4.5752E-05	2
2372	22334	C11H16	1,3-dimethyl-5-propylbenzene	3982-64-7	25	8.5389E+00	1.0376E+00	8.5389E-03	5.7599E-05	2
2373	22335	C11H16	1,4-dimethyl-2-propylbenzene	3042-50-0	25	7.6549E+00	9.3023E-01	7.6550E-03	5.1636E-05	2
2374	22336	C11H16	1,2-dimethyl-3-isopropylbenzene	22539-65-7	25	8.3768E+00	1.0180E+00	8.3769E-03	5.6506E-05	2
2375	22337	C11H16	1,2-dimethyl-4-isopropylbenzene	4132-77-8	25	8.7415E+00	1.0623E+00	8.7415E-03	5.8966E-05	2
2376	22338	C11H16	1,3-dimethyl-2-isopropylbenzene	14411-75-7	25	1.0157E+01	1.2343E+00	1.0157E-02	6.8512E-05	2
2377	22339	C11H16	1,3-dimethyl-4-isopropylbenzene	4706-89-2	25	1.0102E+01	1.2276E+00	1.0102E-02	6.8145E-05	2
2378	22340	C11H16	1,3-dimethyl-5-isopropylbenzene	4706-90-5	25	1.2963E+01	1.5753E+00	1.2963E-02	8.7442E-05	2
2379	22341	C11H16	1,4-dimethyl-2-isopropylbenzene	4132-72-3	25	1.1817E+01	1.4360E+00	1.1817E-02	7.9713E-05	2
2380	22342	C11H16	1-methyl-2,3-diethylbenzene	13632-93-4	25	6.7826E+00	8.2422E-01	6.7826E-03	4.5752E-05	2
2381	22343	C11H16	1-methyl-2,4-diethylbenzene	1758-85-6	25	7.3773E+00	8.9649E-01	7.3774E-03	4.9764E-05	2
2382	22344	C11H16	1-methyl-2,5-diethylbenzene	13632-94-5	25	6.6075E+00	8.0295E-01	6.6076E-03	4.4571E-05	2
2383	22345	C11H16	1-methyl-2,6-diethyl benzene	13632-95-6	25	6.0479E+00	7.3494E-01	6.0479E-03	4.0796E-05	2
2384	22346	C11H16	1-methyl-3,4-diethylbenzene	13732-80-4	25	7.9438E+00	9.6533E-01	7.9438E-03	5.3585E-05	2
2385	22347	C11H16	1-methyl-3,5-diethylbenzene	2050-24-0	25	9.2707E+00	1.1266E+00	9.2708E-03	6.2536E-05	2
2386	22348	C11H16	1,2,3-trimethyl-4-ethylbenzene	61827-86-9	25	3.3704E+00	4.0957E-01	3.3704E-03	2.2735E-05	2
2387	22349	C11H16	1,2,3-trimethyl-5-ethylbenzene	31366-00-4	25	4.2317E+00	5.1424E-01	4.2317E-03	2.8545E-05	2
2388	22350	C11H16	1,2,4-trimethyl-3-ethylbenzene	61827-87-0	25	4.0658E+00	4.9407E-01	4.0658E-03	2.7425E-05	2
2389	22351	C11H16	1,2,4-trimethyl-5-ethylbenzene	---	25	4.8743E+00	5.9232E-01	4.8743E-03	3.2879E-05	2
2390	22352	C11H16	1,2,4-trimethyl-6-ethylbenzene	17851-27-3	25	4.8743E+00	5.9232E-01	4.8743E-03	3.2879E-05	2
2391	22353	C11H16	1,3,5-trimethyl-2-ethylbenzene	3982-67-0	25	5.0255E+00	6.1070E-01	5.0255E-03	3.3899E-05	2
2392	22354	C11H16	pentamethylbenzene	700-12-9	25	1.9726E+00	2.3970E-01	1.9726E-03	1.3306E-05	2
2393	22355	C11H16	1-ethyl-2,3,5-trimethylbenzene	18262-85-6	25	4.8743E+00	5.9232E-01	4.8743E-03	3.2879E-05	2
2394	22356	C11H16	sec-pentylbenzene	29316-05-0	25	4.2209E+00	5.1293E-01	4.2210E-03	2.8472E-05	2
2395	22399	C11H16O	p-tert-amylphenol	80-46-6	25	1.6800E+02	1.8429E+01	1.6803E-01	1.0230E-03	1
2396	22400	C11H16O	2-butyl-4-methylphenol	6891-45-8	25	1.6800E+02	1.8429E+01	1.6803E-01	1.0230E-03	2
2397	22401	C11H16O	2-tert-butyl-4-methylphenol	2409-55-4	25	1.6800E+02	1.8429E+01	1.6803E-01	1.0230E-03	2
2398	22402	C11H16O	2-tert-butyl-5-methylphenol	88-60-8	25	1.6800E+02	1.8429E+01	1.6803E-01	1.0230E-03	2
2399	22403	C11H16O	2-tert-butyl-6-methylphenol	2219-82-1	25	1.6800E+02	1.8429E+01	1.6803E-01	1.0230E-03	2
2400	22404	C11H16O	4-tert-butyl-2-methylphenol	98-27-1	25	1.6800E+02	1.8429E+01	1.6803E-01	1.0230E-03	2

Table 3 Solubility in Water - Organic Compounds

					Solubility in Water, S (ppm - parts per million, wt - weight, g - gram)					
NO	ID	FORMULA	NAME	CAS No	T, C	S@T, ppm(wt)	S@T, ppm(mol)	S@T, g/1000 g of H2O	S@T, mol/1000 g of H2O	code
2401	22405	C11H16O	4-isopentylphenol	1805-61-4	25	1.6800E+02	1.8429E+01	1.6803E-01	1.0230E-03	2
2402	22406	C11H16O	5-methyl-2-sec-butylphenol	29472-95-5	25	1.6800E+02	1.8429E+01	1.6803E-01	1.0230E-03	2
2403	22407	C11H16O	pentamethylphenol	2819-86-5	25	1.6800E+02	1.8429E+01	1.6803E-01	1.0230E-03	2
2404	22408	C11H16O	4-pentylphenol	14938-35-3	25	1.6800E+02	1.8429E+01	1.6803E-01	1.0230E-03	2
2405	22439	C11H16O	2-(1-methylbutyl)phenol	87-26-3	25	1.6800E+02	1.8429E+01	1.6803E-01	1.0230E-03	2
2406	22440	C11H16O	o-amylphenol	136-81-2	25	1.6800E+02	1.8429E+01	1.6803E-01	1.0230E-03	2
2407	22441	C11H16O	2-sec-amylphenol	---	25	1.6800E+02	1.8429E+01	1.6803E-01	1.0230E-03	2
2408	22442	C11H16O	4-sec-amylphenol	25735-67-5	25	1.6800E+02	1.8429E+01	1.6803E-01	1.0230E-03	2
2409	22634	C11H18S	3-heptylthiophene	65016-61-7	25	5.5537E-01	5.4873E-02	5.5537E-04	3.0460E-06	2
2410	22672	C11H20	1-undecyne	2243-98-3	25	6.4282E-01	7.6047E-02	6.4282E-04	4.2213E-06	2
2411	22673	C11H20	2-undecyne	60212-29-5	25	3.8766E-01	4.5861E-02	3.8766E-04	2.5457E-06	2
2412	22674	C11H20	3-undecyne	60212-30-8	25	5.0802E-01	6.0099E-02	5.0802E-04	3.3361E-06	2
2413	22675	C11H20	4-undecyne	60212-31-9	25	5.4473E-01	6.4443E-02	5.4473E-04	3.5772E-06	2
2414	22676	C11H20	5-undecyne	2294-72-6	25	5.5766E-01	6.5972E-02	5.5766E-04	3.6621E-06	2
2415	22677	C11H20	3,3-dimethyl-4-nonyne	29022-31-9	25	3.2384E-01	3.8311E-02	3.2384E-04	2.1266E-06	2
2416	22701	C11H20O2	2-ethylhexyl acrylate	103-11-7	23	1.0000E+02	9.7768E+00	1.0001E-01	5.4271E-04	1
2417	22791	C11H22	hexylcyclopentane	4457-00-5	25	3.7952E-02	4.4312E-03	3.7952E-05	2.4597E-07	2
2418	22792	C11H22	pentylcyclohexane	4292-92-6	25	5.8525E-02	6.8332E-03	5.8525E-05	3.7930E-07	2
2419	22793	C11H22	isopentylcyclohexane	54105-76-9	25	8.1688E-02	9.5376E-03	8.1688E-05	5.2942E-07	2
2420	22795	C11H22	1-undecene	821-95-4	25	4.5957E-02	5.3658E-03	4.5957E-05	2.9785E-07	2
2421	22796	C11H22	cis-2-undecene	821-96-5	25	3.8943E-02	4.5468E-03	3.8943E-05	2.5239E-07	2
2422	22797	C11H22	trans-2-undecene	693-61-8	25	4.6341E-02	5.4106E-03	4.6341E-05	3.0034E-07	2
2423	22798	C11H22	3-undecene	60669-40-1	25	4.4135E-02	5.1531E-03	4.4135E-05	2.8604E-07	2
2424	22799	C11H22	cis-4-undecene	821-98-7	25	4.6115E-02	5.3842E-03	4.6115E-05	2.9887E-07	2
2425	22800	C11H22	trans-4-undecene	693-62-9	25	4.5223E-02	5.2800E-03	4.5223E-05	2.9309E-07	2
2426	22801	C11H22	cis-5-undecene	764-96-5	25	4.6797E-02	5.4638E-03	4.6797E-05	3.0329E-07	2
2427	22802	C11H22	trans-5-undecene	764-97-6	25	4.7491E-02	5.5448E-03	4.7491E-05	3.0779E-07	2
2428	22803	C11H22	2-methyl-1-decene	13151-27-4	25	5.0635E-02	5.9120E-03	5.0635E-05	3.2817E-07	2
2429	22805	C11H22Br2	1,1-dibromoundecane	62168-28-9	25	1.3071E+00	7.4968E-02	1.3071E-03	4.1614E-06	2
2430	22806	C11H22Br2	1,11-dibromoundecane	16696-65-4	25	1.3071E+00	7.4968E-02	1.3071E-03	4.1614E-06	2
2431	22807	C11H22Cl2	1,1-dichloroundecane	822-01-5	25	3.7759E-01	3.0205E-02	3.7759E-04	1.6767E-06	2
2432	22822	C11H22O	undecanal	112-44-7	25	1.9262E+01	2.0377E+00	1.9262E-02	1.1311E-04	2
2433	22823	C11H22O	2-undecanone	112-12-9	25	1.8756E+01	1.9841E+00	1.8756E-02	1.1014E-04	2
2434	22824	C11H22O	3-undecanone	2216-87-7	25	1.9862E+01	2.1011E+00	1.9862E-02	1.1663E-04	2
2435	22825	C11H22O	4-undecanone	14476-37-0	25	1.1756E+01	1.2436E+00	1.1756E-02	6.9033E-05	2
2436	22826	C11H22O	5-undecanone	33083-83-9	25	1.9862E+01	2.1011E+00	1.9862E-02	1.1663E-04	2
2437	22827	C11H22O	6-undecanone	927-49-1	25	1.8536E+01	1.9609E+00	1.8537E-02	1.0885E-04	2
2438	22834	C11H22O	2-methyl-1-decanal	19009-56-4	25	1.7875E+01	1.8909E+00	1.7875E-02	1.0496E-04	2
2439	22835	C11H22O2	undecanoic acid	112-37-8	25	1.0922E+02	1.0563E+01	1.0923E-01	5.8634E-04	2
2440	22836	C11H22O2	decyl formate	5451-52-5	25	5.5425E+00	5.3597E-01	5.5426E-03	2.9752E-05	2
2441	22837	C11H22O2	nonyl acetate	143-13-5	25	2.3813E+01	2.3028E+00	2.3813E-02	1.2783E-04	2
2442	22838	C11H22O2	octyl propanoate	142-60-9	25	1.9147E+01	1.8516E+00	1.9148E-02	1.0278E-04	2
2443	22839	C11H22O2	heptyl butanoate	5870-93-9	25	2.2383E+01	2.1645E+00	2.2384E-02	1.2015E-04	2
2444	22851	C11H22O2	6-methyldecanoic acid	53696-14-3	25	3.3565E+03	3.2557E+02	3.3679E+00	1.8078E-02	2
2445	22853	C11H22O2	hexyl 2,2-dimethylpropanoate	5434-57-1	25	3.5407E+01	3.4240E+00	3.5408E-02	1.9006E-04	2
2446	22856	C11H22O2	3,5,5-trimethylhexyl acetate	58430-94-7	25	5.9792E+01	5.7823E+00	5.9795E-02	3.2097E-04	2
2447	22870	C11H23Br	1-bromoundecane	693-67-4	25	1.4441E-01	1.1061E-02	1.4441E-04	6.1397E-07	2
2448	22872	C11H23Cl	1-chloroundecane	2473-03-2	25	1.4408E-01	1.3607E-02	1.4408E-04	7.5532E-07	2
2449	22890	C11H24	undecane	1120-21-4	25	4.4000E-03	5.0710E-04	4.4000E-06	2.8149E-08	1
2450	22891	C11H24	2-methyldecane	6975-98-0	25	1.3035E-02	1.5022E-03	1.3035E-05	8.3388E-08	2
2451	22892	C11H24	3-methyldecane	13151-34-3	25	1.2518E-02	1.4427E-03	1.2518E-05	8.0085E-08	2
2452	22893	C11H24	4-methyldecane	2847-72-5	25	1.3907E-02	1.6028E-03	1.3907E-05	8.8970E-08	2
2453	22894	C11H24	5-methyldecane	13151-35-4	25	1.5314E-02	1.7649E-03	1.5314E-05	9.7971E-08	2
2454	22895	C11H24	3-ethylnonane	17302-11-3	25	1.3837E-02	1.5947E-03	1.3837E-05	8.8522E-08	2
2455	22896	C11H24	4-ethylnonane	5911-05-7	25	1.6969E-02	1.9557E-03	1.6969E-05	1.0856E-07	2
2456	22897	C11H24	5-ethylnonane	17302-12-4	25	2.0906E-02	2.4095E-03	2.0906E-05	1.3375E-07	2
2457	22898	C11H24	2,2-dimethylnonane	17302-14-6	25	2.0906E-02	2.4095E-03	2.0906E-05	1.3375E-07	2
2458	22899	C11H24	2,3-dimethylnonane	2884-06-2	25	1.5314E-02	1.7649E-03	1.5314E-05	9.7971E-08	2
2459	22900	C11H24	2,4-dimethylnonane	17302-24-8	25	2.3243E-02	2.6788E-03	2.3243E-05	1.4870E-07	2
2460	22901	C11H24	2,5-dimethylnonane	17302-27-1	25	2.2041E-02	2.5402E-03	2.2041E-05	1.4101E-07	2
2461	22902	C11H24	2,6-dimethylnonane	17302-28-2	25	2.2041E-02	2.5402E-03	2.2041E-05	1.4101E-07	2
2462	22903	C11H24	2,7-dimethylnonane	17302-29-3	25	2.0906E-02	2.4095E-03	2.0906E-05	1.3375E-07	2
2463	22904	C11H24	2,8-dimethylnonane	17302-30-6	25	1.7870E-02	2.0595E-03	1.7870E-05	1.1432E-07	2
2464	22905	C11H24	3,3-dimethylnonane	17302-15-7	25	1.8825E-02	2.1695E-03	1.8825E-05	1.2043E-07	2
2465	22906	C11H24	3,4-dimethylnonane	17302-22-6	25	1.6118E-02	1.8576E-03	1.6118E-05	1.0311E-07	2
2466	22907	C11H24	3,5-dimethylnonane	17302-25-9	25	2.0906E-02	2.4095E-03	2.0906E-05	1.3375E-07	2
2467	22908	C11H24	3,6-dimethylnonane	17302-31-7	25	1.8825E-02	2.1695E-03	1.8825E-05	1.2043E-07	2
2468	22909	C11H24	3,7-dimethylnonane	17302-32-8	25	1.6969E-02	1.9557E-03	1.6969E-05	1.0856E-07	2
2469	22910	C11H24	4,4-dimethylnonane	17302-18-0	25	2.3243E-02	2.6788E-03	2.3243E-05	1.4870E-07	2
2470	22911	C11H24	4,5-dimethylnonane	17302-23-7	25	1.8825E-02	2.1695E-03	1.8825E-05	1.2043E-07	2
2471	22912	C11H24	4,6-dimethylnonane	17302-26-0	25	2.2041E-02	2.5402E-03	2.2041E-05	1.4101E-07	2
2472	22913	C11H24	5,5-dimethylnonane	6414-96-6	25	2.4517E-02	2.8256E-03	2.4517E-05	1.5685E-07	2
2473	22914	C11H24	4-propyloctane	17302-13-5	25	2.0906E-02	2.4095E-03	2.0906E-05	1.3375E-07	2
2474	22915	C11H24	4-isopropyloctane	62016-15-3	25	2.3243E-02	2.6788E-03	2.3243E-05	1.4870E-07	2
2475	22916	C11H24	2-methyl-3-ethyloctane	62016-16-4	25	1.8825E-02	2.1695E-03	1.8825E-05	1.2043E-07	2
2476	22917	C11H24	2-methyl-4-ethyloctane	62016-17-5	25	2.5868E-02	2.9813E-03	2.5868E-05	1.6549E-07	2
2477	22918	C11H24	2-methyl-5-ethyloctane	62016-18-6	25	2.3243E-02	2.6788E-03	2.3243E-05	1.4870E-07	2
2478	22919	C11H24	2-methyl-6-ethyloctane	62016-19-7	25	1.8825E-02	2.1695E-03	1.8825E-05	1.2043E-07	2
2479	22920	C11H24	3-methyl-3-ethyloctane	17302-16-8	25	1.5550E-02	1.7922E-03	1.5550E-05	9.9483E-08	2
2480	22921	C11H24	3-methyl-4-ethyloctane	62016-20-0	25	1.8825E-02	2.1695E-03	1.8825E-05	1.2043E-07	2

Table 3 Solubility in Water - Organic Compounds

					Solubility in Water, S (ppm - parts per million, wt - weight, g - gram)					
NO	ID	FORMULA	NAME	CAS No	T, C	S@T, ppm(wt)	S@T, ppm(mol)	S@T, g/1000 g of H2O	S@T, mol/1000 g of H2O	code
2481	22922	C11H24	3-methyl-5-ethyloctane	62016-21-1	25	2.5868E-02	2.9813E-03	2.5868E-05	1.6549E-07	2
2482	22923	C11H24	3-methyl-6-ethyloctane	62016-22-2	25	1.8825E-02	2.1695E-03	1.8825E-05	1.2043E-07	2
2483	22924	C11H24	4-methyl-3-ethyloctane	62016-23-3	25	1.7870E-02	2.0595E-03	1.7870E-05	1.1432E-07	2
2484	22925	C11H24	4-methyl-4-ethyloctane	17302-19-1	25	2.0906E-02	2.4095E-03	2.0906E-05	1.3375E-07	2
2485	22926	C11H24	4-methyl-5-ethyloctane	62016-24-4	25	2.0906E-02	2.4095E-03	2.0906E-05	1.3375E-07	2
2486	22927	C11H24	4-methyl-6-ethyloctane	62016-25-5	25	2.0906E-02	2.4095E-03	2.0906E-05	1.3375E-07	2
2487	22928	C11H24	2,2,3-trimethyloctane	62016-26-6	25	2.2041E-02	2.5402E-03	2.2041E-05	1.4101E-07	2
2488	22929	C11H24	2,2,4-trimethyloctane	18932-14-4	25	3.3023E-02	3.8059E-03	3.3023E-05	2.1126E-07	2
2489	22930	C11H24	2,2,5-trimethyloctane	62016-27-7	25	3.3941E-02	3.9117E-03	3.3941E-05	2.1714E-07	2
2490	22931	C11H24	2,2,6-trimethyloctane	62016-28-8	25	2.8817E-02	3.3211E-03	2.8817E-05	1.8435E-07	2
2491	22932	C11H24	2,2,7-trimethyloctane	62016-29-9	25	3.5860E-02	4.1329E-03	3.5860E-05	2.2941E-07	2
2492	22933	C11H24	2,3,3-trimethyloctane	62016-30-2	25	1.8825E-02	2.1695E-03	1.8825E-05	1.2043E-07	2
2493	22934	C11H24	2,3,4-trimethyloctane	62016-31-3	25	2.0906E-02	2.4095E-03	2.0906E-05	1.3375E-07	2
2494	22935	C11H24	2,3,5-trimethyloctane	62016-32-4	25	2.4517E-02	2.8256E-03	2.4517E-05	1.5685E-07	2
2495	22936	C11H24	2,3,6-trimethyloctane	62016-33-5	25	2.0906E-02	2.4095E-03	2.0906E-05	1.3375E-07	2
2496	22937	C11H24	2,3,7-trimethyloctane	62016-34-6	25	2.2041E-02	2.5402E-03	2.2041E-05	1.4101E-07	2
2497	22938	C11H24	2,4,4-trimethyloctane	62016-35-7	25	3.3941E-02	3.9117E-03	3.3941E-05	2.1714E-07	2
2498	22939	C11H24	2,4,5-trimethyloctane	62016-36-8	25	2.5868E-02	2.9813E-03	2.5868E-05	1.6549E-07	2
2499	22940	C11H24	2,4,6-trimethyloctane	62016-37-9	25	2.8817E-02	3.3211E-03	2.8817E-05	1.8435E-07	2
2500	22941	C11H24	2,4,7-trimethyloctane	62016-38-0	25	3.3941E-02	3.9117E-03	3.3941E-05	2.1714E-07	2
2501	22942	C11H24	2,5,5-trimethyloctane	62016-39-1	25	3.2132E-02	3.7032E-03	3.2132E-05	2.0556E-07	2
2502	22943	C11H24	2,5,6-trimethyloctane	62016-14-2	25	2.3243E-02	2.6788E-03	2.3243E-05	1.4870E-07	2
2503	22944	C11H24	2,6,6-trimethyloctane	54166-32-4	25	2.5868E-02	2.9813E-03	2.5868E-05	1.6549E-07	2
2504	22945	C11H24	3,3,4-trimethyloctane	62016-40-4	25	1.8825E-02	2.1695E-03	1.8825E-05	1.2043E-07	2
2505	22946	C11H24	3,3,5-trimethyloctane	62016-41-5	25	2.8817E-02	3.3211E-03	2.8817E-05	1.8435E-07	2
2506	22947	C11H24	3,3,6-trimethyloctane	62016-42-6	25	2.4517E-02	2.8256E-03	2.4517E-05	1.5685E-07	2
2507	22948	C11H24	3,4,4-trimethyloctane	62016-43-7	25	1.9835E-02	2.2860E-03	1.9835E-05	1.2690E-07	2
2508	22949	C11H24	3,4,5-trimethyloctane	62016-44-8	25	1.9835E-02	2.2860E-03	1.9835E-05	1.2690E-07	2
2509	22950	C11H24	3,4,6-trimethyloctane	62016-45-9	25	2.2041E-02	2.5402E-03	2.2041E-05	1.4101E-07	2
2510	22951	C11H24	3,5,5-trimethyloctane	61868-94-8	25	3.5860E-02	4.1329E-03	3.5860E-05	2.2941E-07	2
2511	22952	C11H24	4,4,5-trimethyloctane	61868-95-9	25	2.0906E-02	2.4095E-03	2.0906E-05	1.3375E-07	2
2512	22953	C11H24	4-tert-butylheptane	60302-21-8	25	2.8817E-02	3.3211E-03	2.8817E-05	1.8435E-07	2
2513	22954	C11H24	2-methyl-4-propylheptane	61868-96-0	25	2.8817E-02	3.3211E-03	2.8817E-05	1.8435E-07	2
2514	22955	C11H24	3-methyl-4-propylheptane	61868-97-1	25	2.2041E-02	2.5402E-03	2.2041E-05	1.4101E-07	2
2515	22956	C11H24	4-methyl-4-propylheptane	17302-20-4	25	2.3243E-02	2.6788E-03	2.3243E-05	1.4870E-07	2
2516	22957	C11H24	2-methyl-3-isopropylheptane	6876-18-2	25	2.2875E-02	2.6364E-03	2.2875E-05	1.4634E-07	2
2517	22958	C11H24	2-methyl-4-isopropylheptane	61868-98-2	25	3.2132E-02	3.7032E-03	3.2132E-05	2.0556E-07	2
2518	22959	C11H24	3-methyl-4-isopropylheptane	61868-99-3	25	2.4517E-02	2.8256E-03	2.4517E-05	1.5685E-07	2
2519	22960	C11H24	4-methyl-4-isopropylheptane	61869-00-9	25	2.0906E-02	2.4095E-03	2.0906E-05	1.3375E-07	2
2520	22961	C11H24	3,3-diethylheptane	17302-17-9	25	1.4629E-02	1.6860E-03	1.4629E-05	9.3586E-08	2
2521	22962	C11H24	3,4-diethylheptane	61869-01-0	25	2.0906E-02	2.4095E-03	2.0906E-05	1.3375E-07	2
2522	22963	C11H24	3,5-diethylheptane	61869-02-1	25	2.2041E-02	2.5402E-03	2.2041E-05	1.4101E-07	2
2523	22964	C11H24	4,4-diethylheptane	17302-21-5	25	2.0579E-02	2.3717E-03	2.0579E-05	1.3165E-07	2
2524	22965	C11H24	2,2-dimethyl-3-ethylheptane	61869-03-2	25	2.5868E-02	2.9813E-03	2.5868E-05	1.6549E-07	2
2525	22966	C11H24	2,2-dimethyl-4-ethylheptane	62016-46-0	25	4.0054E-02	4.6163E-03	4.0054E-05	2.5625E-07	2
2526	22967	C11H24	2,2-dimethyl-5-ethylheptane	62016-47-1	25	3.2132E-02	3.7032E-03	3.2132E-05	2.0556E-07	2
2527	22968	C11H24	2,3-dimethyl-3-ethylheptane	61868-21-1	25	2.0906E-02	2.4095E-03	2.0906E-05	1.3375E-07	2
2528	22969	C11H24	2,3-dimethyl-4-ethylheptane	61868-22-2	25	2.2041E-02	2.5402E-03	2.2041E-05	1.4101E-07	2
2529	22970	C11H24	2,3-dimethyl-5-ethylheptane	61868-23-3	25	2.3243E-02	2.6788E-03	2.3243E-05	1.4870E-07	2
2530	22971	C11H24	2,4-dimethyl-3-ethylheptane	61868-24-4	25	2.2041E-02	2.5402E-03	2.2041E-05	1.4101E-07	2
2531	22972	C11H24	2,4-dimethyl-4-ethylheptane	61868-25-5	25	3.5860E-02	4.1329E-03	3.5860E-05	2.2941E-07	2
2532	22973	C11H24	2,4-dimethyl-5-ethylheptane	61868-26-6	25	2.5868E-02	2.9813E-03	2.5868E-05	1.6549E-07	2
2533	22974	C11H24	2,5-dimethyl-3-ethylheptane	61868-27-7	25	2.5868E-02	2.9813E-03	2.5868E-05	1.6549E-07	2
2534	22975	C11H24	2,5-dimethyl-4-ethylheptane	61868-28-8	25	2.7299E-02	3.1462E-03	2.7299E-05	1.7465E-07	2
2535	22976	C11H24	2,5-dimethyl-5-ethylheptane	61868-29-9	25	2.3243E-02	2.6788E-03	2.3243E-05	1.4870E-07	2
2536	22977	C11H24	2,6-dimethyl-3-ethylheptane	61868-30-2	25	2.5868E-02	2.9813E-03	2.5868E-05	1.6549E-07	2
2537	22978	C11H24	2,6-dimethyl-4-ethylheptane	61868-31-3	25	3.5860E-02	4.1329E-03	3.5860E-05	2.2941E-07	2
2538	22979	C11H24	3,3-dimethyl-4-ethylheptane	61868-32-4	25	2.0906E-02	2.4095E-03	2.0906E-05	1.3375E-07	2
2539	22980	C11H24	3,3-dimethyl-5-ethylheptane	61868-33-5	25	2.7299E-02	3.1462E-03	2.7299E-05	1.7465E-07	2
2540	22981	C11H24	3,4-dimethyl-3-ethylheptane	61868-34-6	25	1.6118E-02	1.8576E-03	1.6118E-05	1.0311E-07	2
2541	22982	C11H24	3,4-dimethyl-4-ethylheptane	61868-35-7	25	1.9835E-02	2.2860E-03	1.9835E-05	1.2690E-07	2
2542	22983	C11H24	3,4-dimethyl-5-ethylheptane	61868-36-8	25	1.8825E-02	2.1695E-03	1.8825E-05	1.2043E-07	2
2543	22984	C11H24	3,5-dimethyl-3-ethylheptane	61868-37-9	25	2.0906E-02	2.4095E-03	2.0906E-05	1.3375E-07	2
2544	22985	C11H24	3,5-dimethyl-4-ethylheptane	61868-38-0	25	1.9835E-02	2.2860E-03	1.9835E-05	1.2690E-07	2
2545	22986	C11H24	4,4-dimethyl-3-ethylheptane	61868-39-1	25	2.5868E-02	2.9813E-03	2.5868E-05	1.6549E-07	2
2546	22987	C11H24	2,2,3,3-tetramethylheptane	61868-40-4	25	2.0906E-02	2.4095E-03	2.0906E-05	1.3375E-07	2
2547	22988	C11H24	2,2,3,4-tetramethylheptane	61868-41-5	25	2.5868E-02	2.9813E-03	2.5868E-05	1.6549E-07	2
2548	22989	C11H24	2,2,3,5-tetramethylheptane	61868-42-6	25	2.8817E-02	3.3211E-03	2.8817E-05	1.8435E-07	2
2549	22990	C11H24	2,2,3,6-tetramethylheptane	61868-43-7	25	3.5860E-02	4.1329E-03	3.5860E-05	2.2941E-07	2
2550	22991	C11H24	2,2,4,4-tetramethylheptane	61868-44-8	25	3.5860E-02	4.1329E-03	3.5860E-05	2.2941E-07	2
2551	22992	C11H24	2,2,4,5-tetramethylheptane	61868-45-9	25	3.7895E-02	4.3675E-03	3.7895E-05	2.4243E-07	2
2552	22993	C11H24	2,2,4,6-tetramethylheptane	61868-46-0	25	5.5444E-02	6.3899E-03	5.5444E-05	3.5470E-07	2
2553	22994	C11H24	2,2,5,5-tetramethylheptane	61868-47-1	25	4.4775E-02	5.1604E-03	4.4775E-05	2.8645E-07	2
2554	22995	C11H24	2,2,5,6-tetramethylheptane	61868-48-2	25	3.7895E-02	4.3675E-03	3.7895E-05	2.4243E-07	2
2555	22996	C11H24	2,2,6,6-tetramethylheptane	40117-45-1	25	5.2993E-02	6.1075E-03	5.2993E-05	3.3902E-07	2
2556	22997	C11H24	2,3,3,4-tetramethylheptane	61868-49-3	25	1.8825E-02	2.1695E-03	1.8825E-05	1.2043E-07	2
2557	22998	C11H24	2,3,3,5-tetramethylheptane	61868-50-6	25	2.5868E-02	2.9813E-03	2.5868E-05	1.6549E-07	2
2558	22999	C11H24	2,3,3,6-tetramethylheptane	61868-51-7	25	2.7299E-02	3.1462E-03	2.7299E-05	1.7465E-07	2
2559	23000	C11H24	2,3,4,4-tetramethylheptane	61868-52-8	25	2.3243E-02	2.6788E-03	2.3243E-05	1.4870E-07	2
2560	23001	C11H24	2,3,4,5-tetramethylheptane	61868-53-9	25	2.2041E-02	2.5402E-03	2.2041E-05	1.4101E-07	2

Table 3 Solubility in Water - Organic Compounds

					Solubility in Water, S (ppm - parts per million, wt - weight, g - gram)					
NO	ID	FORMULA	NAME	CAS No	T, C	S@T, ppm(wt)	S@T, ppm(mol)	S@T, g/1000 g of H2O	S@T, mol/1000 g of H2O	code
2561	23002	C11H24	2,3,4,6-tetramethylheptane	61868-54-0	25	2.8817E-02	3.3211E-03	2.8817E-05	1.8435E-07	2
2562	23003	C11H24	2,3,5,5-tetramethylheptane	61868-55-1	25	3.2132E-02	3.7032E-03	3.2132E-05	2.0556E-07	2
2563	23004	C11H24	2,3,5,6-tetramethylheptane	52670-32-3	25	2.7299E-02	3.1462E-03	2.7299E-05	1.7465E-07	2
2564	23005	C11H24	2,4,4,5-tetramethylheptane	61868-56-2	25	2.8817E-02	3.3211E-03	2.8817E-05	1.8435E-07	2
2565	23006	C11H24	2,4,4,6-tetramethylheptane	61868-57-3	25	5.0090E-02	5.7729E-03	5.0090E-05	3.2045E-07	2
2566	23007	C11H24	2,4,5,5-tetramethylheptane	61868-58-4	25	2.5868E-02	2.9813E-03	2.5868E-05	1.6549E-07	2
2567	23008	C11H24	3,3,4,4-tetramethylheptane	61868-59-5	25	1.6969E-02	1.9557E-03	1.6969E-05	1.0856E-07	2
2568	23009	C11H24	3,3,4,5-tetramethylheptane	61868-60-8	25	1.9835E-02	2.2860E-03	1.9835E-05	1.2690E-07	2
2569	23010	C11H24	3,3,5,5-tetramethylheptane	61868-61-9	25	2.0906E-02	2.4095E-03	2.0906E-05	1.3375E-07	2
2570	23011	C11H24	3,4,4,5-tetramethylheptane	61868-62-0	25	1.6118E-02	1.8576E-03	1.6118E-05	1.0311E-07	2
2571	23012	C11H24	2,2-dimethyl-3-isopropylhexane	61868-63-1	25	3.2132E-02	3.7032E-03	3.2132E-05	2.0556E-07	2
2572	23013	C11H24	2,3-dimethyl-3-isopropylhexane	61868-64-2	25	2.0906E-02	2.4095E-03	2.0906E-05	1.3375E-07	2
2573	23014	C11H24	2,4-dimethyl-3-isopropylhexane	61868-65-3	25	2.7299E-02	3.1462E-03	2.7299E-05	1.7465E-07	2
2574	23015	C11H24	2,5-dimethyl-3-isopropylhexane	61868-66-4	25	3.5860E-02	4.1329E-03	3.5860E-05	2.2941E-07	2
2575	23016	C11H24	2-methyl-3,3-diethylhexane	61868-67-5	25	1.5314E-02	1.7649E-03	1.5314E-05	9.7971E-08	2
2576	23017	C11H24	2-methyl-3,4-diethylhexane	61868-68-6	25	2.2041E-02	2.5402E-03	2.2041E-05	1.4101E-07	2
2577	23018	C11H24	2-methyl-4,4-diethylhexane	61868-69-7	25	2.2041E-02	2.5402E-03	2.2041E-05	1.4101E-07	2
2578	23019	C11H24	3-methyl-3,4-diethylhexane	61868-70-0	25	1.5314E-02	1.7649E-03	1.5314E-05	9.7971E-08	2
2579	23020	C11H24	3-methyl-4,4-diethylhexane	61868-71-1	25	1.2518E-02	1.4427E-03	1.2518E-05	8.0085E-08	2
2580	23021	C11H24	2,2,3-trimethyl-3-ethylhexane	61868-72-2	25	2.0906E-02	2.4095E-03	2.0906E-05	1.3375E-07	2
2581	23022	C11H24	2,2,3-trimethyl-4-ethylhexane	61868-73-3	25	2.5868E-02	2.9813E-03	2.5868E-05	1.6549E-07	2
2582	23023	C11H24	2,2,4-trimethyl-3-ethylhexane	61868-74-4	25	2.7299E-02	3.1462E-03	2.7299E-05	1.7465E-07	2
2583	23024	C11H24	2,2,4-trimethyl-4-ethylhexane	61868-75-5	25	2.3243E-02	2.6788E-03	2.3243E-05	1.4870E-07	2
2584	23025	C11H24	2,2,5-trimethyl-3-ethylhexane	61868-76-6	25	3.5860E-02	4.1329E-03	3.5860E-05	2.2941E-07	2
2585	23026	C11H24	2,2,5-trimethyl-4-ethylhexane	61868-77-7	25	4.4775E-02	5.1604E-03	4.4775E-05	2.8645E-07	2
2586	23027	C11H24	2,3,3-trimethyl-4-ethylhexane	61868-78-8	25	1.7870E-02	2.0595E-03	1.7870E-05	1.1432E-07	2
2587	23028	C11H24	2,3,4-trimethyl-3-ethylhexane	61868-79-9	25	1.4555E-02	1.6774E-03	1.4555E-05	9.3113E-08	2
2588	23029	C11H24	2,3,4-trimethyl-4-ethylhexane	61868-80-2	25	2.0906E-02	2.4095E-03	2.0906E-05	1.3375E-07	2
2589	23030	C11H24	2,3,5-trimethyl-3-ethylhexane	61868-81-3	25	2.5868E-02	2.9813E-03	2.5868E-05	1.6549E-07	2
2590	23031	C11H24	2,3,5-trimethyl-4-ethylhexane	61868-82-4	25	2.4517E-02	2.8256E-03	2.4517E-05	1.5685E-07	2
2591	23032	C11H24	2,4,4-trimethyl-3-ethylhexane	61868-83-5	25	2.3243E-02	2.6788E-03	2.3243E-05	1.4870E-07	2
2592	23033	C11H24	3,3,4-trimethyl-4-ethylhexane	61868-84-6	25	1.2518E-02	1.4427E-03	1.2518E-05	8.0085E-08	2
2593	23034	C11H24	2,2,3,3,4-pentamethylhexane	61868-85-7	25	1.6969E-02	1.9557E-03	1.6969E-05	1.0856E-07	2
2594	23035	C11H24	2,2,3,3,5-pentamethylhexane	61868-86-8	25	3.4886E-02	4.0207E-03	3.4886E-05	2.2318E-07	2
2595	23036	C11H24	2,2,3,4,4-pentamethylhexane	61868-87-9	25	1.5314E-02	1.7649E-03	1.5314E-05	9.7971E-08	2
2596	23037	C11H24	2,2,3,4,5-pentamethylhexane	61868-88-0	25	2.8817E-02	3.3211E-03	2.8817E-05	1.8435E-07	2
2597	23038	C11H24	2,2,3,5,5-pentamethylhexane	14739-73-2	25	4.4775E-02	5.1604E-03	4.4775E-05	2.8645E-07	2
2598	23039	C11H24	2,2,4,4,5-pentamethylhexane	60302-23-0	25	2.5868E-02	2.9813E-03	2.5868E-05	1.6549E-07	2
2599	23040	C11H24	2,3,3,4,4-pentamethylhexane	61868-89-1	25	1.4555E-02	1.6774E-03	1.4555E-05	9.3113E-08	2
2600	23041	C11H24	2,3,3,4,5-pentamethylhexane	52670-33-4	25	1.9835E-02	2.2860E-03	1.9835E-05	1.2690E-07	2
2601	23042	C11H24	2,2,4-trimethyl-3-isopropylpentane	61868-90-4	25	3.5860E-02	4.1329E-03	3.5860E-05	2.2941E-07	2
2602	23043	C11H24	2,3,4-trimethyl-3-isopropylpentane	61868-91-5	25	1.5314E-02	1.7649E-03	1.5314E-05	9.7971E-08	2
2603	23044	C11H24	2,2-dimethyl-3,3-diethylpentane	60302-28-5	25	1.3159E-02	1.5166E-03	1.3159E-05	8.4185E-08	2
2604	23045	C11H24	2,4-dimethyl-3,3-diethylpentane	61868-92-6	25	1.3159E-02	1.5166E-03	1.3159E-05	8.4185E-08	2
2605	23046	C11H24	2,2,3,4-tetramethyl-3-ethylpentane	61868-93-7	25	1.5314E-02	1.7649E-03	1.5314E-05	9.7971E-08	2
2606	23047	C11H24	2,2,4,4-tetramethyl-3-ethylpentane	---	25	2.0906E-02	2.4095E-03	2.0906E-05	1.3375E-07	2
2607	23048	C11H24	2,2,3,3,4,4-hexamethylpentane	60302-27-4	25	1.0286E-02	1.1855E-03	1.0286E-05	6.5805E-08	2
2608	23055	C11H24O	1-undecanol	112-42-5	25	1.4939E+01	1.5619E+00	1.4940E-02	8.6701E-05	2
2609	23056	C11H24O	2-undecanol	1653-30-1	25	3.1172E+01	3.2591E+00	3.1173E-02	1.8091E-04	2
2610	23057	C11H24O	2-undecanol, (±)	113666-64-1	25	5.1030E+01	5.3354E+00	5.1032E-02	2.9616E-04	2
2611	23058	C11H24O	3-undecanol, (R)-	107494-37-1	25	4.7609E+01	4.9777E+00	4.7611E-02	2.7631E-04	2
2612	23059	C11H24O	5-undecanol	37493-70-2	25	4.7609E+01	4.9777E+00	4.7611E-02	2.7631E-04	2
2613	23060	C11H24O	6-undecanol	23708-56-7	25	5.1030E+01	5.3354E+00	5.1032E-02	2.9616E-04	2
2614	23061	C11H24O	2,8-dimethyl-5-nonanol	19780-96-2	25	3.5996E+01	3.7635E+00	3.5998E-02	2.0891E-04	2
2615	23062	C11H24O	undecanol	30207-98-8	25	1.4939E+01	1.5619E+00	1.4940E-02	8.6701E-05	2
2616	23063	C11H24O	5-ethyl-2-nonanol	103-08-2	25	6.3361E+01	6.6247E+00	6.3365E-02	3.6774E-04	2
2617	23064	C11H24O	1-methoxydecane	7289-52-3	25	1.8412E+00	1.9250E-01	1.8413E-03	1.0686E-05	2
2618	23084	C11H24S	1-undecanethiol	5332-52-5	25	2.8701E-01	2.7447E-02	2.8701E-04	1.5236E-06	2
2619	23085	C11H24S	2-undecanethiol	62155-02-6	25	4.6457E-01	4.4428E-02	4.6457E-04	2.4662E-06	2
2620	23086	C11H24S	methyl decyl sulfide	22438-39-7	25	1.0904E+00	1.0427E-01	1.0904E-03	5.7882E-06	2
2621	23087	C11H24S	ethyl nonyl sulfide	59973-08-9	25	1.0904E+00	1.0427E-01	1.0904E-03	5.7882E-06	2
2622	23088	C11H24S	propyl octyl sulfide	3698-93-9	25	1.0904E+00	1.0427E-01	1.0904E-03	5.7882E-06	2
2623	23089	C11H24S	butyl heptyl sulfide	40813-84-1	25	1.0904E+00	1.0427E-01	1.0904E-03	5.7882E-06	2
2624	23090	C11H25N	undecylamine	7307-55-3	25	1.0536E+02	1.1080E+01	1.0537E-01	6.1503E-04	2
2625	23091	C11H25N	methyldecylamine	7516-82-7	25	1.6772E+02	1.7638E+01	1.6774E-01	9.7909E-04	2
2626	23092	C11H25N	ethylnonylamine	66563-84-6	25	2.1025E+02	2.2112E+01	2.1030E-01	1.2275E-03	2
2627	23093	C11H25N	dimethylnonylamine	17373-27-2	25	8.8094E+01	9.2638E+00	8.8102E-02	5.1423E-04	2
2628	23094	C11H25N	diethylheptylamine	26981-81-7	25	2.1979E+02	2.3116E+01	2.1984E-01	1.2832E-03	2
2629	23095	C11H25N	4-ethyl-1-methyloctylamine	10024-78-9	25	2.2381E+02	2.3538E+01	2.2386E-01	1.3066E-03	2
2630	23335	C12H8O	dibenzofuran	132-64-9	25	1.0024E+01	1.0737E+00	1.0024E-02	5.9598E-05	1
2631	23394	C12H9N	dibenzopyrrole	86-74-8	25	9.0750E-01	9.7773E-02	9.0750E-04	5.4273E-06	1
2632	23444	C12H10	acenaphthene	83-32-9	25	3.8861E+00	4.5398E-01	3.8861E-03	2.5200E-05	1
2633	23445	C12H10	phenylbenzene (biphenyl)	92-52-4	25	7.0000E+00	8.1775E-01	7.0000E-03	4.5393E-05	1
2634	23519	C12H10O	diphenyl ether	101-84-8	25	1.8042E+01	1.9096E+00	1.8042E-02	1.0600E-04	1
2635	23609	C12H11N	diphenylamine	122-39-4	20	5.3000E+01	5.6424E+00	5.3003E-02	3.1321E-04	1
2636	23656	C12H11N3	p-aminoazobenzene	60-09-3	18	1.3000E+02	1.1875E+01	1.3002E-01	6.5918E-04	1
2637	23675	C12H12	1,2-dimethylnaphthalene	573-98-8	25	7.9990E+00	9.2239E-01	7.9991E-03	5.1201E-05	1
2638	23676	C12H12	1,3-dimethylnaphthalene	575-41-7	25	7.9990E+00	9.2239E-01	7.9991E-03	5.1201E-05	1
2639	23677	C12H12	1,4-dimethylnaphthalene	571-58-4	25	1.1400E+01	1.3146E+00	1.1400E-02	7.2971E-05	1
2640	23678	C12H12	1,5-dimethylnaphthalene	571-61-9	25	3.3750E+00	3.8918E-01	3.3750E-03	2.1603E-05	1

Table 3 Solubility in Water - Organic Compounds

NO	ID	FORMULA	NAME	CAS No	Solubility in Water, S (ppm - parts per million, wt - weight, g - gram) T, C	S@T, ppm(wt)	S@T, ppm(mol)	S@T, g/1000 g of H2O	S@T, mol/1000 g of H2O	code
2641	23679	C12H12	1,6-dimethylnaphthalene	575-43-9	25	6.8091E+00	7.8518E-01	6.8092E-03	4.3585E-05	2
2642	23680	C12H12	1,7-dimethylnaphthalene	575-37-1	25	6.8091E+00	7.8518E-01	6.8092E-03	4.3585E-05	2
2643	23681	C12H12	1,8-dimethylnaphthalene	569-41-5	25	6.8091E+00	7.8518E-01	6.8092E-03	4.3585E-05	2
2644	23682	C12H12	2,3-dimethylnaphthalene	581-40-8	25	3.0000E+00	3.4594E-01	3.0000E-03	1.9203E-05	1
2645	23683	C12H12	2,6-dimethylnaphthalene	581-42-0	25	2.0000E+00	2.3063E-01	2.0000E-03	1.2802E-05	1
2646	23684	C12H12	2,7-dimethylnaphthalene	582-16-1	25	6.8091E+00	7.8518E-01	6.8092E-03	4.3585E-05	2
2647	23685	C12H12	1-ethylnaphthalene	1127-76-0	25	1.0700E+01	1.2339E+00	1.0700E-02	6.8491E-05	1
2648	23686	C12H12	2-ethylnaphthalene	939-27-5	25	8.0000E+00	9.2251E-01	8.0001E-03	5.1208E-05	1
2649	23687	C12H12	dimethylnaphthalene, isomers	28804-88-8	25	6.8091E+00	7.8518E-01	6.8092E-03	4.3585E-05	2
2650	23718	C12H12N2	p-aminodiphenylamine	101-54-2	25	5.2000E+02	5.0869E+01	5.2027E-01	2.8239E-03	1
2651	24002	C12H14O4	diethyl phthalate	84-66-2	25	1.0000E+03	8.1135E+01	1.0010E+00	4.5041E-03	1
2652	24357	C12H18	hexylbenzene	1077-16-3	25	1.0180E+00	1.1301E-01	1.0180E-03	6.2733E-06	1
2653	24358	C12H18	1,4-dipropylbenzene	4815-57-0	25	6.1061E+00	6.7787E-01	6.1061E-03	3.7628E-05	2
2654	24359	C12H18	diisopropylbenzene	25321-09-9	25	9.3957E+00	1.0431E+00	9.3958E-03	5.7900E-05	2
2655	24360	C12H18	1,2-diisopropylbenzene	577-55-9	25	7.7773E+00	8.6341E-01	7.7774E-03	4.7927E-05	2
2656	24361	C12H18	m-diisopropylbenzene	99-62-7	25	8.1227E+00	9.0175E-01	8.1227E-03	5.0055E-05	2
2657	24362	C12H18	p-diisopropylbenzene	100-18-5	25	5.5393E+00	6.1495E-01	5.5393E-03	3.4136E-05	2
2658	24363	C12H18	triethylbenzene; (mixed isomers)	25340-18-5	25	3.8780E+00	4.3052E-01	3.8780E-03	2.3898E-05	2
2659	24364	C12H18	1,2,3-triethylbenzene	---	25	3.8857E+00	4.3138E-01	3.8857E-03	2.3945E-05	2
2660	24365	C12H18	1,2,4-triethylbenzene	877-44-1	25	3.8857E+00	4.3138E-01	3.8857E-03	2.3945E-05	2
2661	24366	C12H18	1,3,5-triethylbenzene	102-25-0	25	4.1874E+00	4.6487E-01	4.1874E-03	2.5805E-05	2
2662	24367	C12H18	1-tert-butyl-3,5-dimethylbenzene	98-19-1	25	6.6421E+00	7.3738E-01	6.6422E-03	4.0932E-05	2
2663	24368	C12H18	(1,1-dimethylbutyl)benzene	1985-57-5	25	7.1856E+00	7.9772E-01	7.1857E-03	4.4281E-05	2
2664	24369	C12H18	2,4-dimethyl-1-sec-butylbenzene	1483-60-9	25	6.1061E+00	6.7787E-01	6.1061E-03	3.7628E-05	2
2665	24370	C12H18	(1-ethylbutyl)benzene	4468-42-2	25	2.6248E+00	2.9140E-01	2.6248E-03	1.6175E-05	2
2666	24371	C12H18	(1-ethyl-1-methylpropyl)benzene	1985-97-3	25	3.2352E+00	3.5916E-01	3.2352E-03	1.9936E-05	2
2667	24372	C12H18	(1-methylpentyl)benzene	6031-02-3	25	2.7647E+00	3.0693E-01	2.7648E-03	1.7037E-05	2
2668	24373	C12H18	(3-methylpentyl)benzene	54410-69-4	25	1.5072E+00	1.6732E-01	1.5072E-03	9.2879E-06	2
2669	24374	C12H18	(4-methylpentyl)benzene	4215-86-5	25	1.7478E+00	1.9403E-01	1.7478E-03	1.0770E-05	2
2670	24375	C12H18	1,2,4-trimethyl-5-isopropylbenzene	10222-95-4	25	3.2733E+00	3.6339E-01	3.2733E-03	2.0171E-05	2
2671	24376	C12H18	4-tert-butyl-o-xylene	7397-06-0	25	7.3773E+00	8.1900E-01	7.3774E-03	4.5462E-05	2
2672	24377	C12H18	hexamethyldewarbenzene	7641-77-2	25	6.0743E+01	6.7438E+00	6.0747E-02	3.7435E-04	2
2673	24378	C12H18	hexamethylbenzene	87-85-4	25	5.8625E-01	6.5082E-02	5.8625E-04	3.6127E-06	2
2674	24619	C12H20S	3-octylthiophene	65016-62-8	25	2.4878E-01	2.2825E-02	2.4878E-04	1.2670E-06	2
2675	24653	C12H22	1-dodecyne	765-03-7	25	2.6575E-01	2.8787E-02	2.6575E-04	1.5980E-06	2
2676	24654	C12H22	2-dodecyne	629-49-2	25	1.7655E-01	1.9125E-02	1.7655E-04	1.0616E-06	2
2677	24655	C12H22	3-dodecyne	6790-27-8	25	2.2409E-01	2.4274E-02	2.2409E-04	1.3474E-06	2
2678	24656	C12H22	6-dodecyne	6975-99-1	25	3.2639E-01	3.5356E-02	3.2640E-04	1.9626E-06	2
2679	24657	C12H22	3,3-dimethyl-4-decyne	70732-45-5	25	1.6623E-01	1.8006E-02	1.6623E-04	9.9952E-07	2
2680	24669	C12H22	1,11-dodecadiene	5876-87-9	25	6.5375E-02	7.0817E-03	6.5375E-05	3.9310E-07	2
2681	24818	C12H24	heptylcyclopentane	5617-42-5	25	1.6215E-02	1.7354E-03	1.6215E-05	9.6333E-08	2
2682	24819	C12H24	hexylcyclohexane	4292-75-5	25	2.5197E-02	2.6967E-03	2.5197E-05	1.4969E-07	2
2683	24820	C12H24	1-methyl-2-pentylcyclohexane	54411-01-7	25	3.2883E-02	3.5194E-03	3.2883E-05	1.9536E-07	2
2684	24823	C12H24	1-dodecene	112-41-4	25	1.8134E-02	1.9409E-03	1.8134E-05	1.0774E-07	2
2685	24824	C12H24	2-methyl-1-undecene	18516-37-5	25	1.9485E-02	2.0854E-03	1.9485E-05	1.1576E-07	2
2686	24825	C12H24	4,4-dimethyl-2-neopentyl-1-pentene	141-70-8	25	1.6414E-02	1.7567E-03	1.6414E-05	9.7515E-08	2
2687	24826	C12H24	2,2,4,6,6-pentamethyl-3-heptene	123-48-8	25	8.5056E-02	9.1032E-03	8.5056E-05	5.0531E-07	2
2688	24827	C12H24	trans-2,2,4,6,6-pentamethyl-3-heptene	27656-50-4	25	8.5056E-02	9.1032E-03	8.5056E-05	5.0531E-07	2
2689	24828	C12H24	isobutene, trimer	7756-94-7	25	8.7554E-02	9.3706E-03	8.7554E-05	5.2016E-07	2
2690	24831	C12H24Br2	1,1-dibromododecane	62168-29-0	25	7.5129E-01	4.1247E-02	7.5129E-04	2.2896E-06	2
2691	24832	C12H24Br2	1,12-dibromododecane	3344-70-5	25	7.5129E-01	4.1247E-02	7.5129E-04	2.2896E-06	2
2692	24833	C12H24Cl2	1,1-dichlorododecane	62017-17-8	25	1.8277E-01	1.3764E-02	1.8277E-04	7.6401E-07	2
2693	24834	C12H24Cl2	1,12-dichlorododecane	3922-28-9	25	9.9364E-02	7.4826E-03	9.9364E-05	4.1535E-07	2
2694	24852	C12H24O	dodecanal	112-54-9	25	8.1414E+00	7.9572E-01	8.1415E-03	4.4170E-05	2
2695	24853	C12H24O	2-methylundecanal	110-41-8	25	9.9197E+00	9.6953E-01	9.9198E-03	5.3818E-05	2
2696	24854	C12H24O	2-dodecanone	6175-49-1	25	4.7697E+00	4.6618E-01	4.7697E-03	2.5877E-05	2
2697	24855	C12H24O	6-dodecanone	6064-27-3	25	4.9927E+00	4.8797E-01	4.9927E-03	2.7087E-05	2
2698	24860	C12H24O	trimethyl nonanone	1331-50-6	25	4.4422E+01	4.3419E+00	4.4424E-02	2.4102E-04	2
2699	24861	C12H24O	3-dodecanone	1534-27-6	25	4.9927E+00	4.8797E-01	4.9927E-03	2.7087E-05	2
2700	24862	C12H24O	5-dodecanone	19780-10-0	25	4.9927E+00	4.8797E-01	4.9927E-03	2.7087E-05	2
2701	24863	C12H24O	4-dodecanone	6137-26-4	25	4.9927E+00	4.8797E-01	4.9927E-03	2.7087E-05	2
2702	24867	C12H24O2	dodecanoic acid	143-07-7	20	5.5000E+01	4.9464E+00	5.5003E-02	2.7457E-04	1
2703	24868	C12H24O2	undecyl formate	5454-24-0	25	1.4058E+00	1.2643E-01	1.4058E-03	7.0179E-06	2
2704	24869	C12H24O2	decyl acetate	112-17-4	25	8.0909E+00	7.2763E-01	8.0910E-03	4.0390E-05	2
2705	24870	C12H24O2	nonyl propanoate	53184-67-1	25	3.6022E+00	3.2395E-01	3.6022E-03	1.7982E-05	2
2706	24871	C12H24O2	octyl butanoate	110-39-4	25	8.0909E+00	7.2763E-01	8.0910E-03	4.0390E-05	2
2707	24879	C12H24O2	2-butyloctanoic acid	27610-92-0	25	1.7010E+03	1.5321E+02	1.7039E+00	8.5057E-03	2
2708	24880	C12H24O2	3,7-dimethyloctanyl acetate	20780-49-8	25	1.5563E+01	1.3996E+00	1.5563E-02	7.7691E-05	2
2709	24894	C12H25Br	1-bromododecane	143-15-7	25	6.7545E-02	4.8823E-03	6.7545E-05	2.7101E-07	2
2710	24895	C12H25Br	2-bromododecane	13187-99-0	25	6.7545E-02	4.8823E-03	6.7545E-05	2.7101E-07	2
2711	24897	C12H25Cl	1-chlorododecane	112-52-7	25	6.1982E-02	5.4526E-03	6.1982E-05	3.0267E-07	2
2712	24914	C12H26	dodecane	112-40-3	25	3.7000E-03	3.9131E-04	3.7000E-06	2.1721E-08	1
2713	24915	C12H26	3-methylundecane	1002-43-3	25	4.9744E-03	5.2610E-04	4.9744E-06	2.9203E-08	2
2714	24916	C12H26	4-methylundecane	2980-69-0	25	5.1897E-03	5.4886E-04	5.1897E-06	3.0467E-08	2
2715	24917	C12H26	5-methylundecane	1632-70-8	25	5.9082E-03	6.2485E-04	5.9082E-06	3.4685E-08	2
2716	24918	C12H26	6-methylundecane	17302-33-9	25	5.9082E-03	6.2485E-04	5.9082E-06	3.4685E-08	2
2717	24919	C12H26	3-ethyldecane	17085-96-0	25	5.2565E-03	5.5593E-04	5.2565E-06	3.0859E-08	2
2718	24920	C12H26	4-ethyldecane	1636-44-8	25	6.4551E-03	6.8269E-04	6.4551E-06	3.7896E-08	2
2719	24921	C12H26	5-ethyldecane	17302-36-2	25	7.0640E-03	7.4709E-04	7.0640E-06	4.1470E-08	2
2720	24922	C12H26	2,2-dimethyldecane	17302-37-3	25	7.3940E-03	7.8199E-04	7.3940E-06	4.3408E-08	2

Table 3 Solubility in Water - Organic Compounds

NO	ID	FORMULA	NAME	CAS No	Solubility in Water, S (ppm - parts per million, wt - weight, g - gram) T, C	S@T, ppm(wt)	S@T, ppm(mol)	S@T, g/1000 g of H2O	S@T, mol/1000 g of H2O	code
2721	24923	C12H26	2,3-dimethyldecane	17312-44-6	25	5.9082E-03	6.2485E-04	5.9082E-06	3.4685E-08	2
2722	24924	C12H26	2,4-dimethyldecane	2801-84-5	25	7.7424E-03	8.1884E-04	7.7424E-06	4.5453E-08	2
2723	24925	C12H26	2,5-dimethyldecane	17312-50-4	25	8.4990E-03	8.9885E-04	8.4990E-06	4.9895E-08	2
2724	24926	C12H26	2,6-dimethyldecane	13150-81-7	25	8.4990E-03	8.9885E-04	8.4990E-06	4.9895E-08	2
2725	24927	C12H26	2,7-dimethyldecane	17312-51-5	25	7.0640E-03	7.4709E-04	7.0640E-06	4.1470E-08	2
2726	24928	C12H26	2,8-dimethyldecane	17312-52-6	25	6.4551E-03	6.8269E-04	6.4551E-06	3.7896E-08	2
2727	24929	C12H26	2,9-dimethyldecane	1002-17-1	25	6.4551E-03	6.8269E-04	6.4551E-06	3.7896E-08	2
2728	24930	C12H26	3,3-dimethyldecane	17302-38-4	25	6.7513E-03	7.1402E-04	6.7513E-06	3.9635E-08	2
2729	24931	C12H26	3,4-dimethyldecane	17312-45-7	25	6.1743E-03	6.5300E-04	6.1743E-06	3.6247E-08	2
2730	24932	C12H26	3,5-dimethyldecane	17312-48-0	25	7.7424E-03	8.1884E-04	7.7424E-06	4.5453E-08	2
2731	24933	C12H26	3,6-dimethyldecane	17312-53-7	25	7.3940E-03	7.8199E-04	7.3940E-06	4.3408E-08	2
2732	24934	C12H26	3,7-dimethyldecane	17312-54-8	25	7.0640E-03	7.4709E-04	7.0640E-06	4.1470E-08	2
2733	24935	C12H26	3,8-dimethyldecane	17312-55-9	25	6.1743E-03	6.5300E-04	6.1743E-06	3.6247E-08	2
2734	24936	C12H26	4,4-dimethyldecane	17312-39-9	25	8.1104E-03	8.5775E-04	8.1104E-06	4.7613E-08	2
2735	24937	C12H26	4,5-dimethyldecane	17312-46-8	25	7.0640E-03	7.4709E-04	7.0640E-06	4.1470E-08	2
2736	24938	C12H26	4,6-dimethyldecane	17312-49-1	25	8.4990E-03	8.9885E-04	8.4990E-06	4.9895E-08	2
2737	24939	C12H26	4,7-dimethyldecane	17312-56-0	25	8.1104E-03	8.5775E-04	8.1104E-06	4.7613E-08	2
2738	24940	C12H26	5,5-dimethyldecane	17453-92-8	25	8.9095E-03	9.4227E-04	8.9095E-06	5.2305E-08	2
2739	24941	C12H26	5,6-dimethyldecane	1636-43-7	25	7.3940E-03	7.8199E-04	7.3940E-06	4.3408E-08	2
2740	24942	C12H26	4-propylnonane	6165-37-3	25	8.1104E-03	8.5775E-04	8.1104E-06	4.7613E-08	2
2741	24943	C12H26	5-propylnonane	998-35-6	25	8.9095E-03	9.4227E-04	8.9095E-06	5.2305E-08	2
2742	24944	C12H26	4-isopropylnonane	62184-71-8	25	8.4990E-03	8.9885E-04	8.4990E-06	4.9895E-08	2
2743	24945	C12H26	5-isopropylnonane	62184-72-9	25	8.4990E-03	8.9885E-04	8.4990E-06	4.9895E-08	2
2744	24946	C12H26	2-methyl-3-ethylnonane	62184-73-0	25	7.3940E-03	7.8199E-04	7.3940E-06	4.3408E-08	2
2745	24947	C12H26	2-methyl-4-ethylnonane	62184-37-6	25	8.4990E-03	8.9885E-04	8.4990E-06	4.9895E-08	2
2746	24948	C12H26	2-methyl-5-ethylnonane	62184-38-7	25	9.3433E-03	9.8814E-04	9.3433E-06	5.4851E-08	2
2747	24949	C12H26	2-methyl-6-ethylnonane	62184-39-8	25	8.4990E-03	8.9885E-04	8.4990E-06	4.9895E-08	2
2748	24950	C12H26	2-methyl-7-ethylnonane	62184-40-1	25	7.0640E-03	7.4709E-04	7.0640E-06	4.1470E-08	2
2749	24951	C12H26	3-methyl-3-ethylnonane	17302-39-5	25	6.4551E-03	6.8269E-04	6.4551E-06	3.7896E-08	2
2750	24952	C12H26	3-methyl-4-ethylnonane	62184-41-2	25	7.3940E-03	7.8199E-04	7.3940E-06	4.3408E-08	2
2751	24953	C12H26	3-methyl-5-ethylnonane	62184-42-3	25	9.8016E-03	1.0366E-03	9.8016E-06	5.7542E-08	2
2752	24954	C12H26	3-methyl-6-ethylnonane	62184-43-4	25	8.1104E-03	8.5775E-04	8.1104E-06	4.7613E-08	2
2753	24955	C12H26	3-methyl-7-ethylnonane	62184-44-5	25	7.7424E-03	8.1884E-04	7.7424E-06	4.5453E-08	2
2754	24956	C12H26	4-methyl-3-ethylnonane	62184-45-6	25	7.7424E-03	8.1884E-04	7.7424E-06	4.5453E-08	2
2755	24957	C12H26	4-methyl-4-ethylnonane	17312-40-2	25	8.1104E-03	8.5775E-04	8.1104E-06	4.7613E-08	2
2756	24958	C12H26	4-methyl-5-ethylnonane	62184-46-7	25	8.4990E-03	8.9885E-04	8.4990E-06	4.9895E-08	2
2757	24959	C12H26	4-methyl-6-ethylnonane	62184-47-8	25	9.3433E-03	9.8814E-04	9.3433E-06	5.4851E-08	2
2758	24960	C12H26	4-methyl-7-ethylnonane	62184-48-9	25	7.7424E-03	8.1884E-04	7.7424E-06	4.5453E-08	2
2759	24961	C12H26	5-methyl-3-ethylnonane	62184-49-0	25	8.1104E-03	8.5775E-04	8.1104E-06	4.7613E-08	2
2760	24962	C12H26	5-methyl-4-ethylnonane	1632-71-9	25	8.1104E-03	8.5775E-04	8.1104E-06	4.7613E-08	2
2761	24963	C12H26	5-methyl-5-ethylnonane	14531-16-9	25	8.4990E-03	8.9885E-04	8.4990E-06	4.9895E-08	2
2762	24964	C12H26	2,2,3-trimethylnonane	55499-04-2	25	7.7424E-03	8.1884E-04	7.7424E-06	4.5453E-08	2
2763	24965	C12H26	2,2,4-trimethylnonane	62184-50-3	25	1.1913E-02	1.2599E-03	1.1913E-05	6.9935E-08	2
2764	24966	C12H26	2,2,5-trimethylnonane	62184-51-4	25	1.1913E-02	1.2599E-03	1.1913E-05	6.9935E-08	2
2765	24967	C12H26	2,2,6-trimethylnonane	62184-52-5	25	1.2518E-02	1.3239E-03	1.2518E-05	7.3491E-08	2
2766	24968	C12H26	2,2,7-trimethylnonane	62184-53-6	25	9.8016E-03	1.0366E-03	9.8016E-06	5.7542E-08	2
2767	24969	C12H26	2,2,8-trimethylnonane	62184-54-7	25	9.8016E-03	1.0366E-03	9.8016E-06	5.7542E-08	2
2768	24970	C12H26	2,3,3-trimethylnonane	62184-55-8	25	7.0640E-03	7.4709E-04	7.0640E-06	4.1470E-08	2
2769	24971	C12H26	2,3,4-trimethylnonane	62184-56-9	25	7.7424E-03	8.1884E-04	7.7424E-06	4.5453E-08	2
2770	24972	C12H26	2,3,5-trimethylnonane	62184-57-0	25	9.3433E-03	9.8814E-04	9.3433E-06	5.4851E-08	2
2771	24973	C12H26	2,3,6-trimethylnonane	62184-58-1	25	8.4990E-03	8.9885E-04	8.4990E-06	4.9895E-08	2
2772	24974	C12H26	2,3,7-trimethylnonane	62184-59-2	25	7.3940E-03	7.8199E-04	7.3940E-06	4.3408E-08	2
2773	24975	C12H26	2,3,8-trimethylnonane	62184-60-5	25	7.7424E-03	8.1884E-04	7.7424E-06	4.5453E-08	2
2774	24976	C12H26	2,4,4-trimethylnonane	62184-61-6	25	1.2518E-02	1.3239E-03	1.2518E-05	7.3491E-08	2
2775	24977	C12H26	2,4,5-trimethylnonane	62184-62-7	25	9.8016E-03	1.0366E-03	9.8016E-06	5.7542E-08	2
2776	24978	C12H26	2,4,6-trimethylnonane	62184-10-5	25	1.1340E-02	1.1993E-03	1.1340E-05	6.6572E-08	2
2777	24979	C12H26	2,4,7-trimethylnonane	62184-11-6	25	1.0286E-02	1.0879E-03	1.0286E-05	6.0386E-08	2
2778	24980	C12H26	2,4,8-trimethylnonane	49542-74-7	25	1.1913E-02	1.2599E-03	1.1913E-05	6.9935E-08	2
2779	24981	C12H26	2,5,5-trimethylnonane	62184-12-7	25	1.2518E-02	1.3239E-03	1.2518E-05	7.3491E-08	2
2780	24982	C12H26	2,5,6-trimethylnonane	62184-13-8	25	9.3433E-03	9.8814E-04	9.3433E-06	5.4851E-08	2
2781	24983	C12H26	2,5,7-trimethylnonane	62184-14-9	25	1.0286E-02	1.0879E-03	1.0286E-05	6.0386E-08	2
2782	24984	C12H26	2,5,8-trimethylnonane	49557-09-7	25	1.1340E-02	1.1993E-03	1.1340E-05	6.6572E-08	2
2783	24985	C12H26	2,6,6-trimethylnonane	62184-15-0	25	1.2518E-02	1.3239E-03	1.2518E-05	7.3491E-08	2
2784	24986	C12H26	2,6,7-trimethylnonane	62184-16-1	25	8.1104E-03	8.5775E-04	8.1104E-06	4.7613E-08	2
2785	24987	C12H26	2,7,7-trimethylnonane	62184-17-2	25	8.9095E-03	9.4227E-04	8.9095E-06	5.2305E-08	2
2786	24988	C12H26	3,3,4-trimethylnonane	62184-18-3	25	7.0640E-03	7.4709E-04	7.0640E-06	4.1470E-08	2
2787	24989	C12H26	3,3,5-trimethylnonane	62184-19-4	25	1.2518E-02	1.3239E-03	1.2518E-05	7.3491E-08	2
2788	24990	C12H26	3,3,7-trimethylnonane	62184-21-8	25	9.8016E-03	1.0366E-03	9.8016E-06	5.7542E-08	2
2789	24991	C12H26	3,3,6-trimethylnonane	62184-20-7	25	8.4990E-03	8.9885E-04	8.4990E-06	4.9895E-08	2
2790	24992	C12H26	3,4,4-trimethylnonane	62184-22-9	25	7.7424E-03	8.1884E-04	7.7424E-06	4.5453E-08	2
2791	24993	C12H26	3,4,5-trimethylnonane	62184-23-0	25	7.7424E-03	8.1884E-04	7.7424E-06	4.5453E-08	2
2792	24994	C12H26	3,4,6-trimethylnonane	62184-24-1	25	8.9095E-03	9.4227E-04	8.9095E-06	5.2305E-08	2
2793	24995	C12H26	3,4,7-trimethylnonane	27802-85-3	25	7.7424E-03	8.1884E-04	7.7424E-06	4.5453E-08	2
2794	24996	C12H26	3,5,5-trimethylnonane	62184-25-2	25	1.1913E-02	1.2599E-03	1.1913E-05	6.9935E-08	2
2795	24997	C12H26	3,5,6-trimethylnonane	62184-26-3	25	8.9095E-03	9.4227E-04	8.9095E-06	5.2305E-08	2
2796	24998	C12H26	3,5,7-trimethylnonane	62184-27-4	25	9.8016E-03	1.0366E-03	9.8016E-06	5.7542E-08	2
2797	24999	C12H26	3,6,6-trimethylnonane	62184-28-5	25	1.2518E-02	1.3239E-03	1.2518E-05	7.3491E-08	2
2798	25000	C12H26	4,4,5-trimethylnonane	62184-29-6	25	8.4990E-03	8.9885E-04	8.4990E-06	4.9895E-08	2
2799	25001	C12H26	4,4,6-trimethylnonane	62184-30-9	25	1.1913E-02	1.2599E-03	1.1913E-05	6.9935E-08	2
2800	25002	C12H26	4,5,5-trimethylnonane	62184-31-0	25	8.4990E-03	8.9885E-04	8.4990E-06	4.9895E-08	2

Table 3 Solubility in Water - Organic Compounds

					Solubility in Water, S (ppm - parts per million, wt - weight, g - gram)					
NO	ID	FORMULA	NAME	CAS No	T, C	S@T, ppm(wt)	S@T, ppm(mol)	S@T, g/1000 g of H2O	S@T, mol/1000 g of H2O	code
2801	25003	C12H26	4,5,6-trimethylnonane	62211-85-2	25	8.4990E-03	8.9885E-04	8.4990E-06	4.9895E-08	2
2802	25004	C12H26	4-tert-butyloctane	62184-32-1	25	1.1913E-02	1.2599E-03	1.1913E-05	6.9935E-08	2
2803	25005	C12H26	2-methyl-4-propyloctane	62184-33-2	25	1.1913E-02	1.2599E-03	1.1913E-05	6.9935E-08	2
2804	25006	C12H26	2-methyl-5-propyloctane	62184-34-3	25	1.1913E-02	1.2599E-03	1.1913E-05	6.9935E-08	2
2805	25007	C12H26	3-methyl-4-propyloctane	62184-35-4	25	8.9095E-03	9.4227E-04	8.9095E-06	5.2305E-08	2
2806	25008	C12H26	3-methyl-5-propyloctane	62184-36-5	25	1.0286E-02	1.0879E-03	1.0286E-05	6.0386E-08	2
2807	25009	C12H26	4-methyl-4-propyloctane	17312-41-3	25	9.8016E-03	1.0366E-03	9.8016E-06	5.7542E-08	2
2808	25010	C12H26	4-methyl-5-propyloctane	62183-85-1	25	9.3433E-03	9.8814E-04	9.3433E-06	5.4851E-08	2
2809	25011	C12H26	2-methyl-3-isopropyloctane	13287-19-9	25	9.3433E-03	9.8814E-04	9.3433E-06	5.4851E-08	2
2810	25012	C12H26	2-methyl-4-isopropyloctane	62183-86-2	25	1.2518E-02	1.3239E-03	1.2518E-05	7.3491E-08	2
2811	25013	C12H26	2-methyl-5-isopropyloctane	62183-87-3	25	1.1913E-02	1.2599E-03	1.1913E-05	6.9935E-08	2
2812	25014	C12H26	3-methyl-4-isopropyloctane	62183-88-4	25	9.8016E-03	1.0366E-03	9.8016E-06	5.7542E-08	2
2813	25015	C12H26	3-methyl-5-isopropyloctane	62183-89-5	25	1.1340E-02	1.1993E-03	1.1340E-05	6.6572E-08	2
2814	25016	C12H26	4-methyl-4-isopropyloctane	62183-90-8	25	8.4990E-03	8.9885E-04	8.4990E-06	4.9895E-08	2
2815	25017	C12H26	4-methyl-5-isopropyloctane	62183-91-9	25	1.0286E-02	1.0879E-03	1.0286E-05	6.0386E-08	2
2816	25018	C12H26	3,3-diethyloctane	17302-40-8	25	6.1743E-03	6.5300E-04	6.1743E-06	3.6247E-08	2
2817	25019	C12H26	3,4-diethyloctane	62183-92-0	25	8.1104E-03	8.5775E-04	8.1104E-06	4.7613E-08	2
2818	25020	C12H26	3,5-diethyloctane	62183-93-1	25	8.9095E-03	9.4227E-04	8.9095E-06	5.2305E-08	2
2819	25021	C12H26	3,6-diethyloctane	62183-94-2	25	7.0640E-03	7.4709E-04	7.0640E-06	4.1470E-08	2
2820	25022	C12H26	4,4-diethyloctane	17312-42-4	25	7.7424E-03	8.1884E-04	7.7424E-06	4.5453E-08	2
2821	25023	C12H26	4,5-diethyloctane	1636-41-5	25	8.9095E-03	9.4227E-04	8.9095E-06	5.2305E-08	2
2822	25024	C12H26	2,2-dimethyl-3-ethyloctane	62183-95-3	25	9.3433E-03	9.8814E-04	9.3433E-06	5.4851E-08	2
2823	25025	C12H26	2,2-dimethyl-4-ethyloctane	62183-96-4	25	1.4555E-02	1.5393E-03	1.4555E-05	8.5445E-08	2
2824	25026	C12H26	2,2-dimethyl-5-ethyloctane	62183-97-5	25	1.3159E-02	1.3917E-03	1.3159E-05	7.7253E-08	2
2825	25027	C12H26	2,2-dimethyl-6-ethyloctane	62183-98-6	25	1.0798E-02	1.1420E-03	1.0798E-05	6.3393E-08	2
2826	25028	C12H26	2,3-dimethyl-3-ethyloctane	62183-99-7	25	7.0640E-03	7.4709E-04	7.0640E-06	4.1470E-08	2
2827	25029	C12H26	2,3-dimethyl-4-ethyloctane	62184-00-3	25	8.9095E-03	9.4227E-04	8.9095E-06	5.2305E-08	2
2828	25030	C12H26	2,3-dimethyl-5-ethyloctane	62184-01-4	25	1.0286E-02	1.0879E-03	1.0286E-05	6.0386E-08	2
2829	25031	C12H26	2,3-dimethyl-6-ethyloctane	62184-02-5	25	8.1104E-03	8.5775E-04	8.1104E-06	4.7613E-08	2
2830	25032	C12H26	2,4-dimethyl-3-ethyloctane	62184-03-6	25	8.9095E-03	9.4227E-04	8.9095E-06	5.2305E-08	2
2831	25033	C12H26	2,4-dimethyl-4-ethyloctane	62184-04-7	25	1.1913E-02	1.2599E-03	1.1913E-05	6.9935E-08	2
2832	25034	C12H26	2,4-dimethyl-5-ethyloctane	62184-05-8	25	1.0798E-02	1.1420E-03	1.0798E-05	6.3393E-08	2
2833	25035	C12H26	2,4-dimethyl-6-ethyloctane	62184-06-9	25	1.1340E-02	1.1993E-03	1.1340E-05	6.6572E-08	2
2834	25036	C12H26	2,5-dimethyl-3-ethyloctane	62184-07-0	25	1.0286E-02	1.0879E-03	1.0286E-05	6.0386E-08	2
2835	25037	C12H26	2,5-dimethyl-4-ethyloctane	62184-08-1	25	1.1340E-02	1.1993E-03	1.1340E-05	6.6572E-08	2
2836	25038	C12H26	2,5-dimethyl-5-ethyloctane	62184-09-2	25	1.0798E-02	1.1420E-03	1.0798E-05	6.3393E-08	2
2837	25039	C12H26	2,5-dimethyl-6-ethyloctane	62183-50-0	25	8.9095E-03	9.4227E-04	8.9095E-06	5.2305E-08	2
2838	25040	C12H26	2,6-dimethyl-3-ethyloctane	62183-51-1	25	8.9095E-03	9.4227E-04	8.9095E-06	5.2305E-08	2
2839	25041	C12H26	2,6-dimethyl-4-ethyloctane	62183-52-2	25	1.2518E-02	1.3239E-03	1.2518E-05	7.3491E-08	2
2840	25042	C12H26	2,6-dimethyl-5-ethyloctane	62183-53-3	25	9.8016E-03	1.0366E-03	9.8016E-06	5.7542E-08	2
2841	25043	C12H26	2,6-dimethyl-6-ethyloctane	62183-54-4	25	8.4990E-03	8.9885E-04	8.4990E-06	4.9895E-08	2
2842	25044	C12H26	2,7-dimethyl-3-ethyloctane	62183-55-5	25	9.3433E-03	9.8814E-04	9.3433E-06	5.4851E-08	2
2843	25045	C12H26	2,7-dimethyl-4-ethyloctane	62183-56-6	25	1.2518E-02	1.3239E-03	1.2518E-05	7.3491E-08	2
2844	25046	C12H26	3,3-dimethyl-4-ethyloctane	62183-57-7	25	8.4990E-03	8.9885E-04	8.4990E-06	4.9895E-08	2
2845	25047	C12H26	3,3-dimethyl-5-ethyloctane	62183-58-8	25	1.1913E-02	1.2599E-03	1.1913E-05	6.9935E-08	2
2846	25048	C12H26	3,3-dimethyl-6-ethyloctane	62183-59-9	25	9.3433E-03	9.8814E-04	9.3433E-06	5.4851E-08	2
2847	25049	C12H26	3,4-dimethyl-3-ethyloctane	62212-28-6	25	7.7424E-03	8.1884E-04	7.7424E-06	4.5453E-08	2
2848	25050	C12H26	3,4-dimethyl-4-ethyloctane	62183-60-2	25	7.3940E-03	7.8199E-04	7.3940E-06	4.3408E-08	2
2849	25051	C12H26	3,4-dimethyl-5-ethyloctane	62183-61-3	25	8.4990E-03	8.9885E-04	8.4990E-06	4.9895E-08	2
2850	25052	C12H26	3,4-dimethyl-6-ethyloctane	62183-62-4	25	8.4990E-03	8.9885E-04	8.4990E-06	4.9895E-08	2
2851	25053	C12H26	3,5-dimethyl-3-ethyloctane	62183-63-5	25	9.3433E-03	9.8814E-04	9.3433E-06	5.4851E-08	2
2852	25054	C12H26	3,5-dimethyl-4-ethyloctane	62183-64-6	25	8.9095E-03	9.4227E-04	8.9095E-06	5.2305E-08	2
2853	25055	C12H26	3,5-dimethyl-5-ethyloctane	62183-65-7	25	1.2518E-02	1.3239E-03	1.2518E-05	7.3491E-08	2
2854	25056	C12H26	3,5-dimethyl-6-ethyloctane	62183-66-8	25	8.9095E-03	9.4227E-04	8.9095E-06	5.2305E-08	2
2855	25057	C12H26	3,6-dimethyl-3-ethyloctane	62183-67-9	25	8.1104E-03	8.5775E-04	8.1104E-06	4.7613E-08	2
2856	25058	C12H26	3,6-dimethyl-4-ethyloctane	62183-68-0	25	9.3433E-03	9.8814E-04	9.3433E-06	5.4851E-08	2
2857	25059	C12H26	4,4-dimethyl-3-ethyloctane	62183-69-1	25	8.4990E-03	8.9885E-04	8.4990E-06	4.9895E-08	2
2858	25060	C12H26	4,4-dimethyl-5-ethyloctane	62183-70-4	25	9.3433E-03	9.8814E-04	9.3433E-06	5.4851E-08	2
2859	25061	C12H26	4,4-dimethyl-6-ethyloctane	62183-71-5	25	1.1913E-02	1.2599E-03	1.1913E-05	6.9935E-08	2
2860	25062	C12H26	4,5-dimethyl-3-ethyloctane	62183-72-6	25	8.1104E-03	8.5775E-04	8.1104E-06	4.7613E-08	2
2861	25063	C12H26	4,5-dimethyl-4-ethyloctane	62183-73-7	25	7.7424E-03	8.1884E-04	7.7424E-06	4.5453E-08	2
2862	25064	C12H26	2,2,3,3-tetramethyloctane	62183-74-8	25	7.7424E-03	8.1884E-04	7.7424E-06	4.5453E-08	2
2863	25065	C12H26	2,2,3,4-tetramethyloctane	62183-75-9	25	9.8016E-03	1.0366E-03	9.8016E-06	5.7542E-08	2
2864	25066	C12H26	2,2,3,5-tetramethyloctane	62183-76-0	25	1.1340E-02	1.1993E-03	1.1340E-05	6.6572E-08	2
2865	25067	C12H26	2,2,3,6-tetramethyloctane	62183-77-1	25	1.0286E-02	1.0879E-03	1.0286E-05	6.0386E-08	2
2866	25068	C12H26	2,2,3,7-tetramethyloctane	62183-78-2	25	1.0286E-02	1.0879E-03	1.0286E-05	6.0386E-08	2
2867	25069	C12H26	2,2,4,4-tetramethyloctane	62183-79-3	25	1.1340E-02	1.1993E-03	1.1340E-05	6.6572E-08	2
2868	25070	C12H26	2,2,4,5-tetramethyloctane	62183-80-6	25	1.4555E-02	1.5393E-03	1.4555E-05	8.5445E-08	2
2869	25071	C12H26	2,2,4,6-tetramethyloctane	62183-81-7	25	1.6118E-02	1.7046E-03	1.6118E-05	9.4622E-08	2
2870	25072	C12H26	2,2,4,7-tetramethyloctane	62183-82-8	25	1.6284E-02	1.7222E-03	1.6284E-05	9.5599E-08	2
2871	25073	C12H26	2,2,5,5-tetramethyloctane	62183-83-9	25	1.6969E-02	1.7946E-03	1.6969E-05	9.9619E-08	2
2872	25074	C12H26	2,2,5,6-tetramethyloctane	62183-84-0	25	1.2518E-02	1.3239E-03	1.2518E-05	7.3491E-08	2
2873	25075	C12H26	2,2,5,7-tetramethyloctane	62199-19-3	25	1.6118E-02	1.7046E-03	1.6118E-05	9.4622E-08	2
2874	25076	C12H26	2,2,6,6-tetramethyloctane	62199-20-6	25	1.6118E-02	1.7046E-03	1.6118E-05	9.4622E-08	2
2875	25077	C12H26	2,2,6,7-tetramethyloctane	62199-21-7	25	1.1913E-02	1.2599E-03	1.1913E-05	6.9935E-08	2
2876	25078	C12H26	2,2,7,7-tetramethyloctane	1071-31-4	25	1.6118E-02	1.7046E-03	1.6118E-05	9.4622E-08	2
2877	25079	C12H26	2,3,3,4-tetramethyloctane	62199-22-8	25	7.3940E-03	7.8199E-04	7.3940E-06	4.3408E-08	2
2878	25080	C12H26	2,3,3,5-tetramethyloctane	62199-23-9	25	1.0286E-02	1.0879E-03	1.0286E-05	6.0386E-08	2
2879	25081	C12H26	2,3,3,6-tetramethyloctane	62199-24-0	25	9.3433E-03	9.8814E-04	9.3433E-06	5.4851E-08	2
2880	25082	C12H26	2,3,3,7-tetramethyloctane	62199-25-1	25	9.3433E-03	9.8814E-04	9.3433E-06	5.4851E-08	2

Table 3 Solubility in Water - Organic Compounds

NO	ID	FORMULA	NAME	CAS No	Solubility in Water, S (ppm - parts per million, wt - weight, g - gram) T, C	S@T, ppm(wt)	S@T, ppm(mol)	S@T, g/1000 g of H2O	S@T, mol/1000 g of H2O	code
2881	25083	C12H26	2,3,4,4-tetramethyloctane	62199-26-2	25	9.3433E-03	9.8814E-04	9.3433E-06	5.4851E-08	2
2882	25084	C12H26	2,3,4,5-tetramethyloctane	62199-27-3	25	8.9095E-03	9.4227E-04	8.9095E-06	5.2305E-08	2
2883	25085	C12H26	2,3,4,6-tetramethyloctane	62199-28-4	25	9.8016E-03	1.0366E-03	9.8016E-06	5.7542E-08	2
2884	25086	C12H26	2,3,4,7-tetramethyloctane	62199-29-5	25	1.0286E-02	1.0879E-03	1.0286E-05	6.0386E-08	2
2885	25087	C12H26	2,3,5,5-tetramethyloctane	62199-30-8	25	9.8016E-03	1.0366E-03	9.8016E-06	5.7542E-08	2
2886	25088	C12H26	2,3,5,6-tetramethyloctane	62199-31-9	25	9.8016E-03	1.0366E-03	9.8016E-06	5.7542E-08	2
2887	25089	C12H26	2,3,5,7-tetramethyloctane	62199-32-0	25	1.2518E-02	1.3239E-03	1.2518E-05	7.3491E-08	2
2888	25090	C12H26	2,3,6,6-tetramethyloctane	62199-33-1	25	1.0798E-02	1.1420E-03	1.0798E-05	6.3393E-08	2
2889	25091	C12H26	2,3,6,7-tetramethyloctane	52670-34-5	25	9.3433E-03	9.8814E-04	9.3433E-06	5.4851E-08	2
2890	25092	C12H26	2,4,4,5-tetramethyloctane	62199-34-2	25	1.1913E-02	1.2599E-03	1.1913E-05	6.9935E-08	2
2891	25093	C12H26	2,4,4,6-tetramethyloctane	62199-35-3	25	1.6118E-02	1.7046E-03	1.6118E-05	9.4622E-08	2
2892	25094	C12H26	2,4,4,7-tetramethyloctane	35866-96-7	25	1.6969E-02	1.7946E-03	1.6969E-05	9.9619E-08	2
2893	25095	C12H26	2,4,5,5-tetramethyloctane	62199-36-4	25	1.1340E-02	1.1993E-03	1.1340E-05	6.6572E-08	2
2894	25096	C12H26	2,4,5,6-tetramethyloctane	62199-37-5	25	1.0286E-02	1.0879E-03	1.0286E-05	6.0386E-08	2
2895	25097	C12H26	2,4,5,7-tetramethyloctane	2217-17-6	25	1.5082E-02	1.5950E-03	1.5082E-05	8.8539E-08	2
2896	25098	C12H26	2,4,6,6-tetramethyloctane	62199-38-6	25	1.4555E-02	1.5393E-03	1.4555E-05	8.5445E-08	2
2897	25099	C12H26	2,5,5,6-tetramethyloctane	62199-39-7	25	1.0286E-02	1.0879E-03	1.0286E-05	6.0386E-08	2
2898	25100	C12H26	2,5,6,6-tetramethyloctane	62199-40-0	25	9.3433E-03	9.8814E-04	9.3433E-06	5.4851E-08	2
2899	25101	C12H26	3,3,4,4-tetramethyloctane	62199-41-1	25	7.0640E-03	7.4709E-04	7.0640E-06	4.1470E-08	2
2900	25102	C12H26	3,3,4,5-tetramethyloctane	62199-42-2	25	8.4990E-03	8.9885E-04	8.4990E-06	4.9895E-08	2
2901	25103	C12H26	3,3,4,6-tetramethyloctane	62199-43-3	25	9.3433E-03	9.8814E-04	9.3433E-06	5.4851E-08	2
2902	25104	C12H26	3,3,5,5-tetramethyloctane	62199-44-4	25	8.9095E-03	9.4227E-04	8.9095E-06	5.2305E-08	2
2903	25105	C12H26	3,3,5,6-tetramethyloctane	62199-45-5	25	1.0798E-02	1.1420E-03	1.0798E-05	6.3393E-08	2
2904	25106	C12H26	3,3,6,6-tetramethyloctane	62199-46-6	25	1.2707E-02	1.3439E-03	1.2707E-05	7.4597E-08	2
2905	25107	C12H26	3,4,4,5-tetramethyloctane	62199-47-7	25	7.3940E-03	7.8199E-04	7.3940E-06	4.3408E-08	2
2906	25108	C12H26	3,4,4,6-tetramethyloctane	62185-19-7	25	9.8016E-03	1.0366E-03	9.8016E-06	5.7542E-08	2
2907	25109	C12H26	3,4,5,5-tetramethyloctane	62185-20-0	25	8.9095E-03	9.4227E-04	8.9095E-06	5.2305E-08	2
2908	25110	C12H26	3,4,5,6-tetramethyloctane	62185-21-1	25	8.1104E-03	8.5775E-04	8.1104E-06	4.7613E-08	2
2909	25111	C12H26	4,4,5,5-tetramethyloctane	62185-22-2	25	7.7424E-03	8.1884E-04	7.7424E-06	4.5453E-08	2
2910	25112	C12H26	2-methyl-4-tert-butylheptane	62185-23-3	25	1.6118E-02	1.7046E-03	1.6118E-05	9.4622E-08	2
2911	25113	C12H26	3-methyl-4-tert-butylheptane	62185-24-4	25	1.2518E-02	1.3239E-03	1.2518E-05	7.3491E-08	2
2912	25114	C12H26	4-methyl-4-tert-butylheptane	62185-25-5	25	8.4990E-03	8.9885E-04	8.4990E-06	4.9895E-08	2
2913	25115	C12H26	3-ethyl-4-propylheptane	62185-26-6	25	9.3433E-03	9.8814E-04	9.3433E-06	5.4851E-08	2
2914	25116	C12H26	4-ethyl-4-propylheptane	17312-43-5	25	8.9095E-03	9.4227E-04	8.9095E-06	5.2305E-08	2
2915	25117	C12H26	3-ethyl-4-isopropylheptane	62185-27-7	25	9.8016E-03	1.0366E-03	9.8016E-06	5.7542E-08	2
2916	25118	C12H26	4-ethyl-4-isopropylheptane	62185-28-8	25	7.7424E-03	8.1884E-04	7.7424E-06	4.5453E-08	2
2917	25119	C12H26	2,2-dimethyl-4-propylheptane	---	25	1.6969E-02	1.7946E-03	1.6969E-05	9.9619E-08	2
2918	25120	C12H26	2,3-dimethyl-4-propylheptane	62185-30-2	25	1.0286E-02	1.0879E-03	1.0286E-05	6.0386E-08	2
2919	25121	C12H26	2,4-dimethyl-4-propylheptane	62185-31-3	25	1.3159E-02	1.3917E-03	1.3159E-05	7.7253E-08	2
2920	25122	C12H26	2,5-dimethyl-4-propylheptane	62185-32-4	25	1.2518E-02	1.3239E-03	1.2518E-05	7.3491E-08	2
2921	25123	C12H26	2,6-dimethyl-4-propylheptane	62185-33-5	25	1.6452E-02	1.7400E-03	1.6452E-05	9.6587E-08	2
2922	25124	C12H26	3,3-dimethyl-4-propylheptane	62185-34-6	25	9.3433E-03	9.8814E-04	9.3433E-06	5.4851E-08	2
2923	25125	C12H26	3,4-dimethyl-4-propylheptane	62185-35-7	25	8.4990E-03	8.9885E-04	8.4990E-06	4.9895E-08	2
2924	25126	C12H26	3,5-dimethyl-4-propylheptane	62185-36-8	25	9.3433E-03	9.8814E-04	9.3433E-06	5.4851E-08	2
2925	25127	C12H26	2,2-dimethyl-3-isopropylheptane	62185-37-9	25	1.2518E-02	1.3239E-03	1.2518E-05	7.3491E-08	2
2926	25128	C12H26	2,2-dimethyl-4-isopropylheptane	62185-38-0	25	1.8825E-02	1.9909E-03	1.8825E-05	1.1051E-07	2
2927	25129	C12H26	2,3-dimethyl-3-isopropylheptane	62185-39-1	25	7.3940E-03	7.8199E-04	7.3940E-06	4.3408E-08	2
2928	25130	C12H26	2,3-dimethyl-4-isopropylheptane	62185-40-4	25	1.1340E-02	1.1993E-03	1.1340E-05	6.6572E-08	2
2929	25131	C12H26	2,4-dimethyl-3-isopropylheptane	62185-41-5	25	1.1340E-02	1.1993E-03	1.1340E-05	6.6572E-08	2
2930	25132	C12H26	2,4-dimethyl-4-isopropylheptane	62185-42-6	25	1.1340E-02	1.1993E-03	1.1340E-05	6.6572E-08	2
2931	25133	C12H26	2,5-dimethyl-3-isopropylheptane	62185-43-7	25	1.2518E-02	1.3239E-03	1.2518E-05	7.3491E-08	2
2932	25134	C12H26	2,5-dimethyl-4-isopropylheptane	62185-44-8	25	1.3159E-02	1.3917E-03	1.3159E-05	7.7253E-08	2
2933	25135	C12H26	2,6-dimethyl-3-isopropylheptane	62185-45-9	25	1.2031E-02	1.2724E-03	1.2031E-05	7.0630E-08	2
2934	25136	C12H26	2,6-dimethyl-4-isopropylheptane	35866-89-8	25	1.7870E-02	1.8899E-03	1.7870E-05	1.0491E-07	2
2935	25137	C12H26	3,3-dimethyl-4-isopropylheptane	62185-46-0	25	1.0286E-02	1.0879E-03	1.0286E-05	6.0386E-08	2
2936	25138	C12H26	3,4-dimethyl-4-isopropylheptane	62185-47-1	25	7.3940E-03	7.8199E-04	7.3940E-06	4.3408E-08	2
2937	25139	C12H26	3,5-dimethyl-4-isopropylheptane	62198-89-4	25	1.0286E-02	1.0879E-03	1.0286E-05	6.0386E-08	2
2938	25140	C12H26	2-methyl-3,3-diethylheptane	62198-90-7	25	6.7513E-03	7.1402E-04	6.7513E-06	3.9635E-08	2
2939	25141	C12H26	2-methyl-3,4-diethylheptane	62198-91-8	25	9.8016E-03	1.0366E-03	9.8016E-06	5.7542E-08	2
2940	25142	C12H26	2-methyl-3,5-diethylheptane	62198-92-9	25	9.8016E-03	1.0366E-03	9.8016E-06	5.7542E-08	2
2941	25143	C12H26	2-methyl-4,4-diethylheptane	62198-93-0	25	1.0286E-02	1.0879E-03	1.0286E-05	6.0386E-08	2
2942	25144	C12H26	2-methyl-4,5-diethylheptane	62198-94-1	25	1.0798E-02	1.1420E-03	1.0798E-05	6.3393E-08	2
2943	25145	C12H26	2-methyl-5,5-diethylheptane	62198-95-2	25	8.4990E-03	8.9885E-04	8.4990E-06	4.9895E-08	2
2944	25146	C12H26	3-methyl-3,4-diethylheptane	62198-96-3	25	7.3940E-03	7.8199E-04	7.3940E-06	4.3408E-08	2
2945	25147	C12H26	3-methyl-3,5-diethylheptane	62198-97-4	25	8.9095E-03	9.4227E-04	8.9095E-06	5.2305E-08	2
2946	25148	C12H26	3-methyl-4,4-diethylheptane	62198-98-5	25	6.4551E-03	6.8269E-04	6.4551E-06	3.7896E-08	2
2947	25149	C12H26	3-methyl-4,5-diethylheptane	62198-99-6	25	8.4990E-03	8.9885E-04	8.4990E-06	4.9895E-08	2
2948	25150	C12H26	3-methyl-5,5-diethylheptane	62199-00-2	25	8.1104E-03	8.5775E-04	8.1104E-06	4.7613E-08	2
2949	25151	C12H26	4-methyl-3,3-diethylheptane	62199-01-3	25	6.1743E-03	6.5300E-04	6.1743E-06	3.6247E-08	2
2950	25152	C12H26	4-methyl-3,4-diethylheptane	62199-02-4	25	7.3940E-03	7.8199E-04	7.3940E-06	4.3408E-08	2
2951	25153	C12H26	4-methyl-3,5-diethylheptane	62199-03-5	25	7.7424E-03	8.1884E-04	7.7424E-06	4.5453E-08	2
2952	25154	C12H26	2,2,3-trimethyl-3-ethylheptane	62199-04-6	25	7.7424E-03	8.1884E-04	7.7424E-06	4.5453E-08	2
2953	25155	C12H26	2,2,3-trimethyl-4-ethylheptane	62199-05-7	25	1.0798E-02	1.1420E-03	1.0798E-05	6.3393E-08	2
2954	25156	C12H26	2,2,3-trimethyl-5-ethylheptane	62199-06-8	25	1.1340E-02	1.1993E-03	1.1340E-05	6.6572E-08	2
2955	25157	C12H26	2,2,4-trimethyl-3-ethylheptane	62199-07-9	25	1.1340E-02	1.1993E-03	1.1340E-05	6.6572E-08	2
2956	25158	C12H26	2,2,4-trimethyl-4-ethylheptane	62199-08-0	25	1.0286E-02	1.0879E-03	1.0286E-05	6.0386E-08	2
2957	25159	C12H26	2,2,4-trimethyl-5-ethylheptane	62199-09-1	25	1.4555E-02	1.5393E-03	1.4555E-05	8.5445E-08	2
2958	25160	C12H26	2,2,5-trimethyl-3-ethylheptane	62199-10-4	25	1.2518E-02	1.3239E-03	1.2518E-05	7.3491E-08	2
2959	25161	C12H26	2,2,5-trimethyl-4-ethylheptane	62199-11-5	25	1.5314E-02	1.6196E-03	1.5314E-05	8.9903E-08	2
2960	25162	C12H26	2,2,5-trimethyl-5-ethylheptane	62199-12-6	25	1.3159E-02	1.3917E-03	1.3159E-05	7.7253E-08	2

Table 3 Solubility in Water - Organic Compounds

					Solubility in Water, S (ppm - parts per million, wt - weight, g - gram)					
NO	ID	FORMULA	NAME	CAS No	T, C	S@T, ppm(wt)	S@T, ppm(mol)	S@T, g/1000 g of H2O	S@T, mol/1000 g of H2O	code
2961	25163	C12H26	2,2,6-trimethyl-3-ethylheptane	62199-13-7	25	1.2518E-02	1.3239E-03	1.2518E-05	7.3491E-08	2
2962	25164	C12H26	2,2,6-trimethyl-4-ethylheptane	62199-14-8	25	2.0906E-02	2.2111E-03	2.0906E-05	1.2273E-07	2
2963	25165	C12H26	2,2,6-trimethyl-5-ethylheptane	62199-15-9	25	1.4555E-02	1.5393E-03	1.4555E-05	8.5445E-08	2
2964	25166	C12H26	2,3,3-trimethyl-4-ethylheptane	62199-16-0	25	8.1104E-03	8.5775E-04	8.1104E-06	4.7613E-08	2
2965	25167	C12H26	2,3,3-trimethyl-5-ethylheptane	62199-17-1	25	1.0286E-02	1.0879E-03	1.0286E-05	6.0386E-08	2
2966	25168	C12H26	2,3,4-trimethyl-3-ethylheptane	62199-18-2	25	6.7513E-03	7.1402E-04	6.7513E-06	3.9635E-08	2
2967	25169	C12H26	2,3,4-trimethyl-4-ethylheptane	62198-55-4	25	8.4990E-03	8.9885E-04	8.4990E-06	4.9895E-08	2
2968	25170	C12H26	2,3,4-trimethyl-5-ethylheptane	62198-56-5	25	8.9095E-03	9.4227E-04	8.9095E-06	5.2305E-08	2
2969	25171	C12H26	2,3,5-trimethyl-3-ethylheptane	62198-57-6	25	8.7014E-03	9.2026E-04	8.7014E-06	5.1083E-08	2
2970	25172	C12H26	2,3,5-trimethyl-4-ethylheptane	62198-58-7	25	9.3433E-03	9.8814E-04	9.3433E-06	5.4851E-08	2
2971	25173	C12H26	2,3,5-trimethyl-5-ethylheptane	62198-59-8	25	1.0286E-02	1.0879E-03	1.0286E-05	6.0386E-08	2
2972	25174	C12H26	2,3,6-trimethyl-3-ethylheptane	62198-60-1	25	9.3433E-03	9.8814E-04	9.3433E-06	5.4851E-08	2
2973	25175	C12H26	2,3,6-trimethyl-4-ethylheptane	62198-61-2	25	1.2518E-02	1.3239E-03	1.2518E-05	7.3491E-08	2
2974	25176	C12H26	2,3,6-trimethyl-5-ethylheptane	62198-62-3	25	1.0798E-02	1.1420E-03	1.0798E-05	6.3393E-08	2
2975	25177	C12H26	2,4,4-trimethyl-3-ethylheptane	62198-63-4	25	1.0286E-02	1.0879E-03	1.0286E-05	6.0386E-08	2
2976	25178	C12H26	2,4,4-trimethyl-5-ethylheptane	62198-64-5	25	1.1340E-02	1.1993E-03	1.1340E-05	6.6572E-08	2
2977	25179	C12H26	2,4,5-trimethyl-3-ethylheptane	62198-65-6	25	9.3433E-03	9.8814E-04	9.3433E-06	5.4851E-08	2
2978	25180	C12H26	2,4,5-trimethyl-4-ethylheptane	62198-66-7	25	9.8016E-03	1.0366E-03	9.8016E-06	5.7542E-08	2
2979	25181	C12H26	2,4,5-trimethyl-5-ethylheptane	62198-67-8	25	8.9095E-03	9.4227E-04	8.9095E-06	5.2305E-08	2
2980	25182	C12H26	2,4,6-trimethyl-3-ethylheptane	62198-68-9	25	1.1913E-02	1.2599E-03	1.1913E-05	6.9935E-08	2
2981	25183	C12H26	2,4,6-trimethyl-4-ethylheptane	62198-69-0	25	1.6118E-02	1.7046E-03	1.6118E-05	9.4622E-08	2
2982	25184	C12H26	2,5,5-trimethyl-3-ethylheptane	62198-70-3	25	1.2518E-02	1.3239E-03	1.2518E-05	7.3491E-08	2
2983	25185	C12H26	2,5,5-trimethyl-4-ethylheptane	62198-71-4	25	1.1340E-02	1.1993E-03	1.1340E-05	6.6572E-08	2
2984	25186	C12H26	3,3,4-trimethyl-4-ethylheptane	62198-72-5	25	6.1743E-03	6.5300E-04	6.1743E-06	3.6247E-08	2
2985	25187	C12H26	3,3,4-trimethyl-5-ethylheptane	62198-73-6	25	8.1104E-03	8.5775E-04	8.1104E-06	4.7613E-08	2
2986	25188	C12H26	3,3,5-trimethyl-4-ethylheptane	62198-74-7	25	8.9095E-03	9.4227E-04	8.9095E-06	5.2305E-08	2
2987	25189	C12H26	3,3,5-trimethyl-5-ethylheptane	62198-75-8	25	7.0640E-03	7.4709E-04	7.0640E-06	4.1470E-08	2
2988	25190	C12H26	3,4,4-trimethyl-3-ethylheptane	62198-76-9	25	6.1743E-03	6.5300E-04	6.1743E-06	3.6247E-08	2
2989	25191	C12H26	3,4,4-trimethyl-5-ethylheptane	62198-77-0	25	7.0640E-03	7.4709E-04	7.0640E-06	4.1470E-08	2
2990	25192	C12H26	3,4,5-trimethyl-3-ethylheptane	62198-78-1	25	7.0640E-03	7.4709E-04	7.0640E-06	4.1470E-08	2
2991	25193	C12H26	3,4,5-trimethyl-4-ethylheptane	62198-79-2	25	6.1743E-03	6.5300E-04	6.1743E-06	3.6247E-08	2
2992	25194	C12H26	2,2,3,3,4-pentamethylheptane	62198-80-5	25	7.7424E-03	8.1884E-04	7.7424E-06	4.5453E-08	2
2993	25195	C12H26	2,2,3,3,5-pentamethylheptane	62198-81-6	25	1.0286E-02	1.0879E-03	1.0286E-05	6.0386E-08	2
2994	25196	C12H26	2,2,3,3,6-pentamethylheptane	62198-82-7	25	1.0798E-02	1.1420E-03	1.0798E-05	6.3393E-08	2
2995	25197	C12H26	2,2,3,4,4-pentamethylheptane	62198-83-8	25	7.0640E-03	7.4709E-04	7.0640E-06	4.1470E-08	2
2996	25198	C12H26	2,2,3,4,5-pentamethylheptane	62198-84-9	25	1.0286E-02	1.0879E-03	1.0286E-05	6.0386E-08	2
2997	25199	C12H26	2,2,3,4,6-pentamethylheptane	62198-85-0	25	1.3837E-02	1.4634E-03	1.3837E-05	8.1233E-08	2
2998	25200	C12H26	2,2,3,5,5-pentamethylheptane	62198-86-1	25	1.4555E-02	1.5393E-03	1.4555E-05	8.5445E-08	2
2999	25201	C12H26	2,2,3,5,6-pentamethylheptane	62198-87-2	25	1.3292E-02	1.4057E-03	1.3292E-05	7.8031E-08	2
3000	25202	C12H26	2,2,3,6,6-pentamethylheptane	62198-88-3	25	1.6118E-02	1.7046E-03	1.6118E-05	9.4622E-08	2
3001	25203	C12H26	2,2,4,4,5-pentamethylheptane	62199-61-5	25	9.3433E-03	9.8814E-04	9.3433E-06	5.4851E-08	2
3002	25204	C12H26	2,2,4,4,6-pentamethylheptane	62199-62-6	25	1.5630E-02	1.6530E-03	1.5630E-05	9.1759E-08	2
3003	25205	C12H26	2,2,4,5,5-pentamethylheptane	62199-63-7	25	1.5314E-02	1.6196E-03	1.5314E-05	8.9903E-08	2
3004	25206	C12H26	2,2,4,5,6-pentamethylheptane	62199-64-8	25	1.6118E-02	1.7046E-03	1.6118E-05	9.4622E-08	2
3005	25207	C12H26	2,2,4,6,6-pentamethylheptane	13475-82-6	25	2.3305E-02	2.4647E-03	2.3305E-05	1.3682E-07	2
3006	25208	C12H26	2,2,5,5,6-pentamethylheptane	62199-65-9	25	1.5314E-02	1.6196E-03	1.5314E-05	8.9903E-08	2
3007	25209	C12H26	2,3,3,4,4-pentamethylheptane	62199-66-0	25	6.7513E-03	7.1402E-04	6.7513E-06	3.9635E-08	2
3008	25210	C12H26	2,3,3,4,5-pentamethylheptane	62199-67-1	25	7.7424E-03	8.1884E-04	7.7424E-06	4.5453E-08	2
3009	25211	C12H26	2,3,3,4,6-pentamethylheptane	62199-68-2	25	9.8016E-03	1.0366E-03	9.8016E-06	5.7542E-08	2
3010	25212	C12H26	2,3,3,5,5-pentamethylheptane	62199-69-3	25	7.7424E-03	8.1884E-04	7.7424E-06	4.5453E-08	2
3011	25213	C12H26	2,3,3,5,6-pentamethylheptane	52670-35-6	25	1.1340E-02	1.1993E-03	1.1340E-05	6.6572E-08	2
3012	25214	C12H26	2,3,4,4,5-pentamethylheptane	62199-70-6	25	7.7424E-03	8.1884E-04	7.7424E-06	4.5453E-08	2
3013	25215	C12H26	2,3,4,4,6-pentamethylheptane	62199-71-7	25	1.2518E-02	1.3239E-03	1.2518E-05	7.3491E-08	2
3014	25216	C12H26	2,3,4,5,5-pentamethylheptane	62199-72-8	25	8.9095E-03	9.4227E-04	8.9095E-06	5.2305E-08	2
3015	25217	C12H26	2,3,4,5,6-pentamethylheptane	27574-98-7	25	9.8016E-03	1.0366E-03	9.8016E-06	5.7542E-08	2
3016	25218	C12H26	2,4,4,5,5-pentamethylheptane	62199-73-9	25	9.3433E-03	9.8814E-04	9.3433E-06	5.4851E-08	2
3017	25219	C12H26	3,3,4,4,5-pentamethylheptane	62199-74-0	25	5.9082E-03	6.2485E-04	5.9082E-06	3.4685E-08	2
3018	25220	C12H26	3,3,4,5,5-pentamethylheptane	62199-75-1	25	5.1897E-03	5.4886E-04	5.1897E-06	3.0467E-08	2
3019	25221	C12H26	2,2-dimethyl-3-tert-butylhexane	62199-76-2	25	9.3433E-03	9.8814E-04	9.3433E-06	5.4851E-08	2
3020	25222	C12H26	2-methyl-3-ethyl-3-isopropylhexane	62199-77-3	25	7.7424E-03	8.1884E-04	7.7424E-06	4.5453E-08	2
3021	25223	C12H26	2-methyl-4-ethyl-3-isopropylhexane	62199-78-4	25	1.2518E-02	1.3239E-03	1.2518E-05	7.3491E-08	2
3022	25224	C12H26	2,2,3-trimethyl-3-isopropylhexane	62199-79-5	25	7.3940E-03	7.8199E-04	7.3940E-06	4.3408E-08	2
3023	25225	C12H26	2,2,4-trimethyl-3-isopropylhexane	62199-80-8	25	1.3159E-02	1.3917E-03	1.3159E-05	7.7253E-08	2
3024	25226	C12H26	2,2,5-trimethyl-3-isopropylhexane	62199-81-9	25	2.0906E-02	2.2111E-03	2.0906E-05	1.2273E-07	2
3025	25227	C12H26	2,2,5-trimethyl-4-isopropylhexane	62199-82-0	25	2.0906E-02	2.2111E-03	2.0906E-05	1.2273E-07	2
3026	25228	C12H26	2,3,4-trimethyl-3-isopropylhexane	62199-83-1	25	6.4551E-03	6.8269E-04	6.4551E-06	3.7896E-08	2
3027	25229	C12H26	2,3,5-trimethyl-3-isopropylhexane	62199-84-2	25	1.0286E-02	1.0879E-03	1.0286E-05	6.0386E-08	2
3028	25230	C12H26	2,3,5-trimethyl-4-isopropylhexane	62199-85-3	25	1.2518E-02	1.3239E-03	1.2518E-05	7.3491E-08	2
3029	25231	C12H26	2,4,4-trimethyl-3-isopropylhexane	62199-86-4	25	1.1340E-02	1.1993E-03	1.1340E-05	6.6572E-08	2
3030	25232	C12H26	3,3,4-triethylhexane	62199-87-5	25	5.9082E-03	6.2485E-04	5.9082E-06	3.4685E-08	2
3031	25233	C12H26	2,2-dimethyl,3,3-diethylhexane	62199-88-6	25	7.7424E-03	8.1884E-04	7.7424E-06	4.5453E-08	2
3032	25234	C12H26	2,2-dimethyl-3,4-diethylhexane	62199-89-7	25	1.2518E-02	1.3239E-03	1.2518E-05	7.3491E-08	2
3033	25235	C12H26	2,2-dimethyl-4,4-diethylhexane	62184-89-8	25	7.7424E-03	8.1884E-04	7.7424E-06	4.5453E-08	2
3034	25236	C12H26	2,3-dimethyl-3,4-diethylhexane	62184-90-1	25	6.4551E-03	6.8269E-04	6.4551E-06	3.7896E-08	2
3035	25237	C12H26	2,3-dimethyl-4,4-diethylhexane	62184-91-2	25	6.4551E-03	6.8269E-04	6.4551E-06	3.7896E-08	2
3036	25238	C12H26	2,4-dimethyl-3,3-diethylhexane	62184-92-3	25	7.7424E-03	8.1884E-04	7.7424E-06	4.5453E-08	2
3037	25239	C12H26	2,4-dimethyl-3,4-diethylhexane	62184-93-4	25	7.7424E-03	8.1884E-04	7.7424E-06	4.5453E-08	2
3038	25240	C12H26	2,5-dimethyl-3,3-diethylhexane	62184-94-5	25	1.2518E-02	1.3239E-03	1.2518E-05	7.3491E-08	2
3039	25241	C12H26	2,5-dimethyl-3,4-diethylhexane	62184-95-6	25	1.2518E-02	1.3239E-03	1.2518E-05	7.3491E-08	2
3040	25242	C12H26	3,3-dimethyl-4,4-diethylhexane	62184-96-7	25	4.9744E-03	5.2610E-04	4.9744E-06	2.9203E-08	2

Table 3 Solubility in Water - Organic Compounds

					Solubility in Water, S (ppm - parts per million, wt - weight, g - gram)					
NO	ID	FORMULA	NAME	CAS No	T, C	S@T, ppm(wt)	S@T, ppm(mol)	S@T, g/1000 g of H2O	S@T, mol/1000 g of H2O	code
3041	25243	C12H26	3,4-dimethyl-3,4-diethylhexane	62184-97-8	25	4.9744E-03	5.2610E-04	4.9744E-06	2.9203E-08	2
3042	25244	C12H26	2,2,3,3-tetramethyl-4-ethylhexane	62184-98-9	25	7.3940E-03	7.8199E-04	7.3940E-06	4.3408E-08	2
3043	25245	C12H26	2,2,3,4-tetramethyl-3-ethylhexane	62184-99-0	25	6.4551E-03	6.8269E-04	6.4551E-06	3.7896E-08	2
3044	25246	C12H26	2,2,3,4-tetramethyl-4-ethylhexane	62185-00-6	25	5.4166E-03	5.7286E-04	5.4166E-06	3.1799E-08	2
3045	25247	C12H26	2,2,3,5-tetramethyl-3-ethylhexane	62185-01-7	25	1.0286E-02	1.0879E-03	1.0286E-05	6.0386E-08	2
3046	25248	C12H26	2,2,3,5-tetramethyl-4-ethylhexane	62185-02-8	25	1.1913E-02	1.2599E-03	1.1913E-05	6.9935E-08	2
3047	25249	C12H26	2,2,4,4-tetramethyl-3-ethylhexane	62185-03-9	25	7.7424E-03	8.1884E-04	7.7424E-06	4.5453E-08	2
3048	25250	C12H26	2,2,4,5-tetramethyl-3-ethylhexane	62185-04-0	25	1.2518E-02	1.3239E-03	1.2518E-05	7.3491E-08	2
3049	25251	C12H26	2,2,4,5-tetramethyl-4-ethylhexane	62185-05-1	25	8.9095E-03	9.4227E-04	8.9095E-06	5.2305E-08	2
3050	25252	C12H26	2,2,5,5-tetramethyl-3-ethylhexane	62185-06-2	25	2.0906E-02	2.2111E-03	2.0906E-05	1.2273E-07	2
3051	25253	C12H26	2,3,3,4-tetramethyl-4-ethylhexane	62185-07-3	25	5.4166E-03	5.7286E-04	5.4166E-06	3.1799E-08	2
3052	25254	C12H26	2,3,3,5-tetramethyl-4-ethylhexane	62185-08-4	25	8.9095E-03	9.4227E-04	8.9095E-06	5.2305E-08	2
3053	25255	C12H26	2,3,4,4-tetramethyl-3-ethylhexane	62185-09-5	25	5.4166E-03	5.7286E-04	5.4166E-06	3.1799E-08	2
3054	25256	C12H26	2,3,4,5-tetramethyl-3-ethylhexane	62185-10-8	25	7.3940E-03	7.8199E-04	7.3940E-06	4.3408E-08	2
3055	25257	C12H26	2,2,3,3,4,4-hexamethylhexane	62185-11-9	25	3.8943E-03	4.1186E-04	3.8943E-06	2.2862E-08	2
3056	25258	C12H26	2,2,3,3,4,5-hexamethylhexane	62185-12-0	25	8.1104E-03	8.5775E-04	8.1104E-06	4.7613E-08	2
3057	25259	C12H26	2,2,3,3,5,5-hexamethylhexane	60302-24-1	25	9.8016E-03	1.0366E-03	9.8016E-06	5.7542E-08	2
3058	25260	C12H26	2,2,3,4,4,5-hexamethylhexane	62185-13-1	25	5.9082E-03	6.2485E-04	5.9082E-06	3.4685E-08	2
3059	25261	C12H26	2,2,3,4,5,5-hexamethylhexane	62185-14-2	25	1.2518E-02	1.3239E-03	1.2518E-05	7.3491E-08	2
3060	25262	C12H26	2,3,3,4,4,5-hexamethylhexane	52670-36-7	25	5.9082E-03	6.2485E-04	5.9082E-06	3.4685E-08	2
3061	25263	C12H26	2,4-dimethyl-3-ethyl-3-isopropylpentan	62185-16-4	25	5.9082E-03	6.2485E-04	5.9082E-06	3.4685E-08	2
3062	25264	C12H26	2,2,4-trimethyl-3,3-diethylpentane	62185-15-3	25	5.9082E-03	6.2485E-04	5.9082E-06	3.4685E-08	2
3063	25265	C12H26	2,2,3,4-tetramethyl-3-isopropylpentane	62185-17-5	25	6.4551E-03	6.8269E-04	6.4551E-06	3.7896E-08	2
3064	25266	C12H26	2,2,4,4-tetramethyl-3-isopropylpentane	62185-18-6	25	1.0286E-02	1.0879E-03	1.0286E-05	6.0386E-08	2
3065	25267	C12H26	2,2,3,4,4-pentamethyl-3-ethylpentane	66576-21-4	25	4.2177E-03	4.4606E-04	4.2177E-06	2.4760E-08	2
3066	25268	C12H26	2-methylundecane	7045-71-8	25	4.8001E-03	5.0766E-04	4.8001E-06	2.8180E-08	2
3067	25277	C12H26O	1-dodecanol	112-53-8	25	4.0000E+00	3.8672E-01	4.0000E-03	2.1466E-05	1
3068	25278	C12H26O	2-dodecanol	10203-28-8	25	8.7121E+00	8.4228E-01	8.7121E-03	4.6755E-05	2
3069	25279	C12H26O	3-dodecanol	10203-30-2	25	2.2689E+01	2.1936E+00	2.2690E-02	1.2177E-04	2
3070	25280	C12H26O	6-dodecanol	6836-38-0	25	6.2716E+01	6.0637E+00	6.2720E-02	3.3659E-04	2
3071	25281	C12H26O	2-methyl-1-undecanol	10522-26-6	25	2.2689E+01	2.1936E+00	2.2690E-02	1.2177E-04	2
3072	25282	C12H26O	2-methyl-3-undecanol	60671-36-5	25	2.2689E+01	2.1936E+00	2.2690E-02	1.2177E-04	2
3073	25283	C12H26O	2-methyl-5-undecanol	33978-71-1	25	2.2689E+01	2.1936E+00	2.2690E-02	1.2177E-04	2
3074	25284	C12H26O	2-butyl-1-octanol	3913-02-8	25	1.3333E+01	1.2890E+00	1.3333E-02	7.1551E-05	2
3075	25285	C12H26O	dihexyl ether	112-58-3	25	2.6626E+00	2.5742E-01	2.6626E-03	1.4289E-05	2
3076	25286	C12H26O	2,6,8-trimethylnonanol-4	123-17-1	25	6.1654E+01	5.9610E+00	6.1658E-02	3.3089E-04	2
3077	25294	C12H26O3	diethylene glycol dibutyl ether	112-73-2	20	3.0000E+03	2.4821E+02	3.0090E+00	1.3782E-02	1
3078	25310	C12H26S	1-dodecanethiol	112-55-0	25	1.3292E-01	1.1831E-02	1.3292E-04	6.5672E-07	2
3079	25311	C12H26S	2-dodecanethiol	14402-50-7	25	2.0073E-01	1.7866E-02	2.0073E-04	9.9172E-07	2
3080	25312	C12H26S	dihexyl sulfide	6294-31-1	25	2.5364E+00	2.2575E-01	2.5364E-03	1.2531E-05	2
3081	25313	C12H26S	methyl undecyl sulfide	7289-44-3	25	4.8985E-01	4.3599E-02	4.8985E-04	2.4202E-06	2
3082	25314	C12H26S	ethyl decyl sulfide	19313-61-2	25	4.8985E-01	4.3599E-02	4.8985E-04	2.4202E-06	2
3083	25315	C12H26S	propyl nonyl sulfide	62103-66-6	25	4.8985E-01	4.3599E-02	4.8985E-04	2.4202E-06	2
3084	25316	C12H26S	butyl octyl sulfide	16900-07-5	25	4.8985E-01	4.3599E-02	4.8985E-04	2.4202E-06	2
3085	25317	C12H26S	tert-dodecylmercaptan	25103-58-6	25	7.5450E-01	6.7154E-02	7.5450E-04	3.7277E-06	2
3086	25340	C12H27N	dodecylamine	124-22-1	25	7.8000E+01	7.5816E+00	7.8006E-02	4.2085E-04	1
3087	25341	C12H27N	methylundecylamine	66553-53-5	25	6.9709E+01	6.7756E+00	6.9713E-02	3.7611E-04	2
3088	25342	C12H27N	ethyldecylamine	66553-52-4	25	8.1899E+01	7.9605E+00	8.1905E-02	4.4189E-04	2
3089	25343	C12H27N	dihexylamine	143-16-8	25	1.2950E+02	1.2588E+01	1.2952E-01	6.9876E-04	1
3090	25344	C12H27N	dimethyldecylamine	1120-24-7	25	3.2562E+01	3.1649E+00	3.2563E-02	1.7568E-04	2
3091	25345	C12H27N	diethyloctylamine	4088-37-3	25	1.0613E+02	1.0316E+01	1.0615E-01	5.7266E-04	2
3092	25346	C12H27N	tributylamine	102-82-9	25	1.4180E+02	1.3784E+01	1.4182E-01	7.6513E-04	1
3093	25347	C12H27N	triisobutylamine	1116-40-1	25	3.3768E+02	3.2830E+01	3.3780E-01	1.8224E-03	2
3094	25575	C13H10	fluorene	86-73-7	25	1.8949E+00	2.0537E-01	1.8949E-03	1.1400E-05	1
3095	25623	C13H10O	benzophenone	119-61-9	25	1.3670E+02	1.3516E+01	1.3672E-01	7.5029E-04	1
3096	25738	C13H12	diphenylmethane	101-81-5	25	1.4115E+01	1.5115E+00	1.4115E-02	8.3900E-05	1
3097	26283	C13H20	heptylbenzene	1078-71-3	25	4.7273E-01	4.8305E-02	4.7273E-04	2.6814E-06	2
3098	26284	C13H20	1-methyl-2,4-diisopropylbenzene	1460-98-6	25	2.8328E+00	2.8947E-01	2.8329E-03	1.6068E-05	2
3099	26285	C13H20	1-methyl-3,5-diisopropylbenzene	3055-14-9	25	4.0861E+00	4.1753E-01	4.0861E-03	2.3177E-05	2
3100	26286	C13H20	[2-methyl-1-isopropylpropyl]benzene	21777-84-4	25	1.4566E+00	1.4884E-01	1.4566E-03	8.2617E-06	2
3101	26288	C13H20	(2,4-dimethylpentyl)benzene	54518-00-2	25	1.5822E+00	1.6168E-01	1.5822E-03	8.9746E-06	2
3102	26402	C13H24	1-tridecyne	26186-02-7	25	1.3637E-01	1.3623E-02	1.3637E-04	7.5619E-07	2
3103	26403	C13H24	2-tridecyne	28467-75-6	25	1.0164E-01	1.0154E-02	1.0164E-04	5.6364E-07	2
3104	26404	C13H24	3-tridecyne	60186-78-9	25	1.2154E-01	1.2142E-02	1.2154E-04	6.7397E-07	2
3105	26465	C13H26	octylcyclopentane	1795-20-6	25	9.0239E-03	8.9151E-04	9.0239E-06	4.9487E-08	2
3106	26466	C13H26	heptylcyclohexane	5617-41-4	25	1.3899E-02	1.3732E-03	1.3899E-05	7.6224E-08	2
3107	26467	C13H26	2-butyl-1,1,3-trimethylcyclohexane	54676-39-0	25	1.6880E-02	1.6677E-03	1.6880E-05	9.2571E-08	2
3108	26469	C13H26	1-tridecene	2437-56-1	25	9.0349E-03	8.9259E-04	9.0349E-06	4.9547E-08	2
3109	26470	C13H26	2-methyl-1-dodecene	16435-49-7	25	9.4608E-03	9.3467E-04	9.4608E-06	5.1883E-08	2
3110	26471	C13H26	5-butyl-4-nonene	7367-38-6	25	1.6635E-02	1.6434E-03	1.6635E-05	9.1223E-08	2
3111	26472	C13H26Br2	1,1-dibromotridecane	62168-30-3	25	4.8523E-01	2.5548E-02	4.8523E-04	1.4181E-06	2
3112	26473	C13H26Br2	1,13-dibromotridecane	31772-05-1	25	4.8523E-01	2.5548E-02	4.8523E-04	1.4181E-06	2
3113	26474	C13H26Cl2	1,1-dichlorotridecane	821-93-2	25	9.9364E-02	7.0681E-03	9.9364E-05	3.9235E-07	2
3114	26486	C13H26O	1-tridecanal	10486-19-8	25	3.4411E+00	3.1254E-01	3.4411E-03	1.7349E-05	2
3115	26487	C13H26O	2-tridecanone	593-08-8	25	1.3423E+00	1.2191E-01	1.3423E-03	6.7673E-06	2
3116	26488	C13H26O	3-tridecanone	1534-26-5	25	1.0653E+00	9.6758E-02	1.0653E-03	5.3709E-06	2
3117	26489	C13H26O	7-tridecanone	462-18-0	25	1.5310E+00	1.3905E-01	1.5310E-03	7.7186E-06	2
3118	26492	C13H26O2	tridecanoic acid	638-53-9	25	2.6577E+01	2.2337E+00	2.6577E-02	1.2399E-04	2
3119	26493	C13H26O2	2-methyldodecanoic acid	2874-74-0	25	2.3520E+02	1.9772E+01	2.3526E-01	1.0975E-03	2
3120	26494	C13H26O2	dodecyl formate	28303-42-6	25	4.1016E-01	3.4472E-02	4.1016E-04	1.9135E-06	2

Table 3 Solubility in Water - Organic Compounds

					Solubility in Water, S (ppm - parts per million, wt - weight, g - gram)					
NO	ID	FORMULA	NAME	CAS No	T, C	S@T, ppm(wt)	S@T, ppm(mol)	S@T, g/1000 g of H2O	S@T, mol/1000 g of H2O	code
3121	26495	C13H26O2	undecyl acetate	1731-81-3	25	6.9400E+00	5.8328E-01	6.9401E-03	3.2377E-05	2
3122	26496	C13H26O2	decyl propanoate	5454-19-3	25	1.0510E+00	8.8329E-02	1.0510E-03	4.9031E-06	2
3123	26497	C13H26O2	nonyl butanoate	2639-64-7	25	3.0699E+00	2.5801E-01	3.0699E-03	1.4322E-05	2
3124	26515	C13H27Br	1-bromotridecane	765-09-3	25	3.5391E-02	2.4218E-03	3.5391E-05	1.3443E-07	2
3125	26516	C13H27Cl	1-chlorotridecane	822-13-9	25	2.9864E-02	2.4587E-03	2.9864E-05	1.3648E-07	2
3126	26526	C13H28	tridecane	629-50-5	25	1.9904E-03	1.9449E-04	1.9904E-06	1.0796E-08	2
3127	26527	C13H28	2-methyldodecane	1560-97-0	25	2.3932E-03	2.3385E-04	2.3932E-06	1.2981E-08	2
3128	26528	C13H28	3-methyldodecane	17312-57-1	25	2.3321E-03	2.2788E-04	2.3321E-06	1.2649E-08	2
3129	26529	C13H28	2,2-dimethylundecane	17312-64-0	25	3.2169E-03	3.1433E-04	3.2169E-06	1.7448E-08	2
3130	26530	C13H28	2,3-dimethylundecane	17312-77-5	25	2.5144E-03	2.4569E-04	2.5144E-06	1.3638E-08	2
3131	26531	C13H28	2,4-dimethylundecane	17312-80-0	25	3.8943E-03	3.8052E-04	3.8943E-06	2.1123E-08	2
3132	26532	C13H28	5-butylnonane	17312-63-9	25	3.6726E-03	3.5887E-04	3.6726E-06	1.9920E-08	2
3133	26535	C13H28O	1-tridecanol	112-70-9	25	1.3738E+00	1.2352E-01	1.3738E-03	6.8565E-06	2
3134	26536	C13H28O	2-tridecanol	1653-31-2	25	2.3306E+00	2.0955E-01	2.3306E-03	1.1632E-05	2
3135	26537	C13H28O	3-tridecanol	10289-68-6	25	5.3920E+00	4.8480E-01	5.3920E-03	2.6911E-05	2
3136	26538	C13H28O	5-butyl-5-nonanol	597-93-3	25	5.3920E+00	4.8480E-01	5.3920E-03	2.6911E-05	2
3137	26549	C13H28S	1-tridecanethiol	19484-26-5	25	6.8950E-02	5.7391E-03	6.8950E-05	3.1858E-07	2
3138	26550	C13H28S	2-tridecanethiol	62155-03-7	25	9.7619E-02	8.1255E-03	9.7619E-05	4.5104E-07	2
3139	26551	C13H28S	methyl dodecyl sulfide	---	25	2.4481E-01	2.0377E-02	2.4481E-04	1.1311E-06	2
3140	26552	C13H28S	ethyl undecyl sulfide	66577-30-8	25	2.4481E-01	2.0377E-02	2.4481E-04	1.1311E-06	2
3141	26553	C13H28S	propyl decyl sulfide	66577-31-9	25	2.4481E-01	2.0377E-02	2.4481E-04	1.1311E-06	2
3142	26554	C13H28S	butyl nonyl sulfide	66577-32-0	25	2.4481E-01	2.0377E-02	2.4481E-04	1.1311E-06	2
3143	26555	C13H29N	tridecylamine	2869-34-3	25	3.1309E+00	2.8290E-01	3.1310E-03	1.5703E-05	2
3144	26556	C13H29N	methyldodecylamine	7311-30-0	25	3.0101E+01	2.7198E+00	3.0102E-02	1.5098E-04	2
3145	26557	C13H29N	ethylundecylamine	59570-04-6	25	3.3512E+01	3.0281E+00	3.3513E-02	1.6809E-04	2
3146	26558	C13H29N	dimethylundecylamine	17373-28-3	25	1.3331E+01	1.2045E+00	1.3331E-02	6.6861E-05	2
3147	26559	C13H29N	diethylnonylamine	45124-35-4	25	4.3418E+01	3.9232E+00	4.3420E-02	2.1778E-04	2
3148	26560	C13H29N	6-methyl-N-isopentyl-2-heptanamine	502-59-0	25	6.2697E+01	5.6653E+00	6.2701E-02	3.1448E-04	2
3149	26561	C13H29N	tert-amyl-tert-octylamine	150285-07-7	25	8.0456E+02	7.2749E+01	8.0521E-01	4.0385E-03	2
3150	26562	C13H29N	N-methyldihexylamine	37615-53-5	25	1.8068E+01	1.6326E+00	1.8069E-02	9.0624E-05	2
3151	26654	C14H8O2	anthraquinone	84-65-1	25	1.3530E+00	1.1706E-01	1.3530E-03	6.4981E-06	1
3152	26728	C14H10	anthracene	120-12-7	22	5.3470E-02	5.4045E-03	5.3470E-05	3.0000E-07	1
3153	26730	C14H10	phenanthrene	85-01-8	25	1.1817E+00	1.1944E-01	1.1817E-03	6.6301E-06	1
3154	26866	C14H12	cis-stilbene	645-49-8	25	2.9000E-01	2.8984E-02	2.9000E-04	1.6089E-06	2
3155	26867	C14H12	trans-stilbene	103-30-0	25	2.9000E-01	2.8984E-02	2.9000E-04	1.6089E-06	1
3156	26873	C14H12	stilbene	588-59-0	25	2.9000E-01	2.8984E-02	2.9000E-04	1.6089E-06	2
3157	27644	C14H22	octylbenzene	2189-60-8	25	2.4977E-01	2.3641E-02	2.4977E-04	1.3123E-06	2
3158	27645	C14H22	1,2,3,4-tetraethylbenzene	642-32-0	25	8.9385E-01	8.4605E-02	8.9386E-04	4.6964E-06	2
3159	27646	C14H22	1,2,3,5-tetraethylbenzene	38842-05-6	25	9.1060E-01	8.6190E-02	9.1060E-04	4.7844E-06	2
3160	27647	C14H22	1,2,4,5-tetraethylbenzene	635-81-4	25	9.2777E-01	8.7816E-02	9.2777E-04	4.8746E-06	2
3161	27648	C14H22	1,4-di-tert-butylbenzene	1012-72-2	25	1.5027E+00	1.4223E-01	1.5027E-03	7.8951E-06	2
3162	27649	C14H22	(1-methylheptyl)benzene	777-22-0	25	3.9525E-01	3.7411E-02	3.9525E-04	2.0767E-06	2
3163	27650	C14H22	1,3-di-tert-butylbenzene	1014-60-4	25	9.0131E-01	8.5310E-02	9.0131E-04	4.7355E-06	2
3164	27772	C14H24S	3-decylthiophene	65016-55-9	25	6.1440E-02	4.9322E-03	6.1440E-05	2.7378E-07	2
3165	27780	C14H26	1-tetradecyne	765-10-6	25	8.7697E-02	8.1285E-03	8.7697E-05	4.5121E-07	2
3166	27781	C14H26	3-tetradecyne	60212-32-0	25	8.3151E-02	7.7072E-03	8.3151E-05	4.2782E-07	2
3167	27782	C14H26	7-tetradecyne	35216-11-6	25	7.4877E-02	6.9403E-03	7.4877E-05	3.8525E-07	2
3168	27784	C14H26	1,13-tetradecadiene	21964-49-8	25	1.4072E-02	1.3043E-03	1.4072E-05	7.2401E-08	2
3169	27847	C14H28	nonylcyclopentane	2882-98-6	25	6.4718E-03	5.9371E-04	6.4718E-06	3.2956E-08	2
3170	27848	C14H28	octylcyclohexane	1795-15-9	25	1.0088E-02	9.2544E-04	1.0088E-05	5.1371E-08	2
3171	27850	C14H28	1-tetradecene	1120-36-1	25	5.6943E-03	5.2238E-04	5.6943E-06	2.8997E-08	2
3172	27851	C14H28	2-tetradecene	638-60-8	25	4.7496E-03	4.3571E-04	4.7496E-06	2.4186E-08	2
3173	27852	C14H28	2-methyl-1-tridecene	18094-01-4	25	5.8193E-03	5.3384E-04	5.8193E-06	2.9633E-08	2
3174	27853	C14H28	trans-7-tetradecene	41446-63-3	25	5.8193E-03	5.3384E-04	5.8193E-06	2.9633E-08	2
3175	27856	C14H28Cl2	1,1-dichlorotetradecane	4168-40-5	25	5.8679E-02	3.9550E-03	5.8679E-05	2.1954E-07	2
3176	27862	C14H28O	tetradecanal	124-25-4	25	1.5300E+00	1.2979E-01	1.5300E-03	7.2044E-06	2
3177	27863	C14H28O	2-tetradecanone	2345-27-9	25	3.4691E-01	2.9427E-02	3.4691E-04	1.6335E-06	2
3178	27864	C14H28O	7-ethyl-2-methyl-4-undecanone	6976-00-7	25	3.0425E+00	2.5809E-01	3.0425E-03	1.4326E-05	2
3179	27865	C14H28O	2-methyl-3-tridecanone	40239-35-8	25	9.2280E-01	7.8277E-02	9.2280E-04	4.3451E-06	2
3180	27866	C14H28O	3-tetradecanone	629-23-2	25	7.9937E+00	6.7808E-01	7.9938E-03	3.7640E-05	2
3181	27873	C14H28O2	tetradecanoic acid	544-63-8	20	2.0000E+01	1.5777E+00	2.0000E-02	8.7577E-05	1
3182	27874	C14H28O2	tridecyl formate	42875-41-2	25	1.1967E-01	9.4397E-03	1.1967E-04	5.2399E-07	2
3183	27875	C14H28O2	dodecyl acetate	112-66-3	25	2.6120E+00	2.0605E-01	2.6120E-03	1.1437E-05	2
3184	27876	C14H28O2	undecyl propanoate	5458-33-3	25	3.0662E-01	2.4188E-02	3.0662E-04	1.3426E-06	2
3185	27877	C14H28O2	decyl butanoate	5454-09-1	25	1.1648E+00	9.1882E-02	1.1648E-03	5.1003E-06	2
3186	27885	C14H28O2	3,7-dimethyloctanyl butyrate	67874-80-0	25	8.1118E-01	6.3989E-02	8.1119E-04	3.5520E-06	2
3187	27895	C14H29Br	1-bromotetradecane	112-71-0	25	2.0800E-02	1.3513E-03	2.0800E-05	7.5011E-08	2
3188	27896	C14H29Br	1-bromo-12-methyltridecane	111772-90-8	25	2.0800E-02	1.3513E-03	2.0800E-05	7.5011E-08	2
3189	27897	C14H29Cl	1-chlorotetradecane	2425-54-9	25	1.2200E-02	9.4394E-04	1.2200E-05	5.2397E-08	1
3190	27912	C14H30	tetradecane	629-59-4	25	2.2000E-03	1.9977E-04	2.2000E-06	1.1089E-08	1
3191	27913	C14H30	2-methyltridecane	1560-96-9	25	1.4516E-03	1.3181E-04	1.4516E-06	7.3166E-09	2
3192	27914	C14H30	3-methyltridecane	6418-41-3	25	1.4271E-03	1.2959E-04	1.4271E-06	7.1934E-09	2
3193	27915	C14H30	2,2-dimethyldodecane	49598-54-1	25	1.7547E-03	1.5934E-04	1.7547E-06	8.8446E-09	2
3194	27916	C14H30	2,3-dimethyldodecane	6117-98-2	25	1.4803E-03	1.3442E-04	1.4803E-06	7.4615E-09	2
3195	27917	C14H30	2,4-dimethyldodecane	6117-99-3	25	2.0789E-03	1.8878E-04	2.0789E-06	1.0479E-08	2
3196	27918	C14H30	7-methyltridecane	26730-14-3	25	1.5420E-03	1.4002E-04	1.5420E-06	7.7726E-09	2
3197	27922	C14H30O	1-tetradecanol	112-72-1	25	3.0000E-01	2.5209E-02	3.0000E-04	1.3993E-06	1
3198	27923	C14H30O	2-tetradecanol	4706-81-4	25	5.4387E-01	4.5700E-02	5.4387E-04	2.5368E-06	2
3199	27924	C14H30O	3-tetradecanol	1653-32-3	25	8.5479E-01	7.1827E-02	8.5479E-04	3.9870E-06	2
3200	27925	C14H30O	12-methyl-1-tridecanol	21987-21-3	25	8.5479E-01	7.1827E-02	8.5479E-04	3.9870E-06	2

Table 3 Solubility in Water - Organic Compounds

					Solubility in Water, S (ppm - parts per million, wt - weight, g - gram)					
NO	ID	FORMULA	NAME	CAS No	T, C	S@T, ppm(wt)	S@T, ppm(mol)	S@T, g/1000 g of H2O	S@T, mol/1000 g of H2O	code
3201	27926	C14H30O	2-methyl-3-tridecanol	98930-89-3	25	1.3738E+00	1.1544E-01	1.3738E-03	6.4079E-06	2
3202	27927	C14H30O	2-methyl-4-tridecanol	36691-77-7	25	8.5479E-01	7.1827E-02	8.5479E-04	3.9870E-06	2
3203	27928	C14H30O	diheptyl ether	629-64-1	25	4.2335E-01	3.5574E-02	4.2335E-04	1.9747E-06	2
3204	27929	C14H30O	tetradecanol; mixed isomers	27196-00-5	25	3.3449E-01	2.8107E-02	3.3449E-04	1.5602E-06	2
3205	27946	C14H30S	1-tetradecanethiol	2079-95-0	25	3.9756E-02	3.1078E-03	3.9756E-05	1.7251E-07	2
3206	27947	C14H30S	2-tetradecanethiol	62155-04-8	25	5.2816E-02	4.1287E-03	5.2816E-05	2.2918E-07	2
3207	27948	C14H30S	diheptyl sulfide	629-65-2	25	1.2127E-01	9.4801E-03	1.2127E-04	5.2623E-07	2
3208	27949	C14H30S	methyl tridecyl sulfide	62155-09-3	25	1.3111E-01	1.0249E-02	1.3111E-04	5.6893E-07	2
3209	27950	C14H30S	ethyl dodecyl sulfide	2851-83-4	25	1.3111E-01	1.0249E-02	1.3111E-04	5.6893E-07	2
3210	27951	C14H30S	propyl undecyl sulfide	66826-84-4	25	1.3111E-01	1.0249E-02	1.3111E-04	5.6893E-07	2
3211	27952	C14H30S	butyl decyl sulfide	19313-57-6	25	1.3111E-01	1.0249E-02	1.3111E-04	5.6893E-07	2
3212	27956	C14H31N	tetradecylamine	2016-42-4	25	9.8166E+00	8.2869E-01	9.8167E-03	4.6000E-05	2
3213	27957	C14H31N	methyltridecylamine	45165-81-9	25	1.3652E+01	1.1524E+00	1.3652E-02	6.3970E-05	2
3214	27958	C14H31N	ethyldodecylamine	35902-57-9	25	1.4404E+01	1.2160E+00	1.4404E-02	6.7497E-05	2
3215	27959	C14H31N	diheptylamine	2470-68-0	25	1.7853E+01	1.5071E+00	1.7853E-02	8.3659E-05	2
3216	27960	C14H31N	dimethyldodecylamine	112-18-5	25	5.4315E+00	4.5851E-01	5.4315E-03	2.5451E-05	2
3217	27961	C14H31N	diethyldecylamine	6308-94-7	25	1.9689E+01	1.6621E+00	1.9689E-02	9.2262E-05	2
3218	28436	C15H16O2	bisphenol a	80-05-7	20	3.5000E+02	2.7628E+01	3.5012E-01	1.5337E-03	1
3219	28725	C15H24	nonylbenzene	1081-77-2	25	1.6075E-01	1.4171E-02	1.6075E-04	7.8661E-07	2
3220	28726	C15H24	1,2,4-triisopropylbenzene	948-32-3	25	1.1718E+00	1.0330E-01	1.1718E-03	5.7340E-06	2
3221	28727	C15H24	1,3,5-triisopropylbenzene	717-74-8	25	1.5027E+00	1.3247E-01	1.5027E-03	7.3532E-06	2
3222	28728	C15H24	3,5-di-tert-butyltoluene	15181-11-0	25	1.1718E+00	1.0330E-01	1.1718E-03	5.7340E-06	2
3223	28756	C15H24O	2,6-di-tert-butyl-p-cresol	128-37-0	25	7.0000E+00	5.7228E-01	7.0000E-03	3.1767E-05	2
3224	28757	C15H24O	nonylphenol	25154-52-3	25	7.0000E+00	5.7228E-01	7.0000E-03	3.1767E-05	1
3225	28758	C15H24O	2,4-di-tert-butyl-5-methylphenol	497-39-2	25	7.0000E+00	5.7228E-01	7.0000E-03	3.1767E-05	2
3226	28759	C15H24O	2,4-di-tert-butyl-6-methylphenol	616-55-7	25	7.0000E+00	5.7228E-01	7.0000E-03	3.1767E-05	2
3227	28764	C15H24O	4-nonylphenol	104-40-5	25	7.0000E+00	5.7228E-01	7.0000E-03	3.1767E-05	2
3228	28765	C15H24O	4-nonylphenol, isomers	84852-15-3	25	7.0000E+00	5.7228E-01	7.0000E-03	3.1767E-05	2
3229	28820	C15H28	1-pentadecyne	765-13-9	25	7.1094E-02	6.1460E-03	7.1094E-05	3.4116E-07	2
3230	28821	C15H28	2-pentadecyne	52112-25-1	25	6.9012E-02	5.9660E-03	6.9012E-05	3.3117E-07	2
3231	28822	C15H28	3-pentadecyne	61886-61-1	25	6.9818E-02	6.0357E-03	6.9818E-05	3.3504E-07	2
3232	28855	C15H30	decylcyclopentane	1795-21-7	25	5.9341E-03	5.0809E-04	5.9341E-06	2.8204E-08	2
3233	28856	C15H30	nonylcyclohexane	2883-02-5	25	9.4603E-03	8.1001E-04	9.4603E-06	4.4963E-08	2
3234	28857	C15H30	1-ethyl-1-methyl-2,4-diisopropylcyclohe	515-12-8	25	9.3864E-03	8.0367E-04	9.3864E-06	4.4611E-08	2
3235	28859	C15H30	1-pentadecene	13360-61-7	25	4.5233E-03	3.8729E-04	4.5233E-06	2.1498E-08	2
3236	28860	C15H30	2-methyl-1-tetradecene	52254-38-3	25	4.5726E-03	3.9151E-04	4.5726E-06	2.1733E-08	2
3237	28861	C15H30	7-pentadecene	15430-98-5	25	4.3903E-03	3.7591E-04	4.3903E-06	2.0866E-08	2
3238	28862	C15H30	8-pentadecene	---	25	4.3903E-03	3.7591E-04	4.3903E-06	2.0866E-08	2
3239	28873	C15H30O	pentadecanal	2765-11-9	25	7.5284E-01	5.9904E-02	7.5284E-04	3.3252E-06	2
3240	28874	C15H30O	2-pentadecanone	2345-28-0	25	9.0074E-02	7.1672E-03	9.0074E-05	3.9785E-07	2
3241	28875	C15H30O	8-pentadecanone	818-23-5	25	9.9588E-02	7.9243E-03	9.9588E-05	4.3987E-07	2
3242	28879	C15H30O2	pentadecanoic acid	1002-84-2	20	1.2000E+01	8.9183E-01	1.2000E-02	4.9505E-05	1
3243	28880	C15H30O2	tetradecyl formate	5451-63-8	25	3.4914E-02	2.5947E-03	3.4914E-05	1.4403E-07	2
3244	28881	C15H30O2	tridecyl acetate	1072-33-9	25	1.0544E+00	7.8359E-02	1.0544E-03	4.3496E-06	2
3245	28882	C15H30O2	dodecyl propanoate	6221-93-8	25	8.9460E-02	6.6485E-03	8.9460E-05	3.6905E-07	2
3246	28883	C15H30O2	undecyl butanoate	5461-02-9	25	8.9460E-02	6.6485E-03	8.9460E-05	3.6905E-07	2
3247	28898	C15H31Br	1-bromopentadecane	629-72-1	25	1.3212E-02	8.1701E-04	1.3212E-05	4.5352E-08	2
3248	28899	C15H31Cl	1-chloropentadecane	4862-03-1	25	9.2984E-03	6.7855E-04	9.2984E-06	3.7666E-08	2
3249	28904	C15H32	pentadecane	629-62-9	25	1.0658E-03	9.0386E-05	1.0658E-06	5.0173E-09	2
3250	28905	C15H32	2-methyltetradecane	1560-95-8	25	1.1090E-03	9.4055E-05	1.1090E-06	5.2209E-09	2
3251	28906	C15H32	3-methyltetradecane	18435-22-8	25	1.0982E-03	9.3135E-05	1.0982E-06	5.1699E-09	2
3252	28907	C15H32	2,2-dimethyltridecane	61869-04-3	25	1.2273E-03	1.0409E-04	1.2273E-06	5.7779E-09	2
3253	28908	C15H32	2,3-dimethyltridecane	18435-20-6	25	1.1111E-03	9.4232E-05	1.1111E-06	5.2308E-09	2
3254	28909	C15H32	2,4-dimethyltridecane	61868-05-1	25	1.3623E-03	1.1554E-04	1.3623E-06	6.4134E-09	2
3255	28912	C15H32O	1-pentadecanol	629-76-5	25	8.9000E-02	7.0193E-03	8.9000E-05	3.8964E-07	1
3256	28913	C15H32O	2-pentadecanol	1653-34-5	25	1.2050E-01	9.5040E-03	1.2050E-04	5.2756E-07	2
3257	28914	C15H32O	3-pentadecanol	53346-71-7	25	3.5536E-07	2.8027E-08	3.5536E-10	1.5558E-12	2
3258	28915	C15H32O	2,8-dimethyl-6-isobutylnonanol-4	10143-20-1	25	2.9544E-03	2.3300E-04	2.9544E-06	1.2934E-08	2
3259	28921	C15H32S	1-pentadecanethiol	25276-70-4	25	2.5497E-02	1.8787E-03	2.5497E-05	1.0429E-07	2
3260	28922	C15H32S	2-pentadecanethiol	62155-05-9	25	3.1985E-02	2.3568E-03	3.1985E-05	1.3082E-07	2
3261	28923	C15H32S	methyl tetradecyl sulfide	7289-45-4	25	8.1075E-02	5.9740E-03	8.1075E-05	3.3162E-07	2
3262	28924	C15H32S	ethyl tridecyl sulfide	66271-81-6	25	8.1075E-02	5.9740E-03	8.1075E-05	3.3162E-07	2
3263	28925	C15H32S	propyl dodecyl sulfide	66271-82-7	25	8.1075E-02	5.9740E-03	8.1075E-05	3.3162E-07	2
3264	28926	C15H32S	butyl undecyl sulfide	66271-83-8	25	8.1075E-02	5.9740E-03	8.1075E-05	3.3162E-07	2
3265	28931	C15H33N	pentadecylamine	2570-26-5	25	2.1083E+00	1.6700E-01	2.1083E-03	9.2702E-06	2
3266	28932	C15H33N	methyltetradecylamine	29369-63-9	25	6.1655E+00	4.8837E-01	6.1656E-03	2.7109E-05	2
3267	28933	C15H33N	ethyltridecylamine	59570-06-8	25	6.5054E+00	5.1529E-01	6.5054E-03	2.8604E-05	2
3268	28934	C15H33N	dimethyltridecylamine	17373-29-4	25	2.4533E+00	1.9432E-01	2.4533E-03	1.0787E-05	2
3269	28935	C15H33N	diethylundecylamine	54334-64-4	25	8.8918E+00	7.0433E-01	8.8919E-03	3.9097E-05	2
3270	28936	C15H33N	tripentylamine	621-77-2	25	2.1100E+01	1.6714E+00	2.1100E-02	9.2776E-05	1
3271	28937	C15H33N	triisopentylamine	645-41-0	25	3.9960E+01	3.1653E+00	3.9961E-02	1.7570E-04	2
3272	28938	C15H33N	tris(2-methylbutyl)amine	620-43-9	25	4.6943E+01	3.7185E+00	4.6945E-02	2.0641E-04	2
3273	28984	C16H10	fluoranthene	206-44-0	27	2.4068E-01	2.1438E-02	2.4068E-04	1.1900E-06	1
3274	28985	C16H10	pyrene	129-00-0	25	1.2944E-01	1.1529E-02	1.2944E-04	6.3998E-07	1
3275	29518	C16H22O4	dibutyl phthalate	84-74-2	20	1.0800E+01	6.9899E-01	1.0800E-02	3.8801E-05	1
3276	29595	C16H26	decylbenzene	104-72-3	25	1.2861E-01	1.0610E-02	1.2861E-04	5.8894E-07	2
3277	29596	C16H26	pentaethylbenzene	605-01-6	25	4.0774E-01	3.3635E-02	4.0774E-04	1.8671E-06	2
3278	29599	C16H26	(1-methylnonyl)benzene	4537-13-7	25	1.4161E-01	1.1682E-02	1.4161E-04	6.4847E-07	2
3279	29637	C16H28S	3-dodecylthiophene	104934-52-3	25	7.9800E-02	5.6943E-03	7.9800E-05	3.1609E-07	2
3280	29640	C16H30	1-hexadecyne	629-74-3	25	7.0150E-02	5.6820E-03	7.0150E-05	3.1540E-07	2

Table 3 Solubility in Water - Organic Compounds

					Solubility in Water, S (ppm - parts per million, wt - weight, g - gram)					
NO	ID	FORMULA	NAME	CAS No	T, C	S@T, ppm(wt)	S@T, ppm(mol)	S@T, g/1000 g of H2O	S@T, mol/1000 g of H2O	code
3281	29641	C16H30	2-hexadecyne	629-75-4	25	8.0102E-02	6.4881E-03	8.0102E-05	3.6015E-07	2
3282	29642	C16H30	3-hexadecyne	61886-62-2	25	7.1084E-02	5.7576E-03	7.1084E-05	3.1960E-07	2
3283	29704	C16H32	undecylcyclopentane	6785-23-5	25	6.8530E-03	5.5009E-04	6.8530E-06	3.0535E-08	2
3284	29705	C16H32	decylcyclohexane	1795-16-0	25	1.1236E-02	9.0191E-04	1.1236E-05	5.0065E-08	2
3285	29707	C16H32	1-hexadecene	629-73-2	25	4.5012E-03	3.6131E-04	4.5012E-06	2.0056E-08	2
3286	29708	C16H32	2-methyl-1-pentadecene	29833-69-0	25	4.4772E-03	3.5938E-04	4.4772E-06	1.9949E-08	2
3287	29709	C16H32	2-methyl-1-propene, tetramer	15220-85-6	25	6.6430E-03	5.3323E-04	6.6430E-06	2.9599E-08	2
3288	29717	C16H32O	hexadecanal	629-80-1	25	3.7042E-01	2.7755E-02	3.7042E-04	1.5407E-06	2
3289	29718	C16H32O	2-hexadecanone	18787-63-8	25	1.8405E-02	1.3791E-03	1.8405E-05	7.6551E-08	2
3290	29719	C16H32O	3-hexadecanone	18787-64-9	25	5.4922E-02	4.1152E-03	5.4922E-05	2.2843E-07	2
3291	29721	C16H32O2	hexadecanoic acid	57-10-3	20	7.2000E+00	5.0583E-01	7.2001E-03	2.8078E-05	1
3292	29722	C16H32O2	2-hexyldecanoic acid	25354-97-6	25	2.4788E+01	1.7415E+00	2.4789E-02	9.6670E-05	2
3293	29723	C16H32O2	pentadecyl formate	66271-76-9	25	1.0186E-02	7.1562E-04	1.0186E-05	3.9723E-08	2
3294	29724	C16H32O2	tetradecyl acetate	638-59-5	25	4.7018E-01	3.3031E-02	4.7018E-04	1.8336E-06	2
3295	29725	C16H32O2	tridecyl propanoate	66271-77-0	25	2.6100E-02	1.8336E-03	2.6100E-05	1.0178E-07	2
3296	29726	C16H32O2	dodecyl butanoate	3724-61-6	25	2.6100E-02	1.8336E-03	2.6100E-05	1.0178E-07	2
3297	29739	C16H33Br	1-bromohexadecane	112-82-3	25	9.3544E-03	5.5190E-04	9.3544E-06	3.0636E-08	2
3298	29740	C16H33Cl	1-chlorohexadecane	4860-03-1	25	6.1362E-03	4.2372E-04	6.1362E-06	2.3520E-08	2
3299	29754	C16H34	hexadecane	544-76-3	25	9.0000E-04	7.1600E-05	9.0000E-07	3.9745E-09	1
3300	29755	C16H34	2-methylpentadecane	1560-93-6	25	1.0569E-03	8.4085E-05	1.0569E-06	4.6675E-09	2
3301	29756	C16H34	3-methylpentadecane	2882-96-4	25	1.0586E-03	8.4217E-05	1.0586E-06	4.6748E-09	2
3302	29757	C16H34	2,2-dimethyltetradecane	59222-86-5	25	1.0523E-03	8.3715E-05	1.0523E-06	4.6469E-09	2
3303	29758	C16H34	2,3-dimethyltetradecane	18435-23-9	25	1.0586E-03	8.4217E-05	1.0586E-06	4.6748E-09	2
3304	29759	C16H34	2,4-dimethyltetradecane	61868-06-2	25	1.1010E-03	8.7592E-05	1.1010E-06	4.8622E-09	2
3305	29760	C16H34	2,2,4,4,6,8,8-heptamethylnonane	4390-04-9	25	1.5036E-03	1.1962E-04	1.5036E-06	6.6399E-09	2
3306	29766	C16H34O	1-hexadecanol	36653-82-4	25	3.5000E-02	2.6007E-03	3.5000E-05	1.4436E-07	1
3307	29767	C16H34O	2-hexadecanol	14852-31-4	25	2.2846E-02	1.6976E-03	2.2846E-05	9.4232E-08	2
3308	29768	C16H34O	3-hexadecanol	593-03-3	25	1.7583E-01	1.3065E-02	1.7583E-04	7.2523E-07	2
3309	29769	C16H34O	2-methyl-1-pentadecanol	25354-98-7	25	1.7583E-01	1.3065E-02	1.7583E-04	7.2523E-07	2
3310	29770	C16H34O	6-methyl-6-pentadecanol	108836-86-8	25	1.7583E-01	1.3065E-02	1.7583E-04	7.2523E-07	2
3311	29771	C16H34O	dioctyl ether	629-82-3	25	1.2238E-01	9.0933E-03	1.2238E-04	5.0476E-07	2
3312	29772	C16H34O	bis(2-ethylhexyl) ether	10143-60-9	25	2.9696E-01	2.2066E-02	2.9696E-04	1.2249E-06	2
3313	29773	C16H34O	2-hexyl-1-decanol	2425-77-6	25	1.7583E-01	1.3065E-02	1.7583E-04	7.2523E-07	2
3314	29789	C16H34S	1-hexadecanethiol	2917-26-2	25	1.8205E-02	1.2687E-03	1.8205E-05	7.0423E-08	2
3315	29790	C16H34S	2-hexadecanethiol	66271-53-2	25	2.1551E-02	1.5018E-03	2.1551E-05	8.3364E-08	2
3316	29791	C16H34S	dioctyl sulfide	2690-08-6	25	5.3726E-02	3.7440E-03	5.3726E-05	2.0783E-07	2
3317	29792	C16H34S	methyl pentadecyl sulfide	62155-10-6	25	5.3726E-02	3.7440E-03	5.3726E-05	2.0783E-07	2
3318	29793	C16H34S	ethyl tetradecyl sulfide	66271-54-3	25	5.3726E-02	3.7440E-03	5.3726E-05	2.0783E-07	2
3319	29794	C16H34S	propyl tridecyl sulfide	66271-55-4	25	5.3726E-02	3.7440E-03	5.3726E-05	2.0783E-07	2
3320	29795	C16H34S	butyl dodecyl sulfide	16900-08-6	25	5.3726E-02	3.7440E-03	5.3726E-05	2.0783E-07	2
3321	29799	C16H35N	ethyltetradecylamine	66271-56-5	25	8.2661E+00	6.1672E-01	8.2662E-03	3.4234E-05	2
3322	29800	C16H35N	dioctylamine	1120-48-5	25	1.0170E+01	7.5874E-01	1.0170E-02	4.2117E-05	2
3323	29801	C16H35N	methylpentadecylamine	29664-53-7	25	8.2661E+00	6.1672E-01	8.2662E-03	3.4234E-05	2
3324	29802	C16H35N	N-(1-methylheptyl)-2-octanamine	5412-92-0	25	2.2351E+01	1.6676E+00	2.2352E-02	9.2568E-05	2
3325	29803	C16H35N	dimethyltetradecylamine	112-75-4	25	1.1647E+00	8.6899E-02	1.1647E-03	4.8237E-06	2
3326	29804	C16H35N	diethyldodecylamine	4271-27-6	25	4.4540E+00	3.3231E-01	4.4540E-03	1.8446E-05	2
3327	29805	C16H35N	bis(2-ethylhexyl)amine	106-20-7	25	1.0663E+01	7.9554E-01	1.0663E-02	4.4160E-05	2
3328	30352	C17H36O	1-heptadecanol	1454-85-9	25	8.0000E-03	5.6193E-04	8.0000E-06	3.1192E-08	1
3329	30406	C18H12	chrysene	218-01-9	24	1.7122E-02	1.3511E-03	1.7122E-05	7.5000E-08	1
3330	30457	C18H14	m-terphenyl	92-06-8	25	1.5110E+00	1.1819E-01	1.5110E-03	6.5608E-06	1
3331	30458	C18H14	o-terphenyl	84-15-1	25	1.2390E+00	9.6916E-02	1.2390E-03	5.3797E-06	1
3332	30459	C18H14	p-terphenyl	92-94-4	25	1.7960E-02	1.4048E-03	1.7960E-05	7.7982E-08	1
3333	30566	C18H15O4P	triphenyl phosphate	115-86-6	24	7.3000E-01	4.0305E-02	7.3000E-04	2.2373E-06	1
3334	31097	C18H36O2	octadecanoic acid	57-11-4	20	2.9000E+00	1.8364E-01	2.9000E-03	1.0194E-05	1
3335	31131	C18H38	octadecane	593-45-3	25	2.1000E-03	1.4865E-04	2.1000E-06	8.2515E-09	1
3336	31140	C18H38O	1-octadecanol	112-92-5	34	1.1000E-03	7.3259E-05	1.1000E-06	4.0666E-09	1
3337	32346	C20H30O2	abietic acid	514-10-3	20	4.8390E+01	2.8823E+00	4.8392E-02	1.6000E-04	1
3338	32515	C20H42	eicosane	112-95-8	25	1.9000E-03	1.2114E-04	1.9000E-06	6.7244E-09	1
3339	32787	C21H21O4P	tri-o-cresyl phosphate	78-30-8	24	7.4000E-02	3.6189E-03	7.4000E-05	2.0089E-07	1
3340	33925	C24H38O4	dioctyl phthalate	117-84-0	24	2.8500E-01	1.3146E-02	2.8500E-04	7.2972E-07	1
3341	34691	C28H46O4	diisodecyl phthalate	26761-40-0	24	2.8000E-01	1.1293E-02	2.8000E-04	6.2686E-07	1

Code: 1 - data, 2 - estimate

Chapter 4

SOLUBILITY IN WATER – INORGANIC COMPOUNDS

Carl L. Yaws, Meenakshi S. Yadav, and Preeti S. Yadav
Lamar University, Beaumont, Texas

Tabulation Results

The results for solubility in water are given in Table 4. The tabulation is arranged by alphabetical order (Ag, Al, B, ...) to provide ease of use in quickly locating the data by using the chemical formula. The compound name, CAS No (Chemical Abstracts Registry Number), and solubility values are provided in the adjacent columns. The last column provides the code for the tabulation.

In preparing the tabulation, a literature search was conducted to identify data source publications for inorganics (1-13). These data were then keyed into the computer to provide a database of values for compounds for which experimental data are available. The excellent compilations by CRC (4); Krumgalz (6); Sohnel and Novotny (11); and Solubility Data Series of IUPAC (12) were primarily used for experimental data.

Example

In an industrial operation, potassium hydroxide (KOH) comes into contact with a body of water at ambient conditions (25 C). Determine the concentration in water at saturation.

Inspection of the tabulation for solubility in water yields:

S = 54.70 wt %

References

1. Clever, H. L. and F. J. Johnston, J. Phy. Chem. Ref. Data, 9, 751 (1980).
2. Clever, H. L., S. A. Johnston, and M. E. Derrick, J. Phy. Chem. Ref. Data, 14, 631 (1985).
3. Clever, H. L., S. A. Johnston, and M. E. Derrick, J. Phy. Chem. Ref. Data, 21, 941 (1992).
4. CRC HANDBOOK OF CHEMISTRY AND PHYSICS, 75th - 86th eds., CRC Press, Inc., Boca Raton, FL (1994-2006).
5. Knacke, O. and W. Gans, Zeit. Phy. Chem., NF, 104, 41 (1977).
6. Krumgalz, B. S., MINERAL SOLUBILITY IN WATER AT VARIOUS TEMPERATURES, Israel Oceanographic and Limnological Research Ltd., Haifa, Israel (1994).
7. Marcus, Y., J. Phy. Chem. Ref. Data, 9, 1307 (1980).
8. Marshal. W. L. and R. Slusher, J. Phy. Chem., 70, 4015 (1966).
9. Potter, R. W. and M. A. Clynne, J. Phy. Chem. Ref. Data, 24, 338 (1979).
10. Potter, R. W. and M. A. Clynne, J. Research U. S. Geological Survey, 6, 701 (1978).
11. Sohnel, O. and P. Novotny, DENSITIES OF AQUEOUS SOLUTIONS OF INORGANIC SUBSTANCES, Elsevier, Amsterdam, Holland (1985).
12. SOLUBILITY DATA SERIES, IUPAC, Volumes 1 to 53, Pergamon Press, Oxford, England (1979-1994).
13. Stephen, H. and T. Stephen, SOLUBILITIES OF INORGANIC AND ORGANIC COMPOUNDS, Volumes 1, 2, and 3, Macmillan Company, New York, NY (1963-1979).

Table 4 Solubility in Water - Inorganic Compounds

					Solubility in Water, S					
NO	ID	FORMULA	NAME	CAS No	T, C	S@T, wt %	S@T, mol %	S@T, g/1000 g of H2O	S@T, mol/1000 g of H2O	code
1	17	AgBrO3	silver bromate	7783-89-3	25	0.193	1.4773E-02	1.9337E+00	8.2018E-03	1
2	37	AgClO2	silver chlorite	7783-91-7	25	0.550	5.6796E-02	5.5304E+00	3.1545E-02	1
3	38	AgClO3	silver (i) chlorate	7783-92-8	25	15.00	1.6345E+00	1.7647E+02	9.2239E-01	1
4	39	AgClO4	silver (i) perchlorate	7783-93-9	25	84.80	3.2650E+01	5.5789E+03	2.6910E+01	1
5	48	AgNO2	silver (i) nitrite	7783-99-5	25	0.413	4.8530E-02	4.1471E+00	2.6951E-02	1
6	49	AgNO3	silver (i) nitrate	7761-88-8	25	70.10	1.9912E+01	2.3445E+03	1.3801E+01	1
7	68	Ag2SO4	silver (i) sulfate	10294-26-5	25	0.830	4.8333E-02	8.3695E+00	2.6842E-02	1
8	127	AlCl3	aluminum chloride	7446-70-0	25	31.10	5.7479E+00	4.5138E+02	3.3852E+00	1
9	133	AlCl3O12	aluminum perchlorate	14452-39-2	90	64.40	9.1051E+00	1.8090E+03	5.5604E+00	1
10	134	AlF3	aluminum fluoride	7784-18-1	25	0.500	1.0768E-01	5.0251E+00	5.9839E-02	1
11	149	AlN3O9	aluminum nitrate	13473-90-0	25	40.80	5.5080E+00	6.8919E+02	3.2357E+00	1
12	175	Al2S3O12	aluminum sulfate	10043-01-3	25	27.80	1.9870E+00	3.8504E+02	1.1253E+00	1
13	245	As2O3	arsenic trioxide	1327-53-3	25	2.01	1.8643E-01	2.0512E+01	1.0368E-01	1
14	370	BH3O3	boric acid	10043-35-3	25	5.48	1.6611E+00	5.7977E+01	9.3764E-01	1
15	423	BaBr2	barium bromide	10553-31-8	25	50.00	5.7163E+00	1.0000E+03	3.3655E+00	1
16	426	BaBr2O6	barium bromate	13967-90-3	25	0.788	3.6383E-02	7.9426E+00	2.0203E-02	1
17	434	BaC2N2S2	barium thiocyanate	2092-17-3	25	62.60	1.0631E+01	1.6738E+03	6.6029E+00	1
18	437	BaC4H4O6	barium tartrate	5908-81-6	25	44.20	4.7619E+00	7.9211E+02	2.7755E+00	1
19	454	BaCl2	barium chloride	10361-37-2	25	27.03	3.1052E+00	3.7043E+02	1.7789E+00	1
20	458	BaCl2O6	barium chlorate	13477-00-4	25	27.50	2.1968E+00	3.7931E+02	1.2468E+00	1
21	459	BaCl2O8	barium perchlorate	13465-95-7	25	75.75	1.4337E+01	3.1237E+03	9.2905E+00	1
22	464	BaF2	barium fluoride	7787-32-8	25	0.161	1.6567E-02	1.6126E+00	9.1978E-03	1
23	470	BaH2O2	barium hydroxide	17194-00-2	25	4.68	5.1357E-01	4.9098E+01	2.8655E-01	1
24	481	BaI2	barium iodide	13718-50-8	25	68.80	9.2200E+00	2.2051E+03	5.6378E+00	1
25	482	BaI2O6	barium iodate	10567-69-8	25	0.0396	1.4650E-03	3.9616E-01	8.1324E-04	1
26	486	BaN2O4	barium nitrite	13465-94-6	25	44.30	5.8801E+00	7.9533E+02	3.4679E+00	1
27	487	BaN2O6	barium nitrate	10022-31-8	25	9.30	7.0186E-01	1.0254E+02	3.9235E-01	1
28	494	BaO3S	barium sulfite	7787-39-5	25	0.0011	9.1157E-05	1.1000E-02	5.0601E-05	1
29	511	BaS	barium sulfide	21109-95-5	25	8.21	9.4227E-01	8.9443E+01	5.2802E-01	1
30	541	BeCl2	beryllium chloride	7787-47-5	25	41.70	1.3885E+01	7.1527E+02	8.9500E+00	1
31	552	BeO4S	beryllium sulfate	13510-49-1	25	29.22	6.6100E+00	4.1283E+02	3.9289E+00	1
32	682	CaBr2	calcium bromide	7789-41-5	25	61.00	1.2355E+01	1.5641E+03	7.8250E+00	1
33	713	CaCl2	calcium chloride	10043-52-4	25	44.83	1.1653E+01	8.1258E+02	7.3216E+00	1
34	722	CaCl2O6	calcium chlorate	10137-74-3	25	66.30	1.4620E+01	1.9674E+03	9.5051E+00	1
35	723	CaCl2O8	calcium perchlorate	13477-36-6	25	65.30	1.2424E+01	1.8818E+03	7.8745E+00	1
36	728	CaF2	calcium fluoride	7789-75-5	25	0.0016	3.6919E-04	1.6000E-02	2.0493E-04	1
37	746	CaI2	calcium iodide	10102-68-8	25	68.30	1.1667E+01	2.1546E+03	7.3313E+00	1
38	747	CaI2O6	calcium iodate	7789-80-2	25	0.305	1.4134E-02	3.0593E+00	7.8468E-03	1
39	750	CaN2O4	calcium nitrite	13780-06-8	25	48.60	1.1423E+01	9.4553E+02	7.1582E+00	1
40	751	CaN2O6	calcium nitrate	10124-37-5	25	59.00	1.3643E+01	1.4390E+03	8.7698E+00	1
41	757	CaO4S	calcium sulfate	7778-18-9	25	0.205	2.7175E-02	2.0542E+00	1.5089E-02	1
42	777	CdBr2	cadmium bromide	7789-42-6	25	53.40	7.0490E+00	1.1459E+03	4.2096E+00	1
43	783	CdC2O4	cadmium oxalate	814-88-0	25	0.0060	5.3932E-04	6.0004E-02	2.9937E-04	1
44	791	CdCl2	cadmium chloride	10108-64-2	25	54.60	1.0570E+01	1.2026E+03	6.5605E+00	1
45	795	CdCl2O8	cadmium perchlorate	79490-00-9	25	58.70	7.5998E+00	1.4213E+03	4.5655E+00	1
46	798	CdF2	cadmium fluoride	7790-79-6	25	4.18	5.1978E-01	4.3623E+01	2.9003E-01	1
47	804	CdI2	cadmium iodide	7790-80-9	25	46.30	4.0687E+00	8.6220E+02	2.3543E+00	1
48	805	CdI2O6	cadmium iodate	7790-81-0	25	0.0910	3.5499E-03	9.1083E-01	1.9706E-03	1
49	807	CdN2O6	cadmium nitrate	10325-94-7	25	61.00	1.0649E+01	1.5641E+03	6.6158E+00	1
50	816	CdO4S	cadmium sulfate	10124-36-4	25	43.40	6.2143E+00	7.6678E+02	3.6781E+00	1
51	934	CoC2N2S2	cobalt (ii) thiocyanate	3017-60-5	25	50.70	9.5682E+00	1.0284E+03	5.8732E+00	1
52	968	CoCl2	cobalt chloride	7646-79-9	25	35.99	7.2367E+00	5.6226E+02	4.3304E+00	1
53	972	CoCl2O8	cobalt (ii) perchlorate	13455-31-7	25	53.00	7.3036E+00	1.1277E+03	4.3736E+00	1
54	978	CoF2	cobalt (ii) fluoride	10026-17-2	25	1.40	2.6320E-01	1.4199E+01	1.4648E-01	1
55	1000	CoI2	cobalt (ii) iodide	15238-00-3	25	66.99	1.0466E+01	2.0294E+03	6.4890E+00	1
56	1003	CoN2O4	cobalt (ii) nitrite	18488-96-5	25	0.490	5.8734E-02	4.9241E+00	3.2622E-02	1
57	1004	CoN2O6	cobalt (ii) nitrate	10141-05-6	25	50.80	9.2292E+00	1.0325E+03	5.6439E+00	1
58	1010	CoO4S	cobalt (ii) sulfate	10124-43-3	25	27.70	4.2632E+00	3.8313E+02	2.4718E+00	1
59	1100	CrO3	chromium (vi) oxide	1333-82-0	25	62.80	2.3321E+01	1.6882E+03	1.6883E+01	1
60	1128	CsBr	cesium bromide	7787-69-1	25	55.20	9.4453E+00	1.2321E+03	5.7899E+00	1
61	1129	CsBrO3	cesium bromate	13454-75-6	25	3.69	2.6395E-01	3.8314E+01	1.4690E-01	1
62	1137	CsCl	cesium chloride	7647-17-8	25	65.64	1.6972E+01	1.9104E+03	1.1347E+01	1
63	1138	CsClO3	cesium chlorate	13763-67-2	25	7.22	6.4379E-01	7.7818E+01	3.5968E-01	1
64	1139	CsClO4	cesium perchlorate	13454-84-7	25	1.96	1.5476E-01	1.9992E+01	8.6040E-02	1
65	1144	CsI	cesium iodide	7789-17-5	25	45.90	5.5561E+00	8.4843E+02	3.2656E+00	1
66	1145	CsIO3	cesium iodate	13454-81-4	25	2.59	1.5537E-01	2.6589E+01	8.6381E-02	1
67	1147	CsNO3	cesium nitrate	7789-18-6	25	21.80	2.5119E+00	2.7877E+02	1.4303E+00	1
68	1150	CsOH	cesium hydroxide	21351-79-1	30	75.00	2.6498E+01	3.0000E+03	2.0012E+01	1
69	1165	Cs2SO4	cesium sulfate	10294-54-9	25	64.50	8.2947E+00	1.8169E+03	5.0208E+00	1
70	1174	CuBr2	copper (ii) bromide	7789-45-9	25	55.80	9.2414E+00	1.2624E+03	5.6522E+00	1
71	1230	CuCl2	cupric chloride	7447-39-4	25	43.10	9.2141E+00	7.5747E+02	5.6338E+00	1
72	1234	CuCl2O8	copper (ii) perchlorate	13770-18-8	30	59.30	9.0919E+00	1.4570E+03	5.5516E+00	1
73	1240	CuF2	copper (ii) fluoride	7789-19-7	25	0.0750	1.3314E-02	7.5056E-01	7.3916E-03	1
74	1257	CuN2O6	copper (ii) nitrate	3251-23-8	25	59.20	1.2232E+01	1.4510E+03	7.7363E+00	1
75	1265	CuO4S	copper (ii) sulfate	7758-98-7	25	18.00	2.4177E+00	2.1951E+02	1.3753E+00	1
76	1424	FeBr2	iron (ii) bromide	7789-46-0	25	54.60	9.1293E+00	1.2026E+03	5.5768E+00	1
77	1502	FeCl2	ferrous chloride	7758-94-3	25	39.40	8.4591E+00	6.5017E+02	5.1295E+00	1
78	1505	FeCl3	ferric chloride	7705-08-0	25	47.70	9.1979E+00	9.1205E+02	5.6229E+00	1
79	1511	FeF3	iron (iii) fluoride	7783-50-8	25	5.59	9.3644E-01	5.9210E+01	5.2472E-01	1
80	1530	FeN2O6	iron (ii) nitrate	14013-86-6	25	46.67	8.0591E+00	8.7512E+02	4.8657E+00	1

Table 4 Solubility in Water - Inorganic Compounds

					Solubility in Water, S					
NO	ID	FORMULA	NAME	CAS No	T, C	S@T, wt %	S@T, mol %	S@T, g/1000 g of H2O	S@T, mol/1000 g of H2O	code
81	1531	FeN3O9	ferric nitrate	10421-48-4	25	46.57	6.0964E+00	8.7161E+02	3.6038E+00	1
82	1534	FeO4S	iron (ii) sulfate	7720-78-7	25	22.80	3.3839E+00	2.9534E+02	1.9442E+00	1
83	1695	HCl	hydrogen chloride	7647-01-0	25	total	total	total	total	1
84	1698	HF	hydrogen fluoride	7664-39-3	25	total	total	total	total	1
85	1716	H2SO4	sulfuric acid	7664-93-9	25	total	total	total	total	1
86	1773	HgBr2	mercuric bromide	7789-47-1	25	0.610	3.0669E-02	6.1374E+00	1.7030E-02	1
87	1779	HgC2N2	mercury (ii) cyanide	592-04-1	25	10.20	8.0349E-01	1.1359E+02	4.4962E-01	1
88	1781	HgC2N2S2	mercury (ii) thiocyanate	592-85-8	25	0.0700	3.9838E-03	7.0049E-01	2.2114E-03	1
89	1800	HgCl2	mercuric chloride	7487-94-7	25	6.81	4.8256E-01	7.3077E+01	2.6916E-01	1
90	1813	HgI2	mercuric iodide	7774-29-0	25	0.0055	2.1806E-04	5.5003E-02	1.2105E-04	1
91	1829	Hg2Cl2	mercury (i) chloride	10112-91-1	25	0.0004	1.5264E-05	4.0000E-03	8.4731E-06	1
92	1838	Hg2SO4	mercury (i) sulfate	7783-36-0	25	0.0510	1.8486E-03	5.1026E-01	1.0262E-03	1
93	1969	KBF4	potassium fluoroborate	14075-53-7	25	0.550	7.9070E-02	5.5304E+00	4.3926E-02	1
94	1972	KBr	potassium bromide	7758-02-3	25	40.40	9.3066E+00	6.7785E+02	5.6961E+00	1
95	1973	KBrO3	potassium bromate	7758-01-2	25	7.55	8.7327E-01	8.1666E+01	4.8902E-01	1
96	1985	KC2H3O2	potassium acetate	127-08-2	25	72.92	3.3078E+01	2.6928E+03	2.7437E+01	1
97	2007	KCl	potassium chloride	7447-40-7	25	26.22	7.9085E+00	3.5538E+02	4.7669E+00	1
98	2010	KClO3	potassium chlorate	3811-04-9	25	7.93	1.2503E+00	8.6130E+01	7.0282E-01	1
99	2011	KClO4	potassium perchlorate	7778-74-7	25	2.04	2.7005E-01	2.0825E+01	1.5031E-01	1
100	2018	KF	potassium fluoride	7789-23-3	25	50.41	2.3967E+01	1.0165E+03	1.7497E+01	1
101	2023	KHCO3	potassium hydrogen carbonate	298-14-6	25	26.60	6.1219E+00	3.6240E+02	3.6198E+00	1
102	2029	KHSO4	potassium hydrogen sulfate	7646-93-7	25	33.60	6.2745E+00	5.0602E+02	3.7161E+00	1
103	2034	KH2PO4	potassium dihydrogen phosphate	7778-77-0	25	19.97	3.1977E+00	2.4953E+02	1.8336E+00	1
104	2035	KI	potassium iodide	7681-11-0	25	59.70	1.3850E+01	1.4814E+03	8.9239E+00	1
105	2036	KIO3	potassium iodate	7758-05-6	25	8.44	7.7001E-01	9.2180E+01	4.3075E-01	1
106	2037	KIO4	potassium periodate	7790-21-8	25	0.510	4.0135E-02	5.1261E+00	2.2288E-02	1
107	2040	KMnO4	potassium permanganate	7722-64-7	25	7.06	8.5850E-01	7.5963E+01	4.8068E-01	1
108	2041	KNO2	potassium nitrite	7758-09-0	25	75.70	3.9739E+01	3.1152E+03	3.6605E+01	1
109	2042	KNO3	potassium nitrate	7757-79-1	25	27.70	6.3904E+00	3.8313E+02	3.7895E+00	1
110	2047	KOH	potassium hydroxide	1310-58-3	25	54.70	2.7939E+01	1.2075E+03	2.1522E+01	1
111	2053	KSCN	potassium thiocyanate	333-20-0	25	70.40	3.0598E+01	2.3784E+03	2.4474E+01	1
112	2064	K2CO3	potassium carbonate	584-08-7	25	52.70	1.2681E+01	1.1142E+03	8.0617E+00	1
113	2088	K2CrO4	potassium chromate	7789-00-6	25	39.40	5.6885E+00	6.5017E+02	3.3481E+00	1
114	2089	K2Cr2O7	potassium dichromate	7778-50-9	25	13.10	9.1469E-01	1.5075E+02	5.1243E-01	1
115	2099	K2HPO4	potassium hydrogen phosphate	7758-11-4	25	62.70	1.4811E+01	1.6810E+03	9.6510E+00	1
116	2112	K2MoO4	potassium molybdate	13446-49-6	25	64.70	1.2177E+01	1.8329E+03	7.6968E+00	1
117	2116	K2O3S	potassium sulfite	10117-38-1	25	51.55	1.0803E+01	1.0640E+03	6.7230E+00	1
118	2117	K2O3Se	potassium selenite	10431-47-7	25	68.50	1.6034E+01	2.1746E+03	1.0600E+01	1
119	2127	K2SO4	potassium sulfate	7778-80-5	25	10.70	1.2236E+00	1.1982E+02	6.8760E-01	1
120	2128	K2S2O3	potassium thiosulfate	10294-66-3	25	62.30	1.3526E+01	1.6525E+03	8.6825E+00	1
121	2129	K2S2O5	potassium metabisulfite	16731-55-8	25	33.10	3.8546E+00	4.9477E+02	2.2254E+00	1
122	2133	K2SeO4	potassium selenate	7790-59-2	25	53.30	8.5063E+00	1.1413E+03	5.1608E+00	1
123	2141	K3AsO4	potassium arsenate	13464-36-3	25	55.60	8.0924E+00	1.2523E+03	4.8875E+00	1
124	2143	K3C6FeN6	potassium ferricyanide	13746-66-2	25	32.80	2.6012E+00	4.8810E+02	1.4825E+00	1
125	2153	K3PO4	potassium phosphate	7778-53-2	25	51.40	8.2366E+00	1.0576E+03	4.9825E+00	1
126	2177	LaCl3	lanthanum chloride	10099-58-8	25	48.90	6.5673E+00	9.5695E+02	3.9017E+00	1
127	2189	LaN3O9	lanthanum nitrate	10099-59-9	25	60.00	7.6781E+00	1.5000E+03	4.6165E+00	1
128	2213	LiBr	lithium bromide	7550-35-8	25	64.40	2.7286E+01	1.8090E+03	2.0830E+01	1
129	2215	LiBrO3	lithium bromate	13550-28-2	25	65.44	2.0190E+01	1.8935E+03	1.4042E+01	1
130	2223	LiC2H3O2	lithium acetate	546-89-4	25	31.02	1.0935E+01	4.4970E+02	6.8151E+00	1
131	2253	LiCl	lithium chloride	7447-41-8	25	45.81	2.6429E+01	8.4536E+02	1.9941E+01	1
132	2257	LiClO3	lithium chlorate	13453-71-9	25	82.10	4.7756E+01	4.5866E+03	5.0741E+01	1
133	2258	LiClO4	lithium perchlorate	7791-03-9	25	37.00	9.0451E+00	5.8730E+02	5.5202E+00	1
134	2261	LiF	lithium fluoride	7789-24-4	25	0.134	9.3102E-02	1.3418E+00	5.1728E-02	1
135	2267	LiH2O4P	lithium dihydrogen phosphate	13453-80-0	0	55.80	1.7954E+01	1.2624E+03	1.2147E+01	1
136	2270	LiI	lithium iodide	10377-51-2	25	62.30	1.8195E+01	1.6525E+03	1.2346E+01	1
137	2271	LiIO3	lithium iodate	13765-03-2	25	43.80	7.1676E+00	7.7936E+02	4.2859E+00	1
138	2274	LiNO3	lithium nitrate	7790-69-4	25	50.50	2.1047E+01	1.0202E+03	1.4797E+01	1
139	2277	LiOH	lithium hydroxide	1310-65-2	25	11.10	8.5860E+00	1.2486E+02	5.2137E+00	1
140	2283	Li2CO3	lithium carbonate	554-13-2	25	1.28	3.1512E-01	1.2966E+01	1.7547E-01	1
141	2285	Li2C2O4	lithium oxalate	30903-87-8	25	5.87	1.0904E+00	6.2361E+01	6.1197E-01	1
142	2306	Li2SO4	lithium sulfate	10377-48-7	25	25.50	5.3106E+00	3.4228E+02	3.1132E+00	1
143	2313	Li3PO4	lithium phosphate	10377-52-3	25	0.0270	4.2016E-03	2.7007E-01	2.3323E-03	1
144	2341	MgBr2	magnesium bromide	7789-48-2	25	50.60	9.1094E+00	1.0243E+03	5.5634E+00	1
145	2361	MgC4H6O4	magnesium acetate	142-72-3	25	39.61	7.6624E+00	6.5590E+02	4.6063E+00	1
146	2392	MgCl2	magnesium chloride	7786-30-3	25	35.90	9.5817E+00	5.6006E+02	5.8824E+00	1
147	2396	MgCl2O8	magnesium perchlorate	10034-81-8	25	50.10	7.4960E+00	1.0040E+03	4.4981E+00	1
148	2399	MgF2	magnesium fluoride	7783-40-6	25	0.0130	3.7594E-03	1.3002E-01	2.0869E-03	1
149	2424	MgI2	magnesium iodide	10377-58-9	25	59.40	8.6566E+00	1.4631E+03	5.2606E+00	1
150	2427	MgN2O6	magnesium nitrate	10377-60-3	25	41.60	7.9633E+00	7.1233E+02	4.8028E+00	1
151	2431	MgO3S	magnesium sulfite	7757-88-2	25	0.520	9.0144E-02	5.2272E+00	5.0084E-02	1
152	2440	MgSO4	magnesium sulfate	7487-88-9	25	26.30	5.0701E+00	3.5685E+02	2.9647E+00	1
153	2469	MnBr2	manganese (ii) bromide	13446-03-2	25	60.19	1.1256E+01	1.5119E+03	7.0406E+00	1
154	2495	MnCl2	manganese (ii) chloride	7773-01-5	25	43.60	9.9639E+00	7.7305E+02	6.1429E+00	1
155	2499	MnF2	manganese (ii) fluoride	7782-64-1	25	1.01	1.9739E-01	1.0203E+01	1.0979E-01	1
156	2515	MnN2O6	manganese (ii) nitrate	10377-93-2	25	61.70	1.3955E+01	1.6110E+03	9.0024E+00	1
157	2526	MnSO4	manganese (ii) sulfate	7785-87-7	25	38.90	7.0594E+00	6.3666E+02	4.2163E+00	1
158	2619	NH3	ammonia	7664-41-7	25	total	total	total	total	1
159	2625	NH4Br	ammonium bromide	12124-97-9	25	43.90	1.2582E+01	7.8253E+02	7.9897E+00	1
160	2627	NH4Cl	ammonium chloride	12125-02-9	25	28.34	1.1754E+01	3.9548E+02	7.3933E+00	1

Table 4 Solubility in Water - Inorganic Compounds

					Solubility in Water, S					
NO	ID	FORMULA	NAME	CAS No	T, C	S@T, wt %	S@T, mol %	S@T, g/1000 g of H2O	S@T, mol/1000 g of H2O	code
161	2629	NH4ClO4	ammonium perchlorate	7790-98-9	25	19.70	3.6254E+00	2.4533E+02	2.0881E+00	1
162	2631	NH4F	ammonium fluoride	12125-01-8	25	45.50	2.8880E+01	8.3486E+02	2.2541E+01	1
163	2634	NH4I	ammonium iodide	12027-06-4	25	64.00	1.8097E+01	1.7778E+03	1.2265E+01	1
164	2635	NH4IO3	ammonium iodate	13446-09-8	25	3.70	3.5746E-01	3.8422E+01	1.9914E-01	1
165	2641	NH5CO3	ammonium hydrogen carbonate	1066-33-7	25	19.90	5.3581E+00	2.4844E+02	3.1426E+00	1
166	2649	NH6PO4	ammonium dihydrogen phosphate	7722-76-1	25	28.80	5.9577E+00	4.0449E+02	3.5166E+00	1
167	2687	N2H4CS	ammonium thiocyanate	1762-95-4	25	64.40	2.9978E+01	1.8090E+03	2.3764E+01	1
168	2688	N2H4O2	ammonium nitrite	13446-48-5	25	68.80	3.8282E+01	2.2051E+03	3.4431E+01	1
169	2689	N2H4O3	ammonium nitrate	6484-52-2	25	68.00	3.2353E+01	2.1250E+03	2.6548E+01	1
170	2707	N2H8C2O4	ammonium oxalate	1113-38-8	25	4.94	7.4876E-01	5.1967E+01	4.1877E-01	1
171	2721	N2H8O4S	ammonium sulfate	7783-20-2	25	43.30	9.4295E+00	7.6367E+02	5.7792E+00	1
172	2723	N2H8O8S2	ammonium peroxydisulfate	7727-54-0	25	45.49	6.1808E+00	8.3453E+02	3.6569E+00	1
173	2728	N2H8SeO3	ammonium selenite	7783-19-9	25	54.70	1.1772E+01	1.2075E+03	7.4064E+00	1
174	2729	N2H8SeO4	ammonium selenate	7783-21-3	25	54.02	1.0572E+01	1.1749E+03	6.5622E+00	1
175	2734	N2H9PO4	diammonium hydrogen phosphate	7783-28-0	25	41.00	8.6591E+00	6.9492E+02	5.2623E+00	1
176	2800	NaBr	sodium bromide	7647-15-6	25	48.60	1.4203E+01	9.4553E+02	9.1893E+00	1
177	2802	NaBrO3	sodium bromate	7789-38-0	25	28.28	4.4960E+00	3.9431E+02	2.6132E+00	1
178	2805	NaCHO2	sodium formate	141-53-7	25	48.70	2.0094E+01	9.4932E+02	1.3959E+01	1
179	2816	NaC2H2IO2	sodium iodoacetate	305-53-3	25	33.50	4.1820E+00	5.0376E+02	2.4227E+00	1
180	2855	NaCl	sodium chloride	7647-14-5	25	26.45	9.9791E+00	3.5962E+02	6.1534E+00	1
181	2857	NaClH10O6	sodium hypochlorite pentahydrate	7681-52-8	25	44.40	8.0412E+00	7.9856E+02	4.8539E+00	1
182	2858	NaClO	sodium hypochlorite	7681-52-9	25	97.00	8.8668E+01	3.2333E+04	4.3434E+02	1
183	2859	NaClO2	sodium chlorite	7758-19-2	25	50.10	1.6666E+01	1.0040E+03	1.1101E+01	1
184	2860	NaClO3	sodium chlorate	7775-09-9	25	67.20	2.5747E+01	2.0488E+03	1.9248E+01	1
185	2864	NaF	sodium fluoride	7681-49-4	25	3.97	1.7428E+00	4.1341E+01	9.8459E-01	1
186	2870	NaHCO3	sodium hydrogen carbonate	144-55-8	25	9.32	2.1565E+00	1.0278E+02	1.2235E+00	1
187	2875	NaHSO4	sodium hydrogen sulfate	7681-38-1	25	22.20	4.1058E+00	2.8535E+02	2.3767E+00	1
188	2877	NaH2O4P	sodium phosphate, monobasic	7558-80-7	25	48.68	1.2467E+01	9.4856E+02	7.9062E+00	1
189	2888	NaI	sodium iodide	7681-82-5	25	64.80	1.8117E+01	1.8409E+03	1.2281E+01	1
190	2889	NaIO3	sodium iodate	7681-55-2	25	8.65	8.5464E-01	9.4691E+01	4.7850E-01	1
191	2890	NaIO4	sodium periodate	7790-28-5	25	12.62	1.2018E+00	1.4443E+02	6.7523E-01	1
192	2893	NaNO2	sodium nitrite	7632-00-0	25	45.90	1.8135E+01	8.4843E+02	1.2297E+01	1
193	2894	NaNO3	sodium nitrate	7631-99-4	25	47.70	1.6200E+01	9.1205E+02	1.0731E+01	1
194	2897	NaOH	sodium hydroxide	1310-73-2	25	50.00	3.1054E+01	1.0000E+03	2.5002E+01	1
195	2903	NaSCN	sodium thiocyanate	540-72-7	25	60.20	2.5155E+01	1.5126E+03	1.8657E+01	1
196	2916	Na2CO3	sodium carbonate	497-19-8	25	23.50	4.9622E+00	3.0719E+02	2.8983E+00	1
197	2917	Na2C2O4	sodium oxalate	62-76-0	25	3.48	4.8239E-01	3.6055E+01	2.6907E-01	1
198	2930	Na2CrO4	sodium chromate	7775-11-3	25	46.70	8.8797E+00	8.7617E+02	5.4094E+00	1
199	2932	Na2Cr2O7	sodium dichromate	10588-01-9	25	65.20	1.1414E+01	1.8736E+03	7.1519E+00	1
200	2960	Na2MoO4	sodium molybdate	7631-95-0	25	39.40	5.3819E+00	6.5017E+02	3.1574E+00	1
201	2966	Na2O4Se	sodium selenate	13410-01-0	25	36.90	5.2814E+00	5.8479E+02	3.0951E+00	1
202	2968	Na2O4W	sodium tungstate	13472-45-2	25	42.60	4.3524E+00	7.4216E+02	2.5259E+00	1
203	2972	Na2S	sodium sulfide	1313-82-2	25	17.10	4.5449E+00	2.0627E+02	2.6430E+00	1
204	2973	Na2SO3	sodium sulfite	7757-83-7	25	23.50	4.2059E+00	3.0719E+02	2.4372E+00	1
205	2974	Na2SO4	sodium sulfate	7757-82-6	25	21.94	3.4420E+00	2.8107E+02	1.9787E+00	1
206	2975	Na2S2O5	sodium metabisulfite	7681-57-4	25	40.00	5.9421E+00	6.6667E+02	3.5068E+00	1
207	2978	Na2SeO3	sodium selenite	10102-18-8	25	47.30	8.5502E+00	8.9753E+02	5.1899E+00	1
208	3002	Na3O4P	sodium phosphate	7601-54-9	25	12.60	1.5595E+00	1.4416E+02	8.7937E-01	1
209	3013	Na4P2O7	sodium pyrophosphate	7722-88-5	25	6.62	4.7801E-01	7.0893E+01	2.6661E-01	1
210	3070	NdCl3	neodymium trichloride	10024-93-8	25	50.00	6.7067E+00	1.0000E+03	3.9905E+00	1
211	3101	NiC2N2S2	nickel thiocyanate	13689-92-4	25	35.48	5.3617E+00	5.4991E+02	3.1448E+00	1
212	3133	NiCl2	nickel (ii) chloride	7718-54-9	25	40.30	8.5785E+00	6.7504E+02	5.2087E+00	1
213	3139	NiF2	nickel fluoride	10028-18-9	25	2.50	4.7546E-01	2.5641E+01	2.6519E-01	1
214	3154	NiI2	nickel (ii) iodide	13462-90-3	25	60.69	8.1727E+00	1.5439E+03	4.9404E+00	1
215	3155	NiI2O6	nickel iodate	13477-98-0	25	49.80	4.1915E+00	9.9203E+02	2.4285E+00	1
216	3164	NiSO4	nickel (ii) sulfate	7786-81-4	25	28.80	4.4969E+00	4.0449E+02	2.6137E+00	1
217	3487	PbBr2	lead bromide	10031-22-8	25	0.966	4.7857E-02	9.7542E+00	2.6578E-02	1
218	3514	PbCl2	lead chloride	7758-95-4	25	1.07	7.0013E-02	1.0816E+01	3.8891E-02	1
219	3518	PbCl2O8	lead (ii) perchlorate	13453-62-8	25	81.50	1.6348E+01	4.4054E+03	1.0848E+01	1
220	3521	PbF2	lead fluoride	7783-46-2	25	0.0670	4.9257E-03	6.7045E-01	2.7343E-03	1
221	3530	PbI2	lead iodide	10101-63-0	25	0.0760	2.9720E-03	7.6058E-01	1.6498E-03	1
222	3531	PbI2O6	lead (ii) iodate	25659-31-8	25	0.0025	8.0858E-05	2.5001E-02	4.4884E-05	1
223	3533	PbN2O6	lead (ii) nitrate	10099-74-8	25	37.38	3.1447E+00	5.9693E+02	1.8023E+00	1
224	3547	PbSO4	lead (ii) sulfate	7446-14-2	25	0.0044	2.6139E-04	4.4002E-02	1.4509E-04	1
225	3663	PrCl3	praseodymium chloride	10361-79-2	25	49.00	6.5420E+00	9.6078E+02	3.8856E+00	1
226	3786	RbBr	rubidium bromide	7789-39-1	25	53.80	1.1258E+01	1.1645E+03	7.0417E+00	1
227	3787	RbBrO3	rubidium bromate	13446-70-3	25	2.87	2.4886E-01	2.9548E+01	1.3848E-01	1
228	3791	RbCl	rubidium chloride	7791-11-9	25	48.42	1.2270E+01	9.3874E+02	7.7632E+00	1
229	3792	RbClO3	rubidium chlorate	13446-71-4	25	6.22	7.0239E-01	6.6325E+01	3.9265E-01	1
230	3793	RbClO4	rubidium perchlorate	13510-42-4	25	1.50	1.4814E-01	1.5228E+01	8.2352E-02	1
231	3795	RbF	rubidium fluoride	13446-74-7	20	75.00	3.4095E+01	3.0000E+03	2.8717E+01	1
232	3798	RbHCO3	rubidium hydrogen carbonate	19088-74-5	20	53.70	1.2483E+01	1.1598E+03	7.9177E+00	1
233	3801	RbI	rubidium iodide	7790-29-6	25	62.30	1.2294E+01	1.6525E+03	7.7812E+00	1
234	3806	RbOH	rubidium hydroxide	1310-82-3	30	63.40	2.3344E+01	1.7322E+03	1.6904E+01	1
235	3822	Rb2SO4	rubidium sulfate	7488-54-2	25	33.70	3.3159E+00	5.0830E+02	1.9037E+00	1
236	4042	SbCl3	antimony trichloride	10025-91-9	25	90.80	4.3802E+01	9.8696E+03	4.3265E+01	1
237	4045	SbF3	antimony (iii) fluoride	7783-56-4	25	83.10	3.3135E+01	4.9172E+03	2.7508E+01	1
238	4315	SmCl3	samarium (iii) chloride	10361-82-7	25	48.40	6.1757E+00	9.3798E+02	3.6538E+00	1
239	4422	SnCl2	stannous chloride	7772-99-8	10	64.00	1.4450E+01	1.7778E+03	9.3757E+00	1
240	4432	SnI2	tin (ii) iodide	10294-70-9	20	0.970	4.7346E-02	9.7950E+00	2.6294E-02	1

Table 4 Solubility in Water - Inorganic Compounds

					Solubility in Water, S					
NO	ID	FORMULA	NAME	CAS No	T, C	S@T, wt %	S@T, mol %	S@T, g/1000 g of H2O	S@T, mol/1000 g of H2O	code
241	4464	SrBr2	strontium bromide	10476-81-0	25	51.70	7.2300E+00	1.0704E+03	4.3261E+00	1
242	4479	SrCl2	strontium chloride	10476-85-4	25	35.37	5.8551E+00	5.4727E+02	3.4522E+00	1
243	4482	SrCl2O6	strontium chlorate	7791-10-8	25	63.77	1.1078E+01	1.7601E+03	6.9155E+00	1
244	4484	SrF2	strontium fluoride	7783-48-4	25	0.0210	3.0122E-03	2.1004E-01	1.6721E-03	1
245	4487	SrH2O2	strontium hydroxide	18480-07-4	25	2.20	3.3206E-01	2.2495E+01	1.8494E-01	1
246	4492	SrI2	strontium iodide	10476-86-5	25	63.90	8.5418E+00	1.7701E+03	5.1843E+00	1
247	4493	SrI2O6	strontium iodate	13470-01-4	25	0.165	6.8062E-03	1.6527E+00	3.7783E-03	1
248	4495	SrN2O4	strontium nitrite	13470-06-9	30	41.90	6.7447E+00	7.2117E+02	4.0147E+00	1
249	4496	SrN2O6	strontium nitrate	10042-76-9	25	44.50	6.3893E+00	8.0180E+02	3.7887E+00	1
250	4502	SrO4S	strontium sulfate	7759-02-6	25	0.0135	1.3242E-03	1.3502E-01	7.3506E-04	1
251	4744	Tl2SO4	thallium (i) sulfate	7446-18-6	25	5.19	1.9496E-01	5.4741E+01	1.0843E-01	1
252	4799	UN2O8	uranyl nitrate	10102-06-4	25	55.85	5.4673E+00	1.2650E+03	3.2104E+00	1
253	5024	ZnBr2	zinc bromide	7699-45-8	25	83.00	2.8087E+01	4.8824E+03	2.1680E+01	1
254	5062	ZnCl2	zinc chloride	7646-85-7	25	80.30	3.5013E+01	4.0761E+03	2.9907E+01	1
255	5069	ZnF2	zinc fluoride	7783-49-5	25	1.53	2.7001E-01	1.5538E+01	1.5029E-01	1
256	5082	ZnI2	zinc iodide	10139-47-6	25	81.40	1.9807E+01	4.3763E+03	1.3710E+01	1
257	5083	ZnI2O6	zinc iodate	7790-37-6	25	0.640	2.7940E-02	6.4412E+00	1.5514E-02	1
258	5086	ZnN2O6	zinc nitrate	7779-88-6	25	54.60	1.0265E+01	1.2026E+03	6.3498E+00	1
259	5094	ZnSO4	zinc sulfate	7733-02-0	25	36.60	6.0516E+00	5.7729E+02	3.5756E+00	1

Code: 1 - data, 2 - estimate

Chapter 5

SOLUBILITY IN SALT WATER – ORGANIC COMPOUNDS

Carl L. Yaws
Lamar University, Beaumont, Texas

Tabulation Results

The results for solubility of organic compounds in salt water are given in Table 5. The tabulation is applicable for alkanes, cyclopentanes, cyclohexanes, and benzenes. The results are based on a series expansion in salt concentration:

$$\log_{10} S = A + B X + C X^2 \tag{1}$$

where S = solubility in salt water at 25 C, parts per million by weight, ppm (wt)
X = concentration of salt (NaCl) in water, mg of salt/kg of water
A, B, and C = regression coefficients

The tabulation is arranged by carbon number (C, C2, C3, ...) to provide ease of use in quickly locating the data by using the chemical formula. The compound name, CAS No (Chemical Abstracts Registry Number), and regression coefficients are provided in the adjacent columns. The range of application is denoted by minimum and maximum salt concentrations (XMIN and XMAX). Salt concentrations outside the range of application should not be used. The next column provides the code for the tabulation which is based on both experimental data and estimated values.

In preparing the tabulation, a literature search was conducted to identify data source publications for organics (1-22). Both experimental values for the property under consideration and parameter values for estimation of the property are included in the source publications. The publications were screened and copies of appropriate data were made. These data were then keyed into the computer to provide a database of values for compounds for which experimental data are available. The database also served as a basis to check the accuracy of the correlations. The numerous point values were processed using a computer program for minimum deviation.

The excellent compilations by Howard and Meylan (1); Mackay, Shiu, and Ma (2-3); Verschueren (7); Yalkowsky (8); and Yaws (12-21) were primarily used for solubility in water without salt. For solubility in water containing salt, the compilations of API (6), Price (4); Sutton and Calder (5); and Yaws (9-11) were primarily used. Boiling point temperatures were ascertained from data compilations by Yaws (19-22).

For organic compounds, experimental data available in the literature are very limited for solubility in water containing salt. In the absence of data, the estimates for solubility in salt water should be considered as rough approximations useful for initial analysis. If initial analysis is favorable, follow-up experimental determination is recommended.

A comparison of calculated and data values is shown in Figure 5 for a representative compound. The graph discloses favorable agreement of equation and data.

Example

A chemical spill of methylcyclopentane (C6H12) occurs into a body of water containing 34,472 mg of salt (NaCl) / kg of water. Determine the concentration in the salt water at saturation.

Substitution of the regression coefficients from the table and salt concentration into the equation for solubility in water yields:

$$\log_{10} S = 1.6212 - 4.5987\text{E-}06{*}34{,}472 + 2.2993\text{E-}12{*}34{,}472^2$$

$$S = 10^{1.4654} = 29.20 \text{ parts per million by weight, ppm(wt)}$$

References

1. Howard, P. H. and W. M. Meylan, eds., HANDBOOK OF PHYSICAL PROPERTIES OF ORGANIC CHEMICALS, CRC Press, Boca Raton, FL (1997).
2. Mackay, D., W. Y. Shiu, and K. C. Ma, ILLUSTRATED HANDBOOK OF PHYSICAL-CHEMICAL PROPERTIES AND ENVIRONMENTAL FATE FOR ORGANIC CHEMICALS, Volume 1 – Monoaromatic Hydrocarbons, Chlorobenzenes and PCBs, Lewis Publishers, New York, NY (1992).
3. Mackay, D., W. Y. Shiu, and K. C. Ma, ILLUSTRATED HANDBOOK OF PHYSICAL-CHEMICAL PROPERTIES AND ENVIRONMENTAL FATE FOR ORGANIC CHEMICALS, Volume 3 – Volatile Organic Compounds, Lewis Publishers, New York, NY (1993).

4. Price, L. C., Bull. Am. Assoc. Petrol. Geologists, 60(2), 213-244 (1976).
5. Sutton, C. and J. A. Calder, Environ. Sci. Techn., 8(7), 654 (1974).
6. TECHNICAL DATA BOOK - PETROLEUM REFINING, Vols. I and II, American Petroleum Institute, Washington, DC (1972, 1977, 1982).
7. Verschueren, K., HANDBOOK OF ENVIRONMENTAL DATA ON ORGANIC CHEMICALS, 3rd and 4th eds., Van Nostrand Reinhold, New York, NY (1996, 2001).
8. Yalkowsky, S. H., Aquasol Database, 5th edition, College of Pharmacy, University of Arizona, Tuscon, Az (1990).
9. Yaws, C. L., "Hydrocarbon Solubility in Salt Water", [salt effect – alkanes], Pollution Engineering, 24, 46-50 (Sept. 15, 1992).
10. Yaws, C. L. and Xiaoyan Lin, "Solubility of Hydrocarbons in Salt Water", [salt effect – cyclopentanes and cyclohexanes], Pollution Engineering, 26, 70-72 (January, 1994).
11. Yaws, C. L., Li Bu, and S. Nijhawan, "Determining the Solubility of Aromatic Hydrocarbons", [salt effect – benzenes], Pollution Engineering, 27, 78-80 (June, 1995).
12. Yaws, C. L. and others, "Organic Chemicals: Water Solubility Data", Chem. Eng., 97, 115 (July, 1990).
13. Yaws, C. L. and others, "Water Solubility Data for Organic Compounds", Pollution Engineering, 22, 70 (Oct., 1990).
14. Yaws, C. L. and others, "New Correlation Calculates Reliable Paraffin Solubilities", Oil & Gas Journal, 89, 79 (April 8, 1991).
15. Yaws, C. L. and others, "New Correlation Accurately Calculates Naphthene Water Solubilities", Oil & Gas Journal, 89(37), 86 (Sept. 16, 1991).
16. Yaws, C. L. and others, "New Correlation Accurately Calculates Water Solubilities of Aromatics", Oil & Gas Journal, 95 (35), 80 (Aug. 29, 1994).
17. Yaws, C. L. and others, "Solubility of Hydrocarbons and Sulfur Compounds in Water", WATER ENCYCLOPEDIA, Volume 4, ISBN 0-471-73684-8, pages 561-564, John Wiley, New York, NY (2005).
18. Yaws, C. L. and others, "Correlation Measures Properties of Brominated Petchems", [solubility and Henry's law constant in water], Oil & Gas Journal, 104.11, 53-56 (March 20, 2006).
19. Yaws, C. L., CHEMICAL PROPERTIES HANDBOOK, McGraw-Hill, New York, NY (1999).
20. Yaws, C. L., MATHESON GAS DATA BOOK, 7th ed., Matheson Tri-Gas (Parsippany, NJ), McGraw-Hill, New York, NY (2001).
21. Yaws, C. L., YAWS HANDBOOK OF THERMODYNAMIC AND PHYSICAL PROPERTIES OF CHEMICAL COMPOUNDS, electronic edition, www.knovel.com, knovel, Norwich, NY (2003).
22. Yaws, C. L., YAWS HANDBOOK OF PHYSICAL PROPERTIES FOR HYDROCARBONS AND CHEMICALS, Gulf Publishing Co., Houston, TX (2005).

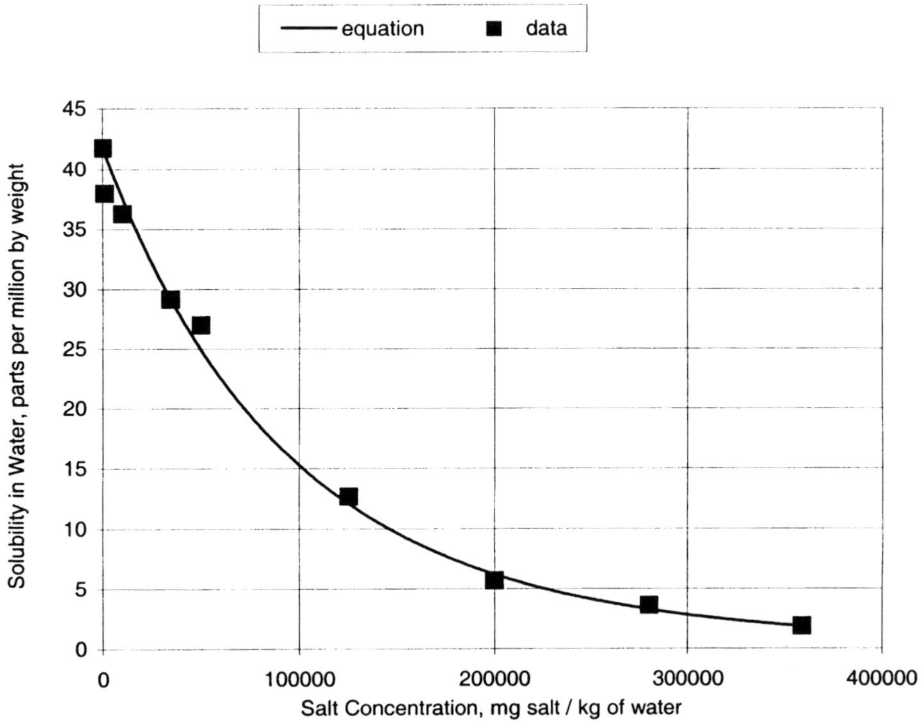

Figure 5 Solubility of Methylcyclopentane in Salt Water

Table 5 Solubility in Salt Water - Organic Compounds

$\log_{10} S = A + B X + C X^2$ (S - solubility, parts per million by weight, ppm (wt), X - mg salt / kg of water)												
NO	ID	FORMULA	NAME	CAS No	A	B	C	XMIN	XMAX	code	S@X=0	S@X=34,472
1	5285	C5H10	cyclopentane	287-92-3	2.1931	-7.1550E-06	3.5775E-12	0	358700	1,2	1.5600E+02	8.9275E+01
2	5772	C5H12	pentane	109-66-0	1.5966	-4.5956E-06	2.2978E-12	0	358700	1,2	3.9500E+01	2.7600E+01
3	5773	C5H12	isopentane	78-78-4	1.6794	-4.4374E-06	2.2187E-12	0	358700	1,2	4.7800E+01	3.3814E+01
4	5774	C5H12	neopentane	463-82-1	1.6175	-4.4374E-06	2.2187E-12	0	358700	1,2	4.1450E+01	2.9322E+01
5	6876	C6H6	benzene	71-43-2	3.2443	-3.2714E-06	1.6357E-12	0	358700	1,2	1.7551E+03	1.3598E+03
6	8214	C6H12	methylcyclopentane	96-37-7	1.6212	-4.5987E-06	2.2993E-12	0	358700	1,2	4.1802E+01	2.9202E+01
7	8215	C6H12	cyclohexane	110-82-7	1.7490	-7.1550E-06	3.5775E-12	0	358700	1,2	5.6100E+01	3.2105E+01
8	8877	C6H14	hexane	110-54-3	0.9763	-4.4374E-06	2.2187E-12	0	358700	1,2	9.4700E+00	6.6992E+00
9	8878	C6H14	2,2-dimethylbutane	75-83-2	1.3769	-4.4374E-06	2.2187E-12	0	358700	1,2	2.3820E+01	1.6850E+01
10	8879	C6H14	2,3-dimethylbutane	79-29-8	1.3126	-4.4374E-06	2.2187E-12	0	358700	1,2	2.0540E+01	1.4530E+01
11	8880	C6H14	2-methylpentane	107-83-5	1.1399	-4.4374E-06	2.2187E-12	0	358700	1,2	1.3800E+01	9.7622E+00
12	8881	C6H14	3-methylpentane	96-14-0	1.2531	-4.4374E-06	2.2187E-12	0	358700	1,2	1.7910E+01	1.2670E+01
13	10725	C7H8	toluene	108-88-3	2.7343	-4.5929E-06	2.2646E-12	0	358700	1,2	5.4240E+02	3.7904E+02
14	11696	C7H14	ethylcyclopentane	1640-89-7	0.9798	-7.1550E-06	3.5775E-12	0	358700	2	9.5450E+00	5.4624E+00
15	11697	C7H14	1,1-dimethylcyclopentane	1638-26-2	1.3528	-7.1550E-06	3.5775E-12	0	358700	2	2.2530E+01	1.2893E+01
16	11698	C7H14	1,2-dimethylcyclopentane	2452-99-5	1.1406	-7.1550E-06	3.5775E-12	0	358700	2	1.3822E+01	7.9099E+00
17	11699	C7H14	cis-1,2-dimethylcyclopentane	1192-18-3	1.0759	-7.1550E-06	3.5775E-12	0	358700	2	1.1910E+01	6.8156E+00
18	11700	C7H14	trans-1,2-dimethylcyclopentane	822-50-4	1.2590	-7.1550E-06	3.5775E-12	0	358700	2	1.8155E+01	1.0389E+01
19	11701	C7H14	1,3-dimethylcyclopentane	2453-00-1	1.2571	-7.1550E-06	3.5775E-12	0	358700	2	1.8076E+01	1.0344E+01
20	11702	C7H14	cis-1,3-dimethylcyclopentane	2532-58-3	1.2848	-7.1550E-06	3.5775E-12	0	358700	2	1.9267E+01	1.1026E+01
21	11703	C7H14	trans-1,3-dimethylcyclopentane	1759-58-6	1.2623	-7.1550E-06	3.5775E-12	0	358700	2	1.8293E+01	1.0469E+01
22	11704	C7H14	methylcyclohexane	108-87-2	1.1467	-7.1550E-06	3.5775E-12	0	358700	1,2	1.4020E+01	8.0233E+00
23	12112	C7H16	heptane	142-82-5	0.3502	-4.4374E-06	2.2187E-12	0	358700	1,2	2.2400E+00	1.5846E+00
24	12113	C7H16	2-methylhexane	591-76-4	0.4048	-4.4374E-06	2.2187E-12	0	358700	1,2	2.5400E+00	1.7968E+00
25	12114	C7H16	3-methylhexane	589-34-4	0.4216	-4.4374E-06	2.2187E-12	0	358700	1,2	2.6400E+00	1.8676E+00
26	12115	C7H16	3-ethylpentane	617-78-7	0.4691	-4.4374E-06	2.2187E-12	0	358700	2	2.9451E+00	2.0834E+00
27	12116	C7H16	2,2-dimethylpentane	590-35-2	0.6435	-4.4374E-06	2.2187E-12	0	358700	1,2	4.4000E+00	3.1126E+00
28	12117	C7H16	2,3-dimethylpentane	565-59-3	0.7202	-4.4374E-06	2.2187E-12	0	358700	1,2	5.2500E+00	3.7139E+00
29	12118	C7H16	2,4-dimethylpentane	108-08-7	0.6444	-4.4374E-06	2.2187E-12	0	358700	1,2	4.4100E+00	3.1197E+00
30	12119	C7H16	3,3-dimethylpentane	562-49-2	0.7738	-4.4374E-06	2.2187E-12	0	358700	1,2	5.9400E+00	4.2020E+00
31	12120	C7H16	2,2,3-trimethylbutane	464-06-2	0.7589	-4.4374E-06	2.2187E-12	0	358700	2	5.7397E+00	4.0603E+00
32	12121	C7H16	3-methylhexane, (S)-	6131-24-4	0.5038	-4.4374E-06	2.2187E-12	0	358700	2	3.1903E+00	2.2568E+00
33	13789	C8H10	ethylbenzene	100-41-4	2.2177	-3.3451E-06	1.6726E-12	0	358700	1,2	1.6510E+02	1.2718E+02
34	13790	C8H10	o-xylene	95-47-6	2.3440	-4.5929E-06	2.2646E-12	0	358700	1,2	2.2080E+02	1.5430E+02
35	13791	C8H10	m-xylene	108-38-3	2.2405	-4.5929E-06	2.2646E-12	0	358700	1,2	1.7400E+02	1.2159E+02
36	13792	C8H10	p-xylene	106-42-3	2.3047	-4.5929E-06	2.2646E-12	0	358700	1,2	2.0170E+02	1.4095E+02
37	13793	C8H10	xylenes	1330-20-7	2.4313	-4.5929E-06	2.2646E-12	0	358700	2	2.6997E+02	1.8866E+02
38	14861	C8H16	propylcyclopentane	2040-96-2	0.3096	-7.1550E-06	3.5775E-12	0	358700	1,2	2.0400E+00	1.1674E+00
39	14862	C8H16	isopropylcyclopentane	3875-51-2	0.4020	-7.1550E-06	3.5775E-12	0	358700	2	2.5232E+00	1.4440E+00
40	14863	C8H16	1-methyl-1-ethylcyclopentane	16747-50-5	0.5273	-7.1550E-06	3.5775E-12	0	358700	2	3.3671E+00	1.9269E+00
41	14864	C8H16	1-methyl-cis-2-ethylcyclopentane	930-89-2	0.3605	-7.1550E-06	3.5775E-12	0	358700	2	2.2936E+00	1.3126E+00
42	14865	C8H16	1-methyl-trans-2-ethylcyclopentane	930-90-5	0.5354	-7.1550E-06	3.5775E-12	0	358700	2	3.4308E+00	1.9634E+00
43	14866	C8H16	1-methyl-cis-3-ethylcyclopentane	2613-66-3	0.5379	-7.1550E-06	3.5775E-12	0	358700	2	3.4510E+00	1.9749E+00
44	14867	C8H16	1-methyl-trans-3-ethylcyclopentane	2613-65-2	0.5379	-7.1550E-06	3.5775E-12	0	358700	2	3.4510E+00	1.9749E+00
45	14868	C8H16	1,1,2-trimethylcyclopentane	4259-00-1	0.7246	-7.1550E-06	3.5775E-12	0	358700	2	5.3034E+00	3.0350E+00
46	14869	C8H16	1,1,3-trimethylcyclopentane	4516-69-2	0.5717	-7.1550E-06	3.5775E-12	0	358700	1,2	3.7300E+00	2.1346E+00
47	14870	C8H16	1,cis-2,cis-3-trimethylcyclopentane	2613-69-6	0.4895	-7.1550E-06	3.5775E-12	0	358700	2	3.0871E+00	1.7666E+00
48	14871	C8H16	1,cis-2,trans-3-trimethylcyclopentane	15890-40-1	0.6293	-7.1550E-06	3.5775E-12	0	358700	2	4.2594E+00	2.4375E+00
49	14872	C8H16	1,trans-2,cis-3-trimethylcyclopentane	19374-46-0	0.8130	-7.1550E-06	3.5775E-12	0	358700	2	6.5020E+00	3.7209E+00
50	14873	C8H16	1,cis-2,cis-4-trimethylcyclopentane	2613-72-1	0.6420	-7.1550E-06	3.5775E-12	0	358700	2	4.3854E+00	2.5096E+00
51	14874	C8H16	1,cis-2,trans-4-trimethylcyclopentane	4850-28-6	0.6488	-7.1550E-06	3.5775E-12	0	358700	2	4.4546E+00	2.5493E+00
52	14875	C8H16	1,trans-2,cis-4-trimethylcyclopentane	16883-48-0	0.8358	-7.1550E-06	3.5775E-12	0	358700	2	6.8511E+00	3.9207E+00
53	14876	C8H16	1-methyl-3-ethylcyclopentane	3726-47-4	0.5392	-7.1550E-06	3.5775E-12	0	358700	2	3.4611E+00	1.9807E+00
54	14877	C8H16	ethylcyclohexane	1678-91-7	0.7993	-7.1550E-06	3.5775E-12	0	358700	1,2	6.3000E+00	3.6053E+00
55	14878	C8H16	1,1-dimethylcyclohexane	590-66-9	0.7774	-7.1550E-06	3.5775E-12	0	358700	2	5.9890E+00	3.4273E+00
56	14879	C8H16	1,2-dimethylcyclohexane, (cis+trans)	583-57-3	0.6155	-7.1550E-06	3.5775E-12	0	358700	2	4.1258E+00	2.3611E+00
57	14880	C8H16	cis-1,2-dimethylcyclohexane	2207-01-4	0.7782	-7.1550E-06	3.5775E-12	0	358700	1,2	6.0000E+00	3.4336E+00
58	14881	C8H16	trans-1,2-dimethylcyclohexane	6876-23-9	0.7160	-7.1550E-06	3.5775E-12	0	358700	1,2	5.2000E+00	2.9758E+00
59	14882	C8H16	1,3-dimethylcyclohexane, cis and trans	591-21-9	0.7672	-7.1550E-06	3.5775E-12	0	358700	2	5.8504E+00	3.3481E+00
60	14883	C8H16	cis-1,3-dimethylcyclohexane	638-04-0	0.7636	-7.1550E-06	3.5775E-12	0	358700	2	5.8027E+00	3.3207E+00
61	14884	C8H16	trans-1,3-dimethylcyclohexane	2207-03-6	0.6523	-7.1550E-06	3.5775E-12	0	358700	2	4.4905E+00	2.5698E+00
62	14885	C8H16	1,4-dimethylcyclohexane, (cis+trans)	589-90-2	0.6959	-7.1550E-06	3.5775E-12	0	358700	2	4.9650E+00	2.8413E+00
63	14886	C8H16	cis-1,4-dimethylcyclohexane	624-29-3	0.6559	-7.1550E-06	3.5775E-12	0	358700	2	4.5276E+00	2.5910E+00
64	14887	C8H16	trans-1,4-dimethylcyclohexane	2207-04-7	0.5843	-7.1550E-06	3.5775E-12	0	358700	1,2	3.8400E+00	2.1975E+00
65	15354	C8H18	octane	111-65-9	-0.3655	-4.4374E-06	2.2187E-12	0	358700	1,2	4.3100E-01	3.0489E-01
66	15355	C8H18	2-methylheptane	592-27-8	-0.1266	-4.4374E-06	2.2187E-12	0	358700	2	7.4722E-01	5.2859E-01
67	15356	C8H18	3-methylheptane	589-81-1	-0.1013	-4.4374E-06	2.2187E-12	0	358700	1,2	7.9200E-01	5.6027E-01
68	15357	C8H18	4-methylheptane	589-53-7	-0.1281	-4.4374E-06	2.2187E-12	0	358700	2	7.4461E-01	5.2674E-01
69	15358	C8H18	3-methyl-3-ethylpentane	1067-08-9	-0.1423	-4.4374E-06	2.2187E-12	0	358700	2	7.2067E-01	5.0981E-01
70	15359	C8H18	3-ethylhexane	619-99-8	-0.1491	-4.4374E-06	2.2187E-12	0	358700	2	7.0939E-01	5.0183E-01
71	15360	C8H18	3-ethyl-2-methylpentane	609-26-7	-0.0760	-4.4374E-06	2.2187E-12	0	358700	2	8.3952E-01	5.9388E-01
72	15361	C8H18	2,2-dimethylhexane	590-73-8	0.1446	-4.4374E-06	2.2187E-12	0	358700	2	1.3949E+00	9.8679E-01
73	15362	C8H18	2,3-dimethylhexane	584-94-1	-0.0750	-4.4374E-06	2.2187E-12	0	358700	2	8.4147E-01	5.9527E-01
74	15363	C8H18	2,4-dimethylhexane	589-43-5	0.0802	-4.4374E-06	2.2187E-12	0	358700	2	1.2027E+00	8.5082E-01
75	15364	C8H18	2,5-dimethylhexane	592-13-2	0.0881	-4.4374E-06	2.2187E-12	0	358700	2	1.2250E+00	8.6659E-01
76	15365	C8H18	3,3-dimethylhexane	563-16-6	0.0166	-4.4374E-06	2.2187E-12	0	358700	2	1.0391E+00	7.3505E-01
77	15366	C8H18	3,4-dimethylhexane	583-48-2	-0.1286	-4.4374E-06	2.2187E-12	0	358700	2	7.4374E-01	5.2613E-01
78	15367	C8H18	2,2,3-trimethylpentane	564-02-3	0.0699	-4.4374E-06	2.2187E-12	0	358700	2	1.1747E+00	8.3101E-01
79	15368	C8H18	2,2,4-trimethylpentane	540-84-1	0.3308	-4.4374E-06	2.2187E-12	0	358700	2	2.1420E+00	1.5153E+00
80	15369	C8H18	2,3,3-trimethylpentane	560-21-4	-0.0538	-4.4374E-06	2.2187E-12	0	358700	2	8.8356E-01	6.2504E-01

Table 5 Solubility in Salt Water - Organic Compounds

			$\log_{10} S = A + B X + C X^2$	(S - solubility, parts per million by weight, ppm (wt),					X - mg salt / kg of water)			
NO	ID	FORMULA	NAME	CAS No	A	B	C	XMIN	XMAX	code	S@X=0	S@X=34,472
81	15370	C8H18	2,3,4-trimethylpentane	565-75-3	0.3621	-4.4374E-06	2.2187E-12	0	358700	1,2	2.3020E+00	1.6285E+00
82	15371	C8H18	2,2,3,3-tetramethylbutane	594-82-1	0.1582	-4.4374E-06	2.2187E-12	0	358700	2	1.4394E+00	1.0182E+00
83	15372	C8H18	octane, mixture	---	0.2375	-4.4374E-06	2.2187E-12	0	358700	2	1.7278E+00	1.2222E+00
84	17004	C9H12	cumene	98-82-8	1.6990	-3.3451E-06	1.6726E-12	0	358700	1,2	5.0000E+01	3.8516E+01
85	17005	C9H12	m-ethyltoluene	620-14-4	1.9868	-4.5929E-06	2.2646E-12	0	358700	1,2	9.7000E+01	6.7785E+01
86	17006	C9H12	o-ethyltoluene	611-14-3	1.9687	-4.5929E-06	2.2646E-12	0	358700	1,2	9.3050E+01	6.5025E+01
87	17007	C9H12	p-ethyltoluene	622-96-8	1.9770	-4.5929E-06	2.2646E-12	0	358700	1,2	9.4850E+01	6.6283E+01
88	17008	C9H12	1,2,3-trimethylbenzene	526-73-8	1.8762	-4.5929E-06	2.2646E-12	0	358700	1,2	7.5200E+01	5.2551E+01
89	17009	C9H12	1,2,4-trimethylbenzene	95-63-6	1.7709	-4.5929E-06	2.2646E-12	0	358700	1,2	5.9000E+01	4.1230E+01
90	17010	C9H12	trimethyl benzene	25551-13-7	1.6610	-4.5929E-06	2.2646E-12	0	358700	2	4.5810E+01	3.2013E+01
91	17011	C9H12	mesitylene	108-67-8	1.9868	-4.5929E-06	2.2646E-12	0	358700	1,2	9.7000E+01	6.7785E+01
92	17012	C9H12	propylbenzene	103-65-1	1.7173	-3.3451E-06	1.6726E-12	0	358700	1,2	5.2160E+01	4.0180E+01
93	17859	C9H18	butylcyclopentane	2040-95-1	-0.3658	-7.1550E-06	3.5775E-12	0	358700	2	4.3077E-01	2.4652E-01
94	17860	C9H18	isobutylcyclopentane	3788-32-7	-0.1479	-7.1550E-06	3.5775E-12	0	358700	2	7.1134E-01	4.0708E-01
95	17861	C9H18	sec-butylcyclopentane	4850-32-2	-0.3093	-7.1550E-06	3.5775E-12	0	358700	2	4.9061E-01	2.8076E-01
96	17862	C9H18	tert-butylcyclopentane	3875-52-3	-0.0692	-7.1550E-06	3.5775E-12	0	358700	2	8.5273E-01	4.8800E-01
97	17863	C9H18	1-methyl-1-propylcyclopentane	16631-63-3	-0.0984	-7.1550E-06	3.5775E-12	0	358700	2	7.9720E-01	4.5622E-01
98	17864	C9H18	1-methyl-cis-2-propylcyclopentane	932-43-4	-0.2648	-7.1550E-06	3.5775E-12	0	358700	2	5.4346E-01	3.1101E-01
99	17865	C9H18	1-methyl-trans-2-propylcyclopentane	932-44-5	-0.1078	-7.1550E-06	3.5775E-12	0	358700	2	7.8013E-01	4.4645E-01
100	17866	C9H18	1-methyl-cis-3-propylcyclopentane	2443-04-1	-0.1492	-7.1550E-06	3.5775E-12	0	358700	2	7.0927E-01	4.0590E-01
101	17867	C9H18	1-methyl-trans-3-propylcyclopentane	2443-03-0	-0.1492	-7.1550E-06	3.5775E-12	0	358700	2	7.0927E-01	4.0590E-01
102	17868	C9H18	1-methyl-1-isopropylcyclopentane	61828-00-0	-0.1492	-7.1550E-06	3.5775E-12	0	358700	2	7.0927E-01	4.0590E-01
103	17869	C9H18	1-methyl-cis-2-isopropylcyclopentane	61868-01-7	-0.1492	-7.1550E-06	3.5775E-12	0	358700	2	7.0927E-01	4.0590E-01
104	17870	C9H18	1-methyl-trans-2-isopropylcyclopentane	61828-01-1	0.0290	-7.1550E-06	3.5775E-12	0	358700	2	1.0690E+00	6.1176E-01
105	17871	C9H18	1-methyl-cis-3-isopropylcyclopentane	61828-02-2	0.0034	-7.1550E-06	3.5775E-12	0	358700	2	1.0080E+00	5.7684E-01
106	17872	C9H18	1-methyl-trans-3-isopropylcyclopentane	61828-03-3	0.0034	-7.1550E-06	3.5775E-12	0	358700	2	1.0080E+00	5.7684E-01
107	17873	C9H18	1,1-diethylcyclopentane	2721-38-2	-0.2124	-7.1550E-06	3.5775E-12	0	358700	2	6.1316E-01	3.5090E-01
108	17874	C9H18	1,cis-2-diethylcyclopentane	932-39-8	-0.2895	-7.1550E-06	3.5775E-12	0	358700	2	5.1351E-01	2.9387E-01
109	17875	C9H18	1,trans-2-diethylcyclopentane	932-40-1	-0.1373	-7.1550E-06	3.5775E-12	0	358700	2	7.2900E-01	4.1719E-01
110	17876	C9H18	1,3-diethylcyclopentane	19398-75-5	-0.1998	-7.1550E-06	3.5775E-12	0	358700	2	6.3125E-01	3.6125E-01
111	17877	C9H18	1,cis-3-diethylcyclopentane	62016-59-5	-0.1998	-7.1550E-06	3.5775E-12	0	358700	2	6.3125E-01	3.6125E-01
112	17878	C9H18	1,trans-3-diethylcyclopentane	62016-60-8	-0.1998	-7.1550E-06	3.5775E-12	0	358700	2	6.3125E-01	3.6125E-01
113	17879	C9H18	1,1-dimethyl-2-ethylcyclopentane	54549-80-3	0.1057	-7.1550E-06	3.5775E-12	0	358700	2	1.2755E+00	7.2994E-01
114	17880	C9H18	1,1-dimethyl-3-ethylcyclopentane	62016-61-9	0.2081	-7.1550E-06	3.5775E-12	0	358700	2	1.6148E+00	9.2411E-01
115	17881	C9H18	1,cis-2-dimethyl-1-ethylcyclopentane	62016-63-1	-0.0475	-7.1550E-06	3.5775E-12	0	358700	2	8.9630E-01	5.1293E-01
116	17882	C9H18	1,trans-2-dimethyl-1-ethylcyclopentane	62016-62-0	-0.0475	-7.1550E-06	3.5775E-12	0	358700	2	8.9630E-01	5.1293E-01
117	17883	C9H18	1,cis-2-dimethyl-cis-3-ethylcyclopentane	---	-0.2502	-7.1550E-06	3.5775E-12	0	358700	2	5.6203E-01	3.2164E-01
118	17884	C9H18	1,cis-2-dimethyl-trans-3-ethylcyclopentane	---	-0.0730	-7.1550E-06	3.5775E-12	0	358700	2	8.4527E-01	4.8373E-01
119	17885	C9H18	1,trans-2-dimethyl-cis-3-ethylcyclopentane	---	0.1057	-7.1550E-06	3.5775E-12	0	358700	2	1.2755E+00	7.2994E-01
120	17886	C9H18	1,trans-2-dimethyl-trans-3-ethylcyclopenta	---	-0.0730	-7.1550E-06	3.5775E-12	0	358700	2	8.4527E-01	4.8373E-01
121	17887	C9H18	1,cis-2-dimethyl-cis-4-ethylcyclopentane	62016-64-2	-0.1238	-7.1550E-06	3.5775E-12	0	358700	2	7.5192E-01	4.3031E-01
122	17888	C9H18	1,cis-2-dimethyl-trans-4-ethylcyclopentane	62016-65-3	-0.1238	-7.1550E-06	3.5775E-12	0	358700	2	7.5192E-01	4.3031E-01
123	17889	C9H18	1,trans-2-dimethyl-cis-4-ethylcyclopentane	62016-66-4	0.0545	-7.1550E-06	3.5775E-12	0	358700	2	1.1338E+00	6.4884E-01
124	17890	C9H18	1,cis-3-dimethyl-1-ethylcyclopentane	62016-68-6	0.1569	-7.1550E-06	3.5775E-12	0	358700	2	1.4351E+00	8.2128E-01
125	17891	C9H18	1,trans-3-dimethyl-1-ethylcyclopentane	62016-67-5	0.1569	-7.1550E-06	3.5775E-12	0	358700	2	1.4351E+00	8.2128E-01
126	17892	C9H18	1,cis-3-dimethyl-cis-2-ethylcyclopentane	19903-00-5	-0.2502	-7.1550E-06	3.5775E-12	0	358700	2	5.6203E-01	3.2164E-01
127	17893	C9H18	1,cis-3-dimethyl-trans-2-ethylcyclopentane	19902-98-8	0.1313	-7.1550E-06	3.5775E-12	0	358700	2	1.3529E+00	7.7426E-01
128	17894	C9H18	1,trans-3-dimethyl-cis-2-ethylcyclopentane	19902-99-9	-0.0475	-7.1550E-06	3.5775E-12	0	358700	2	8.9630E-01	5.1293E-01
129	17895	C9H18	1,cis-3-dimethyl-cis-4-ethylcyclopentane	---	-0.0730	-7.1550E-06	3.5775E-12	0	358700	2	8.4527E-01	4.8373E-01
130	17896	C9H18	1,cis-3-dimethyl-trans-4-ethylcyclopentane	---	0.1313	-7.1550E-06	3.5775E-12	0	358700	2	1.3529E+00	7.7426E-01
131	17897	C9H18	1,trans-3-dimethyl-cis-4-ethylcyclopentane	---	0.0839	-7.1550E-06	3.5775E-12	0	358700	2	1.2132E+00	6.9429E-01
132	17898	C9H18	1,trans-3-dimethyl-trans-4-ethylcyclopenta	---	-0.0730	-7.1550E-06	3.5775E-12	0	358700	2	8.4527E-01	4.8373E-01
133	17899	C9H18	1,1,2,2-tetramethylcyclopentane	52688-89-8	0.2337	-7.1550E-06	3.5775E-12	0	358700	2	1.7129E+00	9.8026E-01
134	17900	C9H18	1,1,cis-2,cis-3-tetramethylcyclopentane	62016-70-0	0.1057	-7.1550E-06	3.5775E-12	0	358700	2	1.2755E+00	7.2994E-01
135	17901	C9H18	1,1,cis-2,trans-3-tetramethylcyclopentane	---	0.3106	-7.1550E-06	3.5775E-12	0	358700	2	2.0445E+00	1.1700E+00
136	17902	C9H18	1,1,cis-2,cis-4-tetramethylcyclopentane	62016-72-2	0.3106	-7.1550E-06	3.5775E-12	0	358700	2	2.0445E+00	1.1700E+00
137	17903	C9H18	1,1,cis-2,trans-4-tetramethylcyclopentane	---	0.3106	-7.1550E-06	3.5775E-12	0	358700	2	2.0445E+00	1.1700E+00
138	17904	C9H18	1,1,3,3-tetramethylcyclopentane	50876-33-0	0.6177	-7.1550E-06	3.5775E-12	0	358700	2	4.1466E+00	2.3730E+00
139	17905	C9H18	1,1,cis-3,cis-4-tetramethylcyclopentane	20309-77-7	0.2337	-7.1550E-06	3.5775E-12	0	358700	2	1.7129E+00	9.8026E-01
140	17906	C9H18	1,1,cis-3,trans-4-tetramethylcyclopentane	---	0.5252	-7.1550E-06	3.5775E-12	0	358700	2	3.3513E+00	1.9179E+00
141	17907	C9H18	1,2,2,cis-3-tetramethylcyclopentane	62016-73-3	0.1057	-7.1550E-06	3.5775E-12	0	358700	2	1.2755E+00	7.2994E-01
142	17908	C9H18	1,2,2,trans-3-tetramethylcyclopentane	---	0.1057	-7.1550E-06	3.5775E-12	0	358700	2	1.2755E+00	7.2994E-01
143	17909	C9H18	1,cis-2,cis-3,cis-4-tetramethylcyclopentane	2532-65-2	-0.0730	-7.1550E-06	3.5775E-12	0	358700	2	8.4527E-01	4.8373E-01
144	17910	C9H18	1,cis-2,cis-3,trans-4-tetramethylcyclopenta	2532-69-6	0.1057	-7.1550E-06	3.5775E-12	0	358700	2	1.2755E+00	7.2994E-01
145	17911	C9H18	1,cis-2,trans-3,cis-4-tetramethylcyclopenta	2532-68-5	0.2850	-7.1550E-06	3.5775E-12	0	358700	2	1.9274E+00	1.1030E+00
146	17912	C9H18	1,cis-2,trans-3,trans-4-tetramethylcyclopen	2532-64-1	0.1057	-7.1550E-06	3.5775E-12	0	358700	2	1.2755E+00	7.2994E-01
147	17913	C9H18	1,trans-2,cis-3,trans-4-tetramethylcyclopen	2532-67-4	0.3771	-7.1550E-06	3.5775E-12	0	358700	2	2.3831E+00	1.3638E+00
148	17914	C9H18	1,trans-2,trans-3,cis-4-tetramethylcyclopen	19907-40-5	0.2850	-7.1550E-06	3.5775E-12	0	358700	2	1.9274E+00	1.1030E+00
149	17915	C9H18	1-methyl-2-isopropylcyclopentane	89223-57-4	-0.0221	-7.1550E-06	3.5775E-12	0	358700	2	9.5047E-01	5.4393E-01
150	17916	C9H18	propylcyclohexane	1678-92-8	-0.1692	-7.1550E-06	3.5775E-12	0	358700	2	6.7726E-01	3.8758E-01
151	17917	C9H18	isopropylcyclohexane	696-29-7	-0.1195	-7.1550E-06	3.5775E-12	0	358700	2	7.5939E-01	4.3458E-01
152	17918	C9H18	1-methyl-1-ethylcyclohexane	4926-90-3	-0.0543	-7.1550E-06	3.5775E-12	0	358700	2	8.8254E-01	5.0506E-01
153	17919	C9H18	1-methyl-cis-2-ethylcyclohexane	4923-77-7	-0.1530	-7.1550E-06	3.5775E-12	0	358700	2	7.0302E-01	4.0232E-01
154	17920	C9H18	1-methyl-trans-2-ethylcyclohexane	4923-78-8	-0.0427	-7.1550E-06	3.5775E-12	0	358700	2	9.0639E-01	5.1871E-01
155	17921	C9H18	1-methyl-cis-3-ethylcyclohexane	19489-10-2	0.0508	-7.1550E-06	3.5775E-12	0	358700	2	1.1241E+00	6.4330E-01
156	17922	C9H18	1-methyl-trans-3-ethylcyclohexane	4926-76-5	-0.0250	-7.1550E-06	3.5775E-12	0	358700	2	9.4397E-01	5.4021E-01
157	17923	C9H18	1-methyl-cis-4-ethylcyclohexane	4926-78-7	-0.0250	-7.1550E-06	3.5775E-12	0	358700	2	9.4397E-01	5.4021E-01
158	17924	C9H18	1-methyl-trans-4-ethylcyclohexane	6236-88-0	0.0508	-7.1550E-06	3.5775E-12	0	358700	2	1.1241E+00	6.4330E-01
159	17925	C9H18	1,1,2-trimethylcyclohexane	7094-26-0	0.1219	-7.1550E-06	3.5775E-12	0	358700	2	1.3241E+00	7.5773E-01
160	17926	C9H18	1,1,3-trimethylcyclohexane	3073-66-3	0.2480	-7.1550E-06	3.5775E-12	0	358700	1,2	1.7700E+00	1.0129E+00

Table 5 Solubility in Salt Water - Organic Compounds

$\log_{10} S = A + B X + C X^2$				(S - solubility, parts per million by weight, ppm (wt), X - mg salt / kg of water)								
NO	ID	FORMULA	NAME	CAS No	A	B	C	XMIN	XMAX	code	S@X=0	S@X=34,472
161	17927	C9H18	1,1,4-trimethylcyclohexane	7094-27-1	0.3825	-7.1550E-06	3.5775E-12	0	358700	2	2.4127E+00	1.3807E+00
162	17928	C9H18	1,2,3-trimethylcyclohexane	1678-97-3	0.2034	-7.1550E-06	3.5775E-12	0	358700	2	1.5975E+00	9.1422E-01
163	17929	C9H18	1,cis-2,cis-3-trimethylcyclohexane	1839-88-9	-0.0754	-7.1550E-06	3.5775E-12	0	358700	2	8.4064E-01	4.8108E-01
164	17930	C9H18	1,cis-2,trans-3-trimethylcyclohexane	7667-55-2	-0.0250	-7.1550E-06	3.5775E-12	0	358700	2	9.4397E-01	5.4021E-01
165	17931	C9H18	1,trans-2,cis-3-trimethylcyclohexane	1678-81-5	0.1525	-7.1550E-06	3.5775E-12	0	358700	2	1.4205E+00	8.1294E-01
166	17932	C9H18	1,2,4-trimethylcyclohexane, isomers	2234-75-5	0.2060	-7.1550E-06	3.5775E-12	0	358700	2	1.6069E+00	9.1961E-01
167	17933	C9H18	1,cis-2,cis-4-trimethylcyclohexane	1678-80-4	0.0508	-7.1550E-06	3.5775E-12	0	358700	2	1.1241E+00	6.4330E-01
168	17934	C9H18	1,cis-2,trans-4-trimethylcyclohexane	7667-58-5	0.0508	-7.1550E-06	3.5775E-12	0	358700	2	1.1241E+00	6.4330E-01
169	17935	C9H18	1,trans-2,cis-4-trimethylcyclohexane	7667-59-6	0.1270	-7.1550E-06	3.5775E-12	0	358700	2	1.3397E+00	7.6666E-01
170	17936	C9H18	1,trans-2,trans-4-trimethylcyclohexane	7667-60-9	0.2234	-7.1550E-06	3.5775E-12	0	358700	2	1.6725E+00	9.5712E-01
171	17937	C9H18	1,3,5,-trimethylcyclohexane	1839-63-0	0.3095	-7.1550E-06	3.5775E-12	0	358700	2	2.0395E+00	1.1672E+00
172	17938	C9H18	cis,cis-1,3,5-trimethylcyclohexane	1795-27-3	0.2926	-7.1550E-06	3.5775E-12	0	358700	2	1.9617E+00	1.1226E+00
173	17939	C9H18	cis,trans-1,3,5-trimethylcyclohexane	1795-26-2	0.2405	-7.1550E-06	3.5775E-12	0	358700	2	1.7397E+00	9.9559E-01
174	17940	C9H18	1-ethyl-2-methylcyclohexane	3728-54-9	-0.0590	-7.1550E-06	3.5775E-12	0	358700	2	8.7288E-01	4.9953E-01
175	18176	C9H20	nonane	111-84-2	-0.9136	-4.4374E-06	2.2187E-12	0	358700	1,2	1.2200E-01	8.6304E-02
176	18177	C9H20	2-methyloctane	3221-61-2	-0.7813	-4.4374E-06	2.2187E-12	0	358700	2	1.6546E-01	1.1705E-01
177	18178	C9H20	3-methyloctane	2216-33-3	-0.8055	-4.4374E-06	2.2187E-12	0	358700	2	1.5649E-01	1.1070E-01
178	18179	C9H20	4-methyloctane	2216-34-4	-0.9393	-4.4374E-06	2.2187E-12	0	358700	1,2	1.1500E-01	8.1352E-02
179	18180	C9H20	3-ethylheptane	15869-80-4	-0.7793	-4.4374E-06	2.2187E-12	0	358700	2	1.6624E-01	1.1760E-01
180	18181	C9H20	4-ethylheptane	2216-32-2	-0.7285	-4.4374E-06	2.2187E-12	0	358700	2	1.8686E-01	1.3219E-01
181	18182	C9H20	2,2-dimethylheptane	1071-26-7	-0.5104	-4.4374E-06	2.2187E-12	0	358700	2	3.0873E-01	2.1840E-01
182	18183	C9H20	2,3-dimethylheptane	3074-71-3	-0.7106	-4.4374E-06	2.2187E-12	0	358700	2	1.9471E-01	1.3774E-01
183	18184	C9H20	2,4-dimethylheptane	2213-23-2	-0.5158	-4.4374E-06	2.2187E-12	0	358700	2	3.0493E-01	2.1571E-01
184	18185	C9H20	2,5-dimethylheptane	2216-30-0	-0.5955	-4.4374E-06	2.2187E-12	0	358700	2	2.5382E-01	1.7956E-01
185	18186	C9H20	2,6-dimethylheptane	1072-05-5	-0.5750	-4.4374E-06	2.2187E-12	0	358700	2	2.6608E-01	1.8823E-01
186	18187	C9H20	3,4-dimethylheptane	922-28-1	-0.7132	-4.4374E-06	2.2187E-12	0	358700	2	1.9357E-01	1.3693E-01
187	18188	C9H20	3,5-dimethylheptane	926-82-9	-0.5955	-4.4374E-06	2.2187E-12	0	358700	2	2.5382E-01	1.7956E-01
188	18189	C9H20	4,4-dimethylheptane	1068-19-5	-0.5750	-4.4374E-06	2.2187E-12	0	358700	2	2.6608E-01	1.8823E-01
189	18190	C9H20	3-ethyl-2-methylhexane	16789-46-1	-0.6467	-4.4374E-06	2.2187E-12	0	358700	2	2.2559E-01	1.5959E-01
190	18191	C9H20	4-ethyl-2-methylhexane	3074-75-7	-0.5391	-4.4374E-06	2.2187E-12	0	358700	2	2.8899E-01	2.0444E-01
191	18192	C9H20	3-ethyl-3-methylhexane	3074-76-8	-0.7132	-4.4374E-06	2.2187E-12	0	358700	2	1.9357E-01	1.3693E-01
192	18193	C9H20	3-ethyl-4-methylhexane	3074-77-9	-0.7080	-4.4374E-06	2.2187E-12	0	358700	2	1.9586E-01	1.3856E-01
193	18194	C9H20	2,2,3-trimethylhexane	16747-25-4	-0.5337	-4.4374E-06	2.2187E-12	0	358700	2	2.9259E-01	2.0698E-01
194	18195	C9H20	2,2,4-trimethylhexane	16747-26-5	-0.3530	-4.4374E-06	2.2187E-12	0	358700	2	4.4365E-01	3.1384E-01
195	18196	C9H20	2,2,5-trimethylhexane	3522-94-9	-0.2904	-4.4374E-06	2.2187E-12	0	358700	2	5.1243E-01	3.6250E-01
196	18197	C9H20	2,3,3-trimethylhexane	16747-28-7	-0.6385	-4.4374E-06	2.2187E-12	0	358700	2	2.2989E-01	1.6263E-01
197	18198	C9H20	2,3,4-trimethylhexane	921-47-1	-0.6733	-4.4374E-06	2.2187E-12	0	358700	2	2.1219E-01	1.5010E-01
198	18199	C9H20	2,3,5-trimethylhexane	1069-53-0	-0.4761	-4.4374E-06	2.2187E-12	0	358700	2	3.3413E-01	2.3636E-01
199	18200	C9H20	2,4,4-trimethylhexane	16747-30-1	-0.4584	-4.4374E-06	2.2187E-12	0	358700	2	3.4801E-01	2.4618E-01
200	18201	C9H20	3,3,4-trimethylhexane	16747-31-2	-0.7096	-4.4374E-06	2.2187E-12	0	358700	2	1.9517E-01	1.3807E-01
201	18202	C9H20	3,3-diethylpentane	1067-20-5	-0.8554	-4.4374E-06	2.2187E-12	0	358700	2	1.3952E-01	9.8699E-02
202	18203	C9H20	2,2-dimethyl-3-ethylpentane	16747-32-3	-0.5399	-4.4374E-06	2.2187E-12	0	358700	2	2.8848E-01	2.0407E-01
203	18204	C9H20	2,3-dimethyl-3-ethylpentane	16747-33-4	-0.8177	-4.4374E-06	2.2187E-12	0	358700	2	1.5215E-01	1.0763E-01
204	18205	C9H20	2,4-dimethyl-3-ethylpentane	1068-87-7	-0.6137	-4.4374E-06	2.2187E-12	0	358700	2	2.4342E-01	1.7219E-01
205	18206	C9H20	2,2,3,3-tetramethylpentane	7154-79-2	-0.7050	-4.4374E-06	2.2187E-12	0	358700	2	1.9725E-01	1.3954E-01
206	18207	C9H20	2,2,3,4-tetramethylpentane	1186-53-4	-0.5191	-4.4374E-06	2.2187E-12	0	358700	2	3.0260E-01	2.1406E-01
207	18208	C9H20	2,2,4,4-tetramethylpentane	1070-87-7	-0.2445	-4.4374E-06	2.2187E-12	0	358700	2	5.6956E-01	4.0291E-01
208	18209	C9H20	2,3,3,4-tetramethylpentane	16747-38-9	-0.7372	-4.4374E-06	2.2187E-12	0	358700	2	1.8316E-01	1.2957E-01
209	18210	C9H20	3,3-dimethylheptane	4032-86-4	-0.6285	-4.4374E-06	2.2187E-12	0	358700	2	2.3523E-01	1.6641E-01
210	19865	C10H14	butylbenzene	104-51-8	1.1405	-3.3451E-06	1.6726E-12	0	358700	1,2	1.3820E+01	1.0646E+01
211	19866	C10H14	isobutylbenzene	538-93-2	1.0043	-3.3451E-06	1.6726E-12	0	358700	1,2	1.0100E+01	7.7803E+00
212	19867	C10H14	sec-butylbenzene	135-98-8	1.2455	-3.3451E-06	1.6726E-12	0	358700	1,2	1.7600E+01	1.3558E+01
213	19868	C10H14	tert-butylbenzene	98-06-6	1.4698	-3.3451E-06	1.6726E-12	0	358700	1,2	2.9500E+01	2.2725E+01
214	19869	C10H14	cymene	25155-15-1	1.5508	-4.5929E-06	2.2646E-12	0	358700	2	3.5549E+01	2.4842E+01
215	19870	C10H14	m-cymene	535-77-3	1.5819	-4.5929E-06	2.2646E-12	0	358700	2	3.8184E+01	2.6684E+01
216	19871	C10H14	o-cymene	527-84-4	1.5061	-4.5929E-06	2.2646E-12	0	358700	2	3.2072E+01	2.2412E+01
217	19872	C10H14	p-cymene	99-87-6	1.5334	-4.5929E-06	2.2646E-12	0	358700	1,2	3.4150E+01	2.3865E+01
218	19873	C10H14	1-methyl-2-propylbenzene	1074-17-5	1.3452	-4.5929E-06	2.2646E-12	0	358700	2	2.2140E+01	1.5472E+01
219	19874	C10H14	1-methyl-3-propylbenzene	1074-43-7	1.4179	-4.5929E-06	2.2646E-12	0	358700	2	2.6179E+01	1.8294E+01
220	19875	C10H14	1-methyl-4-propylbenzene	1074-55-1	1.3815	-4.5929E-06	2.2646E-12	0	358700	2	2.4073E+01	1.6823E+01
221	19876	C10H14	(S)-(1-methylpropyl)benzene	5787-28-0	1.0078	-3.3451E-06	1.6726E-12	0	358700	2	1.0181E+01	7.8428E+00
222	19877	C10H14	diethylbenzene	25340-17-4	1.4155	-4.5929E-06	2.2646E-12	0	358700	2	2.6033E+01	1.8192E+01
223	19878	C10H14	o-diethylbenzene	135-01-3	1.3776	-4.5929E-06	2.2646E-12	0	358700	2	2.3859E+01	1.6673E+01
224	19879	C10H14	m-diethylbenzene	141-93-5	1.4340	-4.5929E-06	2.2646E-12	0	358700	2	2.7164E+01	1.8983E+01
225	19880	C10H14	p-diethylbenzene	105-05-5	1.3950	-4.5929E-06	2.2646E-12	0	358700	1,2	2.4830E+01	1.7352E+01
226	19881	C10H14	3-ethyl-o-xylene	933-98-2	1.1255	-4.5929E-06	2.2646E-12	0	358700	2	1.3351E+01	9.3298E+00
227	19882	C10H14	4-ethyl-o-xylene	934-80-5	1.2252	-4.5929E-06	2.2646E-12	0	358700	2	1.6797E+01	1.1738E+01
228	19883	C10H14	2-ethyl-m-xylene	2870-04-4	1.2190	-4.5929E-06	2.2646E-12	0	358700	2	1.6557E+01	1.1571E+01
229	19884	C10H14	4-ethyl-m-xylene	874-41-9	1.2574	-4.5929E-06	2.2646E-12	0	358700	2	1.8088E+01	1.2640E+01
230	19885	C10H14	5-ethyl-m-xylene	934-74-7	1.3699	-4.5929E-06	2.2646E-12	0	358700	2	2.3436E+01	1.6378E+01
231	19886	C10H14	2-ethyl-p-xylene	1758-88-9	1.2961	-4.5929E-06	2.2646E-12	0	358700	2	1.9776E+01	1.3820E+01
232	19887	C10H14	1,2,3,4-tetramethylbenzene	488-23-3	0.8670	-4.5929E-06	2.2646E-12	0	358700	2	7.3618E+00	5.1445E+00
233	19888	C10H14	1,2,3,5-tetramethylbenzene	527-53-7	1.0302	-4.5929E-06	2.2646E-12	0	358700	2	1.0720E+01	7.4910E+00
234	19889	C10H14	1,2,4,5-tetramethylbenzene	95-93-2	0.5416	-4.5929E-06	2.2646E-12	0	358700	1,2	3.4800E+00	2.4319E+00
235	20922	C10H20	pentylcyclopentane	3741-00-2	-0.9393	-7.1550E-06	3.5775E-12	0	358700	1,2	1.1500E-01	6.5812E-02
236	20923	C10H20	butylcyclohexane	1678-93-9	-0.7501	-7.1550E-06	3.5775E-12	0	358700	2	1.7779E-01	1.0174E-01
237	20924	C10H20	tert-butylcyclohexane	3178-22-1	-0.5292	-7.1550E-06	3.5775E-12	0	358700	2	2.9568E-01	1.6921E-01
238	20925	C10H20	sec-butylcyclohexane	7058-01-7	-0.7117	-7.1550E-06	3.5775E-12	0	358700	2	1.9424E-01	1.1116E-01
239	20926	C10H20	isobutylcyclohexane	1678-98-4	-0.5244	-7.1550E-06	3.5775E-12	0	358700	2	2.9894E-01	1.7107E-01
240	20927	C10H20	1-methyl-1-propylcyclohexane	4258-93-9	-0.5411	-7.1550E-06	3.5775E-12	0	358700	2	2.8770E-01	1.6464E-01

Table 5 Solubility in Salt Water - Organic Compounds

$\log_{10} S = A + B X + C X^2$			(S - solubility, parts per million by weight, ppm (wt), X - mg salt / kg of water)									
NO	ID	FORMULA	NAME	CAS No	A	B	C	XMIN	XMAX	code	S@X=0	S@X=34,472
241	20928	C10H20	1-methyl-2-propylcyclohexane	4291-79-6	-0.6352	-7.1550E-06	3.5775E-12	0	358700	2	2.3161E-01	1.3254E-01
242	20929	C10H20	1-methyl-3-propylcyclohexane	4291-80-9	-0.3605	-7.1550E-06	3.5775E-12	0	358700	2	4.3606E-01	2.4954E-01
243	20930	C10H20	1-methyl-4-propylcyclohexane	4291-81-0	-0.6236	-7.1550E-06	3.5775E-12	0	358700	2	2.3792E-01	1.3616E-01
244	20931	C10H20	1-methyl-2-isopropylcyclohexane	16580-23-7	-0.5173	-7.1550E-06	3.5775E-12	0	358700	2	3.0390E-01	1.7391E-01
245	20932	C10H20	1-methyl-3-isopropylcyclohexane	16580-24-8	-0.4091	-7.1550E-06	3.5775E-12	0	358700	2	3.8986E-01	2.2311E-01
246	20933	C10H20	1-methyl-4-isopropylcyclohexane	99-82-1	-0.5101	-7.1550E-06	3.5775E-12	0	358700	2	3.0894E-01	1.7680E-01
247	20934	C10H20	cis-1-methyl-4-isopropylcyclohexane	6069-98-3	-0.5411	-7.1550E-06	3.5775E-12	0	358700	2	2.8770E-01	1.6464E-01
248	20935	C10H20	trans-1-methyl-4-isopropylcyclohexane	1678-82-6	-0.5077	-7.1550E-06	3.5775E-12	0	358700	2	3.1065E-01	1.7778E-01
249	20936	C10H20	1,1,3,4-tetramethylcyclohexane	24612-75-7	-0.2499	-7.1550E-06	3.5775E-12	0	358700	2	5.6244E-01	3.2187E-01
250	20937	C10H20	trans-1,1,3,5-tetramethylcyclohexane	50876-31-8	-0.1604	-7.1550E-06	3.5775E-12	0	358700	2	6.9120E-01	3.9556E-01
251	20938	C10H20	cis-1,1,3,5-tetramethylcyclohexane	50876-32-9	-0.0628	-7.1550E-06	3.5775E-12	0	358700	2	8.6533E-01	4.9520E-01
252	20939	C10H20	diethylcyclohexane; (mixed isomers)	1331-43-7	-0.5884	-7.1550E-06	3.5775E-12	0	358700	2	2.5801E-01	1.4765E-01
253	21133	C10H22	decane	124-18-5	-1.2840	-4.4374E-06	2.2187E-12	0	358700	1,2	5.2000E-02	3.6785E-02
254	21134	C10H22	2-methylnonane	871-83-0	-1.3733	-4.4374E-06	2.2187E-12	0	358700	2	4.2335E-02	2.9948E-02
255	21135	C10H22	3-methylnonane	5911-04-6	-1.3926	-4.4374E-06	2.2187E-12	0	358700	2	4.0492E-02	2.8645E-02
256	21136	C10H22	4-methylnonane	17301-94-9	-1.3418	-4.4374E-06	2.2187E-12	0	358700	2	4.5522E-02	3.2203E-02
257	21137	C10H22	5-methylnonane	15869-85-9	-1.3284	-4.4374E-06	2.2187E-12	0	358700	2	4.6946E-02	3.3210E-02
258	21138	C10H22	3-ethyloctane	5881-17-4	-1.3614	-4.4374E-06	2.2187E-12	0	358700	2	4.3508E-02	3.0778E-02
259	21139	C10H22	4-ethyloctane	15869-86-0	-1.2918	-4.4374E-06	2.2187E-12	0	358700	2	5.1075E-02	3.6131E-02
260	21140	C10H22	2,2-dimethyloctane	15869-87-1	-1.1251	-4.4374E-06	2.2187E-12	0	358700	2	7.4975E-02	5.3038E-02
261	21141	C10H22	2,3-dimethyloctane	7146-60-3	-1.3082	-4.4374E-06	2.2187E-12	0	358700	2	4.9185E-02	3.4794E-02
262	21142	C10H22	2,4-dimethyloctane	4032-94-4	-1.1001	-4.4374E-06	2.2187E-12	0	358700	2	7.9407E-02	5.6173E-02
263	21143	C10H22	2,5-dimethyloctane	15869-89-3	-1.1649	-4.4374E-06	2.2187E-12	0	358700	2	6.8411E-02	4.8395E-02
264	21144	C10H22	2,6-dimethyloctane	2051-30-1	-1.2114	-4.4374E-06	2.2187E-12	0	358700	2	6.1459E-02	4.3476E-02
265	21145	C10H22	2,7-dimethyloctane	1072-16-8	-1.1991	-4.4374E-06	2.2187E-12	0	358700	2	6.3232E-02	4.4731E-02
266	21146	C10H22	3,3-dimethyloctane	4110-44-5	-1.2319	-4.4374E-06	2.2187E-12	0	358700	2	5.8629E-02	4.1474E-02
267	21147	C10H22	3,4-dimethyloctane	15869-92-8	-1.2859	-4.4374E-06	2.2187E-12	0	358700	2	5.1770E-02	3.6623E-02
268	21148	C10H22	3,5-dimethyloctane	15869-93-9	-1.1874	-4.4374E-06	2.2187E-12	0	358700	2	6.4949E-02	4.5945E-02
269	21149	C10H22	3,6-dimethyloctane	15869-94-0	-1.2220	-4.4374E-06	2.2187E-12	0	358700	2	5.9975E-02	4.2427E-02
270	21150	C10H22	4,4-dimethyloctane	15869-95-1	-1.1403	-4.4374E-06	2.2187E-12	0	358700	2	7.2398E-02	5.1215E-02
271	21151	C10H22	4,5-dimethyloctane	15869-96-2	-1.2548	-4.4374E-06	2.2187E-12	0	358700	2	5.5620E-02	3.9346E-02
272	21152	C10H22	4-propylheptane	3178-29-8	-1.1403	-4.4374E-06	2.2187E-12	0	358700	2	7.2398E-02	5.1215E-02
273	21153	C10H22	4-isopropylheptane	52896-87-4	-1.1750	-4.4374E-06	2.2187E-12	0	358700	2	6.6828E-02	4.7275E-02
274	21154	C10H22	3-ethyl-2-methylheptane	14676-29-0	-1.2319	-4.4374E-06	2.2187E-12	0	358700	2	5.8629E-02	4.1474E-02
275	21155	C10H22	4-ethyl-2-methylheptane	52896-88-5	-1.1079	-4.4374E-06	2.2187E-12	0	358700	2	7.8005E-02	5.5181E-02
276	21156	C10H22	5-ethyl-2-methylheptane	13475-78-0	-1.1949	-4.4374E-06	2.2187E-12	0	358700	2	6.3848E-02	4.5166E-02
277	21157	C10H22	3-ethyl-3-methylheptane	17302-01-1	-1.2957	-4.4374E-06	2.2187E-12	0	358700	2	5.0617E-02	3.5807E-02
278	21158	C10H22	4-ethyl-3-methylheptane	52896-89-6	-1.2565	-4.4374E-06	2.2187E-12	0	358700	2	5.5400E-02	3.9190E-02
279	21159	C10H22	3-ethyl-5-methylheptane	52896-90-9	-1.1577	-4.4374E-06	2.2187E-12	0	358700	2	6.9555E-02	4.9204E-02
280	21160	C10H22	3-ethyl-4-methylheptane	52896-91-0	-1.2761	-4.4374E-06	2.2187E-12	0	358700	2	5.2951E-02	3.7458E-02
281	21161	C10H22	4-ethyl-4-methylheptane	17302-04-4	-1.2220	-4.4374E-06	2.2187E-12	0	358700	2	5.9975E-02	4.2427E-02
282	21162	C10H22	2,2,3-trimethylheptane	52896-92-1	-1.1428	-4.4374E-06	2.2187E-12	0	358700	2	7.1985E-02	5.0923E-02
283	21163	C10H22	2,2,4-trimethylheptane	14720-74-2	-0.9091	-4.4374E-06	2.2187E-12	0	358700	2	1.2327E-01	8.7202E-02
284	21164	C10H22	2,2,5-trimethylheptane	20291-95-6	-0.9723	-4.4374E-06	2.2187E-12	0	358700	2	1.0657E-01	7.5392E-02
285	21165	C10H22	2,2,6-trimethylheptane	1190-83-6	-0.9251	-4.4374E-06	2.2187E-12	0	358700	2	1.1882E-01	8.4057E-02
286	21166	C10H22	2,3,3-trimethylheptane	52896-93-2	-1.2072	-4.4374E-06	2.2187E-12	0	358700	2	6.2056E-02	4.3899E-02
287	21167	C10H22	2,3,4-trimethylheptane	52896-95-4	-1.1998	-4.4374E-06	2.2187E-12	0	358700	2	6.3125E-02	4.4655E-02
288	21168	C10H22	2,3,5-trimethylheptane	20278-85-7	-1.2196	-4.4374E-06	2.2187E-12	0	358700	2	6.0317E-02	4.2669E-02
289	21169	C10H22	2,3,6-trimethylheptane	4032-93-3	-1.1029	-4.4374E-06	2.2187E-12	0	358700	2	7.8907E-02	5.5819E-02
290	21170	C10H22	2,4,4-trimethylheptane	4032-92-2	-0.9774	-4.4374E-06	2.2187E-12	0	358700	2	1.0534E-01	7.4521E-02
291	21171	C10H22	2,4,5-trimethylheptane	20278-84-6	-1.1154	-4.4374E-06	2.2187E-12	0	358700	2	7.6672E-02	5.4239E-02
292	21172	C10H22	2,4,6-trimethylheptane	2613-61-8	-0.8914	-4.4374E-06	2.2187E-12	0	358700	2	1.2841E-01	9.0838E-02
293	21173	C10H22	2,5,5-trimethylheptane	1189-99-7	-1.0227	-4.4374E-06	2.2187E-12	0	358700	2	9.4904E-02	6.7136E-02
294	21174	C10H22	3,3,4-trimethylheptane	20278-87-9	-1.2491	-4.4374E-06	2.2187E-12	0	358700	2	5.6348E-02	3.9861E-02
295	21175	C10H22	3,3,5-trimethylheptane	7154-80-5	-1.0951	-4.4374E-06	2.2187E-12	0	358700	2	8.0326E-02	5.6823E-02
296	21176	C10H22	3,4,4-trimethylheptane	20278-88-0	-1.2294	-4.4374E-06	2.2187E-12	0	358700	2	5.8962E-02	4.1710E-02
297	21177	C10H22	3,4,5-trimethylheptane	20278-89-1	-1.2639	-4.4374E-06	2.2187E-12	0	358700	2	5.4468E-02	3.8531E-02
298	21178	C10H22	3-isopropyl-2-methylhexane	62016-13-1	-1.3663	-4.4374E-06	2.2187E-12	0	358700	2	4.3025E-02	3.0436E-02
299	21179	C10H22	3,3-diethylhexane	17302-02-2	-1.3566	-4.4374E-06	2.2187E-12	0	358700	2	4.3996E-02	3.1123E-02
300	21180	C10H22	3,4-diethylhexane	19398-77-7	-1.2982	-4.4374E-06	2.2187E-12	0	358700	2	5.0333E-02	3.5606E-02
301	21181	C10H22	3-ethyl-2,2-dimethylhexane	20291-91-2	-1.1054	-4.4374E-06	2.2187E-12	0	358700	2	7.8454E-02	5.5499E-02
302	21182	C10H22	4-ethyl-2,2-dimethylhexane	52896-99-8	-0.8762	-4.4374E-06	2.2187E-12	0	358700	2	1.3299E-01	9.4078E-02
303	21183	C10H22	3-ethyl-2,3-dimethylhexane	52897-00-4	-1.2933	-4.4374E-06	2.2187E-12	0	358700	2	5.0902E-02	3.6009E-02
304	21184	C10H22	4-ethyl-2,3-dimethylhexane	52897-01-5	-1.2245	-4.4374E-06	2.2187E-12	0	358700	2	5.9636E-02	4.2187E-02
305	21185	C10H22	3-ethyl-2,4-dimethylhexane	7220-26-0	-1.2047	-4.4374E-06	2.2187E-12	0	358700	2	6.2410E-02	4.4149E-02
306	21186	C10H22	4-ethyl-2,4-dimethylhexane	52897-03-7	-1.2294	-4.4374E-06	2.2187E-12	0	358700	2	5.8962E-02	4.1710E-02
307	21187	C10H22	3-ethyl-2,5-dimethylhexane	52897-04-8	-1.0553	-4.4374E-06	2.2187E-12	0	358700	2	8.8034E-02	6.2276E-02
308	21188	C10H22	4-ethyl-3,3-dimethylhexane	52897-05-9	-1.2737	-4.4374E-06	2.2187E-12	0	358700	2	5.3251E-02	3.7670E-02
309	21189	C10H22	3-ethyl-3,4-dimethylhexane	52897-06-0	-1.2540	-4.4374E-06	2.2187E-12	0	358700	2	5.5714E-02	3.9413E-02
310	21190	C10H22	2,2,3,3-tetramethylhexane	13475-81-5	-1.2102	-4.4374E-06	2.2187E-12	0	358700	2	6.1634E-02	4.3600E-02
311	21191	C10H22	2,2,3,4-tetramethylhexane	52897-08-2	-1.1726	-4.4374E-06	2.2187E-12	0	358700	2	6.7211E-02	4.7545E-02
312	21192	C10H22	2,2,3,5-tetramethylhexane	52897-09-3	-0.9117	-4.4374E-06	2.2187E-12	0	358700	2	1.2255E-01	8.6695E-02
313	21193	C10H22	2,2,4,4-tetramethylhexane	51750-65-3	-1.0478	-4.4374E-06	2.2187E-12	0	358700	2	8.9572E-02	6.3364E-02
314	21194	C10H22	2,2,4,5-tetramethylhexane	16747-42-5	-0.8985	-4.4374E-06	2.2187E-12	0	358700	2	1.2633E-01	8.9366E-02
315	21195	C10H22	2,2,5,5-tetramethylhexane	1071-81-4	-0.6326	-4.4374E-06	2.2187E-12	0	358700	2	2.3303E-01	1.6484E-01
316	21196	C10H22	2,3,3,4-tetramethylhexane	52897-10-6	-1.3150	-4.4374E-06	2.2187E-12	0	358700	2	4.8418E-02	3.4251E-02
317	21197	C10H22	2,3,3,5-tetramethylhexane	52897-11-7	-1.0303	-4.4374E-06	2.2187E-12	0	358700	2	9.3271E-02	6.5981E-02
318	21198	C10H22	2,3,4,4-tetramethylhexane	52897-12-8	-1.2417	-4.4374E-06	2.2187E-12	0	358700	2	5.7314E-02	4.0544E-02
319	21199	C10H22	2,3,4,5-tetramethylhexane	52897-15-1	-1.1079	-4.4374E-06	2.2187E-12	0	358700	2	7.8005E-02	5.5181E-02
320	21200	C10H22	3,3,4,4-tetramethylhexane	5171-84-6	-1.4457	-4.4374E-06	2.2187E-12	0	358700	2	3.5832E-02	2.5348E-02

Table 5 Solubility in Salt Water - Organic Compounds

$\log_{10} S = A + B X + C X^2$ (S - solubility, parts per million by weight, ppm (wt), X - mg salt / kg of water)												
NO	ID	FORMULA	NAME	CAS No	A	B	C	XMIN	XMAX	code	S@X=0	S@X=34,472
321	21201	C10H22	2,4-dimethyl-3-isopropylpentane	13475-79-1	-1.1288	-4.4374E-06	2.2187E-12	0	358700	2	7.4333E-02	5.2584E-02
322	21202	C10H22	3,3-diethyl-2-methylpentane	52897-16-2	-1.4385	-4.4374E-06	2.2187E-12	0	358700	2	3.6429E-02	2.5771E-02
323	21203	C10H22	3-ethyl-2,2,3-trimethylpentane	52897-17-3	-1.4338	-4.4374E-06	2.2187E-12	0	358700	2	3.6834E-02	2.6057E-02
324	21204	C10H22	3-ethyl-2,2,4-trimethylpentane	52897-18-4	-1.0854	-4.4374E-06	2.2187E-12	0	358700	2	8.2149E-02	5.8113E-02
325	21205	C10H22	3-ethyl-2,3,4-trimethylpentane	52897-19-5	-1.4323	-4.4374E-06	2.2187E-12	0	358700	2	3.6956E-02	2.6143E-02
326	21206	C10H22	2,2,3,3,4-pentamethylpentane	16747-44-7	-1.3505	-4.4374E-06	2.2187E-12	0	358700	2	4.4615E-02	3.1561E-02
327	21207	C10H22	2,2,3,4,4-pentamethylpentane	16747-45-8	-1.1847	-4.4374E-06	2.2187E-12	0	358700	2	6.5357E-02	4.6234E-02
328	22304	C11H16	pentylbenzene	538-68-1	0.5843	-3.3451E-06	1.6726E-12	0	358700	1,2	3.8400E+00	2.9581E+00
329	22305	C11H16	2-phenylpentane	2719-52-0	0.8619	-3.3451E-06	1.6726E-12	0	358700	2	7.2769E+00	5.6056E+00
330	22306	C11H16	3-phenylpentane	1196-58-3	0.8380	-3.3451E-06	1.6726E-12	0	358700	2	6.8868E+00	5.3051E+00
331	22307	C11H16	1-phenyl-2-methylbutane	3968-85-2	1.0537	-4.5929E-06	2.2646E-12	0	358700	2	1.1315E+01	7.9074E+00
332	22308	C11H16	1-phenyl-3-methylbutane	2049-94-7	1.0091	-4.5929E-06	2.2646E-12	0	358700	2	1.0212E+01	7.1360E+00
333	22309	C11H16	2-phenyl-2-methylbutane	2049-95-8	1.1631	-4.5929E-06	2.2646E-12	0	358700	2	1.4557E+01	1.0173E+01
334	22310	C11H16	2-phenyl-3-methylbutane	4481-30-5	1.2680	-4.5929E-06	2.2646E-12	0	358700	2	1.8534E+01	1.2952E+01
335	22311	C11H16	1-phenyl-2,2-dimethylpropane	1007-26-7	1.3162	-4.5929E-06	2.2646E-12	0	358700	2	2.0710E+01	1.4472E+01
336	22312	C11H16	1-methyl-2-butylbenzene	1595-11-5	0.7997	-4.5929E-06	2.2646E-12	0	358700	2	6.3046E+00	4.4057E+00
337	22313	C11H16	1-methyl-3-butylbenzene	1595-04-6	0.8679	-4.5929E-06	2.2646E-12	0	358700	2	7.3773E+00	5.1554E+00
338	22314	C11H16	1-methyl-4-butylbenzene	1595-05-7	0.8223	-4.5929E-06	2.2646E-12	0	358700	2	6.6421E+00	4.6416E+00
339	22315	C11H16	1-methyl-2-sec-butylbenzene	1595-06-8	1.0772	-4.5929E-06	2.2646E-12	0	358700	2	1.1946E+01	8.3483E+00
340	22316	C11H16	1-methyl-3-sec-butylbenzene	1772-10-7	1.1246	-4.5929E-06	2.2646E-12	0	358700	2	1.3322E+01	9.3094E+00
341	22317	C11H16	1-methyl-4-sec-butylbenzene	1595-16-0	1.0537	-4.5929E-06	2.2646E-12	0	358700	2	1.1315E+01	7.9074E+00
342	22318	C11H16	1-methyl-2-isobutylbenzene	36301-29-8	1.0772	-4.5929E-06	2.2646E-12	0	358700	2	1.1946E+01	8.3483E+00
343	22319	C11H16	1-methyl-3-isobutylbenzene	5160-99-6	1.1246	-4.5929E-06	2.2646E-12	0	358700	2	1.3322E+01	9.3094E+00
344	22320	C11H16	1-methyl-4-isobutylbenzene	5161-04-6	1.0772	-4.5929E-06	2.2646E-12	0	358700	2	1.1946E+01	8.3483E+00
345	22321	C11H16	1-methyl-2-tert-butyl benzene	1074-92-6	0.9729	-4.5929E-06	2.2646E-12	0	358700	2	9.3957E+00	6.5659E+00
346	22322	C11H16	1-methyl-3-tert-butylbenzene	1075-38-3	1.2377	-4.5929E-06	2.2646E-12	0	358700	2	1.7286E+01	1.2080E+01
347	22323	C11H16	1-methyl-4-tert-butylbenzene	98-51-1	1.1540	-4.5929E-06	2.2646E-12	0	358700	2	1.4257E+01	9.9631E+00
348	22324	C11H16	1-ethyl-2-propylbenzene	16021-20-8	0.9138	-4.5929E-06	2.2646E-12	0	358700	2	8.2007E+00	5.7308E+00
349	22325	C11H16	1-ethyl-3-propylbenzene	20024-91-3	0.9601	-4.5929E-06	2.2646E-12	0	358700	2	9.1231E+00	6.3754E+00
350	22326	C11H16	1-ethyl-4-propylbenzene	20024-90-2	0.8679	-4.5929E-06	2.2646E-12	0	358700	2	7.3773E+00	5.1554E+00
351	22327	C11H16	1-ethyl-2-isopropylbenzene	18970-44-0	1.1483	-4.5929E-06	2.2646E-12	0	358700	2	1.4071E+01	9.8329E+00
352	22328	C11H16	1-ethyl-3-isopropylbenzene	4920-99-4	1.1721	-4.5929E-06	2.2646E-12	0	358700	2	1.4864E+01	1.0387E+01
353	22329	C11H16	1-ethyl-4-isopropylbenzene	4218-48-8	1.0631	-4.5929E-06	2.2646E-12	0	358700	2	1.1563E+01	8.0807E+00
354	22330	C11H16	1,2-dimethyl-3-propylbenzene	17059-44-8	0.7390	-4.5929E-06	2.2646E-12	0	358700	2	5.4826E+00	3.8313E+00
355	22331	C11H16	1,2-dimethyl-4-propylbenzene	3982-66-9	0.7794	-4.5929E-06	2.2646E-12	0	358700	2	6.0166E+00	4.2045E+00
356	22332	C11H16	1,3-dimethyl-2-propylbenzene	17059-45-9	0.8087	-4.5929E-06	2.2646E-12	0	358700	2	6.4373E+00	4.4985E+00
357	22333	C11H16	1,3-dimethyl-4-propylbenzene	61827-85-8	0.8314	-4.5929E-06	2.2646E-12	0	358700	2	6.7826E+00	4.7398E+00
358	22334	C11H16	1,3-dimethyl-5-propylbenzene	3982-64-7	0.9314	-4.5929E-06	2.2646E-12	0	358700	2	8.5389E+00	5.9671E+00
359	22335	C11H16	1,4-dimethyl-2-propylbenzene	3042-50-0	0.8839	-4.5929E-06	2.2646E-12	0	358700	2	7.6549E+00	5.3494E+00
360	22336	C11H16	1,2-dimethyl-3-isopropylbenzene	22539-65-7	0.9231	-4.5929E-06	2.2646E-12	0	358700	2	8.3768E+00	5.8539E+00
361	22337	C11H16	1,2-dimethyl-4-isopropylbenzene	4132-77-8	0.9416	-4.5929E-06	2.2646E-12	0	358700	2	8.7415E+00	6.1087E+00
362	22338	C11H16	1,3-dimethyl-2-isopropylbenzene	14411-75-7	1.0068	-4.5929E-06	2.2646E-12	0	358700	2	1.0157E+01	7.0977E+00
363	22339	C11H16	1,3-dimethyl-4-isopropylbenzene	4706-89-2	1.0044	-4.5929E-06	2.2646E-12	0	358700	2	1.0102E+01	7.0596E+00
364	22340	C11H16	1,3-dimethyl-5-isopropylbenzene	4706-90-5	1.1127	-4.5929E-06	2.2646E-12	0	358700	2	1.2963E+01	9.0587E+00
365	22341	C11H16	1,4-dimethyl-2-isopropylbenzene	4132-72-3	1.0725	-4.5929E-06	2.2646E-12	0	358700	2	1.1817E+01	8.2580E+00
366	22342	C11H16	1-methyl-2,3-diethylbenzene	13632-93-4	0.8314	-4.5929E-06	2.2646E-12	0	358700	2	6.7826E+00	4.7398E+00
367	22343	C11H16	1-methyl-2,4-diethylbenzene	1758-85-6	0.8679	-4.5929E-06	2.2646E-12	0	358700	2	7.3773E+00	5.1554E+00
368	22344	C11H16	1-methyl-2,5-diethylbenzene	13632-94-5	0.8200	-4.5929E-06	2.2646E-12	0	358700	2	6.6075E+00	4.6175E+00
369	22345	C11H16	1-methyl-2,6-diethyl benzene	13632-95-6	0.7816	-4.5929E-06	2.2646E-12	0	358700	2	6.0479E+00	4.2264E+00
370	22346	C11H16	1-methyl-3,4-diethylbenzene	13732-80-4	0.9000	-4.5929E-06	2.2646E-12	0	358700	2	7.9438E+00	5.5512E+00
371	22347	C11H16	1-methyl-3,5-diethylbenzene	2050-24-0	0.9671	-4.5929E-06	2.2646E-12	0	358700	2	9.2707E+00	6.4785E+00
372	22348	C11H16	1,2,3-trimethyl-4-ethylbenzene	61827-86-9	0.5277	-4.5929E-06	2.2646E-12	0	358700	2	3.3704E+00	2.3553E+00
373	22349	C11H16	1,2,3-trimethyl-5-ethylbenzene	31366-00-4	0.6265	-4.5929E-06	2.2646E-12	0	358700	2	4.2317E+00	2.9572E+00
374	22350	C11H16	1,2,4-trimethyl-3-ethylbenzene	61827-87-0	0.6091	-4.5929E-06	2.2646E-12	0	358700	2	4.0658E+00	2.8412E+00
375	22351	C11H16	1,2,4-trimethyl-5-ethylbenzene	---	0.6879	-4.5929E-06	2.2646E-12	0	358700	2	4.8743E+00	3.4062E+00
376	22352	C11H16	1,2,4-trimethyl-6-ethylbenzene	17851-27-3	0.6879	-4.5929E-06	2.2646E-12	0	358700	2	4.8743E+00	3.4062E+00
377	22353	C11H16	1,3,5-trimethyl-2-ethylbenzene	3982-67-0	0.7012	-4.5929E-06	2.2646E-12	0	358700	2	5.0255E+00	3.5119E+00
378	22355	C11H16	1-ethyl-2,3,5-trimethylbenzene	18262-85-6	0.6879	-4.5929E-06	2.2646E-12	0	358700	2	4.8743E+00	3.4062E+00
379	22356	C11H16	sec-pentylbenzene	29316-05-0	0.6254	-3.3451E-06	1.6726E-12	0	358700	2	4.2209E+00	3.2515E+00
380	22791	C11H22	hexylcyclopentane	4457-00-5	-1.4208	-7.1550E-06	3.5775E-12	0	358700	2	3.7952E-02	2.1719E-02
381	22792	C11H22	pentylcyclohexane	4292-92-6	-1.2327	-7.1550E-06	3.5775E-12	0	358700	2	5.8525E-02	3.3492E-02
382	22793	C11H22	isopentylcyclohexane	54105-76-9	-1.0878	-7.1550E-06	3.5775E-12	0	358700	2	8.1688E-02	4.6748E-02
383	22890	C11H24	undecane	1120-21-4	-2.3565	-4.4374E-06	2.2187E-12	0	358700	1,2	4.4000E-03	3.1126E-03
384	22891	C11H24	2-methyldecane	6975-98-0	-1.8850	-4.4374E-06	2.2187E-12	0	358700	2	1.3032E-02	9.2186E-03
385	22892	C11H24	3-methyldecane	13151-34-3	-1.9026	-4.4374E-06	2.2187E-12	0	358700	2	1.2515E-02	8.8535E-03
386	22893	C11H24	4-methyldecane	2847-72-5	-1.8569	-4.4374E-06	2.2187E-12	0	358700	2	1.3904E-02	9.8357E-03
387	22894	C11H24	5-methyldecane	13151-35-4	-1.8150	-4.4374E-06	2.2187E-12	0	358700	2	1.5310E-02	1.0831E-02
388	22895	C11H24	3-ethylnonane	17302-11-3	-1.8591	-4.4374E-06	2.2187E-12	0	358700	2	1.3834E-02	9.7862E-03
389	22896	C11H24	4-ethylnonane	5911-05-7	-1.7704	-4.4374E-06	2.2187E-12	0	358700	2	1.6965E-02	1.2001E-02
390	22897	C11H24	5-ethylnonane	17302-12-4	-1.6798	-4.4374E-06	2.2187E-12	0	358700	2	2.0901E-02	1.4786E-02
391	22898	C11H24	2,2-dimethylnonane	17302-14-6	-1.6798	-4.4374E-06	2.2187E-12	0	358700	2	2.0901E-02	1.4786E-02
392	22899	C11H24	2,3-dimethylnonane	2884-06-2	-1.8150	-4.4374E-06	2.2187E-12	0	358700	2	1.5310E-02	1.0831E-02
393	22900	C11H24	2,4-dimethylnonane	17302-24-8	-1.6338	-4.4374E-06	2.2187E-12	0	358700	2	2.3238E-02	1.6439E-02
394	22901	C11H24	2,5-dimethylnonane	17302-27-1	-1.6569	-4.4374E-06	2.2187E-12	0	358700	2	2.2036E-02	1.5588E-02
395	22902	C11H24	2,6-dimethylnonane	17302-28-2	-1.6569	-4.4374E-06	2.2187E-12	0	358700	2	2.2036E-02	1.5588E-02
396	22903	C11H24	2,7-dimethylnonane	17302-29-3	-1.6798	-4.4374E-06	2.2187E-12	0	358700	2	2.0901E-02	1.4786E-02
397	22904	C11H24	2,8-dimethylnonane	17302-30-6	-1.7480	-4.4374E-06	2.2187E-12	0	358700	2	1.7866E-02	1.2639E-02
398	22905	C11H24	3,3-dimethylnonane	17302-15-7	-1.7254	-4.4374E-06	2.2187E-12	0	358700	2	1.8820E-02	1.3314E-02
399	22906	C11H24	3,4-dimethylnonane	17302-22-6	-1.7928	-4.4374E-06	2.2187E-12	0	358700	2	1.6114E-02	1.1399E-02
400	22907	C11H24	3,5-dimethylnonane	17302-25-9	-1.6798	-4.4374E-06	2.2187E-12	0	358700	2	2.0901E-02	1.4786E-02

Table 5 Solubility in Salt Water - Organic Compounds

$\log_{10} S = A + B X + C X^2$ (S - solubility, parts per million by weight, ppm (wt), X - mg salt / kg of water)												
NO	ID	FORMULA	NAME	CAS No	A	B	C	XMIN	XMAX	code	S@X=0	S@X=34,472
401	22908	C11H24	3,6-dimethylnonane	17302-31-7	-1.7254	-4.4374E-06	2.2187E-12	0	358700	2	1.8820E-02	1.3314E-02
402	22909	C11H24	3,7-dimethylnonane	17302-32-8	-1.7704	-4.4374E-06	2.2187E-12	0	358700	2	1.6965E-02	1.2001E-02
403	22910	C11H24	4,4-dimethylnonane	17302-18-0	-1.6338	-4.4374E-06	2.2187E-12	0	358700	2	2.3238E-02	1.6439E-02
404	22911	C11H24	4,5-dimethylnonane	17302-23-7	-1.7254	-4.4374E-06	2.2187E-12	0	358700	2	1.8820E-02	1.3314E-02
405	22912	C11H24	4,6-dimethylnonane	17302-26-0	-1.6569	-4.4374E-06	2.2187E-12	0	358700	2	2.2036E-02	1.5588E-02
406	22913	C11H24	5,5-dimethylnonane	6414-96-6	-1.6106	-4.4374E-06	2.2187E-12	0	358700	2	2.4512E-02	1.7340E-02
407	22914	C11H24	4-propyloctane	17302-13-5	-1.6798	-4.4374E-06	2.2187E-12	0	358700	2	2.0901E-02	1.4786E-02
408	22915	C11H24	4-isopropyloctane	62016-15-3	-1.6338	-4.4374E-06	2.2187E-12	0	358700	2	2.3238E-02	1.6439E-02
409	22916	C11H24	2-methyl-3-ethyloctane	62016-16-4	-1.7254	-4.4374E-06	2.2187E-12	0	358700	2	1.8820E-02	1.3314E-02
410	22917	C11H24	2-methyl-4-ethyloctane	62016-17-5	-1.5873	-4.4374E-06	2.2187E-12	0	358700	2	2.5862E-02	1.8295E-02
411	22918	C11H24	2-methyl-5-ethyloctane	62016-18-6	-1.6338	-4.4374E-06	2.2187E-12	0	358700	2	2.3238E-02	1.6439E-02
412	22919	C11H24	2-methyl-6-ethyloctane	62016-19-7	-1.7254	-4.4374E-06	2.2187E-12	0	358700	2	1.8820E-02	1.3314E-02
413	22920	C11H24	3-methyl-3-ethyloctane	17302-16-8	-1.8084	-4.4374E-06	2.2187E-12	0	358700	2	1.5547E-02	1.0998E-02
414	22921	C11H24	3-methyl-4-ethyloctane	62016-20-0	-1.7254	-4.4374E-06	2.2187E-12	0	358700	2	1.8820E-02	1.3314E-02
415	22922	C11H24	3-methyl-5-ethyloctane	62016-21-1	-1.5873	-4.4374E-06	2.2187E-12	0	358700	2	2.5862E-02	1.8295E-02
416	22923	C11H24	3-methyl-6-ethyloctane	62016-22-2	-1.7254	-4.4374E-06	2.2187E-12	0	358700	2	1.8820E-02	1.3314E-02
417	22924	C11H24	4-methyl-3-ethyloctane	62016-23-3	-1.7480	-4.4374E-06	2.2187E-12	0	358700	2	1.7866E-02	1.2639E-02
418	22925	C11H24	4-methyl-4-ethyloctane	17302-19-1	-1.6798	-4.4374E-06	2.2187E-12	0	358700	2	2.0901E-02	1.4786E-02
419	22926	C11H24	4-methyl-5-ethyloctane	62016-24-4	-1.6798	-4.4374E-06	2.2187E-12	0	358700	2	2.0901E-02	1.4786E-02
420	22927	C11H24	4-methyl-6-ethyloctane	62016-25-5	-1.6798	-4.4374E-06	2.2187E-12	0	358700	2	2.0901E-02	1.4786E-02
421	22928	C11H24	2,2,3-trimethyloctane	62016-26-6	-1.6569	-4.4374E-06	2.2187E-12	0	358700	2	2.2036E-02	1.5588E-02
422	22929	C11H24	2,2,4-trimethyloctane	18932-14-4	-1.4813	-4.4374E-06	2.2187E-12	0	358700	2	3.3015E-02	2.3355E-02
423	22930	C11H24	2,2,5-trimethyloctane	62016-27-7	-1.4694	-4.4374E-06	2.2187E-12	0	358700	2	3.3933E-02	2.4005E-02
424	22931	C11H24	2,2,6-trimethyloctane	62016-28-8	-1.5405	-4.4374E-06	2.2187E-12	0	358700	2	2.8810E-02	2.0380E-02
425	22932	C11H24	2,2,7-trimethyloctane	62016-29-9	-1.4455	-4.4374E-06	2.2187E-12	0	358700	2	3.5852E-02	2.5362E-02
426	22933	C11H24	2,3,3-trimethyloctane	62016-30-2	-1.7254	-4.4374E-06	2.2187E-12	0	358700	2	1.8820E-02	1.3314E-02
427	22934	C11H24	2,3,4-trimethyloctane	62016-31-3	-1.6798	-4.4374E-06	2.2187E-12	0	358700	2	2.0901E-02	1.4786E-02
428	22935	C11H24	2,3,5-trimethyloctane	62016-32-4	-1.6106	-4.4374E-06	2.2187E-12	0	358700	2	2.4512E-02	1.7340E-02
429	22936	C11H24	2,3,6-trimethyloctane	62016-33-5	-1.6798	-4.4374E-06	2.2187E-12	0	358700	2	2.0901E-02	1.4786E-02
430	22937	C11H24	2,3,7-trimethyloctane	62016-34-6	-1.6569	-4.4374E-06	2.2187E-12	0	358700	2	2.2036E-02	1.5588E-02
431	22938	C11H24	2,4,4-trimethyloctane	62016-35-7	-1.4694	-4.4374E-06	2.2187E-12	0	358700	2	3.3933E-02	2.4005E-02
432	22939	C11H24	2,4,5-trimethyloctane	62016-36-8	-1.5873	-4.4374E-06	2.2187E-12	0	358700	2	2.5862E-02	1.8295E-02
433	22940	C11H24	2,4,6-trimethyloctane	62016-37-9	-1.5405	-4.4374E-06	2.2187E-12	0	358700	2	2.8810E-02	2.0380E-02
434	22941	C11H24	2,4,7-trimethyloctane	62016-38-0	-1.4694	-4.4374E-06	2.2187E-12	0	358700	2	3.3933E-02	2.4005E-02
435	22942	C11H24	2,5,5-trimethyloctane	62016-39-1	-1.4932	-4.4374E-06	2.2187E-12	0	358700	2	3.2124E-02	2.2725E-02
436	22943	C11H24	2,5,6-trimethyloctane	62016-14-2	-1.6338	-4.4374E-06	2.2187E-12	0	358700	2	2.3238E-02	1.6439E-02
437	22944	C11H24	2,6,6-trimethyloctane	54166-32-4	-1.5873	-4.4374E-06	2.2187E-12	0	358700	2	2.5862E-02	1.8295E-02
438	22945	C11H24	3,3,4-trimethyloctane	62016-40-4	-1.7254	-4.4374E-06	2.2187E-12	0	358700	2	1.8820E-02	1.3314E-02
439	22946	C11H24	3,3,5-trimethyloctane	62016-41-5	-1.5405	-4.4374E-06	2.2187E-12	0	358700	2	2.8810E-02	2.0380E-02
440	22947	C11H24	3,3,6-trimethyloctane	62016-42-6	-1.6106	-4.4374E-06	2.2187E-12	0	358700	2	2.4512E-02	1.7340E-02
441	22948	C11H24	3,4,4-trimethyloctane	62016-43-7	-1.7027	-4.4374E-06	2.2187E-12	0	358700	2	1.9831E-02	1.4029E-02
442	22949	C11H24	3,4,5-trimethyloctane	62016-44-8	-1.7027	-4.4374E-06	2.2187E-12	0	358700	2	1.9831E-02	1.4029E-02
443	22950	C11H24	3,4,6-trimethyloctane	62016-45-9	-1.6569	-4.4374E-06	2.2187E-12	0	358700	2	2.2036E-02	1.5588E-02
444	22951	C11H24	3,5,5-trimethyloctane	61868-94-8	-1.4455	-4.4374E-06	2.2187E-12	0	358700	2	3.5852E-02	2.5362E-02
445	22952	C11H24	4,4,5-trimethyloctane	61868-95-9	-1.6798	-4.4374E-06	2.2187E-12	0	358700	2	2.0901E-02	1.4786E-02
446	22953	C11H24	4-tert-butylheptane	60302-21-8	-1.5405	-4.4374E-06	2.2187E-12	0	358700	2	2.8810E-02	2.0380E-02
447	22954	C11H24	2-methyl-4-propylheptane	61868-96-0	-1.5405	-4.4374E-06	2.2187E-12	0	358700	2	2.8810E-02	2.0380E-02
448	22955	C11H24	3-methyl-4-propylheptane	61868-97-1	-1.6569	-4.4374E-06	2.2187E-12	0	358700	2	2.2036E-02	1.5588E-02
449	22956	C11H24	4-methyl-4-propylheptane	17302-20-4	-1.6338	-4.4374E-06	2.2187E-12	0	358700	2	2.3238E-02	1.6439E-02
450	22957	C11H24	2-methyl-3-isopropylheptane	6876-18-2	-1.6407	-4.4374E-06	2.2187E-12	0	358700	2	2.2870E-02	1.6178E-02
451	22958	C11H24	2-methyl-4-isopropylheptane	61868-98-2	-1.4932	-4.4374E-06	2.2187E-12	0	358700	2	3.2124E-02	2.2725E-02
452	22959	C11H24	3-methyl-4-isopropylheptane	61868-99-3	-1.6106	-4.4374E-06	2.2187E-12	0	358700	2	2.4512E-02	1.7340E-02
453	22960	C11H24	4-methyl-4-isopropylheptane	61869-00-9	-1.6798	-4.4374E-06	2.2187E-12	0	358700	2	2.0901E-02	1.4786E-02
454	22961	C11H24	3,3-diethylheptane	17302-17-9	-1.8349	-4.4374E-06	2.2187E-12	0	358700	2	1.4625E-02	1.0346E-02
455	22962	C11H24	3,4-diethylheptane	61869-01-0	-1.6798	-4.4374E-06	2.2187E-12	0	358700	2	2.0901E-02	1.4786E-02
456	22963	C11H24	3,5-diethylheptane	61869-02-1	-1.6569	-4.4374E-06	2.2187E-12	0	358700	2	2.2036E-02	1.5588E-02
457	22964	C11H24	4,4-diethylheptane	17302-21-5	-1.6867	-4.4374E-06	2.2187E-12	0	358700	2	2.0574E-02	1.4554E-02
458	22965	C11H24	2,2-dimethyl-3-ethylheptane	61869-03-2	-1.5873	-4.4374E-06	2.2187E-12	0	358700	2	2.5862E-02	1.8295E-02
459	22966	C11H24	2,2-dimethyl-4-ethylheptane	62016-46-0	-1.3974	-4.4374E-06	2.2187E-12	0	358700	2	4.0045E-02	2.8328E-02
460	22967	C11H24	2,2-dimethyl-5-ethylheptane	62016-47-1	-1.4932	-4.4374E-06	2.2187E-12	0	358700	2	3.2124E-02	2.2725E-02
461	22968	C11H24	2,3-dimethyl-3-ethylheptane	61868-21-1	-1.6798	-4.4374E-06	2.2187E-12	0	358700	2	2.0901E-02	1.4786E-02
462	22969	C11H24	2,3-dimethyl-4-ethylheptane	61868-22-2	-1.6569	-4.4374E-06	2.2187E-12	0	358700	2	2.2036E-02	1.5588E-02
463	22970	C11H24	2,3-dimethyl-5-ethylheptane	61868-23-3	-1.6338	-4.4374E-06	2.2187E-12	0	358700	2	2.3238E-02	1.6439E-02
464	22971	C11H24	2,4-dimethyl-3-ethylheptane	61868-24-4	-1.6569	-4.4374E-06	2.2187E-12	0	358700	2	2.2036E-02	1.5588E-02
465	22972	C11H24	2,4-dimethyl-4-ethylheptane	61868-25-5	-1.4455	-4.4374E-06	2.2187E-12	0	358700	2	3.5852E-02	2.5362E-02
466	22973	C11H24	2,4-dimethyl-5-ethylheptane	61868-26-6	-1.5873	-4.4374E-06	2.2187E-12	0	358700	2	2.5862E-02	1.8295E-02
467	22974	C11H24	2,5-dimethyl-3-ethylheptane	61868-27-7	-1.5873	-4.4374E-06	2.2187E-12	0	358700	2	2.5862E-02	1.8295E-02
468	22975	C11H24	2,5-dimethyl-4-ethylheptane	61868-28-8	-1.5640	-4.4374E-06	2.2187E-12	0	358700	2	2.7293E-02	1.9307E-02
469	22976	C11H24	2,5-dimethyl-5-ethylheptane	61868-29-9	-1.6338	-4.4374E-06	2.2187E-12	0	358700	2	2.3238E-02	1.6439E-02
470	22977	C11H24	2,6-dimethyl-3-ethylheptane	61868-30-2	-1.5873	-4.4374E-06	2.2187E-12	0	358700	2	2.5862E-02	1.8295E-02
471	22978	C11H24	2,6-dimethyl-4-ethylheptane	61868-31-3	-1.4455	-4.4374E-06	2.2187E-12	0	358700	2	3.5852E-02	2.5362E-02
472	22979	C11H24	3,3-dimethyl-4-ethylheptane	61868-32-4	-1.6798	-4.4374E-06	2.2187E-12	0	358700	2	2.0901E-02	1.4786E-02
473	22980	C11H24	3,3-dimethyl-5-ethylheptane	61868-33-5	-1.5640	-4.4374E-06	2.2187E-12	0	358700	2	2.7293E-02	1.9307E-02
474	22981	C11H24	3,4-dimethyl-3-ethylheptane	61868-34-6	-1.7928	-4.4374E-06	2.2187E-12	0	358700	2	1.6114E-02	1.1399E-02
475	22982	C11H24	3,4-dimethyl-4-ethylheptane	61868-35-7	-1.7027	-4.4374E-06	2.2187E-12	0	358700	2	1.9831E-02	1.4029E-02
476	22983	C11H24	3,4-dimethyl-5-ethylheptane	61868-36-8	-1.7254	-4.4374E-06	2.2187E-12	0	358700	2	1.8820E-02	1.3314E-02
477	22984	C11H24	3,5-dimethyl-3-ethylheptane	61868-37-9	-1.6798	-4.4374E-06	2.2187E-12	0	358700	2	2.0901E-02	1.4786E-02
478	22985	C11H24	3,5-dimethyl-4-ethylheptane	61868-38-0	-1.7027	-4.4374E-06	2.2187E-12	0	358700	2	1.9831E-02	1.4029E-02
479	22986	C11H24	4,4-dimethyl-3-ethylheptane	61868-39-1	-1.5873	-4.4374E-06	2.2187E-12	0	358700	2	2.5862E-02	1.8295E-02
480	22987	C11H24	2,2,3,3-tetramethylheptane	61868-40-4	-1.6798	-4.4374E-06	2.2187E-12	0	358700	2	2.0901E-02	1.4786E-02

Table 5 Solubility in Salt Water - Organic Compounds

$\log_{10} S = A + B X + C X^2$ (S - solubility, parts per million by weight, ppm (wt), X - mg salt / kg of water)

NO	ID	FORMULA	NAME	CAS No	A	B	C	XMIN	XMAX	code	S@X=0	S@X=34,472
481	22988	C11H24	2,2,3,4-tetramethylheptane	61868-41-5	-1.5873	-4.4374E-06	2.2187E-12	0	358700	2	2.5862E-02	1.8295E-02
482	22989	C11H24	2,2,3,5-tetramethylheptane	61868-42-6	-1.5405	-4.4374E-06	2.2187E-12	0	358700	2	2.8810E-02	2.0380E-02
483	22990	C11H24	2,2,3,6-tetramethylheptane	61868-43-7	-1.4455	-4.4374E-06	2.2187E-12	0	358700	2	3.5852E-02	2.5362E-02
484	22991	C11H24	2,2,4,4-tetramethylheptane	61868-44-8	-1.4455	-4.4374E-06	2.2187E-12	0	358700	2	3.5852E-02	2.5362E-02
485	22992	C11H24	2,2,4,5-tetramethylheptane	61868-45-9	-1.4215	-4.4374E-06	2.2187E-12	0	358700	2	3.7887E-02	2.6801E-02
486	22993	C11H24	2,2,4,6-tetramethylheptane	61868-46-0	-1.2562	-4.4374E-06	2.2187E-12	0	358700	2	5.5431E-02	3.9212E-02
487	22994	C11H24	2,2,5,5-tetramethylheptane	61868-47-1	-1.3491	-4.4374E-06	2.2187E-12	0	358700	2	4.4765E-02	3.1667E-02
488	22995	C11H24	2,2,5,6-tetramethylheptane	61868-48-2	-1.4215	-4.4374E-06	2.2187E-12	0	358700	2	3.7887E-02	2.6801E-02
489	22996	C11H24	2,2,6,6-tetramethylheptane	40117-45-1	-1.2759	-4.4374E-06	2.2187E-12	0	358700	2	5.2981E-02	3.7479E-02
490	22997	C11H24	2,3,3,4-tetramethylheptane	61868-49-3	-1.7254	-4.4374E-06	2.2187E-12	0	358700	2	1.8820E-02	1.3314E-02
491	22998	C11H24	2,3,3,5-tetramethylheptane	61868-50-6	-1.5873	-4.4374E-06	2.2187E-12	0	358700	2	2.5862E-02	1.8295E-02
492	22999	C11H24	2,3,3,6-tetramethylheptane	61868-51-7	-1.5640	-4.4374E-06	2.2187E-12	0	358700	2	2.7293E-02	1.9307E-02
493	23000	C11H24	2,3,4,4-tetramethylheptane	61868-52-8	-1.6338	-4.4374E-06	2.2187E-12	0	358700	2	2.3238E-02	1.6439E-02
494	23001	C11H24	2,3,4,5-tetramethylheptane	61868-53-9	-1.6569	-4.4374E-06	2.2187E-12	0	358700	2	2.2036E-02	1.5588E-02
495	23002	C11H24	2,3,4,6-tetramethylheptane	61868-54-0	-1.5405	-4.4374E-06	2.2187E-12	0	358700	2	2.8810E-02	2.0380E-02
496	23003	C11H24	2,3,5,5-tetramethylheptane	61868-55-1	-1.4932	-4.4374E-06	2.2187E-12	0	358700	2	3.2124E-02	2.2725E-02
497	23004	C11H24	2,3,5,6-tetramethylheptane	52670-32-3	-1.5640	-4.4374E-06	2.2187E-12	0	358700	2	2.7293E-02	1.9307E-02
498	23005	C11H24	2,4,4,5-tetramethylheptane	61868-56-2	-1.5405	-4.4374E-06	2.2187E-12	0	358700	2	2.8810E-02	2.0380E-02
499	23006	C11H24	2,4,4,6-tetramethylheptane	61868-57-3	-1.3004	-4.4374E-06	2.2187E-12	0	358700	2	5.0078E-02	3.5426E-02
500	23007	C11H24	2,4,5,5-tetramethylheptane	61868-58-4	-1.5873	-4.4374E-06	2.2187E-12	0	358700	2	2.5862E-02	1.8295E-02
501	23008	C11H24	3,3,4,4-tetramethylheptane	61868-59-5	-1.7704	-4.4374E-06	2.2187E-12	0	358700	2	1.6965E-02	1.2001E-02
502	23009	C11H24	3,3,4,5-tetramethylheptane	61868-60-8	-1.7027	-4.4374E-06	2.2187E-12	0	358700	2	1.9831E-02	1.4029E-02
503	23010	C11H24	3,3,5,5-tetramethylheptane	61868-61-9	-1.6798	-4.4374E-06	2.2187E-12	0	358700	2	2.0901E-02	1.4786E-02
504	23011	C11H24	3,4,4,5-tetramethylheptane	61868-62-0	-1.7928	-4.4374E-06	2.2187E-12	0	358700	2	1.6114E-02	1.1399E-02
505	23012	C11H24	2,2-dimethyl-3-isopropylhexane	61868-63-1	-1.4932	-4.4374E-06	2.2187E-12	0	358700	2	3.2124E-02	2.2725E-02
506	23013	C11H24	2,3-dimethyl-3-isopropylhexane	61868-64-2	-1.6798	-4.4374E-06	2.2187E-12	0	358700	2	2.0901E-02	1.4786E-02
507	23014	C11H24	2,4-dimethyl-3-isopropylhexane	61868-65-3	-1.5640	-4.4374E-06	2.2187E-12	0	358700	2	2.7293E-02	1.9307E-02
508	23015	C11H24	2,5-dimethyl-3-isopropylhexane	61868-66-4	-1.4455	-4.4374E-06	2.2187E-12	0	358700	2	3.5852E-02	2.5362E-02
509	23016	C11H24	2-methyl-3,3-diethylhexane	61868-67-5	-1.8150	-4.4374E-06	2.2187E-12	0	358700	2	1.5310E-02	1.0831E-02
510	23017	C11H24	2-methyl-3,4-diethylhexane	61868-68-6	-1.6569	-4.4374E-06	2.2187E-12	0	358700	2	2.2036E-02	1.5588E-02
511	23018	C11H24	2-methyl-4,4-diethylhexane	61868-69-7	-1.6569	-4.4374E-06	2.2187E-12	0	358700	2	2.2036E-02	1.5588E-02
512	23019	C11H24	3-methyl-3,4-diethylhexane	61868-70-0	-1.8150	-4.4374E-06	2.2187E-12	0	358700	2	1.5310E-02	1.0831E-02
513	23020	C11H24	3-methyl-4,4-diethylhexane	61868-71-1	-1.9026	-4.4374E-06	2.2187E-12	0	358700	2	1.2515E-02	8.8535E-03
514	23021	C11H24	2,2,3-trimethyl-3-ethylhexane	61868-72-2	-1.6798	-4.4374E-06	2.2187E-12	0	358700	2	2.0901E-02	1.4786E-02
515	23022	C11H24	2,2,3-trimethyl-4-ethylhexane	61868-73-3	-1.5873	-4.4374E-06	2.2187E-12	0	358700	2	2.5862E-02	1.8295E-02
516	23023	C11H24	2,2,4-trimethyl-3-ethylhexane	61868-74-4	-1.5640	-4.4374E-06	2.2187E-12	0	358700	2	2.7293E-02	1.9307E-02
517	23024	C11H24	2,2,4-trimethyl-4-ethylhexane	61868-75-5	-1.6338	-4.4374E-06	2.2187E-12	0	358700	2	2.3238E-02	1.6439E-02
518	23025	C11H24	2,2,5-trimethyl-3-ethylhexane	61868-76-6	-1.4455	-4.4374E-06	2.2187E-12	0	358700	2	3.5852E-02	2.5362E-02
519	23026	C11H24	2,2,5-trimethyl-4-ethylhexane	61868-77-7	-1.3491	-4.4374E-06	2.2187E-12	0	358700	2	4.4765E-02	3.1667E-02
520	23027	C11H24	2,3,3-trimethyl-4-ethylhexane	61868-78-8	-1.7480	-4.4374E-06	2.2187E-12	0	358700	2	1.7866E-02	1.2639E-02
521	23028	C11H24	2,3,4-trimethyl-3-ethylhexane	61868-79-9	-1.8371	-4.4374E-06	2.2187E-12	0	358700	2	1.4551E-02	1.0294E-02
522	23029	C11H24	2,3,4-trimethyl-4-ethylhexane	61868-80-2	-1.6798	-4.4374E-06	2.2187E-12	0	358700	2	2.0901E-02	1.4786E-02
523	23030	C11H24	2,3,5-trimethyl-3-ethylhexane	61868-81-3	-1.5873	-4.4374E-06	2.2187E-12	0	358700	2	2.5862E-02	1.8295E-02
524	23031	C11H24	2,3,5-trimethyl-4-ethylhexane	61868-82-4	-1.6106	-4.4374E-06	2.2187E-12	0	358700	2	2.4512E-02	1.7340E-02
525	23032	C11H24	2,4,4-trimethyl-3-ethylhexane	61868-83-5	-1.6338	-4.4374E-06	2.2187E-12	0	358700	2	2.3238E-02	1.6439E-02
526	23033	C11H24	3,3,4-trimethyl-4-ethylhexane	61868-84-6	-1.9026	-4.4374E-06	2.2187E-12	0	358700	2	1.2515E-02	8.8535E-03
527	23034	C11H24	2,2,3,3,4-pentamethylhexane	61868-85-7	-1.7704	-4.4374E-06	2.2187E-12	0	358700	2	1.6965E-02	1.2001E-02
528	23035	C11H24	2,2,3,3,5-pentamethylhexane	61868-86-8	-1.4574	-4.4374E-06	2.2187E-12	0	358700	2	3.4878E-02	2.4673E-02
529	23036	C11H24	2,2,3,4,4-pentamethylhexane	61868-87-9	-1.8150	-4.4374E-06	2.2187E-12	0	358700	2	1.5310E-02	1.0831E-02
530	23037	C11H24	2,2,3,4,5-pentamethylhexane	61868-88-0	-1.5405	-4.4374E-06	2.2187E-12	0	358700	2	2.8810E-02	2.0380E-02
531	23038	C11H24	2,2,3,5,5-pentamethylhexane	14739-73-2	-1.3491	-4.4374E-06	2.2187E-12	0	358700	2	4.4765E-02	3.1667E-02
532	23039	C11H24	2,2,4,4,5-pentamethylhexane	60302-23-0	-1.5873	-4.4374E-06	2.2187E-12	0	358700	2	2.5862E-02	1.8295E-02
533	23040	C11H24	2,3,3,4,4-pentamethylhexane	61868-89-1	-1.8371	-4.4374E-06	2.2187E-12	0	358700	2	1.4551E-02	1.0294E-02
534	23041	C11H24	2,3,3,4,5-pentamethylhexane	52670-33-4	-1.7027	-4.4374E-06	2.2187E-12	0	358700	2	1.9831E-02	1.4029E-02
535	23042	C11H24	2,2,4-trimethyl-3-isopropylpentane	61868-90-4	-1.4455	-4.4374E-06	2.2187E-12	0	358700	2	3.5852E-02	2.5362E-02
536	23043	C11H24	2,3,4-trimethyl-3-isopropylpentane	61868-91-5	-1.8150	-4.4374E-06	2.2187E-12	0	358700	2	1.5310E-02	1.0831E-02
537	23044	C11H24	2,2-dimethyl-3,3-diethylpentane	60302-28-5	-1.8809	-4.4374E-06	2.2187E-12	0	358700	2	1.3156E-02	9.3067E-03
538	23045	C11H24	2,4-dimethyl-3,3-diethylpentane	61868-92-6	-1.8809	-4.4374E-06	2.2187E-12	0	358700	2	1.3156E-02	9.3067E-03
539	23046	C11H24	2,2,3,4-tetramethyl-3-ethylpentane	61868-93-7	-1.8150	-4.4374E-06	2.2187E-12	0	358700	2	1.5310E-02	1.0831E-02
540	23047	C11H24	2,2,4,4-tetramethyl-3-ethylpentane	---	-1.6798	-4.4374E-06	2.2187E-12	0	358700	2	2.0901E-02	1.4786E-02
541	23048	C11H24	2,2,3,3,4,4-hexamethylpentane	60302-27-4	-1.9878	-4.4374E-06	2.2187E-12	0	358700	2	1.0284E-02	7.2748E-03
542	24357	C12H18	hexylbenzene	1077-16-3	0.0077	-3.3451E-06	1.6726E-12	0	358700	1,2	1.0180E+00	7.8419E-01
543	24358	C12H18	1,4-dipropylbenzene	4815-57-0	0.7858	-4.5929E-06	2.2646E-12	0	358700	2	6.1061E+00	4.2670E+00
544	24359	C12H18	diisopropylbenzene	25321-09-9	0.9729	-4.5929E-06	2.2646E-12	0	358700	2	9.3957E+00	6.5659E+00
545	24360	C12H18	1,2-diisopropylbenzene	577-55-9	0.8908	-4.5929E-06	2.2646E-12	0	358700	2	7.7773E+00	5.4349E+00
546	24361	C12H18	m-diisopropylbenzene	99-62-7	0.9097	-4.5929E-06	2.2646E-12	0	358700	2	8.1227E+00	5.6763E+00
547	24362	C12H18	p-diisopropylbenzene	100-18-5	0.7435	-4.5929E-06	2.2646E-12	0	358700	2	5.5393E+00	3.8710E+00
548	24363	C12H18	triethylbenzene; (mixed isomers)	25340-18-5	0.5886	-4.5929E-06	2.2646E-12	0	358700	2	3.8780E+00	2.7100E+00
549	24364	C12H18	1,2,3-triethylbenzene	---	0.5895	-4.5929E-06	2.2646E-12	0	358700	2	3.8857E+00	2.7154E+00
550	24365	C12H18	1,2,4-triethylbenzene	877-44-1	0.5789	-4.5929E-06	2.2646E-12	0	358700	2	3.7925E+00	2.6502E+00
551	24366	C12H18	1,3,5-triethylbenzene	102-25-0	0.6243	-4.5929E-06	2.2646E-12	0	358700	2	4.2106E+00	2.9424E+00
552	24367	C12H18	1-tert-butyl-3,5-dimethylbenzene	98-19-1	0.8223	-4.5929E-06	2.2646E-12	0	358700	2	6.6421E+00	4.6416E+00
553	24369	C12H18	2,4-dimethyl-1-sec-butylbenzene	1483-60-9	0.7858	-4.5929E-06	2.2646E-12	0	358700	2	6.1061E+00	4.2670E+00
554	24370	C12H18	(1-ethylbutyl)benzene	4468-42-2	0.7771	-4.5929E-06	2.2646E-12	0	358700	2	5.9855E+00	4.1828E+00
555	24371	C12H18	(1-ethyl-1-methylpropyl)benzene	1985-97-3	0.8679	-4.5929E-06	2.2646E-12	0	358700	2	7.3773E+00	5.1554E+00
556	24372	C12H18	(1-methylpentyl)benzene	6031-02-3	0.7997	-4.5929E-06	2.2646E-12	0	358700	2	6.3046E+00	4.4057E+00
557	24373	C12H18	(3-methylpentyl)benzene	54410-69-4	0.5362	-4.5929E-06	2.2646E-12	0	358700	2	3.4369E+00	2.4018E+00
558	24374	C12H18	(4-methylpentyl)benzene	4215-86-5	0.6005	-4.5929E-06	2.2646E-12	0	358700	2	3.9855E+00	2.7851E+00
559	24375	C12H18	1,2,4-trimethyl-5-isopropylbenzene	10222-95-4	0.5150	-4.5929E-06	2.2646E-12	0	358700	2	3.2733E+00	2.2874E+00
560	24376	C12H18	4-tert-butyl-o-xylene	7397-06-0	0.8679	-4.5929E-06	2.2646E-12	0	358700	2	7.3773E+00	5.1554E+00

Table 5 Solubility in Salt Water - Organic Compounds

			$\log_{10} S = A + B X + C X^2$	(S - solubility, parts per million by weight, ppm (wt),					X - mg salt / kg of water)			
NO	ID	FORMULA	NAME	CAS No	A	B	C	XMIN	XMAX	code	S@X=0	S@X=34,472
561	24818	C12H24	heptylcyclopentane	5617-42-5	-1.7901	-7.1550E-06	3.5775E-12	0	358700	2	1.6215E-02	9.2795E-03
562	24819	C12H24	hexylcyclohexane	4292-75-5	-1.5987	-7.1550E-06	3.5775E-12	0	358700	2	2.5197E-02	1.4420E-02
563	24820	C12H24	1-methyl-2-pentylcyclohexane	54411-01-7	-1.4830	-7.1550E-06	3.5775E-12	0	358700	2	3.2883E-02	1.8818E-02
564	24914	C12H26	dodecane	112-40-3	-2.4318	-4.4374E-06	2.2187E-12	0	358700	1,2	3.7000E-03	2.6174E-03
565	24915	C12H26	3-methylundecane	1002-43-3	-2.3179	-4.4374E-06	2.2187E-12	0	358700	2	4.8091E-03	3.4020E-03
566	24916	C12H26	4-methylundecane	2980-69-0	-2.2850	-4.4374E-06	2.2187E-12	0	358700	2	5.1885E-03	3.6704E-03
567	24917	C12H26	5-methylundecane	1632-70-8	-2.2286	-4.4374E-06	2.2187E-12	0	358700	2	5.9068E-03	4.1786E-03
568	24918	C12H26	6-methylundecane	17302-33-9	-2.2286	-4.4374E-06	2.2187E-12	0	358700	2	5.9068E-03	4.1786E-03
569	24919	C12H26	3-ethyldecane	17085-96-0	-2.2794	-4.4374E-06	2.2187E-12	0	358700	2	5.2553E-03	3.7177E-03
570	24920	C12H26	4-ethyldecane	1636-44-8	-2.1902	-4.4374E-06	2.2187E-12	0	358700	2	6.4536E-03	4.5653E-03
571	24921	C12H26	5-ethyldecane	17302-36-2	-2.1511	-4.4374E-06	2.2187E-12	0	358700	2	7.0624E-03	4.9960E-03
572	24922	C12H26	2,2-dimethyldecane	17302-37-3	-2.1312	-4.4374E-06	2.2187E-12	0	358700	2	7.3923E-03	5.2294E-03
573	24923	C12H26	2,3-dimethyldecane	17312-44-6	-2.2286	-4.4374E-06	2.2187E-12	0	358700	2	5.9068E-03	4.1786E-03
574	24924	C12H26	2,4-dimethyldecane	2801-84-5	-2.1112	-4.4374E-06	2.2187E-12	0	358700	2	7.7407E-03	5.4758E-03
575	24925	C12H26	2,5-dimethyldecane	17312-50-4	-2.0707	-4.4374E-06	2.2187E-12	0	358700	2	8.4970E-03	6.0109E-03
576	24926	C12H26	2,6-dimethyldecane	13150-81-7	-2.0707	-4.4374E-06	2.2187E-12	0	358700	2	8.4970E-03	6.0109E-03
577	24927	C12H26	2,7-dimethyldecane	17312-51-5	-2.1511	-4.4374E-06	2.2187E-12	0	358700	2	7.0624E-03	4.9960E-03
578	24928	C12H26	2,8-dimethyldecane	17312-52-6	-2.1902	-4.4374E-06	2.2187E-12	0	358700	2	6.4536E-03	4.5653E-03
579	24929	C12H26	2,9-dimethyldecane	1002-17-1	-2.1902	-4.4374E-06	2.2187E-12	0	358700	2	6.4536E-03	4.5653E-03
580	24930	C12H26	3,3-dimethyldecane	17302-38-4	-2.1707	-4.4374E-06	2.2187E-12	0	358700	2	6.7498E-03	4.7749E-03
581	24931	C12H26	3,4-dimethyldecane	17312-45-7	-2.2095	-4.4374E-06	2.2187E-12	0	358700	2	6.1729E-03	4.3668E-03
582	24932	C12H26	3,5-dimethyldecane	17312-48-0	-2.1112	-4.4374E-06	2.2187E-12	0	358700	2	7.7407E-03	5.4758E-03
583	24933	C12H26	3,6-dimethyldecane	17312-53-7	-2.1312	-4.4374E-06	2.2187E-12	0	358700	2	7.3923E-03	5.2294E-03
584	24934	C12H26	3,7-dimethyldecane	17312-54-8	-2.1511	-4.4374E-06	2.2187E-12	0	358700	2	7.0624E-03	4.9960E-03
585	24935	C12H26	3,8-dimethyldecane	17312-55-9	-2.2095	-4.4374E-06	2.2187E-12	0	358700	2	6.1729E-03	4.3668E-03
586	24936	C12H26	4,4-dimethyldecane	17312-39-9	-2.0911	-4.4374E-06	2.2187E-12	0	358700	2	8.1085E-03	5.7360E-03
587	24937	C12H26	4,5-dimethyldecane	17312-46-8	-2.1511	-4.4374E-06	2.2187E-12	0	358700	2	7.0624E-03	4.9960E-03
588	24938	C12H26	4,6-dimethyldecane	17312-49-1	-2.0707	-4.4374E-06	2.2187E-12	0	358700	2	8.4970E-03	6.0109E-03
589	24939	C12H26	4,7-dimethyldecane	17312-56-0	-2.0911	-4.4374E-06	2.2187E-12	0	358700	2	8.1085E-03	5.7360E-03
590	24940	C12H26	5,5-dimethyldecane	17453-92-8	-2.0502	-4.4374E-06	2.2187E-12	0	358700	2	8.9075E-03	6.3012E-03
591	24941	C12H26	5,6-dimethyldecane	1636-43-7	-2.1312	-4.4374E-06	2.2187E-12	0	358700	2	7.3923E-03	5.2294E-03
592	24942	C12H26	4-propylnonane	6165-37-3	-2.0911	-4.4374E-06	2.2187E-12	0	358700	2	8.1085E-03	5.7360E-03
593	24943	C12H26	5-propylnonane	998-35-6	-2.0502	-4.4374E-06	2.2187E-12	0	358700	2	8.9075E-03	6.3012E-03
594	24944	C12H26	4-isopropylnonane	62184-71-8	-2.0707	-4.4374E-06	2.2187E-12	0	358700	2	8.4970E-03	6.0109E-03
595	24945	C12H26	5-isopropylnonane	62184-72-9	-2.0707	-4.4374E-06	2.2187E-12	0	358700	2	8.4970E-03	6.0109E-03
596	24946	C12H26	2-methyl-3-ethylnonane	62184-73-0	-2.1312	-4.4374E-06	2.2187E-12	0	358700	2	7.3923E-03	5.2294E-03
597	24947	C12H26	2-methyl-4-ethylnonane	62184-37-6	-2.0707	-4.4374E-06	2.2187E-12	0	358700	2	8.4970E-03	6.0109E-03
598	24948	C12H26	2-methyl-5-ethylnonane	62184-38-7	-2.0296	-4.4374E-06	2.2187E-12	0	358700	2	9.3411E-03	6.6080E-03
599	24949	C12H26	2-methyl-6-ethylnonane	62184-39-8	-2.0707	-4.4374E-06	2.2187E-12	0	358700	2	8.4970E-03	6.0109E-03
600	24950	C12H26	2-methyl-7-ethylnonane	62184-40-1	-2.1511	-4.4374E-06	2.2187E-12	0	358700	2	7.0624E-03	4.9960E-03
601	24951	C12H26	3-methyl-3-ethylnonane	17302-39-5	-2.1902	-4.4374E-06	2.2187E-12	0	358700	2	6.4536E-03	4.5653E-03
602	24952	C12H26	3-methyl-4-ethylnonane	62184-41-2	-2.1312	-4.4374E-06	2.2187E-12	0	358700	2	7.3923E-03	5.2294E-03
603	24953	C12H26	3-methyl-5-ethylnonane	62184-42-3	-2.0088	-4.4374E-06	2.2187E-12	0	358700	2	9.7994E-03	6.9322E-03
604	24954	C12H26	3-methyl-6-ethylnonane	62184-43-4	-2.0911	-4.4374E-06	2.2187E-12	0	358700	2	8.1085E-03	5.7360E-03
605	24955	C12H26	3-methyl-7-ethylnonane	62184-44-5	-2.1112	-4.4374E-06	2.2187E-12	0	358700	2	7.7407E-03	5.4758E-03
606	24956	C12H26	4-methyl-3-ethylnonane	62184-45-6	-2.1112	-4.4374E-06	2.2187E-12	0	358700	2	7.7407E-03	5.4758E-03
607	24957	C12H26	4-methyl-4-ethylnonane	17312-40-2	-2.0911	-4.4374E-06	2.2187E-12	0	358700	2	8.1085E-03	5.7360E-03
608	24958	C12H26	4-methyl-5-ethylnonane	62184-46-7	-2.0707	-4.4374E-06	2.2187E-12	0	358700	2	8.4970E-03	6.0109E-03
609	24959	C12H26	4-methyl-6-ethylnonane	62184-47-8	-2.0296	-4.4374E-06	2.2187E-12	0	358700	2	9.3411E-03	6.6080E-03
610	24960	C12H26	4-methyl-7-ethylnonane	62184-48-9	-2.1112	-4.4374E-06	2.2187E-12	0	358700	2	7.7407E-03	5.4758E-03
611	24961	C12H26	5-methyl-3-ethylnonane	62184-49-0	-2.0911	-4.4374E-06	2.2187E-12	0	358700	2	8.1085E-03	5.7360E-03
612	24962	C12H26	5-methyl-4-ethylnonane	1632-71-9	-2.0911	-4.4374E-06	2.2187E-12	0	358700	2	8.1085E-03	5.7360E-03
613	24963	C12H26	5-methyl-5-ethylnonane	14531-16-9	-2.0707	-4.4374E-06	2.2187E-12	0	358700	2	8.4970E-03	6.0109E-03
614	24964	C12H26	2,2,3-trimethylnonane	55499-04-2	-2.1112	-4.4374E-06	2.2187E-12	0	358700	2	7.7407E-03	5.4758E-03
615	24965	C12H26	2,2,4-trimethylnonane	62184-50-3	-1.9241	-4.4374E-06	2.2187E-12	0	358700	2	1.1910E-02	8.4251E-03
616	24966	C12H26	2,2,5-trimethylnonane	62184-51-4	-1.9241	-4.4374E-06	2.2187E-12	0	358700	2	1.1910E-02	8.4251E-03
617	24967	C12H26	2,2,6-trimethylnonane	62184-52-5	-1.9026	-4.4374E-06	2.2187E-12	0	358700	2	1.2515E-02	8.8535E-03
618	24968	C12H26	2,2,7-trimethylnonane	62184-53-6	-2.0088	-4.4374E-06	2.2187E-12	0	358700	2	9.7994E-03	6.9322E-03
619	24969	C12H26	2,2,8-trimethylnonane	62184-54-7	-2.0088	-4.4374E-06	2.2187E-12	0	358700	2	9.7994E-03	6.9322E-03
620	24970	C12H26	2,3,3-trimethylnonane	62184-55-8	-2.1511	-4.4374E-06	2.2187E-12	0	358700	2	7.0624E-03	4.9960E-03
621	24971	C12H26	2,3,4-trimethylnonane	62184-56-9	-2.1112	-4.4374E-06	2.2187E-12	0	358700	2	7.7407E-03	5.4758E-03
622	24972	C12H26	2,3,5-trimethylnonane	62184-57-0	-2.0296	-4.4374E-06	2.2187E-12	0	358700	2	9.3411E-03	6.6080E-03
623	24973	C12H26	2,3,6-trimethylnonane	62184-58-1	-2.0707	-4.4374E-06	2.2187E-12	0	358700	2	8.4970E-03	6.0109E-03
624	24974	C12H26	2,3,7-trimethylnonane	62184-59-2	-2.1312	-4.4374E-06	2.2187E-12	0	358700	2	7.3923E-03	5.2294E-03
625	24975	C12H26	2,3,8-trimethylnonane	62184-60-5	-2.1112	-4.4374E-06	2.2187E-12	0	358700	2	7.7407E-03	5.4758E-03
626	24976	C12H26	2,4,4-trimethylnonane	62184-61-6	-1.9026	-4.4374E-06	2.2187E-12	0	358700	2	1.2515E-02	8.8535E-03
627	24977	C12H26	2,4,5-trimethylnonane	62184-62-7	-2.0088	-4.4374E-06	2.2187E-12	0	358700	2	9.7994E-03	6.9322E-03
628	24978	C12H26	2,4,6-trimethylnonane	62184-10-5	-1.9455	-4.4374E-06	2.2187E-12	0	358700	2	1.1337E-02	8.0201E-03
629	24979	C12H26	2,4,7-trimethylnonane	62184-11-6	-1.9878	-4.4374E-06	2.2187E-12	0	358700	2	1.0284E-02	7.2748E-03
630	24980	C12H26	2,4,8-trimethylnonane	49542-74-7	-1.9241	-4.4374E-06	2.2187E-12	0	358700	2	1.1910E-02	8.4251E-03
631	24981	C12H26	2,5,5-trimethylnonane	62184-12-7	-1.9026	-4.4374E-06	2.2187E-12	0	358700	2	1.2515E-02	8.8535E-03
632	24982	C12H26	2,5,6-trimethylnonane	62184-13-8	-2.0296	-4.4374E-06	2.2187E-12	0	358700	2	9.3411E-03	6.6080E-03
633	24983	C12H26	2,5,7-trimethylnonane	62184-14-9	-1.9878	-4.4374E-06	2.2187E-12	0	358700	2	1.0284E-02	7.2748E-03
634	24984	C12H26	2,5,8-trimethylnonane	49557-09-7	-1.9455	-4.4374E-06	2.2187E-12	0	358700	2	1.1337E-02	8.0201E-03
635	24985	C12H26	2,6,6-trimethylnonane	62184-15-0	-1.9026	-4.4374E-06	2.2187E-12	0	358700	2	1.2515E-02	8.8535E-03
636	24986	C12H26	2,6,7-trimethylnonane	62184-16-1	-2.0911	-4.4374E-06	2.2187E-12	0	358700	2	8.1085E-03	5.7360E-03
637	24987	C12H26	2,7,7-trimethylnonane	62184-17-2	-2.0502	-4.4374E-06	2.2187E-12	0	358700	2	8.9075E-03	6.3012E-03
638	24988	C12H26	3,3,4-trimethylnonane	62184-18-3	-2.1511	-4.4374E-06	2.2187E-12	0	358700	2	7.0624E-03	4.9960E-03
639	24989	C12H26	3,3,5-trimethylnonane	62184-19-4	-1.9026	-4.4374E-06	2.2187E-12	0	358700	2	1.2515E-02	8.8535E-03
640	24990	C12H26	3,3,7-trimethylnonane	62184-21-8	-2.0088	-4.4374E-06	2.2187E-12	0	358700	2	9.7994E-03	6.9322E-03

Table 5 Solubility in Salt Water - Organic Compounds

$\log_{10} S = A + B X + C X^2$			(S - solubility, parts per million by weight, ppm (wt), X - mg salt / kg of water)									
NO	ID	FORMULA	NAME	CAS No	A	B	C	XMIN	XMAX	code	S@X=0	S@X=34,472
641	24991	C12H26	3,3,6-trimethylnonane	62184-20-7	-2.0707	-4.4374E-06	2.2187E-12	0	358700	2	8.4970E-03	6.0109E-03
642	24992	C12H26	3,4,4-trimethylnonane	62184-22-9	-2.1112	-4.4374E-06	2.2187E-12	0	358700	2	7.7407E-03	5.4758E-03
643	24993	C12H26	3,4,5-trimethylnonane	62184-23-0	-2.1112	-4.4374E-06	2.2187E-12	0	358700	2	7.7407E-03	5.4758E-03
644	24994	C12H26	3,4,6-trimethylnonane	62184-24-1	-2.0502	-4.4374E-06	2.2187E-12	0	358700	2	8.9075E-03	6.3012E-03
645	24995	C12H26	3,4,7-trimethylnonane	27802-85-3	-2.1112	-4.4374E-06	2.2187E-12	0	358700	2	7.7407E-03	5.4758E-03
646	24996	C12H26	3,5,5-trimethylnonane	62184-25-2	-1.9241	-4.4374E-06	2.2187E-12	0	358700	2	1.1910E-02	8.4251E-03
647	24997	C12H26	3,5,6-trimethylnonane	62184-26-3	-2.0502	-4.4374E-06	2.2187E-12	0	358700	2	8.9075E-03	6.3012E-03
648	24998	C12H26	3,5,7-trimethylnonane	62184-27-4	-2.0088	-4.4374E-06	2.2187E-12	0	358700	2	9.7994E-03	6.9322E-03
649	24999	C12H26	3,6,6-trimethylnonane	62184-28-5	-1.9026	-4.4374E-06	2.2187E-12	0	358700	2	1.2515E-02	8.8535E-03
650	25000	C12H26	4,4,5-trimethylnonane	62184-29-6	-2.0707	-4.4374E-06	2.2187E-12	0	358700	2	8.4970E-03	6.0109E-03
651	25001	C12H26	4,4,6-trimethylnonane	62184-30-9	-1.9241	-4.4374E-06	2.2187E-12	0	358700	2	1.1910E-02	8.4251E-03
652	25002	C12H26	4,5,5-trimethylnonane	62184-31-0	-2.0707	-4.4374E-06	2.2187E-12	0	358700	2	8.4970E-03	6.0109E-03
653	25003	C12H26	4,5,6-trimethylnonane	62211-85-2	-2.0707	-4.4374E-06	2.2187E-12	0	358700	2	8.4970E-03	6.0109E-03
654	25004	C12H26	4-tert-butyloctane	62184-32-1	-1.9241	-4.4374E-06	2.2187E-12	0	358700	2	1.1910E-02	8.4251E-03
655	25005	C12H26	2-methyl-4-propyloctane	62184-33-2	-1.9241	-4.4374E-06	2.2187E-12	0	358700	2	1.1910E-02	8.4251E-03
656	25006	C12H26	2-methyl-5-propyloctane	62184-34-3	-1.9241	-4.4374E-06	2.2187E-12	0	358700	2	1.1910E-02	8.4251E-03
657	25007	C12H26	3-methyl-4-propyloctane	62184-35-4	-2.0502	-4.4374E-06	2.2187E-12	0	358700	2	8.9075E-03	6.3012E-03
658	25008	C12H26	3-methyl-5-propyloctane	62184-36-5	-1.9878	-4.4374E-06	2.2187E-12	0	358700	2	1.0284E-02	7.2748E-03
659	25009	C12H26	4-methyl-4-propyloctane	17312-41-3	-2.0088	-4.4374E-06	2.2187E-12	0	358700	2	9.7994E-03	6.9322E-03
660	25010	C12H26	4-methyl-5-propyloctane	62183-85-1	-2.0296	-4.4374E-06	2.2187E-12	0	358700	2	9.3411E-03	6.6080E-03
661	25011	C12H26	2-methyl-3-isopropyloctane	13287-19-9	-2.0296	-4.4374E-06	2.2187E-12	0	358700	2	9.3411E-03	6.6080E-03
662	25012	C12H26	2-methyl-4-isopropyloctane	62183-86-2	-1.9026	-4.4374E-06	2.2187E-12	0	358700	2	1.2515E-02	8.8535E-03
663	25013	C12H26	2-methyl-5-isopropyloctane	62183-87-3	-1.9241	-4.4374E-06	2.2187E-12	0	358700	2	1.1910E-02	8.4251E-03
664	25014	C12H26	3-methyl-4-isopropyloctane	62183-88-4	-2.0088	-4.4374E-06	2.2187E-12	0	358700	2	9.7994E-03	6.9322E-03
665	25015	C12H26	3-methyl-5-isopropyloctane	62183-89-5	-1.9455	-4.4374E-06	2.2187E-12	0	358700	2	1.1337E-02	8.0201E-03
666	25016	C12H26	4-methyl-4-isopropyloctane	62183-90-8	-2.0707	-4.4374E-06	2.2187E-12	0	358700	2	8.4970E-03	6.0109E-03
667	25017	C12H26	4-methyl-5-isopropyloctane	62183-91-9	-1.9878	-4.4374E-06	2.2187E-12	0	358700	2	1.0284E-02	7.2748E-03
668	25018	C12H26	3,3-diethyloctane	17302-40-8	-2.2095	-4.4374E-06	2.2187E-12	0	358700	2	6.1729E-03	4.3668E-03
669	25019	C12H26	3,4-diethyloctane	62183-92-0	-2.0911	-4.4374E-06	2.2187E-12	0	358700	2	8.1085E-03	5.7360E-03
670	25020	C12H26	3,5-diethyloctane	62183-93-1	-2.0502	-4.4374E-06	2.2187E-12	0	358700	2	8.9075E-03	6.3012E-03
671	25021	C12H26	3,6-diethyloctane	62183-94-2	-2.1511	-4.4374E-06	2.2187E-12	0	358700	2	7.0624E-03	4.9960E-03
672	25022	C12H26	4,4-diethyloctane	17312-42-4	-2.1112	-4.4374E-06	2.2187E-12	0	358700	2	7.7407E-03	5.4758E-03
673	25023	C12H26	4,5-diethyloctane	1636-41-5	-2.0502	-4.4374E-06	2.2187E-12	0	358700	2	8.9075E-03	6.3012E-03
674	25024	C12H26	2,2-dimethyl-3-ethyloctane	62183-95-3	-2.0296	-4.4374E-06	2.2187E-12	0	358700	2	9.3411E-03	6.6080E-03
675	25025	C12H26	2,2-dimethyl-4-ethyloctane	62183-96-4	-1.8371	-4.4374E-06	2.2187E-12	0	358700	2	1.4551E-02	1.0294E-02
676	25026	C12H26	2,2-dimethyl-5-ethyloctane	62183-97-5	-1.8809	-4.4374E-06	2.2187E-12	0	358700	2	1.3156E-02	9.3067E-03
677	25027	C12H26	2,2-dimethyl-6-ethyloctane	62183-98-6	-1.9667	-4.4374E-06	2.2187E-12	0	358700	2	1.0796E-02	7.6371E-03
678	25028	C12H26	2,3-dimethyl-3-ethyloctane	62183-99-7	-2.1511	-4.4374E-06	2.2187E-12	0	358700	2	7.0624E-03	4.9960E-03
679	25029	C12H26	2,3-dimethyl-4-ethyloctane	62184-00-3	-2.0502	-4.4374E-06	2.2187E-12	0	358700	2	8.9075E-03	6.3012E-03
680	25030	C12H26	2,3-dimethyl-5-ethyloctane	62184-01-4	-1.9878	-4.4374E-06	2.2187E-12	0	358700	2	1.0284E-02	7.2748E-03
681	25031	C12H26	2,3-dimethyl-6-ethyloctane	62184-02-5	-2.0911	-4.4374E-06	2.2187E-12	0	358700	2	8.1085E-03	5.7360E-03
682	25032	C12H26	2,4-dimethyl-3-ethyloctane	62184-03-6	-2.0502	-4.4374E-06	2.2187E-12	0	358700	2	8.9075E-03	6.3012E-03
683	25033	C12H26	2,4-dimethyl-4-ethyloctane	62184-04-7	-1.9241	-4.4374E-06	2.2187E-12	0	358700	2	1.1910E-02	8.4251E-03
684	25034	C12H26	2,4-dimethyl-5-ethyloctane	62184-05-8	-1.9667	-4.4374E-06	2.2187E-12	0	358700	2	1.0796E-02	7.6371E-03
685	25035	C12H26	2,4-dimethyl-6-ethyloctane	62184-06-9	-1.9455	-4.4374E-06	2.2187E-12	0	358700	2	1.1337E-02	8.0201E-03
686	25036	C12H26	2,5-dimethyl-3-ethyloctane	62184-07-0	-1.9878	-4.4374E-06	2.2187E-12	0	358700	2	1.0284E-02	7.2748E-03
687	25037	C12H26	2,5-dimethyl-4-ethyloctane	62184-08-1	-1.9455	-4.4374E-06	2.2187E-12	0	358700	2	1.1337E-02	8.0201E-03
688	25038	C12H26	2,5-dimethyl-5-ethyloctane	62184-09-2	-1.9667	-4.4374E-06	2.2187E-12	0	358700	2	1.0796E-02	7.6371E-03
689	25039	C12H26	2,5-dimethyl-6-ethyloctane	62183-50-0	-2.0502	-4.4374E-06	2.2187E-12	0	358700	2	8.9075E-03	6.3012E-03
690	25040	C12H26	2,6-dimethyl-3-ethyloctane	62183-51-1	-2.0502	-4.4374E-06	2.2187E-12	0	358700	2	8.9075E-03	6.3012E-03
691	25041	C12H26	2,6-dimethyl-4-ethyloctane	62183-52-2	-1.9026	-4.4374E-06	2.2187E-12	0	358700	2	1.2515E-02	8.8535E-03
692	25042	C12H26	2,6-dimethyl-5-ethyloctane	62183-53-3	-2.0088	-4.4374E-06	2.2187E-12	0	358700	2	9.7994E-03	6.9322E-03
693	25043	C12H26	2,6-dimethyl-6-ethyloctane	62183-54-4	-2.0707	-4.4374E-06	2.2187E-12	0	358700	2	8.4970E-03	6.0109E-03
694	25044	C12H26	2,7-dimethyl-3-ethyloctane	62183-55-5	-2.0296	-4.4374E-06	2.2187E-12	0	358700	2	9.3411E-03	6.6080E-03
695	25045	C12H26	2,7-dimethyl-4-ethyloctane	62183-56-6	-1.9026	-4.4374E-06	2.2187E-12	0	358700	2	1.2515E-02	8.8535E-03
696	25046	C12H26	3,3-dimethyl-4-ethyloctane	62183-57-7	-2.0707	-4.4374E-06	2.2187E-12	0	358700	2	8.4970E-03	6.0109E-03
697	25047	C12H26	3,3-dimethyl-5-ethyloctane	62183-58-8	-1.9241	-4.4374E-06	2.2187E-12	0	358700	2	1.1910E-02	8.4251E-03
698	25048	C12H26	3,3-dimethyl-6-ethyloctane	62183-59-9	-2.0296	-4.4374E-06	2.2187E-12	0	358700	2	9.3411E-03	6.6080E-03
699	25049	C12H26	3,4-dimethyl-3-ethyloctane	62212-28-6	-2.1112	-4.4374E-06	2.2187E-12	0	358700	2	7.7407E-03	5.4758E-03
700	25050	C12H26	3,4-dimethyl-4-ethyloctane	62183-60-2	-2.1312	-4.4374E-06	2.2187E-12	0	358700	2	7.3923E-03	5.2294E-03
701	25051	C12H26	3,4-dimethyl-5-ethyloctane	62183-61-3	-2.0707	-4.4374E-06	2.2187E-12	0	358700	2	8.4970E-03	6.0109E-03
702	25052	C12H26	3,4-dimethyl-6-ethyloctane	62183-62-4	-2.0707	-4.4374E-06	2.2187E-12	0	358700	2	8.4970E-03	6.0109E-03
703	25053	C12H26	3,5-dimethyl-3-ethyloctane	62183-63-5	-2.0296	-4.4374E-06	2.2187E-12	0	358700	2	9.3411E-03	6.6080E-03
704	25054	C12H26	3,5-dimethyl-4-ethyloctane	62183-64-6	-2.0502	-4.4374E-06	2.2187E-12	0	358700	2	8.9075E-03	6.3012E-03
705	25055	C12H26	3,5-dimethyl-5-ethyloctane	62183-65-7	-1.9026	-4.4374E-06	2.2187E-12	0	358700	2	1.2515E-02	8.8535E-03
706	25056	C12H26	3,5-dimethyl-6-ethyloctane	62183-66-8	-2.0502	-4.4374E-06	2.2187E-12	0	358700	2	8.9075E-03	6.3012E-03
707	25057	C12H26	3,6-dimethyl-3-ethyloctane	62183-67-9	-2.0911	-4.4374E-06	2.2187E-12	0	358700	2	8.1085E-03	5.7360E-03
708	25058	C12H26	3,6-dimethyl-4-ethyloctane	62183-68-0	-2.0296	-4.4374E-06	2.2187E-12	0	358700	2	9.3411E-03	6.6080E-03
709	25059	C12H26	4,4-dimethyl-3-ethyloctane	62183-69-1	-2.0707	-4.4374E-06	2.2187E-12	0	358700	2	8.4970E-03	6.0109E-03
710	25060	C12H26	4,4-dimethyl-5-ethyloctane	62183-70-4	-2.0296	-4.4374E-06	2.2187E-12	0	358700	2	9.3411E-03	6.6080E-03
711	25061	C12H26	4,4-dimethyl-6-ethyloctane	62183-71-5	-1.9241	-4.4374E-06	2.2187E-12	0	358700	2	1.1910E-02	8.4251E-03
712	25062	C12H26	4,5-dimethyl-3-ethyloctane	62183-72-6	-2.0911	-4.4374E-06	2.2187E-12	0	358700	2	8.1085E-03	5.7360E-03
713	25063	C12H26	4,5-dimethyl-4-ethyloctane	62183-73-7	-2.1112	-4.4374E-06	2.2187E-12	0	358700	2	7.7407E-03	5.4758E-03
714	25064	C12H26	2,2,3,3-tetramethyloctane	62183-74-8	-2.1112	-4.4374E-06	2.2187E-12	0	358700	2	7.7407E-03	5.4758E-03
715	25065	C12H26	2,2,3,4-tetramethyloctane	62183-75-9	-2.0088	-4.4374E-06	2.2187E-12	0	358700	2	9.7994E-03	6.9322E-03
716	25066	C12H26	2,2,3,5-tetramethyloctane	62183-76-0	-1.9455	-4.4374E-06	2.2187E-12	0	358700	2	1.1337E-02	8.0201E-03
717	25067	C12H26	2,2,3,6-tetramethyloctane	62183-77-1	-1.9878	-4.4374E-06	2.2187E-12	0	358700	2	1.0284E-02	7.2748E-03
718	25068	C12H26	2,2,3,7-tetramethyloctane	62183-78-2	-1.9878	-4.4374E-06	2.2187E-12	0	358700	2	1.0284E-02	7.2748E-03
719	25069	C12H26	2,2,4,4-tetramethyloctane	62183-79-3	-1.9455	-4.4374E-06	2.2187E-12	0	358700	2	1.1337E-02	8.0201E-03
720	25070	C12H26	2,2,4,5-tetramethyloctane	62183-80-6	-1.8371	-4.4374E-06	2.2187E-12	0	358700	2	1.4551E-02	1.0294E-02

Table 5 Solubility in Salt Water - Organic Compounds

$\log_{10} S = A + B X + C X^2$ (S - solubility, parts per million by weight, ppm (wt), X - mg salt / kg of water)												
NO	ID	FORMULA	NAME	CAS No	A	B	C	XMIN	XMAX	code	S@X=0	S@X=34,472
721	25071	C12H26	2,2,4,6-tetramethyloctane	62183-81-7	-1.7928	-4.4374E-06	2.2187E-12	0	358700	2	1.6114E-02	1.1399E-02
722	25072	C12H26	2,2,4,7-tetramethyloctane	62183-82-8	-1.7883	-4.4374E-06	2.2187E-12	0	358700	2	1.6280E-02	1.1517E-02
723	25073	C12H26	2,2,5,5-tetramethyloctane	62183-83-9	-1.7704	-4.4374E-06	2.2187E-12	0	358700	2	1.6965E-02	1.2001E-02
724	25074	C12H26	2,2,5,6-tetramethyloctane	62183-84-0	-1.9026	-4.4374E-06	2.2187E-12	0	358700	2	1.2515E-02	8.8535E-03
725	25075	C12H26	2,2,5,7-tetramethyloctane	62199-19-3	-1.7928	-4.4374E-06	2.2187E-12	0	358700	2	1.6114E-02	1.1399E-02
726	25076	C12H26	2,2,6,6-tetramethyloctane	62199-20-6	-1.7928	-4.4374E-06	2.2187E-12	0	358700	2	1.6114E-02	1.1399E-02
727	25077	C12H26	2,2,6,7-tetramethyloctane	62199-21-7	-1.9241	-4.4374E-06	2.2187E-12	0	358700	2	1.1910E-02	8.4251E-03
728	25078	C12H26	2,2,7,7-tetramethyloctane	1071-31-4	-1.7928	-4.4374E-06	2.2187E-12	0	358700	2	1.6114E-02	1.1399E-02
729	25079	C12H26	2,3,3,4-tetramethyloctane	62199-22-8	-2.1312	-4.4374E-06	2.2187E-12	0	358700	2	7.3923E-03	5.2294E-03
730	25080	C12H26	2,3,3,5-tetramethyloctane	62199-23-9	-1.9878	-4.4374E-06	2.2187E-12	0	358700	2	1.0284E-02	7.2748E-03
731	25081	C12H26	2,3,3,6-tetramethyloctane	62199-24-0	-2.0296	-4.4374E-06	2.2187E-12	0	358700	2	9.3411E-03	6.6080E-03
732	25082	C12H26	2,3,3,7-tetramethyloctane	62199-25-1	-2.0296	-4.4374E-06	2.2187E-12	0	358700	2	9.3411E-03	6.6080E-03
733	25083	C12H26	2,3,4,4-tetramethyloctane	62199-26-2	-2.0296	-4.4374E-06	2.2187E-12	0	358700	2	9.3411E-03	6.6080E-03
734	25084	C12H26	2,3,4,5-tetramethyloctane	62199-27-3	-2.0502	-4.4374E-06	2.2187E-12	0	358700	2	8.9075E-03	6.3012E-03
735	25085	C12H26	2,3,4,6-tetramethyloctane	62199-28-4	-2.0088	-4.4374E-06	2.2187E-12	0	358700	2	9.7994E-03	6.9322E-03
736	25086	C12H26	2,3,4,7-tetramethyloctane	62199-29-5	-1.9878	-4.4374E-06	2.2187E-12	0	358700	2	1.0284E-02	7.2748E-03
737	25087	C12H26	2,3,5,5-tetramethyloctane	62199-30-8	-2.0088	-4.4374E-06	2.2187E-12	0	358700	2	9.7994E-03	6.9322E-03
738	25088	C12H26	2,3,5,6-tetramethyloctane	62199-31-9	-2.0088	-4.4374E-06	2.2187E-12	0	358700	2	9.7994E-03	6.9322E-03
739	25089	C12H26	2,3,5,7-tetramethyloctane	62199-32-0	-1.9026	-4.4374E-06	2.2187E-12	0	358700	2	1.2515E-02	8.8535E-03
740	25090	C12H26	2,3,6,6-tetramethyloctane	62199-33-1	-1.9667	-4.4374E-06	2.2187E-12	0	358700	2	1.0796E-02	7.6371E-03
741	25091	C12H26	2,3,6,7-tetramethyloctane	52670-34-5	-2.0296	-4.4374E-06	2.2187E-12	0	358700	2	9.3411E-03	6.6080E-03
742	25092	C12H26	2,4,4,5-tetramethyloctane	62199-34-2	-1.9241	-4.4374E-06	2.2187E-12	0	358700	2	1.1910E-02	8.4251E-03
743	25093	C12H26	2,4,4,6-tetramethyloctane	62199-35-3	-1.7928	-4.4374E-06	2.2187E-12	0	358700	2	1.6114E-02	1.1399E-02
744	25094	C12H26	2,4,4,7-tetramethyloctane	35866-96-7	-1.7704	-4.4374E-06	2.2187E-12	0	358700	2	1.6965E-02	1.2001E-02
745	25095	C12H26	2,4,5,5-tetramethyloctane	62199-36-4	-1.9455	-4.4374E-06	2.2187E-12	0	358700	2	1.1337E-02	8.0201E-03
746	25096	C12H26	2,4,5,6-tetramethyloctane	62199-37-5	-1.9878	-4.4374E-06	2.2187E-12	0	358700	2	1.0284E-02	7.2748E-03
747	25097	C12H26	2,4,5,7-tetramethyloctane	2217-17-6	-1.8217	-4.4374E-06	2.2187E-12	0	358700	2	1.5078E-02	1.0666E-02
748	25098	C12H26	2,4,6,6-tetramethyloctane	62199-38-6	-1.8371	-4.4374E-06	2.2187E-12	0	358700	2	1.4551E-02	1.0294E-02
749	25099	C12H26	2,5,5,6-tetramethyloctane	62199-39-7	-1.9878	-4.4374E-06	2.2187E-12	0	358700	2	1.0284E-02	7.2748E-03
750	25100	C12H26	2,5,6,6-tetramethyloctane	62199-40-0	-2.0296	-4.4374E-06	2.2187E-12	0	358700	2	9.3411E-03	6.6080E-03
751	25101	C12H26	3,3,4,4-tetramethyloctane	62199-41-1	-2.1511	-4.4374E-06	2.2187E-12	0	358700	2	7.0624E-03	4.9960E-03
752	25102	C12H26	3,3,4,5-tetramethyloctane	62199-42-2	-2.0707	-4.4374E-06	2.2187E-12	0	358700	2	8.4970E-03	6.0109E-03
753	25103	C12H26	3,3,4,6-tetramethyloctane	62199-43-3	-2.0296	-4.4374E-06	2.2187E-12	0	358700	2	9.3411E-03	6.6080E-03
754	25104	C12H26	3,3,5,5-tetramethyloctane	62199-44-4	-2.0502	-4.4374E-06	2.2187E-12	0	358700	2	8.9075E-03	6.3012E-03
755	25105	C12H26	3,3,5,6-tetramethyloctane	62199-45-5	-1.9667	-4.4374E-06	2.2187E-12	0	358700	2	1.0796E-02	7.6371E-03
756	25106	C12H26	3,3,6,6-tetramethyloctane	62199-46-6	-1.8961	-4.4374E-06	2.2187E-12	0	358700	2	1.2704E-02	8.9868E-03
757	25107	C12H26	3,4,4,5-tetramethyloctane	62199-47-7	-2.1312	-4.4374E-06	2.2187E-12	0	358700	2	7.3923E-03	5.2294E-03
758	25108	C12H26	3,4,4,6-tetramethyloctane	62185-19-7	-2.0088	-4.4374E-06	2.2187E-12	0	358700	2	9.7994E-03	6.9322E-03
759	25109	C12H26	3,4,5,5-tetramethyloctane	62185-20-0	-2.0502	-4.4374E-06	2.2187E-12	0	358700	2	8.9075E-03	6.3012E-03
760	25110	C12H26	3,4,5,6-tetramethyloctane	62185-21-1	-2.0911	-4.4374E-06	2.2187E-12	0	358700	2	8.1085E-03	5.7360E-03
761	25111	C12H26	4,4,5,5-tetramethyloctane	62185-22-2	-2.1112	-4.4374E-06	2.2187E-12	0	358700	2	7.7407E-03	5.4758E-03
762	25112	C12H26	2-methyl-4-tert-butylheptane	62185-23-3	-1.7928	-4.4374E-06	2.2187E-12	0	358700	2	1.6114E-02	1.1399E-02
763	25113	C12H26	3-methyl-4-tert-butylheptane	62185-24-4	-1.9026	-4.4374E-06	2.2187E-12	0	358700	2	1.2515E-02	8.8535E-03
764	25114	C12H26	4-methyl-4-tert-butylheptane	62185-25-5	-2.0707	-4.4374E-06	2.2187E-12	0	358700	2	8.4970E-03	6.0109E-03
765	25115	C12H26	3-ethyl-4-propylheptane	62185-26-6	-2.0296	-4.4374E-06	2.2187E-12	0	358700	2	9.3411E-03	6.6080E-03
766	25116	C12H26	4-ethyl-4-propylheptane	17312-43-5	-2.0502	-4.4374E-06	2.2187E-12	0	358700	2	8.9075E-03	6.3012E-03
767	25117	C12H26	3-ethyl-4-isopropylheptane	62185-27-7	-2.0088	-4.4374E-06	2.2187E-12	0	358700	2	9.7994E-03	6.9322E-03
768	25118	C12H26	4-ethyl-4-isopropylheptane	62185-28-8	-2.1112	-4.4374E-06	2.2187E-12	0	358700	2	7.7407E-03	5.4758E-03
769	25119	C12H26	2,2-dimethyl-4-propylheptane	---	-1.7704	-4.4374E-06	2.2187E-12	0	358700	2	1.6965E-02	1.2001E-02
770	25120	C12H26	2,3-dimethyl-4-propylheptane	62185-30-2	-1.9878	-4.4374E-06	2.2187E-12	0	358700	2	1.0284E-02	7.2748E-03
771	25121	C12H26	2,4-dimethyl-4-propylheptane	62185-31-3	-1.8809	-4.4374E-06	2.2187E-12	0	358700	2	1.3156E-02	9.3067E-03
772	25122	C12H26	2,5-dimethyl-4-propylheptane	62185-32-4	-1.9026	-4.4374E-06	2.2187E-12	0	358700	2	1.2515E-02	8.8535E-03
773	25123	C12H26	2,6-dimethyl-4-propylheptane	62185-33-5	-1.7839	-4.4374E-06	2.2187E-12	0	358700	2	1.6449E-02	1.1636E-02
774	25124	C12H26	3,3-dimethyl-4-propylheptane	62185-34-6	-2.0296	-4.4374E-06	2.2187E-12	0	358700	2	9.3411E-03	6.6080E-03
775	25125	C12H26	3,4-dimethyl-4-propylheptane	62185-35-7	-2.0707	-4.4374E-06	2.2187E-12	0	358700	2	8.4970E-03	6.0109E-03
776	25126	C12H26	3,5-dimethyl-4-propylheptane	62185-36-8	-2.0296	-4.4374E-06	2.2187E-12	0	358700	2	9.3411E-03	6.6080E-03
777	25127	C12H26	2,2-dimethyl-3-isopropylheptane	62185-37-9	-1.9026	-4.4374E-06	2.2187E-12	0	358700	2	1.2515E-02	8.8535E-03
778	25128	C12H26	2,2-dimethyl-4-isopropylheptane	62185-38-0	-1.7254	-4.4374E-06	2.2187E-12	0	358700	2	1.8820E-02	1.3314E-02
779	25129	C12H26	2,3-dimethyl-3-isopropylheptane	62185-39-1	-2.1312	-4.4374E-06	2.2187E-12	0	358700	2	7.3923E-03	5.2294E-03
780	25130	C12H26	2,3-dimethyl-4-isopropylheptane	62185-40-4	-1.9455	-4.4374E-06	2.2187E-12	0	358700	2	1.1337E-02	8.0201E-03
781	25131	C12H26	2,4-dimethyl-3-isopropylheptane	62185-41-5	-1.9455	-4.4374E-06	2.2187E-12	0	358700	2	1.1337E-02	8.0201E-03
782	25132	C12H26	2,4-dimethyl-4-isopropylheptane	62185-42-6	-1.9455	-4.4374E-06	2.2187E-12	0	358700	2	1.1337E-02	8.0201E-03
783	25133	C12H26	2,5-dimethyl-3-isopropylheptane	62185-43-7	-1.9026	-4.4374E-06	2.2187E-12	0	358700	2	1.2515E-02	8.8535E-03
784	25134	C12H26	2,5-dimethyl-4-isopropylheptane	62185-44-8	-1.8809	-4.4374E-06	2.2187E-12	0	358700	2	1.3156E-02	9.3067E-03
785	25135	C12H26	2,6-dimethyl-3-isopropylheptane	62185-45-9	-1.9198	-4.4374E-06	2.2187E-12	0	358700	2	1.2028E-02	8.5089E-03
786	25136	C12H26	2,6-dimethyl-4-isopropylheptane	35866-89-8	-1.7480	-4.4374E-06	2.2187E-12	0	358700	2	1.7866E-02	1.2639E-02
787	25137	C12H26	3,3-dimethyl-4-isopropylheptane	62185-46-0	-1.9878	-4.4374E-06	2.2187E-12	0	358700	2	1.0284E-02	7.2748E-03
788	25138	C12H26	3,4-dimethyl-4-isopropylheptane	62185-47-1	-2.1312	-4.4374E-06	2.2187E-12	0	358700	2	7.3923E-03	5.2294E-03
789	25139	C12H26	3,5-dimethyl-4-isopropylheptane	62198-89-4	-1.9878	-4.4374E-06	2.2187E-12	0	358700	2	1.0284E-02	7.2748E-03
790	25140	C12H26	2-methyl-3,3-diethylheptane	62198-90-7	-2.1707	-4.4374E-06	2.2187E-12	0	358700	2	6.7498E-03	4.7749E-03
791	25141	C12H26	2-methyl-3,4-diethylheptane	62198-91-8	-2.0088	-4.4374E-06	2.2187E-12	0	358700	2	9.7994E-03	6.9322E-03
792	25142	C12H26	2-methyl-3,5-diethylheptane	62198-92-9	-2.0088	-4.4374E-06	2.2187E-12	0	358700	2	9.7994E-03	6.9322E-03
793	25143	C12H26	2-methyl-4,4-diethylheptane	62198-93-0	-1.9878	-4.4374E-06	2.2187E-12	0	358700	2	1.0284E-02	7.2748E-03
794	25144	C12H26	2-methyl-4,5-diethylheptane	62198-94-1	-1.9667	-4.4374E-06	2.2187E-12	0	358700	2	1.0796E-02	7.6371E-03
795	25145	C12H26	2-methyl-5,5-diethylheptane	62198-95-2	-2.0707	-4.4374E-06	2.2187E-12	0	358700	2	8.4970E-03	6.0109E-03
796	25146	C12H26	3-methyl-3,4-diethylheptane	62198-96-3	-2.1312	-4.4374E-06	2.2187E-12	0	358700	2	7.3923E-03	5.2294E-03
797	25147	C12H26	3-methyl-3,5-diethylheptane	62198-97-4	-2.0502	-4.4374E-06	2.2187E-12	0	358700	2	8.9075E-03	6.3012E-03
798	25148	C12H26	3-methyl-4,4-diethylheptane	62198-98-5	-2.1902	-4.4374E-06	2.2187E-12	0	358700	2	6.4536E-03	4.5653E-03
799	25149	C12H26	3-methyl-4,5-diethylheptane	62198-99-6	-2.0707	-4.4374E-06	2.2187E-12	0	358700	2	8.4970E-03	6.0109E-03
800	25150	C12H26	3-methyl-5,5-diethylheptane	62199-00-2	-2.0911	-4.4374E-06	2.2187E-12	0	358700	2	8.1085E-03	5.7360E-03

Table 5 Solubility in Salt Water - Organic Compounds

$\log_{10} S = A + B X + C X^2$ (S - solubility, parts per million by weight, ppm (wt), X - mg salt / kg of water)												
NO	ID	FORMULA	NAME	CAS No	A	B	C	XMIN	XMAX	code	S@X=0	S@X=34,472
801	25151	C12H26	4-methyl-3,3-diethylheptane	62199-01-3	-2.2095	-4.4374E-06	2.2187E-12	0	358700	2	6.1729E-03	4.3668E-03
802	25152	C12H26	4-methyl-3,4-diethylheptane	62199-02-4	-2.1312	-4.4374E-06	2.2187E-12	0	358700	2	7.3923E-03	5.2294E-03
803	25153	C12H26	4-methyl-3,5-diethylheptane	62199-03-5	-2.1112	-4.4374E-06	2.2187E-12	0	358700	2	7.7407E-03	5.4758E-03
804	25154	C12H26	2,2,3-trimethyl-3-ethylheptane	62199-04-6	-2.1112	-4.4374E-06	2.2187E-12	0	358700	2	7.7407E-03	5.4758E-03
805	25155	C12H26	2,2,3-trimethyl-4-ethylheptane	62199-05-7	-1.9667	-4.4374E-06	2.2187E-12	0	358700	2	1.0796E-02	7.6371E-03
806	25156	C12H26	2,2,3-trimethyl-5-ethylheptane	62199-06-8	-1.9455	-4.4374E-06	2.2187E-12	0	358700	2	1.1337E-02	8.0201E-03
807	25157	C12H26	2,2,4-trimethyl-3-ethylheptane	62199-07-9	-1.9455	-4.4374E-06	2.2187E-12	0	358700	2	1.1337E-02	8.0201E-03
808	25158	C12H26	2,2,4-trimethyl-4-ethylheptane	62199-08-0	-1.9878	-4.4374E-06	2.2187E-12	0	358700	2	1.0284E-02	7.2748E-03
809	25159	C12H26	2,2,4-trimethyl-5-ethylheptane	62199-09-1	-1.8371	-4.4374E-06	2.2187E-12	0	358700	2	1.4551E-02	1.0294E-02
810	25160	C12H26	2,2,5-trimethyl-3-ethylheptane	62199-10-4	-1.9026	-4.4374E-06	2.2187E-12	0	358700	2	1.2515E-02	8.8535E-03
811	25161	C12H26	2,2,5-trimethyl-4-ethylheptane	62199-11-5	-1.8150	-4.4374E-06	2.2187E-12	0	358700	2	1.5310E-02	1.0831E-02
812	25162	C12H26	2,2,5-trimethyl-5-ethylheptane	62199-12-6	-1.8809	-4.4374E-06	2.2187E-12	0	358700	2	1.3156E-02	9.3067E-03
813	25163	C12H26	2,2,6-trimethyl-3-ethylheptane	62199-13-7	-1.9026	-4.4374E-06	2.2187E-12	0	358700	2	1.2515E-02	8.8535E-03
814	25164	C12H26	2,2,6-trimethyl-4-ethylheptane	62199-14-8	-1.6798	-4.4374E-06	2.2187E-12	0	358700	2	2.0901E-02	1.4786E-02
815	25165	C12H26	2,2,6-trimethyl-5-ethylheptane	62199-15-9	-1.8371	-4.4374E-06	2.2187E-12	0	358700	2	1.4551E-02	1.0294E-02
816	25166	C12H26	2,3,3-trimethyl-4-ethylheptane	62199-16-0	-2.0911	-4.4374E-06	2.2187E-12	0	358700	2	8.1085E-03	5.7360E-03
817	25167	C12H26	2,3,3-trimethyl-5-ethylheptane	62199-17-1	-1.9878	-4.4374E-06	2.2187E-12	0	358700	2	1.0284E-02	7.2748E-03
818	25168	C12H26	2,3,4-trimethyl-3-ethylheptane	62199-18-2	-2.1707	-4.4374E-06	2.2187E-12	0	358700	2	6.7498E-03	4.7749E-03
819	25169	C12H26	2,3,4-trimethyl-4-ethylheptane	62198-55-4	-2.0707	-4.4374E-06	2.2187E-12	0	358700	2	8.4970E-03	6.0109E-03
820	25170	C12H26	2,3,4-trimethyl-5-ethylheptane	62198-56-5	-2.0502	-4.4374E-06	2.2187E-12	0	358700	2	8.9075E-03	6.3012E-03
821	25171	C12H26	2,3,5-trimethyl-3-ethylheptane	62198-57-6	-2.0605	-4.4374E-06	2.2187E-12	0	358700	2	8.6994E-03	6.1541E-03
822	25172	C12H26	2,3,5-trimethyl-4-ethylheptane	62198-58-7	-2.0296	-4.4374E-06	2.2187E-12	0	358700	2	9.3411E-03	6.6080E-03
823	25173	C12H26	2,3,5-trimethyl-5-ethylheptane	62198-59-8	-1.9878	-4.4374E-06	2.2187E-12	0	358700	2	1.0284E-02	7.2748E-03
824	25174	C12H26	2,3,6-trimethyl-3-ethylheptane	62198-60-1	-2.0296	-4.4374E-06	2.2187E-12	0	358700	2	9.3411E-03	6.6080E-03
825	25175	C12H26	2,3,6-trimethyl-4-ethylheptane	62198-61-2	-1.9026	-4.4374E-06	2.2187E-12	0	358700	2	1.2515E-02	8.8535E-03
826	25176	C12H26	2,3,6-trimethyl-5-ethylheptane	62198-62-3	-1.9667	-4.4374E-06	2.2187E-12	0	358700	2	1.0796E-02	7.6371E-03
827	25177	C12H26	2,4,4-trimethyl-3-ethylheptane	62198-63-4	-1.9878	-4.4374E-06	2.2187E-12	0	358700	2	1.0284E-02	7.2748E-03
828	25178	C12H26	2,4,4-trimethyl-5-ethylheptane	62198-64-5	-1.9455	-4.4374E-06	2.2187E-12	0	358700	2	1.1337E-02	8.0201E-03
829	25179	C12H26	2,4,5-trimethyl-3-ethylheptane	62198-65-6	-2.0296	-4.4374E-06	2.2187E-12	0	358700	2	9.3411E-03	6.6080E-03
830	25180	C12H26	2,4,5-trimethyl-4-ethylheptane	62198-66-7	-2.0088	-4.4374E-06	2.2187E-12	0	358700	2	9.7994E-03	6.9322E-03
831	25181	C12H26	2,4,5-trimethyl-5-ethylheptane	62198-67-8	-2.0502	-4.4374E-06	2.2187E-12	0	358700	2	8.9075E-03	6.3012E-03
832	25182	C12H26	2,4,6-trimethyl-3-ethylheptane	62198-68-9	-1.9241	-4.4374E-06	2.2187E-12	0	358700	2	1.1910E-02	8.4251E-03
833	25183	C12H26	2,4,6-trimethyl-4-ethylheptane	62198-69-0	-1.7928	-4.4374E-06	2.2187E-12	0	358700	2	1.6114E-02	1.1399E-02
834	25184	C12H26	2,5,5-trimethyl-3-ethylheptane	62198-70-3	-1.9026	-4.4374E-06	2.2187E-12	0	358700	2	1.2515E-02	8.8535E-03
835	25185	C12H26	2,5,5-trimethyl-4-ethylheptane	62198-71-4	-1.9455	-4.4374E-06	2.2187E-12	0	358700	2	1.1337E-02	8.0201E-03
836	25186	C12H26	3,3,4-trimethyl-4-ethylheptane	62198-72-5	-2.2095	-4.4374E-06	2.2187E-12	0	358700	2	6.1729E-03	4.3668E-03
837	25187	C12H26	3,3,4-trimethyl-5-ethylheptane	62198-73-6	-2.0911	-4.4374E-06	2.2187E-12	0	358700	2	8.1085E-03	5.7360E-03
838	25188	C12H26	3,3,5-trimethyl-4-ethylheptane	62198-74-7	-2.0502	-4.4374E-06	2.2187E-12	0	358700	2	8.9075E-03	6.3012E-03
839	25189	C12H26	3,3,5-trimethyl-5-ethylheptane	62198-75-8	-2.1511	-4.4374E-06	2.2187E-12	0	358700	2	7.0624E-03	4.9960E-03
840	25190	C12H26	3,4,4-trimethyl-3-ethylheptane	62198-76-9	-2.2095	-4.4374E-06	2.2187E-12	0	358700	2	6.1729E-03	4.3668E-03
841	25191	C12H26	3,4,4-trimethyl-5-ethylheptane	62198-77-0	-2.1511	-4.4374E-06	2.2187E-12	0	358700	2	7.0624E-03	4.9960E-03
842	25192	C12H26	3,4,5-trimethyl-3-ethylheptane	62198-78-1	-2.1511	-4.4374E-06	2.2187E-12	0	358700	2	7.0624E-03	4.9960E-03
843	25193	C12H26	3,4,5-trimethyl-4-ethylheptane	62198-79-2	-2.2095	-4.4374E-06	2.2187E-12	0	358700	2	6.1729E-03	4.3668E-03
844	25194	C12H26	2,2,3,3,4-pentamethylheptane	62198-80-5	-2.1112	-4.4374E-06	2.2187E-12	0	358700	2	7.7407E-03	5.4758E-03
845	25195	C12H26	2,2,3,3,5-pentamethylheptane	62198-81-6	-1.9878	-4.4374E-06	2.2187E-12	0	358700	2	1.0284E-02	7.2748E-03
846	25196	C12H26	2,2,3,3,6-pentamethylheptane	62198-82-7	-1.9667	-4.4374E-06	2.2187E-12	0	358700	2	1.0796E-02	7.6371E-03
847	25197	C12H26	2,2,3,4,4-pentamethylheptane	62198-83-8	-2.1511	-4.4374E-06	2.2187E-12	0	358700	2	7.0624E-03	4.9960E-03
848	25198	C12H26	2,2,3,4,5-pentamethylheptane	62198-84-9	-1.9878	-4.4374E-06	2.2187E-12	0	358700	2	1.0284E-02	7.2748E-03
849	25199	C12H26	2,2,3,4,6-pentamethylheptane	62198-85-0	-1.8591	-4.4374E-06	2.2187E-12	0	358700	2	1.3834E-02	9.7862E-03
850	25200	C12H26	2,2,3,5,5-pentamethylheptane	62198-86-1	-1.8371	-4.4374E-06	2.2187E-12	0	358700	2	1.4551E-02	1.0294E-02
851	25201	C12H26	2,2,3,5,6-pentamethylheptane	62198-87-2	-1.8765	-4.4374E-06	2.2187E-12	0	358700	2	1.3289E-02	9.4005E-03
852	25202	C12H26	2,2,3,6,6-pentamethylheptane	62198-88-3	-1.7928	-4.4374E-06	2.2187E-12	0	358700	2	1.6114E-02	1.1399E-02
853	25203	C12H26	2,2,4,4,5-pentamethylheptane	62199-61-5	-2.0296	-4.4374E-06	2.2187E-12	0	358700	2	9.3411E-03	6.6080E-03
854	25204	C12H26	2,2,4,4,6-pentamethylheptane	62199-62-6	-1.8061	-4.4374E-06	2.2187E-12	0	358700	2	1.5626E-02	1.1054E-02
855	25205	C12H26	2,2,4,5,5-pentamethylheptane	62199-63-7	-1.8150	-4.4374E-06	2.2187E-12	0	358700	2	1.5310E-02	1.0831E-02
856	25206	C12H26	2,2,4,5,6-pentamethylheptane	62199-64-8	-1.7928	-4.4374E-06	2.2187E-12	0	358700	2	1.6114E-02	1.1399E-02
857	25207	C12H26	2,2,4,6,6-pentamethylheptane	13475-82-6	-1.6326	-4.4374E-06	2.2187E-12	0	358700	2	2.3300E-02	1.6482E-02
858	25208	C12H26	2,2,5,5,6-pentamethylheptane	62199-65-9	-1.8150	-4.4374E-06	2.2187E-12	0	358700	2	1.5310E-02	1.0831E-02
859	25209	C12H26	2,3,3,4,4-pentamethylheptane	62199-66-0	-2.1707	-4.4374E-06	2.2187E-12	0	358700	2	6.7498E-03	4.7749E-03
860	25210	C12H26	2,3,3,4,5-pentamethylheptane	62199-67-1	-2.1112	-4.4374E-06	2.2187E-12	0	358700	2	7.7407E-03	5.4758E-03
861	25211	C12H26	2,3,3,4,6-pentamethylheptane	62199-68-2	-2.0088	-4.4374E-06	2.2187E-12	0	358700	2	9.7994E-03	6.9322E-03
862	25212	C12H26	2,3,3,5,5-pentamethylheptane	62199-69-3	-2.1112	-4.4374E-06	2.2187E-12	0	358700	2	7.7407E-03	5.4758E-03
863	25213	C12H26	2,3,3,5,6-pentamethylheptane	52670-35-6	-1.9455	-4.4374E-06	2.2187E-12	0	358700	2	1.1337E-02	8.0201E-03
864	25214	C12H26	2,3,4,4,5-pentamethylheptane	62199-70-6	-2.1112	-4.4374E-06	2.2187E-12	0	358700	2	7.7407E-03	5.4758E-03
865	25215	C12H26	2,3,4,4,6-pentamethylheptane	62199-71-7	-1.9026	-4.4374E-06	2.2187E-12	0	358700	2	1.2515E-02	8.8535E-03
866	25216	C12H26	2,3,4,5,5-pentamethylheptane	62199-72-8	-2.0502	-4.4374E-06	2.2187E-12	0	358700	2	8.9075E-03	6.3012E-03
867	25217	C12H26	2,3,4,5,6-pentamethylheptane	27574-98-7	-2.0088	-4.4374E-06	2.2187E-12	0	358700	2	9.7994E-03	6.9322E-03
868	25218	C12H26	2,4,4,5,5-pentamethylheptane	62199-73-9	-2.0296	-4.4374E-06	2.2187E-12	0	358700	2	9.3411E-03	6.6080E-03
869	25219	C12H26	3,3,4,4,5-pentamethylheptane	62199-74-0	-2.2286	-4.4374E-06	2.2187E-12	0	358700	2	5.9068E-03	4.1786E-03
870	25220	C12H26	3,3,4,5,5-pentamethylheptane	62199-75-1	-2.2850	-4.4374E-06	2.2187E-12	0	358700	2	5.1885E-03	3.6704E-03
871	25221	C12H26	2,2-dimethyl-3-tert-butylhexane	62199-76-2	-2.0296	-4.4374E-06	2.2187E-12	0	358700	2	9.3411E-03	6.6080E-03
872	25222	C12H26	2-methyl-3-ethyl-3-isopropylhexane	62199-77-3	-2.1112	-4.4374E-06	2.2187E-12	0	358700	2	7.7407E-03	5.4758E-03
873	25223	C12H26	2-methyl-4-ethyl-3-isopropylhexane	62199-78-4	-1.9026	-4.4374E-06	2.2187E-12	0	358700	2	1.2515E-02	8.8535E-03
874	25224	C12H26	2,2,3-trimethyl-3-isopropylhexane	62199-79-5	-2.1312	-4.4374E-06	2.2187E-12	0	358700	2	7.3923E-03	5.2294E-03
875	25225	C12H26	2,2,4-trimethyl-3-isopropylhexane	62199-80-8	-1.8809	-4.4374E-06	2.2187E-12	0	358700	2	1.3156E-02	9.3067E-03
876	25226	C12H26	2,2,5-trimethyl-3-isopropylhexane	62199-81-9	-1.6798	-4.4374E-06	2.2187E-12	0	358700	2	2.0901E-02	1.4786E-02
877	25227	C12H26	2,2,5-trimethyl-4-isopropylhexane	62199-82-0	-1.6798	-4.4374E-06	2.2187E-12	0	358700	2	2.0901E-02	1.4786E-02
878	25228	C12H26	2,3,4-trimethyl-3-isopropylhexane	62199-83-1	-2.1902	-4.4374E-06	2.2187E-12	0	358700	2	6.4536E-03	4.5653E-03
879	25229	C12H26	2,3,5-trimethyl-3-isopropylhexane	62199-84-2	-1.9878	-4.4374E-06	2.2187E-12	0	358700	2	1.0284E-02	7.2748E-03
880	25230	C12H26	2,3,5-trimethyl-4-isopropylhexane	62199-85-3	-1.9026	-4.4374E-06	2.2187E-12	0	358700	2	1.2515E-02	8.8535E-03

Table 5 Solubility in Salt Water - Organic Compounds

$\log_{10} S = A + B X + C X^2$ (S - solubility, parts per million by weight, ppm (wt), X - mg salt / kg of water)

NO	ID	FORMULA	NAME	CAS No	A	B	C	XMIN	XMAX	code	S@X=0	S@X=34,472
881	25231	C12H26	2,4,4-trimethyl-3-isopropylhexane	62199-86-4	-1.9455	-4.4374E-06	2.2187E-12	0	358700	2	1.1337E-02	8.0201E-03
882	25232	C12H26	3,3,4-triethylhexane	62199-87-5	-2.2286	-4.4374E-06	2.2187E-12	0	358700	2	5.9068E-03	4.1786E-03
883	25233	C12H26	2,2-dimethyl,3,3-diethylhexane	62199-88-6	-2.1112	-4.4374E-06	2.2187E-12	0	358700	2	7.7407E-03	5.4758E-03
884	25234	C12H26	2,2-dimethyl-3,4-diethylhexane	62199-89-7	-1.9026	-4.4374E-06	2.2187E-12	0	358700	2	1.2515E-02	8.8535E-03
885	25235	C12H26	2,2-dimethyl-4,4-diethylhexane	62184-89-8	-2.1112	-4.4374E-06	2.2187E-12	0	358700	2	7.7407E-03	5.4758E-03
886	25236	C12H26	2,3-dimethyl-3,4-diethylhexane	62184-90-1	-2.1902	-4.4374E-06	2.2187E-12	0	358700	2	6.4536E-03	4.5653E-03
887	25237	C12H26	2,3-dimethyl-4,4-diethylhexane	62184-91-2	-2.1902	-4.4374E-06	2.2187E-12	0	358700	2	6.4536E-03	4.5653E-03
888	25238	C12H26	2,4-dimethyl-3,3-diethylhexane	62184-92-3	-2.1112	-4.4374E-06	2.2187E-12	0	358700	2	7.7407E-03	5.4758E-03
889	25239	C12H26	2,4-dimethyl-3,4-diethylhexane	62184-93-4	-2.1112	-4.4374E-06	2.2187E-12	0	358700	2	7.7407E-03	5.4758E-03
890	25240	C12H26	2,5-dimethyl-3,3-diethylhexane	62184-94-5	-1.9026	-4.4374E-06	2.2187E-12	0	358700	2	1.2515E-02	8.8535E-03
891	25241	C12H26	2,5-dimethyl-3,4-diethylhexane	62184-95-6	-1.9026	-4.4374E-06	2.2187E-12	0	358700	2	1.2515E-02	8.8535E-03
892	25242	C12H26	3,3-dimethyl-4,4-diethylhexane	62184-96-7	-2.3034	-4.4374E-06	2.2187E-12	0	358700	2	4.9733E-03	3.5181E-03
893	25243	C12H26	3,4-dimethyl-3,4-diethylhexane	62184-97-8	-2.3034	-4.4374E-06	2.2187E-12	0	358700	2	4.9733E-03	3.5181E-03
894	25244	C12H26	2,2,3,3-tetramethyl-4-ethylhexane	62184-98-9	-2.1312	-4.4374E-06	2.2187E-12	0	358700	2	7.3923E-03	5.2294E-03
895	25245	C12H26	2,2,3,4-tetramethyl-3-ethylhexane	62184-99-0	-2.1902	-4.4374E-06	2.2187E-12	0	358700	2	6.4536E-03	4.5653E-03
896	25246	C12H26	2,2,3,4-tetramethyl-4-ethylhexane	62185-00-6	-2.2664	-4.4374E-06	2.2187E-12	0	358700	2	5.4154E-03	3.8309E-03
897	25247	C12H26	2,2,3,5-tetramethyl-3-ethylhexane	62185-01-7	-1.9878	-4.4374E-06	2.2187E-12	0	358700	2	1.0284E-02	7.2748E-03
898	25248	C12H26	2,2,3,5-tetramethyl-4-ethylhexane	62185-02-8	-1.9241	-4.4374E-06	2.2187E-12	0	358700	2	1.1910E-02	8.4251E-03
899	25249	C12H26	2,2,4,4-tetramethyl-3-ethylhexane	62185-03-9	-2.1112	-4.4374E-06	2.2187E-12	0	358700	2	7.7407E-03	5.4758E-03
900	25250	C12H26	2,2,4,5-tetramethyl-3-ethylhexane	62185-04-0	-1.9026	-4.4374E-06	2.2187E-12	0	358700	2	1.2515E-02	8.8535E-03
901	25251	C12H26	2,2,4,5-tetramethyl-4-ethylhexane	62185-05-1	-2.0502	-4.4374E-06	2.2187E-12	0	358700	2	8.9075E-03	6.3012E-03
902	25252	C12H26	2,2,5,5-tetramethyl-3-ethylhexane	62185-06-2	-1.6798	-4.4374E-06	2.2187E-12	0	358700	2	2.0901E-02	1.4786E-02
903	25253	C12H26	2,3,3,4-tetramethyl-4-ethylhexane	62185-07-3	-2.2664	-4.4374E-06	2.2187E-12	0	358700	2	5.4154E-03	3.8309E-03
904	25254	C12H26	2,3,3,5-tetramethyl-4-ethylhexane	62185-08-4	-2.0502	-4.4374E-06	2.2187E-12	0	358700	2	8.9075E-03	6.3012E-03
905	25255	C12H26	2,3,4,4-tetramethyl-3-ethylhexane	62185-09-5	-2.2664	-4.4374E-06	2.2187E-12	0	358700	2	5.4154E-03	3.8309E-03
906	25256	C12H26	2,3,4,5-tetramethyl-3-ethylhexane	62185-10-8	-2.1312	-4.4374E-06	2.2187E-12	0	358700	2	7.3923E-03	5.2294E-03
907	25257	C12H26	2,2,3,3,4,4-hexamethylhexane	62185-11-9	-2.4097	-4.4374E-06	2.2187E-12	0	358700	2	3.8934E-03	2.7542E-03
908	25258	C12H26	2,2,3,3,4,5-hexamethylhexane	62185-12-0	-2.0911	-4.4374E-06	2.2187E-12	0	358700	2	8.1085E-03	5.7360E-03
909	25259	C12H26	2,2,3,3,5,5-hexamethylhexane	60302-24-1	-2.0088	-4.4374E-06	2.2187E-12	0	358700	2	9.7994E-03	6.9322E-03
910	25260	C12H26	2,2,3,4,4,5-hexamethylhexane	62185-13-1	-2.2286	-4.4374E-06	2.2187E-12	0	358700	2	5.9068E-03	4.1786E-03
911	25261	C12H26	2,2,3,4,5,5-hexamethylhexane	62185-14-2	-1.9026	-4.4374E-06	2.2187E-12	0	358700	2	1.2515E-02	8.8535E-03
912	25262	C12H26	2,3,3,4,4,5-hexamethylhexane	52670-36-7	-2.2286	-4.4374E-06	2.2187E-12	0	358700	2	5.9068E-03	4.1786E-03
913	25263	C12H26	2,4-dimethyl-3-ethyl-3-isopropylpentane	62185-16-4	-2.2286	-4.4374E-06	2.2187E-12	0	358700	2	5.9068E-03	4.1786E-03
914	25264	C12H26	2,2,4-trimethyl-3,3-diethylpentane	62185-15-3	-2.2286	-4.4374E-06	2.2187E-12	0	358700	2	5.9068E-03	4.1786E-03
915	25265	C12H26	2,2,3,4-tetramethyl-3-isopropylpentane	62185-17-5	-2.1902	-4.4374E-06	2.2187E-12	0	358700	2	6.4536E-03	4.5653E-03
916	25266	C12H26	2,2,4,4-tetramethyl-3-isopropylpentane	62185-18-6	-1.9878	-4.4374E-06	2.2187E-12	0	358700	2	1.0284E-02	7.2748E-03
917	25267	C12H26	2,2,3,4,4-pentamethyl-3-ethylpentane	66576-21-4	-2.3750	-4.4374E-06	2.2187E-12	0	358700	2	4.2167E-03	2.9829E-03
918	25268	C12H26	2-methylundecane	7045-71-8	-2.3188	-4.4374E-06	2.2187E-12	0	358700	2	4.7990E-03	3.3949E-03
919	26283	C13H20	heptylbenzene	1078-71-3	-0.3254	-3.3451E-06	1.6726E-12	0	358700	2	4.7273E-01	3.6415E-01
920	26284	C13H20	1-methyl-2,4-diisopropylbenzene	1460-98-6	0.4522	-4.5929E-06	2.2646E-12	0	358700	2	2.8328E+00	1.9796E+00
921	26285	C13H20	1-methyl-3,5-diisopropylbenzene	3055-14-9	0.6113	-4.5929E-06	2.2646E-12	0	358700	2	4.0861E+00	2.8554E+00
922	26286	C13H20	[2-methyl-1-isopropylpropyl]benzene	21777-84-4	0.5213	-4.5929E-06	2.2646E-12	0	358700	2	3.3214E+00	2.3211E+00
923	26288	C13H20	(2,4-dimethylpentyl)benzene	54518-00-2	0.5573	-4.5929E-06	2.2646E-12	0	358700	2	3.6080E+00	2.5214E+00
924	26465	C13H26	octylcyclopentane	1795-20-6	-2.0446	-7.1550E-06	3.5775E-12	0	358700	2	9.0239E-03	5.1642E-03
925	26466	C13H26	heptylcyclohexane	5617-41-4	-1.8570	-7.1550E-06	3.5775E-12	0	358700	2	1.3899E-02	7.9543E-03
926	26467	C13H26	2-butyl-1,1,3-trimethylcyclohexane	54676-39-0	-1.7726	-7.1550E-06	3.5775E-12	0	358700	2	1.6880E-02	9.6602E-03
927	26526	C13H28	tridecane	629-50-5	-2.7012	-4.4374E-06	2.2187E-12	0	358700	2	1.9900E-03	1.4077E-03
928	26527	C13H28	2-methyldodecane	1560-97-0	-2.6211	-4.4374E-06	2.2187E-12	0	358700	2	2.3926E-03	1.6926E-03
929	26528	C13H28	3-methyldodecane	17312-57-1	-2.6324	-4.4374E-06	2.2187E-12	0	358700	2	2.3316E-03	1.6494E-03
930	26529	C13H28	2,2-dimethylundecane	17312-64-0	-2.4927	-4.4374E-06	2.2187E-12	0	358700	2	3.2161E-03	2.2751E-03
931	26530	C13H28	2,3-dimethylundecane	17312-77-5	-2.5997	-4.4374E-06	2.2187E-12	0	358700	2	2.5138E-03	1.7783E-03
932	26531	C13H28	2,4-dimethylundecane	17312-80-0	-2.4097	-4.4374E-06	2.2187E-12	0	358700	2	3.8934E-03	2.7542E-03
933	26532	C13H28	5-butylnonane	17312-63-9	-2.4351	-4.4374E-06	2.2187E-12	0	358700	2	3.6718E-03	2.5974E-03
934	27644	C14H22	octylbenzene	2189-60-8	-0.6025	-3.3451E-06	1.6726E-12	0	358700	2	2.4977E-01	1.9240E-01
935	27645	C14H22	1,2,3,4-tetraethylbenzene	642-32-0	-0.0487	-4.5929E-06	2.2646E-12	0	358700	2	8.9385E-01	6.2464E-01
936	27646	C14H22	1,2,3,5-tetraethylbenzene	38842-05-6	-0.0135	-4.5929E-06	2.2646E-12	0	358700	2	9.6930E-01	6.7736E-01
937	27647	C14H22	1,2,4,5-tetraethylbenzene	635-81-4	-0.0326	-4.5929E-06	2.2646E-12	0	358700	2	9.2777E-01	6.4834E-01
938	27648	C14H22	1,4-di-tert-butylbenzene	1012-72-2	0.1903	-4.5929E-06	2.2646E-12	0	358700	2	1.5497E+00	1.0830E+00
939	27649	C14H22	(1-methylheptyl)benzene	777-22-0	-0.4031	-3.3451E-06	1.6726E-12	0	358700	2	3.9525E-01	3.0447E-01
940	27650	C14H22	1,3-di-tert-butylbenzene	1014-60-4	-0.0451	-4.5929E-06	2.2646E-12	0	358700	2	9.0131E-01	6.2985E-01
941	27847	C14H28	nonylcyclopentane	2882-98-6	-2.1890	-7.1550E-06	3.5775E-12	0	358700	2	6.4718E-03	3.7037E-03
942	27848	C14H28	octylcyclohexane	1795-15-9	-1.9962	-7.1550E-06	3.5775E-12	0	358700	2	1.0088E-02	5.7731E-03
943	27912	C14H30	tetradecane	629-59-4	-2.6576	-4.4374E-06	2.2187E-12	0	358700	1,2	2.2000E-03	1.5563E-03
944	27913	C14H30	2-methyltridecane	1560-96-9	-2.8383	-4.4374E-06	2.2187E-12	0	358700	2	1.4512E-03	1.0266E-03
945	27914	C14H30	3-methyltridecane	6418-41-3	-2.8456	-4.4374E-06	2.2187E-12	0	358700	2	1.4268E-03	1.0093E-03
946	27915	C14H30	2,2-dimethyldodecane	49598-54-1	-2.7559	-4.4374E-06	2.2187E-12	0	358700	2	1.7543E-03	1.2410E-03
947	27916	C14H30	2,3-dimethyldodecane	6117-98-2	-2.8298	-4.4374E-06	2.2187E-12	0	358700	2	1.4800E-03	1.0469E-03
948	27917	C14H30	2,4-dimethyldodecane	6117-99-3	-2.6823	-4.4374E-06	2.2187E-12	0	358700	2	2.0784E-03	1.4703E-03
949	27918	C14H30	7-methyltridecane	26730-14-3	-2.8120	-4.4374E-06	2.2187E-12	0	358700	2	1.5417E-03	1.0906E-03
950	28725	C15H24	nonylbenzene	1081-77-2	-0.7939	-3.3451E-06	1.6726E-12	0	358700	2	1.6075E-01	1.2383E-01
951	28726	C15H24	1,2,4-triisopropylbenzene	948-32-3	0.0688	-4.5929E-06	2.2646E-12	0	358700	2	1.1718E+00	8.1886E-01
952	28727	C15H24	1,3,5-triisopropylbenzene	717-74-8	0.1769	-4.5929E-06	2.2646E-12	0	358700	2	1.5027E+00	1.0501E+00
953	28728	C15H24	3,5-di-tert-butyltoluene	15181-11-0	0.0688	-4.5929E-06	2.2646E-12	0	358700	2	1.1718E+00	8.1886E-01
954	28855	C15H30	decylcyclopentane	1795-21-7	-2.2266	-7.1550E-06	3.5775E-12	0	358700	2	5.9341E-03	3.3959E-03
955	28856	C15H30	nonylcyclohexane	2883-02-5	-2.0241	-7.1550E-06	3.5775E-12	0	358700	2	9.4603E-03	5.4139E-03
956	28857	C15H30	1-ethyl-1-methyl-2,4-diisopropylcyclohexa	515-12-8	-2.0275	-7.1550E-06	3.5775E-12	0	358700	2	9.3864E-03	5.3716E-03
957	28904	C15H32	pentadecane	629-62-9	-2.9724	-4.4374E-06	2.2187E-12	0	358700	2	1.0655E-03	7.5376E-04
958	28905	C15H32	2-methyltetradecane	1560-95-8	-2.9552	-4.4374E-06	2.2187E-12	0	358700	2	1.1088E-03	7.8435E-04
959	28906	C15H32	3-methyltetradecane	18435-22-8	-2.9594	-4.4374E-06	2.2187E-12	0	358700	2	1.0979E-03	7.7668E-04
960	28907	C15H32	2,2-dimethyltridecane	61869-04-3	-2.9111	-4.4374E-06	2.2187E-12	0	358700	2	1.2270E-03	8.6802E-04

Table 5 Solubility in Salt Water - Organic Compounds

			$\log_{10} S = A + B X + C X^2$	(S - solubility, parts per million by weight, ppm (wt), X - mg salt / kg of water)								
NO	ID	FORMULA	NAME	CAS No	A	B	C	XMIN	XMAX	code	S@X=0	S@X=34,472
961	28908	C15H32	2,3-dimethyltridecane	18435-20-6	-2.9543	-4.4374E-06	2.2187E-12	0	358700	2	1.1109E-03	7.8583E-04
962	28909	C15H32	2,4-dimethyltridecane	61868-05-1	-2.8658	-4.4374E-06	2.2187E-12	0	358700	2	1.3620E-03	9.6350E-04
963	29595	C16H26	decylbenzene	104-72-3	-0.8907	-3.3451E-06	1.6726E-12	0	358700	2	1.2861E-01	9.9076E-02
964	29599	C16H26	(1-methylnonyl)benzene	4537-13-7	-0.8489	-3.3451E-06	1.6726E-12	0	358700	2	1.4161E-01	1.0909E-01
965	29704	C16H32	undecylcyclopentane	6785-23-5	-2.1641	-7.1550E-06	3.5775E-12	0	358700	2	6.8530E-03	3.9218E-03
966	29705	C16H32	decylcyclohexane	1795-16-0	-1.9494	-7.1550E-06	3.5775E-12	0	358700	2	1.1236E-02	6.4301E-03
967	29754	C16H34	hexadecane	544-76-3	-3.0458	-4.4374E-06	2.2187E-12	0	358700	1,2	9.0000E-04	6.3667E-04
968	29755	C16H34	2-methylpentadecane	1560-93-6	-2.9760	-4.4374E-06	2.2187E-12	0	358700	2	1.0567E-03	7.4752E-04
969	29756	C16H34	3-methylpentadecane	2882-96-4	-2.9754	-4.4374E-06	2.2187E-12	0	358700	2	1.0583E-03	7.4868E-04
970	29757	C16H34	2,2-dimethyltetradecane	59222-86-5	-2.9780	-4.4374E-06	2.2187E-12	0	358700	2	1.0520E-03	7.4422E-04
971	29758	C16H34	2,3-dimethyltetradecane	18435-23-9	-2.9754	-4.4374E-06	2.2187E-12	0	358700	2	1.0583E-03	7.4868E-04
972	29759	C16H34	2,4-dimethyltetradecane	61868-06-2	-2.9583	-4.4374E-06	2.2187E-12	0	358700	2	1.1008E-03	7.7870E-04
973	29760	C16H34	2,2,4,4,6,8,8-heptamethylnonane	4390-04-9	-2.8235	-4.4374E-06	2.2187E-12	0	358700	2	1.5016E-03	1.0622E-03
974	30264	C17H28	undecylbenzene	6742-54-7	-0.9071	-3.3451E-06	1.6726E-12	0	358700	2	1.2384E-01	9.5395E-02
975	30310	C17H34	dodecylcyclopentane	5634-30-0	-2.0086	-7.1550E-06	3.5775E-12	0	358700	2	9.8042E-03	5.6107E-03
976	30311	C17H34	undecylcyclohexane	54105-66-7	-1.7780	-7.1550E-06	3.5775E-12	0	358700	2	1.6673E-02	9.5417E-03
977	30346	C17H36	heptadecane	629-78-7	-2.8617	-4.4374E-06	2.2187E-12	0	358700	2	1.3749E-03	9.7259E-04
978	30347	C17H36	2-methylhexadecane	1560-92-5	-2.9061	-4.4374E-06	2.2187E-12	0	358700	2	1.2414E-03	8.7816E-04
979	30348	C17H36	3-methylhexadecane	6418-43-5	-2.8983	-4.4374E-06	2.2187E-12	0	358700	2	1.2638E-03	8.9400E-04
980	30349	C17H36	2,2-dimethylpentadecane	61869-05-4	-2.9492	-4.4374E-06	2.2187E-12	0	358700	2	1.1240E-03	7.9510E-04
981	30350	C17H36	2,3-dimethylpentadecane	2882-97-5	-2.8983	-4.4374E-06	2.2187E-12	0	358700	2	1.2638E-03	8.9400E-04
982	30351	C17H36	2,4-dimethylpentadecane	61868-07-3	-2.9770	-4.4374E-06	2.2187E-12	0	358700	2	1.0545E-03	7.4596E-04

code: 1 - data, 2 - estimate

Salt (NaCl) concentration range is X = 0 to 358,700 (saturated) mg of salt / kg of water.

Salt (NaCl) in sea water is X = 34,472 mg of salt / kg of water.

Chapter 6

SOLUBILITY IN WATER AS A FUNCTION OF TEMPERATURE – ORGANIC COMPOUNDS

Carl L. Yaws, Akhila Kodali, and Kayzad J. Vajifdar
Lamar University, Beaumont, Texas

Tabulation Results

The results for solubility in water are given in Table 6. For the tabulation, the correlation for water solubility as a function of temperature is based on a series expansion in temperature:

$$\log_{10} S = A + B\,T + C\,T^2 \qquad (6)$$

where S = solubility in water, parts per million by weight, ppm(wt)
T = temperature, K
A, B, and C = regression coefficients

The tabulation is arranged by carbon number (C, C2, C3, ...) to provide ease of use in quickly locating the data by using the chemical formula. The compound name, CAS No (Chemical Abstracts Registry Number), and regression coefficients are provided in the adjacent columns. The range of application is denoted by minimum and maximum temperatures (TMIN and TMAX). Temperatures outside the range of application should not be used. The next column provides the code for the tabulation which is based on both experimental data and estimated values.

In preparing the tabulation, a literature search was conducted to identify data source publications for organics (1-20). Both experimental values for the property under consideration and parameter values for estimation of the property are included in the source publications. The publications were screened and copies of appropriate data were made. These data were then keyed into the computer to provide a database of values for compounds for which experimental data are available. The database also served as a basis to check the accuracy of the correlations. The numerous point values were processed using a computer program for minimum deviation.

For ambient temperature, the compilations by Howard and Meylan (1); Mackay, Shiu, and Ma (2-3); Verschueren (6); Yalkowsky (7); and Yaws (8-19) were primarily used. For temperatures higher than ambient, the compilations of Price (4), API (5), and Yaws (13-14, 16) were used. Parameter values for the estimates were from data compilations of Yaws (18-20).

For hydrocarbons and chemicals, experimental data at temperatures higher than ambient are very limited for isomers and large molecules (C10 and higher). In the absence of data, the estimates for isomers and large molecules should be considered as rough approximations useful for initial analysis. If initial analysis is favorable, follow-up experimental determination is recommended.

A comparison of calculated and data values is shown in Figure 6 for a representative compound. The graph discloses favorable agreement of equation and data.

Example

A chemical spill of heptane (C_7H_{16}) occurs into a body of water at 55.7 C. Determine the concentration in water at saturation.

Substitution of the regression coefficients from the table and temperature (55.7 C = 328.85 K) into the equation for solubility in water yields:

$$\log_{10} S = 0.6113 - 5.8784\text{E-}03 \times 328.85 + 1.6779\text{E-}05 \times 328.85^2$$

$$S = 10^{0.4928} = 3.11 \text{ parts per million by weight, ppm(wt)}$$

References

1. Howard, P. H. and W. M. Meylan, eds., HANDBOOK OF PHYSICAL PROPERTIES OF ORGANIC CHEMICALS, CRC Press, Boca Raton, FL (1997).
2. Mackay, D., W. Y. Shiu, and K. C. Ma, ILLUSTRATED HANDBOOK OF PHYSICAL-CHEMICAL PROPERTIES AND ENVIRONMENTAL FATE FOR ORGANIC CHEMICALS, Volume 1 – Monoaromatic Hydrocarbons, Chlorobenzenes and PCBs, Lewis Publishers, New York, NY (1992).
3. Mackay, D., W. Y. Shiu, and K. C. Ma, ILLUSTRATED HANDBOOK OF PHYSICAL-CHEMICAL PROPERTIES AND ENVIRONMENTAL FATE FOR ORGANIC CHEMICALS, Volume 3 – Volatile Organic

Compounds, Lewis Publishers, New York, NY (1993).
4. Price, L. C., Bull. Am. Assoc. Petrol. Geologists, 60(2), 213-244 (1976).
5. TECHNICAL DATA BOOK - PETROLEUM REFINING, Vols. I and II, American Petroleum Institute, Washington, DC (1972, 1977, 1982).
6. Verschueren, K., HANDBOOK OF ENVIRONMENTAL DATA ON ORGANIC CHEMICALS, 3rd and 4th eds., Van Nostrand Reinhold, New York, NY (1996, 2001).
7. Yalkowsky, S. H., Aquasol Database, 5th edition, College of Pharmacy, University of Arizona, Tuscon, Az (1990).
8. Yaws, C. L. and others, "Hydrocarbons: Water Solubility Data", Chem. Eng., 97, 177 (April,1990).
9. Yaws, C. L. and others, "Organic Chemicals: Water Solubility Data", Chem. Eng., 97, 115 (July, 1990).
10. Yaws, C. L. and others, "Water Solubility Data for Organic Compounds", Pollution Engineering, 22, 70 (Oct., 1990).
11. Yaws, C. L. and others, "New Correlation Calculates Reliable Paraffin Solubilities", Oil & Gas Journal, 89, 79 (April 8, 1991).
12. Yaws, C. L. and others, "New Correlation Accurately Calculates Naphthene Water Solubilities", Oil & Gas Journal, 89(37), 86 (Sept. 16, 1991).
13. Yaws, C. L. and others, "Water Solubility Data for 151 Hydrocarbons", [temperature effect – alkanes], Chem. Eng., 100, 108 (February, 1993).
14. Yaws, C. L. and others, "The Water Solubility of Naphthenes", [temperature effect – cycloalkanes], Chem. Eng., 100 (10), 122 (Oct., 1993).
15. Yaws, C. L. and others, "New Correlation Accurately Calculates Water Solubilities of Aromatics", Oil & Gas Journal, 95 (35), 80 (Aug. 29, 1994).
16. Yaws, C. L. and others, "Calculate the Solubility of Aromatics", [temperature effect – benzenes], Chem. Eng., 102 (2), 113 (Feb., 1995).
17. Yaws, C. L. and others, "Solubility of Hydrocarbons and Sulfur Compounds in Water", WATER ENCYCLOPEDIA, Volume 4, ISBN 0-471-73684-8, pages 561-564, John Wiley, New York, NY (2005).
18. Yaws, C. L., CHEMICAL PROPERTIES HANDBOOK, McGraw-Hill, New York, NY (1999).
19. Yaws, C. L., YAWS HANDBOOK OF THERMODYNAMIC AND PHYSICAL PROPERTIES OF CHEMICAL COMPOUNDS, electronic edition, www.knovel.com, knovel, Norwich, NY (2003).
20. Yaws, C. L., YAWS HANDBOOK OF PHYSICAL PROPERTIES FOR HYDROCARBONS AND CHEMICALS, Gulf Publishing Co., Houston, TX (2005).

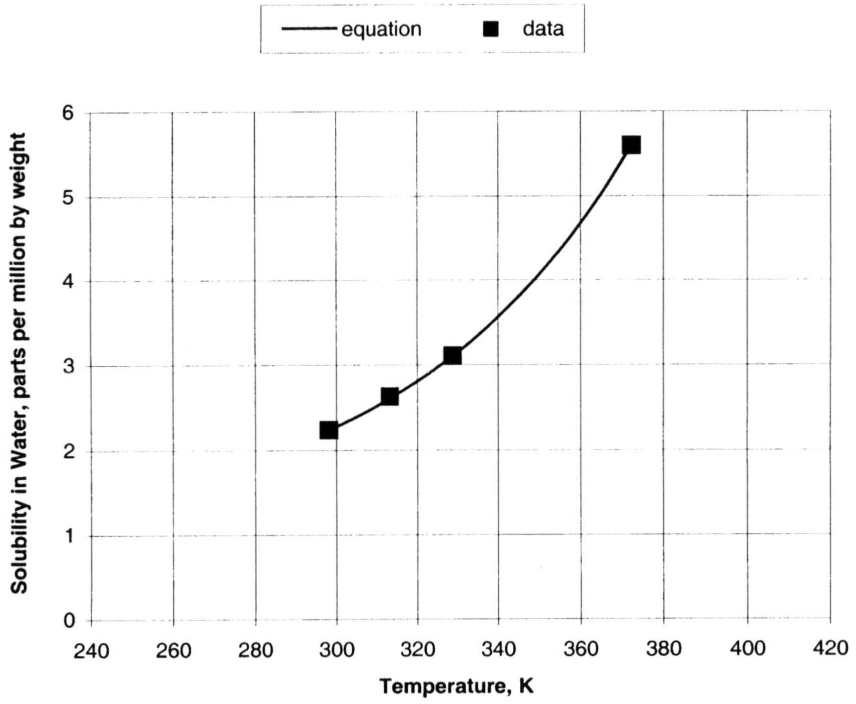

Figure 6 Solubility of Heptane in Water

Table 6 Solubility in Water as a Function of Temperature - Organic Compounds

$\log_{10} S = A + B\,T + C\,T^2$ (S - solubility, parts per million by weight, ppm (wt), T - K)												
NO	ID	FORMULA	NAME	CAS No	A	B	C	TMIN	TMAX	code	S@298, ppm(wt)	S@373, ppm(wt)
1	5285	C5H10	cyclopentane	287-92-3	5.4969	-2.2963E-02	3.9852E-05	298	373	1,2	1.5600E+02	3.0017E+02
2	5772	C5H12	pentane	109-66-0	6.4121	-3.1921E-02	5.2766E-05	298	373	1,2	3.8500E+01	7.0474E+01
3	5773	C5H12	isopentane	78-78-4	-0.7566	1.0284E-02	-7.0903E-06	298	373	1,2	4.7800E+01	1.2410E+02
4	5774	C5H12	neopentane	463-82-1	-13.2611	8.2896E-02	-1.1066E-04	298	373	1,2	4.1450E+01	1.8338E+02
5	6876	C6H6	benzene	71-43-2	5.1955	-1.5691E-02	3.0679E-05	298	373	1,2	1.7550E+03	4.0929E+03
6	8214	C6H12	methylcyclopentane	96-37-7	-3.6532	2.7473E-02	-3.2715E-05	298	373	1,2	4.2640E+01	1.1046E+02
7	8215	C6H12	cyclohexane	110-82-7	2.1052	-6.2114E-03	1.6826E-05	298	373	1,2	5.6100E+01	1.3497E+02
8	8877	C6H14	hexane	110-54-3	0.3619	-3.2984E-04	8.0186E-06	298	373	1,2	9.4700E+00	2.2664E+01
9	8878	C6H14	2,2-dimethylbutane	75-83-2	-0.1217	4.8145E-03	7.1159E-07	298	373	1,2	2.3820E+01	5.9409E+01
10	8879	C6H14	2,3-dimethylbutane	79-29-8	4.9937	-2.5376E-02	4.3701E-05	298	373	1,2	2.0540E+01	4.0703E+01
11	8880	C6H14	2-methylpentane	107-83-5	5.4091	-2.8955E-02	4.9091E-05	298	373	1,2	1.3800E+01	2.7532E+01
12	8881	C6H14	3-methylpentane	96-14-0	6.3631	-3.3752E-02	5.5720E-05	298	373	1,2	1.7910E+01	3.3656E+01
13	10725	C7H8	toluene	108-88-3	1.6230	1.8323E-03	6.3557E-06	298	373	1,2	5.4240E+02	1.5550E+03
14	11696	C7H14	ethylcyclopentane	1640-89-7	-0.2049	3.5563E-03	1.3987E-06	298	373	2	9.5450E+00	2.0745E+01
15	11697	C7H14	1,1-dimethylcyclopentane	1638-26-2	0.1681	3.5563E-03	1.3987E-06	298	373	2	2.2530E+01	4.8967E+01
16	11698	C7H14	1,2-dimethylcyclopentane	2452-99-5	-0.0441	3.5563E-03	1.3987E-06	298	373	2	1.3822E+01	3.0040E+01
17	11699	C7H14	cis-1,2-dimethylcyclopentane	1192-18-3	-0.1087	3.5563E-03	1.3987E-06	298	373	2	1.1910E+01	2.5885E+01
18	11700	C7H14	trans-1,2-dimethylcyclopentane	822-50-4	0.0744	3.5563E-03	1.3987E-06	298	373	2	1.8155E+01	3.9457E+01
19	11701	C7H14	1,3-dimethylcyclopentane	2453-00-1	0.0725	3.5563E-03	1.3987E-06	298	373	2	1.8076E+01	3.9286E+01
20	11702	C7H14	cis-1,3-dimethylcyclopentane	2532-58-3	0.1002	3.5563E-03	1.3987E-06	298	373	2	1.9267E+01	4.1875E+01
21	11703	C7H14	trans-1,3-dimethylcyclopentane	1759-58-6	0.0777	3.5563E-03	1.3987E-06	298	373	2	1.8293E+01	3.9758E+01
22	11704	C7H14	methylcyclohexane	108-87-2	1.9931	-9.2433E-03	2.1481E-05	298	373	1,2	1.4020E+01	3.4277E+01
23	12112	C7H16	heptane	142-82-5	0.6113	-5.8784E-03	1.6779E-05	298	373	1,2	2.2400E+00	5.6775E+00
24	12113	C7H16	2-methylhexane	591-76-4	-8.6851	4.9115E-02	-6.2476E-05	298	373	1,2	2.5400E+00	8.7694E+00
25	12114	C7H16	3-methylhexane	589-34-4	-5.5530	3.0935E-02	-3.6544E-05	298	373	1,2	2.6400E+00	7.9760E+00
26	12115	C7H16	3-ethylpentane	617-78-7	-0.9713	4.4750E-03	1.1958E-06	298	373	2	2.9458E+00	7.3287E+00
27	12116	C7H16	2,2-dimethylpentane	590-35-2	-8.8770	5.1627E-02	-6.6059E-05	298	373	1,2	4.4000E+00	1.5474E+01
28	12117	C7H16	2,3-dimethylpentane	565-59-3	-8.8003	5.1627E-02	-6.6059E-05	298	373	1,2	5.2500E+00	9.0087E+00
29	12118	C7H16	2,4-dimethylpentane	108-08-7	-7.3040	4.2454E-02	-5.2974E-05	298	373	1,2	4.4100E+00	1.4499E+01
30	12119	C7H16	3,3-dimethylpentane	562-49-2	2.1955	-1.3180E-02	2.8213E-05	298	373	1,2	5.9400E+00	1.6059E+01
31	12120	C7H16	2,2,3-trimethylbutane	464-06-2	-0.6815	4.4750E-03	1.1958E-06	298	373	2	5.7410E+00	1.4283E+01
32	12121	C7H16	3-methylhexane, (S)-	6131-24-4	-0.9366	4.4750E-03	1.1958E-06	298	373	2	3.1910E+00	7.9388E+00
33	13789	C8H10	ethylbenzene	100-41-4	-2.2685	2.1518E-02	-2.1706E-05	298	373	1,2	1.6510E+02	5.4803E+02
34	13790	C8H10	o-xylene	95-47-6	2.1884	-2.1165E-03	8.8490E-06	298	373	1,2	2.2080E+02	4.2735E+02
35	13791	C8H10	m-xylene	108-38-3	-9.9885	6.8341E-02	-9.1647E-05	298	373	1,2	1.7400E+02	5.6487E+02
36	13792	C8H10	p-xylene	106-42-3	-7.4735	5.4039E-02	-7.1247E-05	298	373	1,2	2.0170E+02	5.8953E+02
37	13793	C8H10	xylenes	1330-20-7	1.3651	3.1975E-03	1.2695E-06	298	373	2	2.6997E+02	5.4331E+02
38	14861	C8H16	propylcyclopentane	2040-96-2	0.3728	-3.7254E-03	1.1785E-05	298	373	1,2	2.0400E+00	4.2030E+00
39	14862	C8H16	isopropylcyclopentane	3875-51-2	-0.7827	3.5563E-03	1.3987E-06	298	373	2	2.5232E+00	5.4840E+00
40	14863	C8H16	1-methyl-1-ethylcyclopentane	16747-50-5	-0.6574	3.5563E-03	1.3987E-06	298	373	2	3.3671E+00	7.3180E+00
41	14864	C8H16	1-methyl-cis-2-ethylcyclopentane	930-89-2	-0.8241	3.5563E-03	1.3987E-06	298	373	2	2.2936E+00	4.9849E+00
42	14865	C8H16	1-methyl-trans-2-ethylcyclopentane	930-90-5	-0.6492	3.5563E-03	1.3987E-06	298	373	2	3.4308E+00	7.4566E+00
43	14866	C8H16	1-methyl-cis-3-ethylcyclopentane	2613-66-3	-0.6467	3.5563E-03	1.3987E-06	298	373	2	3.4510E+00	7.5004E+00
44	14867	C8H16	1-methyl-trans-3-ethylcyclopentane	2613-65-2	-0.6467	3.5563E-03	1.3987E-06	298	373	2	3.4510E+00	7.5004E+00
45	14868	C8H16	1,1,2-trimethylcyclopentane	4259-00-1	-0.4601	3.5563E-03	1.3987E-06	298	373	2	5.3034E+00	1.1526E+01
46	14869	C8H16	1,1,3-trimethylcyclopentane	4516-69-2	-0.6129	3.5563E-03	1.3987E-06	298	373	1,2	3.7300E+00	1.8853E+01
47	14870	C8H16	1,cis-2,cis-3-trimethylcyclopentane	2613-69-6	-0.6951	3.5563E-03	1.3987E-06	298	373	2	3.0871E+00	6.7094E+00
48	14871	C8H16	1,cis-2,trans-3-trimethylcyclopentane	15890-40-1	-0.5553	3.5563E-03	1.3987E-06	298	373	2	4.2594E+00	9.2573E+00
49	14872	C8H16	1,trans-2,cis-3-trimethylcyclopentane	19374-46-0	-0.3716	3.5563E-03	1.3987E-06	298	373	2	6.5020E+00	1.4131E+01
50	14873	C8H16	1,cis-2,cis-4-trimethylcyclopentane	2613-72-1	-0.5426	3.5563E-03	1.3987E-06	298	373	2	4.3854E+00	9.5311E+00
51	14874	C8H16	1,cis-2,trans-4-trimethylcyclopentane	4850-28-6	-0.5358	3.5563E-03	1.3987E-06	298	373	2	4.4546E+00	9.6816E+00
52	14875	C8H16	1,trans-2,cis-4-trimethylcyclopentane	16883-48-0	-0.3489	3.5563E-03	1.3987E-06	298	373	2	6.8511E+00	1.4890E+01
53	14876	C8H16	1-methyl-3-ethylcyclopentane	3726-47-4	-0.6454	3.5563E-03	1.3987E-06	298	373	2	3.4611E+00	7.5224E+00
54	14877	C8H16	ethylcyclohexane	1678-91-7	3.8282	-2.4252E-02	4.7269E-05	298	373	1,2	6.3000E+00	2.2926E+01
55	14878	C8H16	1,1-dimethylcyclohexane	590-66-9	-0.6837	4.3934E-03	1.7000E-06	298	373	2	5.9890E+00	1.5576E+01
56	14879	C8H16	1,2-dimethylcyclohexane, (cis+trans)	583-57-3	-0.8455	4.3934E-03	1.7000E-06	298	373	2	4.1258E+00	1.0730E+01
57	14880	C8H16	cis-1,2-dimethylcyclohexane	2207-01-4	1.2936	-7.1409E-03	1.8152E-05	298	373	1,2	6.0000E+00	8.6230E+00
58	14881	C8H16	trans-1,2-dimethylcyclohexane	6876-23-9	1.2315	-7.1409E-03	1.8152E-05	298	373	1,2	5.2000E+00	1.2426E+01
59	14882	C8H16	1,3-dimethylcyclohexane, cis and trans	591-21-9	-0.6938	4.3934E-03	1.7000E-06	298	373	2	5.8504E+00	1.5215E+01
60	14883	C8H16	cis-1,3-dimethylcyclohexane	638-04-0	-0.6974	4.3934E-03	1.7000E-06	298	373	2	5.8027E+00	1.5091E+01
61	14884	C8H16	trans-1,3-dimethylcyclohexane	2207-03-6	-0.8087	4.3934E-03	1.7000E-06	298	373	2	4.4905E+00	1.1678E+01
62	14885	C8H16	1,4-dimethylcyclohexane, (cis+trans)	589-90-2	-0.7651	4.3934E-03	1.7000E-06	298	373	2	4.9650E+00	1.2913E+01
63	14886	C8H16	cis-1,4-dimethylcyclohexane	624-29-3	-0.8051	4.3934E-03	1.7000E-06	298	373	2	4.5276E+00	1.1775E+01
64	14887	C8H16	trans-1,4-dimethylcyclohexane	2207-04-7	-11.3252	6.5368E-02	-8.5270E-05	298	373	1,2	3.8400E+00	1.5624E+01
65	15354	C8H18	octane	111-65-9	-1.8060	4.4750E-03	1.1958E-06	298	373	1,2	4.3100E-01	1.1168E+00
66	15355	C8H18	2-methylheptane	592-27-8	-1.5670	4.4750E-03	1.1958E-06	298	373	2	7.4739E-01	1.8594E+00
67	15356	C8H18	3-methylheptane	589-81-1	1.5019	-1.3287E-02	2.6530E-05	298	373	1,2	7.9200E-01	1.7295E+00
68	15357	C8H18	4-methylheptane	589-53-7	-1.5685	4.4750E-03	1.1958E-06	298	373	2	7.4478E-01	1.8529E+00
69	15358	C8H18	3-methyl-3-ethylpentane	1067-08-9	-1.5827	4.4750E-03	1.1958E-06	298	373	2	7.2083E-01	1.7933E+00
70	15359	C8H18	3-ethylhexane	619-99-8	-1.5895	4.4750E-03	1.1958E-06	298	373	2	7.0956E-01	1.7653E+00
71	15360	C8H18	3-ethyl-2-methylpentane	609-26-7	-1.5164	4.4750E-03	1.1958E-06	298	373	2	8.3971E-01	2.0891E+00
72	15361	C8H18	2,2-dimethylhexane	590-73-8	-1.2959	4.4750E-03	1.1958E-06	298	373	2	1.3953E+00	3.4712E+00
73	15362	C8H18	2,3-dimethylhexane	584-94-1	-1.5154	4.4750E-03	1.1958E-06	298	373	2	8.4167E-01	2.0940E+00
74	15363	C8H18	2,4-dimethylhexane	589-43-5	-1.3603	4.4750E-03	1.1958E-06	298	373	2	1.2030E+00	2.9929E+00
75	15364	C8H18	2,5-dimethylhexane	592-13-2	-1.3523	4.4750E-03	1.1958E-06	298	373	2	1.2253E+00	3.0484E+00
76	15365	C8H18	3,3-dimethylhexane	563-16-6	-1.4238	4.4750E-03	1.1958E-06	298	373	2	1.0393E+00	2.5857E+00
77	15366	C8H18	3,4-dimethylhexane	583-48-2	-1.5690	4.4750E-03	1.1958E-06	298	373	2	7.4391E-01	1.8508E+00
78	15367	C8H18	2,2,3-trimethylpentane	564-02-3	-1.3705	4.4750E-03	1.1958E-06	298	373	2	1.1750E+00	2.9233E+00
79	15368	C8H18	2,2,4-trimethylpentane	540-84-1	-1.1096	4.4750E-03	1.1958E-06	298	373	2	2.1425E+00	5.3304E+00
80	15369	C8H18	2,3,3-trimethylpentane	560-21-4	-1.4942	4.4750E-03	1.1958E-06	298	373	2	8.8376E-01	2.1987E+00

Table 6 Solubility in Water as a Function of Temperature - Organic Compounds

$\log_{10} S = A + B\,T + C\,T^2$ (S - solubility, parts per million by weight, ppm (wt), T - K)												
NO	ID	FORMULA	NAME	CAS No	A	B	C	TMIN	TMAX	code	S@298, ppm(wt)	S@373, ppm(wt)
81	15370	C8H18	2,3,4-trimethylpentane	565-75-3	-1.0784	4.4750E-03	1.1958E-06	298	373	1,2	2.3020E+00	2.4078E+00
82	15371	C8H18	2,2,3,3-tetramethylbutane	594-82-1	-1.2822	4.4750E-03	1.1958E-06	298	373	2	1.4397E+00	3.5818E+00
83	15372	C8H18	octane, mixture	---	-1.2029	4.4750E-03	1.1958E-06	298	373	2	1.7282E+00	4.2995E+00
84	17004	C9H12	cumene	98-82-8	-4.1149	2.9210E-02	-3.2570E-05	298	373	1,2	5.0000E+01	1.7780E+02
85	17005	C9H12	m-ethyltoluene	620-14-4	4.5206	-1.7811E-02	3.1235E-05	298	373	1,2	9.7000E+01	1.6731E+02
86	17006	C9H12	o-ethyltoluene	611-14-3	4.5025	-1.7811E-02	3.1235E-05	298	373	1,2	9.3050E+01	1.3510E+02
87	17007	C9H12	p-ethyltoluene	622-96-8	4.8745	-1.9933E-02	3.4262E-05	298	373	1,2	9.4850E+01	1.6108E+02
88	17008	C9H12	1,2,3-trimethylbenzene	526-73-8	4.4100	-1.7811E-02	3.1235E-05	298	373	1,2	7.5200E+01	7.3414E+01
89	17009	C9H12	1,2,4-trimethylbenzene	95-63-6	3.3113	-1.2014E-02	2.2967E-05	298	373	1,2	5.9000E+01	1.0619E+02
90	17010	C9H12	trimethyl benzene	25551-13-7	0.5948	3.1975E-03	1.2695E-06	298	373	2	4.5810E+01	9.2192E+01
91	17011	C9H12	mesitylene	108-67-8	4.5206	-1.7811E-02	3.1235E-05	298	373	1,2	9.7000E+01	1.3854E+02
92	17012	C9H12	propylbenzene	103-65-1	5.6332	-2.7569E-02	4.8416E-05	298	373	1,2	5.2160E+01	1.2226E+02
93	17859	C9H18	butylcyclopentane	2040-95-1	-1.5504	3.5563E-03	1.3987E-06	298	373	2	4.3077E-01	9.3623E-01
94	17860	C9H18	isobutylcyclopentane	3788-32-7	-1.3326	3.5563E-03	1.3987E-06	298	373	2	7.1134E-01	1.5460E+00
95	17861	C9H18	sec-butylcyclopentane	4850-32-2	-1.4939	3.5563E-03	1.3987E-06	298	373	2	4.9061E-01	1.0663E+00
96	17862	C9H18	tert-butylcyclopentane	3875-52-3	-1.2538	3.5563E-03	1.3987E-06	298	373	2	8.5273E-01	1.8533E+00
97	17863	C9H18	1-methyl-1-propylcyclopentane	16631-63-3	-1.2831	3.5563E-03	1.3987E-06	298	373	2	7.9720E-01	1.7326E+00
98	17864	C9H18	1-methyl-cis-2-propylcyclopentane	932-43-4	-1.4495	3.5563E-03	1.3987E-06	298	373	2	5.4346E-01	1.1812E+00
99	17865	C9H18	1-methyl-trans-2-propylcyclopentane	932-44-5	-1.2925	3.5563E-03	1.3987E-06	298	373	2	7.8013E-01	1.6955E+00
100	17866	C9H18	1-methyl-cis-3-propylcyclopentane	2443-04-1	-1.3338	3.5563E-03	1.3987E-06	298	373	2	7.0927E-01	1.5415E+00
101	17867	C9H18	1-methyl-trans-3-propylcyclopentane	2443-03-0	-1.3338	3.5563E-03	1.3987E-06	298	373	2	7.0927E-01	1.5415E+00
102	17868	C9H18	1-methyl-1-isopropylcyclopentane	61828-00-0	-1.3338	3.5563E-03	1.3987E-06	298	373	2	7.0927E-01	1.5415E+00
103	17869	C9H18	1-methyl-cis-2-isopropylcyclopentane	61868-01-7	-1.3338	3.5563E-03	1.3987E-06	298	373	2	7.0927E-01	1.5415E+00
104	17870	C9H18	1-methyl-trans-2-isopropylcyclopentane	61828-01-1	-1.1557	3.5563E-03	1.3987E-06	298	373	2	1.0690E+00	2.3234E+00
105	17871	C9H18	1-methyl-cis-3-isopropylcyclopentane	61828-02-2	-1.1812	3.5563E-03	1.3987E-06	298	373	2	1.0080E+00	2.1907E+00
106	17872	C9H18	1-methyl-trans-3-isopropylcyclopentane	61828-03-3	-1.1812	3.5563E-03	1.3987E-06	298	373	2	1.0080E+00	2.1907E+00
107	17873	C9H18	1,1-diethylcyclopentane	2721-38-2	-1.3971	3.5563E-03	1.3987E-06	298	373	2	6.1316E-01	1.3326E+00
108	17874	C9H18	1,cis-2-diethylcyclopentane	932-39-8	-1.4741	3.5563E-03	1.3987E-06	298	373	2	5.1351E-01	1.1161E+00
109	17875	C9H18	1,trans-2-diethylcyclopentane	932-40-1	-1.3219	3.5563E-03	1.3987E-06	298	373	2	7.2900E-01	1.5844E+00
110	17876	C9H18	1,3-diethylcyclopentane	19398-75-5	-1.3844	3.5563E-03	1.3987E-06	298	373	2	6.3125E-01	1.3720E+00
111	17877	C9H18	1,cis-3-diethylcyclopentane	62016-59-5	-1.3844	3.5563E-03	1.3987E-06	298	373	2	6.3125E-01	1.3720E+00
112	17878	C9H18	1,trans-3-diethylcyclopentane	62016-60-8	-1.3844	3.5563E-03	1.3987E-06	298	373	2	6.3125E-01	1.3720E+00
113	17879	C9H18	1,1-dimethyl-2-ethylcyclopentane	54549-80-3	-1.0789	3.5563E-03	1.3987E-06	298	373	2	1.2755E+00	2.7722E+00
114	17880	C9H18	1,1-dimethyl-3-ethylcyclopentane	62016-61-9	-0.9765	3.5563E-03	1.3987E-06	298	373	2	1.6148E+00	3.5096E+00
115	17881	C9H18	1,cis-2-dimethyl-1-ethylcyclopentane	62016-63-1	-1.2322	3.5563E-03	1.3987E-06	298	373	2	8.9630E-01	1.9480E+00
116	17882	C9H18	1,trans-2-dimethyl-1-ethylcyclopentane	62016-62-0	-1.2322	3.5563E-03	1.3987E-06	298	373	2	8.9630E-01	1.9480E+00
117	17883	C9H18	1,cis-2-dimethyl-cis-3-ethylcyclopentane	---	-1.4349	3.5563E-03	1.3987E-06	298	373	2	5.6203E-01	1.2215E+00
118	17884	C9H18	1,cis-2-dimethyl-trans-3-ethylcyclopentar	---	-1.2576	3.5563E-03	1.3987E-06	298	373	2	8.4527E-01	1.8371E+00
119	17885	C9H18	1,trans-2-dimethyl-cis-3-ethylcyclopentar	---	-1.0789	3.5563E-03	1.3987E-06	298	373	2	1.2755E+00	2.7722E+00
120	17886	C9H18	1,trans-2-dimethyl-trans-3-ethylcyclopen	---	-1.2576	3.5563E-03	1.3987E-06	298	373	2	8.4527E-01	1.8371E+00
121	17887	C9H18	1,cis-2-dimethyl-cis-4-ethylcyclopentane	62016-64-2	-1.3085	3.5563E-03	1.3987E-06	298	373	2	7.5192E-01	1.6342E+00
122	17888	C9H18	1,cis-2-dimethyl-trans-4-ethylcyclopentar	62016-65-3	-1.3085	3.5563E-03	1.3987E-06	298	373	2	7.5192E-01	1.6342E+00
123	17889	C9H18	1,trans-2-dimethyl-cis-4-ethylcyclopentar	62016-66-4	-1.1301	3.5563E-03	1.3987E-06	298	373	2	1.1338E+00	2.4642E+00
124	17890	C9H18	1,cis-3-dimethyl-1-ethylcyclopentane	62016-68-6	-1.0277	3.5563E-03	1.3987E-06	298	373	2	1.4351E+00	3.1191E+00
125	17891	C9H18	1,trans-3-dimethyl-1-ethylcyclopentane	62016-67-5	-1.0277	3.5563E-03	1.3987E-06	298	373	2	1.4351E+00	3.1191E+00
126	17892	C9H18	1,cis-3-dimethyl-cis-2-ethylcyclopentane	19903-00-5	-1.4349	3.5563E-03	1.3987E-06	298	373	2	5.6203E-01	1.2215E+00
127	17893	C9H18	1,cis-3-dimethyl-trans-2-ethylcyclopentar	19902-98-8	-1.0534	3.5563E-03	1.3987E-06	298	373	2	1.3529E+00	2.9405E+00
128	17894	C9H18	1,trans-3-dimethyl-cis-2-ethylcyclopentar	19902-99-9	-1.2322	3.5563E-03	1.3987E-06	298	373	2	8.9630E-01	1.9480E+00
129	17895	C9H18	1,cis-3-dimethyl-cis-4-ethylcyclopentane	---	-1.2576	3.5563E-03	1.3987E-06	298	373	2	8.4527E-01	1.8371E+00
130	17896	C9H18	1,cis-3-dimethyl-trans-4-ethylcyclopentar	---	-1.0534	3.5563E-03	1.3987E-06	298	373	2	1.3529E+00	2.9405E+00
131	17897	C9H18	1,trans-3-dimethyl-cis-4-ethylcyclopentar	---	-1.1007	3.5563E-03	1.3987E-06	298	373	2	1.2132E+00	2.6368E+00
132	17898	C9H18	1,trans-3-dimethyl-trans-4-ethylcyclopen	---	-1.2576	3.5563E-03	1.3987E-06	298	373	2	8.4527E-01	1.8371E+00
133	17899	C9H18	1,1,2,2-tetramethylcyclopentane	52688-89-8	-0.9509	3.5563E-03	1.3987E-06	298	373	2	1.7129E+00	3.7229E+00
134	17900	C9H18	1,1,cis-2,cis-3-tetramethylcyclopentane	62016-70-0	-1.0789	3.5563E-03	1.3987E-06	298	373	2	1.2755E+00	2.7722E+00
135	17901	C9H18	1,1,cis-2,trans-3-tetramethylcyclopentan	---	-0.8740	3.5563E-03	1.3987E-06	298	373	2	2.0445E+00	4.4436E+00
136	17902	C9H18	1,1,cis-2,cis-4-tetramethylcyclopentane	62016-72-2	-0.8740	3.5563E-03	1.3987E-06	298	373	2	2.0445E+00	4.4436E+00
137	17903	C9H18	1,1,cis-2,trans-4-tetramethylcyclopentan	---	-0.8740	3.5563E-03	1.3987E-06	298	373	2	2.0445E+00	4.4436E+00
138	17904	C9H18	1,1,3,3-tetramethylcyclopentane	50876-33-0	-0.5669	3.5563E-03	1.3987E-06	298	373	2	4.1466E+00	9.0122E+00
139	17905	C9H18	1,1,cis-3,cis-4-tetramethylcyclopentane	20309-77-7	-0.9509	3.5563E-03	1.3987E-06	298	373	2	1.7129E+00	3.7229E+00
140	17906	C9H18	1,1,cis-3,trans-4-tetramethylcyclopentan	---	-0.6594	3.5563E-03	1.3987E-06	298	373	2	3.3513E+00	7.2838E+00
141	17907	C9H18	1,2,2,cis-3-tetramethylcyclopentane	62016-73-3	-1.0789	3.5563E-03	1.3987E-06	298	373	2	1.2755E+00	2.7722E+00
142	17908	C9H18	1,2,2,trans-3-tetramethylcyclopentane	---	-1.0789	3.5563E-03	1.3987E-06	298	373	2	1.2755E+00	2.7722E+00
143	17909	C9H18	1,cis-2,cis-3,cis-4-tetramethylcyclopenta	2532-65-2	-1.2576	3.5563E-03	1.3987E-06	298	373	2	8.4527E-01	1.8371E+00
144	17910	C9H18	1,cis-2,cis-3,trans-4-tetramethylcyclopen	2532-69-6	-1.0789	3.5563E-03	1.3987E-06	298	373	2	1.2755E+00	2.7722E+00
145	17911	C9H18	1,cis-2,trans-3,cis-4-tetramethylcyclopen	2532-68-5	-0.8997	3.5563E-03	1.3987E-06	298	373	2	1.9274E+00	4.1891E+00
146	17912	C9H18	1,cis-2,trans-3,trans-4-tetramethylcyclop	2532-64-1	-1.0789	3.5563E-03	1.3987E-06	298	373	2	1.2755E+00	2.7722E+00
147	17913	C9H18	1,trans-2,cis-3,trans-4-tetramethylcyclop	2532-67-4	-0.8075	3.5563E-03	1.3987E-06	298	373	2	2.3831E+00	5.1795E+00
148	17914	C9H18	1,trans-2,trans-3,cis-4-tetramethylcyclop	19907-40-5	-0.8997	3.5563E-03	1.3987E-06	298	373	2	1.9274E+00	4.1891E+00
149	17915	C9H18	1-methyl-2-isopropylcyclopentane	89223-57-4	-1.2067	3.5563E-03	1.3987E-06	298	373	2	9.5047E-01	2.0658E+00
150	17916	C9H18	propylcyclohexane	1678-92-8	-1.6303	4.3934E-03	1.7000E-06	298	373	2	6.7726E-01	1.7614E+00
151	17917	C9H18	isopropylcyclohexane	696-29-7	-1.5805	4.3934E-03	1.7000E-06	298	373	2	7.5939E-01	1.9750E+00
152	17918	C9H18	1-methyl-1-ethylcyclohexane	4926-90-3	-1.5153	4.3934E-03	1.7000E-06	298	373	2	8.8254E-01	2.2952E+00
153	17919	C9H18	1-methyl-cis-2-ethylcyclohexane	4923-77-7	-1.6140	4.3934E-03	1.7000E-06	298	373	2	7.0302E-01	1.8284E+00
154	17920	C9H18	1-methyl-trans-2-ethylcyclohexane	4923-78-8	-1.5037	4.3934E-03	1.7000E-06	298	373	2	9.0639E-01	2.3573E+00
155	17921	C9H18	1-methyl-cis-3-ethylcyclohexane	19489-10-2	-1.4102	4.3934E-03	1.7000E-06	298	373	2	1.1241E+00	2.9235E+00
156	17922	C9H18	1-methyl-trans-3-ethylcyclohexane	4926-76-5	-1.4860	4.3934E-03	1.7000E-06	298	373	2	9.4397E-01	2.4550E+00
157	17923	C9H18	1-methyl-cis-4-ethylcyclohexane	4926-78-7	-1.4860	4.3934E-03	1.7000E-06	298	373	2	9.4397E-01	2.4550E+00
158	17924	C9H18	1-methyl-trans-4-ethylcyclohexane	6236-88-0	-1.4102	4.3934E-03	1.7000E-06	298	373	2	1.1241E+00	2.9235E+00
159	17925	C9H18	1,1,2-trimethylcyclohexane	7094-26-0	-1.3391	4.3934E-03	1.7000E-06	298	373	2	1.3241E+00	3.4435E+00
160	17926	C9H18	1,1,3-trimethylcyclohexane	3073-66-3	-6.1183	3.3019E-02	-3.9130E-05	298	373	1,2	1.7700E+00	5.6797E+00

Table 6 Solubility in Water as a Function of Temperature - Organic Compounds

$\log_{10} S = A + B\,T + C\,T^2$ (S - solubility, parts per million by weight, ppm (wt), T - K)												
NO	ID	FORMULA	NAME	CAS No	A	B	C	TMIN	TMAX	code	S@298, ppm(wt)	S@373, ppm(wt)
161	17927	C9H18	1,1,4-trimethylcyclohexane	7094-27-1	-1.0785	4.3934E-03	1.7000E-06	298	373	2	2.4127E+00	6.2747E+00
162	17928	C9H18	1,2,3-trimethylcyclohexane	1678-97-3	-1.2576	4.3934E-03	1.7000E-06	298	373	2	1.5975E+00	4.1547E+00
163	17929	C9H18	1,cis-2,cis-3-trimethylcyclohexane	1839-88-9	-1.5364	4.3934E-03	1.7000E-06	298	373	2	8.4064E-01	2.1863E+00
164	17930	C9H18	1,cis-2,trans-3-trimethylcyclohexane	7667-55-2	-1.4860	4.3934E-03	1.7000E-06	298	373	2	9.4397E-01	2.4550E+00
165	17931	C9H18	1,trans-2,cis-3-trimethylcyclohexane	1678-81-5	-1.3086	4.3934E-03	1.7000E-06	298	373	2	1.4205E+00	3.6944E+00
166	17932	C9H18	1,2,4-trimethylcyclohexane, isomers	2234-75-5	-1.2550	4.3934E-03	1.7000E-06	298	373	2	1.6069E+00	4.1792E+00
167	17933	C9H18	1,cis-2,cis-4-trimethylcyclohexane	1678-80-4	-1.4102	4.3934E-03	1.7000E-06	298	373	2	1.1241E+00	2.9235E+00
168	17934	C9H18	1,cis-2,trans-4-trimethylcyclohexane	7667-58-5	-1.4102	4.3934E-03	1.7000E-06	298	373	2	1.1241E+00	2.9235E+00
169	17935	C9H18	1,trans-2,cis-4-trimethylcyclohexane	7667-59-6	-1.3340	4.3934E-03	1.7000E-06	298	373	2	1.3397E+00	3.4841E+00
170	17936	C9H18	1,trans-2,trans-4-trimethylcyclohexane	7667-60-9	-1.2376	4.3934E-03	1.7000E-06	298	373	2	1.6725E+00	4.3497E+00
171	17937	C9H18	1,3,5,-trimethylcyclohexane	1839-63-0	-1.1515	4.3934E-03	1.7000E-06	298	373	2	2.0395E+00	5.3042E+00
172	17938	C9H18	cis,cis-1,3,5-trimethylcyclohexane	1795-27-3	-1.1684	4.3934E-03	1.7000E-06	298	373	2	1.9617E+00	5.1019E+00
173	17939	C9H18	cis,trans-1,3,5-trimethylcyclohexane	1795-26-2	-1.2205	4.3934E-03	1.7000E-06	298	373	2	1.7397E+00	4.5245E+00
174	17940	C9H18	1-ethyl-2-methylcyclohexane	3728-54-9	-1.5201	4.3934E-03	1.7000E-06	298	373	2	8.7288E-01	2.2701E+00
175	18176	C9H20	nonane	111-84-2	-2.3542	4.4750E-03	1.1958E-06	298	373	1,2	1.2200E-01	2.5923E-01
176	18177	C9H20	2-methyloctane	3221-61-2	-2.2217	4.4750E-03	1.1958E-06	298	373	2	1.6550E-01	4.1175E-01
177	18178	C9H20	3-methyloctane	2216-33-3	-2.2459	4.4750E-03	1.1958E-06	298	373	2	1.5653E-01	3.8943E-01
178	18179	C9H20	4-methyloctane	2216-34-4	-12.1038	6.1357E-02	-8.0199E-05	298	373	1,2	1.1500E-01	4.2137E-01
179	18180	C9H20	3-ethylheptane	15869-80-4	-2.2197	4.4750E-03	1.1958E-06	298	373	2	1.6628E-01	4.1368E-01
180	18181	C9H20	4-ethylheptane	2216-32-2	-2.1689	4.4750E-03	1.1958E-06	298	373	2	1.8690E-01	4.6499E-01
181	18182	C9H20	2,2-dimethylheptane	1071-26-7	-1.9508	4.4750E-03	1.1958E-06	298	373	2	3.0880E-01	7.6827E-01
182	18183	C9H20	2,3-dimethylheptane	3074-71-3	-2.1510	4.4750E-03	1.1958E-06	298	373	2	1.9476E-01	4.8454E-01
183	18184	C9H20	2,4-dimethylheptane	2213-23-2	-1.9562	4.4750E-03	1.1958E-06	298	373	2	3.0500E-01	7.5881E-01
184	18185	C9H20	2,5-dimethylheptane	2216-30-0	-2.0359	4.4750E-03	1.1958E-06	298	373	2	2.5388E-01	6.3163E-01
185	18186	C9H20	2,6-dimethylheptane	1072-05-5	-2.0154	4.4750E-03	1.1958E-06	298	373	2	2.6615E-01	6.6214E-01
186	18187	C9H20	3,4-dimethylheptane	922-28-1	-2.1536	4.4750E-03	1.1958E-06	298	373	2	1.9362E-01	4.8170E-01
187	18188	C9H20	3,5-dimethylheptane	926-82-9	-2.0359	4.4750E-03	1.1958E-06	298	373	2	2.5388E-01	6.3163E-01
188	18189	C9H20	4,4-dimethylheptane	1068-19-5	-2.0154	4.4750E-03	1.1958E-06	298	373	2	2.6615E-01	6.6214E-01
189	18190	C9H20	3-ethyl-2-methylhexane	16789-46-1	-2.0871	4.4750E-03	1.1958E-06	298	373	2	2.2565E-01	5.6138E-01
190	18191	C9H20	4-ethyl-2-methylhexane	3074-75-7	-1.9795	4.4750E-03	1.1958E-06	298	373	2	2.8906E-01	7.1914E-01
191	18192	C9H20	3-ethyl-3-methylhexane	3074-76-8	-2.1536	4.4750E-03	1.1958E-06	298	373	2	1.9362E-01	4.8170E-01
192	18193	C9H20	3-ethyl-4-methylhexane	3074-77-9	-2.1485	4.4750E-03	1.1958E-06	298	373	2	1.9591E-01	4.8740E-01
193	18194	C9H20	2,2,3-trimethylhexane	16747-25-4	-1.9742	4.4750E-03	1.1958E-06	298	373	2	2.9266E-01	7.2811E-01
194	18195	C9H20	2,2,4-trimethylhexane	16747-26-5	-1.7934	4.4750E-03	1.1958E-06	298	373	2	4.4376E-01	1.1040E+00
195	18196	C9H20	2,2,5-trimethylhexane	3522-94-9	-1.7308	4.4750E-03	1.1958E-06	298	373	2	5.1255E-01	1.2752E+00
196	18197	C9H20	2,3,3-trimethylhexane	16747-28-7	-2.0789	4.4750E-03	1.1958E-06	298	373	2	2.2994E-01	5.7207E-01
197	18198	C9H20	2,3,4-trimethylhexane	921-47-1	-2.1137	4.4750E-03	1.1958E-06	298	373	2	2.1224E-01	5.2802E-01
198	18199	C9H20	2,3,5-trimethylhexane	1069-53-0	-1.9165	4.4750E-03	1.1958E-06	298	373	2	3.3420E-01	8.3146E-01
199	18200	C9H20	2,4,4-trimethylhexane	16747-30-1	-1.8988	4.4750E-03	1.1958E-06	298	373	2	3.4809E-01	8.6600E-01
200	18201	C9H20	3,3,4-trimethylhexane	16747-31-2	-2.1500	4.4750E-03	1.1958E-06	298	373	2	1.9522E-01	4.8568E-01
201	18202	C9H20	3,3-diethylpentane	1067-20-5	-2.2958	4.4750E-03	1.1958E-06	298	373	2	1.3955E-01	3.4719E-01
202	18203	C9H20	2,2-dimethyl-3-ethylpentane	16747-32-3	-1.9803	4.4750E-03	1.1958E-06	298	373	2	2.8855E-01	7.1787E-01
203	18204	C9H20	2,3-dimethyl-3-ethylpentane	16747-33-4	-2.2581	4.4750E-03	1.1958E-06	298	373	2	1.5219E-01	3.7862E-01
204	18205	C9H20	2,4-dimethyl-3-ethylpentane	1068-87-7	-2.0541	4.4750E-03	1.1958E-06	298	373	2	2.4347E-01	6.0573E-01
205	18206	C9H20	2,2,3,3-tetramethylpentane	7154-79-2	-2.1454	4.4750E-03	1.1958E-06	298	373	2	1.9730E-01	4.9085E-01
206	18207	C9H20	2,2,3,4-tetramethylpentane	1186-53-4	-1.9595	4.4750E-03	1.1958E-06	298	373	2	3.0267E-01	7.5301E-01
207	18208	C9H20	2,2,4,4-tetramethylpentane	1070-87-7	-1.6849	4.4750E-03	1.1958E-06	298	373	2	5.6969E-01	1.4173E+00
208	18209	C9H20	2,3,3,4-tetramethylpentane	16747-38-9	-2.1776	4.4750E-03	1.1958E-06	298	373	2	1.8320E-01	4.5579E-01
209	18210	C9H20	3,3-dimethylheptane	4032-86-4	-2.0689	4.4750E-03	1.1958E-06	298	373	2	2.3529E-01	5.8537E-01
210	19865	C10H14	butylbenzene	104-51-8	5.6644	-3.1117E-02	5.3477E-05	298	373	1,2	1.3820E+01	3.1561E+01
211	19866	C10H14	isobutylbenzene	538-93-2	-15.2183	8.9953E-02	-1.1921E-04	298	373	1,2	1.0100E+01	5.6095E+01
212	19867	C10H14	sec-butylbenzene	135-98-8	-1.5410	1.1544E-02	-7.3719E-06	298	373	1,2	1.7600E+01	5.4974E+01
213	19868	C10H14	tert-butylbenzene	98-06-6	5.1205	-2.6022E-02	4.6210E-05	298	373	1,2	2.9500E+01	6.9938E+01
214	19869	C10H14	cymene	25155-15-1	0.4846	3.1975E-03	1.2695E-06	298	373	2	3.5549E+01	7.1541E+01
215	19870	C10H14	m-cymene	535-77-3	0.5157	3.1975E-03	1.2695E-06	298	373	2	3.8184E+01	7.6844E+01
216	19871	C10H14	o-cymene	527-84-4	0.4399	3.1975E-03	1.2695E-06	298	373	2	3.2072E+01	6.4544E+01
217	19872	C10H14	p-cymene	99-87-6	0.5532	2.6959E-03	1.9850E-06	298	373	1,2	3.4150E+01	6.8473E+01
218	19873	C10H14	1-methyl-2-propylbenzene	1074-17-5	0.2790	3.1975E-03	1.2695E-06	298	373	2	2.2140E+01	4.4557E+01
219	19874	C10H14	1-methyl-3-propylbenzene	1074-43-7	0.3518	3.1975E-03	1.2695E-06	298	373	2	2.6179E+01	5.2684E+01
220	19875	C10H14	1-methyl-4-propylbenzene	1074-55-1	0.3153	3.1975E-03	1.2695E-06	298	373	2	2.4073E+01	4.8446E+01
221	19876	C10H14	(S)-(1-methylpropyl)benzene	5787-28-0	0.2996	3.1975E-03	1.2695E-06	298	373	2	2.3216E+01	4.6723E+01
222	19877	C10H14	diethylbenzene	25340-17-4	0.3493	3.1975E-03	1.2695E-06	298	373	2	2.6033E+01	5.2390E+01
223	19878	C10H14	o-diethylbenzene	135-01-3	0.3115	3.1975E-03	1.2695E-06	298	373	2	2.3859E+01	4.8015E+01
224	19879	C10H14	m-diethylbenzene	141-93-5	0.3678	3.1975E-03	1.2695E-06	298	373	2	2.7164E+01	5.4667E+01
225	19880	C10H14	p-diethylbenzene	105-05-5	1.6665	-4.6092E-03	1.2404E-05	298	373	1,2	2.4830E+01	4.7187E+01
226	19881	C10H14	3-ethyl-o-xylene	933-98-2	0.0593	3.1975E-03	1.2695E-06	298	373	2	1.3351E+01	2.6868E+01
227	19882	C10H14	4-ethyl-o-xylene	934-80-5	0.1590	3.1975E-03	1.2695E-06	298	373	2	1.6797E+01	3.3803E+01
228	19883	C10H14	2-ethyl-m-xylene	2870-04-4	0.1528	3.1975E-03	1.2695E-06	298	373	2	1.6557E+01	3.3321E+01
229	19884	C10H14	4-ethyl-m-xylene	874-41-9	0.1912	3.1975E-03	1.2695E-06	298	373	2	1.8088E+01	3.6401E+01
230	19885	C10H14	5-ethyl-m-xylene	934-74-7	0.3037	3.1975E-03	1.2695E-06	298	373	2	2.3436E+01	4.7165E+01
231	19886	C10H14	2-ethyl-p-xylene	1758-88-9	0.2300	3.1975E-03	1.2695E-06	298	373	2	1.9776E+01	3.9799E+01
232	19887	C10H14	1,2,3,4-tetramethylbenzene	488-23-3	-0.1992	3.1975E-03	1.2695E-06	298	373	2	7.3618E+00	1.4815E+01
233	19888	C10H14	1,2,3,5-tetramethylbenzene	527-53-7	-0.0360	3.1975E-03	1.2695E-06	298	373	2	1.0720E+01	2.1573E+01
234	19889	C10H14	1,2,4,5-tetramethylbenzene	95-93-2	-0.5246	3.1975E-03	1.2695E-06	298	373	1,2	3.4800E+00	2.2497E+01
235	20922	C10H20	pentylcyclopentane	3741-00-2	-1.9991	2.8275E-03	2.4381E-06	298	373	1,2	1.1500E-01	2.4861E-01
236	20923	C10H20	butylcyclohexane	1678-93-9	-2.2111	4.3934E-03	1.7000E-06	298	373	2	1.7779E-01	4.6237E-01
237	20924	C10H20	tert-butylcyclohexane	3178-22-1	-1.9902	4.3934E-03	1.7000E-06	298	373	2	2.9568E-01	7.6897E-01
238	20925	C10H20	sec-butylcyclohexane	7058-01-7	-2.1727	4.3934E-03	1.7000E-06	298	373	2	1.9424E-01	5.0516E-01
239	20926	C10H20	isobutylcyclohexane	1678-98-4	-1.9854	4.3934E-03	1.7000E-06	298	373	2	2.9894E-01	7.7745E-01
240	20927	C10H20	1-methyl-1-propylcyclohexane	4258-93-9	-2.0021	4.3934E-03	1.7000E-06	298	373	2	2.8770E-01	7.4822E-01

Table 6 Solubility in Water as a Function of Temperature - Organic Compounds

$\log_{10} S = A + B\,T + C\,T^2$ (S - solubility, parts per million by weight, ppm (wt), T - K)													
NO	ID	FORMULA	NAME	CAS No	A	B	C	TMIN	TMAX	code	S@298, ppm(wt)	S@373, ppm(wt)
241	20928	C10H20	1-methyl-2-propylcyclohexane	4291-79-6	-2.0962	4.3934E-03	1.7000E-06	298	373	2	2.3161E-01	6.0235E-01
242	20929	C10H20	1-methyl-3-propylcyclohexane	4291-80-9	-1.8215	4.3934E-03	1.7000E-06	298	373	2	4.3606E-01	1.1341E+00
243	20930	C10H20	1-methyl-4-propylcyclohexane	4291-81-0	-2.0846	4.3934E-03	1.7000E-06	298	373	2	2.3792E-01	6.1878E-01
244	20931	C10H20	1-methyl-2-isopropylcyclohexane	16580-23-7	-1.9783	4.3934E-03	1.7000E-06	298	373	2	3.0390E-01	7.9035E-01
245	20932	C10H20	1-methyl-3-isopropylcyclohexane	16580-24-8	-1.8701	4.3934E-03	1.7000E-06	298	373	2	3.8986E-01	1.0139E+00
246	20933	C10H20	1-methyl-4-isopropylcyclohexane	99-82-1	-1.9711	4.3934E-03	1.7000E-06	298	373	2	3.0894E-01	8.0348E-01
247	20934	C10H20	cis-1-methyl-4-isopropylcyclohexane	6069-98-3	-2.0021	4.3934E-03	1.7000E-06	298	373	2	2.8770E-01	7.4822E-01
248	20935	C10H20	trans-1-methyl-4-isopropylcyclohexane	1678-82-6	-1.9687	4.3934E-03	1.7000E-06	298	373	2	3.1065E-01	8.0791E-01
249	20936	C10H20	1,1,3,4-tetramethylcyclohexane	24612-75-7	-1.7109	4.3934E-03	1.7000E-06	298	373	2	5.6244E-01	1.4627E+00
250	20937	C10H20	trans-1,1,3,5-tetramethylcyclohexane	50876-31-8	-1.6214	4.3934E-03	1.7000E-06	298	373	2	6.9120E-01	1.7976E+00
251	20938	C10H20	cis-1,1,3,5-tetramethylcyclohexane	50876-32-9	-1.5238	4.3934E-03	1.7000E-06	298	373	2	8.6533E-01	2.2505E+00
252	20939	C10H20	diethylcyclohexane; (mixed isomers)	1331-43-7	-2.0494	4.3934E-03	1.7000E-06	298	373	2	2.5801E-01	6.7102E-01
253	21133	C10H22	decane	124-18-5	11.0008	-7.5622E-02	1.1544E-04	298	373	1,2	5.2000E-02	7.1861E-02
254	21134	C10H22	2-methylnonane	871-83-0	-2.8137	4.4750E-03	1.1958E-06	298	373	2	4.2345E-02	1.0535E-01
255	21135	C10H22	3-methylnonane	5911-04-6	-2.8330	4.4750E-03	1.1958E-06	298	373	2	4.0502E-02	1.0076E-01
256	21136	C10H22	4-methylnonane	17301-94-9	-2.7822	4.4750E-03	1.1958E-06	298	373	2	4.5533E-02	1.1328E-01
257	21137	C10H22	5-methylnonane	15869-85-9	-2.7688	4.4750E-03	1.1958E-06	298	373	2	4.6957E-02	1.1682E-01
258	21138	C10H22	3-ethyloctane	5881-17-4	-2.8019	4.4750E-03	1.1958E-06	298	373	2	4.3518E-02	1.0827E-01
259	21139	C10H22	4-ethyloctane	15869-86-0	-2.7322	4.4750E-03	1.1958E-06	298	373	2	5.1087E-02	1.2710E-01
260	21140	C10H22	2,2-dimethyloctane	15869-87-1	-2.5655	4.4750E-03	1.1958E-06	298	373	2	7.4992E-02	1.8657E-01
261	21141	C10H22	2,3-dimethyloctane	7146-60-3	-2.7486	4.4750E-03	1.1958E-06	298	373	2	4.9197E-02	1.2240E-01
262	21142	C10H22	2,4-dimethyloctane	4032-94-4	-2.5406	4.4750E-03	1.1958E-06	298	373	2	7.9426E-02	1.9760E-01
263	21143	C10H22	2,5-dimethyloctane	15869-89-3	-2.6053	4.4750E-03	1.1958E-06	298	373	2	6.8427E-02	1.7024E-01
264	21144	C10H22	2,6-dimethyloctane	2051-30-1	-2.6518	4.4750E-03	1.1958E-06	298	373	2	6.1473E-02	1.5294E-01
265	21145	C10H22	2,7-dimethyloctane	1072-16-8	-2.6395	4.4750E-03	1.1958E-06	298	373	2	6.3247E-02	1.5735E-01
266	21146	C10H22	3,3-dimethyloctane	4110-44-5	-2.6723	4.4750E-03	1.1958E-06	298	373	2	5.8642E-02	1.4589E-01
267	21147	C10H22	3,4-dimethyloctane	15869-92-8	-2.7263	4.4750E-03	1.1958E-06	298	373	2	5.1782E-02	1.2883E-01
268	21148	C10H22	3,5-dimethyloctane	15869-93-9	-2.6278	4.4750E-03	1.1958E-06	298	373	2	6.4964E-02	1.6162E-01
269	21149	C10H22	3,6-dimethyloctane	15869-94-0	-2.6624	4.4750E-03	1.1958E-06	298	373	2	5.9989E-02	1.4925E-01
270	21150	C10H22	4,4-dimethyloctane	15869-95-1	-2.5807	4.4750E-03	1.1958E-06	298	373	2	7.2415E-02	1.8016E-01
271	21151	C10H22	4,5-dimethyloctane	15869-96-2	-2.6952	4.4750E-03	1.1958E-06	298	373	2	5.5632E-02	1.3841E-01
272	21152	C10H22	4-propylheptane	3178-29-8	-2.5807	4.4750E-03	1.1958E-06	298	373	2	7.2415E-02	1.8016E-01
273	21153	C10H22	4-isopropylheptane	52896-87-4	-2.6155	4.4750E-03	1.1958E-06	298	373	2	6.6843E-02	1.6630E-01
274	21154	C10H22	3-ethyl-2-methylheptane	14676-29-0	-2.6723	4.4750E-03	1.1958E-06	298	373	2	5.8642E-02	1.4589E-01
275	21155	C10H22	4-ethyl-2-methylheptane	52896-88-5	-2.5483	4.4750E-03	1.1958E-06	298	373	2	7.8023E-02	1.9411E-01
276	21156	C10H22	5-ethyl-2-methylheptane	13475-78-0	-2.6353	4.4750E-03	1.1958E-06	298	373	2	6.3862E-02	1.5888E-01
277	21157	C10H22	3-ethyl-3-methylheptane	17302-01-1	-2.7361	4.4750E-03	1.1958E-06	298	373	2	5.0628E-02	1.2596E-01
278	21158	C10H22	4-ethyl-3-methylheptane	52896-89-6	-2.6969	4.4750E-03	1.1958E-06	298	373	2	5.5412E-02	1.3786E-01
279	21159	C10H22	3-ethyl-5-methylheptane	52896-90-9	-2.5981	4.4750E-03	1.1958E-06	298	373	2	6.9571E-02	1.7308E-01
280	21160	C10H22	3-ethyl-4-methylheptane	52896-91-0	-2.7165	4.4750E-03	1.1958E-06	298	373	2	5.2964E-02	1.3177E-01
281	21161	C10H22	4-ethyl-4-methylheptane	17302-04-4	-2.6624	4.4750E-03	1.1958E-06	298	373	2	5.9989E-02	1.4925E-01
282	21162	C10H22	2,2,3-trimethylheptane	52896-92-1	-2.5832	4.4750E-03	1.1958E-06	298	373	2	7.2001E-02	1.7913E-01
283	21163	C10H22	2,2,4-trimethylheptane	14720-74-2	-2.3496	4.4750E-03	1.1958E-06	298	373	2	1.2330E-01	3.0675E-01
284	21164	C10H22	2,2,5-trimethylheptane	20291-95-6	-2.4128	4.4750E-03	1.1958E-06	298	373	2	1.0660E-01	2.6521E-01
285	21165	C10H22	2,2,6-trimethylheptane	1190-83-6	-2.3655	4.4750E-03	1.1958E-06	298	373	2	1.1885E-01	2.9569E-01
286	21166	C10H22	2,3,3-trimethylheptane	52896-93-2	-2.6476	4.4750E-03	1.1958E-06	298	373	2	6.2070E-02	1.5442E-01
287	21167	C10H22	2,3,4-trimethylheptane	52896-95-4	-2.6402	4.4750E-03	1.1958E-06	298	373	2	6.3139E-02	1.5708E-01
288	21168	C10H22	2,3,5-trimethylheptane	20278-85-7	-2.6600	4.4750E-03	1.1958E-06	298	373	2	6.0331E-02	1.5010E-01
289	21169	C10H22	2,3,6-trimethylheptane	4032-93-3	-2.5433	4.4750E-03	1.1958E-06	298	373	2	7.8925E-02	1.9636E-01
290	21170	C10H22	2,4,4-trimethylheptane	4032-92-2	-2.4178	4.4750E-03	1.1958E-06	298	373	2	1.0537E-01	2.6214E-01
291	21171	C10H22	2,4,5-trimethylheptane	20278-84-6	-2.5558	4.4750E-03	1.1958E-06	298	373	2	7.6690E-02	1.9079E-01
292	21172	C10H22	2,4,6-trimethylheptane	2613-61-8	-2.3318	4.4750E-03	1.1958E-06	298	373	2	1.2844E-01	3.1954E-01
293	21173	C10H22	2,5,5-trimethylheptane	1189-99-7	-2.4631	4.4750E-03	1.1958E-06	298	373	2	9.4926E-02	2.3617E-01
294	21174	C10H22	3,3,4-trimethylheptane	20278-87-9	-2.6895	4.4750E-03	1.1958E-06	298	373	2	5.6361E-02	1.4022E-01
295	21175	C10H22	3,3,5-trimethylheptane	7154-80-5	-2.5356	4.4750E-03	1.1958E-06	298	373	2	8.0345E-02	1.9989E-01
296	21176	C10H22	3,4,4-trimethylheptane	20278-88-0	-2.6698	4.4750E-03	1.1958E-06	298	373	2	5.8976E-02	1.4673E-01
297	21177	C10H22	3,4,5-trimethylheptane	20278-89-1	-2.7043	4.4750E-03	1.1958E-06	298	373	2	5.4480E-02	1.3554E-01
298	21178	C10H22	3-isopropyl-2-methylhexane	62016-13-1	-2.8067	4.4750E-03	1.1958E-06	298	373	2	4.3035E-02	1.0707E-01
299	21179	C10H22	3,3-diethylhexane	17302-02-2	-2.7970	4.4750E-03	1.1958E-06	298	373	2	4.4006E-02	1.0948E-01
300	21180	C10H22	3,4-diethylhexane	19398-77-7	-2.7386	4.4750E-03	1.1958E-06	298	373	2	5.0344E-02	1.2525E-01
301	21181	C10H22	3-ethyl-2,2-dimethylhexane	20291-91-2	-2.5458	4.4750E-03	1.1958E-06	298	373	2	7.8472E-02	1.9523E-01
302	21182	C10H22	4-ethyl-2,2-dimethylhexane	52896-99-8	-2.3166	4.4750E-03	1.1958E-06	298	373	2	1.3302E-01	3.3094E-01
303	21183	C10H22	3-ethyl-2,3-dimethylhexane	52897-00-4	-2.7337	4.4750E-03	1.1958E-06	298	373	2	5.0914E-02	1.2667E-01
304	21184	C10H22	4-ethyl-2,3-dimethylhexane	52897-01-5	-2.6649	4.4750E-03	1.1958E-06	298	373	2	5.9649E-02	1.4840E-01
305	21185	C10H22	3-ethyl-2,4-dimethylhexane	7220-26-0	-2.6452	4.4750E-03	1.1958E-06	298	373	2	6.2424E-02	1.5530E-01
306	21186	C10H22	4-ethyl-2,4-dimethylhexane	52897-03-7	-2.6698	4.4750E-03	1.1958E-06	298	373	2	5.8976E-02	1.4673E-01
307	21187	C10H22	3-ethyl-2,5-dimethylhexane	52897-04-8	-2.4958	4.4750E-03	1.1958E-06	298	373	2	8.8054E-02	2.1907E-01
308	21188	C10H22	4-ethyl-3,3-dimethylhexane	52897-05-9	-2.7141	4.4750E-03	1.1958E-06	298	373	2	5.3263E-02	1.3251E-01
309	21189	C10H22	3-ethyl-3,4-dimethylhexane	52897-06-0	-2.6945	4.4750E-03	1.1958E-06	298	373	2	5.5727E-02	1.3864E-01
310	21190	C10H22	2,2,3,3-tetramethylhexane	13475-81-5	-2.6506	4.4750E-03	1.1958E-06	298	373	2	6.1648E-02	1.5337E-01
311	21191	C10H22	2,2,3,4-tetramethylhexane	52897-08-2	-2.6130	4.4750E-03	1.1958E-06	298	373	2	6.7226E-02	1.6725E-01
312	21192	C10H22	2,2,3,5-tetramethylhexane	52897-09-3	-2.3521	4.4750E-03	1.1958E-06	298	373	2	1.2258E-01	3.0497E-01
313	21193	C10H22	2,2,4,4-tetramethylhexane	51750-65-3	-2.4882	4.4750E-03	1.1958E-06	298	373	2	8.9593E-02	2.2290E-01
314	21194	C10H22	2,2,4,5-tetramethylhexane	16747-42-5	-2.3389	4.4750E-03	1.1958E-06	298	373	2	1.2636E-01	3.1436E-01
315	21195	C10H22	2,2,5,5-tetramethylhexane	1071-81-4	-2.0730	4.4750E-03	1.1958E-06	298	373	2	2.3308E-01	5.7987E-01
316	21196	C10H22	2,3,3,4-tetramethylhexane	52897-10-6	-2.7554	4.4750E-03	1.1958E-06	298	373	2	4.8429E-02	1.2048E-01
317	21197	C10H22	2,3,3,5-tetramethylhexane	52897-11-7	-2.4707	4.4750E-03	1.1958E-06	298	373	2	9.3293E-02	2.3210E-01
318	21198	C10H22	2,3,4,4-tetramethylhexane	52897-12-8	-2.6822	4.4750E-03	1.1958E-06	298	373	2	5.7327E-02	1.4262E-01
319	21199	C10H22	2,3,4,5-tetramethylhexane	52897-15-1	-2.5483	4.4750E-03	1.1958E-06	298	373	2	7.8023E-02	1.9411E-01
320	21200	C10H22	3,3,4,4-tetramethylhexane	5171-84-6	-2.8861	4.4750E-03	1.1958E-06	298	373	2	3.5840E-02	8.9166E-02

Table 6 Solubility in Water as a Function of Temperature - Organic Compounds

$\log_{10} S = A + B T + C T^2$ (S - solubility, parts per million by weight, ppm (wt), T - K)												
NO	ID	FORMULA	NAME	CAS No	A	B	C	TMIN	TMAX	code	S@298, ppm(wt)	S@373, ppm(wt)
321	21201	C10H22	2,4-dimethyl-3-isopropylpentane	13475-79-1	-2.5692	4.4750E-03	1.1958E-06	298	373	2	7.4350E-02	1.8497E-01
322	21202	C10H22	3,3-diethyl-2-methylpentane	52897-16-2	-2.8790	4.4750E-03	1.1958E-06	298	373	2	3.6438E-02	9.0653E-02
323	21203	C10H22	3-ethyl-2,2,3-trimethylpentane	52897-17-3	-2.8742	4.4750E-03	1.1958E-06	298	373	2	3.6842E-02	9.1659E-02
324	21204	C10H22	3-ethyl-2,2,4-trimethylpentane	52897-18-4	-2.5258	4.4750E-03	1.1958E-06	298	373	2	8.2168E-02	2.0443E-01
325	21205	C10H22	3-ethyl-2,3,4-trimethylpentane	52897-19-5	-2.8727	4.4750E-03	1.1958E-06	298	373	2	3.6964E-02	9.1963E-02
326	21206	C10H22	2,2,3,3,4-pentamethylpentane	16747-44-7	-2.7909	4.4750E-03	1.1958E-06	298	373	2	4.4625E-02	1.1102E-01
327	21207	C10H22	2,2,3,4,4-pentamethylpentane	16747-45-8	-2.6251	4.4750E-03	1.1958E-06	298	373	2	6.5372E-02	1.6264E-01
328	22304	C11H16	pentylbenzene	538-68-1	3.4038	-2.1171E-02	3.9291E-05	298	373	1,2	3.8400E+00	9.4337E+00
329	22305	C11H16	2-phenylpentane	2719-52-0	-0.8050	5.0102E-03	1.9474E-06	298	373	2	7.2769E+00	2.1665E+01
330	22306	C11H16	3-phenylpentane	1196-58-3	-0.8289	5.0102E-03	1.9474E-06	298	373	2	6.8868E+00	2.0504E+01
331	22307	C11H16	1-phenyl-2-methylbutane	3968-85-2	-0.9712	5.0102E-03	1.9474E-06	298	373	2	4.9621E+00	1.4774E+01
332	22308	C11H16	1-phenyl-3-methylbutane	2049-94-7	-1.0158	5.0102E-03	1.9474E-06	298	373	2	4.4781E+00	1.3332E+01
333	22309	C11H16	2-phenyl-2-methylbutane	2049-95-8	-0.8618	5.0102E-03	1.9474E-06	298	373	2	6.3838E+00	1.9006E+01
334	22310	C11H16	2-phenyl-3-methylbutane	4481-30-5	-0.7569	5.0102E-03	1.9474E-06	298	373	2	8.1277E+00	2.4198E+01
335	22311	C11H16	1-phenyl-2,2-dimethylpropane	1007-26-7	0.2500	3.1975E-03	1.2695E-06	298	373	2	2.0710E+01	4.1678E+01
336	22312	C11H16	1-methyl-2-butylbenzene	1595-11-5	-0.2665	3.1975E-03	1.2695E-06	298	373	2	6.3046E+00	1.2688E+01
337	22313	C11H16	1-methyl-3-butylbenzene	1595-04-6	-0.1983	3.1975E-03	1.2695E-06	298	373	2	7.3773E+00	1.4847E+01
338	22314	C11H16	1-methyl-4-butylbenzene	1595-05-7	-0.2439	3.1975E-03	1.2695E-06	298	373	2	6.6421E+00	1.3367E+01
339	22315	C11H16	1-methyl-2-sec-butylbenzene	1595-06-8	0.0110	3.1975E-03	1.2695E-06	298	373	2	1.1946E+01	2.4041E+01
340	22316	C11H16	1-methyl-3-sec-butylbenzene	1772-10-7	0.0584	3.1975E-03	1.2695E-06	298	373	2	1.3322E+01	2.6810E+01
341	22317	C11H16	1-methyl-4-sec-butylbenzene	1595-16-0	-0.0125	3.1975E-03	1.2695E-06	298	373	2	1.1315E+01	2.2772E+01
342	22318	C11H16	1-methyl-2-isobutylbenzene	36301-29-8	0.0110	3.1975E-03	1.2695E-06	298	373	2	1.1946E+01	2.4041E+01
343	22319	C11H16	1-methyl-3-isobutylbenzene	5160-99-6	0.0584	3.1975E-03	1.2695E-06	298	373	2	1.3322E+01	2.6810E+01
344	22320	C11H16	1-methyl-4-isobutylbenzene	5161-04-6	0.0110	3.1975E-03	1.2695E-06	298	373	2	1.1946E+01	2.4041E+01
345	22321	C11H16	1-methyl-2-tert-butyl benzene	1074-92-6	-0.0933	3.1975E-03	1.2695E-06	298	373	2	9.3957E+00	1.8909E+01
346	22322	C11H16	1-methyl-3-tert-butylbenzene	1075-38-3	0.1715	3.1975E-03	1.2695E-06	298	373	2	1.7286E+01	3.4787E+01
347	22323	C11H16	1-methyl-4-tert-butylbenzene	98-51-1	0.0878	3.1975E-03	1.2695E-06	298	373	2	1.4257E+01	2.8692E+01
348	22324	C11H16	1-ethyl-2-propylbenzene	16021-20-8	-0.1523	3.1975E-03	1.2695E-06	298	373	2	8.2007E+00	1.6504E+01
349	22325	C11H16	1-ethyl-3-propylbenzene	20024-91-3	-0.1060	3.1975E-03	1.2695E-06	298	373	2	9.1231E+00	1.8360E+01
350	22326	C11H16	1-ethyl-4-propylbenzene	20024-90-2	-0.1983	3.1975E-03	1.2695E-06	298	373	2	7.3773E+00	1.4847E+01
351	22327	C11H16	1-ethyl-2-isopropylbenzene	18970-44-0	0.0821	3.1975E-03	1.2695E-06	298	373	2	1.4071E+01	2.8317E+01
352	22328	C11H16	1-ethyl-3-isopropylbenzene	4920-99-4	0.1060	3.1975E-03	1.2695E-06	298	373	2	1.4864E+01	2.9914E+01
353	22329	C11H16	1-ethyl-4-isopropylbenzene	4218-48-8	-0.0031	3.1975E-03	1.2695E-06	298	373	2	1.1563E+01	2.3271E+01
354	22330	C11H16	1,2-dimethyl-3-propylbenzene	17059-44-8	-0.3272	3.1975E-03	1.2695E-06	298	373	2	5.4826E+00	1.1034E+01
355	22331	C11H16	1,2-dimethyl-4-propylbenzene	3982-66-9	-0.2868	3.1975E-03	1.2695E-06	298	373	2	6.0166E+00	1.2108E+01
356	22332	C11H16	1,3-dimethyl-2-propylbenzene	17059-45-9	-0.2575	3.1975E-03	1.2695E-06	298	373	2	6.4373E+00	1.2955E+01
357	22333	C11H16	1,3-dimethyl-4-propylbenzene	61827-85-8	-0.2348	3.1975E-03	1.2695E-06	298	373	2	6.7826E+00	1.3650E+01
358	22334	C11H16	1,3-dimethyl-5-propylbenzene	3982-64-7	-0.1348	3.1975E-03	1.2695E-06	298	373	2	8.5389E+00	1.7184E+01
359	22335	C11H16	1,4-dimethyl-2-propylbenzene	3042-50-0	-0.1822	3.1975E-03	1.2695E-06	298	373	2	7.6549E+00	1.5405E+01
360	22336	C11H16	1,2-dimethyl-3-isopropylbenzene	22539-65-7	-0.1431	3.1975E-03	1.2695E-06	298	373	2	8.3768E+00	1.6858E+01
361	22337	C11H16	1,2-dimethyl-4-isopropylbenzene	4132-77-8	-0.1246	3.1975E-03	1.2695E-06	298	373	2	8.7415E+00	1.7592E+01
362	22338	C11H16	1,3-dimethyl-2-isopropylbenzene	14411-75-7	-0.0594	3.1975E-03	1.2695E-06	298	373	2	1.0157E+01	2.0440E+01
363	22339	C11H16	1,3-dimethyl-4-isopropylbenzene	4706-89-2	-0.0618	3.1975E-03	1.2695E-06	298	373	2	1.0102E+01	2.0330E+01
364	22340	C11H16	1,3-dimethyl-5-isopropylbenzene	4706-90-5	0.0465	3.1975E-03	1.2695E-06	298	373	2	1.2963E+01	2.6088E+01
365	22341	C11H16	1,4-dimethyl-2-isopropylbenzene	4132-72-3	0.0063	3.1975E-03	1.2695E-06	298	373	2	1.1817E+01	2.3782E+01
366	22342	C11H16	1-methyl-2,3-diethylbenzene	13632-93-4	-0.2348	3.1975E-03	1.2695E-06	298	373	2	6.7826E+00	1.3650E+01
367	22343	C11H16	1-methyl-2,4-diethylbenzene	1758-85-6	-0.1983	3.1975E-03	1.2695E-06	298	373	2	7.3773E+00	1.4847E+01
368	22344	C11H16	1-methyl-2,5-diethylbenzene	13632-94-5	-0.2461	3.1975E-03	1.2695E-06	298	373	2	6.6075E+00	1.3297E+01
369	22345	C11H16	1-methyl-2,6-diethyl benzene	13632-95-6	-0.2846	3.1975E-03	1.2695E-06	298	373	2	6.0479E+00	1.2171E+01
370	22346	C11H16	1-methyl-3,4-diethylbenzene	13732-80-4	-0.1662	3.1975E-03	1.2695E-06	298	373	2	7.9438E+00	1.5987E+01
371	22347	C11H16	1-methyl-3,5-diethylbenzene	2050-24-0	-0.0991	3.1975E-03	1.2695E-06	298	373	2	9.2707E+00	1.8657E+01
372	22348	C11H16	1,2,3-trimethyl-4-ethylbenzene	61827-86-9	-0.5385	3.1975E-03	1.2695E-06	298	373	2	3.3704E+00	6.7828E+00
373	22349	C11H16	1,2,3-trimethyl-5-ethylbenzene	31366-00-4	-0.4397	3.1975E-03	1.2695E-06	298	373	2	4.2317E+00	8.5162E+00
374	22350	C11H16	1,2,4-trimethyl-3-ethylbenzene	61827-87-0	-0.4570	3.1975E-03	1.2695E-06	298	373	2	4.0658E+00	8.1822E+00
375	22351	C11H16	1,2,4-trimethyl-5-ethylbenzene	---	-0.3783	3.1975E-03	1.2695E-06	298	373	2	4.8743E+00	9.8093E+00
376	22352	C11H16	1,2,4-trimethyl-6-ethylbenzene	17851-27-3	-0.3783	3.1975E-03	1.2695E-06	298	373	2	4.8743E+00	9.8093E+00
377	22353	C11H16	1,3,5-trimethyl-2-ethylbenzene	3982-67-0	-0.3650	3.1975E-03	1.2695E-06	298	373	2	5.0255E+00	1.0114E+01
378	22354	C11H16	pentamethylbenzene	700-12-9	-0.7633	3.1975E-03	1.2695E-06	298	373	2	2.0085E+00	4.0420E+00
379	22355	C11H16	1-ethyl-2,3,5-trimethylbenzene	18262-85-6	-0.3783	3.1975E-03	1.2695E-06	298	373	2	4.8743E+00	9.8093E+00
380	22356	C11H16	sec-pentylbenzene	29316-05-0	-1.0415	5.0102E-03	1.9474E-06	298	373	2	4.2209E+00	1.2567E+01
381	22791	C11H22	hexylcyclopentane	4457-00-5	-2.6054	3.5563E-03	1.3987E-06	298	373	2	3.7952E-02	8.2485E-02
382	22792	C11H22	pentylcyclohexane	4292-92-6	-2.6937	4.3934E-03	1.7000E-06	298	373	2	5.8525E-02	1.5221E-01
383	22793	C11H22	isopentylcyclohexane	54105-76-9	-2.5488	4.3934E-03	1.7000E-06	298	373	2	8.1688E-02	2.1245E-01
384	22890	C11H24	undecane	1120-21-4	-21.1455	1.0572E-01	-1.4321E-04	298	373	1,2	4.4000E-03	2.3016E-02
385	22891	C11H24	2-methyldecane	6975-98-0	-3.3254	4.4750E-03	1.1958E-06	298	373	2	1.3035E-02	3.2428E-02
386	22892	C11H24	3-methyldecane	13151-34-3	-3.3430	4.4750E-03	1.1958E-06	298	373	2	1.2518E-02	3.1144E-02
387	22893	C11H24	4-methyldecane	2847-72-5	-3.2973	4.4750E-03	1.1958E-06	298	373	2	1.3907E-02	3.4599E-02
388	22894	C11H24	5-methyldecane	13151-35-4	-3.2554	4.4750E-03	1.1958E-06	298	373	2	1.5314E-02	3.8099E-02
389	22895	C11H24	3-ethylnonane	17302-11-3	-3.2995	4.4750E-03	1.1958E-06	298	373	2	1.3837E-02	3.4425E-02
390	22896	C11H24	4-ethylnonane	5911-05-7	-3.2109	4.4750E-03	1.1958E-06	298	373	2	1.6969E-02	4.2217E-02
391	22897	C11H24	5-ethylnonane	17302-12-4	-3.1202	4.4750E-03	1.1958E-06	298	373	2	2.0906E-02	5.2012E-02
392	22898	C11H24	2,2-dimethylnonane	17302-14-6	-3.1202	4.4750E-03	1.1958E-06	298	373	2	2.0906E-02	5.2012E-02
393	22899	C11H24	2,3-dimethylnonane	2884-06-2	-3.2554	4.4750E-03	1.1958E-06	298	373	2	1.5314E-02	3.8099E-02
394	22900	C11H24	2,4-dimethylnonane	17302-24-8	-3.0742	4.4750E-03	1.1958E-06	298	373	2	2.3243E-02	5.7826E-02
395	22901	C11H24	2,5-dimethylnonane	17302-27-1	-3.0973	4.4750E-03	1.1958E-06	298	373	2	2.2041E-02	5.4835E-02
396	22902	C11H24	2,6-dimethylnonane	17302-28-2	-3.0973	4.4750E-03	1.1958E-06	298	373	2	2.2041E-02	5.4835E-02
397	22903	C11H24	2,7-dimethylnonane	17302-29-3	-3.1202	4.4750E-03	1.1958E-06	298	373	2	2.0906E-02	5.2012E-02
398	22904	C11H24	2,8-dimethylnonane	17302-30-6	-3.1884	4.4750E-03	1.1958E-06	298	373	2	1.7870E-02	4.4459E-02
399	22905	C11H24	3,3-dimethylnonane	17302-15-7	-3.1658	4.4750E-03	1.1958E-06	298	373	2	1.8825E-02	4.6833E-02
400	22906	C11H24	3,4-dimethylnonane	17302-22-6	-3.2332	4.4750E-03	1.1958E-06	298	373	2	1.6118E-02	4.0099E-02

Table 6 Solubility in Water as a Function of Temperature - Organic Compounds

$\log_{10} S = A + B\,T + C\,T^2$ (S - solubility, parts per million by weight, ppm (wt), T - K)

NO	ID	FORMULA	NAME	CAS No	A	B	C	TMIN	TMAX	code	S@298, ppm(wt)	S@373, ppm(wt)
401	22907	C11H24	3,5-dimethylnonane	17302-25-9	-3.1202	4.4750E-03	1.1958E-06	298	373	2	2.0906E-02	5.2012E-02
402	22908	C11H24	3,6-dimethylnonane	17302-31-7	-3.1658	4.4750E-03	1.1958E-06	298	373	2	1.8825E-02	4.6833E-02
403	22909	C11H24	3,7-dimethylnonane	17302-32-8	-3.2109	4.4750E-03	1.1958E-06	298	373	2	1.6969E-02	4.2217E-02
404	22910	C11H24	4,4-dimethylnonane	17302-18-0	-3.0742	4.4750E-03	1.1958E-06	298	373	2	2.3243E-02	5.7826E-02
405	22911	C11H24	4,5-dimethylnonane	17302-23-7	-3.1658	4.4750E-03	1.1958E-06	298	373	2	1.8825E-02	4.6833E-02
406	22912	C11H24	4,6-dimethylnonane	17302-26-0	-3.0973	4.4750E-03	1.1958E-06	298	373	2	2.2041E-02	5.4835E-02
407	22913	C11H24	5,5-dimethylnonane	6414-96-6	-3.0510	4.4750E-03	1.1958E-06	298	373	2	2.4517E-02	6.0996E-02
408	22914	C11H24	4-propyloctane	17302-13-5	-3.1202	4.4750E-03	1.1958E-06	298	373	2	2.0906E-02	5.2012E-02
409	22915	C11H24	4-isopropyloctane	62016-15-3	-3.0742	4.4750E-03	1.1958E-06	298	373	2	2.3243E-02	5.7826E-02
410	22916	C11H24	2-methyl-3-ethyloctane	62016-16-4	-3.1658	4.4750E-03	1.1958E-06	298	373	2	1.8825E-02	4.6833E-02
411	22917	C11H24	2-methyl-4-ethyloctane	62016-17-5	-3.0278	4.4750E-03	1.1958E-06	298	373	2	2.5868E-02	6.4356E-02
412	22918	C11H24	2-methyl-5-ethyloctane	62016-18-6	-3.0742	4.4750E-03	1.1958E-06	298	373	2	2.3243E-02	5.7826E-02
413	22919	C11H24	2-methyl-6-ethyloctane	62016-19-7	-3.1658	4.4750E-03	1.1958E-06	298	373	2	1.8825E-02	4.6833E-02
414	22920	C11H24	3-methyl-3-ethyloctane	17302-16-8	-3.2488	4.4750E-03	1.1958E-06	298	373	2	1.5550E-02	3.8687E-02
415	22921	C11H24	3-methyl-4-ethyloctane	62016-20-0	-3.1658	4.4750E-03	1.1958E-06	298	373	2	1.8825E-02	4.6833E-02
416	22922	C11H24	3-methyl-5-ethyloctane	62016-21-1	-3.0278	4.4750E-03	1.1958E-06	298	373	2	2.5868E-02	6.4356E-02
417	22923	C11H24	3-methyl-6-ethyloctane	62016-22-2	-3.1658	4.4750E-03	1.1958E-06	298	373	2	1.8825E-02	4.6833E-02
418	22924	C11H24	4-methyl-3-ethyloctane	62016-23-3	-3.1884	4.4750E-03	1.1958E-06	298	373	2	1.7870E-02	4.4459E-02
419	22925	C11H24	4-methyl-4-ethyloctane	17302-19-1	-3.1202	4.4750E-03	1.1958E-06	298	373	2	2.0906E-02	5.2012E-02
420	22926	C11H24	4-methyl-5-ethyloctane	62016-24-4	-3.1202	4.4750E-03	1.1958E-06	298	373	2	2.0906E-02	5.2012E-02
421	22927	C11H24	4-methyl-6-ethyloctane	62016-25-5	-3.1202	4.4750E-03	1.1958E-06	298	373	2	2.0906E-02	5.2012E-02
422	22928	C11H24	2,2,3-trimethyloctane	62016-26-6	-3.0973	4.4750E-03	1.1958E-06	298	373	2	2.2041E-02	5.4835E-02
423	22929	C11H24	2,2,4-trimethyloctane	18932-14-4	-2.9217	4.4750E-03	1.1958E-06	298	373	2	3.3023E-02	8.2157E-02
424	22930	C11H24	2,2,5-trimethyloctane	62016-27-7	-2.9098	4.4750E-03	1.1958E-06	298	373	2	3.3941E-02	8.4441E-02
425	22931	C11H24	2,2,6-trimethyloctane	62016-28-8	-2.9809	4.4750E-03	1.1958E-06	298	373	2	2.8817E-02	7.1692E-02
426	22932	C11H24	2,2,7-trimethyloctane	62016-29-9	-2.8859	4.4750E-03	1.1958E-06	298	373	2	3.5860E-02	8.9215E-02
427	22933	C11H24	2,3,3-trimethyloctane	62016-30-2	-3.1658	4.4750E-03	1.1958E-06	298	373	2	1.8825E-02	4.6833E-02
428	22934	C11H24	2,3,4-trimethyloctane	62016-31-3	-3.1202	4.4750E-03	1.1958E-06	298	373	2	2.0906E-02	5.2012E-02
429	22935	C11H24	2,3,5-trimethyloctane	62016-32-4	-3.0510	4.4750E-03	1.1958E-06	298	373	2	2.4517E-02	6.0996E-02
430	22936	C11H24	2,3,6-trimethyloctane	62016-33-5	-3.1202	4.4750E-03	1.1958E-06	298	373	2	2.0906E-02	5.2012E-02
431	22937	C11H24	2,3,7-trimethyloctane	62016-34-6	-3.0973	4.4750E-03	1.1958E-06	298	373	2	2.2041E-02	5.4835E-02
432	22938	C11H24	2,4,4-trimethyloctane	62016-35-7	-2.9098	4.4750E-03	1.1958E-06	298	373	2	3.3941E-02	8.4441E-02
433	22939	C11H24	2,4,5-trimethyloctane	62016-36-8	-3.0278	4.4750E-03	1.1958E-06	298	373	2	2.5868E-02	6.4356E-02
434	22940	C11H24	2,4,6-trimethyloctane	62016-37-9	-2.9809	4.4750E-03	1.1958E-06	298	373	2	2.8817E-02	7.1692E-02
435	22941	C11H24	2,4,7-trimethyloctane	62016-38-0	-2.9098	4.4750E-03	1.1958E-06	298	373	2	3.3941E-02	8.4441E-02
436	22942	C11H24	2,5,5-trimethyloctane	62016-39-1	-2.9336	4.4750E-03	1.1958E-06	298	373	2	3.2132E-02	7.9940E-02
437	22943	C11H24	2,5,6-trimethyloctane	62016-14-2	-3.0742	4.4750E-03	1.1958E-06	298	373	2	2.3243E-02	5.7826E-02
438	22944	C11H24	2,6,6-trimethyloctane	54166-32-4	-3.0278	4.4750E-03	1.1958E-06	298	373	2	2.5868E-02	6.4356E-02
439	22945	C11H24	3,3,4-trimethyloctane	62016-40-4	-3.1658	4.4750E-03	1.1958E-06	298	373	2	1.8825E-02	4.6833E-02
440	22946	C11H24	3,3,5-trimethyloctane	62016-41-5	-2.9809	4.4750E-03	1.1958E-06	298	373	2	2.8817E-02	7.1692E-02
441	22947	C11H24	3,3,6-trimethyloctane	62016-42-6	-3.0510	4.4750E-03	1.1958E-06	298	373	2	2.4517E-02	6.0996E-02
442	22948	C11H24	3,4,4-trimethyloctane	62016-43-7	-3.1431	4.4750E-03	1.1958E-06	298	373	2	1.9835E-02	4.9348E-02
443	22949	C11H24	3,4,5-trimethyloctane	62016-44-8	-3.1431	4.4750E-03	1.1958E-06	298	373	2	1.9835E-02	4.9348E-02
444	22950	C11H24	3,4,6-trimethyloctane	62016-45-9	-3.0973	4.4750E-03	1.1958E-06	298	373	2	2.2041E-02	5.4835E-02
445	22951	C11H24	3,5,5-trimethyloctane	61868-94-8	-2.8859	4.4750E-03	1.1958E-06	298	373	2	3.5860E-02	8.9215E-02
446	22952	C11H24	4,4,5-trimethyloctane	61868-95-9	-3.1202	4.4750E-03	1.1958E-06	298	373	2	2.0906E-02	5.2012E-02
447	22953	C11H24	4-tert-butylheptane	60302-21-8	-2.9809	4.4750E-03	1.1958E-06	298	373	2	2.8817E-02	7.1692E-02
448	22954	C11H24	2-methyl-4-propylheptane	61868-96-0	-2.9809	4.4750E-03	1.1958E-06	298	373	2	2.8817E-02	7.1692E-02
449	22955	C11H24	3-methyl-4-propylheptane	61868-97-1	-3.0973	4.4750E-03	1.1958E-06	298	373	2	2.2041E-02	5.4835E-02
450	22956	C11H24	4-methyl-4-propylheptane	17302-20-4	-3.0742	4.4750E-03	1.1958E-06	298	373	2	2.3243E-02	5.7826E-02
451	22957	C11H24	2-methyl-3-isopropylheptane	6876-18-2	-3.0812	4.4750E-03	1.1958E-06	298	373	2	2.2875E-02	5.6911E-02
452	22958	C11H24	2-methyl-4-isopropylheptane	61868-98-2	-2.9336	4.4750E-03	1.1958E-06	298	373	2	3.2132E-02	7.9940E-02
453	22959	C11H24	3-methyl-4-isopropylheptane	61868-99-3	-3.0510	4.4750E-03	1.1958E-06	298	373	2	2.4517E-02	6.0996E-02
454	22960	C11H24	4-methyl-4-isopropylheptane	61869-00-9	-3.1202	4.4750E-03	1.1958E-06	298	373	2	2.0906E-02	5.2012E-02
455	22961	C11H24	3,3-diethylheptane	17302-17-9	-3.2753	4.4750E-03	1.1958E-06	298	373	2	1.4629E-02	3.6394E-02
456	22962	C11H24	3,4-diethylheptane	61869-01-0	-3.1202	4.4750E-03	1.1958E-06	298	373	2	2.0906E-02	5.2012E-02
457	22963	C11H24	3,5-diethylheptane	61869-02-1	-3.0973	4.4750E-03	1.1958E-06	298	373	2	2.2041E-02	5.4835E-02
458	22964	C11H24	4,4-diethylheptane	17302-21-5	-3.1271	4.4750E-03	1.1958E-06	298	373	2	2.0579E-02	5.1197E-02
459	22965	C11H24	2,2-dimethyl-3-ethylheptane	61869-03-2	-3.0278	4.4750E-03	1.1958E-06	298	373	2	2.5868E-02	6.4356E-02
460	22966	C11H24	2,2-dimethyl-4-ethylheptane	62016-46-0	-2.8379	4.4750E-03	1.1958E-06	298	373	2	4.0054E-02	9.9651E-02
461	22967	C11H24	2,2-dimethyl-5-ethylheptane	62016-47-1	-2.9336	4.4750E-03	1.1958E-06	298	373	2	3.2132E-02	7.9940E-02
462	22968	C11H24	2,3-dimethyl-3-ethylheptane	61868-21-1	-3.1202	4.4750E-03	1.1958E-06	298	373	2	2.0906E-02	5.2012E-02
463	22969	C11H24	2,3-dimethyl-4-ethylheptane	61868-22-2	-3.0973	4.4750E-03	1.1958E-06	298	373	2	2.2041E-02	5.4835E-02
464	22970	C11H24	2,3-dimethyl-5-ethylheptane	61868-23-3	-3.0742	4.4750E-03	1.1958E-06	298	373	2	2.3243E-02	5.7826E-02
465	22971	C11H24	2,4-dimethyl-3-ethylheptane	61868-24-4	-3.0973	4.4750E-03	1.1958E-06	298	373	2	2.2041E-02	5.4835E-02
466	22972	C11H24	2,4-dimethyl-4-ethylheptane	61868-25-5	-2.8859	4.4750E-03	1.1958E-06	298	373	2	3.5860E-02	8.9215E-02
467	22973	C11H24	2,4-dimethyl-5-ethylheptane	61868-26-6	-3.0278	4.4750E-03	1.1958E-06	298	373	2	2.5868E-02	6.4356E-02
468	22974	C11H24	2,5-dimethyl-3-ethylheptane	61868-27-7	-3.0278	4.4750E-03	1.1958E-06	298	373	2	2.5868E-02	6.4356E-02
469	22975	C11H24	2,5-dimethyl-4-ethylheptane	61868-28-8	-3.0044	4.4750E-03	1.1958E-06	298	373	2	2.7299E-02	6.7917E-02
470	22976	C11H24	2,5-dimethyl-5-ethylheptane	61868-29-9	-3.0742	4.4750E-03	1.1958E-06	298	373	2	2.3243E-02	5.7826E-02
471	22977	C11H24	2,6-dimethyl-3-ethylheptane	61868-30-2	-3.0278	4.4750E-03	1.1958E-06	298	373	2	2.5868E-02	6.4356E-02
472	22978	C11H24	2,6-dimethyl-4-ethylheptane	61868-31-3	-2.8859	4.4750E-03	1.1958E-06	298	373	2	3.5860E-02	8.9215E-02
473	22979	C11H24	3,3-dimethyl-4-ethylheptane	61868-32-4	-3.1202	4.4750E-03	1.1958E-06	298	373	2	2.0906E-02	5.2012E-02
474	22980	C11H24	3,3-dimethyl-5-ethylheptane	61868-33-5	-3.0044	4.4750E-03	1.1958E-06	298	373	2	2.7299E-02	6.7917E-02
475	22981	C11H24	3,4-dimethyl-3-ethylheptane	61868-34-6	-3.2332	4.4750E-03	1.1958E-06	298	373	2	1.6118E-02	4.0099E-02
476	22982	C11H24	3,4-dimethyl-4-ethylheptane	61868-35-7	-3.1431	4.4750E-03	1.1958E-06	298	373	2	1.9835E-02	4.9348E-02
477	22983	C11H24	3,4-dimethyl-5-ethylheptane	61868-36-8	-3.1658	4.4750E-03	1.1958E-06	298	373	2	1.8825E-02	4.6833E-02
478	22984	C11H24	3,5-dimethyl-3-ethylheptane	61868-37-9	-3.1202	4.4750E-03	1.1958E-06	298	373	2	2.0906E-02	5.2012E-02
479	22985	C11H24	3,5-dimethyl-4-ethylheptane	61868-38-0	-3.1431	4.4750E-03	1.1958E-06	298	373	2	1.9835E-02	4.9348E-02
480	22986	C11H24	4,4-dimethyl-3-ethylheptane	61868-39-1	-3.0278	4.4750E-03	1.1958E-06	298	373	2	2.5868E-02	6.4356E-02

Table 6 Solubility in Water as a Function of Temperature - Organic Compounds

$\log_{10} S = A + B\,T + C\,T^2$ (S - solubility, parts per million by weight, ppm (wt), T - K)

NO	ID	FORMULA	NAME	CAS No	A	B	C	TMIN	TMAX	code	S@298, ppm(wt)	S@373, ppm(wt)
481	22987	C11H24	2,2,3,3-tetramethylheptane	61868-40-4	-3.1202	4.4750E-03	1.1958E-06	298	373	2	2.0906E-02	5.2012E-02
482	22988	C11H24	2,2,3,4-tetramethylheptane	61868-41-5	-3.0278	4.4750E-03	1.1958E-06	298	373	2	2.5868E-02	6.4356E-02
483	22989	C11H24	2,2,3,5-tetramethylheptane	61868-42-6	-2.9809	4.4750E-03	1.1958E-06	298	373	2	2.8817E-02	7.1692E-02
484	22990	C11H24	2,2,3,6-tetramethylheptane	61868-43-7	-2.8859	4.4750E-03	1.1958E-06	298	373	2	3.5860E-02	8.9215E-02
485	22991	C11H24	2,2,4,4-tetramethylheptane	61868-44-8	-2.8859	4.4750E-03	1.1958E-06	298	373	2	3.5860E-02	8.9215E-02
486	22992	C11H24	2,2,4,5-tetramethylheptane	61868-45-9	-2.8619	4.4750E-03	1.1958E-06	298	373	2	3.7895E-02	9.4279E-02
487	22993	C11H24	2,2,4,6-tetramethylheptane	61868-46-0	-2.6967	4.4750E-03	1.1958E-06	298	373	2	5.5444E-02	1.3794E-01
488	22994	C11H24	2,2,5,5-tetramethylheptane	61868-47-1	-2.7895	4.4750E-03	1.1958E-06	298	373	2	4.4775E-02	1.1140E-01
489	22995	C11H24	2,2,5,6-tetramethylheptane	61868-48-2	-2.8619	4.4750E-03	1.1958E-06	298	373	2	3.7895E-02	9.4279E-02
490	22996	C11H24	2,2,6,6-tetramethylheptane	40117-45-1	-2.7163	4.4750E-03	1.1958E-06	298	373	2	5.2993E-02	1.3184E-01
491	22997	C11H24	2,3,3,4-tetramethylheptane	61868-49-3	-3.1658	4.4750E-03	1.1958E-06	298	373	2	1.8825E-02	4.6833E-02
492	22998	C11H24	2,3,3,5-tetramethylheptane	61868-50-6	-3.0278	4.4750E-03	1.1958E-06	298	373	2	2.5868E-02	6.4356E-02
493	22999	C11H24	2,3,3,6-tetramethylheptane	61868-51-7	-3.0044	4.4750E-03	1.1958E-06	298	373	2	2.7299E-02	6.7917E-02
494	23000	C11H24	2,3,4,4-tetramethylheptane	61868-52-8	-3.0742	4.4750E-03	1.1958E-06	298	373	2	2.3243E-02	5.7826E-02
495	23001	C11H24	2,3,4,5-tetramethylheptane	61868-53-9	-3.0973	4.4750E-03	1.1958E-06	298	373	2	2.2041E-02	5.4835E-02
496	23002	C11H24	2,3,4,6-tetramethylheptane	61868-54-0	-2.9809	4.4750E-03	1.1958E-06	298	373	2	2.8817E-02	7.1692E-02
497	23003	C11H24	2,3,5,5-tetramethylheptane	61868-55-1	-2.9336	4.4750E-03	1.1958E-06	298	373	2	3.2132E-02	7.9940E-02
498	23004	C11H24	2,3,5,6-tetramethylheptane	52670-32-3	-3.0044	4.4750E-03	1.1958E-06	298	373	2	2.7299E-02	6.7917E-02
499	23005	C11H24	2,4,4,5-tetramethylheptane	61868-56-2	-2.9809	4.4750E-03	1.1958E-06	298	373	2	2.8817E-02	7.1692E-02
500	23006	C11H24	2,4,4,6-tetramethylheptane	61868-57-3	-2.7408	4.4750E-03	1.1958E-06	298	373	2	5.0090E-02	1.2462E-01
501	23007	C11H24	2,4,5,5-tetramethylheptane	61868-58-4	-3.0278	4.4750E-03	1.1958E-06	298	373	2	2.5868E-02	6.4356E-02
502	23008	C11H24	3,3,4,4-tetramethylheptane	61868-59-5	-3.2109	4.4750E-03	1.1958E-06	298	373	2	1.6969E-02	4.2217E-02
503	23009	C11H24	3,3,4,5-tetramethylheptane	61868-60-8	-3.1431	4.4750E-03	1.1958E-06	298	373	2	1.9835E-02	4.9348E-02
504	23010	C11H24	3,3,5,5-tetramethylheptane	61868-61-9	-3.1202	4.4750E-03	1.1958E-06	298	373	2	2.0906E-02	5.2012E-02
505	23011	C11H24	3,4,4,5-tetramethylheptane	61868-62-0	-3.2332	4.4750E-03	1.1958E-06	298	373	2	1.6118E-02	4.0099E-02
506	23012	C11H24	2,2-dimethyl-3-isopropylhexane	61868-63-1	-2.9336	4.4750E-03	1.1958E-06	298	373	2	3.2132E-02	7.9940E-02
507	23013	C11H24	2,3-dimethyl-3-isopropylhexane	61868-64-2	-3.1202	4.4750E-03	1.1958E-06	298	373	2	2.0906E-02	5.2012E-02
508	23014	C11H24	2,4-dimethyl-3-isopropylhexane	61868-65-3	-3.0044	4.4750E-03	1.1958E-06	298	373	2	2.7299E-02	6.7917E-02
509	23015	C11H24	2,5-dimethyl-3-isopropylhexane	61868-66-4	-2.8859	4.4750E-03	1.1958E-06	298	373	2	3.5860E-02	8.9215E-02
510	23016	C11H24	2-methyl-3,3-diethylhexane	61868-67-5	-3.2554	4.4750E-03	1.1958E-06	298	373	2	1.5314E-02	3.8099E-02
511	23017	C11H24	2-methyl-3,4-diethylhexane	61868-68-6	-3.0973	4.4750E-03	1.1958E-06	298	373	2	2.2041E-02	5.4835E-02
512	23018	C11H24	2-methyl-4,4-diethylhexane	61868-69-7	-3.0973	4.4750E-03	1.1958E-06	298	373	2	2.2041E-02	5.4835E-02
513	23019	C11H24	3-methyl-3,4-diethylhexane	61868-70-0	-3.2554	4.4750E-03	1.1958E-06	298	373	2	1.5314E-02	3.8099E-02
514	23020	C11H24	3-methyl-4,4-diethylhexane	61868-71-1	-3.3430	4.4750E-03	1.1958E-06	298	373	2	1.2518E-02	3.1144E-02
515	23021	C11H24	2,2,3-trimethyl-3-ethylhexane	61868-72-2	-3.1202	4.4750E-03	1.1958E-06	298	373	2	2.0906E-02	5.2012E-02
516	23022	C11H24	2,2,3-trimethyl-4-ethylhexane	61868-73-3	-3.0278	4.4750E-03	1.1958E-06	298	373	2	2.5868E-02	6.4356E-02
517	23023	C11H24	2,2,4-trimethyl-3-ethylhexane	61868-74-4	-3.0044	4.4750E-03	1.1958E-06	298	373	2	2.7299E-02	6.7917E-02
518	23024	C11H24	2,2,4-trimethyl-4-ethylhexane	61868-75-5	-3.0742	4.4750E-03	1.1958E-06	298	373	2	2.3243E-02	5.7826E-02
519	23025	C11H24	2,2,5-trimethyl-3-ethylhexane	61868-76-6	-2.8859	4.4750E-03	1.1958E-06	298	373	2	3.5860E-02	8.9215E-02
520	23026	C11H24	2,2,5-trimethyl-4-ethylhexane	61868-77-7	-2.7895	4.4750E-03	1.1958E-06	298	373	2	4.4775E-02	1.1140E-01
521	23027	C11H24	2,3,3-trimethyl-4-ethylhexane	61868-78-8	-3.1884	4.4750E-03	1.1958E-06	298	373	2	1.7870E-02	4.4459E-02
522	23028	C11H24	2,3,4-trimethyl-3-ethylhexane	61868-79-9	-3.2775	4.4750E-03	1.1958E-06	298	373	2	1.4555E-02	3.6210E-02
523	23029	C11H24	2,3,4-trimethyl-4-ethylhexane	61868-80-2	-3.1202	4.4750E-03	1.1958E-06	298	373	2	2.0906E-02	5.2012E-02
524	23030	C11H24	2,3,5-trimethyl-3-ethylhexane	61868-81-3	-3.0278	4.4750E-03	1.1958E-06	298	373	2	2.5868E-02	6.4356E-02
525	23031	C11H24	2,3,5-trimethyl-4-ethylhexane	61868-82-4	-3.0510	4.4750E-03	1.1958E-06	298	373	2	2.4517E-02	6.0996E-02
526	23032	C11H24	2,4,4-trimethyl-3-ethylhexane	61868-83-5	-3.0742	4.4750E-03	1.1958E-06	298	373	2	2.3243E-02	5.7826E-02
527	23033	C11H24	3,3,4-trimethyl-4-ethylhexane	61868-84-6	-3.3430	4.4750E-03	1.1958E-06	298	373	2	1.2518E-02	3.1144E-02
528	23034	C11H24	2,2,3,3,4-pentamethylhexane	61868-85-7	-3.2109	4.4750E-03	1.1958E-06	298	373	2	1.6969E-02	4.2217E-02
529	23035	C11H24	2,2,3,3,5-pentamethylhexane	61868-86-8	-2.8979	4.4750E-03	1.1958E-06	298	373	2	3.4886E-02	8.6793E-02
530	23036	C11H24	2,2,3,4,4-pentamethylhexane	61868-87-9	-3.2554	4.4750E-03	1.1958E-06	298	373	2	1.5314E-02	3.8099E-02
531	23037	C11H24	2,2,3,4,5-pentamethylhexane	61868-88-0	-2.9809	4.4750E-03	1.1958E-06	298	373	2	2.8817E-02	7.1692E-02
532	23038	C11H24	2,2,3,5,5-pentamethylhexane	14739-73-2	-2.7895	4.4750E-03	1.1958E-06	298	373	2	4.4775E-02	1.1140E-01
533	23039	C11H24	2,2,4,4,5-pentamethylhexane	60302-23-0	-3.0278	4.4750E-03	1.1958E-06	298	373	2	2.5868E-02	6.4356E-02
534	23040	C11H24	2,3,3,4,4-pentamethylhexane	61868-89-1	-3.2775	4.4750E-03	1.1958E-06	298	373	2	1.4555E-02	3.6210E-02
535	23041	C11H24	2,3,3,4,5-pentamethylhexane	52670-33-4	-3.1431	4.4750E-03	1.1958E-06	298	373	2	1.9835E-02	4.9348E-02
536	23042	C11H24	2,2,4-trimethyl-3-isopropylpentane	61868-90-4	-2.8859	4.4750E-03	1.1958E-06	298	373	2	3.5860E-02	8.9215E-02
537	23043	C11H24	2,3,4-trimethyl-3-isopropylpentane	61868-91-5	-3.2554	4.4750E-03	1.1958E-06	298	373	2	1.5314E-02	3.8099E-02
538	23044	C11H24	2,2-dimethyl-3,3-diethylpentane	60302-28-5	-3.3213	4.4750E-03	1.1958E-06	298	373	2	1.3159E-02	3.2738E-02
539	23045	C11H24	2,4-dimethyl-3,3-diethylpentane	61868-92-6	-3.3213	4.4750E-03	1.1958E-06	298	373	2	1.3159E-02	3.2738E-02
540	23046	C11H24	2,2,3,4-tetramethyl-3-ethylpentane	61868-93-7	-3.2554	4.4750E-03	1.1958E-06	298	373	2	1.5314E-02	3.8099E-02
541	23047	C11H24	2,2,4,4-tetramethyl-3-ethylpentane	---	-3.1202	4.4750E-03	1.1958E-06	298	373	2	2.0906E-02	5.2012E-02
542	23048	C11H24	2,2,3,3,4,4-hexamethylpentane	60302-27-4	-3.4283	4.4750E-03	1.1958E-06	298	373	2	1.0286E-02	2.5591E-02
543	24357	C12H18	hexylbenzene	1077-16-3	-3.9340	1.8285E-02	-1.6987E-05	298	373	1,2	1.0180E+00	3.3410E+00
544	24358	C12H18	1,4-dipropylbenzene	4815-57-0	-0.2804	3.1975E-03	1.2695E-06	298	373	2	6.1061E+00	1.2288E+01
545	24359	C12H18	diisopropylbenzene	25321-09-9	-0.0933	3.1975E-03	1.2695E-06	298	373	2	9.3957E+00	1.8909E+01
546	24360	C12H18	1,2-diisopropylbenzene	577-55-9	-0.1754	3.1975E-03	1.2695E-06	298	373	2	7.7773E+00	1.5652E+01
547	24361	C12H18	m-diisopropylbenzene	99-62-7	-0.1565	3.1975E-03	1.2695E-06	298	373	2	8.1227E+00	1.6347E+01
548	24362	C12H18	p-diisopropylbenzene	100-18-5	-0.3227	3.1975E-03	1.2695E-06	298	373	2	5.5393E+00	1.1148E+01
549	24363	C12H18	triethylbenzene; (mixed isomers)	25340-18-5	-0.4776	3.1975E-03	1.2695E-06	298	373	2	3.8780E+00	7.8044E+00
550	24364	C12H18	1,2,3-triethylbenzene	---	-0.4767	3.1975E-03	1.2695E-06	298	373	2	3.8857E+00	7.8199E+00
551	24365	C12H18	1,2,4-triethylbenzene	877-44-1	-0.4873	3.1975E-03	1.2695E-06	298	373	2	3.7925E+00	7.6322E+00
552	24366	C12H18	1,3,5-triethylbenzene	102-25-0	-0.4418	3.1975E-03	1.2695E-06	298	373	2	4.2106E+00	8.4736E+00
553	24367	C12H18	1-tert-butyl-3,5-dimethylbenzene	98-19-1	-0.2439	3.1975E-03	1.2695E-06	298	373	2	6.6421E+00	1.3367E+01
554	24368	C12H18	(1,1-dimethylbutyl)benzene	1985-57-5	-0.2097	3.1975E-03	1.2695E-06	298	373	2	7.1856E+00	1.4461E+01
555	24369	C12H18	2,4-dimethyl-1-sec-butylbenzene	1483-60-9	-0.2804	3.1975E-03	1.2695E-06	298	373	2	6.1061E+00	1.2288E+01
556	24370	C12H18	(1-ethylbutyl)benzene	4468-42-2	-0.2891	3.1975E-03	1.2695E-06	298	373	2	5.9855E+00	1.2046E+01
557	24371	C12H18	(1-ethyl-1-methylpropyl)benzene	1985-97-3	-0.1983	3.1975E-03	1.2695E-06	298	373	2	7.3773E+00	1.4847E+01
558	24372	C12H18	(1-methylpentyl)benzene	6031-02-3	-0.2665	3.1975E-03	1.2695E-06	298	373	2	6.3046E+00	1.2688E+01
559	24373	C12H18	(3-methylpentyl)benzene	54410-69-4	-0.5300	3.1975E-03	1.2695E-06	298	373	2	3.4369E+00	6.9167E+00
560	24374	C12H18	(4-methylpentyl)benzene	4215-86-5	-0.4657	3.1975E-03	1.2695E-06	298	373	2	3.9855E+00	8.0207E+00

Table 6 Solubility in Water as a Function of Temperature - Organic Compounds

			$\log_{10} S = A + B\,T + C\,T^2$ (S - solubility, parts per million by weight, ppm (wt), T - K)									
NO	ID	FORMULA	NAME	CAS No	A	B	C	TMIN	TMAX	code	S@298, ppm(wt)	S@373, ppm(wt)
561	24375	C12H18	1,2,4-trimethyl-5-isopropylbenzene	10222-95-4	-0.5512	3.1975E-03	1.2695E-06	298	373	2	3.2733E+00	6.5874E+00
562	24376	C12H18	4-tert-butyl-o-xylene	7397-06-0	-0.1983	3.1975E-03	1.2695E-06	298	373	2	7.3773E+00	1.4847E+01
563	24377	C12H18	hexamethyldewarbenzene	7641-77-2	0.7173	3.1975E-03	1.2695E-06	298	373	2	6.0743E+01	1.2224E+02
564	24378	C12H18	hexamethylbenzene	87-85-4	-1.2981	3.1975E-03	1.2695E-06	298	373	2	5.8625E-01	1.1798E+00
565	24818	C12H24	heptylcyclopentane	5617-42-5	-2.9747	3.5563E-03	1.3987E-06	298	373	2	1.6215E-02	3.5242E-02
566	24819	C12H24	hexylcyclohexane	4292-75-5	-3.0597	4.3934E-03	1.7000E-06	298	373	2	2.5197E-02	6.5530E-02
567	24820	C12H24	1-methyl-2-pentylcyclohexane	54411-01-7	-2.9440	4.3934E-03	1.7000E-06	298	373	2	3.2883E-02	8.5521E-02
568	24914	C12H26	dodecane	112-40-3	-4.7572	9.6388E-03	-6.1695E-06	298	373	1,2	3.7000E-03	9.5608E-03
569	24915	C12H26	3-methylundecane	1002-43-3	-3.7584	4.4750E-03	1.1958E-06	298	373	2	4.8102E-03	1.1967E-02
570	24916	C12H26	4-methylundecane	2980-69-0	-3.7254	4.4750E-03	1.1958E-06	298	373	2	5.1897E-03	1.2911E-02
571	24917	C12H26	5-methylundecane	1632-70-8	-3.6691	4.4750E-03	1.1958E-06	298	373	2	5.9082E-03	1.4699E-02
572	24918	C12H26	6-methylundecane	17302-33-9	-3.6691	4.4750E-03	1.1958E-06	298	373	2	5.9082E-03	1.4699E-02
573	24919	C12H26	3-ethyldecane	17085-96-0	-3.7198	4.4750E-03	1.1958E-06	298	373	2	5.2565E-03	1.3078E-02
574	24920	C12H26	4-ethyldecane	1636-44-8	-3.6306	4.4750E-03	1.1958E-06	298	373	2	6.4551E-03	1.6059E-02
575	24921	C12H26	5-ethyldecane	17302-36-2	-3.5915	4.4750E-03	1.1958E-06	298	373	2	7.0640E-03	1.7574E-02
576	24922	C12H26	2,2-dimethyldecane	17302-37-3	-3.5716	4.4750E-03	1.1958E-06	298	373	2	7.3940E-03	1.8395E-02
577	24923	C12H26	2,3-dimethyldecane	17312-44-6	-3.6691	4.4750E-03	1.1958E-06	298	373	2	5.9082E-03	1.4699E-02
578	24924	C12H26	2,4-dimethyldecane	2801-84-5	-3.5516	4.4750E-03	1.1958E-06	298	373	2	7.7424E-03	1.9262E-02
579	24925	C12H26	2,5-dimethyldecane	17312-50-4	-3.5111	4.4750E-03	1.1958E-06	298	373	2	8.4990E-03	2.1144E-02
580	24926	C12H26	2,6-dimethyldecane	13150-81-7	-3.5111	4.4750E-03	1.1958E-06	298	373	2	8.4990E-03	2.1144E-02
581	24927	C12H26	2,7-dimethyldecane	17312-51-5	-3.5915	4.4750E-03	1.1958E-06	298	373	2	7.0640E-03	1.7574E-02
582	24928	C12H26	2,8-dimethyldecane	17312-52-6	-3.6306	4.4750E-03	1.1958E-06	298	373	2	6.4551E-03	1.6059E-02
583	24929	C12H26	2,9-dimethyldecane	1002-17-1	-3.6306	4.4750E-03	1.1958E-06	298	373	2	6.4551E-03	1.6059E-02
584	24930	C12H26	3,3-dimethyldecane	17302-38-4	-3.6111	4.4750E-03	1.1958E-06	298	373	2	6.7513E-03	1.6797E-02
585	24931	C12H26	3,4-dimethyldecane	17312-45-7	-3.6499	4.4750E-03	1.1958E-06	298	373	2	6.1743E-03	1.5361E-02
586	24932	C12H26	3,5-dimethyldecane	17312-48-0	-3.5516	4.4750E-03	1.1958E-06	298	373	2	7.7424E-03	1.9262E-02
587	24933	C12H26	3,6-dimethyldecane	17312-53-7	-3.5716	4.4750E-03	1.1958E-06	298	373	2	7.3940E-03	1.8395E-02
588	24934	C12H26	3,7-dimethyldecane	17312-54-8	-3.5915	4.4750E-03	1.1958E-06	298	373	2	7.0640E-03	1.7574E-02
589	24935	C12H26	3,8-dimethyldecane	17312-55-9	-3.6499	4.4750E-03	1.1958E-06	298	373	2	6.1743E-03	1.5361E-02
590	24936	C12H26	4,4-dimethyldecane	17312-39-9	-3.5315	4.4750E-03	1.1958E-06	298	373	2	8.1104E-03	2.0178E-02
591	24937	C12H26	4,5-dimethyldecane	17312-46-8	-3.5915	4.4750E-03	1.1958E-06	298	373	2	7.0640E-03	1.7574E-02
592	24938	C12H26	4,6-dimethyldecane	17312-49-1	-3.5111	4.4750E-03	1.1958E-06	298	373	2	8.4990E-03	2.1144E-02
593	24939	C12H26	4,7-dimethyldecane	17312-56-0	-3.5315	4.4750E-03	1.1958E-06	298	373	2	8.1104E-03	2.0178E-02
594	24940	C12H26	5,5-dimethyldecane	17453-92-8	-3.4907	4.4750E-03	1.1958E-06	298	373	2	8.9095E-03	2.2166E-02
595	24941	C12H26	5,6-dimethyldecane	1636-43-7	-3.5716	4.4750E-03	1.1958E-06	298	373	2	7.3940E-03	1.8395E-02
596	24942	C12H26	4-propylnonane	6165-37-3	-3.5315	4.4750E-03	1.1958E-06	298	373	2	8.1104E-03	2.0178E-02
597	24943	C12H26	5-propylnonane	998-35-6	-3.4907	4.4750E-03	1.1958E-06	298	373	2	8.9095E-03	2.2166E-02
598	24944	C12H26	4-isopropylnonane	62184-71-8	-3.5111	4.4750E-03	1.1958E-06	298	373	2	8.4990E-03	2.1144E-02
599	24945	C12H26	5-isopropylnonane	62184-72-9	-3.5111	4.4750E-03	1.1958E-06	298	373	2	8.4990E-03	2.1144E-02
600	24946	C12H26	2-methyl-3-ethylnonane	62184-73-0	-3.5716	4.4750E-03	1.1958E-06	298	373	2	7.3940E-03	1.8395E-02
601	24947	C12H26	2-methyl-4-ethylnonane	62184-37-6	-3.5111	4.4750E-03	1.1958E-06	298	373	2	8.4990E-03	2.1144E-02
602	24948	C12H26	2-methyl-5-ethylnonane	62184-38-7	-3.4700	4.4750E-03	1.1958E-06	298	373	2	9.3433E-03	2.3245E-02
603	24949	C12H26	2-methyl-6-ethylnonane	62184-39-8	-3.5111	4.4750E-03	1.1958E-06	298	373	2	8.4990E-03	2.1144E-02
604	24950	C12H26	2-methyl-7-ethylnonane	62184-40-1	-3.5915	4.4750E-03	1.1958E-06	298	373	2	7.0640E-03	1.7574E-02
605	24951	C12H26	3-methyl-3-ethylnonane	17302-39-5	-3.6306	4.4750E-03	1.1958E-06	298	373	2	6.4551E-03	1.6059E-02
606	24952	C12H26	3-methyl-4-ethylnonane	62184-41-2	-3.5716	4.4750E-03	1.1958E-06	298	373	2	7.3940E-03	1.8395E-02
607	24953	C12H26	3-methyl-5-ethylnonane	62184-42-3	-3.4492	4.4750E-03	1.1958E-06	298	373	2	9.8016E-03	2.4385E-02
608	24954	C12H26	3-methyl-6-ethylnonane	62184-43-4	-3.5315	4.4750E-03	1.1958E-06	298	373	2	8.1104E-03	2.0178E-02
609	24955	C12H26	3-methyl-7-ethylnonane	62184-44-5	-3.5516	4.4750E-03	1.1958E-06	298	373	2	7.7424E-03	1.9262E-02
610	24956	C12H26	4-methyl-3-ethylnonane	62184-45-6	-3.5516	4.4750E-03	1.1958E-06	298	373	2	7.7424E-03	1.9262E-02
611	24957	C12H26	4-methyl-4-ethylnonane	17312-40-2	-3.5315	4.4750E-03	1.1958E-06	298	373	2	8.1104E-03	2.0178E-02
612	24958	C12H26	4-methyl-5-ethylnonane	62184-46-7	-3.5111	4.4750E-03	1.1958E-06	298	373	2	8.4990E-03	2.1144E-02
613	24959	C12H26	4-methyl-6-ethylnonane	62184-47-8	-3.4700	4.4750E-03	1.1958E-06	298	373	2	9.3433E-03	2.3245E-02
614	24960	C12H26	4-methyl-7-ethylnonane	62184-48-9	-3.5516	4.4750E-03	1.1958E-06	298	373	2	7.7424E-03	1.9262E-02
615	24961	C12H26	5-methyl-3-ethylnonane	62184-49-0	-3.5315	4.4750E-03	1.1958E-06	298	373	2	8.1104E-03	2.0178E-02
616	24962	C12H26	5-methyl-4-ethylnonane	1632-71-9	-3.5315	4.4750E-03	1.1958E-06	298	373	2	8.1104E-03	2.0178E-02
617	24963	C12H26	5-methyl-5-ethylnonane	14531-16-9	-3.5111	4.4750E-03	1.1958E-06	298	373	2	8.4990E-03	2.1144E-02
618	24964	C12H26	2,2,3-trimethylnonane	55499-04-2	-3.5516	4.4750E-03	1.1958E-06	298	373	2	7.7424E-03	1.9262E-02
619	24965	C12H26	2,2,4-trimethylnonane	62184-50-3	-3.3645	4.4750E-03	1.1958E-06	298	373	2	1.1913E-02	2.9637E-02
620	24966	C12H26	2,2,5-trimethylnonane	62184-51-4	-3.3645	4.4750E-03	1.1958E-06	298	373	2	1.1913E-02	2.9637E-02
621	24967	C12H26	2,2,6-trimethylnonane	62184-52-5	-3.3430	4.4750E-03	1.1958E-06	298	373	2	1.2518E-02	3.1144E-02
622	24968	C12H26	2,2,7-trimethylnonane	62184-53-6	-3.4492	4.4750E-03	1.1958E-06	298	373	2	9.8016E-03	2.4385E-02
623	24969	C12H26	2,2,8-trimethylnonane	62184-54-7	-3.4492	4.4750E-03	1.1958E-06	298	373	2	9.8016E-03	2.4385E-02
624	24970	C12H26	2,3,3-trimethylnonane	62184-55-8	-3.5915	4.4750E-03	1.1958E-06	298	373	2	7.0640E-03	1.7574E-02
625	24971	C12H26	2,3,4-trimethylnonane	62184-56-9	-3.5516	4.4750E-03	1.1958E-06	298	373	2	7.7424E-03	1.9262E-02
626	24972	C12H26	2,3,5-trimethylnonane	62184-57-0	-3.4700	4.4750E-03	1.1958E-06	298	373	2	9.3433E-03	2.3245E-02
627	24973	C12H26	2,3,6-trimethylnonane	62184-58-1	-3.5111	4.4750E-03	1.1958E-06	298	373	2	8.4990E-03	2.1144E-02
628	24974	C12H26	2,3,7-trimethylnonane	62184-59-2	-3.5716	4.4750E-03	1.1958E-06	298	373	2	7.3940E-03	1.8395E-02
629	24975	C12H26	2,3,8-trimethylnonane	62184-60-5	-3.5516	4.4750E-03	1.1958E-06	298	373	2	7.7424E-03	1.9262E-02
630	24976	C12H26	2,4,4-trimethylnonane	62184-61-6	-3.3430	4.4750E-03	1.1958E-06	298	373	2	1.2518E-02	3.1144E-02
631	24977	C12H26	2,4,5-trimethylnonane	62184-62-7	-3.4492	4.4750E-03	1.1958E-06	298	373	2	9.8016E-03	2.4385E-02
632	24978	C12H26	2,4,6-trimethylnonane	62184-10-5	-3.3859	4.4750E-03	1.1958E-06	298	373	2	1.1340E-02	2.8212E-02
633	24979	C12H26	2,4,7-trimethylnonane	62184-11-6	-3.4283	4.4750E-03	1.1958E-06	298	373	2	1.0286E-02	2.5591E-02
634	24980	C12H26	2,4,8-trimethylnonane	49542-74-7	-3.3645	4.4750E-03	1.1958E-06	298	373	2	1.1913E-02	2.9637E-02
635	24981	C12H26	2,5,5-trimethylnonane	62184-12-7	-3.3430	4.4750E-03	1.1958E-06	298	373	2	1.2518E-02	3.1144E-02
636	24982	C12H26	2,5,6-trimethylnonane	62184-13-8	-3.4700	4.4750E-03	1.1958E-06	298	373	2	9.3433E-03	2.3245E-02
637	24983	C12H26	2,5,7-trimethylnonane	62184-14-9	-3.4283	4.4750E-03	1.1958E-06	298	373	2	1.0286E-02	2.5591E-02
638	24984	C12H26	2,5,8-trimethylnonane	49557-09-7	-3.3859	4.4750E-03	1.1958E-06	298	373	2	1.1340E-02	2.8212E-02
639	24985	C12H26	2,6,6-trimethylnonane	62184-15-0	-3.3430	4.4750E-03	1.1958E-06	298	373	2	1.2518E-02	3.1144E-02
640	24986	C12H26	2,6,7-trimethylnonane	62184-16-1	-3.5315	4.4750E-03	1.1958E-06	298	373	2	8.1104E-03	2.0178E-02

Table 6 Solubility in Water as a Function of Temperature - Organic Compounds

$\log_{10} S = A + B\,T + C\,T^2$ (S - solubility, parts per million by weight, ppm (wt), T - K)												
NO	ID	FORMULA	NAME	CAS No	A	B	C	TMIN	TMAX	code	S@298, ppm(wt)	S@373, ppm(wt)
641	24987	C12H26	2,7,7-trimethylnonane	62184-17-2	-3.4907	4.4750E-03	1.1958E-06	298	373	2	8.9095E-03	2.2166E-02
642	24988	C12H26	3,3,4-trimethylnonane	62184-18-3	-3.5915	4.4750E-03	1.1958E-06	298	373	2	7.0640E-03	1.7574E-02
643	24989	C12H26	3,3,5-trimethylnonane	62184-19-4	-3.3430	4.4750E-03	1.1958E-06	298	373	2	1.2518E-02	3.1144E-02
644	24990	C12H26	3,3,7-trimethylnonane	62184-21-8	-3.4492	4.4750E-03	1.1958E-06	298	373	2	9.8016E-03	2.4385E-02
645	24991	C12H26	3,3,6-trimethylnonane	62184-20-7	-3.5111	4.4750E-03	1.1958E-06	298	373	2	8.4990E-03	2.1144E-02
646	24992	C12H26	3,4,4-trimethylnonane	62184-22-9	-3.5516	4.4750E-03	1.1958E-06	298	373	2	7.7424E-03	1.9262E-02
647	24993	C12H26	3,4,5-trimethylnonane	62184-23-0	-3.5516	4.4750E-03	1.1958E-06	298	373	2	7.7424E-03	1.9262E-02
648	24994	C12H26	3,4,6-trimethylnonane	62184-24-1	-3.4907	4.4750E-03	1.1958E-06	298	373	2	8.9095E-03	2.2166E-02
649	24995	C12H26	3,4,7-trimethylnonane	27802-85-3	-3.5516	4.4750E-03	1.1958E-06	298	373	2	7.7424E-03	1.9262E-02
650	24996	C12H26	3,5,5-trimethylnonane	62184-25-2	-3.3645	4.4750E-03	1.1958E-06	298	373	2	1.1913E-02	2.9637E-02
651	24997	C12H26	3,5,6-trimethylnonane	62184-26-3	-3.4907	4.4750E-03	1.1958E-06	298	373	2	8.9095E-03	2.2166E-02
652	24998	C12H26	3,5,7-trimethylnonane	62184-27-4	-3.4492	4.4750E-03	1.1958E-06	298	373	2	9.8016E-03	2.4385E-02
653	24999	C12H26	3,6,6-trimethylnonane	62184-28-5	-3.3430	4.4750E-03	1.1958E-06	298	373	2	1.2518E-02	3.1144E-02
654	25000	C12H26	4,4,5-trimethylnonane	62184-29-6	-3.5111	4.4750E-03	1.1958E-06	298	373	2	8.4990E-03	2.1144E-02
655	25001	C12H26	4,4,6-trimethylnonane	62184-30-9	-3.3645	4.4750E-03	1.1958E-06	298	373	2	1.1913E-02	2.9637E-02
656	25002	C12H26	4,5,5-trimethylnonane	62184-31-0	-3.5111	4.4750E-03	1.1958E-06	298	373	2	8.4990E-03	2.1144E-02
657	25003	C12H26	4,5,6-trimethylnonane	62211-85-2	-3.5111	4.4750E-03	1.1958E-06	298	373	2	8.4990E-03	2.1144E-02
658	25004	C12H26	4-tert-butyloctane	62184-32-1	-3.3645	4.4750E-03	1.1958E-06	298	373	2	1.1913E-02	2.9637E-02
659	25005	C12H26	2-methyl-4-propyloctane	62184-33-2	-3.3645	4.4750E-03	1.1958E-06	298	373	2	1.1913E-02	2.9637E-02
660	25006	C12H26	2-methyl-5-propyloctane	62184-34-3	-3.3645	4.4750E-03	1.1958E-06	298	373	2	1.1913E-02	2.9637E-02
661	25007	C12H26	3-methyl-4-propyloctane	62184-35-4	-3.4907	4.4750E-03	1.1958E-06	298	373	2	8.9095E-03	2.2166E-02
662	25008	C12H26	3-methyl-5-propyloctane	62184-36-5	-3.4283	4.4750E-03	1.1958E-06	298	373	2	1.0286E-02	2.5591E-02
663	25009	C12H26	4-methyl-4-propyloctane	17312-41-3	-3.4492	4.4750E-03	1.1958E-06	298	373	2	9.8016E-03	2.4385E-02
664	25010	C12H26	4-methyl-5-propyloctane	62183-85-1	-3.4700	4.4750E-03	1.1958E-06	298	373	2	9.3433E-03	2.3245E-02
665	25011	C12H26	2-methyl-3-isopropyloctane	13287-19-9	-3.4700	4.4750E-03	1.1958E-06	298	373	2	9.3433E-03	2.3245E-02
666	25012	C12H26	2-methyl-4-isopropyloctane	62183-86-2	-3.3430	4.4750E-03	1.1958E-06	298	373	2	1.2518E-02	3.1144E-02
667	25013	C12H26	2-methyl-5-isopropyloctane	62183-87-3	-3.3645	4.4750E-03	1.1958E-06	298	373	2	1.1913E-02	2.9637E-02
668	25014	C12H26	3-methyl-4-isopropyloctane	62183-88-4	-3.4492	4.4750E-03	1.1958E-06	298	373	2	9.8016E-03	2.4385E-02
669	25015	C12H26	3-methyl-5-isopropyloctane	62183-89-5	-3.3859	4.4750E-03	1.1958E-06	298	373	2	1.1340E-02	2.8212E-02
670	25016	C12H26	4-methyl-4-isopropyloctane	62183-90-8	-3.5111	4.4750E-03	1.1958E-06	298	373	2	8.4990E-03	2.1144E-02
671	25017	C12H26	4-methyl-5-isopropyloctane	62183-91-9	-3.4283	4.4750E-03	1.1958E-06	298	373	2	1.0286E-02	2.5591E-02
672	25018	C12H26	3,3-diethyloctane	17302-40-8	-3.6499	4.4750E-03	1.1958E-06	298	373	2	6.1743E-03	1.5361E-02
673	25019	C12H26	3,4-diethyloctane	62183-92-0	-3.5315	4.4750E-03	1.1958E-06	298	373	2	8.1104E-03	2.0178E-02
674	25020	C12H26	3,5-diethyloctane	62183-93-1	-3.4907	4.4750E-03	1.1958E-06	298	373	2	8.9095E-03	2.2166E-02
675	25021	C12H26	3,6-diethyloctane	62183-94-2	-3.5915	4.4750E-03	1.1958E-06	298	373	2	7.0640E-03	1.7574E-02
676	25022	C12H26	4,4-diethyloctane	17312-42-4	-3.5516	4.4750E-03	1.1958E-06	298	373	2	7.7424E-03	1.9262E-02
677	25023	C12H26	4,5-diethyloctane	1636-41-5	-3.4907	4.4750E-03	1.1958E-06	298	373	2	8.9095E-03	2.2166E-02
678	25024	C12H26	2,2-dimethyl-3-ethyloctane	62183-95-3	-3.4700	4.4750E-03	1.1958E-06	298	373	2	9.3433E-03	2.3245E-02
679	25025	C12H26	2,2-dimethyl-4-ethyloctane	62183-96-4	-3.2775	4.4750E-03	1.1958E-06	298	373	2	1.4555E-02	3.6210E-02
680	25026	C12H26	2,2-dimethyl-5-ethyloctane	62183-97-5	-3.3213	4.4750E-03	1.1958E-06	298	373	2	1.3159E-02	3.2738E-02
681	25027	C12H26	2,2-dimethyl-6-ethyloctane	62183-98-6	-3.4072	4.4750E-03	1.1958E-06	298	373	2	1.0798E-02	2.6865E-02
682	25028	C12H26	2,3-dimethyl-3-ethyloctane	62183-99-7	-3.5915	4.4750E-03	1.1958E-06	298	373	2	7.0640E-03	1.7574E-02
683	25029	C12H26	2,3-dimethyl-4-ethyloctane	62184-00-3	-3.4907	4.4750E-03	1.1958E-06	298	373	2	8.9095E-03	2.2166E-02
684	25030	C12H26	2,3-dimethyl-5-ethyloctane	62184-01-4	-3.4283	4.4750E-03	1.1958E-06	298	373	2	1.0286E-02	2.5591E-02
685	25031	C12H26	2,3-dimethyl-6-ethyloctane	62184-02-5	-3.5315	4.4750E-03	1.1958E-06	298	373	2	8.1104E-03	2.0178E-02
686	25032	C12H26	2,4-dimethyl-3-ethyloctane	62184-03-6	-3.4907	4.4750E-03	1.1958E-06	298	373	2	8.9095E-03	2.2166E-02
687	25033	C12H26	2,4-dimethyl-4-ethyloctane	62184-04-7	-3.3645	4.4750E-03	1.1958E-06	298	373	2	1.1913E-02	2.9637E-02
688	25034	C12H26	2,4-dimethyl-5-ethyloctane	62184-05-8	-3.4072	4.4750E-03	1.1958E-06	298	373	2	1.0798E-02	2.6865E-02
689	25035	C12H26	2,4-dimethyl-6-ethyloctane	62184-06-9	-3.3859	4.4750E-03	1.1958E-06	298	373	2	1.1340E-02	2.8212E-02
690	25036	C12H26	2,5-dimethyl-3-ethyloctane	62184-07-0	-3.4283	4.4750E-03	1.1958E-06	298	373	2	1.0286E-02	2.5591E-02
691	25037	C12H26	2,5-dimethyl-4-ethyloctane	62184-08-1	-3.3859	4.4750E-03	1.1958E-06	298	373	2	1.1340E-02	2.8212E-02
692	25038	C12H26	2,5-dimethyl-5-ethyloctane	62184-09-2	-3.4072	4.4750E-03	1.1958E-06	298	373	2	1.0798E-02	2.6865E-02
693	25039	C12H26	2,5-dimethyl-6-ethyloctane	62183-50-0	-3.4907	4.4750E-03	1.1958E-06	298	373	2	8.9095E-03	2.2166E-02
694	25040	C12H26	2,6-dimethyl-3-ethyloctane	62183-51-1	-3.4907	4.4750E-03	1.1958E-06	298	373	2	8.9095E-03	2.2166E-02
695	25041	C12H26	2,6-dimethyl-4-ethyloctane	62183-52-2	-3.3430	4.4750E-03	1.1958E-06	298	373	2	1.2518E-02	3.1144E-02
696	25042	C12H26	2,6-dimethyl-5-ethyloctane	62183-53-3	-3.4492	4.4750E-03	1.1958E-06	298	373	2	9.8016E-03	2.4385E-02
697	25043	C12H26	2,6-dimethyl-6-ethyloctane	62183-54-4	-3.5111	4.4750E-03	1.1958E-06	298	373	2	8.4990E-03	2.1144E-02
698	25044	C12H26	2,7-dimethyl-3-ethyloctane	62183-55-5	-3.4700	4.4750E-03	1.1958E-06	298	373	2	9.3433E-03	2.3245E-02
699	25045	C12H26	2,7-dimethyl-4-ethyloctane	62183-56-6	-3.3430	4.4750E-03	1.1958E-06	298	373	2	1.2518E-02	3.1144E-02
700	25046	C12H26	3,3-dimethyl-4-ethyloctane	62183-57-7	-3.5111	4.4750E-03	1.1958E-06	298	373	2	8.4990E-03	2.1144E-02
701	25047	C12H26	3,3-dimethyl-5-ethyloctane	62183-58-8	-3.3645	4.4750E-03	1.1958E-06	298	373	2	1.1913E-02	2.9637E-02
702	25048	C12H26	3,3-dimethyl-6-ethyloctane	62183-59-9	-3.4700	4.4750E-03	1.1958E-06	298	373	2	9.3433E-03	2.3245E-02
703	25049	C12H26	3,4-dimethyl-3-ethyloctane	62212-28-6	-3.5516	4.4750E-03	1.1958E-06	298	373	2	7.7424E-03	1.9262E-02
704	25050	C12H26	3,4-dimethyl-4-ethyloctane	62183-60-2	-3.5716	4.4750E-03	1.1958E-06	298	373	2	7.3940E-03	1.8395E-02
705	25051	C12H26	3,4-dimethyl-5-ethyloctane	62183-61-3	-3.5111	4.4750E-03	1.1958E-06	298	373	2	8.4990E-03	2.1144E-02
706	25052	C12H26	3,4-dimethyl-6-ethyloctane	62183-62-4	-3.5111	4.4750E-03	1.1958E-06	298	373	2	8.4990E-03	2.1144E-02
707	25053	C12H26	3,5-dimethyl-3-ethyloctane	62183-63-5	-3.4700	4.4750E-03	1.1958E-06	298	373	2	9.3433E-03	2.3245E-02
708	25054	C12H26	3,5-dimethyl-4-ethyloctane	62183-64-6	-3.4907	4.4750E-03	1.1958E-06	298	373	2	8.9095E-03	2.2166E-02
709	25055	C12H26	3,5-dimethyl-5-ethyloctane	62183-65-7	-3.3430	4.4750E-03	1.1958E-06	298	373	2	1.2518E-02	3.1144E-02
710	25056	C12H26	3,5-dimethyl-6-ethyloctane	62183-66-8	-3.4907	4.4750E-03	1.1958E-06	298	373	2	8.9095E-03	2.2166E-02
711	25057	C12H26	3,6-dimethyl-3-ethyloctane	62183-67-9	-3.5315	4.4750E-03	1.1958E-06	298	373	2	8.1104E-03	2.0178E-02
712	25058	C12H26	3,6-dimethyl-4-ethyloctane	62183-68-0	-3.4700	4.4750E-03	1.1958E-06	298	373	2	9.3433E-03	2.3245E-02
713	25059	C12H26	4,4-dimethyl-3-ethyloctane	62183-69-1	-3.5111	4.4750E-03	1.1958E-06	298	373	2	8.4990E-03	2.1144E-02
714	25060	C12H26	4,4-dimethyl-5-ethyloctane	62183-70-4	-3.4700	4.4750E-03	1.1958E-06	298	373	2	9.3433E-03	2.3245E-02
715	25061	C12H26	4,4-dimethyl-6-ethyloctane	62183-71-5	-3.3645	4.4750E-03	1.1958E-06	298	373	2	1.1913E-02	2.9637E-02
716	25062	C12H26	4,5-dimethyl-3-ethyloctane	62183-72-6	-3.5315	4.4750E-03	1.1958E-06	298	373	2	8.1104E-03	2.0178E-02
717	25063	C12H26	4,5-dimethyl-4-ethyloctane	62183-73-7	-3.5516	4.4750E-03	1.1958E-06	298	373	2	7.7424E-03	1.9262E-02
718	25064	C12H26	2,2,3,3-tetramethyloctane	62183-74-8	-3.5516	4.4750E-03	1.1958E-06	298	373	2	7.7424E-03	1.9262E-02
719	25065	C12H26	2,2,3,4-tetramethyloctane	62183-75-9	-3.4492	4.4750E-03	1.1958E-06	298	373	2	9.8016E-03	2.4385E-02
720	25066	C12H26	2,2,3,5-tetramethyloctane	62183-76-0	-3.3859	4.4750E-03	1.1958E-06	298	373	2	1.1340E-02	2.8212E-02

Table 6 Solubility in Water as a Function of Temperature - Organic Compounds

$\log_{10} S = A + B\,T + C\,T^2$ (S - solubility, parts per million by weight, ppm (wt), T - K)

NO	ID	FORMULA	NAME	CAS No	A	B	C	TMIN	TMAX	code	S@298, ppm(wt)	S@373, ppm(wt)
721	25067	C12H26	2,2,3,6-tetramethyloctane	62183-77-1	-3.4283	4.4750E-03	1.1958E-06	298	373	2	1.0286E-02	2.5591E-02
722	25068	C12H26	2,2,3,7-tetramethyloctane	62183-78-2	-3.4283	4.4750E-03	1.1958E-06	298	373	2	1.0286E-02	2.5591E-02
723	25069	C12H26	2,2,4,4-tetramethyloctane	62183-79-3	-3.3859	4.4750E-03	1.1958E-06	298	373	2	1.1340E-02	2.8212E-02
724	25070	C12H26	2,2,4,5-tetramethyloctane	62183-80-6	-3.2775	4.4750E-03	1.1958E-06	298	373	2	1.4555E-02	3.6210E-02
725	25071	C12H26	2,2,4,6-tetramethyloctane	62183-81-7	-3.2332	4.4750E-03	1.1958E-06	298	373	2	1.6118E-02	4.0099E-02
726	25072	C12H26	2,2,4,7-tetramethyloctane	62183-82-8	-3.2288	4.4750E-03	1.1958E-06	298	373	2	1.6284E-02	4.0513E-02
727	25073	C12H26	2,2,5,5-tetramethyloctane	62183-83-9	-3.2109	4.4750E-03	1.1958E-06	298	373	2	1.6969E-02	4.2217E-02
728	25074	C12H26	2,2,5,6-tetramethyloctane	62183-84-0	-3.3430	4.4750E-03	1.1958E-06	298	373	2	1.2518E-02	3.1144E-02
729	25075	C12H26	2,2,5,7-tetramethyloctane	62199-19-3	-3.2332	4.4750E-03	1.1958E-06	298	373	2	1.6118E-02	4.0099E-02
730	25076	C12H26	2,2,6,6-tetramethyloctane	62199-20-6	-3.2332	4.4750E-03	1.1958E-06	298	373	2	1.6118E-02	4.0099E-02
731	25077	C12H26	2,2,6,7-tetramethyloctane	62199-21-7	-3.3645	4.4750E-03	1.1958E-06	298	373	2	1.1913E-02	2.9637E-02
732	25078	C12H26	2,2,7,7-tetramethyloctane	1071-31-4	-3.2332	4.4750E-03	1.1958E-06	298	373	2	1.6118E-02	4.0099E-02
733	25079	C12H26	2,3,3,4-tetramethyloctane	62199-22-8	-3.5716	4.4750E-03	1.1958E-06	298	373	2	7.3940E-03	1.8395E-02
734	25080	C12H26	2,3,3,5-tetramethyloctane	62199-23-9	-3.4283	4.4750E-03	1.1958E-06	298	373	2	1.0286E-02	2.5591E-02
735	25081	C12H26	2,3,3,6-tetramethyloctane	62199-24-0	-3.4700	4.4750E-03	1.1958E-06	298	373	2	9.3433E-03	2.3245E-02
736	25082	C12H26	2,3,3,7-tetramethyloctane	62199-25-1	-3.4700	4.4750E-03	1.1958E-06	298	373	2	9.3433E-03	2.3245E-02
737	25083	C12H26	2,3,4,4-tetramethyloctane	62199-26-2	-3.4700	4.4750E-03	1.1958E-06	298	373	2	9.3433E-03	2.3245E-02
738	25084	C12H26	2,3,4,5-tetramethyloctane	62199-27-3	-3.4907	4.4750E-03	1.1958E-06	298	373	2	8.9095E-03	2.2166E-02
739	25085	C12H26	2,3,4,6-tetramethyloctane	62199-28-4	-3.4492	4.4750E-03	1.1958E-06	298	373	2	9.8016E-03	2.4385E-02
740	25086	C12H26	2,3,4,7-tetramethyloctane	62199-29-5	-3.4283	4.4750E-03	1.1958E-06	298	373	2	1.0286E-02	2.5591E-02
741	25087	C12H26	2,3,5,5-tetramethyloctane	62199-30-8	-3.4492	4.4750E-03	1.1958E-06	298	373	2	9.8016E-03	2.4385E-02
742	25088	C12H26	2,3,5,6-tetramethyloctane	62199-31-9	-3.4492	4.4750E-03	1.1958E-06	298	373	2	9.8016E-03	2.4385E-02
743	25089	C12H26	2,3,5,7-tetramethyloctane	62199-32-0	-3.3430	4.4750E-03	1.1958E-06	298	373	2	1.2518E-02	3.1144E-02
744	25090	C12H26	2,3,6,6-tetramethyloctane	62199-33-1	-3.4072	4.4750E-03	1.1958E-06	298	373	2	1.0798E-02	2.6865E-02
745	25091	C12H26	2,3,6,7-tetramethyloctane	52670-34-5	-3.4700	4.4750E-03	1.1958E-06	298	373	2	9.3433E-03	2.3245E-02
746	25092	C12H26	2,4,4,5-tetramethyloctane	62199-34-2	-3.3645	4.4750E-03	1.1958E-06	298	373	2	1.1913E-02	2.9637E-02
747	25093	C12H26	2,4,4,6-tetramethyloctane	62199-35-3	-3.2332	4.4750E-03	1.1958E-06	298	373	2	1.6118E-02	4.0099E-02
748	25094	C12H26	2,4,4,7-tetramethyloctane	35866-96-7	-3.2109	4.4750E-03	1.1958E-06	298	373	2	1.6969E-02	4.2217E-02
749	25095	C12H26	2,4,5,5-tetramethyloctane	62199-36-4	-3.3859	4.4750E-03	1.1958E-06	298	373	2	1.1340E-02	2.8212E-02
750	25096	C12H26	2,4,5,6-tetramethyloctane	62199-37-5	-3.4283	4.4750E-03	1.1958E-06	298	373	2	1.0286E-02	2.5591E-02
751	25097	C12H26	2,4,5,7-tetramethyloctane	2217-17-6	-3.2621	4.4750E-03	1.1958E-06	298	373	2	1.5082E-02	3.7521E-02
752	25098	C12H26	2,4,6,6-tetramethyloctane	62199-38-6	-3.2775	4.4750E-03	1.1958E-06	298	373	2	1.4555E-02	3.6210E-02
753	25099	C12H26	2,5,5,6-tetramethyloctane	62199-39-7	-3.4283	4.4750E-03	1.1958E-06	298	373	2	1.0286E-02	2.5591E-02
754	25100	C12H26	2,5,6,6-tetramethyloctane	62199-40-0	-3.4700	4.4750E-03	1.1958E-06	298	373	2	9.3433E-03	2.3245E-02
755	25101	C12H26	3,3,4,4-tetramethyloctane	62199-41-1	-3.5915	4.4750E-03	1.1958E-06	298	373	2	7.0640E-03	1.7574E-02
756	25102	C12H26	3,3,4,5-tetramethyloctane	62199-42-2	-3.5111	4.4750E-03	1.1958E-06	298	373	2	8.4990E-03	2.1144E-02
757	25103	C12H26	3,3,4,6-tetramethyloctane	62199-43-3	-3.4700	4.4750E-03	1.1958E-06	298	373	2	9.3433E-03	2.3245E-02
758	25104	C12H26	3,3,5,5-tetramethyloctane	62199-44-4	-3.4907	4.4750E-03	1.1958E-06	298	373	2	8.9095E-03	2.2166E-02
759	25105	C12H26	3,3,5,6-tetramethyloctane	62199-45-5	-3.4072	4.4750E-03	1.1958E-06	298	373	2	1.0798E-02	2.6865E-02
760	25106	C12H26	3,3,6,6-tetramethyloctane	62199-46-6	-3.3365	4.4750E-03	1.1958E-06	298	373	2	1.2707E-02	3.1613E-02
761	25107	C12H26	3,4,4,5-tetramethyloctane	62199-47-7	-3.5716	4.4750E-03	1.1958E-06	298	373	2	7.3940E-03	1.8395E-02
762	25108	C12H26	3,4,4,6-tetramethyloctane	62185-19-7	-3.4492	4.4750E-03	1.1958E-06	298	373	2	9.8016E-03	2.4385E-02
763	25109	C12H26	3,4,5,5-tetramethyloctane	62185-20-0	-3.4907	4.4750E-03	1.1958E-06	298	373	2	8.9095E-03	2.2166E-02
764	25110	C12H26	3,4,5,6-tetramethyloctane	62185-21-1	-3.5315	4.4750E-03	1.1958E-06	298	373	2	8.1104E-03	2.0178E-02
765	25111	C12H26	4,4,5,5-tetramethyloctane	62185-22-2	-3.5516	4.4750E-03	1.1958E-06	298	373	2	7.7424E-03	1.9262E-02
766	25112	C12H26	2-methyl-4-tert-butylheptane	62185-23-3	-3.2332	4.4750E-03	1.1958E-06	298	373	2	1.6118E-02	4.0099E-02
767	25113	C12H26	3-methyl-4-tert-butylheptane	62185-24-4	-3.3430	4.4750E-03	1.1958E-06	298	373	2	1.2518E-02	3.1144E-02
768	25114	C12H26	4-methyl-4-tert-butylheptane	62185-25-5	-3.5111	4.4750E-03	1.1958E-06	298	373	2	8.4990E-03	2.1144E-02
769	25115	C12H26	3-ethyl-4-propylheptane	62185-26-6	-3.4700	4.4750E-03	1.1958E-06	298	373	2	9.3433E-03	2.3245E-02
770	25116	C12H26	4-ethyl-4-propylheptane	17312-43-5	-3.4907	4.4750E-03	1.1958E-06	298	373	2	8.9095E-03	2.2166E-02
771	25117	C12H26	3-ethyl-4-isopropylheptane	62185-27-7	-3.4492	4.4750E-03	1.1958E-06	298	373	2	9.8016E-03	2.4385E-02
772	25118	C12H26	4-ethyl-4-isopropylheptane	62185-28-8	-3.5516	4.4750E-03	1.1958E-06	298	373	2	7.7424E-03	1.9262E-02
773	25119	C12H26	2,2-dimethyl-4-propylheptane	---	-3.2109	4.4750E-03	1.1958E-06	298	373	2	1.6969E-02	4.2217E-02
774	25120	C12H26	2,3-dimethyl-4-propylheptane	62185-30-2	-3.4283	4.4750E-03	1.1958E-06	298	373	2	1.0286E-02	2.5591E-02
775	25121	C12H26	2,4-dimethyl-4-propylheptane	62185-31-3	-3.3213	4.4750E-03	1.1958E-06	298	373	2	1.3159E-02	3.2738E-02
776	25122	C12H26	2,5-dimethyl-4-propylheptane	62185-32-4	-3.3430	4.4750E-03	1.1958E-06	298	373	2	1.2518E-02	3.1144E-02
777	25123	C12H26	2,6-dimethyl-4-propylheptane	62185-33-5	-3.2243	4.4750E-03	1.1958E-06	298	373	2	1.6452E-02	4.0932E-02
778	25124	C12H26	3,3-dimethyl-4-propylheptane	62185-34-6	-3.4700	4.4750E-03	1.1958E-06	298	373	2	9.3433E-03	2.3245E-02
779	25125	C12H26	3,4-dimethyl-4-propylheptane	62185-35-7	-3.5111	4.4750E-03	1.1958E-06	298	373	2	8.4990E-03	2.1144E-02
780	25126	C12H26	3,5-dimethyl-4-propylheptane	62185-36-8	-3.4700	4.4750E-03	1.1958E-06	298	373	2	9.3433E-03	2.3245E-02
781	25127	C12H26	2,2-dimethyl-3-isopropylheptane	62185-37-9	-3.3430	4.4750E-03	1.1958E-06	298	373	2	1.2518E-02	3.1144E-02
782	25128	C12H26	2,2-dimethyl-4-isopropylheptane	62185-38-0	-3.1658	4.4750E-03	1.1958E-06	298	373	2	1.8825E-02	4.6833E-02
783	25129	C12H26	2,3-dimethyl-3-isopropylheptane	62185-39-1	-3.5716	4.4750E-03	1.1958E-06	298	373	2	7.3940E-03	1.8395E-02
784	25130	C12H26	2,3-dimethyl-4-isopropylheptane	62185-40-4	-3.3859	4.4750E-03	1.1958E-06	298	373	2	1.1340E-02	2.8212E-02
785	25131	C12H26	2,4-dimethyl-3-isopropylheptane	62185-41-5	-3.3859	4.4750E-03	1.1958E-06	298	373	2	1.1340E-02	2.8212E-02
786	25132	C12H26	2,4-dimethyl-4-isopropylheptane	62185-42-6	-3.3859	4.4750E-03	1.1958E-06	298	373	2	1.1340E-02	2.8212E-02
787	25133	C12H26	2,5-dimethyl-3-isopropylheptane	62185-43-7	-3.3430	4.4750E-03	1.1958E-06	298	373	2	1.2518E-02	3.1144E-02
788	25134	C12H26	2,5-dimethyl-4-isopropylheptane	62185-44-8	-3.3213	4.4750E-03	1.1958E-06	298	373	2	1.3159E-02	3.2738E-02
789	25135	C12H26	2,6-dimethyl-3-isopropylheptane	62185-45-9	-3.3602	4.4750E-03	1.1958E-06	298	373	2	1.2031E-02	2.9932E-02
790	25136	C12H26	2,6-dimethyl-4-isopropylheptane	35866-89-8	-3.1884	4.4750E-03	1.1958E-06	298	373	2	1.7870E-02	4.4459E-02
791	25137	C12H26	3,3-dimethyl-4-isopropylheptane	62185-46-0	-3.4283	4.4750E-03	1.1958E-06	298	373	2	1.0286E-02	2.5591E-02
792	25138	C12H26	3,4-dimethyl-4-isopropylheptane	62185-47-1	-3.5716	4.4750E-03	1.1958E-06	298	373	2	7.3940E-03	1.8395E-02
793	25139	C12H26	3,5-dimethyl-4-isopropylheptane	62198-89-4	-3.4283	4.4750E-03	1.1958E-06	298	373	2	1.0286E-02	2.5591E-02
794	25140	C12H26	2-methyl-3,3-diethylheptane	62198-90-7	-3.6111	4.4750E-03	1.1958E-06	298	373	2	6.7513E-03	1.6797E-02
795	25141	C12H26	2-methyl-3,4-diethylheptane	62198-91-8	-3.4492	4.4750E-03	1.1958E-06	298	373	2	9.8016E-03	2.4385E-02
796	25142	C12H26	2-methyl-3,5-diethylheptane	62198-92-9	-3.4492	4.4750E-03	1.1958E-06	298	373	2	9.8016E-03	2.4385E-02
797	25143	C12H26	2-methyl-4,4-diethylheptane	62198-93-0	-3.4283	4.4750E-03	1.1958E-06	298	373	2	1.0286E-02	2.5591E-02
798	25144	C12H26	2-methyl-4,5-diethylheptane	62198-94-1	-3.4072	4.4750E-03	1.1958E-06	298	373	2	1.0798E-02	2.6865E-02
799	25145	C12H26	2-methyl-5,5-diethylheptane	62198-95-2	-3.5111	4.4750E-03	1.1958E-06	298	373	2	8.4990E-03	2.1144E-02
800	25146	C12H26	3-methyl-3,4-diethylheptane	62198-96-3	-3.5716	4.4750E-03	1.1958E-06	298	373	2	7.3940E-03	1.8395E-02

Table 6 Solubility in Water as a Function of Temperature - Organic Compounds

$\log_{10} S = A + B\,T + C\,T^2$ (S - solubility, parts per million by weight, ppm (wt), T - K)												
NO	ID	FORMULA	NAME	CAS No	A	B	C	TMIN	TMAX	code	S@298, ppm(wt)	S@373, ppm(wt)
801	25147	C12H26	3-methyl-3,5-diethylheptane	62198-97-4	-3.4907	4.4750E-03	1.1958E-06	298	373	2	8.9095E-03	2.2166E-02
802	25148	C12H26	3-methyl-4,4-diethylheptane	62198-98-5	-3.6306	4.4750E-03	1.1958E-06	298	373	2	6.4551E-03	1.6059E-02
803	25149	C12H26	3-methyl-4,5-diethylheptane	62198-99-6	-3.5111	4.4750E-03	1.1958E-06	298	373	2	8.4990E-03	2.1144E-02
804	25150	C12H26	3-methyl-5,5-diethylheptane	62199-00-2	-3.5315	4.4750E-03	1.1958E-06	298	373	2	8.1104E-03	2.0178E-02
805	25151	C12H26	4-methyl-3,3-diethylheptane	62199-01-3	-3.6499	4.4750E-03	1.1958E-06	298	373	2	6.1743E-03	1.5361E-02
806	25152	C12H26	4-methyl-3,4-diethylheptane	62199-02-4	-3.5716	4.4750E-03	1.1958E-06	298	373	2	7.3940E-03	1.8395E-02
807	25153	C12H26	4-methyl-3,5-diethylheptane	62199-03-5	-3.5516	4.4750E-03	1.1958E-06	298	373	2	7.7424E-03	1.9262E-02
808	25154	C12H26	2,2,3-trimethyl-3-ethylheptane	62199-04-6	-3.5516	4.4750E-03	1.1958E-06	298	373	2	7.7424E-03	1.9262E-02
809	25155	C12H26	2,2,3-trimethyl-4-ethylheptane	62199-05-7	-3.4072	4.4750E-03	1.1958E-06	298	373	2	1.0798E-02	2.6865E-02
810	25156	C12H26	2,2,3-trimethyl-5-ethylheptane	62199-06-8	-3.3859	4.4750E-03	1.1958E-06	298	373	2	1.1340E-02	2.8212E-02
811	25157	C12H26	2,2,4-trimethyl-3-ethylheptane	62199-07-9	-3.3859	4.4750E-03	1.1958E-06	298	373	2	1.1340E-02	2.8212E-02
812	25158	C12H26	2,2,4-trimethyl-4-ethylheptane	62199-08-0	-3.4283	4.4750E-03	1.1958E-06	298	373	2	1.0286E-02	2.5591E-02
813	25159	C12H26	2,2,4-trimethyl-5-ethylheptane	62199-09-1	-3.2775	4.4750E-03	1.1958E-06	298	373	2	1.4555E-02	3.6210E-02
814	25160	C12H26	2,2,5-trimethyl-3-ethylheptane	62199-10-4	-3.3430	4.4750E-03	1.1958E-06	298	373	2	1.2518E-02	3.1144E-02
815	25161	C12H26	2,2,5-trimethyl-4-ethylheptane	62199-11-5	-3.2554	4.4750E-03	1.1958E-06	298	373	2	1.5314E-02	3.8099E-02
816	25162	C12H26	2,2,5-trimethyl-5-ethylheptane	62199-12-6	-3.3213	4.4750E-03	1.1958E-06	298	373	2	1.3159E-02	3.2738E-02
817	25163	C12H26	2,2,6-trimethyl-3-ethylheptane	62199-13-7	-3.3430	4.4750E-03	1.1958E-06	298	373	2	1.2518E-02	3.1144E-02
818	25164	C12H26	2,2,6-trimethyl-4-ethylheptane	62199-14-8	-3.1202	4.4750E-03	1.1958E-06	298	373	2	2.0906E-02	5.2012E-02
819	25165	C12H26	2,2,6-trimethyl-5-ethylheptane	62199-15-9	-3.2775	4.4750E-03	1.1958E-06	298	373	2	1.4555E-02	3.6210E-02
820	25166	C12H26	2,3,3-trimethyl-4-ethylheptane	62199-16-0	-3.5315	4.4750E-03	1.1958E-06	298	373	2	8.1104E-03	2.0178E-02
821	25167	C12H26	2,3,3-trimethyl-5-ethylheptane	62199-17-1	-3.4283	4.4750E-03	1.1958E-06	298	373	2	1.0286E-02	2.5591E-02
822	25168	C12H26	2,3,4-trimethyl-3-ethylheptane	62199-18-2	-3.6111	4.4750E-03	1.1958E-06	298	373	2	6.7513E-03	1.6797E-02
823	25169	C12H26	2,3,4-trimethyl-4-ethylheptane	62198-55-4	-3.5111	4.4750E-03	1.1958E-06	298	373	2	8.4990E-03	2.1144E-02
824	25170	C12H26	2,3,4-trimethyl-5-ethylheptane	62198-56-5	-3.4907	4.4750E-03	1.1958E-06	298	373	2	8.9095E-03	2.2166E-02
825	25171	C12H26	2,3,5-trimethyl-3-ethylheptane	62198-57-6	-3.5009	4.4750E-03	1.1958E-06	298	373	2	8.7014E-03	2.1648E-02
826	25172	C12H26	2,3,5-trimethyl-4-ethylheptane	62198-58-7	-3.4700	4.4750E-03	1.1958E-06	298	373	2	9.3433E-03	2.3245E-02
827	25173	C12H26	2,3,5-trimethyl-5-ethylheptane	62198-59-8	-3.4283	4.4750E-03	1.1958E-06	298	373	2	1.0286E-02	2.5591E-02
828	25174	C12H26	2,3,6-trimethyl-3-ethylheptane	62198-60-1	-3.4700	4.4750E-03	1.1958E-06	298	373	2	9.3433E-03	2.3245E-02
829	25175	C12H26	2,3,6-trimethyl-4-ethylheptane	62198-61-2	-3.3430	4.4750E-03	1.1958E-06	298	373	2	1.2518E-02	3.1144E-02
830	25176	C12H26	2,3,6-trimethyl-5-ethylheptane	62198-62-3	-3.4072	4.4750E-03	1.1958E-06	298	373	2	1.0798E-02	2.6865E-02
831	25177	C12H26	2,4,4-trimethyl-3-ethylheptane	62198-63-4	-3.4283	4.4750E-03	1.1958E-06	298	373	2	1.0286E-02	2.5591E-02
832	25178	C12H26	2,4,4-trimethyl-5-ethylheptane	62198-64-5	-3.3859	4.4750E-03	1.1958E-06	298	373	2	1.1340E-02	2.8212E-02
833	25179	C12H26	2,4,5-trimethyl-3-ethylheptane	62198-65-6	-3.4700	4.4750E-03	1.1958E-06	298	373	2	9.3433E-03	2.3245E-02
834	25180	C12H26	2,4,5-trimethyl-4-ethylheptane	62198-66-7	-3.4492	4.4750E-03	1.1958E-06	298	373	2	9.8016E-03	2.4385E-02
835	25181	C12H26	2,4,5-trimethyl-5-ethylheptane	62198-67-8	-3.4907	4.4750E-03	1.1958E-06	298	373	2	8.9095E-03	2.2166E-02
836	25182	C12H26	2,4,6-trimethyl-3-ethylheptane	62198-68-9	-3.3645	4.4750E-03	1.1958E-06	298	373	2	1.1913E-02	2.9637E-02
837	25183	C12H26	2,4,6-trimethyl-4-ethylheptane	62198-69-0	-3.2332	4.4750E-03	1.1958E-06	298	373	2	1.6118E-02	4.0099E-02
838	25184	C12H26	2,5,5-trimethyl-3-ethylheptane	62198-70-3	-3.3430	4.4750E-03	1.1958E-06	298	373	2	1.2518E-02	3.1144E-02
839	25185	C12H26	2,5,5-trimethyl-4-ethylheptane	62198-71-4	-3.3859	4.4750E-03	1.1958E-06	298	373	2	1.1340E-02	2.8212E-02
840	25186	C12H26	3,3,4-trimethyl-4-ethylheptane	62198-72-5	-3.6499	4.4750E-03	1.1958E-06	298	373	2	6.1743E-03	1.5361E-02
841	25187	C12H26	3,3,4-trimethyl-5-ethylheptane	62198-73-6	-3.5315	4.4750E-03	1.1958E-06	298	373	2	8.1104E-03	2.0178E-02
842	25188	C12H26	3,3,5-trimethyl-4-ethylheptane	62198-74-7	-3.4907	4.4750E-03	1.1958E-06	298	373	2	8.9095E-03	2.2166E-02
843	25189	C12H26	3,3,5-trimethyl-5-ethylheptane	62198-75-8	-3.5915	4.4750E-03	1.1958E-06	298	373	2	7.0640E-03	1.7574E-02
844	25190	C12H26	3,4,4-trimethyl-3-ethylheptane	62198-76-9	-3.6499	4.4750E-03	1.1958E-06	298	373	2	6.1743E-03	1.5361E-02
845	25191	C12H26	3,4,4-trimethyl-5-ethylheptane	62198-77-0	-3.5915	4.4750E-03	1.1958E-06	298	373	2	7.0640E-03	1.7574E-02
846	25192	C12H26	3,4,5-trimethyl-3-ethylheptane	62198-78-1	-3.5915	4.4750E-03	1.1958E-06	298	373	2	7.0640E-03	1.7574E-02
847	25193	C12H26	3,4,5-trimethyl-4-ethylheptane	62198-79-2	-3.6499	4.4750E-03	1.1958E-06	298	373	2	6.1743E-03	1.5361E-02
848	25194	C12H26	2,2,3,3,4-pentamethylheptane	62198-80-5	-3.5516	4.4750E-03	1.1958E-06	298	373	2	7.7424E-03	1.9262E-02
849	25195	C12H26	2,2,3,3,5-pentamethylheptane	62198-81-6	-3.4283	4.4750E-03	1.1958E-06	298	373	2	1.0286E-02	2.5591E-02
850	25196	C12H26	2,2,3,3,6-pentamethylheptane	62198-82-7	-3.4072	4.4750E-03	1.1958E-06	298	373	2	1.0798E-02	2.6865E-02
851	25197	C12H26	2,2,3,4,4-pentamethylheptane	62198-83-8	-3.5915	4.4750E-03	1.1958E-06	298	373	2	7.0640E-03	1.7574E-02
852	25198	C12H26	2,2,3,4,5-pentamethylheptane	62198-84-9	-3.4283	4.4750E-03	1.1958E-06	298	373	2	1.0286E-02	2.5591E-02
853	25199	C12H26	2,2,3,4,6-pentamethylheptane	62198-85-0	-3.2995	4.4750E-03	1.1958E-06	298	373	2	1.3837E-02	3.4425E-02
854	25200	C12H26	2,2,3,5,5-pentamethylheptane	62198-86-1	-3.2775	4.4750E-03	1.1958E-06	298	373	2	1.4555E-02	3.6210E-02
855	25201	C12H26	2,2,3,5,6-pentamethylheptane	62198-87-2	-3.3169	4.4750E-03	1.1958E-06	298	373	2	1.3292E-02	3.3068E-02
856	25202	C12H26	2,2,3,6,6-pentamethylheptane	62198-88-3	-3.2332	4.4750E-03	1.1958E-06	298	373	2	1.6118E-02	4.0099E-02
857	25203	C12H26	2,2,4,4,5-pentamethylheptane	62199-61-5	-3.4700	4.4750E-03	1.1958E-06	298	373	2	9.3433E-03	2.3245E-02
858	25204	C12H26	2,2,4,4,6-pentamethylheptane	62199-62-6	-3.2466	4.4750E-03	1.1958E-06	298	373	2	1.5630E-02	3.8886E-02
859	25205	C12H26	2,2,4,5,5-pentamethylheptane	62199-63-7	-3.2554	4.4750E-03	1.1958E-06	298	373	2	1.5314E-02	3.8099E-02
860	25206	C12H26	2,2,4,5,6-pentamethylheptane	62199-64-8	-3.2332	4.4750E-03	1.1958E-06	298	373	2	1.6118E-02	4.0099E-02
861	25207	C12H26	2,2,4,6,6-pentamethylheptane	13475-82-6	-3.0731	4.4750E-03	1.1958E-06	298	373	2	2.3305E-02	5.7980E-02
862	25208	C12H26	2,2,5,5,6-pentamethylheptane	62199-65-9	-3.2554	4.4750E-03	1.1958E-06	298	373	2	1.5314E-02	3.8099E-02
863	25209	C12H26	2,3,3,4,4-pentamethylheptane	62199-66-0	-3.6111	4.4750E-03	1.1958E-06	298	373	2	6.7513E-03	1.6797E-02
864	25210	C12H26	2,3,3,4,5-pentamethylheptane	62199-67-1	-3.5516	4.4750E-03	1.1958E-06	298	373	2	7.7424E-03	1.9262E-02
865	25211	C12H26	2,3,3,4,6-pentamethylheptane	62199-68-2	-3.4492	4.4750E-03	1.1958E-06	298	373	2	9.8016E-03	2.4385E-02
866	25212	C12H26	2,3,3,5,5-pentamethylheptane	62199-69-3	-3.5516	4.4750E-03	1.1958E-06	298	373	2	7.7424E-03	1.9262E-02
867	25213	C12H26	2,3,3,5,6-pentamethylheptane	52670-35-6	-3.3859	4.4750E-03	1.1958E-06	298	373	2	1.1340E-02	2.8212E-02
868	25214	C12H26	2,3,4,4,5-pentamethylheptane	62199-70-6	-3.5516	4.4750E-03	1.1958E-06	298	373	2	7.7424E-03	1.9262E-02
869	25215	C12H26	2,3,4,4,6-pentamethylheptane	62199-71-7	-3.3430	4.4750E-03	1.1958E-06	298	373	2	1.2518E-02	3.1144E-02
870	25216	C12H26	2,3,4,5,5-pentamethylheptane	62199-72-8	-3.4907	4.4750E-03	1.1958E-06	298	373	2	8.9095E-03	2.2166E-02
871	25217	C12H26	2,3,4,5,6-pentamethylheptane	27574-98-7	-3.4492	4.4750E-03	1.1958E-06	298	373	2	9.8016E-03	2.4385E-02
872	25218	C12H26	2,4,4,5,5-pentamethylheptane	62199-73-9	-3.4700	4.4750E-03	1.1958E-06	298	373	2	9.3433E-03	2.3245E-02
873	25219	C12H26	3,3,4,4,5-pentamethylheptane	62199-74-0	-3.6691	4.4750E-03	1.1958E-06	298	373	2	5.9082E-03	1.4699E-02
874	25220	C12H26	3,3,4,5,5-pentamethylheptane	62199-75-1	-3.7254	4.4750E-03	1.1958E-06	298	373	2	5.1897E-03	1.2911E-02
875	25221	C12H26	2,2-dimethyl-3-tert-butylhexane	62199-76-2	-3.4700	4.4750E-03	1.1958E-06	298	373	2	9.3433E-03	2.3245E-02
876	25222	C12H26	2-methyl-3-ethyl-3-isopropylhexane	62199-77-3	-3.5516	4.4750E-03	1.1958E-06	298	373	2	7.7424E-03	1.9262E-02
877	25223	C12H26	2-methyl-4-ethyl-3-isopropylhexane	62199-78-4	-3.3430	4.4750E-03	1.1958E-06	298	373	2	1.2518E-02	3.1144E-02
878	25224	C12H26	2,2,3-trimethyl-3-isopropylhexane	62199-79-5	-3.5716	4.4750E-03	1.1958E-06	298	373	2	7.3940E-03	1.8395E-02
879	25225	C12H26	2,2,4-trimethyl-3-isopropylhexane	62199-80-8	-3.3213	4.4750E-03	1.1958E-06	298	373	2	1.3159E-02	3.2738E-02
880	25226	C12H26	2,2,5-trimethyl-3-isopropylhexane	62199-81-9	-3.1202	4.4750E-03	1.1958E-06	298	373	2	2.0906E-02	5.2012E-02

Table 6 Solubility in Water as a Function of Temperature - Organic Compounds

$\log_{10} S = A + B\,T + C\,T^2$ (S - solubility, parts per million by weight, ppm (wt), T - K)

NO	ID	FORMULA	NAME	CAS No	A	B	C	TMIN	TMAX	code	S@298, ppm(wt)	S@373, ppm(wt)
881	25227	C12H26	2,2,5-trimethyl-4-isopropylhexane	62199-82-0	-3.1202	4.4750E-03	1.1958E-06	298	373	2	2.0906E-02	5.2012E-02
882	25228	C12H26	2,3,4-trimethyl-3-isopropylhexane	62199-83-1	-3.6306	4.4750E-03	1.1958E-06	298	373	2	6.4551E-03	1.6059E-02
883	25229	C12H26	2,3,5-trimethyl-3-isopropylhexane	62199-84-2	-3.4283	4.4750E-03	1.1958E-06	298	373	2	1.0286E-02	2.5591E-02
884	25230	C12H26	2,3,5-trimethyl-4-isopropylhexane	62199-85-3	-3.3430	4.4750E-03	1.1958E-06	298	373	2	1.2518E-02	3.1144E-02
885	25231	C12H26	2,4,4-trimethyl-3-isopropylhexane	62199-86-4	-3.3859	4.4750E-03	1.1958E-06	298	373	2	1.1340E-02	2.8212E-02
886	25232	C12H26	3,3,4-triethylhexane	62199-87-5	-3.6691	4.4750E-03	1.1958E-06	298	373	2	5.9082E-03	1.4699E-02
887	25233	C12H26	2,2-dimethyl,3,3-diethylhexane	62199-88-6	-3.5516	4.4750E-03	1.1958E-06	298	373	2	7.7424E-03	1.9262E-02
888	25234	C12H26	2,2-dimethyl-3,4-diethylhexane	62199-89-7	-3.3430	4.4750E-03	1.1958E-06	298	373	2	1.2518E-02	3.1144E-02
889	25235	C12H26	2,2-dimethyl-4,4-diethylhexane	62184-89-8	-3.5516	4.4750E-03	1.1958E-06	298	373	2	7.7424E-03	1.9262E-02
890	25236	C12H26	2,3-dimethyl-3,4-diethylhexane	62184-90-1	-3.6306	4.4750E-03	1.1958E-06	298	373	2	6.4551E-03	1.6059E-02
891	25237	C12H26	2,3-dimethyl-4,4-diethylhexane	62184-91-2	-3.6306	4.4750E-03	1.1958E-06	298	373	2	6.4551E-03	1.6059E-02
892	25238	C12H26	2,4-dimethyl-3,3-diethylhexane	62184-92-3	-3.5516	4.4750E-03	1.1958E-06	298	373	2	7.7424E-03	1.9262E-02
893	25239	C12H26	2,4-dimethyl-3,4-diethylhexane	62184-93-4	-3.5516	4.4750E-03	1.1958E-06	298	373	2	7.7424E-03	1.9262E-02
894	25240	C12H26	2,5-dimethyl-3,3-diethylhexane	62184-94-5	-3.3430	4.4750E-03	1.1958E-06	298	373	2	1.2518E-02	3.1144E-02
895	25241	C12H26	2,5-dimethyl-3,4-diethylhexane	62184-95-6	-3.3430	4.4750E-03	1.1958E-06	298	373	2	1.2518E-02	3.1144E-02
896	25242	C12H26	3,3-dimethyl-4,4-diethylhexane	62184-96-7	-3.7438	4.4750E-03	1.1958E-06	298	373	2	4.9744E-03	1.2376E-02
897	25243	C12H26	3,4-dimethyl-3,4-diethylhexane	62184-97-8	-3.7438	4.4750E-03	1.1958E-06	298	373	2	4.9744E-03	1.2376E-02
898	25244	C12H26	2,2,3,3-tetramethyl-4-ethylhexane	62184-98-9	-3.5716	4.4750E-03	1.1958E-06	298	373	2	7.3940E-03	1.8395E-02
899	25245	C12H26	2,2,3,4-tetramethyl-3-ethylhexane	62184-99-0	-3.6306	4.4750E-03	1.1958E-06	298	373	2	6.4551E-03	1.6059E-02
900	25246	C12H26	2,2,3,4-tetramethyl-4-ethylhexane	62185-00-6	-3.7068	4.4750E-03	1.1958E-06	298	373	2	5.4166E-03	1.3476E-02
901	25247	C12H26	2,2,3,5-tetramethyl-3-ethylhexane	62185-01-7	-3.4283	4.4750E-03	1.1958E-06	298	373	2	1.0286E-02	2.5591E-02
902	25248	C12H26	2,2,3,5-tetramethyl-4-ethylhexane	62185-02-8	-3.3645	4.4750E-03	1.1958E-06	298	373	2	1.1913E-02	2.9637E-02
903	25249	C12H26	2,2,4,4-tetramethyl-3-ethylhexane	62185-03-9	-3.5516	4.4750E-03	1.1958E-06	298	373	2	7.7424E-03	1.9262E-02
904	25250	C12H26	2,2,4,5-tetramethyl-3-ethylhexane	62185-04-0	-3.3430	4.4750E-03	1.1958E-06	298	373	2	1.2518E-02	3.1144E-02
905	25251	C12H26	2,2,4,5-tetramethyl-4-ethylhexane	62185-05-1	-3.4907	4.4750E-03	1.1958E-06	298	373	2	8.9095E-03	2.2166E-02
906	25252	C12H26	2,2,5,5-tetramethyl-3-ethylhexane	62185-06-2	-3.1202	4.4750E-03	1.1958E-06	298	373	2	2.0906E-02	5.2012E-02
907	25253	C12H26	2,3,3,4-tetramethyl-4-ethylhexane	62185-07-3	-3.7068	4.4750E-03	1.1958E-06	298	373	2	5.4166E-03	1.3476E-02
908	25254	C12H26	2,3,3,5-tetramethyl-4-ethylhexane	62185-08-4	-3.4907	4.4750E-03	1.1958E-06	298	373	2	8.9095E-03	2.2166E-02
909	25255	C12H26	2,3,4,4-tetramethyl-3-ethylhexane	62185-09-5	-3.7068	4.4750E-03	1.1958E-06	298	373	2	5.4166E-03	1.3476E-02
910	25256	C12H26	2,3,4,5-tetramethyl-3-ethylhexane	62185-10-8	-3.5716	4.4750E-03	1.1958E-06	298	373	2	7.3940E-03	1.8395E-02
911	25257	C12H26	2,2,3,3,4,4-hexamethylhexane	62185-11-9	-3.8501	4.4750E-03	1.1958E-06	298	373	2	3.8943E-03	9.6885E-03
912	25258	C12H26	2,2,3,3,4,5-hexamethylhexane	62185-12-0	-3.5315	4.4750E-03	1.1958E-06	298	373	2	8.1104E-03	2.0178E-02
913	25259	C12H26	2,2,3,3,5,5-hexamethylhexane	60302-24-1	-3.4492	4.4750E-03	1.1958E-06	298	373	2	9.8016E-03	2.4385E-02
914	25260	C12H26	2,2,3,4,4,5-hexamethylhexane	62185-13-1	-3.6691	4.4750E-03	1.1958E-06	298	373	2	5.9082E-03	1.4699E-02
915	25261	C12H26	2,2,3,4,5,5-hexamethylhexane	62185-14-2	-3.3430	4.4750E-03	1.1958E-06	298	373	2	1.2518E-02	3.1144E-02
916	25262	C12H26	2,3,3,4,4,5-hexamethylhexane	52670-36-7	-3.6691	4.4750E-03	1.1958E-06	298	373	2	5.9082E-03	1.4699E-02
917	25263	C12H26	2,4-dimethyl-3-ethyl-3-isopropylpentane	62185-16-4	-3.6691	4.4750E-03	1.1958E-06	298	373	2	5.9082E-03	1.4699E-02
918	25264	C12H26	2,2,4-trimethyl-3,3-diethylpentane	62185-15-3	-3.6691	4.4750E-03	1.1958E-06	298	373	2	5.9082E-03	1.4699E-02
919	25265	C12H26	2,2,3,4-tetramethyl-3-isopropylpentane	62185-17-5	-3.6306	4.4750E-03	1.1958E-06	298	373	2	6.4551E-03	1.6059E-02
920	25266	C12H26	2,2,4,4-tetramethyl-3-isopropylpentane	62185-18-6	-3.4283	4.4750E-03	1.1958E-06	298	373	2	1.0286E-02	2.5591E-02
921	25267	C12H26	2,2,3,4,4-pentamethyl-3-ethylpentane	66576-21-4	-3.8154	4.4750E-03	1.1958E-06	298	373	2	4.2177E-03	1.0493E-02
922	25268	C12H26	2-methylundecane	7045-71-8	-3.7593	4.4750E-03	1.1958E-06	298	373	2	4.8001E-03	1.1942E-02
923	26283	C13H20	heptylbenzene	1078-71-3	-1.9923	5.0102E-03	1.9474E-06	298	373	2	4.7273E-01	1.4074E+00
924	26284	C13H20	1-methyl-2,4-diisopropylbenzene	1460-98-6	-0.6140	3.1975E-03	1.2695E-06	298	373	2	2.8328E+00	5.7010E+00
925	26285	C13H20	1-methyl-3,5-diisopropylbenzene	3055-14-9	-0.4549	3.1975E-03	1.2695E-06	298	373	2	4.0861E+00	8.2232E+00
926	26286	C13H20	[2-methyl-1-isopropylpropyl]benzene	21777-84-4	-0.5449	3.1975E-03	1.2695E-06	298	373	2	3.3214E+00	6.6843E+00
927	26288	C13H20	(2,4-dimethylpentyl)benzene	54518-00-2	-0.5089	3.1975E-03	1.2695E-06	298	373	2	3.6080E+00	7.2610E+00
928	26465	C13H26	octylcyclopentane	1795-20-6	-3.2292	3.5563E-03	1.3987E-06	298	373	2	9.0239E-03	1.9613E-02
929	26466	C13H26	heptylcyclohexane	5617-41-4	-3.3180	4.3934E-03	1.7000E-06	298	373	2	1.3899E-02	3.6148E-02
930	26467	C13H26	2-butyl-1,1,3-trimethylcyclohexane	54676-39-0	-3.2336	4.3934E-03	1.7000E-06	298	373	2	1.6880E-02	4.3901E-02
931	26526	C13H28	tridecane	629-50-5	-4.1416	4.4750E-03	1.1958E-06	298	373	2	1.9904E-03	4.9519E-03
932	26527	C13H28	2-methyldodecane	1560-97-0	-4.0615	4.4750E-03	1.1958E-06	298	373	2	2.3932E-03	5.9539E-03
933	26528	C13H28	3-methyldodecane	17312-57-1	-4.0728	4.4750E-03	1.1958E-06	298	373	2	2.3321E-03	5.8020E-03
934	26529	C13H28	2,2-dimethylundecane	17312-64-0	-3.9331	4.4750E-03	1.1958E-06	298	373	2	3.2169E-03	8.0032E-03
935	26530	C13H28	2,3-dimethylundecane	17312-77-5	-4.0401	4.4750E-03	1.1958E-06	298	373	2	2.5144E-03	6.2556E-03
936	26531	C13H28	2,4-dimethylundecane	17312-80-0	-3.8501	4.4750E-03	1.1958E-06	298	373	2	3.8943E-03	9.6885E-03
937	26532	C13H28	5-butylnonane	17312-63-9	-3.8755	4.4750E-03	1.1958E-06	298	373	2	3.6726E-03	9.1370E-03
938	27644	C14H22	octylbenzene	2189-60-8	-2.2694	5.0102E-03	1.9474E-06	298	373	2	2.4977E-01	7.4362E-01
939	27645	C14H22	1,2,3,4-tetraethylbenzene	642-32-0	-1.1149	3.1975E-03	1.2695E-06	298	373	2	8.9385E-01	1.7989E+00
940	27646	C14H22	1,2,3,5-tetraethylbenzene	38842-05-6	-1.0797	3.1975E-03	1.2695E-06	298	373	2	9.6930E-01	1.9507E+00
941	27647	C14H22	1,2,4,5-tetraethylbenzene	635-81-4	-1.0987	3.1975E-03	1.2695E-06	298	373	2	9.2777E-01	1.8671E+00
942	27648	C14H22	1,4-di-tert-butylbenzene	1012-72-2	-0.8759	3.1975E-03	1.2695E-06	298	373	2	1.5497E+00	3.1188E+00
943	27649	C14H22	(1-methylheptyl)benzene	777-22-0	-1.1113	3.1975E-03	1.2695E-06	298	373	2	9.0131E-01	1.8139E+00
944	27650	C14H22	1,3-di-tert-butylbenzene	1014-60-4	-1.1113	3.1975E-03	1.2695E-06	298	373	2	9.0131E-01	1.8139E+00
945	27847	C14H28	nonylcyclopentane	2882-98-6	-3.3736	3.5563E-03	1.3987E-06	298	373	2	6.4718E-03	1.4066E-02
946	27848	C14H28	octylcyclohexane	1795-15-9	-3.4572	4.3934E-03	1.7000E-06	298	373	2	1.0088E-02	2.6236E-02
947	27912	C14H30	tetradecane	629-59-4	7.9929	-6.6084E-02	1.0184E-04	298	373	1,2	2.2000E-03	3.2608E-03
948	27913	C14H30	2-methyltridecane	1560-96-9	-4.2787	4.4750E-03	1.1958E-06	298	373	2	1.4516E-03	3.6113E-03
949	27914	C14H30	3-methyltridecane	6418-41-3	-4.2861	4.4750E-03	1.1958E-06	298	373	2	1.4271E-03	3.5505E-03
950	27915	C14H30	2,2-dimethyldodecane	49598-54-1	-4.1963	4.4750E-03	1.1958E-06	298	373	2	1.7547E-03	4.3655E-03
951	27916	C14H30	2,3-dimethyldodecane	6117-98-2	-4.2702	4.4750E-03	1.1958E-06	298	373	2	1.4803E-03	3.6828E-03
952	27917	C14H30	2,4-dimethyldodecane	6117-99-3	-4.1227	4.4750E-03	1.1958E-06	298	373	2	2.0789E-03	5.1721E-03
953	27918	C14H30	7-methyltridecane	26730-14-3	-4.2524	4.4750E-03	1.1958E-06	298	373	2	1.5420E-03	3.8364E-03
954	28725	C15H24	nonylbenzene	1081-77-2	-2.4608	5.0102E-03	1.9474E-06	298	373	2	1.6075E-01	4.7859E-01
955	28726	C15H24	1,2,4-triisopropylbenzene	948-32-3	-0.9973	3.1975E-03	1.2695E-06	298	373	2	1.1718E+00	2.3582E+00
956	28727	C15H24	1,3,5-triisopropylbenzene	717-74-8	-0.8893	3.1975E-03	1.2695E-06	298	373	2	1.5027E+00	3.0241E+00
957	28728	C15H24	3,5-di-tert-butyltoluene	15181-11-0	-0.9973	3.1975E-03	1.2695E-06	298	373	2	1.1718E+00	2.3582E+00
958	28855	C15H30	decylcyclopentane	1795-21-7	-3.4113	3.5563E-03	1.3987E-06	298	373	2	5.9341E-03	1.2897E-02
959	28856	C15H30	nonylcyclohexane	2883-02-5	-3.4851	4.3934E-03	1.7000E-06	298	373	2	9.4603E-03	2.4604E-02
960	28857	C15H30	1-ethyl-1-methyl-2,4-diisopropylcyclohex	515-12-8	-3.4885	4.3934E-03	1.7000E-06	298	373	2	9.3864E-03	2.4411E-02

Table 6 Solubility in Water as a Function of Temperature - Organic Compounds

				$\log_{10} S = A + B T + C T^2$		(S - solubility, parts per million by weight, ppm (wt), T - K)						
NO	ID	FORMULA	NAME	CAS No	A	B	C	TMIN	TMAX	code	S@298, ppm(wt)	S@373, ppm(wt)
961	28904	C15H32	pentadecane	629-62-9	-4.4129	4.4750E-03	1.1958E-06	298	373	2	1.0658E-03	2.6515E-03
962	28905	C15H32	2-methyltetradecane	1560-95-8	-4.3956	4.4750E-03	1.1958E-06	298	373	2	1.1090E-03	2.7591E-03
963	28906	C15H32	3-methyltetradecane	18435-22-8	-4.3998	4.4750E-03	1.1958E-06	298	373	2	1.0982E-03	2.7321E-03
964	28907	C15H32	2,2-dimethyltridecane	61869-04-3	-4.3516	4.4750E-03	1.1958E-06	298	373	2	1.2273E-03	3.0534E-03
965	28908	C15H32	2,3-dimethyltridecane	18435-20-6	-4.3948	4.4750E-03	1.1958E-06	298	373	2	1.1111E-03	2.7643E-03
966	28909	C15H32	2,4-dimethyltridecane	61868-05-1	-4.3062	4.4750E-03	1.1958E-06	298	373	2	1.3623E-03	3.3893E-03
967	29595	C16H26	decylbenzene	104-72-3	-2.5576	5.0102E-03	1.9474E-06	298	373	2	1.2861E-01	3.8292E-01
968	29596	C16H26	pentaethylbenzene	605-01-6	-1.4558	3.1975E-03	1.2695E-06	298	373	2	4.0774E-01	8.2056E-01
969	29599	C16H26	(1-methylnonyl)benzene	4537-13-7	-2.5158	5.0102E-03	1.9474E-06	298	373	2	1.4161E-01	4.2162E-01
970	29704	C16H32	undecylcyclopentane	6785-23-5	-3.3487	3.5563E-03	1.3987E-06	298	373	2	6.8530E-03	1.4894E-02
971	29705	C16H32	decylcyclohexane	1795-16-0	-3.4104	4.3934E-03	1.7000E-06	298	373	2	1.1236E-02	2.9222E-02
972	29754	C16H34	hexadecane	544-76-3	-8.8945	3.0200E-02	-3.5497E-05	298	373	1,2	9.0000E-04	2.7045E-03
973	29755	C16H34	2-methylpentadecane	1560-93-6	-4.4165	4.4750E-03	1.1958E-06	298	373	2	1.0569E-03	2.6295E-03
974	29756	C16H34	3-methylpentadecane	2882-96-4	-4.4158	4.4750E-03	1.1958E-06	298	373	2	1.0586E-03	2.6337E-03
975	29757	C16H34	2,2-dimethyltetradecane	59222-86-5	-4.4184	4.4750E-03	1.1958E-06	298	373	2	1.0523E-03	2.6179E-03
976	29758	C16H34	2,3-dimethyltetradecane	18435-23-9	-4.4158	4.4750E-03	1.1958E-06	298	373	2	1.0586E-03	2.6337E-03
977	29759	C16H34	2,4-dimethyltetradecane	61868-06-2	-4.3987	4.4750E-03	1.1958E-06	298	373	2	1.1010E-03	2.7392E-03
978	29760	C16H34	2,2,4,4,6,8,8-heptamethylnonane	4390-04-9	-4.2639	4.4750E-03	1.1958E-06	298	373	2	1.5019E-03	3.7365E-03
979	30264	C17H28	undecylbenzene	6742-54-7	-2.5741	5.0102E-03	1.9474E-06	298	373	---	1.2384E-01	3.6869E-01
980	30310	C17H34	dodecylcyclopentane	5634-30-0	-3.1932	3.5563E-03	1.3987E-06	298	373	---	9.8042E-03	2.1309E-02
981	30311	C17H34	undecylcyclohexane	54105-66-7	-3.2390	4.3934E-03	1.7000E-06	298	373	---	1.6673E-02	4.3363E-02
982	30346	C17H36	heptadecane	629-78-7	-4.3022	4.4750E-03	1.1958E-06	298	373	---	1.3752E-03	3.4213E-03
983	30347	C17H36	2-methylhexadecane	1560-92-5	-4.3465	4.4750E-03	1.1958E-06	298	373	---	1.2417E-03	3.0891E-03
984	30348	C17H36	3-methylhexadecane	6418-43-5	-4.3388	4.4750E-03	1.1958E-06	298	373	---	1.2641E-03	3.1448E-03
985	30349	C17H36	2,2-dimethylpentadecane	61869-05-4	-4.3897	4.4750E-03	1.1958E-06	298	373	---	1.1242E-03	2.7969E-03
986	30350	C17H36	2,3-dimethylpentadecane	2882-97-5	-4.3388	4.4750E-03	1.1958E-06	298	373	---	1.2641E-03	3.1448E-03
987	30351	C17H36	2,4-dimethylpentadecane	61868-07-3	-4.4174	4.4750E-03	1.1958E-06	298	373	---	1.0547E-03	2.6241E-03

code: 1 - data, 2 - estimate

Chapter 7

SOLUBILITY IN WATER AS A FUNCTION OF TEMPERATURE – INORGANIC COMPOUNDS

Carl L. Yaws, Kayzad J. Vajifdar, and Preeti S. Yadav
Lamar University, Beaumont, Texas

Tabulation Results

The results for solubility in water are given in Table 7. For the tabulation, the correlation for water solubility as a function of temperature is based on a series expansion in temperature:

$$\log_{10} S = A + B\,T + C\,T^2 \quad (7)$$

where S = solubility in water, weight percent, %
T = temperature, K
A, B, and C = regression coefficients

The tabulation is arranged by alphabetical order (Ag, Al, B, ...) to provide ease of use in quickly locating the data by using the chemical formula. The compound name, CAS No (Chemical Abstracts Registry Number), and regression coefficients are provided in the adjacent columns. The range of application is denoted by minimum and maximum temperatures (TMIN and TMAX). Temperatures outside the range of application should not be used. The next column provides the code for the tabulation.

In preparing the tabulation, a literature search was conducted to identify data source publications for inorganics (1-13). These data were then keyed into the computer to provide a database of values for compounds for which experimental data are available. The excellent compilations by CRC (4); Krumgalz (6); Sohnel and Novotny (11); and Solubility Data Series of IUPAC (12) were primarily used for experimental data at temperatures below ambient, at ambient, and above ambient.

A comparison of calculated and data values is shown in Figure 7 for a representative compound. The graph discloses favorable agreement of equation and data.

Example

In an industrial operation, potassium chloride (KCl) comes into contact with a body of water at 50 C. Determine the concentration in water at saturation.

Substitution of the regression coefficients from the table and temperature (50 C = 323.15 K) into the equation for solubility in water yields:

$$\log_{10} S = -0.7012 + 1.1318E\text{-}02{*}323.15 - 1.4113E\text{-}05{*}323.15^2$$

$$S = 10^{1.4824} = 30.36 \text{ wt \%}$$

References

1. Clever, H. L. and F. J. Johnston, J. Phy. Chem. Ref. Data, 9, 751 (1980).
2. Clever, H. L., S. A. Johnston, and M. E. Derrick, J. Phy. Chem. Ref. Data, 14, 631 (1985).
3. Clever, H. L., S. A. Johnston, and M. E. Derrick, J. Phy. Chem. Ref. Data, 21, 941 (1992).
4. CRC HANDBOOK OF CHEMISTRY AND PHYSICS, 75th - 86th eds., CRC Press, Inc., Boca Raton, FL (1994-2006).
5. Knacke, O. and W. Gans, Zeit. Phy. Chem., NF, 104, 41 (1977).
6. Krumgalz, B. S., MINERAL SOLUBILITY IN WATER AT VARIOUS TEMPERATURES, Israel Oceanographic and Limnological Research Ltd., Haifa, Israel (1994).
7. Marcus, Y., J. Phy. Chem. Ref. Data, 9, 1307 (1980).
8. Marshal. W. L. and R. Slusher, J. Phy. Chem., 70, 4015 (1966).
9. Potter, R. W. and M. A. Clynne, J. Phy. Chem. Ref. Data, 24, 338 (1979).
10. Potter, R. W. and M. A. Clynne, J. Research U. S. Geological Survey, 6, 701 (1978).
11. Sohnel, O. and P. Novotny, DENSITIES OF AQUEOUS SOLUTIONS OF INORGANIC SUBSTANCES, Elsevier, Amsterdam, Holland (1985).
12. SOLUBILITY DATA SERIES, IUPAC, Volumes 1 to 53, Pergamon Press, Oxford, England (1979-1994).
13. Stephen, H. and T. Stephen, SOLUBILITIES OF INORGANIC AND ORGANIC COMPOUNDS, Volumes 1, 2, and 3, Macmillan Company, New York, NY (1963-1979).

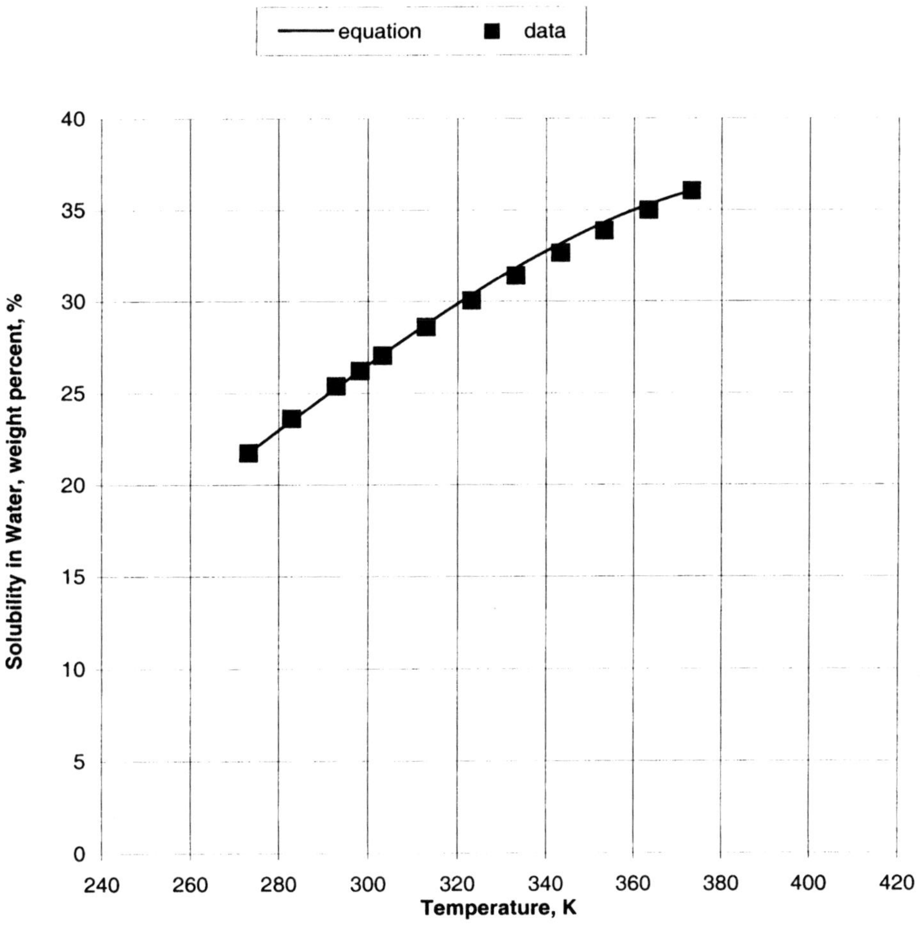

Figure 7 Solubility of Potassium Chloride in Water

Table 7 Solubility in Water as a Function of Temperature - Inorganic Compounds

$\log_{10} S = A + B\,T + C\,T^2$ (S - solubility, weight percent, T - K)

NO	ID	FORMULA	NAME	CAS No	A	B	C	TMIN	TMAX	code	S@TMIN, %	S@298, %	S@TMAX, %
1	37	AgClO2	silver chlorite	7783-91-7	-16.6112	9.2443E-02	-1.2611E-04	273	373	1,2	0.170	0.550	2.110
2	39	AgClO4	silver (i) perchlorate	7783-93-9	1.4023	2.9610E-03	-4.0133E-06	273	373	1,2	81.60	84.80	88.80
3	48	AgNO2	silver (i) nitrite	7783-99-5	-31.6132	2.0049E-01	-3.2114E-04	273	298	1,2	0.155	0.413	0.413
4	49	AgNO3	silver (i) nitrate	7761-88-8	-1.8972	2.1965E-02	-3.1564E-05	273	343	1,2	55.90	70.10	83.80
5	68	Ag2SO4	silver (i) sulfate	10294-26-5	-5.2542	2.8829E-02	-3.8498E-05	273	373	1,2	0.560	0.830	1.390
6	127	AlCl3	aluminum chloride	7446-70-0	1.6557	-1.3020E-03	2.5342E-06	273	373	1,2	30.84	31.10	33.32
7	134	AlF3	aluminum fluoride	7784-18-1	-8.0957	4.1537E-02	-5.1629E-05	273	373	1,2	0.250	0.500	1.640
8	149	AlN3O9	aluminum nitrate	13473-90-0	2.2340	-6.2268E-03	1.3872E-05	273	343	1,2	37.00	40.80	53.80
9	175	Al2S3O12	aluminum sulfate	10043-01-3	3.4210	-1.4074E-02	2.4965E-05	273	373	1,2	27.50	27.80	44.20
10	245	As2O3	arsenic trioxide	1327-53-3	-3.4877	1.6654E-02	-1.3212E-05	273	373	1,2	1.19	2.01	7.71
11	370	BH3O3	boric acid	10043-35-3	-6.0244	3.3379E-02	-3.5872E-05	273	373	1,2	2.61	5.48	27.30
12	423	BaBr2	barium bromide	10553-31-8	1.4156	1.0554E-03	-3.5156E-07	273	373	1,2	47.60	50.00	57.60
13	426	BaBr2O6	barium bromate	13967-90-3	-10.8156	5.5861E-02	-6.6855E-05	273	373	1,2	0.285	0.788	5.250
14	437	BaC4H4O6	barium tartrate	5908-81-6	-0.1689	9.3559E-03	-1.0970E-05	273	298	1,2	37.00	44.20	44.20
15	454	BaCl2	barium chloride	10361-37-2	0.0489	6.8855E-03	-7.5371E-06	273	373	1,2	23.30	27.03	37.05
16	458	BaCl2O6	barium chlorate	13477-00-4	-6.2307	4.4573E-02	-6.3216E-05	273	333	1,2	16.90	27.50	40.05
17	459	BaCl2O8	barium perchlorate	13465-95-7	-0.2118	1.2427E-02	-1.8155E-05	273	323	1,2	67.30	75.75	80.92
18	464	BaF2	barium fluoride	7787-32-8	-0.9844	7.4306E-04	-3.4144E-07	283	298	1,2	0.158	0.161	0.161
19	470	BaH2O2	barium hydroxide	17194-00-2	41.4808	-3.0583E-01	5.6665E-04	273	313	1,2	1.67	4.68	19.00
20	481	BaI2	barium iodide	13718-50-8	2.2410	-4.6506E-03	1.1061E-05	273	298	1,2	62.50	68.80	68.80
21	482	BaI2O6	barium iodate	10567-69-8	-9.2347	4.0203E-02	-4.6733E-05	273	373	1,2	0.0182	0.0396	0.1820
22	486	BaN2O4	barium nitrite	13465-94-6	-3.2787	2.7841E-02	-3.7976E-05	273	333	1,2	31.10	44.30	60.50
23	487	BaN2O6	barium nitrate	10022-31-8	-8.1615	5.1106E-02	-6.8704E-05	273	343	1,2	4.70	9.30	19.30
24	511	BaS	barium sulfide	21109-95-5	-14.3565	8.6659E-02	-1.1887E-04	273	343	1,2	2.79	8.21	24.19
25	541	BeCl2	beryllium chloride	7787-47-5	1.4448	6.7647E-04	-2.9621E-07	273	298	1,2	40.50	41.70	41.70
26	552	BeO4S	beryllium sulfate	13510-49-1	2.0511	-5.8237E-03	1.2948E-05	273	343	1,2	26.69	29.22	37.78
27	682	CaBr2	calcium bromide	7789-41-5	1.8323	-2.2926E-03	7.1614E-06	273	333	1,2	55.00	61.00	73.00
28	713	CaCl2	calcium chloride	10043-52-4	1.2735	-1.1419E-03	8.0834E-06	273	323	1,2	36.70	44.83	56.05
29	722	CaCl2O6	calcium chlorate	10137-74-3	1.9710	-1.9565E-03	4.8807E-06	273	353	1,2	63.20	66.30	77.40
30	728	CaF2	calcium fluoride	7789-75-5	-5.0891	1.2149E-02	-1.4953E-05	273	298	1,2	0.0013	0.0016	0.0016
31	746	CaI2	calcium iodide	10102-68-8	1.5622	8.5329E-04	1.9997E-07	273	373	1,2	64.60	68.30	81.00
32	747	CaI2O6	calcium iodate	7789-80-2	-25.4971	1.5034E-01	-2.2321E-04	273	333	1,2	0.082	0.305	0.652
33	750	CaN2O4	calcium nitrite	13780-06-8	15.3696	-1.0035E-01	1.8266E-04	273	298	1,2	38.60	48.60	48.60
34	751	CaN2O6	calcium nitrate	10124-37-5	1.2113	8.2468E-04	3.5286E-06	273	313	1,2	50.10	59.00	65.40
35	757	CaO4S	calcium sulfate	7778-18-9	-4.9381	2.6704E-02	-4.1757E-05	273	373	1,2	0.174	0.205	0.163
36	777	CdBr2	cadmium bromide	7789-42-6	-7.0968	5.4427E-02	-8.3279E-05	273	313	1,2	36.00	53.40	60.30
37	791	CdCl2	cadmium chloride	10108-64-2	1.3443	-5.1116E-06	4.4375E-06	273	303	1,2	47.20	54.60	56.30
38	798	CdF2	cadmium fluoride	7790-79-6	5.1053	-2.0786E-02	1.9271E-05	283	303	1,2	5.82	4.18	3.76
39	804	CdI2	cadmium iodide	7790-80-9	1.5709	-2.5894E-04	1.9335E-06	273	373	1,2	44.10	46.30	55.40
40	807	CdN2O6	cadmium nitrate	10325-94-7	2.6974	-8.2243E-03	1.7324E-05	273	323	1,2	55.40	61.00	70.60
41	816	CdO4S	cadmium sulfate	10124-36-4	2.2994	-4.7750E-03	8.5690E-06	273	313	1,2	43.10	43.40	44.10
42	968	CoCl2	cobalt chloride	7646-79-9	-0.0800	8.2145E-03	-9.1456E-06	273	373	1,2	30.30	35.99	51.50
43	972	CoCl2O8	cobalt (ii) perchlorate	13455-31-7	1.3264	1.6861E-03	-1.1795E-06	273	298	1,2	50.00	53.00	53.00
44	1000	CoI2	cobalt (ii) iodide	15238-00-3	-0.2851	1.2077E-02	-1.6758E-05	273	343	1,2	58.00	66.99	76.89
45	1003	CoN2O4	cobalt (ii) nitrite	18488-96-5	-96.6308	6.4035E-01	-1.0642E-03	273	298	1,2	0.076	0.490	0.490
46	1004	CoN2O6	cobalt (ii) nitrate	10141-05-6	3.9275	-1.7674E-02	3.4287E-05	273	323	1,2	45.50	50.80	62.60
47	1010	CoO4S	cobalt (ii) sulfate	10124-43-3	-3.7011	2.9811E-02	-4.2125E-05	273	333	1,2	19.90	27.70	35.90
48	1100	CrO3	chromium (vi) oxide	1333-82-0	1.9806	-1.4635E-03	2.8536E-06	273	373	1,2	62.20	62.80	67.90
49	1129	CsBrO3	cesium bromate	13454-75-6	-13.4874	7.6649E-02	-9.8978E-05	273	343	1,2	1.16	3.69	14.45
50	1137	CsCl	cesium chloride	7647-17-8	1.1601	3.4757E-03	-4.2656E-06	273	373	1,2	61.83	65.64	72.96
51	1138	CsClO3	cesium chlorate	13763-67-2	-11.9673	6.9089E-02	-8.7443E-05	273	373	1,2	2.40	7.22	43.42
52	1139	CsClO4	cesium perchlorate	13454-84-7	0.3668	-1.7753E-02	5.8704E-05	273	333	1,2	0.790	1.960	9.290
53	1144	CsI	cesium iodide	7789-17-5	-5.0112	3.9308E-02	-5.6771E-05	273	333	1,2	30.90	45.90	60.70
54	1145	CsIO3	cesium iodate	13454-81-4	-9.0877	5.0064E-02	-6.1034E-05	273	373	1,2	1.08	2.59	12.45
55	1147	CsNO3	cesium nitrate	7789-18-6	-13.4168	8.5560E-02	-1.2098E-04	273	333	1,2	8.46	21.80	45.70
56	1165	Cs2SO4	cesium sulfate	10294-54-9	1.5360	1.3518E-03	-1.4570E-06	273	373	1,2	62.60	64.50	68.80
57	1230	CuCl2	cupric chloride	7447-39-4	1.5229	-2.5710E-04	2.1176E-06	273	373	1,2	40.80	43.10	52.70
58	1234	CuCl2O8	copper (ii) perchlorate	13770-18-8	1.2315	2.3537E-03	-1.8714E-06	273	303	1,2	54.30	58.47	59.30
59	1257	CuN2O6	copper (ii) nitrate	3251-23-8	-4.9009	4.1697E-02	-6.4781E-05	273	303	1,2	45.20	59.20	61.10
60	1265	CuO4S	copper (ii) sulfate	7758-98-7	-1.7862	1.4269E-02	-1.3645E-05	273	373	1,2	12.40	18.00	43.50
61	1502	FeCl2	ferrous chloride	7758-94-3	-0.1198	8.7859E-03	-1.0173E-05	273	298	1,2	33.20	39.40	39.40
62	1505	FeCl3	ferric chloride	7705-08-0	2.9134	-1.1399E-02	2.4535E-05	273	303	1,2	42.70	49.65	51.60
63	1530	FeN2O6	iron (ii) nitrate	14013-86-6	0.6540	4.8670E-03	-4.9052E-06	273	298	1,2	41.44	46.67	46.67
64	1531	FeN3O9	ferric nitrate	10421-48-4	0.2778	6.9402E-03	-7.6376E-06	273	298	1,2	40.15	46.57	46.57
65	1534	FeO4S	iron (ii) sulfate	7720-78-7	-6.2563	4.3477E-02	-6.0166E-05	273	333	1,2	13.50	22.80	35.50
66	1773	HgBr2	mercuric bromide	7789-47-1	-7.8870	3.7651E-02	-3.9974E-05	273	373	1,2	0.260	0.610	3.950
67	1779	HgC2N2	mercury (ii) cyanide	592-04-1	-2.0570	1.3164E-02	-9.6669E-06	273	373	1,2	6.57	10.20	32.30
68	1800	HgCl2	mercuric chloride	7487-94-7	0.7070	-8.1000E-03	2.8586E-05	273	333	1,2	4.24	6.81	15.18
69	1813	HgI2	mercuric iodide	7774-29-0	-19.1206	8.8118E-02	-1.0587E-04	293	323	1,2	0.0041	0.0055	0.0199
70	1838	Hg2SO4	mercury (i) sulfate	7783-36-0	-4.1460	1.4439E-02	-1.6326E-05	273	373	1,2	0.0380	0.0510	0.0930
71	1969	KBF4	potassium fluoroborate	14075-53-7	7.8546	-6.9723E-02	1.4257E-04	273	333	1,2	0.280	0.550	2.820
72	1972	KBr	potassium bromide	7758-02-3	-0.0864	9.1545E-03	-1.1661E-05	273	373	1,2	35.00	40.40	50.80
73	1973	KBrO3	potassium bromate	7758-01-2	-11.1899	6.6965E-02	-8.8846E-05	273	333	1,2	2.97	7.55	18.14
74	1985	KC2H3O2	potassium acetate	127-08-2	1.0954	4.1702E-03	-5.3537E-06	273	373	1,2	68.40	72.92	80.55
75	2007	KCl	potassium chloride	7447-40-7	-0.7012	1.1318E-02	-1.4113E-05	273	373	1,2	21.74	26.22	36.05
76	2010	KClO3	potassium chlorate	3811-04-9	-12.2487	7.3990E-02	-1.0026E-04	273	373	1,2	3.03	7.93	18.78
77	2011	KClO4	potassium perchlorate	7778-74-7	-9.9222	5.1494E-02	-5.7611E-05	273	373	1,2	0.700	2.040	18.670
78	2018	KF	potassium fluoride	7789-23-3	-11.2115	8.1312E-02	-1.2744E-04	273	303	1,2	30.90	50.41	53.20
79	2023	KHCO3	potassium hydrogen carbonate	298-14-6	-2.9248	2.3750E-02	-3.0726E-05	273	343	1,2	18.62	26.60	40.45
80	2029	KHSO4	potassium hydrogen sulfate	7646-93-7	-0.3392	9.0103E-03	-9.2342E-06	273	373	1,2	27.10	33.60	54.60

Table 7 Solubility in Water as a Function of Temperature - Inorganic Compounds

					$\log_{10} S = A + B\,T + C\,T^2$			(S - solubility, weight percent, T - K)					
NO	ID	FORMULA	NAME	CAS No	A	B	C	TMIN	TMAX	code	S@TMIN, %	S@298, %	S@TMAX, %
81	2034	KH2PO4	potassium dihydrogen phosphate	7778-77-0	-5.6405	3.8617E-02	-5.1442E-05	273	333	1,2	11.74	19.97	32.76
82	2035	KI	potassium iodide	7681-11-0	1.1115	3.4479E-03	-4.0898E-06	273	373	1,2	56.00	59.70	67.40
83	2036	KIO3	potassium iodate	7758-05-6	-6.1658	3.7952E-02	-4.7510E-05	273	373	1,2	4.53	8.44	24.03
84	2037	KIO4	potassium periodate	7790-21-8	-4.7692	9.4231E-03	1.8755E-05	273	333	1,2	0.160	0.510	2.830
85	2040	KMnO4	potassium permanganate	7722-64-7	-10.4586	6.1375E-02	-7.8650E-05	273	333	1,2	2.74	7.06	18.16
86	2041	KNO2	potassium nitrite	7758-09-0	1.6280	1.2534E-03	-1.3798E-06	273	373	1,2	73.70	75.70	80.10
87	2042	KNO3	potassium nitrate	7757-79-1	-11.9427	7.8035E-02	-1.1116E-04	273	333	1,2	12.00	27.70	52.20
88	2047	KOH	potassium hydroxide	1310-58-3	0.3783	7.3350E-03	-9.3063E-06	273	313	1,2	48.70	54.70	57.90
89	2053	KSCN	potassium thiocyanate	333-20-0	0.9597	4.3620E-03	-4.6419E-06	273	373	1,2	63.80	70.40	87.30
90	2064	K2CO3	potassium carbonate	584-08-7	1.8912	-1.6986E-03	3.7919E-06	273	373	1,2	51.30	52.70	61.00
91	2088	K2CrO4	potassium chromate	7789-00-6	0.9858	3.1366E-03	-3.6613E-06	273	373	1,2	37.10	39.40	44.30
92	2089	K2Cr2O7	potassium dichromate	7778-50-9	-16.0361	9.9208E-02	-1.3978E-04	273	333	1,2	4.30	13.10	31.70
93	2099	K2HPO4	potassium hydrogen phosphate	7758-11-4	1.5486	-6.2848E-05	3.0081E-06	273	333	1,2	57.00	62.70	72.70
94	2116	K2O3S	potassium sulfite	10117-38-1	1.7544	-3.8818E-04	8.2729E-07	273	373	1,2	51.30	51.55	53.06
95	2117	K2O3Se	potassium selenite	10431-47-7	1.8281	2.5802E-05	-7.4153E-10	273	298	1,2	68.40	68.50	68.50
96	2127	K2SO4	potassium sulfate	7778-80-5	-4.0886	2.8153E-02	-3.6849E-05	273	373	1,2	7.11	10.70	19.30
97	2128	K2S2O3	potassium thiosulfate	10294-66-3	-1.0771	1.5591E-02	-1.9988E-05	273	298	1,2	49.00	62.30	62.30
98	2129	K2S2O5	potassium metabisulfite	16731-55-8	-4.0851	3.1659E-02	-4.3133E-05	273	353	1,2	22.10	33.10	52.00
99	2133	K2SeO4	potassium selenate	7790-59-2	1.7072	-7.7393E-05	4.7971E-07	273	373	1,2	52.70	53.30	55.60
100	2141	K3AsO4	potassium arsenate	13464-36-3	1.1823	2.4951E-03	-2.0382E-06	273	298	1,2	51.50	55.60	55.60
101	2143	K3C6FeN6	potassium ferricyanide	13746-66-2	-3.5443	2.9495E-02	-4.2002E-05	273	333	1,2	23.90	32.80	41.70
102	2153	K3PO4	potassium phosphate	7778-53-2	0.3163	6.9645E-03	-7.6705E-06	273	298	1,2	44.30	51.40	51.40
103	2177	LaCl3	lanthanum chloride	10099-58-8	2.3900	-5.5552E-03	1.0749E-05	298	363	1,2	48.90	48.90	61.70
104	2189	LaN3O9	lanthanum nitrate	10099-59-9	1.9389	-2.7773E-03	7.5072E-06	273	343	1,2	55.00	60.00	74.10
105	2213	LiBr	lithium bromide	7550-35-8	0.8762	4.6887E-03	-5.2333E-06	273	313	1,2	58.40	64.40	67.80
106	2215	LiBrO3	lithium bromate	13550-28-2	2.0240	-2.7833E-03	6.9933E-06	273	313	1,2	61.03	65.44	68.90
107	2223	LiC2H3O2	lithium acetate	546-89-4	-0.1541	6.4890E-03	-3.2508E-06	273	373	1,2	23.76	31.02	65.26
108	2253	LiCl	lithium chloride	7447-41-8	-1.4979	1.9800E-02	-3.0874E-05	273	313	1,2	40.45	45.81	47.30
109	2257	LiClO3	lithium chlorate	13453-71-9	-0.0711	1.1752E-02	-1.7082E-05	273	313	1,2	73.20	82.10	85.90
110	2258	LiClO4	lithium perchlorate	7791-03-9	0.4053	4.2444E-03	-1.1535E-06	273	343	1,2	30.10	37.00	53.20
111	2261	LiF	lithium fluoride	7789-24-4	-2.5411	9.6100E-03	-1.3466E-05	273	298	1,2	0.120	0.134	0.134
112	2270	LiI	lithium iodide	10377-51-2	1.6985	-2.3066E-04	1.8532E-06	273	343	1,2	59.40	62.30	68.80
113	2274	LiNO3	lithium nitrate	7790-69-4	3.1179	-2.1093E-02	5.4831E-05	283	303	1,2	34.80	50.50	57.90
114	2277	LiOH	lithium hydroxide	1310-65-2	2.2695	-9.1070E-03	1.6774E-05	273	373	1,2	10.80	11.10	16.10
115	2283	Li2CO3	lithium carbonate	554-13-2	0.9679	-2.5310E-03	-1.1927E-06	273	373	1,2	1.54	1.28	0.72
116	2306	Li2SO4	lithium sulfate	10377-48-7	1.6384	-1.0408E-03	8.8247E-07	273	373	1,2	26.30	25.50	23.60
117	2341	MgBr2	magnesium bromide	7789-48-2	1.6540	-1.4154E-04	1.0392E-06	273	373	1,2	49.30	50.60	55.70
118	2361	MgC4H6O4	magnesium acetate	142-72-3	0.9371	2.9174E-03	-2.3525E-06	273	298	1,2	36.18	39.61	39.61
119	2392	MgCl2	magnesium chloride	7786-30-3	1.2383	1.1690E-03	-3.5696E-07	273	373	1,2	33.96	35.90	42.15
120	2396	MgCl2O8	magnesium perchlorate	10034-81-8	1.2343	2.3748E-03	-2.7279E-06	273	323	1,2	47.80	50.10	52.10
121	2424	MgI2	magnesium iodide	10377-58-9	2.9525	-9.8314E-03	1.9715E-05	273	303	1,2	54.70	59.40	60.80
122	2427	MgN2O6	magnesium nitrate	10377-60-3	4.1438	-1.9229E-02	3.6092E-05	273	323	1,2	38.40	41.60	50.00
123	2431	MgO3S	magnesium sulfite	7757-88-2	-8.9925	5.1884E-02	-7.6055E-05	273	373	1,2	0.320	0.520	0.600
124	2440	MgSO4	magnesium sulfate	7487-88-9	-4.1250	3.1917E-02	-4.4673E-05	273	343	1,2	18.20	26.30	36.90
125	2469	MnBr2	manganese (ii) bromide	13446-03-2	1.0632	3.6566E-03	-4.2065E-06	273	373	1,2	56.00	60.19	69.50
126	2495	MnCl2	manganese (ii) chloride	7773-01-5	1.1826	9.4445E-04	1.9719E-06	273	323	1,2	38.70	43.60	49.40
127	2499	MnF2	manganese (ii) fluoride	7782-64-1	-2.7368	1.4809E-02	-1.8835E-05	273	298	1,2	0.800	1.010	1.010
128	2515	MnN2O6	manganese (ii) nitrate	10377-93-2	-0.3806	1.1431E-02	-1.3917E-05	273	298	1,2	50.50	61.70	61.70
129	2526	MnSO4	manganese (ii) sulfate	7785-87-7	-5.4140	4.6863E-02	-7.8374E-05	273	303	1,2	34.60	39.02	38.90
130	2625	NH4Br	ammonium bromide	12124-97-9	-0.1385	9.5055E-03	-1.1847E-05	273	373	1,2	37.50	43.90	57.40
131	2627	NH4Cl	ammonium chloride	12125-02-9	-0.6576	1.0777E-02	-1.2409E-05	273	373	1,2	22.92	28.34	43.24
132	2629	NH4ClO4	ammonium perchlorate	7790-98-9	-6.9993	4.6784E-02	-6.3613E-05	273	333	1,2	10.80	19.70	33.60
133	2631	NH4F	ammonium fluoride	12125-01-8	1.0556	2.5719E-03	-1.8500E-06	273	353	1,2	41.70	45.50	54.10
134	2634	NH4I	ammonium iodide	12027-06-4	1.2792	2.6933E-03	-3.1046E-06	273	373	1,2	60.70	64.00	71.10
135	2635	NH4IO3	ammonium iodate	13446-09-8	0.3628	-1.0276E-02	3.6775E-05	298	323	1,2	3.70	3.70	7.63
136	2641	NH5CO3	ammonium hydrogen carbonate	1066-33-7	-4.4327	2.8264E-02	-3.0320E-05	273	373	1,2	10.60	19.90	78.00
137	2649	NH6PO4	ammonium dihydrogen phosphate	7722-76-1	-3.4931	2.5618E-02	-3.0210E-05	273	373	1,2	17.80	28.80	72.40
138	2688	N2H4O2	ammonium nitrite	13446-48-5	7.0940	-4.0879E-02	7.7977E-05	273	298	1,2	55.70	68.80	68.80
139	2689	N2H4O3	ammonium nitrate	6484-52-2	-2.2918	2.4561E-02	-3.5982E-05	273	313	1,2	54.00	68.00	74.30
140	2707	N2H8C2O4	ammonium oxalate	1113-38-8	-6.2200	3.4087E-02	-3.6552E-05	273	373	1,2	2.31	4.94	25.70
141	2721	N2H8O4S	ammonium sulfate	7783-20-2	1.4479	4.2618E-04	6.9199E-07	273	373	1,2	41.30	43.30	50.50
142	2723	N2H8O8S2	ammonium peroxydisulfate	7727-54-0	-0.5761	1.1755E-02	-1.4294E-05	273	353	1,2	37.00	45.49	62.00
143	2728	N2H8SeO3	ammonium selenite	7783-19-9	1.5679	-8.9343E-04	4.9100E-06	273	343	1,2	49.00	54.70	69.10
144	2734	N2H9PO4	diammonium hydrogen phosphate	7783-28-0	0.9971	2.0622E-03	8.8453E-09	273	373	1,2	36.40	41.00	58.60
145	2800	NaBr	sodium bromide	7647-15-6	1.4846	-2.9675E-04	3.2678E-06	273	323	1,2	44.40	48.60	53.70
146	2802	NaBrO3	sodium bromate	7789-38-0	-3.2556	2.6451E-02	-3.5766E-05	273	383	1,2	20.00	28.28	42.52
147	2805	NaCHO2	sodium formate	141-53-7	-13.2674	9.6222E-02	-1.5449E-04	273	303	1,2	30.80	48.70	50.60
148	2816	NaC2H2IO2	sodium iodoacetate	305-53-3	2.0918	-8.4202E-03	2.1866E-05	273	323	1,2	26.50	33.50	45.10
149	2855	NaCl	sodium chloride	7647-14-5	1.5748	-1.1910E-03	2.2808E-06	273	373	1,2	26.28	26.45	28.05
150	2857	NaClH10O6	sodium hypochlorite pentahydrate	7681-52-8	-14.3303	9.9362E-02	-1.5352E-04	273	298	1,2	22.70	44.40	44.40
151	2858	NaClO	sodium hypochlorite	7681-52-9	2.0467	-1.8441E-04	-5.5416E-08	298	333	1,2	97.00	97.00	95.30
152	2859	NaClO2	sodium chlorite	7758-19-2	0.4549	6.3872E-03	-7.4184E-06	273	323	1,2	44.27	50.10	55.50
153	2860	NaClO3	sodium chlorate	7775-09-9	0.8632	5.2061E-03	-6.6146E-06	273	373	1,2	61.90	67.20	76.70
154	2864	NaF	sodium fluoride	7681-49-4	-0.8114	7.6115E-03	-9.6649E-06	273	373	1,2	3.52	3.97	4.82
155	2870	NaHCO3	sodium hydrogen carbonate	144-55-8	-2.6711	1.8646E-02	-2.1587E-05	273	373	1,2	6.48	9.32	19.10
156	2877	NaH2O4P	sodium phosphate, monobasic	7558-80-7	-1.2088	1.4877E-02	-1.7318E-05	273	323	1,2	36.54	48.68	61.70
157	2888	NaI	sodium iodide	7681-82-5	1.9567	-2.1017E-03	5.4169E-06	273	343	1,2	61.20	64.80	74.70
158	2889	NaIO3	sodium iodate	7681-55-2	-32.4704	2.1048E-01	-3.3013E-04	273	313	1,2	2.43	8.65	11.67
159	2893	NaNO2	sodium nitrite	7632-00-0	1.3023	7.9328E-04	1.3840E-06	273	373	1,2	41.90	45.90	61.80
160	2894	NaNO3	sodium nitrate	7631-99-4	0.6822	4.6660E-03	-4.4421E-06	273	373	1,2	42.20	47.70	63.80

Table 7 Solubility in Water as a Function of Temperature - Inorganic Compounds

					$\log_{10} S = A + B\,T + C\,T^2$			(S - solubility, weight percent, T - K)					
NO	ID	FORMULA	NAME	CAS No	A	B	C	TMIN	TMAX	code	S@TMIN, %	S@298, %	S@TMAX, %
161	2897	NaOH	sodium hydroxide	1310-73-2	-11.2972	8.1482E-02	-1.2709E-04	273	303	1,2	30.00	50.00	53.00
162	2903	NaSCN	sodium thiocyanate	540-72-7	-0.2164	9.8031E-03	-1.0426E-05	283	303	1,2	52.90	60.20	62.70
163	2916	Na2CO3	sodium carbonate	497-19-8	-19.2447	1.2007E-01	-1.7082E-04	273	303	1,2	6.44	23.50	28.70
164	2917	Na2C2O4	sodium oxalate	62-76-0	-2.3132	1.4644E-02	-1.7001E-05	273	373	1,2	2.62	3.48	6.08
165	2930	Na2CrO4	sodium chromate	7775-11-3	-17.8407	1.2310E-01	-1.9340E-04	273	298	1,2	22.60	46.70	46.70
166	2932	Na2Cr2O7	sodium dichromate	10588-01-9	2.2339	-3.8679E-03	8.2515E-06	273	323	1,2	62.10	65.20	70.10
167	2960	Na2MoO4	sodium molybdate	7631-95-0	1.8281	-2.0697E-03	4.3248E-06	283	373	1,2	38.80	39.40	45.50
168	2966	Na2O4Se	sodium selenate	13410-01-0	-39.8248	2.6858E-01	-4.3520E-04	273	298	1,2	11.70	36.90	36.90
169	2968	Na2O4W	sodium tungstate	13472-45-2	1.8783	-2.1960E-03	4.5662E-06	273	373	1,2	41.60	42.60	49.50
170	2972	Na2S	sodium sulfide	1313-82-2	-2.9948	2.1464E-02	-2.4431E-05	273	373	1,2	11.10	17.10	41.00
171	2973	Na2SO3	sodium sulfite	7757-83-7	-7.2621	4.7818E-02	-6.3263E-05	273	303	1,2	12.00	23.50	26.30
172	2974	Na2SO4	sodium sulfate	7757-82-6	-22.6353	1.3503E-01	-1.8317E-04	293	303	1,2	16.13	21.94	29.22
173	2975	Na2S2O5	sodium metabisulfite	7681-57-4	1.2985	8.4594E-04	5.7798E-07	283	373	1,2	38.40	40.00	49.50
174	3002	Na3O4P	sodium phosphate	7601-54-9	-26.4290	1.7264E-01	-2.6936E-04	273	313	1,2	4.28	12.60	16.60
175	3013	Na4P2O7	sodium pyrophosphate	7722-88-5	-8.2954	4.4438E-02	-4.6836E-05	273	353	1,2	2.23	6.17	36.03
176	3070	NdCl3	neodymium trichloride	10024-93-8	2.0489	-2.8380E-03	5.5819E-06	273	373	1,2	49.00	50.00	58.50
177	3133	NiCl2	nickel (ii) chloride	7718-54-9	-5.4501	4.4020E-02	-6.8275E-05	283	303	1,2	34.70	40.30	41.70
178	3154	NiI2	nickel (ii) iodide	13462-90-3	0.3198	8.5359E-03	-1.2168E-05	273	363	1,2	55.40	60.69	65.30
179	3155	NiI2O6	nickel iodate	13477-98-0	1.6181	-1.7497E-03	6.7588E-06	273	333	1,2	44.10	49.80	61.00
180	3164	NiSO4	nickel (ii) sulfate	7786-81-4	-2.1124	1.9425E-02	-2.4971E-05	273	303	1,2	21.40	28.80	30.30
181	3487	PbBr2	lead bromide	10031-22-8	-6.6704	3.2160E-02	-3.2997E-05	273	323	1,2	0.449	0.966	1.890
182	3514	PbCl2	lead chloride	7758-95-4	-3.8292	1.7906E-02	-1.6650E-05	273	373	1,2	0.660	1.070	3.420
183	3521	PbF2	lead fluoride	7783-46-2	-2.5849	6.5038E-03	-5.9406E-06	283	303	1,2	0.0603	0.0670	0.0693
184	3530	PbI2	lead iodide	10101-63-0	-6.8748	2.8673E-02	-3.1423E-05	273	383	1,2	0.0410	0.0760	0.3150
185	3533	PbN2O6	lead (ii) nitrate	10099-74-8	-2.0915	2.0534E-02	-2.7652E-05	273	333	1,2	28.46	37.38	47.90
186	3547	PbSO4	lead (ii) sulfate	7446-14-2	-4.1699	7.2660E-03	-3.9707E-06	273	323	1,2	0.0033	0.0044	0.0058
187	3663	PrCl3	praseodymium chloride	10361-79-2	2.6157	-6.8833E-03	1.2675E-05	273	353	1,2	48.00	49.00	58.30
188	3786	RbBr	rubidium bromide	7789-39-1	0.2397	8.0585E-03	-1.0254E-05	273	373	1,2	47.40	53.80	65.90
189	3787	RbBrO3	rubidium bromate	13446-70-3	-11.1761	6.1043E-02	-7.3865E-05	273	373	1,2	0.970	2.870	20.760
190	3791	RbCl	rubidium chloride	7791-11-9	0.5131	6.2238E-03	-7.6917E-06	273	373	1,2	43.58	48.42	58.15
191	3792	RbClO3	rubidium chlorate	13446-71-4	-11.5833	6.6236E-02	-8.2922E-05	273	373	1,2	2.10	6.22	38.60
192	3801	RbI	rubidium iodide	7790-29-6	0.4638	7.2458E-03	-9.3324E-06	273	373	1,2	55.80	62.30	73.80
193	3822	Rb2SO4	rubidium sulfate	7488-54-2	-1.1897	1.5068E-02	-1.9971E-05	273	373	1,2	27.30	33.70	44.90
194	4042	SbCl3	antimony trichloride	10025-91-9	1.5641	1.6674E-03	-1.1608E-06	273	298	1,2	85.70	90.80	90.80
195	4045	SbF3	antimony (iii) fluoride	7783-56-4	1.6250	1.2029E-03	-7.2068E-07	273	298	1,2	79.40	83.10	83.10
196	4315	SmCl3	samarium (iii) chloride	10361-82-7	2.2985	-4.4786E-03	8.1178E-06	283	323	1,2	48.00	48.40	50.00
197	4422	SnCl2	stannous chloride	7772-99-8	-20.4884	1.4549E-01	-2.3575E-04	273	283	1,2	46.00	---	64.00
198	4464	SrBr2	strontium bromide	10476-81-0	0.8170	4.0738E-03	-3.5786E-06	273	373	1,2	46.00	51.70	69.00
199	4479	SrCl2	strontium chloride	10476-85-4	2.8346	-1.0955E-02	2.2278E-05	273	333	1,2	31.94	35.37	45.44
200	4482	SrCl2O6	strontium chlorate	7791-10-8	1.9134	-9.0617E-04	1.8159E-06	273	373	1,2	63.29	63.77	67.31
201	4484	SrF2	strontium fluoride	7783-48-4	-16.6573	9.2820E-02	-1.4281E-04	273	298	1,2	0.0110	0.0210	0.0210
202	4487	SrH2O2	strontium hydroxide	18480-07-4	-26.1775	1.6909E-01	-2.6879E-04	273	298	1,2	0.900	2.200	2.200
203	4492	SrI2	strontium iodide	10476-86-5	2.3896	-4.5178E-03	8.5816E-06	273	373	1,2	62.50	63.90	79.20
204	4493	SrI2O6	strontium iodate	13470-01-4	-6.5862	3.1593E-02	-4.0675E-05	273	373	1,2	0.102	0.165	0.346
205	4495	SrN2O4	strontium nitrite	13470-06-9	0.2495	6.5703E-03	-6.7358E-06	303	313	1,2	41.90	---	44.30
206	4496	SrN2O6	strontium nitrate	10042-76-9	-9.3384	6.8423E-02	-1.0590E-04	273	303	1,2	28.20	44.50	47.00
207	4744	Tl2SO4	thallium (i) sulfate	7446-18-6	-7.1071	4.2128E-02	-5.3301E-05	273	373	1,2	2.65	5.19	15.53
208	4799	UN2O8	uranyl nitrate	10102-06-4	2.9002	-1.0370E-02	2.1810E-05	273	323	1,2	49.52	55.85	67.07
209	5024	ZnBr2	zinc bromide	7699-45-8	2.0927	-2.2801E-03	5.6938E-06	283	303	1,2	80.10	83.00	84.10
210	5082	ZnI2	zinc iodide	10139-47-6	1.9707	-4.9119E-04	9.7205E-07	273	373	1,2	81.10	81.40	83.70
211	5083	ZnI2O6	zinc iodate	7790-37-6	-15.1107	9.2220E-02	-1.4150E-04	293	323	1,2	0.580	0.640	0.820
212	5086	ZnN2O6	zinc nitrate	7779-88-6	1.8389	-3.7444E-03	1.1572E-05	273	303	1,2	47.80	56.38	58.50
213	5094	ZnSO4	zinc sulfate	7733-02-0	-0.6129	1.0919E-02	-1.2140E-05	273	313	1,2	29.10	36.60	41.30

code: 1 - data, 2 - estimate

Chapter 8

SOLUBILITY OF GASES IN WATER – ORGANIC COMPOUNDS

Carl L. Yaws and Ralph W. Pike
Lamar University, Beaumont, Texas / Louisiana State University, Baton Rouge, Louisiana

Tabulation Results

The results for solubility of gases in water are given in Table 8. For the tabulation, the correlation for water solubility as a function of temperature is based on the following equation:

$$\log_{10} S = A + B/T + C \log_{10} T + D\,T \tag{8}$$

where S = solubility in water, parts per million by weight, ppm(wt)
T = temperature, K
A, B, C, and D = regression coefficients

The tabulation is arranged by carbon number (C, C2, C3, ...) to provide ease of use in quickly locating the data by using the chemical formula. The compound name, CAS No (Chemical Abstracts Registry Number), and regression coefficients are provided in the adjacent columns. The range of application is denoted by minimum and maximum temperatures (TMIN and TMAX). Temperatures outside the range of application should not be used. The next column provides the code for the tabulation.

In preparing the tabulation, a literature search was conducted to identify data source publications (1-22). The excellent compilations of CRC (4); Gerrard (9-10); Solubility Data Series of IUPAC (13-15); and Wilhelm, Battino, and Wilcock (19) were primarily used for experimental data.

A comparison of calculated and data values is shown in Figure 8 for a representative compound. The graph discloses favorable agreement of equation and data.

Example

In an industrial operation, propylene (C3H6) comes into contact with a body of water at 37.8 C. Determine the concentration in water at saturation.

Substitution of the regression coefficients from the table and temperature (37.8 C = 310.95 K) into the equation for solubility in water yields:

$$\log_{10} S = -33.4402 + 6.3445E{+}02/310.95 + 1.7542E{+}01 * \log_{10}(310.95) - 3.2416E{-}02 * 310.95$$

$$S = 10^{2.2487} = 177.31 \text{ parts per million by weight, ppm(wt)}$$

References

1. Abraham, M. H., J. A. Haftvan, G. S. Whiting, and A. Leo, J. Chem. Soc. Perkin Trans., 2, 1777 (1994).
2. Azarnoosh, A. and J. J. McKetta, J. Chem. Eng. Data, 4 (3), 211 (1959).
3. Azarnoosh, A. and J. J. McKetta, Petrol. Refiner, 37 (11), 275 (1958).
4. CRC HANDBOOK OF CHEMISTRY AND PHYSICS, 75th - 86th eds., CRC Press, Inc., Boca Raton, FL (1994-2006).
5. Crovetto, R., "Evaluation of Solubility Data for the System CO2-H2O", J. Phys. Chem. Ref. Data, 20, 575 (1991).
6. Davis, J. E. and J. J. McKetta, J. Chem. Eng. Data, 5, 374 (1960).
7. Davis, J. E. and J. J. McKetta, Petrol. Refiner, 39 (3), 205 (1960).
8. Fogg, P. and W. Gerrard, SOLUBILITY OF GASES IN LIQUIDS, John Wiley & Sons, New York, NY (1991).
9. Gerrard, W., GAS SOLUBILITIES, Pergamon Press, New York, NY (1980).
10. Gerrard, W., SOLUBILITY OF GASES AND LIQUIDS, Plenum Press, New York, NY (1976).
11. Horvath, A. L., HALOGENATED HYDROCARBONS, Solubility-Miscibility with Water, Marcel Dekker, Inc., New York, NY (1982).
12. Hayduk, W., ed., IUPAC SOLUBILITY DATA SERIES, Hayduk, W., ed., Vol. 9, Ethane, Pergamon Press, New York, NY (1982).
13. IUPAC SOLUBILITY DATA SERIES, Hayduk, W., ed., Vol. 24, Propane, Butane, and 2-Methylpropane, Pergamon Press, New York, NY (1986).
14. IUPAC SOLUBILITY DATA SERIES, Clever, H. L. and C. L. Young, eds., Vol. 27/28, Methane, Pergamon Press, New York, NY (1987).
15. IUPAC SOLUBILITY DATA SERIES, Cargill, R. W., ed., Vol. 43, Carbon Monoxide, Pergamon Press, New

York, NY (1990).
16. McAuliffe, C., J. Phy. Chem., 70, 1267 (1966).
17. TECHNICAL DATA BOOK - PETROLEUM REFINING, Vols. I and II, American Petroleum Institute, Washington, DC (1972, 1977, 1983).
18. Thibodeaux, L. J., CHEMODYNAMICS, John Wiley & Sons, New York, NY (1979).
19. Wilhelm, E., R. Battino, and R. J. Wilcock, Chemical Reviews, 77, No. 2, 219 (1977).
20. Yaws, C. L., CHEMICAL PROPERTIES HANDBOOK, McGraw-Hill, New York, NY (1999).
21. Yaws, C. L., MATHESON GAS DATA BOOK, 7th ed., Matheson Tri-Gas (Parsippany, NJ), McGraw-Hill, New York, NY (2001).
22. Yaws, C. L., YAWS HANDBOOK OF THERMODYNAMIC AND PHYSICAL PROPERTIES OF CHEMICAL COMPOUNDS, electronic edition, www.knovel.com, knovel, Norwich, NY (2003).

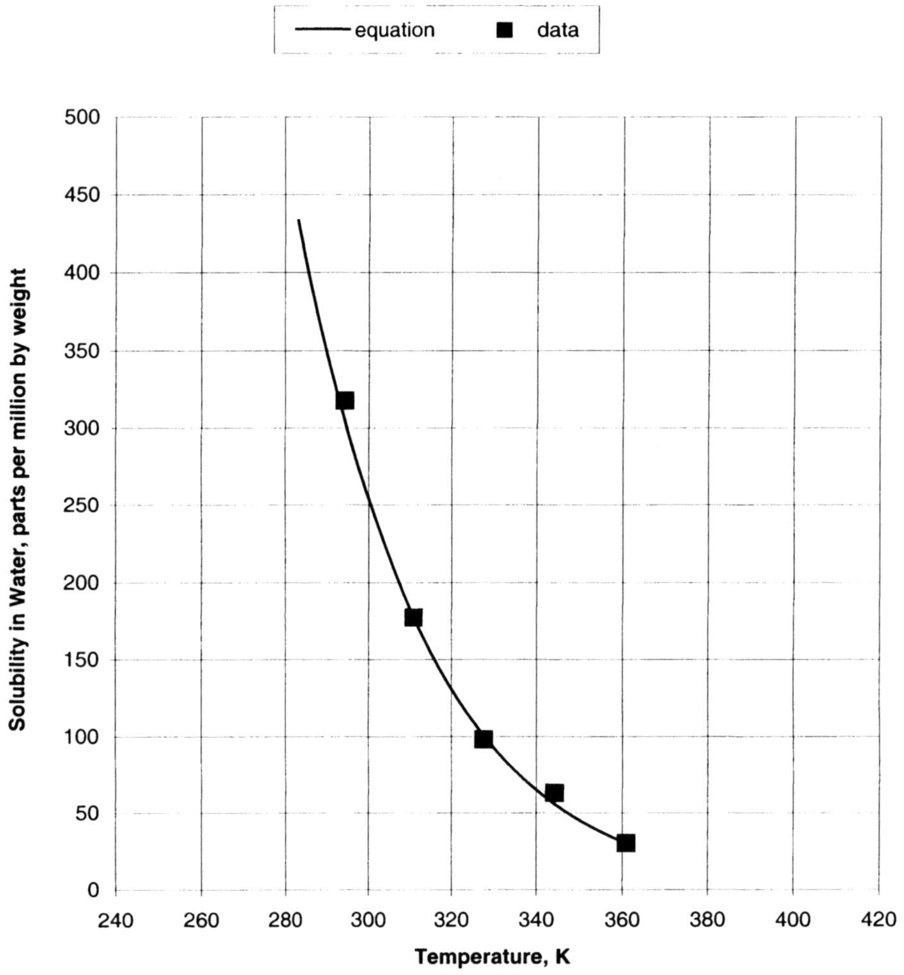

Figure 8 Solubility of Propylene in Water

Table 8 Solubility of Gases in Water - Organic Compounds

$\log_{10} S = A + B/T + C \log_{10}T + D T$ (S - solubility, parts per million by weight, ppm(wt), T - K)

NO	ID	FORMULA	NAME	CAS No	A	B	C	D	TMIN	TMAX	code	S@TMIN,ppm(wt)	S@T298,ppm(wt)	S@TMAX,ppm(wt)
1	17	CBrF3	bromotrifluoromethane	75-63-8	-905.2535	2.3850E+04	3.6800E+02	-2.7781E-01	298	348	1,2	322.42	322.42	80.86
2	34	CClF3	chlorotrifluoromethane	75-72-9	2143.6327	-5.6822E+04	-8.6148E+02	6.0567E-01	298	348	1,2	90.93	90.93	40.56
3	42	CCl2F2	dichlorodifluoromethane	75-71-8	5225.2514	-1.4399E+05	-2.0854E+03	1.4099E+00	298	350	1,2	303.04	303.04	81.67
4	47	CCl3F	trichlorofluoromethane	75-69-4	-476.4021	1.4738E+04	1.8883E+02	-1.2464E-01	303	348	1,2	1091.56	---	313.24
5	74	CF4	carbon tetrafluoride	75-73-0	-1917.2642	5.2308E+04	7.6972E+02	-5.4182E-01	273	323	1,2	38.75	38.75	11.46
6	97	CHClF2	chlorodifluoromethane	75-45-6	-1211.0912	3.4883E+04	4.8301E+02	-3.2745E-01	283	353	1,2	5308.20	5308.20	531.22
7	101	CHCl2F	dichlorofluoromethane	75-43-4	2069.5936	-5.6481E+04	-8.2482E+02	5.5371E-01	298	348	1,2	18690.90	18690.90	4234.71
8	113	CHF3	fluoroform	75-46-7	-2836.7148	7.5870E+04	1.1438E+03	-8.2242E-01	298	348	1,2	720.28	720.28	160.56
9	145	CH2ClF	chlorofluoromethane	593-70-4	-1373.9260	3.7833E+04	5.5299E+02	-3.9346E-01	283	353	1,2	16881.45	16881.45	1806.40
10	191	CH3Br	methyl bromide	74-83-9	-1826.7563	5.0517E+04	7.3363E+02	-5.1597E-01	278	353	1,2	35125.36	35125.36	2190.00
11	196	CH3Cl	methyl chloride	74-87-3	-1424.0488	3.9273E+04	5.7276E+02	-4.0661E-01	273	353	1,2	14075.39	14075.39	881.70
12	214	CH3F	methyl fluoride	593-53-3	-1360.6514	3.7118E+04	5.4811E+02	-3.9183E-01	273	353	1,2	4248.62	4248.62	441.26
13	251	CH4	methane	74-82-8	-1382.5451	3.7802E+04	5.5522E+02	-3.9159E-01	273	361	1,2	41.52	41.52	6.03
14	372	CO	carbon monoxide	630-08-0	-1310.0154	3.5637E+04	5.2664E+02	-3.7309E-01	273	353	1,2	44.74	44.74	9.99
15	373	COS	carbonyl sulfide	463-58-1	-304.8791	1.0397E+04	1.1901E+02	-7.1737E-02	273	303	1,2	3555.82	3555.82	1042.93
16	375	CO2	carbon dioxide	124-38-9	-1301.3028	3.5734E+04	5.2365E+02	-3.7270E-01	273	353	1,2	3388.06	3388.06	313.11
17	427	C2ClF5	chloropentafluoroethane	76-15-3	-209.1559	5.0079E+03	8.8573E+01	-8.3990E-02	298	348	1,2	58.57	58.57	13.17
18	441	C2Cl2F4	1,2-dichloro-1,1,2,2-tetrafluor	76-14-2	2775.8212	-7.4821E+04	-1.1118E+03	7.6594E-01	298	348	1,2	138.71	138.71	41.68
19	488	C2F4	tetrafluoroethylene	116-14-3	-1872.9419	5.2393E+04	7.4850E+02	-5.1221E-01	273	343	1,2	400.02	400.02	60.68
20	571	C2H2	acetylene	74-86-2	-898.6169	2.4867E+04	3.6154E+02	-2.5620E-01	274	343	1,2	1970.33	1970.33	423.60
21	642	C2H2F2	1,1-difluoroethylene	75-38-7	-2058.6482	5.1586E+04	8.4193E+02	-6.5558E-01	293	303	1,2	177.13	177.13	142.13
22	650	C2H2F4	1,1,1,2-tetrafluoroethane	811-97-2	-1825.7631	4.6301E+04	7.4521E+02	-5.7244E-01	293	320	1,2	719.49	719.49	393.73
23	701	C2H3Cl	vinyl chloride	75-01-4	-460.7040	1.3479E+04	1.8523E+02	-1.3236E-01	273	323	1,2	6030.65	6030.65	1120.18
24	702	C2H3ClF2	1-chloro-1,1-difluoroethane	75-68-3	-10320.0434	2.5185E+05	4.2370E+03	-3.3738E+00	286	294	1,2	2107.08	---	1365.59
25	787	C2H4	ethylene	74-85-1	-2711.3688	7.5648E+04	1.0831E+03	-7.3886E-01	283	361	1,2	238.69	238.69	31.70
26	834	C2H4F2	1,1-difluoroethane	75-37-6	-4160.8181	1.0187E+05	1.7088E+03	-1.3608E+00	273	301	1,2	5361.42	5361.42	2408.23
27	939	C2H5F	ethyl fluoride	353-36-6	-25251.3283	6.3590E+05	1.0293E+04	-7.8744E+00	287	298	1,2	3952.25	3952.25	2088.87
28	985	C2H6	ethane	74-84-0	-540.9272	1.6418E+04	2.1380E+02	-1.3899E-01	275	323	1,2	121.88	121.88	30.36
29	1284	C3F6	hexafluoropropylene	116-15-4	-754.4758	2.0425E+04	3.0560E+02	-2.2788E-01	273	343	1,2	419.00	419.00	49.67
30	1299	C3F8	octafluoropropane	76-19-7	-1744.5834	4.2411E+04	7.1792E+02	-5.8037E-01	278	303	1,2	17.78	17.78	10.02
31	1492	C3H4	methylacetylene	74-99-7	-5572.6082	1.4896E+05	2.2447E+03	-1.6028E+00	273	361	1,2	7014.27	7014.27	145.58
32	1809	C3H6	cyclopropane	75-19-4	-3886.0392	1.0415E+05	1.5654E+03	-1.1208E+00	298	361	1,2	478.81	478.81	28.63
33	1810	C3H6	propylene	115-07-1	-33.4402	6.3455E+02	1.7542E+01	-3.2416E-02	283	361	1,2	433.24	433.24	30.29
34	2020	C3H7F	1-fluoropropane	460-13-9	-1845.1003	4.4951E+04	7.6005E+02	-6.1342E-01	278	303	1,2	4503.66	4503.66	2513.11
35	2021	C3H7F	2-fluoropropane	420-26-8	-1742.2778	4.2414E+04	7.1794E+02	-5.8035E-01	278	303	1,2	4352.10	4352.10	2458.07
36	2094	C3H8	propane	74-98-6	-5548.9049	1.5698E+05	2.2079E+03	-1.4727E+00	283	361	1,2	199.05	199.05	10.15
37	2336	C4F8	octafluorocyclobutane	115-25-3	-4993.9175	1.2423E+05	2.0422E+03	-1.5911E+00	273	311	1,2	89.91	89.91	29.98
38	2510	C4H4	vinylacetylene	689-97-4	-587.0277	1.5818E+04	2.3868E+02	-1.7894E-01	273	333	1,2	3417.91	3417.91	894.22
39	2755	C4H6	ethylacetylene	107-00-6	-529.0357	1.3275E+04	2.1887E+02	-1.7931E-01	273	333	1,2	6842.64	6842.64	1579.89
40	2757	C4H6	butadiene (1,3 butadiene)	106-99-0	-587.0277	1.5818E+04	2.3868E+02	-1.7894E-01	273	333	1,2	3417.91	3417.91	894.22
41	3320	C4H8	1-butene	106-98-9	-2985.5526	8.4030E+04	1.1937E+03	-8.2911E-01	311	348	1,2	280.00	---	20.83
42	3324	C4H8	isobutene	115-11-7	-913.4382	2.6306E+04	3.6467E+02	-2.5038E-01	273	343	1,2	906.91	906.91	82.77
43	3783	C4H10	butane	106-97-8	-1154.1487	3.2991E+04	4.6039E+02	-3.1485E-01	273	349	1,2	215.88	215.88	15.79
44	3784	C4H10	isobutane	75-28-5	-560.9083	1.8039E+04	2.1849E+02	-1.2916E-01	278	318	1,2	114.79	114.79	30.81
45	5774	C5H12	neopentane	463-82-1	-2054.5554	5.8998E+04	8.1640E+02	-5.4276E-01	288	353	1,2	65.54	65.54	9.39

code: 1 - data, 2 - estimate

S - solubility in water at pressure of about 1 atm (101.325 kPa), parts per million by weight, ppm(wt)

1. Most compounds in the tabulation are at very low concentration (range of about 0.0001 to 0.000001 mol fraction). At very low concentrations, Henry's law is applicable (p = Hx where x<<<1).

2. For those compounds which react with water, all chemical species of the gas and its reaction products are included (see CRC HANDBOOK OF CHEMISTRY AND PHYSICS, 84th ed., CRC Press (2004)).

Chapter 9

SOLUBILITY OF GASES IN WATER – INORGANIC COMPOUNDS

Carl L. Yaws and Ralph W. Pike
Lamar University, Beaumont, Texas / Louisiana State University, Baton Rouge, Louisiana

Tabulation Results

The results for solubility of gases in water are given in Table 9. For the tabulation, the correlation for water solubility as a function of temperature is based on the following equation:

$$\log_{10} S = A + B/T + C \log_{10} T + D\,T \qquad (9)$$

where
S = solubility in water, parts per million by weight, ppm(wt)
T = temperature, K
A, B, C, and D = regression coefficients

The tabulation is arranged by alphabetical order (Ar, As, C, ...) to provide ease of use in quickly locating the data by using the chemical formula. The compound name, CAS No (Chemical Abstracts Registry Number), and regression coefficients are provided in the adjacent columns. The range of application is denoted by minimum and maximum temperatures (TMIN and TMAX). Temperatures outside the range of application should not be used. The next column provides the code for the tabulation.

In preparing the tabulation, a literature search was conducted to identify data source publications (1-23). The excellent compilations of CRC (1); Gerrard (3-5); Solubility Data Series of IUPAC (7-16); and Wilhelm, Battino, and Wilcock (20) were primarily used for experimental data.

A comparison of calculated and data values is shown in Figure 9 for a representative compound. The graph discloses favorable agreement of equation and data.

Example

In an industrial operation, argon (Ar) comes into contact with a body of water at 35 C. Determine the concentration in water at saturation.

Substitution of the regression coefficients from the table and temperature (35 C = 308.15 K) into the equation for solubility in water yields:

$$\log_{10} S = -1055.3705 + 2.8772E{+}04/308.15 + 4.2467E{+}02 * \log_{10}(308.15) - 3.0260E{-}01 * 308.15$$

$$S = 10^{1.6601} = 45.72 \text{ parts per million by weight, ppm(wt)}$$

References

1. CRC HANDBOOK OF CHEMISTRY AND PHYSICS, 75th - 86th eds., CRC Press, Inc., Boca Raton, FL (1994-2006).
2. Crovetto, R., "Evaluation of Solubility Data for the System CO2-H2O", J. Phys. Chem. Ref. Data, 20, 575 (1991).
3. Fogg, P. and W. Gerrard, SOLUBILITY OF GASES IN LIQUIDS, John Wiley & Sons, New York, NY (1991).
4. Gerrard, W., GAS SOLUBILITIES, Pergamon Press, New York, NY (1980).
5. Gerrard, W., SOLUBILITY OF GASES AND LIQUIDS, Plenum Press, New York, NY (1976).
6. Horne, R. A., THE CHEMISTRY OF OUR ENVIRONMENT, John Wiley & Sons, New York, NY (1978).
7. IUPAC SOLUBILITY DATA SERIES, Clever, H. L., ed., Vol. 1, Helium and Neon, Pergamon Press, New York, NY (1979).
8. IUPAC SOLUBILITY DATA SERIES, Clever, H. L., ed., Vol. 2, Krypton, Xenon, and Radon, Pergamon Press, New York, NY (1979).
9. IUPAC SOLUBILITY DATA SERIES, Clever, H. L., ed., Vol. 4, Argon, Pergamon Press, New York, NY (1980).
10. IUPAC SOLUBILITY DATA SERIES, Young, C. L., ed., Vol. 5/6, Hydrogen and Deuterium, Pergamon Press, New York, NY (1981).
11. IUPAC SOLUBILITY DATA SERIES, Battino, R., ed., Vol. 7, Oxygen and Ozone, Pergamon Press, New York, NY (1981).
12. IUPAC SOLUBILITY DATA SERIES, Young, C. L., ed., Vol. 8, Oxides of Nitrogen, Pergamon Press, New York, NY (1981).
13. IUPAC SOLUBILITY DATA SERIES, Battino, R., ed., Vol. 10, Nitrogen and Air, Pergamon Press, New York, NY (1982).

14. IUPAC SOLUBILITY DATA SERIES, Young, C. L., ed., Vol. 12, Sulfur Dioxide, Chlorine, Fluorine, and Chlorine Oxides, Pergamon Press, New York, NY (1983).
15. IUPAC SOLUBILITY DATA SERIES, P. G. T. Fogg and C. L. Young, eds., Vol. 32, Hydrogen Sulfide, Deuterium Sulfide, and Hydrogen Selenide", Pergamon Press, New York, NY (1988).
16. IUPAC SOLUBILITY DATA SERIES, Cargill, R. W., ed., Vol. 43, Carbon Monoxide, Pergamon Press, New York, NY (1990).
17. Hutchinson, G. E., A TREATISE ON LIMNOLOGY, John Wiley & Sons, New York, NY (1957).
18. TECHNICAL DATA BOOK - PETROLEUM REFINING, Vols. I and II, American Petroleum Institute, Washington, DC (1972, 1977, 1983).
19. Thibodeaux, L. J., CHEMODYNAMICS, John Wiley & Sons, New York, NY (1979).
20. Wilhelm, E., R. Battino, and R. J. Wilcock, Chemical Reviews, Vol. 77, No. 2, 219 (1977).
21. Yaws, C. L., CHEMICAL PROPERTIES HANDBOOK, McGraw-Hill, New York, NY (1999).
22. Yaws, C. L., MATHESON GAS DATA BOOK, 7th ed., Matheson Tri-Gas (Parsippany, NJ), McGraw-Hill, New York, NY (2001).
23. Yaws, C. L., YAWS HANDBOOK OF THERMODYNAMIC AND PHYSICAL PROPERTIES OF CHEMICAL COMPOUNDS, electronic edition, www.knovel.com, knovel, Norwich, NY (2003).

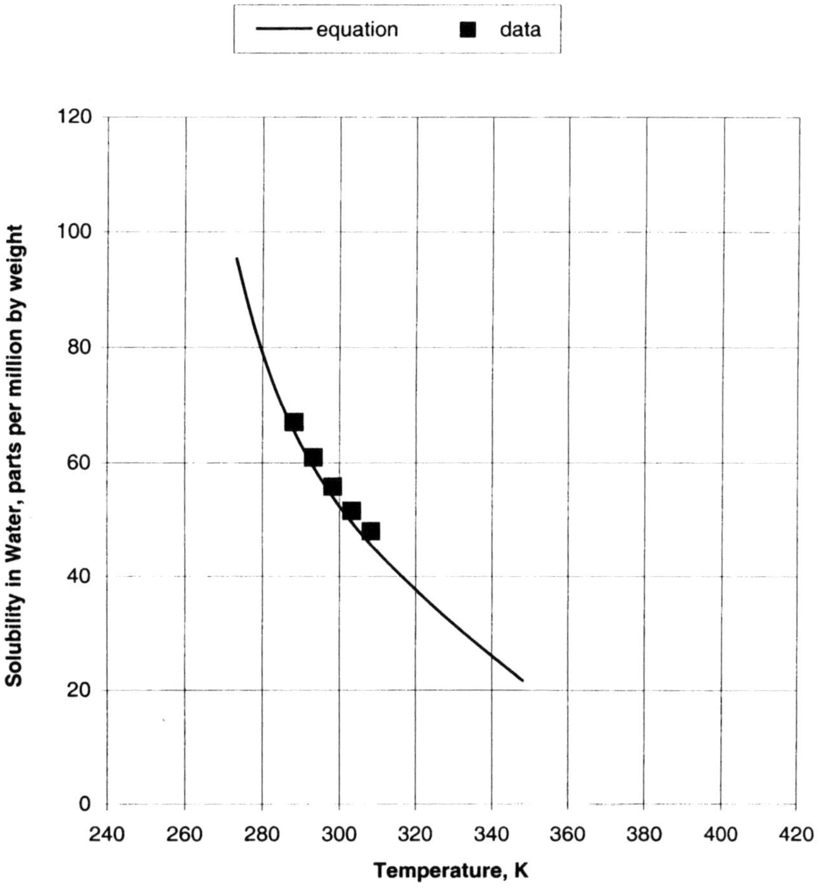

Figure 9 Solubility of Argon in Water

Table 9 Solubility of Gases in Water - Inorganic Compounds

$\log_{10} S = A + B/T + C \log_{10} T + D\,T$ (S - solubility, parts per million by weight, ppm(wt), T - K)

NO	ID	FORMULA	NAME	CAS No	A	B	C	D	TMIN	TMAX	code	S@TMIN,ppm(wt)	S@T298,ppm(wt)	S@TMAX,ppm(wt)
1	200	Ar	argon	7440-37-1	-1055.3705	2.8772E+04	4.2467E+02	-3.0260E-01	273	348	1,2	95.32	54.13	21.76
2	234	AsH3	arsine	7784-42-1	-257.5855	8.2214E+03	1.0220E+02	-6.7230E-02	273	299	1,2	1404.51	676.12	659.53
3	652	CO	carbon monoxide	630-08-0	-1310.0154	3.5637E+04	5.2664E+02	-3.7309E-01	273	353	1,2	44.74	25.94	9.99
4	653	COS	carbonyl sulfide	463-58-1	-304.8791	1.0397E+04	1.1901E+02	-7.1737E-02	273	303	1,2	3555.82	1237.31	1042.93
5	655	CO2	carbon dioxide	124-38-9	-1301.3028	3.5734E+04	5.2365E+02	-3.7270E-01	273	353	1,2	3388.06	1442.91	313.11
6	898	ClO2	chlorine dioxide	10049-04-4	-563.1018	1.4685E+04	2.3242E+02	-1.8939E-01	283	333	1,2	107738.85	62050.35	15611.92
7	899	Cl2	chlorine	7782-50-5	-1841.3759	5.0593E+04	7.3961E+02	-5.1867E-01	283	353	1,2	9680.50	6269.55	1667.96
8	1302	D2	deuterium	7782-39-0	-1613.1148	4.1924E+04	6.5374E+02	-4.8510E-01	283	303	1,2	3.71	3.16	3.00
9	1709	H2	hydrogen	1333-74-0	-932.5889	2.4928E+04	3.7568E+02	-2.6979E-01	273	345	1,2	1.97	1.53	1.01
10	1714	H2S	hydrogen sulfide	7783-06-4	-1241.8348	3.3418E+04	5.0199E+02	-3.6516E-01	273	353	1,2	6644.02	3369.75	787.92
11	1719	H2Se	hydrogen selenide	7783-07-5	-1197.6818	3.1926E+04	4.8524E+02	-3.5650E-01	288	343	1,2	7920.37	6412.65	2217.28
12	1739	He	helium-4	7440-59-7	-1037.1419	2.7377E+04	4.1857E+02	-3.0262E-01	273	348	1,2	1.69	1.51	1.10
13	2162	Kr	krypton	7439-90-9	-1312.1442	3.5977E+04	5.2739E+02	-3.7300E-01	273	353	1,2	414.82	202.97	59.66
14	2613	NF3	nitrogen trifluoride	7783-54-2	-641.1820	1.8922E+04	2.5437E+02	-1.6759E-01	283	323	1,2	81.93	54.95	34.31
15	2674	NO	nitric oxide	10102-43-9	-1265.6006	3.4454E+04	5.0899E+02	-3.6117E-01	273	353	1,2	98.98	56.05	20.43
16	2678	N2	nitrogen	7727-37-9	-1157.3807	3.1611E+04	4.6492E+02	-3.2805E-01	273	350	1,2	28.85	17.28	7.83
17	2684	N2F4	tetrafluorohydrazine	10036-47-2	-668.4338	2.0605E+04	2.6272E+02	-1.6377E-01	288	318	1,2	119.96	86.18	52.90
18	2747	N2O	nitrous oxide	10024-97-2	-365.1999	1.1259E+04	1.4541E+02	-9.8429E-02	273	313	1,2	2512.81	1031.88	673.14
19	3091	Ne	neon	7440-01-9	-1051.1943	2.8179E+04	4.2347E+02	-3.0262E-01	273	348	1,2	11.39	8.85	5.70
20	3198	O2	oxygen	7782-44-7	-1068.0450	2.9325E+04	4.2899E+02	-3.0261E-01	273	348	1,2	70.34	39.45	16.78
21	3389	PH3	phosphine	7803-51-2	-858.2313	2.4113E+04	3.4414E+02	-2.4067E-01	298	323	1,2	269.73	269.73	155.51
22	3940	Rn	radon	10043-92-2	-508.8565	1.5703E+04	2.0136E+02	-1.3000E-01	273	323	1,2	5142.01	1991.01	1034.88
23	4001	SF6	sulfur hexafluoride	2551-62-4	-634.4119	2.0392E+04	2.4651E+02	-1.4224E-01	276	323	1,2	81.32	34.53	20.47
24	4007	SO2	sulfur dioxide	7446-09-5	-353.9136	9.8223E+03	1.4558E+02	-1.1528E-01	283	323	1,2	126867.80	78096.99	35719.54
25	4945	Xe	xenon	7440-63-3	-1075.4609	3.0053E+04	4.3142E+02	-3.0222E-01	273	348	1,2	1314.17	557.14	160.95

code: 1 - data, 2 - estimate

S - solubility in water at pressure of about 1 atm (101.325 kPa), parts per million by weight, ppm(wt)

1. Most compounds in the tabulation are at very low concentration (range of about 0.0001 to 0.000001 mol fraction). At very low concentrations, Henry's law is applicable ($p = Hx$ where $x <<< 1$).

2. A few compounds (ClO2 and SO2) are not at very low concentrations. The tabulated data for these compounds should be considered rough values.

3. For those compounds which react with water, all chemical species of the gas and its reaction products are included (see CRC HANDBOOK OF CHEMISTRY AND PHYSICS, 84th ed., CRC Press (2004)).

Chapter 10

HENRY'S LAW CONSTANT FOR COMPOUND IN WATER – ORGANIC COMPOUNDS

Carl L. Yaws and Prasad K. Narasimhan
Lamar University, Beaumont, Texas

Tabulation Results

The results for Henry's law constant for compound in water are given in Table 10. The tabulation is arranged by carbon number (C, C2, C3, ...) to provide ease of use in quickly locating the data by using the chemical formula. The compound name, CAS No (Chemical Abstracts Registry Number), and Henry's law constant are provided in the adjacent columns. The last column provides the code for the tabulation which is based on both experimental data and estimated values.

In preparing the tabulation, a literature search was conducted to identify data source publications for organics (1-246). Both experimental values for the property under consideration and parameter values for estimation of the property are included in the source publications. The publications were screened and copies of appropriate data were made. These data were then keyed into the computer to provide a database of values for compounds for which experimental data are available. The database also served as a basis to check the accuracy of the estimation methods. Upon completion of data collection, estimation of the values for the remaining compounds was performed. The numerous point values were processed using a computer program for minimum deviation.

Henry's law constant was primarily determined from values for solubility in water, activity coefficient at infinite dilution, and vapor pressure. For solubility in water, the excellent compilations by Horvath (56); Howard and Meylan (57); Mackay, Shiu, and Ma (91-95); Verschueren (141); Yalkowsky (147); and Yaws (149-171) were used for experimental data. Most of the estimates were based on the boiling point method as developed by Yaws (150-169). For activity coefficient at infinite dilution, the vapor-liquid equilibrium data collection of Dechema (173) was used. For vapor pressure, the compilations of Boublik, Fried, and Hala (178); Daubert and Danner (182); Ohe (192); TAMU (198-199); and Yaws (205-228) were utilized.

For water solubility which is used in determining Henry's law constant for hydrocarbons and chemicals, experimental data available in the literature are very limited for isomers, large molecules (C10 and higher), and substituted organics containing halogens, nitrogen, oxygen, and sulfur. In the absence of data, the estimates for isomers, large molecules, and substituted organics should be considered as rough approximations useful for initial analysis. If initial analysis is favorable, follow-up experimental determination is recommended.

Example

A chemical spill of benzene (C6H6) occurs into a body of water. The concentration of benzene in the liquid at the surface of the water is 50 parts per million on a mol basis [x = 50 ppm(mol)]. Calculate the concentration in the air at the surface of the water.

From thermodynamics at low pressure, the partition coefficient is given by $K = H/P_t$. Substitution of Henry's law constant (H = 307.69 atm/mol fraction for benzene) from the table and the total pressure (P_t = 1 atm) into this equation yields K = 307.69/1 = 307.69.

Since the vapor concentration is given by y = K x, substitution of the K value (K = 307.69) and liquid concentration (x = 50 ppm) yields

$$y = K\,x = (307.69)\ (50)$$

$$y = 15{,}385.5 \text{ ppm (mol)} = 1.54\ \%\ \text{(mol)}$$

References

1-172. See **Chapter 3 SOLUBILITY IN WATER – ORGANIC COMPOUNDS**

173. DECHEMA CHEMISTRY DATA SERIES, Volume1, Vapor-Liquid Equilibrium Data Collection, Parts 1, 1a, 1b, 1c, and 1d, Dechema, Frankurt am Main, Germany (1991-2003).

174. Altunin, V. V., V. Z. Geller, E. K. Petrov, D. C. Rasskazov, and G. A. Spiridonov, THERMOPHYSICAL PROPERTIES OF FREONS, Methane Series, Parts 1 and 2, Hemisphere Publishing Corporation, New York, NY (1987).

175. ASHRAE Handbook – Fundamentals, 2001, 1993 and 1989 eds., ASHRAE, Atlanta, GA (2001, 1993, 1989).

176. Bah, A. and N. Dupont-Pavolvsky, J. Phys. Chem. Ref. Data, 40, 869 (1995).

177. Beaton, C. F. and G. F. Hewitt, PHYSICAL PROPERTY DATA FOR THE DESIGN ENGINEER, Hemisphere

Publishing Corporation, New York, NY (1989).
178. Boublick, T., V. Fried and E. Hala, THE VAPOUR PRESSURES OF PURE SUBSTANCES, 1st and 2nd eds., Elsevier, New York, NY (1975, 1984).
179. Chao, J., [Ethane and Propane], J. Phys. Chem. Ref. Data, 2, 427-438 (1973).
180. CRC HANDBOOK OF CHEMISTRY AND PHYSICS, 75th - 86th eds., CRC Press, Inc., Boca Raton, FL (1994-2006).
181. CRC HANDBOOK OF THERMOPHYSICAL AND THERMOCHEMICAL DATA, CRC Press, Boca Raton, FL (1994).
182. Daubert, T. E. and R. P. Danner, DATA COMPILATION OF PROPERTIES OF PURE COMPOUNDS, Parts 1, 2, 3 and 4, Supplements 1 and 2, DIPPR Project 801, AIChE, New York, NY (1985-2003).
183. Driesbach, R. R., PHYSICAL PROPERTIES OF CHEMICAL COMPOUNDS, Vol. I (No. 15), Vol. II (No. 22), Vol. III (No. 29), Advances in Chemistry Series, American Chemical Society, Washington, DC (1955,1959,1961).
184. ENCYCLOPEDIA OF CHEMICAL TECHNOLOGY, 3rd and 4th eds., John Wiley and Sons, Inc., New York, NY (1978-2004).
185. Goodwin, R. D., [Methanol], J. Phys. Chem. Ref. Data, 16, 799-892 (1987).
186. Goodwin, R. D., J. Phys. Chem. Ref. Data, 14, 849 (1985).
187. Jaconsen, R. T., et. Al., INTERNATIONAL THERMODYNAMIC TABLES OF THE FLUID STATE, No. 10, Ethylene, Blackwell Scientific Publications, Oxford (1988).
188. Jahangi, M., R. T. Jacobsen, R. B. Stewart, and R. D. McCarty, [Ethylene], J. Phys. Chem. Ref. Data, 15, 593 (1986).
189. Landolt-Bornstein, ZAHLENWERTE UND FUNKIONEN ANS PHYSIK, CHEMEI, ASTRONOMIE UND TECHNIK, Springer-Verlag, Heidelberg, Germany (1972-2005).
190. LANGE'S HANDBOOK OF CHEMISTRY, 13th, 14th, and 15th eds., McGraw-Hill, New York, NY (1985, 1992, 2005).
191. NIST Chemistry WebBook, webbook.nist.gov (2005).
192. Ohe, S., COMPUTER AIDED DATA BOOK OF VAPOR PRESSURE, Data Book Publishing Company, Tokyo, Japan (1976).
193. Outcalt, S. I. and M. O. McLinden, J. Phys. Chem. Ref. Data, 25, 605 (1996).
194. PERRY'S CHEMICAL ENGINEERING HANDBOOK, 6th and 7th eds., McGraw-Hill, New York, NY (1984, 1997).
195. Poling, B. E., J. M. Prausnitz, and J. P. O'Connell, THE PROPERTIES OF GASES AND LIQUIDS, 5th ed., McGraw-Hill, New York, NY (2000).
196. Reid, R. C., J. M. Prausnitz, and B. E. Poling, THE PROPERTIES OF GASES AND LIQUIDS, 3rd ed. (R. C. Reid and T. K. Sherwood), 4th ed., McGraw-Hill, New York, NY (1977, 1987).
197. Ruzicka, K. and V. Majer, J. Phys. Chem. Ref. Data, 23, 1 (1994).
198. SELECTED VALUES OF PROPERTIES OF CHEMICAL COMPOUNDS, Thermodynamics Research Center, TAMU, College Station, TX (1997).
199. SELECTED VALUES OF PROPERTIES OF HYDROCARBONS AND RELATED COMPOUNDS, Thermodynamics Research Center, TAMU, College Station, TX (1998).
200. Stull, D., Ind. Eng. Chem., 39, 517 (1947).
201. Tillner-Roth, R. and H. D. Baehr, J. Phys. Chem. Ref. Data, 23, 657 (1994).
202. Timmermans, J., PHYSICO-CHEMICAL CONSTANTS OF PURE ORGANIC COMPOUNDS, Vols. 1 and 2, Elsevier, New York, NY (1950, 1965).
203. Vargaftik, N. B., TABLES ON THE THERMOPHYSICAL PROPERTIES OF LIQUIDS AND GASES, 2nd ed., English translation, Hemisphere Publishing Corporation, New York, NY (1975, 1983).
204. Weber, L. A. and D. R. Defibaugh, J. Phys. Chem. Ref. Data, 41, 581 (1996).
205. Yaws, C. L. and Others, "Benzene and Napthalene - Phy. and Thermo. Prop. (13)", Chem. Eng., 82 (18), 107 (Sept. 1, 1975).
206. Yaws, C. L. and Others, "Carbon Oxides (Carbon Monoxide and Dioxide) - Phy. and Thermo. Prop. (4)", Chem. Eng., 81 (20), 115 (Sept. 30, 1974).
207. Yaws, C. L. and Others, "Chlorobenzene, Aniline and Phenol - Phy. and Thermo. Prop. (15)", Chem. Eng., 82, 119 (Oct. 27, 1975).
208. Yaws, C. L. and Others, "Cyclopropane, Cyclobutane, Cyclopentane and Cyclohexane - Phy. and Thermo. Prop. (16)", Chem. Eng., 82 (26), 119 (Dec. 8, 1975).
209. Yaws, C. L. and Others, "Ethylene, Propylene and Butylene Oxides - Phy. and Thermo. Prop. (19)", Chem. Eng., 83, 129 (April 12, 1976).
210. Yaws, C. L. and Others, "Methanol, Ethanol, Propanol and Butanol - Phy. and Thermo. Prop. (20)", Chem. Eng., 83, 119 (June 7, 1976).
211. Yaws, C. L., "Alkanes (Methane, Ethane and Propane) - Phy. and Thermo. Prop. (11)", Chem. Eng., 82 (10), 89 (May 12, 1975).
212. Yaws, C. L., "Diolefins: Butadiene, Isoprene and Chloroprene - Phy. and Thermo. Prop. (18)", Chem. Eng., 83, 107 (March 1, 1976).

213. Yaws, C. L., "Methyl Chloride, Methylene Chloride, Chloroform and Carbon Tetrachloride - Phy. and Thermo. Prop. (21)", Chem. Eng., 83, 81 (July 5, 1976).
214. Yaws, C. L., "Olefin Monomers: Isobutylene and Styrene - Phy. and Thermo. Prop. (17)", Chem. Eng., 83, 107 (Jan. 19, 1976).
215. Yaws, C. L., "Olefins (Ethylene, Propylene and 1-Butene) - Phy. and Thermo. Prop.(10)", Chem. Eng., 82 (7), 101 (March 31, 1975).
216. Yaws, C. L., "Toluene, Ethylbenzene and Cumene - Phy. and Thermo. Prop. (14)", Chem. Eng., 82 (20), 73 (Sept. 29, 1975).
217. Yaws, C. L., "Xylenes (Ortho, Meta and Para) - Phy. and Thermo. Prop. (12)", Chem. Eng., 82 (15), 113 (July 21, 1975).
218. Yaws, C. L., PHYSICAL PROPERTIES, McGraw-Hill, New York, NY (1977).
219. Yaws, C. L. and R. W. Gallant, PHYSICAL PROPERTIES OF HYDROCARBONS, Vols. 1 (2nd ed.), 2 (3rd ed.), 3 and 4, Gulf Publishing Co., Houston, TX (1992,1993,1993,1995).
220. Yaws, C. L., HANDBOOK OF VAPOR PRESSURE, Vols. 1, 2, 3, and 4, Gulf Publishing Co., Houston, TX (1994,1994,1994,1995).
221. Yaws, C. L., HANDBOOK OF THERMODYNAMIC DIAGRAMS, Vols. 1, 2, 3, and 4, Gulf Publishing Co., Houston, TX (1996).
222. Yaws, C. L., HANDBOOK OF CHEMICAL COMPOUND DATA FOR PROCESS SAFETY, Gulf Publishing Co., Houston, TX (1997).
223. Yaws, C. L., CHEMICAL PROPERTIES HANDBOOK, McGraw-Hill, New York, NY (1999).
224. Yaws, C. L., MATHESON GAS DATA BOOK, 7th ed., Matheson Tri-Gas (Parsippany, NJ), McGraw-Hill, New York, NY (2001).
225. Yaws, C. L., YAWS HANDBOOK OF THERMODYNAMIC AND PHYSICAL PROPERTIES FOR CHEMICALS, electronic edition, www.knovel.com, knovel, Norvich, NY (2003).
226. Yaws, C. L., YAWS HANDBOOK OF ANTOINE COEFFICIENTS FOR VAPOR PRESSURE, electronic edition, www.knovel.com, knovel, Norvich, NY (2004).
227. Yaws, C. L., YAWS HANDBOOK OF PHYSICAL PROPERTIES FOR HYDROCARBONS AND CHEMICALS, Gulf Publishing Co., Houston, TX (2005).
228. Yaws, C. L., YAWS HANDBOOK OF ANTOINE COEFFICIENTS FOR VAPOR PRESSURE, Gulf Publishing Co., Houston, TX (2007).
229. Ashworth, R. A., G. B. Howe, M. E. Mullins, and T. N. Rogers, "Air-Water Partitioning Coefficients of Organics in Dilute Aqueous Solutions", 1986 Summer National Meeting, AIChE, Boston, Mass. (August, 1986).
230. Ashworth, R. A., G. B. Howe, M. E. Mullins, and T. N. Rogers, J. Hazardous Materials, 18, 25 (1988).
231. Baker, R. J., B. J. Donelan, L. J. Peterson, W. E. Acree, Jr., and C. C. Tsai, Phys. Chem. Liq., 16, 279 (1987).
232. Burkhard, L. P., D. E. Armstrong, and A. W. Andren, Environ. Sci. Technol., 19, 590 (1985).
233. Byers, W. D. and C. M. Morton, "Removing VOC from Groundwater", 1984 Summer National Meeting, A.I.Ch.E., Phil., PA (August, 1984).
234. Dilling, W. L., Environ. Sci. Technol., 11, 405 (1977).
235. Gossett, J. M., ANAEROBIC DEGRADATION OF C1 AND C2 CHLORINATED HYDROCARBONS, final report, U.S. Air Force, ESL-TR-85-38, Tyndall AFB, FL (Dec., 1985).
236. Gossett, J. M., Environ. Sci. Technol., 21, 202 (1987).
237. Hwang, Y. L., G. E. Keller, II, and J. D. Olson, Ind. Eng. Chem. Res., 31(7), 1753 (1992).
238. Hwang, Y. L., G. E. Keller, II, and J. D. Olson, Ind. Eng. Chem. Res., 31(7), 1759 (1992).
239. Leighton, D. T., Jr. and J. M. Calo, J. Chem. Eng. Data, 26, 382 (1981).
240. Murphy, T. J., M. D. Mullin, and J. A. Meyer, Environ. Sci. Technol., 21, 155 (1987).
241. Preston, G. T. and J. M. Prausnitz, Ind. Eng. Chem. Fundam., 10, 389 (1971).
242. Roberts, P. V. and J. A. Levy, Journal AWWA, 138 (April, 1985).
243. Roberts, P. V., G. D. Hopkins, C. Munz, and A. H. Riojas, Environ. Sci. Technol., 19, 164 (1985).
244. TREATABILITY MANUAL, Vol. I, Treatability Data, EPA-600/2-82-001a, Office of Research and Development, U.S. Environmental Protection Agency, Washington, D.C. (Sept., 1981).
245. Warner, H. P., J. M. Cohen and J. C. Ireland, DETERMINATION OF HENRY'S LAW CONSTANTS OF SELECTED PRIORITY POLLUTANTS, EPA/600/D-87/229 (July, 1987).
246. Yaws, C. L. and others, Chem. Eng. Data, 38(4), 546 (1993).

Table 10 Henry's Law Constant for Compound in Water - Organic Compounds

					Henry's Law Constant, H					
NO	ID	FORMULA	NAME	CAS No	MW, g/mol	T_B, K	T, C	H @ T, atm/mol frac	H @ T, atm/mol/m^3	code
1	13	CBrClF2	bromochlorodifluoromethane	353-59-3	165.365	269.14	2	8.9676E+03	1.6141E-01	1,2
2	17	CBrF3	bromotrifluoromethane	75-63-8	148.911	215.26	25	2.4835E+04	4.4703E-01	1,2
3	29	CBr4	carbon tetrabromide	558-13-4	331.627	462.65	30	1.2603E+02	2.2686E-03	1,2
4	34	CClF3	chlorotrifluoromethane	75-72-9	104.459	191.74	25	6.1787E+04	1.1122E+00	1,2
5	39	CClN	cyanogen chloride	506-77-4	61.471	286.00	25	1.0788E+02	1.9418E-03	1,2
6	42	CCl2F2	dichlorodifluoromethane	75-71-8	120.913	243.36	25	2.1431E+04	3.8575E-01	1,2
7	44	CCl2O	phosgene	75-44-5	98.915	280.71	25	7.7515E+02	1.3953E-02	1,2
8	47	CCl3F	trichlorofluoromethane	75-69-4	137.367	296.97	30	6.8350E+03	1.2303E-01	1,2
9	50	CCl4	carbon tetrachloride	56-23-5	153.821	349.79	25	1.6170E+03	2.9105E-02	1,2
10	74	CF4	carbon tetrafluoride	75-73-0	88.005	145.09	25	2.5970E+05	4.6746E+00	1,2
11	96	CHBr3	bromoform	75-25-2	252.731	422.35	25	2.0521E+01	3.6938E-04	1,2
12	97	CHClF2	chlorodifluoromethane	75-45-6	86.468	232.32	25	1.6718E+03	3.0091E-02	1,2
13	101	CHCl2F	dichlorofluoromethane	75-43-4	102.922	282.05	25	2.9162E+02	5.2491E-03	1,2
14	104	CHCl3	chloroform	67-66-3	119.376	334.33	25	2.1101E+02	3.7982E-03	1,2
15	113	CHF3	fluoroform	75-46-7	70.014	190.99	25	5.2257E+03	9.4062E-02	1,2
16	117	CHI3	iodoform	75-47-8	393.732	491.16	25	2.0623E+02	3.7120E-03	1,2
17	122	CHN	hydrogen cyanide	74-90-8	27.026	298.85	25	4.9370E+00	8.8865E-05	1,2
18	140	CH2BrCl	bromochloromethane	74-97-5	129.383	341.20	25	8.9545E+01	1.6118E-03	1,2
19	144	CH2Br2	dibromomethane	74-95-3	173.835	370.10	25	5.1008E+01	9.1813E-04	1,2
20	145	CH2ClF	chlorofluoromethane	593-70-4	68.478	264.06	25	3.6510E+02	6.5718E-03	1,2
21	149	CH2Cl2	dichloromethane	75-09-2	84.932	312.90	25	1.3689E+02	2.4640E-03	1,2
22	158	CH2F2	difluoromethane	75-10-5	52.024	221.50	25	6.3588E+02	1.1446E-02	1,2
23	161	CH2I2	diiodomethane	75-11-6	267.836	455.15	25	5.1663E+00	9.2992E-05	1,2
24	175	CH2O	formaldehyde	50-00-0	30.026	254.05	25	1.8722E-02	3.3700E-07	1,2
25	176	CH2O2	formic acid	64-18-6	46.026	373.71	100	1.0687E+00	1.9237E-05	1,2
26	191	CH3Br	methyl bromide	74-83-9	94.939	276.71	25	3.4621E+02	6.2317E-03	1,2
27	196	CH3Cl	methyl chloride	74-87-3	50.487	248.93	25	5.2600E+02	9.4680E-03	1,2
28	214	CH3F	methyl fluoride	593-53-3	34.033	194.82	25	9.4931E+02	1.7087E-02	1,2
29	222	CH3I	methyl iodide	74-88-4	141.939	315.58	25	1.5536E+02	2.7965E-03	1,2
30	232	CH3NO	formamide	75-12-7	45.041	493.00	25	7.7223E-05	1.3900E-09	2
31	235	CH3NO2	nitromethane	75-52-5	61.041	374.35	25	1.5389E+01	2.7699E-04	1,2
32	251	CH4	methane	74-82-8	16.043	111.66	25	3.9315E+04	7.0766E-01	1,2
33	272	CH4O	methyl alcohol	67-56-1	32.042	337.85	25	3.7308E-01	6.7153E-06	1,2
34	278	CH4S	methyl mercaptan	74-93-1	48.109	279.11	25	1.0944E+02	1.9699E-03	1,2
35	286	CH5N	methylamine	74-89-5	31.058	266.82	25	6.1667E-01	1.1100E-05	1,2
36	372	CO	carbon monoxide	630-08-0	28.010	81.70	25	5.8088E+04	1.0456E+00	1,2
37	373	COS	carbonyl sulfide	463-58-1	60.076	223.00	25	2.6094E+03	4.6969E-02	1,2
38	375	CO2	carbon dioxide	124-38-9	44.010	194.67	25	1.6392E+03	2.9505E-02	1,2
39	382	CS2	carbon disulfide	75-15-0	76.143	319.37	25	1.0521E+03	1.8937E-02	1,2
40	427	C2ClF5	chloropentafluoroethane	76-15-3	154.467	234.04	25	1.4180E+05	2.5524E+00	1,2
41	441	C2Cl2F4	1,2-dichloro-1,1,2,2-tetrafluoroethane	76-14-2	170.921	276.92	25	6.6271E+04	1.1929E+00	1,2
42	449	C2Cl3F3	1,1,2-trichloro-1,2,2-trifluoroethane	76-13-1	187.375	320.75	25	2.6695E+04	4.8051E-01	1,2
43	455	C2Cl4	tetrachloroethylene	127-18-4	165.832	394.40	25	1.5423E+03	2.7761E-02	1,2
44	456	C2Cl4F2	1,1,2,2-tetrachloro-1,2-difluoroethane	76-12-0	203.829	366.00	25	6.9216E+03	1.2459E-01	1,2
45	463	C2Cl6	hexachloroethane	67-72-1	236.737	460.00	25	1.2811E+03	2.3059E-02	1,2
46	488	C2F4	tetrafluoroethylene	116-14-3	100.016	197.51	25	3.5405E+04	6.3728E-01	1,2
47	494	C2F6	hexafluoroethane	76-16-4	138.013	194.95	25	9.4004E+05	1.6921E+01	1,2
48	511	C2HBrClF3	halothane	151-67-7	197.382	323.35	24	1.1353E+03	2.0436E-02	1,2
49	539	C2HCl3	trichloroethylene	79-01-6	131.387	360.10	25	6.6108E+02	1.1899E-02	1,2
50	551	C2HCl5	pentachloroethane	76-01-7	202.292	433.03	25	1.0635E+02	1.9142E-03	1,2
51	571	C2H2	acetylene	74-86-2	26.038	189.15	25	1.3329E+03	2.3992E-02	1,2
52	592	C2H2Br4	1,1,2,2-tetrabromoethane	79-27-6	345.654	516.65	30	7.3925E-01	1.3306E-05	1,2
53	593	C2H2Br4	1,1,1,2-tetrabromoethane	630-16-0	345.654	523.15	25	1.5803E-02	2.8445E-07	1,2
54	615	C2H2Cl2	1,1-dichloroethylene	75-35-4	96.943	304.71	25	1.2540E+03	2.2572E-02	1,2
55	616	C2H2Cl2	cis-1,2-dichloroethylene	156-59-2	96.943	333.65	25	3.9577E+02	7.1238E-03	1,2
56	617	C2H2Cl2	trans-1,2-dichloroethylene	156-60-5	96.943	320.85	25	3.6096E+02	6.4972E-03	1,2
57	618	C2H2Cl2	dichloroethylene	25323-30-2	96.943	333.15	25	3.0621E+02	5.5118E-03	1,2
58	619	C2H2Cl2	1,2-dichloroethylene, cis and trans	540-59-0	96.943	328.15	25	3.7596E+02	6.7673E-03	1,2
59	627	C2H2Cl2O2	dichloroacetic acid	79-43-6	128.942	467.15	25	1.9556E-02	3.5200E-07	2
60	635	C2H2Cl4	1,1,1,2-tetrachloroethane	630-20-6	167.848	403.65	25	1.4251E+02	2.5652E-03	1,2
61	636	C2H2Cl4	1,1,2,2-tetrachloroethane	79-34-5	167.848	418.25	25	2.1214E+01	3.8185E-04	1,2
62	642	C2H2F2	1,1-difluoroethylene	75-38-7	64.035	190.15	25	2.1568E+04	3.8822E-01	1,2
63	643	C2H2F2	cis-1,2-difluoroethene	1630-77-9	64.035	245.15	25	2.1568E+04	3.8822E-01	1,2
64	644	C2H2F2	trans-1,2-difluoroethene	1630-78-0	64.035	245.15	25	2.1568E+04	3.8822E-01	2
65	650	C2H2F4	1,1,1,2-tetrafluoroethane	811-97-2	102.032	247.15	25	8.3102E+03	1.4958E-01	1,2
66	651	C2H2F4	1,1,2,2-tetrafluoroethane	359-35-3	102.032	253.10	25	8.3102E+03	1.4958E-01	1,2
67	684	C2H2O4	oxalic acid	144-62-7	90.035	569.00	20	1.1655E-07	2.0979E-12	1,2
68	701	C2H3Cl	vinyl chloride	75-01-4	62.498	259.78	25	1.4038E+03	2.5269E-02	1,2
69	702	C2H3ClF2	1-chloro-1,1-difluoroethane	75-68-3	100.495	263.14	25	5.3065E+03	9.5517E-02	1,2
70	720	C2H3Cl3	1,1,1-trichloroethane	71-55-6	133.403	347.23	25	1.1852E+03	2.1333E-02	1,2
71	721	C2H3Cl3	1,1,2-trichloroethane	79-00-5	133.403	387.00	25	5.3212E+01	9.5781E-04	1,2
72	753	C2H3N	acetonitrile	75-05-8	41.053	354.75	25	1.0768E+00	1.9382E-05	1,2
73	787	C2H4	ethylene	74-85-1	28.054	169.47	25	1.1057E+04	1.9903E-01	1,2
74	798	C2H4Br2	1,1-dibromoethane	557-91-5	187.862	381.15	25	2.7509E+01	4.9516E-04	1,2
75	799	C2H4Br2	1,2-dibromoethane	106-93-4	187.862	404.51	25	4.4001E+01	7.9200E-04	1,2
76	812	C2H4Cl2	1,1-dichloroethane	75-34-3	98.959	330.45	25	3.1736E+02	5.7125E-03	1,2
77	813	C2H4Cl2	1,2-dichloroethane	107-06-2	98.959	356.59	20	5.0236E+01	9.0425E-04	1,2
78	814	C2H4Cl2	dichloroethane	1300-21-6	98.959	343.52	25	1.3974E+02	2.5152E-03	2
79	834	C2H4F2	1,1-difluoroethane	75-37-6	66.051	247.35	25	1.3230E+03	2.3813E-02	1,2
80	835	C2H4F2	1,2-difluoroethane	624-72-6	66.051	303.65	25	1.1170E+03	2.0106E-02	1,2

Table 10 Henry's Law Constant for Compound in Water - Organic Compounds

					Henry's Law Constant, H					
NO	ID	FORMULA	NAME	CAS No	MW, g/mol	T_B, K	T, C	H @ T, atm/mol frac	H @ T, atm/mol/m^3	code
81	884	C2H4O	acetaldehyde	75-07-0	44.053	293.55	99	7.0821E+00	1.2748E-04	1,2
82	885	C2H4O	ethylene oxide	75-21-8	44.053	283.85	25	9.8363E+00	1.7705E-04	1,2
83	889	C2H4O2	acetic acid	64-19-7	60.053	391.05	100	1.8382E+00	3.3087E-05	1,2
84	890	C2H4O2	methyl formate	107-31-3	60.053	304.90	25	1.2380E+01	2.2284E-04	1,2
85	910	C2H5Br	bromoethane	74-96-4	108.966	311.50	20	3.2978E+02	5.9359E-03	1,2
86	914	C2H5Cl	ethyl chloride	75-00-3	64.514	285.42	20	6.0629E+02	1.0913E-02	1,2
87	939	C2H5F	ethyl fluoride	353-36-6	48.060	235.45	25	1.2359E+03	2.2246E-02	1,2
88	948	C2H5I	ethyl iodide	75-03-6	155.966	345.45	20	3.0335E+02	5.4602E-03	1,2
89	954	C2H5N	ethyleneimine	151-56-4	43.069	329.00	25	6.7223E-01	1.2100E-05	2
90	956	C2H5NO	acetamide	60-35-5	59.068	494.30	25	9.5273E-05	1.7149E-09	1,2
91	958	C2H5NO2	nitroethane	79-24-3	75.068	387.22	25	2.1990E+00	3.9583E-05	1,2
92	985	C2H6	ethane	74-84-0	30.070	184.55	25	2.9397E+04	5.2914E-01	1,2
93	1065	C2H6O	ethyl alcohol	64-17-5	46.069	351.44	25	3.9981E-01	7.1965E-06	1,2
94	1066	C2H6O	dimethyl ether	115-10-6	46.069	248.31	18	3.6509E+01	6.5715E-04	1,2
95	1072	C2H6O2	ethylene glycol	107-21-1	62.068	470.45	25	7.8494E-05	1.4129E-09	1,2
96	1083	C2H6O4S	dimethyl sulfate	77-78-1	126.133	461.95	18	1.7057E-01	3.0703E-06	1,2
97	1084	C2H6O4S	ethyl sulfate	540-82-9	126.133	461.95	25	2.5959E-01	4.6726E-06	2
98	1092	C2H6S	ethyl mercaptan	75-08-1	62.136	308.15	25	1.5805E+02	2.8448E-03	1,2
99	1093	C2H6S	dimethyl sulfide	75-18-3	62.136	310.48	25	9.8196E+01	1.7675E-03	1,2
100	1115	C2H7N	ethylamine	75-04-7	45.085	289.73	25	6.8334E-01	1.2300E-05	1,2
101	1116	C2H7N	dimethylamine	124-40-3	45.085	280.03	25	9.8334E-01	1.7700E-05	1,2
102	1117	C2H7NO	monoethanolamine	141-43-5	61.084	444.15	25	1.8056E-03	3.2500E-08	2
103	1159	C2H8N2	ethylenediamine	107-15-3	60.100	390.41	25	4.7016E-04	8.4628E-09	1,2
104	1210	C2N2	cyanogen	460-19-5	52.036	252.00	20	2.9320E+02	5.2775E-03	1,2
105	1284	C3F6	hexafluoropropylene	116-15-4	150.024	243.55	25	4.2527E+04	7.6548E-01	1,2
106	1299	C3F8	octafluoropropane	76-19-7	188.021	236.40	25	8.7334E+05	1.5720E+01	1,2
107	1377	C3H2N2	malononitrile	109-77-3	66.063	491.50	20	6.1413E-03	1.1054E-07	1,2
108	1462	C3H3N	acrylonitrile	107-13-1	53.064	350.50	20	4.0637E+00	7.3145E-05	1,2
109	1492	C3H4	methylacetylene	74-99-7	40.065	249.94	25	8.1685E+02	1.4703E-02	1,2
110	1533	C3H4Cl2	2,3-dichloropropene	78-88-6	110.970	365.75	25	2.3953E+02	4.3115E-03	1,2
111	1534	C3H4Cl2	1,1-dichloropropene	563-58-6	110.970	349.65	25	3.5710E+02	6.4277E-03	1,2
112	1535	C3H4Cl2	cis-1,2-dichloropropene	6923-20-2	110.970	366.15	25	1.7321E+02	3.1178E-03	1,2
113	1536	C3H4Cl2	trans-1,2-dichloropropene	7069-38-7	110.970	350.15	25	3.4950E+02	6.2910E-03	1,2
114	1537	C3H4Cl2	cis-1,3-dichloropropene	10061-01-5	110.970	377.45	25	1.0376E+02	1.8676E-03	1,2
115	1538	C3H4Cl2	trans-1,3-dichloropropene	10061-02-6	110.970	385.15	25	7.2581E+01	1.3065E-03	1,2
116	1539	C3H4Cl2	3,3-dichloropropene	563-57-5	110.970	357.55	25	2.5346E+02	4.5623E-03	1,2
117	1540	C3H4Cl2	1,3-dichloropropene; (cis+trans)	542-75-6	110.970	380.65	25	8.9511E+01	1.6112E-03	1,2
118	1541	C3H4Cl2	1,2-dichloropropene	563-54-2	110.970	348.15	25	3.8083E+02	6.8549E-03	1,2
119	1542	C3H4Cl2	dichloropropylene	26952-23-8	110.970	364.52	25	1.8631E+02	3.3536E-03	2
120	1621	C3H4O	acrolein	107-02-8	56.064	325.84	20	3.5356E+00	6.3641E-05	1,2
121	1622	C3H4O	propargyl alcohol	107-19-7	56.064	386.75	25	6.3889E-02	1.1500E-06	2
122	1626	C3H4O2	acrylic acid	79-10-7	72.064	414.15	100	8.0764E-01	1.4537E-05	1,2
123	1628	C3H4O2	vinyl formate	692-45-5	72.064	320.00	25	1.0827E+01	1.9488E-04	1,2
124	1634	C3H4O3	pyruvic acid	127-17-3	88.063	438.15	25	8.3334E-03	1.5000E-07	2
125	1642	C3H5Br	cis-1-bromo-1-propene	590-13-6	120.977	330.95	25	4.4454E+02	8.0016E-03	1,2
126	1643	C3H5Br	trans-1-bromo-1-propene	590-15-8	120.977	336.35	25	3.5082E+02	6.3148E-03	1,2
127	1644	C3H5Br	2-bromo-1-propene	557-93-7	120.977	321.55	25	6.7049E+02	1.2069E-02	1,2
128	1645	C3H5Br	3-bromo-1-propene	106-95-6	120.977	343.16	25	3.2455E+02	5.8419E-03	1,2
129	1647	C3H5Br	1-bromo-1-propene; (cis+trans)	590-14-7	120.977	333.65	25	4.1953E+02	7.5515E-03	1,2
130	1668	C3H5Cl	cis-1-chloro-1-propene	16136-84-8	76.525	305.95	25	7.8215E+02	1.4079E-02	1,2
131	1669	C3H5Cl	trans-1-chloro-1-propene	16136-85-9	76.525	310.55	25	6.5250E+02	1.1745E-02	1,2
132	1670	C3H5Cl	2-chloro-1-propene	557-98-2	76.525	295.80	25	1.0259E+03	1.8466E-02	1,2
133	1671	C3H5Cl	3-chloro-1-propene	107-05-1	76.525	318.11	25	5.2677E+02	9.4818E-03	1,2
134	1672	C3H5Cl	1-chloro-1-propene	590-21-6	76.525	302.50	25	8.9167E+02	1.6050E-02	1,2
135	1703	C3H5Cl3	1,1,1-trichloropropane	7789-89-1	147.430	379.15	25	1.3661E+02	2.4590E-03	1,2
136	1704	C3H5Cl3	1,1,2-trichloropropane	598-77-6	147.430	405.15	25	3.6375E+01	6.5474E-04	1,2
137	1705	C3H5Cl3	1,1,3-trichloropropane	20395-25-9	147.430	418.70	25	1.7522E+01	3.1539E-04	1,2
138	1706	C3H5Cl3	1,2,2-trichloropropane	3175-23-3	147.430	396.15	25	6.1704E+01	1.1107E-03	1,2
139	1707	C3H5Cl3	1,2,3-trichloropropane	96-18-4	147.430	430.00	25	2.3759E+01	4.2766E-04	1,2
140	1752	C3H5N	propionitrile	107-12-0	55.080	370.50	25	1.8792E+00	3.3825E-05	1,2
141	1755	C3H5NO	acrylamide	79-06-1	71.079	465.75	30	8.7865E-07	1.5816E-11	1,2
142	1796	C3H5N3O9	nitroglycerine	55-63-0	227.088	523.00	20	3.2125E-03	5.7825E-08	1,2
143	1809	C3H6	cyclopropane	75-19-4	42.081	240.37	25	4.7260E+03	8.5067E-02	1,2
144	1810	C3H6	propylene	115-07-1	42.081	225.43	25	8.3884E+03	1.5099E-01	1,2
145	1824	C3H6Br2	1,1-dibromopropane	598-17-4	201.889	406.65	25	2.9196E+01	5.2552E-04	1,2
146	1825	C3H6Br2	1,2-dibromopropane	78-75-1	201.889	413.16	25	8.6008E+01	1.5481E-03	1,2
147	1826	C3H6Br2	1,3-dibromopropane	109-64-8	201.889	440.45	20	2.8007E+00	5.0412E-05	1,2
148	1827	C3H6Br2	2,2-dibromopropane	594-16-1	201.889	388.25	25	1.6541E+01	2.9774E-04	1,2
149	1844	C3H6Cl2	1,1-dichloropropane	78-99-9	112.986	361.25	25	1.7544E+02	3.1580E-03	1,2
150	1845	C3H6Cl2	1,2-dichloropropane	78-87-5	112.986	369.52	25	1.5732E+02	2.8317E-03	1,2
151	1846	C3H6Cl2	1,3-dichloropropane	142-28-9	112.986	393.55	25	5.3611E+01	9.6499E-04	1,2
152	1847	C3H6Cl2	2,2-dichloropropane	594-20-7	112.986	342.46	25	1.7680E+02	3.1824E-03	1,2
153	1848	C3H6Cl2	1,2-dichloropropane, (±)	26198-63-0	112.986	369.55	25	1.5168E+02	2.7303E-03	1,2
154	1849	C3H6Cl2	dichloropropane	26638-19-7	112.986	369.75	25	1.5168E+02	2.7303E-03	1,2
155	1920	C3H6O	allyl alcohol	107-18-6	58.080	370.23	25	3.4201E-01	6.1562E-06	1,2
156	1922	C3H6O	propionaldehyde	123-38-6	58.080	321.15	25	2.3930E+00	4.3074E-05	1,2
157	1923	C3H6O	acetone	67-64-1	58.080	329.44	25	2.6896E+00	4.8412E-05	1,2
158	1924	C3H6O	1,2-propylene oxide	75-56-9	58.080	307.05	20	3.3580E+00	6.0443E-05	1,2
159	1925	C3H6O	1,3-propylene oxide	503-30-0	58.080	321.00	25	2.4498E+00	4.4096E-05	1,2
160	1927	C3H6O	(S)-(-)-propylene oxide	16088-62-3	58.080	308.05	25	3.8817E+00	6.9871E-05	1,2

Table 10 Henry's Law Constant for Compound in Water - Organic Compounds

					Henry's Law Constant, H					
NO	ID	FORMULA	NAME	CAS No	MW, g/mol	T_B, K	T, C	H @ T, atm/mol frac	H @ T, atm/mol/m^3	code
161	1928	C3H6O	(R)-(+)-propylene oxide	15448-47-2	58.080	317.97	25	2.6014E+00	4.6825E-05	2
162	1930	C3H6O2	propanoic acid	79-09-4	74.079	414.32	100	3.2300E+00	5.8139E-05	1,2
163	1931	C3H6O2	ethyl formate	109-94-4	74.079	327.46	25	2.1390E+01	3.8502E-04	1,2
164	1932	C3H6O2	methyl acetate	79-20-9	74.079	330.09	25	6.3800E+00	1.1484E-04	1,2
165	1949	C3H6O3	methoxyacetic acid	625-45-6	90.079	478.26	25	3.5667E-04	6.4200E-09	2
166	1950	C3H6O3	trioxane	110-88-3	90.079	387.65	25	5.5830E-01	1.0049E-05	1,2
167	1978	C3H7Br	1-bromopropane	106-94-5	122.993	344.15	20	4.0353E+02	7.2635E-03	1,2
168	1979	C3H7Br	2-bromopropane	75-26-3	122.993	332.56	20	4.7505E+02	8.5509E-03	1,2
169	1986	C3H7Cl	1-chloropropane	540-54-5	78.541	319.67	25	7.8675E+02	1.4161E-02	1,2
170	1987	C3H7Cl	2-chloropropane	75-29-6	78.541	308.85	20	7.8212E+02	1.4078E-02	1,2
171	2020	C3H7F	1-fluoropropane	460-13-9	62.087	269.95	25	1.1441E+03	2.0594E-02	1,2
172	2021	C3H7F	2-fluoropropane	420-26-8	62.087	263.81	25	1.1741E+03	2.1134E-02	1,2
173	2025	C3H7I	1-iodopropane	107-08-4	169.993	375.60	24	4.5883E+02	8.2589E-03	1,2
174	2026	C3H7I	2-iodopropane	75-30-9	169.993	362.65	20	4.7901E+02	8.6222E-03	1,2
175	2030	C3H7N	allylamine	107-11-9	57.096	326.45	25	5.5278E-01	9.9500E-06	2
176	2045	C3H7NO2	1-nitropropane	108-03-2	89.094	404.33	20	3.5844E+00	6.4520E-05	1,2
177	2046	C3H7NO2	2-nitropropane	79-46-9	89.094	393.40	20	5.0395E+00	9.0710E-05	1,2
178	2094	C3H8	propane	74-98-6	44.097	231.11	25	3.6420E+04	6.5556E-01	1,2
179	2129	C3H8O	propyl alcohol	71-23-8	60.096	370.35	25	4.4069E-01	7.9324E-06	1,2
180	2130	C3H8O	isopropyl alcohol	67-63-0	60.096	355.41	25	7.1693E-01	1.2905E-05	1,2
181	2131	C3H8O	methyl ethyl ether	540-67-0	60.096	280.50	25	1.4066E+01	2.5319E-04	1,2
182	2139	C3H8O2	methylal	109-87-5	76.095	315.00	20	6.5722E+00	1.1830E-04	1,2
183	2140	C3H8O2	1,2-propanediol (propylene glycol)	57-55-6	76.095	460.75	25	2.0356E-04	3.6641E-09	1,2
184	2141	C3H8O2	1,3-propanediol	504-63-2	76.095	487.55	25	9.6667E-03	1.7400E-07	2
185	2148	C3H8O3	glycerol	56-81-5	92.095	563.15	25	9.6112E-04	1.7300E-08	1,2
186	2158	C3H8S	propyl mercaptan	107-03-9	76.163	340.87	25	2.3039E+02	4.1470E-03	1,2
187	2159	C3H8S	isopropyl mercaptan	75-33-2	76.163	325.71	25	2.4654E+02	4.4376E-03	1,2
188	2160	C3H8S	methyl ethyl sulfide	624-89-5	76.163	340.15	25	1.0045E+02	1.8082E-03	1,2
189	2200	C3H9N	propylamine	107-10-8	59.111	321.65	25	1.0740E+00	1.9332E-05	1,2
190	2201	C3H9N	isopropylamine	75-31-0	59.111	305.55	20	8.4150E-01	1.5147E-05	1,2
191	2202	C3H9N	methylethylamine	624-78-2	59.111	309.15	25	1.3451E+00	2.4211E-05	1,2
192	2203	C3H9N	trimethylamine	75-50-3	59.111	276.02	25	5.7778E+00	1.0400E-04	1,2
193	2204	C3H9NO	1-amino-2-propanol	78-96-6	75.111	432.61	25	2.7111E-05	4.8800E-10	2
194	2241	C3H9O4P	trimethyl phosphate	512-56-1	140.076	465.85	25	9.6956E-03	1.7452E-07	1,2
195	2318	C4Cl6	hexachloro-1,3-butadiene	87-68-3	260.759	488.15	25	1.7774E+03	3.1993E-02	1,2
196	2336	C4F8	octafluorocyclobutane	115-25-3	200.032	267.17	25	2.0842E+05	3.7516E+00	1,2
197	2374	C4H2	biacetylene	460-12-8	50.060	283.46	25	2.8318E+02	5.0972E-03	1,2
198	2510	C4H4	vinylacetylene	689-97-4	52.076	278.25	25	1.4600E+03	2.6280E-02	1,2
199	2562	C4H4N2	succinonitrile	110-61-2	80.090	540.15	20	1.0581E-04	1.9045E-09	1,2
200	2592	C4H4O	furan	110-00-9	68.075	304.50	25	2.9912E+02	5.3842E-03	1,2
201	2605	C4H4O4	fumaric acid	110-17-8	116.073	563.15	25	1.2330E-10	2.2193E-15	1,2
202	2606	C4H4O4	maleic acid	110-16-7	116.073	565.00	25	1.1052E-07	1.9893E-12	1,2
203	2613	C4H4S	thiophene	110-02-1	84.142	357.31	25	1.5808E+02	2.8454E-03	1,2
204	2754	C4H6	dimethylacetylene	503-17-3	54.092	300.13	25	8.3096E+02	1.4957E-02	1,2
205	2755	C4H6	ethylacetylene	107-00-6	54.092	281.22	25	7.2609E+02	1.3070E-02	1,2
206	2756	C4H6	1,2-butadiene	590-19-2	54.092	284.00	25	4.4953E+03	8.0915E-02	1,2
207	2757	C4H6	butadiene (1,3 butadiene)	106-99-0	54.092	268.74	25	3.9566E+03	7.1218E-02	1,2
208	2940	C4H6O	trans-crotonaldehyde	123-73-9	70.091	377.25	20	7.3798E-01	1.3284E-05	1,2
209	2941	C4H6O	methacrolein	78-85-3	70.091	341.15	25	1.1667E+01	2.1001E-04	1,2
210	2944	C4H6O	divinyl ether	109-93-3	70.091	301.45	37	9.7874E+02	1.7617E-02	1,2
211	2956	C4H6O	crotonaldehyde	4170-30-3	70.091	377.15	25	1.3535E+00	2.4362E-05	1,2
212	2963	C4H6O2	2-butyne-1,4-diol	110-65-6	86.090	511.15	25	8.0001E-04	1.4400E-08	2
213	2965	C4H6O2	gamma-butyrolactone	96-48-0	86.090	477.15	25	2.9278E-03	5.2700E-08	2
214	2966	C4H6O2	cis-2-butenoic acid	503-64-0	86.090	445.05	25	3.7000E-03	6.6600E-08	2
215	2970	C4H6O2	methyl acrylate	96-33-3	86.090	353.35	23	9.5644E+00	1.7216E-04	1,2
216	2971	C4H6O2	vinyl acetate	108-05-4	86.090	345.65	25	2.9013E+01	5.2223E-04	1,2
217	2972	C4H6O2	allyl formate	1838-59-1	86.090	356.75	25	1.0577E+01	1.9039E-04	1,2
218	2990	C4H6O3	acetic anhydride	108-24-7	102.090	411.78	25	1.5538E-01	2.7968E-06	1,2
219	3006	C4H6O4	succinic acid	110-15-6	118.089	591.00	25	7.0236E-08	1.2642E-12	1,2
220	3024	C4H6O5	malic acid	6915-15-7	134.089	602.00	26	2.7127E-11	4.8828E-16	1,2
221	3033	C4H6O6	tartaric acid	87-69-4	150.088	660.00	20	8.4342E-18	1.5182E-22	1,2
222	3238	C4H7N	butyronitrile	109-74-0	69.107	390.75	25	3.0536E+00	5.4964E-05	1,2
223	3239	C4H7N	isobutyronitrile	78-82-0	69.107	376.76	25	5.3034E+00	9.5461E-05	1,2
224	3249	C4H7NO	2-pyrrolidone	616-45-5	85.106	518.15	25	1.2611E-04	2.2700E-09	2
225	3320	C4H8	1-butene	106-98-9	56.108	266.90	25	1.3592E+04	2.4466E-01	1,2
226	3321	C4H8	2-butene; (cis+trans)	107-01-7	56.108	275.50	25	1.3310E+04	2.3957E-01	1,2
227	3322	C4H8	cis-2-butene	590-18-1	56.108	276.87	25	1.3500E+04	2.4299E-01	1,2
228	3323	C4H8	trans-2-butene	624-64-6	56.108	274.03	25	1.3122E+04	2.3620E-01	1,2
229	3324	C4H8	isobutene	115-11-7	56.108	266.25	25	9.7072E+03	1.7473E-01	1,2
230	3333	C4H8Br2	1,1-dibromobutane	62168-25-6	215.916	431.15	25	3.5019E+01	6.3034E-04	1,2
231	3334	C4H8Br2	1,2-dibromobutane	533-98-2	215.916	439.46	25	6.4902E+01	1.1682E-03	1,2
232	3335	C4H8Br2	1,3-dibromobutane	107-80-2	215.916	448.15	25	1.2280E+01	2.2104E-04	1,2
233	3336	C4H8Br2	1,4-dibromobutane	110-52-1	215.916	470.15	25	3.5566E+00	6.4018E-05	1,2
234	3337	C4H8Br2	2,2-dibromobutane	50341-35-0	215.916	417.15	25	1.6473E+01	2.9651E-04	1,2
235	3338	C4H8Br2	DL-2,3-dibromobutane	598-71-0	215.916	433.65	25	1.4370E+01	2.5866E-04	1,2
236	3339	C4H8Br2	meso-2,3-dibromobutane	5780-13-2	215.916	430.45	25	1.4611E+01	2.6300E-04	1,2
237	3340	C4H8Br2	1,1-dibromo-2-methylpropane	33693-78-6	215.916	420.15	25	1.5312E+01	2.7561E-04	1,2
238	3341	C4H8Br2	1,2-dibromo-2-methylpropane	594-34-3	215.916	422.15	25	1.6178E+01	2.9120E-04	1,2
239	3342	C4H8Br2	1,3-dibromo-2-methylpropane	28148-04-1	215.916	448.15	25	1.2280E+01	2.2104E-04	1,2
240	3343	C4H8Br2	2,3-dibromobutane	5408-86-6	215.916	434.15	25	3.0975E+01	5.5755E-04	1,2

Table 10 Henry's Law Constant for Compound in Water - Organic Compounds

				Henry's Law Constant, H						
NO	ID	FORMULA	NAME	CAS No	MW, g/mol	T_B, K	T, C	H @ T, atm/mol frac	H @ T, atm/mol/m³	code
241	3360	C4H8Cl2	1,1-dichlorobutane	541-33-3	127.013	386.95	25	4.2274E+02	7.6093E-03	1,2
242	3361	C4H8Cl2	1,2-dichlorobutane	616-21-7	127.013	397.15	25	1.5018E+02	2.7031E-03	1,2
243	3362	C4H8Cl2	1,3-dichlorobutane	1190-22-3	127.013	406.65	25	1.5057E+02	2.7103E-03	1,2
244	3363	C4H8Cl2	1,4-dichlorobutane	110-56-5	127.013	427.05	25	3.2206E+02	5.7970E-03	1,2
245	3364	C4H8Cl2	2,2-dichlorobutane	4279-22-5	127.013	376.15	25	1.6211E+02	2.9179E-03	1,2
246	3365	C4H8Cl2	DL-2,3-dichlorobutane	2211-67-8	127.013	392.65	20	1.5221E+02	2.7398E-03	1,2
247	3366	C4H8Cl2	meso-2,3-dichlorobutane	4028-56-2	127.013	389.15	20	1.8143E+02	3.2657E-03	1,2
248	3367	C4H8Cl2	1,1-dichloro-2-methylpropane	598-76-5	127.013	378.15	25	1.5600E+02	2.8080E-03	1,2
249	3368	C4H8Cl2	1,2-dichloro-2-methylpropane	594-37-6	127.013	380.15	25	1.6213E+02	2.9184E-03	1,2
250	3369	C4H8Cl2	1,3-dichloro-2-methylpropane	616-19-3	127.013	408.15	25	1.5063E+02	2.7114E-03	1,2
251	3370	C4H8Cl2	2,3-dichlorobutane, dl and meso	7581-97-7	127.013	391.10	25	1.7453E+02	3.1415E-03	1,2
252	3371	C4H8Cl2	mixo-dichlorobutane	26761-81-9	127.013	393.94	25	1.7554E+02	3.1596E-03	2
253	3478	C4H8O	butyraldehyde	123-72-8	72.107	347.95	25	6.3313E+00	1.1396E-04	1,2
254	3479	C4H8O	isobutyraldehyde	78-84-2	72.107	337.25	25	8.9149E+00	1.6047E-04	1,2
255	3480	C4H8O	methyl ethyl ketone	78-93-3	72.107	352.79	25	7.1761E+00	1.2917E-04	1,2
256	3481	C4H8O	tetrahydrofuran	109-99-9	72.107	338.00	20	9.1043E+00	1.6388E-04	1,2
257	3504	C4H8O2	butyric acid	107-92-6	88.106	436.42	100	6.3533E+00	1.1436E-04	1,2
258	3505	C4H8O2	isobutyric acid	79-31-2	88.106	427.85	20	3.8412E-02	6.9142E-07	1,2
259	3506	C4H8O2	propyl formate	110-74-7	88.106	353.97	25	2.2621E+01	4.0717E-04	1,2
260	3507	C4H8O2	isopropyl formate	625-55-8	88.106	341.25	25	3.9679E+01	7.1421E-04	1,2
261	3508	C4H8O2	ethyl acetate	141-78-6	88.106	350.21	25	7.4400E+00	1.3392E-04	1,2
262	3509	C4H8O2	methyl propanoate	554-12-1	88.106	352.60	25	8.5817E+00	1.5447E-04	1,2
263	3510	C4H8O2	1,4-dioxane	123-91-1	88.106	374.47	25	3.0932E-01	5.5677E-06	1,2
264	3511	C4H8O2	cis-2-butene-1,4-diol	6117-80-2	88.106	508.15	25	1.1278E-02	2.0300E-07	2
265	3512	C4H8O2	trans-2-butene-1,4-diol	821-11-4	88.106	510.00	25	1.1278E-02	2.0300E-07	2
266	3581	C4H8S	tetrahydrothiophene	110-01-0	88.174	394.27	25	3.1191E+02	5.6144E-03	1,2
267	3598	C4H9Br	1-bromobutane	109-65-9	137.019	374.75	25	4.3347E+02	7.8024E-03	1,2
268	3599	C4H9Br	2-bromobutane	78-76-2	137.019	364.37	25	7.8582E+02	1.4145E-02	1,2
269	3600	C4H9Br	1-bromo-2-methylpropane	78-77-3	137.019	364.68	18	5.5464E+02	9.9835E-03	1,2
270	3601	C4H9Br	2-bromo-2-methylpropane	507-19-7	137.019	346.40	25	6.4058E+02	1.1530E-02	1,2
271	3602	C4H9Br	2-bromobutane, (±)	5787-31-5	137.019	364.45	25	6.4349E+02	1.1583E-02	1,2
272	3610	C4H9Cl	1-chlorobutane	109-69-3	92.568	351.58	25	7.9161E+02	1.4249E-02	1,2
273	3611	C4H9Cl	1-chloro-2-methylpropane	513-36-0	92.568	342.00	20	7.3923E+02	1.3306E-02	1,2
274	3612	C4H9Cl	2-chlorobutane	78-86-4	92.568	341.25	25	1.0427E+03	1.8768E-02	1,2
275	3613	C4H9Cl	2-chloro-2-methylpropane	507-20-0	92.568	323.75	15	4.7260E+02	8.5067E-03	1,2
276	3614	C4H9Cl	2-chlorobutane	53178-20-4	92.568	341.35	25	8.6872E+02	1.5637E-02	1,2
277	3675	C4H9N	pyrrolidine	123-75-1	71.122	359.72	25	1.3278E-01	2.3900E-06	1,2
278	3681	C4H9NO	N,N-dimethylacetamide	127-19-5	87.122	439.25	25	7.2778E-04	1.3100E-08	1,2
279	3682	C4H9NO	morpholine	110-91-8	87.122	401.15	25	6.4445E-02	1.1600E-06	2
280	3783	C4H10	butane	106-97-8	58.123	272.65	25	4.4573E+04	8.0231E-01	1,2
281	3784	C4H10	isobutane	75-28-5	58.123	261.43	25	6.0286E+04	1.0851E+00	1,2
282	3865	C4H10O	butanol	71-36-3	74.123	390.81	25	6.7826E-01	1.2209E-05	1,2
283	3866	C4H10O	isobutanol	78-83-1	74.123	380.81	25	9.0925E-01	1.6366E-05	1,2
284	3867	C4H10O	sec-butanol	78-92-2	74.123	372.70	25	6.3120E-01	1.1362E-05	1,2
285	3868	C4H10O	tert-butanol	75-65-0	74.123	355.57	25	2.2816E+00	4.1069E-05	1,2
286	3869	C4H10O	diethyl ether	60-29-7	74.123	307.58	25	4.5384E+01	8.1690E-04	1,2
287	3870	C4H10O	methyl propyl ether	557-17-5	74.123	311.72	25	7.8978E+01	1.4216E-03	1,2
288	3871	C4H10O	methyl isopropyl ether	598-53-8	74.123	303.92	25	4.8644E+01	8.7558E-04	1,2
289	3872	C4H10O	(R)-(-)-2-butanol	14898-79-4	74.123	372.65	25	7.2535E-01	1.3056E-05	1,2
290	3873	C4H10O	(S)-(+)-2-butanol	4221-99-2	74.123	372.60	25	7.2433E-01	1.3038E-05	1,2
291	3874	C4H10O	(<+->)-2-butanol	15892-23-6	74.123	372.00	25	7.1205E-01	1.2817E-05	1,2
292	3883	C4H10O2	2-ethoxyethanol	110-80-5	90.122	408.15	25	5.5556E-04	1.0000E-08	2
293	3884	C4H10O2	1,2-butanediol	584-03-2	90.122	464.15	25	3.3000E-04	5.9400E-09	2
294	3885	C4H10O2	1,3-butanediol	107-88-0	90.122	480.15	25	1.3222E-04	2.3800E-09	2
295	3886	C4H10O2	1,4-butanediol	110-63-4	90.122	501.15	25	7.2223E-05	1.3000E-09	2
296	3887	C4H10O2	DL-2,3-butanediol	6982-25-8	90.122	453.85	20	5.6049E-04	1.0089E-08	1,2
297	3888	C4H10O2	meso-2,3-butanediol	5341-95-7	90.122	452.00	20	7.9914E-04	1.4384E-08	2
298	3918	C4H10O3	diethylene glycol	111-46-6	106.122	518.15	25	1.1111E-04	2.0000E-09	2
299	3942	C4H10S	butyl mercaptan	109-79-5	90.189	371.61	25	5.0818E+02	9.1472E-03	1,2
300	3943	C4H10S	isobutyl mercaptan	513-44-0	90.189	361.64	25	2.8995E+02	5.2191E-03	1,2
301	3944	C4H10S	sec-butyl mercaptan	513-53-1	90.189	358.13	25	2.8972E+02	5.2150E-03	1,2
302	3945	C4H10S	tert-butyl mercaptan	75-66-1	90.189	337.37	25	2.8348E+02	5.1026E-03	1,2
303	3946	C4H10S	2-butanethiol	91840-99-2	90.189	358.15	25	2.1801E+02	3.9242E-03	1,2
304	3947	C4H10S	diethyl sulfide	352-93-2	90.189	365.25	25	1.2201E+02	2.1962E-03	1,2
305	3948	C4H10S	methyl propyl sulfide	3877-15-4	90.189	368.71	25	1.2139E+02	2.1851E-03	1,2
306	3949	C4H10S	methyl isopropyl sulfide	1551-21-9	90.189	357.90	25	1.2193E+02	2.1947E-03	1,2
307	3977	C4H11N	butylamine	109-73-9	73.138	350.55	25	1.5715E+00	2.8287E-05	1,2
308	3978	C4H11N	isobutylamine	78-81-9	73.138	340.88	20	1.3711E+00	2.4680E-05	1,2
309	3979	C4H11N	sec-butylamine	13952-84-6	73.138	336.15	25	1.3858E+00	2.4945E-05	1,2
310	3980	C4H11N	tert-butylamine	75-64-9	73.138	317.55	25	1.0371E+00	1.8667E-05	1,2
311	3981	C4H11N	2-butanamine, (.+/-.)-	33966-50-6	73.138	335.88	25	1.3049E+00	2.3489E-05	1,2
312	3982	C4H11N	methyl-propylamine	627-35-0	73.138	336.15	25	1.7628E+00	3.1730E-05	1,2
313	3983	C4H11N	methylisopropylamine	4747-21-1	73.138	323.55	25	1.3645E+00	2.4560E-05	1,2
314	3984	C4H11N	diethylamine	109-89-7	73.138	328.60	25	1.8873E+00	3.3972E-05	1,2
315	3985	C4H11N	dimethylethylamine	598-56-1	73.138	310.15	25	4.9890E+00	8.9802E-05	1,2
316	3986	C4H11N	(R)-(-)-sec-butylamine	13250-12-9	73.138	333.15	25	1.2370E+00	2.2267E-05	1,2
317	3987	C4H11N	(S)-(+)-sec-butylamine	513-49-5	73.138	335.10	25	1.2849E+00	2.3129E-05	1,2
318	3988	C4H11NO	dimethylethanolamine	108-01-0	89.138	407.15	25	2.7389E-03	4.9300E-08	2
319	4006	C4H11NO2	diethanolamine	111-42-2	105.137	542.04	25	2.1500E-06	3.8700E-11	2
320	4118	C4H13N3	diethylene triamine	111-40-0	103.168	480.25	25	1.7500E-02	3.1500E-07	2

Table 10 Henry's Law Constant for Compound in Water - Organic Compounds

								Henry's Law Constant, H		
NO	ID	FORMULA	NAME	CAS No	MW, g/mol	T_B, K	T, C	H @ T, atm/mol frac	H @ T, atm/mol/m^3	code
321	4327	C5H4O2	furfural	98-01-1	96.086	434.85	25	1.8120E-01	3.2615E-06	1,2
322	4391	C5H5N	pyridine	110-86-1	79.102	388.41	25	6.1112E-01	1.1000E-05	1,2
323	4548	C5H6O2	furfuryl alcohol	98-00-0	98.101	443.15	25	2.3358E-02	4.2044E-07	1,2
324	4576	C5H6O4	citraconic acid	498-23-7	130.100	607.00	25	5.8766E-09	1.0578E-13	1,2
325	4577	C5H6O4	itaconic acid	97-65-4	130.100	601.00	20	3.9781E-08	7.1606E-13	1,2
326	4587	C5H6S	2-methylthiophene	554-14-3	98.169	385.71	25	2.8992E+02	5.2186E-03	1,2
327	4588	C5H6S	3-methylthiophene	616-44-4	98.169	388.60	25	3.9458E+02	7.1024E-03	1,2
328	4677	C5H7NO2	ethyl cyanoacetate	105-56-6	113.116	479.15	25	3.8325E-02	6.8984E-07	1,2
329	4737	C5H8	cyclopentene	142-29-0	68.119	317.38	25	3.5229E+03	6.3412E-02	1,2
330	4738	C5H8	isoprene	78-79-5	68.119	307.22	25	4.2648E+03	7.6766E-02	1,2
331	4739	C5H8	3-methyl-1,2-butadiene	598-25-4	68.119	314.00	25	5.2296E+03	9.4132E-02	1,2
332	4740	C5H8	1,2-pentadiene	591-95-7	68.119	318.01	25	5.2888E+03	9.5198E-02	1,2
333	4741	C5H8	cis-1,3-pentadiene	1574-41-0	68.119	317.22	25	5.3039E+03	9.5469E-02	1,2
334	4742	C5H8	trans-1,3-pentadiene	2004-70-8	68.119	315.17	25	5.2906E+03	9.5229E-02	1,2
335	4743	C5H8	1,4-pentadiene	591-93-5	68.119	299.11	25	6.5571E+03	1.1803E-01	1,2
336	4744	C5H8	2,3-pentadiene	591-96-8	68.119	321.40	25	5.1997E+03	9.3593E-02	1,2
337	4745	C5H8	1-pentyne	627-19-0	68.119	313.33	25	1.3654E+03	2.4577E-02	1,2
338	4746	C5H8	2-pentyne	627-21-4	68.119	329.22	25	9.8457E+02	1.7722E-02	1,2
339	4747	C5H8	3-methyl-1-butyne	598-23-2	68.119	302.15	25	1.0245E+03	1.8441E-02	1,2
340	4751	C5H8	1,3-pentadiene; (cis+trans)	504-60-9	68.119	315.05	25	4.8398E+03	8.7115E-02	1,2
341	4845	C5H8N4O12	pentaerythritol tetranitrate	78-11-5	316.139	543.00	25	1.6891E-01	3.0404E-06	1,2
342	4849	C5H8O	methyl isopropenyl ketone	814-78-8	84.118	371.15	25	5.7879E+00	1.0418E-04	1,2
343	4882	C5H8O2	acetylacetone	123-54-6	100.117	413.55	20	1.9398E-01	3.4916E-06	1,2
344	4883	C5H8O2	allyl acetate	591-87-7	100.117	377.15	25	1.2896E+01	2.3213E-04	1,2
345	4884	C5H8O2	isopropenyl acetate	108-22-5	100.117	367.15	25	1.4704E+01	2.6467E-04	1,2
346	4885	C5H8O2	vinyl propionate	105-38-4	100.117	364.35	25	1.3982E+01	2.5167E-04	1,2
347	4886	C5H8O2	methyl cis-2-butenoate	4358-59-2	100.117	391.15	25	4.7417E+00	8.5350E-05	1,2
348	4887	C5H8O2	methyl trans-2-butenoate	623-43-8	100.117	394.15	25	4.0975E+00	7.3754E-05	1,2
349	4888	C5H8O2	ethyl acrylate	140-88-5	100.117	372.65	25	1.3788E+01	2.4818E-04	1,2
350	4889	C5H8O2	methyl methacrylate	80-62-6	100.117	373.45	25	1.3017E+01	2.3430E-04	1,2
351	4918	C5H8O2	methyl 3-butenoate	3724-55-8	100.117	385.15	25	6.3303E+00	1.1395E-04	1,2
352	4945	C5H8O4	glutaric acid	110-94-1	132.116	595.54	65	4.4039E-06	7.9269E-11	1,2
353	5245	C5H9NO4	L-glutamic acid	56-86-0	147.131	670.00	25	1.9848E-11	3.5726E-16	1,2
354	5285	C5H10	cyclopentane	287-92-3	70.134	322.40	25	1.0371E+04	1.8668E-01	1,2
355	5286	C5H10	1-pentene	109-67-1	70.134	303.11	25	2.2044E+04	3.9678E-01	1,2
356	5287	C5H10	2-pentene; (cis+trans)	109-68-2	70.134	306.70	25	1.5992E+04	2.8785E-01	1,2
357	5288	C5H10	cis-2-pentene	627-20-3	70.134	310.08	25	1.2473E+04	2.2451E-01	1,2
358	5289	C5H10	trans-2-pentene	646-04-8	70.134	309.49	25	1.2755E+04	2.2959E-01	1,2
359	5290	C5H10	2-methyl-1-butene	563-46-2	70.134	304.30	25	1.6445E+04	2.9600E-01	1,2
360	5291	C5H10	3-methyl-1-butene	563-45-1	70.134	293.21	25	2.9016E+04	5.2228E-01	1,2
361	5292	C5H10	2-methyl-2-butene	513-35-9	70.134	311.71	25	1.5814E+04	2.8465E-01	1,2
362	5297	C5H10Br2	1,1-dibromopentane	13320-56-4	229.942	453.15	25	4.0829E+01	7.3491E-04	1,2
363	5298	C5H10Br2	1,2-dibromopentane	3234-49-9	229.942	457.15	25	1.2506E+01	2.2511E-04	1,2
364	5299	C5H10Br2	1,3-dibromopentane	42474-20-4	229.942	463.15	25	1.1826E+01	2.1286E-04	1,2
365	5300	C5H10Br2	1,4-dibromopentane	626-87-9	229.942	473.15	25	1.0589E+01	1.9061E-04	1,2
366	5301	C5H10Br2	1,5-dibromopentane	111-24-0	229.942	495.45	25	6.8861E+00	1.2395E-04	1,2
367	5302	C5H10Br2	2,2-dibromopentane	54653-26-8	229.942	443.15	25	1.6044E+01	2.8878E-04	1,2
368	5303	C5H10Br2	2,3-dibromopentane	5398-25-4	229.942	451.65	25	1.4042E+01	2.5275E-04	1,2
369	5304	C5H10Br2	2,4-dibromopentane	19398-53-9	229.942	455.15	25	1.3671E+01	2.4608E-04	1,2
370	5305	C5H10Br2	3,3-dibromopentane	54653-27-9	229.942	440.15	25	1.6302E+01	2.9343E-04	1,2
371	5306	C5H10Br2	1,1-dibromo-2-methylbutane	62127-58-6	229.942	445.24	25	3.3512E+01	6.0321E-04	2
372	5307	C5H10Br2	1,1-dibromo-3-methylbutane	62127-59-7	229.942	445.24	25	3.3512E+01	6.0321E-04	2
373	5308	C5H10Br2	1,2-dibromo-2-methylbutane	10428-64-5	229.942	446.15	25	1.5769E+01	2.8385E-04	1,2
374	5309	C5H10Br2	1,2-dibromo-3-methylbutane	10288-13-8	229.942	445.24	25	3.3512E+01	6.0321E-04	2
375	5310	C5H10Br2	1,3-dibromo-2-methylbutane	49623-50-9	229.942	445.24	25	3.3512E+01	6.0321E-04	2
376	5311	C5H10Br2	1,3-dibromo-3-methylbutane	24443-15-0	229.942	442.89	25	3.3433E+01	6.0180E-04	2
377	5312	C5H10Br2	1,4-dibromo-2-methylbutane	69498-28-8	229.942	445.68	25	3.3524E+01	6.0343E-04	2
378	5313	C5H10Br2	2,2-dibromo-3-methylbutane	62127-60-0	229.942	442.45	25	3.3417E+01	6.0150E-04	2
379	5314	C5H10Br2	2,3-dibromo-2-methylbutane	594-51-4	229.942	444.01	25	3.1016E+01	5.5829E-04	1,2
380	5315	C5H10Br2	1,1-dibromo-2,2-dimethylpropane	2443-91-6	229.942	453.15	25	1.2931E+01	2.3276E-04	1,2
381	5316	C5H10Br2	1,3-dibromo-2,2-dimethylpropane	5434-27-5	229.942	456.15	25	1.5769E+01	2.8384E-04	1,2
382	5317	C5H10Br2	1,3-dibromo-2-ethylpropane	---	229.942	442.89	25	3.3433E+01	6.0180E-04	2
383	5330	C5H10Cl2	1,1-dichloropentane	820-55-3	141.039	412.95	25	2.6980E+02	4.8564E-03	1,2
384	5331	C5H10Cl2	1,2-dichloropentane	1674-33-5	141.039	420.15	25	1.0889E+02	1.9600E-03	1,2
385	5332	C5H10Cl2	1,3-dichloropentane	30122-12-4	141.039	428.15	25	1.6776E+02	3.0196E-03	1,2
386	5333	C5H10Cl2	1,4-dichloropentane	626-92-6	141.039	435.15	25	1.6704E+02	3.0066E-03	1,2
387	5334	C5H10Cl2	1,5-dichloropentane	628-76-2	141.039	453.15	25	4.5844E+02	8.2519E-03	1,2
388	5335	C5H10Cl2	2,2-dichloropentane	34887-14-4	141.039	400.15	25	1.8305E+02	3.2950E-03	1,2
389	5336	C5H10Cl2	2,3-dichloropentane	600-11-3	141.039	412.15	25	1.7368E+02	3.1262E-03	1,2
390	5337	C5H10Cl2	2,4-dichloropentane	625-67-2	141.039	422.15	25	1.7719E+02	3.1893E-03	1,2
391	5338	C5H10Cl2	3,3-dichloropentane	21571-91-5	141.039	405.15	25	1.8433E+02	3.3178E-03	1,2
392	5339	C5H10Cl2	1,1-dichloro-2-methylbutane	23010-05-1	141.039	400.00	25	1.9757E+02	3.5563E-03	2
393	5340	C5H10Cl2	1,1-dichloro-3-methylbutane	625-66-1	141.039	402.75	25	1.7447E+02	3.1405E-03	1,2
394	5341	C5H10Cl2	1,2-dichloro-2-methylbutane	23010-04-0	141.039	407.15	25	1.6959E+02	3.0526E-03	1,2
395	5342	C5H10Cl2	1,2-dichloro-3-methylbutane	600-10-2	141.039	416.15	25	1.7653E+02	3.1776E-03	1,2
396	5343	C5H10Cl2	1,3-dichloro-2-methylbutane	23010-07-3	141.039	418.00	25	2.0651E+02	3.7171E-03	2
397	5344	C5H10Cl2	1,3-dichloro-3-methylbutane	624-96-4	141.039	419.85	25	2.0067E+02	3.6121E-03	1,2
398	5345	C5H10Cl2	1,4-dichloro-2-methylbutane	623-34-7	141.039	441.65	25	4.3425E+02	7.8165E-03	1,2
399	5346	C5H10Cl2	2,2-dichloro-3-methylbutane	17773-66-9	141.039	418.00	25	2.0651E+02	3.7171E-03	2
400	5347	C5H10Cl2	2,3-dichloro-2-methylbutane	507-45-9	141.039	402.15	25	1.9068E+02	3.4322E-03	1,2

Table 10 Henry's Law Constant for Compound in Water - Organic Compounds

					Henry's Law Constant, H					
NO	ID	FORMULA	NAME	CAS No	MW, g/mol	T_B, K	T, C	H @ T, atm/mol frac	H @ T, atm/mol/m^3	code
401	5348	C5H10Cl2	1,3-dichloro-2-ethylpropane	---	141.039	418.00	25	2.0651E+02	3.7171E-03	2
402	5349	C5H10Cl2	1,1-dichloro-2,2-dimethylpropane	29559-54-4	141.039	418.00	25	2.0651E+02	3.7171E-03	2
403	5350	C5H10Cl2	1,3-dichloro-2,2-dimethylpropane	29559-55-5	141.039	419.15	25	1.8770E+02	3.3786E-03	1,2
404	5351	C5H10Cl2	dichloropentane	30586-10-8	141.039	403.15	25	1.9906E+02	3.5831E-03	1,2
405	5445	C5H10O	pentanal	110-62-3	86.134	376.15	25	1.7748E+01	3.1946E-04	1,2
406	5446	C5H10O	2-methylbutanal	96-17-3	86.134	365.15	25	1.2218E+01	2.1992E-04	1,2
407	5447	C5H10O	3-methylbutanal	590-86-3	86.134	365.75	25	1.2288E+01	2.2119E-04	1,2
408	5448	C5H10O	2,2-dimethylpropanal	630-19-3	86.134	347.15	25	1.0694E+01	1.9248E-04	1,2
409	5449	C5H10O	2-methylbutanal, (±)	57456-98-1	86.134	365.65	25	1.0512E+01	1.8922E-04	1,2
410	5450	C5H10O	methyl propyl ketone	107-87-9	86.134	375.46	25	4.2662E+00	7.6790E-05	1,2
411	5451	C5H10O	diethyl ketone	96-22-0	86.134	375.14	25	6.6337E+00	1.1941E-04	1,2
412	5452	C5H10O	methyl isopropyl ketone	563-80-4	86.134	367.55	25	4.3467E+00	7.8239E-05	1,2
413	5492	C5H10O2	pentanoic acid	109-52-4	102.133	458.65	25	9.8571E-02	1.7743E-06	1,2
414	5493	C5H10O2	2-methylbutyric acid	116-53-0	102.133	450.15	25	2.0184E-01	3.6331E-06	1,2
415	5494	C5H10O2	3-methylbutanoic acid	503-74-2	102.133	448.25	20	6.1545E-02	1.1078E-06	1,2
416	5495	C5H10O2	2,2-dimethylpropanoic acid	75-98-9	102.133	436.95	25	6.3587E-02	1.1446E-06	1,2
417	5496	C5H10O2	2-methylbutanoic acid	600-07-7	102.133	450.15	25	1.5835E-01	2.8504E-06	1,2
418	5497	C5H10O2	butyl formate	592-84-7	102.133	379.25	25	3.0889E+01	5.5600E-04	1,2
419	5498	C5H10O2	isobutyl formate	542-55-2	102.133	371.22	25	2.6072E+01	4.6930E-04	1,2
420	5499	C5H10O2	sec-butyl formate	589-40-2	102.133	363.55	25	3.3827E+01	6.0888E-04	1,2
421	5500	C5H10O2	tert-butyl formate	762-75-4	102.133	355.95	25	2.8357E+01	5.1042E-04	1,2
422	5501	C5H10O2	propyl acetate	109-60-4	102.133	374.65	25	1.2722E+01	2.2899E-04	1,2
423	5502	C5H10O2	isopropyl acetate	108-21-4	102.133	361.65	25	1.5053E+01	2.7096E-04	1,2
424	5503	C5H10O2	ethyl propanoate	105-37-3	102.133	372.25	25	1.2526E+01	2.2546E-04	1,2
425	5504	C5H10O2	methyl butanoate	623-42-7	102.133	375.90	25	1.6188E+01	2.9139E-04	1,2
426	5505	C5H10O2	methyl isobutanoate	547-63-7	102.133	365.65	25	1.4373E+01	2.5870E-04	1,2
427	5506	C5H10O2	tetrahydrofurfuryl alcohol	97-99-4	102.133	451.15	25	1.0330E-02	1.8594E-07	1,2
428	5530	C5H10O2	(S)-(+)-2-methylbutyric acid	1730-91-2	102.133	409.77	25	1.7517E+03	3.1530E-02	2
429	5543	C5H10O3	diethyl carbonate	105-58-8	118.133	399.95	20	3.1207E+00	5.6172E-05	1,2
430	5544	C5H10O3	ethyl lactate	97-64-3	118.133	427.65	25	3.2389E-02	5.8300E-07	2
431	5603	C5H11Br	1-bromopentane	110-53-2	151.046	402.74	25	1.1669E+03	2.1003E-02	1,2
432	5604	C5H11Br	2-bromopentane	107-81-3	151.046	390.55	25	5.8082E+02	1.0455E-02	1,2
433	5605	C5H11Br	3-bromopentane	1809-10-5	151.046	391.75	25	5.7993E+02	1.0439E-02	1,2
434	5606	C5H11Br	1-bromo-2-methylbutane	10422-35-2	151.046	393.65	25	5.7860E+02	1.0415E-02	1,2
435	5607	C5H11Br	1-bromo-3-methylbutane	107-82-4	151.046	393.55	17	3.2207E+02	5.7972E-03	1,2
436	5608	C5H11Br	2-bromo-2-methylbutane	507-36-8	151.046	381.15	25	6.3718E+02	1.1469E-02	1,2
437	5609	C5H11Br	2-bromo-3-methylbutane	18295-25-5	151.046	388.45	25	6.0608E+02	1.0909E-02	1,2
438	5610	C5H11Br	1-bromo-2,2-dimethylpropane	630-17-1	151.046	379.15	25	6.3851E+02	1.1493E-02	1,2
439	5611	C5H11Br	1-bromo-2-methylbutane, (S)	534-00-9	151.046	394.75	25	7.1700E+02	1.2906E-02	1,2
440	5620	C5H11Cl	1-chloropentane	543-59-9	106.595	381.54	25	1.2269E+03	2.2084E-02	1,2
441	5621	C5H11Cl	2-chloropentane	625-29-6	106.595	369.67	25	1.5681E+03	2.8226E-02	1,2
442	5622	C5H11Cl	3-chloropentane	616-20-6	106.595	370.94	25	1.1170E+03	2.0107E-02	1,2
443	5623	C5H11Cl	1-chloro-2-methylbutane	616-13-7	106.595	373.69	25	1.2658E+03	2.2784E-02	1,2
444	5624	C5H11Cl	1-chloro-3-methylbutane	107-84-6	106.595	371.66	25	1.4093E+03	2.5366E-02	1,2
445	5625	C5H11Cl	2-chloro-2-methylbutane	594-36-5	106.595	358.76	25	1.8260E+03	3.2867E-02	1,2
446	5626	C5H11Cl	2-chloro-3-methylbutane	631-65-2	106.595	365.95	25	9.5403E+02	1.7172E-02	1,2
447	5627	C5H11Cl	1-chloro-2,2-dimethylpropane	753-89-9	106.595	357.45	25	9.9627E+02	1.7933E-02	1,2
448	5628	C5H11Cl	1-chloro-2-methylbutane, (±)	114180-21-1	106.595	373.05	25	9.0492E+02	1.6288E-02	1,2
449	5629	C5H11Cl	2-chloropentane, (+)	29882-57-3	106.595	370.15	25	9.0750E+02	1.6335E-02	1,2
450	5772	C5H12	pentane	109-66-0	72.150	309.22	25	6.9795E+04	1.2563E+00	1,2
451	5773	C5H12	isopentane	78-78-4	72.150	300.99	25	7.5830E+04	1.3649E+00	1,2
452	5774	C5H12	neopentane	463-82-1	72.150	282.65	25	9.3624E+04	1.6852E+00	1,2
453	5831	C5H12O	1-pentanol	71-41-0	88.150	410.95	25	9.4829E-01	1.7069E-05	1,2
454	5832	C5H12O	2-pentanol	6032-29-7	88.150	392.15	25	1.0792E+00	1.9425E-05	1,2
455	5833	C5H12O	3-pentanol	584-02-1	88.150	388.45	25	1.2586E+00	2.2655E-05	1,2
456	5834	C5H12O	2-methyl-1-butanol	137-32-6	88.150	401.85	25	7.4271E-01	1.3369E-05	1,2
457	5835	C5H12O	2-methyl-1-butanol, (±)	34713-94-5	88.150	400.65	25	1.3991E+00	2.5183E-05	1,2
458	5836	C5H12O	3-methyl-2-butanol, (±)	70116-68-6	88.150	386.05	25	1.1531E+00	2.0756E-05	1,2
459	5837	C5H12O	3-methyl-1-butanol	123-51-3	88.150	404.35	25	1.3115E+00	2.3607E-05	1,2
460	5838	C5H12O	tert-pentyl-alcohol	75-85-4	88.150	375.15	25	9.7119E-01	1.7481E-05	1,2
461	5839	C5H12O	3-methyl-2-butanol	598-75-4	88.150	384.65	25	1.5775E+00	2.8396E-05	1,2
462	5840	C5H12O	2,2-dimethyl-1-propanol	75-84-3	88.150	386.25	25	9.4255E-01	1.6966E-05	1,2
463	5841	C5H12O	methyl butyl ether	628-28-4	88.150	343.35	25	1.0693E+02	1.9247E-03	1,2
464	5842	C5H12O	methyl isobutyl ether	625-44-5	88.150	331.70	25	1.0912E+02	1.9641E-03	1,2
465	5843	C5H12O	methyl sec-butyl ether	6795-87-5	88.150	332.15	25	1.1031E+02	1.9857E-03	1,2
466	5844	C5H12O	methyl tert-butyl ether	1634-04-4	88.150	328.35	25	3.0311E+01	5.4560E-04	1,2
467	5845	C5H12O	ethyl propyl ether	628-32-0	88.150	337.01	25	1.1160E+02	2.0087E-03	1,2
468	5846	C5H12O	ethyl isopropyl ether	625-54-7	88.150	326.15	25	9.6260E+01	1.7327E-03	1,2
469	5847	C5H12O	sec-butyl methyl ether	116783-23-4	88.150	332.25	25	9.9036E+01	1.7826E-03	1,2
470	5848	C5H12O	sec-butyl methyl ether (d)	66610-39-7	88.150	333.85	25	1.0047E+02	1.8084E-03	1,2
471	5849	C5H12O	(S)-(-)-2-methyl-1-butanol	1565-80-6	88.150	401.05	25	1.4046E+00	2.5283E-05	1,2
472	5850	C5H12O	(S)-(+)-2-pentanol	26184-62-3	88.150	391.65	25	1.2572E+00	2.2630E-05	1,2
473	5851	C5H12O	pentanol	30899-19-5	88.150	403.15	25	1.4328E+00	2.5791E-05	1,2
474	5852	C5H12O	(R)-(-)-2-pentanol	31087-44-2	88.150	397.10	25	1.3469E+00	2.4244E-05	1,2
475	5853	C5H12O	(S)-3-methyl-2-butanol	1517-66-4	88.150	372.19	25	8.4566E-01	1.5222E-05	2
476	5863	C5H12O2	1,5-pentanediol	111-29-5	104.149	512.15	25	1.7000E-02	3.0600E-07	2
477	5899	C5H12O3	2-(2-methoxyethoxy)ethanol	111-77-3	120.148	466.75	25	9.1667E-07	1.6500E-11	2
478	5912	C5H12O4	pentaerythritol	115-77-5	136.148	631.00	20	2.2353E-11	4.0236E-16	1,2
479	5921	C5H12S	pentyl mercaptan	110-66-7	104.216	399.79	25	4.0304E+02	7.2546E-03	1,2
480	5922	C5H12S	2-pentanethiol	2084-19-7	104.216	385.55	25	3.6620E+02	6.5916E-03	1,2

Table 10 Henry's Law Constant for Compound in Water - Organic Compounds

			Henry's Law Constant, H							
NO	ID	FORMULA	NAME	CAS No	MW, g/mol	T_B, K	T, C	H @ T, atm/mol frac	H @ T, atm/mol/m^3	code
481	5923	C5H12S	3-pentanethiol	616-31-9	104.216	387.05	25	3.7339E+02	6.7209E-03	1,2
482	5924	C5H12S	2-methyl-1-butanethiol	1878-18-8	104.216	392.15	25	3.7979E+02	6.8362E-03	1,2
483	5925	C5H12S	3-methyl-1-butanethiol	541-31-1	104.216	391.50	25	3.7866E+02	6.8159E-03	1,2
484	5926	C5H12S	2-methyl-2-butanethiol	1679-09-0	104.216	372.28	25	3.5643E+02	6.4157E-03	1,2
485	5927	C5H12S	3-methyl-2-butanethiol	2084-18-6	104.216	382.91	25	3.6992E+02	6.6585E-03	1,2
486	5928	C5H12S	2,2-dimethyl-1-propanethiol	1679-08-9	104.216	376.83	25	3.6109E+02	6.4995E-03	1,2
487	5929	C5H12S	2-methyl-1-butanethiol, (+)	20089-07-0	104.216	392.25	25	2.4412E+02	4.3941E-03	1,2
488	5930	C5H12S	methyl butyl sulfide	628-29-5	104.216	396.58	25	1.6825E+02	3.0284E-03	1,2
489	5931	C5H12S	ethyl propyl sulfide	4110-50-3	104.216	391.65	25	1.6120E+02	2.9015E-03	1,2
490	5932	C5H12S	3-methyl-2-thiapentane	10359-64-5	104.216	385.15	25	1.4614E+02	2.6304E-03	1,2
491	5933	C5H12S	4-methyl-2-thiapentane	5008-69-5	104.216	385.65	25	1.4799E+02	2.6637E-03	1,2
492	5934	C5H12S	2-methyl-3-thiapentane	5145-99-3	104.216	380.53	25	1.4052E+02	2.5294E-03	1,2
493	5935	C5H12S	3,3-dimethyl-2-thiabutane	6163-64-0	104.216	372.05	25	1.4344E+02	2.5818E-03	1,2
494	5947	C5H13N	pentylamine	110-58-7	87.165	377.65	25	2.1570E+00	3.8827E-05	1,2
495	5948	C5H13N	1-methylbutylamine	625-30-9	87.165	364.85	25	1.4518E+00	2.6133E-05	1,2
496	5949	C5H13N	2-methylbutylamine	96-15-1	87.165	368.65	25	1.4694E+00	2.6448E-05	1,2
497	5950	C5H13N	3-methylbutylamine	107-85-7	87.165	370.15	25	1.4831E+00	2.6696E-05	1,2
498	5951	C5H13N	1,1-dimethylpropylamine	594-39-8	87.165	350.15	25	1.3382E+00	2.4088E-05	1,2
499	5952	C5H13N	1,2-dimethylpropylamine	598-74-3	87.165	357.15	25	1.4071E+00	2.5328E-05	1,2
500	5953	C5H13N	2,2-dimethylpropylamine	5813-64-9	87.165	350.15	25	1.3504E+00	2.4306E-05	1,2
501	5954	C5H13N	3-pentanamine	616-24-0	87.165	362.15	25	1.6544E+00	2.9779E-05	1,2
502	5955	C5H13N	methyl-butylamine	110-68-9	87.165	364.25	25	2.1783E+00	3.9209E-05	1,2
503	5956	C5H13N	methylisobutylamine	14610-37-8	87.165	350.15	25	2.0515E+00	3.6927E-05	1,2
504	5957	C5H13N	methyl-sec-butylamine	7713-69-1	87.165	351.65	25	2.0667E+00	3.7200E-05	1,2
505	5958	C5H13N	methyl-tert-butylamine	625-43-4	87.165	346.15	25	2.0128E+00	3.6229E-05	1,2
506	5959	C5H13N	ethylpropylamine	20193-20-8	87.165	353.45	25	2.1433E+00	3.8580E-05	1,2
507	5960	C5H13N	ethylisopropylamine	19961-27-4	87.165	349.15	25	2.0356E+00	3.6640E-05	1,2
508	5961	C5H13N	methyldiethylamine	616-39-7	87.165	339.15	25	6.6503E+00	1.1970E-04	1,2
509	5962	C5H13N	dimethyl-propylamine	926-63-6	87.165	339.15	25	7.5972E+00	1.3675E-04	1,2
510	5963	C5H13N	dimethylisopropylamine	996-35-6	87.165	339.15	25	6.7737E+00	1.2193E-04	1,2
511	5964	C5H13N	(S)-(-)-2-methylbutylamine	20626-52-2	87.165	354.02	25	2.1653E+00	3.8975E-05	1,2
512	5965	C5H13N	N,N-dimethyl-2-propanamine	996-35-0	87.165	339.18	25	5.7652E+00	1.0377E-04	1,2
513	6067	C6Cl6	hexachlorobenzene	118-74-1	284.781	582.55	25	6.7359E-02	1.2125E-06	1,2
514	6281	C6H3ClN2O4	1-chloro-2,4-dinitrobenzene	97-00-7	202.554	588.00	15	3.0052E-02	5.4094E-07	1,2
515	6305	C6H3Cl3	1,2,4-trichlorobenzene	120-82-1	181.447	486.15	25	1.6710E+02	3.0077E-03	1,2
516	6306	C6H3Cl3	1,2,3-trichlorobenzene	87-61-6	181.447	491.75	25	1.2030E+02	2.1654E-03	1,2
517	6307	C6H3Cl3	1,3,5-trichlorobenzene	108-70-3	181.447	481.55	25	1.8608E+02	3.3494E-03	1,2
518	6308	C6H3Cl3	trichlorobenzene	12002-48-1	181.447	486.40	25	1.2951E+02	2.3311E-03	1,2
519	6367	C6H3N3O6	1,3,5-trinitrobenzene	99-35-4	213.107	748.00	15	4.3006E-03	7.7409E-08	1,2
520	6411	C6H4Br2	m-dibromobenzene	108-36-1	235.906	491.15	25	4.6456E+01	8.3620E-04	1,2
521	6412	C6H4Br2	o-dibromobenzene	583-53-9	235.906	498.15	25	2.7170E+01	4.8906E-04	1,2
522	6413	C6H4Br2	p-dibromobenzene	106-37-6	235.906	491.65	25	3.8985E+01	7.0173E-04	1,2
523	6414	C6H4Br2	dibromobenzene	26249-12-7	235.906	493.65	25	3.4911E+01	6.2839E-04	2
524	6443	C6H4ClNO2	m-chloronitrobenzene	121-73-3	157.556	508.75	20	9.1573E-01	1.6483E-05	1,2
525	6444	C6H4ClNO2	o-chloronitrobenzene	88-73-3	157.556	519.00	20	3.3600E-01	6.0479E-06	1,2
526	6445	C6H4ClNO2	p-chloronitrobenzene	100-00-5	157.556	515.15	20	6.6103E-01	1.1898E-05	1,2
527	6469	C6H4Cl2	o-dichlorobenzene	95-50-1	147.003	453.57	25	1.7215E+02	3.0987E-03	1,2
528	6470	C6H4Cl2	m-dichlorobenzene	541-73-1	147.003	446.23	25	1.9863E+02	3.5753E-03	1,2
529	6471	C6H4Cl2	p-dichlorobenzene	106-46-7	147.003	447.21	25	2.3474E+02	4.2254E-03	1,2
530	6524	C6H4F2	m-difluorobenzene	372-18-9	114.095	363.66	25	4.3438E+03	7.8188E-02	1,2
531	6525	C6H4F2	o-difluorobenzene	367-11-3	114.095	364.66	25	3.8634E+02	6.9540E-03	1,2
532	6526	C6H4F2	p-difluorobenzene	540-36-3	114.095	362.00	25	4.2514E+02	7.6525E-03	1,2
533	6580	C6H4N2O4	m-dinitrobenzene	99-65-0	168.109	573.00	25	3.4431E-03	6.1976E-08	1,2
534	6581	C6H4N2O4	o-dinitrobenzene	528-29-0	168.109	592.00	25	3.2411E-03	5.8339E-08	1,2
535	6582	C6H4N2O4	p-dinitrobenzene	100-25-4	168.109	572.00	25	1.8108E-02	3.2593E-07	1,2
536	6656	C6H5Br	bromobenzene	108-86-1	157.010	429.24	25	1.2272E+02	2.2090E-03	1,2
537	6692	C6H5Cl	chlorobenzene	108-90-7	112.558	404.87	25	2.7067E+02	4.8721E-03	1,2
538	6720	C6H5ClO	m-chlorophenol	108-43-0	128.558	487.00	20	8.0319E-02	1.4457E-06	1,2
539	6721	C6H5ClO	o-chlorophenol	95-57-8	128.558	447.53	25	2.0249E+00	3.6448E-05	1,2
540	6722	C6H5ClO	p-chlorophenol	106-48-9	128.558	493.11	20	4.8397E-02	8.7115E-07	1,2
541	6769	C6H5F	fluorobenzene	462-06-6	96.104	357.88	25	3.5729E+02	6.4312E-03	1,2
542	6795	C6H5I	iodobenzene	591-50-4	204.010	461.60	25	8.3726E+01	1.5071E-03	1,2
543	6819	C6H5NO2	nitrobenzene	98-95-3	123.112	483.95	25	1.1267E+00	2.0281E-05	1,2
544	6876	C6H6	benzene	71-43-2	78.114	353.24	25	3.0769E+02	5.5385E-03	1,2
545	6923	C6H6ClN	m-chloroaniline	108-42-9	127.573	501.65	20	7.6946E-02	1.3850E-06	1,2
546	6924	C6H6ClN	o-chloroaniline	95-51-2	127.573	481.99	20	3.1995E-01	5.7590E-06	1,2
547	6925	C6H6ClN	p-chloroaniline	106-47-8	127.573	503.65	20	1.3591E-01	2.4463E-06	1,2
548	6996	C6H6N2O2	m-nitroaniline	99-09-2	138.126	579.00	25	9.4353E-03	1.6983E-07	1,2
549	6997	C6H6N2O2	o-nitroaniline	88-74-4	138.126	558.00	25	2.3496E-02	4.2293E-07	1,2
550	6998	C6H6N2O2	p-nitroaniline	100-01-6	138.126	609.15	25	1.8477E-03	3.3258E-08	1,2
551	7000	C6H6N2O2	N-nitroaniline	645-55-6	138.126	557.33	25	1.1427E-02	2.0568E-07	2
552	7057	C6H6O	phenol	108-95-2	94.113	454.99	25	4.0051E-02	7.2092E-07	1,2
553	7071	C6H6O2	resorcinol	108-46-3	110.112	549.65	30	2.2647E-06	4.0764E-11	1,2
554	7072	C6H6O2	p-hydroquinone	123-31-9	110.112	558.15	25	3.5226E-05	6.3407E-10	1,2
555	7085	C6H6O3	1,2,3-benzenetriol	87-66-1	126.112	581.85	25	8.9301E-06	1.6074E-10	1,2
556	7172	C6H7N	aniline	62-53-3	93.129	457.60	25	8.9141E-02	1.6045E-06	1,2
557	7173	C6H7N	2-methylpyridine	109-06-8	93.129	402.55	25	5.5334E-01	9.9600E-06	1,2
558	7174	C6H7N	3-methylpyridine	108-99-6	93.129	417.29	25	4.2945E-01	7.7300E-06	1,2
559	7175	C6H7N	4-methylpyridine	108-89-4	93.129	418.50	25	3.3334E-01	6.0000E-06	1,2
560	7267	C6H8	1,3-cyclohexadiene	592-57-4	80.130	353.49	25	8.0362E+02	1.4465E-02	1,2

Table 10 Henry's Law Constant for Compound in Water - Organic Compounds

					Henry's Law Constant, H					
NO	ID	FORMULA	NAME	CAS No	MW, g/mol	T_B, K	T, C	H @ T, atm/mol frac	H @ T, atm/mol/m³	code
561	7269	C6H8	1,4-cyclohexadiene	628-41-1	80.130	358.65	25	6.5645E+02	1.1816E-02	1,2
562	7318	C6H8N2	m-phenylenediamine	108-45-2	108.144	560.00	25	7.4113E-05	1.3340E-09	1,2
563	7319	C6H8N2	o-phenylenediamine	95-54-5	108.144	525.00	35	4.0096E-03	7.2173E-08	1,2
564	7320	C6H8N2	p-phenylenediamine	106-50-3	108.144	540.00	24	3.5976E-04	6.4756E-09	1,2
565	7464	C6H8O7	citric acid	77-92-9	192.125	659.00	20	2.7527E-16	4.9549E-21	1,2
566	7467	C6H8S	2-ethylthiophene	872-55-9	112.196	407.15	25	2.6332E+02	4.7398E-03	1,2
567	7468	C6H8S	3-ethylthiophene	1795-01-3	112.196	409.15	25	3.7163E+02	6.6892E-03	1,2
568	7469	C6H8S	2,3-dimethylthiophene	632-16-6	112.196	414.75	25	3.8919E+02	7.0053E-03	1,2
569	7470	C6H8S	2,4-dimethylthiophene	638-00-6	112.196	413.85	25	3.8668E+02	6.9602E-03	1,2
570	7471	C6H8S	2,5-dimethylthiophene	638-02-8	112.196	409.85	25	3.7386E+02	6.7294E-03	1,2
571	7472	C6H8S	3,4-dimethylthiophene	632-15-5	112.196	418.15	25	3.9989E+02	7.1979E-03	1,2
572	7622	C6H10	cyclohexene	110-83-8	82.145	356.12	25	2.4675E+03	4.4414E-02	1,2
573	7623	C6H10	1,2-hexadiene	592-44-9	82.145	349.15	25	7.3796E+03	1.3283E-01	1,2
574	7624	C6H10	1,cis-3-hexadiene	14596-92-0	82.145	346.15	25	7.2020E+03	1.2963E-01	1,2
575	7625	C6H10	1,trans-3-hexadiene	20237-34-7	82.145	346.15	25	7.2020E+03	1.2963E-01	1,2
576	7626	C6H10	1,4-hexadiene	592-45-0	82.145	338.15	25	5.6596E+03	1.0187E-01	1,2
577	7627	C6H10	1,cis-4-hexadiene	7318-67-4	82.145	338.15	25	6.7033E+03	1.2066E-01	1,2
578	7628	C6H10	1,trans-4-hexadiene	7319-00-8	82.145	338.15	25	6.7033E+03	1.2066E-01	1,2
579	7629	C6H10	1,5-hexadiene	592-42-7	82.145	332.61	25	7.9179E+03	1.4252E-01	1,2
580	7630	C6H10	2,3-hexadiene	592-49-4	82.145	341.15	25	6.8678E+03	1.2362E-01	1,2
581	7631	C6H10	2,4-hexadiene	592-46-1	82.145	353.15	25	6.0509E+03	1.0892E-01	1,2
582	7632	C6H10	cis-2,trans-4-hexadiene	5194-50-3	82.145	356.65	25	8.4155E+03	1.5148E-01	1,2
583	7633	C6H10	cis-2,cis-4-hexadiene	6108-61-8	82.145	353.15	25	7.8468E+03	1.4124E-01	1,2
584	7634	C6H10	trans-2,trans-4-hexadiene	5194-51-4	82.145	355.05	25	7.8146E+03	1.4066E-01	1,2
585	7635	C6H10	3-methyl-1,2-pentadiene	7417-48-3	82.145	343.15	25	7.0850E+03	1.2753E-01	1,2
586	7636	C6H10	4-methyl-1,2-pentadiene	13643-05-5	82.145	343.15	25	7.1212E+03	1.2818E-01	1,2
587	7637	C6H10	3-methyl-1,3-pentadiene	4549-74-0	82.145	350.15	25	5.9454E+03	1.0702E-01	1,2
588	7638	C6H10	2-methyl-1,cis-3-pentadiene	1501-60-6	82.145	349.15	25	7.6005E+03	1.3681E-01	1,2
589	7639	C6H10	2-methyl-1,trans-3-pentadiene	926-54-5	82.145	349.15	25	7.6005E+03	1.3681E-01	1,2
590	7640	C6H10	3-methyl-1,cis-3-pentadiene	2787-45-3	82.145	350.15	25	7.6892E+03	1.3841E-01	1,2
591	7641	C6H10	3-methyl-1,trans-3-pentadiene	2787-43-1	82.145	350.15	25	7.6892E+03	1.3841E-01	1,2
592	7642	C6H10	4-methyl-1,3-pentadiene	926-56-7	82.145	349.45	25	7.6268E+03	1.3728E-01	1,2
593	7643	C6H10	2-methyl-1,4-pentadiene	763-30-4	82.145	323.15	25	6.2684E+03	1.1283E-01	1,2
594	7644	C6H10	3-methyl-1,4-pentadiene	1115-08-8	82.145	328.15	25	6.3866E+03	1.1496E-01	1,2
595	7645	C6H10	2-methyl-2,3-pentadiene	3043-33-2	82.145	345.15	25	7.2731E+03	1.3091E-01	1,2
596	7646	C6H10	2-ethyl-1,3-butadiene	3404-63-5	82.145	348.15	25	7.4629E+03	1.3433E-01	1,2
597	7647	C6H10	2,3-dimethyl-1,3-butadiene	513-81-5	82.145	341.93	25	7.1621E+03	1.2892E-01	1,2
598	7648	C6H10	1-hexyne	693-02-7	82.145	344.48	25	2.1599E+03	3.8878E-02	1,2
599	7649	C6H10	2-hexyne	764-35-2	82.145	357.67	25	1.5763E+03	2.8373E-02	1,2
600	7650	C6H10	3-hexyne	928-49-4	82.145	354.35	25	1.0639E+03	1.9151E-02	1,2
601	7651	C6H10	3-methyl-1-pentyne	922-59-8	82.145	330.85	25	1.2037E+03	2.1666E-02	1,2
602	7652	C6H10	4-methyl-1-pentyne	7154-75-8	82.145	334.32	25	1.2278E+03	2.2099E-02	1,2
603	7653	C6H10	4-methyl-2-pentyne	21020-27-9	82.145	346.28	25	1.3718E+03	2.4693E-02	1,2
604	7654	C6H10	3,3-dimethyl-1-butyne	917-92-0	82.145	310.87	25	1.1935E+03	2.1482E-02	1,2
605	7657	C6H10	1,3-hexadiene, cis and trans	592-48-3	82.145	346.55	25	5.8363E+03	1.0505E-01	1,2
606	7658	C6H10	2-methyl-1,3-pentadiene	1118-58-7	82.145	348.55	25	5.8945E+03	1.0610E-01	1,2
607	7659	C6H10	methylpentadiene, isomers	54363-49-4	82.145	349.15	25	5.9132E+03	1.0644E-01	1,2
608	7660	C6H10	hexadiene	42296-74-2	82.145	344.64	25	5.7863E+03	1.0415E-01	2
609	7760	C6H10O	cyclohexanone	108-94-1	98.145	428.90	25	2.6746E-01	4.8142E-06	1,2
610	7761	C6H10O	mesityl oxide	141-79-7	98.145	402.95	20	2.1831E+00	3.9296E-05	1,2
611	7881	C6H10O3	ethylacetoacetate	141-97-9	130.144	453.95	17	3.7185E-02	6.6933E-07	1,2
612	7917	C6H10O4	adipic acid	124-04-9	146.143	611.00	34	5.2933E-06	9.5279E-11	1,2
613	7919	C6H10O4	ethylene glycol diacetate	111-55-7	146.143	463.65	22	5.0680E-03	9.1224E-08	1,2
614	8214	C6H12	methylcyclopentane	96-37-7	84.161	344.96	25	1.9994E+04	3.5988E-01	1,2
615	8215	C6H12	cyclohexane	110-82-7	84.161	353.87	25	1.0718E+04	1.9292E-01	1,2
616	8216	C6H12	1-hexene	592-41-6	84.161	336.63	25	2.2963E+04	4.1333E-01	1,2
617	8217	C6H12	2-hexene; (cis+trans)	592-43-8	84.161	341.15	25	1.9232E+04	3.4618E-01	1,2
618	8218	C6H12	cis-2-hexene	7688-21-3	84.161	342.03	25	2.1892E+04	3.9405E-01	1,2
619	8219	C6H12	trans-2-hexene	4050-45-7	84.161	341.02	25	2.1426E+04	3.8567E-01	1,2
620	8220	C6H12	hex-3-ene	592-47-2	84.161	339.95	25	1.9103E+04	3.4384E-01	1,2
621	8221	C6H12	cis-3-hexene	7642-09-3	84.161	339.60	25	2.1058E+04	3.7905E-01	1,2
622	8222	C6H12	trans-3-hexene	13269-52-8	84.161	340.24	25	2.0881E+04	3.7586E-01	1,2
623	8223	C6H12	2-methyl-1-pentene	763-29-1	84.161	335.25	25	1.5717E+04	2.8291E-01	1,2
624	8224	C6H12	3-methyl-1-pentene	760-20-3	84.161	327.33	25	1.9550E+04	3.5190E-01	1,2
625	8225	C6H12	4-methyl-1-pentene	691-37-2	84.161	327.01	25	3.4724E+04	6.2503E-01	1,2
626	8226	C6H12	4-methyl-2-pentene	4461-48-7	84.161	330.35	25	1.8369E+04	3.3065E-01	1,2
627	8227	C6H12	4-methyl-cis-2-pentene	691-38-3	84.161	329.53	25	1.9775E+04	3.5595E-01	1,2
628	8228	C6H12	4-methyl-trans-2-pentene	674-76-0	84.161	331.75	25	1.9980E+04	3.5964E-01	1,2
629	8229	C6H12	2-methyl-2-pentene	625-27-4	84.161	340.45	25	2.1564E+04	3.8814E-01	1,2
630	8230	C6H12	3-methyl-cis-2-pentene	922-62-3	84.161	340.85	25	2.1505E+04	3.8708E-01	1,2
631	8231	C6H12	3-methyl-2-pentene, cis and trans	922-61-2	84.161	341.05	25	1.9221E+04	3.4598E-01	1,2
632	8232	C6H12	3-methyl-trans-2-pentene	616-12-6	84.161	343.60	25	2.1331E+04	3.8395E-01	1,2
633	8233	C6H12	2,3-dimethyl-1-butene	563-78-0	84.161	328.76	25	2.0043E+04	3.6077E-01	1,2
634	8234	C6H12	2,3-dimethyl-2-butene	563-79-1	84.161	346.35	25	2.2429E+04	4.0372E-01	1,2
635	8235	C6H12	3,3-dimethyl-1-butene	558-37-2	84.161	314.40	25	1.9068E+04	3.4321E-01	1,2
636	8236	C6H12	2-ethyl-1-butene	760-21-4	84.161	337.82	25	2.1306E+04	3.8350E-01	1,2
637	8245	C6H12Br2	1,1-dibromohexane	58133-26-9	243.969	475.15	25	4.7027E+01	8.4648E-04	1,2
638	8246	C6H12Br2	1,2-dibromo-2,3-dimethylbutane	29916-45-8	243.969	486.03	25	3.2379E+01	5.8282E-04	2
639	8247	C6H12Br2	1,4-dibromo-2,3-dimethylbutane	54462-70-3	243.969	486.03	25	3.2379E+01	5.8282E-04	2
640	8248	C6H12Br2	2,3-dibromo-2,3-dimethylbutane	594-81-0	243.969	486.03	25	3.2379E+01	5.8282E-04	2

Table 10 Henry's Law Constant for Compound in Water - Organic Compounds

								Henry's Law Constant, H		
NO	ID	FORMULA	NAME	CAS No	MW, g/mol	T_B, K	T, C	H @ T, atm/mol frac	H @ T, atm/mol/m^3	code
641	8249	C6H12Br2	1,2-dibromohexane	624-20-4	243.969	486.03	25	3.2379E+01	5.8282E-04	2
642	8250	C6H12Br2	1,4-dibromohexane	25118-28-9	243.969	486.03	25	3.2379E+01	5.8282E-04	2
643	8251	C6H12Br2	1,5-dibromohexane	627-96-3	243.969	486.03	25	3.2379E+01	5.8282E-04	2
644	8252	C6H12Br2	1,6-dibromohexane	629-03-8	243.969	518.65	25	2.2685E+01	4.0832E-04	1,2
645	8253	C6H12Br2	2,3-dibromohexane	6423-02-5	243.969	469.15	25	3.4906E+01	6.2830E-04	1,2
646	8254	C6H12Br2	2,5-dibromohexane, (R*,R**)-(±)-	54462-68-9	243.969	486.03	25	3.2379E+01	5.8282E-04	2
647	8255	C6H12Br2	1,2-dibromo-3,3-dimethylbutane	640-21-1	243.969	486.03	25	3.2379E+01	5.8282E-04	2
648	8256	C6H12Br2	3,4-dibromohexane	16230-28-7	243.969	486.03	25	3.2379E+01	5.8282E-04	2
649	8257	C6H12Br2	2,5-dibromohexane	24774-58-1	243.969	481.15	25	3.3303E+01	5.9945E-04	1,2
650	8258	C6H12Br2	1,5-dibromo-3-methylpentane	4457-72-1	243.969	486.03	25	3.2379E+01	5.8282E-04	2
651	8273	C6H12Cl2	1,1-dichlorohexane	62017-16-7	155.066	437.15	25	3.3782E+02	6.0806E-03	1,2
652	8274	C6H12Cl2	1,2-dichlorohexane	2162-92-7	155.066	446.15	25	2.3784E+02	4.2811E-03	1,2
653	8275	C6H12Cl2	1,6-dichlorohexane	2163-00-0	155.066	477.15	25	2.2496E+02	4.0493E-03	1,2
654	8276	C6H12Cl2	2,2-dichlorohexane	42131-89-5	155.066	444.73	25	2.3765E+02	4.2777E-03	2
655	8277	C6H12Cl2	2,3-dichlorohexane	54305-87-2	155.066	436.65	25	2.3567E+02	4.2420E-03	1,2
656	8278	C6H12Cl2	2,5-dichlorohexane, (R*,R*)-(±)-	41761-12-0	155.066	450.15	25	2.3807E+02	4.2852E-03	1,2
657	8279	C6H12Cl2	1,1-dichloro-3,3-dimethylbutane	6130-96-7	155.066	421.10	25	2.2872E+02	4.1169E-03	1,2
658	8280	C6H12Cl2	hexane, 2,5-dichloro-, (R*,S*)-	41761-11-9	155.066	444.73	25	2.3765E+02	4.2777E-03	2
659	8281	C6H12Cl2	2,2-dichloro-3,3-dimethylbutane	594-84-3	155.066	444.73	25	2.3765E+02	4.2777E-03	2
660	8400	C6H12O	cyclohexanol	108-93-0	100.161	434.00	25	1.4139E-01	2.5450E-06	1,2
661	8401	C6H12O	hexanal	66-25-1	100.161	401.45	30	1.9307E+01	3.4752E-04	1,2
662	8402	C6H12O	2-methylpentanal	123-15-9	100.161	390.15	25	1.6119E+01	2.9014E-04	1,2
663	8403	C6H12O	3-methylpentanal	15877-57-3	100.161	395.15	25	1.6697E+01	3.0055E-04	1,2
664	8404	C6H12O	4-methylpentanal	1119-16-0	100.161	395.15	25	1.6697E+01	3.0055E-04	1,2
665	8405	C6H12O	2-ethylbutanal	97-96-1	100.161	389.95	25	1.6096E+01	2.8972E-04	1,2
666	8406	C6H12O	2,2-dimethylbutanal	2094-75-9	100.161	377.15	25	1.6044E+01	2.8878E-04	1,2
667	8407	C6H12O	2,3-dimethylbutanal	2109-98-0	100.161	386.15	25	1.6445E+01	2.9601E-04	1,2
668	8408	C6H12O	3,3-dimethylbutanal	2987-16-8	100.161	377.15	25	1.6044E+01	2.8878E-04	1,2
669	8409	C6H12O	2-hexanone	591-78-6	100.161	400.85	25	5.1126E+00	9.2026E-05	1,2
670	8410	C6H12O	3-hexanone	589-38-8	100.161	396.65	25	6.9177E+00	1.2452E-04	1,2
671	8411	C6H12O	3-methyl-2-pentanone	565-61-7	100.161	390.55	25	8.4488E+00	1.5208E-04	1,2
672	8412	C6H12O	4-methyl-2-pentanone	108-10-1	100.161	389.65	25	6.9048E+00	1.2429E-04	1,2
673	8413	C6H12O	ethyl isopropyl ketone	565-69-5	100.161	386.55	25	5.7269E+00	1.0308E-04	1,2
674	8414	C6H12O	3,3-dimethyl-2-butanone	75-97-8	100.161	379.45	25	7.0658E+00	1.2718E-04	1,2
675	8415	C6H12O	3-methyl-2-pentanone, (±)	55156-16-6	100.161	390.65	25	7.0148E+00	1.2627E-04	1,2
676	8453	C6H12O	2,2-dimethyl-3-butanone	---	100.161	394.21	25	7.4739E+00	1.3453E-04	2
677	8460	C6H12O2	hexanoic acid	142-62-1	116.160	478.85	20	2.7709E-02	4.9876E-07	1,2
678	8461	C6H12O2	2-ethyl butyric acid	88-09-5	116.160	466.95	25	1.7013E-01	3.0624E-06	1,2
679	8462	C6H12O2	tert-butylacetic acid	1070-83-3	116.160	463.15	25	1.6619E-01	2.9913E-06	1,2
680	8463	C6H12O2	2,2-dimethylbutanoic acid	595-37-9	116.160	459.15	25	1.7088E-01	3.0758E-06	1,2
681	8464	C6H12O2	2,3-dimethylbutanoic acid	14287-61-7	116.160	464.85	25	1.6409E-01	2.9536E-06	1,2
682	8465	C6H12O2	2-methylpentanoic acid, (±)	22160-39-0	116.160	468.75	25	1.5909E-01	2.8636E-06	1,2
683	8466	C6H12O2	3-methylpentanoic acid, (±)	22160-40-3	116.160	470.65	25	1.5656E-01	2.8181E-06	1,2
684	8467	C6H12O2	4-methylpentanoic acid	646-07-1	116.160	473.65	25	1.5246E-01	2.7443E-06	1,2
685	8468	C6H12O2	pentyl formate	638-49-3	116.160	406.60	25	5.0052E+01	9.0093E-04	1,2
686	8469	C6H12O2	1-methylbutyl formate	58368-66-4	116.160	393.76	25	4.1379E+01	7.4482E-04	2
687	8470	C6H12O2	2-methylbutyl formate	35073-27-9	116.160	393.76	25	4.1379E+01	7.4482E-04	2
688	8471	C6H12O2	isopentyl formate	110-45-2	116.160	397.25	25	3.2308E+01	5.8154E-04	1,2
689	8472	C6H12O2	1-ethylpropyl formate	58368-67-5	116.160	393.76	25	4.1379E+01	7.4482E-04	2
690	8473	C6H12O2	1,1-dimethylpropyl formate	757-88-0	116.160	385.65	25	4.0293E+01	7.2526E-04	1,2
691	8474	C6H12O2	1,2-dimethylpropyl formate	66794-46-5	116.160	393.32	25	4.1288E+01	7.4318E-04	2
692	8475	C6H12O2	2,2-dimethylpropyl formate	23361-67-3	116.160	390.97	25	4.0790E+01	7.3422E-04	2
693	8476	C6H12O2	butyl acetate	123-86-4	116.160	399.15	25	2.4712E+01	4.4481E-04	1,2
694	8477	C6H12O2	isobutyl acetate	110-19-0	116.160	389.80	25	2.4623E+01	4.4321E-04	1,2
695	8478	C6H12O2	sec-butyl acetate	105-46-4	116.160	385.15	25	3.3565E+01	6.0417E-04	1,2
696	8479	C6H12O2	tert-butyl acetate	540-88-5	116.160	369.15	25	1.6684E+01	3.0030E-04	1,2
697	8480	C6H12O2	propyl propanoate	106-36-5	116.160	395.65	25	2.4995E+01	4.4990E-04	1,2
698	8481	C6H12O2	isopropyl propanoate	637-78-5	116.160	383.15	25	2.5205E+01	4.5368E-04	1,2
699	8482	C6H12O2	ethyl butanoate	105-54-4	116.160	394.65	25	1.5088E+01	2.7159E-04	1,2
700	8483	C6H12O2	ethyl isobutanoate	97-62-1	116.160	383.00	25	2.0361E+01	3.6649E-04	1,2
701	8485	C6H12O2	diacetone alcohol	123-42-2	116.160	441.00	25	1.4500E-02	2.6100E-07	2
702	8497	C6H12O2	methyl 2,2-dimethylpropanoate	598-98-1	116.160	374.25	25	1.5810E+01	2.8458E-04	1,2
703	8514	C6H12O2	(±)-3-methylvaleric acid	105-43-1	116.160	469.10	25	1.5863E-01	2.8553E-06	1,2
704	8515	C6H12O2	2-methylvaleric acid	97-61-0	116.160	435.40	25	1.8920E-01	3.4056E-06	1,2
705	8531	C6H12O3	2-ethoxyethyl acetate	111-15-9	132.159	429.45	20	6.4730E-02	1.1651E-06	1,2
706	8533	C6H12O3	paraldehyde	123-63-7	132.159	397.25	30	1.1659E+00	2.0986E-05	1,2
707	8647	C6H13Br	1-bromohexane	111-25-1	165.073	428.45	25	1.8864E+03	3.3954E-02	1,2
708	8648	C6H13Br	2-bromohexane	3377-86-4	165.073	417.05	25	6.1735E+02	1.1112E-02	1,2
709	8649	C6H13Br	3-bromohexane	3377-87-5	165.073	414.45	25	6.1890E+02	1.1140E-02	1,2
710	8650	C6H13Br	1-bromo-2-methylpentane	25346-33-2	165.073	417.15	25	6.1728E+02	1.1111E-02	1,2
711	8651	C6H13Br	1-bromo-3-methylpentane	51116-73-5	165.073	421.15	25	6.1455E+02	1.1062E-02	1,2
712	8652	C6H13Br	1-bromo-4-methylpentane	626-88-0	165.073	420.15	25	6.1528E+02	1.1075E-02	1,2
713	8653	C6H13Br	2-bromo-2-methylpentane	4283-80-1	165.073	415.65	25	1.1079E+03	1.9942E-02	1,2
714	8654	C6H13Br	2-bromo-3-methylpentane	62168-41-6	165.073	401.96	25	7.9331E+02	1.4279E-02	2
715	8655	C6H13Br	2-bromo-4-methylpentane	30310-22-6	165.073	404.15	25	6.5510E+02	1.1792E-02	1,2
716	8656	C6H13Br	3-bromo-2-methylpentane	4283-83-4	165.073	401.96	25	7.9331E+02	1.4279E-02	2
717	8657	C6H13Br	3-bromo-3-methylpentane	25346-31-0	165.073	403.15	25	7.6147E+02	1.3706E-02	1,2
718	8658	C6H13Br	1-bromo-2,2-dimethylbutane	62168-42-7	165.073	410.00	25	8.0557E+02	1.4500E-02	2
719	8659	C6H13Br	1-bromo-2,3-dimethylbutane	30540-31-9	165.073	413.15	25	6.9225E+02	1.2460E-02	1,2
720	8660	C6H13Br	1-bromo-3,3-dimethylbutane	1647-23-0	165.073	410.15	25	6.5365E+02	1.1766E-02	1,2

Table 10 Henry's Law Constant for Compound in Water - Organic Compounds

				Henry's Law Constant, H						
NO	ID	FORMULA	NAME	CAS No	MW, g/mol	T_B, K	T, C	H @ T, atm/mol frac	H @ T, atm/mol/m^3	code
721	8661	C6H13Br	2-bromo-2,3-dimethylbutane	594-52-5	165.073	406.15	25	8.2918E+02	1.4925E-02	1,2
722	8662	C6H13Br	2-bromo-3,3-dimethylbutane	26356-06-9	165.073	405.15	25	7.2449E+02	1.3041E-02	1,2
723	8663	C6H13Br	1-bromo-2-ethylbutane	3814-34-4	165.073	416.45	25	7.2827E+02	1.3109E-02	1,2
724	8669	C6H13Cl	1-chlorohexane	544-10-5	120.622	408.24	25	1.2926E+03	2.3266E-02	1,2
725	8670	C6H13Cl	2-chlorohexane	638-28-8	120.622	397.15	25	1.0402E+03	1.8724E-02	1,2
726	8671	C6H13Cl	3-chlorohexane	2346-81-8	120.622	396.15	25	1.0391E+03	1.8704E-02	1,2
727	8672	C6H13Cl	1-chloro-2-methylpentane	14753-05-0	120.622	393.15	25	1.0362E+03	1.8651E-02	1,2
728	8673	C6H13Cl	1-chloro-3-methylpentane	62016-93-7	120.622	402.15	25	1.0460E+03	1.8828E-02	1,2
729	8674	C6H13Cl	1-chloro-4-methylpentane	62016-94-8	120.622	398.15	25	1.0413E+03	1.8743E-02	1,2
730	8675	C6H13Cl	2-chloro-2-methylpentane	4325-48-8	120.622	384.65	25	1.1174E+03	2.0112E-02	1,2
731	8676	C6H13Cl	2-chloro-3-methylpentane	24319-09-3	120.622	380.00	25	1.0210E+03	1.8377E-02	2
732	8677	C6H13Cl	2-chloro-4-methylpentane	25346-32-1	120.622	386.15	25	1.0721E+03	1.9297E-02	1,2
733	8678	C6H13Cl	3-chloro-2-methylpentane	38384-05-3	120.622	385.00	25	1.0217E+03	1.8391E-02	2
734	8679	C6H13Cl	3-chloro-3-methylpentane	918-84-3	120.622	389.15	25	1.1246E+03	2.0243E-02	1,2
735	8680	C6H13Cl	1-chloro-2-ethylbutane	4737-41-1	120.622	399.15	25	1.0424E+03	1.8764E-02	1,2
736	8681	C6H13Cl	1-chloro-2,2-dimethylbutane	6366-35-4	120.622	390.15	25	1.1264E+03	2.0276E-02	1,2
737	8682	C6H13Cl	1-chloro-2,3-dimethylbutane	600-06-6	120.622	395.15	25	1.0835E+03	1.9502E-02	1,2
738	8683	C6H13Cl	1-chloro-3,3-dimethylbutane	2855-08-5	120.622	390.15	25	1.1264E+03	2.0276E-02	1,2
739	8684	C6H13Cl	2-chloro-2,3-dimethylbutane	594-57-0	120.622	385.15	25	1.1599E+03	2.0877E-02	1,2
740	8685	C6H13Cl	2-chloro-3,3-dimethylbutane	5750-00-5	120.622	384.15	25	1.1579E+03	2.0842E-02	1,2
741	8877	C6H14	hexane	110-54-3	86.177	341.88	25	9.8105E+04	1.7659E+00	1,2
742	8878	C6H14	2,2-dimethylbutane	75-83-2	86.177	322.88	25	8.3516E+04	1.5033E+00	1,2
743	8879	C6H14	2,3-dimethylbutane	79-29-8	86.177	331.13	25	7.1416E+04	1.2855E+00	1,2
744	8880	C6H14	2-methylpentane	107-83-5	86.177	333.41	25	9.7338E+04	1.7521E+00	1,2
745	8881	C6H14	3-methylpentane	96-14-0	86.177	336.42	25	6.6946E+04	1.2050E+00	1,2
746	8970	C6H14O	1-hexanol	111-27-3	102.177	430.15	25	1.6143E+00	2.9058E-05	1,2
747	8971	C6H14O	2-hexanol	626-93-7	102.177	413.04	25	1.7097E+00	3.0774E-05	1,2
748	8972	C6H14O	3-hexanol	623-37-0	102.177	408.55	25	2.3908E+00	4.3033E-05	1,2
749	8973	C6H14O	2-methyl-1-pentanol	105-30-6	102.177	421.15	25	2.1031E+00	3.7856E-05	1,2
750	8974	C6H14O	3-methyl-1-pentanol	589-35-5	102.177	425.55	25	1.3716E+00	2.4688E-05	1,2
751	8975	C6H14O	4-methyl-1-pentanol	626-89-1	102.177	424.95	25	1.1258E+00	2.0264E-05	1,2
752	8976	C6H14O	4-methyl-2-pentanol	108-11-2	102.177	404.85	27	3.5795E+00	6.4430E-05	1,2
753	8977	C6H14O	2-methyl-2-pentanol	590-36-3	102.177	394.56	25	1.3450E+00	2.4209E-05	1,2
754	8978	C6H14O	3-methyl-2-pentanol	565-60-6	102.177	407.36	25	1.4684E+00	2.6432E-05	1,2
755	8979	C6H14O	2-methyl-3-pentanol	565-67-3	102.177	399.66	25	1.5037E+00	2.7066E-05	1,2
756	8980	C6H14O	3-methyl-3-pentanol	77-74-7	102.177	395.56	25	1.9165E+00	3.4497E-05	1,2
757	8981	C6H14O	2-ethyl-1-butanol	97-95-0	102.177	419.65	25	9.7730E-01	1.7591E-05	1,2
758	8982	C6H14O	2,2-dimethyl-1-butanol	1185-33-7	102.177	409.95	25	1.7177E+00	3.0917E-05	1,2
759	8983	C6H14O	2,3-dimethyl-1-butanol	19550-30-2	102.177	422.15	25	1.7397E+00	3.1315E-05	1,2
760	8984	C6H14O	2,3-dimethyl-1-butanol, (±)-	20281-85-0	102.177	417.65	25	1.8319E+00	3.2975E-05	1,2
761	8985	C6H14O	3,3-dimethyl-1-butanol	624-95-3	102.177	416.15	25	1.7332E+00	3.1198E-05	1,2
762	8986	C6H14O	2,3-dimethyl-2-butanol	594-60-5	102.177	391.75	25	1.4912E+00	2.6842E-05	1,2
763	8987	C6H14O	3,3-dimethyl-2-butanol	464-07-3	102.177	393.15	25	1.3591E+00	2.4464E-05	1,2
764	8988	C6H14O	3,3-dimethyl-2-butanol, (±)	20281-91-8	102.177	393.55	25	1.4934E+00	2.6881E-05	1,2
765	8989	C6H14O	(RS)-2-hexanol	20281-86-1	102.177	413.15	25	1.7884E+00	3.2191E-05	1,2
766	8990	C6H14O	2-hexanol, (R)	26549-24-6	102.177	410.15	25	1.7545E+00	3.1582E-05	1,2
767	8991	C6H14O	3-hexanol, (R)	13471-42-6	102.177	408.15	25	1.7297E+00	3.1135E-05	1,2
768	8992	C6H14O	3-methyl-1-pentanol, (±)	20281-83-8	102.177	426.15	25	1.8918E+00	3.4053E-05	1,2
769	8993	C6H14O	dipropyl ether	111-43-3	102.177	362.79	25	1.2196E+02	2.1952E-03	1,2
770	8994	C6H14O	propyl isopropyl ether	627-08-7	102.177	353.15	25	1.4716E+02	2.6489E-03	1,2
771	8995	C6H14O	diisopropyl ether	108-20-3	102.177	341.45	25	1.0117E+02	1.8210E-03	1,2
772	8996	C6H14O	methyl pentyl ether	628-80-8	102.177	372.15	25	1.7022E+02	3.0640E-03	1,2
773	8997	C6H14O	methyl 1-methylbutyl ether	6795-88-6	102.177	364.15	25	1.6403E+02	2.9526E-03	1,2
774	8998	C6H14O	methyl 2-methylbutyl ether	62016-48-2	102.177	363.15	25	1.6245E+02	2.9240E-03	1,2
775	8999	C6H14O	methyl 3-methylbutyl ether	626-91-5	102.177	363.15	25	1.6245E+02	2.9240E-03	1,2
776	9000	C6H14O	methyl 1-ethylpropyl ether	36839-67-5	102.177	361.15	25	1.5931E+02	2.8676E-03	1,2
777	9001	C6H14O	methyl 1,2-dimethylpropyl ether	62016-49-3	102.177	356.15	25	1.5632E+02	2.8138E-03	1,2
778	9002	C6H14O	methyl tert-pentyl ether	994-05-8	102.177	359.45	25	1.8824E+02	3.3883E-03	1,2
779	9003	C6H14O	methyl 2,2-dimethylpropyl ether	1118-00-9	102.177	356.15	25	1.6160E+02	2.9088E-03	1,2
780	9004	C6H14O	ethyl butyl ether	628-81-9	102.177	365.35	25	1.7839E+02	3.2110E-03	1,2
781	9005	C6H14O	ethyl isobutyl ether	627-02-1	102.177	354.25	25	1.4880E+02	2.6783E-03	1,2
782	9006	C6H14O	ethyl sec-butyl ether	2679-87-0	102.177	354.35	25	1.4895E+02	2.6810E-03	1,2
783	9007	C6H14O	ethyl tert-butyl ether	637-92-3	102.177	345.95	25	1.4407E+02	2.5933E-03	1,2
784	9008	C6H14O	(S)-(+)-2-hexanol	52019-78-0	102.177	411.65	25	1.7720E+00	3.1895E-05	1,2
785	9009	C6H14O	tert-hexyl alcohol	26401-20-7	102.177	390.55	25	1.4327E+00	2.5789E-05	2
786	9020	C6H14O2	acetal	105-57-7	118.176	376.75	25	5.8980E+00	1.0616E-04	1,2
787	9021	C6H14O2	2-butoxyethanol	111-76-2	118.176	444.47	25	1.1556E-03	2.0800E-08	2
788	9077	C6H14O3	2-(2-ethoxyethoxy)ethanol	111-90-0	134.175	475.15	25	1.2389E-03	2.2300E-08	2
789	9092	C6H14O4	triethylene glycol	112-27-6	150.175	551.00	25	1.4500E-05	2.6100E-10	2
790	9100	C6H14O6	sorbitol	50-70-4	182.174	777.00	25	4.0334E-08	7.2600E-13	2
791	9107	C6H14S	hexyl mercaptan	111-31-9	118.243	425.81	25	5.7389E+02	1.0330E-02	1,2
792	9108	C6H14S	2-hexanethiol	1679-06-7	118.243	412.05	25	4.5769E+02	8.2383E-03	1,2
793	9109	C6H14S	2,3-dimethyl-2-butanethiol	1639-01-6	118.243	407.26	25	2.8099E+02	5.0578E-03	2
794	9110	C6H14S	3-hexanethiol	1633-90-5	118.243	407.26	25	2.8099E+02	5.0578E-03	2
795	9111	C6H14S	2-methyl-2-pentanethiol	1633-97-2	118.243	407.26	25	2.8099E+02	5.0578E-03	2
796	9112	C6H14S	dipropyl sulfide	111-47-7	118.243	416.00	25	2.2404E+02	4.0327E-03	1,2
797	9113	C6H14S	diisopropyl sulfide	625-80-9	118.243	393.19	25	1.9900E+02	3.5821E-03	1,2
798	9114	C6H14S	methyl pentyl sulfide	1741-83-9	118.243	401.16	25	1.3898E+02	2.5017E-03	1,2
799	9115	C6H14S	ethyl butyl sulfide	638-46-0	118.243	417.41	25	2.2224E+02	4.0002E-03	1,2
800	9116	C6H14S	3-methyl-2-thiahexane	13286-91-4	118.243	412.15	25	1.8132E+02	3.2637E-03	1,2

Table 10 Henry's Law Constant for Compound in Water - Organic Compounds

				Henry's Law Constant, H						
NO	ID	FORMULA	NAME	CAS No	MW, g/mol	T_B, K	T, C	H @ T, atm/mol frac	H @ T, atm/mol/m^3	code
801	9117	C6H14S	4-methyl-2-thiahexane	15013-37-3	118.243	413.15	25	1.8330E+02	3.2994E-03	1,2
802	9118	C6H14S	5-methyl-2-thiahexane	13286-90-3	118.243	411.15	25	1.7930E+02	3.2273E-03	1,2
803	9119	C6H14S	2-methyl-3-thiahexane	5008-73-1	118.243	405.20	25	1.8621E+02	3.3518E-03	1,2
804	9120	C6H14S	4-methyl-3-thiahexane	5008-72-0	118.243	406.80	25	1.9247E+02	3.4644E-03	1,2
805	9121	C6H14S	5-methyl-3-thiahexane	1613-45-2	118.243	407.37	25	1.9029E+02	3.4251E-03	1,2
806	9122	C6H14S	3-ethyl-2-thiapentane	57093-84-2	118.243	405.02	25	1.6099E+02	2.8979E-03	2
807	9123	C6H14S	3,3-dimethyl-2-thiapentane	13286-92-5	118.243	401.15	25	1.8768E+02	3.3782E-03	1,2
808	9124	C6H14S	3,4-dimethyl-2-thiapentane	53897-51-1	118.243	401.15	25	1.7679E+02	3.1821E-03	1,2
809	9125	C6H14S	4,4-dimethyl-2-thiapentane	6079-57-8	118.243	401.15	25	1.8768E+02	3.3782E-03	1,2
810	9126	C6H14S	2,2-dimethyl-3-thiapentane	14290-92-7	118.243	393.56	25	1.8216E+02	3.2789E-03	1,2
811	9173	C6H15N	hexylamine	111-26-2	101.192	404.65	25	2.7214E+00	4.8985E-05	1,2
812	9174	C6H15N	2-aminohexane	5329-79-3	101.192	403.15	25	1.6463E+00	2.9633E-05	1,2
813	9175	C6H15N	3-aminohexane	16751-58-9	101.192	408.77	25	1.6553E+00	2.9795E-05	2
814	9176	C6H15N	1-amino-2-methylpentane	13364-16-4	101.192	400.00	25	1.6382E+00	2.9487E-05	2
815	9177	C6H15N	1-amino-3-methylpentane	42245-37-4	101.192	395.15	25	1.4697E+00	2.6455E-05	1,2
816	9178	C6H15N	1-amino-4-methylpentane	5344-20-7	101.192	397.05	25	1.6287E+00	2.9317E-05	1,2
817	9179	C6H15N	2-amino-2-methylpentane	53310-02-4	101.192	377.15	25	1.6470E+00	2.9645E-05	1,2
818	9180	C6H15N	2-amino-3-methylpentane	35399-81-6	101.192	408.33	25	1.7429E+00	3.1372E-05	2
819	9181	C6H15N	2-amino-4-methylpentane	108-09-8	101.192	381.65	25	1.6143E+00	2.9057E-05	1,2
820	9182	C6H15N	3-amino-2-methylpentane	54287-41-1	101.192	408.33	25	1.7429E+00	3.1372E-05	2
821	9183	C6H15N	3-amino-3-methylpentane	3495-46-3	101.192	398.45	25	1.8059E+00	3.2506E-05	1,2
822	9184	C6H15N	1-amino-2,2-dimethylbutane	41781-17-3	101.192	386.65	25	1.7283E+00	3.1109E-05	1,2
823	9185	C6H15N	1-amino-2,3-dimethylbutane	66553-05-7	101.192	378.15	25	1.5874E+00	2.8573E-05	1,2
824	9186	C6H15N	1-amino-3,3-dimethylbutane	15673-00-4	101.192	405.98	25	1.8395E+00	3.3112E-05	2
825	9187	C6H15N	2-amino-2,3-dimethylbutane	4358-75-2	101.192	377.65	25	1.7862E+00	3.2152E-05	1,2
826	9188	C6H15N	2-amino-3,3-dimethyl butane	3850-30-4	101.192	375.15	25	1.6862E+00	3.0352E-05	1,2
827	9189	C6H15N	1-amino-2-ethylbutane	617-79-8	101.192	408.77	25	1.6553E+00	2.9795E-05	2
828	9190	C6H15N	2-hexanamine, (±)	68107-05-1	101.192	390.65	25	2.0734E+00	3.7321E-05	1,2
829	9191	C6H15N	methylpentylamine	25419-06-1	101.192	390.55	25	2.4712E+00	4.4481E-05	1,2
830	9192	C6H15N	methyl-1-methylbutylamine	51932-19-5	101.192	386.41	25	2.5844E+00	4.6519E-05	2
831	9193	C6H15N	methyl-2-methylbutylamine	90023-96-4	101.192	386.41	25	2.5844E+00	4.6519E-05	2
832	9194	C6H15N	methyl-3-methylbutylamine	52317-98-3	101.192	381.15	25	2.5235E+00	4.5422E-05	1,2
833	9195	C6H15N	methyl-1,1-dimethylpropylamine	---	101.192	383.62	25	2.7791E+00	5.0023E-05	2
834	9196	C6H15N	methyl-1,2-dimethylpropylamine	34317-39-0	101.192	386.41	25	2.5844E+00	4.6519E-05	2
835	9197	C6H15N	methyl-2,2-dimethylpropylamine	26153-91-3	101.192	383.62	25	2.7791E+00	5.0023E-05	2
836	9198	C6H15N	ethylbutylamine	13360-63-9	101.192	381.15	25	2.4252E+00	4.3654E-05	1,2
837	9199	C6H15N	ethyl-sec-butylamine	21035-44-9	101.192	371.15	25	2.3896E+00	4.3013E-05	1,2
838	9200	C6H15N	ethylisobutylamine	13205-60-2	101.192	371.15	25	2.3896E+00	4.3013E-05	1,2
839	9201	C6H15N	ethyl-tert-butylamine	4432-77-3	101.192	351.15	25	2.1834E+00	3.9302E-05	1,2
840	9202	C6H15N	di-propylamine	142-84-7	101.192	382.00	36	3.0106E+00	5.4190E-05	1,2
841	9203	C6H15N	diisopropylamine	108-18-9	101.192	357.05	25	2.3053E+00	4.1495E-05	1,2
842	9204	C6H15N	propylisopropylamine	21968-17-2	101.192	371.45	25	2.3939E+00	4.3091E-05	1,2
843	9205	C6H15N	N-ethyl-2-butanamine, (±)	116724-10-8	101.192	371.15	25	2.0976E+00	3.7756E-05	1,2
844	9206	C6H15N	methylethylpropylamine	4458-32-6	101.192	364.65	25	8.2528E+00	1.4855E-04	1,2
845	9207	C6H15N	methylethylisopropylamine	39198-07-7	101.192	364.65	25	8.5090E+00	1.5316E-04	1,2
846	9208	C6H15N	dimethylbutylamine	927-62-8	101.192	367.15	25	1.0141E+01	1.8254E-04	1,2
847	9209	C6H15N	dimethyl-sec-butylamine	921-04-0	101.192	367.15	25	8.6730E+00	1.5611E-04	1,2
848	9210	C6H15N	dimethylisobutylamine	7239-24-9	101.192	354.15	25	7.7940E+00	1.4029E-04	1,2
849	9211	C6H15N	dimethyl-tert-butylamine	918-02-5	101.192	363.15	25	8.9464E+00	1.6103E-04	1,2
850	9212	C6H15N	triethylamine	121-44-8	101.192	361.92	25	1.1522E+01	2.0739E-04	1,2
851	9230	C6H15NO2	diisopropanolamine	110-97-4	133.191	521.90	25	3.8389E-06	6.9100E-11	2
852	9239	C6H15NO3	triethanolamine	102-71-6	149.190	613.00	25	3.9167E-08	7.0500E-13	2
853	9268	C6H15O4P	triethyl phosphate	78-40-0	182.157	488.15	25	3.9036E-03	7.0264E-08	1,2
854	9292	C6H16N2	hexamethylenediamine	124-09-4	116.207	475.04	25	1.7833E-04	3.2100E-09	2
855	9359	C6H18N4	triethylene tetramine	112-24-3	146.237	539.65	25	9.2223E-07	1.6600E-11	2
856	9932	C7H5ClO2	o-chlorobenzoic acid	118-91-2	156.568	560.15	25	1.5973E-03	2.8752E-08	1,2
857	10003	C7H5F3	benzotrifluoride	98-08-8	146.112	375.20	25	9.5197E+02	1.7135E-02	1,2
858	10045	C7H5N	benzonitrile	100-47-0	103.124	464.15	100	3.1613E+01	5.6903E-04	1,2
859	10049	C7H5NO	phenyl isocyanate	103-71-9	119.123	438.75	20	2.1881E+04	3.9386E-01	1,2
860	10118	C7H5N3O6	2,4,6-trinitrotoluene	118-96-7	227.134	573.00	20	4.5882E-03	8.2588E-08	1,2
861	10363	C7H6N2O4	2,4-dinitrotoluene	121-14-2	182.136	590.00	22	3.7597E-01	6.7674E-06	1,2
862	10404	C7H6O	benzaldehyde	100-52-7	106.124	451.90	25	1.5220E+00	2.7395E-05	1,2
863	10408	C7H6O2	benzoic acid	65-85-0	122.123	522.40	25	3.8472E-03	6.9250E-08	1,2
864	10409	C7H6O2	p-hydroxybenzaldehyde	123-08-0	122.123	583.15	30	7.0155E-04	1.2628E-08	1,2
865	10410	C7H6O2	salicylaldehyde	90-02-8	122.123	469.65	86	8.1330E+00	1.4639E-04	1,2
866	10420	C7H6O3	salicylic acid	69-72-7	138.123	529.00	25	1.4944E-03	2.6898E-08	1,2
867	10492	C7H7Cl	benzyl chloride	100-44-7	126.585	452.55	25	1.3282E+02	2.3907E-03	1,2
868	10493	C7H7Cl	o-chlorotoluene	95-49-8	126.585	432.30	25	3.1559E+02	5.6805E-03	1,2
869	10494	C7H7Cl	p-chlorotoluene	106-43-4	126.585	435.65	25	2.4057E+02	4.3302E-03	1,2
870	10495	C7H7Cl	m-chlorotoluene	108-41-8	126.585	435.45	25	2.6226E+02	4.7206E-03	1,2
871	10626	C7H7NO2	m-nitrotoluene	99-08-1	137.138	505.00	30	5.8053E+00	1.0449E-04	1,2
872	10628	C7H7NO2	p-nitrotoluene	99-99-0	137.138	511.65	15	2.1216E+01	3.8188E-04	1,2
873	10724	C7H8	1,3,5-cycloheptatriene	544-25-2	92.141	388.65	25	2.4823E+02	4.4682E-03	1,2
874	10725	C7H8	toluene	108-88-3	92.141	383.78	25	3.5934E+02	6.4680E-03	1,2
875	10850	C7H8O	anisole	100-66-3	108.140	426.73	25	2.1296E+02	3.8333E-03	1,2
876	10851	C7H8O	benzyl alcohol	100-51-6	108.140	477.85	25	1.4854E-02	2.6737E-07	1,2
877	10852	C7H8O	m-cresol	108-39-4	108.140	475.43	20	2.5529E-02	4.5952E-07	1,2
878	10853	C7H8O	o-cresol	95-48-7	108.140	464.15	20	6.2138E-02	1.1185E-06	1,2
879	10854	C7H8O	p-cresol	106-44-5	108.140	475.13	20	2.5487E-02	4.5876E-07	1,2
880	10855	C7H8O	cresol mixture	1319-77-3	108.140	470.15	25	7.3485E-02	1.3227E-06	1,2

Table 10 Henry's Law Constant for Compound in Water - Organic Compounds

					Henry's Law Constant, H					
NO	ID	FORMULA	NAME	CAS No	MW, g/mol	T_B, K	T, C	H @ T, atm/mol frac	H @ T, atm/mol/m^3	code
881	10866	C7H8O2	guaiacol	90-05-1	124.139	478.15	37	4.0704E+00	7.3267E-05	1,2
882	10867	C7H8O2	p-methoxyphenol	150-76-5	124.139	516.00	37	7.3039E-01	1.3147E-05	1,2
883	10953	C7H9N	benzylamine	100-46-9	107.155	457.65	25	1.9403E+00	3.4925E-05	1,2
884	10954	C7H9N	N-methylaniline	100-61-8	107.155	469.02	25	6.3060E-01	1.1351E-05	1,2
885	10955	C7H9N	m-toluidine	108-44-1	107.155	476.55	20	1.2462E-01	2.2431E-06	1,2
886	10956	C7H9N	o-toluidine	95-53-4	107.155	473.55	25	1.3314E-01	2.3966E-06	1,2
887	10957	C7H9N	p-toluidine	106-49-0	107.155	473.40	15	1.9361E-01	3.4849E-06	1,2
888	11225	C7H10S	2-propylthiophene	1551-27-5	126.222	431.65	25	4.6883E+02	8.4390E-03	1,2
889	11226	C7H10S	3-propylthiophene	1518-75-8	126.222	434.15	25	4.7836E+02	8.6104E-03	1,2
890	11227	C7H10S	2-isopropylthiophene	4095-22-1	126.222	426.15	25	4.4852E+02	8.0733E-03	1,2
891	11228	C7H10S	3-isopropylthiophene	29488-27-5	126.222	430.15	25	4.9809E+02	8.9656E-03	1,2
892	11229	C7H10S	2-methyl-3-ethylthiophene	53119-51-0	126.222	430.15	25	4.6890E+02	8.4401E-03	1,2
893	11230	C7H10S	2-methyl-4-ethyl thiophene	13678-54-1	126.222	436.15	25	4.9237E+02	8.8625E-03	1,2
894	11231	C7H10S	2-methyl-5-ethyl thiophene	40323-88-4	126.222	433.25	25	4.8090E+02	8.6562E-03	1,2
895	11232	C7H10S	3-methyl-2-ethylthiophene	31805-48-8	126.222	434.15	25	4.8444E+02	8.7198E-03	1,2
896	11233	C7H10S	3-methyl-4-ethyl thiophene	---	126.222	438.65	25	4.0718E+02	7.3293E-03	1,2
897	11234	C7H10S	3-methyl-5-ethylthiophene	---	126.222	427.35	25	3.9370E+02	7.0866E-03	2
898	11235	C7H10S	2,3,4-trimethylthiophene	1795-04-6	126.222	445.85	25	4.1514E+02	7.4725E-03	1,2
899	11236	C7H10S	2,3,5-trimethylthiophene	1795-05-7	126.222	437.65	25	5.0482E+02	9.0867E-03	1,2
900	11340	C7H12	1-heptyne	628-71-7	96.172	372.86	25	3.8437E+03	6.9186E-02	1,2
901	11341	C7H12	2-heptyne	1119-65-9	96.172	385.15	25	1.1149E+03	2.0068E-02	1,2
902	11342	C7H12	3-heptyne	2586-89-2	96.172	380.31	25	2.4836E+03	4.4704E-02	1,2
903	11343	C7H12	3-methyl-1-hexyne	40276-93-5	96.172	358.15	25	1.7874E+03	3.2172E-02	1,2
904	11344	C7H12	4-methyl-1-hexyne	52713-81-2	96.172	364.15	25	1.9386E+03	3.4894E-02	1,2
905	11345	C7H12	5-methyl-1-hexyne	2203-80-7	96.172	365.00	25	1.9624E+03	3.5323E-02	1,2
906	11346	C7H12	4-methyl-2-hexyne	20198-49-6	96.172	372.69	25	2.2366E+03	4.0259E-02	1,2
907	11347	C7H12	5-methyl-2-hexyne	53566-37-3	96.172	375.61	25	2.3466E+03	4.2238E-02	1,2
908	11348	C7H12	2-methyl-3-hexyne	36566-80-0	96.172	368.35	25	2.1264E+03	3.8274E-02	1,2
909	11349	C7H12	3-ethyl-1-pentyne	21020-26-8	96.172	357.15	25	1.7644E+03	3.1759E-02	1,2
910	11350	C7H12	3,3-dimethyl-1-pentyne	918-82-1	96.172	343.15	25	1.5749E+03	2.8348E-02	1,2
911	11351	C7H12	3,4-dimethyl-1-pentyne	61064-08-2	96.172	353.15	25	1.7105E+03	3.0789E-02	1,2
912	11352	C7H12	4,4-dimethyl-1-pentyne	13361-63-2	96.172	349.23	25	1.6843E+03	3.0318E-02	1,2
913	11353	C7H12	4,4-dimethyl-2-pentyne	999-78-0	96.172	356.15	25	1.8364E+03	3.3055E-02	1,2
914	11358	C7H12	1,2-heptadiene	2384-90-9	96.172	378.65	25	8.6992E+03	1.5659E-01	1,2
915	11359	C7H12	1,4-heptadiene	5675-22-9	96.172	366.15	25	7.7886E+03	1.4019E-01	1,2
916	11360	C7H12	1,5-heptadiene	1541-23-7	96.172	367.15	25	7.8536E+03	1.4136E-01	1,2
917	11361	C7H12	1,6-heptadiene	3070-53-9	96.172	370.15	25	8.0569E+03	1.4502E-01	2
918	11362	C7H12	2,4-heptadiene	628-72-8	96.172	381.15	25	8.9048E+03	1.6028E-01	1,2
919	11363	C7H12	2-methyl-1,5-hexadiene	4049-81-4	96.172	361.25	25	7.4919E+03	1.3485E-01	1,2
920	11364	C7H12	2-methyl-2,4-hexadiene	28823-41-8	96.172	384.65	25	9.2035E+03	1.6566E-01	1,2
921	11365	C7H12	2,4-dimethyl-1,3-pentadiene	1000-86-8	96.172	366.35	25	7.8015E+03	1.4043E-01	1,2
922	11369	C7H12	2,4-dimethyl-2,3-pentadiene	1000-87-9	96.172	358.55	25	7.3442E+03	1.3219E-01	1,2
923	11450	C7H12O2	butyl acrylate	141-32-2	128.171	421.00	20	2.4459E+01	4.4026E-04	1,2
924	11451	C7H12O2	isobutyl acrylate	106-63-8	128.171	405.15	23	3.6931E+01	6.6474E-04	1,2
925	11452	C7H12O2	propyl methacrylate	2210-28-8	128.171	414.00	25	3.6181E+01	6.5126E-04	1,2
926	11453	C7H12O2	isopropyl methacrylate	4655-34-9	128.171	398.15	25	4.8618E+01	8.7512E-04	1,2
927	11454	C7H12O2	allyl butanoate	2051-78-7	128.171	415.15	25	2.0662E+01	3.7192E-04	1,2
928	11480	C7H12O2	ethyl 3-methyl-2-butenoate	638-10-8	128.171	426.65	25	1.1351E+01	2.0431E-04	1,2
929	11481	C7H12O2	ethyl trans-2-methyl-2-butenoate	5837-78-5	128.171	429.15	25	9.9426E+00	1.7897E-04	1,2
930	11483	C7H12O2	methyl 3-hexenoate	2396-78-3	128.171	447.69	25	3.6286E+00	6.5315E-05	2
931	11484	C7H12O2	methyl 2-methyl-2-pentenoate	10478-12-3	128.171	447.69	25	3.6286E+00	6.5315E-05	2
932	11485	C7H12O2	methyl 3-methyl-3-pentenoate	2258-58-4	128.171	447.69	25	3.6286E+00	6.5315E-05	2
933	11486	C7H12O2	methyl cis-3-methyl-2-pentenoate	17447-00-6	128.171	447.69	25	3.6286E+00	6.5315E-05	2
934	11487	C7H12O2	methyl trans-3-methyl-2-pentenoate	17447-01-7	128.171	424.15	25	1.2948E+01	2.3306E-04	1,2
935	11488	C7H12O2	tert-butyl acrylate	1663-39-4	128.171	447.69	25	3.6286E+00	6.5315E-05	2
936	11491	C7H12O2	methyl trans-2-methyl-2-pentenoate	1567-14-2	128.171	447.69	25	3.6286E+00	6.5315E-05	2
937	11492	C7H12O2	4-pentenyl acetate	1576-85-8	128.171	418.15	25	1.7702E+01	3.1863E-04	1,2
938	11497	C7H12O2	3,3-dimethylallyl acetate	1191-16-8	128.171	447.69	25	3.6286E+00	6.5315E-05	2
939	11500	C7H12O2	cis-3-hexenyl formate	33467-73-1	128.171	447.69	25	3.6286E+00	6.5315E-05	2
940	11530	C7H12O4	diethyl malonate	105-53-3	160.170	472.05	37	3.4567E-01	6.2221E-06	1,2
941	11696	C7H14	ethylcyclopentane	1640-89-7	98.188	376.62	25	3.0970E+04	5.5745E-01	1,2
942	11697	C7H14	1,1-dimethylcyclopentane	1638-26-2	98.188	361.00	25	2.3170E+04	4.1705E-01	1,2
943	11698	C7H14	1,2-dimethylcyclopentane	2452-99-5	98.188	370.00	25	3.2742E+04	5.8936E-01	1,2
944	11699	C7H14	cis-1,2-dimethylcyclopentane	1192-18-3	98.188	372.68	25	2.7901E+04	5.0221E-01	1,2
945	11700	C7H14	trans-1,2-dimethylcyclopentane	822-50-4	98.188	365.02	25	2.5047E+04	4.5084E-01	1,2
946	11701	C7H14	1,3-dimethylcyclopentane	2453-00-1	98.188	365.10	25	3.0218E+04	5.4392E-01	1,2
947	11702	C7H14	cis-1,3-dimethylcyclopentane	2532-58-3	98.188	363.92	25	2.5402E+04	4.5724E-01	1,2
948	11703	C7H14	trans-1,3-dimethylcyclopentane	1759-58-6	98.188	364.88	25	2.5025E+04	4.5045E-01	1,2
949	11704	C7H14	methylcyclohexane	108-87-2	98.188	374.08	25	2.3780E+04	4.2804E-01	1,2
950	11705	C7H14	cycloheptane	291-64-5	98.188	391.94	25	5.0882E+03	9.1587E-02	1,2
951	11706	C7H14	1-heptene	592-76-7	98.188	366.79	25	2.2531E+04	4.0555E-01	1,2
952	11707	C7H14	2-heptene	592-77-8	98.188	371.40	25	2.9378E+04	5.2881E-01	1,2
953	11708	C7H14	cis-2-heptene	6443-92-1	98.188	371.56	25	3.8927E+04	7.0068E-01	1,2
954	11709	C7H14	trans-2-heptene	14686-13-6	98.188	371.10	25	3.6089E+04	6.4961E-01	1,2
955	11710	C7H14	3-heptene	592-78-9	98.188	369.00	25	2.8612E+04	5.1502E-01	1,2
956	11711	C7H14	3-heptene (mixed isomers)	25339-56-4	98.188	363.69	25	2.7048E+04	4.8685E-01	2
957	11712	C7H14	cis-3-heptene	7642-10-6	98.188	368.90	25	3.6342E+04	6.5415E-01	1,2
958	11713	C7H14	trans-3-heptene	14686-14-7	98.188	368.82	25	3.5816E+04	6.4468E-01	1,2
959	11714	C7H14	2-methyl-1-hexene	6094-02-6	98.188	364.99	25	3.3441E+04	6.0193E-01	1,2
960	11715	C7H14	3-methyl-1-hexene	3404-61-3	98.188	357.05	25	2.9387E+04	5.2896E-01	1,2

Table 10 Henry's Law Constant for Compound in Water - Organic Compounds

								Henry's Law Constant, H		
NO	ID	FORMULA	NAME	CAS No	MW, g/mol	T_B, K	T, C	H @ T, atm/mol frac	H @ T, atm/mol/m^3	code
961	11716	C7H14	4-methyl-1-hexene	3769-23-1	98.188	359.88	25	3.1701E+04	5.7061E-01	1,2
962	11717	C7H14	5-methyl-1-hexene	3524-73-0	98.188	358.46	25	2.9095E+04	5.2371E-01	1,2
963	11718	C7H14	2-methyl-2-hexene	2738-19-4	98.188	368.56	25	3.4554E+04	6.2197E-01	1,2
964	11719	C7H14	3-methyl-cis-2-hexene	10574-36-4	98.188	370.41	25	3.5760E+04	6.4367E-01	1,2
965	11720	C7H14	3-methyl-trans-2-hexene	20710-38-7	98.188	368.33	25	3.4410E+04	6.1938E-01	1,2
966	11721	C7H14	4-methyl-cis-2-hexene	3683-19-0	98.188	359.46	25	2.9622E+04	5.3319E-01	1,2
967	11722	C7H14	4-methyl-trans-2-hexene	3683-22-5	98.188	360.71	25	3.0217E+04	5.4391E-01	1,2
968	11723	C7H14	5-methyl-cis-2-hexene	13151-17-2	98.188	362.65	25	3.1195E+04	5.6150E-01	1,2
969	11724	C7H14	5-methyl-trans-2-hexene	7385-82-2	98.188	361.26	25	3.0490E+04	5.4882E-01	1,2
970	11725	C7H14	2-methyl-cis-3-hexene	15840-60-5	98.188	359.15	25	2.9480E+04	5.3063E-01	1,2
971	11726	C7H14	2-methyl-trans-3-hexene	692-24-0	98.188	359.05	25	2.9432E+04	5.2977E-01	1,2
972	11727	C7H14	3-methyl-cis-3-hexene	4914-89-0	98.188	368.55	25	3.2811E+04	5.9060E-01	1,2
973	11728	C7H14	3-methyl-trans-3-hexene	3899-36-3	98.188	366.69	25	3.2509E+04	5.8516E-01	1,2
974	11729	C7H14	2-ethyl-1-pentene	3404-71-5	98.188	367.15	25	3.3444E+04	6.0199E-01	1,2
975	11730	C7H14	3-ethyl-1-pentene	4038-04-4	98.188	357.26	25	2.9791E+04	5.3623E-01	1,2
976	11731	C7H14	2,3-dimethyl-1-pentene	3404-72-6	98.188	357.43	25	2.8846E+04	5.1923E-01	1,2
977	11732	C7H14	2,4-dimethyl-1-pentene	2213-32-3	98.188	354.76	25	2.8916E+04	5.2048E-01	1,2
978	11733	C7H14	3,3-dimethyl-1-pentene	3404-73-7	98.188	350.63	25	2.6486E+04	4.7674E-01	1,2
979	11734	C7H14	3,4-dimethyl-1-pentene	7385-78-6	98.188	353.95	25	2.7452E+04	4.9414E-01	1,2
980	11735	C7H14	4,4-dimethyl-1-pentene	762-62-9	98.188	345.67	25	2.6741E+04	4.8133E-01	1,2
981	11736	C7H14	3-ethyl-2-pentene	816-79-5	98.188	369.16	25	3.4934E+04	6.2881E-01	1,2
982	11737	C7H14	2,3-dimethyl-2-pentene	10574-37-5	98.188	370.55	25	3.6090E+04	6.4962E-01	1,2
983	11738	C7H14	2,4-dimethyl-2-pentene	625-65-0	98.188	356.45	25	2.8392E+04	5.1106E-01	1,2
984	11739	C7H14	3,4-dimethyl-cis-2-pentene	4914-91-4	98.188	362.40	25	3.1284E+04	5.6310E-01	1,2
985	11740	C7H14	3,4-dimethyl-trans-2-pentene	4914-92-5	98.188	364.65	25	3.2496E+04	5.8491E-01	1,2
986	11741	C7H14	4,4-dimethyl-cis-2-pentene	762-63-0	98.188	353.58	25	2.9199E+04	5.2558E-01	1,2
987	11742	C7H14	4,4-dimethyl-trans-2-pentene	690-08-4	98.188	349.89	25	2.7040E+04	4.8671E-01	1,2
988	11743	C7H14	3-methyl-2-ethyl-1-butene	7357-93-9	98.188	359.52	25	2.9949E+04	5.3908E-01	1,2
989	11744	C7H14	2,3,3-trimethyl-1-butene	594-56-9	98.188	351.04	25	2.8617E+04	5.1510E-01	1,2
990	11745	C7H14	3,4-dimethylpent-2-ene	24910-63-2	98.188	360.80	25	2.6270E+04	4.7286E-01	1,2
991	11751	C7H14Br2	1,1-dibromoheptane	59104-79-9	257.996	495.15	25	4.7571E+01	8.5627E-04	1,2
992	11752	C7H14Br2	1,2-dibromoheptane	42474-21-5	257.996	501.15	25	3.0195E+01	5.4350E-04	1,2
993	11753	C7H14Br2	1,5-dibromoheptane	1622-10-2	257.996	510.82	25	2.6919E+01	4.8455E-04	2
994	11754	C7H14Br2	1,7-dibromoheptane	4549-31-9	257.996	536.15	25	1.7106E+01	3.0790E-04	1,2
995	11755	C7H14Br2	2,3-dibromoheptane	21266-88-6	257.996	510.82	25	2.6919E+01	4.8455E-04	2
996	11756	C7H14Br2	3,4-dibromoheptane	21266-90-0	257.996	510.82	25	2.6919E+01	4.8455E-04	2
997	11763	C7H14Cl2	1,1-dichloroheptane	821-25-0	169.093	460.15	25	3.7629E+02	6.7733E-03	1,2
998	11764	C7H14Cl2	1,2-dichloro-4,4-dimethylpentane	6065-90-3	169.093	448.15	25	2.5954E+02	4.6717E-03	1,2
999	11765	C7H14Cl2	1,5-dichloro-3,3-dimethylpentane	62496-53-1	169.093	454.15	25	2.5933E+02	4.6679E-03	2
1000	11766	C7H14Cl2	2,4-dichloro-2,4-dimethylpentane	33553-93-4	169.093	454.15	25	2.5933E+02	4.6679E-03	2
1001	11767	C7H14Cl2	1,2-dichloroheptane	10575-87-8	169.093	454.15	25	2.5933E+02	4.6679E-03	2
1002	11768	C7H14Cl2	1,7-dichloroheptane	821-76-1	169.093	454.15	25	2.5933E+02	4.6679E-03	2
1003	11769	C7H14Cl2	2,2-dichloroheptane	65786-09-6	169.093	454.15	25	2.5933E+02	4.6679E-03	2
1004	11770	C7H14Cl2	4,4-dichloroheptane	89796-76-9	169.093	454.15	25	2.5933E+02	4.6679E-03	1,2
1005	11827	C7H14O	1-heptanal	111-71-7	114.188	425.95	30	2.5120E+01	4.5216E-04	1,2
1006	11828	C7H14O	3-methylhexanal	19269-28-4	114.188	416.15	25	1.4985E+01	2.6972E-04	1,2
1007	11829	C7H14O	2-heptanone	110-43-0	114.188	424.05	25	7.7328E+00	1.3919E-04	1,2
1008	11830	C7H14O	3-heptanone	106-35-4	114.188	420.55	25	1.1177E+01	2.0119E-04	1,2
1009	11831	C7H14O	4-heptanone	123-19-3	114.188	417.15	25	1.7559E+01	3.1605E-04	1,2
1010	11832	C7H14O	3-methyl-2-hexanone	2550-21-2	114.188	413.15	25	2.0710E+01	3.7278E-04	1,2
1011	11833	C7H14O	4-methyl-2-hexanone	105-42-0	114.188	412.15	25	1.1545E+01	2.0782E-04	1,2
1012	11834	C7H14O	5-methyl-2-hexanone	110-12-3	114.188	417.95	25	1.4353E+01	2.5835E-04	1,2
1013	11835	C7H14O	2-methyl-3-hexanone	7379-12-6	114.188	406.15	25	1.1699E+01	2.1058E-04	1,2
1014	11836	C7H14O	4-methyl-3-hexanone	17042-16-9	114.188	409.15	25	1.8870E+01	3.3965E-04	1,2
1015	11837	C7H14O	5-methyl-3-hexanone	623-56-3	114.188	409.15	25	1.8870E+01	3.3965E-04	1,2
1016	11838	C7H14O	3-ethyl-2-pentanone	6137-03-7	114.188	411.15	25	1.9775E+01	3.5594E-04	1,2
1017	11839	C7H14O	3,3-dimethyl-2-pentanone	20669-04-9	114.188	403.75	25	1.8606E+01	3.3491E-04	1,2
1018	11840	C7H14O	3,4-dimethyl-2-pentanone	565-78-6	114.188	405.15	25	1.8140E+01	3.2652E-04	1,2
1019	11841	C7H14O	4,4-dimethyl-2-pentanone	590-50-1	114.188	398.15	25	1.6065E+01	2.8917E-04	1,2
1020	11842	C7H14O	2,2-dimethyl-3-pentanone	564-04-5	114.188	398.15	25	1.6065E+01	2.8917E-04	1,2
1021	11843	C7H14O	2,4-dimethyl-3-pentanone	565-80-0	114.188	397.55	25	2.0079E+01	3.6141E-04	1,2
1022	11844	C7H14O	1-methylcyclohexanol	590-67-0	114.188	430.15	25	2.4238E-01	4.3628E-06	1,2
1023	11845	C7H14O	cis-2-methylcyclohexanol	7443-70-1	114.188	438.15	25	3.2883E-01	5.9188E-06	1,2
1024	11846	C7H14O	trans-2-methylcyclohexanol	7443-52-9	114.188	439.65	25	2.7516E-01	4.9529E-06	1,2
1025	11847	C7H14O	cis-3-methylcyclohexanol	5454-79-5	114.188	441.15	25	1.3913E-01	2.5043E-06	1,2
1026	11848	C7H14O	trans-3-methylcyclohexanol	7443-55-2	114.188	441.15	25	1.4251E-01	2.5652E-06	1,2
1027	11849	C7H14O	cis-4-methylcyclohexanol	7731-28-4	114.188	444.15	25	7.2286E-02	1.3011E-06	1,2
1028	11850	C7H14O	trans-4-methylcyclohexanol	7731-29-5	114.188	444.15	20	5.3488E-02	9.6278E-07	1,2
1029	11851	C7H14O	trans-2-methylcyclohexanol, (±)	615-39-4	114.188	440.65	25	1.6210E-01	2.9178E-06	1,2
1030	11877	C7H14O	methylcyclohexanol	25639-42-3	114.188	440.65	25	1.6210E-01	2.9178E-06	1,2
1031	11878	C7H14O	2-methylcyclohexanol; (cis+trans)	583-59-5	114.188	438.90	25	1.7910E-01	3.2238E-06	1,2
1032	11879	C7H14O	4-methylcyclohexanol; (cis+trans)	589-91-3	114.188	445.15	25	1.2520E-01	2.2537E-06	1,2
1033	11880	C7H14O	3-methylcyclohexanol; (cis+trans)	591-23-1	114.188	440.25	25	1.6585E-01	2.9852E-06	1,2
1034	11884	C7H14O2	heptanoic acid	111-14-8	130.187	496.15	20	1.3843E-02	2.4917E-07	1,2
1035	11885	C7H14O2	2-ethyl-2-methylbutanoic acid	19889-37-3	130.187	480.15	25	1.6048E-01	2.8886E-06	1,2
1036	11886	C7H14O2	2-ethylpentanoic acid, (±)	116908-83-9	130.187	482.35	25	1.5686E-01	2.8235E-06	1,2
1037	11887	C7H14O2	2-methylhexanoic acid, (±)	22160-12-9	130.187	488.65	25	1.4630E-01	2.6334E-06	1,2
1038	11888	C7H14O2	3-methylhexanoic acid	3780-58-3	130.187	486.15	25	1.5052E-01	2.7094E-06	1,2
1039	11889	C7H14O2	4-methylhexanoic acid, (±)	22160-41-4	130.187	490.65	25	1.4290E-01	2.5722E-06	1,2
1040	11890	C7H14O2	5-methylhexanoic acid	628-46-6	130.187	489.15	25	1.4545E-01	2.6181E-06	1,2

Table 10 Henry's Law Constant for Compound in Water - Organic Compounds

				Henry's Law Constant, H						
NO	ID	FORMULA	NAME	CAS No	MW, g/mol	T_B, K	T, C	H @ T, atm/mol frac	H @ T, atm/mol/m^3	code
1041	11891	C7H14O2	hexyl formate	629-33-4	130.187	428.65	25	4.6148E+01	8.3066E-04	1,2
1042	11892	C7H14O2	pentyl acetate	628-63-7	130.187	422.15	25	3.5248E+01	6.3446E-04	1,2
1043	11893	C7H14O2	1-methylbutyl acetate	626-38-0	130.187	406.15	25	1.6706E+01	3.0070E-04	1,2
1044	11894	C7H14O2	2-methylbutyl acetate	624-41-9	130.187	413.15	25	1.6952E+01	3.0514E-04	1,2
1045	11895	C7H14O2	isopentyl acetate	123-92-2	130.187	415.25	25	3.2285E+01	5.8112E-04	1,2
1046	11896	C7H14O2	1-ethylpropyl acetate	620-11-1	130.187	405.15	25	1.6513E+01	2.9723E-04	1,2
1047	11897	C7H14O2	tert-pentyl acetate	625-16-1	130.187	397.15	25	1.7153E+01	3.0875E-04	1,2
1048	11898	C7H14O2	1,2-dimethylpropyl acetate	5343-96-4	130.187	401.65	25	1.7098E+01	3.0776E-04	1,2
1049	11899	C7H14O2	neopentyl acetate	926-41-0	130.187	401.15	25	1.8567E+01	3.3420E-04	1,2
1050	11900	C7H14O2	sec-pentyl acetate (R)	54638-10-7	130.187	415.15	25	2.1545E+01	3.8781E-04	1,2
1051	11901	C7H14O2	butyl propanoate	590-01-2	130.187	419.75	25	3.2085E+01	5.7753E-04	1,2
1052	11902	C7H14O2	isobutyl propanoate	540-42-1	130.187	409.75	25	1.6216E+01	2.9188E-04	1,2
1053	11903	C7H14O2	sec-butyl propanoate	591-34-4	130.187	406.15	25	2.2434E+01	4.0381E-04	1,2
1054	11904	C7H14O2	tert-butyl propanoate	20487-40-5	130.187	414.55	25	2.3342E+01	4.2016E-04	2
1055	11905	C7H14O2	propyl butanoate	105-66-8	130.187	416.45	25	2.0911E+01	3.7639E-04	1,2
1056	11906	C7H14O2	isopropyl butanoate	638-11-9	130.187	404.15	25	1.6756E+01	3.0160E-04	1,2
1057	11907	C7H14O2	isopropyl isobutanoate	617-50-5	130.187	393.85	25	2.7363E+01	4.9253E-04	1,2
1058	11908	C7H14O2	propyl isobutanoate	644-49-5	130.187	408.55	25	2.1233E+01	3.8219E-04	1,2
1059	11910	C7H14O2	ethyl 2,2-dimethylpropanoate	3938-95-2	130.187	391.55	25	2.0485E+01	3.6873E-04	1,2
1060	11911	C7H14O2	ethyl 2-methylbutanoate, (+)	10307-61-6	130.187	405.15	25	2.3598E+01	4.2475E-04	1,2
1061	11913	C7H14O2	methyl 2-ethylbutanoate	816-11-5	130.187	409.15	25	2.4134E+01	4.3442E-04	1,2
1062	11925	C7H14O2	2-methylhexanoic acid	4536-23-6	130.187	482.65	25	1.5637E-01	2.8146E-06	1,2
1063	11926	C7H14O2	neoheptanoic acid	33113-10-9	130.187	427.81	25	8.6527E+01	1.5575E-02	2
1064	11931	C7H14O3	ethyl-3-ethoxypropionate	763-69-9	146.186	438.15	25	4.0652E+01	7.3172E-04	1,2
1065	11982	C7H15Br	1-bromoheptane	629-04-9	179.100	452.05	25	2.4877E+03	4.4778E-02	1,2
1066	11983	C7H15Br	2-bromoheptane	1974-04-5	179.100	439.15	25	9.1371E+02	1.6447E-02	1,2
1067	11984	C7H15Br	4-bromoheptane	998-93-6	179.100	434.15	25	9.0976E+02	1.6376E-02	1,2
1068	11988	C7H15Cl	1-chloroheptane	629-06-1	134.649	433.59	25	2.1303E+03	3.8345E-02	1,2
1069	11989	C7H15Cl	2-chloro-2,3-dimethylpentane	59889-45-1	134.649	413.42	25	1.1762E+03	2.1171E-02	2
1070	11990	C7H15Cl	2-chloro-2,4-dimethylpentane	35951-33-8	134.649	401.15	25	1.1568E+03	2.0822E-02	1,2
1071	11991	C7H15Cl	3-chloro-2,3-dimethylpentane	595-38-0	134.649	401.15	25	1.1568E+03	2.0822E-02	2
1072	11992	C7H15Cl	4-chloro-2,2-dimethylpentane	33429-72-0	134.649	401.15	25	1.1568E+03	2.0822E-02	1,2
1073	11993	C7H15Cl	3-chloro-3-ethylpentane	994-25-2	134.649	416.65	25	1.1813E+03	2.1264E-02	1,2
1074	11994	C7H15Cl	2-chloroheptane	1001-89-4	134.649	413.42	25	1.1762E+03	2.1171E-02	2
1075	11995	C7H15Cl	3-chloroheptane	999-52-0	134.649	417.15	25	1.1821E+03	2.1278E-02	1,2
1076	11996	C7H15Cl	4-chloroheptane	998-95-8	134.649	417.15	25	1.1821E+03	2.1278E-02	1,2
1077	11997	C7H15Cl	1-chloro-3-methylhexane	101257-63-0	134.649	424.15	25	1.1923E+03	2.1462E-02	1,2
1078	11998	C7H15Cl	2-chloro-2-methylhexane	4398-65-6	134.649	408.15	25	1.1676E+03	2.1016E-02	2
1079	11999	C7H15Cl	2-chloro-5-methylhexane	58766-17-9	134.649	411.15	25	1.1725E+03	2.1104E-02	2
1080	12000	C7H15Cl	3-chloro-3-methylhexane	43197-78-0	134.649	408.15	25	1.1676E+03	2.1016E-02	1,2
1081	12001	C7H15Cl	2-chloro-2,3,3-trimethylbutane	918-07-0	134.649	413.42	25	1.1762E+03	2.1171E-02	2
1082	12112	C7H16	heptane	142-82-5	100.204	371.58	25	1.4634E+05	2.6341E+00	1,2
1083	12113	C7H16	2-methylhexane	591-76-4	100.204	363.20	25	1.8965E+05	3.4136E+00	1,2
1084	12114	C7H16	3-methylhexane	589-34-4	100.204	365.00	25	1.7294E+05	3.1130E+00	1,2
1085	12115	C7H16	3-ethylpentane	617-78-7	100.204	366.62	25	1.4988E+05	2.6978E+00	1,2
1086	12116	C7H16	2,2-dimethylpentane	590-35-2	100.204	352.34	25	1.7718E+05	3.1891E+00	1,2
1087	12117	C7H16	2,3-dimethylpentane	565-59-3	100.204	362.93	25	9.7025E+04	1.7464E+00	1,2
1088	12118	C7H16	2,4-dimethylpentane	108-08-7	100.204	353.64	25	1.6359E+05	2.9446E+00	1,2
1089	12119	C7H16	3,3-dimethylpentane	562-49-2	100.204	359.21	25	1.0345E+05	1.8622E+00	1,2
1090	12120	C7H16	2,2,3-trimethylbutane	464-06-2	100.204	354.03	25	1.2868E+05	2.3163E+00	1,2
1091	12121	C7H16	3-methylhexane, (S)-	6131-24-4	100.204	365.15	25	1.1736E+05	2.1124E+00	1,2
1092	12164	C7H16O	1-heptanol	111-70-6	116.203	449.45	25	1.5406E+00	2.7731E-05	1,2
1093	12165	C7H16O	2-heptanol	543-49-7	116.203	432.35	30	4.4936E+00	8.0884E-05	1,2
1094	12166	C7H16O	3-heptanol	589-82-2	116.203	429.85	25	1.8710E+00	3.3677E-05	1,2
1095	12167	C7H16O	4-heptanol	589-55-9	116.203	427.85	25	2.2895E+00	4.1211E-05	1,2
1096	12168	C7H16O	2-methyl-1-hexanol	624-22-6	116.203	436.15	25	1.5877E+00	2.8578E-05	1,2
1097	12169	C7H16O	3-methyl-1-hexanol	13231-81-7	116.203	445.15	25	9.0371E-01	1.6267E-05	1,2
1098	12170	C7H16O	4-methyl-1-hexanol	818-49-5	116.203	446.15	25	1.0115E+00	1.8206E-05	1,2
1099	12171	C7H16O	5-methyl-1-hexanol	627-98-5	116.203	445.15	25	1.9380E+00	3.4883E-05	1,2
1100	12172	C7H16O	2-methyl-2-hexanol	625-23-0	116.203	415.95	25	1.9963E+00	3.5934E-05	1,2
1101	12173	C7H16O	3-methyl-2-hexanol	2313-65-7	116.203	425.15	25	3.5098E+00	6.3175E-05	1,2
1102	12174	C7H16O	4-methyl-2-hexanol	2313-61-3	116.203	424.15	25	7.8517E-01	1.4133E-05	1,2
1103	12175	C7H16O	5-methyl-2-hexanol	627-59-8	116.203	424.15	25	2.5737E+00	4.6327E-05	1,2
1104	12176	C7H16O	2-methyl-3-hexanol	617-29-8	116.203	418.15	25	1.7032E+00	3.0657E-05	1,2
1105	12177	C7H16O	3-methyl-3-hexanol	597-96-6	116.203	415.95	25	2.1805E+00	3.9249E-05	1,2
1106	12178	C7H16O	4-methyl-3-hexanol	615-29-2	116.203	422.15	25	4.9088E+00	8.8358E-05	1,2
1107	12179	C7H16O	5-methyl-3-hexanol	623-55-2	116.203	421.15	25	3.4484E+00	6.2071E-05	1,2
1108	12180	C7H16O	2-ethyl-1-pentanol	27522-11-8	116.203	439.15	25	5.1840E+00	9.3312E-05	1,2
1109	12181	C7H16O	3-ethyl-1-pentanol	66225-51-2	116.203	439.15	25	5.9953E+00	1.0792E-04	1,2
1110	12182	C7H16O	2,2-dimethyl-1-pentanol	2370-12-9	116.203	426.15	25	5.6894E+00	1.0241E-04	1,2
1111	12183	C7H16O	2,3-dimethyl-1-pentanol	10143-23-4	116.203	437.15	25	1.4450E+00	2.6010E-05	1,2
1112	12184	C7H16O	2,4-dimethyl-1-pentanol	6305-71-1	116.203	432.15	25	5.7971E+00	1.0435E-04	1,2
1113	12185	C7H16O	3,3-dimethyl-1-pentanol	19264-94-9	116.203	438.15	25	1.2290E+00	2.2121E-05	1,2
1114	12186	C7H16O	3,4-dimethyl-1-pentanol	6570-87-2	116.203	438.15	25	3.2722E+00	5.8900E-05	1,2
1115	12187	C7H16O	4,4-dimethyl-1-pentanol	3121-79-7	116.203	433.15	25	6.4128E+00	1.1543E-04	1,2
1116	12188	C7H16O	3-ethyl-2-pentanol	609-27-8	116.203	425.15	25	5.1638E+00	9.2947E-05	1,2
1117	12189	C7H16O	2,3-dimethyl-2-pentanol	4911-70-0	116.203	412.15	25	1.8852E+00	3.3933E-05	1,2
1118	12190	C7H16O	2,4-dimethyl-2-pentanol	625-06-9	116.203	406.15	25	1.9964E+00	3.5934E-05	1,2
1119	12191	C7H16O	3,3-dimethyl-2-pentanol	19781-24-9	116.203	420.15	25	6.0761E+00	1.0937E-04	1,2
1120	12192	C7H16O	3,4-dimethyl-2-pentanol	64502-86-9	116.203	426.15	25	5.6463E+00	1.0163E-04	1,2

Table 10 Henry's Law Constant for Compound in Water - Organic Compounds

NO	ID	FORMULA	NAME	CAS No	MW, g/mol	T_B, K	T, C	H @ T, atm/mol frac	H @ T, atm/mol/m³	code
						Henry's Law Constant, H				
1121	12193	C7H16O	4,4-dimethyl-2-pentanol	6144-93-0	116.203	411.15	25	1.8579E+00	3.3441E-05	1,2
1122	12194	C7H16O	3-ethyl-3-pentanol	597-49-9	116.203	415.65	25	2.5229E+00	4.5412E-05	1,2
1123	12195	C7H16O	2,2-dimethyl-3-pentanol	3970-62-5	116.203	409.15	25	2.3415E+00	4.2147E-05	1,2
1124	12196	C7H16O	2,3-dimethyl-3-pentanol	595-41-5	116.203	413.15	25	2.3224E+00	4.1802E-05	1,2
1125	12197	C7H16O	2,4-dimethyl-3-pentanol	600-36-2	116.203	411.95	25	2.3513E+00	4.2322E-05	1,2
1126	12198	C7H16O	2-methyl-2-ethyl-1-butanol	18371-13-6	116.203	430.15	25	1.0581E+00	1.9045E-05	1,2
1127	12199	C7H16O	3-methyl-2-ethyl-1-butanol	32444-34-1	116.203	435.15	25	2.5937E+00	4.6686E-05	1,2
1128	12200	C7H16O	2,2,3-trimethyl-1-butanol	55505-23-2	116.203	430.15	25	2.1766E+00	3.9178E-05	1,2
1129	12201	C7H16O	2,3,3-trimethyl-1-butanol	36794-64-6	116.203	433.15	25	2.3632E+00	4.2536E-05	1,2
1130	12202	C7H16O	2,3,3-trimethyl-2-butanol	594-83-2	116.203	404.15	25	1.1031E+00	1.9856E-05	1,2
1131	12203	C7H16O	2,4-dimethyl-1-pentanol, (±)	111768-02-6	116.203	432.15	25	2.1816E+00	3.9268E-05	1,2
1132	12204	C7H16O	2-heptanol, (±)	52390-72-4	116.203	432.15	25	2.1816E+00	3.9268E-05	1,2
1133	12205	C7H16O	3-heptanol, (S)	26549-25-7	116.203	430.15	25	2.1730E+00	3.9113E-05	1,2
1134	12206	C7H16O	2-methyl-1-hexanol, (±)	111768-04-8	116.203	437.15	25	2.1970E+00	3.9546E-05	1,2
1135	12207	C7H16O	3-methyl-1-hexanol, (±)	111768-08-2	116.203	443.15	25	2.2057E+00	3.9702E-05	1,2
1136	12208	C7H16O	4-methyl-1-hexanol, (±)	111768-05-9	116.203	446.15	25	2.2067E+00	3.9720E-05	1,2
1137	12209	C7H16O	2-methyl-3-hexanol, (±)	100295-82-7	116.203	416.15	25	2.0676E+00	3.7216E-05	1,2
1138	12210	C7H16O	5-methyl-3-hexanol, (±)	100295-83-8	116.203	420.15	25	2.1065E+00	3.7916E-05	1,2
1139	12211	C7H16O	isopropyl-tert-butyl-ether	17348-59-3	116.203	378.66	25	8.7278E+02	1.5710E-02	1,2
1140	12212	C7H16O	butyl propyl ether	3073-92-5	116.203	391.25	25	1.8264E+02	3.2874E-03	1,2
1141	12213	C7H16O	2-ethoxy-2-methylbutane	919-94-8	116.203	375.15	25	1.7327E+02	3.1189E-03	1,2
1142	12214	C7H16O	ethyl isopentyl ether	628-04-6	116.203	385.65	25	1.7993E+02	3.2387E-03	1,2
1143	12215	C7H16O	ethyl pentyl ether	17952-11-3	116.203	390.75	25	1.8242E+02	3.2835E-03	1,2
1144	12216	C7H16O	hexyl methyl ether	4747-07-3	116.203	399.25	25	1.8539E+02	3.3370E-03	1,2
1145	12217	C7H16O	1-(1-methylethoxy)butane	1860-27-1	116.203	381.15	25	1.7732E+02	3.1917E-03	1,2
1146	12218	C7H16O	2-methyl-1-propoxypropane	15268-49-2	116.203	378.15	25	1.7537E+02	3.1567E-03	1,2
1147	12219	C7H16O	2-methyl-2-propoxypropane	29072-93-3	116.203	373.15	25	1.7179E+02	3.0922E-03	1,2
1148	12220	C7H16O	(S)-(+)-2-heptanol	6033-23-4	116.203	426.80	25	2.1553E+00	3.8794E-05	1,2
1149	12221	C7H16O	(R)-(-)-2-heptanol	6033-24-5	116.203	420.75	25	2.1117E+00	3.8010E-05	2
1150	12300	C7H16S	heptyl mercaptan	1639-09-4	132.270	450.09	25	7.1521E+02	1.2874E-02	1,2
1151	12301	C7H16S	2-heptanethiol	1628-00-2	132.270	436.75	25	6.3351E+02	1.1403E-02	1,2
1152	12302	C7H16S	methyl hexyl sulfide	20291-60-5	132.270	440.00	25	2.0117E+02	3.6211E-03	2
1153	12303	C7H16S	ethyl pentyl sulfide	26158-99-6	132.270	440.00	25	2.0117E+02	3.6211E-03	2
1154	12304	C7H16S	propyl butyl sulfide	1613-46-3	132.270	431.00	25	1.9662E+02	3.5392E-03	1,2
1155	12315	C7H17N	heptylamine	111-68-2	115.219	430.05	25	3.0772E+00	5.5389E-05	1,2
1156	12316	C7H17N	2,4-dimethyl-2-pentanamine	64379-30-2	115.219	395.15	25	2.0652E+00	3.7173E-05	1,2
1157	12317	C7H17N	2-heptanamine	123-82-0	115.219	415.15	25	3.4632E+00	6.2337E-05	1,2
1158	12318	C7H17N	4-heptanamine	16751-59-0	115.219	412.65	25	3.2344E+00	5.8218E-05	1,2
1159	12319	C7H17N	3-methyl-1-hexanamine	65530-93-0	115.219	422.15	25	4.2158E+00	7.5885E-05	1,2
1160	12320	C7H17N	4-methyl-2-hexanamine	105-41-9	115.219	405.65	25	2.6856E+00	4.8341E-05	1,2
1161	12321	C7H17N	4-methyl-1-hexylamine	34263-68-8	115.219	426.15	25	4.7329E+00	8.5192E-05	1,2
1162	12322	C7H17N	methylhexylamine	35161-70-7	115.219	415.15	25	2.5580E+00	4.6043E-05	1,2
1163	12323	C7H17N	ethylpentylamine	17839-26-8	115.219	407.15	25	2.6815E+00	4.8267E-05	1,2
1164	12324	C7H17N	dimethylpentylamine	26153-88-8	115.219	395.15	25	1.2388E+01	2.2298E-04	1,2
1165	12325	C7H17N	dimethyl-2-pentylamine	57303-85-2	115.219	382.15	25	9.4768E+00	1.7058E-04	1,2
1166	12326	C7H17N	dimethyl-3-pentylamine	18636-94-7	115.219	389.15	25	9.8358E+00	1.7704E-04	1,2
1167	12327	C7H17N	dimethyl-2-methylbutylamine	66225-39-6	115.219	388.00	25	9.7799E+00	1.7604E-04	2
1168	12328	C7H17N	dimethyl-3-methylbutylamine	2315-43-7	115.219	386.65	25	9.7126E+00	1.7483E-04	1,2
1169	12329	C7H17N	dimethyl-1,1-dimethylpropylamine	57757-60-5	115.219	391.15	25	1.0873E+01	1.9571E-04	1,2
1170	12330	C7H17N	dimethyl-1,2-dimethylpropylamine	66225-38-5	115.219	390.50	25	9.8999E+00	1.7820E-04	2
1171	12331	C7H17N	dimethyl-2,2-dimethylpropylamine	10076-31-0	115.219	390.50	25	1.0833E+01	1.9500E-04	2
1172	12332	C7H17N	methylethylbutylamine	66225-40-9	115.219	390.15	25	9.4649E+00	1.7037E-04	1,2
1173	12333	C7H17N	methylethyl-sec-butylamine	66225-41-0	115.219	376.15	25	9.1371E+00	1.6447E-04	1,2
1174	12334	C7H17N	methylethylisobutylamine	60247-14-5	115.219	380.15	25	9.3666E+00	1.6860E-04	1,2
1175	12335	C7H17N	methylethyl-tert-butylamine	52841-28-8	115.219	383.15	25	1.0356E+01	1.8640E-04	1,2
1176	12336	C7H17N	methyldipropylamine	3405-42-3	115.219	387.25	25	9.3426E+00	1.6817E-04	1,2
1177	12337	C7H17N	methylpropylisopropylamine	66225-42-1	115.219	386.00	25	9.6796E+00	1.7423E-04	2
1178	12338	C7H17N	methyldiisopropylamine	10342-97-9	115.219	385.15	25	1.0031E+01	1.8056E-04	1,2
1179	12339	C7H17N	diethylpropylamine	4458-31-5	115.219	385.15	25	9.2492E+00	1.6648E-04	1,2
1180	12340	C7H17N	diethylisopropylamine	6006-15-1	115.219	381.65	25	9.4495E+00	1.7009E-04	1,2
1181	12341	C7H17N	N-tert-butylisopropylamine	7515-80-2	115.219	371.15	25	1.9076E+00	3.4337E-05	1,2
1182	12342	C7H17N	(S)-2-heptylamine	44745-29-1	115.219	416.15	25	3.5602E+00	6.4084E-05	1,2
1183	12343	C7H17N	(R)-2-heptylamine	6240-90-0	115.219	416.15	25	3.5602E+00	6.4084E-05	1,2
1184	12344	C7H17N	N-propylbutylamine	20193-21-9	115.219	406.65	25	4.2693E+00	7.6847E-05	1,2
1185	12345	C7H17N	3-aminoheptane	28292-42-4	115.219	414.15	25	3.3694E+00	6.0649E-05	1,2
1186	12346	C7H17N	5-methyl-2-hexylamine	28292-43-5	115.219	397.47	25	2.1847E+00	3.9325E-05	2
1187	12594	C8H4O3	phthalic anhydride	85-44-9	148.118	557.65	25	1.0595E-01	1.9072E-06	1,2
1188	12713	C8H6	ethynylbenzene	536-74-3	102.136	416.20	25	1.0256E+02	1.8461E-03	1,2
1189	12934	C8H6O4	isophthalic acid	121-91-5	166.133	753.00	25	3.5157E-14	6.3283E-19	1,2
1190	12935	C8H6O4	phthalic acid	88-99-3	166.133	598.00	25	2.6498E-06	4.7696E-11	1,2
1191	12936	C8H6O4	terephthalic acid	100-21-0	166.133	832.00	20	3.1205E-21	5.6169E-26	1,2
1192	12947	C8H6S	benzothiophene	95-15-8	134.202	493.05	20	1.2072E+01	2.1730E-04	1,2
1193	13107	C8H7N	indole	120-72-9	117.151	526.15	20	3.5741E-02	6.4334E-07	1,2
1194	13237	C8H8	styrene	100-42-5	104.152	418.31	25	1.5012E+02	2.7022E-03	1,2
1195	13391	C8H8O	acetophenone	98-86-2	120.151	475.15	25	5.0591E-01	9.1063E-06	1,2
1196	13392	C8H8O	p-tolualdehyde	104-87-0	120.151	477.15	25	1.0023E+00	1.8041E-05	1,2
1197	13412	C8H8O2	methyl benzoate	93-58-3	136.150	472.65	25	2.0321E+00	3.6578E-05	1,2
1198	13413	C8H8O2	o-toluic acid	118-90-1	136.150	532.00	25	1.8842E-02	3.3915E-07	1,2
1199	13414	C8H8O2	p-toluic acid	99-94-5	136.150	548.15	25	1.5610E-02	2.8098E-07	1,2
1200	13445	C8H8O3	methyl salicylate	119-36-8	152.150	493.65	30	5.7030E-01	1.0265E-05	1,2

Table 10 Henry's Law Constant for Compound in Water - Organic Compounds

NO	ID	FORMULA	NAME	CAS No	MW, g/mol	T_B, K	T, C	H @ T, atm/mol frac	H @ T, atm/mol/m^3	code
					Henry's Law Constant, H					
1201	13446	C8H8O3	vanillin	121-33-5	152.150	558.00	25	2.5111E-04	4.5200E-09	1,2
1202	13671	C8H9NO	acetanilide	103-84-4	135.166	576.95	25	6.1376E-04	1.1048E-08	1,2
1203	13789	C8H10	ethylbenzene	100-41-4	106.167	409.35	25	4.7617E+02	8.5711E-03	1,2
1204	13790	C8H10	o-xylene	95-47-6	106.167	417.58	25	2.3303E+02	4.1946E-03	1,2
1205	13791	C8H10	m-xylene	108-38-3	106.167	412.27	25	3.7998E+02	6.8395E-03	1,2
1206	13792	C8H10	p-xylene	106-42-3	106.167	411.51	25	3.3838E+02	6.0908E-03	1,2
1207	13793	C8H10	xylenes	1330-20-7	106.167	413.15	25	1.9137E+02	3.4447E-03	1,2
1208	13913	C8H10O	2-phenylethanol	60-12-8	122.167	492.05	25	1.5582E-01	2.8047E-06	1,2
1209	13918	C8H10O	o-ethylphenol	90-00-6	122.167	477.67	25	2.7663E-01	4.9793E-06	1,2
1210	13919	C8H10O	m-ethylphenol	620-17-7	122.167	477.66	25	2.0816E-01	3.7468E-06	1,2
1211	13920	C8H10O	p-ethylphenol	123-07-9	122.167	491.14	25	1.1629E-01	2.0932E-06	1,2
1212	13921	C8H10O	2,3-xylenol	526-75-0	122.167	490.07	25	1.6660E-01	2.9988E-06	1,2
1213	13922	C8H10O	2,4-xylenol	105-67-9	122.167	484.13	25	1.1296E-01	2.0333E-06	1,2
1214	13923	C8H10O	2,5-xylenol	95-87-4	122.167	484.33	25	3.1021E-01	5.5837E-06	1,2
1215	13924	C8H10O	2,6-xylenol	576-26-1	122.167	474.22	25	4.0581E-01	7.3044E-06	1,2
1216	13925	C8H10O	3,4-xylenol	95-65-8	122.167	500.15	25	5.2637E-02	9.4745E-07	1,2
1217	13926	C8H10O	3,5-xylenol	108-68-9	122.167	494.89	25	6.0027E-02	1.0805E-06	1,2
1218	13934	C8H10O	dimethyl phenol	1300-71-6	122.167	411.65	25	8.4056E+00	1.5130E-04	1,2
1219	14082	C8H11N	N,N-dimethylaniline	121-69-7	121.182	466.69	25	5.8860E+00	1.0595E-04	1,2
1220	14083	C8H11N	o-ethylaniline	578-54-1	121.182	482.65	25	1.7622E+00	3.1719E-05	1,2
1221	14084	C8H11N	N,2-dimethylaniline	611-21-2	121.182	480.65	25	1.4852E+00	2.6734E-05	1,2
1222	14085	C8H11N	N,3-dimethylaniline	696-44-6	121.182	479.65	25	1.5729E+00	2.8312E-05	1,2
1223	14086	C8H11N	N,4-dimethylaniline	623-08-5	121.182	483.15	25	1.2859E+00	2.3147E-05	1,2
1224	14087	C8H11N	2,3-dimethylaniline	87-59-2	121.182	494.65	25	6.5443E-01	1.1780E-05	1,2
1225	14088	C8H11N	2,4-dimethylaniline	95-68-1	121.182	487.15	25	1.0191E+00	1.8344E-05	1,2
1226	14089	C8H11N	2,5-dimethylaniline	95-78-3	121.182	487.15	25	1.0191E+00	1.8344E-05	1,2
1227	14090	C8H11N	2,6-dimethylaniline	87-62-7	121.182	488.15	25	9.6115E-01	1.7301E-05	1,2
1228	14091	C8H11N	3,4-dimethylaniline	95-64-7	121.182	501.15	25	4.4254E-01	7.9657E-06	1,2
1229	14092	C8H11N	3,5-dimethylaniline	108-69-0	121.182	493.65	25	6.9460E-01	1.2503E-05	1,2
1230	14093	C8H11N	m-ethylaniline	587-02-0	121.182	487.15	25	1.0191E+00	1.8344E-05	1,2
1231	14094	C8H11N	p-ethylaniline	589-16-2	121.182	490.65	25	8.2973E-01	1.4935E-05	1,2
1232	14095	C8H11N	N-ethylaniline	103-69-5	121.182	476.15	25	1.9203E+00	3.4566E-05	1,2
1233	14231	C8H12	vinylcyclohexene	100-40-3	108.183	401.00	25	2.2046E+03	3.9682E-02	1,2
1234	14379	C8H12O4	1,4-cyclohexanedicarboxylic acid	619-82-9	172.181	669.00	17	2.6295E-10	4.7330E-15	1,2
1235	14410	C8H12S	2-butylthiophene	1455-20-5	140.249	454.65	25	4.6481E+02	8.3664E-03	1,2
1236	14411	C8H12S	3-butylthiophene	34722-01-5	140.249	455.15	25	4.6496E+02	8.3692E-03	1,2
1237	14412	C8H12S	2-(1,1-dimethylethyl)thiophene	1689-78-7	140.249	437.15	25	4.5067E+02	8.1120E-03	1,2
1238	14413	C8H12S	3-(1,1-dimethylethyl)thiophene	1689-79-8	140.249	442.15	25	4.5621E+02	8.2118E-03	1,2
1239	14484	C8H14	1-octyne	629-05-0	110.199	399.35	25	3.0914E+03	5.5645E-02	1,2
1240	14485	C8H14	2-octyne	2809-67-8	110.199	410.88	25	2.7775E+03	4.9994E-02	1,2
1241	14486	C8H14	3-octyne	15232-76-5	110.199	406.29	25	2.6675E+03	4.8014E-02	1,2
1242	14487	C8H14	4-octyne	1942-45-6	110.199	404.75	25	2.6300E+03	4.7339E-02	1,2
1243	14488	C8H14	3-ethyl-3-methyl-1-pentyne	919-12-0	110.199	374.65	25	1.9423E+03	3.4961E-02	1,2
1244	14493	C8H14	1,7-octadiene	3710-30-3	110.199	388.65	25	1.0952E+04	1.9714E-01	1,2
1245	14494	C8H14	2,6-octadiene	4974-27-0	110.199	397.65	25	1.1909E+04	2.1436E-01	1,2
1246	14495	C8H14	3-methyl-1,5-heptadiene	4894-62-6	110.199	384.15	25	1.0496E+04	1.8893E-01	1,2
1247	14496	C8H14	2,5-dimethyl-1,5-hexadiene	627-58-7	110.199	387.45	25	1.0829E+04	1.9492E-01	1,2
1248	14497	C8H14	2,5-dimethyl-2,4-hexadiene	764-13-6	110.199	407.65	25	1.2982E+04	2.3368E-01	1,2
1249	14861	C8H16	propylcyclopentane	2040-96-2	112.215	404.11	25	5.7334E+04	1.0320E+00	1,2
1250	14862	C8H16	isopropylcyclopentane	3875-51-2	112.215	399.57	25	5.2354E+04	9.4237E-01	1,2
1251	14863	C8H16	1-methyl-1-ethylcyclopentane	16747-50-5	112.215	394.67	25	4.9986E+04	8.9974E-01	1,2
1252	14864	C8H16	1-methyl-cis-2-ethylcyclopentane	930-89-2	112.215	401.20	25	5.2345E+04	9.4221E-01	1,2
1253	14865	C8H16	1-methyl-trans-2-ethylcyclopentane	930-90-5	112.215	394.35	25	4.6732E+04	8.4116E-01	1,2
1254	14866	C8H16	1-methyl-cis-3-ethylcyclopentane	2613-66-3	112.215	394.25	25	4.5961E+04	8.2729E-01	1,2
1255	14867	C8H16	1-methyl-trans-3-ethylcyclopentane	2613-65-2	112.215	394.25	25	4.7451E+04	8.5411E-01	1,2
1256	14868	C8H16	1,1,2-trimethylcyclopentane	4259-00-1	112.215	386.88	25	4.2852E+04	7.7134E-01	1,2
1257	14869	C8H16	1,1,3-trimethylcyclopentane	4516-69-2	112.215	378.04	25	8.6656E+04	1.5598E+00	1,2
1258	14870	C8H16	1,cis-2,cis-3-trimethylcyclopentane	2613-69-6	112.215	396.15	25	4.9892E+04	8.9804E-01	1,2
1259	14871	C8H16	1,cis-2,trans-3-trimethylcyclopentane	15890-40-1	112.215	390.65	25	4.5147E+04	8.1264E-01	1,2
1260	14872	C8H16	1,trans-2,cis-3-trimethylcyclopentane	19374-46-0	112.215	383.35	25	4.0240E+04	7.2432E-01	1,2
1261	14873	C8H16	1,cis-2,cis-4-trimethylcyclopentane	2613-72-1	112.215	390.15	25	4.4202E+04	7.9564E-01	1,2
1262	14874	C8H16	1,cis-2,trans-4-trimethylcyclopentane	4850-28-6	112.215	389.88	25	4.4787E+04	8.0615E-01	1,2
1263	14875	C8H16	1,trans-2,cis-4-trimethylcyclopentane	16883-48-0	112.215	382.44	25	3.9049E+04	7.0288E-01	1,2
1264	14876	C8H16	1-methyl-3-ethylcyclopentane	3726-47-4	112.215	394.20	25	5.6623E+04	1.0192E+00	1,2
1265	14877	C8H16	ethylcyclohexane	1678-91-7	112.215	404.95	25	1.8876E+04	3.3976E-01	1,2
1266	14878	C8H16	1,1-dimethylcyclohexane	590-66-9	112.215	392.70	25	3.1333E+04	5.6398E-01	1,2
1267	14879	C8H16	1,2-dimethylcyclohexane, (cis+trans)	583-57-3	112.215	399.05	25	2.8610E+04	5.1497E-01	1,2
1268	14880	C8H16	cis-1,2-dimethylcyclohexane	2207-01-4	112.215	402.94	25	2.0718E+04	3.7292E-01	1,2
1269	14881	C8H16	trans-1,2-dimethylcyclohexane	6876-23-9	112.215	396.58	25	3.2249E+04	5.8048E-01	1,2
1270	14882	C8H16	1,3-dimethylcyclohexane, cis and trans	591-21-9	112.215	393.10	25	2.6319E+04	4.7373E-01	1,2
1271	14883	C8H16	cis-1,3-dimethylcyclohexane	638-04-0	112.215	393.24	25	3.1949E+04	5.7508E-01	1,2
1272	14884	C8H16	trans-1,3-dimethylcyclohexane	2207-03-6	112.215	397.61	25	3.5152E+04	6.3274E-01	1,2
1273	14885	C8H16	1,4-dimethylcyclohexane, (cis+trans)	589-90-2	112.215	395.90	25	2.7381E+04	4.9285E-01	1,2
1274	14886	C8H16	cis-1,4-dimethylcyclohexane	624-29-3	112.215	397.47	25	3.5274E+04	6.3493E-01	1,2
1275	14887	C8H16	trans-1,4-dimethylcyclohexane	2207-04-7	112.215	392.51	25	4.9048E+04	8.8286E-01	1,2
1276	14888	C8H16	cyclooctane	292-64-8	112.215	424.30	25	5.9816E+03	1.0767E-01	1,2
1277	14890	C8H16	1-octene	111-66-0	112.215	394.44	25	3.6056E+04	6.4900E-01	1,2
1278	14891	C8H16	oct-2-ene	111-67-1	112.215	398.05	25	4.5684E+04	8.2230E-01	1,2
1279	14892	C8H16	cis-2-octene	7642-04-8	112.215	398.79	25	6.5999E+04	1.1880E+00	1,2
1280	14893	C8H16	trans-2-octene	13389-42-9	112.215	398.15	25	7.0115E+04	1.2621E+00	1,2

Table 10 Henry's Law Constant for Compound in Water - Organic Compounds

				Henry's Law Constant, H						
NO	ID	FORMULA	NAME	CAS No	MW, g/mol	T_B, K	T, C	H @ T, atm/mol frac	H @ T, atm/mol/m^3	code
1281	14894	C8H16	cis-3-octene	14850-22-7	112.215	396.05	25	6.2111E+04	1.1180E+00	1,2
1282	14895	C8H16	trans-3-octene	14919-01-8	112.215	396.45	25	6.8153E+04	1.2268E+00	1,2
1283	14896	C8H16	oct-4-ene	592-99-4	112.215	398.00	25	4.5658E+04	8.2184E-01	1,2
1284	14897	C8H16	cis-4-octene	7642-15-1	112.215	395.69	25	6.1581E+04	1.1084E+00	1,2
1285	14898	C8H16	trans-4-octene	14850-23-8	112.215	395.41	25	6.4528E+04	1.1615E+00	1,2
1286	14899	C8H16	2-methyl-1-heptene	15870-10-7	112.215	392.45	25	5.7636E+04	1.0374E+00	1,2
1287	14900	C8H16	3-methyl-1-heptene	4810-09-7	112.215	384.15	25	4.8379E+04	8.7081E-01	1,2
1288	14901	C8H16	4-methyl-1-heptene	13151-05-8	112.215	385.95	25	5.0127E+04	9.0227E-01	1,2
1289	14902	C8H16	5-methyl-1-heptene	13151-04-7	112.215	386.45	25	5.0683E+04	9.1229E-01	1,2
1290	14903	C8H16	6-methyl-1-heptene	5026-76-6	112.215	386.35	25	5.0585E+04	9.1052E-01	1,2
1291	14904	C8H16	2-methyl-2-heptene	627-97-4	112.215	395.75	25	6.1839E+04	1.1131E+00	1,2
1292	14905	C8H16	3-methyl-cis-2-heptene	22768-19-0	112.215	395.15	25	6.1141E+04	1.1005E+00	1,2
1293	14906	C8H16	3-methyl-trans-2-heptene	22768-20-3	112.215	395.15	25	6.1141E+04	1.1005E+00	1,2
1294	14907	C8H16	4-methyl-cis-2-heptene	66225-16-9	112.215	387.15	25	5.1553E+04	9.2795E-01	1,2
1295	14908	C8H16	4-methyl-trans-2-heptene	66225-17-0	112.215	387.15	25	5.1553E+04	9.2795E-01	1,2
1296	14909	C8H16	5-methyl-cis-2-heptene	24608-84-2	112.215	391.15	25	5.6097E+04	1.0097E+00	1,2
1297	14910	C8H16	5-methyl-trans-2-heptene	24608-85-3	112.215	391.15	25	5.6097E+04	1.0097E+00	1,2
1298	14911	C8H16	6-methyl-2-heptene	73548-72-8	112.215	390.15	25	4.1712E+04	7.5081E-01	1,2
1299	14912	C8H16	6-methyl-cis-2-heptene	66225-18-1	112.215	390.15	25	5.4909E+04	9.8836E-01	1,2
1300	14913	C8H16	6-methyl-trans-2-heptene	51065-65-7	112.215	390.15	25	5.4909E+04	9.8836E-01	1,2
1301	14914	C8H16	6-methyl-3-heptene	3404-57-7	112.215	388.15	25	4.0747E+04	7.3344E-01	1,2
1302	14915	C8H16	2-methyl-3-heptene	17618-76-7	112.215	385.15	25	3.9338E+04	7.0808E-01	1,2
1303	14916	C8H16	2-methyl-cis-3-heptene	20488-34-0	112.215	385.15	25	4.9462E+04	8.9031E-01	1,2
1304	14917	C8H16	2-methyl-trans-3-heptene	692-96-6	112.215	385.15	25	4.8642E+04	8.7554E-01	1,2
1305	14918	C8H16	3-methyl-3-heptene	7300-03-0	112.215	394.15	25	4.3695E+04	7.8650E-01	1,2
1306	14919	C8H16	3-methyl-cis-3-heptene	22768-17-8	112.215	394.15	25	5.9814E+04	1.0766E+00	1,2
1307	14920	C8H16	3-methyl-trans-3-heptene	22768-18-9	112.215	394.15	25	5.9814E+04	1.0766E+00	1,2
1308	14921	C8H16	4-methyl-3-heptene	4485-16-9	112.215	394.15	25	4.3695E+04	7.8650E-01	1,2
1309	14922	C8H16	4-methyl-cis-3-heptene	14255-24-4	112.215	395.15	25	6.1141E+04	1.1005E+00	1,2
1310	14923	C8H16	4-methyl-trans-3-heptene	13714-85-7	112.215	395.15	25	6.1141E+04	1.1005E+00	1,2
1311	14924	C8H16	5-methyl-3-heptene	13172-91-3	112.215	385.15	25	3.9338E+04	7.0808E-01	1,2
1312	14925	C8H16	5-methyl-cis-3-heptene	50422-80-5	112.215	385.15	25	4.9462E+04	8.9031E-01	1,2
1313	14926	C8H16	5-methyl-trans-3-heptene	53510-18-2	112.215	385.15	25	4.9462E+04	8.9031E-01	1,2
1314	14927	C8H16	6-methyl-cis-3-heptene	66225-19-2	112.215	388.15	25	5.2632E+04	9.4737E-01	1,2
1315	14928	C8H16	6-methyl-trans-3-heptene	66225-20-5	112.215	388.15	25	5.4066E+04	9.7318E-01	1,2
1316	14929	C8H16	2-ethyl-1-hexene	1632-16-2	112.215	393.15	25	5.8151E+04	1.0467E+00	1,2
1317	14930	C8H16	3-ethyl-1-hexene	3404-58-8	112.215	383.45	25	4.7666E+04	8.5798E-01	1,2
1318	14931	C8H16	4-ethyl-1-hexene	16746-85-3	112.215	383.15	25	4.3754E+04	7.8756E-01	1,2
1319	14932	C8H16	2,3-dimethyl-1-hexene	16746-86-4	112.215	383.65	25	4.8104E+04	8.6586E-01	1,2
1320	14933	C8H16	2,4-dimethyl-1-hexene	16746-87-5	112.215	384.35	25	4.8756E+04	8.7760E-01	1,2
1321	14934	C8H16	2,5-dimethyl-1-hexene	6975-92-4	112.215	384.75	25	4.9149E+04	8.8468E-01	1,2
1322	14935	C8H16	3,3-dimethyl-1-hexene	3404-77-1	112.215	377.15	25	4.2760E+04	7.6968E-01	1,2
1323	14936	C8H16	3,4-dimethyl-1-hexene	16745-94-1	112.215	385.15	25	4.9666E+04	8.9397E-01	1,2
1324	14937	C8H16	3,5-dimethyl-1-hexene	7423-69-0	112.215	377.15	25	4.2440E+04	7.6391E-01	1,2
1325	14938	C8H16	4,4-dimethyl-1-hexene	1647-08-1	112.215	380.35	25	4.5380E+04	8.1683E-01	1,2
1326	14939	C8H16	4,5-dimethyl-1-hexene	16106-59-5	112.215	382.15	25	4.6760E+04	8.4167E-01	1,2
1327	14940	C8H16	5,5-dimethyl-1-hexene	7116-86-1	112.215	375.65	25	4.1548E+04	7.4787E-01	1,2
1328	14941	C8H16	3-ethyl-cis-2-hexene	36880-72-5	112.215	394.15	25	5.9814E+04	1.0766E+00	1,2
1329	14942	C8H16	3-ethyl-trans-2-hexene	66225-15-8	112.215	394.15	25	5.9814E+04	1.0766E+00	1,2
1330	14943	C8H16	4-ethyl-cis-2-hexene	54616-49-8	112.215	386.15	25	5.0493E+04	9.0886E-01	1,2
1331	14944	C8H16	4-ethyl-trans-2-hexene	19781-63-6	112.215	386.15	25	5.0493E+04	9.0886E-01	1,2
1332	14945	C8H16	2,3-dimethyl-2-hexene	7145-20-2	112.215	394.92	25	6.0912E+04	1.0964E+00	1,2
1333	14946	C8H16	2,4-dimethyl-2-hexene	14255-23-3	112.215	383.75	25	4.8245E+04	8.6841E-01	1,2
1334	14947	C8H16	2,5-dimethyl-2-hexene	3404-78-2	112.215	385.35	25	4.8250E+04	8.6849E-01	1,2
1335	14948	C8H16	3,4-dimethyl-cis-2-hexene	19550-81-3	112.215	389.15	25	5.3990E+04	9.7182E-01	1,2
1336	14949	C8H16	3,4-dimethyl-trans-2-hexene	19550-82-4	112.215	389.15	25	5.3990E+04	9.7182E-01	1,2
1337	14950	C8H16	3,5-dimethyl-cis-2-hexene	66225-31-8	112.215	385.15	25	4.9707E+04	8.9472E-01	1,2
1338	14951	C8H16	3,5-dimethyl-trans-2-hexene	66225-12-5	112.215	385.15	25	4.9707E+04	8.9472E-01	1,2
1339	14952	C8H16	4,4-dimethyl-cis-2-hexene	19550-83-5	112.215	379.15	25	4.4490E+04	8.0082E-01	1,2
1340	14953	C8H16	4,4-dimethyl-trans-2-hexene	---	112.215	379.15	25	4.4490E+04	8.0082E-01	1,2
1341	14954	C8H16	4,5-dimethyl-2-hexene	73548-71-7	112.215	383.15	25	3.8426E+04	6.9166E-01	1,2
1342	14955	C8H16	4,5-dimethyl-cis-2-hexene	65036-71-7	112.215	383.15	25	4.7776E+04	8.5995E-01	1,2
1343	14956	C8H16	4,5-dimethyl-trans-2-hexene	66225-14-7	112.215	383.15	25	4.7776E+04	8.5995E-01	1,2
1344	14957	C8H16	5,5-dimethyl-cis-2-hexene	39761-61-0	112.215	380.05	25	4.5236E+04	8.1423E-01	1,2
1345	14958	C8H16	5,5-dimethyl-trans-2-hexene	39782-43-9	112.215	377.25	25	4.2893E+04	7.7206E-01	1,2
1346	14959	C8H16	3-ethyl-3-hexene	16789-51-8	112.215	389.15	25	5.3716E+04	9.6689E-01	1,2
1347	14960	C8H16	2,2-dimethyl-cis-3-hexene	690-92-6	112.215	378.58	25	4.7229E+04	8.5012E-01	1,2
1348	14961	C8H16	2,2-dimethyl-trans-3-hexene	690-93-7	112.215	374.00	25	4.1479E+04	7.4661E-01	1,2
1349	14962	C8H16	2,3-dimethyl-cis-3-hexene	59643-75-3	112.215	387.15	25	5.1794E+04	9.3229E-01	1,2
1350	14963	C8H16	2,3-dimethyl-trans-3-hexene	66225-30-7	112.215	387.15	25	5.1794E+04	9.3229E-01	1,2
1351	14964	C8H16	2,4-dimethyl-cis-3-hexene	37549-89-6	112.215	382.15	25	4.6731E+04	8.4116E-01	1,2
1352	14965	C8H16	2,4-dimethyl-trans-3-hexene	61847-78-7	112.215	380.75	25	4.5522E+04	8.1939E-01	1,2
1353	14966	C8H16	2,5-dimethyl-cis-3-hexene	10557-44-5	112.215	375.15	25	4.4208E+04	7.9573E-01	1,2
1354	14967	C8H16	2,5-dimethyl-trans-3-hexene	692-70-6	112.215	375.15	25	4.0838E+04	7.3509E-01	1,2
1355	14968	C8H16	3,4-dimethyl-cis-3-hexene	19550-87-9	112.215	395.15	25	6.1349E+04	1.1043E+00	1,2
1356	14969	C8H16	3,4-dimethyl-trans-3-hexene	19550-88-0	112.215	395.15	25	6.1349E+04	1.1043E+00	1,2
1357	14970	C8H16	2-propyl-1-pentene	15918-08-8	112.215	390.85	25	5.5580E+04	1.0004E+00	1,2
1358	14971	C8H16	2-isopropyl-1-pentene	16746-02-4	112.215	386.15	25	5.2361E+04	9.4249E-01	1,2
1359	14972	C8H16	3-methyl-2-ethyl-1-pentene	3404-67-9	112.215	384.65	25	4.7779E+04	8.6001E-01	1,2
1360	14973	C8H16	4-methyl-2-ethyl-1-pentene	3404-80-6	112.215	383.45	25	4.7892E+04	8.6205E-01	1,2

Table 10 Henry's Law Constant for Compound in Water - Organic Compounds

NO	ID	FORMULA	NAME	CAS No	MW, g/mol	T_B, K	T, C	H @ T, atm/mol frac	H @ T, atm/mol/m³	code
					Henry's Law Constant, H					
1361	14974	C8H16	2-methyl-3-ethyl-1-pentene	19780-66-6	112.215	383.15	25	5.1827E+04	9.3288E-01	1,2
1362	14975	C8H16	3-methyl-3-ethyl-1-pentene	6196-60-7	112.215	385.15	25	4.9981E+04	8.9965E-01	1,2
1363	14976	C8H16	4-methyl-3-ethyl-1-pentene	61847-80-1	112.215	380.65	25	4.5346E+04	8.1623E-01	1,2
1364	14977	C8H16	2,3,3-trimethyl-1-pentene	560-23-6	112.215	381.46	25	4.6641E+04	8.3953E-01	1,2
1365	14978	C8H16	2,3,4-trimethyl-1-pentene	565-76-4	112.215	381.15	25	4.6130E+04	8.3034E-01	1,2
1366	14979	C8H16	2,4,4-trimethyl-1-pentene	107-39-1	112.215	374.59	25	4.7996E+04	8.6392E-01	1,2
1367	14980	C8H16	3,3,4-trimethyl-1-pentene	560-22-5	112.215	378.15	25	4.3860E+04	7.8947E-01	1,2
1368	14981	C8H16	3,4,4-trimethyl-1-pentene	564-03-4	112.215	377.15	25	4.3052E+04	7.7493E-01	1,2
1369	14982	C8H16	2-methyl-3-ethyl-2-pentene	19780-67-7	112.215	390.15	25	5.1223E+04	9.2201E-01	1,2
1370	14983	C8H16	4-methyl-3-ethyl-cis-2-pentene	42067-48-1	112.215	389.15	25	5.3990E+04	9.7182E-01	1,2
1371	14984	C8H16	4-methyl-3-ethyl-trans-2-pentene	42067-49-2	112.215	387.45	25	5.2048E+04	9.3686E-01	1,2
1372	14985	C8H16	2,3,4-trimethyl-2-pentene	565-77-5	112.215	389.41	25	5.4455E+04	9.8019E-01	1,2
1373	14986	C8H16	2,4,4-trimethyl-2-pentene	107-40-4	112.215	378.06	25	4.9966E+04	8.9937E-01	1,2
1374	14987	C8H16	3,4,4-trimethyl-cis-2-pentene	39761-64-3	112.215	385.15	25	5.0191E+04	9.0343E-01	1,2
1375	14988	C8H16	3,4,4-trimethyl-trans-2-pentene	39761-57-4	112.215	385.15	25	5.0310E+04	9.0557E-01	1,2
1376	14989	C8H16	3,4,4-trimethyl-2-pentene	598-96-9	112.215	385.15	25	3.9338E+04	7.0808E-01	1,2
1377	14990	C8H16	2,4,4-trimethylpentene	25167-70-8	112.215	376.15	25	3.5426E+04	6.3767E-01	1,2
1378	14991	C8H16	3-methyl-2-isopropyl-1-butene	111823-35-9	112.215	377.15	25	4.2689E+04	7.6840E-01	1,2
1379	14992	C8H16	3,3-dimethyl-2-ethyl-1-butene	18231-53-3	112.215	383.15	25	4.8262E+04	8.6870E-01	1,2
1380	14994	C8H16Br2	1,1-dibromooctane	62168-26-7	272.023	515.15	25	4.5248E+01	8.1445E-04	1,2
1381	14995	C8H16Br2	1,2-dibromooctane	6269-92-7	272.023	514.15	25	2.7093E+01	4.8767E-04	1,2
1382	14996	C8H16Br2	1,4-dibromooctane	70690-24-3	272.023	524.48	25	2.2890E+01	4.1203E-04	2
1383	14997	C8H16Br2	1,5-dibromooctane	17912-17-3	272.023	524.48	25	2.2890E+01	4.1203E-04	2
1384	14998	C8H16Br2	1,8-dibromooctane	4549-32-0	272.023	544.15	25	1.4831E+01	2.6696E-04	1,2
1385	14999	C8H16Br2	4,5-dibromooctane	61539-75-1	272.023	524.48	25	2.2890E+01	4.1203E-04	2
1386	15005	C8H16Cl2	1,1-dichlorooctane	20395-24-8	183.120	481.15	25	3.8378E+02	6.9080E-03	1,2
1387	15006	C8H16Cl2	2,5-dichloro-2,5-dimethylhexane	6223-78-5	183.120	491.48	25	2.4296E+02	4.3732E-03	2
1388	15007	C8H16Cl2	1,8-dichlorooctane	2162-99-4	183.120	514.15	25	1.8833E+02	3.3899E-03	1,2
1389	15008	C8H16Cl2	2,3-dichlorooctane	21948-47-0	183.120	479.15	25	2.6312E+02	4.7361E-03	1,2
1390	15062	C8H16O	2-ethylhexanal	123-05-7	128.214	433.80	25	2.9497E+01	5.3094E-04	1,2
1391	15063	C8H16O	1-octanal	124-13-0	128.214	447.15	30	3.9575E+01	7.1234E-04	1,2
1392	15064	C8H16O	5-methylheptanal	75579-88-3	128.214	435.14	25	1.5893E+01	2.8607E-04	2
1393	15065	C8H16O	2-propylpentanal	18295-59-5	128.214	433.15	25	1.6023E+01	2.8842E-04	1,2
1394	15066	C8H16O	2-octanone	111-13-7	128.214	445.75	20	4.6188E+00	8.3137E-05	1,2
1395	15067	C8H16O	2,5-dimethyl-3-hexanone	1888-57-9	128.214	420.65	25	1.3422E+01	2.4160E-04	1,2
1396	15068	C8H16O	2,2-dimethyl-3-hexanone	5405-79-8	128.214	419.15	25	1.3234E+01	2.3821E-04	1,2
1397	15069	C8H16O	3,3-dimethyl-2-hexanone	26118-38-7	128.214	420.15	25	1.3360E+01	2.4048E-04	1,2
1398	15070	C8H16O	3,4-dimethyl-2-hexanone	19550-10-8	128.214	431.15	25	1.4627E+01	2.6328E-04	1,2
1399	15071	C8H16O	4,4-dimethyl-3-hexanone	19550-14-2	128.214	421.15	25	1.3484E+01	2.4271E-04	1,2
1400	15072	C8H16O	3-ethyl-4-methyl-2-pentanone	71172-57-1	128.214	427.65	25	1.4247E+01	2.5645E-04	1,2
1401	15073	C8H16O	2-methyl-3-heptanone	13019-20-0	128.214	431.15	25	1.4627E+01	2.6328E-04	1,2
1402	15074	C8H16O	2-methyl-4-heptanone	626-33-5	128.214	427.15	25	1.4191E+01	2.5544E-04	1,2
1403	15075	C8H16O	3-methyl-2-heptanone	2371-19-9	128.214	437.15	25	1.5231E+01	2.7416E-04	1,2
1404	15076	C8H16O	3-methyl-4-heptanone	15726-15-5	128.214	426.15	25	1.4078E+01	2.5340E-04	1,2
1405	15077	C8H16O	5-methyl-3-heptanone	541-85-5	128.214	435.14	25	1.5035E+01	2.7063E-04	2
1406	15078	C8H16O	6-methyl-2-heptanone	928-68-7	128.214	440.15	25	1.5513E+01	2.7923E-04	1,2
1407	15079	C8H16O	6-methyl-3-heptanone	624-42-0	128.214	437.15	25	1.5231E+01	2.7416E-04	1,2
1408	15080	C8H16O	3-octanone	106-68-3	128.214	440.65	25	1.5559E+01	2.8005E-04	1,2
1409	15081	C8H16O	4-octanone	589-63-9	128.214	436.15	25	1.5134E+01	2.7242E-04	1,2
1410	15082	C8H16O	2,2,4-trimethyl-3-pentanone	5857-36-3	128.214	408.25	25	1.1746E+01	2.1143E-04	1,2
1411	15147	C8H16O	3-ethyl-3-methyl-2-pentanone	19780-65-5	128.214	426.60	25	1.4129E+01	2.5432E-04	1,2
1412	15148	C8H16O	5-methyl-2-heptanone	18217-12-4	128.214	437.00	25	1.5217E+01	2.7390E-04	1,2
1413	15156	C8H16O2	octanoic acid	124-07-2	144.214	513.05	20	3.0341E-02	5.4614E-07	1,2
1414	15157	C8H16O2	2-ethylhexanoic acid	149-57-5	144.214	501.15	25	1.3837E-01	2.4906E-06	1,2
1415	15158	C8H16O2	3-ethylhexanoic acid	41065-91-2	144.214	436.35	25	2.3412E-01	4.2142E-06	2
1416	15159	C8H16O2	heptyl formate	112-23-2	144.214	451.25	25	4.9343E+01	8.8817E-04	1,2
1417	15160	C8H16O2	hexyl acetate	142-92-7	144.214	444.65	25	3.1444E+01	5.6599E-04	1,2
1418	15161	C8H16O2	1,1-dimethylbutyl acetate	34859-98-8	144.214	426.15	25	2.4594E+01	4.4269E-04	1,2
1419	15162	C8H16O2	2-ethylbutyl acetate	10031-87-5	144.214	435.65	25	2.4919E+01	4.4854E-04	1,2
1420	15163	C8H16O2	sec-hexyl acetate	108-84-9	144.214	420.65	25	2.7267E+01	4.9080E-04	1,2
1421	15164	C8H16O2	2-methyl-1-pentyl acetate	7789-99-3	144.214	436.15	25	2.4928E+01	4.4870E-04	1,2
1422	15165	C8H16O2	2-methyl-3-pentyl acetate	35897-16-6	144.214	422.15	25	2.4371E+01	4.3868E-04	1,2
1423	15166	C8H16O2	3-methyl-3-pentyl acetate	10250-47-2	144.214	421.15	25	2.4308E+01	4.3754E-04	1,2
1424	15167	C8H16O2	pentyl propanoate	624-54-4	144.214	441.90	25	1.2192E+01	2.1946E-04	1,2
1425	15168	C8H16O2	1-methylbutyl propanoate	54004-43-2	144.214	440.22	25	2.7562E+01	4.9611E-04	2
1426	15169	C8H16O2	2-methylbutyl propanoate	2438-20-2	144.214	440.22	25	2.7562E+01	4.9611E-04	2
1427	15170	C8H16O2	isopentyl propanoate	105-68-0	144.214	432.65	25	4.5347E+01	8.1623E-04	1,2
1428	15171	C8H16O2	1-ethylpropyl propanoate	---	144.214	440.22	25	2.7562E+01	4.9611E-04	2
1429	15172	C8H16O2	1,1-dimethylpropyl propanoate	34949-22-9	144.214	437.43	25	2.7488E+01	4.9478E-04	2
1430	15173	C8H16O2	1,2-dimethylpropyl propanoate	66576-70-3	144.214	439.78	25	2.7552E+01	4.9593E-04	2
1431	15174	C8H16O2	2,2-dimethylpropyl propanoate	3581-69-9	144.214	437.43	25	2.7488E+01	4.9478E-04	2
1432	15175	C8H16O2	butyl butanoate	109-21-7	144.214	438.15	25	4.8475E+01	8.7255E-04	1,2
1433	15176	C8H16O2	isobutyl butanoate	539-90-2	144.214	430.05	25	4.1792E+01	7.5226E-04	1,2
1434	15177	C8H16O2	isobutyl isobutyrate	97-85-8	144.214	420.65	25	1.0049E+02	1.8088E-03	1,2
1435	15178	C8H16O2	sec-butyl butanoate	819-97-6	144.214	425.15	25	1.8711E+01	3.3680E-04	1,2
1436	15179	C8H16O2	tert-butyl butanoate	2308-38-5	144.214	409.15	25	1.9484E+01	3.5071E-04	1,2
1437	15180	C8H16O2	sec-butyl butanoate, (S)	116836-55-6	144.214	425.15	25	2.8646E+01	5.1563E-04	1,2
1438	15181	C8H16O2	tert-butyl 2-methylpropanoate	16889-72-8	144.214	399.85	25	2.3977E+01	4.3157E-04	1,2
1439	15185	C8H16O2	ethyl hexanoate	123-66-0	144.214	440.15	25	3.8402E+01	6.9124E-04	1,2
1440	15186	C8H16O2	ethyl 4-methylpentanoate	25415-67-2	144.214	436.15	25	4.6886E+01	8.4395E-04	1,2

Table 10 Henry's Law Constant for Compound in Water - Organic Compounds

NO	ID	FORMULA	NAME	CAS No	Henry's Law Constant, H					
					MW, g/mol	T_B, K	T, C	H @ T, atm/mol frac	H @ T, atm/mol/m^3	code
1441	15191	C8H16O2	isopropyl 3-methylbutanoate	32665-23-9	144.214	415.15	25	2.7543E+01	4.9578E-04	1,2
1442	15192	C8H16O2	isopropyl pentanoate	18362-97-5	144.214	436.35	25	4.6417E+01	8.3551E-04	2
1443	15193	C8H16O2	propyl 3-methylbutanoate	557-00-6	144.214	429.05	25	4.0729E+01	7.3311E-04	1,2
1444	15194	C8H16O2	propyl pentanoate	141-06-0	144.214	440.65	25	3.7450E+01	6.7410E-04	1,2
1445	15195	C8H16O2	methyl heptanoate	106-73-0	144.214	447.15	25	2.6938E+01	4.8489E-04	1,2
1446	15204	C8H16O2	2-methylheptanoic acid	1188-02-9	144.214	436.35	25	2.3412E-01	4.2142E-06	2
1447	15205	C8H16O2	2-propylpentanoic acid	99-66-1	144.214	494.15	25	1.5166E-01	2.7299E-06	1,2
1448	15207	C8H16O2	butyl isobutyrate	---	144.214	436.35	25	4.6417E+01	8.3551E-04	2
1449	15232	C8H16O4	diethylene glycol ethyl ether acetate	112-15-2	176.213	490.55	25	1.9056E-03	3.4300E-08	2
1450	15248	C8H17Br	1-bromooctane	111-83-1	193.127	473.95	25	3.7213E+03	6.6982E-02	1,2
1451	15249	C8H17Br	2-bromooctane, (±)-	60251-57-2	193.127	461.65	25	9.7127E+02	1.7483E-02	1,2
1452	15250	C8H17Br	2-ethylhexyl bromide	18908-66-2	193.127	467.80	25	9.5517E+02	1.7193E-02	2
1453	15252	C8H17Cl	1-chlorooctane	111-85-3	148.675	456.62	25	2.0318E+03	3.6572E-02	1,2
1454	15253	C8H17Cl	2-chloro-2,5-dimethylhexane	29342-44-7	148.675	442.45	25	1.3296E+03	2.3933E-02	2
1455	15254	C8H17Cl	3-chloro-2,3-dimethylhexane	101654-30-2	148.675	442.45	25	1.3296E+03	2.3933E-02	2
1456	15255	C8H17Cl	2-chloro-2-methylheptane	4325-49-9	148.675	442.45	25	1.3296E+03	2.3933E-02	2
1457	15256	C8H17Cl	2-chloro-6-methylheptane	2350-19-8	148.675	442.45	25	1.3296E+03	2.3933E-02	2
1458	15257	C8H17Cl	3-chloro-3-methylheptane	5272-02-6	148.675	442.45	25	1.3296E+03	2.3933E-02	2
1459	15258	C8H17Cl	4-chloro-4-methylheptane	61764-94-1	148.675	442.45	25	1.3296E+03	2.3933E-02	2
1460	15259	C8H17Cl	3-(chloromethyl)heptane	123-04-6	148.675	445.15	25	1.3284E+03	2.3911E-02	1,2
1461	15260	C8H17Cl	2-chlorooctane	628-61-5	148.675	445.15	25	1.3284E+03	2.3911E-02	1,2
1462	15261	C8H17Cl	2-chloro-2,4,4-trimethylpentane	6111-88-2	148.675	445.15	25	1.3284E+03	2.3911E-02	2
1463	15262	C8H17Cl	(S)-2-chlorooctane	16844-08-9	148.675	442.45	25	1.3296E+03	2.3933E-02	2
1464	15263	C8H17Cl	(<+->)-2-chlorooctane	51261-14-4	148.675	445.20	25	1.3284E+03	2.3910E-02	1,2
1465	15264	C8H17Cl	2-ethylhexyl-6-chloride	2350-24-5	148.675	442.45	25	1.3296E+03	2.3933E-02	2
1466	15354	C8H18	octane	111-65-9	114.231	398.83	25	2.8611E+05	5.1499E+00	1,2
1467	15355	C8H18	2-methylheptane	592-27-8	114.231	390.80	25	2.4121E+05	4.3417E+00	1,2
1468	15356	C8H18	3-methylheptane	589-81-1	114.231	392.08	25	2.1722E+05	3.9099E+00	1,2
1469	15357	C8H18	4-methylheptane	589-53-7	114.231	390.86	25	2.4244E+05	4.3638E+00	1,2
1470	15358	C8H18	3-methyl-3-ethylpentane	1067-08-9	114.231	391.42	25	2.6744E+05	4.8138E+00	1,2
1471	15359	C8H18	3-ethylhexane	619-99-8	114.231	391.69	25	2.3349E+05	4.2027E+00	1,2
1472	15360	C8H18	3-ethyl-2-methylpentane	609-26-7	114.231	388.81	25	2.3526E+05	4.2347E+00	1,2
1473	15361	C8H18	2,2-dimethylhexane	590-73-8	114.231	379.99	25	2.0991E+05	3.7783E+00	1,2
1474	15362	C8H18	2,3-dimethylhexane	584-94-1	114.231	388.76	25	2.2925E+05	4.1264E+00	1,2
1475	15363	C8H18	2,4-dimethylhexane	589-43-5	114.231	382.58	25	2.0804E+05	3.7447E+00	1,2
1476	15364	C8H18	2,5-dimethylhexane	592-13-2	114.231	382.26	25	2.0830E+05	3.7495E+00	1,2
1477	15365	C8H18	3,3-dimethylhexane	563-16-6	114.231	385.12	25	2.3712E+05	4.2682E+00	1,2
1478	15366	C8H18	3,4-dimethylhexane	583-48-2	114.231	390.88	25	2.4065E+05	4.3316E+00	1,2
1479	15367	C8H18	2,2,3-trimethylpentane	564-02-3	114.231	383.00	25	2.3240E+05	4.1832E+00	1,2
1480	15368	C8H18	2,2,4-trimethylpentane	540-84-1	114.231	372.39	25	1.9115E+05	3.4407E+00	1,2
1481	15369	C8H18	2,3,3-trimethylpentane	560-21-4	114.231	387.92	25	2.5784E+05	4.6410E+00	1,2
1482	15370	C8H18	2,3,4-trimethylpentane	565-75-3	114.231	386.62	25	1.0100E+05	1.8180E+00	1,2
1483	15371	C8H18	2,2,3,3-tetramethylbutane	594-82-1	114.231	379.60	25	2.0880E+05	3.7583E+00	1,2
1484	15372	C8H18	octane, mixture	---	114.231	376.22	25	2.0044E+05	3.6079E+00	1,2
1485	15430	C8H18O	1-octanol	111-87-5	130.230	468.35	25	1.8652E+00	3.3573E-05	1,2
1486	15431	C8H18O	2-octanol	123-96-6	130.230	452.95	25	2.5329E+00	4.5592E-05	1,2
1487	15432	C8H18O	2-octanol, (±)	4128-31-8	130.230	453.15	25	2.4680E+00	4.4424E-05	1,2
1488	15433	C8H18O	3-octanol	589-98-0	130.230	447.85	25	4.0419E+00	7.2754E-05	1,2
1489	15434	C8H18O	4-octanol	589-62-8	130.230	449.75	25	3.3890E+00	6.1001E-05	1,2
1490	15435	C8H18O	4-octanol, (±)	74778-22-6	130.230	449.45	25	2.4718E+00	4.4492E-05	1,2
1491	15436	C8H18O	2-methyl-1-heptanol	60435-70-3	130.230	448.75	25	5.8411E+00	1.0514E-04	1,2
1492	15437	C8H18O	3-methyl-1-heptanol	1070-32-2	130.230	459.15	25	5.8615E+00	1.0551E-04	1,2
1493	15438	C8H18O	4-methyl-1-heptanol	817-91-4	130.230	456.35	25	5.6285E+00	1.0131E-04	1,2
1494	15439	C8H18O	5-methyl-1-heptanol	7212-53-5	130.230	459.75	25	3.3653E+00	6.0576E-05	1,2
1495	15440	C8H18O	6-methyl-1-heptanol	1653-40-3	130.230	460.85	25	1.6830E+00	3.0294E-05	1,2
1496	15441	C8H18O	2-methyl-2-heptanol	625-25-2	130.230	429.85	25	1.7848E+00	3.2126E-05	1,2
1497	15442	C8H18O	3-methyl-2-heptanol	31367-46-1	130.230	439.25	25	6.9698E+00	1.2546E-04	1,2
1498	15443	C8H18O	4-methyl-2-heptanol	56298-90-9	130.230	444.75	25	3.4739E+00	6.2530E-05	1,2
1499	15444	C8H18O	5-methyl-2-heptanol	54630-50-1	130.230	445.05	25	5.0873E+00	9.1571E-05	1,2
1500	15445	C8H18O	6-methyl-2-heptanol	4730-22-7	130.230	445.05	25	2.5897E+00	4.6614E-05	1,2
1501	15446	C8H18O	2-methyl-3-heptanol	18720-62-2	130.230	440.75	25	3.0461E+00	5.4830E-05	1,2
1502	15447	C8H18O	3-methyl-3-heptanol	5582-82-1	130.230	434.15	25	2.8707E+00	5.1673E-05	1,2
1503	15448	C8H18O	4-methyl-3-heptanol	14979-39-6	130.230	428.55	25	7.4887E+00	1.3480E-04	1,2
1504	15449	C8H18O	5-methyl-3-heptanol	18720-65-5	130.230	426.75	25	6.2830E+00	1.1309E-04	1,2
1505	15450	C8H18O	6-methyl-3-heptanol	18720-66-6	130.230	433.00	25	2.4483E+00	4.4070E-05	2
1506	15451	C8H18O	2-methyl-4-heptanol	21570-35-4	130.230	439.25	25	2.4684E+00	4.4432E-05	1,2
1507	15452	C8H18O	3-methyl-4-heptanol	1838-73-9	130.230	437.85	25	6.6388E+00	1.1950E-04	1,2
1508	15453	C8H18O	4-methyl-4-heptanol	598-01-6	130.230	434.25	25	2.9256E+00	5.2661E-05	1,2
1509	15454	C8H18O	2-ethyl-1-hexanol	104-76-7	130.230	457.75	20	2.3809E+00	4.2857E-05	1,2
1510	15455	C8H18O	3-ethyl-1-hexanol	41065-95-6	130.230	453.00	25	2.4682E+00	4.4428E-05	2
1511	15456	C8H18O	4-ethyl-1-hexanol	---	130.230	453.00	25	2.4682E+00	4.4428E-05	2
1512	15457	C8H18O	2,2-dimethyl-1-hexanol	2370-13-0	130.230	445.65	25	9.3235E-01	1.6782E-05	1,2
1513	15458	C8H18O	2,3-dimethyl-1-hexanol	---	130.230	447.00	25	2.4729E+00	4.4512E-05	2
1514	15459	C8H18O	2,4-dimethyl-1-hexanol	3965-59-1	130.230	448.15	25	7.9188E+00	1.4254E-04	1,2
1515	15460	C8H18O	2,5-dimethyl-1-hexanol	6886-16-4	130.230	452.15	25	8.4483E+00	1.5207E-04	1,2
1516	15461	C8H18O	3,3-dimethyl-1-hexanol	10524-70-6	130.230	454.00	25	2.4669E+00	4.4403E-05	2
1517	15462	C8H18O	3,4-dimethyl-1-hexanol	---	130.230	454.00	25	2.4669E+00	4.4403E-05	2
1518	15463	C8H18O	3,5-dimethyl-1-hexanol	13501-73-0	130.230	456.15	25	9.0238E+00	1.6243E-04	1,2
1519	15464	C8H18O	4,4-dimethyl-1-hexanol	6481-95-4	130.230	455.00	25	2.4654E+00	4.4376E-05	2
1520	15465	C8H18O	4,5-dimethyl-1-hexanol	---	130.230	455.00	25	2.4654E+00	4.4376E-05	2

Table 10 Henry's Law Constant for Compound in Water - Organic Compounds

								Henry's Law Constant, H		
NO	ID	FORMULA	NAME	CAS No	MW, g/mol	T_B, K	T, C	H @ T, atm/mol frac	H @ T, atm/mol/m³	code
1521	15466	C8H18O	5,5-dimethyl-1-hexanol	---	130.230	455.00	25	2.4654E+00	4.4376E-05	2
1522	15467	C8H18O	3-ethyl-2-hexanol	24448-19-9	130.230	441.15	25	7.0835E+00	1.2750E-04	1,2
1523	15468	C8H18O	4-ethyl-2-hexanol	---	130.230	441.00	25	2.4696E+00	4.4452E-05	2
1524	15469	C8H18O	2,3-dimethyl-2-hexanol	19550-03-9	130.230	433.25	25	1.8191E+00	3.2744E-05	1,2
1525	15470	C8H18O	2,4-dimethyl-2-hexanol	42328-76-7	130.230	423.85	25	2.6277E+00	4.7299E-05	1,2
1526	15471	C8H18O	2,5-dimethyl-2-hexanol	3730-60-7	130.230	425.65	25	1.4471E+00	2.6047E-05	1,2
1527	15472	C8H18O	3,3-dimethyl-2-hexanol	22025-20-3	130.230	433.00	25	2.4483E+00	4.4070E-05	2
1528	15473	C8H18O	3,4-dimethyl-2-hexanol	19550-05-1	130.230	444.15	25	8.0603E+00	1.4508E-04	1,2
1529	15474	C8H18O	3,5-dimethyl-2-hexanol	66576-27-0	130.230	433.15	25	6.7297E+00	1.2113E-04	1,2
1530	15475	C8H18O	4,4-dimethyl-2-hexanol	---	130.230	435.00	25	2.4557E+00	4.4203E-05	2
1531	15476	C8H18O	4,5-dimethyl-2-hexanol	---	130.230	435.00	25	2.4557E+00	4.4203E-05	2
1532	15477	C8H18O	5,5-dimethyl-2-hexanol	31841-77-7	130.230	439.15	25	8.0995E+00	1.4579E-04	1,2
1533	15478	C8H18O	3-ethyl-3-hexanol	597-76-2	130.230	432.15	25	3.6597E+00	6.5873E-05	1,2
1534	15479	C8H18O	4-ethyl-3-hexanol	19780-44-0	130.230	437.15	25	3.8753E+00	6.9755E-05	1,2
1535	15480	C8H18O	2,2-dimethyl-3-hexanol	4209-90-9	130.230	429.25	25	8.8585E+00	1.5945E-04	1,2
1536	15481	C8H18O	2,3-dimethyl-3-hexanol	4166-46-5	130.230	431.35	25	4.8584E+00	8.7450E-05	1,2
1537	15482	C8H18O	2,4-dimethyl-3-hexanol	13432-25-2	130.230	433.15	25	7.9086E+00	1.4235E-04	1,2
1538	15483	C8H18O	2,5-dimethyl-3-hexanol	19550-07-3	130.230	432.15	25	2.7907E+00	5.0233E-05	1,2
1539	15484	C8H18O	3,4-dimethyl-3-hexanol	19550-08-4	130.230	425.15	25	2.9485E+00	5.3073E-05	1,2
1540	15485	C8H18O	3,5-dimethyl-3-hexanol	4209-91-0	130.230	425.15	25	6.1048E+00	1.0989E-04	1,2
1541	15486	C8H18O	4,4-dimethyl-3-hexanol	19550-09-5	130.230	432.15	25	4.1916E+00	7.5449E-05	1,2
1542	15487	C8H18O	4,5-dimethyl-3-hexanol	---	130.230	429.00	25	2.4288E+00	4.3719E-05	2
1543	15488	C8H18O	5,5-dimethyl-3-hexanol	66576-31-6	130.230	426.15	25	3.6891E+00	6.6404E-05	1,2
1544	15489	C8H18O	2-propyl-1-pentanol	58175-57-8	130.230	452.15	25	1.3543E+00	2.4378E-05	1,2
1545	15490	C8H18O	2-isopropyl-1-pentanol	---	130.230	473.74	25	2.4269E+00	4.3684E-05	2
1546	15491	C8H18O	2-methyl-2-ethyl-1-pentanol	5970-63-8	130.230	450.25	25	2.4118E+00	4.3411E-05	1,2
1547	15492	C8H18O	3-methyl-2-ethyl-1-pentanol	---	130.230	450.00	25	2.4714E+00	4.4485E-05	2
1548	15493	C8H18O	4-methyl-2-ethyl-1-pentanol	106-67-2	130.230	449.65	25	2.9567E+00	5.3219E-05	1,2
1549	15494	C8H18O	2-methyl-3-ethyl-1-pentanol	---	130.230	448.00	25	2.4726E+00	4.4506E-05	2
1550	15495	C8H18O	3-methyl-3-ethyl-1-pentanol	---	130.230	448.00	25	2.4726E+00	4.4506E-05	2
1551	15496	C8H18O	4-methyl-3-ethyl-1-pentanol	---	130.230	448.00	25	2.4726E+00	4.4506E-05	2
1552	15497	C8H18O	2,2,3-trimethyl-1-pentanol	57409-53-7	130.230	447.65	25	2.3464E+00	4.2234E-05	1,2
1553	15498	C8H18O	2,2,4-trimethyl-1-pentanol	123-44-4	130.230	441.45	25	3.4271E+00	6.1687E-05	1,2
1554	15499	C8H18O	2,3,3-trimethyl-1-pentanol	---	130.230	449.00	25	2.4721E+00	4.4497E-05	2
1555	15500	C8H18O	2,3,4-trimethyl-1-pentanol	6570-88-3	130.230	456.15	25	9.8569E+00	1.7742E-04	1,2
1556	15501	C8H18O	2,4,4-trimethyl-1-pentanol	16325-63-6	130.230	444.15	25	4.3726E+00	7.8707E-05	1,2
1557	15502	C8H18O	3,3,4-trimethyl-1-pentanol	---	130.230	444.00	25	2.4724E+00	4.4503E-05	2
1558	15503	C8H18O	3,4,4-trimethyl-1-pentanol	16325-64-7	130.230	444.00	25	2.4724E+00	4.4503E-05	2
1559	15504	C8H18O	2-methyl-3-ethyl-2-pentanol	19780-63-3	130.230	430.95	25	2.0337E+00	3.6607E-05	1,2
1560	15505	C8H18O	3-methyl-3-ethyl-2-pentanol	66576-22-5	130.230	433.00	25	2.4483E+00	4.4070E-05	2
1561	15506	C8H18O	4-methyl-3-ethyl-2-pentanol	66576-23-6	130.230	437.15	25	7.1863E+00	1.2935E-04	1,2
1562	15507	C8H18O	2,3,3-trimethyl-2-pentanol	23171-85-9	130.230	433.15	25	6.7715E+00	1.2189E-04	1,2
1563	15508	C8H18O	2,3,4-trimethyl-2-pentanol	66576-26-9	130.230	431.15	25	7.6096E+00	1.3697E-04	1,2
1564	15509	C8H18O	2,4,4-trimethyl-2-pentanol	690-37-9	130.230	419.55	25	3.7458E+00	6.7425E-05	1,2
1565	15510	C8H18O	3,3,4-trimethyl-2-pentanol	19411-41-7	130.230	438.15	25	8.6074E+00	1.5493E-04	1,2
1566	15511	C8H18O	3,4,4-trimethyl-2-pentanol	10575-56-1	130.230	431.15	25	2.5825E+00	4.6485E-05	1,2
1567	15512	C8H18O	2-methyl-3-ethyl-3-pentanol	597-05-7	130.230	433.15	25	7.3134E+00	1.3164E-04	1,2
1568	15513	C8H18O	2,2,3-trimethyl-3-pentanol	7294-05-5	130.230	425.15	25	4.9530E+00	8.9154E-05	1,2
1569	15514	C8H18O	2,2,4-trimethyl-3-pentanol	5162-48-1	130.230	424.15	25	2.9815E+00	5.3666E-05	1,2
1570	15515	C8H18O	2,3,4-trimethyl-3-pentanol	3054-92-0	130.230	430.15	25	8.5148E+00	1.5326E-04	1,2
1571	15516	C8H18O	3-methyl-2-isopropyl-1-butanol	18593-92-5	130.230	446.15	25	8.3306E+00	1.4995E-04	1,2
1572	15517	C8H18O	2,2-diethyl-1-butanol	13023-60-4	130.230	446.00	25	2.4730E+00	4.4513E-05	2
1573	15518	C8H18O	2,3-dimethyl-2-ethyl-1-butanol	---	130.230	446.00	25	2.4730E+00	4.4513E-05	2
1574	15519	C8H18O	3,3-dimethyl-2-ethyl-1-butanol	66576-56-5	130.230	446.00	25	2.4730E+00	4.4513E-05	2
1575	15520	C8H18O	2,2,3,3-tetramethyl-1-butanol	66576-24-7	130.230	446.00	25	2.4730E+00	4.4513E-05	2
1576	15521	C8H18O	3,5-dimethyl-3-hexanol, (±)	19113-78-1	130.230	425.15	25	2.4036E+00	4.3264E-05	1,2
1577	15522	C8H18O	2-methyl-1-heptanol, (±)	111675-77-5	130.230	448.75	25	2.4722E+00	4.4500E-05	1,2
1578	15523	C8H18O	5-methyl-1-heptanol, (±)	111767-95-4	130.230	459.75	25	2.4567E+00	4.4220E-05	1,2
1579	15524	C8H18O	2-methyl-3-heptanol, (±)	100296-26-2	130.230	440.65	25	2.4691E+00	4.4444E-05	1,2
1580	15525	C8H18O	6-methyl-3-heptanol, (±)	100295-85-0	130.230	442.15	25	2.4710E+00	4.4477E-05	1,2
1581	15526	C8H18O	3-methyl-4-heptanol, (R*,S*)-(±)	92737-91-2	130.230	435.15	25	2.4562E+00	4.4212E-05	1,2
1582	15527	C8H18O	dibutyl ether	142-96-1	130.230	413.44	25	2.1936E+02	3.9484E-03	1,2
1583	15528	C8H18O	di-sec-butyl ether	6863-58-7	130.230	394.20	25	2.9829E+02	5.3693E-03	1,2
1584	15529	C8H18O	di-tert-butyl ether	6163-66-2	130.230	380.40	25	3.0063E+02	5.4112E-03	1,2
1585	15530	C8H18O	butyl isobutyl ether	17071-47-5	130.230	424.15	25	2.0716E+02	3.7289E-03	1,2
1586	15531	C8H18O	tert-butyl isobutyl ether	33021-02-2	130.230	439.93	25	1.9853E+02	3.5735E-03	2
1587	15532	C8H18O	di-sec-butyl ether, (±)	17226-28-7	130.230	393.65	25	2.0576E+02	3.7037E-03	1,2
1588	15533	C8H18O	ethyl hexyl ether	5756-43-4	130.230	416.15	25	2.0910E+02	3.7637E-03	1,2
1589	15534	C8H18O	1-methoxyheptane	629-32-3	130.230	424.15	25	2.0716E+02	3.7289E-03	1,2
1590	15535	C8H18O	isooctanol	26952-21-6	130.230	459.15	25	2.4579E+00	4.4242E-05	1,2
1591	15536	C8H18O	(R)-(-)-2-octanol	5978-70-1	130.230	448.90	25	2.4722E+00	4.4498E-05	1,2
1592	15537	C8H18O	(S)-(+)-2-octanol	6169-06-8	130.230	451.05	25	2.4705E+00	4.4468E-05	1,2
1593	15538	C8H18O	octyl alcohol; mixed isomers	29063-28-3	130.230	468.15	25	2.4384E+00	4.3891E-05	1,2
1594	15539	C8H18O	isobutyl ether	628-55-7	130.230	395.85	25	2.0664E+02	3.7195E-03	1,2
1595	15578	C8H18O3	diethylene glycol diethyl ether	112-36-7	162.229	462.15	25	6.1667E-03	1.1100E-07	2
1596	15579	C8H18O3	diethylene glycol monobutyl ether	112-34-5	162.229	504.15	25	7.1112E-04	1.2800E-08	2
1597	15598	C8H18O5	tetraethylene glycol	112-60-7	194.228	581.00	25	3.0556E-14	5.5000E-19	2
1598	15602	C8H18S	octyl mercaptan	111-88-6	146.297	472.19	25	8.4244E+02	1.5164E-02	1,2
1599	15603	C8H18S	2-octanethiol	3001-66-9	146.297	459.55	25	7.2244E+02	1.3004E-02	1,2
1600	15604	C8H18S	tert-octyl mercaptan	141-59-3	146.297	429.00	25	9.9741E+02	1.7953E-02	1,2

Table 10 Henry's Law Constant for Compound in Water - Organic Compounds

								Henry's Law Constant, H		
NO	ID	FORMULA	NAME	CAS No	MW, g/mol	T_B, K	T, C	H @ T, atm/mol frac	H @ T, atm/mol/m^3	code
1601	15605	C8H18S	2-octanethiol, (±)	10435-81-1	146.297	459.55	25	3.4759E+02	6.2565E-03	1,2
1602	15606	C8H18S	dibutyl sulfide	544-40-1	146.297	455.15	25	4.5815E+02	8.2467E-03	1,2
1603	15607	C8H18S	methyl heptyl sulfide	20291-61-6	146.297	461.70	25	2.2684E+02	4.0831E-03	1,2
1604	15608	C8H18S	ethyl hexyl sulfide	7309-44-6	146.297	460.00	25	2.2703E+02	4.0866E-03	2
1605	15609	C8H18S	propyl pentyl sulfide	42841-80-5	146.297	460.00	25	2.2703E+02	4.0866E-03	2
1606	15610	C8H18S	di-sec-butyl sulfide	626-26-6	146.297	438.15	25	2.2158E+02	3.9885E-03	1,2
1607	15611	C8H18S	di-tert-butyl sulfide	107-47-1	146.297	422.25	25	2.1161E+02	3.8089E-03	1,2
1608	15612	C8H18S	diisobutyl sulfide	592-65-4	146.297	444.15	25	2.2432E+02	4.0377E-03	1,2
1609	15631	C8H19N	octylamine	111-86-4	129.246	452.75	25	3.8274E+00	6.8893E-05	1,2
1610	15632	C8H19N	2-ethylhexylamine	104-75-6	129.246	442.35	25	3.5223E+00	6.3402E-05	1,2
1611	15633	C8H19N	6-methyl-2-heptanamine, (±)	5984-58-7	129.246	428.15	25	2.4155E+00	4.3479E-05	1,2
1612	15634	C8H19N	2-octanamine, (±)	44855-57-4	129.246	437.15	25	3.0572E+00	5.5028E-05	1,2
1613	15635	C8H19N	methylheptylamine	36343-05-2	129.246	438.15	25	2.4477E+00	4.4057E-05	1,2
1614	15636	C8H19N	ethylhexylamine	20352-67-4	129.246	431.15	25	2.6363E+00	4.7453E-05	1,2
1615	15637	C8H19N	dibutylamine	111-92-2	129.246	432.00	25	3.5717E+00	6.4291E-05	1,2
1616	15638	C8H19N	diisobutylamine	110-96-3	129.246	412.25	25	5.5443E+00	9.9797E-05	1,2
1617	15639	C8H19N	di-sec-butylamine	626-23-3	129.246	407.15	25	2.2752E+00	4.0954E-05	1,2
1618	15640	C8H19N	N-methyl-2-heptanamine	540-43-2	129.246	428.15	25	3.7412E+00	6.7342E-05	1,2
1619	15641	C8H19N	dimethylhexylamine	4385-04-0	129.246	418.15	25	1.4210E+01	2.5578E-04	1,2
1620	15642	C8H19N	diethylbutylamine	4444-68-2	129.246	409.15	25	9.8785E+00	1.7781E-04	1,2
1621	15643	C8H19N	N-ethyl-N-isopropyl-2-propanamine	7087-68-5	129.246	399.65	25	6.7374E+00	1.2127E-04	1,2
1622	15644	C8H19N	N-ethyl-N-propyl-1-propanamine	20634-92-8	129.246	411.15	25	8.6204E+00	1.5517E-04	1,2
1623	15645	C8H19N	2-amino-6-methylheptane	543-82-8	129.246	428.15	25	3.7412E+00	6.7342E-05	1,2
1624	15646	C8H19N	(S)-2-amino-6-methylheptane	70419-10-2	129.246	428.15	25	3.7412E+00	6.7342E-05	1,2
1625	15647	C8H19N	(R)-2-amino-6-methylheptane	70419-11-3	129.246	428.15	25	3.7412E+00	6.7342E-05	1,2
1626	15648	C8H19N	(S)-2-aminooctane	34566-04-6	129.246	437.15	25	3.0572E+00	5.5028E-05	1,2
1627	15649	C8H19N	(R)-2-aminooctane	34566-05-7	129.246	437.15	25	3.0572E+00	5.5028E-05	1,2
1628	15650	C8H19N	2-aminooctane	693-16-3	129.246	438.15	25	3.1406E+00	5.6530E-05	1,2
1629	15651	C8H19N	N-isopropyl-N-methyl-tert-butylamine	85523-00-8	129.246	400.15	25	6.8062E+00	1.2251E-04	1,2
1630	15652	C8H19N	tert-octylamine	107-45-9	129.246	413.15	25	1.6798E+00	3.0236E-05	1,2
1631	15747	C8H23N5	tetraethylenepentamine	112-57-2	189.306	606.15	25	1.6667E-15	3.0000E-20	2
1632	16010	C9H7N	isoquinoline	119-65-3	129.162	516.40	20	9.5465E-02	1.7184E-06	1,2
1633	16011	C9H7N	quinoline	91-22-5	129.162	510.75	20	5.5870E-02	1.0056E-06	1,2
1634	16018	C9H7NO	8-hydroxyquinoline	148-24-3	145.161	540.00	20	4.4234E-02	7.9620E-07	1,2
1635	16508	C9H10	indane	496-11-7	118.178	451.12	25	1.2761E+02	2.2970E-03	1,2
1636	16512	C9H10	alpha-methylstyrene	98-83-9	118.178	438.65	25	2.7479E+02	4.9462E-03	1,2
1637	16513	C9H10	m-methylstyrene	100-80-1	118.178	444.75	25	2.2060E+02	3.9707E-03	1,2
1638	16514	C9H10	o-methylstyrene	611-15-4	118.178	442.96	25	2.0674E+02	3.7212E-03	1,2
1639	16515	C9H10	p-methylstyrene	622-97-9	118.178	445.93	25	1.8656E+02	3.3580E-03	1,2
1640	16519	C9H10	methylstyrene	25013-15-4	118.178	442.65	25	1.9289E+02	3.4719E-03	1,2
1641	16520	C9H10	3(4)-methylstyrene, isomers	39294-88-7	118.178	443.15	25	1.8812E+02	3.3862E-03	1,2
1642	16676	C9H10O2	ethyl benzoate	93-89-0	150.177	486.55	25	4.2659E+00	7.6786E-05	1,2
1643	17004	C9H12	cumene	98-82-8	120.194	425.56	25	8.6347E+02	1.5542E-02	1,2
1644	17005	C9H12	m-ethyltoluene	620-14-4	120.194	434.48	25	3.4135E+02	6.1443E-03	1,2
1645	17006	C9H12	o-ethyltoluene	611-14-3	120.194	438.33	25	3.1209E+02	5.6177E-03	1,2
1646	17007	C9H12	p-ethyltoluene	622-96-8	120.194	435.16	25	3.0873E+02	5.5572E-03	1,2
1647	17008	C9H12	1,2,3-trimethylbenzene	526-73-8	120.194	449.27	25	2.0907E+02	3.7632E-03	1,2
1648	17009	C9H12	1,2,4-trimethylbenzene	95-63-6	120.194	442.53	25	3.6097E+02	6.4975E-03	1,2
1649	17010	C9H12	trimethyl benzene	25551-13-7	120.194	445.00	25	2.6096E+02	4.6972E-03	2
1650	17011	C9H12	mesitylene	108-67-8	120.194	437.89	25	2.3093E+02	4.1567E-03	1,2
1651	17012	C9H12	propylbenzene	103-65-1	120.194	432.39	25	6.3559E+02	1.1441E-02	1,2
1652	17436	C9H14O	isophorone	78-59-1	138.210	488.35	25	3.5841E-01	6.4513E-06	1,2
1653	17520	C9H14S	2-isopentylthiophene	26963-33-7	154.276	469.15	25	5.0806E+02	9.1450E-03	1,2
1654	17593	C9H16	1-nonyne	3452-09-3	124.226	423.96	25	3.3476E+03	6.0257E-02	1,2
1655	17594	C9H16	2-nonyne	19447-29-1	124.226	435.05	25	3.6267E+03	6.5280E-02	1,2
1656	17595	C9H16	3-nonyne	20184-89-8	124.226	430.25	25	3.5616E+03	6.4108E-02	1,2
1657	17596	C9H16	4-nonyne	20184-91-2	124.226	427.15	25	3.5084E+03	6.3151E-02	1,2
1658	17597	C9H16	7-methyl-3-octyne	37050-06-9	124.226	421.10	25	3.3843E+03	6.0916E-02	2
1659	17598	C9H16	2,6-dimethyl-3-heptyne	19549-97-4	124.226	403.75	25	2.9372E+03	5.2868E-02	1,2
1660	17599	C9H16	5,5-dimethyl-3-heptyne	23097-98-5	124.226	421.10	25	3.3843E+03	6.0916E-02	2
1661	17605	C9H16	1,8-nonadiene	4900-30-5	124.226	415.65	25	1.5540E+04	2.7972E-01	1,2
1662	17606	C9H16	2,7-nonadiene	51333-70-1	124.226	424.65	25	1.6398E+04	2.9517E-01	1,2
1663	17607	C9H16	4-methyl-3,5-octadiene	36903-95-4	124.226	422.65	25	1.6227E+04	2.9209E-01	1,2
1664	17608	C9H16	7-methyl-2,4-octadiene	2216-70-8	124.226	422.15	25	1.6183E+04	2.9129E-01	1,2
1665	17609	C9H16	2,4-dimethyl-2,4-heptadiene	74421-05-9	124.226	411.15	25	1.5042E+04	2.7075E-01	1,2
1666	17610	C9H16	2,6-dimethyl-1,3-heptadiene	2436-84-2	124.226	414.15	25	1.5378E+04	2.7680E-01	1,2
1667	17611	C9H16	2,6-dimethyl-1,5-heptadiene	6709-39-3	124.226	416.15	25	1.5593E+04	2.8067E-01	1,2
1668	17612	C9H16	3,5-dimethyl-2,4-heptadiene	101935-28-8	124.226	413.15	25	1.5268E+04	2.7481E-01	1,2
1669	17613	C9H16	3-ethyl-2-methyl-1,5-hexadiene	73398-15-9	124.226	418.15	25	1.5800E+04	2.8439E-01	1,2
1670	17628	C9H16	7-methyl-1,6-octadiene	42152-47-6	124.226	416.65	25	1.5645E+04	2.8162E-01	1,2
1671	17755	C9H16O4	azelaic acid	123-99-9	188.224	633.36	20	1.0133E-07	1.8240E-12	1,2
1672	17859	C9H18	butylcyclopentane	2040-95-1	126.242	429.76	25	8.4673E+04	1.5241E+00	1,2
1673	17860	C9H18	isobutylcyclopentane	3788-32-7	126.242	421.10	25	7.4179E+04	1.3352E+00	1,2
1674	17861	C9H18	sec-butylcyclopentane	4850-32-2	126.242	427.50	25	8.1397E+04	1.4651E+00	1,2
1675	17862	C9H18	tert-butylcyclopentane	3875-52-3	126.242	418.00	25	8.0786E+04	1.4541E+00	1,2
1676	17863	C9H18	1-methyl-1-propylcyclopentane	16631-63-3	126.242	419.15	25	7.2090E+04	1.2976E+00	1,2
1677	17864	C9H18	1-methyl-cis-2-propylcyclopentane	932-43-4	126.242	425.73	25	8.0499E+04	1.4490E+00	1,2
1678	17865	C9H18	1-methyl-trans-2-propylcyclopentane	932-44-5	126.242	419.52	25	7.3383E+04	1.3209E+00	1,2
1679	17866	C9H18	1-methyl-cis-3-propylcyclopentane	2443-04-1	126.242	421.15	25	7.5231E+04	1.3542E+00	1,2
1680	17867	C9H18	1-methyl-trans-3-propylcyclopentane	2443-03-0	126.242	421.15	25	7.5231E+04	1.3542E+00	1,2

Table 10 Henry's Law Constant for Compound in Water - Organic Compounds

					Henry's Law Constant, H					
NO	ID	FORMULA	NAME	CAS No	MW, g/mol	T_B, K	T, C	H @ T, atm/mol frac	H @ T, atm/mol/m^3	code
1681	17868	C9H18	1-methyl-1-isopropylcyclopentane	61828-00-0	126.242	421.15	25	7.9325E+04	1.4278E+00	1,2
1682	17869	C9H18	1-methyl-cis-2-isopropylcyclopentane	61868-01-7	126.242	421.15	25	8.0293E+04	1.4453E+00	1,2
1683	17870	C9H18	1-methyl-trans-2-isopropylcyclopentane	61828-01-1	126.242	414.15	25	7.1689E+04	1.2904E+00	1,2
1684	17871	C9H18	1-methyl-cis-3-isopropylcyclopentane	61828-02-2	126.242	415.15	25	7.2886E+04	1.3119E+00	1,2
1685	17872	C9H18	1-methyl-trans-3-isopropylcyclopentane	61828-03-3	126.242	415.15	25	7.2886E+04	1.3119E+00	1,2
1686	17873	C9H18	1,1-diethylcyclopentane	2721-38-2	126.242	423.65	25	7.7119E+04	1.3881E+00	1,2
1687	17874	C9H18	1,cis-2-diethylcyclopentane	932-39-8	126.242	426.71	25	8.1635E+04	1.4694E+00	1,2
1688	17875	C9H18	1,trans-2-diethylcyclopentane	932-40-1	126.242	420.68	25	7.4697E+04	1.3445E+00	1,2
1689	17876	C9H18	1,3-diethylcyclopentane	19398-75-5	126.242	423.15	25	9.9552E+04	1.7919E+00	1,2
1690	17877	C9H18	1,cis-3-diethylcyclopentane	62016-59-5	126.242	423.15	25	7.7521E+04	1.3954E+00	1,2
1691	17878	C9H18	1,trans-3-diethylcyclopentane	62016-60-8	126.242	423.15	25	7.7521E+04	1.3954E+00	1,2
1692	17879	C9H18	1,1-dimethyl-2-ethylcyclopentane	54549-80-3	126.242	411.15	25	6.8262E+04	1.2287E+00	1,2
1693	17880	C9H18	1,1-dimethyl-3-ethylcyclopentane	62016-61-9	126.242	407.15	25	6.3750E+04	1.1475E+00	1,2
1694	17881	C9H18	1,cis-2-dimethyl-1-ethylcyclopentane	62016-63-1	126.242	417.15	25	7.5421E+04	1.3576E+00	1,2
1695	17882	C9H18	1,trans-2-dimethyl-1-ethylcyclopentane	62016-62-0	126.242	417.15	25	7.5421E+04	1.3576E+00	1,2
1696	17883	C9H18	1,cis-2-dimethyl-cis-3-ethylcyclopentane	---	126.242	425.15	25	8.6523E+04	1.5574E+00	1,2
1697	17884	C9H18	1,cis-2-dimethyl-trans-3-ethylcyclopentane	---	126.242	418.15	25	7.7495E+04	1.3949E+00	1,2
1698	17885	C9H18	1,trans-2-dimethyl-cis-3-ethylcyclopentane	---	126.242	411.15	25	6.8961E+04	1.2413E+00	1,2
1699	17886	C9H18	1,trans-2-dimethyl-trans-3-ethylcyclopentane	---	126.242	418.15	25	7.7495E+04	1.3949E+00	1,2
1700	17887	C9H18	1,cis-2-dimethyl-cis-4-ethylcyclopentane	62016-64-2	126.242	420.15	25	8.0033E+04	1.4406E+00	1,2
1701	17888	C9H18	1,cis-2-dimethyl-trans-4-ethylcyclopentane	62016-65-3	126.242	420.15	25	8.0033E+04	1.4406E+00	1,2
1702	17889	C9H18	1,trans-2-dimethyl-cis-4-ethylcyclopentane	62016-66-4	126.242	413.15	25	7.1337E+04	1.2841E+00	1,2
1703	17890	C9H18	1,cis-3-dimethyl-1-ethylcyclopentane	62016-68-6	126.242	409.15	25	6.5978E+04	1.1876E+00	1,2
1704	17891	C9H18	1,trans-3-dimethyl-1-ethylcyclopentane	62016-67-5	126.242	409.15	25	6.5978E+04	1.1876E+00	1,2
1705	17892	C9H18	1,cis-3-dimethyl-cis-2-ethylcyclopentane	19903-00-5	126.242	425.15	25	8.6523E+04	1.5574E+00	1,2
1706	17893	C9H18	1,cis-3-dimethyl-trans-2-ethylcyclopentane	19902-98-8	126.242	410.15	25	6.7793E+04	1.2203E+00	1,2
1707	17894	C9H18	1,trans-3-dimethyl-cis-2-ethylcyclopentane	19902-99-9	126.242	417.15	25	7.6240E+04	1.3723E+00	1,2
1708	17895	C9H18	1,cis-3-dimethyl-cis-4-ethylcyclopentane	---	126.242	418.15	25	7.7495E+04	1.3949E+00	1,2
1709	17896	C9H18	1,cis-3-dimethyl-trans-4-ethylcyclopentane	---	126.242	410.15	25	6.7793E+04	1.2203E+00	1,2
1710	17897	C9H18	1,trans-3-dimethyl-cis-4-ethylcyclopentane	---	126.242	412.00	25	7.2152E+04	1.2987E+00	1,2
1711	17898	C9H18	1,trans-3-dimethyl-trans-4-ethylcyclopentane	---	126.242	418.15	25	7.7495E+04	1.3949E+00	1,2
1712	17899	C9H18	1,1,2,2-tetramethylcyclopentane	52688-89-8	126.242	406.15	25	6.6324E+04	1.1938E+00	1,2
1713	17900	C9H18	1,1,cis-2,cis-3-tetramethylcyclopentane	62016-70-0	126.242	411.15	25	7.3246E+04	1.3184E+00	1,2
1714	17901	C9H18	1,1,cis-2,trans-3-tetramethylcyclopentane	---	126.242	403.15	25	6.3473E+04	1.1425E+00	1,2
1715	17902	C9H18	1,1,cis-2,cis-4-tetramethylcyclopentane	62016-72-2	126.242	403.15	25	6.3473E+04	1.1425E+00	1,2
1716	17903	C9H18	1,1,cis-2,trans-4-tetramethylcyclopentane	---	126.242	403.15	25	6.3473E+04	1.1425E+00	1,2
1717	17904	C9H18	1,1,3,3-tetramethylcyclopentane	50876-33-0	126.242	391.11	25	5.0507E+04	9.0912E-01	1,2
1718	17905	C9H18	1,1,cis-3,cis-4-tetramethylcyclopentane	20309-77-7	126.242	406.15	25	6.7007E+04	1.2061E+00	1,2
1719	17906	C9H18	1,1,cis-3,trans-4-tetramethylcyclopentane	---	126.242	394.75	25	5.4450E+04	9.8010E-01	1,2
1720	17907	C9H18	1,2,2,cis-3-tetramethylcyclopentane	62016-73-3	126.242	411.15	25	7.3246E+04	1.3184E+00	1,2
1721	17908	C9H18	1,2,2,trans-3-tetramethylcyclopentane	---	126.242	411.15	25	7.3246E+04	1.3184E+00	1,2
1722	17909	C9H18	1,cis-2,cis-3,cis-4-tetramethylcyclopentane	2532-65-2	126.242	418.15	25	8.3466E+04	1.5024E+00	1,2
1723	17910	C9H18	1,cis-2,cis-3,trans-4-tetramethylcyclopentane	2532-69-6	126.242	411.15	25	7.3928E+04	1.3307E+00	1,2
1724	17911	C9H18	1,cis-2,trans-3,cis-4-tetramethylcyclopentane	2532-68-5	126.242	404.15	25	6.5192E+04	1.1735E+00	1,2
1725	17912	C9H18	1,cis-2,trans-3,trans-4-tetramethylcyclopenta	2532-64-1	126.242	411.15	25	7.3928E+04	1.3307E+00	1,2
1726	17913	C9H18	1,trans-2,cis-3,trans-4-tetramethylcyclopenta	2532-67-4	126.242	400.55	25	6.1045E+04	1.0988E+00	1,2
1727	17914	C9H18	1,trans-2,trans-3,cis-4-tetramethylcyclopenta	19907-40-5	126.242	404.15	25	6.5192E+04	1.1735E+00	1,2
1728	17915	C9H18	1-methyl-2-isopropylcyclopentane	89223-57-4	126.242	416.15	25	9.0437E+04	1.6279E+00	1,2
1729	17916	C9H18	propylcyclohexane	1678-92-8	126.242	429.90	25	7.0458E+04	1.2682E+00	1,2
1730	17917	C9H18	isopropylcyclohexane	696-29-7	126.242	427.91	25	6.0800E+04	1.0944E+00	1,2
1731	17918	C9H18	1-methyl-1-ethylcyclohexane	4926-90-3	126.242	425.31	25	6.6509E+04	1.1972E+00	1,2
1732	17919	C9H18	1-methyl-cis-2-ethylcyclohexane	4923-77-7	126.242	429.25	25	5.3551E+04	9.6392E-01	1,2
1733	17920	C9H18	1-methyl-trans-2-ethylcyclohexane	4923-78-8	126.242	424.85	25	5.0050E+04	9.0088E-01	1,2
1734	17921	C9H18	1-methyl-cis-3-ethylcyclohexane	19489-10-2	126.242	421.15	25	5.0521E+04	9.0937E-01	1,2
1735	17922	C9H18	1-methyl-trans-3-ethylcyclohexane	4926-76-5	126.242	424.15	25	4.7824E+04	8.6082E-01	1,2
1736	17923	C9H18	1-methyl-cis-4-ethylcyclohexane	4926-78-7	126.242	424.15	25	4.1447E+04	7.4604E-01	1,2
1737	17924	C9H18	1-methyl-trans-4-ethylcyclohexane	6236-88-0	126.242	421.15	25	6.3909E+04	1.1504E+00	1,2
1738	17925	C9H18	1,1,2-trimethylcyclohexane	7094-26-0	126.242	418.35	25	5.9655E+04	1.0738E+00	1,2
1739	17926	C9H18	1,1,3-trimethylcyclohexane	3073-66-3	126.242	409.78	25	5.8010E+04	1.0442E+00	1,2
1740	17927	C9H18	1,1,4-trimethylcyclohexane	7094-27-1	126.242	408.15	25	5.0074E+04	9.0133E-01	1,2
1741	17928	C9H18	1,2,3-trimethylcyclohexane	1678-97-3	126.242	415.15	25	3.9618E+04	7.1312E-01	1,2
1742	17929	C9H18	1,cis-2,cis-3-trimethylcyclohexane	1839-88-9	126.242	426.15	25	9.7727E+04	1.7591E+00	1,2
1743	17930	C9H18	1,cis-2,trans-3-trimethylcyclohexane	7667-55-2	126.242	424.15	25	8.7535E+04	1.5756E+00	1,2
1744	17931	C9H18	1,trans-2,cis-3-trimethylcyclohexane	1678-81-5	126.242	417.15	25	7.0074E+04	1.2613E+00	1,2
1745	17932	C9H18	1,2,4-trimethylcyclohexane, isomers	2234-75-5	126.242	415.05	25	3.9572E+04	7.1228E-01	1,2
1746	17933	C9H18	1,cis-2,cis-4-trimethylcyclohexane	1678-80-4	126.242	421.15	25	8.8535E+04	1.5936E+00	1,2
1747	17934	C9H18	1,cis-2,trans-4-trimethylcyclohexane	7667-58-5	126.242	421.15	25	8.1753E+04	1.4715E+00	1,2
1748	17935	C9H18	1,trans-2,cis-4-trimethylcyclohexane	7667-59-6	126.242	418.15	25	7.6301E+04	1.3734E+00	1,2
1749	17936	C9H18	1,trans-2,trans-4-trimethylcyclohexane	7667-60-9	126.242	414.37	25	6.9989E+04	1.2598E+00	1,2
1750	17937	C9H18	1,3,5,-trimethylcyclohexane	1839-63-0	126.242	411.00	25	3.7681E+04	6.7825E-01	1,2
1751	17938	C9H18	cis,cis-1,3,5-trimethylcyclohexane	1795-27-3	126.242	411.66	25	2.5849E+04	4.6527E-01	1,2
1752	17939	C9H18	cis,trans-1,3,5-trimethylcyclohexane	1795-26-2	126.242	413.70	25	3.0153E+04	5.4275E-01	1,2
1753	17940	C9H18	1-ethyl-2-methylcyclohexane	3728-54-9	126.242	425.50	25	4.4257E+04	7.9663E-01	1,2
1754	17942	C9H18	1-nonene	124-11-8	126.242	420.02	25	4.7216E+04	8.4989E-01	1,2
1755	17943	C9H18	trans-2-nonene	6434-78-2	126.242	417.65	25	6.2743E+04	1.1294E+00	1,2
1756	17944	C9H18	trans-3-nonene	20063-92-7	126.242	420.65	25	6.4343E+04	1.1582E+00	1,2
1757	17945	C9H18	4-nonene; (cis+trans)	2198-23-4	126.242	419.05	25	6.3499E+04	1.1430E+00	1,2
1758	17946	C9H18	2-methyl-1-octene	4588-18-5	126.242	417.80	25	8.6040E+04	1.5487E+00	1,2
1759	17947	C9H18	3-methyl-2-octene	113426-22-5	126.242	418.15	25	6.3015E+04	1.1343E+00	1,2
1760	17948	C9H18	7-methyl-3-octene	86668-33-9	126.242	415.15	25	6.1359E+04	1.1045E+00	1,2

Table 10 Henry's Law Constant for Compound in Water - Organic Compounds

NO	ID	FORMULA	NAME	CAS No	MW, g/mol	T_B, K	T, C	Henry's Law Constant, H H @ T, atm/mol frac	H @ T, atm/mol/m³	code
1761	17949	C9H18	2-methyl-4-octene	64501-77-5	126.242	412.15	25	5.9653E+04	1.0737E+00	1,2
1762	17950	C9H18	2,3-dimethyl-2-heptene	3074-64-4	126.242	418.35	25	6.3123E+04	1.1362E+00	1,2
1763	17951	C9H18	2,6-dimethylhept-3-ene	2738-18-3	126.242	394.80	25	4.9526E+04	8.9146E-01	1,2
1764	17955	C9H18Br2	1,1-dibromononane	62168-27-8	286.050	533.15	25	4.0678E+01	7.3220E-04	1,2
1765	17956	C9H18Br2	1,2-dibromononane	73642-91-8	286.050	545.65	25	1.4988E+01	2.6978E-04	2
1766	17957	C9H18Br2	1,9-dibromononane	4549-33-1	286.050	558.15	25	1.0336E+01	1.8604E-04	1,2
1767	17959	C9H18Cl2	1,1-dichlorononane	821-88-5	197.147	501.15	25	3.7911E+02	6.8238E-03	1,2
1768	17960	C9H18Cl2	2,6-dichloro-2,6-dimethylheptane	35951-36-1	197.147	517.15	25	1.9367E+02	3.4859E-03	2
1769	17961	C9H18Cl2	1,9-dichlorononane	821-99-8	197.147	533.15	25	1.4352E+02	2.5833E-03	1,2
1770	17997	C9H18O	1-nonanal	124-19-6	142.241	468.15	30	5.1051E+01	9.1891E-04	1,2
1771	17998	C9H18O	diisobutyl ketone	108-83-8	142.241	441.41	24	6.0586E+00	1.0905E-04	1,2
1772	17999	C9H18O	2-nonanone	821-55-6	142.241	467.15	25	2.3416E+01	4.2148E-04	1,2
1773	18000	C9H18O	di-tert-butyl ketone	815-24-7	142.241	425.15	25	1.5490E+01	2.7881E-04	1,2
1774	18001	C9H18O	3,5-dimethyl-4-heptanone	19549-84-9	142.241	435.15	25	1.6681E+01	3.0025E-04	1,2
1775	18002	C9H18O	3-methyl-2-octanone	6137-08-2	142.241	456.62	25	1.8725E+01	3.3705E-04	2
1776	18003	C9H18O	3-methyl-4-octanone	20754-04-5	142.241	447.15	25	1.7889E+01	3.2199E-04	1,2
1777	18004	C9H18O	4-methyl-3-octanone	6137-15-1	142.241	456.62	25	1.8725E+01	3.3705E-04	2
1778	18005	C9H18O	5-methyl-2-octanone	58654-67-4	142.241	456.62	25	1.8725E+01	3.3705E-04	2
1779	18006	C9H18O	7-methyl-3-octanone	5408-57-1	142.241	455.65	25	1.8642E+01	3.3556E-04	1,2
1780	18007	C9H18O	7-methyl-4-octanone	20809-46-5	142.241	451.15	25	1.8251E+01	3.2851E-04	1,2
1781	18008	C9H18O	3-nonanone	925-78-0	142.241	463.15	25	1.9275E+01	3.4696E-04	1,2
1782	18009	C9H18O	4-nonanone	4485-09-0	142.241	460.65	25	1.9065E+01	3.4318E-04	1,2
1783	18010	C9H18O	5-nonanone	502-56-7	142.241	461.60	25	1.2329E+01	2.2192E-04	1,2
1784	18044	C9H18O	3,5,5-trimethylhexanal	5435-64-3	142.241	456.62	25	1.5597E+01	2.8075E-04	2
1785	18045	C9H18O	2,6-dimethyl-3-heptanone	19549-83-8	142.241	445.00	25	1.7687E+01	3.1836E-04	1,2
1786	18047	C9H18O	2-methyl-3-octanone	923-28-4	142.241	456.00	25	1.8672E+01	3.3610E-04	1,2
1787	18048	C9H18O	2-methyloctanal	7786-29-0	142.241	456.62	25	1.5597E+01	2.8075E-04	2
1788	18052	C9H18O2	nonanoic acid	112-05-0	158.241	528.75	20	2.9280E-02	5.2704E-07	1,2
1789	18053	C9H18O2	2-ethylheptanoic acid	3274-29-1	158.241	456.91	25	2.3597E-01	4.2474E-06	2
1790	18054	C9H18O2	3-ethyl-3-methylhexanoic acid	50902-82-4	158.241	456.91	25	2.3597E-01	4.2474E-06	2
1791	18055	C9H18O2	2-ethyl-3-methylhexanoic acid	74581-94-5	158.241	504.15	25	1.4559E-01	2.6207E-06	1,2
1792	18056	C9H18O2	3-methyloctanoic acid	6061-10-5	158.241	456.91	25	2.3597E-01	4.2474E-06	2
1793	18057	C9H18O2	2-methyloctanoic acid	3004-93-1	158.241	456.91	25	2.3597E-01	4.2474E-06	2
1794	18058	C9H18O2	octyl formate	112-32-3	158.241	471.95	25	1.8960E+02	3.4128E-03	1,2
1795	18059	C9H18O2	heptyl acetate	112-06-1	158.241	465.55	25	3.3828E+01	6.0890E-04	1,2
1796	18060	C9H18O2	4-heptyl acetate	5921-84-6	158.241	444.15	25	2.7382E+01	4.9288E-04	1,2
1797	18061	C9H18O2	hexyl propanoate	2445-76-3	158.241	463.15	25	2.3608E+01	4.2494E-04	1,2
1798	18062	C9H18O2	pentyl butanoate	540-18-1	158.241	458.15	25	8.8513E+01	1.5932E-03	1,2
1799	18063	C9H18O2	1-methylbutyl butanoate	60415-61-4	158.241	455.15	25	1.1291E+01	2.0324E-04	1,2
1800	18064	C9H18O2	2-methylbutyl butanoate	51115-64-1	158.241	452.15	25	3.2685E+01	5.8833E-04	2
1801	18065	C9H18O2	3-methylbutyl butanoate	106-27-4	158.241	452.15	25	5.5734E+01	1.0032E-03	1,2
1802	18066	C9H18O2	1-ethylpropyl butanoate	---	158.241	463.10	25	3.2260E+01	5.8067E-04	2
1803	18067	C9H18O2	1,1-dimethylpropyl butanoate	2050-00-2	158.241	460.31	25	3.2421E+01	5.8357E-04	2
1804	18068	C9H18O2	1,2-dimethylpropyl butanoate	88736-68-9	158.241	462.66	25	3.2288E+01	5.8117E-04	2
1805	18069	C9H18O2	2,2-dimethylpropyl butanoate	23361-69-5	158.241	460.31	25	3.2421E+01	5.8357E-04	2
1806	18070	C9H18O2	isopentyl 2-methylpropanoate	2050-01-3	158.241	441.65	25	3.0271E+01	5.4488E-04	1,2
1807	18072	C9H18O2	butyl 2-methylbutanoate	15706-73-7	158.241	452.15	25	3.2685E+01	5.8833E-04	1,2
1808	18073	C9H18O2	butyl 3-methylbutanoate	109-19-3	158.241	456.91	25	3.2569E+01	5.8623E-04	2
1809	18075	C9H18O2	isobutyl 3-methylbutanoate	589-59-3	158.241	441.65	25	4.7782E+01	8.6007E-04	1,2
1810	18082	C9H18O2	3,5,5-trimethylhexanoic acid	3302-10-1	158.241	456.91	25	2.3597E-01	4.2474E-06	2
1811	18083	C9H18O2	7-methyloctanoic acid	693-19-6	158.241	456.91	25	2.3597E-01	4.2474E-06	2
1812	18087	C9H18O2	4-octanecarboxylic acid	3274-28-0	158.241	358.15	25	3.3163E+04	5.9694E-01	1,2
1813	18115	C9H19Br	1-bromononane	693-58-3	207.154	494.55	25	2.3914E+03	4.3045E-02	1,2
1814	18117	C9H19Cl	1-chlorononane	2473-01-0	162.702	478.37	25	2.9899E+03	5.3817E-02	1,2
1815	18118	C9H19Cl	3-chloro-3-ethyl-2,2-dimethylpentane	86661-53-2	162.702	470.76	25	1.3726E+03	2.4706E-02	2
1816	18119	C9H19Cl	3-chloro-3-ethylheptane	28320-89-0	162.702	470.76	25	1.3726E+03	2.4706E-02	2
1817	18120	C9H19Cl	3-chloro-3-methyloctane	28320-88-9	162.702	470.76	25	1.3726E+03	2.4706E-02	2
1818	18121	C9H19Cl	4-chloro-4-methyloctane	36903-89-6	162.702	470.76	25	1.3726E+03	2.4706E-02	2
1819	18122	C9H19Cl	2-chlorononane	2216-36-6	162.702	463.15	25	1.4112E+03	2.5402E-02	1,2
1820	18123	C9H19Cl	5-chlorononane	28123-70-8	162.702	470.76	25	1.3726E+03	2.4706E-02	2
1821	18124	C9H19Cl	3-chloro-2,2,3-trimethylhexane	102449-95-6	162.702	470.76	25	1.3726E+03	2.4706E-02	2
1822	18176	C9H20	nonane	111-84-2	128.258	423.97	25	3.7628E+05	6.7730E+00	1,2
1823	18177	C9H20	2-methyloctane	3221-61-2	128.258	416.43	25	3.9223E+05	7.0600E+00	1,2
1824	18178	C9H20	3-methyloctane	2216-33-3	128.258	417.38	25	4.2351E+05	7.6230E+00	1,2
1825	18179	C9H20	4-methyloctane	2216-34-4	128.258	415.59	25	6.3196E+05	1.1375E+01	1,2
1826	18180	C9H20	3-ethylheptane	15869-80-4	128.258	416.35	25	4.3295E+05	7.7930E+00	1,2
1827	18181	C9H20	4-ethylheptane	2216-32-2	128.258	414.36	25	3.7381E+05	6.7285E+00	1,2
1828	18182	C9H20	2,2-dimethylheptane	1071-26-7	128.258	405.84	25	3.4862E+05	6.2751E+00	1,2
1829	18183	C9H20	2,3-dimethylheptane	3074-71-3	128.258	413.66	25	3.7887E+05	6.8196E+00	1,2
1830	18184	C9H20	2,4-dimethylheptane	2213-23-2	128.258	406.05	25	3.1733E+05	5.7119E+00	1,2
1831	18185	C9H20	2,5-dimethylheptane	2216-30-0	128.258	409.16	25	3.3679E+05	6.0621E+00	1,2
1832	18186	C9H20	2,6-dimethylheptane	1072-05-5	128.258	408.36	25	3.6598E+05	6.5877E+00	1,2
1833	18187	C9H20	3,4-dimethylheptane	922-28-1	128.258	413.76	25	3.8433E+05	6.9179E+00	1,2
1834	18188	C9H20	3,5-dimethylheptane	926-82-9	128.258	409.16	25	3.4028E+05	6.1250E+00	1,2
1835	18189	C9H20	4,4-dimethylheptane	1068-19-5	128.258	408.36	25	3.6232E+05	6.5217E+00	1,2
1836	18190	C9H20	3-ethyl-2-methylhexane	16789-46-1	128.258	411.16	25	3.6702E+05	6.6063E+00	1,2
1837	18191	C9H20	4-ethyl-2-methylhexane	3074-75-7	128.258	406.96	25	3.3135E+05	5.9643E+00	1,2
1838	18192	C9H20	3-ethyl-3-methylhexane	3074-76-8	128.258	413.76	25	4.0985E+05	7.3772E+00	1,2
1839	18193	C9H20	3-ethyl-4-methylhexane	3074-77-9	128.258	413.56	25	3.8613E+05	6.9503E+00	1,2
1840	18194	C9H20	2,2,3-trimethylhexane	16747-25-4	128.258	406.75	25	3.6070E+05	6.4926E+00	1,2

Table 10 Henry's Law Constant for Compound in Water - Organic Compounds

					Henry's Law Constant, H					
NO	ID	FORMULA	NAME	CAS No	MW, g/mol	T_B, K	T, C	H @ T, atm/mol frac	H @ T, atm/mol/m^3	code
1841	18195	C9H20	2,2,4-trimethylhexane	16747-26-5	128.258	399.69	25	3.1552E+05	5.6793E+00	1,2
1842	18196	C9H20	2,2,5-trimethylhexane	3522-94-9	128.258	397.24	25	3.2713E+05	5.8883E+00	1,2
1843	18197	C9H20	2,3,3-trimethylhexane	16747-28-7	128.258	410.84	25	4.0057E+05	7.2101E+00	1,2
1844	18198	C9H20	2,3,4-trimethylhexane	921-47-1	128.258	412.20	25	3.9862E+05	7.1751E+00	1,2
1845	18199	C9H20	2,3,5-trimethylhexane	1069-53-0	128.258	404.50	25	3.2893E+05	5.9206E+00	1,2
1846	18200	C9H20	2,4,4-trimethylhexane	16747-30-1	128.258	403.81	25	3.5839E+05	6.4510E+00	1,2
1847	18201	C9H20	3,3,4-trimethylhexane	16747-31-2	128.258	413.62	25	4.2631E+05	7.6735E+00	1,2
1848	18202	C9H20	3,3-diethylpentane	1067-20-5	128.258	419.34	25	4.9283E+05	8.8708E+00	1,2
1849	18203	C9H20	2,2-dimethyl-3-ethylpentane	16747-32-3	128.258	406.99	25	4.0677E+05	7.3218E+00	1,2
1850	18204	C9H20	2,3-dimethyl-3-ethylpentane	16747-33-4	128.258	417.86	25	4.7626E+05	8.5725E+00	1,2
1851	18205	C9H20	2,4-dimethyl-3-ethylpentane	1068-87-7	128.258	409.87	25	4.3240E+05	7.7831E+00	1,2
1852	18206	C9H20	2,2,3,3-tetramethylpentane	7154-79-2	128.258	413.44	25	4.5030E+05	8.1053E+00	1,2
1853	18207	C9H20	2,2,3,4-tetramethylpentane	1186-53-4	128.258	406.18	25	4.1190E+05	7.4142E+00	1,2
1854	18208	C9H20	2,2,4,4-tetramethylpentane	1070-87-7	128.258	395.44	25	3.3565E+05	6.0417E+00	1,2
1855	18209	C9H20	2,3,3,4-tetramethylpentane	16747-38-9	128.258	414.72	25	4.5280E+05	8.1503E+00	1,2
1856	18210	C9H20	3,3-dimethylheptane	4032-86-4	128.258	410.45	25	2.4962E+05	4.4932E+00	1,2
1857	18236	C9H20O	1-nonanol	143-08-8	144.257	486.25	25	2.3210E+00	4.1777E-05	1,2
1858	18237	C9H20O	2-nonanol	628-99-9	144.257	471.65	15	1.3167E+00	2.3700E-05	1,2
1859	18238	C9H20O	2-nonanol, (±)	74683-66-2	144.257	466.65	25	2.7047E+00	4.8684E-05	1,2
1860	18239	C9H20O	3-nonanol	624-51-1	144.257	467.85	25	1.0426E+01	1.8767E-04	1,2
1861	18240	C9H20O	3-nonanol, (±)	74742-08-8	144.257	468.15	25	2.7010E+00	4.8618E-05	1,2
1862	18241	C9H20O	4-nonanol	5932-79-6	144.257	466.15	25	1.0129E+01	1.8231E-04	1,2
1863	18242	C9H20O	4-nonanol, (S)	52708-03-9	144.257	465.65	25	2.7072E+00	4.8729E-05	1,2
1864	18243	C9H20O	5-nonanol	623-93-8	144.257	468.25	25	1.0498E+01	1.8897E-04	1,2
1865	18244	C9H20O	2-methyl-1-octanol	818-81-5	144.257	473.00	25	2.6899E+00	4.8417E-05	2
1866	18245	C9H20O	3-methyl-1-octanol	38514-02-2	144.257	473.00	25	2.6899E+00	4.8417E-05	2
1867	18246	C9H20O	4-methyl-1-octanol	38514-03-3	144.257	473.00	25	2.6899E+00	4.8417E-05	2
1868	18247	C9H20O	5-methyl-1-octanol	38514-04-4	144.257	473.00	25	2.6899E+00	4.8417E-05	2
1869	18248	C9H20O	6-methyl-1-octanol	38514-05-5	144.257	479.15	25	1.2802E+01	2.3043E-04	1,2
1870	18249	C9H20O	7-methyl-1-octanol	2430-22-0	144.257	479.15	25	1.2802E+01	2.3043E-04	1,2
1871	18250	C9H20O	2-methyl-2-octanol	628-44-4	144.257	451.15	25	9.4987E+00	1.7098E-04	1,2
1872	18251	C9H20O	3-methyl-2-octanol	27644-49-1	144.257	454.00	25	2.7325E+00	4.9185E-05	2
1873	18252	C9H20O	5-methyl-2-octanol	66793-81-5	144.257	454.00	25	2.7325E+00	4.9185E-05	2
1874	18253	C9H20O	7-methyl-2-octanol	66793-83-7	144.257	454.00	25	2.7325E+00	4.9185E-05	2
1875	18254	C9H20O	2-methyl-3-octanol	26533-34-6	144.257	457.15	25	9.5362E+00	1.7165E-04	1,2
1876	18255	C9H20O	3-methyl-3-octanol	5340-36-3	144.257	462.15	25	1.1443E+01	2.0597E-04	1,2
1877	18256	C9H20O	4-methyl-3-octanol	66793-80-4	144.257	459.00	25	2.7229E+00	4.9013E-05	2
1878	18257	C9H20O	6-methyl-3-octanol	40225-75-0	144.257	459.00	25	2.7229E+00	4.9013E-05	2
1879	18258	C9H20O	7-methyl-3-octanol	66793-84-8	144.257	459.00	25	2.7229E+00	4.9013E-05	2
1880	18259	C9H20O	2-methyl-4-octanol	40575-41-5	144.257	457.15	25	9.5362E+00	1.7165E-04	1,2
1881	18260	C9H20O	3-methyl-4-octanol	26533-35-7	144.257	453.15	25	8.9393E+00	1.6091E-04	1,2
1882	18261	C9H20O	4-methyl-4-octanol	23418-37-3	144.257	454.15	25	9.9825E+00	1.7968E-04	1,2
1883	18262	C9H20O	5-methyl-4-octanol	59734-23-5	144.257	455.00	25	2.7309E+00	4.9155E-05	2
1884	18263	C9H20O	6-methyl-4-octanol	66793-82-6	144.257	455.00	25	2.7309E+00	4.9155E-05	2
1885	18264	C9H20O	7-methyl-4-octanol	33933-77-6	144.257	455.00	25	2.7309E+00	4.9155E-05	2
1886	18265	C9H20O	2-ethyl-1-heptanol	817-60-7	144.257	472.00	25	2.6920E+00	4.8456E-05	2
1887	18266	C9H20O	3-ethyl-1-heptanol	3525-25-5	144.257	480.15	25	1.3052E+01	2.3493E-04	1,2
1888	18267	C9H20O	5-ethyl-1-heptanol	998-65-2	144.257	472.00	25	2.6920E+00	4.8456E-05	2
1889	18268	C9H20O	2,2-dimethyl-1-heptanol	14250-79-4	144.257	465.15	25	1.2067E+01	2.1721E-04	1,2
1890	18269	C9H20O	4,6-dimethyl-1-heptanol	820-05-3	144.257	455.00	25	2.7309E+00	4.9155E-05	2
1891	18270	C9H20O	6,6-dimethyl-1-heptanol	65769-10-0	144.257	455.00	25	2.7309E+00	4.9155E-05	2
1892	18271	C9H20O	2-ethyl-2-heptanol	---	144.257	455.00	25	2.7309E+00	4.9155E-05	2
1893	18272	C9H20O	3-ethyl-2-heptanol	19780-39-3	144.257	455.00	25	2.7309E+00	4.9155E-05	2
1894	18273	C9H20O	2,3-dimethyl-2-heptanol	66794-00-1	144.257	455.00	25	2.7309E+00	4.9155E-05	2
1895	18274	C9H20O	2,4-dimethyl-2-heptanol	65822-93-7	144.257	455.00	25	2.7309E+00	4.9155E-05	2
1896	18275	C9H20O	2,5-dimethyl-2-heptanol	---	144.257	455.00	25	2.7309E+00	4.9155E-05	2
1897	18276	C9H20O	2,6-dimethyl-2-heptanol	13254-34-7	144.257	444.15	25	9.1815E+00	1.6527E-04	1,2
1898	18277	C9H20O	4,6-dimethyl-2-heptanol	51079-52-8	144.257	467.15	25	1.2389E+01	2.2300E-04	1,2
1899	18278	C9H20O	5,6-dimethyl-2-heptanol	58795-24-7	144.257	465.15	25	1.1951E+01	2.1511E-04	1,2
1900	18279	C9H20O	3-ethyl-3-heptanol	19780-41-7	144.257	455.35	25	1.0185E+01	1.8333E-04	1,2
1901	18280	C9H20O	2,2-dimethyl-3-heptanol	19549-70-3	144.257	451.00	25	2.7366E+00	4.9258E-05	2
1902	18281	C9H20O	2,3-dimethyl-3-heptanol	19549-71-4	144.257	447.15	25	9.6533E+00	1.7376E-04	1,2
1903	18282	C9H20O	2,6-dimethyl-3-heptanol	19549-73-6	144.257	448.15	25	8.9652E+00	1.6137E-04	1,2
1904	18283	C9H20O	3,5-dimethyl-3-heptanol	19549-74-7	144.257	450.00	25	2.7375E+00	4.9275E-05	2
1905	18284	C9H20O	3,6-dimethyl-3-heptanol	1573-28-0	144.257	446.15	25	8.0949E+00	1.4571E-04	1,2
1906	18285	C9H20O	4-ethyl-4-heptanol	597-90-0	144.257	452.15	25	9.6566E+00	1.7382E-04	1,2
1907	18286	C9H20O	2,2-dimethyl-4-heptanol	66793-99-5	144.257	447.15	25	9.6533E+00	1.7376E-04	1,2
1908	18287	C9H20O	2,4-dimethyl-4-heptanol	19549-77-0	144.257	444.55	25	9.2429E+00	1.6637E-04	1,2
1909	18288	C9H20O	2,6-dimethyl-4-heptanol	108-82-7	144.257	451.00	25	4.2117E+00	7.5811E-05	1,2
1910	18289	C9H20O	3,3-dimethyl-4-heptanol	19549-78-1	144.257	449.15	25	9.9835E+00	1.7970E-04	1,2
1911	18290	C9H20O	3,5-dimethyl-4-heptanol	19549-79-2	144.257	460.15	25	1.0948E+01	1.9706E-04	1,2
1912	18291	C9H20O	2-propyl-1-hexanol	817-46-9	144.257	462.00	25	2.7161E+00	4.8889E-05	2
1913	18292	C9H20O	3-propyl-1-hexanol	66793-85-9	144.257	462.00	25	2.7161E+00	4.8889E-05	2
1914	18293	C9H20O	1-methyl-2-ethyl-1-hexanol	---	144.257	462.00	25	2.7161E+00	4.8889E-05	2
1915	18294	C9H20O	1-methyl-3-ethyl-1-hexanol	---	144.257	465.15	25	1.0908E+01	1.9635E-04	1,2
1916	18295	C9H20O	3-methyl-2-ethyl-1-hexanol	66794-04-5	144.257	466.15	25	1.1363E+01	2.0454E-04	1,2
1917	18296	C9H20O	4-methyl-2-ethyl-1-hexanol	66794-06-7	144.257	468.15	25	1.1497E+01	2.0695E-04	1,2
1918	18297	C9H20O	5-methyl-2-ethyl-1-hexanol	66794-07-8	144.257	466.00	25	2.7063E+00	4.8713E-05	2
1919	18298	C9H20O	3,3,5-trimethyl-1-hexanol	1484-87-3	144.257	466.00	25	2.7063E+00	4.8713E-05	2
1920	18299	C9H20O	3,4,4-trimethyl-1-hexanol	66793-73-5	144.257	464.15	25	1.2980E+01	2.3364E-04	1,2

								Henry's Law Constant, H		
NO	ID	FORMULA	NAME	CAS No	MW, g/mol	T_B, K	T, C	H @ T, atm/mol frac	H @ T, atm/mol/m^3	code
1921	18300	C9H20O	3,5,5-trimethyl-1-hexanol	3452-97-9	144.257	466.15	25	1.3468E+01	2.4243E-04	1,2
1922	18301	C9H20O	4,5,5-trimethyl-1-hexanol	66793-75-7	144.257	475.15	25	1.6029E+01	2.8852E-04	1,2
1923	18302	C9H20O	2-methyl-3-ethyl-2-hexanol	66794-02-3	144.257	451.15	25	1.0327E+01	1.8589E-04	1,2
1924	18303	C9H20O	2,3,4-trimethyl-2-hexanol	21102-13-6	144.257	466.00	25	2.7063E+00	4.8713E-05	2
1925	18304	C9H20O	2,4,4-trimethyl-2-hexanol	66793-91-7	144.257	466.00	25	2.7063E+00	4.8713E-05	2
1926	18305	C9H20O	2,4,5-trimethyl-2-hexanol	66793-93-9	144.257	466.00	25	2.7063E+00	4.8713E-05	2
1927	18306	C9H20O	2,5,5-trimethyl-2-hexanol	66793-71-3	144.257	466.00	25	2.7063E+00	4.8713E-05	2
1928	18307	C9H20O	2-methyl-3-ethyl-3-hexanol	66794-03-4	144.257	457.25	25	1.1472E+01	2.0650E-04	1,2
1929	18308	C9H20O	2-methyl-4-ethyl-3-hexanol	33943-21-4	144.257	452.00	25	2.7354E+00	4.9236E-05	2
1930	18309	C9H20O	3-methyl-4-ethyl-3-hexanol	66794-05-6	144.257	452.00	25	2.7354E+00	4.9236E-05	2
1931	18310	C9H20O	4-methyl-3-ethyl-3-hexanol	51200-80-7	144.257	452.00	25	2.7354E+00	4.9236E-05	2
1932	18311	C9H20O	5-methyl-3-ethyl-3-hexanol	597-77-3	144.257	445.15	25	9.3358E+00	1.6804E-04	1,2
1933	18312	C9H20O	2,2,3-trimethyl-3-hexanol	5340-41-0	144.257	446.25	25	1.1273E+01	2.0291E-04	1,2
1934	18313	C9H20O	2,2,4-trimethyl-3-hexanol	66793-89-3	144.257	442.15	25	9.6089E+00	1.7296E-04	1,2
1935	18314	C9H20O	2,2,5-trimethyl-3-hexanol	3970-60-3	144.257	433.15	25	8.2356E+00	1.4824E-04	1,2
1936	18315	C9H20O	2,3,4-trimethyl-3-hexanol	66793-90-6	144.257	439.00	25	2.7318E+00	4.9173E-05	2
1937	18316	C9H20O	2,3,5-trimethyl-3-hexanol	65927-60-8	144.257	439.00	25	2.7318E+00	4.9173E-05	2
1938	18317	C9H20O	2,4,4-trimethyl-3-hexanol	66793-92-8	144.257	444.15	25	9.9446E+00	1.7900E-04	1,2
1939	18318	C9H20O	2,5,5-trimethyl-3-hexanol	66793-72-4	144.257	442.00	25	2.7369E+00	4.9263E-05	2
1940	18319	C9H20O	3,4,4-trimethyl-3-hexanol	66793-74-6	144.257	439.15	25	9.9367E+00	1.7886E-04	1,2
1941	18320	C9H20O	3,5,5-trimethyl-3-hexanol	66810-87-5	144.257	439.00	25	2.7318E+00	4.9173E-05	2
1942	18321	C9H20O	4-methyl-2-propyl-1-pentanol	54004-41-0	144.257	465.15	25	1.0821E+01	1.9477E-04	1,2
1943	18322	C9H20O	4-methyl-2-isopropyl-1-pentanol	55505-24-3	144.257	460.15	25	1.1028E+01	1.9850E-04	1,2
1944	18323	C9H20O	2,2-dimethyl-3-ethyl-1-pentanol	66793-95-1	144.257	461.00	25	2.7184E+00	4.8931E-05	2
1945	18324	C9H20O	2,4-dimethyl-2-ethyl-1-pentanol	66793-98-4	144.257	461.15	25	1.2293E+01	2.2127E-04	1,2
1946	18325	C9H20O	3,3-diethyl-2-pentanol	66793-94-0	144.257	461.00	25	2.7184E+00	4.8931E-05	2
1947	18326	C9H20O	2,3-dimethyl-3-ethyl-2-pentanol	66793-97-3	144.257	461.00	25	2.7184E+00	4.8931E-05	2
1948	18327	C9H20O	4,4-dimethyl-3-ethyl-2-pentanol	21102-09-0	144.257	461.00	25	2.7184E+00	4.8931E-05	2
1949	18328	C9H20O	2,3,3,4-tetramethyl-2-pentanol	66793-86-0	144.257	453.00	25	2.7340E+00	4.9212E-05	2
1950	18329	C9H20O	2,3,4,4-tetramethyl-2-pentanol	66793-87-1	144.257	453.00	25	2.7340E+00	4.9212E-05	2
1951	18330	C9H20O	3,3,4,4-tetramethyl-2-pentanol	66793-88-2	144.257	453.15	25	1.3894E+01	2.5009E-04	1,2
1952	18331	C9H20O	2-methyl-3-isopropyl-3-pentanol	---	144.257	451.15	25	1.1296E+01	2.0333E-04	1,2
1953	18332	C9H20O	2,2-dimethyl-3-ethyl-3-pentanol	66793-96-2	144.257	447.15	25	1.1456E+01	2.0621E-04	1,2
1954	18333	C9H20O	2,4-dimethyl-3-ethyl-3-pentanol	3970-59-0	144.257	451.05	25	1.1209E+01	2.0176E-04	1,2
1955	18334	C9H20O	2,2,3,4-tetramethyl-3-pentanol	62185-29-9	144.257	447.35	25	1.2470E+01	2.2446E-04	1,2
1956	18335	C9H20O	2,2,4,4-tetramethyl-3-pentanol	14609-79-1	144.257	440.15	25	1.0927E+01	1.9669E-04	1,2
1957	18336	C9H20O	ethyl heptyl ether	1969-43-3	144.257	439.15	25	2.2054E+02	3.9697E-03	1,2
1958	18365	C9H20S	nonyl mercaptan	1455-21-6	160.324	492.95	25	8.9126E+02	1.6043E-02	1,2
1959	18366	C9H20S	2-nonanethiol	13281-11-3	160.324	481.35	25	8.6790E+02	1.5622E-02	1,2
1960	18367	C9H20S	methyl octyl sulfide	3698-95-1	160.324	480.00	25	2.3831E+02	4.2895E-03	2
1961	18368	C9H20S	ethyl heptyl sulfide	24768-44-3	160.324	480.00	25	2.3831E+02	4.2895E-03	2
1962	18369	C9H20S	propyl hexyl sulfide	24768-43-2	160.324	480.00	25	2.3831E+02	4.2895E-03	2
1963	18370	C9H20S	butyl pentyl sulfide	24768-42-1	160.324	480.00	25	2.3831E+02	4.2895E-03	2
1964	18371	C9H20S	tert-nonyl mercaptan, isomers	25360-10-5	160.324	461.15	25	3.7925E+02	6.8264E-03	1,2
1965	18388	C9H21N	nonylamine	112-20-9	143.273	475.35	25	3.5500E+00	6.3900E-05	1,2
1966	18389	C9H21N	methyloctylamine	2439-54-5	143.273	459.85	25	2.1875E+00	3.9374E-05	1,2
1967	18390	C9H21N	ethylheptylamine	66793-76-8	143.273	453.85	25	2.3997E+00	4.3195E-05	1,2
1968	18391	C9H21N	dimethylheptylamine	5277-11-2	143.273	445.15	25	1.3414E+01	2.4146E-04	1,2
1969	18392	C9H21N	diethylpentylamine	2162-91-6	143.273	429.15	25	9.7901E+00	1.7622E-04	1,2
1970	18393	C9H21N	tripropylamine	102-69-2	143.273	429.65	25	1.2258E+01	2.2064E-04	1,2
1971	18394	C9H21N	tert-amyl-tert-butylamine	2085-66-7	143.273	417.15	25	1.5169E+00	2.7305E-05	1,2
1972	18395	C9H21N	N-methyldibutylamine	3405-45-6	143.273	432.75	25	7.4347E+00	1.3382E-04	1,2
1973	18598	C10H6O8	pyromellitic acid	89-05-4	254.153	722.00	16	1.5913E-20	2.8642E-25	1,2
1974	18601	C10H7Br	1-bromonaphthalene	90-11-9	207.070	554.25	21	1.1927E+01	2.1469E-04	1,2
1975	18606	C10H7Cl	1-chloronaphthalene	90-13-1	162.618	532.45	25	1.0965E+01	1.9736E-04	1,2
1976	18607	C10H7Cl	2-chloronaphthalene	91-58-7	162.618	529.15	25	1.1573E+01	2.0832E-04	1,2
1977	18608	C10H7Cl	monochloronaphthalene	25586-43-0	162.618	532.15	25	9.6012E+00	1.7282E-04	1,2
1978	18669	C10H8	naphthalene	91-20-3	128.174	491.14	25	7.7026E+01	1.3865E-03	1,2
1979	19195	C10H10O4	dimethyl phthalate	131-11-3	194.187	556.85	20	6.9585E-03	1.2525E-07	1,2
1980	19359	C10H12	1,2,3,4-tetrahydronaphthalene	119-64-2	132.205	480.77	28	1.0260E+02	1.8468E-03	1,2
1981	19484	C10H12O	anethole	104-46-1	148.205	508.45	25	5.5306E+00	9.9550E-05	1,2
1982	19865	C10H14	butylbenzene	104-51-8	134.221	456.46	25	8.1157E+02	1.4608E-02	1,2
1983	19866	C10H14	isobutylbenzene	538-93-2	134.221	445.94	25	1.9434E+03	3.4982E-02	1,2
1984	19867	C10H14	sec-butylbenzene	135-98-8	134.221	446.48	25	1.1654E+03	2.0976E-02	1,2
1985	19868	C10H14	tert-butylbenzene	98-06-6	134.221	442.30	25	8.3200E+02	1.4976E-02	1,2
1986	19869	C10H14	cymene	25155-15-1	134.221	449.50	25	2.9679E+02	5.3421E-03	1,2
1987	19870	C10H14	m-cymene	535-77-3	134.221	448.23	25	5.0917E+02	9.1649E-03	1,2
1988	19871	C10H14	o-cymene	527-84-4	134.221	451.33	25	5.4274E+02	9.7692E-03	1,2
1989	19872	C10H14	p-cymene	99-87-6	134.221	450.28	25	4.5201E+02	8.1361E-03	1,2
1990	19873	C10H14	1-methyl-2-propylbenzene	1074-17-5	134.221	458.08	25	7.3115E+02	1.3161E-02	1,2
1991	19874	C10H14	1-methyl-3-propylbenzene	1074-43-7	134.221	455.13	25	6.9084E+02	1.2435E-02	1,2
1992	19875	C10H14	1-methyl-4-propylbenzene	1074-55-1	134.221	453.53	25	5.3480E+02	9.6263E-03	1,2
1993	19876	C10H14	(S)-(1-methylpropyl)benzene	5787-28-0	134.221	457.10	25	3.0345E+02	5.4620E-03	2
1994	19877	C10H14	diethylbenzene	25340-17-4	134.221	455.05	25	3.0200E+02	5.4359E-03	1,2
1995	19878	C10H14	o-diethylbenzene	135-01-3	134.221	456.61	25	4.9294E+02	8.8729E-03	1,2
1996	19879	C10H14	m-diethylbenzene	141-93-5	134.221	454.29	25	5.1999E+02	9.3597E-03	1,2
1997	19880	C10H14	p-diethylbenzene	105-05-5	134.221	456.94	25	4.5430E+02	8.1773E-03	1,2
1998	19881	C10H14	3-ethyl-o-xylene	933-98-2	134.221	467.11	25	5.4210E+02	9.7578E-03	1,2
1999	19882	C10H14	4-ethyl-o-xylene	934-80-5	134.221	462.93	25	4.7740E+02	8.5930E-03	1,2
2000	19883	C10H14	2-ethyl-m-xylene	2870-04-4	134.221	463.19	25	4.3868E+02	7.8961E-03	1,2

Table 10 Henry's Law Constant for Compound in Water - Organic Compounds

				Henry's Law Constant, H						
NO	ID	FORMULA	NAME	CAS No	MW, g/mol	T_B, K	T, C	H @ T, atm/mol frac	H @ T, atm/mol/m³	code
2001	19884	C10H14	4-ethyl-m-xylene	874-41-9	134.221	461.59	25	5.6547E+02	1.0178E-02	1,2
2002	19885	C10H14	5-ethyl-m-xylene	934-74-7	134.221	456.93	25	4.6726E+02	8.4105E-03	1,2
2003	19886	C10H14	2-ethyl-p-xylene	1758-88-9	134.221	459.98	25	5.4760E+02	9.8567E-03	1,2
2004	19887	C10H14	1,2,3,4-tetramethylbenzene	488-23-3	134.221	478.25	25	4.4280E+02	7.9703E-03	1,2
2005	19888	C10H14	1,2,3,5-tetramethylbenzene	527-53-7	134.221	471.15	25	4.9285E+02	8.8712E-03	1,2
2006	19889	C10H14	1,2,4,5-tetramethylbenzene	95-93-2	134.221	469.99	25	1.5141E+03	2.7254E-02	1,2
2007	20089	C10H14O2	p-tert-butylcatechol	98-29-3	166.220	558.00	25	2.3580E+07	4.2444E+02	1,2
2008	20195	C10H15N	N,N-dietyhlaniline	91-66-7	149.236	489.42	12	4.6991E-02	8.4582E-07	1,2
2009	20196	C10H15N	2,6-diethylaniline	579-66-8	149.236	508.65	27	6.6481E-01	1.1966E-05	1,2
2010	20197	C10H15N	2-sec-butylaniline	55751-54-7	149.236	502.03	25	7.5325E-02	1.3558E-06	2
2011	20198	C10H15N	2-tert-butylaniline	6310-21-0	149.236	507.15	25	5.5037E-02	9.9066E-07	1,2
2012	20199	C10H15N	4-butylaniline	104-13-2	149.236	534.15	25	9.7523E-03	1.7554E-07	1,2
2013	20200	C10H15N	4-sec-butylaniline	30273-11-1	149.236	511.15	25	4.2937E-02	7.7285E-07	1,2
2014	20201	C10H15N	4-tert-butylaniline	769-92-6	149.236	514.15	25	3.5578E-02	6.4039E-07	1,2
2015	20202	C10H15N	N-sec-butylaniline	6068-69-5	149.236	498.15	25	9.5276E-02	1.7150E-06	1,2
2016	20203	C10H15N	N-tert-butylaniline	937-33-7	149.236	488.15	25	1.7259E-01	3.1067E-06	1,2
2017	20204	C10H15N	N-butylaniline	1126-78-9	149.236	516.65	25	3.0382E-02	5.4687E-07	1,2
2018	20205	C10H15N	4-isobutylaniline	30090-17-6	149.236	510.15	25	4.5698E-02	8.2256E-07	1,2
2019	20206	C10H15N	N-isobutylaniline	588-47-6	149.236	505.15	25	6.2248E-02	1.1205E-06	1,2
2020	20207	C10H15N	4-methyl-N-isopropylaniline	10436-75-6	149.236	493.15	25	1.2849E-01	2.3129E-06	1,2
2021	20208	C10H15N	2-methyl-5-isopropylaniline	2051-53-8	149.236	514.15	25	3.5578E-02	6.4039E-07	1,2
2022	20209	C10H15N	4-methyl-N-propylaniline	54837-90-0	149.236	508.15	25	5.1738E-02	9.3128E-07	1,2
2023	20210	C10H15N	2,3,4,5-tetramethylaniline	2217-45-0	149.236	532.65	25	1.0774E-02	1.9392E-07	1,2
2024	20211	C10H15N	2,3,4,6-tetramethylaniline	488-71-1	149.236	528.15	25	1.4489E-02	2.6080E-07	1,2
2025	20212	C10H15N	N,N,2,6-tetramethylaniline	769-06-2	149.236	469.15	25	5.1113E-01	9.2002E-06	1,2
2026	20230	C10H15N	3,5,N,N-tetramethylaniline	4913-13-7	149.236	500.15	25	8.4436E-02	1.5198E-06	1,2
2027	20231	C10H15N	aniline, p-isopropyl-N-methyl-,	6950-79-4	149.236	502.03	25	7.5325E-02	1.3558E-06	2
2028	20316	C10H16	D-limonene	5989-27-5	136.237	449.65	25	1.6835E+03	3.0303E-02	1,2
2029	20317	C10H16	limonene	138-86-3	136.237	451.15	25	1.4944E+03	2.6898E-02	1,2
2030	20318	C10H16	L-limonene	5989-54-8	136.237	451.15	25	1.4944E+03	2.6898E-02	1,2
2031	20423	C10H16O	camphor	76-22-2	152.236	480.57	25	2.3819E+00	4.2873E-05	1,2
2032	20540	C10H16S	3-hexylthiophene	1693-86-3	168.303	502.15	25	4.6418E+02	8.3552E-03	2
2033	20593	C10H18	1-decyne	764-93-2	138.253	447.16	25	4.0925E+03	7.3664E-02	1,2
2034	20594	C10H18	2-decyne	2384-70-5	138.253	457.75	25	3.9631E+03	7.1335E-02	1,2
2035	20595	C10H18	3-decyne	2384-85-2	138.253	452.45	25	4.0502E+03	7.2902E-02	1,2
2036	20596	C10H18	4-decyne	2384-86-3	138.253	443.65	25	4.0970E+03	7.3745E-02	2
2037	20597	C10H18	5-decyne	1942-46-7	138.253	450.15	25	4.0739E+03	7.3330E-02	1,2
2038	20598	C10H18	2,2-dimethyl-3-octyne	19482-57-6	138.253	443.65	25	4.0970E+03	7.3745E-02	2
2039	20599	C10H18	8-methyl-4-nonyne	70732-43-3	138.253	443.65	25	4.0970E+03	7.3745E-02	2
2040	20600	C10H18	5-ethyl-5-methyl-3-heptyne	61228-10-2	138.253	443.65	25	4.0970E+03	7.3745E-02	2
2041	20601	C10H18	cis-decahydronaphthalene	493-01-6	138.253	468.97	25	9.2282E+03	1.6611E-01	1,2
2042	20602	C10H18	trans-decahydronaphthalene	493-02-7	138.253	460.46	25	1.3797E+04	2.4835E-01	1,2
2043	20608	C10H18	1,3-decadiene	2051-25-4	138.253	442.15	25	1.9121E+04	3.4417E-01	1,2
2044	20609	C10H18	1,9-decadiene	1647-16-1	138.253	440.15	25	1.9108E+04	3.4393E-01	1,2
2045	20610	C10H18	2,6-dimethyl-2,5-octadiene	116668-48-5	138.253	443.65	25	1.9113E+04	3.4404E-01	2
2046	20611	C10H18	2,7-dimethyl-2,6-octadiene	16736-42-8	138.253	439.15	25	1.9092E+04	3.4365E-01	1,2
2047	20612	C10H18	3,6-dimethyl-2,6-octadiene	116668-49-6	138.253	427.15	25	1.8466E+04	3.3239E-01	1,2
2048	20613	C10H18	3,7-dimethyl-2,4-octadiene	56523-26-3	138.253	438.65	25	1.9081E+04	3.4346E-01	1,2
2049	20614	C10H18	4,5-dimethyl-2,6-octadiene	18476-57-8	138.253	426.65	25	1.8425E+04	3.3165E-01	1,2
2050	20615	C10H18	cis-2,6-dimethyl-2,6-octadiene	2492-22-0	138.253	441.15	25	1.9117E+04	3.4411E-01	1,2
2051	20636	C10H18	decahydronaphthalene; (cis+trans)	91-17-8	138.253	446.40	25	2.6660E+04	4.7988E-01	1,2
2052	20637	C10H18	3,7-dimethyl-1,6-octadiene	2436-90-0	138.253	429.60	25	1.8653E+04	3.3575E-01	1,2
2053	20820	C10H18O4	sebacic acid	111-20-6	202.251	642.09	20	2.7153E-07	4.8875E-12	1,2
2054	20922	C10H20	pentylcyclopentane	3741-00-2	140.269	453.76	25	1.0470E+05	1.8846E+00	1,2
2055	20923	C10H20	butylcyclohexane	1678-93-9	140.269	454.13	25	1.0269E+05	1.8484E+00	1,2
2056	20924	C10H20	tert-butylcyclohexane	3178-22-1	140.269	444.65	25	5.6109E+04	1.0100E+00	1,2
2057	20925	C10H20	sec-butylcyclohexane	7058-01-7	140.269	452.45	25	5.7165E+04	1.0290E+00	1,2
2058	20926	C10H20	isobutylcyclohexane	1678-98-4	140.269	444.45	25	5.6065E+04	1.0092E+00	1,2
2059	20927	C10H20	1-methyl-1-propylcyclohexane	4258-93-9	140.269	445.15	25	5.6214E+04	1.0119E+00	1,2
2060	20928	C10H20	1-methyl-2-propylcyclohexane	4291-79-6	140.269	449.15	25	5.6877E+04	1.0238E+00	1,2
2061	20929	C10H20	1-methyl-3-propylcyclohexane	4291-80-9	140.269	437.65	25	5.4163E+04	9.7492E-01	1,2
2062	20930	C10H20	1-methyl-4-propylcyclohexane	4291-81-0	140.269	448.65	25	5.6812E+04	1.0226E+00	1,2
2063	20931	C10H20	1-methyl-2-isopropylcyclohexane	16580-23-7	140.269	444.15	25	5.5999E+04	1.0080E+00	1,2
2064	20932	C10H20	1-methyl-3-isopropylcyclohexane	16580-24-8	140.269	439.65	25	5.4803E+04	9.8644E-01	1,2
2065	20933	C10H20	1-methyl-4-isopropylcyclohexane	99-82-1	140.269	443.85	25	5.5930E+04	1.0067E+00	1,2
2066	20934	C10H20	cis-1-methyl-4-isopropylcyclohexane	6069-98-3	140.269	445.15	25	5.6214E+04	1.0119E+00	1,2
2067	20935	C10H20	trans-1-methyl-4-isopropylcyclohexane	1678-82-6	140.269	443.75	25	5.5907E+04	1.0063E+00	1,2
2068	20936	C10H20	1,1,3,4-tetramethylcyclohexane	24612-75-7	140.269	433.15	25	5.2512E+04	9.4521E-01	1,2
2069	20937	C10H20	trans-1,1,3,5-tetramethylcyclohexane	50876-31-8	140.269	429.55	25	5.1010E+04	9.1818E-01	1,2
2070	20938	C10H20	cis-1,1,3,5-tetramethylcyclohexane	50876-32-9	140.269	425.65	25	4.9245E+04	8.8641E-01	1,2
2071	20939	C10H20	diethylcyclohexane; (mixed isomers)	1331-43-7	140.269	447.15	25	5.6587E+04	1.0186E+00	1,2
2072	20943	C10H20	1-decene	872-05-9	140.269	443.75	25	1.3433E+05	2.4179E+00	1,2
2073	20944	C10H20	4-decene	19689-18-0	140.269	443.75	25	8.0467E+04	1.4484E+00	1,2
2074	20945	C10H20	cis-5-decene	7433-78-5	140.269	444.15	25	8.0516E+04	1.4493E+00	1,2
2075	20946	C10H20	trans-5-decene	7433-56-9	140.269	444.15	25	8.0516E+04	1.4493E+00	1,2
2076	20947	C10H20	2-methyl-1-nonene	2980-71-4	140.269	441.55	25	1.2039E+05	2.1670E+00	1,2
2077	20948	C10H20	2-methyl-3-nonene	53966-53-3	140.269	434.15	25	7.8031E+04	1.4045E+00	1,2
2078	20949	C10H20	5-methyl-4-nonene	15918-07-7	140.269	439.96	25	7.9779E+04	1.4360E+00	2
2079	20950	C10H20	3,7-dimethyl-1-octene	4984-01-4	140.269	427.15	25	7.4995E+04	1.3499E+00	1,2
2080	20951	C10H20	2,6-dimethyl-2-octene	4057-42-5	140.269	436.15	25	7.8721E+04	1.4170E+00	1,2

Table 10 Henry's Law Constant for Compound in Water - Organic Compounds

NO	ID	FORMULA	NAME	CAS No	MW, g/mol	T_B, K	T, C	H @ T, atm/mol frac	H @ T, atm/mol/m^3	code
					Henry's Law Constant, H					
2081	20952	C10H20	4-propyl-3-heptene	4485-13-6	140.269	433.65	25	7.7845E+04	1.4012E+00	1,2
2082	20953	C10H20	2,4,6-trimethyl-3-heptene, isomer	123434-07-1	140.269	439.96	25	7.9779E+04	1.4360E+00	2
2083	20954	C10H20	2,2,5,5-tetramethylhex-3-ene	22808-06-6	140.269	397.45	25	5.6719E+04	1.0209E+00	1,2
2084	20955	C10H20	cis-1,2-di-tert-butylethene	692-47-7	140.269	439.96	25	7.9779E+04	1.4360E+00	2
2085	20956	C10H20	(E)-2,2,5,5-tetramethylhex-3-ene	692-48-8	140.269	398.35	25	5.7298E+04	1.0314E+00	1,2
2086	20958	C10H20Br2	1,1-dibromodecane	59104-80-2	300.077	550.15	25	2.9632E+01	5.3337E-04	1,2
2087	20959	C10H20Br2	1,10-dibromodecane	4101-68-2	300.077	550.15	25	1.3870E+01	2.4966E-04	2
2088	20960	C10H20Br2	5,6-dibromodecane	77928-86-0	300.077	550.15	25	1.3870E+01	2.4966E-04	2
2089	20962	C10H20Cl2	1,1-dichlorodecane	3162-62-7	211.174	520.15	25	3.0038E+02	5.4067E-03	1,2
2090	20963	C10H20Cl2	1,10-dichlorodecane	2162-98-3	211.174	520.15	25	1.9753E+02	3.5555E-03	2
2091	20992	C10H20O	decanal	112-31-2	156.268	488.15	25	2.3063E+01	4.1514E-04	1,2
2092	20993	C10H20O	2-decanone	693-54-9	156.268	483.35	25	2.6652E+01	4.7973E-04	1,2
2093	20994	C10H20O	3-decanone	928-80-3	156.268	476.15	25	2.2425E+01	4.0364E-04	1,2
2094	20995	C10H20O	4-decanone	624-16-8	156.268	479.65	25	2.2794E+01	4.1030E-04	1,2
2095	20996	C10H20O	2-methyl-5-nonanone	22287-02-1	156.268	476.65	25	2.2476E+01	4.0457E-04	1,2
2096	21019	C10H20O	5-decanone	820-29-1	156.268	477.05	25	2.2518E+01	4.0531E-04	1,2
2097	21025	C10H20O	isodecanal	1321-89-7	156.268	470.15	25	1.5341E+01	2.7614E-04	1,2
2098	21034	C10H20O	2-methyl-3-nonanone	5445-31-8	156.268	473.00	25	2.2108E+01	3.9794E-04	1,2
2099	21040	C10H20O2	decanoic acid	334-48-5	172.268	543.15	20	1.0277E-03	1.8499E-08	1,2
2100	21041	C10H20O2	2,2-dimethyloctanoic acid	29662-90-6	172.268	473.73	25	2.2585E-01	4.0653E-06	2
2101	21042	C10H20O2	3-methylnonanoic acid	35205-79-9	172.268	473.73	25	2.2585E-01	4.0653E-06	2
2102	21043	C10H20O2	nonyl formate	5451-92-3	172.268	485.72	25	5.7932E+01	1.0428E-03	2
2103	21044	C10H20O2	2-ethylhexyl acetate	103-09-3	172.268	471.75	25	2.2308E+02	4.0155E-03	1,2
2104	21045	C10H20O2	octyl acetate	112-14-1	172.268	484.45	25	2.8099E+02	5.0578E-03	1,2
2105	21046	C10H20O2	6-methyl-2-heptyl acetate	67952-57-2	172.268	460.15	25	2.9089E+01	5.2360E-04	1,2
2106	21047	C10H20O2	6-methyl-3-heptyl acetate	32764-34-4	172.268	457.65	25	2.9270E+01	5.2686E-04	1,2
2107	21048	C10H20O2	2-methylheptyl acetate, (±)	74112-36-0	172.268	468.15	25	2.8342E+01	5.1016E-04	1,2
2108	21049	C10H20O2	heptyl propanoate	2216-81-1	172.268	483.15	25	2.4587E+01	4.4256E-04	1,2
2109	21050	C10H20O2	hexyl butanoate	2639-63-6	172.268	479.15	25	2.8677E+01	5.1618E-04	1,2
2110	21051	C10H20O2	1-methylpentyl butanoate, (S)	116723-96-7	172.268	473.73	25	3.4099E+01	6.1377E-04	2
2111	21053	C10H20O2	1,1-dimethylpropyl 3-methylbutanoate	542-37-0	172.268	446.65	25	3.5588E+01	6.4058E-04	1,2
2112	21066	C10H20O2	8-methylnonanoic acid	5963-14-4	172.268	473.73	25	2.2585E-01	4.0653E-06	2
2113	21067	C10H20O2	neodecanoic acid	26896-20-8	172.268	528.15	25	1.0709E-01	1.9276E-06	1,2
2114	21068	C10H20O2	tert-decanoic acid	52627-73-3	172.268	473.73	25	2.2585E-01	4.0653E-06	2
2115	21074	C10H20O2	2-(4-isobutylphenyl)butyric acid	55837-18-8	172.268	473.73	25	2.2585E-01	4.0653E-06	2
2116	21077	C10H20O2	4-nonanecarboxylic acid	31080-39-4	172.268	473.73	25	2.2585E-01	4.0653E-06	2
2117	21090	C10H21Br	1-bromodecane	112-29-8	221.181	513.75	25	2.2917E+03	4.1251E-02	1,2
2118	21091	C10H21Br	2-bromodecane	39563-53-6	221.181	513.75	25	7.3582E+02	1.3245E-02	2
2119	21093	C10H21Cl	1-chlorodecane	1002-69-3	176.729	499.02	25	3.2415E+03	5.8347E-02	1,2
2120	21094	C10H21Cl	decyl chloride (mixed isomers)	28519-06-4	176.729	499.02	25	1.2056E+03	2.1700E-02	2
2121	21133	C10H22	decane	124-18-5	142.285	447.30	25	3.1825E+05	5.7285E+00	1,2
2122	21134	C10H22	2-methylnonane	871-83-0	142.285	440.15	25	4.9971E+05	8.9947E+00	1,2
2123	21135	C10H22	3-methylnonane	5911-04-6	142.285	440.95	25	5.5958E+05	1.0072E+01	1,2
2124	21136	C10H22	4-methylnonane	17301-94-9	142.285	438.85	25	6.3127E+05	1.1363E+01	1,2
2125	21137	C10H22	5-methylnonane	15869-85-9	142.285	438.30	25	5.5740E+05	1.0033E+01	1,2
2126	21138	C10H22	3-ethyloctane	5881-17-4	142.285	439.66	25	5.3194E+05	9.5748E+00	1,2
2127	21139	C10H22	4-ethyloctane	15869-86-0	142.285	436.80	25	5.2986E+05	9.5374E+00	1,2
2128	21140	C10H22	2,2-dimethyloctane	15869-87-1	142.285	430.05	25	5.5173E+05	9.9311E+00	1,2
2129	21141	C10H22	2,3-dimethyloctane	7146-60-3	142.285	437.47	25	5.3798E+05	9.6836E+00	1,2
2130	21142	C10H22	2,4-dimethyloctane	4032-94-4	142.285	429.06	25	4.5275E+05	8.1495E+00	1,2
2131	21143	C10H22	2,5-dimethyloctane	15869-89-3	142.285	431.66	25	4.8626E+05	8.7526E+00	1,2
2132	21144	C10H22	2,6-dimethyloctane	2051-30-1	142.285	433.54	25	5.2244E+05	9.4038E+00	1,2
2133	21145	C10H22	2,7-dimethyloctane	1072-16-8	142.285	433.03	25	4.6491E+05	8.3683E+00	1,2
2134	21146	C10H22	3,3-dimethyloctane	4110-44-5	142.285	434.36	25	5.4728E+05	9.8509E+00	1,2
2135	21147	C10H22	3,4-dimethyloctane	15869-92-8	142.285	436.56	25	5.4550E+05	9.8190E+00	1,2
2136	21148	C10H22	3,5-dimethyloctane	15869-93-9	142.285	432.56	25	5.1021E+05	9.1836E+00	1,2
2137	21149	C10H22	3,6-dimethyloctane	15869-94-0	142.285	433.96	25	4.9889E+05	8.9800E+00	1,2
2138	21150	C10H22	4,4-dimethyloctane	15869-95-1	142.285	430.66	25	5.2442E+05	9.4394E+00	1,2
2139	21151	C10H22	4,5-dimethyloctane	15869-96-2	142.285	435.29	25	5.4618E+05	9.8311E+00	1,2
2140	21152	C10H22	4-propylheptane	3178-29-8	142.285	430.66	25	4.5921E+05	8.2656E+00	1,2
2141	21153	C10H22	4-isopropylheptane	52896-87-4	142.285	432.06	25	5.3200E+05	9.5759E+00	1,2
2142	21154	C10H22	3-ethyl-2-methylheptane	14676-29-0	142.285	434.36	25	5.4173E+05	9.7510E+00	1,2
2143	21155	C10H22	4-ethyl-2-methylheptane	52896-88-5	142.285	429.36	25	4.9723E+05	8.9501E+00	1,2
2144	21156	C10H22	5-ethyl-2-methylheptane	13475-78-0	142.285	432.86	25	5.1539E+05	9.2769E+00	1,2
2145	21157	C10H22	3-ethyl-3-methylheptane	17302-01-1	142.285	436.96	25	6.0883E+05	1.0959E+01	1,2
2146	21158	C10H22	4-ethyl-3-methylheptane	52896-89-6	142.285	435.36	25	5.7088E+05	1.0276E+01	1,2
2147	21159	C10H22	3-ethyl-5-methylheptane	52896-90-9	142.285	431.36	25	4.9815E+05	8.9666E+00	1,2
2148	21160	C10H22	3-ethyl-4-methylheptane	52896-91-0	142.285	436.16	25	5.7261E+05	1.0307E+01	1,2
2149	21161	C10H22	4-ethyl-4-methylheptane	17302-04-4	142.285	433.96	25	5.8326E+05	1.0499E+01	1,2
2150	21162	C10H22	2,2,3-trimethylheptane	52896-92-1	142.285	430.76	25	5.4374E+05	9.7872E+00	1,2
2151	21163	C10H22	2,2,4-trimethylheptane	14720-74-2	142.285	421.46	25	4.5964E+05	8.2735E+00	1,2
2152	21164	C10H22	2,2,5-trimethylheptane	20291-95-6	142.285	423.96	25	4.7404E+05	8.5326E+00	1,2
2153	21165	C10H22	2,2,6-trimethylheptane	1190-83-6	142.285	422.09	25	4.3593E+05	7.8467E+00	1,2
2154	21166	C10H22	2,3,3-trimethylheptane	52896-93-2	142.285	433.36	25	5.7959E+05	1.0433E+01	1,2
2155	21167	C10H22	2,3,4-trimethylheptane	52896-95-4	142.285	433.06	25	5.5939E+05	1.0069E+01	1,2
2156	21168	C10H22	2,3,5-trimethylheptane	20278-85-7	142.285	433.86	25	5.8914E+05	1.0604E+01	1,2
2157	21169	C10H22	2,3,6-trimethylheptane	4032-93-3	142.285	429.16	25	4.9293E+05	8.8727E+00	1,2
2158	21170	C10H22	2,4,4-trimethylheptane	4032-92-2	142.285	424.16	25	5.0400E+05	9.0720E+00	1,2
2159	21171	C10H22	2,4,5-trimethylheptane	20278-84-6	142.285	429.66	25	5.2272E+05	9.4089E+00	1,2
2160	21172	C10H22	2,4,6-trimethylheptane	2613-61-8	142.285	420.76	25	4.1240E+05	7.4231E+00	1,2

Table 10 Henry's Law Constant for Compound in Water - Organic Compounds

				Henry's Law Constant, H						
NO	ID	FORMULA	NAME	CAS No	MW, g/mol	T_B, K	T, C	H @ T, atm/mol frac	H @ T, atm/mol/m^3	code
2161	21173	C10H22	2,5,5-trimethylheptane	1189-99-7	142.285	425.96	25	4.9441E+05	8.8994E+00	1,2
2162	21174	C10H22	3,3,4-trimethylheptane	20278-87-9	142.285	435.06	25	6.1432E+05	1.1058E+01	1,2
2163	21175	C10H22	3,3,5-trimethylheptane	7154-80-5	142.285	428.85	25	5.5720E+05	1.0030E+01	1,2
2164	21176	C10H22	3,4,4-trimethylheptane	20278-88-0	142.285	434.26	25	6.1282E+05	1.1031E+01	1,2
2165	21177	C10H22	3,4,5-trimethylheptane	20278-89-1	142.285	435.66	25	6.0067E+05	1.0812E+01	1,2
2166	21178	C10H22	3-isopropyl-2-methylhexane	62016-13-1	142.285	439.86	25	7.7343E+05	1.3922E+01	1,2
2167	21179	C10H22	3,3-diethylhexane	17302-02-2	142.285	439.46	25	6.5816E+05	1.1847E+01	1,2
2168	21180	C10H22	3,4-diethylhexane	19398-77-7	142.285	437.06	25	6.0555E+05	1.0900E+01	1,2
2169	21181	C10H22	3-ethyl-2,2-dimethylhexane	20291-91-2	142.285	429.26	25	5.6385E+05	1.0149E+01	1,2
2170	21182	C10H22	4-ethyl-2,2-dimethylhexane	52896-99-8	142.285	420.16	25	4.5834E+05	8.2501E+00	1,2
2171	21183	C10H22	3-ethyl-2,3-dimethylhexane	52897-00-4	142.285	436.86	25	6.4693E+05	1.1645E+01	1,2
2172	21184	C10H22	4-ethyl-2,3-dimethylhexane	52897-01-5	142.285	434.06	25	5.8891E+05	1.0600E+01	1,2
2173	21185	C10H22	3-ethyl-2,4-dimethylhexane	7220-26-0	142.285	433.26	25	5.8738E+05	1.0573E+01	1,2
2174	21186	C10H22	4-ethyl-2,4-dimethylhexane	52897-03-7	142.285	434.26	25	6.1761E+05	1.1117E+01	1,2
2175	21187	C10H22	3-ethyl-2,5-dimethylhexane	52897-04-8	142.285	427.26	25	5.1338E+05	9.2408E+00	1,2
2176	21188	C10H22	4-ethyl-3,3-dimethylhexane	52897-05-9	142.285	436.06	25	6.4690E+05	1.1644E+01	1,2
2177	21189	C10H22	3-ethyl-3,4-dimethylhexane	52897-06-0	142.285	435.26	25	6.4577E+05	1.1624E+01	1,2
2178	21190	C10H22	2,2,3,3-tetramethylhexane	13475-81-5	142.285	433.48	25	6.6919E+05	1.2045E+01	1,2
2179	21191	C10H22	2,2,3,4-tetramethylhexane	52897-08-2	142.285	431.96	25	6.3833E+05	1.1490E+01	1,2
2180	21192	C10H22	2,2,3,5-tetramethylhexane	52897-09-3	142.285	421.56	25	4.7634E+05	8.5740E+00	1,2
2181	21193	C10H22	2,2,4,4-tetramethylhexane	51750-65-3	142.285	426.96	25	6.7287E+05	1.2112E+01	1,2
2182	21194	C10H22	2,2,4,5-tetramethylhexane	16747-42-5	142.285	421.04	25	4.9801E+05	8.9642E+00	1,2
2183	21195	C10H22	2,2,5,5-tetramethylhexane	1071-81-4	142.285	410.63	25	3.9119E+05	7.0413E+00	1,2
2184	21196	C10H22	2,3,3,4-tetramethylhexane	52897-10-6	142.285	437.75	25	6.9704E+05	1.2547E+01	1,2
2185	21197	C10H22	2,3,3,5-tetramethylhexane	52897-11-7	142.285	426.26	25	5.4628E+05	9.8330E+00	1,2
2186	21198	C10H22	2,3,4,4-tetramethylhexane	52897-12-8	142.285	434.76	25	6.7016E+05	1.2063E+01	1,2
2187	21199	C10H22	2,3,4,5-tetramethylhexane	52897-15-1	142.285	429.36	25	5.4352E+05	9.7832E+00	1,2
2188	21200	C10H22	3,3,4,4-tetramethylhexane	5171-84-6	142.285	443.16	25	8.3769E+05	1.5078E+01	1,2
2189	21201	C10H22	2,4-dimethyl-3-isopropylpentane	13475-79-1	142.285	430.20	25	6.0878E+05	1.0958E+01	1,2
2190	21202	C10H22	3,3-diethyl-2-methylpentane	52897-16-2	142.285	442.86	25	7.4426E+05	1.3397E+01	1,2
2191	21203	C10H22	3-ethyl-2,2,3-trimethylpentane	52897-17-3	142.285	442.66	25	8.4393E+05	1.5191E+01	1,2
2192	21204	C10H22	3-ethyl-2,2,4-trimethylpentane	52897-18-4	142.285	428.46	25	6.0183E+05	1.0833E+01	1,2
2193	21205	C10H22	3-ethyl-2,3,4-trimethylpentane	52897-19-5	142.285	442.60	25	8.0251E+05	1.4445E+01	1,2
2194	21206	C10H22	2,2,3,3,4-pentamethylpentane	16747-44-7	142.285	439.21	25	8.1350E+05	1.4643E+01	1,2
2195	21207	C10H22	2,2,3,4,4-pentamethylpentane	16747-45-8	142.285	432.45	25	7.7745E+05	1.3994E+01	1,2
2196	21224	C10H22O	1-decanol	112-30-1	158.284	503.35	25	3.1444E+00	5.6598E-05	1,2
2197	21225	C10H22O	isodecanol	25339-17-7	158.284	493.00	25	3.9467E+00	7.1040E-05	1,2
2198	21226	C10H22O	2-decanol	1120-06-5	158.284	490.15	25	7.8637E-01	1.4155E-05	1,2
2199	21227	C10H22O	2-decanol, (±)	74742-10-2	158.284	484.15	25	2.9346E+00	5.2823E-05	1,2
2200	21228	C10H22O	3-decanol	1565-81-7	158.284	483.00	25	2.9353E+00	5.2834E-05	2
2201	21229	C10H22O	4-decanol	2051-31-2	158.284	483.15	25	1.4242E+01	2.5635E-04	1,2
2202	21230	C10H22O	5-decanol	5205-34-5	158.284	474.15	25	1.2022E+01	2.1640E-04	1,2
2203	21231	C10H22O	2-methyl-1-nonanol	40589-14-8	158.284	495.15	25	1.8400E+01	3.3120E-04	1,2
2204	21232	C10H22O	4-methyl-1-nonanol	1489-47-0	158.284	489.15	25	1.8900E+01	3.4020E-04	1,2
2205	21233	C10H22O	5-methyl-1-nonanol	2768-16-3	158.284	487.00	25	2.9346E+00	5.2822E-05	2
2206	21234	C10H22O	7-methyl-1-nonanol	33234-93-4	158.284	487.00	25	2.9346E+00	5.2822E-05	2
2207	21235	C10H22O	3-methyl-1-nonanol	22663-64-5	158.284	487.00	25	2.9346E+00	5.2822E-05	2
2208	21236	C10H22O	2-methyl-2-nonanol	10297-57-1	158.284	487.00	25	2.9346E+00	5.2822E-05	2
2209	21237	C10H22O	3-methyl-2-nonanol	60671-32-1	158.284	487.00	25	2.9346E+00	5.2822E-05	2
2210	21238	C10H22O	5-methyl-2-nonanol	66731-95-1	158.284	487.00	25	2.9346E+00	5.2822E-05	2
2211	21239	C10H22O	6-methyl-2-nonanol	66256-60-8	158.284	487.00	25	2.9346E+00	5.2822E-05	2
2212	21240	C10H22O	7-methyl-2-nonanol	66256-61-9	158.284	487.00	25	2.9346E+00	5.2822E-05	2
2213	21241	C10H22O	8-methyl-2-nonanol	14779-92-1	158.284	487.00	25	2.9346E+00	5.2822E-05	2
2214	21242	C10H22O	2-methyl-3-nonanol	26533-33-5	158.284	481.55	25	1.5199E+01	2.7358E-04	1,2
2215	21243	C10H22O	3-methyl-3-nonanol	21078-72-8	158.284	482.00	25	2.9360E+00	5.2848E-05	2
2216	21244	C10H22O	5-methyl-3-nonanol	66719-43-5	158.284	482.00	25	2.9360E+00	5.2848E-05	2
2217	21245	C10H22O	2-methyl-4-nonanol	26533-31-3	158.284	482.00	25	2.9360E+00	5.2848E-05	2
2218	21246	C10H22O	4-methyl-4-nonanol	23418-38-4	158.284	482.00	25	2.9360E+00	5.2848E-05	2
2219	21247	C10H22O	5-methyl-4-nonanol	66719-44-6	158.284	482.00	25	2.9360E+00	5.2848E-05	2
2220	21248	C10H22O	7-methyl-4-nonanol	26981-98-6	158.284	482.00	25	2.9360E+00	5.2848E-05	2
2221	21249	C10H22O	2-methyl-5-nonanol	29843-62-7	158.284	482.00	25	2.9360E+00	5.2848E-05	2
2222	21250	C10H22O	3-methyl-5-nonanol	66779-42-4	158.284	482.00	25	2.9360E+00	5.2848E-05	2
2223	21251	C10H22O	5-methyl-5-nonanol	33933-78-7	158.284	475.15	25	9.4084E+00	1.6935E-04	1,2
2224	21252	C10H22O	2-ethyl-1-octanol	20592-10-3	158.284	482.00	25	2.9360E+00	5.2848E-05	2
2225	21253	C10H22O	3-ethyl-1-octanol	66719-36-6	158.284	482.00	25	2.9360E+00	5.2848E-05	2
2226	21254	C10H22O	2,2-dimethyl-1-octanol	2370-14-1	158.284	481.15	25	1.6790E+01	3.0221E-04	1,2
2227	21255	C10H22O	2,6-dimethyl-1-octanol	62417-08-7	158.284	483.00	25	2.9353E+00	5.2834E-05	2
2228	21256	C10H22O	3,7-dimethyl-1-octanol	106-21-8	158.284	485.15	25	1.6356E+01	2.9441E-04	1,2
2229	21257	C10H22O	4,5-dimethyl-1-octanol	66719-32-2	158.284	485.00	25	2.9344E+00	5.2819E-05	2
2230	21258	C10H22O	4,6-dimethyl-1-octanol	66719-33-3	158.284	485.00	25	2.9344E+00	5.2819E-05	2
2231	21259	C10H22O	4,7-dimethyl-1-octanol	66719-34-4	158.284	485.00	25	2.9344E+00	5.2819E-05	2
2232	21260	C10H22O	7,7-dimethyl-1-octanol	66719-35-5	158.284	485.00	25	2.9344E+00	5.2819E-05	2
2233	21261	C10H22O	2,4-dimethyl-2-octanol	18675-20-2	158.284	485.00	25	2.9344E+00	5.2819E-05	2
2234	21262	C10H22O	2,6-dimethyt-2-octanol	18479-57-7	158.284	485.00	25	2.9344E+00	5.2819E-05	2
2235	21263	C10H22O	2,7-dimethyl-2-octanol	42007-73-8	158.284	485.00	25	2.9344E+00	5.2819E-05	2
2236	21264	C10H22O	3,7-dimethyl-2-octanol	15340-96-2	158.284	485.00	25	2.9344E+00	5.2819E-05	2
2237	21265	C10H22O	3-ethyl-3-octanol	2051-32-3	158.284	472.15	25	1.4084E+01	2.5351E-04	1,2
2238	21266	C10H22O	6-ethyl-3-octanol	19781-27-2	158.284	472.00	25	2.9538E+00	5.3168E-05	2
2239	21267	C10H22O	2,2-dimethyl-3-octanol	19841-72-6	158.284	467.00	25	2.9667E+00	5.3401E-05	2
2240	21268	C10H22O	2,3-dimethyl-3-octanol	19781-10-3	158.284	462.25	25	1.2878E+01	2.3181E-04	1,2

Table 10 Henry's Law Constant for Compound in Water - Organic Compounds

					Henry's Law Constant, H					
NO	ID	FORMULA	NAME	CAS No	MW, g/mol	T_B, K	T, C	H @ T, atm/mol frac	H @ T, atm/mol/m³	code
2241	21269	C10H22O	2,7-dimethyl-3-octanol	66719-55-9	158.284	467.15	25	1.2732E+01	2.2917E-04	1,2
2242	21270	C10H22O	3,5-dimethyl-3-octanol	56065-42-0	158.284	466.00	25	2.9694E+00	5.3449E-05	2
2243	21271	C10H22O	3,6-dimethyl-3-octanol	151-19-9	158.284	465.15	25	1.3563E+01	2.4413E-04	1,2
2244	21272	C10H22O	3,7-dimethyl-3-octanol	78-69-3	158.284	469.15	25	1.4597E+01	2.6274E-04	1,2
2245	21273	C10H22O	3-ethyl-4-octanol	63126-48-7	158.284	466.00	25	2.9694E+00	5.3449E-05	2
2246	21274	C10H22O	4-ethyl-4-octanol	38395-42-5	158.284	466.00	25	2.9694E+00	5.3449E-05	2
2247	21275	C10H22O	2,2-dimethyl-4-octanol	66719-52-6	158.284	463.15	25	1.3085E+01	2.3554E-04	1,2
2248	21276	C10H22O	2,4-dimethyl-4-octanol	33933-79-8	158.284	464.00	25	2.9748E+00	5.3547E-05	2
2249	21277	C10H22O	2,5-dimethyl-4-octanol	66719-53-7	158.284	464.00	25	2.9748E+00	5.3547E-05	2
2250	21278	C10H22O	2,6-dimethyl-4-octanol	66719-54-8	158.284	468.15	25	1.6239E+01	2.9230E-04	1,2
2251	21279	C10H22O	2,7-dimethyl-4-octanol	19781-11-4	158.284	475.15	25	2.6912E+01	4.8441E-04	1,2
2252	21280	C10H22O	3,4-dimethyl-4-octanol	66719-30-0	158.284	464.00	25	2.9748E+00	5.3547E-05	2
2253	21281	C10H22O	3,6-dimethyl-4-octanol	66719-31-1	158.284	464.00	25	2.9748E+00	5.3547E-05	2
2254	21282	C10H22O	4,6-dimethyl-4-octanol	56065-43-1	158.284	464.00	25	2.9748E+00	5.3547E-05	2
2255	21283	C10H22O	4,7-dimethyl-4-octanol	19781-13-6	158.284	465.15	25	1.3563E+01	2.4413E-04	1,2
2256	21284	C10H22O	2-propyl-1-heptanol	10042-59-8	158.284	491.05	25	1.6785E+01	3.0212E-04	1,2
2257	21285	C10H22O	3-isopropyl-1-heptanol	38514-15-7	158.284	488.35	25	1.7506E+01	3.1510E-04	1,2
2258	21286	C10H22O	2,4,6-trimethyl-2-heptanol	66256-47-1	158.284	466.00	25	2.9694E+00	5.3449E-05	2
2259	21287	C10H22O	2,5,6-trimethyl-2-heptanol	66256-48-2	158.284	466.15	25	1.5107E+01	2.7192E-04	1,2
2260	21288	C10H22O	4,6,6-trimethyl-2-heptanol	51079-79-9	158.284	466.00	25	2.9694E+00	5.3449E-05	2
2261	21289	C10H22O	2-methyl-3-ethyl-3-heptanol	66719-37-7	158.284	466.15	25	1.3811E+01	2.4860E-04	1,2
2262	21290	C10H22O	4-methyl-5-ethyl-3-heptanol	66731-94-0	158.284	466.00	25	2.9694E+00	5.3449E-05	2
2263	21291	C10H22O	5-methyl-3-ethyl-3-heptanol	---	158.284	466.00	25	2.9694E+00	5.3449E-05	2
2264	21292	C10H22O	2,2,3-trimethyl-3-heptanol	29772-40-5	158.284	459.15	25	1.4619E+01	2.6315E-04	1,2
2265	21293	C10H22O	2,2,6-trimethyl-3-heptanol	66256-43-7	158.284	463.00	25	2.9775E+00	5.3595E-05	2
2266	21294	C10H22O	2,3,6-trimethyl-3-heptanol	58046-40-5	158.284	463.00	25	2.9775E+00	5.3595E-05	2
2267	21295	C10H22O	3,5,5-trimethyl-3-heptanol	66256-50-6	158.284	468.75	25	1.7552E+01	3.1592E-04	1,2
2268	21296	C10H22O	4-propyl-4-heptanol	2198-72-3	158.284	467.15	25	1.2857E+01	2.3143E-04	1,2
2269	21297	C10H22O	4-isopropyl-4-heptanol	51200-82-9	158.284	463.15	25	1.3085E+01	2.3554E-04	1,2
2270	21298	C10H22O	2,2,4-trimethyl-4-heptanol	57233-31-5	158.284	454.15	25	1.3349E+01	2.4029E-04	1,2
2271	21299	C10H22O	2,2,5-trimethyl-4-heptanol	66256-42-6	158.284	455.00	25	2.9964E+00	5.3934E-05	2
2272	21300	C10H22O	2,2,6-trimethyl-4-heptanol	66256-44-8	158.284	455.00	25	2.9964E+00	5.3934E-05	2
2273	21301	C10H22O	2,4,5-trimethyl-4-heptanol	66256-46-0	158.284	455.00	25	2.9964E+00	5.3934E-05	2
2274	21302	C10H22O	2,4,6-trimethyl-4-heptanol	60836-07-9	158.284	455.15	25	1.2388E+01	2.2298E-04	1,2
2275	21303	C10H22O	3,3,6-trimethyl-4-heptanol	---	158.284	455.00	25	2.9964E+00	5.3934E-05	2
2276	21304	C10H22O	4-methyl-2-propyl-1-hexanol	66256-62-0	158.284	481.15	25	1.5078E+01	2.7141E-04	1,2
2277	21305	C10H22O	4-methyl-2-isopropyl-1-hexanol	66719-41-3	158.284	474.15	25	1.4601E+01	2.6281E-04	1,2
2278	21306	C10H22O	5-methyl-2-isopropyl-1-hexanol	2051-33-4	158.284	486.15	25	1.8451E+01	3.3212E-04	1,2
2279	21307	C10H22O	5,5-dimethyl-2-ethyl-1-hexanol	---	158.284	486.00	25	2.9343E+00	5.2818E-05	2
2280	21308	C10H22O	2,3,4,4-tetramethyl-2-hexanol	66256-66-4	158.284	463.15	25	1.7206E+01	3.0971E-04	1,2
2281	21309	C10H22O	3,3,5,5-tetramethyl-2-hexanol	66256-69-7	158.284	463.00	25	2.9775E+00	5.3595E-05	2
2282	21310	C10H22O	2-methyl-3-isopropyl-3-hexanol	51200-81-8	158.284	465.15	25	1.4931E+01	2.6876E-04	1,2
2283	21311	C10H22O	2,2-dimethyl-4-ethyl-3-hexanol	66719-47-9	158.284	460.15	25	1.3533E+01	2.4360E-04	1,2
2284	21312	C10H22O	2,4-dimethyl-4-ethyl-3-hexanol	66719-48-0	158.284	460.00	25	2.9852E+00	5.3734E-05	2
2285	21313	C10H22O	5,5-dimethyl-3-ethyl-3-hexanol	5340-62-5	158.284	460.00	25	2.9852E+00	5.3734E-05	2
2286	21314	C10H22O	2,2,3,4-tetramethyl-3-hexanol	66256-63-1	158.284	465.15	25	1.7893E+01	3.2208E-04	1,2
2287	21315	C10H22O	2,2,3,5-tetramethyl-3-hexanol	66256-64-2	158.284	513.48	25	3.0890E+00	5.5601E-05	2
2288	21316	C10H22O	2,2,4,4-tetramethyl-3-hexanol	66256-65-3	158.284	463.15	25	1.7206E+01	3.0971E-04	1,2
2289	21317	C10H22O	2,2,5,5-tetramethyl-3-hexanol	55073-86-4	158.284	443.15	25	1.1888E+01	2.1398E-04	1,2
2290	21318	C10H22O	2,3,4,4-tetramethyl-3-hexanol	66256-67-5	158.284	474.15	25	2.1505E+01	3.8709E-04	1,2
2291	21319	C10H22O	2,3,5,5-tetramethyl-3-hexanol	5396-09-8	158.284	475.00	25	2.9470E+00	5.3045E-05	2
2292	21320	C10H22O	2,4,4,5-tetramethyl-3-hexanol	66256-68-6	158.284	475.00	25	2.9470E+00	5.3045E-05	2
2293	21321	C10H22O	3,4,4,5-tetramethyl-3-hexanol	66256-39-1	158.284	475.15	25	2.1968E+01	3.9543E-04	1,2
2294	21322	C10H22O	3,4,5,5-tetramethyl-3-hexanol	66256-40-4	158.284	468.15	25	1.8997E+01	3.4193E-04	1,2
2295	21323	C10H22O	4-methyl-2-isobutyl-1-pentanol	22417-45-4	158.284	477.15	25	1.5471E+01	2.7848E-04	1,2
2296	21324	C10H22O	4,4-dimethyl-3-isopropyl-1-pentanol	66719-51-5	158.284	456.00	25	2.9944E+00	5.3898E-05	2
2297	21325	C10H22O	3,4-dimethyl-3-isopropyl-2-pentanol	66719-50-4	158.284	456.00	25	2.9944E+00	5.3898E-05	2
2298	21326	C10H22O	2,4-dimethyl-3-propyl-3-pentanol	---	158.284	457.15	25	1.2830E+01	2.3094E-04	1,2
2299	21327	C10H22O	2,4-dimethyl-3-isopropyl-3-pentanol	51200-83-0	158.284	467.65	25	1.7128E+01	3.0830E-04	1,2
2300	21328	C10H22O	2,2,4-trimethyl-3-ethyl-3-pentanol	66256-41-5	158.284	464.15	25	1.7545E+01	3.1581E-04	1,2
2301	21329	C10H22O	2,2,3,4,4-pentamethyl-3-pentanol	5857-69-2	158.284	467.55	25	2.2698E+01	4.0855E-04	1,2
2302	21330	C10H22O	2,3-dimethyl-2-tert-butyl-1-butanol	81931-81-9	158.284	466.00	25	2.9694E+00	5.3449E-05	2
2303	21331	C10H22O	3,7-dimethyl-3-octanol, (±)	57706-88-4	158.284	469.65	25	2.9597E+00	5.3274E-05	1,2
2304	21332	C10H22O	3,7-dimethyl-1-octanol, (±)	59204-02-3	158.284	485.65	25	2.9343E+00	5.2817E-05	1,2
2305	21333	C10H22O	3-ethyl-6-methyl-3-heptanol	66719-40-2	158.284	472.78	25	2.9519E+00	5.3134E-05	2
2306	21334	C10H22O	8-methyl-1-nonanol	55505-26-5	158.284	472.78	25	2.9519E+00	5.3134E-05	2
2307	21335	C10H22O	di-pentyl ether	693-65-2	158.284	459.90	25	5.3082E+02	9.5546E-03	1,2
2308	21336	C10H22O	diisopentyl ether	544-01-4	158.284	445.65	25	2.3566E+02	4.2418E-03	1,2
2309	21337	C10H22O	ethyl octyl ether	929-61-3	158.284	459.45	25	2.1849E+02	3.9328E-03	1,2
2310	21338	C10H22O	(S)-(+)-2-decanol	33758-16-6	158.284	484.15	25	2.9346E+00	5.2823E-05	1,2
2311	21339	C10H22O	(R)-3,7-dimethyl-1-octanol	1117-60-8	158.284	485.70	25	2.9343E+00	5.2817E-05	1,2
2312	21340	C10H22O	s-3,7-dimethyl-1-octanol	68680-98-8	158.284	485.00	25	2.8615E+00	5.1507E-05	2
2313	21341	C10H22O	decyl alcohol (mixed isomers)	85566-12-7	158.284	472.78	25	2.9519E+00	5.3134E-05	2
2314	21385	C10H22S	decyl mercaptan	143-10-2	174.351	512.35	25	7.0928E+02	1.2767E-02	1,2
2315	21386	C10H22S	2-decanethiol	13402-60-3	174.351	501.45	25	9.0753E+02	1.6335E-02	1,2
2316	21387	C10H22S	dipentyl sulfide	872-10-6	174.351	500.00	25	2.2702E+02	4.0864E-03	2
2317	21388	C10H22S	methyl nonyl sulfide	59973-07-8	174.351	500.00	25	2.2702E+02	4.0864E-03	2
2318	21389	C10H22S	ethyl octyl sulfide	3698-94-0	174.351	500.00	25	2.2702E+02	4.0864E-03	2
2319	21390	C10H22S	propyl heptyl sulfide	24768-46-5	174.351	500.00	25	2.2702E+02	4.0864E-03	2
2320	21391	C10H22S	butyl hexyl sulfide	16967-04-7	174.351	500.00	25	2.2702E+02	4.0864E-03	2

Table 10 Henry's Law Constant for Compound in Water - Organic Compounds

					Henry's Law Constant, H					
NO	ID	FORMULA	NAME	CAS No	MW, g/mol	T_B, K	T, C	H @ T, atm/mol frac	H @ T, atm/mol/m^3	code
2321	21392	C10H22S	bis(2-methylbutyl) sulfide	96034-00-3	174.351	438.15	25	2.6407E+02	4.7533E-03	1,2
2322	21393	C10H22S	diisopentyl sulfide	544-02-5	174.351	489.20	25	2.4696E+02	4.4452E-03	1,2
2323	21400	C10H23N	decylamine	2016-57-1	157.300	493.65	25	3.7224E+00	6.7003E-05	1,2
2324	21401	C10H23N	methylnonylamine	39093-27-1	157.300	480.15	25	1.8432E+00	3.3178E-05	1,2
2325	21402	C10H23N	dimethyloctylamine	7378-99-6	157.300	464.15	25	1.3925E+01	2.5064E-04	1,2
2326	21403	C10H23N	diethylhexylamine	44979-90-0	157.300	452.15	25	9.0545E+00	1.6298E-04	1,2
2327	21404	C10H23N	ethyloctylamine	4088-36-2	157.300	475.15	25	2.0200E+00	3.6359E-05	1,2
2328	21405	C10H23N	dipentylamine	2050-92-2	157.300	476.15	25	1.9352E+00	3.4833E-05	1,2
2329	21406	C10H23N	diisopentylamine	544-00-3	157.300	461.15	25	1.9500E+00	3.5099E-05	1,2
2330	21407	C10H23N	N,N-diisopropylisobutylamine	44976-81-0	157.300	445.00	25	1.5434E+00	2.7781E-05	1,2
2331	21601	C11H10	1-methylnaphthalene	90-12-0	142.200	517.83	25	2.9586E+01	5.3254E-04	1,2
2332	21602	C11H10	2-methylnaphthalene	91-57-6	142.200	514.20	25	2.3932E+01	4.3077E-04	1,2
2333	21603	C11H10	methylnaphthalene	1321-94-4	142.200	515.15	25	2.2692E+01	4.0845E-04	1,2
2334	22304	C11H16	pentylbenzene	538-68-1	148.248	478.61	25	1.4801E+03	2.6642E-02	1,2
2335	22305	C11H16	2-phenylpentane	2719-52-0	148.248	463.15	25	1.0195E+03	1.8350E-02	1,2
2336	22306	C11H16	3-phenylpentane	1196-58-3	148.248	464.15	25	1.0268E+03	1.8482E-02	1,2
2337	22307	C11H16	1-phenyl-2-methylbutane	3968-85-2	148.248	470.15	25	4.6801E+02	8.4241E-03	1,2
2338	22308	C11H16	1-phenyl-3-methylbutane	2049-94-7	148.248	472.05	25	4.7295E+02	8.5130E-03	1,2
2339	22309	C11H16	2-phenyl-2-methylbutane	2049-95-8	148.248	465.53	25	5.5357E+02	9.9642E-03	1,2
2340	22310	C11H16	2-phenyl-3-methylbutane	4481-30-5	148.248	461.15	25	4.8260E+02	8.6867E-03	1,2
2341	22311	C11H16	1-phenyl-2,2-dimethylpropane	1007-26-7	148.248	459.15	25	5.2353E+02	9.4235E-03	1,2
2342	22312	C11H16	1-methyl-2-butylbenzene	1595-11-5	148.248	481.15	25	4.4860E+02	8.0747E-03	1,2
2343	22313	C11H16	1-methyl-3-butylbenzene	1595-04-6	148.248	478.15	25	4.4505E+02	8.0109E-03	1,2
2344	22314	C11H16	1-methyl-4-butylbenzene	1595-05-7	148.248	480.15	25	4.4755E+02	8.0558E-03	1,2
2345	22315	C11H16	1-methyl-2-sec-butylbenzene	1595-06-8	148.248	469.15	25	4.7182E+02	8.4927E-03	1,2
2346	22316	C11H16	1-methyl-3-sec-butylbenzene	1772-10-7	148.248	467.15	25	4.6590E+02	8.3862E-03	1,2
2347	22317	C11H16	1-methyl-4-sec-butylbenzene	1595-16-0	148.248	470.15	25	4.7465E+02	8.5435E-03	1,2
2348	22318	C11H16	1-methyl-2-isobutylbenzene	36301-29-8	148.248	469.15	25	4.7182E+02	8.4927E-03	1,2
2349	22319	C11H16	1-methyl-3-isobutylbenzene	5160-99-6	148.248	467.15	25	4.6590E+02	8.3862E-03	1,2
2350	22320	C11H16	1-methyl-4-isobutylbenzene	5161-04-6	148.248	469.15	25	4.7182E+02	8.4927E-03	1,2
2351	22321	C11H16	1-methyl-2-tert-butyl benzene	1074-92-6	148.248	473.60	25	5.9495E+02	1.0709E-02	1,2
2352	22322	C11H16	1-methyl-3-tert-butylbenzene	1075-38-3	148.248	462.41	25	5.4622E+02	9.8319E-03	1,2
2353	22323	C11H16	1-methyl-4-tert-butylbenzene	98-51-1	148.248	465.91	25	5.6268E+02	1.0128E-02	1,2
2354	22324	C11H16	1-ethyl-2-propylbenzene	16021-20-8	148.248	476.15	25	4.4206E+02	7.9570E-03	1,2
2355	22325	C11H16	1-ethyl-3-propylbenzene	20024-91-3	148.248	474.15	25	4.3860E+02	7.8947E-03	1,2
2356	22326	C11H16	1-ethyl-4-propylbenzene	20024-90-2	148.248	478.15	25	4.4505E+02	8.0109E-03	1,2
2357	22327	C11H16	1-ethyl-2-isopropylbenzene	18970-44-0	148.248	463.15	25	4.5311E+02	8.1558E-03	1,2
2358	22328	C11H16	1-ethyl-3-isopropylbenzene	4920-99-4	148.248	465.15	25	4.5965E+02	8.2737E-03	1,2
2359	22329	C11H16	1-ethyl-4-isopropylbenzene	4218-48-8	148.248	469.75	25	4.7353E+02	8.5234E-03	1,2
2360	22330	C11H16	1,2-dimethyl-3-propylbenzene	17059-44-8	148.248	483.85	25	4.5779E+02	8.2402E-03	1,2
2361	22331	C11H16	1,2-dimethyl-4-propylbenzene	3982-66-9	148.248	482.05	25	4.5637E+02	8.2145E-03	1,2
2362	22332	C11H16	1,3-dimethyl-2-propylbenzene	17059-45-9	148.248	480.75	25	4.5505E+02	8.1909E-03	1,2
2363	22333	C11H16	1,3-dimethyl-4-propylbenzene	61827-85-8	148.248	479.75	25	4.5389E+02	8.1699E-03	1,2
2364	22334	C11H16	1,3-dimethyl-5-propylbenzene	3982-64-7	148.248	475.39	25	4.4731E+02	8.0516E-03	1,2
2365	22335	C11H16	1,4-dimethyl-2-propylbenzene	3042-50-0	148.248	477.45	25	4.5071E+02	8.1127E-03	1,2
2366	22336	C11H16	1,2-dimethyl-3-isopropylbenzene	22539-65-7	148.248	475.75	25	4.9564E+02	8.9215E-03	1,2
2367	22337	C11H16	1,2-dimethyl-4-isopropylbenzene	4132-77-8	148.248	474.95	25	4.9382E+02	8.8887E-03	1,2
2368	22338	C11H16	1,3-dimethyl-2-isopropylbenzene	14411-75-7	148.248	472.15	25	4.8683E+02	8.7629E-03	1,2
2369	22339	C11H16	1,3-dimethyl-4-isopropylbenzene	4706-89-2	148.248	472.25	25	4.8710E+02	8.7677E-03	1,2
2370	22340	C11H16	1,3-dimethyl-5-isopropylbenzene	4706-90-5	148.248	467.65	25	4.7388E+02	8.5297E-03	1,2
2371	22341	C11H16	1,4-dimethyl-2-isopropylbenzene	4132-72-3	148.248	469.35	25	4.7900E+02	8.6219E-03	1,2
2372	22342	C11H16	1-methyl-2,3-diethylbenzene	13632-93-4	148.248	479.75	25	4.5389E+02	8.1699E-03	1,2
2373	22343	C11H16	1-methyl-2,4-diethylbenzene	1758-85-6	148.248	478.15	25	4.5175E+02	8.1314E-03	1,2
2374	22344	C11H16	1-methyl-2,5-diethylbenzene	13632-94-5	148.248	480.25	25	4.5449E+02	8.1807E-03	1,2
2375	22345	C11H16	1-methyl-2,6-diethyl benzene	13632-95-6	148.248	481.95	25	4.5627E+02	8.2128E-03	1,2
2376	22346	C11H16	1-methyl-3,4-diethylbenzene	13732-80-4	148.248	476.75	25	4.4961E+02	8.0930E-03	1,2
2377	22347	C11H16	1-methyl-3,5-diethylbenzene	2050-24-0	148.248	473.85	25	4.4445E+02	8.0001E-03	1,2
2378	22348	C11H16	1,2,3-trimethyl-4-ethylbenzene	61827-86-9	148.248	493.55	25	4.6434E+02	8.3581E-03	1,2
2379	22349	C11H16	1,2,3-trimethyl-5-ethylbenzene	31366-00-4	148.248	488.95	25	4.6652E+02	8.3973E-03	1,2
2380	22350	C11H16	1,2,4-trimethyl-3-ethylbenzene	61827-87-0	148.248	489.75	25	4.6640E+02	8.3951E-03	1,2
2381	22351	C11H16	1,2,4-trimethyl-5-ethylbenzene	---	148.248	486.15	25	4.6612E+02	8.3901E-03	1,2
2382	22352	C11H16	1,2,4-trimethyl-6-ethylbenzene	17851-27-3	148.248	486.15	25	4.6612E+02	8.3901E-03	1,2
2383	22353	C11H16	1,3,5-trimethyl-2-ethylbenzene	3982-67-0	148.248	485.55	25	4.6587E+02	8.3857E-03	1,2
2384	22354	C11H16	pentamethylbenzene	700-12-9	148.248	504.95	25	4.5050E+02	8.1089E-03	1,2
2385	22355	C11H16	1-ethyl-2,3,5-trimethylbenzene	18262-85-6	148.248	486.15	25	3.1532E+02	5.6757E-03	1,2
2386	22356	C11H16	sec-pentylbenzene	29316-05-0	148.248	473.15	25	7.6431E+02	1.3757E-02	1,2
2387	22399	C11H16O	p-tert-amylphenol	80-46-6	164.247	535.15	25	4.2394E-01	7.6308E-06	1,2
2388	22400	C11H16O	2-butyl-4-methylphenol	6891-45-8	164.247	501.15	25	2.2458E+00	4.0425E-05	1,2
2389	22401	C11H16O	2-tert-butyl-4-methylphenol	2409-55-4	164.247	510.15	25	1.2612E+00	2.2702E-05	1,2
2390	22402	C11H16O	2-tert-butyl-5-methylphenol	88-60-8	164.247	486.71	25	5.5121E+00	9.9217E-05	2
2391	22403	C11H16O	2-tert-butyl-6-methylphenol	2219-82-1	164.247	503.15	25	1.9779E+00	3.5602E-05	1,2
2392	22404	C11H16O	4-tert-butyl-2-methylphenol	98-27-1	164.247	510.15	25	1.2612E+00	2.2702E-05	1,2
2393	22405	C11H16O	4-isopentylphenol	1805-61-4	164.247	528.15	25	3.8126E-01	6.8627E-06	1,2
2394	22406	C11H16O	5-methyl-2-sec-butylphenol	29472-95-5	164.247	519.15	25	6.9847E-01	1.2572E-05	1,2
2395	22407	C11H16O	pentamethylphenol	2819-86-5	164.247	540.15	25	1.6616E-01	2.9908E-06	1,2
2396	22408	C11H16O	4-pentylphenol	14938-35-3	164.247	523.65	25	5.1699E-01	9.3058E-06	1,2
2397	22439	C11H16O	2-(1-methylbutyl)phenol	87-26-3	164.247	515.74	25	8.7521E-01	1.5754E-05	1,2
2398	22440	C11H16O	o-amylphenol	136-81-2	164.247	486.71	25	5.5121E+00	9.9217E-05	2
2399	22441	C11H16O	2-sec-amylphenol	---	164.247	407.65	25	4.3620E+02	7.8516E-03	1,2
2400	22442	C11H16O	4-sec-amylphenol	25735-67-5	164.247	409.65	25	3.9455E+02	7.1019E-03	1,2

Table 10 Henry's Law Constant for Compound in Water - Organic Compounds

							Henry's Law Constant, H			
NO	ID	FORMULA	NAME	CAS No	MW, g/mol	T_B, K	T, C	H @ T, atm/mol frac	H @ T, atm/mol/m^3	code
2401	22634	C11H18S	3-heptylthiophene	65016-61-7	182.330	519.15	25	4.0767E+02	7.3381E-03	2
2402	22672	C11H20	1-undecyne	2243-98-3	152.280	468.16	25	4.0328E+03	7.2590E-02	1,2
2403	22673	C11H20	2-undecyne	60212-29-5	152.280	479.15	25	3.4968E+03	6.2943E-02	1,2
2404	22674	C11H20	3-undecyne	60212-30-8	152.280	473.15	25	3.8103E+03	6.8584E-02	1,2
2405	22675	C11H20	4-undecyne	60212-31-9	152.280	471.65	25	3.8811E+03	6.9859E-02	1,2
2406	22676	C11H20	5-undecyne	2294-72-6	152.280	471.15	25	3.9040E+03	7.0271E-02	1,2
2407	22677	C11H20	3,3-dimethyl-4-nonyne	29022-31-9	152.280	483.34	25	3.2541E+03	5.8574E-02	2
2408	22701	C11H20O2	2-ethylhexyl acrylate	103-11-7	184.279	489.15	23	2.4415E+01	4.3947E-04	1,2
2409	22791	C11H22	hexylcyclopentane	4457-00-5	154.296	476.26	25	1.0697E+05	1.9254E+00	1,2
2410	22792	C11H22	pentylcyclohexane	4292-92-6	154.296	476.87	25	7.4872E+04	1.3477E+00	1,2
2411	22793	C11H22	isopentylcyclohexane	54105-76-9	154.296	469.65	25	5.9820E+04	1.0768E+00	1,2
2412	22795	C11H22	1-undecene	821-95-4	154.296	465.82	25	1.3013E+05	2.3424E+00	1,2
2413	22796	C11H22	cis-2-undecene	821-96-5	154.296	469.25	25	8.0912E+04	1.4564E+00	1,2
2414	22797	C11H22	trans-2-undecene	693-61-8	154.296	465.65	25	8.3403E+04	1.5012E+00	1,2
2415	22798	C11H22	3-undecene	60669-40-1	154.296	466.65	25	8.2756E+04	1.4896E+00	1,2
2416	22799	C11H22	cis-4-undecene	821-98-7	154.296	465.75	25	8.3340E+04	1.5001E+00	1,2
2417	22800	C11H22	trans-4-undecene	693-62-9	154.296	466.15	25	8.3084E+04	1.4955E+00	1,2
2418	22801	C11H22	cis-5-undecene	764-96-5	154.296	465.45	25	8.3528E+04	1.5035E+00	1,2
2419	22802	C11H22	trans-5-undecene	764-97-6	154.296	465.15	25	8.3713E+04	1.5068E+00	1,2
2420	22803	C11H22	2-methyl-1-decene	13151-27-4	154.296	463.85	25	1.3948E+05	2.5107E+00	1,2
2421	22805	C11H22Br2	1,1-dibromoundecane	62168-28-9	314.104	567.15	25	9.3031E+00	1.6745E-04	1,2
2422	22806	C11H22Br2	1,11-dibromoundecane	16696-65-4	314.104	567.15	25	8.2853E+00	1.4913E-04	2
2423	22807	C11H22Cl2	1,1-dichloroundecane	822-01-5	225.201	539.15	25	1.4277E+02	2.5699E-03	1,2
2424	22822	C11H22O	undecanal	112-44-7	170.295	506.15	25	3.0863E+01	5.5552E-04	1,2
2425	22823	C11H22O	2-undecanone	112-12-9	170.295	500.98	25	1.8400E+01	3.3120E-04	1,2
2426	22824	C11H22O	3-undecanone	2216-87-7	170.295	500.15	25	2.7948E+01	5.0306E-04	1,2
2427	22825	C11H22O	4-undecanone	14476-37-0	170.295	507.63	25	2.9598E+01	5.3276E-04	2
2428	22826	C11H22O	5-undecanone	33083-83-9	170.295	500.15	25	2.7948E+01	5.0306E-04	1,2
2429	22827	C11H22O	6-undecanone	927-49-1	170.295	501.15	25	2.8147E+01	5.0665E-04	1,2
2430	22834	C11H22O	2-methyl-1-decanal	19009-56-4	170.295	507.63	25	1.0625E+01	1.9126E-04	2
2431	22835	C11H22O2	undecanoic acid	112-37-8	186.294	557.35	25	7.2710E-03	1.3088E-07	1,2
2432	22836	C11H22O2	decyl formate	5451-52-5	186.294	508.60	25	4.5365E+01	8.1656E-04	2
2433	22837	C11H22O2	nonyl acetate	143-13-5	186.294	483.15	25	1.7746E-01	3.1943E-06	1,2
2434	22838	C11H22O2	octyl propanoate	142-60-9	186.294	501.15	25	4.5597E+00	8.2074E-05	1,2
2435	22839	C11H22O2	heptyl butanoate	5870-93-9	186.294	498.25	25	4.8042E+01	8.6476E-04	1,2
2436	22851	C11H22O2	6-methyldecanoic acid	53696-14-3	186.294	489.73	25	2.0669E-01	3.7204E-06	2
2437	22853	C11H22O2	hexyl 2,2-dimethylpropanoate	5434-57-1	186.294	489.73	25	3.1021E+01	5.5837E-04	2
2438	22856	C11H22O2	3,5,5-trimethylhexyl acetate	58430-94-7	186.294	480.00	25	2.8979E+01	5.2162E-04	2
2439	22870	C11H23Br	1-bromoundecane	693-67-4	235.208	531.95	25	1.9532E+03	3.5157E-02	1,2
2440	22872	C11H23Cl	1-chloroundecane	2473-03-2	190.756	518.49	25	3.0211E+03	5.4379E-02	1,2
2441	22890	C11H24	undecane	1120-21-4	156.312	469.08	25	1.1000E+06	1.9799E+01	1,2
2442	22891	C11H24	2-methyldecane	6975-98-0	156.312	462.34	25	4.2433E+05	7.6379E+00	1,2
2443	22892	C11H24	3-methyldecane	13151-34-3	156.312	463.15	25	4.2522E+05	7.6539E+00	1,2
2444	22893	C11H24	4-methyldecane	2847-72-5	156.312	461.05	25	4.2273E+05	7.6090E+00	1,2
2445	22894	C11H24	5-methyldecane	13151-35-4	156.312	459.15	25	4.1995E+05	7.5591E+00	1,2
2446	22895	C11H24	3-ethylnonane	17302-11-3	156.312	461.15	25	4.2286E+05	7.6114E+00	1,2
2447	22896	C11H24	4-ethylnonane	5911-05-7	156.312	457.15	25	4.1653E+05	7.4976E+00	1,2
2448	22897	C11H24	5-ethylnonane	17302-12-4	156.312	453.15	25	4.0827E+05	7.3489E+00	1,2
2449	22898	C11H24	2,2-dimethylnonane	17302-14-6	156.312	453.15	25	4.8970E+05	8.8146E+00	1,2
2450	22899	C11H24	2,3-dimethylnonane	2884-06-2	156.312	459.15	25	4.5969E+05	8.2744E+00	1,2
2451	22900	C11H24	2,4-dimethylnonane	17302-24-8	156.312	451.15	25	4.3954E+05	7.9116E+00	1,2
2452	22901	C11H24	2,5-dimethylnonane	17302-27-1	156.312	452.15	25	4.4246E+05	7.9642E+00	1,2
2453	22902	C11H24	2,6-dimethylnonane	17302-28-2	156.312	452.15	25	4.4246E+05	7.9642E+00	1,2
2454	22903	C11H24	2,7-dimethylnonane	17302-29-3	156.312	453.15	25	4.4528E+05	8.0149E+00	1,2
2455	22904	C11H24	2,8-dimethylnonane	17302-30-6	156.312	456.15	25	4.5305E+05	8.1548E+00	1,2
2456	22905	C11H24	3,3-dimethylnonane	17302-15-7	156.312	455.15	25	4.9619E+05	8.9313E+00	1,2
2457	22906	C11H24	3,4-dimethylnonane	17302-22-6	156.312	458.15	25	4.5761E+05	8.2369E+00	1,2
2458	22907	C11H24	3,5-dimethylnonane	17302-25-9	156.312	453.15	25	4.4528E+05	8.0149E+00	1,2
2459	22908	C11H24	3,6-dimethylnonane	17302-31-7	156.312	455.15	25	4.5058E+05	8.1103E+00	1,2
2460	22909	C11H24	3,7-dimethylnonane	17302-32-8	156.312	457.15	25	4.5539E+05	8.1970E+00	1,2
2461	22910	C11H24	4,4-dimethylnonane	17302-18-0	156.312	451.15	25	4.8275E+05	8.6894E+00	1,2
2462	22911	C11H24	4,5-dimethylnonane	17302-23-7	156.312	455.15	25	4.5058E+05	8.1103E+00	1,2
2463	22912	C11H24	4,6-dimethylnonane	17302-26-0	156.312	452.15	25	4.4246E+05	7.9642E+00	1,2
2464	22913	C11H24	5,5-dimethylnonane	6414-96-6	156.312	450.15	25	4.7911E+05	8.6239E+00	1,2
2465	22914	C11H24	4-propyloctane	17302-13-5	156.312	453.15	25	4.0827E+05	7.3489E+00	1,2
2466	22915	C11H24	4-isopropyloctane	62016-15-3	156.312	451.15	25	4.3954E+05	7.9116E+00	1,2
2467	22916	C11H24	2-methyl-3-ethyloctane	62016-16-4	156.312	455.15	25	4.5058E+05	8.1103E+00	1,2
2468	22917	C11H24	2-methyl-4-ethyloctane	62016-17-5	156.312	449.15	25	4.3339E+05	7.8010E+00	1,2
2469	22918	C11H24	2-methyl-5-ethyloctane	62016-18-6	156.312	451.15	25	4.3954E+05	7.9116E+00	1,2
2470	22919	C11H24	2-methyl-6-ethyloctane	62016-19-7	156.312	455.15	25	4.5058E+05	8.1103E+00	1,2
2471	22920	C11H24	3-methyl-3-ethyloctane	17302-16-8	156.312	458.85	25	5.0680E+05	9.1223E+00	1,2
2472	22921	C11H24	3-methyl-4-ethyloctane	62016-20-0	156.312	455.15	25	4.5058E+05	8.1103E+00	1,2
2473	22922	C11H24	3-methyl-5-ethyloctane	62016-21-1	156.312	449.15	25	4.3339E+05	7.8010E+00	1,2
2474	22923	C11H24	3-methyl-6-ethyloctane	62016-22-2	156.312	455.15	25	4.5058E+05	8.1103E+00	1,2
2475	22924	C11H24	4-methyl-3-ethyloctane	62016-23-3	156.312	456.15	25	4.5305E+05	8.1548E+00	1,2
2476	22925	C11H24	4-methyl-4-ethyloctane	17302-19-1	156.312	453.15	25	4.8970E+05	8.8146E+00	1,2
2477	22926	C11H24	4-methyl-5-ethyloctane	62016-24-4	156.312	453.15	25	4.4528E+05	8.0149E+00	1,2
2478	22927	C11H24	4-methyl-6-ethyloctane	62016-25-5	156.312	453.15	25	4.4528E+05	8.0149E+00	1,2
2479	22928	C11H24	2,2,3-trimethyloctane	62016-26-6	156.312	452.15	25	5.2957E+05	9.5322E+00	1,2
2480	22929	C11H24	2,2,4-trimethyloctane	18932-14-4	156.312	444.65	25	4.9592E+05	8.9264E+00	1,2

Table 10 Henry's Law Constant for Compound in Water - Organic Compounds

					Henry's Law Constant, H					
NO	ID	FORMULA	NAME	CAS No	MW, g/mol	T_B, K	T, C	H @ T, atm/mol frac	H @ T, atm/mol/m^3	code
2481	22930	C11H24	2,2,5-trimethyloctane	62016-27-7	156.312	444.15	25	4.9350E+05	8.8829E+00	1,2
2482	22931	C11H24	2,2,6-trimethyloctane	62016-28-8	156.312	447.15	25	5.0771E+05	9.1387E+00	1,2
2483	22932	C11H24	2,2,7-trimethyloctane	62016-29-9	156.312	443.15	25	4.8862E+05	8.7951E+00	1,2
2484	22933	C11H24	2,3,3-trimethyloctane	62016-30-2	156.312	455.15	25	5.4133E+05	9.7439E+00	1,2
2485	22934	C11H24	2,3,4-trimethyloctane	62016-31-3	156.312	453.15	25	4.8547E+05	8.7383E+00	1,2
2486	22935	C11H24	2,3,5-trimethyloctane	62016-32-4	156.312	450.15	25	4.7505E+05	8.5508E+00	1,2
2487	22936	C11H24	2,3,6-trimethyloctane	62016-33-5	156.312	453.15	25	4.8547E+05	8.7383E+00	1,2
2488	22937	C11H24	2,3,7-trimethyloctane	62016-34-6	156.312	452.15	25	4.8210E+05	8.6778E+00	1,2
2489	22938	C11H24	2,4,4-trimethyloctane	62016-35-7	156.312	444.15	25	4.9350E+05	8.8829E+00	1,2
2490	22939	C11H24	2,4,5-trimethyloctane	62016-36-8	156.312	449.15	25	4.7137E+05	8.4845E+00	1,2
2491	22940	C11H24	2,4,6-trimethyloctane	62016-37-9	156.312	447.15	25	4.6372E+05	8.3470E+00	1,2
2492	22941	C11H24	2,4,7-trimethyloctane	62016-38-0	156.312	444.15	25	4.5164E+05	8.1294E+00	1,2
2493	22942	C11H24	2,5,5-trimethyloctane	62016-39-1	156.312	445.15	25	4.9831E+05	8.9696E+00	1,2
2494	22943	C11H24	2,5,6-trimethyloctane	62016-14-2	156.312	451.15	25	4.7863E+05	8.6153E+00	1,2
2495	22944	C11H24	2,6,6-trimethyloctane	54166-32-4	156.312	449.15	25	5.1675E+05	9.3015E+00	1,2
2496	22945	C11H24	3,3,4-trimethyloctane	62016-40-4	156.312	455.15	25	5.4133E+05	9.7439E+00	1,2
2497	22946	C11H24	3,3,5-trimethyloctane	62016-41-5	156.312	447.15	25	5.0771E+05	9.1387E+00	1,2
2498	22947	C11H24	3,3,6-trimethyloctane	62016-42-6	156.312	450.15	25	5.2113E+05	9.3803E+00	1,2
2499	22948	C11H24	3,4,4-trimethyloctane	62016-43-7	156.312	454.15	25	5.3754E+05	9.6756E+00	1,2
2500	22949	C11H24	3,4,5-trimethyloctane	62016-44-8	156.312	454.15	25	4.8871E+05	8.7968E+00	1,2
2501	22950	C11H24	3,4,6-trimethyloctane	62016-45-9	156.312	452.15	25	4.8210E+05	8.6778E+00	1,2
2502	22951	C11H24	3,5,5-trimethyloctane	61868-94-8	156.312	443.15	25	4.8862E+05	8.7951E+00	1,2
2503	22952	C11H24	4,4,5-trimethyloctane	61868-95-9	156.312	453.15	25	5.3362E+05	9.6050E+00	1,2
2504	22953	C11H24	4-tert-butylheptane	60302-21-8	156.312	447.15	25	5.0771E+05	9.1387E+00	1,2
2505	22954	C11H24	2-methyl-4-propylheptane	61868-96-0	156.312	447.15	25	4.2688E+05	7.6837E+00	1,2
2506	22955	C11H24	3-methyl-4-propylheptane	61868-97-1	156.312	452.15	25	4.4246E+05	7.9642E+00	1,2
2507	22956	C11H24	4-methyl-4-propylheptane	17302-20-4	156.312	451.15	25	4.8275E+05	8.6894E+00	1,2
2508	22957	C11H24	2-methyl-3-isopropylheptane	6876-18-2	156.312	451.45	25	4.7968E+05	8.6342E+00	1,2
2509	22958	C11H24	2-methyl-4-isopropylheptane	61868-98-2	156.312	445.15	25	4.5574E+05	8.2033E+00	1,2
2510	22959	C11H24	3-methyl-4-isopropylheptane	61868-99-3	156.312	450.15	25	4.7505E+05	8.5508E+00	1,2
2511	22960	C11H24	4-methyl-4-isopropylheptane	61869-00-9	156.312	453.15	25	5.3362E+05	9.6050E+00	1,2
2512	22961	C11H24	3,3-diethylheptane	17302-17-9	156.312	460.05	25	5.0982E+05	9.1767E+00	1,2
2513	22962	C11H24	3,4-diethylheptane	61869-01-0	156.312	453.15	25	4.4528E+05	8.0149E+00	1,2
2514	22963	C11H24	3,5-diethylheptane	61869-02-1	156.312	452.15	25	4.4246E+05	7.9642E+00	1,2
2515	22964	C11H24	4,4-diethylheptane	17302-21-5	156.312	453.45	25	4.9071E+05	8.8327E+00	1,2
2516	22965	C11H24	2,2-dimethyl-3-ethylheptane	61869-03-2	156.312	449.15	25	5.1675E+05	9.3015E+00	1,2
2517	22966	C11H24	2,2-dimethyl-4-ethylheptane	62016-46-0	156.312	441.15	25	4.7867E+05	8.6160E+00	1,2
2518	22967	C11H24	2,2-dimethyl-5-ethylheptane	62016-47-1	156.312	445.15	25	4.9831E+05	8.9696E+00	1,2
2519	22968	C11H24	2,3-dimethyl-3-ethylheptane	61868-21-1	156.312	453.15	25	5.3362E+05	9.6050E+00	1,2
2520	22969	C11H24	2,3-dimethyl-4-ethylheptane	61868-22-2	156.312	452.15	25	4.8210E+05	8.6778E+00	1,2
2521	22970	C11H24	2,3-dimethyl-5-ethylheptane	61868-23-3	156.312	451.15	25	4.7863E+05	8.6153E+00	1,2
2522	22971	C11H24	2,4-dimethyl-3-ethylheptane	61868-24-4	156.312	452.15	25	4.8210E+05	8.6778E+00	1,2
2523	22972	C11H24	2,4-dimethyl-4-ethylheptane	61868-25-5	156.312	443.15	25	4.8862E+05	8.7951E+00	1,2
2524	22973	C11H24	2,4-dimethyl-5-ethylheptane	61868-26-6	156.312	449.15	25	4.7137E+05	8.4845E+00	1,2
2525	22974	C11H24	2,5-dimethyl-3-ethylheptane	61868-27-7	156.312	449.15	25	4.7137E+05	8.4845E+00	1,2
2526	22975	C11H24	2,5-dimethyl-4-ethylheptane	61868-28-8	156.312	448.15	25	4.6759E+05	8.4166E+00	1,2
2527	22976	C11H24	2,5-dimethyl-5-ethylheptane	61868-29-9	156.312	451.15	25	5.2541E+05	9.4573E+00	1,2
2528	22977	C11H24	2,6-dimethyl-3-ethylheptane	61868-30-2	156.312	449.15	25	4.7137E+05	8.4845E+00	1,2
2529	22978	C11H24	2,6-dimethyl-4-ethylheptane	61868-31-3	156.312	443.15	25	4.4746E+05	8.0542E+00	1,2
2530	22979	C11H24	3,3-dimethyl-4-ethylheptane	61868-32-4	156.312	453.15	25	5.3362E+05	9.6050E+00	1,2
2531	22980	C11H24	3,3-dimethyl-5-ethylheptane	61868-33-5	156.312	448.15	25	5.1228E+05	9.2209E+00	1,2
2532	22981	C11H24	3,4-dimethyl-3-ethylheptane	61868-34-6	156.312	458.15	25	5.5188E+05	9.9337E+00	1,2
2533	22982	C11H24	3,4-dimethyl-4-ethylheptane	61868-35-7	156.312	454.15	25	5.3754E+05	9.6756E+00	1,2
2534	22983	C11H24	3,4-dimethyl-5-ethylheptane	61868-36-8	156.312	455.15	25	4.9184E+05	8.8530E+00	1,2
2535	22984	C11H24	3,5-dimethyl-3-ethylheptane	61868-37-9	156.312	453.15	25	5.3362E+05	9.6050E+00	1,2
2536	22985	C11H24	3,5-dimethyl-4-ethylheptane	61868-38-0	156.312	454.15	25	4.8871E+05	8.7968E+00	1,2
2537	22986	C11H24	4,4-dimethyl-3-ethylheptane	61868-39-1	156.312	449.15	25	5.1675E+05	9.3015E+00	1,2
2538	22987	C11H24	2,2,3,3-tetramethylheptane	61868-40-4	156.312	453.15	25	6.3791E+05	1.1482E+01	1,2
2539	22988	C11H24	2,2,3,4-tetramethylheptane	61868-41-5	156.312	449.15	25	5.6147E+05	1.0106E+01	1,2
2540	22989	C11H24	2,2,3,5-tetramethylheptane	61868-42-6	156.312	447.15	25	5.5098E+05	9.9176E+00	1,2
2541	22990	C11H24	2,2,3,6-tetramethylheptane	61868-43-7	156.312	443.15	25	5.2901E+05	9.5220E+00	1,2
2542	22991	C11H24	2,2,4,4-tetramethylheptane	61868-44-8	156.312	443.15	25	5.7690E+05	1.0384E+01	1,2
2543	22992	C11H24	2,2,4,5-tetramethylheptane	61868-45-9	156.312	442.15	25	5.2334E+05	9.4201E+00	1,2
2544	22993	C11H24	2,2,4,6-tetramethylheptane	61868-46-0	156.312	435.35	25	4.8364E+05	8.7054E+00	1,2
2545	22994	C11H24	2,2,5,5-tetramethylheptane	61868-47-1	156.312	439.15	25	5.5043E+05	9.9077E+00	1,2
2546	22995	C11H24	2,2,5,6-tetramethylheptane	61868-48-2	156.312	442.15	25	5.2334E+05	9.4201E+00	1,2
2547	22996	C11H24	2,2,6,6-tetramethylheptane	40117-45-1	156.312	436.15	25	5.3021E+05	9.5437E+00	1,2
2548	22997	C11H24	2,3,3,4-tetramethylheptane	61868-49-3	156.312	455.15	25	5.9028E+05	1.0625E+01	1,2
2549	22998	C11H24	2,3,3,5-tetramethylheptane	61868-50-6	156.312	449.15	25	5.6147E+05	1.0106E+01	1,2
2550	22999	C11H24	2,3,3,6-tetramethylheptane	61868-51-7	156.312	448.15	25	5.5627E+05	1.0013E+01	1,2
2551	23000	C11H24	2,3,4,4-tetramethylheptane	61868-52-8	156.312	451.15	25	5.7155E+05	1.0288E+01	1,2
2552	23001	C11H24	2,3,4,5-tetramethylheptane	61868-53-9	156.312	452.15	25	5.2508E+05	9.4513E+00	1,2
2553	23002	C11H24	2,3,4,6-tetramethylheptane	61868-54-0	156.312	447.15	25	5.0354E+05	9.0637E+00	1,2
2554	23003	C11H24	2,3,5,5-tetramethylheptane	61868-55-1	156.312	445.15	25	5.4014E+05	9.7225E+00	1,2
2555	23004	C11H24	2,3,5,6-tetramethylheptane	52670-32-3	156.312	448.15	25	5.0805E+05	9.1448E+00	1,2
2556	23005	C11H24	2,4,4,5-tetramethylheptane	61868-56-2	156.312	447.15	25	5.5098E+05	9.9176E+00	1,2
2557	23006	C11H24	2,4,4,6-tetramethylheptane	61868-57-3	156.312	437.15	25	4.9430E+05	8.8973E+00	1,2
2558	23007	C11H24	2,4,5,5-tetramethylheptane	61868-58-4	156.312	449.15	25	5.6147E+05	1.0106E+01	1,2
2559	23008	C11H24	3,3,4,4-tetramethylheptane	61868-59-5	156.312	457.15	25	6.5899E+05	1.1862E+01	1,2
2560	23009	C11H24	3,3,4,5-tetramethylheptane	61868-60-8	156.312	454.15	25	5.8579E+05	1.0544E+01	1,2

Table 10 Henry's Law Constant for Compound in Water - Organic Compounds

					Henry's Law Constant, H					
NO	ID	FORMULA	NAME	CAS No	MW, g/mol	T_B, K	T, C	H @ T, atm/mol frac	H @ T, atm/mol/m^3	code
2561	23010	C11H24	3,3,5,5-tetramethylheptane	61868-61-9	156.312	453.15	25	6.3791E+05	1.1482E+01	1,2
2562	23011	C11H24	3,4,4,5-tetramethylheptane	61868-62-0	156.312	458.15	25	6.0286E+05	1.0851E+01	1,2
2563	23012	C11H24	2,2-dimethyl-3-isopropylhexane	61868-63-1	156.312	445.15	25	5.4014E+05	9.7225E+00	1,2
2564	23013	C11H24	2,3-dimethyl-3-isopropylhexane	61868-64-2	156.312	453.15	25	5.8117E+05	1.0461E+01	1,2
2565	23014	C11H24	2,4-dimethyl-3-isopropylhexane	61868-65-3	156.312	448.15	25	5.0805E+05	9.1448E+00	1,2
2566	23015	C11H24	2,5-dimethyl-3-isopropylhexane	61868-66-4	156.312	443.15	25	4.8472E+05	8.7250E+00	1,2
2567	23016	C11H24	2-methyl-3,3-diethylhexane	61868-67-5	156.312	459.15	25	5.5509E+05	9.9915E+00	1,2
2568	23017	C11H24	2-methyl-3,4-diethylhexane	61868-68-6	156.312	452.15	25	4.8210E+05	8.6778E+00	1,2
2569	23018	C11H24	2-methyl-4,4-diethylhexane	61868-69-7	156.312	452.15	25	5.2957E+05	9.5322E+00	1,2
2570	23019	C11H24	3-methyl-3,4-diethylhexane	61868-70-0	156.312	459.15	25	5.5509E+05	9.9915E+00	1,2
2571	23020	C11H24	3-methyl-4,4-diethylhexane	61868-71-1	156.312	463.15	25	5.6629E+05	1.0193E+01	1,2
2572	23021	C11H24	2,2,3-trimethyl-3-ethylhexane	61868-72-2	156.312	453.15	25	6.3791E+05	1.1482E+01	1,2
2573	23022	C11H24	2,2,3-trimethyl-4-ethylhexane	61868-73-3	156.312	449.15	25	5.6147E+05	1.0106E+01	1,2
2574	23023	C11H24	2,2,4-trimethyl-3-ethylhexane	61868-74-4	156.312	448.15	25	5.5627E+05	1.0013E+01	1,2
2575	23024	C11H24	2,2,4-trimethyl-4-ethylhexane	61868-75-5	156.312	451.15	25	6.2653E+05	1.1277E+01	1,2
2576	23025	C11H24	2,2,5-trimethyl-3-ethylhexane	61868-76-6	156.312	443.15	25	5.2901E+05	9.5220E+00	1,2
2577	23026	C11H24	2,2,5-trimethyl-4-ethylhexane	61868-77-7	156.312	439.15	25	5.0604E+05	9.1086E+00	1,2
2578	23027	C11H24	2,3,3-trimethyl-4-ethylhexane	61868-78-8	156.312	456.15	25	5.9463E+05	1.0703E+01	1,2
2579	23028	C11H24	2,3,4-trimethyl-3-ethylhexane	61868-79-9	156.312	460.15	25	6.1044E+05	1.0988E+01	1,2
2580	23029	C11H24	2,3,4-trimethyl-4-ethylhexane	61868-80-2	156.312	453.15	25	5.8117E+05	1.0461E+01	1,2
2581	23030	C11H24	2,3,5-trimethyl-3-ethylhexane	61868-81-3	156.312	449.15	25	5.6147E+05	1.0106E+01	1,2
2582	23031	C11H24	2,3,5-trimethyl-4-ethylhexane	61868-82-4	156.312	450.15	25	5.1677E+05	9.3018E+00	1,2
2583	23032	C11H24	2,4,4-trimethyl-3-ethylhexane	61868-83-5	156.312	451.15	25	5.7155E+05	1.0288E+01	1,2
2584	23033	C11H24	3,3,4-trimethyl-4-ethylhexane	61868-84-6	156.312	463.15	25	6.8549E+05	1.2339E+01	1,2
2585	23034	C11H24	2,2,3,3,4-pentamethylhexane	61868-85-7	156.312	457.15	25	7.1846E+05	1.2932E+01	1,2
2586	23035	C11H24	2,2,3,3,5-pentamethylhexane	61868-86-8	156.312	443.65	25	6.2751E+05	1.1295E+01	1,2
2587	23036	C11H24	2,2,3,4,4-pentamethylhexane	61868-87-9	156.312	459.15	25	7.2977E+05	1.3136E+01	1,2
2588	23037	C11H24	2,2,3,4,5-pentamethylhexane	61868-88-0	156.312	447.15	25	5.9761E+05	1.0757E+01	1,2
2589	23038	C11H24	2,2,3,5,5-pentamethylhexane	14739-73-2	156.312	439.15	25	5.9381E+05	1.0688E+01	1,2
2590	23039	C11H24	2,2,4,4,5-pentamethylhexane	60302-23-0	156.312	449.15	25	6.6701E+05	1.2006E+01	1,2
2591	23040	C11H24	2,3,3,4,4-pentamethylhexane	61868-89-1	156.312	460.15	25	7.3513E+05	1.3232E+01	1,2
2592	23041	C11H24	2,3,3,4,5-pentamethylhexane	52670-33-4	156.312	454.15	25	6.3801E+05	1.1484E+01	1,2
2593	23042	C11H24	2,2,4-trimethyl-3-isopropylpentane	61868-90-4	156.312	443.15	25	5.7242E+05	1.0304E+01	1,2
2594	23043	C11H24	2,3,4-trimethyl-3-isopropylpentane	61868-91-5	156.312	459.15	25	6.6278E+05	1.1930E+01	1,2
2595	23044	C11H24	2,2-dimethyl-3,3-diethylpentane	60302-28-5	156.312	462.15	25	6.8156E+05	1.2268E+01	1,2
2596	23045	C11H24	2,4-dimethyl-3,3-diethylpentane	61868-92-6	156.312	462.15	25	6.1731E+05	1.1112E+01	1,2
2597	23046	C11H24	2,2,3,4-tetramethyl-3-ethylpentane	61868-93-7	156.312	459.15	25	7.2977E+05	1.3136E+01	1,2
2598	23047	C11H24	2,2,4,4-tetramethyl-3-ethylpentane	---	156.312	453.15	25	6.9384E+05	1.2489E+01	1,2
2599	23048	C11H24	2,2,3,3,4,4-hexamethylpentane	60302-27-4	156.312	467.15	25	9.2800E+05	1.6704E+01	1,2
2600	23055	C11H24O	1-undecanol	112-42-5	172.311	518.15	25	2.6853E+00	4.8336E-05	1,2
2601	23056	C11H24O	2-undecanol	1653-30-1	172.311	508.15	25	7.3601E-01	1.3248E-05	1,2
2602	23057	C11H24O	2-undecanol, (±)	113666-64-1	172.311	501.15	25	3.2384E+00	5.8290E-05	1,2
2603	23058	C11H24O	3-undecanol, (R)-	107494-37-1	172.311	502.15	25	3.2449E+00	5.8409E-05	1,2
2604	23059	C11H24O	5-undecanol	37493-70-2	172.311	502.15	25	3.2449E+00	5.8409E-05	1,2
2605	23060	C11H24O	6-undecanol	23708-56-7	172.311	501.15	25	3.2384E+00	5.8290E-05	1,2
2606	23061	C11H24O	2,8-dimethyl-5-nonanol	19780-96-2	172.311	506.13	25	3.2768E+00	5.8982E-05	2
2607	23062	C11H24O	undecanol	30207-98-8	172.311	518.15	25	3.4389E+00	6.1900E-05	1,2
2608	23063	C11H24O	5-ethyl-2-nonanol	103-08-2	172.311	498.00	25	3.2210E+00	5.7978E-05	1,2
2609	23064	C11H24O	1-methoxydecane	7289-52-3	172.311	506.13	25	1.4912E+02	2.6842E-03	2
2610	23084	C11H24S	1-undecanethiol	5332-52-5	188.378	530.55	25	2.0341E+02	3.6614E-03	1,2
2611	23085	C11H24S	2-undecanethiol	62155-02-6	188.378	520.45	25	3.6421E+02	6.5557E-03	1,2
2612	23086	C11H24S	methyl decyl sulfide	22438-39-7	188.378	520.00	25	1.9138E+02	3.4448E-03	2
2613	23087	C11H24S	ethyl nonyl sulfide	59973-08-9	188.378	520.00	25	1.9138E+02	3.4448E-03	2
2614	23088	C11H24S	propyl octyl sulfide	3698-93-9	188.378	520.00	25	1.9138E+02	3.4448E-03	2
2615	23089	C11H24S	butyl heptyl sulfide	40813-84-1	188.378	520.00	25	1.9138E+02	3.4448E-03	2
2616	23090	C11H25N	undecylamine	7307-55-3	171.327	516.15	25	9.1055E-01	1.6390E-05	1,2
2617	23091	C11H25N	methyldecylamine	7516-82-7	171.327	499.35	25	2.5809E-02	4.6456E-07	1,2
2618	23092	C11H25N	ethylnonylamine	66563-84-6	171.327	495.15	25	2.0800E-02	3.7439E-07	1,2
2619	23093	C11H25N	dimethylnonylamine	17373-27-2	171.327	488.15	25	9.6204E+00	1.7317E-04	1,2
2620	23094	C11H25N	diethylheptylamine	26981-81-7	171.327	471.15	25	6.3836E+00	1.1490E-04	1,2
2621	23095	C11H25N	4-ethyl-1-methyloctylamine	10024-78-9	171.327	493.99	25	5.1167E-01	9.2099E-06	2
2622	23335	C12H8O	dibenzofuran	132-64-9	168.195	557.86	25	6.0122E+01	1.0822E-03	1,2
2623	23394	C12H9N	dibenzopyrrole	86-74-8	167.210	627.86	25	1.8895E+00	3.4012E-05	1,2
2624	23444	C12H10	acenaphthene	83-32-9	154.211	550.54	25	3.1108E+01	5.5994E-04	1,2
2625	23445	C12H10	phenylbenzene (biphenyl)	92-52-4	154.211	528.15	25	3.0240E+01	5.4431E-04	1,2
2626	23519	C12H10O	diphenyl ether	101-84-8	170.211	531.46	25	2.9525E+01	5.3144E-04	1,2
2627	23609	C12H11N	diphenylamine	122-39-4	169.226	575.15	20	2.9527E-01	5.3149E-06	1,2
2628	23656	C12H11N3	p-aminoazobenzene	60-09-3	197.240	633.00	18	2.5230E-04	4.5414E-09	1,2
2629	23675	C12H12	1,2-dimethylnaphthalene	573-98-8	156.227	539.46	25	1.1343E+01	2.0416E-04	1,2
2630	23676	C12H12	1,3-dimethylnaphthalene	575-41-7	156.227	538.36	25	5.9293E+01	1.0673E-03	1,2
2631	23677	C12H12	1,4-dimethylnaphthalene	571-58-4	156.227	540.46	25	5.4136E+01	9.7444E-04	1,2
2632	23678	C12H12	1,5-dimethylnaphthalene	571-61-9	156.227	538.16	25	4.4211E+01	7.9579E-04	1,2
2633	23679	C12H12	1,6-dimethylnaphthalene	575-43-9	156.227	536.16	25	5.2036E+01	9.3665E-04	1,2
2634	23680	C12H12	1,7-dimethylnaphthalene	575-37-1	156.227	536.16	25	4.8503E+00	8.7306E-05	1,2
2635	23681	C12H12	1,8-dimethylnaphthalene	569-41-5	156.227	543.15	25	1.7634E+01	3.1741E-04	1,2
2636	23682	C12H12	2,3-dimethylnaphthalene	581-40-8	156.227	541.16	25	4.3533E+01	7.8359E-04	1,2
2637	23683	C12H12	2,6-dimethylnaphthalene	581-42-0	156.227	535.15	25	1.3181E+02	2.3726E-03	1,2
2638	23684	C12H12	2,7-dimethylnaphthalene	582-16-1	156.227	536.15	25	3.8117E+01	6.8611E-04	1,2
2639	23685	C12H12	1-ethylnaphthalene	1127-76-0	156.227	531.48	25	3.3124E+01	5.9622E-04	1,2
2640	23686	C12H12	2-ethylnaphthalene	939-27-5	156.227	531.49	25	4.2939E+01	7.7290E-04	1,2

Table 10 Henry's Law Constant for Compound in Water - Organic Compounds

					Henry's Law Constant, H					
NO	ID	FORMULA	NAME	CAS No	MW, g/mol	T_B, K	T, C	H @ T, atm/mol frac	H @ T, atm/mol/m^3	code
2641	23687	C12H12	dimethylnaphthalene, isomers	28804-88-8	156.227	537.28	25	2.5447E+01	4.5805E-04	2
2642	23718	C12H12N2	p-aminodiphenylamine	101-54-2	184.241	627.15	25	2.0939E-04	3.7689E-09	1,2
2643	24002	C12H14O4	diethyl phthalate	84-66-2	222.241	567.15	25	9.1793E-03	1.6523E-07	1,2
2644	24357	C12H18	hexylbenzene	1077-16-3	162.275	499.26	25	1.7783E+03	3.2009E-02	1,2
2645	24358	C12H18	1,4-dipropylbenzene	4815-57-0	162.275	481.77	25	3.5484E+02	6.3871E-03	2
2646	24359	C12H18	diisopropylbenzene	25321-09-9	162.275	473.60	25	3.6648E+02	6.5966E-03	1,2
2647	24360	C12H18	1,2-diisopropylbenzene	577-55-9	162.275	477.15	25	3.6243E+02	6.5237E-03	1,2
2648	24361	C12H18	m-diisopropylbenzene	99-62-7	162.275	476.33	25	8.1748E+02	1.4715E-02	1,2
2649	24362	C12H18	p-diisopropylbenzene	100-18-5	162.275	483.65	25	6.1782E+02	1.1121E-02	1,2
2650	24363	C12H18	triethylbenzene; (mixed isomers)	25340-18-5	162.275	490.70	25	3.3258E+02	5.9865E-03	1,2
2651	24364	C12H18	1,2,3-triethylbenzene	---	162.275	490.66	25	5.5851E+02	1.0053E-02	1,2
2652	24365	C12H18	1,2,4-triethylbenzene	877-44-1	162.275	490.66	25	7.1995E+02	1.2959E-02	1,2
2653	24366	C12H18	1,3,5-triethylbenzene	102-25-0	162.275	489.16	25	5.2106E+02	9.3790E-03	1,2
2654	24367	C12H18	1-tert-butyl-3,5-dimethylbenzene	98-19-1	162.275	480.15	25	3.5780E+02	6.4404E-03	1,2
2655	24368	C12H18	(1,1-dimethylbutyl)benzene	1985-57-5	162.275	478.65	25	3.6026E+02	6.4846E-03	1,2
2656	24369	C12H18	2,4-dimethyl-1-sec-butylbenzene	1483-60-9	162.275	481.77	25	3.5484E+02	6.3871E-03	2
2657	24370	C12H18	(1-ethylbutyl)benzene	4468-42-2	162.275	482.15	25	8.0744E+02	1.4534E-02	1,2
2658	24371	C12H18	(1-ethyl-1-methylpropyl)benzene	1985-97-3	162.275	478.15	25	8.2324E+02	1.4818E-02	1,2
2659	24372	C12H18	(1-methylpentyl)benzene	6031-02-3	162.275	481.15	25	8.1182E+02	1.4613E-02	1,2
2660	24373	C12H18	(3-methylpentyl)benzene	54410-69-4	162.275	493.15	25	7.4063E+02	1.3331E-02	1,2
2661	24374	C12H18	(4-methylpentyl)benzene	4215-86-5	162.275	490.15	25	7.6218E+02	1.3719E-02	1,2
2662	24375	C12H18	1,2,4-trimethyl-5-isopropylbenzene	10222-95-4	162.275	494.15	25	3.2141E+02	5.7853E-03	1,2
2663	24376	C12H18	4-tert-butyl-o-xylene	7397-06-0	162.275	478.15	25	3.6101E+02	6.4982E-03	1,2
2664	24377	C12H18	hexamethyldewarbenzene	7641-77-2	162.275	440.00	25	6.1399E+02	1.1052E-02	2
2665	24378	C12H18	hexamethylbenzene	87-85-4	162.275	536.60	25	2.4742E+03	4.4535E-02	1,2
2666	24619	C12H20S	3-octylthiophene	65016-62-8	196.357	536.15	25	3.2131E+02	5.7835E-03	2
2667	24653	C12H22	1-dodecyne	765-03-7	166.307	488.16	25	3.2291E+03	5.8123E-02	1,2
2668	24654	C12H22	2-dodecyne	629-49-2	166.307	499.15	25	2.4557E+03	4.4202E-02	1,2
2669	24655	C12H22	3-dodecyne	6790-27-8	166.307	492.55	25	2.9222E+03	5.2600E-02	1,2
2670	24656	C12H22	6-dodecyne	6975-99-1	166.307	483.15	25	3.5660E+03	6.4188E-02	1,2
2671	24657	C12H22	3,3-dimethyl-4-decyne	70732-45-5	166.307	500.93	25	2.3316E+03	4.1968E-02	2
2672	24669	C12H22	1,11-dodecadiene	5876-87-9	166.307	481.65	25	1.6565E+04	2.9817E-01	1,2
2673	24818	C12H24	heptylcyclopentane	5617-42-5	168.323	497.30	25	8.1841E+04	1.4731E+00	1,2
2674	24819	C12H24	hexylcyclohexane	4292-75-5	168.323	497.86	25	6.0214E+04	1.0838E+00	1,2
2675	24820	C12H24	1-methyl-2-pentylcyclohexane	54411-01-7	168.323	490.65	25	4.9889E+04	8.9800E-01	1,2
2676	24823	C12H24	1-dodecene	112-41-4	168.323	486.50	25	1.2407E+05	2.2332E+00	1,2
2677	24824	C12H24	2-methyl-1-undecene	18516-37-5	168.323	484.75	25	1.3006E+05	2.3411E+00	1,2
2678	24825	C12H24	4,4-dimethyl-2-neopentyl-1-pentene	141-70-8	168.323	488.99	25	6.5998E+04	1.1880E+00	2
2679	24826	C12H24	2,2,4,6,6-pentamethyl-3-heptene	123-48-8	168.323	453.70	25	9.6337E+04	1.7341E+00	1,2
2680	24827	C12H24	trans-2,2,4,6,6-pentamethyl-3-heptene	27656-50-4	168.323	453.70	25	9.6337E+04	1.7341E+00	2
2681	24828	C12H24	isobutene, trimer	7756-94-7	168.323	453.15	25	9.6447E+04	1.7360E+00	1,2
2682	24831	C12H24Br2	1,1-dibromododecane	62168-29-0	328.131	583.15	25	4.9931E+00	8.9875E-05	1,2
2683	24832	C12H24Br2	1,12-dibromododecane	3344-70-5	328.131	583.15	25	4.4474E+00	8.0052E-05	2
2684	24833	C12H24Cl2	1,1-dichlorododecane	62017-17-8	239.228	556.15	25	9.4128E+01	1.6943E-03	1,2
2685	24834	C12H24Cl2	1,12-dichlorododecane	3922-28-9	239.228	572.15	25	5.2887E+01	9.5196E-04	1,2
2686	24852	C12H24O	dodecanal	112-54-9	184.322	523.15	25	2.5264E+01	4.5475E-04	1,2
2687	24853	C12H24O	2-methylundecanal	110-41-8	184.322	519.25	25	9.5304E+00	1.7155E-04	2
2688	24854	C12H24O	2-dodecanone	6175-49-1	184.322	519.85	25	2.1300E+01	3.8340E-04	1,2
2689	24855	C12H24O	6-dodecanone	6064-27-3	184.322	519.25	25	3.5801E+01	6.4441E-04	2
2690	24860	C12H24O	trimethyl nonanone	1331-50-6	184.322	488.15	25	2.8078E+01	5.0540E-04	1,2
2691	24861	C12H24O	3-dodecanone	1534-27-6	184.322	519.25	25	3.5801E+01	6.4441E-04	2
2692	24862	C12H24O	5-dodecanone	19780-10-0	184.322	519.25	25	3.5801E+01	6.4441E-04	2
2693	24863	C12H24O	4-dodecanone	6137-26-4	184.322	519.25	25	3.5801E+01	6.4441E-04	2
2694	24867	C12H24O2	dodecanoic acid	143-07-7	200.321	571.85	20	2.2451E-03	4.0412E-08	1,2
2695	24868	C12H24O2	undecyl formate	5454-24-0	200.321	531.48	25	3.8665E+01	6.9597E-04	2
2696	24869	C12H24O2	decyl acetate	112-17-4	200.321	517.15	25	4.6878E+01	8.4380E-04	1,2
2697	24870	C12H24O2	nonyl propanoate	53184-67-1	200.321	532.18	25	2.1726E+01	3.9107E-04	2
2698	24871	C12H24O2	octyl butanoate	110-39-4	200.321	517.15	25	2.6168E+01	4.7102E-04	1,2
2699	24879	C12H24O2	2-butyloctanoic acid	27610-92-0	200.321	503.15	25	1.8693E-01	3.3647E-06	1,2
2700	24880	C12H24O2	3,7-dimethyloctanyl acetate	20780-49-8	200.321	505.00	25	2.5753E+01	4.6355E-04	2
2701	24894	C12H25Br	1-bromododecane	143-15-7	249.235	549.05	25	1.4966E+03	2.6938E-02	1,2
2702	24895	C12H25Br	2-bromododecane	13187-99-0	249.235	549.05	25	3.8373E+02	6.9071E-03	2
2703	24897	C12H25Cl	1-chlorododecane	112-52-7	204.783	536.33	25	2.3305E+03	4.1948E-02	1,2
2704	24914	C12H26	dodecane	112-40-3	170.338	489.47	25	5.0131E+05	9.0236E+00	1,2
2705	24915	C12H26	3-methylundecane	1002-43-3	170.338	483.15	25	3.5744E+05	6.4338E+00	1,2
2706	24916	C12H26	4-methylundecane	2980-69-0	170.338	482.15	25	3.5991E+05	6.4783E+00	1,2
2707	24917	C12H26	5-methylundecane	1632-70-8	170.338	479.15	25	3.6645E+05	6.5960E+00	1,2
2708	24918	C12H26	6-methylundecane	17302-33-9	170.338	479.15	25	3.6645E+05	6.5960E+00	1,2
2709	24919	C12H26	3-ethyldecane	17085-96-0	170.338	481.85	25	3.6062E+05	6.4912E+00	1,2
2710	24920	C12H26	4-ethyldecane	1636-44-8	170.338	477.15	25	3.7006E+05	6.6611E+00	1,2
2711	24921	C12H26	5-ethyldecane	17302-36-2	170.338	475.15	25	3.7308E+05	6.7154E+00	1,2
2712	24922	C12H26	2,2-dimethyldecane	17302-37-3	170.338	474.15	25	4.6117E+05	8.3010E+00	1,2
2713	24923	C12H26	2,3-dimethyldecane	17312-44-6	170.338	479.15	25	4.0594E+05	7.3069E+00	1,2
2714	24924	C12H26	2,4-dimethyldecane	2801-84-5	170.338	473.15	25	4.1444E+05	7.4599E+00	1,2
2715	24925	C12H26	2,5-dimethyldecane	17312-50-4	170.338	471.15	25	4.1595E+05	7.4870E+00	1,2
2716	24926	C12H26	2,6-dimethyldecane	13150-81-7	170.338	471.15	25	4.1595E+05	7.4870E+00	1,2
2717	24927	C12H26	2,7-dimethyldecane	17312-51-5	170.338	475.15	25	4.1227E+05	7.4209E+00	1,2
2718	24928	C12H26	2,8-dimethyldecane	17312-52-6	170.338	477.15	25	4.0944E+05	7.3699E+00	1,2
2719	24929	C12H26	2,9-dimethyldecane	1002-17-1	170.338	477.15	25	4.0944E+05	7.3699E+00	1,2
2720	24930	C12H26	3,3-dimethyldecane	17302-38-4	170.338	476.15	25	4.5900E+05	8.2620E+00	1,2

Table 10 Henry's Law Constant for Compound in Water - Organic Compounds

								Henry's Law Constant, H		
NO	ID	FORMULA	NAME	CAS No	MW, g/mol	T_B, K	T, C	H @ T, atm/mol frac	H @ T, atm/mol/m^3	code
2721	24931	C12H26	3,4-dimethyldecane	17312-45-7	170.338	478.15	25	4.0778E+05	7.3399E+00	1,2
2722	24932	C12H26	3,5-dimethyldecane	17312-48-0	170.338	473.15	25	4.1444E+05	7.4599E+00	1,2
2723	24933	C12H26	3,6-dimethyldecane	17312-53-7	170.338	474.15	25	4.1344E+05	7.4418E+00	1,2
2724	24934	C12H26	3,7-dimethyldecane	17312-54-8	170.338	475.15	25	4.1227E+05	7.4209E+00	1,2
2725	24935	C12H26	3,8-dimethyldecane	17312-55-9	170.338	478.15	25	4.0778E+05	7.3399E+00	1,2
2726	24936	C12H26	4,4-dimethyldecane	17312-39-9	170.338	472.15	25	4.6259E+05	8.3266E+00	1,2
2727	24937	C12H26	4,5-dimethyldecane	17312-46-8	170.338	475.15	25	4.1227E+05	7.4209E+00	1,2
2728	24938	C12H26	4,6-dimethyldecane	17312-49-1	170.338	471.15	25	4.1595E+05	7.4870E+00	1,2
2729	24939	C12H26	4,7-dimethyldecane	17312-56-0	170.338	472.15	25	4.1528E+05	7.4749E+00	1,2
2730	24940	C12H26	5,5-dimethyldecane	17453-92-8	170.338	470.15	25	4.6329E+05	8.3391E+00	1,2
2731	24941	C12H26	5,6-dimethyldecane	1636-43-7	170.338	474.15	25	4.1344E+05	7.4418E+00	1,2
2732	24942	C12H26	4-propylnonane	6165-37-3	170.338	472.15	25	3.7649E+05	6.7768E+00	1,2
2733	24943	C12H26	5-propylnonane	998-35-6	170.338	470.15	25	3.7804E+05	6.8046E+00	1,2
2734	24944	C12H26	4-isopropylnonane	62184-71-8	170.338	471.15	25	4.1595E+05	7.4870E+00	1,2
2735	24945	C12H26	5-isopropylnonane	62184-72-9	170.338	471.15	25	4.1595E+05	7.4870E+00	1,2
2736	24946	C12H26	2-methyl-3-ethylnonane	62184-73-0	170.338	474.15	25	4.1344E+05	7.4418E+00	1,2
2737	24947	C12H26	2-methyl-4-ethylnonane	62184-37-6	170.338	471.15	25	4.1595E+05	7.4870E+00	1,2
2738	24948	C12H26	2-methyl-5-ethylnonane	62184-38-7	170.338	469.15	25	4.1682E+05	7.5026E+00	1,2
2739	24949	C12H26	2-methyl-6-ethylnonane	62184-39-8	170.338	471.15	25	4.1595E+05	7.4870E+00	1,2
2740	24950	C12H26	2-methyl-7-ethylnonane	62184-40-1	170.338	475.15	25	4.1227E+05	7.4209E+00	1,2
2741	24951	C12H26	3-methyl-3-ethylnonane	17302-39-5	170.338	477.15	25	4.5764E+05	8.2375E+00	1,2
2742	24952	C12H26	3-methyl-4-ethylnonane	62184-41-2	170.338	474.15	25	4.1344E+05	7.4418E+00	1,2
2743	24953	C12H26	3-methyl-5-ethylnonane	62184-42-3	170.338	468.15	25	4.1701E+05	7.5061E+00	1,2
2744	24954	C12H26	3-methyl-6-ethylnonane	62184-43-4	170.338	472.15	25	4.1528E+05	7.4749E+00	1,2
2745	24955	C12H26	3-methyl-7-ethylnonane	62184-44-5	170.338	473.15	25	4.1444E+05	7.4599E+00	1,2
2746	24956	C12H26	4-methyl-3-ethylnonane	62184-45-6	170.338	473.15	25	4.1444E+05	7.4599E+00	1,2
2747	24957	C12H26	4-methyl-4-ethylnonane	17312-40-2	170.338	472.15	25	4.6259E+05	8.3266E+00	1,2
2748	24958	C12H26	4-methyl-5-ethylnonane	62184-46-7	170.338	471.15	25	4.1595E+05	7.4870E+00	1,2
2749	24959	C12H26	4-methyl-6-ethylnonane	62184-47-8	170.338	469.15	25	4.1682E+05	7.5026E+00	1,2
2750	24960	C12H26	4-methyl-7-ethylnonane	62184-48-9	170.338	473.15	25	4.1444E+05	7.4599E+00	1,2
2751	24961	C12H26	5-methyl-3-ethylnonane	62184-49-0	170.338	472.15	25	4.1528E+05	7.4749E+00	1,2
2752	24962	C12H26	5-methyl-4-ethylnonane	1632-71-9	170.338	472.15	25	4.1528E+05	7.4749E+00	1,2
2753	24963	C12H26	5-methyl-5-ethylnonane	14531-16-9	170.338	471.15	25	4.6303E+05	8.3345E+00	1,2
2754	24964	C12H26	2,2,3-trimethylnonane	55499-04-2	170.338	473.15	25	5.0963E+05	9.1733E+00	1,2
2755	24965	C12H26	2,2,4-trimethylnonane	62184-50-3	170.338	464.15	25	5.0598E+05	9.1075E+00	1,2
2756	24966	C12H26	2,2,5-trimethylnonane	62184-51-4	170.338	464.15	25	5.0598E+05	9.1075E+00	1,2
2757	24967	C12H26	2,2,6-trimethylnonane	62184-52-5	170.338	463.15	25	5.0465E+05	9.0837E+00	1,2
2758	24968	C12H26	2,2,7-trimethylnonane	62184-53-6	170.338	468.15	25	5.0950E+05	9.1710E+00	1,2
2759	24969	C12H26	2,2,8-trimethylnonane	62184-54-7	170.338	468.15	25	5.0950E+05	9.1710E+00	1,2
2760	24970	C12H26	2,3,3-trimethylnonane	62184-55-8	170.338	475.15	25	5.0828E+05	9.1489E+00	1,2
2761	24971	C12H26	2,3,4-trimethylnonane	62184-56-9	170.338	473.15	25	4.5736E+05	8.2324E+00	1,2
2762	24972	C12H26	2,3,5-trimethylnonane	62184-57-0	170.338	469.15	25	4.5885E+05	8.2592E+00	1,2
2763	24973	C12H26	2,3,6-trimethylnonane	62184-58-1	170.338	471.15	25	4.5846E+05	8.2522E+00	1,2
2764	24974	C12H26	2,3,7-trimethylnonane	62184-59-2	170.338	474.15	25	4.5654E+05	8.2176E+00	1,2
2765	24975	C12H26	2,3,8-trimethylnonane	62184-60-5	170.338	473.15	25	4.5736E+05	8.2324E+00	1,2
2766	24976	C12H26	2,4,4-trimethylnonane	62184-61-6	170.338	463.15	25	5.0465E+05	9.0837E+00	1,2
2767	24977	C12H26	2,4,5-trimethylnonane	62184-62-7	170.338	468.15	25	4.5878E+05	8.2580E+00	1,2
2768	24978	C12H26	2,4,6-trimethylnonane	62184-10-5	170.338	465.15	25	4.5757E+05	8.2362E+00	1,2
2769	24979	C12H26	2,4,7-trimethylnonane	62184-11-6	170.338	467.15	25	4.5855E+05	8.2537E+00	1,2
2770	24980	C12H26	2,4,8-trimethylnonane	49542-74-7	170.338	464.15	25	4.5683E+05	8.2230E+00	1,2
2771	24981	C12H26	2,5,5-trimethylnonane	62184-12-7	170.338	463.15	25	5.0465E+05	9.0837E+00	1,2
2772	24982	C12H26	2,5,6-trimethylnonane	62184-13-8	170.338	469.15	25	4.5885E+05	8.2592E+00	1,2
2773	24983	C12H26	2,5,7-trimethylnonane	62184-14-9	170.338	467.15	25	4.5855E+05	8.2537E+00	1,2
2774	24984	C12H26	2,5,8-trimethylnonane	49557-09-7	170.338	465.15	25	4.5757E+05	8.2362E+00	1,2
2775	24985	C12H26	2,6,6-trimethylnonane	62184-15-0	170.338	463.15	25	5.0465E+05	9.0837E+00	1,2
2776	24986	C12H26	2,6,7-trimethylnonane	62184-16-1	170.338	472.15	25	4.5800E+05	8.2439E+00	1,2
2777	24987	C12H26	2,7,7-trimethylnonane	62184-17-2	170.338	470.15	25	5.1014E+05	9.1825E+00	1,2
2778	24988	C12H26	3,3,4-trimethylnonane	62184-18-3	170.338	475.15	25	5.0828E+05	9.1489E+00	1,2
2779	24989	C12H26	3,3,5-trimethylnonane	62184-19-4	170.338	463.15	25	5.0465E+05	9.0837E+00	1,2
2780	24990	C12H26	3,3,7-trimethylnonane	62184-21-8	170.338	468.15	25	5.0950E+05	9.1710E+00	1,2
2781	24991	C12H26	3,3,6-trimethylnonane	62184-20-7	170.338	471.15	25	5.1017E+05	9.1830E+00	1,2
2782	24992	C12H26	3,4,4-trimethylnonane	62184-22-9	170.338	473.15	25	5.0963E+05	9.1733E+00	1,2
2783	24993	C12H26	3,4,5-trimethylnonane	62184-23-0	170.338	473.15	25	4.5736E+05	8.2324E+00	1,2
2784	24994	C12H26	3,4,6-trimethylnonane	62184-24-1	170.338	470.15	25	4.5874E+05	8.2573E+00	1,2
2785	24995	C12H26	3,4,7-trimethylnonane	27802-85-3	170.338	473.15	25	4.5736E+05	8.2324E+00	1,2
2786	24996	C12H26	3,5,5-trimethylnonane	62184-25-2	170.338	464.15	25	5.0598E+05	9.1075E+00	1,2
2787	24997	C12H26	3,5,6-trimethylnonane	62184-26-3	170.338	470.15	25	4.5874E+05	8.2573E+00	1,2
2788	24998	C12H26	3,5,7-trimethylnonane	62184-27-4	170.338	468.15	25	4.5878E+05	8.2580E+00	1,2
2789	24999	C12H26	3,6,6-trimethylnonane	62184-28-5	170.338	463.15	25	5.0465E+05	9.0837E+00	1,2
2790	25000	C12H26	4,4,5-trimethylnonane	62184-29-6	170.338	471.15	25	5.1017E+05	9.1830E+00	1,2
2791	25001	C12H26	4,4,6-trimethylnonane	62184-30-9	170.338	464.15	25	5.0598E+05	9.1075E+00	1,2
2792	25002	C12H26	4,5,5-trimethylnonane	62184-31-0	170.338	471.15	25	5.1017E+05	9.1830E+00	1,2
2793	25003	C12H26	4,5,6-trimethylnonane	62211-85-2	170.338	471.15	25	4.5846E+05	8.2522E+00	1,2
2794	25004	C12H26	4-tert-butyloctane	62184-32-1	170.338	464.15	25	5.0598E+05	9.1075E+00	1,2
2795	25005	C12H26	2-methyl-4-propyloctane	62184-33-2	170.338	464.15	25	4.1626E+05	7.4926E+00	1,2
2796	25006	C12H26	2-methyl-5-propyloctane	62184-34-3	170.338	464.15	25	4.1626E+05	7.4926E+00	1,2
2797	25007	C12H26	3-methyl-4-propyloctane	62184-35-4	170.338	470.15	25	4.1646E+05	7.4963E+00	1,2
2798	25008	C12H26	3-methyl-5-propyloctane	62184-36-5	170.338	467.15	25	4.1705E+05	7.5068E+00	1,2
2799	25009	C12H26	4-methyl-4-propyloctane	17312-41-3	170.338	468.15	25	4.6327E+05	8.3388E+00	1,2
2800	25010	C12H26	4-methyl-5-propyloctane	62183-85-1	170.338	469.15	25	4.1682E+05	7.5026E+00	1,2

Table 10 Henry's Law Constant for Compound in Water - Organic Compounds

					Henry's Law Constant, H					
NO	ID	FORMULA	NAME	CAS No	MW, g/mol	T_B, K	T, C	H @ T, atm/mol frac	H @ T, atm/mol/m^3	code
2801	25011	C12H26	2-methyl-3-isopropyloctane	13287-19-9	170.338	469.15	25	4.5885E+05	8.2592E+00	1,2
2802	25012	C12H26	2-methyl-4-isopropyloctane	62183-86-2	170.338	463.15	25	4.5594E+05	8.2069E+00	1,2
2803	25013	C12H26	2-methyl-5-isopropyloctane	62183-87-3	170.338	464.15	25	4.5683E+05	8.2230E+00	1,2
2804	25014	C12H26	3-methyl-4-isopropyloctane	62183-88-4	170.338	468.15	25	4.5878E+05	8.2580E+00	1,2
2805	25015	C12H26	3-methyl-5-isopropyloctane	62183-89-5	170.338	465.15	25	4.5757E+05	8.2362E+00	1,2
2806	25016	C12H26	4-methyl-4-isopropyloctane	62183-90-8	170.338	471.15	25	5.1017E+05	9.1830E+00	1,2
2807	25017	C12H26	4-methyl-5-isopropyloctane	62183-91-9	170.338	467.15	25	4.5855E+05	8.2537E+00	1,2
2808	25018	C12H26	3,3-diethyloctane	17302-40-8	170.338	478.15	25	4.5609E+05	8.2095E+00	1,2
2809	25019	C12H26	3,4-diethyloctane	62183-92-0	170.338	472.15	25	4.1528E+05	7.4749E+00	1,2
2810	25020	C12H26	3,5-diethyloctane	62183-93-1	170.338	470.15	25	4.1646E+05	7.4963E+00	1,2
2811	25021	C12H26	3,6-diethyloctane	62183-94-2	170.338	475.15	25	4.1227E+05	7.4209E+00	1,2
2812	25022	C12H26	4,4-diethyloctane	17312-42-4	170.338	473.15	25	4.6197E+05	8.3154E+00	1,2
2813	25023	C12H26	4,5-diethyloctane	1636-41-5	170.338	470.15	25	4.1646E+05	7.4963E+00	1,2
2814	25024	C12H26	2,2-dimethyl-3-ethyloctane	62183-95-3	170.338	469.15	25	5.0992E+05	9.1785E+00	1,2
2815	25025	C12H26	2,2-dimethyl-4-ethyloctane	62183-96-4	170.338	460.15	25	4.9967E+05	8.9940E+00	1,2
2816	25026	C12H26	2,2-dimethyl-5-ethyloctane	62183-97-5	170.338	462.15	25	5.0316E+05	9.0567E+00	1,2
2817	25027	C12H26	2,2-dimethyl-6-ethyloctane	62183-98-6	170.338	466.15	25	5.0810E+05	9.1458E+00	1,2
2818	25028	C12H26	2,3-dimethyl-3-ethyloctane	62183-99-7	170.338	475.15	25	5.0828E+05	9.1489E+00	1,2
2819	25029	C12H26	2,3-dimethyl-4-ethyloctane	62184-00-3	170.338	470.15	25	4.5874E+05	8.2573E+00	1,2
2820	25030	C12H26	2,3-dimethyl-5-ethyloctane	62184-01-4	170.338	467.15	25	4.5855E+05	8.2537E+00	1,2
2821	25031	C12H26	2,3-dimethyl-6-ethyloctane	62184-02-5	170.338	472.15	25	4.5800E+05	8.2439E+00	1,2
2822	25032	C12H26	2,4-dimethyl-3-ethyloctane	62184-03-6	170.338	470.15	25	4.5874E+05	8.2573E+00	1,2
2823	25033	C12H26	2,4-dimethyl-4-ethyloctane	62184-04-7	170.338	464.15	25	5.0598E+05	9.1075E+00	1,2
2824	25034	C12H26	2,4-dimethyl-5-ethyloctane	62184-05-8	170.338	466.15	25	4.5814E+05	8.2464E+00	1,2
2825	25035	C12H26	2,4-dimethyl-6-ethyloctane	62184-06-9	170.338	465.15	25	4.5757E+05	8.2362E+00	1,2
2826	25036	C12H26	2,5-dimethyl-3-ethyloctane	62184-07-0	170.338	467.15	25	4.5855E+05	8.2537E+00	1,2
2827	25037	C12H26	2,5-dimethyl-4-ethyloctane	62184-08-1	170.338	465.15	25	4.5757E+05	8.2362E+00	1,2
2828	25038	C12H26	2,5-dimethyl-5-ethyloctane	62184-09-2	170.338	466.15	25	5.0810E+05	9.1458E+00	1,2
2829	25039	C12H26	2,5-dimethyl-6-ethyloctane	62183-50-0	170.338	470.15	25	4.5874E+05	8.2573E+00	1,2
2830	25040	C12H26	2,6-dimethyl-3-ethyloctane	62183-51-1	170.338	470.15	25	4.5874E+05	8.2573E+00	1,2
2831	25041	C12H26	2,6-dimethyl-4-ethyloctane	62183-52-2	170.338	463.15	25	4.5594E+05	8.2069E+00	1,2
2832	25042	C12H26	2,6-dimethyl-5-ethyloctane	62183-53-3	170.338	468.15	25	4.5878E+05	8.2580E+00	1,2
2833	25043	C12H26	2,6-dimethyl-6-ethyloctane	62183-54-4	170.338	471.15	25	5.1017E+05	9.1830E+00	1,2
2834	25044	C12H26	2,7-dimethyl-3-ethyloctane	62183-55-5	170.338	469.15	25	4.5885E+05	8.2592E+00	1,2
2835	25045	C12H26	2,7-dimethyl-4-ethyloctane	62183-56-6	170.338	463.15	25	4.5594E+05	8.2069E+00	1,2
2836	25046	C12H26	3,3-dimethyl-4-ethyloctane	62183-57-7	170.338	471.15	25	5.1017E+05	9.1830E+00	1,2
2837	25047	C12H26	3,3-dimethyl-5-ethyloctane	62183-58-8	170.338	464.15	25	5.0598E+05	9.1075E+00	1,2
2838	25048	C12H26	3,3-dimethyl-6-ethyloctane	62183-59-9	170.338	469.15	25	5.0992E+05	9.1785E+00	1,2
2839	25049	C12H26	3,4-dimethyl-3-ethyloctane	62212-28-6	170.338	473.15	25	5.0963E+05	9.1733E+00	1,2
2840	25050	C12H26	3,4-dimethyl-4-ethyloctane	62183-60-2	170.338	474.15	25	5.0906E+05	9.1630E+00	1,2
2841	25051	C12H26	3,4-dimethyl-5-ethyloctane	62183-61-3	170.338	471.15	25	4.5846E+05	8.2522E+00	1,2
2842	25052	C12H26	3,4-dimethyl-6-ethyloctane	62183-62-4	170.338	471.15	25	4.5846E+05	8.2522E+00	1,2
2843	25053	C12H26	3,5-dimethyl-3-ethyloctane	62183-63-5	170.338	469.15	25	5.0992E+05	9.1785E+00	1,2
2844	25054	C12H26	3,5-dimethyl-4-ethyloctane	62183-64-6	170.338	470.15	25	4.5874E+05	8.2573E+00	1,2
2845	25055	C12H26	3,5-dimethyl-5-ethyloctane	62183-65-7	170.338	463.15	25	5.0465E+05	9.0837E+00	1,2
2846	25056	C12H26	3,5-dimethyl-6-ethyloctane	62183-66-8	170.338	470.15	25	4.5874E+05	8.2573E+00	1,2
2847	25057	C12H26	3,6-dimethyl-3-ethyloctane	62183-67-9	170.338	472.15	25	5.1000E+05	9.1800E+00	1,2
2848	25058	C12H26	3,6-dimethyl-4-ethyloctane	62183-68-0	170.338	469.15	25	4.5885E+05	8.2592E+00	1,2
2849	25059	C12H26	4,4-dimethyl-3-ethyloctane	62183-69-1	170.338	471.15	25	5.1017E+05	9.1830E+00	1,2
2850	25060	C12H26	4,4-dimethyl-5-ethyloctane	62183-70-4	170.338	469.15	25	5.0992E+05	9.1785E+00	1,2
2851	25061	C12H26	4,4-dimethyl-6-ethyloctane	62183-71-5	170.338	464.15	25	5.0598E+05	9.1075E+00	1,2
2852	25062	C12H26	4,5-dimethyl-3-ethyloctane	62183-72-6	170.338	472.15	25	4.5800E+05	8.2439E+00	1,2
2853	25063	C12H26	4,5-dimethyl-4-ethyloctane	62183-73-7	170.338	473.15	25	5.0963E+05	9.1733E+00	1,2
2854	25064	C12H26	2,2,3,3-tetramethyloctane	62183-74-8	170.338	473.15	25	6.2553E+05	1.1259E+01	1,2
2855	25065	C12H26	2,2,3,4-tetramethyloctane	62183-75-9	170.338	468.15	25	5.6016E+05	1.0083E+01	1,2
2856	25066	C12H26	2,2,3,5-tetramethyloctane	62183-76-0	170.338	465.15	25	5.5654E+05	1.0018E+01	1,2
2857	25067	C12H26	2,2,3,6-tetramethyloctane	62183-77-1	170.338	467.15	25	5.5916E+05	1.0065E+01	1,2
2858	25068	C12H26	2,2,3,7-tetramethyloctane	62183-78-2	170.338	467.15	25	5.5916E+05	1.0065E+01	1,2
2859	25069	C12H26	2,2,4,4-tetramethyloctane	62183-79-3	170.338	465.15	25	6.1614E+05	1.1090E+01	1,2
2860	25070	C12H26	2,2,4,5-tetramethyloctane	62183-80-6	170.338	460.15	25	5.4669E+05	9.8404E+00	1,2
2861	25071	C12H26	2,2,4,6-tetramethyloctane	62183-81-7	170.338	458.15	25	5.4154E+05	9.7476E+00	1,2
2862	25072	C12H26	2,2,4,7-tetramethyloctane	62183-82-8	170.338	457.95	25	5.4099E+05	9.7377E+00	1,2
2863	25073	C12H26	2,2,5,5-tetramethyloctane	62183-83-9	170.338	457.15	25	5.9326E+05	1.0679E+01	1,2
2864	25074	C12H26	2,2,5,6-tetramethyloctane	62183-84-0	170.338	463.15	25	5.5315E+05	9.9566E+00	1,2
2865	25075	C12H26	2,2,5,7-tetramethyloctane	62199-19-3	170.338	458.15	25	5.4154E+05	9.7476E+00	1,2
2866	25076	C12H26	2,2,6,6-tetramethyloctane	62199-20-6	170.338	458.15	25	5.9676E+05	1.0742E+01	1,2
2867	25077	C12H26	2,2,6,7-tetramethyloctane	62199-21-7	170.338	464.15	25	5.5494E+05	9.9888E+00	1,2
2868	25078	C12H26	2,2,7,7-tetramethyloctane	1071-31-4	170.338	458.15	25	5.9676E+05	1.0742E+01	1,2
2869	25079	C12H26	2,3,3,4-tetramethyloctane	62199-22-8	170.338	474.15	25	5.6173E+05	1.0111E+01	1,2
2870	25080	C12H26	2,3,3,5-tetramethyloctane	62199-23-9	170.338	467.15	25	5.5916E+05	1.0065E+01	1,2
2871	25081	C12H26	2,3,3,6-tetramethyloctane	62199-24-0	170.338	469.15	25	5.6097E+05	1.0097E+01	1,2
2872	25082	C12H26	2,3,3,7-tetramethyloctane	62199-25-1	170.338	469.15	25	5.6097E+05	1.0097E+01	1,2
2873	25083	C12H26	2,3,4,4-tetramethyloctane	62199-26-2	170.338	469.15	25	5.6097E+05	1.0097E+01	1,2
2874	25084	C12H26	2,3,4,5-tetramethyloctane	62199-27-3	170.338	470.15	25	5.0520E+05	9.0936E+00	1,2
2875	25085	C12H26	2,3,4,6-tetramethyloctane	62199-28-4	170.338	468.15	25	5.0463E+05	9.0832E+00	1,2
2876	25086	C12H26	2,3,4,7-tetramethyloctane	62199-29-5	170.338	467.15	25	5.0406E+05	9.0730E+00	1,2
2877	25087	C12H26	2,3,5,5-tetramethyloctane	62199-30-8	170.338	468.15	25	5.6016E+05	1.0083E+01	1,2
2878	25088	C12H26	2,3,5,6-tetramethyloctane	62199-31-9	170.338	468.15	25	5.0463E+05	9.0832E+00	1,2
2879	25089	C12H26	2,3,5,7-tetramethyloctane	62199-32-0	170.338	463.15	25	4.9997E+05	8.9994E+00	1,2
2880	25090	C12H26	2,3,6,6-tetramethyloctane	62199-33-1	170.338	466.15	25	5.5795E+05	1.0043E+01	1,2

Table 10 Henry's Law Constant for Compound in Water - Organic Compounds

					Henry's Law Constant, H					
NO	ID	FORMULA	NAME	CAS No	MW, g/mol	T_B, K	T, C	H @ T, atm/mol frac	H @ T, atm/mol/m^3	code
2881	25091	C12H26	2,3,6,7-tetramethyloctane	52670-34-5	170.338	469.15	25	5.0501E+05	9.0901E+00	1,2
2882	25092	C12H26	2,4,4,5-tetramethyloctane	62199-34-2	170.338	464.15	25	5.5494E+05	9.9888E+00	1,2
2883	25093	C12H26	2,4,4,6-tetramethyloctane	62199-35-3	170.338	458.15	25	5.4154E+05	9.7476E+00	1,2
2884	25094	C12H26	2,4,4,7-tetramethyloctane	35866-96-7	170.338	457.15	25	5.3872E+05	9.6969E+00	1,2
2885	25095	C12H26	2,4,5,5-tetramethyloctane	62199-36-4	170.338	465.15	25	5.5654E+05	1.0018E+01	1,2
2886	25096	C12H26	2,4,5,6-tetramethyloctane	62199-37-5	170.338	467.15	25	5.0406E+05	9.0730E+00	1,2
2887	25097	C12H26	2,4,5,7-tetramethyloctane	2217-17-6	170.338	459.45	25	4.9379E+05	8.8882E+00	1,2
2888	25098	C12H26	2,4,6,6-tetramethyloctane	62199-38-6	170.338	460.15	25	5.4669E+05	9.8404E+00	1,2
2889	25099	C12H26	2,5,5,6-tetramethyloctane	62199-39-7	170.338	467.15	25	5.5916E+05	1.0065E+01	1,2
2890	25100	C12H26	2,5,6,6-tetramethyloctane	62199-40-0	170.338	469.15	25	5.6097E+05	1.0097E+01	1,2
2891	25101	C12H26	3,3,4,4-tetramethyloctane	62199-41-1	170.338	475.15	25	6.2547E+05	1.1258E+01	1,2
2892	25102	C12H26	3,3,4,5-tetramethyloctane	62199-42-2	170.338	471.15	25	5.6193E+05	1.0115E+01	1,2
2893	25103	C12H26	3,3,4,6-tetramethyloctane	62199-43-3	170.338	469.15	25	5.6097E+05	1.0097E+01	1,2
2894	25104	C12H26	3,3,5,5-tetramethyloctane	62199-44-4	170.338	470.15	25	6.2376E+05	1.1228E+01	1,2
2895	25105	C12H26	3,3,5,6-tetramethyloctane	62199-45-5	170.338	466.15	25	5.5795E+05	1.0043E+01	1,2
2896	25106	C12H26	3,3,6,6-tetramethyloctane	62199-46-6	170.338	462.85	25	6.1082E+05	1.0995E+01	1,2
2897	25107	C12H26	3,4,4,5-tetramethyloctane	62199-47-7	170.338	474.15	25	5.6173E+05	1.0111E+01	1,2
2898	25108	C12H26	3,4,4,6-tetramethyloctane	62185-19-7	170.338	468.15	25	5.6016E+05	1.0083E+01	1,2
2899	25109	C12H26	3,4,5,5-tetramethyloctane	62185-20-0	170.338	470.15	25	5.6155E+05	1.0108E+01	1,2
2900	25110	C12H26	3,4,5,6-tetramethyloctane	62185-21-1	170.338	472.15	25	5.0500E+05	9.0900E+00	1,2
2901	25111	C12H26	4,4,5,5-tetramethyloctane	62185-22-2	170.338	473.15	25	6.2553E+05	1.1259E+01	1,2
2902	25112	C12H26	2-methyl-4-tert-butylheptane	62185-23-3	170.338	458.15	25	5.4154E+05	9.7476E+00	1,2
2903	25113	C12H26	3-methyl-4-tert-butylheptane	62185-24-4	170.338	463.15	25	5.5315E+05	9.9566E+00	1,2
2904	25114	C12H26	4-methyl-4-tert-butylheptane	62185-25-5	170.338	471.15	25	6.2460E+05	1.1243E+01	1,2
2905	25115	C12H26	3-ethyl-4-propylheptane	62185-26-6	170.338	469.15	25	4.1682E+05	7.5026E+00	1,2
2906	25116	C12H26	4-ethyl-4-propylheptane	17312-43-5	170.338	470.15	25	4.6329E+05	8.3391E+00	1,2
2907	25117	C12H26	3-ethyl-4-isopropylheptane	62185-27-7	170.338	468.15	25	4.5878E+05	8.2580E+00	1,2
2908	25118	C12H26	4-ethyl-4-isopropylheptane	62185-28-8	170.338	473.15	25	5.0963E+05	9.1733E+00	1,2
2909	25119	C12H26	2,2-dimethyl-4-propylheptane	---	170.338	457.15	25	4.9328E+05	8.8790E+00	1,2
2910	25120	C12H26	2,3-dimethyl-4-propylheptane	62185-30-2	170.338	467.15	25	4.5855E+05	8.2537E+00	1,2
2911	25121	C12H26	2,4-dimethyl-4-propylheptane	62185-31-3	170.338	462.15	25	5.0316E+05	9.0567E+00	1,2
2912	25122	C12H26	2,5-dimethyl-4-propylheptane	62185-32-4	170.338	463.15	25	4.5594E+05	8.2069E+00	1,2
2913	25123	C12H26	2,6-dimethyl-4-propylheptane	62185-33-5	170.338	457.75	25	4.4854E+05	8.0736E+00	1,2
2914	25124	C12H26	3,3-dimethyl-4-propylheptane	62185-34-6	170.338	469.15	25	5.0992E+05	9.1785E+00	1,2
2915	25125	C12H26	3,4-dimethyl-4-propylheptane	62185-35-7	170.338	471.15	25	5.1017E+05	9.1830E+00	1,2
2916	25126	C12H26	3,5-dimethyl-4-propylheptane	62185-36-8	170.338	469.15	25	4.5885E+05	8.2592E+00	1,2
2917	25127	C12H26	2,2-dimethyl-3-isopropylheptane	62185-37-9	170.338	463.15	25	5.5315E+05	9.9566E+00	1,2
2918	25128	C12H26	2,2-dimethyl-4-isopropylheptane	62185-38-0	170.338	455.15	25	5.3264E+05	9.5874E+00	1,2
2919	25129	C12H26	2,3-dimethyl-3-isopropylheptane	62185-39-1	170.338	474.15	25	5.6173E+05	1.0111E+01	1,2
2920	25130	C12H26	2,3-dimethyl-4-isopropylheptane	62185-40-4	170.338	465.15	25	5.0237E+05	9.0426E+00	1,2
2921	25131	C12H26	2,4-dimethyl-3-isopropylheptane	62185-41-5	170.338	465.15	25	5.0237E+05	9.0426E+00	1,2
2922	25132	C12H26	2,4-dimethyl-4-isopropylheptane	62185-42-6	170.338	465.15	25	5.5654E+05	1.0018E+01	1,2
2923	25133	C12H26	2,5-dimethyl-3-isopropylheptane	62185-43-7	170.338	463.15	25	4.9997E+05	8.9994E+00	1,2
2924	25134	C12H26	2,5-dimethyl-4-isopropylheptane	62185-44-8	170.338	462.15	25	4.9852E+05	8.9733E+00	1,2
2925	25135	C12H26	2,6-dimethyl-3-isopropylheptane	62185-45-9	170.338	463.95	25	5.0101E+05	9.0182E+00	1,2
2926	25136	C12H26	2,6-dimethyl-4-isopropylheptane	35866-89-8	170.338	456.15	25	4.8651E+05	8.7571E+00	1,2
2927	25137	C12H26	3,3-dimethyl-4-isopropylheptane	62185-46-0	170.338	467.15	25	5.5916E+05	1.0065E+01	1,2
2928	25138	C12H26	3,4-dimethyl-4-isopropylheptane	62185-47-1	170.338	474.15	25	5.6173E+05	1.0111E+01	1,2
2929	25139	C12H26	3,5-dimethyl-4-isopropylheptane	62198-89-4	170.338	467.15	25	5.0406E+05	9.0730E+00	1,2
2930	25140	C12H26	2-methyl-3,3-diethylheptane	62198-90-7	170.338	476.15	25	5.0729E+05	9.1312E+00	1,2
2931	25141	C12H26	2-methyl-3,4-diethylheptane	62198-91-8	170.338	468.15	25	4.5878E+05	8.2580E+00	1,2
2932	25142	C12H26	2-methyl-3,5-diethylheptane	62198-92-9	170.338	468.15	25	4.5878E+05	8.2580E+00	1,2
2933	25143	C12H26	2-methyl-4,4-diethylheptane	62198-93-0	170.338	467.15	25	5.0890E+05	9.1600E+00	1,2
2934	25144	C12H26	2-methyl-4,5-diethylheptane	62198-94-1	170.338	466.15	25	4.5814E+05	8.2464E+00	1,2
2935	25145	C12H26	2-methyl-5,5-diethylheptane	62198-95-2	170.338	471.15	25	5.1017E+05	9.1830E+00	1,2
2936	25146	C12H26	3-methyl-3,4-diethylheptane	62198-96-3	170.338	474.15	25	5.0906E+05	9.1630E+00	1,2
2937	25147	C12H26	3-methyl-3,5-diethylheptane	62198-97-4	170.338	470.15	25	5.1014E+05	9.1825E+00	1,2
2938	25148	C12H26	3-methyl-4,4-diethylheptane	62198-98-5	170.338	477.15	25	5.0610E+05	9.1097E+00	1,2
2939	25149	C12H26	3-methyl-4,5-diethylheptane	62198-99-6	170.338	471.15	25	4.5846E+05	8.2522E+00	1,2
2940	25150	C12H26	3-methyl-5,5-diethylheptane	62199-00-2	170.338	472.15	25	5.1000E+05	9.1800E+00	1,2
2941	25151	C12H26	4-methyl-3,3-diethylheptane	62199-01-3	170.338	478.15	25	5.0469E+05	9.0843E+00	1,2
2942	25152	C12H26	4-methyl-3,4-diethylheptane	62199-02-4	170.338	474.15	25	5.0906E+05	9.1630E+00	1,2
2943	25153	C12H26	4-methyl-3,5-diethylheptane	62199-03-5	170.338	473.15	25	4.5736E+05	8.2324E+00	1,2
2944	25154	C12H26	2,2,3-trimethyl-3-ethylheptane	62199-04-6	170.338	473.15	25	6.2553E+05	1.1259E+01	1,2
2945	25155	C12H26	2,2,3-trimethyl-4-ethylheptane	62199-05-7	170.338	466.15	25	5.5795E+05	1.0043E+01	1,2
2946	25156	C12H26	2,2,3-trimethyl-5-ethylheptane	62199-06-8	170.338	465.15	25	5.5654E+05	1.0018E+01	1,2
2947	25157	C12H26	2,2,4-trimethyl-3-ethylheptane	62199-07-9	170.338	465.15	25	5.5654E+05	1.0018E+01	1,2
2948	25158	C12H26	2,2,4-trimethyl-4-ethylheptane	62199-08-0	170.338	467.15	25	6.1986E+05	1.1157E+01	1,2
2949	25159	C12H26	2,2,4-trimethyl-5-ethylheptane	62199-09-1	170.338	460.15	25	5.4669E+05	9.8404E+00	1,2
2950	25160	C12H26	2,2,5-trimethyl-3-ethylheptane	62199-10-4	170.338	463.15	25	5.5315E+05	9.9566E+00	1,2
2951	25161	C12H26	2,2,5-trimethyl-4-ethylheptane	62199-11-5	170.338	459.15	25	5.4419E+05	9.7954E+00	1,2
2952	25162	C12H26	2,2,5-trimethyl-5-ethylheptane	62199-12-6	170.338	462.15	25	6.0899E+05	1.0962E+01	1,2
2953	25163	C12H26	2,2,6-trimethyl-3-ethylheptane	62199-13-7	170.338	463.15	25	5.5315E+05	9.9566E+00	1,2
2954	25164	C12H26	2,2,6-trimethyl-4-ethylheptane	62199-14-8	170.338	453.15	25	5.2599E+05	9.4678E+00	1,2
2955	25165	C12H26	2,2,6-trimethyl-5-ethylheptane	62199-15-9	170.338	460.15	25	5.4669E+05	9.8404E+00	1,2
2956	25166	C12H26	2,3,3-trimethyl-4-ethylheptane	62199-16-0	170.338	472.15	25	5.6209E+05	1.0117E+01	1,2
2957	25167	C12H26	2,3,3-trimethyl-5-ethylheptane	62199-17-1	170.338	467.15	25	5.5916E+05	1.0065E+01	1,2
2958	25168	C12H26	2,3,4-trimethyl-3-ethylheptane	62199-18-2	170.338	476.15	25	5.6047E+05	1.0088E+01	1,2
2959	25169	C12H26	2,3,4-trimethyl-4-ethylheptane	62198-55-4	170.338	471.15	25	5.6193E+05	1.0115E+01	1,2
2960	25170	C12H26	2,3,4-trimethyl-5-ethylheptane	62198-56-5	170.338	470.15	25	5.0520E+05	9.0936E+00	1,2

Table 10 Henry's Law Constant for Compound in Water - Organic Compounds

					Henry's Law Constant, H					
NO	ID	FORMULA	NAME	CAS No	MW, g/mol	T_B, K	T, C	H @ T, atm/mol frac	H @ T, atm/mol/m^3	code
2961	25171	C12H26	2,3,5-trimethyl-3-ethylheptane	62198-57-6	170.338	470.65	25	5.6177E+05	1.0112E+01	1,2
2962	25172	C12H26	2,3,5-trimethyl-4-ethylheptane	62198-58-7	170.338	469.15	25	5.0501E+05	9.0901E+00	1,2
2963	25173	C12H26	2,3,5-trimethyl-5-ethylheptane	62198-59-8	170.338	467.15	25	5.5916E+05	1.0065E+01	1,2
2964	25174	C12H26	2,3,6-trimethyl-3-ethylheptane	62198-60-1	170.338	469.15	25	5.6097E+05	1.0097E+01	1,2
2965	25175	C12H26	2,3,6-trimethyl-4-ethylheptane	62198-61-2	170.338	463.15	25	4.9997E+05	8.9994E+00	1,2
2966	25176	C12H26	2,3,6-trimethyl-5-ethylheptane	62198-62-3	170.338	466.15	25	5.0330E+05	9.0594E+00	1,2
2967	25177	C12H26	2,4,4-trimethyl-3-ethylheptane	62198-63-4	170.338	467.15	25	5.5916E+05	1.0065E+01	1,2
2968	25178	C12H26	2,4,4-trimethyl-5-ethylheptane	62198-64-5	170.338	465.15	25	5.5654E+05	1.0018E+01	1,2
2969	25179	C12H26	2,4,5-trimethyl-3-ethylheptane	62198-65-6	170.338	469.15	25	5.0501E+05	9.0901E+00	1,2
2970	25180	C12H26	2,4,5-trimethyl-4-ethylheptane	62198-66-7	170.338	468.15	25	5.6016E+05	1.0083E+01	1,2
2971	25181	C12H26	2,4,5-trimethyl-5-ethylheptane	62198-67-8	170.338	470.15	25	5.6155E+05	1.0108E+01	1,2
2972	25182	C12H26	2,4,6-trimethyl-3-ethylheptane	62198-68-9	170.338	464.15	25	5.0126E+05	9.0226E+00	1,2
2973	25183	C12H26	2,4,6-trimethyl-4-ethylheptane	62198-69-0	170.338	458.15	25	5.4154E+05	9.7476E+00	1,2
2974	25184	C12H26	2,5,5-trimethyl-3-ethylheptane	62198-70-3	170.338	463.15	25	5.5315E+05	9.9566E+00	1,2
2975	25185	C12H26	2,5,5-trimethyl-4-ethylheptane	62198-71-4	170.338	465.15	25	5.5654E+05	1.0018E+01	1,2
2976	25186	C12H26	3,3,4-trimethyl-4-ethylheptane	62198-72-5	170.338	478.15	25	6.2344E+05	1.1222E+01	1,2
2977	25187	C12H26	3,3,4-trimethyl-5-ethylheptane	62198-73-6	170.338	472.15	25	5.6209E+05	1.0117E+01	1,2
2978	25188	C12H26	3,3,5-trimethyl-4-ethylheptane	62198-74-7	170.338	470.15	25	5.6155E+05	1.0108E+01	1,2
2979	25189	C12H26	3,3,5-trimethyl-5-ethylheptane	62198-75-8	170.338	475.15	25	6.2547E+05	1.1258E+01	1,2
2980	25190	C12H26	3,4,4-trimethyl-3-ethylheptane	62198-76-9	170.338	478.15	25	6.2344E+05	1.1222E+01	1,2
2981	25191	C12H26	3,4,4-trimethyl-5-ethylheptane	62198-77-0	170.338	475.15	25	5.6122E+05	1.0102E+01	1,2
2982	25192	C12H26	3,4,5-trimethyl-3-ethylheptane	62198-78-1	170.338	475.15	25	5.6122E+05	1.0102E+01	1,2
2983	25193	C12H26	3,4,5-trimethyl-4-ethylheptane	62198-79-2	170.338	478.15	25	5.5828E+05	1.0049E+01	1,2
2984	25194	C12H26	2,2,3,3,4-pentamethylheptane	62198-80-5	170.338	473.15	25	6.8909E+05	1.2404E+01	1,2
2985	25195	C12H26	2,2,3,3,5-pentamethylheptane	62198-81-6	170.338	467.15	25	6.8038E+05	1.2247E+01	1,2
2986	25196	C12H26	2,2,3,3,6-pentamethylheptane	62198-82-7	170.338	466.15	25	6.7805E+05	1.2205E+01	1,2
2987	25197	C12H26	2,2,3,4,4-pentamethylheptane	62198-83-8	170.338	475.15	25	6.8986E+05	1.2417E+01	1,2
2988	25198	C12H26	2,2,3,4,5-pentamethylheptane	62198-84-9	170.338	467.15	25	6.1413E+05	1.1054E+01	1,2
2989	25199	C12H26	2,2,3,4,6-pentamethylheptane	62198-85-0	170.338	461.15	25	6.0081E+05	1.0814E+01	1,2
2990	25200	C12H26	2,2,3,5,5-pentamethylheptane	62198-86-1	170.338	460.15	25	6.5935E+05	1.1868E+01	1,2
2991	25201	C12H26	2,2,3,5,6-pentamethylheptane	62198-87-2	170.338	461.95	25	6.0300E+05	1.0854E+01	1,2
2992	25202	C12H26	2,2,3,6,6-pentamethylheptane	62198-88-3	170.338	458.15	25	6.5149E+05	1.1727E+01	1,2
2993	25203	C12H26	2,2,4,4,5-pentamethylheptane	62199-61-5	170.338	469.15	25	6.8431E+05	1.2318E+01	1,2
2994	25204	C12H26	2,2,4,4,6-pentamethylheptane	62199-62-6	170.338	458.75	25	6.5392E+05	1.1771E+01	1,2
2995	25205	C12H26	2,2,4,5,5-pentamethylheptane	62199-63-7	170.338	459.15	25	6.5551E+05	1.1799E+01	1,2
2996	25206	C12H26	2,2,4,5,6-pentamethylheptane	62199-64-8	170.338	458.15	25	5.9155E+05	1.0648E+01	1,2
2997	25207	C12H26	2,2,4,6,6-pentamethylheptane	13475-82-6	170.338	451.10	25	6.1844E+05	1.1132E+01	1,2
2998	25208	C12H26	2,2,5,5,6-pentamethylheptane	62199-65-9	170.338	459.15	25	6.5551E+05	1.1799E+01	1,2
2999	25209	C12H26	2,3,3,4,4-pentamethylheptane	62199-66-0	170.338	476.15	25	6.8982E+05	1.2417E+01	1,2
3000	25210	C12H26	2,3,3,4,5-pentamethylheptane	62199-67-1	170.338	473.15	25	6.1953E+05	1.1151E+01	1,2
3001	25211	C12H26	2,3,3,4,6-pentamethylheptane	62199-68-2	170.338	468.15	25	6.1561E+05	1.1081E+01	1,2
3002	25212	C12H26	2,3,3,5,5-pentamethylheptane	62199-69-3	170.338	473.15	25	6.8909E+05	1.2404E+01	1,2
3003	25213	C12H26	2,3,3,5,6-pentamethylheptane	52670-35-6	170.338	465.15	25	6.1051E+05	1.0989E+01	1,2
3004	25214	C12H26	2,3,4,4,5-pentamethylheptane	62199-70-6	170.338	473.15	25	6.1953E+05	1.1151E+01	1,2
3005	25215	C12H26	2,3,4,4,6-pentamethylheptane	62199-71-7	170.338	463.15	25	6.0606E+05	1.0909E+01	1,2
3006	25216	C12H26	2,3,4,5,5-pentamethylheptane	62199-72-8	170.338	470.15	25	6.1789E+05	1.1122E+01	1,2
3007	25217	C12H26	2,3,4,5,6-pentamethylheptane	27574-98-7	170.338	468.15	25	5.5487E+05	9.9876E+00	1,2
3008	25218	C12H26	2,4,4,5,5-pentamethylheptane	62199-73-9	170.338	469.15	25	6.8431E+05	1.2318E+01	1,2
3009	25219	C12H26	3,3,4,4,5-pentamethylheptane	62199-74-0	170.338	479.15	25	6.8797E+05	1.2383E+01	1,2
3010	25220	C12H26	3,3,4,5,5-pentamethylheptane	62199-75-1	170.338	482.15	25	6.8346E+05	1.2302E+01	1,2
3011	25221	C12H26	2,2-dimethyl-3-tert-butylhexane	62199-76-2	170.338	469.15	25	6.8431E+05	1.2318E+01	1,2
3012	25222	C12H26	2-methyl-3-ethyl-3-isopropylhexane	62199-77-3	170.338	473.15	25	5.0963E+05	9.1733E+00	1,2
3013	25223	C12H26	2-methyl-4-ethyl-3-isopropylhexane	62199-78-4	170.338	463.15	25	4.5594E+05	8.2069E+00	1,2
3014	25224	C12H26	2,2,3-trimethyl-3-isopropylhexane	62199-79-5	170.338	474.15	25	6.8962E+05	1.2413E+01	1,2
3015	25225	C12H26	2,2,4-trimethyl-3-isopropylhexane	62199-80-8	170.338	462.15	25	6.0353E+05	1.0863E+01	1,2
3016	25226	C12H26	2,2,5-trimethyl-3-isopropylhexane	62199-81-9	170.338	453.15	25	5.7284E+05	1.0311E+01	1,2
3017	25227	C12H26	2,2,5-trimethyl-4-isopropylhexane	62199-82-0	170.338	453.15	25	5.7284E+05	1.0311E+01	1,2
3018	25228	C12H26	2,3,4-trimethyl-3-isopropylhexane	62199-83-1	170.338	477.15	25	6.1824E+05	1.1128E+01	1,2
3019	25229	C12H26	2,3,5-trimethyl-3-isopropylhexane	62199-84-2	170.338	467.15	25	6.1413E+05	1.1054E+01	1,2
3020	25230	C12H26	2,3,5-trimethyl-4-isopropylhexane	62199-85-3	170.338	463.15	25	5.4808E+05	9.8654E+00	1,2
3021	25231	C12H26	2,4,4-trimethyl-3-isopropylhexane	62199-86-4	170.338	465.15	25	6.1051E+05	1.0989E+01	1,2
3022	25232	C12H26	3,3,4-triethylhexane	62199-87-5	170.338	479.15	25	5.0307E+05	9.0552E+00	1,2
3023	25233	C12H26	2,2-dimethyl,3,3-diethylhexane	62199-88-6	170.338	473.15	25	6.2553E+05	1.1259E+01	1,2
3024	25234	C12H26	2,2-dimethyl-3,4-diethylhexane	62199-89-7	170.338	463.15	25	5.5315E+05	9.9566E+00	1,2
3025	25235	C12H26	2,2-dimethyl-4,4-diethylhexane	62184-89-8	170.338	473.15	25	6.2553E+05	1.1259E+01	1,2
3026	25236	C12H26	2,3-dimethyl-3,4-diethylhexane	62184-90-1	170.338	477.15	25	5.5949E+05	1.0071E+01	1,2
3027	25237	C12H26	2,3-dimethyl-4,4-diethylhexane	62184-91-2	170.338	477.15	25	5.5949E+05	1.0071E+01	1,2
3028	25238	C12H26	2,4-dimethyl-3,3-diethylhexane	62184-92-3	170.338	473.15	25	5.6202E+05	1.0116E+01	1,2
3029	25239	C12H26	2,4-dimethyl-3,4-diethylhexane	62184-93-4	170.338	473.15	25	5.6202E+05	1.0116E+01	1,2
3030	25240	C12H26	2,5-dimethyl-3,3-diethylhexane	62184-94-5	170.338	463.15	25	5.5315E+05	9.9566E+00	1,2
3031	25241	C12H26	2,5-dimethyl-3,4-diethylhexane	62184-95-6	170.338	463.15	25	4.9997E+05	8.9994E+00	1,2
3032	25242	C12H26	3,3-dimethyl-4,4-diethylhexane	62184-96-7	170.338	483.15	25	6.1476E+05	1.1066E+01	1,2
3033	25243	C12H26	3,4-dimethyl-3,4-diethylhexane	62184-97-8	170.338	483.15	25	6.1476E+05	1.1066E+01	1,2
3034	25244	C12H26	2,2,3,3-tetramethyl-4-ethylhexane	62184-98-9	170.338	474.15	25	6.8962E+05	1.2413E+01	1,2
3035	25245	C12H26	2,2,3,4-tetramethyl-3-ethylhexane	62184-99-0	170.338	477.15	25	6.8949E+05	1.2411E+01	1,2
3036	25246	C12H26	2,2,3,4-tetramethyl-4-ethylhexane	62185-00-6	170.338	481.15	25	6.8526E+05	1.2335E+01	1,2
3037	25247	C12H26	2,2,3,5-tetramethyl-3-ethylhexane	62185-01-7	170.338	467.15	25	6.8038E+05	1.2247E+01	1,2
3038	25248	C12H26	2,2,3,5-tetramethyl-4-ethylhexane	62185-02-8	170.338	464.15	25	6.0839E+05	1.0951E+01	1,2
3039	25249	C12H26	2,2,4,4-tetramethyl-3-ethylhexane	62185-03-9	170.338	473.15	25	6.8909E+05	1.2404E+01	1,2
3040	25250	C12H26	2,2,4,5-tetramethyl-3-ethylhexane	62185-04-0	170.338	463.15	25	6.0606E+05	1.0909E+01	1,2

Table 10 Henry's Law Constant for Compound in Water - Organic Compounds

					Henry's Law Constant, H					
NO	ID	FORMULA	NAME	CAS No	MW, g/mol	T_B, K	T, C	H @ T, atm/mol frac	H @ T, atm/mol/m³	code
3041	25251	C12H26	2,2,4,5-tetramethyl-4-ethylhexane	62185-05-1	170.338	470.15	25	6.8590E+05	1.2346E+01	1,2
3042	25252	C12H26	2,2,5,5-tetramethyl-3-ethylhexane	62185-06-2	170.338	453.15	25	6.2882E+05	1.1319E+01	1,2
3043	25253	C12H26	2,3,3,4-tetramethyl-4-ethylhexane	62185-07-3	170.338	481.15	25	6.8526E+05	1.2335E+01	1,2
3044	25254	C12H26	2,3,3,5-tetramethyl-4-ethylhexane	62185-08-4	170.338	470.15	25	6.1789E+05	1.1122E+01	1,2
3045	25255	C12H26	2,3,4,4-tetramethyl-3-ethylhexane	62185-09-5	170.338	481.15	25	6.8526E+05	1.2335E+01	1,2
3046	25256	C12H26	2,3,4,5-tetramethyl-3-ethylhexane	62185-10-8	170.338	474.15	25	6.1959E+05	1.1152E+01	1,2
3047	25257	C12H26	2,2,3,3,4,4-hexamethylhexane	62185-11-9	170.338	489.15	25	8.2626E+05	1.4872E+01	1,2
3048	25258	C12H26	2,2,3,3,4,5-hexamethylhexane	62185-12-0	170.338	472.15	25	7.5732E+05	1.3632E+01	1,2
3049	25259	C12H26	2,2,3,3,5,5-hexamethylhexane	60302-24-1	170.338	468.15	25	8.2913E+05	1.4924E+01	1,2
3050	25260	C12H26	2,2,3,4,4,5-hexamethylhexane	62185-13-1	170.338	479.15	25	7.6015E+05	1.3683E+01	1,2
3051	25261	C12H26	2,2,3,4,5,5-hexamethylhexane	62185-14-2	170.338	463.15	25	7.3286E+05	1.3191E+01	1,2
3052	25262	C12H26	2,3,3,4,4,5-hexamethylhexane	52670-36-7	170.338	479.15	25	7.6015E+05	1.3683E+01	1,2
3053	25263	C12H26	2,4-dimethyl-3-ethyl-3-isopropylpentane	62185-16-4	170.338	479.15	25	6.1606E+05	1.1089E+01	1,2
3054	25264	C12H26	2,2,4-trimethyl-3,3-diethylpentane	62185-15-3	170.338	479.15	25	6.8797E+05	1.2383E+01	1,2
3055	25265	C12H26	2,2,3,4-tetramethyl-3-isopropylpentane	62185-17-5	170.338	477.15	25	7.6092E+05	1.3696E+01	1,2
3056	25266	C12H26	2,2,4,4-tetramethyl-3-isopropylpentane	62185-18-6	170.338	467.15	25	7.4637E+05	1.3435E+01	1,2
3057	25267	C12H26	2,2,3,4,4-pentamethyl-3-ethylpentane	66576-21-4	170.338	487.15	25	8.3354E+05	1.5004E+01	1,2
3058	25268	C12H26	2-methylundecane	7045-71-8	170.338	484.00	25	3.0750E+05	5.5350E+00	2
3059	25277	C12H26O	1-dodecanol	112-53-8	186.338	535.00	25	2.9343E+00	5.2817E-05	1,2
3060	25278	C12H26O	2-dodecanol	10203-28-8	186.338	525.15	25	5.0915E-01	9.1646E-06	1,2
3061	25279	C12H26O	3-dodecanol	10203-30-2	186.338	512.53	25	3.6221E+00	6.5198E-05	2
3062	25280	C12H26O	6-dodecanol	6836-38-0	186.338	498.15	25	3.4840E+00	6.2711E-05	1,2
3063	25281	C12H26O	2-methyl-1-undecanol	10522-26-6	186.338	512.53	25	3.6221E+00	6.5198E-05	2
3064	25282	C12H26O	2-methyl-3-undecanol	60671-36-5	186.338	512.53	25	3.6221E+00	6.5198E-05	2
3065	25283	C12H26O	2-methyl-5-undecanol	33978-71-1	186.338	512.53	25	3.6221E+00	6.5198E-05	2
3066	25284	C12H26O	2-butyl-1-octanol	3913-02-8	186.338	519.65	25	3.7497E+00	6.7494E-05	1,2
3067	25285	C12H26O	dihexyl ether	112-58-3	186.338	498.85	25	3.1815E+02	5.7267E-03	1,2
3068	25286	C12H26O	2,6,8-trimethylnonanol-4	123-17-1	186.338	498.40	25	3.4853E+00	6.2735E-05	1,2
3069	25294	C12H26O3	diethylene glycol dibutyl ether	112-73-2	218.337	529.15	20	6.9729E-02	1.2551E-06	1,2
3070	25310	C12H26S	1-dodecanethiol	112-55-0	202.404	547.75	25	8.8716E+02	1.5969E-02	1,2
3071	25311	C12H26S	2-dodecanethiol	14402-50-7	202.404	538.35	25	2.5682E+02	4.6227E-03	1,2
3072	25312	C12H26S	dihexyl sulfide	6294-31-1	202.404	503.15	25	2.5558E+02	4.6004E-03	1,2
3073	25313	C12H26S	methyl undecyl sulfide	7289-44-3	202.404	537.00	25	1.4899E+02	2.6818E-03	2
3074	25314	C12H26S	ethyl decyl sulfide	19313-61-2	202.404	537.00	25	1.4899E+02	2.6818E-03	2
3075	25315	C12H26S	propyl nonyl sulfide	62103-66-6	202.404	537.00	25	1.4899E+02	2.6818E-03	2
3076	25316	C12H26S	butyl octyl sulfide	16900-07-5	202.404	537.00	25	1.4899E+02	2.6818E-03	2
3077	25317	C12H26S	tert-dodecylmercaptan	25103-58-6	202.404	510.65	25	3.1897E+02	5.7415E-03	1,2
3078	25340	C12H27N	dodecylamine	124-22-1	185.353	532.35	25	4.3051E+00	7.7491E-05	1,2
3079	25341	C12H27N	methylundecylamine	66553-53-5	185.353	517.15	25	4.0466E-02	7.2839E-07	1,2
3080	25342	C12H27N	ethyldecylamine	66553-52-4	185.353	514.15	25	3.3311E-02	5.9959E-07	1,2
3081	25343	C12H27N	dihexylamine	143-16-8	185.353	512.95	25	3.7294E-02	6.7129E-07	1,2
3082	25344	C12H27N	dimethyldecylamine	1120-24-7	185.353	508.15	25	9.5288E+00	1.7152E-04	1,2
3083	25345	C12H27N	diethyloctylamine	4088-37-3	185.353	486.15	25	5.0315E+00	9.0567E-05	1,2
3084	25346	C12H27N	tributylamine	102-82-9	185.353	487.15	25	3.0202E+01	5.4363E-04	1,2
3085	25347	C12H27N	triisobutylamine	1116-40-1	185.353	464.65	25	1.0575E+00	1.9036E-05	1,2
3086	25575	C13H10	fluorene	86-73-7	166.222	570.44	25	2.1841E+01	3.9314E-04	1,2
3087	25623	C13H10O	benzophenone	119-61-9	182.222	579.24	25	2.9188E-01	5.2537E-06	1,2
3088	25738	C13H12	diphenylmethane	101-81-5	168.238	537.42	25	1.2620E+01	2.2715E-04	1,2
3089	26283	C13H20	heptylbenzene	1078-71-3	176.302	519.25	25	9.7442E+02	1.7539E-02	1,2
3090	26284	C13H20	1-methyl-2,4-diisopropylbenzene	1460-98-6	176.302	497.15	25	3.3746E+02	6.0742E-03	1,2
3091	26285	C13H20	1-methyl-3,5-diisopropylbenzene	3055-14-9	176.302	489.65	25	3.6473E+02	6.5650E-03	1,2
3092	26286	C13H20	[2-methyl-1-isopropylpropyl]benzene	21777-84-4	176.302	493.85	25	7.9882E+02	1.4379E-02	1,2
3093	26288	C13H20	(2,4-dimethylpentyl)benzene	54518-00-2	176.302	492.16	25	8.1267E+02	1.4628E-02	1,2
3094	26402	C13H24	1-tridecyne	26186-02-7	180.334	507.16	25	2.0699E+03	3.7258E-02	1,2
3095	26403	C13H24	2-tridecyne	28467-75-6	180.334	518.15	25	1.3547E+03	2.4384E-02	1,2
3096	26404	C13H24	3-tridecyne	60186-78-9	180.334	511.15	25	1.7940E+03	3.2291E-02	1,2
3097	26465	C13H26	octylcyclopentane	1795-20-6	182.349	516.86	25	4.6724E+04	8.4103E-01	1,2
3098	26466	C13H26	heptylcyclohexane	5617-41-4	182.349	518.06	25	3.5171E+04	6.3307E-01	1,2
3099	26467	C13H26	2-butyl-1,1,3-trimethylcyclohexane	54676-39-0	182.349	510.52	25	3.2365E+04	5.8256E-01	2
3100	26469	C13H26	1-tridecene	2437-56-1	182.349	505.93	25	1.1149E+05	2.0069E+00	1,2
3101	26470	C13H26	2-methyl-1-dodecene	16435-49-7	182.349	504.45	25	9.6442E+04	1.7359E+00	1,2
3102	26471	C13H26	5-butyl-4-nonene	7367-38-6	182.349	488.65	25	7.1993E+04	1.2959E+00	1,2
3103	26472	C13H26Br2	1,1-dibromotridecane	62168-30-3	342.157	598.15	25	2.1702E+00	3.9064E-05	1,2
3104	26473	C13H26Br2	1,13-dibromotridecane	31772-05-1	342.157	598.15	25	2.1702E+00	3.9064E-05	2
3105	26474	C13H26Cl2	1,1-dichlorotridecane	821-93-2	253.254	572.15	25	5.5988E+01	1.0078E-03	1,2
3106	26486	C13H26O	1-tridecanal	10486-19-8	198.349	540.15	25	1.1794E+01	2.1229E-04	1,2
3107	26487	C13H26O	2-tridecanone	593-08-8	198.349	535.73	25	3.9858E+01	7.1743E-04	1,2
3108	26488	C13H26O	3-tridecanone	1534-26-5	198.349	538.47	25	4.9926E+01	8.9866E-04	2
3109	26489	C13H26O	7-tridecanone	462-18-0	198.349	534.15	25	4.6656E+01	8.3981E-04	1,2
3110	26492	C13H26O2	tridecanoic acid	638-53-9	214.348	585.25	25	5.5441E-03	9.9794E-08	1,2
3111	26493	C13H26O2	2-methyldodecanoic acid	2874-74-0	214.348	542.21	25	1.0082E-01	1.8147E-06	2
3112	26494	C13H26O2	dodecyl formate	28303-42-6	214.348	554.36	25	2.7644E+01	4.9758E-04	2
3113	26495	C13H26O2	undecyl acetate	1731-81-3	214.348	520.00	25	2.3377E+01	4.2078E-04	2
3114	26496	C13H26O2	decyl propanoate	5454-19-3	214.348	555.06	25	1.6098E+01	2.8977E-04	2
3115	26497	C13H26O2	nonyl butanoate	2639-64-7	214.348	535.15	25	2.1598E+01	3.8876E-04	1,2
3116	26515	C13H27Br	1-bromotridecane	765-09-3	263.261	565.15	25	1.1357E+03	2.0443E-02	1,2
3117	26516	C13H27Cl	1-chlorotridecane	822-13-9	218.810	553.15	25	1.7664E+03	3.1795E-02	1,2
3118	26526	C13H28	tridecane	629-50-5	184.365	508.62	25	3.9870E+05	7.1765E+00	1,2
3119	26527	C13H28	2-methyldodecane	1560-97-0	184.365	502.65	25	2.2171E+05	3.9908E+00	1,2
3120	26528	C13H28	3-methyldodecane	17312-57-1	184.365	503.45	25	2.1830E+05	3.9294E+00	1,2

Table 10 Henry's Law Constant for Compound in Water - Organic Compounds

					Henry's Law Constant, H					
NO	ID	FORMULA	NAME	CAS No	MW, g/mol	T_B, K	T, C	H @ T, atm/mol frac	H @ T, atm/mol/m^3	code
3121	26529	C13H28	2,2-dimethylundecane	17312-64-0	184.365	494.15	25	3.2309E+05	5.8155E+00	1,2
3122	26530	C13H28	2,3-dimethylundecane	17312-77-5	184.365	501.15	25	2.5592E+05	4.6065E+00	1,2
3123	26531	C13H28	2,4-dimethylundecane	17312-80-0	184.365	489.15	25	3.0430E+05	5.4773E+00	1,2
3124	26532	C13H28	5-butylnonane	17312-63-9	184.365	490.65	25	2.9232E+05	5.2617E+00	1,2
3125	26535	C13H28O	1-tridecanol	112-70-9	200.365	547.15	25	4.3117E+00	7.7610E-05	1,2
3126	26536	C13H28O	2-tridecanol	1653-31-2	200.365	541.15	25	3.2951E-01	5.9312E-06	1,2
3127	26537	C13H28O	3-tridecanol	10289-68-6	200.365	531.15	25	4.3794E+00	7.8829E-05	2
3128	26538	C13H28O	5-butyl-5-nonanol	597-93-3	200.365	531.15	25	4.3794E+00	7.8829E-05	2
3129	26549	C13H28S	1-tridecanethiol	19484-26-5	216.431	563.96	25	3.3371E+02	6.0067E-03	1,2
3130	26550	C13H28S	2-tridecanethiol	62155-03-7	216.431	555.15	25	1.3692E+02	2.4645E-03	1,2
3131	26551	C13H28S	methyl dodecyl sulfide	---	216.431	553.00	25	1.0545E+02	1.8980E-03	2
3132	26552	C13H28S	ethyl undecyl sulfide	66577-30-8	216.431	553.00	25	1.0545E+02	1.8980E-03	2
3133	26553	C13H28S	propyl decyl sulfide	66577-31-9	216.431	553.00	25	1.0545E+02	1.8980E-03	2
3134	26554	C13H28S	butyl nonyl sulfide	66577-32-0	216.431	553.00	25	1.0545E+02	1.8980E-03	2
3135	26555	C13H29N	tridecylamine	2869-34-3	199.380	553.15	25	3.9813E+00	7.1663E-05	1,2
3136	26556	C13H29N	methyldodecylamine	7311-30-0	199.380	534.15	25	7.2195E-02	1.2995E-06	1,2
3137	26557	C13H29N	ethylundecylamine	59570-04-6	199.380	532.15	25	8.5770E-02	1.5438E-06	1,2
3138	26558	C13H29N	dimethylundecylamine	17373-28-3	199.380	526.15	25	4.6095E+00	8.2970E-05	1,2
3139	26559	C13H29N	diethylnonylamine	45124-35-4	199.380	504.15	25	2.5764E+00	4.6375E-05	1,2
3140	26560	C13H29N	6-methyl-N-isopentyl-2-heptanamine	502-59-0	199.380	520.48	25	3.1446E-01	5.6603E-06	2
3141	26561	C13H29N	tert-amyl-tert-octylamine	150285-07-7	199.380	473.15	25	9.7173E-02	1.7491E-06	1,2
3142	26562	C13H29N	N-methyldihexylamine	37615-53-5	199.380	520.48	25	1.0904E+00	1.9626E-05	2
3143	26654	C14H8O2	anthraquinone	84-65-1	208.216	653.05	25	6.7479E+00	1.2146E-04	1,2
3144	26728	C14H10	anthracene	120-12-7	178.233	615.18	22	2.9329E+02	5.2791E-03	1,2
3145	26730	C14H10	phenanthrene	85-01-8	178.233	613.45	25	7.1931E+00	1.2947E-04	1,2
3146	26866	C14H12	cis-stilbene	645-49-8	180.249	535.00	25	5.0307E+02	9.0551E-03	1,2
3147	26867	C14H12	trans-stilbene	103-30-0	180.249	579.65	25	2.2701E+01	4.0861E-04	1,2
3148	26873	C14H12	stilbene	588-59-0	180.249	579.65	25	2.1747E+01	3.9144E-04	1,2
3149	27644	C14H22	octylbenzene	2189-60-8	190.329	537.55	25	8.2507E+02	1.4851E-02	1,2
3150	27645	C14H22	1,2,3,4-tetraethylbenzene	642-32-0	190.329	524.16	25	9.9931E+02	1.7987E-02	1,2
3151	27646	C14H22	1,2,3,5-tetraethylbenzene	38842-05-6	190.329	523.66	25	7.1851E+02	1.2933E-02	1,2
3152	27647	C14H22	1,2,4,5-tetraethylbenzene	635-81-4	190.329	523.16	25	7.0191E+02	1.2634E-02	1,2
3153	27648	C14H22	1,4-di-tert-butylbenzene	1012-72-2	190.329	511.15	25	2.9249E+02	5.2648E-03	1,2
3154	27649	C14H22	(1-methylheptyl)benzene	777-22-0	190.329	523.94	25	4.9518E+02	8.9133E-03	2
3155	27650	C14H22	1,3-di-tert-butylbenzene	1014-60-4	190.329	523.94	25	2.1715E+02	3.9087E-03	2
3156	27772	C14H24S	3-decylthiophene	65016-55-9	224.411	570.15	25	1.3518E+02	2.4332E-03	2
3157	27780	C14H26	1-tetradecyne	765-10-6	194.360	525.15	25	1.0594E+03	1.9070E-02	1,2
3158	27781	C14H26	3-tetradecyne	60212-32-0	194.360	528.15	25	9.1167E+02	1.6410E-02	1,2
3159	27782	C14H26	7-tetradecyne	35216-11-6	194.360	535.60	25	6.0669E+02	1.0920E-02	2
3160	27784	C14H26	1,13-tetradecadiene	21964-49-8	194.360	535.60	25	2.6251E+03	4.7252E-02	2
3161	27847	C14H28	nonylcyclopentane	2882-98-6	196.376	535.26	25	2.0884E+04	3.7591E-01	1,2
3162	27848	C14H28	octylcyclohexane	1795-15-9	196.376	536.76	25	1.5675E+04	2.8214E-01	1,2
3163	27850	C14H28	1-tetradecene	1120-36-1	196.376	524.25	25	4.1095E+04	7.3970E-01	1,2
3164	27851	C14H28	2-tetradecene	638-60-8	196.376	536.15	25	4.0194E+04	7.2349E-01	1,2
3165	27852	C14H28	2-methyl-1-tridecene	18094-01-4	196.376	523.15	25	3.8109E+04	6.8595E-01	1,2
3166	27853	C14H28	trans-7-tetradecene	41446-63-3	196.376	523.15	25	2.5349E+04	4.5628E-01	1,2
3167	27856	C14H28Cl2	1,1-dichlorotetradecane	4168-40-5	267.281	588.15	25	2.8898E+01	5.2016E-04	1,2
3168	27862	C14H28O	tetradecanal	124-25-4	212.376	556.15	25	2.6387E+01	4.7496E-04	1,2
3169	27863	C14H28O	2-tetradecanone	2345-27-9	212.376	551.15	25	4.8275E+01	8.6895E-04	1,2
3170	27864	C14H28O	7-ethyl-2-methyl-4-undecanone	6976-00-7	212.376	525.65	25	4.4452E+01	8.0013E-04	1,2
3171	27865	C14H28O	2-methyl-3-tridecanone	40239-35-8	212.376	540.15	25	5.4973E+01	9.8951E-04	1,2
3172	27866	C14H28O	3-tetradecanone	629-23-2	212.376	512.95	25	3.8698E+01	6.9656E-04	2
3173	27873	C14H28O2	tetradecanoic acid	544-63-8	228.375	599.35	20	1.9018E-03	3.4233E-08	1,2
3174	27874	C14H28O2	tridecyl formate	42875-41-2	228.375	577.24	25	1.7674E+01	3.1813E-04	2
3175	27875	C14H28O2	dodecyl acetate	112-66-3	228.375	538.15	25	2.5641E-02	4.6154E-07	1,2
3176	27876	C14H28O2	undecyl propanoate	5458-33-3	228.375	577.94	25	1.0672E+01	1.9210E-04	2
3177	27877	C14H28O2	decyl butanoate	5454-09-1	228.375	553.15	25	8.1082E-04	1.4595E-08	1,2
3178	27885	C14H28O2	3,7-dimethyloctanyl butyrate	67874-80-0	228.375	559.87	25	1.7836E+01	3.2104E-04	2
3179	27895	C14H29Br	1-bromotetradecane	112-71-0	277.288	580.15	25	6.3890E+02	1.1500E-02	1,2
3180	27896	C14H29Br	1-bromo-12-methyltridecane	111772-90-8	277.288	580.15	25	1.3541E+02	2.4373E-03	2
3181	27897	C14H29Cl	1-chlorotetradecane	2425-54-9	232.837	569.99	25	1.3530E+03	2.4353E-02	1,2
3182	27912	C14H30	tetradecane	629-59-4	198.392	526.73	25	7.8914E+04	1.4204E+00	1,2
3183	27913	C14H30	2-methyltridecane	1560-96-9	198.392	521.05	25	1.0882E+05	1.9587E+00	1,2
3184	27914	C14H30	3-methyltridecane	6418-41-3	198.392	521.85	25	1.0601E+05	1.9081E+00	1,2
3185	27915	C14H30	2,2-dimethyldodecane	49598-54-1	198.392	513.15	25	1.7792E+05	3.2026E+00	1,2
3186	27916	C14H30	2,3-dimethyldodecane	6117-98-2	198.392	520.15	25	1.2698E+05	2.2856E+00	1,2
3187	27917	C14H30	2,4-dimethyldodecane	6117-99-3	198.392	507.15	25	1.8044E+05	3.2479E+00	1,2
3188	27918	C14H30	7-methyltridecane	26730-14-3	198.392	518.35	25	1.3231E+05	2.3816E+00	2
3189	27922	C14H30O	1-tetradecanol	112-72-1	214.392	560.15	25	4.6730E+00	8.4113E-05	1,2
3190	27923	C14H30O	2-tetradecanol	4706-81-4	214.392	557.15	25	2.0348E-01	3.6627E-06	1,2
3191	27924	C14H30O	3-tetradecanol	1653-32-3	214.392	552.35	25	6.0716E+00	1.0929E-04	2
3192	27925	C14H30O	12-methyl-1-tridecanol	21987-21-3	214.392	552.35	25	6.0716E+00	1.0929E-04	2
3193	27926	C14H30O	2-methyl-3-tridecanol	98930-89-3	214.392	547.15	25	5.6160E+00	1.0109E-04	1,2
3194	27927	C14H30O	2-methyl-4-tridecanol	36691-77-7	214.392	552.35	25	6.0716E+00	1.0929E-04	2
3195	27928	C14H30O	diheptyl ether	629-64-1	214.392	535.15	25	5.3861E+02	9.6948E-03	1,2
3196	27929	C14H30O	tetradecanol; mixed isomers	27196-00-5	214.392	562.15	25	7.2383E+00	1.3029E-04	1,2
3197	27946	C14H30S	1-tetradecanethiol	2079-95-0	230.458	579.36	25	1.9294E+02	3.4728E-03	1,2
3198	27947	C14H30S	2-tetradecanethiol	62155-04-8	230.458	571.15	25	5.9713E+01	1.0748E-03	1,2
3199	27948	C14H30S	diheptyl sulfide	629-65-2	230.458	571.15	25	6.0690E+01	1.0924E-03	1,2
3200	27949	C14H30S	methyl tridecyl sulfide	62155-09-3	230.458	569.00	25	6.5856E+01	1.1854E-03	2

Table 10 Henry's Law Constant for Compound in Water - Organic Compounds

NO	ID	FORMULA	NAME	CAS No	MW, g/mol	T_B, K	T, C	H @ T, atm/mol frac	H @ T, atm/mol/m^3	code
					Henry's Law Constant, H					
3201	27950	C14H30S	ethyl dodecyl sulfide	2851-83-4	230.458	569.00	25	6.5856E+01	1.1854E-03	2
3202	27951	C14H30S	propyl undecyl sulfide	66826-84-4	230.458	569.00	25	6.5856E+01	1.1854E-03	2
3203	27952	C14H30S	butyl decyl sulfide	19313-57-6	230.458	569.00	25	6.5856E+01	1.1854E-03	2
3204	27956	C14H31N	tetradecylamine	2016-42-4	213.407	564.45	25	4.8722E+00	8.7699E-05	1,2
3205	27957	C14H31N	methyltridecylamine	45165-81-9	213.407	550.15	25	1.6770E-01	3.0185E-06	1,2
3206	27958	C14H31N	ethyldodecylamine	35902-57-9	213.407	549.15	25	2.3397E-01	4.2114E-06	1,2
3207	27959	C14H31N	diheptylamine	2470-68-0	213.407	545.15	25	1.6226E-01	2.9207E-06	1,2
3208	27960	C14H31N	dimethyldodecylamine	112-18-5	213.407	544.15	25	3.6570E+00	6.5825E-05	1,2
3209	27961	C14H31N	diethyldecylamine	6308-94-7	213.407	520.15	25	1.7474E+00	3.1454E-05	1,2
3210	28436	C15H16O2	bisphenol a	80-05-7	228.291	633.65	20	2.4204E-16	4.3567E-21	1,2
3211	28725	C15H24	nonylbenzene	1081-77-2	204.356	555.20	25	4.6381E+02	8.3485E-03	1,2
3212	28726	C15H24	1,2,4-triisopropylbenzene	948-32-3	204.356	517.15	25	2.7650E+02	4.9770E-03	1,2
3213	28727	C15H24	1,3,5-triisopropylbenzene	717-74-8	204.356	511.15	25	3.1404E+02	5.6528E-03	1,2
3214	28728	C15H24	3,5-di-tert-butyltoluene	15181-11-0	204.356	517.15	25	2.7650E+02	4.9770E-03	1,2
3215	28756	C15H24O	2,6-di-tert-butyl-p-cresol	128-37-0	220.355	538.00	25	8.0563E+00	1.4501E-04	1,2
3216	28757	C15H24O	nonylphenol	25154-52-3	220.355	581.00	25	2.6819E-01	4.8275E-06	1,2
3217	28758	C15H24O	2,4-di-tert-butyl-5-methylphenol	497-39-2	220.355	555.15	25	1.8217E+00	3.2790E-05	1,2
3218	28759	C15H24O	2,4-di-tert-butyl-6-methylphenol	616-55-7	220.355	542.15	25	4.6467E+00	8.3640E-05	1,2
3219	28764	C15H24O	4-nonylphenol	104-40-5	220.355	568.15	25	6.8980E-01	1.2416E-05	1,2
3220	28765	C15H24O	4-nonylphenol, isomers	84852-15-3	220.355	568.15	25	6.8980E-01	1.2416E-05	1,2
3221	28820	C15H28	1-pentadecyne	765-13-9	208.387	541.16	25	4.6402E+02	8.3523E-03	1,2
3222	28821	C15H28	2-pentadecyne	52112-25-1	208.387	553.15	25	2.0227E+02	3.6408E-03	1,2
3223	28822	C15H28	3-pentadecyne	61886-61-1	208.387	544.15	25	3.8231E+02	6.8815E-03	1,2
3224	28855	C15H30	decylcyclopentane	1795-21-7	210.403	552.54	25	7.2899E+03	1.3122E-01	1,2
3225	28856	C15H30	nonylcyclohexane	2883-02-5	210.403	554.66	25	5.2541E+03	9.4573E-02	1,2
3226	28857	C15H30	1-ethyl-1-methyl-2,4-diisopropylcyclohexane,	515-12-8	210.403	549.62	25	5.3835E+03	9.6902E-02	2
3227	28859	C15H30	1-pentadecene	13360-61-7	210.403	541.61	25	1.6583E+04	2.9849E-01	1,2
3228	28860	C15H30	2-methyl-1-tetradecene	52254-38-3	210.403	540.15	25	2.7425E+04	4.9364E-01	1,2
3229	28861	C15H30	7-pentadecene	15430-98-5	210.403	549.62	25	5.9032E+03	1.0626E-01	2
3230	28862	C15H30	8-pentadecene	---	210.403	549.62	25	5.9032E+03	1.0626E-01	2
3231	28873	C15H30O	pentadecanal	2765-11-9	226.403	570.15	25	1.5796E+01	2.8433E-04	1,2
3232	28874	C15H30O	2-pentadecanone	2345-28-0	226.403	565.15	25	1.5256E+02	2.7461E-03	1,2
3233	28875	C15H30O	8-pentadecanone	818-23-5	226.403	564.15	25	9.7912E+01	1.7624E-03	1,2
3234	28879	C15H30O2	pentadecanoic acid	1002-84-2	242.402	612.05	20	1.4751E-05	2.6551E-10	1,2
3235	28880	C15H30O2	tetradecyl formate	5451-63-8	242.402	600.12	25	1.0004E+01	1.8006E-04	2
3236	28881	C15H30O2	tridecyl acetate	1072-33-9	242.402	555.00	25	1.5340E+01	2.7611E-04	2
3237	28882	C15H30O2	dodecyl propanoate	6221-93-8	242.402	600.82	25	6.2674E+00	1.1281E-04	2
3238	28883	C15H30O2	undecyl butanoate	5461-02-9	242.402	600.82	25	7.2832E+00	1.3110E-04	2
3239	28898	C15H31Br	1-bromopentadecane	629-72-1	291.315	595.15	25	3.5212E+02	6.3381E-03	1,2
3240	28899	C15H31Cl	1-chloropentadecane	4862-03-1	246.864	585.15	25	5.2769E+02	9.4984E-03	1,2
3241	28904	C15H32	pentadecane	629-62-9	212.419	543.83	25	5.1704E+04	9.3066E-01	1,2
3242	28905	C15H32	2-methyltetradecane	1560-95-8	212.419	538.35	25	4.1595E+04	7.4871E-01	1,2
3243	28906	C15H32	3-methyltetradecane	18435-22-8	212.419	539.45	25	3.9486E+04	7.1074E-01	1,2
3244	28907	C15H32	2,2-dimethyltridecane	61869-04-3	212.419	530.15	25	7.8238E+04	1.4083E+00	1,2
3245	28908	C15H32	2,3-dimethyltridecane	18435-20-6	212.419	538.15	25	4.7975E+04	8.6354E-01	1,2
3246	28909	C15H32	2,4-dimethyltridecane	61868-05-1	212.419	524.15	25	8.5063E+04	1.5311E+00	1,2
3247	28912	C15H32O	1-pentadecanol	629-76-5	228.418	578.01	25	3.0616E+01	5.5108E-04	1,2
3248	28913	C15H32O	2-pentadecanol	1653-34-5	228.418	572.15	25	1.2005E-01	2.1609E-06	1,2
3249	28914	C15H32O	3-pentadecanol	53346-71-7	228.418	661.15	25	8.7211E+02	1.5698E-02	1,2
3250	28915	C15H32O	2,8-dimethyl-6-isobutylnonanol-4	10143-20-1	228.418	603.77	25	2.6933E+01	4.8478E-04	2
3251	28921	C15H32S	1-pentadecanethiol	25276-70-4	244.485	593.86	25	9.2681E+01	1.6682E-03	1,2
3252	28922	C15H32S	2-pentadecanethiol	62155-05-9	244.485	586.15	25	2.2956E+01	4.1320E-04	1,2
3253	28923	C15H32S	methyl tetradecyl sulfide	7289-45-4	244.485	583.00	25	3.9215E+01	7.0586E-04	2
3254	28924	C15H32S	ethyl tridecyl sulfide	66271-81-6	244.485	583.00	25	3.9215E+01	7.0586E-04	2
3255	28925	C15H32S	propyl dodecyl sulfide	66271-82-7	244.485	583.00	25	3.9215E+01	7.0586E-04	2
3256	28926	C15H32S	butyl undecyl sulfide	66271-83-8	244.485	583.00	25	3.9215E+01	7.0586E-04	2
3257	28931	C15H33N	pentadecylamine	2570-26-5	227.434	586.15	25	8.7828E-01	1.5809E-05	1,2
3258	28932	C15H33N	methyltetradecylamine	29369-63-9	227.434	566.15	25	5.8958E-01	1.0612E-05	1,2
3259	28933	C15H33N	ethyltridecylamine	59570-06-8	227.434	565.15	25	6.2060E-01	1.1171E-05	1,2
3260	28934	C15H33N	dimethyltridecylamine	17373-29-4	227.434	560.15	25	2.3356E+00	4.2040E-05	1,2
3261	28935	C15H33N	diethylundecylamine	54334-64-4	227.434	536.15	25	1.0896E+00	1.9612E-05	1,2
3262	28936	C15H33N	tripentylamine	621-77-2	227.434	516.15	25	2.5357E-01	4.5642E-06	1,2
3263	28937	C15H33N	triisopentylamine	645-41-0	227.434	508.15	25	3.7418E-01	6.7352E-06	1,2
3264	28938	C15H33N	tris(2-methylbutyl)amine	620-43-9	227.434	505.15	25	3.4799E-01	6.2637E-06	1,2
3265	28984	C16H10	fluoranthene	206-44-0	202.255	655.95	27	1.5094E+01	2.7169E-04	1,2
3266	28985	C16H10	pyrene	129-00-0	202.255	667.95	25	2.9012E+00	5.2221E-05	1,2
3267	29518	C16H22O4	dibutyl phthalate	84-74-2	278.348	613.15	20	1.2812E-02	2.3061E-07	1,2
3268	29595	C16H26	decylbenzene	104-72-3	218.382	571.04	25	1.9364E+02	3.4854E-03	1,2
3269	29596	C16H26	pentaethylbenzene	605-01-6	218.382	550.16	25	5.4299E+02	9.7737E-03	1,2
3270	29599	C16H26	(1-methylnonyl)benzene	4537-13-7	218.382	562.65	25	1.1375E+02	2.0474E-03	1,2
3271	29637	C16H28S	3-dodecylthiophene	104934-52-3	252.464	563.15	25	1.9557E+02	3.5202E-03	1,2
3272	29640	C16H30	1-hexadecyne	629-74-3	222.414	557.16	25	1.5826E+02	2.8487E-03	1,2
3273	29641	C16H30	2-hexadecyne	629-75-4	222.414	569.15	25	5.6381E+01	1.0148E-03	1,2
3274	29642	C16H30	3-hexadecyne	61886-62-2	222.414	559.15	25	1.3481E+02	2.4265E-03	1,2
3275	29704	C16H32	undecylcyclopentane	6785-23-5	224.430	568.76	25	1.6504E+03	2.9708E-02	1,2
3276	29705	C16H32	decylcyclohexane	1795-16-0	224.430	570.75	25	1.6368E+03	2.9463E-02	1,2
3277	29707	C16H32	1-hexadecene	629-73-2	224.430	558.02	25	1.0980E+04	1.9763E-01	1,2
3278	29708	C16H32	2-methyl-1-pentadecene	29833-69-0	224.430	557.15	25	1.0333E+04	1.8599E-01	1,2
3279	29709	C16H32	2-methyl-1-propene, tetramer	15220-85-6	224.430	517.15	25	3.7549E+04	6.7588E-01	1,2
3280	29717	C16H32O	hexadecanal	629-80-1	240.429	584.15	25	9.5596E+00	1.7207E-04	1,2

Table 10 Henry's Law Constant for Compound in Water - Organic Compounds

								Henry's Law Constant, H		
NO	ID	FORMULA	NAME	CAS No	MW, g/mol	T_B, K	T, C	H @ T, atm/mol frac	H @ T, atm/mol/m³	code
3281	29718	C16H32O	2-hexadecanone	18787-63-8	240.429	580.15	25	2.5779E+02	4.6401E-03	1,2
3282	29719	C16H32O	3-hexadecanone	18787-64-9	240.429	569.98	25	1.2212E+02	2.1981E-03	2
3283	29721	C16H32O2	hexadecanoic acid	57-10-3	256.429	624.15	20	7.9093E-02	1.4237E-06	1,2
3284	29722	C16H32O2	2-hexyldecanoic acid	25354-97-6	256.429	586.63	25	3.8304E-02	6.8947E-07	2
3285	29723	C16H32O2	pentadecyl formate	66271-76-9	256.429	623.00	25	4.9543E+00	8.9177E-05	2
3286	29724	C16H32O2	tetradecyl acetate	638-59-5	256.429	570.00	25	1.1924E+01	2.1464E-04	2
3287	29725	C16H32O2	tridecyl propanoate	66271-77-0	256.429	623.70	25	3.2235E+00	5.8022E-05	2
3288	29726	C16H32O2	dodecyl butanoate	3724-61-6	256.429	623.70	25	3.7911E+00	6.8239E-05	2
3289	29739	C16H33Br	1-bromohexadecane	112-82-3	305.342	609.15	25	1.4490E+02	2.6081E-03	1,2
3290	29740	C16H33Cl	1-chlorohexadecane	4860-03-1	260.891	599.75	25	2.0919E+02	3.7654E-03	1,2
3291	29754	C16H34	hexadecane	544-76-3	226.446	560.01	25	2.7031E+04	4.8655E-01	1,2
3292	29755	C16H34	2-methylpentadecane	1560-93-6	226.446	554.75	25	1.1994E+04	2.1590E-01	1,2
3293	29756	C16H34	3-methylpentadecane	2882-96-4	226.446	555.15	25	1.1697E+04	2.1055E-01	1,2
3294	29757	C16H34	2,2-dimethyltetradecane	59222-86-5	226.446	547.15	25	2.5062E+04	4.5112E-01	1,2
3295	29758	C16H34	2,3-dimethyltetradecane	18435-23-9	226.446	555.15	25	1.3440E+04	2.4192E-01	1,2
3296	29759	C16H34	2,4-dimethyltetradecane	61868-06-2	226.446	539.15	25	3.2688E+04	5.8838E-01	1,2
3297	29760	C16H34	2,2,4,4,6,8,8-heptamethylnonane	4390-04-9	226.446	519.45	25	1.4415E+05	2.5947E+00	1,2
3298	29766	C16H34O	1-hexadecanol	36653-82-4	242.445	585.15	25	3.3769E+00	6.0784E-05	1,2
3299	29767	C16H34O	2-hexadecanol	14852-31-4	242.445	587.15	25	6.4547E-02	1.1618E-06	1,2
3300	29768	C16H34O	3-hexadecanol	593-03-3	242.445	568.53	25	9.3789E+00	1.6882E-04	2
3301	29769	C16H34O	2-methyl-1-pentadecanol	25354-98-7	242.445	568.53	25	9.3789E+00	1.6882E-04	2
3302	29770	C16H34O	6-methyl-6-pentadecanol	108836-86-8	242.445	568.53	25	9.3789E+00	1.6882E-04	2
3303	29771	C16H34O	dioctyl ether	629-82-3	242.445	559.65	25	9.9896E+01	1.7981E-03	1,2
3304	29772	C16H34O	bis(2-ethylhexyl) ether	10143-60-9	242.445	542.15	25	1.1394E+02	2.0509E-03	1,2
3305	29773	C16H34O	2-hexyl-1-decanol	2425-77-6	242.445	568.53	25	9.3789E+00	1.6882E-04	2
3306	29789	C16H34S	1-hexadecanethiol	2917-26-2	258.512	607.16	25	3.0952E+01	5.5714E-04	1,2
3307	29790	C16H34S	2-hexadecanethiol	66271-53-2	258.512	600.15	25	8.4865E+00	1.5276E-04	1,2
3308	29791	C16H34S	dioctyl sulfide	2690-08-6	258.512	597.00	25	2.0770E+01	3.7385E-04	2
3309	29792	C16H34S	methyl pentadecyl sulfide	62155-10-6	258.512	597.00	25	2.0770E+01	3.7385E-04	2
3310	29793	C16H34S	ethyl tetradecyl sulfide	66271-54-3	258.512	597.00	25	2.0770E+01	3.7385E-04	2
3311	29794	C16H34S	propyl tridecyl sulfide	66271-55-4	258.512	597.00	25	2.0770E+01	3.7385E-04	2
3312	29795	C16H34S	butyl dodecyl sulfide	16900-08-6	258.512	597.00	25	2.0770E+01	3.7385E-04	2
3313	29803	C16H35N	dimethyltetradecylamine	112-75-4	241.461	575.15	25	5.0872E-01	9.1569E-06	1,2
3314	29804	C16H35N	diethyldodecylamine	4271-27-6	241.461	550.15	25	2.4297E-01	4.3734E-06	1,2
3315	30352	C17H36O	1-heptadecanol	1454-85-9	256.472	597.15	25	4.2744E+00	7.6939E-05	1,2
3316	30372	C17H37N	dimethylpentadecylamine	17678-60-3	255.488	590.15	25	3.2137E-01	5.7847E-06	1,2
3317	30373	C17H37N	diethyltridecylamine	66577-48-8	255.488	564.15	25	1.1855E-01	2.1339E-06	1,2
3318	30374	C17H37N	methyldioctylamine	4455-26-9	255.488	591.95	25	1.4231E+00	2.5615E-05	2
3319	30406	C18H12	chrysene	218-01-9	228.293	714.15	24	1.4525E-02	2.6144E-07	1,2
3320	30457	C18H14	m-terphenyl	92-06-8	230.309	650.00	25	1.1896E-01	2.1412E-06	1,2
3321	30458	C18H14	o-terphenyl	84-15-1	230.309	609.00	25	2.7579E+00	4.9641E-05	1,2
3322	30459	C18H14	p-terphenyl	92-94-4	230.309	649.15	25	3.2587E+00	5.8656E-05	1,2
3323	30566	C18H15O4P	triphenyl phosphate	115-86-6	326.289	686.65	24	4.8864E-02	8.7955E-07	1,2
3324	31097	C18H36O2	octadecanoic acid	57-11-4	284.483	648.35	20	4.9465E-05	8.9036E-10	1,2
3325	31131	C18H38	octadecane	593-45-3	254.500	589.86	25	1.4912E+03	2.6842E-02	1,2
3326	31140	C18H38O	1-octadecanol	112-92-5	270.499	608.15	34	7.7140E+01	1.3885E-03	1,2
3327	31166	C18H39N	dimethylhexadecylamine	112-69-6	269.515	603.15	25	6.1223E-02	1.1020E-06	1,2
3328	31167	C18H39N	diethyltetradecylamine	7307-58-6	269.515	577.15	25	1.3251E-01	2.3852E-06	1,2
3329	31168	C18H39N	trihexylamine	102-86-3	269.515	537.15	25	5.0431E-03	9.0775E-08	1,2
3330	32346	C20H30O2	abietic acid	514-10-3	302.457	649.70	20	1.7054E-12	3.0697E-17	1,2
3331	32515	C20H42	eicosane	112-95-8	282.553	616.93	25	1.9254E+02	3.4657E-03	1,2
3332	32787	C21H21O4P	tri-o-cresyl phosphate	78-30-8	368.369	683.15	24	3.6956E+00	6.6520E-05	1,2
3333	33925	C24H38O4	dioctyl phthalate	117-84-0	390.563	657.15	24	5.9814E-03	1.0766E-07	1,2
3334	34691	C28H46O4	diisodecyl phthalate	26761-40-0	446.671	723.00	24	1.0799E-01	1.9438E-06	1,2

Code: 1 - data, 2 - estimate
H - Henry's law constant for compound in water
T - temperature, C
MW - molecular weight of compound, g/mol
T_B - boiling point temperature of compound, K

Chapter 11

HENRY'S LAW CONSTANT FOR GASES IN WATER – ORGANIC COMPOUNDS

Carl L. Yaws and Prasad K. Narasimhan
Lamar University, Beaumont, Texas

Tabulation Results

The results for Henry's law constant of gases in water are given in Table 11. For the tabulation, the correlation for Henry's law constant as a function of temperature is based on the following equation:

$$\log_{10} H = A + B/T + C \log_{10} T + D T \qquad (11)$$

where H = Henry's law constant in water, atm/mol fraction
T = temperature, K
A, B, C, and D = regression coefficients

The tabulation is arranged by carbon number (C, C2, C3, ...) to provide ease of use in quickly locating the data by using the chemical formula. The compound name, CAS No (Chemical Abstracts Registry Number), and regression coefficients are provided in the adjacent columns. The range of application is denoted by minimum and maximum temperatures (TMIN and TMAX). Temperatures outside the range of application should not be used. The next column provides the code for the tabulation.

In preparing the tabulation, a literature search was conducted to identify data source publications (1-22). The excellent compilations of CRC (4); Gerrard (9-10); Solubility Data Series of IUPAC (13-15); and Wilhelm, Battino, and Wilcock (19) were primarily used for experimental data for water solubility.

A graph of Henry's law constant as a function of temperature is shown in Figure 11 for a representative compound.

Example

In an industrial operation, a body of water at 37.8 C is saturated with propylene (C3H6). Determine Henry's law constant.

Substitution of the regression coefficients from the table and temperature (37.8 C = 310.95 K) into the equation for Henry's law consant yields:

$$\log_{10} H = -2570.0227 + 6.86740E+04/310.95 + 1.03384E+03+01*\log_{10}(310.95) - 7.19694E-01*310.95$$

$$H = 10^{4.0929} = 12{,}384.5 \text{ atm/mol fraction}$$

References

1. Abraham, M. H., J. A. Haftvan, G. S. Whiting, and A. Leo, J. Chem. Soc. Perkin Trans. 2, 1777 (1994).
2. Azarnoosh, A. and J. J. McKetta, J. Chem. Eng. Data, 4 (3), 211 (1959).
3. Azarnoosh, A. and J. J. McKetta, Petrol. Refiner, 37 (11), 275 (1958).
4. CRC HANDBOOK OF CHEMISTRY AND PHYSICS, 75th - 86th eds., CRC Press, Inc., Boca Raton, FL (1994-2006).
5. Crovetto, R., "Evaluation of Solubility Data for the System CO2-H2O", J. Phys. Chem. Ref. Data, 20, 575 (1991).
6. Davis, J. E. and J. J. McKetta, J. Chem. Eng. Data, 5, 374 (1960).
7. Davis, J. E. and J. J. McKetta, Petrol. Refiner, 39 (3), 205 (1960).
8. Fogg, P. and W. Gerrard, SOLUBILITY OF GASES IN LIQUIDS, John Wiley & Sons, New York, NY (1991).
9. Gerrard, W., GAS SOLUBILITIES, Pergamon Press, New York, NY (1980).
10. Gerrard, W., SOLUBILITY OF GASES AND LIQUIDS, Plenum Press, New York, NY (1976).
11. Horvath, A. L., HALOGENATED HYDROCARBONS, Solubility-Miscibility with Water, Marcel Dekker, Inc., New York, NY (1982).
12. Hayduk, W., ed., IUPAC SOLUBILITY DATA SERIES, Hayduk, W., ed., Vol. 9, Ethane, Pergamon Press, New York, NY (1982).
13. IUPAC SOLUBILITY DATA SERIES, Hayduk, W., ed., Vol. 24, Propane, Butane, and 2-Methylpropane, Pergamon Press, New York, NY (1986).
14. IUPAC SOLUBILITY DATA SERIES, Clever, H. L. and C. L. Young, eds., Vol. 27/28, Methane, Pergamon Press, New York, NY (1987).
15. IUPAC SOLUBILITY DATA SERIES, Cargill, R. W., ed., Vol. 43, Carbon Monoxide, Pergamon Press, New

York, NY (1990).
16. McAuliffe, C., J. Phy. Chem., 70, 1267 (1966).
17. TECHNICAL DATA BOOK - PETROLEUM REFINING, Vols. I and II, American Petroleum Institute, Washington, DC (1972, 1977, 1983).
18. Thibodeaux, L. J., CHEMODYNAMICS, John Wiley & Sons, New York, NY (1979).
19. Wilhelm, E., R. Battino, and R. J. Wilcock, Chemical Reviews, Vol. 77, No. 2, 219 (1977).
20. Yaws, C. L., CHEMICAL PROPERTIES HANDBOOK, McGraw-Hill, New York, NY (1999).
21. Yaws, C. L., MATHESON GAS DATA BOOK, 7th ed., Matheson Tri-Gas (Parsippany, NJ), McGraw-Hill, New York, NY (2001).
22. Yaws, C. L., YAWS HANDBOOK OF THERMODYNAMIC AND PHYSICAL PROPERTIES OF CHEMICAL COMPOUNDS, electronic edition, www.knovel.com, knovel, Norwich, NY (2003).

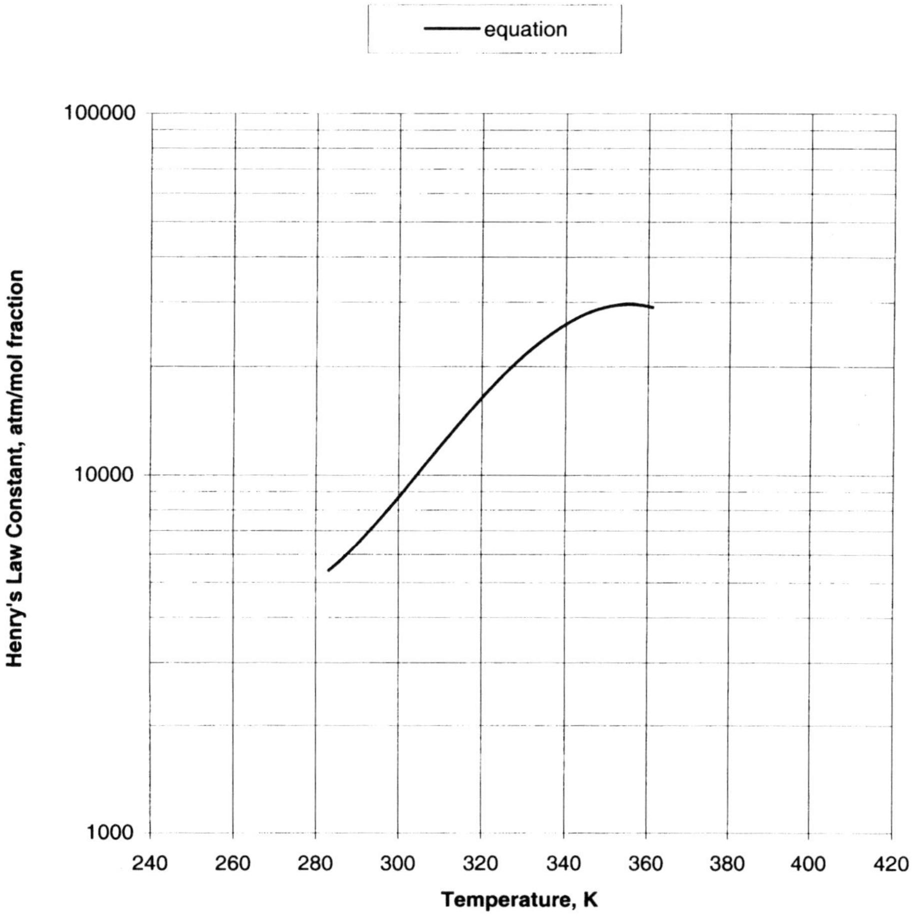

Figure 11 Henry's Law Constant for Propylene in Water

Table 11 Henry's Law Constant for Gases in Water - Organic Compounds

$\log_{10} H = A + B/T + C \log_{10}T + D\,T$ (H - atm/mol fraction, T - K)

NO	ID	FORMULA	NAME	CAS No	A	B	C	D	TMIN	TMAX	code	H@TMIN	H@T298	H@TMAX
1	17	CBrF3	bromotrifluoromethane	75-63-8	-1069.7640	2.90784E+04	4.29670E+02	-2.90325E-01	298	348	1,2	24919.6	24919.6	63664.4
2	34	CClF3	chlorotrifluoromethane	75-72-9	-4118.9177	1.09754E+05	1.65920E+03	-1.17383E+00	298	348	1,2	61999.8	61999.8	89041.2
3	42	CCl2F2	dichlorodifluoromethane	75-71-8	-7360.4488	2.01337E+05	2.94702E+03	-2.02153E+00	298	350	1,2	21548.0	21548.0	48950.2
4	47	CCl3F	trichlorofluoromethane	75-69-4	-1821.2555	4.73031E+04	7.37118E+02	-5.28555E-01	303	348	1,2	6704.9	---	15136.1
5	74	CF4	carbon tetrafluoride	75-73-0	1507.2737	-4.19725E+04	-5.99350E+02	4.09087E-01	273	323	1,2	125414.3	259777.3	374833.0
6	97	CHClF2	chlorodifluoromethane	75-45-6	-399.3369	7.53560E+03	1.70238E+02	-1.47443E-01	283	353	1,2	898.9	1657.9	4903.9
7	101	CHCl2F	dichlorofluoromethane	75-43-4	-4039.6476	1.09206E+05	1.62061E+03	-1.12114E+00	298	348	1,2	291.6	291.6	836.7
8	113	CHF3	fluoroform	75-46-7	861.4687	-2.29451E+04	-3.46185E+02	2.54308E-01	298	348	1,2	5242.7	5242.7	15073.0
9	145	CH2ClF	chlorofluoromethane	593-70-4	-227.2523	4.30705E+03	9.65677E+01	-7.91087E-02	283	353	1,2	221.3	361.3	1140.7
10	191	CH3Br	methyl bromide	74-83-9	470.4908	-1.51981E+04	-1.81666E+02	1.09130E-01	278	353	1,2	145.6	341.0	1308.7
11	196	CH3Cl	methyl chloride	74-87-3	206.3333	-7.74012E+03	-7.63810E+01	3.80554E-02	273	353	1,2	198.1	524.0	1736.3
12	214	CH3F	methyl fluoride	593-53-3	129.1531	-5.19490E+03	-4.63291E+01	1.97394E-02	273	353	1,2	446.7	946.6	2339.3
13	251	CH4	methane	74-82-8	146.8858	-5.76834E+03	-5.19144E+01	1.84936E-02	273	361	1,2	21634.2	39240.1	64820.5
14	372	CO	carbon monoxide	630-08-0	74.5962	-3.60330E+03	-2.33376E+01	0.00000E+00	273	353	1,2	35055.4	57978.5	85034.9
15	373	COS	carbonyl sulfide	463-58-1	96.0704	-5.22240E+03	-3.03658E+01	0.00000E+00	273	303	1,2	928.9	2607.5	3060.5
16	375	CO2	carbon dioxide	124-38-9	69.4237	-3.79646E+03	-2.16694E+01	4.78857E-04	273	353	1,2	725.0	1635.2	4263.7
17	427	C2ClF5	chloropentafluoroethane	76-15-3	-1765.8838	4.79217E+04	7.09115E+02	-4.84156E-01	298	348	1,2	142338.6	142338.6	405400.9
18	441	C2Cl2F4	1,2-dichloro-1,1,2,2-tetrafluor	76-14-2	-4750.8798	1.27752E+05	1.90952E+03	-1.33411E+00	298	348	1,2	66499.9	66499.9	141752.8
19	488	C2F4	tetrafluoroethylene	116-14-3	1062.9876	-3.16137E+04	-4.16423E+02	2.61631E-01	273	343	1,2	13870.0	35434.3	63896.6
20	571	C2H2	acetylene	74-86-2	67.9714	-3.54393E+03	-2.14030E+01	0.00000E+00	274	343	1,2	731.5	1332.6	2381.4
21	642	C2H2F2	1,1-difluoroethylene	75-38-7	1530.6900	-3.80908E+04	-6.24187E+02	4.89392E-01	293	303	1,2	19600.7	21560.9	23956.9
22	650	C2H2F4	1,1,1,2-tetrafluoroethane	811-97-2	1194.6622	-3.00480E+04	-4.85879E+02	3.76706E-01	293	320	1,2	7685.3	8306.0	12872.5
23	701	C2H3Cl	vinyl chloride	75-01-4	61.1868	-3.43853E+03	-1.90984E+01	2.51906E-03	273	323	1,2	568.9	1403.3	2719.5
24	702	C2H3ClF2	1-chloro-1,1-difluoroethane	75-68-3	11425.9117	-2.80274E+05	-4.68224E+03	3.70203E+00	286	294	1,2	2602.6	---	3979.6
25	787	C2H4	ethylene	74-85-1	107.7298	-6.33990E+03	-3.17169E+01	-1.32533E-02	283	361	1,2	6548.7	10784.4	18502.9
26	834	C2H4F2	1,1-difluoroethane	75-37-6	3959.7025	-9.68702E+04	-1.62333E+03	1.29182E+00	273	301	1,2	676.1	1321.6	1463.8
27	939	C2H5F	ethyl fluoride	353-36-6	24250.8144	-6.10318E+05	-9.88353E+03	7.56482E+00	287	298	1,2	661.8	1234.5	1234.5
28	985	C2H6	ethane	74-84-0	108.9263	-5.51363E+03	-3.47413E+01	0.00000E+00	275	323	1,2	13608.8	29402.4	48324.7
29	1284	C3F6	hexafluoropropylene	116-15-4	-55.7390	3.65553E+02	2.66503E+01	-2.28158E-02	273	343	1,2	19861.7	42561.1	117084.4
30	1299	C3F8	octafluoropropane	76-19-7	1494.0570	-3.61382E+04	-6.12187E+02	4.96092E-01	278	303	1,2	581900.5	873006.6	997557.6
31	1492	C3H4	methylacetylene	74-99-7	3681.1929	-9.94148E+04	-1.47741E+03	1.04279E+00	273	361	1,2	323.9	807.0	5888.8
32	1809	C3H6	cyclopropane	75-19-4	-141.9730	5.87466E+03	5.09015E+01	0.00000E+00	298	361	1,2	4820.7	4820.7	30453.0
33	1810	C3H6	propylene	115-07-1	-2570.0227	6.86740E+04	1.03384E+03	-7.19694E-01	283	361	1,2	5412.0	8182.3	29043.9
34	2020	C3H7F	1-fluoropropane	460-13-9	1594.2867	-3.86828E+04	-6.54404E+02	5.29246E-01	278	303	1,2	755.3	1142.7	1310.6
35	2021	C3H7F	2-fluoropropane	420-26-8	1491.1894	-3.61383E+04	-6.12189E+02	4.96093E-01	278	303	1,2	781.7	1172.7	1340.1
36	2094	C3H8	propane	74-98-6	2874.1130	-8.56732E+04	-1.12809E+03	7.01580E-01	283	361	1,2	12386.2	35420.1	90695.7
37	2336	C4F8	octafluorocyclobutane	115-25-3	4711.6007	-1.17163E+05	-1.92351E+03	1.49687E+00	273	311	1,2	122771.6	208361.3	346491.0
38	2510	C4H4	vinylacetylene	689-97-4	21.6379	-1.46228E+03	-5.48363E+00	0.00000E+00	273	333	1,2	839.7	1460.3	2602.4
39	2755	C4H6	ethylacetylene	107-00-6	213.3233	-1.10111E+04	-6.98370E+01	0.00000E+00	298	363	1,2	3843.8	3843.8	16347.9
40	2757	C4H6	butadiene (1,3 butadiene)	106-99-0	-37.5760	1.11123E+03	1.48356E+01	0.00000E+00	273	333	1,2	434.2	725.8	1528.9
41	3320	C4H8	1-butene	106-98-9	12.9593	-2.78226E+03	0.00000E+00	0.00000E+00	311	378	1,2	10410.5	---	399742.0
42	3324	C4H8	isobutene	115-11-7	103.9814	-5.54785E+03	-3.28909E+01	0.00000E+00	273	343	1,2	3430.2	9714.4	26279.4
43	3783	C4H10	butane	106-97-8	121.8305	-6.34244E+03	-3.87599E+01	0.00000E+00	273	349	1,2	15003.1	44563.3	125372.9
44	3784	C4H10	isobutane	75-28-5	161.2644	-7.94950E+03	-5.24651E+01	0.00000E+00	278	318	1,2	27870.9	60280.3	94825.0
45	5774	C5H12	neopentane	463-82-1	189.8679	-9.46836E+03	-6.18900E+01	0.00000E+00	288	353	1,2	60670.3	92918.6	230969.7

code: 1 - data, 2 - estimate

1. Most compounds in the tabulation are at very low concentration (range of about 0.0001 to 0.000001 mol fraction). At very low concentrations, Henry's law is applicable (p = Hx where x<<<1).

2. For those compounds which react with water, all chemical species of the gas and its reaction products are included (see CRC HANDBOOK OF CHEMISTRY AND PHYSICS, 84th ed., CRC Press (2004)).

Chapter 12

HENRY'S LAW CONSTANT FOR GASES IN WATER – INORGANIC COMPOUNDS

Carl L. Yaws
Lamar University, Beaumont, Texas

Tabulation Results

The results for Henry's law constant of gases in water are given in Table 12. For the tabulation, the correlation for Henry's law constant as a function of temperature is based on the following equation:

$$\log_{10} H = A + B/T + C \log_{10} T + D\,T \qquad (12)$$

where H = Henry's law constant in water, atm/mol fraction
T = temperature, K
A, B, C, and D = regression coefficients

The tabulation is arranged by alphabetical order (Ar, As, C, ...) to provide ease of use in quickly locating the data by using the chemical formula. The compound name, CAS No (Chemical Abstracts Registry Number), and regression coefficients are provided in the adjacent columns. The range of application is denoted by minimum and maximum temperatures (TMIN and TMAX). Temperatures outside the range of application should not be used. The next column provides the code for the tabulation.

In preparing the tabulation, a literature search was conducted to identify data source publications (1-23). The excellent compilations of CRC (1); Gerrard (3-5); Solubility Data Series of IUPAC (7-16); and Wilhelm, Battino, and Wilcock (20) were primarily used for experimental data for water solubility.

A graph of Henry's law constant as a function of temperature is shown in Figure 12 for a representative compound.

Example

In an industrial operation, a body of water at 35 is saturated with argon (Ar). Determine Henry's law constant.

Substitution of the regression coefficients from the table and temperature (35 C = 308.15 K) into the equation for Henry's law consant yields:

$$\log_{10} H = -1055.3705 + 2.8772E{+}04/308.15 + 4.2467E{+}02 * \log_{10}(308.15) - 3.0260E{-}01 * 308.15$$

$$H = 10^{4.6636} = 46{,}089.6 \text{ atm/mol fraction}$$

References

1. CRC HANDBOOK OF CHEMISTRY AND PHYSICS, 75th - 86th eds., CRC Press, Inc., Boca Raton, FL (1994-2006).
2. Crovetto, R., "Evaluation of Solubility Data for the System CO2-H2O", J. Phys. Chem. Ref. Data, 20, 575 (1991).
3. Fogg, P. and W. Gerrard, SOLUBILITY OF GASES IN LIQUIDS, John Wiley & Sons, New York, NY (1991).
4. Gerrard, W., GAS SOLUBILITIES, Pergamon Press, New York, NY (1980).
5. Gerrard, W., SOLUBILITY OF GASES AND LIQUIDS, Plenum Press, New York, NY (1976).
6. Horne, R. A., THE CHEMISTRY OF OUR ENVIRONMENT, John Wiley & Sons, New York, NY (1978).
7. IUPAC SOLUBILITY DATA SERIES, Clever, H. L., ed., Vol. 1, Helium and Neon, Pergamon Press, New York, NY (1979).
8. IUPAC SOLUBILITY DATA SERIES, Clever, H. L., ed., Vol. 2, Krypton, Xenon, and Radon, Pergamon Press, New York, NY (1979).
9. IUPAC SOLUBILITY DATA SERIES, Clever, H. L., ed., Vol. 4, Argon, Pergamon Press, New York, NY (1980).
10. IUPAC SOLUBILITY DATA SERIES, Young, C. L., ed., Vol. 5/6, Hydrogen and Deuterium, Pergamon Press, New York, NY (1981).
11. IUPAC SOLUBILITY DATA SERIES, Battino, R., ed., Vol. 7, Oxygen and Ozone, Pergamon Press, New York, NY (1981).
12. IUPAC SOLUBILITY DATA SERIES, Young, C. L., ed., Vol. 8, Oxides of Nitrogen, Pergamon Press, New York, NY (1981).
13. IUPAC SOLUBILITY DATA SERIES, Battino, R., ed., Vol. 10, Nitrogen and Air, Pergamon Press, New York, NY (1982).

14. IUPAC SOLUBILITY DATA SERIES, Young, C. L., ed., Vol. 12, Sulfur Dioxide, Chlorine, Fluorine, and Chlorine Oxides, Pergamon Press, New York, NY (1983).
15. IUPAC SOLUBILITY DATA SERIES, P. G. T. Fogg and C. L. Young, eds., Vol. 32, Hydrogen Sulfide, Deuterium Sulfide, and Hydrogen Selenide", Pergamon Press, New York, NY (1988).
16. IUPAC SOLUBILITY DATA SERIES, Cargill, R. W., ed., Vol. 43, Carbon Monoxide, Pergamon Press, New York, NY (1990).
17. Hutchinson, G. E., A TREATISE ON LIMNOLOGY, John Wiley & Sons, New York, NY (1957).
18. TECHNICAL DATA BOOK - PETROLEUM REFINING, Vols. I and II, American Petroleum Institute, Washington, DC (1972, 1977, 1983).
19. Thibodeaux, L. J., CHEMODYNAMICS, John Wiley & Sons, New York, NY (1979).
20. Wilhelm, E., R. Battino, and R. J. Wilcock, Chemical Reviews, Vol. 77, No. 2, 219 (1977).
21. Yaws, C. L., CHEMICAL PROPERTIES HANDBOOK, McGraw-Hill, New York, NY (1999).
22. Yaws, C. L., MATHESON GAS DATA BOOK, 7th ed., Matheson Tri-Gas (Parsippany, NJ), McGraw-Hill, New York, NY (2001).
23. Yaws, C. L., YAWS HANDBOOK OF THERMODYNAMIC AND PHYSICAL PROPERTIES OF CHEMICAL COMPOUNDS, electronic edition, www.knovel.com, knovel, Norwich, NY (2003).

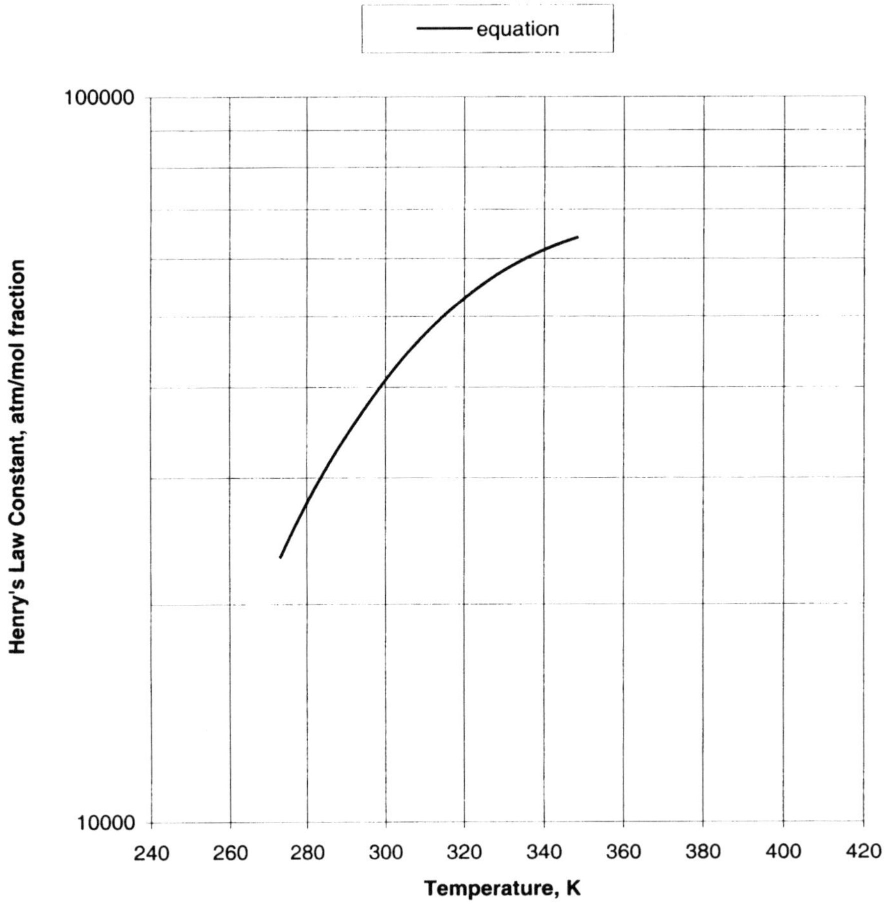

Figure 12 Henry's Law Constant for Argon in Water

Table 12 Henry's Law Constant for Gases in Water - Inorganic Compounds

$\log_{10} H = A + B/T + C \log_{10}T + D\,T$ (H - atm/mol fraction, T - K)

NO	ID	FORMULA	NAME	CAS No	A	B	C	D	TMIN	TMAX	code	H@TMIN	H@T298	H@TMAX
1	200	Ar	argon	7440-37-1	65.3235	-3.24690E+03	-2.01398E+01	0.00000E+00	273	348	1,2	23335.5	39696.1	64077.9
2	234	AsH3	arsine	7784-42-1	62.5415	-3.37389E+03	-1.91694E+01	0.00000E+00	273	299	1,2	3057.8	6194.9	6338.5
3	652	CO	carbon monoxide	630-08-0	74.5962	-3.60330E+03	-2.33376E+01	0.00000E+00	273	353	1,2	35055.4	57978.5	85034.9
4	653	COS	carbonyl sulfide	463-58-1	96.0704	-5.22240E+03	-3.03658E+01	0.00000E+00	273	303	1,2	928.9	2607.5	3060.5
5	655	CO2	carbon dioxide	124-38-9	69.4237	-3.79646E+03	-2.16694E+01	4.78857E-04	273	353	1,2	725.0	1635.2	4263.7
6	898	ClO2	chlorine dioxide	10049-04-4	-24.6413	-6.21813E+01	1.07454E+01	0.00000E+00	283	333	1,2	30.7	54.8	190.0
7	899	Cl2	chlorine	7782-50-5	232.4396	-8.21980E+03	-8.69997E+01	4.42155E-02	283	353	1,2	402.4	599.8	1279.3
8	1302	D2	deuterium	7782-39-0	1322.7872	-3.46668E+04	-5.32554E+02	3.89375E-01	283	303	1,2	59549.5	68580.5	71501.8
9	1709	H2	hydrogen	1333-74-0	54.6946	-2.40098E+03	-1.68893E+01	0.00000E+00	273	345	1,2	56957.4	70832.1	74664.1
10	1714	H2S	hydrogen sulfide	7783-06-4	10.8191	-1.51009E+03	-3.99300E-01	-6.81842E-03	273	353	1,2	285.4	541.2	1311.5
11	1719	H2Se	hydrogen selenide	7783-07-5	-3.9738	-4.23003E+02	3.54200E+00	-1.82404E-03	288	343	1,2	555.4	673.1	1406.9
12	1739	He	helium-4	7440-59-7	46.0252	-1.84993E+03	-1.40094E+01	0.00000E+00	273	348	1,2	131865.2	142950.4	126766.7
13	2162	Kr	krypton	7439-90-9	77.5359	-3.95280E+03	-2.42207E+01	0.00000E+00	273	353	1,2	11307.9	22161.3	42614.9
14	2613	NF3	nitrogen trifluoride	7783-54-2	110.0934	-5.37525E+03	-3.52496E+01	0.00000E+00	283	323	1,2	47539.0	69485.3	100924.6
15	2674	NO	nitric oxide	10102-43-9	30.2512	-2.42215E+03	-5.70490E+00	-1.19149E-02	273	353	1,2	16974.0	28739.1	44559.0
16	2678	N2	nitrogen	7727-37-9	78.8622	-3.74498E+03	-2.47981E+01	0.00000E+00	273	350	1,2	54172.3	87143.1	118634.5
17	2684	N2F4	tetrafluorohydrazine	10036-47-2	149.1454	-7.33702E+03	-4.83847E+01	0.00000E+00	288	318	1,2	47318.4	64889.0	98818.4
18	2747	N2O	nitrous oxide	10024-97-2	68.8882	-3.85775E+03	-2.12530E+01	0.00000E+00	273	313	1,2	964.3	2291.4	3363.7
19	3091	Ne	neon	7440-01-9	60.7869	-2.65134E+03	-1.89157E+01	0.00000E+00	273	348	1,2	98655.2	122626.8	123620.2
20	3198	O2	oxygen	7782-44-7	77.8881	-3.79901E+03	-2.44526E+01	0.00000E+00	273	348	1,2	25328.5	43630.9	66575.2
21	3389	PH3	phosphine	7803-51-2	67.5831	-3.57648E+03	-2.09165E+01	0.00000E+00	298	323	1,2	6776.6	6776.6	10656.0
22	3940	Rn	radon	10043-92-2	109.3341	-5.64696E+03	-3.50047E+01	0.00000E+00	273	323	1,2	2372.0	5987.3	10449.8
23	4001	SF6	sulfur hexafluoride	2551-62-4	191.8514	-9.19008E+03	-6.29116E+01	0.00000E+00	276	323	1,2	99017.3	227521.1	348071.1
24	4007	SO2	sulfur dioxide	7446-09-5	22.3423	-1.98711E+03	-5.68540E+00	0.00000E+00	283	323	1,2	24.2	40.7	84.4
25	4945	Xe	xenon	7440-63-3	87.3918	-4.56921E+03	-2.74664E+01	0.00000E+00	273	348	1,2	5555.9	12669.1	28466.9

code: 1 - data, 2 - estimate

1. Most compounds in the tabulation are at very low concentration (range of about 0.0001 to 0.000001 mol fraction). At very low concentrations, Henry's law is applicable (p = Hx where x<<<1).

2. A few compounds (ClO2 and SO2) are not at very low concentrations. The tabulated data for these compounds should be considered rough values.

3. For those compounds which react with water, all chemical species of the gas and its reaction products are included (see CRC HANDBOOK OF CHEMISTRY AND PHYSICS, 84th ed., CRC Press (2004)).

Chapter 13

OCTANOL-WATER PARTITION COEFFICIENT – ORGANIC COMPOUNDS

Carl L. Yaws
Lamar University, Beaumont, Texas

Tabulation Results

The results are given in Table 13. The tabulation is arranged by carbon number (C, C2, C3, ...) to provide ease of use in quickly locating the data by using the chemical formula. The compound name and CAS No (Chemical Abstracts Registry Number) are next provided. Values for octanol-water partition coefficient are given in the adjacent columns. The next column provides the code for the tabulation.

The octanol-water partition coefficient is the ratio of a compound's concentration in octanol and its concentration in water:

$$Kow = \text{concentration in octanol phase} / \text{concentration in aqueous phase} \qquad (13)$$

The concentration in the octanol phase and water phase should be in the same units (mg/kg). The octanol-water partition coefficient is not the same as the ratio of solubility of compound in octanol and water because the phases are not pure octanol and pure water. At equilibrium, the octanol phase contains 2.3 mol/liter of water, and the aqueous phase contains 0.0045 mol/liter of octanol (35).

The octanol-water partition coefficient is an important parameter in ascertaining the environmental fate of substances. It has special significance in indicating the tendency of a compound to partition between an organic phase (such as fish or organic content of soils) and an aqueous phase. Compounds with low Kow values tend to have high solubility in water and small bioconcentration factors for aquatic life. On the other hand, if a compound tends to partition into the organic phase, then Kow values will be high. For this case, the compound will partition into the fatty tissue of fish and will also bioaccumulate in animals that consume fish.

In preparing the tabulation, a literature search was conducted to identify data source publications (1-76). Both experimental values for the property under consideration and parameter values for estimation of the property are included in the source publications. The publications were screened and copies of appropriate data were made. These data were then keyed into the computer to provide a database of values for compounds for which experimental data are available. The database also served as a basis to check the accuracy of the estimation methods. Upon completion of data collection, estimation of the values for the remaining compounds was performed.

The excellent compilations by CRC (7); Hansch and Leo (22); Howard and Meylan (30); Lyman, Reehl, and Rosenblatt (35); Mackay, Shiu, and Ma (37-41); Montgomery (49); Sangster (58); Verschueren (71); and Yaws (75-76) were primarily used for experimental data. Most of the estimates were based on the group contribution method as developed by Meylan and Howard (1).

Example

Determine the octanol-water partition coefficient for p-dichlorobenzene ($C_6H_4Cl_2$) at ambient conditions.

Inspection of the tabulation for octanol-water partition coefficient yields:

$$\log_{10} Kow = 3.38$$

$$Kow = 10^{3.38} = 2400$$

References

1. Allen, D. T. and D. R. Shonnard, GREEN ENGINEERING, Prentice Hall PTR, Upper Saddle River, NJ (2002).
2. Carlson, R., R. Carlson and H. Kopperman, J. Chromatogr., 107, 210 (1975).
3. Catz, P. and D. R. Friend, Int. J. Pharm., 55(1), 17 (1989).
4. Chem. Inspect. Test Inst., Japan Chemical Industry Ecology – Toxicol. and Inform. Center, ISBN 4 - 89074-101-1 (1992).
5. Chiou, C. T., V. H. Freed, D. W. Schmedding, and R. L. Kohnert, Environ. Sci. Technol., 11, 475 (1977).
6. Chou, J. T. and P. C. Jurs, J. Chem. Inf. Comput. Sci., 19, 172 (1979).

7. CRC HANDBOOK OF CHEMISTRY AND PHYSICS, 84th - 87th eds., CRC Press, Inc., Boca Raton, FL (2004-2007).
8. Currie, D. J., C. E. Lough, R. F. Silver, and H. L. Holmes, Can. J. Chem., 44, 1035 (1966).
9. Debnath, A. K., G. Debnath, A. J. Shusterman, and C. Hansch, *Environ. Mol. Mutagen.*, 19(1), 37 (1992).
10. Debruijn, J. and others, Environ. Toxicol. Chem., 8, 499 (1989).
11. Deneer, J. W. and others, Aquatic Toxicol., 13(3), 195 (1988).
12. Dunn, W. J. and others, Quant Struct-Act. Relat., 2, 156 (1983).
13. Dyott T. M., A. J. Stuper, and G. S. Zander, J.Chem. Inf. Comput. Sci., 20, 28-35 (1980).
14. Fujita, T., J. Iwasa, and C. Hansch, J. Am. Chem. Soc., 86, 5175 (1964).
15. Funasaki , N., S. Hada, and S. Neya, J. Chromatogr., 361, 33 (1986).
16. Funasaki N., S. Hada, S. Neya, and K. Machida, J. Phys. Chem., 88(24), 5786 (1984).
17. Hall, L. H. and L. B. Kier, J. Pharm. Sci., 66, 642 (1977).
18. Hansch, C., J. E. Quinlan, and G. L. Lawrence, J. Org. Chem., 33, 347 (1968).
19. Hansch, C. and A. Leo, J. Org. Chem., 36, 1539 (1971).
20. Hansch, C., A. Leo, and D. Elkins, Chem. Rev., 71, 525 (1971).
21. Hansch, C. and others, J. Med. Chem., 16, 1207 (1973).
22. Hansch, C. and A. J. Leo, SUBSTITUENT CONSTANTS FOR CORRELATION ANALYSIS IN CHEMISTRY AND BIOLOGY, John Wiley and Sons, New York, NY (1979).
23. Hansch, C., A. J. Leo, Medchem Project, Claremont, CA, Pomona College. Issue # 26 (1985).
24. Hansch C., A. Leo, and A.J. Pomona, Pomona College Medicinal Project, Claremont, CA 91711, Log P Database (July, 1987).
25. Hansch, C. and others , Med. Chem., 34(2), 786 (1991).
26. Hanch, C., A. J. Leo, and D. Hoekman, ACS Professional Reference Book, "Exploring QSAR. Hydrophobic, Electronic and Steric Factors", American Chemical Society, Washington, D C (1995).
27. Holmes, H. L., STRUCTURE-ACTIVITY RELATIONSHIPS FOR SOME CONJUGATED HETEROENOID COMPOUNDS, CATECHOL MONOETHERS AND MORPHINE ALKALOIDS, Vols. 1 and 2, Defence Research Establishment, Suffield, Ralston, Alberta, Canada (1975).
28. Holmes, H. L. and C. E. Lough, (NTIS AD A030683), Suffield Technical Note No. 365, Defence Research Establishment, Suffield, Ralston, Alberta (July, 1976).
29. Hopfinger. A. J. and R. D. Battershell, J. Med. Chem., 19, 569 (1976).
30. Howard, P. H. and W. M. Meylan, eds., HANDBOOK OF PHYSICAL PROPERTIES OF ORGANIC CHEMICALS, CRC Press, Boca Raton, FL (1997).
31. Jaber, H. M. and others, EPA-600/6-84/009, Menlo Park, CA (1984).
32. Karickhoff, S. W. and D. S. Brown, Report No. EPA-600/4-79-032, U.S. Environmental Protection Agency, Athens, GA (1979).
33. Karickhoff, S. W., D. S. Brown, and T. A. Scott, Water Res., 13, 241 (1979).
34. Kenaga, E. E. and C. A. I. Goring, "RELATIONSHIP BETWEEN WATER SOLUBILITY, SOIL SORPTION, OCTANOL-WATER PARTITIONING, AND BIOCONCENTRATION OF CHEMICALS IN BIOTA", American Society for Testing and Materials, Third Aquatic toxicology Symposium, New Orleans, LA (October 17-18, 1978).
35. Lyman, W. J., W. F. Reehl, and D. A. Rosenblatt, HANDBOOK OF CHEMICAL PROPERTY ESTIMATION METHODS, American Chemical Society, Washington, DC (1990).
36. Mackay, D. and W. Y. Shiu, J. Phys. Chem. Ref. Data, 15, 911 (1986).
37. Mackay, D., W. Y. Shiu, and K. C. Ma, ILLUSTRATED HANDBOOK OF PHYSICAL-CHEMICAL PROPERTIES AND ENVIRONMENTAL FATE FOR ORGANIC CHEMICALS, Volume 1 – Monoaromatic Hydrocarbons, Chlorobenzenes and PCBs, Lewis Publishers, New York, NY (1992).
38. Mackay, D., W. Y. Shiu, and K. C. Ma, ILLUSTRATED HANDBOOK OF PHYSICAL-CHEMICAL PROPERTIES AND ENVIRONMENTAL FATE FOR ORGANIC CHEMICALS, Volume 2 – Polynuclear Aromatic Hydrocarbons, Polychlorinated Dioxins, and Dibenzofurans, Lewis Publishers, New York, NY (1992).
39. Mackay, D., W. Y. Shiu, and K. C. Ma, ILLUSTRATED HANDBOOK OF PHYSICAL-CHEMICAL PROPERTIES AND ENVIRONMENTAL FATE FOR ORGANIC CHEMICALS, Volume 3 – Volatile Organic Compounds, Lewis Publishers, New York, NY (1993).
40. Mackay, D., W. Y. Shiu, and K. C. Ma, ILLUSTRATED HANDBOOK OF PHYSICAL-CHEMICAL PROPERTIES AND ENVIRONMENTAL FATE FOR ORGANIC CHEMICALS, Volume 4 – Oxygen, Nitrogen, and Sulfur Containing Compounds, Lewis Publishers, New York, NY (1995).
41. Mackay, D., W. Y. Shiu, and K. C. Ma, ILLUSTRATED HANDBOOK OF PHYSICAL-CHEMICAL PROPERTIES AND ENVIRONMENTAL FATE FOR ORGANIC CHEMICALS, Volume 5 – Pesticide Chemicals, Lewis Publishers, New York, NY (1997).
42. Martiska, A. and V. Bekarek, ACTA Univ. Palacki Olomuc., Fac. Rerum Nat. 97 (Chem. 29) 63-7 (1990).
43. Meylan, W.M. and P. H. Howard, Environ. Toxicol. Chem., 10, 1283 (1991).
44. Meylan, W. M. and P. H. Howard, J. Pham. Sci. 84, 83 (1995).
45. Meylan, W.M. and P. H. Howard, J. Pharm. Sci., 84, 83 (1995).

46. Meylan, W.M. and others, Environ. Toxicol. Chem., 15, 100 (1996).
47. Miller, M. M. and others, J. Chem. Eng. Data, 29, 184 (1984).
48. Miller, M. M., Wasik, S. P., Huang, G., Shiu, W., and Mackay, D., Environ. Sci. Technol., 19, 522 (1985).
49. Montgomery, J. H., GROUNDWATER CHEMICALS, Lewis Publishers, New York, NY (2000).
50. Nakagawa, Y., K. Izumi, N. Oikawa, T. Sotomatsu, M. Shigemura, and T. Fujita, Environ. Toxicol. Chem., 11, 901 (1992).
51. Neely, W. B. and G. E. Blau, ENVIRONMENTAL EXPOSURE FROM CHEMICALS, Vol. 1, CRC Press, Baca Raton, FL (1985).
52. Pinsuwan, S., L. Li, and S. H. Yalkowsky, J. Chem. Eng. Data, 40, 623 (1995).
53. Rao, P. S. C. and J.M. Davidson, ENVIRONMENTAL IMPACT OF NONPOINT SOURCE POLLUTION, M R. Overcash and J. M. Davidson, eds., Ann Arbor Science Publishers, Inc., Ann Arbor, MI (1980).
54. Rekker, R. F. and G. G. Nys, Chim. Ther., 8, 521 (1973).
55. Rekker, R. F. and G. G. Nys, Eur. J. Med. Chem. – Chim. Ther., 9, 361 (1974).
56. Rekker, R. F., THE HYDROPHOBIC FRAGMENT CONSTANT, Elsevier Scientific Publishing Co., New York, NY (1977).
57. Saeger, V. W., O. Hicks, R. G. Kaley, P. R. Michael, J. P. Mieure and E. S. Tucker, Environ. Sci. Technol., 13, 840 (1979).
58. Sangster, J., J. Phys. Chem. Ref. Data, 18, 1111-1229 (1989).
59. Sangster, J., LogKow Databank, Sangster Research Laboratories, Montreal, Quebec, Canada, (1994).
60. Sasaki, H. and others, J. Med. Chem., 34(2), 628 (1991).
61. Sasaki, H., and others, Int. J. Pharm. 44, 15-24 (1988).
62. Seiler, P., Eur. J. Med. Chem. – Chim. Ther., 9, 473 (1974).
63. Suzuki, T., J. Computer Aided Molecular Design, 5, 149 (1991).
64. Tanii, H. and K. Hashimoto, Arch. Toxicol., 55(1), 47 (1984).
65. Tanii, H. and K. Hashimoto, Arch. Toxicol., 57(2), 88 (1985).
66. Tanii, H., H. Tsuji, and K. Hashimoto, Toxicol. Lett., 30(1), 13 (1986).
67. Unger, S. H., J. Cook, and J. Hollenberg, J. Pharm. Sci., 67, 1374 (1978).
68. Valvani, S. C. and others, J. Pharm. Sci., 70, 502 (1981).
69. Veith, G. D. and R. T. Morris, Report EPA-600/3-78-049, U.S. Environmental Protection Agency, Duluth, MN (1978).
70. Veith, G. D., N. M. Austin, and R. T. Morris, Water Res., 13, 43 (1979).
71. Verschueren, K., HANDBOOK OF ENVIRONMENTAL DATA ON ORGANIC CHEMICALS, 3rd and 4th eds., Van Nostrand Reinhold, New York, NY (1996, 2001).
72. Yalkowsky, S. H. and S. C.Valvani, J. Chem. Eng. Data, 24, 127 (1979).
73. Yalkowsky, S. H., R. J. Orr, and S. C. Valvani, Ind. Eng. Chem. Fundam., 18, 351 (1979).
74. Yamana, T., J. Pharm. Sci., 66, 747 (1977).
75. Yaws, C. L., CHEMICAL PROPERTIES HANDBOOK, McGraw-Hill, New York, NY (1999).
76. Yaws, C. L., YAWS HANDBOOK OF THERMODYNAMIC AND PHYSICAL PROPERTIES FOR CHEMICALS, electronic edition, www.knovel.com, knovel, Norvich, NY (2005).

Table 13 Octanol-Water Partition Coefficient - Oganic Compounds

Octanol-Water Partition Coefficient								Octanol-Water Partition Coefficient							
NO	ID	FORMULA	NAME	CAS No	$\log_{10}$ Kow	Kow	code	NO	ID	FORMULA	NAME	CAS No	$\log_{10}$ Kow	Kow	code
1	13	CBrClF2	bromochlorodifluoromethane	353-59-3	1.90	7.94E+01	2	81	635	C2H2Cl4	1,1,1,2-tetrachloroethane	630-20-6	2.93	8.51E+02	2
2	16	CBrCl3	bromotrichloromethane	75-62-7	2.53	3.39E+02	2	82	636	C2H2Cl4	1,1,2,2-tetrachloroethane	79-34-5	2.39	2.45E+02	1
3	17	CBrF3	bromotrifluoromethane	75-63-8	1.86	7.24E+01	1	83	642	C2H2F2	1,1-difluoroethylene	75-38-7	1.24	1.74E+01	1
4	23	CBr2F2	dibromodifluoromethane	75-61-6	1.99	9.77E+01	2	84	643	C2H2F2	cis-1,2-difluoroethene	1630-77-9	1.24	1.74E+01	2
5	29	CBr4	carbon tetrabromide	558-13-4	3.42	2.63E+03	1	85	644	C2H2F2	trans-1,2-difluoroethene	1630-78-0	1.24	1.74E+01	2
6	34	CClF3	chlorotrifluoromethane	75-72-9	1.65	4.47E+01	1	86	648	C2H2F3NO	trifluoroacetamide	354-38-1	0.12	1.32E+00	1
7	39	CClN	cyanogen chloride	506-77-4	0.79	6.13E+00	2	87	650	C2H2F4	1,1,1,2-tetrafluoroethane	811-97-2	1.68	4.79E+01	2
8	42	CCl2F2	dichlorodifluoromethane	75-71-8	2.16	1.45E+02	1	88	651	C2H2F4	1,1,2,2-tetrafluoroethane	359-35-3	1.68	4.79E+01	2
9	44	CCl2O	phosgene	75-44-5	-0.71	1.95E-01	2	89	681	C2H2O	ketene	463-51-4	-0.52	3.02E-01	2
10	47	CCl3F	trichlorofluoromethane	75-69-4	2.53	3.39E+02	1	90	684	C2H2O4	oxalic acid	144-62-7	-2.22	6.03E-03	2
11	50	CCl4	carbon tetrachloride	56-23-5	2.64	4.37E+02	1	91	688	C2H3Br	vinyl bromide	593-60-2	1.57	3.72E+01	1
12	68	CF2O	carbonyl fluoride	353-50-4	-1.34	4.57E-02	2	92	693	C2H3BrO2	bromoacetic acid	79-08-3	0.41	2.57E+00	1
13	74	CF4	carbon tetrafluoride	75-73-0	1.18	1.51E+01	1	93	701	C2H3Cl	vinyl chloride	75-01-4	1.38	2.40E+01	1
14	96	CHBr3	bromoform	75-25-2	2.38	2.40E+02	1	94	702	C2H3ClF2	1-chloro-1,1-difluoroethane	75-68-3	2.05	1.12E+02	2
15	97	CHClF2	chlorodifluoromethane	75-45-6	1.08	1.20E+01	1	95	707	C2H3ClO	acetyl chloride	75-36-5	-0.47	3.39E-01	2
16	101	CHCl2F	dichlorofluoromethane	75-43-4	1.55	3.55E+01	1	96	708	C2H3ClO	chloroacetaldehyde	107-20-0	0.09	1.23E+00	2
17	104	CHCl3	chloroform	67-66-3	1.97	9.33E+01	1	97	710	C2H3ClO2	chloroacetic acid	79-11-8	0.22	1.66E+00	1
18	113	CHF3	fluoroform	75-46-7	0.64	4.37E+00	1	98	711	C2H3ClO2	methyl chloroformate	79-22-1	0.14	1.38E+00	2
19	117	CHI3	iodoform	75-47-8	3.03	1.07E+03	2	99	720	C2H3Cl3	1,1,1-trichloroethane	71-55-6	2.49	3.09E+02	1
20	140	CH2BrCl	bromochloromethane	74-97-5	1.41	2.57E+01	1	100	721	C2H3Cl3	1,1,2-trichloroethane	79-00-5	2.38	2.40E+02	1
21	144	CH2Br2	dibromomethane	74-95-3	2.30	2.00E+02	1	101	723	C2H3Cl3O	2,2,2-trichloroethanol	115-20-8	1.35	2.24E+01	1
22	145	CH2ClF	chlorofluoromethane	593-70-4	0.51	3.24E+00	1	102	728	C2H3F	vinyl fluoride	75-02-5	1.19	1.55E+01	2
23	149	CH2Cl2	dichloromethane	75-09-2	1.25	1.78E+01	1	103	735	C2H3F3	1,1,1-trifluoroethane	420-46-2	1.74	5.50E+01	2
24	158	CH2F2	difluoromethane	75-10-5	0.20	1.58E+00	1	104	736	C2H3F3	1,1,2-trifluoroethane	430-66-0	1.74	5.50E+01	2
25	161	CH2I2	diiodomethane	75-11-6	2.50	3.16E+02	1	105	738	C2H3F3O	2,2,2-trifluoroethanol	75-89-8	0.41	2.57E+00	1
26	175	CH2O	formaldehyde	50-00-0	0.35	2.24E+00	1	106	753	C2H3N	acetonitrile	75-05-8	-0.34	4.57E-01	1
27	176	CH2O2	formic acid	64-18-6	-0.54	2.88E-01	1	107	755	C2H3NO	methyl isocyanate	624-83-9	0.79	6.17E+00	2
28	191	CH3Br	methyl bromide	74-83-9	1.19	1.55E+01	1	108	787	C2H4	ethylene	74-85-1	1.13	1.35E+01	1
29	196	CH3Cl	methyl chloride	74-87-3	0.91	8.13E+00	1	109	798	C2H4Br2	1,1-dibromoethane	557-91-5	1.94	8.71E+01	2
30	210	CH3Cl3Si	methyl trichlorosilane	75-79-6	2.01	1.02E+02	2	110	799	C2H4Br2	1,2-dibromoethane	106-93-4	1.96	9.12E+01	1
31	214	CH3F	methyl fluoride	593-53-3	0.51	3.24E+00	1	111	812	C2H4Cl2	1,1-dichloroethane	75-34-3	1.79	6.17E+01	1
32	222	CH3I	methyl iodide	74-88-4	1.50	3.16E+01	1	112	813	C2H4Cl2	1,2-dichloroethane	107-06-2	1.48	3.02E+01	1
33	232	CH3NO	formamide	75-12-7	-1.51	3.09E-02	1	113	814	C2H4Cl2	dichloroethane	1300-21-6	1.64	4.32E+01	2
34	235	CH3NO2	nitromethane	75-52-5	-0.33	4.68E-01	1	114	816	C2H4Cl2O	bis(chloromethyl) ether	542-88-1	0.57	3.72E+00	2
35	236	CH3NO2	methyl nitrite	624-91-9	0.88	7.59E+00	2	115	834	C2H4F2	1,1-difluoroethane	75-37-6	0.75	5.62E+00	1
36	238	CH3NO3	methyl nitrate	598-58-3	0.76	5.75E+00	2	116	835	C2H4F2	1,2-difluoroethane	624-72-6	1.21	1.62E+01	2
37	251	CH4	methane	74-82-8	1.09	1.23E+01	1	117	844	C2H4I2	1,1-diiodoethane	594-02-5	2.71	5.13E+02	2
38	254	CH4Cl2Si	methyl dichlorosilane	75-54-7	1.70	5.01E+01	2	118	845	C2H4I2	1,2-diiodoethane	624-73-7	2.71	5.13E+02	1
39	272	CH4O	methyl alcohol	67-56-1	-0.74	1.82E-01	1	119	884	C2H4O	acetaldehyde	75-07-0	0.45	2.82E+00	1
40	275	CH4O3S	methanesulfonic acid	75-75-2	-2.38	4.17E-03	2	120	885	C2H4O	ethylene oxide	75-21-8	-0.30	5.01E-01	1
41	278	CH4S	methyl mercaptan	74-93-1	0.78	6.03E+00	2	121	889	C2H4O2	acetic acid	64-19-7	-0.17	6.76E-01	1
42	286	CH5N	methylamine	74-89-5	-0.57	2.69E-01	1	122	890	C2H4O2	methyl formate	107-31-3	0.03	1.07E+00	1
43	363	CN4O8	tetranitromethane	509-14-8	-2.05	8.91E-03	2	123	910	C2H5Br	bromoethane	74-96-4	1.60	3.98E+01	1
44	372	CO	carbon monoxide	630-08-0	3.42	2.60E+03	2	124	914	C2H5Cl	ethyl chloride	75-00-3	1.43	2.69E+01	1
45	373	COS	carbonyl sulfide	463-58-1	-1.33	4.68E-02	2	125	917	C2H5ClO	2-chloroethanol	107-07-3	0.03	1.07E+00	1
46	375	CO2	carbon dioxide	124-38-9	-1.33	4.68E-02	2	126	939	C2H5F	ethyl fluoride	353-36-6	1.26	1.82E+01	2
47	382	CS2	carbon disulfide	75-15-0	2.14	1.38E+02	1	127	948	C2H5I	ethyl iodide	75-03-6	2.00	1.00E+02	1
48	404	C2BrF3	bromotrifluoroethylene	598-73-2	1.55	3.55E+01	2	128	954	C2H5N	ethyleneimine	151-56-4	-0.28	5.25E-01	2
49	413	C2Br2F4	1,2-dibromotetrafluoroethane	124-73-2	2.96	9.12E+02	2	129	956	C2H5NO	acetamide	60-35-5	-1.26	5.50E-02	1
50	423	C2ClF3	chlorotrifluoroethylene	79-38-9	1.65	4.47E+01	2	130	957	C2H5NO	N-methylformamide	123-39-7	-0.97	1.07E-01	1
51	427	C2ClF5	chloropentafluoroethane	76-15-3	2.47	2.95E+02	2	131	958	C2H5NO2	nitroethane	79-24-3	0.18	1.51E+00	1
52	441	C2Cl2F4	1,2-dichloro-1,1,2,2-tetrafluor	76-14-2	2.82	6.61E+02	1	132	964	C2H5NO3	ethyl nitrate	625-58-1	1.25	1.78E+01	2
53	442	C2Cl2F4	1,1-dichloro-1,2,2,2-tetrafluor	374-07-2	2.82	6.61E+02	2	133	985	C2H6	ethane	74-84-0	1.81	6.46E+01	1
54	443	C2Cl2F4	dichlorotetrafluoroethane	1320-37-2	2.82	6.61E+02	2	134	1065	C2H6O	ethyl alcohol	64-17-5	-0.30	5.01E-01	1
55	449	C2Cl3F3	1,1,2-trichloro-1,2,2-trifluoroe	76-13-1	3.16	1.45E+03	1	135	1066	C2H6O	dimethyl ether	115-10-6	0.10	1.26E+00	1
56	455	C2Cl4	tetrachloroethylene	127-18-4	2.88	7.59E+02	1	136	1067	C2H6OS	dimethyl sulfoxide	67-68-5	-1.35	4.47E-02	1
57	456	C2Cl4F2	1,1,2,2-tetrachloro-1,2-difluor	76-12-0	3.41	2.57E+03	2	137	1072	C2H6O2	ethylene glycol	107-21-1	-1.36	4.37E-02	1
58	459	C2Cl4O	trichloroacetyl chloride	76-02-8	0.88	7.59E+00	2	138	1075	C2H6O2S	dimethyl sulfone	67-71-0	-1.41	3.89E-02	1
59	463	C2Cl6	hexachloroethane	67-72-1	4.00	1.00E+04	1	139	1083	C2H6O4S	dimethyl sulfate	77-78-1	1.16	1.45E+01	1
60	488	C2F4	tetrafluoroethylene	116-14-3	1.21	1.62E+01	2	140	1084	C2H6O4S	ethyl sulfate	540-82-9	1.16	1.45E+01	2
61	494	C2F6	hexafluoroethane	76-16-4	2.00	1.00E+02	1	141	1092	C2H6S	ethyl mercaptan	75-08-1	1.27	1.86E+01	2
62	511	C2HBrClF3	halothane	151-67-7	2.30	2.00E+02	1	142	1093	C2H6S	dimethyl sulfide	75-18-3	0.92	8.32E+00	2
63	525	C2HClF2	2-chloro-1,1-difluoroethylene	359-10-4	1.60	3.98E+01	2	143	1094	C2H6S2	dimethyl disulfide	624-92-0	1.77	5.89E+01	1
64	539	C2HCl3	trichloroethylene	79-01-6	2.53	3.39E+02	1	144	1115	C2H7N	ethylamine	75-04-7	-0.13	7.41E-01	1
65	543	C2HCl3O	dichloroacetyl chloride	79-36-7	-0.04	9.12E-01	2	145	1116	C2H7N	dimethylamine	124-40-3	-0.38	4.17E-01	1
66	544	C2HCl3O	trichloroacetaldehyde	75-87-6	0.99	9.77E+00	1	146	1117	C2H7NO	monoethanolamine	141-43-5	-1.31	4.90E-02	1
67	551	C2HCl5	pentachloroethane	76-01-7	2.89	7.76E+02	1	147	1118	C2H7NO	1-aminoethanol	75-39-8	-1.31	4.90E-02	2
68	556	C2HF3O2	trifluoroacetic acid	76-05-1	0.50	3.16E+00	2	148	1159	C2H8N2	ethylenediamine	107-15-3	-2.04	9.12E-03	1
69	558	C2HF5	pentafluoroethane	354-33-6	1.55	3.55E+01	2	149	1210	C2N2	cyanogen	460-19-5	0.07	1.17E+00	1
70	571	C2H2	acetylene	74-86-2	0.37	2.34E+00	1	150	1284	C3F6	hexafluoropropylene	116-15-4	2.12	1.32E+02	2
71	592	C2H2Br4	1,1,2,2-tetrabromoethane	79-27-6	2.55	3.55E+02	2	151	1287	C3F6O	hexafluoroacetone	684-16-2	1.46	2.88E+01	1
72	593	C2H2Br4	1,1,1,2-tetrabromoethane	630-16-0	2.55	3.55E+02	2	152	1299	C3F8	octafluoropropane	76-19-7	3.12	1.32E+03	2
73	615	C2H2Cl2	1,1-dichloroethylene	75-35-4	2.13	1.35E+02	1	153	1377	C3H2N2	malononitrile	109-77-3	-0.60	2.51E-01	1
74	616	C2H2Cl2	cis-1,2-dichloroethylene	156-59-2	1.86	7.24E+01	1	154	1462	C3H3N	acrylonitrile	107-13-1	0.25	1.78E+00	1
75	617	C2H2Cl2	trans-1,2-dichloroethylene	156-60-5	1.93	8.51E+01	1	155	1463	C3H3NO	oxazole	288-42-6	0.12	1.32E+00	1
76	618	C2H2Cl2	dichloroethylene	25323-30-2	2.03	1.06E+02	2	156	1473	C3H3NS	thiazole	288-47-1	0.44	2.75E+00	1
77	619	C2H2Cl2	1,2-dichloroethylene, cis and	540-59-0	2.03	1.06E+02	2	157	1492	C3H4	methylacetylene	74-99-7	0.94	8.71E+00	1
78	625	C2H2Cl2O	chloroacetyl chloride	79-04-9	-0.22	6.03E-01	2	158	1493	C3H4	allene	463-49-0	1.45	2.82E+01	1
79	627	C2H2Cl2O2	dichloroacetic acid	79-43-6	0.92	8.32E+00	1	159	1533	C3H4Cl2	2,3-dichloropropene	78-88-6	2.03	1.07E+02	2
80	633	C2H2Cl3NO	2,2,2-trichloroacetamide	594-65-0	1.04	1.10E+01	1	160	1534	C3H4Cl2	1,1-dichloropropene	563-58-6	2.03	1.07E+02	2

Table 13 Octanol-Water Partition Coefficient - Oganic Compounds

Octanol-Water Partition Coefficient								
NO	ID	FORMULA	NAME	CAS No	log_{10} Kow	Kow	code
161	1535	C3H4Cl2	cis-1,2-dichloropropene	6923-20-2	2.03	1.07E+02	1
162	1536	C3H4Cl2	trans-1,2-dichloropropene	7069-38-7	2.03	1.07E+02	2
163	1537	C3H4Cl2	cis-1,3-dichloropropene	10061-01-5	2.03	1.07E+02	2
164	1538	C3H4Cl2	trans-1,3-dichloropropene	10061-02-6	2.03	1.07E+02	2
165	1539	C3H4Cl2	3,3-dichloropropene	563-57-5	2.03	1.07E+02	2
166	1540	C3H4Cl2	1,3-dichloropropene; (cis+tra	542-75-6	2.03	1.07E+02	2
167	1541	C3H4Cl2	1,2-dichloropropene	563-54-2	2.03	1.07E+02	2
168	1542	C3H4Cl2	dichloropropylene	26952-23-8	2.03	1.07E+02	2
169	1592	C3H4N2	imidazole	288-32-4	-0.08	8.32E-01	1
170	1594	C3H4N2	1H-pyrazole	288-13-1	0.13	1.35E+00	1
171	1621	C3H4O	acrolein	107-02-8	-0.01	9.77E-01	1
172	1622	C3H4O	propargyl alcohol	107-19-7	-0.38	4.17E-01	1
173	1626	C3H4O2	acrylic acid	79-10-7	0.35	2.24E+00	1
174	1627	C3H4O2	beta-propiolactone	57-57-8	-0.80	1.58E-01	2
175	1628	C3H4O2	vinyl formate	692-45-5	0.18	1.51E+00	2
176	1633	C3H4O3	ethylene carbonate	96-49-1	0.12	1.32E+00	2
177	1634	C3H4O3	pyruvic acid	127-17-3	-1.24	5.75E-02	2
178	1642	C3H5Br	cis-1-bromo-1-propene	590-13-6	1.79	6.17E+01	2
179	1643	C3H5Br	trans-1-bromo-1-propene	590-15-8	1.79	6.17E+01	2
180	1644	C3H5Br	2-bromo-1-propene	557-93-7	1.79	6.17E+01	2
181	1645	C3H5Br	3-bromo-1-propene	106-95-6	1.79	6.17E+01	1
182	1647	C3H5Br	1-bromo-1-propene; (cis+tran	590-14-7	1.79	6.17E+01	2
183	1668	C3H5Cl	cis-1-chloro-1-propene	16136-84-8	1.97	9.23E+01	2
184	1669	C3H5Cl	trans-1-chloro-1-propene	16136-85-9	1.97	9.23E+01	2
185	1670	C3H5Cl	2-chloro-1-propene	557-98-2	2.00	1.00E+02	1
186	1671	C3H5Cl	3-chloro-1-propene	107-05-1	1.93	8.51E+01	2
187	1672	C3H5Cl	1-chloro-1-propene	590-21-6	1.97	9.23E+01	2
188	1676	C3H5ClO	alpha-epichlorohydrin	106-89-8	0.30	2.00E+00	1
189	1688	C3H5ClO2	methyl chloroacetate	96-34-4	0.63	4.27E+00	2
190	1689	C3H5ClO2	ethyl chloroformate	541-41-3	0.63	4.27E+00	2
191	1703	C3H5Cl3	1,1,1-trichloropropane	7789-89-1	2.27	1.86E+02	2
192	1704	C3H5Cl3	1,1,2-trichloropropane	598-77-6	2.27	1.86E+02	2
193	1705	C3H5Cl3	1,1,3-trichloropropane	20395-25-9	2.27	1.86E+02	2
194	1706	C3H5Cl3	1,2,2-trichloropropane	3175-23-3	2.27	1.86E+02	2
195	1707	C3H5Cl3	1,2,3-trichloropropane	96-18-4	2.27	1.86E+02	1
196	1752	C3H5N	propionitrile	107-12-0	0.16	1.45E+00	1
197	1755	C3H5NO	acrylamide	79-06-1	-0.78	1.66E-01	1
198	1756	C3H5NO	hydracrylonitrile	109-78-4	-1.12	7.59E-02	2
199	1757	C3H5NO	lactonitrile	78-97-7	-0.94	1.15E-01	1
200	1796	C3H5N3O9	nitroglycerine	55-63-0	1.62	4.17E+01	1
201	1809	C3H6	cyclopropane	75-19-4	1.72	5.25E+01	1
202	1810	C3H6	propylene	115-07-1	1.77	5.89E+01	1
203	1824	C3H6Br2	1,1-dibromopropane	598-17-4	2.43	2.69E+02	2
204	1825	C3H6Br2	1,2-dibromopropane	78-75-1	2.43	2.69E+02	2
205	1826	C3H6Br2	1,3-dibromopropane	109-64-8	2.43	2.69E+02	2
206	1827	C3H6Br2	2,2-dibromopropane	594-16-1	2.43	2.69E+02	2
207	1844	C3H6Cl2	1,1-dichloropropane	78-99-9	2.00	1.00E+02	2
208	1845	C3H6Cl2	1,2-dichloropropane	78-87-5	2.00	1.00E+02	1
209	1846	C3H6Cl2	1,3-dichloropropane	142-28-9	2.00	1.00E+02	2
210	1847	C3H6Cl2	2,2-dichloropropane	594-20-7	2.00	1.00E+02	2
211	1848	C3H6Cl2	1,2-dichloropropane, (±)	26198-63-0	2.00	1.00E+02	2
212	1849	C3H6Cl2	dichloropropane	26638-19-7	2.00	1.00E+02	2
213	1920	C3H6O	allyl alcohol	107-18-6	0.17	1.48E+00	1
214	1921	C3H6O	methyl vinyl ether	107-25-5	0.42	2.63E+00	2
215	1922	C3H6O	propionaldehyde	123-38-6	0.59	3.89E+00	1
216	1923	C3H6O	acetone	67-64-1	-0.24	5.75E-01	1
217	1924	C3H6O	1,2-propylene oxide	75-56-9	0.03	1.07E+00	1
218	1925	C3H6O	1,3-propylene oxide	503-30-0	-0.14	7.24E-01	1
219	1926	C3H6O	methyloxirane	16033-71-9	0.03	1.07E+00	1
220	1927	C3H6O	(S)-(-)-propylene oxide	16088-62-3	-0.06	8.81E-01	2
221	1928	C3H6O	(R)-(+)-propylene oxide	15448-47-2	-0.06	8.81E-01	2
222	1930	C3H6O2	propanoic acid	79-09-4	0.33	2.14E+00	1
223	1931	C3H6O2	ethyl formate	109-94-4	0.23	1.70E+00	1
224	1932	C3H6O2	methyl acetate	79-20-9	0.18	1.51E+00	1
225	1942	C3H6O2S	3-mercaptopropionic acid	107-96-0	0.43	2.69E+00	1
226	1948	C3H6O3	lactic acid	50-21-5	-0.72	1.91E-01	1
227	1949	C3H6O3	methoxyacetic acid	625-45-6	-0.68	2.09E-01	2
228	1950	C3H6O3	trioxane	110-88-3	-0.43	3.72E-01	1
229	1967	C3H6S	thiacyclobutane	287-27-4	1.30	2.00E+01	2
230	1978	C3H7Br	1-bromopropane	106-94-5	2.10	1.26E+02	1
231	1979	C3H7Br	2-bromopropane	75-26-3	1.90	7.94E+01	1
232	1986	C3H7Cl	1-chloropropane	540-54-5	2.04	1.10E+02	1
233	1987	C3H7Cl	2-chloropropane	75-29-6	1.90	7.94E+01	1
234	2020	C3H7F	1-fluoropropane	460-13-9	1.76	5.75E+01	2
235	2021	C3H7F	2-fluoropropane	420-26-8	1.66	4.58E+01	2
236	2025	C3H7I	1-iodopropane	107-08-4	2.50	3.16E+02	1
237	2026	C3H7I	2-iodopropane	75-30-9	2.89	7.76E+02	1
238	2030	C3H7N	allylamine	107-11-9	0.03	1.07E+00	1
239	2031	C3H7N	propyleneimine	75-55-8	0.13	1.35E+00	2
240	2036	C3H7NO	N,N-dimethylformamide	68-12-2	-1.01	9.77E-02	1

Octanol-Water Partition Coefficient							
NO	ID	FORMULA	NAME	CAS No	log_{10} Kow	Kow	code
241	2037	C3H7NO	N-methylacetamide	79-16-3	-1.05	8.91E-02	1
242	2038	C3H7NO	N-ethylformamide	627-45-2	-1.01	9.77E-02	2
243	2045	C3H7NO2	1-nitropropane	108-03-2	0.87	7.41E+00	1
244	2046	C3H7NO2	2-nitropropane	79-46-9	0.93	8.51E+00	1
245	2061	C3H7NO3	propyl nitrate	627-13-4	1.74	5.50E+01	2
246	2062	C3H7NO3	isopropyl nitrate	1712-64-7	1.74	5.50E+01	2
247	2094	C3H8	propane	74-98-6	2.36	2.29E+02	1
248	2129	C3H8O	propyl alcohol	71-23-8	0.25	1.78E+00	1
249	2130	C3H8O	isopropyl alcohol	67-63-0	0.05	1.12E+00	1
250	2131	C3H8O	methyl ethyl ether	540-67-0	0.56	3.63E+00	2
251	2138	C3H8O2	2-methoxyethanol	109-86-4	-0.77	1.70E-01	1
252	2140	C3H8O2	1,2-propanediol (propylene g	57-55-6	-0.92	1.20E-01	1
253	2141	C3H8O2	1,3-propanediol	504-63-2	-1.04	9.12E-02	1
254	2143	C3H8O2	(R)-(-)-1,2-propanediol	4254-14-2	-0.98	1.05E-01	2
255	2144	C3H8O2	(S)-(+)-1,2-propanediol	4254-15-3	-0.98	1.05E-01	2
256	2148	C3H8O3	glycerol	56-81-5	-1.76	1.74E-02	1
257	2158	C3H8S	propyl mercaptan	107-03-9	1.81	6.46E+01	1
258	2159	C3H8S	isopropyl mercaptan	75-33-2	1.68	4.79E+01	2
259	2160	C3H8S	methyl ethyl sulfide	624-89-5	1.54	3.47E+01	1
260	2200	C3H9N	propylamine	107-10-8	0.48	3.02E+00	1
261	2201	C3H9N	isopropylamine	75-31-0	0.26	1.82E+00	1
262	2202	C3H9N	methylethylamine	624-78-2	0.15	1.41E+00	1
263	2203	C3H9N	trimethylamine	75-50-3	0.16	1.45E+00	1
264	2204	C3H9NO	1-amino-2-propanol	78-96-6	-0.96	1.10E-01	1
265	2205	C3H9NO	3-amino-1-propanol	156-87-6	-1.12	7.59E-02	1
266	2206	C3H9NO	methylethanolamine	109-83-1	-0.94	1.15E-01	1
267	2207	C3H9NO	2-amino-1-propanol, (±)	6168-72-5	-1.01	9.85E-02	2
268	2210	C3H9NO	s(+)-2-amino-1-propanol	2749-11-3	-1.01	9.85E-02	2
269	2211	C3H9NO	(R)-(-)-1-amino-2-propanol	2799-16-8	-1.01	9.85E-02	2
270	2212	C3H9NO	(S)-(+)-1-amino-2-propanol	2799-17-9	-1.01	9.85E-02	2
271	2213	C3H9NO	(R)-(-)-2-amino-1-propanol	35320-23-1	-1.01	9.85E-02	2
272	2214	C3H9NO	2-aminopropanol	78-91-1	-1.01	9.85E-02	2
273	2241	C3H9O4P	trimethyl phosphate	512-56-1	-0.65	2.24E-01	1
274	2255	C3H10N2	1,2-propanediamine	78-90-0	-1.20	6.31E-02	2
275	2257	C3H10N2	1,2-propanediamine, (±)	10424-38-1	-1.20	6.31E-02	2
276	2258	C3H10N2	1,3-propanediamine	109-76-2	-1.20	6.31E-02	2
277	2318	C4Cl6	hexachloro-1,3-butadiene	87-68-3	4.78	6.03E+04	1
278	2334	C4F8	octafluoro-2-butene	360-89-4	3.03	1.07E+03	2
279	2335	C4F8	perfluoroisobutene	382-21-8	3.03	1.07E+03	2
280	2336	C4F8	octafluorocyclobutane	115-25-3	1.29	1.95E+01	2
281	2337	C4F8	1,1,2,3,3,4,4,4-octafluoro-1-b	357-26-6	3.03	1.07E+03	2
282	2344	C4F10	decafluorobutane	355-25-9	4.09	1.23E+04	2
283	2345	C4F10	perfluoroisobutane	354-92-7	4.09	1.23E+04	2
284	2374	C4H2	biacetylene	460-12-8	1.30	2.00E+01	2
285	2430	C4H2O3	maleic anhydride	108-31-6	1.62	4.17E+01	2
286	2510	C4H4	vinylacetylene	689-97-4	1.40	2.51E+01	2
287	2562	C4H4N2	succinonitrile	110-61-2	-0.99	1.02E-01	1
288	2565	C4H4N2	pyrimidine	289-95-2	-0.40	3.98E-01	1
289	2592	C4H4O	furan	110-00-9	1.34	2.19E+01	1
290	2598	C4H4O2	diketene	674-82-8	-0.39	4.07E-01	2
291	2603	C4H4O3	succinic anhydride	108-30-5	0.81	6.46E+00	2
292	2605	C4H4O4	fumaric acid	110-17-8	0.46	2.88E+00	1
293	2606	C4H4O4	maleic acid	110-16-7	-0.48	3.31E-01	1
294	2613	C4H4S	thiophene	110-02-1	1.81	6.46E+01	1
295	2634	C4H5Cl	chloroprene	126-99-8	2.53	3.39E+02	2
296	2635	C4H5Cl	4-chloro-1,2-butadiene	25790-55-0	2.53	3.39E+02	2
297	2636	C4H5Cl	1-chloro-1,3-butadiene	627-22-5	2.53	3.39E+02	2
298	2695	C4H5N	trans-crotonitrile	627-26-9	0.54	3.47E+00	2
299	2696	C4H5N	cis-crotonitrile	1190-76-7	0.54	3.47E+00	2
300	2697	C4H5N	methacrylonitrile	126-98-7	0.68	4.79E+00	1
301	2698	C4H5N	vinylacetonitrile	109-75-1	0.40	2.51E+00	1
302	2699	C4H5N	2-butenenitrile	4786-20-3	0.54	3.47E+00	2
303	2701	C4H5N	pyrrole	109-97-7	0.75	5.62E+00	1
304	2713	C4H5NO2	methyl cyanoacetate	105-34-0	-0.47	3.39E-01	1
305	2754	C4H6	dimethylacetylene	503-17-3	1.46	2.88E+01	1
306	2755	C4H6	ethylacetylene	107-00-6	1.53	3.39E+01	2
307	2756	C4H6	1,2-butadiene	590-19-2	2.20	1.58E+02	2
308	2757	C4H6	butadiene (1,3 butadiene)	106-99-0	1.99	9.77E+01	1
309	2806	C4H6Cl2	1,3-dichloro-trans-2-butene	7415-31-8	2.60	3.98E+02	2
310	2807	C4H6Cl2	1,4-dichloro-cis-2-butene	1476-11-5	2.60	3.98E+02	2
311	2808	C4H6Cl2	1,4-dichloro-trans-2-butene	110-57-6	2.60	3.98E+02	2
312	2809	C4H6Cl2	3,4-dichloro-1-butene	760-23-6	2.60	3.98E+02	2
313	2810	C4H6Cl2	3-chloro-2-(chloromethyl)-1-p	1871-57-4	2.60	3.98E+02	2
314	2811	C4H6Cl2	1,3-dichloro-1-butene	52497-07-1	2.60	3.98E+02	2
315	2812	C4H6Cl2	2,3-dichloro-1-butene	7013-11-8	2.60	3.98E+02	2
316	2813	C4H6Cl2	1,1-dichloro-2-butene	56800-09-0	2.60	3.98E+02	2
317	2814	C4H6Cl2	1,2-dichloro-2-butene	13602-13-6	2.60	3.98E+02	2
318	2815	C4H6Cl2	cis-1,3-dichloro-2-butene	10075-38-4	2.60	3.98E+02	2
319	2816	C4H6Cl2	cis-2,3-dichloro-2-butene	1587-26-4	2.60	3.98E+02	2
320	2817	C4H6Cl2	trans-2,3-dichloro-2-butene	1587-29-7	2.60	3.98E+02	2

Table 13 Octanol-Water Partition Coefficient - Oganic Compounds

Octanol-Water Partition Coefficient							
NO	ID	FORMULA	NAME	CAS No	$\log_{10}$ Kow	Kow	code
321	2818	C4H6Cl2	1,1-dichloro-2-methyl-1-prope	6065-93-6	2.60	3.98E+02	2
322	2819	C4H6Cl2	3,3-dichloro-2-methyl-1-prope	22227-75-4	2.60	3.98E+02	2
323	2820	C4H6Cl2	1,4-dichloro-2-butene	764-41-0	2.60	3.98E+02	2
324	2821	C4H6Cl2	1,3-dichloro-2-butene, cis and	926-57-8	2.60	3.98E+02	2
325	2822	C4H6Cl2	dichlorobutene	11069-19-5	2.60	3.98E+02	2
326	2940	C4H6O	trans-crotonaldehyde	123-73-9	0.60	3.98E+00	2
327	2941	C4H6O	methacrolein	78-85-3	0.74	5.50E+00	2
328	2942	C4H6O	2,5-dihydrofuran	1708-29-8	0.46	2.88E+00	1
329	2943	C4H6O	2,3-dihydrofuran	1191-99-7	0.46	2.88E+00	2
330	2944	C4H6O	divinyl ether	109-93-3	0.78	6.03E+00	2
331	2956	C4H6O	crotonaldehyde	4170-30-3	0.67	4.68E+00	2
332	2963	C4H6O2	2-butyne-1,4-diol	110-65-6	-0.93	1.17E-01	2
333	2965	C4H6O2	gamma-butyrolactone	96-48-0	-0.64	2.29E-01	1
334	2966	C4H6O2	cis-2-butenoic acid	503-64-0	0.85	7.08E+00	2
335	2967	C4H6O2	trans-2-butenoic acid	107-93-7	0.85	7.08E+00	2
336	2968	C4H6O2	vinylacetic acid	625-38-7	0.88	7.53E+00	2
337	2969	C4H6O2	methacrylic acid	79-41-4	0.93	8.51E+00	1
338	2970	C4H6O2	methyl acrylate	96-33-3	0.80	6.31E+00	1
339	2971	C4H6O2	vinyl acetate	108-05-4	0.73	5.37E+00	1
340	2972	C4H6O2	allyl formate	1838-59-1	0.77	5.82E+00	2
341	2978	C4H6O2	cyclopropanecarboxylic acid	1759-53-1	0.88	7.53E+00	2
342	2981	C4H6O2	crotonic acid	3724-65-0	0.72	5.25E+00	2
343	2990	C4H6O3	acetic anhydride	108-24-7	-0.58	2.63E-01	2
344	3006	C4H6O4	succinic acid	110-15-6	-0.59	2.57E-01	1
345	3024	C4H6O5	malic acid	6915-15-7	-0.12	7.52E-01	2
346	3033	C4H6O6	tartaric acid	87-69-4	0.44	2.73E+00	2
347	3238	C4H7N	butyronitrile	109-74-0	0.60	3.98E+00	1
348	3239	C4H7N	isobutyronitrile	78-82-0	0.46	2.88E+00	1
349	3246	C4H7NO	acetone cyanohydrin	75-86-5	-0.03	9.33E-01	2
350	3249	C4H7NO	2-pyrrolidone	616-45-5	-0.85	1.41E-01	1
351	3319	C4H8	cyclobutane	287-23-0	2.19	1.55E+02	2
352	3320	C4H8	1-butene	106-98-9	2.40	2.51E+02	1
353	3321	C4H8	2-butene; (cis+trans)	107-01-7	2.35	2.21E+02	2
354	3322	C4H8	cis-2-butene	590-18-1	2.33	2.14E+02	1
355	3323	C4H8	trans-2-butene	624-64-6	2.31	2.04E+02	1
356	3324	C4H8	isobutene	115-11-7	2.35	2.24E+02	1
357	3333	C4H8Br2	1,1-dibromobutane	62168-25-6	2.92	8.32E+02	2
358	3334	C4H8Br2	1,2-dibromobutane	533-98-2	2.92	8.32E+02	2
359	3335	C4H8Br2	1,3-dibromobutane	107-80-2	2.92	8.32E+02	2
360	3336	C4H8Br2	1,4-dibromobutane	110-52-1	2.92	8.32E+02	2
361	3337	C4H8Br2	2,2-dibromobutane	50341-35-0	2.92	8.32E+02	2
362	3338	C4H8Br2	DL-2,3-dibromobutane	598-71-0	2.92	8.32E+02	2
363	3339	C4H8Br2	meso-2,3-dibromobutane	5780-13-2	2.92	8.32E+02	2
364	3340	C4H8Br2	1,1-dibromo-2-methylpropane	33693-78-6	2.92	8.32E+02	2
365	3341	C4H8Br2	1,2-dibromo-2-methylpropane	594-34-3	2.92	8.32E+02	2
366	3342	C4H8Br2	1,3-dibromo-2-methylpropane	28148-04-1	2.92	8.32E+02	2
367	3343	C4H8Br2	2,3-dibromobutane	5408-86-6	2.92	8.32E+02	2
368	3360	C4H8Cl2	1,1-dichlorobutane	541-33-3	2.81	6.46E+02	2
369	3361	C4H8Cl2	1,2-dichlorobutane	616-21-7	2.81	6.46E+02	2
370	3362	C4H8Cl2	1,3-dichlorobutane	1190-22-3	2.81	6.46E+02	2
371	3363	C4H8Cl2	1,4-dichlorobutane	110-56-5	2.81	6.46E+02	2
372	3364	C4H8Cl2	2,2-dichlorobutane	4279-22-5	2.81	6.46E+02	2
373	3365	C4H8Cl2	DL-2,3-dichlorobutane	2211-67-8	2.81	6.46E+02	2
374	3366	C4H8Cl2	meso-2,3-dichlorobutane	4028-56-2	2.81	6.46E+02	2
375	3367	C4H8Cl2	1,1-dichloro-2-methylpropane	598-76-5	2.81	6.46E+02	2
376	3368	C4H8Cl2	1,2-dichloro-2-methylpropane	594-37-6	2.81	6.46E+02	2
377	3369	C4H8Cl2	1,3-dichloro-2-methylpropane	616-19-3	2.81	6.46E+02	2
378	3370	C4H8Cl2	2,3-dichlorobutane, dl and me	7581-97-7	2.81	6.46E+02	2
379	3371	C4H8Cl2	mixo-dichlorobutane	26761-81-9	2.81	6.46E+02	2
380	3374	C4H8Cl2O	bis(1-chloroethyl) ether	6986-48-7	1.12	1.32E+01	2
381	3375	C4H8Cl2O	bis(2-chloroethyl) ether	111-44-4	1.12	1.32E+01	1
382	3474	C4H8O	1,2-epoxybutane	106-88-7	0.86	7.24E+00	2
383	3476	C4H8O	ethyl vinyl ether	109-92-2	1.04	1.10E+01	1
384	3477	C4H8O	3-methoxy-1-propene	627-40-7	1.04	1.10E+01	2
385	3478	C4H8O	butyraldehyde	123-72-8	0.88	7.59E+00	1
386	3479	C4H8O	isobutyraldehyde	78-84-2	0.74	5.50E+00	2
387	3480	C4H8O	methyl ethyl ketone	78-93-3	0.29	1.95E+00	1
388	3481	C4H8O	tetrahydrofuran	109-99-9	0.46	2.88E+00	1
389	3504	C4H8O2	butyric acid	107-92-6	0.79	6.17E+00	1
390	3505	C4H8O2	isobutyric acid	79-31-2	0.94	8.71E+00	1
391	3506	C4H8O2	propyl formate	110-74-7	0.83	6.76E+00	1
392	3507	C4H8O2	isopropyl formate	625-55-8	0.74	5.51E+00	2
393	3508	C4H8O2	ethyl acetate	141-78-6	0.73	5.37E+00	1
394	3509	C4H8O2	methyl propanoate	554-12-1	0.84	6.92E+00	1
395	3510	C4H8O2	1,4-dioxane	123-91-1	-0.42	3.80E-01	1
396	3511	C4H8O2	cis-2-butene-1,4-diol	6117-80-2	-0.43	3.72E-01	2
397	3512	C4H8O2	trans-2-butene-1,4-diol	821-11-4	-0.43	3.72E-01	2
398	3513	C4H8O2	3-butene-1,2-diol	497-06-3	-0.43	3.72E-01	2
399	3514	C4H8O2	1,3-dioxane	505-22-6	-0.42	3.80E-01	2
400	3526	C4H8O2	2-butene-1,4-diol	110-64-5	-0.43	3.72E-01	2

Octanol-Water Partition Coefficient							
NO	ID	FORMULA	NAME	CAS No	$\log_{10}$ Kow	Kow	code
401	3533	C4H8O2	R,S-3-butene-1,2-diol	86161-40-2	-0.43	3.72E-01	2
402	3536	C4H8O2S	sulfolane	126-33-0	-0.77	1.70E-01	1
403	3581	C4H8S	tetrahydrothiophene	110-01-0	1.79	6.17E+01	2
404	3598	C4H9Br	1-bromobutane	109-65-9	2.75	5.62E+02	1
405	3599	C4H9Br	2-bromobutane	78-76-2	2.58	3.80E+02	2
406	3600	C4H9Br	1-bromo-2-methylpropane	78-77-3	2.67	4.62E+02	2
407	3601	C4H9Br	2-bromo-2-methylpropane	507-19-7	2.67	4.62E+02	2
408	3602	C4H9Br	2-bromobutane, (±)	5787-31-5	2.67	4.62E+02	2
409	3610	C4H9Cl	1-chlorobutane	109-69-3	2.64	4.37E+02	1
410	3611	C4H9Cl	1-chloro-2-methylpropane	513-36-0	2.47	2.97E+02	2
411	3612	C4H9Cl	2-chlorobutane	78-86-4	2.33	2.14E+02	1
412	3613	C4H9Cl	2-chloro-2-methylpropane	507-20-0	2.45	2.82E+02	2
413	3614	C4H9Cl	2-chlorobutane	53178-20-4	2.47	2.97E+02	2
414	3657	C4H9F	1-fluorobutane	2366-52-1	2.58	3.80E+02	1
415	3666	C4H9I	1-iodobutane	542-69-8	3.00	1.00E+03	1
416	3675	C4H9N	pyrrolidine	123-75-1	0.46	2.88E+00	1
417	3681	C4H9NO	N,N-dimethylacetamide	127-19-5	-0.77	1.70E-01	1
418	3682	C4H9NO	morpholine	110-91-8	-1.80	1.58E-02	1
419	3685	C4H9NO	butanamide	541-35-5	-0.21	6.17E-01	1
420	3697	C4H9NO2	1-nitrobutane	627-05-4	1.47	2.95E+01	1
421	3698	C4H9NO2	2-nitrobutane	600-24-8	1.47	2.95E+01	2
422	3699	C4H9NO2	1-nitro-2-methylpropane	625-74-1	1.47	2.95E+01	2
423	3700	C4H9NO2	2-nitro-2-methylpropane	594-70-7	1.47	2.95E+01	2
424	3715	C4H9NO2	2-nitrobutane, (±)	116781-85-2	1.47	2.95E+01	2
425	3783	C4H10	butane	106-97-8	2.89	7.76E+02	1
426	3784	C4H10	isobutane	75-28-5	2.80	6.31E+02	2
427	3830	C4H10N2	piperazine	110-85-0	-1.17	6.76E-02	1
428	3865	C4H10O	butanol	71-36-3	0.84	6.92E+00	1
429	3866	C4H10O	isobutanol	78-83-1	0.76	5.75E+00	1
430	3867	C4H10O	sec-butanol	78-92-2	0.65	4.47E+00	1
431	3868	C4H10O	tert-butanol	75-65-0	0.35	2.24E+00	1
432	3869	C4H10O	diethyl ether	60-29-7	0.89	7.76E+00	1
433	3870	C4H10O	methyl propyl ether	557-17-5	1.21	1.62E+01	1
434	3871	C4H10O	methyl isopropyl ether	598-53-8	0.98	9.55E+00	2
435	3872	C4H10O	(R)-(-)-2-butanol	14898-79-4	0.65	4.47E+00	2
436	3873	C4H10O	(S)-(+)-2-butanol	4221-99-2	0.65	4.47E+00	2
437	3874	C4H10O	(<+>)-2-butanol	15892-23-6	0.65	4.47E+00	2
438	3881	C4H10O2	t-butyl hydroperoxide	75-91-2	0.94	8.71E+00	2
439	3882	C4H10O2	1,2-dimethoxyethane	110-71-4	-0.21	6.17E-01	1
440	3883	C4H10O2	2-ethoxyethanol	110-80-5	-0.32	4.79E-01	1
441	3884	C4H10O2	1,2-butanediol	584-03-2	-0.56	2.75E-01	2
442	3885	C4H10O2	1,3-butanediol	107-88-0	-0.29	5.13E-01	2
443	3886	C4H10O2	1,4-butanediol	110-63-4	-0.83	1.48E-01	1
444	3887	C4H10O2	DL-2,3-butanediol	6982-25-8	-0.56	2.75E-01	2
445	3888	C4H10O2	meso-2,3-butanediol	5341-95-7	-0.56	2.75E-01	2
446	3889	C4H10O2	2-methyl-1,2-propanediol	558-43-0	-0.56	2.75E-01	2
447	3890	C4H10O2	1-methyl-1,3-propanediol	---	-0.56	2.75E-01	2
448	3891	C4H10O2	1,2-butanediol, (±)	26171-83-5	-0.56	2.75E-01	2
449	3894	C4H10O2	2-methoxy-1-propanol	1589-47-5	-0.32	4.79E-01	2
450	3895	C4H10O2	propylene glycol monomethyl	107-98-2	-0.32	4.79E-01	2
451	3896	C4H10O2	(2S,3S)-(+)-2,3-butanediol	19132-06-0	-0.56	2.75E-01	2
452	3897	C4H10O2	(2R,3R)-(-)-2,3-butanediol	24347-58-8	-0.56	2.75E-01	2
453	3898	C4H10O2	(S)-(+)-1,3-butanediol	24621-61-2	-0.56	2.75E-01	2
454	3899	C4H10O2	2,3-butanediol	513-85-9	-0.56	2.75E-01	2
455	3900	C4H10O2	(R)-(-)-1,3-butanediol	6290-03-5	-0.56	2.75E-01	2
456	3901	C4H10O2	(S)-(+)-2-methoxypropanol	116422-39-0	-0.32	4.79E-01	2
457	3902	C4H10O2	methoxypropanol	1320-67-8	-0.32	4.79E-01	2
458	3903	C4H10O2	(S)-(+)-1-methoxy-2-propanol	26550-55-0	-0.32	4.79E-01	2
459	3904	C4H10O2	(R)-(-)-1-methoxy-2-propanol	4984-22-9	-0.32	4.79E-01	2
460	3905	C4H10O2	2-methyl-1,3-propanediol	2163-42-0	-0.56	2.75E-01	2
461	3906	C4H10O2	3-methoxy-1-propanol	1589-49-7	-0.32	4.79E-01	2
462	3918	C4H10O3	diethylene glycol	111-46-6	-1.47	3.39E-02	2
463	3937	C4H10O4S	diethyl sulfate	64-67-5	1.14	1.38E+01	1
464	3938	C4H10O4S	methyl propyl sulfate	5867-91-4	1.14	1.38E+01	2
465	3942	C4H10S	butyl mercaptan	109-79-5	2.28	1.91E+02	1
466	3943	C4H10S	isobutyl mercaptan	513-44-0	2.18	1.51E+02	2
467	3944	C4H10S	sec-butyl mercaptan	513-53-1	2.18	1.51E+02	2
468	3945	C4H10S	tert-butyl mercaptan	75-66-1	2.14	1.38E+02	2
469	3946	C4H10S	2-butanethiol	91840-99-2	2.20	1.57E+02	2
470	3947	C4H10S	diethyl sulfide	352-93-2	1.95	8.91E+01	1
471	3948	C4H10S	methyl propyl sulfide	3877-15-4	1.90	7.94E+01	2
472	3949	C4H10S	methyl isopropyl sulfide	1551-21-9	1.83	6.76E+01	2
473	3950	C4H10S2	diethyl disulfide	110-81-6	2.86	7.24E+02	2
474	3951	C4H10S2	1,2-bis(methylthio)ethane	6628-18-8	2.86	7.24E+02	2
475	3954	C4H10S2	methyl propyl disulfide	2179-60-4	2.86	7.24E+02	2
476	3977	C4H11N	butylamine	109-73-9	0.86	7.24E+00	1
477	3978	C4H11N	isobutylamine	78-81-9	0.73	5.37E+00	1
478	3979	C4H11N	sec-butylamine	13952-84-6	0.76	5.75E+00	2
479	3980	C4H11N	tert-butylamine	75-64-9	0.40	2.51E+00	1
480	3981	C4H11N	2-butanamine, (.+/-.)-	33966-50-6	0.72	5.19E+00	2

Table 13 Octanol-Water Partition Coefficient - Oganic Compounds

Octanol-Water Partition Coefficient							
NO	ID	FORMULA	NAME	CAS No	log_{10} Kow	Kow	code
481	3982	C4H11N	methyl-propylamine	627-35-0	0.58	3.80E+00	2
482	3983	C4H11N	methylisopropylamine	4747-21-1	0.58	3.80E+00	2
483	3984	C4H11N	diethylamine	109-89-7	0.58	3.80E+00	1
484	3985	C4H11N	dimethylethylamine	598-56-1	0.65	4.42E+00	2
485	3986	C4H11N	(R)-(-)-sec-butylamine	13250-12-9	0.72	5.19E+00	2
486	3987	C4H11N	(S)-(+)-sec-butylamine	513-49-5	0.72	5.19E+00	2
487	3988	C4H11NO	dimethylethanolamine	108-01-0	-0.94	1.15E-01	2
488	3989	C4H11NO	3-amino-1-butanol	2867-59-6	-0.94	1.15E-01	2
489	3990	C4H11NO	3-amino-2-butanol	42551-55-3	-0.94	1.15E-01	2
490	3991	C4H11NO	4-amino-1-butanol	13325-10-5	-0.94	1.15E-01	2
491	3992	C4H11NO	2-amino-1-butanol, (±)-	13054-87-0	-0.94	1.15E-01	2
492	3993	C4H11NO	2-amino-2-methyl-1-propanol	124-68-5	-0.94	1.15E-01	2
493	3995	C4H11NO	2-(ethylamino)ethanol	110-73-6	-0.94	1.15E-01	2
494	3998	C4H11NO	4-amino-2-butanol	---	-0.94	1.15E-01	2
495	3999	C4H11NO	1-amino-2-butanol	13552-21-1	-0.94	1.15E-01	2
496	4000	C4H11NO	4-aminobutan-2-ol	39884-48-5	-0.94	1.15E-01	2
497	4001	C4H11NO	(S)-(+)-2-amino-1-butanol	5856-62-2	-0.94	1.15E-01	2
498	4002	C4H11NO	(R)-(-)-2-amino-1-butanol	5856-63-3	-0.94	1.15E-01	2
499	4003	C4H11NO	2-aminobutan-1-ol	96-20-8	-0.94	1.15E-01	2
500	4005	C4H11NO	1-amino-2-methyl-2-propanol	2854-16-2	-0.94	1.15E-01	2
501	4006	C4H11NO2	diethanolamine	111-42-2	-1.43	3.72E-02	1
502	4008	C4H11NO2	2-amino-2-methyl-1,3-propan	115-69-5	-1.43	3.72E-02	2
503	4010	C4H11NO2	3-methylamino-1,2-propanedi	40137-22-2	-1.43	3.72E-02	2
504	4080	C4H12N2O	N-aminoethyl ethanolamine	111-41-1	-2.13	7.41E-03	2
505	4110	C4H12Si	tetramethylsilane	75-76-3	2.72	5.25E+02	2
506	4118	C4H13N3	diethylene triamine	111-40-0	-2.13	7.41E-03	2
507	4169	C5Cl6	hexachlorocyclopentadiene	77-47-4	5.04	1.10E+05	1
508	4327	C5H4O2	furfural	98-01-1	0.41	2.57E+00	1
509	4391	C5H5N	pyridine	110-86-1	0.65	4.47E+00	1
510	4447	C5H6	cyclopentadiene	542-92-7	2.25	1.78E+02	2
511	4448	C5H6	2-methyl-1-butene-3-yne	78-80-8	2.06	1.16E+02	2
512	4449	C5H6	1-pentene-3-yne	646-05-9	2.20	1.59E+02	2
513	4450	C5H6	1-pentene-4-yne	871-28-3	1.93	8.43E+01	2
514	4451	C5H6	cis-3-penten-1-yne	1574-40-9	2.06	1.16E+02	2
515	4452	C5H6	trans-3-penten-1-yne	2004-69-5	2.06	1.16E+02	2
516	4490	C5H6N2	glutaronitrile	544-13-8	-0.72	1.91E-01	1
517	4534	C5H6O	2-methylfuran	534-22-5	1.85	7.08E+01	1
518	4548	C5H6O2	furfuryl alcohol	98-00-0	0.28	1.91E+00	1
519	4567	C5H6O3	glutaric anhydride	108-55-4	1.30	2.00E+01	2
520	4576	C5H6O4	citraconic acid	498-23-7	-0.25	5.60E-01	2
521	4577	C5H6O4	itaconic acid	97-65-4	0.71	5.11E+00	2
522	4587	C5H6S	2-methylthiophene	554-14-3	2.33	2.14E+02	1
523	4588	C5H6S	3-methylthiophene	616-44-4	2.34	2.19E+02	1
524	4658	C5H7N	N-methylpyrrole	96-54-8	1.21	1.62E+01	1
525	4659	C5H7N	2-methylpyrrole	636-41-9	1.21	1.62E+01	2
526	4660	C5H7N	3-methylpyrrole	616-43-3	1.21	1.62E+01	2
527	4677	C5H7NO2	ethyl cyanoacetate	105-56-6	1.01	1.02E+01	2
528	4736	C5H8	spiropentane	157-40-4	2.46	2.88E+02	2
529	4737	C5H8	cyclopentene	142-29-0	2.47	2.95E+02	2
530	4738	C5H8	isoprene	78-79-5	2.42	2.63E+02	1
531	4739	C5H8	3-methyl-1,2-butadiene	598-25-4	2.61	4.07E+02	2
532	4740	C5H8	1,2-pentadiene	591-95-7	2.55	3.55E+02	2
533	4741	C5H8	cis-1,3-pentadiene	1574-41-0	2.40	2.51E+02	1
534	4742	C5H8	trans-1,3-pentadiene	2004-70-8	2.44	2.75E+02	1
535	4743	C5H8	1,4-pentadiene	591-93-5	2.48	3.02E+02	1
536	4744	C5H8	2,3-pentadiene	591-96-8	2.72	5.23E+02	2
537	4745	C5H8	1-pentyne	627-19-0	1.98	9.55E+01	1
538	4746	C5H8	2-pentyne	627-21-4	2.08	1.20E+02	2
539	4747	C5H8	3-methyl-1-butyne	598-23-2	1.76	5.74E+01	2
540	4751	C5H8	1,3-pentadiene; (cis+trans)	504-60-9	2.52	3.29E+02	2
541	4845	C5H8N4O12	pentaerythritol tetranitrate	78-11-5	2.38	2.40E+02	2
542	4848	C5H8O	cyclopentanone	120-92-3	0.63	4.27E+00	2
543	4849	C5H8O	methyl isopropenyl ketone	814-78-8	0.73	5.39E+00	2
544	4882	C5H8O2	acetylacetone	123-54-6	0.40	2.51E+00	1
545	4883	C5H8O2	allyl acetate	591-87-7	0.97	9.33E+00	1
546	4884	C5H8O2	isopropenyl acetate	108-22-5	1.22	1.67E+01	2
547	4885	C5H8O2	vinyl propionate	105-38-4	1.22	1.66E+01	2
548	4886	C5H8O2	methyl cis-2-butenoate	4358-59-2	1.22	1.67E+01	2
549	4887	C5H8O2	methyl trans-2-butenoate	623-43-8	1.22	1.67E+01	2
550	4888	C5H8O2	ethyl acrylate	140-88-5	1.32	2.09E+01	1
551	4889	C5H8O2	methyl methacrylate	80-62-6	1.38	2.40E+01	1
552	4918	C5H8O2	methyl 3-butenoate	3724-55-8	1.22	1.67E+01	2
553	4927	C5H8O3	2-hydroxyethyl acrylate	818-61-1	-0.21	6.17E-01	1
554	4928	C5H8O3	levulinic acid	123-76-2	-0.49	3.24E-01	1
555	4929	C5H8O3	methyl acetoacetate	105-45-3	-0.69	2.04E-01	2
556	4945	C5H8O4	glutaric acid	110-94-1	-0.29	5.13E-01	1
557	5183	C5H9N	valeronitrile	110-59-8	0.94	8.71E+00	1
558	5184	C5H9N	2-methylbutanenitrile	18936-17-9	1.12	1.32E+01	2
559	5185	C5H9N	3-methylbutanenitrile	625-28-5	1.12	1.32E+01	2
560	5186	C5H9N	2,2-dimethylpropanenitrile	630-18-2	1.12	1.32E+01	2

Octanol-Water Partition Coefficient							
NO	ID	FORMULA	NAME	CAS No	log_{10} Kow	Kow	code
561	5195	C5H9N	(S)-(+)-2-methylbutyronitrile	25570-03-0	1.12	1.32E+01	2
562	5197	C5H9NO	butyl isocyanate	111-36-4	2.26	1.82E+02	2
563	5198	C5H9NO	N-methyl-2-pyrrolidone	872-50-4	-0.38	4.17E-01	1
564	5203	C5H9NO	isobutyl isocyanate	1873-29-6	2.26	1.82E+02	2
565	5204	C5H9NO	2-isocyanato-2-methylpropan	1609-86-5	2.26	1.82E+02	2
566	5245	C5H9NO4	L-glutamic acid	56-86-0	-3.69	2.04E-04	1
567	5285	C5H10	cyclopentane	287-92-3	3.00	1.00E+03	1
568	5286	C5H10	1-pentene	109-67-1	2.20	1.58E+02	2
569	5287	C5H10	2-pentene; (cis+trans)	109-68-2	2.63	4.25E+02	2
570	5288	C5H10	cis-2-pentene	627-20-3	2.58	3.80E+02	2
571	5289	C5H10	trans-2-pentene	646-04-8	2.58	3.80E+02	2
572	5290	C5H10	2-methyl-1-butene	563-46-2	2.72	5.25E+02	2
573	5291	C5H10	3-methyl-1-butene	563-45-1	2.59	3.89E+02	2
574	5292	C5H10	2-methyl-2-butene	513-35-9	2.64	4.37E+02	2
575	5330	C5H10Cl2	1,1-dichloropentane	820-55-3	3.30	2.00E+03	2
576	5331	C5H10Cl2	1,2-dichloropentane	1674-33-5	3.30	2.00E+03	2
577	5332	C5H10Cl2	1,3-dichloropentane	30122-12-4	3.30	2.00E+03	2
578	5333	C5H10Cl2	1,4-dichloropentane	626-92-6	3.30	2.00E+03	2
579	5334	C5H10Cl2	1,5-dichloropentane	628-76-2	3.30	2.00E+03	2
580	5335	C5H10Cl2	2,2-dichloropentane	34887-14-4	3.30	2.00E+03	2
581	5336	C5H10Cl2	2,3-dichloropentane	600-11-3	3.30	2.00E+03	2
582	5337	C5H10Cl2	2,4-dichloropentane	625-67-2	3.30	2.00E+03	2
583	5338	C5H10Cl2	3,3-dichloropentane	21571-91-5	3.30	2.00E+03	2
584	5339	C5H10Cl2	1,1-dichloro-2-methylbutane	23010-05-1	3.30	2.00E+03	2
585	5340	C5H10Cl2	1,1-dichloro-3-methylbutane	625-66-1	3.30	2.00E+03	2
586	5341	C5H10Cl2	1,2-dichloro-2-methylbutane	23010-04-0	3.30	2.00E+03	2
587	5342	C5H10Cl2	1,2-dichloro-3-methylbutane	600-10-2	3.30	2.00E+03	2
588	5343	C5H10Cl2	1,3-dichloro-2-methylbutane	23010-07-3	3.30	2.00E+03	2
589	5344	C5H10Cl2	1,3-dichloro-3-methylbutane	624-96-4	3.30	2.00E+03	2
590	5345	C5H10Cl2	1,4-dichloro-2-methylbutane	623-34-7	3.30	2.00E+03	2
591	5346	C5H10Cl2	2,2-dichloro-3-methylbutane	17773-66-9	3.30	2.00E+03	2
592	5347	C5H10Cl2	2,3-dichloro-2-methylbutane	507-45-9	3.30	2.00E+03	2
593	5348	C5H10Cl2	1,3-dichloro-2-ethylpropane	---	3.30	2.00E+03	2
594	5349	C5H10Cl2	1,1-dichloro-2,2-dimethylprop	29559-54-4	3.30	2.00E+03	2
595	5350	C5H10Cl2	1,3-dichloro-2,2-dimethylprop	29559-55-5	3.30	2.00E+03	2
596	5351	C5H10Cl2	dichloropentane	30586-10-8	3.30	2.00E+03	2
597	5445	C5H10O	pentanal	110-62-3	1.31	2.04E+01	2
598	5446	C5H10O	2-methylbutanal	96-17-3	0.95	8.96E+00	2
599	5447	C5H10O	3-methylbutanal	590-86-3	0.96	9.21E+00	2
600	5448	C5H10O	2,2-dimethylpropanal	630-19-3	0.59	3.86E+00	2
601	5449	C5H10O	2-methylbutanal, (±)	57456-98-1	0.95	8.98E+00	2
602	5450	C5H10O	methyl propyl ketone	107-87-9	0.84	6.92E+00	1
603	5451	C5H10O	diethyl ketone	96-22-0	0.82	6.61E+00	1
604	5452	C5H10O	methyl isopropyl ketone	563-80-4	0.56	3.63E+00	1
605	5453	C5H10O	2-methyltetrahydrofuran	96-47-9	1.85	7.08E+01	1
606	5454	C5H10O	3-methyltetrahydrofuran	13423-15-9	1.85	7.08E+01	2
607	5471	C5H10O	tetrahydropyran	142-68-7	0.82	6.61E+00	1
608	5492	C5H10O2	pentanoic acid	109-52-4	1.39	2.45E+01	1
609	5493	C5H10O2	2-methylbutyric acid	116-53-0	0.97	9.31E+00	2
610	5494	C5H10O2	3-methylbutanoic acid	503-74-2	1.16	1.45E+01	1
611	5495	C5H10O2	2,2-dimethylpropanoic acid	75-98-9	0.67	4.69E+00	2
612	5496	C5H10O2	2-methylbutanoic acid	600-07-7	1.05	1.12E+01	2
613	5497	C5H10O2	butyl formate	592-84-7	1.30	2.00E+01	2
614	5498	C5H10O2	isobutyl formate	542-55-2	1.23	1.70E+01	2
615	5499	C5H10O2	sec-butyl formate	589-40-2	1.28	1.91E+01	2
616	5500	C5H10O2	tert-butyl formate	762-75-4	1.12	1.30E+01	2
617	5501	C5H10O2	propyl acetate	109-60-4	1.24	1.74E+01	1
618	5502	C5H10O2	isopropyl acetate	108-21-4	1.03	1.07E+01	2
619	5503	C5H10O2	ethyl propanoate	105-37-3	1.21	1.62E+01	1
620	5504	C5H10O2	methyl butanoate	623-42-7	1.29	1.95E+01	1
621	5505	C5H10O2	methyl isobutanoate	547-63-7	1.29	1.95E+01	2
622	5506	C5H10O2	tetrahydrofurfuryl alcohol	97-99-4	-0.11	7.76E-01	2
623	5530	C5H10O2	(S)-(+)-2-methylbutyric acid	1730-91-2	1.05	1.12E+01	2
624	5537	C5H10O2S	3-methyl sulfolane	872-93-5	0.18	1.51E+00	2
625	5543	C5H10O3	diethyl carbonate	105-58-8	1.21	1.62E+01	1
626	5544	C5H10O3	ethyl lactate	97-64-3	-0.18	6.61E-01	2
627	5593	C5H10S	thiacyclohexane	1613-51-0	2.28	1.91E+02	2
628	5597	C5H10S	cyclopentanethiol	1679-07-8	2.55	3.55E+02	2
629	5603	C5H11Br	1-bromopentane	110-53-2	3.37	2.34E+03	1
630	5604	C5H11Br	2-bromopentane	107-81-3	3.37	2.34E+03	2
631	5605	C5H11Br	3-bromopentane	1809-10-5	3.37	2.34E+03	2
632	5606	C5H11Br	1-bromo-2-methylbutane	10422-35-2	3.37	2.34E+03	2
633	5607	C5H11Br	1-bromo-3-methylbutane	107-82-4	3.37	2.34E+03	2
634	5608	C5H11Br	2-bromo-2-methylbutane	507-36-8	3.37	2.34E+03	2
635	5609	C5H11Br	2-bromo-3-methylbutane	18295-25-5	3.37	2.34E+03	2
636	5610	C5H11Br	1-bromo-2,2-dimethylpropane	630-17-1	3.37	2.34E+03	2
637	5611	C5H11Br	1-bromo-2-methylbutane, (S)	534-00-9	3.37	2.34E+03	2
638	5620	C5H11Cl	1-chloropentane	543-59-9	2.73	5.37E+02	1
639	5621	C5H11Cl	2-chloropentane	625-29-6	2.63	4.22E+02	2
640	5622	C5H11Cl	3-chloropentane	616-20-6	2.63	4.22E+02	2

Table 13 Octanol-Water Partition Coefficient - Oganic Compounds

Octanol-Water Partition Coefficient								Octanol-Water Partition Coefficient							
NO	ID	FORMULA	NAME	CAS No	$\log_{10}$ Kow	Kow	code	NO	ID	FORMULA	NAME	CAS No	$\log_{10}$ Kow	Kow	code
641	5623	C5H11Cl	1-chloro-2-methylbutane	616-13-7	2.63	4.22E+02	2	721	5929	C5H12S	2-methyl-1-butanethiol, (+)	20089-07-0	2.67	4.72E+02	2
642	5624	C5H11Cl	1-chloro-3-methylbutane	107-84-6	2.63	4.22E+02	2	722	5930	C5H12S	methyl butyl sulfide	628-29-5	2.39	2.45E+02	2
643	5625	C5H11Cl	2-chloro-2-methylbutane	594-36-5	2.52	3.31E+02	1	723	5931	C5H12S	ethyl propyl sulfide	4110-50-3	2.39	2.45E+02	2
644	5626	C5H11Cl	2-chloro-3-methylbutane	631-65-2	2.63	4.22E+02	2	724	5932	C5H12S	3-methyl-2-thiapentane	10359-64-5	2.31	2.04E+02	2
645	5627	C5H11Cl	1-chloro-2,2-dimethylpropane	753-89-9	2.63	4.22E+02	2	725	5933	C5H12S	4-methyl-2-thiapentane	5008-69-5	2.32	2.08E+02	2
646	5628	C5H11Cl	1-chloro-2-methylbutane, (±)	114180-21-1	2.63	4.22E+02	2	726	5934	C5H12S	2-methyl-3-thiapentane	5145-99-3	2.22	1.65E+02	2
647	5629	C5H11Cl	2-chloropentane, (+)	29882-57-3	2.63	4.22E+02	2	727	5935	C5H12S	3,3-dimethyl-2-thiabutane	6163-64-0	2.06	1.14E+02	2
648	5649	C5H11F	1-fluoropentane	592-50-7	2.33	2.14E+02	1	728	5947	C5H13N	pentylamine	110-58-7	1.49	3.09E+01	1
649	5669	C5H11N	N-methylpyrrolidine	120-94-5	0.92	8.32E+00	1	729	5948	C5H13N	1-methylbutylamine	625-30-9	1.49	3.09E+01	2
650	5670	C5H11N	piperidine	110-89-4	0.84	6.92E+00	1	730	5949	C5H13N	2-methylbutylamine	96-15-1	1.49	3.09E+01	2
651	5673	C5H11N	2-methylpyrrolidine	765-38-8	0.92	8.32E+00	2	731	5950	C5H13N	3-methylbutylamine	107-85-7	1.49	3.09E+01	2
652	5674	C5H11N	3-methylpyrrolidine	34375-89-8	0.92	8.32E+00	2	732	5951	C5H13N	1,1-dimethylpropylamine	594-39-8	1.49	3.09E+01	2
653	5696	C5H11NO2	1-nitropentane	628-05-7	2.01	1.02E+02	1	733	5952	C5H13N	1,2-dimethylpropylamine	598-74-3	1.49	3.09E+01	2
654	5697	C5H11NO2	2-nitropentane	4609-89-6	2.01	1.02E+02	2	734	5953	C5H13N	2,2-dimethylpropylamine	5813-64-9	1.49	3.09E+01	2
655	5698	C5H11NO2	3-nitropentane	551-88-2	2.01	1.02E+02	2	735	5954	C5H13N	3-pentanamine	616-24-0	1.49	3.09E+01	2
656	5772	C5H12	pentane	109-66-0	3.45	2.82E+03	1	736	5955	C5H13N	methyl-butylamine	110-68-9	1.18	1.51E+01	2
657	5773	C5H12	isopentane	78-78-4	2.30	2.00E+02	1	737	5956	C5H13N	methylisobutylamine	14610-37-8	1.18	1.51E+01	2
658	5774	C5H12	neopentane	463-82-1	3.11	1.29E+03	1	738	5957	C5H13N	methyl-sec-butylamine	7713-69-1	1.18	1.51E+01	2
659	5831	C5H12O	1-pentanol	71-41-0	1.51	3.24E+01	1	739	5958	C5H13N	methyl-tert-butylamine	625-43-4	1.18	1.51E+01	2
660	5832	C5H12O	2-pentanol	6032-29-7	1.25	1.78E+01	1	740	5959	C5H13N	ethylpropylamine	20193-20-8	1.18	1.51E+01	2
661	5833	C5H12O	3-pentanol	584-02-1	1.21	1.62E+01	1	741	5960	C5H13N	ethylisopropylamine	19961-27-4	1.18	1.51E+01	2
662	5834	C5H12O	2-methyl-1-butanol	137-32-6	1.29	1.95E+01	1	742	5961	C5H13N	methyldiethylamine	616-39-7	0.97	9.37E+00	2
663	5835	C5H12O	2-methyl-1-butanol, (±)	34713-94-5	1.23	1.70E+01	2	743	5962	C5H13N	dimethyl-propylamine	926-63-6	0.97	9.37E+00	2
664	5836	C5H12O	3-methyl-2-butanol, (±)	70116-68-6	1.28	1.91E+01	2	744	5963	C5H13N	dimethylisopropylamine	996-35-6	0.97	9.37E+00	2
665	5837	C5H12O	3-methyl-1-butanol	123-51-3	1.28	1.91E+01	1	745	5964	C5H13N	(S)-(-)-2-methylbutylamine	20626-52-2	1.18	1.51E+01	2
666	5838	C5H12O	tert-pentyl-alcohol	75-85-4	0.89	7.76E+00	1	746	5965	C5H13N	N,N-dimethyl-2-propanamine	996-35-0	0.97	9.37E+00	2
667	5839	C5H12O	3-methyl-2-butanol	598-75-4	1.28	1.91E+01	1	747	5986	C5H13NO2	methyl diethanolamine	105-59-9	-1.50	3.16E-02	2
668	5840	C5H12O	2,2-dimethyl-1-propanol	75-84-3	1.31	2.04E+01	1	748	5987	C5H13NO2	2-amino-2-ethyl-1,3-propaned	115-70-8	-1.50	3.16E-02	2
669	5841	C5H12O	methyl butyl ether	628-28-4	1.59	3.87E+01	2	749	6067	C6Cl6	hexachlorobenzene	118-74-1	5.47	2.95E+05	1
670	5842	C5H12O	methyl isobutyl ether	625-44-5	1.47	2.95E+01	2	750	6080	C6F6	hexafluorobenzene	392-56-3	2.55	3.55E+02	1
671	5843	C5H12O	methyl sec-butyl ether	6795-87-5	1.47	2.95E+01	2	751	6115	C6HCl5	pentachlorobenzene	608-93-5	5.03	1.07E+05	1
672	5844	C5H12O	methyl tert-butyl ether	1634-04-4	0.94	8.71E+00	1	752	6116	C6HCl5O	pentachlorophenol	87-86-5	5.07	1.17E+05	1
673	5845	C5H12O	ethyl propyl ether	628-32-0	1.54	3.47E+01	2	753	6174	C6H2Cl4	1,2,3,4-tetrachlorobenzene	634-66-2	4.55	3.55E+04	1
674	5846	C5H12O	ethyl isopropyl ether	625-54-7	1.22	1.65E+01	2	754	6175	C6H2Cl4	1,2,3,5-tetrachlorobenzene	634-90-2	4.65	4.47E+04	1
675	5847	C5H12O	sec-butyl methyl ether	116783-23-4	1.37	2.35E+01	2	755	6176	C6H2Cl4	1,2,4,5-tetrachlorobenzene	95-94-3	4.51	3.24E+04	1
676	5848	C5H12O	sec-butyl methyl ether (d)	66610-39-7	1.37	2.35E+01	2	756	6281	C6H3ClN2O4	1-chloro-2,4-dinitrobenzene	97-00-7	2.17	1.48E+02	1
677	5849	C5H12O	(S)-(-)-2-methyl-1-butanol	1565-80-6	1.23	1.70E+01	2	757	6296	C6H3Cl2NO2	1,2-dichloro-4-nitrobenzene	99-54-7	3.12	1.32E+03	1
678	5850	C5H12O	(S)-(+)-2-pentanol	26184-62-3	1.23	1.70E+01	2	758	6305	C6H3Cl3	1,2,4-trichlorobenzene	120-82-1	3.98	9.55E+03	1
679	5851	C5H12O	pentanol	30899-19-5	1.23	1.70E+01	2	759	6306	C6H3Cl3	1,2,3-trichlorobenzene	87-61-6	4.04	1.10E+04	1
680	5852	C5H12O	(R)-(-)-2-pentanol	31087-44-2	1.23	1.70E+01	2	760	6307	C6H3Cl3	1,3,5-trichlorobenzene	108-70-3	4.02	1.05E+04	1
681	5853	C5H12O	(S)-3-methyl-2-butanol	1517-66-4	1.23	1.70E+01	2	761	6308	C6H3Cl3	trichlorobenzene	12002-48-1	4.02	1.05E+04	2
682	5857	C5H12O2	ethylene glycol monopropyl e	2807-30-9	0.08	1.20E+00	2	762	6367	C6H3N3O6	1,3,5-trinitrobenzene	99-35-4	1.18	1.51E+01	1
683	5858	C5H12O2	neopentyl glycol	126-30-7	0.27	1.86E+00	2	763	6411	C6H4Br2	m-dibromobenzene	108-36-1	3.75	5.62E+03	1
684	5859	C5H12O2	1,2-pentanediol	5343-92-0	0.27	1.86E+00	2	764	6412	C6H4Br2	o-dibromobenzene	583-53-9	3.75	5.62E+03	2
685	5860	C5H12O2	1,2-pentanediol, (±)	91049-43-3	0.27	1.86E+00	2	765	6413	C6H4Br2	p-dibromobenzene	106-37-6	3.75	5.62E+03	2
686	5861	C5H12O2	1,3-pentanediol	3174-67-2	0.27	1.86E+00	2	766	6414	C6H4Br2	dibromobenzene	26249-12-7	3.75	5.62E+03	2
687	5862	C5H12O2	1,4-pentanediol	626-95-9	0.27	1.86E+00	2	767	6443	C6H4ClNO2	m-chloronitrobenzene	121-73-3	2.46	2.88E+02	1
688	5863	C5H12O2	1,5-pentanediol	111-29-5	0.27	1.86E+00	2	768	6444	C6H4ClNO2	o-chloronitrobenzene	88-73-3	2.24	1.74E+02	1
689	5864	C5H12O2	2,3-pentanediol	42027-23-6	0.27	1.86E+00	2	769	6445	C6H4ClNO2	p-chloronitrobenzene	100-00-5	2.39	2.45E+02	1
690	5865	C5H12O2	2,4-pentanediol	625-69-4	0.27	1.86E+00	2	770	6469	C6H4Cl2	o-dichlorobenzene	95-50-1	3.38	2.40E+03	1
691	5866	C5H12O2	DL-erythro-2,3-pentanediol	61828-35-1	0.27	1.86E+00	2	771	6470	C6H4Cl2	m-dichlorobenzene	541-73-1	3.48	3.02E+03	1
692	5867	C5H12O2	DL-threo-2,3-pentanediol	61828-36-2	0.27	1.86E+00	2	772	6471	C6H4Cl2	p-dichlorobenzene	106-46-7	3.38	2.40E+03	1
693	5868	C5H12O2	DL-2,4-pentanediol	1825-14-5	0.27	1.86E+00	2	773	6480	C6H4Cl2O	2,4-dichlorophenol	120-83-2	3.23	1.70E+03	1
694	5869	C5H12O2	meso-2,4-pentanediol	3950-21-8	0.27	1.86E+00	2	774	6524	C6H4F2	m-difluorobenzene	372-18-9	2.21	1.62E+02	1
695	5870	C5H12O2	2-methyl-1,2-butanediol	41051-72-3	0.27	1.86E+00	2	775	6525	C6H4F2	o-difluorobenzene	367-11-3	2.37	2.34E+02	1
696	5871	C5H12O2	2-methyl-1,3-butanediol	684-84-4	0.27	1.86E+00	2	776	6526	C6H4F2	p-difluorobenzene	540-36-3	2.13	1.35E+02	1
697	5872	C5H12O2	2-methyl-1,4-butanediol	2938-98-9	0.27	1.86E+00	2	777	6580	C6H4N2O4	m-dinitrobenzene	99-65-0	1.49	3.09E+01	1
698	5873	C5H12O2	2-methyl-2,3-butanediol	5396-58-7	0.27	1.86E+00	2	778	6581	C6H4N2O4	o-dinitrobenzene	528-29-0	1.69	4.90E+01	1
699	5874	C5H12O2	3-methyl-1,2-butanediol	50468-22-9	0.27	1.86E+00	2	779	6582	C6H4N2O4	p-dinitrobenzene	100-25-4	1.46	2.88E+01	1
700	5875	C5H12O2	3-methyl-1,3-butanediol	2568-33-4	0.27	1.86E+00	2	780	6656	C6H5Br	bromobenzene	108-86-1	2.99	9.77E+02	1
701	5876	C5H12O2	2-ethyl-1,3-propanediol	2612-29-5	0.27	1.86E+00	2	781	6692	C6H5Cl	chlorobenzene	108-90-7	2.84	6.92E+02	1
702	5881	C5H12O2	1-ethoxy-2-propanol	1569-02-4	0.08	1.20E+00	2	782	6720	C6H5ClO	m-chlorophenol	108-43-0	2.50	3.16E+02	1
703	5882	C5H12O2	2-ethoxy-1-propanol	19089-47-5	0.08	1.20E+00	2	783	6721	C6H5ClO	o-chlorophenol	95-57-8	2.15	1.41E+02	1
704	5883	C5H12O2	2-isopropoxyethanol	109-59-1	0.08	1.20E+00	2	784	6722	C6H5ClO	p-chlorophenol	106-48-9	2.39	2.45E+02	1
705	5884	C5H12O2	3-methoxy-1-butanol	2517-43-3	0.08	1.20E+00	2	785	6740	C6H5Cl2N	3,4-dichloroaniline	95-76-1	2.69	4.90E+02	1
706	5886	C5H12O2	3-ethoxy-1-propanol	111-35-3	0.08	1.20E+00	2	786	6769	C6H5F	fluorobenzene	462-06-6	2.27	1.86E+02	1
707	5887	C5H12O2	1-methoxy-2-butanol	53778-73-7	0.08	1.20E+00	2	787	6795	C6H5I	iodobenzene	591-50-4	3.28	1.91E+03	1
708	5888	C5H12O2	(R,R)-(-)-2,4-pentanediol	42075-32-1	0.27	1.86E+00	2	788	6819	C6H5NO2	nitrobenzene	98-95-3	1.85	7.08E+01	1
709	5889	C5H12O2	(2S,4S)-(+)-pentanediol	72345-23-4	0.27	1.86E+00	2	789	6830	C6H5NO3	o-nitrophenol	88-75-5	1.91	8.13E+01	1
710	5890	C5H12O2	ethyl ether of propylene glycc	52125-53-8	0.08	1.20E+00	2	790	6831	C6H5NO3	m-nitrophenol	554-84-7	1.91	8.13E+01	2
711	5899	C5H12O3	2-(2-methoxyethoxy)ethanol	111-77-3	-1.18	6.61E-02	2	791	6832	C6H5NO3	p-nitrophenol	100-02-7	1.91	8.13E+01	1
712	5912	C5H12O4	pentaerythritol	115-77-5	-1.69	2.04E-02	1	792	6876	C6H6	benzene	71-43-2	2.13	1.35E+02	1
713	5921	C5H12S	pentyl mercaptan	110-66-7	2.74	5.50E+02	2	793	6923	C6H6ClN	m-chloroaniline	108-42-9	1.88	7.59E+01	1
714	5922	C5H12S	2-pentanethiol	2084-19-7	2.66	4.58E+02	2	794	6924	C6H6ClN	o-chloroaniline	95-51-2	1.90	7.94E+01	1
715	5923	C5H12S	3-pentanethiol	616-31-9	2.69	4.90E+02	2	795	6925	C6H6ClN	p-chloroaniline	106-47-8	1.83	6.76E+01	1
716	5924	C5H12S	2-methyl-1-butanethiol	1878-18-8	2.79	6.21E+02	2	796	6996	C6H6N2O2	m-nitroaniline	99-09-2	1.37	2.34E+01	1
717	5925	C5H12S	3-methyl-1-butanethiol	541-31-1	2.78	6.02E+02	2	797	6997	C6H6N2O2	o-nitroaniline	88-74-4	1.85	7.08E+01	1
718	5926	C5H12S	2-methyl-2-butanethiol	1679-09-0	2.63	4.27E+02	2	798	6998	C6H6N2O2	p-nitroaniline	100-01-6	1.39	2.45E+01	1
719	5927	C5H12S	3-methyl-2-butanethiol	2084-18-6	2.61	4.06E+02	2	799	7000	C6H6N2O2	N-nitroaniline	645-55-6	1.54	3.44E+01	2
720	5928	C5H12S	2,2-dimethyl-1-propanethiol	1679-08-9	2.49	3.10E+02	2	800	7057	C6H6O	phenol	108-95-2	1.48	3.02E+01	1

Table 13 Octanol-Water Partition Coefficient - Oganic Compounds

Octanol-Water Partition Coefficient							
NO	ID	FORMULA	NAME	CAS No	log_{10} Kow	Kow	code
801	7070	C6H6O2	pyrocatechol	120-80-9	0.88	7.59E+00	1
802	7071	C6H6O2	resorcinol	108-46-3	0.80	6.31E+00	1
803	7072	C6H6O2	p-hydroquinone	123-31-9	0.59	3.89E+00	1
804	7085	C6H6O3	1,2,3-benzenetriol	87-66-1	0.03	1.08E+00	2
805	7121	C6H6S	phenyl mercaptan	108-98-5	2.52	3.31E+02	1
806	7172	C6H7N	aniline	62-53-3	0.90	7.94E+00	1
807	7173	C6H7N	2-methylpyridine	109-06-8	1.11	1.29E+01	1
808	7174	C6H7N	3-methylpyridine	108-99-6	1.20	1.58E+01	1
809	7175	C6H7N	4-methylpyridine	108-89-4	1.22	1.66E+01	1
810	7178	C6H7N	methylpyridine	1333-41-1	1.18	1.50E+01	2
811	7267	C6H8	1,3-cyclohexadiene	592-57-4	2.47	2.95E+02	2
812	7269	C6H8	1,4-cyclohexadiene	628-41-1	2.47	2.95E+02	1
813	7316	C6H8N2	adiponitrile	111-69-3	-0.32	4.79E-01	1
814	7317	C6H8N2	methylglutaronitrile	4553-62-2	0.28	1.91E+00	2
815	7318	C6H8N2	m-phenylenediamine	108-45-2	-0.33	4.68E-01	1
816	7319	C6H8N2	o-phenylenediamine	95-54-5	0.15	1.41E+00	1
817	7320	C6H8N2	p-phenylenediamine	106-50-3	-0.30	5.01E-01	1
818	7321	C6H8N2	phenylhydrazine	100-63-0	1.25	1.78E+01	1
819	7381	C6H8O	2-ethylfuran	3208-16-0	2.40	2.51E+02	1
820	7382	C6H8O	3-ethylfuran	67363-95-5	2.40	2.51E+02	2
821	7387	C6H8O	2-cyclohexen-1-one	930-68-7	0.61	4.07E+00	1
822	7388	C6H8O	3-cyclohexen-1-one	4096-34-8	0.61	4.07E+00	2
823	7395	C6H8O	3-hexyn-2-one	1679-36-3	0.58	3.80E+00	2
824	7396	C6H8O	5-hexyn-2-one	2550-28-9	0.58	3.80E+00	1
825	7459	C6H8O6	ascorbic acid	50-81-7	-1.64	2.29E-02	1
826	7464	C6H8O7	citric acid	77-92-9	-1.72	1.91E-02	1
827	7467	C6H8S	2-ethylthiophene	872-55-9	3.09	1.22E+03	2
828	7468	C6H8S	3-ethylthiophene	1795-01-3	3.13	1.35E+03	2
829	7469	C6H8S	2,3-dimethylthiophene	632-16-6	3.25	1.78E+03	2
830	7470	C6H8S	2,4-dimethylthiophene	638-00-6	3.23	1.70E+03	2
831	7471	C6H8S	2,5-dimethylthiophene	638-02-8	3.15	1.40E+03	2
832	7472	C6H8S	3,4-dimethylthiophene	632-15-5	3.32	2.11E+03	2
833	7622	C6H10	cyclohexene	110-83-8	2.86	7.24E+02	1
834	7623	C6H10	1,2-hexadiene	592-44-9	3.31	2.03E+03	2
835	7624	C6H10	1,cis-3-hexadiene	14596-92-0	3.25	1.76E+03	2
836	7625	C6H10	1,trans-3-hexadiene	20237-34-7	3.25	1.76E+03	2
837	7626	C6H10	1,4-hexadiene	592-45-0	3.16	1.45E+03	2
838	7627	C6H10	1,cis-4-hexadiene	7318-67-4	3.09	1.23E+03	2
839	7628	C6H10	1,trans-4-hexadiene	7319-00-8	3.09	1.23E+03	2
840	7629	C6H10	1,5-hexadiene	592-42-7	2.80	6.31E+02	1
841	7630	C6H10	2,3-hexadiene	592-49-4	3.15	1.40E+03	2
842	7631	C6H10	2,4-hexadiene	592-46-1	3.16	1.45E+03	2
843	7632	C6H10	cis-2,trans-4-hexadiene	5194-50-3	2.80	6.31E+02	1
844	7633	C6H10	cis-2,cis-4-hexadiene	6108-61-8	3.39	2.46E+03	2
845	7634	C6H10	trans-2,trans-4-hexadiene	5194-51-4	3.01	1.02E+03	1
846	7635	C6H10	3-methyl-1,2-pentadiene	7417-48-3	3.19	1.53E+03	2
847	7636	C6H10	4-methyl-1,2-pentadiene	13643-05-5	3.19	1.53E+03	2
848	7637	C6H10	3-methyl-1,3-pentadiene	4549-74-0	3.16	1.45E+03	2
849	7638	C6H10	2-methyl-1,cis-3-pentadiene	1501-60-6	3.31	2.03E+03	2
850	7639	C6H10	2-methyl-1,trans-3-pentadien	926-54-5	3.31	2.03E+03	2
851	7640	C6H10	3-methyl-1,cis-3-pentadiene	2787-45-3	3.33	2.13E+03	2
852	7641	C6H10	3-methyl-1,trans-3-pentadien	2787-43-1	3.33	2.13E+03	2
853	7642	C6H10	4-methyl-1,3-pentadiene	926-56-7	3.31	2.06E+03	2
854	7643	C6H10	2-methyl-1,4-pentadiene	763-30-4	2.82	6.67E+02	2
855	7644	C6H10	3-methyl-1,4-pentadiene	1115-08-8	2.91	8.09E+02	2
856	7645	C6H10	2-methyl-2,3-pentadiene	3043-33-2	3.23	1.68E+03	2
857	7646	C6H10	2-ethyl-1,3-butadiene	3404-63-5	3.29	1.93E+03	2
858	7647	C6H10	2,3-dimethyl-1,3-butadiene	513-81-5	3.13	1.35E+03	2
859	7648	C6H10	1-hexyne	693-02-7	2.73	5.37E+02	1
860	7649	C6H10	2-hexyne	764-35-2	2.80	6.25E+02	2
861	7650	C6H10	3-hexyne	928-49-4	2.72	5.30E+02	2
862	7651	C6H10	3-methyl-1-pentyne	922-59-8	2.26	1.84E+02	2
863	7652	C6H10	4-methyl-1-pentyne	7154-75-8	2.33	2.12E+02	2
864	7653	C6H10	4-methyl-2-pentyne	21020-27-9	2.56	3.61E+02	2
865	7654	C6H10	3,3-dimethyl-1-butyne	917-92-0	1.95	8.86E+01	2
866	7657	C6H10	1,3-hexadiene, cis and trans	592-48-3	3.16	1.45E+03	2
867	7658	C6H10	2-methyl-1,3-pentadiene	1118-58-7	3.16	1.45E+03	2
868	7659	C6H10	methylpentadiene, isomers	54363-49-4	3.16	1.45E+03	2
869	7660	C6H10	hexadiene	42296-74-2	2.80	6.31E+02	2
870	7760	C6H10O	cyclohexanone	108-94-1	0.81	6.46E+00	1
871	7761	C6H10O	mesityl oxide	141-79-7	1.37	2.34E+01	2
872	7776	C6H10O	5-hexen-2-one	109-49-9	1.02	1.05E+01	1
873	7813	C6H10O2	epsilon-caprolactone	502-44-3	0.68	4.79E+00	2
874	7814	C6H10O2	ethyl methacrylate	97-63-2	1.94	8.71E+01	1
875	7815	C6H10O2	propyl acrylate	925-60-0	1.71	5.13E+01	2
876	7817	C6H10O2	2-buten-1-yl acetate	628-08-0	1.71	5.13E+01	2
877	7818	C6H10O2	allyl propanoate	2408-20-0	1.71	5.13E+01	2
878	7829	C6H10O2	ethyl 3-butenoate	1617-18-1	1.71	5.13E+01	2
879	7830	C6H10O2	ethyl cis-2-butenoate	6776-19-8	1.71	5.13E+01	2
880	7831	C6H10O2	ethyl 2-butenoate	10544-63-5	1.71	5.13E+01	2

Octanol-Water Partition Coefficient							
NO	ID	FORMULA	NAME	CAS No	log_{10} Kow	Kow	code
881	7832	C6H10O2	ethyl trans-2-butenoate	623-70-1	1.71	5.13E+01	2
882	7833	C6H10O2	vinyl butanoate	123-20-6	1.71	5.13E+01	2
883	7834	C6H10O2	vinyl 2-methylpropanoate	2424-98-8	1.71	5.13E+01	2
884	7845	C6H10O2	methyl 2-methyl-2-butenoate	6622-76-0	1.71	5.13E+01	2
885	7846	C6H10O2	methyl 3-methyl-2-butenoate	924-50-5	1.71	5.13E+01	2
886	7859	C6H10O2	methyl trans-2-pentenoate	15790-88-2	1.71	5.13E+01	2
887	7860	C6H10O2	methyl trans-3-pentenoate	20515-19-9	1.71	5.13E+01	2
888	7881	C6H10O3	ethylacetoacetate	141-97-9	0.25	1.78E+00	1
889	7882	C6H10O3	propionic anhydride	123-62-6	0.40	2.51E+00	2
890	7917	C6H10O4	adipic acid	124-04-9	0.08	1.20E+00	1
891	7918	C6H10O4	diethyl oxalate	95-92-1	0.56	3.63E+00	1
892	7919	C6H10O4	ethylene glycol diacetate	111-55-7	0.40	2.51E+00	2
893	7981	C6H11Br	bromocyclohexane	108-85-0	3.20	1.58E+03	1
894	8088	C6H11N	hexanenitrile	628-73-9	1.60	3.98E+01	1
895	8089	C6H11N	2-methylpentanenitrile	6339-13-5	1.60	3.98E+01	2
896	8090	C6H11N	2-ethylbutanenitrile	617-80-1	1.60	3.98E+01	2
897	8091	C6H11N	3-methylpentanenitrile	21101-88-2	1.60	3.98E+01	2
898	8092	C6H11N	4-methylpentanenitrile	542-54-1	1.60	3.98E+01	2
899	8093	C6H11N	2,2-dimethylbutanenitrile	20654-46-0	1.60	3.98E+01	2
900	8094	C6H11N	2,3-dimethylbutanenitrile	20654-44-8	1.60	3.98E+01	2
901	8095	C6H11N	3,3-dimethylbutanenitrile	3302-16-7	1.60	3.98E+01	2
902	8101	C6H11NO	epsilon-caprolactam	105-60-2	-0.19	6.46E-01	1
903	8102	C6H11NO	cyclohexanone oxime	100-64-1	0.84	6.92E+00	1
904	8214	C6H12	methylcyclopentane	96-37-7	3.37	2.34E+03	1
905	8215	C6H12	cyclohexane	110-82-7	3.44	2.75E+03	1
906	8216	C6H12	1-hexene	592-41-6	3.40	2.51E+03	1
907	8217	C6H12	2-hexene; (cis+trans)	592-43-8	3.24	1.72E+03	2
908	8218	C6H12	cis-2-hexene	7688-21-3	3.07	1.17E+03	2
909	8219	C6H12	trans-2-hexene	4050-45-7	3.07	1.17E+03	2
910	8220	C6H12	hex-3-ene	592-47-2	3.24	1.72E+03	2
911	8221	C6H12	cis-3-hexene	7642-09-3	3.57	3.75E+03	2
912	8222	C6H12	trans-3-hexene	13269-52-8	3.59	3.85E+03	2
913	8223	C6H12	2-methyl-1-pentene	763-29-1	3.21	1.62E+03	2
914	8224	C6H12	3-methyl-1-pentene	760-20-3	3.08	1.20E+03	2
915	8225	C6H12	4-methyl-1-pentene	691-37-2	2.50	3.16E+02	1
916	8226	C6H12	4-methyl-2-pentene	4461-48-7	3.24	1.72E+03	2
917	8227	C6H12	4-methyl-cis-2-pentene	691-38-3	3.00	1.00E+03	2
918	8228	C6H12	4-methyl-trans-2-pentene	674-76-0	3.00	1.00E+03	2
919	8229	C6H12	2-methyl-2-pentene	625-27-4	3.13	1.35E+03	2
920	8230	C6H12	3-methyl-cis-2-pentene	922-62-3	3.60	3.96E+03	2
921	8231	C6H12	3-methyl-2-pentene, cis and t	922-61-2	3.24	1.72E+03	2
922	8232	C6H12	3-methyl-trans-2-pentene	616-12-6	3.65	4.48E+03	2
923	8233	C6H12	2,3-dimethyl-1-butene	563-78-0	3.13	1.35E+03	2
924	8234	C6H12	2,3-dimethyl-2-butene	563-79-1	3.19	1.55E+03	2
925	8235	C6H12	3,3-dimethyl-1-butene	558-37-2	3.04	1.10E+03	2
926	8236	C6H12	2-ethyl-1-butene	760-21-4	3.21	1.62E+03	2
927	8315	C6H12N2	triethylenediamine	280-57-9	-0.83	1.48E-01	2
928	8394	C6H12O	butyl vinyl ether	111-34-2	1.89	7.76E+01	2
929	8395	C6H12O	sec-butyl vinyl ether	1888-85-3	1.89	7.76E+01	2
930	8396	C6H12O	tert-butyl vinyl ether	926-02-3	1.89	7.76E+01	2
931	8397	C6H12O	isobutyl vinyl ether	109-53-5	1.89	7.76E+01	2
932	8398	C6H12O	3-(1-methylethoxy)-1-propene	6140-80-3	1.89	7.76E+01	2
933	8399	C6H12O	3-propoxy-1-propene	1471-03-0	1.89	7.76E+01	2
934	8400	C6H12O	cyclohexanol	108-93-0	1.23	1.70E+01	1
935	8401	C6H12O	hexanal	66-25-1	1.78	6.03E+01	1
936	8402	C6H12O	2-methylpentanal	123-15-9	1.53	3.37E+01	2
937	8403	C6H12O	3-methylpentanal	15877-57-3	1.63	4.29E+01	2
938	8404	C6H12O	4-methylpentanal	1119-16-0	1.63	4.29E+01	2
939	8405	C6H12O	2-ethylbutanal	97-96-1	1.52	3.34E+01	2
940	8406	C6H12O	2,2-dimethylbutanal	2094-75-9	1.26	1.82E+01	2
941	8407	C6H12O	2,3-dimethylbutanal	2109-98-0	1.44	2.78E+01	2
942	8408	C6H12O	3,3-dimethylbutanal	2987-16-8	1.26	1.82E+01	2
943	8409	C6H12O	2-hexanone	591-78-6	1.38	2.40E+01	1
944	8410	C6H12O	3-hexanone	589-38-8	1.24	1.74E+01	2
945	8411	C6H12O	3-methyl-2-pentanone	565-61-7	1.21	1.62E+01	2
946	8412	C6H12O	4-methyl-2-pentanone	108-10-1	1.31	2.04E+01	1
947	8413	C6H12O	ethyl isopropyl ketone	565-69-5	1.16	1.45E+01	2
948	8414	C6H12O	3,3-dimethyl-2-butanone	75-97-8	0.91	8.07E+00	2
949	8415	C6H12O	3-methyl-2-pentanone, (±)	55156-16-6	1.20	1.59E+01	2
950	8453	C6H12O	2,2-dimethyl-3-butanone	---	1.20	1.59E+01	2
951	8460	C6H12O2	hexanoic acid	142-62-1	1.92	8.32E+01	1
952	8461	C6H12O2	2-ethyl butyric acid	88-09-5	1.68	4.79E+01	1
953	8462	C6H12O2	tert-butylacetic acid	1070-83-3	1.80	6.31E+01	2
954	8463	C6H12O2	2,2-dimethylbutanoic acid	595-37-9	1.80	6.31E+01	2
955	8464	C6H12O2	2,3-dimethylbutanoic acid	14287-61-7	1.80	6.31E+01	2
956	8465	C6H12O2	2-methylpentanoic acid, (±)	22160-39-0	1.80	6.31E+01	2
957	8466	C6H12O2	3-methylpentanoic acid, (±)	22160-40-3	1.80	6.31E+01	2
958	8467	C6H12O2	4-methylpentanoic acid	646-07-1	1.80	6.31E+01	2
959	8468	C6H12O2	pentyl formate	638-49-3	2.32	2.08E+02	2
960	8469	C6H12O2	1-methylbutyl formate	58368-66-4	2.01	1.03E+02	2

Table 13 Octanol-Water Partition Coefficient - Oganic Compounds

			Octanol-Water Partition Coefficient				
NO	ID	FORMULA	NAME	CAS No	log_{10} Kow	Kow	code
961	8470	C6H12O2	2-methylbutyl formate	35073-27-9	2.01	1.03E+02	2
962	8471	C6H12O2	isopentyl formate	110-45-2	2.10	1.25E+02	2
963	8472	C6H12O2	1-ethylpropyl formate	58368-67-5	2.01	1.03E+02	2
964	8473	C6H12O2	1,1-dimethylpropyl formate	757-88-0	1.83	6.72E+01	2
965	8474	C6H12O2	1,2-dimethylpropyl formate	66794-46-5	2.00	1.01E+02	2
966	8475	C6H12O2	2,2-dimethylpropyl formate	23361-67-3	1.95	8.91E+01	2
967	8476	C6H12O2	butyl acetate	123-86-4	1.82	6.61E+01	1
968	8477	C6H12O2	isobutyl acetate	110-19-0	1.78	6.03E+01	1
969	8478	C6H12O2	sec-butyl acetate	105-46-4	1.72	5.25E+01	1
970	8479	C6H12O2	tert-butyl acetate	540-88-5	1.76	5.75E+01	1
971	8480	C6H12O2	propyl propanoate	106-36-5	1.85	7.08E+01	2
972	8481	C6H12O2	isopropyl propanoate	637-78-5	1.85	7.08E+01	2
973	8482	C6H12O2	ethyl butanoate	105-54-4	1.85	7.08E+01	2
974	8483	C6H12O2	ethyl isobutanoate	97-62-1	1.77	5.89E+01	2
975	8485	C6H12O2	diacetone alcohol	123-42-2	-0.34	4.57E-01	2
976	8497	C6H12O2	methyl 2,2-dimethylpropanoa	598-98-1	1.85	7.08E+01	2
977	8514	C6H12O2	(±)-3-methylvaleric acid	105-43-1	1.80	6.31E+01	2
978	8515	C6H12O2	2-methylvaleric acid	97-61-0	1.80	6.31E+01	2
979	8531	C6H12O3	2-ethoxyethyl acetate	111-15-9	0.59	3.89E+00	2
980	8532	C6H12O3	hydroxycaproic acid	1191-25-9	0.81	6.46E+00	1
981	8533	C6H12O3	paraldehyde	123-63-7	0.67	4.68E+00	1
982	8609	C6H12O6	glucose	50-99-7	-3.24	5.75E-04	1
983	8610	C6H12O6	L(-)-glucose	921-60-8	-3.24	5.75E-04	2
984	8611	C6H12O6	a-D(+)-glucose, anhydrous	492-62-6	-3.24	5.75E-04	2
985	8612	C6H12O6	L(+)-gulose	6027-89-0	-3.24	5.75E-04	2
986	8647	C6H13Br	1-bromohexane	111-25-1	3.80	6.31E+03	1
987	8648	C6H13Br	2-bromohexane	3377-86-4	3.80	6.31E+03	2
988	8649	C6H13Br	3-bromohexane	3377-87-5	3.80	6.31E+03	2
989	8650	C6H13Br	1-bromo-2-methylpentane	25346-33-2	3.80	6.31E+03	2
990	8651	C6H13Br	1-bromo-3-methylpentane	51116-73-5	3.80	6.31E+03	2
991	8652	C6H13Br	1-bromo-4-methylpentane	626-88-0	3.80	6.31E+03	2
992	8653	C6H13Br	2-bromo-2-methylpentane	4283-80-1	3.80	6.31E+03	2
993	8654	C6H13Br	2-bromo-3-methylpentane	62168-41-6	3.80	6.31E+03	2
994	8655	C6H13Br	2-bromo-4-methylpentane	30310-22-6	3.80	6.31E+03	2
995	8656	C6H13Br	3-bromo-2-methylpentane	4283-83-4	3.80	6.31E+03	2
996	8657	C6H13Br	3-bromo-3-methylpentane	25346-31-0	3.80	6.31E+03	2
997	8658	C6H13Br	1-bromo-2,2-dimethylbutane	62168-42-7	3.80	6.31E+03	2
998	8659	C6H13Br	1-bromo-2,3-dimethylbutane	30540-31-9	3.80	6.31E+03	2
999	8660	C6H13Br	1-bromo-3,3-dimethylbutane	1647-23-0	3.80	6.31E+03	2
1000	8661	C6H13Br	2-bromo-2,3-dimethylbutane	594-52-5	3.80	6.31E+03	2
1001	8662	C6H13Br	2-bromo-3,3-dimethylbutane	26356-06-9	3.80	6.31E+03	2
1002	8663	C6H13Br	1-bromo-2-ethylbutane	3814-34-4	3.80	6.31E+03	2
1003	8669	C6H13Cl	1-chlorohexane	544-10-5	3.54	3.47E+03	2
1004	8670	C6H13Cl	2-chlorohexane	638-28-8	3.47	2.95E+03	2
1005	8671	C6H13Cl	3-chlorohexane	2346-81-8	3.50	3.16E+03	2
1006	8672	C6H13Cl	1-chloro-2-methylpentane	14753-05-0	3.50	3.16E+03	2
1007	8673	C6H13Cl	1-chloro-3-methylpentane	62016-93-7	3.50	3.16E+03	2
1008	8674	C6H13Cl	1-chloro-4-methylpentane	62016-94-8	3.50	3.16E+03	2
1009	8675	C6H13Cl	2-chloro-2-methylpentane	4325-48-8	3.50	3.16E+03	2
1010	8676	C6H13Cl	2-chloro-3-methylpentane	24319-09-3	3.50	3.16E+03	2
1011	8677	C6H13Cl	2-chloro-4-methylpentane	25346-32-1	3.50	3.16E+03	2
1012	8678	C6H13Cl	3-chloro-2-methylpentane	38384-05-3	3.50	3.16E+03	2
1013	8679	C6H13Cl	3-chloro-3-methylpentane	918-84-3	3.50	3.16E+03	2
1014	8680	C6H13Cl	1-chloro-2-ethylbutane	4737-41-1	3.50	3.16E+03	2
1015	8681	C6H13Cl	1-chloro-2,2-dimethylbutane	6366-35-4	3.50	3.16E+03	2
1016	8682	C6H13Cl	1-chloro-2,3-dimethylbutane	600-06-6	3.50	3.16E+03	2
1017	8683	C6H13Cl	1-chloro-3,3-dimethylbutane	2855-08-5	3.50	3.16E+03	2
1018	8684	C6H13Cl	2-chloro-2,3-dimethylbutane	594-57-0	3.50	3.16E+03	2
1019	8685	C6H13Cl	2-chloro-3,3-dimethylbutane	5750-00-5	3.50	3.16E+03	2
1020	8742	C6H13N	cyclohexylamine	108-91-8	1.49	3.09E+01	1
1021	8745	C6H13N	hexamethyleneimine	111-49-9	1.68	4.79E+01	2
1022	8877	C6H14	hexane	110-54-3	4.00	1.00E+04	1
1023	8878	C6H14	2,2-dimethylbutane	75-83-2	3.82	6.61E+03	1
1024	8879	C6H14	2,3-dimethylbutane	79-29-8	3.85	7.08E+03	1
1025	8880	C6H14	2-methylpentane	107-83-5	3.21	1.62E+03	2
1026	8881	C6H14	3-methylpentane	96-14-0	3.60	3.98E+03	1
1027	8944	C6H14N2O2	lysine	56-87-1	-3.05	8.91E-04	1
1028	8970	C6H14O	1-hexanol	111-27-3	2.03	1.07E+02	1
1029	8971	C6H14O	2-hexanol	626-93-7	1.76	5.75E+01	1
1030	8972	C6H14O	3-hexanol	623-37-0	1.65	4.47E+01	1
1031	8973	C6H14O	2-methyl-1-pentanol	105-30-6	1.75	5.62E+01	2
1032	8974	C6H14O	3-methyl-1-pentanol	589-35-5	2.07	1.16E+02	2
1033	8975	C6H14O	4-methyl-1-pentanol	626-89-1	2.05	1.12E+02	2
1034	8976	C6H14O	4-methyl-2-pentanol	108-11-2	1.43	2.69E+01	2
1035	8977	C6H14O	2-methyl-2-pentanol	590-36-3	1.29	1.93E+01	2
1036	8978	C6H14O	3-methyl-2-pentanol	565-60-6	1.61	4.12E+01	2
1037	8979	C6H14O	2-methyl-3-pentanol	565-67-3	1.42	2.62E+01	2
1038	8980	C6H14O	3-methyl-3-pentanol	77-74-7	1.31	2.05E+01	2
1039	8981	C6H14O	2-ethyl-1-butanol	97-95-0	1.75	5.62E+01	2
1040	8982	C6H14O	2,2-dimethyl-1-butanol	1185-33-7	1.68	4.79E+01	2

			Octanol-Water Partition Coefficient				
NO	ID	FORMULA	NAME	CAS No	log_{10} Kow	Kow	code
1041	8983	C6H14O	2,3-dimethyl-1-butanol	19550-30-2	1.98	9.60E+01	2
1042	8984	C6H14O	2,3-dimethyl-1-butanol, (±)-	20281-85-0	1.65	4.48E+01	2
1043	8985	C6H14O	3,3-dimethyl-1-butanol	624-95-3	1.83	6.83E+01	2
1044	8986	C6H14O	2,3-dimethyl-2-butanol	594-60-5	1.21	1.63E+01	2
1045	8987	C6H14O	3,3-dimethyl-2-butanol	464-07-3	1.48	3.02E+01	1
1046	8988	C6H14O	3,3-dimethyl-2-butanol, (±)	20281-91-8	1.65	4.48E+01	2
1047	8989	C6H14O	(RS)-2-hexanol	20281-86-1	1.65	4.48E+01	2
1048	8990	C6H14O	2-hexanol, (R)	26549-24-6	1.65	4.48E+01	2
1049	8991	C6H14O	3-hexanol, (R)	13471-42-6	1.65	4.48E+01	2
1050	8992	C6H14O	3-methyl-1-pentanol, (±)	20281-83-8	1.65	4.48E+01	2
1051	8993	C6H14O	dipropyl ether	111-43-3	2.03	1.07E+02	1
1052	8994	C6H14O	propyl isopropyl ether	627-08-7	1.99	9.75E+01	2
1053	8995	C6H14O	diisopropyl ether	108-20-3	1.52	3.31E+01	1
1054	8996	C6H14O	methyl pentyl ether	628-80-8	2.43	2.71E+02	2
1055	8997	C6H14O	methyl 1-methylbutyl ether	6795-88-6	2.24	1.75E+02	2
1056	8998	C6H14O	methyl 2-methylbutyl ether	62016-48-2	2.22	1.66E+02	2
1057	8999	C6H14O	methyl 3-methylbutyl ether	626-91-5	2.22	1.66E+02	2
1058	9000	C6H14O	methyl 1-ethylpropyl ether	36839-67-5	2.17	1.49E+02	2
1059	9001	C6H14O	methyl 1,2-dimethylpropyl eth	62016-49-3	2.06	1.14E+02	2
1060	9002	C6H14O	methyl tert-pentyl ether	994-05-8	1.92	8.32E+01	2
1061	9003	C6H14O	methyl 2,2-dimethylpropyl eth	1118-00-9	2.06	1.14E+02	2
1062	9004	C6H14O	ethyl butyl ether	628-81-9	2.03	1.07E+02	1
1063	9005	C6H14O	ethyl isobutyl ether	627-02-1	2.01	1.03E+02	2
1064	9006	C6H14O	ethyl sec-butyl ether	2679-87-0	2.02	1.04E+02	2
1065	9007	C6H14O	ethyl tert-butyl ether	637-92-3	1.83	6.71E+01	2
1066	9008	C6H14O	(S)-(+)-2-hexanol	52019-78-0	1.65	4.48E+01	2
1067	9009	C6H14O	tert-hexyl alcohol	26401-20-7	1.65	4.48E+01	2
1068	9020	C6H14O2	acetal	105-57-7	0.84	6.92E+00	1
1069	9021	C6H14O2	2-butoxyethanol	111-76-2	0.83	6.76E+00	1
1070	9022	C6H14O2	hexylene glycol	107-41-5	0.58	3.80E+00	2
1071	9023	C6H14O2	1,2-hexanediol	6920-22-5	0.67	4.68E+00	2
1072	9024	C6H14O2	1,3-hexanediol	21531-91-9	0.67	4.68E+00	2
1073	9025	C6H14O2	1,4-hexanediol	16432-53-4	0.67	4.68E+00	2
1074	9026	C6H14O2	1,5-hexanediol	928-40-5	0.67	4.68E+00	2
1075	9027	C6H14O2	1,6-hexanediol	629-11-8	0.76	5.75E+00	2
1076	9028	C6H14O2	2,3-hexanediol	617-30-1	0.67	4.68E+00	2
1077	9029	C6H14O2	2,4-hexanediol	19780-90-6	0.67	4.68E+00	2
1078	9030	C6H14O2	2,5-hexanediol	2935-44-6	0.67	4.68E+00	2
1079	9031	C6H14O2	3,4-hexanediol	922-17-8	0.67	4.68E+00	2
1080	9032	C6H14O2	2-methyl-1,2-pentanediol	20667-05-4	0.67	4.68E+00	2
1081	9033	C6H14O2	2-methyl-1,3-pentanediol	149-31-5	0.67	4.68E+00	2
1082	9034	C6H14O2	2-methyl-1,5-pentanediol	42856-62-2	0.67	4.68E+00	2
1083	9035	C6H14O2	2-methyl-2,3-pentanediol	7795-80-4	0.67	4.68E+00	2
1084	9036	C6H14O2	3-methyl-1,3-pentanediol	33879-72-0	0.67	4.68E+00	2
1085	9037	C6H14O2	3-methyl-1,5-pentanediol	4457-71-0	0.67	4.68E+00	2
1086	9038	C6H14O2	3-methyl-2,3-pentanediol	63521-37-9	0.67	4.68E+00	2
1087	9039	C6H14O2	3-methyl-2,4-pentanediol	5683-44-3	0.67	4.68E+00	2
1088	9040	C6H14O2	4-methyl-1,4-pentanediol	1462-10-8	0.67	4.68E+00	2
1089	9041	C6H14O2	DL-erythro-4-methyl-2,3-pent	6702-10-9	0.67	4.68E+00	2
1090	9042	C6H14O2	DL-threo-4-methyl-2,3-pentar	6464-40-0	0.67	4.68E+00	2
1091	9043	C6H14O2	2-ethyl-1,2-butanediol	66553-16-0	0.67	4.68E+00	2
1092	9044	C6H14O2	2-ethyl-1,3-butanediol	66553-17-1	0.67	4.68E+00	2
1093	9045	C6H14O2	2-ethyl-1,4-butanediol	57716-79-7	0.67	4.68E+00	2
1094	9046	C6H14O2	2,2-dimethyl-1,3-butanediol	76-35-7	0.67	4.68E+00	2
1095	9047	C6H14O2	2,2-dimethyl-1,4-butanediol	32812-23-0	0.67	4.68E+00	2
1096	9048	C6H14O2	2,3-dimethyl-1,2-butanediol	66553-15-9	0.67	4.68E+00	2
1097	9049	C6H14O2	2,3-dimethyl-1,3-butanediol	24893-35-4	0.67	4.68E+00	2
1098	9050	C6H14O2	DL-2,3 -dimethyl-1,4-butaned	---	0.67	4.68E+00	2
1099	9051	C6H14O2	meso-2,3-dimethyl-1,4-butan	---	0.67	4.68E+00	2
1100	9052	C6H14O2	2,3-dimethyl-2,3-butanediol	76-09-5	0.67	4.68E+00	2
1101	9053	C6H14O2	3,3-dimethyl-1,2-butanediol	59562-82-2	0.67	4.68E+00	2
1102	9054	C6H14O2	2-propyl-1,3-propanediol	2612-28-4	0.67	4.68E+00	2
1103	9055	C6H14O2	2-isopropyl-1,3-propanediol	2612-27-3	0.67	4.68E+00	2
1104	9056	C6H14O2	2-methyl-2-ethyl-1,3-propane	77-84-9	0.67	4.68E+00	2
1105	9057	C6H14O2	1-(1-methylethoxy)-2-propanc	3944-36-3	0.83	6.76E+00	2
1106	9058	C6H14O2	2-(2-methylpropoxy)ethanol	4439-24-1	0.83	6.76E+00	2
1107	9059	C6H14O2	1-propoxy-2-propanol	1569-01-3	0.83	6.76E+00	2
1108	9060	C6H14O2	ethylene glycol diethyl ether	629-14-1	0.83	6.76E+00	2
1109	9061	C6H14O2	(2R,5R)-2,5-hexanediol	17299-07-9	0.67	4.68E+00	2
1110	9062	C6H14O2	(2S,5S)-2,5-hexanediol	34338-96-0	0.67	4.68E+00	2
1111	9063	C6H14O2	3-methoxy-3-methylbutanol	56539-66-3	0.83	6.76E+00	2
1112	9064	C6H14O2	(R)-(-)-2-methyl-2,4-pentanec	99210-90-9	0.67	4.68E+00	2
1113	9067	C6H14O2	ethylene glycol mono-sec-but	7795-91-7	0.83	6.76E+00	2
1114	9068	C6H14O2	n-propoxypropanol (mixed isc	30136-13-1	0.83	6.76E+00	2
1115	9069	C6H14O2S	dipropyl sulfone	598-03-8	0.39	2.45E+00	1
1116	9075	C6H14O3	diethylene glycol dimethyl eth	111-96-6	-0.36	4.37E-01	1
1117	9077	C6H14O3	2-(2-ethoxyethoxy)ethanol	111-90-0	-0.54	2.88E-01	1
1118	9078	C6H14O3	trimethylolpropane	77-99-6	-1.48	3.31E-02	1
1119	9092	C6H14O4	triethylene glycol	112-27-6	-1.98	1.05E-02	2
1120	9100	C6H14O6	sorbitol	50-70-4	-2.20	6.31E-03	1

Table 13 Octanol-Water Partition Coefficient - Oganic Compounds

Octanol-Water Partition Coefficient							
NO	ID	FORMULA	NAME	CAS No	log_{10} Kow	Kow	code
1121	9107	C6H14S	hexyl mercaptan	111-31-9	3.23	1.70E+03	2
1122	9108	C6H14S	2-hexanethiol	1679-06-7	3.26	1.83E+03	2
1123	9109	C6H14S	2,3-dimethyl-2-butanethiol	1639-01-6	3.25	1.76E+03	2
1124	9110	C6H14S	3-hexanethiol	1633-90-5	3.25	1.76E+03	2
1125	9111	C6H14S	2-methyl-2-pentanethiol	1633-97-2	3.25	1.76E+03	2
1126	9112	C6H14S	dipropyl sulfide	111-47-7	2.88	7.59E+02	2
1127	9113	C6H14S	diisopropyl sulfide	625-80-9	2.52	3.33E+02	2
1128	9114	C6H14S	methyl pentyl sulfide	1741-83-9	2.88	7.59E+02	2
1129	9115	C6H14S	ethyl butyl sulfide	638-46-0	2.88	7.59E+02	2
1130	9116	C6H14S	3-methyl-2-thiahexane	13286-91-4	2.92	8.32E+02	2
1131	9117	C6H14S	4-methyl-2-thiahexane	15013-37-3	2.94	8.74E+02	2
1132	9118	C6H14S	5-methyl-2-thiahexane	13286-90-3	2.90	7.92E+02	2
1133	9119	C6H14S	2-methyl-3-thiahexane	5008-73-1	2.77	5.91E+02	2
1134	9120	C6H14S	4-methyl-3-thiahexane	5008-72-0	2.81	6.39E+02	2
1135	9121	C6H14S	5-methyl-3-thiahexane	1613-45-2	2.82	6.57E+02	2
1136	9122	C6H14S	3-ethyl-2-thiapentane	57093-84-2	2.77	5.86E+02	2
1137	9123	C6H14S	3,3-dimethyl-2-thiapentane	13286-92-5	2.69	4.86E+02	2
1138	9124	C6H14S	3,4-dimethyl-2-thiapentane	53897-51-1	2.69	4.86E+02	2
1139	9125	C6H14S	4,4-dimethyl-2-thiapentane	6079-57-8	2.69	4.86E+02	2
1140	9126	C6H14S	2,2-dimethyl-3-thiapentane	14290-92-7	2.53	3.39E+02	2
1141	9127	C6H14S2	1,1-bis(ethylthio)ethane	14252-42-7	3.84	6.92E+03	2
1142	9128	C6H14S2	1,2-bis(ethylthio)ethane	5395-75-5	3.84	6.92E+03	2
1143	9129	C6H14S2	diisopropyl disulfide	4253-89-8	3.84	6.92E+03	2
1144	9130	C6H14S2	dipropyl disulfide	629-19-6	3.84	6.92E+03	2
1145	9173	C6H15N	hexylamine	111-26-2	2.06	1.15E+02	1
1146	9174	C6H15N	2-aminohexane	5329-79-3	1.85	7.03E+01	2
1147	9175	C6H15N	3-aminohexane	16751-58-9	1.98	9.65E+01	2
1148	9176	C6H15N	1-amino-2-methylpentane	13364-16-4	1.77	5.89E+01	2
1149	9177	C6H15N	1-amino-3-methylpentane	42245-37-4	1.77	5.89E+01	2
1150	9178	C6H15N	1-amino-4-methylpentane	5344-20-7	1.70	5.00E+01	2
1151	9179	C6H15N	2-amino-2-methylpentane	53310-02-4	1.22	1.67E+01	2
1152	9180	C6H15N	2-amino-3-methylpentane	35399-81-6	1.97	9.41E+01	2
1153	9181	C6H15N	2-amino-4-methylpentane	108-09-8	1.33	2.13E+01	2
1154	9182	C6H15N	3-amino-2-methylpentane	54287-41-1	1.97	9.41E+01	2
1155	9183	C6H15N	3-amino-3-methylpentane	3495-46-3	1.73	5.40E+01	2
1156	9184	C6H15N	1-amino-2,2-dimethylbutane	41781-17-3	1.45	2.80E+01	2
1157	9185	C6H15N	1-amino-2,3-dimethylbutane	66553-05-7	1.25	1.76E+01	2
1158	9186	C6H15N	1-amino-3,3-dimethylbutane	15673-00-4	1.92	8.25E+01	2
1159	9187	C6H15N	2-amino-2,3-dimethylbutane	4358-75-2	1.23	1.71E+01	2
1160	9188	C6H15N	2-amino-3,3-dimethyl butane	3850-30-4	1.18	1.50E+01	2
1161	9189	C6H15N	1-amino-2-ethylbutane	617-79-8	1.98	9.65E+01	2
1162	9190	C6H15N	2-hexanamine, (±)	68107-05-1	1.67	4.66E+01	2
1163	9191	C6H15N	methylpentylamine	25419-06-1	1.78	6.02E+01	2
1164	9192	C6H15N	methyl-1-methylbutylamine	51932-19-5	1.68	4.78E+01	2
1165	9193	C6H15N	methyl-2-methylbutylamine	90023-96-4	1.68	4.78E+01	2
1166	9194	C6H15N	methyl-3-methylbutylamine	52317-98-3	1.55	3.59E+01	2
1167	9195	C6H15N	methyl-1,1-dimethylpropylami	---	1.61	4.10E+01	2
1168	9196	C6H15N	methyl-1,2-dimethylpropylami	34317-39-0	1.68	4.78E+01	2
1169	9197	C6H15N	methyl-2,2-dimethylpropylami	26153-91-3	1.61	4.10E+01	2
1170	9198	C6H15N	ethylbutylamine	13360-63-9	1.55	3.59E+01	2
1171	9199	C6H15N	ethyl-sec-butylamine	21035-44-9	1.32	2.10E+01	2
1172	9200	C6H15N	ethylisobutylamine	13205-60-2	1.32	2.10E+01	2
1173	9201	C6H15N	ethyl-tert-butylamine	4432-77-3	0.88	7.61E+00	2
1174	9202	C6H15N	di-propylamine	142-84-7	1.67	4.68E+01	1
1175	9203	C6H15N	diisopropylamine	108-18-9	1.40	2.51E+01	1
1176	9204	C6H15N	propylisopropylamine	21968-17-2	1.33	2.13E+01	2
1177	9205	C6H15N	N-ethyl-2-butanamine, (±)	116724-10-8	1.51	3.20E+01	2
1178	9206	C6H15N	methylethylpropylamine	4458-32-6	1.30	1.98E+01	2
1179	9207	C6H15N	methylethylisopropylamine	39198-07-7	1.30	1.98E+01	2
1180	9208	C6H15N	dimethylbutylamine	927-62-8	1.35	2.26E+01	2
1181	9209	C6H15N	dimethyl-sec-butylamine	921-04-0	1.35	2.26E+01	2
1182	9210	C6H15N	dimethylisobutylamine	7239-24-9	1.07	1.17E+01	2
1183	9211	C6H15N	dimethyl-tert-butylamine	918-02-5	1.26	1.84E+01	2
1184	9212	C6H15N	triethylamine	121-44-8	1.45	2.82E+01	1
1185	9230	C6H15NO2	diisopropanolamine	110-97-4	-0.82	1.51E-01	1
1186	9239	C6H15NO3	triethanolamine	102-71-6	-1.00	1.00E-01	1
1187	9268	C6H15O4P	triethyl phosphate	78-40-0	0.80	6.31E+00	1
1188	9292	C6H16N2	hexamethylenediamine	124-09-4	0.35	2.24E+00	2
1189	9293	C6H16N2	N,N-diethyl-1,2-ethanediamin	100-36-7	0.35	2.24E+00	2
1190	9294	C6H16N2	N,N'-diethyl-1,2-ethanediamir	111-74-0	0.35	2.24E+00	2
1191	9295	C6H16N2	N,N,N',N'-tetramethyl-1,2-eth	110-18-9	0.35	2.24E+00	2
1192	9297	C6H16N2	N-butylethylenediamine	19522-69-1	0.35	2.24E+00	2
1193	9298	C6H16N2	1,5-diamino-2-methylpentane	15520-10-2	0.35	2.24E+00	2
1194	9299	C6H16N2	N-isopropyl-1,3-propanediam	3360-16-5	0.35	2.24E+00	2
1195	9300	C6H16N2	N-propyl-1,3-propanediamine	23764-31-0	0.35	2.24E+00	2
1196	9301	C6H16N2	N,N,N'-trimethyl-1,3-propane	4543-96-8	0.35	2.24E+00	2
1197	9302	C6H16N2	2-methyl-2,4-pentanediamine	21586-21-0	0.35	2.24E+00	2
1198	9356	C6H18N3OP	hexamethyl phosphoramide	680-31-9	0.28	1.91E+00	1
1199	9359	C6H18N4	triethylene tetramine	112-24-3	-2.65	2.24E-03	2
1200	9362	C6H18OSi2	hexamethyldisiloxane	107-46-0	4.20	1.58E+04	1

Octanol-Water Partition Coefficient							
NO	ID	FORMULA	NAME	CAS No	log_{10} Kow	Kow	code
1201	9366	C6H18O3Si3	hexamethylcyclotrisiloxane	541-05-9	4.47	2.95E+04	2
1202	9378	C6H19NSi2	hexamethyldisilazane	999-97-3	2.62	4.17E+02	2
1203	9519	C7H3Cl2F3	2,4-dichlorobenzotrifluoride	320-60-5	4.24	1.74E+04	2
1204	9527	C7H3Cl2NO	3,4-dichlorophenyl isocyanat	102-36-3	3.88	7.59E+03	2
1205	9684	C7H4ClF3	p-chlorobenzotrifluoride	98-56-6	3.60	3.98E+03	2
1206	9736	C7H4Cl2O	m-chlorobenzoyl chloride	618-46-2	2.08	1.20E+02	2
1207	9799	C7H4F3NO2	3-nitrobenzotrifluoride	98-46-4	2.62	4.17E+02	1
1208	9890	C7H5BrO2	2-bromobenzoic acid	88-65-3	2.20	1.58E+02	1
1209	9891	C7H5BrO2	3-bromobenzoic acid	585-76-2	2.87	7.41E+02	1
1210	9892	C7H5BrO2	4-bromobenzoic acid	586-76-5	2.86	7.24E+02	1
1211	9927	C7H5ClO	benzoyl chloride	98-88-4	1.44	2.75E+01	2
1212	9932	C7H5ClO2	o-chlorobenzoic acid	118-91-2	2.05	1.12E+02	1
1213	9967	C7H5Cl3	benzotrichloride	98-07-7	3.90	7.94E+03	2
1214	9968	C7H5Cl3	1,2-dichloro-4-(chloromethyl)	102-47-6	3.90	7.94E+03	2
1215	9969	C7H5Cl3	1,3-dichloro-5-(chloromethyl)	3290-06-0	3.90	7.94E+03	2
1216	9970	C7H5Cl3	1,2,3-trichloro-4-methylbenze	7359-72-0	3.90	7.94E+03	2
1217	9971	C7H5Cl3	1,2,3-trichloro-5-methylbenze	21472-86-6	3.90	7.94E+03	2
1218	9972	C7H5Cl3	1,2,4-trichloro-5-methylbenze	6639-30-1	3.90	7.94E+03	2
1219	9973	C7H5Cl3	1,2,5-trichloro-3-methylbenze	56961-86-5	3.90	7.94E+03	2
1220	9974	C7H5Cl3	3-chlorobenzal chloride	---	3.90	7.94E+03	2
1221	9975	C7H5Cl3	2,6-dichlorobenzyl chloride	2014-83-7	3.90	7.94E+03	2
1222	9976	C7H5Cl3	dichlorobenzyl chloride	38721-71-0	3.90	7.94E+03	2
1223	9977	C7H5Cl3	2,4-dichlorobenzyl chloride	94-99-5	3.90	7.94E+03	2
1224	9978	C7H5Cl3	2,3,6-trichlorotoluene	2077-46-5	3.90	7.94E+03	2
1225	9979	C7H5Cl3	a-a-p-trichlorotoluene	13940-94-8	3.90	7.94E+03	2
1226	10003	C7H5F3	benzotrifluoride	98-08-8	3.05	1.12E+03	1
1227	10045	C7H5N	benzonitrile	100-47-0	1.56	3.63E+01	1
1228	10049	C7H5NO	phenyl isocyanate	103-71-9	2.59	3.89E+02	2
1229	10094	C7H5NS	benzothiazole	95-16-9	2.01	1.02E+02	1
1230	10118	C7H5N3O6	2,4,6-trinitrotoluene	118-96-7	1.60	3.98E+01	1
1231	10252	C7H6Cl2	benzyl dichloride	98-87-3	2.97	9.33E+02	2
1232	10253	C7H6Cl2	2,4-dichlorotoluene	95-73-8	4.24	1.74E+04	1
1233	10254	C7H6Cl2	1-chloro-2-(chloromethyl)ben	611-19-8	3.61	4.03E+03	2
1234	10255	C7H6Cl2	1-chloro-3-(chloromethyl)ben	620-20-2	3.61	4.03E+03	2
1235	10256	C7H6Cl2	1-chloro-4-(chloromethyl)ben	104-83-6	3.61	4.03E+03	2
1236	10257	C7H6Cl2	2,3-dichlorotoluene	32768-54-0	3.61	4.03E+03	2
1237	10258	C7H6Cl2	2,5-dichlorotoluene	19398-61-9	3.61	4.03E+03	2
1238	10259	C7H6Cl2	2,6-dichlorotoluene	118-69-4	3.61	4.03E+03	2
1239	10260	C7H6Cl2	3,4-dichlorotoluene	95-75-0	3.61	4.03E+03	2
1240	10261	C7H6Cl2	3,5-dichlorotoluene	25186-47-4	3.61	4.03E+03	2
1241	10341	C7H6N2	1H-benzimidazole	51-17-2	1.34	2.19E+01	1
1242	10352	C7H6N2O	2-hydroxybenzimidazole	615-16-7	1.12	1.32E+01	1
1243	10363	C7H6N2O4	2,4-dinitrotoluene	121-14-2	1.98	9.55E+01	1
1244	10364	C7H6N2O4	2,5-dinitrotoluene	619-15-8	2.18	1.51E+02	2
1245	10365	C7H6N2O4	2,6-dinitrotoluene	606-20-2	2.10	1.26E+02	1
1246	10366	C7H6N2O4	3,4-dinitrotoluene	610-39-9	2.08	1.20E+02	1
1247	10404	C7H6O	benzaldehyde	100-52-7	1.48	3.02E+01	1
1248	10408	C7H6O2	benzoic acid	65-85-0	1.88	7.59E+01	1
1249	10409	C7H6O2	p-hydroxybenzaldehyde	123-08-0	1.35	2.24E+01	1
1250	10410	C7H6O2	salicylaldehyde	90-02-8	1.81	6.46E+01	1
1251	10411	C7H6O2	1,3-benzodioxole	274-09-9	2.08	1.20E+02	1
1252	10417	C7H6O2	phenyl formate	1864-94-4	1.26	1.82E+01	1
1253	10420	C7H6O3	salicylic acid	69-72-7	2.20	1.58E+02	1
1254	10462	C7H7Br	p-bromotoluene	106-38-7	3.42	2.63E+03	1
1255	10463	C7H7Br	(bromomethyl)benzene	100-39-0	2.92	8.32E+02	1
1256	10464	C7H7Br	o-bromotoluene	95-46-5	3.42	2.63E+03	2
1257	10465	C7H7Br	m-bromotoluene	591-17-3	3.42	2.63E+03	2
1258	10492	C7H7Cl	benzyl chloride	100-44-7	2.30	2.00E+02	1
1259	10493	C7H7Cl	o-chlorotoluene	95-49-8	3.42	2.63E+03	1
1260	10494	C7H7Cl	p-chlorotoluene	106-43-4	3.33	2.14E+03	1
1261	10495	C7H7Cl	m-chlorotoluene	108-41-8	3.28	1.91E+03	2
1262	10555	C7H7F	p-fluorotoluene	352-32-9	2.58	3.80E+02	1
1263	10556	C7H7F	(fluoromethyl)benzene	350-50-5	2.58	3.80E+02	2
1264	10557	C7H7F	o-fluorotoluene	95-52-3	2.58	3.80E+02	2
1265	10558	C7H7F	m-fluorotoluene	352-70-5	2.58	3.80E+02	2
1266	10611	C7H7NO	formanilide	103-70-8	1.15	1.41E+01	1
1267	10626	C7H7NO2	m-nitrotoluene	99-08-1	2.45	2.82E+02	1
1268	10627	C7H7NO2	o-nitrotoluene	88-72-2	2.30	2.00E+02	1
1269	10628	C7H7NO2	p-nitrotoluene	99-99-0	2.42	2.63E+02	1
1270	10660	C7H7NO3	o-nitroanisole	91-23-6	1.73	5.37E+01	1
1271	10724	C7H8	1,3,5-cycloheptatriene	544-25-2	2.63	4.27E+02	2
1272	10725	C7H8	toluene	108-88-3	2.73	5.37E+02	1
1273	10850	C7H8O	anisole	100-66-3	2.11	1.29E+02	1
1274	10851	C7H8O	benzyl alcohol	100-51-6	1.05	1.12E+01	1
1275	10852	C7H8O	m-cresol	108-39-4	1.98	9.55E+01	1
1276	10853	C7H8O	o-cresol	95-48-7	1.98	9.55E+01	1
1277	10854	C7H8O	p-cresol	106-44-5	1.97	9.33E+01	1
1278	10855	C7H8O	cresol mixture	1319-77-3	1.97	9.33E+01	2
1279	10866	C7H8O2	guaiacol	90-05-1	1.32	2.09E+01	1
1280	10867	C7H8O2	p-methoxyphenol	150-76-5	1.34	2.19E+01	1

Table 13 Octanol-Water Partition Coefficient - Oganic Compounds

Octanol-Water Partition Coefficient								Octanol-Water Partition Coefficient							
NO	ID	FORMULA	NAME	CAS No	$\log_{10}$ Kow	Kow	code	NO	ID	FORMULA	NAME	CAS No	$\log_{10}$ Kow	Kow	code
1281	10953	C7H9N	benzylamine	100-46-9	1.09	1.23E+01	1	1361	11715	C7H14	3-methyl-1-hexene	3404-61-3	3.57	3.72E+03	2
1282	10954	C7H9N	N-methylaniline	100-61-8	1.66	4.57E+01	1	1362	11716	C7H14	4-methyl-1-hexene	3769-23-1	3.57	3.72E+03	2
1283	10955	C7H9N	m-toluidine	108-44-1	1.40	2.51E+01	1	1363	11717	C7H14	5-methyl-1-hexene	3524-73-0	4.02	1.06E+04	2
1284	10956	C7H9N	o-toluidine	95-53-4	1.32	2.09E+01	1	1364	11718	C7H14	2-methyl-2-hexene	2738-19-4	4.25	1.78E+04	2
1285	10957	C7H9N	p-toluidine	106-49-0	1.39	2.45E+01	1	1365	11719	C7H14	3-methyl-cis-2-hexene	10574-36-4	4.29	1.96E+04	2
1286	10958	C7H9N	2,3-dimethylpyridine	583-61-9	1.68	4.79E+01	2	1366	11720	C7H14	3-methyl-trans-2-hexene	20710-38-7	4.24	1.76E+04	2
1287	10959	C7H9N	2,4-dimethylpyridine	108-47-4	1.68	4.79E+01	2	1367	11721	C7H14	4-methyl-cis-2-hexene	3683-19-0	4.05	1.11E+04	2
1288	10960	C7H9N	2,5-dimethylpyridine	589-93-5	1.68	4.79E+01	2	1368	11722	C7H14	4-methyl-trans-2-hexene	3683-22-5	4.07	1.19E+04	2
1289	10961	C7H9N	2,6-dimethylpyridine	108-48-5	1.68	4.79E+01	1	1369	11723	C7H14	5-methyl-cis-2-hexene	13151-17-2	4.12	1.31E+04	2
1290	10962	C7H9N	3,4-dimethylpyridine	583-58-4	1.68	4.79E+01	2	1370	11724	C7H14	5-methyl-trans-2-hexene	7385-82-2	4.09	1.22E+04	2
1291	10963	C7H9N	3,5-dimethylpyridine	591-22-0	1.68	4.79E+01	2	1371	11725	C7H14	2-methyl-cis-3-hexene	15840-60-5	4.04	1.10E+04	2
1292	10964	C7H9N	2-ethylpyridine	100-71-0	1.68	4.79E+01	2	1372	11726	C7H14	2-methyl-trans-3-hexene	692-24-0	4.04	1.09E+04	2
1293	10965	C7H9N	3-ethylpyridine	536-78-7	1.68	4.79E+01	2	1373	11727	C7H14	3-methyl-cis-3-hexene	4914-89-0	4.25	1.78E+04	2
1294	10966	C7H9N	4-ethylpyridine	536-75-4	1.68	4.79E+01	2	1374	11728	C7H14	3-methyl-trans-3-hexene	3899-36-3	4.21	1.61E+04	2
1295	11047	C7H10	2-norbornene	498-66-8	2.85	7.08E+02	2	1375	11729	C7H14	2-ethyl-1-pentene	3404-71-5	4.22	1.65E+04	2
1296	11079	C7H10N2	m-toluenediamine	95-80-7	0.14	1.38E+00	1	1376	11730	C7H14	3-ethyl-1-pentene	4038-04-4	3.57	3.72E+03	2
1297	11225	C7H10S	2-propylthiophene	1551-27-5	3.67	4.68E+03	2	1377	11731	C7H14	2,3-dimethyl-1-pentene	3404-72-6	4.00	1.01E+04	2
1298	11226	C7H10S	3-propylthiophene	1518-75-8	3.73	5.32E+03	2	1378	11732	C7H14	2,4-dimethyl-1-pentene	2213-32-3	3.95	8.81E+03	2
1299	11227	C7H10S	2-isopropylthiophene	4095-22-1	3.55	3.54E+03	2	1379	11733	C7H14	3,3-dimethyl-1-pentene	3404-73-7	3.86	7.21E+03	2
1300	11228	C7H10S	3-isopropylthiophene	29488-27-5	3.64	4.34E+03	2	1380	11734	C7H14	3,4-dimethyl-1-pentene	7385-78-6	3.93	8.47E+03	2
1301	11229	C7H10S	2-methyl-3-ethylthiophene	53119-51-0	3.64	4.34E+03	2	1381	11735	C7H14	4,4-dimethyl-1-pentene	762-62-9	3.76	5.71E+03	2
1302	11230	C7H10S	2-methyl-4-ethyl thiophene	13678-54-1	3.77	5.90E+03	2	1382	11736	C7H14	3-ethyl-2-pentene	816-79-5	4.26	1.83E+04	2
1303	11231	C7H10S	2-methyl-5-ethyl thiophene	40323-88-4	3.71	5.08E+03	2	1383	11737	C7H14	2,3-dimethyl-2-pentene	10574-37-5	4.29	1.97E+04	2
1304	11232	C7H10S	3-methyl-2-ethylthiophene	31805-48-8	3.73	5.32E+03	2	1384	11738	C7H14	2,4-dimethyl-2-pentene	625-65-0	3.98	9.58E+03	2
1305	11233	C7H10S	3-methyl-4-ethyl thiophene	---	3.58	3.76E+03	2	1385	11739	C7H14	3,4-dimethyl-cis-2-pentene	4914-91-4	4.11	1.29E+04	2
1306	11234	C7H10S	3-methyl-5-ethylthiophene	---	3.58	3.76E+03	2	1386	11740	C7H14	3,4-dimethyl-trans-2-pentene	4914-92-5	4.16	1.45E+04	2
1307	11235	C7H10S	2,3,4-trimethylthiophene	1795-04-6	3.99	9.74E+03	2	1387	11741	C7H14	4,4-dimethyl-cis-2-pentene	762-63-0	3.92	8.32E+03	2
1308	11236	C7H10S	2,3,5-trimethylthiophene	1795-05-7	3.80	6.37E+03	2	1388	11742	C7H14	4,4-dimethyl-trans-2-pentene	690-08-4	3.84	6.96E+03	2
1309	11275	C7H11NO	cyclohexyl isocyanate	3173-53-3	3.06	1.15E+03	2	1389	11743	C7H14	3-methyl-2-ethyl-1-butene	7357-93-9	4.05	1.12E+04	2
1310	11340	C7H12	1-heptyne	628-71-7	3.01	1.02E+03	2	1390	11744	C7H14	2,3,3-trimethyl-1-butene	594-56-9	3.59	3.89E+03	2
1311	11341	C7H12	2-heptyne	1119-65-9	3.49	3.09E+03	2	1391	11745	C7H14	3,4-dimethylpent-2-ene	24910-63-2	3.95	8.93E+03	2
1312	11342	C7H12	3-heptyne	2586-89-2	3.38	2.37E+03	2	1392	11827	C7H14O	1-heptanal	111-71-7	2.29	1.95E+02	2
1313	11343	C7H12	3-methyl-1-hexyne	40276-93-5	2.87	7.44E+02	2	1393	11828	C7H14O	3-methylhexanal	19269-28-4	2.29	1.95E+02	2
1314	11344	C7H12	4-methyl-1-hexyne	52713-81-2	3.00	1.01E+03	2	1394	11829	C7H14O	2-heptanone	110-43-0	1.98	9.55E+01	1
1315	11345	C7H12	5-methyl-1-hexyne	2203-80-7	3.02	1.05E+03	2	1395	11830	C7H14O	3-heptanone	106-35-4	2.03	1.06E+02	2
1316	11346	C7H12	4-methyl-2-hexyne	20198-49-6	3.20	1.58E+03	2	1396	11831	C7H14O	4-heptanone	123-19-3	1.94	8.75E+01	2
1317	11347	C7H12	5-methyl-2-hexyne	53566-37-3	3.27	1.84E+03	2	1397	11832	C7H14O	3-methyl-2-hexanone	2550-21-2	1.84	6.96E+01	2
1318	11348	C7H12	2-methyl-3-hexyne	36566-80-0	3.10	1.25E+03	2	1398	11833	C7H14O	4-methyl-2-hexanone	105-42-0	1.82	6.57E+01	2
1319	11349	C7H12	3-ethyl-1-pentyne	21020-26-8	2.85	7.07E+02	2	1399	11834	C7H14O	5-methyl-2-hexanone	110-12-3	1.88	7.59E+01	1
1320	11350	C7H12	3,3-dimethyl-1-pentyne	918-82-1	2.56	3.63E+02	2	1400	11835	C7H14O	2-methyl-3-hexanone	7379-12-6	1.67	4.65E+01	2
1321	11351	C7H12	3,4-dimethyl-1-pentyne	61064-08-2	2.76	5.81E+02	2	1401	11836	C7H14O	4-methyl-3-hexanone	17042-16-9	1.74	5.53E+01	2
1322	11352	C7H12	4,4-dimethyl-1-pentyne	13361-63-2	2.68	4.81E+02	2	1402	11837	C7H14O	5-methyl-3-hexanone	623-56-3	1.74	5.53E+01	2
1323	11353	C7H12	4,4-dimethyl-2-pentyne	999-78-0	2.83	6.73E+02	2	1403	11838	C7H14O	3-ethyl-2-pentanone	6137-03-7	1.79	6.21E+01	2
1324	11450	C7H12O2	butyl acrylate	141-32-2	2.36	2.29E+02	1	1404	11839	C7H14O	3,3-dimethyl-2-pentanone	20669-04-9	1.61	4.04E+01	2
1325	11451	C7H12O2	isobutyl acrylate	106-63-8	2.22	1.66E+02	1	1405	11840	C7H14O	3,4-dimethyl-2-pentanone	565-78-6	1.64	4.38E+01	2
1326	11452	C7H12O2	propyl methacrylate	2210-28-8	2.26	1.82E+02	2	1406	11841	C7H14O	4,4-dimethyl-2-pentanone	590-50-1	1.46	2.90E+01	2
1327	11453	C7H12O2	isopropyl methacrylate	4655-34-9	2.28	1.91E+02	2	1407	11842	C7H14O	2,2-dimethyl-3-pentanone	564-04-5	1.46	2.90E+01	2
1328	11454	C7H12O2	allyl butanoate	2051-78-7	2.28	1.91E+02	2	1408	11843	C7H14O	2,4-dimethyl-3-pentanone	565-80-0	1.86	7.24E+01	1
1329	11480	C7H12O2	ethyl 3-methyl-2-butenoate	638-10-8	2.28	1.91E+02	2	1409	11844	C7H14O	1-methylcyclohexanol	590-67-0	2.09	1.23E+02	2
1330	11481	C7H12O2	ethyl trans-2-methyl-2-butenc	5837-78-5	2.28	1.91E+02	2	1410	11845	C7H14O	cis-2-methylcyclohexanol	7443-70-1	1.84	6.92E+01	1
1331	11483	C7H12O2	methyl 3-hexenoate	2396-78-3	2.28	1.91E+02	2	1411	11846	C7H14O	trans-2-methylcyclohexanol	7443-52-9	1.82	6.61E+01	1
1332	11484	C7H12O2	methyl 2-methyl-2-pentenoate	10478-12-3	2.28	1.91E+02	2	1412	11847	C7H14O	cis-3-methylcyclohexanol	5454-79-5	2.05	1.12E+02	2
1333	11485	C7H12O2	methyl 3-methyl-3-pentenoate	2258-58-4	2.28	1.91E+02	2	1413	11848	C7H14O	trans-3-methylcyclohexanol	7443-55-2	2.05	1.12E+02	2
1334	11486	C7H12O2	methyl cis-3-methyl-2-penten	17447-00-6	2.28	1.91E+02	2	1414	11849	C7H14O	cis-4-methylcyclohexanol	7731-28-4	2.05	1.12E+02	2
1335	11487	C7H12O2	methyl trans-3-methyl-2-pent	17447-01-7	2.28	1.91E+02	2	1415	11850	C7H14O	trans-4-methylcyclohexanol	7731-29-5	2.05	1.12E+02	2
1336	11488	C7H12O2	tert-butyl acrylate	1663-39-4	2.28	1.91E+02	2	1416	11851	C7H14O	trans-2-methylcyclohexanol, (	615-39-4	1.99	9.84E+01	2
1337	11491	C7H12O2	methyl trans-2-methyl-2-pent	1567-14-2	2.28	1.91E+02	2	1417	11877	C7H14O	methylcyclohexanol	25639-42-3	1.99	9.84E+01	2
1338	11492	C7H12O2	4-pentenyl acetate	1576-85-8	2.28	1.91E+02	2	1418	11878	C7H14O	2-methylcyclohexanol; (cis+tr	583-59-5	1.99	9.84E+01	2
1339	11497	C7H12O2	3,3-dimethylallyl acetate	1191-16-8	2.28	1.91E+02	2	1419	11879	C7H14O	4-methylcyclohexanol; (cis+tr	589-91-3	1.99	9.84E+01	2
1340	11500	C7H12O2	cis-3-hexenyl formate	33467-73-1	2.28	1.91E+02	2	1420	11880	C7H14O	3-methylcyclohexanol; (cis+tr	591-23-1	1.99	9.84E+01	2
1341	11530	C7H12O4	diethyl malonate	105-53-3	0.96	9.12E+00	1	1421	11884	C7H14O2	heptanoic acid	111-14-8	2.42	2.63E+02	1
1342	11696	C7H14	ethylcyclopentane	1640-89-7	3.59	3.89E+03	2	1422	11885	C7H14O2	2-ethyl-2-methylbutanoic acic	19889-37-3	2.42	2.63E+02	2
1343	11697	C7H14	1,1-dimethylcyclopentane	1638-26-2	3.56	3.63E+03	2	1423	11886	C7H14O2	2-ethylpentanoic acid, (±)	116908-83-9	2.42	2.63E+02	2
1344	11698	C7H14	1,2-dimethylcyclopentane	2452-99-5	3.54	3.45E+03	2	1424	11887	C7H14O2	2-methylhexanoic acid, (±)	22160-12-9	2.42	2.63E+02	2
1345	11699	C7H14	cis-1,2-dimethylcyclopentane	1192-18-3	3.52	3.31E+03	2	1425	11888	C7H14O2	3-methylhexanoic acid	3780-58-3	2.42	2.63E+02	2
1346	11700	C7H14	trans-1,2-dimethylcyclopenta	822-50-4	3.52	3.31E+03	2	1426	11889	C7H14O2	4-methylhexanoic acid, (±)	22160-41-4	2.42	2.63E+02	2
1347	11701	C7H14	1,3-dimethylcyclopentane	2453-00-1	3.54	3.45E+03	2	1427	11890	C7H14O2	5-methylhexanoic acid	628-46-6	2.42	2.63E+02	2
1348	11702	C7H14	cis-1,3-dimethylcyclopentane	2532-58-3	3.52	3.31E+03	2	1428	11891	C7H14O2	hexyl formate	629-33-4	2.91	8.15E+02	2
1349	11703	C7H14	trans-1,3-dimethylcyclopenta	1759-58-6	3.52	3.31E+03	2	1429	11892	C7H14O2	pentyl acetate	628-63-7	2.34	2.19E+02	2
1350	11704	C7H14	methylcyclohexane	108-87-2	3.88	7.59E+03	1	1430	11893	C7H14O2	1-methylbutyl acetate	626-38-0	2.04	1.11E+02	2
1351	11705	C7H14	cycloheptane	291-64-5	4.00	1.00E+04	1	1431	11894	C7H14O2	2-methylbutyl acetate	624-41-9	2.20	1.58E+02	2
1352	11706	C7H14	1-heptene	592-76-7	3.99	9.77E+03	1	1432	11895	C7H14O2	isopentyl acetate	123-92-2	2.26	1.82E+02	2
1353	11707	C7H14	2-heptene	592-77-8	3.95	8.93E+03	2	1433	11896	C7H14O2	1-ethylpropyl acetate	620-11-1	2.02	1.05E+02	2
1354	11708	C7H14	cis-2-heptene	6443-92-1	3.56	3.63E+03	2	1434	11897	C7H14O2	tert-pentyl acetate	625-16-1	1.84	6.97E+01	2
1355	11709	C7H14	trans-2-heptene	14686-13-6	3.56	3.63E+03	2	1435	11898	C7H14O2	1,2-dimethylpropyl acetate	5343-96-4	1.94	8.78E+01	2
1356	11710	C7H14	3-heptene	592-78-9	3.95	8.93E+03	2	1436	11899	C7H14O2	neopentyl acetate	926-41-0	1.93	8.56E+01	2
1357	11711	C7H14	3-heptene (mixed isomers)	25339-56-4	3.95	8.93E+03	2	1437	11900	C7H14O2	sec-pentyl acetate (R)	54638-10-7	2.07	1.18E+02	2
1358	11712	C7H14	cis-3-heptene	7642-10-6	3.56	3.63E+03	2	1438	11901	C7H14O2	butyl propanoate	590-01-2	2.34	2.19E+02	2
1359	11713	C7H14	trans-3-heptene	14686-14-7	3.56	3.63E+03	2	1439	11902	C7H14O2	isobutyl propanoate	540-42-1	2.34	2.19E+02	2
1360	11714	C7H14	2-methyl-1-hexene	6094-02-6	3.70	5.01E+03	2	1440	11903	C7H14O2	sec-butyl propanoate	591-34-4	2.34	2.19E+02	2

Table 13 Octanol-Water Partition Coefficient - Oganic Compounds

Octanol-Water Partition Coefficient							
NO	ID	FORMULA	NAME	CAS No	$\log_{10}$ Kow	Kow	code
1441	11904	C7H14O2	tert-butyl propanoate	20487-40-5	2.34	2.19E+02	2
1442	11905	C7H14O2	propyl butanoate	105-66-8	2.34	2.19E+02	2
1443	11906	C7H14O2	isopropyl butanoate	638-11-9	2.34	2.19E+02	2
1444	11907	C7H14O2	isopropyl isobutanoate	617-50-5	2.34	2.19E+02	2
1445	11908	C7H14O2	propyl isobutanoate	644-49-5	2.34	2.19E+02	2
1446	11909	C7H14O2	ethyl isovalerate	108-64-5	2.26	1.82E+02	2
1447	11910	C7H14O2	ethyl 2,2-dimethylpropanoate	3938-95-2	2.34	2.19E+02	2
1448	11911	C7H14O2	ethyl 2-methylbutanoate, (+)	10307-61-6	2.34	2.19E+02	2
1449	11912	C7H14O2	ethyl pentanoate	539-82-2	2.26	1.82E+02	2
1450	11913	C7H14O2	methyl 2-ethylbutanoate	816-11-5	2.34	2.19E+02	2
1451	11914	C7H14O2	methyl hexanoate	106-70-7	2.26	1.82E+02	2
1452	11923	C7H14O2	ethyl 2-methylbutyrate	7452-79-1	2.26	1.82E+02	2
1453	11925	C7H14O2	2-methylhexanoic acid	4536-23-6	2.42	2.63E+02	2
1454	11926	C7H14O2	neoheptanoic acid	33113-10-9	2.42	2.63E+02	2
1455	11931	C7H14O3	ethyl-3-ethoxypropionate	763-69-9	2.95	8.97E+02	2
1456	11982	C7H15Br	1-bromoheptane	629-04-9	4.36	2.29E+04	1
1457	11983	C7H15Br	2-bromoheptane	1974-04-5	4.36	2.29E+04	2
1458	11984	C7H15Br	4-bromoheptane	998-93-6	4.36	2.29E+04	2
1459	11988	C7H15Cl	1-chloroheptane	629-06-1	4.15	1.41E+04	1
1460	11989	C7H15Cl	2-chloro-2,3-dimethylpentane	59889-45-1	4.15	1.41E+04	2
1461	11990	C7H15Cl	2-chloro-2,4-dimethylpentane	35951-33-8	4.15	1.41E+04	2
1462	11991	C7H15Cl	3-chloro-2,3-dimethylpentane	595-38-0	4.15	1.41E+04	2
1463	11992	C7H15Cl	4-chloro-2,2-dimethylpentane	33429-72-0	4.15	1.41E+04	2
1464	11993	C7H15Cl	3-chloro-3-ethylpentane	994-25-2	4.15	1.41E+04	2
1465	11994	C7H15Cl	2-chloroheptane	1001-89-4	4.15	1.41E+04	2
1466	11995	C7H15Cl	3-chloroheptane	999-52-0	4.15	1.41E+04	2
1467	11996	C7H15Cl	4-chloroheptane	998-95-8	4.15	1.41E+04	2
1468	11997	C7H15Cl	1-chloro-3-methylhexane	101257-63-0	4.15	1.41E+04	2
1469	11998	C7H15Cl	2-chloro-2-methylhexane	4398-65-6	4.15	1.41E+04	2
1470	11999	C7H15Cl	2-chloro-5-methylhexane	58766-17-9	4.15	1.41E+04	2
1471	12000	C7H15Cl	3-chloro-3-methylhexane	43197-78-0	4.15	1.41E+04	2
1472	12001	C7H15Cl	2-chloro-2,3,3-trimethylbutan	918-07-0	4.15	1.41E+04	2
1473	12020	C7H15I	1-iodoheptane	4282-40-0	4.70	5.01E+04	---
1474	12112	C7H16	heptane	142-82-5	4.50	3.16E+04	1
1475	12113	C7H16	2-methylhexane	591-76-4	3.71	5.13E+03	2
1476	12114	C7H16	3-methylhexane	589-34-4	3.71	5.13E+03	2
1477	12115	C7H16	3-ethylpentane	617-78-7	3.71	5.13E+03	2
1478	12116	C7H16	2,2-dimethylpentane	590-35-2	3.67	4.68E+03	2
1479	12117	C7H16	2,3-dimethylpentane	565-59-3	3.63	4.27E+03	2
1480	12118	C7H16	2,4-dimethylpentane	108-08-7	3.63	4.27E+03	2
1481	12119	C7H16	3,3-dimethylpentane	562-49-2	3.67	4.68E+03	2
1482	12120	C7H16	2,2,3-trimethylbutane	464-06-2	3.59	3.89E+03	2
1483	12121	C7H16	3-methylhexane, (S)-	6131-24-4	3.71	5.13E+03	2
1484	12164	C7H16O	1-heptanol	111-70-6	2.62	4.17E+02	1
1485	12165	C7H16O	2-heptanol	543-49-7	2.31	2.04E+02	1
1486	12166	C7H16O	3-heptanol	589-82-2	2.24	1.74E+02	1
1487	12167	C7H16O	4-heptanol	589-55-9	2.22	1.66E+02	1
1488	12168	C7H16O	2-methyl-1-hexanol	624-22-6	2.38	2.38E+02	2
1489	12169	C7H16O	3-methyl-1-hexanol	13231-81-7	2.60	3.94E+02	2
1490	12170	C7H16O	4-methyl-1-hexanol	818-49-5	2.62	4.17E+02	2
1491	12171	C7H16O	5-methyl-1-hexanol	627-98-5	2.60	3.94E+02	2
1492	12172	C7H16O	2-methyl-2-hexanol	625-23-0	1.88	7.63E+01	2
1493	12173	C7H16O	3-methyl-2-hexanol	2313-65-7	2.11	1.28E+02	2
1494	12174	C7H16O	4-methyl-2-hexanol	2313-61-3	2.08	1.21E+02	2
1495	12175	C7H16O	5-methyl-2-hexanol	627-59-8	2.08	1.21E+02	2
1496	12176	C7H16O	2-methyl-3-hexanol	617-29-8	1.94	8.65E+01	2
1497	12177	C7H16O	3-methyl-3-hexanol	597-96-6	1.88	7.63E+01	2
1498	12178	C7H16O	4-methyl-3-hexanol	615-29-2	2.04	1.08E+02	2
1499	12179	C7H16O	5-methyl-3-hexanol	623-55-2	2.01	1.03E+02	2
1500	12180	C7H16O	2-ethyl-1-pentanol	27522-11-8	2.45	2.82E+02	2
1501	12181	C7H16O	3-ethyl-1-pentanol	66225-51-2	2.45	2.82E+02	2
1502	12182	C7H16O	2,2-dimethyl-1-pentanol	2370-12-9	2.13	1.36E+02	2
1503	12183	C7H16O	2,3-dimethyl-1-pentanol	10143-23-4	2.40	2.52E+02	2
1504	12184	C7H16O	2,4-dimethyl-1-pentanol	6305-71-1	2.28	1.90E+02	2
1505	12185	C7H16O	3,3-dimethyl-1-pentanol	19264-94-9	2.43	2.66E+02	2
1506	12186	C7H16O	3,4-dimethyl-1-pentanol	6570-87-2	2.43	2.66E+02	2
1507	12187	C7H16O	4,4-dimethyl-1-pentanol	3121-79-7	2.30	2.01E+02	2
1508	12188	C7H16O	3-ethyl-2-pentanol	609-27-8	2.11	1.28E+02	2
1509	12189	C7H16O	2,3-dimethyl-2-pentanol	4911-70-0	1.79	6.14E+01	2
1510	12190	C7H16O	2,4-dimethyl-2-pentanol	625-06-9	1.64	4.34E+01	2
1511	12191	C7H16O	3,3-dimethyl-2-pentanol	19781-24-9	1.99	9.69E+01	2
1512	12192	C7H16O	3,4-dimethyl-2-pentanol	64502-86-9	2.13	1.36E+02	2
1513	12193	C7H16O	4,4-dimethyl-2-pentanol	6144-93-0	1.76	5.80E+01	2
1514	12194	C7H16O	3-ethyl-3-pentanol	597-49-9	1.88	7.50E+01	2
1515	12195	C7H16O	2,2-dimethyl-3-pentanol	3970-62-5	1.71	5.16E+01	2
1516	12196	C7H16O	2,3-dimethyl-3-pentanol	595-41-5	1.81	6.50E+01	2
1517	12197	C7H16O	2,4-dimethyl-3-pentanol	600-36-2	1.78	6.07E+01	2
1518	12198	C7H16O	2-methyl-2-ethyl-1-butanol	18371-13-6	2.23	1.70E+02	2
1519	12199	C7H16O	3-methyl-2-ethyl-1-butanol	32444-34-1	2.35	2.25E+02	2
1520	12200	C7H16O	2,2,3-trimethyl-1-butanol	55505-23-2	2.23	1.70E+02	2

Octanol-Water Partition Coefficient							
NO	ID	FORMULA	NAME	CAS No	$\log_{10}$ Kow	Kow	code
1521	12201	C7H16O	2,3,3-trimethyl-1-butanol	36794-64-6	2.30	2.01E+02	2
1522	12202	C7H16O	2,3,3-trimethyl-2-butanol	594-83-2	1.59	3.86E+01	2
1523	12203	C7H16O	2,4-dimethyl-1-pentanol, (±)	111768-02-6	2.15	1.40E+02	2
1524	12204	C7H16O	2-heptanol, (±)	52390-72-4	2.15	1.40E+02	2
1525	12205	C7H16O	3-heptanol, (S)	26549-25-7	2.15	1.40E+02	2
1526	12206	C7H16O	2-methyl-1-hexanol, (±)	111768-04-8	2.15	1.40E+02	2
1527	12207	C7H16O	3-methyl-1-hexanol, (±)	111768-08-2	2.15	1.40E+02	2
1528	12208	C7H16O	4-methyl-1-hexanol, (±)	111768-05-9	2.15	1.40E+02	2
1529	12209	C7H16O	2-methyl-3-hexanol, (±)	100295-82-7	2.15	1.40E+02	2
1530	12210	C7H16O	5-methyl-3-hexanol, (±)	100295-83-8	2.15	1.40E+02	2
1531	12211	C7H16O	isopropyl-tert-butyl-ether	17348-59-3	3.04	1.11E+03	2
1532	12212	C7H16O	butyl propyl ether	3073-92-5	3.04	1.11E+03	2
1533	12213	C7H16O	2-ethoxy-2-methylbutane	919-94-8	3.04	1.11E+03	2
1534	12214	C7H16O	ethyl isopentyl ether	628-04-6	3.04	1.11E+03	2
1535	12215	C7H16O	ethyl pentyl ether	17952-11-3	3.04	1.11E+03	2
1536	12216	C7H16O	hexyl methyl ether	4747-07-3	3.04	1.11E+03	2
1537	12217	C7H16O	1-(1-methylethoxy)butane	1860-27-1	3.04	1.11E+03	2
1538	12218	C7H16O	2-methyl-1-propoxypropane	15268-49-2	3.04	1.11E+03	2
1539	12219	C7H16O	2-methyl-2-propoxypropane	29072-93-3	3.04	1.11E+03	2
1540	12220	C7H16O	(S)-(+)-2-heptanol	6033-23-4	2.15	1.40E+02	2
1541	12221	C7H16O	(R)-(-)-2-heptanol	6033-24-5	2.15	1.40E+02	2
1542	12300	C7H16S	heptyl mercaptan	1639-09-4	3.72	5.25E+03	2
1543	12301	C7H16S	2-heptanethiol	1628-00-2	3.85	7.10E+03	2
1544	12302	C7H16S	methyl hexyl sulfide	20291-60-5	3.67	4.72E+03	2
1545	12303	C7H16S	ethyl pentyl sulfide	26158-99-6	3.67	4.72E+03	2
1546	12304	C7H16S	propyl butyl sulfide	1613-46-3	3.67	4.72E+03	2
1547	12315	C7H17N	heptylamine	111-68-2	2.57	3.72E+02	1
1548	12316	C7H17N	2,4-dimethyl-2-pentanamine	64379-30-2	2.57	3.72E+02	2
1549	12317	C7H17N	2-heptanamine	123-82-0	2.57	3.72E+02	2
1550	12318	C7H17N	4-heptanamine	16751-59-0	2.57	3.72E+02	2
1551	12319	C7H17N	3-methyl-1-hexanamine	65530-93-0	2.57	3.72E+02	2
1552	12320	C7H17N	4-methyl-2-hexanamine	105-41-9	2.57	3.72E+02	2
1553	12321	C7H17N	4-methyl-1-hexylamine	34263-68-8	2.57	3.72E+02	2
1554	12322	C7H17N	methylhexylamine	35161-70-7	2.43	2.70E+02	2
1555	12323	C7H17N	ethylpentylamine	17839-26-8	2.24	1.72E+02	2
1556	12324	C7H17N	dimethylpentylamine	26153-88-8	2.07	1.17E+02	2
1557	12325	C7H17N	dimethyl-2-pentylamine	57303-85-2	1.76	5.70E+01	2
1558	12326	C7H17N	dimethyl-3-pentylamine	18636-94-7	1.92	8.38E+01	2
1559	12327	C7H17N	dimethyl-2-methylbutylamine	66225-39-6	1.90	7.86E+01	2
1560	12328	C7H17N	dimethyl-3-methylbutylamine	2315-43-7	1.86	7.30E+01	2
1561	12329	C7H17N	dimethyl-1,1-dimethylpropylai	57757-60-5	1.97	9.36E+01	2
1562	12330	C7H17N	dimethyl-1,2-dimethylpropylai	66225-38-5	1.96	9.03E+01	2
1563	12331	C7H17N	dimethyl-2,2-dimethylpropylai	10076-31-0	1.96	9.03E+01	2
1564	12332	C7H17N	methylethylbutylamine	66225-40-9	1.95	8.85E+01	2
1565	12333	C7H17N	methylethyl-sec-butylamine	66225-41-0	1.61	4.12E+01	2
1566	12334	C7H17N	methylethylisobutylamine	60247-14-5	1.71	5.11E+01	2
1567	12335	C7H17N	methylethyl-tert-butylamine	52841-28-8	1.78	6.02E+01	2
1568	12336	C7H17N	methyldipropylamine	3405-42-3	1.88	7.54E+01	2
1569	12337	C7H17N	methylpropylisopropylamine	66225-42-1	1.85	7.04E+01	2
1570	12338	C7H17N	methyldiisopropylamine	10342-97-9	1.83	6.72E+01	2
1571	12339	C7H17N	diethylpropylamine	4458-31-5	1.83	6.72E+01	2
1572	12340	C7H17N	diethylisopropylamine	6006-15-1	1.74	5.54E+01	2
1573	12341	C7H17N	N-tert-butylisopropylamine	7515-80-2	2.33	2.16E+02	2
1574	12342	C7H17N	(S)-2-heptylamine	44745-29-1	2.57	3.72E+02	2
1575	12343	C7H17N	(R)-2-heptylamine	6240-90-0	2.57	3.72E+02	2
1576	12344	C7H17N	N-propylbutylamine	20193-21-9	2.33	2.16E+02	2
1577	12345	C7H17N	3-aminoheptane	28292-42-4	2.57	3.72E+02	2
1578	12346	C7H17N	5-methyl-2-hexylamine	28292-43-5	2.57	3.72E+02	2
1579	12594	C8H4O3	phthalic anhydride	85-44-9	1.60	3.98E+01	1
1580	12713	C8H6	ethynylbenzene	536-74-3	2.40	2.51E+02	1
1581	12868	C8H6N2	quinoxaline	91-19-0	1.08	1.20E+01	1
1582	12914	C8H6O	benzofuran	271-89-6	2.67	4.68E+02	1
1583	12934	C8H6O4	isophthalic acid	121-91-5	1.66	4.57E+01	1
1584	12935	C8H6O4	phthalic acid	88-99-3	0.73	5.37E+00	1
1585	12936	C8H6O4	terephthalic acid	100-21-0	2.00	1.00E+02	1
1586	12947	C8H6S	benzothiophene	95-15-8	3.12	1.32E+03	1
1587	13107	C8H7N	indole	120-72-9	2.14	1.38E+02	1
1588	13108	C8H7N	benzeneacetonitrile	140-29-4	1.56	3.63E+01	1
1589	13237	C8H8	styrene	100-42-5	3.05	1.12E+03	1
1590	13238	C8H8	1,3,5,7-cyclooctatetraene	629-20-9	3.08	1.20E+03	1
1591	13391	C8H8O	acetophenone	98-86-2	1.63	4.27E+01	1
1592	13392	C8H8O	p-tolualdehyde	104-87-0	2.26	1.82E+02	2
1593	13393	C8H8O	benzeneacetaldehyde	122-78-1	1.78	6.03E+01	1
1594	13394	C8H8O	2-methylbenzaldehyde	529-20-4	2.26	1.82E+02	1
1595	13395	C8H8O	3-methylbenzaldehyde	620-23-5	2.26	1.82E+02	2
1596	13396	C8H8O	2,3-dihydrobenzofuran	496-16-2	2.14	1.38E+02	1
1597	13397	C8H8O	phenyloxirane	96-09-3	1.61	4.07E+01	1
1598	13412	C8H8O2	methyl benzoate	93-58-3	2.20	1.58E+02	1
1599	13413	C8H8O2	o-toluic acid	118-90-1	2.32	2.09E+02	1
1600	13414	C8H8O2	p-toluic acid	99-94-5	2.34	2.19E+02	1

Table 13 Octanol-Water Partition Coefficient - Oganic Compounds

Octanol-Water Partition Coefficient							
NO	ID	FORMULA	NAME	CAS No	$\log_{10}$ Kow	Kow	code
1601	13415	C8H8O2	benzeneacetic acid	103-82-2	1.41	2.57E+01	1
1602	13417	C8H8O2	phenyl acetate	122-79-2	1.49	3.09E+01	1
1603	13433	C8H8O2	m-toluic acid	99-04-7	2.37	2.34E+02	1
1604	13445	C8H8O3	methyl salicylate	119-36-8	2.55	3.55E+02	1
1605	13446	C8H8O3	vanillin	121-33-5	1.21	1.62E+01	1
1606	13671	C8H9NO	acetanilide	103-84-4	1.16	1.45E+01	1
1607	13789	C8H10	ethylbenzene	100-41-4	3.15	1.41E+03	1
1608	13790	C8H10	o-xylene	95-47-6	3.12	1.32E+03	1
1609	13791	C8H10	m-xylene	108-38-3	3.20	1.58E+03	1
1610	13792	C8H10	p-xylene	106-42-3	3.15	1.41E+03	1
1611	13793	C8H10	xylenes	1330-20-7	3.16	1.43E+03	2
1612	13912	C8H10O	phenetole	103-73-1	2.51	3.24E+02	1
1613	13913	C8H10O	2-phenylethanol	60-12-8	1.36	2.29E+01	1
1614	13915	C8H10O	3-methylbenzenemethanol	587-03-1	1.60	3.98E+01	1
1615	13916	C8H10O	4-methylbenzenemethanol	589-18-4	1.58	3.80E+01	1
1616	13917	C8H10O	1-phenylethanol	98-85-1	1.42	2.63E+01	1
1617	13918	C8H10O	o-ethylphenol	90-00-6	2.47	2.95E+02	1
1618	13919	C8H10O	m-ethylphenol	620-17-7	2.50	3.16E+02	1
1619	13920	C8H10O	p-ethylphenol	123-07-9	2.50	3.16E+02	1
1620	13921	C8H10O	2,3-xylenol	526-75-0	2.61	4.07E+02	2
1621	13922	C8H10O	2,4-xylenol	105-67-9	2.35	2.24E+02	1
1622	13923	C8H10O	2,5-xylenol	95-87-4	2.34	2.19E+02	1
1623	13924	C8H10O	2,6-xylenol	576-26-1	2.36	2.29E+02	1
1624	13925	C8H10O	3,4-xylenol	95-65-8	2.23	1.70E+02	1
1625	13926	C8H10O	3,5-xylenol	108-68-9	2.35	2.24E+02	1
1626	13927	C8H10O	benzyl methyl ether	538-86-3	1.35	2.24E+01	1
1627	13929	C8H10O	2-methylanisole	578-58-5	2.74	5.50E+02	1
1628	13930	C8H10O	3-methylanisole	100-84-5	2.66	4.57E+02	1
1629	13931	C8H10O	4-methylanisole	104-93-8	2.81	6.46E+02	1
1630	13934	C8H10O	dimethyl phenol	1300-71-6	2.40	2.48E+02	2
1631	13951	C8H10O2	1,2-dimethoxybenzene	91-16-7	2.08	1.20E+02	1
1632	13952	C8H10O2	1,3-dimethoxybenzene	151-10-0	2.08	1.20E+02	2
1633	13953	C8H10O2	1,4-dimethoxybenzene	150-78-7	2.08	1.20E+02	2
1634	14082	C8H11N	N,N-dimethylaniline	121-69-7	2.31	2.04E+02	1
1635	14083	C8H11N	o-ethylaniline	578-54-1	1.74	5.50E+01	1
1636	14084	C8H11N	N,2-dimethylaniline	611-21-2	2.03	1.06E+02	2
1637	14085	C8H11N	N,3-dimethylaniline	696-44-6	2.03	1.06E+02	2
1638	14086	C8H11N	N,4-dimethylaniline	623-08-5	2.03	1.06E+02	2
1639	14087	C8H11N	2,3-dimethylaniline	87-59-2	2.03	1.06E+02	2
1640	14088	C8H11N	2,4-dimethylaniline	95-68-1	2.03	1.06E+02	2
1641	14089	C8H11N	2,5-dimethylaniline	95-78-3	2.03	1.06E+02	2
1642	14090	C8H11N	2,6-dimethylaniline	87-62-7	2.03	1.06E+02	2
1643	14091	C8H11N	3,4-dimethylaniline	95-64-7	2.03	1.06E+02	2
1644	14092	C8H11N	3,5-dimethylaniline	108-69-0	2.03	1.06E+02	2
1645	14093	C8H11N	m-ethylaniline	587-02-0	2.03	1.06E+02	2
1646	14094	C8H11N	p-ethylaniline	589-16-2	1.96	9.12E+01	2
1647	14095	C8H11N	N-ethylaniline	103-69-5	2.03	1.06E+02	2
1648	14096	C8H11N	2,4,6-trimethylpyridine	108-75-8	1.88	7.59E+01	1
1649	14097	C8H11N	2-ethyl-4-methylpyridine	2150-18-7	1.88	7.59E+01	2
1650	14098	C8H11N	2-ethyl-6-methylpyridine	1122-69-6	1.88	7.59E+01	2
1651	14099	C8H11N	3-ethyl-4-methylpyridine	529-21-5	1.88	7.59E+01	2
1652	14100	C8H11N	4-ethyl-2-methylpyridine	536-88-9	1.88	7.59E+01	2
1653	14101	C8H11N	2-isopropylpyridine	644-98-4	1.88	7.59E+01	2
1654	14102	C8H11N	3-isopropylpyridine	6304-18-3	1.88	7.59E+01	2
1655	14103	C8H11N	4-isopropylpyridine	696-30-0	1.88	7.59E+01	2
1656	14104	C8H11N	4-propylpyridine	1122-81-2	1.88	7.59E+01	2
1657	14105	C8H11N	2-propylpyridine	622-39-9	1.88	7.59E+01	2
1658	14106	C8H11N	2,3,4-trimethylpyridine	2233-29-6	1.88	7.59E+01	2
1659	14107	C8H11N	2,3,5-trimethylpyridine	695-98-7	1.88	7.59E+01	2
1660	14108	C8H11N	2,3,6-trimethylpyridine	1462-84-6	1.88	7.59E+01	2
1661	14109	C8H11N	2,4,5-trimethylpyridine	1122-39-0	1.88	7.59E+01	2
1662	14110	C8H11N	benzeneethanamine	64-04-0	1.41	2.57E+01	1
1663	14122	C8H11N	1-phenylethylamine	98-84-0	1.41	2.57E+01	2
1664	14228	C8H12	1,5-cyclooctadiene	111-78-4	3.16	1.45E+03	1
1665	14229	C8H12	1,4-cyclooctadiene	1073-07-0	3.16	1.45E+03	2
1666	14230	C8H12	cis,cis-1,5-cyclooctadiene	1552-12-1	3.16	1.45E+03	2
1667	14231	C8H12	vinylcyclohexene	100-40-3	3.93	8.51E+03	1
1668	14247	C8H12	1,3-cyclooctadiene	1700-10-3	3.16	1.45E+03	2
1669	14248	C8H12	cis,cis-1,3-cyclooctadiene	3806-59-5	3.16	1.45E+03	2
1670	14252	C8H12	(Z,E)-1,3-cyclooctadiene	3806-60-8	3.16	1.45E+03	2
1671	14379	C8H12O4	1,4-cyclohexanedicarboxylic a	619-82-9	2.71	5.16E+02	2
1672	14380	C8H12O4	diethyl maleate	141-05-9	2.20	1.58E+02	2
1673	14484	C8H14	1-octyne	629-05-0	3.89	7.76E+03	2
1674	14485	C8H14	2-octyne	2809-67-8	4.17	1.48E+04	2
1675	14486	C8H14	3-octyne	15232-76-5	4.06	1.15E+04	2
1676	14487	C8H14	4-octyne	1942-45-6	4.04	1.10E+04	2
1677	14488	C8H14	3-ethyl-3-methyl-1-pentyne	919-12-0	4.04	1.10E+04	2
1678	14496	C8H14	2,5-dimethyl-1,5-hexadiene	627-58-7	4.11	1.29E+04	2
1679	14497	C8H14	2,5-dimethyl-2,4-hexadiene	764-13-6	3.50	3.16E+03	1
1680	14599	C8H14O2	butyl methacrylate	97-88-1	2.88	7.59E+02	1
1681	14659	C8H14O3	butyric anhydride	106-31-0	1.39	2.45E+01	2
1682	14688	C8H14O4	diethyl succinate	123-25-1	1.20	1.58E+01	1
1683	14775	C8H15N	octanenitrile	124-12-9	2.75	5.62E+02	1
1684	14861	C8H16	propylcyclopentane	2040-96-2	4.08	1.20E+04	2
1685	14862	C8H16	isopropylcyclopentane	3875-51-2	4.92	8.26E+04	2
1686	14863	C8H16	1-methyl-1-ethylcyclopentane	16747-50-5	4.80	6.28E+04	2
1687	14864	C8H16	1-methyl-cis-2-ethylcyclopent	930-89-2	4.96	9.05E+04	2
1688	14865	C8H16	1-methyl-trans-2-ethylcyclope	930-90-5	4.79	6.17E+04	2
1689	14866	C8H16	1-methyl-cis-3-ethylcyclopent	2613-66-3	4.79	6.13E+04	2
1690	14867	C8H16	1-methyl-trans-3-ethylcyclope	2613-65-2	4.79	6.13E+04	2
1691	14868	C8H16	1,1,2-trimethylcyclopentane	4259-00-1	4.61	4.07E+04	2
1692	14869	C8H16	1,1,3-trimethylcyclopentane	4516-69-2	4.40	2.51E+04	2
1693	14870	C8H16	1,cis-2,cis-3-trimethylcyclope	2613-69-6	4.83	6.82E+04	2
1694	14871	C8H16	1,cis-2,trans-3-trimethylcyclo	15890-40-1	4.70	5.02E+04	2
1695	14872	C8H16	1,trans-2,cis-3-trimethylcyclo	19374-46-0	4.53	3.36E+04	2
1696	14873	C8H16	1,cis-2,cis-4-trimethylcyclope	2613-72-1	4.69	4.88E+04	2
1697	14874	C8H16	1,cis-2,trans-4-trimethylcyclo	4850-28-6	4.68	4.81E+04	2
1698	14875	C8H16	1,trans-2,cis-4-trimethylcyclo	16883-48-0	4.50	3.19E+04	2
1699	14876	C8H16	1-methyl-3-ethylcyclopentane	3726-47-4	4.67	4.69E+04	2
1700	14877	C8H16	ethylcyclohexane	1678-91-7	4.08	1.20E+04	2
1701	14878	C8H16	1,1-dimethylcyclohexane	590-66-9	4.05	1.12E+04	2
1702	14879	C8H16	1,2-dimethylcyclohexane, (cis	583-57-3	4.02	1.06E+04	2
1703	14880	C8H16	cis-1,2-dimethylcyclohexane	2207-01-4	4.01	1.02E+04	2
1704	14881	C8H16	trans-1,2-dimethylcyclohexan	6876-23-9	4.01	1.02E+04	2
1705	14882	C8H16	1,3-dimethylcyclohexane, cis	591-21-9	4.02	1.06E+04	2
1706	14883	C8H16	cis-1,3-dimethylcyclohexane	638-04-0	4.01	1.02E+04	2
1707	14884	C8H16	trans-1,3-dimethylcyclohexan	2207-03-6	4.01	1.02E+04	2
1708	14885	C8H16	1,4-dimethylcyclohexane, (cis	589-90-2	4.02	1.06E+04	2
1709	14886	C8H16	cis-1,4-dimethylcyclohexane	624-29-3	4.01	1.02E+04	2
1710	14887	C8H16	trans-1,4-dimethylcyclohexan	2207-04-7	4.01	1.02E+04	2
1711	14888	C8H16	cyclooctane	292-64-8	4.45	2.82E+04	1
1712	14890	C8H16	1-octene	111-66-0	4.57	3.72E+04	1
1713	14891	C8H16	oct-2-ene	111-67-1	4.69	4.86E+04	2
1714	14892	C8H16	cis-2-octene	7642-04-8	5.02	1.05E+05	2
1715	14893	C8H16	trans-2-octene	13389-42-9	4.06	1.15E+04	2
1716	14894	C8H16	cis-3-octene	14850-22-7	4.96	9.02E+04	2
1717	14895	C8H16	trans-3-octene	14919-01-8	4.96	9.22E+04	2
1718	14896	C8H16	oct-4-ene	592-99-4	4.69	4.86E+04	2
1719	14897	C8H16	cis-4-octene	7642-15-1	4.95	8.84E+04	2
1720	14898	C8H16	trans-4-octene	14850-23-8	4.06	1.15E+04	2
1721	14899	C8H16	2-methyl-1-heptene	15870-10-7	4.87	7.38E+04	2
1722	14900	C8H16	3-methyl-1-heptene	4810-09-7	4.67	4.66E+04	2
1723	14901	C8H16	4-methyl-1-heptene	13151-05-8	4.71	5.15E+04	2
1724	14902	C8H16	5-methyl-1-heptene	13151-04-7	4.72	5.29E+04	2
1725	14903	C8H16	6-methyl-1-heptene	5026-76-6	4.72	5.26E+04	2
1726	14904	C8H16	2-methyl-2-heptene	627-97-4	4.95	8.87E+04	2
1727	14905	C8H16	3-methyl-cis-2-heptene	22768-19-0	4.93	8.58E+04	2
1728	14906	C8H16	3-methyl-trans-2-heptene	22768-20-3	4.93	8.58E+04	2
1729	14907	C8H16	4-methyl-cis-2-heptene	66225-16-9	4.74	5.50E+04	2
1730	14908	C8H16	4-methyl-trans-2-heptene	66225-17-0	4.74	5.50E+04	2
1731	14909	C8H16	5-methyl-cis-2-heptene	24608-84-2	4.84	6.86E+04	2
1732	14910	C8H16	5-methyl-trans-2-heptene	24608-85-3	4.84	6.86E+04	2
1733	14911	C8H16	6-methyl-2-heptene	73548-72-8	4.69	4.86E+04	2
1734	14912	C8H16	6-methyl-cis-2-heptene	66225-18-1	4.81	6.49E+04	2
1735	14913	C8H16	6-methyl-trans-2-heptene	51065-65-7	4.81	6.49E+04	2
1736	14914	C8H16	6-methyl-3-heptene	3404-57-7	4.69	4.86E+04	2
1737	14915	C8H16	2-methyl-3-heptene	17618-76-7	4.69	4.86E+04	2
1738	14916	C8H16	2-methyl-cis-3-heptene	20488-34-0	4.69	4.93E+04	2
1739	14917	C8H16	2-methyl-trans-3-heptene	692-96-6	4.69	4.93E+04	2
1740	14918	C8H16	3-methyl-3-heptene	7300-03-0	4.69	4.86E+04	2
1741	14919	C8H16	3-methyl-cis-3-heptene	22768-17-8	4.91	8.11E+04	2
1742	14920	C8H16	3-methyl-trans-3-heptene	22768-18-9	4.91	8.11E+04	2
1743	14921	C8H16	4-methyl-3-heptene	4485-16-9	4.69	4.86E+04	2
1744	14922	C8H16	4-methyl-cis-3-heptene	14255-24-4	4.93	8.58E+04	2
1745	14923	C8H16	4-methyl-trans-3-heptene	13714-85-7	4.93	8.58E+04	2
1746	14924	C8H16	5-methyl-3-heptene	13172-91-3	4.69	4.86E+04	2
1747	14925	C8H16	5-methyl-cis-3-heptene	50422-80-5	4.69	4.93E+04	2
1748	14926	C8H16	5-methyl-trans-3-heptene	53510-18-2	4.69	4.93E+04	2
1749	14927	C8H16	6-methyl-cis-3-heptene	66225-19-2	4.76	5.81E+04	2
1750	14928	C8H16	6-methyl-trans-3-heptene	66225-20-5	4.76	5.81E+04	2
1751	14929	C8H16	2-ethyl-1-hexene	1632-16-2	4.19	1.55E+04	2
1752	14930	C8H16	3-ethyl-1-hexene	3404-58-8	4.65	4.49E+04	2
1753	14931	C8H16	4-ethyl-1-hexene	16746-85-3	4.64	4.41E+04	2
1754	14932	C8H16	2,3-dimethyl-1-hexene	16746-86-4	4.66	4.54E+04	2
1755	14933	C8H16	2,4-dimethyl-1-hexene	16746-87-5	4.67	4.71E+04	2
1756	14934	C8H16	2,5-dimethyl-1-hexene	6975-92-4	4.68	4.82E+04	2
1757	14935	C8H16	3,3-dimethyl-1-hexene	3404-77-1	4.50	3.19E+04	2
1758	14936	C8H16	3,4-dimethyl-1-hexene	16745-94-1	4.69	4.93E+04	2
1759	14937	C8H16	3,5-dimethyl-1-hexene	7423-69-0	4.50	3.19E+04	2
1760	14938	C8H16	4,4-dimethyl-1-hexene	1647-08-1	4.58	3.79E+04	2

Table 13 Octanol-Water Partition Coefficient - Oganic Compounds

Octanol-Water Partition Coefficient								Octanol-Water Partition Coefficient							
NO	ID	FORMULA	NAME	CAS No	log_{10} Kow	Kow	code	NO	ID	FORMULA	NAME	CAS No	log_{10} Kow	Kow	code
1761	14939	C8H16	4,5-dimethyl-1-hexene	16106-59-5	4.62	4.18E+04	2	1841	15159	C8H16O2	heptyl formate	112-23-2	3.55	3.54E+03	2
1762	14940	C8H16	5,5-dimethyl-1-hexene	7116-86-1	4.47	2.94E+04	2	1842	15160	C8H16O2	hexyl acetate	142-92-7	2.83	6.76E+02	2
1763	14941	C8H16	3-ethyl-cis-2-hexene	36880-72-5	4.91	8.11E+04	2	1843	15161	C8H16O2	1,1-dimethylbutyl acetate	34859-98-8	2.83	6.76E+02	2
1764	14942	C8H16	3-ethyl-trans-2-hexene	66225-15-8	4.91	8.11E+04	2	1844	15162	C8H16O2	2-ethylbutyl acetate	10031-87-5	2.83	6.76E+02	2
1765	14943	C8H16	4-ethyl-cis-2-hexene	54616-49-8	4.72	5.20E+04	2	1845	15163	C8H16O2	sec-hexyl acetate	108-84-9	2.83	6.76E+02	2
1766	14944	C8H16	4-ethyl-trans-2-hexene	19781-63-6	4.72	5.20E+04	2	1846	15164	C8H16O2	2-methyl-1-pentyl acetate	7789-99-3	2.83	6.76E+02	2
1767	14945	C8H16	2,3-dimethyl-2-hexene	7145-20-2	4.93	8.47E+04	2	1847	15165	C8H16O2	2-methyl-3-pentyl acetate	35897-16-6	2.83	6.76E+02	2
1768	14946	C8H16	2,4-dimethyl-2-hexene	14255-23-3	4.66	4.56E+04	2	1848	15166	C8H16O2	3-methyl-3-pentyl acetate	10250-47-2	2.83	6.76E+02	2
1769	14947	C8H16	2,5-dimethyl-2-hexene	3404-78-2	4.70	4.98E+04	2	1849	15175	C8H16O2	butyl butanoate	109-21-7	2.83	6.76E+02	2
1770	14948	C8H16	3,4-dimethyl-cis-2-hexene	19550-81-3	4.79	6.14E+04	2	1850	15176	C8H16O2	isobutyl butanoate	539-90-2	2.83	6.76E+02	2
1771	14949	C8H16	3,4-dimethyl-trans-2-hexene	19550-82-4	4.79	6.14E+04	2	1851	15177	C8H16O2	isobutyl isobutyrate	97-85-8	2.68	4.79E+02	2
1772	14950	C8H16	3,5-dimethyl-cis-2-hexene	66225-31-8	4.69	4.93E+04	2	1852	15178	C8H16O2	sec-butyl butanoate	819-97-6	2.83	6.76E+02	2
1773	14951	C8H16	3,5-dimethyl-trans-2-hexene	66225-12-5	4.69	4.93E+04	2	1853	15179	C8H16O2	tert-butyl butanoate	2308-38-5	2.83	6.76E+02	2
1774	14952	C8H16	4,4-dimethyl-cis-2-hexene	19550-83-5	4.55	3.55E+04	2	1854	15180	C8H16O2	sec-butyl butanoate, (S)	116836-55-6	2.83	6.76E+02	2
1775	14953	C8H16	4,4-dimethyl-trans-2-hexene	---	4.55	3.55E+04	2	1855	15185	C8H16O2	ethyl hexanoate	123-66-0	2.68	4.79E+02	2
1776	14954	C8H16	4,5-dimethyl-2-hexene	73548-71-7	4.69	4.86E+04	2	1856	15186	C8H16O2	ethyl 4-methylpentanoate	25415-67-2	2.68	4.79E+02	2
1777	14955	C8H16	4,5-dimethyl-cis-2-hexene	65036-71-7	4.64	4.41E+04	2	1857	15191	C8H16O2	isopropyl 3-methylbutanoate	32665-23-9	2.83	6.76E+02	2
1778	14956	C8H16	4,5-dimethyl-trans-2-hexene	66225-14-7	4.64	4.41E+04	2	1858	15192	C8H16O2	isopropyl pentanoate	18362-97-5	2.68	4.79E+02	2
1779	14957	C8H16	5,5-dimethyl-cis-2-hexene	39761-61-0	4.57	3.73E+04	2	1859	15193	C8H16O2	propyl 3-methylbutanoate	557-00-6	2.83	6.76E+02	2
1780	14958	C8H16	5,5-dimethyl-trans-2-hexene	39782-43-9	4.51	3.20E+04	2	1860	15194	C8H16O2	propyl pentanoate	141-06-0	2.68	4.79E+02	2
1781	14959	C8H16	3-ethyl-3-hexene	16789-51-8	4.79	6.14E+04	2	1861	15195	C8H16O2	methyl heptanoate	106-73-0	2.68	4.79E+02	2
1782	14960	C8H16	2,2-dimethyl-cis-3-hexene	690-92-6	4.54	3.44E+04	2	1862	15204	C8H16O2	2-methylheptanoic acid	1188-02-9	3.05	1.12E+03	2
1783	14961	C8H16	2,2-dimethyl-trans-3-hexene	690-93-7	4.43	2.69E+04	2	1863	15205	C8H16O2	2-propylpentanoic acid	99-66-1	3.05	1.12E+03	2
1784	14962	C8H16	2,3-dimethyl-cis-3-hexene	59643-75-3	4.74	5.50E+04	2	1864	15207	C8H16O2	butyl isobutyrate	---	2.68	4.79E+02	2
1785	14963	C8H16	2,3-dimethyl-trans-3-hexene	66225-30-7	4.74	5.50E+04	2	1865	15232	C8H16O4	diethylene glycol ethyl ether a	112-15-2	0.32	2.09E+00	2
1786	14964	C8H16	2,4-dimethyl-cis-3-hexene	37549-89-6	4.62	4.18E+04	2	1866	15248	C8H17Br	1-bromooctane	111-83-1	4.89	7.76E+04	1
1787	14965	C8H16	2,4-dimethyl-trans-3-hexene	61847-78-7	4.59	3.87E+04	2	1867	15249	C8H17Br	2-bromooctane, (±)-	60251-57-2	4.89	7.76E+04	2
1788	14966	C8H16	2,5-dimethyl-cis-3-hexene	10557-44-5	4.46	2.86E+04	2	1868	15250	C8H17Br	2-ethylhexyl bromide	18908-66-2	4.89	7.76E+04	2
1789	14967	C8H16	2,5-dimethyl-trans-3-hexene	692-70-6	4.46	2.86E+04	2	1869	15252	C8H17Cl	1-chlorooctane	111-85-3	4.52	3.31E+04	2
1790	14968	C8H16	3,4-dimethyl-cis-3-hexene	19550-87-9	4.93	8.58E+04	2	1870	15253	C8H17Cl	2-chloro-2,5-dimethylhexane	29342-44-7	4.45	2.82E+04	2
1791	14969	C8H16	3,4-dimethyl-trans-3-hexene	19550-88-0	4.93	8.58E+04	2	1871	15254	C8H17Cl	3-chloro-2,3-dimethylhexane	101654-30-2	4.45	2.82E+04	2
1792	14970	C8H16	2-propyl-1-pentene	15918-08-8	4.83	6.75E+04	2	1872	15255	C8H17Cl	2-chloro-2-methylheptane	4325-49-9	4.45	2.82E+04	2
1793	14971	C8H16	2-isopropyl-1-pentene	16746-02-4	4.72	5.20E+04	2	1873	15256	C8H17Cl	2-chloro-6-methylheptane	2350-19-8	4.45	2.82E+04	2
1794	14972	C8H16	3-methyl-2-ethyl-1-pentene	3404-67-9	4.68	4.79E+04	2	1874	15257	C8H17Cl	3-chloro-3-methylheptane	5272-02-6	4.45	2.82E+04	2
1795	14973	C8H16	4-methyl-2-ethyl-1-pentene	3404-80-6	4.65	4.49E+04	2	1875	15258	C8H17Cl	4-chloro-4-methylheptane	61764-94-1	4.45	2.82E+04	2
1796	14974	C8H16	2-methyl-3-ethyl-1-pentene	19780-66-6	4.64	4.41E+04	2	1876	15259	C8H17Cl	3-(chloromethyl)heptane	123-04-6	4.45	2.82E+04	2
1797	14975	C8H16	3-methyl-3-ethyl-1-pentene	6196-60-7	4.69	4.93E+04	2	1877	15260	C8H17Cl	2-chlorooctane	628-61-5	4.45	2.82E+04	2
1798	14976	C8H16	4-methyl-3-ethyl-1-pentene	61847-80-1	4.59	3.85E+04	2	1878	15261	C8H17Cl	2-chloro-2,4,4-trimethylpenta	6111-88-2	4.45	2.82E+04	2
1799	14977	C8H16	2,3,3-trimethyl-1-pentene	560-23-6	4.60	4.02E+04	2	1879	15262	C8H17Cl	(S)-2-chlorooctane	16844-08-9	4.45	2.82E+04	2
1800	14978	C8H16	2,3,4-trimethyl-1-pentene	565-76-4	4.60	3.96E+04	2	1880	15263	C8H17Cl	(<+->)-2-chlorooctane	51261-14-4	4.45	2.82E+04	2
1801	14979	C8H16	2,4,4-trimethyl-1-pentene	107-39-1	4.55	3.55E+04	1	1881	15264	C8H17Cl	2-ethylhexyl-6-chloride	2350-24-5	4.45	2.82E+04	2
1802	14980	C8H16	3,3,4-trimethyl-1-pentene	560-22-5	4.53	3.36E+04	2	1882	15354	C8H18	octane	111-65-9	5.15	1.41E+05	1
1803	14981	C8H16	3,4,4-trimethyl-1-pentene	564-03-4	4.50	3.19E+04	2	1883	15355	C8H18	2-methylheptane	592-27-8	4.20	1.58E+04	2
1804	14982	C8H16	2-methyl-3-ethyl-2-pentene	19780-67-7	4.81	6.49E+04	2	1884	15356	C8H18	3-methylheptane	589-81-1	4.20	1.58E+04	2
1805	14983	C8H16	4-methyl-3-ethyl-cis-2-penten	42067-48-1	4.79	6.14E+04	2	1885	15357	C8H18	4-methylheptane	589-53-7	4.20	1.58E+04	2
1806	14984	C8H16	4-methyl-3-ethyl-trans-2-pent	42067-49-2	4.75	5.59E+04	2	1886	15358	C8H18	3-methyl-3-ethylpentane	1067-08-9	4.16	1.45E+04	2
1807	14985	C8H16	2,3,4-trimethyl-2-pentene	565-77-5	4.79	6.23E+04	2	1887	15359	C8H18	3-ethylhexane	619-99-8	4.20	1.58E+04	2
1808	14986	C8H16	2,4,4-trimethyl-2-pentene	107-40-4	4.00	1.00E+04	2	1888	15360	C8H18	3-ethyl-2-methylpentane	609-26-7	4.12	1.32E+04	2
1809	14987	C8H16	3,4,4-trimethyl-cis-2-pentene	39761-64-3	4.69	4.93E+04	2	1889	15361	C8H18	2,2-dimethylhexane	590-73-8	4.16	1.45E+04	2
1810	14988	C8H16	3,4,4-trimethyl-trans-2-penten	39761-57-4	4.69	4.93E+04	2	1890	15362	C8H18	2,3-dimethylhexane	584-94-1	4.12	1.32E+04	2
1811	14989	C8H16	3,4,4-trimethyl-2-pentene	598-96-9	4.69	4.86E+04	2	1891	15363	C8H18	2,4-dimethylhexane	589-43-5	1.90	7.94E+01	2
1812	14990	C8H16	2,4,4-trimethylpentene	25167-70-8	4.69	4.86E+04	2	1892	15364	C8H18	2,5-dimethylhexane	592-13-2	4.12	1.32E+04	2
1813	14991	C8H16	3-methyl-2-isopropyl-1-buten	111823-35-9	4.50	3.19E+04	2	1893	15365	C8H18	3,3-dimethylhexane	563-16-6	4.16	1.45E+04	2
1814	14992	C8H16	3,3-dimethyl-2-ethyl-1-butene	18231-53-3	4.64	4.41E+04	2	1894	15366	C8H18	3,4-dimethylhexane	583-48-2	4.12	1.32E+04	2
1815	15062	C8H16O	2-ethylhexanal	123-05-7	2.71	5.13E+02	2	1895	15367	C8H18	2,2,3-trimethylpentane	564-02-3	4.09	1.23E+04	2
1816	15063	C8H16O	1-octanal	124-13-0	2.78	6.03E+02	2	1896	15368	C8H18	2,2,4-trimethylpentane	540-84-1	4.09	1.23E+04	2
1817	15064	C8H16O	5-methylheptanal	75579-88-3	2.75	5.56E+02	2	1897	15369	C8H18	2,3,3-trimethylpentane	560-21-4	4.09	1.23E+04	2
1818	15065	C8H16O	2-propylpentanal	18295-59-5	2.75	5.56E+02	2	1898	15370	C8H18	2,3,4-trimethylpentane	565-75-3	4.05	1.12E+04	2
1819	15066	C8H16O	2-octanone	111-13-7	2.37	2.34E+02	1	1899	15371	C8H18	2,2,3,3-tetramethylbutane	594-82-1	4.05	1.12E+04	2
1820	15067	C8H16O	2,5-dimethyl-3-hexanone	1888-57-9	2.37	2.34E+02	2	1900	15372	C8H18	octane, mixture	---	4.09	1.23E+04	2
1821	15068	C8H16O	2,2-dimethyl-3-hexanone	5405-79-8	2.37	2.34E+02	2	1901	15430	C8H18O	1-octanol	111-87-5	3.07	1.17E+03	1
1822	15069	C8H16O	3,3-dimethyl-2-hexanone	26118-38-7	2.37	2.34E+02	2	1902	15431	C8H18O	2-octanol	123-96-6	2.90	7.94E+02	1
1823	15070	C8H16O	3,4-dimethyl-2-hexanone	19550-10-8	2.37	2.34E+02	2	1903	15432	C8H18O	2-octanol, (±)	4128-31-8	2.44	2.78E+02	2
1824	15071	C8H16O	4,4-dimethyl-3-hexanone	19550-14-2	2.37	2.34E+02	2	1904	15433	C8H18O	3-octanol	589-98-0	2.59	3.88E+02	2
1825	15072	C8H16O	3-ethyl-4-methyl-2-pentanone	71172-57-1	2.37	2.34E+02	2	1905	15434	C8H18O	4-octanol	589-62-8	2.68	4.79E+02	1
1826	15073	C8H16O	2-methyl-3-heptanone	13019-20-0	2.37	2.34E+02	2	1906	15435	C8H18O	4-octanol, (±)	74778-22-6	2.44	2.78E+02	2
1827	15074	C8H16O	2-methyl-4-heptanone	626-33-5	2.37	2.34E+02	2	1907	15436	C8H18O	2-methyl-1-heptanol	60435-70-3	2.61	4.08E+02	2
1828	15075	C8H16O	3-methyl-2-heptanone	2371-19-9	2.37	2.34E+02	2	1908	15437	C8H18O	3-methyl-1-heptanol	1070-32-2	2.87	7.35E+02	2
1829	15076	C8H16O	3-methyl-4-heptanone	15726-15-5	2.37	2.34E+02	2	1909	15438	C8H18O	4-methyl-1-heptanol	817-91-4	2.80	6.26E+02	2
1830	15077	C8H16O	5-methyl-3-heptanone	541-85-5	2.37	2.34E+02	2	1910	15439	C8H18O	5-methyl-1-heptanol	7212-53-5	2.88	7.60E+02	2
1831	15078	C8H16O	6-methyl-2-heptanone	928-68-7	2.37	2.34E+02	2	1911	15440	C8H18O	6-methyl-1-heptanol	1653-40-3	2.91	8.10E+02	2
1832	15079	C8H16O	6-methyl-3-heptanone	624-42-0	2.37	2.34E+02	2	1912	15441	C8H18O	2-methyl-2-heptanol	625-25-2	2.15	1.41E+02	2
1833	15080	C8H16O	3-octanone	106-68-3	2.37	2.34E+02	2	1913	15442	C8H18O	3-methyl-2-heptanol	31367-46-1	2.38	2.39E+02	2
1834	15081	C8H16O	4-octanone	589-63-9	2.37	2.34E+02	2	1914	15443	C8H18O	4-methyl-2-heptanol	56298-90-9	2.51	3.26E+02	2
1835	15082	C8H16O	2,2,4-trimethyl-3-pentanone	5857-36-3	2.37	2.34E+02	2	1915	15444	C8H18O	5-methyl-2-heptanol	54630-50-1	2.52	3.31E+02	2
1836	15147	C8H16O	3-ethyl-3-methyl-2-pentanone	19780-65-5	2.37	2.34E+02	2	1916	15445	C8H18O	6-methyl-2-heptanol	4730-22-7	2.52	3.31E+02	2
1837	15148	C8H16O	5-methyl-2-heptanone	18217-12-4	2.37	2.34E+02	2	1917	15446	C8H18O	2-methyl-3-heptanol	18720-62-2	2.42	2.60E+02	2
1838	15156	C8H16O2	octanoic acid	124-07-2	3.05	1.12E+03	1	1918	15447	C8H18O	3-methyl-3-heptanol	5582-82-1	2.26	1.80E+02	2
1839	15157	C8H16O2	2-ethylhexanoic acid	149-57-5	3.05	1.12E+03	2	1919	15448	C8H18O	4-methyl-3-heptanol	14979-39-6	2.12	1.32E+02	2
1840	15158	C8H16O2	3-ethylhexanoic acid	41065-91-2	3.05	1.12E+03	2	1920	15449	C8H18O	5-methyl-3-heptanol	18720-65-5	2.08	1.19E+02	2

Octanol-Water Partition Coefficient								Octanol-Water Partition Coefficient							
NO	ID	FORMULA	NAME	CAS No	$\log_{10}$ Kow	Kow	code	NO	ID	FORMULA	NAME	CAS No	$\log_{10}$ Kow	Kow	code
1921	15450	C8H18O	6-methyl-3-heptanol	18720-66-6	2.23	1.69E+02	2	2001	15530	C8H18O	butyl isobutyl ether	17071-47-5	2.96	9.05E+02	2
1922	15451	C8H18O	2-methyl-4-heptanol	21570-35-4	2.38	2.39E+02	2	2002	15531	C8H18O	tert-butyl isobutyl ether	33021-02-2	2.96	9.05E+02	2
1923	15452	C8H18O	3-methyl-4-heptanol	1838-73-9	2.35	2.21E+02	2	2003	15532	C8H18O	di-sec-butyl ether, (±)	17226-28-7	2.96	9.05E+02	2
1924	15453	C8H18O	4-methyl-4-heptanol	598-01-6	2.26	1.81E+02	2	2004	15533	C8H18O	ethyl hexyl ether	5756-43-4	2.96	9.05E+02	2
1925	15454	C8H18O	2-ethyl-1-hexanol	104-76-7	2.73	5.37E+02	2	2005	15534	C8H18O	1-methoxyheptane	629-32-3	2.96	9.05E+02	2
1926	15455	C8H18O	3-ethyl-1-hexanol	41065-95-6	2.71	5.18E+02	2	2006	15535	C8H18O	isooctanol	26952-21-6	2.44	2.78E+02	2
1927	15456	C8H18O	4-ethyl-1-hexanol	---	2.71	5.18E+02	2	2007	15536	C8H18O	(R)-(-)-2-octanol	5978-70-1	2.44	2.78E+02	2
1928	15457	C8H18O	2,2-dimethyl-1-hexanol	2370-13-0	2.53	3.43E+02	2	2008	15537	C8H18O	(S)-(+)-2-octanol	6169-06-8	2.44	2.78E+02	2
1929	15458	C8H18O	2,3-dimethyl-1-hexanol	---	2.57	3.70E+02	2	2009	15538	C8H18O	octyl alcohol; mixed isomers	29063-28-3	2.44	2.78E+02	2
1930	15459	C8H18O	2,4-dimethyl-1-hexanol	3965-59-1	2.60	3.94E+02	2	2010	15539	C8H18O	isobutyl ether	628-55-7	2.96	9.05E+02	2
1931	15460	C8H18O	2,5-dimethyl-1-hexanol	6886-16-4	2.69	4.94E+02	2	2011	15578	C8H18O3	diethylene glycol diethyl ethe	112-36-7	0.39	2.45E+00	1
1932	15461	C8H18O	3,3-dimethyl-1-hexanol	10524-70-6	2.74	5.48E+02	2	2012	15579	C8H18O3	diethylene glycol monobutyl e	112-34-5	0.56	3.63E+00	1
1933	15462	C8H18O	3,4-dimethyl-1-hexanol	---	2.74	5.48E+02	2	2013	15591	C8H18O4	triethylene glycol dimethyl eth	112-49-2	-0.76	1.74E-01	2
1934	15463	C8H18O	3,5-dimethyl-1-hexanol	13501-73-0	2.79	6.19E+02	2	2014	15598	C8H18O5	tetraethylene glycol	112-60-7	-2.02	9.55E-03	2
1935	15464	C8H18O	4,4-dimethyl-1-hexanol	6481-95-4	2.76	5.80E+02	2	2015	15602	C8H18S	octyl mercaptan	111-88-6	4.21	1.62E+04	2
1936	15465	C8H18O	4,5-dimethyl-1-hexanol	---	2.76	5.80E+02	2	2016	15603	C8H18S	2-octanethiol	3001-66-9	4.41	2.55E+04	2
1937	15466	C8H18O	5,5-dimethyl-1-hexanol	---	2.76	5.80E+02	2	2017	15604	C8H18S	tert-octyl mercaptan	141-59-3	3.99	9.77E+03	2
1938	15467	C8H18O	3-ethyl-2-hexanol	24448-19-9	2.43	2.66E+02	2	2018	15605	C8H18S	2-octanethiol, (±)	10435-81-1	4.20	1.59E+04	2
1939	15468	C8H18O	4-ethyl-2-hexanol	---	2.42	2.64E+02	2	2019	15606	C8H18S	dibutyl sulfide	544-40-1	3.87	7.41E+03	2
1940	15469	C8H18O	2,3-dimethyl-2-hexanol	19550-03-9	2.23	1.71E+02	2	2020	15607	C8H18S	methyl heptyl sulfide	20291-61-6	3.87	7.41E+03	2
1941	15470	C8H18O	2,4-dimethyl-2-hexanol	42328-76-7	2.00	1.01E+02	2	2021	15608	C8H18S	ethyl hexyl sulfide	7309-44-6	3.87	7.41E+03	2
1942	15471	C8H18O	2,5-dimethyl-2-hexanol	3730-60-7	2.05	1.12E+02	2	2022	15609	C8H18S	propyl pentyl sulfide	42841-80-5	3.87	7.41E+03	2
1943	15472	C8H18O	3,3-dimethyl-2-hexanol	22025-20-3	2.23	1.69E+02	2	2023	15610	C8H18S	di-sec-butyl sulfide	626-26-6	3.87	7.41E+03	2
1944	15473	C8H18O	3,4-dimethyl-2-hexanol	19550-05-1	2.50	3.15E+02	2	2024	15611	C8H18S	di-tert-butyl sulfide	107-47-1	3.87	7.41E+03	2
1945	15474	C8H18O	3,5-dimethyl-2-hexanol	66576-27-0	2.23	1.70E+02	2	2025	15612	C8H18S	diisobutyl sulfide	592-65-4	3.87	7.41E+03	2
1946	15475	C8H18O	4,4-dimethyl-2-hexanol	---	2.28	1.89E+02	2	2026	15614	C8H18S2	dibutyl disulfide	629-45-8	4.82	6.61E+04	2
1947	15476	C8H18O	4,5-dimethyl-2-hexanol	---	2.28	1.89E+02	2	2027	15615	C8H18S2	di-tert-butyl disulfide	110-06-5	4.82	6.61E+04	2
1948	15477	C8H18O	5,5-dimethyl-2-hexanol	31841-77-7	2.38	2.38E+02	2	2028	15617	C8H18S2	disulfide, bis(2-methylpropyl)	1518-72-5	4.82	6.61E+04	2
1949	15478	C8H18O	3-ethyl-3-hexanol	597-76-2	2.21	1.61E+02	2	2029	15618	C8H18S2	disulfide, bis(1-methylpropyl)	5943-30-6	4.82	6.61E+04	2
1950	15479	C8H18O	4-ethyl-3-hexanol	19780-44-0	2.33	2.13E+02	2	2030	15631	C8H19N	octylamine	111-86-4	2.90	7.94E+02	1
1951	15480	C8H18O	2,2-dimethyl-3-hexanol	4209-90-9	2.14	1.37E+02	2	2031	15632	C8H19N	2-ethylhexylamine	104-75-6	2.90	7.94E+02	2
1952	15481	C8H18O	2,3-dimethyl-3-hexanol	4166-46-5	2.19	1.54E+02	2	2032	15633	C8H19N	6-methyl-2-heptanamine, (±)	5984-58-7	2.90	7.94E+02	2
1953	15482	C8H18O	2,4-dimethyl-3-hexanol	13432-25-2	2.23	1.70E+02	2	2033	15634	C8H19N	2-octanamine, (±)	44855-57-4	2.90	7.94E+02	2
1954	15483	C8H18O	2,5-dimethyl-3-hexanol	19550-07-3	2.21	1.61E+02	2	2034	15635	C8H19N	methylheptylamine	36343-05-2	3.03	1.07E+03	2
1955	15484	C8H18O	3,4-dimethyl-3-hexanol	19550-08-4	2.04	1.09E+02	2	2035	15636	C8H19N	ethylhexylamine	20352-67-4	2.86	7.30E+02	2
1956	15485	C8H18O	3,5-dimethyl-3-hexanol	4209-91-0	2.04	1.09E+02	2	2036	15637	C8H19N	dibutylamine	111-92-2	2.83	6.76E+02	1
1957	15486	C8H18O	4,4-dimethyl-3-hexanol	19550-09-5	2.21	1.61E+02	2	2037	15638	C8H19N	diisobutylamine	110-96-3	2.63	4.27E+02	2
1958	15487	C8H18O	4,5-dimethyl-3-hexanol	---	2.13	1.35E+02	2	2038	15639	C8H19N	di-sec-butylamine	626-23-3	2.84	6.88E+02	2
1959	15488	C8H18O	5,5-dimethyl-3-hexanol	66576-31-6	2.06	1.15E+02	2	2039	15640	C8H19N	N-methyl-2-heptanamine	540-43-2	2.84	6.88E+02	2
1960	15489	C8H18O	2-propyl-1-pentanol	58175-57-8	2.69	4.94E+02	2	2040	15641	C8H19N	dimethylhexylamine	4385-04-0	2.68	4.74E+02	2
1961	15490	C8H18O	2-isopropyl-1-pentanol	---	3.23	1.71E+03	2	2041	15642	C8H19N	diethylbutylamine	4444-68-2	2.46	2.86E+02	2
1962	15491	C8H18O	2-methyl-2-ethyl-1-pentanol	5970-63-8	2.65	4.44E+02	2	2042	15643	C8H19N	N-ethyl-N-isopropyl-2-propan	7087-68-5	2.57	3.68E+02	2
1963	15492	C8H18O	3-methyl-2-ethyl-1-pentanol	---	2.64	4.37E+02	2	2043	15644	C8H19N	N-ethyl-N-propyl-1-propanam	20634-92-8	2.57	3.68E+02	2
1964	15493	C8H18O	4-methyl-2-ethyl-1-pentanol	106-67-2	2.63	4.29E+02	2	2044	15645	C8H19N	2-amino-6-methylheptane	543-82-8	2.84	6.88E+02	2
1965	15494	C8H18O	2-methyl-3-ethyl-1-pentanol	---	2.59	3.91E+02	2	2045	15646	C8H19N	(S)-2-amino-6-methylheptane	70419-10-2	2.84	6.88E+02	2
1966	15495	C8H18O	3-methyl-3-ethyl-1-pentanol	---	2.59	3.91E+02	2	2046	15647	C8H19N	(R)-2-amino-6-methylheptane	70419-11-3	2.84	6.88E+02	2
1967	15496	C8H18O	4-methyl-3-ethyl-1-pentanol	---	2.59	3.91E+02	2	2047	15648	C8H19N	(S)-2-aminooctane	34566-04-6	2.90	7.94E+02	2
1968	15497	C8H18O	2,2,3-trimethyl-1-pentanol	57409-53-7	2.58	3.83E+02	2	2048	15649	C8H19N	(R)-2-aminooctane	34566-05-7	2.90	7.94E+02	2
1969	15498	C8H18O	2,2,4-trimethyl-1-pentanol	123-44-4	2.43	2.71E+02	2	2049	15650	C8H19N	2-aminooctane	693-16-3	2.90	7.94E+02	2
1970	15499	C8H18O	2,3,3-trimethyl-1-pentanol	---	2.62	4.14E+02	2	2050	15651	C8H19N	N-isopropyl-N-methyl-tert-but	85523-00-8	2.57	3.68E+02	2
1971	15500	C8H18O	2,3,4-trimethyl-1-pentanol	6570-88-3	2.79	6.19E+02	2	2051	15652	C8H19N	tert-octylamine	107-45-9	2.90	7.94E+02	2
1972	15501	C8H18O	2,4,4-trimethyl-1-pentanol	16325-63-6	2.50	3.15E+02	2	2052	15747	C8H23N5	tetraethylenepentamine	112-57-2	-3.16	6.92E-04	2
1973	15502	C8H18O	3,3,4-trimethyl-1-pentanol	---	2.49	3.12E+02	2	2053	15756	C8H24O4Si4	octamethylcyclotetrasiloxane	556-67-2	5.10	1.26E+05	1
1974	15503	C8H18O	3,4,4-trimethyl-1-pentanol	16325-64-7	2.49	3.12E+02	2	2054	15803	C9H4O5	trimellitic anhydride	552-30-7	1.95	8.91E+01	2
1975	15504	C8H18O	2-methyl-3-ethyl-2-pentanol	19780-63-3	2.18	1.50E+02	2	2055	15925	C9H6O2	1H-2-benzopyran-1-one	491-31-6	1.39	2.45E+01	2
1976	15505	C8H18O	3-methyl-3-ethyl-2-pentanol	66576-22-5	2.23	1.69E+02	2	2056	15926	C9H6O2	2H-1-benzopyran-2-one	91-64-5	1.39	2.45E+01	1
1977	15506	C8H18O	4-methyl-3-ethyl-2-pentanol	66576-23-6	2.33	2.13E+02	2	2057	15927	C9H6O2	4H-1-benzopyran-4-one	491-38-3	1.39	2.45E+01	2
1978	15507	C8H18O	2,3,3-trimethyl-2-pentanol	23171-85-9	2.23	1.70E+02	2	2058	15928	C9H6O2	1H-indene-1,3(2H)-dione	606-23-5	0.61	4.07E+00	1
1979	15508	C8H18O	2,3,4-trimethyl-2-pentanol	66576-26-9	2.18	1.52E+02	2	2059	16010	C9H7N	isoquinoline	119-65-3	2.08	1.20E+02	1
1980	15509	C8H18O	2,4,4-trimethyl-2-pentanol	690-37-9	1.90	7.92E+01	2	2060	16011	C9H7N	quinoline	91-22-5	2.03	1.07E+02	1
1981	15510	C8H18O	3,3,4-trimethyl-2-pentanol	19411-41-7	2.35	2.25E+02	2	2061	16017	C9H7N	cinnamonitrile	4360-47-8	1.96	9.12E+01	1
1982	15511	C8H18O	3,4,4-trimethyl-2-pentanol	10575-56-1	2.18	1.52E+02	2	2062	16018	C9H7NO	8-hydroxyquinoline	148-24-3	2.02	1.05E+02	1
1983	15512	C8H18O	2-methyl-3-ethyl-3-pentanol	597-05-7	2.23	1.70E+02	2	2063	16090	C9H8	indene	95-13-6	2.92	8.32E+02	1
1984	15513	C8H18O	2,2,3-trimethyl-3-pentanol	7294-05-5	2.04	1.09E+02	2	2064	16185	C9H8O	2-methylbenzofuran	4265-25-2	3.22	1.66E+03	1
1985	15514	C8H18O	2,2,4-trimethyl-3-pentanol	5162-48-1	2.01	1.03E+02	2	2065	16200	C9H8O	2-propenophenone	768-03-6	1.88	7.59E+01	1
1986	15515	C8H18O	2,3,4-trimethyl-3-pentanol	3054-92-0	2.16	1.44E+02	2	2066	16203	C9H8O2	cis-cinnamic acid	102-94-3	2.13	1.35E+02	2
1987	15516	C8H18O	3-methyl-2-isopropyl-1-butan	18593-92-5	2.55	3.52E+02	2	2067	16204	C9H8O2	trans-cinnamic acid	140-10-3	2.13	1.35E+02	2
1988	15517	C8H18O	2,2-diethyl-1-butanol	13023-60-4	2.54	3.49E+02	2	2068	16212	C9H8O2	cinnamic acid	621-82-9	2.13	1.35E+02	1
1989	15518	C8H18O	2,3-dimethyl-2-ethyl-1-butano	---	2.54	3.49E+02	2	2069	16364	C9H9N	benzenepropanenitrile	645-59-0	1.72	5.25E+01	1
1990	15519	C8H18O	3,3-dimethyl-2-ethyl-1-butano	66576-56-5	2.54	3.49E+02	2	2070	16388	C9H9NO	cinnamamide, predominantly	621-79-4	1.41	2.57E+01	1
1991	15520	C8H18O	2,2,3,3-tetramethyl-1-butanol	66576-24-7	2.54	3.49E+02	2	2071	16508	C9H10	indane	496-11-7	3.33	2.14E+03	1
1992	15521	C8H18O	3,5-dimethyl-3-hexanol, (±)	19113-78-1	2.44	2.78E+02	2	2072	16509	C9H10	cis-1-propenylbenzene	766-90-5	3.31	2.04E+03	2
1993	15522	C8H18O	2-methyl-1-heptanol, (±)	111675-77-5	2.44	2.78E+02	2	2073	16510	C9H10	trans-1-propenylbenzene	873-66-5	3.31	2.04E+03	2
1994	15523	C8H18O	5-methyl-1-heptanol, (±)	111767-95-4	2.44	2.78E+02	2	2074	16511	C9H10	2-propenylbenzene	300-57-2	3.31	2.04E+03	2
1995	15524	C8H18O	2-methyl-3-heptanol, (±)	100296-26-2	2.44	2.78E+02	2	2075	16512	C9H10	alpha-methylstyrene	98-83-9	3.48	3.02E+03	1
1996	15525	C8H18O	6-methyl-3-heptanol, (±)	100295-85-0	2.44	2.78E+02	2	2076	16513	C9H10	m-methylstyrene	100-80-1	3.44	2.75E+03	2
1997	15526	C8H18O	3-methyl-4-heptanol, (R*,S*)-	92737-91-2	2.44	2.78E+02	2	2077	16514	C9H10	o-methylstyrene	611-15-4	3.44	2.75E+03	2
1998	15527	C8H18O	dibutyl ether	142-96-1	3.21	1.62E+03	1	2078	16515	C9H10	p-methylstyrene	622-97-9	3.35	2.24E+03	2
1999	15528	C8H18O	di-sec-butyl ether	6863-58-7	2.87	7.41E+02	2	2079	16516	C9H10	1-propenylbenzene	637-50-3	3.31	2.04E+03	2
2000	15529	C8H18O	di-tert-butyl ether	6163-66-2	2.79	6.17E+02	2	2080	16519	C9H10	methylstyrene	25013-15-4	3.43	2.68E+03	2

Table 13 Octanol-Water Partition Coefficient - Oganic Compounds

Octanol-Water Partition Coefficient								Octanol-Water Partition Coefficient							
NO	ID	FORMULA	NAME	CAS No	$\log_{10}$ Kow	Kow	code	NO	ID	FORMULA	NAME	CAS No	$\log_{10}$ Kow	Kow	code
2081	16520	C9H10	3(4)-methylstyrene, isomers	39294-88-7	3.43	2.68E+03	2	2161	17888	C9H18	1,cis-2-dimethyl-trans-4-ethyl	62016-65-3	5.47	2.93E+05	2
2082	16631	C9H10O	4-methylacetophenone	122-00-9	2.19	1.55E+02	1	2162	17889	C9H18	1,trans-2-dimethyl-cis-4-ethyl	62016-66-4	5.30	1.98E+05	2
2083	16634	C9H10O	1-phenyl-1-propanone	93-55-0	2.19	1.55E+02	1	2163	17890	C9H18	1,cis-3-dimethyl-1-ethylcyclop	62016-68-6	5.20	1.58E+05	2
2084	16635	C9H10O	1-phenyl-2-propanone	103-79-7	1.44	2.75E+01	1	2164	17891	C9H18	1,trans-3-dimethyl-1-ethylcyc	62016-67-5	5.20	1.58E+05	2
2085	16676	C9H10O2	ethyl benzoate	93-89-0	2.64	4.37E+02	1	2165	17892	C9H18	1,cis-3-dimethyl-cis-2-ethylcy	19903-00-5	5.59	3.86E+05	2
2086	16680	C9H10O2	benzyl acetate	140-11-4	1.96	9.12E+01	1	2166	17893	C9H18	1,cis-3-dimethyl-trans-2-ethyl	19902-98-8	5.22	1.67E+05	2
2087	16681	C9H10O2	2-methylphenyl acetate	533-18-6	2.11	1.29E+02	2	2167	17894	C9H18	1,trans-3-dimethyl-cis-2-ethyl	19902-99-9	5.39	2.48E+05	2
2088	16682	C9H10O2	3-methylphenyl acetate	122-46-3	2.11	1.29E+02	2	2168	17895	C9H18	1,cis-3-dimethyl-cis-4-ethylcy	---	5.42	2.62E+05	2
2089	16683	C9H10O2	4-methylphenyl acetate	140-39-6	2.11	1.29E+02	1	2169	17896	C9H18	1,cis-3-dimethyl-trans-4-ethyl	---	5.22	1.67E+05	2
2090	16701	C9H10O2	(±)-2-phenylpropionic acid	492-37-5	1.80	6.31E+01	1	2170	17897	C9H18	1,trans-3-dimethyl-cis-4-ethyl	---	6.46	2.92E+06	2
2091	16702	C9H10O2	(S)-(+)-2-phenylpropionic aci	7782-24-3	1.80	6.31E+01	2	2171	17898	C9H18	1,trans-3-dimethyl-trans-4-eth	---	5.42	2.62E+05	2
2092	16703	C9H10O2	(R)-(-)-2-phenylpropionic acid	7782-26-5	1.80	6.31E+01	2	2172	17899	C9H18	1,1,2,2-tetramethylcyclopenta	52688-89-8	5.13	1.34E+05	2
2093	16711	C9H10O3	ethyl vanillin	121-32-4	1.58	3.80E+01	1	2173	17900	C9H18	1,1,cis-2,cis-3-tetramethylcyc	62016-70-0	5.25	1.77E+05	2
2094	16893	C9H11N	1,2,3,4-tetrahydroisoquinoline	91-21-4	2.29	1.95E+02	2	2174	17901	C9H18	1,1,cis-2,trans-3-tetramethylc	---	5.05	1.13E+05	2
2095	16894	C9H11N	5,6,7,8-tetrahydroisoquinoline	36556-06-6	2.29	1.95E+02	2	2175	17902	C9H18	1,1,cis-2,cis-4-tetramethylcyc	62016-72-2	5.05	1.13E+05	2
2096	16895	C9H11N	1,2,3,4-tetrahydroquinoline	635-46-1	2.29	1.95E+02	1	2176	17903	C9H18	1,1,cis-2,trans-4-tetramethylc	---	5.05	1.13E+05	2
2097	16896	C9H11N	5,6,7,8-tetrahydroquinoline	10500-57-9	2.29	1.95E+02	2	2177	17904	C9H18	1,1,3,3-tetramethylcyclopenta	50876-33-0	4.76	5.76E+04	2
2098	16905	C9H11NO	p-dimethylaminobenzaldehyd	100-10-7	1.81	6.46E+01	1	2178	17905	C9H18	1,1,cis-3,cis-4-tetramethylcyc	20309-77-7	5.13	1.34E+05	2
2099	17004	C9H12	cumene	98-82-8	3.66	4.57E+03	1	2179	17906	C9H18	1,1,cis-3,trans-4-tetramethylc	---	4.85	7.06E+04	2
2100	17005	C9H12	m-ethyltoluene	620-14-4	3.98	9.55E+03	1	2180	17907	C9H18	1,2,2,cis-3-tetramethylcyclope	62016-73-3	5.25	1.77E+05	2
2101	17006	C9H12	o-ethyltoluene	611-14-3	3.53	3.39E+03	1	2181	17908	C9H18	1,2,2,trans-3-tetramethylcycl	---	5.25	1.77E+05	2
2102	17007	C9H12	p-ethyltoluene	622-96-8	3.63	4.27E+03	1	2182	17909	C9H18	1,cis-2,cis-3,cis-4-tetramethyl	2532-65-2	5.42	2.62E+05	2
2103	17008	C9H12	1,2,3-trimethylbenzene	526-73-8	3.60	3.98E+03	1	2183	17910	C9H18	1,cis-2,cis-3,trans-4-tetramet	2532-69-6	5.25	1.77E+05	2
2104	17009	C9H12	1,2,4-trimethylbenzene	95-63-6	3.63	4.27E+03	1	2184	17911	C9H18	1,cis-2,trans-3,cis-4-tetramet	2532-68-5	5.08	1.19E+05	2
2105	17010	C9H12	trimethyl benzene	25551-13-7	3.67	4.66E+03	2	2185	17912	C9H18	1,cis-2,trans-3,trans-4-tetram	2532-64-1	5.25	1.77E+05	2
2106	17011	C9H12	mesitylene	108-67-8	3.42	2.63E+03	1	2186	17913	C9H18	1,trans-2,cis-3,trans-4-tetram	2532-67-4	4.99	9.76E+04	2
2107	17012	C9H12	propylbenzene	103-65-1	3.69	4.90E+03	1	2187	17914	C9H18	1,trans-2,trans-3,cis-4-tetram	19907-40-5	5.08	1.19E+05	2
2108	17077	C9H12O	benzyl ethyl ether	539-30-0	2.16	1.45E+02	1	2188	17915	C9H18	1-methyl-2-isopropylcyclopen	89223-57-4	5.35	2.26E+05	2
2109	17090	C9H12O	2-phenyl-2-propanol	617-94-7	1.95	8.91E+01	2	2189	17916	C9H18	propylcyclohexane	1678-92-8	4.58	3.80E+04	2
2110	17091	C9H12O	benzenepropanol	122-97-4	1.88	7.59E+01	1	2190	17917	C9H18	isopropylcyclohexane	696-29-7	4.50	3.16E+04	2
2111	17101	C9H12O	2-propylphenol	644-35-9	2.93	8.51E+02	1	2191	17918	C9H18	1-methyl-1-ethylcyclohexane	4926-90-3	5.40	2.51E+05	2
2112	17102	C9H12O	3-propylphenol	621-27-2	3.00	1.00E+03	2	2192	17919	C9H18	1-methyl-cis-2-ethylcyclohexa	4923-77-7	5.49	3.12E+05	2
2113	17103	C9H12O	4-propylphenol	645-56-7	3.20	1.58E+03	1	2193	17920	C9H18	1-methyl-trans-2-ethylcyclohe	4923-78-8	5.39	2.45E+05	2
2114	17116	C9H12O	2,3,4-trimethylphenol	526-85-2	2.50	3.16E+02	2	2194	17921	C9H18	1-methyl-cis-3-ethylcyclohexa	19489-10-2	5.30	2.00E+05	2
2115	17117	C9H12O	2,3,5-trimethylphenol	697-82-5	2.50	3.16E+02	2	2195	17922	C9H18	1-methyl-trans-3-ethylcyclohe	4926-76-5	5.37	2.36E+05	2
2116	17118	C9H12O	2,3,6-trimethylphenol	2416-94-6	2.67	4.68E+02	1	2196	17923	C9H18	1-methyl-cis-4-ethylcyclohexa	4926-78-7	5.37	2.36E+05	2
2117	17119	C9H12O	2,4,5-trimethylphenol	496-78-6	2.50	3.16E+02	2	2197	17924	C9H18	1-methyl-trans-4-ethylcyclohe	6236-88-0	5.30	2.00E+05	2
2118	17120	C9H12O	2,4,6-trimethylphenol	527-60-6	2.46	2.88E+02	1	2198	17925	C9H18	1,1,2-trimethylcyclohexane	7094-26-0	5.23	1.71E+05	2
2119	17138	C9H12O2	cumene hydroperoxide	80-15-9	1.55	3.55E+01	2	2199	17926	C9H18	1,1,3-trimethylcyclohexane	3073-66-3	5.02	1.06E+05	2
2120	17250	C9H13N	amphetamine	300-62-9	1.76	5.75E+01	1	2200	17927	C9H18	1,1,4-trimethylcyclohexane	7094-27-1	4.98	9.65E+04	2
2121	17253	C9H13N	N,N-dimethylbenzylamine	103-83-3	1.98	9.55E+01	1	2201	17928	C9H18	1,2,3-trimethylcyclohexane	1678-97-3	5.20	1.57E+05	2
2122	17436	C9H14O	isophorone	78-59-1	1.70	5.01E+01	1	2202	17929	C9H18	1,cis-2,cis-3-trimethylcyclohe	1839-88-9	5.42	2.63E+05	2
2123	17513	C9H14O6	glyceryl triacetate	102-76-1	0.25	1.78E+00	1	2203	17930	C9H18	1,cis-2,trans-3-trimethylcyclol	7667-55-2	5.37	2.36E+05	2
2124	17593	C9H16	1-nonyne	3452-09-3	3.99	9.77E+03	2	2204	17931	C9H18	1,trans-2,cis-3-trimethylcyclol	1678-81-5	5.20	1.60E+05	2
2125	17594	C9H16	2-nonyne	19447-29-1	4.80	6.32E+04	2	2205	17932	C9H18	1,2,4-trimethylcyclohexane, is	2234-75-5	5.20	1.57E+05	2
2126	17595	C9H16	3-nonyne	20184-89-8	4.69	4.87E+04	2	2206	17933	C9H18	1,cis-2,cis-4-trimethylcyclohe	1678-80-4	5.30	2.00E+05	2
2127	17596	C9H16	4-nonyne	20184-91-2	4.49	3.11E+04	2	2207	17934	C9H18	1,cis-2,trans-4-trimethylcyclol	7667-58-5	5.30	2.00E+05	2
2128	17597	C9H16	7-methyl-3-octyne	37050-06-9	4.49	3.11E+04	2	2208	17935	C9H18	1,trans-2,cis-4-trimethylcyclol	7667-59-6	5.23	1.69E+05	2
2129	17598	C9H16	2,6-dimethyl-3-heptyne	19549-97-4	4.49	3.11E+04	2	2209	17936	C9H18	1,trans-2,trans-4-trimethylcyc	7667-60-9	5.14	1.37E+05	2
2130	17599	C9H16	5,5-dimethyl-3-heptyne	23097-98-5	4.49	3.11E+04	2	2210	17937	C9H18	1,3,5,-trimethylcyclohexane	1839-63-0	5.20	1.57E+05	2
2131	17755	C9H16O4	azelaic acid	123-99-9	1.57	3.72E+01	1	2211	17938	C9H18	cis,cis-1,3,5-trimethylcyclohe:	1795-27-3	5.07	1.18E+05	2
2132	17859	C9H18	butylcyclopentane	2040-95-1	5.70	4.98E+05	2	2212	17939	C9H18	cis,trans-1,3,5-trimethylcyclol	1795-26-2	5.12	1.32E+05	2
2133	17860	C9H18	isobutylcyclopentane	3788-32-7	5.49	3.09E+05	2	2213	17940	C9H18	1-ethyl-2-methylcyclohexane	3728-54-9	5.20	1.57E+05	2
2134	17861	C9H18	sec-butylcyclopentane	4850-32-2	5.64	4.40E+05	2	2214	17942	C9H18	1-nonene	124-11-8	5.15	1.41E+05	1
2135	17862	C9H18	tert-butylcyclopentane	3875-52-3	5.41	2.60E+05	2	2215	17943	C9H18	trans-2-nonene	6434-78-2	5.34	2.20E+05	2
2136	17863	C9H18	1-methyl-1-propylcyclopentar	16631-63-3	5.44	2.77E+05	2	2216	17944	C9H18	trans-3-nonene	20063-92-7	5.34	2.20E+05	2
2137	17864	C9H18	1-methyl-cis-2-propylcyclope	932-43-4	5.60	3.99E+05	2	2217	17945	C9H18	4-nonene; (cis+trans)	2198-23-4	5.34	2.20E+05	2
2138	17865	C9H18	1-methyl-trans-2-propylcyclop	932-44-5	5.45	2.83E+05	2	2218	17946	C9H18	2-methyl-1-octene	4588-18-5	5.53	3.42E+05	2
2139	17866	C9H18	1-methyl-cis-3-propylcyclope	2443-04-1	5.49	3.10E+05	2	2219	17947	C9H18	3-methyl-2-octene	113426-22-5	5.34	2.20E+05	2
2140	17867	C9H18	1-methyl-trans-3-propylcyclop	2443-03-0	5.49	3.10E+05	2	2220	17948	C9H18	7-methyl-3-octene	86668-33-9	5.34	2.20E+05	2
2141	17868	C9H18	1-methyl-1-isopropylcyclopen	61828-00-0	5.49	3.10E+05	2	2221	17949	C9H18	2-methyl-4-octene	64501-77-5	5.34	2.20E+05	2
2142	17869	C9H18	1-methyl-cis-2-isopropylcyclo	61868-01-7	5.49	3.10E+05	2	2222	17950	C9H18	2,3-dimethyl-2-heptene	3074-64-4	5.34	2.20E+05	2
2143	17870	C9H18	1-methyl-trans-2-isopropylcyc	61828-01-1	5.32	2.09E+05	2	2223	17951	C9H18	2,6-dimethylhept-3-ene	2738-18-3	5.34	2.20E+05	2
2144	17871	C9H18	1-methyl-cis-3-isopropylcyclo	61828-02-2	5.35	2.22E+05	2	2224	17997	C9H18O	1-nonanal	124-19-6	3.27	1.86E+03	2
2145	17872	C9H18	1-methyl-trans-3-isopropylcyc	61828-03-3	5.35	2.22E+05	2	2225	17998	C9H18O	diisobutyl ketone	108-83-8	2.56	3.63E+02	2
2146	17873	C9H18	1,1-diethylcyclopentane	2721-38-2	5.55	3.56E+05	2	2226	17999	C9H18O	2-nonanone	821-55-6	3.16	1.45E+03	1
2147	17874	C9H18	1,cis-2-diethylcyclopentane	932-39-8	5.62	4.21E+05	2	2227	18000	C9H18O	di-tert-butyl ketone	815-24-7	2.91	8.13E+02	2
2148	17875	C9H18	1,trans-2-diethylcyclopentane	932-40-1	5.48	3.02E+05	2	2228	18001	C9H18O	3,5-dimethyl-4-heptanone	19549-84-9	2.91	8.13E+02	2
2149	17876	C9H18	1,3-diethylcyclopentane	19398-75-5	5.35	2.26E+05	2	2229	18002	C9H18O	3-methyl-2-octanone	6137-08-2	2.91	8.13E+02	2
2150	17877	C9H18	1,cis-3-diethylcyclopentane	62016-59-5	5.54	3.46E+05	2	2230	18003	C9H18O	3-methyl-4-octanone	20754-04-5	2.91	8.13E+02	2
2151	17878	C9H18	1,trans-3-diethylcyclopentane	62016-60-8	5.54	3.46E+05	2	2231	18004	C9H18O	4-methyl-3-octanone	6137-15-1	2.91	8.13E+02	2
2152	17879	C9H18	1,1-dimethyl-2-ethylcyclopent	54549-80-3	5.25	1.77E+05	2	2232	18005	C9H18O	5-methyl-2-octanone	58654-67-4	2.92	8.32E+02	1
2153	17880	C9H18	1,1-dimethyl-3-ethylcyclopent	62016-61-9	5.15	1.41E+05	2	2233	18006	C9H18O	7-methyl-3-octanone	5408-57-1	2.91	8.13E+02	2
2154	17881	C9H18	1,cis-2-dimethyl-1-ethylcyclop	62016-63-1	5.39	2.48E+05	2	2234	18007	C9H18O	7-methyl-4-octanone	20809-46-5	2.91	8.13E+02	2
2155	17882	C9H18	1,trans-2-dimethyl-1-ethylcyc	62016-62-0	5.39	2.48E+05	2	2235	18008	C9H18O	3-nonanone	925-78-0	2.91	8.13E+02	2
2156	17883	C9H18	1,cis-2-dimethyl-cis-3-ethylcy	---	5.59	3.86E+05	2	2236	18009	C9H18O	4-nonanone	4485-09-0	2.91	8.13E+02	2
2157	17884	C9H18	1,cis-2-dimethyl-trans-3-ethyl	---	5.42	2.62E+05	2	2237	18010	C9H18O	5-nonanone	502-56-7	2.91	8.13E+02	2
2158	17885	C9H18	1,trans-2-dimethyl-cis-3-ethyl	---	5.25	1.77E+05	2	2238	18044	C9H18O	3,5,5-trimethylhexanal	5435-64-3	3.27	1.86E+03	2
2159	17886	C9H18	1,trans-2-dimethyl-trans-3-eth	---	5.42	2.62E+05	2	2239	18045	C9H18O	2,6-dimethyl-3-heptanone	19549-83-8	2.91	8.13E+02	2
2160	17887	C9H18	1,cis-2-dimethyl-cis-4-ethylcy	62016-64-2	5.47	2.93E+05	2	2240	18047	C9H18O	2-methyl-3-octanone	923-28-4	2.91	8.13E+02	2

Table 13 Octanol-Water Partition Coefficient - Oganic Compounds

Octanol-Water Partition Coefficient							
NO	ID	FORMULA	NAME	CAS No	$\log_{10}$ Kow	Kow	code
2241	18048	C9H18O	2-methyloctanal	7786-29-0	3.27	1.86E+03	2
2242	18052	C9H18O2	nonanoic acid	112-05-0	3.42	2.63E+03	1
2243	18053	C9H18O2	2-ethylheptanoic acid	3274-29-1	3.42	2.63E+03	2
2244	18054	C9H18O2	3-ethyl-3-methylhexanoic aci	50902-82-4	3.42	2.63E+03	2
2245	18055	C9H18O2	2-ethyl-3-methylhexanoic aci	74581-94-5	3.42	2.63E+03	2
2246	18056	C9H18O2	3-methyloctanoic acid	6061-10-5	3.42	2.63E+03	2
2247	18057	C9H18O2	2-methyloctanoic acid	3004-93-1	3.42	2.63E+03	2
2248	18058	C9H18O2	octyl formate	112-32-3	4.17	1.48E+04	2
2249	18059	C9H18O2	heptyl acetate	112-06-1	3.52	3.34E+03	2
2250	18060	C9H18O2	4-heptyl acetate	5921-84-6	3.52	3.34E+03	2
2251	18082	C9H18O2	3,5,5-trimethylhexanoic acid	3302-10-1	3.42	2.63E+03	2
2252	18083	C9H18O2	7-methyloctanoic acid	693-19-6	3.42	2.63E+03	2
2253	18087	C9H18O2	4-octanecarboxylic acid	3274-28-0	3.42	2.63E+03	2
2254	18115	C9H19Br	1-bromononane	693-58-3	5.10	1.26E+05	2
2255	18117	C9H19Cl	1-chlorononane	2473-01-0	5.02	1.05E+05	2
2256	18118	C9H19Cl	3-chloro-3-ethyl-2,2-dimethyl	86661-53-2	5.02	1.05E+05	2
2257	18119	C9H19Cl	3-chloro-3-ethylheptane	28320-89-0	5.02	1.05E+05	2
2258	18120	C9H19Cl	3-chloro-3-methyloctane	28320-88-9	5.02	1.05E+05	2
2259	18121	C9H19Cl	4-chloro-4-methyloctane	36903-89-6	5.02	1.05E+05	2
2260	18122	C9H19Cl	2-chlorononane	2216-36-6	5.02	1.05E+05	2
2261	18123	C9H19Cl	5-chlorononane	28123-70-8	5.02	1.05E+05	2
2262	18124	C9H19Cl	3-chloro-2,2,3-trimethylhexan	102449-95-6	5.02	1.05E+05	2
2263	18176	C9H20	nonane	111-84-2	5.65	4.47E+05	1
2264	18177	C9H20	2-methyloctane	3221-61-2	4.69	4.90E+04	2
2265	18178	C9H20	3-methyloctane	2216-33-3	4.69	4.90E+04	2
2266	18179	C9H20	4-methyloctane	2216-34-4	4.69	4.90E+04	2
2267	18180	C9H20	3-ethylheptane	15869-80-4	4.69	4.90E+04	2
2268	18181	C9H20	4-ethylheptane	2216-32-2	4.69	4.90E+04	2
2269	18182	C9H20	2,2-dimethylheptane	1071-26-7	4.65	4.47E+04	2
2270	18183	C9H20	2,3-dimethylheptane	3074-71-3	4.65	4.47E+04	2
2271	18184	C9H20	2,4-dimethylheptane	2213-23-2	4.65	4.47E+04	2
2272	18185	C9H20	2,5-dimethylheptane	2216-30-0	4.65	4.47E+04	2
2273	18186	C9H20	2,6-dimethylheptane	1072-05-5	4.65	4.47E+04	2
2274	18187	C9H20	3,4-dimethylheptane	922-28-1	4.65	4.47E+04	2
2275	18188	C9H20	3,5-dimethylheptane	926-82-9	4.65	4.47E+04	2
2276	18189	C9H20	4,4-dimethylheptane	1068-19-5	4.65	4.47E+04	2
2277	18190	C9H20	3-ethyl-2-methylhexane	16789-46-1	4.65	4.47E+04	2
2278	18191	C9H20	4-ethyl-2-methylhexane	3074-75-7	4.65	4.47E+04	2
2279	18192	C9H20	3-ethyl-3-methylhexane	3074-76-8	4.65	4.47E+04	2
2280	18193	C9H20	3-ethyl-4-methylhexane	3074-77-9	4.65	4.47E+04	2
2281	18194	C9H20	2,2,3-trimethylhexane	16747-25-4	4.58	3.80E+04	2
2282	18195	C9H20	2,2,4-trimethylhexane	16747-26-5	4.58	3.80E+04	2
2283	18196	C9H20	2,2,5-trimethylhexane	3522-94-9	4.58	3.80E+04	2
2284	18197	C9H20	2,3,3-trimethylhexane	16747-28-7	4.58	3.80E+04	2
2285	18198	C9H20	2,3,4-trimethylhexane	921-47-1	4.54	3.47E+04	2
2286	18199	C9H20	2,3,5-trimethylhexane	1069-53-0	4.54	3.47E+04	2
2287	18200	C9H20	2,4,4-trimethylhexane	16747-30-1	4.58	3.80E+04	2
2288	18201	C9H20	3,3,4-trimethylhexane	16747-31-2	4.58	3.80E+04	2
2289	18202	C9H20	3,3-diethylpentane	1067-20-5	4.65	4.47E+04	2
2290	18203	C9H20	2,2-dimethyl-3-ethylpentane	16747-32-3	4.58	3.80E+04	2
2291	18204	C9H20	2,3-dimethyl-3-ethylpentane	16747-33-4	5.13	1.35E+05	2
2292	18205	C9H20	2,4-dimethyl-3-ethylpentane	1068-87-7	4.54	3.47E+04	2
2293	18206	C9H20	2,2,3,3-tetramethylpentane	7154-79-2	4.54	3.47E+04	2
2294	18207	C9H20	2,2,3,4-tetramethylpentane	1186-53-4	4.50	3.16E+04	2
2295	18208	C9H20	2,2,4,4-tetramethylpentane	1070-87-7	4.54	3.47E+04	2
2296	18209	C9H20	2,3,3,4-tetramethylpentane	16747-38-9	4.50	3.16E+04	2
2297	18210	C9H20	3,3-dimethylheptane	4032-86-4	4.70	5.01E+04	2
2298	18236	C9H20O	1-nonanol	143-08-8	4.02	1.05E+04	1
2299	18237	C9H20O	2-nonanol	628-99-9	3.48	3.05E+03	2
2300	18238	C9H20O	2-nonanol, (±)	74683-66-2	3.02	1.04E+03	2
2301	18239	C9H20O	3-nonanol	624-51-1	3.27	1.85E+03	2
2302	18240	C9H20O	3-nonanol, (±)	74742-08-8	3.02	1.04E+03	2
2303	18241	C9H20O	4-nonanol	5932-79-6	3.22	1.67E+03	2
2304	18242	C9H20O	4-nonanol, (S)	52708-03-9	3.02	1.04E+03	2
2305	18243	C9H20O	5-nonanol	623-93-8	3.28	1.89E+03	2
2306	18244	C9H20O	2-methyl-1-octanol	818-81-5	3.40	2.50E+03	2
2307	18245	C9H20O	3-methyl-1-octanol	38514-02-2	3.40	2.50E+03	2
2308	18246	C9H20O	4-methyl-1-octanol	38514-03-3	3.40	2.50E+03	2
2309	18247	C9H20O	5-methyl-1-octanol	38514-04-4	3.40	2.50E+03	2
2310	18248	C9H20O	6-methyl-1-octanol	38514-05-5	3.56	3.60E+03	2
2311	18249	C9H20O	7-methyl-1-octanol	2430-22-0	3.56	3.60E+03	2
2312	18250	C9H20O	2-methyl-2-octanol	628-44-4	2.85	7.10E+02	2
2313	18251	C9H20O	3-methyl-2-octanol	27644-49-1	2.92	8.34E+02	2
2314	18252	C9H20O	5-methyl-2-octanol	66793-81-5	2.92	8.34E+02	2
2315	18253	C9H20O	7-methyl-2-octanol	66793-83-7	2.92	8.34E+02	2
2316	18254	C9H20O	2-methyl-3-octanol	26533-34-6	3.00	9.98E+02	2
2317	18255	C9H20O	3-methyl-3-octanol	5340-36-3	3.12	1.33E+03	2
2318	18256	C9H20O	4-methyl-3-octanol	66793-80-4	3.04	1.11E+03	2
2319	18257	C9H20O	6-methyl-3-octanol	40225-75-0	3.04	1.11E+03	2
2320	18258	C9H20O	7-methyl-3-octanol	66793-84-8	3.04	1.11E+03	2
2321	18259	C9H20O	2-methyl-4-octanol	40575-41-5	3.00	9.98E+02	2
2322	18260	C9H20O	3-methyl-4-octanol	26533-35-7	2.90	7.95E+02	2
2323	18261	C9H20O	4-methyl-4-octanol	23418-37-3	2.93	8.41E+02	2
2324	18262	C9H20O	5-methyl-4-octanol	59734-23-5	2.95	8.83E+02	2
2325	18263	C9H20O	6-methyl-4-octanol	66793-82-6	2.95	8.83E+02	2
2326	18264	C9H20O	7-methyl-4-octanol	33933-77-6	2.95	8.83E+02	2
2327	18265	C9H20O	2-ethyl-1-heptanol	817-60-7	3.37	2.35E+03	2
2328	18266	C9H20O	3-ethyl-1-heptanol	3525-25-5	3.58	3.83E+03	2
2329	18267	C9H20O	5-ethyl-1-heptanol	998-65-2	3.37	2.35E+03	2
2330	18268	C9H20O	2,2-dimethyl-1-heptanol	14250-79-4	3.20	1.58E+03	2
2331	18269	C9H20O	4,6-dimethyl-1-heptanol	820-05-3	2.95	8.83E+02	2
2332	18270	C9H20O	6,6-dimethyl-1-heptanol	65769-10-0	2.95	8.83E+02	2
2333	18271	C9H20O	2-ethyl-2-heptanol	---	2.95	8.83E+02	2
2334	18272	C9H20O	3-ethyl-2-heptanol	19780-39-3	2.95	8.83E+02	2
2335	18273	C9H20O	2,3-dimethyl-2-heptanol	66794-00-1	2.95	8.83E+02	2
2336	18274	C9H20O	2,4-dimethyl-2-heptanol	65822-93-7	2.95	8.83E+02	2
2337	18275	C9H20O	2,5-dimethyl-2-heptanol	---	2.95	8.83E+02	2
2338	18276	C9H20O	2,6-dimethyl-2-heptanol	13254-34-7	2.68	4.79E+02	2
2339	18277	C9H20O	4,6-dimethyl-2-heptanol	51079-52-8	3.25	1.77E+03	2
2340	18278	C9H20O	5,6-dimethyl-2-heptanol	58795-24-7	3.20	1.58E+03	2
2341	18279	C9H20O	3-ethyl-3-heptanol	19780-41-7	2.95	9.01E+02	2
2342	18280	C9H20O	2,2-dimethyl-3-heptanol	19549-70-3	2.85	7.04E+02	2
2343	18281	C9H20O	2,3-dimethyl-3-heptanol	19549-71-4	2.75	5.67E+02	2
2344	18282	C9H20O	2,6-dimethyl-3-heptanol	19549-73-6	2.78	6.00E+02	2
2345	18283	C9H20O	3,5-dimethyl-3-heptanol	19549-74-7	2.82	6.66E+02	2
2346	18284	C9H20O	3,6-dimethyl-3-heptanol	1573-28-0	2.73	5.36E+02	2
2347	18285	C9H20O	4-ethyl-4-heptanol	597-90-0	2.88	7.51E+02	2
2348	18286	C9H20O	2,2-dimethyl-4-heptanol	66793-99-5	2.75	5.67E+02	2
2349	18287	C9H20O	2,4-dimethyl-4-heptanol	19549-77-0	2.69	4.90E+02	2
2350	18288	C9H20O	2,6-dimethyl-4-heptanol	108-82-7	3.08	1.20E+03	1
2351	18289	C9H20O	3,3-dimethyl-4-heptanol	19549-78-1	2.80	6.35E+02	2
2352	18290	C9H20O	3,5-dimethyl-4-heptanol	19549-79-2	3.07	1.18E+03	2
2353	18291	C9H20O	2-propyl-1-hexanol	817-46-9	3.12	1.32E+03	2
2354	18292	C9H20O	3-propyl-1-hexanol	66793-85-9	3.12	1.32E+03	2
2355	18293	C9H20O	1-methyl-2-ethyl-1-hexanol	---	3.12	1.32E+03	2
2356	18294	C9H20O	1-methyl-3-ethyl-1-hexanol	---	3.20	1.58E+03	2
2357	18295	C9H20O	3-methyl-2-ethyl-1-hexanol	66794-04-5	3.22	1.67E+03	2
2358	18296	C9H20O	4-methyl-2-ethyl-1-hexanol	66794-06-7	3.27	1.88E+03	2
2359	18297	C9H20O	5-methyl-2-ethyl-1-hexanol	66794-07-8	3.22	1.66E+03	2
2360	18298	C9H20O	3,3,5-trimethyl-1-hexanol	1484-87-3	3.22	1.66E+03	2
2361	18299	C9H20O	3,4,4-trimethyl-1-hexanol	66793-73-5	3.17	1.49E+03	2
2362	18300	C9H20O	3,5,5-trimethyl-1-hexanol	3452-97-9	3.22	1.67E+03	2
2363	18301	C9H20O	4,5,5-trimethyl-1-hexanol	66793-75-7	3.45	2.84E+03	2
2364	18302	C9H20O	2-methyl-3-ethyl-2-hexanol	66794-02-3	2.85	7.10E+02	2
2365	18303	C9H20O	2,3,4-trimethyl-2-hexanol	21102-13-6	3.22	1.66E+03	2
2366	18304	C9H20O	2,4,4-trimethyl-2-hexanol	66793-91-7	3.22	1.66E+03	2
2367	18305	C9H20O	2,4,5-trimethyl-2-hexanol	66793-93-9	3.22	1.66E+03	2
2368	18306	C9H20O	2,5,5-trimethyl-2-hexanol	66793-71-3	3.22	1.66E+03	2
2369	18307	C9H20O	2-methyl-3-ethyl-3-hexanol	66794-03-4	3.00	1.00E+03	2
2370	18308	C9H20O	2-methyl-4-ethyl-3-hexanol	33943-21-4	2.87	7.45E+02	2
2371	18309	C9H20O	3-methyl-4-ethyl-3-hexanol	66794-05-6	2.87	7.45E+02	2
2372	18310	C9H20O	4-methyl-3-ethyl-3-hexanol	51200-80-7	2.87	7.45E+02	2
2373	18311	C9H20O	5-methyl-3-ethyl-3-hexanol	597-77-3	2.70	5.07E+02	2
2374	18312	C9H20O	2,2,3-trimethyl-3-hexanol	5340-41-0	2.73	5.39E+02	2
2375	18313	C9H20O	2,2,4-trimethyl-3-hexanol	66793-89-3	2.63	4.28E+02	2
2376	18314	C9H20O	2,2,5-trimethyl-3-hexanol	3970-60-3	2.41	2.59E+02	2
2377	18315	C9H20O	2,3,4-trimethyl-3-hexanol	66793-90-6	2.56	3.59E+02	2
2378	18316	C9H20O	2,3,5-trimethyl-3-hexanol	65927-60-8	2.56	3.59E+02	2
2379	18317	C9H20O	2,4,4-trimethyl-3-hexanol	66793-92-8	2.68	4.79E+02	2
2380	18318	C9H20O	2,5,5-trimethyl-3-hexanol	66793-72-4	2.63	4.25E+02	2
2381	18319	C9H20O	3,4,4-trimethyl-3-hexanol	66793-74-6	2.56	3.62E+02	2
2382	18320	C9H20O	3,5,5-trimethyl-3-hexanol	66810-87-5	2.56	3.59E+02	2
2383	18321	C9H20O	4-methyl-2-propyl-1-pentanol	54004-41-0	3.20	1.58E+03	2
2384	18322	C9H20O	4-methyl-2-isopropyl-1-penta	55505-24-3	3.07	1.18E+03	2
2385	18323	C9H20O	2,2-dimethyl-3-ethyl-1-pentan	66793-95-1	3.09	1.24E+03	2
2386	18324	C9H20O	2,4-dimethyl-2-ethyl-1-pentan	66793-98-4	3.10	1.25E+03	2
2387	18325	C9H20O	3,3-diethyl-2-pentanol	66793-94-0	3.09	1.24E+03	2
2388	18326	C9H20O	2,3-dimethyl-3-ethyl-2-pentan	66793-97-3	3.09	1.24E+03	2
2389	18327	C9H20O	4,4-dimethyl-3-ethyl-2-pentan	21102-09-0	3.09	1.24E+03	2
2390	18328	C9H20O	2,3,3,4-tetramethyl-2-pentano	66793-86-0	2.90	7.88E+02	2
2391	18329	C9H20O	2,3,4,4-tetramethyl-2-pentano	66793-87-1	2.90	7.88E+02	2
2392	18330	C9H20O	3,3,4,4-tetramethyl-2-pentano	66793-88-2	2.90	7.95E+02	2
2393	18331	C9H20O	2-methyl-3-isopropyl-3-penta	---	2.85	7.10E+02	2
2394	18332	C9H20O	2,2-dimethyl-3-ethyl-3-pentan	66793-96-2	2.75	5.67E+02	2
2395	18333	C9H20O	2,4-dimethyl-3-ethyl-3-pentan	3970-59-0	2.85	7.06E+02	2
2396	18334	C9H20O	2,2,3,4-tetramethyl-3-pentano	62185-29-9	2.76	5.73E+02	2
2397	18335	C9H20O	2,2,4,4-tetramethyl-3-pentano	14609-79-1	2.58	3.83E+02	2
2398	18365	C9H20S	nonyl mercaptan	1455-21-6	4.70	5.01E+04	2
2399	18366	C9H20S	2-nonanethiol	13281-11-3	4.93	8.56E+04	2
2400	18367	C9H20S	methyl octyl sulfide	3698-95-1	4.37	2.34E+04	2

Table 13 Octanol-Water Partition Coefficient - Oganic Compounds

Octanol-Water Partition Coefficient								
NO	ID	FORMULA	NAME	CAS No	log_{10} Kow	Kow	code
2401	18368	C9H20S	ethyl heptyl sulfide	24768-44-3	4.37	2.34E+04	2
2402	18369	C9H20S	propyl hexyl sulfide	24768-43-2	4.37	2.34E+04	2
2403	18370	C9H20S	butyl pentyl sulfide	24768-42-1	4.37	2.34E+04	2
2404	18371	C9H20S	tert-nonyl mercaptan, isomers	25360-10-5	4.82	6.55E+04	2
2405	18388	C9H21N	nonylamine	112-20-9	3.29	1.95E+03	2
2406	18389	C9H21N	methyloctylamine	2439-54-5	3.55	3.56E+03	2
2407	18390	C9H21N	ethylheptylamine	66793-76-8	3.42	2.65E+03	2
2408	18391	C9H21N	dimethylheptylamine	5277-11-2	2.75	5.68E+02	2
2409	18392	C9H21N	diethylpentylamine	2162-91-6	2.38	2.41E+02	2
2410	18393	C9H21N	tripropylamine	102-69-2	2.79	6.17E+02	1
2411	18394	C9H21N	tert-amyl-tert-butylamine	2085-66-7	3.49	3.07E+03	2
2412	18395	C9H21N	N-methyldibutylamine	3405-45-6	2.64	4.39E+02	2
2413	18586	C10H6O3	2-hydroxy-1,4-naphthalenedi	83-72-7	1.46	2.88E+01	1
2414	18587	C10H6O3	5-hydroxy-1,4-naphthalenedi	481-39-0	1.46	2.88E+01	2
2415	18598	C10H6O8	pyromellitic acid	89-05-4	0.15	1.41E+00	2
2416	18601	C10H7Br	1-bromonaphthalene	90-11-9	4.06	1.15E+04	2
2417	18606	C10H7Cl	1-chloronaphthalene	90-13-1	3.90	7.94E+03	1
2418	18607	C10H7Cl	2-chloronaphthalene	91-58-7	3.98	9.55E+03	1
2419	18608	C10H7Cl	monochloronaphthalene	25586-43-0	3.90	7.94E+03	2
2420	18669	C10H8	naphthalene	91-20-3	3.34	2.19E+03	1
2421	18670	C10H8	azulene	275-51-4	3.22	1.66E+03	1
2422	18740	C10H8O	1-naphthol	90-15-3	2.84	6.92E+02	1
2423	18741	C10H8O	2-naphthol	135-19-3	2.70	5.01E+02	1
2424	18849	C10H9N	2-methylquinoline	91-63-4	2.59	3.89E+02	1
2425	18850	C10H9N	3-methylquinoline	612-58-8	2.59	3.89E+02	2
2426	18851	C10H9N	4-methylquinoline	491-35-0	2.59	3.89E+02	1
2427	18852	C10H9N	5-methylquinoline	7661-55-4	2.59	3.89E+02	2
2428	18853	C10H9N	6-methylquinoline	91-62-3	2.59	3.89E+02	2
2429	18854	C10H9N	7-methylquinoline	612-60-2	2.59	3.89E+02	2
2430	18855	C10H9N	8-methylquinoline	611-32-5	2.59	3.89E+02	2
2431	18856	C10H9N	1-methylisoquinoline	1721-93-3	2.59	3.89E+02	2
2432	18857	C10H9N	3-methylisoquinoline	1125-80-0	2.59	3.89E+02	2
2433	18858	C10H9N	4-methylisoquinoline	1196-39-0	2.59	3.89E+02	2
2434	18859	C10H9N	6-methylisoquinoline	42398-73-2	2.59	3.89E+02	2
2435	18860	C10H9N	8-methylisoquinoline	62882-00-2	2.59	3.89E+02	2
2436	18958	C10H10	m-divinylbenzene	108-57-6	3.80	6.31E+03	2
2437	19106	C10H10O	benzalacetone	122-57-6	2.07	1.17E+02	1
2438	19119	C10H10O2	methyl cinnamate	1754-62-7	2.62	4.17E+02	1
2439	19195	C10H10O4	dimethyl phthalate	131-11-3	1.56	3.63E+01	1
2440	19196	C10H10O4	dimethyl terephthalate	120-61-6	2.25	1.78E+02	1
2441	19358	C10H12	1,3-dicyclopentadiene	77-73-6	3.51	3.24E+03	2
2442	19359	C10H12	1,2,3,4-tetrahydronaphthalen	119-64-2	3.49	3.09E+03	1
2443	19484	C10H12O	anethole	104-46-1	3.39	2.45E+03	2
2444	19551	C10H12O2	isopropyl benzoate	939-48-0	3.18	1.51E+03	1
2445	19675	C10H12O4	diallyl maleate	999-21-3	2.90	7.94E+02	2
2446	19865	C10H14	butylbenzene	104-51-8	4.26	1.82E+04	1
2447	19866	C10H14	isobutylbenzene	538-93-2	4.01	1.02E+04	1
2448	19867	C10H14	sec-butylbenzene	135-98-8	4.57	3.72E+04	1
2449	19868	C10H14	tert-butylbenzene	98-06-6	4.11	1.29E+04	1
2450	19869	C10H14	cymene	25155-15-1	4.32	2.10E+04	2
2451	19870	C10H14	m-cymene	535-77-3	4.50	3.16E+04	1
2452	19871	C10H14	o-cymene	527-84-4	4.38	2.40E+04	1
2453	19872	C10H14	p-cymene	99-87-6	4.10	1.26E+04	1
2454	19873	C10H14	1-methyl-2-propylbenzene	1074-17-5	4.10	1.25E+04	2
2455	19874	C10H14	1-methyl-3-propylbenzene	1074-43-7	4.03	1.07E+04	2
2456	19875	C10H14	1-methyl-4-propylbenzene	1074-55-1	3.99	9.79E+03	2
2457	19876	C10H14	(S)-(1-methylpropyl)benzene	5787-28-0	4.32	2.10E+04	2
2458	19877	C10H14	diethylbenzene	25340-17-4	4.32	2.10E+04	2
2459	19878	C10H14	o-diethylbenzene	135-01-3	4.42	2.63E+04	1
2460	19879	C10H14	m-diethylbenzene	141-93-5	4.57	3.72E+04	1
2461	19880	C10H14	p-diethylbenzene	105-05-5	4.58	3.80E+04	1
2462	19881	C10H14	3-ethyl-o-xylene	933-98-2	4.34	2.19E+04	1
2463	19882	C10H14	4-ethyl-o-xylene	934-80-5	4.50	3.16E+04	1
2464	19883	C10H14	2-ethyl-m-xylene	2870-04-4	4.28	1.91E+04	1
2465	19884	C10H14	4-ethyl-m-xylene	874-41-9	4.47	2.95E+04	1
2466	19885	C10H14	5-ethyl-m-xylene	934-74-7	4.55	3.55E+04	1
2467	19886	C10H14	2-ethyl-p-xylene	1758-88-9	4.43	2.69E+04	1
2468	19887	C10H14	1,2,3,4-tetramethylbenzene	488-23-3	4.00	1.00E+04	1
2469	19888	C10H14	1,2,3,5-tetramethylbenzene	527-53-7	4.10	1.26E+04	1
2470	19889	C10H14	1,2,4,5-tetramethylbenzene	95-93-2	4.10	1.26E+04	1
2471	19982	C10H14O	3-butylphenol	4074-43-5	3.65	4.47E+03	2
2472	19983	C10H14O	4-butylphenol	1638-22-8	3.65	4.47E+03	1
2473	19984	C10H14O	2-isobutylphenol	4167-75-3	3.65	4.47E+03	2
2474	19985	C10H14O	3-isobutylphenol	30749-25-8	3.65	4.47E+03	2
2475	19986	C10H14O	4-isobutylphenol	4167-74-2	3.65	4.47E+03	2
2476	20037	C10H14O	2-butylphenol	3180-09-4	3.65	4.47E+03	2
2477	20089	C10H14O2	p-tert-butylcatechol	98-29-3	11.84	6.96E+11	2
2478	20195	C10H15N	N,N-dietyhlaniline	91-66-7	3.31	2.04E+03	1
2479	20196	C10H15N	2,6-diethylaniline	579-66-8	3.15	1.41E+03	2
2480	20197	C10H15N	2-sec-butylaniline	55751-54-7	3.23	1.70E+03	2

Octanol-Water Partition Coefficient								
NO	ID	FORMULA	NAME	CAS No	log_{10} Kow	Kow	code
2481	20198	C10H15N	2-tert-butylaniline	6310-21-0	3.23	1.70E+03	2
2482	20199	C10H15N	4-butylaniline	104-13-2	3.23	1.70E+03	2
2483	20200	C10H15N	4-sec-butylaniline	30273-11-1	3.23	1.70E+03	2
2484	20201	C10H15N	4-tert-butylaniline	769-92-6	3.23	1.70E+03	2
2485	20202	C10H15N	N-sec-butylaniline	6068-69-5	3.23	1.70E+03	2
2486	20203	C10H15N	N-tert-butylaniline	937-33-7	3.23	1.70E+03	2
2487	20204	C10H15N	N-butylaniline	1126-78-9	3.23	1.70E+03	2
2488	20205	C10H15N	4-isobutylaniline	30090-17-6	3.23	1.70E+03	2
2489	20206	C10H15N	N-isobutylaniline	588-47-6	3.23	1.70E+03	2
2490	20207	C10H15N	4-methyl-N-isopropylaniline	10436-75-6	3.23	1.70E+03	2
2491	20208	C10H15N	2-methyl-5-isopropylaniline	2051-53-8	3.23	1.70E+03	2
2492	20209	C10H15N	4-methyl-N-propylaniline	54837-90-0	3.23	1.70E+03	2
2493	20210	C10H15N	2,3,4,5-tetramethylaniline	2217-45-0	3.23	1.70E+03	2
2494	20211	C10H15N	2,3,4,6-tetramethylaniline	488-71-1	3.23	1.70E+03	2
2495	20212	C10H15N	N,N,2,6-tetramethylaniline	769-06-2	3.23	1.70E+03	2
2496	20230	C10H15N	3,5,N,N-tetramethylaniline	4913-13-7	3.23	1.70E+03	2
2497	20231	C10H15N	aniline, p-isopropyl-N-methyl-	6950-79-4	3.23	1.70E+03	2
2498	20315	C10H16	camphene	79-92-5	4.35	2.24E+04	2
2499	20316	C10H16	D-limonene	5989-27-5	4.83	6.76E+04	2
2500	20317	C10H16	limonene	138-86-3	4.83	6.76E+04	2
2501	20318	C10H16	L-limonene	5989-54-8	4.83	6.76E+04	2
2502	20319	C10H16	alpha-phellandrene	99-83-2	4.62	4.17E+04	2
2503	20320	C10H16	beta-phellandrene	555-10-2	4.70	5.01E+04	2
2504	20321	C10H16	alpha-pinene	80-56-8	4.83	6.76E+04	1
2505	20322	C10H16	alpha-pinene, (-)	7785-26-4	4.56	3.63E+04	2
2506	20323	C10H16	beta-pinene	127-91-3	4.35	2.24E+04	2
2507	20324	C10H16	beta-pinene, (1R)	19902-08-0	4.56	3.63E+04	2
2508	20325	C10H16	alpha-terpinene	99-86-5	4.56	3.63E+04	2
2509	20326	C10H16	gamma-terpinene	99-85-4	4.50	3.16E+04	1
2510	20327	C10H16	terpinolene	586-62-9	4.47	2.95E+04	1
2511	20356	C10H16	(+)-camphene	5794-03-6	4.35	2.24E+04	2
2512	20357	C10H16	(-)-camphene	5794-04-7	4.35	2.24E+04	2
2513	20365	C10H16	(-)-a-pinene	1330-16-1	4.56	3.63E+04	2
2514	20366	C10H16	(1S)-(-)-b-pinene	18172-67-3	4.56	3.63E+04	2
2515	20367	C10H16	(±)-a-pinene	2437-95-8	4.56	3.63E+04	2
2516	20368	C10H16	(+)-a-pinene	7785-70-8	4.56	3.63E+04	2
2517	20423	C10H16O	camphor	76-22-2	2.34	2.19E+02	2
2518	20593	C10H18	1-decyne	764-93-2	5.12	1.33E+05	2
2519	20594	C10H18	2-decyne	2384-70-5	5.35	2.26E+05	2
2520	20595	C10H18	3-decyne	2384-85-2	5.24	1.74E+05	2
2521	20596	C10H18	4-decyne	2384-86-3	5.24	1.74E+05	2
2522	20597	C10H18	5-decyne	1942-46-7	5.24	1.74E+05	2
2523	20598	C10H18	2,2-dimethyl-3-octyne	19482-57-6	5.24	1.74E+05	2
2524	20599	C10H18	8-methyl-4-nonyne	70732-43-3	5.24	1.74E+05	2
2525	20600	C10H18	5-ethyl-5-methyl-3-heptyne	61228-10-2	5.24	1.74E+05	2
2526	20601	C10H18	cis-decahydronaphthalene	493-01-6	4.20	1.58E+04	2
2527	20602	C10H18	trans-decahydronaphthalene	493-02-7	4.20	1.58E+04	2
2528	20636	C10H18	decahydronaphthalene; (cis+	91-17-8	4.20	1.58E+04	2
2529	20820	C10H18O4	sebacic acid	111-20-6	2.19	1.55E+02	2
2530	20922	C10H20	pentylcyclopentane	3741-00-2	6.29	1.94E+06	2
2531	20923	C10H20	butylcyclohexane	1678-93-9	5.07	1.17E+05	2
2532	20924	C10H20	tert-butylcyclohexane	3178-22-1	5.07	1.17E+05	2
2533	20925	C10H20	sec-butylcyclohexane	7058-01-7	5.07	1.17E+05	2
2534	20926	C10H20	isobutylcyclohexane	1678-98-4	5.07	1.17E+05	2
2535	20927	C10H20	1-methyl-1-propylcyclohexan	4258-93-9	5.07	1.17E+05	2
2536	20928	C10H20	1-methyl-2-propylcyclohexan	4291-79-6	5.07	1.17E+05	2
2537	20929	C10H20	1-methyl-3-propylcyclohexan	4291-80-9	5.07	1.17E+05	2
2538	20930	C10H20	1-methyl-4-propylcyclohexan	4291-81-0	5.07	1.17E+05	2
2539	20931	C10H20	1-methyl-2-isopropylcyclohex	16580-23-7	5.07	1.17E+05	2
2540	20932	C10H20	1-methyl-3-isopropylcyclohex	16580-24-8	5.07	1.17E+05	2
2541	20933	C10H20	1-methyl-4-isopropylcyclohex	99-82-1	5.07	1.17E+05	2
2542	20934	C10H20	cis-1-methyl-4-isopropylcyclo	6069-98-3	5.07	1.17E+05	2
2543	20935	C10H20	trans-1-methyl-4-isopropylcyc	1678-82-6	5.07	1.17E+05	2
2544	20936	C10H20	1,1,3,4-tetramethylcyclohexa	24612-75-7	5.07	1.17E+05	2
2545	20937	C10H20	trans-1,1,3,5-tetramethylcyclo	50876-31-8	5.07	1.17E+05	2
2546	20938	C10H20	cis-1,1,3,5-tetramethylcyclohe	50876-32-9	5.07	1.17E+05	2
2547	20939	C10H20	diethylcyclohexane; (mixed is	1331-43-7	5.07	1.17E+05	2
2548	20943	C10H20	1-decene	872-05-9	5.12	1.32E+05	2
2549	20944	C10H20	4-decene	19689-18-0	5.63	4.27E+05	2
2550	20945	C10H20	cis-5-decene	7433-78-5	5.63	4.27E+05	2
2551	20946	C10H20	trans-5-decene	7433-56-9	5.63	4.27E+05	2
2552	20947	C10H20	2-methyl-1-nonene	2980-71-4	6.14	1.38E+06	2
2553	20948	C10H20	2-methyl-3-nonene	53966-53-3	5.63	4.27E+05	2
2554	20949	C10H20	5-methyl-4-nonene	15918-07-7	5.63	4.27E+05	2
2555	20950	C10H20	3,7-dimethyl-1-octene	4984-01-4	5.63	4.27E+05	2
2556	20951	C10H20	2,6-dimethyl-2-octene	4057-42-5	5.63	4.27E+05	2
2557	20952	C10H20	4-propyl-3-heptene	4485-13-6	5.63	4.27E+05	2
2558	20953	C10H20	2,4,6-trimethyl-3-heptene, iso	123434-07-1	5.63	4.27E+05	2
2559	20954	C10H20	2,2,5,5-tetramethylhex-3-ene	22808-06-6	5.63	4.27E+05	2
2560	20955	C10H20	cis-1,2-di-tert-butylethene	692-47-7	5.63	4.27E+05	2

Table 13 Octanol-Water Partition Coefficient - Oganic Compounds

Octanol-Water Partition Coefficient							
NO	ID	FORMULA	NAME	CAS No	log_{10} Kow	Kow	code
2561	20956	C10H20	(E)-2,2,5,5-tetramethylhex-3-(	692-48-8	5.63	4.27E+05	2
2562	20992	C10H20O	decanal	112-31-2	3.76	5.75E+03	2
2563	20993	C10H20O	2-decanone	693-54-9	3.77	5.89E+03	1
2564	20994	C10H20O	3-decanone	928-80-3	3.72	5.22E+03	2
2565	20995	C10H20O	4-decanone	624-16-8	3.72	5.22E+03	2
2566	20996	C10H20O	2-methyl-5-nonanone	22287-02-1	3.72	5.22E+03	2
2567	21019	C10H20O	5-decanone	820-29-1	3.72	5.22E+03	2
2568	21025	C10H20O	isodecanal	1321-89-7	3.76	5.75E+03	2
2569	21034	C10H20O	2-methyl-3-nonanone	5445-31-8	3.72	5.22E+03	2
2570	21040	C10H20O2	decanoic acid	334-48-5	4.09	1.23E+04	1
2571	21041	C10H20O2	2,2-dimethyloctanoic acid	29662-90-6	4.09	1.23E+04	2
2572	21042	C10H20O2	3-methylnonanoic acid	35205-79-9	4.09	1.23E+04	2
2573	21043	C10H20O2	nonyl formate	5451-92-3	4.61	4.12E+04	2
2574	21044	C10H20O2	2-ethylhexyl acetate	103-09-3	3.74	5.50E+03	2
2575	21045	C10H20O2	octyl acetate	112-14-1	4.07	1.18E+04	2
2576	21046	C10H20O2	6-methyl-2-heptyl acetate	67952-57-2	3.91	8.04E+03	2
2577	21047	C10H20O2	6-methyl-3-heptyl acetate	32764-34-4	3.91	8.04E+03	2
2578	21048	C10H20O2	2-methylheptyl acetate, (±)	74112-36-0	3.91	8.04E+03	2
2579	21052	C10H20O2	isopentyl isovalerate	659-70-1	3.66	4.57E+03	2
2580	21054	C10H20O2	isopentyl pentanoate	2050-09-1	3.66	4.57E+03	2
2581	21055	C10H20O2	pentyl pentanoate	2173-56-0	3.66	4.57E+03	2
2582	21056	C10H20O2	butyl hexanoate	626-82-4	3.66	4.57E+03	2
2583	21057	C10H20O2	isobutyl 4-methylpentanoate	25415-70-7	3.66	4.57E+03	2
2584	21058	C10H20O2	propyl heptanoate	7778-87-2	3.66	4.57E+03	2
2585	21059	C10H20O2	ethyl 2-ethylhexanoate	2983-37-1	3.66	4.57E+03	2
2586	21060	C10H20O2	ethyl octanoate	106-32-1	3.66	4.57E+03	2
2587	21063	C10H20O2	methyl nonanoate	1731-84-6	3.66	4.57E+03	2
2588	21066	C10H20O2	8-methylnonanoic acid	5963-14-4	4.09	1.23E+04	2
2589	21067	C10H20O2	neodecanoic acid	26896-20-8	4.09	1.23E+04	2
2590	21068	C10H20O2	tert-decanoic acid	52627-73-3	4.09	1.23E+04	2
2591	21071	C10H20O2	n-hexyl isobutyrate	2349-07-7	3.66	4.57E+03	2
2592	21073	C10H20O2	isobutyl hexanoate	105-79-3	3.66	4.57E+03	2
2593	21074	C10H20O2	2-(4-isobutylphenyl)butyric ac	55837-18-8	4.09	1.23E+04	2
2594	21077	C10H20O2	4-nonanecarboxylic acid	31080-39-4	4.09	1.23E+04	2
2595	21090	C10H21Br	1-bromodecane	112-29-8	5.60	3.98E+05	2
2596	21091	C10H21Br	2-bromodecane	39563-53-6	5.60	3.98E+05	2
2597	21093	C10H21Cl	1-chlorodecane	1002-69-3	5.51	3.24E+05	2
2598	21094	C10H21Cl	decyl chloride (mixed isomers	28519-06-4	5.51	3.24E+05	2
2599	21133	C10H22	decane	124-18-5	6.25	1.78E+06	1
2600	21134	C10H22	2-methylnonane	871-83-0	5.18	1.51E+05	2
2601	21135	C10H22	3-methylnonane	5911-04-6	5.18	1.51E+05	2
2602	21136	C10H22	4-methylnonane	17301-94-9	5.18	1.51E+05	2
2603	21137	C10H22	5-methylnonane	15869-85-9	5.18	1.51E+05	2
2604	21138	C10H22	3-ethyloctane	5881-17-4	5.18	1.51E+05	2
2605	21139	C10H22	4-ethyloctane	15869-86-0	5.18	1.51E+05	2
2606	21140	C10H22	2,2-dimethyloctane	15869-87-1	5.11	1.29E+05	2
2607	21141	C10H22	2,3-dimethyloctane	7146-60-3	5.11	1.29E+05	2
2608	21142	C10H22	2,4-dimethyloctane	4032-94-4	5.11	1.29E+05	2
2609	21143	C10H22	2,5-dimethyloctane	15869-89-3	5.11	1.29E+05	2
2610	21144	C10H22	2,6-dimethyloctane	2051-30-1	5.11	1.29E+05	2
2611	21145	C10H22	2,7-dimethyloctane	1072-16-8	5.11	1.29E+05	2
2612	21146	C10H22	3,3-dimethyloctane	4110-44-5	5.11	1.29E+05	2
2613	21147	C10H22	3,4-dimethyloctane	15869-92-8	5.11	1.29E+05	2
2614	21148	C10H22	3,5-dimethyloctane	15869-93-9	5.11	1.29E+05	2
2615	21149	C10H22	3,6-dimethyloctane	15869-94-0	5.11	1.29E+05	2
2616	21150	C10H22	4,4-dimethyloctane	15869-95-1	5.11	1.29E+05	2
2617	21151	C10H22	4,5-dimethyloctane	15869-96-2	5.11	1.29E+05	2
2618	21152	C10H22	4-propylheptane	3178-29-8	5.18	1.51E+05	2
2619	21153	C10H22	4-isopropylheptane	52896-87-4	5.11	1.29E+05	2
2620	21154	C10H22	3-ethyl-2-methylheptane	14676-29-0	5.11	1.29E+05	2
2621	21155	C10H22	4-ethyl-2-methylheptane	52896-88-5	5.11	1.29E+05	2
2622	21156	C10H22	5-ethyl-2-methylheptane	13475-78-0	5.11	1.29E+05	2
2623	21157	C10H22	3-ethyl-3-methylheptane	17302-01-1	5.11	1.29E+05	2
2624	21158	C10H22	4-ethyl-3-methylheptane	52896-89-6	5.11	1.29E+05	2
2625	21159	C10H22	3-ethyl-5-methylheptane	52896-90-9	5.11	1.29E+05	2
2626	21160	C10H22	3-ethyl-4-methylheptane	52896-91-0	5.11	1.29E+05	2
2627	21161	C10H22	4-ethyl-4-methylheptane	17302-04-4	5.11	1.29E+05	2
2628	21162	C10H22	2,2,3-trimethylheptane	52896-92-1	5.07	1.17E+05	2
2629	21163	C10H22	2,2,4-trimethylheptane	14720-74-2	5.07	1.17E+05	2
2630	21164	C10H22	2,2,5-trimethylheptane	20291-95-6	5.07	1.17E+05	2
2631	21165	C10H22	2,2,6-trimethylheptane	1190-83-6	5.07	1.17E+05	2
2632	21166	C10H22	2,3,3-trimethylheptane	52896-93-2	5.07	1.17E+05	2
2633	21167	C10H22	2,3,4-trimethylheptane	52896-95-4	5.07	1.17E+05	2
2634	21168	C10H22	2,3,5-trimethylheptane	20278-85-7	5.07	1.17E+05	2
2635	21169	C10H22	2,3,6-trimethylheptane	4032-93-3	5.07	1.17E+05	2
2636	21170	C10H22	2,4,4-trimethylheptane	4032-92-2	5.07	1.17E+05	2
2637	21171	C10H22	2,4,5-trimethylheptane	20278-84-6	5.07	1.17E+05	2
2638	21172	C10H22	2,4,6-trimethylheptane	2613-61-8	5.07	1.17E+05	2
2639	21173	C10H22	2,5,5-trimethylheptane	1189-99-7	5.07	1.17E+05	2
2640	21174	C10H22	3,3,4-trimethylheptane	20278-87-9	5.07	1.17E+05	2
2641	21175	C10H22	3,3,5-trimethylheptane	7154-80-5	5.07	1.17E+05	2
2642	21176	C10H22	3,4,4-trimethylheptane	20278-88-0	5.07	1.17E+05	2
2643	21177	C10H22	3,4,5-trimethylheptane	20278-89-1	5.07	1.17E+05	2
2644	21178	C10H22	3-isopropyl-2-methylhexane	62016-13-1	5.11	1.29E+05	2
2645	21179	C10H22	3,3-diethylhexane	17302-02-2	5.11	1.29E+05	2
2646	21180	C10H22	3,4-diethylhexane	19398-77-7	5.11	1.29E+05	2
2647	21181	C10H22	3-ethyl-2,2-dimethylhexane	20291-91-2	5.08	1.20E+05	2
2648	21182	C10H22	4-ethyl-2,2-dimethylhexane	52896-99-8	5.08	1.20E+05	2
2649	21183	C10H22	3-ethyl-2,3-dimethylhexane	52897-00-4	5.08	1.20E+05	2
2650	21184	C10H22	4-ethyl-2,3-dimethylhexane	52897-01-5	5.08	1.20E+05	2
2651	21185	C10H22	3-ethyl-2,4-dimethylhexane	7220-26-0	5.08	1.20E+05	2
2652	21186	C10H22	4-ethyl-2,4-dimethylhexane	52897-03-7	5.08	1.20E+05	2
2653	21187	C10H22	3-ethyl-2,5-dimethylhexane	52897-04-8	5.08	1.20E+05	2
2654	21188	C10H22	4-ethyl-3,3-dimethylhexane	52897-05-9	5.08	1.20E+05	2
2655	21189	C10H22	3-ethyl-3,4-dimethylhexane	52897-06-0	5.08	1.20E+05	2
2656	21190	C10H22	2,2,3,3-tetramethylhexane	13475-81-5	5.03	1.07E+05	1
2657	21191	C10H22	2,2,3,4-tetramethylhexane	52897-08-2	5.03	1.07E+05	2
2658	21192	C10H22	2,2,3,5-tetramethylhexane	52897-09-3	5.03	1.07E+05	2
2659	21193	C10H22	2,2,4,4-tetramethylhexane	51750-65-3	5.03	1.07E+05	2
2660	21194	C10H22	2,2,4,5-tetramethylhexane	16747-42-5	4.99	9.77E+04	2
2661	21195	C10H22	2,2,5,5-tetramethylhexane	1071-81-4	5.03	1.07E+05	2
2662	21196	C10H22	2,3,3,4-tetramethylhexane	52897-10-6	5.03	1.07E+05	2
2663	21197	C10H22	2,3,3,5-tetramethylhexane	52897-11-7	5.03	1.07E+05	2
2664	21198	C10H22	2,3,4,4-tetramethylhexane	52897-12-8	5.03	1.07E+05	2
2665	21199	C10H22	2,3,4,5-tetramethylhexane	52897-15-1	5.03	1.07E+05	2
2666	21200	C10H22	3,3,4,4-tetramethylhexane	5171-84-6	5.03	1.07E+05	2
2667	21201	C10H22	2,4-dimethyl-3-isopropylpenta	13475-79-1	5.07	1.17E+05	2
2668	21202	C10H22	3,3-diethyl-2-methylpentane	52897-16-2	5.07	1.17E+05	2
2669	21203	C10H22	3-ethyl-2,2,3-trimethylpentan	52897-17-3	5.03	1.07E+05	2
2670	21204	C10H22	3-ethyl-2,2,4-trimethylpentan	52897-18-4	5.03	1.07E+05	2
2671	21205	C10H22	3-ethyl-2,3,4-trimethylpentan	52897-19-5	5.03	1.07E+05	2
2672	21206	C10H22	2,2,3,3,4-pentamethylpentan	16747-44-7	4.97	9.33E+04	2
2673	21207	C10H22	2,2,3,4,4-pentamethylpentan	16747-45-8	4.97	9.33E+04	2
2674	21224	C10H22O	1-decanol	112-30-1	4.57	3.72E+04	1
2675	21225	C10H22O	isodecanol	25339-17-7	3.71	5.13E+03	2
2676	21226	C10H22O	2-decanol	1120-06-5	4.13	1.34E+04	2
2677	21227	C10H22O	2-decanol, (±)	74742-10-2	3.70	4.98E+03	2
2678	21228	C10H22O	3-decanol	1565-81-7	3.94	8.64E+03	2
2679	21229	C10H22O	4-decanol	2051-31-2	3.94	8.72E+03	2
2680	21230	C10H22O	5-decanol	5205-34-5	3.71	5.07E+03	2
2681	21231	C10H22O	2-methyl-1-nonanol	40589-14-8	4.27	1.85E+04	2
2682	21232	C10H22O	4-methyl-1-nonanol	1489-47-0	4.10	1.26E+04	2
2683	21233	C10H22O	5-methyl-1-nonanol	2768-16-3	4.04	1.10E+04	2
2684	21234	C10H22O	7-methyl-1-nonanol	33234-93-4	4.04	1.10E+04	2
2685	21235	C10H22O	3-methyl-1-nonanol	22663-64-5	4.04	1.10E+04	2
2686	21236	C10H22O	2-methyl-2-nonanol	10297-57-1	4.04	1.10E+04	2
2687	21237	C10H22O	3-methyl-2-nonanol	60671-32-1	4.04	1.10E+04	2
2688	21238	C10H22O	5-methyl-2-nonanol	66731-95-1	4.04	1.10E+04	2
2689	21239	C10H22O	6-methyl-2-nonanol	66256-60-8	4.04	1.10E+04	2
2690	21240	C10H22O	7-methyl-2-nonanol	66256-61-9	4.04	1.10E+04	2
2691	21241	C10H22O	8-methyl-2-nonanol	14779-92-1	4.04	1.10E+04	2
2692	21242	C10H22O	2-methyl-3-nonanol	26533-33-5	3.90	7.91E+03	2
2693	21243	C10H22O	3-methyl-3-nonanol	21078-72-8	3.91	8.13E+03	2
2694	21244	C10H22O	5-methyl-3-nonanol	66719-43-5	3.91	8.13E+03	2
2695	21245	C10H22O	2-methyl-4-nonanol	26533-31-3	3.91	8.13E+03	2
2696	21246	C10H22O	4-methyl-4-nonanol	23418-38-4	3.91	8.13E+03	2
2697	21247	C10H22O	5-methyl-4-nonanol	66719-44-6	3.91	8.13E+03	2
2698	21248	C10H22O	7-methyl-4-nonanol	26981-98-6	3.91	8.13E+03	2
2699	21249	C10H22O	2-methyl-5-nonanol	29843-62-7	3.91	8.13E+03	2
2700	21250	C10H22O	3-methyl-5-nonanol	66779-42-4	3.91	8.13E+03	2
2701	21251	C10H22O	5-methyl-5-nonanol	33933-78-7	3.73	5.38E+03	2
2702	21252	C10H22O	2-ethyl-1-octanol	20592-10-3	3.91	8.13E+03	2
2703	21253	C10H22O	3-ethyl-1-octanol	66719-36-6	3.91	8.13E+03	2
2704	21254	C10H22O	2,2-dimethyl-1-octanol	2370-14-1	3.89	7.72E+03	2
2705	21255	C10H22O	2,6-dimethyl-1-octanol	62417-08-7	3.94	8.64E+03	2
2706	21256	C10H22O	3,7-dimethyl-1-octanol	106-21-8	3.99	9.86E+03	2
2707	21257	C10H22O	4,5-dimethyl-1-octanol	66719-32-2	3.99	9.76E+03	2
2708	21258	C10H22O	4,6-dimethyl-1-octanol	66719-33-3	3.99	9.76E+03	2
2709	21259	C10H22O	4,7-dimethyl-1-octanol	66719-34-4	3.99	9.76E+03	2
2710	21260	C10H22O	7,7-dimethyl-1-octanol	66719-35-5	3.99	9.76E+03	2
2711	21261	C10H22O	2,4-dimethyl-2-octanol	18675-20-2	3.99	9.76E+03	2
2712	21262	C10H22O	2,6-dimethyl-2-octanol	18479-57-7	3.99	9.76E+03	2
2713	21263	C10H22O	2,7-dimethyl-2-octanol	42007-73-8	3.99	9.76E+03	2
2714	21264	C10H22O	3,7-dimethyl-2-octanol	15340-96-2	3.99	9.76E+03	2
2715	21265	C10H22O	3-ethyl-3-octanol	2051-32-3	3.65	4.51E+03	2
2716	21266	C10H22O	6-ethyl-3-octanol	19781-27-2	3.65	4.47E+03	2
2717	21267	C10H22O	2,2-dimethyl-3-octanol	19841-72-6	3.52	3.34E+03	2
2718	21268	C10H22O	2,3-dimethyl-3-octanol	19781-10-3	3.40	2.54E+03	2
2719	21269	C10H22O	2,7-dimethyl-3-octanol	66719-55-9	3.53	3.36E+03	2
2720	21270	C10H22O	3,5-dimethyl-3-octanol	56065-42-0	3.50	3.15E+03	2

Table 13 Octanol-Water Partition Coefficient - Oganic Compounds

Octanol-Water Partition Coefficient							
NO	ID	FORMULA	NAME	CAS No	log_{10} Kow	Kow	code
2721	21271	C10H22O	3,6-dimethyl-3-octanol	151-19-9	3.48	3.00E+03	2
2722	21272	C10H22O	3,7-dimethyl-3-octanol	78-69-3	3.58	3.78E+03	2
2723	21273	C10H22O	3-ethyl-4-octanol	63126-48-7	3.50	3.15E+03	2
2724	21274	C10H22O	4-ethyl-4-octanol	38395-42-5	3.50	3.15E+03	2
2725	21275	C10H22O	2,2-dimethyl-4-octanol	66719-52-6	3.43	2.67E+03	2
2726	21276	C10H22O	2,4-dimethyl-4-octanol	33933-79-8	3.45	2.80E+03	2
2727	21277	C10H22O	2,5-dimethyl-4-octanol	66719-53-7	3.45	2.80E+03	2
2728	21278	C10H22O	2,6-dimethyl-4-octanol	66719-54-8	3.55	3.57E+03	2
2729	21279	C10H22O	2,7-dimethyl-4-octanol	19781-11-4	3.73	5.38E+03	2
2730	21280	C10H22O	3,4-dimethyl-4-octanol	66719-30-0	3.45	2.80E+03	2
2731	21281	C10H22O	3,6-dimethyl-4-octanol	66719-31-1	3.45	2.80E+03	2
2732	21282	C10H22O	4,6-dimethyl-4-octanol	56065-43-1	3.45	2.80E+03	2
2733	21283	C10H22O	4,7-dimethyl-4-octanol	19781-13-6	3.48	3.00E+03	2
2734	21284	C10H22O	2-propyl-1-heptanol	10042-59-8	4.15	1.42E+04	2
2735	21285	C10H22O	3-isopropyl-1-heptanol	38514-15-7	4.08	1.20E+04	2
2736	21286	C10H22O	2,4,6-trimethyl-2-heptanol	66256-47-1	3.50	3.15E+03	2
2737	21287	C10H22O	2,5,6-trimethyl-2-heptanol	66256-48-2	3.50	3.17E+03	2
2738	21288	C10H22O	4,6,6-trimethyl-2-heptanol	51079-79-9	3.50	3.15E+03	2
2739	21289	C10H22O	2-methyl-3-ethyl-3-heptanol	66719-37-7	3.50	3.17E+03	2
2740	21290	C10H22O	4-methyl-5-ethyl-3-heptanol	66731-94-0	3.50	3.15E+03	2
2741	21291	C10H22O	5-methyl-3-ethyl-3-heptanol	---	3.50	3.15E+03	2
2742	21292	C10H22O	2,2,3-trimethyl-3-heptanol	29772-40-5	3.33	2.12E+03	2
2743	21293	C10H22O	2,2,6-trimethyl-3-heptanol	66256-43-7	3.42	2.65E+03	2
2744	21294	C10H22O	2,3,6-trimethyl-3-heptanol	58046-40-5	3.42	2.65E+03	2
2745	21295	C10H22O	3,5,5-trimethyl-3-heptanol	66256-50-6	3.57	3.69E+03	2
2746	21296	C10H22O	4-propyl-4-heptanol	2198-72-3	3.53	3.36E+03	2
2747	21297	C10H22O	4-isopropyl-4-heptanol	51200-82-9	3.43	2.67E+03	2
2748	21298	C10H22O	2,2,4-trimethyl-4-heptanol	57233-31-5	3.20	1.60E+03	2
2749	21299	C10H22O	2,2,5-trimethyl-4-heptanol	66256-42-6	3.22	1.68E+03	2
2750	21300	C10H22O	2,2,6-trimethyl-4-heptanol	66256-44-8	3.22	1.68E+03	2
2751	21301	C10H22O	2,4,5-trimethyl-4-heptanol	66256-46-0	3.22	1.68E+03	2
2752	21302	C10H22O	2,4,6-trimethyl-4-heptanol	60836-07-9	3.23	1.69E+03	2
2753	21303	C10H22O	3,3,6-trimethyl-4-heptanol	---	3.22	1.68E+03	2
2754	21304	C10H22O	4-methyl-2-propyl-1-hexanol	66256-62-0	3.89	7.72E+03	2
2755	21305	C10H22O	4-methyl-2-isopropyl-1-hexan	66719-41-3	3.71	5.07E+03	2
2756	21306	C10H22O	5-methyl-2-isopropyl-1-hexan	2051-33-4	4.02	1.05E+04	2
2757	21307	C10H22O	5,5-dimethyl-2-ethyl-1-hexan	---	4.02	1.04E+04	2
2758	21308	C10H22O	2,3,4,4-tetramethyl-2-hexanol	66256-66-4	3.43	2.67E+03	2
2759	21309	C10H22O	3,3,5,5-tetramethyl-2-hexanol	66256-69-7	3.42	2.65E+03	2
2760	21310	C10H22O	2-methyl-3-isopropyl-3-hexan	51200-81-8	3.48	3.00E+03	2
2761	21311	C10H22O	2,2-dimethyl-4-ethyl-3-hexan	66719-47-9	3.35	2.25E+03	2
2762	21312	C10H22O	2,4-dimethyl-4-ethyl-3-hexan	66719-48-0	3.35	2.23E+03	2
2763	21313	C10H22O	5,5-dimethyl-3-ethyl-3-hexan	5340-62-5	3.35	2.23E+03	2
2764	21314	C10H22O	2,2,3,4-tetramethyl-3-hexanol	66256-63-1	3.48	3.00E+03	2
2765	21315	C10H22O	2,2,3,5-tetramethyl-3-hexanol	66256-64-2	4.80	6.31E+04	2
2766	21316	C10H22O	2,2,4,4-tetramethyl-3-hexanol	66256-65-3	3.43	2.67E+03	2
2767	21317	C10H22O	2,2,5,5-tetramethyl-3-hexanol	55073-86-4	2.93	8.60E+02	2
2768	21318	C10H22O	2,3,4,4-tetramethyl-3-hexanol	66256-67-5	3.71	5.07E+03	2
2769	21319	C10H22O	2,3,5,5-tetramethyl-3-hexanol	5396-09-8	3.73	5.34E+03	2
2770	21320	C10H22O	2,4,4,5-tetramethyl-3-hexanol	66256-68-6	3.73	5.34E+03	2
2771	21321	C10H22O	3,4,4,5-tetramethyl-3-hexanol	66256-39-1	3.73	5.38E+03	2
2772	21322	C10H22O	3,4,5,5-tetramethyl-3-hexanol	66256-40-4	3.55	3.57E+03	2
2773	21323	C10H22O	4-methyl-2-isobutyl-1-pentan	22417-45-4	3.78	6.07E+03	2
2774	21324	C10H22O	4,4-dimethyl-3-isopropyl-1-pe	66719-51-5	3.25	1.77E+03	2
2775	21325	C10H22O	3,4-dimethyl-3-isopropyl-2-pe	66719-50-4	3.25	1.77E+03	2
2776	21326	C10H22O	2,4-dimethyl-3-propyl-3-penta	---	3.28	1.89E+03	2
2777	21327	C10H22O	2,4-dimethyl-3-isopropyl-3-pe	51200-83-0	3.54	3.46E+03	2
2778	21328	C10H22O	2,2,4-trimethyl-3-ethyl-3-pent	66256-41-5	3.45	2.83E+03	2
2779	21329	C10H22O	2,2,3,4,4-pentamethyl-3-pent	5857-69-2	3.54	3.44E+03	2
2780	21330	C10H22O	2,3-dimethyl-2-tert-butyl-1-bu	81931-81-9	3.50	3.15E+03	2
2781	21331	C10H22O	3,7-dimethyl-3-octanol, (±)	57706-88-4	3.70	4.98E+03	2
2782	21332	C10H22O	3,7-dimethyl-1-octanol, (±)	59204-02-3	3.70	4.98E+03	2
2783	21333	C10H22O	3-ethyl-6-methyl-3-heptanol	66719-40-2	3.70	4.98E+03	2
2784	21334	C10H22O	8-methyl-1-nonanol	55505-26-5	3.70	4.98E+03	2
2785	21335	C10H22O	di-pentyl ether	693-65-2	4.00	1.00E+04	2
2786	21336	C10H22O	diisopentyl ether	544-01-4	4.00	1.00E+04	2
2787	21337	C10H22O	ethyl octyl ether	929-61-3	4.00	1.00E+04	2
2788	21338	C10H22O	(S)-(+)-2-decanol	33758-16-6	3.70	4.98E+03	2
2789	21339	C10H22O	(R)-3,7-dimethyl-1-octanol	1117-60-8	3.70	4.98E+03	2
2790	21340	C10H22O	s-3,7-dimethyl-1-octanol	68680-98-8	3.70	4.98E+03	2
2791	21341	C10H22O	decyl alcohol (mixed isomers)	85566-12-7	3.70	4.98E+03	2
2792	21385	C10H22S	decyl mercaptan	143-10-2	5.20	1.58E+05	2
2793	21386	C10H22S	2-decanethiol	13402-60-3	5.40	2.54E+05	2
2794	21387	C10H22S	dipentyl sulfide	872-10-6	4.86	7.24E+04	2
2795	21388	C10H22S	methyl nonyl sulfide	59973-07-8	4.86	7.24E+04	2
2796	21389	C10H22S	ethyl octyl sulfide	3698-94-0	4.86	7.24E+04	2
2797	21390	C10H22S	propyl heptyl sulfide	24768-46-5	4.86	7.24E+04	2
2798	21391	C10H22S	butyl hexyl sulfide	16967-04-7	4.86	7.24E+04	2
2799	21392	C10H22S	bis(2-methylbutyl) sulfide	96034-00-3	4.86	7.24E+04	2
2800	21393	C10H22S	diisopentyl sulfide	544-02-5	4.86	7.24E+04	2

Octanol-Water Partition Coefficient							
NO	ID	FORMULA	NAME	CAS No	log_{10} Kow	Kow	code
2801	21394	C10H22S2	dipentyl disulfide	112-51-6	5.80	6.31E+05	2
2802	21395	C10H22S2	diisopentyl disulfide	2051-04-9	5.80	6.31E+05	2
2803	21400	C10H23N	decylamine	2016-57-1	3.78	6.03E+03	2
2804	21401	C10H23N	methylnonylamine	39093-27-1	3.99	9.72E+03	2
2805	21402	C10H23N	dimethyloctylamine	7378-99-6	3.20	1.60E+03	2
2806	21403	C10H23N	diethylhexylamine	44979-90-0	2.95	8.89E+02	2
2807	21404	C10H23N	ethyloctylamine	4088-36-2	3.90	7.87E+03	2
2808	21405	C10H23N	dipentylamine	2050-92-2	3.91	8.22E+03	2
2809	21406	C10H23N	diisopentylamine	544-00-3	3.93	8.57E+03	2
2810	21407	C10H23N	N,N-diisopropylisobutylamine	44976-81-0	3.08	1.19E+03	2
2811	21531	C11H8O2	2-methyl-1,4-naphthalenedio	58-27-5	2.20	1.58E+02	1
2812	21535	C11H8O3	2-hydroxy-3-methyl-1,4-naph	483-55-6	1.20	1.58E+01	1
2813	21536	C11H8O3	5-hydroxy-2-methyl-1,4-naph	481-42-5	1.20	1.58E+01	2
2814	21543	C11H8O3	2-methoxy-1,4-naphthalenedi	2348-82-5	1.35	2.24E+01	1
2815	21563	C11H9N	2-phenylpyridine	1008-89-5	2.59	3.89E+02	2
2816	21564	C11H9N	3-phenylpyridine	1008-88-4	2.59	3.89E+02	2
2817	21565	C11H9N	4-phenylpyridine	939-23-1	2.59	3.89E+02	1
2818	21601	C11H10	1-methylnaphthalene	90-12-0	3.87	7.41E+03	1
2819	21602	C11H10	2-methylnaphthalene	91-57-6	4.00	1.00E+04	1
2820	21603	C11H10	methylnaphthalene	1321-94-4	3.87	7.33E+03	2
2821	21801	C11H12Cl2N2O5	chloramphenicol	56-75-7	1.14	1.38E+01	1
2822	22113	C11H14O2	butyl benzoate	136-60-7	3.84	6.92E+03	1
2823	22304	C11H16	pentylbenzene	538-68-1	4.90	7.94E+04	1
2824	22305	C11H16	2-phenylpentane	2719-52-0	4.59	3.93E+04	2
2825	22306	C11H16	3-phenylpentane	1196-58-3	4.62	4.14E+04	2
2826	22307	C11H16	1-phenyl-2-methylbutane	3968-85-2	4.75	5.66E+04	2
2827	22308	C11H16	1-phenyl-3-methylbutane	2049-94-7	4.80	6.24E+04	2
2828	22309	C11H16	2-phenyl-2-methylbutane	2049-95-8	4.65	4.45E+04	2
2829	22310	C11H16	2-phenyl-3-methylbutane	4481-30-5	4.55	3.54E+04	2
2830	22311	C11H16	1-phenyl-2,2-dimethylpropan	1007-26-7	4.50	3.18E+04	2
2831	22312	C11H16	1-methyl-2-butylbenzene	1595-11-5	4.65	4.50E+04	2
2832	22313	C11H16	1-methyl-3-butylbenzene	1595-04-6	4.59	3.88E+04	2
2833	22314	C11H16	1-methyl-4-butylbenzene	1595-05-7	4.63	4.29E+04	2
2834	22315	C11H16	1-methyl-2-sec-butylbenzene	1595-06-8	4.39	2.45E+04	2
2835	22316	C11H16	1-methyl-3-sec-butylbenzene	1772-10-7	4.34	2.21E+04	2
2836	22317	C11H16	1-methyl-4-sec-butylbenzene	1595-16-0	4.41	2.58E+04	2
2837	22318	C11H16	1-methyl-2-isobutylbenzene	36301-29-8	4.39	2.45E+04	2
2838	22319	C11H16	1-methyl-3-isobutylbenzene	5160-99-6	4.34	2.21E+04	2
2839	22320	C11H16	1-methyl-4-isobutylbenzene	5161-04-6	4.39	2.45E+04	2
2840	22321	C11H16	1-methyl-2-tert-butyl benzene	1074-92-6	4.49	3.08E+04	2
2841	22322	C11H16	1-methyl-3-tert-butylbenzene	1075-38-3	4.24	1.72E+04	2
2842	22323	C11H16	1-methyl-4-tert-butylbenzene	98-51-1	4.32	2.07E+04	2
2843	22324	C11H16	1-ethyl-2-propylbenzene	16021-20-8	4.54	3.51E+04	2
2844	22325	C11H16	1-ethyl-3-propylbenzene	20024-91-3	4.50	3.17E+04	2
2845	22326	C11H16	1-ethyl-4-propylbenzene	20024-90-2	4.59	3.88E+04	2
2846	22327	C11H16	1-ethyl-2-isopropylbenzene	18970-44-0	4.25	1.79E+04	2
2847	22328	C11H16	1-ethyl-3-isopropylbenzene	4920-99-4	4.30	1.99E+04	2
2848	22329	C11H16	1-ethyl-4-isopropylbenzene	4218-48-8	4.40	2.53E+04	2
2849	22330	C11H16	1,2-dimethyl-3-propylbenzene	17059-44-8	4.71	5.15E+04	2
2850	22331	C11H16	1,2-dimethyl-4-propylbenzene	3982-66-9	4.67	4.71E+04	2
2851	22332	C11H16	1,3-dimethyl-2-propylbenzene	17059-45-9	4.65	4.42E+04	2
2852	22333	C11H16	1,3-dimethyl-4-propylbenzene	61827-85-8	4.62	4.20E+04	2
2853	22334	C11H16	1,3-dimethyl-5-propylbenzene	3982-64-7	4.53	3.37E+04	2
2854	22335	C11H16	1,4-dimethyl-2-propylbenzene	3042-50-0	4.57	3.74E+04	2
2855	22336	C11H16	1,2-dimethyl-3-isopropylbenz	22539-65-7	4.54	3.44E+04	2
2856	22337	C11H16	1,2-dimethyl-4-isopropylbenz	4132-77-8	4.52	3.30E+04	2
2857	22338	C11H16	1,3-dimethyl-2-isopropylbenz	14411-75-7	4.46	2.86E+04	2
2858	22339	C11H16	1,3-dimethyl-4-isopropylbenz	4706-89-2	4.46	2.88E+04	2
2859	22340	C11H16	1,3-dimethyl-5-isopropylbenz	4706-90-5	4.36	2.27E+04	2
2860	22341	C11H16	1,4-dimethyl-2-isopropylbenz	4132-72-3	4.39	2.48E+04	2
2861	22342	C11H16	1-methyl-2,3-diethylbenzene	13632-93-4	4.62	4.20E+04	2
2862	22343	C11H16	1-methyl-2,4-diethylbenzene	1758-85-6	4.59	3.88E+04	2
2863	22344	C11H16	1-methyl-2,5-diethylbenzene	13632-94-5	4.63	4.31E+04	2
2864	22345	C11H16	1-methyl-2,6-diethyl benzene	13632-95-6	4.67	4.69E+04	2
2865	22346	C11H16	1-methyl-3,4-diethylbenzene	13732-80-4	4.56	3.61E+04	2
2866	22347	C11H16	1-methyl-3,5-diethylbenzene	2050-24-0	4.49	3.12E+04	2
2867	22348	C11H16	1,2,3-trimethyl-4-ethylbenzen	61827-86-9	4.53	3.36E+04	2
2868	22349	C11H16	1,2,3-trimethyl-5-ethylbenzen	31366-00-4	4.53	3.36E+04	2
2869	22350	C11H16	1,2,4-trimethyl-3-ethylbenzen	61827-87-0	4.53	3.36E+04	2
2870	22351	C11H16	1,2,4-trimethyl-5-ethylbenzen	---	4.53	3.36E+04	2
2871	22352	C11H16	1,2,4-trimethyl-6-ethylbenzen	17851-27-3	4.53	3.36E+04	2
2872	22353	C11H16	1,3,5-trimethyl-2-ethylbenzen	3982-67-0	4.53	3.36E+04	2
2873	22354	C11H16	pentamethylbenzene	700-12-9	4.56	3.63E+04	1
2874	22355	C11H16	1-ethyl-2,3,5-trimethylbenzen	18262-85-6	4.53	3.36E+04	2
2875	22356	C11H16	sec-pentylbenzene	29316-05-0	4.53	3.36E+04	2
2876	22399	C11H16O	p-tert-amylphenol	80-46-6	3.91	8.13E+03	2
2877	22400	C11H16O	2-butyl-4-methylphenol	6891-45-8	3.91	8.13E+03	2
2878	22401	C11H16O	2-tert-butyl-4-methylphenol	2409-55-4	3.91	8.13E+03	2
2879	22402	C11H16O	2-tert-butyl-5-methylphenol	88-60-8	3.91	8.13E+03	2
2880	22403	C11H16O	2-tert-butyl-6-methylphenol	2219-82-1	3.91	8.13E+03	2

Table 13 Octanol-Water Partition Coefficient - Oganic Compounds

Octanol-Water Partition Coefficient							
NO	ID	FORMULA	NAME	CAS No	$\log_{10}$ Kow	Kow	code
2881	22404	C11H16O	4-tert-butyl-2-methylphenol	98-27-1	3.91	8.13E+03	2
2882	22405	C11H16O	4-isopentylphenol	1805-61-4	3.91	8.13E+03	2
2883	22406	C11H16O	5-methyl-2-sec-butylphenol	29472-95-5	3.91	8.13E+03	2
2884	22407	C11H16O	pentamethylphenol	2819-86-5	3.91	8.13E+03	2
2885	22408	C11H16O	4-pentylphenol	14938-35-3	3.91	8.13E+03	2
2886	22439	C11H16O	2-(1-methylbutyl)phenol	87-26-3	3.91	8.13E+03	2
2887	22440	C11H16O	o-amylphenol	136-81-2	3.91	8.13E+03	2
2888	22441	C11H16O	2-sec-amylphenol	---	3.91	8.13E+03	2
2889	22442	C11H16O	4-sec-amylphenol	25735-67-5	3.91	8.13E+03	2
2890	22672	C11H20	1-undecyne	2243-98-3	5.61	4.07E+05	2
2891	22673	C11H20	2-undecyne	60212-29-5	5.82	6.58E+05	2
2892	22674	C11H20	3-undecyne	60212-30-8	5.71	5.09E+05	2
2893	22675	C11H20	4-undecyne	60212-31-9	5.71	5.14E+05	2
2894	22676	C11H20	5-undecyne	2294-72-6	5.71	5.14E+05	2
2895	22677	C11H20	3,3-dimethyl-4-nonyne	29022-31-9	5.71	5.14E+05	2
2896	22701	C11H20O2	2-ethylhexyl acrylate	103-11-7	4.09	1.23E+04	2
2897	22791	C11H22	hexylcyclopentane	4457-00-5	6.78	6.09E+06	2
2898	22792	C11H22	pentylcyclohexane	4292-92-6	5.56	3.63E+05	2
2899	22793	C11H22	isopentylcyclohexane	54105-76-9	5.56	3.63E+05	2
2900	22795	C11H22	1-undecene	821-95-4	5.61	4.07E+05	2
2901	22796	C11H22	cis-2-undecene	821-96-5	6.14	1.37E+06	2
2902	22797	C11H22	trans-2-undecene	693-61-8	6.14	1.37E+06	2
2903	22798	C11H22	3-undecene	60669-40-1	6.14	1.37E+06	2
2904	22799	C11H22	cis-4-undecene	821-98-7	6.14	1.37E+06	2
2905	22800	C11H22	trans-4-undecene	693-62-9	6.14	1.37E+06	2
2906	22801	C11H22	cis-5-undecene	764-96-5	6.14	1.37E+06	2
2907	22802	C11H22	trans-5-undecene	764-97-6	6.14	1.37E+06	2
2908	22803	C11H22	2-methyl-1-decene	13151-27-4	6.67	4.63E+06	2
2909	22822	C11H22O	undecanal	112-44-7	4.64	4.38E+04	2
2910	22823	C11H22O	2-undecanone	112-12-9	4.09	1.23E+04	1
2911	22824	C11H22O	3-undecanone	2216-87-7	4.24	1.73E+04	2
2912	22825	C11H22O	4-undecanone	14476-37-0	4.24	1.73E+04	2
2913	22826	C11H22O	5-undecanone	33083-83-9	4.24	1.73E+04	2
2914	22827	C11H22O	6-undecanone	927-49-1	4.24	1.73E+04	2
2915	22834	C11H22O	2-methyl-1-decanal	19009-56-4	4.64	4.38E+04	2
2916	22835	C11H22O2	undecanoic acid	112-37-8	3.72	5.21E+03	2
2917	22836	C11H22O2	decyl formate	5451-52-5	5.37	2.34E+05	2
2918	22837	C11H22O2	nonyl acetate	143-13-5	4.07	1.16E+04	2
2919	22848	C11H22O2	methyl decanoate	110-42-9	4.41	2.57E+04	1
2920	22851	C11H22O2	6-methyldecanoic acid	53696-14-3	3.72	5.21E+03	2
2921	22856	C11H22O2	3,5,5-trimethylhexyl acetate	58430-94-7	4.07	1.16E+04	2
2922	22870	C11H23Br	1-bromoundecane	693-67-4	6.09	1.23E+06	2
2923	22872	C11H23Cl	1-chloroundecane	2473-03-2	5.02	1.05E+05	2
2924	22890	C11H24	undecane	1120-21-4	5.74	5.50E+05	2
2925	22891	C11H24	2-methyldecane	6975-98-0	7.23	1.71E+07	2
2926	22892	C11H24	3-methyldecane	13151-34-3	7.25	1.77E+07	2
2927	22893	C11H24	4-methyldecane	2847-72-5	7.21	1.61E+07	2
2928	22894	C11H24	5-methyldecane	13151-35-4	7.17	1.46E+07	2
2929	22895	C11H24	3-ethylnonane	17302-11-3	7.21	1.61E+07	2
2930	22896	C11H24	4-ethylnonane	5911-05-7	7.12	1.33E+07	2
2931	22897	C11H24	5-ethylnonane	17302-12-4	7.04	1.09E+07	2
2932	22898	C11H24	2,2-dimethylnonane	17302-14-6	7.04	1.09E+07	2
2933	22899	C11H24	2,3-dimethylnonane	2884-06-2	7.17	1.46E+07	2
2934	22900	C11H24	2,4-dimethylnonane	17302-24-8	6.99	9.84E+06	2
2935	22901	C11H24	2,5-dimethylnonane	17302-27-1	7.02	1.04E+07	2
2936	22902	C11H24	2,6-dimethylnonane	17302-28-2	7.02	1.04E+07	2
2937	22903	C11H24	2,7-dimethylnonane	17302-29-3	7.04	1.09E+07	2
2938	22904	C11H24	2,8-dimethylnonane	17302-30-6	7.10	1.26E+07	2
2939	22905	C11H24	3,3-dimethylnonane	17302-15-7	7.08	1.20E+07	2
2940	22906	C11H24	3,4-dimethylnonane	17302-22-6	7.14	1.39E+07	2
2941	22907	C11H24	3,5-dimethylnonane	17302-25-9	7.04	1.09E+07	2
2942	22908	C11H24	3,6-dimethylnonane	17302-31-7	7.08	1.20E+07	2
2943	22909	C11H24	3,7-dimethylnonane	17302-32-8	7.12	1.33E+07	2
2944	22910	C11H24	4,4-dimethylnonane	17302-18-0	6.99	9.84E+06	2
2945	22911	C11H24	4,5-dimethylnonane	17302-23-7	7.08	1.20E+07	2
2946	22912	C11H24	4,6-dimethylnonane	17302-26-0	7.02	1.04E+07	2
2947	22913	C11H24	5,5-dimethylnonane	6414-96-6	6.97	9.35E+06	2
2948	22914	C11H24	4-propyloctane	17302-13-5	7.04	1.09E+07	2
2949	22915	C11H24	4-isopropyloctane	62016-15-3	6.99	9.84E+06	2
2950	22916	C11H24	2-methyl-3-ethyloctane	62016-16-4	7.08	1.20E+07	2
2951	22917	C11H24	2-methyl-4-ethyloctane	62016-17-5	6.95	8.89E+06	2
2952	22918	C11H24	2-methyl-5-ethyloctane	62016-18-6	6.99	9.84E+06	2
2953	22919	C11H24	2-methyl-6-ethyloctane	62016-19-7	7.08	1.20E+07	2
2954	22920	C11H24	3-methyl-3-ethyloctane	17302-16-8	7.16	1.44E+07	2
2955	22921	C11H24	3-methyl-4-ethyloctane	62016-20-0	7.08	1.20E+07	2
2956	22922	C11H24	3-methyl-5-ethyloctane	62016-21-1	6.95	8.89E+06	2
2957	22923	C11H24	3-methyl-6-ethyloctane	62016-22-2	7.08	1.20E+07	2
2958	22924	C11H24	4-methyl-3-ethyloctane	62016-23-3	7.10	1.26E+07	2
2959	22925	C11H24	4-methyl-4-ethyloctane	17302-19-1	7.04	1.09E+07	2
2960	22926	C11H24	4-methyl-5-ethyloctane	62016-24-4	7.04	1.09E+07	2

Octanol-Water Partition Coefficient							
NO	ID	FORMULA	NAME	CAS No	$\log_{10}$ Kow	Kow	code
2961	22927	C11H24	4-methyl-6-ethyloctane	62016-25-5	7.04	1.09E+07	2
2962	22928	C11H24	2,2,3-trimethyloctane	62016-26-6	7.02	1.04E+07	2
2963	22929	C11H24	2,2,4-trimethyloctane	18932-14-4	6.85	7.04E+06	2
2964	22930	C11H24	2,2,5-trimethyloctane	62016-27-7	6.84	6.86E+06	2
2965	22931	C11H24	2,2,6-trimethyloctane	62016-28-8	6.90	8.02E+06	2
2966	22932	C11H24	2,2,7-trimethyloctane	62016-29-9	6.81	6.51E+06	2
2967	22933	C11H24	2,3,3-trimethyloctane	62016-30-2	7.08	1.20E+07	2
2968	22934	C11H24	2,3,4-trimethyloctane	62016-31-3	7.04	1.09E+07	2
2969	22935	C11H24	2,3,5-trimethyloctane	62016-32-4	6.97	9.35E+06	2
2970	22936	C11H24	2,3,6-trimethyloctane	62016-33-5	7.04	1.09E+07	2
2971	22937	C11H24	2,3,7-trimethyloctane	62016-34-6	7.02	1.04E+07	2
2972	22938	C11H24	2,4,4-trimethyloctane	62016-35-7	6.84	6.86E+06	2
2973	22939	C11H24	2,4,5-trimethyloctane	62016-36-8	6.95	8.89E+06	2
2974	22940	C11H24	2,4,6-trimethyloctane	62016-37-9	6.90	8.02E+06	2
2975	22941	C11H24	2,4,7-trimethyloctane	62016-38-0	6.84	6.86E+06	2
2976	22942	C11H24	2,5,5-trimethyloctane	62016-39-1	6.86	7.23E+06	2
2977	22943	C11H24	2,5,6-trimethyloctane	62016-14-2	6.99	9.84E+06	2
2978	22944	C11H24	2,6,6-trimethyloctane	54166-32-4	6.95	8.89E+06	2
2979	22945	C11H24	3,3,4-trimethyloctane	62016-40-4	7.08	1.20E+07	2
2980	22946	C11H24	3,3,5-trimethyloctane	62016-41-5	6.90	8.02E+06	2
2981	22947	C11H24	3,3,6-trimethyloctane	62016-42-6	6.97	9.35E+06	2
2982	22948	C11H24	3,4,4-trimethyloctane	62016-43-7	7.06	1.14E+07	2
2983	22949	C11H24	3,4,5-trimethyloctane	62016-44-8	7.06	1.14E+07	2
2984	22950	C11H24	3,4,6-trimethyloctane	62016-45-9	7.02	1.04E+07	2
2985	22951	C11H24	3,5,5-trimethyloctane	61868-94-8	6.81	6.51E+06	2
2986	22952	C11H24	4,4,5-trimethyloctane	61868-95-9	7.04	1.09E+07	2
2987	22953	C11H24	4-tert-butylheptane	60302-21-8	6.90	8.02E+06	2
2988	22954	C11H24	2-methyl-4-propylheptane	61868-96-0	6.90	8.02E+06	2
2989	22955	C11H24	3-methyl-4-propylheptane	61868-97-1	7.02	1.04E+07	2
2990	22956	C11H24	4-methyl-4-propylheptane	17302-20-4	6.99	9.84E+06	2
2991	22957	C11H24	2-methyl-3-isopropylheptane	6876-18-2	7.00	9.99E+06	2
2992	22958	C11H24	2-methyl-4-isopropylheptane	61868-98-2	6.86	7.23E+06	2
2993	22959	C11H24	3-methyl-4-isopropylheptane	61868-99-3	6.97	9.35E+06	2
2994	22960	C11H24	4-methyl-4-isopropylheptane	61869-00-9	7.04	1.09E+07	2
2995	22961	C11H24	3,3-diethylheptane	17302-17-9	7.18	1.53E+07	2
2996	22962	C11H24	3,4-diethylheptane	61869-01-0	7.04	1.09E+07	2
2997	22963	C11H24	3,5-diethylheptane	61869-02-1	7.02	1.04E+07	2
2998	22964	C11H24	4,4-diethylheptane	17302-21-5	7.04	1.11E+07	2
2999	22965	C11H24	2,2-dimethyl-3-ethylheptane	61869-03-2	6.95	8.89E+06	2
3000	22966	C11H24	2,2-dimethyl-4-ethylheptane	62016-46-0	6.77	5.86E+06	2
3001	22967	C11H24	2,2-dimethyl-5-ethylheptane	62016-47-1	6.86	7.23E+06	2
3002	22968	C11H24	2,3-dimethyl-3-ethylheptane	61868-21-1	7.04	1.09E+07	2
3003	22969	C11H24	2,3-dimethyl-4-ethylheptane	61868-22-2	7.02	1.04E+07	2
3004	22970	C11H24	2,3-dimethyl-5-ethylheptane	61868-23-3	6.99	9.84E+06	2
3005	22971	C11H24	2,4-dimethyl-3-ethylheptane	61868-24-4	7.02	1.04E+07	2
3006	22972	C11H24	2,4-dimethyl-4-ethylheptane	61868-25-5	6.81	6.51E+06	2
3007	22973	C11H24	2,4-dimethyl-5-ethylheptane	61868-26-6	6.95	8.89E+06	2
3008	22974	C11H24	2,5-dimethyl-3-ethylheptane	61868-27-7	6.95	8.89E+06	2
3009	22975	C11H24	2,5-dimethyl-4-ethylheptane	61868-28-8	6.93	8.44E+06	2
3010	22976	C11H24	2,5-dimethyl-5-ethylheptane	61868-29-9	6.99	9.84E+06	2
3011	22977	C11H24	2,6-dimethyl-3-ethylheptane	61868-30-2	6.95	8.89E+06	2
3012	22978	C11H24	2,6-dimethyl-4-ethylheptane	61868-31-3	6.81	6.51E+06	2
3013	22979	C11H24	3,3-dimethyl-4-ethylheptane	61868-32-4	7.04	1.09E+07	2
3014	22980	C11H24	3,3-dimethyl-5-ethylheptane	61868-33-5	6.93	8.44E+06	2
3015	22981	C11H24	3,4-dimethyl-3-ethylheptane	61868-34-6	7.14	1.39E+07	2
3016	22982	C11H24	3,4-dimethyl-4-ethylheptane	61868-35-7	7.06	1.14E+07	2
3017	22983	C11H24	3,4-dimethyl-5-ethylheptane	61868-36-8	7.08	1.20E+07	2
3018	22984	C11H24	3,5-dimethyl-3-ethylheptane	61868-37-9	7.04	1.09E+07	2
3019	22985	C11H24	3,5-dimethyl-4-ethylheptane	61868-38-0	7.06	1.14E+07	2
3020	22986	C11H24	4,4-dimethyl-3-ethylheptane	61868-39-1	6.95	8.89E+06	2
3021	22987	C11H24	2,2,3,3-tetramethylheptane	61868-40-4	7.04	1.09E+07	2
3022	22988	C11H24	2,2,3,4-tetramethylheptane	61868-41-5	6.95	8.89E+06	2
3023	22989	C11H24	2,2,3,5-tetramethylheptane	61868-42-6	6.90	8.02E+06	2
3024	22990	C11H24	2,2,3,6-tetramethylheptane	61868-43-7	6.81	6.51E+06	2
3025	22991	C11H24	2,2,4,4-tetramethylheptane	61868-44-8	6.81	6.51E+06	2
3026	22992	C11H24	2,2,4,5-tetramethylheptane	61868-45-9	6.79	6.18E+06	2
3027	22993	C11H24	2,2,4,6-tetramethylheptane	61868-46-0	6.63	4.30E+06	2
3028	22994	C11H24	2,2,5,5-tetramethylheptane	61868-47-1	6.72	5.27E+06	2
3029	22995	C11H24	2,2,5,6-tetramethylheptane	61868-48-2	6.79	6.18E+06	2
3030	22996	C11H24	2,2,6,6-tetramethylheptane	40117-45-1	6.65	4.49E+06	2
3031	22997	C11H24	2,3,3,4-tetramethylheptane	61868-49-3	7.08	1.20E+07	2
3032	22998	C11H24	2,3,3,5-tetramethylheptane	61868-50-6	6.95	8.89E+06	2
3033	22999	C11H24	2,3,3,6-tetramethylheptane	61868-51-7	6.93	8.44E+06	2
3034	23000	C11H24	2,3,4,4-tetramethylheptane	61868-52-8	6.99	9.84E+06	2
3035	23001	C11H24	2,3,4,5-tetramethylheptane	61868-53-9	7.02	1.04E+07	2
3036	23002	C11H24	2,3,4,6-tetramethylheptane	61868-54-0	6.90	8.02E+06	2
3037	23003	C11H24	2,3,5,5-tetramethylheptane	61868-55-1	6.86	7.23E+06	2
3038	23004	C11H24	2,3,5,6-tetramethylheptane	52670-32-3	6.93	8.44E+06	2
3039	23005	C11H24	2,4,4,5-tetramethylheptane	61868-56-2	6.90	8.02E+06	2
3040	23006	C11H24	2,4,4,6-tetramethylheptane	61868-57-3	6.68	4.74E+06	2

Table 13 Octanol-Water Partition Coefficient - Oganic Compounds

Octanol-Water Partition Coefficient								Octanol-Water Partition Coefficient							
NO	ID	FORMULA	NAME	CAS No	log_{10} Kow	Kow	code	NO	ID	FORMULA	NAME	CAS No	log_{10} Kow	Kow	code
3041	23007	C11H24	2,4,5,5-tetramethylheptane	61868-58-4	6.95	8.89E+06	2	3121	23204	C12H6Cl4	2,3,4,5-tetrachlorobiphenyl	33284-53-6	5.72	5.25E+05	1
3042	23008	C11H24	3,3,4,4-tetramethylheptane	61868-59-5	7.12	1.33E+07	2	3122	23205	C12H6Cl4	2,2',4',5-tetrachlorobiphenyl	41464-40-8	5.73	5.37E+05	1
3043	23009	C11H24	3,3,4,5-tetramethylheptane	61868-60-8	7.06	1.14E+07	2	3123	23207	C12H6Cl4	3,3',4,4'-tetrachlorobiphenyl	32598-13-3	5.72	5.25E+05	2
3044	23010	C11H24	3,3,5,5-tetramethylheptane	61868-61-9	7.04	1.09E+07	2	3124	23208	C12H6Cl4	2,2',3,3'-tetrachlorobiphenyl	38444-93-8	5.72	5.25E+05	2
3045	23011	C11H24	3,4,4,5-tetramethylheptane	61868-62-0	7.14	1.39E+07	2	3125	23209	C12H6Cl4	2,2',5,5'-tetrachlorobiphenyl	35693-99-3	5.72	5.25E+05	2
3046	23012	C11H24	2,2-dimethyl-3-isopropylhexa	61868-63-1	6.86	7.23E+06	2	3126	23210	C12H6Cl4	2,2',6,6'-tetrachlorobiphenyl	15968-05-5	5.72	5.25E+05	2
3047	23013	C11H24	2,3-dimethyl-3-isopropylhexa	61868-64-2	7.04	1.09E+07	2	3127	23247	C12H7Cl3	2,4,5-trichlorobiphenyl	15862-07-4	5.60	3.98E+05	1
3048	23014	C11H24	2,4-dimethyl-3-isopropylhexa	61868-65-3	6.93	8.44E+06	2	3128	23248	C12H7Cl3	2,4,6-trichlorobiphenyl	35693-92-6	5.47	2.95E+05	1
3049	23015	C11H24	2,5-dimethyl-3-isopropylhexa	61868-66-4	6.81	6.51E+06	2	3129	23249	C12H7Cl3	2,2',5-trichlorobiphenyl	37680-65-2	5.54	3.47E+05	2
3050	23016	C11H24	2-methyl-3,3-diethylhexane	61868-67-5	7.17	1.46E+07	2	3130	23250	C12H7Cl3	2,3',4'-trichlorobiphenyl	38444-86-9	5.54	3.47E+05	2
3051	23017	C11H24	2-methyl-3,4-diethylhexane	61868-68-6	7.02	1.04E+07	2	3131	23251	C12H7Cl3	2,3,5-trichlorobiphenyl	38444-81-4	5.54	3.47E+05	2
3052	23018	C11H24	2-methyl-4,4-diethylhexane	61868-69-7	7.02	1.04E+07	2	3132	23252	C12H7Cl3	2,4,4'-trichlorobiphenyl	7012-37-5	5.54	3.47E+05	2
3053	23019	C11H24	3-methyl-3,4-diethylhexane	61868-70-0	7.17	1.46E+07	2	3133	23253	C12H7Cl3	2,4',5-trichlorobiphenyl	16606-02-3	5.54	3.47E+05	2
3054	23020	C11H24	3-methyl-4,4-diethylhexane	61868-71-1	7.25	1.77E+07	2	3134	23254	C12H7Cl3	trichlorodiphenyl	25323-68-6	5.54	3.47E+05	2
3055	23021	C11H24	2,2,3-trimethyl-3-ethylhexane	61868-72-2	7.04	1.09E+07	2	3135	23272	C12H8Cl2	2,5-dichlorobiphenyl	34883-39-1	5.10	1.26E+05	1
3056	23022	C11H24	2,2,3-trimethyl-4-ethylhexane	61868-73-3	6.95	8.89E+06	2	3136	23273	C12H8Cl2	2,6-dichlorobiphenyl	33146-45-1	5.00	1.00E+05	1
3057	23023	C11H24	2,2,4-trimethyl-3-ethylhexane	61868-74-4	6.93	8.44E+06	2	3137	23305	C12H8N2	1,10-phenanthroline	66-71-7	1.83	6.76E+01	1
3058	23024	C11H24	2,2,4-trimethyl-4-ethylhexane	61868-75-5	6.99	9.84E+06	2	3138	23306	C12H8N2	1,7-phenanthroline	230-46-6	1.83	6.76E+01	2
3059	23025	C11H24	2,2,5-trimethyl-3-ethylhexane	61868-76-6	6.81	6.51E+06	2	3139	23307	C12H8N2	4,7-phenanthroline	230-07-9	1.83	6.76E+01	2
3060	23026	C11H24	2,2,5-trimethyl-4-ethylhexane	61868-77-7	6.72	5.27E+06	2	3140	23308	C12H8N2	phenazine	92-82-0	2.84	6.92E+02	1
3061	23027	C11H24	2,3,3-trimethyl-4-ethylhexane	61868-78-8	7.10	1.26E+07	2	3141	23335	C12H8O	dibenzofuran	132-64-9	4.12	1.32E+04	1
3062	23028	C11H24	2,3,4-trimethyl-3-ethylhexane	61868-79-9	7.19	1.54E+07	2	3142	23365	C12H9Cl	2-chlorobiphenyl	2051-60-7	4.52	3.31E+04	1
3063	23029	C11H24	2,3,4-trimethyl-4-ethylhexane	61868-80-2	7.04	1.09E+07	2	3143	23366	C12H9Cl	3-chlorobiphenyl	2051-61-8	4.58	3.80E+04	1
3064	23030	C11H24	2,3,5-trimethyl-3-ethylhexane	61868-81-3	6.95	8.89E+06	2	3144	23367	C12H9Cl	4-chlorobiphenyl	2051-62-9	4.61	4.07E+04	1
3065	23031	C11H24	2,3,5-trimethyl-4-ethylhexane	61868-82-4	6.97	9.35E+06	2	3145	23394	C12H9N	dibenzopyrrole	86-74-8	3.72	5.25E+03	1
3066	23032	C11H24	2,4,4-trimethyl-3-ethylhexane	61868-83-5	6.99	9.84E+06	2	3146	23444	C12H10	acenaphthene	83-32-9	3.96	9.12E+03	1
3067	23033	C11H24	3,3,4-trimethyl-4-ethylhexane	61868-84-6	7.25	1.77E+07	2	3147	23445	C12H10	phenylbenzene (biphenyl)	92-52-4	3.76	5.75E+03	1
3068	23034	C11H24	2,2,3,3,4-pentamethylhexane	61868-85-7	7.12	1.33E+07	2	3148	23488	C12H10N2	azobenzene	17082-12-1	3.82	6.61E+03	1
3069	23035	C11H24	2,2,3,3,5-pentamethylhexane	61868-86-8	6.83	6.68E+06	2	3149	23489	C12H10N2	cis-azobenzene	1080-16-6	3.82	6.61E+03	2
3070	23036	C11H24	2,2,3,4,4-pentamethylhexane	61868-87-9	7.17	1.46E+07	2	3150	23519	C12H10O	diphenyl ether	101-84-8	4.21	1.62E+04	1
3071	23037	C11H24	2,2,3,4,5-pentamethylhexane	61868-88-0	6.90	8.02E+06	2	3151	23576	C12H10S	diphenyl sulfide	139-66-2	4.45	2.82E+04	1
3072	23038	C11H24	2,2,3,5,5-pentamethylhexane	14739-73-2	6.72	5.27E+06	2	3152	23607	C12H11N	p-aminodiphenyl	92-67-1	2.86	7.24E+02	1
3073	23039	C11H24	2,2,4,4,5-pentamethylhexane	60302-23-0	6.95	8.89E+06	2	3153	23609	C12H11N	diphenylamine	122-39-4	3.44	2.75E+03	1
3074	23040	C11H24	2,3,3,4,4-pentamethylhexane	61868-89-1	7.19	1.54E+07	2	3154	23656	C12H11N3	p-aminoazobenzene	60-09-3	3.41	2.57E+03	1
3075	23041	C11H24	2,3,3,4,5-pentamethylhexane	52670-33-4	7.06	1.14E+07	2	3155	23657	C12H11N3	1,3-diphenyltriazene	136-35-6	3.99	9.77E+03	2
3076	23042	C11H24	2,2,4-trimethyl-3-isopropylpe	61868-90-4	6.81	6.51E+06	2	3156	23675	C12H12	1,2-dimethylnaphthalene	573-98-8	4.31	2.04E+04	1
3077	23043	C11H24	2,3,4-trimethyl-3-isopropylpe	61868-91-5	7.17	1.46E+07	2	3157	23676	C12H12	1,3-dimethylnaphthalene	575-41-7	4.42	2.63E+04	1
3078	23044	C11H24	2,2-dimethyl-3,3-diethylpenta	60302-28-5	7.23	1.69E+07	2	3158	23677	C12H12	1,4-dimethylnaphthalene	571-58-4	4.37	2.34E+04	1
3079	23045	C11H24	2,4-dimethyl-3,3-diethylpenta	61868-92-6	7.23	1.69E+07	2	3159	23678	C12H12	1,5-dimethylnaphthalene	571-61-9	4.38	2.40E+04	1
3080	23046	C11H24	2,2,3,4-tetramethyl-3-ethylpe	61868-93-7	7.17	1.46E+07	2	3160	23679	C12H12	1,6-dimethylnaphthalene	575-43-9	4.26	1.82E+04	2
3081	23047	C11H24	2,2,4,4-tetramethyl-3-ethylpe	---	7.04	1.09E+07	2	3161	23680	C12H12	1,7-dimethylnaphthalene	575-37-1	4.44	2.75E+04	1
3082	23048	C11H24	2,2,3,3,4,4-hexamethylpentar	60302-27-4	7.33	2.14E+07	2	3162	23681	C12H12	1,8-dimethylnaphthalene	569-41-5	4.36	2.28E+04	2
3083	23055	C11H24O	1-undecanol	112-42-5	4.72	5.25E+04	1	3163	23682	C12H12	2,3-dimethylnaphthalene	581-40-8	4.40	2.51E+04	1
3084	23056	C11H24O	2-undecanol	1653-30-1	4.41	2.60E+04	2	3164	23683	C12H12	2,6-dimethylnaphthalene	581-42-0	4.31	2.04E+04	1
3085	23057	C11H24O	2-undecanol, (±)	113666-64-1	4.57	3.69E+04	2	3165	23684	C12H12	2,7-dimethylnaphthalene	582-16-1	4.26	1.82E+04	2
3086	23058	C11H24O	3-undecanol, (R)-	107494-37-1	4.57	3.69E+04	2	3166	23685	C12H12	1-ethylnaphthalene	1127-76-0	4.40	2.51E+04	1
3087	23059	C11H24O	5-undecanol	37493-70-2	4.57	3.69E+04	2	3167	23686	C12H12	2-ethylnaphthalene	939-27-5	4.38	2.40E+04	1
3088	23060	C11H24O	6-undecanol	23708-56-7	4.57	3.69E+04	2	3168	23687	C12H12	dimethylnaphthalene, isomers	28804-88-8	4.36	2.28E+04	2
3089	23061	C11H24O	2,8-dimethyl-5-nonanol	19780-96-2	4.57	3.69E+04	2	3169	23718	C12H12N2	p-aminodiphenylamine	101-54-2	1.82	6.61E+01	2
3090	23062	C11H24O	undecanol	30207-98-8	4.57	3.69E+04	2	3170	23719	C12H12N2	hydrazobenzene	122-66-7	2.94	8.71E+02	1
3091	23063	C11H24O	5-ethyl-2-nonanol	103-08-2	4.57	3.69E+04	2	3171	23965	C12H14O	4-phenylcyclohexanone	4894-75-1	2.45	2.82E+02	1
3092	23084	C11H24S	1-undecanethiol	5332-52-5	5.69	4.90E+05	2	3172	24002	C12H14O4	diethyl phthalate	84-66-2	2.47	2.95E+02	1
3093	23085	C11H24S	2-undecanethiol	62155-02-6	5.83	6.78E+05	2	3173	24111	C12H16	cyclohexylbenzene	827-52-1	4.81	6.46E+04	2
3094	23086	C11H24S	methyl decyl sulfide	22438-39-7	5.36	2.29E+05	2	3174	24357	C12H18	hexylbenzene	1077-16-3	5.52	3.31E+05	1
3095	23087	C11H24S	ethyl nonyl sulfide	59973-08-9	5.36	2.29E+05	2	3175	24358	C12H18	1,4-dipropylbenzene	4815-57-0	5.05	1.11E+05	2
3096	23088	C11H24S	propyl octyl sulfide	3698-93-9	5.36	2.29E+05	2	3176	24359	C12H18	diisopropylbenzene	25321-09-9	5.05	1.11E+05	2
3097	23089	C11H24S	butyl heptyl sulfide	40813-84-1	5.36	2.29E+05	2	3177	24360	C12H18	1,2-diisopropylbenzene	577-55-9	5.05	1.11E+05	2
3098	23090	C11H25N	undecylamine	7307-55-3	4.30	2.00E+04	2	3178	24361	C12H18	m-diisopropylbenzene	99-62-7	4.90	7.94E+04	2
3099	23091	C11H25N	methyldecylamine	7516-82-7	4.33	2.15E+04	2	3179	24362	C12H18	p-diisopropylbenzene	100-18-5	4.90	7.94E+04	2
3100	23092	C11H25N	ethylnonylamine	66563-84-6	4.27	1.87E+04	2	3180	24363	C12H18	triethylbenzene; (mixed isome	25340-18-5	5.05	1.11E+05	2
3101	23093	C11H25N	dimethylnonylamine	17373-27-2	3.68	4.84E+03	2	3181	24364	C12H18	1,2,3-triethylbenzene	---	4.89	7.78E+04	2
3102	23094	C11H25N	diethylheptylamine	26981-81-7	3.38	2.39E+03	2	3182	24365	C12H18	1,2,4-triethylbenzene	877-44-1	4.89	7.78E+04	2
3103	23095	C11H25N	4-ethyl-1-methyloctylamine	10024-78-9	4.30	2.01E+04	2	3183	24366	C12H18	1,3,5-triethylbenzene	102-25-0	5.11	1.29E+05	2
3104	23119	C12Cl10	decachlorobiphenyl	2051-24-3	8.26	1.82E+08	1	3184	24367	C12H18	1-tert-butyl-3,5-dimethylbenz	98-19-1	5.05	1.11E+05	2
3105	23134	C12HCl9	2,2',3,3',4,5,5',6,6'-nonachlor	52663-77-1	8.16	1.45E+08	1	3185	24368	C12H18	(1,1-dimethylbutyl)benzene	1985-57-5	5.05	1.11E+05	2
3106	23145	C12H2Cl8	2,2',3,3',5,5',6,6'-octachlorobi	2136-99-4	7.10	1.26E+07	1	3186	24369	C12H18	2,4-dimethyl-1-sec-butylbenz	1483-60-9	5.05	1.11E+05	2
3107	23152	C12H3Cl7	2,2',3,3',4,4',6-heptachlorobip	52663-71-5	6.70	5.01E+06	1	3187	24370	C12H18	(1-ethylbutyl)benzene	4468-42-2	5.05	1.11E+05	2
3108	23159	C12H4Cl6	2,2',3,3',4,4'-hexachlorobiphe	38380-07-3	7.00	1.00E+07	1	3188	24371	C12H18	(1-ethyl-1-methylpropyl)benz	1985-97-3	5.05	1.11E+05	2
3109	23160	C12H4Cl6	2,2',4,4',6,6'-hexachlorobiphe	33979-03-2	7.00	1.00E+07	1	3189	24372	C12H18	(1-methylpentyl)benzene	6031-02-3	5.05	1.11E+05	2
3110	23161	C12H4Cl6	2,2',3,3',6,6'-hexachlorobiphe	38411-22-2	6.70	5.01E+06	1	3190	24373	C12H18	(3-methylpentyl)benzene	54410-69-4	5.05	1.11E+05	2
3111	23162	C12H4Cl6	2,3,3',4,4',5-hexachlorobiphe	38380-08-4	6.85	7.08E+06	2	3191	24374	C12H18	(4-methylpentyl)benzene	4215-86-5	5.05	1.11E+05	2
3112	23163	C12H4Cl6	2,2',4,4',5'5'-hexachloro-1,1'-	35065-27-1	6.85	7.08E+06	2	3192	24375	C12H18	1,2,4-trimethyl-5-isopropylbe	10222-95-4	5.05	1.11E+05	2
3113	23164	C12H4Cl6	3,3',4,4',5,5'-hexachlorobiphe	32774-16-6	6.85	7.08E+06	2	3193	24376	C12H18	4-tert-butyl-o-xylene	7397-06-0	5.05	1.11E+05	2
3114	23177	C12H5Cl5	2,3,4,5,6-pentachlorobipheny	18259-05-7	6.30	2.00E+06	1	3194	24377	C12H18	hexamethyldewarbenzene	7641-77-2	5.05	1.11E+05	2
3115	23178	C12H5Cl5	2,2',4,5,5'-pentachlorobiphen	37680-73-2	6.40	2.51E+06	1	3195	24378	C12H18	hexamethylbenzene	87-85-4	4.69	4.90E+04	1
3116	23180	C12H5Cl5	pentachlorobiphenyl	25429-29-2	6.35	2.24E+06	2	3196	24594	C12H20O4	dibutyl maleate	105-76-0	4.16	1.45E+04	2
3117	23181	C12H5Cl5	2,2',3,4,5'-pentachlorobiphen	38380-02-8	6.35	2.24E+06	2	3197	24652	C12H22	bicyclohexyl	92-51-3	5.86	7.24E+05	2
3118	23182	C12H5Cl5	2,3',4,4',5-pentachlorobiphen	31508-00-6	6.35	2.24E+06	2	3198	24653	C12H22	1-dodecyne	765-03-7	6.01	1.03E+06	2
3119	23183	C12H5Cl5	2,3,3',4,4'-pentachlorobiphen	32598-14-4	6.35	2.24E+06	2	3199	24654	C12H22	2-dodecyne	629-49-2	6.18	1.51E+06	2
3120	23184	C12H5Cl5	3,3',4,4',5-pentachlorobiphen	57465-28-8	6.35	2.24E+06	2	3200	24655	C12H22	3-dodecyne	6790-27-8	6.08	1.21E+06	2

Table 13 Octanol-Water Partition Coefficient - Oganic Compounds

Octanol-Water Partition Coefficient								Octanol-Water Partition Coefficient							
NO	ID	FORMULA	NAME	CAS No	log_{10} Kow	Kow	code	NO	ID	FORMULA	NAME	CAS No	log_{10} Kow	Kow	code
3201	24656	C12H22	6-dodecyne	6975-99-1	6.09	1.23E+06	2	3281	24966	C12H26	2,2,5-trimethylnonane	62184-51-4	7.31	2.02E+07	2
3202	24657	C12H22	3,3-dimethyl-4-decyne	70732-45-5	6.09	1.23E+06	2	3282	24967	C12H26	2,2,6-trimethylnonane	62184-52-5	7.28	1.93E+07	2
3203	24690	C12H22O	cyclododecanone	830-13-7	4.10	1.26E+04	1	3283	24968	C12H26	2,2,7-trimethylnonane	62184-53-6	7.39	2.43E+07	2
3204	24818	C12H24	heptylcyclopentane	5617-42-5	7.17	1.49E+07	2	3284	24969	C12H26	2,2,8-trimethylnonane	62184-54-7	7.39	2.43E+07	2
3205	24819	C12H24	hexylcyclohexane	4292-75-5	6.05	1.12E+06	2	3285	24970	C12H26	2,3,3-trimethylnonane	62184-55-8	7.52	3.32E+07	2
3206	24820	C12H24	1-methyl-2-pentylcyclohexane	54411-01-7	6.05	1.12E+06	2	3286	24971	C12H26	2,3,4-trimethylnonane	62184-56-9	7.48	3.04E+07	2
3207	24823	C12H24	1-dodecene	112-41-4	6.10	1.26E+06	2	3287	24972	C12H26	2,3,5-trimethylnonane	62184-57-0	7.41	2.54E+07	2
3208	24824	C12H24	2-methyl-1-undecene	18516-37-5	7.10	1.25E+07	2	3288	24973	C12H26	2,3,6-trimethylnonane	62184-58-1	7.44	2.78E+07	2
3209	24825	C12H24	4,4-dimethyl-2-neopentyl-1-p	141-70-8	6.60	3.97E+06	2	3289	24974	C12H26	2,3,7-trimethylnonane	62184-59-2	7.50	3.18E+07	2
3210	24826	C12H24	2,2,4,6,6-pentamethyl-3-hept	123-48-8	6.60	3.97E+06	2	3290	24975	C12H26	2,3,8-trimethylnonane	62184-60-5	7.48	3.04E+07	2
3211	24827	C12H24	trans-2,2,4,6,6-pentamethyl-3	27656-50-4	6.60	3.97E+06	2	3291	24976	C12H26	2,4,4-trimethylnonane	62184-61-6	7.28	1.93E+07	2
3212	24828	C12H24	isobutene, trimer	7756-94-7	6.60	3.97E+06	2	3292	24977	C12H26	2,4,5-trimethylnonane	62184-62-7	7.39	2.43E+07	2
3213	24852	C12H24O	dodecanal	112-54-9	4.75	5.62E+04	2	3293	24978	C12H26	2,4,6-trimethylnonane	62184-10-5	7.33	2.12E+07	2
3214	24853	C12H24O	2-methylundecanal	110-41-8	4.75	5.62E+04	2	3294	24979	C12H26	2,4,7-trimethylnonane	62184-11-6	7.37	2.32E+07	2
3215	24854	C12H24O	2-dodecanone	6175-49-1	4.84	6.87E+04	2	3295	24980	C12H26	2,4,8-trimethylnonane	49542-74-7	7.31	2.02E+07	2
3216	24855	C12H24O	6-dodecanone	6064-27-3	4.84	6.87E+04	2	3296	24981	C12H26	2,5,5-trimethylnonane	62184-12-7	7.28	1.93E+07	2
3217	24860	C12H24O	trimethyl nonanone	1331-50-6	4.84	6.87E+04	2	3297	24982	C12H26	2,5,6-trimethylnonane	62184-13-8	7.41	2.54E+07	2
3218	24861	C12H24O	3-dodecanone	1534-27-6	4.84	6.87E+04	2	3298	24983	C12H26	2,5,7-trimethylnonane	62184-14-9	7.37	2.32E+07	2
3219	24862	C12H24O	5-dodecanone	19780-10-0	4.84	6.87E+04	2	3299	24984	C12H26	2,5,8-trimethylnonane	49557-09-7	7.33	2.12E+07	2
3220	24863	C12H24O	4-dodecanone	6137-26-4	4.84	6.87E+04	2	3300	24985	C12H26	2,6,6-trimethylnonane	62184-15-0	7.28	1.93E+07	2
3221	24867	C12H24O2	dodecanoic acid	143-07-7	4.60	3.98E+04	1	3301	24986	C12H26	2,6,7-trimethylnonane	62184-16-1	7.46	2.91E+07	2
3222	24868	C12H24O2	undecyl formate	5454-24-0	6.18	1.51E+06	2	3302	24987	C12H26	2,7,7-trimethylnonane	62184-17-2	7.43	2.66E+07	2
3223	24869	C12H24O2	decyl acetate	112-17-4	5.15	1.41E+05	2	3303	24988	C12H26	3,3,4-trimethylnonane	62184-18-3	7.52	3.32E+07	2
3224	24879	C12H24O2	2-butyloctanoic acid	27610-92-0	4.60	3.98E+04	2	3304	24989	C12H26	3,3,5-trimethylnonane	62184-19-4	7.28	1.93E+07	2
3225	24880	C12H24O2	3,7-dimethyloctanyl acetate	20780-49-8	5.15	1.41E+05	2	3305	24990	C12H26	3,3,7-trimethylnonane	62184-21-8	7.39	2.43E+07	2
3226	24894	C12H25Br	1-bromododecane	143-15-7	6.58	3.80E+06	2	3306	24991	C12H26	3,3,6-trimethylnonane	62184-20-7	7.44	2.78E+07	2
3227	24895	C12H25Br	2-bromododecane	13187-99-0	6.58	3.80E+06	2	3307	24992	C12H26	3,4,4-trimethylnonane	62184-22-9	7.48	3.04E+07	2
3228	24897	C12H25Cl	1-chlorododecane	112-52-7	6.49	3.09E+06	2	3308	24993	C12H26	3,4,5-trimethylnonane	62184-23-0	7.48	3.04E+07	2
3229	24914	C12H26	dodecane	112-40-3	6.10	1.26E+06	1	3309	24994	C12H26	3,4,6-trimethylnonane	62184-24-1	7.43	2.66E+07	2
3230	24915	C12H26	3-methylundecane	1002-43-3	7.67	4.64E+07	2	3310	24995	C12H26	3,4,7-trimethylnonane	27802-85-3	7.48	3.04E+07	2
3231	24916	C12H26	4-methylundecane	2980-69-0	7.65	4.45E+07	2	3311	24996	C12H26	3,5,5-trimethylnonane	62184-25-2	7.31	2.02E+07	2
3232	24917	C12H26	5-methylundecane	1632-70-8	7.60	3.94E+07	2	3312	24997	C12H26	3,5,6-trimethylnonane	62184-26-3	7.43	2.66E+07	2
3233	24918	C12H26	6-methylundecane	17302-33-9	7.60	3.94E+07	2	3313	24998	C12H26	3,5,7-trimethylnonane	62184-27-4	7.39	2.43E+07	2
3234	24919	C12H26	3-ethyldecane	17085-96-0	7.64	4.40E+07	2	3314	24999	C12H26	3,6,6-trimethylnonane	62184-28-5	7.28	1.93E+07	2
3235	24920	C12H26	4-ethyldecane	1636-44-8	7.56	3.62E+07	2	3315	25000	C12H26	4,4,5-trimethylnonane	62184-29-6	7.44	2.78E+07	2
3236	24921	C12H26	5-ethyldecane	17302-36-2	7.52	3.32E+07	2	3316	25001	C12H26	4,4,6-trimethylnonane	62184-30-9	7.31	2.02E+07	2
3237	24922	C12H26	2,2-dimethyldecane	17302-37-3	7.50	3.18E+07	2	3317	25002	C12H26	4,5,5-trimethylnonane	62184-31-0	7.44	2.78E+07	2
3238	24923	C12H26	2,3-dimethyldecane	17312-44-6	7.60	3.94E+07	2	3318	25003	C12H26	4,5,6-trimethylnonane	62211-85-2	7.44	2.78E+07	2
3239	24924	C12H26	2,4-dimethyldecane	2801-84-5	7.48	3.04E+07	2	3319	25004	C12H26	4-tert-butyloctane	62184-32-1	7.31	2.02E+07	2
3240	24925	C12H26	2,5-dimethyldecane	17312-50-4	7.44	2.78E+07	2	3320	25005	C12H26	2-methyl-4-propyloctane	62184-33-2	7.31	2.02E+07	2
3241	24926	C12H26	2,6-dimethyldecane	13150-81-7	7.44	2.78E+07	2	3321	25006	C12H26	2-methyl-5-propyloctane	62184-34-3	7.31	2.02E+07	2
3242	24927	C12H26	2,7-dimethyldecane	17312-51-5	7.52	3.32E+07	2	3322	25007	C12H26	3-methyl-4-propyloctane	62184-35-4	7.43	2.66E+07	2
3243	24928	C12H26	2,8-dimethyldecane	17312-52-6	7.56	3.62E+07	2	3323	25008	C12H26	3-methyl-5-propyloctane	62184-36-5	7.37	2.32E+07	2
3244	24929	C12H26	2,9-dimethyldecane	1002-17-1	7.56	3.62E+07	2	3324	25009	C12H26	4-methyl-4-propyloctane	17312-41-3	7.39	2.43E+07	2
3245	24930	C12H26	3,3-dimethyldecane	17302-38-4	7.54	3.47E+07	2	3325	25010	C12H26	4-methyl-5-propyloctane	62183-85-1	7.41	2.54E+07	2
3246	24931	C12H26	3,4-dimethyldecane	17312-45-7	7.58	3.77E+07	2	3326	25011	C12H26	2-methyl-3-isopropyloctane	13287-19-9	7.41	2.54E+07	2
3247	24932	C12H26	3,5-dimethyldecane	17312-48-0	7.48	3.04E+07	2	3327	25012	C12H26	2-methyl-4-isopropyloctane	62183-86-2	7.28	1.93E+07	2
3248	24933	C12H26	3,6-dimethyldecane	17312-53-7	7.50	3.18E+07	2	3328	25013	C12H26	2-methyl-5-isopropyloctane	62183-87-3	7.31	2.02E+07	2
3249	24934	C12H26	3,7-dimethyldecane	17312-54-8	7.52	3.32E+07	2	3329	25014	C12H26	3-methyl-4-isopropyloctane	62183-88-4	7.39	2.43E+07	2
3250	24935	C12H26	3,8-dimethyldecane	17312-55-9	7.58	3.77E+07	2	3330	25015	C12H26	3-methyl-5-isopropyloctane	62183-89-5	7.33	2.12E+07	2
3251	24936	C12H26	4,4-dimethyldecane	17312-39-9	7.46	2.91E+07	2	3331	25016	C12H26	4-methyl-4-isopropyloctane	62183-90-8	7.44	2.78E+07	2
3252	24937	C12H26	4,5-dimethyldecane	17312-46-8	7.52	3.32E+07	2	3332	25017	C12H26	4-methyl-5-isopropyloctane	62183-91-9	7.37	2.32E+07	2
3253	24938	C12H26	4,6-dimethyldecane	17312-49-1	7.44	2.78E+07	2	3333	25018	C12H26	3,3-diethyloctane	17302-40-8	7.58	3.77E+07	2
3254	24939	C12H26	4,7-dimethyldecane	17312-56-0	7.46	2.91E+07	2	3334	25019	C12H26	3,4-diethyloctane	62183-92-0	7.46	2.91E+07	2
3255	24940	C12H26	5,5-dimethyldecane	17453-92-8	7.43	2.66E+07	2	3335	25020	C12H26	3,5-diethyloctane	62183-93-1	7.43	2.66E+07	2
3256	24941	C12H26	5,6-dimethyldecane	1636-43-7	7.50	3.18E+07	2	3336	25021	C12H26	3,6-diethyloctane	62183-94-2	7.52	3.32E+07	2
3257	24942	C12H26	4-propylnonane	6165-37-3	7.46	2.91E+07	2	3337	25022	C12H26	4,4-diethyloctane	17312-42-4	7.48	3.04E+07	2
3258	24943	C12H26	5-propylnonane	998-35-6	7.43	2.66E+07	2	3338	25023	C12H26	4,5-diethyloctane	1636-41-5	7.43	2.66E+07	2
3259	24944	C12H26	4-isopropylnonane	62184-71-8	7.44	2.78E+07	2	3339	25024	C12H26	2,2-dimethyl-3-ethyloctane	62183-95-3	7.41	2.54E+07	2
3260	24945	C12H26	5-isopropylnonane	62184-72-9	7.44	2.78E+07	2	3340	25025	C12H26	2,2-dimethyl-4-ethyloctane	62183-96-4	7.22	1.67E+07	2
3261	24946	C12H26	2-methyl-3-ethylnonane	62184-73-0	7.50	3.18E+07	2	3341	25026	C12H26	2,2-dimethyl-5-ethyloctane	62183-97-5	7.26	1.84E+07	2
3262	24947	C12H26	2-methyl-4-ethylnonane	62184-37-6	7.44	2.78E+07	2	3342	25027	C12H26	2,2-dimethyl-6-ethyloctane	62183-98-6	7.35	2.22E+07	2
3263	24948	C12H26	2-methyl-5-ethylnonane	62184-38-7	7.41	2.54E+07	2	3343	25028	C12H26	2,3-dimethyl-3-ethyloctane	62183-99-7	7.52	3.32E+07	2
3264	24949	C12H26	2-methyl-6-ethylnonane	62184-39-8	7.44	2.78E+07	2	3344	25029	C12H26	2,3-dimethyl-4-ethyloctane	62184-00-3	7.43	2.66E+07	2
3265	24950	C12H26	2-methyl-7-ethylnonane	62184-40-1	7.52	3.32E+07	2	3345	25030	C12H26	2,3-dimethyl-5-ethyloctane	62184-01-4	7.37	2.32E+07	2
3266	24951	C12H26	3-methyl-3-ethylnonane	17302-39-5	7.56	3.62E+07	2	3346	25031	C12H26	2,3-dimethyl-6-ethyloctane	62184-02-5	7.46	2.91E+07	2
3267	24952	C12H26	3-methyl-4-ethylnonane	62184-41-2	7.50	3.18E+07	2	3347	25032	C12H26	2,4-dimethyl-3-ethyloctane	62184-03-6	7.43	2.66E+07	2
3268	24953	C12H26	3-methyl-5-ethylnonane	62184-42-3	7.39	2.43E+07	2	3348	25033	C12H26	2,4-dimethyl-4-ethyloctane	62184-04-7	7.31	2.02E+07	2
3269	24954	C12H26	3-methyl-6-ethylnonane	62184-43-4	7.46	2.91E+07	2	3349	25034	C12H26	2,4-dimethyl-5-ethyloctane	62184-05-8	7.35	2.22E+07	2
3270	24955	C12H26	3-methyl-7-ethylnonane	62184-44-5	7.48	3.04E+07	2	3350	25035	C12H26	2,4-dimethyl-6-ethyloctane	62184-06-9	7.33	2.12E+07	2
3271	24956	C12H26	4-methyl-3-ethylnonane	62184-45-6	7.48	3.04E+07	2	3351	25036	C12H26	2,5-dimethyl-3-ethyloctane	62184-07-0	7.37	2.32E+07	2
3272	24957	C12H26	4-methyl-4-ethylnonane	17312-40-2	7.46	2.91E+07	2	3352	25037	C12H26	2,5-dimethyl-4-ethyloctane	62184-08-1	7.33	2.12E+07	2
3273	24958	C12H26	4-methyl-5-ethylnonane	62184-46-7	7.44	2.78E+07	2	3353	25038	C12H26	2,5-dimethyl-5-ethyloctane	62184-09-2	7.35	2.22E+07	2
3274	24959	C12H26	4-methyl-6-ethylnonane	62184-47-8	7.41	2.54E+07	2	3354	25039	C12H26	2,5-dimethyl-6-ethyloctane	62183-50-0	7.43	2.66E+07	2
3275	24960	C12H26	4-methyl-7-ethylnonane	62184-48-9	7.48	3.04E+07	2	3355	25040	C12H26	2,6-dimethyl-3-ethyloctane	62183-51-1	7.43	2.66E+07	2
3276	24961	C12H26	5-methyl-3-ethylnonane	62184-49-0	7.46	2.91E+07	2	3356	25041	C12H26	2,6-dimethyl-4-ethyloctane	62183-52-2	7.28	1.93E+07	2
3277	24962	C12H26	5-methyl-4-ethylnonane	1632-71-9	7.46	2.91E+07	2	3357	25042	C12H26	2,6-dimethyl-5-ethyloctane	62183-53-3	7.39	2.43E+07	2
3278	24963	C12H26	5-methyl-5-ethylnonane	14531-16-9	7.44	2.78E+07	2	3358	25043	C12H26	2,6-dimethyl-6-ethyloctane	62183-54-4	7.44	2.78E+07	2
3279	24964	C12H26	2,2,3-trimethylnonane	55499-04-2	7.48	3.04E+07	2	3359	25044	C12H26	2,7-dimethyl-3-ethyloctane	62183-55-5	7.41	2.54E+07	2
3280	24965	C12H26	2,2,4-trimethylnonane	62184-50-3	7.31	2.02E+07	2	3360	25045	C12H26	2,7-dimethyl-4-ethyloctane	62183-56-6	7.28	1.93E+07	2

Table 13 Octanol-Water Partition Coefficient - Oganic Compounds

NO	ID	FORMULA	NAME	CAS No	log_{10} Kow	Kow	code
3361	25046	C12H26	3,3-dimethyl-4-ethyloctane	62183-57-7	7.44	2.78E+07	2
3362	25047	C12H26	3,3-dimethyl-5-ethyloctane	62183-58-8	7.31	2.02E+07	2
3363	25048	C12H26	3,3-dimethyl-6-ethyloctane	62183-59-9	7.41	2.54E+07	2
3364	25049	C12H26	3,4-dimethyl-3-ethyloctane	62212-28-6	7.48	3.04E+07	2
3365	25050	C12H26	3,4-dimethyl-4-ethyloctane	62183-60-2	7.50	3.18E+07	2
3366	25051	C12H26	3,4-dimethyl-5-ethyloctane	62183-61-3	7.44	2.78E+07	2
3367	25052	C12H26	3,4-dimethyl-6-ethyloctane	62183-62-4	7.44	2.78E+07	2
3368	25053	C12H26	3,5-dimethyl-3-ethyloctane	62183-63-5	7.41	2.54E+07	2
3369	25054	C12H26	3,5-dimethyl-4-ethyloctane	62183-64-6	7.43	2.66E+07	2
3370	25055	C12H26	3,5-dimethyl-5-ethyloctane	62183-65-7	7.28	1.93E+07	2
3371	25056	C12H26	3,5-dimethyl-6-ethyloctane	62183-66-8	7.43	2.66E+07	2
3372	25057	C12H26	3,6-dimethyl-3-ethyloctane	62183-67-9	7.46	2.91E+07	2
3373	25058	C12H26	3,6-dimethyl-4-ethyloctane	62183-68-0	7.41	2.54E+07	2
3374	25059	C12H26	4,4-dimethyl-3-ethyloctane	62183-69-1	7.44	2.78E+07	2
3375	25060	C12H26	4,4-dimethyl-5-ethyloctane	62183-70-4	7.41	2.54E+07	2
3376	25061	C12H26	4,4-dimethyl-6-ethyloctane	62183-71-5	7.31	2.02E+07	2
3377	25062	C12H26	4,5-dimethyl-3-ethyloctane	62183-72-6	7.46	2.91E+07	2
3378	25063	C12H26	4,5-dimethyl-4-ethyloctane	62183-73-7	7.48	3.04E+07	2
3379	25064	C12H26	2,2,3,3-tetramethyloctane	62183-74-8	7.48	3.04E+07	2
3380	25065	C12H26	2,2,3,4-tetramethyloctane	62183-75-9	7.39	2.43E+07	2
3381	25066	C12H26	2,2,3,5-tetramethyloctane	62183-76-0	7.33	2.12E+07	2
3382	25067	C12H26	2,2,3,6-tetramethyloctane	62183-77-1	7.37	2.32E+07	2
3383	25068	C12H26	2,2,3,7-tetramethyloctane	62183-78-2	7.37	2.32E+07	2
3384	25069	C12H26	2,2,4,4-tetramethyloctane	62183-79-3	7.33	2.12E+07	2
3385	25070	C12H26	2,2,4,5-tetramethyloctane	62183-80-6	7.22	1.67E+07	2
3386	25071	C12H26	2,2,4,6-tetramethyloctane	62183-81-7	7.18	1.51E+07	2
3387	25072	C12H26	2,2,4,7-tetramethyloctane	62183-82-8	7.18	1.50E+07	2
3388	25073	C12H26	2,2,5,5-tetramethyloctane	62183-83-9	7.16	1.44E+07	2
3389	25074	C12H26	2,2,5,6-tetramethyloctane	62183-84-0	7.28	1.93E+07	2
3390	25075	C12H26	2,2,5,7-tetramethyloctane	62199-19-3	7.18	1.51E+07	2
3391	25076	C12H26	2,2,6,6-tetramethyloctane	62199-20-6	7.18	1.51E+07	2
3392	25077	C12H26	2,2,6,7-tetramethyloctane	62199-21-7	7.31	2.02E+07	2
3393	25078	C12H26	2,2,7,7-tetramethyloctane	1071-31-4	7.18	1.51E+07	2
3394	25079	C12H26	2,3,3,4-tetramethyloctane	62199-22-8	7.50	3.18E+07	2
3395	25080	C12H26	2,3,3,5-tetramethyloctane	62199-23-9	7.37	2.32E+07	2
3396	25081	C12H26	2,3,3,6-tetramethyloctane	62199-24-0	7.41	2.54E+07	2
3397	25082	C12H26	2,3,3,7-tetramethyloctane	62199-25-1	7.41	2.54E+07	2
3398	25083	C12H26	2,3,4,4-tetramethyloctane	62199-26-2	7.41	2.54E+07	2
3399	25084	C12H26	2,3,4,5-tetramethyloctane	62199-27-3	7.43	2.66E+07	2
3400	25085	C12H26	2,3,4,6-tetramethyloctane	62199-28-4	7.39	2.43E+07	2
3401	25086	C12H26	2,3,4,7-tetramethyloctane	62199-29-5	7.37	2.32E+07	2
3402	25087	C12H26	2,3,5,5-tetramethyloctane	62199-30-8	7.39	2.43E+07	2
3403	25088	C12H26	2,3,5,6-tetramethyloctane	62199-31-9	7.39	2.43E+07	2
3404	25089	C12H26	2,3,5,7-tetramethyloctane	62199-32-0	7.28	1.93E+07	2
3405	25090	C12H26	2,3,6,6-tetramethyloctane	62199-33-1	7.35	2.22E+07	2
3406	25091	C12H26	2,3,6,7-tetramethyloctane	52670-34-5	7.41	2.54E+07	2
3407	25092	C12H26	2,4,4,5-tetramethyloctane	62199-34-2	7.31	2.02E+07	2
3408	25093	C12H26	2,4,4,6-tetramethyloctane	62199-35-3	7.18	1.51E+07	2
3409	25094	C12H26	2,4,4,7-tetramethyloctane	35866-96-7	7.16	1.44E+07	2
3410	25095	C12H26	2,4,5,5-tetramethyloctane	62199-36-4	7.33	2.12E+07	2
3411	25096	C12H26	2,4,5,6-tetramethyloctane	62199-37-5	7.37	2.32E+07	2
3412	25097	C12H26	2,4,5,7-tetramethyloctane	2217-17-6	7.21	1.61E+07	2
3413	25098	C12H26	2,4,6,6-tetramethyloctane	62199-38-6	7.22	1.67E+07	2
3414	25099	C12H26	2,5,5,6-tetramethyloctane	62199-39-7	7.37	2.32E+07	2
3415	25100	C12H26	2,5,6,6-tetramethyloctane	62199-40-0	7.41	2.54E+07	2
3416	25101	C12H26	3,3,4,4-tetramethyloctane	62199-41-1	7.52	3.32E+07	2
3417	25102	C12H26	3,3,4,5-tetramethyloctane	62199-42-2	7.44	2.78E+07	2
3418	25103	C12H26	3,3,4,6-tetramethyloctane	62199-43-3	7.41	2.54E+07	2
3419	25104	C12H26	3,3,5,5-tetramethyloctane	62199-44-4	7.43	2.66E+07	2
3420	25105	C12H26	3,3,5,6-tetramethyloctane	62199-45-5	7.35	2.22E+07	2
3421	25106	C12H26	3,3,6,6-tetramethyloctane	62199-46-6	7.28	1.90E+07	2
3422	25107	C12H26	3,4,4,5-tetramethyloctane	62199-47-7	7.50	3.18E+07	2
3423	25108	C12H26	3,4,4,6-tetramethyloctane	62185-19-7	7.39	2.43E+07	2
3424	25109	C12H26	3,4,5,5-tetramethyloctane	62185-20-0	7.43	2.66E+07	2
3425	25110	C12H26	3,4,5,6-tetramethyloctane	62185-21-1	7.46	2.91E+07	2
3426	25111	C12H26	4,4,5,5-tetramethyloctane	62185-22-2	7.48	3.04E+07	2
3427	25112	C12H26	2-methyl-4-tert-butylheptane	62185-23-3	7.18	1.51E+07	2
3428	25113	C12H26	3-methyl-4-tert-butylheptane	62185-24-4	7.28	1.93E+07	2
3429	25114	C12H26	4-methyl-4-tert-butylheptane	62185-25-5	7.44	2.78E+07	2
3430	25115	C12H26	3-ethyl-4-propylheptane	62185-26-6	7.41	2.54E+07	2
3431	25116	C12H26	4-ethyl-4-propylheptane	17312-43-5	7.43	2.66E+07	2
3432	25117	C12H26	3-ethyl-4-isopropylheptane	62185-27-7	7.39	2.43E+07	2
3433	25118	C12H26	4-ethyl-4-isopropylheptane	62185-28-8	7.48	3.04E+07	2
3434	25119	C12H26	2,2-dimethyl-4-propylheptane	---	7.16	1.44E+07	2
3435	25120	C12H26	2,3-dimethyl-4-propylheptane	62185-30-2	7.37	2.32E+07	2
3436	25121	C12H26	2,4-dimethyl-4-propylheptane	62185-31-3	7.26	1.84E+07	2
3437	25122	C12H26	2,5-dimethyl-4-propylheptane	62185-32-4	7.28	1.93E+07	2
3438	25123	C12H26	2,6-dimethyl-4-propylheptane	62185-33-5	7.17	1.48E+07	2
3439	25124	C12H26	3,3-dimethyl-4-propylheptane	62185-34-6	7.41	2.54E+07	2
3440	25125	C12H26	3,4-dimethyl-4-propylheptane	62185-35-7	7.44	2.78E+07	2
3441	25126	C12H26	3,5-dimethyl-4-propylheptane	62185-36-8	7.41	2.54E+07	2
3442	25127	C12H26	2,2-dimethyl-3-isopropylhepta	62185-37-9	7.28	1.93E+07	2
3443	25128	C12H26	2,2-dimethyl-4-isopropylhepta	62185-38-0	7.12	1.31E+07	2
3444	25129	C12H26	2,3-dimethyl-3-isopropylhepta	62185-39-1	7.50	3.18E+07	2
3445	25130	C12H26	2,3-dimethyl-4-isopropylhepta	62185-40-4	7.33	2.12E+07	2
3446	25131	C12H26	2,4-dimethyl-3-isopropylhepta	62185-41-5	7.33	2.12E+07	2
3447	25132	C12H26	2,4-dimethyl-4-isopropylhepta	62185-42-6	7.33	2.12E+07	2
3448	25133	C12H26	2,5-dimethyl-3-isopropylhepta	62185-43-7	7.28	1.93E+07	2
3449	25134	C12H26	2,5-dimethyl-4-isopropylhepta	62185-44-8	7.26	1.84E+07	2
3450	25135	C12H26	2,6-dimethyl-3-isopropylhepta	62185-45-9	7.30	2.00E+07	2
3451	25136	C12H26	2,6-dimethyl-4-isopropylhepta	35866-89-8	7.14	1.37E+07	2
3452	25137	C12H26	3,3-dimethyl-4-isopropylhepta	62185-46-0	7.37	2.32E+07	2
3453	25138	C12H26	3,4-dimethyl-4-isopropylhepta	62185-47-1	7.50	3.18E+07	2
3454	25139	C12H26	3,5-dimethyl-4-isopropylhepta	62198-89-4	7.37	2.32E+07	2
3455	25140	C12H26	2-methyl-3,3-diethylheptane	62198-90-7	7.54	3.47E+07	2
3456	25141	C12H26	2-methyl-3,4-diethylheptane	62198-91-8	7.39	2.43E+07	2
3457	25142	C12H26	2-methyl-3,5-diethylheptane	62198-92-9	7.39	2.43E+07	2
3458	25143	C12H26	2-methyl-4,4-diethylheptane	62198-93-0	7.37	2.32E+07	2
3459	25144	C12H26	2-methyl-4,5-diethylheptane	62198-94-1	7.35	2.22E+07	2
3460	25145	C12H26	2-methyl-5,5-diethylheptane	62198-95-2	7.44	2.78E+07	2
3461	25146	C12H26	3-methyl-3,4-diethylheptane	62198-96-3	7.50	3.18E+07	2
3462	25147	C12H26	3-methyl-3,5-diethylheptane	62198-97-4	7.43	2.66E+07	2
3463	25148	C12H26	3-methyl-4,4-diethylheptane	62198-98-5	7.56	3.62E+07	2
3464	25149	C12H26	3-methyl-4,5-diethylheptane	62198-99-6	7.44	2.78E+07	2
3465	25150	C12H26	3-methyl-5,5-diethylheptane	62199-00-2	7.46	2.91E+07	2
3466	25151	C12H26	4-methyl-3,3-diethylheptane	62199-01-3	7.58	3.77E+07	2
3467	25152	C12H26	4-methyl-3,4-diethylheptane	62199-02-4	7.50	3.18E+07	2
3468	25153	C12H26	4-methyl-3,5-diethylheptane	62199-03-5	7.48	3.04E+07	2
3469	25154	C12H26	2,2,3-trimethyl-3-ethylheptane	62199-04-6	7.48	3.04E+07	2
3470	25155	C12H26	2,2,3-trimethyl-4-ethylheptane	62199-05-7	7.35	2.22E+07	2
3471	25156	C12H26	2,2,3-trimethyl-5-ethylheptane	62199-06-8	7.33	2.12E+07	2
3472	25157	C12H26	2,2,4-trimethyl-3-ethylheptane	62199-07-9	7.33	2.12E+07	2
3473	25158	C12H26	2,2,4-trimethyl-4-ethylheptane	62199-08-0	7.37	2.32E+07	2
3474	25159	C12H26	2,2,4-trimethyl-5-ethylheptane	62199-09-1	7.22	1.67E+07	2
3475	25160	C12H26	2,2,5-trimethyl-3-ethylheptane	62199-10-4	7.28	1.93E+07	2
3476	25161	C12H26	2,2,5-trimethyl-4-ethylheptane	62199-11-5	7.20	1.59E+07	2
3477	25162	C12H26	2,2,5-trimethyl-5-ethylheptane	62199-12-6	7.26	1.84E+07	2
3478	25163	C12H26	2,2,6-trimethyl-3-ethylheptane	62199-13-7	7.28	1.93E+07	2
3479	25164	C12H26	2,2,6-trimethyl-4-ethylheptane	62199-14-8	7.07	1.18E+07	2
3480	25165	C12H26	2,2,6-trimethyl-5-ethylheptane	62199-15-9	7.22	1.67E+07	2
3481	25166	C12H26	2,3,3-trimethyl-4-ethylheptane	62199-16-0	7.46	2.91E+07	2
3482	25167	C12H26	2,3,3-trimethyl-5-ethylheptane	62199-17-1	7.37	2.32E+07	2
3483	25168	C12H26	2,3,4-trimethyl-3-ethylheptane	62199-18-2	7.54	3.47E+07	2
3484	25169	C12H26	2,3,4-trimethyl-4-ethylheptane	62198-55-4	7.44	2.78E+07	2
3485	25170	C12H26	2,3,4-trimethyl-5-ethylheptane	62198-56-5	7.43	2.66E+07	2
3486	25171	C12H26	2,3,5-trimethyl-3-ethylheptane	62198-57-6	7.43	2.72E+07	2
3487	25172	C12H26	2,3,5-trimethyl-4-ethylheptane	62198-58-7	7.41	2.54E+07	2
3488	25173	C12H26	2,3,5-trimethyl-5-ethylheptane	62198-59-8	7.37	2.32E+07	2
3489	25174	C12H26	2,3,6-trimethyl-3-ethylheptane	62198-60-1	7.41	2.54E+07	2
3490	25175	C12H26	2,3,6-trimethyl-4-ethylheptane	62198-61-2	7.28	1.93E+07	2
3491	25176	C12H26	2,3,6-trimethyl-5-ethylheptane	62198-62-3	7.35	2.22E+07	2
3492	25177	C12H26	2,4,4-trimethyl-3-ethylheptane	62198-63-4	7.37	2.32E+07	2
3493	25178	C12H26	2,4,4-trimethyl-5-ethylheptane	62198-64-5	7.33	2.12E+07	2
3494	25179	C12H26	2,4,5-trimethyl-3-ethylheptane	62198-65-6	7.41	2.54E+07	2
3495	25180	C12H26	2,4,5-trimethyl-4-ethylheptane	62198-66-7	7.39	2.43E+07	2
3496	25181	C12H26	2,4,5-trimethyl-5-ethylheptane	62198-67-8	7.43	2.66E+07	2
3497	25182	C12H26	2,4,6-trimethyl-3-ethylheptane	62198-68-9	7.31	2.02E+07	2
3498	25183	C12H26	2,4,6-trimethyl-4-ethylheptane	62198-69-0	7.18	1.51E+07	2
3499	25184	C12H26	2,5,5-trimethyl-3-ethylheptane	62198-70-3	7.28	1.93E+07	2
3500	25185	C12H26	2,5,5-trimethyl-4-ethylheptane	62198-71-4	7.33	2.12E+07	2
3501	25186	C12H26	3,3,4-trimethyl-4-ethylheptane	62198-72-5	7.58	3.77E+07	2
3502	25187	C12H26	3,3,4-trimethyl-5-ethylheptane	62198-73-6	7.46	2.91E+07	2
3503	25188	C12H26	3,3,5-trimethyl-4-ethylheptane	62198-74-7	7.43	2.66E+07	2
3504	25189	C12H26	3,3,5-trimethyl-5-ethylheptane	62198-75-8	7.52	3.32E+07	2
3505	25190	C12H26	3,4,4-trimethyl-3-ethylheptane	62198-76-9	7.58	3.77E+07	2
3506	25191	C12H26	3,4,4-trimethyl-5-ethylheptane	62198-77-0	7.52	3.32E+07	2
3507	25192	C12H26	3,4,5-trimethyl-3-ethylheptane	62198-78-1	7.52	3.32E+07	2
3508	25193	C12H26	3,4,5-trimethyl-4-ethylheptane	62198-79-2	7.58	3.77E+07	2
3509	25194	C12H26	2,2,3,3,4-pentamethylheptane	62198-80-5	7.48	3.04E+07	2
3510	25195	C12H26	2,2,3,3,5-pentamethylheptane	62198-81-6	7.37	2.32E+07	2
3511	25196	C12H26	2,2,3,3,6-pentamethylheptane	62198-82-7	7.35	2.22E+07	2
3512	25197	C12H26	2,2,3,4,4-pentamethylheptane	62198-83-8	7.52	3.32E+07	2
3513	25198	C12H26	2,2,3,4,5-pentamethylheptane	62198-84-9	7.37	2.32E+07	2
3514	25199	C12H26	2,2,3,4,6-pentamethylheptane	62198-85-0	7.24	1.75E+07	2
3515	25200	C12H26	2,2,3,5,5-pentamethylheptane	62198-86-1	7.22	1.67E+07	2
3516	25201	C12H26	2,2,3,5,6-pentamethylheptane	62198-87-2	7.26	1.82E+07	2
3517	25202	C12H26	2,2,3,6,6-pentamethylheptane	62198-88-3	7.18	1.51E+07	2
3518	25203	C12H26	2,2,4,4,5-pentamethylheptane	62199-61-5	7.41	2.54E+07	2
3519	25204	C12H26	2,2,4,4,6-pentamethylheptane	62199-62-6	7.19	1.56E+07	2
3520	25205	C12H26	2,2,4,5,5-pentamethylheptane	62199-63-7	7.20	1.59E+07	2

Table 13 Octanol-Water Partition Coefficient - Oganic Compounds

Octanol-Water Partition Coefficient								Octanol-Water Partition Coefficient							
NO	ID	FORMULA	NAME	CAS No	log_{10} Kow	Kow	code	NO	ID	FORMULA	NAME	CAS No	log_{10} Kow	Kow	code
3521	25206	C12H26	2,2,4,5,6-pentamethylheptane	62199-64-8	7.18	1.51E+07	2	3601	25316	C12H26S	butyl octyl sulfide	16900-07-5	5.85	7.08E+05	2
3522	25207	C12H26	2,2,4,6,6-pentamethylheptane	13475-82-6	7.03	1.07E+07	2	3602	25317	C12H26S	tert-dodecylmercaptan	25103-58-6	6.19	1.56E+06	2
3523	25208	C12H26	2,2,5,5,6-pentamethylheptane	62199-65-9	7.20	1.59E+07	2	3603	25318	C12H26S2	dihexyl disulfide	10496-15-8	6.78	6.03E+06	2
3524	25209	C12H26	2,3,3,4,4-pentamethylheptane	62199-66-0	7.54	3.47E+07	2	3604	25340	C12H27N	dodecylamine	124-22-1	4.46	2.90E+04	2
3525	25210	C12H26	2,3,3,4,5-pentamethylheptane	62199-67-1	7.48	3.04E+07	2	3605	25341	C12H27N	methylundecylamine	66553-53-5	4.58	3.82E+04	2
3526	25211	C12H26	2,3,3,4,6-pentamethylheptane	62199-68-2	7.39	2.43E+07	2	3606	25342	C12H27N	ethyldecylamine	66553-52-4	4.55	3.55E+04	2
3527	25212	C12H26	2,3,3,5,5-pentamethylheptane	62199-69-3	7.48	3.04E+07	2	3607	25343	C12H27N	dihexylamine	143-16-8	4.54	3.45E+04	2
3528	25213	C12H26	2,3,3,5,6-pentamethylheptane	52670-35-6	7.33	2.12E+07	2	3608	25344	C12H27N	dimethyldecylamine	1120-24-7	4.01	1.01E+04	2
3529	25214	C12H26	2,3,4,4,5-pentamethylheptane	62199-70-6	7.48	3.04E+07	2	3609	25345	C12H27N	diethyloctylamine	4088-37-3	3.68	4.83E+03	2
3530	25215	C12H26	2,3,4,4,6-pentamethylheptane	62199-71-7	7.28	1.93E+07	2	3610	25346	C12H27N	tributylamine	102-82-9	4.46	2.88E+04	2
3531	25216	C12H26	2,3,4,5,5-pentamethylheptane	62199-72-8	7.43	2.66E+07	2	3611	25347	C12H27N	triisobutylamine	1116-40-1	4.05	1.12E+04	2
3532	25217	C12H26	2,3,4,5,6-pentamethylheptane	27574-98-7	7.39	2.43E+07	2	3612	25446	C13H6Cl6O2	hexachlorophene	70-30-4	3.93	8.51E+03	1
3533	25218	C12H26	2,4,4,5,5-pentamethylheptane	62199-73-9	7.41	2.54E+07	2	3613	25496	C13H8O	9H-fluoren-9-one	486-25-9	3.58	3.80E+03	1
3534	25219	C12H26	3,3,4,4,5-pentamethylheptane	62199-74-0	7.60	3.94E+07	2	3614	25537	C13H9N	acridine	260-94-6	3.40	2.51E+03	1
3535	25220	C12H26	3,3,4,5,5-pentamethylheptane	62199-75-1	7.65	4.45E+07	2	3615	25575	C13H10	fluorene	86-73-7	4.20	1.58E+04	1
3536	25221	C12H26	2,2-dimethyl-3-tert-butylhexan	62199-76-2	7.41	2.54E+07	2	3616	25623	C13H10O	benzophenone	119-61-9	3.18	1.51E+03	1
3537	25222	C12H26	2-methyl-3-ethyl-3-isopropylh	62199-77-3	7.48	3.04E+07	2	3617	25631	C13H10O2	phenyl benzoate	93-99-2	3.59	3.89E+03	1
3538	25223	C12H26	2-methyl-4-ethyl-3-isopropylh	62199-78-4	7.28	1.93E+07	2	3618	25694	C13H11NO	N-phenylbenzamide	93-98-1	2.62	4.17E+02	1
3539	25224	C12H26	2,2,3-trimethyl-3-isopropylhex	62199-79-5	7.50	3.18E+07	2	3619	25738	C13H12	diphenylmethane	101-81-5	4.14	1.38E+04	1
3540	25225	C12H26	2,2,4-trimethyl-3-isopropylhex	62199-80-8	7.26	1.84E+07	2	3620	25741	C13H12	4-methylbiphenyl	644-08-6	4.63	4.27E+04	1
3541	25226	C12H26	2,2,5-trimethyl-3-isopropylhex	62199-81-9	7.07	1.18E+07	2	3621	25793	C13H12O	benzyl phenyl ether	946-80-5	3.79	6.17E+03	1
3542	25227	C12H26	2,2,5-trimethyl-4-isopropylhex	62199-82-0	7.07	1.18E+07	2	3622	25798	C13H12O	diphenylmethanol	91-01-0	2.67	4.68E+02	1
3543	25228	C12H26	2,3,4-trimethyl-3-isopropylhex	62199-83-1	7.56	3.62E+07	2	3623	26283	C13H20	heptylbenzene	1078-71-3	5.49	3.09E+05	2
3544	25229	C12H26	2,3,5-trimethyl-3-isopropylhex	62199-84-2	7.37	2.32E+07	2	3624	26284	C13H20	1-methyl-2,4-diisopropylbenz	1460-98-6	5.49	3.09E+05	2
3545	25230	C12H26	2,3,5-trimethyl-4-isopropylhex	62199-85-3	7.28	1.93E+07	2	3625	26285	C13H20	1-methyl-3,5-diisopropylbenz	3055-14-9	5.49	3.09E+05	2
3546	25231	C12H26	2,4,4-trimethyl-3-isopropylhex	62199-86-4	7.33	2.12E+07	2	3626	26286	C13H20	[2-methyl-1-isopropylpropyl]b	21777-84-4	5.49	3.09E+05	2
3547	25232	C12H26	3,3,4-triethylhexane	62199-87-5	7.60	3.94E+07	2	3627	26288	C13H20	(2,4-dimethylpentyl)benzene	54518-00-2	5.49	3.09E+05	2
3548	25233	C12H26	2,2-dimethyl,3,3-diethylhexan	62199-88-6	7.48	3.04E+07	2	3628	26402	C13H24	1-tridecyne	26186-02-7	6.32	2.09E+06	2
3549	25234	C12H26	2,2-dimethyl-3,4-diethylhexan	62199-89-7	7.28	1.93E+07	2	3629	26403	C13H24	2-tridecyne	28467-75-6	6.44	2.77E+06	2
3550	25235	C12H26	2,2-dimethyl-4,4-diethylhexan	62184-89-8	7.48	3.04E+07	2	3630	26404	C13H24	3-tridecyne	60186-78-9	6.37	2.33E+06	2
3551	25236	C12H26	2,3-dimethyl-3,4-diethylhexan	62184-90-1	7.56	3.62E+07	2	3631	26465	C13H26	octylcyclopentane	1795-20-6	7.45	2.81E+07	2
3552	25237	C12H26	2,3-dimethyl-4,4-diethylhexan	62184-91-2	7.56	3.62E+07	2	3632	26466	C13H26	heptylcyclohexane	5617-41-4	6.54	3.47E+06	2
3553	25238	C12H26	2,4-dimethyl-3,3-diethylhexan	62184-92-3	7.48	3.04E+07	2	3633	26467	C13H26	2-butyl-1,1,3-trimethylcyclohe	54676-39-0	6.54	3.47E+06	2
3554	25239	C12H26	2,4-dimethyl-3,4-diethylhexan	62184-93-4	7.48	3.04E+07	2	3634	26469	C13H26	1-tridecene	2437-56-1	6.59	3.89E+06	2
3555	25240	C12H26	2,5-dimethyl-3,3-diethylhexan	62184-94-5	7.28	1.93E+07	2	3635	26470	C13H26	2-methyl-1-dodecene	16435-49-7	7.43	2.68E+07	2
3556	25241	C12H26	2,5-dimethyl-3,4-diethylhexan	62184-95-6	7.28	1.93E+07	2	3636	26471	C13H26	5-butyl-4-nonene	7367-38-6	7.01	1.02E+07	2
3557	25242	C12H26	3,3-dimethyl-4,4-diethylhexan	62184-96-7	7.67	4.64E+07	2	3637	26486	C13H26O	1-tridecanal	10486-19-8	5.87	7.33E+05	2
3558	25243	C12H26	3,4-dimethyl-3,4-diethylhexan	62184-97-8	7.67	4.64E+07	2	3638	26487	C13H26O	2-tridecanone	593-08-8	5.39	2.47E+05	2
3559	25244	C12H26	2,2,3,3-tetramethyl-4-ethylhex	62184-98-9	7.50	3.18E+07	2	3639	26488	C13H26O	3-tridecanone	1534-26-5	5.39	2.47E+05	2
3560	25245	C12H26	2,2,3,4-tetramethyl-3-ethylhex	62184-99-0	7.56	3.62E+07	2	3640	26489	C13H26O	7-tridecanone	462-18-0	5.39	2.47E+05	2
3561	25246	C12H26	2,2,3,4-tetramethyl-4-ethylhex	62185-00-6	7.63	4.28E+07	2	3641	26492	C13H26O2	tridecanoic acid	638-53-9	4.32	2.09E+04	2
3562	25247	C12H26	2,2,3,5-tetramethyl-3-ethylhex	62185-01-7	7.37	2.32E+07	2	3642	26493	C13H26O2	2-methyldodecanoic acid	2874-74-0	4.32	2.09E+04	2
3563	25248	C12H26	2,2,3,5-tetramethyl-4-ethylhex	62185-02-8	7.31	2.02E+07	2	3643	26494	C13H26O2	dodecyl formate	28303-42-6	7.05	1.13E+07	2
3564	25249	C12H26	2,2,4,4-tetramethyl-3-ethylhex	62185-03-9	7.48	3.04E+07	2	3644	26495	C13H26O2	undecyl acetate	1731-81-3	6.63	4.27E+06	2
3565	25250	C12H26	2,2,4,5-tetramethyl-3-ethylhex	62185-04-0	7.28	1.93E+07	2	3645	26498	C13H26O2	octyl pentanoate	5451-85-4	5.28	1.91E+05	2
3566	25251	C12H26	2,2,4,5-tetramethyl-4-ethylhex	62185-05-1	7.43	2.66E+07	2	3646	26499	C13H26O2	heptyl hexanoate	6976-72-3	5.28	1.91E+05	2
3567	25252	C12H26	2,2,5,5-tetramethyl-3-ethylhex	62185-06-2	7.07	1.18E+07	2	3647	26500	C13H26O2	hexyl heptanoate	1119-06-8	5.28	1.91E+05	2
3568	25253	C12H26	2,3,3,4-tetramethyl-4-ethylhex	62185-07-3	7.63	4.28E+07	2	3648	26501	C13H26O2	pentyl octanoate	638-25-5	5.28	1.91E+05	2
3569	25254	C12H26	2,3,3,5-tetramethyl-4-ethylhex	62185-08-4	7.43	2.66E+07	2	3649	26502	C13H26O2	butyl nonanoate	50623-57-9	5.28	1.91E+05	2
3570	25255	C12H26	2,3,4,4-tetramethyl-3-ethylhex	62185-09-5	7.63	4.28E+07	2	3650	26503	C13H26O2	propyl decanoate	30673-60-0	5.28	1.91E+05	2
3571	25256	C12H26	2,3,4,5-tetramethyl-3-ethylhex	62185-10-8	7.50	3.18E+07	2	3651	26504	C13H26O2	isopropyl decanoate	2311-59-3	5.28	1.91E+05	2
3572	25257	C12H26	2,2,3,3,4,4-hexamethylhexane	62185-11-9	7.77	5.85E+07	2	3652	26505	C13H26O2	methyl dodecanoate	111-82-0	5.28	1.91E+05	2
3573	25258	C12H26	2,2,3,3,4,5-hexamethylhexane	62185-12-0	7.46	2.91E+07	2	3653	26506	C13H26O2	ethyl undecanoate	627-90-7	5.28	1.91E+05	2
3574	25259	C12H26	2,2,3,3,5,5-hexamethylhexane	60302-24-1	7.39	2.43E+07	2	3654	26507	C13H26O2	isoamyl caprylate	2035-99-6	5.28	1.91E+05	2
3575	25260	C12H26	2,2,3,4,4,5-hexamethylhexane	62185-13-1	7.60	3.94E+07	2	3655	26515	C13H27Br	1-bromotridecane	765-09-3	7.07	1.17E+07	2
3576	25261	C12H26	2,2,3,4,5,5-hexamethylhexane	62185-14-2	7.28	1.93E+07	2	3656	26516	C13H27Cl	1-chlorotridecane	822-13-9	6.97	9.33E+06	2
3577	25262	C12H26	2,3,3,4,4,5-hexamethylhexane	52670-36-7	7.60	3.94E+07	2	3657	26526	C13H28	tridecane	629-50-5	6.73	5.37E+06	2
3578	25263	C12H26	2,4-dimethyl-3-ethyl-3-isoprop	62185-16-4	7.60	3.94E+07	2	3658	26527	C13H28	2-methyldodecane	1560-97-0	8.00	1.00E+08	2
3579	25264	C12H26	2,2,4-trimethyl-3,3-diethylpen	62185-15-3	7.60	3.94E+07	2	3659	26528	C13H28	3-methyldodecane	17312-57-1	8.01	1.03E+08	2
3580	25265	C12H26	2,2,3,4-tetramethyl-3-isoprop	62185-17-5	7.56	3.62E+07	2	3660	26529	C13H28	2,2-dimethylundecane	17312-64-0	7.88	7.57E+07	2
3581	25266	C12H26	2,2,4,4-tetramethyl-3-isoprop	62185-18-6	7.37	2.32E+07	2	3661	26530	C13H28	2,3-dimethylundecane	17312-77-5	7.98	9.58E+07	2
3582	25267	C12H26	2,2,3,4,4-pentamethyl-3-ethyl	66576-21-4	7.73	5.43E+07	2	3662	26531	C13H28	2,4-dimethylundecane	17312-80-0	7.80	6.31E+07	2
3583	25268	C12H26	2-methylundecane	7045-71-8	7.67	4.64E+07	2	3663	26532	C13H28	5-butylnonane	17312-63-9	7.73	5.42E+07	2
3584	25277	C12H26O	1-dodecanol	112-53-8	5.13	1.35E+05	1	3664	26535	C13H28O	1-tridecanol	112-70-9	5.82	6.61E+05	1
3585	25278	C12H26O	2-dodecanol	10203-28-8	5.06	1.16E+05	2	3665	26536	C13H28O	2-tridecanol	1653-31-2	5.64	4.36E+05	2
3586	25279	C12H26O	3-dodecanol	10203-30-2	5.10	1.25E+05	2	3666	26537	C13H28O	3-tridecanol	10289-68-6	5.73	5.37E+05	2
3587	25280	C12H26O	6-dodecanol	6836-38-0	5.10	1.25E+05	2	3667	26538	C13H28O	5-butyl-5-nonanol	597-93-3	5.73	5.37E+05	2
3588	25281	C12H26O	2-methyl-1-undecanol	10522-26-6	5.10	1.25E+05	2	3668	26549	C13H28S	1-tridecanethiol	19484-26-5	6.68	4.76E+06	2
3589	25282	C12H26O	2-methyl-3-undecanol	60671-36-5	5.10	1.25E+05	2	3669	26550	C13H28S	2-tridecanethiol	62155-03-7	6.53	3.42E+06	2
3590	25283	C12H26O	2-methyl-5-undecanol	33978-71-1	5.10	1.25E+05	2	3670	26551	C13H28S	methyl dodecyl sulfide	---	6.35	2.24E+06	2
3591	25284	C12H26O	2-butyl-1-octanol	3913-02-8	5.10	1.25E+05	2	3671	26552	C13H28S	ethyl undecyl sulfide	66577-30-8	6.35	2.24E+06	2
3592	25285	C12H26O	dihexyl ether	112-58-3	4.98	9.55E+04	2	3672	26553	C13H28S	propyl decyl sulfide	66577-31-9	6.35	2.24E+06	2
3593	25286	C12H26O	2,6,8-trimethylnonanol-4	123-17-1	5.10	1.25E+05	2	3673	26554	C13H28S	butyl nonyl sulfide	66577-32-0	6.35	2.24E+06	2
3594	25294	C12H26O3	diethylene glycol dibutyl ethe	112-73-2	1.92	8.32E+01	1	3674	26555	C13H29N	tridecylamine	2869-34-3	4.54	3.50E+04	2
3595	25310	C12H26S	1-dodecanethiol	112-55-0	6.18	1.51E+06	2	3675	26556	C13H29N	methyldodecylamine	7311-30-0	4.74	5.51E+04	2
3596	25311	C12H26S	2-dodecanethiol	14402-50-7	6.21	1.62E+06	2	3676	26557	C13H29N	ethylundecylamine	59570-04-6	4.73	5.37E+04	2
3597	25312	C12H26S	dihexyl sulfide	6294-31-1	5.85	7.08E+05	2	3677	26558	C13H29N	dimethylundecylamine	17373-28-3	4.21	1.64E+04	2
3598	25313	C12H26S	methyl undecyl sulfide	7289-44-3	5.85	7.08E+05	2	3678	26559	C13H29N	diethylnonylamine	45124-35-4	3.99	9.67E+03	2
3599	25314	C12H26S	ethyl decyl sulfide	19313-61-2	5.85	7.08E+05	2	3679	26560	C13H29N	6-methyl-N-isopentyl-2-hepta	502-59-0	4.74	5.44E+04	2
3600	25315	C12H26S	propyl nonyl sulfide	62103-66-6	5.85	7.08E+05	2	3680	26561	C13H29N	tert-amyl-tert-octylamine	150285-07-7	4.74	5.44E+04	2

Table 13 Octanol-Water Partition Coefficient - Oganic Compounds

Octanol-Water Partition Coefficient								Octanol-Water Partition Coefficient							
NO	ID	FORMULA	NAME	CAS No	log_{10} Kow	Kow	code	NO	ID	FORMULA	NAME	CAS No	log_{10} Kow	Kow	code
3681	26562	C13H29N	N-methyldihexylamine	37615-53-5	4.25	1.77E+04	2	3761	27956	C14H31N	tetradecylamine	2016-42-4	5.75	5.62E+05	2
3682	26654	C14H8O2	anthraquinone	84-65-1	3.39	2.45E+03	1	3762	27957	C14H31N	methyltridecylamine	45165-81-9	4.81	6.47E+04	2
3683	26728	C14H10	anthracene	120-12-7	4.56	3.63E+04	1	3763	27958	C14H31N	ethyldodecylamine	35902-57-9	4.81	6.47E+04	2
3684	26729	C14H10	diphenylacetylene	501-65-5	4.78	6.03E+04	1	3764	27959	C14H31N	diheptylamine	2470-68-0	4.81	6.41E+04	2
3685	26730	C14H10	phenanthrene	85-01-8	4.52	3.31E+04	1	3765	27960	C14H31N	dimethyldodecylamine	112-18-5	4.33	2.13E+04	2
3686	26866	C14H12	cis-stilbene	645-49-8	4.52	3.31E+04	2	3766	27961	C14H31N	diethyldecylamine	6308-94-7	4.19	1.56E+04	2
3687	26867	C14H12	trans-stilbene	103-30-0	4.81	6.46E+04	1	3767	28029	C15H10N2O2	diphenylmethane-4,4'-diisocy	101-68-8	5.22	1.66E+05	2
3688	26871	C14H12	1-methylfluorene	1730-37-6	4.97	9.33E+04	1	3768	28044	C15H10O2	2-phenyl-1H-indene-1,3(2H)-	83-12-5	2.90	7.94E+02	1
3689	26873	C14H12	stilbene	588-59-0	4.67	4.62E+04	2	3769	28129	C15H12	1-methylanthracene	610-48-0	5.08	1.20E+05	2
3690	26933	C14H12O	2-phenylacetophenone	451-40-1	3.18	1.51E+03	1	3770	28130	C15H12	2-methylanthracene	613-12-7	5.15	1.41E+05	1
3691	26942	C14H12O2	benzyl benzoate	120-51-4	3.97	9.33E+03	1	3771	28131	C15H12	9-methylanthracene	779-02-2	5.07	1.17E+05	1
3692	27052	C14H14	1,1-diphenylethane	612-00-0	4.15	1.41E+04	2	3772	28132	C15H12	1-methylphenanthrene	832-69-9	5.14	1.38E+05	1
3693	27053	C14H14	1,2-diphenylethane	103-29-7	4.70	5.01E+04	1	3773	28436	C15H16O2	bisphenol a	80-05-7	3.32	2.09E+03	1
3694	27054	C14H14	2-methyldiphenylmethane	713-36-0	4.47	2.95E+04	2	3774	28725	C15H24	nonylbenzene	1081-77-2	7.11	1.29E+07	1
3695	27055	C14H14	3-methyldiphenylmethane	620-47-3	4.47	2.95E+04	2	3775	28726	C15H24	1,2,4-triisopropylbenzene	948-32-3	7.11	1.29E+07	2
3696	27056	C14H14	4-methyldiphenylmethane	620-83-7	4.47	2.95E+04	2	3776	28727	C15H24	1,3,5-triisopropylbenzene	717-74-8	7.11	1.29E+07	2
3697	27057	C14H14	2,2'-dimethylbiphenyl	605-39-0	5.09	1.23E+05	2	3777	28728	C15H24	3,5-di-tert-butyltoluene	15181-11-0	7.11	1.29E+07	2
3698	27058	C14H14	2,3'-trimethylbiphenyl	611-43-8	5.09	1.23E+05	2	3778	28756	C15H24O	2,6-di-tert-butyl-p-cresol	128-37-0	5.10	1.26E+05	1
3699	27059	C14H14	2,4'-dimethlbiphenyl	611-61-0	5.09	1.23E+05	2	3779	28757	C15H24O	nonylphenol	25154-52-3	4.77	5.95E+04	2
3700	27060	C14H14	3,3'-dimethylbiphenyl	612-75-9	5.09	1.23E+05	2	3780	28758	C15H24O	2,4-di-tert-butyl-5-methylpher	497-39-2	4.94	8.65E+04	2
3701	27061	C14H14	3,4'-dimethylbiphenyl	7383-90-6	5.09	1.23E+05	2	3781	28759	C15H24O	2,4-di-tert-butyl-6-methylpher	616-55-7	4.94	8.65E+04	2
3702	27062	C14H14	4,4'-dimethylbiphenyl	613-33-2	5.09	1.23E+05	1	3782	28764	C15H24O	4-nonylphenol	104-40-5	4.94	8.65E+04	2
3703	27138	C14H14O	dibenzyl ether	103-50-4	3.31	2.04E+03	1	3783	28765	C15H24O	4-nonylphenol, isomers	84852-15-3	4.94	8.65E+04	2
3704	27644	C14H22	octylbenzene	2189-60-8	6.30	2.00E+06	1	3784	28820	C15H28	1-pentadecyne	765-13-9	6.65	4.46E+06	2
3705	27645	C14H22	1,2,3,4-tetraethylbenzene	642-32-0	6.30	2.00E+06	2	3785	28821	C15H28	2-pentadecyne	52112-25-1	6.66	4.59E+06	2
3706	27646	C14H22	1,2,3,5-tetraethylbenzene	38842-05-6	6.30	2.00E+06	2	3786	28822	C15H28	3-pentadecyne	61886-61-1	6.66	4.54E+06	2
3707	27647	C14H22	1,2,4,5-tetraethylbenzene	635-81-4	6.30	2.00E+06	2	3787	28855	C15H30	decylcyclopentane	1795-21-7	7.68	4.79E+07	2
3708	27648	C14H22	1,4-di-tert-butylbenzene	1012-72-2	6.30	2.00E+06	2	3788	28856	C15H30	nonylcyclohexane	2883-02-5	7.52	3.31E+07	2
3709	27649	C14H22	(1-methylheptyl)benzene	777-22-0	6.30	2.00E+06	2	3789	28857	C15H30	1-ethyl-1-methyl-2,4-diisopro	515-12-8	7.52	3.31E+07	2
3710	27650	C14H22	1,3-di-tert-butylbenzene	1014-60-4	6.30	2.00E+06	2	3790	28859	C15H30	1-pentadecene	13360-61-7	7.57	3.72E+07	2
3711	27679	C14H22O	p-tert-octylphenol	140-66-9	5.28	1.91E+05	2	3791	28860	C15H30	2-methyl-1-tetradecene	52254-38-3	7.79	6.14E+07	2
3712	27680	C14H22O	2,4-di-tert-butylphenol	96-76-4	5.28	1.91E+05	2	3792	28861	C15H30	7-pentadecene	15430-98-5	7.68	4.78E+07	2
3713	27681	C14H22O	2,6-di-sec-butylphenol	5510-99-6	5.28	1.91E+05	2	3793	28862	C15H30	8-pentadecene	---	7.68	4.78E+07	2
3714	27682	C14H22O	2,6-di-tert-butylphenol	128-39-2	5.28	1.91E+05	2	3794	28875	C15H30O	8-pentadecanone	818-23-5	5.66	4.57E+05	2
3715	27687	C14H22O	4-octylphenol	1806-26-4	5.28	1.91E+05	2	3795	28879	C15H30O2	pentadecanoic acid	1002-84-2	6.47	2.95E+06	2
3716	27688	C14H22O	tert-octylphenol	27193-28-8	5.28	1.91E+05	2	3796	28898	C15H31Br	1-bromopentadecane	629-72-1	8.05	1.12E+08	2
3717	27689	C14H22O	3,5-bis(1,1-dimethylethyl)phe	1138-52-9	5.28	1.91E+05	2	3797	28899	C15H31Cl	1-chloropentadecane	4862-03-1	7.95	8.91E+07	2
3718	27691	C14H22O	2-isopropyl-6-(1-methylbutyl)	74926-98-0	5.28	1.91E+05	2	3798	28904	C15H32	pentadecane	629-62-9	7.71	5.13E+07	2
3719	27780	C14H26	1-tetradecyne	765-10-6	6.53	3.42E+06	2	3799	28905	C15H32	2-methyltetradecane	1560-95-8	8.38	2.39E+08	2
3720	27781	C14H26	3-tetradecyne	60212-32-0	6.56	3.60E+06	2	3800	28906	C15H32	3-methyltetradecane	18435-22-8	8.38	2.41E+08	2
3721	27782	C14H26	7-tetradecyne	35216-11-6	6.54	3.51E+06	2	3801	28907	C15H32	2,2-dimethyltridecane	61869-04-3	8.34	2.17E+08	2
3722	27847	C14H28	nonylcyclopentane	2882-98-6	7.62	4.13E+07	2	3802	28908	C15H32	2,3-dimethyltridecane	18435-20-6	8.38	2.39E+08	2
3723	27848	C14H28	octylcyclohexane	1795-15-9	7.03	1.07E+07	2	3803	28909	C15H32	2,4-dimethyltridecane	61868-05-1	8.29	1.96E+08	2
3724	27850	C14H28	1-tetradecene	1120-36-1	7.08	1.20E+07	2	3804	28912	C15H32O	1-pentadecanol	629-76-5	6.24	1.74E+06	2
3725	27851	C14H28	2-tetradecene	638-60-8	7.74	5.55E+07	2	3805	28913	C15H32O	2-pentadecanol	1653-34-5	6.47	2.95E+06	2
3726	27852	C14H28	2-methyl-1-tridecene	18094-01-4	7.66	4.57E+07	2	3806	28914	C15H32O	3-pentadecanol	53346-71-7	6.35	2.26E+06	2
3727	27853	C14H28	trans-7-tetradecene	41446-63-3	7.49	3.12E+07	2	3807	28915	C15H32O	2,8-dimethyl-6-isobutylnonan	10143-20-1	6.35	2.26E+06	2
3728	27862	C14H28O	tetradecanal	124-25-4	6.51	3.24E+06	2	3808	28923	C15H32S	methyl tetradecyl sulfide	7289-45-4	7.34	2.19E+07	2
3729	27863	C14H28O	2-tetradecanone	2345-27-9	5.98	9.55E+05	2	3809	28924	C15H32S	ethyl tridecyl sulfide	66271-81-6	7.34	2.19E+07	2
3730	27864	C14H28O	7-ethyl-2-methyl-4-undecanol	6976-00-7	5.98	9.55E+05	2	3810	28925	C15H32S	propyl dodecyl sulfide	66271-82-7	7.34	2.19E+07	2
3731	27865	C14H28O	2-methyl-3-tridecanone	40239-35-8	5.98	9.55E+05	2	3811	28926	C15H32S	butyl undecyl sulfide	66271-83-8	7.34	2.19E+07	2
3732	27866	C14H28O	3-tetradecanone	629-23-2	5.98	9.55E+05	2	3812	28931	C15H33N	pentadecylamine	2570-26-5	4.91	8.07E+04	2
3733	27873	C14H28O2	tetradecanoic acid	544-63-8	6.10	1.26E+06	1	3813	28932	C15H33N	methyltetradecylamine	29369-63-9	4.91	8.07E+04	2
3734	27875	C14H28O2	dodecyl acetate	112-66-3	5.97	9.24E+05	2	3814	28933	C15H33N	ethyltridecylamine	59570-06-8	4.91	8.07E+04	2
3735	27895	C14H29Br	1-bromotetradecane	112-71-0	7.56	3.63E+07	2	3815	28934	C15H33N	dimethyltridecylamine	17373-29-4	4.33	2.13E+04	2
3736	27896	C14H29Br	1-bromo-12-methyltridecane	111772-90-8	7.56	3.63E+07	2	3816	28935	C15H33N	diethylundecylamine	54334-64-4	4.33	2.13E+04	2
3737	27897	C14H29Cl	1-chlorotetradecane	2425-54-9	7.47	2.95E+07	2	3817	28936	C15H33N	tripentylamine	621-77-2	4.18	1.51E+04	2
3738	27912	C14H30	tetradecane	629-59-4	7.20	1.58E+07	1	3818	28937	C15H33N	triisopentylamine	645-41-0	4.33	2.13E+04	2
3739	27913	C14H30	2-methyltridecane	1560-96-9	8.24	1.73E+08	2	3819	28938	C15H33N	tris(2-methylbutyl)amine	620-43-9	4.33	2.13E+04	2
3740	27914	C14H30	3-methyltridecane	6418-41-3	8.25	1.76E+08	2	3820	28984	C16H10	fluoranthene	206-44-0	5.07	1.17E+05	1
3741	27915	C14H30	2,2-dimethyldodecane	49598-54-1	8.16	1.45E+08	2	3821	28985	C16H10	pyrene	129-00-0	5.08	1.20E+05	1
3742	27916	C14H30	2,3-dimethyldodecane	6117-98-2	8.23	1.70E+08	2	3822	29111	C16H14	9,10-dimethylphenanthrene	604-83-1	5.69	4.90E+05	1
3743	27917	C14H30	2,4-dimethyldodecane	6117-99-3	8.09	1.23E+08	2	3823	29518	C16H22O4	dibutyl phthalate	84-74-2	4.72	5.25E+04	1
3744	27918	C14H30	7-methyltridecane	26730-14-3	8.03	1.07E+08	2	3824	29595	C16H26	decylbenzene	104-72-3	7.35	2.24E+07	1
3745	27922	C14H30O	1-tetradecanol	112-72-1	6.03	1.07E+06	1	3825	29596	C16H26	pentaethylbenzene	605-01-6	7.35	2.24E+07	2
3746	27923	C14H30O	2-tetradecanol	4706-81-4	5.82	6.60E+05	2	3826	29599	C16H26	(1-methylnonyl)benzene	4537-13-7	7.35	2.24E+07	2
3747	27924	C14H30O	3-tetradecanol	1653-32-3	5.92	8.41E+05	2	3827	29640	C16H30	1-hexadecyne	629-74-3	6.68	4.81E+06	2
3748	27925	C14H30O	12-methyl-1-tridecanol	21987-21-3	5.92	8.41E+05	2	3828	29641	C16H30	2-hexadecyne	629-75-4	6.68	4.78E+06	2
3749	27926	C14H30O	2-methyl-3-tridecanol	98930-89-3	5.92	8.41E+05	2	3829	29642	C16H30	3-hexadecyne	61886-62-2	6.68	4.75E+06	2
3750	27927	C14H30O	2-methyl-4-tridecanol	36691-77-7	5.92	8.41E+05	2	3830	29705	C16H32	decylcyclohexane	1795-16-0	8.01	1.02E+08	2
3751	27928	C14H30O	diheptyl ether	629-64-1	7.54	3.49E+07	2	3831	29707	C16H32	1-hexadecene	629-73-2	8.06	1.15E+08	2
3752	27929	C14H30O	tetradecanol; mixed isomers	27196-00-5	5.92	8.41E+05	2	3832	29708	C16H32	2-methyl-1-pentadecene	29833-69-0	7.82	6.67E+07	2
3753	27946	C14H30S	1-tetradecanethiol	2079-95-0	6.93	8.54E+06	2	3833	29709	C16H32	2-methyl-1-propene, tetramer	15220-85-6	7.94	8.75E+07	2
3754	27947	C14H30S	2-tetradecanethiol	62155-04-8	6.81	6.52E+06	2	3834	29721	C16H32O2	hexadecanoic acid	57-10-3	7.17	1.48E+07	1
3755	27948	C14H30S	diheptyl sulfide	629-65-2	6.84	6.92E+06	2	3835	29722	C16H32O2	2-hexyldecanoic acid	25354-97-6	7.17	1.48E+07	2
3756	27949	C14H30S	methyl tridecyl sulfide	62155-09-3	6.84	6.92E+06	2	3836	29754	C16H34	hexadecane	544-76-3	8.25	1.78E+08	2
3757	27950	C14H30S	ethyl dodecyl sulfide	2851-83-4	6.84	6.92E+06	2	3837	29755	C16H34	2-methylpentadecane	1560-93-6	8.42	2.66E+08	2
3758	27951	C14H30S	propyl undecyl sulfide	66826-84-4	6.84	6.92E+06	2	3838	29756	C16H34	3-methylpentadecane	2882-96-4	8.42	2.65E+08	2
3759	27952	C14H30S	butyl decyl sulfide	19313-57-6	6.84	6.92E+06	2	3839	29757	C16H34	2,2-dimethyltetradecane	59222-86-5	8.43	2.67E+08	2
3760	27953	C14H30S2	diheptyl disulfide	10496-16-9	7.76	5.75E+07	2	3840	29758	C16H34	2,3-dimethyltetradecane	18435-23-9	8.42	2.65E+08	2

Table 13 Octanol-Water Partition Coefficient - Oganic Compounds

Octanol-Water Partition Coefficient								Octanol-Water Partition Coefficient							
NO	ID	FORMULA	NAME	CAS No	log_{10} Kow	Kow	code	NO	ID	FORMULA	NAME	CAS No	log_{10} Kow	Kow	code
3841	29759	C16H34	2,4-dimethyltetradecane	61868-06-2	8.41	2.56E+08	2	3921	31132	C18H38	9-methylheptadecane	18869-72-2	9.18	1.51E+09	2
3842	29760	C16H34	2,2,4,4,6,8,8-heptamethylnon	4390-04-9	8.39	2.47E+08	2	3922	31133	C18H38	2-methylheptadecane	1560-89-0	9.18	1.51E+09	2
3843	29766	C16H34O	1-hexadecanol	36653-82-4	6.73	5.37E+06	2	3923	31134	C18H38	3-methylheptadecane	6418-44-6	9.18	1.51E+09	2
3844	29767	C16H34O	2-hexadecanol	14852-31-4	7.18	1.52E+07	2	3924	31135	C18H38	2,2-dimethylhexadecane	19486-08-9	9.18	1.51E+09	2
3845	29768	C16H34O	3-hexadecanol	593-03-3	6.96	9.03E+06	2	3925	31136	C18H38	2,3-dimethylhexadecane	61868-02-8	9.18	1.51E+09	2
3846	29769	C16H34O	2-methyl-1-pentadecanol	25354-98-7	6.96	9.03E+06	2	3926	31137	C18H38	2,4-dimethylhexadecane	61868-08-4	9.18	1.51E+09	2
3847	29770	C16H34O	6-methyl-6-pentadecanol	108836-86-8	6.96	9.03E+06	2	3927	31140	C18H38O	1-octadecanol	112-92-5	7.70	5.01E+07	2
3848	29771	C16H34O	dioctyl ether	629-82-3	8.46	2.91E+08	2	3928	31141	C18H38O	2-octadecanol	593-32-8	8.65	4.47E+08	2
3849	29772	C16H34O	bis(2-ethylhexyl) ether	10143-60-9	8.46	2.91E+08	2	3929	31142	C18H38O	3-octadecanol, (±)	111897-18-8	8.18	1.50E+08	2
3850	29773	C16H34O	2-hexyl-1-decanol	2425-77-6	6.96	9.03E+06	2	3930	31143	C18H38O	dinonyl ether	2456-27-1	7.92	8.32E+07	2
3851	29789	C16H34S	1-hexadecanethiol	2917-26-2	8.14	1.38E+08	2	3931	31149	C18H38S	1-octadecanethiol	2885-00-9	9.12	1.32E+09	2
3852	29790	C16H34S	2-hexadecanethiol	66271-53-2	8.14	1.38E+08	2	3932	31150	C18H38S	2-octadecanethiol	62155-06-0	9.12	1.32E+09	2
3853	29791	C16H34S	dioctyl sulfide	2690-08-6	7.83	6.76E+07	2	3933	31151	C18H38S	2-thianonadecane	62155-11-7	9.12	1.32E+09	2
3854	29792	C16H34S	methyl pentadecyl sulfide	62155-10-6	7.83	6.76E+07	2	3934	31152	C18H38S	dinonyl sulfide	929-98-6	8.82	6.61E+08	2
3855	29793	C16H34S	ethyl tetradecyl sulfide	66271-54-3	7.83	6.76E+07	2	3935	31153	C18H38S	methyl heptadecyl sulfide	---	8.82	6.61E+08	2
3856	29794	C16H34S	propyl tridecyl sulfide	66271-55-4	7.83	6.76E+07	2	3936	31154	C18H38S	ethyl hexadecyl sulfide	66292-31-7	8.82	6.61E+08	2
3857	29795	C16H34S	butyl dodecyl sulfide	16900-08-6	7.83	6.76E+07	2	3937	31155	C18H38S	propyl pentadecyl sulfide	66292-32-8	8.82	6.61E+08	2
3858	29797	C16H34S2	dioctyl disulfide	822-27-5	8.74	5.50E+08	2	3938	31156	C18H38S	butyl tetradecyl sulfide	66292-33-9	8.82	6.61E+08	2
3859	29803	C16H35N	dimethyltetradecylamine	112-75-4	4.39	2.43E+04	2	3939	31157	C18H38S2	dinonyl disulfide	4485-77-2	9.72	5.25E+09	2
3860	29804	C16H35N	diethyldodecylamine	4271-27-6	4.39	2.43E+04	2	3940	31166	C18H39N	dimethylhexadecylamine	112-69-6	4.39	2.43E+04	2
3861	29862	C17H12	11H-benzo[a]fluorene	238-84-6	5.40	2.51E+05	1	3941	31167	C18H39N	diethyltetradecylamine	7307-58-6	4.39	2.43E+04	2
3862	29863	C17H12	11H-benzo[b]fluorene	243-17-4	5.75	5.62E+05	1	3942	31168	C18H39N	trihexylamine	102-86-3	4.40	2.53E+04	2
3863	30097	C17H19NO3	morphine	57-27-2	0.83	6.76E+00	1	3943	31344	C19H16O	triphenylmethanol	76-84-6	3.68	4.79E+03	1
3864	30264	C17H28	undecylbenzene	6742-54-7	8.14	1.38E+08	1	3944	31646	C19H32	tridecylbenzene	123-02-4	9.36	2.29E+09	1
3865	30284	C17H32	1-heptadecyne	26186-00-5	8.55	3.55E+08	2	3945	31647	C19H32	(1-hexylheptyl)benzene	2400-01-3	9.36	2.29E+09	2
3866	30285	C17H32	2-heptadecyne	61847-96-9	8.55	3.55E+08	2	3946	31675	C19H36O2	methyl oleate	112-62-9	8.02	1.05E+08	2
3867	30286	C17H32	3-heptadecyne	61886-63-3	8.55	3.55E+08	2	3947	31696	C19H38	tridecylcyclohexane	6006-33-3	9.49	3.09E+09	2
3868	30313	C17H34	1-heptadecene	6765-39-5	8.55	3.55E+08	2	3948	31698	C19H38	1-nonadecene	18435-45-5	9.54	3.47E+09	2
3869	30314	C17H34	2-methyl-1-hexadecene	61868-19-7	8.55	3.55E+08	2	3949	31699	C19H38	2-methyl-1-octadecene	61868-20-0	9.54	3.47E+09	2
3870	30325	C17H34O2	heptadecanoic acid	506-12-7	7.63	4.27E+07	2	3950	31708	C19H38O2	nonadecanoic acid	646-30-0	8.34	2.19E+08	2
3871	30346	C17H36	heptadecane	629-78-7	8.69	4.90E+08	2	3951	31709	C19H38O2	14-methyloctadecanoic acid	94434-64-7	8.34	2.19E+08	2
3872	30347	C17H36	2-methylhexadecane	1560-92-5	8.57	3.72E+08	2	3952	31710	C19H38O2	17-methyloctadecanoic acid	2724-59-6	8.34	2.19E+08	2
3873	30348	C17H36	3-methylhexadecane	6418-43-5	8.57	3.72E+08	2	3953	31711	C19H38O2	9-methyloctadecanoic acid	86073-38-3	8.34	2.19E+08	2
3874	30349	C17H36	2,2-dimethylpentadecane	61869-05-4	8.57	3.72E+08	2	3954	31733	C19H40	7-hexyltridecane	7225-66-3	9.67	4.68E+09	2
3875	30350	C17H36	2,3-dimethylpentadecane	2882-97-5	8.57	3.72E+08	2	3955	31734	C19H40	2,6,10,14-tetramethylpentade	1921-70-6	9.67	4.68E+09	2
3876	30351	C17H36	2,4-dimethylpentadecane	61868-07-3	8.45	2.82E+08	2	3956	31735	C19H40	nonadecane	629-92-5	9.67	4.68E+09	2
3877	30352	C17H36O	1-heptadecanol	1454-85-9	7.23	1.70E+07	2	3957	31736	C19H40	2-methyloctadecane	1560-88-9	9.67	4.68E+09	2
3878	30353	C17H36O	2-heptadecanol	16813-18-6	7.91	8.15E+07	2	3958	31737	C19H40	3-methyloctadecane	6561-44-0	9.67	4.68E+09	2
3879	30354	C17H36O	9-heptadecanol	624-08-8	7.57	3.72E+07	2	3959	31738	C19H40	2,2-dimethylheptadecane	53594-82-4	9.67	4.68E+09	2
3880	30355	C17H36O	heptadecanol (mixed primary	52783-44-5	7.57	3.72E+07	2	3960	31739	C19H40	2,3-dimethylheptadecane	61868-03-9	9.67	4.68E+09	2
3881	30364	C17H36S	methyl hexadecyl sulfide	27563-68-4	8.33	2.14E+08	2	3961	31740	C19H40	2,4-dimethylheptadecane	61868-09-5	9.67	4.68E+09	2
3882	30365	C17H36S	ethyl pentadecyl sulfide	64919-20-6	8.33	2.14E+08	2	3962	31742	C19H40O	1-nonadecanol	1454-84-8	9.45	2.83E+09	2
3883	30366	C17H36S	propyl tetradecyl sulfide	66577-61-5	8.33	2.14E+08	2	3963	31743	C19H40O	2-nonadecanol	26533-36-8	9.45	2.83E+09	2
3884	30367	C17H36S	butyl tridecyl sulfide	66577-62-6	8.33	2.14E+08	2	3964	31752	C19H40S	methyl octadecyl sulfide	40289-98-3	9.32	2.09E+09	2
3885	30372	C17H37N	dimethylpentadecylamine	17678-60-3	4.39	2.43E+04	2	3965	31753	C19H40S	ethyl heptadecyl sulfide	66359-40-8	9.32	2.09E+09	2
3886	30373	C17H37N	diethyltridecylamine	66577-48-8	4.39	2.43E+04	2	3966	31754	C19H40S	propyl hexadecyl sulfide	66359-41-9	9.32	2.09E+09	2
3887	30374	C17H37N	methyldioctylamine	4455-26-9	4.39	2.43E+04	2	3967	31755	C19H40S	butyl pentadecyl sulfide	66359-42-0	9.32	2.09E+09	2
3888	30406	C18H12	chrysene	218-01-9	5.73	5.37E+05	1	3968	31800	C20H12	perylene	198-55-0	6.25	1.78E+06	1
3889	30407	C18H12	benz[a]anthracene	56-55-3	5.91	8.13E+05	1	3969	31804	C20H12	benzo[a]pyrene	50-32-8	6.20	1.58E+06	1
3890	30409	C18H12	naphthacene	92-24-0	5.76	5.75E+05	1	3970	32346	C20H30O2	abietic acid	514-10-3	4.11	1.28E+04	2
3891	30410	C18H12	triphenylene	217-59-4	5.49	3.09E+05	1	3971	32387	C20H32O2	5,8,11,14-eicosatetraenoic ac	7771-44-0	6.98	9.55E+06	2
3892	30457	C18H14	m-terphenyl	92-06-8	5.52	3.31E+05	2	3972	32388	C20H32O2	5,8,11,14-eicosatetraenoic ac	506-32-1	6.98	9.55E+06	1
3893	30458	C18H14	o-terphenyl	84-15-1	5.52	3.31E+05	2	3973	32474	C20H40	tetradecylcyclohexane	1795-18-2	9.98	9.55E+09	2
3894	30459	C18H14	p-terphenyl	92-94-4	6.03	1.07E+06	1	3974	32476	C20H40	1-eicosene	3452-07-1	10.03	1.07E+10	2
3895	30548	C18H15N	triphenylamine	603-34-9	5.74	5.50E+05	1	3975	32477	C20H40	2-methyl-1-nonadecene	52254-50-9	10.03	1.07E+10	2
3896	30566	C18H15O4P	triphenyl phosphate	115-86-6	4.59	3.89E+04	1	3976	32491	C20H40O2	eicosanoic acid	506-30-9	9.29	1.95E+09	1
3897	30567	C18H15P	triphenylphosphine	603-35-0	5.69	4.90E+05	1	3977	32515	C20H42	eicosane	112-95-8	10.16	1.45E+10	2
3898	30588	C18H16N2	N,N'-diphenyl-p-phenylenedia	74-31-7	4.04	1.10E+04	2	3978	32516	C20H42	2-methylnonadecane	1560-86-7	10.16	1.45E+10	2
3899	30806	C18H22O2	dicumyl peroxide	80-43-3	5.50	3.16E+05	1	3979	32517	C20H42	3-methylnonadecane	6418-45-7	10.16	1.45E+10	2
3900	30947	C18H30	dodecylbenzene	123-01-3	8.65	4.47E+08	1	3980	32518	C20H42	2,2-dimethyloctadecane	61869-06-5	10.16	1.45E+10	2
3901	30948	C18H30	hexaethylbenzene	604-88-6	8.65	4.47E+08	2	3981	32519	C20H42	2,3-dimethyloctadecane	61868-04-0	10.16	1.45E+10	2
3902	30949	C18H30	1,2,4,5-tetraisopropylbenzene	635-11-0	8.65	4.47E+08	2	3982	32520	C20H42	2,4-dimethyloctadecane	61868-10-8	10.16	1.45E+10	2
3903	30952	C18H30	1,3,5-tri-tert-butylbenzene	1460-02-2	8.65	4.47E+08	2	3983	32522	C20H42O	1-eicosanol	629-96-9	8.70	5.01E+08	2
3904	30953	C18H30	1,2,4-tri-tert-butylbenzene	1459-11-6	8.65	4.47E+08	2	3984	32523	C20H42O	2-eicosanol	4340-76-5	8.70	5.01E+08	2
3905	30964	C18H30O2	linolenic acid	463-40-1	6.46	2.88E+06	1	3985	32525	C20H42O	2-octyl-1-dodecanol	5333-42-6	8.70	5.01E+08	2
3906	30991	C18H32O2	linoleic acid	60-33-3	7.05	1.12E+07	1	3986	32539	C20H42S	didecyl sulfide	693-83-4	9.81	6.46E+09	2
3907	31030	C18H34O2	oleic acid	112-80-1	7.64	4.37E+07	1	3987	32540	C20H42S	methyl nonadecyl sulfide	62155-12-8	9.81	6.46E+09	2
3908	31031	C18H34O2	elaidic acid	112-79-8	7.73	5.37E+07	2	3988	32541	C20H42S	ethyl octadecyl sulfide	41947-84-6	9.81	6.46E+09	2
3909	31032	C18H34O2	trans-2-octadecenoic acid	2825-79-8	7.73	5.37E+07	2	3989	32542	C20H42S	propyl heptadecyl sulfide	66455-35-4	9.81	6.46E+09	2
3910	31033	C18H34O2	trans-11-octadecenoic acid	693-72-1	7.73	5.37E+07	2	3990	32543	C20H42S	butyl hexadecyl sulfide	18437-89-3	9.81	6.46E+09	2
3911	31034	C18H34O2	petroselinic acid	593-39-5	7.73	5.37E+07	2	3991	32544	C20H42S2	didecyl disulfide	10496-18-1	10.70	5.01E+10	2
3912	31041	C18H34O4	dibutyl sebacate	109-43-3	6.30	2.00E+06	2	3992	32632	C21H16	1,2-dihydro-3-methylbenz[j]ac	56-49-5	6.75	5.62E+06	1
3913	31042	C18H34O4	dihexyl adipate	110-33-8	6.30	2.00E+06	2	3993	32787	C21H21O4P	tri-o-cresyl phosphate	78-30-8	6.87	7.39E+06	2
3914	31071	C18H36	1-octadecene	112-88-9	9.04	1.10E+09	2	3994	32998	C21H36	pentadecylbenzene	2131-18-2	9.42	2.63E+09	2
3915	31072	C18H36	9-octadecene	5557-31-3	9.04	1.10E+09	2	3995	33080	C22H12	benzo[ghi]perylene	191-24-2	6.90	7.94E+06	1
3916	31073	C18H36	2-methyl-1-heptadecene	42764-74-9	9.04	1.10E+09	2	3996	33401	C22H38	1-phenylhexadecane	1459-09-2	9.91	8.13E+09	2
3917	31075	C18H36	dodecylcyclohexane	1795-17-1	9.00	1.00E+09	2	3997	33466	C22H44O2	butyl stearate	123-95-5	9.70	5.01E+09	2
3918	31097	C18H36O2	octadecanoic acid	57-11-4	8.23	1.70E+08	1	3998	33467	C22H44O2	isobutyl stearate	646-13-9	9.70	5.01E+09	2
3919	31098	C18H36O2	2-octyldecanoic acid	619-39-6	8.23	1.70E+08	2	3999	33469	C22H44O2	ethyl eicosanoate	18281-05-5	9.70	5.01E+09	2
3920	31131	C18H38	octadecane	593-45-3	9.18	1.51E+09	2	4000	33470	C22H44O2	methyl heneicosanoate	6064-90-0	9.70	5.01E+09	2

Table 13 Octanol-Water Partition Coefficient - Oganic Compounds

Octanol-Water Partition Coefficient								Octanol-Water Partition Coefficient							
NO	ID	FORMULA	NAME	CAS No	$\log_{10}$ Kow	Kow	code	NO	ID	FORMULA	NAME	CAS No	$\log_{10}$ Kow	Kow	code
4001	33717	C24H12	coronene	191-07-1	6.05	1.12E+06	1	4003	33925	C24H38O4	dioctyl phthalate	117-84-0	7.60	3.98E+07	1
4002	33924	C24H38O4	diisooctyl phthalate	27554-26-3	8.39	2.45E+08	2	4004	34691	C28H46O4	diisodecyl phthalate	26761-40-0	10.36	2.29E+10	2

Code: 1 - data, 2 - estimate

Kow = (concentration in octanol phase, mg/kg) / (concentration in aqueous phase, mg/kg)

Values apply at ambient conditions (1 atm, 25 C)

Chapter 14

BIOCONCENTRATION FACTOR – ORGANIC COMPOUNDS

Carl L. Yaws and Kayzad J. Vajifdar
Lamar University, Beaumont, Texas

Tabulation Results

The results are given in Table 14. The tabulation is arranged by carbon number (C, C2, C3, ...) to provide ease of use in quickly locating the data by using the chemical formula. The compound name and CAS No (Chemical Abstracts Registry Number) are next provided. Values for bioconcentration factor are given in the adjacent columns. The next column provides the code for the tabulation.

The biconcentration factor is the ratio of a compound's concentration at equilibrium in an organism and its concentration in water:

$$\text{BCF} = \text{concentration in living organism} / \text{concentration in water} \qquad (14)$$

The concentration in the organism tissue and water should be the same units (mg/kg). The measurements for concentration must be made over a sufficient time to make sure that equilibrium is attained. Flow-through systems should be used to insure that the mean concentration is water is constant during the test (18).

The bioconcentration factor has special environmental significance since it represents the tendency of a compound to partition into living aquatic organism. High values for the bioconcentration factor indicate that the living organism will tend to extract the compound from the aqueous phase and concentrate it in the fatty tissue. Health hazards can result if the aquatic organism (such as fish containing the fatty tissue) is consumed.

In preparing the tabulation, a literature search was conducted to identify data source publications (1-36). Both experimental values for the property under consideration and parameter values for estimation of the property are included in the source publications. The publications were screened and copies of appropriate data were made. These data were then keyed into the computer to provide a database of values for compounds for which experimental data are available. The database also served as a basis to check the accuracy of the estimation methods. Upon completion of data collection, estimation of the values for the remaining compounds was performed.

The excellent compilations by Lyman, Reehl, and Rosenblatt (18); Mackay, Shiu, and Ma (19-23); Montgomery (25); Veight et. al. (33-35); and Verschueren (36) were primarily used for experimental data. Most of the estimates were based on the method of Veith and Kosian (1) which correlates bioconcentration factor as a function of octanol-water partition coefficient.

Example

In an industrial operation, small amounts of p-dichlorobenzene (C6H4Cl2) are leaked into water at ambient conditions. Determine the bioconcentration factor for p-dichlorobenzene.

Inspection of the tabulation for bioconcentration factor yields:

$$\log_{10} \text{BCF} = 1.78$$

$$\text{Kow} = 10^{1.78} = 60$$

The above bioconcentration factor applies for bluegill sunfish in equilibrium with aqueous phase.

References

1. Allen, D. T. and D. R. Shonnard, GREEN ENGINEERING, Prentice Hall PTR, Upper Saddle River, NJ (2002).
2. Appleton, H. T. and H. C. Sikka, "Accumulation, Elimination, Metabolism of Dichlorobenzidine in the Bluegill Sunfish, Environ. Sci. Technol., 14(1), 50-54 (1980).
3. Barrows, M. E. and others, "Bioconcentration and Elimination of Selected Water Pollutants by Bluegill Sunfish", DYNAMIC, EXPOSURE, HAZARD ASSESSMENT TOXIC CHEMICALS, Haque, R., editor, Ann Arbor Science Publishers, Ann Arbor, MI (1980).
4. Carlberg, G. E. and others, "Influence of Aquatic Humus on the Bioavailability of Chlorinated Micropollutants

in Atlantic Salmon", Arch. Environ. Contam. Toxicol., 15(5), 543-548 (1986).
5. Dauble, D. D. and others, "Bioaccumulation of Fossil Fuel Components during Single-Compound and Complex-Mixture Exposures of Daphnia Magna", Bull. Environ. Contam. Toxicol., 37(1), 125-132 (1986).
6. Davies, R. P. and A. J. Dobbs, "The Prediction of Bioconcentration in Fish", *Water Res.*, 18(10), 1253-1262 (1984).
7. Devillers, J. and others, "Comparison of BCF Models Based on log P", Chemosphere, 33(6), 1047-1065 (1996).
8. Freitag, D. and others, "Environmental Hazard Profile of Organic Chemicals", Chemosphere, 14(10), 1589-1616 (1985).
9. Garten, C. T. and J. R. Trabalka, "Evaluation of Models for Predicting Terrestrial Fod Chain Behavior of Xenobiotics", Environ. Sci. Technol., 17(10), 590-595 (1983).
10. Geyer, H. and others, "Prediction of Ecotoxicological Behavior of Chemicals – Relationship between Octanol-Water Partition Coefficient and Bioaccumulation of Organic Chemicals by Algae Clorella, Chemosphere, 13(2), 269-284 (1984).
11. Isnard, S. and S. Lambert, "Estimating Biocentration Factors from Octanol-Water Partition Coefficient and Aqueous Solubility", 17(1), 21-34 (1988).
12. Kalsch, W. and others, "Uptake, Elimination, and Bioconcentration of Ten Anilines in Zebrafish", Chemosphere, 22(3-4), 351-363 (1991).
13. Kenaga, E. E. and C. A. I. Goring, "Reltionship between Water Solubility, Soil Sorption, Octanol-Water Partition, and Concentration of Chemicals in Biota", AQUATIC TOXICOLOGY, ASTM STP 707, 78-115, ASTM, Philadelphia, PA (1980).
14. Khan, M. A. Q. and others, "Metabolism of Cyclodiene Insecticides by Fish", PESTICIDE AND XENOBIOTIC METABOLISM IN AQUATIC ORGANISMS, ACS Symposium Series 99, American Chemical Society, Washington, DC (1979).
15. Kobayshi, H., "Metabolism of Pentachlorophenol in Fish", PESTICIDE AND XENOBIOTIC METABOLISM IN AQUATIC ORGANISMS, ACS Symposium Series 99, American Chemical Society, Washington, DC (1979).
16. Lu, P. Y. and others, "Evaluation of Environmental Distribution and Fate of Hexachlorocyclopentadiene, Chlordene, Heptachlor, and Heptachlor Epoxide in a Laboratory Model Ecosystem", J. Agric. Food Chem., 23(5), 967-973 (1975).
17. Lu, P. Y. and R. L. Metcalf, "Environmental Fate and Biodegradability of Benzene Derivatives as Studied in a Model Ecosystem", Environ. Health Perspect., 10, 269-284 (1975).
18. Lyman, W. J., W. F. Reehl, and D. A. Rosenblatt, HANDBOOK OF CHEMICAL PROPERTY ESTIMATION METHODS, American Chemical Society, Washington, DC (1990).
19. Mackay, D., W. Y. Shiu, and K. C. Ma, ILLUSTRATED HANDBOOK OF PHYSICAL-CHEMICAL PROPERTIES AND ENVIRONMENTAL FATE FOR ORGANIC CHEMICALS, Volume 1 – Monoaromatic Hydrocarbons, Chlorobenzenes and PCBs, Lewis Publishers, New York, NY (1992).
20. Mackay, D., W. Y. Shiu, and K. C. Ma, ILLUSTRATED HANDBOOK OF PHYSICAL-CHEMICAL PROPERTIES AND ENVIRONMENTAL FATE FOR ORGANIC CHEMICALS, Volume 2 – Polynuclear Aromatic Hydrocarbons, Polychlorinated Dioxins, and Dibenzofurans, Lewis Publishers, New York, NY (1992).
21. Mackay, D., W. Y. Shiu, and K. C. Ma, ILLUSTRATED HANDBOOK OF PHYSICAL-CHEMICAL PROPERTIES AND ENVIRONMENTAL FATE FOR ORGANIC CHEMICALS, Volume 3 – Volatile Organic Compounds, Lewis Publishers, New York, NY (1993).
22. Mackay, D., W. Y. Shiu, and K. C. Ma, ILLUSTRATED HANDBOOK OF PHYSICAL-CHEMICAL PROPERTIES AND ENVIRONMENTAL FATE FOR ORGANIC CHEMICALS, Volume 4 – Oxygen, Nitrogen, and Sulfur Containing Compounds, Lewis Publishers, New York, NY (1995).
23. Mackay, D., W. Y. Shiu, and K. C. Ma, ILLUSTRATED HANDBOOK OF PHYSICAL-CHEMICAL PROPERTIES AND ENVIRONMENTAL FATE FOR ORGANIC CHEMICALS, Volume 5 – Pesticide Chemicals, Lewis Publishers, New York, NY (1997).
24. Mayer, F. L. and H. O. Sanders, "Toxicology of Phthalate Acid Esters in Aquatic Organisms", Environ. Health Perspect., 3, 153-157 (1973).
25. Montgomery, J. H., GROUNDWATER CHEMICALS, Lewis Publishers, New York, NY (2000).
26. Neely, W. B. and others, "Partition Coefficients to Measure Biocentration Potential of Organic Chemicals in Fish", Environ. Sci. Technol., 8, 1113-1115 (1974).
27. Ogata, M. and others, "Partition Coefficients as a Measure of Biocentration Potential of Crude Oil Compounds in Fish and Shellfish", Bull. Environ. Contam. Toxicol., 33(4), 561-567 (1984).
28. Oliver, B. G. and A. J. Niimi, "Bioconcentration of Chlorobenzenes from Water by Rainbow Trout", Environ. Sci. 17(5), 287-291 (1983).
29. Pereira, W. E. and others, "contamination of Esturine Water, Biota, and Sediment by Halogenated Organic Compounds – A Field Study", Environ. Sci. Technol., 22(7), 772-778 (1988).
30. Spacie, A. and others, "Uptake, Depuration, and Biotransformation of Anthracene and Bexzo[a]pyrene in Bluegill Sunfish", Ecotoxicol. Environ. Saf., 7(3), 330-341 (1983).
31. Stalling, D. L. and F. L. Meyer, "Toxicities of PCBs to Fish and Environmental Residues", Environ. Health

Perspect., 1, 159-164 (1972).

32. van Hattum, B. and others, "Polycyclic Aromatic Hydrocarbons in Freshwater Isopods and Field-Partitioning between Abiotic Phases", Arch. Environ. Contam. Toxicol., 35(2), 257-267 (1998).
33. Veith, G. D. and others, "Measuring and Estimating the *Bioconcentration Factor of Chemicals in Fish*", prepublication copy, J. Fish. Res. Board Can., 36, 1040-48 (1979).
34. Veith, G. D. and others, "An Evaluation of Using Parition Coefficients and Water Solubility to Estimate Bioconcentration Factors for Organic Chemicals in Fish", prepublication copy, J. Fish. Res. Board Can. (1980).
35. Veith, G. D. and others, "An Evaluation of Using Parition Coefficients and Water Solubility to Estimate Bioconcentration Factors for Organic Chemicals in Fish", AQUATIC TOXICOLOGY, ASTM STP 707, 116-129, ASTM, Philadelphia, PA (1980).
36. Verschueren, K., HANDBOOK OF ENVIRONMENTAL DATA ON ORGANIC CHEMICALS, 3rd and 4th eds., Van Nostrand Reinhold, New York, NY (1996, 2001).

Table 14 Bioconcentration Factor - Oganic Compounds

NO	ID	FORMULA	NAME	CAS No	$\log_{10}$ BCF	BCF	code
Bioconcentration Factor							
1	17	CBrF3	bromotrifluoromethane	75-63-8	1.07	1.17E+01	2
2	29	CBr4	carbon tetrabromide	558-13-4	2.30	2.00E+02	2
3	34	CClF3	chlorotrifluoromethane	75-72-9	0.90	8.01E+00	2
4	42	CCl2F2	dichlorodifluoromethane	75-71-8	1.31	2.02E+01	2
5	47	CCl3F	trichlorofluoromethane	75-69-4	1.60	3.97E+01	2
6	50	CCl4	carbon tetrachloride	56-23-5	1.48	3.00E+01	1
7	74	CF4	carbon tetrafluoride	75-73-0	0.53	3.41E+00	2
8	96	CHBr3	bromoform	75-25-2	1.48	3.02E+01	2
9	97	CHClF2	chlorodifluoromethane	75-45-6	0.45	2.84E+00	2
10	101	CHCl2F	dichlorofluoromethane	75-43-4	0.82	6.68E+00	2
11	104	CHCl3	chloroform	67-66-3	0.78	6.00E+00	1
12	113	CHF3	fluoroform	75-46-7	0.11	1.28E+00	2
13	140	CH2BrCl	bromochloromethane	74-97-5	0.71	5.17E+00	2
14	144	CH2Br2	dibromomethane	74-95-3	1.42	2.61E+01	2
15	145	CH2ClF	chlorofluoromethane	593-70-4	0.00	1.01E+00	2
16	149	CH2Cl2	dichloromethane	75-09-2	0.59	3.87E+00	2
17	158	CH2F2	difluoromethane	75-10-5	-0.24	5.73E-01	2
18	161	CH2I2	diiodomethane	75-11-6	1.58	3.76E+01	2
19	175	CH2O	formaldehyde	50-00-0	-0.12	7.52E-01	2
20	176	CH2O2	formic acid	64-18-6	-0.83	1.49E-01	2
21	191	CH3Br	methyl bromide	74-83-9	0.54	3.47E+00	2
22	196	CH3Cl	methyl chloride	74-87-3	0.32	2.08E+00	2
23	214	CH3F	methyl fluoride	593-53-3	0.00	1.01E+00	2
24	222	CH3I	methyl iodide	74-88-4	0.79	6.10E+00	2
25	232	CH3NO	formamide	75-12-7	-1.59	2.55E-02	2
26	235	CH3NO2	nitromethane	75-52-5	-0.66	2.18E-01	2
27	251	CH4	methane	74-82-8	0.46	2.89E+00	2
28	272	CH4O	methyl alcohol	67-56-1	-0.98	1.04E-01	2
29	286	CH5N	methylamine	74-89-5	-0.85	1.41E-01	2
30	382	CS2	carbon disulfide	75-15-0	1.29	1.95E+01	2
31	441	C2Cl2F4	1,2-dichloro-1,1,2,2-tetrafluor	76-14-2	1.83	6.73E+01	2
32	449	C2Cl3F3	1,1,2-trichloro-1,2,2-trifluoroe	76-13-1	2.10	1.25E+02	2
33	455	C2Cl4	tetrachloroethylene	127-18-4	1.69	4.90E+01	1
34	463	C2Cl6	hexachloroethane	67-72-1	2.14	1.39E+02	1
35	494	C2F6	hexafluoroethane	76-16-4	1.18	1.51E+01	2
36	511	C2HBrClF3	halothane	151-67-7	1.42	2.61E+01	2
37	539	C2HCl3	trichloroethylene	79-01-6	1.23	1.70E+01	1
38	544	C2HCl3O	trichloroacetaldehyde	75-87-6	0.38	2.41E+00	2
39	551	C2HCl5	pentachloroethane	76-01-7	1.83	6.70E+01	1
40	571	C2H2	acetylene	74-86-2	-0.11	7.80E-01	2
41	615	C2H2Cl2	1,1-dichloroethylene	75-35-4	1.28	1.92E+01	2
42	616	C2H2Cl2	cis-1,2-dichloroethylene	156-59-2	1.07	1.17E+01	2
43	617	C2H2Cl2	trans-1,2-dichloroethylene	156-60-5	1.12	1.33E+01	2
44	627	C2H2Cl2O2	dichloroacetic acid	79-43-6	0.33	2.12E+00	2
45	633	C2H2Cl3NO	2,2,2-trichloroacetamide	594-65-0	0.42	2.64E+00	2
46	636	C2H2Cl4	1,1,2,2-tetrachloroethane	79-34-5	0.90	7.94E+00	1
47	642	C2H2F2	1,1-difluoroethylene	75-38-7	0.58	3.80E+00	2
48	648	C2H2F3NO	trifluoroacetamide	354-38-1	-0.31	4.95E-01	2
49	688	C2H3Br	vinyl bromide	593-60-2	0.84	6.92E+00	2
50	693	C2H3BrO2	bromoacetic acid	79-08-3	-0.08	8.39E-01	2
51	701	C2H3Cl	vinyl chloride	75-01-4	0.69	4.90E+00	2
52	710	C2H3ClO2	chloroacetic acid	79-11-8	-0.23	5.94E-01	2
53	720	C2H3Cl3	1,1,1-trichloroethane	71-55-6	0.95	9.00E+00	1
54	721	C2H3Cl3	1,1,2-trichloroethane	79-00-5	1.48	3.02E+01	2
55	723	C2H3Cl3O	2,2,2-trichloroethanol	115-20-8	0.67	4.64E+00	2
56	738	C2H3F3O	2,2,2-trifluoroethanol	75-89-8	-0.08	8.39E-01	2
57	753	C2H3N	acetonitrile	75-05-8	-0.67	2.14E-01	2
58	787	C2H4	ethylene	74-85-1	0.49	3.11E+00	2
59	799	C2H4Br2	1,2-dibromoethane	106-93-4	1.15	1.41E+01	2
60	812	C2H4Cl2	1,1-dichloroethane	75-34-3	1.01	1.03E+01	2
61	813	C2H4Cl2	1,2-dichloroethane	107-06-2	0.30	2.00E+00	1
62	834	C2H4F2	1,1-difluoroethane	75-37-6	0.19	1.56E+00	2
63	845	C2H4I2	1,2-diiodoethane	624-73-7	1.74	5.51E+01	2
64	884	C2H4O	acetaldehyde	75-07-0	-0.04	9.03E-01	2
65	885	C2H4O	ethylene oxide	75-21-8	-0.64	2.31E-01	2
66	889	C2H4O2	acetic acid	64-19-7	-0.53	2.92E-01	2
67	890	C2H4O2	methyl formate	107-31-3	-0.38	4.20E-01	2
68	910	C2H5Br	bromoethane	74-96-4	0.86	7.31E+00	2
69	914	C2H5Cl	ethyl chloride	75-00-3	0.73	5.37E+00	2
70	917	C2H5ClO	2-chloroethanol	107-07-3	-0.38	4.20E-01	2
71	948	C2H5I	ethyl iodide	75-03-6	1.18	1.51E+01	2
72	956	C2H5NO	acetamide	60-35-5	-1.40	4.02E-02	2
73	957	C2H5NO	N-methylformamide	123-39-7	-1.17	6.82E-02	2
74	958	C2H5NO2	nitroethane	79-24-3	-0.26	5.52E-01	2
75	985	C2H6	ethane	74-84-0	1.03	1.07E+01	2
76	1065	C2H6O	ethyl alcohol	64-17-5	-0.64	2.31E-01	2
77	1066	C2H6O	dimethyl ether	115-10-6	-0.32	4.78E-01	2
78	1067	C2H6OS	dimethyl sulfoxide	67-68-5	-1.47	3.42E-02	2
79	1072	C2H6O2	ethylene glycol	107-21-1	-1.47	3.35E-02	2
80	1075	C2H6O2S	dimethyl sulfone	67-71-0	-1.51	3.06E-02	2

NO	ID	FORMULA	NAME	CAS No	$\log_{10}$ BCF	BCF	code
Bioconcentration Factor							
81	1083	C2H6O4S	dimethyl sulfate	77-78-1	0.52	3.28E+00	2
82	1094	C2H6S2	dimethyl disulfide	624-92-0	1.00	9.96E+00	2
83	1115	C2H7N	ethylamine	75-04-7	-0.50	3.14E-01	2
84	1116	C2H7N	dimethylamine	124-40-3	-0.70	1.99E-01	2
85	1117	C2H7NO	monoethanolamine	141-43-5	-1.43	3.67E-02	2
86	1159	C2H8N2	ethylenediamine	107-15-3	-2.01	9.74E-03	2
87	1210	C2N2	cyanogen	460-19-5	-0.34	4.52E-01	2
88	1287	C3F6O	hexafluoroacetone	684-16-2	0.75	5.67E+00	2
89	1377	C3H2N2	malononitrile	109-77-3	-0.87	1.34E-01	2
90	1462	C3H3N	acrylonitrile	107-13-1	1.68	4.80E+01	1
91	1463	C3H3NO	oxazole	288-42-6	-0.31	4.95E-01	2
92	1473	C3H3NS	thiazole	288-47-1	-0.05	8.86E-01	2
93	1492	C3H4	methylacetylene	74-99-7	0.34	2.20E+00	2
94	1493	C3H4	allene	463-49-0	0.75	5.57E+00	2
95	1535	C3H4Cl2	cis-1,2-dichloropropene	6923-20-2	1.20	1.60E+01	2
96	1592	C3H4N2	imidazole	288-32-4	-0.46	3.44E-01	2
97	1594	C3H4N2	1H-pyrazole	288-13-1	-0.30	5.04E-01	2
98	1621	C3H4O	acrolein	107-02-8	2.54	3.44E+02	1
99	1622	C3H4O	propargyl alcohol	107-19-7	-0.70	1.99E-01	2
100	1626	C3H4O2	acrylic acid	79-10-7	-0.12	7.52E-01	2
101	1645	C3H5Br	3-bromo-1-propene	106-95-6	1.01	1.03E+01	2
102	1670	C3H5Cl	2-chloro-1-propene	557-98-2	1.18	1.51E+01	2
103	1676	C3H5ClO	alpha-epichlorohydrin	106-89-8	-0.16	6.87E-01	2
104	1707	C3H5Cl3	1,2,3-trichloropropane	96-18-4	1.39	2.47E+01	2
105	1752	C3H5N	propionitrile	107-12-0	-0.27	5.33E-01	2
106	1755	C3H5NO	acrylamide	79-06-1	-1.02	9.63E-02	2
107	1757	C3H5NO	lactonitrile	78-97-7	-1.14	7.20E-02	2
108	1796	C3H5N3O9	nitroglycerine	55-63-0	0.88	7.58E+00	2
109	1809	C3H6	cyclopropane	75-19-4	0.96	9.09E+00	2
110	1810	C3H6	propylene	115-07-1	1.00	9.96E+00	2
111	1845	C3H6Cl2	1,2-dichloropropane	78-87-5	1.18	1.51E+01	2
112	1920	C3H6O	allyl alcohol	107-18-6	-0.27	5.42E-01	2
113	1922	C3H6O	propionaldehyde	123-38-6	0.07	1.16E+00	2
114	1923	C3H6O	acetone	67-64-1	-0.59	2.57E-01	2
115	1924	C3H6O	1,2-propylene oxide	75-56-9	-0.38	4.20E-01	2
116	1925	C3H6O	1,3-propylene oxide	503-30-0	-0.51	3.09E-01	2
117	1926	C3H6O	methyloxirane	16033-71-9	-0.38	4.20E-01	2
118	1930	C3H6O2	propanoic acid	79-09-4	-0.14	7.26E-01	2
119	1931	C3H6O2	ethyl formate	109-94-4	-0.22	6.05E-01	2
120	1932	C3H6O2	methyl acetate	79-20-9	-0.26	5.52E-01	2
121	1942	C3H6O2S	3-mercaptopropionic acid	107-96-0	-0.06	8.70E-01	2
122	1948	C3H6O3	lactic acid	50-21-5	-0.97	1.07E-01	2
123	1950	C3H6O3	trioxane	110-88-3	-0.74	1.82E-01	2
124	1978	C3H7Br	1-bromopropane	106-94-5	1.26	1.82E+01	2
125	1979	C3H7Br	2-bromopropane	75-26-3	1.10	1.26E+01	2
126	1986	C3H7Cl	1-chloropropane	540-54-5	1.21	1.63E+01	2
127	1987	C3H7Cl	2-chloropropane	75-29-6	1.10	1.26E+01	2
128	2025	C3H7I	1-iodopropane	107-08-4	1.58	3.76E+01	2
129	2026	C3H7I	2-iodopropane	75-30-9	1.88	7.64E+01	2
130	2030	C3H7N	allylamine	107-11-9	-0.38	4.20E-01	2
131	2036	C3H7NO	N,N-dimethylformamide	68-12-2	-1.20	6.34E-02	2
132	2037	C3H7NO	N-methylacetamide	79-16-3	-1.23	5.90E-02	2
133	2045	C3H7NO2	1-nitropropane	108-03-2	0.29	1.94E+00	2
134	2046	C3H7NO2	2-nitropropane	79-46-9	0.33	2.16E+00	2
135	2094	C3H8	propane	74-98-6	1.46	2.91E+01	2
136	2129	C3H8O	propyl alcohol	71-23-8	-0.20	6.27E-01	2
137	2130	C3H8O	isopropyl alcohol	67-63-0	-0.36	4.36E-01	2
138	2138	C3H8O2	2-methoxyethanol	109-86-4	-1.01	9.81E-02	2
139	2140	C3H8O2	1,2-propanediol (propylene g	57-55-6	-1.13	7.47E-02	2
140	2141	C3H8O2	1,3-propanediol	504-63-2	-1.22	6.00E-02	2
141	2148	C3H8O3	glycerol	56-81-5	-1.79	1.62E-02	2
142	2158	C3H8S	propyl mercaptan	107-03-9	1.03	1.07E+01	2
143	2160	C3H8S	methyl ethyl sulfide	624-89-5	0.82	6.56E+00	2
144	2200	C3H9N	propylamine	107-10-8	-0.02	9.53E-01	2
145	2201	C3H9N	isopropylamine	75-31-0	-0.19	6.39E-01	2
146	2202	C3H9N	methylethylamine	624-78-2	-0.28	5.23E-01	2
147	2203	C3H9N	trimethylamine	75-50-3	-0.27	5.33E-01	2
148	2204	C3H9NO	1-amino-2-propanol	78-96-6	-1.16	6.94E-02	2
149	2205	C3H9NO	3-amino-1-propanol	156-87-6	-1.28	5.19E-02	2
150	2206	C3H9NO	methylethanolamine	109-83-1	-1.14	7.20E-02	2
151	2241	C3H9O4P	trimethyl phosphate	512-56-1	-0.91	1.22E-01	2
152	2318	C4Cl6	hexachloro-1,3-butadiene	87-68-3	4.55	3.55E+04	1
153	2562	C4H4N2	succinonitrile	110-61-2	-1.18	6.58E-02	2
154	2565	C4H4N2	pyrimidine	289-95-2	-0.72	1.92E-01	2
155	2592	C4H4O	furan	110-00-9	0.66	4.56E+00	2
156	2605	C4H4O4	fumaric acid	110-17-8	-0.04	9.19E-01	2
157	2606	C4H4O4	maleic acid	110-16-7	-0.78	1.66E-01	2
158	2613	C4H4S	thiophene	110-02-1	1.03	1.07E+01	2
159	2697	C4H5N	methacrylonitrile	126-98-7	0.14	1.37E+00	2
160	2698	C4H5N	vinylacetonitrile	109-75-1	-0.08	8.24E-01	2

Table 14 Bioconcentration Factor - Oganic Compounds

Bioconcentration Factor								Bioconcentration Factor							
NO	ID	FORMULA	NAME	CAS No	$\log_{10}$ BCF	BCF	code	NO	ID	FORMULA	NAME	CAS No	$\log_{10}$ BCF	BCF	code
161	2701	C4H5N	pyrrole	109-97-7	0.19	1.56E+00	2	241	5183	C5H9N	valeronitrile	110-59-8	0.34	2.20E+00	2
162	2713	C4H5NO2	methyl cyanoacetate	105-34-0	-0.77	1.69E-01	2	242	5198	C5H9NO	N-methyl-2-pyrrolidone	872-50-4	-0.70	1.99E-01	2
163	2754	C4H6	dimethylacetylene	503-17-3	0.75	5.67E+00	2	243	5245	C5H9NO4	L-glutamic acid	56-86-0	-3.32	4.84E-04	2
164	2757	C4H6	butadiene (1,3 butadiene)	106-99-0	1.17	1.49E+01	2	244	5285	C5H10	cyclopentane	287-92-3	1.97	9.33E+01	2
165	2942	C4H6O	2,5-dihydrofuran	1708-29-8	-0.04	9.19E-01	2	245	5450	C5H10O	methyl propyl ketone	107-87-9	0.26	1.83E+00	2
166	2965	C4H6O2	gamma-butyrolactone	96-48-0	-0.91	1.24E-01	2	246	5451	C5H10O	diethyl ketone	96-22-0	0.25	1.77E+00	2
167	2969	C4H6O2	methacrylic acid	79-41-4	0.33	2.16E+00	2	247	5452	C5H10O	methyl isopropyl ketone	563-80-4	0.04	1.10E+00	2
168	2970	C4H6O2	methyl acrylate	96-33-3	0.23	1.71E+00	2	248	5453	C5H10O	2-methyltetrahydrofuran	96-47-9	1.06	1.15E+01	2
169	2971	C4H6O2	vinyl acetate	108-05-4	0.18	1.50E+00	2	249	5471	C5H10O	tetrahydropyran	142-68-7	0.25	1.77E+00	2
170	3006	C4H6O4	succinic acid	110-15-6	-0.87	1.36E-01	2	250	5492	C5H10O2	pentanoic acid	109-52-4	0.70	4.99E+00	2
171	3155	C4H7Cl2O4P	dichlorvos	62-73-7	-0.10	7.94E-01	1	251	5494	C5H10O2	3-methylbutanoic acid	503-74-2	0.52	3.28E+00	2
172	3238	C4H7N	butyronitrile	109-74-0	0.07	1.19E+00	2	252	5501	C5H10O2	propyl acetate	109-60-4	0.58	3.80E+00	2
173	3239	C4H7N	isobutyronitrile	78-82-0	-0.04	9.19E-01	2	253	5503	C5H10O2	ethyl propanoate	105-37-3	0.56	3.60E+00	2
174	3249	C4H7NO	2-pyrrolidone	616-45-5	-1.07	8.48E-02	2	254	5504	C5H10O2	methyl butanoate	623-42-7	0.62	4.16E+00	2
175	3320	C4H8	1-butene	106-98-9	1.50	3.13E+01	2	255	5543	C5H10O3	diethyl carbonate	105-58-8	0.56	3.60E+00	2
176	3322	C4H8	cis-2-butene	590-18-1	1.44	2.76E+01	2	256	5603	C5H11Br	1-bromopentane	110-53-2	2.26	1.83E+02	2
177	3323	C4H8	trans-2-butene	624-64-6	1.42	2.66E+01	2	257	5620	C5H11Cl	1-chloropentane	543-59-9	1.76	5.71E+01	2
178	3324	C4H8	isobutene	115-11-7	1.46	2.86E+01	2	258	5625	C5H11Cl	2-chloro-2-methylbutane	594-36-5	1.59	3.90E+01	2
179	3375	C4H8Cl2O	bis(2-chloroethyl) ether	111-44-4	1.04	1.10E+01	1	259	5649	C5H11F	1-fluoropentane	592-50-7	1.44	2.76E+01	2
180	3476	C4H8O	ethyl vinyl ether	109-92-2	0.42	2.64E+00	2	260	5669	C5H11N	N-methylpyrrolidine	120-94-5	0.33	2.12E+00	2
181	3478	C4H8O	butyraldehyde	123-72-8	0.30	1.97E+00	2	261	5670	C5H11N	piperidine	110-89-4	0.26	1.83E+00	2
182	3480	C4H8O	methyl ethyl ketone	78-93-3	-0.17	6.75E-01	2	262	5696	C5H11NO2	1-nitropentane	628-05-7	1.19	1.54E+01	2
183	3481	C4H8O	tetrahydrofuran	109-99-9	-0.04	9.19E-01	2	263	5772	C5H12	pentane	109-66-0	2.33	2.12E+02	2
184	3504	C4H8O2	butyric acid	107-92-6	0.22	1.68E+00	2	264	5773	C5H12	isopentane	78-78-4	1.42	2.61E+01	2
185	3505	C4H8O2	isobutyric acid	79-31-2	0.34	2.20E+00	2	265	5774	C5H12	neopentane	463-82-1	2.06	1.14E+02	2
186	3506	C4H8O2	propyl formate	110-74-7	0.26	1.80E+00	2	266	5831	C5H12O	1-pentanol	71-41-0	0.79	6.21E+00	2
187	3508	C4H8O2	ethyl acetate	141-78-6	1.48	3.02E+01	1	267	5832	C5H12O	2-pentanol	6032-29-7	0.59	3.87E+00	2
188	3509	C4H8O2	methyl propanoate	554-12-1	0.26	1.83E+00	2	268	5833	C5H12O	3-pentanol	584-02-1	0.56	3.60E+00	2
189	3510	C4H8O2	1,4-dioxane	123-91-1	-0.73	1.85E-01	2	269	5834	C5H12O	2-methyl-1-butanol	137-32-6	0.62	4.16E+00	2
190	3536	C4H8O2S	sulfolane	126-33-0	-1.01	9.81E-02	2	270	5837	C5H12O	3-methyl-1-butanol	123-51-3	0.61	4.09E+00	2
191	3598	C4H9Br	1-bromobutane	109-65-9	1.77	5.92E+01	2	271	5838	C5H12O	tert-pentyl-alcohol	75-85-4	0.30	2.01E+00	2
192	3610	C4H9Cl	1-chlorobutane	109-69-3	1.69	4.85E+01	2	272	5839	C5H12O	3-methyl-2-butanol	598-75-4	0.61	4.09E+00	2
193	3612	C4H9Cl	2-chlorobutane	78-86-4	1.44	2.76E+01	2	273	5840	C5H12O	2,2-dimethyl-1-propanol	75-84-3	0.63	4.31E+00	2
194	3657	C4H9F	1-fluorobutane	2366-52-1	1.64	4.35E+01	2	274	5844	C5H12O	methyl tert-butyl ether	1634-04-4	0.34	2.20E+00	2
195	3666	C4H9I	1-iodobutane	542-69-8	1.97	9.33E+01	2	275	5912	C5H12O4	pentaerythritol	115-77-5	-1.74	1.84E-02	2
196	3675	C4H9N	pyrrolidine	123-75-1	-0.04	9.19E-01	2	276	5947	C5H13N	pentylamine	110-58-7	0.78	5.99E+00	2
197	3681	C4H9NO	N,N-dimethylacetamide	127-19-5	-1.01	9.81E-02	2	277	6067	C6Cl6	hexachlorobenzene	118-74-1	3.73	5.37E+03	1
198	3682	C4H9NO	morpholine	110-91-8	-1.82	1.51E-02	2	278	6080	C6F6	hexafluorobenzene	392-56-3	1.61	4.12E+01	2
199	3685	C4H9NO	butanamide	541-35-5	-0.57	2.72E-01	2	279	6115	C6HCl5	pentachlorobenzene	608-93-5	3.53	3.40E+03	1
200	3697	C4H9NO2	1-nitrobutane	627-05-4	0.76	5.77E+00	2	280	6116	C6HCl5O	pentachlorophenol	87-86-5	2.89	7.70E+02	1
201	3783	C4H10	butane	106-97-8	1.88	7.64E+01	2	281	6150	C6H2Br4	1,2,4,5-tetrabromobenzene	636-28-2	3.57	3.72E+03	1
202	3830	C4H10N2	piperazine	110-85-0	-1.32	4.74E-02	2	282	6174	C6H2Cl4	1,2,3,4-tetrachlorobenzene	634-66-2	3.89	7.76E+03	1
203	3865	C4H10O	butanol	71-36-3	0.26	1.83E+00	2	283	6175	C6H2Cl4	1,2,3,5-tetrachlorobenzene	634-90-2	3.26	1.80E+03	1
204	3866	C4H10O	isobutanol	78-83-1	0.20	1.59E+00	2	284	6176	C6H2Cl4	1,2,4,5-tetrachlorobenzene	95-94-3	4.90	7.94E+04	1
205	3867	C4H10O	sec-butanol	78-92-2	0.11	1.30E+00	2	285	6259	C6H3Br3	1,3,5-tribromobenzene	626-39-1	3.97	9.33E+03	1
206	3868	C4H10O	tert-butanol	75-65-0	-0.12	7.52E-01	2	286	6281	C6H3ClN2O4	1-chloro-2,4-dinitrobenzene	97-00-7	1.31	2.06E+01	2
207	3869	C4H10O	diethyl ether	60-29-7	0.30	2.01E+00	2	287	6296	C6H3Cl2NO2	1,2-dichloro-4-nitrobenzene	99-54-7	2.06	1.16E+02	2
208	3870	C4H10O	methyl propyl ether	557-17-5	0.56	3.60E+00	2	288	6305	C6H3Cl3	1,2,4-trichlorobenzene	120-82-1	2.95	8.91E+02	1
209	3882	C4H10O2	1,2-dimethoxyethane	110-71-4	-0.57	2.72E-01	2	289	6306	C6H3Cl3	1,2,3-trichlorobenzene	87-61-6	4.49	3.09E+04	1
210	3883	C4H10O2	2-ethoxyethanol	110-80-5	-0.65	2.22E-01	2	290	6307	C6H3Cl3	1,3,5-trichlorobenzene	108-70-3	4.22	1.66E+04	1
211	3886	C4H10O2	1,4-butanediol	110-63-4	-1.06	8.80E-02	2	291	6313	C6H3Cl3O	2,4,5-trichlorophenol	95-95-4	3.27	1.86E+03	1
212	3937	C4H10O4S	diethyl sulfate	64-67-5	0.50	3.17E+00	2	292	6314	C6H3Cl3O	2,4,6-trichlorophenol	88-06-2	2.75	5.62E+02	1
213	3942	C4H10S	butyl mercaptan	109-79-5	1.40	2.52E+01	2	293	6367	C6H3N3O6	1,3,5-trinitrobenzene	99-35-4	0.53	3.41E+00	2
214	3947	C4H10S	diethyl sulfide	352-93-2	1.14	1.38E+01	2	294	6411	C6H4Br2	m-dibromobenzene	108-36-1	2.56	3.65E+02	2
215	3977	C4H11N	butylamine	109-73-9	0.28	1.90E+00	2	295	6443	C6H4ClNO2	m-chloronitrobenzene	121-73-3	1.54	3.49E+01	2
216	3978	C4H11N	isobutylamine	78-81-9	0.18	1.50E+00	2	296	6444	C6H4ClNO2	o-chloronitrobenzene	88-73-3	1.37	2.34E+01	2
217	3980	C4H11N	tert-butylamine	75-64-9	-0.08	8.24E-01	2	297	6445	C6H4ClNO2	p-chloronitrobenzene	100-00-5	1.49	3.08E+01	2
218	3984	C4H11N	diethylamine	109-89-7	0.06	1.14E+00	2	298	6469	C6H4Cl2	o-dichlorobenzene	95-50-1	1.95	8.90E+01	1
219	4006	C4H11NO2	diethanolamine	111-42-2	-1.53	2.95E-02	2	299	6470	C6H4Cl2	m-dichlorobenzene	541-73-1	1.82	6.60E+01	1
220	4169	C5Cl6	hexachlorocyclopentadiene	77-47-4	1.47	2.95E+01	1	300	6471	C6H4Cl2	p-dichlorobenzene	106-46-7	1.78	6.00E+01	1
221	4327	C5H4O2	furfural	98-01-1	-0.08	8.39E-01	2	301	6480	C6H4Cl2O	2,4-dichlorophenol	120-83-2	1.53	3.39E+01	1
222	4391	C5H5N	pyridine	110-86-1	0.11	1.30E+00	2	302	6524	C6H4F2	m-difluorobenzene	372-18-9	1.35	2.22E+01	2
223	4490	C5H6N2	glutaronitrile	544-13-8	-0.97	1.07E-01	2	303	6525	C6H4F2	o-difluorobenzene	367-11-3	1.47	2.97E+01	2
224	4534	C5H6O	2-methylfuran	534-22-5	1.06	1.15E+01	2	304	6526	C6H4F2	p-difluorobenzene	540-36-3	1.28	1.92E+01	2
225	4548	C5H6O2	furfuryl alcohol	98-00-0	-0.18	6.63E-01	2	305	6580	C6H4N2O4	m-dinitrobenzene	99-65-0	0.78	5.99E+00	2
226	4587	C5H6S	2-methylthiophene	554-14-3	1.44	2.76E+01	2	306	6581	C6H4N2O4	o-dinitrobenzene	528-29-0	0.94	8.61E+00	2
227	4588	C5H6S	3-methylthiophene	616-44-4	1.45	2.81E+01	2	307	6582	C6H4N2O4	p-dinitrobenzene	100-25-4	0.75	5.67E+00	2
228	4658	C5H7N	N-methylpyrrole	96-54-8	0.56	3.60E+00	2	308	6656	C6H5Br	bromobenzene	108-86-1	1.96	9.16E+01	2
229	4738	C5H8	isoprene	78-79-5	1.51	3.25E+01	2	309	6692	C6H5Cl	chlorobenzene	108-90-7	2.65	4.50E+02	1
230	4741	C5H8	cis-1,3-pentadiene	1574-41-0	1.50	3.13E+01	2	310	6720	C6H5ClO	m-chlorophenol	108-43-0	1.58	3.76E+01	2
231	4742	C5H8	trans-1,3-pentadiene	2004-70-8	1.53	3.37E+01	2	311	6721	C6H5ClO	o-chlorophenol	95-57-8	2.33	2.14E+02	1
232	4743	C5H8	1,4-pentadiene	591-93-5	1.56	3.62E+01	2	312	6722	C6H5ClO	p-chlorophenol	106-48-9	1.49	3.08E+01	2
233	4745	C5H8	1-pentyne	627-19-0	1.16	1.46E+01	2	313	6740	C6H5Cl2N	3,4-dichloroaniline	95-76-1	1.73	5.31E+01	2
234	4882	C5H8O2	acetylacetone	123-54-6	-0.08	8.24E-01	2	314	6769	C6H5F	fluorobenzene	462-06-6	1.39	2.47E+01	2
235	4883	C5H8O2	allyl acetate	591-87-7	0.37	2.32E+00	2	315	6795	C6H5I	iodobenzene	591-50-4	2.19	1.55E+02	2
236	4888	C5H8O2	ethyl acrylate	140-88-5	0.64	4.39E+00	2	316	6819	C6H5NO2	nitrobenzene	98-95-3	1.18	1.50E+01	1
237	4889	C5H8O2	methyl methacrylate	80-62-6	0.69	4.90E+00	2	317	6830	C6H5NO3	o-nitrophenol	88-75-5	1.60	3.98E+01	1
238	4927	C5H8O3	2-hydroxyethyl acrylate	818-61-1	-0.57	2.72E-01	2	318	6832	C6H5NO3	p-nitrophenol	100-02-7	1.60	3.98E+01	1
239	4928	C5H8O3	levulinic acid	123-76-2	-0.79	1.63E-01	2	319	6876	C6H6	benzene	71-43-2	0.63	4.27E+00	1
240	4945	C5H8O4	glutaric acid	110-94-1	-0.63	2.35E-01	2	320	6923	C6H6ClN	m-chloroaniline	108-42-9	1.09	1.22E+01	2

Table 14 Bioconcentration Factor - Oganic Compounds

Bioconcentration Factor							
NO	ID	FORMULA	NAME	CAS No	log_{10} BCF	BCF	code
321	6924	C6H6ClN	o-chloroaniline	95-51-2	1.10	1.26E+01	2
322	6925	C6H6ClN	p-chloroaniline	106-47-8	0.91	8.13E+00	1
323	6948	C6H6Cl6	1,2,3,4,5,6-hexachlorocycloh	58-89-9	2.65	4.47E+02	1
324	6949	C6H6Cl6	1,2,3,4,5,6-hexachlorocycloh	319-85-7	2.86	7.24E+02	1
325	6953	C6H6Cl6	a-hexachlorocyclohexane	319-84-6	2.79	6.17E+02	1
326	6996	C6H6N2O2	m-nitroaniline	99-09-2	0.68	4.81E+00	2
327	6997	C6H6N2O2	o-nitroaniline	88-74-4	0.91	8.13E+00	1
328	6998	C6H6N2O2	p-nitroaniline	100-01-6	0.64	4.37E+00	1
329	7057	C6H6O	phenol	108-95-2	1.30	2.00E+01	1
330	7070	C6H6O2	pyrocatechol	120-80-9	0.30	1.97E+00	2
331	7071	C6H6O2	resorcinol	108-46-3	0.23	1.71E+00	2
332	7072	C6H6O2	p-hydroquinone	123-31-9	1.60	3.98E+01	1
333	7121	C6H6S	phenyl mercaptan	108-98-5	1.59	3.90E+01	2
334	7172	C6H7N	aniline	62-53-3	0.78	6.03E+00	1
335	7173	C6H7N	2-methylpyridine	109-06-8	0.48	3.00E+00	2
336	7174	C6H7N	3-methylpyridine	108-99-6	0.55	3.53E+00	2
337	7175	C6H7N	4-methylpyridine	108-89-4	0.56	3.66E+00	2
338	7269	C6H8	1,4-cyclohexadiene	628-41-1	1.55	3.56E+01	2
339	7316	C6H8N2	adiponitrile	111-69-3	-0.65	2.22E-01	2
340	7318	C6H8N2	m-phenylenediamine	108-45-2	-0.66	2.18E-01	2
341	7319	C6H8N2	o-phenylenediamine	95-54-5	-0.28	5.23E-01	2
342	7320	C6H8N2	p-phenylenediamine	106-50-3	-0.64	2.31E-01	2
343	7321	C6H8N2	phenylhydrazine	100-63-0	0.59	3.87E+00	2
344	7381	C6H8O	2-ethylfuran	3208-16-0	1.50	3.13E+01	2
345	7387	C6H8O	2-cyclohexen-1-one	930-68-7	0.08	1.21E+00	2
346	7396	C6H8O	5-hexyn-2-one	2550-28-9	0.06	1.14E+00	2
347	7459	C6H8O6	ascorbic acid	50-81-7	-1.70	2.02E-02	2
348	7464	C6H8O7	citric acid	77-92-9	-1.76	1.74E-02	2
349	7622	C6H10	cyclohexene	110-83-8	1.86	7.23E+01	2
350	7629	C6H10	1,5-hexadiene	592-42-7	1.81	6.49E+01	2
351	7632	C6H10	cis-2,trans-4-hexadiene	5194-50-3	1.81	6.49E+01	2
352	7634	C6H10	trans-2,trans-4-hexadiene	5194-51-4	1.98	9.50E+01	2
353	7648	C6H10	1-hexyne	693-02-7	1.76	5.71E+01	2
354	7760	C6H10O	cyclohexanone	108-94-1	0.24	1.74E+00	2
355	7776	C6H10O	5-hexen-2-one	109-49-9	0.41	2.55E+00	2
356	7814	C6H10O2	ethyl methacrylate	97-63-2	1.13	1.36E+01	2
357	7881	C6H10O3	ethylacetoacetate	141-97-9	-0.20	6.27E-01	2
358	7917	C6H10O4	adipic acid	124-04-9	-0.34	4.60E-01	2
359	7918	C6H10O4	diethyl oxalate	95-92-1	0.04	1.10E+00	2
360	7981	C6H11Br	bromocyclohexane	108-85-0	2.13	1.34E+02	2
361	8088	C6H11N	hexanenitrile	628-73-9	0.86	7.31E+00	2
362	8101	C6H11NO	epsilon-caprolactam	105-60-2	-0.55	2.82E-01	2
363	8102	C6H11NO	cyclohexanone oxime	100-64-1	0.26	1.83E+00	2
364	8214	C6H12	methylcyclopentane	96-37-7	2.26	1.83E+02	2
365	8215	C6H12	cyclohexane	110-82-7	2.32	2.08E+02	2
366	8216	C6H12	1-hexene	592-41-6	2.29	1.93E+02	2
367	8225	C6H12	4-methyl-1-pentene	691-37-2	1.58	3.76E+01	2
368	8400	C6H12O	cyclohexanol	108-93-0	0.57	3.73E+00	2
369	8401	C6H12O	hexanal	66-25-1	1.01	1.01E+01	2
370	8409	C6H12O	2-hexanone	591-78-6	0.69	4.90E+00	2
371	8412	C6H12O	4-methyl-2-pentanone	108-10-1	0.63	4.31E+00	2
372	8460	C6H12O2	hexanoic acid	142-62-1	1.12	1.31E+01	2
373	8461	C6H12O2	2-ethyl butyric acid	88-09-5	0.93	8.46E+00	2
374	8476	C6H12O2	butyl acetate	123-86-4	1.04	1.09E+01	2
375	8477	C6H12O2	isobutyl acetate	110-19-0	1.01	1.01E+01	2
376	8478	C6H12O2	sec-butyl acetate	105-46-4	0.96	9.09E+00	2
377	8479	C6H12O2	tert-butyl acetate	540-88-5	0.99	9.78E+00	2
378	8532	C6H12O3	hydroxycaproic acid	1191-25-9	0.24	1.74E+00	2
379	8533	C6H12O3	paraldehyde	123-63-7	0.13	1.35E+00	2
380	8609	C6H12O6	glucose	50-99-7	-2.96	1.10E-03	2
381	8647	C6H13Br	1-bromohexane	111-25-1	2.60	4.00E+02	2
382	8742	C6H13N	cyclohexylamine	108-91-8	0.78	5.99E+00	2
383	8877	C6H14	hexane	110-54-3	2.76	5.75E+02	2
384	8878	C6H14	2,2-dimethylbutane	75-83-2	2.62	4.15E+02	2
385	8879	C6H14	2,3-dimethylbutane	79-29-8	2.64	4.38E+02	2
386	8881	C6H14	3-methylpentane	96-14-0	2.44	2.78E+02	2
387	8944	C6H14N2O2	lysine	56-87-1	-2.81	1.55E-03	2
388	8970	C6H14O	1-hexanol	111-27-3	1.20	1.60E+01	2
389	8971	C6H14O	2-hexanol	626-93-7	0.99	9.78E+00	2
390	8972	C6H14O	3-hexanol	623-37-0	0.90	8.01E+00	2
391	8987	C6H14O	3,3-dimethyl-2-butanol	464-07-3	0.77	5.88E+00	2
392	8993	C6H14O	dipropyl ether	111-43-3	1.20	1.60E+01	2
393	8995	C6H14O	diisopropyl ether	108-20-3	0.80	6.32E+00	2
394	9004	C6H14O	ethyl butyl ether	628-81-9	1.20	1.60E+01	2
395	9020	C6H14O2	acetal	105-57-7	0.26	1.83E+00	2
396	9021	C6H14O2	2-butoxyethanol	111-76-2	0.26	1.80E+00	2
397	9069	C6H14O2S	dipropyl sulfone	598-03-8	-0.09	8.09E-01	2
398	9075	C6H14O3	diethylene glycol dimethyl eth	111-96-6	-0.68	2.07E-01	2
399	9077	C6H14O3	2-(2-ethoxyethoxy)ethanol	111-90-0	-0.83	1.49E-01	2
400	9078	C6H14O3	trimethylolpropane	77-99-6	-1.57	2.70E-02	2

Bioconcentration Factor							
NO	ID	FORMULA	NAME	CAS No	log_{10} BCF	BCF	code
401	9100	C6H14O6	sorbitol	50-70-4	-2.14	7.28E-03	2
402	9173	C6H15N	hexylamine	111-26-2	1.23	1.69E+01	2
403	9202	C6H15N	di-propylamine	142-84-7	0.92	8.30E+00	2
404	9203	C6H15N	diisopropylamine	108-18-9	0.71	5.08E+00	2
405	9212	C6H15N	triethylamine	121-44-8	0.75	5.57E+00	2
406	9230	C6H15NO2	diisopropanolamine	110-97-4	-1.05	8.96E-02	2
407	9239	C6H15NO3	triethanolamine	102-71-6	-1.19	6.46E-02	2
408	9268	C6H15O4P	triethyl phosphate	78-40-0	0.23	1.71E+00	2
409	9356	C6H18N3OP	hexamethyl phosphoramide	680-31-9	-0.18	6.63E-01	2
410	9362	C6H18OSi2	hexamethyldisiloxane	107-46-0	2.92	8.28E+02	2
411	9799	C7H4F3NO2	3-nitrobenzotrifluoride	98-46-4	1.67	4.68E+01	2
412	9890	C7H5BrO2	2-bromobenzoic acid	88-65-3	1.34	2.18E+01	2
413	9891	C7H5BrO2	3-bromobenzoic acid	585-76-2	1.87	7.37E+01	2
414	9892	C7H5BrO2	4-bromobenzoic acid	586-76-5	1.86	7.23E+01	2
415	9932	C7H5ClO2	o-chlorobenzoic acid	118-91-2	1.22	1.66E+01	2
416	10003	C7H5F3	benzotrifluoride	98-08-8	2.01	1.02E+02	2
417	10045	C7H5N	benzonitrile	100-47-0	0.83	6.80E+00	2
418	10094	C7H5NS	benzothiazole	95-16-9	1.19	1.54E+01	2
419	10118	C7H5N3O6	2,4,6-trinitrotoluene	118-96-7	0.86	7.31E+00	2
420	10253	C7H6Cl2	2,4-dichlorotoluene	95-73-8	2.95	8.90E+02	2
421	10341	C7H6N2	1H-benzimidazole	51-17-2	0.66	4.56E+00	2
422	10352	C7H6N2O	2-hydroxybenzimidazole	615-16-7	0.48	3.05E+00	2
423	10363	C7H6N2O4	2,4-dinitrotoluene	121-14-2	1.16	1.46E+01	2
424	10365	C7H6N2O4	2,6-dinitrotoluene	606-20-2	1.26	1.82E+01	2
425	10366	C7H6N2O4	3,4-dinitrotoluene	610-39-9	1.24	1.75E+01	2
426	10404	C7H6O	benzaldehyde	100-52-7	0.77	5.88E+00	2
427	10408	C7H6O2	benzoic acid	65-85-0	2.00	1.00E+02	1
428	10409	C7H6O2	p-hydroxybenzaldehyde	123-08-0	0.67	4.64E+00	2
429	10410	C7H6O2	salicylaldehyde	90-02-8	1.03	1.07E+01	2
430	10411	C7H6O2	1,3-benzodioxole	274-09-9	1.24	1.75E+01	2
431	10417	C7H6O2	phenyl formate	1864-94-4	0.60	3.94E+00	2
432	10420	C7H6O3	salicylic acid	69-72-7	1.34	2.18E+01	2
433	10462	C7H7Br	p-bromotoluene	106-38-7	2.30	2.00E+02	2
434	10463	C7H7Br	(bromomethyl)benzene	100-39-0	1.91	8.07E+01	2
435	10492	C7H7Cl	benzyl chloride	100-44-7	1.42	2.61E+01	2
436	10493	C7H7Cl	o-chlorotoluene	95-49-8	2.30	2.00E+02	2
437	10494	C7H7Cl	p-chlorotoluene	106-43-4	2.23	1.70E+02	2
438	10555	C7H7F	p-fluorotoluene	352-32-9	1.64	4.35E+01	2
439	10611	C7H7NO	formanilide	103-70-8	0.51	3.22E+00	2
440	10626	C7H7NO2	m-nitrotoluene	99-08-1	1.54	3.43E+01	2
441	10627	C7H7NO2	o-nitrotoluene	88-72-2	1.42	2.61E+01	2
442	10628	C7H7NO2	p-nitrotoluene	99-99-0	1.51	3.25E+01	2
443	10660	C7H7NO3	o-nitroanisole	91-23-6	0.97	9.26E+00	2
444	10725	C7H8	toluene	108-88-3	0.92	8.32E+00	1
445	10850	C7H8O	anisole	100-66-3	1.27	1.85E+01	2
446	10851	C7H8O	benzyl alcohol	100-51-6	0.43	2.69E+00	2
447	10852	C7H8O	m-cresol	108-39-4	1.16	1.46E+01	2
448	10853	C7H8O	o-cresol	95-48-7	1.16	1.46E+01	2
449	10854	C7H8O	p-cresol	106-44-5	1.16	1.43E+01	2
450	10866	C7H8O2	guaiacol	90-05-1	0.64	4.39E+00	2
451	10867	C7H8O2	p-methoxyphenol	150-76-5	0.66	4.56E+00	2
452	10953	C7H9N	benzylamine	100-46-9	0.46	2.89E+00	2
453	10954	C7H9N	N-methylaniline	100-61-8	0.91	8.15E+00	2
454	10955	C7H9N	m-toluidine	108-44-1	0.71	5.08E+00	2
455	10956	C7H9N	o-toluidine	95-53-4	0.64	4.39E+00	2
456	10957	C7H9N	p-toluidine	106-49-0	0.70	4.99E+00	2
457	10961	C7H9N	2,6-dimethylpyridine	108-48-5	0.93	8.46E+00	2
458	11079	C7H10N2	m-toluenediamine	95-80-7	-0.29	5.14E-01	2
459	11107	C7H10N2	toluenediamine	25376-45-8	1.96	9.10E+01	1
460	11450	C7H12O2	butyl acrylate	141-32-2	1.46	2.91E+01	2
461	11451	C7H12O2	isobutyl acrylate	106-63-8	1.35	2.26E+01	2
462	11530	C7H12O4	diethyl malonate	105-53-3	0.36	2.28E+00	2
463	11704	C7H14	methylcyclohexane	108-87-2	2.67	4.63E+02	2
464	11705	C7H14	cycloheptane	291-64-5	2.76	5.75E+02	2
465	11706	C7H14	1-heptene	592-76-7	2.75	5.65E+02	2
466	11829	C7H14O	2-heptanone	110-43-0	1.16	1.46E+01	2
467	11834	C7H14O	5-methyl-2-hexanone	110-12-3	1.09	1.22E+01	2
468	11843	C7H14O	2,4-dimethyl-3-pentanone	565-80-0	1.07	1.17E+01	2
469	11845	C7H14O	cis-2-methylcyclohexanol	7443-70-1	1.05	1.13E+01	2
470	11846	C7H14O	trans-2-methylcyclohexanol	7443-52-9	1.04	1.09E+01	2
471	11884	C7H14O2	heptanoic acid	111-14-8	1.51	3.25E+01	2
472	11982	C7H15Br	1-bromoheptane	629-04-9	3.04	1.11E+03	2
473	11988	C7H15Cl	1-chloroheptane	629-06-1	2.88	7.56E+02	2
474	12112	C7H16	heptane	142-82-5	3.16	1.43E+03	2
475	12164	C7H16O	1-heptanol	111-70-6	1.67	4.68E+01	2
476	12165	C7H16O	2-heptanol	543-49-7	1.42	2.66E+01	2
477	12166	C7H16O	3-heptanol	589-82-2	1.37	2.34E+01	2
478	12167	C7H16O	4-heptanol	589-55-9	1.35	2.26E+01	2
479	12315	C7H17N	heptylamine	111-68-2	1.63	4.27E+01	2
480	12419	C8Cl8	octachlorostyrene	29082-74-4	4.52	3.30E+04	1

Table 14 Bioconcentration Factor - Oganic Compounds

Bioconcentration Factor							
NO	ID	FORMULA	NAME	CAS No	$\log_{10}$ BCF	BCF	code
481	12594	C8H4O3	phthalic anhydride	85-44-9	0.86	7.31E+00	2
482	12639	C8H5Cl3O3	2,4,5-trichlorophenoxyacetic	93-76-5	1.36	2.29E+01	1
483	12713	C8H6	ethynylbenzene	536-74-3	1.50	3.13E+01	2
484	12725	C8H6BrN	5-bromoindole	10075-50-0	1.15	1.40E+01	1
485	12868	C8H6N2	quinoxaline	91-19-0	0.45	2.84E+00	2
486	12914	C8H6O	benzofuran	271-89-6	1.71	5.12E+01	2
487	12934	C8H6O4	isophthalic acid	121-91-5	0.91	8.15E+00	2
488	12935	C8H6O4	phthalic acid	88-99-3	0.18	1.50E+00	2
489	12936	C8H6O4	terephthalic acid	100-21-0	1.18	1.51E+01	2
490	12947	C8H6S	benzothiophene	95-15-8	2.06	1.16E+02	2
491	13107	C8H7N	indole	120-72-9	1.29	1.95E+01	2
492	13108	C8H7N	benzeneacetonitrile	140-29-4	0.83	6.80E+00	2
493	13237	C8H8	styrene	100-42-5	1.13	1.35E+01	1
494	13238	C8H8	1,3,5,7-cyclooctatetraene	629-20-9	2.03	1.08E+02	2
495	13309	C8H8Cl3O3PS	ronnel	299-84-3	4.64	4.37E+04	1
496	13391	C8H8O	acetophenone	98-86-2	0.89	7.72E+00	2
497	13393	C8H8O	benzeneacetaldehyde	122-78-1	1.01	1.01E+01	2
498	13394	C8H8O	2-methylbenzaldehyde	529-20-4	1.39	2.43E+01	2
499	13396	C8H8O	2,3-dihydrobenzofuran	496-16-2	1.29	1.95E+01	2
500	13397	C8H8O	phenyloxirane	96-09-3	0.87	7.45E+00	2
501	13412	C8H8O2	methyl benzoate	93-58-3	1.34	2.18E+01	2
502	13413	C8H8O2	o-toluic acid	118-90-1	1.43	2.71E+01	2
503	13414	C8H8O2	p-toluic acid	99-94-5	1.45	2.81E+01	2
504	13415	C8H8O2	benzeneacetic acid	103-82-2	0.71	5.17E+00	2
505	13417	C8H8O2	phenyl acetate	122-79-2	0.78	5.99E+00	2
506	13433	C8H8O2	m-toluic acid	99-04-7	1.47	2.97E+01	2
507	13445	C8H8O3	methyl salicylate	119-36-8	1.61	4.12E+01	2
508	13446	C8H8O3	vanillin	121-33-5	0.56	3.60E+00	2
509	13671	C8H9NO	acetanilide	103-84-4	0.52	3.28E+00	2
510	13789	C8H10	ethylbenzene	100-41-4	1.19	1.55E+01	1
511	13790	C8H10	o-xylene	95-47-6	1.15	1.41E+01	1
512	13791	C8H10	m-xylene	108-38-3	2.13	1.34E+02	2
513	13792	C8H10	p-xylene	106-42-3	1.17	1.48E+01	1
514	13912	C8H10O	phenetole	103-73-1	1.58	3.83E+01	2
515	13913	C8H10O	2-phenylethanol	60-12-8	0.67	4.72E+00	2
516	13915	C8H10O	3-methylbenzenemethanol	587-03-1	0.86	7.31E+00	2
517	13916	C8H10O	4-methylbenzenemethanol	589-18-4	0.85	7.05E+00	2
518	13917	C8H10O	1-phenylethanol	98-85-1	0.72	5.27E+00	2
519	13918	C8H10O	o-ethylphenol	90-00-6	1.55	3.56E+01	2
520	13919	C8H10O	m-ethylphenol	620-17-7	1.58	3.76E+01	2
521	13920	C8H10O	p-ethylphenol	123-07-9	1.58	3.76E+01	2
522	13922	C8H10O	2,4-xylenol	105-67-9	2.18	1.51E+02	1
523	13923	C8H10O	2,5-xylenol	95-87-4	1.45	2.81E+01	2
524	13924	C8H10O	2,6-xylenol	576-26-1	1.46	2.91E+01	2
525	13925	C8H10O	3,4-xylenol	95-65-8	1.36	2.30E+01	2
526	13926	C8H10O	3,5-xylenol	108-68-9	1.46	2.86E+01	2
527	13927	C8H10O	benzyl methyl ether	538-86-3	0.67	4.64E+00	2
528	13929	C8H10O	2-methylanisole	578-58-5	1.76	5.82E+01	2
529	13930	C8H10O	3-methylanisole	100-84-5	1.70	5.03E+01	2
530	13931	C8H10O	4-methylanisole	104-93-8	1.82	6.61E+01	2
531	13934	C8H10O	dimethyl phenol	1300-71-6	2.18	1.50E+02	1
532	13951	C8H10O2	1,2-dimethoxybenzene	91-16-7	1.24	1.75E+01	2
533	14082	C8H11N	N,N-dimethylaniline	121-69-7	1.42	2.66E+01	2
534	14083	C8H11N	o-ethylaniline	578-54-1	0.97	9.43E+00	2
535	14096	C8H11N	2,4,6-trimethylpyridine	108-75-8	1.09	1.22E+01	2
536	14110	C8H11N	benzeneethanamine	64-04-0	0.71	5.17E+00	2
537	14228	C8H12	1,5-cyclooctadiene	111-78-4	2.10	1.25E+02	2
538	14231	C8H12	vinylcyclohexene	100-40-3	2.70	5.07E+02	2
539	14497	C8H14	2,5-dimethyl-2,4-hexadiene	764-13-6	2.37	2.32E+02	2
540	14599	C8H14O2	butyl methacrylate	97-88-1	1.88	7.50E+01	2
541	14688	C8H14O4	diethyl succinate	123-25-1	0.55	3.53E+00	2
542	14775	C8H15N	octanenitrile	124-12-9	1.77	5.92E+01	2
543	14888	C8H16	cyclooctane	292-64-8	3.12	1.30E+03	2
544	14890	C8H16	1-octene	111-66-0	3.21	1.62E+03	2
545	14979	C8H16	2,4,4-trimethyl-1-pentene	107-39-1	3.19	1.56E+03	2
546	15066	C8H16O	2-octanone	111-13-7	1.47	2.97E+01	2
547	15156	C8H16O2	octanoic acid	124-07-2	2.01	1.02E+02	2
548	15248	C8H17Br	1-bromooctane	111-83-1	3.46	2.90E+03	2
549	15354	C8H18	octane	111-65-9	3.67	4.66E+03	2
550	15430	C8H18O	1-octanol	111-87-5	2.03	1.06E+02	2
551	15431	C8H18O	2-octanol	123-96-6	1.89	7.78E+01	2
552	15434	C8H18O	4-octanol	589-62-8	1.72	5.21E+01	2
553	15527	C8H18O	dibutyl ether	142-96-1	2.14	1.37E+02	2
554	15578	C8H18O3	diethylene glycol diethyl ethe	112-36-7	-0.09	8.09E-01	2
555	15579	C8H18O3	diethylene glycol monobutyl e	112-34-5	0.04	1.10E+00	2
556	15631	C8H19N	octylamine	111-86-4	1.89	7.78E+01	2
557	15637	C8H19N	dibutylamine	111-92-2	1.84	6.85E+01	2
558	15756	C8H24O4Si4	octamethylcyclotetrasiloxane	556-67-2	3.63	4.26E+03	2
559	15877	C9H6Cl6O3S	endosulfan	115-29-7	3.35	2.24E+03	1
560	15878	C9H6Cl6O3S	endosulfan 2	33213-65-9	3.35	2.24E+03	1
561	15926	C9H6O2	2H-1-benzopyran-2-one	91-64-5	0.70	4.99E+00	2
562	15928	C9H6O2	1H-indene-1,3(2H)-dione	606-23-5	0.08	1.21E+00	2
563	16010	C9H7N	isoquinoline	119-65-3	1.24	1.75E+01	2
564	16011	C9H7N	quinoline	91-22-5	1.20	1.60E+01	2
565	16017	C9H7N	cinnamonitrile	4360-47-8	1.15	1.41E+01	2
566	16018	C9H7NO	8-hydroxyquinoline	148-24-3	1.20	1.57E+01	2
567	16090	C9H8	indene	95-13-6	1.91	8.07E+01	2
568	16185	C9H8O	2-methylbenzofuran	4265-25-2	2.14	1.39E+02	2
569	16200	C9H8O	2-propenophenone	768-03-6	1.09	1.22E+01	2
570	16212	C9H8O2	cinnamic acid	621-82-9	1.28	1.92E+01	2
571	16364	C9H9N	benzenepropanenitrile	645-59-0	0.96	9.09E+00	2
572	16388	C9H9NO	cinnamamide, predominantly	621-79-4	0.71	5.17E+00	2
573	16508	C9H10	indane	496-11-7	2.23	1.70E+02	2
574	16512	C9H10	alpha-methylstyrene	98-83-9	2.35	2.23E+02	2
575	16631	C9H10O	4-methylacetophenone	122-00-9	1.33	2.14E+01	2
576	16634	C9H10O	1-phenyl-1-propanone	93-55-0	1.33	2.14E+01	2
577	16635	C9H10O	1-phenyl-2-propanone	103-79-7	0.74	5.47E+00	2
578	16676	C9H10O2	ethyl benzoate	93-89-0	1.69	4.85E+01	2
579	16680	C9H10O2	benzyl acetate	140-11-4	1.15	1.41E+01	2
580	16683	C9H10O2	4-methylphenyl acetate	140-39-6	1.27	1.85E+01	2
581	16701	C9H10O2	(±)-2-phenylpropionic acid	492-37-5	1.02	1.05E+01	2
582	16711	C9H10O3	ethyl vanillin	121-32-4	0.85	7.05E+00	2
583	16876	C9H11Cl3NO3PS	chlorpyrifos	2921-88-2	2.67	4.70E+02	1
584	16895	C9H11N	1,2,3,4-tetrahydroquinoline	635-46-1	1.41	2.57E+01	2
585	16905	C9H11NO	p-dimethylaminobenzaldehyd	100-10-7	1.03	1.07E+01	2
586	17004	C9H12	cumene	98-82-8	1.55	3.55E+01	1
587	17005	C9H12	m-ethyltoluene	620-14-4	2.74	5.55E+02	2
588	17006	C9H12	o-ethyltoluene	611-14-3	2.39	2.45E+02	2
589	17007	C9H12	p-ethyltoluene	622-96-8	2.47	2.94E+02	2
590	17008	C9H12	1,2,3-trimethylbenzene	526-73-8	2.44	2.78E+02	2
591	17009	C9H12	1,2,4-trimethylbenzene	95-63-6	2.47	2.94E+02	2
592	17011	C9H12	mesitylene	108-67-8	2.30	2.00E+02	2
593	17012	C9H12	propylbenzene	103-65-1	2.52	3.27E+02	2
594	17077	C9H12O	benzyl ethyl ether	539-30-0	1.31	2.02E+01	2
595	17091	C9H12O	benzenepropanol	122-97-4	1.09	1.22E+01	2
596	17101	C9H12O	2-propylphenol	644-35-9	1.91	8.22E+01	2
597	17103	C9H12O	4-propylphenol	645-56-7	2.13	1.34E+02	2
598	17118	C9H12O	2,3,6-trimethylphenol	2416-94-6	1.71	5.12E+01	2
599	17120	C9H12O	2,4,6-trimethylphenol	527-60-6	1.54	3.49E+01	2
600	17250	C9H13N	amphetamine	300-62-9	0.99	9.78E+00	2
601	17253	C9H13N	N,N-dimethylbenzylamine	103-83-3	1.16	1.46E+01	2
602	17436	C9H14O	isophorone	78-59-1	0.85	7.00E+00	1
603	17513	C9H14O6	glyceryl triacetate	102-76-1	-0.20	6.27E-01	2
604	17533	C9H15Br6O4P	2,3-dibromo-1-propanol, phos	126-72-7	0.43	2.70E+00	1
605	17755	C9H16O4	azelaic acid	123-99-9	0.84	6.92E+00	2
606	17942	C9H18	1-nonene	124-11-8	3.67	4.66E+03	2
607	17999	C9H18O	2-nonanone	821-55-6	2.10	1.25E+02	2
608	18005	C9H18O	5-methyl-2-octanone	58654-67-4	1.91	8.07E+01	2
609	18052	C9H18O2	nonanoic acid	112-05-0	2.30	2.00E+02	2
610	18176	C9H20	nonane	111-84-2	4.06	1.16E+04	2
611	18236	C9H20O	1-nonanol	143-08-8	2.78	5.97E+02	2
612	18288	C9H20O	2,6-dimethyl-4-heptanol	108-82-7	2.03	1.08E+02	2
613	18393	C9H21N	tripropylamine	102-69-2	1.80	6.37E+01	2
614	18458	C10Cl10O	kepone	143-50-0	2.76	5.75E+02	1
615	18459	C10Cl12	mirex	2385-85-5	4.26	1.81E+04	1
616	18507	C10H5Cl7	heptachlor	76-44-8	4.14	1.38E+04	1
617	18508	C10H5Cl7O	heptachlor epoxide	1024-57-3	4.16	1.44E+04	1
618	18551	C10H6Cl8	chlordane	57-74-9	4.58	3.78E+04	1
619	18553	C10H6Cl8	a-chlordan	5103-71-9	4.18	1.51E+04	1
620	18554	C10H6Cl8	trans-chlordane	5103-74-2	3.62	4.17E+03	1
621	18586	C10H6O3	2-hydroxy-1,4-naphthalenedi	83-72-7	0.75	5.67E+00	2
622	18606	C10H7Cl	1-chloronaphthalene	90-13-1	2.68	4.80E+02	2
623	18607	C10H7Cl	2-chloronaphthalene	91-58-7	2.74	5.55E+02	2
624	18669	C10H8	naphthalene	91-20-3	2.63	4.30E+02	1
625	18670	C10H8	azulene	275-51-4	2.14	1.39E+02	2
626	18740	C10H8O	1-naphthol	90-15-3	1.84	6.98E+01	2
627	18741	C10H8O	2-naphthol	135-19-3	1.73	5.41E+01	2
628	18836	C10H9Cl9	toxaphene	8001-35-2	3.81	6.46E+03	1
629	18849	C10H9N	2-methylquinoline	91-63-4	1.65	4.43E+01	2
630	19106	C10H10O	benzalacetone	122-57-6	1.24	1.72E+01	2
631	19119	C10H10O2	methyl cinnamate	1754-62-7	1.67	4.68E+01	2
632	19195	C10H10O4	dimethyl phthalate	131-11-3	1.76	5.70E+01	1
633	19196	C10H10O4	dimethyl terephthalate	120-61-6	1.38	2.39E+01	2
634	19359	C10H12	1,2,3,4-tetrahydronaphthalen	119-64-2	2.36	2.28E+02	2
635	19551	C10H12O2	isopropyl benzoate	939-48-0	2.11	1.29E+02	2
636	19865	C10H14	butylbenzene	104-51-8	2.97	9.23E+02	2
637	19866	C10H14	isobutylbenzene	538-93-2	2.77	5.86E+02	2
638	19867	C10H14	sec-butylbenzene	135-98-8	3.21	1.62E+03	2
639	19868	C10H14	tert-butylbenzene	98-06-6	2.85	7.03E+02	2
640	19870	C10H14	m-cymene	535-77-3	3.16	1.43E+03	2

Table 14 Bioconcentration Factor - Oganic Compounds

Bioconcentration Factor								Bioconcentration Factor							
NO	ID	FORMULA	NAME	CAS No	log_{10} BCF	BCF	code	NO	ID	FORMULA	NAME	CAS No	log_{10} BCF	BCF	code
641	19871	C10H14	o-cymene	527-84-4	3.06	1.15E+03	2	721	23678	C12H12	1,5-dimethylnaphthalene	571-61-9	3.06	1.15E+03	2
642	19872	C10H14	p-cymene	99-87-6	2.84	6.90E+02	2	722	23680	C12H12	1,7-dimethylnaphthalene	575-37-1	3.11	1.28E+03	2
643	19878	C10H14	o-diethylbenzene	135-01-3	3.09	1.24E+03	2	723	23682	C12H12	2,3-dimethylnaphthalene	581-40-8	3.08	1.19E+03	2
644	19879	C10H14	m-diethylbenzene	141-93-5	3.21	1.62E+03	2	724	23683	C12H12	2,6-dimethylnaphthalene	581-42-0	3.00	1.01E+03	2
645	19880	C10H14	p-diethylbenzene	105-05-5	3.22	1.65E+03	2	725	23685	C12H12	1-ethylnaphthalene	1127-76-0	3.08	1.19E+03	2
646	19881	C10H14	3-ethyl-o-xylene	933-98-2	3.03	1.07E+03	2	726	23686	C12H12	2-ethylnaphthalene	939-27-5	3.06	1.15E+03	2
647	19882	C10H14	4-ethyl-o-xylene	934-80-5	3.16	1.43E+03	2	727	23719	C12H12N2	hydrazobenzene	122-66-7	1.92	8.37E+01	2
648	19883	C10H14	2-ethyl-m-xylene	2870-04-4	2.98	9.58E+02	2	728	23965	C12H14O	4-phenylcyclohexanone	4894-75-1	1.54	3.43E+01	2
649	19884	C10H14	4-ethyl-m-xylene	874-41-9	3.13	1.35E+03	2	729	24002	C12H14O4	diethyl phthalate	84-66-2	2.07	1.17E+02	1
650	19885	C10H14	5-ethyl-m-xylene	934-74-7	3.19	1.56E+03	2	730	24357	C12H18	hexylbenzene	1077-16-3	3.96	9.14E+03	2
651	19886	C10H14	2-ethyl-p-xylene	1758-88-9	3.10	1.26E+03	2	731	24378	C12H18	hexamethylbenzene	87-85-4	3.31	2.02E+03	2
652	19887	C10H14	1,2,3,4-tetramethylbenzene	488-23-3	2.76	5.75E+02	2	732	24690	C12H22O	cyclododecanone	830-13-7	2.84	6.90E+02	2
653	19888	C10H14	1,2,3,5-tetramethylbenzene	527-53-7	2.84	6.90E+02	2	733	24867	C12H24O2	dodecanoic acid	143-07-7	3.23	1.71E+03	2
654	19889	C10H14	1,2,4,5-tetramethylbenzene	95-93-2	2.84	6.90E+02	2	734	24914	C12H26	dodecane	112-40-3	4.42	2.62E+04	2
655	19983	C10H14O	4-butylphenol	1638-22-8	2.48	3.04E+02	2	735	25277	C12H26O	1-dodecanol	112-53-8	3.65	4.49E+03	2
656	20195	C10H15N	N,N-dietyhlaniline	91-66-7	2.21	1.64E+02	2	736	25294	C12H26O3	diethylene glycol dibutyl ethe	112-73-2	1.12	1.31E+01	2
657	20321	C10H16	alpha-pinene	80-56-8	3.42	2.60E+03	2	737	25427	C12H4Br6	hexabromobiphenyl	36355-01-8	4.26	1.81E+04	1
658	20326	C10H16	gamma-terpinene	99-85-4	3.16	1.43E+03	2	738	25446	C13H6Cl6O2	hexachlorophene	70-30-4	2.70	5.07E+02	2
659	20327	C10H16	terpinolene	586-62-9	3.13	1.35E+03	2	739	25496	C13H8O	9H-fluoren-9-one	486-25-9	2.43	2.68E+02	2
660	20919	C10H19O6PS2	malathion	121-75-5	1.58	3.80E+01	1	740	25537	C13H9N	acridine	260-94-6	2.29	1.93E+02	2
661	20993	C10H20O	2-decanone	693-54-9	2.58	3.79E+02	2	741	25575	C13H10	fluorene	86-73-7	3.11	1.30E+03	1
662	21040	C10H20O2	decanoic acid	334-48-5	2.83	6.78E+02	2	742	25623	C13H10O	benzophenone	119-61-9	2.11	1.29E+02	2
663	21133	C10H22	decane	124-18-5	4.54	3.45E+04	2	743	25631	C13H10O2	phenyl benzoate	93-99-2	2.44	2.73E+02	2
664	21190	C10H22	2,2,3,3-tetramethylhexane	13475-81-5	3.57	3.75E+03	2	744	25694	C13H11NO	N-phenylbenzamide	93-98-1	1.67	4.68E+01	2
665	21224	C10H22O	1-decanol	112-30-1	3.21	1.62E+03	2	745	25738	C13H12	diphenylmethane	101-81-5	2.87	7.42E+02	2
666	21531	C11H8O2	2-methyl-1,4-naphthalenedio	58-27-5	1.34	2.18E+01	2	746	25741	C13H12	4-methylbiphenyl	644-08-6	3.26	1.81E+03	2
667	21535	C11H8O3	2-hydroxy-3-methyl-1,4-napht	483-55-6	0.55	3.53E+00	2	747	25793	C13H12O	benzyl phenyl ether	946-80-5	2.59	3.93E+02	2
668	21543	C11H8O3	2-methoxy-1,4-naphthalenedi	2348-82-5	0.67	4.64E+00	2	748	25798	C13H12O	diphenylmethanol	91-01-0	1.71	5.12E+01	2
669	21565	C11H9N	4-phenylpyridine	939-23-1	1.65	4.43E+01	2	749	26049	C13H16F3N3O4	trifluralin	1582-09-8	3.03	1.07E+03	1
670	21601	C11H10	1-methylnaphthalene	90-12-0	2.66	4.54E+02	2	750	26535	C13H28O	1-tridecanol	112-70-9	4.20	1.58E+04	2
671	21602	C11H10	2-methylnaphthalene	91-57-6	2.76	5.75E+02	2	751	26637	C14H8Cl4	2,2-dichloro-1,1-bis(4-chlorop	72-55-9	4.71	5.10E+04	1
672	21801	C11H12Cl2N2O5	chloramphenicol	56-75-7	0.50	3.17E+00	2	752	26654	C14H8O2	anthraquinone	84-65-1	2.28	1.90E+02	2
673	22113	C11H14O2	butyl benzoate	136-60-7	2.63	4.30E+02	2	753	26689	C14H9Cl	2-chloroanthracene	17135-78-3	3.63	4.27E+03	1
674	22304	C11H16	pentylbenzene	538-68-1	3.47	2.96E+03	2	754	26699	C14H9Cl5	1,1,1-trichloro-2,2-bis(4-chlor	50-29-3	4.53	3.39E+04	1
675	22354	C11H16	pentamethylbenzene	700-12-9	3.20	1.59E+03	2	755	26700	C14H9Cl5	o,p'-DDT	789-02-6	4.57	3.70E+04	1
676	22823	C11H22O	2-undecanone	112-12-9	2.83	6.78E+02	2	756	26728	C14H10	anthracene	120-12-7	2.95	8.91E+02	1
677	22848	C11H22O2	methyl decanoate	110-42-9	3.08	1.21E+03	2	757	26729	C14H10	diphenylacetylene	501-65-5	3.38	2.38E+03	2
678	23055	C11H24O	1-undecanol	112-42-5	3.33	2.13E+03	2	758	26730	C14H10	phenanthrene	85-01-8	3.42	2.63E+03	1
679	23119	C12Cl10	decachlorobiphenyl	2051-24-3	6.13	1.33E+06	2	759	26739	C14H10Cl4	1,1-dichloro-2,2-bis(p-chlorop	72-54-8	4.92	8.32E+04	1
680	23134	C12HCl9	2,2',3,3',4,5,5',6,6'-nonachlor	52663-77-1	6.05	1.11E+06	2	760	26867	C14H12	trans-stilbene	103-30-0	3.40	2.51E+03	2
681	23145	C12H2Cl8	2,2',3,3',5,5',6,6'-octachlorobi	2136-99-4	5.21	1.62E+05	2	761	26871	C14H12	1-methylfluorene	1730-37-6	3.53	3.36E+03	2
682	23152	C12H3Cl7	2,2',3,3',4,4',6-heptachlorobip	52663-71-5	4.89	7.82E+04	2	762	26933	C14H12O	2-phenylacetophenone	451-40-1	2.11	1.29E+02	2
683	23153	C12H4Cl4O2	2,3,7,8-tetrachloro-dibenzo-p	1746-01-6	3.72	5.25E+03	1	763	26942	C14H12O2	benzyl benzoate	120-51-4	2.74	5.45E+02	2
684	23159	C12H4Cl6	2,2',3,3',4,4'-hexachlorobiphe	38380-07-3	5.13	1.35E+05	2	764	27053	C14H14	1,2-diphenylethane	103-29-7	3.31	2.06E+03	2
685	23160	C12H4Cl6	2,2',4,4',6,6'-hexachlorobiphe	33979-03-2	5.13	1.35E+05	2	765	27062	C14H14	4,4'-dimethylbiphenyl	613-33-2	3.62	4.18E+03	2
686	23161	C12H4Cl6	2,2',3,3',6,6'-hexachlorobiphe	38411-22-2	4.89	7.82E+04	2	766	27096	C14H14NO4PS	ethyl p-nitrophenyl benzeneth	2104-64-5	3.05	1.12E+03	1
687	23177	C12H5Cl5	2,3,4,5,6-pentachlorobipheny	18259-05-7	4.58	3.78E+04	2	767	27138	C14H14O	dibenzyl ether	103-50-4	2.21	1.64E+02	2
688	23178	C12H5Cl5	2,2',4,5,5'-pentachlorobiphen	37680-73-2	4.66	4.53E+04	2	768	27644	C14H22	octylbenzene	2189-60-8	4.58	3.78E+04	2
689	23179	C12H5Cl5	alochlor 1254	11097-69-1	4.70	5.01E+04	1	769	27873	C14H28O2	tetradecanoic acid	544-63-8	4.42	2.62E+04	2
690	23204	C12H6Cl4	2,3,4,5-tetrachlorobiphenyl	33284-53-6	4.12	1.31E+04	2	770	27912	C14H30	tetradecane	629-59-4	5.29	1.94E+05	2
691	23205	C12H6Cl4	2,2',4',5-tetrachlorobiphenyl	41464-40-8	4.13	1.34E+04	2	771	27922	C14H30O	1-tetradecanol	112-72-1	4.36	2.31E+04	2
692	23206	C12H6Cl4	chlorodiphenyl	53469-21-9	4.51	3.24E+04	1	772	28044	C15H10O2	2-phenyl-1H-indene-1,3(2H)-	83-12-5	1.89	7.78E+01	2
693	23247	C12H7Cl3	2,4,5-trichlorobiphenyl	15862-07-4	4.02	1.06E+04	2	773	28130	C15H12	2-methylanthracene	613-12-7	3.67	4.66E+03	2
694	23248	C12H7Cl3	2,4,6-trichlorobiphenyl	35693-92-6	3.92	8.34E+03	2	774	28131	C15H12	9-methylanthracene	779-02-2	3.61	4.03E+03	2
695	23272	C12H8Cl2	2,5-dichlorobiphenyl	34883-39-1	3.63	4.26E+03	2	775	28132	C15H12	1-methylphenanthrene	832-69-9	3.66	4.58E+03	2
696	23273	C12H8Cl2	2,6-dichlorobiphenyl	33146-45-1	3.55	3.55E+03	2	776	28135	C15H12	2-methylphenanthrene	2531-84-2	3.48	3.00E+03	1
697	23291	C12H8Cl6	aldrin	309-00-2	3.50	3.16E+03	1	777	28436	C15H16O2	bisphenol a	80-05-7	2.22	1.67E+02	2
698	23293	C12H8Cl6O	dieldrin	60-57-1	4.15	1.41E+04	1	778	28725	C15H24	nonylbenzene	1081-77-2	5.22	1.65E+05	2
699	23294	C12H8Cl6O	endrin	72-20-8	3.42	2.63E+03	1	779	28756	C15H24O	2,6-di-tert-butyl-p-cresol	128-37-0	3.63	4.26E+03	2
700	23305	C12H8N2	1,10-phenanthroline	66-71-7	1.05	1.11E+01	2	780	28984	C16H10	fluoranthene	206-44-0	4.00	1.00E+04	1
701	23308	C12H8N2	phenazine	92-82-0	1.84	6.98E+01	2	781	28985	C16H10	pyrene	129-00-0	4.20	1.58E+04	1
702	23335	C12H8O	dibenzofuran	132-64-9	3.13	1.35E+03	1	782	29088	C16H13N	N-phenyl-2-naphthalenamine	135-88-6	2.17	1.47E+02	1
703	23365	C12H9Cl	2-chlorobiphenyl	2051-60-7	3.17	1.48E+03	2	783	29111	C16H14	9,10-dimethylphenanthrene	604-83-1	4.10	1.24E+04	2
704	23366	C12H9Cl	3-chlorobiphenyl	2051-61-8	3.22	1.65E+03	2	784	29198	C16H15Cl3O2	methoxychlor	72-43-5	3.92	8.30E+03	1
705	23367	C12H9Cl	4-chlorobiphenyl	2051-62-9	3.24	1.75E+03	2	785	29518	C16H22O4	dibutyl phthalate	84-74-2	3.15	1.41E+03	1
706	23394	C12H9N	dibenzopyrrole	86-74-8	2.54	3.46E+02	2	786	29595	C16H26	decylbenzene	104-72-3	5.41	2.55E+05	2
707	23444	C12H10	acenaphthene	83-32-9	2.59	3.87E+02	1	787	29721	C16H32O2	hexadecanoic acid	57-10-3	5.26	1.84E+05	2
708	23445	C12H10	phenylbenzene (biphenyl)	92-52-4	2.45	2.82E+02	1	788	29862	C17H12	11H-benzo[a]fluorene	238-84-6	3.87	7.35E+03	2
709	23469	C12H10Cl2N2	3,3'-dichloro-p-benzidine	91-94-1	2.70	5.01E+02	1	789	29863	C17H12	11H-benzo[b]fluorene	243-17-4	4.14	1.39E+04	2
710	23488	C12H10N2	azobenzene	17082-12-1	2.62	4.15E+02	2	790	30097	C17H19NO3	morphine	57-27-2	0.26	1.80E+00	2
711	23494	C12H10N2O	N-nitrosodiphenylamine	86-30-6	2.34	2.17E+02	1	791	30264	C17H28	undecylbenzene	6742-54-7	6.03	1.07E+06	2
712	23519	C12H10O	diphenyl ether	101-84-8	2.29	1.95E+02	1	792	30406	C18H12	chrysene	218-01-9	4.30	2.00E+04	1
713	23576	C12H10S	diphenyl sulfide	139-66-2	3.12	1.30E+03	2	793	30407	C18H12	benz[a]anthracene	56-55-3	4.10	1.26E+04	1
714	23607	C12H11N	p-aminodiphenyl	92-67-1	1.86	7.23E+01	2	794	30409	C18H12	naphthacene	92-24-0	4.15	1.41E+04	2
715	23609	C12H11N	diphenylamine	122-39-4	1.48	3.00E+01	1	795	30410	C18H12	triphenylene	217-59-4	3.94	8.65E+03	2
716	23632	C12H11NO2	carbaryl	63-25-2	1.86	7.24E+01	1	796	30459	C18H14	p-terphenyl	92-94-4	4.36	2.31E+04	2
717	23656	C12H11N3	p-aminoazobenzene	60-09-3	2.29	1.97E+02	2	797	30548	C18H15N	triphenylamine	603-34-9	4.13	1.36E+04	2
718	23675	C12H12	1,2-dimethylnaphthalene	573-98-8	3.00	1.01E+03	2	798	30566	C18H15O4P	triphenyl phosphate	115-86-6	2.60	3.98E+02	1
719	23676	C12H12	1,3-dimethylnaphthalene	575-41-7	3.09	1.24E+03	2	799	30567	C18H15P	triphenylphosphine	603-35-0	4.10	1.24E+04	2
720	23677	C12H12	1,4-dimethylnaphthalene	571-58-4	3.05	1.13E+03	2	800	30806	C18H22O2	dicumyl peroxide	80-43-3	3.95	8.81E+03	2

Table 14 Bioconcentration Factor - Oganic Compounds

Bioconcentration Factor								Bioconcentration Factor							
NO	ID	FORMULA	NAME	CAS No	$\log_{10}$ BCF	BCF	code	NO	ID	FORMULA	NAME	CAS No	$\log_{10}$ BCF	BCF	code
801	30947	C18H30	dodecylbenzene	123-01-3	6.43	2.71E+06	2	813	31805	C20H12	benzo[e]pyrene	192-97-2	4.56	3.63E+04	1
802	30964	C18H30O2	linolenic acid	463-40-1	4.70	5.05E+04	2	814	32388	C20H32O2	5,8,11,14-eicosatetraenoic ac	506-32-1	5.11	1.30E+05	2
803	30991	C18H32O2	linoleic acid	60-33-3	5.17	1.48E+05	2	815	32491	C20H40O2	eicosanoic acid	506-30-9	6.94	8.69E+06	2
804	31030	C18H34O2	oleic acid	112-80-1	5.64	4.32E+05	2	816	32632	C21H16	1,2-dihydro-3-methylbenz[j]ac	56-49-5	4.93	8.56E+04	2
805	31097	C18H36O2	octadecanoic acid	57-11-4	6.10	1.26E+06	2	817	32787	C21H21O4P	tri-o-cresyl phosphate	78-30-8	2.22	1.66E+02	1
806	31344	C19H16O	triphenylmethanol	76-84-6	2.51	3.22E+02	2	818	32790	C21H21O4P	tricresyl phosphate	1330-78-5	2.22	1.65E+02	1
807	31452	C19H20O4	benzyl butyl phthalate	85-68-7	2.89	7.72E+02	1	819	33080	C22H12	benzo[ghi]perylene	191-24-2	3.70	5.01E+03	1
808	31646	C19H32	tridecylbenzene	123-02-4	6.99	9.87E+06	2	820	33081	C22H12	indeno[1,2,3-cd]pyrene	193-39-5	3.40	2.51E+03	1
809	31800	C20H12	perylene	198-55-0	4.54	3.45E+04	2	821	33098	C22H14	dibenz[a,h]anthracene	53-70-3	3.70	5.01E+03	1
810	31802	C20H12	benzo[k]fluoranthene	207-08-9	4.40	2.51E+04	1	822	33717	C24H12	coronene	191-07-1	4.38	2.40E+04	2
811	31803	C20H12	benzo[b]fluoranthene	205-99-2	4.30	2.00E+04	1	823	33925	C24H38O4	dioctyl phthalate	117-84-0	5.60	4.02E+05	2
812	31804	C20H12	benzo[a]pyrene	50-32-8	3.69	4.90E+03	1	824	33926	C24H38O4	bis(2-ethylhexyl) phthalate	117-81-7	2.49	3.09E+02	1

Code: 1 - data, 2 - estimate

BCF = (concentration in organism tissue, mg/kg) / (concentration in water, mg/kg)

Values apply at ambient conditions (1 atm, 25 C)

Most of the BCF values are for fish - bluegill sunfish, rainbow trout, mosquito fish, ide fish, ...

Chapter 15

SOIL SORPTION COEFFICIENT – ORGANIC COMPOUNDS

Carl L. Yaws and Daniel H. Chen
Lamar University, Beaumont, Texas

Tabulation Results

The results are given in Table 15. The tabulation is arranged by carbon number (C, C2, C3, ...) to provide ease of use in quickly locating the data by using the chemical formula. The compound name and CAS No (Chemical Abstracts Registry Number) are next provided. Values for soil sorption coefficient are given in the adjacent columns. The next column provides the code for the tabulation.

The soil sorption coefficient is the ratio of a compound's concentration in solid phase (normalized for organic carbon content) to its concentration in water:

$$Koc = (\text{mg sorbed/kg organic carbon}) / (\text{mg/L aqueous solution}) \quad (15\text{-}1)$$

By basing the sorption coefficient on organic carbon in soil, it is possible to eliminate much of the variation between soils. For organic content, values of 1% to 8% are recommended in EPA's tests (49) for soils.

In instances where the coefficient is expressed in terms of soil organic matter (Kom), the following relationship is often used:

$$Koc = 1.724\ Kom \quad (15\text{-}2)$$

This relationship assumes organic matter contains about 58% carbon (61).

The amount of compound sorbed onto soil sediment depends on the concentration in aqueous phase and its soil sorption coefficient. For dilute aqueous solutions, the relationship can be expressed by the following equation which utilitizes the Freundlich equation with 1/n equal to unity:

$$x / m = Koc\ C \quad (15\text{-}3)$$

where x = weight of compound sorbed, mg
m = weight of organic carbon in soil, kg
Koc = soil sorption coefficient, L/kg
C = equilibrium concentration of compound in aqueous solution, mg/L

In preparing the tabulation, a literature search was conducted to identify data source publications (1-86). Both experimental values for the property under consideration and parameter values for estimation of the property are included in the source publications. The publications were screened and copies of appropriate data were made. These data were then keyed into the computer to provide a database of values for compounds for which experimental data are available. The database also served as a basis to check the accuracy of the estimation methods. Upon completion of data collection, estimation of the values for the remaining compounds was performed.

The excellent compilations by Lyman, Reehl, and Rosenblatt (49); Mackay, Shiu, and Ma (50-54); Montgomery (61); and Yaws (86) were primarily used for experimental data. Most of the estimates were based on the method given by Lyman, Reehl, and Rosenblatt (49) in which soil sorption coefficient is a function of octanol-water coefficient.

Example

River water at 20 C is contaminated with toluene (C_7H_8) at a concentration of 10 ppm (weight). The sediment in the river bottom contains about 3% carbon. Determine the uptake of toluene per metric ton (1000 kg) of bottom sediment.

First, the amount of organic carbon per metric ton of bottom sediment is calculated: m = 1000 kg (3%) = 30 kg of organic carbon. Inspection of the tabulated values indicates that Koc = 93.3 for toluene. Substitution of these values and C = 10 ppm = 10 mg/L into the equation x / m = Koc C yields:

$$x / 30\ \text{kg} = 93.3\ (\text{mg sorbed/kg organic carbon})/(\text{mg/L aqueous solution})\ (10\ \text{mg/L}) = 933\ \text{mg/kg}$$

$$x = 30\ \text{kg}\ (933\ \text{mg/kg}) = 27{,}990\ \text{mg} = 27.99\ \text{g}$$

References

1. Allen, D. T. and D. R. Shonnard, GREEN ENGINEERING, Prentice Hall PTR, Upper Saddle River, NJ (2002).
2. Baarschers, W. H., J. Elvish, and S. P. Ryan, "Adsorption of Fenitrothion and 3-Methyl-4-Nitrophenol on Soils and Sediment", Bull. Environ. Contam. Toxicol., 30(5), 621-627 (1983).
3. Banwart, W. L., J. J. Hassett, S. G. Wood, and J. C. Means, "Sorption of Nitrogen-Hetrocyclic Compounds by Soils and Sediments", Soil Sci., 133(1), 42-47 (1982).
4. Beyerle-Pfnur, R. and J. P. Lay, "Adsorption and Desorption of 3,4-Dichloroaniline on Soil", Chemosphere, 21(9), 1087-1094 (1990).
5. Bilkert, J. N. and P. S. C. Rao, "Sorption and Leaching of Three Nonfumigant Nematicides in Soils", J. Environ. Sci. Health, B20(1), 1-26 (1985).
6. Boyd, S. A., "Adsorption of Substituted Phenols by Soil", Soil Sci., 134(5), 337-343. (1982).
7. Boyd, S. A., M. D. Mikesell, and J. F. Lee, "Chlorophenols in Soil", Sawhney, B. L. and K. Brown (eds.), REACTION AND MOVEMENT OF ORGANIC CHEMICALS IN SOILS , Soil Science Society of American and American Society Agronomy, SSSA Special Publication No. 22, 209 (1989).
8. Briggs, G. G., J. Agric. Food Chem., 29(5), 1050 (1981).
9. Briggs, G. G., "Theoretical and Experimental Relationships between Soil Adsoption, Octanol-Water Partition Coefficients, Water Solubilities, Bioconcentration Factors, and the Parachor", J. Agric. Food Chem., 29(5), 1050-1059 (1981).
10. Brown., D. S. and E. W. Flagg, "Emprical Prediction of Oranic Pollutant Sorption in Natural Sediments", J. Environ. Qual., 10(3), 382-386 (1981).
11. Buchter, B., B. Davidoff, M. C. Amacher, C. Hinz, K. Iskandar, and H. M. Selim, "Correlation of Freundlich Kd and n Retention Parameters with Soils and Elements", Soil Sci., 148(5), 370-379 (1989).
12. Carton, A., T., Isla, and J. Alcazer-Benedi, "Sorption-Desorption of Imazamethabenz on Three Spanish Soils", J. Agric. Food Chem., 45(4), 1454-1458 (1997).
13. Chen, T. L., D. H. Chen, and C. L. Yaws, "Predicting Soil Adsorption with Molecular Structure", Paper 54b, AIChE National Meeting, San Diego, CA (August 19-22, 1990).
14. Chin, Y. P., C. S. Peven, and W. J. Weber, "Estimating Soil/Sediment Partition Coefficients for Organic Compounds by High Performance Reverse Liquid Chromatography", Water Res., 22(7), 873-881 (1988).
15. Chiou, C. T., L. J. Peters, and V. H. Freed, Science, 206, 831 (1979).
16. Chiou, C. T., Soil Science Society of American and American Society Agronomy, SSSA Special Publication No. 22, 1 (1989).
17. de Jonge, H. and L. W. de Jonge, "Influence of pH and Solution Composition on the Sorption of Glyphosate and Prochloraz to a Sandy Loam Soil", Chemosphere, 39(5), 753-763. (1999).
18. Deeley, G. M., M. Reinhard, and S. M. Stearns, "Transformation and Sorption of 1,2-Dibromo-2-Chloropropane in Subsurface Samples Collected at Fresno, California", J. Environ. Qual., 2(3), 547-556 (1991).
19. Dewulf, J., H., H. Van Langenhove, and S. Grare, "Sediment/Water and Octanol/Water Equilibrium Partitioning of Volatile Organic Compounds: Temperature Dependence in the 2-25 °C Range", Water Res., 33(10), 2424-2436 (1999a).
20. Dickson, K. I. (ed.), MODELING THE FATE OF CHEMICALS IN THE AQUATIC ENVIRONMENT, Ann Arbor Sci. Pub., Ann Arbor, MI (1982).
21. Fairbanks, B. C. and G. A. O'Connor, "Effect of Sewage on the Adsoption of Polychlorinated Biphenyls by Three New Mexico Soils", J. Environ. Qual., 13(2), 297-300 (1984).
22. Felsot, A. and P. A. Dahm, "Sorption of Organophosphorus and Carbamate Insecticides by Soil", J. Agric. Food Chem., 27(3), 557-563 (1979).
23. Gan, J. and S. R. Yates, "Degradation and Phase Partition of Methyl Iodide in Soil", J. Agric. Food Chem., 44(12), 4001-4008 (1996).
24. Gerstl, Z. and B. Yaron, "Behavior of Bromacil and Napropamide in Soils: I. Adsorption and Degradation", Soil Sci. Soc. Am. J., 47(3), 474-478 (1983).
25. Girvin, D. C. and A. J. Scott, "Polychlorinated Biphenyl Sorption by Soils: Measurement of Soil-Water Partition Coefficients at Equilibrium", Chemosphere, 35(9), 2007-2025 (1997).
26. Gonzalez, J. and L. Ukraineczyk, "Transport of Nicosulfuron in Soil Columns", J. Environ. Qual., 28(1), 101-107 (1999).
27. Graveel, J. G., L. E. Sommers, and D. W. Nelson, "Decomposition of Benzidine, α-Naphthylamine, and *p*-Toluidine in Soils", J. Environ. Qual., 15(1), 53-59 (1986).
28. Green, R. E. and S. R. Obien, Weed Sci., 17, 514 (1969).
29. Grover, R. and A. E. Smith, "Adsorption Studies with the Acid and Dimethylamine Forms of 2,4-D and Dicamba", Can, J. Soil Sci., 54, 179-186 (1974).
30. Gschwend, P. M. and S. C. Wu, Environ. Sci. Technol., 19 90 (1985).
31. Hance, R. J., J. Agric. Food Chem., 17, 667 (1969).
32. Hodson, J. and N. A. Williams, Chemosphere, 17, No. 1, 67 (1988).

33. Horzempa, L. M. and D. M. Di Toro, "The Extent of Reversibility of Polychlorinated Biphenyl Adsorption", Water Res., 17(8), 851-859 (1983).
34. Karickhoff, S. W. and D. S. Brown, J. Environ. Qual., 7, 246 (1978).
35. Karickhoff, S. W., Chemosphere, 10(8), 833 (1981).
36. Karickhoff, S. W., J. of Hydraulic Engineering, ASCE, 110, 707 (1984).
37. Karickhoff, S. W., D.S. Brown, and T.A. Scott, Water Res., 13, 241 (1979).
38. Kenaga, E. E. and C. A. I. Goring, AQUATIC TOXICOLOGY, Eaton, J. C., P. R. Parrish, A. C. Hendricks, (eds), American Society for Testing and Materials, Philadelphia, PA, 78 (1980).
39. Khan, S. V., J. of Environmental Quality, 3 ,202 (1974).
40. Kishi, H., N. Kogure, and Y. Hashimoto, "Contribution of Soil Constituents in Adsorption Coefficient of Aromatic Compounds, Halogenated Alicyclic and Aromatic Compounds to Soil", Chemosphere, 21(7), 867-876 (1990).
41. Kishk, F. M., T. M. Abu-Shara, N. Bakry, and M. B. Abou-Donia, "Adsorption of Methyl Parathion by Soils", Bull. Environ, Contam, Toxicol., 22(6), 733-738 (1979).
42. Kozak, J., J. B. Weber, and T. J. Sheets, "Adsorption of Prometryn and Metolachlor by Selected Soil Organic Matter Fractions", Soil Sci., 136(2), 94-101 (1983).
43. Ladlie, J. S., W.F. Meggitt, and D. Penner, Weed Science, 24, 477 (1976).
44. Lagas, P., "Sorption of Chlorophenols in the Soil", Chemosphere, 17(2), 205-216 (1988).
45. Lai, W. Y., D. H. Chen, and R. N. Maddox, Ind. Eng. Chem. Res., 26, 1072 (1987).
46. Lambert, S. M., J. Agric. Food Chem., 15, 572 (1967).
47. Liu, L. C., M. Cibes-Viade, and F. K. S. Koo, "Adsorption of Ametryne and Diuron by Soils", Weed Sci., 18, 470-474 (1970).
48. Lokke, H., "Sorption of Selected Organic Pollutants in Danish Soils", Ecotoxicol. Environ. Saf., 8(5), 395-409 (1984).
49. Lyman, W. J., W. F. Reehl, and D. A. Rosenblatt, HANDBOOK OF CHEMICAL PROPERTY ESTIMATION METHODS, American Chemical Society, Washington, DC (1990).
50. Mackay, D., W. Y. Shiu, and K. C. Ma, ILLUSTRATED HANDBOOK OF PHYSICAL-CHEMICAL PROPERTIES AND ENVIRONMENTAL FATE FOR ORGANIC CHEMICALS, Volume 1 – Monoaromatic Hydrocarbons, Chlorobenzenes and PCBs, Lewis Publishers, New York, NY (1992).
51. Mackay, D., W. Y. Shiu, and K. C. Ma, ILLUSTRATED HANDBOOK OF PHYSICAL-CHEMICAL PROPERTIES AND ENVIRONMENTAL FATE FOR ORGANIC CHEMICALS, Volume 2 – Polynuclear Aromatic Hydrocarbons, Polychlorinated Dioxins, and Dibenzofurans, Lewis Publishers, New York, NY (1992).
52. Mackay, D., W. Y. Shiu, and K. C. Ma, ILLUSTRATED HANDBOOK OF PHYSICAL-CHEMICAL PROPERTIES AND ENVIRONMENTAL FATE FOR ORGANIC CHEMICALS, Volume 3 – Volatile Organic Compounds, Lewis Publishers, New York, NY (1993).
53. Mackay, D., W. Y. Shiu, and K. C. Ma, ILLUSTRATED HANDBOOK OF PHYSICAL-CHEMICAL PROPERTIES AND ENVIRONMENTAL FATE FOR ORGANIC CHEMICALS, Volume 4 – Oxygen, Nitrogen, and Sulfur Containing Compounds, Lewis Publishers, New York, NY (1995).
54. Mackay, D., W. Y. Shiu, and K. C. Ma, ILLUSTRATED HANDBOOK OF PHYSICAL-CHEMICAL PROPERTIES AND ENVIRONMENTAL FATE FOR ORGANIC CHEMICALS, Volume 5 – Pesticide Chemicals, Lewis Publishers, New York, NY (1997).
55. Madhun, Y. A., V. H. Freed, J. L. Young, and S. C. Fang, "Sorption of Bromacil, Chlortoluron, and Diuron by Soils", Soil Sci. Soc. Am. J., 50, 1467-1471 (1986).
56. Means, J. C., S. G. Wood, J. J. Hasset, and W. L. Banwart, "Sorption of Polynuclear Aromatic Hydrocarbons by Sediments and Soils", Environ. Sci. Technol., 14(2), 1524-1528 (1980).
57. Means, J. C., S. G. Wood, J. J. Hasset, and W. L. Banwart, "Sorption of Amino- and Carboxy-Substituted Polynuclear Aromatic Hydrocarbons by Sediments and Soils", Environ. Sci. Technol., 16(2), 93-98 (1982).
58. Means, J. C., S. G. Wood, J. J. Hassett, and W. L. Banwart, Environ. Sci. Technol., 14, 1524 (1980).
59. Mills, A. C. and J. W. Biggar, "Solubility-Temperature Effect on the Adsorption of gamma- and beta-BHC from Aqueous and Hexane Solutions by Soil Materials", Soil Sci. Soc. Am. Proc., 33, 210-216 (1969).
60. Mills, A. L. and M. Alexander, "Factors Affecting Dimethylnitrosamine Formation in Samples of Soil and Water", J. Environ, Qual., 5(4), 437-440 (1976).
61. Montgomery, J. H., GROUNDWATER CHEMICALS, Lewis Publishers, New York, NY (2000).
62. Moreale, A. and R. Van Bladel, "Behavior of 2,4-D in Belgian Soils", J. Environ. Qual., 9(4), 627-633 (1980).
63. Nearpass, D. C., Soil Sci., 103, 177 (1967).
64. Nowara, A., J. Burhenne, and M. Spiteller, "Binding of Fluoroquinoline Carboxylic Acid Derivatives to Clay Minerals", J. Agric. Food Chem., 45(4), 1459-1463 (1997).
65. Pickens, J. and W. C. Lennox, Water Resources Research, 12, No. 2, 171 (1976).
66. Pierce, R. H., Jr., C. E. Olney, and G. T. Felbeck, Jr., Geochem. Cosmochim. Acta, 38, 1061 (1974)
67. Poinke, H. B. and G. Chesters, J. Environ. Qual., 2 , 29 (1973).
68. Reddy, K. N., M. Singh, and A. K. Alva, "Sorption and Leaching of Bromacil and Simazine in Florida Flatwoods Soils", Bull. Environ. Contam Toxicol., 48(5), 662-670 (1992).

69. Sabljic, A., J. Argic. Food Chem., 32, 243 (1984).
70. Sabljic, A., Environ. Sci. Technol. 21(4), 358 (1987).
71. Schleibe, K. A., O. C. Burnside, and T. L. Lavy, "Dissipation of Amiben", Weeds, 13, 321-325 (1965).
72. Schwarzenbach, R. P. and J. Westall, "Transport of Nonpolar *Organic Compounds from Surface* Water to Groundwater: Laboratory Sorption Studies", Environ. Sci. Technol., 22(1), 1360-1367 (1981).
73. Seip, H. M., J. Alstad, G. E. Crlberg, K. Martinsen, and R. Skaane, "Measurement of Mobility of Organic Compounds in Soils", Sci. Total Environ., 50, 87-101 (1986).
74. Shang, C. and M. A. Arshad, "Sorption of Clopyralid, Dicamba and MCPA by two Soils with Conventional and No-Till Management", Can. J. Soil. Sci., 78, 181-186 (1998).
75. Sharom, M. S. and L. V. Edgington, "The Adsorption, Mobility, and Persistence of Metalaxyl in Soil and Aqueous Systems", Can. J. Plant Pathol., 4, 334-340 (1982).
76. Sharom, M. S., J. R. W. Miles, C. R. Harris, and F. L. McEwen, "Behavior of 12 Pesticides in Soil and Aqueous Suspension of Soil and Sediment", Water Res., 14(8), 1095-1100 (1980a).
77. Sharom, M. S., J. R. W. Miles, J. W. Harris, and F. L. McEwen, "Persistence of 12 Insecticides in Water", Water Res., 14(8), 1089-1093 (1980).
78. Southworth, G. R. and J. L. Keller, "Hydrophobic Sorption of Polar Organics by Low Organic Carbon Soils", Water, Air, Soil Pollut., 28(3-4), 239-248 (1986).
79. Starr, R. I., R. W. Trim, K. G. Doxtader, D. B. Hurlbut, S. A. Volz, and M. Goodall, "Sorption and Aerobic Biodegradation of Strychnine Alkaloid in Various Soil Systems", J. Agric. Food Chem., 44(6), 1603-1608 (1996).
80. Steen, W. C., D. F. Paris, and G. L. Baughman, "Partitioning of Selected Polychlorinated Biphenyls to Natural Sediments", Water Res., 12(9), 655-657 (1978).
81. Suba, J. D. and M. E. Essington, "Adsorption of Fluometuron and Norflurazon: Effect of Tillage and Dissolved Organic Carbon", Soil Sci., 164(3), 145-155 (1999).
82. Swann, R. L., D. A. Laskowski, P. J. McCall, K. Vander Kuy, and H. J. Dishburger, Residue Reviews, 85, 17 (1983).
83. Talbert, R. E. and O. H. Fletchall, "The Adsoption of s-Triazines in Soils", Weeds, 46-52 (1965).
84. Valverde-Gracia, A., E. Gonzalez-Pradas, M. Villafranca-Sanchez, F. del Rey-Bueno, and A. Gracia-Rodriguez, "Adsorption of Thiram and Dimethoate on Almeria Soils", Soil Sci. Soc. Am. J., 52(6), 1571-1574 (1988).
85. Xu, F., X. M. M. Liang, F. Su, Q. Zhang, B. C. Lin, W. Z. Wu, A. Yediler, and A. Kettrup, "A Column Method for Determination of Soil Organic Partition Coefficients of Eight Pesticides", Chemosphere, 39(5), 787-794 (1999).
86. Yaws, C. L., CHEMICAL PROPERTIES HANDBOOK, McGraw-Hill, New York, NY (1999).

Table 15 Soil Sorption Coefficient - Oganic Compounds

Soil Sorption Coefficient							
NO	ID	FORMULA	NAME	CAS No	log_{10} Koc	Koc	code
1	17	CBrF3	bromotrifluoromethane	75-63-8	2.39	2.45E+02	2
2	29	CBr4	carbon tetrabromide	558-13-4	3.24	1.73E+03	2
3	34	CClF3	chlorotrifluoromethane	75-72-9	2.27	1.88E+02	2
4	42	CCl2F2	dichlorodifluoromethane	75-71-8	2.55	3.56E+02	2
5	47	CCl3F	trichlorofluoromethane	75-69-4	2.75	5.67E+02	2
6	50	CCl4	carbon tetrachloride	56-23-5	2.27	1.86E+02	1
7	74	CF4	carbon tetrafluoride	75-73-0	2.02	1.04E+02	2
8	96	CHBr3	bromoform	75-25-2	2.67	4.70E+02	2
9	97	CHClF2	chlorodifluoromethane	75-45-6	1.96	9.22E+01	2
10	101	CHCl2F	dichlorofluoromethane	75-43-4	2.22	1.66E+02	2
11	104	CHCl3	chloroform	67-66-3	1.65	4.47E+01	1
12	113	CHF3	fluoroform	75-46-7	1.73	5.31E+01	2
13	140	CH2BrCl	bromochloromethane	74-97-5	2.14	1.39E+02	2
14	144	CH2Br2	dibromomethane	74-95-3	2.63	4.25E+02	2
15	145	CH2ClF	chlorofluoromethane	593-70-4	1.65	4.51E+01	2
16	149	CH2Cl2	dichloromethane	75-09-2	2.06	1.14E+02	2
17	158	CH2F2	difluoromethane	75-10-5	1.49	3.06E+01	2
18	161	CH2I2	diiodomethane	75-11-6	2.74	5.46E+02	2
19	175	CH2O	formaldehyde	50-00-0	1.57	3.69E+01	2
20	176	CH2O2	formic acid	64-18-6	1.08	1.21E+01	2
21	191	CH3Br	methyl bromide	74-83-9	0.79	6.17E+00	1
22	196	CH3Cl	methyl chloride	74-87-3	1.87	7.45E+01	2
23	214	CH3F	methyl fluoride	593-53-3	1.65	4.51E+01	2
24	222	CH3I	methyl iodide	74-88-4	1.04	1.10E+01	1
25	232	CH3NO	formamide	75-12-7	0.56	3.59E+00	2
26	235	CH3NO2	nitromethane	75-52-5	1.20	1.58E+01	2
27	251	CH4	methane	74-82-8	1.97	9.33E+01	2
28	272	CH4O	methyl alcohol	67-56-1	0.97	9.43E+00	2
29	286	CH5N	methylamine	74-89-5	1.07	1.17E+01	2
30	382	CS2	carbon disulfide	75-15-0	2.54	3.48E+02	2
31	441	C2Cl2F4	1,2-dichloro-1,1,2,2-tetrafluor	76-14-2	2.91	8.15E+02	2
32	449	C2Cl3F3	1,1,2-trichloro-1,2,2-trifluoroe	76-13-1	3.10	1.25E+03	2
33	455	C2Cl4	tetrachloroethylene	127-18-4	2.31	2.04E+02	1
34	463	C2Cl6	hexachloroethane	67-72-1	3.55	3.57E+03	2
35	494	C2F6	hexafluoroethane	76-16-4	2.47	2.92E+02	2
36	511	C2HBrClF3	halothane	151-67-7	2.63	4.25E+02	2
37	539	C2HCl3	trichloroethylene	79-01-6	2.15	1.41E+02	1
38	544	C2HCl3O	trichloroacetaldehyde	75-87-6	1.92	8.23E+01	2
39	551	C2HCl5	pentachloroethane	76-01-7	2.95	8.90E+02	2
40	571	C2H2	acetylene	74-86-2	1.58	3.79E+01	2
41	615	C2H2Cl2	1,1-dichloroethylene	75-35-4	2.54	3.43E+02	2
42	616	C2H2Cl2	cis-1,2-dichloroethylene	156-59-2	2.39	2.45E+02	2
43	617	C2H2Cl2	trans-1,2-dichloroethylene	156-60-5	2.43	2.67E+02	2
44	627	C2H2Cl2O2	dichloroacetic acid	79-43-6	1.88	7.54E+01	2
45	633	C2H2Cl3NO	2,2,2-trichloroacetamide	594-65-0	1.94	8.77E+01	2
46	636	C2H2Cl4	1,1,2,2-tetrachloroethane	79-34-5	2.68	4.76E+02	2
47	642	C2H2F2	1,1-difluoroethylene	75-38-7	2.05	1.13E+02	2
48	648	C2H2F3NO	trifluoroacetamide	354-38-1	1.44	2.77E+01	2
49	688	C2H3Br	vinyl bromide	593-60-2	2.23	1.70E+02	2
50	693	C2H3BrO2	bromoacetic acid	79-08-3	1.60	3.98E+01	2
51	701	C2H3Cl	vinyl chloride	75-01-4	2.13	1.34E+02	2
52	710	C2H3ClO2	chloroacetic acid	79-11-8	1.50	3.14E+01	2
53	720	C2H3Cl3	1,1,1-trichloroethane	71-55-6	2.01	1.02E+02	1
54	721	C2H3Cl3	1,1,2-trichloroethane	79-00-5	1.80	6.31E+01	1
55	723	C2H3Cl3O	2,2,2-trichloroethanol	115-20-8	2.11	1.29E+02	2
56	738	C2H3F3O	2,2,2-trifluoroethanol	75-89-8	1.60	3.98E+01	2
57	753	C2H3N	acetonitrile	75-05-8	1.19	1.56E+01	2
58	787	C2H4	ethylene	74-85-1	1.99	9.81E+01	2
59	799	C2H4Br2	1,2-dibromoethane	106-93-4	2.44	2.77E+02	2
60	812	C2H4Cl2	1,1-dichloroethane	75-34-3	1.49	3.09E+01	1
61	813	C2H4Cl2	1,2-dichloroethane	107-06-2	1.65	4.47E+01	1
62	814	C2H4Cl2	dichloroethane	1300-21-6	1.55	3.55E+01	2
63	834	C2H4F2	1,1-difluoroethane	75-37-6	1.79	6.10E+01	2
64	845	C2H4I2	1,2-diiodoethane	624-73-7	2.85	7.10E+02	2
65	884	C2H4O	acetaldehyde	75-07-0	1.62	4.19E+01	2
66	885	C2H4O	ethylene oxide	75-21-8	1.21	1.64E+01	2
67	889	C2H4O2	acetic acid	64-19-7	1.28	1.93E+01	2
68	890	C2H4O2	methyl formate	107-31-3	1.39	2.47E+01	2
69	910	C2H5Br	bromoethane	74-96-4	2.25	1.77E+02	2
70	914	C2H5Cl	ethyl chloride	75-00-3	2.15	1.43E+02	2
71	917	C2H5ClO	2-chloroethanol	107-07-3	1.39	2.47E+01	2
72	948	C2H5I	ethyl iodide	75-03-6	2.47	2.92E+02	2
73	956	C2H5NO	acetamide	60-35-5	0.69	4.92E+00	2
74	957	C2H5NO	N-methylformamide	123-39-7	0.85	7.07E+00	2
75	958	C2H5NO2	nitroethane	79-24-3	1.47	2.98E+01	2
76	985	C2H6	ethane	74-84-0	2.36	2.30E+02	2
77	1065	C2H6O	ethyl alcohol	64-17-5	1.21	1.64E+01	2
78	1066	C2H6O	dimethyl ether	115-10-6	1.43	2.70E+01	2
79	1067	C2H6OS	dimethyl sulfoxide	67-68-5	0.64	4.39E+00	2
80	1072	C2H6O2	ethylene glycol	107-21-1	0.64	4.34E+00	2
81	1075	C2H6O2S	dimethyl sulfone	67-71-0	0.61	4.07E+00	2
82	1083	C2H6O4S	dimethyl sulfate	77-78-1	2.01	1.02E+02	2
83	1094	C2H6S2	dimethyl disulfide	624-92-0	2.34	2.19E+02	2
84	1115	C2H7N	ethylamine	75-04-7	1.31	2.02E+01	2
85	1116	C2H7N	dimethylamine	124-40-3	2.72	5.25E+02	1
86	1117	C2H7NO	monoethanolamine	141-43-5	0.66	4.62E+00	2
87	1159	C2H8N2	ethylenediamine	107-15-3	0.27	1.85E+00	2
88	1210	C2N2	cyanogen	460-19-5	1.42	2.60E+01	2
89	1287	C3F6O	hexafluoroacetone	684-16-2	2.17	1.48E+02	2
90	1377	C3H2N2	malononitrile	109-77-3	1.05	1.12E+01	2
91	1462	C3H3N	acrylonitrile	107-13-1	1.51	3.26E+01	2
92	1463	C3H3NO	oxazole	288-42-6	1.44	2.77E+01	2
93	1473	C3H3NS	thiazole	288-47-1	1.62	4.13E+01	2
94	1492	C3H4	methylacetylene	74-99-7	1.89	7.73E+01	2
95	1493	C3H4	allene	463-49-0	2.17	1.46E+02	2
96	1535	C3H4Cl2	cis-1,2-dichloropropene	6923-20-2	2.48	3.03E+02	2
97	1592	C3H4N2	imidazole	288-32-4	1.33	2.16E+01	2
98	1594	C3H4N2	1H-pyrazole	288-13-1	1.45	2.80E+01	2
99	1621	C3H4O	acrolein	107-02-8	1.37	2.35E+01	2
100	1622	C3H4O	propargyl alcohol	107-19-7	1.17	1.48E+01	2
101	1626	C3H4O2	acrylic acid	79-10-7	1.57	3.69E+01	2
102	1645	C3H5Br	3-bromo-1-propene	106-95-6	2.35	2.24E+02	2
103	1670	C3H5Cl	2-chloro-1-propene	557-98-2	2.47	2.92E+02	2
104	1676	C3H5ClO	alpha-epichlorohydrin	106-89-8	1.54	3.47E+01	2
105	1707	C3H5Cl3	1,2,3-trichloropropane	96-18-4	2.61	4.09E+02	2
106	1752	C3H5N	propionitrile	107-12-0	1.46	2.91E+01	2
107	1755	C3H5NO	acrylamide	79-06-1	0.95	8.97E+00	2
108	1757	C3H5NO	lactonitrile	78-97-7	0.87	7.34E+00	2
109	1796	C3H5N3O9	nitroglycerine	55-63-0	2.26	1.81E+02	2
110	1809	C3H6	cyclopropane	75-19-4	2.31	2.05E+02	2
111	1810	C3H6	propylene	115-07-1	2.34	2.19E+02	2
112	1845	C3H6Cl2	1,2-dichloropropane	78-87-5	2.47	2.92E+02	2
113	1920	C3H6O	allyl alcohol	107-18-6	1.47	2.95E+01	2
114	1922	C3H6O	propionaldehyde	123-38-6	1.70	4.99E+01	2
115	1923	C3H6O	acetone	67-64-1	1.25	1.76E+01	2
116	1924	C3H6O	1,2-propylene oxide	75-56-9	1.39	2.47E+01	2
117	1925	C3H6O	1,3-propylene oxide	503-30-0	1.30	2.00E+01	2
118	1926	C3H6O	methyloxirane	16033-71-9	1.39	2.47E+01	2
119	1930	C3H6O2	propanoic acid	79-09-4	1.56	3.60E+01	2
120	1931	C3H6O2	ethyl formate	109-94-4	1.50	3.18E+01	2
121	1932	C3H6O2	methyl acetate	79-20-9	1.47	2.98E+01	2
122	1942	C3H6O2S	3-mercaptopropionic acid	107-96-0	1.61	4.08E+01	2
123	1948	C3H6O3	lactic acid	50-21-5	0.99	9.67E+00	2
124	1950	C3H6O3	trioxane	110-88-3	1.14	1.39E+01	2
125	1978	C3H7Br	1-bromopropane	106-94-5	2.52	3.31E+02	2
126	1979	C3H7Br	2-bromopropane	75-26-3	2.41	2.57E+02	2
127	1986	C3H7Cl	1-chloropropane	540-54-5	2.49	3.07E+02	2
128	1987	C3H7Cl	2-chloropropane	75-29-6	2.41	2.57E+02	2
129	2025	C3H7I	1-iodopropane	107-08-4	2.74	5.46E+02	2
130	2026	C3H7I	2-iodopropane	75-30-9	2.95	8.90E+02	2
131	2030	C3H7N	allylamine	107-11-9	1.39	2.47E+01	2
132	2036	C3H7NO	N,N-dimethylformamide	68-12-2	0.83	6.72E+00	2
133	2037	C3H7NO	N-methylacetamide	79-16-3	0.81	6.39E+00	2
134	2045	C3H7NO2	1-nitropropane	108-03-2	1.85	7.08E+01	2
135	2046	C3H7NO2	2-nitropropane	79-46-9	1.88	7.64E+01	2
136	2094	C3H8	propane	74-98-6	2.66	4.58E+02	2
137	2103	C3H8NO5P	glyphosate	1071-83-6	3.46	2.88E+03	1
138	2129	C3H8O	propyl alcohol	71-23-8	1.51	3.26E+01	2
139	2130	C3H8O	isopropyl alcohol	67-63-0	1.40	2.54E+01	2
140	2138	C3H8O2	2-methoxyethanol	109-86-4	0.96	9.08E+00	2
141	2140	C3H8O2	1,2-propanediol (propylene g	57-55-6	0.88	7.53E+00	2
142	2141	C3H8O2	1,3-propanediol	504-63-2	0.81	6.48E+00	2
143	2148	C3H8O3	glycerol	56-81-5	0.42	2.63E+00	2
144	2158	C3H8S	propyl mercaptan	107-03-9	2.36	2.30E+02	2
145	2160	C3H8S	methyl ethyl sulfide	624-89-5	2.21	1.64E+02	2
146	2200	C3H9N	propylamine	107-10-8	1.64	4.35E+01	2
147	2201	C3H9N	isopropylamine	75-31-0	1.52	3.30E+01	2
148	2202	C3H9N	methylethylamine	624-78-2	1.46	2.87E+01	2
149	2203	C3H9N	trimethylamine	75-50-3	1.46	2.91E+01	2
150	2204	C3H9NO	1-amino-2-propanol	78-96-6	0.85	7.16E+00	2
151	2205	C3H9NO	3-amino-1-propanol	156-87-6	0.77	5.86E+00	2
152	2206	C3H9NO	methylethanolamine	109-83-1	0.87	7.34E+00	2
153	2241	C3H9O4P	trimethyl phosphate	512-56-1	1.02	1.06E+01	2
154	2318	C4Cl6	hexachloro-1,3-butadiene	87-68-3	3.98	9.49E+03	2
155	2562	C4H4N2	succinonitrile	110-61-2	0.84	6.89E+00	2
156	2565	C4H4N2	pyrimidine	289-95-2	1.16	1.44E+01	2
157	2592	C4H4O	furan	110-00-9	2.11	1.28E+02	2
158	2605	C4H4O4	fumaric acid	110-17-8	1.63	4.24E+01	2
159	2606	C4H4O4	maleic acid	110-16-7	1.12	1.31E+01	2
160	2613	C4H4S	thiophene	110-02-1	2.36	2.30E+02	2

Table 15 Soil Sorption Coefficient - Oganic Compounds

Soil Sorption Coefficient							
NO	ID	FORMULA	NAME	CAS No	log_{10} Koc	Koc	code
161	2697	C4H5N	methacrylonitrile	126-98-7	1.75	5.58E+01	2
162	2698	C4H5N	vinylacetonitrile	109-75-1	1.59	3.93E+01	2
163	2701	C4H5N	pyrrole	109-97-7	1.79	6.10E+01	2
164	2713	C4H5NO2	methyl cyanoacetate	105-34-0	1.12	1.32E+01	2
165	2754	C4H6	dimethylacetylene	503-17-3	2.17	1.48E+02	2
166	2757	C4H6	butadiene (1,3 butadiene)	106-99-0	2.46	2.88E+02	2
167	2942	C4H6O	2,5-dihydrofuran	1708-29-8	1.63	4.24E+01	2
168	2965	C4H6O2	gamma-butyrolactone	96-48-0	1.03	1.07E+01	2
169	2969	C4H6O2	methacrylic acid	79-41-4	1.88	7.64E+01	2
170	2970	C4H6O2	methyl acrylate	96-33-3	1.81	6.49E+01	2
171	2971	C4H6O2	vinyl acetate	108-05-4	1.77	5.94E+01	2
172	3006	C4H6O4	succinic acid	110-15-6	1.06	1.14E+01	2
173	3238	C4H7N	butyronitrile	109-74-0	1.70	5.05E+01	2
174	3239	C4H7N	isobutyronitrile	78-82-0	1.63	4.24E+01	2
175	3249	C4H7NO	2-pyrrolidone	616-45-5	0.91	8.21E+00	2
176	3320	C4H8	1-butene	106-98-9	2.68	4.82E+02	2
177	3322	C4H8	cis-2-butene	590-18-1	2.64	4.41E+02	2
178	3323	C4H8	trans-2-butene	624-64-6	2.63	4.30E+02	2
179	3324	C4H8	isobutene	115-11-7	2.66	4.52E+02	2
180	3375	C4H8Cl2O	bis(2-chloroethyl) ether	111-44-4	1.99	9.69E+01	2
181	3476	C4H8O	ethyl vinyl ether	109-92-2	1.94	8.77E+01	2
182	3478	C4H8O	butyraldehyde	123-72-8	1.86	7.17E+01	2
183	3480	C4H8O	methyl ethyl ketone	78-93-3	1.53	3.43E+01	2
184	3481	C4H8O	tetrahydrofuran	109-99-9	1.63	4.24E+01	2
185	3504	C4H8O2	butyric acid	107-92-6	1.81	6.41E+01	2
186	3505	C4H8O2	isobutyric acid	79-31-2	1.89	7.73E+01	2
187	3506	C4H8O2	propyl formate	110-74-7	1.83	6.74E+01	2
188	3508	C4H8O2	ethyl acetate	141-78-6	1.77	5.94E+01	2
189	3509	C4H8O2	methyl propanoate	554-12-1	1.83	6.82E+01	2
190	3510	C4H8O2	1,4-dioxane	123-91-1	1.15	1.41E+01	2
191	3536	C4H8O2S	sulfolane	126-33-0	0.96	9.08E+00	2
192	3598	C4H9Br	1-bromobutane	109-65-9	2.87	7.46E+02	2
193	3610	C4H9Cl	1-chlorobutane	109-69-3	2.81	6.50E+02	2
194	3612	C4H9Cl	2-chlorobutane	78-86-4	2.64	4.41E+02	2
195	3657	C4H9F	1-fluorobutane	2366-52-1	2.78	6.03E+02	2
196	3666	C4H9I	1-iodobutane	542-69-8	3.01	1.02E+03	2
197	3675	C4H9N	pyrrolidine	123-75-1	1.63	4.24E+01	2
198	3681	C4H9NO	N,N-dimethylacetamide	127-19-5	0.96	9.08E+00	2
199	3682	C4H9NO	morpholine	110-91-8	0.40	2.50E+00	2
200	3685	C4H9NO	butanamide	541-35-5	1.26	1.83E+01	2
201	3697	C4H9NO2	1-nitrobutane	627-05-4	2.18	1.50E+02	2
202	3783	C4H10	butane	106-97-8	2.95	8.90E+02	2
203	3830	C4H10N2	piperazine	110-85-0	0.74	5.50E+00	2
204	3865	C4H10O	butanol	71-36-3	1.83	6.82E+01	2
205	3866	C4H10O	isobutanol	78-83-1	1.79	6.17E+01	2
206	3867	C4H10O	sec-butanol	78-92-2	1.73	5.38E+01	2
207	3868	C4H10O	tert-butanol	75-65-0	1.57	3.69E+01	2
208	3869	C4H10O	diethyl ether	60-29-7	1.86	7.26E+01	2
209	3870	C4H10O	methyl propyl ether	557-17-5	2.04	1.08E+02	2
210	3882	C4H10O2	1,2-dimethoxyethane	110-71-4	1.26	1.83E+01	2
211	3883	C4H10O2	2-ethoxyethanol	110-80-5	1.20	1.60E+01	2
212	3886	C4H10O2	1,4-butanediol	110-63-4	0.93	8.42E+00	2
213	3937	C4H10O4S	diethyl sulfate	64-67-5	2.00	9.93E+01	2
214	3942	C4H10S	butyl mercaptan	109-79-5	2.62	4.14E+02	2
215	3947	C4H10S	diethyl sulfide	352-93-2	2.44	2.74E+02	2
216	3977	C4H11N	butylamine	109-73-9	1.84	7.00E+01	2
217	3978	C4H11N	isobutylamine	78-81-9	1.77	5.94E+01	2
218	3980	C4H11N	tert-butylamine	75-64-9	1.59	3.93E+01	2
219	3984	C4H11N	diethylamine	109-89-7	1.69	4.93E+01	2
220	4006	C4H11NO2	diethanolamine	111-42-2	0.60	3.97E+00	2
221	4169	C5Cl6	hexachlorocyclopentadiene	77-47-4	4.12	1.31E+04	2
222	4327	C5H4O2	furfural	98-01-1	1.60	3.98E+01	2
223	4391	C5H5N	pyridine	110-86-1	1.73	5.38E+01	2
224	4490	C5H6N2	glutaronitrile	544-13-8	0.99	9.67E+00	2
225	4534	C5H6O	2-methylfuran	534-22-5	2.38	2.42E+02	2
226	4548	C5H6O2	furfuryl alcohol	98-00-0	1.53	3.38E+01	2
227	4587	C5H6S	2-methylthiophene	554-14-3	2.64	4.41E+02	2
228	4588	C5H6S	3-methylthiophene	616-44-4	2.65	4.47E+02	2
229	4658	C5H7N	N-methylpyrrole	96-54-8	2.04	1.08E+02	2
230	4738	C5H8	isoprene	78-79-5	2.69	4.94E+02	2
231	4741	C5H8	cis-1,3-pentadiene	1574-41-0	2.68	4.82E+02	2
232	4742	C5H8	trans-1,3-pentadiene	2004-70-8	2.70	5.06E+02	2
233	4743	C5H8	1,4-pentadiene	591-93-5	2.73	5.32E+02	2
234	4745	C5H8	1-pentyne	627-19-0	2.45	2.85E+02	2
235	4882	C5H8O2	acetylacetone	123-54-6	1.59	3.93E+01	2
236	4883	C5H8O2	allyl acetate	591-87-7	1.90	8.03E+01	2
237	4888	C5H8O2	ethyl acrylate	140-88-5	2.10	1.24E+02	2
238	4889	C5H8O2	methyl methacrylate	80-62-6	2.13	1.34E+02	2
239	4927	C5H8O3	2-hydroxyethyl acrylate	818-61-1	1.26	1.83E+01	2
240	4928	C5H8O3	levulinic acid	123-76-2	1.11	1.29E+01	2

Soil Sorption Coefficient							
NO	ID	FORMULA	NAME	CAS No	log_{10} Koc	Koc	code
241	4945	C5H8O4	glutaric acid	110-94-1	1.22	1.66E+01	2
242	5183	C5H9N	valeronitrile	110-59-8	1.89	7.73E+01	2
243	5198	C5H9NO	N-methyl-2-pyrrolidone	872-50-4	1.17	1.48E+01	2
244	5245	C5H9NO4	L-glutamic acid	56-86-0	-0.63	2.34E-01	2
245	5285	C5H10	cyclopentane	287-92-3	3.01	1.02E+03	2
246	5450	C5H10O	methyl propyl ketone	107-87-9	1.83	6.82E+01	2
247	5451	C5H10O	diethyl ketone	96-22-0	1.82	6.65E+01	2
248	5452	C5H10O	methyl isopropyl ketone	563-80-4	1.68	4.80E+01	2
249	5453	C5H10O	2-methyltetrahydrofuran	96-47-9	2.38	2.42E+02	2
250	5471	C5H10O	tetrahydropyran	142-68-7	1.82	6.65E+01	2
251	5492	C5H10O2	pentanoic acid	109-52-4	2.13	1.36E+02	2
252	5494	C5H10O2	3-methylbutanoic acid	503-74-2	2.01	1.02E+02	2
253	5501	C5H10O2	propyl acetate	109-60-4	2.05	1.13E+02	2
254	5503	C5H10O2	ethyl propanoate	105-37-3	2.04	1.08E+02	2
255	5504	C5H10O2	methyl butanoate	623-42-7	2.08	1.20E+02	2
256	5543	C5H10O3	diethyl carbonate	105-58-8	2.04	1.08E+02	2
257	5603	C5H11Br	1-bromopentane	110-53-2	3.21	1.62E+03	2
258	5620	C5H11Cl	1-chloropentane	543-59-9	2.86	7.28E+02	2
259	5625	C5H11Cl	2-chloro-2-methylbutane	594-36-5	2.75	5.60E+02	2
260	5649	C5H11F	1-fluoropentane	592-50-7	2.64	4.41E+02	2
261	5669	C5H11N	N-methylpyrrolidine	120-94-5	1.88	7.54E+01	2
262	5670	C5H11N	piperidine	110-89-4	1.83	6.82E+01	2
263	5696	C5H11NO2	1-nitropentane	628-05-7	2.47	2.95E+02	2
264	5772	C5H12	pentane	109-66-0	3.25	1.79E+03	2
265	5773	C5H12	isopentane	78-78-4	2.63	4.25E+02	2
266	5774	C5H12	neopentane	463-82-1	3.07	1.17E+03	2
267	5790	C5H12NO3PS2	dimethoate	60-51-5	2.56	3.63E+02	1
268	5831	C5H12O	1-pentanol	71-41-0	2.20	1.58E+02	2
269	5832	C5H12O	2-pentanol	6032-29-7	2.06	1.14E+02	2
270	5833	C5H12O	3-pentanol	584-02-1	2.04	1.08E+02	2
271	5834	C5H12O	2-methyl-1-butanol	137-32-6	2.08	1.20E+02	2
272	5837	C5H12O	3-methyl-1-butanol	123-51-3	2.07	1.18E+02	2
273	5838	C5H12O	tert-pentyl-alcohol	75-85-4	1.86	7.26E+01	2
274	5839	C5H12O	3-methyl-2-butanol	598-75-4	2.07	1.18E+02	2
275	5840	C5H12O	2,2-dimethyl-1-propanol	75-84-3	2.09	1.23E+02	2
276	5844	C5H12O	methyl tert-butyl ether	1634-04-4	1.89	7.73E+01	2
277	5912	C5H12O4	pentaerythritol	115-77-5	0.46	2.87E+00	2
278	5947	C5H13N	pentylamine	110-58-7	2.19	1.54E+02	2
279	6067	C6Cl6	hexachlorobenzene	118-74-1	4.49	3.09E+04	1
280	6080	C6F6	hexafluorobenzene	392-56-3	2.76	5.81E+02	2
281	6115	C6HCl5	pentachlorobenzene	608-93-5	4.11	1.30E+04	2
282	6116	C6HCl5O	pentachlorophenol	87-86-5	2.47	2.95E+02	1
283	6174	C6H2Cl4	1,2,3,4-tetrachlorobenzene	634-66-2	3.52	3.31E+03	1
284	6175	C6H2Cl4	1,2,3,5-tetrachlorobenzene	634-90-2	3.52	3.31E+03	1
285	6176	C6H2Cl4	1,2,4,5-tetrachlorobenzene	95-94-3	3.72	5.25E+03	1
286	6177	C6H2Cl4O	2,3,4,5-tetrachlorophenol	4901-51-3	2.88	7.59E+02	2
287	6178	C6H2Cl4O	2,3,4,6-tetrachlorophenol	58-90-2	2.88	7.59E+02	1
288	6179	C6H2Cl4O	2,3,5,6-tetrachlorophenol	935-95-5	2.88	7.59E+02	2
289	6281	C6H3ClN2O4	1-chloro-2,4-dinitrobenzene	97-00-7	2.56	3.61E+02	2
290	6296	C6H3Cl2NO2	1,2-dichloro-4-nitrobenzene	99-54-7	2.53	3.39E+02	1
291	6305	C6H3Cl3	1,2,4-trichlorobenzene	120-82-1	3.11	1.29E+03	1
292	6306	C6H3Cl3	1,2,3-trichlorobenzene	87-61-6	3.23	1.70E+03	1
293	6307	C6H3Cl3	1,3,5-trichlorobenzene	108-70-3	2.85	7.08E+02	1
294	6308	C6H3Cl3	trichlorobenzene	12002-48-1	3.11	1.29E+03	2
295	6310	C6H3Cl3O	2,3,4-trichlorophenol	15950-66-0	1.96	9.12E+01	2
296	6311	C6H3Cl3O	2,3,5-trichlorophenol	933-78-8	1.96	9.12E+01	2
297	6312	C6H3Cl3O	2,3,6-trichlorophenol	933-75-5	1.96	9.12E+01	2
298	6313	C6H3Cl3O	2,4,5-trichlorophenol	95-95-4	1.96	9.12E+01	2
299	6314	C6H3Cl3O	2,4,6-trichlorophenol	88-06-2	1.96	9.12E+01	1
300	6315	C6H3Cl3O	3,4,5-trichlorophenol	609-19-8	1.96	9.12E+01	2
301	6325	C6H3Cl4N	nitrapyrin	1929-82-4	2.24	1.74E+02	1
302	6367	C6H3N3O6	1,3,5-trinitrobenzene	99-35-4	2.02	1.04E+02	2
303	6383	C6H4BrCl	1-bromo-2-chlorobenzene	694-80-4	2.60	3.98E+02	2
304	6384	C6H4BrCl	1-bromo-3-chlorobenzene	108-37-2	2.60	3.98E+02	2
305	6385	C6H4BrCl	1-bromo-4-chlorobenzene	106-39-8	2.60	3.98E+02	2
306	6385.1	C6H4BrCl	3-chloro-4-bromobenzene	29682-39-1	2.60	3.98E+02	1
307	6386	C6H4BrClO	3-bromo-5-chlorophenol	56962-04-0	2.60	3.98E+02	2
308	6387	C6H4BrClO	4-bromo-2-chlorophenol	3964-56-5	2.60	3.98E+02	2
309	6388	C6H4BrClO	2-bromo-4-chlorophenol	695-96-5	2.60	3.98E+02	2
310	6403	C6H4BrNO2	1-bromo-2-nitrobenzene	577-19-5	2.42	2.63E+02	2
311	6404	C6H4BrNO2	1-bromo-3-nitrobenzene	585-79-5	2.42	2.63E+02	2
312	6405	C6H4BrNO2	1-bromo-4-nitrobenzene	586-78-7	2.42	2.63E+02	1
313	6411	C6H4Br2	m-dibromobenzene	108-36-1	3.42	2.61E+03	2
314	6443	C6H4ClNO2	m-chloronitrobenzene	121-73-3	2.72	5.19E+02	2
315	6444	C6H4ClNO2	o-chloronitrobenzene	88-73-3	2.60	3.94E+02	2
316	6445	C6H4ClNO2	p-chloronitrobenzene	100-00-5	2.68	4.76E+02	2
317	6469	C6H4Cl2	o-dichlorobenzene	95-50-1	2.78	6.03E+02	1
318	6470	C6H4Cl2	m-dichlorobenzene	541-73-1	2.78	6.03E+02	1
319	6471	C6H4Cl2	p-dichlorobenzene	106-46-7	2.78	6.03E+02	1
320	6479	C6H4Cl2O	2,3-dichlorophenol	576-24-9	2.55	3.55E+02	2

Table 15 Soil Sorption Coefficient - Oganic Compounds

Soil Sorption Coefficient							
NO	ID	FORMULA	NAME	CAS No	log_{10} Koc	Koc	code
321	6480	C6H4Cl2O	2,4-dichlorophenol	120-83-2	2.55	3.55E+02	1
322	6481	C6H4Cl2O	2,5-dichlorophenol	583-78-8	2.55	3.55E+02	2
323	6482	C6H4Cl2O	2,6-dichlorophenol	87-65-0	2.55	3.55E+02	2
324	6483	C6H4Cl2O	3,4-dichlorophenol	95-77-2	2.55	3.55E+02	2
325	6484	C6H4Cl2O	3,5-dichlorophenol	591-35-5	2.55	3.55E+02	2
326	6524	C6H4F2	m-difluorobenzene	372-18-9	2.58	3.80E+02	2
327	6525	C6H4F2	o-difluorobenzene	367-11-3	2.67	4.64E+02	2
328	6526	C6H4F2	p-difluorobenzene	540-36-3	2.54	3.43E+02	2
329	6580	C6H4N2O4	m-dinitrobenzene	99-65-0	2.19	1.54E+02	2
330	6581	C6H4N2O4	o-dinitrobenzene	528-29-0	2.30	1.98E+02	2
331	6582	C6H4N2O4	p-dinitrobenzene	100-25-4	2.17	1.48E+02	2
332	6656	C6H5Br	bromobenzene	108-86-1	3.00	1.01E+03	2
333	6671	C6H5BrO	o-bromophenol	95-56-7	2.41	2.57E+02	2
334	6672	C6H5BrO	m-bromophenol	591-20-8	2.41	2.57E+02	2
335	6673	C6H5BrO	p-bromophenol	106-41-2	2.41	2.57E+02	1
336	6674	C6H5BrO	bromophenol	32762-51-9	2.41	2.57E+02	2
337	6692	C6H5Cl	chlorobenzene	108-90-7	2.22	1.66E+02	1
338	6720	C6H5ClO	m-chlorophenol	108-43-0	1.82	6.61E+01	1
339	6721	C6H5ClO	o-chlorophenol	95-57-8	1.71	5.13E+01	1
340	6722	C6H5ClO	p-chlorophenol	106-48-9	1.85	7.08E+01	1
341	6740	C6H5Cl2N	3,4-dichloroaniline	95-76-1	0.67	4.68E+00	1
342	6741	C6H5Cl2N	2,3-dichloroaniline	608-27-5	0.67	4.68E+00	2
343	6742	C6H5Cl2N	2,4-dichloroaniline	554-00-7	0.67	4.68E+00	2
344	6743	C6H5Cl2N	2,5-dichloroaniline	95-82-9	0.67	4.68E+00	2
345	6744	C6H5Cl2N	2,6-dichloroaniline	608-31-1	0.67	4.68E+00	2
346	6745	C6H5Cl2N	3,5-dichloroaniline	626-43-7	0.67	4.68E+00	2
347	6769	C6H5F	fluorobenzene	462-06-6	2.61	4.09E+02	2
348	6795	C6H5I	iodobenzene	591-50-4	3.16	1.45E+03	2
349	6819	C6H5NO2	nitrobenzene	98-95-3	2.01	1.02E+02	1
350	6830	C6H5NO3	o-nitrophenol	88-75-5	2.06	1.15E+02	1
351	6831	C6H5NO3	m-nitrophenol	554-84-7	1.72	5.25E+01	1
352	6832	C6H5NO3	p-nitrophenol	100-02-7	2.72	5.25E+02	1
353	6876	C6H6	benzene	71-43-2	1.87	7.41E+01	1
354	6905	C6H6BrN	o-bromoaniline	615-36-1	1.96	9.12E+01	2
355	6906	C6H6BrN	m-bromoaniline	591-19-5	1.96	9.12E+01	2
356	6907	C6H6BrN	p-bromoaniline	106-40-1	1.96	9.12E+01	1
357	6923	C6H6ClN	m-chloroaniline	108-42-9	3.13	1.35E+03	1
358	6924	C6H6ClN	o-chloroaniline	95-51-2	3.13	1.35E+03	1
359	6925	C6H6ClN	p-chloroaniline	106-47-8	3.13	1.35E+03	1
360	6948	C6H6Cl6	1,2,3,4,5,6-hexachlorocyclohe	58-89-9	3.41	2.57E+03	1
361	6949	C6H6Cl6	1,2,3,4,5,6-hexachlorocyclohe	319-85-7	3.12	1.32E+03	1
362	6950	C6H6Cl6	1,2,3,4,5,6-hexachlorocyclohe	319-86-8	3.30	2.00E+03	2
363	6951	C6H6Cl6	1,2,3,4,5,6-hexachlorocyclohe	60291-32-9	3.30	2.00E+03	2
364	6953	C6H6Cl6	a-hexachlorocyclohexane	319-84-6	3.30	2.00E+03	2
365	6954	C6H6Cl6	hexachlorocyclohexane; (mix	608-73-1	3.30	2.00E+03	2
366	6996	C6H6N2O2	m-nitroaniline	99-09-2	2.12	1.33E+02	2
367	6997	C6H6N2O2	o-nitroaniline	88-74-4	2.38	2.42E+02	2
368	6998	C6H6N2O2	p-nitroaniline	100-01-6	2.13	1.36E+02	2
369	7057	C6H6O	phenol	108-95-2	1.74	5.50E+01	1
370	7070	C6H6O2	pyrocatechol	120-80-9	2.07	1.17E+02	1
371	7071	C6H6O2	resorcinol	108-46-3	1.02	1.05E+01	1
372	7072	C6H6O2	p-hydroquinone	123-31-9	1.70	4.99E+01	2
373	7121	C6H6S	phenyl mercaptan	108-98-5	2.75	5.60E+02	2
374	7172	C6H7N	aniline	62-53-3	1.87	7.36E+01	1
375	7173	C6H7N	2-methylpyridine	109-06-8	1.98	9.57E+01	2
376	7174	C6H7N	3-methylpyridine	108-99-6	2.03	1.07E+02	2
377	7175	C6H7N	4-methylpyridine	108-89-4	2.04	1.10E+02	2
378	7269	C6H8	1,4-cyclohexadiene	628-41-1	2.72	5.26E+02	2
379	7316	C6H8N2	adiponitrile	111-69-3	1.20	1.60E+01	2
380	7318	C6H8N2	m-phenylenediamine	108-45-2	1.20	1.58E+01	2
381	7319	C6H8N2	o-phenylenediamine	95-54-5	1.46	2.87E+01	2
382	7320	C6H8N2	p-phenylenediamine	106-50-3	1.21	1.64E+01	2
383	7321	C6H8N2	phenylhydrazine	100-63-0	2.06	1.14E+02	2
384	7381	C6H8O	2-ethylfuran	3208-16-0	2.68	4.82E+02	2
385	7387	C6H8O	2-cyclohexen-1-one	930-68-7	1.71	5.11E+01	2
386	7396	C6H8O	5-hexyn-2-one	2550-28-9	1.69	4.93E+01	2
387	7459	C6H8O6	ascorbic acid	50-81-7	0.48	3.05E+00	2
388	7464	C6H8O7	citric acid	77-92-9	0.44	2.76E+00	2
389	7622	C6H10	cyclohexene	110-83-8	2.93	8.57E+02	2
390	7629	C6H10	1,5-hexadiene	592-42-7	2.90	7.95E+02	2
391	7632	C6H10	cis-2,trans-4-hexadiene	5194-50-3	2.90	7.95E+02	2
392	7634	C6H10	trans-2,trans-4-hexadiene	5194-51-4	3.01	1.03E+03	2
393	7648	C6H10	1-hexyne	693-02-7	2.86	7.28E+02	2
394	7760	C6H10O	cyclohexanone	108-94-1	1.82	6.57E+01	2
395	7776	C6H10O	5-hexen-2-one	109-49-9	1.93	8.55E+01	2
396	7814	C6H10O2	ethyl methacrylate	97-63-2	2.43	2.71E+02	2
397	7881	C6H10O3	ethylacetoacetate	141-97-9	1.51	3.26E+01	2
398	7917	C6H10O4	adipic acid	124-04-9	1.42	2.63E+01	2
399	7918	C6H10O4	diethyl oxalate	95-92-1	1.68	4.80E+01	2
400	7981	C6H11Br	bromocyclohexane	108-85-0	3.12	1.31E+03	2

Soil Sorption Coefficient							
NO	ID	FORMULA	NAME	CAS No	log_{10} Koc	Koc	code
401	8088	C6H11N	hexanenitrile	628-73-9	2.25	1.77E+02	2
402	8101	C6H11NO	epsilon-caprolactam	105-60-2	1.27	1.88E+01	2
403	8102	C6H11NO	cyclohexanone oxime	100-64-1	1.83	6.82E+01	2
404	8214	C6H12	methylcyclopentane	96-37-7	3.21	1.62E+03	2
405	8215	C6H12	cyclohexane	110-82-7	3.25	1.77E+03	2
406	8216	C6H12	1-hexene	592-41-6	3.23	1.69E+03	2
407	8225	C6H12	4-methyl-1-pentene	691-37-2	2.74	5.46E+02	2
408	8380	C6H12N2S4	thiram	137-26-8	3.01	1.02E+03	1
409	8400	C6H12O	cyclohexanol	108-93-0	2.05	1.11E+02	2
410	8401	C6H12O	hexanal	66-25-1	2.35	2.21E+02	2
411	8409	C6H12O	2-hexanone	591-78-6	2.13	1.34E+02	2
412	8411	C6H12O	3-methyl-2-pentanone	565-61-7	0.63	4.27E+00	2
413	8412	C6H12O	4-methyl-2-pentanone	108-10-1	0.63	4.27E+00	1
414	8460	C6H12O2	hexanoic acid	142-62-1	2.42	2.64E+02	2
415	8461	C6H12O2	2-ethyl butyric acid	88-09-5	2.29	1.95E+02	2
416	8476	C6H12O2	butyl acetate	123-86-4	2.37	2.33E+02	2
417	8477	C6H12O2	isobutyl acetate	110-19-0	2.35	2.21E+02	2
418	8478	C6H12O2	sec-butyl acetate	105-46-4	2.31	2.05E+02	2
419	8479	C6H12O2	tert-butyl acetate	540-88-5	2.33	2.16E+02	2
420	8532	C6H12O3	hydroxycaproic acid	1191-25-9	1.82	6.57E+01	2
421	8533	C6H12O3	paraldehyde	123-63-7	1.74	5.51E+01	2
422	8609	C6H12O6	glucose	50-99-7	-0.39	4.12E-01	2
423	8647	C6H13Br	1-bromohexane	111-25-1	3.44	2.78E+03	2
424	8742	C6H13N	cyclohexylamine	108-91-8	2.19	1.54E+02	2
425	8877	C6H14	hexane	110-54-3	3.55	3.57E+03	2
426	8878	C6H14	2,2-dimethylbutane	75-83-2	3.46	2.85E+03	2
427	8879	C6H14	2,3-dimethylbutane	79-29-8	3.47	2.96E+03	2
428	8881	C6H14	3-methylpentane	96-14-0	3.34	2.16E+03	2
429	8944	C6H14N2O2	lysine	56-87-1	-0.28	5.22E-01	2
430	8970	C6H14O	1-hexanol	111-27-3	2.48	3.03E+02	2
431	8971	C6H14O	2-hexanol	626-93-7	2.33	2.16E+02	2
432	8972	C6H14O	3-hexanol	623-37-0	2.27	1.88E+02	2
433	8987	C6H14O	3,3-dimethyl-2-butanol	464-07-3	2.18	1.52E+02	2
434	8993	C6H14O	dipropyl ether	111-43-3	2.48	3.03E+02	2
435	8995	C6H14O	diisopropyl ether	108-20-3	2.20	1.60E+02	2
436	9004	C6H14O	ethyl butyl ether	628-81-9	2.48	3.03E+02	2
437	9020	C6H14O2	acetal	105-57-7	1.83	6.82E+01	2
438	9021	C6H14O2	2-butoxyethanol	111-76-2	1.83	6.74E+01	2
439	9069	C6H14O2S	dipropyl sulfone	598-03-8	1.59	3.88E+01	2
440	9075	C6H14O3	diethylene glycol dimethyl eth	111-96-6	1.18	1.52E+01	2
441	9077	C6H14O3	2-(2-ethoxyethoxy)ethanol	111-90-0	1.08	1.21E+01	2
442	9078	C6H14O3	trimethylolpropane	77-99-6	0.57	3.73E+00	2
443	9100	C6H14O6	sorbitol	50-70-4	0.18	1.51E+00	2
444	9173	C6H15N	hexylamine	111-26-2	2.50	3.15E+02	2
445	9202	C6H15N	di-propylamine	142-84-7	2.29	1.93E+02	2
446	9203	C6H15N	diisopropylamine	108-18-9	2.14	1.38E+02	2
447	9212	C6H15N	triethylamine	121-44-8	2.17	1.46E+02	2
448	9230	C6H15NO2	diisopropanolamine	110-97-4	0.93	8.53E+00	2
449	9239	C6H15NO3	triethanolamine	102-71-6	0.83	6.81E+00	2
450	9268	C6H15O4P	triethyl phosphate	78-40-0	1.81	6.49E+01	2
451	9356	C6H18N3OP	hexamethyl phosphoramide	680-31-9	1.53	3.38E+01	2
452	9362	C6H18OSi2	hexamethyldisiloxane	107-46-0	3.66	4.59E+03	2
453	9799	C7H4F3NO2	3-nitrobenzotrifluoride	98-46-4	2.80	6.34E+02	2
454	9890	C7H5BrO2	2-bromobenzoic acid	88-65-3	2.57	3.75E+02	2
455	9891	C7H5BrO2	3-bromobenzoic acid	585-76-2	2.94	8.68E+02	2
456	9892	C7H5BrO2	4-bromobenzoic acid	586-76-5	2.93	8.57E+02	2
457	9932	C7H5ClO2	o-chlorobenzoic acid	118-91-2	2.49	3.11E+02	2
458	9957	C7H5Cl2NO2	chloramben	133-90-4	1.25	1.78E+01	1
459	9987	C7H5Cl3O2	4,5,6-trichloroguaiacol	2668-24-8	2.99	9.77E+02	1
460	10003	C7H5F3	benzotrifluoride	98-08-8	3.04	1.09E+03	2
461	10045	C7H5N	benzonitrile	100-47-0	2.23	1.68E+02	2
462	10094	C7H5NS	benzothiazole	95-16-9	2.47	2.95E+02	2
463	10118	C7H5N3O6	2,4,6-trinitrotoluene	118-96-7	2.25	1.77E+02	2
464	10253	C7H6Cl2	2,4-dichlorotoluene	95-73-8	3.68	4.83E+03	2
465	10261.1	C7H6Cl2N2O	3,4-dichlorophenyl urea	2327-02-8	2.49	3.09E+02	1
466	10320	C7H6F3N	2-(trifluoromethyl)aniline	88-17-5	2.36	2.29E+02	2
467	10321	C7H6F3N	3-(trifluoromethyl)aniline	98-16-8	2.36	2.29E+02	1
468	10322	C7H6F3N	4-(trifluoromethyl)aniline	455-14-1	2.36	2.29E+02	2
469	10323	C7H6F3NO	3-(trifluoromethoxy)aniline	1535-73-5	2.36	2.29E+02	2
470	10324	C7H6F3NO	2-(trifluoromethoxy)aniline	1535-75-7	2.36	2.29E+02	2
471	10325	C7H6F3NO	4-(trifluoromethoxy)aniline	461-82-5	2.36	2.29E+02	2
472	10326	C7H6F3NS	3-(trifluoromethylthio)aniline	369-68-6	2.36	2.29E+02	2
473	10327	C7H6F3NS	4-(trifluoromethylthio)aniline	372-16-7	2.36	2.29E+02	2
474	10341	C7H6N2	1H-benzimidazole	51-17-2	2.11	1.28E+02	2
475	10352	C7H6N2O	2-hydroxybenzimidazole	615-16-7	1.99	9.69E+01	2
476	10363	C7H6N2O4	2,4-dinitrotoluene	121-14-2	2.45	2.85E+02	2
477	10365	C7H6N2O4	2,6-dinitrotoluene	606-20-2	2.52	3.31E+02	2
478	10366	C7H6N2O4	3,4-dinitrotoluene	610-39-9	2.51	3.22E+02	2
479	10404	C7H6O	benzaldehyde	100-52-7	2.18	1.52E+02	2
480	10408	C7H6O2	benzoic acid	65-85-0	1.95	8.91E+01	1

Table 15 Soil Sorption Coefficient - Oganic Compounds

Soil Sorption Coefficient								Soil Sorption Coefficient							
NO	ID	FORMULA	NAME	CAS No	log_{10} Koc	Koc	code	NO	ID	FORMULA	NAME	CAS No	log_{10} Koc	Koc	code
481	10409	C7H6O2	p-hydroxybenzaldehyde	123-08-0	2.11	1.29E+02	2	561	12713	C8H6	ethynylbenzene	536-74-3	2.68	4.82E+02	2
482	10410	C7H6O2	salicylaldehyde	90-02-8	2.36	2.30E+02	2	562	12820	C8H6Cl2O3	3,6-dichloro-2-methoxybenzo	1918-00-9	0.99	9.77E+00	1
483	10411	C7H6O2	1,3-benzodioxole	274-09-9	2.51	3.22E+02	2	563	12821	C8H6Cl2O3	(2,4-dichlorophenoxy)acetic a	94-75-7	2.11	1.29E+02	1
484	10417	C7H6O2	phenyl formate	1864-94-4	2.06	1.15E+02	2	564	12868	C8H6N2	quinoxaline	91-19-0	1.96	9.22E+01	2
485	10420	C7H6O3	salicylic acid	69-72-7	2.57	3.75E+02	2	565	12914	C8H6O	benzofuran	271-89-6	2.83	6.75E+02	2
486	10462	C7H7Br	p-bromotoluene	106-38-7	3.24	1.73E+03	2	566	12934	C8H6O4	isophthalic acid	121-91-5	2.28	1.91E+02	2
487	10463	C7H7Br	(bromomethyl)benzene	100-39-0	2.97	9.24E+02	2	567	12935	C8H6O4	phthalic acid	88-99-3	1.77	5.94E+01	2
488	10469	C7H7BrN2O	(4-bromophenyl)urea	1967-25-5	2.12	1.32E+02	1	568	12936	C8H6O4	terephthalic acid	100-21-0	2.47	2.92E+02	2
489	10492	C7H7Cl	benzyl chloride	100-44-7	2.63	4.25E+02	2	569	12947	C8H6S	benzothiophene	95-15-8	3.07	1.19E+03	2
490	10493	C7H7Cl	o-chlorotoluene	95-49-8	3.24	1.73E+03	2	570	13080.1	C8H7F3N2O	3-(trifluoromethylphenyl) urea	13114-87-9	1.60	3.98E+01	1
491	10494	C7H7Cl	p-chlorotoluene	106-43-4	3.19	1.54E+03	2	571	13107	C8H7N	indole	120-72-9	2.54	3.48E+02	2
492	10502.1	C7H7ClN2O	2-chlorophenyl urea	114-38-5	1.61	4.07E+01	1	572	13108	C8H7N	benzeneacetonitrile	140-29-4	2.23	1.68E+02	2
493	10502.2	C7H7ClN2O	3-chlorophenyl urea	1967-27-7	2.01	1.02E+02	1	573	13237	C8H8	styrene	100-42-5	3.04	1.09E+03	2
494	10555	C7H7F	p-fluorotoluene	352-32-9	2.78	6.03E+02	2	574	13238	C8H8	1,3,5,7-cyclooctatetraene	629-20-9	3.05	1.13E+03	2
495	10559.1	C7H7FN2O	2-fluorophenyl urea	656-31-5	1.31	2.04E+01	1	575	13391	C8H8O	acetophenone	98-86-2	2.26	1.84E+02	2
496	10559.2	C7H7FN2O	3-fluorophenyl urea	770-19-4	1.77	5.89E+01	1	576	13393	C8H8O	benzeneacetaldehyde	122-78-1	2.35	2.21E+02	2
497	10559.3	C7H7FN2O	4-fluorophenyl urea	659-30-3	1.52	3.31E+01	1	577	13394	C8H8O	2-methylbenzaldehyde	529-20-4	2.61	4.04E+02	2
498	10611	C7H7NO	formanilide	103-70-8	2.00	1.01E+02	2	578	13396	C8H8O	2,3-dihydrobenzofuran	496-16-2	2.54	3.48E+02	2
499	10626	C7H7NO2	m-nitrotoluene	99-08-1	2.71	5.13E+02	2	579	13397	C8H8O	phenyloxirane	96-09-3	2.25	1.79E+02	2
500	10627	C7H7NO2	o-nitrotoluene	88-72-2	2.63	4.25E+02	2	580	13412	C8H8O2	methyl benzoate	93-58-3	2.57	3.75E+02	2
501	10628	C7H7NO2	p-nitrotoluene	99-99-0	2.69	4.94E+02	2	581	13413	C8H8O2	o-toluic acid	118-90-1	2.64	4.36E+02	2
502	10660	C7H7NO3	o-nitroanisole	91-23-6	2.32	2.08E+02	2	582	13414	C8H8O2	p-toluic acid	99-94-5	2.65	4.47E+02	2
503	10668	C7H7NO3	4-methyl-3-nitrophenol	2042-14-0	2.61	4.07E+02	2	583	13415	C8H8O2	benzeneacetic acid	103-82-2	2.14	1.39E+02	2
504	10669	C7H7NO3	3-methyl-4-nitrophenol	2581-34-2	2.61	4.07E+02	1	584	13417	C8H8O2	phenyl acetate	122-79-2	2.19	1.54E+02	2
505	10670	C7H7NO3	3-methyl-2-nitrophenol	4920-77-8	2.61	4.07E+02	2	585	13433	C8H8O2	m-toluic acid	99-04-7	2.67	4.64E+02	2
506	10671	C7H7NO3	2-methyl-3-nitrophenol	5460-31-1	2.61	4.07E+02	2	586	13445	C8H8O3	methyl salicylate	119-36-8	2.76	5.81E+02	2
507	10672	C7H7NO3	5-methyl-2-nitrophenol	700-38-9	2.61	4.07E+02	2	587	13446	C8H8O3	vanillin	121-33-5	2.04	1.08E+02	2
508	10725	C7H8	toluene	108-88-3	1.97	9.33E+01	1	588	13671	C8H9NO	acetanilide	103-84-4	2.01	1.02E+02	2
509	10735	C7H8BrN	2-bromo-4-methylaniline	583-68-6	1.96	9.12E+01	2	589	13789	C8H10	ethylbenzene	100-41-4	2.73	5.37E+02	1
510	10736	C7H8BrN	2-bromo-5-methylaniline	53078-85-6	1.96	9.12E+01	2	590	13790	C8H10	o-xylene	95-47-6	2.70	5.01E+02	1
511	10737	C7H8BrN	3-bromo-4-methylaniline	7745-91-7	1.96	9.12E+01	2	591	13791	C8H10	m-xylene	108-38-3	2.46	2.88E+02	1
512	10738	C7H8BrN	4-bromo-2-methylaniline	583-75-5	1.96	9.12E+01	2	592	13792	C8H10	p-xylene	106-42-3	2.77	5.89E+02	1
513	10739	C7H8BrN	4-bromo-3-methylaniline	6933-10-4	1.96	9.12E+01	1	593	13793	C8H10	xylenes	1330-20-7	2.60	3.98E+02	2
514	10740	C7H8BrN	5-bromo-2-methylaniline	39478-78-9	1.96	9.12E+01	2	594	13854	C8H10NO5PS	methyl parathion	298-00-0	2.64	4.37E+02	1
515	10741	C7H8BrN	3-bromo-2-methylaniline	55289-36-6	1.96	9.12E+01	2	595	13911	C8H10Na2O5	endothall disodium	145-73-3	2.14	1.38E+02	1
516	10760	C7H8ClNO	3-chloroanisidine	5345-54-0	1.93	8.51E+01	1	596	13912	C8H10O	phenetole	103-73-1	2.74	5.53E+02	2
517	10794	C7H8N2O	phenylurea	64-10-8	1.35	2.24E+01	1	597	13913	C8H10O	2-phenylethanol	60-12-8	2.12	1.31E+02	2
518	10850	C7H8O	anisole	100-66-3	2.52	3.35E+02	2	598	13915	C8H10O	3-methylbenzenemethanol	587-03-1	2.25	1.77E+02	2
519	10851	C7H8O	benzyl alcohol	100-51-6	1.95	8.88E+01	2	599	13916	C8H10O	4-methylbenzenemethanol	589-18-4	2.24	1.72E+02	2
520	10852	C7H8O	m-cresol	108-39-4	1.54	3.47E+01	1	600	13917	C8H10O	1-phenylethanol	98-85-1	1.57	3.72E+01	1
521	10853	C7H8O	o-cresol	95-48-7	1.34	2.19E+01	1	601	13918	C8H10O	o-ethylphenol	90-00-6	2.72	5.26E+02	2
522	10854	C7H8O	p-cresol	106-44-5	1.69	4.90E+01	1	602	13919	C8H10O	m-ethylphenol	620-17-7	2.74	5.46E+02	2
523	10855	C7H8O	cresol mixture	1319-77-3	1.45	2.82E+01	2	603	13920	C8H10O	p-ethylphenol	123-07-9	2.74	5.46E+02	2
524	10866	C7H8O2	guaiacol	90-05-1	1.60	3.98E+01	1	604	13921	C8H10O	2,3-xylenol	526-75-0	2.66	4.57E+02	2
525	10867	C7H8O2	p-methoxyphenol	150-76-5	1.75	5.62E+01	1	605	13922	C8H10O	2,4-xylenol	105-67-9	2.66	4.57E+02	1
526	10881	C7H8O2	3-methoxyphenol	150-19-6	1.55	3.55E+01	1	606	13923	C8H10O	2,5-xylenol	95-87-4	2.66	4.57E+02	1
527	10953	C7H9N	benzylamine	100-46-9	1.97	9.33E+01	2	607	13924	C8H10O	2,6-xylenol	576-26-1	2.66	4.57E+02	1
528	10954	C7H9N	N-methylaniline	100-61-8	2.28	1.91E+02	2	608	13925	C8H10O	3,4-xylenol	95-65-8	2.66	4.57E+02	1
529	10955	C7H9N	m-toluidine	108-44-1	1.74	5.50E+01	1	609	13926	C8H10O	3,5-xylenol	108-68-9	2.66	4.57E+02	1
530	10956	C7H9N	o-toluidine	95-53-4	1.74	5.50E+01	1	610	13927	C8H10O	benzyl methyl ether	538-86-3	2.11	1.29E+02	2
531	10957	C7H9N	p-toluidine	106-49-0	1.90	7.94E+01	1	611	13929	C8H10O	2-methylanisole	578-58-5	2.87	7.37E+02	2
532	10961	C7H9N	2,6-dimethylpyridine	108-48-5	2.29	1.95E+02	2	612	13930	C8H10O	3-methylanisole	100-84-5	2.82	6.67E+02	2
533	11079	C7H10N2	m-toluenediamine	95-80-7	1.45	2.84E+01	2	613	13931	C8H10O	4-methylanisole	104-93-8	2.91	8.05E+02	2
534	11376	C7H12ClN5	simazine	122-34-9	2.08	1.20E+02	1	614	13951	C8H10O2	1,2-dimethoxybenzene	91-16-7	2.51	3.22E+02	2
535	11450	C7H12O2	butyl acrylate	141-32-2	2.66	4.58E+02	2	615	14033.1	C8H10O5	endothall	145-73-3	2.14	1.38E+02	1
536	11451	C7H12O2	isobutyl acrylate	106-63-8	2.58	3.84E+02	2	616	14082	C8H11N	N,N-dimethylaniline	121-69-7	2.63	4.30E+02	2
537	11530	C7H12O4	diethyl malonate	105-53-3	1.90	7.93E+01	2	617	14083	C8H11N	o-ethylaniline	578-54-1	2.32	2.11E+02	2
538	11688	C7H13N3O3S	oxamyl	23135-22-0	0.90	7.94E+00	1	618	14096	C8H11N	2,4,6-trimethylpyridine	108-75-8	2.40	2.51E+02	2
539	11695	C7H13O6P	mevinphos	7786-34-7	1.64	4.37E+01	1	619	14110	C8H11N	benzeneethanamine	64-04-0	2.14	1.39E+02	2
540	11704	C7H14	methylcyclohexane	108-87-2	3.49	3.07E+03	2	620	14228	C8H12	1,5-cyclooctadiene	111-78-4	3.10	1.25E+03	2
541	11705	C7H14	cycloheptane	291-64-5	3.55	3.57E+03	2	621	14231	C8H12	vinylcyclohexene	100-40-3	3.51	3.27E+03	2
542	11706	C7H14	1-heptene	592-76-7	3.55	3.53E+03	2	622	14497	C8H14	2,5-dimethyl-2,4-hexadiene	764-13-6	3.28	1.91E+03	2
543	11803	C7H14N2O2S	aldicarb	116-06-3	1.30	2.00E+01	1	623	14599	C8H14O2	butyl methacrylate	97-88-1	2.94	8.78E+02	2
544	11829	C7H14O	2-heptanone	110-43-0	2.45	2.85E+02	2	624	14688	C8H14O4	diethyl succinate	123-25-1	2.03	1.07E+02	2
545	11834	C7H14O	5-methyl-2-hexanone	110-12-3	2.40	2.51E+02	2	625	14775	C8H15N	octanenitrile	124-12-9	2.87	7.46E+02	2
546	11843	C7H14O	2,4-dimethyl-3-pentanone	565-80-0	2.39	2.45E+02	2	626	14888	C8H16	cyclooctane	292-64-8	3.80	6.28E+03	2
547	11845	C7H14O	cis-2-methylcyclohexanol	7443-70-1	2.38	2.39E+02	2	627	14890	C8H16	1-octene	111-66-0	3.86	7.30E+03	2
548	11846	C7H14O	trans-2-methylcyclohexanol	7443-52-9	2.37	2.33E+02	2	628	14979	C8H16	2,4,4-trimethyl-1-pentene	107-39-1	3.85	7.12E+03	2
549	11884	C7H14O2	heptanoic acid	111-14-8	2.69	4.94E+02	2	629	15066	C8H16O	2-octanone	111-13-7	2.67	4.64E+02	2
550	11982	C7H15Br	1-bromoheptane	629-04-9	3.75	5.61E+03	2	630	15156	C8H16O2	octanoic acid	124-07-2	3.04	1.09E+03	2
551	11988	C7H15Cl	1-chloroheptane	629-06-1	3.63	4.31E+03	2	631	15248	C8H17Br	1-bromooctane	111-83-1	4.04	1.09E+04	2
552	12112	C7H16	heptane	142-82-5	3.83	6.68E+03	2	632	15354	C8H18	octane	111-65-9	4.18	1.51E+04	2
553	12164	C7H16O	1-heptanol	111-70-6	2.80	6.34E+02	2	633	15430	C8H18O	1-octanol	111-87-5	3.05	1.11E+03	2
554	12165	C7H16O	2-heptanol	543-49-7	2.63	4.30E+02	2	634	15431	C8H18O	2-octanol	123-96-6	2.95	9.01E+02	2
555	12166	C7H16O	3-heptanol	589-82-2	2.60	3.94E+02	2	635	15434	C8H18O	4-octanol	589-62-8	2.83	6.84E+02	2
556	12167	C7H16O	4-heptanol	589-55-9	2.58	3.84E+02	2	636	15527	C8H18O	dibutyl ether	142-96-1	3.12	1.33E+03	2
557	12315	C7H17N	heptylamine	111-68-2	2.78	5.96E+02	2	637	15578	C8H18O3	diethylene glycol diethyl ethe	112-36-7	1.59	3.88E+01	2
558	12364	C7H17O2PS3	phorate	298-02-2	2.82	6.61E+02	1	638	15579	C8H18O3	diethylene glycol monobutyl e	112-34-5	1.68	4.80E+01	2
559	12415	C8Cl4N2	chlorothalonil	1897-45-6	2.98	9.55E+02	1	639	15631	C8H19N	octylamine	111-86-4	2.95	9.01E+02	2
560	12594	C8H4O3	phthalic anhydride	85-44-9	2.25	1.77E+02	2	640	15637	C8H19N	dibutylamine	111-92-2	2.92	8.25E+02	2

Table 15 Soil Sorption Coefficient - Oganic Compounds

Soil Sorption Coefficient								Soil Sorption Coefficient							
NO	ID	FORMULA	NAME	CAS No	log_{10} Koc	Koc	code	NO	ID	FORMULA	NAME	CAS No	log_{10} Koc	Koc	code
641	15756	C8H24O4Si4	octamethylcyclotetrasiloxane	556-67-2	4.15	1.42E+04	2	721	18432	C9H22O4P2S4	ethion	563-12-2	3.94	8.71E+03	1
642	15788	C9H4Cl3NO2S	folpet	133-07-3	3.27	1.86E+03	1	722	18551	C10H6Cl8	chlordane	57-74-9	5.15	1.41E+05	2
643	15926	C9H6O2	2H-1-benzopyran-2-one	91-64-5	2.13	1.36E+02	2	723	18552	C10H6Cl8	trans-chlordan	5566-34-7	5.15	1.41E+05	2
644	15928	C9H6O2	1H-indene-1,3(2H)-dione	606-23-5	1.71	5.11E+01	2	724	18553	C10H6Cl8	a-chlordan	5103-71-9	5.15	1.41E+05	2
645	16010	C9H7N	isoquinoline	119-65-3	2.51	3.22E+02	2	725	18554	C10H6Cl8	trans-chlordane	5103-74-2	5.15	1.41E+05	1
646	16011	C9H7N	quinoline	91-22-5	2.48	3.03E+02	2	726	18586	C10H6O3	2-hydroxy-1,4-naphthalenedi	83-72-7	2.17	1.48E+02	2
647	16017	C9H7N	cinnamonitrile	4360-47-8	2.44	2.77E+02	2	727	18606	C10H7Cl	1-chloronaphthalene	90-13-1	3.50	3.15E+03	2
648	16018	C9H7NO	8-hydroxyquinoline	148-24-3	2.48	2.99E+02	2	728	18607	C10H7Cl	2-chloronaphthalene	91-58-7	3.54	3.48E+03	2
649	16090	C9H8	indene	95-13-6	2.97	9.24E+02	2	729	18669	C10H8	naphthalene	91-20-3	3.00	1.00E+03	1
650	16128	C9H8Cl3NO2S	captan	133-06-2	2.30	2.00E+02	1	730	18670	C10H8	azulene	275-51-4	3.13	1.34E+03	2
651	16185	C9H8O	2-methylbenzofuran	4265-25-2	3.13	1.34E+03	2	731	18740	C10H8O	1-naphthol	90-15-3	2.92	8.36E+02	2
652	16200	C9H8O	2-propenophenone	768-03-6	2.40	2.51E+02	2	732	18741	C10H8O	2-naphthol	135-19-3	2.85	7.01E+02	2
653	16212	C9H8O2	cinnamic acid	621-82-9	2.54	3.43E+02	2	733	18832	C10H9Cl4NO2S	captafol	2425-06-1	3.32	2.09E+03	1
654	16324	C9H9ClO3	(4-chloro-2-methylphenoxy)a	94-74-6	3.86	7.24E+03	1	734	18849	C10H9N	2-methylquinoline	91-63-4	2.79	6.11E+02	2
655	16331	C9H9Cl2NO	propanil	709-98-8	2.48	3.02E+02	1	735	18861	C10H9N	1-naphthylamine	134-32-7	3.58	3.80E+03	1
656	16364	C9H9N	benzenepropanenitrile	645-59-0	2.31	2.05E+02	2	736	18862	C10H9N	2-naphthylamine	91-59-8	3.58	3.80E+03	2
657	16388	C9H9NO	cinnamamide, predominantly	621-79-4	2.14	1.39E+02	2	737	19076	C10H10N4O	4-amino-3-methyl-6-phenyl-1	41394-05-2	1.55	3.55E+01	1
658	16508	C9H10	indane	496-11-7	3.19	1.54E+03	2	738	19106	C10H10O	benzalacetone	122-57-6	2.50	3.18E+02	2
659	16512	C9H10	alpha-methylstyrene	98-83-9	3.27	1.86E+03	2	739	19119	C10H10O2	methyl cinnamate	1754-62-7	2.80	6.34E+02	2
660	16522	C9H10BrClN2O2	chlorbromuron	13360-45-7	2.58	3.80E+02	1	740	19195	C10H10O4	dimethyl phthalate	131-11-3	1.63	4.27E+01	1
661	16537.1	C9H10ClN5O2	imidacloprid	105827-78-9	2.64	4.37E+02	1	741	19196	C10H10O4	dimethyl terephthalate	120-61-6	2.60	3.99E+02	2
662	16540	C9H10Cl2N2O	diuron	330-54-1	2.82	6.61E+02	1	742	19275	C10H11F3N2O	N,N-dimethyl-N'-[3-(trifluorom	2164-17-2	1.96	9.12E+01	1
663	16541	C9H10Cl2N2O2	linuron	330-55-2	2.43	2.69E+02	1	743	19359	C10H12	1,2,3,4-tetrahydronaphthalen	119-64-2	3.28	1.89E+03	2
664	16627	C9H10O	2,3-dihydro-1H-inden-1-ol	6351-10-6	4.06	1.15E+04	2	744	19441.1	C10H12N2O	3-phenyl-1-cyclopropyl urea	13140-86-8	1.72	5.25E+01	1
665	16628	C9H10O	2,3-dihydro-1H-inden-5-ol	1470-94-6	4.06	1.15E+04	1	745	19468	C10H12N3O3PS2	azinphos-methyl	86-50-0	2.69	4.90E+02	1
666	16631	C9H10O	4-methylacetophenone	122-00-9	2.57	3.70E+02	2	746	19551	C10H12O2	isopropyl benzoate	939-48-0	3.11	1.28E+03	2
667	16634	C9H10O	1-phenyl-1-propanone	93-55-0	2.57	3.70E+02	2	747	19729	C10H13ClN2O	N-(3-chloro-4-methylphenyl)-	15545-48-9	2.43	2.69E+02	1
668	16635	C9H10O	1-phenyl-2-propanone	103-79-7	2.16	1.45E+02	2	748	19745.1	C10H13Cl2NO4	2,4-D dimethylammonium	3599-58-4	1.21	1.62E+01	1
669	16676	C9H10O2	ethyl benzoate	93-89-0	2.81	6.50E+02	2	749	19799	C10H13NO2	isopropyl phenylcarbamate	122-42-9	1.95	8.91E+01	1
670	16680	C9H10O2	benzyl acetate	140-11-4	2.44	2.77E+02	2	750	19865	C10H14	butylbenzene	104-51-8	3.69	4.95E+03	2
671	16683	C9H10O2	4-methylphenyl acetate	140-39-6	2.52	3.35E+02	2	751	19866	C10H14	isobutylbenzene	538-93-2	3.56	3.62E+03	2
672	16701	C9H10O2	(±)-2-phenylpropionic acid	492-37-5	2.36	2.27E+02	2	752	19867	C10H14	sec-butylbenzene	135-98-8	3.86	7.30E+03	2
673	16711	C9H10O3	ethyl vanillin	121-32-4	2.24	1.72E+02	2	753	19868	C10H14	tert-butylbenzene	98-06-6	3.61	4.10E+03	2
674	16834	C9H11BrN2O2	metobromuron	3060-89-7	2.02	1.05E+02	1	754	19870	C10H14	m-cymene	535-77-3	3.83	6.68E+03	2
675	16854	C9H11ClN2O	N'-(4-chlorophenyl)-N,N-dime	150-68-5	1.70	5.01E+01	1	755	19871	C10H14	o-cymene	527-84-4	3.76	5.75E+03	2
676	16855	C9H11ClN2O2	3-(4-chlorophenyl)-1-methoxy	1746-81-2	1.84	6.92E+01	1	756	19872	C10H14	p-cymene	99-87-6	3.61	4.05E+03	2
677	16876	C9H11Cl3NO3PS	chlorpyrifos	2921-88-2	3.79	6.17E+03	1	757	19878	C10H14	o-diethylbenzene	135-01-3	3.78	6.05E+03	2
678	16877.1	C9H11FN2O	1,1-dimethyl-3-(3-fluoropheny	330-39-2	1.73	5.37E+01	1	758	19879	C10H14	m-diethylbenzene	141-93-5	3.86	7.30E+03	2
679	16877.2	C9H11FN2O	1,1-dimethyl-3-(4-fluoropheny	332-33-2	1.43	2.69E+01	1	759	19880	C10H14	p-diethylbenzene	105-05-5	3.87	7.39E+03	2
680	16895	C9H11N	1,2,3,4-tetrahydroquinoline	635-46-1	2.62	4.20E+02	2	760	19881	C10H14	3-ethyl-o-xylene	933-98-2	3.74	5.47E+03	2
681	16905	C9H11NO	p-dimethylaminobenzaldehyd	100-10-7	2.36	2.30E+02	2	761	19882	C10H14	4-ethyl-o-xylene	934-80-5	3.83	6.68E+03	2
682	17004	C9H12	cumene	98-82-8	3.37	2.33E+03	2	762	19883	C10H14	2-ethyl-m-xylene	2870-04-4	3.71	5.07E+03	2
683	17005	C9H12	m-ethyltoluene	620-14-4	3.54	3.48E+03	2	763	19884	C10H14	4-ethyl-m-xylene	874-41-9	3.81	6.44E+03	2
684	17006	C9H12	o-ethyltoluene	611-14-3	3.30	1.98E+03	2	764	19885	C10H14	5-ethyl-m-xylene	934-74-7	3.85	7.12E+03	2
685	17007	C9H12	p-ethyltoluene	622-96-8	3.35	2.25E+03	2	765	19886	C10H14	2-ethyl-p-xylene	1758-88-9	3.79	6.12E+03	2
686	17008	C9H12	1,2,3-trimethylbenzene	526-73-8	3.34	2.16E+03	2	766	19887	C10H14	1,2,3,4-tetramethylbenzene	488-23-3	3.55	3.57E+03	2
687	17009	C9H12	1,2,4-trimethylbenzene	95-63-6	3.35	2.25E+03	2	767	19888	C10H14	1,2,3,5-tetramethylbenzene	527-53-7	3.61	4.05E+03	2
688	17011	C9H12	mesitylene	108-67-8	3.24	1.73E+03	2	768	19889	C10H14	1,2,4,5-tetramethylbenzene	95-93-2	3.61	4.05E+03	2
689	17012	C9H12	propylbenzene	103-65-1	3.38	2.42E+03	2	769	19927	C10H14NO5PS	parathion	56-38-2	3.02	1.05E+03	1
690	17035	C9H12NO5PS	fenitrothion	122-14-5	3.51	3.24E+03	1	770	19950.1	C10H14N2O2	1,1-dimethyl-3-(4-methoxy-ph	28170-54-9	1.72	5.25E+01	1
691	17077	C9H12O	benzyl ethyl ether	539-30-0	2.55	3.56E+02	2	771	19983	C10H14O	4-butylphenol	1638-22-8	3.36	2.30E+03	2
692	17091	C9H12O	benzenepropanol	122-97-4	2.40	2.51E+02	2	772	20195	C10H15N	N,N-dietyhlaniline	91-66-7	3.18	1.51E+03	2
693	17101	C9H12O	2-propylphenol	644-35-9	2.97	9.35E+02	2	773	20321	C10H16	alpha-pinene	80-56-8	4.00	1.01E+04	2
694	17103	C9H12O	4-propylphenol	645-56-7	3.12	1.31E+03	2	774	20326	C10H16	gamma-terpinene	99-85-4	3.83	6.68E+03	2
695	17116	C9H12O	2,3,4-trimethylphenol	526-85-2	3.76	5.75E+03	2	775	20327	C10H16	terpinolene	586-62-9	3.81	6.44E+03	2
696	17117	C9H12O	2,3,5-trimethylphenol	697-82-5	3.76	5.75E+03	1	776	20647.1	C10H18ClN5	ipazine	1912-25-0	3.39	2.45E+03	1
697	17118	C9H12O	2,3,6-trimethylphenol	2416-94-6	3.76	5.75E+03	1	777	20915	C10H19N5O	2-methoxy-4,6-bis(isopropyla	1610-18-0	2.43	2.69E+02	1
698	17119	C9H12O	2,4,5-trimethylphenol	496-78-6	3.76	5.75E+03	2	778	20917	C10H19N5S	prometryn	7287-19-6	2.80	6.31E+02	1
699	17120	C9H12O	2,4,6-trimethylphenol	527-60-6	3.76	5.75E+03	1	779	20993	C10H20O	2-decanone	693-54-9	3.43	2.68E+03	2
700	17121	C9H12O	3,4,5-trimethylphenol	527-54-8	3.76	5.75E+03	2	780	21040	C10H20O2	decanoic acid	334-48-5	3.60	4.00E+03	2
701	17234	C9H13BrN2O2	bromacil	314-40-9	1.97	9.33E+01	1	781	21133	C10H22	decane	124-18-5	4.78	5.98E+04	2
702	17237	C9H13ClN2O2	terbacil	5902-51-2	1.63	4.27E+01	1	782	21190	C10H22	2,2,3,3-tetramethylhexane	13475-81-5	4.11	1.30E+04	2
703	17238	C9H13ClN6	cyanazine	21725-46-2	2.26	1.82E+02	1	783	21224	C10H22O	1-decanol	112-30-1	3.86	7.30E+03	2
704	17250	C9H13N	amphetamine	300-62-9	2.33	2.16E+02	2	784	21520	C11H8N2O	2-(2-furyl)benzimidazole	3878-19-1	2.55	3.55E+02	1
705	17253	C9H13N	N,N-dimethylbenzylamine	103-83-3	2.45	2.85E+02	2	785	21531	C11H8O2	2-methyl-1,4-naphthalenedio	58-27-5	2.57	3.75E+02	2
706	17436	C9H14O	isophorone	78-59-1	2.30	2.00E+02	2	786	21535	C11H8O3	2-hydroxy-3-methyl-1,4-napht	483-55-6	2.03	1.07E+02	2
707	17513	C9H14O6	glyceryl triacetate	102-76-1	1.51	3.26E+01	2	787	21543	C11H8O3	2-methoxy-1,4-naphthalenedi	2348-82-5	2.11	1.29E+02	2
708	17637	C9H16ClN5	propazine	139-40-2	2.19	1.55E+02	1	788	21565	C11H9N	4-phenylpyridine	939-23-1	2.79	6.11E+02	2
709	17640	C9H16ClN5	triethazine	1912-26-1	2.74	5.50E+02	1	789	21601	C11H10	1-methylnaphthalene	90-12-0	3.48	3.04E+03	2
710	17755	C9H16O4	azelaic acid	123-99-9	2.23	1.70E+02	2	790	21602	C11H10	2-methylnaphthalene	91-57-6	3.55	3.57E+03	2
711	17857	C9H17N5S	ametryn	834-12-8	2.13	1.35E+02	1	791	21650	C11H10O	1-naphthalenemethanol	4780-79-4	2.22	1.66E+02	1
712	17942	C9H18	1-nonene	124-11-8	4.18	1.51E+04	2	792	21651	C11H10O	2-naphthalenemethanol	1592-38-7	2.22	1.66E+02	2
713	17999	C9H18O	2-nonanone	821-55-6	3.10	1.25E+03	2	793	21801	C11H12Cl2N2O5	chloramphenicol	56-75-7	2.00	9.93E+01	2
714	18005	C9H18O	5-methyl-2-octanone	58654-67-4	2.97	9.24E+02	2	794	22113	C11H14O2	butyl benzoate	136-60-7	3.47	2.92E+03	2
715	18052	C9H18O2	nonanoic acid	112-05-0	3.24	1.73E+03	2	795	22278	C11H15NO2S	methiocarb	2032-65-7	2.25	1.78E+02	1
716	18176	C9H20	nonane	111-84-2	4.45	2.82E+04	2	796	22304	C11H16	pentylbenzene	538-68-1	4.04	1.10E+04	2
717	18236	C9H20O	1-nonanol	143-08-8	3.56	3.66E+03	2	797	22354	C11H16	pentamethylbenzene	700-12-9	3.86	7.21E+03	2
718	18288	C9H20O	2,6-dimethyl-4-heptanol	108-82-7	3.05	1.13E+03	2	798	22823	C11H22O	2-undecanone	112-12-9	3.60	4.00E+03	2
719	18393	C9H21N	tripropylamine	102-69-2	2.89	7.85E+02	2	799	22848	C11H22O2	methyl decanoate	110-42-9	3.78	5.97E+03	2
720	18408	C9H21O2PS3	terbufos	13071-79-9	2.50	3.16E+02	1	800	23055	C11H24O	1-undecanol	112-42-5	3.94	8.80E+03	2

Table 15 Soil Sorption Coefficient - Oganic Compounds

Soil Sorption Coefficient								Soil Sorption Coefficient							
NO	ID	FORMULA	NAME	CAS No	$\log_{10}$ Koc	Koc	code	NO	ID	FORMULA	NAME	CAS No	$\log_{10}$ Koc	Koc	code
801	23119	C12Cl10	decachlorobiphenyl	2051-24-3	5.87	7.42E+05	2	881	25277	C12H26O	1-dodecanol	112-53-8	4.17	1.47E+04	2
802	23134	C12HCl9	2,2',3,3',4,5,5',6,6'-nonachlor	52663-77-1	5.82	6.55E+05	2	882	25294	C12H26O3	diethylene glycol dibutyl ethe	112-73-2	2.42	2.64E+02	2
803	23145	C12H2Cl8	2,2',3,3',5,5',6,6'-octachlorobi	2136-99-4	5.24	1.74E+05	2	883	25446	C13H6Cl6O2	hexachlorophene	70-30-4	3.51	3.27E+03	2
804	23152	C12H3Cl7	2,2',3,3',4,4',6-heptachlorobip	52663-71-5	5.02	1.05E+05	2	884	25496	C13H8O	9H-fluoren-9-one	486-25-9	3.32	2.11E+03	2
805	23152.1	C12H4ClF6N4OS	fiponil	120068-37-3	3.08	1.20E+03	1	885	25537	C13H9N	acridine	260-94-6	3.23	1.69E+03	2
806	23159	C12H4Cl6	2,2',3,3',4,4'-hexachlorobiphe	38380-07-3	3.83	6.76E+03	1	886	25575	C13H10	fluorene	86-73-7	3.66	4.59E+03	2
807	23160	C12H4Cl6	2,2',4,4',6,6'-hexachlorobiphe	33979-03-2	3.83	6.76E+03	1	887	25576	C13H10BrCl2O2PS	o-(4-bromo-2,5-dichloropheny	21609-90-5	5.07	1.17E+05	1
808	23161	C12H4Cl6	2,2',3,3',6,6'-hexachlorobiphe	38411-22-2	3.83	6.76E+03	1	888	25623	C13H10O	benzophenone	119-61-9	2.64	4.37E+02	1
809	23162	C12H4Cl6	2,3,3',4,4',5-hexachlorobiphe	38380-08-4	3.83	6.76E+03	2	889	25631	C13H10O2	phenyl benzoate	93-99-2	3.33	2.14E+03	2
810	23163	C12H4Cl6	2,2',4,4',5'5'-hexachloro-1,1'-l	35065-27-1	3.83	6.76E+03	1	890	25694	C13H11NO	N-phenylbenzamide	93-98-1	2.80	6.34E+02	2
811	23164	C12H4Cl6	3,3',4,4',5,5'-hexachlorobiphe	---	3.83	6.76E+03	2	891	25738	C13H12	diphenylmethane	101-81-5	3.63	4.26E+03	2
812	23177	C12H5Cl5	2,3,4,5,6-pentachlorobipheny	18259-05-7	4.80	6.37E+04	2	892	25741	C13H12	4-methylbiphenyl	644-08-6	3.90	7.87E+03	2
813	23178	C12H5Cl5	2,2',4,5,5'-pentachlorobiphen	37680-73-2	4.86	7.22E+04	2	893	25774.1	C13H12N2O2	4-phenoxyphenyl urea	78508-44-8	2.56	3.63E+02	1
814	23179	C12H5Cl5	alochlor 1254	11097-69-1	4.47	2.95E+04	1	894	25793	C13H12O	benzyl phenyl ether	946-80-5	3.44	2.75E+03	2
815	23204	C12H6Cl4	2,3,4,5-tetrachlorobiphenyl	33284-53-6	5.64	4.37E+05	1	895	25798	C13H12O	diphenylmethanol	91-01-0	2.83	6.75E+02	2
816	23205	C12H6Cl4	2,2',4',5-tetrachlorobiphenyl	41464-40-8	5.64	4.37E+05	1	896	25803	C13H12O	4-biphenylmethanol	3597-91-9	2.69	4.90E+02	1
817	23206	C12H6Cl4	chlorodiphenyl	53469-21-9	5.13	1.35E+05	1	897	26032	C13H15N3O3	imazapyr	81334-34-1	2.35	2.24E+02	1
818	23207	C12H6Cl4	3,3',4,4'-tetrachlorobiphenyl	32598-13-3	5.00	1.00E+05	2	898	26033.1	C13H15N3O3	imazapyr	81334-34-1	2.35	2.24E+02	1
819	23208	C12H6Cl4	2,2',3,3'-tetrachlorobiphenyl	38444-93-8	5.00	1.00E+05	2	899	26049	C13H16F3N3O4	trifluralin	1582-09-8	4.49	3.09E+04	1
820	23209	C12H6Cl4	2,2',5,5'-tetrachlorobiphenyl	35693-99-3	5.37	2.34E+05	1	900	26183.1	C13H18N2O	3-phenyl-1-cyclohexyl urea	886-59-9	2.07	1.17E+02	1
821	23210	C12H6Cl4	2,2',6,6'-tetrachlorobiphenyl	15968-05-5	4.89	7.76E+04	1	901	26367	C13H22NO3PS	fenamiphos	22224-92-6	2.52	3.31E+02	1
822	23210.1	C12H6Cl4	2,3',4',5-tetrachlorobiphenyl	32598-11-1	4.85	7.08E+04	1	902	26535	C13H28O	1-tridecanol	112-70-9	4.54	3.49E+04	2
823	23247	C12H7Cl3	2,4,5-trichlorobiphenyl	15862-07-4	5.21	1.62E+05	1	903	26654	C14H8O2	anthraquinone	84-65-1	3.22	1.66E+03	2
824	23248	C12H7Cl3	2,4,6-trichlorobiphenyl	35693-92-6	5.21	1.62E+05	1	904	26728	C14H10	anthracene	120-12-7	3.86	7.21E+03	2
825	23249	C12H7Cl3	2,2',5-trichlorobiphenyl	37680-65-2	5.21	1.62E+05	1	905	26729	C14H10	diphenylacetylene	501-65-5	3.98	9.49E+03	2
826	23250	C12H7Cl3	2,3',4'-trichlorobiphenyl	38444-86-9	5.21	1.62E+05	2	906	26730	C14H10	phenanthrene	85-01-8	3.77	5.89E+03	1
827	23251	C12H7Cl3	2,3,5-trichlorobiphenyl	38444-81-4	5.21	1.62E+05	2	907	26821	C14H11N	2-anthracenamine	613-13-8	4.48	3.02E+04	1
828	23252	C12H7Cl3	2,4,4'-trichlorobiphenyl	7012-37-5	5.21	1.62E+05	2	908	26867	C14H12	trans-stilbene	103-30-0	3.99	9.85E+03	2
829	23253	C12H7Cl3	2,4',5-trichlorobiphenyl	16606-02-3	5.21	1.62E+05	2	909	26871	C14H12	1-methylfluorene	1730-37-6	4.08	1.20E+04	2
830	23254	C12H7Cl3	trichlorodiphenyl	25323-68-6	5.21	1.62E+05	2	910	26933	C14H12O	2-phenylacetophenone	451-40-1	3.11	1.28E+03	2
831	23272	C12H8Cl2	2,5-dichlorobiphenyl	34883-39-1	4.70	5.01E+04	1	911	26942	C14H12O2	benzyl benzoate	120-51-4	3.54	3.44E+03	2
832	23273	C12H8Cl2	2,6-dichlorobiphenyl	33146-45-1	4.70	5.01E+04	1	912	27053	C14H14	1,2-diphenylethane	103-29-7	3.93	8.59E+03	2
833	23274	C12H8Cl2	3,3'-dichloro-1,1'-biphenyl	2050-67-1	4.70	5.01E+04	2	913	27062	C14H14	4,4'-dimethylbiphenyl	613-33-2	4.15	1.40E+04	2
834	23275	C12H8Cl2	4,4'-dichloro-1,1'-biphenyl	2050-68-2	5.27	1.86E+05	1	914	27138	C14H14O	dibenzyl ether	103-50-4	3.18	1.51E+03	2
835	23276	C12H8Cl2	2,2'-dichloro-1,1'-biphenyl	13029-08-8	4.70	5.01E+04	2	915	27432	C14H18ClN3O2	2-(4-chlorophenoxy)-1-tert-bu	55219-65-3	1.95	8.91E+01	1
836	23277	C12H8Cl2	3,4-dichloro-1,1'-biphenyl	2974-92-7	4.70	5.01E+04	2	916	27572	C14H20ClNO2	alachlor	15972-60-8	2.48	3.02E+02	1
837	23278	C12H8Cl2	dichloro-1,1'-biphenyl	25512-42-9	4.70	5.01E+04	2	917	27644	C14H22	octylbenzene	2189-60-8	4.80	6.37E+04	2
838	23279	C12H8Cl2	2,4'-dichloro-1,1'-biphenyl	34883-43-7	4.55	3.55E+04	1	918	27873	C14H28O2	tetradecanoic acid	544-63-8	4.70	4.96E+04	2
839	23293	C12H8Cl6O	dieldrin	60-57-1	4.11	1.29E+04	1	919	27912	C14H30	tetradecane	629-59-4	5.29	1.97E+05	2
840	23294	C12H8Cl6O	endrin	72-20-8	4.20	1.58E+04	1	920	27922	C14H30O	1-tetradecanol	112-72-1	4.66	4.54E+04	2
841	23305	C12H8N2	1,10-phenanthroline	66-71-7	2.37	2.36E+02	2	921	28038	C15H10O2	9-anthracenecarboxylic acid	723-62-6	2.54	3.47E+02	1
842	23308	C12H8N2	phenazine	92-82-0	2.92	8.36E+02	2	922	28044	C15H10O2	2-phenyl-1H-indene-1,3(2H)-(	83-12-5	2.95	9.01E+02	2
843	23335	C12H8O	dibenzofuran	132-64-9	3.62	4.15E+03	2	923	28130	C15H12	2-methylanthracene	613-12-7	4.18	1.51E+04	2
844	23365	C12H9Cl	2-chlorobiphenyl	2051-60-7	3.84	6.85E+03	2	924	28131	C15H12	9-methylanthracene	779-02-2	4.14	1.36E+04	2
845	23366	C12H9Cl	3-chlorobiphenyl	2051-61-8	3.87	7.39E+03	2	925	28132	C15H12	1-methylphenanthrene	832-69-9	4.17	1.49E+04	2
846	23367	C12H9Cl	4-chlorobiphenyl	2051-62-9	3.88	7.67E+03	2	926	28179	C15H12O	9-anthracenemethanol	1468-95-7	3.43	2.69E+03	1
847	23370	C12H9ClF3N3O	norflurazon	27314-13-2	2.66	4.57E+02	1	927	28414	C15H16Cl3N3O2	N-propyl-N-(2-(2,4,6-trichloro	67747-09-5	4.13	1.35E+04	1
848	23394	C12H9N	dibenzopyrrole	86-74-8	3.40	2.52E+03	2	928	28436	C15H16O2	bisphenol a	80-05-7	3.18	1.52E+03	2
849	23444	C12H10	acenaphthene	83-32-9	3.53	3.40E+03	2	929	28512.1	C15H18N2O3	imazamethabenz	100728-84-5	2.04	1.10E+02	1
850	23445	C12H10	phenylbenzene (biphenyl)	92-52-4	3.04	1.10E+03	1	930	28531	C15H18N6O6S	nicosulfuron	111991-09-4	2.26	1.82E+02	1
851	23488	C12H10N2	azobenzene	17082-12-1	3.46	2.85E+03	2	931	28623	C15H21NO4	metalaxyl	57837-19-1	1.66	4.57E+01	1
852	23491.1	C12H10N2	azobenzene	103-33-3	3.13	1.35E+03	1	932	28653	C15H22ClNO2	metolachlor	51218-45-2	2.20	1.58E+02	1
853	23519	C12H10O	diphenyl ether	101-84-8	3.67	4.65E+03	2	933	28725	C15H24	nonylbenzene	1081-77-2	5.24	1.76E+05	2
854	23576	C12H10S	diphenyl sulfide	139-66-2	3.80	6.28E+03	2	934	28756	C15H24O	2,6-di-tert-butyl-p-cresol	128-37-0	4.15	1.42E+04	2
855	23607	C12H11N	p-aminodiphenyl	92-67-1	2.93	8.57E+02	2	935	28984	C16H10	fluoranthene	206-44-0	4.14	1.36E+04	2
856	23609	C12H11N	diphenylamine	122-39-4	2.78	6.03E+02	1	936	28985	C16H10	pyrene	129-00-0	4.80	6.31E+04	1
857	23632	C12H11NO2	carbaryl	63-25-2	2.02	1.05E+02	1	937	29111	C16H14	9,10-dimethylphenanthrene	604-83-1	4.47	2.97E+04	2
858	23656	C12H11N3	p-aminoazobenzene	60-09-3	3.23	1.71E+03	2	938	29518	C16H22O4	dibutyl phthalate	84-74-2	3.94	8.80E+03	2
859	23675	C12H12	1,2-dimethylnaphthalene	573-98-8	3.72	5.27E+03	2	939	29538	C16H23ClN3O	tebuconazole	107534-96-3	2.67	4.68E+02	1
860	23676	C12H12	1,3-dimethylnaphthalene	575-41-7	3.78	6.05E+03	2	940	29595	C16H26	decylbenzene	104-72-3	5.38	2.37E+05	2
861	23677	C12H12	1,4-dimethylnaphthalene	571-58-4	3.75	5.68E+03	2	941	29721	C16H32O2	hexadecanoic acid	57-10-3	5.28	1.89E+05	2
862	23678	C12H12	1,5-dimethylnaphthalene	571-61-9	3.76	5.75E+03	2	942	29862	C17H12	11H-benzo[a]fluorene	238-84-6	4.31	2.06E+04	2
863	23680	C12H12	1,7-dimethylnaphthalene	575-37-1	3.79	6.20E+03	2	943	29863	C17H12	11H-benzo[b]fluorene	243-17-4	4.51	3.20E+04	2
864	23682	C12H12	2,3-dimethylnaphthalene	581-40-8	3.77	5.90E+03	2	944	30058.1	C17H18FN3O3	ciprofloxacin	85721-33-1	4.79	6.17E+04	1
865	23683	C12H12	2,6-dimethylnaphthalene	581-42-0	3.72	5.27E+03	2	945	30097	C17H19NO3	morphine	57-27-2	1.83	6.74E+01	2
866	23685	C12H12	1-ethylnaphthalene	1127-76-0	3.77	5.90E+03	2	946	30161.1	C17H21NO2	napropamide	15299-99-7	2.54	3.47E+02	1
867	23686	C12H12	2-ethylnaphthalene	939-27-5	3.76	5.75E+03	2	947	30264	C17H28	undecylbenzene	6742-54-7	5.81	6.38E+05	2
868	23719	C12H12N2	hydrazobenzene	122-66-7	2.98	9.47E+02	2	948	30406	C18H12	chrysene	218-01-9	4.49	3.12E+04	2
869	23720	C12H12N2	p-benzidine	92-87-5	5.36	2.29E+05	1	949	30407	C18H12	benz[a]anthracene	56-55-3	4.59	3.91E+04	2
870	23911	C12H14Cl3O4P	chlorfenvinphos	470-90-6	2.47	2.95E+02	1	950	30409	C18H12	naphthacene	92-24-0	4.51	3.24E+04	2
871	23965	C12H14O	4-phenylcyclohexanone	4894-75-1	2.71	5.13E+02	2	951	30410	C18H12	triphenylene	217-59-4	4.36	2.31E+04	2
872	24002	C12H14O4	diethyl phthalate	84-66-2	2.72	5.26E+02	2	952	30424	C18H12N2	2,2'-biquinoline	119-91-5	4.26	1.82E+04	1
873	24076	C12H15NO3	carbofuran	1563-66-2	1.79	6.17E+01	1	953	30445	C18H13N	6-chrysenamine	2642-98-0	5.58	3.80E+05	1
874	24357	C12H18	hexylbenzene	1077-16-3	4.38	2.40E+04	2	954	30459	C18H14	p-terphenyl	92-94-4	4.66	4.54E+04	2
875	24378	C12H18	hexamethylbenzene	87-85-4	3.93	8.48E+03	2	955	30548	C18H15N	triphenylamine	603-34-9	4.50	3.16E+04	2
876	24411.1	C12H18N2O	isoproturon	34123-59-6	2.35	2.24E+02	1	956	30566	C18H15O4P	triphenyl phosphate	115-86-6	3.87	7.48E+03	2
877	24649	C12H21N2O3PS	diazinon	333-41-5	2.36	2.29E+02	1	957	30567	C18H15P	triphenylphosphine	603-35-0	4.47	2.97E+04	2
878	24690	C12H22O	cyclododecanone	830-13-7	3.61	4.05E+03	2	958	30806	C18H22O2	dicumyl peroxide	80-43-3	4.37	2.34E+04	2
879	24867	C12H24O2	dodecanoic acid	143-07-7	3.88	7.58E+03	2	959	30947	C18H30	dodecylbenzene	123-01-3	6.08	1.21E+06	2
880	24914	C12H26	dodecane	112-40-3	4.70	4.96E+04	2	960	30964	C18H30O2	linolenic acid	463-40-1	4.89	7.78E+04	2

Table 15 Soil Sorption Coefficient - Oganic Compounds

Soil Sorption Coefficient								Soil Sorption Coefficient							
NO	ID	FORMULA	NAME	CAS No	log_{10} Koc	Koc	code	NO	ID	FORMULA	NAME	CAS No	log_{10} Koc	Koc	code
961	30991	C18H32O2	linoleic acid	60-33-3	5.21	1.63E+05	2	976	32632	C21H16	1,2-dihydro-3-methylbenz[j]ac	56-49-5	6.10	1.26E+06	1
962	31030	C18H34O2	oleic acid	112-80-1	5.53	3.41E+05	2	977	32738	C21H20Cl2O3	permethrin	52645-53-1	3.19	1.55E+03	1
963	31097	C18H36O2	octadecanoic acid	57-11-4	5.85	7.15E+05	2	978	32739	C21H20Cl2O3	cis-permethrin	61949-76-6	3.19	1.55E+03	2
964	31279	C19H14F3NO	fluridone	59756-60-4	3.01	1.02E+03	1	979	32740	C21H20Cl2O3	(-)-cis-permethrin	54774-46-8	3.19	1.55E+03	2
965	31344	C19H16O	triphenylmethanol	76-84-6	3.38	2.39E+03	2	980	32741	C21H20Cl2O3	(±)-cis-permethrin	52341-33-0	3.19	1.55E+03	2
966	31455	C19H21ClN2O	N-((4-chlorophenyl)methyl)-N	66063-05-6	3.33	2.14E+03	1	981	32742	C21H20Cl2O3	(+)-cis-permethrin	54774-45-7	3.19	1.55E+03	2
967	31481.1	C19H22FN3O3	enrofloxacin	93106-60-6	4.85	7.08E+04	1	982	32743	C21H20Cl2O3	trans-permethrin	61949-77-7	3.19	1.55E+03	2
968	31646	C19H32	tridecylbenzene	123-02-4	6.47	2.94E+06	2	983	32744	C21H20Cl2O3	(+)-trans-permethrin	51877-74-8	3.19	1.55E+03	2
969	31800	C20H12	perylene	198-55-0	4.78	5.98E+04	2	984	32745	C21H20Cl2O3	trans-(±)-permethrin	52341-32-9	3.19	1.55E+03	2
970	31804	C20H12	benzo[a]pyrene	50-32-8	4.75	5.62E+04	2	985	32805	C21H22N2O2	strychnine	57-24-9	4.14	1.38E+04	1
971	31847	C20H13N	13H-dibenzo[a,i]carbazole	239-64-5	6.10	1.26E+06	1	986	33080	C22H12	benzo[ghi]perylene	191-24-2	5.13	1.35E+05	2
972	31852.1	C20H13N	1,2,7,8-dibenzocarbazole	239-64-5	6.10	1.26E+06	1	987	33098	C22H14	dibenz[a,h]anthracene	53-70-3	6.07	1.17E+06	1
973	31977	C20H16	7,12-dimethylbenz[a]anthrace	57-97-6	5.29	1.95E+05	1	988	33717	C24H12	coronene	191-07-1	4.67	4.66E+04	2
974	32388	C20H32O2	5,8,11,14-eicosatetraenoic ac	506-32-1	5.17	1.49E+05	2	989	33925	C24H38O4	dioctyl phthalate	117-84-0	5.51	3.25E+05	2
975	32491	C20H40O2	eicosanoic acid	506-30-9	6.43	2.70E+06	2								

Code: 1 - data, 2 - estimate

Koc = (mg sorbed/kg organic carbon) / (mg/L aqueous solution)

Values apply at ambient conditions (1 atm, 20 C)

Chapter 16

THRESHOLD LIMIT VALUE AND PERMISSIBLE EXPOSURE LIMIT – ORGANIC COMPOUNDS

Carl L. Yaws and Prasad K. Narasimhan
Lamar University, Beaumont, Texas

Tabulation Results

The results are given in Table 16. The tabulation is arranged by carbon number (C, C2, C3, ...) to provide ease of use in quickly locating the data by using the chemical formula. The compound name, CAS No (Chemical Abstracts Registry Number), and molecular weight are next provided. Values for threshold limit value and permissible exposure limit are given in the adjacent columns.

The threshold limit value and permissible exposure value are important in establishing the exposure limits for safeguarding health in the workplace. The threshold limit value (TLV) is given by American Conference of Governmental Industrial Hygienists (ACGIH). The permissible exposure limit (PEL) is provided by the Occupational Safety and Health Administration (OSHA). The values which represent exposure limits for safeguarding health apply to exposure in the workplace for a 8 hour day or 40 hour week.

In preparing the tabulation, a literature search was conducted to identify data source publications (1-15). The publications were screened and copies of appropriate data were made. These data were then keyed into the computer to provide a database of values for the compounds. The excellent compilations by ACGIH (1-2), CRC (4), Lewis (10), Montgomery (11), and Yaws (15) were primarily used for the values.

Example

Determine the threshold limit value (TLV) and permissible exposure limit (PEL) for propyl alcohol (C3H8O) in the workplace.

Inspection of the tabulated values yields:

TLV = 200 ppm (vol)

PEL = 200 ppm (vol)

References

1. ACGIH Publication, 1997 TLVs and BEIs, American Conference of Governmental Industrial Hygienists, ACGIH, Inc., Cincinnati, OH (1997).
2. ACGIH Publication, GUIDE TO OCCUPATIONAL EXPOSURE VALUES – 1997, American Conference of Governmental Industrial Hygienists, ACGIH, Inc., Cincinnati, OH (1997).
3. Allen, D. T. and D. R. Shonnard, GREEN ENGINEERING, Prentice Hall PTR, Upper Saddle River, NJ (2002).
4. CRC HANDBOOK OF CHEMISTRY AND PHYSICS, 84th - 87th ed., CRC Press, Inc., Boca Raton, FL (2004-2007).
5. Crowl, D. A. and J. F. Louvar, CHEMICAL PROCESS SAFETY, 2nd edition, Prentice Hall, Inc., Englewood Cliffs, NJ (2002).
6. de la Cruz, P. L. and D. G. Sarvadi, Am. Ind. Hyg. Assoc. J., 55(10), 894 (1994).
7. Fawcett, H. H. and W. C. Wood, eds., SAFETY AND ACCIDENT PREVENTION IN CHEMICAL OPERATIONS, 2nd ed., John Wiley and Sons, New York, NY (1982).
8. June 1993 Air Contaminants Final Rule, specified in Tables Z-1, Z-2 and Z-3 (Federal Register, 58:35388-35351, June 30, 1993; corrected in Federal Register, 58:40191, July 27, 1993; and subsequent amendments).
9. Lees, F. P., LOSS PREVENTION IN THE PROCESS INDUSTRIES, Vols. 1 and 2, Butterworth-Heinemann, London, England (1992).
10. Lewis, R. J., SAX'S DANGEROUS PROPERTIES OF INDUSTRIAL MATERIALS, 9th ed., Vols. 1, 2 and 3, Van Nostrand Reinhold Company, New York, NY (1996).
11. Montgomery, J. H., GROUNDWATER CHEMICALS, Lewis Publishers, New York, NY (2000).
12. NIOSH POCKET GUIDE TO CHEMICAL HAZARDS, U. S. Dept. of Health and Human Services, Superintendent of Documents, Washington, DC (June, 1994).
13. Springer, C. and J. R. Welker, INDUSTRIAL HYGIENE: AN INTRODUCTION FOR CHEMICAL ENGINEERS, American Institute of Chemical Engineers, New York, NY (1995).
14. Williams, P. L. and J. L. Burson, eds., INDUSTRIAL TOXICOLOGY, SAFETY AND HEALTH APPLICATIONS IN THE WORKPLACE, Van Nostrand Reinhold Company, New York, NY (1985).
15. Yaws, C. L., CHEMICAL PROPERTIES HANDBOOK, McGraw-Hill, New York, NY (1999).

Table 16 Threshold Limit Value and Permissible Exposure Level - Oganic Compounds

						Threshold Limit Value - TLV		Permissible Exposure Limit - PEL	
NO	ID	FORMULA	NAME	CAS No	MW, g/mol	ppm (vol)	mg/m3	ppm (vol)	mg/m3
1	17	CBrF3	bromotrifluoromethane	75-63-8	148.911	1000	6090	1000	6100
2	23	CBr2F2	dibromodifluoromethane	75-61-6	209.816	100	858	100	860
3	29	CBr4	carbon tetrabromide	558-13-4	331.627	0.1	1.4	---	---
4	42	CCl2F2	dichlorodifluoromethane	75-71-8	120.913	1000	4950	1000	4950
5	44	CCl2O	phosgene	75-44-5	98.915	0.1	0.4	0.1	0.4
6	47	CCl3F	trichlorofluoromethane	75-69-4	137.367	---	---	1000	5600
7	49	CCl3NO2	trichloronitromethane	76-06-2	164.374	0.1	0.67	0.1	0.7
8	50	CCl4	carbon tetrachloride	56-23-5	153.821	5	31	10	---
9	54	CCl4S	perchloromethyl mercaptan	594-42-3	185.887	0.1	0.76	0.1	0.8
10	68	CF2O	carbonyl fluoride	353-50-4	66.007	2	5.4	---	---
11	96	CHBr3	bromoform	75-25-2	252.731	0.5	5.2	0.5	5
12	97	CHClF2	chlorodifluoromethane	75-45-6	86.468	1000	3540	---	---
13	101	CHCl2F	dichlorofluoromethane	75-43-4	102.922	10	42	1000	4200
14	104	CHCl3	chloroform	67-66-3	119.376	10	49	---	---
15	117	CHI3	iodoform	75-47-8	393.732	0.6	10	---	---
16	122	CHN	hydrogen cyanide	74-90-8	27.026	---	---	10	11
17	140	CH2BrCl	bromochloromethane	74-97-5	129.383	200	1060	200	1050
18	149	CH2Cl2	dichloromethane	75-09-2	84.932	50	174	25	---
19	165	CH2N2	cyanamide	420-04-2	42.041	---	2	---	---
20	166	CH2N2	diazomethane	334-88-3	42.041	0.2	0.34	0.2	0.4
21	175	CH2O	formaldehyde	50-00-0	30.026	---	---	0.75	---
22	176	CH2O2	formic acid	64-18-6	46.026	5	9.4	5	9
23	191	CH3Br	methyl bromide	74-83-9	94.939	1	3.9	---	---
24	196	CH3Cl	methyl chloride	74-87-3	50.487	50	103	100	205
25	220	CH3Hg	methylmercury	22967-92-6	215.625	---	0.01	---	0.01
26	222	CH3I	methyl iodide	74-88-4	141.939	2	12	5	28
27	232	CH3NO	formamide	75-12-7	45.041	10	18	---	---
28	235	CH3NO2	nitromethane	75-52-5	61.041	20	50	100	250
29	272	CH4O	methyl alcohol	67-56-1	32.042	200	262	200	260
30	278	CH4S	methyl mercaptan	74-93-1	48.109	0.5	0.98	---	---
31	286	CH5N	methylamine	74-89-5	31.058	5	6.4	10	12
32	312	CH6N2	methylhydrazine	60-34-4	46.073	0.01	0.019	---	---
33	338	CKN	potassium cyanide	151-50-8	65.116	---	---	---	5
34	351	CNNa	sodium cyanide	143-33-9	49.008	---	---	---	5
35	363	CN4O8	tetranitromethane	509-14-8	196.034	0.005	0.04	1	8
36	372	CO	carbon monoxide	630-08-0	28.010	25	29	50	55
37	375	CO2	carbon dioxide	124-38-9	44.010	5000	9000	5000	9000
38	382	CS2	carbon disulfide	75-15-0	76.143	10	31	20	---
39	418	C2CaN2	calcium cyanide	592-01-8	92.114	---	---	---	5
40	427	C2ClF5	chloropentafluoroethane	76-15-3	154.467	1000	6320	---	---
41	441	C2Cl2F4	1,2-dichloro-1,1,2,2-tetrafluoroethane	76-14-2	170.921	1000	6990	1000	7000
42	449	C2Cl3F3	1,1,2-trichloro-1,2,2-trifluoroethane	76-13-1	187.375	1000	7670	1000	7600
43	455	C2Cl4	tetrachloroethylene	127-18-4	165.832	25	170	100	---
44	456	C2Cl4F2	1,1,2,2-tetrachloro-1,2-difluoroethane	76-12-0	203.829	500	4170	500	4170
45	457	C2Cl4F2	1,1,1,2-tetrachloro-2,2-difluoroethane	76-11-9	203.829	500	4170	500	4170
46	463	C2Cl6	hexachloroethane	67-72-1	236.737	1	9.7	1	10
47	511	C2HBrClF3	halothane	151-67-7	197.382	50	404	---	---
48	539	C2HCl3	trichloroethylene	79-01-6	131.387	50	269	100	---
49	547	C2HCl3O2	trichloroacetic acid	76-03-9	163.386	1	6.7	---	---
50	592	C2H2Br4	1,1,2,2-tetrabromoethane	79-27-6	345.654	1	14	1	14
51	615	C2H2Cl2	1,1-dichloroethylene	75-35-4	96.943	5	20	---	---
52	616	C2H2Cl2	cis-1,2-dichloroethylene	156-59-2	96.943	200	793	200	790
53	617	C2H2Cl2	trans-1,2-dichloroethylene	156-60-5	96.943	200	793	200	790
54	619	C2H2Cl2	1,2-dichloroethylene, cis and trans	540-59-0	96.943	200	793	200	790
55	625	C2H2Cl2O	chloroacetyl chloride	79-04-9	112.942	0.05	0.23	---	---
56	636	C2H2Cl4	1,1,2,2-tetrachloroethane	79-34-5	167.848	1	6.9	5	35
57	641	C2H2FNaO2	sodium fluoroacetate	62-74-8	100.025	---	0.05		0.05
58	681	C2H2O	ketene	463-51-4	42.037	0.5	0.86	0.5	0.9
59	684	C2H2O4	oxalic acid	144-62-7	90.035	---	1	---	1
60	688	C2H3Br	vinyl bromide	593-60-2	106.950	5	22	---	---
61	701	C2H3Cl	vinyl chloride	75-01-4	62.498	5	13	1	---
62	718	C2H3Cl2NO2	1,1-dichloro-1-nitroethane	594-72-9	143.957	2	12	---	---
63	720	C2H3Cl3	1,1,1-trichloroethane	71-55-6	133.403	350	1910	350	1900
64	721	C2H3Cl3	1,1,2-trichloroethane	79-00-5	133.403	10	55	10	45
65	753	C2H3N	acetonitrile	75-05-8	41.053	40	67	40	70
66	755	C2H3NO	methyl isocyanate	624-83-9	57.052	0.02	0.047	0.02	0.05
67	799	C2H4Br2	1,2-dibromoethane	106-93-4	187.862	---	---	20	---
68	812	C2H4Cl2	1,1-dichloroethane	75-34-3	98.959	100	405	100	400
69	813	C2H4Cl2	1,2-dichloroethane	107-06-2	98.959	10	40	50	---
70	816	C2H4Cl2O	bis(chloromethyl) ether	542-88-1	114.958	0.001	0.0047	---	---
71	865	C2H4N2O6	ethylene glycol dinitrate	628-96-6	152.064	0.05	0.31	---	---
72	871	C2H4N4	1H-1,2,4-triazol-3-amine	61-82-5	84.082	---	0.2	---	---
73	884	C2H4O	acetaldehyde	75-07-0	44.053	---	---	200	360
74	885	C2H4O	ethylene oxide	75-21-8	44.053	1	1.8	1	---
75	889	C2H4O2	acetic acid	64-19-7	60.053	10	25	10	25
76	890	C2H4O2	methyl formate	107-31-3	60.053	100	246	100	250
77	892	C2H4O2S	thioglycolic acid	68-11-1	92.119	1	3.8	---	---
78	910	C2H5Br	bromoethane	74-96-4	108.966	5	22	200	890
79	914	C2H5Cl	ethyl chloride	75-00-3	64.514	100	264	1000	2600
80	917	C2H5ClO	2-chloroethanol	107-07-3	80.514	---	---	5	16

Table 16 Threshold Limit Value and Permissible Exposure Level - Oganic Compounds

NO	ID	FORMULA	NAME	CAS No	MW, g/mol	Threshold Limit Value - TLV		Permissible Exposure Limit - PEL	
						ppm (vol)	mg/m3	ppm (vol)	mg/m3
81	954	C2H5N	ethyleneimine	151-56-4	43.069	0.5	0.88	---	---
82	958	C2H5NO2	nitroethane	79-24-3	75.068	100	307	100	310
83	1065	C2H6O	ethyl alcohol	64-17-5	46.069	1000	1880	1000	1900
84	1083	C2H6O4S	dimethyl sulfate	77-78-1	126.133	0.1	0.52	1	5
85	1092	C2H6S	ethyl mercaptan	75-08-1	62.136	0.5	1.3	---	---
86	1115	C2H7N	ethylamine	75-04-7	45.085	5	9.2	10	18
87	1116	C2H7N	dimethylamine	124-40-3	45.085	5	9.2	10	18
88	1117	C2H7NO	monoethanolamine	141-43-5	61.084	3	7.5	3	6
89	1159	C2H8N2	ethylenediamine	107-15-3	60.100	10	25	10	25
90	1160	C2H8N2	1,1-dimethylhydrazine	57-14-7	60.100	0.01	0.025	0.5	1
91	1210	C2N2	cyanogen	460-19-5	52.036	10	21	---	---
92	1287	C3F6O	hexafluoroacetone	684-16-2	166.023	0.1	0.68	---	---
93	1346	C3H2ClF5O	enflurane	13838-16-9	184.493	75	566	---	---
94	1462	C3H3N	acrylonitrile	107-13-1	53.064	2	4.3	2	---
95	1492	C3H4	methylacetylene	74-99-7	40.065	1000	1640	1000	1650
96	1540	C3H4Cl2	1,3-dichloropropene; (cis+trans)	542-75-6	110.970	1	4.5	---	---
97	1554	C3H4Cl2O2	2,2-dichloropropanoic acid	75-99-0	142.969	1	5.8	---	---
98	1621	C3H4O	acrolein	107-02-8	56.064	0.1	0.23	0.1	0.25
99	1622	C3H4O	propargyl alcohol	107-19-7	56.064	1	2.3	---	---
100	1626	C3H4O2	acrylic acid	79-10-7	72.064	2	5.9	---	---
101	1627	C3H4O2	beta-propiolactone	57-57-8	72.064	0.5	1.5	---	---
102	1662	C3H5Br2Cl	1,2-dibromo-3-chloropropane	96-12-8	236.333	---	---	0.001	---
103	1671	C3H5Cl	3-chloro-1-propene	107-05-1	76.525	1	3	1	3
104	1676	C3H5ClO	alpha-epichlorohydrin	106-89-8	92.525	0.5	1.9	5	19
105	1691	C3H5ClO2	2-chloropropanoic acid	598-78-7	108.524	0.1	0.44	---	---
106	1707	C3H5Cl3	1,2,3-trichloropropane	96-18-4	147.430	10	60	50	300
107	1755	C3H5NO	acrylamide	79-06-1	71.079	---	0.03	---	0.3
108	1796	C3H5N3O9	nitroglycerine	55-63-0	227.088	0.05	0.46	---	---
109	1837	C3H6ClNO2	1-chloro-1-nitropropane	600-25-9	123.539	2	10	20	100
110	1845	C3H6Cl2	1,2-dichloropropane	78-87-5	112.986	75	347	75	350
111	1902	C3H6N2O6	1,2-propylene glycol dinitrate	6423-43-4	166.091	0.05	0.34	---	---
112	1919	C3H6N6O6	hexahydro-1,3,5-trinitro-1,3,5-triazine	121-82-4	222.119	---	0.5	---	---
113	1920	C3H6O	allyl alcohol	107-18-6	58.080	. 2	4.8	2	5
114	1923	C3H6O	acetone	67-64-1	58.080	500	1188	1000	2400
115	1924	C3H6O	1,2-propylene oxide	75-56-9	58.080	20	48	100	240
116	1930	C3H6O2	propanoic acid	79-09-4	74.079	10	30	---	---
117	1931	C3H6O2	ethyl formate	109-94-4	74.079	100	303	100	300
118	1932	C3H6O2	methyl acetate	79-20-9	74.079	200	606	200	610
119	1937	C3H6O2	(±)-glycidol	556-52-5	74.079	2	6.1	50	150
120	2031	C3H7N	propyleneimine	75-55-8	57.096	2	4.7	2	5
121	2036	C3H7NO	N,N-dimethylformamide	68-12-2	73.095	10	30	10	30
122	2045	C3H7NO2	1-nitropropane	108-03-2	89.094	25	91	25	90
123	2046	C3H7NO2	2-nitropropane	79-46-9	89.094	10	36	25	90
124	2061	C3H7NO3	propyl nitrate	627-13-4	105.094	25	107	25	110
125	2094	C3H8	propane	74-98-6	44.097	2500	4508	1000	1800
126	2129	C3H8O	propyl alcohol	71-23-8	60.096	200	492	200	500
127	2130	C3H8O	isopropyl alcohol	67-63-0	60.096	400	983	400	980
128	2138	C3H8O2	2-methoxyethanol	109-86-4	76.095	5	16	25	80
129	2139	C3H8O2	methylal	109-87-5	76.095	1000	3110	1000	3100
130	2148	C3H8O3	glycerol	56-81-5	92.095	10	---	---	15*/5**
131	2201	C3H9N	isopropylamine	75-31-0	59.111	5	12	5	12
132	2203	C3H9N	trimethylamine	75-50-3	59.111	5	12	---	---
133	2236	C3H9O3P	trimethyl phosphite	121-45-9	124.077	2	10	---	---
134	2318	C4Cl6	hexachloro-1,3-butadiene	87-68-3	260.759	0.02	0.21	---	---
135	2365	C4HCoO4	cobalt hydrocarbonyl	16842-03-8	171.983	---	0.1	---	---
136	2430	C4H2O3	maleic anhydride	108-31-6	98.058	0.25	1	0.25	1
137	2634	C4H5Cl	chloroprene	126-99-8	88.536	10	36	25	90
138	2697	C4H5N	methacrylonitrile	126-98-7	67.091	1	2.7	---	---
139	2757	C4H6	butadiene (1,3 butadiene)	106-99-0	54.092	2	4.4	1	5
140	2820	C4H6Cl2	1,4-dichloro-2-butene	764-41-0	124.997	0.005	0.025	---	---
141	2956	C4H6O	crotonaldehyde	4170-30-3	70.091	2	5.7	2	6
142	2969	C4H6O2	methacrylic acid	79-41-4	86.090	20	70	---	---
143	2970	C4H6O2	methyl acrylate	96-33-3	86.090	2	7	10	35
144	2971	C4H6O2	vinyl acetate	108-05-4	86.090	10	35	---	---
145	2990	C4H6O3	acetic anhydride	108-24-7	102.090	5	21	5	20
146	3082	C4H7Br2Cl2O4P	naled	300-76-5	380.784	---	3	---	3
147	3155	C4H7Cl2O4P	dichlorvos	62-73-7	220.976	0.1	0.9	---	1
148	3375	C4H8Cl2O	bis(2-chloroethyl) ether	111-44-4	143.012	5	29	---	---
149	3480	C4H8O	methyl ethyl ketone	78-93-3	72.107	200	590	200	590
150	3481	C4H8O	tetrahydrofuran	109-99-9	72.107	200	590	200	590
151	3508	C4H8O2	ethyl acetate	141-78-6	88.106	400	1440	400	1400
152	3510	C4H8O2	1,4-dioxane	123-91-1	88.106	25	90	100	360
153	3681	C4H9NO	N,N-dimethylacetamide	127-19-5	87.122	10	36	10	35
154	3682	C4H9NO	morpholine	110-91-8	87.122	20	71	20	70
155	3783	C4H10	butane	106-97-8	58.123	800	1900	---	---
156	3865	C4H10O	butanol	71-36-3	74.123	---	---	100	300
157	3866	C4H10O	isobutanol	78-83-1	74.123	50	152	100	300
158	3867	C4H10O	sec-butanol	78-92-2	74.123	100	300	150	450
159	3868	C4H10O	tert-butanol	75-65-0	74.123	100	303	100	300
160	3869	C4H10O	diethyl ether	60-29-7	74.123	400	1210	400	1200

Table 16 Threshold Limit Value and Permissible Exposure Level - Oganic Compounds

						Threshold Limit Value - TLV		Permissible Exposure Limit - PEL	
NO	ID	FORMULA	NAME	CAS No	MW, g/mol	ppm (vol)	mg/m3	ppm (vol)	mg/m3
161	3883	C4H10O2	2-ethoxyethanol	110-80-5	90.122	5	18	200	740
162	3942	C4H10S	butyl mercaptan	109-79-5	90.189	0.5	1.8	10	35
163	3984	C4H11N	diethylamine	109-89-7	73.138	5	15	25	75
164	4006	C4H11NO2	diethanolamine	111-42-2	105.137	0.46	2	---	---
165	4049	C4H12Cl2N2	piperazine dihydrochloride	142-64-3	159.058	---	5	---	---
166	4095	C4H12OSi	ethoxydimethylsilane	14857-34-2	104.224	0.5	2.1	---	---
167	4101	C4H12O4Si	tetramethyl silicate	681-84-5	152.222	1	6	---	---
168	4106	C4H12Pb	tetramethyl lead	75-74-1	267.339	---	0.15	---	0.075
169	4118	C4H13N3	diethylene triamine	111-40-0	103.168	1	4.2	---	---
170	4156	C4NiO4	nickel carbonyl	13463-39-3	170.735	0.05	0.12	0.001	0.007
171	4169	C5Cl6	hexachlorocyclopentadiene	77-47-4	272.770	0.01	0.11	---	---
172	4183	C5FeO5	iron pentacarbonyl	13463-40-6	195.897	0.1	0.8	---	---
173	4327	C5H4O2	furfural	98-01-1	96.086	2	7.9	5	20
174	4391	C5H5N	pyridine	110-86-1	79.102	5	16	5	15
175	4408	C5H5NO2	methyl 2-cyanoacrylate	137-05-3	111.101	2	9.1	---	---
176	4447	C5H6	cyclopentadiene	542-92-7	66.103	75	203	75	200
177	4476	C5H6Cl2N2O2	1,3-dichloro-5,5-dimethyl hydantoin	118-52-5	197.020	---	0.2	---	0.2
178	4497	C5H6N2	2-pyridinamine	504-29-0	94.117	0.5	2	0.5	2
179	4548	C5H6O2	furfuryl alcohol	98-00-0	98.101	10	40	50	200
180	4888	C5H8O2	ethyl acrylate	140-88-5	100.117	5	20	25	100
181	4889	C5H8O2	methyl methacrylate	80-62-6	100.117	100	410	100	410
182	5285	C5H10	cyclopentane	287-92-3	70.134	600	1720	---	---
183	5426	C5H10N2O2S	methomyl	16752-77-5	162.213	---	2.5	---	---
184	5445	C5H10O	pentanal	110-62-3	86.134	50	176	---	---
185	5450	C5H10O	methyl propyl ketone	107-87-9	86.134	200	705	200	700
186	5451	C5H10O	diethyl ketone	96-22-0	86.134	200	705	---	---
187	5452	C5H10O	methyl isopropyl ketone	563-80-4	86.134	200	705	---	---
188	5501	C5H10O2	propyl acetate	109-60-4	102.133	200	835	200	840
189	5502	C5H10O2	isopropyl acetate	108-21-4	102.133	250	1040	250	950
190	5545	C5H10O3	ethylene glycol monomethyl ether acetate	110-49-6	118.133	5	24	25	120
191	5772	C5H12	pentane	109-66-0	72.150	600	1770	1000	2950
192	5773	C5H12	isopentane	78-78-4	72.150	600	1770	1000	2950
193	5774	C5H12	neopentane	463-82-1	72.150	600	1770	1000	2950
194	5837	C5H12O	3-methyl-1-butanol	123-51-3	88.150	100	361	100	360
195	5844	C5H12O	methyl tert-butyl ether	1634-04-4	88.150	40	144	---	---
196	5883	C5H12O2	2-isopropoxyethanol	109-59-1	104.149	25	106	---	---
197	5912	C5H12O4	pentaerythritol	115-77-5	136.148	---	10	---	15*/5**
198	6065	C6Cl5NO2	pentachloronitrobenzene	82-68-8	295.334	---	0.5	---	---
199	6067	C6Cl6	hexachlorobenzene	118-74-1	284.781	---	0.002	---	---
200	6116	C6HCl5O	pentachlorophenol	87-86-5	266.336	---	0.5	---	0.5
201	6309	C6H3Cl3N2O2	4-amino-3,5,6-trichloropyridinecarboxlic acid	1918-02-1	241.460	---	10	---	15*/5**
202	6325	C6H3Cl4N	nitrapyrin	1929-82-4	230.907	---	10	---	15*/5**
203	6368	C6H3N3O7	2,4,6-trinitrophenol	88-89-1	229.107	---	0.1	---	0.1
204	6445	C6H4ClNO2	p-chloronitrobenzene	100-00-5	157.556	0.1	0.64	---	1
205	6469	C6H4Cl2	o-dichlorobenzene	95-50-1	147.003	25	150	---	---
206	6471	C6H4Cl2	p-dichlorobenzene	106-46-7	147.003	10	60	75	450
207	6580	C6H4N2O4	m-dinitrobenzene	99-65-0	168.109	0.15	1	---	1
208	6581	C6H4N2O4	o-dinitrobenzene	528-29-0	168.109	0.15	1	---	1
209	6582	C6H4N2O4	p-dinitrobenzene	100-25-4	168.109	0.15	1	---	1
210	6616	C6H4O2	p-benzoquinone	106-51-4	108.097	0.1	0.44	0.1	0.4
211	6692	C6H5Cl	chlorobenzene	108-90-7	112.558	10	46	75	350
212	6819	C6H5NO2	nitrobenzene	98-95-3	123.112	1	5	1	5
213	6876	C6H6	benzene	71-43-2	78.114	0.5	1.6	1	3
214	6948	C6H6Cl6	1,2,3,4,5,6-hexachlorocyclohexane, (1alpha,2	58-89-9	290.829	---	0.5	---	0.5
215	6998	C6H6N2O2	p-nitroaniline	100-01-6	138.126	---	3	1	6
216	7057	C6H6O	phenol	108-95-2	94.113	5	19	5	19
217	7070	C6H6O2	pyrocatechol	120-80-9	110.112	5	23	---	---
218	7071	C6H6O2	resorcinol	108-46-3	110.112	10	45	---	---
219	7072	C6H6O2	p-hydroquinone	123-31-9	110.112	---	2	---	2
220	7121	C6H6S	phenyl mercaptan	108-98-5	110.180	0.5	2.3	---	---
221	7172	C6H7N	aniline	62-53-3	93.129	2	7.6	5	19
222	7212	C6H7NO2	ethyl cyanoacrylate	7085-85-0	125.127	0.2	1	---	---
223	7316	C6H8N2	adiponitrile	111-69-3	108.144	2	8.8	---	---
224	7318	C6H8N2	m-phenylenediamine	108-45-2	108.144	---	0.1	---	---
225	7319	C6H8N2	o-phenylenediamine	95-54-5	108.144	---	0.1	---	---
226	7320	C6H8N2	p-phenylenediamine	106-50-3	108.144	---	0.1	---	0.1
227	7321	C6H8N2	phenylhydrazine	100-63-0	108.144	0.1	0.44	5	22
228	7622	C6H10	cyclohexene	110-83-8	82.145	300	1010	300	1015
229	7760	C6H10O	cyclohexanone	108-94-1	98.145	25	100	50	200
230	7816	C6H10O2	allyl glycidyl ether	106-92-3	114.144	5	23	---	---
231	7884	C6H10O3	diglycidyl ether	2238-07-5	130.144	0.1	0.53	---	---
232	7892	C6H10O3	2-hydroxypropyl acrylate	999-61-1	130.144	0.5	2.8	---	---
233	7917	C6H10O4	adipic acid	124-04-9	146.143	---	5	---	---
234	8101	C6H11NO	epsilon-caprolactam	105-60-2	113.160	---	1	---	---
235	8215	C6H12	cyclohexane	110-82-7	84.161	300	1030	300	1050
236	8216	C6H12	1-hexene	592-41-6	84.161	30	130	---	---
237	8380	C6H12N2S4	thiram	137-26-8	240.439	---	1	---	5
238	8400	C6H12O	cyclohexanol	108-93-0	100.161	50	206	50	200
239	8409	C6H12O	2-hexanone	591-78-6	100.161	5	20	100	410
240	8412	C6H12O	4-methyl-2-pentanone	108-10-1	100.161	50	205	100	410

Table 16 Threshold Limit Value and Permissible Exposure Level - Oganic Compounds

						Threshold Limit Value - TLV		Permissible Exposure Limit - PEL	
NO	ID	FORMULA	NAME	CAS No	MW, g/mol	ppm (vol)	mg/m3	ppm (vol)	mg/m3
241	8476	C6H12O2	butyl acetate	123-86-4	116.160	150	713	150	710
242	8477	C6H12O2	isobutyl acetate	110-19-0	116.160	150	713	150	700
243	8478	C6H12O2	sec-butyl acetate	105-46-4	116.160	200	950	200	950
244	8479	C6H12O2	tert-butyl acetate	540-88-5	116.160	200	950	200	950
245	8485	C6H12O2	diacetone alcohol	123-42-2	116.160	50	238	50	240
246	8496	C6H12O2	isopropyl glycidyl ether	4016-14-2	116.160	50	238	50	240
247	8531	C6H12O3	2-ethoxyethyl acetate	111-15-9	132.159	5	27	100	540
248	8639	C6H12S2	allyl propyl disulfide	2179-59-1	148.293	2	12	2	12
249	8742	C6H13N	cyclohexylamine	108-91-8	99.176	10	41	---	---
250	8769	C6H13NO	N-ethylmorpholine	100-74-3	115.176	5	24	20	94
251	8877	C6H14	hexane	110-54-3	86.177	50	176	500	1800
252	8878	C6H14	2,2-dimethylbutane	75-83-2	86.177	500	1760	---	---
253	8879	C6H14	2,3-dimethylbutane	79-29-8	86.177	500	1760	---	---
254	8880	C6H14	2-methylpentane	107-83-5	86.177	500	1760	---	---
255	8881	C6H14	3-methylpentane	96-14-0	86.177	500	1760	---	---
256	8976	C6H14O	4-methyl-2-pentanol	108-11-2	102.177	25	104	25	100
257	8995	C6H14O	diisopropyl ether	108-20-3	102.177	250	1040	500	2100
258	9021	C6H14O2	2-butoxyethanol	111-76-2	118.176	25	121	50	240
259	9203	C6H15N	diisopropylamine	108-18-9	101.192	5	21	5	20
260	9212	C6H15N	triethylamine	121-44-8	101.192	1	4.1	25	100
261	9217	C6H15NO	2-diethylaminoethanol	100-37-8	117.192	2	9.6	10	50
262	9239	C6H15NO3	triethanolamine	102-71-6	149.190	---	5	---	---
263	9267	C6H15O3PS2	methyl demeton	8022-00-2	230.289	---	0.5	---	---
264	9292	C6H16N2	hexamethylenediamine	124-09-4	116.207	0.5	2.3	---	---
265	10118	C7H5N3O6	2,4,6-trinitrotoluene	118-96-7	227.134	---	0.1	---	1.5
266	10127	C7H5N5O8	N-methyl-N,2,4,6-tetranitroaniline	479-45-8	287.147	---	1.5	---	1.5
267	10374	C7H6N2O4	dinitrotoluene	25321-14-6	182.136	---	0.2	---	1.5
268	10383	C7H6N2O5	2-methyl-4,6-dinitrophenol	534-52-1	198.136	---	0.2	---	0.2
269	10492	C7H7Cl	benzyl chloride	100-44-7	126.585	1	5.2	1	5
270	10493	C7H7Cl	o-chlorotoluene	95-49-8	126.585	50	259	---	---
271	10547	C7H7Cl2NO	clopidol	2971-90-6	192.044	---	10	---	15*/5**
272	10626	C7H7NO2	m-nitrotoluene	99-08-1	137.138	2	11	5	30
273	10627	C7H7NO2	o-nitrotoluene	88-72-2	137.138	2	11	5	30
274	10628	C7H7NO2	p-nitrotoluene	99-99-0	137.138	2	11	5	30
275	10725	C7H8	toluene	108-88-3	92.141	50	188	200	---
276	10855	C7H8O	cresol mixture	1319-77-3	108.140	5	22	5	22
277	10867	C7H8O2	p-methoxyphenol	150-76-5	124.139	---	5	---	---
278	10954	C7H9N	N-methylaniline	100-61-8	107.155	0.5	2.2	2	9
279	10955	C7H9N	m-toluidine	108-44-1	107.155	2	8.8	---	---
280	10956	C7H9N	o-toluidine	95-53-4	107.155	2	8.8	5	22
281	10957	C7H9N	p-toluidine	106-49-0	107.155	2	8.8	---	---
282	10979	C7H9NO	2-methoxyaniline	90-04-0	123.155	0.1	0.5	---	0.5
283	10981	C7H9NO	4-methoxyaniline	104-94-9	123.155	0.1	0.5	---	0.5
284	11441	C7H12O	2-methylcyclohexanone	583-60-8	112.172	50	229	100	460
285	11450	C7H12O2	butyl acrylate	141-32-2	128.171	10	52	---	---
286	11695	C7H13O6P	mevinphos	7786-34-7	224.151	0.01	0.092	---	0.1
287	11704	C7H14	methylcyclohexane	108-87-2	98.188	400	1610	500	2000
288	11781	C7H14NO5P	monocrotophos	6923-22-4	223.166	---	0.25	---	---
289	11829	C7H14O	2-heptanone	110-43-0	114.188	50	233	100	465
290	11830	C7H14O	3-heptanone	106-35-4	114.188	50	---	50	230
291	11831	C7H14O	4-heptanone	123-19-3	114.188	50	233	---	---
292	11834	C7H14O	5-methyl-2-hexanone	110-12-3	114.188	50	234	100	475
293	11877	C7H14O	methylcyclohexanol	25639-42-3	114.188	50	234	100	470
294	11892	C7H14O2	pentyl acetate	628-63-7	130.187	100	532	100	525
295	11893	C7H14O2	1-methylbutyl acetate	626-38-0	130.187	125	665	125	650
296	11895	C7H14O2	isopentyl acetate	123-92-2	130.187	100	532	100	525
297	11915	C7H14O2	butyl glycidyl ether	2426-08-6	130.187	25	133	50	270
298	11937	C7H14O3	butyl lactate	138-22-7	146.186	5	30	---	---
299	12112	C7H16	heptane	142-82-5	100.204	400	1640	500	2000
300	12282	C7H16O3	dipropylene glycol monomethyl ether	34590-94-8	148.202	100	606	100	600
301	12364	C7H17O2PS3	phorate	298-02-2	260.383	---	0.05	---	---
302	12421	C8Co2O8	cobalt carbonyl	10210-68-1	341.950	---	0.1	---	---
303	12570	C8H4F15NO2	ammonium perfluorooctanoate	3825-26-1	431.103	---	0.01	---	---
304	12575	C8H4N2	m-dicyanobenzene	626-17-5	128.134	---	5	---	---
305	12594	C8H4O3	phthalic anhydride	85-44-9	148.118	1	6.1	2	12
306	12639	C8H5Cl3O3	2,4,5-trichlorophenoxyacetic acid	93-76-5	255.483	---	10	---	10
307	12821	C8H6Cl2O3	(2,4-dichlorophenoxy)acetic acid	94-75-7	221.039	---	10	---	10
308	12936	C8H6O4	terephthalic acid	100-21-0	166.133	---	10	---	---
309	12985	C8H7Cl	2-chlorostyrene	2039-87-4	138.596	50	283	---	---
310	13000	C8H7ClO	alpha-chloroacetophenone	532-27-4	154.595	0.05	0.32	0.05	0.3
311	13047	C8H7Cl2NaO5S	sesone	136-78-7	309.101	---	10	---	15*/5**
312	13226	C8H7N3O5	2-methyl-3,5-dinitrobenzamide	148-01-6	225.162	---	5	---	---
313	13237	C8H8	styrene	100-42-5	104.152	20	85	100	---
314	13391	C8H8O	acetophenone	98-86-2	120.151	10	49	---	---
315	13789	C8H10	ethylbenzene	100-41-4	106.167	100	434	100	435
316	13790	C8H10	o-xylene	95-47-6	106.167	100	434	100	435
317	13791	C8H10	m-xylene	108-38-3	106.167	100	434	100	435
318	13792	C8H10	p-xylene	106-42-3	106.167	100	434	100	435
319	13793	C8H10	xylenes	1330-20-7	106.167	100	434	100	435
320	13854	C8H10NO5PS	methyl parathion	298-00-0	263.211	---	0.2	---	---

Table 16 Threshold Limit Value and Permissible Exposure Level - Oganic Compounds

						Threshold Limit Value - TLV		Permissible Exposure Limit - PEL	
NO	ID	FORMULA	NAME	CAS No	MW, g/mol	ppm (vol)	mg/m3	ppm (vol)	mg/m3
321	14082	C8H11N	N,N-dimethylaniline	121-69-7	121.182	5	25	5	25
322	14123	C8H11N	xylidine, isomers	1300-73-8	121.182	0.5	2.5	5	25
323	14231	C8H12	vinylcyclohexene	100-40-3	108.183	0.1	0.44	---	---
324	14286	C8H12N2	tetramethyl succinonitrile	3333-52-6	136.197	0.5	2.8	0.5	3
325	14306	C8H12N2O2	hexamethylene diisocyanate	822-06-0	168.196	0.005	0.034	---	---
326	14348	C8H12O2	1,2-epoxy-4-(epoxyethyl)cyclohexane	106-87-6	140.182	0.1	0.57	---	---
327	14510	C8H14ClN5	atrazine	1912-24-9	215.687	---	5	---	---
328	14538	C8H14N4OS	metribuzin	21087-64-9	214.293	---	5	---	---
329	15023	C8H16NO5P	dicrotophos	141-66-2	237.193	---	0.25	---	---
330	15077	C8H16O	5-methyl-3-heptanone	541-85-5	128.214	25	131	25	130
331	15163	C8H16O2	sec-hexyl acetate	108-84-9	144.214	50	295	50	300
332	15354	C8H18	octane	111-65-9	114.231	300	1400	500	2350
333	15535	C8H18O	isooctanol	26952-21-6	130.230	50	266	---	---
334	15667	C8H19O2PS3	disulfoton	298-04-4	274.410	---	0.1	---	---
335	15676	C8H19O4P	dibutyl phosphate	107-66-4	210.210	1	8.6	1	5
336	15719	C8H20O4Si	ethyl silicate	78-10-4	208.329	10	85	100	850
337	15724	C8H20O5P2S2	sulfotep	3689-24-5	322.324	---	0.2	---	0.2
338	15727	C8H20O7P2	tetraethyl pyrophosphate	107-49-3	290.191	0.004	0.047	---	0.05
339	15729	C8H20Pb	tetraethyl lead	78-00-2	323.447	---	0.1	---	0.075
340	15803	C9H4O5	trimellitic anhydride	552-30-7	192.128	---	5	---	---
341	15903	C9H6N2O2	toluene diisocyanate	584-84-9	174.159	---	0.1	---	---
342	15908	C9H6N2O2	toluene-2,6-diisocyanate	91-08-7	174.159	0.005	0.036	---	---
343	16233	C9H8O4	2-(acetyloxy)benzoic acid	50-78-2	180.160	---	5	---	---
344	16512	C9H10	alpha-methylstyrene	98-83-9	118.178	50	242	---	---
345	16519	C9H10	methylstyrene	25013-15-4	118.178	50	242	100	480
346	16540	C9H10Cl2N2O	diuron	330-54-1	233.097	---	10	---	---
347	16680	C9H10O2	benzyl acetate	140-11-4	150.177	10	61	---	---
348	16686	C9H10O2	phenyl glycidyl ether	122-60-1	150.177	0.1	0.6	10	60
349	16876	C9H11Cl3NO3PS	chlorpyrifos	2921-88-2	350.589	---	0.2	---	---
350	17004	C9H12	cumene	98-82-8	120.194	50	---	50	245
351	17010	C9H12	trimethyl benzene	25551-13-7	120.194	25	123	---	---
352	17234	C9H13BrN2O2	bromacil	314-40-9	261.119	---	10	---	---
353	17275	C9H13N	N-isopropylaniline	768-52-5	135.209	2	---	---	---
354	17436	C9H14O	isophorone	78-59-1	138.210	---	---	25	140
355	17966	C9H18FeN3S6	ferbam	14484-64-1	416.504	---	10	---	15
356	17998	C9H18O	diisobutyl ketone	108-83-8	142.241	25	145	50	290
357	18176	C9H20	nonane	111-84-2	128.258	200	1050	---	---
358	18432	C9H22O4P2S4	ethion	563-12-2	384.483	---	0.4	---	---
359	18455	C10Cl8	octachloronaphthalene	2234-13-1	403.730	0.1	0.3	---	0.1
360	18474	C10H2Cl6	hexachloronaphthalene	1335-87-1	334.841	---	0.2	---	0.2
361	18482	C10H3Cl5	pentachloronaphthalene	1321-64-8	300.396	---	0.5	---	0.5
362	18489	C10H4Cl4	tetrachloronaphthalene	1335-88-2	265.952	---	2	---	2
363	18498	C10H5ClN2	o-chlorobenzylidene malononitrile	2698-41-1	188.616	---	---	0.05	0.4
364	18505	C10H5Cl3	trichloronaphthalene	1321-65-9	231.507	---	5	---	5
365	18507	C10H5Cl7	heptachlor	76-44-8	373.317	---	0.5	---	0.5
366	18508	C10H5Cl7O	heptachlor epoxide	1024-57-3	389.317	---	0.5	---	0.5
367	18551	C10H6Cl8	chlordane	57-74-9	409.778	---	0.5	---	0.5
368	18669	C10H8	naphthalene	91-20-3	128.174	10	52	10	50
369	18832	C10H9Cl4NO2S	captafol	2425-06-1	349.063	---	0.1	---	---
370	18836	C10H9Cl9	toxaphene	8001-35-2	448.254	---	0.5	---	0.5
371	18973	C10H10	divinylbenzene	1321-74-0	130.189	10	53	---	---
372	19015	C10H10Fe	ferrocene	102-54-5	186.034	---	10	---	15
373	19195	C10H10O4	dimethyl phthalate	131-11-3	194.187	---	5	---	5
374	19358	C10H12	1,3-dicyclopentadiene	77-73-6	132.205	5	27	---	---
375	19468	C10H12N3O3PS2	azinphos-methyl	86-50-0	317.330	---	0.2	---	0.2
376	19927	C10H14NO5PS	parathion	56-38-2	291.265	---	0.1	---	0.1
377	19933	C10H14N2	L-nicotine	54-11-5	162.235	---	0.5	---	0.5
378	19987	C10H14O	2-sec-butylphenol	89-72-5	150.221	5	31	---	---
379	20304	C10H15OPS2	fonofos	944-22-9	246.335	---	0.1	---	---
380	20311	C10H15O3PS2	fenthion	55-38-9	278.333	---	0.2	---	---
381	20423	C10H16O	camphor	76-22-2	152.236	2	12	---	2
382	20919	C10H19O6PS2	malathion	121-75-5	330.363	---	10	---	15
383	20989	C10H20N2S4	disulfiram	97-77-8	296.547	---	2	---	---
384	21409	C10H23NO	2-dibutylaminoethanol	102-81-8	173.299	0.5	3.5	---	---
385	21638	C11H10N2S	1-naphthalenylthiourea	86-88-4	202.280	---	0.3	---	0.3
386	22282	C11H15NO3	propoxur	114-26-1	209.245	---	0.5	---	---
387	22323	C11H16	1-methyl-4-tert-butylbenzene	98-51-1	148.248	1	6.1	10	60
388	22552	C11H17O4PS2	fensulfothion	115-90-2	308.360	---	0.1	---	---
389	23166	C12H4Cl6O	chlorinated diphenyl oxide	31242-93-0	376.878	---	0.5	---	0.5
390	23291	C12H8Cl6	aldrin	309-00-2	364.911	---	0.25	---	0.25
391	23293	C12H8Cl6O	dieldrin	60-57-1	380.910	---	0.25	---	0.25
392	23294	C12H8Cl6O	endrin	72-20-8	380.910	---	0.1	---	0.1
393	23425	C12H9NS	10H-phenothiazine	92-84-2	199.276	---	5	---	---
394	23445	C12H10	phenylbenzene (biphenyl)	92-52-4	154.211	0.2	1.3	0.2	1
395	23519	C12H10O	diphenyl ether	101-84-8	170.211	1	7	1	7
396	23609	C12H11N	diphenylamine	122-39-4	169.226	---	10	---	---
397	23632	C12H11NO2	carbaryl	63-25-2	201.225	---	5	---	5
398	23729.1	C12H12N2	diquat	2764-72-9	184.241	---	0.5	---	---
399	23917	C12H14N2	paraquat	4685-14-7	186.257	---	0.5	---	0.5
400	24002	C12H14O4	diethyl phthalate	84-66-2	222.241	---	5	---	---

Table 16 Threshold Limit Value and Permissible Exposure Level - Oganic Compounds

NO	ID	FORMULA	NAME	CAS No	MW, g/mol	Threshold Limit Value - TLV ppm (vol)	Threshold Limit Value - TLV mg/m3	Permissible Exposure Limit - PEL ppm (vol)	Permissible Exposure Limit - PEL mg/m3
401	24076	C12H15NO3	carbofuran	1563-66-2	221.256	---	0.1	---	---
402	24103	C12H15N3O6	glycidyl isocyanurate	2451-62-9	297.269	---	0.05	---	---
403	24412	C12H18N2O2	isophorone diisocyanate	4098-71-9	222.288	0.005	0.045	---	---
404	24503	C12H19ClNO3P	crufomate	299-86-5	291.715	---	5	---	---
405	24527	C12H19O2PS3	sulprofos	35400-43-2	322.454	---	1	---	---
406	24649	C12H21N2O3PS	diazinon	333-41-5	304.351	---	0.1	---	---
407	24774	C12H22O11	sucrose	57-50-1	342.300	---	10	---	15*/5**
408	25307	C12H26O6P2S4	dioxathion	78-34-2	456.547	---	0.2	---	---
409	25355	C12H27O4P	tributyl phosphate	126-73-8	266.318	0.2	2.2	---	5
410	25748	C13H12Cl2N2	bis(4-amino-3-chlorophenyl)methane	101-14-4	267.157	0.01	0.11	---	---
411	25950	C13H14N2	4,4'-diaminodiphenylmethane	101-77-9	198.268	0.1	0.81	0.01	---
412	26367	C13H22NO3PS	fenamiphos	22224-92-6	303.363	---	0.1	---	---
413	26699	C14H9Cl5	1,1,1-trichloro-2,2-bis(4-chlorophenyl)ethane	50-29-3	354.488	---	1	---	1
414	26788	C14H10O4	benzoyl peroxide	94-36-0	242.231	---	5	---	5
415	27096	C14H14NO4PS	ethyl p-nitrophenyl benzenethiophosphate	2104-64-5	323.310	---	0.1	---	0.5
416	27180	C14H14O3	2-pivaloyl-1,3-indandione	83-26-1	230.263	---	0.1	---	0.1
417	27476	C14H18N4O3	benomyl	17804-35-2	290.323	0.84	10	---	15*/5**
418	27742	C14H23O4P	dibutylphenyl phosphate	2528-36-1	286.308	0.3	3.5	---	---
419	28029	C15H10N2O2	diphenylmethane-4,4'-diisocyanate	101-68-8	250.257	0.005	0.051	---	---
420	28428	C15H16N4O5S	sulfometuron methyl	74222-97-2	364.383	---	5	---	---
421	28658	C15H22N2O2	methylene bis(4-cyclohexylisocyanate)	5124-30-1	262.352	0.005	0.054	---	---
422	28756	C15H24O	2,6-di-tert-butyl-p-cresol	128-37-0	220.355	---	10	---	---
423	29198	C16H15Cl3O2	methoxychlor	72-43-5	345.651	---	10	---	15
424	29470	C16H20O6P2S3	abate	3383-96-8	466.477	---	10	---	15*/5**
425	29518	C16H22O4	dibutyl phthalate	84-74-2	278.348	---	5	---	5
426	29834.1	C16H38O6P2S4	demeton	8065-48-3	516.680	0.01	0.11	---	0.1
427	30406	C18H12	chrysene	218-01-9	228.293	---	---	---	0.2
428	30548	C18H15N	triphenylamine	603-34-9	245.324	---	5	---	---
429	30566	C18H15O4P	triphenyl phosphate	115-86-6	326.289	---	3	---	3
430	31029	C18H34OSn	cyhexatin	13121-70-5	385.177	---	5	---	---
431	31366	C19H16O4	warfarin	81-81-2	308.334	---	0.1	---	0.1
432	31804	C20H12	benzo[a]pyrene	50-32-8	252.315	---	---	---	0.2
433	32787	C21H21O4P	tri-o-cresyl phosphate	78-30-8	368.369	---	0.1	---	0.1
434	32805	C21H22N2O2	strychnine	57-24-9	334.418	---	0.15	---	0.15
435	33356	C22H30O2S	4,4'-thiobis(6-tert-butyl-m-cresol)	96-69-5	358.545	---	10	---	15*/5**
436	33546	C23H22O6	rotenone	83-79-4	394.424	---	5	---	5
437	33926	C24H38O4	bis(2-ethylhexyl) phthalate	117-81-7	390.563	---	5	---	5
438	35415	C36H70O4Zn	zinc stearate	557-05-1	632.339	---	10	---	15*/5**

* - total dust, ** - respirable dust

Threshold Limit Value of ACGIH - exposure for time weighted average of 8-hour workday (or 40-hour work week) without adverse effects on health.

ACGIH - American Conference of Governmental Industrial Hygienists.

Permissible Exposure Level of OSHA.

OSHA - Occupational Safety and Health Adminstration.

ppm (vol) - parts per million by volume in air

mg/m3 - milligrams per cubic meter of air

Chapter 17

THRESHOLD LIMIT VALUE AND PERMISSIBLE EXPOSURE LIMIT – INORGANIC COMPOUNDS

Carl L. Yaws and Kayzad J. Vajifdar
Lamar University, Beaumont, Texas

Tabulation Results

The results are given in Table 17. The tabulation is arranged by alphabetical order (Ag, Al, As, C, ...) to provide ease of use in quickly locating the data by using the chemical formula. The compound name, CAS No (Chemical Abstracts Registry Number), and molecular weight are next provided. Values for threshold limit value and permissible exposure limit are given in the adjacent columns.

The threshold limit value and permissible exposure value are important in establishing the exposure limits for safeguarding health in the workplace. The threshold limit value (TLV) is given by American Conference of Governmental Industrial Hygienists (ACGIH). The permissible exposure limit (PEL) is provided by the Occupational Safety and Health Administration (OSHA). The values which represent exposure limits for safeguarding health apply to exposure in the workplace for a 8 hour day or 40 hour week.

In preparing the tabulation, a literature search was conducted to identify data source publications (1-15). The publications were screened and copies of appropriate data were made. These data were then keyed into the computer to provide a database of values for the compounds. The excellent compilations by ACGIH (1-2), CRC (4), Lewis (10), Montgomery (11), and Yaws (15) were primarily used for the values.

Example

Determine the threshold limit value (TLV) and permissible exposure limit (PEL) for diborane (B2H6) in the workplace.

Inspection of the tabulated values yields:

TLV = 0.1 ppm (vol)

PEL = 0.1 ppm (vol)

References

1. ACGIH Publication, 1997 TLVs and BEIs, American Conference of Governmental Industrial Hygienists, ACGIH, Inc., Cincinnati, OH (1997).
2. ACGIH Publication, GUIDE TO OCCUPATIONAL EXPOSURE VALUES – 1997, American Conference of Governmental Industrial Hygienists, ACGIH, Inc., Cincinnati, OH (1997).
3. Allen, D. T. and D. R. Shonnard, GREEN ENGINEERING, Prentice Hall PTR, Upper Saddle River, NJ (2002).
4. CRC HANDBOOK OF CHEMISTRY AND PHYSICS, 84th - 87th ed., CRC Press, Inc., Boca Raton, FL (2004-2007).
5. Crowl, D. A. and J. F. Louvar, CHEMICAL PROCESS SAFETY, 2nd edition, Prentice Hall, Inc., Englewood Cliffs, NJ (2002).
6. de la Cruz, P. L. and D. G. Sarvadi, Am. Ind. Hyg. Assoc. J., 55(10), 894 (1994).
7. Fawcett, H. H. and W. C. Wood, eds., SAFETY AND ACCIDENT PREVENTION IN CHEMICAL OPERATIONS, 2nd ed., John Wiley and Sons, New York, NY (1982).
8. June 1993 Air Contaminants Final Rule, specified in Tables Z-1, Z-2 and Z-3 (Federal Register, 58:35388-35351, June 30, 1993; corrected in Federal Register, 58:40191, July 27, 1993; and subsequent amendments).
9. Lees, F. P., LOSS PREVENTION IN THE PROCESS INDUSTRIES, Vols. 1 and 2, Butterworth-Heinemann, London, England (1992).
10. Lewis, R. J., SAX'S DANGEROUS PROPERTIES OF INDUSTRIAL MATERIALS, 9th ed., Vols. 1, 2 and 3, Van Nostrand Reinhold Company, New York, NY (1996).
11. Montgomery, J. H., GROUNDWATER CHEMICALS, Lewis Publishers, New York, NY (2000).
12. NIOSH POCKET GUIDE TO CHEMICAL HAZARDS, U. S. Dept. of Health and Human Services, Superintendent of Documents, Washington, DC (June, 1994).
13. Springer, C. and J. R. Welker, INDUSTRIAL HYGIENE: AN INTRODUCTION FOR CHEMICAL ENGINEERS, American Institute of Chemical Engineers, New York, NY (1995).
14. Williams, P. L. and J. L. Burson, eds., INDUSTRIAL TOXICOLOGY, SAFETY AND HEALTH APPLICATIONS IN THE WORKPLACE, Van Nostrand Reinhold Company, New York, NY (1985).
15. Yaws, C. L., CHEMICAL PROPERTIES HANDBOOK, McGraw-Hill, New York, NY (1999).

Table 17 Threshold Limit Value and Permissible Exposure Level - Inoganic Compounds

NO	ID	FORMULA	NAME	CAS No	MW, g/mol	Threshold Limit Value - TLV ppm (vol)	Threshold Limit Value - TLV mg/m3	Permissible Exposure Limit - PEL ppm (vol)	Permissible Exposure Limit - PEL mg/m3
1	13	Ag	silver	7440-22-4	107.868	---	0.1	---	0.01
2	77	Al	aluminum	7429-90-5	26.982	---	10	---	15*/5**
3	167	Al2O3	aluminum oxide	1344-28-1	101.961	---	10	---	15*/5**
4	168	Al2O3	aluminum oxide (alpha)	1302-74-5	101.961	---	10	---	15
5	202	As	arsenic	7440-38-3	74.922	---	0.01	---	0.01
6	235	AsH3O4	arsenic acid	7778-39-4	141.943	---	0.01	---	---
7	238	AsI3	arsenic triiodide	7784-45-4	455.635	0.05	0.16	0.05	0.2
8	394	B2H6	diborane	19287-45-7	27.670	0.1	0.11	0.1	0.1
9	395	B2O3	boron oxide	1303-86-2	69.620	---	10	---	15*/5**
10	405	B5H9	pentaborane	19624-22-7	63.126	0.005	0.013	0.005	0.01
11	413	B10H14	decaborane	177702-41-9	122.221	0.05	0.25	0.05	0.3
12	416	Ba	barium	7440-39-3	137.327	---	0.5	---	0.5
13	501	BaO4S	barium sulfate	7727-43-7	233.391	---	10	---	15*/5**
14	530	Be	beryllium	7440-41-7	9.012	---	0.002	---	0.002
15	604	Bi2Te3	bismuth telluride	1304-82-1	800.761	---	10	---	15*/5**
16	604.1	Bi2Te3	bismuth telluride, Se-doped	---	800.761	---	5	---	---
17	628	BrF5	bromine pentafluoride	7789-30-2	174.896	0.1	0.72	---	---
18	633	Br2	bromine	7726-95-6	159.808	0.1	0.66	0.1	0.7
19	639	C	graphite	7782-42-5	12.011	---	2	---	---
20	641	CCl2O	carbonyl chloride	75-44-5	98.916	0.1	0.4	0.1	0.4
21	652	CO	carbon monoxide	630-08-0	28.010	25	29	50	55
22	655	CO2	carbon dioxide	124-38-9	44.010	5000	9000	5000	9000
23	656	CS2	carbon disulfide	75-15-0	76.143	10	31	20	---
24	687	CaCN2	calcium cyanamide	156-62-7	80.102	---	0.5	---	---
25	688	CaCO3	calcium carbonate (calcite)	471-34-1	100.087	---	10	---	15*/5**
26	688.1	CaCO3	calcium carbonate (limestone)	1317-65-3	100.087	---	10	---	15*/5**
27	724	CaCrO4	calcium chromate	13756-19-0	156.072	---	0.001	---	---
28	733	CaH2O2	calcium hydroxide	1305-62-0	74.093	---	5	---	15*/5**
29	738	CaH4O6S	calcium sulfate dihydrate	10101-41-4	172.172	---	10	---	15*/5**
30	752	CaO	calcium oxide	1305-78-8	56.077	---	2	---	5
31	754	CaO3Si	calcium metasilicate	1344-95-2	116.162	---	10	---	15*/5**
32	757	CaO4S	calcium sulfate	7778-18-9	136.142	---	10	---	15*/5**
33	776	Cd	cadmium	7440-43-9	112.411	---	0.01	---	0.005
34	809	CdO	cadmium oxide	1306-19-0	128.410	---	---	---	0.005
35	892	ClFO3	perchloryl fluoride	7616-94-6	102.449	3	13	3	13.5
36	898	ClO2	chlorine dioxide	10049-04-4	67.452	0.1	0.28	0.1	0.3
37	899	Cl2	chlorine	7782-50-5	70.905	0.5	1.5	---	---
38	916	Co	cobalt	7440-48-4	58.933	---	0.02	---	0.1
39	1035	Cr	chromium	7440-47-3	51.996	---	0.5	---	1
40	1049.1	CrC6H9O6	chromic acetate (as Cr for TLV)	1066-30-4	226.057	---	0.05	---	---
41	1099	CrO2Cl2	chromium oxychloride	14977-61-8	154.900	0.025	0.16	---	---
42	1100	CrO3	chromium (vi) oxide	1333-82-0	99.994	---	0.05	---	---
43	1150	CsOH	cesium hydroxide	21351-79-1	149.913	---	2	---	---
44	1169	Cu	copper, fume	7440-50-8	63.546	---	0.2	---	0.1
45	1169.1	Cu	copper, dusts and mists	7440-50-8	63.546	---	---	---	1
46	1409	F2	fluorine	7782-41-4	37.997	1	1.6	0.1	0.2
47	1412	F2O	fluorine oxide	7783-41-7	53.996	---	---	0.05	0.1
48	1434	FeC5O5	iron pentacarbonyl	13463-40-6	195.896	0.1	0.8	---	---
49	1558	Fe2O3	iron (iii) oxide	1309-37-1	159.688	---	5	---	10
50	1675	GeH4	germane	7782-65-2	76.642	0.2	0.63	---	---
51	1691	HBr	hydrogen bromide	10035-10-6	80.912	---	---	3	10
52	1694	HCN	hydrogen cyanide	74-90-8	27.025	---	---	10	11
53	1698	HF	hydrogen fluoride	7664-39-3	20.006	---	---	3	---
54	1702	HNO3	nitric acid	7697-37-2	63.013	2	5.2	2	5
55	1713	H2O2	hydrogen peroxide	7722-84-1	34.015	1	1.4	1	1.4
56	1714	H2S	hydrogen sulfide	7783-06-4	34.082	10	14	---	---
57	1716	H2SO4	sulfuric acid	7664-93-9	98.079	---	1	---	1
58	1719	H2Se	hydrogen selenide	7783-07-5	80.976	0.05	0.16	0.05	0.2
59	1725	H3PO4	phosphoric acid	7664-38-2	97.995	---	1	---	1
60	1740	Hf	hafnium	7440-58-6	178.490	---	0.5	---	0.5
61	1877	In	indium	7440-74-6	114.818	---	0.1	---	---
62	2013	KCN	potassium cyanide	151-50-8	65.116	---	---	---	5
63	2131	K2S2O8	potassium persulfate	7727-21-1	270.324	---	0.1	---	---
64	2265	LiH	lithium hydride	7580-67-8	7.949	---	0.025	---	0.025
65	2335	Mg	magnesium	7439-95-4	24.305	---	10	---	15*/5**
66	2350	MgCO3	magnesium carbonate	546-93-0	84.314	---	10	---	15
67	2429	MgO	magnesium oxide	1309-48-4	40.304	---	0.2	---	---
68	2464	Mn	manganese	7439-96-5	54.938	---	0.2	---	---
69	2548	Mo	molybdenum (soluble compounds, as Mo)	7439-98-7	95.940	---	5	---	5
70	2548.1	Mo	molybdenum (insoluble compounds, as Mo)	7439-98-7	95.940	---	10	---	15
71	2613	NF3	nitrogen trifluoride	7783-54-2	71.002	10	29	10	29
72	2619	NH3	ammonia	7664-41-7	17.031	25	17	50	35
73	2627	NH4Cl	ammonium chloride	12125-02-9	53.491	---	10	---	---
74	2674	NO	nitric oxide	10102-43-9	30.006	25	---	25	30
75	2676	NO2	nitrogen dioxide	10102-44-0	46.006	3	5.6	---	---
76	2685	N2H4	hydrazine	302-01-2	32.045	0.01	0.013	1	1.3
77	2702	N2H6O3S	ammonium sulfamate	7773-06-0	114.125	---	10	---	15*/5*
78	2723	N2H8O8S2	ammonium peroxydisulfate	7727-54-0	228.204	---	0.1	---	---
79	2747	N2O	nitrous oxide	10024-97-2	44.013	50	90	---	---
80	2810	NaCN	sodium cyanide	143-33-9	49.007	---	---	---	5
81	2864	NaF	sodium fluoride	7681-49-4	41.988	---	---	---	2.5
82	2874	NaHSO3	sodium hydrogen sulfite	7631-90-5	104.062	---	5	---	---

Table 17 Threshold Limit Value and Permissible Exposure Level - Inoganic Compounds

						Threshold Limit Value - TLV		Permissible Exposure Limit - PEL	
NO	ID	FORMULA	NAME	CAS No	MW, g/mol	ppm (vol)	mg/m3	ppm (vol)	mg/m3
83	2897	NaOH	sodium hydroxide	1310-73-2	39.997	---	---	---	2
84	2909	Na2B4H20O17	sodium tetraborate decahydrate	1303-96-4	381.372	---	5	---	---
85	2971	Na2O8S2	sodium persulfate	7775-27-1	238.107	---	0.1	---	---
86	2975	Na2S2O5	sodium metabisulfite	7681-57-4	190.109	---	5	---	---
87	3013	Na4P2O7	sodium pyrophosphate	7722-88-5	265.902	---	5	---	---
88	3092	Ni	nickel	7440-02-0	58.693	---	1	---	1
89	3105	NiC4O4	nickel carbonyl (as Ni)	13463-39-3	170.734	---	---	0.001	0.007
90	3199	O3	ozone	10028-15-6	47.998	---	---	0.1	0.2
91	3222	OsO4	osmium tetroxide (as Os)	20816-12-0	254.228	0.0002	0.0016	---	0.002
92	3234	P	phosphorus (yellow)	7723-14-0	30.974	0.02	0.1	---	0.1
93	3381	PCl3	phosphorus (iii) chloride	7719-12-2	137.332	0.2	1.1	0.5	3
94	3382	PCl5	phosphorus pentachloride	10026-13-8	208.237	0.1	0.85	---	1
95	3389	PH3	phosphine	7803-51-2	33.998	0.3	0.42	0.3	0.4
96	3396	POCl3	phosphorus oxychloride	10025-87-3	153.331	0.1	0.63	---	---
97	3451	P2S5	phosphorus (v) sulfide	1314-80-3	222.278	---	1	---	1
98	3483	Pb	lead	7439-92-1	207.200	---	0.05	---	0.05
99	3520	PbCrO4	lead (ii) chromate	7758-97-6	323.194	---	0.012	---	---
100	3564	Pb3As2O8	lead (ii) arsenate	3687-31-8	899.438	---	0.15	---	0.01
101	3569	Pb3O8P2	lead (ii) phosphate	7446-27-7	811.543	---	0.05	---	0.05
102	3691	Pt	platinum	7440-06-4	195.078	---	1	---	---
103	3880	Rh	rhodium	7440-16-6	102.906	---	1	---	0.1
104	4001	SF6	sulfur hexafluoride	2551-62-4	146.056	1000	---	1000	6000
105	4007	SO2	sulfur dioxide	7446-09-5	64.065	2	5.2	5	13
106	4009	SO2F2	sulfuryl fluoride	2699-79-8	102.062	5	21	5	20
107	4013	S2Cl2	sulfur monochloride	10025-67-9	135.037	---	---	1	6
108	4017	S2F10	sulfur decafluoride	5714-22-7	254.116	---	---	0.025	0.25
109	4023	Sb	antimony	7440-36-0	121.760	---	0.5	---	0.5
110	4047	SbH3	stibine	7803-52-3	124.784	0.1	0.51	0.1	0.5
111	4056	Sb2O3	antimony (iii) oxide	1309-64-4	291.518	---	0.5	---	0.5
112	4083	Se	selenium	7782-49-2	78.960	---	0.2	---	0.2
113	4110	SeF6	selenium hexafluoride	7783-79-1	192.950	0.05	0.16	0.05	0.16
114	4140	Si	silicon	7440-21-3	28.086	---	10	---	15*/5**
115	4147	SiC	silicon carbide (hexagonal)	409-21-2	40.096	---	10	---	15*/5**
116	4197	SiC8H20O4	silane, tetraethoxy	78-10-4	208.328	10	85	100	850
117	4261	SiH4	silane	7803-62-5	32.117	5	---	---	---
118	4265	SiO2	silicon dioxide (quartz)	14808-60-7	60.084	---	0.1	---	---
119	4266	SiO2	silicon dioxide (tridymite)	15468-32-3	60.084	---	0.05	---	---
120	4267	SiO2	silicon dioxide (cristobalite)	14464-46-1	60.084	---	0.05	---	---
121	4268	SiO2	silicon dioxide (vitreous)	60676-86-0	60.084	---	0.1	---	---
122	4269.1	SiO2	silica, amprphous silica gel	112926-00-8	60.084	---	10	---	---
123	4269.2	SiO2	silica, crystalline tripoli	1317-95-9	60.084	---	0.1	---	---
124	4269.3	SiO2	silica, fume	69012-64-2	60.084	---	2	---	---
125	4332	Sn	tin	7440-31-5	118.710	---	2	---	2
126	4435	SnO2	tin (iv) oxide	18282-10-5	150.709	---	2	---	---
127	4483	SrCrO4	strontium chromate	7789-06-2	203.614	---	0.0005	---	---
128	4517	Ta	tantalum	7440-25-7	180.948	---	5	---	5
129	4548	Ta2O5	tantalum (v) oxide	1314-61-0	441.893	---	5	---	5
130	4583	Te	tellurium	13494-80-9	127.600	---	0.1	---	0.1
131	4590	TeF6	tellurium hexafluoride	7783-80-4	241.590	0.02	0.2	0.02	0.2
132	4689	TiO2	titanium (iv) oxide	13463-67-7	79.866	---	10	---	15
133	4703	Tl	thallium	7440-28-0	204.383	---	0.1	---	0.1
134	4766	U	uranium	7440-61-1	238.029	---	0.2	---	0.05
135	4891	V2O5	vanadium (v) oxide	1314-62-1	181.880	---	0.05	---	---
136	4899	W	tungsten	7440-33-7	183.840	---	5	---	---
137	4995	Yb	ytterbium	7440-64-4	173.040	---	1	---	1
138	5062	ZnCl2	zinc chloride	7646-85-7	136.295	---	1	---	1
139	5087	ZnO	zinc oxide, dust	1314-13-2	81.389	---	10	---	15*/5**
140	5087.1	ZnO	zinc oxide, fume	1314-13-2	81.389	---	5	---	5
141	5113	Zr	zirconium	7440-67-7	91.224	---	---	---	5

* - total dust, ** - respirable dust

Threshold Limit Value of ACGIH - exposure for time weighted average of 8-hour workday (or 40-hour work week) without adverse effects on health.

ACGIH - American Conference of Governmental Industrial Hygienists.

Permissible Exposure Level of OSHA.

OSHA - Occupational Safety and Health Adminstration.

ppm (vol) - parts per million by volume in air

mg/m3 - milligrams per cubic meter of air

Appendix A

Conversion Table

1. **Temperature**
   To convert from Centigrade to:
   Kelvin, add 273.15
   Rankine, multiply Kelvin by 1.8
   Fahrenheit, multiply Centigrade by 1.8 and add 32

2. **Pressure**
   To convert from psia to:
   kPa, multiply by 6.895
   psig, subtract 14.7
   mm Hg, multiply by 51.71
   atmospheres, divide by 14.7
   bars, divide by 14.508

3. **Heat of Vaporization**
   To convert from kJ/kg to:
   BTU/lb, multiply by 0.43
   cal/gram, multiply by 0.239

4. **Density**
   To convert from g/ml to:
   lb/ft^3, multiply by 62.43
   lb/gallon, multiply by 8.345

5. **Surface Tension**
   To convert from dynes/cm to:
   N/m, multiply by 0.001

6. **Heat Capacity**
   To convert from J/g K to:
   BTU/lb R, multiply by 0.239
   cal/gram K, multiply by 0.239

7. **Viscosity**
   To convert from micropoise to:
   lb/ft s, multiply by 0.0672E-06
   centipoise, multiply by 1.0E-04
   poise, multiply by 1.0E-06
   Pa s (Pascal seconds), multiply by 1.0E-07

   To convert from centipoise to:
   lb/ft s, multiply by 0.000672
   micropoise, multiply by 10,000
   poise, multiply by 0.01
   Pa s (Pascal seconds), multiply by 0.001

8. **Thermal Conductivity**
   To convert from W/m K to:
   BTU/hr ft R, multiply by 0.5770
   calorie/cm s K, multiply by .002388

9. **Enthalpy of Formation**
   To convert from kJ/mol to:
   kcal/mol, multiply by 0.239

10. **Gibbs Energy of Formation**
    To convert from kJ/mol to:
    kcal/mol, multiply by 0.239

11. **Henry's Law Constant for Compound in Water**
    To convert from atm/mol fraction to:
    atm/(mol/m^3), divide by 55,556
    kPa/(mol/m^3), divide by 548.295

Appendix B

Compound List by CAS Registry Number – Organic Compounds

Organics The compilation for organics provides the compound list by CAS registry number, name, chemical formula, and compound number.

50-00-0 formaldehyde CH2O 175
50-21-5 lactic acid C3H6O3 1948
50-29-3 1,1,1-trichloro-2,2-bis(4-chlorophenyl)ethane C14H9Cl5 26699
50-32-8 benzo[a]pyrene C20H12 31804
50-70-4 sorbitol C6H14O6 9100
50-78-2 2-(acetyloxy)benzoic acid C9H8O4 16233
50-81-7 ascorbic acid C6H8O6 7459
50-99-7 glucose C6H12O6 8609
51-17-2 1H-benzimidazole C7H6N2 10341
53-70-3 dibenz[a,h]anthracene C22H14 33098
54-11-5 L-nicotine C10H14N2 19933
55-38-9 fenthion C10H15O3PS2 20311
55-63-0 nitroglycerine C3H5N3O9 1796
56-23-5 carbon tetrachloride CCl4 50
56-38-2 parathion C10H14NO5PS 19927
56-49-5 1,2-dihydro-3-methylbenz[j]aceanthrylene C21H16 32632
56-55-3 benz[a]anthracene C18H12 30407
56-75-7 chloramphenicol C11H12Cl2N2O5 21801
56-81-5 glycerol C3H8O3 2148
56-86-0 L-glutamic acid C5H9NO4 5245
56-87-1 lysine C6H14N2O2 8944
57-10-3 hexadecanoic acid C16H32O2 29721
57-11-4 octadecanoic acid C18H36O2 31097
57-14-7 1,1-dimethylhydrazine C2H8N2 1160
57-24-9 strychnine C21H22N2O2 32805
57-27-2 morphine C17H19NO3 30097
57-50-1 sucrose C12H22O11 24774
57-55-6 1,2-propanediol (propylene glycol) C3H8O2 2140
57-57-8 beta-propiolactone C3H4O2 1627
57-74-9 chlordane C10H6Cl8 18551
57-97-6 7,12-dimethylbenz[a]anthracene C20H16 31977
58-27-5 2-methyl-1,4-naphthalenedione C11H8O2 21531
58-89-9 1,2,3,4,5,6-hexachlorocyclohexane, (1alpha,2alpha, C6H6Cl6 6948
58-90-2 2,3,4,6-tetrachlorophenol C6H2Cl4O 6178
60-09-3 p-aminoazobenzene C12H11N3 23656
60-12-8 2-phenylethanol C8H10O 13913
60-29-7 diethyl ether C4H10O 3869
60-33-3 linoleic acid C18H32O2 30991
60-34-4 methylhydrazine CH6N2 312
60-35-5 acetamide C2H5NO 956
60-51-5 dimethoate C5H12NO3PS2 5790
60-57-1 dieldrin C12H8Cl6O 23293
61-82-5 1H-1,2,4-triazol-3-amine C2H4N4 871
62-53-3 aniline C6H7N 7172
62-73-7 dichlorvos C4H7Cl2O4P 3155
62-74-8 sodium fluoroacetate C2H2FNaO2 641
63-25-2 carbaryl C12H11NO2 23632
64-04-0 benzeneethanamine C8H11N 14110
64-10-8 phenylurea C7H8N2O 10794
64-17-5 ethyl alcohol C2H6O 1065
64-18-6 formic acid CH2O2 176
64-19-7 acetic acid C2H4O2 889
64-67-5 diethyl sulfate C4H10O4S 3937
65-85-0 benzoic acid C7H6O2 10408
66-25-1 hexanal C6H12O 8401
66-71-7 1,10-phenanthroline C12H8N2 23305
67-56-1 methyl alcohol CH4O 272
67-63-0 isopropyl alcohol C3H8O 2130
67-64-1 acetone C3H6O 1923
67-66-3 chloroform CHCl3 104
67-68-5 dimethyl sulfoxide C2H6OS 1067
67-71-0 dimethyl sulfone C2H6O2S 1075
67-72-1 hexachloroethane C2Cl6 463
68-11-1 thioglycolic acid C2H4O2S 892
68-12-2 N,N-dimethylformamide C3H7NO 2036
69-72-7 salicylic acid C7H6O3 10420
70-30-4 hexachlorophene C13H6Cl6O2 25446
71-23-8 propyl alcohol C3H8O 2129
71-36-3 butanol C4H10O 3865
71-41-0 1-pentanol C5H12O 5831
71-43-2 benzene C6H6 6876
71-55-6 1,1,1-trichloroethane C2H3Cl3 720
72-20-8 endrin C12H8Cl6O 23294
72-43-5 methoxychlor C16H15Cl3O2 29198
72-54-8 1,1-dichloro-2,2-bis(p-chlorophenyl)ethane C14H10Cl4 26739
72-55-9 2,2-dichloro-1,1-bis(4-chlorophenyl)ethene C14H8Cl4 26637
74-31-7 N,N'-diphenyl-p-phenylenediamine C18H16N2 30588
74-82-8 methane CH4 251
74-83-9 methyl bromide CH3Br 191
74-84-0 ethane C2H6 985
74-85-1 ethylene C2H4 787
74-86-2 acetylene C2H2 571
74-87-3 methyl chloride CH3Cl 196
74-88-4 methyl iodide CH3I 222
74-89-5 methylamine CH5N 286
74-90-8 hydrogen cyanide CHN 122
74-93-1 methyl mercaptan CH4S 278
74-95-3 dibromomethane CH2Br2 144
74-96-4 bromoethane C2H5Br 910
74-97-5 bromochloromethane CH2BrCl 140
74-98-6 propane C3H8 2094
74-99-7 methylacetylene C3H4 1492
75-00-3 ethyl chloride C2H5Cl 914
75-01-4 vinyl chloride C2H3Cl 701
75-02-5 vinyl fluoride C2H3F 728
75-03-6 ethyl iodide C2H5I 948
75-04-7 ethylamine C2H7N 1115
75-05-8 acetonitrile C2H3N 753
75-07-0 acetaldehyde C2H4O 884
75-08-1 ethyl mercaptan C2H6S 1092
75-09-2 dichloromethane CH2Cl2 149
75-10-5 difluoromethane CH2F2 158
75-11-6 diiodomethane CH2I2 161
75-12-7 formamide CH3NO 232
75-15-0 carbon disulfide CS2 382
75-18-3 dimethyl sulfide C2H6S 1093
75-19-4 cyclopropane C3H6 1809
75-21-8 ethylene oxide C2H4O 885
75-25-2 bromoform CHBr3 96
75-26-3 2-bromopropane C3H7Br 1979
75-28-5 isobutane C4H10 3784
75-29-6 2-chloropropane C3H7Cl 1987
75-30-9 2-iodopropane C3H7I 2026
75-31-0 isopropylamine C3H9N 2201
75-33-2 isopropyl mercaptan C3H8S 2159
75-34-3 1,1-dichloroethane C2H4Cl2 812
75-35-4 1,1-dichloroethylene C2H2Cl2 615
75-36-5 acetyl chloride C2H3ClO 707
75-37-6 1,1-difluoroethane C2H4F2 834
75-38-7 1,1-difluoroethylene C2H2F2 642
75-39-8 1-aminoethanol C2H7NO 1118
75-43-4 dichlorofluoromethane CHCl2F 101
75-44-5 phosgene CCl2O 44
75-45-6 chlorodifluoromethane CHClF2 97
75-46-7 fluoroform CHF3 113
75-47-8 iodoform CHI3 117
75-50-3 trimethylamine C3H9N 2203
75-52-5 nitromethane CH3NO2 235
75-54-7 methyl dichlorosilane CH4Cl2Si 254
75-55-8 propyleneimine C3H7N 2031
75-56-9 1,2-propylene oxide C3H6O 1924
75-61-6 dibromodifluoromethane CBr2F2 23
75-62-7 bromotrichloromethane CBrCl3 16
75-63-8 bromotrifluoromethane CBrF3 17
75-64-9 tert-butylamine C4H11N 3980
75-65-0 tert-butanol C4H10O 3868
75-66-1 tert-butyl mercaptan C4H10S 3945
75-68-3 1-chloro-1,1-difluoroethane C2H3ClF2 702
75-69-4 trichlorofluoromethane CCl3F 47
75-71-8 dichlorodifluoromethane CCl2F2 42
75-72-9 chlorotrifluoromethane CClF3 34
75-73-0 carbon tetrafluoride CF4 74
75-74-1 tetramethyl lead C4H12Pb 4106
75-75-2 methanesulfonic acid CH4O3S 275
75-76-3 tetramethylsilane C4H12Si 4110
75-79-6 methyl trichlorosilane CH3Cl3Si 210
75-83-2 2,2-dimethylbutane C6H14 8878
75-84-3 2,2-dimethyl-1-propanol C5H12O 5840
75-85-4 tert-pentyl-alcohol C5H12O 5838
75-86-5 acetone cyanohydrin C4H7NO 3246
75-87-6 trichloroacetaldehyde C2HCl3O 544
75-88-7 2-chloro-1,1,1-trifluoroethane C2H2ClF3 606
75-89-8 2,2,2-trifluoroethanol C2H3F3O 738
75-91-2 t-butyl hydroperoxide C4H10O2 3881
75-97-8 3,3-dimethyl-2-butanone C6H12O 8414
75-98-9 2,2-dimethylpropanoic acid C5H10O2 5495
75-99-0 2,2-dichloropropanoic acid C3H4Cl2O2 1554
76-01-7 pentachloroethane C2HCl5 551
76-02-8 trichloroacetyl chloride C2Cl4O 459
76-03-9 trichloroacetic acid C2HCl3O2 547
76-05-1 trifluoroacetic acid C2HF3O2 556
76-06-2 trichloronitromethane CCl3NO2 49
76-09-5 2,3-dimethyl-2,3-butanediol C6H14O2 9052
76-11-9 1,1,1,2-tetrachloro-2,2-difluoroethane C2Cl4F2 457
76-12-0 1,1,2,2-tetrachloro-1,2-difluoroethane C2Cl4F2 456
76-13-1 1,1,2-trichloro-1,2,2-trifluoroethane C2Cl3F3 449
76-14-2 1,2-dichloro-1,1,2,2-tetrafluoroethane C2Cl2F4 441
76-15-3 chloropentafluoroethane C2ClF5 427
76-16-4 hexafluoroethane C2F6 494
76-19-7 octafluoropropane C3F8 1299
76-22-2 camphor C10H16O 20423
76-35-7 2,2-dimethyl-1,3-butanediol C6H14O2 9046
76-44-8 heptachlor C10H5Cl7 18507
76-84-6 triphenylmethanol C19H16O 31344
77-47-4 hexachlorocyclopentadiene C5Cl6 4169
77-73-6 1,3-dicyclopentadiene C10H12 19358
77-74-7 3-methyl-3-pentanol C6H14O 8980
77-78-1 dimethyl sulfate C2H6O4S 1083
77-84-9 2-methyl-2-ethyl-1,3-propanediol C6H14O2 9056
77-92-9 citric acid C6H8O7 7464
77-99-6 trimethylolpropane C6H14O3 9078
78-00-2 tetraethyl lead C8H20Pb 15729
78-10-4 ethyl silicate C8H20O4Si 15719
78-11-5 pentaerythritol tetranitrate C5H8N4O12 4845
78-26-2 2-methyl-2-propyl-1,3-propanediol C7H16O2 12264
78-30-8 tri-o-cresyl phosphate C21H21O4P 32787
78-34-2 dioxathion C12H26O6P2S4 25307
78-40-0 triethyl phosphate C6H15O4P 9268
78-59-1 isophorone C9H14O 17436
78-69-3 3,7-dimethyl-3-octanol C10H22O 21272
78-74-0 1,1,2-tribromoethane C2H3Br3 698
78-75-1 1,2-dibromopropane C3H6Br2 1825
78-76-2 2-bromobutane C4H9Br 3599
78-77-3 1-bromo-2-methylpropane C4H9Br 3600
78-78-4 isopentane C5H12 5773
78-79-5 isoprene C5H8 4738
78-80-8 2-methyl-1-butene-3-yne C5H6 4448
78-81-9 isobutylamine C4H11N 3978
78-82-0 isobutyronitrile C4H7N 3239
78-83-1 isobutanol C4H10O 3866
78-84-2 isobutyraldehyde C4H8O 3479
78-85-3 methacrolein C4H6O 2941
78-86-4 2-chlorobutane C4H9Cl 3612
78-87-5 1,2-dichloropropane C3H6Cl2 1845
78-88-6 2,3-dichloropropene C3H4Cl2 1533
78-90-0 1,2-propanediamine C3H10N2 2255
78-91-1 2-aminopropanol C3H9NO 2214
78-92-2 sec-butanol C4H10O 3867
78-93-3 methyl ethyl ketone C4H8O 3480
78-96-6 1-amino-2-propanol C3H9NO 2204
78-97-7 lactonitrile C3H5NO 1757
78-99-9 1,1-dichloropropane C3H6Cl2 1844
79-00-5 1,1,2-trichloroethane C2H3Cl3 721
79-01-6 trichloroethylene C2HCl3 539
79-04-9 chloroacetyl chloride C2H2Cl2O 625
79-06-1 acrylamide C3H5NO 1755
79-08-3 bromoacetic acid C2H3BrO2 693
79-09-4 propanoic acid C3H6O2 1930
79-10-7 acrylic acid C3H4O2 1626
79-11-8 chloroacetic acid C2H3ClO2 710
79-16-3 N-methylacetamide C3H7NO 2037
79-20-9 methyl acetate C3H6O2 1932
79-22-1 methyl chloroformate C2H3ClO2 711
79-24-3 nitroethane C2H5NO2 958
79-27-6 1,1,2,2-tetrabromoethane C2H2Br4 592
79-29-8 2,3-dimethylbutane C6H14 8879
79-31-2 isobutyric acid C4H8O2 3505
79-34-5 1,1,2,2-tetrachloroethane C2H2Cl4 636
79-36-7 dichloroacetyl chloride C2HCl3O 543
79-38-9 chlorotrifluoroethylene C2ClF3 423
79-41-4 methacrylic acid C4H6O2 2969

79-43-6 dichloroacetic acid C2H2Cl2O2 627
79-46-9 2-nitropropane C3H7NO2 2046
79-92-5 camphene C10H16 20315
80-05-7 bisphenol a C15H16O2 28436
80-15-9 cumene hydroperoxide C9H12O2 17138
80-43-3 dicumyl peroxide C18H22O2 30806
80-46-6 p-tert-amylphenol C11H16O 22399
80-56-8 alpha-pinene C10H16 20321
80-59-1 trans-2-methyl-2-butenoic acid C5H8O2 4891
80-62-6 methyl methacrylate C5H8O2 4889
81-81-2 warfarin C19H16O4 31366
82-68-8 pentachloronitrobenzene C6Cl5NO2 6065
83-12-5 2-phenyl-1H-indene-1,3(2H)-dione C15H10O2 28044
83-26-1 2-pivaloyl-1,3-indandione C14H14O3 27180
83-32-9 acenaphthene C12H10 23444
83-72-7 2-hydroxy-1,4-naphthalenedione C10H6O3 18586
83-79-4 rotenone C23H22O6 33546
84-15-1 o-terphenyl C18H14 30458
84-65-1 anthraquinone C14H8O2 26654
84-66-2 diethyl phthalate C12H14O4 24002
84-74-2 dibutyl phthalate C16H22O4 29518
85-01-8 phenanthrene C14H10 26730
85-44-9 phthalic anhydride C8H4O3 12594
85-68-7 benzyl butyl phthalate C19H20O4 31452
86-30-6 N-nitrosodiphenylamine C12H10N2O 23494
86-50-0 azinphos-methyl C10H12N3O3PS2 19468
86-73-7 fluorene C13H10 25575
86-74-8 dibenzopyrrole C12H9N 23394
86-88-4 1-naphthalenylthiourea C11H10N2S 21638
87-26-3 2-(1-methylbutyl)phenol C11H16O 22439
87-59-2 2,3-dimethylaniline C8H11N 14087
87-61-6 1,2,3-trichlorobenzene C6H3Cl3 6306
87-62-7 2,6-dimethylaniline C8H11N 14090
87-65-0 2,6-dichlorophenol C6H4Cl2O 6482
87-66-1 1,2,3-benzenetriol C6H6O3 7085
87-68-3 hexachloro-1,3-butadiene C4Cl6 2318
87-69-4 tartaric acid C4H6O6 3033
87-85-4 hexamethylbenzene C12H18 24378
87-86-5 pentachlorophenol C6HCl5O 6116
88-06-2 2,4,6-trichlorophenol C6H3Cl3O 6314
88-09-5 2-ethyl butyric acid C6H12O2 8461
88-17-5 2-(trifluoromethyl)aniline C7H6F3N 10320
88-18-6 2-tert-butylphenol C10H14O 19990
88-60-8 2-tert-butyl-5-methylphenol C11H16O 22402
88-65-3 2-bromobenzoic acid C7H5BrO2 9890
88-69-7 2-isopropylphenol C9H12O 17104
88-72-2 o-nitrotoluene C7H7NO2 10627
88-73-3 o-chloronitrobenzene C6H4ClNO2 6444
88-74-4 o-nitroaniline C6H6N2O2 6997
88-75-5 o-nitrophenol C6H5NO3 6830
88-89-1 2,4,6-trinitrophenol C6H3N3O7 6368
88-99-3 phthalic acid C8H6O4 12935
89-05-4 pyromellitic acid C10H6O8 18598
89-72-5 2-sec-butylphenol C10H14O 19987
89-83-8 5-methyl-2-isopropylphenol C10H14O 20011
90-00-6 o-ethylphenol C8H10O 13918
90-02-8 salicylaldehyde C7H6O2 10410
90-04-0 2-methoxyaniline C7H9NO 10979
90-05-1 guaiacol C7H8O2 10866
90-11-9 1-bromonaphthalene C10H7Br 18601
90-12-0 1-methylnaphthalene C11H10 21601
90-13-1 1-chloronaphthalene C10H7Cl 18606
90-15-3 1-naphthol C10H8O 18740
91-01-0 diphenylmethanol C13H12O 25798
91-08-7 toluene-2,6-diisocyanate C9H6N2O2 15908
91-16-7 1,2-dimethoxybenzene C8H10O2 13951
91-17-8 decahydronaphthalene; (cis+trans) C10H18 20636
91-19-0 quinoxaline C8H6N2 12868
91-20-3 naphthalene C10H8 18669
91-21-4 1,2,3,4-tetrahydroisoquinoline C9H11N 16893
91-22-5 quinoline C9H7N 16011
91-23-6 o-nitroanisole C7H7NO3 10660
91-57-6 2-methylnaphthalene C11H10 21602
91-58-7 2-chloronaphthalene C10H7Cl 18607
91-59-8 2-naphthylamine C10H9N 18862
91-62-3 6-methylquinoline C10H9N 18853
91-63-4 2-methylquinoline C10H9N 18849
91-64-5 2H-1-benzopyran-2-one C9H6O2 15926
91-66-7 N,N-diethylaniline C10H15N 20195
91-94-1 3,3'-dichloro-p-benzidine C12H10Cl2N2 23469
92-06-8 m-terphenyl C18H14 30457
92-24-0 naphthacene C18H12 30409
92-51-3 bicyclohexyl C12H22 24652
92-52-4 phenylbenzene (biphenyl) C12H10 23445
92-67-1 p-aminodiphenyl C12H11N 23607
92-82-0 phenazine C12H8N2 23308
92-84-2 10H-phenothiazine C12H9NS 23425
92-87-5 p-benzidine C12H12N2 23720
92-94-4 p-terphenyl C18H14 30459
93-37-8 2,7-dimethylquinoline C11H11N 21708
93-55-0 1-phenyl-1-propanone C9H10O 16634
93-58-3 methyl benzoate C8H8O2 13412
93-76-5 2,4,5-trichlorophenoxyacetic acid C8H5Cl3O3 12639
93-89-0 ethyl benzoate C9H10O2 16676
93-98-1 N-phenylbenzamide C13H11NO 25694
93-99-2 phenyl benzoate C13H10O2 25631
94-36-0 benzoyl peroxide C14H10O4 26788
94-74-6 (4-chloro-2-methylphenoxy)acetic acid C9H9ClO3 16324
94-75-7 (2,4-dichlorophenoxy)acetic acid C8H6Cl2O3 12821
94-99-5 2,4-dichlorobenzyl chloride C7H5Cl3 9977
95-13-6 indene C9H8 16090
95-15-8 benzothiophene C8H6S 12947
95-16-9 benzothiazole C7H5NS 10094
95-46-5 o-bromotoluene C7H7Br 10464
95-47-6 o-xylene C8H10 13790
95-48-7 o-cresol C7H8O 10853
95-49-8 o-chlorotoluene C7H7Cl 10493
95-50-1 o-dichlorobenzene C6H4Cl2 6469
95-51-2 o-chloroaniline C6H6ClN 6924
95-52-3 o-fluorotoluene C7H7F 10557
95-53-4 o-toluidine C7H9N 10956
95-54-5 o-phenylenediamine C6H8N2 7319
95-56-7 o-bromophenol C6H5BrO 6671
95-57-8 o-chlorophenol C6H5ClO 6721
95-63-6 1,2,4-trimethylbenzene C9H12 17009
95-64-7 3,4-dimethylaniline C8H11N 14091
95-65-8 3,4-xylenol C8H10O 13925
95-68-1 2,4-dimethylaniline C8H11N 14088
95-73-8 2,4-dichlorotoluene C7H6Cl2 10253
95-75-0 3,4-dichlorotoluene C7H6Cl2 10260
95-76-1 3,4-dichloroaniline C6H5Cl2N 6740
95-77-2 3,4-dichlorophenol C6H4Cl2O 6483
95-78-3 2,5-dimethylaniline C8H11N 14089
95-80-7 m-toluenediamine C7H10N2 11079
95-82-9 2,5-dichloroaniline C6H5Cl2N 6743
95-87-4 2,5-xylenol C8H10O 13923
95-92-1 diethyl oxalate C6H10O4 7918
95-93-2 1,2,4,5-tetramethylbenzene C10H14 19889
95-94-3 1,2,4,5-tetrachlorobenzene C6H2Cl4 6176
95-95-4 2,4,5-trichlorophenol C6H3Cl3O 6313
96-09-3 phenyloxirane C8H8O 13397
96-11-7 1,2,3-tribromopropane C3H5Br3 1667
96-12-8 1,2-dibromo-3-chloropropane C3H5Br2Cl 1662
96-14-0 3-methylpentane C6H14 8881
96-15-1 2-methylbutylamine C5H13N 5949
96-17-3 2-methylbutanal C5H10O 5446
96-18-4 1,2,3-trichloropropane C3H5Cl3 1707
96-20-8 2-aminobutan-1-ol C4H11NO 4003
96-22-0 diethyl ketone C5H10O 5451
96-33-3 methyl acrylate C4H6O2 2970
96-34-4 methyl chloroacetate C3H5ClO2 1688
96-37-7 methylcyclopentane C6H12 8214
96-47-9 2-methyltetrahydrofuran C5H10O 5453
96-48-0 gamma-butyrolactone C4H6O2 2965
96-49-1 ethylene carbonate C3H4O3 1633
96-54-8 N-methylpyrrole C5H7N 4658
96-69-5 4,4'-thiobis(6-tert-butyl-m-cresol) C22H30O2S 33356
96-76-4 2,4-di-tert-butylphenol C14H22O 27680
97-00-7 1-chloro-2,4-dinitrobenzene C6H3ClN2O4 6281
97-61-0 2-methylvaleric acid C6H12O2 8515
97-62-1 ethyl isobutanoate C6H12O2 8483
97-63-2 ethyl methacrylate C6H10O2 7814
97-64-3 ethyl lactate C5H10O3 5544
97-65-4 itaconic acid C5H6O4 4577
97-77-8 disulfiram C10H20N2S4 20989
97-85-8 isobutyl isobutyrate C8H16O2 15177
97-88-1 butyl methacrylate C8H14O2 14599
97-93-8 triethyl aluminum C6H15Al 9135
97-95-0 2-ethyl-1-butanol C6H14O 8981
97-96-1 2-ethylbutanal C6H12O 8405
97-99-4 tetrahydrofurfuryl alcohol C5H10O2 5506
98-00-0 furfuryl alcohol C5H6O2 4548
98-01-1 furfural C5H4O2 4327
98-06-6 tert-butylbenzene C10H14 19868
98-07-7 benzotrichloride C7H5Cl3 9967
98-08-8 benzotrifluoride C7H5F3 10003
98-16-8 3-(trifluoromethyl)aniline C7H6F3N 10321
98-19-1 1-tert-butyl-3,5-dimethylbenzene C12H18 24367
98-27-1 4-tert-butyl-2-methylphenol C11H16O 22404
98-29-3 p-tert-butylcatechol C10H14O2 20089
98-46-4 3-nitrobenzotrifluoride C7H4F3NO2 9799
98-51-1 1-methyl-4-tert-butylbenzene C11H16 22323
98-56-6 p-chlorobenzotrifluoride C7H4ClF3 9684
98-82-8 cumene C9H12 17004
98-83-9 alpha-methylstyrene C9H10 16512
98-84-0 1-phenylethylamine C8H11N 14122
98-85-1 1-phenylethanol C8H10O 13917
98-86-2 acetophenone C8H8O 13391
98-87-3 benzyl dichloride C7H6Cl2 10252
98-88-4 benzoyl chloride C7H5ClO 9927
98-95-3 nitrobenzene C6H5NO2 6819
99-04-7 m-toluic acid C8H8O2 13433
99-08-1 m-nitrotoluene C7H7NO2 10626
99-09-2 m-nitroaniline C6H6N2O2 6996
99-35-4 1,3,5-trinitrobenzene C6H3N3O6 6367
99-54-7 1,2-dichloro-4-nitrobenzene C6H3Cl2NO2 6296
99-62-7 m-diisopropylbenzene C12H18 24361
99-65-0 m-dinitrobenzene C6H4N2O4 6580
99-66-1 2-propylpentanoic acid C8H16O2 15205
99-71-8 4-sec-butylphenol C10H14O 19989
99-82-1 1-methyl-4-isopropylcyclohexane C10H20 20933
99-83-2 alpha-phellandrene C10H16 20319
99-85-4 gamma-terpinene C10H16 20326
99-86-5 alpha-terpinene C10H16 20325
99-87-6 p-cymene C10H14 19872
99-89-8 4-isopropylphenol C9H12O 17106
99-94-5 p-toluic acid C8H8O2 13414
99-99-0 p-nitrotoluene C7H7NO2 10628
100-00-5 p-chloronitrobenzene C6H4ClNO2 6445
100-01-6 p-nitroaniline C6H6N2O2 6998
100-02-7 p-nitrophenol C6H5NO3 6832
100-10-7 p-dimethylaminobenzaldehyde C9H11NO 16905
100-18-5 p-diisopropylbenzene C12H18 24362
100-21-0 terephthalic acid C8H6O4 12936
100-25-4 p-dinitrobenzene C6H4N2O4 6582
100-36-7 N,N-diethyl-1,2-ethanediamine C6H16N2 9293
100-37-8 2-diethylaminoethanol C6H15NO 9217
100-39-0 (bromomethyl)benzene C7H7Br 10463
100-40-3 vinylcyclohexene C8H12 14231
100-41-4 ethylbenzene C8H10 13789
100-42-5 styrene C8H8 13237
100-44-7 benzyl chloride C7H7Cl 10492
100-46-9 benzylamine C7H9N 10953
100-47-0 benzonitrile C7H5N 10045
100-51-6 benzyl alcohol C7H8O 10851
100-52-7 benzaldehyde C7H6O 10404
100-60-7 N-methylcyclohexylamine C7H15N 12024
100-61-8 N-methylaniline C7H9N 10954
100-63-0 phenylhydrazine C6H8N2 7321
100-64-1 cyclohexanone oxime C6H11NO 8102
100-66-3 anisole C7H8O 10850
100-68-5 methyl phenyl sulfide C7H8S 10921
100-71-0 2-ethylpyridine C7H9N 10964
100-74-3 N-ethylmorpholine C6H13NO 8769
100-80-1 m-methylstyrene C9H10 16513
100-84-5 3-methylanisole C8H10O 13930
101-14-4 bis(4-amino-3-chlorophenyl)methane C13H12Cl2N2 25748
101-54-2 p-aminodiphenylamine C12H12N2 23718
101-68-8 diphenylmethane-4,4'-diisocyanate C15H10N2O2 28029
101-77-9 4,4'-diaminodiphenylmethane C13H14N2 25950
101-81-5 diphenylmethane C13H12 25738
101-83-7 dicyclohexylamine C12H23N 24804
101-84-8 diphenyl ether C12H10O 23519
102-25-0 1,3,5-triethylbenzene C12H18 24366
102-36-3 3,4-dichlorophenyl isocyanate C7H3Cl2NO 9527
102-47-6 1,2-dichloro-4-(chloromethyl)benzene C7H5Cl3 9968
102-54-5 ferrocene C10H10Fe 19015
102-69-2 tripropylamine C9H21N 18393
102-71-6 triethanolamine C6H15NO3 9239
102-76-1 glyceryl triacetate C9H14O6 17513
102-81-8 2-dibutylaminoethanol C10H23NO 21409
102-82-9 tributylamine C12H27N 25346
102-86-3 trihexylamine C18H39N 31168
102-94-3 cis-cinnamic acid C9H8O2 16203
103-08-2 5-ethyl-2-nonanol C11H24O 23063
103-09-3 2-ethylhexyl acetate C10H20O2 21044
103-11-7 2-ethylhexyl acrylate C11H20O2 22701
103-29-7 1,2-diphenylethane C14H14 27053
103-30-0 trans-stilbene C14H12 26867
103-33-3 azobenzene C12H10N2 23491
103-50-4 dibenzyl ether C14H14O 27138
103-65-1 propylbenzene C9H12 17012
103-69-5 N-ethylaniline C8H11N 14095
103-70-8 formanilide C7H7NO 10611

103-71-9 phenyl isocyanate C7H5NO 10049
103-73-1 phenetole C8H10O 13912
103-79-7 1-phenyl-2-propanone C9H10O 16635
103-82-2 benzeneacetic acid C8H8O2 13415
103-83-3 N,N-dimethylbenzylamine C9H13N 17253
103-84-4 acetanilide C8H9NO 13671
104-13-2 4-butylaniline C10H15N 20199
104-40-5 4-nonylphenol C15H24O 28764
104-46-1 anethole C10H12O 19484
104-51-8 butylbenzene C10H14 19865
104-72-3 decylbenzene C16H26 29595
104-75-6 2-ethylhexylamine C8H19N 15632
104-76-7 2-ethyl-1-hexanol C8H18O 15454
104-83-6 1-chloro-4-(chloromethyl)benzene C7H6Cl2 10256
104-87-0 p-tolualdehyde C8H8O 13392
104-93-8 4-methylanisole C8H10O 13931
104-94-9 4-methoxyaniline C7H9NO 10981
105-05-5 p-diethylbenzene C10H14 19880
105-30-6 2-methyl-1-pentanol C6H14O 8973
105-34-0 methyl cyanoacetate C4H5NO2 2713
105-37-3 ethyl propanoate C5H10O2 5503
105-38-4 vinyl propionate C5H8O2 4885
105-41-9 4-methyl-2-hexanamine C7H17N 12320
105-42-0 4-methyl-2-hexanone C7H14O 11833
105-43-1 (±)-3-methylvaleric acid C6H12O2 8514
105-45-3 methyl acetoacetate C5H8O3 4929
105-46-4 sec-butyl acetate C6H12O2 8478
105-53-3 diethyl malonate C7H12O4 11530
105-54-4 ethyl butanoate C6H12O2 8482
105-56-6 ethyl cyanoacetate C5H7NO2 4677
105-57-7 acetal C6H14O2 9020
105-58-8 diethyl carbonate C5H10O3 5543
105-59-9 methyl diethanolamine C5H13NO2 5986
105-60-2 epsilon-caprolactam C6H11NO 8101
105-66-8 propyl butanoate C7H14O2 11905
105-67-9 2,4-xylenol C8H10O 13922
105-68-0 isopentyl propanoate C8H16O2 15170
105-76-0 dibutyl maleate C12H20O4 24594
105-79-3 isobutyl hexanoate C10H20O2 21073
106-20-7 bis(2-ethylhexyl)amine C16H35N 29805
106-21-8 3,7-dimethyl-1-octanol C10H22O 21256
106-27-4 3-methylbutyl butanoate C9H18O2 18065
106-31-0 butyric anhydride C8H14O3 14659
106-32-1 ethyl octanoate C10H20O2 21060
106-35-4 3-heptanone C7H14O 11830
106-36-5 propyl propanoate C6H12O2 8480
106-37-6 p-dibromobenzene C6H4Br2 6413
106-38-7 p-bromotoluene C7H7Br 10462
106-39-8 1-bromo-4-chlorobenzene C6H4BrCl 6385
106-40-1 p-bromoaniline C6H6BrN 6907
106-41-2 p-bromophenol C6H5BrO 6673
106-42-3 p-xylene C8H10 13792
106-43-4 p-chlorotoluene C7H7Cl 10494
106-44-5 p-cresol C7H8O 10854
106-45-6 p-methylthiophenol C7H8S 10920
106-46-7 p-dichlorobenzene C6H4Cl2 6471
106-47-8 p-chloroaniline C6H6ClN 6925
106-48-9 p-chlorophenol C6H5ClO 6722
106-49-0 p-toluidine C7H9N 10957
106-50-3 p-phenylenediamine C6H8N2 7320
106-51-4 p-benzoquinone C6H4O2 6616
106-63-8 isobutyl acrylate C7H12O2 11451
106-67-2 4-methyl-2-ethyl-1-pentanol C8H18O 15493
106-68-3 3-octanone C8H16O 15080
106-70-7 methyl hexanoate C7H14O2 11914
106-73-0 methyl heptanoate C8H16O2 15195
106-87-6 1,2-epoxy-4-(epoxyethyl)cyclohexane C8H12O2 14348
106-88-7 1,2-epoxybutane C4H8O 3474
106-89-8 alpha-epichlorohydrin C3H5ClO 1676
106-92-3 allyl glycidyl ether C6H10O2 7816
106-93-4 1,2-dibromoethane C2H4Br2 799
106-94-5 1-bromopropane C3H7Br 1978
106-95-6 3-bromo-1-propene C3H5Br 1645
106-97-8 butane C4H10 3783
106-98-9 1-butene C4H8 3320
106-99-0 butadiene (1,3 butadiene) C4H6 2757
107-00-6 ethylacetylene C4H6 2755
107-01-7 2-butene; (cis+trans) C4H8 3321
107-02-8 acrolein C3H4O 1621
107-03-9 propyl mercaptan C3H8S 2158
107-05-1 3-chloro-1-propene C3H5Cl 1671
107-06-2 1,2-dichloroethane C2H4Cl2 813
107-07-3 2-chloroethanol C2H5ClO 917
107-08-4 1-iodopropane C3H7I 2025
107-10-8 propylamine C3H9N 2200
107-11-9 allylamine C3H7N 2030
107-12-0 propionitrile C3H5N 1752
107-13-1 acrylonitrile C3H3N 1462
107-15-3 ethylenediamine C2H8N2 1159
107-18-6 allyl alcohol C3H6O 1920
107-19-7 propargyl alcohol C3H4O 1622
107-20-0 chloroacetaldehyde C2H3ClO 708
107-21-1 ethylene glycol C2H6O2 1072
107-25-5 methyl vinyl ether C3H6O 1921
107-31-3 methyl formate C2H4O2 890
107-39-1 2,4,4-trimethyl-1-pentene C8H16 14979
107-40-4 2,4,4-trimethyl-2-pentene C8H16 14986
107-41-5 hexylene glycol C6H14O2 9022
107-45-9 tert-octylamine C8H19N 15652
107-46-0 hexamethyldisiloxane C6H18OSi2 9362
107-47-1 di-tert-butyl sulfide C8H18S 15611
107-49-3 tetraethyl pyrophosphate C8H20O7P2 15727
107-66-4 dibutyl phosphate C8H19O4P 15676
107-80-2 1,3-dibromobutane C4H8Br2 3335
107-81-3 2-bromopentane C5H11Br 5604
107-82-4 1-bromo-3-methylbutane C5H11Br 5607
107-83-5 2-methylpentane C6H14 8880
107-84-6 1-chloro-3-methylbutane C5H11Cl 5624
107-85-7 3-methylbutylamine C5H13N 5950
107-87-9 methyl propyl ketone C5H10O 5450
107-88-0 1,3-butanediol C4H10O2 3885
107-92-6 butyric acid C4H8O2 3504
107-93-7 trans-2-butenoic acid C4H6O2 2967
107-96-0 3-mercaptopropionic acid C3H6O2S 1942
107-98-2 propylene glycol monomethyl ether C4H10O2 3895
108-01-0 dimethylethanolamine C4H11NO 3988
108-03-2 1-nitropropane C3H7NO2 2045
108-05-4 vinyl acetate C4H6O2 2971
108-08-7 2,4-dimethylpentane C7H16 12118
108-09-8 2-amino-4-methylpentane C6H15N 9181
108-10-1 4-methyl-2-pentanone C6H12O 8412
108-11-2 4-methyl-2-pentanol C6H14O 8976
108-18-9 diisopropylamine C6H15N 9203
108-20-3 diisopropyl ether C6H14O 8995
108-21-4 isopropyl acetate C5H10O2 5502
108-22-5 isopropenyl acetate C5H8O2 4884
108-24-7 acetic anhydride C4H6O3 2990
108-30-5 succinic anhydride C4H4O3 2603
108-31-6 maleic anhydride C4H2O3 2430
108-36-1 m-dibromobenzene C6H4Br2 6411
108-37-2 1-bromo-3-chlorobenzene C6H4BrCl 6384
108-38-3 m-xylene C8H10 13791
108-39-4 m-cresol C7H8O 10852
108-40-7 m-methylthiophenol C7H8S 10919
108-41-8 m-chlorotoluene C7H7Cl 10495
108-42-9 m-chloroaniline C6H6ClN 6923
108-43-0 m-chlorophenol C6H5ClO 6720
108-44-1 m-toluidine C7H9N 10955
108-45-2 m-phenylenediamine C6H8N2 7318
108-46-3 resorcinol C6H6O2 7071
108-47-4 2,4-dimethylpyridine C7H9N 10959
108-48-5 2,6-dimethylpyridine C7H9N 10961
108-55-4 glutaric anhydride C5H6O3 4567
108-57-6 m-divinylbenzene C10H10 18958
108-64-5 ethyl isovalerate C7H14O2 11909
108-67-8 mesitylene C9H12 17011
108-68-9 3,5-xylenol C8H10O 13926
108-69-0 3,5-dimethylaniline C8H11N 14092
108-70-3 1,3,5-trichlorobenzene C6H3Cl3 6307
108-75-8 2,4,6-trimethylpyridine C8H11N 14096
108-82-7 2,6-dimethyl-4-heptanol C9H20O 18288
108-83-8 diisobutyl ketone C9H18O 17998
108-84-9 sec-hexyl acetate C8H16O2 15163
108-85-0 bromocyclohexane C6H11Br 7981
108-86-1 bromobenzene C6H5Br 6656
108-87-2 methylcyclohexane C7H14 11704
108-88-3 toluene C7H8 10725
108-89-4 4-methylpyridine C6H7N 7175
108-90-7 chlorobenzene C6H5Cl 6692
108-91-8 cyclohexylamine C6H13N 8742
108-93-0 cyclohexanol C6H12O 8400
108-94-1 cyclohexanone C6H10O 7760
108-95-2 phenol C6H6O 7057
108-98-5 phenyl mercaptan C6H6S 7121
108-99-6 3-methylpyridine C6H7N 7174
109-06-8 2-methylpyridine C6H7N 7173
109-19-3 butyl 3-methylbutanoate C9H18O2 18073
109-21-7 butyl butanoate C8H16O2 15175
109-43-3 dibutyl sebacate C18H34O4 31041
109-49-9 5-hexen-2-one C6H10O 7776
109-52-4 pentanoic acid C5H10O2 5492
109-53-5 isobutyl vinyl ether C6H12O 8397
109-59-1 2-isopropoxyethanol C5H12O2 5883
109-60-4 propyl acetate C5H10O2 5501
109-64-8 1,3-dibromopropane C3H6Br2 1826
109-65-9 1-bromobutane C4H9Br 3598
109-66-0 pentane C5H12 5772
109-67-1 1-pentene C5H10 5286
109-68-2 2-pentene; (cis+trans) C5H10 5287
109-69-3 1-chlorobutane C4H9Cl 3610
109-73-9 butylamine C4H11N 3977
109-74-0 butyronitrile C4H7N 3238
109-75-1 vinylacetonitrile C4H5N 2698
109-76-2 1,3-propanediamine C3H10N2 2258
109-77-3 malononitrile C3H2N2 1377
109-78-4 hydracrylonitrile C3H5NO 1756
109-79-5 butyl mercaptan C4H10S 3942
109-83-1 methylethanolamine C3H9NO 2206
109-86-4 2-methoxyethanol C3H8O2 2138
109-87-5 methylal C3H8O2 2139
109-89-7 diethylamine C4H11N 3984
109-92-2 ethyl vinyl ether C4H8O 3476
109-93-3 divinyl ether C4H6O 2944
109-94-4 ethyl formate C3H6O2 1931
109-97-7 pyrrole C4H5N 2701
109-99-9 tetrahydrofuran C4H8O 3481
110-00-9 furan C4H4O 2592
110-01-0 tetrahydrothiophene C4H8S 3581
110-02-1 thiophene C4H4S 2613
110-05-4 di-t-butyl peroxide C8H18O2 15556
110-06-5 di-tert-butyl disulfide C8H18S2 15615
110-12-3 5-methyl-2-hexanone C7H14O 11834
110-15-6 succinic acid C4H6O4 3006
110-16-7 maleic acid C4H4O4 2606
110-17-8 fumaric acid C4H4O4 2605
110-18-9 N,N,N',N'-tetramethyl-1,2-ethanediamine C6H16N2 9295
110-19-0 isobutyl acetate C6H12O2 8477
110-33-8 dihexyl adipate C18H34O4 31042
110-39-4 octyl butanoate C12H24O2 24871
110-41-8 2-methylundecanal C12H24O 24853
110-42-9 methyl decanoate C11H22O2 22848
110-43-0 2-heptanone C7H14O 11829
110-45-2 isopentyl formate C6H12O2 8471
110-49-6 ethylene glycol monomethyl ether acetate C5H10O3 5545
110-52-1 1,4-dibromobutane C4H8Br2 3336
110-53-2 1-bromopentane C5H11Br 5603
110-54-3 hexane C6H14 8877
110-56-5 1,4-dichlorobutane C4H8Cl2 3363
110-57-6 1,4-dichloro-trans-2-butene C4H6Cl2 2808
110-58-7 pentylamine C5H13N 5947
110-59-8 valeronitrile C5H9N 5183
110-61-2 succinonitrile C4H4N2 2562
110-62-3 pentanal C5H10O 5445
110-63-4 1,4-butanediol C4H10O2 3886
110-64-5 2-butene-1,4-diol C4H8O2 3526
110-65-6 2-butyne-1,4-diol C4H6O2 2963
110-66-7 pentyl mercaptan C5H12S 5921
110-67-8 3-methoxypropionitrile C4H7NO 3248
110-68-9 methyl-butylamine C5H13N 5955
110-71-4 1,2-dimethoxyethane C4H10O2 3882
110-74-7 propyl formate C4H8O2 3506
110-80-5 2-ethoxyethanol C4H10O2 3883
110-81-6 diethyl disulfide C4H10S2 3950
110-82-7 cyclohexane C6H12 8215
110-83-8 cyclohexene C6H10 7622
110-85-0 piperazine C4H10N2 3830
110-86-1 pyridine C5H5N 4391
110-88-3 trioxane C3H6O3 1950
110-89-4 piperidine C5H11N 5670
110-91-8 morpholine C4H9NO 3682
110-94-1 glutaric acid C5H8O4 4945
110-96-3 diisobutylamine C8H19N 15638
110-97-4 diisopropanolamine C6H15NO2 9230
111-13-7 2-octanone C8H16O 15066
111-14-8 heptanoic acid C7H14O2 11884
111-15-9 2-ethoxyethyl acetate C6H12O3 8531
111-20-6 sebacic acid C10H18O4 20820
111-24-0 1,5-dibromopentane C5H10Br2 5301
111-25-1 1-bromohexane C6H13Br 8647
111-26-2 hexylamine C6H15N 9173
111-27-3 1-hexanol C6H14O 8970
111-29-5 1,5-pentanediol C5H12O2 5863
111-31-9 hexyl mercaptan C6H14S 9107
111-34-2 butyl vinyl ether C6H12O 8394

111-35-3 3-ethoxy-1-propanol C5H12O2 5886
111-36-4 butyl isocyanate C5H9NO 5197
111-40-0 diethylene triamine C4H13N3 4118
111-41-1 N-aminoethyl ethanolamine C4H12N2O 4080
111-42-2 diethanolamine C4H11NO2 4006
111-43-3 dipropyl ether C6H14O 8993
111-44-4 bis(2-chloroethyl) ether C4H8Cl2O 3375
111-46-6 diethylene glycol C4H10O3 3918
111-47-7 dipropyl sulfide C6H14S 9112
111-49-9 hexamethyleneimine C6H13N 8745
111-55-7 ethylene glycol diacetate C6H10O4 7919
111-65-9 octane C8H18 15354
111-66-0 1-octene C8H16 14890
111-67-1 oct-2-ene C8H16 14891
111-68-2 heptylamine C7H17N 12315
111-69-3 adiponitrile C6H8N2 7316
111-70-6 1-heptanol C7H16O 12164
111-71-7 1-heptanal C7H14O 11827
111-74-0 N,N'-diethyl-1,2-ethanediamine C6H16N2 9294
111-76-2 2-butoxyethanol C6H14O2 9021
111-77-3 2-(2-methoxyethoxy)ethanol C5H12O3 5899
111-78-4 1,5-cyclooctadiene C8H12 14228
111-82-0 methyl dodecanoate C13H26O2 26505
111-83-1 1-bromooctane C8H17Br 15248
111-84-2 nonane C9H20 18176
111-85-3 1-chlorooctane C8H17Cl 15252
111-86-4 octylamine C8H19N 15631
111-87-5 1-octanol C8H18O 15430
111-88-6 octyl mercaptan C8H18S 15602
111-90-0 2-(2-ethoxyethoxy)ethanol C6H14O3 9077
111-92-2 dibutylamine C8H19N 15637
111-96-6 diethylene glycol dimethyl ether C6H14O3 9075
112-05-0 nonanoic acid C9H18O2 18052
112-06-1 heptyl acetate C9H18O2 18059
112-12-9 2-undecanone C11H22O 22823
112-14-1 octyl acetate C10H20O2 21045
112-15-2 diethylene glycol ethyl ether acetate C8H16O4 15232
112-17-4 decyl acetate C12H24O2 24869
112-18-5 dimethyldodecylamine C14H31N 27960
112-20-9 nonylamine C9H21N 18388
112-23-2 heptyl formate C8H16O2 15159
112-24-3 triethylene tetramine C6H18N4 9359
112-27-6 triethylene glycol C6H14O4 9092
112-29-8 1-bromodecane C10H21Br 21090
112-30-1 1-decanol C10H22O 21224
112-31-2 decanal C10H20O 20992
112-32-3 octyl formate C9H18O2 18058
112-34-5 diethylene glycol monobutyl ether C8H18O3 15579
112-36-7 diethylene glycol diethyl ether C8H18O3 15578
112-37-8 undecanoic acid C11H22O2 22835
112-40-3 dodecane C12H26 24914
112-41-4 1-dodecene C12H24 24823
112-42-5 1-undecanol C11H24O 23055
112-44-7 undecanal C11H22O 22822
112-49-2 triethylene glycol dimethyl ether C8H18O4 15591
112-51-6 dipentyl disulfide C10H22S2 21394
112-52-7 1-chlorododecane C12H25Cl 24897
112-53-8 1-dodecanol C12H26O 25277
112-54-9 dodecanal C12H24O 24852
112-55-0 1-dodecanethiol C12H26S 25310
112-57-2 tetraethylenepentamine C8H23N5 15747
112-58-3 dihexyl ether C12H26O 25285
112-60-7 tetraethylene glycol C8H18O5 15598
112-62-9 methyl oleate C19H36O2 31675
112-66-3 dodecyl acetate C14H28O2 27875
112-69-6 dimethylhexadecylamine C18H39N 31166
112-70-9 1-tridecanol C13H28O 26535
112-71-0 1-bromotetradecane C14H29Br 27895
112-72-1 1-tetradecanol C14H30O 27922
112-73-2 diethylene glycol dibutyl ether C12H26O3 25294
112-75-4 dimethyltetradecylamine C16H35N 29803
112-79-8 elaidic acid C18H34O2 31031
112-80-1 oleic acid C18H34O2 31030
112-82-3 1-bromohexadecane C16H33Br 29739
112-88-9 1-octadecene C18H36 31071
112-92-5 1-octadecanol C18H38O 31140
112-95-8 eicosane C20H42 32515
114-26-1 propoxur C11H15NO3 22282
114-38-5 2-chlorophenyl urea C7H7ClN2O 10502
115-07-1 propylene C3H6 1810
115-10-6 dimethyl ether C2H6O 1066
115-11-7 isobutene C4H8 3324
115-20-8 2,2,2-trichloroethanol C2H3Cl3O 723
115-25-3 octafluorocyclobutane C4F8 2336
115-29-7 endosulfan C9H6Cl6O3S 15877
115-69-5 2-amino-2-methyl-1,3-propanediol C4H11NO2 4008
115-70-8 2-amino-2-ethyl-1,3-propanediol C5H13NO2 5987
115-76-4 2,2-diethyl-1,3-propanediol C7H16O2 12265
115-77-5 pentaerythritol C5H12O4 5912
115-86-6 triphenyl phosphate C18H15O4P 30566
115-90-2 fensulfothion C11H17O4PS2 22552
116-06-3 aldicarb C7H14N2O2S 11803
116-14-3 tetrafluoroethylene C2F4 488
116-15-4 hexafluoropropylene C3F6 1284
116-53-0 2-methylbutyric acid C5H10O2 5493
117-81-7 bis(2-ethylhexyl) phthalate C24H38O4 33926
117-84-0 dioctyl phthalate C24H38O4 33925
118-52-5 1,3-dichloro-5,5-dimethyl hydantoin C5H6Cl2N2O2 4476
118-69-4 2,6-dichlorotoluene C7H6Cl2 10259
118-72-9 2,6-dimethylthiophenol C8H10S 14045
118-74-1 hexachlorobenzene C6Cl6 6067
118-90-1 o-toluic acid C8H8O2 13413
118-91-2 o-chlorobenzoic acid C7H5ClO2 9932
118-96-7 2,4,6-trinitrotoluene C7H5N3O6 10118
119-36-8 methyl salicylate C8H8O3 13445
119-61-9 benzophenone C13H10O 25623
119-64-2 1,2,3,4-tetrahydronaphthalene C10H12 19359
119-65-3 isoquinoline C9H7N 16010
119-91-5 2,2'-biquinoline C18H12N2 30424
120-12-7 anthracene C14H10 26728
120-51-4 benzyl benzoate C14H12O2 26942
120-61-6 dimethyl terephthalate C10H10O4 19196
120-72-9 indole C8H7N 13107
120-80-9 pyrocatechol C6H6O2 7070
120-82-1 1,2,4-trichlorobenzene C6H3Cl3 6305
120-83-2 2,4-dichlorophenol C6H4Cl2O 6480
120-92-3 cyclopentanone C5H8O 4848
120-94-5 N-methylpyrrolidine C5H11N 5669
121-14-2 2,4-dinitrotoluene C7H6N2O4 10363
121-17-5 4-chloro-3-nitrobenzotrifluoride C7H3ClF3NO2 9508
121-32-4 ethyl vanillin C9H10O3 16711
121-33-5 vanillin C8H8O3 13446
121-44-8 triethylamine C6H15N 9212
121-45-9 trimethyl phosphite C3H9O3P 2236
121-69-7 N,N-dimethylaniline C8H11N 14082
121-73-3 m-chloronitrobenzene C6H4ClNO2 6443
121-75-5 malathion C10H19O6PS2 20919
121-82-4 hexahydro-1,3,5-trinitro-1,3,5-triazine C3H6N6O6 1919
121-91-5 isophthalic acid C8H6O4 12934
122-00-9 4-methylacetophenone C9H10O 16631
122-14-5 fenitrothion C9H12NO5PS 17035
122-34-9 simazine C7H12ClN5 11376
122-39-4 diphenylamine C12H11N 23609
122-42-9 isopropyl phenylcarbamate C10H13NO2 19799
122-46-3 3-methylphenyl acetate C9H10O2 16682
122-57-6 benzalacetone C10H10O 19106
122-60-1 phenyl glycidyl ether C9H10O2 16686
122-66-7 hydrazobenzene C12H12N2 23719
122-78-1 benzeneacetaldehyde C8H8O 13393
122-79-2 phenyl acetate C8H8O2 13417
122-97-4 benzenepropanol C9H12O 17091
123-01-3 dodecylbenzene C18H30 30947
123-02-4 tridecylbenzene C19H32 31646
123-04-6 3-(chloromethyl)heptane C8H17Cl 15259
123-05-7 2-ethylhexanal C8H16O 15062
123-07-9 p-ethylphenol C8H10O 13920
123-08-0 p-hydroxybenzaldehyde C7H6O2 10409
123-15-9 2-methylpentanal C6H12O 8402
123-17-1 2,6,8-trimethylnonanol-4 C12H26O 25286
123-19-3 4-heptanone C7H14O 11831
123-20-6 vinyl butanoate C6H10O2 7833
123-25-1 diethyl succinate C8H14O4 14688
123-31-9 p-hydroquinone C6H6O2 7072
123-38-6 propionaldehyde C3H6O 1922
123-39-7 N-methylformamide C2H5NO 957
123-42-2 diacetone alcohol C6H12O2 8485
123-44-4 2,2,4-trimethyl-1-pentanol C8H18O 15498
123-48-8 2,2,4,6,6-pentamethyl-3-heptene C12H24 24826
123-51-3 3-methyl-1-butanol C5H12O 5837
123-54-6 acetylacetone C5H8O2 4882
123-62-6 propionic anhydride C6H10O3 7882
123-63-7 paraldehyde C6H12O3 8533
123-66-0 ethyl hexanoate C8H16O2 15185
123-72-8 butyraldehyde C4H8O 3478
123-73-9 trans-crotonaldehyde C4H6O 2940
123-75-1 pyrrolidine C4H9N 3675
123-76-2 levulinic acid C5H8O3 4928
123-82-0 2-heptanamine C7H17N 12317
123-86-4 butyl acetate C6H12O2 8476
123-91-1 1,4-dioxane C4H8O2 3510
123-92-2 isopentyl acetate C7H14O2 11895
123-95-5 butyl stearate C22H44O2 33466
123-96-6 2-octanol C8H18O 15431
123-99-9 azelaic acid C9H16O4 17755
124-04-9 adipic acid C6H10O4 7917
124-07-2 octanoic acid C8H16O2 15156
124-09-4 hexamethylenediamine C6H16N2 9292
124-11-8 1-nonene C9H18 17942
124-12-9 octanenitrile C8H15N 14775
124-13-0 1-octanal C8H16O 15063
124-18-5 decane C10H22 21133
124-19-6 1-nonanal C9H18O 17997
124-22-1 dodecylamine C12H27N 25340
124-25-4 tetradecanal C14H28O 27862
124-38-9 carbon dioxide CO2 375
124-40-3 dimethylamine C2H7N 1116
124-68-5 2-amino-2-methyl-1-propanol C4H11NO 3993
124-73-2 1,2-dibromotetrafluoroethane C2Br2F4 413
126-30-7 neopentyl glycol C5H12O2 5858
126-33-0 sulfolane C4H8O2S 3536
126-72-7 2,3-dibromo-1-propanol, phosphate (3:1) C9H15Br6O4P 17533
126-73-8 tributyl phosphate C12H27O4P 25355
126-98-7 methacrylonitrile C4H5N 2697
126-99-8 chloroprene C4H5Cl 2634
127-17-3 pyruvic acid C3H4O3 1634
127-18-4 tetrachloroethylene C2Cl4 455
127-19-5 N,N-dimethylacetamide C4H9NO 3681
127-91-3 beta-pinene C10H16 20323
128-37-0 2,6-di-tert-butyl-p-cresol C15H24O 28756
128-39-2 2,6-di-tert-butylphenol C14H22O 27682
129-00-0 pyrene C16H10 28985
131-11-3 dimethyl phthalate C10H10O4 19195
132-64-9 dibenzofuran C12H8O 23335
133-06-2 captan C9H8Cl3NO2S 16128
133-07-3 folpet C9H4Cl3NO2S 15788
133-90-4 chloramben C7H5Cl2NO2 9957
134-32-7 1-naphthylamine C10H9N 18861
135-01-3 o-diethylbenzene C10H14 19878
135-19-3 2-naphthol C10H8O 18741
135-88-6 N-phenyl-2-naphthalenamine C16H13N 29088
135-98-8 sec-butylbenzene C10H14 19867
136-35-6 1,3-diphenyltriazene C12H11N3 23657
136-60-7 butyl benzoate C11H14O2 22113
136-78-7 sesone C8H7Cl2NaO5S 13047
136-81-2 o-amylphenol C11H16O 22440
137-05-3 methyl 2-cyanoacrylate C5H5NO2 4408
137-06-4 o-methylthiophenol C7H8S 10918
137-26-8 thiram C6H12N2S4 8380
137-32-6 2-methyl-1-butanol C5H12O 5834
138-22-7 butyl lactate C7H14O3 11937
138-86-3 limonene C10H16 20317
139-40-2 propazine C9H16ClN5 17637
139-66-2 diphenyl sulfide C12H10S 23576
140-10-3 trans-cinnamic acid C9H8O2 16204
140-11-4 benzyl acetate C9H10O2 16680
140-29-4 benzeneacetonitrile C8H7N 13108
140-31-8 N-aminoethyl piperazine C6H15N3 9247
140-39-6 4-methylphenyl acetate C9H10O2 16683
140-66-9 p-tert-octylphenol C14H22O 27679
140-88-5 ethyl acrylate C5H8O2 4888
141-05-9 diethyl maleate C8H12O4 14380
141-06-0 propyl pentanoate C8H16O2 15194
141-32-2 butyl acrylate C7H12O2 11450
141-43-5 monoethanolamine C2H7NO 1117
141-59-3 tert-octyl mercaptan C8H18S 15604
141-66-2 dicrotophos C8H16NO5P 15023
141-70-8 4,4-dimethyl-2-neopentyl-1-pentene C12H24 24825
141-78-6 ethyl acetate C4H8O2 3508
141-79-7 mesityl oxide C6H10O 7761
141-93-5 m-diethylbenzene C10H14 19879
141-97-9 ethylacetoacetate C6H10O3 7881
142-28-9 1,3-dichloropropane C3H6Cl2 1846
142-29-0 cyclopentene C5H8 4737
142-60-9 octyl propanoate C11H22O2 22838
142-62-1 hexanoic acid C6H12O2 8460
142-64-3 piperazine dihydrochloride C4H12Cl2N2 4049
142-68-7 tetrahydropyran C5H10O 5471
142-82-5 heptane C7H16 12112
142-84-7 di-propylamine C6H15N 9202
142-92-7 hexyl acetate C8H16O2 15160
142-96-1 dibutyl ether C8H18O 15527
143-07-7 dodecanoic acid C12H24O2 24867
143-08-8 1-nonanol C9H20O 18236
143-10-2 decyl mercaptan C10H22S 21385

143-13-5 nonyl acetate C11H22O2 22837
143-15-7 1-bromododecane C12H25Br 24894
143-16-8 dihexylamine C12H27N 25343
143-24-8 tetraethylene glycol dimethyl ether C10H22O5 21380
143-33-9 sodium cyanide CNNa 351
143-50-0 kepone C10Cl10O 18458
144-62-7 oxalic acid C2H2O4 684
145-73-3 endothall disodium C8H10Na2O5 13911
145-73-3 endothall C8H10O5 14033
148-01-6 2-methyl-3,5-dinitrobenzamide C8H7N3O5 13226
148-24-3 8-hydroxyquinoline C9H7NO 16018
149-31-5 2-methyl-1,3-pentanediol C6H14O2 9033
149-57-5 2-ethylhexanoic acid C8H16O2 15157
150-19-6 3-methoxyphenol C7H8O2 10881
150-68-5 N'-(4-chlorophenyl)-N,N-dimethylurea C9H11ClN2O 16854
150-76-5 p-methoxyphenol C7H8O2 10867
150-78-7 1,4-dimethoxybenzene C8H10O2 13953
151-10-0 1,3-dimethoxybenzene C8H10O2 13952
151-19-9 3,6-dimethyl-3-octanol C10H22O 21271
151-50-8 potassium cyanide CKN 338
151-56-4 ethyleneimine C2H5N 954
151-67-7 halothane C2HBrClF3 511
156-43-4 p-phenetidine C8H11NO 14124
156-59-2 cis-1,2-dichloroethylene C2H2Cl2 616
156-60-5 trans-1,2-dichloroethylene C2H2Cl2 617
156-87-6 3-amino-1-propanol C3H9NO 2205
157-40-4 spiropentane C5H8 4736
191-07-1 coronene C24H12 33717
191-24-2 benzo[ghi]perylene C22H12 33080
192-97-2 benzo[e]pyrene C20H12 31805
193-39-5 indeno[1,2,3-cd]pyrene C22H12 33081
198-55-0 perylene C20H12 31800
205-99-2 benzo[b]fluoranthene C20H12 31803
206-44-0 fluoranthene C16H10 28984
207-08-9 benzo[k]fluoranthene C20H12 31802
217-59-4 triphenylene C18H12 30410
218-01-9 chrysene C18H12 30406
230-07-9 4,7-phenanthroline C12H8N2 23307
230-46-6 1,7-phenanthroline C12H8N2 23306
238-84-6 11H-benzo[a]fluorene C17H12 29862
239-64-5 13H-dibenzo[a,i]carbazole C20H13N 31847
239-64-5 1,2,7,8-dibenzocarbazole C20H13N 31852
243-17-4 11H-benzo[b]fluorene C17H12 29863
260-94-6 acridine C13H9N 25537
271-89-6 benzofuran C8H6O 12914
274-09-9 1,3-benzodioxole C7H6O2 10411
275-51-4 azulene C10H8 18670
280-57-9 triethylenediamine C6H12N2 8315
287-23-0 cyclobutane C4H8 3319
287-27-4 thiacyclobutane C3H6S 1967
287-92-3 cyclopentane C5H10 5285
288-13-1 1H-pyrazole C3H4N2 1594
288-32-4 imidazole C3H4N2 1592
288-42-6 oxazole C3H3NO 1463
288-47-1 thiazole C3H3NS 1473
289-95-2 pyrimidine C4H4N2 2565
291-64-5 cycloheptane C7H14 11705
292-64-8 cyclooctane C8H16 14888
293-55-0 cyclononane C9H18 17941
293-96-9 cyclodecane C10H20 20942
294-41-7 cycloundecane C11H22 22794
294-65-5 thiacyclododecane C11H22S 22866
295-02-3 cyclotridecane C13H26 26468
295-05-6 thiacyclotridecane C12H24S 24892
298-00-0 methyl parathion C8H10NO5PS 13854
298-02-2 phorate C7H17O2PS3 12364
298-04-4 disulfoton C8H19O2PS3 15667
299-84-3 ronnel C8H8Cl3O3PS 13309
299-86-5 crufomate C12H19ClNO3P 24503
300-57-2 2-propenylbenzene C9H10 16511
300-62-9 amphetamine C9H13N 17250
300-76-5 naled C4H7Br2Cl2O4P 3082
309-00-2 aldrin C12H8Cl6 23291
314-40-9 bromacil C9H13BrN2O2 17234
319-84-6 a-hexachlorocyclohexane C6H6Cl6 6953
319-85-7 C6H6Cl6 1,2,3,4,5,6-hexachlorocyclohexane 6949
319-86-8 C6H6Cl6 1,2,3,4,5,6-hexachlorocyclohexane 6950
320-60-5 2,4-dichlorobenzotrifluoride C7H3Cl2F3 9519
330-39-2 1,1-dimethyl-3-(3-fluorophenyl) urea C9H11FN2O 16877
330-54-1 diuron C9H10Cl2N2O 16540
330-55-2 linuron C9H10Cl2N2O2 16541
332-33-2 1,1-dimethyl-3-(4-fluorophenyl) urea C9H11FN2O 16877
333-41-5 diazinon C12H21N2O3PS 24649
334-48-5 decanoic acid C10H20O2 21040
334-56-5 1-fluorodecane C10H21F 21097
334-68-9 1-fluorododecane C12H25F 24899
334-88-3 diazomethane CH2N2 166
350-50-5 (fluoromethyl)benzene C7H7F 10556
352-32-9 p-fluorotoluene C7H7F 10555
352-70-5 m-fluorotoluene C7H7F 10558
352-93-2 diethyl sulfide C4H10S 3947
353-36-6 ethyl fluoride C2H5F 939
353-50-4 carbonyl fluoride CF2O 68
353-59-3 bromochlorodifluoromethane CBrClF2 13
353-61-7 2-fluoro-2-methylpropane C4H9F 3660
353-81-1 2,2-difluorobutane C4H8F2 3393
354-09-6 2-fluoro-2,3-dimethylbutane C6H13F 8719
354-11-0 1,1,1,2-tetrachloro-2-fluoroethane C2HCl4F 549
354-12-1 1,1,1-trichloro-2,2-difluoroethane C2HCl3F2 540
354-14-3 1,1,2,2-tetrachloro-1-fluoroethane C2HCl4F 550
354-15-4 1,1,2-trichloro-1,2-difluoroethane C2HCl3F2 541
354-21-2 1,2,2-trichloro-1,1-difluoroethane C2HCl3F2 542
354-33-6 pentafluoroethane C2HF5 558
354-38-1 trifluoroacetamide C2H2F3NO 648
354-58-5 1,1,1-trichloro-2,2,2-trifluoroethane C2Cl3F3 450
354-92-7 perfluoroisobutane C4F10 2345
355-25-9 decafluorobutane C4F10 2344
357-26-6 1,1,2,3,3,4,4,4-octafluoro-1-butene C4F8 2337
358-03-2 3,3-difluoropentane C5H10F2 5366
359-00-2 1-fluoro-2-methylpropane C4H9F 3659
359-01-3 2-fluorobutane C4H9F 3658
359-10-4 2-chloro-1,1-difluoroethylene C2HClF2 525
359-28-4 1,1,2-trichloro-2-fluoroethane C2H2Cl3F 632
359-35-3 1,1,2,2-tetrafluoroethane C2H2F4 651
360-89-4 octafluoro-2-butene C4F8 2334
367-11-3 o-difluorobenzene C6H4F2 6525
369-68-6 3-(trifluoromethylthio)aniline C7H6F3NS 10326
371-64-2 1-fluoro-3,3-dimethylbutane C6H13F 8718
371-65-3 2,2-difluoropentane C5H10F2 5363
372-16-7 4-(trifluoromethylthio)aniline C7H6F3NS 10327
372-18-9 m-difluorobenzene C6H4F2 6524
372-54-3 2-fluorohexane C6H13F 8706
372-90-7 1,4-difluorobutane C4H8F2 3392
373-14-8 1-fluorohexane C6H13F 8705
373-17-1 1,5-difluoropentane C5H10F2 5362
374-07-2 1,1-dichloro-1,2,2,2-tetrafluoroethane C2Cl2F4 442
382-21-8 perfluoroisobutene C4F8 2335
392-56-3 hexafluorobenzene C6F6 6080
406-82-6 1,1,1-trifluoropentane C5H9F3 5144
407-06-7 1-fluoro-3-methylbutane C5H11F 5653
407-79-4 5-fluoro-1-pentene C5H9F 5114
407-96-5 1,1-difluoroheptane C7H14F2 11772
408-32-2 thiacycloundecane C10H20S 21088
408-38-8 1-fluorohexadecane C16H33F 29744
420-04-2 cyanamide CH2N2 165
420-12-2 ethylene sulfide C2H4S 901
420-26-8 2-fluoropropane C3H7F 2021
420-45-1 2,2-difluoropropane C3H6F2 1868
420-46-2 1,1,1-trifluoroethane C2H3F3 735
421-04-5 1-chloro-1,1,2-trifluoroethane C2H2ClF3 604
421-48-7 1,1,1,2-tetrafluoropropane C3H4F4 1578
430-53-5 1,1-dichloro-2-fluoroethane C2H3Cl2F 715
430-57-9 1,2-dichloro-1-fluoroethane C2H3Cl2F 716
430-61-5 1,1-difluoropropane C3H6F2 1865
430-66-0 1,1,2-trifluoroethane C2H3F3 736
431-06-1 1,2-dichloro-1,2-difluoroethane C2H2Cl2F2 623
431-07-2 1-chloro-1,2,2-trifluoroethane C2H2ClF3 605
451-40-1 2-phenylacetophenone C14H12O 26933
455-14-1 4-(trifluoromethyl)aniline C7H6F3N 10322
460-12-8 biacetylene C4H2 2374
460-13-9 1-fluoropropane C3H7F 2020
460-19-5 cyanogen C2N2 1210
460-36-6 1,1,1,3-tetrafluoropropane C3H4F4 1579
461-82-5 4-(trifluoromethoxy)aniline C7H6F3NO 10325
462-06-6 fluorobenzene C6H5F 6769
462-18-0 7-tridecanone C13H26O 26489
462-39-5 1,3-difluoropropane C3H6F2 1867
463-11-6 1-fluorooctane C8H17F 15273
463-18-3 1-fluorononane C9H19F 18127
463-40-1 linolenic acid C18H30O2 30964
463-49-0 allene C3H4 1493
463-51-4 ketene C2H2O 681
463-58-1 carbonyl sulfide COS 373
463-82-1 neopentane C5H12 5774
464-06-2 2,2,3-trimethylbutane C7H16 12120
464-07-3 3,3-dimethyl-2-butanol C6H14O 8987
470-90-6 chlorfenvinphos C12H14Cl3O4P 23911
471-43-2 1,1-dichloro-2,2-difluoroethane C2H2Cl2F2 621
479-45-8 N-methyl-N,2,4,6-tetranitroaniline C7H5N5O8 10127
481-39-0 5-hydroxy-1,4-naphthalenedione C10H6O3 18587
481-42-5 5-hydroxy-2-methyl-1,4-naphthalenedione C11H8O3 21536
483-55-6 2-hydroxy-3-methyl-1,4-naphthalenedione C11H8O3 21535
486-25-9 9H-fluoren-9-one C13H8O 25496
488-23-3 1,2,3,4-tetramethylbenzene C10H14 19887
488-71-1 2,3,4,6-tetramethylaniline C10H15N 20211
491-31-6 1H-2-benzopyran-1-one C9H6O2 15925
491-35-0 4-methylquinoline C10H9N 18851
491-38-3 4H-1-benzopyran-4-one C9H6O2 15927
492-37-5 (±)-2-phenylpropionic acid C9H10O2 16701
492-62-6 a-D(+)-glucose, anhydrous C6H12O6 8611
493-01-6 cis-decahydronaphthalene C10H18 20601
493-02-7 trans-decahydronaphthalene C10H18 20602
496-11-7 indane C9H10 16508
496-16-2 2,3-dihydrobenzofuran C8H8O 13396
496-78-6 2,4,5-trimethylphenol C9H12O 17119
497-06-3 3-butene-1,2-diol C4H8O2 3513
497-39-2 2,4-di-tert-butyl-5-methylphenol C15H24O 28758
498-23-7 citraconic acid C5H6O4 4576
498-66-8 2-norbornene C7H10 11047
499-75-2 2-methyl-5-isopropylphenol C10H14O 20005
501-65-5 diphenylacetylene C14H10 26729
502-44-3 epsilon-caprolactone C6H10O2 7813
502-56-7 5-nonanone C9H18O 18010
502-59-0 6-methyl-N-isopentyl-2-heptanamine C13H29N 26560
503-17-3 dimethylacetylene C4H6 2754
503-30-0 1,3-propylene oxide C3H6O 1925
503-60-6 1-chloro-3-methyl-2-butene C5H9Cl 5050
503-64-0 cis-2-butenoic acid C4H6O2 2966
503-74-2 3-methylbutanoic acid C5H10O2 5494
504-29-0 2-pyridinamine C5H6N2 4497
504-60-9 1,3-pentadiene; (cis+trans) C5H8 4751
504-63-2 1,3-propanediol C3H8O2 2141
505-22-6 1,3-dioxane C4H8O2 3514
506-05-8 1-fluoroundecane C11H23F 22875
506-12-7 heptadecanoic acid C17H34O2 30325
506-30-9 eicosanoic acid C20H40O2 32491
506-32-1 5,8,11,14-eicosatetraenoic acid, (all-trans) C20H32O2 32388
506-77-4 cyanogen chloride CClN 39
507-09-5 thioacetic-acid C2H4OS 888
507-19-7 2-bromo-2-methylpropane C4H9Br 3601
507-20-0 2-chloro-2-methylpropane C4H9Cl 3613
507-36-8 2-bromo-2-methylbutane C5H11Br 5608
507-45-9 2,3-dichloro-2-methylbutane C5H10Cl2 5347
509-14-8 tetranitromethane CN4O8 363
512-56-1 trimethyl phosphate C3H9O4P 2241
513-35-9 2-methyl-2-butene C5H10 5292
513-36-0 1-chloro-2-methylpropane C4H9Cl 3611
513-37-1 1-chloro-2-methyl-1-propene C4H7Cl 3107
513-38-2 1-iodo-2-methylpropane C4H9I 3668
513-44-0 isobutyl mercaptan C4H10S 3943
513-48-4 2-iodobutane C4H9I 3667
513-49-5 (S)-(+)-sec-butylamine C4H11N 3987
513-53-1 sec-butyl mercaptan C4H10S 3944
513-81-5 2,3-dimethyl-1,3-butadiene C6H10 7647
513-85-9 2,3-butanediol C4H10O2 3899
514-10-3 abietic acid C20H30O2 32346
515-12-8 C15H30 1-ethyl-1-methyl-2,4-diisopropylcyclo 28857
526-73-8 1,2,3-trimethylbenzene C9H12 17008
526-75-0 2,3-xylenol C8H10O 13921
526-85-2 2,3,4-trimethylphenol C9H12O 17116
527-53-7 1,2,3,5-tetramethylbenzene C10H14 19888
527-54-8 3,4,5-trimethylphenol C9H12O 17121
527-60-6 2,4,6-trimethylphenol C9H12O 17120
527-84-4 o-cymene C10H14 19871
528-29-0 o-dinitrobenzene C6H4N2O4 6581
529-20-4 2-methylbenzaldehyde C8H8O 13394
529-21-5 3-ethyl-4-methylpyridine C8H11N 14099
530-48-3 1,1-diphenylethene C14H12 26868
532-27-4 alpha-chloroacetophenone C8H7ClO 13000
533-18-6 2-methylphenyl acetate C9H10O2 16681
533-98-2 1,2-dibromobutane C4H8Br2 3334
534-00-9 1-bromo-2-methylbutane, (S) C5H11Br 5611
534-22-5 2-methylfuran C5H6O 4534
534-52-1 2-methyl-4,6-dinitrophenol C7H6N2O5 10383
535-77-3 m-cymene C10H14 19870
536-74-3 ethynylbenzene C8H6 12713
536-75-4 4-ethylpyridine C7H9N 10966
536-78-7 3-ethylpyridine C7H9N 10965
536-88-9 4-ethyl-2-methylpyridine C8H11N 14100
538-68-1 pentylbenzene C11H16 22304
538-86-3 benzyl methyl ether C8H10O 13927
538-93-2 isobutylbenzene C10H14 19866
539-30-0 benzyl ethyl ether C9H12O 17077
539-82-2 ethyl pentanoate C7H14O2 11912

598-29-8 1,2-diiodopropane C3H6I2 1872
598-45-8 isopropyl isocyanide C4H7N 3241
598-53-8 methyl isopropyl ether C4H10O 3871
598-56-1 dimethylethylamine C4H11N 3985
598-58-3 methyl nitrate CH3NO3 238
598-61-8 methylcyclobutane C5H10 5284
598-71-0 DL-2,3-dibromobutane C4H8Br2 3338
598-73-2 bromotrifluoroethylene C2BrF3 404
598-74-3 1,2-dimethylpropylamine C5H13N 5952
598-75-4 3-methyl-2-butanol C5H12O 5839
598-76-5 1,1-dichloro-2-methylpropane C4H8Cl2 3367
598-77-6 1,1,2-trichloropropane C3H5Cl3 1704
598-78-7 2-chloropropanoic acid C3H5ClO2 1691
598-96-9 3,4,4-trimethyl-2-pentene C8H16 14989
598-98-1 methyl 2,2-dimethylpropanoate C6H12O2 8497
600-06-6 1-chloro-2,3-dimethylbutane C6H13Cl 8682
600-07-7 2-methylbutanoic acid C5H10O2 5496
600-10-2 1,2-dichloro-3-methylbutane C5H10Cl2 5342
600-11-3 2,3-dichloropentane C5H10Cl2 5336
600-24-8 2-nitrobutane C4H9NO2 3698
600-25-9 1-chloro-1-nitropropane C3H6ClNO2 1837
600-36-2 2,4-dimethyl-3-pentanol C7H16O 12197
603-34-9 triphenylamine C18H15N 30548
603-35-0 triphenylphosphine C18H15P 30567
604-83-1 9,10-dimethylphenanthrene C16H14 29111
604-88-6 hexaethylbenzene C18H30 30948
605-01-6 pentaethylbenzene C16H26 29596
605-39-0 2,2'-dimethylbiphenyl C14H14 27057
606-20-2 2,6-dinitrotoluene C7H6N2O4 10365
606-23-5 1H-indene-1,3(2H)-dione C9H6O2 15928
608-27-5 2,3-dichloroaniline C6H5Cl2N 6741
608-31-1 2,6-dichloroaniline C6H5Cl2N 6744
608-73-1 hexachlorocyclohexane; (mixed isomers) C6H6Cl6 6954
608-93-5 pentachlorobenzene C6HCl5 6115
609-19-8 3,4,5-trichlorophenol C6H3Cl3O 6315
609-26-7 3-ethyl-2-methylpentane C8H18 15360
609-27-8 3-ethyl-2-pentanol C7H16O 12188
610-39-9 3,4-dinitrotoluene C7H6N2O4 10366
610-48-0 1-methylanthracene C15H12 28129
611-14-3 o-ethyltoluene C9H12 17006
611-15-4 o-methylstyrene C9H10 16514
611-19-8 1-chloro-2-(chloromethyl)benzene C7H6Cl2 10254
611-21-2 N,2-dimethylaniline C8H11N 14084
611-32-5 8-methylquinoline C10H9N 18855
611-43-8 2,3'-trimethylbiphenyl C14H14 27058
611-61-0 2,4'-dimethlbiphenyl C14H14 27059
612-00-0 1,1-diphenylethane C14H14 27052
612-58-8 3-methylquinoline C10H9N 18850
612-60-2 7-methylquinoline C10H9N 18854
612-75-9 3,3'-dimethylbiphenyl C14H14 27060
613-12-7 2-methylanthracene C15H12 28130
613-13-8 2-anthracenamine C14H11N 26821
613-33-2 4,4'-dimethylbiphenyl C14H14 27062
615-16-7 2-hydroxybenzimidazole C7H6N2O 10352
615-29-2 4-methyl-3-hexanol C7H16O 12178
615-36-1 o-bromoaniline C6H6BrN 6905
615-39-4 trans-2-methylcyclohexanol, (±) C7H14O 11851
616-12-6 3-methyl-trans-2-pentene C6H12 8232
616-13-7 1-chloro-2-methylbutane C5H11Cl 5623
616-19-3 1,3-dichloro-2-methylpropane C4H8Cl2 3369
616-20-6 3-chloropentane C5H11Cl 5622
616-21-7 1,2-dichlorobutane C4H8Cl2 3361
616-24-0 3-pentanamine C5H13N 5954
616-31-9 3-pentanethiol C5H12S 5923
616-39-7 methyldiethylamine C5H13N 5961
616-43-3 3-methylpyrrole C5H7N 4660
616-44-4 3-methylthiophene C5H6S 4588
616-45-5 2-pyrrolidone C4H7NO 3249
616-55-7 2,4-di-tert-butyl-6-methylphenol C15H24O 28759
617-29-8 2-methyl-3-hexanol C7H16O 12176
617-30-1 2,3-hexanediol C6H14O2 9028
617-50-5 isopropyl isobutanoate C7H14O2 11907
617-78-7 3-ethylpentane C7H16 12115
617-79-8 1-amino-2-ethylbutane C6H15N 9189
617-80-1 2-ethylbutanenitrile C6H11N 8090
617-94-7 2-phenyl-2-propanol C9H12O 17090
618-45-1 3-isopropylphenol C9H12O 17105
618-46-2 m-chlorobenzoyl chloride C7H4Cl2O 9736
619-15-8 2,5-dinitrotoluene C7H6N2O4 10364
619-39-6 2-octyldecanoic acid C18H36O2 31098
619-82-9 1,4-cyclohexanedicarboxylic acid C8H12O4 14379
619-99-8 3-ethylhexane C8H18 15359
620-11-1 1-ethylpropyl acetate C7H14O2 11896
620-14-4 m-ethyltoluene C9H12 17005
620-17-7 m-ethylphenol C8H10O 13919
620-20-2 1-chloro-3-(chloromethyl)benzene C7H6Cl2 10255
620-23-5 3-methylbenzaldehyde C8H8O 13395
620-43-9 tris(2-methylbutyl)amine C15H33N 28938
620-47-3 3-methyldiphenylmethane C14H14 27055
620-83-7 4-methyldiphenylmethane C14H14 27056
620-85-9 4-ethyldiphenylmethane C15H16 28391
621-27-2 3-propylphenol C9H12O 17102
621-77-2 tripentylamine C15H33N 28936
621-79-4 cinnamamide, predominantly trans C9H9NO 16388
621-82-9 cinnamic acid C9H8O2 16212
622-39-9 2-propylpyridine C8H11N 14105
622-63-9 4-methyl-(1-thiapropyl)-benzene C9H12S 17216
622-96-8 p-ethyltoluene C9H12 17007
622-97-9 p-methylstyrene C9H10 16515
623-08-5 N,4-dimethylaniline C8H11N 14086
623-13-2 4-methyl-(1-thiaethyl)-benzene C8H10S 14051
623-34-7 1,4-dichloro-2-methylbutane C5H10Cl2 5345
623-37-0 3-hexanol C6H14O 8972
623-42-7 methyl butanoate C5H10O2 5504
623-43-8 methyl trans-2-butenoate C5H8O2 4887
623-55-2 5-methyl-3-hexanol C7H16O 12179
623-56-3 5-methyl-3-hexanone C7H14O 11837
623-70-1 ethyl trans-2-butenoate C6H10O2 7832
623-93-8 5-nonanol C9H20O 18243
624-08-8 9-heptadecanol C17H36O 30354
624-16-8 4-decanone C10H20O 20995
624-20-4 1,2-dibromohexane C6H12Br2 8249
624-22-6 2-methyl-1-hexanol C7H16O 12168
624-29-3 cis-1,4-dimethylcyclohexane C8H16 14886
624-41-9 2-methylbutyl acetate C7H14O2 11894
624-42-0 6-methyl-3-heptanone C8H16O 15079
624-48-6 dimethyl maleate C6H8O4 7446
624-51-1 3-nonanol C9H20O 18239
624-54-4 pentyl propanoate C8H16O2 15167
624-64-6 trans-2-butene C4H8 3323
624-65-7 propargyl chloride C3H3Cl 1402
624-72-6 1,2-difluoroethane C2H4F2 835
624-73-7 1,2-diiodoethane C2H4I2 845
624-78-2 methylethylamine C3H9N 2202
624-79-3 ethyl isocyanide C3H5N 1753
624-83-9 methyl isocyanate C2H3NO 755
624-89-5 methyl ethyl sulfide C3H8S 2160
624-91-9 methyl nitrite CH3NO2 236
624-92-0 dimethyl disulfide C2H6S2 1094
624-95-3 3,3-dimethyl-1-butanol C6H14O 8985
624-96-4 1,3-dichloro-3-methylbutane C5H10Cl2 5344
625-06-9 2,4-dimethyl-2-pentanol C7H16O 12190
625-16-1 tert-pentyl acetate C7H14O2 11897
625-23-0 2-methyl-2-hexanol C7H16O 12172
625-25-2 2-methyl-2-heptanol C8H18O 15441
625-27-4 2-methyl-2-pentene C6H12 8229
625-28-5 3-methylbutanenitrile C5H9N 5185
625-29-6 2-chloropentane C5H11Cl 5621
625-30-9 1-methylbutylamine C5H13N 5948
625-38-7 vinylacetic acid C4H6O2 2968
625-43-4 methyl-tert-butylamine C5H13N 5958
625-44-5 methyl isobutyl ether C5H12O 5842
625-45-6 methoxyacetic acid C3H6O3 1949
625-54-7 ethyl isopropyl ether C5H12O 5846
625-55-8 isopropyl formate C4H8O2 3507
625-58-1 ethyl nitrate C2H5NO3 964
625-65-0 2,4-dimethyl-2-pentene C7H14 11738
625-66-1 1,1-dichloro-3-methylbutane C5H10Cl2 5340
625-67-2 2,4-dichloropentane C5H10Cl2 5337
625-69-4 2,4-pentanediol C5H12O2 5865
625-74-1 1-nitro-2-methylpropane C4H9NO2 3699
625-80-9 diisopropyl sulfide C6H14S 9113
625-86-5 2,5-dimethylfuran C6H8O 7385
626-17-5 m-dicyanobenzene C8H4N2 12575
626-23-3 di-sec-butylamine C8H19N 15639
626-26-6 di-sec-butyl sulfide C8H18S 15610
626-33-5 2-methyl-4-heptanone C8H16O 15074
626-38-0 1-methylbutyl acetate C7H14O2 11893
626-39-1 1,3,5-tribromobenzene C6H3Br3 6259
626-43-7 3,5-dichloroaniline C6H5Cl2N 6745
626-82-4 butyl hexanoate C10H20O2 21056
626-87-9 1,4-dibromopentane C5H10Br2 5300
626-88-0 1-bromo-4-methylpentane C6H13Br 8652
626-89-1 4-methyl-1-pentanol C6H14O 8975
626-91-5 methyl 3-methylbutyl ether C6H14O 8999
626-92-6 1,4-dichloropentane C5H10Cl2 5333
626-93-7 2-hexanol C6H14O 8971
626-95-9 1,4-pentanediol C5H12O2 5862
627-02-1 ethyl isobutyl ether C6H14O 9005
627-05-4 1-nitrobutane C4H9NO2 3697
627-08-7 propyl isopropyl ether C6H14O 8994
627-13-4 propyl nitrate C3H7NO3 2061
627-19-0 1-pentyne C5H8 4745
627-20-3 cis-2-pentene C5H10 5288
627-21-4 2-pentyne C5H8 4746
627-22-5 1-chloro-1,3-butadiene C4H5Cl 2636
627-26-9 trans-crotonitrile C4H5N 2695
627-31-6 1,3-diiodopropane C3H6I2 1873
627-35-0 methyl-propylamine C4H11N 3982
627-36-1 propyl isocyanide C4H7N 3240
627-40-7 3-methoxy-1-propene C4H8O 3477
627-45-2 N-ethylformamide C3H7NO 2038
627-58-7 2,5-dimethyl-1,5-hexadiene C8H14 14496
627-59-8 5-methyl-2-hexanol C7H16O 12175
627-67-8 1-nitro-3-methylbutane C5H11NO2 5700
627-90-7 ethyl undecanoate C13H26O2 26506
627-96-3 1,5-dibromohexane C6H12Br2 8251
627-97-4 2-methyl-2-heptene C8H16 14904
627-98-5 5-methyl-1-hexanol C7H16O 12171
628-04-6 ethyl isopentyl ether C7H16O 12214
628-05-7 1-nitropentane C5H11NO2 5696
628-08-0 2-buten-1-yl acetate C6H10O2 7817
628-17-1 1-iodopentane C5H11I 5660
628-21-7 1,4-diiodobutane C4H8I2 3404
628-28-4 methyl butyl ether C5H12O 5841
628-29-5 methyl butyl sulfide C5H12S 5930
628-32-0 ethyl propyl ether C5H12O 5845
628-41-1 1,4-cyclohexadiene C6H8 7269
628-44-4 2-methyl-2-octanol C9H20O 18250
628-46-6 5-methylhexanoic acid C7H14O2 11890
628-55-7 isobutyl ether C8H18O 15539
628-61-5 2-chlorooctane C8H17Cl 15260
628-63-7 pentyl acetate C7H14O2 11892
628-71-7 1-heptyne C7H12 11340
628-72-8 2,4-heptadiene C7H12 11362
628-73-9 hexanenitrile C6H11N 8088
628-76-2 1,5-dichloropentane C5H10Cl2 5334
628-77-3 1,5-diiodopentane C5H10I2 5384
628-80-8 methyl pentyl ether C6H14O 8996
628-81-9 ethyl butyl ether C6H14O 9004
628-96-6 ethylene glycol dinitrate C2H4N2O6 865
628-99-9 2-nonanol C9H20O 18237
629-03-8 1,6-dibromohexane C6H12Br2 8252
629-04-9 1-bromoheptane C7H15Br 11982
629-05-0 1-octyne C8H14 14484
629-06-1 1-chloroheptane C7H15Cl 11988
629-08-3 heptanenitrile C7H13N 11622
629-11-8 1,6-hexanediol C6H14O2 9027
629-14-1 ethylene glycol diethyl ether C6H14O2 9060
629-19-6 dipropyl disulfide C6H14S2 9130
629-20-9 1,3,5,7-cyclooctatetraene C8H8 13238
629-23-2 3-tetradecanone C14H28O 27866
629-27-6 1-iodooctane C8H17I 15276
629-30-1 1,7-heptanediol C7H16O2 12229
629-32-3 1-methoxyheptane C8H18O 15534
629-33-4 hexyl formate C7H14O2 11891
629-41-4 1,8-octanediol C8H18O2 15547
629-45-8 dibutyl disulfide C8H18S2 15614
629-49-2 2-dodecyne C12H22 24654
629-50-5 tridecane C13H28 26526
629-59-4 tetradecane C14H30 27912
629-62-9 pentadecane C15H32 28904
629-64-1 diheptyl ether C14H30O 27928
629-65-2 diheptyl sulfide C14H30S 27948
629-72-1 1-bromopentadecane C15H31Br 28898
629-73-2 1-hexadecene C16H32 29707
629-74-3 1-hexadecyne C16H30 29640
629-75-4 2-hexadecyne C16H30 29641
629-76-5 1-pentadecanol C15H32O 28912
629-78-7 heptadecane C17H36 30346
629-80-1 hexadecanal C16H32O 29717
629-82-3 dioctyl ether C16H34O 29771
629-92-5 nonadecane C19H40 31735
629-96-9 1-eicosanol C20H42O 32522
630-08-0 carbon monoxide CO 372
630-13-7 2,2-diiodopropane C3H6I2 1874
630-16-0 1,1,1,2-tetrabromoethane C2H2Br4 593
630-17-1 1-bromo-2,2-dimethylpropane C5H11Br 5610
630-18-2 2,2-dimethylpropanenitrile C5H9N 5186
630-19-3 2,2-dimethylpropanal C5H10O 5448
630-20-6 1,1,1,2-tetrachloroethane C2H2Cl4 635
631-28-7 1,2,3-tribromo-2-methylpropane C4H7Br3 3095
631-65-2 2-chloro-3-methylbutane C5H11Cl 5626
632-05-3 1,2,3-tribromobutane C4H7Br3 3088

632-15-5 3,4-dimethylthiophene C6H8S 7472
632-16-6 2,3-dimethylthiophene C6H8S 7469
634-66-2 1,2,3,4-tetrachlorobenzene C6H2Cl4 6174
634-90-2 1,2,3,5-tetrachlorobenzene C6H2Cl4 6175
635-11-0 1,2,4,5-tetraisopropylbenzene C18H30 30949
635-46-1 1,2,3,4-tetrahydroquinoline C9H11N 16895
635-81-4 1,2,4,5-tetraethylbenzene C14H22 27647
636-28-2 1,2,4,5-tetrabromobenzene C6H2Br4 6150
636-41-9 2-methylpyrrole C5H7N 4659
637-50-3 1-propenylbenzene C9H10 16516
637-78-5 isopropyl propanoate C6H12O2 8481
637-92-3 ethyl tert-butyl ether C6H14O 9007
637-97-8 2-iodopentane C5H11I 5661
638-00-6 2,4-dimethylthiophene C6H8S 7470
638-02-8 2,5-dimethylthiophene C6H8S 7471
638-04-0 cis-1,3-dimethylcyclohexane C8H16 14883
638-10-8 ethyl 3-methyl-2-butenoate C7H12O2 11480
638-11-9 isopropyl butanoate C7H14O2 11906
638-25-5 pentyl octanoate C13H26O2 26501
638-28-8 2-chlorohexane C6H13Cl 8670
638-45-9 1-iodohexane C6H13I 8724
638-46-0 ethyl butyl sulfide C6H14S 9115
638-49-3 pentyl formate C6H12O2 8468
638-53-9 tridecanoic acid C13H26O2 26492
638-59-5 tetradecyl acetate C16H32O2 29724
638-60-8 2-tetradecene C14H28 27851
640-21-1 1,2-dibromo-3,3-dimethylbutane C6H12Br2 8255
642-32-0 1,2,3,4-tetraethylbenzene C14H22 27645
643-93-6 3-methylbiphenyl C13H12 25740
644-08-6 4-methylbiphenyl C13H12 25741
644-35-9 2-propylphenol C9H12O 17101
644-49-5 propyl isobutanoate C7H14O2 11908
644-98-4 2-isopropylpyridine C8H11N 14101
645-41-0 triisopentylamine C15H33N 28937
645-49-8 cis-stilbene C14H12 26866
645-55-6 N-nitroaniline C6H6N2O2 7000
645-56-7 4-propylphenol C9H12O 17103
645-59-0 benzenepropanenitrile C9H9N 16364
646-04-8 trans-2-pentene C5H10 5289
646-05-9 1-pentene-3-yne C5H6 4449
646-07-1 4-methylpentanoic acid C6H12O2 8467
646-13-9 isobutyl stearate C22H44O2 33467
646-14-0 1-nitrohexane C6H13NO2 8790
646-30-0 nonadecanoic acid C19H38O2 31708
656-31-5 2-fluorophenyl urea C7H7FN2O 10559
659-30-3 4-fluorophenyl urea C7H7FN2O 10559
659-70-1 isopentyl isovalerate C10H20O2 21052
661-11-0 1-fluoroheptane C7H15F 12015
661-53-0 2-fluoro-2-methylbutane C5H11F 5654
666-21-7 2,3-difluorobutane C4H8F2 3394
674-76-0 4-methyl-trans-2-pentene C6H12 8228
674-82-8 diketene C4H4O2 2598
680-31-9 hexamethyl phosphoramide C6H18N3OP 9356
681-84-5 tetramethyl silicate C4H12O4Si 4101
684-16-2 hexafluoroacetone C3F6O 1287
684-84-4 2-methyl-1,3-butanediol C5H12O2 5871
686-65-7 1,2-difluorobutane C4H8F2 3390
688-74-4 tributyl borate C12H27BO3 25327
689-97-4 vinylacetylene C4H4 2510
690-08-4 4,4-dimethyl-trans-2-pentene C7H14 11742
690-37-9 2,4,4-trimethyl-2-pentanol C8H18O 15509
690-92-6 2,2-dimethyl-cis-3-hexene C8H16 14960
690-93-7 2,2-dimethyl-trans-3-hexene C8H16 14961
691-37-2 4-methyl-1-pentene C6H12 8225
691-38-3 4-methyl-cis-2-pentene C6H12 8227
691-42-9 1,3-difluorobutane C4H8F2 3391
692-24-0 2-methyl-trans-3-hexene C7H14 11726
692-45-5 vinyl formate C3H4O2 1628
692-47-7 cis-1,2-di-tert-butylethene C10H20 20955
692-48-8 (E)-2,2,5,5-tetramethylhex-3-ene C10H20 20956
692-70-6 2,5-dimethyl-trans-3-hexene C8H16 14967
692-96-6 2-methyl-trans-3-heptene C8H16 14917
693-02-7 1-hexyne C6H10 7648
693-16-3 2-aminooctane C8H19N 15650
693-19-6 7-methyloctanoic acid C9H18O2 18083
693-54-9 2-decanone C10H20O 20993
693-58-3 1-bromononane C9H19Br 18115
693-61-8 trans-2-undecene C11H22 22797
693-62-9 trans-4-undecene C11H22 22800
693-65-2 di-pentyl ether C10H22O 21335
693-67-4 1-bromoundecane C11H23Br 22870
693-72-1 trans-11-octadecenoic acid C18H34O2 31033
693-83-4 didecyl sulfide C20H42S 32539
693-89-0 1-methylcyclopentene C6H10 7619
694-35-9 3-ethylcyclopentene C7H12 11324
694-80-4 1-bromo-2-chlorobenzene C6H4BrCl 6383
695-96-5 2-bromo-4-chlorophenol C6H4BrClO 6388
695-98-7 2,3,5-trimethylpyridine C8H11N 14107
696-29-7 isopropylcyclohexane C9H18 17917
696-30-0 4-isopropylpyridine C8H11N 14103
696-44-6 N,3-dimethylaniline C8H11N 14085
697-82-5 2,3,5-trimethylphenol C9H12O 17117
698-71-5 3-methyl-5-ethylphenol C9H12O 17112
700-12-9 pentamethylbenzene C11H16 22354
700-38-9 5-methyl-2-nitrophenol C7H7NO3 10672
709-98-8 propanil C9H9Cl2NO 16331
713-36-0 2-methyldiphenylmethane C14H14 27054
717-74-8 1,3,5-triisopropylbenzene C15H24 28727
719-79-9 1,1-diphenylbutane C16H18 29342
723-62-6 9-anthracenecarboxylic acid C15H10O2 28038
753-89-9 1-chloro-2,2-dimethylpropane C5H11Cl 5627
757-88-0 1,1-dimethylpropyl formate C6H12O2 8473
760-20-3 3-methyl-1-pentene C6H12 8224
760-21-4 2-ethyl-1-butene C6H12 8236
760-23-6 3,4-dichloro-1-butene C4H6Cl2 2809
762-50-5 1-chloro-2-fluoroethane C2H4ClF 802
762-62-9 4,4-dimethyl-1-pentene C7H14 11735
762-63-0 4,4-dimethyl-cis-2-pentene C7H14 11741
762-75-4 tert-butyl formate C5H10O2 5500
763-29-1 2-methyl-1-pentene C6H12 8223
763-30-4 2-methyl-1,4-pentadiene C6H10 7643
763-69-9 ethyl-3-ethoxypropionate C7H14O3 11931
764-13-6 2,5-dimethyl-2,4-hexadiene C8H14 14497
764-35-2 2-hexyne C6H10 7649
764-41-0 1,4-dichloro-2-butene C4H6Cl2 2820
764-93-2 1-decyne C10H18 20593
764-96-5 cis-5-undecene C11H22 22801
764-97-6 trans-5-undecene C11H22 22802
765-03-7 1-dodecyne C12H22 24653
765-09-3 1-bromotridecane C13H27Br 26515
765-10-6 1-tetradecyne C14H26 27780
765-13-9 1-pentadecyne C15H28 28820
765-38-8 2-methylpyrrolidine C5H11N 5673
765-47-9 1,2-dimethylcyclopentene C7H12 11326
766-07-4 cyclohexyl peroxide C6H12O2 8484
766-90-5 cis-1-propenylbenzene C9H10 16509
767-99-7 cis-(1-methyl-1-propenyl)benzene C10H12 19362
768-00-3 trans-(1-methyl-1-propenyl)benzene C10H12 19363
768-03-6 2-propenophenone C9H8O 16200
768-49-0 (2-methyl-1-propenyl)benzene C10H12 19364
768-52-5 N-isopropylaniline C9H13N 17275
769-06-2 N,N,2,6-tetramethylaniline C10H15N 20212
769-92-6 4-tert-butylaniline C10H15N 20201
770-19-4 3-fluorophenyl urea C7H7FN2O 10559
777-22-0 (1-methylheptyl)benzene C14H22 27649
778-22-3 2,2-diphenylpropane C15H16 28382
779-02-2 9-methylanthracene C15H12 28131
789-02-6 o,p'-DDT C14H9Cl5 26700
811-95-0 1,1,2-trichloro-1-fluoroethane C2H2Cl3F 631
811-97-2 1,1,1,2-tetrafluoroethane C2H2F4 650
812-03-3 1,1,1,2-tetrachloropropane C3H4Cl4 1565
813-75-2 1,2,2,3-tetrafluoropropane C3H4F4 1583
814-78-8 methyl isopropenyl ketone C5H8O 4849
815-24-7 di-tert-butyl ketone C9H18O 18000
816-11-5 methyl 2-ethylbutanoate C7H14O2 11913
816-79-5 3-ethyl-2-pentene C7H14 11736
817-46-9 2-propyl-1-hexanol C9H20O 18291
817-60-7 2-ethyl-1-heptanol C9H20O 18265
817-91-4 4-methyl-1-heptanol C8H18O 15438
818-23-5 8-pentadecanone C15H30O 28875
818-49-5 4-methyl-1-hexanol C7H16O 12170
818-61-1 2-hydroxyethyl acrylate C5H8O3 4927
818-81-5 2-methyl-1-octanol C9H20O 18244
818-92-8 3-fluoro-1-propene C3H5F 1714
819-97-6 sec-butyl butanoate C8H16O2 15178
820-05-3 4,6-dimethyl-1-heptanol C9H20O 18269
820-29-1 5-decanone C10H20O 21019
820-55-3 1,1-dichloropentane C5H10Cl2 5330
821-11-4 trans-2-butene-1,4-diol C4H8O2 3512
821-25-0 1,1-dichloroheptane C7H14Cl2 11763
821-55-6 2-nonanone C9H18O 17999
821-76-1 1,7-dichloroheptane C7H14Cl2 11768
821-88-5 1,1-dichlorononane C9H18Cl2 17959
821-93-2 1,1-dichlorotridecane C13H26Cl2 26474
821-95-4 1-undecene C11H22 22795
821-96-5 cis-2-undecene C11H22 22796
821-98-7 cis-4-undecene C11H22 22799
821-99-8 1,9-dichlorononane C9H18Cl2 17961
822-01-5 1,1-dichloroundecane C11H22Cl2 22807
822-06-0 hexamethylene diisocyanate C8H12N2O2 14306
822-13-9 1-chlorotridecane C13H27Cl 26516
822-27-5 dioctyl disulfide C16H34S2 29797
822-50-4 trans-1,2-dimethylcyclopentane C7H14 11700
826-77-7 4,6-dimethylquinoline C11H11N 21715
827-52-1 cyclohexylbenzene C12H16 24111
830-13-7 cyclododecanone C12H22O 24690
832-69-9 1-methylphenanthrene C15H12 28132
834-12-8 ametryn C9H17N5S 17857
865-58-7 3-bromo-3-methyl-1-butene C5H9Br 4987
870-63-3 1-bromo-3-methyl-2-butene C5H9Br 4991
871-28-3 1-pentene-4-yne C5H6 4450
871-83-0 2-methylnonane C10H22 21134
872-05-9 1-decene C10H20 20943
872-10-6 dipentyl sulfide C10H22S 21387
872-50-4 N-methyl-2-pyrrolidone C5H9NO 5198
872-55-9 2-ethylthiophene C6H8S 7467
872-93-5 3-methyl sulfolane C5H10O2S 5537
873-66-5 trans-1-propenylbenzene C9H10 16510
874-41-9 4-ethyl-m-xylene C10H14 19884
874-79-3 propyl phenyl sulfide C9H12S 17213
877-43-0 2,6-dimethylquinoline C11H11N 21707
877-44-1 1,2,4-triethylbenzene C12H18 24365
886-59-9 3-phenyl-1-cyclohexyl urea C13H18N2O 26183
917-92-0 3,3-dimethyl-1-butyne C6H10 7654
918-02-5 dimethyl-tert-butylamine C6H15N 9211
918-07-0 2-chloro-2,3,3-trimethylbutane C7H15Cl 12001
918-82-1 3,3-dimethyl-1-pentyne C7H12 11350
918-84-3 3-chloro-3-methylpentane C6H13Cl 8679
919-12-0 3-ethyl-3-methyl-1-pentyne C8H14 14488
919-94-8 2-ethoxy-2-methylbutane C7H16O 12213
921-04-0 dimethyl-sec-butylamine C6H15N 9209
921-47-1 2,3,4-trimethylhexane C9H20 18198
921-60-8 L(-)-glucose C6H12O6 8610
922-17-8 3,4-hexanediol C6H14O2 9031
922-28-1 3,4-dimethylheptane C9H20 18187
922-59-8 3-methyl-1-pentyne C6H10 7651
922-61-2 3-methyl-2-pentene, cis and trans C6H12 8231
922-62-3 3-methyl-cis-2-pentene C6H12 8230
923-28-4 2-methyl-3-octanone C9H18O 18047
924-50-5 methyl 3-methyl-2-butenoate C6H10O2 7846
925-60-0 propyl acrylate C6H10O2 7815
925-78-0 3-nonanone C9H18O 18008
926-02-3 tert-butyl vinyl ether C6H12O 8396
926-41-0 neopentyl acetate C7H14O2 11899
926-54-5 2-methyl-1,trans-3-pentadiene C6H10 7639
926-56-7 4-methyl-1,3-pentadiene C6H10 7642
926-57-8 1,3-dichloro-2-butene, cis and trans C4H6Cl2 2821
926-63-6 dimethyl-propylamine C5H13N 5962
926-82-9 3,5-dimethylheptane C9H20 18188
927-49-1 6-undecanone C11H22O 22827
927-62-8 dimethylbutylamine C6H15N 9208
927-73-1 4-chloro-1-butene C4H7Cl 3102
928-40-5 1,5-hexanediol C6H14O2 9026
928-49-4 3-hexyne C6H10 7650
928-50-7 5-chloro-1-pentene C5H9Cl 5028
928-68-7 6-methyl-2-heptanone C8H16O 15078
928-80-3 3-decanone C10H20O 20994
929-61-3 ethyl octyl ether C10H22O 21337
930-18-7 1,cis-2-dimethylcyclopropane C5H10 5281
930-27-8 3-methylfuran C5H6O 4535
930-68-7 2-cyclohexen-1-one C6H8O 7387
930-89-2 1-methyl-cis-2-ethylcyclopentane C8H16 14864
930-90-5 1-methyl-trans-2-ethylcyclopentane C8H16 14865
932-39-8 1,cis-2-diethylcyclopentane C9H18 17874
932-40-1 1,trans-2-diethylcyclopentane C9H18 17875
932-43-4 1-methyl-cis-2-propylcyclopentane C9H18 17864
932-44-5 1-methyl-trans-2-propylcyclopentane C9H18 17865
933-75-5 2,3,6-trichlorophenol C6H3Cl3O 6312
933-78-8 2,3,5-trichlorophenol C6H3Cl3O 6311
933-98-2 3-ethyl-o-xylene C10H14 19881
934-74-7 5-ethyl-m-xylene C10H14 19885
934-80-5 4-ethyl-o-xylene C10H14 19882
935-95-5 2,3,5,6-tetrachlorophenol C6H2Cl4O 6179
937-33-7 N-tert-butylaniline C10H15N 20203
939-23-1 4-phenylpyridine C11H9N 21565
939-27-5 2-ethylnaphthalene C12H12 23686
939-48-0 isopropyl benzoate C10H12O2 19551
944-22-9 fonofos C10H15OPS2 20304
946-80-5 benzyl phenyl ether C13H12O 25793
948-32-3 1,2,4-triisopropylbenzene C15H24 28726
993-00-0 methyl chlorosilane CH5ClSi 284
994-05-8 methyl tert-pentyl ether C6H14O 9002
994-25-2 3-chloro-3-ethylpentane C7H15Cl 11993
996-35-0 N,N-dimethyl-2-propanamine C5H13N 5965

996-35-6 dimethylisopropylamine C5H13N 5963
998-35-6 5-propylnonane C12H26 24943
998-65-2 5-ethyl-1-heptanol C9H20O 18267
998-76-5 1,2-diiodopentane C5H10I2 5381
998-93-6 4-bromoheptane C7H15Br 11984
998-95-8 4-chloroheptane C7H15Cl 11996
999-21-3 diallyl maleate C10H12O4 19675
999-52-0 3-chloroheptane C7H15Cl 11995
999-61-1 2-hydroxypropyl acrylate C6H10O3 7892
999-78-0 4,4-dimethyl-2-pentyne C7H12 11353
999-97-3 hexamethyldisilazane C6H19NSi2 9378
1000-86-8 2,4-dimethyl-1,3-pentadiene C7H12 11365
1000-87-9 2,4-dimethyl-2,3-pentadiene C7H12 11369
1001-89-4 2-chloroheptane C7H15Cl 11994
1002-17-1 2,9-dimethyldecane C12H26 24929
1002-43-3 3-methylundecane C12H26 24915
1002-69-3 1-chlorodecane C10H21Cl 21093
1002-84-2 pentadecanoic acid C15H30O2 28879
1003-30-1 2-ethyltetrahydrofuran C6H12O 8416
1003-38-9 2,5-dimethyltetrahydrofuran C6H12O 8421
1005-64-7 trans-(1-butenyl)benzene C10H12 19361
1007-26-7 1-phenyl-2,2-dimethylpropane C11H16 22311
1008-88-4 3-phenylpyridine C11H9N 21564
1008-89-5 2-phenylpyridine C11H9N 21563
1012-72-2 1,4-di-tert-butylbenzene C14H22 27648
1014-60-4 1,3-di-tert-butylbenzene C14H22 27650
1024-57-3 heptachlor epoxide C10H5Cl7O 18508
1067-08-9 3-methyl-3-ethylpentane C8H18 15358
1067-20-5 3,3-diethylpentane C9H20 18202
1068-19-5 4,4-dimethylheptane C9H20 18189
1068-87-7 2,4-dimethyl-3-ethylpentane C9H20 18205
1069-53-0 2,3,5-trimethylhexane C9H20 18199
1070-32-2 3-methyl-1-heptanol C8H18O 15437
1070-78-6 1,1,1,3-tetrachloropropane C3H4Cl4 1566
1070-83-3 tert-butylacetic acid C6H12O2 8462
1070-87-7 2,2,4,4-tetramethylpentane C9H20 18208
1071-26-7 2,2-dimethylheptane C9H20 18182
1071-31-4 2,2,7,7-tetramethyloctane C12H26 25078
1071-81-4 2,2,5,5-tetramethylhexane C10H22 21195
1071-83-6 glyphosate C3H8NO5P 2103
1071-84-7 1,1,1-trichlorononane C9H17Cl3 17804
1072-05-5 2,6-dimethylheptane C9H20 18186
1072-16-8 2,7-dimethyloctane C10H22 21145
1072-33-9 tridecyl acetate C15H30O2 28881
1072-43-1 2-methylthiacyclopropane C3H6S 1968
1073-07-0 1,4-cyclooctadiene C8H12 14229
1074-17-5 1-methyl-2-propylbenzene C10H14 19873
1074-43-7 1-methyl-3-propylbenzene C10H14 19874
1074-55-1 1-methyl-4-propylbenzene C10H14 19875
1074-92-6 1-methyl-2-tert-butyl benzene C11H16 22321
1075-38-3 1-methyl-3-tert-butylbenzene C11H16 22322
1077-16-3 hexylbenzene C12H18 24357
1078-71-3 heptylbenzene C13H20 26283
1080-16-6 cis-azobenzene C12H10N2 23489
1081-75-0 1,3-diphenylpropane C15H16 28381
1081-77-2 nonylbenzene C15H24 28725
1115-08-8 3-methyl-1,4-pentadiene C6H10 7644
1116-40-1 triisobutylamine C12H27N 25347
1117-60-8 (R)-3,7-dimethyl-1-octanol C10H22O 21339
1117-86-8 1,2-octanediol C8H18O2 15544
1118-00-9 methyl 2,2-dimethylpropyl ether C6H14O 9003
1118-58-7 2-methyl-1,3-pentadiene C6H10 7658
1119-06-8 hexyl heptanoate C13H26O2 26500
1119-16-0 4-methylpentanal C6H12O 8404
1119-51-3 5-bromo-1-pentene C5H9Br 4969
1119-65-9 2-heptyne C7H12 11341
1119-85-3 trans-1,4-dicyano-2-butene C6H6N2 6981
1120-06-5 2-decanol C10H22O 21226
1120-21-4 undecane C11H24 22890
1120-24-7 dimethyldecylamine C12H27N 25344
1120-36-1 1-tetradecene C14H28 27850
1120-48-5 dioctylamine C16H35N 29800
1120-62-3 3-methylcyclopentene C6H10 7620
1122-39-0 2,4,5-trimethylpyridine C8H11N 14109
1122-69-6 2-ethyl-6-methylpyridine C8H11N 14098
1122-81-2 4-propylpyridine C8H11N 14104
1123-73-5 2-methyl-3-ethylphenol C9H12O 17107
1124-94-0 3-methyl-4-ethylphenol C9H12O 17111
1124-20-5 1-methyl-3-(1-methylethenyl)benzene C10H12 19370
1125-80-0 3-methylisoquinoline C10H9N 18857
1126-78-9 N-butylaniline C10H15N 20204
1127-76-0 1-ethylnaphthalene C12H12 23685
1134-40-3 2,3,6,7-tetramethylnaphthalene C14H16 27311
1134-62-9 2-butylnaphthalene C14H16 27268
1138-52-9 3,5-bis(1,1-dimethylethyl)phenol C14H22O 27689
1184-60-7 2-fluoro-1-propene C3H5F 1713
1185-33-7 2,2-dimethyl-1-butanol C6H14O 8982
1186-53-4 2,2,3,4-tetramethylpentane C9H20 18207
1188-02-9 2-methylheptanoic acid C8H16O2 15204
1189-99-7 2,5,5-trimethylheptane C10H22 21173
1190-22-3 1,3-dichlorobutane C4H8Cl2 3362
1190-76-7 cis-crotonitrile C4H5N 2696
1190-83-6 2,2,6-trimethylheptane C10H22 21165
1191-16-8 3,3-dimethylallyl acetate C7H12O2 11497
1191-25-9 hydroxycaproic acid C6H12O3 8532
1191-96-4 ethylcyclopropane C5H10 5283
1191-99-7 2,3-dihydrofuran C4H6O 2943
1192-18-3 cis-1,2-dimethylcyclopentane C7H14 11699
1195-32-0 1-methyl-4-(1-methylethenyl)benzene C10H12 19371
1196-39-0 4-methylisoquinoline C10H9N 18858
1196-58-3 3-phenylpentane C11H16 22306
1198-37-4 2,4-dimethylquinoline C11H11N 21706
1300-21-6 dichloroethane C2H4Cl2 814
1300-71-6 dimethyl phenol C8H10O 13934
1300-73-8 xylidine, isomers C8H11N 14123
1319-77-3 cresol mixture C7H8O 10855
1320-37-2 dichlorotetrafluoroethane C2Cl2F4 443
1320-67-8 methoxypropanol C4H10O2 3902
1321-64-8 pentachloronaphthalene C10H3Cl5 18482
1321-65-9 trichloronaphthalene C10H5Cl3 18505
1321-74-0 divinylbenzene C10H10 18973
1321-89-7 isodecanal C10H20O 21025
1321-94-4 methylnaphthalene C11H10 21603
1330-16-1 (-)-a-pinene C10H16 20365
1330-20-7 xylenes C8H10 13793
1330-78-5 tricresyl phosphate C21H21O4P 32790
1331-43-7 diethylcyclohexane; (mixed isomers) C10H20 20939
1331-50-6 trimethyl nonanone C12H24O 24860
1333-41-1 methylpyridine C6H7N 7178
1335-87-1 hexachloronaphthalene C10H2Cl6 18474
1335-88-2 tetrachloronaphthalene C10H4Cl4 18489
1453-24-3 1-ethylcyclohexene C8H14 14466
1454-84-8 1-nonadecanol C19H40O 31742
1454-85-9 1-heptadecanol C17H36O 30352
1455-20-5 2-butylthiophene C8H12S 14410
1455-21-6 nonyl mercaptan C9H20S 18365
1458-98-6 3-bromo-2-methyl-1-propene C4H7Br 3057
1459-09-2 1-phenylhexadecane C22H38 33401
1459-11-6 1,2,4-tri-tert-butylbenzene C18H30 30953
1460-02-2 1,3,5-tri-tert-butylbenzene C18H30 30952
1460-98-6 1-methyl-2,4-diisopropylbenzene C13H20 26284
1462-10-8 4-methyl-1,4-pentanediol C6H14O2 9040
1462-11-9 5-methyl-1,5-hexanediol C7H16O2 12241
1462-84-6 2,3,6-trimethylpyridine C8H11N 14108
1463-17-8 2,8-dimethylquinoline C11H11N 21709
1468-95-7 9-anthracenemethanol C15H12O 28179
1470-94-6 2,3-dihydro-1H-inden-5-ol C9H10O 16628
1471-03-0 3-propoxy-1-propene C6H12O 8399
1476-11-5 1,4-dichloro-cis-2-butene C4H6Cl2 2807
1483-60-9 2,4-dimethyl-1-sec-butylbenzene C12H18 24369
1484-87-3 3,3,5-trimethyl-1-hexanol C9H20O 18298
1489-47-0 4-methyl-1-nonanol C10H22O 21232
1501-60-6 2-methyl-1,cis-3-pentadiene C6H10 7638
1517-66-4 (S)-3-methyl-2-butanol C5H12O 5853
1518-72-5 disulfide, bis(2-methylpropyl) C8H18S2 15617
1518-75-8 3-propylthiophene C7H10S 11226
1530-03-6 1,1-diphenylpropane C15H16 28379
1534-26-5 3-tridecanone C13H26O 26488
1534-27-6 3-dodecanone C12H24O 24861
1535-73-5 3-(trifluoromethoxy)aniline C7H6F3NO 10323
1535-75-7 2-(trifluoromethoxy)aniline C7H6F3NO 10324
1536-21-6 1-fluorotridecane C13H27F 26517
1541-23-7 1,5-heptadiene C7H12 11360
1551-21-9 methyl isopropyl sulfide C4H10S 3949
1551-27-5 2-propylthiophene C7H10S 11225
1551-32-2 2-ethylthiacyclopentane C6H12S 8622
1552-12-1 cis,cis-1,5-cyclooctadiene C8H12 14230
1555-17-5 1-fluoropentadecane C15H31F 28900
1559-81-5 1-methyl-[1,2,3,4-tetrahydronaphthalene] C11H14 22025
1560-09-4 cis-(1-butenyl)benzene C10H12 19360
1560-86-7 2-methylnonadecane C20H42 32516
1560-88-9 2-methyloctadecane C19H40 31736
1560-89-0 2-methylheptadecane C18H38 31133
1560-92-5 2-methylhexadecane C17H36 30347
1560-93-6 2-methylpentadecane C16H34 29755
1560-95-8 2-methyltetradecane C15H32 28905
1560-96-9 2-methyltridecane C14H30 27913
1560-97-0 2-methyldodecane C13H28 26527
1563-66-2 carbofuran C12H15NO3 24076
1565-80-6 (S)-(-)-2-methyl-1-butanol C5H12O 5849
1565-81-7 3-decanol C10H22O 21228
1567-14-2 methyl trans-2-methyl-2-pentenoate C7H12O2 11491
1569-01-3 1-propoxy-2-propanol C6H14O2 9059
1569-02-4 1-ethoxy-2-propanol C5H12O2 5881
1573-28-0 3,6-dimethyl-3-heptanol C9H20O 18284
1574-40-9 cis-3-penten-1-yne C5H6 4451
1574-41-0 cis-1,3-pentadiene C5H8 4741
1576-85-8 4-pentenyl acetate C7H12O2 11492
1582-09-8 trifluralin C13H16F3N3O4 26049
1587-26-4 cis-2,3-dichloro-2-butene C4H6Cl2 2816
1587-29-7 trans-2,3-dichloro-2-butene C4H6Cl2 2817
1589-47-5 2-methoxy-1-propanol C4H10O2 3894
1589-49-7 3-methoxy-1-propanol C4H10O2 3906
1592-38-7 2-naphthalenemethanol C11H10O 21651
1595-04-6 1-methyl-3-butylbenzene C11H16 22313
1595-05-7 1-methyl-4-butylbenzene C11H16 22314
1595-06-8 1-methyl-2-sec-butylbenzene C11H16 22315
1595-11-5 1-methyl-2-butylbenzene C11H16 22312
1595-16-0 1-methyl-4-sec-butylbenzene C11H16 22317
1609-86-5 2-isocyanato-2-methylpropane C5H9NO 5204
1610-18-0 2-methoxy-4,6-bis(isopropylamino)-1,3,5-triaz C10H19N5O 20915
1613-45-2 5-methyl-3-thiahexane C6H14S 9121
1613-46-3 propyl butyl sulfide C7H16S 12304
1613-51-0 thiacyclohexane C5H10S 5593
1615-75-4 1-chloro-1-fluoroethane C2H4ClF 801
1617-18-1 ethyl 3-butenoate C6H10O2 7829
1617-31-8 3-methyl-3-butenoic acid C5H8O2 4895
1617-32-9 trans-3-pentenoic acid C5H8O2 4898
1622-10-2 1,5-dibromoheptane C7H14Br2 11753
1628-00-2 2-heptanethiol C7H16S 12301
1630-77-9 cis-1,2-difluoroethene C2H2F2 643
1630-78-0 trans-1,2-difluoroethene C2H2F2 644
1630-94-0 1,1-dimethylcyclopropane C5H10 5280
1632-16-2 2-ethyl-1-hexene C8H16 14929
1632-70-8 5-methylundecane C12H26 24917
1632-71-9 5-methyl-4-ethylnonane C12H26 24962
1633-90-5 3-hexanethiol C6H14S 9110
1633-97-2 2-methyl-2-pentanethiol C6H14S 9111
1634-04-4 methyl tert-butyl ether C5H12O 5844
1634-09-9 1-butylnaphthalene C14H16 27267
1636-41-5 4,5-diethyloctane C12H26 25023
1636-43-7 5,6-dimethyldecane C12H26 24941
1636-44-8 4-ethyldecane C12H26 24920
1638-22-8 4-butylphenol C10H14O 19983
1638-26-2 1,1-dimethylcyclopentane C7H14 11697
1639-01-6 2,3-dimethyl-2-butanethiol C6H14S 9109
1639-09-4 heptyl mercaptan C7H16S 12300
1640-89-7 ethylcyclopentane C7H14 11696
1647-08-1 4,4-dimethyl-1-hexene C8H16 14938
1647-16-1 1,9-decadiene C10H18 20609
1647-23-0 1-bromo-3,3-dimethylbutane C6H13Br 8660
1649-08-7 1,2-dichloro-1,1-difluoroethane C2H2Cl2F2 622
1653-17-4 1,1,3,3-tetrachloropropane C3H4Cl4 1569
1653-30-1 2-undecanol C11H24O 23056
1653-31-2 2-tridecanol C13H28O 26536
1653-32-3 3-tetradecanol C14H30O 27924
1653-34-5 2-pentadecanol C15H32O 28913
1653-40-3 6-methyl-1-heptanol C8H18O 15440
1663-39-4 tert-butyl acrylate C7H12O2 11488
1674-10-8 1,2-dimethylcyclohexene C8H14 14469
1674-33-5 1,2-dichloropentane C5H10Cl2 5331
1678-80-4 1,cis-2,cis-4-trimethylcyclohexane C9H18 17933
1678-81-5 1,trans-2,cis-3-trimethylcyclohexane C9H18 17931
1678-82-6 trans-1-methyl-4-isopropylcyclohexane C10H20 20935
1678-91-7 ethylcyclohexane C8H16 14877
1678-92-8 propylcyclohexane C9H18 17916
1678-93-9 butylcyclohexane C10H20 20923
1678-97-3 1,2,3-trimethylcyclohexane C9H18 17928
1678-98-4 isobutylcyclohexane C10H20 20926
1679-06-7 2-hexanethiol C6H14S 9108
1679-07-8 cyclopentanethiol C5H10S 5597
1679-08-9 2,2-dimethyl-1-propanethiol C5H12S 5928
1679-09-0 2-methyl-2-butanethiol C5H12S 5926
1679-36-3 3-hexyn-2-one C6H8O 7395
1680-58-6 1-sec-butylnaphthalene C14H16 27269
1687-61-2 3-methyl-6-ethylphenol C9H12O 17113
1687-64-5 2-methyl-6-ethylphenol C9H12O 17110
1687-65-6 2-methyl-5-ethylphenol C9H12O 17109
1689-78-7 2-(1,1-dimethylethyl)thiophene C8H12S 14412
1689-79-8 3-(1,1-dimethylethyl)thiophene C8H12S 14413
1693-86-3 3-hexylthiophene C10H16S 20540
1700-10-3 1,3-cyclooctadiene C8H12 14247
1708-29-8 2,5-dihydrofuran C4H6O 2942
1712-64-7 isopropyl nitrate C3H7NO3 2062

1717-00-6 1,1-dichloro-1-fluoroethane C2H3Cl2F 714
1721-89-7 2,3-dimethylquinoline C11H11N 21705
1721-93-3 1-methylisoquinoline C10H9N 18856
1726-14-3 1,1-diphenyl-1-butene C16H16 29232
1730-37-6 1-methylfluorene C14H12 26871
1730-91-2 (S)-(+)-2-methylbutyric acid C5H10O2 5530
1731-81-3 undecyl acetate C13H26O2 26495
1731-84-6 methyl nonanoate C10H20O2 21063
1740-97-2 2-methyl-4-isopropylphenol C10H14O 20004
1741-83-9 methyl pentyl sulfide C6H14S 9114
1746-01-6 2,3,7,8-tetrachloro-dibenzo-p-dioxin C12H4Cl4O2 23153
1746-81-2 3-(4-chlorophenyl)-1-methoxy-1-methylurea C9H11ClN2O2 16855
1754-62-7 methyl cinnamate C10H10O2 19119
1758-85-6 1-methyl-2,4-diethylbenzene C11H16 22343
1758-88-9 2-ethyl-p-xylene C10H14 19886
1759-53-1 cyclopropanecarboxylic acid C4H6O2 2978
1759-58-6 trans-1,3-dimethylcyclopentane C7H14 11703
1759-64-4 1,6-dimethylcyclohexene C8H14 14473
1759-81-5 4-methylcyclopentene C6H10 7621
1772-10-7 1-methyl-3-sec-butylbenzene C11H16 22316
1795-01-3 3-ethylthiophene C6H8S 7468
1795-04-6 2,3,4-trimethylthiophene C7H10S 11235
1795-05-7 2,3,5-trimethylthiophene C7H10S 11236
1795-09-1 2-methylthiacyclopentane C5H10S 5595
1795-15-9 octylcyclohexane C14H28 27848
1795-16-0 decylcyclohexane C16H32 29705
1795-17-1 dodecylcyclohexane C18H36 31075
1795-18-2 tetradecylcyclohexane C20H40 32474
1795-20-6 octylcyclopentane C13H26 26465
1795-21-7 decylcyclopentane C15H30 28855
1795-26-2 cis,trans-1,3,5-trimethylcyclohexane C9H18 17939
1795-27-3 cis,cis-1,3,5-trimethylcyclohexane C9H18 17938
1805-61-4 4-isopentylphenol C11H16O 22405
1806-26-4 4-octylphenol C14H22O 27687
1809-05-8 3-iodopentane C5H11I 5662
1809-10-5 3-bromopentane C5H11Br 5605
1812-51-7 2-ethylbiphenyl C14H14 27063
1825-14-5 DL-2,4-pentanediol C5H12O2 5868
1838-59-1 allyl formate C4H6O2 2972
1838-73-9 3-methyl-4-heptanol C8H18O 15452
1839-63-0 1,3,5,-trimethylcyclohexane C9H18 17937
1839-88-9 1,cis-2,cis-3-trimethylcyclohexane C9H18 17929
1842-05-3 1,1-dichloro-1,2-difluoroethane C2H2Cl2F2 620
1860-27-1 1-(1-methylethoxy)butane C7H16O 12217
1864-94-4 phenyl formate C7H6O2 10417
1871-57-4 3-chloro-2-(chloromethyl)-1-propene C4H6Cl2 2810
1873-29-6 isobutyl isocyanate C5H9NO 5203
1878-18-8 2-methyl-1-butanethiol C5H12S 5924
1888-57-9 2,5-dimethyl-3-hexanone C8H16O 15067
1888-85-3 sec-butyl vinyl ether C6H12O 8395
1897-45-6 chlorothalonil C8Cl4N2 12415
1912-24-9 atrazine C8H14ClN5 14510
1912-25-0 ipazine C10H18ClN5 20647
1912-26-1 triethazine C9H16ClN5 17640
1918-00-9 3,6-dichloro-2-methoxybenzoic acid C8H6Cl2O3 12820
1918-02-1 4-amino-3,5,6-trichloropyridinecarboxlic aci C6H3Cl3N2O2 6309
1921-70-6 2,6,10,14-tetramethylpentadecane C19H40 31734
1929-82-4 nitrapyrin C6H3Cl4N 6325
1942-45-6 4-octyne C8H14 14487
1942-46-7 5-decyne C10H18 20597
1967-25-5 (4-bromophenyl)urea C7H7BrN2O 10469
1967-27-7 3-chlorophenyl urea C7H7ClN2O 10502
1969-43-3 ethyl heptyl ether C9H20O 18336
1974-04-5 2-bromoheptane C7H15Br 11983
1975-78-6 decanenitrile C10H19N 20886
1985-57-5 (1,1-dimethylbutyl)benzene C12H18 24368
1985-97-3 (1-ethyl-1-methylpropyl)benzene C12H18 24371
2004-69-5 trans-3-penten-1-yne C5H6 4452
2004-70-8 trans-1,3-pentadiene C5H8 4742
2014-83-7 2,6-dichlorobenzyl chloride C7H5Cl3 9975
2016-42-4 tetradecylamine C14H31N 27956
2016-57-1 decylamine C10H23N 21400
2027-17-0 2-isopropylnaphthalene C13H14 25906
2027-19-2 2-propylnaphthalene C13H14 25904
2032-65-7 methiocarb C11H15NO2S 22278
2035-99-6 isoamyl caprylate C13H26O2 26507
2039-87-4 2-chlorostyrene C8H7Cl 12985
2039-89-6 2-ethenyl-1,4-dimethylbenzene C10H12 19378
2039-90-9 2-ethenyl-1,3-dimethylbenzene C10H12 19377
2039-93-2 (1-methylenepropyl)benzene C10H12 19387
2040-95-1 butylcyclopentane C9H18 17859
2040-96-2 propylcyclopentane C8H16 14861
2042-14-0 4-methyl-3-nitrophenol C7H7NO3 10668
2049-94-7 1-phenyl-3-methylbutane C11H16 22308
2049-95-8 2-phenyl-2-methylbutane C11H16 22309
2050-00-2 1,1-dimethylpropyl butanoate C9H18O2 18067
2050-01-3 isopentyl 2-methylpropanoate C9H18O2 18070
2050-09-1 isopentyl pentanoate C10H20O2 21054
2050-24-0 1-methyl-3,5-diethylbenzene C11H16 22347
2050-67-1 3,3'-dichloro-1,1'-biphenyl C12H8Cl2 23274
2050-68-2 4,4'-dichloro-1,1'-biphenyl C12H8Cl2 23275
2050-77-3 1-iododecane C10H21I 21100
2050-92-2 dipentylamine C10H23N 21405
2051-04-9 diisopentyl disulfide C10H22S2 21395
2051-24-3 decachlorobiphenyl C12Cl10 23119
2051-25-4 1,3-decadiene C10H18 20608
2051-30-1 2,6-dimethyloctane C10H22 21144
2051-31-2 4-decanol C10H22O 21229
2051-32-3 3-ethyl-3-octanol C10H22O 21265
2051-33-4 5-methyl-2-isopropyl-1-hexanol C10H22O 21306
2051-53-8 2-methyl-5-isopropylaniline C10H15N 20208
2051-60-7 2-chlorobiphenyl C12H9Cl 23365
2051-61-8 3-chlorobiphenyl C12H9Cl 23366
2051-62-9 4-chlorobiphenyl C12H9Cl 23367
2051-78-7 allyl butanoate C7H12O2 11454
2077-29-4 cis-1-methyl-4-(1-propenyl)benzene C10H12 19367
2077-30-7 trans-1-methyl-4-(1-propenyl)benzene C10H12 19368
2077-31-8 cis-1-methyl-3-(1-propenyl)benzene C10H12 19366
2077-33-0 cis-1-methyl-2-(1-propenyl)benzene C10H12 19365
2077-46-5 2,3,6-trichlorotoluene C7H5Cl3 9978
2079-95-0 1-tetradecanethiol C14H30S 27946
2084-18-6 3-methyl-2-butanethiol C5H12S 5927
2084-19-7 2-pentanethiol C5H12S 5922
2085-66-7 tert-amyl-tert-butylamine C9H21N 18394
2094-75-9 2,2-dimethylbutanal C6H12O 8406
2104-64-5 ethyl p-nitrophenyl benzenethiophosphate C14H14NO4PS 27096
2109-98-0 2,3-dimethylbutanal C6H12O 8407
2121-69-9 meso-2,4-dimethyl-1,5-pentanediol C7H16O2 12252
2131-18-2 pentadecylbenzene C21H36 32998
2131-38-6 1,3,7-trimethylnaphthalene C13H14 25929
2131-43-3 1,2,5,6-tetramethylnaphthalene C14H16 27304
2136-99-4 2,2',3,3',5,5',6,6'-octachlorobiphenyl C12H2Cl8 23145
2141-58-4 cis-dicyano-1-butene C6H6N2 6979
2141-59-5 trans-dicyano-1-butene C6H6N2 6980
2146-38-5 1-ethylcyclopentene C7H12 11323
2150-18-7 2-ethyl-4-methylpyridine C8H11N 14097
2157-31-5 2,2-dimethyl-1,3-pentanediol C7H16O2 12247
2162-91-6 diethylpentylamine C9H21N 18392
2162-92-7 1,2-dichlorohexane C6H12Cl2 8274
2162-98-3 1,10-dichlorodecane C10H20Cl2 20963
2162-99-4 1,8-dichlorooctane C8H16Cl2 15007
2163-00-0 1,6-dichlorohexane C6H12Cl2 8275
2163-42-0 2-methyl-1,3-propanediol C4H10O2 3905
2164-17-2 N,N-dimethyl-N'-[3-(trifluoromethyl)phenyl] C10H11F3N2O 19275
2173-56-0 pentyl pentanoate C10H20O2 21055
2179-59-1 allyl propyl disulfide C6H12S2 8639
2179-60-4 methyl propyl disulfide C4H10S2 3954
2189-60-8 octylbenzene C14H22 27644
2190-48-9 3-chloro-3-methyl-1-butene C5H9Cl 5046
2198-23-4 4-nonene; (cis+trans) C9H18 17945
2198-72-3 4-propyl-4-heptanol C10H22O 21296
2203-80-7 5-methyl-1-hexyne C7H12 11345
2207-01-4 cis-1,2-dimethylcyclohexane C8H16 14880
2207-03-6 trans-1,3-dimethylcyclohexane C8H16 14884
2207-04-7 trans-1,4-dimethylcyclohexane C8H16 14887
2210-28-8 propyl methacrylate C7H12O2 11452
2211-67-8 DL-2,3-dichlorobutane C4H8Cl2 3365
2211-68-9 2-chloro-trans-2-butene C4H7Cl 3106
2211-69-0 2-chloro-cis-2-butene C4H7Cl 3105
2211-70-3 2-chloro-1-butene C4H7Cl 3100
2213-23-2 2,4-dimethylheptane C9H20 18184
2213-32-3 2,4-dimethyl-1-pentene C7H14 11732
2216-30-0 2,5-dimethylheptane C9H20 18185
2216-32-2 4-ethylheptane C9H20 18181
2216-33-3 3-methyloctane C9H20 18178
2216-34-4 4-methyloctane C9H20 18179
2216-36-6 2-chlorononane C9H19Cl 18122
2216-70-8 7-methyl-2,4-octadiene C9H16 17608
2216-81-1 heptyl propanoate C10H20O2 21049
2216-87-7 3-undecanone C11H22O 22824
2217-17-6 2,4,5,7-tetramethyloctane C12H26 25097
2217-45-0 2,3,4,5-tetramethylaniline C10H15N 20210
2219-73-0 2-methyl-4-ethylphenol C9H12O 17108
2219-82-1 2-tert-butyl-6-methylphenol C11H16O 22403
2233-29-6 2,3,4-trimethylpyridine C8H11N 14106
2234-13-1 octachloronaphthalene C10Cl8 18455
2234-20-0 1-ethenyl-2,4-dimethylbenzene C10H12 19375
2234-75-5 1,2,4-trimethylcyclohexane, isomers C9H18 17932
2238-07-5 diglycidyl ether C6H10O3 7884
2243-27-8 nonanenitrile C9H17N 17813
2243-98-3 1-undecyne C11H20 22672
2244-07-7 undecanenitrile C11H21N 22771
2258-58-4 methyl 3-methyl-3-pentenoate C7H12O2 11485
2294-72-6 5-undecyne C11H20 22676
2308-38-5 tert-butyl butanoate C8H16O2 15179
2311-14-0 1,1,1-tribromoethane C2H3Br3 697
2311-59-3 isopropyl decanoate C13H26O2 26504
2313-61-3 4-methyl-2-hexanol C7H16O 12174
2313-65-7 3-methyl-2-hexanol C7H16O 12173
2315-43-7 dimethyl-3-methylbutylamine C7H17N 12328
2319-61-1 1,1-dicyclohexylethane C14H26 27777
2327-02-8 3,4-dichlorophenyl urea C7H6Cl2N2O 10261
2345-27-9 2-tetradecanone C14H28O 27863
2345-28-0 2-pentadecanone C15H30O 28874
2346-81-8 3-chlorohexane C6H13Cl 8671
2348-82-5 2-methoxy-1,4-naphthalenedione C11H8O3 21543
2349-07-7 n-hexyl isobutyrate C10H20O2 21071
2350-19-8 2-chloro-6-methylheptane C8H17Cl 15256
2350-24-5 2-ethylhexyl-6-chloride C8H17Cl 15264
2358-38-5 1,1-difluorobutane C4H8F2 3389
2366-36-1 1,1,1-trichloro-2-fluoroethane C2H2Cl3F 630
2366-52-1 1-fluorobutane C4H9F 3657
2370-12-9 2,2-dimethyl-1-pentanol C7H16O 12182
2370-13-0 2,2-dimethyl-1-hexanol C8H18O 15457
2370-14-1 2,2-dimethyl-1-octanol C10H22O 21254
2371-19-9 3-methyl-2-heptanone C8H16O 15075
2384-70-5 2-decyne C10H18 20594
2384-85-2 3-decyne C10H18 20595
2384-86-3 4-decyne C10H18 20596
2384-90-9 1,2-heptadiene C7H12 11358
2385-85-5 mirex C10Cl12 18459
2396-78-3 methyl 3-hexenoate C7H12O2 11483
2398-09-6 1,cis-3-dimethylcyclobutane C6H12 8211
2398-10-9 1,trans-3-dimethylcyclobutane C6H12 8212
2400-01-3 (1-hexylheptyl)benzene C19H32 31647
2402-06-4 1,trans-2-dimethylcyclopropane C5H10 5282
2408-20-0 allyl propanoate C6H10O2 7818
2409-55-4 2-tert-butyl-4-methylphenol C11H16O 22401
2415-72-7 propylcyclopropane C6H12 8206
2416-94-6 2,3,6-trimethylphenol C9H12O 17118
2423-01-0 1-butylcyclopentene C9H16 17590
2424-98-8 vinyl 2-methylpropanoate C6H10O2 7834
2425-06-1 captafol C10H9Cl4NO2S 18832
2425-54-9 1-chlorotetradecane C14H29Cl 27897
2425-74-3 tert-butylformamide C5H11NO 5675
2425-77-6 2-hexyl-1-decanol C16H34O 29773
2426-08-6 butyl glycidyl ether C7H14O2 11915
2430-22-0 7-methyl-1-octanol C9H20O 18249
2430-73-1 1,4-nonanediol C9H20O2 18341
2436-84-2 2,6-dimethyl-1,3-heptadiene C9H16 17610
2436-90-0 3,7-dimethyl-1,6-octadiene C10H18 20637
2436-92-2 3,4-dimethylquinoline C11H11N 21710
2437-25-4 dodecanenitrile C12H23N 24805
2437-56-1 1-tridecene C13H26 26469
2437-95-8 (±)-a-pinene C10H16 20367
2438-20-2 2-methylbutyl propanoate C8H16O2 15169
2439-54-5 methyloctylamine C9H21N 18389
2443-03-0 1-methyl-trans-3-propylcyclopentane C9H18 17867
2443-04-1 1-methyl-cis-3-propylcyclopentane C9H18 17866
2443-89-2 1,1-diiodo-2,2-dimethylpropane C5H10I2 5399
2443-91-6 1,1-dibromo-2,2-dimethylpropane C5H10Br2 5315
2445-76-3 hexyl propanoate C9H18O2 18061
2451-62-9 glycidyl isocyanurate C12H15N3O6 24103
2452-99-5 1,2-dimethylcyclopentane C7H14 11698
2453-00-1 1,3-dimethylcyclopentane C7H14 11701
2456-27-1 dinonyl ether C18H38O 31143
2470-68-0 diheptylamine C14H31N 27959
2473-01-0 1-chlorononane C9H19Cl 18117
2473-03-2 1-chloroundecane C11H23Cl 22872
2492-22-0 cis-2,6-dimethyl-2,6-octadiene C10H18 20615
2517-43-3 3-methoxy-1-butanol C5H12O2 5884
2528-36-1 dibutylphenyl phosphate C14H23O4P 27742
2531-84-2 2-methylphenanthrene C15H12 28135
2532-58-3 cis-1,3-dimethylcyclopentane C7H14 11702
2532-64-1 1,cis-2,trans-3,trans-4-tetramethylcyclopentane C9H18 17912
2532-65-2 1,cis-2,cis-3,cis-4-tetramethylcyclopentane C9H18 17909
2532-67-4 1,trans-2,cis-3,trans-4-tetramethylcyclopentane C9H18 17913
2532-68-5 1,cis-2,trans-3,cis-4-tetramethylcyclopentane C9H18 17911
2532-69-6 1,cis-2,cis-3,trans-4-tetramethylcyclopentane C9H18 17910
2547-26-4 4a-methyl-cis-decahydronaphthalene C11H20 22669
2547-27-5 4a-methyl-trans-decahydronaphthalene C11H20 22670
2550-21-2 3-methyl-2-hexanone C7H14O 11832
2550-28-9 5-hexyn-2-one C6H8O 7396
2568-33-4 3-methyl-1,3-butanediol C5H12O2 5875

2570-26-5	pentadecylamine	C15H33N	28931
2581-34-2	3-methyl-4-nitrophenol	C7H7NO3	10669
2586-89-2	3-heptyne	C7H12	11342
2612-27-3	2-isopropyl-1,3-propanediol	C6H14O2	9055
2612-28-4	2-propyl-1,3-propanediol	C6H14O2	9054
2612-29-5	2-ethyl-1,3-propanediol	C5H12O2	5876
2613-61-8	2,4,6-trimethylheptane	C10H22	21172
2613-65-2	1-methyl-trans-3-ethylcyclopentane	C8H16	14867
2613-66-3	1-methyl-cis-3-ethylcyclopentane	C8H16	14866
2613-69-6	1,cis-2,cis-3-trimethylcyclopentane	C8H16	14870
2613-72-1	1,cis-2,cis-4-trimethylcyclopentane	C8H16	14873
2625-29-8	1-nitro-2,2-dimethylbutane	C6H13NO2	8801
2625-35-6	2-nitro-3-methylbutane	C5H11NO2	5702
2639-63-6	hexyl butanoate	C10H20O2	21050
2639-64-7	nonyl butanoate	C13H26O2	26497
2642-98-0	6-chrysenamine	C18H13N	30445
2668-24-8	4,5,6-trichloroguaiacol	C7H5Cl3O2	9987
2679-87-0	ethyl sec-butyl ether	C6H14O	9006
2690-08-6	dioctyl sulfide	C16H34S	29791
2698-41-1	o-chlorobenzylidene malononitrile	C10H5ClN2	18498
2717-39-7	1,4,5,8-tetramethylnaphthalene	C14H16	27309
2719-52-0	2-phenylpentane	C11H16	22305
2721-38-2	1,1-diethylcyclopentane	C9H18	17873
2724-59-6	17-methyloctadecanoic acid	C19H38O2	31710
2738-18-3	2,6-dimethylhept-3-ene	C9H18	17951
2738-19-4	2-methyl-2-hexene	C7H14	11718
2749-11-3	s(+)-2-amino-1-propanol	C3H9NO	2210
2764-72-9	diquat	C12H12N2	23729
2765-11-9	pentadecanal	C15H30O	28873
2765-18-6	1-propylnaphthalene	C13H14	25903
2768-16-3	5-methyl-1-nonanol	C10H22O	21233
2769-64-4	butyl isocyanide	C5H9N	5187
2787-43-1	3-methyl-1,trans-3-pentadiene	C6H10	7641
2787-45-3	3-methyl-1,cis-3-pentadiene	C6H10	7640
2799-16-8	(R)-(-)-1-amino-2-propanol	C3H9NO	2211
2799-17-9	(S)-(+)-1-amino-2-propanol	C3H9NO	2212
2801-84-5	2,4-dimethyldecane	C12H26	24924
2807-30-9	ethylene glycol monopropyl ether	C5H12O2	5857
2808-71-1	3-ethylcyclohexene	C8H14	14467
2808-76-6	1,3-dimethylcyclohexene	C8H14	14470
2808-77-7	1,5-dimethylcyclohexene	C8H14	14472
2808-79-9	1,4-dimethylcyclohexene	C8H14	14471
2809-67-8	2-octyne	C8H14	14485
2819-86-5	pentamethylphenol	C11H16O	22407
2825-79-8	trans-2-octadecenoic acid	C18H34O2	31032
2847-72-5	4-methyldecane	C11H24	22893
2851-83-4	ethyl dodecyl sulfide	C14H30S	27950
2854-16-2	1-amino-2-methyl-2-propanol	C4H11NO	4005
2855-08-5	1-chloro-3,3-dimethylbutane	C6H13Cl	8683
2867-59-6	3-amino-1-butanol	C4H11NO	3989
2869-34-3	tridecylamine	C13H29N	26555
2870-04-4	2-ethyl-m-xylene	C10H14	19883
2874-74-0	2-methyldodecanoic acid	C13H26O2	26493
2876-35-9	2-tert-butylnaphthalene	C14H16	27274
2882-96-4	3-methylpentadecane	C16H34	29756
2882-97-5	2,3-dimethylpentadecane	C17H36	30350
2882-98-6	nonylcyclopentane	C14H28	27847
2883-02-5	nonylcyclohexane	C15H30	28856
2884-06-2	2,3-dimethylnonane	C11H24	22899
2885-00-9	1-octadecanethiol	C18H38S	31149
2917-26-2	1-hexadecanethiol	C16H34S	29789
2921-88-2	chlorpyrifos	C9H11Cl3NO3PS	16876
2935-44-6	2,5-hexanediol	C6H14O2	9030
2938-98-9	2-methyl-1,4-butanediol	C5H12O2	5872
2971-90-6	clopidol	C7H7Cl2NO	10547
2974-92-7	3,4-dichloro-1,1'-biphenyl	C12H8Cl2	23277
2980-69-0	4-methylundecane	C12H26	24916
2980-71-4	2-methyl-1-nonene	C10H20	20947
2983-37-1	ethyl 2-ethylhexanoate	C10H20O2	21059
2987-16-8	3,3-dimethylbutanal	C6H12O	8408
3001-66-9	2-octanethiol	C8H18S	15603
3004-93-1	2-methyloctanoic acid	C9H18O2	18057
3017-68-3	2-bromo-cis-2-butene	C4H7Br	3054
3017-69-4	1-bromo-2-methyl-1-propene	C4H7Br	3056
3017-71-8	2-bromo-trans-2-butene	C4H7Br	3055
3019-20-3	isopropyl phenyl sulfide	C9H12S	17214
3031-05-8	1,2,6-trimethylnaphthalene	C13H14	25924
3031-15-0	1,2,3,4-tetramethylnaphthalene	C14H16	27299
3031-16-1	1,2,5,8-tetramethylnaphthalene	C14H16	27305
3042-50-0	1,4-dimethyl-2-propylbenzene	C11H16	22335
3043-33-2	2-methyl-2,3-pentadiene	C6H10	7645
3054-92-0	2,3,4-trimethyl-3-pentanol	C8H18O	15515
3055-14-9	1-methyl-3,5-diisopropylbenzene	C13H20	26285
3060-89-7	metobromuron	C9H11BrN2O2	16834
3070-53-9	1,6-heptadiene	C7H12	11361
3073-66-3	1,1,3-trimethylcyclohexane	C9H18	17926
3073-92-5	butyl propyl ether	C7H16O	12212
3074-61-1	1-propylcyclopentene	C8H14	14463
3074-64-4	2,3-dimethyl-2-heptene	C9H18	17950
3074-71-3	2,3-dimethylheptane	C9H20	18183
3074-75-7	4-ethyl-2-methylhexane	C9H20	18191
3074-76-8	3-ethyl-3-methylhexane	C9H20	18192
3074-77-9	3-ethyl-4-methylhexane	C9H20	18193
3121-79-7	4,4-dimethyl-1-pentanol	C7H16O	12187
3121-82-2	2,2-dimethyl-1,5-pentanediol	C7H16O2	12248
3162-62-7	1,1-dichlorodecane	C10H20Cl2	20962
3173-53-3	cyclohexyl isocyanate	C7H11NO	11275
3174-67-2	1,3-pentanediol	C5H12O2	5861
3175-23-3	1,2,2-trichloropropane	C3H5Cl3	1706
3178-22-1	tert-butylcyclohexane	C10H20	20924
3178-23-2	dicyclohexylmethane	C13H24	26405
3178-29-8	4-propylheptane	C10H22	21152
3180-09-4	2-butylphenol	C10H14O	20037
3195-86-6	2-ethylthiacyclopropane	C4H8S	3583
3208-16-0	2-ethylfuran	C6H8O	7381
3221-61-2	2-methyloctane	C9H20	18177
3228-01-1	3-methyl-2-isopropylphenol	C10H14O	20007
3228-02-2	3-methyl-4-isopropylphenol	C10H14O	20008
3228-03-3	3-methyl-5-isopropylphenol	C10H14O	20009
3228-04-4	2-methyl-6-isopropylphenol	C10H14O	20006
3234-49-9	1,2-dibromopentane	C5H10Br2	5298
3274-28-0	4-octanecarboxylic acid	C9H18O2	18087
3274-29-1	2-ethylheptanoic acid	C9H18O2	18053
3290-06-0	1,3-dichloro-5-(chloromethyl)benzene	C7H5Cl3	9969
3302-10-1	3,5,5-trimethylhexanoic acid	C9H18O2	18082
3302-16-7	3,3-dimethylbutanenitrile	C6H11N	8095
3333-52-6	tetramethyl succinonitrile	C8H12N2	14286
3344-70-5	1,12-dibromododecane	C12H24Br2	24832
3360-16-5	N-isopropyl-1,3-propanediamine	C6H16N2	9299
3377-86-4	2-bromohexane	C6H13Br	8648
3377-87-5	3-bromohexane	C6H13Br	8649
3383-96-8	abate	C16H20O6P2S3	29470
3404-57-7	6-methyl-3-heptene	C8H16	14914
3404-58-8	3-ethyl-1-hexene	C8H16	14930
3404-61-3	3-methyl-1-hexene	C7H14	11715
3404-63-5	2-ethyl-1,3-butadiene	C6H10	7646
3404-67-9	3-methyl-2-ethyl-1-pentene	C8H16	14972
3404-71-5	2-ethyl-1-pentene	C7H14	11729
3404-72-6	2,3-dimethyl-1-pentene	C7H14	11731
3404-73-7	3,3-dimethyl-1-pentene	C7H14	11733
3404-77-1	3,3-dimethyl-1-hexene	C8H16	14935
3404-78-2	2,5-dimethyl-2-hexene	C8H16	14947
3404-80-6	4-methyl-2-ethyl-1-pentene	C8H16	14973
3405-42-3	methyldipropylamine	C7H17N	12336
3405-45-6	N-methyldibutylamine	C9H21N	18395
3452-07-1	1-eicosene	C20H40	32476
3452-09-3	1-nonyne	C9H16	17593
3452-97-9	3,5,5-trimethyl-1-hexanol	C9H20O	18300
3454-07-7	p-ethylstyrene	C10H12	19374
3457-57-6	1-nitro-2-methylbutane	C5H11NO2	5699
3495-46-3	3-amino-3-methylpentane	C6H15N	9183
3522-86-9	3-sec-butylphenol	C10H14O	19988
3522-94-9	2,2,5-trimethylhexane	C9H20	18196
3524-73-0	5-methyl-1-hexene	C7H14	11717
3525-25-5	3-ethyl-1-heptanol	C9H20O	18266
3581-69-9	2,2-dimethylpropyl propanoate	C8H16O2	15174
3586-58-1	2-methylenebutanoic acid	C5H8O2	4890
3597-91-9	4-biphenylmethanol	C13H12O	25803
3599-58-4	2,4-D dimethylammonium	C10H13Cl2NO4	19745
3675-69-2	1,2,2-tribromobutane	C4H7Br3	3087
3683-19-0	4-methyl-cis-2-hexene	C7H14	11721
3683-22-5	4-methyl-trans-2-hexene	C7H14	11722
3685-00-5	3,trans-6-dimethylcyclohexene	C8H14	14480
3689-24-5	sulfotep	C8H20O5P2S2	15724
3698-93-9	propyl octyl sulfide	C11H24S	23088
3698-94-0	ethyl octyl sulfide	C10H22S	21389
3698-95-1	methyl octyl sulfide	C9H20S	18367
3710-30-3	1,7-octadiene	C8H14	14493
3710-31-4	1,2-heptanediol	C7H16O2	12224
3710-43-8	2,4-dimethylfuran	C6H8O	7384
3724-55-8	methyl 3-butenoate	C5H8O2	4918
3724-61-6	dodecyl butanoate	C16H32O2	29726
3724-65-0	crotonic acid	C4H6O2	2981
3726-47-4	1-methyl-3-ethylcyclopentane	C8H16	14876
3728-54-9	1-ethyl-2-methylcyclohexane	C9H18	17940
3730-60-7	2,5-dimethyl-2-hexanol	C8H18O	15471
3741-00-2	pentylcyclopentane	C10H20	20922
3742-38-9	4-ethylcyclopentene	C7H12	11325
3742-42-5	4-ethylcyclohexene	C8H14	14468
3756-30-7	3-iodo-2-methyl-1-propene	C4H7I	3219
3769-23-1	4-methyl-1-hexene	C7H14	11716
3772-13-2	2,2-dimethylthiacyclopropane	C4H8S	3584
3780-58-3	3-methylhexanoic acid	C7H14O2	11888
3788-32-7	isobutylcyclopentane	C9H18	17860
3806-59-5	cis,cis-1,3-cyclooctadiene	C8H12	14248
3806-60-8	(Z,E)-1,3-cyclooctadiene	C8H12	14252
3814-34-4	1-bromo-2-ethylbutane	C6H13Br	8663
3825-26-1	ammonium perfluorooctanoate	C8H4F15NO2	12570
3850-30-4	2-amino-3,3-dimethyl butane	C6H15N	9188
3855-26-3	4-methyl-2-ethylphenol	C9H12O	17114
3875-51-2	isopropylcyclopentane	C8H16	14862
3875-52-3	tert-butylcyclopentane	C9H18	17862
3877-15-4	methyl propyl sulfide	C4H10S	3948
3877-19-8	2-methyl-[1,2,3,4-tetrahydronaphthalene]	C11H14	22026
3878-19-1	2-(2-furyl)benzimidazole	C11H8N2O	21520
3899-36-3	3-methyl-trans-3-hexene	C7H14	11728
3913-02-8	2-butyl-1-octanol	C12H26O	25284
3922-26-7	1,1,1-trichloroheptane	C7H13Cl3	11612
3922-27-8	1,1,1-trichloropentane	C5H9Cl3	5100
3922-28-9	1,12-dichlorododecane	C12H24Cl2	24834
3937-56-2	1,9-nonanediol	C9H20O2	18342
3938-95-2	ethyl 2,2-dimethylpropanoate	C7H14O2	11910
3944-36-3	1-(1-methylethoxy)-2-propanol	C6H14O2	9057
3950-21-8	meso-2,4-pentanediol	C5H12O2	5869
3964-56-5	4-bromo-2-chlorophenol	C6H4BrClO	6387
3965-59-1	2,4-dimethyl-1-hexanol	C8H18O	15459
3968-85-2	1-phenyl-2-methylbutane	C11H16	22307
3970-59-0	2,4-dimethyl-3-ethyl-3-pentanol	C9H20O	18333
3970-60-3	2,2,5-trimethyl-3-hexanol	C9H20O	18314
3970-62-5	2,2-dimethyl-3-pentanol	C7H16O	12195
3982-64-7	1,3-dimethyl-5-propylbenzene	C11H16	22334
3982-66-9	1,2-dimethyl-4-propylbenzene	C11H16	22331
3982-67-0	1,3,5-trimethyl-2-ethylbenzene	C11H16	22353
4001-61-0	2,5-dimethylthiophenol	C8H10S	14044
4016-14-2	isopropyl glycidyl ether	C6H12O2	8496
4028-56-2	meso-2,3-dichlorobutane	C4H8Cl2	3366
4032-86-4	3,3-dimethylheptane	C9H20	18210
4032-92-2	2,4,4-trimethylheptane	C10H22	21170
4032-93-3	2,3,6-trimethylheptane	C10H22	21169
4032-94-4	2,4-dimethyloctane	C10H22	21142
4038-04-4	3-ethyl-1-pentene	C7H14	11730
4049-81-4	2-methyl-1,5-hexadiene	C7H12	11363
4050-45-7	trans-2-hexene	C6H12	8219
4057-42-5	2,6-dimethyl-2-octene	C10H20	20951
4074-43-5	3-butylphenol	C10H14O	19982
4088-36-2	ethyloctylamine	C10H23N	21404
4088-37-3	diethyloctylamine	C12H27N	25345
4095-22-1	2-isopropylthiophene	C7H10S	11227
4096-34-8	3-cyclohexen-1-one	C6H8O	7388
4098-71-9	isophorone diisocyanate	C12H18N2O2	24412
4101-68-2	1,10-dibromodecane	C10H20Br2	20959
4110-44-5	3,3-dimethyloctane	C10H22	21146
4110-50-3	ethyl propyl sulfide	C5H12S	5931
4127-45-1	1,1,2-trimethylcyclopropane	C6H12	8203
4128-31-8	2-octanol, (±)	C8H18O	15432
4132-72-3	1,4-dimethyl-2-isopropylbenzene	C11H16	22341
4132-77-8	1,2-dimethyl-4-isopropylbenzene	C11H16	22337
4166-46-5	2,3-dimethyl-3-hexanol	C8H18O	15481
4167-74-2	4-isobutylphenol	C10H14O	19986
4167-75-3	2-isobutylphenol	C10H14O	19984
4168-40-5	1,1-dichlorotetradecane	C14H28Cl2	27856
4170-30-3	crotonaldehyde	C4H6O	2956
4209-90-9	2,2-dimethyl-3-hexanol	C8H18O	15480
4209-91-0	3,5-dimethyl-3-hexanol	C8H18O	15485
4215-86-5	(4-methylpentyl)benzene	C12H18	24374
4218-48-8	1-ethyl-4-isopropylbenzene	C11H16	22329
4221-99-2	(S)-(+)-2-butanol	C4H10O	3873
4253-89-8	diisopropyl disulfide	C6H14S2	9129
4254-14-2	(R)-(-)-1,2-propanediol	C3H8O2	2143
4254-15-3	(S)-(+)-1,2-propanediol	C3H8O2	2144
4258-93-9	1-methyl-1-propylcyclohexane	C10H20	20927
4259-00-1	1,1,2-trimethylcyclopentane	C8H16	14868
4265-25-2	2-methylbenzofuran	C9H8O	16185
4271-27-6	diethyldodecylamine	C16H35N	29804
4279-22-5	2,2-dichlorobutane	C4H8Cl2	3364
4282-40-0	1-iodoheptane	C7H15I	12020
4282-42-2	1-iodononane	C9H19I	18129
4282-44-4	1-iodoundecane	C11H23I	22876
4283-80-1	2-bromo-2-methylpentane	C6H13Br	8653
4283-83-4	3-bromo-2-methylpentane	C6H13Br	8656
4291-79-6	1-methyl-2-propylcyclohexane	C10H20	20928
4291-80-9	1-methyl-3-propylcyclohexane	C10H20	20929
4291-81-0	1-methyl-4-propylcyclohexane	C10H20	20930
4291-98-9	1-pentylcyclopentene	C10H18	20603

4291-99-0 1-hexylcyclopentene $C_{11}H_{20}$ 22671
4292-00-6 1-heptylcyclopentene $C_{12}H_{22}$ 24666
4292-19-7 1-iodododecane $C_{12}H_{25}I$ 24900
4292-75-5 hexylcyclohexane $C_{12}H_{24}$ 24819
4292-92-6 pentylcyclohexane $C_{11}H_{22}$ 22792
4325-48-8 2-chloro-2-methylpentane $C_6H_{13}Cl$ 8675
4325-49-9 2-chloro-2-methylheptane $C_8H_{17}Cl$ 15255
4340-76-5 2-eicosanol $C_{20}H_{42}O$ 32523
4358-59-2 methyl cis-2-butenoate $C_5H_8O_2$ 4886
4358-75-2 2-amino-2,3-dimethylbutane $C_6H_{15}N$ 9187
4360-47-8 cinnamonitrile C_9H_7N 16017
4371-48-6 2-methyl-3-isopropylphenol $C_{10}H_{14}O$ 20003
4375-96-6 2-iodo-1-propene C_3H_5I 1736
4385-04-0 dimethylhexylamine $C_8H_{19}N$ 15641
4390-04-9 2,2,4,4,6,8,8-heptamethylnonane $C_{16}H_{34}$ 29760
4398-65-6 2-chloro-2-methylhexane $C_7H_{15}Cl$ 11998
4427-56-9 4-methyl-2-isopropylphenol $C_{10}H_{14}O$ 20010
4432-77-3 ethyl-tert-butylamine $C_6H_{15}N$ 9201
4439-24-1 2-(2-methylpropoxy)ethanol $C_6H_{14}O_2$ 9058
4444-68-2 diethylbutylamine $C_8H_{19}N$ 15642
4455-26-9 methyldioctylamine $C_{17}H_{37}N$ 30374
4457-00-5 hexylcyclopentane $C_{11}H_{22}$ 22791
4457-71-0 3-methyl-1,5-pentanediol $C_6H_{14}O_2$ 9037
4457-72-1 1,5-dibromo-3-methylpentane $C_6H_{12}Br_2$ 8258
4458-31-5 diethylpropylamine $C_7H_{17}N$ 12339
4458-32-6 methylethylpropylamine $C_6H_{15}N$ 9206
4461-48-7 4-methyl-2-pentene C_6H_{12} 8226
4468-42-2 (1-ethylbutyl)benzene $C_{12}H_{18}$ 24370
4481-30-5 2-phenyl-3-methylbutane $C_{11}H_{16}$ 22310
4485-09-0 4-nonanone $C_9H_{18}O$ 18009
4485-13-6 4-propyl-3-heptene $C_{10}H_{20}$ 20952
4485-16-9 4-methyl-3-heptene C_8H_{16} 14921
4485-77-2 dinonyl disulfide $C_{18}H_{38}S_2$ 31157
4500-58-7 o-ethylthiophenol $C_8H_{10}S$ 14039
4516-69-2 1,1,3-trimethylcyclopentane C_8H_{16} 14869
4536-23-6 2-methylhexanoic acid $C_7H_{14}O_2$ 11925
4537-13-7 (1-methylnonyl)benzene $C_{16}H_{26}$ 29599
4543-96-8 N,N,N'-trimethyl-1,3-propanediamine $C_6H_{16}N_2$ 9301
4549-31-9 1,7-dibromoheptane $C_7H_{14}Br_2$ 11754
4549-32-0 1,8-dibromooctane $C_8H_{16}Br_2$ 14998
4549-33-1 1,9-dibromononane $C_9H_{18}Br_2$ 17957
4549-74-0 3-methyl-1,3-pentadiene C_6H_{10} 7637
4553-62-2 methylglutaronitrile $C_6H_8N_2$ 7317
4588-18-5 2-methyl-1-octene C_9H_{18} 17946
4609-89-6 2-nitropentane $C_5H_{11}NO_2$ 5697
4614-03-3 2-sec-butylnaphthalene $C_{14}H_{16}$ 27270
4628-21-1 1-chloro-cis-2-butene C_4H_7Cl 3103
4655-34-9 isopropyl methacrylate $C_7H_{12}O_2$ 11453
4683-94-7 2-methyl-trans-decahydronaphthalene $C_{11}H_{20}$ 22668
4685-14-7 paraquat $C_{12}H_{14}N_2$ 23917
4706-81-4 2-tetradecanol $C_{14}H_{30}O$ 27923
4706-89-2 1,3-dimethyl-4-isopropylbenzene $C_{11}H_{16}$ 22339
4706-90-5 1,3-dimethyl-5-isopropylbenzene $C_{11}H_{16}$ 22340
4730-22-7 6-methyl-2-heptanol $C_8H_{18}O$ 15445
4737-41-1 1-chloro-2-ethylbutane $C_6H_{13}Cl$ 8680
4740-00-5 3-methylthiacyclopentane $C_5H_{10}S$ 5596
4747-07-3 hexyl methyl ether $C_7H_{16}O$ 12216
4747-21-1 methylisopropylamine $C_4H_{11}N$ 3983
4753-80-4 thiacycloheptane $C_6H_{12}S$ 8621
4780-79-4 1-naphthalenemethanol $C_{11}H_{10}O$ 21650
4786-20-3 2-butenenitrile C_4H_5N 2699
4806-58-0 1,cis-2,cis-3-trimethylcyclopropane C_6H_{12} 8204
4806-59-1 1,cis-2,trans-3-trimethylcyclopropane C_6H_{12} 8205
4806-61-5 ethylcyclobutane C_6H_{12} 8213
4810-09-7 3-methyl-1-heptene C_8H_{16} 14900
4815-57-0 1,4-dipropylbenzene $C_{12}H_{18}$ 24358
4850-28-6 1,cis-2,trans-4-trimethylcyclopentane C_8H_{16} 14874
4850-32-2 sec-butylcyclopentane C_9H_{18} 17861
4860-03-1 1-chlorohexadecane $C_{16}H_{33}Cl$ 29740
4862-03-1 1-chloropentadecane $C_{15}H_{31}Cl$ 28899
4886-77-5 3-methyl-(1-thiaethyl)-benzene $C_8H_{10}S$ 14050
4894-61-5 1-chloro-trans-2-butene C_4H_7Cl 3104
4894-62-6 3-methyl-1,5-heptadiene C_8H_{14} 14495
4894-75-1 4-phenylcyclohexanone $C_{12}H_{14}O$ 23965
4900-30-5 1,8-nonadiene C_9H_{16} 17605
4901-51-3 2,3,4,5-tetrachlorophenol $C_6H_2Cl_4O$ 6177
4905-79-7 1,1,1-trichlorooctane $C_8H_{15}Cl_3$ 14765
4911-70-0 2,3-dimethyl-2-pentanol $C_7H_{16}O$ 12189
4913-13-7 3,5,N,N-tetramethylaniline $C_{10}H_{15}N$ 20230
4914-89-0 3-methyl-cis-3-hexene C_7H_{14} 11727
4914-91-4 3,4-dimethyl-cis-2-pentene C_7H_{14} 11739
4914-92-5 3,4-dimethyl-trans-2-pentene C_7H_{14} 11740
4920-77-8 3-methyl-2-nitrophenol $C_7H_7NO_3$ 10670
4920-99-4 1-ethyl-3-isopropylbenzene $C_{11}H_{16}$ 22328
4923-77-7 1-methyl-cis-2-ethylcyclohexane C_9H_{18} 17919
4923-78-8 1-methyl-trans-2-ethylcyclohexane C_9H_{18} 17920
4926-76-5 1-methyl-trans-3-ethylcyclohexane C_9H_{18} 17922
4926-78-7 1-methyl-cis-4-ethylcyclohexane C_9H_{18} 17923
4926-90-3 1-methyl-1-ethylcyclohexane C_9H_{18} 17918
4974-27-0 2,6-octadiene C_8H_{14} 14494
4984-01-4 3,7-dimethyl-1-octene $C_{10}H_{20}$ 20950
4984-22-9 (R)-(-)-1-methoxy-2-propanol $C_4H_{10}O_2$ 3904
5008-69-5 4-methyl-2-thiapentane $C_5H_{12}S$ 5933
5008-72-0 4-methyl-3-thiahexane $C_6H_{14}S$ 9120
5008-73-1 2-methyl-3-thiahexane $C_6H_{14}S$ 9119
5026-76-6 6-methyl-1-heptene C_8H_{16} 14903
5103-71-9 a-chlordan $C_{10}H_6Cl_8$ 18553
5103-74-2 trans-chlordane $C_{10}H_6Cl_8$ 18554
5124-30-1 methylene bis(4-cyclohexylisocyanate) $C_{15}H_{22}N_2O_2$ 28658
5145-99-3 2-methyl-3-thiapentane $C_5H_{12}S$ 5934
5160-99-6 1-methyl-3-isobutylbenzene $C_{11}H_{16}$ 22319
5161-04-6 1-methyl-4-isobutylbenzene $C_{11}H_{16}$ 22320
5161-11-5 3,cis-4-dimethylthiacyclopentane $C_6H_{12}S$ 8632
5161-12-6 3,trans-4-dimethylthiacyclopentane $C_6H_{12}S$ 8633
5161-13-7 2,cis-5-dimethylthiacyclopentane $C_6H_{12}S$ 8629
5161-14-8 2,trans-5-dimethylthiacyclopentane $C_6H_{12}S$ 8630
5161-16-0 2-methylthiacyclohexane $C_6H_{12}S$ 8634
5161-17-1 4-methylthiacyclohexane $C_6H_{12}S$ 8636
5161-75-1 2,2-dimethylthiacyclopentane $C_6H_{12}S$ 8624
5161-76-2 3,3-dimethylthiacyclopentane $C_6H_{12}S$ 8631
5161-77-3 2,cis-3-dimethylthiacyclopentane $C_6H_{12}S$ 8625
5161-78-4 2,trans-3-dimethylthiacyclopentane $C_6H_{12}S$ 8626
5161-79-5 2,cis-4-dimethylthiacyclopentane $C_6H_{12}S$ 8627
5161-80-8 2,trans-4-dimethylthiacyclopentane $C_6H_{12}S$ 8628
5162-44-7 4-bromo-1-butene C_4H_7Br 3051
5162-48-1 2,2,4-trimethyl-3-pentanol $C_8H_{18}O$ 15514
5166-35-8 3-chloro-2-methyl-1-butene C_5H_9Cl 5041
5171-84-6 3,3,4,4-tetramethylhexane $C_{10}H_{22}$ 21200
5194-50-3 cis-2,trans-4-hexadiene C_6H_{10} 7632
5194-51-4 trans-2,trans-4-hexadiene C_6H_{10} 7634
5205-34-5 5-decanol $C_{10}H_{22}O$ 21230
5258-50-4 3-methylthiacyclohexane $C_6H_{12}S$ 8635
5272-02-6 3-chloro-3-methylheptane $C_8H_{17}Cl$ 15257
5277-11-2 dimethylheptylamine $C_9H_{21}N$ 18391
5329-79-3 2-aminohexane $C_6H_{15}N$ 9174
5332-52-5 1-undecanethiol $C_{11}H_{24}S$ 23084
5333-42-6 2-octyl-1-dodecanol $C_{20}H_{42}O$ 32525
5340-36-3 3-methyl-3-octanol $C_9H_{20}O$ 18255
5340-41-0 2,2,3-trimethyl-3-hexanol $C_9H_{20}O$ 18312
5340-62-5 5,5-dimethyl-3-ethyl-3-hexanol $C_{10}H_{22}O$ 21313
5341-95-7 meso-2,3-butanediol $C_4H_{10}O_2$ 3888
5343-92-0 1,2-pentanediol $C_5H_{12}O_2$ 5859
5343-96-4 1,2-dimethylpropyl acetate $C_7H_{14}O_2$ 11898
5344-20-7 1-amino-4-methylpentane $C_6H_{15}N$ 9178
5345-54-0 3-chloroanisidine C_7H_8ClNO 10760
5379-20-4 1-ethenyl-3,5-dimethylbenzene $C_{10}H_{12}$ 19376
5395-75-5 1,2-bis(ethylthio)ethane $C_6H_{14}S_2$ 9128
5396-09-8 2,3,5,5-tetramethyl-3-hexanol $C_{10}H_{22}O$ 21319
5396-58-7 2-methyl-2,3-butanediol $C_5H_{12}O_2$ 5873
5398-25-4 2,3-dibromopentane $C_5H_{10}Br_2$ 5303
5405-79-8 2,2-dimethyl-3-hexanone $C_8H_{16}O$ 15068
5408-57-1 7-methyl-3-octanone $C_9H_{18}O$ 18006
5408-86-6 2,3-dibromobutane $C_4H_8Br_2$ 3343
5412-92-0 N-(1-methylheptyl)-2-octanamine $C_{16}H_{35}N$ 29802
5434-27-5 1,3-dibromo-2,2-dimethylpropane $C_5H_{10}Br_2$ 5316
5434-57-1 hexyl 2,2-dimethylpropanoate $C_{11}H_{22}O_2$ 22853
5435-64-3 3,5,5-trimethylhexanal $C_9H_{18}O$ 18044
5445-31-8 2-methyl-3-nonanone $C_{10}H_{20}O$ 21034
5451-52-5 decyl formate $C_{11}H_{22}O_2$ 22836
5451-63-8 tetradecyl formate $C_{15}H_{30}O_2$ 28880
5451-85-4 octyl pentanoate $C_{13}H_{26}O_2$ 26498
5451-92-3 nonyl formate $C_{10}H_{20}O_2$ 21043
5454-09-1 decyl butanoate $C_{14}H_{28}O_2$ 27877
5454-19-3 decyl propanoate $C_{13}H_{26}O_2$ 26496
5454-24-0 undecyl formate $C_{12}H_{24}O_2$ 24868
5454-79-5 cis-3-methylcyclohexanol $C_7H_{14}O$ 11847
5458-33-3 undecyl propanoate $C_{14}H_{28}O_2$ 27876
5460-31-1 2-methyl-3-nitrophenol $C_7H_7NO_3$ 10671
5461-02-9 undecyl butanoate $C_{15}H_{30}O_2$ 28883
5510-99-6 2,6-di-sec-butylphenol $C_{14}H_{22}O$ 27681
5557-31-3 9-octadecene $C_{18}H_{36}$ 31072
5566-34-7 trans-chlordan $C_{10}H_6Cl_8$ 18552
5582-82-1 3-methyl-3-heptanol $C_8H_{18}O$ 15447
5617-41-4 heptylcyclohexane $C_{13}H_{26}$ 26466
5617-42-5 heptylcyclopentane $C_{12}H_{24}$ 24818
5634-30-0 dodecylcyclopentane $C_{17}H_{34}$ 30310
5668-93-9 3-ethylbiphenyl $C_{14}H_{14}$ 27064
5675-22-9 1,4-heptadiene C_7H_{12} 11359
5683-44-3 3-methyl-2,4-pentanediol $C_6H_{14}O_2$ 9039
5707-44-8 4-ethylbiphenyl $C_{14}H_{14}$ 27065
5750-00-5 2-chloro-3,3-dimethylbutane $C_6H_{13}Cl$ 8685
5756-43-4 ethyl hexyl ether $C_8H_{18}O$ 15533
5780-13-2 meso-2,3-dibromobutane $C_4H_8Br_2$ 3339
5787-28-0 (S)-(1-methylpropyl)benzene $C_{10}H_{14}$ 19876
5787-31-5 2-bromobutane, (±) C_4H_9Br 3602
5794-03-6 (+)-camphene $C_{10}H_{16}$ 20356
5794-04-7 (-)-camphene $C_{10}H_{16}$ 20357
5813-64-9 2,2-dimethylpropylamine $C_5H_{13}N$ 5953
5814-85-7 1,2-diphenylpropane $C_{15}H_{16}$ 28380
5837-78-5 ethyl trans-2-methyl-2-butenoate $C_7H_{12}O_2$ 11481
5856-62-2 (S)-(+)-2-amino-1-butanol $C_4H_{11}NO$ 4001
5856-63-3 (R)-(-)-2-amino-1-butanol $C_4H_{11}NO$ 4002
5857-36-3 2,2,4-trimethyl-3-pentanone $C_8H_{16}O$ 15082
5857-69-2 2,2,3,4,4-pentamethyl-3-pentanol $C_{10}H_{22}O$ 21329
5867-91-4 methyl propyl sulfate $C_4H_{10}O_4S$ 3938
5870-93-9 heptyl butanoate $C_{11}H_{22}O_2$ 22839
5876-87-9 1,11-dodecadiene $C_{12}H_{22}$ 24669
5881-17-4 3-ethyloctane $C_{10}H_{22}$ 21138
5902-51-2 terbacil $C_9H_{13}ClN_2O_2$ 17237
5911-04-6 3-methylnonane $C_{10}H_{22}$ 21135
5911-05-7 4-ethylnonane $C_{11}H_{24}$ 22896
5921-84-6 4-heptyl acetate $C_9H_{18}O_2$ 18060
5932-79-6 4-nonanol $C_9H_{20}O$ 18241
5943-30-6 disulfide, bis(1-methylpropyl) $C_8H_{18}S_2$ 15618
5954-71-2 2,cis-3-dimethylthiacyclopropane C_4H_8S 3585
5955-98-6 2,trans-3-dimethylthiacyclopropane C_4H_8S 3586
5963-14-4 8-methylnonanoic acid $C_{10}H_{20}O_2$ 21066
5969-12-0 2,6-heptanediol $C_7H_{16}O_2$ 12232
5970-63-8 2-methyl-2-ethyl-1-pentanol $C_8H_{18}O$ 15491
5978-70-1 (R)-(-)-2-octanol $C_8H_{18}O$ 15536
5984-58-7 6-methyl-2-heptanamine, (±) $C_8H_{19}N$ 15633
5989-27-5 D-limonene $C_{10}H_{16}$ 20316
5989-54-8 L-limonene $C_{10}H_{16}$ 20318
6006-15-1 diethylisopropylamine $C_7H_{17}N$ 12340
6006-33-3 tridecylcyclohexane $C_{19}H_{38}$ 31696
6007-54-1 thiacyclononane $C_8H_{16}S$ 15243
6027-89-0 L(+)-gulose $C_6H_{12}O_6$ 8612
6031-02-3 (1-methylpentyl)benzene $C_{12}H_{18}$ 24372
6032-29-7 2-pentanol $C_5H_{12}O$ 5832
6033-23-4 (S)-(+)-2-heptanol $C_7H_{16}O$ 12220
6033-24-5 (R)-(-)-2-heptanol $C_7H_{16}O$ 12221
6048-83-5 thiacyclodecane $C_9H_{18}S$ 18111
6061-10-5 3-methyloctanoic acid $C_9H_{18}O_2$ 18056
6064-27-3 6-dodecanone $C_{12}H_{24}O$ 24855
6064-90-0 methyl heneicosanoate $C_{22}H_{44}O_2$ 33470
6065-90-3 1,2-dichloro-4,4-dimethylpentane $C_7H_{14}Cl_2$ 11764
6065-93-6 1,1-dichloro-2-methyl-1-propene $C_4H_6Cl_2$ 2818
6068-69-5 N-sec-butylaniline $C_{10}H_{15}N$ 20202
6069-98-3 cis-1-methyl-4-isopropylcyclohexane $C_{10}H_{20}$ 20934
6079-57-8 4,4-dimethyl-2-thiapentane $C_6H_{14}S$ 9125
6094-02-6 2-methyl-1-hexene C_7H_{14} 11714
6108-61-8 cis-2,cis-4-hexadiene C_6H_{10} 7633
6111-88-2 2-chloro-2,4,4-trimethylpentane $C_8H_{17}Cl$ 15261
6117-80-2 cis-2-butene-1,4-diol $C_4H_8O_2$ 3511
6117-98-2 2,3-dimethyldodecane $C_{14}H_{30}$ 27916
6117-99-3 2,4-dimethyldodecane $C_{14}H_{30}$ 27917
6130-96-7 1,1-dichloro-3,3-dimethylbutane $C_6H_{12}Cl_2$ 8279
6131-24-4 3-methylhexane, (S)- C_7H_{16} 12121
6137-03-7 3-ethyl-2-pentanone $C_7H_{14}O$ 11838
6137-08-2 3-methyl-2-octanone $C_9H_{18}O$ 18002
6137-15-1 4-methyl-3-octanone $C_9H_{18}O$ 18004
6137-26-4 4-dodecanone $C_{12}H_{24}O$ 24863
6140-80-3 3-(1-methylethoxy)-1-propene $C_6H_{12}O$ 8398
6144-93-0 4,4-dimethyl-2-pentanol $C_7H_{16}O$ 12193
6158-45-8 1-isopropylnaphthalene $C_{13}H_{14}$ 25905
6161-67-7 4-methyl-3-ethylphenol $C_9H_{12}O$ 17115
6163-64-0 3,3-dimethyl-2-thiabutane $C_5H_{12}S$ 5935
6163-66-2 di-tert-butyl ether $C_8H_{18}O$ 15529
6165-37-3 4-propylnonane $C_{12}H_{26}$ 24942
6168-72-5 2-amino-1-propanol, (±) C_3H_9NO 2207
6169-06-8 (S)-(+)-2-octanol $C_8H_{18}O$ 15537
6175-49-1 2-dodecanone $C_{12}H_{24}O$ 24854
6196-60-7 3-methyl-3-ethyl-1-pentene C_8H_{16} 14975
6196-80-1 1-iodo-4-methylpentane $C_6H_{13}I$ 8729
6221-93-8 dodecyl propanoate $C_{15}H_{30}O_2$ 28882
6223-78-5 2,5-dichloro-2,5-dimethylhexane $C_8H_{16}Cl_2$ 15006
6236-88-0 1-methyl-trans-4-ethylcyclohexane C_9H_{18} 17924
6240-90-0 (R)-2-heptylamine $C_7H_{17}N$ 12343
6261-19-4 1-chloro-cis-2-pentene C_5H_9Cl 5029
6261-25-2 1-chloro-trans-2-pentene C_5H_9Cl 5030
6269-92-7 1,2-dibromooctane $C_8H_{16}Br_2$ 14995
6290-03-5 (R)-(-)-1,3-butanediol $C_4H_{10}O_2$ 3900
6294-31-1 dihexyl sulfide $C_{12}H_{26}S$ 25312
6304-18-3 3-isopropylpyridine $C_8H_{11}N$ 14102

6305-71-1 2,4-dimethyl-1-pentanol C7H16O 12184
6308-94-7 diethyldecylamine C14H31N 27961
6310-21-0 2-tert-butylaniline C10H15N 20198
6339-13-5 2-methylpentanenitrile C6H11N 8089
6351-10-6 2,3-dihydro-1H-inden-1-ol C9H10O 16627
6366-35-4 1-chloro-2,2-dimethylbutane C6H13Cl 8681
6414-96-6 5,5-dimethylnonane C11H24 22913
6418-41-3 3-methyltridecane C14H30 27914
6418-43-5 3-methylhexadecane C17H36 30348
6418-44-6 3-methylheptadecane C18H38 31134
6418-45-7 3-methylnonadecane C20H42 32517
6423-02-5 2,3-dibromohexane C6H12Br2 8253
6423-43-4 1,2-propylene glycol dinitrate C3H6N2O6 1902
6434-78-2 trans-2-nonene C9H18 17943
6443-92-1 cis-2-heptene C7H14 11708
6464-40-0 DL-threo-4-methyl-2,3-pentanediol C6H14O2 9042
6481-95-4 4,4-dimethyl-1-hexanol C8H18O 15464
6561-44-0 3-methyloctadecane C19H40 31737
6570-87-2 3,4-dimethyl-1-pentanol C7H16O 12186
6570-88-3 2,3,4-trimethyl-1-pentanol C8H18O 15500
6622-76-0 methyl 2-methyl-2-butenoate C6H10O2 7845
6628-18-8 1,2-bis(methylthio)ethane C4H10S2 3951
6639-30-1 1,2,4-trichloro-5-methylbenzene C7H5Cl3 9972
6702-10-9 DL-erythro-4-methyl-2,3-pentanediol C6H14O2 9041
6709-39-3 2,6-dimethyl-1,5-heptadiene C9H16 17611
6742-54-7 undecylbenzene C17H28 30264
6765-39-5 1-heptadecene C17H34 30313
6776-19-8 ethyl cis-2-butenoate C6H10O2 7830
6785-23-5 undecylcyclopentane C16H32 29704
6790-27-8 3-dodecyne C12H22 24655
6795-87-5 methyl sec-butyl ether C5H12O 5843
6795-88-6 methyl 1-methylbutyl ether C6H14O 8997
6836-38-0 6-dodecanol C12H26O 25280
6863-58-7 di-sec-butyl ether C8H18O 15528
6876-18-2 2-methyl-3-isopropylheptane C11H24 22957
6876-23-9 trans-1,2-dimethylcyclohexane C8H16 14881
6886-16-4 2,5-dimethyl-1-hexanol C8H18O 15460
6891-45-8 2-butyl-4-methylphenol C11H16O 22400
6915-15-7 malic acid C4H6O5 3024
6920-22-5 1,2-hexanediol C6H14O2 9023
6923-20-2 cis-1,2-dichloropropene C3H4Cl2 1535
6923-22-4 monocrotophos C7H14NO5P 11781
6931-70-0 2,3-dimethyl-2,3-pentanediol C7H16O2 12250
6933-10-4 4-bromo-3-methylaniline C7H8BrN 10739
6950-79-4 aniline, p-isopropyl-N-methyl- C10H15N 20231
6975-92-4 2,5-dimethyl-1-hexene C8H16 14934
6975-98-0 2-methyldecane C11H24 22891
6975-99-1 6-dodecyne C12H22 24656
6976-00-7 7-ethyl-2-methyl-4-undecanone C14H28O 27864
6976-72-3 heptyl hexanoate C13H26O2 26499
6982-25-8 DL-2,3-butanediol C4H10O2 3887
6986-48-7 bis(1-chloroethyl) ether C4H8Cl2O 3374
7012-37-5 2,4,4'-trichlorobiphenyl C12H7Cl3 23252
7013-11-8 2,3-dichloro-1-butene C4H6Cl2 2812
7045-71-8 2-methylundecane C12H26 25268
7058-01-7 sec-butylcyclohexane C10H20 20925
7069-38-7 trans-1,2-dichloropropene C3H4Cl2 1536
7085-85-0 ethyl cyanoacrylate C6H7NO2 7212
7087-68-5 N-ethyl-N-isopropyl-2-propanamine C8H19N 15643
7094-26-0 1,1,2-trimethylcyclohexane C9H18 17925
7094-27-1 1,1,4-trimethylcyclohexane C9H18 17927
7116-86-1 5,5-dimethyl-1-hexene C8H16 14940
7145-20-2 2,3-dimethyl-2-hexene C8H16 14945
7146-60-3 2,3-dimethyloctane C10H22 21141
7154-75-8 4-methyl-1-pentyne C6H10 7652
7154-79-2 2,2,3,3-tetramethylpentane C9H20 18206
7154-80-5 3,3,5-trimethylheptane C10H22 21175
7188-38-7 tert-butyl isocyanide C5H9N 5190
7212-53-5 5-methyl-1-heptanol C8H18O 15439
7220-26-0 3-ethyl-2,4-dimethylhexane C10H22 21185
7225-66-3 7-hexyltridecane C19H40 31733
7239-24-9 dimethylisobutylamine C6H15N 9210
7287-19-6 prometryn C10H19N5S 20917
7289-44-3 methyl undecyl sulfide C12H26S 25313
7289-45-4 methyl tetradecyl sulfide C15H32S 28923
7289-52-3 1-methoxydecane C11H24O 23064
7294-05-5 2,2,3-trimethyl-3-pentanol C8H18O 15513
7300-03-0 3-methyl-3-heptene C8H16 14918
7307-55-3 undecylamine C11H25N 23090
7307-58-6 diethyltetradecylamine C18H39N 31167
7309-44-6 ethyl hexyl sulfide C8H18S 15608
7311-30-0 methyldodecylamine C13H29N 26556
7318-67-4 1,cis-4-hexadiene C6H10 7627
7319-00-8 1,trans-4-hexadiene C6H10 7628
7348-71-2 1-bromo-trans-2-pentene C5H9Br 4971
7348-78-9 1-bromo-cis-2-pentene C5H9Br 4970
7357-93-9 3-methyl-2-ethyl-1-butene C7H14 11743
7359-72-0 1,2,3-trichloro-4-methylbenzene C7H5Cl3 9970
7367-38-6 5-butyl-4-nonene C13H26 26471
7378-99-6 dimethyloctylamine C10H23N 21402
7379-12-6 2-methyl-3-hexanone C7H14O 11835
7383-90-6 3,4'-dimethylbiphenyl C14H14 27061
7385-78-6 3,4-dimethyl-1-pentene C7H14 11734
7385-82-2 5-methyl-trans-2-hexene C7H14 11724
7397-06-0 4-tert-butyl-o-xylene C12H18 24376
7399-49-7 1-methyl-2-(1-methylethenyl)benzene C10H12 19369
7415-31-8 1,3-dichloro-trans-2-butene C4H6Cl2 2806
7417-48-3 3-methyl-1,2-pentadiene C6H10 7635
7423-69-0 3,5-dimethyl-1-hexene C8H16 14937
7433-56-9 trans-5-decene C10H20 20946
7433-78-5 cis-5-decene C10H20 20945
7435-50-9 1,3,6,7-tetramethylnaphthalene C14H16 27308
7443-52-9 trans-2-methylcyclohexanol C7H14O 11846
7443-55-2 trans-3-methylcyclohexanol C7H14O 11848
7443-70-1 cis-2-methylcyclohexanol C7H14O 11845
7452-79-1 ethyl 2-methylbutyrate C7H14O2 11923
7515-62-0 5-bromo-trans-2-pentene C5H9Br 4979
7515-80-2 N-tert-butylisopropylamine C7H17N 12341
7516-82-7 methyldecylamine C11H25N 23091
7525-62-4 m-ethylstyrene C10H12 19373
7564-63-8 o-ethylstyrene C10H12 19372
7581-97-7 2,3-dichlorobutane, dl and meso C4H8Cl2 3370
7611-86-1 cis-1-chloro-1-butene C4H7Cl 3098
7611-87-2 trans-1-chloro-1-butene C4H7Cl 3099
7611-88-3 cis-1-chloro-2-methyl-1-butene C5H9Cl 5039
7611-89-4 trans-1-chloro-2-methyl-1-butene C5H9Cl 5040
7641-77-2 hexamethyldewarbenzene C12H18 24377
7642-04-8 cis-2-octene C8H16 14892
7642-09-3 cis-3-hexene C6H12 8221
7642-10-6 cis-3-heptene C7H14 11712
7642-15-1 cis-4-octene C8H16 14897
7661-55-4 5-methylquinoline C10H9N 18852
7667-55-2 1,cis-2,trans-3-trimethylcyclohexane C9H18 17930
7667-58-5 1,cis-2,trans-4-trimethylcyclohexane C9H18 17934
7667-59-6 1,trans-2,cis-4-trimethylcyclohexane C9H18 17935
7667-60-9 1,trans-2,trans-4-trimethylcyclohexane C9H18 17936
7688-21-3 cis-2-hexene C6H12 8218
7713-69-1 methyl-sec-butylamine C5H13N 5957
7731-28-4 cis-4-methylcyclohexanol C7H14O 11849
7731-29-5 trans-4-methylcyclohexanol C7H14O 11850
7745-91-7 3-bromo-4-methylaniline C7H8BrN 10737
7756-94-7 isobutene, trimer C12H24 24828
7766-48-5 5-iodo-1-pentene C5H9I 5152
7766-51-0 4-iodo-1-butene C4H7I 3213
7771-44-0 5,8,11,14-eicosatetraenoic acid C20H32O2 32387
7778-87-2 propyl heptanoate C10H20O2 21058
7782-24-3 (S)-(+)-2-phenylpropionic acid C9H10O2 16702
7782-26-5 (R)-(-)-2-phenylpropionic acid C9H10O2 16703
7785-26-4 alpha-pinene, (-) C10H16 20322
7785-70-8 (+)-a-pinene C10H16 20368
7786-29-0 2-methyloctanal C9H18O 18048
7786-34-7 mevinphos C7H13O6P 11695
7789-89-1 1,1,1-trichloropropane C3H5Cl3 1703
7789-99-3 2-methyl-1-pentyl acetate C8H16O2 15164
7795-80-4 2-methyl-2,3-pentanediol C6H14O2 9035
7795-91-7 ethylene glycol mono-sec-butyl ether C6H14O2 9067
7796-36-3 cis-1-iodo-1-propene C3H5I 1734
7796-54-5 trans-1-iodo-1-propene C3H5I 1735
8001-35-2 toxaphene C10H9Cl9 18836
8022-00-2 methyl demeton C6H15O3PS2 9267
8065-48-3 demeton C16H38O6P2S4 29834
10024-78-9 4-ethyl-1-methyloctylamine C11H25N 23095
10031-87-5 2-ethylbutyl acetate C8H16O2 15162
10042-59-8 2-propyl-1-heptanol C10H22O 21284
10061-01-5 cis-1,3-dichloropropene C3H4Cl2 1537
10061-02-6 trans-1,3-dichloropropene C3H4Cl2 1538
10075-38-4 cis-1,3-dichloro-2-butene C4H6Cl2 2815
10075-50-0 5-bromoindole C8H6BrN 12725
10076-31-0 dimethyl-2,2-dimethylpropylamine C7H17N 12331
10086-64-3 1-fluoro-2-methylbutane C5H11F 5652
10143-20-1 2,8-dimethyl-6-isobutylnonanol-4 C15H32O 28915
10143-23-4 2,3-dimethyl-1-pentanol C7H16O 12183
10143-60-9 bis(2-ethylhexyl) ether C16H34O 29772
10203-28-8 2-dodecanol C12H26O 25278
10203-30-2 3-dodecanol C12H26O 25279
10210-68-1 cobalt carbonyl C8Co2O8 12421
10222-95-4 1,2,4-trimethyl-5-isopropylbenzene C12H18 24375
10250-47-2 3-methyl-3-pentyl acetate C8H16O2 15166
10250-52-9 1,1-diiodopropane C3H6I2 1871
10250-55-2 1,1-diiodo-2-methylpropane C4H8I2 3407
10288-13-8 1,2-dibromo-3-methylbutane C5H10Br2 5309
10289-68-6 3-tridecanol C13H28O 26537
10297-57-1 2-methyl-2-nonanol C10H22O 21236
10307-61-6 ethyl 2-methylbutanoate, (+) C7H14O2 11911
10342-97-9 methyldiisopropylamine C7H17N 12338
10359-64-5 3-methyl-2-thiapentane C5H12S 5932
10422-35-2 1-bromo-2-methylbutane C5H11Br 5606
10424-38-1 1,2-propanediamine, (±) C3H10N2 2257
10428-64-5 1,2-dibromo-2-methylbutane C5H10Br2 5308
10435-81-1 2-octanethiol, (±) C8H18S 15605
10436-75-6 4-methyl-N-isopropylaniline C10H15N 20207
10478-12-3 methyl 2-methyl-2-pentenoate C7H12O2 11484
10486-19-8 1-tridecanal C13H26O 26486
10496-15-8 dihexyl disulfide C12H26S2 25318
10496-16-9 diheptyl disulfide C14H30S2 27953
10496-18-1 didecyl disulfide C20H42S2 32544
10500-57-9 5,6,7,8-tetrahydroquinoline C9H11N 16896
10522-26-6 2-methyl-1-undecanol C12H26O 25281
10523-96-3 4-chloro-2-methyl-1-butene C5H9Cl 5042
10524-01-3 4-chloro-3-methyl-1-butene C5H9Cl 5047
10524-07-9 5-chloro-trans-2-pentene C5H9Cl 5038
10524-08-0 4-chloro-1-pentene C5H9Cl 5027
10524-70-6 3,3-dimethyl-1-hexanol C8H18O 15461
10544-63-5 ethyl 2-butenoate C6H10O2 7831
10557-44-5 2,5-dimethyl-cis-3-hexene C8H16 14966
10574-36-4 3-methyl-cis-2-hexene C7H14 11719
10574-37-5 2,3-dimethyl-2-pentene C7H14 11737
10575-56-1 3,4,4-trimethyl-2-pentanol C8H18O 15511
10575-87-8 1,2-dichloroheptane C7H14Cl2 11767
11069-19-5 dichlorobutene C4H6Cl2 2822
11097-69-1 alochlor 1254 C12H5Cl5 23179
12002-48-1 trichlorobenzene C6H3Cl3 6308
13019-20-0 2-methyl-3-heptanone C8H16O 15073
13023-60-4 2,2-diethyl-1-butanol C8H18O 15517
13029-08-8 2,2'-dichloro-1,1'-biphenyl C12H8Cl2 23276
13054-87-0 2-amino-1-butanol, (±)- C4H11NO 3992
13071-79-9 terbufos C9H21O2PS3 18408
13114-87-9 3-(trifluoromethylphenyl) urea C8H7F3N2O 13080
13116-53-5 1,2,2,3-tetrachloropropane C3H4Cl4 1570
13116-60-4 1,1,2,2-tetrachloropropane C3H4Cl4 1567
13121-70-5 cyhexatin C18H34OSn 31029
13140-86-8 3-phenyl-1-cyclopropyl urea C10H12N2O 19441
13150-81-7 2,6-dimethyldecane C12H26 24926
13151-04-7 5-methyl-1-heptene C8H16 14902
13151-05-8 4-methyl-1-heptene C8H16 14901
13151-17-2 5-methyl-cis-2-hexene C7H14 11723
13151-27-4 2-methyl-1-decene C11H22 22803
13151-34-3 3-methyldecane C11H24 22892
13151-35-4 5-methyldecane C11H24 22894
13154-12-6 cis-1-bromo-1-hexene C6H11Br 7979
13154-13-7 trans-1-bromo-1-hexene C6H11Br 7980
13172-91-3 5-methyl-3-heptene C8H16 14924
13175-27-4 1,6-heptanediol C7H16O2 12228
13187-99-0 2-bromododecane C12H25Br 24895
13205-60-2 ethylisobutylamine C6H15N 9200
13231-81-7 3-methyl-1-hexanol C7H16O 12169
13250-12-9 (R)-(-)-sec-butylamine C4H11N 3986
13254-34-7 2,6-dimethyl-2-heptanol C9H20O 18276
13269-52-8 trans-3-hexene C6H12 8222
13281-11-3 2-nonanethiol C9H20S 18366
13286-90-3 5-methyl-2-thiahexane C6H14S 9118
13286-91-4 3-methyl-2-thiahexane C6H14S 9116
13286-92-5 3,3-dimethyl-2-thiapentane C6H14S 9123
13287-19-9 2-methyl-3-isopropyloctane C12H26 25011
13320-56-4 1,1-dibromopentane C5H10Br2 5297
13325-10-5 4-amino-1-butanol C4H11NO 3991
13360-45-7 chlorbromuron C9H10BrClN2O2 16522
13360-61-7 1-pentadecene C15H30 28859
13360-63-9 ethylbutylamine C6H15N 9198
13361-63-2 4,4-dimethyl-1-pentyne C7H12 11352
13362-80-6 4,8-dimethylquinoline C11H11N 21716
13364-16-4 1-amino-2-methylpentane C6H15N 9176
13389-42-9 trans-2-octene C8H16 14893
13402-60-3 2-decanethiol C10H22S 21386
13423-15-9 3-methyltetrahydrofuran C5H10O 5454
13432-25-2 2,4-dimethyl-3-hexanol C8H18O 15482
13463-39-3 nickel carbonyl C4NiO4 4156
13463-40-6 iron pentacarbonyl C5FeO5 4183
13471-42-6 3-hexanol, (R) C6H14O 8991
13475-78-0 5-ethyl-2-methylheptane C10H22 21156
13475-79-1 2,4-dimethyl-3-isopropylpentane C10H22 21201
13475-81-5 2,2,3,3-tetramethylhexane C10H22 21190
13475-82-6 2,2,4,6,6-pentamethylheptane C12H26 25207
13501-73-0 3,5-dimethyl-1-hexanol C8H18O 15463

13552-21-1 1-amino-2-butanol C4H11NO 3999
13556-58-6 1-ethyl-[1,2,3,4-tetrahydronaphthalene] C12H16 24112
13602-13-6 1,2-dichloro-2-butene C4H6Cl2 2814
13616-82-5 2,4-dimethylthiophenol C8H10S 14043
13632-93-4 1-methyl-2,3-diethylbenzene C11H16 22342
13632-94-5 1-methyl-2,5-diethylbenzene C11H16 22344
13632-95-6 1-methyl-2,6-diethyl benzene C11H16 22345
13643-05-5 4-methyl-1,2-pentadiene C6H10 7636
13678-54-1 2-methyl-4-ethyl thiophene C7H10S 11230
13714-85-7 4-methyl-trans-3-heptene C8H16 14923
13732-80-4 1-methyl-3,4-diethylbenzene C11H16 22346
13764-18-6 1,4,6,7-tetramethylnaphthalene C14H16 27310
13838-16-9 enflurane C3H2ClF5O 1346
13940-94-8 a-a-p-trichlorotoluene C7H5Cl3 9979
13952-84-6 sec-butylamine C4H11N 3979
13991-37-2 trans-2-pentenoic acid C5H8O2 4896
14069-89-7 sec-butyl isocyanide C5H9N 5189
14189-13-0 2-ethyl-1,5-pentanediol C7H16O2 12244
14250-79-4 2,2-dimethyl-1-heptanol C9H20O 18268
14252-42-7 1,1-bis(ethylthio)ethane C6H14S2 9127
14255-23-3 2,4-dimethyl-2-hexene C8H16 14946
14255-24-4 4-methyl-cis-3-heptene C8H16 14922
14255-44-8 2-nitrohexane C6H13NO2 8791
14287-61-7 2,3-dimethylbutanoic acid C6H12O2 8464
14290-92-7 2,2-dimethyl-3-thiapentane C6H14S 9126
14402-50-7 2-dodecanethiol C12H26S 25311
14411-75-7 1,3-dimethyl-2-isopropylbenzene C11H16 22338
14476-30-3 1,2,2-tribromopropane C3H5Br3 1666
14476-37-0 4-undecanone C11H22O 22825
14484-64-1 ferbam C9H18FeN3S6 17966
14531-16-9 5-methyl-5-ethylnonane C12H26 24963
14558-12-4 1,3,5,8-tetramethylnaphthalene C14H16 27307
14596-92-0 1,cis-3-hexadiene C6H10 7624
14602-62-1 1,1,2-tribromopropane C3H5Br3 1664
14609-79-1 2,2,4,4-tetramethyl-3-pentanol C9H20O 18335
14610-37-8 methylisobutylamine C5H13N 5956
14676-29-0 3-ethyl-2-methylheptane C10H22 21154
14686-13-6 trans-2-heptene C7H14 11709
14686-14-7 trans-3-heptene C7H14 11713
14720-74-2 2,2,4-trimethylheptane C10H22 21163
14739-73-2 2,2,3,5,5-pentamethylhexane C11H24 23038
14753-05-0 1-chloro-2-methylpentane C6H13Cl 8672
14779-92-1 8-methyl-2-nonanol C10H22O 21241
14850-22-7 cis-3-octene C8H16 14894
14850-23-8 trans-4-octene C8H16 14898
14852-31-4 2-hexadecanol C16H34O 29767
14857-34-2 ethoxydimethylsilane C4H12OSi 4095
14898-79-4 (R)-(-)-2-butanol C4H10O 3872
14919-01-8 trans-3-octene C8H16 14895
14920-89-9 2,3-dimethylfuran C6H8O 7383
14938-35-3 4-pentylphenol C11H16O 22408
14979-39-6 4-methyl-3-heptanol C8H18O 15448
15013-37-3 4-methyl-2-thiahexane C6H14S 9117
15181-11-0 3,5-di-tert-butyltoluene C15H24 28728
15220-85-6 2-methyl-1-propene, tetramer C16H32 29709
15232-76-5 3-octyne C8H14 14486
15268-49-2 2-methyl-1-propoxypropane C7H16O 12218
15299-99-7 napropamide C17H21NO2 30161
15340-96-2 3,7-dimethyl-2-octanol C10H22O 21264
15430-98-5 7-pentadecene C15H30 28861
15448-47-2 (R)-(+)-propylene oxide C3H6O 1928
15501-33-4 1-iodo-2,2-dimethylpropane C5H11I 5667
15520-10-2 1,5-diamino-2-methylpentane C6H16N2 9298
15545-48-9 N-(3-chloro-4-methylphenyl)-N',N'-dimethyl C10H13ClN2O 19729
15672-88-5 1-iodo-3,3-dimethylbutane C6H13I 8737
15673-00-4 1-amino-3,3-dimethylbutane C6H15N 9186
15679-01-3 1,cis-2-dimethylcyclobutane C6H12 8209
15679-02-4 1,trans-2-dimethylcyclobutane C6H12 8210
15706-73-7 butyl 2-methylbutanoate C9H18O2 18072
15726-15-5 3-methyl-4-heptanone C8H16O 15076
15790-88-2 methyl trans-2-pentenoate C6H10O2 7859
15833-75-7 3,3-dimethyltetrahydrofuran C6H12O 8422
15840-60-5 2-methyl-cis-3-hexene C7H14 11725
15862-07-4 2,4,5-trichlorobiphenyl C12H7Cl3 23247
15869-80-4 3-ethylheptane C9H20 18180
15869-85-9 5-methylnonane C10H22 21137
15869-86-0 4-ethyloctane C10H22 21139
15869-87-1 2,2-dimethyloctane C10H22 21140
15869-89-3 2,5-dimethyloctane C10H22 21143
15869-92-8 3,4-dimethyloctane C10H22 21147
15869-93-9 3,5-dimethyloctane C10H22 21148
15869-94-0 3,6-dimethyloctane C10H22 21149
15869-95-1 4,4-dimethyloctane C10H22 21150
15869-96-2 4,5-dimethyloctane C10H22 21151
15870-10-7 2-methyl-1-heptene C8H16 14899
15877-57-3 3-methylpentanal C6H12O 8403
15890-40-1 1,cis-2,trans-3-trimethylcyclopentane C8H16 14871
15892-23-6 (<+->)-2-butanol C4H10O 3874
15918-07-7 5-methyl-4-nonene C10H20 20949
15918-08-8 2-propyl-1-pentene C8H16 14970
15950-66-0 2,3,4-trichlorophenol C6H3Cl3O 6310
15968-05-5 2,2',6,6'-tetrachlorobiphenyl C12H6Cl4 23210
15972-60-8 alachlor C14H20ClNO2 27572
16020-17-0 1,2,4,7-tetramethylnaphthalene C14H16 27302
16021-20-8 1-ethyl-2-propylbenzene C11H16 22324
16033-71-9 methyloxirane C3H6O 1926
16088-62-3 (S)-(-)-propylene oxide C3H6O 1927
16106-59-5 4,5-dimethyl-1-hexene C8H16 14939
16136-84-8 cis-1-chloro-1-propene C3H5Cl 1668
16136-85-9 trans-1-chloro-1-propene C3H5Cl 1669
16230-28-7 3,4-dibromohexane C6H12Br2 8256
16325-63-6 2,4,4-trimethyl-1-pentanol C8H18O 15501
16325-64-7 3,4,4-trimethyl-1-pentanol C8H18O 15503
16416-41-4 trans-1-bromo-3-methyl-1-butene C5H9Br 4985
16416-44-7 cis-1-bromo-3-methyl-1-butene C5H9Br 4984
16432-53-4 1,4-hexanediol C6H14O2 9025
16435-49-7 2-methyl-1-dodecene C13H26 26470
16491-15-9 1,5-dimethylcyclopentene C7H12 11329
16538-47-9 cis-1-iodo-1-hexene C6H11I 8080
16580-23-7 1-methyl-2-isopropylcyclohexane C10H20 20931
16580-24-8 1-methyl-3-isopropylcyclohexane C10H20 20932
16606-02-3 2,4',5-trichlorobiphenyl C12H7Cl3 23253
16631-63-3 1-methyl-1-propylcyclopentane C9H18 17863
16644-98-7 trans-1-iodo-1-hexene C6H11I 8081
16666-42-5 cis-2-pentenoic acid C5H8O2 4897
16696-65-4 1,11-dibromoundecane C11H22Br2 22806
16727-91-6 1-isobutylnaphthalene C14H16 27271
16736-42-8 2,7-dimethyl-2,6-octadiene C10H18 20611
16745-94-1 3,4-dimethyl-1-hexene C8H16 14936
16746-02-4 2-isopropyl-1-pentene C8H16 14971
16746-85-3 4-ethyl-1-hexene C8H16 14931
16746-86-4 2,3-dimethyl-1-hexene C8H16 14932
16746-87-5 2,4-dimethyl-1-hexene C8H16 14933
16747-25-4 2,2,3-trimethylhexane C9H20 18194
16747-26-5 2,2,4-trimethylhexane C9H20 18195
16747-28-7 2,3,3-trimethylhexane C9H20 18197
16747-30-1 2,4,4-trimethylhexane C9H20 18200
16747-31-2 3,3,4-trimethylhexane C9H20 18201
16747-32-3 2,2-dimethyl-3-ethylpentane C9H20 18203
16747-33-4 2,3-dimethyl-3-ethylpentane C9H20 18204
16747-38-9 2,3,3,4-tetramethylpentane C9H20 18209
16747-42-5 2,2,4,5-tetramethylhexane C10H22 21194
16747-44-7 2,2,3,3,4-pentamethylpentane C10H22 21206
16747-45-8 2,2,3,4,4-pentamethylpentane C10H22 21207
16747-50-5 1-methyl-1-ethylcyclopentane C8H16 14863
16751-58-9 3-aminohexane C6H15N 9175
16751-59-0 4-heptanamine C7H17N 12318
16752-77-5 methomyl C5H10N2O2S 5426
16789-46-1 3-ethyl-2-methylhexane C9H20 18190
16789-51-8 3-ethyl-3-hexene C8H16 14959
16813-18-6 2-heptadecanol C17H36O 30353
16842-03-8 cobalt hydrocarbonyl C4HCoO4 2365
16844-08-9 (S)-2-chlorooctane C8H17Cl 15262
16883-48-0 1,trans-2,cis-4-trimethylcyclopentane C8H16 14875
16889-72-8 tert-butyl 2-methylpropanoate C8H16O2 15181
16900-07-5 butyl octyl sulfide C12H26S 25316
16900-08-6 butyl dodecyl sulfide C16H34S 29795
16967-04-7 butyl hexyl sulfide C10H22S 21391
17042-16-9 4-methyl-3-hexanone C7H14O 11836
17059-44-8 1,2-dimethyl-3-propylbenzene C11H16 22330
17059-45-9 1,3-dimethyl-2-propylbenzene C11H16 22332
17071-47-5 butyl isobutyl ether C8H18O 15530
17082-12-1 azobenzene C12H10N2 23488
17085-91-5 1-tert-butylnaphthalene C14H16 27273
17085-96-0 3-ethyldecane C12H26 24919
17135-78-3 2-chloroanthracene C14H9Cl 26689
17226-28-7 di-sec-butyl ether, (±) C8H18O 15532
17299-07-9 (2R,5R)-2,5-hexanediol C6H14O2 9061
17301-94-9 4-methylnonane C10H22 21136
17302-01-1 3-ethyl-3-methylheptane C10H22 21157
17302-02-2 3,3-diethylhexane C10H22 21179
17302-04-4 4-ethyl-4-methylheptane C10H22 21161
17302-11-3 3-ethylnonane C11H24 22895
17302-12-4 5-ethylnonane C11H24 22897
17302-13-5 4-propyloctane C11H24 22914
17302-14-6 2,2-dimethylnonane C11H24 22898
17302-15-7 3,3-dimethylnonane C11H24 22905
17302-16-8 3-methyl-3-ethyloctane C11H24 22920
17302-17-9 3,3-diethylheptane C11H24 22961
17302-18-0 4,4-dimethylnonane C11H24 22910
17302-19-1 4-methyl-4-ethyloctane C11H24 22925
17302-20-4 4-methyl-4-propylheptane C11H24 22956
17302-21-5 4,4-diethylheptane C11H24 22964
17302-22-6 3,4-dimethylnonane C11H24 22906
17302-23-7 4,5-dimethylnonane C11H24 22911
17302-24-8 2,4-dimethylnonane C11H24 22900
17302-25-9 3,5-dimethylnonane C11H24 22907
17302-26-0 4,6-dimethylnonane C11H24 22912
17302-27-1 2,5-dimethylnonane C11H24 22901
17302-28-2 2,6-dimethylnonane C11H24 22902
17302-29-3 2,7-dimethylnonane C11H24 22903
17302-30-6 2,8-dimethylnonane C11H24 22904
17302-31-7 3,6-dimethylnonane C11H24 22908
17302-32-8 3,7-dimethylnonane C11H24 22909
17302-33-9 6-methylundecane C12H26 24918
17302-36-2 5-ethyldecane C12H26 24921
17302-37-3 2,2-dimethyldecane C12H26 24922
17302-38-4 3,3-dimethyldecane C12H26 24930
17302-39-5 3-methyl-3-ethylnonane C12H26 24951
17302-40-8 3,3-diethyloctane C12H26 25018
17312-39-9 4,4-dimethyldecane C12H26 24936
17312-40-2 4-methyl-4-ethylnonane C12H26 24957
17312-41-3 4-methyl-4-propyloctane C12H26 25009
17312-42-4 4,4-diethyloctane C12H26 25022
17312-43-5 4-ethyl-4-propylheptane C12H26 25116
17312-44-6 2,3-dimethyldecane C12H26 24923
17312-45-7 3,4-dimethyldecane C12H26 24931
17312-46-8 4,5-dimethyldecane C12H26 24937
17312-48-0 3,5-dimethyldecane C12H26 24932
17312-49-1 4,6-dimethyldecane C12H26 24938
17312-50-4 2,5-dimethyldecane C12H26 24925
17312-51-5 2,7-dimethyldecane C12H26 24927
17312-52-6 2,8-dimethyldecane C12H26 24928
17312-53-7 3,6-dimethyldecane C12H26 24933
17312-54-8 3,7-dimethyldecane C12H26 24934
17312-55-9 3,8-dimethyldecane C12H26 24935
17312-56-0 4,7-dimethyldecane C12H26 24939
17312-57-1 3-methyldodecane C13H28 26528
17312-63-9 5-butylnonane C13H28 26532
17312-64-0 2,2-dimethylundecane C13H28 26529
17312-77-5 2,3-dimethylundecane C13H28 26530
17312-80-0 2,4-dimethylundecane C13H28 26531
17348-59-3 isopropyl-tert-butyl-ether C7H16O 12211
17373-27-2 dimethylnonylamine C11H25N 23093
17373-28-3 dimethylundecylamine C13H29N 26558
17373-29-4 dimethyltridecylamine C15H33N 28934
17447-00-6 methyl cis-3-methyl-2-pentenoate C7H12O2 11486
17447-01-7 methyl trans-3-methyl-2-pentenoate C7H12O2 11487
17453-92-8 5,5-dimethyldecane C12H26 24940
17616-44-3 1,3-diiodo-2-methylpropane C4H8I2 3409
17618-76-7 2-methyl-3-heptene C8H16 14915
17678-60-3 dimethylpentadecylamine C17H37N 30372
17760-40-6 1,1,1-trichlorohexane C6H11Cl3 8066
17773-64-7 2-chloro-3-methyl-1-butene C5H9Cl 5045
17773-66-9 2,2-dichloro-3-methylbutane C5H10Cl2 5346
17804-35-2 benomyl C14H18N4O3 27476
17839-26-8 ethylpentylamine C7H17N 12323
17851-27-3 1,2,4-trimethyl-6-ethylbenzene C11H16 22352
17912-17-3 1,5-dibromooctane C8H16Br2 14997
17952-11-3 ethyl pentyl ether C7H16O 12215
18094-01-4 2-methyl-1-tridecene C14H28 27852
18172-67-3 (1S)-(-)-b-pinene C10H16 20366
18217-12-4 5-methyl-2-heptanone C8H16O 15148
18231-53-3 3,3-dimethyl-2-ethyl-1-butene C8H16 14992
18259-05-7 2,3,4,5,6-pentachlorobiphenyl C12H5Cl5 23177
18262-85-6 1-ethyl-2,3,5-trimethylbenzene C11H16 22355
18281-05-5 ethyl eicosanoate C22H44O2 33469
18295-25-5 2-bromo-3-methylbutane C5H11Br 5609
18295-27-7 2-iodo-3-methylbutane C5H11I 5666
18295-59-5 2-propylpentanal C8H16O 15065
18338-40-4 1,2,3-trichlorobutane C4H7Cl3 3161
18362-97-5 isopropyl pentanoate C8H16O2 15192
18371-13-6 2-methyl-2-ethyl-1-butanol C7H16O 12198
18371-24-9 1,3-diiodobutane C4H8I2 3403
18435-20-6 2,3-dimethyltridecane C15H32 28908
18435-22-8 3-methyltetradecane C15H32 28906
18435-23-9 2,3-dimethyltetradecane C16H34 29758
18435-45-5 1-nonadecene C19H38 31698
18437-89-3 butyl hexadecyl sulfide C20H42S 32543
18476-57-8 4,5-dimethyl-2,6-octadiene C10H18 20614
18479-57-7 2,6-dimethyl-2-octanol C10H22O 21262
18495-30-2 1,1,2,3-tetrachloropropane C3H4Cl4 1568
18516-37-5 2-methyl-1-undecene C12H24 24824
18593-92-5 3-methyl-2-isopropyl-1-butanol C8H18O 15516

18610-33-8 4-chloro-trans-2-pentene C5H9Cl 5036
18636-94-7 dimethyl-3-pentylamine C7H17N 12326
18675-20-2 2,4-dimethyl-2-octanol C10H22O 21261
18720-62-2 2-methyl-3-heptanol C8H18O 15446
18720-65-5 5-methyl-3-heptanol C8H18O 15449
18720-66-6 6-methyl-3-heptanol C8H18O 15450
18787-63-8 2-hexadecanone C16H32O 29718
18787-64-9 3-hexadecanone C16H32O 29719
18800-53-8 3,4-dimethylthiophenol C8H10S 14046
18869-72-2 9-methylheptadecane C18H38 31132
18908-66-2 2-ethylhexyl bromide C8H17Br 15250
18932-14-4 2,2,4-trimethyloctane C11H24 22929
18936-17-9 2-methylbutanenitrile C5H9N 5184
18938-47-1 3-methyl-3,4-hexanediol C7H16O2 12238
18970-44-0 1-ethyl-2-isopropylbenzene C11H16 22327
19009-56-4 2-methyl-1-decanal C11H22O 22834
19031-61-9 3-fluoro-3-methylpentane C6H13F 8715
19037-72-0 4,4-dimethylcyclopentene C7H12 11335
19089-47-5 2-ethoxy-1-propanol C5H12O2 5882
19113-78-1 3,5-dimethyl-3-hexanol, (±) C8H18O 15521
19132-06-0 (2S,3S)-(+)-2,3-butanediol C4H10O2 3896
19182-11-7 1,2-diethylnaphthalene C14H16 27285
19182-13-9 1,6-diethylnaphthalene C14H16 27287
19184-10-2 cis-1-fluoro-1-propene C3H5F 1711
19218-94-1 1-iodotetradecane C14H29I 27901
19264-94-9 3,3-dimethyl-1-pentanol C7H16O 12185
19269-28-4 3-methylhexanal C7H14O 11828
19313-57-6 butyl decyl sulfide C14H30S 27952
19313-61-2 ethyl decyl sulfide C12H26S 25314
19374-46-0 1,trans-2,cis-3-trimethylcyclopentane C8H16 14872
19398-53-9 2,4-dibromopentane C5H10Br2 5304
19398-61-9 2,5-dichlorotoluene C7H6Cl2 10258
19398-75-5 1,3-diethylcyclopentane C9H18 17876
19398-77-7 3,4-diethylhexane C10H22 21180
19411-41-7 3,3,4-trimethyl-2-pentanol C8H18O 15510
19447-29-1 2-nonyne C9H16 17594
19482-57-6 2,2-dimethyl-3-octyne C10H18 20598
19484-26-5 1-tridecanethiol C13H28S 26549
19486-08-9 2,2-dimethylhexadecane C18H38 31135
19489-10-2 1-methyl-cis-3-ethylcyclohexane C9H18 17921
19522-69-1 N-butylethylenediamine C6H16N2 9297
19549-70-3 2,2-dimethyl-3-heptanol C9H20O 18280
19549-71-4 2,3-dimethyl-3-heptanol C9H20O 18281
19549-73-6 2,6-dimethyl-3-heptanol C9H20O 18282
19549-74-7 3,5-dimethyl-3-heptanol C9H20O 18283
19549-77-0 2,4-dimethyl-4-heptanol C9H20O 18287
19549-78-1 3,3-dimethyl-4-heptanol C9H20O 18289
19549-79-2 3,5-dimethyl-4-heptanol C9H20O 18290
19549-83-8 2,6-dimethyl-3-heptanone C9H18O 18045
19549-84-9 3,5-dimethyl-4-heptanone C9H18O 18001
19549-97-4 2,6-dimethyl-3-heptyne C9H16 17598
19550-03-9 2,3-dimethyl-2-hexanol C8H18O 15469
19550-05-1 3,4-dimethyl-2-hexanol C8H18O 15473
19550-07-3 2,5-dimethyl-3-hexanol C8H18O 15483
19550-08-4 3,4-dimethyl-3-hexanol C8H18O 15484
19550-09-5 4,4-dimethyl-3-hexanol C8H18O 15486
19550-10-8 3,4-dimethyl-2-hexanone C8H16O 15070
19550-14-2 4,4-dimethyl-3-hexanone C8H16O 15071
19550-30-2 2,3-dimethyl-1-butanol C6H14O 8983
19550-48-2 1,4-dimethylcyclopentene C7H12 11328
19550-81-3 3,4-dimethyl-cis-2-hexene C8H16 14948
19550-82-4 3,4-dimethyl-trans-2-hexene C8H16 14949
19550-83-5 4,4-dimethyl-cis-2-hexene C8H16 14952
19550-87-9 3,4-dimethyl-cis-3-hexene C8H16 14968
19550-88-0 3,4-dimethyl-trans-3-hexene C8H16 14969
19689-18-0 4-decene C10H20 20944
19780-10-0 5-dodecanone C12H24O 24862
19780-39-3 3-ethyl-2-heptanol C9H20O 18272
19780-41-7 3-ethyl-3-heptanol C9H20O 18279
19780-44-0 4-ethyl-3-hexanol C8H18O 15479
19780-63-3 2-methyl-3-ethyl-2-pentanol C8H18O 15504
19780-65-5 3-ethyl-3-methyl-2-pentanone C8H16O 15147
19780-66-6 2-methyl-3-ethyl-1-pentene C8H16 14974
19780-67-7 2-methyl-3-ethyl-2-pentene C8H16 14982
19780-90-6 2,4-hexanediol C6H14O2 9029
19780-96-2 2,8-dimethyl-5-nonanol C11H24O 23061
19781-10-3 2,3-dimethyl-3-octanol C10H22O 21268
19781-11-4 2,7-dimethyl-4-octanol C10H22O 21279
19781-13-6 4,7-dimethyl-4-octanol C10H22O 21283
19781-24-9 3,3-dimethyl-2-pentanol C7H16O 12191
19781-27-2 6-ethyl-3-octanol C10H22O 21266
19781-63-6 4-ethyl-trans-2-hexene C8H16 14944
19781-68-1 1-methyl-cis-2-ethylcyclopropane C6H12 8201
19781-69-2 1-methyl-trans-2-ethylcyclopropane C6H12 8202
19841-72-6 2,2-dimethyl-3-octanol C10H22O 21267
19889-37-3 2-ethyl-2-methylbutanoic acid C7H14O2 11885
19902-08-0 beta-pinene, (1R) C10H16 20324
19902-98-8 1,cis-3-dimethyl-trans-2-ethylcyclopentane C9H18 17893
19902-99-9 1,trans-3-dimethyl-cis-2-ethylcyclopentane C9H18 17894
19903-00-5 1,cis-3-dimethyl-cis-2-ethylcyclopentane C9H18 17892
19907-40-5 1,trans-2,trans-3,cis-4-tetramethylcyclopentane C9H18 17914
19961-27-4 ethylisopropylamine C5H13N 5960
20024-90-2 1-ethyl-4-propylbenzene C11H16 22326
20024-91-3 1-ethyl-3-propylbenzene C11H16 22325
20038-12-4 4-bromo-2-methyl-1-butene C5H9Br 4983
20063-92-7 trans-3-nonene C9H18 17944
20089-07-0 2-methyl-1-butanethiol, (+) C5H12S 5929
20150-89-4 5,7-dimethylquinoline C11H11N 21718
20184-89-8 3-nonyne C9H16 17595
20184-91-2 4-nonyne C9H16 17596
20193-20-8 ethylpropylamine C5H13N 5959
20193-21-9 N-propylbutylamine C7H17N 12344
20198-49-6 4-methyl-2-hexyne C7H12 11346
20237-34-7 1,trans-3-hexadiene C6H10 7625
20278-84-6 2,4,5-trimethylheptane C10H22 21171
20278-85-7 2,3,5-trimethylheptane C10H22 21168
20278-87-9 3,3,4-trimethylheptane C10H22 21174
20278-88-0 3,4,4-trimethylheptane C10H22 21176
20278-89-1 3,4,5-trimethylheptane C10H22 21177
20281-83-8 3-methyl-1-pentanol, (±) C6H14O 8992
20281-85-0 2,3-dimethyl-1-butanol, (±)- C6H14O 8984
20281-86-1 (RS)-2-hexanol C6H14O 8989
20281-91-8 3,3-dimethyl-2-butanol, (±) C6H14O 8988
20291-60-5 methyl hexyl sulfide C7H16S 12302
20291-61-6 methyl heptyl sulfide C8H18S 15607
20291-91-2 3-ethyl-2,2-dimethylhexane C10H22 21181
20291-95-6 2,2,5-trimethylheptane C10H22 21164
20309-77-7 1,1,cis-3,cis-4-tetramethylcyclopentane C9H18 17905
20327-65-5 trans-1-fluoro-1-propene C3H5F 1712
20352-67-4 ethylhexylamine C8H19N 15636
20395-24-8 1,1-dichlorooctane C8H16Cl2 15005
20395-25-9 1,1,3-trichloropropane C3H5Cl3 1705
20487-40-5 tert-butyl propanoate C7H14O2 11904
20488-34-0 2-methyl-cis-3-heptene C8H16 14916
20515-19-9 methyl trans-3-pentenoate C6H10O2 7860
20592-10-3 2-ethyl-1-octanol C10H22O 21252
20626-52-2 (S)-(-)-2-methylbutylamine C5H13N 5964
20634-92-8 N-ethyl-N-propyl-1-propanamine C8H19N 15644
20654-44-8 2,3-dimethylbutanenitrile C6H11N 8094
20654-46-0 2,2-dimethylbutanenitrile C6H11N 8093
20667-05-4 2-methyl-1,2-pentanediol C6H14O2 9032
20668-26-2 3,6-dimethylquinoline C11H11N 21711
20668-28-4 3,7-dimethylquinoline C11H11N 21712
20668-29-5 3,8-dimethylquinoline C11H11N 21713
20668-30-8 5,6-dimethylquinoline C11H11N 21717
20668-33-1 6,7-dimethylquinoline C11H11N 21719
20669-04-9 3,3-dimethyl-2-pentanone C7H14O 11839
20687-01-8 1-iodo-2-methyl-1-propene C4H7I 3218
20710-38-7 3-methyl-trans-2-hexene C7H14 11720
20748-86-1 2,4-heptanediol C7H16O2 12231
20754-04-5 3-methyl-4-octanone C9H18O 18003
20780-49-8 3,7-dimethyloctanyl acetate C12H24O2 24880
20809-46-5 7-methyl-4-octanone C9H18O 18007
20843-07-6 3,4-dimethylfuran C6H8O 7386
21020-26-8 3-ethyl-1-pentyne C7H12 11349
21020-27-9 4-methyl-2-pentyne C6H10 7653
21035-44-9 ethyl-sec-butylamine C6H15N 9199
21078-72-8 3-methyl-3-nonanol C10H22O 21243
21087-64-9 metribuzin C8H14N4OS 14538
21101-88-2 3-methylpentanenitrile C6H11N 8091
21102-09-0 4,4-dimethyl-3-ethyl-2-pentanol C9H20O 18327
21102-13-6 2,3,4-trimethyl-2-hexanol C9H20O 18303
21266-88-6 2,3-dibromoheptane C7H14Br2 11755
21266-90-0 3,4-dibromoheptane C7H14Br2 11756
21472-86-6 1,2,3-trichloro-5-methylbenzene C7H5Cl3 9971
21508-07-6 2,3-heptanediol C7H16O2 12230
21531-91-9 1,3-hexanediol C6H14O2 9024
21570-35-4 2-methyl-4-heptanol C8H18O 15451
21571-91-5 3,3-dichloropentane C5H10Cl2 5338
21586-21-0 2-methyl-2,4-pentanediamine C6H16N2 9302
21609-90-5 C13H10BrCl2O2PS o-(4-bromo-2,5-dichloroph 25576
21725-46-2 cyanazine C9H13ClN6 17238
21777-84-4 [2-methyl-1-isopropylpropyl]benzene C13H20 26286
21948-47-0 2,3-dichlorooctane C8H16Cl2 15008
21964-49-8 1,13-tetradecadiene C14H26 27784
21968-17-2 propylisopropylamine C6H15N 9204
21987-21-3 12-methyl-1-tridecanol C14H30O 27925
22025-20-3 3,3-dimethyl-2-hexanol C8H18O 15472
22037-73-6 3-bromo-1-butene C4H7Br 3050
22160-12-9 2-methylhexanoic acid, (±) C7H14O2 11887
22160-39-0 2-methylpentanoic acid, (±) C6H12O2 8465
22160-40-3 3-methylpentanoic acid, (±) C6H12O2 8466
22160-41-4 4-methylhexanoic acid, (±) C7H14O2 11889
22224-92-6 fenamiphos C13H22NO3PS 26367
22227-75-4 3,3-dichloro-2-methyl-1-propene C4H6Cl2 2819
22287-02-1 2-methyl-5-nonanone C10H20O 20996
22417-45-4 4-methyl-2-isobutyl-1-pentanol C10H22O 21323
22438-39-7 methyl decyl sulfide C11H24S 23086
22539-65-7 1,2-dimethyl-3-isopropylbenzene C11H16 22336
22663-64-5 3-methyl-1-nonanol C10H22O 21235
22768-17-8 3-methyl-cis-3-heptene C8H16 14919
22768-18-9 3-methyl-trans-3-heptene C8H16 14920
22768-19-0 3-methyl-cis-2-heptene C8H16 14905
22768-20-3 3-methyl-trans-2-heptene C8H16 14906
22808-06-6 2,2,5,5-tetramethylhex-3-ene C10H20 20954
22967-92-6 methylmercury CH3Hg 220
23009-73-6 1-chloro-2-methyl-trans-2-butene C5H9Cl 5049
23009-74-7 1-chloro-2-methyl-cis-2-butene C5H9Cl 5048
23010-04-0 1,2-dichloro-2-methylbutane C5H10Cl2 5341
23010-05-1 1,1-dichloro-2-methylbutane C5H10Cl2 5339
23010-07-3 1,3-dichloro-2-methylbutane C5H10Cl2 5343
23068-94-2 3-bromo-cis-2-pentene C5H9Br 4974
23068-95-3 4-bromo-trans-2-pentene C5H9Br 4977
23074-36-4 2-bromo-1-butene C4H7Br 3049
23097-98-5 5,5-dimethyl-3-heptyne C9H16 17599
23135-22-0 oxamyl C7H13N3O3S 11688
23171-85-9 2,3,3-trimethyl-2-pentanol C8H18O 15507
23361-67-3 2,2-dimethylpropyl formate C6H12O2 8475
23361-69-5 2,2-dimethylpropyl butanoate C9H18O2 18069
23418-37-3 4-methyl-4-octanol C9H20O 18261
23418-38-4 4-methyl-4-nonanol C10H22O 21246
23425-81-2 1-fluoro-3-methyl-2-butene C5H9F 5136
23433-04-7 1,3-heptanediol C7H16O2 12225
23511-78-6 1,1,3-tribromopropane C3H5Br3 1665
23646-57-3 erythro-4,4-dimethyl-2,3-pentanediol C7H16O2 12258
23646-58-4 threo-4,4-dimethyl-2,3-pentanediol C7H16O2 12259
23708-56-7 6-undecanol C11H24O 23060
23764-31-0 N-propyl-1,3-propanediamine C6H16N2 9300
24298-08-6 2-iodo-trans-2-butene C4H7I 3217
24298-09-7 2-iodo-cis-2-butene C4H7I 3216
24308-61-0 2-iodo-1-butene C4H7I 3211
24319-05-2 3-iodo-3-methylpentane C6H13I 8734
24319-07-1 2-iodo-3-methylpentane C6H13I 8731
24319-09-3 2-chloro-3-methylpentane C6H13Cl 8676
24346-53-0 1-iodo-3-methylpentane C6H13I 8728
24347-58-8 (2R,3R)-(-)-2,3-butanediol C4H10O2 3897
24356-00-1 3-chloro-1-pentene C5H9Cl 5026
24443-15-0 1,3-dibromo-3-methylbutane C5H10Br2 5311
24448-19-9 3-ethyl-2-hexanol C8H18O 15467
24608-84-2 5-methyl-cis-2-heptene C8H16 14909
24608-85-3 5-methyl-trans-2-heptene C8H16 14910
24612-75-7 1,1,3,4-tetramethylcyclohexane C10H20 20936
24621-61-2 (S)-(+)-1,3-butanediol C4H10O2 3898
24768-42-1 butyl pentyl sulfide C9H20S 18370
24768-43-2 propyl hexyl sulfide C9H20S 18369
24768-44-3 ethyl heptyl sulfide C9H20S 18368
24768-46-5 propyl heptyl sulfide C10H22S 21390
24774-58-1 2,5-dibromohexane C6H12Br2 8257
24892-49-7 2,4-dimethyl-2,4-pentanediol C7H16O2 12254
24893-35-4 2,3-dimethyl-1,3-butanediol C6H14O2 9049
24910-63-2 3,4-dimethylpent-2-ene C7H14 11745
25013-15-4 methylstyrene C9H10 16519
25103-58-6 tert-dodecylmercaptan C12H26S 25317
25118-28-9 1,4-dibromohexane C6H12Br2 8250
25154-52-3 nonylphenol C15H24O 28757
25155-15-1 cymene C10H14 19869
25167-70-8 2,4,4-trimethylpentene C8H16 14990
25186-47-4 3,5-dichlorotoluene C7H6Cl2 10261
25265-71-8 dipropylene glycol C6H14O3 9076
25276-70-4 1-pentadecanethiol C15H32S 28921
25321-09-9 diisopropylbenzene C12H18 24359
25321-14-6 dinitrotoluene C7H6N2O4 10374
25323-30-2 dichloroethylene C2H2Cl2 618
25323-68-6 trichlorodiphenyl C12H7Cl3 23254
25339-17-7 isodecanol C10H22O 21225
25339-56-4 3-heptene (mixed isomers) C7H14 11711
25340-17-4 diethylbenzene C10H14 19877
25340-18-5 triethylbenzene; (mixed isomers) C12H18 24363
25346-31-0 3-bromo-3-methylpentane C6H13Br 8657
25346-32-1 2-chloro-4-methylpentane C6H13Cl 8677
25346-33-2 1-bromo-2-methylpentane C6H13Br 8650
25354-97-6 2-hexyldecanoic acid C16H32O2 29722
25354-98-7 2-methyl-1-pentadecanol C16H34O 29769
25360-10-5 tert-nonyl mercaptan, isomers C9H20S 18371

25376-45-8 toluenediamine C7H10N2 11107
25415-67-2 ethyl 4-methylpentanoate C8H16O2 15186
25415-70-7 isobutyl 4-methylpentanoate C10H20O2 21057
25419-06-1 methylpentylamine C6H15N 9191
25429-29-2 pentachlorobiphenyl C12H5Cl5 23180
25512-42-9 dichloro-1,1'-biphenyl C12H8Cl2 23278
25551-13-7 trimethyl benzene C9H12 17010
25570-03-0 (S)-(+)-2-methylbutyronitrile C5H9N 5195
25586-43-0 monochloronaphthalene C10H7Cl 18608
25639-42-3 methylcyclohexanol C7H14O 11877
25735-67-5 4-sec-amylphenol C11H16O 22442
25790-55-0 4-chloro-1,2-butadiene C4H5Cl 2635
26118-38-7 3,3-dimethyl-2-hexanone C8H16O 15069
26143-08-8 trans-1-iodo-1-nonene C9H17I 17810
26153-88-8 dimethylpentylamine C7H17N 12324
26153-91-3 methyl-2,2-dimethylpropylamine C6H15N 9197
26158-99-6 ethyl pentyl sulfide C7H16S 12303
26171-83-5 1,2-butanediol, (±) C4H10O2 3891
26184-62-3 (S)-(+)-2-pentanol C5H12O 5850
26186-00-5 1-heptadecyne C17H32 30284
26186-02-7 1-tridecyne C13H24 26402
26198-63-0 1,2-dichloropropane, (±) C3H6Cl2 1848
26249-12-7 dibromobenzene C6H4Br2 6414
26356-06-9 2-bromo-3,3-dimethylbutane C6H13Br 8662
26401-20-7 tert-hexyl alcohol C6H14O 9009
26423-60-9 3-chloro-cis-2-pentene C5H9Cl 5033
26423-61-0 3-chloro-trans-2-pentene C5H9Cl 5034
26423-63-2 4-chloro-cis-2-pentene C5H9Cl 5035
26490-07-3 2-isobutylnaphthalene C14H16 27272
26519-91-5 methylcyclopentadiene C6H8 7268
26533-31-3 2-methyl-4-nonanol C10H22O 21245
26533-33-5 2-methyl-3-nonanol C10H22O 21242
26533-34-6 2-methyl-3-octanol C9H20O 18254
26533-35-7 3-methyl-4-octanol C9H20O 18260
26533-36-8 2-nonadecanol C19H40O 31743
26549-24-6 2-hexanol, (R) C6H14O 8990
26549-25-7 3-heptanol, (S) C7H16O 12205
26550-55-0 (S)-(+)-1-methoxy-2-propanol C4H10O2 3903
26638-19-7 dichloropropane C3H6Cl2 1849
26730-14-3 7-methyltridecane C14H30 27918
26761-40-0 diisodecyl phthalate C28H46O4 34691
26761-81-9 mixo-dichlorobutane C4H8Cl2 3371
26896-20-8 neodecanoic acid C10H20O2 21067
26952-21-6 isooctanol C8H18O 15535
26952-23-8 dichloropropylene C3H4Cl2 1542
26963-33-7 2-isopentylthiophene C9H14S 17520
26981-81-7 diethylheptylamine C11H25N 23094
26981-98-6 7-methyl-4-nonanol C10H22O 21248
27193-28-8 tert-octylphenol C14H22O 27688
27196-00-5 tetradecanol; mixed isomers C14H30O 27929
27314-13-2 norflurazon C12H9ClF3N3O 23370
27370-94-1 cis-1-iodo-1-nonene C9H17I 17809
27522-11-8 2-ethyl-1-pentanol C7H16O 12180
27554-26-3 diisooctyl phthalate C24H38O4 33924
27563-68-4 methyl hexadecyl sulfide C17H36S 30364
27574-98-7 2,3,4,5,6-pentamethylheptane C12H26 25217
27610-92-0 2-butyloctanoic acid C12H24O2 24879
27644-49-1 3-methyl-2-octanol C9H20O 18251
27656-50-4 trans-2,2,4,6,6-pentamethyl-3-heptene C12H24 24827
27802-85-3 3,4,7-trimethylnonane C12H26 24995
27831-13-6 4-ethenyl-1,2-dimethylbenzene C10H12 19379
28028-91-3 cis-1-fluoro-1-heptene C7H13F 11613
28122-24-9 3-ethyldiphenylmethane C15H16 28390
28122-25-0 2-ethyldiphenylmethane C15H16 28389
28123-70-8 5-chlorononane C9H19Cl 18123
28148-04-1 1,3-dibromo-2-methylpropane C4H8Br2 3342
28170-54-9 1,1-dimethyl-3-(4-methoxy-phenyl) urea C10H14N2O2 19950
28292-42-4 3-aminoheptane C7H17N 12345
28292-43-5 5-methyl-2-hexylamine C7H17N 12346
28303-42-6 dodecyl formate C13H26O2 26494
28320-88-9 3-chloro-3-methyloctane C9H19Cl 18120
28320-89-0 3-chloro-3-ethylheptane C9H19Cl 18119
28467-75-6 2-tridecyne C13H24 26403
28519-06-4 decyl chloride (mixed isomers) C10H21Cl 21094
28804-88-8 dimethylnaphthalene, isomers C12H12 23687
28823-41-8 2-methyl-2,4-hexadiene C7H12 11364
29022-31-9 3,3-dimethyl-4-nonyne C11H20 22677
29044-06-2 2-methyl-2,5-hexanediol C7H16O2 12235
29063-28-3 octyl alcohol; mixed isomers C8H18O 15538
29072-93-3 2-methyl-2-propoxypropane C7H16O 12219
29082-74-4 octachlorostyrene C8Cl8 12419
29316-05-0 sec-pentylbenzene C11H16 22356
29342-44-7 2-chloro-2,5-dimethylhexane C8H17Cl 15253
29369-63-9 methyltetradecylamine C15H33N 28932
29394-58-9 1-iodo-2-methylbutane C5H11I 5663
29443-50-3 2,2-diiodobutane C4H8I2 3405
29472-95-5 5-methyl-2-sec-butylphenol C11H16O 22406
29488-27-5 3-isopropylthiophene C7H10S 11228
29559-52-2 1,1,2-trichloro-2-methylpropane C4H7Cl3 3166
29559-54-4 1,1-dichloro-2,2-dimethylpropane C5H10Cl2 5349
29559-55-5 1,3-dichloro-2,2-dimethylpropane C5H10Cl2 5350
29576-14-5 1-bromo-trans-2-butene C4H7Br 3053
29662-90-6 2,2-dimethyloctanoic acid C10H20O2 21041
29664-53-7 methylpentadecylamine C16H35N 29801
29682-39-1 3-chloro-4-bromobenzene C6H4BrCl 6385
29772-40-5 2,2,3-trimethyl-3-heptanol C10H22O 21292
29833-69-0 2-methyl-1-pentadecene C16H32 29708
29843-62-7 2-methyl-5-nonanol C10H22O 21249
29882-57-3 2-chloropentane, (+) C5H11Cl 5629
29887-11-4 2-ethyl-1,3-pentanediol C7H16O2 12243
29916-45-8 1,2-dibromo-2,3-dimethylbutane C6H12Br2 8246
30090-17-6 4-isobutylaniline C10H15N 20205
30122-12-4 1,3-dichloropentane C5H10Cl2 5332
30136-13-1 n-propoxypropanol (mixed isomers) C6H14O2 9068
30207-98-8 undecanol C11H24O 23062
30213-29-7 3,cis-5-dimethylcyclopentene C7H12 11333
30273-11-1 4-sec-butylaniline C10H15N 20200
30310-22-6 2-bromo-4-methylpentane C6H13Br 8655
30388-40-0 3-methyl-4-propylphenol C10H14O 19998
30540-31-9 1-bromo-2,3-dimethylbutane C6H13Br 8659
30586-10-8 dichloropentane C5H10Cl2 5351
30673-60-0 propyl decanoate C13H26O2 26503
30749-25-8 3-isobutylphenol C10H14O 19985
30899-19-5 pentanol C5H12O 5851
31080-39-4 4-nonanecarboxylic acid C10H20O2 21077
31087-44-2 (R)-(-)-2-pentanol C5H12O 5852
31242-93-0 chlorinated diphenyl oxide C12H4Cl6O 23166
31294-91-4 3-iodohexane C6H13I 8726
31294-94-7 1-iodo-2-methylpentane C6H13I 8727
31294-95-8 2-iodo-2-methylpentane C6H13I 8730
31294-96-9 2-iodo-4-methylpentane C6H13I 8732
31294-97-0 3-iodo-2-methylpentane C6H13I 8733
31294-98-1 1-iodo-2,2-dimethylbutane C6H13I 8735
31295-00-8 1-iodo-2,3-dimethylbutane C6H13I 8736
31366-00-4 1,2,3-trimethyl-5-ethylbenzene C11H16 22349
31367-46-1 3-methyl-2-heptanol C8H18O 15442
31508-00-6 2,3',4,4',5-pentachlorobiphenyl C12H5Cl5 23182
31772-05-1 1,13-dibromotridecane C13H26Br2 26473
31805-48-8 3-methyl-2-ethylthiophene C7H10S 11232
31841-77-7 5,5-dimethyl-2-hexanol C8H18O 15477
31844-95-8 2-bromo-1-pentene C5H9Br 4966
31844-96-9 2-bromo-3-methyl-1-butene C5H9Br 4986
31849-75-9 cis-1-bromo-1-pentene C5H9Br 4964
31849-76-0 trans-1-bromo-1-pentene C5H9Br 4965
31849-78-2 cis-1-bromo-1-butene C4H7Br 3047
31950-55-7 4-bromo-3-methyl-1-butene C5H9Br 4988
31950-56-8 4-bromo-1-pentene C5H9Br 4968
32367-54-7 2-ethyl-[1,2,3,4-tetrahydronaphthalene] C12H16 24113
32442-47-0 2-iodo-3-methyl-1-butene C5H9I 5169
32444-34-1 3-methyl-2-ethyl-1-butanol C7H16O 12199
32598-11-1 2,3',4',5-tetrachlorobiphenyl C12H6Cl4 23210
32598-13-3 3,3',4,4'-tetrachlorobiphenyl C12H6Cl4 23207
32598-14-4 2,3,3',4,4'-pentachlorobiphenyl C12H5Cl5 23183
32620-08-9 trans-1-bromo-1-butene C4H7Br 3048
32665-23-9 isopropyl 3-methylbutanoate C8H16O2 15191
32762-51-9 bromophenol C6H5BrO 6674
32764-34-4 6-methyl-3-heptyl acetate C10H20O2 21047
32768-54-0 2,3-dichlorotoluene C7H6Cl2 10257
32812-23-0 2,2-dimethyl-1,4-butanediol C6H14O2 9047
32814-16-7 cis-1-fluoro-1-octene C8H15F 14766
32814-17-8 trans-1-fluoro-1-octene C8H15F 14767
33021-02-2 tert-butyl isobutyl ether C8H18O 15531
33083-83-9 5-undecanone C11H22O 22826
33113-10-9 neoheptanoic acid C7H14O2 11926
33146-45-1 2,6-dichlorobiphenyl C12H8Cl2 23273
33213-65-9 endosulfan 2 C9H6Cl6O3S 15878
33234-93-4 7-methyl-1-nonanol C10H22O 21234
33284-53-6 2,3,4,5-tetrachlorobiphenyl C12H6Cl4 23204
33429-72-0 4-chloro-2,2-dimethylpentane C7H15Cl 11992
33467-73-1 cis-3-hexenyl formate C7H12O2 11500
33553-93-4 2,4-dichloro-2,4-dimethylpentane C7H14Cl2 11766
33673-01-7 2-sec-butyl-1,3-propanediol C7H16O2 12263
33693-78-6 1,1-dibromo-2-methylpropane C4H8Br2 3340
33698-87-2 cis-3-pentenoic acid C5H8O2 4899
33758-16-6 (S)-(+)-2-decanol C10H22O 21338
33879-72-0 3-methyl-1,3-pentanediol C6H14O2 9036
33933-77-6 7-methyl-4-octanol C9H20O 18264
33933-78-7 5-methyl-5-nonanol C10H22O 21251
33933-79-8 2,4-dimethyl-4-octanol C10H22O 21276
33943-21-4 2-methyl-4-ethyl-3-hexanol C9H20O 18308
33966-50-6 2-butanamine, (.+/-.)- C4H11N 3981
33978-71-1 2-methyl-5-undecanol C12H26O 25283
33979-03-2 2,2',4,4',6,6'-hexachlorobiphenyl C12H4Cl6 23160
34075-28-0 2-nitro-2,3-dimethylbutane C6H13NO2 8804
34123-59-6 isoproturon C12H18N2O 24411
34263-68-8 4-methyl-1-hexylamine C7H17N 12321
34317-39-0 methyl-1,2-dimethylpropylamine C6H15N 9196
34338-96-0 (2S,5S)-2,5-hexanediol C6H14O2 9062
34375-89-8 3-methylpyrrolidine C5H11N 5674
34566-04-6 (S)-2-aminooctane C8H19N 15648
34566-05-7 (R)-2-aminooctane C8H19N 15649
34590-94-8 dipropylene glycol monomethyl ether C7H16O3 12282
34713-94-5 2-methyl-1-butanol, (±) C5H12O 5835
34715-98-5 1-nitro-2,2-dimethylpropane C5H11NO2 5703
34722-01-5 3-butylthiophene C8H12S 14411
34786-24-8 3-methyl-(1-thiapropyl)-benzene C9H12S 17215
34859-98-8 1,1-dimethylbutyl acetate C8H16O2 15161
34883-39-1 2,5-dichlorobiphenyl C12H8Cl2 23272
34883-43-7 2,4'-dichloro-1,1'-biphenyl C12H8Cl2 23279
34887-14-4 2,2-dichloropentane C5H10Cl2 5335
34949-22-9 1,1-dimethylpropyl propanoate C8H16O2 15172
35065-27-1 2,2',4,4',5,5'-hexachloro-1,1'-biphenyl C12H4Cl6 23163
35073-27-9 2-methylbutyl formate C6H12O2 8470
35161-70-7 methylhexylamine C7H17N 12322
35205-79-9 3-methylnonanoic acid C10H20O2 21042
35216-11-6 7-tetradecyne C14H26 27782
35320-23-1 (R)-(-)-2-amino-1-propanol C3H9NO 2213
35399-81-6 2-amino-3-methylpentane C6H15N 9180
35400-43-2 sulprofos C12H19O2PS3 24527
35599-77-0 1-iodotridecane C13H27I 26518
35693-92-6 2,4,6-trichlorobiphenyl C12H7Cl3 23248
35693-99-3 2,2',5,5'-tetrachlorobiphenyl C12H6Cl4 23209
35866-89-8 2,6-dimethyl-4-isopropylheptane C12H26 25136
35866-96-7 2,4,4,7-tetramethyloctane C12H26 25094
35895-39-7 3-iodo-cis-2-pentene C5H9I 5157
35897-16-6 2-methyl-3-pentyl acetate C8H16O2 15165
35902-57-9 ethyldodecylamine C14H31N 27958
35951-33-8 2-chloro-2,4-dimethylpentane C7H15Cl 11990
35951-36-1 2,6-dichloro-2,6-dimethylheptane C9H18Cl2 17960
36186-96-6 3-methyl-5-propylphenol C10H14O 19999
36230-28-1 2-butyl-[1,2,3,4-tetrahydronaphthalene] C14H20 27561
36301-29-8 1-methyl-2-isobutylbenzene C11H16 22318
36343-05-2 methylheptylamine C8H19N 15635
36355-01-8 hexabromobiphenyl C12H4Br6 25427
36556-06-6 5,6,7,8-tetrahydroisoquinoline C9H11N 16894
36566-80-0 2-methyl-3-hexyne C7H12 11348
36653-82-4 1-hexadecanol C16H34O 29766
36668-55-0 cis-1-bromo-2-methyl-1-butene C5H9Br 4980
36691-77-7 2-methyl-4-tridecanol C14H30O 27927
36794-64-6 2,3,3-trimethyl-1-butanol C7H16O 12201
36839-67-5 methyl 1-ethylpropyl ether C6H14O 9000
36880-72-5 3-ethyl-cis-2-hexene C8H16 14941
36903-89-6 4-chloro-4-methyloctane C9H19Cl 18121
36903-95-4 4-methyl-3,5-octadiene C9H16 17607
37050-06-9 7-methyl-3-octyne C9H16 17597
37493-70-2 5-undecanol C11H24O 23059
37549-89-6 2,4-dimethyl-cis-3-hexene C8H16 14964
37615-53-5 N-methyldihexylamine C13H29N 26562
37680-65-2 2,2',5-trichlorobiphenyl C12H7Cl3 23249
37680-73-2 2,2',4,5,5'-pentachlorobiphenyl C12H5Cl5 23178
37796-58-0 1,4-diethylnaphthalene C14H16 27286
38169-04-9 1-iodo-trans-2-butene C4H7I 3215
38300-67-3 1,2,4-tribromobutane C4H7Br3 3089
38360-81-5 3,5-dimethylthiophenol C8H10S 14047
38380-02-8 2,2',3,4,5'-pentachlorobiphenyl C12H5Cl5 23181
38380-07-3 2,2',3,3',4,4'-hexachlorobiphenyl C12H4Cl6 23159
38380-08-4 2,3,3',4,4',5-hexachlorobiphenyl C12H4Cl6 23162
38384-05-3 3-chloro-2-methylpentane C6H13Cl 8678
38395-42-5 4-ethyl-4-octanol C10H22O 21274
38411-22-2 2,2',3,3',6,6'-hexachlorobiphenyl C12H4Cl6 23161
38444-81-4 2,3,5-trichlorobiphenyl C12H7Cl3 23251
38444-86-9 2,3',4'-trichlorobiphenyl C12H7Cl3 23250
38444-93-8 2,2',3,3'-tetrachlorobiphenyl C12H6Cl4 23208
38514-02-2 3-methyl-1-octanol C9H20O 18245
38514-03-3 4-methyl-1-octanol C9H20O 18246
38514-04-4 5-methyl-1-octanol C9H20O 18247
38514-05-5 6-methyl-1-octanol C9H20O 18248
38514-15-7 3-isopropyl-1-heptanol C10H22O 21285
38721-71-0 dichlorobenzyl chloride C7H5Cl3 9976
38836-25-8 4-methyl-2,4-hexanediol C7H16O2 12240
38842-05-6 1,2,3,5-tetraethylbenzene C14H22 27646
38857-76-0 1-butyl-[1,2,3,4-tetrahydronaphthalene] C14H20 27560
39093-27-1 methylnonylamine C10H23N 21401
39198-07-7 methylethylisopropylamine C6H15N 9207

39294-88-7 3(4)-methylstyrene, isomers C9H10 16520
39478-78-9 5-bromo-2-methylaniline C7H8BrN 10740
39563-53-6 2-bromodecane C10H21Br 21091
39616-19-8 1-bromo-cis-2-butene C4H7Br 3052
39622-45-2 1,3-dimethyl-6-ethylnaphthalene C14H16 27296
39761-57-4 3,4,4-trimethyl-trans-2-pentene C8H16 14988
39761-61-0 5,5-dimethyl-cis-2-hexene C8H16 14957
39761-64-3 3,4,4-trimethyl-cis-2-pentene C8H16 14987
39782-43-9 5,5-dimethyl-trans-2-hexene C8H16 14958
39884-48-5 4-aminobutan-2-ol C4H11NO 4000
39924-57-7 cis-1-bromo-1-heptene C7H13Br 11569
39924-58-8 cis-1-bromo-1-nonene C9H17Br 17785
40117-45-1 2,2,6,6-tetramethylheptane C11H24 22996
40137-22-2 3-methylamino-1,2-propanediol C4H11NO2 4010
40225-75-0 6-methyl-3-octanol C9H20O 18257
40239-35-8 2-methyl-3-tridecanone C14H28O 27865
40276-93-5 3-methyl-1-hexyne C7H12 11343
40289-98-3 methyl octadecyl sulfide C19H40S 31752
40323-88-4 2-methyl-5-ethyl thiophene C7H10S 11231
40575-41-5 2-methyl-4-octanol C9H20O 18259
40589-14-8 2-methyl-1-nonanol C10H22O 21231
40646-07-9 1,4-heptanediol C7H16O2 12226
40813-84-1 butyl heptyl sulfide C11H24S 23089
41004-19-7 1-iodo-3-methyl-2-butene C5H9I 5174
41051-72-3 2-methyl-1,2-butanediol C5H12O2 5870
41065-91-2 3-ethylhexanoic acid C8H16O2 15158
41065-95-6 3-ethyl-1-hexanol C8H18O 15455
41394-05-2 4-amino-3-methyl-6-phenyl-1,2,4-triazin-5-on C10H10N4O 19076
41446-63-3 trans-7-tetradecene C14H28 27853
41464-40-8 2,2',4',5-tetrachlorobiphenyl C12H6Cl4 23205
41761-11-9 hexane, 2,5-dichloro-, (R*,S*)- C6H12Cl2 8280
41761-12-0 2,5-dichlorohexane, (R*,R*)-(±)- C6H12Cl2 8278
41781-17-3 1-amino-2,2-dimethylbutane C6H15N 9184
41909-29-9 3-fluoropentane C5H11F 5651
41947-84-6 ethyl octadecyl sulfide C20H42S 32541
42007-73-8 2,7-dimethyl-2-octanol C10H22O 21263
42027-23-6 2,3-pentanediol C5H12O2 5864
42067-48-1 4-methyl-3-ethyl-cis-2-pentene C8H16 14983
42067-49-2 4-methyl-3-ethyl-trans-2-pentene C8H16 14984
42075-32-1 (R,R)-(-)-2,4-pentanediol C5H12O2 5888
42131-85-1 2-chloro-1-pentene C5H9Cl 5025
42131-89-5 2,2-dichlorohexane C6H12Cl2 8276
42131-98-6 2-chloro-cis-2-pentene C5H9Cl 5031
42132-00-3 2-chloro-trans-2-pentene C5H9Cl 5032
42152-47-6 7-methyl-1,6-octadiene C9H16 17628
42245-37-4 1-amino-3-methylpentane C6H15N 9177
42296-74-2 hexadiene C6H10 7660
42328-76-7 2,4-dimethyl-2-hexanol C8H18O 15470
42398-73-2 6-methylisoquinoline C10H9N 18859
42474-20-4 1,3-dibromopentane C5H10Br2 5299
42474-21-5 1,2-dibromoheptane C7H14Br2 11752
42551-55-3 3-amino-2-butanol C4H11NO 3990
42599-17-7 trans-1-iodo-1-octene C8H15I 14773
42764-74-9 2-methyl-1-heptadecene C18H36 31073
42841-80-5 propyl pentyl sulfide C8H18S 15609
42843-49-2 cis-1-bromo-1-octene C8H15Br 14736
42856-62-2 2-methyl-1,5-pentanediol C6H14O2 9034
42875-41-2 tridecyl formate C14H28O2 27874
43197-78-0 3-chloro-3-methylhexane C7H15Cl 12000
44745-29-1 (S)-2-heptylamine C7H17N 12342
44855-57-4 2-octanamine, (±) C8H19N 15634
44976-81-0 N,N-diisopropylisobutylamine C10H23N 21407
44979-90-0 diethylhexylamine C10H23N 21403
45124-35-4 diethylnonylamine C13H29N 26559
45165-81-9 methyltridecylamine C14H31N 27957
49542-74-7 2,4,8-trimethylnonane C12H26 24980
49557-09-7 2,5,8-trimethylnonane C12H26 24984
49598-54-1 2,2-dimethyldodecane C14H30 27915
49623-50-9 1,3-dibromo-2-methylbutane C5H10Br2 5310
50273-84-2 5-bromo-cis-2-pentene C5H9Br 4978
50341-35-0 2,2-dibromobutane C4H8Br2 3337
50422-80-5 5-methyl-cis-3-heptene C8H16 14925
50468-22-9 3-methyl-1,2-butanediol C5H12O2 5874
50586-18-0 cis-1-chloro-1-hexene C6H11Cl 8006
50586-19-1 trans-1-chloro-1-hexene C6H11Cl 8007
50623-57-9 butyl nonanoate C13H26O2 26502
50876-31-8 trans-1,1,3,5-tetramethylcyclohexane C10H20 20937
50876-32-9 cis-1,1,3,5-tetramethylcyclohexane C10H20 20938
50876-33-0 1,1,3,3-tetramethylcyclopentane C9H18 17904
50902-82-4 3-ethyl-3-methylhexanoic acid C9H18O2 18054
51065-65-7 6-methyl-trans-2-heptene C8H16 14913
51079-52-8 4,6-dimethyl-2-heptanol C9H20O 18277
51079-79-9 4,6,6-trimethyl-2-heptanol C10H22O 21288
51115-64-1 2-methylbutyl butanoate C9H18O2 18064
51116-73-5 1-bromo-3-methylpentane C6H13Br 8651
51200-80-7 4-methyl-3-ethyl-3-hexanol C9H20O 18310
51200-81-8 2-methyl-3-isopropyl-3-hexanol C10H22O 21310
51200-82-9 4-isopropyl-4-heptanol C10H22O 21297
51200-83-0 2,4-dimethyl-3-isopropyl-3-pentanol C10H22O 21327
51218-45-2 metolachlor C15H22ClNO2 28653
51261-14-4 (<+->)-2-chlorooctane C8H17Cl 15263
51333-70-1 2,7-nonadiene C9H16 17606
51750-65-3 2,2,4,4-tetramethylhexane C10H22 21193
51751-87-2 trans-1-bromo-1-octene C8H15Br 14737
51872-48-1 3-bromo-2-methyl-1-butene C5H9Br 4982
51877-74-8 (+)-trans-permethrin C21H20Cl2O3 32744
51891-58-8 2,2-difluoro-3-methylbutane C5H10F2 5374
51932-19-5 methyl-1-methylbutylamine C6H15N 9192
52019-78-0 (S)-(+)-2-hexanol C6H14O 9008
52112-25-1 2-pentadecyne C15H28 28821
52125-53-8 ethyl ether of propylene glycol C5H12O2 5890
52254-38-3 2-methyl-1-tetradecene C15H30 28860
52254-50-9 2-methyl-1-nonadecene C20H40 32477
52315-44-1 1-octylcyclopentene C13H24 26406
52317-98-3 methyl-3-methylbutylamine C6H15N 9194
52341-32-9 trans-(±)-permethrin C21H20Cl2O3 32745
52341-33-0 (±)-cis-permethrin C21H20Cl2O3 32741
52356-93-1 cis-1-iodo-1-octene C8H15I 14772
52390-72-4 2-heptanol, (±) C7H16O 12204
52497-07-1 1,3-dichloro-1-butene C4H6Cl2 2811
52627-73-3 tert-decanoic acid C10H20O2 21068
52645-53-1 permethrin C21H20Cl2O3 32738
52663-71-5 2,2',3,3',4,4',6-heptachlorobiphenyl C12H3Cl7 23152
52663-77-1 2,2',3,3',4,5,5',6,6'-nonachlorobiphenyl C12HCl9 23134
52670-32-3 2,3,5,6-tetramethylheptane C11H24 23004
52670-33-4 2,3,3,4,5-pentamethylhexane C11H24 23041
52670-34-5 2,3,6,7-tetramethyloctane C12H26 25091
52670-35-6 2,3,3,5,6-pentamethylheptane C12H26 25213
52670-36-7 2,3,3,4,4,5-hexamethylhexane C12H26 25262
52688-75-2 3-fluorohexane C6H13F 8707
52688-89-8 1,1,2,2-tetramethylcyclopentane C9H18 17899
52708-03-9 4-nonanol, (S) C9H20O 18242
52713-81-2 4-methyl-1-hexyne C7H12 11344
52783-44-5 heptadecanol (mixed primary isomers) C17H36O 30355
52812-57-4 cis-1-iodo-2-methyl-1-butene C5H9I 5163
52841-28-8 methylethyl-tert-butylamine C7H17N 12335
52896-87-4 4-isopropylheptane C10H22 21153
52896-88-5 4-ethyl-2-methylheptane C10H22 21155
52896-89-6 4-ethyl-3-methylheptane C10H22 21158
52896-90-9 3-ethyl-5-methylheptane C10H22 21159
52896-91-0 3-ethyl-4-methylheptane C10H22 21160
52896-92-1 2,2,3-trimethylheptane C10H22 21162
52896-93-2 2,3,3-trimethylheptane C10H22 21166
52896-95-4 2,3,4-trimethylheptane C10H22 21167
52896-99-8 4-ethyl-2,2-dimethylhexane C10H22 21182
52897-00-4 3-ethyl-2,3-dimethylhexane C10H22 21183
52897-01-5 4-ethyl-2,3-dimethylhexane C10H22 21184
52897-03-7 4-ethyl-2,4-dimethylhexane C10H22 21186
52897-04-8 3-ethyl-2,5-dimethylhexane C10H22 21187
52897-05-9 4-ethyl-3,3-dimethylhexane C10H22 21188
52897-06-0 3-ethyl-3,4-dimethylhexane C10H22 21189
52897-08-2 2,2,3,4-tetramethylhexane C10H22 21191
52897-09-3 2,2,3,5-tetramethylhexane C10H22 21192
52897-10-6 2,3,3,4-tetramethylhexane C10H22 21196
52897-11-7 2,3,3,5-tetramethylhexane C10H22 21197
52897-12-8 2,3,4,4-tetramethylhexane C10H22 21198
52897-15-1 2,3,4,5-tetramethylhexane C10H22 21199
52897-16-2 3,3-diethyl-2-methylpentane C10H22 21202
52897-17-3 3-ethyl-2,2,3-trimethylpentane C10H22 21203
52897-18-4 3-ethyl-2,2,4-trimethylpentane C10H22 21204
52897-19-5 3-ethyl-2,3,4-trimethylpentane C10H22 21205
53045-71-9 3-bromo-1-pentene C5H9Br 4967
53078-85-6 2-bromo-5-methylaniline C7H8BrN 10736
53119-51-0 2-methyl-3-ethylthiophene C7H10S 11229
53120-74-4 3,3-dimethyl-1,5-pentanediol C7H16O2 12255
53121-23-6 1-iodo-cis-2-butene C4H7I 3214
53161-72-1 1,2-diiodobutane C4H8I2 3402
53178-20-4 2-chlorobutane C4H9Cl 3614
53184-67-1 nonyl propanoate C12H24O2 24870
53225-40-4 3,trans-4-dimethylcyclopentene C7H12 11332
53268-66-9 cis-1-chloro-1-heptene C7H13Cl 11587
53268-67-0 trans-1-chloro-1-heptene C7H13Cl 11588
53310-02-4 2-amino-2-methylpentane C6H15N 9179
53346-71-7 3-pentadecanol C15H32O 28914
53392-86-2 1,1,1-trifluorooctane C8H15F3 14770
53434-74-5 trans-1-bromo-1-heptene C7H13Br 11570
53434-75-6 trans-1-bromo-1-nonene C9H17Br 17786
53469-21-9 chlorodiphenyl C12H6Cl4 23206
53510-18-2 5-methyl-trans-3-heptene C8H16 14926
53543-44-5 5-chloro-cis-2-pentene C5H9Cl 5037
53566-37-3 5-methyl-2-hexyne C7H12 11347
53594-82-4 2,2-dimethylheptadecane C19H40 31738
53696-14-3 6-methyldecanoic acid C11H22O2 22851
53731-22-9 1,1-difluoro-3-methylbutane C5H10F2 5368
53731-23-0 1,1-difluoro-2,2-dimethylpropane C5H10F2 5377
53731-24-1 3-fluoro-2-methyl-1-butene C5H9F 5127
53750-52-0 4-iodo-2-methyl-1-butene C5H9I 5166
53774-20-2 2-methyl-3-butenoic acid C5H8O2 4893
53778-43-1 1-methyl-1-ethylcyclopropane C6H12 8200
53778-73-7 1-methoxy-2-butanol C5H12O2 5887
53897-51-1 3,4-dimethyl-2-thiapentane C6H14S 9124
53966-53-3 2-methyl-3-nonene C10H20 20948
53971-47-4 2,2,3-trimethylthiacyclopropane C5H10S 5594
54004-41-0 4-methyl-2-propyl-1-pentanol C9H20O 18321
54004-43-2 1-methylbutyl propanoate C8H16O2 15168
54068-75-6 cis-1-iodo-1-butene C4H7I 3209
54105-66-7 undecylcyclohexane C17H34 30311
54105-76-9 isopentylcyclohexane C11H22 22793
54166-32-4 2,6,6-trimethyloctane C11H24 22944
54265-17-7 trans-1-bromo-2-methyl-1-butene C5H9Br 4981
54268-02-9 1,2,2,3-tetrabromopropane C3H4Br4 1519
54287-41-1 3-amino-2-methylpentane C6H15N 9182
54305-87-2 2,3-dichlorohexane C6H12Cl2 8277
54334-64-4 diethylundecylamine C15H33N 28935
54363-49-4 methylpentadiene, isomers C6H10 7659
54410-69-4 (3-methylpentyl)benzene C12H18 24373
54411-01-7 1-methyl-2-pentylcyclohexane C12H24 24820
54462-68-9 2,5-dibromohexane, (R*,R**)-(±)- C6H12Br2 8254
54462-70-3 1,4-dibromo-2,3-dimethylbutane C6H12Br2 8247
54518-00-2 (2,4-dimethylpentyl)benzene C13H20 26288
54549-80-3 1,1-dimethyl-2-ethylcyclopentane C9H18 17879
54616-49-8 4-ethyl-cis-2-hexene C8H16 14943
54630-50-1 5-methyl-2-heptanol C8H18O 15444
54630-82-9 DL-2,4-dimethyl-1,5-pentanediol C7H16O2 12251
54638-10-7 sec-pentyl acetate (R) C7H14O2 11900
54653-26-8 2,2-dibromopentane C5H10Br2 5302
54653-27-9 3,3-dibromopentane C5H10Br2 5305
54653-28-0 3-bromo-trans-2-pentene C5H9Br 4975
54653-29-1 2-bromo-cis-2-pentene C5H9Br 4972
54653-30-4 2-bromo-trans-2-pentene C5H9Br 4973
54676-39-0 2-butyl-1,1,3-trimethylcyclohexane C13H26 26467
54774-45-7 (+)-cis-permethrin C21H20Cl2O3 32742
54774-46-8 (-)-cis-permethrin C21H20Cl2O3 32740
54837-90-0 4-methyl-N-propylaniline C10H15N 20209
54877-00-8 5-methyl-2,4-hexanediol C7H16O2 12242
54890-00-5 1,1-dicyclohexylbutane C16H30 29643
54934-91-7 1,1-dicyclohexylpropane C15H28 28815
55073-86-4 2,2,5,5-tetramethyl-3-hexanol C10H22O 21317
55156-16-6 3-methyl-2-pentanone, (±) C6H12O 8415
55219-65-3 C14H18ClN3O2 2-(4-chlorophenoxy)-1-tert- 27432
55289-36-6 3-bromo-2-methylaniline C7H8BrN 10741
55499-04-2 2,2,3-trimethylnonane C12H26 24964
55505-23-2 2,2,3-trimethyl-1-butanol C7H16O 12200
55505-24-3 4-methyl-2-isopropyl-1-pentanol C9H20O 18322
55505-26-5 8-methyl-1-nonanol C10H22O 21334
55751-54-7 2-sec-butylaniline C10H15N 20197
55757-34-1 1,1,1-trifluorononane C9H17F3 17808
55837-18-8 2-(4-isobutylphenyl)butyric acid C10H20O2 21074
55930-45-5 1,4-diiodopentane C5H10I2 5383
56039-55-5 3,cis-4-dimethylcyclopentene C7H12 11331
56065-42-0 3,5-dimethyl-3-octanol C10H22O 21270
56065-43-1 4,6-dimethyl-4-octanol C10H22O 21282
56255-50-6 2-methyl-1,2-hexanediol C7H16O2 12233
56298-90-9 4-methyl-2-heptanol C8H18O 15443
56399-98-5 5-iodo-trans-2-pentene C5H9I 5162
56523-26-3 3,7-dimethyl-2,4-octadiene C10H18 20613
56535-63-8 4-bromo-cis-2-pentene C5H9Br 4976
56539-66-3 3-methoxy-3-methylbutanol C6H14O2 9063
56800-09-0 1,1-dichloro-2-butene C4H6Cl2 2813
56962-04-0 3-bromo-5-chlorophenol C6H4BrClO 6386
57093-84-2 3-ethyl-2-thiapentane C6H14S 9122
57233-31-5 2,2,4-trimethyl-4-heptanol C10H22O 21298
57253-29-9 1-bromo-2-methyl-cis-2-butene C5H9Br 4989
57253-30-2 1-bromo-2-methyl-trans-2-butene C5H9Br 4990
57303-85-2 dimethyl-2-pentylamine C7H17N 12325
57409-53-7 2,2,3-trimethyl-1-pentanol C8H18O 15497
57456-98-1 2-methylbutanal, (±) C5H10O 5449
57465-28-8 3,3',4,4',5-pentachlorobiphenyl C12H5Cl5 23184
57706-88-4 3,7-dimethyl-3-octanol, (±) C10H22O 21331
57716-79-7 2-ethyl-1,4-butanediol C6H14O2 9045
57757-60-5 dimethyl-1,1-dimethylpropylamine C7H17N 12329
57837-19-1 metalaxyl C15H21NO4 28623
58046-40-5 2,3,6-trimethyl-3-heptanol C10H22O 21294
58049-91-5 3,3-dimethylcyclopentene C7H12 11330

58133-26-9 1,1-dibromohexane C6H12Br2 8245
58175-57-8 2-propyl-1-pentanol C8H18O 15489
58368-66-4 1-methylbutyl formate C6H12O2 8469
58368-67-5 1-ethylpropyl formate C6H12O2 8472
58430-94-7 3,5,5-trimethylhexyl acetate C11H22O2 22856
58654-67-4 5-methyl-2-octanone C9H18O 18005
58766-17-9 2-chloro-5-methylhexane C7H15Cl 11999
58795-24-7 5,6-dimethyl-2-heptanol C9H20O 18278
59006-05-2 1-fluoro-2,2-dimethylpropane C5H11F 5656
59104-79-9 1,1-dibromoheptane C7H14Br2 11751
59104-80-2 1,1-dibromodecane C10H20Br2 20958
59128-91-5 3-iodo-3-methyl-1-butene C5H9I 5170
59129-74-7 4,5-dimethylquinoline C11H11N 21714
59204-02-3 3,7-dimethyl-1-octanol, (±) C10H22O 21332
59222-86-5 2,2-dimethyltetradecane C16H34 29757
59562-82-2 3,3-dimethyl-1,2-butanediol C6H14O2 9053
59570-04-6 ethylundecylamine C13H29N 26557
59570-06-8 ethyltridecylamine C15H33N 28933
59643-75-3 2,3-dimethyl-cis-3-hexene C8H16 14962
59654-13-6 2-iodohexane C6H13I 8725
59734-23-5 5-methyl-4-octanol C9H20O 18262
59756-60-4 fluridone C19H14F3NO 31279
59871-24-8 trans-1-chloro-1-octene C8H15Cl 14750
59889-45-1 2-chloro-2,3-dimethylpentane C7H15Cl 11989
59967-14-5 4-iodo-1-pentene C5H9I 5151
59973-07-8 methyl nonyl sulfide C10H22S 21388
59973-08-9 ethyl nonyl sulfide C11H24S 23087
60186-78-9 3-tridecyne C13H24 26404
60212-29-5 2-undecyne C11H20 22673
60212-30-8 3-undecyne C11H20 22674
60212-31-9 4-undecyne C11H20 22675
60212-32-0 3-tetradecyne C14H26 27781
60247-14-5 methylethylisobutylamine C7H17N 12334
60251-57-2 2-bromooctane, (±)- C8H17Br 15249
60291-32-9 C6H6Cl6 1,2,3,4,5,6-hexachlorocyclohexan 6951
60302-21-8 4-tert-butylheptane C11H24 22953
60302-23-0 2,2,4,4,5-pentamethylhexane C11H24 23039
60302-24-1 2,2,3,3,5,5-hexamethylhexane C12H26 25259
60302-27-4 2,2,3,3,4,4-hexamethylpentane C11H24 23048
60302-28-5 2,2-dimethyl-3,3-diethylpentane C11H24 23044
60415-61-4 1-methylbutyl butanoate C9H18O2 18063
60435-70-3 2-methyl-1-heptanol C8H18O 15436
60595-37-1 trans-1-iodo-1-heptene C7H13I 11620
60669-40-1 3-undecene C11H22 22798
60671-32-1 3-methyl-2-nonanol C10H22O 21237
60671-36-5 2-methyl-3-undecanol C12H26O 25282
60836-07-9 2,4,6-trimethyl-4-heptanol C10H22O 21302
61064-08-2 3,4-dimethyl-1-pentyne C7H12 11351
61228-10-2 5-ethyl-5-methyl-3-heptyne C10H18 20600
61350-03-6 1,1-difluorooctane C8H16F2 15016
61394-27-2 3,trans-5-dimethylcyclopentene C7H12 11334
61539-75-1 4,5-dibromooctane C8H16Br2 14999
61764-94-1 4-chloro-4-methylheptane C8H17Cl 15258
61827-85-8 1,3-dimethyl-4-propylbenzene C11H16 22333
61827-86-9 1,2,3-trimethyl-4-ethylbenzene C11H16 22348
61827-87-0 1,2,4-trimethyl-3-ethylbenzene C11H16 22350
61828-00-0 1-methyl-1-isopropylcyclopentane C9H18 17868
61828-01-1 1-methyl-trans-2-isopropylcyclopentane C9H18 17870
61828-02-2 1-methyl-cis-3-isopropylcyclopentane C9H18 17871
61828-03-3 1-methyl-trans-3-isopropylcyclopentane C9H18 17872
61828-35-1 DL-erythro-2,3-pentanediol C5H12O2 5866
61828-36-2 DL-threo-2,3-pentanediol C5H12O2 5867
61847-78-7 2,4-dimethyl-trans-3-hexene C8H16 14965
61847-80-1 4-methyl-3-ethyl-1-pentene C8H16 14976
61847-96-9 2-heptadecyne C17H32 30285
61868-01-7 1-methyl-cis-2-isopropylcyclopentane C9H18 17869
61868-02-8 2,3-dimethylhexadecane C18H38 31136
61868-03-9 2,3-dimethylheptadecane C19H40 31739
61868-04-0 2,3-dimethyloctadecane C20H42 32519
61868-05-1 2,4-dimethyltridecane C15H32 28909
61868-06-2 2,4-dimethyltetradecane C16H34 29759
61868-07-3 2,4-dimethylpentadecane C17H36 30351
61868-08-4 2,4-dimethylhexadecane C18H38 31137
61868-09-5 2,4-dimethylheptadecane C19H40 31740
61868-10-8 2,4-dimethyloctadecane C20H42 32520
61868-19-7 2-methyl-1-hexadecene C17H34 30314
61868-20-0 2-methyl-1-octadecene C19H38 31699
61868-21-1 2,3-dimethyl-3-ethylheptane C11H24 22968
61868-22-2 2,3-dimethyl-4-ethylheptane C11H24 22969
61868-23-3 2,3-dimethyl-5-ethylheptane C11H24 22970
61868-24-4 2,4-dimethyl-3-ethylheptane C11H24 22971
61868-25-5 2,4-dimethyl-4-ethylheptane C11H24 22972
61868-26-6 2,4-dimethyl-5-ethylheptane C11H24 22973
61868-27-7 2,5-dimethyl-3-ethylheptane C11H24 22974
61868-28-8 2,5-dimethyl-4-ethylheptane C11H24 22975
61868-29-9 2,5-dimethyl-5-ethylheptane C11H24 22976
61868-30-2 2,6-dimethyl-3-ethylheptane C11H24 22977
61868-31-3 2,6-dimethyl-4-ethylheptane C11H24 22978
61868-32-4 3,3-dimethyl-4-ethylheptane C11H24 22979
61868-33-5 3,3-dimethyl-5-ethylheptane C11H24 22980
61868-34-6 3,4-dimethyl-3-ethylheptane C11H24 22981
61868-35-7 3,4-dimethyl-4-ethylheptane C11H24 22982
61868-36-8 3,4-dimethyl-5-ethylheptane C11H24 22983
61868-37-9 3,5-dimethyl-3-ethylheptane C11H24 22984
61868-38-0 3,5-dimethyl-4-ethylheptane C11H24 22985
61868-39-1 4,4-dimethyl-3-ethylheptane C11H24 22986
61868-40-4 2,2,3,3-tetramethylheptane C11H24 22987
61868-41-5 2,2,3,4-tetramethylheptane C11H24 22988
61868-42-6 2,2,3,5-tetramethylheptane C11H24 22989
61868-43-7 2,2,3,6-tetramethylheptane C11H24 22990
61868-44-8 2,2,4,4-tetramethylheptane C11H24 22991
61868-45-9 2,2,4,5-tetramethylheptane C11H24 22992
61868-46-0 2,2,4,6-tetramethylheptane C11H24 22993
61868-47-1 2,2,5,5-tetramethylheptane C11H24 22994
61868-48-2 2,2,5,6-tetramethylheptane C11H24 22995
61868-49-3 2,3,3,4-tetramethylheptane C11H24 22997
61868-50-6 2,3,3,5-tetramethylheptane C11H24 22998
61868-51-7 2,3,3,6-tetramethylheptane C11H24 22999
61868-52-8 2,3,4,4-tetramethylheptane C11H24 23000
61868-53-9 2,3,4,5-tetramethylheptane C11H24 23001
61868-54-0 2,3,4,6-tetramethylheptane C11H24 23002
61868-55-1 2,3,5,5-tetramethylheptane C11H24 23003
61868-56-2 2,4,4,5-tetramethylheptane C11H24 23005
61868-57-3 2,4,4,6-tetramethylheptane C11H24 23006
61868-58-4 2,4,5,5-tetramethylheptane C11H24 23007
61868-59-5 3,3,4,4-tetramethylheptane C11H24 23008
61868-60-8 3,3,4,5-tetramethylheptane C11H24 23009
61868-61-9 3,3,5,5-tetramethylheptane C11H24 23010
61868-62-0 3,4,4,5-tetramethylheptane C11H24 23011
61868-63-1 2,2-dimethyl-3-isopropylhexane C11H24 23012
61868-64-2 2,3-dimethyl-3-isopropylhexane C11H24 23013
61868-65-3 2,4-dimethyl-3-isopropylhexane C11H24 23014
61868-66-4 2,5-dimethyl-3-isopropylhexane C11H24 23015
61868-67-5 2-methyl-3,3-diethylhexane C11H24 23016
61868-68-6 2-methyl-3,4-diethylhexane C11H24 23017
61868-69-7 2-methyl-4,4-diethylhexane C11H24 23018
61868-70-0 3-methyl-3,4-diethylhexane C11H24 23019
61868-71-1 3-methyl-4,4-diethylhexane C11H24 23020
61868-72-2 2,2,3-trimethyl-3-ethylhexane C11H24 23021
61868-73-3 2,2,3-trimethyl-4-ethylhexane C11H24 23022
61868-74-4 2,2,4-trimethyl-3-ethylhexane C11H24 23023
61868-75-5 2,2,4-trimethyl-4-ethylhexane C11H24 23024
61868-76-6 2,2,5-trimethyl-3-ethylhexane C11H24 23025
61868-77-7 2,2,5-trimethyl-4-ethylhexane C11H24 23026
61868-78-8 2,3,3-trimethyl-4-ethylhexane C11H24 23027
61868-79-9 2,3,4-trimethyl-3-ethylhexane C11H24 23028
61868-80-2 2,3,4-trimethyl-4-ethylhexane C11H24 23029
61868-81-3 2,3,5-trimethyl-3-ethylhexane C11H24 23030
61868-82-4 2,3,5-trimethyl-4-ethylhexane C11H24 23031
61868-83-5 2,4,4-trimethyl-3-ethylhexane C11H24 23032
61868-84-6 3,3,4-trimethyl-4-ethylhexane C11H24 23033
61868-85-7 2,2,3,3,4-pentamethylhexane C11H24 23034
61868-86-8 2,2,3,3,5-pentamethylhexane C11H24 23035
61868-87-9 2,2,3,4,4-pentamethylhexane C11H24 23036
61868-88-0 2,2,3,4,5-pentamethylhexane C11H24 23037
61868-89-1 2,3,3,4,4-pentamethylhexane C11H24 23040
61868-90-4 2,2,4-trimethyl-3-isopropylpentane C11H24 23042
61868-91-5 2,3,4-trimethyl-3-isopropylpentane C11H24 23043
61868-92-6 2,4-dimethyl-3,3-diethylpentane C11H24 23045
61868-93-7 2,2,3,4-tetramethyl-3-ethylpentane C11H24 23046
61868-94-8 3,5,5-trimethyloctane C11H24 22951
61868-95-9 4,4,5-trimethyloctane C11H24 22952
61868-96-0 2-methyl-4-propylheptane C11H24 22954
61868-97-1 3-methyl-4-propylheptane C11H24 22955
61868-98-2 2-methyl-4-isopropylheptane C11H24 22958
61868-99-3 3-methyl-4-isopropylheptane C11H24 22959
61869-00-9 4-methyl-4-isopropylheptane C11H24 22960
61869-01-0 3,4-diethylheptane C11H24 22962
61869-02-1 3,5-diethylheptane C11H24 22963
61869-03-2 2,2-dimethyl-3-ethylheptane C11H24 22965
61869-04-3 2,2-dimethyltridecane C15H32 28907
61869-05-4 2,2-dimethylpentadecane C17H36 30349
61869-06-5 2,2-dimethyloctadecane C20H42 32518
61886-61-1 3-pentadecyne C15H28 28822
61886-62-2 3-hexadecyne C16H30 29642
61886-63-3 3-heptadecyne C17H32 30286
61949-76-6 cis-permethrin C21H20Cl2O3 32739
61949-77-7 trans-permethrin C21H20Cl2O3 32743
62016-13-1 3-isopropyl-2-methylhexane C10H22 21178
62016-14-2 2,5,6-trimethyloctane C11H24 22943
62016-15-3 4-isopropyloctane C11H24 22915
62016-16-4 2-methyl-3-ethyloctane C11H24 22916
62016-17-5 2-methyl-4-ethyloctane C11H24 22917
62016-18-6 2-methyl-5-ethyloctane C11H24 22918
62016-19-7 2-methyl-6-ethyloctane C11H24 22919
62016-20-0 3-methyl-4-ethyloctane C11H24 22921
62016-21-1 3-methyl-5-ethyloctane C11H24 22922
62016-22-2 3-methyl-6-ethyloctane C11H24 22923
62016-23-3 4-methyl-3-ethyloctane C11H24 22924
62016-24-4 4-methyl-5-ethyloctane C11H24 22926
62016-25-5 4-methyl-6-ethyloctane C11H24 22927
62016-26-6 2,2,3-trimethyloctane C11H24 22928
62016-27-7 2,2,5-trimethyloctane C11H24 22930
62016-28-8 2,2,6-trimethyloctane C11H24 22931
62016-29-9 2,2,7-trimethyloctane C11H24 22932
62016-30-2 2,3,3-trimethyloctane C11H24 22933
62016-31-3 2,3,4-trimethyloctane C11H24 22934
62016-32-4 2,3,5-trimethyloctane C11H24 22935
62016-33-5 2,3,6-trimethyloctane C11H24 22936
62016-34-6 2,3,7-trimethyloctane C11H24 22937
62016-35-7 2,4,4-trimethyloctane C11H24 22938
62016-36-8 2,4,5-trimethyloctane C11H24 22939
62016-37-9 2,4,6-trimethyloctane C11H24 22940
62016-38-0 2,4,7-trimethyloctane C11H24 22941
62016-39-1 2,5,5-trimethyloctane C11H24 22942
62016-40-4 3,3,4-trimethyloctane C11H24 22945
62016-41-5 3,3,5-trimethyloctane C11H24 22946
62016-42-6 3,3,6-trimethyloctane C11H24 22947
62016-43-7 3,4,4-trimethyloctane C11H24 22948
62016-44-8 3,4,5-trimethyloctane C11H24 22949
62016-45-9 3,4,6-trimethyloctane C11H24 22950
62016-46-0 2,2-dimethyl-4-ethylheptane C11H24 22966
62016-47-1 2,2-dimethyl-5-ethylheptane C11H24 22967
62016-48-2 methyl 2-methylbutyl ether C6H14O 8998
62016-49-3 methyl 1,2-dimethylpropyl ether C6H14O 9001
62016-59-5 1,cis-3-diethylcyclopentane C9H18 17877
62016-60-8 1,trans-3-diethylcyclopentane C9H18 17878
62016-61-9 1,1-dimethyl-3-ethylcyclopentane C9H18 17880
62016-62-0 1,trans-2-dimethyl-1-ethylcyclopentane C9H18 17882
62016-63-1 1,cis-2-dimethyl-1-ethylcyclopentane C9H18 17881
62016-64-2 1,cis-2-dimethyl-cis-4-ethylcyclopentane C9H18 17887
62016-65-3 1,cis-2-dimethyl-trans-4-ethylcyclopentane C9H18 17888
62016-66-4 1,trans-2-dimethyl-cis-4-ethylcyclopentane C9H18 17889
62016-67-5 1,trans-3-dimethyl-1-ethylcyclopentane C9H18 17891
62016-68-6 1,cis-3-dimethyl-1-ethylcyclopentane C9H18 17890
62016-70-0 1,1,cis-2,cis-3-tetramethylcyclopentane C9H18 17900
62016-72-2 1,1,cis-2,cis-4-tetramethylcyclopentane C9H18 17902
62016-73-3 1,2,2,cis-3-tetramethylcyclopentane C9H18 17907
62016-93-7 1-chloro-3-methylpentane C6H13Cl 8673
62016-94-8 1-chloro-4-methylpentane C6H13Cl 8674
62017-16-7 1,1-dichlorohexane C6H12Cl2 8273
62017-17-8 1,1-dichlorododecane C12H24Cl2 24833
62103-66-6 propyl nonyl sulfide C12H26S 25315
62108-56-9 1,1,1-trichlorodecane C10H19Cl3 20876
62108-65-0 1,1,3-trichloro-2-methylpropane C4H7Cl3 3167
62108-95-6 2-fluoro-3-methylbutane C5H11F 5655
62126-90-3 1,2-difluoropropane C3H6F2 1866
62126-91-4 1,1-difluoro-2-methylpropane C4H8F2 3395
62126-92-5 1,2-difluoro-2-methylpropane C4H8F2 3396
62126-93-6 1,3-difluoro-2-methylpropane C4H8F2 3397
62126-94-7 1,2-difluoropentane C5H10F2 5359
62126-95-8 1,3-difluoropentane C5H10F2 5360
62126-96-9 1,4-difluoropentane C5H10F2 5361
62127-02-0 1,1-difluorotridecane C13H26F2 26475
62127-03-1 1,1-difluorotetradecane C14H28F2 27857
62127-27-9 1-fluoro-2-methylpentane C6H13F 8708
62127-28-0 1-fluoro-3-methylpentane C6H13F 8709
62127-29-1 1-fluoro-4-methylpentane C6H13F 8710
62127-30-4 2-fluoro-2-methylpentane C6H13F 8711
62127-31-5 2-fluoro-3-methylpentane C6H13F 8712
62127-32-6 2-fluoro-4-methylpentane C6H13F 8713
62127-33-7 3-fluoro-2-methylpentane C6H13F 8714
62127-34-8 1-fluoro-2,2-dimethylbutane C6H13F 8716
62127-35-9 1-fluoro-2,3-dimethylbutane C6H13F 8717
62127-40-6 1,1-difluoropentane C5H10F2 5358
62127-41-7 1,1-difluorohexane C6H12F2 8298
62127-42-8 1,1-difluorononane C9H18F2 17964
62127-43-9 1,1-difluorodecane C10H20F2 20966
62127-44-0 1,1-difluoroundecane C11H22F2 22809
62127-45-1 1,1-difluorododecane C12H24F2 24838
62127-46-2 1,3,3-tribromobutane C4H7Br3 3090
62127-48-4 1,3-dibromo-2-(bromomethyl)propane C4H7Br3 3096
62127-58-6 1,1-dibromo-2-methylbutane C5H10Br2 5306
62127-59-7 1,1-dibromo-3-methylbutane C5H10Br2 5307

62127-60-0 2,2-dibromo-3-methylbutane C5H10Br2 5313
62127-61-1 1,1,1-tribromopropane C3H5Br3 1663
62127-62-2 1,1,1-tribromobutane C4H7Br3 3083
62154-74-9 3-iodo-1-butene C4H7I 3212
62154-77-2 m-ethylthiophenol C8H10S 14040
62154-92-1 trans-1-iodo-1-butene C4H7I 3210
62155-02-6 2-undecanethiol C11H24S 23085
62155-03-7 2-tridecanethiol C13H28S 26550
62155-04-8 2-tetradecanethiol C14H30S 27947
62155-05-9 2-pentadecanethiol C15H32S 28922
62155-06-0 2-octadecanethiol C18H38S 31150
62155-09-3 methyl tridecyl sulfide C14H30S 27949
62155-10-6 methyl pentadecyl sulfide C16H34S 29792
62155-11-7 2-thianonadecane C18H38S 31151
62155-12-8 methyl nonadecyl sulfide C20H42S 32540
62168-25-6 1,1-dibromobutane C4H8Br2 3333
62168-26-7 1,1-dibromooctane C8H16Br2 14994
62168-27-8 1,1-dibromononane C9H18Br2 17955
62168-28-9 1,1-dibromoundecane C11H22Br2 22805
62168-29-0 1,1-dibromododecane C12H24Br2 24831
62168-30-3 1,1-dibromotridecane C13H26Br2 26472
62168-41-6 2-bromo-3-methylpentane C6H13Br 8654
62168-42-7 1-bromo-2,2-dimethylbutane C6H13Br 8658
62183-50-0 2,5-dimethyl-6-ethyloctane C12H26 25039
62183-51-1 2,6-dimethyl-3-ethyloctane C12H26 25040
62183-52-2 2,6-dimethyl-4-ethyloctane C12H26 25041
62183-53-3 2,6-dimethyl-5-ethyloctane C12H26 25042
62183-54-4 2,6-dimethyl-6-ethyloctane C12H26 25043
62183-55-5 2,7-dimethyl-3-ethyloctane C12H26 25044
62183-56-6 2,7-dimethyl-4-ethyloctane C12H26 25045
62183-57-7 3,3-dimethyl-4-ethyloctane C12H26 25046
62183-58-8 3,3-dimethyl-5-ethyloctane C12H26 25047
62183-59-9 3,3-dimethyl-6-ethyloctane C12H26 25048
62183-60-2 3,4-dimethyl-4-ethyloctane C12H26 25050
62183-61-3 3,4-dimethyl-5-ethyloctane C12H26 25051
62183-62-4 3,4-dimethyl-6-ethyloctane C12H26 25052
62183-63-5 3,5-dimethyl-3-ethyloctane C12H26 25053
62183-64-6 3,5-dimethyl-4-ethyloctane C12H26 25054
62183-65-7 3,5-dimethyl-5-ethyloctane C12H26 25055
62183-66-8 3,5-dimethyl-6-ethyloctane C12H26 25056
62183-67-9 3,6-dimethyl-3-ethyloctane C12H26 25057
62183-68-0 3,6-dimethyl-4-ethyloctane C12H26 25058
62183-69-1 4,4-dimethyl-3-ethyloctane C12H26 25059
62183-70-4 4,4-dimethyl-5-ethyloctane C12H26 25060
62183-71-5 4,4-dimethyl-6-ethyloctane C12H26 25061
62183-72-6 4,5-dimethyl-3-ethyloctane C12H26 25062
62183-73-7 4,5-dimethyl-4-ethyloctane C12H26 25063
62183-74-8 2,2,3,3-tetramethyloctane C12H26 25064
62183-75-9 2,2,3,4-tetramethyloctane C12H26 25065
62183-76-0 2,2,3,5-tetramethyloctane C12H26 25066
62183-77-1 2,2,3,6-tetramethyloctane C12H26 25067
62183-78-2 2,2,3,7-tetramethyloctane C12H26 25068
62183-79-3 2,2,4,4-tetramethyloctane C12H26 25069
62183-80-6 2,2,4,5-tetramethyloctane C12H26 25070
62183-81-7 2,2,4,6-tetramethyloctane C12H26 25071
62183-82-8 2,2,4,7-tetramethyloctane C12H26 25072
62183-83-9 2,2,5,5-tetramethyloctane C12H26 25073
62183-84-0 2,2,5,6-tetramethyloctane C12H26 25074
62183-85-1 4-methyl-5-propyloctane C12H26 25010
62183-86-2 2-methyl-4-isopropyloctane C12H26 25012
62183-87-3 2-methyl-5-isopropyloctane C12H26 25013
62183-88-4 3-methyl-4-isopropyloctane C12H26 25014
62183-89-5 3-methyl-5-isopropyloctane C12H26 25015
62183-90-8 4-methyl-4-isopropyloctane C12H26 25016
62183-91-9 4-methyl-5-isopropyloctane C12H26 25017
62183-92-0 3,4-diethyloctane C12H26 25019
62183-93-1 3,5-diethyloctane C12H26 25020
62183-94-2 3,6-diethyloctane C12H26 25021
62183-95-3 2,2-dimethyl-3-ethyloctane C12H26 25024
62183-96-4 2,2-dimethyl-4-ethyloctane C12H26 25025
62183-97-5 2,2-dimethyl-5-ethyloctane C12H26 25026
62183-98-6 2,2-dimethyl-6-ethyloctane C12H26 25027
62183-99-7 2,3-dimethyl-3-ethyloctane C12H26 25028
62184-00-3 2,3-dimethyl-4-ethyloctane C12H26 25029
62184-01-4 2,3-dimethyl-5-ethyloctane C12H26 25030
62184-02-5 2,3-dimethyl-6-ethyloctane C12H26 25031
62184-03-6 2,4-dimethyl-3-ethyloctane C12H26 25032
62184-04-7 2,4-dimethyl-4-ethyloctane C12H26 25033
62184-05-8 2,4-dimethyl-5-ethyloctane C12H26 25034
62184-06-9 2,4-dimethyl-6-ethyloctane C12H26 25035
62184-07-0 2,5-dimethyl-3-ethyloctane C12H26 25036
62184-08-1 2,5-dimethyl-4-ethyloctane C12H26 25037
62184-09-2 2,5-dimethyl-5-ethyloctane C12H26 25038
62184-10-5 2,4,6-trimethylnonane C12H26 24978
62184-11-6 2,4,7-trimethylnonane C12H26 24979
62184-12-7 2,5,5-trimethylnonane C12H26 24981
62184-13-8 2,5,6-trimethylnonane C12H26 24982
62184-14-9 2,5,7-trimethylnonane C12H26 24983
62184-15-0 2,6,6-trimethylnonane C12H26 24985
62184-16-1 2,6,7-trimethylnonane C12H26 24986
62184-17-2 2,7,7-trimethylnonane C12H26 24987
62184-18-3 3,3,4-trimethylnonane C12H26 24988
62184-19-4 3,3,5-trimethylnonane C12H26 24989
62184-20-7 3,3,6-trimethylnonane C12H26 24991
62184-21-8 3,3,7-trimethylnonane C12H26 24990
62184-22-9 3,4,4-trimethylnonane C12H26 24992
62184-23-0 3,4,5-trimethylnonane C12H26 24993
62184-24-1 3,4,6-trimethylnonane C12H26 24994
62184-25-2 3,5,5-trimethylnonane C12H26 24996
62184-26-3 3,5,6-trimethylnonane C12H26 24997
62184-27-4 3,5,7-trimethylnonane C12H26 24998
62184-28-5 3,6,6-trimethylnonane C12H26 24999
62184-29-6 4,4,5-trimethylnonane C12H26 25000
62184-30-9 4,4,6-trimethylnonane C12H26 25001
62184-31-0 4,5,5-trimethylnonane C12H26 25002
62184-32-1 4-tert-butyloctane C12H26 25004
62184-33-2 2-methyl-4-propyloctane C12H26 25005
62184-34-3 2-methyl-5-propyloctane C12H26 25006
62184-35-4 3-methyl-4-propyloctane C12H26 25007
62184-36-5 3-methyl-5-propyloctane C12H26 25008
62184-37-6 2-methyl-4-ethylnonane C12H26 24947
62184-38-7 2-methyl-5-ethylnonane C12H26 24948
62184-39-8 2-methyl-6-ethylnonane C12H26 24949
62184-40-1 2-methyl-7-ethylnonane C12H26 24950
62184-41-2 3-methyl-4-ethylnonane C12H26 24952
62184-42-3 3-methyl-5-ethylnonane C12H26 24953
62184-43-4 3-methyl-6-ethylnonane C12H26 24954
62184-44-5 3-methyl-7-ethylnonane C12H26 24955
62184-45-6 4-methyl-3-ethylnonane C12H26 24956
62184-46-7 4-methyl-5-ethylnonane C12H26 24958
62184-47-8 4-methyl-6-ethylnonane C12H26 24959
62184-48-9 4-methyl-7-ethylnonane C12H26 24960
62184-49-0 5-methyl-3-ethylnonane C12H26 24961
62184-50-3 2,2,4-trimethylnonane C12H26 24965
62184-51-4 2,2,5-trimethylnonane C12H26 24966
62184-52-5 2,2,6-trimethylnonane C12H26 24967
62184-53-6 2,2,7-trimethylnonane C12H26 24968
62184-54-7 2,2,8-trimethylnonane C12H26 24969
62184-55-8 2,3,3-trimethylnonane C12H26 24970
62184-56-9 2,3,4-trimethylnonane C12H26 24971
62184-57-0 2,3,5-trimethylnonane C12H26 24972
62184-58-1 2,3,6-trimethylnonane C12H26 24973
62184-59-2 2,3,7-trimethylnonane C12H26 24974
62184-60-5 2,3,8-trimethylnonane C12H26 24975
62184-61-6 2,4,4-trimethylnonane C12H26 24976
62184-62-7 2,4,5-trimethylnonane C12H26 24977
62184-67-2 3-ethylthiacyclopentane C6H12S 8623
62184-71-8 4-isopropylnonane C12H26 24944
62184-72-9 5-isopropylnonane C12H26 24945
62184-73-0 2-methyl-3-ethylnonane C12H26 24946
62184-74-1 1-nonylcyclopentene C14H26 27779
62184-75-2 1-decylcyclopentene C15H28 28819
62184-76-3 1-undecylcyclopentene C16H30 29646
62184-82-1 1,3-dimethylcyclopentene C7H12 11327
62184-89-8 2,2-dimethyl-4,4-diethylhexane C12H26 25235
62184-90-1 2,3-dimethyl-3,4-diethylhexane C12H26 25236
62184-91-2 2,3-dimethyl-4,4-diethylhexane C12H26 25237
62184-92-3 2,4-dimethyl-3,3-diethylhexane C12H26 25238
62184-93-4 2,4-dimethyl-3,4-diethylhexane C12H26 25239
62184-94-5 2,5-dimethyl-3,3-diethylhexane C12H26 25240
62184-95-6 2,5-dimethyl-3,4-diethylhexane C12H26 25241
62184-96-7 3,3-dimethyl-4,4-diethylhexane C12H26 25242
62184-97-8 3,4-dimethyl-3,4-diethylhexane C12H26 25243
62184-98-9 2,2,3,3-tetramethyl-4-ethylhexane C12H26 25244
62184-99-0 2,2,3,4-tetramethyl-3-ethylhexane C12H26 25245
62185-00-6 2,2,3,4-tetramethyl-4-ethylhexane C12H26 25246
62185-01-7 2,2,3,5-tetramethyl-3-ethylhexane C12H26 25247
62185-02-8 2,2,3,5-tetramethyl-4-ethylhexane C12H26 25248
62185-03-9 2,2,4,4-tetramethyl-3-ethylhexane C12H26 25249
62185-04-0 2,2,4,5-tetramethyl-3-ethylhexane C12H26 25250
62185-05-1 2,2,4,5-tetramethyl-4-ethylhexane C12H26 25251
62185-06-2 2,2,5,5-tetramethyl-3-ethylhexane C12H26 25252
62185-07-3 2,3,3,4-tetramethyl-4-ethylhexane C12H26 25253
62185-08-4 2,3,3,5-tetramethyl-4-ethylhexane C12H26 25254
62185-09-5 2,3,4,4-tetramethyl-3-ethylhexane C12H26 25255
62185-10-8 2,3,4,5-tetramethyl-3-ethylhexane C12H26 25256
62185-11-9 2,2,3,3,4,4-hexamethylhexane C12H26 25257
62185-12-0 2,2,3,3,4,5-hexamethylhexane C12H26 25258
62185-13-1 2,2,3,4,4,5-hexamethylhexane C12H26 25260
62185-14-2 2,2,3,4,5,5-hexamethylhexane C12H26 25261
62185-15-3 2,2,4-trimethyl-3,3-diethylpentane C12H26 25264
62185-16-4 2,4-dimethyl-3-ethyl-3-isopropylpentane C12H26 25263
62185-17-5 2,2,3,4-tetramethyl-3-isopropylpentane C12H26 25265
62185-18-6 2,2,4,4-tetramethyl-3-isopropylpentane C12H26 25266
62185-19-7 3,4,4,6-tetramethyloctane C12H26 25108
62185-20-0 3,4,5,5-tetramethyloctane C12H26 25109
62185-21-1 3,4,5,6-tetramethyloctane C12H26 25110
62185-22-2 4,4,5,5-tetramethyloctane C12H26 25111
62185-23-3 2-methyl-4-tert-butylheptane C12H26 25112
62185-24-4 3-methyl-4-tert-butylheptane C12H26 25113
62185-25-5 4-methyl-4-tert-butylheptane C12H26 25114
62185-26-6 3-ethyl-4-propylheptane C12H26 25115
62185-27-7 3-ethyl-4-isopropylheptane C12H26 25117
62185-28-8 4-ethyl-4-isopropylheptane C12H26 25118
62185-29-9 2,2,3,4-tetramethyl-3-pentanol C9H20O 18334
62185-30-2 2,3-dimethyl-4-propylheptane C12H26 25120
62185-31-3 2,4-dimethyl-4-propylheptane C12H26 25121
62185-32-4 2,5-dimethyl-4-propylheptane C12H26 25122
62185-33-5 2,6-dimethyl-4-propylheptane C12H26 25123
62185-34-6 3,3-dimethyl-4-propylheptane C12H26 25124
62185-35-7 3,4-dimethyl-4-propylheptane C12H26 25125
62185-36-8 3,5-dimethyl-4-propylheptane C12H26 25126
62185-37-9 2,2-dimethyl-3-isopropylheptane C12H26 25127
62185-38-0 2,2-dimethyl-4-isopropylheptane C12H26 25128
62185-39-1 2,3-dimethyl-3-isopropylheptane C12H26 25129
62185-40-4 2,3-dimethyl-4-isopropylheptane C12H26 25130
62185-41-5 2,4-dimethyl-3-isopropylheptane C12H26 25131
62185-42-6 2,4-dimethyl-4-isopropylheptane C12H26 25132
62185-43-7 2,5-dimethyl-3-isopropylheptane C12H26 25133
62185-44-8 2,5-dimethyl-4-isopropylheptane C12H26 25134
62185-45-9 2,6-dimethyl-3-isopropylheptane C12H26 25135
62185-46-0 3,3-dimethyl-4-isopropylheptane C12H26 25137
62185-47-1 3,4-dimethyl-4-isopropylheptane C12H26 25138
62198-55-4 2,3,4-trimethyl-4-ethylheptane C12H26 25169
62198-56-5 2,3,4-trimethyl-5-ethylheptane C12H26 25170
62198-57-6 2,3,5-trimethyl-3-ethylheptane C12H26 25171
62198-58-7 2,3,5-trimethyl-4-ethylheptane C12H26 25172
62198-59-8 2,3,5-trimethyl-5-ethylheptane C12H26 25173
62198-60-1 2,3,6-trimethyl-3-ethylheptane C12H26 25174
62198-61-2 2,3,6-trimethyl-4-ethylheptane C12H26 25175
62198-62-3 2,3,6-trimethyl-5-ethylheptane C12H26 25176
62198-63-4 2,4,4-trimethyl-3-ethylheptane C12H26 25177
62198-64-5 2,4,4-trimethyl-5-ethylheptane C12H26 25178
62198-65-6 2,4,5-trimethyl-3-ethylheptane C12H26 25179
62198-66-7 2,4,5-trimethyl-4-ethylheptane C12H26 25180
62198-67-8 2,4,5-trimethyl-5-ethylheptane C12H26 25181
62198-68-9 2,4,6-trimethyl-3-ethylheptane C12H26 25182
62198-69-0 2,4,6-trimethyl-4-ethylheptane C12H26 25183
62198-70-3 2,5,5-trimethyl-3-ethylheptane C12H26 25184
62198-71-4 2,5,5-trimethyl-4-ethylheptane C12H26 25185
62198-72-5 3,3,4-trimethyl-4-ethylheptane C12H26 25186
62198-73-6 3,3,4-trimethyl-5-ethylheptane C12H26 25187
62198-74-7 3,3,5-trimethyl-4-ethylheptane C12H26 25188
62198-75-8 3,3,5-trimethyl-5-ethylheptane C12H26 25189
62198-76-9 3,4,4-trimethyl-3-ethylheptane C12H26 25190
62198-77-0 3,4,4-trimethyl-5-ethylheptane C12H26 25191
62198-78-1 3,4,5-trimethyl-3-ethylheptane C12H26 25192
62198-79-2 3,4,5-trimethyl-4-ethylheptane C12H26 25193
62198-80-5 2,2,3,3,4-pentamethylheptane C12H26 25194
62198-81-6 2,2,3,3,5-pentamethylheptane C12H26 25195
62198-82-7 2,2,3,3,6-pentamethylheptane C12H26 25196
62198-83-8 2,2,3,4,4-pentamethylheptane C12H26 25197
62198-84-9 2,2,3,4,5-pentamethylheptane C12H26 25198
62198-85-0 2,2,3,4,6-pentamethylheptane C12H26 25199
62198-86-1 2,2,3,5,5-pentamethylheptane C12H26 25200
62198-87-2 2,2,3,5,6-pentamethylheptane C12H26 25201
62198-88-3 2,2,3,6,6-pentamethylheptane C12H26 25202
62198-89-4 3,5-dimethyl-4-isopropylheptane C12H26 25139
62198-90-7 2-methyl-3,3-diethylheptane C12H26 25140
62198-91-8 2-methyl-3,4-diethylheptane C12H26 25141
62198-92-9 2-methyl-3,5-diethylheptane C12H26 25142
62198-93-0 2-methyl-4,4-diethylheptane C12H26 25143
62198-94-1 2-methyl-4,5-diethylheptane C12H26 25144
62198-95-2 2-methyl-5,5-diethylheptane C12H26 25145
62198-96-3 3-methyl-3,4-diethylheptane C12H26 25146
62198-97-4 3-methyl-3,5-diethylheptane C12H26 25147
62198-98-5 3-methyl-4,4-diethylheptane C12H26 25148
62198-99-6 3-methyl-4,5-diethylheptane C12H26 25149
62199-00-2 3-methyl-5,5-diethylheptane C12H26 25150
62199-01-3 4-methyl-3,3-diethylheptane C12H26 25151
62199-02-4 4-methyl-3,4-diethylheptane C12H26 25152
62199-03-5 4-methyl-3,5-diethylheptane C12H26 25153
62199-04-6 2,2,3-trimethyl-3-ethylheptane C12H26 25154
62199-05-7 2,2,3-trimethyl-4-ethylheptane C12H26 25155

62199-06-8 2,2,3-trimethyl-5-ethylheptane C12H26 25156
62199-07-9 2,2,4-trimethyl-3-ethylheptane C12H26 25157
62199-08-0 2,2,4-trimethyl-4-ethylheptane C12H26 25158
62199-09-1 2,2,4-trimethyl-5-ethylheptane C12H26 25159
62199-10-4 2,2,5-trimethyl-3-ethylheptane C12H26 25160
62199-11-5 2,2,5-trimethyl-4-ethylheptane C12H26 25161
62199-12-6 2,2,5-trimethyl-5-ethylheptane C12H26 25162
62199-13-7 2,2,6-trimethyl-3-ethylheptane C12H26 25163
62199-14-8 2,2,6-trimethyl-4-ethylheptane C12H26 25164
62199-15-9 2,2,6-trimethyl-5-ethylheptane C12H26 25165
62199-16-0 2,3,3-trimethyl-4-ethylheptane C12H26 25166
62199-17-1 2,3,3-trimethyl-5-ethylheptane C12H26 25167
62199-18-2 2,3,4-trimethyl-3-ethylheptane C12H26 25168
62199-19-3 2,2,5,7-tetramethyloctane C12H26 25075
62199-20-6 2,2,6,6-tetramethyloctane C12H26 25076
62199-21-7 2,2,6,7-tetramethyloctane C12H26 25077
62199-22-8 2,3,3,4-tetramethyloctane C12H26 25079
62199-23-9 2,3,3,5-tetramethyloctane C12H26 25080
62199-24-0 2,3,3,6-tetramethyloctane C12H26 25081
62199-25-1 2,3,3,7-tetramethyloctane C12H26 25082
62199-26-2 2,3,4,4-tetramethyloctane C12H26 25083
62199-27-3 2,3,4,5-tetramethyloctane C12H26 25084
62199-28-4 2,3,4,6-tetramethyloctane C12H26 25085
62199-29-5 2,3,4,7-tetramethyloctane C12H26 25086
62199-30-8 2,3,5,5-tetramethyloctane C12H26 25087
62199-31-9 2,3,5,6-tetramethyloctane C12H26 25088
62199-32-0 2,3,5,7-tetramethyloctane C12H26 25089
62199-33-1 2,3,6,6-tetramethyloctane C12H26 25090
62199-34-2 2,4,4,5-tetramethyloctane C12H26 25092
62199-35-3 2,4,4,6-tetramethyloctane C12H26 25093
62199-36-4 2,4,5,5-tetramethyloctane C12H26 25095
62199-37-5 2,4,5,6-tetramethyloctane C12H26 25096
62199-38-6 2,4,6,6-tetramethyloctane C12H26 25098
62199-39-7 2,5,5,6-tetramethyloctane C12H26 25099
62199-40-0 2,5,6,6-tetramethyloctane C12H26 25100
62199-41-1 3,3,4,4-tetramethyloctane C12H26 25101
62199-42-2 3,3,4,5-tetramethyloctane C12H26 25102
62199-43-3 3,3,4,6-tetramethyloctane C12H26 25103
62199-44-4 3,3,5,5-tetramethyloctane C12H26 25104
62199-45-5 3,3,5,6-tetramethyloctane C12H26 25105
62199-46-6 3,3,6,6-tetramethyloctane C12H26 25106
62199-47-7 3,4,4,5-tetramethyloctane C12H26 25107
62199-61-5 2,2,4,4,5-pentamethylheptane C12H26 25203
62199-62-6 2,2,4,4,6-pentamethylheptane C12H26 25204
62199-63-7 2,2,4,5,5-pentamethylheptane C12H26 25205
62199-64-8 2,2,4,5,6-pentamethylheptane C12H26 25206
62199-65-9 2,2,5,5,6-pentamethylheptane C12H26 25208
62199-66-0 2,3,3,4,4-pentamethylheptane C12H26 25209
62199-67-1 2,3,3,4,5-pentamethylheptane C12H26 25210
62199-68-2 2,3,3,4,6-pentamethylheptane C12H26 25211
62199-69-3 2,3,3,5,5-pentamethylheptane C12H26 25212
62199-70-6 2,3,4,4,5-pentamethylheptane C12H26 25214
62199-71-7 2,3,4,4,6-pentamethylheptane C12H26 25215
62199-72-8 2,3,4,5,5-pentamethylheptane C12H26 25216
62199-73-9 2,4,4,5,5-pentamethylheptane C12H26 25218
62199-74-0 3,3,4,4,5-pentamethylheptane C12H26 25219
62199-75-1 3,3,4,5,5-pentamethylheptane C12H26 25220
62199-76-2 2,2-dimethyl-3-tert-butylhexane C12H26 25221
62199-77-3 2-methyl-3-ethyl-3-isopropylhexane C12H26 25222
62199-78-4 2-methyl-4-ethyl-3-isopropylhexane C12H26 25223
62199-79-5 2,2,3-trimethyl-3-isopropylhexane C12H26 25224
62199-80-8 2,2,4-trimethyl-3-isopropylhexane C12H26 25225
62199-81-9 2,2,5-trimethyl-3-isopropylhexane C12H26 25226
62199-82-0 2,2,5-trimethyl-4-isopropylhexane C12H26 25227
62199-83-1 2,3,4-trimethyl-3-isopropylhexane C12H26 25228
62199-84-2 2,3,5-trimethyl-3-isopropylhexane C12H26 25229
62199-85-3 2,3,5-trimethyl-4-isopropylhexane C12H26 25230
62199-86-4 2,4,4-trimethyl-3-isopropylhexane C12H26 25231
62199-87-5 3,3,4-triethylhexane C12H26 25232
62199-88-6 2,2-dimethyl,3,3-diethylhexane C12H26 25233
62199-89-7 2,2-dimethyl-3,4-diethylhexane C12H26 25234
62211-85-2 4,5,6-trimethylnonane C12H26 25003
62212-28-6 3,4-dimethyl-3-ethyloctane C12H26 25049
62417-08-7 2,6-dimethyl-1-octanol C10H22O 21255
62496-53-1 1,5-dichloro-3,3-dimethylpentane C7H14Cl2 11765
62882-00-2 8-methylisoquinoline C10H9N 18860
62946-68-3 2-propyl-1,4-butanediol C7H16O2 12260
63126-48-7 3-ethyl-4-octanol C10H22O 21273
63318-29-6 cis-1-iodo-1-heptene C7H13I 11619
63521-36-8 3,4-dimethyl-1,4-pentanediol C7H16O2 12256
63521-37-9 3-methyl-2,3-pentanediol C6H14O2 9038
64245-25-6 cis-1-iodo-3-methyl-1-butene C5H9I 5167
64265-26-5 2,4-dimethyltetrahydrofuran C6H12O 8420
64379-30-2 2,4-dimethyl-2-pentanamine C7H17N 12316
64501-77-5 2-methyl-4-octene C9H18 17949
64502-86-9 3,4-dimethyl-2-pentanol C7H16O 12192
64531-23-3 cis-1-chloro-1-octene C8H15Cl 14749
64919-20-6 ethyl pentadecyl sulfide C17H36S 30365
65016-55-9 3-decylthiophene C14H24S 27772
65016-61-7 3-heptylthiophene C11H18S 22634
65016-62-8 3-octylthiophene C12H20S 24619
65036-71-7 4,5-dimethyl-cis-2-hexene C8H16 14955
65530-93-0 3-methyl-1-hexanamine C7H17N 12319
65769-10-0 6,6-dimethyl-1-heptanol C9H20O 18270
65786-09-6 2,2-dichloroheptane C7H14Cl2 11769
65822-93-7 2,4-dimethyl-2-heptanol C9H20O 18274
65927-60-8 2,3,5-trimethyl-3-hexanol C9H20O 18316
66063-05-6 C19H21ClN2O N-((4-chlorophenyl)methyl 31455
66142-59-4 cis-1-bromo-1-undecene C11H21Br 22749
66142-60-7 trans-1-bromo-1-undecene C11H21Br 22750
66142-61-8 trans-1-chloro-1-undecene C11H21Cl 22757
66142-62-9 trans-1-fluoro-1-undecene C11H21F 22765
66142-63-0 cis-1-iodo-1-undecene C11H21I 22768
66142-64-1 trans-1-iodo-1-undecene C11H21I 22769
66172-73-4 cis-1-chloro-1-undecene C11H21Cl 22756
66172-74-5 cis-1-fluoro-1-undecene C11H21F 22764
66213-67-0 cis-1-chloro-3-methyl-1-butene C5H9Cl 5043
66213-68-1 trans-1-chloro-3-methyl-1-butene C5H9Cl 5044
66213-69-2 cis-1-chloro-1-pentene C5H9Cl 5023
66213-70-5 trans-1-chloro-1-pentene C5H9Cl 5024
66213-72-7 cis-1-fluoro-2-methyl-1-butene C5H9F 5125
66213-73-8 cis-1-fluoro-3-methyl-1-butene C5H9F 5129
66213-74-9 trans-1-fluoro-2-methyl-1-butene C5H9F 5126
66213-75-0 trans-1-fluoro-3-methyl-1-butene C5H9F 5130
66213-76-1 1-fluoro-2-methyl-cis-2-butene C5H9F 5134
66213-77-2 1-fluoro-2-methyl-trans-2-butene C5H9F 5135
66213-78-3 2-fluoro-3-methyl-1-butene C5H9F 5131
66213-80-7 3-fluoro-3-methyl-1-butene C5H9F 5132
66213-81-8 4-fluoro-2-methyl-1-butene C5H9F 5128
66213-82-9 4-fluoro-3-methyl-1-butene C5H9F 5133
66213-83-0 cis-1-fluoro-1-pentene C5H9F 5109
66213-84-1 trans-1-fluoro-1-pentene C5H9F 5110
66213-85-2 1-fluoro-cis-2-pentene C5H9F 5115
66213-86-3 1-fluoro-trans-2-pentene C5H9F 5116
66213-87-4 2-fluoro-cis-2-pentene C5H9F 5117
66213-88-5 2-fluoro-trans-2-pentene C5H9F 5118
66213-89-6 2-fluoro-1-pentene C5H9F 5111
66213-90-9 3-fluoro-cis-2-pentene C5H9F 5119
66225-12-5 3,5-dimethyl-trans-2-hexene C8H16 14951
66225-14-7 4,5-dimethyl-trans-2-hexene C8H16 14956
66225-15-8 3-ethyl-trans-2-hexene C8H16 14942
66225-16-9 4-methyl-cis-2-heptene C8H16 14907
66225-17-0 4-methyl-trans-2-heptene C8H16 14908
66225-18-1 6-methyl-cis-2-heptene C8H16 14912
66225-19-2 6-methyl-cis-3-heptene C8H16 14927
66225-20-5 6-methyl-trans-3-heptene C8H16 14928
66225-22-7 1,1-diiodooctane C8H16I2 15017
66225-30-7 2,3-dimethyl-trans-3-hexene C8H16 14963
66225-31-8 3,5-dimethyl-cis-2-hexene C8H16 14950
66225-32-9 3-ethyl-2,3-pentanediol C7H16O2 12245
66225-33-0 3-ethyl-2,4-pentanediol C7H16O2 12246
66225-35-2 2-methyl-2,4-hexanediol C7H16O2 12234
66225-37-4 4-methyl-1,5-hexanediol C7H16O2 12239
66225-38-5 dimethyl-1,2-dimethylpropylamine C7H17N 12330
66225-39-6 dimethyl-2-methylbutylamine C7H17N 12327
66225-40-9 methylethylbutylamine C7H17N 12332
66225-41-0 methylethyl-sec-butylamine C7H17N 12333
66225-42-1 methylpropylisopropylamine C7H17N 12337
66225-46-5 cis-1-fluoro-1-hexene C6H11F 8071
66225-50-1 1,1-diiodohexane C6H12I2 8305
66225-51-2 3-ethyl-1-pentanol C7H16O 12181
66225-52-3 2,3-dimethyl-1,3-pentanediol C7H16O2 12249
66225-53-4 2,4-dimethyl-2,3-pentanediol C7H16O2 12253
66225-54-5 4,4-dimethyl-1,2-pentanediol C7H16O2 12257
66256-39-1 3,4,4,5-tetramethyl-3-hexanol C10H22O 21321
66256-40-4 3,4,5,5-tetramethyl-3-hexanol C10H22O 21322
66256-41-5 2,2,4-trimethyl-3-ethyl-3-pentanol C10H22O 21328
66256-42-6 2,2,5-trimethyl-4-heptanol C10H22O 21299
66256-43-7 2,2,6-trimethyl-3-heptanol C10H22O 21293
66256-44-8 2,2,6-trimethyl-4-heptanol C10H22O 21300
66256-46-0 2,4,5-trimethyl-4-heptanol C10H22O 21301
66256-47-1 2,4,6-trimethyl-2-heptanol C10H22O 21286
66256-48-2 2,5,6-trimethyl-2-heptanol C10H22O 21287
66256-50-6 3,5,5-trimethyl-3-heptanol C10H22O 21295
66256-60-8 6-methyl-2-nonanol C10H22O 21239
66256-61-9 7-methyl-2-nonanol C10H22O 21240
66256-62-0 4-methyl-2-propyl-1-hexanol C10H22O 21304
66256-63-1 2,2,3,4-tetramethyl-3-hexanol C10H22O 21314
66256-64-2 2,2,3,5-tetramethyl-3-hexanol C10H22O 21315
66256-65-3 2,2,4,4-tetramethyl-3-hexanol C10H22O 21316
66256-66-4 2,3,4,4-tetramethyl-2-hexanol C10H22O 21308
66256-67-5 2,3,4,4-tetramethyl-3-hexanol C10H22O 21318
66256-68-6 2,4,4,5-tetramethyl-3-hexanol C10H22O 21320
66256-69-7 3,3,5,5-tetramethyl-2-hexanol C10H22O 21309
66271-53-2 2-hexadecanethiol C16H34S 29790
66271-54-3 ethyl tetradecyl sulfide C16H34S 29793
66271-55-4 propyl tridecyl sulfide C16H34S 29794
66271-56-5 ethyltetradecylamine C16H35N 29799
66271-71-4 cis-1-fluoro-1-hexadecene C16H31F 29693
66271-72-5 trans-1-fluoro-1-hexadecene C16H31F 29694
66271-76-9 pentadecyl formate C16H32O2 29723
66271-77-0 tridecyl propanoate C16H32O2 29725
66271-81-6 ethyl tridecyl sulfide C15H32S 28924
66271-82-7 propyl dodecyl sulfide C15H32S 28925
66271-83-8 butyl undecyl sulfide C15H32S 28926
66291-42-7 cis-1-fluoro-1-pentadecene C15H29F 28843
66291-43-8 trans-1-fluoro-1-pentadecene C15H29F 28844
66291-47-2 cis-1-chloro-1-decene C10H19Cl 20868
66291-48-3 trans-1-chloro-1-decene C10H19Cl 20869
66291-49-4 cis-1-fluoro-1-decene C10H19F 20877
66291-50-7 trans-1-fluoro-1-decene C10H19F 20878
66291-51-8 cis-1-iodo-1-decene C10H19I 20883
66291-52-9 trans-1-iodo-1-decene C10H19I 20884
66291-72-3 cis-1-bromo-1-decene C10H19Br 20856
66291-73-4 trans-1-bromo-1-decene C10H19Br 20857
66292-31-7 ethyl hexadecyl sulfide C18H38S 31154
66292-32-8 propyl pentadecyl sulfide C18H38S 31155
66292-33-9 butyl tetradecyl sulfide C18H38S 31156
66309-90-8 1,4-dimethyl-5-ethylnaphthalene C14H16 27295
66324-50-3 cis-1-bromo-1-tridecene C13H25Br 26439
66324-51-4 trans-1-bromo-1-tridecene C13H25Br 26440
66324-52-5 cis-1-chloro-1-tridecene C13H25Cl 26443
66324-53-6 trans-1-chloro-1-tridecene C13H25Cl 26444
66324-54-7 cis-1-fluoro-1-tridecene C13H25F 26446
66324-55-8 trans-1-fluoro-1-tridecene C13H25F 26447
66324-56-9 cis-1-iodo-1-tridecene C13H25I 26449
66324-57-0 trans-1-iodo-1-tridecene C13H25I 26450
66324-83-2 1-propyl-[1,2,3,4-tetrahydronaphthalene] C13H18 26158
66324-84-3 2-propyl-[1,2,3,4-tetrahydronaphthalene] C13H18 26159
66325-09-5 2-hexyl-[1,2,3,4-tetrahydronaphthalene] C16H24 29560
66325-11-9 1-hexyl-[1,2,3,4-tetrahydronaphthalene] C16H24 29559
66359-06-6 1-pentyl-[1,2,3,4-tetrahydronaphthalene] C15H22 28647
66359-07-7 2-pentyl-[1,2,3,4-tetrahydronaphthalene] C15H22 28648
66359-40-8 ethyl heptadecyl sulfide C19H40S 31753
66359-41-9 propyl hexadecyl sulfide C19H40S 31754
66359-42-0 butyl pentadecyl sulfide C19H40S 31755
66455-35-4 propyl heptadecyl sulfide C20H42S 32542
66553-05-7 1-amino-2,3-dimethylbutane C6H15N 9185
66553-15-9 2,3-dimethyl-1,2-butanediol C6H14O2 9048
66553-16-0 2-ethyl-1,2-butanediol C6H14O2 9043
66553-17-1 2-ethyl-1,3-butanediol C6H14O2 9044
66553-39-7 cis-1-bromo-1-dodecene C12H23Br 24787
66553-40-0 trans-1-bromo-1-dodecene C12H23Br 24788
66553-41-1 cis-1-chloro-1-dodecene C12H23Cl 24792
66553-42-2 trans-1-chloro-1-dodecene C12H23Cl 24793
66553-43-3 cis-1-fluoro-1-dodecene C12H23F 24796
66553-44-4 trans-1-fluoro-1-dodecene C12H23F 24797
66553-45-5 cis-1-iodo-1-dodecene C12H23I 24800
66553-46-6 trans-1-iodo-1-dodecene C12H23I 24801
66553-52-4 ethyldecylamine C12H27N 25342
66553-53-5 methylundecylamine C12H27N 25341
66553-61-5 4a-ethyl-cis-decahydronaphthalene C12H22 24658
66563-84-6 ethylnonylamine C11H25N 23092
66576-21-4 2,2,3,4,4-pentamethyl-3-ethylpentane C12H26 25267
66576-22-5 3-methyl-3-ethyl-2-pentanol C8H18O 15505
66576-23-6 4-methyl-3-ethyl-2-pentanol C8H18O 15506
66576-24-7 2,2,3,3-tetramethyl-1-butanol C8H18O 15520
66576-26-9 2,3,4-trimethyl-2-pentanol C8H18O 15508
66576-27-0 3,5-dimethyl-2-hexanol C8H18O 15474
66576-31-6 5,5-dimethyl-3-hexanol C8H18O 15488
66576-56-5 3,3-dimethyl-2-ethyl-1-butanol C8H18O 15519
66576-70-3 1,2-dimethylpropyl propanoate C8H16O2 15173
66577-00-2 1,2,6,8-tetramethylnaphthalene C14H16 27306
66577-14-8 2,4-dimethyl-1-ethylnaphthalene C14H16 27290
66577-20-6 1,2,3,6-tetramethylnaphthalene C14H16 27300
66577-21-7 1,2,4,6-tetramethylnaphthalene C14H16 27301
66577-22-8 1,2,4,8-tetramethylnaphthalene C14H16 27303
66577-30-8 ethyl undecyl sulfide C13H28S 26552
66577-31-9 propyl decyl sulfide C13H28S 26553
66577-32-0 butyl nonyl sulfide C13H28S 26554
66577-37-5 1,7-diethylnaphthalene C14H16 27288
66577-38-6 1,2-dimethyl-4-ethylnaphthalene C14H16 27293
66577-39-7 1,2-dimethyl-7-ethylnaphthalene C14H16 27298
66577-40-0 1,3-dimethyl-5-ethylnaphthalene C14H16 27294

Appendix C

Compound List by CAS Registry Number – Inorganic Compounds

Inorganics The compilation for inorganics provides the compound list by CAS registry number, name, chemical formula, and compound number.

62-76-0 sodium oxalate Na2C2O4 2917
74-90-8 hydrogen cyanide HCN 1694
75-15-0 carbon disulfide CS2 656
75-44-5 carbonyl chloride CCl2O 641
78-10-4 silane, tetraethoxy SiC8H20O4 4197
124-38-9 carbon dioxide CO2 655
127-08-2 potassium acetate KC2H3O2 1985
141-53-7 sodium formate NaCHO2 2805
142-72-3 magnesium acetate MgC4H6O4 2361
143-33-9 sodium cyanide NaCN 2810
144-55-8 sodium hydrogen carbonate NaHCO3 2870
151-50-8 potassium cyanide KCN 2013
156-62-7 calcium cyanamide CaCN2 687
298-14-6 potassium hydrogen carbonate KHCO3 2023
302-01-2 hydrazine N2H4 2685
305-53-3 sodium iodoacetate NaC2H2IO2 2816
333-20-0 potassium thiocyanate KSCN 2053
409-21-2 silicon carbide (hexagonal) SiC 4147
463-58-1 carbonyl sulfide COS 653
471-34-1 calcium carbonate (calcite) CaCO3 688
497-19-8 sodium carbonate Na2CO3 2916
540-72-7 sodium thiocyanate NaSCN 2903
546-89-4 lithium acetate LiC2H3O2 2223
546-93-0 magnesium carbonate MgCO3 2350
554-13-2 lithium carbonate Li2CO3 2283
584-08-7 potassium carbonate K2CO3 2064
592-04-1 mercury (ii) cyanide HgC2N2 1779
592-85-8 mercury (ii) thiocyanate HgC2N2S2 1781
630-08-0 carbon monoxide CO 652
814-88-0 cadmium oxalate CdC2O4 783
1066-30-4 chromic acetate (as Cr for TLV) CrC6H9O6 1049
1066-33-7 ammonium hydrogen carbonate NH5CO3 2641
1113-38-8 ammonium oxalate N2H8C2O4 2707
1302-74-5 aluminum oxide (alpha) Al2O3 168
1303-86-2 boron oxide B2O3 395
1303-96-4 sodium tetraborate decahydrate Na2B4H20O17 2909
1305-62-0 calcium hydroxide CaH2O2 733
1305-78-8 calcium oxide CaO 752
1306-19-0 cadmium oxide CdO 809
1309-37-1 iron (iii) oxide Fe2O3 1558
1309-48-4 magnesium oxide MgO 2429
1309-64-4 antimony (iii) oxide Sb2O3 4056
1310-58-3 potassium hydroxide KOH 2047
1310-65-2 lithium hydroxide LiOH 2277
1310-73-2 sodium hydroxide NaOH 2897
1310-82-3 rubidium hydroxide RbOH 3806
1313-82-2 sodium sulfide Na2S 2972
1314-13-2 zinc oxide, dust ZnO 5087
1314-13-2 zinc oxide, fume ZnO 5087
1314-61-0 tantalum (v) oxide Ta2O5 4548
1314-62-1 vanadium (v) oxide V2O5 4891
1314-80-3 phosphorus (v) sulfide P2S5 3451
1317-65-3 calcium carbonate (limestone) CaCO3 688
1317-95-9 silica, crystalline tripoli SiO2 4269
1327-53-3 arsenic trioxide As2O3 245
1333-74-0 hydrogen H2 1709
1333-82-0 chromium (vi) oxide CrO3 1100
1344-28-1 aluminum oxide Al2O3 167
1344-95-2 calcium metasilicate CaO3Si 754
1762-95-4 ammonium thiocyanate N2H4CS 2687
2092-17-3 barium thiocyanate BaC2N2S2 434
2551-62-4 sulfur hexafluoride SF6 4001
2699-79-8 sulfuryl fluoride SO2F2 4009
3017-60-5 cobalt (ii) thiocyanate CoC2N2S2 934
3251-23-8 copper (ii) nitrate CuN2O6 1257
3687-31-8 lead (ii) arsenate Pb3As2O8 3564
3811-04-9 potassium chlorate KClO3 2010
5714-22-7 sulfur decafluoride S2F10 4017
5908-81-6 barium tartrate BaC4H4O6 437
6484-52-2 ammonium nitrate N2H4O3 2689
7429-90-5 aluminum Al 77
7439-90-9 krypton Kr 2162
7439-92-1 lead Pb 3483
7439-95-4 magnesium Mg 2335
7439-96-5 manganese Mn 2464
7439-98-7 molybdenum (soluble compounds, as Mo) Mo 2548
7439-98-7 molybdenum (insoluble compounds, as Mo) Mo 2548
7440-01-9 neon Ne 3091
7440-02-0 nickel Ni 3092
7440-06-4 platinum Pt 3691
7440-16-6 rhodium Rh 3880
7440-21-3 silicon Si 4140
7440-22-4 silver Ag 13
7440-25-7 tantalum Ta 4517
7440-28-0 thallium Tl 4703
7440-31-5 tin Sn 4332
7440-33-7 tungsten W 4899
7440-36-0 antimony Sb 4023
7440-37-1 argon Ar 200
7440-38-2 arsenic As 202
7440-39-3 barium Ba 416
7440-41-7 beryllium Be 530
7440-43-9 cadmium Cd 776
7440-47-3 chromium Cr 1035
7440-48-4 cobalt Co 916
7440-50-8 copper, fume Cu 1169
7440-50-8 copper, dusts and mists Cu 1169
7440-58-6 hafnium Hf 1740
7440-59-7 helium-4 He 1739
7440-61-1 uranium U 4766
7440-63-3 xenon Xe 4945
7440-64-4 ytterbium Yb 4995
7440-67-7 zirconium Zr 5113
7440-74-6 indium In 1877
7446-09-5 sulfur dioxide SO2 4007
7446-14-2 lead (ii) sulfate PbSO4 3547
7446-18-6 thallium (i) sulfate Tl2SO4 4744
7446-27-7 lead (ii) phosphate Pb3O8P2 3569
7446-70-0 aluminum chloride AlCl3 127
7447-39-4 cupric chloride CuCl2 1230
7447-40-7 potassium chloride KCl 2007
7447-41-8 lithium chloride LiCl 2253
7487-88-9 magnesium sulfate MgSO4 2440
7487-94-7 mercuric chloride HgCl2 1800
7488-54-2 rubidium sulfate Rb2SO4 3822
7550-35-8 lithium bromide LiBr 2213
7558-80-7 sodium phosphate, monobasic NaH2O4P 2877
7580-67-8 lithium hydride LiH 2265
7601-54-9 sodium phosphate Na3O4P 3002
7616-94-6 perchloryl fluoride ClFO3 892
7631-90-5 sodium hydrogen sulfite NaHSO3 2874
7631-95-0 sodium molybdate Na2MoO4 2960
7631-99-4 sodium nitrate NaNO3 2894
7632-00-0 sodium nitrite NaNO2 2893
7646-79-9 cobalt chloride CoCl2 968
7646-85-7 zinc chloride ZnCl2 5062
7646-93-7 potassium hydrogen sulfate KHSO4 2029
7647-01-0 hydrogen chloride HCl 1695
7647-14-5 sodium chloride NaCl 2855
7647-15-6 sodium bromide NaBr 2800
7647-17-8 cesium chloride CsCl 1137
7664-38-2 phosphoric acid H3PO4 1725
7664-39-3 hydrogen fluoride HF 1698
7664-41-7 ammonia NH3 2619
7664-93-9 sulfuric acid H2SO4 1716
7681-11-0 potassium iodide KI 2035
7681-38-1 sodium hydrogen sulfate NaHSO4 2875
7681-49-4 sodium fluoride NaF 2864
7681-52-8 sodium hypochlorite pentahydrate NaClH10O6 2857
7681-52-9 sodium hypochlorite NaClO 2858
7681-55-2 sodium iodate NaIO3 2889
7681-57-4 sodium metabisulfite Na2S2O5 2975
7681-82-5 sodium iodide NaI 2888
7697-37-2 nitric acid HNO3 1702
7699-45-8 zinc bromide ZnBr2 5024
7705-08-0 ferric chloride FeCl3 1505
7718-54-9 nickel (ii) chloride NiCl2 3133
7719-12-2 phosphorus (iii) chloride PCl3 3381
7720-78-7 iron (ii) sulfate FeO4S 1534
7722-64-7 potassium permanganate KMnO4 2040
7722-76-1 ammonium dihydrogen phosphate NH6PO4 2649
7722-84-1 hydrogen peroxide H2O2 1713
7722-88-5 sodium pyrophosphate Na4P2O7 3013
7723-14-0 phosphorus (yellow) P 3234
7726-95-6 bromine Br2 633
7727-21-1 potassium persulfate K2S2O8 2131
7727-37-9 nitrogen N2 2678
7727-43-7 barium sulfate BaO4S 501
7727-54-0 ammonium peroxydisulfate N2H8O8S2 2723
7733-02-0 zinc sulfate ZnSO4 5094
7757-79-1 potassium nitrate KNO3 2042
7757-82-6 sodium sulfate Na2SO4 2974
7757-83-7 sodium sulfite Na2SO3 2973
7757-88-2 magnesium sulfite MgO3S 2431
7758-01-2 potassium bromate KBrO3 1973
7758-02-3 potassium bromide KBr 1972
7758-05-6 potassium iodate KIO3 2036
7758-09-0 potassium nitrite KNO2 2041
7758-11-4 potassium hydrogen phosphate K2HPO4 2099
7758-19-2 sodium chlorite NaClO2 2859
7758-94-3 ferrous chloride FeCl2 1502
7758-95-4 lead chloride PbCl2 3514
7758-97-6 lead (ii) chromate PbCrO4 3520
7758-98-7 copper (ii) sulfate CuO4S 1265
7759-02-6 strontium sulfate SrO4S 4502
7761-88-8 silver (i) nitrate AgNO3 49
7772-99-8 stannous chloride SnCl2 4422
7773-01-5 manganese (ii) chloride MnCl2 2495
7773-06-0 ammonium sulfamate N2H6O3S 2702
7774-29-0 mercuric iodide HgI2 1813
7775-09-9 sodium chlorate NaClO3 2860
7775-11-3 sodium chromate Na2CrO4 2930
7775-27-1 sodium persulfate Na2O8S2 2971
7778-18-9 calcium sulfate CaO4S 757
7778-39-4 arsenic acid AsH3O4 235
7778-50-9 potassium dichromate K2Cr2O7 2089
7778-53-2 potassium phosphate K3PO4 2153
7778-74-7 potassium perchlorate KClO4 2011
7778-77-0 potassium dihydrogen phosphate KH2PO4 2034
7778-80-5 potassium sulfate K2SO4 2127
7779-88-6 zinc nitrate ZnN2O6 5086
7782-39-0 deuterium D2 1302
7782-41-4 fluorine F2 1409
7782-42-5 graphite C 639
7782-44-7 oxygen O2 3198
7782-49-2 selenium Se 4083
7782-50-5 chlorine Cl2 899
7782-64-1 manganese (ii) fluoride MnF2 2499
7782-65-2 germane GeH4 1675
7783-06-4 hydrogen sulfide H2S 1714
7783-07-5 hydrogen selenide H2Se 1719
7783-19-9 ammonium selenite N2H8SeO3 2728
7783-20-2 ammonium sulfate N2H8O4S 2721
7783-21-3 ammonium selenate N2H8SeO4 2729
7783-28-0 diammonium hydrogen phosphate N2H9PO4 2734
7783-36-0 mercury (i) sulfate Hg2SO4 1838
7783-40-6 magnesium fluoride MgF2 2399
7783-41-7 fluorine oxide F2O 1412
7783-46-2 lead fluoride PbF2 3521
7783-48-4 strontium fluoride SrF2 4484
7783-49-5 zinc fluoride ZnF2 5069
7783-50-8 iron (iii) fluoride FeF3 1511
7783-54-2 nitrogen trifluoride NF3 2613
7783-56-4 antimony (iii) fluoride SbF3 4045
7783-79-1 selenium hexafluoride SeF6 4110
7783-80-4 tellurium hexafluoride TeF6 4590
7783-89-3 silver bromate AgBrO3 17
7783-91-7 silver chlorite AgClO2 37
7783-92-8 silver (i) chlorate AgClO3 38
7783-93-9 silver (i) perchlorate AgClO4 39
7783-99-5 silver (i) nitrite AgNO2 48
7784-18-1 aluminum fluoride AlF3 134
7784-42-1 arsine AsH3 234
7784-45-4 arsenic triiodide AsI3 238
7785-87-7 manganese (ii) sulfate MnSO4 2526
7786-30-3 magnesium chloride MgCl2 2392
7786-81-4 nickel (ii) sulfate NiSO4 3164
7787-32-8 barium fluoride BaF2 464
7787-39-5 barium sulfite BaO3S 494
7787-47-5 beryllium chloride BeCl2 541
7787-69-1 cesium bromide CsBr 1128

Appendix D

Compound List by Name – Organic Compounds

ORGANICS The compilation for organics provides the compound list by name and compound number.

abate 29470
abate 29470
abietic acid 32346
acenaphthene 23444
acetal 9020
acetaldehyde 884
acetamide 956
acetanilide 13671
acetic acid 889
acetic anhydride 2990
acetone 1923
acetone cyanohydrin 3246
acetonitrile 753
acetophenone 13391
acetyl chloride 707
acetylacetone 4882
acetylene 571
2-(acetyloxy)benzoic acid 16233
acridine 25537
acrolein 1621
acrylamide 1755
acrylic acid 1626
acrylonitrile 1462
adipic acid 7917
adiponitrile 7316
alachlor 27572
aldicarb 11803
aldrin 23291
allene 1493
allyl acetate 4883
allyl alcohol 1920
allyl butanoate 11454
allyl formate 2972
allyl glycidyl ether 7816
allyl propanoate 7818
allyl propyl disulfide 8639
allylamine 2030
alochlor 1254 23179
ametryn 17857
3-amino-1-butanol 3989
3-amino-2-butanol 3990
4-amino-1-butanol 3991
4-amino-2-butanol 3998
1-amino-2-butanol 3999
(S)-(+)-2-amino-1-butanol 4001
(R)-(-)-2-amino-1-butanol 4002
2-amino-1-butanol, (±)- 3992
4-aminobutan-2-ol 4000
2-aminobutan-1-ol 4003
bis(4-amino-3-chlorophenyl)methane 25748
chloramben 9957
2-amino-3,3-dimethyl butane 9188
1-amino-2,2-dimethylbutane 9184
1-amino-2,3-dimethylbutane 9185
1-amino-3,3-dimethylbutane 9186
2-amino-2,3-dimethylbutane 9187
1-aminoethanol 1118
N-aminoethyl ethanolamine 4080
N-aminoethyl piperazine 9247
1-amino-2-ethylbutane 9189
2-amino-2-ethyl-1,3-propanediol 5987
3-aminoheptane 12345
2-aminohexane 9174
3-aminohexane 9175
2-amino-6-methylheptane 15645
(S)-2-amino-6-methylheptane 15646
(R)-2-amino-6-methylheptane 15647
1-amino-2-methylpentane 9176
1-amino-3-methylpentane 9177
1-amino-4-methylpentane 9178
2-amino-2-methylpentane 9179
2-amino-3-methylpentane 9180
2-amino-4-methylpentane 9181
3-amino-2-methylpentane 9182
3-amino-3-methylpentane 9183
4-amino-3-methyl-6-phenyl-1,2,4-triazin- 19076
2-amino-2-methyl-1,3-propanediol 4008
2-amino-2-methyl-1-propanol 3993
1-amino-2-methyl-2-propanol 4005
(S)-2-aminooctane 15648
(R)-2-aminooctane 15649
2-aminooctane 15650
1-amino-2-propanol 2204
3-amino-1-propanol 2205
(R)-(-)-1-amino-2-propanol 2211
(S)-(+)-1-amino-2-propanol 2212
(R)-(-)-2-amino-1-propanol 2213
2-aminopropanol 2214
2-amino-1-propanol, (±) 2207
4-amino-3,5,6-trichloropyridinecarboxlic 6309
ammonium perfluorooctanoate 12570
amphetamine 17250
tert-amyl-tert-butylamine 18394
tert-amyl-tert-octylamine 26561
p-tert-amylphenol 22399
o-amylphenol 22440
4-sec-amylphenol 22442
anethole 19484
aniline 7172
aniline, p-isopropyl-N-methyl-, 20231
anisole 10850
2-anthracenamine 26821
anthracene 26728
9-anthracenecarboxylic acid 28038
9-anthracenemethanol 28179
anthraquinone 26654
ascorbic acid 7459
atrazine 14510
azelaic acid 17755
azinphos-methyl 19468
azobenzene 23488
cis-azobenzene 23489
azulene 18670
benomyl 27476
benz[a]anthracene 30407
benzalacetone 19106
benzaldehyde 10404
benzene 6876
benzeneacetaldehyde 13393
benzeneacetic acid 13415
benzeneacetonitrile 13108
benzeneethanamine 14110
benzenepropanenitrile 16364
benzenepropanol 17091
1,2,3-benzenetriol 7085
p-benzidine 23720
1H-benzimidazole 10341
11H-benzo[a]fluorene 29862
benzo[a]pyrene 31804
benzo[b]fluoranthene 31803
11H-benzo[b]fluorene 29863
1,3-benzodioxole 10411
benzo[e]pyrene 31805
benzofuran 12914
benzo[ghi]perylene 33080
benzoic acid 10408
benzo[k]fluoranthene 31802
benzonitrile 10045
benzophenone 25623
1H-2-benzopyran-1-one 15925
2H-1-benzopyran-2-one 15926
4H-1-benzopyran-4-one 15927
p-benzoquinone 6616
benzothiazole 10094
benzothiophene 12947
benzotrichloride 9967
benzotrifluoride 10003
benzoyl chloride 9927
benzoyl peroxide 26788
benzyl acetate 16680
benzyl alcohol 10851
benzyl benzoate 26942
benzyl butyl phthalate 31452
benzyl chloride 10492
benzyl dichloride 10252
benzyl ethyl ether 17077
benzyl methyl ether 13927
benzyl phenyl ether 25793
benzylamine 10953
biacetylene 2374
bicyclohexyl 24652
4-biphenylmethanol 25803
2,2'-biquinoline 30424
bromacil 17234
bromoacetic acid 693
o-bromoaniline 6905
m-bromoaniline 6906
p-bromoaniline 6907
bromobenzene 6656
2-bromobenzoic acid 9890
3-bromobenzoic acid 9891
4-bromobenzoic acid 9892
1-bromobutane 3598
2-bromobutane 3599
2-bromobutane, (±) 3602
cis-1-bromo-1-butene 3047
trans-1-bromo-1-butene 3048
2-bromo-1-butene 3049
3-bromo-1-butene 3050
4-bromo-1-butene 3051
1-bromo-cis-2-butene 3052
1-bromo-trans-2-butene 3053
2-bromo-cis-2-butene 3054
2-bromo-trans-2-butene 3055
1-bromo-2-chlorobenzene 6383
1-bromo-3-chlorobenzene 6384
1-bromo-4-chlorobenzene 6385
bromochlorodifluoromethane 13
bromochloromethane 140
3-bromo-5-chlorophenol 6386
4-bromo-2-chlorophenol 6387
2-bromo-4-chlorophenol 6388
bromocyclohexane 7981
1-bromodecane 21090
2-bromodecane 21091
cis-1-bromo-1-decene 20856
trans-1-bromo-1-decene 20857
o-(4-bromo-2,5-dichlorophenyl) o-methyl 25576
1-bromo-2,2-dimethylbutane 8658
1-bromo-2,3-dimethylbutane 8659
1-bromo-3,3-dimethylbutane 8660
2-bromo-2,3-dimethylbutane 8661
2-bromo-3,3-dimethylbutane 8662
1-bromo-2,2-dimethylpropane 5610
1-bromododecane 24894
2-bromododecane 24895
cis-1-bromo-1-dodecene 24787
trans-1-bromo-1-dodecene 24788
bromoethane 910
1-bromo-2-ethylbutane 8663
3-bromo-2-ethyl-1-propene 4993
bromoform 96
1-bromoheptane 11982
2-bromoheptane 11983
4-bromoheptane 11984
cis-1-bromo-1-heptene 11569
trans-1-bromo-1-heptene 11570
1-bromohexadecane 29739
1-bromohexane 8647
2-bromohexane 8648
3-bromohexane 8649
cis-1-bromo-1-hexene 7979
trans-1-bromo-1-hexene 7980
5-bromoindole 12725
2-bromo-4-methylaniline 10735
2-bromo-5-methylaniline 10736
3-bromo-4-methylaniline 10737
4-bromo-2-methylaniline 10738
4-bromo-3-methylaniline 10739
5-bromo-2-methylaniline 10740
3-bromo-2-methylaniline 10741
(bromomethyl)benzene 10463
1-bromo-2-methylbutane 5606
1-bromo-3-methylbutane 5607
2-bromo-2-methylbutane 5608
2-bromo-3-methylbutane 5609
1-bromo-2-methylbutane, (S) 5611
cis-1-bromo-2-methyl-1-butene 4980
trans-1-bromo-2-methyl-1-butene 4981
3-bromo-2-methyl-1-butene 4982
4-bromo-2-methyl-1-butene 4983
cis-1-bromo-3-methyl-1-butene 4984
trans-1-bromo-3-methyl-1-butene 4985
2-bromo-3-methyl-1-butene 4986
3-bromo-3-methyl-1-butene 4987
4-bromo-3-methyl-1-butene 4988
1-bromo-2-methyl-cis-2-butene 4989
1-bromo-2-methyl-trans-2-butene 4990
1-bromo-3-methyl-2-butene 4991
1-bromo-2-methylpentane 8650
1-bromo-3-methylpentane 8651
1-bromo-4-methylpentane 8652
2-bromo-2-methylpentane 8653
2-bromo-3-methylpentane 8654
2-bromo-4-methylpentane 8655
3-bromo-2-methylpentane 8656
3-bromo-3-methylpentane 8657
1-bromo-2-methylpropane 3600
2-bromo-2-methylpropane 3601
1-bromo-2-methyl-1-propene 3056
3-bromo-2-methyl-1-propene 3057
1-bromo-12-methyltridecane 27896
1-bromonaphthalene 18601
1-bromo-2-nitrobenzene 6403
1-bromo-3-nitrobenzene 6404
1-bromo-4-nitrobenzene 6405
1-bromononane 18115
cis-1-bromo-1-nonene 17785
trans-1-bromo-1-nonene 17786
1-bromooctane 15248
2-bromooctane, (±)- 15249
cis-1-bromo-1-octene 14736
trans-1-bromo-1-octene 14737
1-bromopentadecane 28898
1-bromopentane 5603
2-bromopentane 5604
3-bromopentane 5605
cis-1-bromo-1-pentene 4964
trans-1-bromo-1-pentene 4965
2-bromo-1-pentene 4966
3-bromo-1-pentene 4967
4-bromo-1-pentene 4968
5-bromo-1-pentene 4969
1-bromo-cis-2-pentene 4970
1-bromo-trans-2-pentene 4971
2-bromo-cis-2-pentene 4972
2-bromo-trans-2-pentene 4973
3-bromo-cis-2-pentene 4974
3-bromo-trans-2-pentene 4975
4-bromo-cis-2-pentene 4976
4-bromo-trans-2-pentene 4977
5-bromo-cis-2-pentene 4978
5-bromo-trans-2-pentene 4979
o-bromophenol 6671
m-bromophenol 6672
p-bromophenol 6673
bromophenol 6674
(4-bromophenyl)urea 10469
1-bromopropane 1978
2-bromopropane 1979
cis-1-bromo-1-propene 1642
trans-1-bromo-1-propene 1643
2-bromo-1-propene 1644
3-bromo-1-propene 1645
1-bromo-1-propene; (cis+trans) 1647
1-bromotetradecane 27895
p-bromotoluene 10462
o-bromotoluene 10464
m-bromotoluene 10465
bromotrichloromethane 16
1-bromotridecane 26515
cis-1-bromo-1-tridecene 26439
trans-1-bromo-1-tridecene 26440
bromotrifluoroethylene 404
bromotrifluoromethane 17
1-bromoundecane 22870
cis-1-bromo-1-undecene 22749
trans-1-bromo-1-undecene 22750
1,2-butadiene 2756
butadiene (1,3 butadiene) 2757
butanamide 3685
2-butanamine, (.+/-.)- 3981
butane 3783
1,2-butanediol 3884
1,3-butanediol 3885
1,4-butanediol 3886

Appendix E

Compound List by Name – Inorganic Compounds

INORGANICS The compilation for inorganics provides the compound list by name and compound number.

aluminum 77
aluminum chloride 127
aluminum fluoride 134
aluminum nitrate 149
aluminum oxide 167
aluminum oxide (alpha) 168
aluminum perchlorate 133
aluminum sulfate 175
ammonia 2619
ammonium bromide 2625
ammonium chloride 2627
ammonium dihydrogen phosphate 2649
ammonium fluoride 2631
ammonium hydrogen carbonate 2641
ammonium iodate 2635
ammonium iodide 2634
ammonium nitrate 2689
ammonium nitrite 2688
ammonium oxalate 2707
ammonium perchlorate 2629
ammonium peroxydisulfate 2723
ammonium selenate 2729
ammonium selenite 2728
ammonium sulfamate 2702
ammonium sulfate 2721
ammonium thiocyanate 2687
antimony 4023
antimony (iii) fluoride 4045
antimony (iii) oxide 4056
antimony trichloride 4042
argon 200
arsenic 202
arsenic acid 235
arsenic triiodide 238
arsenic trioxide 245
arsine 234
barium 416
barium bromate 426
barium bromide 423
barium chlorate 458
barium chloride 454
barium fluoride 464
barium hydroxide 470
barium iodate 482
barium iodide 481
barium nitrate 487
barium nitrite 486
barium perchlorate 459
barium sulfate 501
barium sulfide 511
barium sulfite 494
barium tartrate 437
barium thiocyanate 434
beryllium 530
beryllium chloride 541
beryllium sulfate 552
boric acid 370
boron oxide 395
bromine 633
bromine pentafluoride 628
cadmium 776
cadmium bromide 777
cadmium chloride 791
cadmium fluoride 798
cadmium iodate 805
cadmium iodide 804
cadmium nitrate 807
cadmium oxalate 783
cadmium oxide 809
cadmium perchlorate 795
cadmium sulfate 816
calcium bromide 682
calcium carbonate (calcite) 688
calcium carbonate (limestone) 688
calcium chlorate 722
calcium chloride 713
calcium chromate 724
calcium cyanamide 687
calcium fluoride 728
calcium hydroxide 733
calcium iodate 747
calcium iodide 746
calcium metasilicate 754
calcium nitrate 751
calcium nitrite 750
calcium oxide 752
calcium perchlorate 723
calcium sulfate 757
calcium sulfate dihydrate 738
carbon dioxide 655
carbon disulfide 656
carbon monoxide 652
carbonyl chloride 641
carbonyl sulfide 653
cesium bromate 1129
cesium bromide 1128
cesium chlorate 1138
cesium chloride 1137
cesium hydroxide 1150
cesium iodate 1145
cesium iodide 1144
cesium nitrate 1147
cesium perchlorate 1139
cesium sulfate 1165
chlorine 899
chlorine dioxide 898
chromic acetate (as Cr for TLV) 1049
chromium 1035
chromium (vi) oxide 1100
chromium oxychloride 1099
cobalt 916
cobalt chloride 968
cobalt (ii) fluoride 978
cobalt (ii) iodide 1000
cobalt (ii) nitrate 1004
cobalt (ii) nitrite 1003
cobalt (ii) perchlorate 972
cobalt (ii) sulfate 1010
cobalt (ii) thiocyanate 934
copper, fume 1169
copper (ii) bromide 1174
copper (ii) fluoride 1240
copper (ii) nitrate 1257
copper (ii) perchlorate 1234
copper (ii) sulfate 1265
cupric chloride 1230
decaborane 413
deuterium 1302
diammonium hydrogen phosphate 2734
diborane 394
ferric chloride 1505
ferric nitrate 1531
ferrous chloride 1502
fluorine 1409
fluorine oxide 1412
germane 1675
graphite 639
hafnium 1740
helium-4 1739
hydrazine 2685
hydrogen 1709
hydrogen bromide 1691
hydrogen chloride 1695
hydrogen cyanide 1694
hydrogen fluoride 1698
hydrogen peroxide 1713
hydrogen selenide 1719
hydrogen sulfide 1714
indium 1877
iron (ii) bromide 1424
iron (iii) fluoride 1511
iron (ii) nitrate 1530
iron (iii) oxide 1558
iron pentacarbonyl 1434
iron (ii) sulfate 1534
krypton 2162
lanthanum chloride 2177
lanthanum nitrate 2189
lead 3483
lead (ii) arsenate 3564
lead bromide 3487
lead chloride 3514
lead (ii) chromate 3520
lead fluoride 3521
lead (ii) iodate 3531
lead iodide 3530
lead (ii) nitrate 3533
lead (ii) perchlorate 3518
lead (ii) phosphate 3569
lead (ii) sulfate 3547
lithium acetate 2223
lithium bromate 2215
lithium bromide 2213
lithium carbonate 2283
lithium chlorate 2257
lithium chloride 2253
lithium dihydrogen phosphate 2267
lithium fluoride 2261
lithium hydride 2265
lithium hydroxide 2277
lithium iodate 2271
lithium iodide 2270
lithium nitrate 2274
lithium oxalate 2285
lithium perchlorate 2258
lithium phosphate 2313
lithium sulfate 2306
magnesium 2335
magnesium acetate 2361
magnesium bromide 2341
magnesium carbonate 2350
magnesium chloride 2392
magnesium fluoride 2399
magnesium iodide 2424
magnesium nitrate 2427
magnesium oxide 2429
magnesium perchlorate 2396
magnesium sulfate 2440
magnesium sulfite 2431
manganese 2464
manganese (ii) bromide 2469
manganese (ii) chloride 2495
manganese (ii) fluoride 2499
manganese (ii) nitrate 2515
manganese (ii) sulfate 2526
mercuric bromide 1773
mercuric chloride 1800
mercuric iodide 1813
mercury (i) chloride 1829
mercury (ii) cyanide 1779
mercury (i) sulfate 1838
mercury (ii) thiocyanate 1781
molybdenum (soluble compounds, as Mo) 2548
neodymium trichloride 3070
neon 3091
nickel 3092
nickel carbonyl (as Ni) 3105
nickel (ii) chloride 3133
nickel fluoride 3139
nickel iodate 3155
nickel (ii) iodide 3154
nickel (ii) sulfate 3164
nickel thiocyanate 3101
nitric acid 1702
nitric oxide 2674
nitrogen 2678
nitrogen dioxide 2676
nitrogen trifluoride 2613
nitrous oxide 2747
osmium tetroxide (as Os) 3222
oxygen 3198
ozone 3199
pentaborane 405
perchloryl fluoride 892
phosphine 3389
phosphoric acid 1725
phosphorus (iii) chloride 3381
phosphorus oxychloride 3396
phosphorus pentachloride 3382
phosphorus (v) sulfide 3451
phosphorus (yellow) 3234
platinum 3691
potassium acetate 1985
potassium arsenate 2141
potassium bromate 1973
potassium bromide 1972
potassium carbonate 2064
potassium chlorate 2010
potassium chloride 2007
potassium chromate 2088
potassium cyanide 2013
potassium dichromate 2089
potassium dihydrogen phosphate 2034
potassium ferricyanide 2143
potassium fluoride 2018
potassium fluoroborate 1969
potassium hydrogen carbonate 2023
potassium hydrogen phosphate 2099
potassium hydrogen sulfate 2029
potassium hydroxide 2047
potassium iodate 2036
potassium iodide 2035
potassium metabisulfite 2129
potassium molybdate 2112
potassium nitrate 2042
potassium nitrite 2041
potassium perchlorate 2011
potassium periodate 2037
potassium permanganate 2040
potassium persulfate 2131
potassium phosphate 2153
potassium selenate 2133
potassium selenite 2117
potassium sulfate 2127
potassium sulfite 2116
potassium thiocyanate 2053
potassium thiosulfate 2128
praseodymium chloride 3663
radon 3940
rhodium 3880
rubidium bromate 3787
rubidium bromide 3786
rubidium chlorate 3792
rubidium chloride 3791
rubidium fluoride 3795
rubidium hydrogen carbonate 3798
rubidium hydroxide 3806
rubidium iodide 3801
rubidium perchlorate 3793
rubidium sulfate 3822
samarium (iii) chloride 4315
selenium 4083
selenium hexafluoride 4110
silane 4261
silane, tetraethoxy 4197
silica, amprphous silica gel 4269
silica, colloidal 4269
silica, crystalline tripoli 4269
silica, fume 4269
silicon 4140
silicon carbide (hexagonal) 4147
silicon dioxide (quartz) 4265
silicon dioxide (cristobalite) 4267
silicon dioxide (tridymite) 4266
silicon dioxide (vitreous) 4268
silver 13
silver bromate 17
silver (i) chlorate 38
silver chlorite 37
silver (i) nitrate 49
silver (i) nitrite 48
silver (i) perchlorate 39
silver (i) sulfate 68
sodium bromate 2802
sodium bromide 2800
sodium carbonate 2916
sodium chlorate 2860
sodium chloride 2855
sodium chlorite 2859
sodium chromate 2930
sodium cyanide 2810
sodium dichromate 2932
sodium fluoride 2864
sodium formate 2805
sodium hydrogen carbonate 2870
sodium hydrogen sulfate 2875

Appendix F

Adsorption Capacity of Activated Carbon for Aqueous System - Correlations

Carl L. Yaws
Lamar University, Beaumont, Texas

The results for adsorption capacity of activated carbon for removal of organic chemicals in water are given in Table F. The results are applicable for a wide variety of substances – alcohols, ketones, aldehydes, acetates, benzenes, chloro alkanes, chloro olefins, and bromo alkanes. The results utilize the adsorption capacity as a function of the adsorption potential of the chemical in the aqueous liquid.

For the tabulation, the adsorption capacity was correlated as a function of adsorption potential (Polyani theory) of the chemical in the aqueous phase using the following equation:

$$\log Q = A + B\, T/V \log (x_s/x) \qquad \text{(F-1)}$$

where
Q = adsorption capacity at equilibrium, g of compound/100 g of carbon
A and B = regression coefficients
T = temperature, K
V = molar volume of liquid, cm^3/mol
x_s = solubility in water, parts per million by weight, ppm(wt)
x = concentration in water, parts per million by weight, ppm(wt)

The results were determined from regression of the available data. In preparing the correlation, a literature search was conducted to identify data source data publications (1-11). The excellent compilations by Arbuckle (1); Sonthier, Crittenden and Summer (2); and Speth (3,4) were utilized for adsorption capacity. The compilations by Howard and Meylan (5); Mackay, Shiu, and Ma (6); Verschueren (7); Yalkowsky (8); and Yaws (9) were used for water solubility. Liquid molar volume was ascertained from data compilations by Yaws (9-11). The publications were screened and copies of appropriate data were made. These data were then keyed into the computer to provide a database of adsorption capacity values at different adsorption potentials in aqueous phase. The database also served as a basis to check the accuracy of the correlation.

In general, experimental data available in the literature are very limited for adsorption capacity of activated carbon for organic compounds in aqueous liquid at low concentrations (parts per million range). In view of the very limited experimental data, estimates in the tabulation should be considered rough approximations. Also, adsorption capacity is complicated by differences in carbon, lots of carbon, source of water, and concentration ranges (Speth and Miller, 4). If initial feasibility studies using the correlation are favorable, follow-up experimental determination of equilibrium adsorption capacity is recommended.

Representative graphs of adsorption capacity as a function of the adsorption potential are shown in Figures F-1, F-2, F-3, F-4 and F-5 for the substances. The graphs discloses favorable agreement of correlation results and experimental data.

References

1. Arbuckle, W. B., Environ. Sci. Technol., 15(7), 812 (1981).
2. Sontheimer, H., Crittenden, J. C. and Summers, R. S., ACTIVATED CARBON FOR WATER TREATMENT, 2nd edition, DVGW – Forschungsstelle, FRG (1988).
3. Speth, T. F., Masters Thesis, Michigan Technological University, Houghton, MI (1986).
4. Speth, T. F. and Miller, J. R., Journal AWWA, 72 (February, 1990).
5. Howard, P. H. and W. M. Meylan, eds., HANDBOOK OF PHYSICAL PROPERTIES OF ORGANIC CHEMICALS, CRC Press, Boca Raton, FL (1997).
6. Mackay, D., W. Y. Shiu, and K. C. Ma, ILLUSTRATED HANDBOOK OF PHYSICAL-CHEMICAL PROPERTIES AND ENVIRONMENTAL FATE FOR ORGANIC CHEMICALS, Vols. 1, 2, 3, 4 and 5, Lewis Publishers, New York, NY (1992, 1992, 1993, 1995, 1997).
7. Verschueren, K., HANDBOOK OF ENVIRONMENTAL DATA ON ORGANIC CHEMICALS, 3rd and 4th eds., Van Nostrand Reinhold, New York, NY (1996, 2001).
8. Yalkowsky, S. H., AQUASOL DATABASE, University of Arizona, Tucson, AZ (1990-2004).
9. Yaws, C. L., CHEMICAL PROPERTIES HANDBOOK, McGraw-Hill, New York, NY (1999).
10. Yaws, C. L., YAWS HANDBOOK OF THERMODYNAMIC AND PYYSICAL PROPERTIES OF CHEMICAL COMPOUNDS, electronic edition, www.knovel.com, knovel Corporation, Norwich, NY (2003).
11. Yaws, C. L., YAWS HANDBOOK OF PHYSICAL PROPERTIES FOR HYDROCARBONS AND CHEMICALS, Gulf Publishing Co., Houston, TX (2005).

Table F Adsorption Capacity of Activated Carbon – Aqueous System

$\log_{10} Q = A + B\ T/V \log_{10}(xs/x)$ (Q - g of compound/100 g of carbon, x - ppm(wt))					
No	**Chemical Family**	**A**	**B**	**Formula**	**Example Compound**
1	alcohols	1.7250	-0.1210	$C_nH_{2n+2}O$	1-butanol
2	ketones	1.7250	-0.1210	$C_nH_{2n}O$	diethyl ketone
3	aldehydes	1.7250	-0.1210	$C_nH_{2n}O$	butyraldehyde
4	acetates	1.7250	-0.1210	$C_nH_{2n}O_2$	butyl acetate
5	benzenes	1.8893	-0.1199	$C_{n+6}H_{2n+6}$	ethyl benzene
6	chloro alkanes	2.6021	-0.1931	$C_nH_{2n+2-x}Cl_x$	chloroform
7	chloro olefins	2.2175	-0.1599	$C_nH_{2n-x}Cl_x$	trichloroethylene
8	bromo alkanes	3.0607	-0.1991	$C_nH_{2n+2-x}Br_x$	bromoform
F 400 Activated Carbon, Calgon					
Q - adsorption capacity at equilibrium, g of compound/100 g of carbon					
xs - solubility in water at 25 C, parts per million by weight, ppm(wt)					
x - concentration in water at 25 C, parts per million by weight, ppm(wt)					
T - temperature, K					
V - molar volume of liquid, cm3/mole					
A and B - regression coefficients					

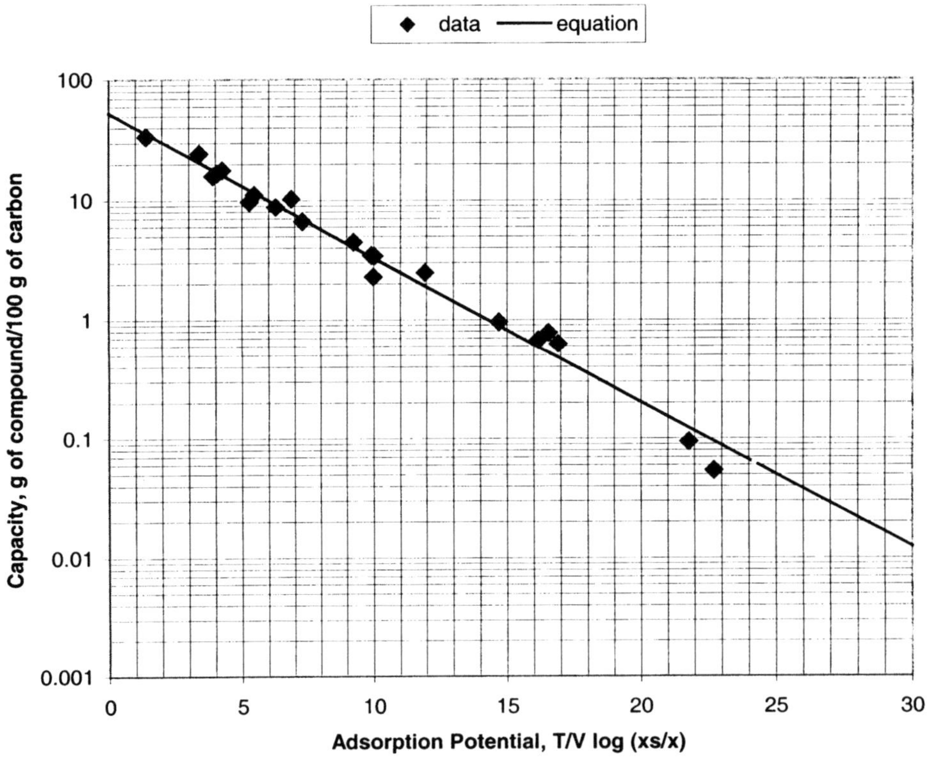

Figure F-1 Adsorption Capacity of Activated Carbon for Organic Oxygen Chemicals (Alcohols, Ketones, Aldehydes and Acetates) in Water

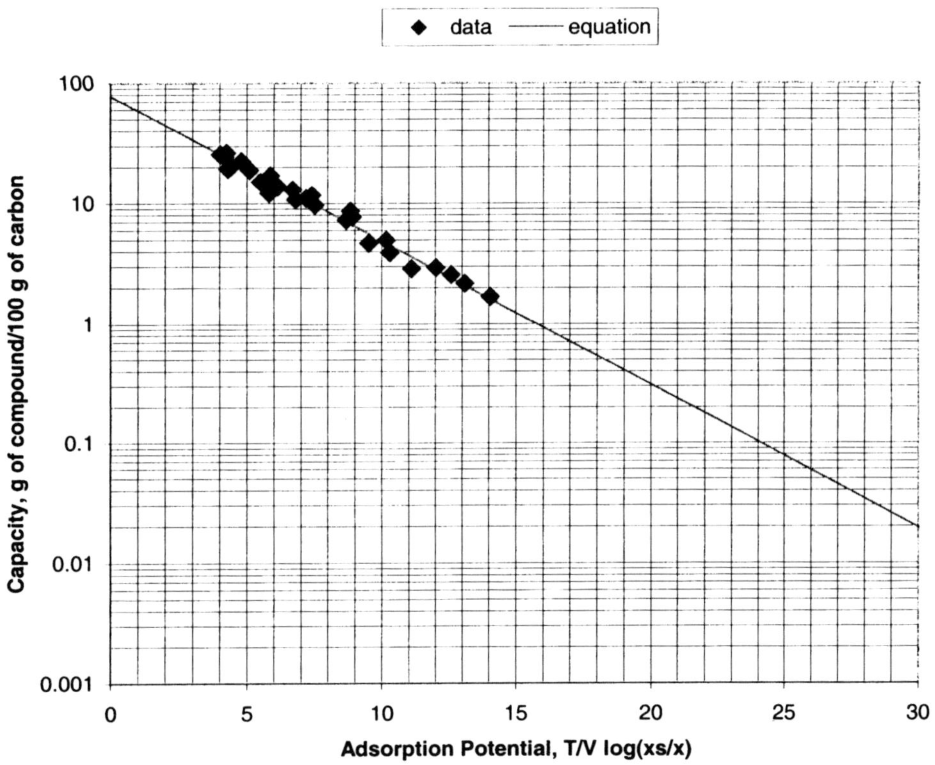

Figure F-2 Adsorption Capacity of Activated Carbon for Benzenes in Water

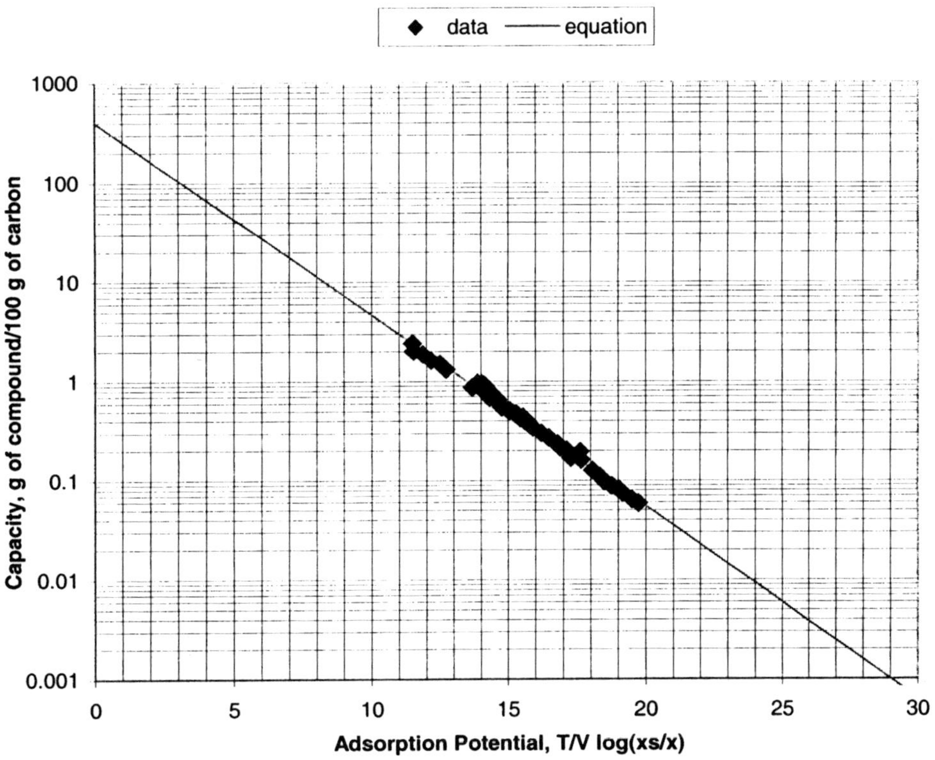

Figure F- 3 Adsorption Capacity of Activated Carbon for Chloro Alkanes in Water

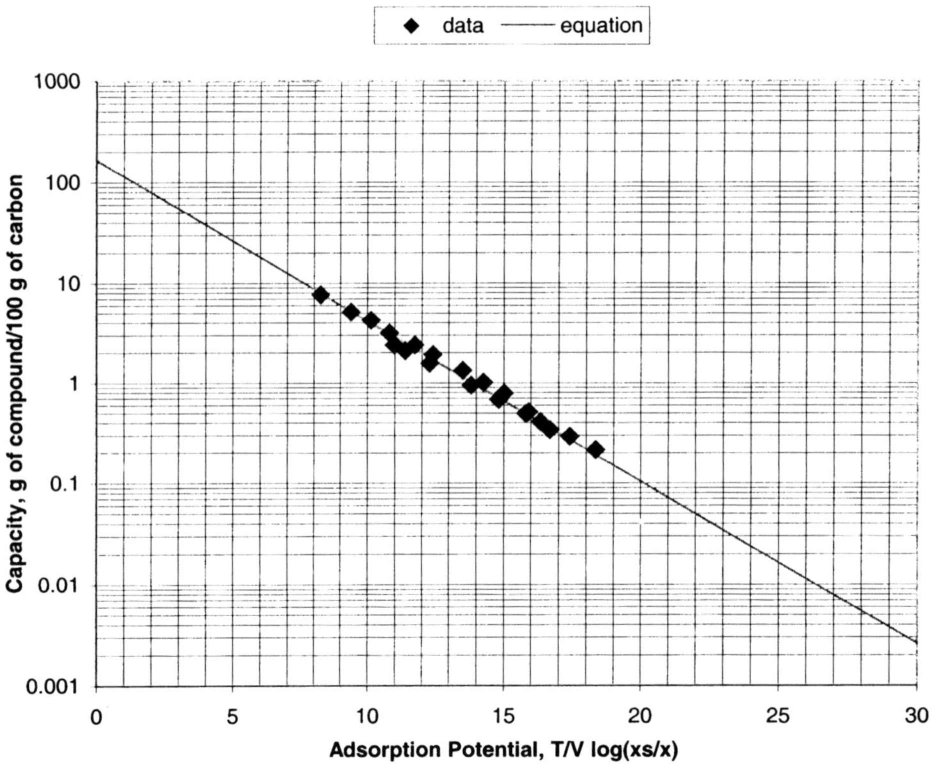

Figure F-4 Adsorption Capacity of Activated Carbon for Chloro Olefins in Water

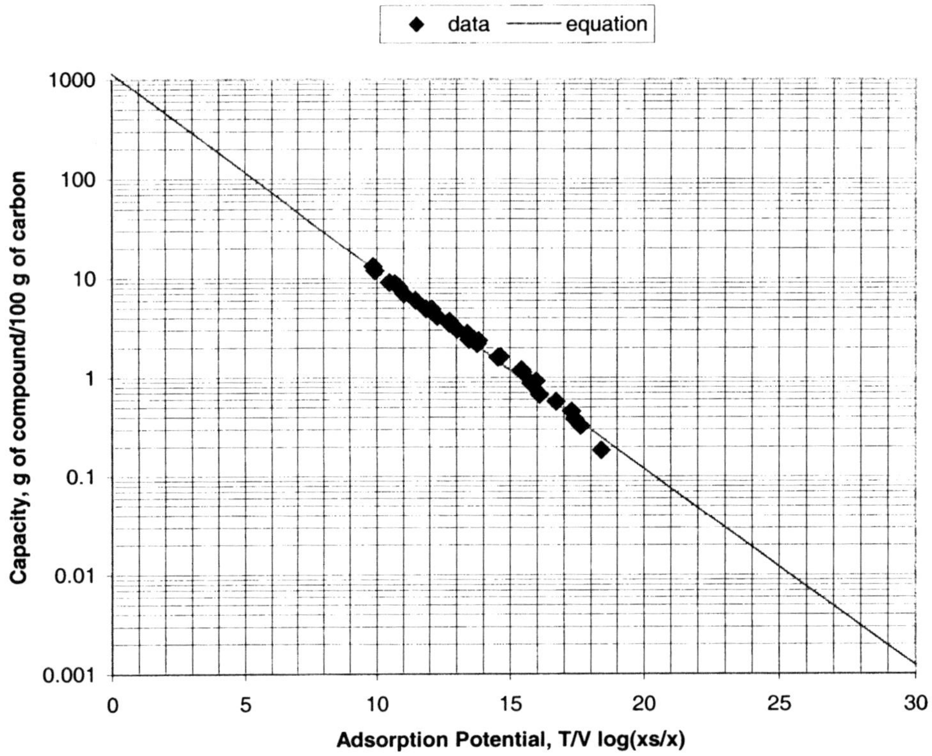

Figure F-5 Adsorption Capacity of Activated Carbon for Bromo Alkanes in Water

Appendix G

Solubility of Organic Compounds in Water - Correlations

Carl L. Yaws
Lamar University, Beaumont, Texas

The results for solubility of organic compounds in water are given in Table G. The results are applicable for a wide variety of substances – aetates, acetylenes, acids, alcohols, aldehydes, alkanes, amines, benzenes, butanoates, cyclohexanes, cyclopentanes, diolefins, ethers, formats, ketones, mercaptans, monobromo alkanes, monochloro alkanes, olefins, propanoates, suldides, and thiophenes.

The results are based on the boiling point method as developed by Yaws for water solubility as a function of the boiling point temperature of the compound using the following equation:

$$\log_{10} S = A + B\,T_B + C\,T_B^2 + D\,T_B^3 \qquad \text{(G-1)}$$

where S = solubility in water at 25 C, parts per million by weight, ppm (wt)
T_B = boiling point temperature of compound, K
A, B, C, and D = regression coefficients

In preparing the tabulation, a literature search was conducted to identify data source publications for organics (1-172). Both experimental values for the property under consideration and parameter values for estimation of the property are included in the source publications. The publications were screened and copies of appropriate data were made. These data were then keyed into the computer to provide a database of values for compounds for which experimental data are available. The database also served as a basis to check the accuracy of the correlations. The numerous point values were processed using a computer program for minimum deviation.

The excellent compilations by Horvath (56); Howard and Meylan (57); Mackay, Shiu, and Ma (91-95); Verschueren (141); Yalkowsky (147); and Yaws (149-171) were primarily used for experimental data for water solubility. Boiling point temperatures were ascertained from data compilations by Yaws (169-172).

For hydrocarbons and chemicals, experimental data available in the literature are very limited for isomers, large molecules (C10 and higher), and substituted organics containing halogens, nitrogen, oxygen, and sulfur. In the absence of data, the estimates for isomers, large molecules, and substituted organics should be considered as rough approximations useful for initial analysis. If initial analysis is favorable, follow-up experimental determination is recommended.

References

1-172. See **References** in **Chapter 3 – SOLUBILITY IN WATER - ORGANIC COMPOUNDS**

Table G Solubility of Organic Compounds in Water - Correlations

$$\log_{10} S = A + B\,T_B + C\,T_B^2 + D\,T_B^3 \qquad (S\text{ - ppm(wt)},\ T_B\text{ - K})$$

No	Chemical Family	A	B	C	D	TBMIN	TBMAX	Formula	Example Compound
1	acetates	13.0000	-2.3382E-02	0.0000E+00	0.0000E+00	330	600	$C_nH_{2n}O_2$	butyl acetate
2	acetylenes	-15.8350	1.77811E-01	-5.00907E-04	4.11124E-07	298	560	$C_{n+2}H_{2n+2}$	1-hexyne
3	acids	14.3000	-2.2000E-02	0.0000E+00	0.0000E+00	440	560	$C_nH_{2n}O_2$	n-hexanoic acid
4	alcohols	45.6398	-2.3859E-01	4.8739E-04	-3.7160E-07	370	620	$C_nH_{2n+2}O$	n-pentanol
5	aldehydes	12.4200	-2.2000E-02	0.0000E+00	0.0000E+00	320	560	$C_nH_{2n}O$	1-hexanal
6	alkanes	-17.6520	1.77811E-01	-5.00907E-04	4.11124E-07	298	560	C_nH_{2n+2}	n-hexane
7	amines-primary	11.4300	-1.8000E-02	0.0000E+00	0.0000E+00	400	560	$C_nH_{2n+3}N$	n-hexyl amine
8	amines-secondary	11.3600	-1.8000E-02	0.0000E+00	0.0000E+00	360	560	$C_nH_{2n+3}N$	dipropyl amine
9	amines-tertiary	11.8200	-2.0000E-02	0.0000E+00	0.0000E+00	360	560	$C_nH_{2n+3}N$	triethyl amine
10	benzenes (multiple substitution)	-23.6500	2.21196E-01	-5.55632E-04	4.18830E-07	298	560	$C_{n+6}H_{2n+6}$	p-xylene
11	benzenes (single substitution)	-24.0080	2.21196E-01	-5.55632E-04	4.18830E-07	298	560	$C_{n+6}H_{2n+6}$	toluene
12	butanoates	13.0000	-2.3382E-02	0.0000E+00	0.0000E+00	330	600	$C_nH_{2n}O_2$	propyl butanoate
13	cyclohexanes	-16.7000	1.77811E-01	-5.00907E-04	4.11124E-07	298	560	C_nH_{2n}	methyl cyclohexane
14	cyclopentanes	-16.9000	1.77811E-01	-5.00907E-04	4.11124E-07	298	560	C_nH_{2n}	methyl cyclopentane
15	diolefins	-16.5610	1.77811E-01	-5.00907E-04	4.11124E-07	298	560	C_nH_{2n-2}	1,5 hexadiene
16	ethers	11.4000	-2.2000E-02	0.0000E+00	0.0000E+00	300	560	$C_nH_{2n+2}O$	dipropyl ether
17	formates	12.5750	-2.3382E-02	0.0000E+00	0.0000E+00	330	600	$C_nH_{2n}O_2$	hexyl formate
18	ketones	45.2000	-2.3859E-01	4.8739E-04	-3.7160E-07	350	590	$C_nH_{2n}O$	diethyl ketone
19	mercaptans	-6.9000	1.00500E-01	-2.72880E-04	1.99870E-07	298	560	$C_nH_{2n+2}S$	propyl mercaptan
20	monobromo alkanes	-7.1700	1.0050E-01	-2.7288E-04	1.9987E-07	300	590	$C_nH_{2n+1}Br$	1-bromobutane
21	monochloro alkanes	-7.4500	1.0050E-01	-2.7288E-04	1.9987E-07	300	590	$C_nH_{2n+1}Cl$	2-chlorobutane
22	olefins	-17.0300	1.77811E-01	-5.00907E-04	4.11124E-07	298	560	C_nH_{2n}	1-hexene
23	propanoates	13.0000	-2.3382E-02	0.0000E+00	0.0000E+00	330	600	$C_nH_{2n}O_2$	propyl propanoate
24	sulfides	-6.5390	1.00500E-01	-2.72880E-04	1.99870E-07	298	560	$C_nH_{2n+2}S$	ethyl methyl sulfide
25	thiophenes	-6.8500	1.00500E-01	-2.72880E-04	1.99870E-07	298	560	$C_nH_{2n-4}S$	2-methyl thiophene

S - solubility in water at 25 C, parts per million by weight, ppm(wt)

A, B, C, and D - regression coefficients

T_B - boiling point temperature of compound, K

TBMAX - maximum temperature, K

TBMIN - minimum temperature, K

Appendix H

Solubility of Organic Compounds in Salt Water - Correlations

Carl L. Yaws
Lamar University, Beaumont, Texas

The results for solubility of organic compounds in salt water are given in Table H. The tabulation is applicable for alkanes, cyclopentanes, cyclohexanes, and benzenes. The results for solubility in water containing salt are based on a series expansion in salt concentration:

$$\log_{10} S = A + B X + C X^2 \tag{H-1}$$

where S = solubility in salt water at 25 C, parts per million by weight, ppm (wt)
X = concentration of salt (NaCl) in water, mg/kg of water
A, B, and C = regression coefficients

In preparing the tabulation, a literature search was conducted to identify data source publications for organics (1-21). Both experimental values for the property under consideration and parameter values for estimation of the property are included in the source publications. The publications were screened and copies of appropriate data were made. These data were then keyed into the computer to provide a database of values for compounds for which experimental data are available. The database also served as a basis to check the accuracy of the correlations. The numerous point values were processed using a computer program for minimum deviation.

The excellent compilations by Howard and Meylan (1); Mackay, Shiu, and Ma (2-3); Verschueren (7); Yalkowsky (8); and Yaws (12-21) were primarily used for solubility in water without salt. For solubility in water containing salt, the compilations of API (6), Price (4); Sutton and Calder (5); and Yaws (9-11) were primarily used. Boiling point temperatures were ascertained from data compilations by Yaws (19-21).

For organic compounds, experimental data available in the literature are very limited for solubility in water containing salt. In the absence of data, the estimates for solubility in salt water should be considered as rough approximations useful for initial analysis. If initial analysis is favorable, follow-up experimental determination is recommended.

References

1. Howard, P. H. and W. M. Meylan, eds., HANDBOOK OF PHYSICAL PROPERTIES OF ORGANIC CHEMICALS, CRC Press, Boca Raton, FL (1997).
2. Mackay, D., W. Y. Shiu, and K. C. Ma, ILLUSTRATED HANDBOOK OF PHYSICAL-CHEMICAL PROPERTIES AND ENVIRONMENTAL FATE FOR ORGANIC CHEMICALS, Volume 1 – Monoaromatic Hydrocarbons, Chlorobenzenes and PCBs, Lewis Publishers, New York, NY (1992).
3. Mackay, D., W. Y. Shiu, and K. C. Ma, ILLUSTRATED HANDBOOK OF PHYSICAL-CHEMICAL PROPERTIES AND ENVIRONMENTAL FATE FOR ORGANIC CHEMICALS, Volume 3 – Volatile Organic Compounds, Lewis Publishers, New York, NY (1993).
4. Price, L. C., Bull. Am. Assoc. Petrol. Geologists, 60(2), 213-244 (1976).
5. Sutton, C. and J. A. Calder, Environ. Sci. Techn., 8(7), 654 (1974).
6. TECHNICAL DATA BOOK - PETROLEUM REFINING, Vols. I and II, American Petroleum Institute, Washington, DC (1972, 1977, 1982).
7. Verschueren, K., HANDBOOK OF ENVIRONMENTAL DATA ON ORGANIC CHEMICALS, 3rd and 4th eds., Van Nostrand Reinhold, New York, NY (1996, 2001).
8. Yalkowsky, S. H., Aquasol Database, 5th edition, College of Pharmacy, University of Arizona, Tuscon, Az (1990).
9. Yaws, C. L., "Hydrocarbon Solubility in Salt Water", [salt effect – alkanes], Pollution Engineering, 24, 46-50 (Sept. 15, 1992).
10. Yaws, C. L. and Xiaoyan Lin, "Solubility of Hydrocarbons in Salt Water", [salt effect – cyclopentanes and cyclohexanes], Pollution Engineering, 26, 70-72 (January, 1994).
11. Yaws, C. L., Li Bu, and S. Nijhawan, "Determining the Solubility of Aromatic Hydrocarbons", [salt effect – benzenes], Pollution Engineering, 27, 78-80 (June, 1995).
12. Yaws, C. L. and others, "Organic Chemicals: Water Solubility Data", Chem. Eng., 97, 115 (July, 1990).
13. Yaws, C. L. and others, "Water Solubility Data for Organic Compounds", Pollution Engineering, 22, 70 (Oct., 1990).
14. Yaws, C. L. and others, "New Correlation Calculates Reliable Paraffin Solubilities", Oil & Gas Journal, 89, 79 (April 8, 1991).

15. Yaws, C. L. and others, "New Correlation Accurately Calculates Naphthene Water Solubilities", Oil & Gas Journal, 89(37), 86 (Sept. 16, 1991).
16. Yaws, C. L. and others, "New Correlation Accurately Calculates Water Solubilities of Aromatics", Oil & Gas Journal, 95 (35), 80 (Aug. 29, 1994).
17. Yaws, C. L. and others, "Solubility of Hydrocarbons and Sulfur Compounds in Water", WATER ENCYCLOPEDIA, Volume 4, ISBN 0-471-73684-8, pages 561-564, John Wiley, New York, NY (2005).
18. Yaws, C. L., CHEMICAL PROPERTIES HANDBOOK, McGraw-Hill, New York, NY (1999).
19. Yaws, C. L., MATHESON GAS DATA BOOK, 7th ed., Matheson Tri-Gas (Parsippany, NJ), McGraw-Hill, New York, NY (2001).
20. Yaws, C. L., YAWS HANDBOOK OF THERMODYNAMIC AND PHYSICAL PROPERTIES OF CHEMICAL COMPOUNDS, electronic edition, www.knovel.com, knovel, Norwich, NY (2003).
21. Yaws, C. L., YAWS HANDBOOK OF PHYSICAL PROPERTIES FOR HYDROCARBONS AND CHEMICALS, Gulf Publishing Co., Houston, TX (2005).

Table H Solubility of Organic Compounds in Salt Water – Correlations

$\log_{10} S = A + B X + C X^2$ S - solubility, ppm (wt), X(salt) - mg of salt / kg of water

No	Chemical Family	A, B and C	Formula	Example Compound
1	alkanes	$A = -17.6521 + 1.77811E\text{-}01\ T_B - 5.00907E\text{-}04\ T_B^2 + 4.11124E\text{-}07\ T_B^3$	C_nH_{2n+2}	n-pentane
		B = -4.4374E-06		
		C = 2.2187E-12		
2	cyclopentanes	$A = -16.9000 + 1.77811E\text{-}01\ T_B - 5.00907E\text{-}04\ T_B^2 + 4.11124E\text{-}07\ T_B^3$	C_nH_{2n}	methyl cyclopentane
		B = -7.1550E-06		
		C = 3.5775E-12		
3	cyclohexanes	$A = -16.7000 + 1.77811E\text{-}01\ T_B - 5.00907E\text{-}04\ T_B^2 + 4.11124E\text{-}07\ T_B^3$	C_nH_{2n}	cyclohexane
		B = -7.1550E-06		
		C = 3.5775E-12		
4	benzenes	$A = -24.0080 + 2.21196E\text{-}01\ T_B - 5.55632E\text{-}04\ T_B^2 + 4.18830E\text{-}07\ T_B^3$	$C_{n+6}H_{2n+6}$	benzene
	(no and single substitution)	B = -3.3451E-06		
		C = 1.6726E-12		
5	benzenes	$A = -23.6500 + 2.21196E\text{-}01\ T_B - 5.55632E\text{-}04\ T_B^2 + 4.18830E\text{-}07\ T_B^3$	$C_{n+6}H_{2n+6}$	o-xylene
	(double and triple substitution)	B = -4.5292E-06		
		C = 2.2646E-12		

S - solubility in salt water at 25 C, parts per million by weight, ppm(wt)
X - salt (NaCl) concentration in water, mg of salt/kg of water
Salt concentration range is X = 0 to 358,700 (saturated).
Salt in sea water is 34,472 mg/kg of water.
A, B, and C - regression coefficients
T_B - boiling point temperature of compound, K
TBMAX - maximum temperature, K = 560
TBMIN - minimum temperature, K = 300
Correlation should not be used for compounds having boiling points outside TBMAX and TBMIN.
Range of application for correlation is about C5 to C16.

Appendix I

Solubility in Water as a Function of Temperature for Organic Compounds - Correlations

Carl L. Yaws
Lamar University, Beaumont, Texas

The results for solubility of organic compounds in water are given in Table I. The tabulation is applicable for alkanes, cyclopentanes, cyclohexanes, and benzenes. Solubilities at three different temperatures are provided in the tabulation.

The results are useable for water solubility as a function of temperature using a series expansion in temperature:

$$\log_{10} S = A + B\,T + C\,T^2 \qquad \text{(I-1)}$$

where S = solubility in water, parts per million by weight, ppm(wt)
T = temperature, K
A, B, and C = regression coefficients

The coefficients (A, B, and C) maybe ascertained from the solubilities at the three different temperatures (25, 99.1, and 121.3 C) in the tabulation.

In preparing the tabulation, a literature search was conducted to identify data source publications for organics (1-20). Both experimental values for the property under consideration and parameter values for estimation of the property are included in the source publications. The publications were screened and copies of appropriate data were made. These data were then keyed into the computer to provide a database of values for compounds for which experimental data are available. The database also served as a basis to check the accuracy of the correlations. The numerous point values were processed using a computer program for minimum deviation.

For ambient temperature, the compilations by Howard and Meylan (1); Mackay, Shiu, and Ma (2-3); Verschueren (6); Yalkowsky (7); and Yaws (8-19) were primarily used. For temperatures higher than ambient, the compilations of Price (4), API (5), and Yaws (13-14, 16) were used. Parameter values for the estimates were from data compilations of Yaws (18-20).

For hydrocarbons and chemicals, experimental data at temperatures higher than ambient are very limited for isomers and large molecules (C10 and higher). In the absence of data, the estimates for isomers and large molecules should be considered as rough approximations useful for initial analysis. If initial analysis is favorable, follow-up experimental determination is recommended.

References

1. Howard, P. H. and W. M. Meylan, eds., HANDBOOK OF PHYSICAL PROPERTIES OF ORGANIC CHEMICALS, CRC Press, Boca Raton, FL (1997).
2. Mackay, D., W. Y. Shiu, and K. C. Ma, ILLUSTRATED HANDBOOK OF PHYSICAL-CHEMICAL PROPERTIES AND ENVIRONMENTAL FATE FOR ORGANIC CHEMICALS, Volume 1 – Monoaromatic Hydrocarbons, Chlorobenzenes and PCBs, Lewis Publishers, New York, NY (1992).
3. Mackay, D., W. Y. Shiu, and K. C. Ma, ILLUSTRATED HANDBOOK OF PHYSICAL-CHEMICAL PROPERTIES AND ENVIRONMENTAL FATE FOR ORGANIC CHEMICALS, Volume 3 – Volatile Organic Compounds, Lewis Publishers, New York, NY (1993).
4. Price, L. C., Bull. Am. Assoc. Petrol. Geologists, 60(2), 213-244 (1976).
5. TECHNICAL DATA BOOK - PETROLEUM REFINING, Vols. I and II, American Petroleum Institute, Washington, DC (1972, 1977, 1982).
6. Verschueren, K., HANDBOOK OF ENVIRONMENTAL DATA ON ORGANIC CHEMICALS, 3rd and 4th eds., Van Nostrand Reinhold, New York, NY (1996, 2001).
7. Yalkowsky, S. H., Aquasol Database, 5th edition, College of Pharmacy, University of Arizona, Tuscon, Az (1990).
8. Yaws, C. L. and others, "Hydrocarbons: Water Solubility Data", Chem. Eng., 97, 177 (April,1990).
9. Yaws, C. L. and others, "Organic Chemicals: Water Solubility Data", Chem. Eng., 97, 115 (July, 1990).
10. Yaws, C. L. and others, "Water Solubility Data for Organic Compounds", Pollution Engineering, 22, 70 (Oct., 1990).
11. Yaws, C. L. and others, "New Correlation Calculates Reliable Paraffin Solubilities", Oil & Gas Journal, 89, 79 (April 8, 1991).
12. Yaws, C. L. and others, "New Correlation Accurately Calculates Naphthene Water Solubilities", Oil & Gas Journal, 89(37), 86 (Sept. 16, 1991).
13. Yaws, C. L. and others, "Water Solubility Data for 151 Hydrocarbons", [temperature effect – alkanes], Chem. Eng., 100, 108 (February, 1993).
14. Yaws, C. L. and others, "The Water Solubility of Naphthenes", [temperature effect – cycloalkanes], Chem. Eng.,

100 (10), 122 (Oct., 1993).
15. Yaws, C. L. and others, "New Correlation Accurately Calculates Water Solubilities of Aromatics", Oil & Gas Journal, 95 (35), 80 (Aug. 29, 1994).
16. Yaws, C. L. and others, "Calculate the Solubility of Aromatics", [temperature effect – benzenes], Chem. Eng., 102 (2), 113 (Feb., 1995).
17. Yaws, C. L. and others, "Solubility of Hydrocarbons and Sulfur Compounds in Water", WATER ENCYCLOPEDIA, Volume 4, ISBN 0-471-73684-8, pages 561-564, John Wiley, New York, NY (2005).
18. Yaws, C. L., CHEMICAL PROPERTIES HANDBOOK, McGraw-Hill, New York, NY (1999).
19. Yaws, C. L., YAWS HANDBOOK OF THERMODYNAMIC AND PHYSICAL PROPERTIES OF CHEMICAL COMPOUNDS, electronic edition, www.knovel.com, knovel, Norwich, NY (2003).
20. Yaws, C. L., YAWS HANDBOOK OF PHYSICAL PROPERTIES FOR HYDROCARBONS AND CHEMICALS, Gulf Publishing Co., Houston, TX (2005).

Table I Solubility in Water as a Function of Temperature for Organic Compounds – Correlations

$\log_{10} S = A + B\,T_B + C\,T_B^2 + D\,T_B^3$ S - solubility, ppm (wt), T_B - K

No	Chemical Family	A, B, C, and D	Formula	Example Compound
1	alkanes	A = -17.6520 for 25 C, -17.2610 for 99.1 C, -0.7360 for 121.3 C	C_nH_{2n+2}	n-pentane
		B = 1.7781E-01 for 25 C, 1.7781E-01 for 99.1 C, 4.1114E-02 for 121.3 C		
		C = -5.0091E-08 for 25 C, -5.0091E-08 for 99.1 C, -1.3698E-08 for 121.3 C		
		D = 4.1112E-11 for 25 C, 4.1112E-11 for 99.1 C, 1.7002E-11 for 121.3 C		
2	cyclopentanes	A = -16.9000 for 25 C, -16.5670 for 99.1 C, -0.0330 for 121.3 C	C_nH_{2n}	cyclopentane
		B = 1.7781E-01 for 25 C, 1.7781E-01 for 99.1 C, 4.1114E-02 for 121.3 C		
		C = -5.0091E-08 for 25 C, -5.0091E-08 for 99.1 C, -1.3698E-08 for 121.3 C		
		D = 4.1112E-11 for 25 C, 4.1112E-11 for 99.1 C, 1.7002E-11 for 121.3 C		
3	cyclohexanes	A = -16.7000 for 25 C, -16.2900 for 99.1 C, -0.0850 for 121.3 C	C_nH_{2n}	cyclohexane
		B = 1.7781E-01 for 25 C, 1.7781E-01 for 99.1 C, 4.1114E-02 for 121.3 C		
		C = -5.0091E-08 for 25 C, -5.0091E-08 for 99.1 C, -1.3698E-08 for 121.3 C		
		D = 4.1112E-11 for 25 C, 4.1112E-11 for 99.1 C, 1.7002E-11 for 121.3 C		
4	benzenes	A = -24.0080 for 25 C, -23.5400 for 99.1 C, -23.2900 for 121.3 C	$C_{n+6}H_{2n+6}$	benzene
	(no and single substitution)	B = 2.2120E-01 for 25 C, 2.2120E-01 for 99.1 C, 2.2120E-03 for 121.3 C		
		C = -5.5563E-08 for 25 C, -5.5563E-08 for 99.1 C, -5.5563E-08 for 121.3 C		
		D = 4.1883E-11 for 25 C, 4.1883E-11 for 99.1 C, 4.1883E-11 for 121.3 C		
5	benzenes	A = -23.6500 for 25 C, -23.3500 for 99.1 C, -23.0800 for 121.3 C	$C_{n+6}H_{2n+6}$	o-xylene
	(double and triple substitution)	B = 2.2120E-01 for 25 C, 2.2120E-01 for 99.1 C, 2.2120E-03 for 121.3 C		
		C = -5.5563E-08 for 25 C, -5.5563E-08 for 99.1 C, -5.5563E-08 for 121.3 C		
		D = 4.1883E-11 for 25 C, 4.1883E-11 for 99.1 C, 4.1883E-11 for 121.3 C		

S - solubility in water, parts per million by weight, ppm(wt)

TB - boiling point temperature of compound, K

TBMAX - maximum temperature, K = 560

TBMIN - minimum temperature, K = 300

Correlation should not be used for compounds having boiling points outside TBMAX and TBMIN.

Index

ABOUT THE AUTHOR

Carl L. Yaws, Ph. D., is professor of chemical engineering at Lamar University, Beaumont, Texas. He has industrial experience in process engineering, research, development and design at Exxon, Ethyl and Texas Instruments. He is the author of 33 books and has published more than 690 technical papers in process engineering, property data and pollution prevention.